Machine Design with CAD and Optimization

Machine Design with CAD and Optimization

Sayed M. Metwalli
Faculty of Engineering, Cairo University
Egypt

WILEY

This edition first published 2021

Registered Offices
John Wiley & Sons, Inc., 111 River Street, Hoboken, NJ 07030, USA
John Wiley & Sons Ltd, The Atrium, Southern Gate, Chichester, West Sussex, PO19 8SQ, UK

Editorial Office
The Atrium, Southern Gate, Chichester, West Sussex, PO19 8SQ, UK

For details of our global editorial offices, customer services, and more information about Wiley products visit us at www.wiley.com.

Wiley also publishes its books in a variety of electronic formats and by print-on-demand. Some content that appears in standard print versions of this book may not be available in other formats.

Library of Congress Cataloging-in-Publication Data

Names: Metwalli, Sayed, author. | John Wiley & Sons, Ltd., publisher.
Title: Machine design with CAD & optimization / Sayed Metwalli.
Description: First edition. | Hoboken, NJ : Wiley, 2021. | Includes index.
Identifiers: LCCN 2020025973 (print) | LCCN 2020025974 (ebook) | ISBN 9781119156642 (cloth) | ISBN 9781119156659 (adobe pdf) | ISBN 9781119156666 (epub)
Subjects: LCSH: Machine design–Computer-aided design. | Computer-aided design.
Classification: LCC TJ233 .M48 2021 (print) | LCC TJ233 (ebook) | DDC 621.8/150285–dc23
LC record available at https://lccn.loc.gov/2020025973
LC ebook record available at https://lccn.loc.gov/2020025974

Cover Design: Wiley
Cover Image: © adventtr/Getty Images

Set in 9.5/12.5pt STIXTwoText by SPi Global, Chennai, India
Printed and bound by CPI Group (UK) Ltd, Croydon, CR0 4YY

C9781119156642_050421

To all members of my family for their understanding, patience, and endless support.
To all my teachers and colleagues for their nourishing knowledge, inspiration, and backing.
To all my students for their perseverance, tolerance, efforts, and dedication.

Contents

Preface

This book intends to provide the tools to "*really design*" or "*synthesize*" machine elements and assembly of prospective machine elements in systems or products. It employs knowledge base, computer-aided design (*CAD*), and *optimization* tools to directly define appropriate geometry and material selection of machine elements. The treatment is set in both US and international (SI) units.

For each machine element, there is a chart, a simple Excel sheet or a *Tablet*, a MATLAB program, or an interactive program to calculate the element geometry and to help in selecting suitable material.

The book is divided into three main parts:

1. *Introduction and Design Considerations.*
2. *Knowledge-Based Design*: Introduction to the new machine element *synthesis* for a *real design* intent, which includes bases for computer-aided techniques. Overviews are dedicated for **CAD** and **optimization** fields. Bases for stresses, deformations, and deflections in addition to materials static and dynamic strength are given as cornerstones to synthesis of machine elements. These knowledge-based tools are utilized in the *initial synthesis* of machine elements and system.
3. *Detailed Design of Machine Elements*: Rigorous traditional *detailed design* requirements are given for basic joints and machine elements such as screws fasteners and permanent joins, springs, rolling bearings, and journal bearings. This part of the text also covers power transmission elements such as spur gears; helical, bevel, and worm gears; flexible elements; shafts; and clutches, breaks, and flywheel.

These parts include the following outline of chapters and enlightening pivotal details:

- The first part covers a basic introduction to machine design and presents several design factors to consider. It introduces an overview of the design process and the necessary background knowledge and information that is needed to make sound judgments in decision making.
  The introduction discusses the main goal of the text and argues that it is intended to generate or create designs directly through *synthesis* rather than designs by repeated analysis. This process is implemented in designing machine elements. For system or product design, conceptual considerations and innovations are cited to generate better products or systems that include assembly of some of those machine elements. It necessitates exposure to the necessary phases of design, basic mechanical functions, codes and standards, design factors, and proper design approach to end up with a sound design assembly. It is also important to realize that for any product there is a product life cycle. The design should consider important business measures, economics of scale, technology adoption, and the research and development process in the product cycle. Teamwork for complex product or large system design is indispensable to get the job done in an effective way and on schedule. Design and development case studies are used to demonstrate the concepts covered in this introduction.
  The designer will need to use mathematical modeling tools. Appropriate calculation resources must be at hand to accomplish a good design. Manufacturing means and needed tolerance limits and surface quality must be matched to effectively produce the intended design and guarantee its performance. The designer should be aware

of available standards, sets of original equipment manufacturer (OEM), and standard components to effectively use them in the total construction of the design. The design procedure should also account for possible uses of reverse engineering tools as exemplar knowledge so as not to reinvent the wheel particularly for starters or design modifications of products or systems. The designer should also be aware of units, codes, and standards particularly if the design is intended for a foreign client or a certain field with specific operating codes to abide by.

- In the knowledge-based design part of the text, certain introduction to *CAD* and *optimization* is presented. It is necessary to have a thorough overview of these indispensable fields since they are utilized throughout the book and for being common practice to design engineers nowadays. A thorough understanding of the subject and techniques provides a solid foundation to use the tools efficiently and without ambiguity or pitfalls.

  The ability of using CAD software is very essential to completing the design process of products or systems. It increases the usage proficiency, allows the assembly of elements in ***3D***, avoiding interference of geometry, and allows the utility of exporting elements for finite element (***FE***) analysis, ***3D*** printing, and CNC code generation. CAD is used effectively in photo realistic design viewing and presentation. The applications of *virtual reality* render even more help in that regard particularly if more complex geometry variations exist. A demonstration to this effect is also presented.

  Optimization understanding is an important factor in the last stage of redesign. It allows the best design to be implemented rather than the last iteration without a specific objective. The application of optimization tools provides a strong performance of the iteration process without reaching unrealistic solutions. The proper use of optimization successfully allows the optimum design of numerous machine elements for some specific objective or multiple objectives. These optimum elements are offered as a better datum, initial design synthesis, and further reference to other objectives.

  Detailed coverage in those fields with sufficient mathematical treatments is provided in the defined chapters. Please refer to the table of contents for characterization of these details.

- The real *design synthesis* approach is introduced in the **Knowledge-Based Design** part of the book. The necessary knowledge in design foundation is presented. Knowledge about induced stresses and material failure due to applied loading during intended performance is of a paramount importance. This knowledge is used to synthesize the element geometry and select its material. These are the necessary characteristics of element design synthesis.

  Stress, *deformation*, and *deflection* knowledge are essential for full understanding of machine element withstanding of real loading conditions and defining geometry to do that. Different means of loading calculations and estimations are introduced including random load variations to appropriately find representative internal shear and moments inside elements. Stress and deflection considerations are then accounted for and properly evaluated. Simple and combined stresses are considered in evaluating the right state of stress in the element. Simple straight members representing elements or more complex curved members are considered to evaluate stresses and deflections in some machine elements. Strain energy means are also used to find deflections in more complex elements. Some special loading considerations are covered in elements such as columns, variable section members, and thermal effects due to temperature variations. Stress concentration ensued by existing holes, notches, fillets, shoulders, and grooves must be considered in effectively evaluating the rise of internal stresses inside the machine elements. The ability of using ***FE*** programs is then very essential in evaluating internal stresses and different deformations due to complex geometry variations and complex loading conditions. Effective coverage of the finite element technique guarantees the suitable evaluation of machine element responses to loading conditions. A demonstration to this effect is also provided. The finite element process can be used effectively after defining both the geometry and the material that are explained in the subsequent chapters for different machine elements.

  Knowledge of *material properties* facilitates the appropriate selection of the material, which is a significant characteristic of element design synthesis. Materials are defined by specific designation depending on established standards. Knowledge of designation allows the identification of material properties that are presented including structure and failure modes. A large scope of material variations is provided to allow the consideration of a wide variety of applications. Special groups of materials that are more suitable for specific machine elements are predetermined in categories. With this knowledge, the selection of the material for that element is then focused and narrowed down to a few possible choices. Hardness and strength relations provide support to an even narrower selection of the material. Distinction between static and fatigue failure theories allows the distinct definition of the course of action to home in more on the most suitable material to be used. If the case requires contemplation for toughness due to inevitable notch existence, a fracture mechanics approach is provided to account for appropriate safety. Cases of safety consideration and strength variations are examined for failure prevention.

With the knowledge of generated stress and the material selection process, the initial *design synthesis* approach of major machine elements is introduced. The intended initial design encompasses the definition of ***geometry*** and the appropriate ***material*** of the machine element. *CAD* and optimization means are employed with this focused knowledge to directly generate tools (charts, codes, and programs) to define the necessary ***geometry*** (dimensions) of the element. This is the first significant identified characteristic of the initial design synthesis of the element. The initial design synthesis family of elements covers most machine components such as beams, columns, fasteners, springs, shafts, gears, bearings, belts, chains, ropes, clutches, etc. The appropriate selection of ***material*** for each of these elements provides the second significant identified characteristic of the initial design synthesis of the element. For product and system design, reverse engineering tools may provide insight for proper design innovation particularly for inception or design modifications. Components in these products or systems are handled specifically as previously synthesized.

- The traditional rigorous *detailed design* of machine elements can then be enacted after the initial design is synthesized. This process strengthens the concept of the appropriateness of the initial design synthesis. It further defines the more specific external and internal loading conditions and allows the definite evaluation of safety and reliability if needed. The process is straightforward since the geometry and material have been decided upon earlier in the initial synthesis of the element. The treatment is divided into two sections: (A) basic Joints and Machine Elements and (B) Power Transmitting and Controlling Elements. The first section covers screws, fasteners, and permanent joints, springs, rolling bearings, and journal bearings. The second section covers spur, helical, bevel, and worm gears, flexible elements, shafts, clutches, breaks, couplings, and flywheels.

  Detailed coverage of particulars of those elements with extensive mathematical derivations and calculation treatments are provided in the subsequent chapters. Please refer to the table of contents for details.

  It should be noted that the coverage includes *CAD* and optimization of the incorporated machine elements. This is also provided through numerous computer tools in the form of *Excel*© ***Synthesis Tablets*** and MATLAB© *CAD* or ***CAD codes***. Available also is a specialized interactive software (*PanDesign*) that implements the rigorous procedures for some basic elements, applying detailed design iterations and further recursion for optimization in addition to providing assembly of the elements for mostly SI units. This software has been used extensively by students and by few industries in the past.

The approach and means presented in the book intend to accomplish the following:

- Provides the tools to perform a new *direct design synthesis* rather than design by a process of repeated analysis.
- Knowledge-based design utilization of *CAD* tools, software, and optimum component design for the new *direct design synthesis* of machine elements.
- Allows initial suitable design synthesis in a very short time (five minutes for a machine element).
- Utility of *CAD* and optimization to attain better designs.

The synthesis approach has evolved through its implementation in my machine design courses at Cairo University ever since 1988. It has also been implemented for a short time at The American University in Cairo. Other traditional machine design courses have also been taught at several other universities and organizations during the span of my career.

It is expected that if other fields are treated with the same provided concepts and practices, much better designs would materialize.

This text is intended for a wide scope of courses. It can be used at the introductory as well as at the advance level machine design. The book reviews several bases and prerequisites of machine design courses. Prerequisite knowledge is of most regular capstone machine design courses in most engineering schools around the world. The book can also cover more than one semester course. The selection of topics for each course depends on the prerequisite coverage, scope, and objectives of each course. Advanced sections and adequately covered prerequisite materials can be skipped.

Machine design has been my great interest, hobby, profession, research, and educational field since even before my graduation with an engineering degree from Cairo University, Egypt in 1965. Since my graduation, it has continued to be my passion in teaching, consultation, design of mechanical products, developing *CAD* tools, and in design optimization applications. During that time, the concept of *real design synthesis* has evolved and implemented in the machine design courses taught since the 1980s.

Undergraduate engineering major students and junior graduate engineers have limited to little design experience. That is why they needed a simple tool to generate reasonable designs in a short time with no or very little iteration. The design should provide synthesis to geometry as well as suitable material selection. The design has thus far been obtained by utilizing previous knowledge, developed selection charts, *CAD*, and optimization procedures. Through this process, the participants have been able to generate component design in a short time. Almost all machine elements have been designed directly by this knowledge-based concept. Very rigorous checks are performed afterwards (detailed calculations) to guarantee the suitability and robustness of the original initially synthesized design. *CAD* software has also been developed to implement the new concept to several machine elements and system assembly. It also allowed the users to export the ***3D*** geometry and details to other geometric modeling (CAD) software.

It is envisioned that the book can provide students and engineers with the new *CAD* and optimization tools and skills to generate real design synthesis of machine elements and systems on solid ground for better development of products and systems. Many of the engineering graduates had this experience and grasped the introduced concepts and tools. Their feedbacks indicate that they still use their lecture notes as an indispensable reference in their careers.

Cairo, October 2020 *Sayed M. Metwalli*

Acknowledgments

This text has been the culmination of nourishing knowledge, efforts, responses, and support of my teachers, colleagues, and students. It is difficult to name all, but one has at least to remember some and most probably inadvertently miss to mention others. I am indebted to my late teachers H. Fahmy, G. Shawki, S Bayoumi, and others of Cairo University. In my graduate work, it is my pleasure to have had Roger Mayne as my PhD advisor and friend at the University at Buffalo. During my professional career, I would like to acknowledge Douglas Wild of Stanford University, Salah Elmaghraby of NC State, Glen Johnson of Tennessee Tech, Steve Velinsky of UC Davis, Daniel Inman of the University of Michigan, Panos Papalambros of the University of Michigan, Yasser Hosni of UCF, Ahmed Shabana of UIC, Judy Vance of Iowa State, Waguih and Hoda El Maraghy of Windsor University, Mohamed Trabia of UNLV, Georges Fadel of Clemson University, Amr Baz of the University of Maryland, Elsayed Elsayed of Rutgers University, Ashwani Gupta of the University of Maryland, H Ezzat Khalifa of Syracuse University, Shaker Meguid of the University of Toronto, Farrokh Mistree of the University of Oklahoma, Singiresu Rao of the University of Miami, David Rosen of Georgia Tech, Jami Shah of Ohio State, Osman Shinaishin of NSF, Mohamed Zikry of NC State, Mahmoud El Sherif of Drexel University, and many others for inspiring my research and endeavors. Of my colleagues, I would like to cite M. Younan of AUC and I. Fawzi, M. Elaraby, A. Mostafa, H. Arafa, A. Ragab, A. Radwan, M. Mokhtar, M. Said, A. Wifi, and the rest of the department faculty members at Cairo University and so many others for being there. Of my former students and present colleagues, I would like to mention and thank H. Ghoneim of RIT, S. Megahed of Cairo University, A. Elzoghby of Cairo University, M. Sharobeam of Stockton University, A. Nassef of AUC, A. Bastawros of Iowa State, A. El Danaf of Cairo University, H. Hegazi of Cairo University, M. Shalaby of GE, and so many others of my graduate students for their perseverance, tolerance, efforts, and dedication.

I would like to thank Cairo University for providing means and tools to complete the endeavors of this job. Thanks are also due for the Mechanical Design and Production Engineering Department and its members for the understanding of the undertaking efforts needed to complete this work. Thanks are also due for the John Wiley & Sons staff for their patience, tolerance, and dedication in support of producing a proper, consistent, and clear text.

About the Companion Website

This book is accompanied by a companion website:

https://www.wiley.com/go/metwalli/machine

The website includes:

- Supplementary materials for classroom use
- Solutions Manual

Part I

Introduction and Design Considerations

1

Introduction to Design

A design is understood to be an object that fulfils a function or performs a job. Different designs can be of objects other than machines. In engineering, design can be associated with different fields of engineering such as bridges and highways in civil engineering; ship design in marine engineering; electric motors, computers, and communication equipment in electrical and electronic engineering; and so on. In mechanical engineering, machine design, engines, turbines, pumps, and heating or cooling systems are some of the main areas of mechanical design. *Machine design*, however, can be defined as the *process* of getting the design of a machine and its components. The design of a machine can be understood as the machine form and construction as an object that is made to perform the *function* of the machine. The machine is generally a device composed of moving parts to perform a function that consumes or transforms power or energy. A *motorcycle* or a *vehicle* is a machine that transforms energy into moving a person or persons from one point to the other and consumes energy generated by the engine or the motor to do so. The consumed energy is spent mainly in overcoming air and ground resistances, acceleration and deceleration, and potential energy for different elevations up or down if no regenerative mean is present. Figure 1.1 shows an example of a *trans-mixer* design that transports concrete from a mixing station to a construction site. The truck carrying the mixer consumes power in continuous rotation and in transporting the concrete charge over the roads to the construction site. A separate power is usually used to keep the mixer drum continuously rotating during transit, accelerations and decelerations, and pouring the concrete into the desired location (Badawy et al. 1994). The *machine tool* is another design of a complex device consuming energy in producing components by machining or other manufacturing processes. A *mechanism* is also a machine that is used in many equipment and other contraptions.

Machine design in this text is considered as the design process of the machine or the *design*. It involves the process of *assembly* of mechanical components to form a design of the machine. In this text the process to get the component design is done mainly by *synthesis* rather than repeated analysis. The resulting component is the design of the component. The process of getting the design is the *synthesis* of the component. The adopted synthesis process is simple, but it is so sophisticated in the sense of using the maximum utilization of *knowledge-base* and previous results of *optimization*. The handling of different machine elements will demonstrate such a paradigm throughout the text. Assembly of synthesized optimum components into machine systems should evidently produce better machines. This is the obvious expected strategy of a better machine design. In general, this chapter gives an overview of the design process with relations to general machine design and its larger associated fields. It also covers the *standard units* used throughout the text and the fundamentals necessarily needed for the successful understanding through the course of study.

Machine Design with CAD and Optimization, First Edition. Sayed M. Metwalli.

Companion website: www.wiley.com/go/metwalli/machine

Figure 1.1 A trans-mixer design that transports concrete from a mixing station to a construction site. Source: Sayed Metwalli.

Symbols and Abbreviations

The adopted units are [in, lb, psi] or [m, kg, N, Pa], others given at each symbol definition. [k...] is 10^3, [M...] is 10^6 and [G...] is 10^9.

Symbol	Quantity, units (adopted)
$	US dollar
°C	Temperature in Celsius, [°C]
°F	Temperature in Fahrenheit, [°F]
°K	Temperature in Kelvin, [°K]
3D	Three-dimensional
$\boldsymbol{a}$	Acceleration, [m/s^2] or [in/s^2]
A	Area, [m^2] or [in^2]
A	Electric current in ampere [A]
AC	Alternating electric current
CAD	Computer-aided design
CAD	Computer aided drafting
CAE	Computer-aided engineering
CAM	Computer-aided manufacturing
CAS	Computer-aided synthesis
cd	Luminous intensity in candela [cd]
Cell	Cellular phone
CIP	Concurrent idea to product
CNC	Computerized numerical control machine

Symbol	Quantity, units (adopted)
d	Diameter, [m] or [in]
DC	Direct electric current
E	Energy or work, [lb in] or joule [J] ≡ [N m]
$\boldsymbol{F}$, F	Force or force magnitude, [lb] or newton [N]
$\boldsymbol{F}_i$	Force vector i with components F_{ix}, F_{iy}, F_{iz}
$\boldsymbol{F}'$ **or** $\boldsymbol{F}^T$	Transpose of force vector or matrix $\boldsymbol{F}$
FBD	Free body diagram
FE	Finite element
ft	Length in foot, 1 ft = 12 [in]
g	Gravitational acceleration, 9.806 65 [m/s^2] or 386.088 [in/s^2]
H	Power, [lb in/s] or [N m/s] ≡ watt [W]
H_{hp}	Power H in horsepower (hp), [hp] (1 [hp] = 6600 [in lb/s])
hp	Power unit in horsepower, [hp] = 6600 [in lb/s]
Hz	Hertz, one cycle per second, [Hz] or [rev/s] or [s^{-1}]
i	Electric current, ampere [A]
in or ′	Length in inch [in]
ISO	International Organization for Standardization
J	Energy or work unit "joule" [J] ≡ [N m]
lb	Force in pounds of US set of units, pound [lb]
lb$_m$	Mass, US pound mass [lb$_m$]
LED	Light-emitting diode
m	Mass, SI, kilogram [kg], US pound mass [lb$_\text{m}$]
m	Length in meter of SI units, meter [m]
$\boldsymbol{M}_0$	Moment vector at origin point 0
mol	Amount of substance in SI system [mol]
N	Force unit in SI system, newton [N]
N	Rotational speed, [rad/s]
N_{rpm}	Rotational speed in revolutions per minute, [rpm]
N_{rps}	Rotational speed in revolutions per second, [rps]
NBS	National Bureau of Standards
NIST	National Institute of Standards and Technology
OEM	Original equipment manufacturer
p	Pressure, force per unit area ($p = F/A$), [Pa = 1 N/m^2] or [psi = 1 lb/in^2]
Pa	Pressure unit of pascal [Pa] or (1 N/1 m/s^2)
PC	Personal computers
psi	Pressure unit in US system defined as [lb/in^2], pound per square inch [psi]
r	Twisting arm length, [m]
$\boldsymbol{r}_1$	Position vector of application point "**1**"
$\boldsymbol{R}_F$	Resultant of i force vectors $\boldsymbol{F}_i$
R&D	Research and development

Symbol	Quantity, units (adopted)
RE	Reverse engineering
rad	Radian angle (2π rad = 360°) [rad]
s	Time (US or SI), second [s]
SAE	Society of Automotive Engineers
SI	International System of Units, [m, kg, s]
slug	A mass in US set of units, [lb s^2/ft]
STL	Stereolithography or standard tessellation language
t	Time in seconds, [s]
$\boldsymbol{T}$	Torque, [in lb] or [m N], ([lb in] or [N m])
US	United States (customary US units), [in lb s]
v	Velocity, [in/s] or [m/s]
w_{kg}	Weight of a [kg] mass in Newton
w_{lbm}	Weight of a [lb_m] mass in pound [lb]
w_{slug}	Weight of a [slug] mass in pound [lb]
W	Work or energy E, [lb in] or joule [J] ≡ [N m]
Wi-Fi	Wireless fidelity communication network
δ	Displacement, [m] or [mm]
θ	Angle of rotation in radians, [rad]
ω	Angular velocity, [rad/s]
ω_H	Frequency [rev/s] or [s^{-1}], hertz [Hz] or [s^{-1}]

1.1 Introduction

Machine design is an eventual termination field of mechanical engineering. It should utilize all previous knowledge and innovations in mechanical engineering and other related fields. Some contraptions and inventions, however, might result from clever minds without significant knowledge of mechanical engineering subjects. These cases are the exceptions and not the overwhelming rule. The generation of new designs is usually stemming from individuals with ample knowledge of mechanical engineering, in addition to some capacity for inventiveness and critical mind for upgrading existing designs. The main constructing blocks of machine design are the *machine components* or *machine elements*. These are covered herein and in most of general machine design textbooks and references. The thorough analysis of these *machine elements* is the foundation for their proper utilization in a whole machine, a product, or a mechanical system. The machine, or the design, or the system is an assembly of some machine elements to perform certain function or task. The assembly of these *elements* is created from scratch or from an imitation to some comparable assemblies that would perform the same function or task. This *conceptual* and *cognitive* process would require experience and knowledge-based methodology. The intention of this text is not focused on generating products, systems, or machines as such. It focuses on the thorough consideration of the appropriate design or *synthesis* of the basic machine elements comprising those machines. Better design of machine elements should normally produce better assembly of components in machines. These are also provided in some useful applications in here. The goal of this book is to *synthesize* and employ *optimization* in a new procedure of a *real* computer-aided design (*CAD*). This generates methods that exploit knowledge-base and optimization for a *computer-aided synthesis* (CAS) of machine elements. This would almost surely produce better assembly of mechanical systems or designs.

The traditional design process usually employs *repeated analysis* to home into the geometry of the element and the material selection of that element. The repeated analysis is found necessary since most representative analytical models generate expressions that are *implicit* and not usually *explicit* in form. The implicit mathematical models do not allow the design parameters to be explicitly defined in terms of the other input data or parameters. As an example for designing a shaft connecting an engine or a motor to an impeller or a fan, the shaft diameter is not mathematically defined explicitly in terms of all other complex input loads and other essential parameters. This indicates that it is not simple or possible to have the diameter in an equation as equal to a complex expression in all other inputs and parameters. Some of these parameters are usually function of the diameter and cannot be explicitly separated or extracted. Without any previous knowledge of the appropriate solution, iterations are necessary to reach the proper diameter. Our approach resolves this dilemma by employing *preliminary synthesis* that converges close to the proper diameter by employing knowledge-based expectation and simple or formalized optimization. The process is dubbed here as an *initial synthesis*. Refinements of results may need a minimum of further iterations.

Machines, mechanical systems, or products are usually assemblies of some basic machine elements. Standard machine components are some of these machine elements such as *bolts* and *rivets*, standard *beams* or *plates*, and *keys*, *pins*, and *retaining rings*. Some combinations of other basic components form standard machine elements such as *seals*, *ball screws*, *couplings*, *rolling bearings*, and *chains*. Other machine components are mass produced in specific dimensions to help designer use such elements economically such as *springs*, *linings of journal bearings*, some *gears*, *wires*, and *ropes*. In a mechanical system or product design, one can use already available subsystems in a total design configuration such as *engines*, *motors*, *valves*, *actuators*, and even a full *transmission* or a *differential* of a vehicle. Specialized manufacturers provide such subsystems as components of "*original equipment manufacturer* (OEM)" to be assembled in other complex products. Inventions are new systems or components that have patents protections. The use of these inventions should be done with the approval of the inventor or as a whole, which is produced under the inventor's consent.

Design of machines or machine components requires ample knowledge and experience. To offset that demanding prerequisites, *knowledge-base*, *CAD*, *expert systems*, and *optimum design* are utilized to ease the design process. Few prerequisites are necessary to have sufficient background and some creativity to attain good designs. All previous studies of subjects and courses in engineering, mechanical, and related fields are indispensable background knowledge that is very beneficially employed in the design process. Traditionally mechanical engineering curriculum including mathematics, physics, statics, dynamics, graphics, mechanism's kinematics, material science, strength of materials, and others are basics for machine design. Thermodynamics, heat transfer, and fluid mechanics are also needed for the design of such interacting systems and designs for these fields. The underlying tools in these fields are applied mathematics and physics. The need for modeling, calculating, and simulating and other tools are essential to guarantee, verify, and validate such tools and their results. CAD packages are needed to accomplish some of such tasks. Several programs are available to perform some of these needs such as ***3D** geometric modeling*, *finite elements* (***FEs***), *flexible body dynamics*, and other packages. Most of these are, however, analysis rather than design tools. The design is usually attained by repeated analysis then the result is delivered to the ***3D*** geometric modeling programs for further processing such as adjusting or checking by some connected *finite element* program. If one starts with a guess of the ***3D*** model, it will take several finite element reruns to iterate or reach an optimum solution. It would be better to start synthesizing the components and use these as *optimum designs* before the ***3D*** modeling and finite element verifications. It is the intention of this text to adopt this approach and deliver sufficient tools and definite procedures to achieve a significant fraction of this task. Reruns are therefore minimized to only adjust and fine-tune the design to accommodate other special secondary requirements.

1.2 Phases of Design

Usually, the design process intends to end up with a hardware production of a system, a machine, or a product for a certain market. In the serious attempts to produce for a market, the design process should go through three phases. To have a viable competitive product, the first phase of design should be a *feasibility study* to secure the potential market success of intentionally produced design. The second phase is the *initial synthesis* or the preliminary design to select available alternatives and different configurations as initial design proposals. The third phase is the *detailed design* that should generate a final construction of the design and assure its production success. The following details each of these phases in a general sense. The suggested details are not rigid and should be adjustable in specifics according to the field of the intended manufacture of a design. Some of these details may be shifted from one phase to the other or present in more phases with different emphasis. Steps in all phases usually execute evaluation checks and implement feedbacks to home into step outcome satisfaction.

Feasibility study is a crucial phase with envisioned steps as suggested in Figure 1.2. The objective of this phase is to develop a set of beneficial designs that are producible and economically feasible. At first, one is to consider the *market* need for such a design. Therefore, one must perform *needs analysis* via market information and investigation. If there is a need, some desired products can be developed. Using technical expertise and creativity, few credible and appropriate *design concepts* and *system identification* should materialize. Alternative solutions are thus generated, and various engineering statements of the cases are developed. Applying technical skills, the *physical analysis* is accomplished to check if the alternatives are possibly achievable. Realizable alternative solutions are the defined outcome. The *economic* and *financial* feasibilities of the alternatives require economic and financial considerations. This defines if the production is worthwhile. It should also identify the *required capital* for each fabrication line. The outcome conclusion is a set of useful *design alternatives*. If not, one must go back and iterate from the start of the feasibility study.

Initial synthesis is the second phase of the process with steps as illustrated in Figure 1.3. The target is to generate acceptable design proposals satisfying the stipulated needs. Further scrutiny for the selection of best *design concepts* requires experience and technical knowledge. The outcome is some tentative selections of most suitable design concepts to use. Engineering science is then used to generate valid *mathematical models* to adopt for the analytical formulation of the performance and *synthesis* of the designs. This entails generating suitable geometries and materials for the designs of mechanical components and other parameters for other components. Synthesis is to generate or create designs directly rather than designs by repeated analysis. This step is the *major paradigm* in this text as intended throughout. Further mathematical analysis and adjustments of parameters are needed to

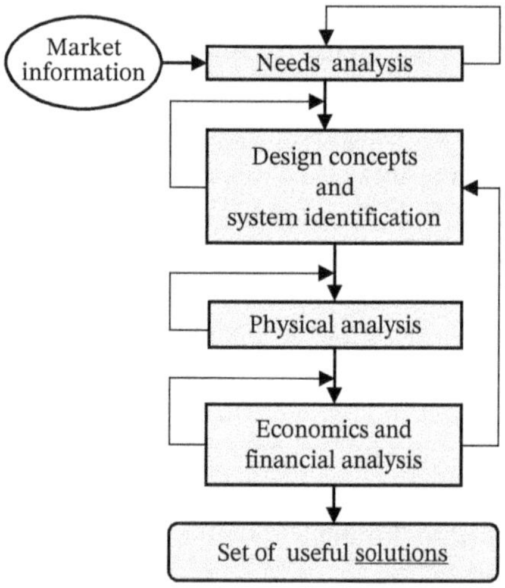

Figure 1.2 Suggested feasibility study steps to control and reiterate feedbacks.

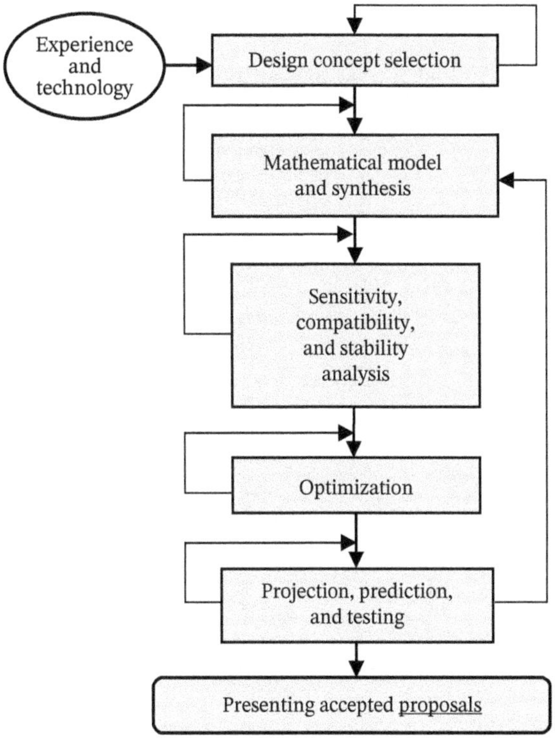

Figure 1.3 Suggested initial synthesis steps to control and reiterate feedbacks.

perform analysis of *compatibility, sensitivity, and stability* of design parameters. This is to define the proper fit between components and find sensitive and stable parameters for more appropriately adjusted design. Furthermore, math and computers are also needed to *optimize* the design for some other objectives in cost or performance. Applying mathematical trend analysis and conducting tests in labs can provide *projection, prediction,* and *assessment* of the design performance into the *future*. If not so satisfactory, one would feedback to reiterate from the beginning of the synthesis. With that, a better performance and expectations are attained for the accepted *design proposals*.

Detailed design is the last phase of the process to develop improved optimum design that can be produced and introduced into the market as shown in Figure 1.4. *Preparations* for such a design necessitate experience and produce the necessary *budget and organization*. By means of that and technical knowledge, *subsystems, components, and parts descriptions* are realized to guarantee a good fit. Specifications, ***3D*** *models*, or *drawings* are subsequently the outcome. The ***3D*** models or the assembly drawings are defined by utilizing technical experience and software. These create complete engineering description of the design that fits the requirements. The *construction* of the design and its *testing* should then need workshops and labs to produce a prototype. *Redesign and optimization* are further required to generate a better and improved *optimum design* by employing more mathematical and technical knowledge. If outcome is not satisfactory, one would feedback to reiterate from the beginning of the detailed design process. Some aspects of redesign might also require going back to the initial phase of synthesis if some new features have not been already accounted for or covered in that phase.

1.3 Basic Mechanical Functions

So many mechanical functions are studied throughout the course structure of mechanical engineering curriculum. Several if not all are necessarily considered in machine design. Depending on the type of design at hand, basic

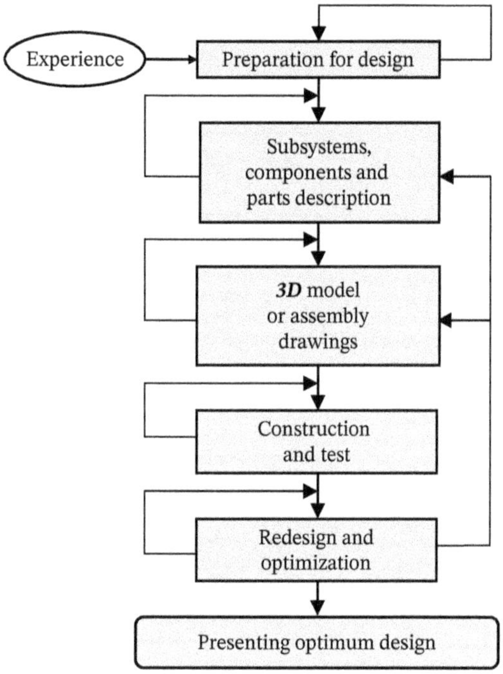

Figure 1.4 Suggested detailed design steps to control and reiterate feedbacks.

Table 1.1 Basic mechanical functions and some of their detailed elemental functions for a complex machine.

Basic function	Elemental functions
Supporting	Attaching, motion constraining, pivoting, removable fastening, limiting, continuous rolling, latching, fastening, load distributing, force limiting, flexible supporting, and sliding
Power transmitting	Force or torque or motion transmitting, friction reduction, energy transforming or absorbing, linking, coupling, clutching, liquid or gas transferring, pumping, deflecting, lighting, and electrical or signal conducting
Enclosing	Protective covering, shielding, pressure supporting, covering, flexible spacing, and shape constraining
Sealing	Sealing, liquid constraining, contaminant constraining, gas constraining, filtering, variable position maintenance, and vibrations or sound and thermal insulation
Sensing	Force sensing and indicating, displacement or position sensing and indicating, pressure and temperature sensing and indicating, information indicating, and viewing
Controlling	Guiding, stabilizing, pressure increasing or limiting, motion reduction or damping or limiting, partitioning, position restoring, disconnecting, gas switching, power absorbing, sound absorbing, damping, gas guiding, force or position maintaining, energy storing, liquid or gas storing, torque limiting, electric switching, electrical amplifying or limiting, and electrical reducing or insulating
Form aesthetics	Appearance, streamlining, artistic shape and color, viewing, and spacing

mechanical functions and other few related non-mechanical functions can be observed as shown in Table 1.1 as one example of a complex mechanical system such as a helicopter, an airplane, or a vehicle. Categorizing these elemental functions into basic functions simplifies tackling of complex machines and gives insight into the basic treatments of these functions. The basic mechanical functions that are of concern in this text are the supporting, power transmission, enclosing, and sealing functions. The machine components that are considered herein serve these functions. Beams, columns, plates, rolling and journal bearings, and even shafts are used as supporting elements. Screws; bolts, rivets, keys, splines, pins, retaining rings, permanent welded or bonded joints, and springs are also types of supporting elements. They join other elements and cause one to be supported by the other and carry loads. Springs, in addition, can also be used as part of sensing and controlling elements. Power transmission components that are the major thrust of this text are elements such as power and ball screws, couplings, gears, belts and chains, wires and ropes, shafts, flywheels, clutches, and breaks. Flywheels, clutches, and breaks, however, are also considered as power controlling elements. Enclosing and sealing is extremely important in machine design. Most of power transmission elements and other machines require housing on frames or in some enclosure. With components connected to these housings or enclosures, one usually need sealing to prevent fluids or unwanted particles or materials to be exchanged with either internal or external environment. Even though sensing and other controlling elements would have mechanical functions, they are usually treated in the measurements and control courses. Few of our basic machine elements are used extensively in the design of these systems. The treatment presented herein can be very useful in that regard. Form, texture, and color aesthetics are very essential in so many designs of products or machines that are used by regular consumers. In designing teams of these products, there should be some artistic talents to satisfy the aesthetics requirements.

1.4 Design Factors

Of a paramount importance in the design process are factors to be observed, considered, or must be satisfied. Design performance is an objective for the designer to fulfill and one of the most important and indispensable factors. The goal to satisfy is that the product design should perform the intended task. Constructional details are necessary to execute the intended task with the required form, dimensions, and component materials. The design should also be produced and assembled to perform the task. The product design must also be within a reasonable cost for intended users. It, therefore, should also be economical to operate and execute the intended task. Some details of these factors and concerns are as follows:

A. **Performance**:
   It is imperative that performance should be guaranteed first and foremost. The intended performance utility in terms of productivity, efficiency, accuracy, and so on is the main objective of the design. Safety and reliability of the produced design should warrant the continuous performance for the expected life of the product. Full control over performance is also expected. Environmental aspects and factors such as noise level or emission should be adhered to. The impact of the design on the environment raised concerns and warrants consideration for an added specific objective of design for the environment (Kutz 2007). Other considerations of specific performance are emphasized such as design for reliability, maintainability, sustainability, and life-cycle optimization (Kutz 2007; Metwalli 2009).
B. **Constructional Details**:
   In constructing the design assembly, important details of the design factors are in order. The main factors are as follows:
   i. Form and dimensions such as shape and geometry, size, and weight are important details that some might be optimized for space, lightness, or cost. Styling (aesthetics) is of paramount significance for market suitability, acceptability, and likeness.
   ii. Appropriate selection of materials is vital to withstand different loading conditions such as forces or heat. The strength, stiffness, or rigidity of the system or its components should be carefully calculated with safety and reliability evaluated. Appropriate material selection for intended machine elements are given in Chapter 7. An introduction to safety and reliability is further scrutinized in Section 1.4C.
   iii. To manufacture the design, *production processing* of each component is selected during the design procedure. Surface finish and dimensional tolerances are defined to suite the function and the mating fit of

other interacting elements. The surface finish and tolerance would necessitate the selection of the suitable production process and the proper machine tool. Details of fits, tolerance assignment, surface finish, and related manufacturing processes and methods such as casting, sheet metal work, machining, etc. are covered in Section 2.4.

iv. *Assemblies* of components or subsystems to complete the design require the considerations of proper fits between components and needed fittings to assemble. Assembly procedure and disassembly for maintenance should be carefully accounted for during the design process. Most geometric modeling software has this check as available options; See Section 2.2.

C. **Safety and Reliability**:
The design should perform the intended function safely and reliably. The safety implies that the maximum conditions of loading are within the allowable limits. The adoption of an appropriate *safety factor* guarantees that all possible loading conditions are accounted for. All loads can be tolerated and should be within the allowable boundaries. The *safety factor* is the ratio of the design failure limit and the allowable loading the design is exposed to. For multiple components, it is expected that the optimum safety factor should be about the same for all components.
Reliability is a measure that has a value related to the randomness of both loading conditions and properties of selected materials. It is the higher probability percentage of design survival relative to a lower probability of failure occurrence. To quantify the reliability, *probabilistic design* procedure is applied. A limited treatment in that field is presented later on. Usually a higher safety factor increases the reliability. An extensive treatment of probabilistic design is beyond the scope of this text; see also Metwalli et al. (1983, 1989).

D. **Cost and Economy**:
The total cost of the product or the machine including running cost is an important factor for the successful entrenchment in the market. Reliability can reduce the need for maintenance. Friction, wear, and corrosion affect running and maintenance cost with the need for lubrication replacement, scheduling, or concern. These affect product economy and the perception of quality.

1.5 Synthesis Approach to Design

This section presents an introductory *synthesis* approach to the design of machines that also conforms to the phases of design in Section 1.2. The main target and objective is to have the intended *function* or functions as the persistent primary goal. Without achieving the objective tasks and functions, the design is considered a failure and useless. The common rules and procedure to synthesize components or systems are general ones pertaining to attitude, perceptions, and reason. The goal is to synthesize a design for a better *performance* and other objectives such as *cost*. All brain storming tools, imagination, previous knowledge, *reverse engineering* (RE), dissecting with sketching of *free body diagrams* (***FBDs***), and suitable analysis such as *finite element* (***FE***) codes are geared toward achieving the design objectives.

General procedure to implement *synthesis* tasks involves the study of function or functions to be attained by the design. The function is usually dissected into sub-functions or steps, if the faction cannot be attained directly in one step. All concepts, previous knowledge of similar options, and alternatives should be scanned and scrutinized by tools such as reverse engineering (RE) to home into the most appropriate ones. To synthesize, one has to account for the external loading and the model representing the system to be designed. All standard parts, components, and subsystems are exploited to attain reasonable cost. *CAD* or – better off – CAS tools are used to find appropriate geometry, topology, and material for each component of the design. Verification tools such as *finite element* (***FE***) codes are to validate the safety of the synthesized design. The manufacturability and maintainability of the design has a priority during the synthesis process to guarantee that the design is producible and maintainable. Computer-aided manufacturing (CAM) and computer numerical control (CNC) machine codes or standard tessellation language (STL) code for ***3D** printing* of components can generate prototypes to check on those and other concerns of design form and performance.

1.6 Product Life Cycle

The usual product life cycle is shown in Figure 1.5. This cycle is valid for any product with the product life cycle curve varying according to the specific product and may not necessarily be smooth. The curve mostly represents the sales of the product in the market. The scales of time, sales, and returns are only indicative, relative, and even nonlinear. The regions for introduction and growth due to development can be steeper or slower depending on the relative need and reception of the market to product. For instance, growth can be fast for new attractive products such as movies, personal computers (PC) or tablets, cell phones, and the Internet. The maturity duration can be very long if the product is a satisfaction of general basic needs such as transportation using internal combustion engines for vehicles, alternating current (AC) electricity for filament light bulb illumination, and cable phones for communications. The decline occurs when a new alternative is introduced such as solar-electric vehicles, direct current (DC) light-emitting diode (LED) lamps, cell phones, and Wi-Fi or satellite Internet, maybe solar-electric (DC) grid or solar-lighting grid from daylight regions to the night regions on planet Earth, or a closed loop *force generator* that may replace internal combustion, jet, and rocket engines. This is when those innovative designs create new markets and starts a new product life cycle. At this stage the returns per product are usually high due to the new innovation. The development of the product generates more competitive outfits that add to the growth of product sales in the market to reach maturity. During this growth period, the returns are usually declining due to the competition. The stage of maturity is very competitive and requires extensive optimization to reduce cost and improve sales margins by having more reliable and robust products to increase market share. The returns are usually modest, and the outfits should be of very large and of massive production size to have worthwhile returns. Usually outfits in this case would have departments of *research and development* (***R&D***) that introduce several product developments or new product innovations to create new markets or sustain and improve market share.

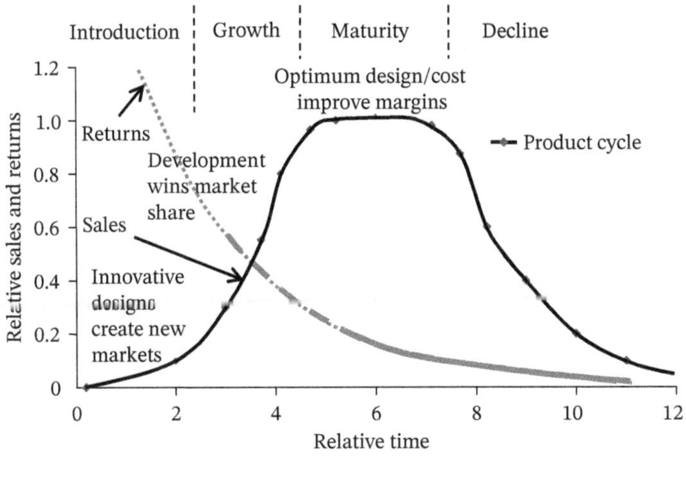

Figure 1.5 Sketch of the usual product life cycle indicating stages of innovative introduction, growth due to developments, maturity, and decline.

1.7 Business Measures

Economics of scale indicates that as the size of production increases, the cost per unit decreases. That may entice a decline in product price, which may also increase sales worth. This interaction is postulated in Figure 1.6 for prospective production of a *new product* where unit cost scale is obviously different from the other scale of costs and sales. As the number of manufactured units increases, the unit cost approaches that of its material cost. The total percentage of ***R&D***, design, labor, and overhead cost to the material cost is reduced to a few percentages as the numbers of units greatly increase. Figure 1.6 shows the *break-even point* where sales worth is equal to the total cost without any return. This indicates incurring losses below that size of production. Gains occur beyond the break-even point and that should define the appropriate size of production far beyond the break-even point. As that size of production increases, the gain increases almost exponentially.

The cost of the product affects the proper business measures as it defines the sales margin, which is the percentage difference between the price and cost. The price is usually defined by the competitors' perceptive ranking. The ranking of the producer usually defines the market share it holds in similar products. With the world as a *global village*, the total market size is the global size and not the local one. The returns or profit of the business can be defined by the following relation of the main three components (Marks and Riley 1995):

$$\text{Global Market Size} \times \text{Market Share} \times \text{Sales Margin} = \text{Profit} \tag{1.1}$$

As an illustrative example of a global market size of 200 million units, a large market share of 5% and a sales margin beyond cost of 15% gives a profit of $1.5 million for each dollar of product cost with the sales of 10 million units and a 15% markup over cost. For a product to be successful, trend in each of the three business measures provide strength, weaknesses, opportunities, and threats. A producer should maintain or improve at least one of the three business measures. A better product should be competitive in price and quality. This can only be achieved by a better or optimum design. This should boost both *market share* and *sales margin* in due time.

To have a better or an optimum successful product, one should use the most advanced technology to reach that goal. The productivity of such advanced technologies is much higher than contemporary or conventional technology. It is therefore imperative to employ innovations and more advance CAS, *computer-aided engineering* (CAE), and concurrent idea to product (CIP) technology with the utilization of ***3D*** printing and advance nanocomposite smart materials. These advancements are used in a ***R&D*** environment to access the highest industry returns due to advanced products or moderate industry returns due to regular product developments.

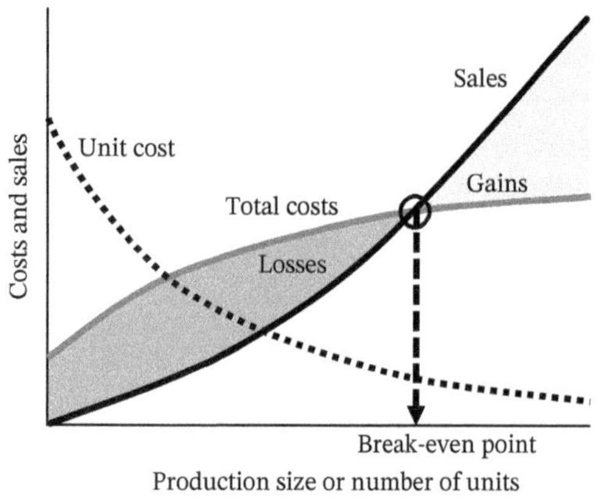

Figure 1.6 The interaction between the number of produced units and unit cost, total cost, and total sales value for prospective production of a new product. Unit cost scale is different from other costs and sales scale.

1.8 Research and Development Process in Product Cycle

Product cycle is different than product life cycle (Section 1.6). The product cycle pertains to the process of getting the product to the market. The product life cycle is the duration in time a product is needed in the market from the time of its market introduction to the time the market does not ask for it as in Figure 1.5. The product cycle is depicted in Figure 1.7 with the start at the market needs of customers. Naturally, the process of getting the product to the market goes through design, detailing, production process planning, and scheduling before the production with a suitable quality control as shown in the inner loop of Figure 1.7 (Groover and Zimmers 1984). The design or synthesis step requires product concept selection or innovations to have the design synthesized in addition to the infrastructure deemed necessary for that.

Figure 1.7 shows the necessary infrastructure needed particularly the ***R&D*** in the design stage and experts backed by software in all cycle infrastructure steps. Without these infrastructures, it is not possible to have a real and viable product delivered to the market. This starts with research of market needs, cost, economics, and financial experts to satisfy consumer needs with least cost for overwhelming competitors. ***R&D***s are essential to secure innovations and advanced design synthesis. The ample spending on ***R&D*** is compensated for and rewarded by great returns particularly for huge production size. Further, it is important to use advanced technology, computer software, and experts to deliver first-rate engineering details, superior manufacturing process planning and scheduling, precise production, and impeccable quality control. Details of these infrastructures are shown in Figure 1.7 and further detailed information belongs to *industrial* and *manufacturing* engineering fields with ***CAD/CAM*** interaction.

The design synthesis, detailing, and production planning and scheduling are usually accounts for licenses provided by original produces to some other small outfits or franchisee to assemble the product. The main beneficiary is usually the original producer who did the innovative design, the engineering, and the ***R&D*** of the product. This provides the experience and develops the experts to generate the next developed or innovated product. Seldom,

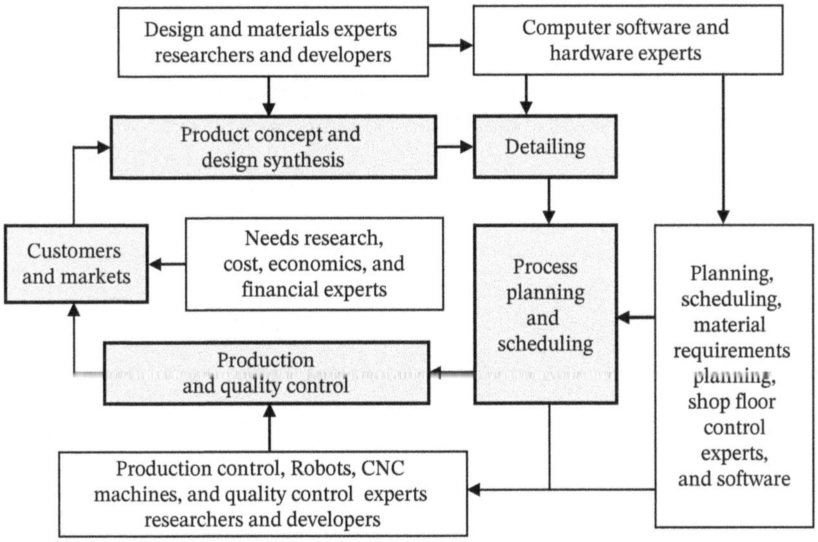

Figure 1.7 The infrastructures required for each step of the product cycle from the market needs of consumers to delivering the product to the market. Source: Adapted from Groover and Zimmers (1984).

these innovations or inventions sprout or get to the market without the support of the ***R&D*** provided by strong producers or sponsors. That is why governments and large companies spend heavily in ***R&D*** to entice the creation of advanced new products for the market with a technological superiority to capture the high introduction returns, larger market share, and sales margin (see Figure 1.5).

1.9 Teamwork for Product or System Design

Few jobs are possibly done by one individual. Most jobs are too large for anyone to finish by himself in a reasonable time if ever. More than one or few individuals should then cooperate to accomplish larger jobs. These few individuals constitute a team that work together in a *teamwork* style with an objective to get the job done in an effective way in due time. Each one of the team should do a part of the job and provides his expertise in conjunction with others without overlap but with integration and matching to complete the job. For large jobs, this might need a teamwork leader to organize the different tasks and coordinate among members to ensure amalgamation and matching without overlap. Products or system design or synthesis typically demands teamwork. Referring to Figure 1.4, the preparation for design requires budget and organization. The organization is a conglomerate teamwork consisting of many teams. Each team performs a specific task to accomplish such as parts of the steps in Figure 1.4 or Figures 1.2 and 1.3. Teamwork procedure starts with *goal analysis and specification* to formulate the *strategy* to accomplish the *goal* or task. Team members render their knowledge and capabilities to support teamwork. Sharing these traits improves teamwork *productivity*. During work, monitoring and coordination is continuous to manage conflicts, motivate, and build confidence (Marks et al. 2001; LePine et al. 2008).

1.10 Design and Development Case Study

A very interesting case study by Done (1989) reveals a real-world situation for the ***R&D*** of a design project. The case is some years ago but it demonstrates the concepts covered in this introduction, and it is somewhat generic with respect to time. The case reports about *Rover*, a company that went through the development process in a relatively long period of time to conclude few important outcomes of the (***R&D***) practice. The objective was to develop an engine for a class of vehicles the company intended to produce. After spending 11 years in ***R&D*** at an investment program cost of 240 million British pounds ($377 million at that time), the company decided to retain its ***R&D*** capability to develop its own engine. However, the other outcome of the ***R&D*** is the company's decision to use another company's transmission gearbox for the new class of vehicles. The case study demonstrated the need for ***R&D*** organization and teamwork with ample budget and time to have calculated decisions about self-design or procurement of other more established subsystems as OEM. The time and money were spent to carry out the design process with all its phases, factors, considerations, and business measurers. Considering moving into market through competition and cooperation with leading manufacturers; the company created a special management, developed a product offensive with large investment program, and invested in a highly automated engine plant.

Nowadays, it might not need such a lengthy time with the availability of a great deal of software for the analysis and simulations. Nowadays the development of a new car model may take about 18 months rather than the usual 5 years. Now, the software applications extend to so many calculations of components and subsystems and simulations and render help in visualization and decision making, which reduces the time and cost needed for a decision. With the *synthesis* procedure adopter herein, the time and money to design optimum components and systems should be even less.

1.11 Units and Fundamentals

Units are extremely important means to consistently define variables and parameters. In machine design and other fields, they have to be unified to be able to exchange components, to manage maintenance, and to have presence in the *global market*. That is why the world has been keen to establish a unified system of units, which is the SI system. Few countries are still using their own system of units (the classical English system). The relations between units are tied to fundamental relations in mechanics, thermal, fluid, chemical, electrical, physics, and all other fields. The importance of units has been further recognized when a $125 million NASA space probe was lost because of a mishap in converting US to metric units (The Boston Globe 1999). Here the system of units of the US, however, and the international SI units are carefully covered. Basic fundamental relations are presented and verified in both systems.

Fundamentals of design also relate to units in ***3D*** space variables, where they are better defined by *vectors* and *matrices*. The fundamentals of vectors and matrices are reviewed herein, since they are dependent on the units defined in this section and extensively used in this text.

1.11.1 Units

The common set of *US units* is used extensively in the US even though there have been early attempts to use the international *SI units*. The SI abbreviation is coined from the French name "Le **S**ystème **I**nternational d'Unités" (ISO 2009). This International System of Units (**SI**) is used worldwide and has developed over the years from the metric units of *meter* [m] for length, *kilogram* [kg] for mass, and the *second* [s] for time. The common **US** set of units adopted in this text are the *inch* [in] for length, *pound* [lb] for force, and *second* [s] for time. Both systems are reviewed in this section and will be alternatively used throughout the text; see also NBS (1968) and NIST (2008). The basic quantities, units, and ratio of both systems are given in Table 1.2. The ratio is simply the division of the magnitude of the SI unit over that for the US unit. If this ratio is greater than 1, the SI unit is then larger than the US unit. In *conversion* of units between US and SI, one can use this ratio to get the desired unit from the other known value. For engineering calculations and particularly for design purposes, only five digits are usually used in the ratio of SI to US units. Values of these ratios described in more digits can be found in references, Taylor and Thompson (2008) and SAE (1999). In this text the *units* are always given *between brackets*. Other quantities or variables are usually given in Italic symbols. This can be observed in the *Symbols* section at the beginning of this chapter.

Other basic quantities such as the *ampere* [A] for *electric current*, *Kelvin* [°K] for *absolute temperature*, *candela* [cd] for *luminous intensity*, and the *amount of substance mole* [mol] are similar in both systems of units. For more details, one can consult further references, NBS (1968) and NIST (2008, 2010). For temperature [°K], both systems have the same scale for *Celsius* [°C] with scale shift of 273.15 such that

$$[°C] = [°K] - 273.15 \tag{1.2}$$

For the US temperature unit of *Fahrenheit,* the following relation is used:

$$[°F] = (9/5)[°K] - 459.67 \tag{1.3}$$

Table 1.2 Basic quantities, units, ratio of US to SI systems, and conversion ratio.

Quantity	US set	SI set	Ratio SI/US or (conversion ratio = 1)
Length	inch [in or ′]	meter [m]	39.370, 1 [m] = 39.37 [in] or (0.0254 [m]/[in])
Mass	slug [1 lb s^2/ft] [lb$_m$]	kilogram [kg]	0.068 522, 1 [kg] = 0.0685 [slug] or (14.594 [kg]/[slug])
			2.2046, 1 [kg] = 2.2056 [lb$_m$] or (0.4536 [kg]/[lb$_m$])
Time	second [s]	second [s]	1
Force	pound [lb]	Newton [N]	0.224 81, 1 [N] = 0.224 81 [lb] or (4.4482 [N]/[lb])

Table 1.3 Conversion ratios of selected fundamental quantities (continued from Table 1.2).

Quantity	US system	SI system	Conversion ratio = 1
Acceleration a	[in/s^2]	[m/s^2]	0.025 40 [m/s^2]/[in/s^2]
Area second moment I_A	[in^4]	[m^4]	0.416 23 × 10^{-6} [m^4]/[in^4]
Density ρ	[lb$_m$/in^3]	[kg/m^3]	27 680 [kg/m^3]/[lb$_m$/in^3]
Heat energy	[Btu]	[N · m] ≡ [J]	1055.056 [J]/[Btu]
Mass m	Slug, [lb s^2/ft]	[kg]	14.594 [kg]/[slug]
Mass m	[lb s^2/in]	[kg]	175 [kg]/[lb s^2/in]
Mass m	[lb$_m$]	[kg]	0.453 592 37 [kg]/[lb$_m$]
Mass moment of inertia I_m	[lb$_m$ in^2]	[kg m^2]	2.9264 × 10^{-4} [kg m^2]/[lb$_m$ in^2]
Power H	[lb in/s] or [hp]	[N m/s] ≡ [W]	745.7 [W]/[hp]
Pressure p or stress σ, τ	[psi]	[Pa]	6894.8 [Pa]/[psi]
Section modulus Z	[in^3]	[m^3]	16.387 × 10^{-6} [m^3]/[in^3]
Spring rate or stiffness k_s	[lb/in]	[N/m]	175.13 [N/m]/[lb/in]
Torque T	[in lb] or [lb in]	[m N] or [N m]	0.112 985 [m N]/[in lb]
Velocity v	[in/s]	[m/s]	0.0254 [m/s]/[in/s]
Weight w	[lb]	[N]	0.453 592 37 [N]/[lb]
Work W and energy E	[lb in] or [lb in]	[N m] ≡ [J]	0.112 985 [J]/[lb in]

Derivative quantities such as area, volume, torque, pressure, work, energy, power, etc. and their respective SI/US ratios are obtained from basic units defined in Table 1.3. These basic quantities are of vital importance to machine design field. It is essential to review these quantities, their definitions, and the specific relations between these different quantities. The units of these quantities are ***derived*** from their definitions and relations as given next.

1.11.1.1 Force and Mass

The main *basic/derivative* quantity is the *force* ***F***. It is a basic *pound* force [lb] in the **US** units, but in the **SI** units, it is a derivative unit given the abbreviation name of "*newton*" [N] in appreciation to Sir Isaac Newton (1642–1726/7). The *newton* [N] is defined as the force required to accelerate 1 [kg] of mass m a 1 [m/s^2] of the acceleration $\boldsymbol{a}$. The addition of the unit of *newton* [N] is also due to the *Newton's* law relating the force $\boldsymbol{F}$, the mass m, and the acceleration $\boldsymbol{a}$ (with $\boldsymbol{a}$ = standard gravitational acceleration "g" of 9.806 65 ≅ 9.81 [m/s^2]) as defined in Eq. (1.4):

$$\boldsymbol{F} = m\boldsymbol{a} \tag{1.4}$$

The customary **US** units usually use the *pound* force [lb] as a basic unit and in some cases the "*slug*" unit as the mass. The "*slug*" therefore has the units of $(\boldsymbol{F}/\boldsymbol{a})$, where $\boldsymbol{a}$ is in [ft/s^2] and the *slug* is then in [lb s^2/ft] as indicated by the application of Eq. (1.4). Therefore, the weight of *slug* is equal to 32.17 [lb]. In that case the acceleration $\boldsymbol{a}$ is depicted as [ft/s^2] rather than [in/s^2]. To use the units of [in/s^2] for the acceleration, the mass will then be different than a *slug*. However, one may encounter the name of *pound mass* or [lb$_m$] in some applications. The *pound mass* [lb$_m$] is the amount of substance that has the weight of 1 pound [lb] at a standard gravitational acceleration g of 32.174 049 [ft/s^2]. Care and consistency should be exercised when using units in the customary US system of units. For engineering calculations, it is suggested to stick with one US system of units. The adoption of [lb] for the *pound force* is usually designated as "p" in the known [psi], which is [lb/in^2]. The force [lb] is sometimes abbreviated as a pound force [lbf] to be distinguished from the pound mass [lb$_\mathrm{m}$]. Herein, we ***only use*** [lb] for the *force* and [lbm] or better [lb_m] for the *mass*, if we need to. Adoption of the *inch* [in] as the unit of *length* should cause all other quantities to be in inches or converted to inches. The acceleration units should be [in/s^2], and the gravitational acceleration should be g = 386.088 [in/s^2] instead of "g" = 32.1740 (or 32.2) [ft/s^2]. The adoption of the *basic* units of ***pound–inch–second*** [lb, in, s] as the *basis* of the customary US units would eliminate the need for the "slug" unit of the mass. Therefore *one ought to change all of the variables or quantities to the* ***adopted US units*** of [lb, in, s] before proceeding into any further calculations. This can make life much easier, cautious, and consistent.

For the **SI**, the [kg] defines the unit of the *mass m* and the "newton" [N] defines the unit of the *force* $\boldsymbol{F}$. The force is considered as a derivative defined off the *basic* units of ***meter–kilogram–second*** [m, kg, s] for length, mass, and time. The force $\boldsymbol{F}$ is obtained from Newton's law as in Eq. (1.4). Since forces are used and encountered extensively, the derived value of [kg s^2/m] is given the abbreviation "*newton*" [N] as indicated before and shown in Table 1.2.

1.11.1.2 Pressure

The *pressure p*, or the normal *stress*, which is simply defined as the force per unit area can be obtained as

$$p = \boldsymbol{F}/A \tag{1.5}$$

where A is the area.

The **US** system usually defines the pressure unit as [psi], which is pound per square inch. Applying Eq. (1.5), the units of pressure p is then [lb/in^2] or pound/in^2, thus the name [psi]. The unit of [psi] will be kept in this text since it is widely used and its definition conforms to the basic units used throughout the text. The same unit of [psi] can also be used for the normal *stress*, which will be further defined later in the text (see Chapter 6).

The pressure p in the **SI** units is a simple substitution in Eq. (1.5) to get [N/m^2]. This is given the name "*pascal*" or [Pa] as an abbreviation and in appreciation to Blaise Pascal (1623–1662).

1.11.1.3 Velocity, Acceleration, and Rotational Speed

The essential quantities that are used extensively in mechanics are the *velocity* $\boldsymbol{v}$, *acceleration* $\boldsymbol{a}$, rotational speed in revolutions per minutes N_{rpm}, and *angular velocity* $\boldsymbol{\omega}$. The velocity $\boldsymbol{v}$ is simply the distance traveled over time. The velocity $\boldsymbol{v}$ has then the US units of [in/s]. The acceleration $\boldsymbol{a}$ is simply the rate of change of velocity $\boldsymbol{v}$ over time. The acceleration $\boldsymbol{a}$ has then the US units of [in/s^2]. The *angle* of rotation θ is usually given in *radian* [rad], which is about 57.296°. If degrees are used, one should use the factor $(2\pi/180)$ to convert degrees to radians. The angular velocity $\boldsymbol{\omega}$ is defined as the rate of change of rotation θ with time. It is an important quantity that is frequently used. It is obtained from the usually known rotational speed N_{rpm} (revolution per minute) through the relation shown in Eq. (1.6):

$$\omega = 2\pi N_{rpm}/60 \tag{1.6}$$

where $\boldsymbol{\omega}$ is defined in radians per second [rad/s] or [rad s^{-1}] for both US and SI systems. In other applications such as vibratory systems, the quantity ω_H is the frequency defined as revolution per second [rev/s] or [s^{-1}] that is given the name "*hertz*" [Hz] in appreciation to Heinrich Hertz (1857–1894).

The velocity $\boldsymbol{v}$ in SI units is just [m/s], the acceleration $\boldsymbol{a}$ is [m/s^2], the speed N_{rpm} is [rpm], and the speed N_{rps} is revolution per second or [rev/s] usually [rps], and both are defined by the angular velocity $\boldsymbol{\omega}$ as [rad/s] as shown in Eq. (1.6) for N_{rpm} and multiplying N_{rps} by 2π to get $\boldsymbol{\omega}$.

1.11.1.4 Moments, Work, and Power

The units of other *derived* quantities such as *torque* $\boldsymbol{T}$ or moment $\boldsymbol{M}$, work W or energy E, and power H are evaluated using the basic units and the equations defining each quantity. The power H has been assigned a specific symbol of "*watt*" [W], in appreciation to James Watt (1736–1819), or the classical *horsepower* [hp]. It is obvious that the work W is not the same as the power H, even though the units of power H can have the unit of "watt"

[W]. The units, again, are non-italic characters given between brackets, while the symbols of the quantities are given in *italic*.

The ***torque T***, or ***moment M***, is defined by the twisting action or moment of the force $\boldsymbol{F}$ due to its arm $\boldsymbol{r}$ from its point of application to the point of its action as defined by Eq. (1.7):

$$\boldsymbol{T} = \boldsymbol{r} \times \boldsymbol{F} \tag{1.7}$$

where × is the *cross product* as presented in Section 1.11.3. Similarly, the moment $\boldsymbol{M}$ is given by Eq. (1.8):

$$\boldsymbol{M} = \boldsymbol{r} \times \boldsymbol{F} \tag{1.8}$$

The torques and the moments are then having the units of [in lb] (or [lb in]) in the US system.

The work W is simply defined by the amount of energy E exerted by the force $\boldsymbol{F}$ when it moves a displacement δ as shown in Eq. (1.9):

$$W = \boldsymbol{F} \cdot \delta \tag{1.9}$$

where the "·" is the *dot product* as presented in Section 1.11.3. The ***work*** W and the ***energy*** E are then having the units of [lb in] in the **US** system. Comparing the work to the torque, one finds that both are defined in a similar form and thus both would have the same units [lb in]. If need be, to differentiate between the units of the torque [in lb] and the units of the work W [lb in], one may have a space between the force in pound [lb] and the length [in] for the units of the work W as [lb in] instead of the dot in-between [in · lb] and [lb · in] for the torque $\boldsymbol{T}$. The use of [in lb] for torque $\boldsymbol{T}$ against the [lb in] for work may provide some distinction between $\boldsymbol{T}$ and W; however, both have essentially the same units [lb in]. Then, one may also use the customary [lb · in] for torque with the dot rather than a space for work or energy if one needs to differentiate.

In the **SI** system, the units of the *torque* $\boldsymbol{T}$, or *moment* $\boldsymbol{M}$, is obtained as [m N] through Eq. (1.7) or (1.8). The *work* W or *energy* E has the units [N m] from Eq. (1.9). Also, one may then use the ***customary*** [N · m] for torque with the dot or the space for work or energy. The units of energy E or work W [N m] is distinctively given an abbreviated name "*joule*" [J] in appreciation to James Joule (1818–1889). This *joule* [J] is equal to [N m] or [N · m].

The ***power*** H is defined by the rate of dispensing energy or doing work with time. It can be obtained simply from Eq. (1.10), defining the rate of change of work W with time.

$$H(t) = \frac{dW}{dt} \tag{1.10}$$

For a constant force $\boldsymbol{F}$, substituting Eq. (1.9) into Eq. (1.10) gives the following alternative form that can be easily utilized:

$$H(t) = \boldsymbol{F} \cdot \frac{d\delta}{dt} = \boldsymbol{F} \cdot \boldsymbol{v}(t) \tag{1.11}$$

where the "·" is the *dot product* as presented in Section 1.11.3.

The **US** system of units for *power* H is then derived from the units of force $\boldsymbol{F}$ [lb] and the units of velocity $\boldsymbol{v}$ [in/s]. This gives the units of power H as [lb in/s]. For rotating elements, one can use the torque $\boldsymbol{T}$ [in lb] or [lb in] and the angular velocity or rate of rotation $\boldsymbol{\omega} = (d\theta/dt)$ [rad/s] in the power expression of H to get Eq. (1.12):

$$H = \boldsymbol{T} \cdot \boldsymbol{\omega} \tag{1.12}$$

where the "·" is the *dot product* as presented in Section 1.11.3. The power H will then have the same units of [in lb/s], which is the same as [lb in/s]. A special unit for the power H is the *horsepower* [hp] that is extensively used in the US as well as SI units. The definition of this quantity involves some history that might have been given in earlier subjects and can be checked in numerous references; see also Internet list at the References section. The relation used for the power H in horsepower [hp] is defined as the work rate of 33 000 [ft lb/min], 550 foot-pounds per second, or 6600 [in lb/s], which is also given by Eq. (1.13):

$$H_{\text{hp}} = \frac{2\pi N_{\text{rpm}} \boldsymbol{T}_{\text{lb ft}}}{33\,000} = \frac{N_{\text{rpm}} \boldsymbol{T}_{\text{lb in.}}}{63\,025} = \frac{N_{\text{rps}} \boldsymbol{T}_{\text{lb in.}}}{1050} = \frac{\boldsymbol{T}_{\text{lb in.}} \cdot \boldsymbol{\omega}}{6600} \tag{1.13}$$

The units of the horsepower H_{hp} can then be given in [lb in/s], which is the same units of the power.

The units of *power H* in the **SI** system of units is [N m/s] as derived from Eq. (1.11) or (1.12). A unit of "*watt*" [W] ≡ [N m/s] is assigned to the power unit as an abbreviation and in appreciation to James Watt (1736–1819) as previously indicated. One *horsepower* [hp] is equal to 745.7 watts [W], which is also obtained through converting US [hp] to SI system in watts [W]; see Table 1.3. Also, the *horsepower* [hp] is then equal to 0.7457 kilowatt [kW], which is 1000 [W].

1.11.1.5 Weight

The gravitational acceleration in the **SI** system ($g = 9.806\,65$ [m/s^2]) is not used, per se, in defining the unit of force [N]. The *newton* [N] is defined as the force required to accelerate 1 [kg] of mass m a 1 [m/s^2] of the acceleration $\boldsymbol{a}$. In that case the weight $\boldsymbol{w}_{\text{kg}}$ of 1 [kg] mass is given by Eq. (1.14), where the magnitude $\boldsymbol{a}$ is the standard gravitational acceleration g (taken as 9.806 65 [m/s^2]):

$$\boldsymbol{w}_{\text{kg}} = m\boldsymbol{a} = 9.806\,65\ [\text{N}] \tag{1.14}$$

In that case, 9.806 65 [N] = 1 [kg m/s^2].

The **US** *weight* $\boldsymbol{w}$ of 1 [slug] of mass is given by an equation similar to Eq. (1.14), where $\boldsymbol{a}$ is the gravitational acceleration g (taken as 32.1740 (or ≈32.2) [ft/s^2] instead of $g = 386.088$ (or ≈386) [in/s^2]). The weight of 1 [slug] is then

$$\boldsymbol{w}_{\text{slug}} = m\boldsymbol{a} = 32.1740\ [\text{lb}] \tag{1.15}$$

The *weight* $\boldsymbol{w}_{\text{lbm}}$ of 1 [lb$_m$] mass is given by a similar to Eq. (1.15), where $\boldsymbol{a}$ is the gravitational acceleration g (taken as 386.088 588 (or 386) [in/s^2] instead of $g = 32.174\,049$ (or 32.2) [ft/s^2]). The *pound mass* [lb$_m$] is the amount of substance that has the weight of 1 pound [lb] at a standard gravitational acceleration. The weight of 1 [lb$_m$] is then

$$\boldsymbol{w}_{\text{lb}_m} = m\boldsymbol{a} = \frac{[\text{lb}]}{386.088\,588}386.088\,588 = [\text{lb}] \tag{1.16}$$

That is why one should be very careful when using the mass in the US system of units. Table 1.3 devotes few entries for the mass conversion in the US to SI system of units.

1.11.1.6 Prefixes

The prefixes are used in both SI and US systems of units to account for very large and very small values of the quantities. The major ones are the value prefixed by G for *giga* or (10^9), M for *mega* or (10^6), k for *kilo* or one thousand (1000) times the value, m for milli or one thousandth (0.001) the value, and the *micro* μ or one millionth (10^{-6}) the value. These are used in this text, and it is recommended to convert other values to those indicated previously, namely, G, M, k, m, and μ. Other prefixes exist and can be found in literature and abbreviated in Appendix A.2. Note that the **non-italic** form is used for these prefixes of units inside the brackets such as [MPa] for *mega*-pascal and [mm] for *milli* meter. All other quantities or variables are usually specified in *italic*.

All derived quantities in SI system of units are straightforward substitution in their definitions or relations. That is why the SI system is regarded an *absolute system of units*. This has been previously demonstrated for developing the SI units of several *derived* quantities.

1.11.2 Unit Conversion

To convert from US to SI system of units or from SI to US system of units, one is advised to use the basic units for each and the ratio of SI/US depicted in Table 1.2. In addition to that, one should use the relations between units in each system; see SAE (1999) and NIST (2010). Table 1.3 provides conversion factors for the selected basic quantities in the column of SI/US conversion ratio = 1.

The number of significant figures in the quantity is a factor to consider when converting units between systems. Using approximate conversion factor is useful but the resulting truncation may not be acceptable particularly if several conversions are implemented at the same time. For design purposes the use of four or five significant figures is generally sufficient when defining geometry or dimensions. These would usually be rounded or changed to closest standard value after calculations anyway.

The conversion is performed by sequential multiplication of ratio factors between quantities where each ratio factor has the value of 1. Table 1.2 shows the ratio SI/US for length as (1 [m] = 39.37 [in]). The ratio factors having the value of 1 for length conversions are then (39.37 [in]/[m]) and ([m]/39.37 [in]). These are used in the conversion process with either the numerator or the denominator cancelled out sequentially to get the sought units. Similar conversion factors are derived from Table 1.2 and from quantities in either SI or US system of units such as (12 [in]/[ft]) = 1 or (1000 [mm]/[m]) = 1. The process is shown in Examples 1.1 and 1.2.

Conversion ratios of some frequently used quantities are given in Table 1.3. Other conversion factors related to the subjects covered herein are given latter in the text and are also available in literature; see NIST (2010), SAE (1999), and the Internet references (http://www.iso.org/iso/home.html; http://standards.iso.org/ittf/PubliclyAvailableStandards/; https://en.wikipedia.org/wiki/Conversion_of_units; https://en.wikipedia.org/wiki/Horsepower; https://www.nist.gov/pml/weights-and-measures/metric-si/unit-conversion).

Example 1.1 In a design problem, the quantities given are as follows: the drive power H is 25 [kW] and the rotational speed N_{rpm} is 1575 [rpm]. It is required to evaluate the torque $\boldsymbol{T}$ in both SI and customary US units. What is the equivalent power H in horsepower [hp] and in pound–inch–second units?

Solution

Data: $H_{kW} = 25$ [kW], $N_{rpm} = 1575$ [rpm], conversion Tables 1.2 and 1.3

1. Convert the given quantities to the adopted set units:

$$H = 25\,[\text{kW}] = 25\,[\cancel{\text{kW}}][1000\,[\text{W}]/[\cancel{\text{kW}}]] = 25\,000\,[\text{W}] \tag{a}$$

The strikethrough units are cancelled out.

$$N_{rpm} = 1575\,[\text{rev/minute}] = 1575\left[\frac{\text{rev}}{\cancel{\text{minute}}}\right]\left[\frac{1\,\cancel{\text{minute}}}{60\text{ seconds}}\right] = 1575/60[\text{rev/s}] = 26.25\,[\text{rps}] \tag{b}$$

$$= 26.25\left[\frac{\cancel{\text{rev}}}{\text{s}}\right]\left[\frac{2\pi\text{ rad}}{\cancel{\text{rev}}}\right] = 164.9\,[\text{rad/s}] \tag{b'}$$

The strikethrough units are cancelled out.
Alternatively, from Eq. (1.4),

$$\omega = 2\pi N_{rpm}/60 = 2\pi(1575)\,[\text{rad}/\cancel{\text{min}}]/[60\,[\text{s}]/[\cancel{\text{min}}]] = 164.9\,[\text{rad/s}] \tag{c}$$

which is the same as converting N_{rpm} to [rad/s] with the factor $(2\pi/60)$. This factor is about equal to (0.105). To quickly check the calculations, one can divide N_{rpm} by 10 to get a rough estimate of $\approx\boldsymbol{\omega}$. The value of the angular velocity $\boldsymbol{\omega}$ should be a little more than that.

2. The torque is obtained from Eq. (1.12) and substituting the respective values from Eqs. (a) and (c) to get

$$\boldsymbol{T} = H/\boldsymbol{\omega} = 25\,000/164.9 = 151.6\,[\text{m N}] \text{ or } [\text{N m}] \tag{d}$$

To convert the torque to the US units, substitute the ratio for each unit from Table 1.2 such that

$$\boldsymbol{T} = 150.6\,[\text{m N}] = 151.6\,[[\cancel{\text{m}}]\,(39.37\,[\text{in}]/[\cancel{\text{m}}])\cdot[\cancel{\text{N}}]\,(0.2248\,[\text{lb}]/[\cancel{\text{N}}])] = 1341.7\,[\text{in lb}] \text{ or } [\text{lb in}] \tag{e}$$

The strikethrough units are cancelled out.

3. Use Eqs. (1.13), (b), and (e) to convert the power H from [kW] to horsepower [hp],

$$H_{\mathrm{hp}} = N_{\mathrm{rps}}\, \boldsymbol{T}_{\mathrm{lbin}}/1050 = 26.25\,[\mathrm{rps}]\,1341.7\,[\mathrm{in\ lb}]/1050 = 33.54\,[\mathrm{hp}] \tag{f}$$

Alternatively, one can use the information that 1 [hp] = 745.7 [W] such that

$$H_{\mathrm{hp}} = H\,[[\cancel{\mathrm{W}}]\,([\mathrm{hp}]/745.7\,[\cancel{\mathrm{W}}])] = 25\,000/745.7\,[\mathrm{hp}] = 33.53\,[\mathrm{hp}] \tag{g}$$

This is about the same as the value in Eq. (f).
Use Eqs. (1.12), (e), and (c) to get the power H in pound–inch–second units:

$$H = \boldsymbol{T} \cdot \boldsymbol{\omega} = 1341.7\,[\mathrm{in\ lb}] \cdot 164.9\,[\mathrm{rad/s}] = 221\,246\,[\mathrm{in\ lb/s}] \tag{h}$$

From Eqs. (g) and (h), one can get the [hp] in terms of [in lb/s], or

$$[\mathrm{hp}] = [[\cancel{\mathrm{hp}}]\,(221\,246\,[\mathrm{in\ lb/s}]/33.54\,[\cancel{\mathrm{hp}}])] = 6598.5\,[\mathrm{in\ lb/s}]$$

This is about the same as the 6600 [in lb/s] in the definition of the horsepower [hp] in Eq. (1.13). Divide Eq. (h) by 6600 to get H_{hp} such that

$$H_{\mathrm{hp}} = 221\,246\,[\cancel{\mathrm{in\ lb/s}}]/6600\,([\cancel{\mathrm{in\ lb/s}}]/[\mathrm{hp}]) = 33.52\,[\mathrm{hp}] \tag{i}$$

One should note the deviations in the calculated values. The variation is in the last digit that can be an acceptable value in engineering calculations. For more accuracy, more significant figures should be used in quantities and conversion factors.

Example 1.2 A pipe has an internal diameter of 150 [mm]. The internal pressure is 0.35 [MPa]. Find the total force $\boldsymbol{F}$ at its flange covered end. Evaluate the quantities in both SI and US system of units. Estimate the conversion factor of the pressure p between SI and US units.

Solution
Data: $d = 150$ [mm], $p = 0.35$ [MPa], conversion Tables 1.2 and 1.3

1. Convert the given quantities to the adopted units:

$$d = 150\,[\cancel{\mathrm{mm}}]\,([\mathrm{m}]/1000\,[\cancel{\mathrm{mm}}]) = 0.15\,[\mathrm{m}], \quad p = 0.35\,[\cancel{\mathrm{MPa}}]\,(10^6\,[\mathrm{Pa}]/[\cancel{\mathrm{MPa}}]) = 350\,000\,[\mathrm{Pa}] \tag{a}$$

The strikethrough units are cancelled out. In terms of the customary US units these values are

$$d = 0.15\,[\cancel{\mathrm{m}}]\,(39.37\,[\mathrm{in}]/[\cancel{\mathrm{m}}]) = 5.91\,[\mathrm{in}] \tag{b}$$

$$p = 350\,000\,[\cancel{\mathrm{N}}/\cancel{\mathrm{m}}^2]\,(0.2248\,[\mathrm{lb}]/[\cancel{\mathrm{N}}])/(39.37\,[\mathrm{in}]/[\cancel{\mathrm{m}}])^2 = 350\,000\,(0.2248)/(39.37)^2 = 50.76\,[\mathrm{psi}] \tag{c}$$

2. Find the internal cross-sectional area A of the pipe at the flange covered end:

$$A = \pi d^2/4 = \pi(0.15)^2/4 = 0.017\,67\,[\mathrm{m}^2] \tag{d}$$

$$A = \pi d^2/4 = \pi(5.91)^2/4 = 27.43\,[\mathrm{in}^2] \tag{e}$$

3. The total force $\boldsymbol{F}$ at the flange covered end of the pipe is obtained by the following:

$$\boldsymbol{F} = pA = 350\,000\,(0.017\,67) = 6184.5\,[\mathrm{N}] \tag{f}$$

$$\boldsymbol{F} = pA = 50.76\,(27.43) = 1392.35\,[\mathrm{lb}] \tag{g}$$

4. The conversion factor between SI and US units for the pressure p is estimated from the ratio between the pressures in both cases as found in Eqs. (a) and (c).

$$\text{Ratio of SI/US units for } p = 350\,000/50.76 = 6895\ [\text{Pa}]/[\text{psi}] \tag{h}$$

This would indicate that 1 [psi] = 6895 [Pa], which is about the same ratio found in Table 1.3 and shows that the [psi] unit is much larger than the [Pa].
If prefixes are used, one can find that the ratio 6895 [MPa]/1000 [kpsi] is 6.895 [MPa]/[kpsi] or 1 [MPa] = 6.895 [kpsi] (or [ksi] in some references). In conversion ratio, this gives (6.985 [MPa]/[kpsi]); see Table 1.3.

1.11.3 Vectors and Matrices

In ***3D*** space it is more convenient to define variables by their components in *Cartesian coordinates* rather than their *magnitude* and *direction*. Figure 1.8 represents a diagram for a door-handle model subjected to the maximum opening load $\boldsymbol{F}_1$ acting at point "1" as shown. The lock reached its maximum position as the handle ends up in the shown location. The lock is fixed in the fixed door that is not shown but the origin 0 of the coordinate system is considered fixed at that point and assumed as the ground or support. The applied force is better defined by a vector $\boldsymbol{F}_1$ with its *components* $[F_{1x} \quad F_{1y} \quad F_{1z}]^T$. The symbol "^{T}" on top right corner of the force vector indicates that we are considering the vector to be a *column vector* and we are writing its *Transpose*. The transpose of a column vector sets the vector in a *row vector* configuration rather than the usual column vector form. From Figure 1.8, it is obvious that only the y component of the force $\boldsymbol{F}_1$ is present or $\boldsymbol{F}_1 = [0 \quad -F_{1y} \quad 0]^T$. This generalizes the definition of the applied force in its component values and also indicates that the component in the y direction is opposite to the positive reference coordinate direction y. It also indicates that no components of the force $\boldsymbol{F}_1$ are present in the x or z directions. Figure 1.8 also shows the *position* of point "1" as a *ray vector* $\boldsymbol{r}_1$, which is defined by its components $[r_{1x} \quad r_{1y} \quad r_{1z}]^T$. Location of any ***3D*** point "***P***" in space are usually defined by its *ray* vector $\boldsymbol{r}_i$ starting from the origin 0 of the coordinate system (x, y, z) and ends at the coordinates of the position point (x_P, y_P, z_P). The position vector of point "1", i.e. $\boldsymbol{r}_1$, is then $[x_1 \quad y_1 \quad z_1]^T$. In equation form the applied force vector $\boldsymbol{F}_1$ and the application position vector $\boldsymbol{r}_1$ of the applied force at point "1" are defined by the following expressions:

$$\boldsymbol{F}_1 = \begin{bmatrix} F_{1x} \\ F_{1y} \\ F_{1z} \end{bmatrix} \text{ and } \boldsymbol{r}_1 = \begin{bmatrix} r_{1x} \\ r_{1y} \\ r_{1y} \end{bmatrix} \tag{1.17}$$

Any of these can be set in the transpose form such as

$$\boldsymbol{F}_1 = [F_{1x}\ F_{1y}\ F_{1z}]^T = \begin{bmatrix} F_{1x} \\ F_{1y} \\ F_{1z} \end{bmatrix} \text{ and } \boldsymbol{r}_1^T = \begin{bmatrix} r_{1x} \\ r_{1y} \\ r_{1y} \end{bmatrix}^T \tag{1.18}$$

The adopted convention in our text is to usually use ***bold*** face for ***vectors*** and lowercase font for the *components* of the vector with subscript indicating the direction of each component. Since these components are also vectors in the direction of the coordinates, one may use a ***bold*** face for these components such that $\boldsymbol{F}_{1x} = F_{1x}\,\boldsymbol{i}$, $\boldsymbol{F}_{1y} = F_{1y}\,\boldsymbol{j}$, and $\boldsymbol{F}_{1z} = F_{1z}\,\boldsymbol{k}$, where $\boldsymbol{i}, \boldsymbol{j}, \boldsymbol{k}$ are the unit vectors in x, y, z directions. The location of these components in the vector can then eliminate the need to specify the $\boldsymbol{i}, \boldsymbol{j}, \boldsymbol{k}$ unit-vectors in x, y, z directions.

At the handle support in Figure 1.8, the unknown reaction force vector $\boldsymbol{R}_0$ is better defined by its components $[R_{0x} \quad R_{0y} \quad R_{0z}]^T$. The components of the reaction vector $\boldsymbol{R}_0$ are shown rather than the vector itself. The moment vector $\boldsymbol{M}_0$ is also indicated but without showing its components $[M_{0x} \quad M_{0y} \quad M_{0z}]^T$.

The ***matrices*** are usually two-dimensional arrays. Arrays, however, can have several dimensions or coordinates. A load *matrix* $[\boldsymbol{F}]$ can be defined by its indexed components F_{ij}, where i is the row index, and j is the column index. All the loads acting on the body can be set in one matrix such that the index i represents each of the different n

force vectors $\boldsymbol{F}_1, \boldsymbol{F}_2, \ldots, \boldsymbol{F}_n$, and the j index represents the x, y, z components of each force. In a matrix form and for three loads ($n = 3$), this loading case $[\boldsymbol{F}]$ can be defined as the following *force matrix*:

$$[\boldsymbol{F}] = \begin{bmatrix} F_{1x} & F_{1y} & F_{1z} \\ F_{2x} & F_{2y} & F_{2z} \\ F_{3x} & F_{3y} & F_{3z} \end{bmatrix} \quad \text{or} \quad [\boldsymbol{F}]' = [\boldsymbol{F}]^T = \begin{bmatrix} F_{1x} & F_{2x} & F_{3x} \\ F_{1y} & F_{2y} & F_{3y} \\ F_{1z} & F_{2z} & F_{3z} \end{bmatrix} = \begin{bmatrix} \boldsymbol{F}_1 & \boldsymbol{F}_2 & \boldsymbol{F}_3 \end{bmatrix} \tag{1.19}$$

If more convenient, the forces can be specified as columns rather than rows in the force matrix $[\boldsymbol{F}]$. This form uses the matrix $[\boldsymbol{F}]$ in its *transpose* configuration $[\boldsymbol{F}]'$ or $[\boldsymbol{F}]^T$. The resultant of these forces' $\boldsymbol{R}_F$ is simply obtained by summing the components of these forces (Eq. (1.19)) in x, y, and z coordinates. In a matrix or vector form, one can get

$$\boldsymbol{R}_F = \sum_{i=1}^{3} \boldsymbol{F}_i = \boldsymbol{F}_1 + \boldsymbol{F}_2 + \boldsymbol{F}_3 = \sum_{\text{over rows}} \boldsymbol{F}' = \begin{bmatrix} \sum_{i=1}^{3} F_{ix} \\ \sum_{i=1}^{3} F_{iy} \\ \sum_{i=1}^{3} F_{iz} \end{bmatrix} \tag{1.20}$$

This is an *element-by-element* addition over each row for the force-matrix set as columns of forces, i.e. the matrix $\boldsymbol{F}'$.

The ***dot product*** of two vectors produces the *projection* of one vector onto the other. It is the *transpose* of one vector multiplied by the other vector. The result is a scalar quantity. In a vector form, a dot product of vector $\boldsymbol{F}_1$ onto itself produces the square of its *magnitude* such that

$$\boldsymbol{F}_1 \cdot \boldsymbol{F}_1 = \boldsymbol{F}_1^T\ \boldsymbol{F}_1 = \begin{bmatrix} F_{1x} & F_{1y} & F_{1z} \end{bmatrix}^T \begin{bmatrix} F_{1x} \\ F_{1y} \\ F_{1z} \end{bmatrix} = F_{1x}{}^2 + F_{1y}{}^2 + F_{1z}{}^2 = |\boldsymbol{F}_1|^2 \tag{1.21}$$

The product is obtained by multiplying each column component of the row by the corresponding component of the column and adds each of the multiplicity to the others as shown in Eq. (1.21). The number of columns in the first vector should be equal to the number of rows in the second vector. This is to have a ***conformable*** inner matrix dimension for multiplication to be carried out.

The *dot product* of two *unit vectors* provides the *direction cosine* between the two vectors. The direction cosines of a vector are then the component of the vector in each coordinate x, y, z divided by its magnitude. The direction cosines of the force vector $\boldsymbol{F}_1$ are then $F_{1x}/|\boldsymbol{F}_1|$, $F_{1y}/|\boldsymbol{F}_1|$, and $F_{1z}/|\boldsymbol{F}_1|$. These are the cosine of the angle between the vector and each of the coordinates x, y, z.

The ***cross product*** of two vectors produces a vector normal (or perpendicular ⊥) to the plane of the original two vectors. In Figure 1.8 the moment of the force $\boldsymbol{F}_1$ about the origin 0 is obtained by the cross product of its

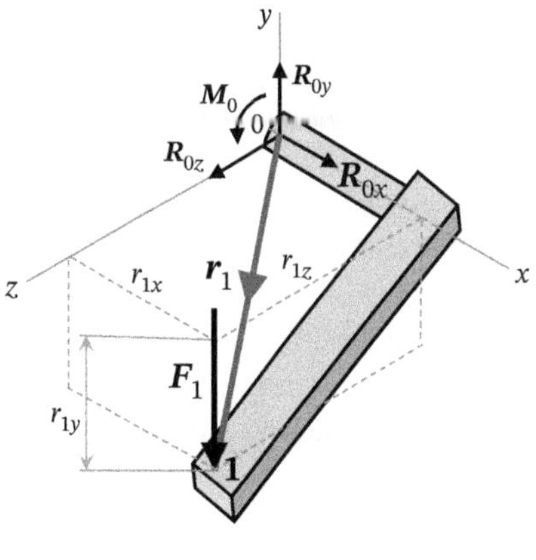

Figure 1.8 Door handle under the maximum opening load $\boldsymbol{F}_1$. The applied load is acting at point "1." The position vector of point "1" is the vector $\boldsymbol{r}_1$. Support reaction vectors $\boldsymbol{R}_0$ and $\boldsymbol{M}_0$ are shown at the origin 0.

position vector $\boldsymbol{r}_1$ times the force. This *cross product* is best handled by the following matrix product of the *special equivalent position matrix* and the force vector such that

$$\boldsymbol{r}_1 \times \boldsymbol{F}_1 \equiv [\boldsymbol{r}_1 \times]\boldsymbol{F}_1 = \begin{bmatrix} 0 & -r_{1z} & r_{1y} \\ r_{1z} & 0 & -r_{1x} \\ -r_{1y} & r_{1x} & 0 \end{bmatrix} \begin{bmatrix} F_{1x} \\ F_{1y} \\ F_{1z} \end{bmatrix} = \begin{bmatrix} -r_{1z}F_{1y} + r_{1y}F_{1z} \\ r_{1z}F_{1x} - r_{1x}F_{1z} \\ -r_{1y}F_{1x} + r_{1x}F_{1y} \end{bmatrix} \qquad \textbf{(1.22)}$$

where the matrix $[\boldsymbol{r}_1 \times]$ is the *special equivalent matrix* or ***equivalent cross product matrix*** for the cross-product of the *position vector* $\boldsymbol{r}_1$ as defined in Eq. (1.22). The same multiplication procedure of the two vectors of Eq. (1.21) is used for every row of the special $[\boldsymbol{r}_1 \times]$ matrix in Eq. (1.22). Note that a ***conformable*** or *consistent* inner matrix dimension is satisfied for multiplication to be carried out. The outcome produced in Eq. (1.22) is verified by the usual *determinant* expansion process regularly used to evaluate the cross products, which is

$$\boldsymbol{r}_1 \times \boldsymbol{F}_1 = \begin{vmatrix} \boldsymbol{i} & \boldsymbol{j} & \boldsymbol{k} \\ r_{1x} & r_{1y} & r_{1z} \\ F_{1x} & F_{1y} & F_{1z} \end{vmatrix} = (r_{1y}F_{1z} - r_{1z}F_{1y})\boldsymbol{i} + (r_{1z}F_{1x} - r_{1x}F_{1z})\boldsymbol{j} + (r_{1x}F_{1y} - r_{1y}F_{1x})\boldsymbol{k} \qquad (1.23)$$

where $\boldsymbol{i}$ is the unit vector in the x direction, $\boldsymbol{j}$ is the unit vector in the y direction, and $\boldsymbol{k}$ is the unit vector in the z direction. The cross product in Eq. (1.22) is much easier to evaluate and manipulate in a computer program.

During the previous review, the product of vectors and matrices is inclusively demonstrated. Further treatments of vectors and matrices can be checked off prerequisite subjects or other references. In this text, however, more applications and utilities will be given in due course.

1.12 Summary

This chapter introduced an outline of design with accent on machine design and its broader related fields. More focused consideration is given to the initiation of machine design or synthesis. Common phases of design are presented including feasibility study, initial synthesis, and detailed design. The *initial synthesis* aims at generating or creating the designs directly rather than design by repeated analysis. This step is the major paradigm in this text as intended throughout. Basic mechanical functions are characterized, and some detailed elemental functions are classified accordingly. Main design factors defined the necessary concerns that are significantly studied while designing. The approach and the procedure of design are presented with an eye on the full scope of the design cycle from the idea to the final production and the introduction into the market.

The chapter also covered the wide extent and critically related fields to machine design. The background knowledge of these fields is needed to make sound judgments in design achievement. It is vital to situate the design importance in the product life cycle and the product cycle. Understanding product life cycle gives impetus, reflection, and feedback into the design process and requirements. Business measures are essential to the success of the design in satisfying needs and providing deserving stature to the design. The role of ***R&D*** in innovative design outcome is highlighted. Necessary infrastructure means in the product cycle support the design process and provide indispensable tools to good design accomplishment. Teamwork is emphasized in the course of developing complex designs or large systems. A real-world case of ***R&D*** process to design such systems is presented to demonstrate the applicability of the concepts and treatment presented in this chapter.

As a significant and basic topic, the units and basic fundamentals are covered in this chapter. The units are reviewed for both US and SI systems. Converting between US and SI systems is clearly established. The utility of vectors and matrices is also demonstrated as an introduction and basic definitions. They are extensively used in the course of the text due to their power in computer applications and coding.

Problems

1.1 What is the difference between *conceptual* and *cognitive* design process? Which is used in each design process?

1.2 Find the necessary information to consider in the design of a motorcycle.

1.3 A trans-mixer design consideration is defined in Badawy et al. (1994) and Figure 1.1. Discuss the main problem considered and the method of solution used in that reference. Can the case of that reference be considered as the main design problem of this equipment? What should be the primary performance goal of the trans-mixer design?

1.4 Define the expected subsystems and components in the design of the trans-mixer of Figure 1.1. Suggest the main components for a manufacturer of the trans-mixer to produce and the other components to acquire as standard or as OEM. Are there other alternatives for the trans-mixer configurations and alternatives? What are the powers driving the trans-mixer drum and the usual truck carrying the drum for different sizes of the concrete loads?

1.5 What type of loading conditions you expect the trans-mixer to be subjected to on its journey from the initial unloaded state to the loading state at the mixing station and from the mixing station to the construction site?

1.6 Select a machine tool and define its main subsystems and components. Identify the expected loading types and variations.

1.7 Few mechanisms are usually used in dump trucks. Select one of these mechanisms and define its main subsystems and components. What are the expected loading variations and extremes?

1.8 Define the expected basic mechanical functions in the systems and subsystems of the trans-mixer of Figure 1.1.

1.9 Identify the basic mechanical functions in each of the subsystems in a motorcycle.

1.10 Select a machine tool in the workshop or lab and study the subsystems defining the basic mechanical functions in each subsystem and its connections.

1.11 Which are the major performance objectives in the design of motorcycles, trans-mixers, machine tools, and dump trucks? Can you suggest a categorized hierarchy from the necessary to the required objectives if there are more than one?

1.12 What are the safety and reliability measures anticipated in each of the designs of motorcycles, trans-mixers, machine tools, and dump trucks?

1.13 Is it possible to guess or find out the cost and economy of the motorcycles, trans-mixers, machine tools, and dump trucks? What is the usual price range for each excluding off-range values of special products if any?

1.14 Would you prefer design by repeated analysis or by synthesized means to aid in defining geometry and selection of materials? Identify your reasons for your selection.

1.15 Select a very old technology of a design that is not used nowadays and a very advanced technology of a design that you expect to be used in the future suggesting each product life cycle with past and future time.

1.16 Study one of the following products, namely, motorcycles, trans-mixers, machine tools, and dump trucks (or any other similar product). What is the global market size last year or the year before? What are the market shares of the top producers? Is it possible to guess the sales margin from the profit of the top producer's annual budget? What is the amount of research and development (***R&D***) spending by the producer in its budget? Is it possible to identify the (***R&D***) teams in the producer's employee structure?

1.17 What do you suggest to make sure that no confusion or errors in the applications of mass, weight, and force in the US or SI systems of units? Did you experience a problem or problems in applications?

1.18 Define the relations to the adopted ones (US [lb, in, s] and SI [m, kg, s]) among the following units:
- Ton (long), ton (short), ounce [oz] to [lb].
- Mile, yard, foot, to [in].
- Gallon, quart, pint to [in^3].
- Yard^3, ft^3 to [in^3].
- Hour, minute to second [s].
- Degree, minute, second to radian [rad].
- Liter to meter^3 [m^3].
- Bar of pressure to pascal [Pa] and [psi].

1.19 What is the factor of the following related to the basic unit: tera, deci, centi, pico, and femto?

1.20 A vehicle is traveling at 70 miles per hour. What is the vehicle speed in [in/s], [m/s], and [km/s]?

1.21 The trunk space of a vehicle is 50 ft^3. What is the volume in [in^3], liter, and [m^3]?

1.22 The drive power H in a power station is 15 000 [kW], and the rotational speed N_{rpm} is 3000 [rpm]. It is required to evaluate the torque $\boldsymbol{T}$ in both SI and customary US units. What is the equivalent power H in horsepower [hp] and in pound–inch–second units?

1.23 The power H of a fan is 0.075 [kW], and the rotational speed N_{rpm} is 1000 [rpm]. Evaluate the torque $\boldsymbol{T}$ in both SI and customary US units. What is the power H in horsepower [hp] and in pound–inch–second units?

1.24 A cylinder with an internal piston has a diameter of 6 [in]. The internal pressure can have a maximum value of 3000 [psi]. What is the maximum force the piston rod can deliver in US and SI units?

1.25 A manual air pump cylinder has an internal diameter of 1.5 [in]. If the maximum force that can be exerted on the piston rod is 60 [lb], what is the maximum pressure that can be delivered by the pump? If the pump is to be used to pressurize a tire to 35 [psi], what should be the maximum necessary force exerted manually to do the job?

1.26 Solve Problem 1.25 using the SI units.

1.27 Use the determinant expansion of the cross product in Eq. (1.23) to verify the equivalent *position matrix* multiplication of Eq. (1.22) of the text.

1.28 A body is subjected to the three force vectors of $[3 \quad 0 \quad 0]'$, $[0 \quad 4 \quad 0]'$, and $[0 \quad 0 \quad 5]'$ in [lb]. Find the force matrix and the resultant of the combined force vectors on the body in US and SI units. What is the magnitude of the resultant force in US and SI units? Find the direction cosines of the resultant.

1.29 The maximum applied force on the door handle of Figure 1.8 is 20 [N], and the applied force position vector is $[0.05, 0.05, 0.12]^T$ in [m]. What is the maximum moment at the base of the handle?

1.30 Solve Problem 1.29 using the US system of units.

References

Badawy, H.S., Mourad, S.A., and Metwalli, S.M. (1994). Finite element analysis for a CAD of a concrete mixer drum. In: *Proceedings of the ASME International Computers in Engineering Conference* (11–14 September 1994) (ed. K. Ishii), 741–747. Minneapolis, MN: American Society of Mechanical Engineers (ASME).

Done, K. (1989) Rover unveils first new engine for 11 years. *Financial Times* (30 August), p. 16.

Groover, M.P. and Zimmers, E.W. (1984). *CAD/CAM Computer Aided Design and Manufacturing*. Prentice Hall.

ISO 80000 (2009). *Quantities and units*. International Organization for Standardization (ISO). Geneva, Switzerland.

Kutz, M. (ed.) (2007). *Environmentally Conscious Mechanical Design*. Wiley.

LePine, J.A., Piccolo, R.F., Jackson, C.L. et al. (2008). A meta-analysis of teamwork processes: tests of a multidimensional model and relationships with team effectiveness criteria. *Personnel Psychology* 61 (2): 273–307.

Marks, M.A., Mathieu, J.E., and Zaccaro, S.J. (2001). A temporally based framework and taxonomy of team processes. *The Academy of Management Review* 26 (3): 356–376.

Marks, P. and Riley, K. (1995). *Aligning Technology for Best Business Results*. Santa Cruz, CA: Design Insight.

Metwalli, S.M. (2009). Life cycle optimization in the design for environment (Keynote Paper). In: *Proceedings of 16th CIRP International Conference on Life Cycle Engineering (LCE 2009)* (4–6 May 2009) (ed. W. ElMaraghy), 48–53. International Academy for Production Engineering (CIRP).

Metwalli, S.M., Salama, M.S., and Taher, R.A. (1989). Computer aided reliability for optimum maintenance planning. *Computers & Industrial Engineering* 35 (3): 603–606.

Metwalli, S.M., Younan, M.Y., and El-Hebery, M.R. (1983). Application of fracture mechanics and reliability in fail safe prediction of fire tube boilers. In: *Proceedings of the 5th ASME Failure Prevention and Reliability Conference* (11–14 September 1983) (ed. G.M. Kurajian), 1–7. ASME.

NBS (1968). Appendix 8. *Federal Register* 33 (146): 1068. National Bureau of Standards, U.S. Department of Commerce.

NIST (2008). *The International System of Units (SI)* (Special Publication 330). U.S. Department of Commerce.

NIST (2010). *Specifications, Tolerances, and Other Technical Requirements for Weighing and Measuring Devices, NIST Handbook 44*. U.S. Department of Commerce.

SAE (1999). *Rules for SAE Use of SI (Metric) Units*. TSB 003 Society of Automotive Engineers.

Taylor, B.N. and Thompson, A. (eds.) (2008). *The International System of Units (SI)* (NIST Special Publication 330). U.S. Department of Commerce.

The Boston Globe (1999). Thank you, NASA. *The Boston Globe* (5 October), p. A16.

Internet Sites

http://www.iso.org/iso/home.html ISO: Standards
http://standards.iso.org/ittf/PubliclyAvailableStandards/ ISO: Available Standards
https://en.wikipedia.org/wiki/Conversion_of_units C
https://en.wikipedia.org/wiki/Horsepower Wikipedia: Horsepower
https://www.nist.gov/pml/weights-and-measures/metric-si/unit-conversion NIST: Unit conversion
http://www.sae.org/standardsdev/tsb/tsb003.pdf SAE: Standards
https://www.nist.gov/pml/weights-and-measures/approximate-conversions-us-customary-measures-metric NIST: Approximate unit conversion

2

Design Considerations

Machine design involves several factors to consider in addition to the previous broader associated fields discussed in Chapter 1. Tools and procedures are essential to appropriately design machine components and systems. To perform appropriate calculations, an understanding of the adopted mathematical model is essential. The model closeness to the real world causes the calculations to be representative to what is expected. If the model is loosely representative, the expected behavior of the design is not necessarily assured. The mathematical model includes approximating system structure and the form of applied loading. After decisions on the mathematical model, the available tools of calculations are chosen. Design procedures and tools helping to develop the synthesis should be appropriately defined. These tools are to provide the synthesis of each component and the consideration of its suitable manufacturing or procurement process. This constitutes a paradigm shift in design process. Selected standard sets and components are to be defined in detail during the course of synthesis. Codes and standards involved in the process should be adhered to. These are the topics considered in this chapter to attain appropriate machine component synthesis and proper machine system assembly.

Symbols and Abbreviations

The adopted units are [in, lb, psi] or [m, kg, N, Pa], others given at each symbol definition. [k…] is 10^3, [M…] is 10^6 and [G…] is 10^9.

Symbol	Quantity, units (adopted)
$[\boldsymbol{r}\times]$	Cross-product matrix (Eq. (2.6))
2D	Two-dimensional
3D	Three-dimensional
$\boldsymbol{a}$	Acceleration vector $[a_x \quad a_y \quad a_z]^T$
A–Z	Fundamental deviation categories for holes
a–z	Fundamental deviation categories for shafts
A_1, A_2	Areas of side 1 and 2 of hydraulic transformer
AC	Alternating electric current
A_B	Billows area
$\boldsymbol{A}_1$	Reduced node matrix
$\boldsymbol{A}_{\text{in}}$	Input matrix or vector
$\boldsymbol{A}_n$	Node matrix $(n_{v-1} \times n_e)$
A_p	Piston area of hydraulic cylinder

Machine Design with CAD and Optimization, First Edition. Sayed M. Metwalli.

Companion website: www.wiley.com/go/metwalli/machine

Symbol	Quantity, units (adopted)
$\boldsymbol{A}_S$	System matrix
CAD	Computer-aided design
CAE	Computer-aided engineering
CAS	Computer-aided synthesis
CNC	Computer numerical control
CNT	Carbon nanotubes
C	Across-type component or electric capacitance
c_0	Fundamental deviation in tolerances and fits
c_1	Centrifugal motor transformation ratio
c_T	Translational damping coefficient
c_R	Rotational damping coefficient
DC	Direct current
d_C	Cylindrical diameter
EDM	Electric discharge machine
E	Effort or *across* variable
e_D	Effort or across *driver*
e_i	Effort variable in component i
$\boldsymbol{F}$	Force vector, $[F_x \quad F_y \quad F_z]^T$
$\boldsymbol{F}_i, \ldots, \boldsymbol{F}_n$	Force vectors at points $1, 2, \ldots, i, \ldots n$
$\boldsymbol{F}_{\text{in}}$	Input force vector
FBD	Free body diagram
FE	Finite element
FN	Force interference
F	Flow or *through* variable
f_1, f_2	Flow or *through* variable at terminals 1 and 2
f_D	Flow or through *driver*
f_i	Flow or *through* variable in component i
$\boldsymbol{f}_c$	Flow vector of components n
$\boldsymbol{f}_n$	Flow vector of nodes n
GY	Gyrator
g	Gravitational acceleration, 9.81 [m/s^2] or 386 [in/s^2]
h_1, h_2	Heads at terminal 1 and 2
H	Heat flux
I	Through-type component or electric inductance
I_F	Fluid inductance
I_T	Thermal inductance (not identified)
I	Electric current or a counting index
i_g	Tolerance grade number
i_1	Current at terminal 1 of DC motor

Symbol	Quantity, units (adopted)
ITi	International tolerance grade i
J_G	Gyroscope polar moment of inertia
k	Translation spring stiffness
k_1, k_2	Stiffness of springs 1 and 2
k_S	Combined spring stiffness
k_T	Rotational or torsional spring stiffness
k_A	Amplifier gain coefficient
L	Electric inductor
LC, LT, LN	Location: clearance, transition, and interference
l_1, l_2	Lever lengths for terminals 1 and 2
$\boldsymbol{M}$	Moment vector, $[M_x \quad M_y \quad M_z]^T$
$\boldsymbol{M}_j, \ldots, \boldsymbol{M}_k$	Moment vectors at points 1, 2, …, j, …, k
$\boldsymbol{M}_T$	Torsional moment or torque
m	Mass
N_{rpm}	Rotational speed in revolutions per minute, [rpm]
N_G	Gyrator ratio
N_T	Transformer ratio
N_1, N_2	Number of teeth of input and output gears
N_1, N_2	Number of turns of input and output coils
n_e	Number of elements or components
n_v	Number of vertices or nodes
OEM	Original equipment manufacturer
p	Pressure
p_1	Pressure of component at terminal 1
q	Flow rate of fluid
q_1	Flow rate of component at terminal 1
$\boldsymbol{R}$	Reaction vector $[R_x \quad R_y \quad R_z]^T$
$\boldsymbol{R}_1, \boldsymbol{R}_2$	Reaction vectors at supports 1, 2
R_D, R_E	Dissipative components or electric resistance
R_F	Fluid resistance, tube, porous plug
RC	Running clearance
RP	Rapid prototyping
R_T	Thermal insulation
$\boldsymbol{r}$	Position vector $[r_x \quad r_y \quad r_z]^T$
$\boldsymbol{r}_i$	Position vector at points 1, 2, …, i, …, n
r_P	Pulley radius
s	Shaft
S	Amplifier component name
SLS	Selective laser sintering

Symbol	Quantity, units (adopted)
STL	Stereolithography or standard tessellation language
t	Time
$\boldsymbol{T}$	Torque
T_1, T_2	Torques at input–output sides
$\boldsymbol{T}_c$	Component flow (torque) vector
T_{id}	Tolerance for grade number i_g and diameter d_C
$\boldsymbol{T}_{\text{in}}$	Input flow (torque) vector
$\boldsymbol{T}_n$	Node flow (torque) vector
TF	Transformer
v_1	Voltage at terminal 1 of DC motor
v_1, v_2	Voltages at terminals 1 and 2
$\boldsymbol{W}_c$	Component admittance matrix
$\boldsymbol{W}_L$	Lever component matrix
$\boldsymbol{W}_{\text{HC}}$	Hydraulic cylinder component matrix
$\boldsymbol{W}_{\text{TG}}$	Transformer to gyrator component matrix
δ	Displacement
$\boldsymbol{\delta}_c$	Component displacement vector
$\boldsymbol{\delta}_{\text{in}}$	Input displacement vector
$\boldsymbol{\delta}_n$	Node displacement vector
δ_1, δ_2	Displacements at terminals 1 and 2
θ	Angle of rotation
$\boldsymbol{\theta}_c$	Component across (rotation) vector
$\boldsymbol{\theta}_{\text{in}}$	Input across (rotation) vector
$\boldsymbol{\theta}_n$	Node across (rotation) vector
$\dot{\theta}$	The time derivative of θ.
$\dot{\delta}$	The time derivative of the displacement δ
$\boldsymbol{\omega}$	Angular velocity
ω_G	Gyroscope rotational speed

2.1 Mathematical Modeling

The fundamental concepts defining basic models are crucial in the true representation of the physical or real system. Employing a three-dimensional (***3D***) space generalizes the treatment of the systems, subsystems, and elements connected to form the design. Inertial frame of reference or *inertial coordinates* are a good choice to start with, even though the analysis in a two-dimensional (***2D***) plane might be applicable and accurate. It should be known that forces are generated by the action of one body on the other body in contact or in a connection. The *inertia* of a body is present to resist any variation of motion if the motion is sufficiently changing with a sizably noticeable rate. The *mass* is the quantity that characterizes the inertia. In mechanical design it is seldom that one needs to consider the point mass as a particle. Usually rigid or flexible bodies are adopted. Adopted *scalar* quantities are only the magnitudes of time, volume, density, speed, energy, and mass. *Vectors* are defined by magnitude and direction or better by the components of the vector in the ***3D*** region of space. Vectors are any of the displacement, velocity, acceleration, force, moment, and momentum. Generally *bold face* characters are denoting vectors, while lightface characters are used for the scalar quantities. These generalities help in the clear understanding and development of the mathematical relations used or derived herein. All previous knowledge pertaining to these variables and their interrelations are highly useful, and some are briefly covered as an essential review. It is recommended that this necessary background should be retrieved by all who need to closely follow the course of the material presented herein and make further use of it.

Analytical relations in mathematical modeling employ ***Newton's laws***. The first law asserts that "a particle remains at rest or continue to move with a uniform velocity in a straight line if no unbalanced force acting on it." The second law states that "the particle *acceleration* is proportional to the resultant *force* acting on it and is in the direction of this force." This law is the basis of mechanics. In a mathematical form, a particle of mass m subjected to a force $\boldsymbol{F}$ will have acceleration $\boldsymbol{a}$ defined from the well-known relation

$$\boldsymbol{F} = m\boldsymbol{a} \tag{2.1}$$

In Eq. (2.1), the force $\boldsymbol{F}$ and the acceleration $\boldsymbol{a}$ are both vectors; however, the mass m is a scalar. The third law of ***Newton*** indicates that "the forces between interacting bodies be it action or reaction are equal in magnitude, but opposite in direction. They are also collinear." This is widely utilized in the contact between mechanical components. For analysis, the third law is applied through detaching the bodies from one another to form the *free body diagram* (***FBD***) for each body.

2.1.1 Mathematical Model Initiation and Adoption

A mathematical model represents an ideal and restraining approximation for the natural physical and real situation. The use of this model requires few assumptions to simplify the treatment of elements and their assembly. One may neglect small distances, angles, or forces compared to large ones. Effect of bearing friction may be neglected if friction forces are small. If velocity variation is nearly uniform, one can assume constant acceleration. The degree of a simple assumption depends on the desired accuracy. The following sample application demonstrates the relevance of some of these assumptions.

Figure 2.1 shows an assembly of a shaft carrying an element through its hub and suspended by two bearings. The clearance between the hub and the shaft is exaggerated. If the clearance is smaller than the shaft *deflection*, the distribution of the pressure between them might be as shown in Figure 2.1. Under load, the hub will first contact the shaft at its right and left ends, and the deflection of the hub causes the interior to start transferring the load accordingly. That is why the pressure distribution between the hub and the shaft may give the shown nonlinear form. To mathematically model the loading on the shaft, one can simply assume the load $\boldsymbol{F}$ to be concentrated

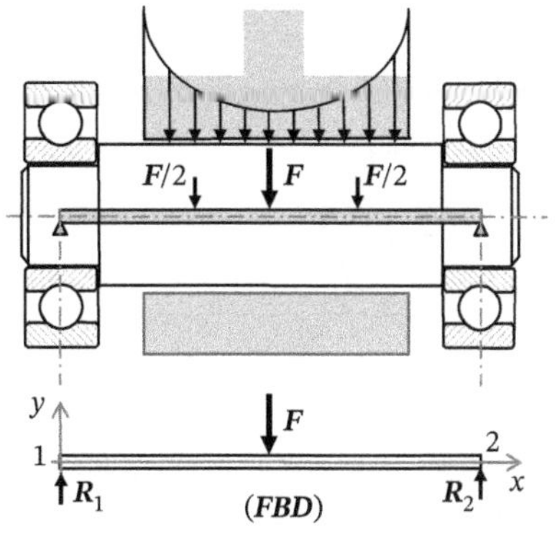

Figure 2.1 An assembly of a shaft carrying an element through its hub and suspended by two bearings. One simplified mathematical model is identified in the lower part of this multipart design assembly.

about the middle of the hub or assume it as two loads each as $F/2$ and at some distance apart with a maximum at the ends of the hub. The first assumption is more conservative and can be adopted for the initial or preliminary synthesis. The second assumption might be valid if the clearance is much larger than the anticipated deflection of the shaft and hub. The assumption that the reaction at the bearings are at the bearing centerline is not grossly in error if the angular deflection of the shaft is very small at the bearing. In general, the ***mathematical model*** of this assembly can then be represented by a *simply supported beam* with the load $\boldsymbol{F}$ at about the middle as shown in Figure 2.1 as the grayish bar on the assembly axis. This assumption is conservative and is adopted for the treatment of similar problems in this text. The simplified mathematical model is then identified in the detached lower part of Figure 2.1 for this multipart design assembly.

The mathematical model shown in Figure 2.1 represents the first stage of the ***FBD***. The second stage is to assume reactions $\boldsymbol{R}_1$ at support number 1 and $\boldsymbol{R}_2$ at the support number 2 and to identify the frame of reference or coordinates (x, y) on that diagram. The representations of the supports are eliminated, and the reactions are substituted in their place. This is shown in Figure 2.1 below the assembly with the assumed origin set at the left where $\boldsymbol{R}_1$ is. The x-axis is the horizontal one, the y-axis is the vertical one, and the z-axis is normal to the plane of the page. This is the usual procedure and the adopted coordinates in this text. Support 1 is always located at the origin.

To calculate stresses and deflections, few other mathematical modeling tools such as *finite element* (***FE***) codes can handle such a connection of Figure 2.1 with assumed dimensions, clearances, and different contact loading and shape conditions. In real engineering product development, such a situation should be checked for validation of appropriate adoption of conditions that are closer to the physical or real situation. It should be pointed out, however, that ***FE*** codes are also simplified mathematical models to the actual physical situations. They may be more closely representing the real loading and design behavior conditions, if the loading and boundary conditions are close to the real case. The ***FE*** analysis is a tool used after the dimensions are specifically defined. The mathematical model adopted for the problem in Figure 2.1, however, may be convenient and conservative with close and acceptably evaluated internal stresses that can be used in initial synthesis. Results are more or less a close average rather than the ones of the concentrated loads or connections at a single location or *node* in the ***FE***s to be discussed later on in Sections 2.2.4 and 6.11. The discrepancies are taken care of by the appropriate modeling and the *safety factor* considered later in Sections 2.1.3 and 2.3.

Other similar assumptions can be made in the adoption of the mathematical model. Cases such as those shown in Figure 2.2 are some of the encountered situations. Several bearing supports are shown where the bearing in (a) can be mathematically modeled as a support at the bearing centerline as previously adopted. The mathematical models of the tandem bearings in (b) and the journal bearing in (c) are not apparent. The location of the mathematically modeled equivalent is not obviously definite and can be adopted as shown in Figure 2.2 with approximated dimensions depending on the support of the bearings and their configurations. The pressure distribution of the sleeve bearing in Figure 2.2c is nonlinear and depends on the fluid flow, which is more restrictive to the left and

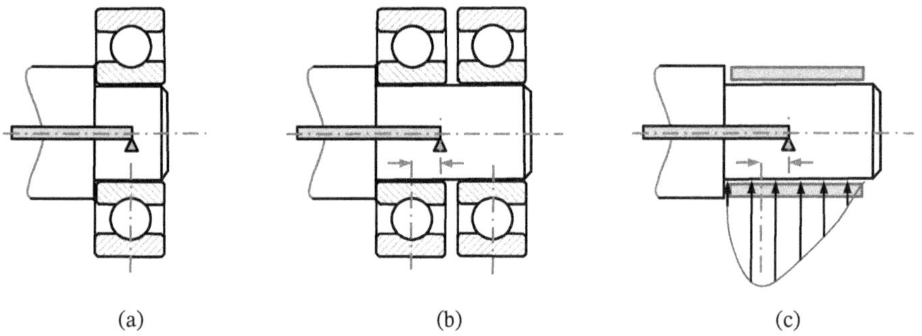

Figure 2.2 Several bearing supports with the mathematical model of the bearing in (a) at the bearing centerline. The mathematical models of the tandem bearings in (b) and the journal bearing in (c) are suggested.

accounts for the shift of the maximum pressure to the left. Conservative adoptions take the reactions in these cases at the middle of the support. This will be the adopted mathematical model that is favored in this text. More representative considerations require more involved modeling than the scope of this course. Most of statically undetermined problems, complex contact cases, and *elastohydrodynamic* considerations are beyond the intended treatment presented herein. As an example, you may consult Abbas et al. (2010) and Abbas and Metwalli (2011).

In general, the mathematical model formulation procedure is to start with identifying the given data and the desired result accuracy for performance and expected due loading case. From that, one should define necessary diagrams such as ***FBDs*** for the whole system and the composing constituent components. The process also needs to identify necessary relations between components. By performing calculations, one finds the answers to the required solution specifically the required synthesis. One should then check and verify intermediate and final results in addition to evaluating the conclusions.

An isolated ***FBD*** representing a whole system is denoted as an *external **FBD***. For every internal member of the system, one can construct an ***FBD***. The relation between each of these members is defined by Newton's third law, where forces and moments of connection will have the same magnitude but opposite in direction for each member. It is recommended first to have an external ***FBD*** and then dissect the system into more of its internal members (or elements) to evaluate the forces and moment on each member. This allows the evaluation of internal shear forces and moments for each member. For the whole system, it is recommended to set the origin of the coordinate system at the main support of the system. A right-hand coordinate system is used in all cases. All reactions are assumed in the right positive direction. If they are not, they will come out as negative. The equilibrium of the ***FBD*** will be assumed static, which is also valid for bodies moving at constant velocity or the acceleration is small compared to the applied loads, i.e. the acceleration $\boldsymbol{a}$ is assumed equal to 0. The ***equilibrium*** of the ***FBD*** is then governed by the following relations:

$$\sum_{\text{FBD}} \boldsymbol{F} = \boldsymbol{0} \tag{2.2}$$

$$\sum_{\text{About any Point}} \boldsymbol{M} = \boldsymbol{0} \tag{2.3}$$

These relations provide six equations in the ***3D*** space, which allows the evaluation of all three components of both *reaction forces* at two support locations or reactions of three force components and three moment components at a single support position. In a general ***FBD*** subjected to external n force vectors $\boldsymbol{F}_1, \boldsymbol{F}_2, \ldots, \boldsymbol{F}_n$ acting at position vectors $\boldsymbol{r}_1, \boldsymbol{r}_2, \ldots, \boldsymbol{r}_n$, respectively, and other external k moment vectors $\boldsymbol{M}_1, \boldsymbol{M}_2, \ldots, \boldsymbol{M}_k$ with reaction vectors at supports $\boldsymbol{R}_1$ and $\boldsymbol{R}_2$ or $\boldsymbol{R}_1$ and $\boldsymbol{M}_1$, the equilibrium equations of (2.2) and (2.3) will be as follows:

$$\left(\sum_1^n \boldsymbol{F}_i\right) + \boldsymbol{R}_1 + \boldsymbol{R}_2 = \boldsymbol{0} \tag{2.4}$$

$$\sum_1^k \boldsymbol{M}_j + \left(\sum_1^n \boldsymbol{r}_i \times \boldsymbol{F}_i\right) + (\boldsymbol{r}_{R_2} \times \boldsymbol{R}_2) + \boldsymbol{M}_1 = \boldsymbol{0} \tag{2.5}$$

Locating the coordinate origin at support 1 and with moment $\boldsymbol{M}_1 = 0$, the moment Eq. (2.5) produces $\boldsymbol{R}_2$. Substitution of $\boldsymbol{R}_2$ into Eq. (2.4) allows the evaluation of $\boldsymbol{R}_1$. If support 1 carries a reaction moment $\boldsymbol{M}_1$, support 2 is assumed free with $\boldsymbol{R}_2 = 0$. Equation (2.5) will then produce the moment $\boldsymbol{M}_1$, and Eq. (2.4) generates $\boldsymbol{R}_1$ at support number 1. The *cross product* in Eq. (2.5) is best handled by the following matrix product:

$$\boldsymbol{r} \times \boldsymbol{F} = \begin{bmatrix} 0 & -r_z & r_y \\ r_z & 0 & -r_x \\ -r_y & r_x & 0 \end{bmatrix} \begin{bmatrix} F_x \\ F_y \\ F_z \end{bmatrix} = [\boldsymbol{r}\times]\boldsymbol{F} \tag{2.6}$$

where (r_x, r_y, r_z) are the components of the position vector $\boldsymbol{r}$ that comprise the ***cross-product matrix*** $[\boldsymbol{r}\times]$ and (F_x, F_y, F_z) are the components of the force vector $\boldsymbol{F}$. This expression is the same as getting the solution of the cross product by any other means. It is, however, very useful as will be demonstrated in Example 2.1 and in computer coding.

Example 2.1 Figure 2.3 represents an ***FBD*** for a model of a door handle under maximum opening load $\boldsymbol{F}_1$. The lock reached its maximum position as the handle ends up in the shown position. The lock is fixed in the fixed door that is not shown, but the origin of the coordinate system is considered fixed at that point presumed as the ground or support 1. Assume the maximum applied load is 4 [lb] or about 18 [N] and its location vector is $[3 \quad -2 \quad 4]^T$ [in] or $[0.08 \quad -0.05 \quad 0.1]^T$ [m]. It is required to get the reaction at the door handle support.

Solution

Data: $\boldsymbol{F}_1 = -4$ [lb] or -18 [N] and $\boldsymbol{r}_1 = [3 \quad -2 \quad 4]^T$ [in] or $[0.08 \quad -0.05 \quad 0.1]^T$ [m].
At the support, the reaction force vector $\boldsymbol{R}_1$ is defined by its components $[R_{1x} \quad R_{1y} \quad R_{1z}]^T$. The moment vector $\boldsymbol{M}_1$ is indicated but without showing its components $[M_{1x} \quad M_{1y} \quad M_{1z}]^T$. It should be noted that no external moments are present. Applying Eqs. (2.4), (2.5), and (2.6) gives

$$\begin{bmatrix} 0 \\ -F_{1y} \\ 0 \end{bmatrix} + \begin{bmatrix} R_{1x} \\ R_{1y} \\ R_{1z} \end{bmatrix} = \mathbf{0} \tag{a}$$

$$\begin{bmatrix} 0 & -r_{1z} & (-r_{1y}) \\ r_{1z} & 0 & -r_{1x} \\ -(-r_{1y}) & r_{1x} & 0 \end{bmatrix} \begin{bmatrix} 0 \\ -F_{1y} \\ 0 \end{bmatrix} + \begin{bmatrix} M_{1x} \\ M_{1y} \\ M_{1z} \end{bmatrix} = \mathbf{0} \tag{b}$$

From Eq. (a), one gets

$$R_{1x} = 0, \; R_{1y} = F_{1y}, \; R_{1z} = 0 \tag{c}$$

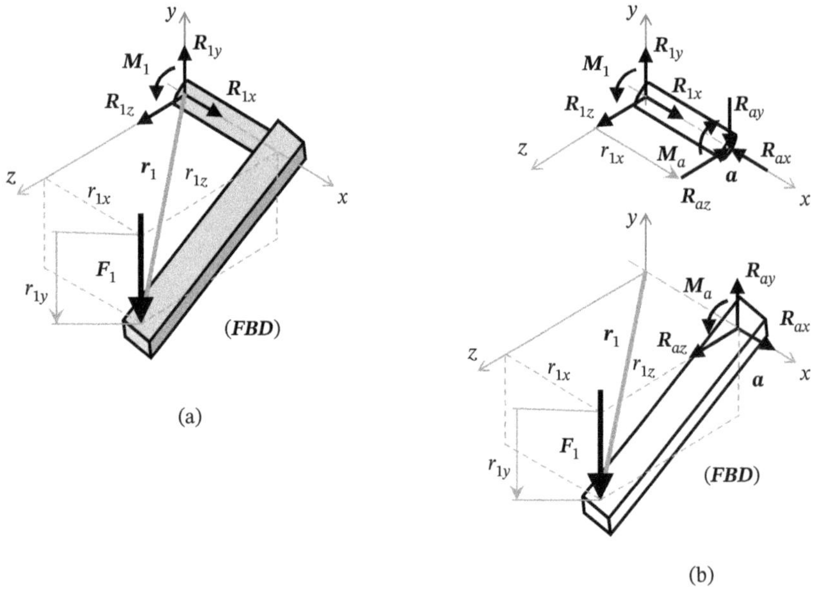

Figure 2.3 External free body diagrams (***FBDs***): (a) door handle under maximum opening load $\boldsymbol{F}_1$ and (b) handle components having separate ***FBD***.

This indicates that the components of the reaction vector $\boldsymbol{R}_1$ are positive (upward) with a magnitude of the applied force vector $\boldsymbol{F}_1$. Processing Eq. (b) in details shows that

$$\begin{bmatrix} r_{1z}F_{1y} \\ 0 \\ -r_{1x}F_{1y} \end{bmatrix} + \begin{bmatrix} M_{1x} \\ M_{1y} \\ M_{1z} \end{bmatrix} = \mathbf{0} \tag{d}$$

or

$$\begin{bmatrix} M_{1x} \\ M_{1y} \\ M_{1z} \end{bmatrix} = \begin{bmatrix} -r_{1z}F_{1y} \\ 0 \\ r_{1x}F_{1y} \end{bmatrix} \tag{e}$$

This states that we have generated a reaction moment of magnitude $(-r_{1z}F_{1y})$ about the x-axis (i.e. a torque) and a moment about the z-axis of magnitude $(r_{1x}F_{1y})$. It should be noted that the component z of the position vector of the applied force did not affect the reaction moment at the support. The inclination of the handle affects only the x and y components of its position vector and thus the magnitude of the moment components.

By substitution of the numerical values of the force $\boldsymbol{F}_1$ and its position vector $\boldsymbol{r}_1$, one gets for the US and SI systems: $R_{1x} = R_{1z} = 0$, $R_{1y} = 4$ [lb] or 18 [N], and $\boldsymbol{M}_1 = [-16 \quad 0 \quad 12]^T$ [lb in] or $[-1.8 \quad 0 \quad 1.44]^T$ [N m].

Figure 2.3b shows the handle dissected into the hand part and the lock axel part. Each part can better be related to the same coordinate system. Therefore, location or position vectors are the same. Applying the same equilibrium equations to each part gives the following for the hand:

$$\begin{bmatrix} 0 \\ -F_{1y} \\ 0 \end{bmatrix} + \begin{bmatrix} R_{ax} \\ R_{ay} \\ R_{az} \end{bmatrix} = \mathbf{0} \tag{f}$$

$$\begin{bmatrix} 0 & -r_{1z} & (-r_{1y}) \\ r_{1z} & 0 & -r_{1x} \\ -(-r_{1y}) & r_{1x} & 0 \end{bmatrix} \begin{bmatrix} 0 \\ -F_{1y} \\ 0 \end{bmatrix} + \begin{bmatrix} 0 & 0 & 0 \\ 0 & 0 & -r_{ax} \\ 0 & r_{ax} & 0 \end{bmatrix} \begin{bmatrix} R_{ax} \\ R_{ay} \\ R_{az} \end{bmatrix} + \begin{bmatrix} M_{ax} \\ M_{ay} \\ M_{az} \end{bmatrix} = \mathbf{0} \tag{g}$$

From that

$$R_{ax} = 0, \quad R_{ay} = F_{1y}, \quad R_{az} = 0 \tag{h}$$

And as $r_{ax} = r_{1x}$, then

$$\begin{bmatrix} r_{1z}F_{1y} \\ 0 \\ -r_{1x}F_{1y} \end{bmatrix} + \begin{bmatrix} 0 & 0 & 0 \\ 0 & 0 & -r_{1x} \\ 0 & r_{1x} & 0 \end{bmatrix} \begin{bmatrix} 0 \\ F_{1y} \\ 0 \end{bmatrix} + \begin{bmatrix} M_{ax} \\ M_{ay} \\ M_{az} \end{bmatrix} = \mathbf{0} \tag{i}$$

or

$$\begin{bmatrix} r_{1z}F_{1y} \\ 0 \\ -r_{1x}F_{1y} \end{bmatrix} + \begin{bmatrix} 0 \\ 0 \\ r_{1x}F_{1y} \end{bmatrix} + \begin{bmatrix} M_{ax} \\ M_{ay} \\ M_{az} \end{bmatrix} = \mathbf{0}$$

Then,

$$\begin{bmatrix} M_{ax} \\ M_{ay} \\ M_{az} \end{bmatrix} = \begin{bmatrix} -r_{1z}F_{1y} \\ 0 \\ 0 \end{bmatrix} \tag{i'}$$

```
% Cross Product Matrix for reaction position "rR2", Force_Analysis_3D_Short.m
    a=[0 -rR2(3) rR2(2);rR2(3) 0 -rR2(1); -rR2(2) rR2(1) 0];
    ainv=pinv(a);                                   % Find inverse of matrix "a"
% Calculate Moments of each F as = (r × F), f and r are Input vectors components
    for i=1:n                                       % n = Input force number
      af=[0 -r(i,3) r(i,2); r(i,3) 0 -r(i,1); -r(i,2) r(i,1) 0];
      rXf(i,:)=af*f(i,:)';
    end
% Calculate Sum of Moments (r × F) and Sum of forces
    sumM=[0; 0; 0];                                 % Initialize the sum
    sumf=[0; 0; 0];                                 % Initialize the sum
   for i=1:n
      sumM = sumM - rXf(i,:)' - m(i,:)';       % m = Input moment components
      sumf = sumf - f(i,:)';
   end
R2=[ainv*[sumM] -ainv*M1]'                     % Row vector for display
R1= (sumf -R2')'                               % Row vector for display
M1=[sumM - a*R2']'                             % Row vector for display
```

Figure 2.4 Main statements of a MATLAB© code for n applied loads and moments. Reactions R_1, R_2, and M_1 are obtained at the two supports 1 and 2. External operational loading f = F_i and m = M_i are inputs that must be provided by the user, not shown.

Taking the opposite direction for these forces and moment to be applied to the lock axel as shown in Figure 2.3, one clearly finds the same solution for $\boldsymbol{R}_1$ and $\boldsymbol{M}_1$ by inspection. Similar relations to Eqs. (f) and (g) can be obtained for the lock axel, which produce the same results for $\boldsymbol{R}_1$ and $\boldsymbol{M}_1$. This demonstrates the systematic nature of the procedure that is adopted in this text.

By substituting for the numerical values of the force $\boldsymbol{F}_1$ and its position vector $\boldsymbol{r}_1$, one gets for the US and SI systems: $R_{ax} = R_{az} = 0$, $R_{ay} = 4$ [lb] or 18 [N], and $\boldsymbol{M}_a = [-16 \quad 0 \quad 0]^T$ [lb in] or $[-1.8 \quad 0 \quad 0]^T$ [N m].

The treatment of this example seems lengthy relative to others. This is due to its limited number of applied loads, having one fixed support and applying generalized procedure. When the number of applied loads and moments increases, the procedure is still the same with marked savings particularly if a computer code is implemented. The main statements of a MATLAB code are depicted in Figure 2.4 with the program available through the **Wiley website** of this textbook under the name ***Force_Analysis_3D_Short.m***. Note that external operational loading f = $\boldsymbol{F}_i$ and m = $\boldsymbol{M}_i$ are inputs that must be provided by the user in column vector form, not shown in Figure 2.4. The code uses ***pseudo inverse*** (pinv) of the *cross-product* matrix since it is singular. This pseudo inverse is, however, useful in finding solutions of sparsely populated and non-square matrix systems. The code should be clear for users having MATLAB experience. For other users, it could be clearer after reading Section 2.2.2 and practicing further with the use of MATLAB.

2.1.2 Generalized System Modeling

This section covers a modeling process for systems that are composed of different elements and can have mechanical, fluid, thermal, and electrical components. It is a systematic process that makes the connection of such elements in different fields an easy and formalized job particularly for linear and *dynamic systems*. The main applications are for *lumped parameters* dynamic rather than static systems. However, it can also be used for systems that act more statically than dynamically. Since it is general, the process can be seen as a lengthy one for simple systems. It should be much easier when the system is complex and has many elements with few fields involved, i.e. mechanical, fluid, and electrical at the same time. It is also very useful in linking translational and rotational systems as will be seen in Example 2.4. The possibility of designing across different fields, and in dynamic cases where loads are variable with time such as impact, is presented herein for *generalization*. It is also very useful in demonstrating important interactions between drivers and systems particularly the effect of the systems on the drivers.

The system components and links of this system model depend on the ***power*** interaction through the system. The selections of the *generalized variables* attest to that. In static cases, we can use ***energy*** rather than power and these are most of the cases considered in this text. For more information on different methods, one needs to refer to the classical treatments such as Martens and Allen (1969) and Karnopp and Rosenberg (1968). These references treat systems using *system graphs* or *bond graph*. The treatment herein utilizes ***system graph***, but the variables and relations describing components are about the same for bond graphs (Borutzky 2010):

A. ***Generalized Variables:***

(a) Effort or ***across*** variables *"e"*:

These variables designate a "*potential.*" They may be selected as velocity $\dot{\delta}$, angular velocity $\dot{\theta}$, pressure p (or head h), temperature T for mechanical systems, and voltage v for the electrical systems.

These variables should satisfy the "*compatibility*" laws:

$$\sum_{\text{around closed paths}} e_i = 0 \tag{2.7}$$

(b) Flow or ***through*** variables *"f"*:

These variables designate a "*flow.*" They may be selected as force $\boldsymbol{F}$, torque $\boldsymbol{M}_T$, flow rate q, and heat flux H, for mechanical systems and the current i for the electrical systems.

These variables should satisfy the "*continuity*" laws:

$$\sum_{\text{at points of junction}} f_i = 0 \tag{2.8}$$

The product of the ***across*** variable and the ***flow*** variable is the ***power***.

B. ***Two Terminal Components***

The components in this category have only two connections or terminals to other components (see Tables 2.1–2.4). The terminals of the component are the ends "a" and "b." Some of these ends

Table 2.1 The schematic and system graph symbol for each element of the *through-type* components (storage in time via the through variable f).

Component	Schematic and system graph symbol	Relation
Translation spring k	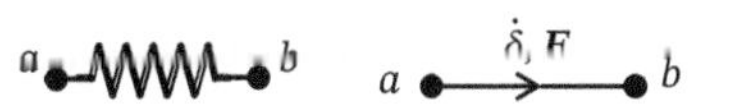	$\frac{d}{dt}F = k\ \dot{\delta}$
Torsional spring k_T	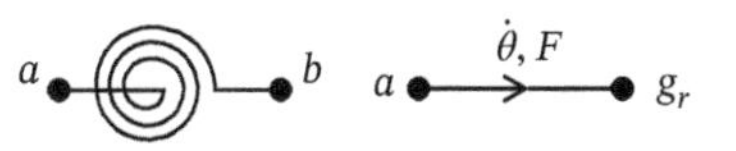	$\frac{d}{dt}\dot{\theta} = k_T\ \boldsymbol{T}$
Flow inductance I	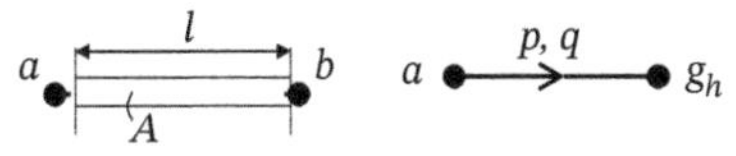	$\frac{d}{dt}p = \frac{1}{I_F}\ q$
Electric inductance L	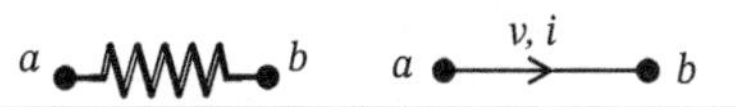	$\frac{d}{dt}v = \frac{1}{L}\ i$

would be referenced to the ground. Some of these components store *energy* through one of the generalized variables, others dissipate energy, and two are *sources* of energy through a generalized variable. The characteristic of each type is as follows:

(a) ***Storage or Semi-passive Components***

i. ***Through-type components "I"***: The form of these components is explicit in the ***through*** variable (or storage in time via the through variable) f. The characteristic coefficient is the symbol "*I*." This symbol is borrowed from electrical inductance I. In mechanical engineering, the symbol is the potential storage in the spring stiffness k.

$$f(t) = \frac{1}{I}\int_0^t e(t)\,dt + f(0)$$
$$\frac{d}{dt}f(t) = \frac{1}{I}e(t), \quad f(0) \tag{2.9}$$

where $f(0)$ is the value of the through variable at the initial condition of time at 0. The translation spring stiffness k, the rotational spring stiffness k_T, the fluid inductance I_F, and the electric inductor L are given in Table 2.1. The *differential relation* of each element is also defined. Table 2.1 presents the schematic and system graph symbol for each element. For mechanical applications where mainly static analysis is of the main concern, it is feasible to use ***energy*** rather than power in the relations among components. This gives the following relations in addition to or in place of Eq. (2.8) for springs:

$$F = k\delta \tag{2.10}$$

and

$$T = k_T\theta \tag{2.10'}$$

For thermal systems, the thermal inductance "I_T" is not identified yet.

ii. ***Across-type components "c"***: Their form is explicit in ***across*** variable (or storage via an across variable) e. The characteristic coefficient is the symbol "*c*." This is borrowed from electrical capacitance "*c*." In mechanical engineering, the symbol is the inertial capacitance in the mass m.

$$e(t) = \frac{1}{c}\int_0^t f(t)\,dt + e(0)$$
$$\frac{d}{dt}e(t) = \frac{1}{c}f(t), \quad e(0) \tag{2.11}$$

where $e(0)$ is the value of the across variable at the initial condition of time at 0. The translation mass m, the rotational inertia J, the flow capacitance c, the thermal capacitance c_p, and the electric capacitor c are shown in Table 2.2 with their *differential relations* and the schematic and system graph symbol.

(b) ***Dissipative- or Passive-Type Components***

The dissipative-type components can have the symbol "R_D," which is borrowed from the electrical resistance. Their form is a mathematical relation between ***across*** variable and ***flow*** variable as follows:

$$e(t) = \Im_1[f(t)], \text{ i.e. } e = R_D f$$
$$f(t) = \Im_2[e(t)], \text{ i.e. } f = e/R_D \tag{2.12}$$

The translation dash pot c_T, the rotational dash pot c_R, the long tube (or porous plug) R_F, the heat conduction or thermal insulation R_T, and the electrical resistance R_E are presented in Table 2.3. The relations and the schematics and system graph symbols for these components are also shown in Table 2.3.

(c) ***Active Components: Drivers or Sources***

The drivers or sources of power are active components that deliver power explicit in ***across*** variable or ***flow*** variable. The ***ideal*** value of the variable is a fixed function with time irrespective of the effect of the

Table 2.2 The schematic and system graph symbol for each element of the *effort-type* components (storage in time via the across variable *e*).

Component	Schematic and system graph symbol	Relation
Translation mass m	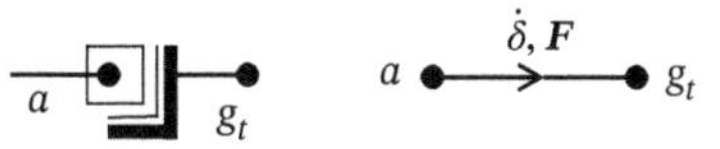	$\frac{d}{dt}\dot{\delta} = \frac{1}{m}F$
Rotational inertia J	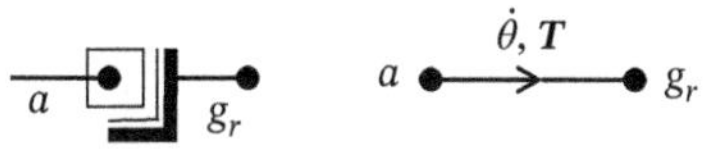	$\frac{d}{dt}\dot{\theta} = \frac{1}{J}\boldsymbol{T}$
Flow capacitance c	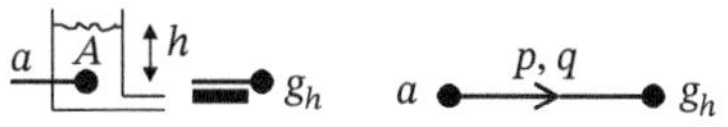	$\frac{d}{dt}p = \frac{1}{c}q$
Thermal capacitance c_p	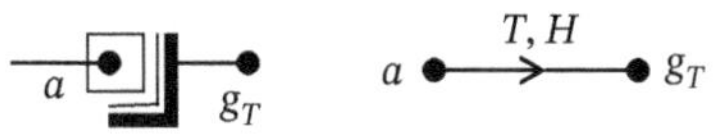	$\frac{d}{dt}T = \frac{1}{c_p m}H$
Electric capacitor c	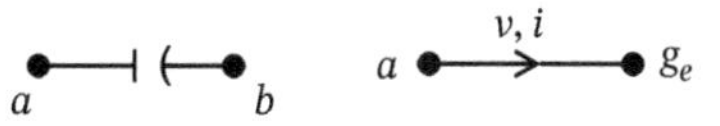	$\frac{d}{dt}v = \frac{1}{c}i$

Table 2.3 The schematic and system graph symbol for each element of the *dissipative-type* components.

Component	Schematic and system graph symbol	Relation
Translation dash pot c	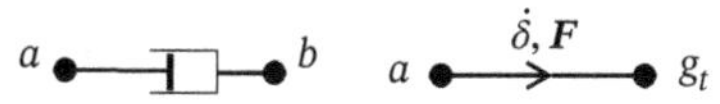	$F = c_T\ \dot{\delta}$
Rotational dash pot c_T	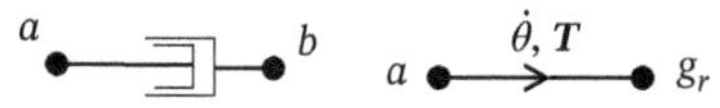	$\boldsymbol{T} = c_R\ \dot{\theta}$
Flow resistance R_F	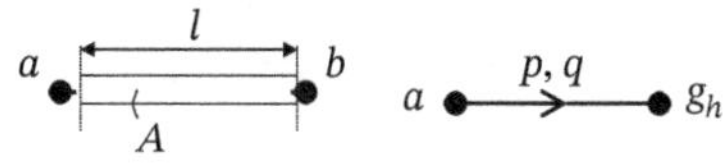	$p = R_F\ q$
Thermal insulation R_T	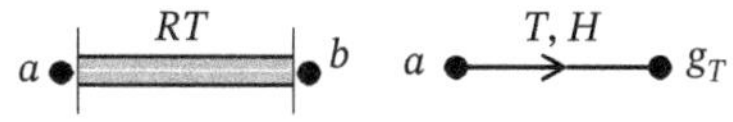	$H = \frac{1}{R_T}T$
Electric resistance R_E	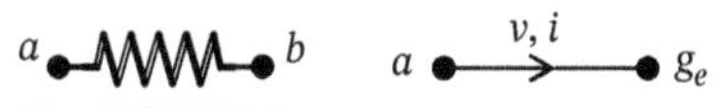	$v = R_E\ i$

system on the driver. The effect of the system on the driver will be the other generalized variable that can vary according to the power needed by the system from the driver. Types of drivers or sources are shown in Table 2.4 with their relations and schematic and system graph symbol for each.

i. *Across drivers* (e_D)

$$e_D(t) = \Im_e(t) \tag{2.13}$$

This means that the input is – for *ideal case* – a constant angular velocity $\omega = d\theta/dt$ irrespective of the torque $\boldsymbol{T}$, which is required to drive the system. Usually electric motors would have a constant angular velocity rating. The ***power***, however, is equal to $\boldsymbol{T}\omega$, which the driver should be able to deliver within its

Table 2.4 The schematic and system graph symbol for each element of *active components* that deliver power explicit in *across* variable or *flow* variable.

Component	Schematic and system graph symbol	Relation
Across drivers (e_D)	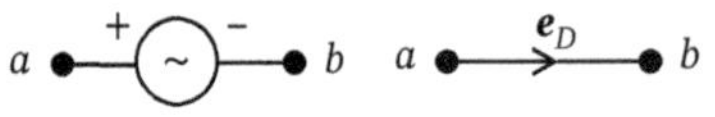	$e_D(t) = \Im_e(t)$
Through drivers (f_D)	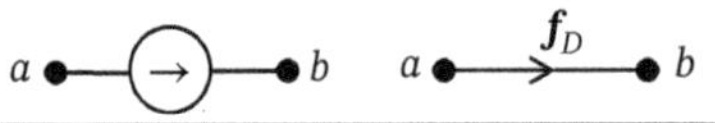	$f_D(t) = \Im_f(t)$

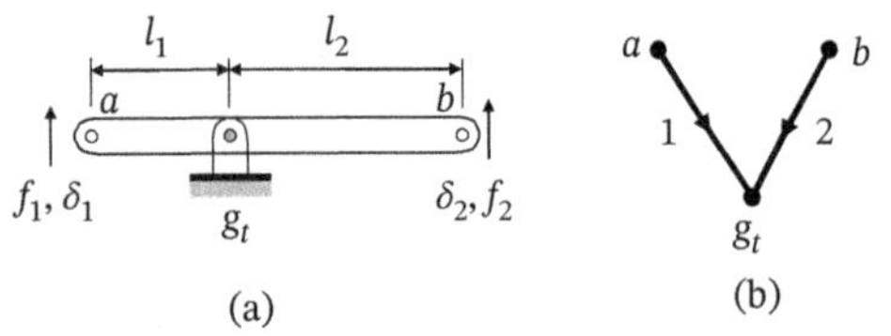

Figure 2.5 A multi-terminal *transformer* component: (a) schematic of an ideal lever and (b) system graph of the ideal lever.

maximum power capacity. The angular velocity ω is independent, and the torque $\boldsymbol{T}$ is system dependent. Another example is a cam driver with a specific displacement irrespective of the force that is the reaction of the connected system onto the cam.

ii. *Through drivers (f_D)*

$$f_D(t) = \Im_f(t) \tag{2.14}$$

This means that the input is – for *ideal case* – a specific time varying flow rate $q(t)$ irrespective of the pressure p developed in the system. This driver type is, for example, a positive displacement pump. The ***power***, however, is equal to qp, which the driver should be able to deliver within its maximum power capacity. Another simple example is an eccentric mass rotating at some specific speed that generates a dynamic force variation dependent only on the mass, the eccentricity, and rotating speed.

C. ***Multi-terminal Components***

(a) ***Transformers (TF)***: The transformer name is loaned out from electrical engineering. However, many mechanical engineering components or systems operate like transformers such as *levers*, *pulley systems*, *gear sets*, and *centrifugal pumps* or *motors*. Figure 2.5a shows a diagram of an ideal pivoted lever (a), and its system graph representation is shown in Figure 2.5b. Either (f_1, δ_2) or (δ_1, f_2) may be arbitrary specified, but each of the couples (f_1, f_2), (δ_1, δ_2), (δ_1, f_1), or (δ_2, f_2) cannot be independently specified. This is especially the case if the lever is ideal and rigid with no account for its *flexibility* or *mass*. The terminal equation of the lever transformer is then as follows:

$$\begin{bmatrix} f_1 \\ \delta_2 \end{bmatrix} = \begin{bmatrix} 0 & l_2/l_1 \\ -l_2/l_1 & 0 \end{bmatrix} \begin{bmatrix} \delta_1 \\ f_2 \end{bmatrix} \tag{2.15}$$

$$\begin{bmatrix} \text{Measurements} \\ \text{or “effect”} \end{bmatrix} = [\boldsymbol{W}_L]_{2\times 2} \begin{bmatrix} \text{Driver or} \\ \text{“Cause”} \end{bmatrix}$$

where $\boldsymbol{W}_L$ is the lever component matrix and the left vector is the force f_1 of terminal 1 and the displacement δ_2 of terminal 2. The right vector is the across and flow variables of the terminals 1 and 2, respectively. The equations are obtained by setting $f_2 = 0$ (*no load* or “open circuit”), then $\delta_1 = 0$ (*locked position* or “short circuit”). The process of setting is tantamount to using a driver or a “*cause*.” The left-hand vector of Eq. (2.15) is the measurements or the “*effect*” of setting the right-hand vector.

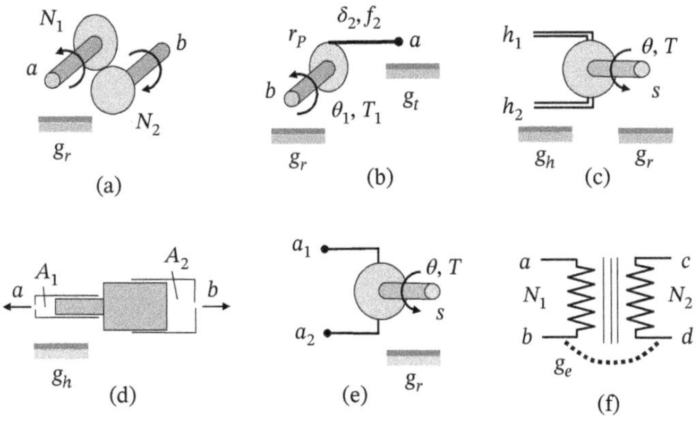

Figure 2.6 Other *transformer* components: (a) a gearbox, (b) a shaft and a pulley, (c) a hydraulic motor (centrifugal), (d) a hydraulic transformer, (e) a direct current (DC) motor (permanent magnet field), and (f) an (AC) transformer.

In general, a ***transformer*** is then

$$\begin{bmatrix} e_1 \\ f_2 \end{bmatrix} = \begin{bmatrix} 0 & N_T \\ -N_T & 0 \end{bmatrix} \begin{bmatrix} f_1 \\ e_2 \end{bmatrix} \tag{2.16}$$

where N_T is the *transformer ratio*, which is l_2/l_1 for the lever. The transformer equation can also be inverted to the following form:

$$\begin{bmatrix} f_1 \\ e_2 \end{bmatrix} = \begin{bmatrix} 0 & -1/N_T \\ 1/N_T & 0 \end{bmatrix} \begin{bmatrix} e_1 \\ f_2 \end{bmatrix} \tag{2.17}$$

A *gearbox* relating the input–output sides of the rotational speeds (θ_1, θ_2) and the torques (T_1, T_2) should have the transformer ratio N_T of N_1/N_2. The parameters N_1 and N_2 are the number of teeth in the input gear number 1 and the output gear number 2, respectively, as shown in Figure 2.6a. The left vector is $[\theta_1 \quad T_2]^T$ and the right vector is $[T_1 \quad \theta_2]^T$. For a shaft and pulley system shown in Figure 2.6b, the transformer ratio N_T is $1/r_P$. The variable r_P is the pulley radius. The left vector is $[\theta_1 \quad f_2]^T$ and the right vector is $[T_1 \quad \delta_2]^T$. The hydraulic centrifugal motor in Figure 2.6c has the transformation ratio $N_T = c_1$ for the flow rate to torque relation. The parameter c_2 is for the relation between the rotational speed and the pressure. The left vector is $[q_1 \quad \dot{\theta}_2]^T$, and the right vector is $[p_1 \quad T_2]^T$. The pressure p_1 is the head difference $h_1 - h_2$. A hydraulic transformer shown in Figure 2.6d would have the transformation ratio $N^T = A_2/A_1$. The variables A_1 and A_2 are the areas of side 1 and side 2, respectively. The left vector is $[p_1 \quad q_2]^T$, and the right vector is $[q_1 \quad p_2]^T$.

A direct current (DC) motor (permanent magnet field) is shown in Figure 2.6e. The transformation ratio $N_T = k_1$ for the volt to rotational speed relation and k_2 for the torque to the current relation. The left vector is $[v_1 \quad T_2]^T$, and the right vector is $[i_1 \quad \dot{\theta}_2]^T$. This can also be used as a generator of a DC current with the mechanical shaft s as the input side. Figure 2.6f shows the alternating electric current (AC) transformer that has the transformer ratio N_T of N_1/N_2, where N_1 and N_2 are the number of coil turns in the input coil 1 and the output coil 2, respectively. The left vector is $[v_1 \quad i_2]^T$, and the right vector is $[i_1 \quad v_2]^T$.

(b) ***Gyrator (GY)***: The gyrator designation is borrowed from the mechanical gyroscope even though few other mechanical engineering components or systems perform gyration such as hydraulic cylinders, hydraulic positive displacement motors, and air billows. The main characteristic is the composition of the left and right vectors solely out of either across or flow variables. Figure 2.7 shows a schematic of an ideal hydraulic cylinder (a) and its system graph (b). Either (q_1, f_2) or (p_1, δ_2) may be arbitrary specified, but not any other pair combinations can be arbitrarily set.

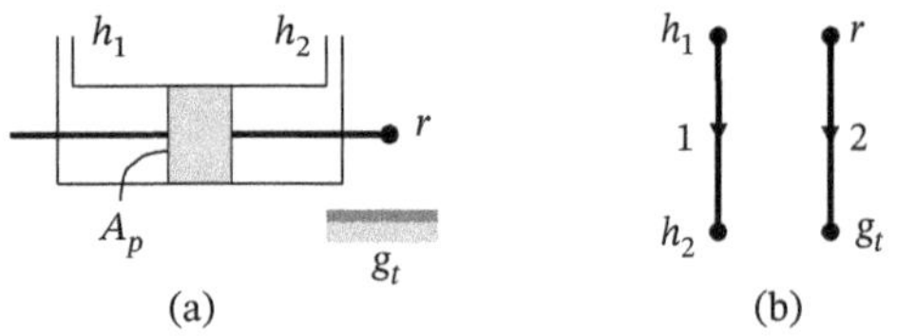

Figure 2.7 A multi-terminal *gyrator* component: (a) schematic of an ideal hydraulic cylinder; the pressure p_1 is the head difference $h_1 - h_2$; (b) system graph of the hydraulic cylinder.

The terminal equation of this hydraulic cylinder is then

$$\begin{bmatrix} q_1 \\ f_2 \end{bmatrix} = \begin{bmatrix} 0 & A_p \\ -A_p & 0 \end{bmatrix} \begin{bmatrix} p_1 \\ \dot{\delta}_2 \end{bmatrix} \tag{2.18}$$

$$\begin{bmatrix} \text{Measurement} \\ \text{or "effect"} \end{bmatrix} = [\boldsymbol{W}_{\text{HC}}]_{2\text{x}2} \begin{bmatrix} \text{Driver or} \\ \text{"Cause"} \end{bmatrix}$$

where A_p is the piston area, $\boldsymbol{W}_{\text{HC}}$ is the hydraulic cylinder component matrix, and the left vector is the flow rate q_1 of terminal 1 and the force f_2 of terminal 2. The right vector is all of the across variables of terminals 1 and 2, respectively, that is, the pressure p_1 and the velocity $\dot{\delta}_2$. The pressure p_1 is the head difference $h_1 - h_2$. The equations are obtained by setting $p_1 = 0$ (*no load* or "open circuit"), then $\delta_2 = 0$ (*locked position* or "short circuit"). The process of setting is equivalent to using a driver or a "*cause*." The left-hand vector of Eq. (2.15) is the measurements or the "*effect*" of setting the right-hand vector.

In general, a ***Gyrator*** is then

$$\begin{bmatrix} e_1 \\ e_2 \end{bmatrix} = \begin{bmatrix} 0 & N_G \\ -N_G & 0 \end{bmatrix} \begin{bmatrix} f_1 \\ f_2 \end{bmatrix} \tag{2.19}$$

where N_G is the gyrator ratio, which is A_P for the hydraulic cylinder. The gyrator equation can also be inverted to the following form:

$$\begin{bmatrix} f_1 \\ f_2 \end{bmatrix} = \begin{bmatrix} 0 & -1/N_G \\ 1/N_G & 0 \end{bmatrix} \begin{bmatrix} e_1 \\ e_2 \end{bmatrix} \tag{2.20}$$

A hydraulic positive displacement motor relating the flow rate to the rotational speed $(q_1, \dot{\theta}_2)$ and the torque to the pressure (T_1, p_2) of input–output sides will have the gyrator ratio N_G of V_H, where V_H is the volumetric displacement of the motor (Figure 2.8a). The pressure p_1 is the head difference $h_1 - h_2$. The left vector is $[q_1 \quad T_2]^T$, and the right vector is $[p_1 \quad \dot{\theta}_2]^T$. For a gyroscope (Figure 2.8b), the gyrator ratio N_G is $J_G\omega_G$, where J_G and ω_G are the polar moment of inertia and the rotational speed of the gyroscope rotor, respectively. The left vector is $[T_1 \quad T_2]^T$ and the right vector is $[\dot{\theta}_1 \quad \dot{\theta}_2]^T$. The air billows in Figure 2.8c has the gyrator ratio $N_G = A_B$, where A_B is the air billow's area. The left vector is $[f_1 \quad q_2]^T$ and the right vector is $[\dot{\delta}_1 \quad p_2]^T$.

(c) ***<u>Amplifier (S)</u>:*** The amplifier component is mainly used in measurements and control systems. Figure 2.9 shows a schematic of an ideal amplifier (a), its system graph (b), and its block representation (c). It is

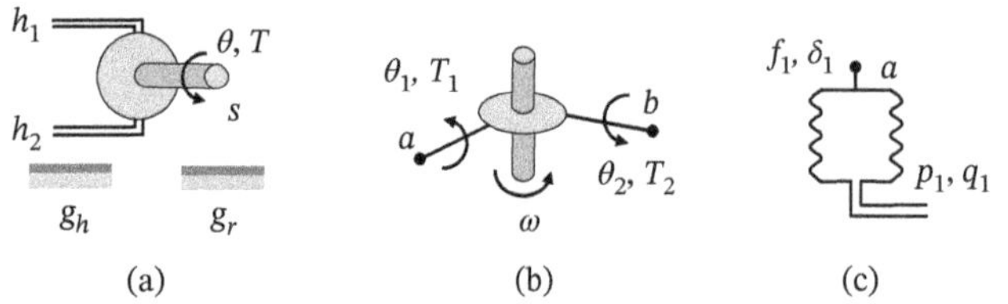

Figure 2.8 Other *gyrator* components: (a) hydraulic positive displacement motor, (b) a gyroscope (or a top), and (c) air billows.

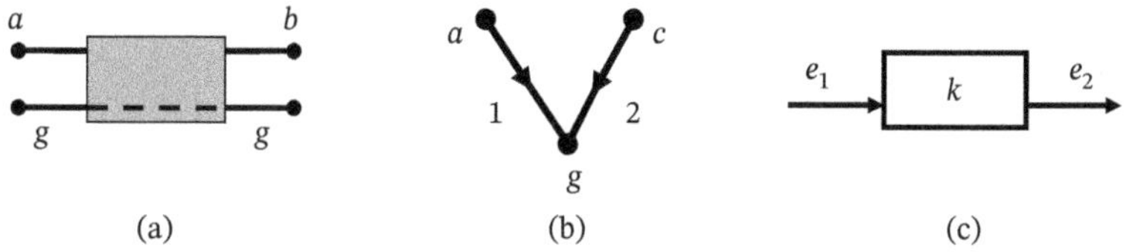

Figure 2.9 Schematic of an ideal *amplifier* (a), its system graph (b), and its block representation (c).

reviewed herein for some completion. It usually separates the systems into unconnected subsystems in the sense of transfer power and dependencies. The power consumption of the amplifier is typically coming from an outside source to the system. It might be used if one knows that some subsystems of a design are minutely affecting the design. Therefore, we have a decoupling between the design and subsystem, and both can be synthesized separately. In general, the amplifier can take the form

$$\begin{bmatrix} f_1 \\ e_2 \end{bmatrix} = \begin{bmatrix} \overset{\infty\text{ input impedance}}{0} & \overset{\text{Unilateral component}}{0} \\ \underset{\text{Gain coefficient}}{k_A} & \underset{\text{Zero output impedance}}{0} \end{bmatrix} \begin{bmatrix} e_1 \\ f_2 \end{bmatrix} \tag{2.21}$$

where k_A is the amplifier gain coefficient and the rest of zero terms attest to the fact that there is no coupling or other effect of terminal 2 on terminal 1 and vice versa. The amplifier is then indicated by the simple block diagram in Figure 2.9c. Equation (2.21) is also reduced to the following relation:

$$e_2 = k_A e_1 \tag{2.21$'$}$$

Examples of amplifiers are all measuring instruments that should not load the system to any appreciable measure. A hydraulic amplifier is another example. A *claw-lever* with a tremendous ratio may be considered as an amplifier.

The following examples are introduced mainly to demonstrate the systematic nature of formulation procedure. Much simpler procedures can be used to get similar results. However, the demonstrated procedure herein can be applied to multitude of components and much more complex dynamic systems with very little added effort.

Example 2.2 The simple system of two springs assembled in parallel is shown in Figure 2.10a. The assembly is to be modeled by a system graph to show the procedure and define the equivalent spring to the system. The springs are acted upon by a flow or across driver. Find the difference in performance of each driver.

Solution

The standard sign convention used in system graphs is depicted in Figure 2.10b. Any element assigned to the direction exiting a node (or a junction) is positive (+), and it is negative (−) if the assigned direction is entering the node. The system graph is shown in Figure 2.10c indicating the links between the input at node (a), and the

Figure 2.10 A simple system of two springs assembled in parallel: (a) schematic diagram, (b) the standard sign convention, and (c) the system graph representation.

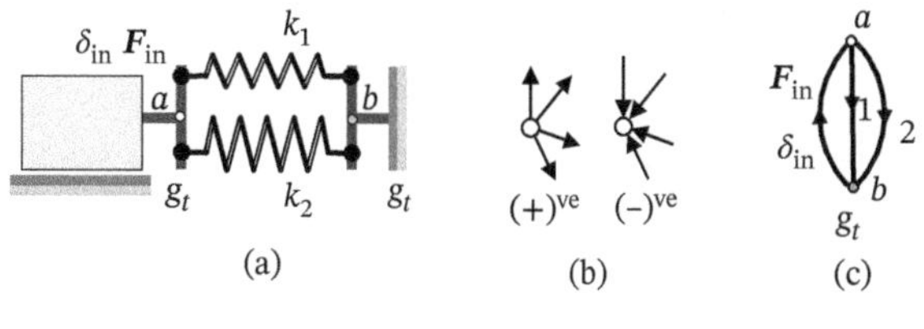

springs 1 and 2 are connecting (a) and (b), which is the ground of translation g_t. Usually *drivers* are assigned to a direction that is *entering* the node (a). The spring components are assigned to the direction from (a) to (g_t). The solution, however, is not affected by these assigned directions. Any assignment usually works.

The node equations are written with reference to Eq. (2.8) as

$$\sum_{\text{at points of junction}} f_i = 0 \tag{a}$$

The node equation at node (a) is then (exit is + and enter is −) $F_1 + F_2 - F_{\text{in}} = 0$ or in a *vector* form

$$\begin{bmatrix} 1 & 1 & -1 \end{bmatrix} \begin{bmatrix} F_1 \\ F_2 \\ F_{\text{in}} \end{bmatrix} = 0 \tag{b}$$

or in a general *matrix* form,

$$\boldsymbol{A}_n \boldsymbol{f}_n = 0 \tag{b$'$}$$

where $\boldsymbol{A}_n$ is the *node matrix* ($n_{v-1} \times n_e$) and $\boldsymbol{f}_n$ is the *flow vector* (n_e), with n_e as the number of components, and n_{v-1} is the number of vertices or nodes −1. By partitioning Eq. (b) to separate the input driver term, one gets

$$\begin{bmatrix} 1 & 1 \end{bmatrix} \begin{bmatrix} F_1 \\ F_2 \end{bmatrix} + [-1]F_{\text{in}} = 0 \tag{c}$$

or in a general matrix form,

$$\boldsymbol{A}_1 \boldsymbol{f}_c + \boldsymbol{A}_{\text{in}} \boldsymbol{f}_{\text{in}} = 0 \tag{c$'$}$$

where $\boldsymbol{A}_1$ is the reduced node matrix (reduced by the input flow or across driver effect), $\boldsymbol{f}_c$ is the component flow vector, and $\boldsymbol{A}_{\text{in}}$ is the *input matrix* (or vector). The component equations are

$$\begin{bmatrix} F_1 \\ F_2 \end{bmatrix} = \begin{bmatrix} k_1 & 0 \\ 0 & k_2 \end{bmatrix} \begin{bmatrix} \delta_1 \\ \delta_2 \end{bmatrix} \tag{d}$$

or in a matrix form,

$$\boldsymbol{f}_c = \boldsymbol{W}_c \boldsymbol{\delta}_c \tag{d$'$}$$

Here, $\boldsymbol{W}_c$ is the *components admittance* matrix, and $\boldsymbol{\delta}_c$ is the component displacement vector. Substituting into the partitioned Eq. (c) gives

$$\begin{bmatrix} 1 & 1 \end{bmatrix} \boldsymbol{W}_c \begin{bmatrix} \delta_1 \\ \delta_2 \end{bmatrix} + [-1]F_{\text{in}} = \boldsymbol{0} \tag{e}$$

or in a matrix form,

$$\boldsymbol{A}_1 \boldsymbol{W}_c \boldsymbol{\delta}_c + \boldsymbol{A}_{\text{in}} F_{\text{in}} = 0 \tag{e$'$}$$

The node variable transformation relates the *element* displacements to the *node* displacements using the same sign convention of Figure 2.10b such that

$$\begin{bmatrix} \delta_1 \\ \delta_2 \end{bmatrix} = \begin{bmatrix} 1 \\ 1 \end{bmatrix} \delta_a \tag{f}$$

or in a matrix form,

$$\boldsymbol{\delta}_c = \boldsymbol{A}_1^T \delta_n \tag{f$'$}$$

A_1^T is the *node variable transformation* matrix. This is known from graph theory that the node variable transformation matrix is the transpose of the reduced node matrix. Substituting Eq. (f) into Eq. (e) gives

$$\begin{bmatrix} 1 & 1 \end{bmatrix} \begin{bmatrix} k_1 & \\ & k_2 \end{bmatrix} \begin{bmatrix} 1 \\ 1 \end{bmatrix} \delta_a + [-1]F_{\text{in}} = 0 \tag{g}$$

or in a matrix form,

$$\boldsymbol{A}_1 \boldsymbol{W}_c \boldsymbol{A}_1^T \delta_n + \boldsymbol{A}_{\text{in}} F_{\text{in}} = 0 \tag{g$'$}$$

This is the ***general node equation*** that can be used irrespective of the number of components and how they are connected in any system graph. Multiplying the constituents of Eq. (g) provides the relation

$$[k_1 + k_2]\delta_a + [-1]F_{\text{in}} = 0 \tag{h}$$

This is the ***system equation*** as a function of the *node across variables*, which simply provides the following sought relation:

$$[k_1 + k_2]\delta_a = F_{\text{in}} \tag{i}$$

But as δ_a is the same as δ_{in}, then Eq. (i) gives

$$\frac{F_{\text{in}}}{\delta_{\text{in}}} = (k_1 + k_2) = k_S \tag{j}$$

The solution indicates that the parallel spring combined stiffness k_S is equal to the sum of both stiffnesses. The performance will be the same if the input is a force or a displacement since we have only one node (a) relative to the ground.

The solution can simply be obtained by summing the spring forces F_1 and F_2 as equal to the input F_{in} (***FBD*** at a and Eqs. (b) and (c)) and realizing that the deflection of either is the same as the deflection of the input δ_{in} (Eq. (f)). One has also to account for the spring relations (Eq. (g)) to substitute in Eq. (a).

Example 2.3 The simple system of two springs assembled in series is shown in Figure 2.11a as two connected flexible shafts. The assembly is to be modeled by a system graph to show the procedure and define the equivalent spring to the system. The system is acted upon by a flow or across driver. Find the difference in the performance of the system relative to other modeling techniques.

Solution

Again, the standard sign convention that is used in system graphs is depicted in Figure 2.11b. Any element assigned to the direction exiting a node is positive (+), and it is negative (−) if the assigned direction is entering the node. The system graph is shown in Figure 2.11c, indicating the links between the input at node (a). The spring 1 is connecting (a) and (b), while spring 2 is connecting node (b) and (c), which is the ground of rotation g_r. Usually drivers are assigned to a direction that is entering the node (a). The spring components are assigned to the direction

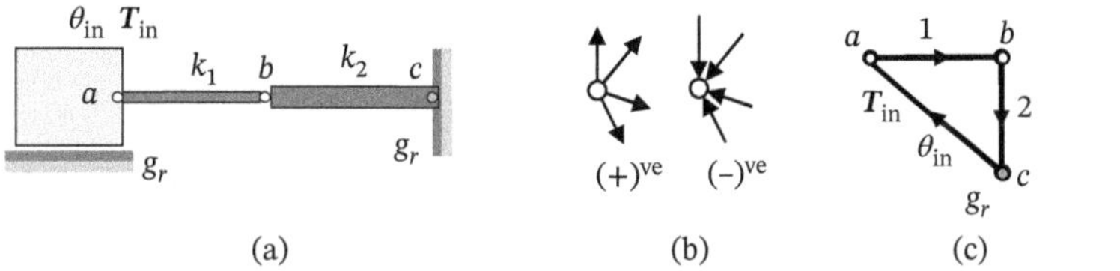

Figure 2.11 A simple system of two springs assembled in series: (a) schematic diagram, (b) the standard sign convention, and (c) the system graph representation.

from (a) to (b) and from (b) to the ground (g_r). The solution, however, is not affected by these assigned directions. Any assignment usually works.

The node equations are written with reference to Eq. (2.8) as

$$\sum_{\text{at points of junction}} f_i = 0 \tag{a}$$

The node equations from Figure 2.11c at nodes (a) and (b) are then

$$\begin{matrix}\text{at } a: \\ \text{at } b:\end{matrix} \begin{bmatrix} 1 & 0 & -1 \\ -1 & 1 & 0 \end{bmatrix} \begin{bmatrix} T_1 \\ T_2 \\ T_{\text{in}} \end{bmatrix} = \mathbf{0} \tag{b}$$

or in a general *matrix* form,

$$\boldsymbol{A}_n \boldsymbol{T}_n = 0 \tag{b$'$}$$

where $\boldsymbol{A}_n$ is the *node matrix* $(n_{v-1} \times n_e)$ and $\boldsymbol{T}_n$ is the *flow vector* (n_e), with n_e as the number of components, and n_{v-1} is the number of vertices or nodes −1. By partitioning Eq. (b) to separate the input driver term, one gets

$$\begin{bmatrix} 1 & 0 \\ -1 & 1 \end{bmatrix} \begin{bmatrix} T_1 \\ T_2 \end{bmatrix} + \begin{bmatrix} -1 \\ 0 \end{bmatrix} T_{\text{in}} = 0 \tag{c}$$

or in a general matrix form,

$$\boldsymbol{A}_1 \boldsymbol{T}_c + \boldsymbol{A}_{\text{in}} \boldsymbol{T}_{\text{in}} = \mathbf{0} \tag{c$'$}$$

where $\boldsymbol{A}_1$ is the reduced node matrix (reduced by the input flow or across driver effect), $\boldsymbol{T}_c$ is the component flow vector, and $\boldsymbol{A}_{\text{in}}$ is the *input matrix* (or vector). The component equations are

$$\begin{bmatrix} T_1 \\ T_2 \end{bmatrix} = \begin{bmatrix} k_1 & \\ & k_2 \end{bmatrix} \begin{bmatrix} \theta_1 \\ \theta_2 \end{bmatrix} \tag{d}$$

or in a matrix form,

$$\boldsymbol{T}_c = \boldsymbol{W}_c \boldsymbol{\theta}_c \tag{d$'$}$$

Here, $\boldsymbol{W}_c$ is the *components admittance* matrix, and $\boldsymbol{\theta}_c$ is the component across vector. Substituting into the partitioned Eq. (c) gives

$$\begin{bmatrix} 1 & 0 \\ -1 & 1 \end{bmatrix} \boldsymbol{W} \begin{bmatrix} \theta_1 \\ \theta_2 \end{bmatrix} + \begin{bmatrix} -1 \\ 0 \end{bmatrix} T_{\text{in}} = \mathbf{0} \tag{e}$$

or in a matrix form,

$$\boldsymbol{A}_1 \boldsymbol{W}_c \boldsymbol{\theta}_c + \boldsymbol{A}_{\text{in}} T_{\text{in}} = \mathbf{0} \tag{e$'$}$$

The node variable transformation relates the *element* angular displacements to the *node* angular displacements using the same sign convention of Figure 2.11b such that

$$\begin{bmatrix} \theta_1 \\ \theta_2 \end{bmatrix} = \begin{bmatrix} 1 & -1 \\ 0 & 1 \end{bmatrix} \begin{bmatrix} \theta_a \\ \theta_b \end{bmatrix} \tag{f}$$

or in a matrix form,

$$\boldsymbol{\theta}_c = \boldsymbol{A}_1^T \boldsymbol{\theta}_n \tag{f'}$$

Again, $\boldsymbol{A}_1^T$ is the node variable transformation matrix. This is known from graph theory that the node variable transformation matrix is the transpose of the reduced node matrix. Substituting Eq. (f) into Eq. (e) gives

$$\begin{bmatrix} 1 & 0 \\ -1 & 1 \end{bmatrix} \begin{bmatrix} k_1 & \\ & k_2 \end{bmatrix} \begin{bmatrix} 1 & -1 \\ 0 & 1 \end{bmatrix} \begin{bmatrix} \theta_a \\ \theta_b \end{bmatrix} + \begin{bmatrix} -1 \\ 0 \end{bmatrix} T_{\text{in}} = 0 \tag{g}$$

or in a matrix form,

$$\boldsymbol{A}_1 \boldsymbol{W}_c \boldsymbol{A}_1^T \boldsymbol{\theta}_n + \boldsymbol{A}_{\text{in}} T_{\text{in}} = 0 \tag{g'}$$

This is the ***general node equation*** that can be used irrespective of the number of components and how they are connected in any system graph. Multiplying the constituents of Eq. (g) provides the relation

$$\begin{bmatrix} k_1 & -k_1 \\ -k_1 & k_1 + k_2 \end{bmatrix} \begin{bmatrix} \theta_a \\ \theta_b \end{bmatrix} + \begin{bmatrix} -1 \\ 0 \end{bmatrix} T_{\text{in}} = 0 \tag{h}$$

This is the ***system equation*** as a function of the *node across variables*. It is observed that the ***system matrix*** on the left side of Eq. (h) is *always symmetric*. Expanding the system equation gives

$$\begin{aligned} k_1\theta_a - k_1\theta_b - T_{\text{in}} &= 0 \\ -k_1\theta_a + (k_1 + k_2)\theta_b &= 0 \end{aligned} \tag{i}$$

But as θ_a is the same as θ_{in}, then substitute into the first equation of (i) for θ_b from the second equation and with little manipulation to get the following:

$$\frac{T_{\text{in}}}{\theta_{\text{in}}} = \frac{k_1 k_2}{(k_1 + k_2)} = k_S \tag{j}$$

This solution indicates that the series springs combined stiffness k_S is a combination of both stiffness's. The performance will be the same if the input is a torque or an angular displacement since we have only one expression (*j*) that can be used for either case.

This solution may regularly be obtained by summing the spring deflections θ_1 and θ_2 as equal to the input θ_{in} and realizing that the torque on either is the same as the torque of the input T_{in}. Some manipulation similar to solving Eq. (i) is also needed to get the same relation as Eq. (j). The performance here is elaborating on the system response at each of the nodes or connections (a), (b), and (g_r). This might not be apparent unless we consider more than one (***FBD***), i.e. one for each of the components of the system.

Example 2.4 Figure 2.12a shows a schematic diagram of a constant angular velocity motor driving a hoist "3" that is lifting a mass m_5 through a flexible cable "4." The motor and the hoist are connected by a flexible shaft "1," which is defined by its stiffness k_1. The flexibility of the cable is defined by its stiffness k_4. It is required to model the system during its uniform velocity state of use. No dynamics of its start or stop are needed for now. It is also required to find the reactions and the torque considered necessary by the motor to drive the system. The angular deflections of the shaft and the total deformation of the cable are to be obtained.

Solution

The ***FBD*** of the system is shown in Figure 2.12b where the cable mass is detached only to demonstrate the transmitted load to the hoist drum. At that support, the reaction force vector $\boldsymbol{R}_1$ is defined by its expected component

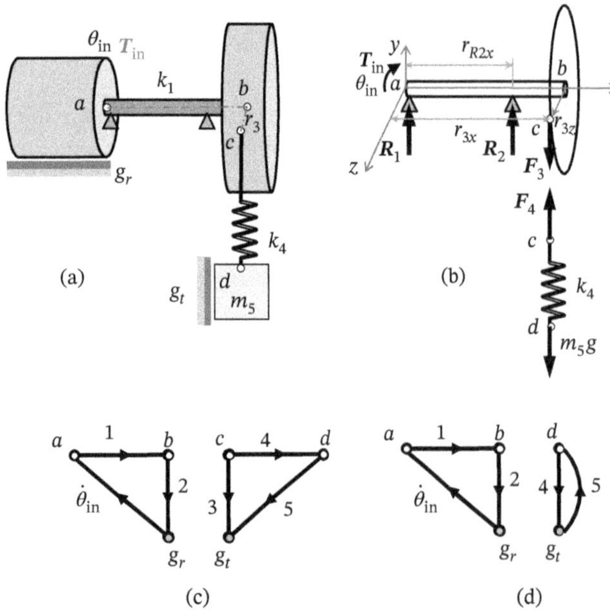

Figure 2.12 A schematic diagram in (a), ***FBDs*** in (b), and system graphs in (c) and (d) showing a constant angular velocity motor using a flexible shaft "1" to drive a hoist "3" that is lifting a mass "5" through a flexible cable "4."

in the y direction. The reaction force vector $\boldsymbol{R}_2$ is treated the same way. The torque vector $\boldsymbol{T}_{\text{in}}$ is also indicated but without showing its components. It is understood that the motor will only provide a torque about the x-axis or only $T_{\text{in},x}$ is not zero. It should be noted that no other external moment is present for the whole ***FBD*** particularly at the applied force location. Applying Eqs. (2.4), (2.5), and (2.6) gives

$$\begin{bmatrix} 0 \\ -m_5 g \\ 0 \end{bmatrix} + \begin{bmatrix} R_{1x} \\ R_{1y} \\ R_{1z} \end{bmatrix} + \begin{bmatrix} R_{2x} \\ R_{2y} \\ R_{2z} \end{bmatrix} = 0 \tag{a}$$

$$\begin{bmatrix} 0 & -r_{3z} & r_{3y} \\ r_{3z} & 0 & -r_{3x} \\ -r_{3y} & r_{3x} & 0 \end{bmatrix} \begin{bmatrix} 0 \\ -m_5 g \\ 0 \end{bmatrix} + \begin{bmatrix} 0 & -r_{R2z} & r_{R2y} \\ r_{R2z} & 0 & -r_{R2x} \\ -r_{R2y} & r_{R2x} & 0 \end{bmatrix} \begin{bmatrix} R_{2x} \\ R_{2y} \\ R_{2z} \end{bmatrix} + \begin{bmatrix} T_{\text{in},x} \\ T_{\text{in},y} \\ T_{\text{in},z} \end{bmatrix} = \mathbf{0} \tag{b}$$

From Eq. (a) and since there are no applied forces in the x or z directions, one gets

$$R_{1x} = R_{2x} = 0, \quad R_{1y} = m_5 g - R_{2y}, \quad R_{1z} = R_{2z} = 0 \tag{c}$$

Processing Eq. (b) in details and taking the values in Eq. (c) into account shows that

$$\begin{bmatrix} 0 & -r_{3z} & 0 \\ r_{3z} & 0 & -r_{3x} \\ 0 & r_{3x} & 0 \end{bmatrix} \begin{bmatrix} 0 \\ -m_5 g \\ 0 \end{bmatrix} + \begin{bmatrix} 0 & 0 & 0 \\ 0 & 0 & -r_{R2x} \\ 0 & r_{R2x} & 0 \end{bmatrix} \begin{bmatrix} 0 \\ R_{2y} \\ 0 \end{bmatrix} + \begin{bmatrix} T_{\text{in},x} \\ T_{\text{in},y} \\ T_{\text{in},z} \end{bmatrix} = \mathbf{0} \tag{d}$$

or

$$\begin{bmatrix} T_{\text{in},x} \\ T_{\text{in},y} \\ T_{\text{in},z} \end{bmatrix} = \begin{bmatrix} -r_{3z} m_5 g \\ 0 \\ r_{3x} m_5 g - r_{R2x} R_{2y} \end{bmatrix} \tag{d$'$}$$

This states that the motor must generate a moment of magnitude $(-r_{3z}m_5g)$ about the x-axis (i.e. a torque). Since no torque is possibly applied by the motor about z, the value of R_{2y} is obtained to balance the system and verify that $T_{in,z} = 0$ in Eq. (d) and R_{1y} is thereafter obtained from Eq. (c).

It should be noted that the conditions leading to Eq. (c) have been intuitively decided upon. Another intuitive observation is applied to Eq. (d) to justify the results obtained for R_{2y} and R_{1y}. Mathematically, iterations of Eqs. (a) and (b) to solve simultaneously should produce the same results.

System Model

Again, the standard sign convention in system graphs is used. Any element assigned to the direction exiting a node is positive (+), and it is negative (−) if the assigned direction is entering the node. The system graph is shown in Figure 2.12c, indicating the links between the input at node (a) and the spring "1" connecting (a) and (b), while the cable spring "4" is connecting node (c) and (d), which is acted upon by the gravitational force of mass m_5 relative to the ground of translation g_t. The cable drum system is a ***transformer*** relating the drum rotation to the cable translation. Usually *drivers* are assigned to a direction that is entering the node (a). The springs (shaft) and cable components are assigned to the direction from (a) to (b) and from (c) to (d) with the weight of m_5 from (d) to the ground of translation (g_t). The solution, however, is not affected by these assigned directions. Any assignment usually works.

The node equations are written with reference to Eq. (2.8) as

$$\sum_{\text{at points of junction}} f_i = 0 \tag{e}$$

The node equations are then

$$\begin{bmatrix} 1 & 0 & -1 & 0 & 0 & 0 \\ -1 & 1 & 0 & 0 & 0 & 0 \\ 0 & 0 & 0 & 1 & 1 & 0 \\ 0 & 0 & 0 & 0 & -1 & 1 \end{bmatrix} \begin{bmatrix} T_1 \\ T_2 \\ T_{in} \\ f_3 \\ f_4 \\ f_5 \end{bmatrix} = 0 \tag{f}$$

or in a matrix form,

$$\boldsymbol{A}_n \boldsymbol{f}_n = \boldsymbol{0} \tag{2.22}$$

where $\boldsymbol{A}_n$ is the *node matrix* $(n_{v-1} \times n_e)$ and $\boldsymbol{f}_n$ is the *flow vector* (n_e), with n_e as the number of components and n_{v-1} is the number of vertices or nodes −1. By partitioning Eq. (f) to separate the input driver term, one gets

$$\begin{bmatrix} 1 & & & & \\ -1 & 1 & & & \\ & & 1 & 1 & \\ & & & -1 & 1 \end{bmatrix} \begin{bmatrix} T_1 \\ T_2 \\ f_3 \\ f_4 \\ f_5 \end{bmatrix} + \begin{bmatrix} -1 \\ \\ \\ \\ \end{bmatrix} T_{in} = 0 \tag{g}$$

or in a matrix form,

$$\boldsymbol{A}_1 \boldsymbol{f}_c + \boldsymbol{A}_{in} \boldsymbol{f}_{in} = \boldsymbol{0} \tag{2.23}$$

where $\boldsymbol{A}_1$ is the reduced node matrix (reduced by the input flow or across driver effect), $\boldsymbol{f}_c$ is the component flow vector, $\boldsymbol{A}_{in}$ is the input matrix (or vector), and $\boldsymbol{f}_{in}$ is the input flow vector. The zeros in the matrices $\boldsymbol{A}_1$ and $\boldsymbol{A}_{in}$ are not shown and left as empty entries. The component equations are

$$\text{for the transformer,}\quad \begin{bmatrix}\theta_2\\ f_3\end{bmatrix}=\begin{bmatrix}0 & {}^{1}/_{r_3}\\ {}^{-1}/_{r_3} & 0\end{bmatrix}\begin{bmatrix}T_2\\ \delta_3\end{bmatrix}\quad\text{or}\quad\begin{bmatrix}T_2\\ \delta_3\end{bmatrix}=\begin{bmatrix} & -r_3\\ r_3 & \end{bmatrix}\begin{bmatrix}\theta_2\\ f_3\end{bmatrix}\tag{h}$$

$$\text{for the two “springs,”}\quad \begin{bmatrix}T_1\\ f_4\end{bmatrix}=\begin{bmatrix}k_1 & \\ & k_4\end{bmatrix}\begin{bmatrix}\theta_1\\ \delta_4\end{bmatrix}\tag{i}$$

Apparently, it is not possible or straightforward to substitute the transformer equation in (h) into Eq. (g). The ***transformer*** equation (h) should be put in a different form similar to a ***gyrator*** to be able to substitute for the component relations with only flow variables to the left as Eq. (i). This can be done by including the stiffness of the shaft k_1 or the cable k_4 into the hoist drum model. Regenerating the drum–cable relations in Eqs. (h) and (i), one gets

$$T_2=-r_3f_3,\quad \theta_2=\left({}^{-1}/_{r_3}\right)\delta_3,\quad\text{and}\quad f_4=k_4\delta_4\tag{j}$$

Using a setting of $\theta_2=0$ (*locked position*-short circuit), then $f_4=0$ (*no load*-open circuit), one gets

$$T_2=-r_3f_3=-r_3f_4=-r_3k_4\delta_4$$

$$\theta_2=\left({}^{-1}/_{r_3}\right)\delta_3=\left({}^{-1}/_{r_3}\right)\left(\delta_5-\frac{f_4}{k_4}\right)\quad\text{or}\quad\frac{f_4}{k_4}=-r_3\theta_2+\delta_5\tag{k}$$

Or from Eq. (k), a *non-ideal* ***transformer*** is turned into a *non-ideal* ***gyrator*** as

$$\begin{bmatrix}T_2\\ f_4\end{bmatrix}=\begin{bmatrix}0 & -r_3k_4\\ -r_3k_4 & k_4\end{bmatrix}\begin{bmatrix}\theta_2\\ \delta_5\end{bmatrix}\tag{l}$$

In a general matrix form,

$$\boldsymbol{f}_c=\boldsymbol{W}_{\mathrm{TG}}\boldsymbol{e}_c\tag{2.24}$$

where $\boldsymbol{W}_{\mathrm{TG}}$ is the component matrix of the transformer turned into a non-ideal gyrator. The form of Eq. (l) encompasses element 3 into the hoist cable model defined in Eq. (l). The system graph becomes as defined in Figure 2.12d. Equation (g) therefore becomes

$$\begin{bmatrix}1 & & & \\ -1 & 1 & & \\ & & 1 & -1\end{bmatrix}\begin{bmatrix}T_1\\ T_2\\ f_4\\ m_5g\end{bmatrix}+\begin{bmatrix}-1\\ \\ \end{bmatrix}T_{\mathrm{in}}=0\tag{m}$$

By separating the constant force input of m_5g and substituting for the component Eqs. (i) and (l), we get

$$\begin{bmatrix}1 & & \\ -1 & 1 & \\ & & 1\end{bmatrix}\begin{bmatrix}k_1 & & \\ & & -r_3k_4\\ & -r_3k_4 & k_4\end{bmatrix}\begin{bmatrix}\theta_1\\ \theta_2\\ \delta_5\end{bmatrix}+\begin{bmatrix}-1 & \\ & \\ & -1\end{bmatrix}\begin{bmatrix}T_{\mathrm{in}}\\ m_5g\end{bmatrix}=\mathbf{0}\tag{n}$$

In a general matrix form,

$$\boldsymbol{A}_1\boldsymbol{W}_c\boldsymbol{e}_c+\boldsymbol{A}_{\mathrm{in}}\boldsymbol{f}_{\mathrm{in}}=\mathbf{0}\tag{2.25}$$

where $\boldsymbol{e}_c$ is the vector of the effort variables of all components. The node variable transformation relates the *element* displacements to the *node* displacements using the same sign convention of Figure 2.10b such that

$$\begin{bmatrix} \theta_1 \\ \theta_2 \\ \delta_5 \end{bmatrix} = \begin{bmatrix} 1 & -1 & \\ & 1 & \\ & & 1 \end{bmatrix} \begin{bmatrix} \theta_a \\ \theta_b \\ \delta_d \end{bmatrix} \tag{o}$$

In a matrix form,

$$\boldsymbol{e}_c = \boldsymbol{A}_1^T \boldsymbol{e}_n \tag{2.26}$$

where $\boldsymbol{e}_n$ is the vector of the effort variables of the nodes. Again, $\boldsymbol{A}_1^T$ is the node variable transformation matrix. This is known from graph theory that the *node variable transformation matrix* is the *transpose* of the *reduced node matrix*. Substituting Eq. (o) into Eq. (n) gives

$$\begin{bmatrix} 1 & & \\ -1 & 1 & \\ & & 1 \end{bmatrix} \begin{bmatrix} k_1 & & \\ & & -r_3k_4 \\ & -r_3k_4 & k_4 \end{bmatrix} \begin{bmatrix} 1 & -1 & \\ & 1 & \\ & & 1 \end{bmatrix} \begin{bmatrix} \theta_a \\ \theta_b \\ \delta_d \end{bmatrix} + \begin{bmatrix} -1 & \\ & \\ & -1 \end{bmatrix} \begin{bmatrix} T_{\text{in}} \\ m_5 g \end{bmatrix} = \boldsymbol{0} \tag{p}$$

Or in a general matrix form,

$$\boldsymbol{A}_1 \boldsymbol{W}_c \boldsymbol{A}_1^T \boldsymbol{e}_n + \boldsymbol{A}_{\text{in}} \boldsymbol{f}_{\text{in}} = \boldsymbol{0} \tag{2.27}$$

This is the ***general node equation*** that can be used irrespective of the number of components and how they are connected in any system graph. Multiplying the constituents of Eq. (p) provides the relation

$$\begin{bmatrix} k_1 & -k_1 & 0 \\ -k_1 & k_1 & -r_3k_4 \\ 0 & -r_3k_4 & k_4 \end{bmatrix} \begin{bmatrix} \theta_a \\ \theta_b \\ \delta_d \end{bmatrix} + \begin{bmatrix} -1 & 0 \\ 0 & 0 \\ 0 & -1 \end{bmatrix} \begin{bmatrix} T_{\text{in}} \\ m_5 g \end{bmatrix} = \boldsymbol{0} \tag{q}$$

or

$$\begin{bmatrix} k_1 & -k_1 & 0 \\ -k_1 & k_1 & -r_3k_4 \\ 0 & -r_3k_4 & k_4 \end{bmatrix} \begin{bmatrix} \theta_{\text{in}} \\ \theta_b \\ \delta_d \end{bmatrix} = \begin{bmatrix} T_{\text{in}} \\ 0 \\ m_5 g \end{bmatrix} \tag{r}$$

$$\boldsymbol{A}_S \boldsymbol{e}_n = -\boldsymbol{A}_{\text{in}} \boldsymbol{f}_{\text{in}} \tag{2.28}$$

This is the ***system equation*** as a function of the *node across variables*. It is observed that the ***system matrix*** $\boldsymbol{A}_S$ on the left side of Eq. (r) and Eq. (2.28) is *always symmetric*. Usually the system parameters k_1, k_4, and r_3 are known. The angles θ_{in} and θ_b should be in radians. Separating the first equation in (r), the next two equations can be solved for θ_b and δ_d for the input values of θ_{in} and m_5g. The input is moved to the right-hand side of Eq. (r). The effect of the system on the driver is the torque T_{in}, which is obtained from the first equation in (r). These simple manipulations give

$$\begin{bmatrix} k_1 & -r_3k_4 \\ -r_3k_4 & k_4 \end{bmatrix} \begin{bmatrix} \theta_b \\ \delta_d \end{bmatrix} = \begin{bmatrix} k_1 \\ 0 \end{bmatrix} \theta_{\text{in}} + \begin{bmatrix} 0 \\ m_5 g \end{bmatrix}$$

$$T_{\text{in}} = k_1\theta_{\text{in}} - k_1\theta_b \tag{s}$$

Inverting the matrix on the left-hand side of Eq. (s) provides the solution of θ_b and δ_d. With that, the torque T_{in} that should be provided by the input motor is obtained from Eq. (s). The steady-state energy provided by the motor is $T_{in}\theta_{in}$. This should help in providing the rated power of the motor if the rotational speed is used instead of the input angular displacement.

It is obvious that the system model provided herein gives a complete picture of the system performance. All system behavioral variables are derived and the effect of the system on the driver is also inherently an obtained output. Getting an equivalent output by other means of modeling might take about the same effort particularly mental. The systematic procedure using system modeling generalizes the approach and guarantees results with reasonable and less mental exercise. It is, however, left to the reader to assess the utility of this modeling technique relative to other traditional ways according to the system under his deliberation.

2.1.3 Modeling of Loads and Material Variations

Natural loading conditions are generally variable and *random* in time and value. However, these conditions are usually having some limits. Material properties in dynamic loading cases are usually not specified deterministically. It is unusual to find a material property data that is presented in a proper statistical manner by a *mean* value and a *standard deviation*. For critical applications where *reliability* is of a great significance, it is necessary to have the *frequency distribution* for the expected situations of loading variations (e.g. stresses) and material characteristic deviations (e.g. strength) as schematically represented in Figure 2.13. The loading is usually dynamic and random in nature with a wide disparity in magnitude. The frequency of loading occurrence may also be random. An example of that is a vehicle with different number of occupants and subjected to different road undulations at different times of operation during its expected life. The frequency of occurrence is little for very high occupants' weight in [lb] or [N] and very low occupants' weight in [lb] or [N]. The frequency of occurrence for *average* occupants' weight is usually high. The statistical model may be accepted as the customary bell-shaped *normal distribution* as illustrated in Figure 2.13. Population of many physical quantities follows the usual "*normal distribution*." The material properties are expected to behave as believed in Figure 2.13. However, the material property variation in strength is usually much narrower than indicated. The exaggeration is only used to clarify the intersection between the load and the material property distributions. The gray intersection area indicates that there is a probable occurrence of a higher load value than the property value, and *failure* is then expected to occur. The grayish area of intersection is then the probability of *failure*.

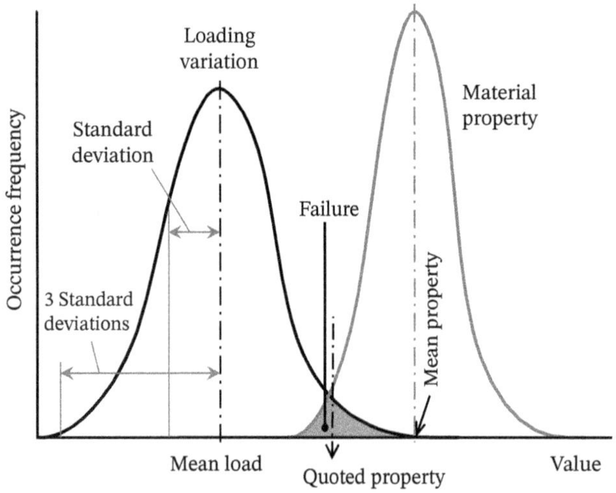

Figure 2.13 Schematic diagram of frequency distributions for loading variations and material property variation as randomly expected.

The ***reliability*** percent (=100% − failure probability %) increases as the area of failure probability decreases; see Figure 2.13. This can be achieved if loading variation and material property variations are limited or the mean values of both are further apart and the *standard deviation* of either is small. More than 99.7% of occurrences happen within ±3 *standard deviations*; see Figure 2.13. If the value of the maximum loading is 3 standard deviations above the average and if it happens at the minimum strength property of 3 standard deviations below the average, the reliability would be about 99.7%. Material-*quoted properties* are usually *cited* by manufacturers or suppliers at about −2 standard deviations to guarantee about 97.75% *reliability* of the quoted property value. It is known that 95.5% of occurrences happen within ±2 standard deviations; see Figure 2.13 and any statistical textbook. Optimum 100% *reliability* of a system requires that all components and subsystems of the design would fail at the same essentially specified extent of lifetime. This is almost probabilistically impossible. *Probabilistic design* is the approach to adopt in this specific reliability requested cases, which is beyond the scope of the text.

The classical and conservative design approach is habitually known to use the maximum load (i.e. stress) and the material property (i.e. strength). The classical *safety factor* is then the value of the material property (i.e. *strength*) divided by the value of the maximum load (i.e. *stress*). Usually for extra precaution or safety, one might use the maximum load and the minimum strength in place of the mean strength. Usually the material strength provided by the manufacturers is generally the mean strength property minus 2 standard deviations to be sure that strength tests would fall above their specified or quoted values. If all loading conditions and all material characteristic behaviors are exactly known, the classical factor of safety would be close to unity, and the material is then loaded to its maximum *possible, permissible,* or *allowable* limit. This is seldom done, but in aviation cases, the safety factor is surprisingly close to 1 (1.1–1.15). Regular *safety factor* in machine design cases would be around 1.5–2.5 depending on the sophistication of synthesis. Sometimes the factor of safety is dubbed as a *factor of ignorance and fear* of failure.

2.2 Calculation Tools

In the preceding examples few tools are used. *Vector–matrix* equations are utilized. Further review of these algebra tools is useful. This part of mathematics is usually a prerequisite to a machine design course. The basic matrix operations mostly needed are multiplication, transpose, and inverse. Some of these are reviewed in Section 1.11.3. The following tools are current, and some information might be checked since the software market is fast changing with mergers, acquisitions, and liquidations. These tools are used in the course of the text, and the need for a review is useful. Other similar or more useful tools may also be utilized.

2.2.1 Excel©

Excel© is one of the *spreadsheets* copyrighted by Microsoft©, Excel (2016). It can be used extensively, just like any other spreadsheet programs, for calculations such as iterations, curve fitting, statistics, simulations, synthesis, and some optimization. It is using a template of cells in rows (1, 2, 3, etc.) and columns (A, B, C, etc.) to perform calculations for a parameter with its name usually stated by the user in a cell to the left or top of the calculation cell. The calculation uses an equation or a function *fx* typed in the ***formula bar*** of say a `B5` cell (`B` is the column, and `5` is the row). Here, we highlight the cell formula or entered text for further clarity. The equation or formula is equal to a function of other variables defined numerically in other referenced cells such as a formula **(=B4*(PI()*B3^2)/4)** defining the volume of a cylinder that has a diameter in cell **B3** and a length in cell **B4**. One ought to use the enter symbol ✓ to the left of the function formula bar to admit his typed formula. The function **PI()** returns the value of π (i.e. 3.141 592 654) into the formula. For clarity, you could type the text "**Cylinder Diameter**" in cell **A3** (to the left of B3) and the text "**Cylinder Length**" in cell **A4** (to the left of B4). You could also type "**Cylinder Volume**" in **A5** cell (to the left of B5). All calculations of cylinder volumes in cell `B5` are then updated after cell values in `B3` and B4 are changed. In cell A1, you could have typed the title: "**Cylinder Volume Calculations.**" Cells A2, B2, etc. are left blank as line spacing. The formulas accommodate most of the known algebraic operations, geometric functions, statistics, and more.

One can also use cells C3–C5 to find the volume for another system of units. Click on the cell B5, and drag the lower right corner marked **(+)** to cell **C5**. The formula is copied in cell C5, but the B3 and B4 in the formula are changed to C3 and C4 where the new units' values should be entered. Using other cells, it is simple to calculate the mass of the cylinder or its mass moment of inertia or any other property related to the diameter and length of the cylinder. The proficiency of the user provides a useful background to follow and appreciate the utility of such a tool. *Excel* spreadsheets are utilized here in the synthesis of several machine components. With ample user proficiency, similar synthesis tools can be developed to the satisfaction of further requirements beyond what is developed in this text. One can also add more power to Excel by developing Macros and Visual Basic Applications (VBA) into Excel. The procedure and utility are left for the reader to peruse; if not familiar with, see Chapra and Canale (2010).

2.2.2 MATLAB©

MATLAB©, a copyright of MathWorks©, MATLAB (2014), is a known software package dealing with so many applications with an extensive staunch utility of matrix operations. It is similar to a programming language, if an ".m" extension MATLAB file is created rather than using the software "*command window*." The ".m" file (or *M-file*) can be executed from the MATLAB *interactive editor*, which is activated when the ".m" file is opened in any directory. Many calculations, however, can be processed using the command window of MATLAB. At the prompt "≫," you can enter your variables in name and value and use them – after that – in a formula defining other variables. The following statements demonstrate the process of calculating a cylinder volume. The statements can be written line by line or in one line separated by a comma or a semicolon: ≫**d=3; l=5;volume=l*pi*d∧2/4**. When pressing the computer *enter key*, the result is displayed in the next line as: volume = 35.3429. Note that the variable **pi** is the regular π value. When a comma rather than a semicolon is typed after an input, the input will also be rewritten after pressing the enter key. Using a semicolon will suppress rewriting of the variable. One can also open a new file in the MATLAB *Editor* and type the previous statement (or statements) without the prompt, i.e. **d=3; l=5; volume=l*pi*d∧2/4**. Running this program produces the answer that is displayed in the *command window* as: volume = 35.3429. Saving this as "CylinderVolume.m" allows its editing and running again at different values of *d* and *l*. The name of the file may not have a space or inadmissible characters.

Arrays are used in MATLAB for a one-dimensional *vector* such as a row vector $[x \;\; y \;\; z]$ and for a two-dimensional *matrix* such as a 2×2 $[x1 \;\; y1;\, x2 \;\; y2]$. The number of rows and columns is specified by the user. The semicolon implies the next row. The number of columns is specified by the number of entries with a single space or few spaces between each entry. It is feasible to calculate several cylinder volumes simultaneously by constructing a diameter array or vector, say **da = [3 5 7 9]**, and a length array, say **la = [5 10 15 20]**. The volume of each of the four cylinders can then be the components of the volume array obtained by the formula of **volume =pi*da.∧2.*la/4**. When clicking the *enter key*, the result is displayed in the *command window* as volume = 1.0e+003 *[0.0353 0.1963 0.5773 1.2723]. Note that the power operation (.∧) is an element-by-element power calculation. The multiplication operation denoted (.*) is an element-by-element multiplication for the two arrays of (da.∧2) and (la). The element-by-element multiplication or element-by-element powers of arrays are not familiar to regular users of vector or matrix multiplications. This operation, however, is available in MATLAB. It has been usefully applied in our cases. Other element-by-element operations are performed similarly by adding *a dot* (.) before the operation.

The data can also be entered in the command window when running a ".m" file using the line **d = input ('Diameter :')** in the M file. The ".m" file can then have the line: **d = input ('Cylinder Diameter :'); l = input ('Cylinder Length: '); Volume = l*pi*d∧2/4**. One can also use the string **'Input Cylinder Diameter'** or the string **'Cylinder Diameter =?'** for further clarity of the request. Running this program requires the response to each request for input with the desired value in the MATLAB *command window*, and the value of the "`Volume =`" will be displayed accordingly in the *command window*.

Conditional statement such as "**`if-then-else`**" can be used to advantage in MATLAB programming. For the cylinder volume problem, one can use the `if-then-else` on one line with parts separated by commas or semicolons if no displayed results are needed. The conditional statement can be typed such as **`if l==d; volume = pi*d^3/4, else, volume = l*pi*d^2/4; end`**. This procedure demonstrates the calculation expressions used when `l=d` and when `l` is not equal to `d`. Other statement that can be used is the "loop" control that takes the form **`while`** ***`expression, statements,`*** **`end`**. This loop control is substituting the known "DO" loop in other languages.

Components of arrays, matrices, or vectors can be specified by their location in the array such as the reaction *position vector* `rR2` in Figure 2.4. The components of this vector are `rR2(1)`, `rR2(2)`, and `rR2(3)`, which, respectively, are the x, y, and z components of the position vector `rR2` (of the reaction R2). The matrix **`a`** in Figure 2.4 is defined by the components of the reaction position vector `rR2`. The colon (:) used in the statement "**`for i=1:n`**" in Figure 2.4 denotes running the steps to the **`end`** statement for `i=1` to `i=n`. The colon (:) used in the array **`rXf(i,:)`** in Figure 2.4 denotes all the `i`-th row of `rXf` (r *cross* f) while executing the steps to the **`end`** statement. The same is for the `i`-th row of **`f(i,:)'`** in Figure 2.4. However, the code uses the *transpose* of `f(i,:)`, which indicates that `f(i,:)'` is the transpose of `f(i,:)`. The code in Figure 2.4 also uses the ***pseudo inverse*** (**`pinv`**) of the cross-product matrix since it is singular. This pseudo inverse is, however, valuable in finding solutions of *sparsely* populated and *non-square* matrix systems. The regular inverse of a matrix is calculated by the function (**`inv`**). The pseudo inverse is useful in many respects and seldom fails. The reader is advised to get more familiar with the pseudo inverse, which is also called the *minimum norm inverse*.

More capabilities are available in MATLAB. An extensive library and *Toolboxes* of built-in functions are accessible and can be applied equally to vectors or matrices. You can use MATLAB help to get more information and how each function can be utilized. Some of these will be used and explained later on. Similar to these, the own development of a "**`function`**" can be very useful in multiple applications of compound expressions to be evaluated and checked throughout your personal program. It has the form: **`function [out1, out2, out3 ... etc] = funname (in1, in2, in3 ... etc)`** with formulas defining the `out1, out2 ... etc` for the `in1, in2, in3 ... etc` and ending with the word **`end`**. This must be saved as "**`funname.m`**" where funname can be such as CylVol from our previous example. The "`CylVol.m`" will have the form of the starting line as **`function [volume] = CylVol (d, l)`**. The following lines are **`volume = l*pi*d^2/4`**, then the last line should be the key word **`end`**. When the function is called from the *command window* at the prompt such as ≫ `[volume] = CylVol (3, 5)`, the function will be executed, and the value of the volume will be returned in the command window as: `volume = 35.3429`. For more details, refer to other references that utilize more MATLAB codes such as Chapra and Canale (2010).

2.2.3 Computer-Aided Design (CAD)

Computer-aided design (CAD) is supposedly a computer-aided program to consummate a design. Most of the packages that are called "CAD" programs are mainly ***geometric modeling*** packages but has been customary nicknamed "CAD" packages. Most of these require the specifications of *geometry* and *topology* to generate the ***3D*** model of the design. To continue on with the completion of the design, the user has to specify the material and may transfer to an ***FE*** package to apply the load and calculate the deflections and the stresses or the safety factor. This is far from a *CAD* tool for calculations or an aid for design, since numerous and may be exhausting ***iterations*** are needed to reach a reasonable design. If an *expert* uses these packages, it is his experience that generates the design, and the package only helps in setting the design geometry in a computer ***3D*** model. Some ***3D*** geometric modeling (or "CAD") packages are *free* and available to download such as *FreeCAD, Fusion 360, Sketchup, OpenCasCade,*

BRL-CAD, etc. However, general *verification*, *validation*, and connectivity to others might be checked to allow for practical and accurate use of the software. Available commercial ***3D*** "CAD" packages such as *SolidWorks*©, *Pro Creo*© (formerly *Pro/Engineer*), *Unigraphics NX*©, *CATIA*©, *Inventor*©, *Solid Edge*©, etc. are useful particularly when additional computer programs are present to help in defining geometry and material selection. Few packages are accessible through some of these "CAD" packages to help in that endeavor. Available ***3D*** models of many standard components such as bolts, nuts, pins, keys, rolling bearings, etc. are also ready to be inserted into the ***3D*** model of the design. In our textbook, several tools are provided to perform a real *CAD* of machine components that provide the synthesis of these components defining the geometry and suitable material selection. These tools and programs are dubbed "***computer-aided synthesis***" (CAS) rather than CAD. The outcome can be transferred to other ***3D*** geometric programming packages for completing the full ***3D*** *assembly* of these components into the final design and further processing. The CAS process would constitute the paradigm shift in the real design process.

2.2.4 Finite Element (*FE*)

FE analysis is a numerical method to solve *partial differential equation* representing the ***3D*** material continuum of a component under boundary loads and support restraints. Other means are also available to numerically solve partial differential equations such as *boundary element*, *finite difference*, and other methods. Usually the more widely spread ***FE*** packages produce the deformations and stress distributions in a solid component under load. Some ***FE*** packages consider other special cases such as *contact* of components, existing *cracks*, nonlinear problems, and others. Several ***3D*** geometric modeling (CAD) packages have a selected ***FE*** package as an integral part to perform the analysis and allow geometry modification in a repetitive process to generate more suitable design by ***FE*** iterations or even some optimization in the ***FE*** package. Traditionally, ***FE*** analysis has been considered as part of the CAD, since geometric modeling does not produce calculations of stresses or deformations. The possible calculations produced by most ***3D*** geometric modeling packages are related to the ***3D*** encompassing volume such as *mass*, *center of gravity*, *mass moment of inertia*, and possible geometrical interference between components. Other ***3D*** model interaction with loading conditions and support restraints are handled by ***FE*** packages. Available commercial ***FE*** packages are standing alone or imbedded with the ***3D*** modeling packages such as *ANSYS*, *Nastran*, *Abaqus*, *ADINA*, etc. Other ***FE*** packages are *free* and available to download such as *FreeFem++*, *Deal.II*, *Elmer*, *GetFEM++*, etc. However, general *verification* and *validation* of these might be confirmed and should be needed to allow for accurate use of the software. In all cases, the successful use of any ***FE*** package is dependent on the user defining the right model, load, and boundary conditions. Any misrepresentation of these would produce a solution that can be totally unrepresentative and can be grossly inaccurate and misleading. An introduction to the ***FE*** method is covered in Section 6.11, which presents some basics and guidelines for additional understanding of this analysis and calculation tool.

2.3 Design Procedure

The initial process of a design requires some previous *knowledge* of similar or close designs, unless an invention is the intention. To gain knowledge about similar designs, one needs the technique of *reverse engineering*. This process entails a thorough examination of performance, geometry, and articulation of existing designs. If the design is to be developed from an existing one that has no ***3D*** model, you might need to create one. Reverse engineering procedure intends to generate a ***3D*** model of any component of a design by any measuring means such as a *vernier* or a *micrometer* or others. Also, a ***3D*** scanning using ***3D*** imaging, laser scanning or point-by-point generation of the body is a first step to generate the geometry; see, for example, Metwalli et al. (1999) for more details. The massive *cloud point data* generated by imaging or laser scanning is used to fit accurate surfaces to these data (Nassef et al. 1999). The generated surfaces encompass the ***3D*** solid representing the model. Several softwares are available in some geometric modeling (CAD) packages that do the ***3D*** model from the cloud point data or from the surfaces generated thereof. Evaluation of the ***3D*** model and its interaction with other components of the design provides a valuable knowledge to perform the modification and verification process of generating the new design.

To better design or synthesize components or systems, the following provides general guidelines. These include but are not limited to the following:

- One should always attempt to *synthesize* and *optimize* the design from the head start.
- The process is commenced with the utilization of *freehand sketch* to materialize the innovations and imagination throughout the intended synthesis.
- An ***FBD*** is always essential to include the *loading* conditions and general *geometry* and *topology* or configuration.

The common procedural steps to design or synthesize components or systems are defined hereafter. One is expected to follow the general steps. These steps, however, are supposed to be altered, adjusted, and scrutinized to suit the design at hand before proceeding. The procedure is further taken into account and assimilated after rather than before the completion of the design course:

1. Study the objective and the function that the *design* should perform.
2. Separate an action or a function into steps or sections and subsections.
   - Example: Lathe gearbox has housing body, shafts, gears, bearings, levers, fasteners, couplings, etc.
3. Consider ALL possible *alternative* forms and concept solutions (even unconventional or seemingly bad ones), e.g. mechanical, electrical, hydraulic, pneumatic, magnetic, etc.
4. Consider composition or inversion such as holding a bar or a pin from the middle with bearings on either end (such as Figure 2.1) or holding at both ends with bearings in the middle.
5. Decide on "best" *conceptual* solution or alternative solutions and complete a detailed sketch.
6. Start ***synthesis*** (*real design*) such as performing the following steps with possibly switching steps.
   - Define loading conditions and mathematical model utilizing ***FBD*** (loading such as forces, heat, torques, dynamic, static, deterministic, random or stochastic, etc.).
   - Define input–output relations (kinematics, deflections, compatibility, etc.).
   - Select standard parts and purchased subassemblies for specific characteristics (bolts, motors, control elements, ball bearings, etc.).
   - Define *geometry, dimensions,* and *topology* (initial ***synthesis***) with the assignments of appropriate *material* or materials and their alternatives (initial or preliminary selection) using CAS and appropriate ***safety factor*** to guarantee that the maximum *stress* state is lower than the minimum available *strength*. Produce a ***3D*** model of the design utilizing the defined geometry and topology.
   - *Verify* stresses and *factors of safety* (detailed design) using computer-aided engineering (CAE) tools such as ***FE*** analysis.
   - Consider *manufacturing* factors and requirements (this might have been considered during synthesis).
   - Consider *assembly* and disassembly problems (tolerances and fits, Section 2.4.7).
7. Redesign for a "best" or an *optimum* solution or other alternative solutions.
   - Repeat from step 5 to adjust for more accurate design or other optimum considerations.
8. Produce a complete ***3D*** constructional *assembly* and machine shop models or drawings (final design) using computer-aided manufacturing (CAM), computer numerical control (CNC) machine code generation or standard tessellation language (STL) code for ***3D*** printing or additive manufacturing of components (Section 2.4).

The previous procedure seems to be similar to the usual design process. However, the *synthesis* adoption eliminates or greatly reduces the iterations present in the regular design process. The redesign operations are minimized or eliminated altogether when the synthesis is optimized. The CAS process would constitute the paradigm shift in this real design procedure.

2.4 Manufacturing Processes

This section covers an introduction to the usual manufacturing processes that are carried out in workshops or factories to produce parts that can also be even further processed. It is important to think of how to produce the component of a design while designing rather than after the design is finished. In fact, the design is not finished until the manufacturing process is defined. For a serious designer, a prerequisite of a ***manufacturing engineering*** course is extremely essential. The short review herein may not be sufficient by itself to have a good design of a system or a product. Machine component synthesis is, however, possible with such a short review.

Many manufacturing processing methods are available, and some would be competing alternative to others. The manufacturing processes of concern here are not those needed to transform oars into ingots, blooms, billets, or even cold-rolled sheets and hot rolled bars. These are used as raw materials to be further processed in the manufacturing processes of concern to form machine design components and systems. Manufacturing a part can be processed by melting and then solidifying or setting in shape such as ***casting*** and some types of ***3D*** printing. Manufacturing a part by pressing a sheet or a bar into the desired shape is termed a ***deformation*** process. Shaping the part by removing metal from an original bar or a body is called ***machining*** operation. Parts can also be produced by permanently ***joining*** other members by welding or bonding. Other manufacturing tools use ***particulates*** processing, where materials as powder undergo some temperature and pressure to be sintered. After the part is manufactured, further processing may be applied such as ***surface and heat treatment*** to improve properties or aesthetics. More details of these and others are described as follows, and one can get extensive and detailed coverage in Groover (2010).

2.4.1 Casting or Molding

Casting or molding is one of the less expensive means of producing machine housings, enclosures, and blocks. Other components such as cylinder heads, brake drums, clutch plates, flywheels, cylinder liners, and even some camshafts are produced by a casting process. These and others are manufactured by the widespread conventional sand casting using a ***pattern*** – close in shape to the finished part – in a sand ***mold*** with a sand ***core*** for the hollow space in the part as shown in Figure 2.14. The mold is usually in two sections, the lower section (drag) and the top section (cope), with a parting line to remove the pattern after packing the two sections. Removing the pattern provides the part cavity to be filled with molten material poured into the pouring sprue. The molten material hits the well and flow through runner or runners to fill up the part cavity and comes out through the riser, which is also used as a reservoir to replenish the part due to its shrinking while solidification.

Patterns are usually made of wood, metal, plastics, or other materials. The pattern size is a little larger to account for cooling shrinkage and further machining. Also each pattern has a small draft to allow for easy removal after sand packing. Polystyrene (styrofoam) might also be used as patterns that are not removed but consumed inside the mold under the high molten material temperature. Other expendable mold castings are shell molding, vacuum molding, investment casting, and plaster and ceramic molding. Instead of using the expendable molds to remove the part after solidification, permanent molds are also used. They are made of metals or ceramic refractory material

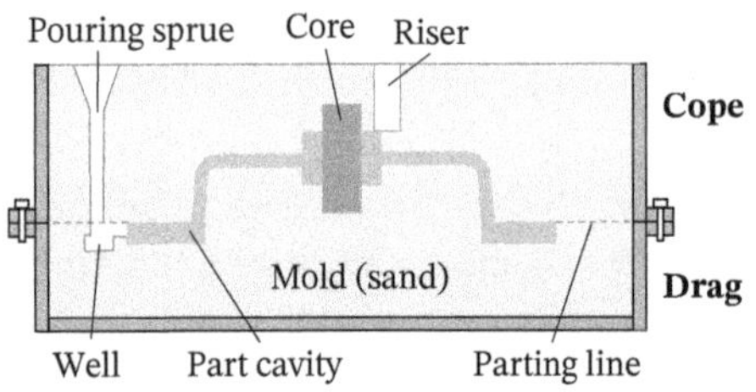

Figure 2.14 Conventional sand casting using a *pattern* – close in shape to the finished part – in the sand mold with a sand *core* for the hollow space in the part.

and have two sections and maybe ejectors to remove the part. Die casting and centrifugal casting are some of these types of permanent molds. Further information and details are found in Groover (2010).

Design for casting accounts for the process limitations and recommendations. Minimum thicknesses are material and general size constraints. One should consult the knowledge base or the rule of thumb in such matters. The minimum thicknesses in sand castings of some materials are about: cast irons and copper 0.1 [in] or 2.5 [mm], aluminum 0.2 [in] or 5 [mm], and steel 0.25 [in] or 6 [mm]. The variation of thicknesses should be gradual with no sharp corners and large intersecting sections. Machining allowances are 0.1–0.2 [in] or 2.5–5 [mm] for cast iron, 0.125–0.25 [in] or 3–6 [mm] for steel, and 0.06–0.2 [in] or 1.5–5 [mm] for nonferrous materials.

2.4.2 Deformation

Deformation processes involves applying larger forces and pressures to yield materials into plastic deformation to the required shapes. That includes ***bending*** or ***drawing*** of sheets into shapes as frames or cups (Figure 2.15a,b). Bending is usually on straight edges or lines to produce some sheet metal works. Blank drawing of a sheet is usually on circular on nearly circular forms. ***Rolling*** off a sheet deforms it into a partial cylindrical or a cylindrical form such as a pipe. Pressing a sheet into a die will deform it into the shape imprinted in the die such as an auto fender. Spinning a blank over a form using a roller deforms the blank to conform to the shape of the form such as some dishes and funnels. ***Forging*** and ***extrusion*** of blocks, slabs, or billets are extreme deformation process that may utilize high temperature to soften materials into the desired shape; see Figure 2.15c,d. Forging is used to produce sections of elements that are highly stresses. The extra flash protrusion in the forged section is usually trimmed. Extrusion is usually used to produce special sections of aluminum girders generally used in construction of windows and doors. Extrusion is also used in some polymer section production such as plastic filaments, pipes, tubes, and hoses.

After deformation, the produced part will experience a spring back where dimensions will be different from the tool used in deformation. This factor is considered in designing the production tool to have the spring back accounted for and the final product as initially required. Most deformation processes are employed to produce basic forms, parts, and sections. The main deformation process affecting machine component design is the forging, which can be used to produce some special shafts and gear blanks. Important factors are the draft angles that are

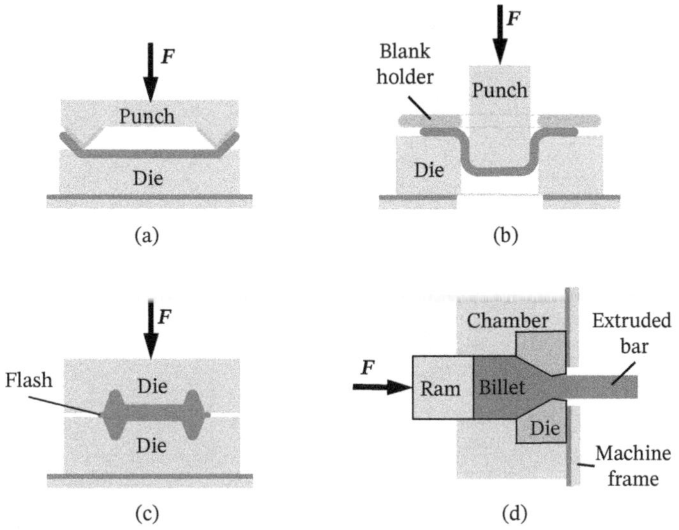

Figure 2.15 Deformation processes schematics: (a) sheet double bending, (b) blank drawing, (c) slab forging, and (d) bar extrusion.

about 7° on the outside and 10° on the inside and the fillets that are about 0.1–0.25 [in] or 2.5–6 [mm]. For these and others, it is recommended to confer with manufacturing experts for more proper considerations. The reference by Groover (2010) is a valuable background to consult with.

2.4.3 Machining

Machining operations involve some sorts of metal cutting or material removal. Conventional machining includes mostly ***turning***, ***milling***, ***drilling***, ***sawing***, ***grinding***, and ***broaching***. Schematic diagrams of these operations are given in Figure 2.16. Other machining operations utilize about the same tools where the machining would be internally in such an operation termed as *boring* instead of turning. The tool can have one cutting edge or multi-cutting edges. The tool can translate or rotate in place or also translate. The work piece can rotate or translate. These options would generate other operations such as *shaping* and *planning* where the tool and the work piece translate. When the work piece rotates while the tool translates on the flat face, the operation is termed *facing*.

The ***turning*** process is operated by the lathe machine, where the chuck grips the work piece, rotates it, and the tool translates with a specific feed to cut the work piece (Figure 2.16a). A hot chip is continuously generated with a cross-section of the feed and the depth of cut. The heat can be reduced by using a cutting fluid, which may also improve cutting operation. The cutting speed depends on the strength of work piece and tool materials. Generated surface quality varies depending on these values of feed and depth of cut providing appropriate cutting speed. This is also the case for most other machining operations.

On a milling machine, the ***milling*** operation uses tools of multi-cutting edges to straighten a surface and cut a groove, a keyway, or a gear tooth, etc. The tool can have many shapes and rotates about either a vertical axis or a horizontal axis. The tool rotates and may translate but the work piece usually translates with the machine table. Figure 2.16b shows two milling operations, the vertical milling and the horizontal milling, depending on the tool's axis of rotation.

The ***drilling*** operation in Figure 2.16c is well known and widespread to the point of the availability of portable hand drills to all persons in many department stores. The tools are drill bits that come in many sizes conventionally 0.014–2.362 [in] or 0.5–60 [mm]. In a drilling machine, the drill bit tool rotates and usually translates

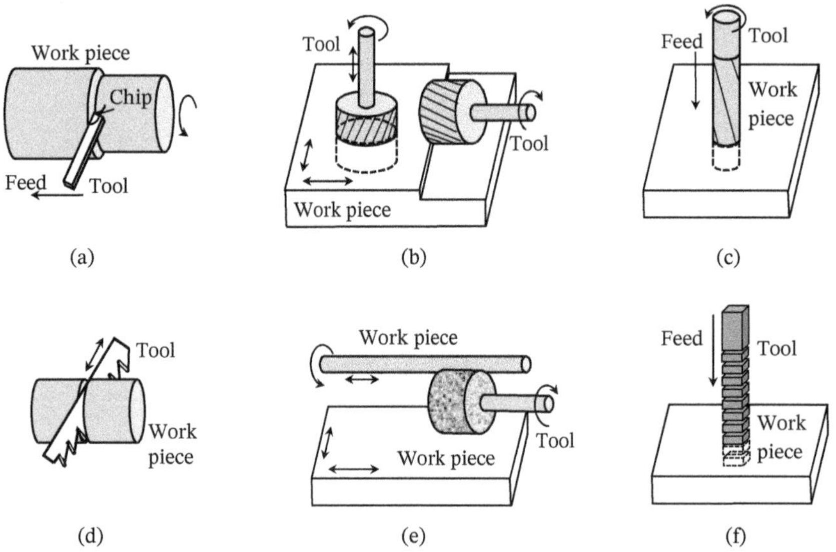

Figure 2.16 Machining processes schematics: (a) turning, (b) two milling operations shown, (c) drilling, (d) sawing, (e) two grinding operations shown, and (f) broaching.

as feeding to cut a circular hole. The quality of holes, however, is less than the milling operation, but reamers can be used to improve quality.

Sawing is another operation used to cut bars into suitable lengths as shown in Figure 2.16d for further processing on other machines. It can also be used to cut a shape off a sheet by *contouring* while the saw is translating vertically, and the work piece is rotating or moving horizontally in a closed path. The sawing blade has many small teeth that each will cut a small chip while feeding. The teeth have few forms, spacing, and set to allow different applications and a little wider cut than the blade width.

Grinding is usually a finishing process by which a bar, or a surface, is ground to a suitable finish as indicated in Figure 2.16e for both operations. The grinding wheel cuts very small chips or pieces of material proportional to the gain (grit) size of the abrasive alumina, silicon carbide, or even industrial diamond particles bonded in the grinding wheel.

Broaching is a process similar to sawing but in ***3D*** starting in a hole where the progression of the tool provides a progression of increasing size cutters all around the tool as schematically shown in Figure 2.16f for a rectangular section tool. The broach design starts with the size of the initial hole and ends with the intended size and shape of the intended hole. The tool is especially designed for a particular initial and final shape of the hole. The number of progressive cutters and their shape depend on the amount each cutter is required to shave or saw.

Blanking or punching is a shearing process, where a die with a specific shape punches to separate the entire shape from a strip or a sheet. The *cutting process* is a direct shear through the thickness of the stock or sheet. Usually presses are used to blank a piece or a group of pieces every stroke with specially designed dies.

Nonconventional machining use other means instead of hard tools that cut in direct contact. Some of these are electric discharge machines (EDMs), electrochemical wire cutting, chemical erosion, lasers, plasma, water, and electron beams. These use erosions or very high vaporizing temperature to separate material continuum along the progression of this unconventional tool.

So many *computer numerically controlled* (CNC) machine tools are performing one or more of the machining operations in usually several coordinates or axes. They are also combined to form *flexible manufacturing* systems to produce several components at any specific time demand. The manufacturing processes allow the flexibility and mass production capability at the same time.

2.4.4 Joining

Joining operation might be considered an assembly operation. However, many machine components can be produced by joining before further machining such as machine frames, some shafts, and supports. The main permanent joining operations are *welding*, *brazing*, *soldering*, *riveting*, and *adhesive bonding*. Other joining operations that can be disassembled are threaded *fasteners*, *screws*, *bolts*, *snap fasteners*, press, and expansion fittings. Some of these are machine components that are treated in Chapters 8 and 9 of the text.

Particulates processing may be considered as a joining process in the micro level than the macro joining process such as welding. In particulate process such as *powder metallurgy*, small particles or powder are sintered together under high pressure and temperature. The particles are joined to form the part shape enclosure under pressure then bonded together under temperature. This process might also be considered a shaping or a deformation process since the particulates would deform under the pressure and may be fused under temperature to get the final shape. The final product can be used as insert or special component in the design of machines.

2.4.5 Surface and Heat Treatment

Surface and heat treatment are manufacturing processes that intend to improve surface properties or gives an aesthetic impression such as *polishing*, *lapping*, *honing*, sand *blasting*, *plating*, *hot dipping*, *anodizing*, *vapor deposition*, or a coat of *paint*, *etc*. Classically known *electroplating* thickness is about 0.002 [in] or 0.05 [mm]. In machine design, however, the *heat treatment* is used to improve component strength to withstand different loading conditions. The heat treatment process is sophisticated in which the component is heated to a specified temperature and cooled with a specified rate and may be reheated to another specified temperature and cooled with another cooling rate. The process affects the material structure and/or composition to render specified surface or core properties. The main heat treatment processes are further discussed in Chapter 7.

2.4.6 *3D* Printing or Additive Manufacturing

3D printing **or** ***additive manufacturing*** is a process that is gaining an attention to produce parts particularly models or spare parts and recent endeavors to produce regular machine parts by this process. Its original name is ***rapid prototyping*** (RP) since it was slow to produce a part, but not to produce a prototype; see samples in Figure 2.17. It was usually lacking strength and regular surface quality due the nature of production in layers with a specific thickness. The lack of strength was due to the initial use of the photosensitive polymer that is cured (hardened) by ultraviolet laser in layers to form a solid plastic part in the *stereolithography* (STL) process; see samples in Figure 2.17. The process starts from the bottom, where a platform near the top of the liquid basin is subjected to the laser light that scans the section of the part at the bottom position. The platform then dips down a small distance, and the laser solidifies the section at that position as another layer on top of the one before. Continuing to the top of the object produces the solidified part out of these layers. The layer thickness has been in the range of 0.003–0.02 [in] or 0.08–0.5 [mm].

Other ***3D*** printing techniques involve *droplets deposition manufacturing (DDM)* or *fused deposition manufacturing* (FDM) using heated hard wax or thermoplastic to be deposited and cooled to form a layer on top of a layer to the last top layer of the ***3D*** object. Layer thicknesses are in the range of about 0.0035–0.03 [in] or 0.09–0.75 [mm]. Another method of *laminated objects manufacturing* (LOM) uses layers of sheets that are laid on top of each other's and laser cut to fit the section at that position and bond down to the previous layer. Thickness of each layer is in the order of 0.002–0.02 [in] or 0.05–0.5 [mm].

A powder-based ***3D*** printing is a process that uses a bed of special metal powder that is selectively sintered by laser and thus heat-fused into solid. The coined name given is *selective laser sintering* (SLS). It also starts from the bottom with the laser heat fuses the cross-section at that bottom level, and the platform then goes down to

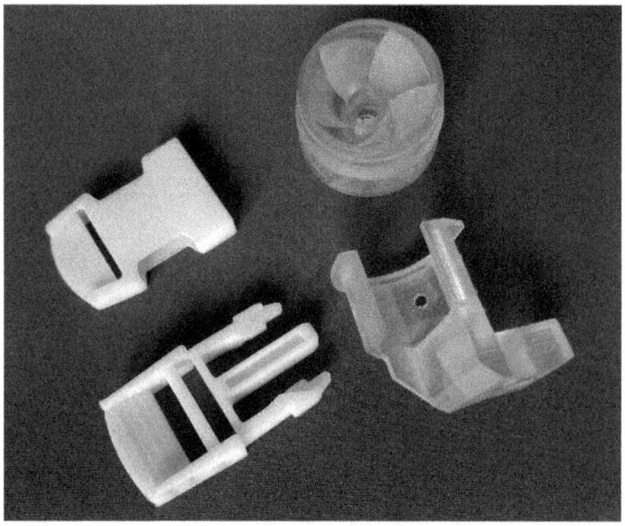

Figure 2.17 Samples of ***3D*** printing or additive manufacturing of few parts or models. Source: Sayed Metwalli.

fuse the next layer and so on till it reaches the top layer. Layer thickness is in the range of 0.003–0.020 [in] or 0.075–0.50 [mm]. *Plasma **3D*** printing is also an alternative.

Extensive research is underway to improve the speed and the strength of the ***3D*** printed objects to make them suitable as usable machine elements. The challenge is to use regularly known materials such as cast iron, steel, and alloys of other metals that gives comparable strength, accuracy, and manufacturing speed or economics. Nanocomposites such as *carbon nanotubes* (CNTs) composites might be a potential alternative, but these need extensive research and development to implement; see Hassan et al. (2014) as a material improvement attempt. Traditional materials with a low melting temperature can, however, be used for very soft rubbery like components or parts that do not experience extensive loading. One such material has been used to produce complex prosthetic finger as defined in El-Sheikh et al. (2012).

2.4.7 Tolerances, Surface Finish, and Fits

Manufacturing processes produce components in an imperfect dimensions and geometry. The variation is usually random but within some expected limits or range. The probability distribution of the imperfection is usually defined as a bell-shaped "*normal distribution*." The distribution presents the probability of dimension imperfection occurrence, which is similar to material property variation in Figure 2.13. More than 99.7% of occurrences happen within ±3 standard deviations. To follow standards, however, the tolerance limits or grades are defined, and the manufacturing processes would be capable of producing within one or few grades of these tolerance ranges according to the accuracy of the process.

2.4.7.1 Tolerances

The limits (about ±3 standard deviations) range is defined as the produced *tolerance* variety occurring during regular production and are given the symbol IT (International Tolerance) with *grades i* from 0 to16, ISO 286 (1962) (R87). The smaller grade numbers represent smaller gap variation of the tolerance zone or a better accuracy of producing dimensions. Table 2.5 shows the range of manufacturing processes IT *tolerance grades* and the *surface roughness* in (usually average or *rms*), [μm] (i.e. 10^{-6} [m]) and [μin] (i.e. 10^{-6} [in]). General operations have a wider span than the usual common operationss since they can include either more accurate or less accurate manufacturing means than the usual common operations. Surface roughness is, however, not generally correlated with tolerances, and it depends more on the type and nature of manufacturing process. The span of the tolerance range for the practical manufacturing processes is IT5–IT16. High tolerance or low *accuracy* is expected from *casting* (IT16), and very low tolerance or more accuracy is expected from *polishing* (IT5). Casting and deformation manufacturing processes has less accuracy (IT12–IT16) than machining (IT5–IT13). The value of the tolerance T_{id} is dependent on the *grade number* i_g in (IT*i*) and the size or diameter d_C of the cylindrical surface. These might be obtained from standard tables in the proper standards or using the following equation (2.29):

$$T_{id} = 10^{0.2(i_g-1)}(0.45\sqrt[3]{d_C} + 0.001d_C) \tag{2.29}$$

where T_{id} is in [μm], i_g is the tolerance grade number, and d_C is the diameter of the cylindrical shaft or hole in [mm]. In standard tables, the diameter d_C is the *geometric mean* of the diameter range in the table. Using the diameter d_C as the basic diameter of the design might give a slightly different value than the standard tables. From that it is simple to find the tolerance of any basic diameter in [mm] at any tolerance grade. As an example, a 25-[mm] (about 1 [in]) basic shaft diameter produced by *fine turning* can reach a tolerance grade of IT8; see Table 2.5, which gives a tolerance of $T_{id} = 10^{0.2(8-1)}(0.45\sqrt[3]{25} + 0.001(25)) = 33.68\ [\mu m]$ according to Eq. (2.29). For small diameters, the term of $(0.001d_c)$ in the estimation of T_{id} may be omitted. Checking the value with the available table of the standards, it was found that IT8 gives 33 [μm] for the range of diameters 18–30 [mm] with a *geometric mean* of 23.24 [mm], ANSI B4.2-(1978). The use of the geometric mean produces a value of T_{id} = 32.84 [μm], which is rounded to 33 [μm]. Only for small diameters and small IT grades that a value less than 1 [μm] is added that is

Table 2.5 Some manufacturing processes IT tolerance grades and surface roughness range in [μm] and [μin] for selected general (including less frequent) and average common operations.

Manufacturing process		IT grade	Surface roughness [μm] ([μin])	
			General	Average (common)
Casting	Sand casting	16	6.3–50 (250–2000)	12.5–25 (500–1000)
	Investment casting	15	0.4–63 (16–250)	1.6–3.2 (63–125)
	Die casting	14	0.4–3.2 (16–125)	0.8–1.6 (32–63)
Deformation	Forging	15	1.6–25 (63–1000)	3.2–12.5 (125–500)
	Extruding	12–13	0.4–12.5 (16–500)	0.8–3.2 (32–125)
	Hot rolling	13	6.3–50 (250–2000)	12.5–25 (500–1000)
Machining	Turning, rough	9–12	6.3–50 (250–2000)	12.5–25 (500–1000)
	Turning, finish-fine	6–9	0.4–6.3 (16–250)	0.8–3.2 (32–125)
	Milling	9–10	0.2–25 (8–1000)	0.8–6.3 (32–250)
	Drilling	10–13	0.8–12.5 (32–500)	1.6–6.3 (63–250)
	Sawing	10–13	0.8–50 (32–2000)	1.6–25 (63–1000)
	Grinding	5–8	0.025–6.3 (1–250)	0.1–1.6 (4–63)
	Broaching	6–8	0.4–6.3 (16–250)	0.8–3.2 (32–125)
	Nonconventional (EDM)	10–13	0.8–12.5 (32–500)	1.6–6.3 (63–250)
Joining	Powder metallurgy	9	1.6–12.5 (63–500)	3.2–6.3 (125–250)
Surface treatment	Polishing	5	0.012–0.8 (0.5–32)	0.1–0.4 (4–16)
	Lapping	4–5	0.025–0.2 (1–8)	0.05–0.1(2–4)

also why the IT8 grade is quoted as 33 [μm]. In addition to the tolerance grades, Table 2.5 projects the surface roughness of the fine turning to be in the average range of 0.8–3.2 [μm], which is a low to a medium roughness. For a better surface roughness, one might need to use grinding after turning.

2.4.7.2 Surface Finish

Surface finish is a function of the surface roughness. In manufacturing processes, excellent *surface finish* has a roughness up to about 0.2 [μm] or 8 [μin]. Very good surface finish has a roughness up to 0.8 [μm] or 32 [μin]. Good surface finish has a roughness up to 3.2 [μm] or 125 [μin]. Ordinary surface finish has a roughness up to 12.5 [μm] or 500 [μin]. More roughness than that is considered "as produced" surface. These assessments are conceptual evaluations. However, proper definition of surface finish is done by a specific roughness indicating the value of the acceptable limit of the (usually average or root-mean-square *rms*) magnitude of the roughness in a specific ***standard*** *surface finish symbol* inserted on the surface. Table 2.5 shows the range of manufacturing processes IT *tolerance grades* and the *surface roughness* in (usually average or *rms*) [μm] (i.e. 10^{-6} [m]) and [μin] (i.e. 10^{-6} [in]) for selected general (including less frequent occurrence) and average common operations. A surface finish symbol is usually used on components *working drawings* used for production. The form of the surface finish symbol is dependent on the standard used by the designer; see ISO 1302 (2002) and ANSI/ASME (1996). Surface finish is very critical in contacts of parts that have relative motion between them such as seals on rotating shafts in gearboxes and in journal bearings. The surface roughness should be as fine as possible (at least in the excellent surface finish range) not to have intolerable ware on the seal material or bearing surfaces.

2.4.7.3 Fits

A schematic diagram of fits for "Hole" basis mating of cylindrical parts is given in Figure 2.18 with an *actual* shaft section of "d9" tolerance indicating its possible location within the IT9 tolerance zone and a clearance mating with any hole of any tolerance grade of "H" *fundamental deviation*. The spaces are enormously exaggerated relative to the shaft dimensions so as to demonstrate the idea. The *basic size* is the preferred nominal (round figure) design size. It might be the size of the shaft where a bearing will mate at that segment of the shaft. *Preferred sizes* are defined according to a geometric series progression, and their standards can be found in Appendix A.3, see ISO 3 (1973), ANSI B4.2 (1978), and ANSI B4.1 (1967). In Figure 2.18, categories of *clearances*, *locations*, and *interferences* between the two mating components depend on the gap between the two mating cylindrical surfaces. The gap is a positive clearance for running and sliding *fits* between the shaft and the hole. The location of the IT tolerance zone for both hole and shaft defines the mating case. The location of the IT tolerance zone is given by the *fundamental deviation categories* from A–Z for holes and a–z for shafts, ISO 286 (1962). The H and h location categories cause the IT zone to snuggle the *basic size* as shown at the end of the clearance zone of Figure 2.18 for H7/h6. The shaft will have a maximum diameter of the *basic size* and the minimum diameter of the basic size – (minus) the IT tolerance zone. As an example, the 25 [mm] shaft of h8 tolerance will have a diameter variation of 0.0 for h and −33 [μm] for the IT8 as calculated in the previous paragraph. The shaft diameter can be quoted as 25 + 0.0 and 25 − 0.033 [mm] or 25.000 and 24.967 [mm]. For shaft manufactured by *turning* process, it is better to give the *maximum metal condition* first (i.e. 25.000 [mm]) so the operator attempts to produce this value, and if he misses, he most probably will be above the 24.967 [mm] diameter.

Figure 2.18 shows the fundamental deviations for holes of only H categories due to ***hole base*** system of *fits* between mating shafts and holes. Categories H11, H9, H8, and H7 have the letter H as the *fundamental deviation category*, and the number is for the tolerance IT value. The *fundamental deviations* for shafts of c11, d9, f7, g6, h6, k6, n6, p6, s6, and u6 have the letter for the deviation category and the number for the tolerance IT value. Shaft fundamental deviations of categories c11–g9 are negative or the maximum shaft diameter should be smaller than the basic size for all these categories. The mating of these shafts with holes of category H will always have a ***clearance fit***. Categories of k6 and n6 will cause some cases of clearance or some cases of interference between the shaft and the hole due to the tolerance zone of the IT6. These are then causing ***transition fit*** between clearance

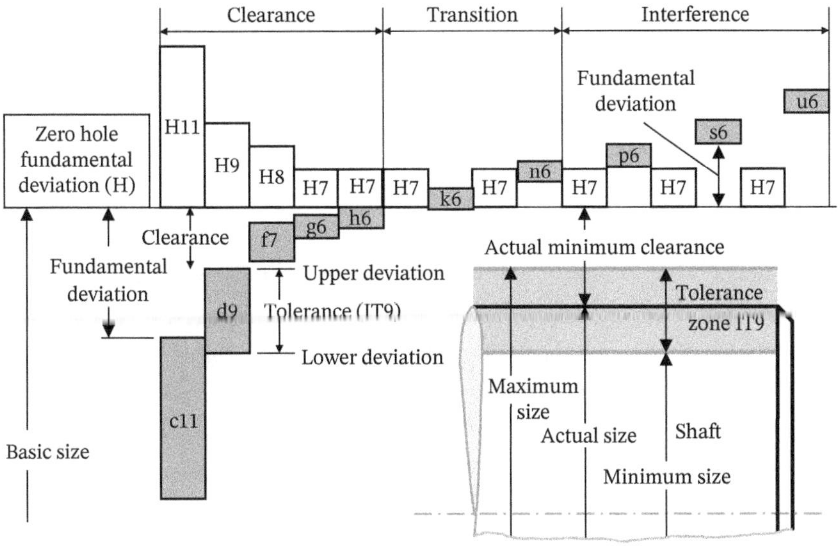

Figure 2.18 Schematic diagram of fits for hole basis cylindrical parts with an actual shaft section for d9 tolerance indicating its possible location within the IT9 tolerance zone.

and interference. Categories of p6, s6, and u6 will always cause ***interference fit*** between the shafts for categories H7 of the holes. The fundamental deviation is positive, which means that the maximum shaft diameter should be larger than the basic size for all these categories and larger than the largest H7 hole diameter. The discussion herein is confined to the hole base system of fits since it is more widely used. The ***shaft base*** system of fits can, however, be easily deduced from the hole base system by just vertically flipping Figure 2.18 and capitalizing the shaft deviation letters to be the hole deviation letters and by changing H to the lower case h for the shaft.

2.4.7.4 Fundamental Deviations

The *actual* shaft section for d9 tolerance in Figure 2.18 indicates its location within the IT9 tolerance zone. The actual shaft diameter is one of the produced shafts to be accepted. It has to be within the specified tolerance zone. Its maximum possibly accepted diameter (max. size at *upper deviation*) is the basic diameter minus the fundamental deviation of the deviation letter d. Its minimum possibly accepted diameter (min. size at *lower deviation*) is the basic diameter minus the fundamental deviation of the letter d minus the IT9 tolerance zone. The *fundamental deviation* c_0 for each deviation letter is a function of the *basic size diameter* d_C. These might be obtained from standard tables in the proper standards or using the following relations for the preferred letters:

- For the deviation letter c, the fundamental deviation is $52{d_C}^{0.2}$, $d_C \leq 40$ or $(95 + 0.8d_C)$, $d_C > 40$ [mm].
- For the deviation letters d, e, f, and g, the fundamental deviations are $16{d_C}^{0.44}$, $11{d_C}^{0.41}$, $5.5{d_C}^{0.41}$, and $2.5{d_C}^{0.34}$, respectively.
- For the deviation letter k, the fundamental deviation is $(0.6{d_C}^{0.333})$, $3 < \mathrm{IT} < 8$ else, 0.
- For the deviation letter n, the fundamental deviation is $(+5{d_C}^{0.34})$.
- For the deviation letter p, the fundamental deviation is almost $(+\mathrm{IT7}+(d_c/100))$.
- For the deviation letter s, the fundamental deviation is almost $(\mathrm{IT8}+0.08d_C)$, $d_C \leq 50$, or $(\mathrm{IT7}+0.4d_C)$, $d_C > 50$.
- For the deviation letter u, the fundamental deviation is $(\mathrm{IT8}+d_C)$.

These relations are very useful in developing computer-aided tolerance software as defined later on.

The span of the tolerance range for practical manufacturing processes (IT5–IT16) in Table 2.5 is also the range for the preferred fits suggested for mating the cylindrical shaft–hole connections. These preferred fits, their performance, anticipated applications, and the selected ANSI 4.2 (1978) similar and comparable fits are defined in

Table 2.6 Basic selected fits for regular cylindrical hole base, typical applications, and selected ANSI similar and comparable fits.

Fit	ISO fit hole base	Performance	Applications	ANSI fit (similar)
Clearance	H11/c11	Loose running	Wide commercial uses	RC 9
	H9/d9	Free running	Heavy journals, high speed, and temperature	RC 7
	H8/f7	Close running	Accurate machines, sensible speed, and pressure	RC 4
	H7/g6	Sliding location	Move and locate accurately	RC 2
Location	H7/h6	Location clearance	Snug fit, freely assembled, and disassembled	LC 2
	H7/k6	Location transition	Accurate location	LT 3
	H7/n6	Location transition	More accurate location	LT 5
	H7/p6	Location interference	Rigid alignment	LN 2
Interference	H7/s6	Medium drive	Shrink fit on light sections	FN 2
	H7/u6	Force fit	Highly stressed shrink fit	FN 4

Table 2.6. The ISO symbols for preferred fits are given for the ***hole base*** system. Similar preferred fits for *shaft base* system are obtained by just switching to the h for H and capitalizing all shaft deviation letters to become for holes. The preferred *clearance* for loose running fit (H11/c11) is for wide commercial applications where IT11 (for both hole and shaft) specifies drilling for the hole and rough turning for the shaft as Table 2.5 indicates. Free running (H9/d9) and close running (H8/f7) are suitable for journal bearings depending on the speed and load of operation. Higher loads and temperatures suit free running, where close running is for accurate and moderate speeds and loads. For other different conditions, one may select other suitable fits such as H8/e7 for average *journal bearings*; see Chapter 12. The sliding location (H7/g6) is preferred for fits that requires some movement and location accuracy. All previous *clearance* fits have the equivalent ANSI B4.1 (1967) fits that uses the symbol RC (running clearance) with a number indicating the clearance and tolerance magnitudes.

The next class of fit is the *location* or transition between clearance and interference (Table 2.5). The applications are mainly for location but with some degree of play or tightness such as the location clearance (H7/h6) for snug fit, the location transition (H7/k6 and H7/n6) for accurate locations, and the location interference (H7/p6) for rigid alignments. The *location* equivalent ANSI B4.1 (1967) fits are the symbols (LC, LT, and LN) for location clearance, location transition, and location interference, respectively.

The preferred *interference* fits are the medium drive (H7/s6), which is suitable for shrink fit on light sections, and the force fit (H7/u6) that is for highly stressed shrink fit. One may also use other fits, but the preferred ones are more in general use. The ANSI/ASME B4.1 (R2009) (1967) equivalent to these *interference* fits has the symbol FN for force interference with a number indicating the magnitude of the interference.

More utility and elaborations are provided in the trial-version of ***computer-aided fits and tolerances*** software provided through the **Wiley website**. Figure 2.19 provides the interface of the available computer-aided fits and tolerances program. Figure 2.19 shows hole and shaft base options, the interactive pop-up help to select appropriate fits, and the calculated limits of shaft and hole diameters.

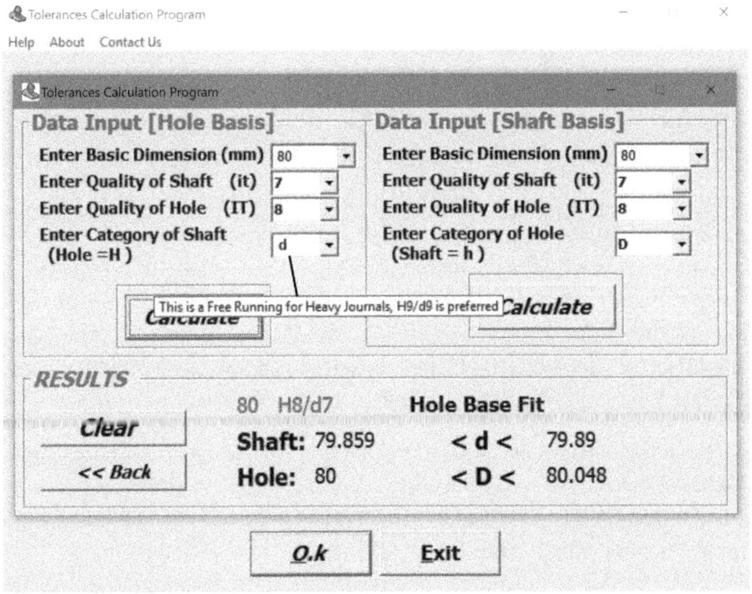

Figure 2.19 Interface of the available fits and tolerances computer-aided program showing hole and shaft base options, the interactive pop-up help to select appropriate fits, and the calculated limits of shaft and hole diameters. Source: Sayed Metwalli.

2.5 Standard Sets and Components

It is difficult to have machines without the inclusion of some standard components or standard sets. It is imperative to use such components or sets since it is futile to produce these while producing the machine. Standard components or sets are mass produced by specialized outfits that result in extreme savings compared with self-produced ones. For machine design, bolts, nuts, keys, pins, retaining rings, rods, beams, plates, pipes, belts, chains, ropes, seals, rolling bearings, couplings, rivets, etc. are frequently used. These are basic standard components. Other standard sets, which are also frequently used, are motors, gear motors, pumps, hydraulic cylinders, valves, control systems, etc. In fact, many auto manufacturers employ standard transmissions for their drive trains, let alone starters, generators, batteries, air-conditioning compressors, lamps, locks, measuring sensors, micro processing and control elements, motors, etc. that are supplied by other manufacturers as (original equipment manufacturer (OEM)) components.

It is important for a machine designer to have all information about standard components and systems. The engineer should have full documentations about all the standards for the countries he or she is doing the design for. Internet access to these standards is also an option other than documentations. With some details, standard components covered in this text are the following:

- Screws, bolts, rivets, power screws, and ball screws.
- Keys, pins, retaining rings, splines, and couplings.
- Seals and springs.
- Rolling bearings and journal bearings.
- Belts and chains.

Almost all these components have standards; however, *springs* and *journal bearings* are available in specific dimensions from so many manufacturers such that they may not necessarily be produced to order unless essentially needed to be specific. If they are of limited numbers, the manufacturer may be willing to produce them. If they are of large numbers, usually manufacturers would produce them.

2.6 Codes and Standards

With the global village the world has become, it is imperative that design engineers and others should be aware of the codes and standards of worldwide nature. The products are usually crossing borders and have to conform to the rules and standards of the country they land in. That is why the design engineer should be aware of international codes and standards generated by effective organizations and societies. To design for a country or countries, the rules of their codes should be adhered to, and the design should not violate any of these standards. The codes and standards are developed by governmental bodies or societies. Common selected bodies, organizations, and societies that generate codes or standards related to machine design are categorized as follows:

International Standards: These are the standards, which so many nation members (162) agreed to. The official organization that generates many standards is as follows:

- International Organization for Standardization (ISO), the website link is http://www.iso.org/iso/home.html

American Codes and Standards: This includes some North American organizations and societies:

- Aluminum Association (AA), the website link is www.aluminum.org/resources/industry-standards
- American Bearing Manufacturers Association (ABMA), the website link is www.americanbearings.org/
- American Gear Manufacturers Association (AGMA), the website link is https://www.agma.org/
- American Iron and Steel Institute (AISI), the website link is www.steel.org/
- American National Standards Institute (ANSI), the website link is www.ansi.org/

- American Society of Heating, Refrigerating and Air-Conditioning Engineers (ASHRAE), the website link is www.ashrae.org/
- American Society of Mechanical Engineers (ASME), the website link is www.asme.org/
- American Society of Testing and Materials (ASTM), the website link is www.astm.org/
- American Welding Society (AWS), the website link is www.aws.org/
- ASM International (American Society for Metals) (ASM), the website link is www.asminternational.org/
- Canadian Standards Association Group (CSA), the website link is www.csagroup.org/
- Industrial Fasteners Institute (IFI), the website link is www.indfast.org/
- National Institute for Standards and Technology (NIST), the website link is www.nist.gov/
- Society of Automotive Engineers (SAE), the website link is www.sae.org/

European Codes and Standards

- Association Française de Normalisation (AFNOR), the website link is www.afnor.org/
- British Standards Institution (BSI), the website link is www.bsigroup.com/
- Deutsches Institut für Normung "German Institute for Standardization" (DIN), the website link is www.din.de/
- Euro-Asian Council for Standardization, Metrology and Certification "EASC" (GOST), the website link is www.runorm.com/
- European Committee for Standardization (CEN), the website link is www.cen.eu/
- Institution of Mechanical Engineers (IMechE), the website link is www.imeche.org/
- International Bureau of Weights and Measures (BIPM) – Patron of (SI) Standards, the website link is www.bipm.org/

Asia Pacific Standards

- Bureau of Indian Standards (IS), the website link is www.bis.org.in/
- Indonesian National Standard (SNI), Badan Nasional Standardisasi "BSN," the website link is www.bsn.go.id/
- Japanese Standards Association "JSA" (JIS), the website link is www.jisc.go.jp/
- Korean Standards Association (KS), the website link is www.ksa.or.kr/
- Standardization Administration of China "SAC," Guobiao standards (GB), the website link is www.sac.gov.cn/
- Standards Australia (AS), the website link is www.standards.org.au/

It should be noticed that usually standards and codes are the responsibility of governments while codes can be developed by societies or organizations. Following codes and standards of countries, the design is exported to, is important to meet the customers' requirements.

2.7 Summary

This chapter presents the basic considerations for machine design. Necessary tools in mathematical modeling and calculations are covered. Mathematical modeling allows the representation of mechanical systems to be formulated in the closest way to the intended physical system. Force analysis program is also provided to calculate reactions for ***3D*** models of externally loaded systems. The adoption of generalized mathematical model provides means to calculate mechanical behavior in *steady-state* and in *dynamic interaction* with loading conditions. The steady-state or static behavior is mainly handled in this chapter. Dynamic considerations can simply be taken care of, but this has been postponed for latter developments. The usual random loading conditions and variations of the material strength have been introduced to be aware of the statistical nature of failure and design reliability and safety.

To implement mathematical treatments of the mathematical models of a design, calculation tools are presented. Excel as a spreadsheet program, MATLAB software, geometric modeling, or CAD and ***FE*** packages are reviewed. Introduction to the implementation of Excel and MATLAB is given. The random nature of loading and strength may be modeled by such tools to account for a realistic physical model of the design conditions.

A design procedure is introduced that can be used as a template for modifications according to the design at hand. The concept of *synthesis* rather than design by iteration is emphasized. The tools to perform synthesis of the components have not been introduced since that will be the task of the following Chapters 5–18 of the text. The necessary tools of the knowledge base and reverse engineering process are introduced. These tools are used to advantage in the synthesis process and optimization, which constitutes the paradigm shift in the real design procedure.

An introduction to manufacturing processes is vital to identify the way to produce the component of a design while designing. This is a better way rather than defining the manufacturing processes after the design is finalized. The design is not really finalized until the manufacturing process is identified. It also includes the implementation of a successful assembly of the finalized design. Casting and molding, welding and bonding, metal forming, metal cutting, surface and heat treatment, ***3D*** printing, or RP are main manufacturing processes. These are reviewed in this chapter. Successful assembly for proper operation is dependent on tolerances, surface finish, and fits. These are some of the consequences of the manufacturing processes used in producing the components of the design. The selection of the appropriate assembly fits and associated manufacturing processes to properly execute the design performance is introduced. A computer-aided program is also available to aid in the appropriate definition of fits and the calculation of tolerances to define the manufacturing bounds.

The use of standard sets and components are usually indispensable in any design. Several of these are introduced and recommended to be acquired rather than self-produced such as nuts, keys, pins, retaining rings, rods, beams, plates, pipes, belts, chains, ropes, seals, rolling bearings, couplings, and rivets. Also, other frequently subsystems are used such as motors, gear motors, pumps, hydraulic cylinders, valves, control systems, etc. These and others are supplied by other manufacturers as (OEM) components to be included in the product design. They are designed and mass produced by specialized outfits that result in extreme savings compared with self-designed and self-produced ones.

To design for a country or countries, the design engineer should be aware of international codes and standards generated by effective organizations and societies. Common selected bodies, organizations, and societies that generate codes or standards related to machine design are categorized in this chapter. The designer should follow codes and standards of countries where the design is exported to. This is important to meet the customers' requirements and the regulations of the receiving countries.

Problems

2.1 What is the *inertial* coordinate system? What is the difference between inertial and *Cartesian*, polar, cylindrical, and spherical coordinate systems? Which is the most used in each design process and why? Did you know that there is a *homogeneous* coordinate system that is very useful in *projection*?

2.2 A door is pivoted to the left by two hinges into the wall. The door handle is to the right edge of the door similar to Figure 2.3. To open the door, one needs to unlock the door using the handle and then pull the door open. What mathematical model can be adopted? Is the origin better being at the handle base or one of the door hinges? Draw a free body diagram (***FBD***) of the door and handle. When do you consider friction at the hinges or handle lock?

2.3 A very heavy steel door has a handle similar to that in Figure 2.3. The door is homogeneous and weighs 500 [lb]. To open the door, the handle is first pushed down by a vertical force to its maximum of 5 [lb] to unlock and simultaneously pulled horizontally by a necessary force of 9 [lb] to move the door. It is required to model the door and its handle to find the reaction at the lock support and door hinges. If the door is 4 ft wide, find the forces and moments acting on the door at its inception of movement. The door has two vertical hinges. Suggest the locations of the hinges and handle. Assume any missing data. Do vertical locations of the hinges affect the calculations?

2.4 Solve Problem 2.3 in the SI units if the forces and dimensions are the same. Resolve the same problem for other assumed values that suits a different dimensions and operators. Use available software to verify your solutions.

2.5 Consider the shaft assembly of Figure 2.1. If the total force $\boldsymbol{F}$ on the middle of the shaft is 1000 [N] and the distance between the bearings centerlines is 0.1 [m], what are the values of the reactions at the bearings? If one assumes a uniformly distributed model for the load over the middle 0.08 [m], what would be the values of the reactions at the bearings? If one assumes an equivalent model for the load as two forces $\boldsymbol{F}/2$ at the 0.02 and 0.06 [m] over the middle 0.08 [m], what should be the values of the reactions at the bearings? Compare and discuss the different models. Which is the conservative and which is closer to reality? Verify your suggestions by the available force analysis software and divide the distributed load into eight equal loads with each applied at the middle of the 0.08 [m] sections. Redo the solution for the US system of units if the forces and dimensions are the same. What conclusion would you draw?

2.6 A motorcycle is climbing a steep uphill of an angle of 15° at a constant velocity of 25 [km/h]. The motorcycle wheels are 0.5 [m] in diameter with a 1-[m] centerline distance apart. The maximum power of the motorcycle is 15 [hp]. The curb weight of the motorcycle is 800 [N] acting in the middle between the wheel centerlines but above that centerline by 0.15 [m]. The driver weight is 750 [N] acting on the rear quarter between the wheel centerline but above that centerline by 0.50 [m]. Draw the ***FBD*** of the motorcycle and the driver. Find the reactions at the wheels assuming the driving friction at the rear wheel only. Use available force analysis software to verify your solution. What is the power needed at that situation, which the motorcycle should deliver?

2.7 Solve Problem 2.6 in the US system of units if the forces and dimensions are the same. Use available force analysis software to verify your solution.

2.8 A simple gearbox has an input and an output parallel shafts that lay in the horizontal plane. Both are 4 [in] apart and 5 [in] higher than the horizontal base plane that the gearbox is fixed to by two vertical bolts. The gearbox is transmitting 2.5 [hp] at input shaft speed of 1200 [rpm] and output shaft speed of 300 [rpm]. It is required to model the system and find the reaction forces at the two ground bolts. Where should you select the support bolts locations? If the gearbox is 3 [in] in maximum thickness and 10 [in] in maximum width, suggest other locations of the fixation bolts. Verify your suggestions by the available force analysis software.

2.9 Solve Problem 2.8 in the SI system of units if the information and dimensions are the same. Use available force analysis software to verify your solution and suggestions.

2.10 Remodel the system defined in Example 2.2 using ***FBD*** and classical force analysis to verify Eq. (j) of the example.

2.11 Remodel the system defined in Example 2.3 using ***FBDs*** and classical force analysis to verify Eq. (j) of the example.

2.12 Remodel the system defined in Example 2.4 using ***FBDs*** and classical force analysis to verify Eqs. (d) and (s) of the example. Compare the classical method and the system modeling procedure with respect to simplicity, methodical process, and generality.

2.13 The system in Example 2.4 has a *driver* running at 30 [rpm], and the drum diameter of the hoist is 0.5 [m]. The system is lifting a mass of 1500 [kg] connected to the drum by a cable that has a stiffness of 1000 [N/mm]. The shaft connecting the driver and the drum has a stiffness of 10^6 [m N/rad]. Find the twist angle of the connecting shaft, the cable deflection, and the velocity of the mass at steady lifting. What is the input torque the driver should provide at that steady-state running?

2.14 Solve problem number 2.13 in the US system of units if the forces and dimensions are the same.

2.15 Suggest a load occurrence frequency distribution model for a regular private car in the US [lb] or SI [N] system of units. If the car is used as a regular cab or a taxi, what model of load occurrence frequency-distribution you should expect? Would the distribution be affected by the country in question? To consider a load on a car of your design, would you use the average load or the maximum expected load and why?

2.16 Properties of produced materials are controlled to fall in a specific standard. The standard deviation of strength is usually less than 5% of the mean. Search references to find out the variation and qualify a similar statement for some specific material and some specific strength.

2.17 If the value of the maximum loading is 3 standard deviations above the average and if it happens at the minimum strength property of 3 standard deviations below the average, verify that the reliability would be about 99.7%.

2.18 Develop a program using spreadsheet (e.g. Excel) to calculate the weight and some other selected properties at each step of a general stepped shaft.

2.19 Develop a program using MATLAB to calculate the weight and some other selected properties at each step of a general stepped shaft.

2.20 From the developed stepped shaft program, develop a mean to transfer the stepped shaft geometry to a geometric modeling (CAD) program of your choice.

2.21 Suggest a geometric stepped model for the shaft of Problem 2.5, and find the diameters of the shaft at each step using the suggested loading models and any of the ***FE*** programs at your disposal. Use any material of your choice for now.

2.22 Select a manufacturing process (machining) to produce the stepped shaft of Problem 2.5 noting that the bearings and the middle element should be located on the shaft with location clearance fit for proper assembly. Find the tolerance of each region of the stepped shaft assuming maximum diameter value of 25 [mm]. Check your values using the available *fits and tolerance software*. What is the expected surface finish value for a fine turning process to produce the stepped shaft?

2.23 Consider the inner diameter of both bearings in Figure 2.1 as 20 [mm]. If they need to be located on the shaft using a location interference fit, find the proper fit knowing that the bearings are hole base system. What is the expected tolerance value of the bearings and the mating shaft? Also find the fundamental deviations for bearings and mating shaft. Check your values using the available fits and tolerance software.

2.24 Develop a ***shaft base*** system of fits that can be deduced from the hole base system by just vertically flipping Figure 2.18. To consider holes and shafts, capitalize the shaft deviation letters to be the hole deviation letters and change H to the lowercase h for the shaft.

2.25 Find the tolerance limits of H11, H8, and H6 for holes of diameters 10, 70, and 400 [mm]. Define the upper and lower deviation for each case. What is the *fundamental deviation?* Check your values using the available fits and tolerance software.

2.26 Find the tolerance limits of H9, H7, and H5 for holes of diameters 5, 30, and 150 [mm]. Define the upper and lower deviation for each case. What is the *fundamental deviation?* Check your values using the available fits and tolerance software.

2.27 Find the *fundamental deviations* for shafts of c11, g6, and p6 if the shafts are 5, 50, and 300 [mm] in diameter. Define the tolerance for each diameter. Check the values against accessible standards using the available fits and tolerance software.

2.28 Find the *fundamental deviations* for shafts of e7, k6, and u6 if the shafts are 15, 70, and 350 [mm] in diameter. Define the tolerance for each diameter. Check the values against accessible standards using the available fits and tolerance software.

2.29 Find the *fundamental deviations* for shafts of d9, and n6 if the shafts are 20, 100, and 500 [mm] in diameter. Define the tolerance for each diameter. Check the values against accessible standards using the available fits and tolerance software.

2.30 Find the *fundamental deviations* for shafts of f7, and s6 if the shafts are 10, 80, and 450 [mm] in diameter. Define the tolerance for each diameter. Check the values against accessible standards using the available fits and tolerance software.

References

Abbas, M.H. and Metwalli, S.M. (2011). Elastohydrodynamic ball bearing optimization using genetic algorithm and heuristic gradient projection. *Proceeding of the ASME IDETC/CIE 2011*, Washington, DC, USA (29–31 August 2011), Paper no: DETC2011-47624. NY, USA: ASME (American Society of Mechanical Engineers).

Abbas, M.H., Youssef, A.M.A., and Metwalli, S.M. (2010). Ball bearing fatigue and wear life optimization using elastohydrodynamic and genetic algorithm. *Proceedings of the ASME IDETC/CIE 2010*, Montreal, Quebec, Canada (15–18 August 2010), Paper no: DETC2010-28849. NY, USA: ASME (American Society of Mechanical Engineers).

ANSI B4.1 (R1987) (1967). *Preferred limits and fits for cylindrical parts*. American National Standards Institute.

ANSI B4.2 (R1994) (1978). *Preferred metric limits and fits*. American National Standards Institute.

ANSI/ASME (R2002) (1996). *Surface texture symbols – metric version*, Y14.36M-1996. American National Standards Institute/American Society of Mechanical Engineers.

ANSI/ASME B4.1 (R2009) (1967). *Preferred limits and fits for cylindrical parts*. American National Standards Institute.

Borutzky, W. (2010). *Bond Graph Methodology: Development and Analysis of Multidisciplinary Dynamic System Models*. Springer.

Chapra, S.C. and Canale, R.P. (2010). *Numerical Methods for Engineers*, 6e. McGraw Hill.

El-Sheikh, M.A., Taher, M.F., and Metwalli, S.M. (2012). New optimum humanoid hand design for prosthetic applications. *International Journal of Artificial Organs* 35 (4): 251–262.

Excel (2016). *Microsoft Office Excel*. Microsoft Corporation.

Groover, M.P. (2010). *Fundamentals of Modern Manufacturing: Materials, Processes, and Systems*, 4e. Wiley.

Hassan, M.T.Z., Esawi, A.M.K., and Metwalli, S. (2014). Effect of carbon nanotube damage on the mechanical properties of aluminium – carbon nanotube composites. *Journal of Alloys and Compounds* 607: 215–222.

ISO 1302:(2002E) (2002). *Geometrical Product Specifications (GPS) – Indication of surface texture in technical product documentation*. International Organization for Standardization.

ISO 286 (1962). *ISO system of limits and fits – Part 1: General, tolerances and deviations* (ISO/R 286). International Organization for Standardization.

ISO 3 (1973). *Preferred numbers – Series of preferred numbers*. International Organization for Standardization.

Karnopp, D.C. and Rosenberg, R.C. (1968). *Analysis and Simulation of Multiport Systems – The Bond Graph Approach to Physical System Dynamics*. MIT Press.

Martens, H.R. and Allen, D.R. (1969). *Introduction to Systems Theory*. Merrill.

MATLAB (2014). *Matrix Laboratory, The Language of Technical Computing*. MathWorks Inc.

Metwalli, S.M., Radwan, A., Abdel-Wahab, O., et al. (1999). Maintenance and parts fabrication by reverse engineering. *Proceedings of the ASME DETC99/CIE*, Las Vegas, Nevada (12–15 September 1999), Paper no: DETC99/CIE-9134. NY, USA: ASME (American Society of Mechanical Engineers).

Nassef, A.O., Ashraf, A.M., and Metwalli, S.M. (1999). Accuracy and fitting-time minimization in the reverse engineering of prismatic features. *Proceedings of the ASME DETC99/CIE*, Las Vegas, Nevada (12–15 September 1999), Paper no: DETC99/CIE-9131. NY, USA: ASME (American Society of Mechanical Engineers).

Internet Links (Selected)

Abaqus, SIMULIA (Abaqus and CATIA), www.3ds.com/products/simulia/ Dassault Systèmes Simulia Corp

ANSYS, www.ansys.com/ ANSYS, Inc, TA Associates

Excel, office.microsoft.com/en-us/excel, Microsoft

ISO, http://www.iso.org/iso/home.html and free available at http://standards.iso.org/ittf/PubliclyAvailableStandards/

MATLAB, www.mathworks.com MathWorks

Nastran, http://www.mscsoftware.com/product/msc-nastran, The MacNeal-Schwendler Corp. (MSC)

SolidWorks, www.solidworks.com/ Dassault Systèmes

Part II

Knowledge-Based Design

3

Introduction to Computer-Aided Techniques

With the availability of computers and other advent gadgets that are gaining power, the utility of these tools should be very valuable in machine design. With appropriate tools, software, and codes, the optimum synthesis can be guaranteed. This should implement the best characteristics of human and machine. Coupling of these beneficial and advantageous characteristics is the utmost goal of the efficient computer-aided techniques. The computer-aided characteristics of tools should capitalize on the most efficient interaction of human and machine to generate an effective computer-aided technique. This chapter probes some of these computer-aided techniques that are mostly related to machine design.

Designers are using codes to develop mechanical systems or products. The codes are usually commercial and sometimes self-developed. The widespread codes are the currently called *computer-aided design* (CAD) software packages. These are usually helping users to develop a three-dimensional (***3D***) *solid model*, and they are thus *computer-aided geometric (CAG) modeling* packages. The codes do not commonly provide aides to suggest geometry and material for any component or part thereof. The task is usually kept to the repeated numerical analysis by *finite element* (***FE***) codes; see Sections 2.2 and 6.11. It would be useful for the designer to develop his codes to define geometry and material before resorting to the geometric modeling packages. One might employ a spreadsheet package such as Microsoft *Excel*© (2016), a matrix-manipulation package such as the MathWorks *MATLAB*© (2014), programming packages such as the simple legacy Microsoft *Visual Studio 6*© (1998), and any recent packages thereof. To develop such codes, one would be advised to follow the prospective *CAD-software* characteristics, if other than a personal usage is intended. Section 2.2, provides an overview of some basic capabilities of both *Excel* and *MATLAB*. The prospective *CAD-software* characteristics are discussed in this chapter.

Symbols and Abbreviations

Symbol	Quantity, Units (adopted)
3D	Three-dimensional
AI	Artificial intelligence
APIs	Application programming interfaces
CAD	Computer-aided design
CAD	Computer-aided geometric modeling or Computer-aided drafting
CAE	Computer-aided engineering
CAG	Computer-aided geometric modeling or “CAD”
CAI	Computer-aided inspection

Machine Design with CAD and Optimization, First Edition. Sayed M. Metwalli.

Companion website: www.wiley.com/go/metwalli/machine

Symbol	Quantity, Units (adopted)
CAM	Computer-aided manufacturing
CAPP	Computer-aided process planning
CAQC	Computer-aided quality control
CAS	Computer-aided synthesis – real *CAD*
CAVE	Cave automatic virtual environment
CNC	Computer numerical control
FE	Finite element
GPS	Global Positioning System
HMD	Head mounted display
NURBS	Non-uniform rational B-splines
STL	Stereolithography (**S**tandard **T**essellation **L**anguage or **S**tandard **T**riangle **L**anguage)
VR	Virtual reality

3.1 CAD and Geometric Modeling

The *CAD* software is expected to aid the designer in the component and system synthesis. One would expect that the computer should at least suggest the appropriate geometry and material of each element.

3.1.1 Classical Design Process

The available commonly called CAD packages are providing means to construct ***3D*** solid models for the machine elements. The user must provide the dimensions and the material to proceed with the analysis to define the stresses and deformations of the machine element under some specific loading and deformation constraints. The user in the *classical design* process with the accustomed "CAD" systems requires heavy iterations via the ***FE*** analysis as shown in Figure 3.1; see references such as Groover and Zimmers (1984). Design through repeated iterations are usually extensive to further acceptance or optimization of the design. Final ***3D*** solid model of the design assembly is rendered for presentation. The ***3D*** solid model is necessary for proceeding to the *computer-aided manufacturing* (*CAM*) such as the *computer numerical control* (*CNC*) code generation.

3.1.2 Synthesis Design Process

The designer in the prospective design synthesis paradigm or *computer-aided synthesis* (*CAS*), i.e. the *real CAD*, is shown in Figure 3.2. Design iterations are limited to the optimization of further objectives or special tuning. The properly evaluated final design assembly is rendered for *virtual reality* (***VR***) presentation. The geometric modeling should be replaced by *geometric synthesis* of the ***3D*** *solid model*. This should provide the appropriate *geometry* and *material* of each machine element. The result ought to be a direct *synthesis* of the machine element. The *geometric synthesis* directly provides the needed *CAS*. The further *optimization* is limited to additional objectives or special tuning requirements. The synthesis is therefore realizable with few iterations, if needed. This should be the philosophy and objective of a real *CAD*.

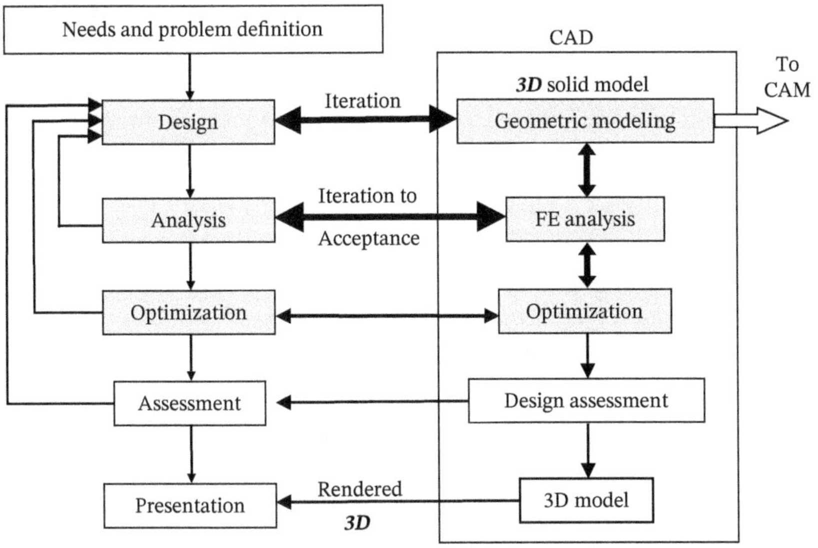

Figure 3.1 Classical design process with the accustomed "CAD" systems requiring heavy iterations for analysis. Design iterations are usually extensive to further acceptance or optimization. The ***3D*** model of design is rendered for presentation.

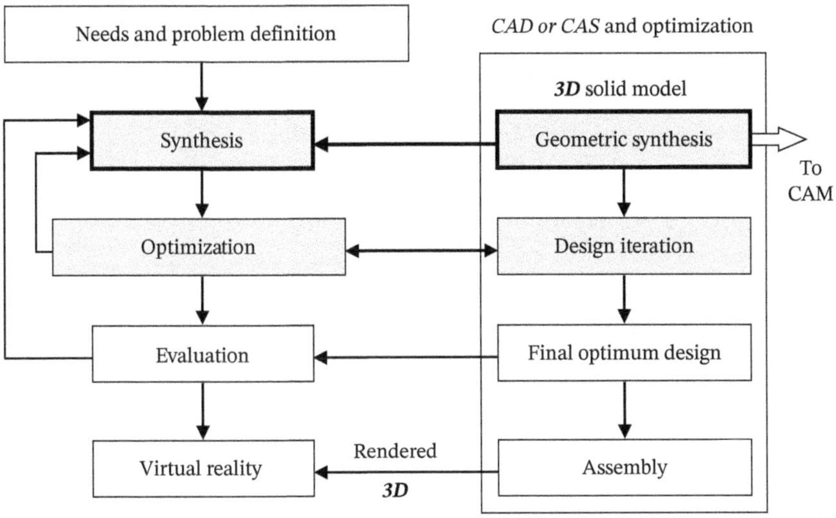

Figure 3.2 Design synthesis paradigm in real *CAD* or computer-aided synthesis (*CAS*). Design iterations are limited to the optimization of further objectives or special tuning. Final evaluated design assembly is rendered for virtual reality (***VR***) presentation.

3.1.3 *Human–Machine Characteristics*

The realistic *CAD* program should capitalize on the advantages of both human and machine to generate an efficient *CAD* program. The pitfalls or disadvantages of both should be avoidable. The following categories are the attributes of human–machine in several functions:

- *Information input*: The human is capable of *large input* information that occurs at once (sight and sound). The computer needs sequential and formal process to input information (data or images).

- *Information output*: The human is *slow* and sequential in information output (speech, gesturing). The computer is rapid and delivers sequentially (graphics, simulation).
- *Information organization*: The human is informal and intuitive. The computer is formal and detailed.
- *Information editing*: The human edits information in an easy and instantaneous manner. On the other hand, the computer procedure is usually difficult and involved.
- *Information extraction*: The human is good in extracting information. The computer is poor in extracting information.
- *Erroneous information and error detection*: The human have intuitive error detection and correction. The computer is highly intolerant to erroneous information but can be systematic in error detection.
- *Error production*: The human frequently produces errors, while computer rarely makes errors.
- *Tolerance to repetitions*: The human poorly handles repetitive tasks. The computer is excellent for repetitive tasks.
- *Analysis capabilities*: The human has a good intuitive analysis but is poor in numerical analysis. The computer does not have intuitive analysis but is good in numerical analysis.
- *Logic, reason, and intelligence*: The human is usually intuitive, gain experience, has imagination, provide judgment, learns rapidly but is sequential and unreliable in intelligence. The computer is systematic and stylized, has little learning but has reliable abilities. *Artificial intelligence* (***AI***) is working to improve computer intelligence.
- *Upgrading*: Human upgrading is rare and slow. The computer is usually possible to have speedy upgrades.

It is then better to adopt the positive attributes for human and machine in computer codes, procedures, and applications. It is also advisable to refrain from depending on the negative attributes of both human and machine. This should provide efficient human–machine interaction.

The characteristics of *CAD software* should then at least be a user-friendly, self-explanatory, easy to use, interactive, and verifying and should tolerate misuse. These characteristics guarantee a software that does not really need a manual. The characteristics may not necessarily be easy to realize, but it would be good if materialized.

3.2 Geometric Construction and *FE* Analysis

To do a full ***3D*** construction or an assembly of a design, each component should be in a ***3D*** solid model form. The absolute coordinates of each element allow the easy assembly of elements in a design. Most geometric modeling or customary "CAD" packages can demonstrate the *interference* of components, if geometry and dimensions are not fitting the mating components. The local coordinate of each element should ease the numerical analysis in a ***FE*** software. In most cases, the ***FE*** package should handle the absolute coordinates just as good as the local coordinates.

For numerical analysis in a ***FE*** package, the software usually performs a *preprocessing* operation on the ***3D*** solid model to devise the suitable meshing of the massive number of needed ***FEs***. Then the ***FE*** package performs the analysis on these massive number of *elements* to define the stresses and deformations of the machine component under some specific loading, deformation, and boundary constraints. The *postprocessing* in the ***FE*** package provides the values of the *stresses* and *deformations* at any point in the ***3D*** *solid* model of the machine component.

The ***3D*** *solid model* is essential to define the extent of material in a component. Figure 3.3 shows different geometric modeling of a bolt. Figure 3.3a presents the *wireframe* model of the bolt. Figure 3.3b gives a rendered ***3D*** solid model of the same bolt. To calculate *stresses* or write a *CNC code* to machine this bolt, the ***3D*** *solid* model clearly defines the boundaries of the bolt. These boundaries are used to bound the *elements* in a ***FE*** package or would be the final target for manufacturing or machining on a *CNC* machine. The wire frame model in Figure 3.3a does not really define boundaries for the computer to comprehend and handle.

The ***3D*** *surface model* shown in Figure 3.4 for a bolt can be directly used in ***3D*** *printing* or *layered manufacturing* since it defines the triangular outer-boundary. The outer-boundary of the solid is defined as the positive normal

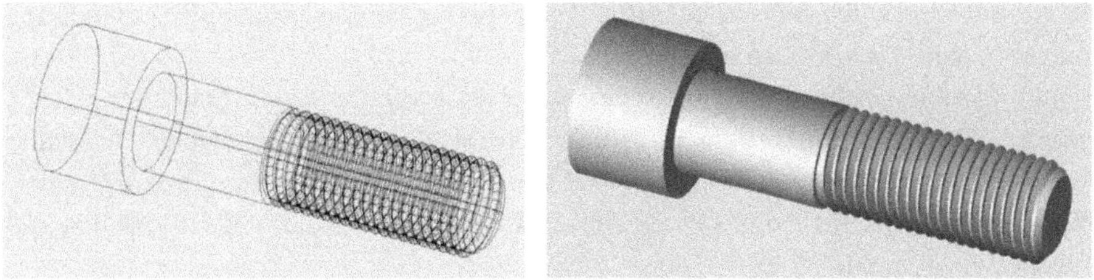

(a) (b)

Figure 3.3 Different geometric modeling of a bolt: (a) the ***3D*** wireframe model and (b) a rendered ***3D*** model.

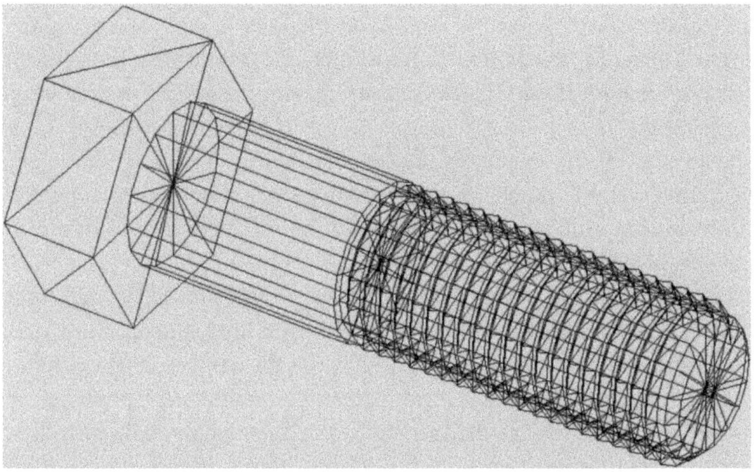

Figure 3.4 A ***3D*** surface model of a bolt in triangles and polygons. This model can be used in ***3D*** printing or layered manufacturing since it defines the triangular outer-boundary. The back triangles of the bolt head are not shown.

to the face triangle when its sides are set in the *counter clockwise direction* (*right-hand rule*). The back triangles of the bolt head are not shown in Figure 3.4, and the straight cylindrical sides are given in polygonal form that are ordinarily split into their triangles. Similarly, all other polygons are split into triangles. The triangles between the head and the bolt stem are omitted in the corrected *stereolithography* (***STL***) file for ***3D*** printing.

3.3 CAD/CAM/CAE and Advanced Systems

In product cycle, the integration of computer-aided design, computer-aided manufacturing, computer-aided engineering (*CAD/CAM/CAE*), and other advanced tools should save a great deal of time, money, and effort. It should also produce better or optimum designs. The *CAD* should produce a real synthesis to name it a *CAS*. In that case, the design process is much faster and should even render an optimum design. With a ***3D*** *solid model*, the boundaries of the machine component are defined, and therefore the numerical analysis is possible through any ***FE*** package as defined earlier in Section 3.2. With a ***3D*** *solid model*, the boundaries of the machine component are defined, and therefore a computer code can also be developed to manufacture that component on a *CNC* machine or by any other means of production. A ***3D*** solid model provides information to perform a real-time *quality control* on the machine component while manufacturing. The product dimension is measured during manufacturing thru cutting on a *CNC* machine and compared to the ***3D*** solid dimension-value. A code is used to decide on continued cutting or termination when the dimension reaches the target value. This is part of the computer-aided engineering (*CAE*) process that can be integrated with *CAM* or can be a part of *CAM*.

CAE is more amenable and feasible with a ***3D** solid*. This *involves computer-aided inspection* (*CAI*), *computer-aided process planning* (*CAPP*), computer-aided quality control (*CAQC*), computer-aided testing (*CAT*), and computer-aided maintenance, Metwalli et al. (1998). *Robotics* is also a field that can utilize ***3D*** models particularly to assemble components into products or systems. These areas are beyond the scope of this text and can be further studied in their specialized fields.

The expansion of technology in ***3D** printing* or *layered manufacturing* and *rapid-prototyping* depends on the ***3D*** *solid* or ***3D** surface* models. They can be part of *CAE*. The software of the ***3D*** printers slices the ***3D** solid* model into layers to define the boundaries-to-fill with printing material. The ***3D** surface* model directly provides the boundaries and therefore easy to slice. A ***3D** surface* format such as ***STL*** (***stereolithography***, or ***Standard Triangle Language*** or ***Standard Tessellation Language***) directly provides the *right-hand-triangles* that are wrapping the outer surface of the ***3D*** solid. This ***STL*** format is generally the preferable file format for ***3D** printers*. If the ***3D** solid* is bound or wrapped by a single *non-uniform rational B-splines* (NURBS) surface, the process is then just as easy; see Section 4.2.5 and El-Komy and Metwalli (2014).

The ***3D*** printing is gaining popularity and attractiveness particularly in producing lightweight structures. Optimizing ***3D*** solid models to have voids for lighter weight and constant or uniform stresses across the solid can be produced by ***3D*** printing or layered manufacturing. Hollow honeycomb configurations are attractive forms for such a light solid. A hollowed ***3D** frame* that is reduced from a solid can also be a solution, El-Rahman et al. (2017). Figure 3.5 shows a couple of optimum ***3D*** cantilever beams with a fixed left-support that is loaded at mid-right end. Figure 3.5a presents the original cantilever beam of a constant thickness. Figure 3.5b shows an optimum cantilever beam of variable thickness in ***3D***, Saad and Metwalli (2013). Figure 3.5c shows the stresses in an optimum cantilever beam using a ***FE*** package, Abd El-Rahman (2017). These optimum configurations can be produced by ***3D*** printing. Extensive research in ***3D** printing* is geared toward the suitable material, geometry, and form that can be utilized to meet *the competing objectives* of lighter, stronger, and cheaper products. *Geometry* and *topology optimization* would be useful in that regard. This is beyond the scope of this text.

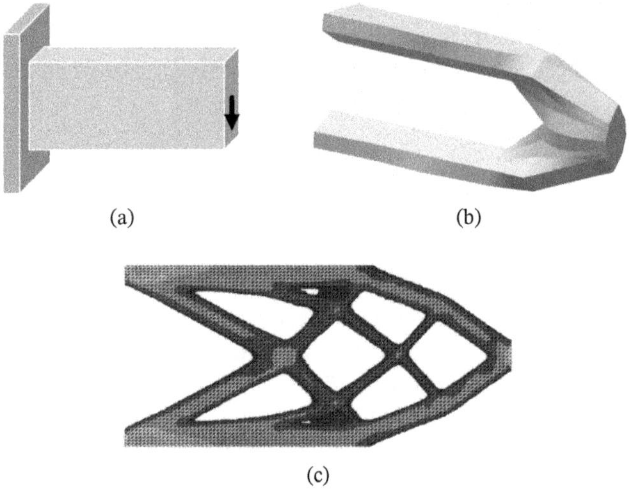

Figure 3.5 ***3D*** cantilever beam with a fixed left support and loaded at mid-right end: (a) original cantilever beam of constant thickness, (b) optimum cantilever beam of variable thickness in ***3D*** (Saad and Metwalli 2013), and (c) stresses in an optimum cantilever beam using a ***FE*** package (Abd El-Rahman 2017).

3.4 Virtual Reality

A *CAE* field is the ***VR***. It allows users to view ***3D*** *solids* in a *stereographic* and *photorealistic* environment. The user is then virtually immersed in a perceptual reality as if the objects are literary produced and are a front. Figure 3.6 shows two views of a ***3D*** bolt in a stereographic projection that represent the left-eye view of the ***3D*** bolt in Figure 3.6a and the right-eye view of ***3D*** bolt in Figure 3.6b. The distance between the two views (a) and (b) should be about the same or less than the distance between the two eyes. If not, one needs to scan Figure 3.6 and insert it on any page, slide, or a picture screen and zoom till the two views are one ***3D*** stereo image. If one places a cardboard between the two views and one's two eyes, he/she should see a single stereographic view of the bolt, just as *stereoscope* with a simple material implementation as shown in Figure 3.7. A facial board is cut off a cardboard or a thin balsa-wood, and two holes are cut for the left and right eyes. A nose slit is also cut to allow the board to be as close as possible to the eyes. Another spacer-board is glued normal to the facial board. This simple *stereoscope* is placed normal to the page of Figure 3.6 with the spacer-board situated in-between the two bolt-views

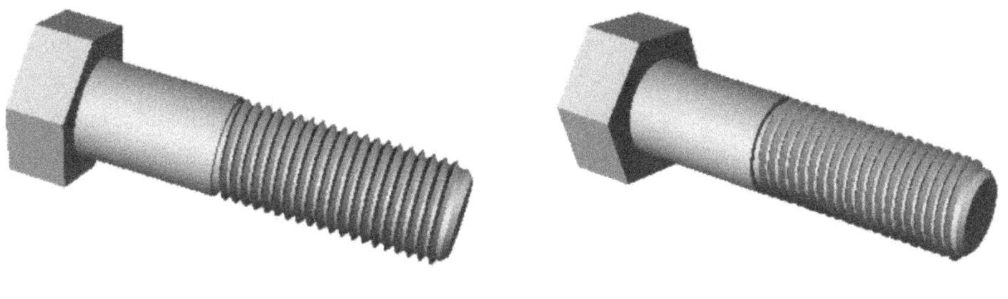

(a) (b)

Figure 3.6 Two views of a ***3D*** bolt in a stereographic projection that represents the left-eye view of the ***3D*** bolt in (a) and the right-eye view of the ***3D*** bolt in (b). The distance between the two views should be about the same or less than the distance between the two eyes.

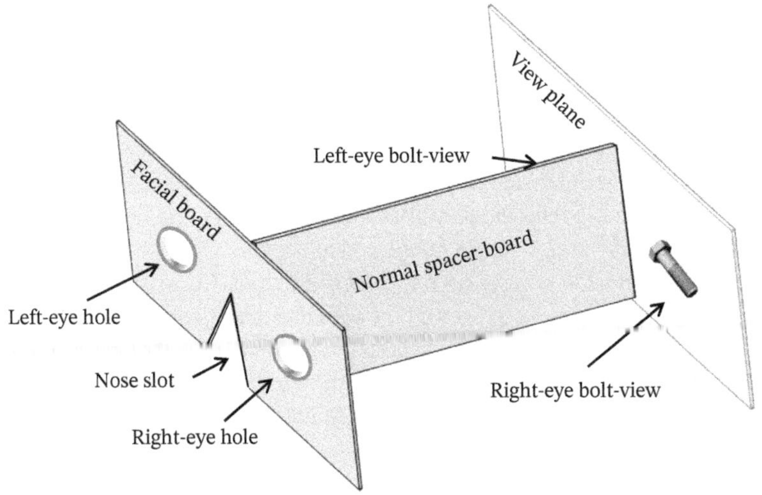

Figure 3.7 An assembly of a *stereoscope* with a simple material implementation. A facial board is cut off a cardboard or thin balsa wood, and two holes are cut for the left and right eyes. A nose slit is also cut to allow the board to be as close as possible to the eyes. Another board is glued normal to the facial board. This *stereoscope* is placed normal to the view plane such as the page of Figure 3.6 and situated in-between the two bolt views.

in (a) and (b). One might need to be a little far or close to the two views to be able to see them as one bolt. The proper distance between the two views would be about the same or less than the distance between the two eyes; See Section 4.1.

3.4.1 Virtual Reality Process

VR tools provide a large screen that may subtend some 180° or so around viewers to immerse one into the scene. In some options, the projection into the screen is dual: left and right eye views – one view at a time – of all ***3D*** objects in the space. Of course, the front objects obscure the back ones and the *rendering* of the objects should be *photorealistic*. The dual viewing is synchronized with the viewer goggles to let him/her see only the left view by the left eye and the right view by only the right eye. The viewer then sees the ***3D*** objects in a *photorealistic stereographic* form as if they are real. The objects are then virtually real and ***VR*** figuratively materializes.

3.4.2 Virtual Reality Hardware Requirements

The expected hardware for ***VR*** would incorporate good graphics capabilities of the computer to perform photorealistic simulations; large high-definition screen or screens (may be curved) or *CAVE*© (cave automatic virtual environment), *HMD* (head mounted display), *Powerwall*, or others; sensation tools such as synchronized goggles, sound, touch (haptic feedback); and may be smell and taste. The *CAVE* was invented by a group at the University of Illinois Chicago, Cruz-Neira et al. (1992). Other low-cost systems such as *stereo televisions*, *Oculus Rift*© HMD, and *Google Cardboard*© are available. *Google Cardboard* is a more involved material implementation of Figure 3.7 with lenses placed in the eye holes to allow the close viewing of ***VR*** apps on cell phones. These available or envisioned tools should cause ***VR*** to be more realistic for the user to visualize, think it is real, and get immersed into the action. If the screens have active polarized film to view the left-eye scene following the right-eye scene (or interlaced), the goggles can just be two passive polarized films in a perpendicular polarization of one eyepiece to the other. This would cause many spectators to view the scene in an easier way. If the goggles are made to be the screens without being bulky (as *stereoscopes*), that would immerse the viewer even more. The viewer interacting with the scene would use haptic gloves to feel the touching of the object in a feedback manner to move it or interact with it. These gloves may need to be long and anchoring to the shoulders. They may even need to be a whole chest, arm, hands, and fingers. The seat might also be involved just like flight simulators.

3.4.3 Virtual Reality Interactive-Process Tools

To prepare the ***3D*** scenes, a powerful graphics software (imbedded in a hardware) is needed to calculate the right and left eye views in their rendering-effect state. These capabilities are available in such *application programming interfaces* (APIs) software such as OpenGL© or DirectX©; see Internet references in Section 3.6. Both capabilities are available in the graphics hardware of most computers, laptops, or even smart cell phones (most cells by now). This should also be capable of performing real-time simulation graphics. The synchronized goggles should also be tracked as to the head position and orientation to dynamically affect the looking angles and field of view. If the goggles are not synchronized, the viewer head and eyes should also be tracked by some hardware and software system. The same applies for tracking the interactive haptic gloves to dynamically affect the interaction intent of fingers and hand.

3.4.4 Virtual Reality Applications

With ***VR*** tools, many applications are possible for humans or robotic interaction, such as computer-aided *assembly* of components, subsystems and products, CAI, VR *presentation* for design evaluation (Figure 3.2), etc. (see Seth et al. 2011; Berg and Vance 2017). If fact, assembly by computers, robots, or machines is feasible. Other applications in the design field are for *ergonomics* and *reachability* of utility components in designs. *Simulation* of factory layout before implementation is one other application of ***VR***. Autonomous vehicles should also use ***VR*** to effectively guide the vehicle in addition to ***AI*** and a position tracking system such as Global Positioning System (*GPS*).

3.5 Summary

This chapter presented an overview of computer-aided techniques that are mostly related to machine design. It is an introduction to differentiate between real *CAD* process and the geometric modeling that is commonly known as *CAD*. In real *CAD*, one should expect that the computer would at least suggest the appropriate geometry and material of each element. The *geometric modeling* packages known as *CAD* help users to develop a ***3D*** *solid model* but do not normally provide aides to suggest geometry and material for any component or part thereof. They usually depend on the repeated numerical analysis by ***FE*** codes to perform design. It would be useful for the designer to develop his codes to define geometry and material or have a real *CAS* (real *CAD*) to do so. This text intends to help in this synthesis process before resorting to the CAG-modeling packages or to have a numerical analysis by ***FE*** to validate the synthesized parts or components.

The chapter introduces some computer-aided tools that are adopted throughout the text and are potentially useful in CAS and optimization. Of these are the spreadsheet of Microsoft *Excel* (2016) and the matrix-manipulation package of MathWorks *MATLAB* (2014). Section 2.2, provides an overview of some basic capabilities of both *Excel* and *MATLAB*. The prospective *CAD*-software characteristics are also discussed to provide help in utilizing computers in effective and real *CAD* software. Capitalizing on the advantages of human and machine characteristics would be very valuable and efficient in that regard.

The integration of *CAD/CAM/CAE* and other advanced tools should save a great deal of time, money, and effort in the synthesis and production of designs, systems, or products. The utility of the ***3D*** solid modeling or surface modeling dramatically shortens the time of this integration. This chapter demonstrates some applications of these *CAD/CAM/CAE* tools such as the ***3D*** *printing*.

VR coverage in this chapter is only a short or abridged review. It presents the basics and some envisioned tools. ***VR*** is essential in design, manufacturing, or production processes, and particularly in design presentation and evaluation. The covered material is hopefully useful in the understanding of this potentially vital tool in many applications.

Problems

3.1 Identify the computer-aided design (CAD) software packages or the *CAG* modeling packages available at your school and those available in the world market. Define the mostly used package worldwide and find its capabilities comparing its market share relative to others.

3.2 Define the available real *CAD* software, i.e. the computer-aided synthesis (*CAS*) software. Find the machine elements that can be synthesized by these packages.

3.3 Identify the used characteristics of human and machine in the interaction of the CAD or *CAS* package you use. How often did you need the help or manual of the package?

3.4 How simple can you transfer the CAD file to the provided or available ***FE*** package to perform analysis?

3.5 Use CAD and ***FE*** to model a cantilever beam of a unit length and a rectangular cross-section loaded by a unit load at its free end. What is the maximum stress and deflection for an assumed material strength and cross-sectional dimensions? Try to synthesize the cross-sectional dimensions.

3.6 Use your CAD software to export an ***STL*** file for a cantilever, and get a ***3D*** printed model for it. Check the format of the ***STL*** file.

3.7 Use your CAD software to generate a *CNC* code to machine the cantilever off a cylindrical bar.

3.8 Produce a stereoscope as envisioned in Figure 3.7, and use it to observe the stereo image of the ***3D*** bolt in Figure 3.6.

3.9 Identify the ***VR*** facility at your school or organization and transfer the ***3D*** model of the cantilever or any ***3D*** solid model to it. Use the ***VR*** tools to visualize the ***3D*** object and utilize the available capabilities to move the object in space.

References

Abd El-Rahman, M.S. (2017). *A generalization of the heuristic gradient projection for 2D and 3D frame optimization*. MS Thesis. Cairo University.

Berg, L.P. and Vance, J.M. (2017). Industry use of virtual reality in product design and manufacturing: a survey. *Virtual Reality* 21 (1): 1–17.

Cruz-Neira, C., Sandin, D.J., DeFanti, T.A. (1992). The CAVE: audio visual experience automatic virtual environment. *Communications of the ACM* 35 (6): 64–72.

El-Komy, M. A., Metwalli, S.M. (2014). Optimum 3D wrapping of CAD models using single NURBS. Proceedings of the ASME IMECE 2014, Montreal, Quebec, Canada (8–13 November 2014), Paper # IMECE2014-36736.

El-Rahman, M.S., Abd El-Aziz, K.M. and Metwalli, S.M. (2017). A generalization of the heuristic gradient projection for 2D and 3D frame optimization. *Proceedings of the ASME IMECE 2017*, Paper # IMECE2017-70560. November 3–9 2017, Tampa, Florida, USA.

Groover, M.P. and Zimmers, E.W. (1984). *CAD/CAM Computer-Aided Design and Manufacturing*. Prentice Hall.

MathWorks (2014). *MATLAB©*. MathWorks.

Metwalli, S.M., Salama, M.S., and Taher, R.A. (1998). Computer aided reliability for optimum maintenance planning. *Computers and Industrial Engineering* 35 (3–4): 603–606.

Microsoft (1998). *Visual Studio-6©*. Microsoft Corporation.

Microsoft (2016). *Excel©*. Microsoft Corporation.

Saad, M. R., and Metwalli, S. M. (2013). Heuristic GP finite element method for structural thickness optimization. *Proceedings of the ASME International Mechanical Engineering Congress & Exposition IMECE 2013*, San Diego, CA (15–21 November 2013), Paper # IMECE2013-66364

Seth, A., Vance, J.M., and Oliver, J.H. (2011). Virtual reality for assembly methods prototyping: a review. *Virtual Reality* 15 (5): 5–20.

Internet Links

http://opengl.org/ OpenGL free download graphics APIs.

https://www.microsoft.com/en-us/download/search.aspx?q=directx DirectX free download graphics APIs.

4

Computer-Aided Design

This chapter intends to present a brief synopsis of the computer-aided design field as mainly as geometric modeling. The coverage is geared toward the applications in machine design to develop a three-dimensional (***3D***) model. This necessitates the coverage of *geometric modeling* or ***3D*** *computer-aided drafting* known as "CAD" rather than the traditional *computer-aided design* commonly known as "*CAD*." The computer-aided drafting is related to *computer graphics* or geometric modeling. The real computer-aided design in machine element selection or adoption should be related to having the computer define proper geometry and material, and not only helping in ***3D*** modeling or ***3D*** drafting. CAD in the geometric modeling sense is just a *computer-**a**ided **d**rafting* (CAD) rather than *computer-aided design (CAD).* Since a great deal of *geometric modeling* – known as "CAD" – is employed in the design process, the coverage is focused on the ***3D*** modeling as a necessary part in the design process; see Section 3.1. The understanding of the technology of producing a ***3D*** geometric model is therefore of a profound effect in efficient production and utilization of available *geometric modeling* packages traditionally known as "CAD" tools. The use of computers to design synthesis of components is the objective of this textbook. The synthesis in terms of letting the computer aids in the definition of dimensions or geometry and material selection is what one should call a real *computer-aided design* or a *computer-aided synthesis* (CAS). This task is the purpose of the text.

The ***3D*** geometric modeling is essential in design *presentation* and product *manufacturing*. The ***3D*** modeling is made easier, if the basics of the *computer graphics* are understood. This chapter intends to help in that regard. For further coverage, one can refer to texts in *computer graphics* such as Newman and Sproull (1979), Foley et al. (1990), and Hill (2001).

Symbols and Abbreviations

Symbol	Quantity, Units (adopted)
$\Im$	Conic or quadric surface
(h_C, k_C)	Circle center location
(X, Y, Z)	Non-visible primary *CIE* colors
$(x, y, z, 1)$	Normalized homogeneous coordinate
$[\boldsymbol{P}_B]$	Body matrix in homogeneous coordinates
3D	Three-dimensional
a, b, c	Scaling quantities in x, y, and z directions
AI	Artificial intelligence
a_L, b_L, c_L	Components of line direction vector

Machine Design with CAD and Optimization, First Edition. Sayed M. Metwalli.

Companion website: www.wiley.com/go/metwalli/machine

Symbol	Quantity, Units (adopted)
a_N, b_N, c_N	Components of unit normal vector of face F
a_P, b_P, c_P	Components of plane-normal vector
A_P, B_P, C_P	Explicit plane coefficients
APIs	Application programming interfaces
a_q, … k_q	Coefficients of conic or quadric surface
$B_1, \ldots, B_5$	Body solid components
$\boldsymbol{b}_1$ and $\boldsymbol{b}_2$	Blending functions of parameter t
$\boldsymbol{b}_1, \boldsymbol{b}_2, \boldsymbol{b}_3$	***2D** blending functions*
$\boldsymbol{b}_B$	Blending function vector
$\boldsymbol{B}_C$	Body matrix with vertices as column vectors
$\boldsymbol{B}_H$	Body matrix in homogeneous coordinates
$\boldsymbol{b}_{i,k}$	Spline blending functions
$\boldsymbol{b}_{i,n}(u)$	Blending functions of parameter u, $i = 0$ to n
$\boldsymbol{b}_{i,n}(u)$	Blending functions in the u direction
$\boldsymbol{b}_{j,m}(v)$	Blending functions in the v direction
$\boldsymbol{B}_L$	Lines matrix
***B**-rep*	Boundary representation
CAD	Computer-aided drafting
CAD	Computer-aided design
CAM	Computer-aided manufacturing
CAS	Computer-aided synthesis – real *CAD*
CCW	Counterclockwise
CIE	Commission Internationale L'Eclairage
C-loops	Child loops or boundaries of holes in a face
CMY	Cyan, magenta, and yellow colors
CNC	Computer numerical control
***CSG**-rep*	Constructive solid geometry
d, e, k, m, n, p	Skewing amounts in x, y, and z
d_F	Length of normal vector of face F
$E_1, \ldots, E_m$	Body edges from *1* to m connecting vertices
E_B	Number of the body edges
$F_1, \ldots, F_k$	Body boundaries of faces ($F_1, \ldots, F_k$)
F_B	Number of the body faces
FE	Finite Element method
$\boldsymbol{f}_z$	Eye z focal distance to projection plane
h, g, f	Perspective amounts in x, y, and z
H_B	Number of holes that pass through the solid
h_H	Homogeneous extra coordinate

Symbol	Quantity, Units (adopted)
HSL	Hue, saturation, and luminance
I_d	Incident diffuse illumination
I_{Pd}	Diffuse illumination from a point
I_{Ps}	Reflected diffuse illumination
I_{Pt}	Transmitted illumination due to translucency
I_S	Incident specular illumination
I_T	Total rendered illumination
k	Spline curve order
$\boldsymbol{K}_H$	Skewing matrix about *x*, *y*, or *z*
K_{Ps}	Reflectance coefficient at surface point $\boldsymbol{P}_s$
K_{PsT}	Transmission coefficient
k_S	Light source constant
$K_S(\theta_s)$	Specular reflection coefficient
$\boldsymbol{L}_{\mathrm{H}}$	Line in homogeneous coordinates
$\boldsymbol{L}_I$	Intersecting-line of two planes
$\boldsymbol{L}_P$	Parametric line defined by $\boldsymbol{P}_L(t)$ on it
LUT	Look-up tables
m	Number of edges in a body
$\boldsymbol{M}_{\mathrm{BB}}$	Bicubic blending coefficients matrix
n	Number of vertices or points in a body
$\boldsymbol{N}_F, \ldots, N_{F1}$	Unit normal vector of face F or F_1
$\boldsymbol{N}_P$	Plane normal vector
np	Phong *cos* exponent varies from 1 to 200
np	*Phong* exponent ranges from 1 to 10
N_{Ps}	Surface normal at point $\boldsymbol{P}_s$
$N_{Px}\, N_{Py}\, N_{Pz}$	Components of unit normal vector
NURBS	Non-uniform rational B-splines
$\boldsymbol{P}_1, \ldots, \boldsymbol{P}_n$	Body points $\boldsymbol{P}_1$ to $\boldsymbol{P}_n$ (or vertices) from 1 to n
$\boldsymbol{P}_B(u, v)$	Boundary curves $u, v = 0$ or 1
$\boldsymbol{P}_{BC}(u)$	Point on *Bezier* curve at u
$\boldsymbol{P}_{\mathrm{BS}}(u, v)$	A point on the *Bezier* surface
PC	Personal computers
$\boldsymbol{P}_C$	Matrix of position vectors of corner points
$\boldsymbol{P}_{\boldsymbol{c}}$	Position vectors of corner points
$\boldsymbol{P}_C(u)$	Point on a parametric curve in u
$\boldsymbol{P}_{Ci}$	Control points on *Bezier* curve, $i = 0$ to n
$\boldsymbol{P}_{Cij}$	Control points in ***3D***
$\boldsymbol{P}_H$	Perspective matrix in z direction
$\boldsymbol{P}_I$	Intersection point
$\boldsymbol{P}_L(t), \boldsymbol{L}_P$	Parametric point on the line $\boldsymbol{L}_P$

Symbol	Quantity, Units (adopted)
P-loops	Parent loops or boundaries of faces
$\boldsymbol{P}_{\mathrm{NC}}$	Point on NURBS curve
$\boldsymbol{P}_q$	Quadric surface point
$\boldsymbol{P}_S\,(u,v)$	Point on a parametric surface in *u and v*
$\boldsymbol{P}_{\mathrm{SC}}\,(u)$	Point on spline curve at *u*
$\boldsymbol{Q}_q$	Quadric surface matrix in implicit form
R_B	Number of rings or cavities in body faces
r_C	Radius of a circle
RGB	Red, green, and blue colors
r_S	Distance from light source to the surface
$\boldsymbol{R}_S$	Reflected ray off a surface
$\boldsymbol{R}_{xyz}$	Rotation matrix about *x, y*, or *z*
s	Parameter defines location of a point
S_B	Number of shells or body wrapping surfaces
$\boldsymbol{S}_B\,(u,\,v)$	Bilinear parametric surface
$\boldsymbol{S}_{\mathrm{BS}}$	Bezier surface
$\boldsymbol{S}_{\mathrm{BS}}\,(u,\,v)$	Bicubic Bezier patch or surface
$\boldsymbol{S}_C$	Coons surface
$\boldsymbol{S}_e$	Eye line-of-sight direction
$\boldsymbol{S}_H$	Scaling matrix in *x, y*, or *z*
s_j	Spline *knots* for the *v* direction
$\boldsymbol{S}_P$	Plane "surface"
$\boldsymbol{S}_R$	Ruled or lofted surface
STL	**S**tereo**l**ithography (**S**tandard **T**essellation **L**anguage or **S**tandard **T**riangle **L**anguage)
S_x, S_y, S_z	Scales in *x, y*, and *z* directions
t	Parameter defines location of a point
t, u, v	Translation amounts in *x, y*, and *z*
$\boldsymbol{T}_{\mathrm{GH}}$	Total homogeneous transformation matrix
t_i	Spline *knot* values
t_I	Intersection parameter of line with a plane
$\boldsymbol{t}_K$	NURBS knot vector
t_x, t_y, t_z	Translation amounts in *x, y*, and *z*
$\boldsymbol{T}_{xyz}$	Translation matrix in *x, y*, or *z*
u and *v*	Parametric surface or plane coordinates
$V_1, \ldots, V_n$	Body vertices or points $\boldsymbol{P}_1$ to $\boldsymbol{P}_n$ from *1* to *n*
$\boldsymbol{V}_1\boldsymbol{V}_2\boldsymbol{V}_n$	Body vertices
V_B	Number of the body vertices or points
$\boldsymbol{V}_{C1}$	Cartesian vector 1
$\boldsymbol{V}_{H1}$	Homogeneous coordinate vector 1

Symbol	Quantity, Units (adopted)
$\boldsymbol{V}_L$	Line direction vector
VR	Virtual reality
w_i	NURBS weights
$x_C y_C z_C$	*CIE* visible chromaticity primates
x_i, y_i, z_i	Coordinates of vertex or point $\boldsymbol{P}_i$
z_f	Zooming factor in z direction
$\boldsymbol{Z}_H$	Zooming matrix in x, y, and z
θ_C	Angle on the circle
θ_i	Incident angle
θ_r	Reflectance angle
θ_s	Reflected angle between ray and observer
θ_v	Surface normals to light source angle
θ_x	Rotation angle about x
θ_y	Rotation angle about y
θ_z	Rotation angle about z

4.1 *3D* Geometric Modeling and Viewing Transformation

Developing a ***3D*** model of a design is made easy through the availability of extensive packages that competitively do the job. Understanding the process by which this is accomplished supports the proficiency needed to do that. Basic presented concepts, relations, and definitions are useful in understanding and developing geometric models.

4.1.1 *3D* Geometric Modeling

Some time ago, the ***3D*** *geometric modeling* of solids has been represented by a *wire frame* model as discussed in Section 3.2. It defined main body points, corners or *vertices*, and the *edges* connecting these points or vertices. Edges can be a straight or a curved line. Further developments generated a *surface model*, which defines *vertices*, *edges*, and *faces* representing the outer enclosure of the body. Again edges can be a straight line or a curved line attached to a face. Further to that, ***boundary*** *representation* dobbed (***B***-*rep*) has the same faces, edges, and vertices as in surface model. Topology, however, defines relationship between faces, edges, and vertices. A *solid* is then an enclosed space in ***3D***. In a parallel development, the *constructive solid geometry* dobbed (***CSG***-*rep*) depended on a set of ***3D*** primitive solids (e.g. a block, cylinder, cone, sphere, a torus, etc.) to be used as building blocks to construct a solid. Figure 4.1 shows some of the primitives that can be used. *Boolean* operations such as *union*, *intersection*, and *difference* manipulate these primitives to construct the solid. A body is therefore made of several intersecting primitive solids by adding or subtracting primitives as shown in Figure 4.2. Figure 4.2a shows a body created by adding a cylinder to a block (a *union*). Subtracting a smaller cylinder from the large one generates a hollow cylinder (a *difference*). *Shelling* the block produces a hollow block with a specific thickness. Other local *operations* can also be similarly employed such as *filleting, chamfering,* and *drafting* or *tapering* as defined in Figure 4.2b,c. In shelling operation, the solid (or part of it) is hollowed by the same but smaller shape to form a shelled object that has a specific shell thickness through *subtracting* a smaller object from the original one.

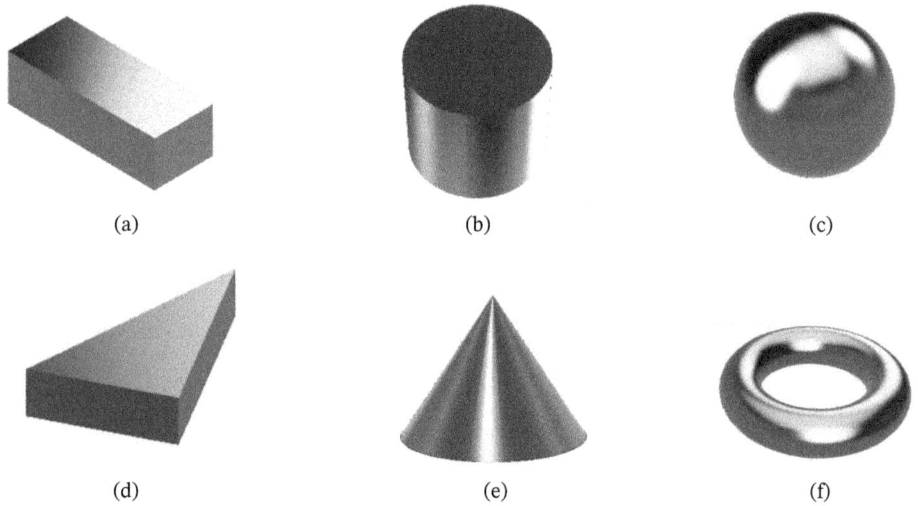

Figure 4.1 Basic ***3D*** primitive solids, e.g. a block or parallelepiped (a), cylinder (b), sphere (c), wedge (d) cone (e), and torus (f) to be used as building blocks to construct a solid.

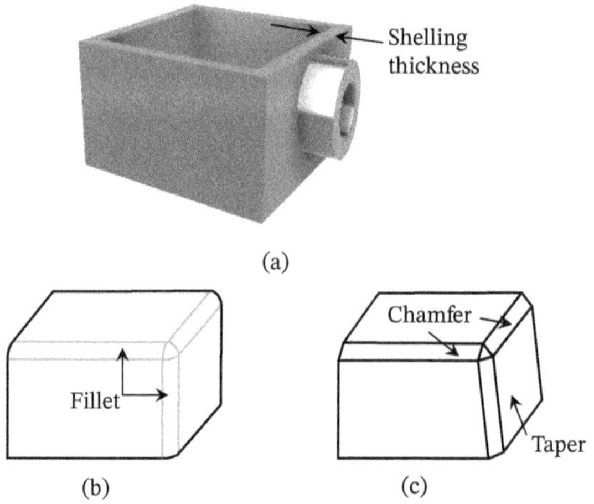

Figure 4.2 A body is made of several intersecting primitive solids in (a). Other ***3D*** local operations can be applied such as shelling (a), filleting (b), chamfering (c), or drafting and tapering as in (c).

Recently, most systems employ ***boundary*** representation even though some would still employ primitives as simplified means to build solids. The generation of a solid of revolution starts by a *sketch* or a section of the object and then *revolving* the section perimeter about a centerline to generate the object *boundary*. The generated ***3D*** body is encompassed by the generated revolved *boundary* as demonstrated in Figure 4.3. The shaft section in Figure 4.3a is revolved about the centerline to produce the shaft in Figure 4.3b. Numerous other objects are similarly generated. A torus is generated by *revolving* the circular section about a distant centerline. Another operation is to *loft, sweep,* or *extrude* a sketch or a section in a specific direction as given in Figure 4.4. The desired section is sketched by a combination of a rectangle and a circle at its edge. The sketch is then *lofted, swept,* or *extruded* in a direction normal (or inclined) to the plane of the sketch to generate the object shown in Figure 4.4a. This figure also shows another inscribed circular sketch on its face. The inscribed circular sketch on the face in Figure 4.4a is used to generate the solid in Figure 4.4b. The extrusion of the inscribed circular sketch is in a *cutting* direction to create the hole in Figure 4.4b. The cylindrical hole in Figure 4.4b may have also been created by using a cylinder to *subtract*, *drill*, or *cut* a hole in a generated "blind" object (as the sketch in Figure 4.4a but without the inscribed circle). The lofting of a circular section can be done on a curve such as a helix or spiral to generate a helical or a spiral spring.

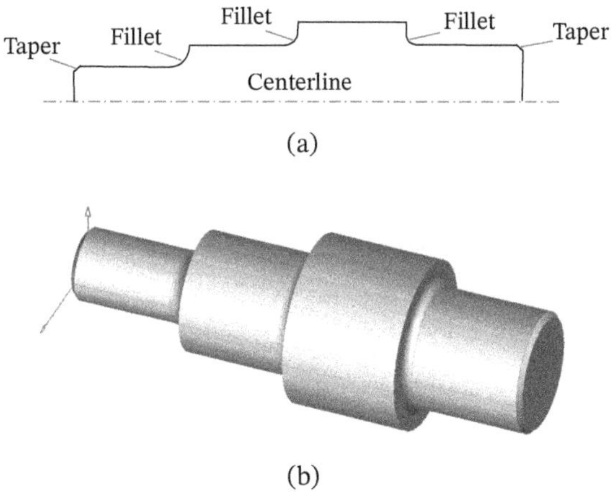

Figure 4.3 Generating a ***3D*** solid by *revolving* about the centerline of a *wire frame* section or a shaft profile sketch such as in (a) to produce the ***3D*** shaft shown in (b).

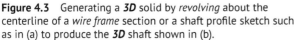

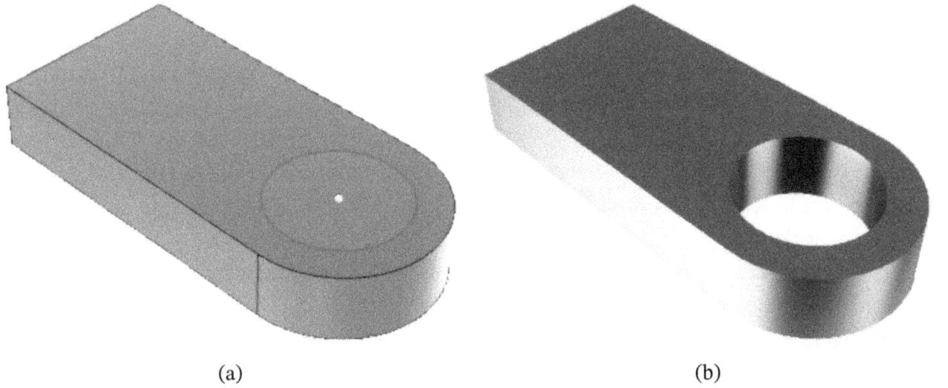

Figure 4.4 Generating a ***3D*** solid by *lofting* or *extruding* a sketch as shown in (a). An inscribed circular sketch on the face in (a) is used to generate the solid in (b). The extrusion of the inscribed circular sketch is in a cutting direction to create the hole in (b).

Solid modeling system utility depends on the type of system at hand. Several solid modeling systems are available as commercial packages from several vendors. The field is volatile such that several acquisitions and mergers occur, and packages change hands and names. Some vendors upgrade packages and thus change names. Current solid modeling systems are such as the commercial *SolidWorks*© and *CATIA*© from Dassault Systems©, *Creo*© (formerly *Pro/Engineer*) from PTC©, *Inventor*©, and *Fusion 360*© from Autodesk©, *UnigraphicsNX*©, and *Solid Edge*© from Siemens, etc. They use several modeling *kernels* or graphics software *engines* such as *ACIS*, Proprietary "*Granite*," *Parasolid, Creo*, etc. Now most kernels and systems depend on and employ **OpenGL,** *Direct3D,* or ***DirectX*** for Windows or emulate these tools. These *OpenGL, Direct3D,* or *DirectX* are *application programming interfaces* (APIs) that typically *bound* or *wrap* the object with all its necessary surface *polygons* or *triangles* and operate on these as the representation of the object. Intersecting lines between subsections or sub-objects are the boundaries of these subsections. The total body is then treated as the conglomeration of these triangles. All geometric modeling systems strive to achieve better representation, response time, wider range of solid shapes generated, ease of use, and system reliability.

Internal representation of systems may have advantages of *CSG-rep* such as simple data structure and size, which result in easy internal management of data. It is easy to modify a solid shape because it defines both the shape and the process of modeling the solid. The internal representation of solids may include the advantages of a *B-rep* since edges and faces are explicitly represented and found from *intersections* of objects. Faces are easily *tessellated* to generate the *triangles* needed for *OpenGL, Direct3D*, or *DirectX* processing of objects. It is also easy and quick to ascertain topological relationships for the integrity of the ***3D*** object. The drawbacks of *B-rep* are inherent in the complex data structure and may require a large memory space. Procedures for manipulating the internal data may therefore be complex. To correct such an error, the user may need to reconstruct parts of the original solid. The software *kernel/engine* is usually guarding against such problems.

The first method for internal *data handling* is to save all data on a solid as a file at each stage of the design process. However, when complex solids are constructed using many building commands, a vast amount of storage space is required. This method is therefore economically taxing. Recomputing building commands is another alternative, where it needs to store all the commands entered by the user when generating a solid. This is time consuming if a great number of commands were used. The problem of reconstructing solids may use undo and redo for primitive operations. Each primitive operation may have a corresponding inverse operation for undoing changes. The primitive operations are stored in ***tree*** structures. These *tree* structures represent the solid as it evolves in the design process, and usually support undo and redo operations.

A ***3D*** model can be represented by sets of *faces, edges*, and *vertices*. For example, a triangular tetrahedron pyramid is shown in Figure 4.5 with assigned symbols for vertices (points), edges, and faces. For clarity, symbols and dashed lines defining back edges or faces are in gray color. Edges (E_1 to E_6) connect vertices (or points $\boldsymbol{P}_1$ to $\boldsymbol{P}_4$) and form the boundaries of *faces* (F_1 to F_3). Citing a face with *counterclockwise* (CCW) of edges indicates the orientation of the face outward direction similar to the treatments of vectors. The hierarchy of boundaries in an involved process of object creation defines the dependency or belonging of each to an original object or not. The ***loops of faces*** are useful in the hierarchy of body construction. Figure 4.6 points up the boundaries of faces as *P-loops* (or parent loops) and the boundaries of holes in a face as *C-loops* (or child loops). A *P-loop* is an outside boundary of a face and corresponds to a face. For every face there is a single *P-loop*. Therefore, the number of faces in a solid is the same as the number of *P-loops*. A *C-loop* is a boundary indicating a *hole* in a face. A face may have several *C-loops*. Since there are fewer topological elements, the internal management of a solid's data structure becomes easier. One can delete a *C-loop* (child) without affecting the *P-loop* (parent). Deleting a *P-loop* would eliminate

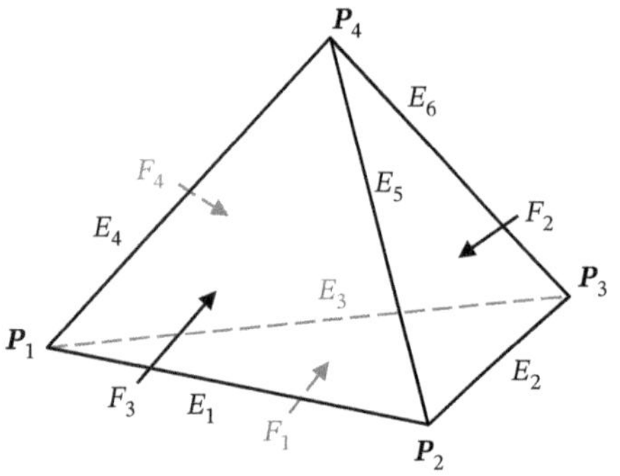

Figure 4.5 A tetrahedron showing *edges* (E_1 to E_6) connecting *vertices* ($\boldsymbol{P}_1$ to $\boldsymbol{P}_4$) and forming the boundaries of *faces* (F_1 to F_4).

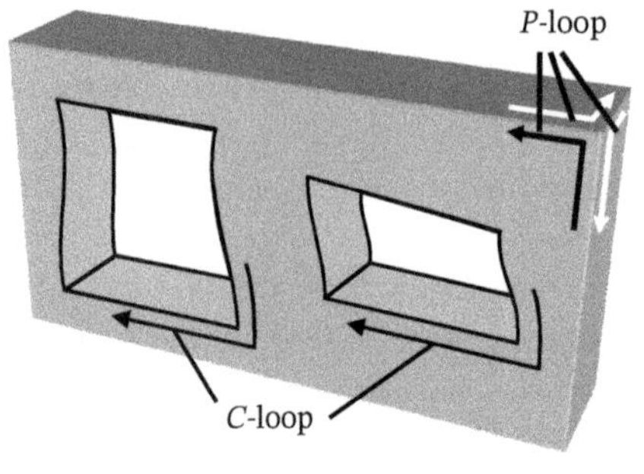

Figure 4.6 A simple ***3D*** object where the boundaries of faces are defined as *P*-loops (or parent loops) and the boundaries of holes in a face are defined as *C*-loops (or child loops).

the associated offspring *C-loops*. These loops are generated while the body tree is constructed. Deleting a hole in a body should delete the associated *C-loops* created by the intersection of the hole with the face surface or surfaces of a body.

4.1.1.1 Geometric Computations

Normal vector of a loop or a surface is very important in defining the loop or surface outer orientation. A plane *surface* polygon or a *P-loop* is defined by m edges E_j ($j = 1, \ldots, m$) and n vertices V_i or body points $\boldsymbol{P}_i$ ($i = 1, \ldots, n$). The coordinates of vertex $\boldsymbol{P}_i$ are (x_i, y_i, z_i). The *unit normal vector* $\boldsymbol{N}_F$ (a_N, b_N, c_N) of the face F or the *P-loop* polygon is calculated from the following expressions (Nowell et al. (1972)):

$$\left\{\begin{aligned} a_F &= \sum_{i=1}^{n}(y_i - y_j)(z_i + z_j) \\ b_F &= \sum_{i=1}^{n}(z_i - z_j)(x_i + x_j) \\ c_F &= \sum_{i=1}^{n}(x_i - x_j)(y_i + y_j) \end{aligned}\right\} \quad j = 1 \quad \text{if} \quad i = n, \quad \text{else} \quad j = i + 1$$

$$d_F = \sqrt{a_F^2 + b_F^2 + c_F^2}$$

$$a_N = a_F/d_F, \quad b_N = b_F/d_F, \quad c_N = c_F/d_F \tag{4.1}$$

where d_F is the length of the normal vector of the face or loop. If the *polygon* is not in a flat plane, the normal is the average value of the boundary normals. When the *vertices* are traversed in CCW direction, the normal will point outward of the face. Equation (4.1) is derivable for a triangular face F_1 in Figure 4.5 by considering the vertices or points $\boldsymbol{P}_1$, $\boldsymbol{P}_2$, and $\boldsymbol{P}_3$ forming vectors $(\boldsymbol{P}_2 - \boldsymbol{P}_1)$, $(\boldsymbol{P}_3 - \boldsymbol{P}_2)$, and $(\boldsymbol{P}_1 - \boldsymbol{P}_3)$ that are traversed in CCW direction. The normal N_{F1} of the F_1 face is obtained by the cross product of any consecutive two vectors of these. The three cross products should produce the same normal N_{F1}. The cross product $(\boldsymbol{P}_2 - \boldsymbol{P}_1) \times (\boldsymbol{P}_3 - \boldsymbol{P}_2)$ of the first two vectors can be evaluated by an equation similar to Eq. (2.6) of Chapter 2. The same is done for the other two cross products. The three cross products should produce the same result as of Eq. (4.1).

The evaluation of the face or polygon normal is very useful in defining the outward direction of the face. This is essential to calculate the proper color for face rendering; see Section 4.4.

4.1.1.2 Topological Operations and the Euler Formula

To check the topological integrity of the solid, the *Euler–Poincaré* formula for its topology ought to be satisfied (Leonhard Euler (1707–1783) and Henri Poincaré (1854–1912)). This check may be performed by the software to alert the system or the user of the appropriateness of the constructional operation. After several developments, the formula for an arbitrary *solid body* is as follows (see original works of Euler (1758) and Poincaré (1895)):

$$F_B + V_B - E_B - R_B = 2\,(S_B - H_B) \tag{4.2}$$

where F_B, E_B, and V_B are the number of the body *faces*, *edges*, and *vertices*, respectively. R_B is the number of *rings* that are cavities in the faces. S_B is the number of *shells*, or disconnected body wrapping surfaces, and H_B is the number of *holes*, which pass all the way through the solid. Figure 4.7a shows a ring R_{B1} in a face F_{B1} and a ring R_{B2} in a face F_{B2}. Counting the values of the parameters in the solid shown in Figure 4.7a, one gets $E_B = 24$, $F_B = 10$, $V_B = 16$, $S_B = 1$, $H_B = 1$, and $R_B = 2$. Substituting in Eq. (4.2), the Euler–Poincaré formula is found to be satisfied. One should note that F_B and R_B correspond to the number of *P-loops* and *C-loops*, respectively.

Another topology example is depicted in Figure 4.7b for a pyramid with an internal compartment and a passage to it. Examining the pyramid, one can count the different parameters. Faces, vertices, and edges including internal ones get $F_B = 14$, $V_B = 17$, and $E_B = 28$, respectively. The ring $R_B = 1$, which is the entrance of the passage on the face of the pyramid. One can also wrap the whole body and its cavity with one shell or $S_B = 1$. There is no through

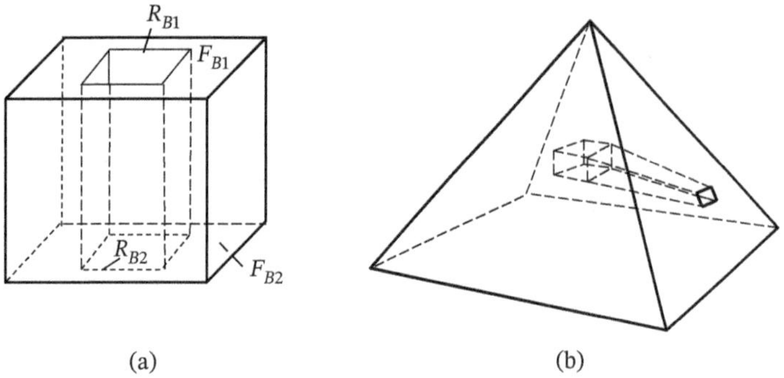

Figure 4.7 Object drawing (a) shows a ring R_{B1} in a face F_{B1} and a ring R_{B2} in a face F_{B2}. Object drawing (b) is a pyramid with an internal compartment and a passage to it.

hole in the pyramid and thus $H_B = 0$. Substituting these values in Eq. (4.2), one gets $14 + 17 - 28 - 1 = 2\,(1 - 0) = 2$. The Euler–Poincaré formula is thus satisfied.

Note that a circle can be modeled as few or many sides and vertices on its perimeter. The cylinder is modeled as two end circles with edges connecting the vertices along the cylinder length.

4.1.1.3 Geometric and Global Operations

In geometric modeling operations, one might need to translate a vertex, change a line to a curve, change a curve to a line, or translate a point on a curve. Globally, one might need to translate a solid, rotate a solid, negate a solid (where all the faces of solid are turned over so that their normal vectors are reversed), combine two solids, or separate a solid into its original components. These operations would cause the geometric modeling package to be more useful.

4.1.1.4 Procedures for Constructing a Single or a Compound Solid

A single solid construction can be produced as demonstrated in Figure 4.8. The first action produces the initial solid body B_1 using a solid primitive such as a parallelepiped or extruding, sweeping, or lofting a sketch of a face F_1. Subsequent action is to modify it to step B_2 by identifying a face F_2, sweep F_2 to get the B_3 body, then select edge E_1 to round and get the body B_4. A data *tree* is usually constructed to correspond to the design process. Each node of the tree represents the solid at a specific stage of construction. This procedure in the tree may support *undo* and *redo* operations. After saving, however, undo and redo operations may not be conceivable in some modeling systems.

Constructing a compound solid is depicted in Figure 4.9 as one alternative solution. New solid is generated from several developed primitive solids. This includes *Boolean* operations. The design process using a compound solid

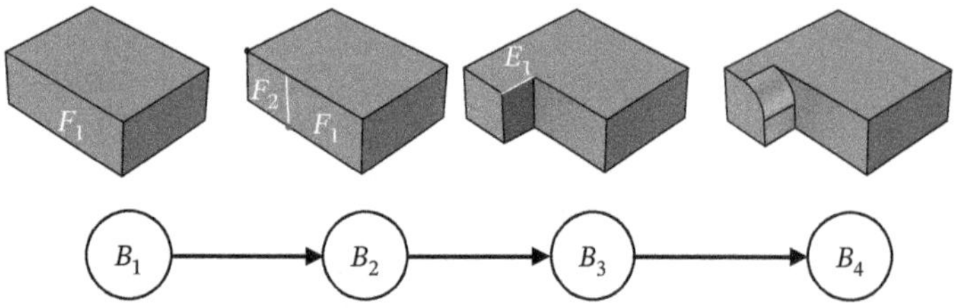

Figure 4.8 A single solid construction of a solid body. Each node of the tree to the bottom represents the solid at a specific stage of construction.

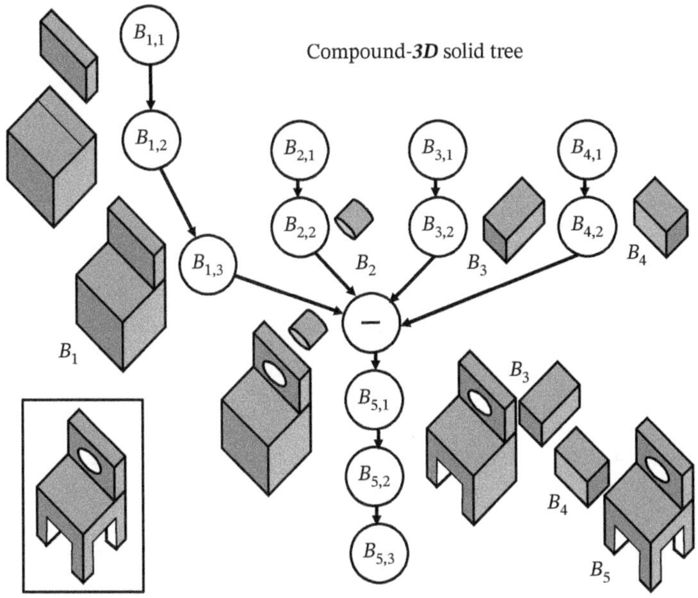

Figure 4.9 Constructing compound ***3D*** solid generated from several developed primitive solids.

operations is utilizing solids $B_{1,1}$ and $B_{1,2}$ to produce a single solid body B_1 through the $B_{1,3}$ *union* process. Similarly solid bodies B_2, B_3, and B_4 are created. Solid B_5 is then generated from solids B_1, B_2, B_3, and B_4 using a *difference Boolean* operation. After the *Boolean* operation, several edges of solid B_5 may be changed to fillet surfaces using a rounding operation (not shown in Figure 4.9). Other procedures are possible to produce a compound solid object. These procedures depend on the proficiency of the user in building ***3D*** models in different software packages. The chair-like object in Figure 4.9 can be produced in fewer steps such as a seat, a back, and four legs. The seat and the back can also be a one free-form surface with a certain thickness.

Trees representing the solid design process are usually displayed on a side screen to enable the user to document and edit the process. A single solid tree is representing the design process of a single solid (Figure 4.8). A compound solid tree is representing the relationships of solid bodies, which are combined using *Boolean* or other operations; see Figure 4.9. Using such trees, a designer can enter undo or redo commands to regenerate a previous solid if it is allowed. To generate or modify a solid, the user gives commands to the geometric modeling module. The *tree processor* stores the operations of all trees in some geometric modeling packages.

4.1.2 Homogeneous Coordinates Versus Cartesian Coordinates

Homogeneous coordinates have been devised to provide means to solve several problems such as solving sets of simultaneous *homogeneous* equations. They are also described or defined as *affine transformation*. Figure 4.10 defines the homogeneous coordinates in the two-dimensional (***2D***) Cartesian space. The homogeneous coordinates introduce an extra coordinate h_H. In that, the *homogeneous* coordinate of a *Cartesian* point of (x, y) is defined as (x, y, h_H) with the extra h_H added to the definition. However, we usually *normalize* the coordinates to have all coordinates lie on the homogenous plane of $(x, y, 1)$. All points (x, y, h_H) in the homogeneous coordinates will map into $(x/h_H, y/h_H, 1)$ as apparent in Figure 4.10. Homogenization operation in R^3 affirms that point (x, y, h_H) is homogenized to the point $(x/h_H, y/h_H, 1)$. In ***3D***, the *normalized homogeneous coordinate* $(x, y, z, 1)$ represents the *Cartesian* point (x, y, z). The homogeneous coordinates of (x, y, z, h_H) represent the *Cartesian* coordinates of $(x/h_H, y/h_H, z/h_H)$. Homogeneous coordinates of $(h_H x, h_H y, h_H z, h_H)$ should represent the *Cartesian* coordinates of

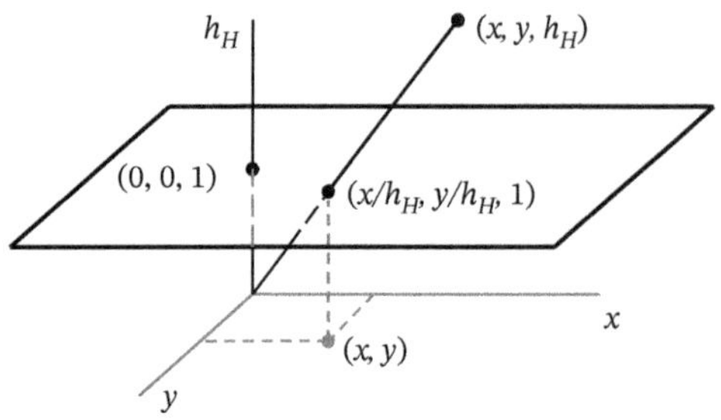

Figure 4.10 Definition of the homogeneous coordinates in the ***2D*** *Cartesian* space.

(x, y, z). In homogeneous coordinates, each (unique) *Cartesian* coordinate represents an unlimited or (∞) numbers of *homogeneous* coordinates, i.e. homogeneous coordinates are redundant. The value of h_H is an arbitrary constant.

4.1.2.1 Point in Space

In *Cartesian* coordinates a point is defined by its Cartesian coordinate values $\boldsymbol{P}_{C1}\ (x_1, y_1, z_1)$ and can also take the form of a *position vector* from the origin to the point.

$$\boldsymbol{P}_{C1} = [x_1\ y_1\ z_1]^T \tag{4.3}$$

The vectors are usually defined as *column arrays* but may also be defined as *row arrays*.
In *homogeneous* coordinates,

$$\boldsymbol{P}_{H1} = \lfloor x_1\ y_1\ z_1\ 1 \rfloor \tag{4.4}$$

In this text, it was observed that points in homogeneous coordinates are better defined as *rows*. This allows the definition of body geometry by its main characteristic points, vertices, or corners as a set in a column array of these distinctive points. It can also be assumed that one might use $\boldsymbol{P}_1 = \boldsymbol{P}_{H1}$.
In *geometric modeling*, points $\boldsymbol{P}_1, \boldsymbol{P}_2, \ldots, \boldsymbol{P}_n$ may also be referred to as vertices $\boldsymbol{V}_1, \boldsymbol{V}_2, \ldots, \boldsymbol{V}_n$.

4.1.2.2 Vectors

In *Cartesian* coordinates a vector $\boldsymbol{V}_{C1}$ is defined by its coordinate components. It is usually set in a *column* array form.

$$\boldsymbol{V}_{C1} = \lfloor \boldsymbol{V}_{1x}\ \boldsymbol{V}_{1y}\ \boldsymbol{V}_{1z} \rfloor^T \tag{4.5}$$

In *homogeneous* coordinates, a vector can be defined as the subtraction of two position arrays.

$$\boldsymbol{V}_{H1} = \lfloor \boldsymbol{V}_{1x}\ \boldsymbol{V}_{1y}\ \boldsymbol{V}_{1z}\ 0 \rfloor^T \tag{4.6}$$

It should be noted that the *0* in the fourth location should be carefully handled. The components of the vector are given the names $\boldsymbol{V}_{1x}$, $\boldsymbol{V}_{1y}$, and $\boldsymbol{V}_{1z}$ rather than x_1, y_1, and z_1 for location of point "*1*."

4.1.2.3 Lines

In homogeneous coordinates, a line $\boldsymbol{L}_{\mathrm{H}}$ between two points $\boldsymbol{P}_1$ and $\boldsymbol{P}_2$ is given by

$$\boldsymbol{L}_H = \begin{bmatrix} \boldsymbol{P}_1 \\ \boldsymbol{P}_2 \end{bmatrix} = \begin{bmatrix} x_1 & y_1 & z_1 & 1 \\ x_2 & y_2 & z_2 & 1 \end{bmatrix}, (2 \times j),\ j = 4 \tag{4.7}$$

The number of rows is 2, which is the number of end points $\boldsymbol{P}_1$ and $\boldsymbol{P}_2$ defining the line. The number of the columns *j* is the number of homogeneous coordinates, which is 4. Note that $\boldsymbol{P}_1$ and $\boldsymbol{P}_2$ are used to represent $\boldsymbol{P}_{H1}$ and $\boldsymbol{P}_{H2}$.

4.1.2.4 Body Geometry and Vertices

Define the geometry of the body $\boldsymbol{B}_H$ in homogeneous coordinates by the matrix $[\boldsymbol{P}_B]$ representing the object *vertices* or *points* $\boldsymbol{P}_i$, each of which is a row $(x_i, y_i, z_i, 1)$ in this matrix. The body $\boldsymbol{B}_H$ can then have the following definition:

$$\boldsymbol{B}_H = \begin{bmatrix} \boldsymbol{P}_1 \\ \boldsymbol{P}_2 \\ \cdots \\ \boldsymbol{P}_n \end{bmatrix} = \begin{bmatrix} x_1 & y_1 & z_1 & 1 \\ x_2 & y_2 & z_2 & 1 \\ \cdots & \cdots & \cdots & \cdots \\ x_n & y_n & z_n & 1 \end{bmatrix}, (i \times j), \quad i = n, \quad j = 4 \tag{4.8}$$

Note that $\boldsymbol{P}_1, \ldots, \boldsymbol{P}_n$ are used to represent $\boldsymbol{P}_{H1}, \ldots, \boldsymbol{P}_{Hn}$.

4.1.3 Body Transformation

The body $\boldsymbol{B}_H$ can be transformed in the homogeneous space by *translation, rotation, scaling, zooming, skewing*, or *perspective viewing*. A simple original ***2D*** shape of a body in a plane is shown in Figure 4.11a, and each of its particular transformation is indicated as in Figure 4.11b for translation, in Figure 4.11c for scaling, in Figure 4.11d for rotation, in Figure 4.11e for skewing, and in Figure 4.11f for perspective. The transformed body matrix $\boldsymbol{B}_H'$ is obtained collectively or by any one of the transformations in any other order by the following matrix multiplication:

$$\begin{aligned} \boldsymbol{B}_H' &= \boldsymbol{B}_H \ [\boldsymbol{T}_{xyz}][\boldsymbol{R}_{xyz}][\boldsymbol{S}_H][\boldsymbol{Z}_H][\boldsymbol{K}_H] \ldots [\boldsymbol{P}_H] \\ &= \boldsymbol{B}_H[\boldsymbol{T}_{xyz}\, \boldsymbol{R}_{xyz}\, \boldsymbol{S}_H\, \boldsymbol{Z}_H\, \boldsymbol{K}_H \ldots \boldsymbol{P}_H] \\ &= \boldsymbol{B}_H \ \boldsymbol{T}_{\text{GH}} \end{aligned} \tag{4.9}$$

where $\boldsymbol{T}_{xyz}$ is the translation matrix; $\boldsymbol{R}_{xyz}$ is the rotation matrix about x, y, or z; $\boldsymbol{S}_H$ is the scaling matrix; $\boldsymbol{Z}_H$ is the zooming matrix; $\boldsymbol{K}_H$ is the skewing matrix; $\boldsymbol{P}_H$ is the perspective matrix; and $\boldsymbol{T}_{\text{GH}}$ is the general total homogeneous

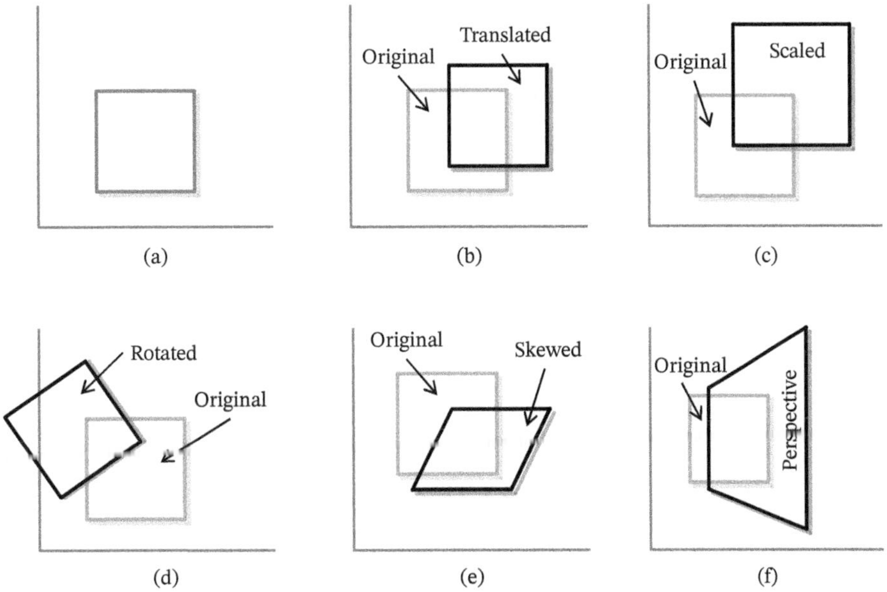

Figure 4.11 A simple original shape of a body (a) in a plane and each of its transformation is indicated in (b) translation, (c) scaling, (d) rotation, (e) skewing, and (f) perspective.

transformation matrix. The order of multiplication is important in defining the intended "motion" or transformation of the body in space from its original position to its final intended location and defined shape. Note that matrix multiplication is *associative*, but not *commutative*. The matrices will be defined later on, but their final multiplication form represents the general transformation matrix, which is post-multiplied by the body matrix $\boldsymbol{B}_H$ to get the shape of the transformed body matrix $\boldsymbol{B}_H'$. The general homogeneous transformation matrix $\boldsymbol{T}_{\mathrm{GH}}$ can then have the following form:

$$\boldsymbol{T}_{\mathrm{GH}} = [\boldsymbol{T}_{\mathrm{GH}}] = \begin{bmatrix} a & k & n & h \\ d & b & p & g \\ e & m & c & f \\ t & u & v & z \end{bmatrix} \tag{4.10}$$

The components of the matrix are representing different transformations. Components a, b, and c are the *scaling* amounts in x, y, and z, respectively. The components t, u, and v are the *translation* amounts in x, y, and z, respectively. The components h, g, and f are the *perspective* amounts in x, y, and z, respectively. The components d, e, k, m, n, and p are the *skewing* amounts in x, y, and z. The component z is the *inverse zooming* amount in all x, y, and z. Each of the respective transformation matrixes is defined separately as follows:

4.1.3.1 Translation

The *translation matrix* can have the form

$$\boldsymbol{T}_{xyx} = \begin{bmatrix} 1 & & & \\ & 1 & & \\ & & 1 & \\ t_x & t_y & t_z & 1 \end{bmatrix} \tag{4.11}$$

The components t_x, t_y, and t_z are the translation amounts in x, y, and z, respectively. All empty locations in the $\boldsymbol{T}_T$ matrix are zeros.

4.1.3.2 Rotation

The *rotation matrix* about z, which can be checked with that defined in mechanics subject, has the following form:

$$\boldsymbol{R}_z = \begin{bmatrix} \cos(\theta_z) & \sin(\theta_z) & 0 & 0 \\ -\sin(\theta_z) & \cos(\theta_z) & 0 & 0 \\ 0 & 0 & 1 & 0 \\ 0 & 0 & 0 & 1 \end{bmatrix} \tag{4.12}$$

where θ_z is the rotation angle about z.

The *rotation matrix* about y has the form

$$\boldsymbol{R}_y = \begin{bmatrix} \cos(\theta_y) & 0 & -\sin(\theta_y) & 0 \\ 0 & 1 & 0 & 0 \\ \sin(\theta_y) & 0 & \cos(\theta_y) & 0 \\ 0 & 0 & 0 & 1 \end{bmatrix} \tag{4.13}$$

where θ_y is the rotation angle about y.

The *rotation matrix* about x has the form

$$\boldsymbol{R}_x = \begin{bmatrix} 0 & 0 & 0 & 0 \\ 0 & \cos(\theta_x) & \sin(\theta_x) & 0 \\ 0 & -\sin(\theta_x) & \cos(\theta_x) & 0 \\ 0 & 0 & 0 & 1 \end{bmatrix} \tag{4.14}$$

where θ_x is the rotation angle about x.

4.1.3.3 Scaling

The *scaling matrix* is given by the following form:

$$S_H = \begin{bmatrix} S_x & 0 & 0 & 0 \\ 0 & S_y & 0 & 0 \\ 0 & 0 & S_z & 0 \\ 0 & 0 & 0 & 1 \end{bmatrix} \tag{4.15}$$

The components S_x, S_y, S_z are the scales in x, y, and z, respectively.

4.1.3.4 Zooming

The *zooming matrix* is given by the following form:

$$Z_H = \begin{bmatrix} 1 & & & \\ & 1 & & \\ & & 1 & \\ & & & 1/z_f \end{bmatrix} \tag{4.16}$$

where z_f is the zooming factor and all empty locations are zeros.

4.1.3.5 Skewing

The *skewing matrix* is given by the following form:

$$K_H = \begin{bmatrix} 1 & k & n & \\ d & 1 & p & \\ e & m & 1 & \\ & & & 1 \end{bmatrix} \tag{4.17}$$

where d, e, k, m, n, and p are the skewing amounts in x, y, and z, where all empty locations in the matrix K_H are zeros.

4.1.3.6 Perspective

The *perspective matrix* is defined on the condition that the eye of the observer is on the z-coordinate axis and at a distance of f_z from the body.

$$P_H = \begin{bmatrix} 1 & 0 & 0 & 0 \\ 0 & 1 & 0 & 0 \\ 0 & 0 & 1 & -1/f_z \\ 0 & 0 & 0 & 1 \end{bmatrix} \tag{4.18}$$

where f_z is the distance from the eye to the projection plane when the *eye* is on the z-axis as previously required by this procedure. Usually the projection plane is set to be the x–y plane (or z plane). The distance f_z may be called the *focal distance* along the z-axis. This is usually performed after rotating the body and the eye to set the looking *eye* on the z direction at a f_z distance off the body location.

4.1.3.7 Orthographic Projection

After all transformations and considering the z-axis to be normal to the plane of the screen (the +ve behind the screen), the simplest method of projection is to ignore the z-coordinate of every point in space. This would render the ***3D*** points (x, y, z) to be only ***2D*** points (x, y).

4.1.3.8 Body Transformation Systems

Note that since matrix multiplication is *associative*, but not *commutative*, the order of multiplication is important.

If the body matrix B_C representing the object vertices or points each of which as a ***column*** vector $[x_i \quad y_i \quad z_i \quad 1]^T$ in this matrix, the transformed body matrix B_C' is obtained by any one of the transformations or collectively in any other proper order by

$$\begin{aligned} \boldsymbol{B}_C' &= [\boldsymbol{T}^T_{xyz}][\boldsymbol{R}^T_{xyz}][\boldsymbol{S}^T_H][\boldsymbol{Z}^T_H][\boldsymbol{K}^T_H]\cdots[\boldsymbol{P}^T_H]\,\boldsymbol{B}_C \\ &= \boldsymbol{T}^T_{\mathrm{GH}}\,\boldsymbol{B}_C \end{aligned} \tag{4.19}$$

This process is used by many computer graphics texts and by most *OpenGL* systems. For the computer programming, however, it is usually easier to represent and save a body n points or vertices in an $(n \times 4)$ matrix instead of the $(4 \times n)$ matrix. Several graphics texts are also using the formulation procedure adopted herein rather than the usual *OpenGL* Systems. Note that using the transpose will cause the multiplication to be in a *post-multiplication* rather a *pre-multiplication* way. In providing *OpenGL* systems with vertices, this is not an issue since the order of the internal calculations is not observed or be concerned with.

Example 4.1 A line is defined by its two ***3D*** end points $\boldsymbol{P}_1$ (1, 0, 0) and $\boldsymbol{P}_2$ (2, 2, 0) in *Cartesian* coordinates. It is required to translate the line 2 units in the x direction and 1 unit in the y direction simultaneously. After translation, it is required to rotate the line by −10° about the z-axis.

Solution

The line $\boldsymbol{L}$ can be identified in *homogeneous* coordinate by applying Eq. (4.7) such that

$$\boldsymbol{L} = \boldsymbol{L}_H = \begin{bmatrix} \boldsymbol{P}_1 \\ \boldsymbol{P}_2 \end{bmatrix} = \begin{bmatrix} 1 & 0 & 0 & 1 \\ 2 & 2 & 0 & 1 \end{bmatrix} \tag{a}$$

To translate in x and y by 2 and 1, respectively, the transformation matrix for translation is

$$\boldsymbol{T}_{xyz} = \begin{bmatrix} 1 & & & \\ & 1 & & \\ & & 1 & \\ t_x & t_y & t_z & 1 \end{bmatrix} = \begin{bmatrix} 1 & & & \\ & 1 & & \\ & & 1 & \\ 2 & 1 & 0 & 1 \end{bmatrix} \tag{b}$$

Considering the line to be our ***3D*** body, the new translated line $\boldsymbol{L}_1$ is obtained by applying Eq. (4.9) to get the following:

$$\boldsymbol{L}_1 = \boldsymbol{L}_H\ \boldsymbol{T}_{xyz} = \begin{bmatrix} 1 & 0 & 0 & 1 \\ 2 & 2 & 0 & 1 \end{bmatrix} \begin{bmatrix} 1 & & & \\ & 1 & & \\ & & 1 & \\ 2 & 1 & 0 & 1 \end{bmatrix} = \begin{bmatrix} 3 & 1 & 0 & 1 \\ 4 & 3 & 0 & 1 \end{bmatrix} \tag{c}$$

The solution is simply the multiplication of the two matrices. Figure 4.12 shows the original line $\boldsymbol{L}$ and the translated line $\boldsymbol{L}_1$.

The following rotation is −10° about the z-axis with the rotation matrix defined by Eq. (4.11):

$$\boldsymbol{R}_z = \begin{bmatrix} \cos(\theta_z) & \sin(\theta_z) & 0 & 0 \\ -\sin(\theta_z) & \cos(\theta_z) & 0 & 0 \\ 0 & 0 & 1 & 0 \\ 0 & 0 & 0 & 1 \end{bmatrix} = \begin{bmatrix} \cos(-10) & \sin(-10) & 0 & 0 \\ -\sin(-10) & \cos(-10) & 0 & 0 \\ 0 & 0 & 1 & 0 \\ 0 & 0 & 0 & 1 \end{bmatrix} = \begin{bmatrix} 0.9848 & -0.1736 & 0 & 0 \\ -(-0.1736) & 0.9848 & 0 & 0 \\ 0 & 0 & 1 & 0 \\ 0 & 0 & 0 & 1 \end{bmatrix} \tag{d}$$

$$\boldsymbol{L}_2 = \boldsymbol{L}_1\ \boldsymbol{R}_z = \begin{bmatrix} 3 & 1 & 0 & 1 \\ 4 & 3 & 0 & 1 \end{bmatrix} \begin{bmatrix} 0.985 & -0.174 & 0 & 0 \\ 0.174 & 0.985 & 0 & 0 \\ 0 & 0 & 1 & 0 \\ 0 & 0 & 0 & 1 \end{bmatrix} = \begin{bmatrix} 3.129 & 0.463 & 0 & 1 \\ 4.462 & 2.259 & 0 & 1 \end{bmatrix} \tag{e}$$

Figure 4.12 shows the original line $\boldsymbol{L}$, the translated line $\boldsymbol{L}_1$, and the rotated line $\boldsymbol{L}_2$.

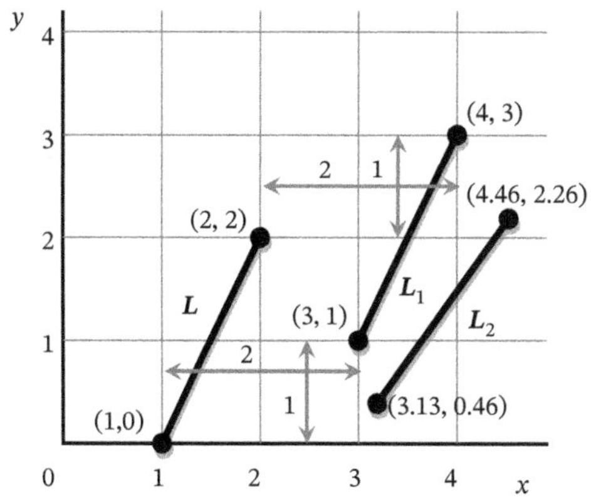

Figure 4.12 A line $\boldsymbol{L}$ defined by its two end points $\boldsymbol{P}_1$ and $\boldsymbol{P}_2$ in *Cartesian* coordinates. It is translated to line $\boldsymbol{L}_1$ by *2* units in the x direction and *1* unit in the y direction. It is then rotated to line $\boldsymbol{L}_2$ by −10°.

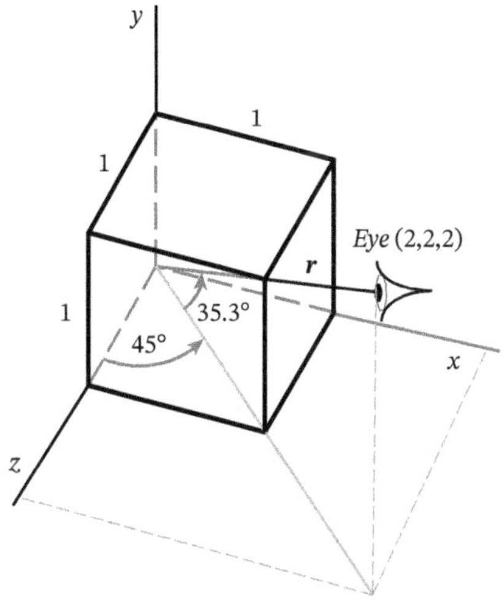

Figure 4.13 A unit cube body attached to the origin of the coordinate system *xyz*. The observer eye is located at the position (2, 2, 2) and is looking along the cube diagonal toward the origin.

Example 4.2 A unit cube body has its bottom corner attached to the origin of the coordinate system *xyz* as shown in Figure 4.13. The observer eye is located at the position (2, 2, 2) and is looking along the cube diagonal toward the origin. It is required to find the projection of the cube on the normal plane to the line of sight and passing by the origin. Does this projection produce the perceived ***perspective*** of the cube by the observer?

Solution

The body matrix $\boldsymbol{B}_H$ is defined by the following:

$$\boldsymbol{B}_H = \begin{bmatrix} 0 & 0 & 0 & 1 \\ 1 & 0 & 0 & 1 \\ 1 & 1 & 0 & 1 \\ 0 & 1 & 0 & 1 \\ 0 & 0 & 1 & 1 \\ 1 & 0 & 1 & 1 \\ 1 & 1 & 1 & 1 \\ 0 & 1 & 1 & 1 \end{bmatrix} = \begin{bmatrix} 0 & 1 & 1 & 0 & 0 & 1 & 1 & 0 \\ 0 & 0 & 1 & 1 & 0 & 0 & 1 & 1 \\ 0 & 0 & 0 & 0 & 1 & 1 & 1 & 1 \\ 1 & 1 & 1 & 1 & 1 & 1 & 1 & 1 \end{bmatrix}^T \tag{a}$$

Note that the cube corners, vertices, or points $\boldsymbol{P}_i$ are considered as row vectors. The *eye (2, 2, 2)*, or in general location vector $\boldsymbol{r}\,(a_E, b_E, c_E)$, makes the following angles to transform the *eye* to be located on the *z*-axis in preparation for applying the perspective transformation. The radial distance r_{eye} from the origin to the eye is then

$$r_{\text{eye}} = (a_E{}^2 + b_E{}^2 + c_E{}^2)^{1/2} \quad \text{Or} \quad r_{2,2,2} = (12)^{1/2} = 3.46 = f_z \tag{b}$$

The angle θ_y between the projected $\boldsymbol{r}$ vector on the x–z plane and the y–z plane is

$$\theta_y = \tan^{-1} a_E/c_E = \tan^{-1} 1 = 45° \tag{c}$$

Then the *eye* and the body should be rotated in the opposite direction, i.e. rotated by $-\theta_y$ so that the *eye* would be located in the y–z plane. Therefore, the rotation should be $\theta_y = -45°$.

The angle between the *eye* – while on the y–z plane after θ_y rotation – and the x–z plane is given by

$$\beta_x = -\sin^{-1} (b_E/(a_E{}^2 + b_E{}^2 + c_E{}^2)^{1/2}) = -\sin^{-1} (2/3.46) = -35.3° \tag{d}$$

Here, the *eye* and the body should be rotated in the opposite direction, i.e. by $-\beta_x$ so that the eye would be located on the z-axis. Therefore, the rotation should be $\beta_x = 35.3°$.

For the rotation matrices, one calculates $\cos\theta_y = 0.707$, $\sin\theta_y = -0.707$, $\cos\beta_x = 0.816$, and $\sin\beta_x = 0.578$. The rotation matrices are then

$$\boldsymbol{R}_y = \begin{bmatrix} \cos(\theta_y) & 0 & -\sin(\theta_y) & 0 \\ 0 & 1 & 0 & 0 \\ \sin(\theta_y) & 0 & \cos(\theta_y) & 0 \\ 0 & 0 & 0 & 1 \end{bmatrix} = \begin{bmatrix} 0.707 & 0 & 0.707 & 0 \\ 0 & 1 & 0 & 0 \\ -0.707 & 0 & 0.707 & 0 \\ 0 & 0 & 0 & 1 \end{bmatrix} \tag{e}$$

$$\boldsymbol{R}_x = \begin{bmatrix} 0 & 0 & 0 & 0 \\ 0 & \cos(\beta_x) & \sin(\beta_x) & 0 \\ 0 & -\sin(\beta_x) & \cos(\beta_x) & 0 \\ 0 & 0 & 0 & 1 \end{bmatrix} = \begin{bmatrix} 0 & 0 & 0 & 0 \\ 0 & 0.816 & 0.578 & 0 \\ 0 & -0.578 & 0.816 & 0 \\ 0 & 0 & 0 & 1 \end{bmatrix} \tag{f}$$

The order of rotation is important and should be $\boldsymbol{R}_y$ then $\boldsymbol{R}_x$, i.e.,

$$\boldsymbol{R}_{xyz} = \boldsymbol{R}_y \boldsymbol{R}_x = \begin{bmatrix} 0.707 & 0 & 0.707 & 0 \\ 0 & 1 & 0 & 0 \\ -0.707 & 0 & 0.707 & 0 \\ 0 & 0 & 0 & 1 \end{bmatrix} \begin{bmatrix} 0 & 0 & 0 & 0 \\ 0 & 0.816 & 0.578 & 0 \\ 0 & -0.578 & 0.816 & 0 \\ 0 & 0 & 0 & 1 \end{bmatrix} = \begin{bmatrix} 0.707 & -0.409 & 0.577 & 0 \\ 0 & 0.816 & 0.578 & 0 \\ -0.707 & -0.409 & 0.577 & 0 \\ 0 & 0 & 0 & 1 \end{bmatrix} \tag{g}$$

Transformation of rotation then *perspective* $\boldsymbol{P}_H$, where $\boldsymbol{P}_H$ is given by

$$\boldsymbol{P}_H = \begin{bmatrix} 1 & 0 & 0 & 0 \\ 0 & 1 & 0 & 0 \\ 0 & 0 & 1 & -1/3.464 \\ 0 & 0 & 0 & 1 \end{bmatrix} = \begin{bmatrix} 1 & 0 & 0 & 0 \\ 0 & 1 & 0 & 0 \\ 0 & 0 & 1 & -0.289 \\ 0 & 0 & 0 & 1 \end{bmatrix} \tag{h}$$

The total transformation matrix $\boldsymbol{T}_{\text{GH}}$ is then (after multiplication)

$$\boldsymbol{T}_{\text{GH}} = \boldsymbol{R}_{xyz}\ \boldsymbol{P}_H = \begin{bmatrix} 0.707 & -0.409 & 0.577 & 0 \\ 0 & 0.816 & 0.578 & 0 \\ -0.707 & -0.409 & 0.577 & 0 \\ 0 & 0 & 0 & 1 \end{bmatrix} \begin{bmatrix} 1 & 0 & 0 & 0 \\ 0 & 1 & 0 & 0 \\ 0 & 0 & 1 & -0.289 \\ 0 & 0 & 0 & 1 \end{bmatrix} = \begin{bmatrix} 0.707 & -0.409 & 0.577 & -0.167 \\ 0 & 0.816 & 0.578 & -0.167 \\ -0.707 & -0.409 & 0.577 & -0.167 \\ 0 & 0 & 0 & 1 \end{bmatrix} \tag{i}$$

Note that the first three columns are exactly the same as the total rotation matrix $\boldsymbol{R}_{xyz}$.

The body after total transformation is then given by (after multiplication)

$$B'_H = B_H\ T_{GH} = \begin{bmatrix} 0 & 0 & 0 & 1.0000 \\ 0.7071 & -0.4082 & 0.5774 & 0.8333 \\ 0.7071 & 0.4082 & 1.547 & 0.6667 \\ 0 & 0.8165 & 0.5774 & 0.8333 \\ -0.7071 & -0.4082 & 0.5774 & 0.8333 \\ 0.0000 & -0.8165 & 1.1547 & 0.6667 \\ 0.0000 & 0.0000 & 1.7321 & 0.5000 \\ -0.7071 & 0.4082 & 1.1547 & 0.6667 \end{bmatrix} \qquad \text{(j)}$$

To project the body, a *homogeneous normalization* is necessary. It means that the fourth column must have the value of one for each point or row. This can be implemented by dividing each component of a row by the fourth column value of that row, which normalizes the fourth component to the value of *1*. Performing the process for all the rows causes the fourth column to a unity. After *homogeneous normalization*, the perspective projection body is then

$$B'_H = \begin{bmatrix} 0 & 0 & 0 & 1.0000 \\ 0.8486 & -0.4899 & 0.6929 & 1.0000 \\ 1.0606 & 0.6123 & 1.7320 & 1.0000 \\ 0 & 0.9798 & 0.6929 & 1.0000 \\ -0.8486 & -0.4899 & 0.6929 & 1.0000 \\ 0 & -1.2247 & 1.7320 & 1.0000 \\ 0 & 0 & 3.4642 & 1.0000 \\ -1.0606 & 0.6123 & 1.7320 & 1.0000 \end{bmatrix} \qquad \text{(k)}$$

The plotting of the body on x–y plane is shown in Figure 4.14. This projection produces the perceived *perspective* of the cube by the observer. If all lines are extended, they will meet at the three *vanishing points* known in regular graphics or drawing subjects.

4.1.4 Stereo Viewing

To produce a stereo view for a body, we need to generate two views of the body, one for each eye of the viewer. The two views are looked at simultaneously by each eye to generate the ***3D*** stereo of the body. The right eye should only see the right body view, and the left eye should only see the left view of the body. This is similar to the ***3D*** movies where one would use eyeglasses with right and left piece of different polarization planes. Usually one piece is of a vertical polarization, and the other is of a horizontal polarization. The movie is produced with two cameras for the right and left eyes and projects both simultaneously through the two polarized planes onto the theater screen. For LCD or LED screens, the process is different so far. It requires that the eyeglasses should synchronize with the projection of the right and left views; see also Section 3.4 about virtual reality (VR).

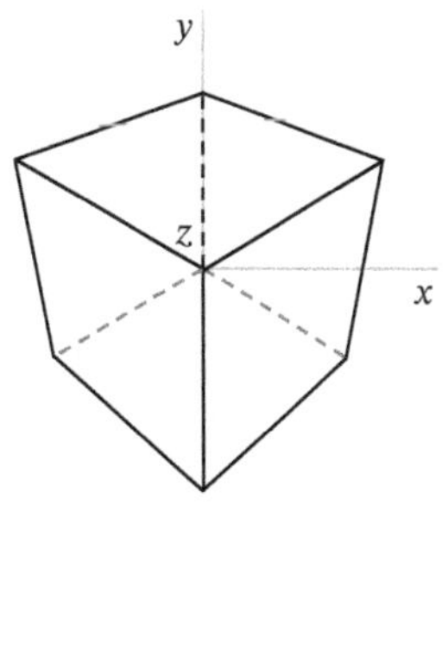

Figure 4.14 Plotting of the body onto the projection x–y plane. The projection produces the perceived perspective of the cube by the observer.

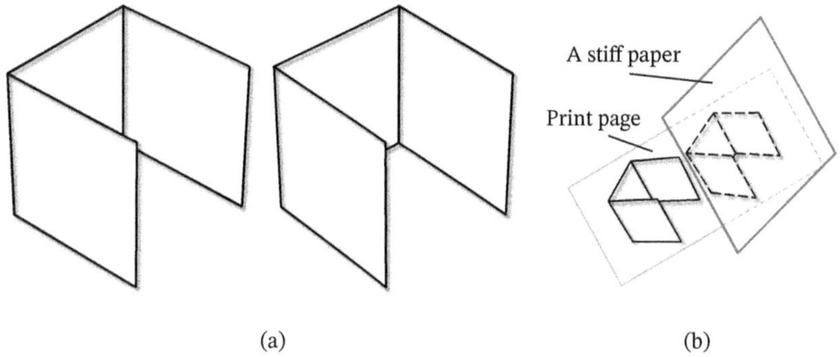

Figure 4.15 An example of open cube perspectives with 6° angle between the two left and right eyes (a). One should look at the printed page with the stiff paper separating the left-eye view from the right-eye view as shown in (b).

The procedure to generate stereo viewing is then as follows:

A. The "left-eye" and the "right-eye" projections are made by having two views of the body at two different rotations for each eye. If the viewer will be away from the viewing print page by some distance and his left- and right-eye spacing is defined, the angle difference between the left eye and the right eye would be calculated.
B. The projections are built using the same *look-at point*, but at two different eye position angles.
C. The two projections are viewed side by side where the left-eye projection is displayed in the left side and the right-eye projection is displayed in the right side as shown in Figure 4.15a.
D. The distance between the two sides is usually less than or the same as the distance between the two eyes.
E. To view as a stereo, the left eye should look at the left projection only, and the right eye should look at the right projection only. This can be done by using a stiff paper placed normal to the print page in-between the two views as shown in Figure 4.15b. One should look at the printed page with the stiff paper separating the left-eye view from the right-eye view; see also Section 3.4.
F. When done properly, the two images should fuse into a single image that appears to have the stereo depth or a ***3D*** perception of the body.

4.1.5 *3D* Graphics

So far, ***3D*** bodies or objects are defined by their characteristic points as corners, points, or vertices. Graphics is implemented by viewing or projecting bodies through these points and the connecting edges or lines. This has been done as a demonstration of different viewing procedures; see Figures 4.7, 4.13, 4.14, and 4.15. Hidden lines are either removed or drawn as dashed for demonstration clarity. This process has been done in the early developments of graphics and has been replaced by using ***3D*** graphics viewing by the generation of ***3D*** shapes as defined previously in Section 4.1.1. These ***3D*** bodies are defined by proprietary software engines or by *OpenGL*, *DirexX*, and other graphics languages. These languages use several API software to perform several operations or drive the hardware to process the necessary static or dynamic viewing scenes. Each integral body is usually treated as one encompassing boundary, which is divided into *triangles*. The screen viewing is dependent on the orientation and color of each *triangle*.

A plane surface is *tessellated* into some *triangles* such as the face of a cube can be tessellated into two *triangles*. A cylindrical part can be tessellated into several rectangular sides, which can each be divided into two *triangles*. The order of giving the *triangle* nodes or vertices defines the direction of the *triangle* surface normal. Right-hand order of vertices defines the *normal* direction of the surface to be facing the outside; see Section 4.1.1. This information is used to perform graphics positioning and lighting or coloring of each *triangle*. All vertices of body surface *triangles* are also used to perform transformations of the ***3D*** body in every viewing move. These operations are performed by the proprietary software engines or by *OpenGL*, *DirexX*, and other graphics languages.

4.2 Parametric Modeling

Parametric modeling provides generalization and control over all geometric interties such as lines, curves, planes, and surfaces. It helps in identifying intersections of lines, planes, and surfaces. It is also very useful in editing any of the ***2D*** or ***3D*** geometric interties and their interactions.

4.2.1 Parametric Lines

The parametric equation of a straight line joining $\boldsymbol{P}_1$ and $\boldsymbol{P}_2$ is defined by the parametric point $\boldsymbol{P}_L(t)$ on the *parametric line* $\boldsymbol{L}_P$ such that

$$\boldsymbol{L}_P = \boldsymbol{P}_L(t) = \begin{bmatrix} t & 1 \end{bmatrix} \begin{bmatrix} \boldsymbol{P}_2 - \boldsymbol{P}_1 \\ \boldsymbol{P}_1 \end{bmatrix} = t(\boldsymbol{P}_2 - \boldsymbol{P}_1) + \boldsymbol{P}_1 \quad 0 < t < 1 \tag{4.20}$$

where $t = 0$ at $\boldsymbol{P}_1$ and $t = 1$ at $\boldsymbol{P}_2$, and the interval describes the line segment $\boldsymbol{P}_1\boldsymbol{P}_2$. The parameter t defines the location of a point on the line $\boldsymbol{L}_P$ connecting $\boldsymbol{P}_1$ and $\boldsymbol{P}_2$.

The parametric form of a line joining the two points $\boldsymbol{P}_1$ and $\boldsymbol{P}_2$ is identified in Figure 4.16a by the point $\boldsymbol{P}_L(t)$. This point is representing the *parametric line* $\boldsymbol{L}_P$ such that $t = 0$ at $\boldsymbol{P}_1$ and $t = 1$ at $\boldsymbol{P}_2$. The line $\boldsymbol{L}_P$ increasingly changes color from darker gray at $\boldsymbol{P}_1$ to a lighter gray at $\boldsymbol{P}_2$. This is compliant with the triangle color on top of the line at $\boldsymbol{P}_1$ and the lower triangle color below the line at $\boldsymbol{P}_2$. The interval $0 < t < 1$ defines the line segment $\boldsymbol{P}_1\boldsymbol{P}_2$. The parameter t identifies the location of $\boldsymbol{P}_L(t)$ on the line $\boldsymbol{P}_1\boldsymbol{P}_2$.

4.2.1.1 Alternative Parametric Form

Another parametric form is deduced from previous Eq. (4.20) and provides a consistent start to the definition of future lines and curves formulation. This parametric form utilizes the concept of ***blending functions*** such that

$$\boldsymbol{L}_P = \boldsymbol{P}_L(t) = [(1 - t)\boldsymbol{P}_1 + t\boldsymbol{P}_2] = \begin{bmatrix} 1 - t & t \end{bmatrix} \begin{bmatrix} P_1 \\ P_2 \end{bmatrix} \quad 0 < t < 1 \tag{4.21}$$

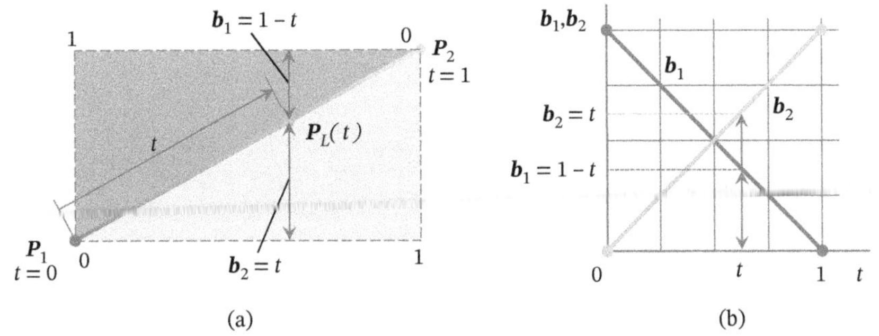

Figure 4.16 Parametric representation of the line in (a) indicating the amount of blending function $\boldsymbol{b}_1$ along t that constitutes the blending proportion of $\boldsymbol{P}_1$ and the amount of blending function $\boldsymbol{b}_2$ along t that constitutes the blending proportion of $\boldsymbol{P}_2$ that compose the value of $\boldsymbol{P}_L(t)$. Blending functions $\boldsymbol{b}_1(t)$ and $\boldsymbol{b}_2(t)$ versus t are plotted in (b) with their values indicated at (t).

or

$$L_P = P_L(t) = \begin{bmatrix} b_1 & b_2 \end{bmatrix} \begin{bmatrix} P_1 \\ P_2 \end{bmatrix}, \quad b_1 = (1-t), b_2 = (t) \quad 0 < t < 1 \tag{4.22}$$

where $\boldsymbol{b}_1$ and $\boldsymbol{b}_2$ are the ***blending functions*** that indicate how $\boldsymbol{P}_L(t)$ is obtained by *blending* the effect of $\boldsymbol{P}_1$ (i.e. with an amount of $\boldsymbol{b}_1$ at (t)) and with the effect of $\boldsymbol{P}_2$ (i.e. with an amount of $\boldsymbol{b}_2$ at (t)). This alternative parametric representation is shown in Figure 4.16. Observe again that the line $\boldsymbol{L}_P$ changes color from darker gray at $\boldsymbol{P}_1$ to a lighter gray at $\boldsymbol{P}_2$. This conforms with the color of the *blending function* $\boldsymbol{b}_1$ at $\boldsymbol{P}_1$ and the color of the *blending function* $\boldsymbol{b}_2$ at $\boldsymbol{P}_2$. The color of the *blending functions* $\boldsymbol{b}_1$ and $\boldsymbol{b}_2$ are shown in Figure 4.16b, indicating the amount of blending function $\boldsymbol{b}_1$ along t that constitutes the blending proportion of $\boldsymbol{P}_1$ and the amount of blending function $\boldsymbol{b}_2$ along t that constitutes the blending proportion of $\boldsymbol{P}_2$ that compose the value of $\boldsymbol{P}_L(t)$. Blending functions $\boldsymbol{b}_1(t)$ and $\boldsymbol{b}_2(t)$ are plotted versus t in Figure 4.16b with their values indicated at (t).

Classically, a straight *line* is defined by a vector $\boldsymbol{V}_L$ (a_L, b_L, c_L) – the line direction vector – and a point $\boldsymbol{P}_1$. Any point $\boldsymbol{P}_L$ on the line has the same direction as $\boldsymbol{V}_L$ and therefore is a scalar multiple of it, i.e.,

$$\boldsymbol{P}_1\boldsymbol{P}_L = t\ \boldsymbol{V}_L \tag{4.23}$$

But $\boldsymbol{P}_1\ \boldsymbol{P}_L = \boldsymbol{P}_L - \boldsymbol{P}_1$ and $\boldsymbol{V}_L = \boldsymbol{P}_2 - \boldsymbol{P}_1$. From that, one gets the same parametric form as Eq. (4.20). In expanded form, one has $\boldsymbol{P}_1 = [x_1\ \ y_1\ \ z_1\ \ 1]$, $\boldsymbol{P}_L = [x\ \ y\ \ z\ \ 1]$, $\boldsymbol{P}_2 = [x_2\ \ y_2\ \ z_2\ \ 1]$, and $\boldsymbol{V}_L = \boldsymbol{V}_{HL} = [a_L\ \ b_L\ \ c_L\ \ 0]^T$, where a_L, b_L, and c_L are the components of the line direction vector. In a *column form*, Eq. (4.23) becomes

$$\begin{bmatrix} x - x_1 \\ y - y_1 \\ z - z_1 \\ 1 - 1 \end{bmatrix} = t \begin{bmatrix} x_2 - x_1 \\ y_2 - y_1 \\ z_2 - z_1 \\ 1 - 1 \end{bmatrix} = t \begin{bmatrix} a_L \\ b_L \\ c_L \\ 0 \end{bmatrix} \tag{4.24}$$

Separating the components, one gets

$$x - x_1 = t\,a_L, \quad y - y_1 = t\,b_L, \quad z - z_1 = t\,c_L \tag{4.25}$$

The *Cartesian* equations are obtained by eliminating t from Eq. (4.25), thereby one gets the *straight line* form known as

$$\frac{x - x_1}{a_L} = \frac{y - y_1}{b_L} = \frac{z - z_1}{c_L} \tag{4.26}$$

If any of the coefficients a_L, b_L, or c_L is zero in a denominator, the corresponding numerator must also be zero. That is, if $a_L = 0$, $x - x_1 = 0$. Note that in *Cartesian* form, the line in ***2D*** is defined by one equation and in ***3D*** by two equations.

4.2.2 Parametric Planes

In *explicit* form, the plane equation is given by the following form:

$$z = A_P\,x + B_P\,y + C_P \tag{4.27}$$

where A_P, B_P, and C_P are the explicit plane coefficients or components.

In *implicit* form, the plane is given by

$$a_P\,x + b_P\,y + c_P\,z + d_P = 0 \tag{4.28}$$

where a_P, b_P, and c_P are the components of $\boldsymbol{N}_P$, which is the *normal vector* to the plane given by

$$\boldsymbol{N}_P = \begin{bmatrix} a_P & b_P & c_P \end{bmatrix}^T \tag{4.29}$$

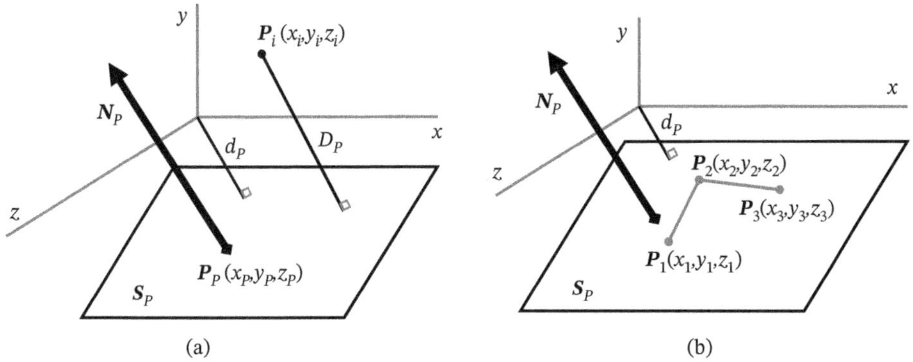

Figure 4.17 A plane surface $\boldsymbol{S}_P$, its normal vector $\boldsymbol{N}_P$, its normal distance from the origin d_P, and the normal distance D_P of any point $\boldsymbol{P}_i$ from the plane are shown in (a). A plane defined by three in-plane points is shown in (b).

The *implicit* form is also the result of the following *dot product*:

$$\boldsymbol{P}_i \cdot \boldsymbol{S}_P = \boldsymbol{0}_H \tag{4.30}$$

where $\boldsymbol{0}_H$ is the homogeneous coordinate origin and $\boldsymbol{S}_P$ defines the plane "*surface*" by

$$\boldsymbol{S}_P = (\,a_P, b_P, c_P, d_P) = \begin{bmatrix} a_P & b_P & c_P & d_P \end{bmatrix}^T \tag{4.31}$$

The point $\boldsymbol{P}_i$ is any *homogeneous* point $[x, y, z, h_H]$ or $[x, y, z, 1]$ satisfying the *dot product* of Eq. (4.30).

If the normal vector to the plane $\boldsymbol{N}_P$ is made to be a unit vector by scaling $\boldsymbol{S}_P$, then d_P is the signed distance to the origin from the plane. Thus, the *dot product* of the plane coefficients (or component) vector and any point $\boldsymbol{P}_i$ (in homogeneous coordinates) gives the distance D_P of the point to the plane, i.e.,

$$D_P = \boldsymbol{S}_P \cdot \boldsymbol{P}_i \tag{4.32}$$

That is why $D_P = 0$ when $\boldsymbol{P}_i$ lies in the plane.

The plane $\boldsymbol{S}_P$ can be determined for the three non-collinear points $\boldsymbol{P}_1$, $\boldsymbol{P}_2$, and $\boldsymbol{P}_3$ from Eq. (4.30) such that

$$\begin{bmatrix} & \boldsymbol{P}_1 & & \\ & \boldsymbol{P}_2 & & \\ & \boldsymbol{P}_3 & & \\ 0 & 0 & 0 & 1 \end{bmatrix} \boldsymbol{S}_P = \begin{bmatrix} 0 \\ 0 \\ 0 \\ 1 \end{bmatrix} \quad \text{or} \quad \begin{bmatrix} P_{1x} & P_{1y} & P_{1z} & 1 \\ P_{2x} & P_{2y} & P_{2z} & 1 \\ P_{3x} & P_{3y} & P_{3z} & 1 \\ 0 & 0 & 0 & 1 \end{bmatrix} \boldsymbol{S}_P = \begin{bmatrix} 0 \\ 0 \\ 0 \\ 1 \end{bmatrix} \tag{4.33}$$

where each of the points is substituted as a raw in the matrix as shown in Eq. (4.33). If the homogeneous $\boldsymbol{P}_1$, $\boldsymbol{P}_2$, and $\boldsymbol{P}_3$ matrix in Eq. (4.33) is inverted, the plane $\boldsymbol{S}_P$ is thus the *fourth column* of the inverse. Figure 4.17a shows a plane surface $\boldsymbol{S}_P$, its normal vector $\boldsymbol{N}_P$, its normal distance from the origin d_P, and the normal distance D_P of any point $\boldsymbol{P}_i$ from the plane. Figure 4.17b shows a plane defined by three *in-plane* points with some plane parameters indicated.

The *parametric* equation of a *plane* encompassing or containing points $\boldsymbol{P}_1$, $\boldsymbol{P}_2$, and $\boldsymbol{P}_3$ is then defined by the point $\boldsymbol{P}_P\,(u, v)$ on the *plane surface* $\boldsymbol{S}_P$ such that

$$\boldsymbol{S}_P = \boldsymbol{P}_P(u, v) = \boldsymbol{P}_1 + u(\boldsymbol{P}_2 - \boldsymbol{P}_1) + v(\boldsymbol{P}_3 - \boldsymbol{P}_2) \tag{4.34}$$

where the u and v are the parameters representing the parametric *surface or plane coordinates.* Figure 4.18 shows a plane defined by three *in-plane* points with indicated directions of u and v parameters.

In a matrix form, the plane surface is given by

$$\boldsymbol{S}_P = \boldsymbol{P}_P(u, v) = \begin{bmatrix} (1 - u) & (u - v) & v \end{bmatrix} \begin{bmatrix} \boldsymbol{P}_1 \\ \boldsymbol{P}_2 \\ \boldsymbol{P}_3 \end{bmatrix} \tag{4.35}$$

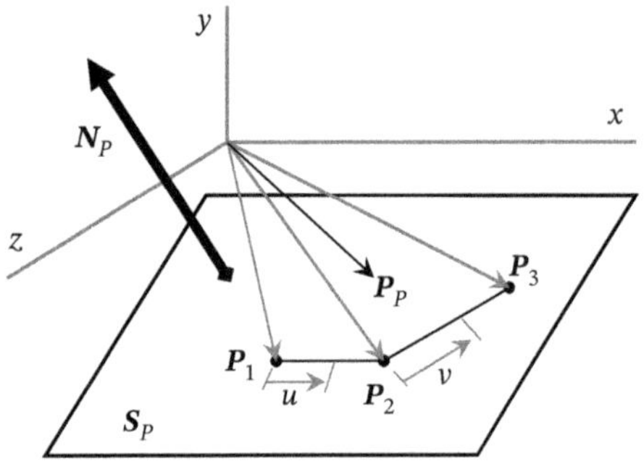

Figure 4.18 A plane defined by three in-plane points with indicated directions of u and v parameters.

Or

$$S_P = P_P(u, v) = (1 - u)P_1 + (u - v)P_2 + vP_3 \tag{4.36}$$

The parametric equation of a plane containing points P_1, P_2, and P_3 can then be given in terms of *blending functions* $\mathbf{b}_i$, $i = 1,2,3$, such that

$$S_P = P_P(u, v) = \mathbf{b}_1 \; P_1 + \mathbf{b}_2 \; P_2 + \mathbf{b}_3 \; P_3, \tag{4.37}$$

where $\mathbf{b}_1 = (1 - u)$, $\mathbf{b}_2 = (u - v)$, and $\mathbf{b}_3 = v$. The parameters $\mathbf{b}_1$, $\mathbf{b}_2$, and $\mathbf{b}_3$ are the ***2D** blending functions*, which indicate how P_P will be obtained by blending the effect of P_1 (i.e. $\mathbf{b}_1$) with the effect of P_2 (i.e. $\mathbf{b}_2$) and P_3 (i.e. $\mathbf{b}_3$). The *normal vector* to the plane N_P is given by the following *cross product*:

$$N_P = (P_2 - P_1) \times (P_3 - P_2) \tag{4.38}$$

The cross product can be evaluated by the matrix product in the form given in Eq. (2.6).

Example 4.3 Draw the two planes that have the intersection points of (2,0,0), (0,1,0), and (0,0,1) for the first plane and the intersection points of (−2,0,0), (0,−1,0), and (0,0,−1) for the second plane.

Solution

Since the points of coordinate intersections are given, the plotting task is greatly simplified as shown in Figure 4.19a.

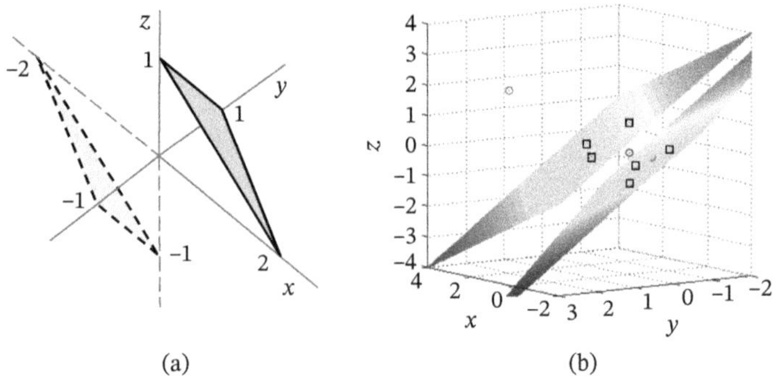

Figure 4.19 Two planes that have the intersection points of (2,0,0), (0,1,0), and (0,0,1) for the first plane and the intersection points of (−2,0,0), (0,−1,0), and (0,0,−1) for the second plane are shown in (a). The two planes are plotted using a MATLAB code in (b) showing the intersection points as squares in addition to other points in space.

According to Eq. (4.33), the first plane can be obtained from

$$\begin{bmatrix} 2 & 0 & 0 & 1 \\ 0 & 1 & 0 & 1 \\ 0 & 0 & 1 & 1 \\ 0 & 0 & 0 & 1 \end{bmatrix} \boldsymbol{S}_P = \begin{bmatrix} 0 \\ 0 \\ 0 \\ 1 \end{bmatrix} \tag{a}$$

The inverse of the three points matrix is simply transforming Eq. (a) to

$$\boldsymbol{S}_P = \begin{bmatrix} 0.5 & 0 & 0 & -0.5 \\ 0 & 1 & 0 & -1 \\ 0 & 0 & 1 & -1 \\ 0 & 0 & 0 & 1 \end{bmatrix} \begin{bmatrix} 0 \\ 0 \\ 0 \\ 1 \end{bmatrix} = \begin{bmatrix} -0.5 \\ -1 \\ -1 \\ 1 \end{bmatrix} \tag{b}$$

From Eq. (b), one can find the homogeneous plane form, the plane normal, the implicit plane form, and the normal distance of the plane to the origin.

According to Eq. (4.33), the second plane can be obtained from

$$\begin{bmatrix} -2 & 0 & 0 & 1 \\ 0 & -1 & 0 & 1 \\ 0 & 0 & -1 & 1 \\ 0 & 0 & 0 & 1 \end{bmatrix} \boldsymbol{S}_P = \begin{bmatrix} 0 \\ 0 \\ 0 \\ 1 \end{bmatrix} \tag{c}$$

The inverse of the three points matrix is simply transforming Eq. (a) to

$$\boldsymbol{S}_P = \begin{bmatrix} -0.5 & 0 & 0 & 0.5 \\ 0 & -1 & 0 & 1 \\ 0 & 0 & -1 & 1 \\ 0 & 0 & 0 & 1 \end{bmatrix} \begin{bmatrix} 0 \\ 0 \\ 0 \\ 1 \end{bmatrix} = \begin{bmatrix} 0.5 \\ 1 \\ 1 \\ 1 \end{bmatrix} \tag{d}$$

From Eq. (d), one can find the homogeneous plane form, the plane normal, the implicit plane form, and the normal distance of the plane to the origin.

For any other three points on a plane, the task is more involved, and a MATLAB code has been developed for that. Figure 4.19b shows the two planes in question plotted using the MATLAB code in (b) and indicating the intersection points as squares in addition to other points in space to demonstrate the extent of the two planes. Eq. (4.33) has been used to advantage to get the plane surface $\boldsymbol{S}_P$ by the inverse of the three points matrix in the equation. The plane $\boldsymbol{S}_P$ is thus the *fourth column* of the inverse.

The MATLAB code is available in the **Wiley website** under the name `CAD_Planes.m`. The code provides many other parameter and options such as calculated plane normal, implicit plane form, normal distance of plane to origin, homogeneous plane form, normal distance of any fourth point to the plane, and plane *x*, *y*, and *z* coordinates intersects.

4.2.3 Parametric Bilinear Surfaces

A *bilinear parametric surface* $\boldsymbol{S}_B$ *(u, v),* where the parameters ($u, v \in [0, 1]$), allows the linear interpolation of any point on the surface such that

$$\boldsymbol{S}_B(u, v) = \begin{bmatrix} (1-u) & u \end{bmatrix} \begin{bmatrix} \boldsymbol{P}_1(0,0) & \boldsymbol{P}_2(0,1) \\ \boldsymbol{P}_3(1,0) & \boldsymbol{P}_4(1,1) \end{bmatrix} \begin{bmatrix} 1-v \\ v \end{bmatrix} \tag{4.39}$$

Note that $\boldsymbol{S}_B\ (0, 0) = \boldsymbol{P}_1\ (0, 0)$ and the center of the surface is at the parameter values ($u = v = 0.5$) or

$$\boldsymbol{S}_B(0.5, 0.5) = [\boldsymbol{P}_c(0,0) + \boldsymbol{P}_c(0,1) + \boldsymbol{P}_c(1,0) + \boldsymbol{P}_c(1,1)]/4 \tag{4.40}$$

In a matrix form, the bilinear surface is given by

$$\boldsymbol{S}_B(x(u,v), y(u,v), z(u,v)) = \boldsymbol{b}_B\, \boldsymbol{P}_c \tag{4.41}$$

where $\boldsymbol{S}_B$ is a position vector on the interpolated surface, $\boldsymbol{P}_c$ is the matrix of position vectors of ***corner*** *points* and the *blending function vector* $\boldsymbol{b}_B$ is given by

$$\boldsymbol{b}_B = [(1-u)(1-v) \quad (1-u)v \quad u(1-v) \quad uv] \tag{4.42}$$

Note that the matrix of ***corner*** *points* of the bilinear surface is given by

$$\boldsymbol{P}_c = \begin{bmatrix} \boldsymbol{P}_{c1} \\ \boldsymbol{P}_{c2} \\ \boldsymbol{P}_{c3} \\ \boldsymbol{P}_{c4} \end{bmatrix} = \begin{bmatrix} P_{1x} & P_{1y} & P_{1z} & 1 \\ P_{2x} & P_{2y} & P_{2z} & 1 \\ P_{3x} & P_{3y} & P_{3z} & 1 \\ P_{4x} & P_{4y} & P_{4z} & 1 \end{bmatrix} \tag{4.43}$$

Figure 4.20 shows a bilinear surface defined by four corner points (0,0,0), (3,0,0), (3,3,3), and (0,3,0). The center of the surface is shown at the parameter values ($u = v = 0.5$), where – according to Eq. (4.42) – the *blending function vector* $\boldsymbol{b}_B = [0.25 \quad 0.25 \quad 0.25 \quad 0.25]$.

This process can also generate ***lofted*** or ***ruled*** surfaces $\boldsymbol{S}_R$ between two opposite *boundary curves*, say $\boldsymbol{P}_B\,(u, 0)$ and $\boldsymbol{P}_B\,(u, 1)$.

$$\boldsymbol{S}_R(u,v) = \left[\boldsymbol{P}_B(u,0)\ \boldsymbol{P}_B(u,1)\right] \begin{bmatrix} 1-v \\ v \end{bmatrix} \tag{4.44}$$

Another extension is the generation of the linear ***Coons*** surface by the following linear interpolation:

$$\boldsymbol{S}_C(u,v) = \left[(1-u) \quad u \quad 1\right] \begin{bmatrix} -\boldsymbol{P}(0,0) & -\boldsymbol{P}(0,1) & -\boldsymbol{P}(0,v) \\ -\boldsymbol{P}(1,0) & -\boldsymbol{P}(1,1) & -\boldsymbol{P}(1,v) \\ -\boldsymbol{P}(u,0) & -\boldsymbol{P}(u,1) & 0 \end{bmatrix} \begin{bmatrix} 1-v \\ v \\ 1 \end{bmatrix} \tag{4.45}$$

The surface is linearly connecting the four curves at the boundaries $\boldsymbol{P}\,(u, 0)$, $\boldsymbol{P}\,(u, 1)$, $\boldsymbol{P}\,(0, v)$, and $\boldsymbol{P}\,(1, v)$.

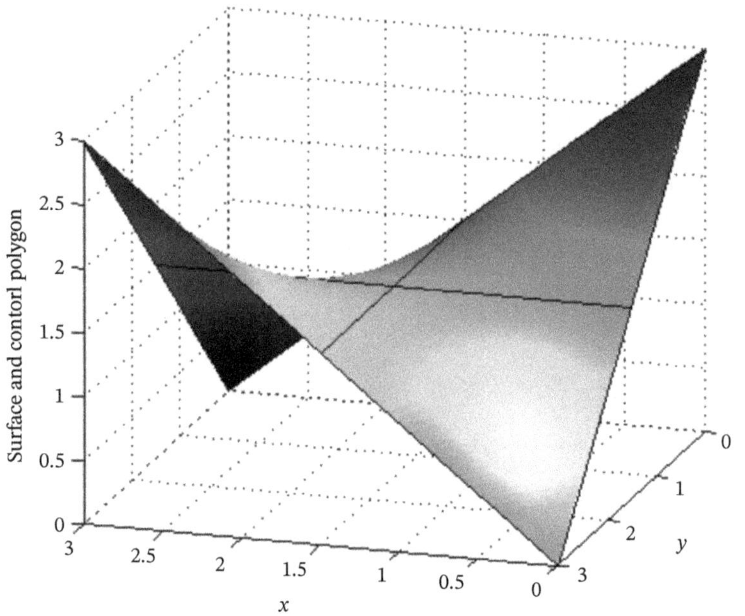

Figure 4.20 Bilinear surface defined by four corner points and the center of the surface is at the parameter values ($u = v = 0.5$).

4.2.4 Parametric Curves and Surfaces

Some frequently used curves and surfaces are presented herein by their conventional and parametric equations. The conventional forms are usually used as input parameters or coefficients by the users of geometric molding packages. The *parametric* form is usually used by the geometric modeling graphics to generalize the code and to easily use it as a parametric form by the users for editing afterward. This is applied to *circles, conics, spheres, cylinders, cones*, and *quadric surfaces*.

A ***circle*** of radius r_C and its center at point (h_C, k_C) is conventionally given by

$$(x - h_C)^2 + (y - k_C)^2 = r_C^2 \tag{4.46}$$

where x and y are the coordinates of any point $\boldsymbol{P}_c\ (x, y)$ on the circle. In geometric modeling packages, the required input is usually r_C and (h_C, k_C). These parameters may be associated with other entities or objects in the model.

The *parametric* representation of an origin-centered circle is given by

$$x = r_C \cos\theta_C, \qquad \text{and} \qquad y = r_C \sin\theta_C. \tag{4.47}$$

where θ_C is the angle on the circle. Or for generic unit *circle*, we have the parametric equation represented by

$$\boldsymbol{P}_c(u) = \begin{bmatrix}\cos\theta_C & \sin\theta_C\end{bmatrix} \tag{4.48}$$

For n fixed number of points on the *circle*, the angle between each point is $d\theta_C = 2\pi/n$, and any point on the circle can then be obtained recursively as

$$\begin{bmatrix}x_{n+1} & y_{n+1}\end{bmatrix} = \begin{bmatrix}x_n & y_n\end{bmatrix}\begin{bmatrix}\cos(d\theta_C) & \sin(d\theta_C)\\ -\sin(d\theta_C) & \cos(d\theta_C)\end{bmatrix} \tag{4.49}$$

or

$$P_{c,n+1} = \boldsymbol{P}_{c,n}\boldsymbol{R}_z \tag{4.50}$$

where $\boldsymbol{R}_z$ is the rotation transformation matrix in ***2D***; see Eq. (4.11) for ***3D***. Equation (4.49) draws a circle much faster than Eq. (4.48) since cos $(d\theta_C)$ and sin $(d\theta_C)$ are calculated only once. In Eq. (4.48), cos (θ_C) and sin (θ_C) are calculated n times.

The family of ***conic*** or ***quadric surfaces*** $\Im$ in implicit form is conventionally defined by

$$\Im(x, y, z) = a_q x^2 + b_q y^2 + c_q z^2 + 2d_q xy + 2e_q yz + 2f_q xz + 2g_q x + 2h_q y + 2j_q z + k_q = 0 \tag{4.51}$$

where $a_q, \ldots, k_q$ are the coefficients of the *conic* or *quadric surface*. If the coefficients a_q through f_q are zero, a *plane* is the outcome, i.e.,

$$\Im(x, y, z) = 2g_q x + 2h_q y + 2j_q z + k_q = 0 \tag{4.52}$$

If the coefficients $a_q = b_q = c_q = -k_q = 1$ and the remaining coefficients are zero, a generic *unit sphere* is conventionally defined with its center at the origin such that

$$\Im(x, y, z) = x^2 + y^2 + z^2 - 1 \tag{4.53}$$

The parametric form of a ***unit sphere*** is given by

$$\boldsymbol{P}_s(u, v) = (\cos(v)\ \cos(u), \cos(v)\ \sin(u), \sin(v)) \tag{4.54}$$

where u corresponds to the *azimuth* and v corresponds to the *latitude* with u vary over $(0, 2\pi)$ and v vary over $(-\pi/2, \pi/2)$. For geographical symbol u-contours are called *meridians*, and v-contours are known as *parallels*.

The generic or ***unit cylinder*** in z direction is conventionally given by

$$\Im(x, y, z) = x^2 + y^2 - 1 \tag{4.55}$$

This equation is valid for all values of z. The parametric form is given by

$$\boldsymbol{P}_{cl}(u, v) = (\cos(u), \sin(u), 0) \tag{4.56}$$

where for *azimuth*, the parameter u is in the range $[0, 2\pi]$ for all values of v.

The generic or ***unit cone*** is conventionally given by

$$\Im(x, y, z) = x^2 + y^2 - (1 - z)^2 = 0, \quad \text{for} \quad 0 < z < 1. \tag{4.57}$$

The parametric form of the *unit cone* is then given by

$$\boldsymbol{P}_{\text{cone}}(u, v) = ((1 - v)\cos(u), (1 - v)\sin(u), v) \tag{4.58}$$

where, for *azimuth*, the parameter u is in the range $[0, 2\pi]$ and the parameter v is in the range $[0, 1]$.

In a matrix form, the *general quadric surface* is conventionally given by

$$\boldsymbol{P}_q \ \boldsymbol{Q}_q \ \boldsymbol{P}_q^T = \boldsymbol{0}_H \tag{4.59}$$

with

$$\boldsymbol{Q}_q = \begin{bmatrix} a & d & f & g \\ d & b & e & h \\ f & e & c & j \\ g & h & j & k \end{bmatrix} \quad \text{and} \quad \boldsymbol{P}_q{}^T = \begin{bmatrix} P_x \\ P_y \\ P_z \\ 1 \end{bmatrix}. \tag{4.60}$$

where $a_q, b_q, c_q, \ldots, k_q$ are of the *quadric* surfaces in *implicit* form. This form of a matrix can easily be transformed.

As before, generic *unit sphere* is given by

$$\Im(x, y, z) = x^2 + y^2 + z^2 - 1 \tag{4.61}$$

The matrix form of a *unit sphere* is then given by $\boldsymbol{P}_q\boldsymbol{Q}_q\boldsymbol{P}_q{}^T = \boldsymbol{0}$ with

$$\boldsymbol{Q}_q = \begin{bmatrix} 1 & 0 & 0 & 0 \\ 0 & 1 & 0 & 0 \\ 0 & 0 & 1 & 0 \\ 0 & 0 & 0 & -1 \end{bmatrix} \quad \text{and} \quad \boldsymbol{P}_q{}^T = \begin{bmatrix} P_x \\ P_y \\ P_z \\ 1 \end{bmatrix}. \tag{4.62}$$

4.2.5 Free-Form Parametric Curves and Surfaces

Mathematical representation of *free-form curves* and *free-form surfaces* is indispensable in computer implementation and applications particularly in *geometric modeling* (CAD or "computer-aided drafting"). Polynomials and exponentials are traditionally used for fitting or regression analysis of input data. *Splines* are piecewise defined polynomials. They are lately and extensively been used to model free-form curves and surfaces in design. They are attractive in many aspects particularly simplicity, ease of use, flexibility, smoothness, and accuracy.

4.2.5.1 Surface Patches and Curves

The selection of mathematical modeling of curves and surface patches depends on shape description requirements. The user requires that curves and surfaces must be mathematically tractable, computationally convenient, and economical to store. Some important properties for designing curves are related to entering data points that can be used to control the shapes of curves or surfaces. These points can then be named as ***control points***. A curve is said to *interpolate* the control points if it passes through them Figure 4.21a. One regularly connects the control points with straight lines to show the *"polygon;"* see Figure 4.21a. Multiple coordinate values of points on the curve as depicted in Figure 4.21b might cause problems, and it is better to have the curve defined to be axis independent. It is also desired to have a local control rather than global control over the curve shape when

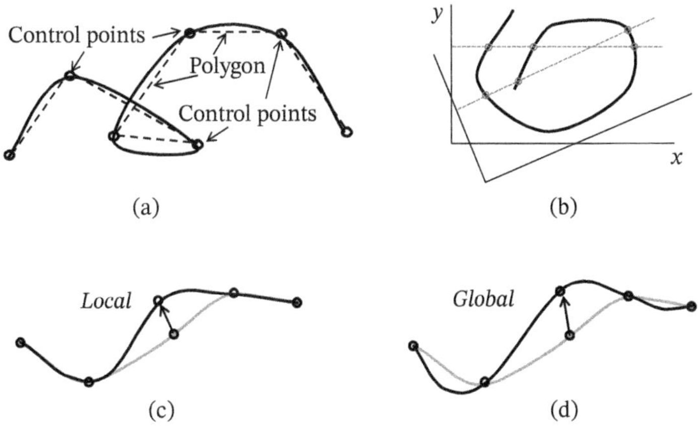

Figure 4.21 A curve interpolates the control points passing through them in (a), and the "polygon" connects the control points with straight lines in (a). Multiple coordinate values of points on the curve are depicted in (b). When moving the control points, local control in (c) rather than global control in (d) is desirable.

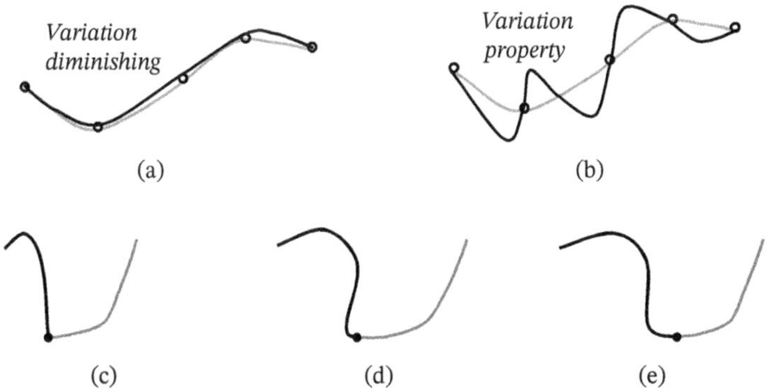

Figure 4.22 The curve having a variation-diminishing property in (a) rather than variation property in (b). Zero-order continuity of two curves just meeting at connection point in (c). First-order continuity of curves to have the same tangent at the point of intersection as in (d). Second-order continuity requires curvatures of both curves to be the same at connection point as in (e).

moving the control points as shown in Figure 4.21c,d. The devised curve should also have a variation-diminishing property rather than variation property as shown in Figure 4.22a,b. Versatility of the mathematical model would allow the representation and sculpting of several desired shapes. Order of continuity of curves and surfaces when connected to others are essential to the sought geometric model. Zero-order continuity simply means that the two curves meet at connection point as demonstrated in Figure 4.22c. First-order continuity requires the curves to be tangent at the point of intersection as defined in Figure 4.22d. Second-order continuity requires that curvatures of both curves be the same at connection point as shown in Figure 4.22e.

Curves representing input data or control points can either *interpolate* the data or control points as defined in Figure 4.23a or *approximate* the location of the data or control points as shown in Figure 4.23b. Control points are connected by gray straight lines to form the *polygon* encompassing the data or control points. The adopted mathematical model of curves may then either interpolate or approximate the control points. Mathematically, the approximation is attractive. For the users, however, the interpolation is more utilizable and easier to edit and manipulate. The user interface may show an interpolated curve (or a surface), but the internal system might use approximation model with some way to figure out how to move control points when approximated curve (or

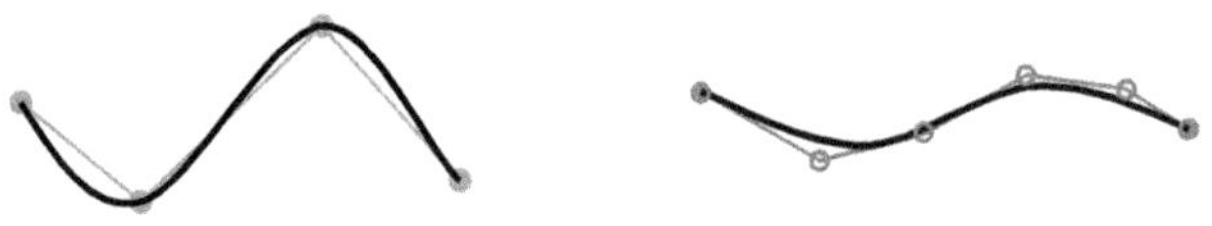

(a) (b)

Figure 4.23 A curve representing input data or control points interpolates the data or control points in (a). A curve approximating the location of data or control points is shown in (b).

surface) points are moved. The user may be interested to find out the direct relation from the information presented in this text or other related references. The treatment presented herein is the more attractive mathematical approximation using parametric representation.

Parametric or vector-valued function where a point on a *parametric curve* $\boldsymbol{P}_C$ is given by a vector in the following form:

$$\boldsymbol{P}_C(u) = \begin{bmatrix} x(u) & y(u) & z(u) \end{bmatrix} \tag{4.63}$$

The parameter u takes on values in a specified range, usually 0–1.

For surfaces, two parameters are required, which renders the point on a *parametric surface* $\boldsymbol{P}_S$ as

$$\boldsymbol{P}_S(u, v) = \begin{bmatrix} x(u, v) & y(u, v) & z(u, v) \end{bmatrix} \tag{4.64}$$

The parameters u and v take on values in a specified range, usually 0–1. Both ***Bezier*** and ***B-spline*** curve formulations use polynomial parametric functions. The desirable local control property is obtained by defining a piece in terms of *control points* near it and of the continuity requirements at its joints. *Bezier* and *B-spline* formulations use control points that lie off the curve but nevertheless provide remarkably effective control of the curve shape. An interpolation variant of *B-spline* uses control points that lie on the curve. The approximating representation is provided herein for *Bezier* and *B-spline* curves and patches or surfaces.

4.2.5.2 Bezier Curves

Bezier defines the *parametric curve* $\boldsymbol{P}_{BC}$ (u) of *n degree* in terms of the parameter u and the locations of $(n + 1)$ control points $\boldsymbol{P}_{Ci}$ such as for the four ***3D*** control points in Figure 4.24 (Bezier 1986). The parametric curve is given by

$$\boldsymbol{P}_{BC}(u) = \sum_{i=0}^{n} \boldsymbol{P}_{Ci}\ \boldsymbol{b}_{i,n}(u), u = (0 \text{ to } 1) \tag{4.65}$$

where $\boldsymbol{P}_{BC}$ (u) is the point on the *Bezier* curve at u, $\boldsymbol{P}_{Ci}$ are the $(n + 1)$ control points, and $\boldsymbol{b}_{i,n}$ (u) are the $(n + 1)$ *blending functions* of the parameter u. The blending functions are defined as follows (see Figure 4.24b):

$$\boldsymbol{b}_{i,n}(u) = \left(\frac{n!}{i!(n-1)!}\right) u^i (1-u)^{n-i} \tag{4.66}$$

The first term in the value of $\boldsymbol{b}_{i,n}(u)$ in Eq. (4.66) is the *binomial coefficient*. The ***3D*** location of the *control point* $\boldsymbol{P}_{Ci}$ is $[x_i \ \ y_i \ \ z_i]$ or $[x_i \ \ y_i \ \ z_i \ \ 1]$, and any point $\boldsymbol{P}_{BC}(u)$ on the *Bezier* curve has the following location coordinates:

$$\boldsymbol{P}_{BC,x}(u) = \sum_{i=0}^{n} x_i\ \boldsymbol{b}_{i,n}(u), \quad \boldsymbol{P}_{BC,y}(u) = \sum_{i=0}^{n} y_i\ \boldsymbol{b}_{i,n}(u), \quad \boldsymbol{P}_{BC,z}(u) = \sum_{i=0}^{n} z_i\ \boldsymbol{b}_{i,n}(u) \tag{4.67}$$

Note again that the $x_i\ y_i\ z_i$ in Eq. (4.67) are for the *control points coordinates*. The ***3D*** *Bezier* curve defined by four control points (just happened in a plane) is shown in Figure 4.24a. The four blending functions are plotted on top of each other in Figure 4.24b. It should be noted that the *four control points* generate a *cubic Bezier curve*. They are usually the most commonly used curves in addition to the *quadratic* one, which is generated by only three control points.

(a)

(b)

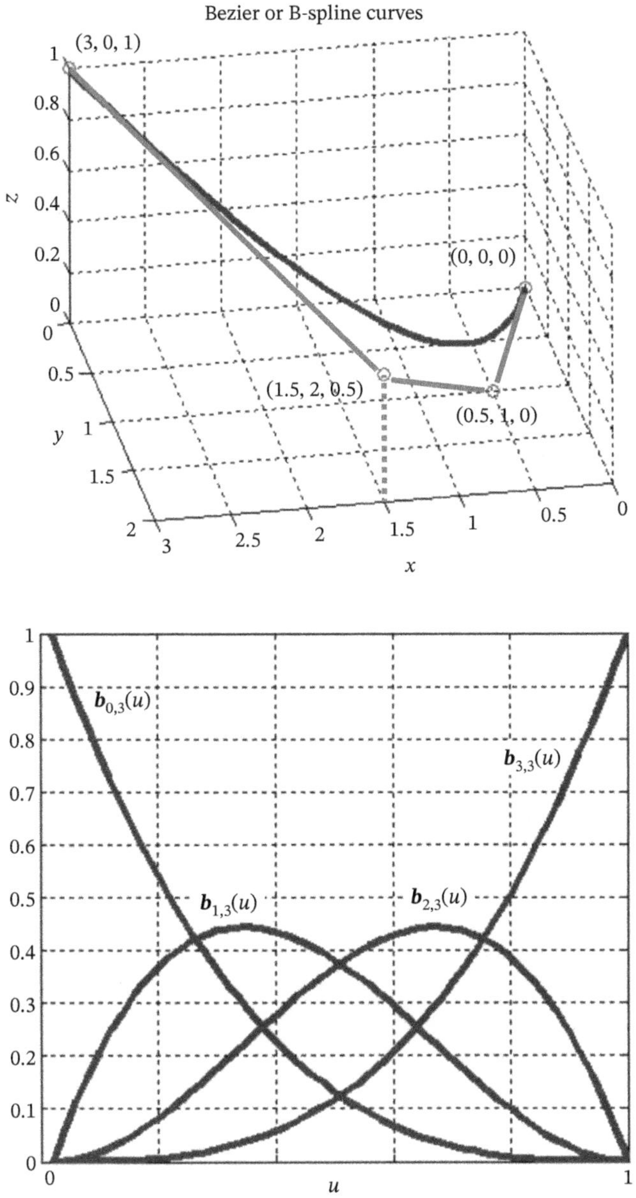

Figure 4.24 The *3D Bezier* curve defined by four control points (in a plane) is shown in (a). The four blending functions are plotted on top of each other in (b).

The *Bezier* curve has several properties and advantages. The curve is related to the locations of *control points* – each exerts more "pull" on the portion of the curve near it. The curve does pass through the two end points ($\boldsymbol{P}_0$ and $\boldsymbol{P}_n$), and it is tangent at the end points to the corresponding edge of the polygon of control points. The curve solves the problem of multiple values of Figure 4.21b since points are a parametric function of u, which is the distance from the original point $\boldsymbol{P}_0$. It is axis independent of the coordinate system. It, then, represents multiple-valued shapes. If the first and last control points coincide, the curve is closed. The *Bezier* curves do not provide localized control. They have global control, which means that moving any control point will change the shape of every part of the curve. This is apparent since all *blending functions* have values between the two end points ($u = 0$ to 1), which specify that any control point movement will change all of the points on the curve. The *Bezier* curve has also a variation-diminishing property since it never oscillates wildly away from its defining control points. Also adding more control points describes more complex shapes, which contributes to its versatility. To achieve *first-order continuity*, the edges of the two polygons adjacent to the common end points must lie in a line (i.e. control points $\boldsymbol{P}_{n-1}$ and $\boldsymbol{P}_n$ of one curve and $\boldsymbol{P}_0$ and $\boldsymbol{P}_1$ of the next connecting curve must all be collinear).

A simple MATLAB code is available in the ***Wiley website*** to get the ***3D*** *Bezier* curve for any *four control points* defined in ***3D***. The name of the code is ***Bezier_B_Spline_3.m***. The same code can be used for any three control points by using two identical middle control points.

4.2.5.3 Bezier Surfaces or Patches

By generating the Cartesian product of two curves, two similar blending functions are used; one for each parameter. Any point $\boldsymbol{P}_{BS}$ (u, v) on the *Bezier surface* $\boldsymbol{S}_{BS}$ has the following location:

$$\boldsymbol{S}_{\mathrm{BS}} = \boldsymbol{P}_{\mathrm{BS}}(u,v) = \sum_{i=0}^{n}\sum_{j=0}^{m} \boldsymbol{P}_{Cij}\ \mathbf{b}_{i,n}(u)\ \ \mathbf{b}_{j,m}(v) \tag{4.68}$$

where $\boldsymbol{P}_{Cij}$ are the *control points* in ***3D***, $\mathbf{b}_{i,n}$ (u) are the *blending functions* in the u direction, and, $\mathbf{b}_{j,m}$ (v) are the *blending functions* in the v direction. The blending functions in u and v directions are defined similar to the one in the definition of curves. *Bezier* surfaces are generated with $(n+1)\times(m+1)$ control points, arranged in a mesh. Figure 4.25 shows specific *Bezier* surfaces with $(n = 2)\times(m = 3)$ in v and u, respectively. The control points are arranged in a mesh of four control points in u direction and three control points in v direction. The polygons are the gray lines connecting the control points in Figure 4.25. The surface is produced by drawing two sets of curves. A set is drawn by holding the u parameter constant and allowing v to range from 0 to 1. The other set is drawn by holding v constant and varying u from 0 to 1. Only the set of *Bezier* curves in u direction (i.e. of constant v) are drawn in Figure 4.25 to reduce confusion. As $\mathbf{b}_{j,m}(0) = 1$ and $\mathbf{b}_{j,m}(1) = 1$, the boundary curve of u at $v = 0$ and the boundary curve of u at $v = 1$ are the *Bezier* curves given by

$$\boldsymbol{P}_{\mathrm{BS}}(u,0) = \sum_{i=0}^{n} \boldsymbol{P}_{Ci,0}\ \mathbf{b}_{i,n}(u) \quad \text{and} \quad \boldsymbol{P}_{\mathrm{BS}}(u,1) = \sum_{i=0}^{n} \boldsymbol{P}_{Ci,1}\ \mathbf{b}_{i,n}(u) \tag{4.69}$$

Other curves are generated similarly.

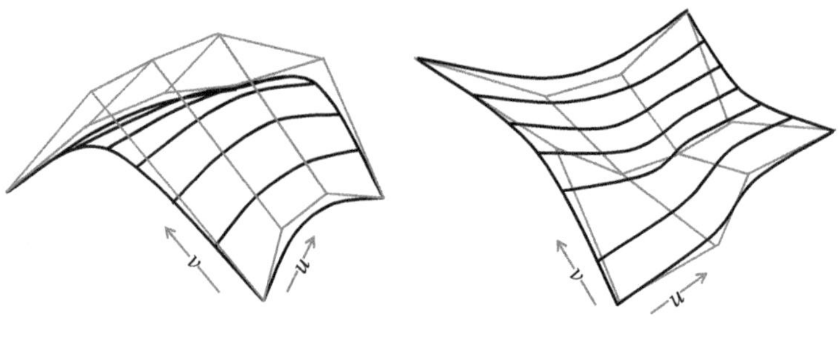

(a) (b)

Figure 4.25 Two *Bezier* surfaces in (a) and (b) with $(n = 2)\times(m = 3)$ in v and u. The control points are arranged in a mesh of $(n+1)$ or three control points in v direction and $(m+1)$ or four control points in u direction. The polygons are the gray lines connecting the control points. Only sets of *Bezier* curves in u direction (i.e. of constant v) are shown.

Large surfaces can be mended together from separate patches. To achieve first-order continuity at a boundary requires that first and last edges of the joining polygons be collinear. It also requires that the ratio of the lengths of all these edge pairs be constant.

A *bicubic Bezier* patch defined by $\boldsymbol{S}_{\text{BS}}(u, v)$ can be formulated using matrices such as the following relation:

$$\boldsymbol{S}_{\text{BS}}(u,v) = \begin{bmatrix}1 & u & u^2 & u^3\end{bmatrix} \boldsymbol{M}_{\text{BB}}\, \boldsymbol{P}_C\, \boldsymbol{M}_{\text{BB}}^T \begin{bmatrix}1 & v & v^2 & v^3\end{bmatrix}^T, \qquad 0 < u, \quad v < 1 \tag{4.70}$$

where

$$\boldsymbol{M}_{\text{BB}} = \begin{bmatrix} 1 & 0 & 0 & 0 \\ -3 & 3 & 0 & 0 \\ 3 & -6 & 3 & 0 \\ -1 & 3 & -3 & 1 \end{bmatrix} \quad \text{and} \quad \boldsymbol{P}_C = \begin{bmatrix} P_{00} & P_{01} & P_{02} & P_{03} \\ P_{10} & P_{11} & P_{12} & P_{13} \\ P_{20} & P_{21} & P_{22} & P_{23} \\ P_{30} & P_{31} & P_{32} & P_{33} \end{bmatrix} \tag{4.71}$$

This *bicubic Bezier patch* is characterized by the depicted 16 control points $\boldsymbol{P}_{\text{C}}$ ($i = 0, 1, 2, 3$ and $j = 0, 1, 2, 3$) of what is called the *characteristic polyhedron*. The coefficients in the *blending coefficients matrix* $\boldsymbol{M}_{\text{BB}}$ are found from the expansion of the *blending functions* designation in Eq. (4.66). This matrix form is simpler to formulate on computer programs or software with available matrix operations and graphing such as MATLAB. Figure 4.26 illustrates a *Bezier* surface patch, where the locations of the control points are directed to be collinear to produce the cylindrical surface effect. The boundary curves at $x = 0$ and $x = 3$ are about the same as the *Bezier* curve in Figure 4.24a.

The *Bezier* patch has similar advantages to the *Bezier* curve. Any point on a *Bezier* patch is included in a *convex hull* defined by 16 control points. This is used in the interference check on two *Bezier* patches. We can quickly find the case when two *Bezier* patches do not intersect. The characteristic polyhedron is approximately the same shape as the patch. In addition, the shape of the patch is smoother than the characteristic polyhedron. The designer can therefore easily understand the relationship between the two and define or modify the shape of the patch by manipulating the control points. There is a simple relationship between a polyhedron and derivative vectors at the patch four corners defined by control points. These can be found in the literature.

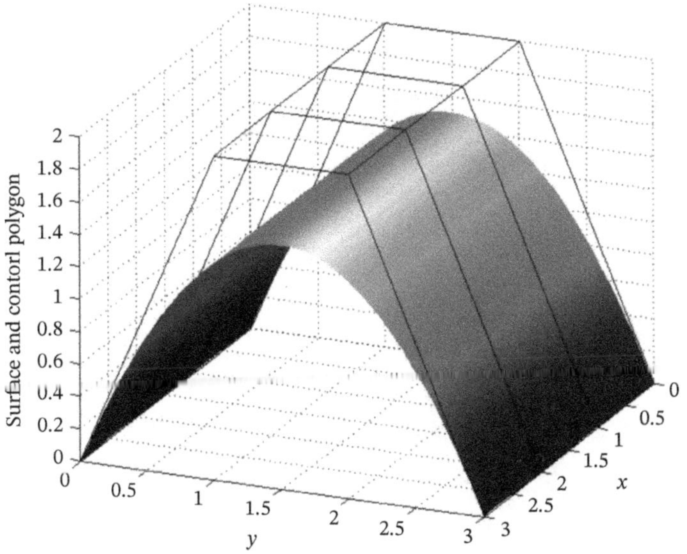

Figure 4.26 *Bezier* surface patch, where the location of the control points are manipulated to be collinear to produce the cylindrical surface effect.

4.2.5.4 B-Spline Curves

The *B-spline* curve is a piecewise polynomial function. The parametric ***non-periodic B-spline*** (basis spline) curve is expressed as follows (see, e.g. Lee 1999; Hill 2001):

$$\boldsymbol{P}_{SC}(u) = \sum_{i=0}^{n} \boldsymbol{P}_{Ci}\ \mathbf{b}_{i,k}(u) \quad (t_{k-1} \leq u \leq t_{n-1}) \tag{4.72}$$

where $\boldsymbol{P}_{SC}(u)$ is any point on the *spline curve* at u, $(n+1)$ is the number of *control points* $(\boldsymbol{P}_{Ci})$, k is the ***curve order*** that controls the order of *continuity* with $(k-1)$ ***curve degree***, t_i are the *knot* values, and the *blending functions* $\mathbf{b}_{i,k}$ are of a degree $(k-1)$ and given by

$$\boldsymbol{b}_{i,1}(u) = \begin{cases} 1 & \text{if } t_i \leq u \leq t_{i+1} \\ 0 & \text{otherwise} \end{cases} \tag{4.73}$$

$$\boldsymbol{b}_{i,k}(u) = \frac{u - t_i}{t_{i+k-1} - t_i}\boldsymbol{b}_{i,k-1}(u) + \frac{t_{i+k} - u}{t_{i+k} - t_{i+1}}\boldsymbol{b}_{i+1,k-1}(u) \tag{4.74}$$

In the previous Eq. (4.74), 0/0 is presumed to be zero. The *knot values* t_i are the parameter values limiting the finite intervals over which the blending functions have nonzero values. There are $(n+k-1)$ knot values from t_0 to t_{n+k} that need to be specified to define the $(n+1)$ blending functions from $\mathbf{b}_{0,k}$ to $\mathbf{b}_{n,k}$. The parameter u has the range (0 to $n-k+2$). For the *non-periodic B-spline*, which is used to model *open curves* (Figure 4.27a), the *non-periodic knot* values t_i are determined from

$$t_i = \begin{cases} 0 & 0 \leq i < k \\ i - k + 1 & k \leq i \leq n \\ n - k + 2 & n < i \leq n + k \end{cases} \tag{4.75}$$

Figure 4.27a shows *B-spline* curves for curve orders k of 3, 4, and 5. It is apparent that the lower the degree $(k-1)$ the closer the curve to the control points. It is also important to point out that the curve order k can take any value, and it is not tied to the number of control points as the case for the *Bezier* curves. It is important also to point out that the *blending functions* $\mathbf{b}_{i,k}$ in Eqs. (4.73) and (4.74) have been typically given the symbol $(N_{i,k})$ in many computer graphics and other texts. This notation is in conflict with the notations used extensively in this text particularly for the rotational speed N. The symbol $\mathbf{b}_{i,k}$ as a ***blending function*** seems to be more representative.

According to previous Eqs. (4.72–4.75), the six non-periodic blending functions for $n = 5$ and $k = 3$ are shown on top of each other in Figure 4.27b. The knot vector is [0 0 0 1 2 3 4 4 4], which represents the knot values for t_0–t_{n+k} or t_0–t_8. The parameter u has then the range (0 to $n-k+2$), i.e. the range is (0–4). The *blending functions* indicate that moving the control point $\boldsymbol{P}_{Ci}$ changes the *B-spline* curve only in the vicinity of the control point since each would only affect the region preceding and following the control point as mainly shown in Figure 4.27b.

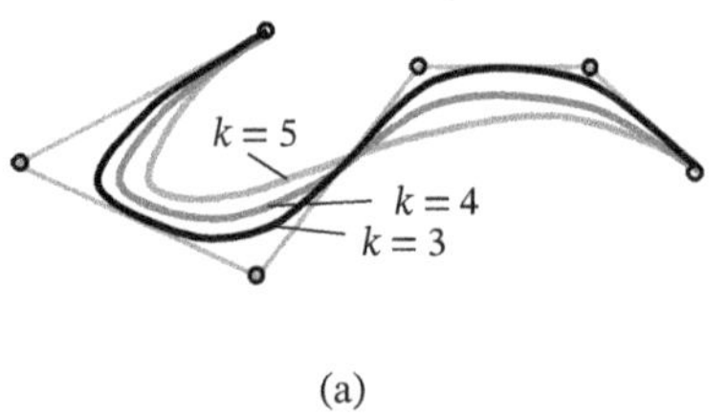

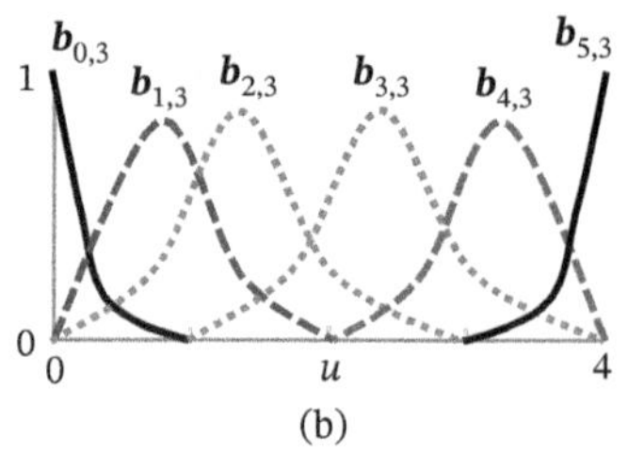

Figure 4.27 *B*-spline curves for curve orders k of 3, 4, and 5 in (a), where the lower the degree $(k-1)$, the closer the curve to the control points. The six non-periodic blending functions for $n = 5$ and $k = 3$ are shown on top of each other in (b).

Periodic B-Spline Curve

For the periodic *B-spline* that is typically used to model *closed curves* (Figure 4.28), the periodic *knot* values t_i are determined from

$$t_i = i - k \quad 0 \leq i \leq n + k \tag{4.76}$$

The *periodic blending functions* are reduced to one expression determined from

$$\mathbf{b}_{i,k}(u) = \mathbf{b}_{0,k}((u - i + n - 1) \text{ mod}(n + 1)) \tag{4.77}$$

The parameter u takes the values in the range of $0 < u < n + i$. The "*mod*" is the remaindering operator (i.e. 3 mod 7 = 3, 9 mod 7 = 2, 7, mod 7 = 0, etc.). Figure 4.28 shows a periodic *B-spline* curve for the curve orders k of 4 with its 4 control points, where the end control point (5) is the same as the start control point (1). It should be noted that the periodic *B-spline* curve does not pass by the start or the end control points; see Figure 4.28.

B-splines have good properties. The local control of shape using *B-splines* makes it possible to generate "corners" as shown in Figure 4.29a. This is achieved by adding multiple control points, i.e. by using several repeated control points at the required corner. As the number of repeated control points increases, the sharper is the intended corner as demonstrated in Figure 4.29a. Local shape control is also demonstrated by moving one control point as shown in Figure 4.29b with only a local change in the original gray curve to the adjusted black one. This is valid particularly if the numbers of control points are much larger than the order k or one is using a *periodic B-spline*. *B-splines* also diminish the necessity to add many curves to describe a shape. Control points can be added with no reasonable bound to the numbers without needing to increase the degree of the curve. Increasing the degree of a curve would make the curve more difficult to control and to accurately calculate. As a consequence, ***cubic B-splines*** ($k = 4$) suffice for a large number of applications by employing more control points to tweak or generate the desired shape.

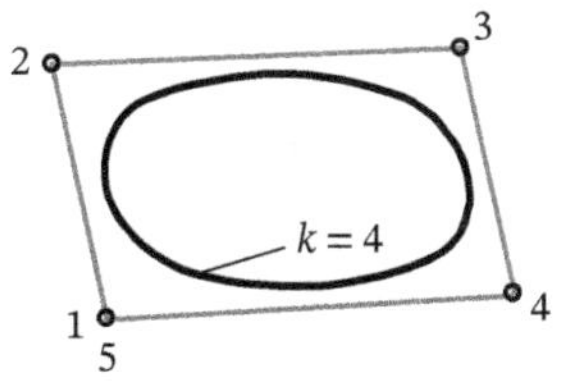

Figure 4.28 Periodic *B-spline* curve for the curve orders k of 4 with its four control points, where the end control point (5) is the same as the start control point (1). The periodic *B-spline* curve does not pass by the start control point (1) or the end control point (5, i.e. 1).

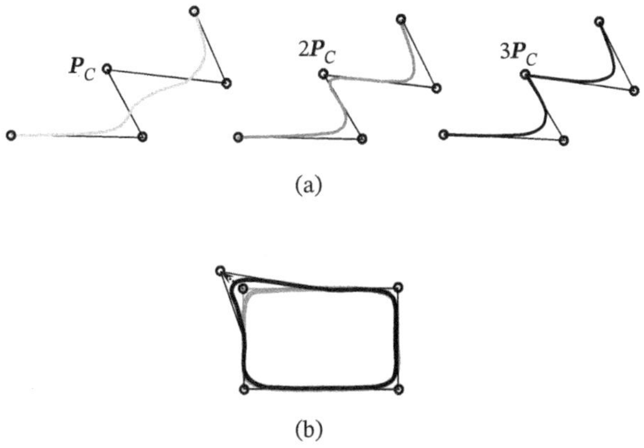

Figure 4.29 Local control of shape using *B-splines*. Part (a) develops sharp corner by using multiple control points at the intended corner, and part (b) demonstrates local control by moving a control point to a new location.

Example 4.4 demonstrates the procedure of developing the blending functions through the implementation of Eqs. (4.73–4.75) to be able to apply Eq. (4.72) of the curve formula.

Example 4.4 Find the equation of a non-periodic *B-spline* curve of order 3 in a polynomial structure. Assume that the control points to develop the curve are $\boldsymbol{P}_{C0}$, $\boldsymbol{P}_{C1}$, and $\boldsymbol{P}_{C2}$. Plot the curve for the ***2D*** control points of (4, 0), (0, 4), and (−4, 0).

Solution

Since we have $\boldsymbol{P}_{C0}$, $\boldsymbol{P}_{C1}$, and $\boldsymbol{P}_{C2}$ control points and the curve order $k = 3$, then $n = k - 1 = 2$. Also, since the required order is 3, the curve degree is $n = k - 1 = 2$. The non-periodic knot values t_i are (Eq. (4.75))

$$t_i = \begin{cases} 0 & 0 \le i < 3 \\ 2 - 3 + 1 & 3 \le i \le 2 \\ 2 - 3 + 2 & 2 < i \le 5 \end{cases} \tag{a}$$

Or $t_0 = 0$, $t_1 = 0$, $t_2 = 0$, $t_3 = 1$, $t_4 = 1$, and $t_5 = 1$. The parameter u ranges from 0 to 1 (Eq. (4.72)).

The blending functions of order 1, $\boldsymbol{b}_{i,1}(u)$ are

$$\boldsymbol{b}_{0,1}(u) = \begin{cases} 1 & \text{if } t_0 \le u \le t_1 \\ 0 & \text{otherwise} \end{cases} \quad (u = 0) \tag{b}$$

$$\boldsymbol{b}_{1,1}(u) = \begin{cases} 1 & \text{if } t_1 \le u \le t_2 \\ 0 & \text{otherwise} \end{cases} \quad (u = 0) \tag{c}$$

$$\boldsymbol{b}_{2,1}(u) = \begin{cases} 1 & \text{if } t_2 \le u \le t_3 \\ 0 & \text{otherwise} \end{cases} \quad (u \le 1) \tag{d}$$

$$\boldsymbol{b}_{3,1}(u) = \begin{cases} 1 & \text{if } t_3 \le u \le t_4 \\ 0 & \text{otherwise} \end{cases} \quad (u = 1) \tag{e}$$

$$\boldsymbol{b}_{4,1}(u) = \begin{cases} 1 & \text{if } t_4 \le u \le t_5 \\ 0 & \text{otherwise} \end{cases} \quad (u = 1) \tag{f}$$

Only $\boldsymbol{b}_{2,1}(u)$ has nonzero value at $u = 0$ and at $u = 1$. Therefore $\boldsymbol{b}_{2,1}(u)$ is the only nonzero blending function of order 1 in the range of u (0–1) and has a constant value of 1 over the entire range. The blending function $\boldsymbol{b}_{2,1}(u)$ is then the only admissible blending function to be of use.

The nontrivial blending functions of order 2, ($k = 2$), are then developed as follows:

$$\boldsymbol{b}_{1,2}(u) = \frac{u - t_1}{t_{1+2-1} - t_1}\boldsymbol{b}_{1,2-1}(u) + \frac{t_{1+2} - u}{t_{1+2} - t_{1+1}}\boldsymbol{b}_{1+1,2-1}(u) \tag{g}$$

i.e.

$$\boldsymbol{b}_{1,2}(u) = \cancel{\frac{u - t_1}{t_2 - t_1}\boldsymbol{b}_{1,1}(u)} + \frac{t_3 - u}{t_3 - t_2}\boldsymbol{b}_{2,1}(u) = \frac{1 - u}{1}\boldsymbol{b}_{2,1}(u) \tag{h}$$

Or

$$\boldsymbol{b}_{1,2}(u) = 1 - u \tag{i}$$

And

$$\boldsymbol{b}_{2,2}(u) = \frac{u - t_2}{t_3 - t_2}\boldsymbol{b}_{2,1}(u) + \cancel{\frac{t_4 - u}{t_4 - t_3}\boldsymbol{b}_{3,1}(u)} = \frac{u}{1}\boldsymbol{b}_{2,1}(u) = u \tag{j}$$

Similarly, we get the blending functions of order 3, $\boldsymbol{b}_{i,3}(u)$

$$\boldsymbol{b}_{0,3}(u) = \frac{u - t_0}{t_3 - t_0}\boldsymbol{b}_{0,2}(u) + \frac{t_3 - u}{t_4 - t_1}\boldsymbol{b}_{1,2}(u) = \frac{1 - u}{1}\boldsymbol{b}_{1,2}(u) = (1 - u)^2 \tag{k}$$

$$\boldsymbol{b}_{1,3}(u) = \frac{u - t_1}{t_3 - t_1}\boldsymbol{b}_{1,2}(u) + \frac{t_4 - u}{t_4 - t_2}\boldsymbol{b}_{2,2}(u) = u(1 - u) + (1 - u)u = 2u(1 - u) \tag{l}$$

$$\boldsymbol{b}_{2,3}(u) = \frac{u - t_2}{t_4 - t_2}\boldsymbol{b}_{2,2}(u) + \frac{t_5 - u}{t_5 - t_3}\boldsymbol{b}_{3,2}(u) = u^2 \tag{m}$$

The expanded equation of the *B-spline* is then

$$\boldsymbol{P}_{\mathrm{SC}}(u) = (1 - u)^2\,\boldsymbol{P}_{C0} + 2u(1 - u)\,\boldsymbol{P}_{C1} + u^2\boldsymbol{P}_{C2} \tag{n}$$

However, the equation of the *Bezier* curve specified by the control points $\boldsymbol{P}_{C0}$, $\boldsymbol{P}_{C1}$, and $\boldsymbol{P}_{C2}$ can be defined by (Eqs. (4.65) and (4.66))

$$\boldsymbol{P}_{\mathrm{BC}}(u) = \binom{2}{0}u^0(1 - u)^2\boldsymbol{P}_{C0} + \binom{2}{1}u^1(1 - u)^1\boldsymbol{P}_{C1} + \binom{2}{2}u^2(1 - u)^0\boldsymbol{P}_{C2} \tag{o}$$

Or

$$\boldsymbol{P}_{\mathrm{BC}}(u) = (1 - u)^2\boldsymbol{P}_{C0} + 2u(1 - u)\boldsymbol{P}_{C1} + u^2\boldsymbol{P}_{C2} \tag{p}$$

This is exactly the same as Eq. (n) of the *B-spline*.

Then, the non-periodic uniform *B-spline* curve of order 3 defined by three *control points* is the same as the *Bezier* curve defined by the same number of *control points*. Therefore, the *non-periodic* and *uniform B-spline* curve is the same as the *Bezier* curve identified by the same control points if its order k is the same as the number of control points ($n + 1$). Therefore, a *Bezier* curve is just a special state of a *B-spline* curve.

The equation of the *Bezier* curve (or the *B-spline* curve) defined by the control points $\boldsymbol{P}_{C0}$, $\boldsymbol{P}_{C1}$, and $\boldsymbol{P}_{C2}$ can then be defined by the components of any point on it $\boldsymbol{P}_{\mathrm{BC}}$ (x, y, z) in an expanded form as a function of the parameter u as follows:

$$\begin{bmatrix}\boldsymbol{P}_{\mathrm{BC},x}\\ \boldsymbol{P}_{\mathrm{BC},y}\\ \boldsymbol{P}_{\mathrm{BC},z}\end{bmatrix}(u) = (1 - u)^2\begin{bmatrix}\boldsymbol{P}_{C0,x}\\ \boldsymbol{P}_{C0,y}\\ \boldsymbol{P}_{C0,z}\end{bmatrix} + 2u(1 - u)\begin{bmatrix}\boldsymbol{P}_{C1,x}\\ \boldsymbol{P}_{C1,y}\\ \boldsymbol{P}_{C1,z}\end{bmatrix} + u^2\begin{bmatrix}\boldsymbol{P}_{C2,x}\\ \boldsymbol{P}_{C2,y}\\ \boldsymbol{P}_{C2,z}\end{bmatrix} \tag{q}$$

To plot the *Bezier* (or *B-spline*) curve for the **2D** control points (4, 0), (0, 4), and (−4, 0), we have only x and y components, and Eq. (q) becomes

$$\begin{bmatrix}\boldsymbol{P}_{\mathrm{BC},x}\\ \boldsymbol{P}_{\mathrm{BC},y}\end{bmatrix}(u) = (1 - u)^2\begin{bmatrix}4\\ 0\end{bmatrix} + 2u(1 - u)\begin{bmatrix}0\\ 4\end{bmatrix} + u^2\begin{bmatrix}-4\\ 0\end{bmatrix} \tag{r}$$

Plotting for parameter $u = 0.25$, 0.5, and 0.75, one needs only to substitute for u in Eq. (r). For $u = 0.25$, $\boldsymbol{P}_{\mathrm{BC}}$ (0.25) = (2.0, 1.5); for $u = 0.5$, $\boldsymbol{P}_{\mathrm{BC}}$ (0.5) = (0, 2), and for $u = 0.75$, $\boldsymbol{P}_{\mathrm{BC}}$ (0.75) = (−2.0, 1.5). Figure 4.30 shows the control points and the plot of the curve passing by the points $\boldsymbol{P}_{C0}$, $\boldsymbol{P}_{\mathrm{BC}}$ (0.25), $\boldsymbol{P}_{\mathrm{BC}}$ (0.5), $\boldsymbol{P}_{\mathrm{BC}}$ (0.75), and $\boldsymbol{P}_{C2}$.

A simple MATLAB code is available in the ***Wiley website*** to get the ***3D*** *cubic B-spline* curve for any *four control points* defined in ***3D***, which is the same as ***3D*** *cubic Bezier* curve for any *four control points*. The code plots the ***3D*** curve and the blending functions (code name is `Bezier_B_Spline_3.m`). The code is for demonstrating approximation rather than interpolation fitting. It is also a good tool to understand the relations between the curves and the control points. The available MATLAB codes have the names of `Bezier_B_Spline_3.m` or ***`CAD_Bezier_B_Spline.m`***.

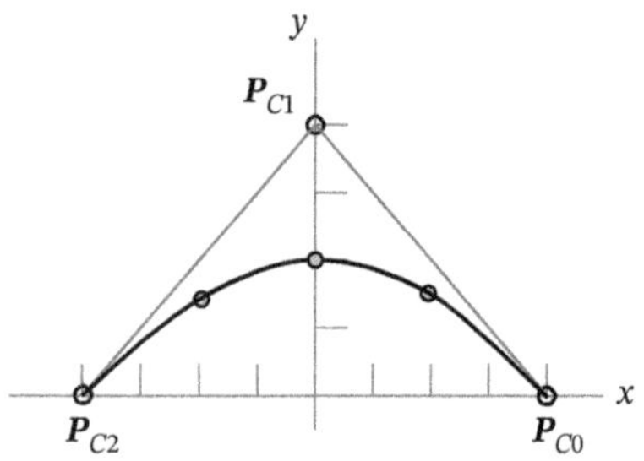

Figure 4.30 Plot of the *B-spline* or *Bezier* curve for the control points (4, 0), (0, 4), and (−4, 0).

4.2.5.5 B-Spline Surfaces

The *B-spline* curves are overlaid to describe the *B-spline surface* $\boldsymbol{S}_{SS}$ such that the point on the surface $\boldsymbol{P}_{SS}$ (Piegl and Tiller 1997)

$$\boldsymbol{S}_{SS} = \boldsymbol{P}_{SS}(u, v) = \sum_{i=0}^{n}\sum_{j=0}^{m} \boldsymbol{P}_{Cij} \quad \boldsymbol{b}_{i,k}(u) \quad \boldsymbol{b}_{j,l}(v) \tag{4.78}$$

where n is ***curve degree*** in the direction u and m is ***curve degree*** in the direction v. The $\boldsymbol{P}_{Cij}$ are the $(n + 1) \times (m + 1)$ control points. The order of the spline surfaces can look similar to the *Bezier* surfaces shown in Figure 4.25. These are examples of special *B-spline* surfaces and their mesh of control points. The knot vectors $\boldsymbol{t}_k$ and $\boldsymbol{s}_k$ that are used in developing the blending functions $\boldsymbol{b}_{i,n}$ (u) and $\boldsymbol{b}_{j,m}$ (v) are defined similar to Eq. (4.75) for the knots t_i for u direction and the knots s_j for the v direction with m in place of n such that

$$s_i = \begin{cases} 0 & 0 \le i < l \\ j - l + 1 & l \le i \le m \\ m - l + 2 & m < i \le m + l \end{cases} \tag{4.79}$$

The $\boldsymbol{t}_k$ vector has $(n + 1)$ knots t_i. The $\boldsymbol{s}_k$ vector has $(m + 1)$ knots s_j.
The *blending functions* $\boldsymbol{b}_{i,n}$ (u) and $\boldsymbol{b}_{j,m}$ (v) are defined like Eq. (4.74).

$$\boldsymbol{b}_{i,k}(u) = \frac{u - t_i}{t_{i+k-1} - t_i}\boldsymbol{b}_{i,k-1}(u) + \frac{t_{i+k} - u}{t_{i+k} - t_{i+1}}\boldsymbol{b}_{i+1,k-1}(u) \tag{4.80}$$

$$\boldsymbol{b}_{i,l}(v) = \frac{v - s_i}{s_{i+l-1} - s_i}\boldsymbol{b}_{i,l-1}(v) + \frac{s_{i+l} - v}{s_{i+l} - s_{i+1}}\boldsymbol{b}_{i+1,l-1}(v) \tag{4.81}$$

The knot values are evaluated similar to Eq. (4.75) for the parameter u with the same k and n. For the parameter v, Eq. (4.75) is modified to replace k by l and replace n by m as depicted in Eq. (4.79).

The *B-spline surfaces* have the same advantages as the *B-spline* curves. They have smooth joining since the continuity of the derivative is implicitly represented in the formulation. When the order k equals the number of control points $(n + 1)$ and the order l equals the number of control points $(m + 1)$, one gets the same patch as *Bezier* patch shown in Figure 4.25a,b. It shows a mesh of control points $\boldsymbol{P}_{Cij}$ $(i = 0, \ldots, n, j = 0, \ldots, m)$ given by the designer and *quadratic B-spline* in one direction and *bicubic B-spline* in the other direction of the patch. For local control, the designer can make local changes to the shape of the surface patch by manipulating control points.

A simple MATLAB code is provided in Figure 4.31 to calculate the ***3D*** *bicubic B-spline* surface for any *16 control points* (4×4) defined in ***3D***. This is the same as the ***3D*** *cubic Bezier* surface for the same *16 control points*. One can then plot the surface and change control points at will. The code is for demonstrating approximation rather than interpolation fitting. In Figure 4.31, an extra line defines the z-coordinates of the control points at the boundaries. This line can be commented to let the previous line as the active boundaries. The MATLAB code is also a good tool to understand the relations between the surfaces and the control points. Figure 4.32 shows the *bicubic B-spline* or *bicubic Bezier* surface and the control polygon for 16 control points, where the 16 control points are marked by

```
clear all; clc; format compact                    % CAD_3D_Spline_Bezier_Color.m
MB=[1 0 0 0;-3 3 0 0;3 -6 3 0;-1 3 -3 1];         % Matrix of Blending coefficients
Pcx=[0 1 2 3; 0 1 2 3; 0 1 2 3; 0 1 2 3]          % Control Points x coordinates
Pcy=[0 0 0 0; 1 1 1 1; 2 2 2 2; 3 3 3 3]          % Control Points y coordinates
Pcz=[0 0 0 0; 0 2 2 0; 0 2 2 0; 0 0 0 0];         % z Boundaries are all zeros
Pcz=[2 -2 2 -2; 0 2 2 0; 0 2 2 0; -1 -2 2 -2]     % z Variations at boundaries
for i=1:1:41
  u=(i-1)/40;                                            % 40 steps in u
  for j=1:1:41
    v=(j-1)/40;                                          % 40 steps in v
    Sx(i,j)=[1 u u^2 u^3]*MB*Pcx*MB'*[1 v v^2 v^3]';
    Sy(i,j)=[1 u u^2 u^3]*MB*Pcy*MB'*[1 v v^2 v^3]';
    Sz(i,j)=[1 u u^2 u^3]*MB*Pcz*MB'*[1 v v^2 v^3]';
  end
end
```

Figure 4.31 A simple MATLAB code to calculate the ***3D*** bicubic *B-spline* surface for any 16 control points (4 × 4) defined in ***3D***. It is the same as the ***3D*** cubic *Bezier* surface for the same 16 control points.

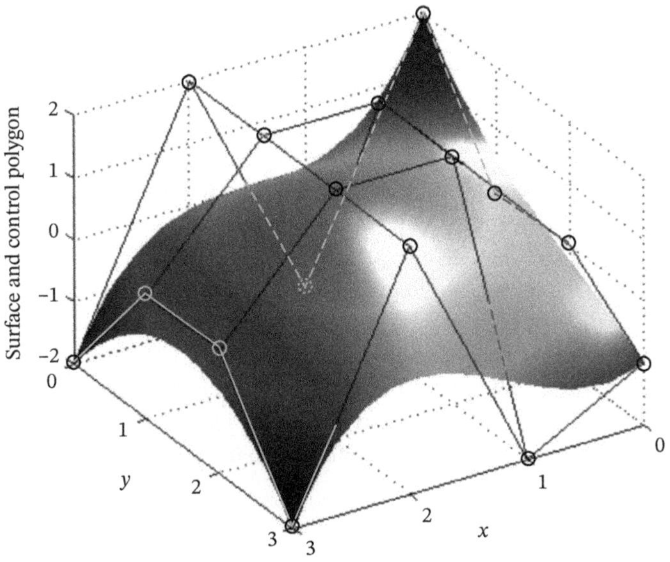

Figure 4.32 Plot of the *bicubic B-spline* or *bicubic Bezier* surface and the control polygon for the 16 control points shown in the MATLAB code of Figure 4.31.

small circles. The control points are those given in the MATLAB code of Figure 4.31. The base *x*–*y* grid values are the same (*x*, *y*) components for all 16 control points. One can change that and observe the variation in the shape of the surface. In fact all points can be changed at will. It should be noted that using the *bicubic B-spline* with 16 control points would necessitate the same conditions of patch joining as the *Bezier* patches. Achieving first-order continuity at a boundary necessitates that first and last edges of the joining polygons be collinear. It also requires that the ratio of the lengths of all these edge pairs be constant.

A simple MATLAB code is available in the ***Wiley website*** to get the ***3D*** *cubic B-spline* surface for any *16 control points* defined in ***3D***, which is the same as ***3D*** *cubic Bezier* surface for the same *16 control points*. The code plots the ***3D*** surface. The code is for demonstrating approximation rather than interpolation fitting. It is also a good tool to understand the relations between the surfaces and the control points. The available MATLAB code has the name of ***CAD_3D_Spline_Bezier_Color.m***.

4.2.5.6 NURBS

NURBS stands for **non-uniform rational B-splines**. They are much widely used since they can exactly represent known curves and surfaces such as a circle, an ellipse, a sphere, a cylinder, a cones, and a torus. They have also been used in several design applications. *NURBS* were used effectively to optimally design ***3D*** C-frame constructions using real-coded and hybrid genetic algorithms, Nassef et al. (1999, 2000) and Hegazi et al. (2002).

NURBS can represent curves, surfaces, and multivariate objects (Piegl and Tiller 1997). The mathematical form that computes the *NURBS curve's* point $\boldsymbol{P}_{\mathrm{NC}}$ at any value of the parameter u is

$$\boldsymbol{P}_{\mathrm{NC}}(u) = \frac{\sum_{i=0}^{n} w_i \boldsymbol{P}_{Ci} \mathbf{b}_{i,k}(u)}{\sum_{i=0}^{n} w_i \mathbf{b}_{i,k}(u)} \quad (t_{k-1} \le u \le t_{n-1}) \tag{4.82}$$

where P_{Ci} are the control points, w_i is the weight, t_K is the knot value, and $\mathbf{b}_{i,k}$ are the kth degree *B-spline* basis functions defined by the Cox (1972) and de Boor (1972) recursion formulas. Again, k is the degree, and n is the number of control points-1. The knot vector components are the same as for *B-splines* and are defined in Eq. (4.75). The number of control points P_{Ci} is $(n + 1)$, and the number of knots t_i in the knot vector is $(k + 1)$. The weights w_i are defined to adjust the curve to exactly represent the specific known curves or any other requirements.

A *NURBS surface* of degree k in the u direction and degree l in the v direction is a bivariate rational function of the form

$$\boldsymbol{P}_{\mathrm{NC}}(u) = \frac{\sum_{i=0}^{n} \sum_{j=0}^{m} w_{i,j}\ \mathbf{b}_{i,k}(u)\ \boldsymbol{P}_{Cij}\ \mathbf{b}_{j,l}(v)}{\sum_{i=0}^{n} \sum_{j=0}^{m} w_{i,j}\ \mathbf{b}_{i,k}(u)\ \mathbf{b}_{j,l}(v)}, \quad (t_{k-1} \le u \le t_{n-1}), (s_{l-1} \le v \le s_{m-1}) \tag{4.83}$$

where the $\boldsymbol{P}_{Cij}$ form a control points net, the $w_{i,j}$ are the weights, and the $\mathbf{b}_{i,k}$ (u) and $\mathbf{b}_{j,l}$ (v) are the *B-spline* basis functions defined on the knot by the vectors $\boldsymbol{t}_i$ and $\boldsymbol{s}_j$. The $\boldsymbol{t}_k$ vector has $(n + 1)$ knots t_i. The $\boldsymbol{s}_k$ vector has $(m + 1)$ knots s_j. The knot values are evaluated similar to Eq. (4.75) for the parameter u with the same k and n. For the parameter v, Eq. (4.75) is modified to replace k by l and replace n by m as depicted in Eq. (4.79).

NURBS curves and *surfaces* have been utilized extensively in so many applications and endeavors. They have also been used to model singular points in connecting different curves and generating ***3D*** models (El-Komy and Metwalli 2014). It is concluded that the singular corner point can be identified with *cubic NURBS*. The optimization identifies the parameters and geometry to ensure any required level of accuracy to represent singular corner solid models and allow a single *cubic* or other *NURBS* representing the whole solid as shown in Figure 4.33. Figure 4.33 shows a single *NURBS* wrapping a shaft (a) having fillets and chamfers and a cube (b) that has the middle point

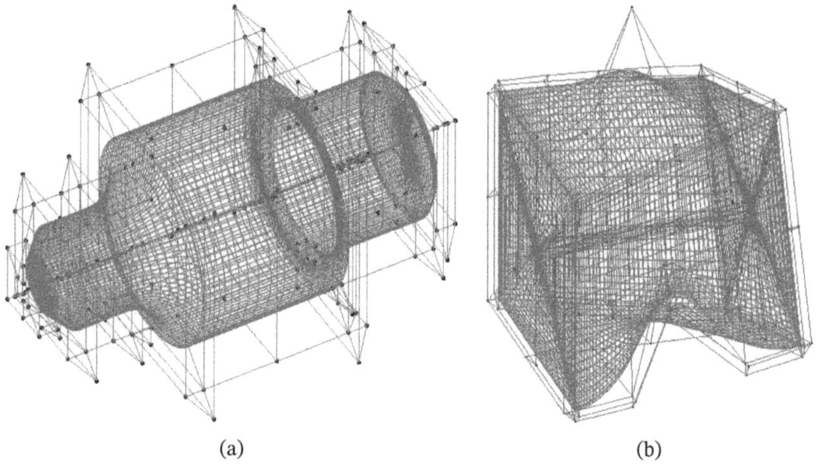

(a) (b)

Figure 4.33 A single *NURBS* wrapping of ***3D*** objects with surface polygons shown: (a) a shaft that has fillets and tapered chamfers and (b) a cube that has the middle point on the top surface pulled out, a middle point on one of its sides pushed up, and one of its faces rotated with certain angle (El-Komy and Metwalli 2014).

on top surface pulled out, a middle point on one of its sides pushed up, and one of its faces rotated with certain angle. Other applications to several ***3D*** solid models are also used to verify such a technique.

4.2.6 Intersections

While modeling compound designs, interaction between the basic sub-objects depends on the intersection between these objects. The intersections create the boundaries between the objects. These boundaries are defined by the intersection of any two lines on the two interacting basic objects, the intersection of a line on one with a plane on the other, intersection of two planes on both objects, and may be intersection of three planes on three interacting objects. Finding these intersections is, therefore, of paramount importance in defining the boundaries between sub-objects in a compound design. This section is dedicated to identifying these basic intersections to create shared boundaries.

4.2.6.1 Intersection of Two Lines

Intersection of two *lines* should always occur in a plane containing these lines. If one considers the plane to be coincident within the *x–y* plane, one can find the intersection by a procedure defined as follows. Consider the equations of the two lines in an *implicit* form as follows:

$$\begin{aligned} a_1x + b_1y + c_1 &= 0 \\ a_2x + b_2y + c_2 &= 0 \end{aligned} \tag{4.84}$$

In a matrix form, one can write that as the simultaneous homogeneous linear equations:

$$[x\ y\ 1]\begin{bmatrix} a_1 & a_2 \\ b_1 & b_2 \\ c_1 & c_2 \end{bmatrix} = [0\ 0] \tag{4.85}$$

Using the *homogeneous* coordinates, however, gives a square matrix that is easier to invert and to find the solution. The matrix equation of the two lines can be rewritten in *homogeneous* coordinates as

$$[x\ y\ 1]\begin{bmatrix} a_1 & a_2 & 0 \\ b_1 & b_2 & 0 \\ c_1 & c_2 & 1 \end{bmatrix} = [0\ \ 0\ \ 1] \tag{4.86}$$

Or

$$\boldsymbol{P}_i\,\boldsymbol{B}_L = \boldsymbol{0}_H \tag{4.87}$$

By inverting the lines matrix $\boldsymbol{B}_L$ and multiplying, one can get the intersection point $\boldsymbol{P}_i$ or

$$\boldsymbol{P}_i = \boldsymbol{0}_H\,\boldsymbol{B}_L^{-1} \tag{4.88}$$

Since the third term in $\boldsymbol{0}_H$ is *1*, the intersection point can be directly obtained as the third row of $\boldsymbol{B}_L^{-1}$. For other cases of ***3D*** line intersections, the intersection point will then be the last or fourth row of $\boldsymbol{B}_L^{-1}$.

Example 4.5 The two lines are given by $x+y-1=0$ and $2x-3y=0$. Show that the two lines have a point of intersection $x=3/5$, $y=2/5$.

Solution

The solution can be directly obtained from applying the homogeneous Eqs. (4.86) and (4.87) for the two equations such that

$$[x\ y\ 1]\begin{bmatrix} 1 & 2 & 0 \\ 1 & -3 & 0 \\ 1 & 0 & 1 \end{bmatrix} = [0\ 0\ 1] \tag{a}$$

The inverse of the matrix in (a) gives

$$B_L^{-1} = \begin{bmatrix} 3/5 & 2/5 & 0 \\ 1/5 & -1/5 & 0 \\ 3/5 & 2/5 & 1 \end{bmatrix} \tag{b}$$

The third row of this inverted matrix is the point of intersection *(3/5, 2/5).*

Intersection of 3D Parametric Lines

The intersection of any two ***3D*** or ***2D*** lines can also be obtained by using the *parametric* form of the two lines represented by

$$\begin{aligned} \boldsymbol{P}_{L1}(t) &= (t\,(\boldsymbol{P}_2 - \boldsymbol{P}_1) + \boldsymbol{P}_1) \\ \boldsymbol{P}_{L2}(s) &= (s\,(\boldsymbol{P}_4 - \boldsymbol{P}_3) + \boldsymbol{P}_3) \end{aligned} \tag{4.89}$$

where t and s are parameters defining the location of points on the lines. At the point of intersection, one has $\boldsymbol{P}_{L1}(t) = \boldsymbol{P}_{L2}(s)$. Therefore, equating the equations of the two lines, one gets

$$t\,(\boldsymbol{P}_2 - \boldsymbol{P}_1) - s\,(\boldsymbol{P}_4 - \boldsymbol{P}_3) = (\boldsymbol{P}_3 - \boldsymbol{P}_1) \tag{4.90}$$

In a matrix form, one can rewrite Eq. (4.90) to get

$$\begin{bmatrix} t & s \end{bmatrix} \begin{bmatrix} \boldsymbol{P}_2 - \boldsymbol{P}_1 \\ \boldsymbol{P}_4 - \boldsymbol{P}_3 \end{bmatrix} = [\boldsymbol{P}_3 - \boldsymbol{P}_1] \tag{4.91}$$

Using the *homogeneous* coordinates, one should get a square matrix that is easier to invert.

$$\begin{bmatrix} t & s & 1 \end{bmatrix} \begin{bmatrix} \boldsymbol{P}_2 & -\boldsymbol{P}_1 & 0 \\ \boldsymbol{P}_4 & -\boldsymbol{P}_3 & 0 \\ 0 & 0 & 1 \end{bmatrix} = \begin{bmatrix} \boldsymbol{P}_3 & -\boldsymbol{P}_1 & 1 \end{bmatrix} \tag{4.92}$$

It should be noted that in ***2D***, all the points $\boldsymbol{P}_1, \boldsymbol{P}_2, \boldsymbol{P}_3,$ and $\boldsymbol{P}_4$ are ***2D*** position vectors in the *x*–*y* plane that means

$$\begin{bmatrix} t & s & 1 \end{bmatrix} \begin{bmatrix} x_2 - x_1 & y_2 - y_1 & 0 \\ x_4 - x_3 & y_4 - y_3 & 0 \\ 0 & 0 & 1 \end{bmatrix} = \begin{bmatrix} x_3 - x_1 & y_3 - y_1 & 1 \end{bmatrix} \tag{4.93}$$

By inverting the matrix and multiplying, one gets the intersection values for t and s. Substituting the t value into $\boldsymbol{P}_{L1}(t)$ and s value into $\boldsymbol{P}_{L2}(s)$, if all corresponding values of x and y are the same, the two lines intersect. If not, the lines do not physically intersect. Next, if both t and s have a value from *0* to *1*, the intersection point is between $\boldsymbol{P}_1$ and $\boldsymbol{P}_2$, and $\boldsymbol{P}_3$ and $\boldsymbol{P}_4$.

If all points $\boldsymbol{P}_1, \boldsymbol{P}_2, \boldsymbol{P}_3,$ and $\boldsymbol{P}_4$ are in ***3D*** space, one can use only x and y components to find t and s. The values of both z-coordinates after substituting t and s in the two lines should be the same for the intersection to be a point. If not, the location of the two points on $\boldsymbol{P}_{L1}$ and $\boldsymbol{P}_{L2}$ gives the location of the minimum distance between the two points. A simple MATLAB code is available in the ***Wiley website*** to get the intersection of any two ***3D*** nonparallel lines. The name of this MATLAB code is ***`CAD_Line_Line_Intersection.m`***.

Example 4.6 Two lines are defined by two points each. Line one is between $\boldsymbol{P}_1$ = (100, 60, 60) and $\boldsymbol{P}_2$ = (40 80 30). Line two is between $\boldsymbol{P}_3$ = (80, 50, 40) and $\boldsymbol{P}_4$ = (−10, 20, 50). Use the available MATLAB code (`CAD_Line_Line_Intersection.m`) to find the intersection of the two lines and their plotting figure.

Solution

Figure 4.34a shows the MATLAB code plot of the two lines and the shortest distance between them. The two lines are not in the same plane, and the solution provides the minimum normal distance between them. The solution also shows that the shortest distance is off the second line.

4.2.6.2 Intersection of a Line with a Plane

The intersection point of a line defined by two points $\boldsymbol{P}_1$ and $\boldsymbol{P}_2$ and a given plane that is defined by a point $\boldsymbol{P}_0$ and a normal $\boldsymbol{N}_P$, is given by

$$t_I = -\frac{(\boldsymbol{N}_P \cdot (\boldsymbol{P}_0\boldsymbol{P}_1))}{(\boldsymbol{N}_P \cdot (\boldsymbol{P}_1\boldsymbol{P}_2))} \tag{4.94}$$

where t_I is the intersection parameter. The vectors $(\boldsymbol{P}_0\ \boldsymbol{P}_1)$ and $(\boldsymbol{P}_1\ \boldsymbol{P}_2)$ are simply obtained by

$$\begin{aligned}(\boldsymbol{P}_0\boldsymbol{P}_1) &= \boldsymbol{P}_1 - \boldsymbol{P}_0 \\ (\boldsymbol{P}_1\boldsymbol{P}_2) &= \boldsymbol{P}_2 - \boldsymbol{P}_1\end{aligned} \tag{4.95}$$

The intersecting point $\boldsymbol{P}_I$ can be found from the *parametric* equation of the line by substituting t_I value. If the parameter value is found in the interval $0 \le t_I \le 1$, the intersection point on the plane is on the segment from $\boldsymbol{P}_1$ to $\boldsymbol{P}_2$. If not, the intersection point is on the extended line. A simple MATLAB code is available in the ***Wiley website*** to get the intersection of any ***3D*** line with any nonparallel ***3D*** plane. The name of this MATLAB code is ***`CAD_Plane_Line_Intersection.m`***.

Example 4.7 A line and a plane are to be investigated as to their intersection. The line one is defined to be between $\boldsymbol{P}_1 = [5\quad 4\quad 4\quad 1]$ and $\boldsymbol{P}_2 = [2\quad 3\quad -2\quad 1]$. The plane is defined by the three points $\boldsymbol{P}_3 = [4\quad -1\quad 2\quad 1]$; $\boldsymbol{P}_4 = [-2\quad 3\quad 1\quad 1]$; and $\boldsymbol{P}_5 = [3\quad 1\quad -2\quad 1]$.

Use the available MATLAB code (`CAD_Plane_Line_Intersection.m`) to find the intersection of the line with the plane and their plotting figure.

Solution

Figure 4.34b shows the plot of the MATLAB code for the intersection of the line with the plane. The plot shows that the defined intersection point (a square) on the plane has to be on the line extension. The three points on the plane are also indicated.

4.2.6.3 Intersection of Two Planes

The intersecting line $\boldsymbol{L}_I$ of two planes $\boldsymbol{S}_1$ and $\boldsymbol{S}_2$ must be normal to both plane normals $\boldsymbol{N}_1$ and $\boldsymbol{N}_2$ or it should be parallel to

$$\boldsymbol{L}_I = \boldsymbol{N}_1 \times \boldsymbol{N}_2 \tag{4.96}$$

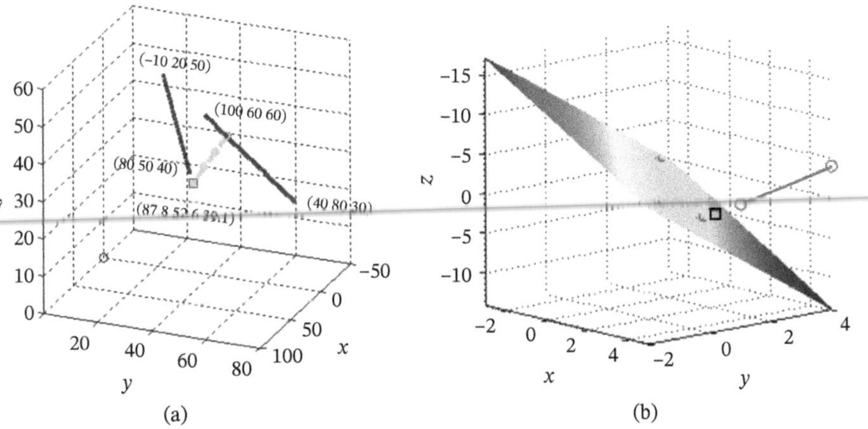

Figure 4.34 Intersection of the two lines of Example 4.6 and their plotting in (a) using the MATLAB code. In (b), the intersection of a line with a plane of Example 4.7 and the plotting showing the intersection point on the plane to be on the line extension.

To uniquely specify the line, it is also necessary to find a point that is simultaneously located on both planes, i.e. a point $\boldsymbol{P}_0$ that also simultaneously satisfies

$$\boldsymbol{P}_0 \cdot \boldsymbol{S}_1 = 0 \quad \text{and} \quad \boldsymbol{P}_0 \cdot \boldsymbol{S}_2 = 0 \tag{4.97}$$

It can be obtained by the intersections of any of the two plane edges (lines) with the line $\boldsymbol{L}_I$.

In parametric form, the intersecting point $\boldsymbol{P}_L$ on the line is then

$$\boldsymbol{P}_L = \boldsymbol{P}_0 + t\ \boldsymbol{L}_I \tag{4.98}$$

The *cross product* in Eq. (4.96) can be expressed as the product of the skew-symmetric matrix of components of $\boldsymbol{N}_1$ with $\boldsymbol{N}_2$ or as Eq. (2.6).

$$\boldsymbol{L}_I = \begin{bmatrix} 0 & -N_{1,z} & N_{1,y} \\ N_{1,z} & 0 & -N_{1,x} \\ -N_{1,y} & N_{1,x} & 0 \end{bmatrix} \begin{bmatrix} N_{2,x} \\ N_{2,y} \\ N_{2,z} \end{bmatrix} \tag{4.99}$$

The dot product of $\boldsymbol{P}_0$ and $\boldsymbol{S}_1$ and $\boldsymbol{P}_0$ and $\boldsymbol{S}_2$ is

$$\boldsymbol{P}_0 \cdot \boldsymbol{S}_1 = \boldsymbol{P}_0{}^T \boldsymbol{S}_1 \ \text{and}\ \boldsymbol{P}_0 \cdot \boldsymbol{S}_2 = \boldsymbol{P}_0{}^T \boldsymbol{S}_2 \tag{4.100}$$

With these relations, one can identify the point $\boldsymbol{P}_0$. A simple MATLAB code is available in the ***Wiley website*** to get the intersection of any two ***3D*** nonparallel planes. The name of this MATLAB code is ***`CAD_Plane_Plane_Intersection.m`***.

Example 4.8 Two planes are each defined by three points. Plane one is defined by $\boldsymbol{P}_1 = [4\ -1\ 2\ 1]$, $\boldsymbol{P}_2 = [-2\ 3\ 1\ 1]$, and $\boldsymbol{P}_3 = [3\ 1\ -2\ 1]$. Plane two is defined by $\boldsymbol{P}_4 = [5\ 4\ 4\ 1]$, $\boldsymbol{P}_5 = [2\ 3\ -2\ 1]$, and $\boldsymbol{P}_6 = [3\ 1\ -2\ 1]$.

Use the MATLAB code (`CAD_Plane_Plane_Intersection.m`) to find the line intersection of the two planes and their plotting figure.

Solution

Figure 4.35 shows the plot of the MATLAB code for the two planes and their intersecting line.

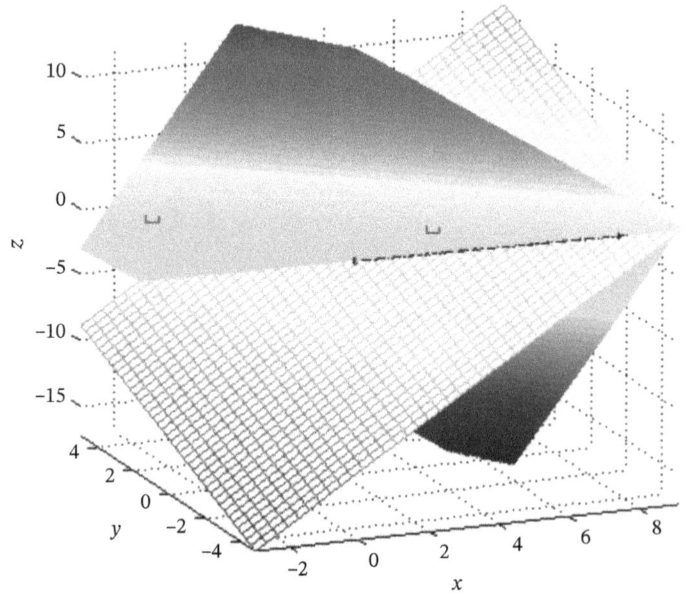

Figure 4.35 Intersection of the two planes of Example 4.8 and their plotting using the MATLAB code.

4.2.6.4 Intersection of Three Planes

Three planes $\boldsymbol{S}_1$, $\boldsymbol{S}_2$, and $\boldsymbol{S}_3$ may intersect at a point $\boldsymbol{P}_I$. To find it, one can write

$$\boldsymbol{P}_I \begin{bmatrix} S_{1,x} & S_{2,x} & S_{3,x} & 0 \\ S_{1,y} & S_{2,y} & S_{3,y} & 0 \\ S_{1,z} & S_{2,z} & S_{3,z} & 0 \\ d_{N1} & d_{N2} & d_{N3} & 1 \end{bmatrix} = [0 \quad 0 \quad 0 \quad 1] \tag{4.101}$$

The components d_{N1}, d_{N2}, and d_{N3} are the normalized parameters with respect to each plane normal. If the matrix containing the plane vectors can be inverted, the desired point $\boldsymbol{P}_I$ is given by the fourth row of the inverse. If the planes do not intersect in a point, the inverse does not exist.

4.3 CAD Hardware and Software

From Sections 4.1 and 4.2, it is apparent that ***3D*** modeling and manipulation requires so much of processing power to do the calculations and viewing. In the initial stage of this process, special hardware and software has been required. Dedicated computers such as *Workstations* or even *Mainframes* have been needed for that. Special graphics cards were required to do viewing and processing of objects, their needed simulation, and animation. Specially dedicated graphics software languages have been developed to perform such tasks. With further advancements of *PC* hardware and needed requirements of computer games, APIs such as *OpenGL* and *ActiveX* have been developed with dedicated and able graphics cards. These capable hardware and software have migrated to laptops, tablets, and even cell phones. The introduction of ***VR*** to such common equipment allowed more capabilities to be utilized. This can deliver interactive ease, construction efficiency, and presentation proficiency; see ***VR*** Section 3.4. Cursors in ***3D***, tactile gloves, voice, and possible facial or eye tracking can further advance these capabilities.

4.4 Rendering and Animation

Rendering and *animation* of bodies intends to generate "*photo-realistic*" or photograph-like quality for each scene of the body and environment. This task would attempt to mimic the photographic objects such as those shown in Figure 4.36 for three shiny balls in (a) and shiny and translucent balls in (b) with some different environment. The *specular*, *diffused*, *reflections*, *transparency*, and *shadows* are evident. In attempting to mimic that, complexity due to many surface identities, textures, color shifts, shadows, reflections, and slight irregularities presents the

(a) (b)

Figure 4.36 Pictures of three shiny balls in (a) and shiny and translucent balls in (b) with some different environment. The *specular*, *diffused*, *reflections*, *transparency*, and *shadows* are evident. Source: Sayed Metwalli.

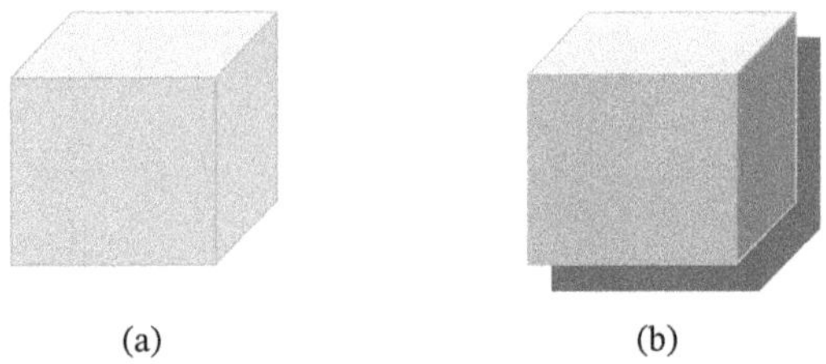

Figure 4.37 Light from all directions giving *constant illumination* as shown in (a), where cube edges are only highlighted to differentiate between surfaces. The light source illuminating the surface depends on its orientation to incident light rays, where surfaces facing the light are brightly illuminated as in (b). However, the shading in (b) is *constant shading* as the simplest model of shading excluding the shadow.

many difficulties faced by the graphics system to produce proper *rendering*. Computational costs for each frame or scene can therefore be high. It requires the computation of color and intensity for millions of pixels. Therefore, the high-quality-like pictures would usually take a lot of computer processing time for each frame or scene.

The perception of a surface hinges on the types of light sources illuminating the object. It also depends on the properties of the surface in terms of color, texture, and reflectance. The other factors are the position and orientation of the surface with respect to the light source or sources, and other surfaces on or near the body. Light source can be the *ambient light* from all directions. This gives a *constant illumination* as shown in Figure 4.37a, where cube edges are only highlighted in this figure to differentiate between surfaces. Alternatively, a light source that illuminates the surface should produce surface illumination that hinges on its orientation to incident light rays. The surfaces that face the light rays are supposed to be brightly illuminated as demonstrated in Figure 4.37b. However, the shading in Figure 4.37b – excluding the shadow – is a *constant shading*. This is the simplest model of shading, where each surface has a single tone or illumination intensity. To approach a little realism, constant shading is much better than constant illumination. This realism is somewhat improved by reproducing shadows created by the light source as the case in Figure 4.37b. The shadow in Figure 4.37b is nevertheless not realistically evaluated.

Objects illuminated with only a point light source tend to look unrealistically harsh. Light illumination energy drops off as the inverse square of the distance r_S from the source to the surface and back to the eye. More realistic effects are achieved by replacing the $1/(r_S{}^2)$ light source with $1/(r_S + k_S)$, where k_S is a constant and r_S is the distance from the perspective viewpoint to the surface. *Diffuse reflection* of light is a subtractive process. For a yellow surface, all rays of *yellow* light are reflected, while all incident *magenta* and *cyan* light rays are absorbed. *Specular reflection* is observed on shiny surfaces such as the *marbles* represented in Figure 4.36. For round surfaces, as a marble, the intensity of the reflected light falls off as the angle increases. ***Phong*** approximates this fall off with a ($\cos^{np}$) function. The value of *np* varies from 1 to 200 depending on the surface (*Phong* 1975). For a perfect reflector, *np* has an infinite value.

The *specular* transmission of light occurs through transparent materials, such as glass or polished Lucite; see Figure 4.36. *Diffuse* transmission occurs in *translucent* materials like frosted glass or plastic. In general, a ray of light striking a surface breaks up into three parts, the *diffused reflected* rays, *specularly* reflected rays, and *transmitted* (or *refracted*) rays as shown in the photos of Figure 4.36. Light leaving the surface of an object is the sum of contributions from the three sources. Further treatment of that is presented in Section 4.4.5.

To enhance realism one should also reproduce the surface texturing properties. Some surfaces are dull and disperse reflected light in many directions, while others reflect light only in certain directions. Shadows for point light sources are also adding to realism. The details of the main factors of defining shading and rendering of bodies are discussed in the following sections.

4.4.1 Realistic Presentations

Realistic graphics started with *hidden* surface (or line) removal. Alternatively, it may include the hidden line as a dashed or different lightly colored style such as demonstrated in Figures 4.7, 4.13, and 4.14. This practice was used by early draftsmen for better appearance and perception. *Shading* is applied after computers were able to do more graphics. *Rendering* is comprehensively used nowadays in addition to ***VR*** for more realistic appearance.

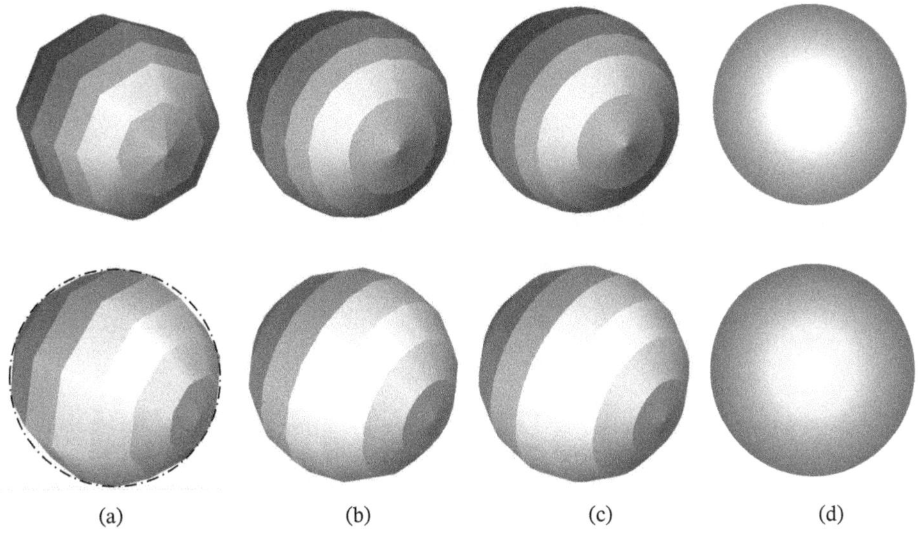

Figure 4.38 Gouraud-shaded bodies with fewer divisions produce poor behavior of the specular highlight shown at two orientation angles. An original sphere in (a) has eight longitudinal slices with each slice radially divided into eight polygons. In (b) and (c), higher division count (16 and 32) produces better rendered behavior in each division. The last sphere in (d) has much more slices and divisions and better smoothing.

Shading and *rendering* hinge on light sources illuminating the ***3D*** object. This is implemented by using a single light or a group of light sources. The surface properties such as color, texture, and reflectance define object shading and rendering. Object location and orientation with respect to light sources and relative to other objects affect the realism of shading and rendering; see, for instance, Foley et al. (1990) or Hill (2001).

For *polygon* or *triangular* mesh *shading*, three main techniques have been used. Constant or *flat shading* is one that has a single intensity value used for the entire polygon, *triangle,* or flat surface; see Figure 4.37b. Intensity interpolation shading or *Gouraud* shading uses linear interpolation of *vertex intensities* along each edge and then between edges along the scan lines inside each *triangle* of the surface (Gouraud 1971). *Triangle* vertex intensities are found from *vertex normals*, which are found by averaging *polygon normals* (Eq. (4.1)) on each side of an edge.

Rendering, however, is widely spread with stronger computer graphics capabilities that are currently available. The main factors affecting realistic rendering procedures are the *specular* reflections, the *diffused* reflections, the *transmitted* refraction (transparent and translucent materials), and the *shadows* calculations. *Gouraud*-shaded bodies such as the sphere in Figure 4.38 with fewer divisions produce poor behavior of the specular highlight. An original sphere has eight longitudinal slices with each slice radially divided into eight polygons at two orientation angles is shown in Figure 4.38a. To the right of the original sphere, a higher division count (16 and 32) produces better rendered behavior in each division as shown in Figure 4.38b,c. The last sphere to the right, Figure 4.38d, has much more slices and divisions and better smoothing.

Interpolating *polygon* surface normal vectors between starting and ending normals is used in *normal vector shading*. These normals are obtained from vertex normals. The utility of normal vector technique provides some improvement over intensity interpolation. The technique reduces the apparent peaks between polygons and slices of *Gouraud* shading demonstrated in Figure 4.38. The technique, however, increases the cost of calculation time to apply this shading model. When the light is white, the ambient and diffuse colors are both the same, and the specular color is white, *Phong* rendering should then produce far better realistic scenes. In *Phong* shading, one linearly interpolates a normal vector across the surface of the polygon from the polygon's vertex normals (Phong 1975). This is usually more involved than *Gouraud* shading. Most graphics engines and APIs, however, are using *Phong* shading to generate better rendering.

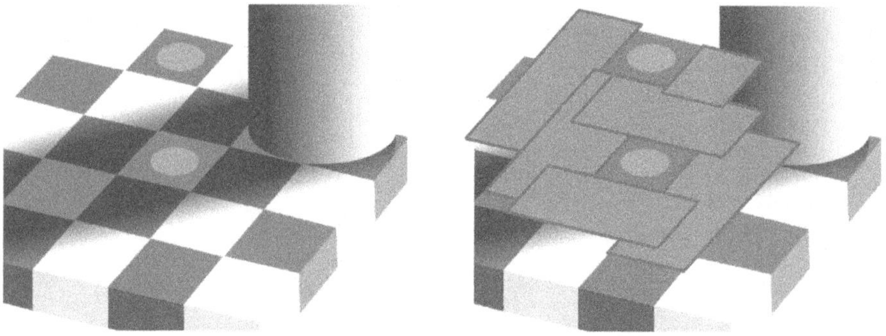

(a) (b)

Figure 4.39 The upper small gray-filled circle and the lower gray-filled circle in the checker bed in (a) have the same color and are in identical gray checker surroundings. One can perceive the colors differently. Isolating the two squares from the rest of the figure using gray strips as in (b), it is clearly apparent that the colors are the same.

Perception is extensively present in the interpretation of graphics. The upper small gray-filled circle and the lower gray-filled circle in the checker bed of Figure 4.39a have the same color and are in identical gray checker surroundings, adapted by Pingstone (2004) from the original work by Adelson (1995). Viewing Figure 4.39a, individuals can perceive the squares and circles as having different interpretation and may comprehend the colors differently. By isolating the two squares and circles from the rest of the figure using gray strips as done in Figure 4.39b, it is clearly apparent that the colors are the same. This highlights the importance of colors, lightings, and shadows in addition to understanding their interrelations, perception, and interpretations.

4.4.2 Color Use

Colors are associated with the visible electromagnetic radiation spectrum. The visible region is very small (wavelength of about 400–700 [nm]). Main *wavelength* ranges of some colors are about: *violet* (380–450), *blue* (450–495), *green* (495–570), *yellow* (570–590), *orange* (590–620), and *red* (620–750) with all values in nanometers [nm]. Gamma rays of electromagnetic radiation have much shorter wavelength in the range of 10^{-16}–10^{-11} [m]. Long radio waves have a much longer wavelength in the range 10^3–10^8 [m]. The wavelength is a physics quantity, but the color is the *eye* perception of the wavelength. The *visibility* is a physiological human interpretation through the eye photoreceptors or cone cells. Figure 4.40 shows a sketch of roughly normalized human *eye response spectrum* and the fraction of light absorbed by the eye for each color spectrum at different visible wavelengths.

4.4.2.1 Visual Color Description

Color description is defined by several means. It can relate the spectral characteristics of the eye to various colors as shown in Figure 4.40, which is known as the *luminosity* curve. This suggests that colors may be defined by weighted sums of ***red, green, and blue*** (***RGB***). Computer displays the range for color defined by the ***RGB*** primary colors. Few color models are available to produce the desired visual color. Most of these are presently based on ***RGB*** primaries. The major ones are as follows.

RGB model is using a *Cartesian* coordinate space of a cube as shown in Figure 4.41. Each of the primaries represents a coordinate and can be added to contribute to the color outcome. The cube main diagonal has equal magnitude of each primary. The diagonal stands for the gray levels from black to white. Figure 4.41a represents the main ***RGB*** coordinates with the *black* origin, while Figure 4.41b represents the ***complementary RGB*** colors as the coordinates with the *white* origin. The ***complementary RGB*** colors are the ***cyan, magenta, and yellow*** (***CMY***) colors. Any two colors are *complementary* if they produce *white* light when mixed or added.

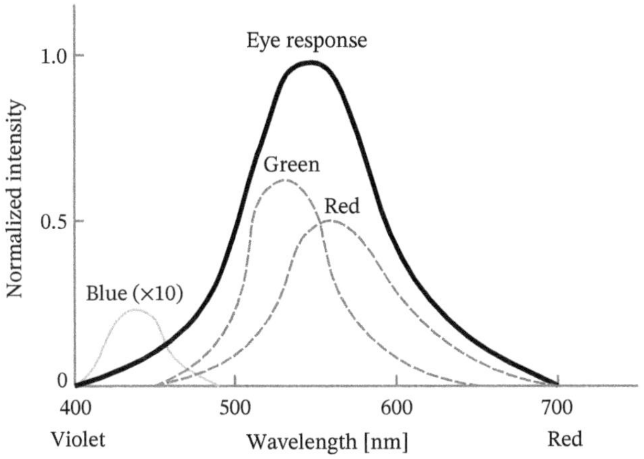

Figure 4.40 A sketch of roughly normalized eye response spectrum and fraction of light absorbed by the eye for each indicated color spectrum at different visible wavelengths (color spectrum in light and dark gray colors identified by the color name).

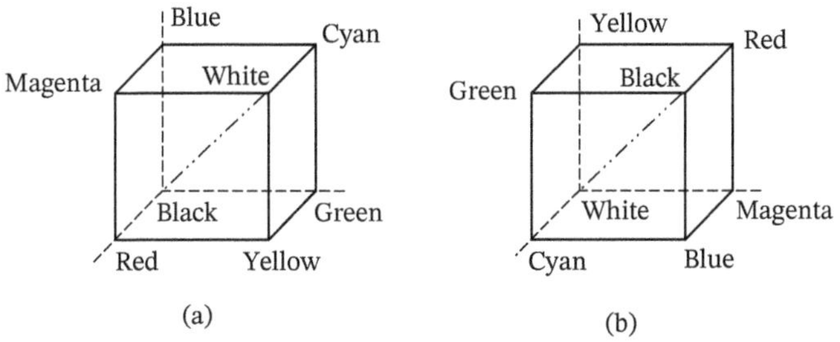

Figure 4.41 The main ***RGB*** (*red, green,* and *blue*) coordinates with the *black* origin is shown in (a). The complementary ***RGB*** colors, which is ***CMY*** (*cyan, magenta,* and *yellow*) as the coordinates with the *white* origin is shown in (b).

Both *RGB* and *CMY* are hardware color. The *RGB* represents an ***additive*** light waves system, where adding or mixing *red* light beams to *blue* light beams produces *magenta* color beam as shown in Figure 4.42a (in light and dark gray colors identified by letters R, G, B, or C, M, and Y). The *yellow* light is formed of *red* plus *green* rays of light. *Cyan* color is obtained by mixing *green* and *blue* light waves. The *white* color is obtained by mixing the *RGB* rays. On the other hand, the *CMY* represents a ***subtractive*** color system as shown in Figure 4.42b. *CMY* is used in hard copy devices. The ink jet printers deposit colored inks or pigmentation onto surfaces. The pigment color comes from absorbing all light waves except the color of the pigment. That is why mixing pigments would be a subtractive process. The *red* color is produced by mixing *yellow* pigments with *magenta* pigments as shown in Figure 4.42b. The *green* color is generated by mixing *yellow* and *cyan* pigments. The *blue* color is the result of mixing *magenta* and *cyan* pigments. The black color is expected to be generated by mixing the *CMY* pigments as shown in Figure 4.42b.

In digital evaluations for 0–255 (8 bits), the *RGB* of (255, 0, 0) is for the *red* with *green* and *blue* values of zero. Similarly, the values of *RGB* of (0, 255, 0) and (0, 0, 255) are for *green* and *blue* colors, respectively. *Cyan*, which is the *complement* of the red, has an *RGB* of (0, 255, 255). This means that red (255, 0, 0) plus cyan (0, 255, 255) gives (255, 255, 255), which is white. *Magenta* as the *complement* of green has an *RGB* of (255, 0, 255). *Yellow* has an *RGB* of (255, 255, 0), which is the *complement* of blue. Any other color can be generated from values of *RGB* with values from 0 to 255 for each of *RGB*. The *gray* is about (127, 127, 127) or (128, 128, 128), which is halfway between the white (255, 255, 255) and the black (0, 0, 0). The distinctive *silver* is about (192, 192, 192). *Gold* is about *RGB* of

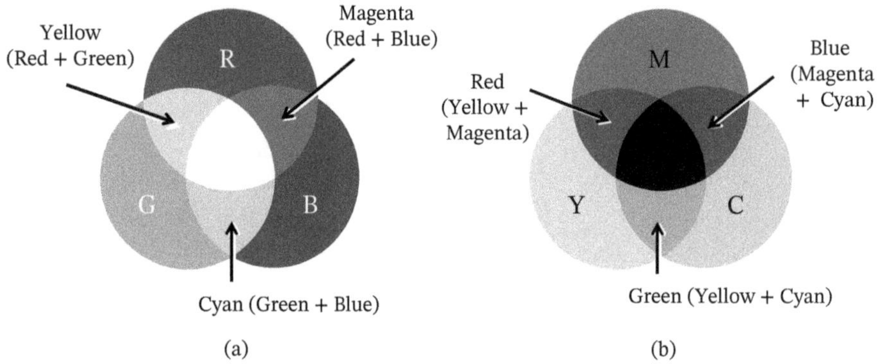

Figure 4.42 The ***RGB*** is an *additive* light waves system as shown in (a). Adding or mixing *red* light beams to *blue* light beams produces *magenta* color beam, and adding ***RGB*** rays produces the white color. ***CMY*** in (b) is a *subtractive* color system. It is used in hard copy devices, e.g. *red* color is produced by mixing *yellow* pigments with *magenta* pigments (colors in light and dark Gray colors identified by letters R, G, B, or C, M, and Y).

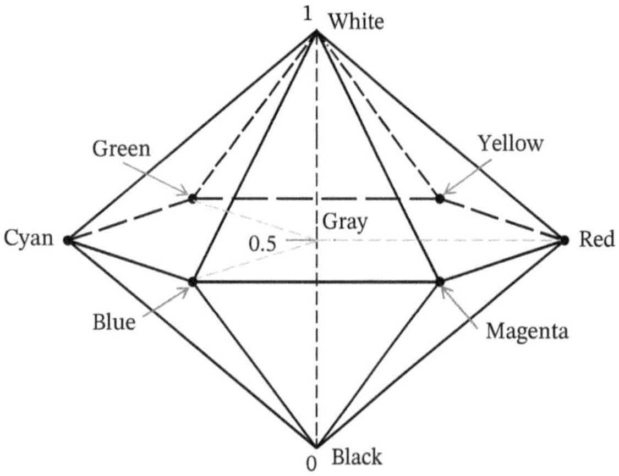

Figure 4.43 Hue, saturation, and luminance (HSL) as properties that generally provide the color descriptions of bodies. It is modeled as a double hex-cone.

(255, 215, 0) while *metallic gold* can be of about (212, 175, 55). However, the realism of these materials is greatly affected by light sources, shininess, and reflectivity.

Hue, saturation, and luminance (HSL) are characteristics that commonly describe the color of bodies. It is described by a double hex-cone, as demonstrated in Figure 4.43. It may be imagined as looking at the *RGB* cube of Figure 4.41a along its diagonal line from white to black corners. This is also forcing or projecting the *RGB* and the *CMY* to be on one plane. That plane lays normal to the line (0–1) linking the black at (0) to the white at (1). The plane is crossing the line at midpoint, i.e. at the *gray* color (0.5). On that plane, *hue* uses a right-hand angle rotation for pure colors from red to yellow, to green to cyan, to blue to magenta, and back to red again. *Saturation* is increasing radially from the gray to the pure color. The amount of white in the color relative to a zero for black is the *luminance* for any color, i.e. moving the mid-plane up for whiter color or down for blackish color.

Hue differentiates between different colors such as actual *RGB*. Hue uses the right-hand angle on the plane where hue at the angle zero starts as *red*. Hue of 60° is the yellow. Hue of 120° is the green color. Hue of 180° is the *cyan*, which is the complement of red. Hue value increases to represent blue at (240°), magenta at (300°), and goes back to (360° or 0) for the red color. The *hue range* of zero to the 360° is taken by some systems as a range of 0–239, which is in steps of 1.5°. Cyan is then (180°/1.5) or digitally has a *hue* of 120 for that system. The usual *hue* range is normally 0–255.

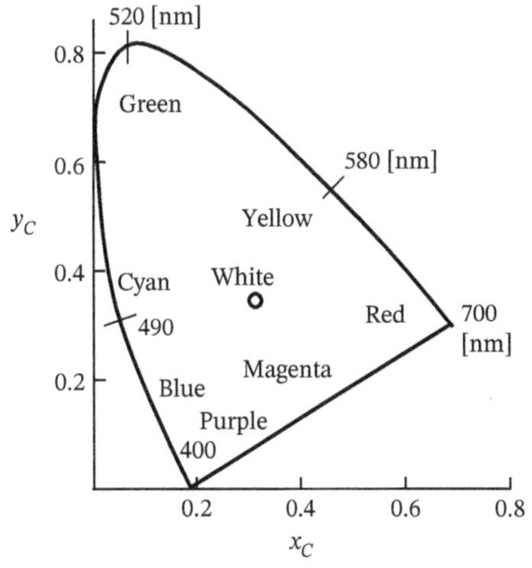

Figure 4.44 *CIE* chromaticity diagram of x_C and y_C for visible chromaticity region. Pure colors fall on the curved part of the boundary, where the wavelengths are indicated in [nm]. White light is marked by the center circular mark close to $x_C = y_C = z_C = 1/3$.

Saturation is the purity or vividness from dilution by white (or intense versus dull). Pastel colors have low saturation or more diluted by white. In digital values for some systems, the saturation value starts radially at the *gray* saturation value of 0 to the vivid color of 240 for full red or green or blue colors at the corners of the hexagonal shown in Figure 4.43 or any other color at the edges of the hexagonal. The usual *saturation* range is normally 0–255.

Luminance, *lightness*, or brightness refers to the intensity of color and how much white or black in the color. Lightness refers to reflecting objects and *brightness* refers to self-luminous objects or sources. In digital values for some systems, the *luminance* value starts at the *black* color value of 0 to the white color of 240, instead of 1 in Figure 4.43. The midpoint of *gray* for such a system has a luminance of 120, and therefore all primaries *RGB*, *CYM*, or any color on the midpoint plane has a luminance of 120. A pure *red* has a *hue* of (0), a *saturation* of (240), and a *luminance* of (120). A very whitish red (a pinkish color) has a hue of (0), a saturation of (240), and a luminance of, say, (230). This sets the *RGB* of the pinkish color to (255, 234, 234), which indicates the addition of a large green and blue close to the 255 for white. The usual luminance range is normally 0–255.

4.4.2.2 Color Specification System

As indicated previously, Figure 4.40 for luminosity shows eye's response to various colors at constant luminance light as dominant wavelength varied. Since the human eye is less sensitive to blue, the matching of colors may require some negative weight on the sensitivity axis in the addition of the primary *RGB* components. Commission Internationale L'Eclairage (*CIE*) (1931) defined non-visible primary colors (X, Y, Z) that can be combined to define all experienced light sensations. The primary colors (X, Y, Z) are spectral energy distributions that depend only on wavelength and saturation. Y is defined with matched energy distribution to the luminosity response of the eye. *CIE* chromaticity diagram of x_C and y_C is shown in Figure 4.44 for visible chromaticity region. The x_c and y_c are functions of primary (X, Y, Z) colors. Since by definition, $x_C = X/(X+Y+Z)$, $y_C = Y/(X+Y+Z)$, $z_C = Z/(X+Y+Z)$, and $x_C + y_C + z_C = 1$, one needs to plot only the projection on $x_C - y_C$ plane, which is shown in Figure 4.44.

In *CIE* chromaticity, the 100% pure colors of the spectrum fall on the curved part of the boundary (Figure 4.44), where the wavelengths are indicated in [nm]. *White* light is marked by the center circular mark close to $x_C = y_C = z_C = 1/3$. When two colors are added together, the new color lies on a straight line in the chromaticity diagram connecting the two colors being added. Because the diagram factors out luminance, color sensations that are luminance related are excluded. Therefore, brown, which is an orange red chromaticity at very low luminance, is not shown. As indicated previously, *complementary* colors are those that can be mixed to produce *white* light. The chromaticity in the mixture is the ratio of line lengths. The line has to pass by the white color circular mark shown in Figure 4.44. *CIE* improved chromaticity in 1976 provided a more accurate chromaticity

(ISO/CIE (2016)). Chromaticity is usually used to define and compare color displays and hard copy devices. Since no great involvement with that is intended in this text, the details of this matter and more related ones can be found in *computer graphics* and other pertinent references.

4.4.3 Shading and Rendering Technique

This section presents a treatment of few technical details of shading and rendering process. It intends to provide some mathematical treatment to grasp the extent of the calculation performed in the graphics software or better hardware. Calculations are usually performed in hardware of the graphics cards. If not, software is utilized but with slower results.

4.4.3.1 Methods of Shading a Polygon or a Triangle

Constant shading has no interpolation, as each *pixel* contained in a *polygon* has the same color value. It is the fastest shading technique, but lacks in realism perceptiveness. *Gouraud* or intensity interpolation varies the pixel values and smooths out the polygons from vertices intensity so that they appear as a part of the surface. It takes about 10% longer than constant shading. *Phong*, or normal vector interpolation, breaks the pixel value into two parts, each of which is linearly interpolated across the face of the *polygon*. It takes much longer than *Gouraud* shading.

Look-up tables (LUT) may be used to speed up the shading or coloring of each *pixel* of the viewing screen. As an example, *LUT* linearly interpolates a 12-bit intensity *pixel* value to change only the *shade* when interpolating across an object of one color. To achieve a *smooth* ***3D*** effect, 64 shades would be needed for each of the colors. The *LUT* uses a 12 bits wide by 8 (bit planes in frame buffer), which has 2^{12} or 4096 simultaneous colors. Dividing 4096 by 64 shades yields the 64-word-wide blocks in the table with the first used for the background. The choice of 64 shades of 64 colors forced the *bits* of each *pixel* to be essentially divided in half, with six bits denoting the *color* and six bits denoting the *shade*. The six *least significant bits* (*lsb's*) of each pixel correspond to the *shade*, and the six *most significant bits* (*msb's*) correspond to the *color*. To interpolate, only the six *lsb's* change, and the six *msb's* remain the same. The six bits corresponding to the *shade* are encoded to be the angle θ_v between the normal of the surface at the *vertex* and the normal to the light source unit direction vector; see Figure 4.45. Note that this angle should determine the shade.

If the *light source* is placed at the viewer, or in the positive z direction as shown in Figure 4.45, the angle θ_v between the vectors ranges from 0° to 90°. This is broken up into 64 different parts, each $90°/64 = 1.4°$ wide. The *LUT* then decoded the six color bits and six shade bits of each pixel into a mix of *RGB* to create the selected color and the correct shade.

Each *polygon* is described by its set of *vertices* at its corners. Any change in orientation of objects, sectioning, or making an object translucent requires that the whole picture be redrawn. The system keeps only the visible surfaces, and the rest of an object is erased or not viewed.

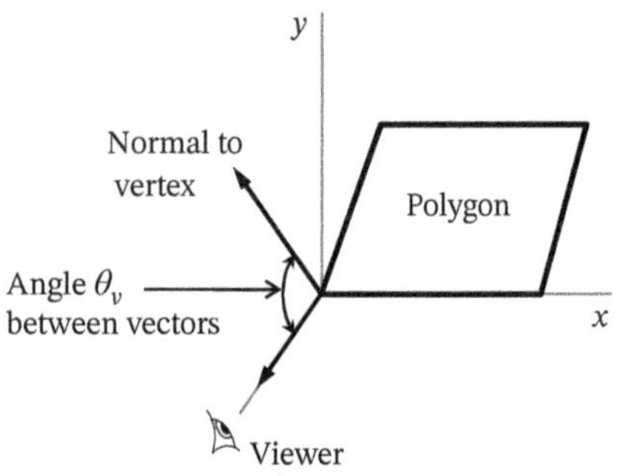

Figure 4.45 If the light source is placed at the viewer, or in the positive z direction as shown, the angle θ_v between the vectors ranges from 0° to 90°.

4.4.4 Computing Vertex and Surface Normals

A *polygon* surface normal is a vector perpendicular to the surface of a planar polygon. A *polygon vertex* normal is a vector perpendicular to the surface of the polygon at that vertex; see Figure 4.46. The polygon vertex normals can be different from the surface normals, as would occur in the case of approximating the surface of a sphere with polygons. Each vertex of a polygon should have the true normal of the sphere at that point. Thus, the true vertex normal can be approximated by averaging all the surface normals of the polygons containing the vertex.

Two vectors in the plane of a *polygon* (x_1, y_1, z_1) and (x_2, y_2, z_2) can be obtained by subtracting two adjacent vertices of the polygon, provided that the vertices 1, 2, and 3 are not all collinear as indicated in Figure 4.46. One then has

$$\begin{aligned}(x_1, \quad y_1, \quad z_1) &= (P_{1x} - P_{2x}, \quad P_{1y} - P_{2y}, \quad P_{1z} - P_{2z}) \\ (x_2, \quad y_2, \quad z_2) &= (P_{2x} - P_{3x}, \quad P_{2y} - P_{3y}, \quad P_{2z} - P_{3z})\end{aligned} \tag{4.102}$$

The ordering of the vertices of the polygon determines the signs of the components of the vectors (x_1, y_1, z_1) and (x_2, y_2, z_2), which in turn determine the signs of the components of the *unit normal vector* (N_{Px}, N_{Py}, N_{Pz}). The signs of the vector components indicate the front- or back-facing surfaces and the different lighting values if the light source is not parallel to the z-axis. A *CCW* polygon order with positive z normal components is facing outward.

Given the calculated vectors within a polygon (x_1, y_1, z_1) and (x_2, y_2, z_2), Figure 4.46, the normal for that polygon can be computed using the following *cross product* formula:

$$\boldsymbol{N}_{P2} = \begin{bmatrix} 0 & -z_1 & y_1 \\ z_1 & 0 & -x_1 \\ -y_1 & x_1 & 0 \end{bmatrix} \begin{bmatrix} x_2 \\ y_2 \\ z_2 \end{bmatrix} \tag{4.103}$$

As indicated previously, the vertex polygon normal for a sphere can be approximated by averaging the surface normals for all polygons containing that vertex. If a cylinder is approximated by polygons, the cylinder should have sharp distinguishable edges at the top and bottom but have a smooth rounded side. If all polygons were averaged to compute the vertex normals, the sharp edges for the top and bottom would be rounded. Thus, the angle between the polygons is used to determine whether polygons sharing a vertex also lie in the same surface. The *dot product* between the surface normals can be used to yield a cosine value that can be compared with a test value. If the magnitude of the cosine is less than the test value, one should average the surface normals.

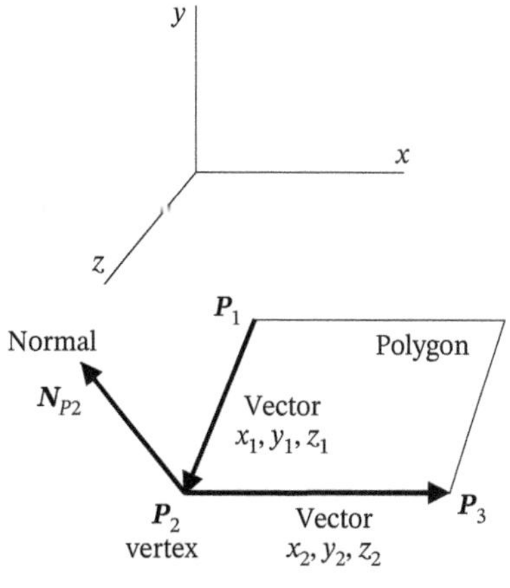

Figure 4.46 Two vectors in the plane of a polygon (x_1, y_1, z_1) and (x_2, y_2, z_2) can be obtained by subtracting two adjacent vertices of the polygon, provided that the vertices 1, 2, and 3 are not all collinear. The normal $\boldsymbol{N}_{P2}$ is the cross product of the two vectors.

Because the normals are usually unit normals, the cosine between the normals is the *dot product* obtained as $(N_{P1,x}\, N_{P2,x} + N_{P1,y}\, N_{P2,y} + N_{P1,z}\, N_{P2,z})$, where $(N_{P1,x}, N_{P1,y}, N_{P1,z})$ and $(N_{P2,x}\, N_{P2,y}\, N_{P2,z})$ are the respective unit normals of the two polygons. For the cylinder, if the arbitrary cosine value is 0 for a 90°, the top and bottom polygon normals would not be averaged with the side normals. The resulting shaded figure would have a smooth shaded side with sharply defined edges at the top and bottom of the cylinder.

4.4.5 Rendering Process

The objective is to create a computer image of the object with the effect of multiple *light sources*, *shadows*, *specular* reflection, and *transparent* surfaces to look realistic. Accurate rendering process takes a lot of computer time due to the required massive calculations of every point (*pixel*) on the surface of the object. Presently, few moments or minutes may be needed for every screen to be *photo-realistically* rendered. The rendering model has two main ingredients, the properties of the surface and properties of the *incident illumination* falling on it. The principal surface property is its reflectance. The surface will appear to be colored if it has different reflectance for light of different wavelengths. If a surface is textured or has a pattern on it, the reflectance will vary with position on the surface. Another surface property is uniform reflection for all directions, called *diffuse illumination*. There may be also a *point source of light* in the scene, and *specular reflections*, or highlights, appear on the surfaces; see Figures 4.36 and 4.47. The body may be *transparent* or *translucent* and would partially show objects that are behind. Finally, the incidence illumination on an object may be partially blocked due to *shadows* or *reflecting* images of neighboring objects.

Rendering models should determine the shade of each point on the surface of an object in terms of a number of attributes. The main attributes reviewed herein are the *diffuse* illumination, the *specular* contributions from one or more light sources, and the *transparency* effect. Other attributes such as *shadows* and more refined considerations are the realm of computer graphics and their references are to be consulted.

4.4.5.1 Diffuse Illumination

The diffuse illumination I_{Pd} coming from a surface point $\boldsymbol{P}_S$ can be given by the form

$$I_{Pd} = K_{Ps}\, I_d \tag{4.104}$$

where I_d is the *incident diffuse illumination* falling on the scene and K_{Ps} is the *reflectance coefficient* at $\boldsymbol{P}_S$. The *reflectance coefficient* takes the range from 0 to 1 depending on the type of surface at that point. For colored surfaces one can use the common *RGB* system, i.e. for each color one should have

$$I_{Pd,R} = K_{Ps,R}\, I_{d,R}, \quad I_{Pd,G} = K_{Ps,G}\, I_{d,G}, \quad I_{Pd,B} = K_{Ps,B}\, I_{d,B} \tag{4.105}$$

It should be noted that diffuse rendering alone does not look very realistic because changing the orientation of a surface does not change its shade.

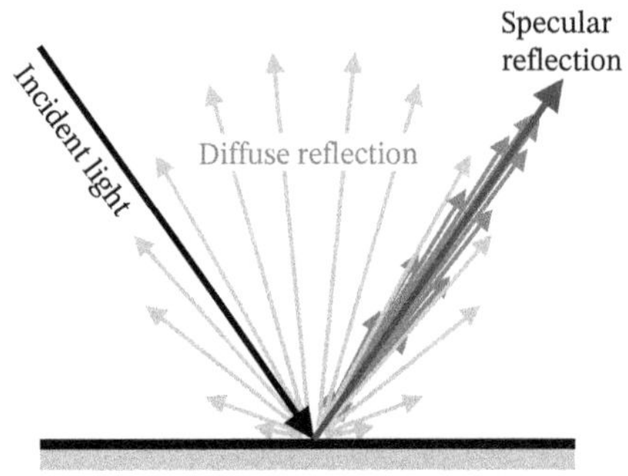

Figure 4.47 Diffuse and specular reflection from a glossy surface for a point source of light.

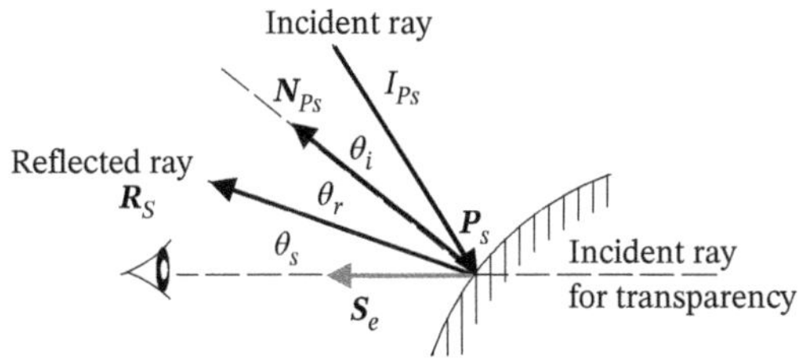

Figure 4.48 A reflected ray $\boldsymbol{R}_S$ varies as the cosine of the angle of incidence in-between the surface normal $\boldsymbol{N}_{Ps}$ at point $\boldsymbol{P}_s$ and the ray of light source I_{Ps}. The angle θ_s between the reflected ray and the observer is shown.

4.4.5.2 Specular Reflection

Rendering due to a specific light source differs according to surface orientation with reference to light source variations. *Lambert's* law, Lambert (1760), states that the illumination falling off a surface in a reflected ray $\boldsymbol{R}_S$ varies as the cosine of the *angle of incidence* in-between the surface normal $\boldsymbol{N}_{Ps}$ at point $\boldsymbol{P}_s$ and the ray of light source (Figure 4.48). If the *incident specular illumination* I_S arriving from the light source to point $\boldsymbol{P}_s$ is reflected in all directions (*diffused*), one has

$$I_{Ps} = (K_{Ps} \cos \theta_i) I_S \tag{4.106}$$

When the *incident angle* θ_i exceeds 90° (i.e. the surface is hidden from light source), one should set I_{Ps} to zero. One should note that the *reflectance angle* θ_r is the same as incident angle θ_i.

Specular reflection requires the calculation of relationship between observer, light source, and the surface as shown in Figure 4.48. The amount of light seen by the observer is determined by a *specular reflection coefficient* $K_S(\theta_s)$ and the angle θ_s between the reflected ray and the observer as defined in Figure 4.48. Then, Eq. (4.106) becomes

$$I_{Ps} = [K_{Ps} \cos \theta_i + K_S(\theta_s) (\cos \theta_s)^{np}] \; I_S \tag{4.107}$$

It should be noted that *specular reflections* have the same color as the illumination. The reflection coefficient $K_S(\theta_s)$ is a function of the incidence angle, as in Figure 4.48, and the *Phong* exponent *np*, which may simply range from 1 to 10. This controls how "*shiny*" the surface would appear. The large value of *np* is for shiny metals. When θ_s is small, the eye should have a *specular reflection* as a strong shiny spot (for highly curved surface) or a large shiny zone (for flatter surface).

For several light sources, Eq. (4.107) would be the summation of the effect of each of the light sources. It should also be clear that I_{Ps} from a point source should decrease as the distance between the light source and the surface point $\boldsymbol{P}_S$ increases. For simplicity, one would assume the source illumination to be constant throughout the scene.

4.4.5.3 Transparency

Transparency or translucency adds a contribution to the scene. If the illumination behind the body I_{Pb} arrives at a point $\boldsymbol{P}_S$ from behind due to its transparency or translucency, the *transmitted illumination* I_{Pt} in the eye line-of-sight direction $\boldsymbol{S}_e$ has the form

$$I_{Pt} = K_{PsT} \, I_{Pb} \tag{4.108}$$

The *transmission coefficient* K_{PsT} (which lies between 0 and 1) governs how much light is transmitted at surface point $\boldsymbol{P}_S$ from behind.

4.4.5.4 Total Rendering Effect

The *total rendered illumination* I_T that arrives at the eye from a surface point $\boldsymbol{P}_S$ is the sum of the illuminations from the previous individual effects. The total rendered illumination is then

$$I_T = I_{Pd} + I_{Ps} + I_{Pt} \tag{4.109}$$

To calculate color lighting, the resulting illumination is computed as color vectors with *RGB* components.

Although the required calculations are simple, they should be performed at every point so many times in the process of generating a rendered image, objects, and scene. The evaluation of cos θ_i and cos θ_s can be carried out efficiently by using the *dot product* of surface normal and light source directions ($\boldsymbol{N}_{Ps}\cdot\boldsymbol{I}_S = \boldsymbol{N}_{Ps}{}^T\boldsymbol{I}_S$) for (cos θ_i) and the *dot product* of the reflected ray and the line-of-sight directions ($\boldsymbol{R}_S\cdot\boldsymbol{S}_e = \boldsymbol{R}_S{}^T\boldsymbol{S}_e$) for (cos θ_s).

4.4.6 *3D* Cursor and Picking

A ***3D*** *cursor* can be smoothly moved around the screen or the scene using some interactive device. Unlike a ***2D*** cursor, the ***3D*** cursor has a depth or *z* value associated with it. This may mean that any portion of the cursor that is behind or inside an object will not be visible. It can be used for defining viewing points and cutting planes. It can be implemented by the identification of the ***3D*** object boundaries (edges and surfaces) projected on the screen and highlighting the object for selection or picking.

4.5 Data Structure

Data structure is essential in communication between systems and software packages. The *CAD* data about a ***3D*** model created in a software package is to be possibly communicated to other software packages. For instance, one needs to communicate the ***3D*** model to a finite element (***FE)*** package to perform stress and deformation calculations. Communicating the ***3D*** model may also be needed in other packages for developing a *CNC* code to produce the ***3D*** model. The data structure can involve a drawing, a wire frame model, a surface model, or a solid model. Other data may also be involved such as material and tolerances. Several data structure file formats have been devised particularly for specific solid modeling or "CAD" geometric modeling software. Many geometric modeling "CAD" packages have their own *proprietary* data structure format for generated ***2D*** models, ***3D*** models, and assemblies. They may change the *graphics kernel/engine* and thus may change the data file structure and its extension. They may also discontinue the use of some of their data formats and the associated extensions. ***AutoCAD***© of Autodesk has (.dwg) and (.dxf) file format extensions. ***Inventor*** of Autodesk previously used (.sat) file format extension ACIS 3D modeling kernel/engine from Spatial Corporation (2017) or used (.ipt) for parts and (.iam) for assembly from ShapeManager© kernel of Autodesk. ShapeManager kernel is a fork of ACIS. ***Fusion 360*** of Autodesk uses (.smt) and (.fbx) with the change of kernel to its newer *Core*© kernel. ***SolidWorks*** from Dassault Systems has the extensions (.sldprt) for part file, (.sldasm) for the assembly, and (.slddrw) for drawing. ***CATIA*** from Dassault Systems has an extension (.CATPart) for parts, (.CATProduct) for assembly, and (.3DXML). ***NX***© or ***Unigraphics****NX* by Siemens© PLM Software used the Parasolid© kernel by Siemens (.x_t) for ASCII or (.x_b) for binary and (.prt) for NX part. ***Solid Edge*** by Siemens used (.prt) for part or (.SLDPRT) and (.asm) for assembly. ***Creo*** (formerly Pro/ENGINEER©) from PTC used the extension (.neu) for neutral files; (.prt), (.XPR), (.ASM), (.XAS), and (.DEX) for design exploration; (.ISDX) for interactive surface design extension; and other extensions for other functions. These file extensions of different packages are mainly to be for the package utility while using the software. They do not usually intend to transfer to other packages.

Some of the "CAD" packages open or exports ***3D*** or ***2D*** data files from or to other packages. The full translation is the responsibility of the "CAD" package. Most packages also open or export the widespread or standard formats of DXF (.dxf), IGES (.igs), STL (.stl), and STEP (.step or .stp). Some of these formats are generated as standard for the community of computer-aided design (CAD) packages to use and follow. Information about and reference to some of the common and standard data files are as follows:

4.5.1 Drawing Exchange Format (DXF)

Very early in the exploitation of computers in drafting and modeling, the data structure has been the concern of many *geometric modeling* "CAD" packages. This is for the package to be useful in communicating with other packages. In 1982, Autodesk developed the DXF data file with the extension (.dxf). This format had a widespread dissemination of ***3D*** modeling for *AutoCAD* package (Autodesk (2011)). The DXF format has been mainly for ***3D*** *wire frame* similar to the wire frame shaft profile in Figure 4.3a. The DXF file structure is simple entities with each entity on a separate line (Autodesk (2011)). This causes the DXF file to be very long. A cylinder is made of two end circles and four lines as six entities. Each of these entities should be specified in location and dimensions of x, y, and z components with each component or entity on a separate line preceded with an identifier number or so (Autodesk 2011). The DXF file may still be used to communicate ***2D*** sketches, even though it is possible to draw ***2D*** sketches in all geometric modeling "CAD" packages. The DXF file also communicates a ***3D*** surface model or the wire frame of the ***3D*** model. Many "CAD" packages may import or export DXF file format.

4.5.2 STL File Format

The **stereolithography** or **s**tandard **t**essellation **l**anguage (STL) is a *boundary surface* model composed of external triangles wrapping the object. The triangles are given in a clockwise order to indicate the surface normal pointing to the outside of the ***3D*** object (Hull (1986)) of 3D Systems; see Internet Links section of the Reference "https://www.3dsystems.com/". The collection of triangles encompasses or wraps the ***3D*** model. The STL files are used extensively in ***3D*** printing. The format of the STL ASCII data file is given in Figure 4.49 for the ***3D*** tetrahedron object in Figure 4.5. The notations used in Figure 4.49 are for the components of the normal vector $\boldsymbol{N}_{Fi}$ of each face F_i in the x, y, and z directions. The vertices components are the points $\boldsymbol{P}_i$ in the x, y, and z directions. The citation of the vertices is in a *clockwise* order. The same treatment is performed for as many triangles as the model requires to produce smoother surface generation.

4.5.3 IGES File Format

The IGES data file is a platform-independent format that stands for ***I***nitial ***G***raphics ***E***xchange ***S***pecification. This format is first introduced by the US National Bureau of Standard as NBSIR 80-1978 (Nagel et al. (1980)). It intended to be a *standard* for ***3D*** modeling packages to adopt. After its initiation, many updates have been devised. With being a standard, most modeling packages import and export IGES file format. The file has the extension (.igs).

The IGES file has five section, namely, start (S), global (g), directory entry (D), parameter data (P), and Terminate (T). The letter representing the section is placed in the column 73 of the ASCII file of IGES. The letter is repeated on each line of the file for that section. The characteristics of an entity are given in fixed length format in the D section. The geometry information of each entity is given next in a coma delimited format in the P section. For more details and sample files, one can consult Nagel et al. (1980), ASME Y14.26M (1989) and IGES 5.3 (1996).

Entity names are specified by numbers such as the circle is assigned the number 100, the face or a plane is number 108, the line is 110, the parametric spline is 112, the parametric bicubic spline surface is 114, the point is 116, surface of revolution is 120, the tabulated cylinder is 122, etc. (Nagel et al. (1980)). For each of these entities, definite parameters are specified to sufficiently define the geometry, position, orientation, properties, etc. Dimensioning is also included as entities. Functions, statements, and macros are used for complex, repeated entities, or other tasks.

After the STEP file format was introduced, ISO 10303 (1994), the further development of IGES has stopped. Geometric modeling packages had the IGES file format to open and save for a long time, and still, they keep the option. The last IGES version 5.3 was in 1996 (IGES 5.3 (1996)).

4.5.4 STEP File Format

The STEP file name is short for ***S****tandard for* ***t****he* ***E****xchange of* ***P****roduct model data* (ISO 10303 (1994) and Pratt (2001)). The standard is vendor free, i.e. independent from any particular package. The file format is not propitiatory and is considered as a successor to IGES. Most of the "CAD" software packages import and export STEP files,

```
solid Tetrahedron
    facet normal N_F1i N_F1j N_F1k
        outer loop
                    vertex P1_x P1_y P1_z
                    vertex P2_x P2_y P2_z
                    vertex P3_x P3_y P3_z
        endloop
    endfacet
    facet normal N_F2i N_F2j N_F2k
        outer loop
                    vertex P3_x P3_y P3_z
                    vertex P4_x P4_y P4_z
                    vertex P2_x P2_y P2_z
        endloop
    Endfacet
    facet normal N_F3i N_F3j N_F3k
        outer loop
                    vertex P1_x P1_y P1_z
                    vertex P2_x P2_y P2_z
                    vertex P4_x P4_y P4_z
        endloop
    Endfacet
    facet normal N_F4i N_F4j N_F4k
        outer loop
                    vertex P1_x P1_y P1_z
                    vertex P4_x P4_y P4_z
                    vertex P3_x P3_y P3_z
        endloop
    Endfacet
endsolid Tetrahedron
```

Figure 4.49 The STL data file format for the ***3D*** tetrahedron object in Figure 4.5.

but they still use their native propitiatory file formats. Software vendors may need to adopt the STEP file as their native data structure format. Migration between packages would then be easier.

The STEP file is extensive and has so many attributes to guarantee a possible general applicability in exchanging ***3D*** models and other product data. The exchange of product data necessitates the encompassing of so many applications in mechanical, manufacturing, life cycle, electrical, electronics, dimensional or geometrical tolerances, etc. It is also tailored to specific industries such as automotive, construction, naval, aerospace, petroleum, etc. Each of the applications would have a special STEP treatment in addition to the general geometrical modeling attributes like IGES. This generated many *Application Protocols (***APs)** as parts of STEP. The later AP AP242 combined and replaced some earlier protocols to manage model-based ***3D*** engineering (ISO 10303–242 (2014)). It includes ***2D*** and ***3D*** *explicit* and *associative draughting*, ***3D*** mechanical *design* of parts and assemblies including *boundary representation* (***B***-rep), and other core data for *tolerancing*, *dimensioning*, *machining* features, *assembly* constraints, and *kinematics*. Other modifications are handled or underway. The main areas of the APs of concern are in design, manufacturing, and life cycles including items such as materials, composites, sheet metal, product verification, inspection, casting, machining, CNC, process planning, models for life cycle support, etc.

There are many parts in groups of the STEP file. The groups define the *environment*, *integrated data*, and *top* parts. The *top* part for APs refers to product kind, structure, geometry, and others in the *integrated data* group. The file format specifics are produced by the ***3D*** modeling software packages and verified by their successful import and export of such STEP files. Specific implementation and data file format are thus beyond the scope of the text.

4.6 Using CAD in *3D* Modeling and CAM

This section emphasizes the objective of this text in the development process toward the realization of a design. The real *computer-aided design* (*CAD*) in machine element selection or adoption should be related to having the computer define proper geometry and material, and not only helping in ***3D*** modeling or ***3D*** drafting. However, ***3D*** modeling is an essential part of the design process after defining proper geometry and material. Applications transferring geometry and material data to geometric modeling software or commonly known CAD packages are not readily available in a seamless or widely accessible way. One might have to do the design geometry and material data generation for machine elements (real *CAD*) and manually enter this information into the geometric modeling software or CAD packages. This would facilitate the generation of the ***3D*** models into the geometric modeling packages to generate the *CAM* codes for CNC, ***3D*** printing, rapid prototyping, and other manufacturing functions.

4.7 Summary

This chapter was mainly dedicated to the *geometric modeling* or ***3D*** modeling process and a glimpse of basic *computer graphics* field. The *computer-aided design (CAD)* in the sense of defining proper geometry and material would have to be transferred to ***3D*** models for the integration of CAD/CAM/CAE. This process has not been readily available in a seamless or widely accessible way in ***3D*** modeling packages.

The presented *geometric modeling* or ***3D*** modeling process is extended to its basic *computer graphics* technology to provide the user with the profound understanding of the process and the capability of proper adoption of details. This process involved the ***3D*** geometric modeling and viewing details including mathematical relations to achieve that. The underlying basics of the new *homogeneous* coordinates and transformation have been introduced including stereo viewing for ***VR***. Details of *parametric modeling* are laid out, and its utility of defining intersecting boundaries between ***3D*** entities of the objects is provided. *Rendering* techniques and process of ***3D*** models including color usage provide the basic means to understand the need for photorealistic generation of scenes. Needed graphics *data structures* to transfer these ***3D*** models to other fields have been presented to allow the integration and usefulness of the generated ***3D*** models.

The information provided in this chapter may be used as a preliminary foundation to develop geometric modeling tools particularly to transfer the generated *computer-aided design (CAD)* geometry and material to other CAD packages for the integration of CAD/CAM/CAE.

Problems

4.1 For the geometric modeling CAD package available to you, identify if it provides *3D* primitives for utility, or you are to generate any of these primitives.

4.2 Does the geometric modeling CAD package available to you provide any means to define the geometry and material of the machine elements you try to model in *3D*?

4.3 Does the geometric modeling CAD package available to you provide lofting, or sweeping, or extrusion in a direction normal (or inclined) to the plane of the sketch to generate the object?

4.4 Does the geometric modeling CAD package available to you provide *3D* model generation from a free-form surface model you generate?

4.5 For any *2D* four-sided planar polygon, use Eq. (4.1) to calculate the unit normal vector of the polygon. Find the unit normals of the polygon at its four corners by the cross product of the two vectors of the polygon sides at each corner. Compare the values of the unit normal vector components for all cases.

4.6 Examine the Euler–Poincaré formula for the solid torus model in Figure 4.1f.

4.7 Examine the Euler–Poincaré formula for the solid model in Figure 4.2a.

4.8 Examine the Euler–Poincaré formula for the solid model in Figure 4.6.

4.9 Examine the Euler–Poincaré formula for the final solid model in Figure 4.9.

4.10 Construct a *3D* model for the shaft of Figure 4.3 using the suggested method of revolving section. Use any other method to generate the *3D* shaft. Assume any proportional dimensions to those in Figure 4.3.

4.11 Construct a *3D* model for such a seat like Figure 4.9 using different methods of 3D modeling in the available geometric modeling CAD package available to you. Which method was the easiest? Can you edit any of the sub-object in the model? Can you use parametric characteristics in some sub-objects in the model to change geometry? Do you have available tree for your model?

4.12 Develop a *3D* model for the *Jib* crane constructed in the text.

4.13 Are all vectors considered projections on the zero plane of the homogeneous coordinates? How is that related to the conception that a vector is the same as any parallel vector in *Cartesian* coordinates?

4.14 The line in the homogeneous coordinates is defined by its two end points. Is that more suitable for implementing computer graphics in *3D* modeling of objects and why?

4.15 If the z-coordinates of the two points in Example 4.1 are 2 and 3, respectively, what should be the translated line in *3D*? Sketch the original and translated lines on a *3D* Cartesian coordinates.

4.16 If the eye in Example 4.2 is moved to the location of (3, 3, 3), what should be the perspective view of the cube? Compare with the perspective view of the said example.

4.17 Write a MATLAB code to perform a similar perspective view of Example 4.2 on other bodies defined by few corner points or vertices such as the *3D* object of Figure 4.6 assuming suitable dimensions.

4.18 Generate the stereo views of Figure 4.15 of the cube of Example 4.2.

4.19 Use the available MATLAB code to plot the plane passing by any three points of your selection. Verify the results and the other output parameters.

4.20 Use the geometric modeling CAD package available to you so as to develop the bilinear surface in Figure 4.20. Compare results.

4.21 Use the geometric modeling CAD package available to you so as to join two free-form surfaces similar to that in Figure 4.20.

4.22 Use the available MATLAB code of `CAD_3D_Spline_Bezier_Color.m` to change control points to your interest, and observe the shape variation capabilities. Compare the output with Figures 4.26 and 4.32.

4.23 Use the available MATLAB code of `CAD_3D_Spline_Bezier_Color.m` to duplicate some control points, and observe the shape variation capabilities.

4.24 Use the available MATLAB code (`CAD_Line_Line_Intersection.m`) to find the intersection of any two lines of your choice, and observe their plotting figure.

4.25 Use the available MATLAB code (`CAD_Plane_Line_Intersection.m`) to find the intersection of any line with any plane of your choice, and observe their plotting figure.

4.26 Use the available MATLAB code (`CAD_Plane_Plane_Intersection.m`) to find the intersection of any two planes of your selection, and observe their plotting figure.

4.27 Use the geometric modeling CAD package available to you so as to attempt to generate the *3D* rendered model of the three shiny balls in Figure 4.36.

4.28 Copy Figure 4.39a and generate a copy similar to Figure 4.39b to confirm the observed result.

4.29 Use the geometric modeling CAD package available to you so as to model different metals such as steel or aluminum, and identify their RGB values. Check the values of silver and gold provided in the text.

4.30 Search for the CIE zones of the PC monitors and laptops screens on the CIE chromaticity diagram.

4.31 Use the geometric modeling CAD package available to you so as to develop a simple cube of any dimensions, and study the form of the STL and STEP files exported by the package.

4.32 Use the geometric modeling CAD package available to you so as to develop a simple cylinder of any dimensions, and study the form of the STL and STEP files exported by the package.

References

Adelson, E.H. (1995). Checker shadow illusion.svg http://persci.mit.edu/gallery/checkershadow

ASME Y14.26M (1989). *Digital representation for communication of product definition data*. American Society of Mechanical Engineers.

Autodesk (2011). *AutoCAD 2012 DXF-reference*. Autodesk.

Bezier, P. (1986). *The Mathematical Basis of the UNISURF CAD System*. Butterworths.

CIE (1931). *Commission internationale de l'Eclairage proceedings, 1931*. Cambridge: Cambridge University Press.

Cox, M.G. (1972). The numerical evaluation of B-splines. *Journal of the Institute of Mathematics and its Applications* 15: 95–108.

De Boor, C. (1972). On calculating with B-spline. *Journal of Approximation Theory* 6: 52–60.

El-Komy, M.A., and Metwalli, S.M. (2014) "Optimum 3D ***W***rapping of CAD models using single NURBS", ASME IMECE2014, Montreal, Quebec, Canada (November 8–13 2014), Paper # IMECE2014-36736, Volume 11: Systems, Design, and Complexity, pp. V011T14A001 (8 pp.).

Euler, L. (1758) "Elementa doctrinae solidorum – Demonstratio nonnullarum insignium proprietatum, quibus solida hedris planis inclusa sunt praedita" Novi commentarii academiae scientiarum imp.

Foley, J.D., van Dam, A., Feiner, S.K., and Hughes, J.F. (1990). *Computer Graphics: Principles and Practice*. Addison Wesley.

Gouraud, H. (1971). Continuous shading of curved surfaces. *IEEE Transactions on Computers* C–20 (6): 623–629.

Hegazi, H.A., Nassef, A.O. and Metwalli, S.M. (2002). Shape optimization of NURBS modeled 3D C-frames using hybrid genetic algorithm. *Proceedings of ASME DETC & CIE Conference*, Montreal, Canada (29 September–2 October 2002), Paper # DETC2002/DAC-34107.

Hill, F.S. (2001). *Computer Graphics Using Open GL*, 2e. Prentice Hall: Pearson.

Hull, C.W. (1986). *Apparatus for production of three-dimensional objects by stereolithography 3D systems*. US Patent 4, 575,330.

IGES 5.3 (1996). *Initial graphics exchange specification*. US PRO, U.S. Product Data Association.

ISO 10303 (1994). *Industrial automation systems and integration – Product data representation and exchange*. International Organization for Standardization.

ISO 10303-242 (2014). *Industrial automation systems and integration – Product data representation and exchange – Part 242: Application protocol: Managed model-based 3D engineering*. International Organization for Standardization.

ISO/CIE (2016). *Colorimetry – Part 5: CIE 1976 L*u*v* colour space and u', v' uniform chromaticity scale diagram*. ISO/CIE 11664-5:2016(E). International Organization for Standardization/International Commission on Illumination.

Lambert, J.H. (1760). *Photometria, sive de Mensura et gradibus luminis, colorum et umbrae*. Verlag von Wilhelm Engelmann, Leipzig 1892.

Lee, K. (1999). *Principles of CAD/CAM/CAE Systems*. Addison Wesley.

Nagel, R.N., Braithwaite, W.W., and Kennicott, P.R. (1980). *Initial graphics exchange specification IGES, Version 1.0*, NBSIR 80-1978. U.S. National Bureau of Standard.

Nassef, A.O., Hegazi, H.A. and Metwalli, S.M. (1999). Design of C-frame using real-coded genetic optimization algorithm and NURBS. *Proceedings of ASME DETC*, Las Vegas, Nevada, (12–15 September 1999), Paper No. DETC99/CIE-9138.

Nassef, A.O., Hegazi, H.A. and Metwalli, S.M. (2000). A hybrid genetic-direct search algorithm for the shape optimization of solid C-frame cross-section. *Proceedings of ASME DETC*, Baltimore, MD (10–14 September 2000), Paper No. DETC2000/DAC-14295.

Newman, W.M. and Sproull, R.F. (1979). *Principles of Interactive Computer Graphics*, 2e. McGraw-Hill.

Nowell, M.E., Nowell, R.G., and Sancha, T.L. (1972). A solution to the hidden surface problem J.J. Donovan and R. Shields (eds.). *Proceedings of the ACM National Conference*, pp 443–450, Volume 1. New York, NY, United States: Association for Computing Machinery.

Phong, B.T. (1975). Illumination for computer generated pictures. *Communications of ACM* 18 (6): 311–317.

Piegl, L. and Tiller, W. (1997). *The NURBS Book*. Springer Verlag.

Pingstone, A. (2004) *Optical grey squares orange brown.png*.

Poincaré, H. (1895). Analysis situs. *Journal de l'École Polytechnique ser* 2 (1): 1–123.

Pratt, M.J. (2001). Introduction to ISO 10303 – the STEP standard for product sata exchange. *Journal of Computing and Information Science in Engineering* 1 (1): 102–103.

Spatial Corp (2017). *Spatial Release 2017 1.0, Delivering Technology Enhancements Aimed at Innovation and Industrialization*. Dassault Systèmes.

Internet Links

http://support.ptc.com/products/granite/gplugs PTC (Parametric Technology) Granite kernel/engine.
https://opengl.org/ **OpenG**raphics **L**ibrary: Industry's Foundation for High Performance Graphics.
https://www.3ds.com/products-services/ Dassault Systems: SolidWorks, CATIA, and others.
https://www.3dsystems.com/ 3D Systems: 3D Printers and 3D Printing.
https://www.autodesk.com/products/ Autodesk: Inventor, Fusion 360, 3DS Max, AutoCAD, and others.
https://www.plm.automation.siemens.com/global/en/products/ Siemens: NX (Unigraphics), Solid Edge,and others.
https://www.ptc.com/en/products/cad/ PTC: *Creo*© (formerly *Pro/Engineer*), and others.
www.astm.org/ ASTM: ASTM E284-13b (2013) *Standard Terminology of Appearance*, Similar standard to CIE.

5

Optimization

In this chapter, the concepts, formulation, and definitions of the field of optimization are introduced. For generalization of problems, a useful tool of nondimensionalization is presented to set problems in more wide-ranging form. Problem classification with respect to form, methods of optimization, and optimization fields is also introduced. This overview might be useful for newcomers to the field of optimum design.

Symbols and Abbreviations

The adopted units are [in, lb, psi] or [m, kg, N, Pa], others given at each symbol definition. [k…] is 10^3, [M…] is 10^6 and [G…] is 10^9.

Symbol	Quantity, units (adapted)
$\check{f}$	The minimum objective function
$\bar{f}$	*Augmented* function
$^*\boldsymbol{X}_i$	Step optimum vector
$^*x_j$	A *step optimum*
:	In set theory, (:) means "provided that."
2D	Two-dimensional
3D	Three-dimensional
$\boldsymbol{A}$	Quadratic function constant matrix
a, b, c	Coefficients of a function or distances
a_{ij}	Components of $\boldsymbol{A}$ matrix
$\boldsymbol{B}$	Quadratic function constant vector
BFGS	Broyden–Fletcher–Goldfarb–Shanno method
b_j	Components of $\boldsymbol{B}$ vector x
b_t	Breadth of the truss base
B_t	Fixed breadth of the truss base
C	Quadratic function constant
C_{ls}	Cost of seam length to cost of surface area
$\boldsymbol{D}$	Design vector
$\underline{D}$	Nondimensional tube diameter

Machine Design with CAD and Optimization, First Edition. Sayed M. Metwalli.

Companion website: www.wiley.com/go/metwalli/machine

Symbol	Quantity, units (adapted)
$\boldsymbol{D}^*$	Unconstrained optimum design vector
d^*, H^*	Truss optimum diameter and height
d_c	Can diameter
$\boldsymbol{D}_c^*$	Constrained optimum design vector
d_c^*	Optimum can diameter
DFP	Davidon–Fletcher–Powell optimization
$D_{H,n}$	The *nth* determinant of $\boldsymbol{H}$
$\boldsymbol{d}_T$	Tray design vector
d_t	Cylindrical tubes diameter
$\boldsymbol{D}_t$	Tube design vector
E	Modulus of elasticity
$\boldsymbol{f}$	Objective function
f_1, f_2, f_3	Function values at three different s values
f_i	First derivative of a function with respect to i
$\mathcal{F}_i(\ldots)$	Function i of some variables in parenthesis
f_{ij}	Second function derivatives with respect to i,j
$\boldsymbol{f}_p$	Projection vector
$\mathcal{F}_s'$	First derivative of the function at s
$\mathcal{F}_s''$	Second derivative of the function at s
$\boldsymbol{g}$	Inequality constraint vector, each as $\boldsymbol{g}_i$
$\boldsymbol{G}$	constraint gradient matrix
GA	Genetic algorithms
$\boldsymbol{h}$	Equality constraints vector, each as $\boldsymbol{h}_i$
$\underline{H}$	Nondimensional truss height
$\boldsymbol{H}$	The *Hessian* matrix of second derivatives
h_c	Can height
h_c^*	Optimum can height
HGP	Heuristic gradient projection
h_T	Tray height
H_T	Tray specific height
h_t	Truss height
k	Arbitrary constant
K–T	Kuhn–Tucker
L	A length or interval of uncertainty
$l_{\max1}, l_{\max2}$	Tray length constraints maximum 1 and 2
l_T	Tray length
l_t	Tube length
l_T^*	Optimum tray length

Symbol	Quantity, units (adapted)
l_{Tc}	Tray length constraint
l^*_{Tc}	Optimum constrained tray length
m	Number of constraints
n	Number of variables
n_T	Number of trials
$\boldsymbol{P}$	Truss load or top force
Pi (π)	*Buckingham* theorem
PSO	Particle swarm optimization
Q	Quadratic function
q	Search step
r	A multiplier factor
$\boldsymbol{R}$	The residual vector
R^n	The set of real numbers of dimension n
s	Distance on a line or slackness to constraint
s^*	Optimum location of the minimum
$\mathbf{S}, \boldsymbol{S}_i$	Search vector
s_i	A slack variable
S_y	Yield strength
t	Independent variable such as time
t	Step in the direction s
t_T	Tray material thickness
t_t	Tube thickness
T_T	Tray specific material thickness
T_t	Fixed tube thickness
$\boldsymbol{U}$	Uncontrollable design or behavior variable
$\boldsymbol{u}$	Constraint gradient multiplier vector
V_c	Can volume
V_T	Tray inner volume
v_T	Tray material volume
W_t	Truss weight
$\underline{W}_t$	Nondimensional truss weight
w^*_T	Optimum tray width
w_T	Tray width
X	A normalized or nondimensional variable
x and y	Variables
$\boldsymbol{X}_0, \boldsymbol{X}_1, \ldots$	Location or position vectors at 0, 1, …, step
$x_1, x_2, \ldots, x_n$	Independent design parameters

Symbol	Quantity, units (adapted)
α^*, α_i^*	Optimum distance in the direction $\boldsymbol{S}_i$
β	Conjugate gradient factor
δs	Small distance on s
Δ	Step distance
ε	Small distance in a direction x_i
ε_e	Acceptable error value
$\in$	In set theory, meaning "in"
$\boldsymbol{\lambda}$	*Lagrange* multipliers vector
λ, λ_i	Lagrange multiplier
∇	First derivative operator "del"
$\boldsymbol{\phi}$	Penalty function
$\boldsymbol{\varphi}$	Constraint vector
φ, φ_i	Constraints
ρ	Material density
σ_B	Buckling stress
$\underline{\sigma}_B$	Nondimensional buckling stress in tube
σ_t	Compressive stress in tube
$\underline{\sigma}_t$	Nondimensional compressive stress in tube
τ_G	The *golden ratio* constant or *golden* number

5.1 Introduction

The optimization problem needs the definition of associated terminology and particular formulation process. These definitions, dimensional considerations, classification of optimization problems, and optimization methods are introduced. A good introduction that sheds considerable overview is by scrutinizing an example for simple application as follows.

Example 5.1 ***A sample demonstration problem***
Consider building a plastic tray to hold a given volume of liquid V_T. The tray has a specific height $h_T = H_T$ and a specific thickness $t_T = T_T$. The tray length is l_T and the tray width is w_T. The tray is shown in Figure 5.1. The specified variables are given uppercase fonts. The subscript is for the tray object T. For a workable or adequate design, the volume of the tray is then

$$V_T = w_T \; l_T \; H_T \tag{a}$$

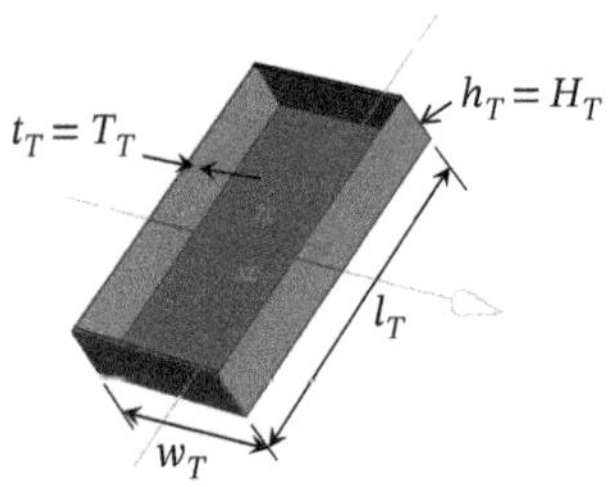

Figure 5.1 A plastic tray that holds a volume of liquid V_T with a specific height $h_T = H_T$ and a specific thickness $t_T = T_T$.

Rearrange the problem variables so as to move the specified values to the right-hand side and the design variables to the left side such that

$$w_T\ l_T = V_T/H_T \tag{b}$$

To design such a tray, one can then pick any w_T or any l_T and solve for the other. Therefore, no optimization is attempted or performed.

For an *optimum design* let us minimize the tray cost. Wall thickness is specified as T_T to give decent structural properties and to allow ready manufacture. Material is specified as plastic, so we minimize cost by minimizing the volume v_T of the material.

$$v_T = T_T\ (2H_T\, w_T + 2H_T\, l_T + w_T\, l_T) \tag{c}$$

The optimization process thinks of the volume v_T as the ***objective function*** $\boldsymbol{f}$, and the quantities w_T and l_T as the ***design variables***. The tray ***design vector*** $\mathbf{d}_T$ is given as follows:

$$\mathbf{d}_T = \begin{bmatrix} w_T \\ l_T \end{bmatrix} \tag{d}$$

The mathematical form of the *optimization* problem is usually stated as to find

$$\min_{\mathbf{d}_T}\ \boldsymbol{f} = T_T(2H_T\, w_T + 2H_T\, l_T + w_T\, l_T) \tag{e}$$

This min expression with underneath $\mathbf{d}_T$ states that the minimum should be performed for all components of the design vector $\mathbf{d}_T$. Solution may be obtained by the usual mathematical process to find the minimum of a mathematical function in the w_T and l_T variables such as

$$\frac{\partial \boldsymbol{f}}{\partial w_T} = 0 \quad \text{and} \quad \frac{\partial \boldsymbol{f}}{\partial l_T} = 0 \tag{f}$$

This gives

$$\begin{aligned} \frac{\partial \boldsymbol{f}}{\partial w_T} &= T_T(2H_T + l_T) = 0 \quad \text{or} \quad w_T = -2H_T \\ \frac{\partial \boldsymbol{f}}{\partial l_T} &= T_T(2H_T + w_T) = 0 \quad \text{or} \quad l_T = -2H_T \end{aligned} \tag{g}$$

This result is not acceptable due to the fact that there was an oversight of the dependent and ***independent parameters***. As recognized out of Eq. (b), there is only one independent design variable. It was even stated that one would pick any w_T or any l_T and solve for the other. This is a volume ***constraint*** tying the two design variables. The problem has only one independent parameter, and the other parameter is dependent on the independent one through the volume constraining relation of Eq. (a). From that constraint, one can define the dependency of w_T on l_T from the following solution of the constraint relation in Eq. (a) to get the following:

$$w_T = \frac{V_T}{l_T\, H_T} \tag{h}$$

Substituting this *constraint solution* into the *objective function* of Eq. (e), one gets the adjusted *objective function*:

$$\min_{\mathbf{d}_T}\ \boldsymbol{f} = T_T\left(2H_T\, \frac{V_T}{l_T\, H_T} + 2H_T\, l_T + \frac{V_T}{l_T\, H_T}\, l_T\right) \tag{i}$$

The problem is then a function of one variable and to get the *extremum* of this function, one can equate the derivative to zero such that

$$\frac{\partial \boldsymbol{f}}{\partial l_T} = T_T(-2V_T\, l_T^{-2} + 2H_T) = 0 \tag{j}$$

This relation can be easily manipulated to find the optimum tray length l_T^* as

$$l_T^* = \sqrt{\frac{V_T}{H_T}} \tag{k}$$

Substituting into the constraining equation, one gets the optimum tray width w_T^* as

$$w_T^* = \frac{V_T}{H_T\sqrt{V_T/H_T}} \quad \text{or} \quad w_T^* = \sqrt{\frac{V_T}{H_T}} \tag{l}$$

It is observed that the optimum width is equal to the optimum length, or the optimum is a square based tray. The total volume at the optimum is obtained by substituting the optimum into the objective function of Eq. (e) to get

$$\begin{aligned} \min_{\boldsymbol{d}_T} \ \boldsymbol{f} = \check{\boldsymbol{f}} &= T_T\left(2H_T\sqrt{\frac{V_T}{H_T}} + 2H_T\sqrt{\frac{V_T}{H_T}} + \sqrt{\frac{V_T}{H_T}}\sqrt{\frac{V_T}{H_T}}\right) \\ &= T_T\left(4\sqrt{V_T H_T} + \frac{V_T}{H_T}\right) \end{aligned} \tag{m}$$

Considering the existence of an additional length ***constraint*** l_{Tc} that is imposed on the problem such that $l_T < l_{Tc}$. Figure 5.2a shows a demonstrative schematic of the problem without adherence to the exact shape of the objective function. It also indicates the probable locations of the constraint l_{Tc}, which takes two possible values l_{max1} and l_{max2}. It shows the main objective function as black dashed line, which has only one minimum at l_T^*. This function shape is called a ***unimodal*** function with a single unique optimum. Figure 5.2a indicates that the value of the constraint affects the outcome of the ***constrained optimization*** problem. It is suggested that if the length constraint l_{max1} is less than l_T^*, the constrained optimum of the length will be $l_{Tc}^* = l_{\text{max1}}$. If the length constraint l_{max2} is larger than l_T^*, the constrained optimum of the length will be $l_{Tc}^* = l_{\text{max2}}$. It is clear that the optimum is dependent on the constraint value where the acceptable solution should be in the ***feasible*** region of the design, and not in the ***infeasible*** region of the design as indicated in Figure 5.2.

Figure 5.2b presents a parallel case of another objective function that can be constrained and might be shaped differently than the problem at hand. It shows the other fictitious objective function as dashed line of a grayish color. If the function in the Figure 5.2b shows two minima (or more), the function is called a ***multimodal*** function. The value of the constraint greatly affects the outcome of the ***constrained optimization*** problem. If $l_T < l_{\text{max1}}$, the constrained optimum is at the first minimum of the function, and not at the constraint value of l_{max1}. If $l_T < l_{\text{max2}}$, the constrained optimum is at the second minimum of the function, and also not at the constraint value of l_{max2}.

5.1.1 Formulation of Optimization Problem

In the optimization problem formulation, the ***objective*** should be clearly defined. This can be easily attained when the problem is sufficiently understood. Therefore, we should first understand the problem and its physical

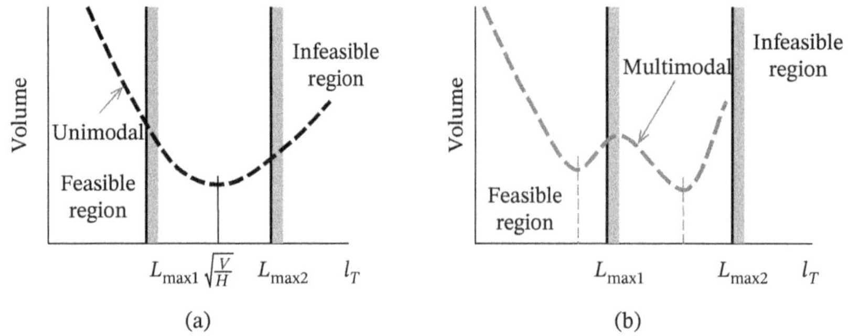

Figure 5.2 A schematic of a one-dimensional problem indicating the probable locations of the constraint: (a) unimodal function with one minimum and (b) a multimodal function with more than one minimum.

interpretations and laws with completely defining *dependent* and *independent* parameters and input and output relations.

After understanding the problem domain, the objective of optimization is then usually to find the "*best*" design or system with respect to some specific requirement or aim, which is the ***objective function***. Sometimes, it is also necessary to specify, in addition, some other restrictions or ***constraints*** on the system performance or design variables. The following are some related definitions and notations:

5.1.1.1 Design Vector *D*

This vector defines all independent design parameters $x_1, x_2, \ldots, x_n$ that directly or indirectly affect the system or design objective and would completely define the system or design configuration. In a mathematical form the design vector $\boldsymbol{D}$ is

$$\boldsymbol{D} = [x_1 \; x_2 \; \cdots \; x_n]^T \tag{5.1}$$

The design vector, in its form, defines the size of the system or the design problem implicitly as n.

5.1.1.2 Objective Function *f*

The objective function $\boldsymbol{f}$ is mathematically defining the aim to be minimized or maximized and which in doing so would result in the "best" *extremum* design or system *with respect to* this defined objective. The objective is thus formulated mathematically as a dependent *function* of the design parameters or design vector $\boldsymbol{D}$.

$$\boldsymbol{f} = \mathcal{F}(\boldsymbol{D}) \tag{5.2}$$

In a more general form, it can be defined as a *function* of many other variables such that

$$\boldsymbol{f} = \mathcal{F}(\boldsymbol{D}, \boldsymbol{U}, t, \ldots) \tag{5.3}$$

In this function, $\boldsymbol{U}$ is an *uncontrollable* design or behavioral variable or vector. The variable t is an independent variable such as time and any other independent variable or variables.

5.1.1.3 Constraints

The constraints define the conditions or restrictions that should be satisfied or physically restrict parameters, response behavior, properties, etc. Two types of constraints are common. The first type is the ***equality constraints*** and may take the form

$$\boldsymbol{h}(\boldsymbol{D}) = \left[\boldsymbol{h}_1 \; \boldsymbol{h}_2 \; \ldots \; \boldsymbol{h}_r\right]^T = \boldsymbol{0} \tag{5.4}$$

where the vector component $\boldsymbol{h}_i = \mathcal{F}_i(\boldsymbol{D}) = 0, i = 1, 2, 3, \ldots, r$.

The second type of constraints is the ***inequality constraints*** that can be formulated as follows:

$$\boldsymbol{g}(\boldsymbol{D}) = \left[g_1 \; g_2 \; \ldots \; g_m\right]^T \leq \boldsymbol{0} \tag{5.5}$$

where $\boldsymbol{g}_j = \mathcal{F}_i(\boldsymbol{D}) = 0, j = 1, 2, 3, \ldots, m$. The form of the constraint is assumed always as ≤ 0 to have a unified treatment. Any other form such as $>$ or so can be transformed to the $<$ by pre-multiplication by -1 or any other means such as using reciprocals.

5.1.1.4 Problem Statement

In a mathematical form, the optimization problem may then be stated as to find

$$\min_{\boldsymbol{D}} \; \boldsymbol{f}(\boldsymbol{D}) = ? \tag{5.6}$$

$$\begin{aligned} \text{Subject to } \; \boldsymbol{h}(\boldsymbol{D}) &= \boldsymbol{0} \quad (r \times 1) \\ \boldsymbol{g}(\boldsymbol{D}) &\leq \boldsymbol{0} \quad (m \times 1) \end{aligned} \tag{5.7}$$

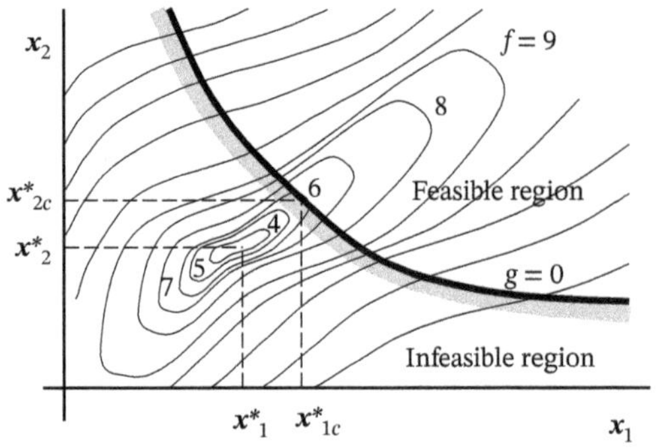

Figure 5.3 Two-dimensional space demonstrating the objective function contours; the constraint boundary is the black line $g = 0$ with the infeasible region to the left of the line. The unconstrained optimum in the feasible region and the unconstrained optimum on the constraint boundary.

In Eq. (5.6), the min expression with the underneath vector $\boldsymbol{D}$ states that the minimum should be performed for the span of $\boldsymbol{D}$ or for all components of the design vector $\boldsymbol{D}$. In another *set theory* form, this can be expressed as

$$\min_{\boldsymbol{D} \in R^n} \boldsymbol{f}(\boldsymbol{D}) : \boldsymbol{h}(\boldsymbol{D}) = \boldsymbol{0}, \quad \boldsymbol{g}(\boldsymbol{D}) < \boldsymbol{0} \tag{5.8}$$

In Eq. (5.8), R defines the set of real numbers, $\boldsymbol{D} \in R^n$ is the design vector $\boldsymbol{D}$ in an n dimensional space (or set) of real numbers R, and the colon mark (:) means "provided that." The form $\{\boldsymbol{D}: \boldsymbol{D} \in R^n : \boldsymbol{h}(\boldsymbol{D}) = 0, \boldsymbol{g}(\boldsymbol{D}) < 0\}$ defines the *feasible* domain. In this textbook, we use the mathematical form in Eqs. (5.6) and (5.7) rather than the *set theory* form of Eq. (5.8).

In two-dimensional (**2D**) space, Figure 5.3 demonstrates the previous definitions and notations. The objective function *contours* $f = 9, 8, 7, 6, 5$, and 4 are shown with the value of each contour stated close to it. The constraint boundary is represented by the black line $g = 0$ with the *infeasible* region to the left of the line and denoted by the gray band to the left of the black line. The solution is then to find the "*optimum*," which should be the *constraint optimum* $\boldsymbol{D}_c^*$ since the *unconstraint optimum* $\boldsymbol{D}^*$ is in the *infeasible* region. Note that the design vector $\boldsymbol{D}$ is defined by its components x_1 and x_2.

5.1.1.5 Dimensional Considerations in Analytical Design "Nondimensionalization"

(a) If there are two variables x and y and one redefines x so that $X = x/y$, then one can say that X is a normalized variable or x has been normalized by y, i.e. with respect to y. If y has the dimensions of x, then X is a nondimensional variable, and we say that x has been nondimensionalized.
(b) This process is performed to generalize the solution to a problem by solving the problem once we essentially get a family of solutions to similar problems. The dimensionless form reduces the number of variables by at least the number of independent variables (for mechanics this means mass, length, and time). There have been so many nondimensional variables that were derived by *Buckingham* Pi (π) theorem. The theorem finds the relevant dimensionless parameters (or variables) without knowing the relevant governing equations. All variables, however, must be known in a problem. In this text, we shall use another procedure that employs the governing equations and the independent parameters in the *nondimensionalization*. This alternative process provided more insight and direct physical interpretation of the *nondimensional parameters*. The process is demonstrated better by an example as follows:

Example 5.2 ***Nondimensionalization from governing equations***
A two-bar truss is shown in Figure 5.4 (Fox 1971). It is assumed that the bars are hollow circular cylindrical tubes, each with the same diameter d_t and a fixed thickness $t_t = T_t$. The two-bar truss is of a height h_t. It is hinged to the ground through two pivots b_t apart and is loaded by a force $2\boldsymbol{P}$ at the top hinge. The fixed breadth of the truss base $b_t = 2B_t$. The material has the known parameters such as the density ρ, the yield strength S_y, and the modulus

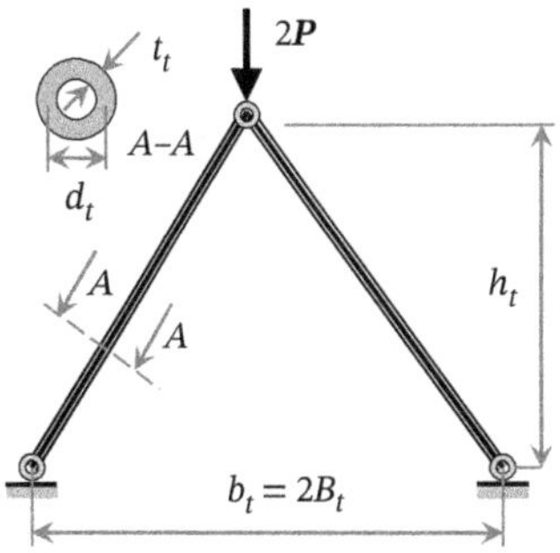

Figure 5.4 A two-bar truss of a height h_t is supported to the ground through two pivots b_t apart and loaded by a force $2\boldsymbol{P}$ at the top hinge. The base is fixed at $b_t = 2B_t$.

of elasticity E. It is required to analyze the problem, define the design vector, and find the optimum design for a minimum truss weight objective. The tubes must neither buckle nor yield. Set the problem in a nondimensional form to generalize the solution to a possible multitude of applications.

Solution

Data: Tube diameter, assumed as the average, d_t, and fixed thickness $t_t = T_t$. Height h_t and fixed base $b_t = 2B_t$. Top load $2\boldsymbol{P}$, material ρ, S_y, and E are known. Tubes must neither buckle nor yield. The d_t^* and H_t^* are to minimize the weight. *Nondimensionalization* is sought.

The truss has a design vector $\boldsymbol{D}_t = [d_t \quad h_t]^T$. The objective function can be simplified if $t_t \ll d_t$ or alternatively more accurately defined. Both are given as follows:

$$\min_{d_t,h_t} f = W_t = \rho V_t = 2\rho\,\pi\, d_t\, T_t (B_t^2 + h_t^2)^{1/2}$$

$$\min_{d_t,h_t} f = W_t = \rho V_t = 2\rho\frac{\pi}{4}\,((d_t + T_t)^2 - (d_t - T_t)^2)(B_t^2 + h_t^2)^{1/2} \tag{a}$$

In the tube, the simplified and the more accurate compressive stresses σ_t are given by

$$\sigma_t = \frac{\boldsymbol{P}\sqrt{B_t^2 + h_t^2}}{\frac{\pi}{4}\,((d_T + T_t)^2 - (d_T - T_t)^2)h_t} \cong \frac{\boldsymbol{P}}{\pi T_t}\,\frac{\sqrt{B_t^2 + h_t^2}}{h_t\, d_t} \tag{b}$$

The simple and more accurate buckling stresses σ_B are given as follows:

$$\sigma_B = \frac{\pi^2 E\, I_t}{l_t^2 A_t} = \frac{\pi^2 E(\pi((d_T + T_t)^4 - (d_T - T_t)^4)/64)}{\pi\ (B_t^2 + h_t^2)\ ((d_T + T_t)^2 - (d_T - T_t)^2)/4} \approx \frac{\pi^2 E}{8}\,\frac{d_t^2 + T_t^2}{(B_t^2 + h_t^2)} \tag{c}$$

The constraints are then expected to be as follows:

$$\sigma_t < S_y$$

$$\sigma_t < \sigma_B \tag{d}$$

One can develop three tables for σ_t, σ_B, and W_t at various values of d_t and h_t as shown in Figure 5.5A. The major gridlines are only shown in each table. The constraints of Eq. (d) are checked in each cell of the table, and the *infeasible* region is highlighted with a grayish color (Fox 1971). The darker color is for violating the yield strength S_y as shown in Figure 5.5a. The lighter gray in Figure 5.5b is for violating the buckling constraint σ_B. An examination of the last table of calculated weights in Figure 5.5c, with overlapping highlighted regions, reveals an approximation to the desired optimum design for minimum weight W_t. This method of optimizing is called ***gridding***. It is convenient for problems of few design variables and if each analysis can be performed quickly. It is given here as the first intuitive *method of optimization*. This is also defined as *enumeration* of parameters between different bounds for each variable.

The nondimensionalization from the previous governing Eqs. (a)–(d) is an involved process that needs insight into the problem and the prospective use of the formulation in real applications. It is not a unique formulation,

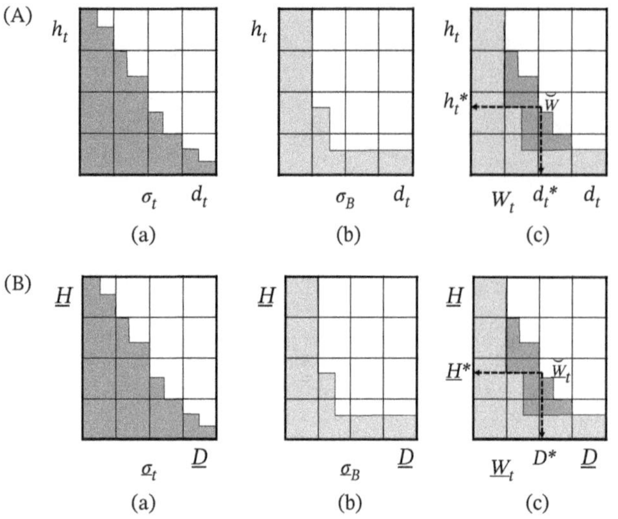

Figure 5.5 (A) Three response tables: (a) for the tube compressive stresses σ_t, (b) for the buckling stresses σ_B, and (c) for the truss weights W_t at various values of tube diameter d_t and truss height h_t. (B) Three nondimensional response tables: (a) for the nondimensional compressive stresses $\underline{\sigma}_t$, (b) for the nondimensional buckling stresses $\underline{\sigma}_B$, and (c) for the nondimensional weights $\underline{W}_t$ at various values of nondimensional diameter $\underline{D}_t$ and nondimensional height $\underline{H}_t$.

but the more understanding of the problem and its utility, the more useful and unambiguous the formulation is. For this example, one would consider a micro look at the tube cross section, which suggests normalizing with respect to the fixed tube thickness T_t. The other concept is to consider the total macro geometry of the truss and normalize the height with respect to the fixed breadth B_t. If that approach is logical, the nondimensional design variables are as follows:

$$\underline{D} = \frac{d_t}{T_t}, \quad d_t = T_t\,\underline{D}$$
$$\underline{H} = \frac{h_t}{B_t}, \quad h_t = B_t\,\underline{H} \tag{e}$$

To show the procedure of nondimensionalization, we will confine the derivation to the simplified relations rather than the more accurate relations. Substituting for the dimensional parameters in Eq. (e) into Eqs. (a–c) and moving all dimensional parameters to the left-hand side, one gets the following equations:

$$\sigma_t = \frac{\boldsymbol{P}}{\pi T_t}\frac{\sqrt{B_t^2 + h_t^2}}{h_t\,d_t} = \frac{\boldsymbol{P}}{\pi T_t}\frac{\sqrt{B_t^2 + (B_t\underline{H})^2}}{B_t\underline{H}\,T_t\,\underline{D}} = \frac{\boldsymbol{P}}{\pi T_t^2}\frac{B_t\sqrt{1 + \underline{H}^2}}{B_t\underline{H}\,\underline{D}}$$
$$\frac{\sigma_t}{P/T_t^2} = \frac{1}{\pi}\frac{\sqrt{1 + \underline{H}^2}}{\underline{H}\,\underline{D}} \equiv \underline{\sigma}_t \tag{f}$$

In Eq. (f), the term $\underline{\sigma}_t$ is the nondimensional compressive stress. Applying the same process to the buckling stress gives the following:

$$\sigma_B = \frac{\pi^2 E}{8}\frac{d_t^2 + T_t^2}{(B_t^2 + h_t^2)} = \frac{\pi^2 E}{8}\frac{T_t^2\underline{D}^2 + T_t^2}{(B_t^2 + (B_t\underline{H})^2)}$$
$$\frac{\sigma_B}{ET_t^2/B_t^2} = \frac{\pi^2}{8}\frac{\underline{D}^2 + 1}{(1 + \underline{H}^2)} \equiv \underline{\sigma}_B \tag{g}$$

In Eq. (g), the term $\underline{\sigma}_B$ is the nondimensional buckling stress.

The constraints in Eq. (d) are set in nondimensional form by dividing each by the same terms of nondimensional stresses. This gives the following nondimensional constraints:

$$\frac{\sigma_t}{P/T_r^{\,2}} < \frac{S_y}{P/T_r^{\,2}}$$

$$\underline{\sigma}_t < \underline{\sigma}_B \rightarrow \frac{\sigma_t/(\boldsymbol{P}/T_t^{\,2})}{\sigma_B/(ET_t^{\,2}/B_t^{\,2})} < \frac{E}{\boldsymbol{P}}\frac{T_t^{\,4}}{B_t^{\,2}} \tag{h}$$

Applying the same process to the truss weight gives the following:

$$W_t = 2\rho\,\pi\;\; d_T\, T_t(B_t^2 + h_t^2)^{1/2} = 2\rho\,\pi\;\;(T_t\;\; \underline{D})\, T_t(B_t^2 + (B_t\underline{H})^2)^{1/2}$$

$$\frac{W_t}{\rho\;\, T_t^{\,2} B_t} = 2\,\pi \underline{D}(1 + \underline{H}^2)^{1/2} \equiv \underline{W}_t \tag{i}$$

In Eq. (i), the term $\underline{W}_t$ is the nondimensional truss weight.

Similar to the dimensional case, one can develop three tables for $\underline{\sigma}_t$, $\underline{\sigma}_B$, and $\underline{W}_t$ as shown in Figure 5.5B with the nondimensional coordinates $\underline{D}$ and $\underline{H}$ for a general solution of any t_t, B_t, P, E, σ_y, and ρ. One should note the ratio between the nondimensional normal stress and the nondimensional buckling stress in Eq. (h). The difference stems from the nondimensionalization and should be carefully observed for different design cases. Noting that difference, the examination of the last table in Figure 5.5B with overlapping highlighted regions, reveals an approximation to the desired nondimensional optimum design for minimum nondimensional weight $\underline{W}_t$. This is again a ***gridding*** method of optimizing, which may be convenient for problems of few design variables. It is stated here, again, as a first *method of optimization*.

5.1.2 Classification of Optimization

The optimization problem can be classified depending on few situations. Classification can be according to the problem type or number of design variables. It can be according to the existence of constraints or not. If the problem is static, dynamic, deterministic, or stochastic, the approach to the optimization formulation can be different. Methods of optimization may be dependent on the classification and optimization fields. These are presented in the following sections:

5.1.2.1 Problem Classification

Classification of optimization can be according to the following classes:

i. *According to the number of independent variables*: The problem is either one dimension, multidimensional, or infinite dimensions as in the case of the number of points on an optimum trajectory. The classification of the problem depends on the number of independent variables in the optimization case. The method or approach to the problem is different accordingly as discussed in this chapter.
ii. *According to the constrained or unconstrained conditions*: The optimization method of solution is different if the problem is constrained or unconstrained. Both cases are discussed in this chapter.
iii. *According to the problem state: Static* (i.e. time independent) or *dynamic* (time dependent with or without feedback control).
iv. *According to the problem case: Deterministic* when no unknown factors or random variables are present or *stochastic* for a system containing random variables or experimental errors and/or other random factors.

5.1.2.2 Methods of Optimization

A classification of the general methods of optimization is shown in Figure 5.6. It shows a classification according to constrained or unconstrained methods in addition to the problem dimensions. Figure 5.6 also presents classification according to two branches; one is a ***direct*** approach, and the other is an ***indirect*** approach to optimization.

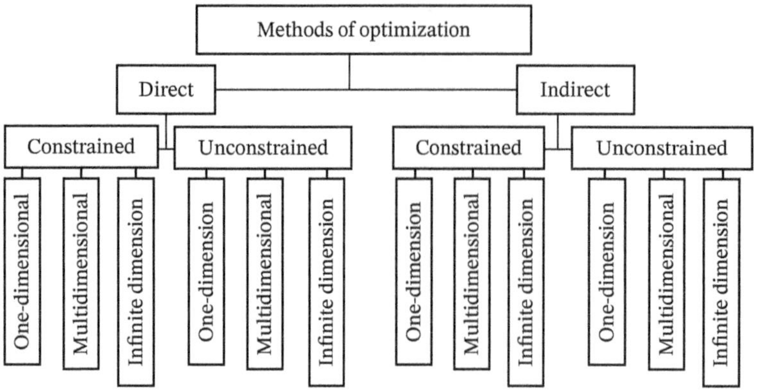

Figure 5.6 A classification of the general methods of optimization.

Indirect method of optimization is one that is based on a *necessary condition* for that optimization such as vanishing derivatives. Direct method, however, depends on a direct *comparison of function* values at two or more points. The discussions of direct and indirect methods are best presented by a simple example as follows:

Example 5.3 ***Direct versus indirect methods***

Consider the optimization problem of only one variable.

$$\min_{x} f = \mathcal{F}(x) \tag{5.9}$$

An *indirect* method would be to determine the optimum x^* for which $f_x = 0$, where

$$f_x = \frac{\partial \mathcal{F}}{\partial x} \tag{5.10}$$

A *direct* method would use one of the search methods of Section 5.2. Classical or *indirect* methods are covered in Section 5.3.

Comparing direct versus indirect methods reveals the following observations:

1. When analytical solutions are desired, indirect methods are preferred.
2. Direct methods have the advantages:
   a. End points and points of discontinuity in the function value present no additional computational difficulties.
   b. As the number of independent variables increases, the indirect method will require the evaluation of at least "n" derivative equations whereas the direct method will require the successive evaluation of only one equation, which is the objective function.

5.1.2.3 Optimization Fields

Several fields are concerned with optimization and may have special means, methods, and considerations for solution methods. Some of these are as follows:

- *Linear programming:* Where the problem has a linear objective function and linear constraints.
- *Quadratic programming:* The problem has a quadratic objective function and linear constraints.
- *Nonlinear programming*: Where problems have nonlinear objective functions and nonlinear constraints.
- *Dynamic programming:* Where the concern is for *multistage* decision process.
- *Geometric programming:* Where the problem has geometric objective function.

- *Integer programming:* When independent design variables have only integer values. Different cases are as follows:
  - Quadratic integer programming.
  - Nonlinear integer programming.
  - Mixed nonlinear integer programming (i.e. both integer and continuous variables).

In this chapter, we will be confined to the relevant nonlinear programming approaches to solve most of the machine design optimization problems. The other fields are covered in specialized literature on optimization where some of these areas are presented such as Reklaitis et al. (1983), Wilde (1978), Wilde and Beightler (1979), Walsh (1975), Arora (2004), and Rao (2009).

5.2 Searches in One Direction

In this section we introduce different ***indirect*** methods to find the optimum in *one dimension* or *one direction*. These methods will be used extensively in the multidimensional optimization methods. This is due to the fact that usually different searches are made in specific *direction* in each step. The first few methods (Sections 5.2.1 and 5.2.2) are *zero order*, which means no need for evaluating derivatives of the objective function with respect to the variable. The methods that need the evaluation of the first derivatives of the objective function with respect to the variable are called *first-order* methods. The method presented in Section 5.2.3, however, requires the evaluation of the *first* and *second derivatives* that makes it a *second-order* method, since it needs the second derivatives. All searches assume the objective function to be a unimodal in the range of interest as shown in Figure 5.2a. Special handling of multimodal functions is not generally considered in this chapter; see Elseddik (2005).

5.2.1 Quadratic Interpolation

This method approximates the objective function f with a *quadratic* function Q. Figure 5.7 shows the unknown objective function f as a thick, grayish, and dashed curve. If the objective function is a function of the distance s, i.e. $f = \mathcal{F}(s)$, then the optimization problem is to find

$$\min_{s} f = \mathcal{F}(s) \tag{5.11}$$

Let a quadratic function $Q(s)$ approximates the function or $Q(s) \approx \mathcal{F}(s)$. The quadratic function $Q(s)$ is shown in Figure 5.7 by the black curve. The quadratic function $Q(s)$ is generally defined as follows:

$$Q(s) = a + b\,s + c\,s^2 \tag{5.12}$$

The minimum of the quadratic function occurs at the first derivative $Q' = 0$, i.e. when (see Figure 5.7)

$$Q' = \frac{dQ}{ds} = 0 \quad \text{and} \quad \frac{d^2Q}{ds^2} > 0 \tag{5.13}$$

Applying the condition in Eq. (5.13) to Eq. (5.12), one gets

$$0 = b + 2cs \tag{5.14}$$

Equation (5.14) is then used to define the optimum location s^* of the minimum of the quadratic function as follows:

$$s^* = -\frac{b}{2c} \tag{5.15}$$

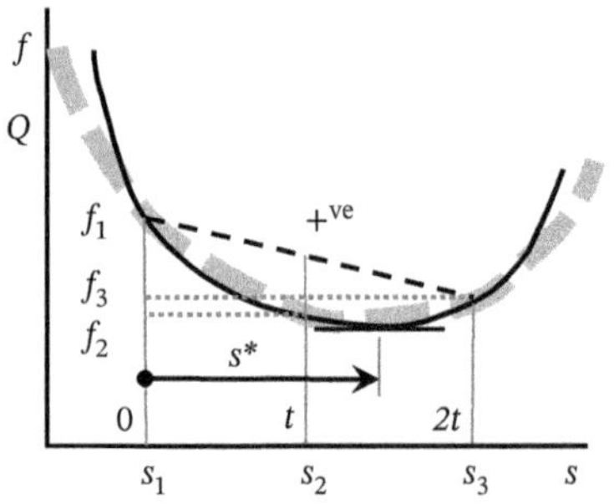

Figure 5.7 The unknown objective function f is shown as a thick, grayish, and dashed curve. The approximate quadratic function $Q(s)$ is shown as the black curve. The optimum location s^* at $Q' = 0$ is also indicated.

To find the value of optimum location for the objective function, one needs to sample the function at three different s values, s_1, s_2, and s_3, and solving these equations can determine the constants a, b, and c. This process gives the following three equations to be solved simultaneously:

$$\begin{aligned} f_1 &= a + b\,s_1 + c\,s_1^2 \\ f_2 &= a + b\,s_2 + c\,s_2^2 \\ f_3 &= a + b\,s_3 + c\,s_3^2 \end{aligned} \tag{5.16}$$

If we use $s_1 = 0$, $s_2 = t$, and $s_3 = 2t$, as shown in Figure 5.7, we get the following by substituting these values in Eq. (5.16):

$$\begin{aligned} f_1 &= a \\ f_2 &= a + b\,t + c\,t^2 \\ f_3 &= a + 2b\,t + 4c\,t^2 \end{aligned} \tag{5.17}$$

Solving Eq. (5.17) for the coefficients of the quadratic function and after minor manipulations, one gets the following:

$$\begin{aligned} a &= f_1 \\ b &= \frac{4f_2 - 3f_1 - f_3}{2t} \\ c &= \frac{f_1 + f_3 - 2f_2}{2t^2} \end{aligned} \tag{5.18}$$

Substituting the values in Eq. (5.18) into Eq. (5.15) gives the optimum location s^* of the minimum of the quadratic function as given in Eq. (5.19).

$$s^* = \frac{4f_2 - 3f_1 - f_3}{4f_2 - 2f_1 - 2f_3}t \tag{5.19}$$

The case in which Q is quadratic requires that the second derivative is positive ($+^{ve}$) for a minimum; see Figure 5.7, i.e. $c > 0$, or

$$f_3 + f_1 > 2f_2 \tag{5.20}$$

Therefore, f_2 must be below the line connecting f_1 and f_3 as shown in Figure 5.7, where the optimum location s^* is also indicated. The two final acceptable situations are suggested in Figure 5.8.

A good alternative to terminate the search is to compare $F(s^*)$ to $Q(s^*)$ and consider s^* a sufficiently good approximation if $F(s^*)$ and $Q(s^*)$ differ by a small amount. It can be shown that

$$Q(s^*) = a - \frac{b^2}{4c} = f_1 - \frac{(4f_2 - 3f_1 - f_3)^2}{8(f_1 + f_3 - 2f_2)} \tag{5.21}$$

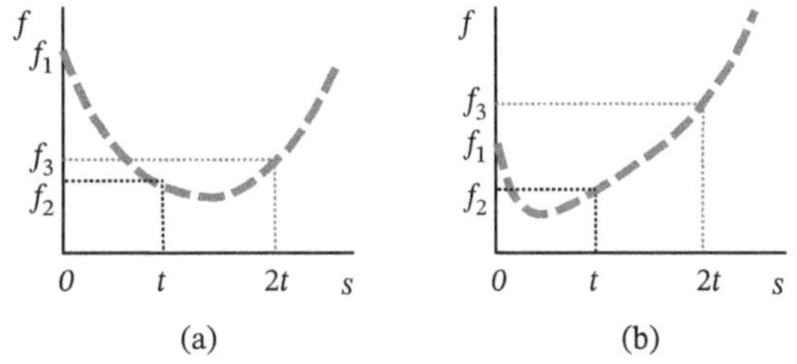

Figure 5.8 Two final acceptable situations for quadratic interpolation are suggested in (a) and (b). For a quadratic objective function, any three points should work for an expected minimum.

We might then require that the following relation holds:

$$\frac{|Q(s^*) - \mathcal{F}(s^*)|}{|Q(s^*)|} < \varepsilon_e \tag{5.22}$$

where ε_e is the acceptable error value, say, 0.01 (1%) or less.

An efficient logic for the algorithm is to double the step t if the situation in Figure 5.8a is not satisfied and f_3 is less than f_2. This case may be repeated until either situation in Figure 5.8 is satisfied. When f_2 is larger than f_1, one may use $t/2$ instead of t or use $-t$. One might use the acceptable error ε_e as an initial step to help in defining the positive or negative step t to start with.

A one-dimensional (***1D***) computer-aided program is available in ***Wiley website*** to apply *quadratic interpolation* and golden section methods to find the optimum of any ***1D*** function. Few inputs are needed as defined at the end of Section 5.2.2.

5.2.2 Golden Section (Euclid)

This method is dated back to Euclid (300 BCE). It has been utilized lately for the optimization procedure. The principle of the *golden section* can be stated as follows: "When dividing a length L into two parts a and b, the ratio of the smaller segment b to the larger segment a should be the same as the ratio of the larger segment a to total length L." The principle is shown in Figure 5.9a. This *golden ratio* has been used in the proportion of ancient Greek's temples. It has been considered as a *beauty* ratio. Beauty may be considered as an optimum look. The golden ratio is used in the optimization procedure provided that the objective function is *unimodal* in the range of search or interest, which is also called the *interval of uncertainty* as shown in Figure 5.9b. Applying the *golden section* principle to the segments in Figure 5.9a, one gets the following relation:

$$\frac{b}{a} = \frac{a}{L} = \tau_G \tag{5.23}$$

where τ_G is the *golden ratio* constant. But the next relation holds as observed from Figure 5.9a.

$$a + b = L \tag{5.24}$$

Substituting the relations in Eq. (5.23) into Eq. (5.24) gives

$$\tau_G L + \tau_G a = \tau_G L + \tau_G(\tau_G L) = L$$

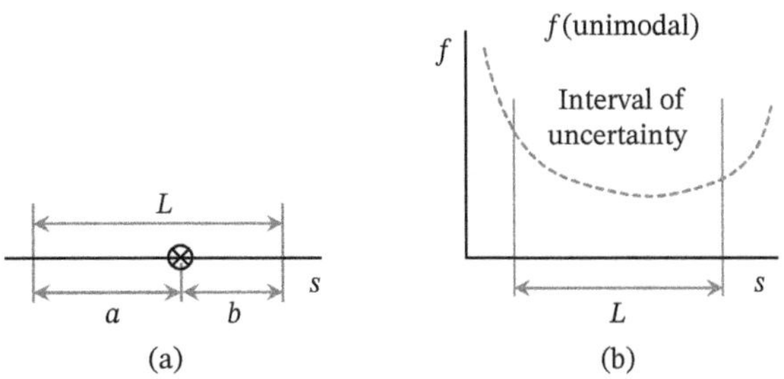

Figure 5.9 The golden section over an interval L that is divided by the golden ratio of a and b as shown in (a). (b) The interval of uncertainty that may be selected for a unimodal function.

or

$$\tau_G^2 + \tau_G - 1 = 0 \tag{5.25}$$

The single positive root of this equation is given by the following *golden ratio*:

$$\tau_G = 0.618\,034 \cong 0.618 \tag{5.26}$$

From that golden ratio, the segments a and b are found by substitution in Eq. (5.23) so that

$$a = 0.618\,034\,L$$
$$b = a\ \tau_G = \tau_G^2\ L = 0.381\,966\,L \tag{5.27}$$

Optimization procedure utilizes the *golden ratio* by stepping from either ends of the interval of uncertainty as defined in the first and second trials in Figure 5.10. The interval of uncertainty is normalized to 1.0, which should not affect the procedure. In Figure 2.8, we have used the following key to simplify the tracking of the procedure from one trial to the next. The gray-filled circle denotes the present trial point. The unfilled circle denotes the previous trial point. The optimum sign * denotes the smaller value of the objective function at that location. From that logic, one can devise the software to perform the optimization. It should be noted that the procedure in Figure 5.10 converges to the right side of the figure. This is just to keep track of the stepping values in obvious clear manner. The convergence can happen in any direction, and the adopted assumption should not affect the logic of the procedure. One should continue the trials till the satisfaction of some stopping criterion such as checking the size of the step to be less than an acceptable error fraction ε_e, say, 0.01 (1%) or less. Table 5.1 summarizes the intervals of uncertainty corresponding to the number of *golden section* trials as a ratio of L. After about 10 trials, the interval of uncertainty is reduced to about 1%. It is also clear that the interval of uncertainty is reduced by the *golden ratio* at each trial, and therefore the interval of uncertainty is reduced by $(\tau_G)^n$ for n trials.

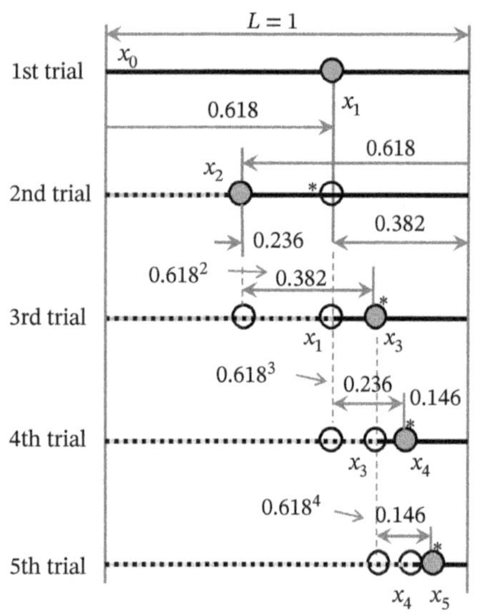

Figure 5.10 The golden ratio is used in the stepping from either ends of the interval of uncertainty as defined in the first and second trials. The gray-filled circle denotes the present trial point. The unfilled circle denotes the previous trial point. The optimum sign * denotes the smaller value of the objective function at that location.

Table 5.1 Interval of uncertainty for the number of golden ratio trials $(0.618)^n$.

Number of trials n	0	1	2	3	4	5	6	7	8	9	10	11	12
Interval of uncertainty $(0.618)^n$	1	1	0.618	0.382	0.236	0.146	0.090	0.056	0.0345	0.0213	0.0132	0.0081	0.0050

From experience, however, the *golden section* procedure is usually less efficient than the *quadratic interpolation*. An optimization program (***1D Optimization.exe***) is provided through the ***Wiley website*** that can be used to find the optimum of any closed form functions using either method. Any ***1D*** problem can be rewritten as a function of x to be typed in the *text box*, or use the *dropdown menu* to select from few imbedded functions to be optimized. Notes are attached to each function between *brackets* to provide information such as the location of the minimum or minima or the maximum or maxima and their location interval. The note should be ***deleted*** before running any of the functions except the first default, just to make life easier. Few other inputs can be used to control the optimization search. The starting step of the search "t," the accepted accuracy to terminate the search, and the initial value of the starting point location "x" can be changed to accommodate the problem at hand. All functions have to be entered in the x variable form. The *interval of uncertainty* should be specified if one should define the range of interest particularly for *golden section* method.

5.2.3 Newton–Raphson

Again, for an optimization problem (and as a reiteration), it is required to find

$$\min_{s} f = \mathcal{F}(s) \tag{5.11}$$

Expanding the function of s specified in Eq. (*5.11*) in terms of Taylor series about a point s_0, one gets

$$\mathcal{F}(s_o + \delta s) = \mathcal{F}(s_o) + \delta s \left.\frac{\partial \mathcal{F}}{\partial s}\right|_{s_o} + \frac{1}{2}(\delta s)^2 \left.\frac{\partial^2 \mathcal{F}}{\partial s^2}\right|_{s_o} + \cdots \tag{5.28}$$

where δs is a small distance from point s_o on the direction s. Taking the derivatives of Eq. (5.28) and ignoring the higher order terms generates the following equation:

$$\left.\frac{\partial \mathcal{F}}{\partial s}\right|_{s_o+\delta s} = \left.\frac{\partial \mathcal{F}}{\partial s}\right|_{s_o} + \delta s \left.\frac{\partial^2 \mathcal{F}}{\partial s^2}\right|_{s_o} \tag{5.29}$$

Since the first derivative at the extremum location of $(s_o + \delta s)$ should be zero, one should have

$$\left.\frac{\partial \mathcal{F}}{\partial s}\right|_{s_o+\delta s} = 0 \tag{5.30}$$

The substitution of Eq. (5.30) into Eq. (5.29) gives the condition of the extremum as

$$\delta s \left.\frac{\partial^2 \mathcal{F}}{\partial s^2}\right|_{s_o} = -\left.\frac{\partial \mathcal{F}}{\partial s}\right|_{s_o} \tag{5.31}$$

This relation defines the *Newton–Raphson* division rule as

$$\delta s = -\frac{\left({}^{\partial \mathcal{F}}/_{\partial s}\right)_{s_o}}{\left({}^{\partial^2 \mathcal{F}}/_{\partial s^2}\right)_{s_o}} = -\frac{\mathcal{F}'_{s_o}}{\mathcal{F}''_{s_o}} \tag{\textbf{\textit{5.32}}}$$

Therefore, we can search for an extremum by stepping from any point s_i to the next point s_{i+1} using the following recursive formula:

$$s_{i+1} = s_i + \delta s \tag{5.33}$$

The *Newton–Raphson* divisor δs is defined by Eq. (5.32). Repeating the recursive process as necessary till the satisfaction of some stopping criterion such as the first derivative or the size of the step being less than acceptable error fraction ε_e, say, 0.01 (1%) or less. The derivatives of the objective function have to be evaluated from a closed form expression, if available, or by numerical approximations; see Chapra and Canale (2010). Numerical evaluation of derivatives, however, needs more evaluations of function values.

It should be noted that *Newton–Raphson* technique needs the evaluation of the first and second derivatives of the objective function. It is then a *second-order* method as stated previously.

5.2.4 Other Methods

Other methods for ***1D*** search can be found in the literature. For zero-order methods where derivatives are not needed or available, other methods may be used such as *dichotomous*, *Fibonacci*, and other possible methods in the literature. If the gradient or derivative of the function being minimized is easily obtained, some of the other searches can be used such as *cubic interpolation*, *direct root method*, etc.; see Fox (1971), Reklaitis et al. (1983), and Rao (2009).

Example 5.4 For Example 5.1, find the optimum length and width of the tray, if the volume is 1.0, the height is 0.5, and the thickness is 0.01. Use quadratic interpolation, and golden section searches. Compare results.

Solution

Data: $V = 1.0$. $H = 0.5$, and $T = 0.01$.

From Eq. (i) of Example 5.1, the function to be minimized is then

$$f = T_T\left(2H_T\frac{V_T}{l_T H_T} + 2H_T l_T + \frac{V_T}{l_T H_T} l_T\right) = 0.01\left(2\frac{1}{l_T} + 2(0.5)l_T + \frac{1}{0.5 l_T} l_T\right)$$

$$= \frac{0.02}{l_T} + 0.01\ l_T + \frac{0.01}{0.5} = \frac{0.02}{l_T} + 0.01\ l_T + 0.02 \tag{a}$$

Since the thickness does not affect the optimum, we could have eliminated it from the objective function. However, the objective function would not have the direct representation of the tray volume per se. The optimization problem is then to find l_T for

$$\min_{l_T}\ f = \frac{0.02}{l_T} + 0.01\ l_T + 0.02 \tag{b}$$

Table 5.2 presents the value of the objective function f or tray material volume v_T at different lengths l_T. It is clear that the function is highly nonlinear with a singularity at $l_T = 0.0$, where the value of the function $f = \infty$. Using any optimization procedure with a starting point at $l_T = 0.0$ will not work. Defining an interval of uncertainty between 1.0 and 2.0 would work for all methods. As a comparison, consider an initial step of 0.3 for *quadratic interpolation*. The three values of the objective functions are as follows:

$$f_1\ (1) = 0.02/1 + 0.01\ \ (1) + 0.02 = 0.050\,000$$

$$f_2\ (1.3) = 0.02/1.3 + 0.01\ \ (1.3) + 0.02 = 0.048\,384\,6$$

$$f_3\ (1.6) = 0.02/1.6 + 0.01\ \ (1.6) + 0.02 = 0.048\,500\,0 \tag{c}$$

Using $s_o = 1.0$ as a starting point, the value of the optimum location s^* is given by Eq. (5.19) that gives

$$s^* = \frac{4f_2 - 3f_1 - f_3}{4f_2 - 2f_1 - 2f_3}t = \frac{4(0.048\,384\,6) - 3(0.05) - 0.0485)}{4(0.048\,384\,6) - 2(0.05) - 2(0.0485)}(0.3)$$

$$= \frac{-0.004\,961\,6}{-0.003\,461\,6}(0.3) = 0.429\,998 \tag{d}$$

The new point is then obtained from the following relation:

$$x_1 = x_0 + s^* = 1.0 + 0.429\,998 = 1.429\,998 \tag{e}$$

Table 5.2 Values of objective function of the tray (Example 5.1) at different lengths.

Variable l_T	−0.5	0	0.5	1	1.5	2	2.5
Objective function value f	−0.025	∞	0.065	0.05	0.0483	0.05	0.053

Table 5.3 Values of objective function of the tray (Example 5.1) at consecutive golden section lengths.

Variable $l_{T1\ to\ T7}$	1.618	1.381 92	1.2361	1.4721	1.3262	1.4164	1.4377
Objective function $f \times 10^2$	4.8541	4.8292	4.8541	4.8307	4.8343	4.8284	4.8288

This value is compared with Eq. (k) in Example 5.1 for the real optimum of $l_T^* = \sqrt{(V_T/H_T)} = \sqrt{2} = 1.414\,214$ or a difference of 0.015 784, i.e. 1.1%.

Defining an interval of uncertainty between 1.0 and 2.0 generates an interval length of 1.0. The *golden section* would then require stepping 0.618 from point $l_T = 1.0$ and another 0.618 back from point $l_T = 2.0$; see Figure 5.10. The first point is then $l_{T1} = 1.0 + 0.618 = 1.618$. The following point is $l_{T2} = 2.0 - 0.618 = 1.382$, or $l_{T2} = 1.0 + (0.618)^2 = 1.0 + 0.3819 = 1.3819$; see Figure 5.10. Tables 5.2 and 5.3 lists these points and the following points for the next point l_{T3} … etc. and the values of the objective function at these points.

Since the value of the function at l_{T2} is less than the value of the function at l_{T1}, the interval of uncertainty becomes 1.0–1.618, and the following point should be $l_{T3} = 1.618 - (0.618)^3 = 1.2361$ as seen in Table 5.3. Since the value of the function at l_{T2} is less than the value of the function at l_{T3}, the interval of uncertainty becomes 1.2361–1.618, and the following point should be $l_{T4} = 1.618 - (0.618)^4 = 1.4721$ as seen in Table 5.3. Since the value of the function at l_{T4} is less than the value of the function at l_{T3}, the interval of uncertainty becomes 1.2361–1.4721, and the following point should be $l_{T5} = 1.2361 + (0.618)^5 = 1.3262$ as seen in Table 5.3. Since the value of the function at l_{T5} is less than the value of the function at l_{T4}, the interval of uncertainty becomes 1.3262–1.4721, and the following point should be $l_{T5} = 1.3262 + (0.618)^5 = 1.4164$ as seen in Table 5.3. Since the value of the function at l_{T6} is less than the value of the function at l_{T5}, the interval of uncertainty becomes 1.3819–1.4721, and the following point should be $l_{T7} = 1.3819 + (0.618)^5 = 1.4377$ as seen in Table 5.3. Since the value of the function at l_{T6} is less than the value of the function at l_{T7}, the interval of uncertainty becomes 1.3819–1.4377.

The values of the objective function at the last two tray lengths are very close, and one can either terminate the trials or continue for more accurate values. The smallest objective function in Table 5.3 is 0.048 284 at $l_{T6} = 1.4164$. The value of the optimum tray length l^* is 1.414 214 as indicated previously. After seven golden section trials, the length at the minimum objective function differs by about 0.0022 from the optimum, which is about 0.2%. The interval of uncertainty after seven trials, however, is (1.4377 − 1.3819) or 0.0558, which is about 5.6% of original interval of uncertainty. This is about the same value observed in Table 5.1 after seven trials.

5.3 Multidimensional: Classical Indirect Approach

In this section we consider the *classical indirect* approach to the solution of *multidimensional* design problem. This approach is familiar to the reader if he has been introduced to the subject of finding the minimum or the maximum of a function in general courses of calculus. Here we expand and formalize the information with respect to our needs in design problems of unconstrained and constrained optimization.

5.3.1 Unconstrained Problem

If we have an objective function f defined as a function $\mathcal{F}$ of several design variables arranged in the design vector $\boldsymbol{X}$ such that

$$f(\boldsymbol{X}) = \mathcal{F}(x_1, x_2, \ldots, x_n) = \mathcal{F}(\boldsymbol{X}) \tag{5.34}$$

The optimization problem will then be to find

$$\min_{\boldsymbol{X}} \ f(\boldsymbol{X}) = ? \tag{5.35}$$

This will occur at a point where

$$f_i = \frac{\partial \mathcal{F}}{\partial X_i} = 0 \tag{5.36}$$

Provided that the *Hessian matrix* $\boldsymbol{H}$, after Ludwig Otto Hesse (1811–1874),

$$\boldsymbol{H} = \begin{bmatrix} f_{11} & f_{12} & \cdots & f_{1n} \\ f_{21} & f_{22} & \cdots & f_{2n} \\ \cdots & \cdots & \cdots & \cdots \\ f_{n1} & \cdots & \cdots & f_{nn} \end{bmatrix} \tag{5.37}$$

is a "*positive definite*" matrix, where

$$f_{jk} = \frac{\partial^2 \mathcal{F}}{\partial x_j \partial x_k} \tag{5.38}$$

It should be of interest to note that the matrix will be positive definite *iff* (if and only if) all its *eigenvalues* are positive.

Using *second partial derivative test*, a local *minimum* will then be realized if (Abramowitz and Stegun 1972)

$$D_{H,n} > 0, \quad n = 1, 2, \ldots \tag{5.39}$$

where $D_{H,n}$ is the determinant of $\boldsymbol{H}$ and

$$D_{H,n} = \begin{vmatrix} f_{11} & f_{12} & \cdots & f_{1n} \\ f_{21} & f_{22} & \cdots & f_{2n} \\ \cdots & \cdots & \cdots & \cdots \\ f_{n1} & \cdots & \cdots & f_{nn} \end{vmatrix} \tag{5.40}$$

For other extremum cases:

- If $\boldsymbol{H}$ is negative definite, f is a *maximum*.
- If $\boldsymbol{H}$ is indefinite we are at a *saddle* point.
- If $\boldsymbol{H}$ is positive or negative semidefinite, this test fails (we have a valley).

By this method, a local *maximum* will be realized if

$$\begin{aligned} D_{H,n} &> 0, \quad n = 2, 4, 6, \ldots \\ D_{H,n} &< 0, \quad n = 1, 3, 5, \ldots \end{aligned} \tag{5.41}$$

There will be a *saddle* point if any of

$$D_{H,n} = 0, \quad n = 1, 2, 3, \ldots \tag{5.42}$$

One can also check the following condition for a saddle (Stewart 2005):

$$D_{H,n} < 0, \quad n = 2, 4, 6, \ldots$$

Example 5.5 For a ***2D*** case (i.e. $n = 2$), an *extremum* occurs when $f_1 = f_2 = 0$:

- A minimum occurs when

$$\begin{aligned} & f_{11} > 0 \\ & f_{11} f_{22} - f_{12}^2 > 0 \end{aligned} \tag{a}$$

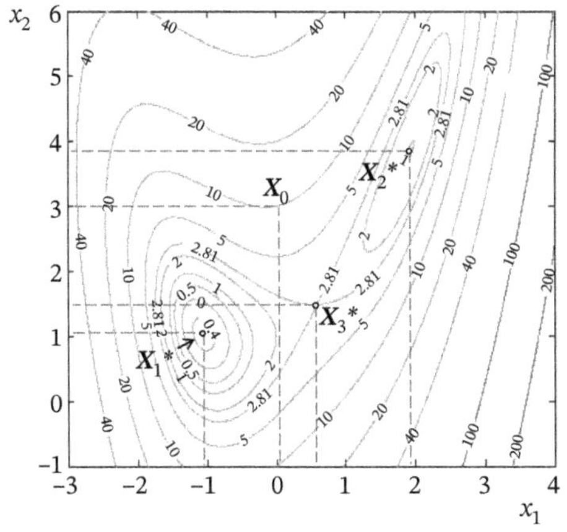

Figure 5.11 The contours of the objective function of Example 5.6. Two function minima are at $\boldsymbol{X}_1^*$ and $\boldsymbol{X}_2^*$ A saddle point is at $\boldsymbol{X}_3^*$ Starting the search at $\boldsymbol{X}_0$ may end at any of the minima or at the saddle point.

- A maximum occurs when

$$\begin{aligned} &f_{11} < 0 \\ &f_{11}f_{22} - f_{12}^2 > 0 \end{aligned} \tag{b}$$

- For a "saddle point," check

$$\begin{aligned} &f_{11}f_{22} - f_{12}^2 = 0 \\ &\text{or } f_{11} = 0, \quad \text{or } f_{22} = 0 \end{aligned} \tag{c}$$

Example 5.6 Let us consider the following objective function and find its extremum (Fox 1971):

$$f(\boldsymbol{X}) = 4 + 4.5x_1 - 4x_2 + x_1{}^2 + 2x_2^2 - 2x_1x_2 + x_1^4 - 2x_1^2x_2 \tag{a}$$

Solution

This function has continuous derivatives of all orders and hence at the minimum.

$$\begin{aligned} \frac{\partial f}{\partial x_1} &= 4.5 + 2x_1 - 2x_2 + 4x_1^3 - 4x_1x_2 = 0 \\ \frac{\partial f}{\partial x_2} &= -4 + 4x_2 - 2x_1 - 2x_1^2 = 0 \end{aligned} \tag{b}$$

Solving these equations simultaneously gives relative function minima of 0.9855 at $\boldsymbol{X}_1^* = (1.941, 3.854)$ and -0.5134 at $\boldsymbol{X}_2^* = (-1.053, 1.028)$. The contours of the objective function are shown in Figure 5.11. Note that there is a saddle point at $\boldsymbol{X}_3^* = (0.557\,469, 1.4909)$ where $f_1 = f_2 = 0$, and $\boldsymbol{H} = [f_{jk}]$ is not positive definite. To verify these results, we can easily find that

$$\boldsymbol{H} = \begin{bmatrix} 2 + 12x_1^2 - 4x_2 & -2 - 4x_1 \\ -2 - 4x_1 & 4 \end{bmatrix} \tag{c}$$

At $\boldsymbol{X}_1^* = (-1.053, 1.028)$, the *Hessian* matrix is

$$\boldsymbol{H}_{X_1} = \begin{bmatrix} 19.4177 & -6.212 \\ -6.212 & 4 \end{bmatrix} \tag{d}$$

From Eq. (d), $D_{H1} = f_{11} = 19.4177 > 0$ and $D_{H2} = (19.4177*4) - (-6.212)^2 = 39.0819 > 0$. Then the function at $\boldsymbol{X_1}$ is a minimum.

At $X_2^* = (1.941, 3.854)$, the *Hessian* matrix is

$$H_{X_2} = \begin{bmatrix} 31.7938 & -9.764 \\ -9.764 & 4 \end{bmatrix} \tag{e}$$

From Eq. (e), $D_{H1} = f_{11} = 31.7938 > 0$ and $D_{H2} = (31.7938*4) - (-9.764)^2 = 31.8395 > 0$. Then the function at X_2 is a minimum.

At $X_3^* = (0.557\,469, 1.4909)$, the *Hessian* matrix is

$$H_{X_3} = \begin{bmatrix} 0.000\,023 & -4.298 \\ -4.298 & 4 \end{bmatrix} \tag{f}$$

From Eq. (f), $D_{H1} = 0$ and $D_{H2} = (0.000\,023*4) - (-4.298)^2 = -18.4727 < 0$. Then X_3 is a saddle point.

Figure 5.11 shows an arbitrary starting point X_0 of some optimization search. It is close to both minima X_1 and X_2. Many optimization techniques may, however, converge to the saddle point X_3 as shown later on in Example 5.12.

5.3.2 Equality Constrained Problem

In this case and for a ***2D*** example, the optimization problem is to find

$$\min_X \; f(X) = \mathcal{F}(x_1, x_2) = ? \tag{5.43}$$

Subject to the following equality constraint

$$\varphi(x_1, x_2) = 0 \tag{5.44}$$

Such a case is illustrated as shown previously in Figure 5.3. One straightforward method for resolving the problem would be to solve the equation $\varphi\,(x_1, x_2) = 0$ and obtain a relation between x_1 and x_2.

Substitution of this relationship into $\mathcal{F}\,(x_1, x_2)$ will reduce the minimization problem to one of a single independent variable either x_1 or x_2.

In general, the optimization problem can be stated to find

$$\min_X \quad f(X) = \mathcal{F}(x_1, x_2, \ldots,\; x_n) = ? \tag{5.45}$$

Subject to the m equality constraints defined by

$$\varphi_i(x_1, x_2, \ldots\; , x_n) = 0 \quad i = 1, 2, \ldots, m \tag{5.46}$$

If the "m" simultaneous Eq. (5.46) can be solved for "m" of the independent variables, these can be substituted into Eq. (5.45), thereby reducing the problem to one of "$n - m$" independent variables. It is, however, extremely difficult, if not impossible, to obtain explicit expressions for "m" of the variables in Eq. (5.46). For these types of situations, we may use the method of "*Lagrange multipliers*," which is described in the following segment:

5.3.2.1 Lagrange Multipliers

Again the constrained optimization problem is to find

$$\min_X \; f(X) = ?, \quad X = (n \times 1) \tag{5.45}$$

subject to the equality constraints

$$\boldsymbol{\varphi}(X) = \mathbf{0}, \quad \boldsymbol{\varphi} = (m \times 1) \tag{5.46$'$}$$

To solve this problem, we define a new *augmented function*

$$\bar{f} = f + \lambda^T \boldsymbol{\varphi}, \quad \lambda = (m \times 1) \tag{5.47}$$

where the "m" undetermined multipliers in the vector λ are the *Lagrange multipliers*.

For $\bar{\boldsymbol{f}}$ to have a minimum, which minimizes $\boldsymbol{f}$, we should have

$$\frac{\partial \bar{\boldsymbol{f}}}{\partial x_i} = 0 \tag{5.48}$$

In expanded form, these are

$$\nabla \boldsymbol{f} + [\nabla \boldsymbol{\varphi}^T]\lambda = \mathbf{0} \tag{5.49}$$

where

$$\nabla \boldsymbol{f} = \text{"del"}\boldsymbol{f} = \begin{bmatrix} f_1 & f_2 & \ldots & \ldots & f_n \end{bmatrix}^T, \quad \nabla \boldsymbol{f} = (n \times 1) \tag{5.50}$$

and

$$[\nabla \boldsymbol{\varphi}^T] = \begin{bmatrix} \nabla \boldsymbol{\varphi}_1 & \nabla \boldsymbol{\varphi}_2 & \ldots & \ldots & \nabla \boldsymbol{\varphi}_m \end{bmatrix} = (n \times m) \tag{5.51}$$

The "n" Eq. (5.49) together with the "m" constraining Eq. (5.46′) allows the solution of the "$n + m$" unknowns in $\boldsymbol{X}$ and λ.

Example 5.7 Right circular cylindrical cans are to hold a given volume V_c. In can manufacturing, there is no waste in cutting the material of the vertical side of the can. Each end piece, however, is to be cut from a square. The corners of the square are wasted. The cost of the unit seam length to the cost of unit surface area is C_{1s}. It is required to find the most economical ratio of the height h_c to the diameter d_c.

Solution

To formulate the objective function, we note that the cost is proportional to the total surface area and the total seam length. The objective function can then take the form

$$f = \pi\, d_c\, h_c + 2d_c^2 + C_{1s}(h_c + 2\,\pi\, d_c) \tag{a}$$

where d_c is the can diameter and h_c is the height of the can. The volume constraint V_c, however, gives

$$h_c = 4V_c/\pi d_c^2 \tag{b}$$

By direct substitution, we "may not" be able to solve explicitly for the optimum d_c for some other problems. Therefore *Lagrange multipliers* are used in this problem for demonstration. For that, the following augmented function is formed:

$$\bar{f} = f + \lambda((\pi\, d_c^2\ h_c/4) - V_c) \tag{c}$$

Taking the derives with respect to d_c, h_c, and λ then equate each to zero gives

$$\begin{aligned} \bar{f}_{d_c} &= \pi\, h_c + 4d_c + 2\pi C_{1c} + \lambda\pi\, d_c h_c/2 = 0 \\ \bar{f}_{h_c} &= \pi d_c + C_{1c} + \lambda\, \pi\, d_c^2/4 = 0 \\ \bar{f}_{\lambda} &= (\pi\, d_c^2\, h_c/4) - V_c = \varphi = 0 \end{aligned} \tag{d}$$

From the second equation in (d), one can get

$$\lambda = (-4/d_c) - (4C_{1c}/\pi d_c^2) \tag{e}$$

Substituting into the first equation in (d), we get the optimum h_c^* as

$$h_c^* = d_c(4d_c + 2\pi\, C_{1c})/(\pi\, d_c + 2C_{1c}) \tag{f}$$

Substituting into Eq. (d) gives the optimum d_c^* as

$$d_c^* = (4V_c/\pi\, h_c^*)^{0.5} \tag{g}$$

With the defined value of the volume V_c and the cost of the unit seam length to the cost of unit surface area C_{ls}, Eqs. (f) and (g) are solved iteratively to get the optimum can diameter and height. Dividing Eq. (f) by Eq. (g), we can get the optimum ratio of the can geometry.

$$\frac{h_c^*}{d_c^*} = \left(\frac{d_c(4d_c + 2\pi\, C_{lc})}{(\pi\, d_c + 2C_{lc})(4V_c/\pi\, h_c^*)^{0.5}}\right) \tag{h}$$

For a volume of 1.0 and an equal cost of the unit seam length to the cost of unit surface area, the optimum values are almost $d_c^* = 0.849\,99$ and $h_c^* = 1.7623$. The ratio of optimum height to optimum length is about 2.0733.

5.3.3 Inequality Constraints Problem

In this case *inequality constraints* exist in an optimization problem defined as in Eq. (5.45). These constraints take the following form:

$$g_j(\mathbf{X}) \le 0, \quad j = 1, 2, \ldots, m \tag{5.52}$$

We can still apply the method of *Lagrange multipliers*. This can be achieved by using what we call "*slack*" variables. A new augmented function will then be defined as

$$\bar{\boldsymbol{f}} = \boldsymbol{f} + \lambda^T(\mathbf{g} + \mathbf{s}^2), \quad \mathbf{s} = (m \times 1) \tag{5.53}$$

where s_i is the *slack* variable denoting the amount by which the constraint g_i is ineffective. The slack variable is used squared in Eq. (5.53) to ensure that the constraints are always <0. For $\bar{f}$ to have a minimum, we should have

$$\nabla\bar{\boldsymbol{f}} = \mathbf{0} \tag{5.54}$$

which gives, due to the *slack* variables in the vector $\mathbf{s}$, the following additional conditions to the constrained problem:

$$f_{sj} = 2\lambda_j\, s_j = 0, \quad j = 1, 2, \ldots, m \tag{5.55}$$

These conditions are given the name of the ***Kuhn–Tucker (K–T) slackness*** conditions.

Example 5.8 Find the minimum of the objective function

$$f = 2x_1^2 - 2x_1x_2 + 2x_2^2 - 6x_1 \tag{a}$$

$$\text{Subject to}: \quad \begin{aligned} 3x_1 + 4x_2 &\le 6 \\ -x_1 + 4x_2 &\le 2 \end{aligned} \tag{b}$$

Solution

The augmented function becomes

$$\begin{aligned}\bar{f} &= 2x_1^2 - 2x_1x_2 + 2x_2^2 - 6x_1 \\ &\quad + \lambda_1(3x_1 + 4x_2 - 6 + s_1^2) + \lambda_2(-x_1 + 4x_2 - 2 + s_2^2)\end{aligned} \tag{c}$$

Taking the derivatives and equate each one to zero, we get

$$\overline{f_1} = 4x_1 - 2x_2 - 6 - 3\lambda_1 - \lambda_2 = 0, \quad \overline{f_2} = 2x_1 + 4x_2 + 4\lambda_1 + 4\lambda_2 = 0$$
$$\overline{f_{\lambda_1}} = 3x_1 + 4x_2 - 6 + s_1^2 = 0, \quad \overline{f_{\lambda_2}} = -x_1 + 4x_2 - 2 + s_2^2 = 0$$
$$\overline{f_{s_1}} = 2\lambda_1 s_1 = 0, \quad \overline{f_{s_2}} = 2\lambda_2 s_2 = 0 \tag{d}$$

Solving these equations simultaneously gives the values $x_1 = 1.4594, x_2 = 0.4054, \lambda_1 = -0.3245, \lambda_2 = 0, s_1 = 0$, and $s_2 = 1.355\,65$. This locates the constrained optimum and indicates that it lies on the first constraint ($\lambda_1 = -0.3245$, and $s_1 = 0$), which indicates that the constraint is *in effect*, or it is an *effective constraint*. It also confirms that the second constraint is an *ineffective constraint*, i.e. it is not affecting the solution. The constrained optimum is not on the second constraint ($\lambda_2 = 0$, and $s_2 = 1.355\,65$). Note that there is *slackness* of s_2 value between the second constraint and the constrained optimum.

5.4 Multidimensional Unconstrained Problem

In this section an overview is presented for the prospective *direct methods* of finding the optimum of multidimensional objective functions. Few of these methods are given as first intuitive means, even though they may not necessarily be efficient. Techniques that are most suited for numerical implementation are of main interest. The cases of unconstrained and constrained optimization problems are provided in Sections 5.4 and 5.5, respectively.

The solution of unconstrained optimization problem involves techniques that are most suited for numerical implementation. Procedures that do not require derivative evaluations are considered *zero-order* methods such as *univariate* and *Powell's conjugate directions*. *First-order* methods require the evaluation of first derivatives of the objective function such as *steepest decent*, *conjugate gradients*, and *quasi-Newton methods*. Methods involving the evaluation of second derivatives are *second-order* techniques such as *Newton–Raphson* method.

5.4.1 Univariate Method

The idea is to take steps in well-defined directions (e.g. coordinates), which reduce the objective function $\boldsymbol{f}$. This is a first intuitive scheme of searches. This method, however, is usually time consuming, may be inefficient, and can bug down and terminate prematurely for complex, highly nonlinear, and high dimensionality. It may be used for limited number of dimensions such as two or three design variables with caution and concern for possible premature termination. Few means to implement the search are such as the following alternatives:

i. ***First Option***

   The procedure in that option is to follow the steps as shown in Figure 5.12a for a ***2D*** function:

   1. Hold all x_j constant except x_i.
   2. Step a small ε in the direction x_i using the recursive equation defining the next or new point X_{q+1} stepped from the current or old point X_q.

   $$\boldsymbol{X}_{q+1} = \boldsymbol{X}_q + \varepsilon\ \boldsymbol{S}_i \tag{5.56}$$

   where $q = i$ and $\boldsymbol{S}_i = [0 \quad 0 \quad 0 \quad \dots \quad s_i \quad \dots \quad 0]^T$ with $s_i = 1$. If the new function $\boldsymbol{f}_{q+1}$ is less than old $\boldsymbol{f}_q$, one would accept the new $\boldsymbol{X}_{i+1}$. If not, one should try the other direction and/or reduce the step size (Figure 5.12a, as the dashed directions).
   3. Advance i to consider another coordinate, and repeat until changes in $\boldsymbol{f}$ satisfy some stopping rule.

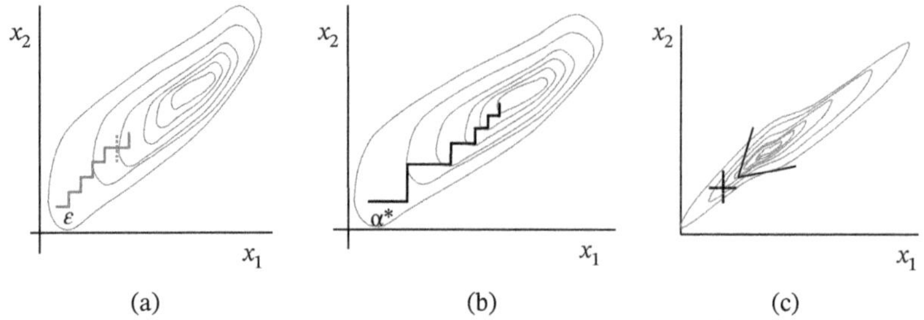

Figure 5.12 Univariate method in ***2D***: (a) first option of stepping ε in each direction, (b) second option of performing ***1D*** optimization in each direction where α^* is locating the minimum in each direction, and (c) indicates condition of univariate search failure.

ii. ***Second Option***

The steps in that option are as follows:

1. Hold all x_j constant except x_i.
2. Perform ***1D*** search optimization in the direction $\boldsymbol{S}_i$ to minimize $\boldsymbol{f}$ (Figure 5.12b) using the following general recursive equation:

$$\boldsymbol{X}_{q+1} = \boldsymbol{X}_q + \alpha^* \ \boldsymbol{S}_i \tag{5.57}$$

where α^* is the optimum distance in the direction $\boldsymbol{S}_i$.

3. Increment i to consider another coordinate and perform a minimization along that coordinate.
4. Continue until some stopping rule is satisfied.

The *univariate* method fails when none of the univariate directions are "*downhill*" as shown (Figure 5.12c) when a hypothetical function with a sharp "ridge" exists.

As shown in Figure 5.12, one may recognize a trend of the search envelope. This trend or pattern is obtained from connecting the search ends or envelopes after all directions are searched in each loop. It is used in another development of an optimization method that is called the "*pattern move;*" see Fox (1971), Reklaitis et al. (1983), or Rao (2009).

5.4.2 Powell's Method of Conjugate Directions

This method improves the convergence of the *pattern move* method. The *Powell's* procedure is as follows (Powell 1964):

1. Start with *univariate* search minimizing in each i of n coordinate directions; see Figure 5.13 for ***2D*** case as steps 1 and 2.
2. Minimizing in the associated pattern direction $\boldsymbol{S}_{n+1}$; see Figure 5.13 for step 3 as a dashed direction with $n = 2$ in the next general equation for n dimensions.

$$\boldsymbol{S}_{n+1} = \boldsymbol{X}_n - \boldsymbol{X}_0 \tag{5.58}$$

3. Discard one of the coordinate directions in favor of the pattern direction $\boldsymbol{S}_{n+1}$ for inclusion in the next m minimizations; see Figure 5.13 for ***2D*** case as step 2 after step 3.
4. Generate a new pattern direction and again replace one of the coordinate directions; see Figure 5.13 for ***2D*** case as step 3 after 2. In general form, we have the following relation for the pattern directions:

$$\boldsymbol{S}_{q+1} = \boldsymbol{X}_q - \boldsymbol{X}_{q-m} \tag{5.59}$$

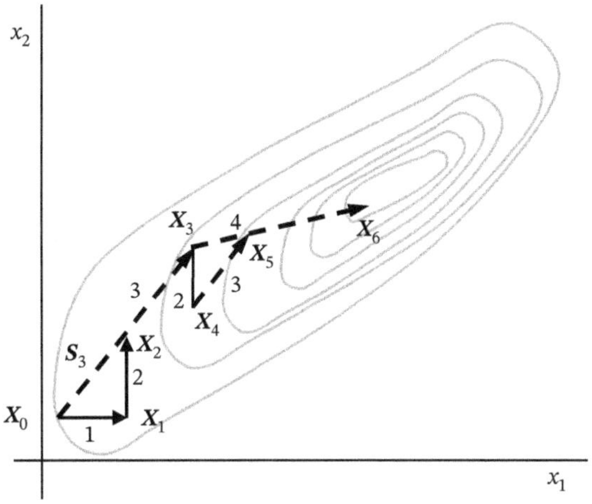

Figure 5.13 The Powell's procedure in ***2D***. Steps 1 and 2 are univariate searches. Direction 3 is A conjugate to direction 4.

These parallel maneuvers result in the *conjugate directions*, which are defined shortly.

This *Powell's* method converges to the minimum in $n(n+1)$ minimizations on a *quadratic function*, which is known as the *quadratic convergence*. A *quadratic function* is defined as follows:

$$f(\boldsymbol{X}) = \frac{1}{2}\boldsymbol{X}^T\boldsymbol{A}\boldsymbol{X} + \boldsymbol{X}^T\boldsymbol{B} + C \tag{5.60}$$

where $\boldsymbol{A}$ is an $(n \times n)$ matrix, which is symmetric. The $\boldsymbol{A}$ matrix in a quadratic function is a matrix of constants, and each term of the matrix is the second derivative of the function f. It is, therefore, equal to the *Hessian* matrix $\boldsymbol{H}$.

Conjugate directions are defined as follows where the ***2D*** case is used as a simplified treatment.

For n vectors $\boldsymbol{S}_i$, $i = 1, 2, \ldots, n$ to be $\boldsymbol{A}$-conjugate, they should satisfy the following relations:

$$\boldsymbol{S}_i^T\boldsymbol{A}\ \boldsymbol{S}_j = 0 \quad \text{for all } i \neq j \tag{5.61}$$

As a simplified demonstration of ***2D*** treatment and for $\boldsymbol{S}_1$ and $\boldsymbol{S}_2$ to be $\boldsymbol{A}$-conjugate, one should have the following:

$$\begin{aligned} \boldsymbol{S}_1^T\boldsymbol{A}\ \boldsymbol{S}_2 &= 0 \\ \boldsymbol{S}_2^T\boldsymbol{A}\ \boldsymbol{S}_1 &= 0 \end{aligned} \tag{5.61$'$}$$

If we start at a point $\boldsymbol{X}_0$ and perform ***1D*** searches in $\boldsymbol{S}_1$ and $\boldsymbol{S}_2$ directions, we get (see Figure 5.13)

$$\boldsymbol{X}_1 = \boldsymbol{X}_0 + \alpha_1^*\boldsymbol{S}_1 \tag{5.62}$$

$$\boldsymbol{X}_2 = \boldsymbol{X}_0 + \alpha_1^*\boldsymbol{S}_1 + \alpha_2^*\boldsymbol{S}_2 \tag{5.63}$$

For a *quadratic* function $f(\boldsymbol{X})$, the minimum is located where

$$\frac{\partial f}{\partial \alpha_1} = 0 \text{ and } \frac{\partial f}{\partial \alpha_2} = 0, \text{ or } \nabla \boldsymbol{f} = \boldsymbol{A}\ \boldsymbol{X} + \boldsymbol{B} = \boldsymbol{0} \tag{5.64}$$

From that and performing the differentiation of the quadratic function at $\boldsymbol{X}_1$ and $\boldsymbol{X}_2$, and solving for the optimum α_1^* and α_2^*, one finds that

$$\begin{aligned} \alpha_1^* &= \frac{-\boldsymbol{S}_1^T(\boldsymbol{A}\boldsymbol{X}_0 + \boldsymbol{B})}{\boldsymbol{S}_1^T\boldsymbol{A}\boldsymbol{S}_1} \\ \alpha_2^* &= \frac{-\boldsymbol{S}_2^T(\boldsymbol{A}\boldsymbol{X}_0 + \boldsymbol{B})}{\boldsymbol{S}_2^T\boldsymbol{A}\boldsymbol{S}_2} \end{aligned} \tag{5.65}$$

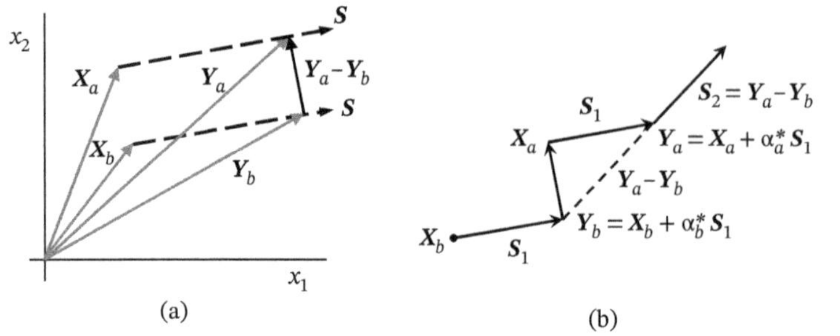

Figure 5.14 Powell's method generates conjugate directions; (a) two optimum points on directions S generates a conjugate direction and (b) the conjugate direction S_2 is used as a search direction to its A conjugate direction S_1.

Note that α_1^* does not depend on $\boldsymbol{S}_2$ and α_2^* does not depend on $\boldsymbol{S}_1$. Thus, the sequential stepping produces the same point, which is the minimum in the plane.

In general, the sequential minimization along the *conjugate vectors* $\boldsymbol{S}_i$, $i = 1, 2, 3, \ldots, n$ produces the minimum point of the function f (in the subspace spanned by the vectors $S_1, \ldots, S_j$). Thus, at or before the nth step, the global minimum of the function f will be reached.

To show that *Powell's* method generates conjugate directions, refer to Figure 5.14. In Figure 5.14a, take two vectors $\boldsymbol{X}_a$ and $\boldsymbol{X}_b$, and also consider two directions $\boldsymbol{S}$ each from the ends of the two vectors. If $\boldsymbol{Y}_a$ is a minimum of the function f from $\boldsymbol{X}_a$ along $\boldsymbol{S}$, and $\boldsymbol{Y}_b$ is a minimum from $\boldsymbol{X}_b$ along $\boldsymbol{S}$, i.e. if

$$\boldsymbol{Y}_a = \boldsymbol{X}_a + \alpha_a^* \boldsymbol{S}$$
$$\boldsymbol{Y}_b = \boldsymbol{X}_b + \alpha_b^* \boldsymbol{S} \tag{5.66}$$

Then $\boldsymbol{Y}_a - \boldsymbol{Y}_b$ and $\boldsymbol{S}$ are $\boldsymbol{A}$-conjugate. To proceed with the proof, by definition

$$\frac{d}{d\alpha}\{f(\boldsymbol{Y}_\text{a} + \alpha\ \boldsymbol{S})\} = 0, \quad \text{at} \quad \alpha = 0$$
$$\frac{d}{d\alpha}\{f(\boldsymbol{Y}_\text{b} + \alpha\ \boldsymbol{S})\} = 0, \quad \text{at} \quad \alpha = 0 \tag{5.67}$$

By the substituting $\boldsymbol{Y}_a$ and $\boldsymbol{Y}_b$ into the equation of the quadratic $f(\boldsymbol{X})$, differentiating, and then setting $\alpha = 0$, we obtain the following:

$$\boldsymbol{S}^T(\boldsymbol{A}\boldsymbol{Y}_a + \boldsymbol{B}) = 0$$
$$\boldsymbol{S}^T(\boldsymbol{A}\boldsymbol{Y}_b + \boldsymbol{B}) = 0 \tag{5.68}$$

Subtracting the two equations in (5.68), one gets

$$\boldsymbol{S}^T\boldsymbol{A}(\boldsymbol{Y}_a - \boldsymbol{Y}_b) = 0 \tag{5.69}$$

This demonstrates the *conjugacy* of $\boldsymbol{S}$ and $\boldsymbol{Y}_a - \boldsymbol{Y}_b$.

In Powell's method we have the process shown in the Figure 5.14b. From this figure, one can conclude that

$$\boldsymbol{S}_1^T\boldsymbol{A}\ \boldsymbol{S}_2 = 0 \tag{5.61'}$$

This indicates that S_1 and $\boldsymbol{S}_2$ are $\boldsymbol{A}$-conjugate.

Example 5.9 A ***2D*** quadratic objective function has the following form:

$$f = 3x_1^2 - 4x_1x_2 + 3x_2^2 \tag{a}$$

It is required to find the optimum point by *Powell's conjugate directions* if one starts at (4, 4).

(a)

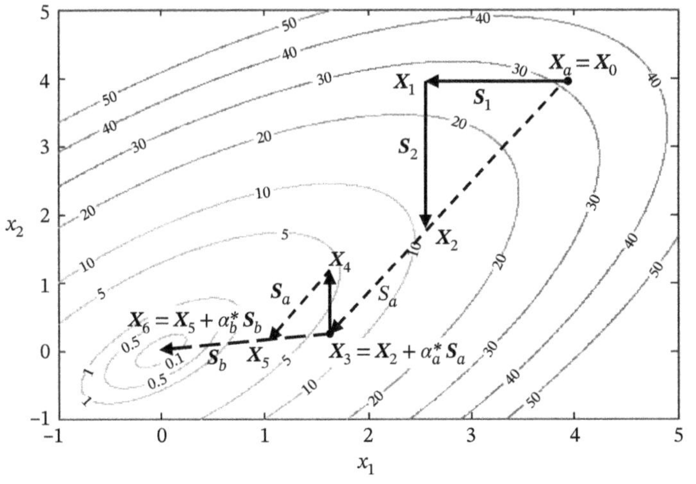

(b)

```
x10=4; x20=4; X0=[x10;x20]                 % Starting point
C=0;B=[0;0];                               % Quadratic Function defined
A=[6 -4; -4 6];
s1=[1;0]; s2=[0;1];                        % Search directions
alfa1=(-s1'*(A*X0+B))/(s1'*A*s1);          % Optimum on search s1
X1=X0+ alfa1*s1                            % Next point X1
alfa2=(-s2'*(A*X1+B))/(s2'*A*s2);          % Optimum on search s2
X2=X1+ alfa2*s2                            % Next point X2
Sa=X2-X0;                                  % Conjugate direction 1
alfaSa=(-Sa'*(A*X2+B))/(Sa'*A*Sa);          % Optimum on search s3
X3=X2+ alfaSa*Sa                           % Next point X3
alfa2=(-s2'*(A*X3+B))/(s2'*A*s2);          % Optimum on search s2
X4=X3+ alfa2*s2%;                          % Next point X4
Sa=X2-X0;                                  % Conjugate direction 1
alfaSa=(-Sa'*(A*X4+B))/(Sa'*A*Sa);         % Optimum on search Sa
X5=X4+ alfaSa*Sa %;                        % Next point X5
Sb=X5-X3;                                  % Conjugate direction 2
alfaSb=(-Sb'*(A*X5+B))/(Sb'*A*Sb);         % Optimum in search Sa2
X6_Optimum=X5+ alfaSb*Sb                   % Next point X6 the OPTIMUM
```

Figure 5.15 (a) Plot of the objective function $3x_1^{\,2} - 4x_1x_2 + 3x_2^{\,2}$ contours and Powell's conjugate directions to the minimum starting at (4, 4). (b) The MATLAB code to optimize the objective function $3x_1^{\,2} - 4x_1x_2 + 3x_2^{\,2}$ using Powell's conjugate directions, a start at (4, 4), and the quadratic optimum step.

Solution

Figure 5.15a gives the contours of the objective function and shows the starting point $\boldsymbol{X}_a = \boldsymbol{X}_0 = [4 \quad 4]^T$ and the following search directions. As a quadratic function, the objective function in Eq. (a) has the following components; see Eq. (5.60). The constant $C = 0$. Since there are no first order or linear terms in x_1 or x_2, the vector $\boldsymbol{B} = \boldsymbol{0}$. The gradient is given in Eq. (b) to be able to calculate the second derivatives in the matrix $\boldsymbol{A}$.

$$\nabla \boldsymbol{f} = \begin{bmatrix} \partial f/\partial x_1 \\ \partial f/\partial x_2 \end{bmatrix} = \begin{bmatrix} 6x_1 - 4x_2 \\ -4x_1 + 6x_2 \end{bmatrix} \tag{b}$$

The $\boldsymbol{A}$ matrix is the Hessian matrix $\boldsymbol{H}$, which is given by

$$\boldsymbol{A} = \begin{bmatrix} \dfrac{\partial^2 f}{\partial x_1^2} & \dfrac{\partial^2 f}{\partial x_1 \partial x_2} \\ \dfrac{\partial^2 f}{\partial x_2 \partial x_1} & \dfrac{\partial^2 f}{\partial x_2^2} \end{bmatrix} = \begin{bmatrix} 6 & -4 \\ -4 & 6 \end{bmatrix} \tag{c}$$

Figure 5.15b shows the basic MATLAB code for solving the example. After the main data are defined as before, the directions `s1` $= [1 \quad 0]^T$ and `s2` $= [0 \quad 1]^T$ are defined in the coordinate directions x_1 and x_2. We will present the first few steps, and the rest of the procedure is produced by the code in Figure 5.15b and the search details shown in Figure 5.15a.

In the direction of x_1, we minimize to reach $\boldsymbol{X}_1$ where according to Eq. (5.65),

$$\alpha_1^* = \frac{-\boldsymbol{S}_1^T(\boldsymbol{A}\boldsymbol{X}_0 + \boldsymbol{B})}{\boldsymbol{S}_1^T\boldsymbol{A}\boldsymbol{S}_1} = \frac{-[1 \quad 0]\left(\begin{bmatrix} 6 & -4 \\ -4 & 6 \end{bmatrix}\begin{bmatrix} 4 \\ 4 \end{bmatrix} + \begin{bmatrix} 0 \\ 0 \end{bmatrix}\right)}{[1 \quad 0]\left(\begin{bmatrix} 6 & -4 \\ -4 & 6 \end{bmatrix}\begin{bmatrix} 1 \\ 0 \end{bmatrix}\right)} = \frac{-[1 \quad 0]\begin{bmatrix} 8 \\ 8 \end{bmatrix}}{[1 \quad 0]\begin{bmatrix} 6 \\ -4 \end{bmatrix}} = \frac{-8}{6} = -1.3333 \tag{d}$$

The next point $\boldsymbol{X}_1$ will then be

$$\boldsymbol{X}_1 = \boldsymbol{X}_0 + \alpha_1^*\boldsymbol{S}_1 = \begin{bmatrix} 4 \\ 4 \end{bmatrix} - 1.3333\begin{bmatrix} 1 \\ 0 \end{bmatrix} = \begin{bmatrix} 2.6667 \\ 4 \end{bmatrix} \tag{e}$$

The next search will be in the x_2 direction, i.e. we use $\boldsymbol{S}_2$ vector. We minimize to reach $\boldsymbol{X}_2$ according to Eq. (5.65)

$$\alpha_2^* = \frac{-\boldsymbol{S}_2^T(\boldsymbol{A}\boldsymbol{X}_1 + \boldsymbol{B})}{\boldsymbol{S}_2^T\boldsymbol{A}\boldsymbol{S}_2} = \frac{-[0 \quad 1]\left(\begin{bmatrix} 6 & -4 \\ -4 & 6 \end{bmatrix}\begin{bmatrix} 2.6667 \\ 4 \end{bmatrix} + \begin{bmatrix} 0 \\ 0 \end{bmatrix}\right)}{[0 \quad 1]\left(\begin{bmatrix} 6 & -4 \\ -4 & 6 \end{bmatrix}\begin{bmatrix} 0 \\ 1 \end{bmatrix}\right)} = \frac{-[0 \quad 1]\begin{bmatrix} 0.0002 \\ 13.3332 \end{bmatrix}}{[0 \quad 1]\begin{bmatrix} -4 \\ 6 \end{bmatrix}} = \frac{-13.3332}{6} = -2.2222 \tag{f}$$

The next point $\boldsymbol{X}_2$ will then be

$$\boldsymbol{X}_2 = \boldsymbol{X}_1 + \alpha_2^*\boldsymbol{S}_2 = \begin{bmatrix} 2.6667 \\ 4 \end{bmatrix} - 2.2222\begin{bmatrix} 0 \\ 1 \end{bmatrix} = \begin{bmatrix} 2.6667 \\ 1.7778 \end{bmatrix} \tag{g}$$

The first conjugate direction $\boldsymbol{S}_a$ is obtained from Eq. (5.58), which gives

$$\boldsymbol{S}_a = \boldsymbol{S}_3 = \boldsymbol{X}_2 - \boldsymbol{X}_0 = \begin{bmatrix} 2.6667 \\ 1.7778 \end{bmatrix} - \begin{bmatrix} 4 \\ 4 \end{bmatrix} = \begin{bmatrix} -1.3333 \\ -2.2222 \end{bmatrix} \tag{h}$$

Note that one does not need to normalize the direction $\boldsymbol{S}_a$, since it is used as is in all equations.

The next point of search is obtained after we get the optimum distance on $\boldsymbol{S}_a$ by applying Eq. (5.65).

$$\alpha_a^* = \frac{-\boldsymbol{S}_a^T(\boldsymbol{A}\boldsymbol{X}_2 + \boldsymbol{B})}{\boldsymbol{S}_a^T\boldsymbol{A}\boldsymbol{S}_a} = \frac{-[-1.3333 \quad -2.2222]\left(\begin{bmatrix} 6 & -4 \\ -4 & 6 \end{bmatrix}\begin{bmatrix} 2.6667 \\ 1.7778 \end{bmatrix} + \begin{bmatrix} 0 \\ 0 \end{bmatrix}\right)}{[-1.3333 \quad -2.2222]\left(\begin{bmatrix} 6 & -4 \\ -4 & 6 \end{bmatrix}\begin{bmatrix} -1.3333 \\ -2.2222 \end{bmatrix}\right)}$$

$$= \frac{-[-1.3333 \quad -2.2222]\begin{bmatrix} 8.889 \\ 0 \end{bmatrix}}{[-1.3333 \quad -2.2222]\begin{bmatrix} 0.4445 \\ -8 \end{bmatrix}} = \frac{11.852}{16.593} = 0.714\,28 \tag{i}$$

The next point $\boldsymbol{X}_3$ will then be

$$\boldsymbol{X}_3 = \boldsymbol{X}_2 + \alpha_a^* \ \boldsymbol{S}_a = \begin{bmatrix} 2.6667 \\ 1.7778 \end{bmatrix} + 0.714\,28\begin{bmatrix} -1.3333 \\ -2.2222 \end{bmatrix} = \begin{bmatrix} 2.6667 - 0.952\,35 \\ 1.7778 - 1.5873 \end{bmatrix} = \begin{bmatrix} 1.7144 \\ 0.1905 \end{bmatrix} \tag{j}$$

Following similarly for the next process, one gets $\alpha_3^* = 0.952\,38$, $\boldsymbol{X}_4 = [1.7143 \quad 1.1429]^T$, $\boldsymbol{S}_a = [-1.3333 \quad -2.2222]^T$, $\alpha_4^* = 0.459\,18$, $\boldsymbol{X}_5 = [1.1020 \quad 0.1225]^T$, $\boldsymbol{S}_b = [-0.612\,25 \quad -0.068\,027]^T$, and $\boldsymbol{X}_6 = [0 \quad 0]^T$.

Note that the solution of Example 5.9 took six steps to get to the optimum of the quadratic function. This is in agreement with the earlier statement that *Powell's* method converges to the minimum in $n(n+1)$ minimizations on a *quadratic* function. Since we have a ***2D*** problem, then $n = 2$. The convergence would be in $2(2 + 1) = 6$ minimizations, which is exactly what it took to reach the optimum.

5.4.3 Linearized Ridge Path Method

This method is developed to be independent of consecutive optimization in each coordinate system (Elzoghby et al. 1980). It depends on linearity following the *ridge path*, or *valley bed*, of the function to reach the optimum as shown in Figure 5.16 (Metwalli and Mayne 1977). The figure shows a function contours represented as a mountain that we seek to find the top. This is maximization rather than a minimization of the height. If one looks at the mountain in the x_1 direction (black arrow), the *ridge* of the mountain is the black line defining the edge or line of sight. This ridge is tangent to the contours of the mountain. If one looks at the mountain in the x_2 direction (gray dashed arrow), the *ridge* of the mountain is the gray dashed line defining the edge or line of sight.

The *linearized ridge path* method is implemented as defined in the following procedure (see Figure 5.16 for ***2D*** case):

1. The search starts by optimizing in the x_1 direction starting at $\boldsymbol{X}_0$ and ending at $\boldsymbol{X}_1$, which is the minimum of the function in the x_1 direction as shown in Figure 5.16 for ***2D***. This $\boldsymbol{X}_1$ point is on the ridge path or literary the valley bed of the function, which is obtained by

$$\boldsymbol{X}_1 = \boldsymbol{X}_0 + \alpha_1^* \ \boldsymbol{S}_1, \quad \boldsymbol{S}_1 = [1 \ 0 \ 0 \ \dots \ 0]^T \tag{5.70}$$

   where α_1^* is the optimum distance on $\boldsymbol{S}_1$ and is given by Eq. (5.65) for a quadratic function.
2. Step a distance Δx_2 in the x_2 direction to point $\boldsymbol{X}_2$ as shown in Figure 5.16, and use the following relation:

$$\boldsymbol{X}_2 = \boldsymbol{X}_1 + \Delta x_2 \boldsymbol{S}_2, \quad \boldsymbol{S}_2 = [0 \ 1 \ 0 \ \dots \ 0]^T \tag{5.71}$$

3. Perform a ***1D*** optimization in the x_1 direction or $\boldsymbol{S}_a$ direction starting at $\boldsymbol{X}_2$ and ending at $\boldsymbol{X}_3$, which is the minimum of the function on x_1 direction as shown in Figure 5.16. The direction $\boldsymbol{S}_a$ is the first conjugate direction since it is parallel to $\boldsymbol{S}_1$. The optimum end point $\boldsymbol{X}_3$ is then

$$\boldsymbol{X}_3 = \boldsymbol{X}_2 + \alpha_a^* \ \boldsymbol{S}_a, \quad \boldsymbol{S}_a = \boldsymbol{S}_1 \tag{5.72}$$

   where α_a^* is the optimum distance on $\boldsymbol{S}_a$ and is given by Eq. (5.73), similar to Eq. (5.65), for a quadratic function.

$$\alpha_a^* = \frac{-\boldsymbol{S}_a^T(\boldsymbol{A}\boldsymbol{X}_2 + \boldsymbol{B})}{\boldsymbol{S}_a^T \boldsymbol{A} \boldsymbol{S}_a} \tag{5.73}$$

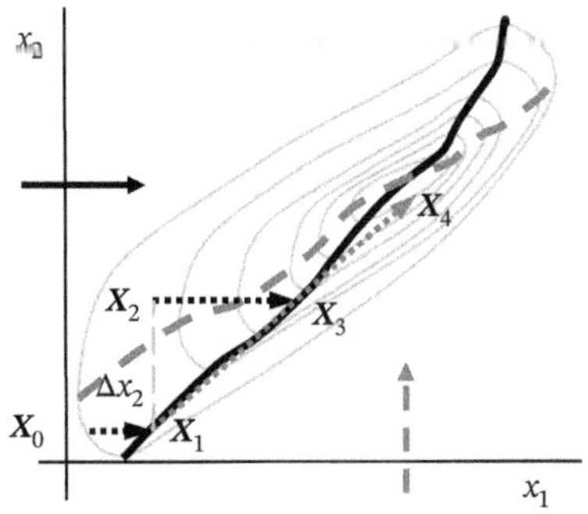

Figure 5.16 The ridge path of the function to reach the optimum locking in the x_1 or x_2 directions and the linearized ridge path starting in the x_1 direction.

4. Generate the next *conjugate direction* $\boldsymbol{S}_b$ by applying the following general relation:

$$\boldsymbol{S}_b = \boldsymbol{X}_3 - \boldsymbol{X}_1 \tag{5.74}$$

5. Perform a ***1D*** optimization in the $\boldsymbol{S}_b$ direction starting at $\boldsymbol{X}_3$ and ending at $\boldsymbol{X}_4$, which is the minimum of the function on $\boldsymbol{S}_b$ direction as shown in Figure 5.16. The end point is given by

$$\boldsymbol{X}_4 = \boldsymbol{X}_3 + \alpha_b^* \ \boldsymbol{S}_b \tag{5.75}$$

where α_b^* is the optimum distance on $\boldsymbol{S}_b$ and is given by Eq. (5.76), similar to Eq. (5.65), for a quadratic function.

$$\alpha_b^* = \frac{-\boldsymbol{S}_b^T(\boldsymbol{A}\boldsymbol{X}_3 + \boldsymbol{B})}{\boldsymbol{S}_b^T\boldsymbol{A}\boldsymbol{S}_b} \tag{5.76}$$

6. For a ***2D*** quadratic function, $\boldsymbol{X}_4$ should be the minimum. If the function is not quadratic, S_b is a linearized direction of the ridge path. A further iteration is needed to find the true minimum on the ridge path as clearly shown in Figure 5.16.

For *n-dimensional* functions, the procedure is repeated by stepping a distance in the third dimension and repeating the previous ***2D*** search (steps 1–6) to reach the minimum in that parallel plane in the third dimension. Minimizing in the conjugate direction connecting the two minima in the two parallel planes will get the minimum in the third dimension. Repeating the process for the subsequent dimensions will reach the minimum of the multidimensional problem; see Elzoghby et al. (1980). This method has a *quadratic converges* to the minimum in $n(n+1)/2$ minimizations, which is twice as efficient as *Powell's conjugate directions.*

Example 5.10 Solve Example 5.9 using the *linearized ridge path* method, and compare convergence efficiency with *Powell's* method.

Solution

Figure 5.17a gives the same contours of the objective function and shows the starting point $\boldsymbol{X}_a = \boldsymbol{X}_0 = [4 \quad 4]^T$ and the following search directions. As a quadratic function, the objective function in Eq. (a) has the following components; see Eq. (5.60). The constant $C = 0$. Since there are no first order or linear terms in x_1 or x_2, the vector $\boldsymbol{B} = \boldsymbol{0}$. The gradient is also given in Eq. (b) of Example 5.9 to be able to calculate the second derivatives in the matrix $\boldsymbol{A}$, which is given by Eq. (c) of Example 5.9.

Figure 5.17b shows the basic MATLAB code for solving this example by the *linearized ridge path* method. After the main data are defined as before, the directions `s1` $= [1 \quad 0]^T$ and `s2` $= [0 \quad 1]^T$ are explicit in the coordinate directions x_1 and x_2, respectively. We will present the steps, and the procedure can be verified by the code in Figure 5.17b. The search details are depicted in Figure 5.17a.

In the direction of x_1, we minimize to reach $\boldsymbol{X}_1$ where according to Eq. (5.65) and the same steps of Example 5.9, $\alpha_1^* = 1.3333$ and $\boldsymbol{X}_1 = [2.6667 \quad 4]^T$.

To have comparable points similar to Example 5.9, we step a distance $\Delta x_2 = -2.22$ in the x_2 direction to point $\boldsymbol{X}_2$. For a quadratic function, the value of Δx_2 should not affect the solution or convergence. The next point $\boldsymbol{X}_2$ will then be, due to Eq. (5.71)

$$\boldsymbol{X}_2 = \boldsymbol{X}_1 + \Delta x_2 \boldsymbol{S}_2 = \begin{bmatrix} 2.6667 \\ 4 \end{bmatrix} - 2.22 \begin{bmatrix} 0 \\ 1 \end{bmatrix} = \begin{bmatrix} 2.6667 \\ 1.78 \end{bmatrix} \tag{a}$$

(a)

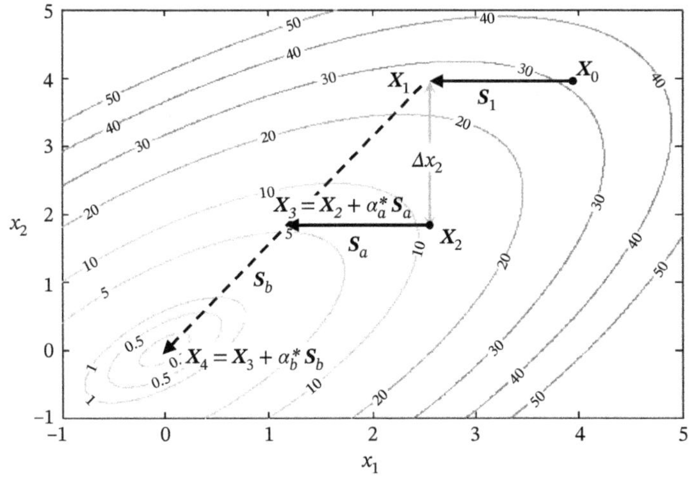

(b)

```
x10=4; x20=4; X0=[x10;x20]                    % Starting point
Deltax2=[0; -2.22]                            % Step in x2
C=0;B=[0;0];                                  % Quadratic Function defined
A=[6 -4; -4 6];
s1=[1;0]; s2=[0;1];                           % Initial search directions
alfa1=(-s1'*(A*X0+B))/(s1'*A*s1);             % Optimum on search s1
X1=X0+ alfa1*s1                               % Next point X1
X2=X1+Deltax2                                 % Next point X2
alfa2=(-s1'*(A*X2+B))/(s1'*A*s1);             % Optimum on search s1=Sa
X3=X2+ alfa2*s1                               % Next point X3
Sb=X3-X1;                                     % Conjugate direction 2
alfaSb=(-Sb'*(A*X3+B))/(Sb'*A*Sb);            % Optimum on search Sb
X4_Optimum=X3+ alfaSb*Sb                      % Next point X4, the OPTIMUM
```

Figure 5.17 (a) The contours of the objective function $3x_1^{\,2} - 4x_1x_2 + 3x_2^{\,2}$ and the linearized ridge path conjugate directions to the minimum starting at (4, 4). (b) The MATLAB code to optimize the quadratic objective function $3x_1^{\,2} - 4x_1x_2 + 3x_2^{\,2}$ using linearized ridge path conjugate directions, a start at (4, 4), and the quadratic optimum step α^*.

We get the optimum distance in the x_1 direction or $\boldsymbol{S}_a$ direction by applying Eq. (5.73).

$$\alpha_a^* = \frac{-\boldsymbol{S}_a^T(\boldsymbol{A}\boldsymbol{X}_2 + \boldsymbol{B})}{\boldsymbol{S}_a^T\boldsymbol{A}\boldsymbol{S}_a} = \frac{-\begin{bmatrix}1 & 0\end{bmatrix}\left(\begin{bmatrix}6 & -4\\ -4 & 6\end{bmatrix}\begin{bmatrix}2.6667\\ 1.78\end{bmatrix} + \begin{bmatrix}0\\ 0\end{bmatrix}\right)}{\begin{bmatrix}1 & 0\end{bmatrix}\left(\begin{bmatrix}6 & -4\\ -4 & 6\end{bmatrix}\begin{bmatrix}1\\ 0\end{bmatrix}\right)} = \frac{-\begin{bmatrix}1 & 0\end{bmatrix}\begin{bmatrix}8.8802\\ 0\end{bmatrix}}{\begin{bmatrix}1 & 0\end{bmatrix}\begin{bmatrix}6\\ -4\end{bmatrix}} = \frac{-8.8802}{6} = -1.4800 \tag{b}$$

The next point $\boldsymbol{X}_3$ will then be, due to Eq. (5.72)

$$\boldsymbol{X}_3 = \boldsymbol{X}_2 + \alpha_a^*\ \boldsymbol{S}_a = \begin{bmatrix}2.6667\\ 1.78\end{bmatrix} - 1.48\begin{bmatrix}1\\ 0\end{bmatrix} = \begin{bmatrix}2.6667 - 1.48\\ 1.78\end{bmatrix} = \begin{bmatrix}1.1867\\ 1.78\end{bmatrix} \tag{c}$$

The second conjugate direction S_b is obtained from Eq. (5.74), which gives

$$S_b = X_3 - X_1 = \begin{bmatrix} 1.1867 \\ 1.78 \end{bmatrix} - \begin{bmatrix} 2.6667 \\ 4 \end{bmatrix} = \begin{bmatrix} -1.48 \\ -2.22 \end{bmatrix} \tag{d}$$

The next point of search is obtained after we get the optimum distance α_b^* on S_b by applying Eq. (5.76).

$$\alpha_b^* = \frac{-S_b^T(AX_3 + B)}{S_b^T A S_b} = \frac{-\begin{bmatrix} -1.48 & -2.22 \end{bmatrix}\left(\begin{bmatrix} 6 & -4 \\ -4 & 6 \end{bmatrix}\begin{bmatrix} 1.1867 \\ 1.78 \end{bmatrix}\right)}{\begin{bmatrix} -1.48 & -2.22 \end{bmatrix}\left(\begin{bmatrix} 6 & -4 \\ -4 & 6 \end{bmatrix}\begin{bmatrix} -1.48 \\ -2.22 \end{bmatrix}\right)}$$
$$= \frac{-\begin{bmatrix} -1.48 & -2.22 \end{bmatrix}\begin{bmatrix} 0.0002 \\ 5.9332 \end{bmatrix}}{\begin{bmatrix} -1.48 & -2.22 \end{bmatrix}\begin{bmatrix} 0 \\ -3.7 \end{bmatrix}} = \frac{13.172}{16.428} = 0.801\,80 \tag{e}$$

The next optimum point X_4 will then be obtained by Eq. (5.75) such that

$$X_4 = X_3 + \alpha_b^* \; S_b = \begin{bmatrix} 1.1867 \\ 1.78 \end{bmatrix} + 0.8018\begin{bmatrix} -1.48 \\ -2.22 \end{bmatrix} = \begin{bmatrix} 1.1867 - 1.1867 \\ 1.78 - 1.779\,996 \end{bmatrix} = \begin{bmatrix} 0 \\ 0 \end{bmatrix} \tag{f}$$

Note that the solution of this example took three minimization steps to get to the optimum of the quadratic function. This is in agreement with the earlier statement that *linearized ridge path* method is twice as efficient as *Powell's* method. It converges to the minimum in $n(n+1)/2$ minimizations on a quadratic function. Since we have a ***2D*** problem, then $n = 2$. The convergence would be in $2(2 + 1)/2 = 3$ minimizations, which is exactly what it took to reach the optimum.

5.4.4 Random Search Methods

Random searches are *direct methods* that use direct comparison of function values, and do not need any derivative condition to satisfy. They are usually used particularly for *global optimization* where the objective function can be *multimodal*. In that case, most classical *direct methods* would converge to any local minima. Random search techniques usually converge to the global minima; however, there is no absolute guarantee for that.

The present widely used techniques are mainly the *genetic algorithms* (*GA*), *simulated annealing*, *particle swarm optimization* (*PSO*), *ant colony* optimization, and the *tabu* (*taboo*) *search*. These are also called *evolutionary algorithms* since they evolve with the search advances. Some of these are covered in the literature, optimization applications, and specialized work on optimization such as Pham and Karaboga (2000), Youssef et al. (2007), El-Mahdy et al. (2010), Badran et al. (2009), and Elmoselhy et al. (2006).

5.4.5 Steepest Descent Method

Consider the ***2D*** problem of taking a small step toward minimizing $f(x_1, x_2)$, and it is required to get the most efficient way to do so. Take a step of a length Δ on the direction vector S so that

$$\Delta = \sqrt{(\Delta x_1)^2 + (\Delta x_2)^2}, \quad S = \begin{bmatrix} \Delta x_1 & \Delta x_2 \end{bmatrix}^T \tag{5.77}$$

We want to maximize the decrease in f, thus the problem is to find

$$\min_{\Delta x_1, \Delta x_2} \; f = ?$$

$$\text{subject to the constraint } \; \Delta^2 = (\Delta x_1)^2 + (\Delta x_2)^2 \tag{5.78}$$

Using *Lagrange multipliers* for the following optimization problem:

$$\Delta f = \frac{\partial f}{\partial x_1}\Delta x_1 + \frac{\partial f}{\partial x_2}\Delta x_2 \tag{5.79}$$

subject to the constraints $\phi = \Delta^2$ (5.80)

This will give

$$\frac{\partial \Delta f}{\partial \Delta x_1} - \lambda\frac{\partial \phi}{\partial \Delta x_1} = 0$$
$$\frac{\partial \Delta f}{\partial \Delta x_2} - \lambda\frac{\partial \phi}{\partial \Delta x_2} = 0 \tag{5.81}$$

Solving these two equations will give

$$\lambda = \pm\frac{1}{2\Delta}\left[\sqrt{\left(\frac{\partial f}{\partial x_1}\right)^2 + \left(\frac{\partial f}{\partial x_2}\right)^2}\right] \tag{5.82}$$

$$\Delta x_1 = \pm\Delta\frac{{}^{\partial f}/_{\partial x_1}}{\sqrt{\left({}^{\partial f}/_{\partial x_1}\right)^2 + \left({}^{\partial f}/_{\partial x_2}\right)^2}} = \pm\Delta\frac{{}^{\partial f}/_{\partial x_1}}{|\nabla f|}$$
$$\Delta x_2 = \pm\Delta\frac{{}^{\partial f}/_{\partial x_2}}{\sqrt{\left({}^{\partial f}/_{\partial x_1}\right)^2 + \left({}^{\partial f}/_{\partial x_2}\right)^2}} = \pm\ \Delta\frac{{}^{\partial f}/_{\partial x_2}}{|\nabla f|} \tag{5.83}$$

From Eq. (5.83), one can get

$$\boldsymbol{S} = \pm\frac{\Delta}{|\nabla \boldsymbol{f}|}\left[{}^{\partial f}/_{\partial x_1} \quad {}^{\partial f}/_{\partial x_2}\right]^T = \pm\ \Delta\frac{\nabla \boldsymbol{f}}{|\nabla \boldsymbol{f}|} \tag{5.84}$$

Looking at the second derivatives for the minimum, we find the *steepest decent*

$$\boldsymbol{S} = -\ \Delta\frac{\nabla \boldsymbol{f}}{|\nabla \boldsymbol{f}|} \tag{5.85}$$

Looking at the second derivatives for the maximum, we find the *steepest ascent*

$$\boldsymbol{S} = +\ \Delta\frac{\nabla \boldsymbol{f}}{|\nabla \boldsymbol{f}|} \tag{5.86}$$

This means that we take the "small" step in the direction of the *gradient* of the function $\boldsymbol{f}$. Note that Δ is the length of $\boldsymbol{S}$.

It should be noted that the *steepest decent* method and steepest ascend method are depending on the evaluation of the first derivatives; therefore they are ***first-order*** methods.

5.4.5.1 Implementation

There are two ways to implement the steepest decent or steepest assent method;

1. First method
   The procedure for minimization is as follows (see Figure 5.18a):
   (a) Compute $\nabla \boldsymbol{f}$ form $\partial \boldsymbol{f}/\partial x_1$, $\partial \boldsymbol{f}/\partial x_2$, or by numerical approximation.
   (b) Let k be a small distance and

$$\Delta \boldsymbol{X} = -k\frac{\nabla \boldsymbol{f}}{|\nabla \boldsymbol{f}|} \tag{5.87}$$

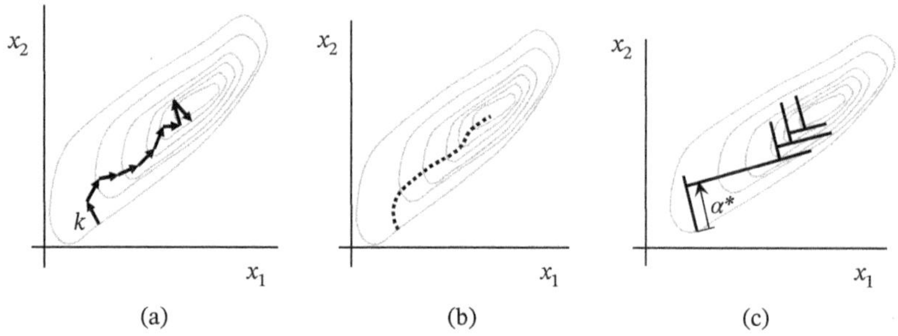

Figure 5.18 Steepest descent method for ***2D*** problem: (a) first method using some distance k, (b) first method using a very small k, and (c) second method by consecutively minimizing in the steepest decent direction.

Therefore,

$$\boldsymbol{X}_{j+1} = \boldsymbol{X}_j - k\frac{\nabla \boldsymbol{f}}{|\nabla \boldsymbol{f}|} \tag{5.88}$$

where k is an arbitrary constant chosen to give a suitably small step.

(c) Adjust k to keep the search moving along the steepest slope.

Objection to large steps (k) are as follows:

1. Only the initial movement is steepest descent.
2. May overshoot the minimum as shown in Figure 5.18a.

On the other hand a small k results in slow convergence as evident from Figure 5.18b.

2. Second method

The procedure for minimization is as follows (see Figure 5.18c):

(a) Evaluate direction $\boldsymbol{S} = \nabla \boldsymbol{f}/|\nabla \boldsymbol{f}|$.

(b) Let

$$\boldsymbol{X}_{j+1} = \boldsymbol{X}_j - \alpha\frac{\nabla \boldsymbol{f}}{|\nabla \boldsymbol{f}|} \tag{5.89}$$

(c) Search in one direction (such as *quadratic interpolation*, *golden section*, etc.) for α^* that gives the minimum $\boldsymbol{f}$ in the direction $\boldsymbol{S}$.

(d) At the new point evaluate new $\boldsymbol{S} = \nabla \boldsymbol{f}/|\nabla \boldsymbol{f}|$ and repeat from step (b).

5.4.6 Fletcher–Reeves Conjugate Gradient

This method utilizes the *gradient* of the objective function, and, therefore, it is a *first-order* method. All first-order methods of minimization usually start by a *steepest decent* step. To present the method, we use a ***2D*** case to develop an understanding of the concept.

For ***2D*** (quadratic), one can start using a steepest decent search for the first step or

$$\boldsymbol{S}_1 = -\nabla \boldsymbol{f}_1 \tag{5.90}$$

As before, the general recursive equation that can be applied at any step or for the start is

$$\boldsymbol{X}_{q+1} = \boldsymbol{X}_q + \alpha_q\,\boldsymbol{S}_q \tag{5.91}$$

As a result of the first minimization, the gradient vector resolved in the direction $\boldsymbol{S}_1$ must be zero thus

$$\boldsymbol{S}_1^T\ \nabla \boldsymbol{f}_2 = 0 \tag{5.92}$$

Similarly, for the second minimization

$$\begin{aligned} &\boldsymbol{S}_2 = -\nabla \boldsymbol{f}_2 + \beta\ \boldsymbol{S}_1 \\ &\boldsymbol{S}_2^T\ \nabla \boldsymbol{f}_3 = 0 \end{aligned} \tag{5.93}$$

From Eqs. (5.90) and (5.92), one should note that

$$\nabla \boldsymbol{f}_1^T\ \nabla \boldsymbol{f}_2 = 0 \tag{5.94}$$

Assuming a quadratic function defined by Eq. (5.60) and performing the differentiation, one gets

$$\begin{aligned} \nabla \boldsymbol{f}_1 &= \boldsymbol{B} + \boldsymbol{A}\,(\boldsymbol{X}_1) \\ \nabla \boldsymbol{f}_2 &= \boldsymbol{B} + \boldsymbol{A}\,(\boldsymbol{X}_2). \end{aligned} \tag{5.95}$$

Therefore, using Eq. (5.91) gives

$$\nabla \boldsymbol{f}_2 - \nabla \boldsymbol{f}_1 = \boldsymbol{A}\,(\boldsymbol{X}_2 - \boldsymbol{X}_1) = \boldsymbol{A}\,((\boldsymbol{X}_1 + \alpha_1\,\boldsymbol{S}_1) - \boldsymbol{X}_1)) = \alpha_1 \boldsymbol{A}\,\boldsymbol{S}_1 \tag{5.96}$$

Thus, to cause $\boldsymbol{S}_1$ and $\boldsymbol{S}_2$ to be $\boldsymbol{A}$ conjugate, we proceed with Eq. (5.61) and substituting Eq. (5.96) to get

$$\boldsymbol{S}_2^T \boldsymbol{A}\ \boldsymbol{S}_1 = \boldsymbol{S}_2^T \frac{\nabla \boldsymbol{f}_2 - \nabla \boldsymbol{f}_1}{\alpha_1} = 0 \tag{5.97}$$

Substitute Eqs. (5.93) and (5.92) into Eq. (5.97) to get

$$\boldsymbol{S}_2^T \boldsymbol{A}\ \boldsymbol{S}_1 = \boldsymbol{S}_2^T \frac{\nabla \boldsymbol{f}_2 - \nabla \boldsymbol{f}_1}{\alpha_1} = [-\nabla \boldsymbol{f}_2 + \beta\ \boldsymbol{S}_1]^T \frac{\nabla \boldsymbol{f}_2 - \nabla \boldsymbol{f}_1}{\alpha_1} = \frac{[-\nabla \boldsymbol{f}_2 + \beta\ (-\nabla \boldsymbol{f}_1)]^T (\nabla \boldsymbol{f}_2 - \nabla \boldsymbol{f}_1)}{\alpha_1} = 0 \tag{5.98}$$

Since α_1 may not necessarily be zero, the value of β that causes the nominator to be zero should in general be

$$\beta = \frac{\nabla \boldsymbol{f}_2^T\ \nabla \boldsymbol{f}_2}{\nabla \boldsymbol{f}_1^T\ \nabla \boldsymbol{f}_1} = \frac{\nabla \boldsymbol{f}_q^T\ \nabla \boldsymbol{f}_q}{\nabla \boldsymbol{f}_{q-1}^T\ \nabla \boldsymbol{f}_{q-1}} \tag{5.99}$$

The procedure is then to use the general recursive Eq. (5.91) with the direction $\boldsymbol{S}_q$ defined by the following formula as similar to Eq. (5.93), i.e.

$$\boldsymbol{S}_q = -\nabla \boldsymbol{f}_q + \beta\ \boldsymbol{S}_{q-1} \tag{5.100}$$

Note that the value of β is chosen so that $\boldsymbol{S}_q$'s ($\boldsymbol{S}_1, \boldsymbol{S}_2, \ldots, \boldsymbol{S}_n$) are *conjugate directions* as seen from the derivation. Thus a *quadratic function* will be minimized in only "n" ***1D*** minimizations. For a quadratic function defined by Eq. (5.60), the optimum α_q^* is obtained as follows:

$$\alpha_q^* = -\frac{\nabla \boldsymbol{f}_q^T\ \nabla \boldsymbol{f}_q}{\nabla \boldsymbol{f}_q^T \boldsymbol{A}\ \boldsymbol{S}_q} \tag{5.101}$$

Note that Eq. (5.101) is used in one step in $\boldsymbol{S}_q$ direction in Eq. (5.91), but Eq. (5.99) is used for the *conjugate gradient* direction defined by Eq. (5.100).

For non-quadratic function it would take more iterations than the n directions due to the following reasons:

(1) Approximation of ***1D*** minimizations of a function.
(2) Region near the optimum is different than away from it, and, the $\boldsymbol{S}_q$'s need not depend on all pervious $\boldsymbol{S}_{q-1}$'s.

To eliminate these difficulties, restart with a new $\boldsymbol{S}_q = -\nabla \boldsymbol{f}$; and maybe after "$2n$" ***1D*** minimizations, we reach the minimum.

5.4.7 Newton–Raphson Method

This method utilizes the first and second derivatives of the objective function, and, therefore, it is a *second-order* method. The development of the method depends on *Taylor* series, after *Brook Taylor* (1685–1731).

The *Taylor* series approximation of the objective function $\boldsymbol{f}(\boldsymbol{X}_{q+1})$ at $\boldsymbol{X}_{q+1}$ as a function of values at $\boldsymbol{X}_q$ is

$$\boldsymbol{f}(\boldsymbol{X}_{q+1}) = \boldsymbol{f}(\boldsymbol{X}_q) + \nabla \boldsymbol{f}_q{}^T(\boldsymbol{X}_{q+1} - \boldsymbol{X}_q) + \frac{1}{2}(\boldsymbol{X}_{q+1} - \boldsymbol{X}_q)^T \boldsymbol{H}_q(\boldsymbol{X}_{q+1} - \boldsymbol{X}_q) + \cdots \tag{5.102}$$

$$\boldsymbol{H}_q = \begin{bmatrix} \frac{\partial^2 f}{\partial x_1^2} & \cdots\cdots & \frac{\partial^2 f}{\partial x_1 \partial x_n} \\ \vdots & \ddots & \vdots \\ \vdots & \ddots & \vdots \\ \frac{\partial^2 f}{\partial x_n \partial x_1} & \cdots\cdots & \frac{\partial^2 f}{\partial x_2^2} \end{bmatrix} \tag{5.103}$$

Also, the *gradient* can be written as *Taylor* series

$$\nabla \boldsymbol{f}_{q+1} = \nabla \boldsymbol{f}_q + \boldsymbol{H}_q(\boldsymbol{X}_{q+1} - \boldsymbol{X}_q) + \cdots \tag{5.104}$$

For an *extremum*, the first derivatives $\nabla \boldsymbol{f}_{q+1}$ at $\boldsymbol{X}_{q+1}$ should be zero, which renders Eq. (5.104) as

$$\begin{aligned} \nabla \boldsymbol{f}_{q+1} &= \boldsymbol{0} = \nabla \boldsymbol{f}_q + \boldsymbol{H}_q(\boldsymbol{X}_{q+1} - \boldsymbol{X}_q) + \cdots \\ \boldsymbol{H}_q(\boldsymbol{X}_{q+1} - \boldsymbol{X}_q) &= -\nabla \boldsymbol{f}_q \end{aligned} \tag{5.105}$$

From Eq. (5.105) and pre-multiplying by $\boldsymbol{H}^{-1}$, the *Newton–Raphson* rule is then

$$\boldsymbol{X}_{q+1} = \boldsymbol{X}_q - \boldsymbol{H}_q^{-1}\ \nabla \boldsymbol{f}_q \tag{5.106}$$

So this jumps to the predicted location of the optimum directly. For a *quadratic function*, $\boldsymbol{H}_q$ is a matrix of constants, and we move to the optimum in one step, or the *quadratic convergence* is equal to one. If the function is not quadratic, $\boldsymbol{H}_q$ depends on $\boldsymbol{X}_q$, and we must *iterate* to find out the optimum. Note that for a *quadratic function* defined by Eq. (5.60), the *Hessian* matrix $\boldsymbol{H}$ is the $\boldsymbol{A}$ matrix or $\boldsymbol{H} = \boldsymbol{A}$.

Disadvantages of this powerful method are as follows:

(1) It is not easy to find the *second derivatives* that make up $\boldsymbol{H}_q$, and it is not easy or costly to approximate it by *finite difference*.
(2) The method converges to a maximum, or a minimum, also might go to a saddle point.
(3) The *inverse matrix* $\boldsymbol{H}_q{}^{-1}$ has to be computed.

Example 5.11 Solve Example 5.9 using the *Newton–Raphson* method. The quadratic objective function in that example is $\boldsymbol{f} = 3x_1^2 - 4x_1x_2 + 3x_2^2$.

Solution

Figure 5.17 gives the same contours of the objective function and shows the starting point $\boldsymbol{X}_0 = [4 \quad 4]^T$. The subsequent search directions of the ridge path method are also defined by the black vectors. As a quadratic function, the objective function in Example 5.9 has the following components; see Eq. (5.60). The constant $C = 0$. Since there are no first order or linear terms in x_1 or x_2, the vector $\boldsymbol{B} = \boldsymbol{0}$. The gradient is also given in Eq. (b) of Example 5.9 to be able to calculate the second derivatives in the matrix $\boldsymbol{A}$, which is given by Eq. (c) of Example 5.9. These are repeated and calculated again herein for an easier procedure to follow and for the same starting point of $\boldsymbol{X}_0 = (4, 4)$.

$$\nabla \boldsymbol{f}_0 = \begin{bmatrix} 6x_1 - 4x_2 \\ -4x_1 + 6x_2 \end{bmatrix} = \begin{bmatrix} 6(4) - 4(4) \\ -4(4) + 6(4) \end{bmatrix} = \begin{bmatrix} 8 \\ 8 \end{bmatrix} \tag{a}$$

$$\boldsymbol{A} = \boldsymbol{H} = \boldsymbol{H}_0 = \begin{bmatrix} 6 & -4 \\ -4 & 6 \end{bmatrix} \tag{b}$$

To use Newton–Raphson method (Eq. 5.106), we need to invert the *Hessian* matrix $\boldsymbol{H}$. This gives the following:

$$\boldsymbol{H}_q^{-1} = \boldsymbol{H}_0^{-1} = \frac{1}{20}\begin{bmatrix} 6 & 4 \\ 4 & 6 \end{bmatrix} \tag{c}$$

Using Newton–Raphson method defined by Eq. (5.106) and noting that the index $q = 0$, we move from point $\boldsymbol{X}_0$ to $\boldsymbol{X}_1$ such that

$$\boldsymbol{X}_{q+1} = \boldsymbol{X}_q - \boldsymbol{H}_q^{-1}\ \nabla \boldsymbol{f}_q, \text{ or } \boldsymbol{X}_1 = \boldsymbol{X}_0 - \boldsymbol{H}_0^{-1}\ \nabla \boldsymbol{f}_0 = \begin{bmatrix} 4 \\ 4 \end{bmatrix} - \begin{bmatrix} 0.3 & 0.2 \\ 0.2 & 0.3 \end{bmatrix}\begin{bmatrix} 8 \\ 8 \end{bmatrix} = \begin{bmatrix} 4 \\ 4 \end{bmatrix} - \begin{bmatrix} 4 \\ 4 \end{bmatrix} = \begin{bmatrix} 0 \\ 0 \end{bmatrix} \tag{d}$$

It is clear from Eq. (d) that we have reached the optimum in *one step* of implementing Eq. (5.106). As seen, it is powerful, but we needed to calculate the inverse of the *Hessian* matrix. This was simple for this ***2D*** example; however, it might be involving for a much larger dimensionality and a non-quadratic objective function.

5.4.8 Quasi-Newton Methods

Since the Newton or Newton–Raphson method is quadratically converging in one step, many quasi-Newton methods have been developed (Rao 2009). Other widely used methods have been utilized in MATLAB function minimizations such as "`fminunc`." These other methods include variable metric Davidon–Fletcher–Powell (DFP) method or the Broyden–Fletcher–Goldfarb–Shanno (BFGS) method. In this section, we present few other methods that have been used successfully.

5.4.8.1 A Quadratic Optimization Technique

This method is based on approximating the objective function as a quadratic function and iteratively finding the *step optimum* solution (Metwalli and Mayne 1977; Metwalli and Elmi 1989).

The objective function could be approximated as before and reiterated here such that

$$f(X) = \frac{1}{2}\boldsymbol{X}^T\boldsymbol{A}\boldsymbol{X} + \boldsymbol{X}^T\boldsymbol{B} + C \tag{5.60}$$

At an optimum point $\boldsymbol{X}^*$, the values of the components of the gradient vector $\nabla \boldsymbol{f}$ should be zero, such that

$$\nabla \boldsymbol{f} = \boldsymbol{A}\boldsymbol{X}^* + \boldsymbol{B} = 0 \tag{5.107}$$

where $\boldsymbol{X}^*$ is the value of the design vector at the optimum. The matrix $\boldsymbol{A}$ is the design system matrix that is represented by the second derivatives *Hessian* matrix $\boldsymbol{H}$.

The procedure of the quadratic optimization technique is defined in the following steps (Metwalli and Elmi 1989):

1. Identify the design matrix $\boldsymbol{A}$. The objective function is differentiated using closed form mathematical derivatives or finite difference formulas. The matrix $\boldsymbol{A}$ is calculated at a starting point $\boldsymbol{X}_0$.
2. The vector $\boldsymbol{B}$ in Eq. (5.107) may be identified by using the relation

   $$\boldsymbol{B} = \nabla \boldsymbol{f} - \boldsymbol{A}\boldsymbol{X} \tag{5.108}$$

   where $\boldsymbol{B}$ and $\boldsymbol{A}$ are evaluated at $\boldsymbol{X}_0$. This gives exact optimum solution for *quadratic objective function* in only one iteration.

3. The solution of Eq. (5.107) is obtained by the utilization of the *ridge path iterative method* (Metwalli and Mayne 1977; Elzoghby et al. 1980). This step is implemented by using the following cycle:
   - Solving Eq. (5.107) for first variable as a *step optimum* ${}^{*}x_1$, one gets

$$ {}^{*}x_1 = \left(-\sum_{k=1}^{n} a_{1k}x_k - b_1\right) \Big/ a_{11} \quad k \neq 1 \tag{5.109}$$

   where ${}^{*}x_1$ is the *step optimum* after the full cycle is settled and the solution is *stationary*.
   - Solving Eq. (5.107) for the second variable, one similarly obtains

$$ {}^{*}x_2 = \left(-\sum_{k=1}^{n} a_{2k}x_k - b_2\right) \Big/ a_{22} \quad k \neq 2 \tag{5.110}$$

   where ${}^{*}x_2$ is the *step optimum* and after the full cycle is settled and the solution is *stationary*. However, ${}^{*}x_1$ should be used for the solution that generates ${}^{*}x_2$. Similarly, for the solution of ${}^{*}x_3$, the values of ${}^{*}x_1$ and ${}^{*}x_2$ are always used.
   - The general formula is then

$$ {}^{*}x_j = \frac{\left(-\sum_{k=1}^{n} a_{jk}x_k - b_j\right)}{a_{jj}} \quad k \neq j \tag{5.111}$$

   where ${}^{*}x_j$ is the *step optimum* after the full cycle is settled and the solution is *stationary*. Again, all ${}^{*}x_{j-1}$ should be used for the solution that generates ${}^{*}x_j$.
4. Reiterate from Eq. (5.109) till the solution is ***stationary*** for all n variables. The stationary solution is the step optimum vector ${}^{*}\boldsymbol{X}_1$.
5. The new value of the vector ${}^{*}\boldsymbol{X}_1$ is used in order to calculate the value of the objective function. This value is compared to the value obtained in the previous loop. If the difference is less than the specified error, then the optimum point is reached, and the search is terminated. If not, one should continue with another cycle from step 1.
6. The new design vector ${}^{*}\boldsymbol{X}_1$ is used in identifying the $\boldsymbol{A}$ matrix and the vector $\boldsymbol{B}$. Both are used in calculating a new value of ${}^{*}\boldsymbol{X}_i$ and so on till one reaches the final acceptable optimum $\boldsymbol{X}^{*}$.

It should be noted that the previously mentioned procedure provides the solution of the set of linear Eq. (5.108). It is similar to *Gauss–Seidel* procedure (Chapra and Canale 2010). The procedure presented herein is, however, based on the *ridge path* method of function optimization, which significantly affects the aforementioned iteration procedure and as shown in Example 5.11. It is dependent on the $\boldsymbol{A}$ matrix. If $\boldsymbol{A}$ is not *positive definite*, the procedure would converge to a maximum or a saddle.

Example 5.12 Solve the renowned *Rosenbrock's banana* test function defined by (Rosenbrock 1960)

$$ f(\boldsymbol{X}) = 100\,(x_2 - x_1^2)^2 + (x_1 - 1)^2 \tag{a}$$

Start your search at the usual initial point (−1.2, 1).

Solution

Figure 5.19a gives the contours of the *Rosenbrock's* function and shows the starting point $\boldsymbol{X}_0 = [-1.2 \quad 1]^T$. The subsequent search directions are also shown. These searches are obtained by the use of numerical evaluation of derivatives. The solution of this example, however, utilizes the closed form evaluation of derivatives. The objective function in Eq. (a) has the following components. The gradient is obtained as

$$\nabla \boldsymbol{f} = \begin{bmatrix} -400(x_2 x_1 - x_1^3) - 2(1 - x_1) \\ 200(x_2 - x_1^2) \end{bmatrix} \tag{b}$$

The *Hessian* matrix is derived as

$$\boldsymbol{H} = \begin{bmatrix} -400x_2 + 1200x_1^2 + 2 & -400x_1 \\ -400x_1 & 200 \end{bmatrix} \tag{c}$$

At start, where $\boldsymbol{X}_0 = [-1.2\ 1]$, the gradient and the *Hessian* matrix are

$$\nabla \boldsymbol{f} = \begin{bmatrix} -400(x_2 x_1 - x_1^3) - 2(1 - x_1) \\ 200(x_2 - x_1^2) \end{bmatrix} = \begin{bmatrix} -400(1.0(-1.2) - (-1.2)^3) - 2(1 + 1.2) \\ 200(1 - (-1.2)^2) \end{bmatrix} = \begin{bmatrix} -215.6 \\ -88 \end{bmatrix} \tag{d}$$

$$\boldsymbol{H} = \begin{bmatrix} -400x_2 + 1200x_1^2 + 2 & -400x_1 \\ -400x_1 & 200 \end{bmatrix} = \begin{bmatrix} 1330 & 480 \\ 480 & 200 \end{bmatrix} \tag{e}$$

Then

$$\boldsymbol{B} = \nabla \boldsymbol{f} - \boldsymbol{A}\boldsymbol{X} = \begin{bmatrix} -215.6 \\ -88 \end{bmatrix} - \begin{bmatrix} 1330 & 480 \\ 480 & 200 \end{bmatrix} \begin{bmatrix} -1.2 \\ 1 \end{bmatrix} = \begin{bmatrix} -215.6 \\ -88 \end{bmatrix} - \begin{bmatrix} -1116 \\ -376 \end{bmatrix} = \begin{bmatrix} 900.4 \\ 288 \end{bmatrix} \tag{f}$$

Ridge path iteration starts at

$$x_1 = \left(-\sum_{k=1}^{n} a_{1k} x_k - b_1 \right) \Big/ a_{11} = (-a_{12}(x_{2,0}) - b_1)/a_{11} = (-480(1.0) - 900.4)/1330 = -1.0379$$

$$x_2 = \left(-\sum_{k=1}^{n} a_{2k} x_k - b_2 \right) \Big/ a_{22} = (-a_{21}(x_{1,1}) - b_2)/a_{22} = (-480(-1.0379) - 288)/200 = 1.050\,96 \tag{g}$$

Note that to get x_2, the prior calculated value of x_1 is used. Equation (g) are used recursively with the calculated value of x_2 replaces the values of $x_{2,0}$ and so on for the next calculations. After many recursive implementation of Eq. (g) by setting $\boldsymbol{X}_m$ to the calculated values and recalculating x_1 and x_2, the values of x_1 and x_2, will be *stationary* at about $x_1 = -1.1753$ and $x_2 = 1.3807$. These values depend on either the number of recalculation cycles or a termination criterion on the acceptable error in the values of x_1 and x_2. The MATLAB code of Figure 5.19b has been used to get these stationary values or the step optimum $^*\boldsymbol{X}_1$ after a set recalculation value of $m = 300$. The procedure to reach the stationary point or step optimum $^*\boldsymbol{X}_1$ is only *one step optimization cycle* since there is only one evaluation process for $\nabla \boldsymbol{f}$ and $\boldsymbol{A}$ at the start point $\boldsymbol{X}_0$ to reach the step optimum $^*\boldsymbol{X}_1$. This point is shown on Figure 5.19a as $^*\boldsymbol{X}_1$.

Each of the subsequent cycles requires the calculations of $\nabla \boldsymbol{f}$ and $\boldsymbol{A}$ at the step point $^*\boldsymbol{X}_i$. The subsequent optimization cycles and their end points $^*\boldsymbol{X}_2, {}^*\boldsymbol{X}_3, \ldots, {}^*\boldsymbol{X}_7$ are shown in Figure 5.19a. For this problem and the intended accuracy, it took only seven cycles to reach a point $^*\boldsymbol{X}_7 \approx \boldsymbol{X}_7^* = (0.9963, 0.9925)$, which is very close to the theoretical optimum of (1, 1). If eight cycles are used, the result would be $\boldsymbol{X}_8^* = (0.9982, 0.9965)$. If the recalculation in each cycle is 500 instead of 300, the end point would be $\boldsymbol{X}_8^* = (1.0000, 0.9999)$. If an error controller is used instead of the 500 recalculations, the results would be around those values.

5.4.8.2 Identified Quadratic Optimization Technique

This method is based on approximating the objective function as a *quadratic function* and iteratively identifying the optimum solution (Metwalli et al. 1989). It can be considered as a *quasi-Newton* technique since it approximates the *Hessian* matrix by identification from the first derivatives only. It can be applied to non-quadratic functions; however, more iteration cycles are usually needed.

(a)

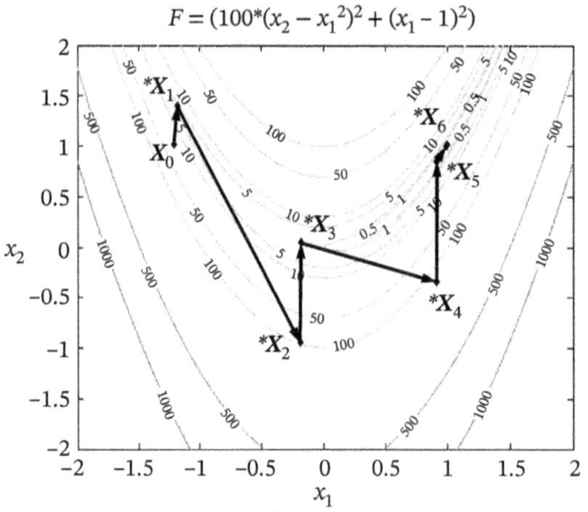

(b)

```
for n=1:7
  X0=[x1; x2];                                              % Start X0=(-1.2  1.0)
   f=100*(x2-x1^2)^2+(1-x1)^2;
   GRAD=[100*(4*x1^3-4*x1*x2)+2*x1-2; 100*(2*x2-2*x1^2)];
   AH=[-400*x2+1200*x1^2+2  -400*x1;
     -400*x1   200];
   B=GRAD-AH*X0;
   b1=B(1); b2=B(2); a11=AH(1,1); a12=AH(1,2); a22=AH(2,2);
  for m=1:300
   x11=(-a12*x20-b1)/a11;
   x22=(-a12*x11-b2)/a22;                                   % Note prior x11
   x1d=x11-x10;      x2d=x22-x20; Xerror=x1d+x2d;
   x10=x11;          x20=x22;
   x1=x11;
   x2=x22;
   m=m+1;
  end
   Xerror=x1d+x2d
   x1=x11; x2=x22;
   X=[x1; x2]
   n=n+1;
end
```

(c)

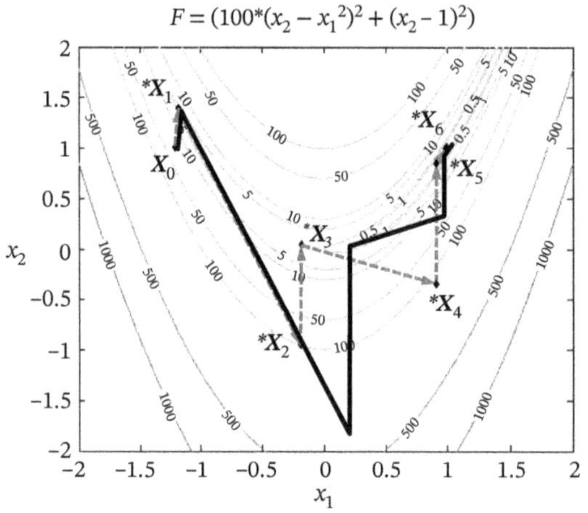

Figure 5.19 (a) The contours of the Rosenbrock's function and the starting point $\boldsymbol{X}_0 = [-1.2 \;\; 1]^T$. The vectors are for the output of the subsequent search directions and their end points. (b) The MATLAB code for the Rosenbrock's function optimization using a quadratic optimization technique and a starting point $\boldsymbol{X}_0 = [-1.2 \;\; 1]^T$. (c) The contours of the Rosenbrock's function and the starting point $\boldsymbol{X}_0 = [-1.2 \;\; 1]^T$. The vectors are for the output of the subsequent search directions and their end points. (d) The MATLAB code to optimize the Rosenbrock's banana test function using the identified quadratic optimization technique. The code calls the function "FBanan" given in (e). (e) The privately written function "Fbanan" needed for the identified quadratic optimization technique shown in Figure 5.19d. This function evaluates the value of the banana function and its derivatives at any point x_1, x_2 using randomly generated differences n.

(d)
```
for i=1:N                                    % Defining the search points
  x1(i+1)=x1(i)+(err/(nit+1))*(rand-0.5);
  x2(i+1)=x2(i)+(err/(nit+1))*(rand-0.5);
end
for i=1:N                                    % Defining Derivatives at search points
  X(:,i)=[x1(i); x2(i)];                     % Matrix of search points
  [fn,dfx1(i),dfx2(i)]=FBanan(x1(i),x2(i),err/(100*(nit+1)));
  nfeval=nfeval+3;                           % nfeval=nfeval+1  the +3 is for derivatives;
  dfX(:,i)=[dfx1(i);dfx2(i)];                % Derivative Deference
end
Points=X                                     % Display Matrix of search points
for i=1:(N-1)                                % Search points and Derivative differencMatrices
  DX(:,i)=[x1(i+1)-x1(i);x2(i+1)-x2(i)];
  Ddf(:,i)=[dfx1(i+1)-dfx1(i);dfx2(i+1)-dfx2(i)];
end
DeltaX_Matrix=DX                             % To Display position difference Matrix
Delta_dfX_Matrix=Ddf                         % To Display derivatives difference Matrix
DXinv=pinv(DX);
Ddfinv=pinv(Ddf);
A=Ddf*DXinv                                  % Display identified A Matrix
Ainv=DX*Ddfinv                               % Display identified inverse A Matrix
                           %================= Newton Raphson Optimization STEP ===========
XOpt= X(:,1)-Ainv*dfX(:,1);
Function_at_Optimum_X= 100*(XOpt(2)-XOpt(1)^2)^2+(XOpt(1)-1)^2
```

(e)
```
function [fB,f1,f2] = FBanan(x1,x2,n)
                    % Evaluates Banana function value fBand derivatives f1 & f2
format long
fB=100*((x2-x1^2)^2)+(x1-1)^2;        % function value at x1,x2
fBn1=100*((x2-(x1+n)^2)^2)+((x1+n)-1)^2;
fBn2=100*(((x2+n)-x1^2)^2)+(x1-1)^2;
if n== 0
f1=100*(4*x1^3-4*x1*x2)+2*x1-2;       % closed form derivative w.r.tx1
f2=100*((2*x2)-2*x1^2);               % closed form derivative w.r.tx2
else
f1=(fBn1-fB)/n;
f2=(fBn2-fB)/n;
end
end
```

Figure 5.19 *(Continued)*

The objective function could be approximated by a *quadratic* function (Eq. 5.60) as follows:

$$f(X) = \frac{1}{2}X^T AX + X^T B + C \tag{5.60}$$

At an optimum point, again, values of the components of the vector $\nabla \boldsymbol{f}$ should be zero such that

$$\nabla \boldsymbol{f} = \boldsymbol{A}\boldsymbol{X}^* + \boldsymbol{B} = 0 \tag{5.112}$$

where $\boldsymbol{X}^*$ is the optimum value of the design vector. The matrix $\boldsymbol{A}$ is the system matrix that has been represented previously by the second derivatives matrix or ***identified*** in this current technique as follows.

At any point, $\boldsymbol{X}_1, \boldsymbol{X}_2, \ldots, \boldsymbol{X}_i$, the values of the components of the vector $\nabla \boldsymbol{f}_i$ should be such that

$$\nabla \boldsymbol{f}_i = \boldsymbol{A}\boldsymbol{X}_i + \boldsymbol{B} \tag{5.113}$$

where $\boldsymbol{X}_i$ is the value of the design vector at any point 1, 2, 3, etc. The matrix $\boldsymbol{A}$ or its inverse $\boldsymbol{A}^{-1}$ can be ***identified*** as follows:

At any two points, $\boldsymbol{X}_1$ and $\boldsymbol{X}_2$, … or $\boldsymbol{X}_i$ and $\boldsymbol{X}_{i+1}$, the values of the components of the *difference vector* $\nabla \boldsymbol{f}_i - \nabla \boldsymbol{f}_{i+1}$ can obtained by subtracting (5.113) for $\boldsymbol{X}_i$ and $\boldsymbol{X}_{i+1}$ to get

$$\nabla \boldsymbol{f}_i - \nabla \boldsymbol{f}_{i+1} = \boldsymbol{A}\ [\boldsymbol{X}_i - \boldsymbol{X}_{i+1}] \tag{5.114}$$

At another two points, X_3 and X_4 … or X_{i+2} and X_{i+3}, we get similar difference vector $\nabla f_{i+2} - \nabla f_{i+3}$. Stacking the *difference vector* equations gives

$$[\Delta \nabla f_{i,i+1} \quad \Delta \nabla f_{i+2,i+3}] = A[\Delta X_{i,i+1} \quad \Delta X_{i+2,i+3}] \tag{5.115}$$

Stacking more equations (n or $m > n$) gives

$$[\Delta \nabla f] = A \ [\Delta X] \tag{5.116}$$

where $[\Delta \nabla f]$ is the *difference gradient matrix* of at least $[n \times n]$ and the *difference position matrix* $[\Delta X\ ]$ is $[n \times n]$. More columns can be used to have a better estimate of A or its inverse A^{-1} as follows.

The estimate of A or its inverse A^{-1} is then obtained from Eq. (5.116) as

$$A = [\Delta \nabla f][\ \Delta X]^{-1} \tag{5.117}$$

$$\mathbf{A}^{-1} = [\ \Delta \mathbf{X}] \ [\Delta \nabla \mathbf{f}]^{-1} \tag{5.118}$$

It should be noted that $[\Delta \nabla f]$ can be a non-square matrix of $[n \times m]$ with $m > n$ and the matrix $[\Delta X]$ is also $[n \times m]$. The inverse should then be implemented as a ***pseudo inverse***; see Section 5.5.4.

Note that $[\Delta \nabla f]$ can be a *singular matrix* if the difference matrix $[\Delta X]$ is singular. This will occur if the difference is the same between the positions to calculate derivatives, which produce duplicate columns. That is why the position components should be different in values with no repetition or being ***unequally spaced*** in any dimension. If not, the ***pseudo inverse*** will be very large or very small, and no reasonable result is attained. To implement this idea, the variation of the search point derivatives is a *random error* apart and is reduced as the number of iteration steps increases.

The matrix A or its inverse A^{-1} is then used either in a *quadratic optimization* method or as in *Newton–Raphson* $H^{-1} = A^{-1}$ to get the optimum X^*. If the function is quadratic, one should reach the optimum in one step just like *Newton–Raphson*. If the function is not quadratic, more steps or cycles are needed to reach the optimum similar to Example 5.11.

If the gradient ∇f is not available, one can estimate the required number of function evaluations by $(n + 1)$ calculations to get ∇f. From that, the total number of function evaluations is obtained as $n\,(n + 1)$. This is expected to be just as efficient as the *conjugate gradient* method.

A program is devised to optimize the *Rosenbrock's* banana test function using the technique of this section. It is written in MATLAB code and can be run at different parametric conditions; see Figure 5.19c,d for the output and the code, respectively. Figure 5.19e defines the separately written "`Fbanan`" function needed for the MATLAB code in Figure 5.19d. The *Rosenbrock's* function can be optimized in 5–8 iterations of this code. Since it depends on random generation of the differences "n," it might not converge as fast some of the times. This does not happen so often. Only one of the six runs is shown in Figure 5.19c on top of the solution of Example 5.11 in Figure 5.19a. Each run has exactly six iterations converging close to the optimum. The solution of Example 5.11 is shown in gray dashed lines and the solution of the *identification* is in black solid lines. The maximum and minimum deviations in x_1 and x_2 for the six runs are (0.432 48 0.007 855 4) for x_1 and (1.0369 0.017 112) for x_2. The maximum and the minimum values of the optimum result x_1^* and x_2^* for the six runs are (1.023 16 0.9989) for x_1^* and (1.042 989 0.9978) for x_2^* relative to the exact optimum of (1.0 1.0). The maximum and the minimum of the optimized function value for the 6 runs are (2.04E−03 3.02E−07) relative to the exact optimum of (0.0). Note that these values are deviations due to the randomness and expected uncertainty of the differences to identify the Hessian H or A matrix. If only seven iterations are conducted instead of six in each run, the values would be much closer to the exact optimum. The abridged MATLAB code is shown in Figure 5.19d, which can be used to verify the results or modified for other design problems. The error used "`err`" is 0.001, and the number of iterations "`nit`" is 6. These could be changed to examine their effect on the solution efficiency. The number of function evaluations "`nfeval`" is 4 for both numerical derivatives calculated during each single iteration. The total number of function evaluations has always been 54 for all of the 6 iterations in every run. The banana function and its derivatives are evaluated by the MATLAB function of "`FBanan.m`" privately written as shown in Figure 5.19e.

Another executable optimization program (***Multi-D Optimization.exe***) is provided through the ***Wiley website*** that can be used to find the optimum of closed form functions using either quadratic optimization or Newton–Raphson methods. A ***2D*** Rosenbrock's banana test function is imbedded as a function of x_1 *and* x_2 to be examined and run for different optimization conditions.

Example 5.13 Again let us consider the objective function in Example 5.6, and find its extremum using Newton–Raphson and starting at $\boldsymbol{X}_0 = (0, 3)$.

$$f(X) = 4 + 4.5x_1 - 4x_2 + {x_1}^2 + 2x_2^2 - 2x_1x_2 + x_1^4 - 2x_1^2x_2 \tag{a}$$

Solution

Data: $\boldsymbol{X}_0 = (0, 3)$ and the value of the objective point at $\boldsymbol{X}_0$ is $\boldsymbol{f}_0 = 10$.

As defined in Example 5.6, the derivative vector and the *Hessian* matrix are as follows:

$$\nabla f = \begin{bmatrix} 4.5 + 2x_1 - 2x_2 + 4x_1^3 - 4x_1x_2 \\ -4 + 4x_2 - 2x_1 - 2x_1^2 \end{bmatrix} \tag{b}$$

$$\boldsymbol{H} = \begin{bmatrix} 2 + 12x_1^2 - 4x_2 & -2 - 4x_1 \\ -2 - 4x_1 & 4 \end{bmatrix} \tag{c}$$

The procedure for solution is to use the recursive Eq. (5.106), which gives

$$\boldsymbol{X}_{q+1} = \boldsymbol{X}_q - \boldsymbol{H}_q^{-1}\ \nabla \boldsymbol{f}_q, \text{ or } \boldsymbol{X}_1 = \boldsymbol{X}_0 - \boldsymbol{H}_0^{-1}\ \nabla \boldsymbol{f}_0 \tag{d}$$

To get the next point, we need to evaluate the derivative vector and the Hessian matrix at $\boldsymbol{X}_0 = (0, 3)$, which gives

$$\nabla \boldsymbol{f}_0 = \begin{bmatrix} -3/2 \\ 8 \end{bmatrix}, \quad \boldsymbol{H}_0 = \begin{bmatrix} -10 & -2 \\ -2 & 4 \end{bmatrix} \quad \text{and} \quad \boldsymbol{H}_0^{-1} = \frac{-1}{44}\begin{bmatrix} 4 & 2 \\ 2 & -10 \end{bmatrix} \tag{e}$$

Substituting the values in Eq. (e) into Eq. (d) gives the next point

$$\boldsymbol{X}_1 = \boldsymbol{X}_0 - \boldsymbol{H}_0^{-1}\ \nabla \boldsymbol{f}_0 = \begin{bmatrix} 0 \\ 3 \end{bmatrix} - \frac{-1}{44}\begin{bmatrix} 4 & 2 \\ 2 & -10 \end{bmatrix}\begin{bmatrix} -3/2 \\ 8 \end{bmatrix} = \begin{bmatrix} 0.227 \\ 1.115 \end{bmatrix} \tag{f}$$

The value of the objective function at point $\boldsymbol{X}_1$ is 2.4812. To proceed further, we need to evaluate the derivative vector and the Hessian matrix at $\boldsymbol{X}_1$ of Eq. (f), which gives

$$\nabla f_1 = \begin{bmatrix} 1.759 \\ 1.115 \end{bmatrix}, \quad \boldsymbol{H}_1 = \begin{bmatrix} -1.842 & -2.908 \\ -2.908 & 4 \end{bmatrix} \quad \text{and} \quad \boldsymbol{H}_1^{-1} = \frac{-1}{16.88}\begin{bmatrix} 4 & 2.908 \\ 2.908 & -1.842 \end{bmatrix} \tag{g}$$

Substituting the values in Eq. (g) into Eq. (d) gives the next point

$$\boldsymbol{X}_2 = X_1 - \boldsymbol{H}_1^{-1}\ \nabla f_1 = \begin{bmatrix} 0.227 \\ 1.115 \end{bmatrix} + \frac{1}{16.88}\begin{bmatrix} 4 & 2.91 \\ 2.91 & -1.84 \end{bmatrix}\begin{bmatrix} 1.759 \\ -0.097 \end{bmatrix} = \begin{bmatrix} 0.627 \\ 1.428 \end{bmatrix} \tag{h}$$

This point is very close to the saddle point (0.557 469, 1.4909) in Example 5.6. The value of the objective function at point $\boldsymbol{X}_2$ is 2.818, which is very close to the function contour of 2.18 in Figure 5.11. Figure 5.11 presents the main contours of the function and the initial start point $\boldsymbol{X}_0$. One can locate the search points $\boldsymbol{X}_1$ and $\boldsymbol{X}_2$ on Figure 5.11. It is clear that the convergence occurs toward the saddle point. Any further searches would get closer to the saddle point.

5.4.9 Comparison of Unconstrained Optimization Methods

In mechanical design of machine elements, the number of design variables is usually limited. If the objective function is differentiable once or twice, the *first* or *second-order methods* would be advantageous. If the objective function is highly nonlinear, and not easily approximated by quadratic function, *quasi-Newton* methods would be efficiently used. A shallow nonlinear objective function with difficult to attain derivatives would be better solved by *ridge path* or *linearized ridge path* method. *Powell's conjugate gradient* method can also be used in such cases. Other efficient techniques such as the *simplex* method are available (Rao 2009). In MATLAB, the optimization toolbox has a function "`fminsearch`" that utilizes the *Nelder–Mead* simplex direct search method to find the minimum of unconstrained multivariable function.

An executable program (***multi-D Optimization.exe***) and a text file (***objective.txt***) are provided through the ***Wiley website*** that can be used to find the optimum of closed form functions using new *quadratic optimization* techniques and *quasi-Newton* technique. It uses the text file (***objective.txt***) that must be included in the same directory having the ***multi-D Optimization.exe*** program. The ***objective.txt*** file has the number of functions in the first line of the file and each function is defined in a separate line by its number of variables (n), the function written in x1, x2, …, xn. Any multidimensional problem can be rewritten as a function of x1, x2, …, xn to be typed in the *text box*, or use the *dropdown menu* to select from few imbedded functions to be optimized.

5.5 Multidimensional Constrained Problem

Optimization problems are usually constrained by equality or inequality constraints. Previously treated cases of indirect methods include *Lagrange multipliers* and *slackness* variables; see Sections 5.3 and 5.3. Several direct methods are used to treat such constraints. Some selected ones are covered in the following sections:

5.5.1 Eliminating Constraints by Transformation

The optimization problem is again to find min $\boldsymbol{f}$ =?, subject to the inequality constraints $\mathbf{g}_i(\boldsymbol{X}) \leq 0$ $i = 1, 2, \ldots, n$. If the constraints are in a simple form, the substitution of a new variable can guarantee the satisfaction of the constraint. This means performing *variables transformation* and conducting the minimization with respect to the new variable. After the optimization, an *inverse transformation* to the original variable is performed. For a demonstration, few examples are given as follows:

1. If $g_1 \equiv -x_1 \leq 0$, one can use the transformation $x_1 = y_1^2$.
2. If $g_2 \equiv -x_2 \leq 0$, and $g_3 \equiv -x_2 \leq 1$, one can use the transformation $x_2 = \sin^2(y_2)$.
3. If $g_4 \equiv -x_3 \leq 1$, and $g_5 \equiv x_3 \leq 1$, one can use the transformation of $x_3 = \sin(y_2)$.

5.5.2 Exterior Penalty Functions

The strategy of the *exterior penalty function* is to penalize by redefining the "objective function" outside the constraints. The new *penalty function* may be

$$\phi(\boldsymbol{X}, r) = f(\boldsymbol{X}) + r\langle g(\boldsymbol{X})\rangle^z, \quad \text{for } g(\boldsymbol{X}) \leq 0 \tag{5.119}$$

where

$$\begin{aligned} \langle g(\boldsymbol{X})\rangle &= g(\boldsymbol{X}), \quad \text{for } g(\boldsymbol{X}) > 0 \\ \langle g(\boldsymbol{X})\rangle &= 0, \qquad\;\; \text{for } g(\boldsymbol{X}) < 0 \end{aligned} \tag{5.120}$$

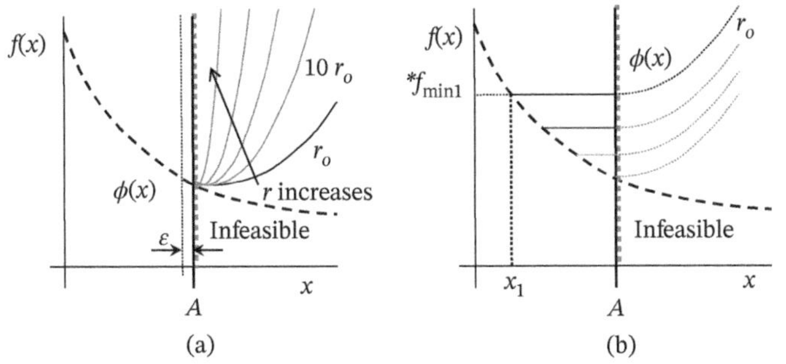

Figure 5.20 A sequence of external penalty searches with increasing r is performed to find the constrained optimum A as shown in (a). Another exterior penalty function is shown in (b), where $^*f_{\min 1}$ is lowest value obtained for f at any step of the search and no need to increase r.

where r is a multiplier factor. Usually the exponent z in Eq. (5.119) is taken as 2.0, which works well because for $z = 2$ we have continuous first derivative as the constraint comes into effect. This is evident from the following partial derivatives:

$$\frac{\partial \phi}{\partial x_i} = \frac{\partial f}{\partial x_i} + 2rg\frac{\partial g}{\partial x_i}, \quad i = 1, 2 \ \ldots n \tag{5.121}$$

A rule that might be followed is that if $\mathbf{X}_0$ is a conservative design, one can pick r_0 (initial value of r) such that

$$r_0 \langle g(\mathbf{X}_0)\rangle^2 \cong f(\mathbf{X}_0) \tag{5.122}$$

The value of r in Eq. (5.119) is kept constant for a given search. It is then increased after locating the minimum of the penalty function defined in Eq. (5.119). A sequence of searches with increasing r is performed to find the constrained optimum as shown in Figure 5.20a. As the sequence progresses we converge to the constrained minimum from the infeasible region. Thus using a constrained of $g + \varepsilon \leq 0$ may be desirable as demonstrated in Figure 5.20a.

For more than one constraint, the penalty function can take the following form:

$$\phi(\mathbf{X}, r) = f(\mathbf{X}) + r\sum_{1}^{m}\langle g(\mathbf{X})\rangle^2, \quad \text{for } g(\mathbf{X}) \leq 0 \tag{5.123}$$

Another exterior penalty function has been devised and used successfully as described next; see Metwalli and Mayne (1977). The penalty function takes the following form:

$$\phi(\mathbf{X}, r) = \begin{cases} ^*f(\mathbf{X}) + r\sum_{i=1}^{m}\langle g_i(\mathbf{X})\rangle^2, & \sum_{i=1}^{m}\langle g_i(\mathbf{X})\rangle^2 > 0 \\ f(\mathbf{X}), & \sum_{i=1}^{m}\langle g_i(\mathbf{X})\rangle^2 = 0 \end{cases} \tag{5.124}$$

where $^*f$ is the lowest value obtained for f at any step of the search as shown in Figure 5.20b. For that penalty function, no need to use $g + \varepsilon \leq 0$, and no need to increase r. It should be noted, however, that this function is only continuous in the first order at the constraint boundary, but it causes no problem for numerical search means with simple controls.

5.5.3 Interior Penalty Functions

In this *interior penalty function*, convergence takes place from inside the feasible region. The interior penalty function may take the following form:

$$\phi(\mathbf{X}, r) = f(\mathbf{X}) + r\sum_{1}^{m}\frac{1}{\langle g(\mathbf{X})\rangle^2}, \quad \text{for } g(\mathbf{X}) \leq 0 \tag{5.125}$$

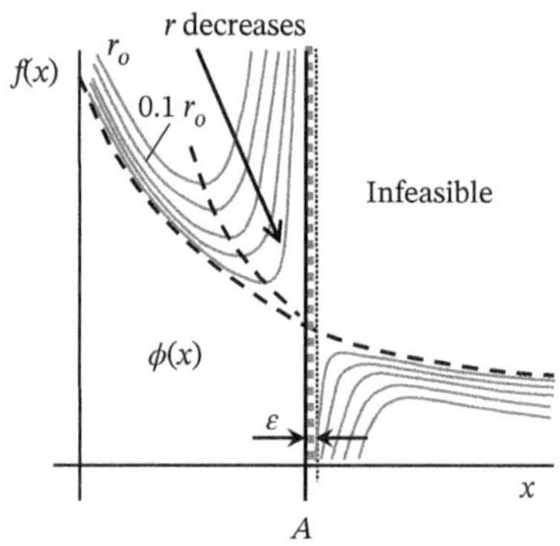

Figure 5.21 Interior penalty function showing that convergence takes place from inside the feasible region. The value of r is decreased from minimization to minimization. A transformed "mirror" image of the penalty function is on the other side of the constraint A inside the infeasible region.

The pointed brackets of $g(\mathbf{X})$ in Eq. (5.125) are defined as given in Eq. (5.120). Similar to the exterior penalty function, the initial value of r can be defined so that

$$r_0 \sum_{1}^{m} \frac{1}{\langle g(\mathbf{X}_0)\rangle} \cong f(\mathbf{X}_0) \tag{5.126}$$

where $\mathbf{X}_0$ is the initial design vector. The value of r is then decreased from minimization to minimization as shown in Figure 5.21. Since a transformed "mirror" image of the penalty function is on the other side of the constraint, we have to be careful in the searching process so that not to jump over the constraint and find an infeasible minimum.

5.5.4 Direct Methods for Constrained Problems

Constrained problems can be handled in several ways. Some of these are covered herein as basic understanding and usable means. The basic properties of convex–concave shed light onto the suitability of the applied method of solution. *K–T* conditions are frequently used to check if the constrained optimum has been reached or not. *Gradient projection* (GP) methods are usually used to guide the search when constraints are active at some search points. These methods are presented in the following segments.

5.5.4.1 Convex–Concave Property

The function is *convex* over a region if between any two points within the region the function value is less than (or equal) to the value of a straight line drawn between the two points. This is shown in Figure 5.22a and defined by the following relations:

$$\begin{aligned} f - \alpha(f_1 - f_2) &\geq f[(\mathbf{X}_1) + \alpha(\mathbf{X}_2 - \mathbf{X}_1)] \\ (1-\alpha)f_1 + \alpha f_2 &\geq f[(1-\alpha)\mathbf{X}_1 + \alpha\mathbf{X}_2] \end{aligned} \tag{5.127}$$

The function is *concave* if

$$(1-\alpha)f_1 + \alpha f_2 < f[(1-\alpha)\mathbf{X}_1 + \alpha\mathbf{X}_2] \tag{5.128}$$

for any two points within the region of the constraints.

Figure 5.22b depicts the *convex* and *concave* state for few feasible internal regions. The constraints are convex if within the feasible region, no straight line can be drawn between two points and can go outside the feasible region.

5.5.4.2 Kuhn–Tucker Conditions

Recall *Lagrange multipliers*, Eq. (5.49) is rewritten next as follows:

$$\nabla \boldsymbol{f} + [\nabla \boldsymbol{\varphi}^T]\lambda = \mathbf{0}, \quad \text{or} \quad \nabla \boldsymbol{f} + \sum_{i=1}^{m} \lambda_i \nabla \boldsymbol{\varphi}_i = \mathbf{0} \tag{5.129}$$

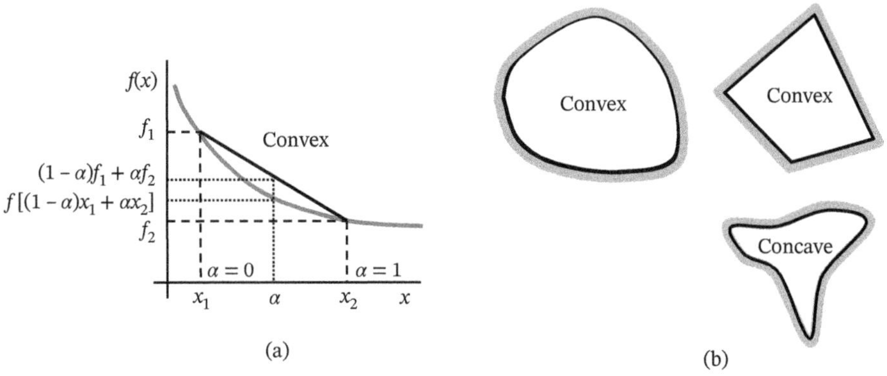

Figure 5.22 The function is convex over a region, where between any two points within the region the function value is less than (or equal) to the value of a straight line between the two points as shown in (a). The convex and concave state for few feasible internal regions are depicted in (b).

Taking *slack variables* to change inequality constraints to equality constraints, i.e. $\boldsymbol{\varphi}_i = \mathbf{g}_i + s_i^2$ (allows us to deal with $g_i \leq 0$), one can have

$$\nabla \boldsymbol{f} + \sum_{i=1}^{m} \lambda_i \nabla(\mathbf{g}_i + s_i^2) = \mathbf{0} \qquad (5.130)$$

Looking at one component of these vectors with respect to a slack variable s_i

$$\frac{\partial \boldsymbol{f}}{\partial s_i} + \lambda_i \left(\frac{\partial \mathbf{g}_i}{\partial s_i} + 2s_i \right) = 0 \qquad (5.131)$$

From that, one gets

$$2\lambda_i \ s_i = 0 \qquad (5.132)$$

Therefore, either $s_i = 0$ or $\lambda_i = 0$. As discussed previously in Section 5.3.3, the following is indicating that *iff*

$$\begin{aligned} s_i &= 0, \quad \text{constraint is in effect} \\ \lambda_i &= 0, \quad \text{constraint is not in effect} \end{aligned} \qquad (5.133)$$

Consider the "*I*" *constraints* that are in effect. Therefore

$$\nabla \boldsymbol{f} + \sum_{I} \lambda_i \nabla \mathbf{g}_i = \mathbf{0} \qquad (5.134)$$

Also, for those constraints that are not in effect, the λ_i's are zeros. Therefore, the constrained minimum should have

$$\begin{aligned} &\nabla \boldsymbol{f} + \sum_{i=1}^{m} \lambda_i \nabla \mathbf{g}_i = \mathbf{0} \\ &\lambda_i \geq 0 \end{aligned} \qquad (5.135)$$

These are the ***K–T*** conditions. If ***K–T*** conditions are not satisfied, we are not at a constraints minimum. On the other hand ***K–T*** conditions may be used to give a point that in constraints minimum.

For the two constraints shown in Figure 5.23, the ***K–T*** conditions give

$$-\nabla \boldsymbol{f} = \lambda_1 \nabla \mathbf{g}_1 + \lambda_2 \nabla \mathbf{g}_2 \qquad (5.136)$$

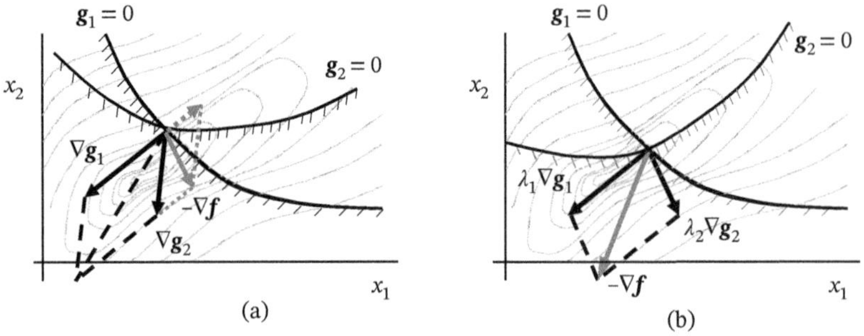

Figure 5.23 Two constraints showing that the ***K–T*** conditions are not satisfied in (a). The ***K–T*** conditions are satisfied in (b).

For the case in Figure 5.23a, at least one of the constraints λ_1 or λ_2 will be negative to have a resultant of $-\nabla \boldsymbol{f}$. According to the ***K–T*** conditions of Eq. (5.134), both λ_1 and λ_2 have to be positive.

For the case in Figure 5.23b, the gradient vector $-\nabla \boldsymbol{f}$ lies in the sector defined by the two vectors $\lambda_1 \nabla \boldsymbol{g}_1$ and $\lambda_2 \nabla \boldsymbol{g}_2$, and the point is thus a constrained local optimum.

To evaluate the λ_i's without solving for the *Lagrange multipliers*, one can proceed as follows:

Let the *Lagrange multipliers* vector λ and the constraint gradient matrix $\boldsymbol{G}$ be defined as

$$\lambda = \begin{bmatrix} \lambda_1 \\ \lambda_2 \\ \vdots \\ \lambda_m \end{bmatrix} \text{ and } \boldsymbol{G} = \begin{bmatrix} \dfrac{\partial g_1}{\partial x_1} & \dfrac{\partial g_2}{\partial x_1} & \cdots \\ \dfrac{\partial g_1}{\partial x_2} & \dfrac{\partial g_2}{\partial x_2} & \cdots \\ \vdots & \vdots & \ddots \end{bmatrix} = [\nabla g_1 \ \nabla g_2 \ \cdots \ \nabla g_m] \tag{5.137}$$

Equation (5.135) become

$$\nabla \boldsymbol{f} + \boldsymbol{G}\ \lambda = \boldsymbol{0} \text{ and } \lambda_i \geq 0 \tag{5.138}$$

The first expression in Eq. (5.138) can be manipulated and solved for λ such that

$$\boldsymbol{G}\lambda = -\nabla \boldsymbol{f}$$
$$\lambda = -\ \boldsymbol{G}^{-1}\ \nabla \boldsymbol{f} \tag{5.139}$$

However, the constraint gradient matrix $\boldsymbol{G}$ may not be a *square matrix* since n is usually not equal to m. Therefore, one can solve this problem by *minimizing a residual* $\boldsymbol{R}$ defined as follows:

$$\boldsymbol{R} = \boldsymbol{G}\lambda + \nabla \boldsymbol{f} \tag{5.140}$$

Squaring $\boldsymbol{R}$ and differentiating with respect to λ gives the following expressions:

$$|\boldsymbol{R}|^2 = (\boldsymbol{G}\lambda + \nabla \boldsymbol{f})^T(\boldsymbol{G}\lambda + \nabla \boldsymbol{f}) = \lambda^T \boldsymbol{G}^T \boldsymbol{G}\lambda + 2\lambda^T \boldsymbol{G}^T \nabla \boldsymbol{f} + \nabla \boldsymbol{f}^T \nabla \boldsymbol{f}$$
$$\frac{\partial |\boldsymbol{R}|^2}{\partial \lambda} = 2\boldsymbol{G}^T[\boldsymbol{G}\lambda + \nabla \boldsymbol{f}] = \boldsymbol{0} = 2\boldsymbol{G}^T\boldsymbol{G}\lambda + 2\boldsymbol{G}^T \nabla \boldsymbol{f} \tag{5.141}$$

Equation (5.141) gives the following relations:

$$2\boldsymbol{G}^T\boldsymbol{G}\lambda + 2\boldsymbol{G}^T\nabla \boldsymbol{f} = \boldsymbol{0}$$
$$\boldsymbol{G}^T\boldsymbol{G}\lambda = -\ \boldsymbol{G}^T\nabla \boldsymbol{f} \tag{5.142}$$

The last expression in Eq. (5.142) provides the solution of the *Lagrange multipliers* as follows:

$$\lambda = -[\boldsymbol{G}^T\boldsymbol{G}]^{-1}\boldsymbol{G}^T\ \nabla \boldsymbol{f} \tag{5.143}$$

Now compute λ at the point of interest. If all λ_i's are ≥ 0 and $\boldsymbol{G}\ \lambda = -\nabla \boldsymbol{f}$, then the point satisfies ***K–T*** conditions.

It should be of interest to note that by comparing Eq. (5.139) to Eq. (5.143), one can get the *inverse* of the *non-square matrix* as follows:

$$G^{-1} = [G^T G]^{-1} G^T \tag{5.144}$$

This is called the ***pseudo inverse*** or *minimum norm inverse*. MATLAB provide a *pseudo inverse function* "`pinv`" to carry such an inverse.

5.5.4.3 Gradient Projection Method

The method of ***GP*** is concerned about moving along the constraint boundary. If we hit the constraints, we may like to move on the constraints by projecting the gradient vector onto the constraints tangent as shown in Figure 5.24. The projected gradient vector consists of the gradient vector itself minus the component u_i normal to the tangent (i.e. parallel to $\nabla \mathbf{g}$). This is translated to the following relation for r constraints *in effect* at the point in question:

$$\nabla f_p = \nabla f - u_1 \nabla g_1 - u_2 \nabla g_2 \ldots \; u_r \nabla g_r \tag{5.145}$$

Note again that only few "r" of the "m" constraints are active. Equation (5.145) can be rewritten as

$$\nabla \boldsymbol{f}_p = \nabla \boldsymbol{f} - \begin{bmatrix} \nabla \mathbf{g}_1 & \nabla \mathbf{g}_2 & \ldots & \nabla \mathbf{g}_r \end{bmatrix} \begin{bmatrix} u_1 \\ u_2 \\ \ldots \\ u_r \end{bmatrix} = \nabla \boldsymbol{f} - \boldsymbol{G}_r \boldsymbol{u} \tag{5.146}$$

It should be noted that, generally, $\boldsymbol{G}_r \neq \boldsymbol{G}$. But the rejection vector must be *orthogonal* to the constraints gradient, which requires that

$$\nabla \mathbf{g}_1^T \nabla \boldsymbol{f}_p = \nabla \mathbf{g}_2^T \nabla \boldsymbol{f}_p = \cdots = \nabla \mathbf{g}_r^T \nabla \boldsymbol{f}_p = 0 \tag{5.147}$$

This is rewritten as

$$\boldsymbol{G}_r^T \nabla \boldsymbol{f}_p = \mathbf{0} \tag{5.148}$$

Substituting Eq. (5.148) into Eq. (5.145) by the through multiplication by G_r^T, one obtains the following:

$$\boldsymbol{G}_r^T [\nabla \boldsymbol{f} - \boldsymbol{G}_r \boldsymbol{u}] = \mathbf{0}$$
$$\boldsymbol{G}_r^T \nabla \boldsymbol{f} = \boldsymbol{G}_r^T \boldsymbol{G}_r \boldsymbol{u} \tag{5.149}$$

Or

$$\boldsymbol{u} = [\boldsymbol{G}_r^T \boldsymbol{G}_r]^{-1} \boldsymbol{G}_r^T \nabla \boldsymbol{f} \tag{5.150}$$

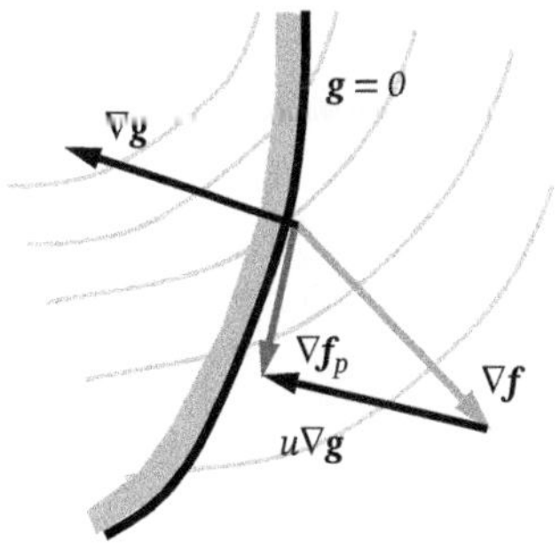

Figure 5.24 Projecting the gradient vector $\nabla \boldsymbol{f}$ onto the constraints tangent to get the gradient projection $\nabla \boldsymbol{f}_p$.

This relation is in a similar form to the *Lagrange multipliers* in Eq. (5.143), but without the negative sign. Substituting Eq. (5.150) into Eq. (5.145) gives the projection vector.

$$\nabla \boldsymbol{f}_p = \nabla \boldsymbol{f} - \boldsymbol{G}_r\,[\boldsymbol{G}_r^T\boldsymbol{G}_r]^{-1}\boldsymbol{G}_r^T\nabla \boldsymbol{f}$$
$$\nabla \boldsymbol{f}_p = \{I - \boldsymbol{G}_r\,[\boldsymbol{G}_r^T\boldsymbol{G}_r]^{-1}\boldsymbol{G}_r^T\nabla \boldsymbol{f}\}\nabla \boldsymbol{f} \tag{5.151}$$

Therefore, for minimization we can proceed in the $-\nabla \boldsymbol{f}_p$ direction. In that situation, what might happen? Three cases are expected to occur; see Figure 5.25 for a hypothetical constraint that has *convex*, *linear*, and *concave* sections:

- At point (a) on the *convex constraint* section, obviously we go into the infeasible region. If we want to continue, we have to step in the $-\nabla \mathbf{g}$ direction to converge. Thus, method is not appropriate for convex constraints.
- For *concave constraint* section at point (b), we move into the feasible region, and we can continue in the normal search procedure.
- For *linear constraint* section at (c), we move in just the right direction. The method in most useful and suitable. This is one way of handling constraints in *linear programming*.

5.5.4.4 *Heuristic Gradient Projection Method (HGP)*

As indicated in the previous ***GP*** section and for convex constraints, the search direction should depart the constraint boundary into the infeasible region if these constraints were active. To continue, one must step into the $-\nabla \mathbf{g}_i$ to converge. If the component of $-\nabla \mathbf{g}_i$ is included in the new combined ***GP*** search direction $-\nabla \mathbf{g}_i{}^T\nabla \boldsymbol{f}$, the process should be automatic. Therefore, this method is suitable for convex constraints. For concave constraints, the search direction also steps into the feasible region. To continue, one may need to step into the $\nabla \boldsymbol{g}_i$ to converge. If the component of $\nabla \boldsymbol{g}_i$ is also included in the new combined search direction $\nabla \boldsymbol{g}_i{}^T\nabla \boldsymbol{f}$, the process should also be automatic. Figure 5.26 shows the projection of the gradient vector ∇f onto the constraints tangent to get the *heuristic gradient projection* (***HGP***) ∇f_{pH}, which is away from the constraint. In many cases, however, the constraint gradient may be in the same direction as the gradient of the objective function. The search can thus continue in the normal procedure; see, e.g. Abd El Malek (2005) and Metwalli (2012).

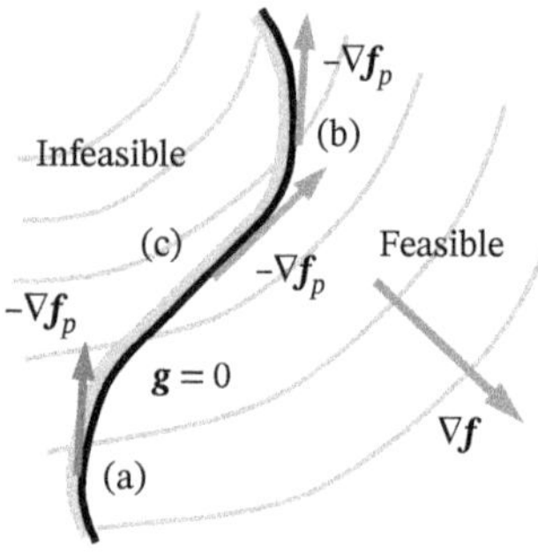

Figure 5.25 A hypothetical constraint having convex, linear, and concave sections.

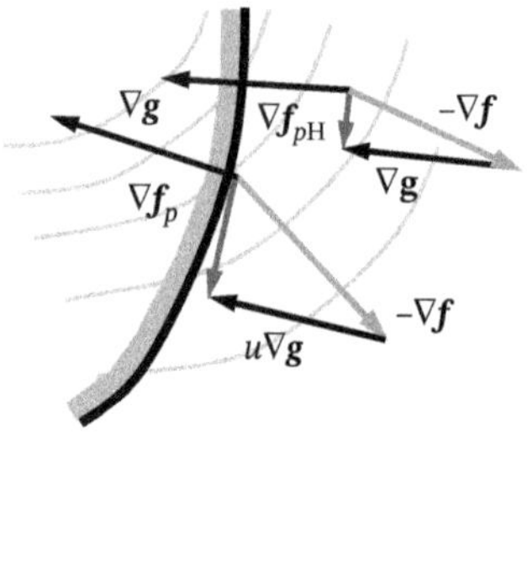

Figure 5.26 Projecting the gradient vector $\nabla \boldsymbol{f}$ onto the constraints tangent to get the heuristic gradient projection $\nabla \boldsymbol{f}_{\mathrm{pH}}$ away from the constraint.

The ***HGP*** method is most suitable to apply for problems that are expected to have a constrained minimum with defined formulation for both objective function and constraints. The optimization of ***3D*** space frames is a problem where the objective function is defined as the total volume and the constraints are restricted to satisfy allowable stress values. The optimization of ***3D*** space frames is then most suitable to demonstrate the ***HGP*** method. The objective function for ***3D*** space frames is the total volume of the n members of circular cross section each of a diameter d_i and a length l_i. The total volume is therefore the objective function defined as

$$f(d) = \sum_{i=1}^{n} \pi d_i^2 l_i/4 \tag{5.152}$$

Without any loss of generality, if the ***3D*** frame will not be *topologically optimized* except by removing trivially loaded members, the objective function will take the form

$$f(d) = \sum_{i=1}^{n} a_i d_i^2 \tag{5.153}$$

where a_i is a constant that can represent $\pi l_i/4$. In the local coordinates, the stress in each member is evaluated numerically (Abd El Malek et al. 2005). Without key loss of generality, the stress constraints may be initially defined by the following expression:

$$g_i(d) \cong b_i d_i^{-3} - c_i \leq 0 \tag{5.154}$$

where b_i is a constant that includes the *stress* constants and c_i is a constant representing the *allowable strength*. It is assumed initially that the coupling between members and its feedback can be automatically accounted for during iterations. If, however, the members were mainly subjected to axial loading, Eq. (5.154) would instead be mainly in (d^{-2}). This is one other alternative, which can be used. Other options have also been used; see, e.g. Senousy et al. (2005), Hanafy and Metwalli (2009), and Abd El-Aziz et al. (2016).

The components of the gradient vectors for the objective function and the constraints are obtained from Eqs. (5.153) and (5.154) as follows:

$$\begin{aligned} \nabla f(d) &= \alpha_i d_i \\ \nabla g_i(d) &= -\beta_i d_i^{-4} \end{aligned} \tag{5.155}$$

The components of *projected gradient vector* can be obtained as follows:

$$\nabla g_i^T \nabla f = -\gamma_i \, d_i^{-3} \tag{5.156}$$

To implement the procedure and considering the *slackness variables* along the ***HGP*** direction, we use the recursive heuristic direction components for constraints and minimize the objective function at the same time such that

$$s_{\text{pi}}(d_i) = \sigma_{i,\text{equiv}} \; (d_{i,\text{new}})^{-3} - S_{i,\text{const}} \; (d_i)^{-3} \tag{5.157}$$

where d_i is the *member diameter*, $d_{i,\text{new}}$ is the *new diameter* of the member, s_{pi} is the component of slackness on the projection vector, $\sigma_{i,\text{equiv}}$ is the *equivalent stress* (e.g. *von Mises*), and $S_{i,\text{const}}$ is the *allowable strength* constraint.

This can be implemented by *recursive* use of the following relation:

$$d_{i,\text{new}}^3 = d_i^3 \frac{\sigma_{i,\text{equiv}}}{S_{i,\text{const}}} \tag{5.158}$$

The relation defined in Eq. (5.158) satisfies the constraint; in addition, it reduces the objective function. Therefore, these steps chase the constraint boundary while reducing the objective function. This is tantamount to the process of ***GP*** method, but does not require following the constraint tangents. This is the reason of applying the name ***HGP*** on this method.

5.5.4.5 Constrained Optimization Sample

This section presents a simple constrained problem to clarify concepts and demonstrate constraint optimization handling.

Example 5.14 Consider the simple objective function defined by

$$\boldsymbol{f} = -(x_1 + x_2) \tag{a}$$

The constraints for this problem are

$$\begin{aligned} g_1 &= -x_1 \le 0 \\ g_2 &= x_1^2 + x_2^2 - 1 \le 0 \end{aligned} \tag{b}$$

Start at $\boldsymbol{X}_o = (0, -1/2)$, use the first direction of any first-order method (*gradient*) to reach the next point. Check if the point $x_1 = \sqrt{2}/2,\ x_2 = \sqrt{2}/2$ is a constraint optimum.

Solution

Data: $\boldsymbol{X}_0 = (0, -1/2)$ and check *K–T* conditions at (0.707, 0.707). The value of the objective point at $\boldsymbol{X}_0$ is $\boldsymbol{f}_0 = 1/2$.

The plot of the function contours and constraints is given in Figure 5.27 showing the infeasible and feasible regions. The objective function gradient is

$$\nabla f = \begin{bmatrix} \dfrac{\partial f}{\partial x_1} \\ \dfrac{\partial f}{\partial x_2} \end{bmatrix} = \begin{bmatrix} -1 \\ -1 \end{bmatrix} \tag{c}$$

First search direction $\boldsymbol{S}$ of any first-order method is in the negative direction of the gradient, *i.e.* $-\nabla \boldsymbol{f}$. Get the intersection of $\boldsymbol{S}$ line with the constraint such as

$$\begin{aligned} x_2 &= x_1 - 0.5, \\ x_1^2 &+ (x_1 - 0.5)^2 - 1 = 0 \end{aligned} \tag{d}$$

which gives a point at $x_1 = 0.9114$ and -0.4114. At $x_1 = 0.9114$, the next point is then $\boldsymbol{X}_1 = (0.9114, 0.4115)$. At $x_1 = -0.4114$, the point is in the infeasible region and thus will be discarded. The required next point $\boldsymbol{X}_1 = (0.9114, 0.4115)$ is then shown in Figure 5.13.

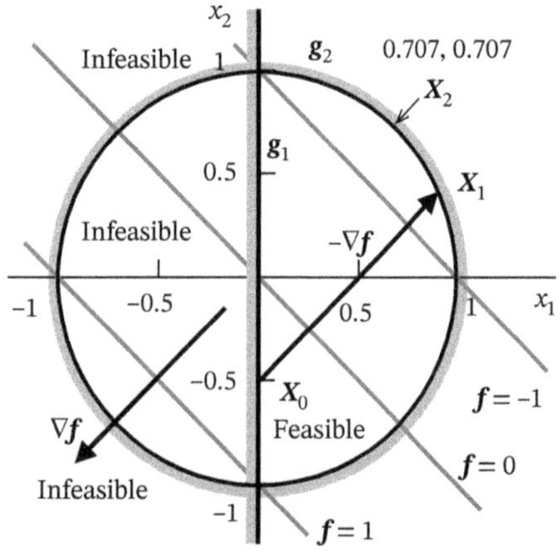

Figure 5.27 The plot of the function contours and constraints of Example 5.14 showing the infeasible and feasible regions.

To check if the point $x_1 = \sqrt{2}/2,\ x_2 = \sqrt{2}/2$ or $\boldsymbol{X}_2 = (0.707, 0.707)$ is a constraint optimum, we use the ***K–T*** conditions. The procedure is to evaluate the Lagrange multipliers vector λ and check Eq. (5.135) at $\boldsymbol{X}_2$. It is then required to calculate the following components of Eq. (5.135) and Eq. (5.143) at $\boldsymbol{X}_2$ such that

$$\boldsymbol{G} = [\nabla \boldsymbol{g}_1 \quad \nabla \boldsymbol{g}_2] = \begin{bmatrix} -1 & 2x_1 \\ 0 & 2x_2 \end{bmatrix}$$

$$\boldsymbol{G}|_{\boldsymbol{X}_2} = \begin{bmatrix} -1 & \sqrt{2} \\ 0 & \sqrt{2} \end{bmatrix} \tag{e}$$

and

$$\boldsymbol{G}^T\boldsymbol{G} = \begin{bmatrix} -1 & 0 \\ \sqrt{2} & \sqrt{2} \end{bmatrix} \begin{bmatrix} -1 & \sqrt{2} \\ 0 & \sqrt{2} \end{bmatrix} = \begin{bmatrix} 1 & -\sqrt{2} \\ -\sqrt{2} & 4 \end{bmatrix}$$

$$[\boldsymbol{G}^T\boldsymbol{G}]^{-1} = \begin{bmatrix} 2 & \frac{\sqrt{2}}{2} \\ \frac{\sqrt{2}}{2} & 1/2 \end{bmatrix} \tag{f}$$

Substituting Eqs. (e) and (f) into Eq. (5.135), we get

$$\lambda = -[\boldsymbol{G}^T\boldsymbol{G}]^{-1}\boldsymbol{G}^T\nabla \boldsymbol{f}$$

$$\lambda = \begin{bmatrix} 2 & \frac{\sqrt{2}}{2} \\ \frac{\sqrt{2}}{2} & \frac{1}{2} \end{bmatrix} \begin{bmatrix} -1 & 0 \\ \sqrt{2} & \sqrt{2} \end{bmatrix} \begin{bmatrix} -1 \\ -1 \end{bmatrix} = \begin{bmatrix} 0 \\ \frac{\sqrt{2}}{2} \end{bmatrix} \geq 0 \tag{g}$$

Checking the second equation in K–T conditions of Eq. (5.135) at $\boldsymbol{X}_2$, we get

$$\nabla \boldsymbol{f} + \boldsymbol{G}\lambda = 0$$

$$\begin{bmatrix} -1 \\ -1 \end{bmatrix} + \begin{bmatrix} \sqrt{2} \\ \sqrt{2} \end{bmatrix} \frac{\sqrt{2}}{2} = \begin{bmatrix} -1 \\ -1 \end{bmatrix} + \begin{bmatrix} 1 \\ 1 \end{bmatrix} = \begin{bmatrix} 0 \\ 0 \end{bmatrix} \tag{h}$$

Therefore, K–T condition is satisfied. Point $\boldsymbol{X}_2$ is then a constrained minimum, which is clearly recognized from Figure 5.27. Graphically, the objective function gradient $-\nabla \boldsymbol{f}(\boldsymbol{X}_2)$ is in the same direction as the constraint gradient $\nabla \boldsymbol{g}_1(\boldsymbol{X}_2)$. The Lagrange multipliers λ_1 should then be positive, and point $\boldsymbol{X}_2$ should be a constrained optimum.

5.5.5 Comparison of Optimum Constrained Methods

Usually problems in mechanical design are constrained by allowable stresses or buckling constraints in addition to other possible constraints of geometry or behavioral responses. The number of design variables is usually limited. If the objective function is highly nonlinear, and not easily approximated by a quadratic function, penalty function methods could be efficiently utilized. A shallow nonlinear objective function with difficult to attain reliable derivatives would be better solved by ***GP*** or ***HGP*** if the optimum is expected to be on the constraints, and the slackness variables can lead to recursive relations to adjust design variables. If the optimum is not expected to be on the constraints boundaries, hybrid methods can also be used to find the optimum design of mechanical components and systems.

5.6 Applications to Machine Elements and Systems

So many applications are there for optimum design of machine elements, mechanical systems, and other fields. This textbook utilizes the optimum designs of many machine elements to synthesize mechanical components.

From these optimum results, it is easy to define the optimum machine elements as the synthesized design of the element. The synthesis process is then more efficient, and does not require repeated analysis to reach a reasonable design. Adjusting the initial optimum design to accommodate different objectives and constraints should be easier than starting from scratch.

Many machine elements have been optimized, and it is useful to search the literature for the available optimized machine components and systems before the synthesis process. The samples used herein are just to consider few of the available optimized machine components by the author and his colleagues and collaborators. A review of some of these is presented next.

Optimum designs of helical compression, tension, and torsion springs have been performed by Metwalli et al. (1993, 1994). Metwalli and El Danaf (1996) have worked on optimization of spur and helical gear sets. Disk brakes have been optimized for multi-objective as presented in Metwalli and Hegazi (1999, 2001). Optimization of ball bearings has been conducted including the effects of elastohydrodynamics (Abbas and Metwalli 2011). Optimum configuration for isotropic rotor is given by Shawki et al. (1984). Solid C-Frame cross-section optimization has been presented in Nassef et al. (2000), and Hegazi et al. (2002). Optimum design of uniformly stresses rotors under angular acceleration effects has been presented by Metwalli (1978). Optimum design of variable-material flywheels has been given by Metwalli et al. (1983). Multiple design objectives in hydrodynamic journal bearing optimization have been considered in Metwalli et al. (1984). Optimal design of disk springs is performed by Metwalli et al. (1985). Flywheel optimization under speed fluctuation effects is performed by Metwalli (1986a, b). Y-stiffened panel multi-objective optimization has been given by Badran et al. (2009). Elmoselhy et al. (2006) presented the shape optimization of laminated fibrous composite E-springs. Optimum design of pressure vessels has also been attained by the use of hybrid ***HGP*** and GA (Abd El Aziz and Metwalli 2017).

Optimum designs of some mechanical systems have also been conducted. Automobile gear train design optimization has been given by Elmaghraby et al. (1979). The design of vehicle suspensions have been optimized by Metwalli et al. (1973); Ghoneim and Metwalli (1984); and Metwalli (1986a,b). MEMS, ***3D*** frames, elutriators, and compliant grippers are some of the systems been optimized (Metwalli and Afify 1981; Hegazi et al. 2002; Senousy et al. 2005, 2008; Shalaby et al. 2003a, b; Abd El Malek et al. 2005; Hanafy and Metwalli 2009). Solar desalination systems are also optimized for remote communities (Abd El-Aziz et al. 2016).

The aforementioned references present several applications with multitude of objectives, competing objectives, and multi-objective cases of optimization. Several realistic constraints are considered. These treatments would be useful in the formulation of other optimization applications or studying other optimization objectives. So many other optimization applications are available in the literatures that cover a very large scope in machine element applications. The space and scope of this text does not allow enough treatment of these other vast applications.

An optimization example is presented next to treat one simple mechanical application that involves realistic considerations.

Example 5.15 Consider Example 5.2 of the two-bar truss in Figure 5.4. It is required to carry an optimization study to minimize the truss weight subject to satisfying the normal and buckling stress constraints. Use dimensional state with the accurate equations to obtain the solution. Discuss the results relative to those in Fox (1971).

Solution

Data: Tube average diameter d_t and fixed thickness $t_t = T_t$. Height h_t and fixed base $b_t = 2B_t$. Top load $2\boldsymbol{P}$, material ρ, S_y, and E are known. Tubes must neither buckle nor yield. The d_t^* and h_t^* are to satisfy both constraints and minimize weight. The truss has a design vector $\boldsymbol{D}_t = [d_t \;\; h_t]^T$. The objective function can be more accurately defined. Both accurate and simplified cases are given for the dimensional case in Example 5.2. Accurate relations, except for the weight, are rewritten as follows:

$$\underline{D} = \frac{d_t}{T_t}, \quad \underline{H} = \frac{h_t}{B_t} \tag{a}$$

$$\boldsymbol{f} = \underline{\boldsymbol{W}}_t \equiv \frac{W_t}{\rho T_t^{\,2} B_t} = 2\,\pi \underline{D}(1 + \underline{H}^2)^{1/2} \tag{b}$$

$$\sigma_t = \frac{\boldsymbol{P}\sqrt{B_t^2 + h_t^2}}{\frac{\pi}{4}\ ((d_T + T_t)^2 - (d_T - T_t)^2)h_t} \tag{c}$$

$$\sigma_B = \frac{\pi^2 E(\pi((d_T + T_t)^4 - (d_T - T_t)^4)/64)}{\pi\ (B_t^2 + h_t^2)\ ((d_T + T_t)^2 - (d_T - T_t)^2)/4} \tag{d}$$

Figure 5.28 shows the MATLAB code for the iteration to get the optimum truss design. The optimum values are approximately given in Table 5.4 to check that the stress constraints are about the same and the variable "`diff`" in the MATLAB code of Figure 5.28 is very small. The iteration is performed by adjusting h to have the variable "`diff`" as close to zero as possible. The solution of Fox (1971) is also available to calculate; however, it is in the US system of units as noted. The values in Table 5.4 are to indicate the differences between the present accurate solution and the approximate solution provided by Fox (1971). The values of h^* are approximated in Table 5.4 but has to be further refined with more digits to get almost equal normal and buckling stresses. The closer value of $h^* = 10.081\,12$ for $d = 2.0$ is shown in Figure 5.29 of the MATLAB code. Figure 5.29 shows the plots of the optimum truss design (d^*, h^*) in [in], the objective function W in [lb], and the stress constraint, which is about the same for both normal and buckling stresses. Running the MATLAB code in Figure 5.28 for different parameters can generate these different values. It was observed, however, that the accurate relation gives very close values to the approximate relation for the normal stresses, but not the same for the buckling stresses.

The constrained optimization performed on the two-bar truss assumed that activating the two constraints would minimize the weight. Even though this provides the maximum utilization of the material, it may not necessarily reduce the weight for all design requirements. This is evident from Figure 5.29 where the weight for lower d^* produced a weight that is lower than the initial solution of Fox (1971) as depicted in bold letters in Table 5.4. The use of approximate relation for the buckling stress may be the reason. The more accurate relation for the buckling

```
clc; clear; clear all; format long; format compact
t=0.1; B=30; d=2; h=19;                                      % Fox Dimensional
P=33*10^3; Sa=100*10^3; E=3*10^7; ro=0.3;                    % Fox Dimensional
d=2; h=10.08112;                                         % Optimum iteration
D=d/t; H=h/B;                                                % Nondimensional
        % Dimensional Calculations
SigN=4*P*sqrt(B^2+h^2)/(pi*((d+t)^2-(d-t)^2)*h)
SigNa=P*sqrt(B^2+h^2)/(pi*t*(d*h))                           % Approximate
SigB=(pi^2*E*((d+t)^4-(d-t)^4))/(16*(B^2+h^2)*((d+t)^2-(d-t)^2))
SigBa=(pi^2*E*(d^2+t^2))/(8*(B^2+h^2))                       % Approximate
Wd=2*ro*pi*d*t*(sqrt(B^2+h^2))
diff=SigN-SigB                                               % diff should ≈0.0
diffa=SigNa-SigBa                                            % Approximate
```

Figure 5.28 MATLAB code for nondimensional optimum truss design. The optimum values are given to check that the stress constraints of Example 5.15 are the same and the variable "`diff`" is very small. The solution of Fox (1971) is also available to calculate.

Table 5.4 Values of optimum truss design (d^*, h^*), objective function W [lb], stress constraints of Example 5.15, and the nondimensional height $\underline{H}$ and nondimensional diameter $\underline{D}$. The thickness $t = 0.1$ [in] and breadth $B = 30$ [in].

	Diameter d^* [in]								Fox (1971)
	1.4	1.5	1.6	1.7	1.8	1.9	2	2.5	2
Height h^* [in]	77.82	34.499	23.807	18.183	14.565	12.002	10.081	4.9632	19
Weight W [lb]	22.01	12.93	11.55	**11.24**	11.31	11.57	**11.93**	14.33	**13.4**
Stress [kpsi]	80.41	**92.80**	**105.6**	119.2	133.6	148.4	164.9	257.4	98[a]
$\underline{H}$	2.594	1.15	0.794	0.606	0.486	0.4001	0.336	0.1654	0.667
$\underline{D}$	14	15	16	17	18	19	20	25	20

a) The buckling stress for accurate relation gives 164.8 [kpsi], which indicates the possibility of adjusting some design variables and thus reducing the weight.

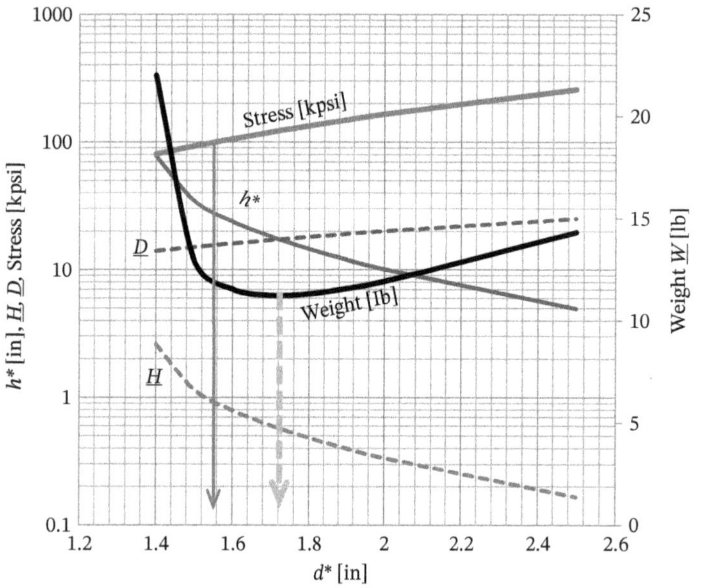

Figure 5.29 Plots of the optimum truss design (d^*, h^*) in [in], objective function W in [lb] with the scale to the right, and the stress constraint, which is about the same for both normal and buckling stresses. Extending the search to lower values of d^*, the weight is reduced.

stress produced a smaller height and a lower weight, which is also depicted in bold letters in Table 5.4. When extending the search to lower values of d^*, the weight is reduced as shown in Table 5.4 and Figure 5.29. The stress is, however, increased. The initial low stress was due to larger design variables, particularly the height h. This height produced a lower approximate normal and buckling stresses. The accurate buckling stress is evaluated according to Eqs. (d) and (c) of Example 5.2. To guarantee an accurate normal and buckling stress of about 100 [kpsi], the optimum constrained design vector (d^*, h^*) in [in] is about (1.52, 29) as indicated by the grayish solid arrow in Figure 5.29. The weight in that case is about 12 [lb], which is still smaller than the original optimum design of Fox (1971); see Table 5.4. According to Figure 5.29, the minimum weight occurs at a design vector (d^*, h^*) in [in] of about (1.62, 18) as deduced from the grayish dashed arrow in the figure. However, the stress would be about 120 [kpsi], which is higher than the constraint value. Note that t is fixed at 0.1 [in]. Another optimization can be performed including t as an additional design variable. The optimum t can additionally be sought with (d^*, h^*) to minimize the weight and satisfy the constraint. One should be aware that the thickness might tend to zero; however, the stresses should increase as the thickness decreases.

5.7 Summary

This chapter is an introduction to the field of optimization. It presents the basic concepts and some optimization tools to formulate and solve basic optimization problems. Formulating optimization problems in terms of defining the independent design variables, the objective function, and the constraints is essential in handling the tools to find the optimum solution. Classical and numerical techniques to find the optimum solution are introduced. This would help in selecting the appropriate technique to reach the optimum in a suitable way in terms of accuracy and speed.

Few optimization tools are provided to solve simple closed form expressions that would simulate the objective functions. Some of these tools have been developed as personal codes that cannot be construed as perfect. They may, however, demonstrate some of the concepts and methods developed in this chapter. Few sample problems have been solved by these tools as presented in this chapter. Other optimization packages are available to solve wide ranges of problems such as the optimization toolbox in MATLAB. These packages can be used to solve many problems, even though no guarantee to have a successful outcome unless the problem is well formulated. Hopefully, this chapter would help in that regard. A more dedicated and advanced courses, textbooks, and research work should be used for a more real and complex optimization problems.

Few applications in design are presented in this chapter. More applications are given in each chapter of the text about each machine element covered. One might look at those applications and extract the method used in the solution of the applied optimization case.

Problems

5.1 Change the tray problem of Section 5.1.1 to minimize cost with a specified volume V_T, but no constraint on h_T. Why is the optimum height not the same as the optimum length and the optimum width?

5.2 Change the tray problem of Section 5.1.1 to a cylindrical tray, and minimize cost with a specified volume V_T with and without constraint on h_T. Why is the optimum height not the same as the optimum diameter and the optimum width?

5.3 Formulate the tray problem of Section 5.1.1 using nondimensionalization, and minimize cost with a specified nondimensional volume considering constraint on the height and no height constraint.

5.4 Use MATLAB to plot the contours of the objective function and the constraints of Example 5.2 using the more accurate relations. Use the same values of Fox (1971), which are $t = 0.1$ [in]; $B = 30$ [in]; $d = 2$ [in]; $h = 19$ [in]; $\boldsymbol{P} = 33\,(10^3)$ [lb]; $S_a = 100\,(10)^3$ [psi]; $E = 3\,(10)^7$ [psi]; and $\rho = 0.3$ [lb/in^3]. Identify the feasible and unfeasible regions. From the plots, find the optimum solution.

5.5 Use MATLAB to plot the contours of the objective function and the constraints of Example 5.2 using the more accurate relations and the SI units. Use about the same values as Fox (1971), which are $t = 2.54$ [mm]; $B = 0.762$ [m]; $d = 50.8$ [mm]; $h = 0.4826$ [m]; $\boldsymbol{P} = 146.8\,(10^3)$ [N]; $S_a = 690$ [MPa]; $E = 207\,(10)^3$ [MPa]; and $\rho = 7800$ [kg/m^3]. Identify the feasible and unfeasible regions.

5.6 Use MATLAB to plot the contours of the objective function and the constraints of Example 5.2 using the more accurate relations and the selected nondimensionalization. Identify the feasible and unfeasible regions.

5.7 Use MATLAB to plot the contours of the objective function and the constraints of Example 5.2 using only one vertical tube fixed to the ground instead of the two-bar truss. Use the more accurate relations. Use the same values of Fox (1971), which are $t = 0.1$ [in]; $d = 2$ [in]; $h = 19$ [in]; $\boldsymbol{P} = 33\ (10^3)$ [psi]; $S_a = 100\ (10)^3$ [kpsi]; $E = 3\ (10)^7$ [kpsi]; and $\rho = 0.3$ [lb/in^3]. Identify the feasible and unfeasible regions.

5.8 Use MATLAB to plot the contours of the objective function and the constraints of Example 5.2 using only one vertical tube fixed to the ground instead of the two-bar truss. Use the more accurate relations. Use about the same values as Fox (1971), which are $t = 2.54$ [mm]; $B = 0.762$ [m]; $d = 50.8$ [mm]; $h = 0.4826$ [m]; $\boldsymbol{P} = 146.8\ (10^3)$ [N]; $S_a = 690$ [MPa]; $E = 207\ (10)^3$ [MPa]; and $\rho = 7800$ [kg/m^3]. Identify the feasible and unfeasible regions.

5.9 Develop a MATLAB code to optimize any ***1D*** function using quadratic interpolation and a special MATLAB function to be changed for every objective function. Compare results to Example 5.4.

5.10 Solve Problem 5.2 using your developed ***1D*** code of Problem 5.9.

5.11 Solve Problem 5.3 using your developed ***1D*** code of Problem 5.9.

5.12 Solve Example 5.4 using Newton–Raphson method starting at the same starting point of the example. Compare results to quadratic interpolation and golden section methods.

5.13 Rework Problem 5.7 by direct substitution of the constraints solution, and compare result to the solution provided by the problem.

5.14 Construct some optimum design curves for different parameters in Example 5.7. Use different ratios of the cost of the unit seam length to the cost of unit surface area. Is it useful to formulate the problem in nondimensional design variables?

5.15 A heat exchanger is to be optimized for minimum cost. A 200 straight unit lengths of tubes must be used. This is to provide the required surface area for heat transfer. The cost particulars of the heat exchanger are as follows. The tubes cost 35 units. The shell cost is a function of the design parameters and is equal to $1.25\ d^{2.5}l$. The heat exchanger outer diameter is d, and its length is l. The heat exchanger floor space cost is equal to dl. A specified number of 200 tubes should fit in each unit area of heat exchanger cross section. In addition, present a nondimensional formulation and results.

5.16 A tubular column of length l, a diameter d, and a thickness t is subjected to a top compressive load $\boldsymbol{P}$. The column is fixed to the ground. Find the optimum tubular column, which is stressed to its allowable strength without buckling. The column is assumed free at the top. The load causing buckling is assumed as $\pi^2\ EI/4l^2$. The cross-sectional second area moment is I, and the modulus of elasticity is E. The column is subject to the compressive stress given by $\boldsymbol{P}/A$. The column cross-sectional area is A. Consider an allowable compressive strength of the material. Find the optimum column for a load $\boldsymbol{P} = 1$ [MN] or 225 [klb] and a length of 50 [m] or 2000 [in]. Use steel of $E = 207$ [GPa] or 30 [Mpsi] and allowable strength of 255 [MPa] or 37 [kpsi]. Reformulate the problem in a nondimensional form.

5.17 Design a minimum cost reinforced concrete beam that can carry a bending moment of $2\,(10)^5$ [N m]. The costs of concrete, steel, and formwork are given by $C_c = 800/\text{m}^{-3}$, $C_s = 200\,00/\text{m}^{-3}$, and $C_f = 160/\text{m}^{-2}$ of surface area. The strength of concrete and steel are limited to $S_c = 25$ [MPa] and $S_s = 500$ [MPa], respectively. Assume the area of reinforcing steel is $\boldsymbol{A_s}$. The cross-sectional dimensions of the beam are the width b and the depth d. The resisting moment of the beam is given by $\boldsymbol{M}_R = A_s S_s\,(d - (0.59A_sS_s/S_c b))$. The area of steel A_s is bounded by the balanced steel area $A_S^b = (0.542)bd(600S_c/(600S_s + S_s{}^2))$.

5.18 A cylindrical pressure vessel with hemispherical ends is required to hold at least 100 [m^3] or 3530 [ft^3] of a fluid under a presser of 20 [MPa] or 2.9 [kpsi]. The length of the cylindrical part is l, its radius is R, and its thickness is t_c. The thickness of each hemispherical part is t_h. Each thickness should be at least as that recommended by ASME code. These are given by $t_c = pR/(Se + 0.4p)$ and $t_h = pR/(Se + 0.8p)$, where S is the yield strength, e the joint efficiency, p the pressure, and R the radius. Find the optimum structural material assuming $S = 200$ [MPa] or 29 [kpsi], and $e = 1.0$. Suggest a nondimensional formulation to generalize the solution.

5.19 A cylindrical pressure vessel with hemispherical ends is required to hold at least 0.06 [m^3] or 3660 [in^3] of a fluid under a presser of 30 [MPa] or 4.35 [kpsi]. The length of the cylindrical part is l, its radius is R, and its thickness is t_c. The thickness of each hemispherical part is t_h. Each thickness should be at least as that recommended by ASME code. These are given by $t_c = pR/(Se + 0.4p)$ and $t_h = pR/(Se + 0.8p)$, where S is the yield strength, e the joint efficiency, p the pressure, and R the radius. Find the optimum structural material assuming $S = 400$ [MPa] or 58 [kpsi] and $e = 0.8$. Suggest a nondimensional formulation to generalize the solution.

5.20 Use a first-order optimization method to search for the maximum of a quadratic function $\mathcal{F}(\boldsymbol{x}) = \boldsymbol{x}^T\boldsymbol{A}\boldsymbol{x}$, where the matrix $\boldsymbol{A} = [1 \quad 0 \quad 1; 0 \quad -1 \quad 1; 1 \quad 1 \quad -5]$. Start at $\boldsymbol{x} = (4, 3, -3)$.

5.21 Use the process defined in Section 5.4.2 to derive Eq. (5.65).

5.22 Use the process defined in Section 5.4.6 to derive Eq. (5.101).

5.23 The design function $\mathcal{F}(\boldsymbol{x}) = (2x_1 - 2)^2 + (3x_2 - 6)^2$ is subject to the two constraints: $2x_2 - x_1 \le 3$ and $x_1 + x_2 \le 6$.
(a) Sketch the design space indicating the function contours and constraints.
(b) Starting from (3,1), find unconstrained minimum by second-order method.
(c) Starting from (1,0), proceed to the next point using conjugate gradient method.
(d) From the last point, use a feasible direction to find the next point.
(e) Check the Kuhn–Tucker conditions at the last point.

5.24 The design function $\mathcal{F}(\boldsymbol{x}) = (x_1 - 1)^2 + (x_2 - 2)^2$ is subject to the two constraints: $2x_2 - x_1 < 2$ and $x_1 + x_2 < 4$.
(a) Sketch the design space indicating the function contours and constraints.
(b) Starting from (2,1), find unconstrained minimum by a second-order method.
(c) Starting from (1,1), proceed to the next point using conjugate gradient method.
(d) From the last point, use a feasible direction to find the next point.
(e) Check the Kuhn–Tucker conditions at the last point.

5.25 A design optimization problem is defined by the objective function $\boldsymbol{f}(\boldsymbol{x}) = 9.82x_1x_2 + 2x_1$, which is constrained by $g_1 = [2500/(\pi x_1x_2)] - 500 \leq 0$, and $g_2 = x_1\, x_2\, ({x_1}^2 + {x_2}^2) \leq 47.3$.

(a) Sketch the design space {x_1 (0 to 12) and x_2 (0 to 0.8)} indicating the function values of 20, 30, 40, and 50 and the constraints g_1, g_2.

(b) Use the sketch to check whether the constrained minimum is at (5.44, 0.293).

5.26 The design function $F(\boldsymbol{x}) = 0.1x_1 + 0.057\,73x_2$ is subject to the three constraints: $(0.6/x_1) + (0.3464/x_2) - 1 \leq 0$, $6 - x_1 \leq 0$, and $7 - x_2 \leq 0$.

(a) Sketch the design space indicating the function contours and constraints.

(b) Starting from (11.8765, 7), determine the search direction using a gradient method, and show the next stopping point on your sketch.

(c) Use the Kuhn–Tucker conditions to show that (8.7657, 8.8719) is close to a constrained minimum. Can you improve on that?

5.27 A design optimization problem is defined by the objective function $\boldsymbol{f}(\boldsymbol{x}) = x_1 - x_2$, which is constrained by $g_1 = [2x_1 - x_2^2 - 1] \geq 0$, and $g_2 = [9 - 0.8x_1^2 - 2x_2] \geq 0$. (Note the $\geq$.)

(a) Sketch the design space {x_1 (0 to 4) and x_2 (0 to 3)} indicating the proper function values and the constraints g_1, g_2.

(b) Starting from (2.0, 0.0), determine the search direction using any gradient method and show the next stopping point on your sketch.

(c) Use the Kuhn–Tucker conditions to check whether or not the constrained minimum is at (2.5, 2.0)

5.28 As an optimization of a multi-variable functions, use a quadratic optimization method to find the minimum of the function $f(\boldsymbol{x}) = (x_1^2 + x_2 - 11)^2 + (x_1 + x_2^2 - 7)^2$. Start at an initial point of (4.0, 6.0). Use few full cycles, and find the value of the function $f(\boldsymbol{x})$ at the end of each cycle. Compare result to some available optimization program.

5.29 Use a similar MATLAB code to that in Figure 5.19b to solve Problem 5.28. Compare results.

5.30 Use a similar MATLAB code to that in Figures 5.19d,e to solve Problem 5.28. Compare results.

5.31 Use the developed code of Problem 5.29 to solve Example 5.13. Discuss results.

5.32 Use the developed code of Problem 5.30 to solve Example 5.13. Discuss results.

5.33 Use any optimization code to solve Example 5.15 using the diameter, the height, and the thickness as the design variables. Compare results to previous optimization cases.

References

Abbas, M.H. and Metwalli, S.M. (2011). Elastohydrodynamic ball bearing optimization using genetic algorithm and heuristic gradient projection. *Proceedings of the ASME 2011 IDETC/CIE*, Washington, DC, USA (29–31 August), Paper # DETC2011-47624.

Abd El Aziz, K.M. and Metwalli, S.M. (2017). Optimum design of pressure vessels using hybrid HGP and genetic algorithm. *Proceedings of the ASME 2017 Pressure Vessels and Piping Conference, PVP2017*, Waikoloa, HI, USA (16–20 July 2017), Paper No. PVP2017-65538.

Abd El Malek, M.R., Senousy, M.S., Hegazi, H.A., and Metwalli, S.M. (2005). Heuristic gradient projection For 3D space frame optimization. *Proceedings of ASME 2005 International Design Engineering Technical Conferences and Computers and Information in Engineering Conference (DETC/CIE2005)*, Long Beach, CA, USA (24–28 September), Paper # DETC2005-85348.

Abd El-Aziz, K.M., Hamza, K., El-Morsi, M.A.O., Metwalli, S.M. and Saitou, K. (2016). Optimum solar humidification–dehumidification desalination for micro grids and remote area communities. *Journal of Solar Energy Engineering* 138 (2): 21005-8.

Abramowitz, M. and Stegun, I.A. (eds.) (1972). *Handbook of Mathematical Functions with Formulas, Graphs, and Mathematical Tables, 9th Printing*, 14. Dover.

Arora, J.S. (2004). *Introduction to Optimum Design*. Academic Press.

Badran, S.F., Nassef, A.O., and Metwalli, S.M. (2009). Y-stiffened panel multi-objective optimization using genetic algorithm. *Thin-Walled Structures* 47 (11): 1331–1342.

Chapra, S.C. and Canale, R.P. (2010). *Numerical Methods for Engineer*, 6e. McGraw-Hill.

Elmaghraby, S.E., Metwalli, S.M., and Zorowski, C.F. (1979). Some operations research approaches to automobile gear train design. *Mathematical Programming Study* II: 150–175.

El-Mahdy, O.F., Ahmed, M.E.H., and Metwalli, S. (2010). Computer aided optimization of natural gas pipe networks using genetic algorithm. *Applied Soft Computing* 10 (4): 1141–1150.

Elmoselhy, S.A., Azzam, B.S., Metwalli, S.M., and Dadoura, H.H. (2006). Hybrid shape optimization and failure analysis of laminated fibrous composite E-springs for vehicle suspension. *SAE Transactions, Journal of Commercial Vehicles* 115-2: 226–236.

Elseddik, B.A.E. (2005). *Fuzzy nonlinear indirect optimization technique for closed form engineering applications*. MS thesis. Cairo University.

Elzoghby, A.A., Metwalli, S.M., and Shawki, G.S.A. (1980). Linearized ridge-path method for function minimization. *Journal of Optimization Theory and Applications* 30 (2): 161–179.

Fox, R.L. (1971). *Optimization Methods for Engineering Design*. Addison-Wesley.

Ghoneim, H. and Metwalli, S.M. (1984). Optimum vehicle suspension with a damped absorber. *Journal of Mechanisms, Transmission and Automation in Design* 106 (2): 148–155.

Hanafy, M.M.A. and Metwalli, S.M. (2009). Hybrid general heuristic gradient projection for frame optimization of micro and macro applications. *Proceedings of the ASME 2009 International Design Engineering Technical Conferences & Computers and Information in Engineering Conference (IDETC/CIE 2009)*, San Diego, CA, USA (August 30–September 2), Paper # DETC2009-87012.

Hegazi, H.A., Nassef, A.O., and Metwalli, S.M. (2002). Shape optimization of NURBS modeled 3D C-frames using hybrid genetic algorithm. *Proceedings of the 2002 ASME Design Engineering Technical Conferences & Computers and Information in Engineering Conference*, Montreal, Canada (September 29–October 2), Paper # DETC2002/DAC-34107.

Metwalli, S.M. (1978). Effect of angular acceleration on optimum design of uniformly stressed rotors. *Proceedings of the Design Engineering Conference*, Rhode, S.M., Allair, P.E. and Maday, C.J. (eds.), *Topics in Fluid Film Bearing and Rotor Bearing System Design and Optimization*, ASME, pp. 231–244.

Metwalli, S.M. (1986). Flywheel optimization under speed fluctuation effects. *Proceedings of the 11th ASME Design Automation Conference*, Columbus, OH, USA (5–8 October), ASME Paper No. 86-DET-129.

Metwalli, S.M. (1986b). Optimum nonlinear suspension systems. *Journal of Mechanisms, Transmissions, and Automation in Design* 108 (2): 197–203.

Metwalli, S.M. (2012). Heuristic GP optimization technique for design synthesis of monotonic objectives, World Scientific Proceedings Series on Computer Engineering and Information Science: Volume 7 *Uncertainty Modeling in*

Knowledge Engineering and Decision Making, Proceedings of 10th FLINS Conference, August 26–29, Istanbul, Turkey, pp. 1233–1238.

Metwalli, S.M. and Afify, E.M. (1981). Analysis, design and optimization of a new horizontal elutriator – part II: design and optimization. *Proceedings of the First Cairo University Conference on Mechanical Design and Production*, Vol. 1, No. MDP-79-DES-15 (December 1979), and in Shawki, G.S.A. and Metwalli, S.M. (Eds.), *Current Advances in Mechanical Design and Production*. Cairo, Egypt: Pergamon Press, pp. 41–47.

Metwalli, S.M., El Danaf, E.A. (1996). CAD and optimization of spur and helical gear sets. *Proceeding of the 21st ASME Design Automation Conference*, Irvine, CA, USA (18–22 August), Paper # 96-DETC/DAC-1433.

Metwalli, S.M. and Elmi, S. (1989). A quadratic optimization technique. *Proceedings of the 4th Cairo University MDP Conference* (27–29 December 1988) in Kabil, Y.H and Saiid, M.E. (eds). *Current Advances in Mechanical Design and Production*. Cairo, Egypt: Pergamon Press, pp. 369–376.

Metwalli, S.M. and Hegazi, H.A. (1999). CAD of disc brakes by multi-objective optimization. *Proceedings of the 1999 ASME Design Engineering Technical Conference*, Las Vegas, NV, USA (12–15 September), Paper No. DETC99/CIE-9139.

Metwalli, S.M. and Hegazi, H.A. (2001). Computer-based design of disc brakes by multi-objective form optimization. *Proceedings of the ASME 2001 Design Engineering Technical Conference and Computers and Information in Engineering Conference*, Pittsburgh, PA (9–12 September), Paper No. DETC2001/CIE-21680

Metwalli, S.M. and Mayne, R.W. (1977). New optimization techniques. *The 4th ASME Design Automation Conference*, Chicago, IL, USA (September), ASME Paper No. 77-DAC-9.

Metwalli, S.M., Afimiwalla, K.A., and Mayne, R.W. (1973). Optimum design of vehicle suspensions. *Proceedings of the Inter society Conference on Transportation*, Denver, CO, USA (September), ASME Paper No. 73-ICT-48.

Metwalli, S.M., Shawki, G.S.A., and Sharobeam, M.H. (1983). Optimum design of variable-material flywheels. *Journal of Mechanisms, Transmission and Automation in Design* 105 (2): 249–253.

Metwalli, S.M., Shawki, G.S.A., Mokhtar, M.O.A., and Seif, M.A.A. (1984). Multiple design objectives in hydrodynamic journal bearing optimization. *Journal of Mechanisms, Transmissions and Automation in Design* 106 (1): 54–61.

Metwalli, S.M., Shawki, G.S.A., and Elzoghby, A. (1985). Optimal design of disk springs. *Journal of Mechanisms, Transmissions, and Automation in Design* 107 (4): 477–481.

Metwalli, S.M., Radwan, M.A., and Elmeligy, A.A. (1993). Optimization of helical compression and tension springs. *Proceeding of the 19th ASME Design Automation Conference*, Albuquerque, NM, USA (19–22 September).

Metwalli, S.M., Radwan, M.A., and Elmeligy, A.A.M. (1994). CAD and optimization of helical torsion springs. *Proceedings of the ASME International Computers in Engineering Conference*, Minneapolis, MN, USA (11–14 September), pp. 767–773.

Nassef, A.O., Hegazi, H.A., and Metwalli, S.M. (2000). A hybrid genetic-direct search algorithm for the shape optimization of solid C-frame cross-section. *Proceedings of the ASME 2000 Design Engineering Technical Conference*, Baltimore, MD, USA (10–14 September), Paper No. DETC2000/DAC-14295.

Pham, D.T. and Karaboga, D. (2000). *Intelligent Optimization Technique*. Springer.

Powell, M.J.D. (1964). An efficient method for finding the minimum of a function of several variables without calculating derivatives. *Computer Journal* 7 (4): 303–307.

Rao, S.S. (2009). *Engineering Optimization: Theory and Practice*, 4e. Wiley.

Reklaitis, G.V., Ravindran, A., and Ragsdell, K.M. (1983). *Engineering Optimization Methods and Applications*. Wiley.

Rosenbrock, H.H. (1960). An automatic method for finding the greatest or least value of a function. *Computer Journal* 3 (3): 175–184.

Senousy, M.S., Hegazi, H.A., and Metwalli, S.M. (2005). Fuzzy heuristic gradient projection for frame topology optimization. *Proceedings of ASME 2005 International Design Engineering Technical Conferences and Computers and Information in Engineering Conference (DETC/CIE2005)*, Long Beach, CA, USA (24–28 September), Paper # DETC2005-85353.

Senousy, M.S., Hegazi, H.A., and Metwalli, S.M. (2008). Topology optimization of a MEMS resonator using hybrid fuzzy techniques. *Proceedings of the ASME 2008 International Design Engineering TechnicalConferences & Computers and Information in Engineering Conference (IDETC/CIE 2008)*, Brooklyn, NY, USA (3–6 August), Paper # DETC2008-49563.

Shalaby, M.M., Hegazi, H.A., Nassef, A.O., and Metwalli, S.M. (2003). Topology optimization of a compliant gripper using hybrid simulated annealing and direct search. *Proceedings of ASME 2003 International Design Engineering Technical Conferences and the Computers and Information in Engineering Conference (DETC'03)*, Chicago, IL, USA (2–6 September), Paper # DETC2003/DAC-48770.

Shalaby, M.M., Hegazi, H.A., Nassef, A.O., and Metwalli, S.M. (2003b). Topology optimization of auxetic structures using hybrid simulated annealing. *Proceedings of the 7th International Conference on Intelligent Engineering Systems (IEEE-INES 2003)*, Assiut-Luxor, Egypt (4–6 March), pp. 774–779.

Shawki, G.S.A., Metwalli, S.M., and Sharobeam, M.H. (1984). Optimum configuration for an isotropic rotor. *Journal of Mechanisms, Transmissions, and Automation in Design* 106 (3): 376–379.

Stewart, J. (2005). *Multivariable Calculus: Concepts & Contexts*. Thomson Brooks/Cole.

Walsh, G.R. (1975). *Methods of Optimization*. Wiley.

Wilde, D.J. (1978). *Globally Optimal Design*. Wiley.

Wilde, D.J. and Beightler, C.S. (1979). *Foundation of Optimization*, 2e. Prentice-Hall.

Youssef, A.M., Nassef, A.O., and Metwalli, S.M. (2007). Reverse engineering of geometric surfaces using tabu search optimization technique. *Proceedings of the ASME 2007 International Design Engineering Technical Conferences & Computers and Information in Engineering Conference (IDETC/CIE 2007)*, Las Vegas, NV, USA (4–7 September), Paper # DETC2007/DAC-34970.

6

Stresses, Deformations, and Deflections

When bodies are loaded by external loads such as forces or moments, internal forces and moments are developed, and consequently *internal stresses* exist. Stresses are conceptually loads per unit area. Due to these internal loads and ensued stresses, the body deforms. As a result of these *deformations*, one gets the deflections that may be observed by naked eye. The *deflections* are observed by the *displacements* of body's individual points relative to their original positions or locations. This displacement may, however, include rigid body motion. *Deflections* are distances or angular rotations that may not include rigid body motion. *Deformations* are the different alterations occurring in the body due to the developed internal stresses. The deformations can be either elastic or plastic. *Elastic deformations* rebound after the loads are removed. *Plastic deformations* remain after the loads are removed. Machine designs of mechanical elements are mainly concerned with *elastic deformations* and avoid any failures due to plastic deformations or any permanent degradation of geometry.

In this chapter, *stresses* generated by different loading conditions are studied. Generated deformations or *strains* are defined. The resulting displacements or deflections are determined. The main concern of mechanical design is the *linear elastic* range of material that should withstand the maximum stresses generated due to the applied loading.

Symbols and Abbreviations

The adopted units are [in, lb, psi] [m, kg, N, Pa]; others given at each symbol definition ([k...] is 10^3, [M...] is 10^6, and [G...] is 10^9).

Symbols	Quantity, Units (adopted)
$(s_r)_T$	Slenderness ratio at *Euler* and parabolic tangent
$\langle\rangle$	Pointed brackets for singularity function
1D	One-dimensional
2D	Two-dimensional
3D	Three-dimensional
A	Cross-sectional area normal to the x direction
$\mathbf{A}$	Matrix of the eigenvalue problem
$a_{0,1,2,3}$	Shape or blending function coefficients
$A_{1,2,3}$	Subtriangle areas
a_c	Contact diameter of the circular contact

Machine Design with CAD and Optimization, First Edition. Sayed M. Metwalli.

Companion website: www.wiley.com/go/metwalli/machine

Symbol	Quantity, units (adopted)
A_C	Column cross-sectional area
A_e	Element area
A_i	Area of element i
$\boldsymbol{A}_n$	Node matrix
a_P, b_P	*Johnson's* parabolic equation constants
A_R	Area of the rectangular cross-section
A_s	Cross-sectional area
$\boldsymbol{B}$	Strain-displacement matrix
b_c	Breadth of contact area
b_h, a_h	Principal dimensions of the elliptic hole
b_R	Rectangular section breadth or width
b_T	T-section breadth or width
C	Circular cross section
c_1, c_2	Neutral axis location from outer fibers
$C_{1,2,3,4}$	Constants of integration
CAD	Computer-aided design
CAS	Computer-aided synthesis
C_E	Column end condition constant
C_{EB}	Equivalent beam factor or coefficient
c_S	Outer fiber distance from neutral axis
C_S	Least distance from neutral axis to outer fiber
CST	Constant strain triangle
$\boldsymbol{D}_E$	Stress–strain constitutive or elasticity matrix
dA	Incremental area in cross section
d_C	Diameter of circular cross section
d_c	Diameter of pressure cylinder
d_C	Cylindrical column diameter
d_{ci}, d_i	Internal cylinder diameter
d_{co}, d_o	Outside cylinder diameter
d_h	Cylindrical hole diameter
d_i	Diameter of element i
$d_{i,\mathrm{old}}$	Old diameter of element i
$d\boldsymbol{M}$	Incremental bending moment
ds	Incremental element length
dU	Incremental strain energy
$d\boldsymbol{V}$	Incremental shear force
dW	Incremental work
dx	Incremental length in x direction
$d\delta$	Incremental deformation

Symbol	Quantity, units (adopted)
$d\theta$	Infinitesimal rotation angle
E	Modulus of elasticity, or *Young*'s modulus
e	Load eccentricity
E_1, E_2	Moduli of elasticity for material 1 and material 2
E_i	Modulus of elasticity of element i
e_n	Eccentricity of neutral axis
e_{Sc}	Shear center location distance off cross section
$\boldsymbol{F}$	Force, load, or action vector
$\mathcal{F}(\,)$	Function of ( )
$\boldsymbol{F}, \boldsymbol{F}_i$	Concentrated force
$\boldsymbol{F}_e$	Equivalent force
FE	Finite element method
f_i	Flow variable in element i
$\boldsymbol{F}_x$	Applied force is in x direction
G	Shear modulus of elasticity or modulus of rigidity
H, H_W	Power in watt [W]
$\boldsymbol{h}_e(x)$	*Hermite* blending or shape function
H_{hp}	Power in horsepower [hp]
H_{kW}	Power in kilowatt [kW]
h_R	Rectangular section height or depth
h_T	T-section height or depth
$\boldsymbol{I}$	The identity matrix
I	I shaped cross section
I_1, I_2	Stress invariants
I_C	Column second area moment
I_x, I_y, I_z	Second area moments about x, y, or z
J_x	Polar second area moment about the x-axis
$\boldsymbol{K}$	Stiffness assembly or property matrix
K_{ci}	Curved beam inner fiber stress factor
K_{co}	Curved beam outer fiber stress factor
$\boldsymbol{K}_e$	Element stiffness matrix
k_i	Stiffness of element i
K_{SC}	Stress concentration factor
K_{SF}	Safety factor or factor of safety
l, l_B	Length or beam length
$L_{1,2,3}$	Area coordinates
l_2', l_3'	Equivalent lengths
l_c	Length of contact surface or pressure cylinder
l_C	Column length

Symbol	Quantity, units (adopted)
l_e	Beam element length
l_i	Length of element *i*
$\boldsymbol{L}_P$	Applied load (= $\boldsymbol{F}_P$, $\boldsymbol{M}_P$, or $\boldsymbol{T}_P$)
$\boldsymbol{M}$	Internal bending moment
$\boldsymbol{M}_1, \boldsymbol{M}_2$	Moments at support 1 or 2
$\boldsymbol{M}_i$	Concentrated moment
$\boldsymbol{M}_z$	Moment about *z* coordinate
n	Exponent of singularity function
$N_{1,2,3,4}$	Weight effect of each node displacement
$N_{31,\,32\ldots}$	Components of the ***3D*** shape function
n_B	Buckling integer for critical buckling modes
n_e	Number of elements or components
$\boldsymbol{N}_e$	Element shape function matrix
$\boldsymbol{N}_i$	Principal directions, $i = 1,2,3$
N_{rpm}	Rotational speed in revolution per minute [rpm]
$\boldsymbol{N}_T$	Plane normal of transformed plane
n_{v-1}	Number of vertices or *nodes* −1
$P(x,y)$	A local point at any (x_i,y_i)
p_{avg}	Average pressure or average stress σ_{av}
$\boldsymbol{P}_{cr}$	Critical buckling load
PDE	Partial differential equation
p_F	Shrink fitting pressure
p_i	Internal pressure
p_{max}	Maximum pressure or compressive stress σ_{max}
p_o	Outer or external pressure
Q_A	First area moment of the area *dA*, or *A*
$\boldsymbol{q}_i$	Generalized loading
q_S	Shear flow
r	Radius
R	Rectangular cross section
R_C	Radius of curvature
R_c	Radius of centroidal axis
r_F	Interfacial shrink fit radius
r_{GC}	Radius of gyration of cross section
R_M	Radius of the *Mohr's* circle
r_n	Neutral axis radius
$\boldsymbol{R}_z$	Rotation transformation matrix about *z*
s_r	Column slenderness ratio = l_C/r_{GC}
S_y, S_{yt}	Yield strength of material in tension

Symbol	Quantity, units (adopted)
S_{yc}	Compressive yield strength of material
$\boldsymbol{T}$	External boundary forces or ***traction*** forces
$\boldsymbol{T}, \boldsymbol{M}_x$	Torsion or moment around torsional x axis
t_B	Bimetal strip thickness
t_C	Channel outer thickness
t_c	Pressure cylinder thickness
t_e	Constant element thickness
$T_{\mathrm{N\,m,lb\,in}}$	Torque magnitude in [N m] or [lb in]
t_P	Plate thickness
t_T	T-section top, bottom, and web thickness
$\boldsymbol{T}_x, \boldsymbol{T}_y$	Traction forces in x and y directions
u	Deflection or displacement in the x direction
u_ε	Strain energy density
U, U_ε	Strain energy
u, v, w	Displacements in the x, y, z directions
$\boldsymbol{u}_e$	Element displacement vector
U_e	Element strain energy
u_e	Element energy density
u_i	Uniform displacement in element i
$\boldsymbol{u}_n$	Node displacement vector
$\boldsymbol{u}_T$	Thermal deformation in x direction
$\boldsymbol{V}$	Internal transverse shear force
v	Deflection or displacement in the y direction
v', v''	Slope dv/dx and curvature of v
$v_{1,2\ldots}$	Deflection or displacement at points 1, 2…
$v_{\max}$	Maximum deflection or displacement
$\boldsymbol{v}_T$	Thermal deflect in the y direction
w	Deflection or displacement in the z direction.
W_F	Work done by the force $\boldsymbol{F}$
$\boldsymbol{w}_i$	Distributed load per unit area or length
w_P	Plate width
$\boldsymbol{x}$	Eigenvector of the matrix $\boldsymbol{A}$
x_0	Position at which the load is applied
XYZ	Global coordinates
xyz	Local coordinates
y	Distance from the neutral axis
y–z	Plane of yz is also the x *plane*
Z_R	Rectangular section modulus
Z_y	Section modulus (I_y/c_S) about the y coordinate

Symbol	Quantity, units (adopted)
Z_z	Section modulus (I_z/c_S) about the z coordinate
α	Coefficient of thermal expansion
α_T	Thermal expansion coefficient
δ	Deflection or deformation
$\boldsymbol{\delta}$	Displacement or behavior vector
δ_{1o}	Outer diameter deformation of inner cylinder 1
δ_{2i}	Inner diameter deformation of outer cylinder 2
δ_F	Interfacial shrinkage or fundamental deviation
δ_i	Displacement in the direction of force vector $\boldsymbol{F}_i$
Δ_L	Displacement (δ, θ, or ϕ...) for load $\boldsymbol{L}_P$
ΔT_T	Temperature difference
ε	Strain
$\boldsymbol{\varepsilon}$	Strain tensor or strain matrix
ε_e	Element strain vector
ε_T	Thermal strain
ε_{Tx}	Thermal strain in x direction
ε_x	Normal elastic strain in the x direction
ε_x	Strain in the x direction
$\boldsymbol{\varepsilon}_x$	Strain vector in the x direction
γ	Angular rotation of the element along the length l
γ	Shear strain
γ_1, γ_2	Angles of the principal directions
γ_{xy}	Shearing strain between y and x axes
κ	Curvature ($1/R_C$)
λ	Eigenvalue
ν	Poisson's ratio
ν_1, ν_2	*Poisson's* ratios for material 1 and material 2
ω, $\omega_{\text{rad/s}}$	Rotational speed in [rad/s]
ϕ	Angular rotation or twist of the cross section
$\boldsymbol{\sigma}$	Matrix of stress state or stress tensor
σ	Stress value
$\sigma_{1,2,3}$	Principal stresses in principal directions 1,2,3
σ_a	Axial or longitudinal stress
σ_{all}	Allowable stress limit or target stress
σ_{av}	Average stress
σ_{av}	Average or hydrostatic stress
σ_{cr}	Critical stress or critical unit load
σ_d	Distortion or deviatoric stress
$\boldsymbol{\sigma}_e$	Element stress vector

Symbol	Quantity, units (adopted)
σ_i	Current stress in element i
σ_r	Radial stress
σ_t	Tangential or hoop stress
σ_{t1}	Tangential stress in outer surface of inner cylinder
σ_{t2}	Tangential stress in inner surface of outer cylinder
$\boldsymbol{\sigma}_{Tx}$	Longitudinal thermal stress in x direction
$\boldsymbol{\sigma}_x$	Normal stress vector in x direction
σ_x	Normal stress in x direction
$\boldsymbol{\sigma}_x$	Stress in x direction
$\sigma_{x,\max}$	Maximum normal stress in x direction
$\sigma_{x,\min}$	Minimum normal stress in x direction
$\boldsymbol{\sigma}_y$	Normal stress vector in y direction
τ	Shear stress or direct shear
τ_1, τ_2	Maximum and minimum shear stresses
τ_{ij}	Stress matrix or stress tensor
$\tau_{\max}$	Maximum shear stress
$\tau_{\max,\min}$	Maximum and minimum shear stresses
τ_T	Transverse shear stress
τ_{xy}	Shear stress on the x plane in the y direction
τ_{xz}	Shear stress on the x plane in the z direction
θ	Slope or an angle
$\theta_{1,2,\ldots}$	Slope or angle at points 1, 2, …
θ_i, ϕ_i	Deformations in $\boldsymbol{M}_i$ or $\boldsymbol{T}_i$ directions
$\theta_{\max}$	Maximum slope or angle

6.1 Loads, Shear, Moment, Slope, and Deflection

Externally applied loads such as concentrated forces $\boldsymbol{F}_i$, distributed loads ($\boldsymbol{w}_i$), moments $\boldsymbol{M}$, and torques $\boldsymbol{T}$ produce internal normal and shear stresses generated by internal bending moments and shear forces. These loads are applied on a cylindrical machine element as shown in Figure 6.1a. In this section the relations of the external loads to the generated internal shear forces and moments are developed. Ensued internal stresses thereof are defined, and the deflections are found particularly for beam like machine elements.

6.1.1 External and Internal Loads

In this section the developed internal loads in the form of internal *bending moment (M)* and *shear force (V)* are developed due to the assumed externally applied loads ($\boldsymbol{w}_i$) as shown in Figure 6.1a. This case is assumed first, and other external loadings will be considered latter on such as concentrated forces $\boldsymbol{F}_i$ or moments $\boldsymbol{M}_i$, not shown in Figure 6.1a. For power transmission H, the external torque $\boldsymbol{T}$ generates internal shear stress, which is discussed in Section 6.3.

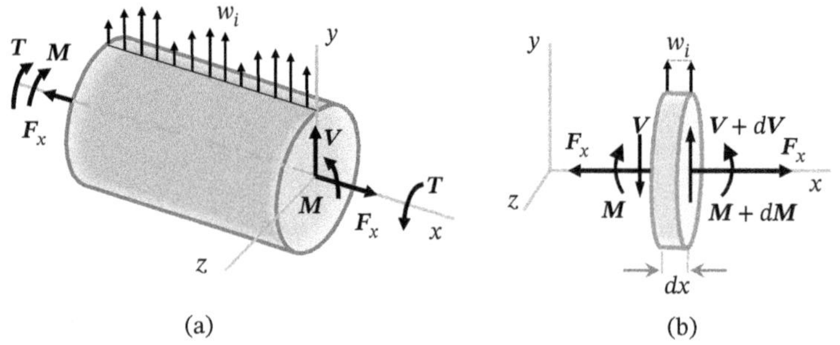

Figure 6.1 External and internal loading of beams: (a) externally loaded beam of circular cross section and (b) a free body diagram of a beam slice of an infinitesimal element with the adapted positive sign conventions for all loads.

The externally applied distributed loads (w_i) is alternatively called *transverse* loading just like the concentrated forces $\boldsymbol{F}_i$ that are applied transversally to the element axis. The axial forces $\boldsymbol{F}_x$ are the axially applied loads in the assigned x centerline axis of the element as shown in Figure 6.1a. The produced internal *normal stresses* and *shear stresses* due to internal bending moments and shear are developed using Figure 6.1b. ***Sign convention*** is adhering to the rule of the *right-hand coordinates* and positive coordinate directions and that the *positive plane* has a positive *normal direction.*

Figure 6.1b shows a free body diagram of a beam slice of an infinitesimal element that has a length dx. The adapted ***sign convention*** is the positive direction for all loads shown in Figure 6.1b. The seemingly negative loads at the left side of the infinitesimal element are adapted as positive since they act on a *negative x plane*. A negative x plane has its *normal* in the negative x direction. A positive tensile force $\boldsymbol{F}_x$ on the *positive x plane* to the right of the element should be in opposite direction on the *negative x plane* on the element left side. The *distributed transverse load* per unit length $\boldsymbol{w}_i$ is assumed constant on the infinitesimal beam element. The external load $\boldsymbol{w}_i$ is usually downward, i.e. negative. For consistency, it is assumed positive so that after the development of the relations, one can then substitute the correct negative direction for $\boldsymbol{w}_i$ or any other external load. In Figure 6.1b, the indicated *internal transverse shear force* $\boldsymbol{V}$ and *internal bending moment* $\boldsymbol{M}$ on left side of the element are also classified as positive since they are on a *negative x plane*. Due to the distributed load $\boldsymbol{w}_i$, the shear force $\boldsymbol{V}$ and the bending moment $\boldsymbol{M}$ on the element should vary by $d\boldsymbol{V}$ and $d\boldsymbol{M}$, respectively, as shown on the right side of the element in Figure 6.1b. Considering the *equilibrium of forces* in the y direction of the element provides the equality: $-\boldsymbol{V} + \boldsymbol{w}_i dx + (\boldsymbol{V} + d\boldsymbol{V}) = 0$. This gives the following relation (see Popov (1968 and 1990)):

$$\frac{d\boldsymbol{V}}{dx} = -\boldsymbol{w}_i \tag{6.1}$$

The *equilibrium of moments* in Figure 6.1b about the z-axis and considering moments about the left end of the element provide the equality: $-\boldsymbol{M} + \boldsymbol{w}_i\, dx\,(dx/2) + (\boldsymbol{V} + d\boldsymbol{V})\, dx + (\boldsymbol{M} + d\boldsymbol{M}) = 0$. In the limits and as $dx \to 0$, one can neglect the higher order differentials of $dx^2/2$ and $d\boldsymbol{V}dx$. In doing so, one gets the following equation:

$$\frac{d\boldsymbol{M}}{dx} = -\boldsymbol{V} \tag{6.2}$$

This indicates that the maximum bending moment will occur where the *internal shear* force $\boldsymbol{V} = 0$. Combining Eqs. (6.1) and (6.2) gives

$$-\boldsymbol{w}_i = \frac{d\boldsymbol{V}}{dx} = \frac{d}{dx}\left(-\frac{d\boldsymbol{M}}{dx}\right) \quad \text{or} \quad \frac{d^2\boldsymbol{M}}{dx^2} = \boldsymbol{w}_i \tag{6.3}$$

These relations are important in the construction of *shear force diagram* and *bending moment diagram* of loaded beams. These are treated in conjunction with beam deflection and slopes in Section 6.1.3, which requires further prerequisite treatments as follows.

The *sign convention* used herein is consistent with the coordinate directions and the sign convention used in the elasticity theory. Some references might use different sign convention, but the one used here is found to be more consistent for generalized treatments and should pose no ambiguity or for the need to consistently check the sign convention.

6.1.2 Pure Bending

Figure 6.2a shows an extremely exaggerated *deformation* of the beam under ***pure bending M*** about the *z*-axis. It is assumed that the coordinate *x* is kept as *undeformed* centerline or what is called the *neutral axis* of the beam. That is why the *x* centerline is with the beam. This process has been adapted just to magnify the deformation of the beam and thus be able to derive the intended concepts and relations. The beam cross section is in the normal *yz* plane. It has a projection that is shown as the *A* black colored line, which is coincident with the *y*-axis in Figure 6.2a. Under *pure bending*, a similar section *B–B* at an infinitesimally rotated angle $d\theta$ is shown. Originally, the *B–B* section (the grayish dashed line) was parallel to *A–A*, but under bending it is rotated by the angle $d\theta$ about the *neutral axis x*. The *curvature radius* is R_C as shown in Figure 6.2a. The original length *ds* of the fiber at any distant *y* from the *neutral axis* is deformed by a *dx* distance due to the $d\theta$ rotation. These deformations produce the following relations:

$$
\begin{aligned}
ds &= R_c\, d\theta \\
dx &= y\, d\theta
\end{aligned}
\tag{6.4}
$$

where R_C is the ***radius of curvature***. It is to be known that the *curvature* κ is defined as the reciprocal of the radius of curvature or $\kappa = 1/R_C$. The curvature of a straight bar is zero since it has an infinite radius of curvature.

The ***strain*** ε_x in the *x* direction is defined as the deformation relative to the original length, or

$$\varepsilon_x = -\frac{dx}{ds} \tag{6.5}$$

It should be noted that the negative sign is due to the fact that *dx* is in the negative direction, which is also reducing *ds* as shown in Figure 6.2a. Solving Eqs. (6.4) and (6.5) give the following *strain* relation:

$$\varepsilon_x = -\frac{y}{R_C} \tag{6.6}$$

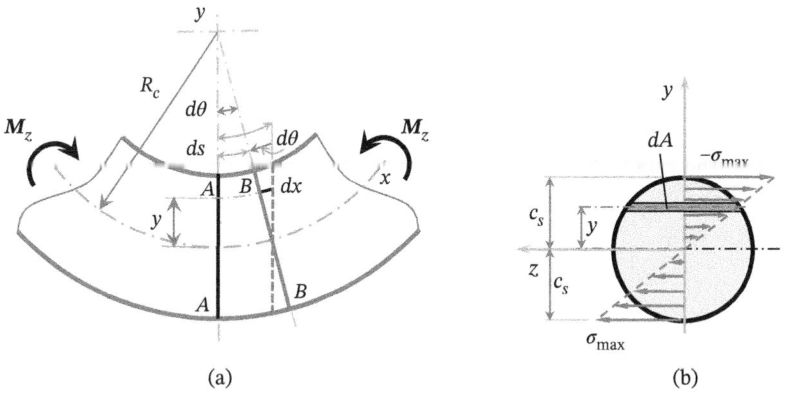

Figure 6.2 Deformation and beam stresses: (a) extremely exaggerated deformation of the beam under pure bending moment M_z and (b) developed stress distribution due to bending.

Utilizing the ***Hook's law*** for linearly elastic materials (i.e. $\sigma_x = E\varepsilon_x$, *Robert Hooke* (1635–1703); see Section 7.1.3), one gets

$$\sigma_x = -\frac{Ey}{R_c} \tag{6.7}$$

where σ_x is the stress in x direction and E is the ***modulus of elasticity***, or the traditional *Young*'s modulus (Thomas Young (1773–1829)), which is the slope of the elastic *stress–strain* curve of the material; see Section 7.1.3.

Figure 6.2b defines the means to develop the stress distribution due to bending. A circular cross section is shown in the y–z plane. The y–z plane is usually called the *x plane* since its normal is in the x direction. An element dA in the cross section is shown parallel to the z-axis and at a distance y from the neutral axis. Due to the normal stress σ_x generated by the moment $\boldsymbol{M}_z$, the internal force on the element is $\sigma_x\, dA$. The moments of these forces ($\int y\, \sigma_x\, dA$) about the *neutral axis* must be balanced with the applied moment $\boldsymbol{M}_z$. Thus, using Eq. (6.7) gives

$$\boldsymbol{M}_z = -\int_{-c_S}^{c_S} y\sigma_x\, dA = \frac{E}{R_c}\int_{-c_S}^{c_S} y^2\, dA \tag{6.8}$$

The term c_S is the cross-section's outer fiber distance from the *neutral axis*. It should be observed that the second integral in Eq. (6.8) is the ***second area moment*** I_z of the section about the z-axis, i.e.,

$$I_z = \int_{-c_S}^{c_S} y^2\, dA \tag{6.9}$$

This relation is valid for any cross section other than the circular cross section, provided that the limits could be different from c_S and $-c_S$. These limits of integration are obtained so that the integral of the forces about the *neutral axis* should vanish. The integration of internal forces about the *neutral axis* should be zero when or where limits c_1 and c_2 should provide

$$\int_{-c_1}^{c_2} \sigma_x\, dA = -\frac{E}{R_c}\int_{-c_1}^{c_2} y\, dA = 0 \tag{6.10}$$

This can define the *neutral axis* location of other cross sections as the values of c_1 and c_2, which are the distances of the outer fibers of the cross section from the neutral axis; see Example 6.1 and Figure 6.3 for a *T section*. For symmetrical cross sections, the *neutral axis* is the axis of symmetry; see Example 6.1 and Figure 6.3 for a *rectangular section*.

Manipulating Eqs. (6.8) and (6.9) gives the following curvature κ and radius of curvature R_C of beams under pure bending.

$$\kappa = \frac{1}{R_C} = \frac{\boldsymbol{M}_z}{EI_z} \tag{6.11}$$

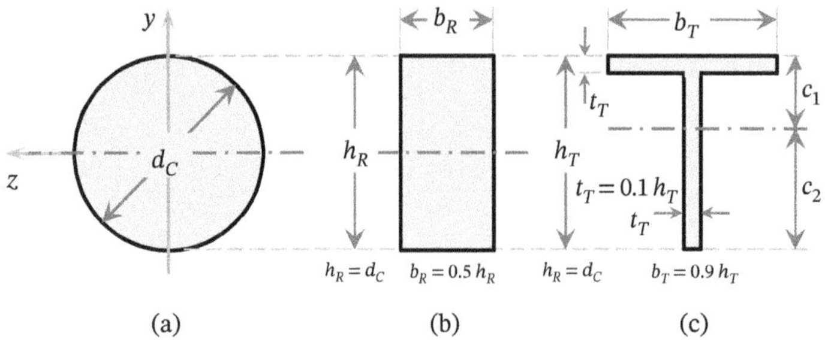

Figure 6.3 Different beam cross sections: (a) circular, (b) rectangular, and (c) T-section. The sections are dimensioned with the indicated predetermined proportions.

Eliminating R_c from Eqs. (6.7) and (6.11) gives

$$\sigma_x = -\frac{\boldsymbol{M}_z y}{I_z} \tag{6.12}$$

The negative sign indicates that the *normal stress* at the outer positive y distance is a ***compression stress*** for a positive moment $\boldsymbol{M}_z$. At the outer negative y distance, the stress is positive, which is a ***tensile stress***. Equation (6.12) indicates a linear distribution of normal stresses with respect to the distance y as shown in Figure 6.2b. The maximum stress occurs at the maximum (or minimum) value of y, which is when $y = \pm\, c_S$. The term c_S is the distance of the outer fibers of the cross section from the *neutral axis*, which is the same for symmetrical sections such as circular and rectangular cross sections. The ***maximum normal stress*** $\sigma_{x,\max}$ is then

$$\sigma_{x,\max} = \pm\frac{\boldsymbol{M}_z y_{\max}}{I_z} = \pm\frac{\boldsymbol{M}_z c_S}{I_z} \tag{6.13}$$

The positive sign is for tension and the negative sign is for compression, where $c_S = y_{\max}$ is the magnitude of the distance or dimension. On the other hand, one may use $\sigma_{x,\min}$ for the *compressive stress* and $c_S = y_{\min}$ as may be being negative. It is obvious from Figure 6.2a that the top fibers are compacted or compressed, and the bottom fibers are elongated or stretched; see Section 6.1.3 for an experimental demonstration.

Equation (6.13) is also written in a usual customary form such that

$$\sigma_{x,\max} = \pm\frac{\boldsymbol{M}_z}{I_z/c_S} = \pm\frac{\boldsymbol{M}_z}{Z_z} \tag{6.14}$$

where Z_z is the ***section modulus*** about the z coordinate and is given by $Z_z = I_z/c_S$, which is useful for *symmetrical sections* only. Many *standard cross sections* are available in Appendix A.8 where section properties, the ***second area moments***, and ***section moduli*** are provided about some commonly used coordinates. The sections are assumed to be in the yz plane (x plane); see Figure 6.2b. A frequently used cross section is the circular section of diameter d_C where $I_y = I_z = \pi d_C{}^4/64$ and $Z_y = Z_z = \pi d_C{}^3/32$. Another regularly used cross section is the rectangular section of a width b_R and a depth h_R, where $I_z = b_R h_R{}^3/12$ and $Z_z = b_R h_R{}^2/6$.

Example 6.1 A circular, rectangular, and T-section profiles are to be selected to withstand a specific bending moment $\boldsymbol{M}_z$. The sections are shown in Figure 6.3 with the adopted predetermined proportions. Find the second area moment of the circular and rectangular sections. Derive the second area moment of the T-section and compare these sections with respect to the most suitable one to withstand the moment $\boldsymbol{M}_z$. If the moment $\boldsymbol{M}_z$ is 10 [kN m] or 88.5 [klb in], what are the maximum normal stresses in the three cross-sectional profiles for the same value of the depth $d_C = 0.1$ [m] or 4 [in]?

Solution

Data: $M_z = +10\,000$ [N m] or +88 500 [lb in], $d_C = 0.1$ [m] or 4 [in].

Circular section:

The second area moments are obtained as follows:

$$\begin{aligned} I_z &= \pi d_C{}^4/64 = \pi(0.1)^4/64 = 4.91E-06\,[\text{m}^4] \\ I_z &= \pi d_C{}^4/64 = \pi(4)^4/64 = 12.566\,[\text{in}^4] \end{aligned} \tag{a}$$

The maximum stresses are obtained by applying Eq. (6.14) as follows;

$$\begin{aligned} \sigma_{x,\max} &= \pm\frac{\boldsymbol{M}_z\,(d_C/2)}{I_z} = \pm\frac{10(10^3)(0.1/2)}{4.91(10^{-6})} = \pm 101.8\,(10^6)\,[\text{Pa}] \\ \sigma_{x,\max} &= \pm\frac{\boldsymbol{M}_z\,(d_C/2)}{I_z} = \pm\frac{88.5(10^3)(4/2)}{12.566} = \pm 14.13\,(10^3)\,[\text{psi}] \end{aligned} \tag{b}$$

Rectangular section:

The second area moments are obtained as follows:

$$I_z = b_R d_C{}^3/12 = (0.5(0.1))(0.1)^3/12 = 4.17E-06\ [\text{m}^4]$$
$$I_z = b_R d_C{}^3/12 = (0.5(4))(4)^3/12 = 10.67[\text{in}^4] \tag{c}$$

The maximum stresses are obtained by applying Eq. (6.14) as follows:

$$\sigma_{x,\max} = \pm\frac{\boldsymbol{M}_z(d_C/2)}{I_z} = \pm\frac{10(10^3)(0.1/2)}{4.17(10^{-6})} = \pm 119.9(10^6)\ [\text{Pa}]$$
$$\sigma_{x,\max} = \pm\frac{\boldsymbol{M}_z(d_C/2)}{I_z} = \pm\frac{88.5(10^3)(4/2)}{10.67} = \pm 16.64(10^3)\ [\text{psi}] \tag{d}$$

T-section:

To get the second area moment, we need to locate the *neutral axis* or the values of c_1 and c_2 in Eq. (6.10). The neutral axis is the dashed dot line in Figure 6.3c. The process of finding the neutral axis necessitates the evaluation of the top and the web rectangular areas of thickness t_T and the location of each of their centers with respect to the section top fiber. These relations are as follows:

$$\text{Top-section area } A_t = b_T\ t_T = (0.9(0.1))(0.1(0.1)) = 0.0009\ [\text{m}^2]$$
$$\text{Top-section area } A_t = b_T\ t_T = (0.9(4))(0.1(4)) = 1.44\ [\text{in}^2] \tag{e}$$

$$\text{Web-section area } A_w = (h_T - t_T)\, t_T = (0.1 - (0.1(0.1)))(0.1(0.1)) = 0.0009\ [\text{m}^2]$$
$$\text{Web-section area } A_w = (h_T - t_T)\, t_T = (4 - (0.1(4)))(0.1(4)) = 1.44\ [\text{in}^2] \tag{f}$$

$$\text{Top-section center } c_t = t_T/2 = (0.1(0.1))/2 = 0.005\ [\text{m}]$$
$$\text{Top-section area } c_t = t_T/2 = (0.1(4))/2 = 0.2\ [\text{in}] \tag{g}$$

$$\text{Web-section center } c_w = ((h_T - t_T)/2) + t_T = ((0.1 - (0.1(0.1))/2) + (0.1(0.1)) = 0.055\ \ [\text{m}]$$
$$\text{Web-section center } c_w = ((h_T - t_T)/2) + t_T = ((4 - (0.1(4))/2) + (0.1(4)) = 2.2\ \ [\text{in}] \tag{h}$$

The location of the neutral axis relative to the top fiber c_1 is thus

$$\begin{aligned}\text{Neutral axix position } c_1 &= ((A_t(c_t) + A_w(c_w))/(A_t + A_w)\\ &= ((0.0009(0.005) + 0.0009(0.055))/(0.0009 + 0.0009) = 0.03\ \ [\text{m}]\\ \text{Neutral axix position } c_1 &= ((A_t(c_t) + A_w(c_w))/(A_t + A_w) = ((1.44(0.2) + 1.44(2.2))/(1.44 + 1.44) = 1.2\ \ [\text{in}]\end{aligned} \tag{i}$$

$$\text{Neutral axix position } c_2 = h_T - (c_1) = 0.1 - 0.03 = 0.07\ \ [\text{m}]$$
$$\text{Neutral axix position } c_2 = h_T - (c_1) = 4 - 1.2 = 2.8\ \ [\text{in}] \tag{j}$$

The second area moments are obtained using the *parallel axis theorem* as follows. It, however, needs the evaluation of each second area moment about each center and its center distance to the neutral axis as follows:

$$\text{Top-section } I_{zt} = b_T\ t_T{}^3/12 = (0.9(0.1))(0.1(0.1))^3/12 = 7.5E-09\ [\text{m}^4]$$
$$\text{Top-section } I_{zt} = b_T\ t_T{}^3/12 = (0.9(4))(0.1(4))^3/12 = 0.0192\ [\text{in}^4]$$
$$\text{Web-section } I_{zw} = t_T\ (h_T - t_T)^3/12 = (0.1(0.1))(0.1 - (0.1(0.1))^3/12 = 6.075E-7\ [\text{m}^4]$$
$$\text{Web-section } I_{zw} = t_T\ (h_T - t_T)^3/12 = (0.1(4))(4 - (0.1(4))^3/12 = 1.5552\ [\text{in}^4] \tag{k}$$

$$\text{Top-section distance } d_t = c_1 - (t_T/2) = 0.03 - (0.1(0.1)/2) = 0.025\ [\text{m}]$$
$$\text{Top-section distance } d_t = c_1 - (t_T/2) = 1.2 - (0.1(4)/2) = 1.0\ [\text{in}]$$
$$\text{Web-section distance } d_w = c_w - c_1 = 0.055 - (0.03) = 0.025\ [\text{m}]$$
$$\text{Web-section distance } d_w = c_w - c_1 = 2.2 - (1.2) = 1.0\ [\text{in}] \tag{l}$$

Applying the parallel axis theorem gives the following second area moment of the T-section:

$$\begin{aligned}\text{T-section } I_z &= (I_{zt} + A_t(d_t^2)) + (I_{zw} + A_w(d_w^2))\\ &= (7.5(10^{-9}) + 0.0009(0.025)^2) + (6.075(10^{-7}) + 0.0009(0.025)^2) = 0.000\,001\,74\ [\text{m}^4]\\ \text{T-section } I_z &= (I_{zt} + A_t(d_t^2)) + (I_{zw} + A_w(d_w^2))\\ &= (0.0192 + 1.44(1.0)^2) + (1.5552 + 1.44(1.0)^2) = 4.4544\ [\text{in}^4]\end{aligned} \tag{m}$$

The maximum stresses are obtained by applying Eq. (6.14) as follows:

$$\begin{aligned}\sigma_{x,\max 1} &= -\frac{\boldsymbol{M}_z\ c_1}{I_z} = -\frac{10(10^3)(0.03)}{0.000\,001\,74} = -172.4\ (10^6)\ [\text{Pa}]\\ \sigma_{x,\max 1} &= -\frac{\boldsymbol{M}_z\ c_1}{I_z} = -\frac{88.5(10^3)(1.2)}{4.4544} = -23.8\ (10^3)\ [\text{psi}]\\ \sigma_{x,\max 2} &= +\frac{\boldsymbol{M}_z\ c_2}{I_z} = +\frac{10(10^3)(0.07)}{0.000\,001\,74} = 402.3\ (10^6)\ [\text{Pa}]\\ \sigma_{x,\max 2} &= +\frac{\boldsymbol{M}_z\ c_2}{I_z} = +\frac{88.5(10^3)(2.8)}{4.4544} = 55.6\ (10^3)\ [\text{psi}]\end{aligned} \tag{n}$$

These stresses indicate that there is a smaller compression (at the top fiber) than the tension (at the bottom fiber). If the moment is in the opposite direction, we get an opposite result. This can be more suitable for some materials that have a *compressive strength* higher that their *tensile strength* such as some cast irons; see Section 7.1.3. The opposite is not usually the case.

It should be noted that the cross-sectional areas of the three profiles are as follows:

$$\begin{aligned}&\text{Circular area } A_C = \pi d_C{}^2/4 = \pi(0.1)^2/4 = 0.007854\ [\text{m}^2]\\ &\text{Circular area } A_C = \pi d_C{}^2/4 = \pi(4)^2/4 = 12.566\ [\text{in}^4]\\ &\text{Rectangular area } A_R = h_R\ b_R = (0.1)(0.5(0.1)) = 0.005\ [\text{m}^2]\\ &\text{Rectangular area } A_R = h_R\ b_R = (4)(0.5(4)) = 8\ [\text{in}^2]\\ &\text{T-section area } A_T = A_t + A_w = (0.0009) + (0.0009) = 0.0018\ [\text{m}^2]\\ &\text{T-section area } A_T = A_t + A_w = (1.44) + (1.44) = 2.88\ [\text{in}^2]\end{aligned} \tag{o}$$

It is clear that the circular cross-sectional area is larger than the rectangular cross-sectional area by about 57%. Maximum stress in the circular profile is less than the maximum stress in the rectangular profile by about 18%. This might suggest that the rectangular section could be favorable in such a loading case and proportional conditions. On the other hand, the T-section has a much lower area than the rectangular section. However, the stresses in the T-section are much higher than the rectangular section particularly in the tension side. These competing results suggest that an *optimization* problem is potentially needed. It should also be noted that there is a constraint on the section depth and only few profiles are considered. No consideration for *deflection* is also accounted for.

6.1.3 Beam Deflection

In the preceding section, externally applied load has been considered in developing internally ensued loads such as shear $\boldsymbol{V}$ and bending $\boldsymbol{M}$. The deformation of the beam under *pure bending* has been demonstrated by beam curvature of a radius R_C as demonstrated by Eqs. (6.6) and (6.11). In this section, the deflection is obtained by few different means. The generalized loading $\boldsymbol{q}_i$ and its differential relations with shear and bending is used to find the beam *slope* and *deflection* by successive *integration* as shown in Section 6.1.3.1. Afterward, another method is presented, which utilizes *singularity function* for the successive integration. A simple mean of available solution *superposition* is also useful in finding deflections of beams if the sub-regions or segments already having definite closed form solution. With the available software such as *finite element* (***FE***), *numerical* solutions are also attractive and right at hand.

6.1.3.1 Deflection by Integration

Figure 6.4 shows an extremely exaggerated *deformation* of the beam under some loading including mainly the bending moment $\boldsymbol{M}_z$ about the z-axis. The loadings are not shown in Figure 6.4a since they are away from the shown part of the beam. Figure 6.4b shows an extreme exaggeration of the expected three-dimensional ***(3D)*** deformation that is not considered in the present analysis. However, it can be roughly demonstrated experimentally by a straight piece of thick but very low density sponge or foam rubber. The straight sponge of a rectangular section is bent from both ends by two hefty straws or balsa-wood rods glued to the sponge as shown in Figure 6.4b. See Problems 6.5–6.7 for experiments in that regard.

It is assumed that the coordinate x is kept as fixed and the *deformed neutral axis* of the beam is named the ***elastic curve*** of the beam as illustrated in Figure 6.4a. To differentiate between the location y of a point on the beam cross section and the deflection of the beam in the y direction due to its deformation, the symbol v is used to represent the beam deflection or the displacement in the y direction. This deflection v as a function of x defines the *elastic curve* of the beam as identified by the dashed line in Figure 6.4a. The presentation herein is considering the *Euler–Bernoulli beam* theory (after Leonhard Euler 1707–1783 and Daniel Bernoulli 1700–1782), which is concerned with small deformations and lateral loads. It is a simplified case of the *Timoshenko* beam theory (after Stephen Timoshenko 1878–1972), which includes the effect of *transverse shear*, Reisman and Pawlik (1980, 1991). For machine design and synthesis, the engineering *Euler–Bernoulli beam* theory is sufficiently accurate and simpler to implement. The ***3D*** deformation in Figure 6.4b will not then be considered due to its special case of very large deformations that are rarely the case in usual machine design applications.

In *analytic geometry*, the curvature at any point (x,v) on the curve is defined as

$$\frac{1}{R_C} = \frac{\dfrac{d^2v}{dx^2}}{\left[1 + \left(\dfrac{dv}{dx}\right)^2\right]^{3/2}} = \frac{v''}{[1 + (v')^2]^{3/2}} \tag{6.15}$$

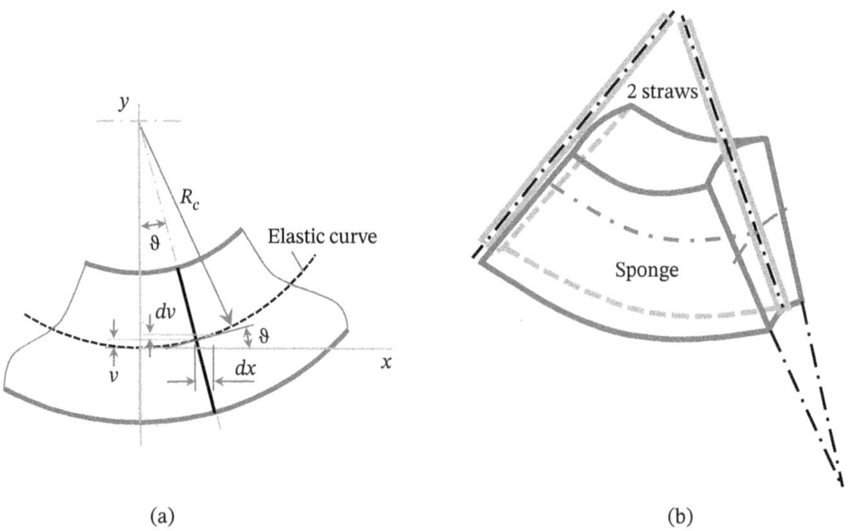

Figure 6.4 Exaggerated deformation of a beam (a) showing the elastic curve and its radius of curvature and (b) showing the expected ***3D*** deformation (extremely exaggerated), which is not considered in the present analysis. The loadings are away from the shown part of the beam in (a).

This expression is used to define the *curvature* of any point on the elastic curve. In practice, the slope dv/dx of the elastic curve is usually very small. Therefore, $(v')^2$ can be neglected – relative to 1 – in Eq. (6.15) and the equation can be reduced to the following:

$$\frac{1}{R_C} \cong \frac{d^2v}{dx^2} = v'' \tag{6.16}$$

Substituting the curvature of Eq. (6.16) into Eq. (6.11), one can write

$$\frac{1}{R_c} = \frac{\boldsymbol{M}_z}{EI_z} = \frac{d^2v}{dx^2} = v'' \tag{6.17}$$

However, from Figure 6.4a, the **slope** θ or v' of the ***elastic curve*** v at any point x is given by

$$\theta(x) = \frac{dv}{dx} = v' \quad \left(\text{note, } \frac{dv}{dx} = \tan\theta \cong \theta\right) \tag{6.18}$$

Rewriting Eq. (6.6), one gets the moment at any point x such that

$$\boldsymbol{M}_z = EI_z\frac{d^2v}{dx^2} = EI_z v'' \tag{6.19}$$

Considering Eq. (6.2) relating the moment $\boldsymbol{M}_z$ to the shear force $\boldsymbol{V}$, we can define the shear force at any point x upon the application of Eq. (6.19) as follows:

$$V(x) = -\frac{d\boldsymbol{M}_z}{dx} = -\frac{d}{dx}\left(EI_z\frac{d^2v}{dx^2}\right) = -EI_z\frac{d^3v}{dx^3} = -EI_z v''' \tag{6.20}$$

One can also apply Eqs. (6.1) and (6.20) to get the relation between the distributed load $\boldsymbol{w}_i$ and the shear force $\boldsymbol{V}$ at any point x on the elastic curve as

$$\boldsymbol{w}_i(x) = -\frac{d\boldsymbol{V}}{dx} = \frac{d}{dx}\left(EI_z\frac{d^3v}{dx^3}\right) = EI_z\frac{d^4v}{dx^4} = EI_z v'''' \tag{6.21}$$

It should be noted that these relations from Eq. (6.18) to Eq. (6.21) are valid for beams with constant *flexural rigidity* EI_z. This means any region of a beam with a constant or uniform cross section or a non-varying form of a cross section.

It is usually the case that the load $\boldsymbol{w}_i(x)$ is known and the deflection $v(x)$ is the sought definition of the elastic curve. In that common case, we use successive integration of Eq. (6.21) through Eq. (6.18) to get the deflection $v(x)$. The *boundary conditions* should be observed throughout the process as is given next.

To consider more general case than the distributed load $\boldsymbol{w}_i$, a more general representation of the distributed load $\boldsymbol{q}_i$ can include concentrated forces $\boldsymbol{F}_i$, moments $\boldsymbol{M}_i$, and others in addition to the distributed load $\boldsymbol{w}_i$. The general load can then take the following form of a function $\mathcal{F}$:

$$\boldsymbol{q}_i(x) = \mathcal{F}(\boldsymbol{F}_i, \boldsymbol{M}_i, \boldsymbol{w}_i, \ldots) \tag{6.22}$$

By substituting the general loading $\boldsymbol{q}_i$ into Eq. (6.21) and performing the integration from any x_1 to x_2, we get the shear force $\boldsymbol{V}$, and continue the process through Eqs. (6.18, 6.20).

$$V(x) = -EI_z v''' = -\int_{x_1}^{x_1} \boldsymbol{q}_i(x)\ dx + C_1, \quad \boldsymbol{q}_i = \mathcal{F}(\boldsymbol{F}_i, \boldsymbol{M}_i, \boldsymbol{w}_i, \ldots) \tag{6.23}$$

$$\boldsymbol{M}_z(x) = EI_z v'' = -\int_{x_1}^{x_2} V(x)\ dx + C_2 \tag{6.24}$$

$$\theta(x) = v' = \int_{x_1}^{x_2} \frac{\boldsymbol{M}_z(x)}{EI_z} dx + C_3 \tag{6.25}$$

$$v(x) = \int_{x_1}^{x_2} \theta(x)\ dx + C_4 \tag{6.26}$$

It is obvious that the constants C_1, C_2, C_3, and C_4 are the *boundary conditions* on the forces, moments, slopes, and deflections at both x_1 and x_2, respectively. Constants C_1 and C_2 represent boundary *reactions* and *moments;* C_3 and C_4 are obtained from boundary conditions on *slope* and *deflection* at supports or beam segments.

Furthermore, from Eq. (6.1), one can write $d\boldsymbol{V} = -\boldsymbol{w}_i\ dx$ or more general $d\boldsymbol{V} = -\boldsymbol{q}_i\ dx$. Integrating between specific limits of x_1 and x_2 including the effect of every one of *singular forces* $\boldsymbol{F}_i$ in-between x_1 and x_2, produces the following:

$$\begin{aligned}\int_{V_1}^{V_2} dV &= -\int_{x_1}^{x_2} \boldsymbol{q}_i\, dx = -\int_{x_1}^{x_2} \boldsymbol{w}_i\, dx - \sum_{x_1}^{x_2} \boldsymbol{F}_i, \quad \text{or } (\boldsymbol{V}_2 - \boldsymbol{V}_1) = -\int_{x_1}^{x_2} \boldsymbol{w}_i\, dx - \sum_{x_1}^{x_2} \boldsymbol{F}_i \\ \boldsymbol{V}_2 &= \boldsymbol{V}_1 - \int_{x_1}^{x_2} \boldsymbol{w}_i\, dx - \sum_{x_1}^{x_2} \boldsymbol{F}_i\end{aligned} \tag{6.27}$$

This indicates that the change in the internal shear force between x_1 and x_2 is the negative of the area under the load diagram between x_1 and x_2. Similarly, from Eq. (6.2), one can write $d\boldsymbol{M} = -\boldsymbol{V}\,dx$. Integrating between the same specific limits of x_1 and x_2 including the effect of every *singular moment* $\boldsymbol{M}_i$ in-between x_1 and x_2 produces the following:

$$\begin{aligned}\int_{M_1}^{M_2} dM &= -\int_{x_1}^{x_2} \boldsymbol{V}_i\, dx = -\int_{x_1}^{x_2} \boldsymbol{V}_i\ dx - \sum_{x_1}^{x_2} \boldsymbol{M}_i \quad \text{or } (\boldsymbol{M}_2 - \boldsymbol{M}_1) = -\int_{x_1}^{x_2} \boldsymbol{V}_i\ dx - \sum_{x_1}^{x_2} \boldsymbol{M}_i \\ \boldsymbol{M}_2 &= \boldsymbol{M}_1 - \int_{x_1}^{x_2} \boldsymbol{V}_i\ dx - \sum_{x_1}^{x_2} \boldsymbol{M}_i\end{aligned} \tag{6.28}$$

Therefore, the change in the internal bending moments between x_1 and x_2 is the negative of the area under the *shear force diagram* between x_1 and x_2.

The relations in Eqs. (6.27) and (6.28) are very useful in checking the results of integration and the evaluation of the boundary conditions.

It is also useful to *rewrite* and *integrate* the previous generalized Eqs. (6.21)–(6.26) in an orderly, systematic, and complete manner to have the following:

$$EI_z \frac{d^4v}{dx^4} = \boldsymbol{q}_i(x) \tag{6.21$'$}$$

$$EI_z \frac{d^3v}{dx^3} = -\boldsymbol{V}(x) = \int_{x_1}^{x_2} q_i(x)dx + C_1 \tag{6.23$'$}$$

$$EI_z \frac{d^2v}{dx^2} = \boldsymbol{M}(x) = \int_{x_1}^{x_2} dx \int_{x_1}^{x_2} q_i(x)dx + C_1x + C_2 \tag{6.24$'$}$$

$$EI_z \frac{dv}{dx} = \theta(x) = \int_{x_1}^{x_2} dx \int_{x_1}^{x_2} dx \int_{x_1}^{x_2} q_i(x)dx + C_1x^2/2! + C_2x + C_3 \tag{6.25$'$}$$

$$EI_z v(x) = \int_{x_1}^{x_2} dx \int_{x_1}^{x_2} dx \int_{x_1}^{x_2} dx \int_{x_1}^{x_2} q_i(x)dx + C_1x^3/3! + C_2x^2/2! + C_3x + C_4 \tag{6.26$'$}$$

The integration process has been consecutively applied to the constants C_1, C_2, C_3, and C_4. These relations are used to advantage particularly Eq. (6.26) in the procedure to assume a displacement *shape function* for *prismatic beam element* in ***FE*** formulation; see Section 6.11.2.

Figure 6.5 shows *shear force diagram* and *bending moment diagram* of beams under some externally concentrated and other uniformly distributed loading conditions. The concentrated load $\boldsymbol{F}$ and the distributed load $\boldsymbol{w}$ are assumed in the usual *negative direction* to conform to the *gravitational direction.* Other beams with deflection and slope expressions are covered next, but several more can be found in numerous references about solid mechanics, strength of materials, and handbooks such as Young and Budynas (2002). The deflections and slopes are obtained by successively applying Eqs. (6.23)–(6.26) and considering Eqs. (6.27) and (6.28) for verifying shear and moment diagrams. The following example provides the integration procedure for the simply supported beam with a concentrated force $\boldsymbol{F}$ at mid-span, Figure 6.5a, and considering $\boldsymbol{F}$ in its vectorial negative direction for consistency with coordinate adoption. The conventional beams of Figure 6.5 use loads in the usual *negative direction*. Extreme care should be exercised not to use the negative direction more than once.

Example 6.2 A simply supported beam of length l is subjected to a mid-span load $-\boldsymbol{F}$. It is required to find the reactions. Use the integration method to find the shear force and bending moment diagram. Derive the expressions for the slope and deflection along the beam. If the force $\boldsymbol{F} = -40$ [kN] or -9 [klb], the length $l = 1.0$ [m] or 40 [in], and the modulus of elasticity $E = 207$ [GPa] or 30 [Mpsi], what are the maximum slope, deflection, and normal stresses in the circular section profile of a diameter $d_C = 0.1$ [m] or 4 [in]? What is the beam radius of curvature?

Solution

Data: $\boldsymbol{F} = -40$ [kN] or -9 [klb] and the length $l = 1.0$ [m] or 40 [in], $d_C = 0.1$ [m] or 4 [in].

The maximum bending moment is calculated according to the value depicted from Figure 6.5a such that $\boldsymbol{M}_z = ¼\,Fl = ¼\,(40)(1) = 10$ [kN m] or $\boldsymbol{M}_z = ¼\,(9)(40) = 90$ [klb in]. It should be noted that the positive bending moment magnitude has included the negative sign of the applied force $-\boldsymbol{F}$. No need to reapply the negative sign. That is why it should have been more appropriate to use Figure 6.5 with all loads as positive. That would have caused the shear force diagrams and bending moment diagrams to flip vertically causing the maximum moment to be negative. In that case, one would use the real negative $\boldsymbol{F}$ in its vectorial negative direction, and a positive bending moment would commence. The same applies for the shear force diagrams.

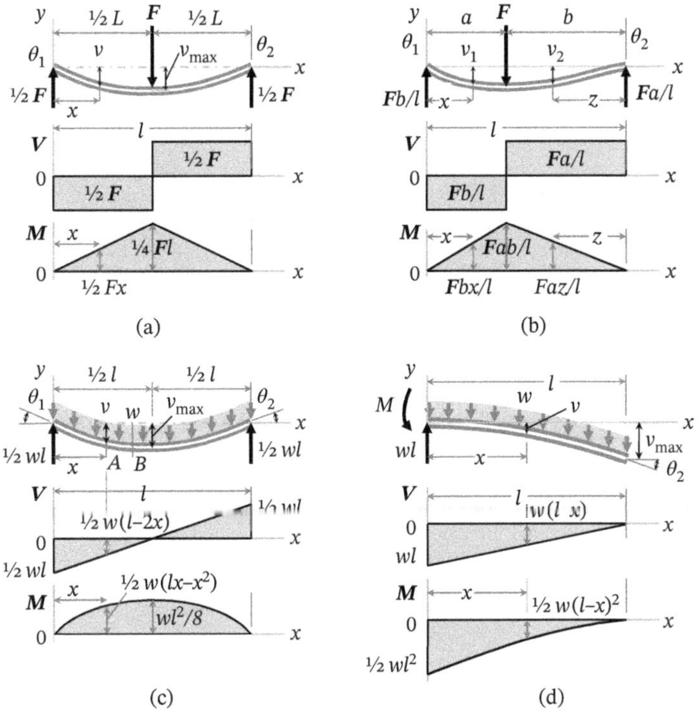

Figure 6.5 Different bem loading and associated shear and bending diagrams: (a) simply supported with a mid-span concentrated force, (b) simply supported and asymmetric concentrated force, (c) simply supported and a uniformly distributed load, and (d) cantilever with a uniformly distributed load.

Reactions: Taking the moment about both ends 1 and 2 of the beam in Figure 6.5a, we get

$$\begin{aligned} \boldsymbol{M}_{z1}\big|_{x=0} &= \boldsymbol{R}_2 l - \boldsymbol{F}l/2 = 0 \quad \text{or} \quad \boldsymbol{R}_2 = {}^1\!/\!_2\boldsymbol{F} \\ \boldsymbol{M}_{z2}\big|_{x=l} &= -\boldsymbol{R}_1 l + \boldsymbol{F}l/2 = 0 \quad \text{or} \quad \boldsymbol{R}_1 = {}^1\!/\!_2\boldsymbol{F} \end{aligned} \tag{a}$$

Note that the negative direction of $\boldsymbol{F}$ has been taken into effect to confirm that the reactions are in the positive direction. The generalized load is then

$$\boldsymbol{q}_1 = \boldsymbol{R}_1 = {}^1\!/\!_2\boldsymbol{F},\ x = 0, \quad \boldsymbol{q}_2 = -\boldsymbol{F},\ x = {}^1\!/\!_2 l, \quad \boldsymbol{q}_3 = \boldsymbol{R}_2 = {}^1\!/\!_2\boldsymbol{F},\ x = l \tag{b}$$

Note that the generalized loads are in the usual *positive coordinate direction.*

Shear force: The shear force is obtained by Eq. (6.23) or (6.27) for both regions of the beam

$$\begin{aligned} &\boldsymbol{V}_1(x) = -\int_0^{(l/2)^-} 0\,dx + C_1, \quad \text{at } x = 0,\ \boldsymbol{V}_1(0) = -{}^1\!/\!_2\boldsymbol{F} \to C_1 = -{}^1\!/\!_2\boldsymbol{F} \\ &\boldsymbol{V}_1(x)\big|_0^{(l/2)^-} = -{}^1\!/\!_2 F \end{aligned} \tag{c}$$

or

$$\boldsymbol{V}_1(x) = -\int_0^{(l/2)^-} 0\ dx - \sum_0^{(l/2)^-} \boldsymbol{F}_i = -({}^1\!/\!_2\boldsymbol{F}) \tag{d}$$

$$\begin{aligned} &\boldsymbol{V}_2(x) = -\int_{l/2}^{x} 0\,dx - \sum_{l/2}^{x} \boldsymbol{F}_i + C_1 = -({}^1\!/\!_2\boldsymbol{F} - \boldsymbol{F}) + C_1 = {}^1\!/\!_2\boldsymbol{F} + C_1 \\ &\text{At } x = l,\ \boldsymbol{V}_2(l) = {}^1\!/\!_2\boldsymbol{F} \to C_1 = 0 \quad \to \quad \boldsymbol{V}_2(x)\big|_{l/2}^{x} = {}^1\!/\!_2\boldsymbol{F} \end{aligned} \tag{e}$$

As a verifying check, use Eq. (6.27) to get

$$(\boldsymbol{V}_2 - \boldsymbol{V}_1) = -\int_0^l 0\ dx - \sum_0^l \boldsymbol{F}_i = -({}^1\!/\!_2\boldsymbol{F} - \boldsymbol{F} + {}^1\!/\!_2\boldsymbol{F}) = 0 \tag{f}$$

The *shear force diagram* is shown in Figure 6.5a.

Bending moment: The bending moment diagram is obtained by Eq. (6.24) or (6.28) for both regions of the beam:

$$\begin{aligned} &\boldsymbol{M}_{z1}(x) = -\int_0^{(l/2)} V\,dx + C_1 = -\int_0^{(l/2)} -{}^1\!/\!_2\boldsymbol{F}\ dx + C_1 = {}^1\!/\!_2\boldsymbol{F}x + C_1 \\ &\text{At } x = 0,\ M_{z1} = 0 \to C_1 = 0 \quad \to \quad \boldsymbol{M}_{z1}(x)\big|_0^{(l/2)} = {}^1\!/\!_2\boldsymbol{F}x \end{aligned} \tag{g}$$

$$\begin{aligned} &M_{z2}(x) = -\int_{(l/2)}^{x} V_2\,dx - \sum_{1/2}^{x} \boldsymbol{M}_i = -\int_{l/2}^{x} {}^1\!/\!_2\boldsymbol{F}dx + C_2 = -{}^1\!/\!_2\boldsymbol{F}x + C_2 \\ &\text{At } x = l,\ M_{z2} = 0 \to C_2 = {}^1\!/\!_2\boldsymbol{F}l \to \boldsymbol{M}_2(x)\big|_{l/2}^{x} = -{}^1\!/\!_2\boldsymbol{F}x + {}^1\!/\!_2\boldsymbol{F}l \end{aligned} \tag{h}$$

As a verifying check, use Eq. (6.28) to get

$$(M_2 - M_1) = -\int_0^l V\,dx - \sum_0^l M_i = 0 \tag{i}$$

The *bending moment diagram* is shown in Figure 6.5a.

Common free body diagram: The previous lengthy procedure was for the *verification* of the integration procedure applicability. A more common procedure is to have a section at a point in the interval between x_1 and x_2, develop a *free body diagram*, and perform equilibrium of forces and moments. The sign convention is the same as that adapted and shown in Figure 6.1b. Applying the common *free body diagram* procedure produces the following for the section from 0 to x and just before $x = \tfrac{1}{2}\, l$.

$$\sum_{0}^{x<l/2} \boldsymbol{F}_i = 0 \rightarrow \tfrac{1}{2}\boldsymbol{F} + \boldsymbol{V}_1(x) = 0 \rightarrow \boldsymbol{V}_1(x) = -\tfrac{1}{2}\boldsymbol{F} \tag{j}$$

$$\sum_{0}^{x<l/2} \boldsymbol{M}_i\big|_x = 0 \rightarrow -\tfrac{1}{2}\boldsymbol{F}x + \boldsymbol{M}_{z1}(x) = 0 \rightarrow M_{z1}(x) = \tfrac{1}{2}\boldsymbol{F}x \tag{k}$$

Applying the common *free body diagram* procedure produce the following for $\tfrac{1}{2}\, l$ to x just to l:

$$\sum_{x>l/2}^{x<l} \boldsymbol{F}_i = 0 \rightarrow \tfrac{1}{2}\boldsymbol{F} - \boldsymbol{F} + \boldsymbol{V}_2(x) = 0 \rightarrow \boldsymbol{V}_2(x) = \tfrac{1}{2}\boldsymbol{F} \tag{l}$$

$$\sum_{x>l/2}^{x<l} \boldsymbol{M}_i\big|_x = 0 \rightarrow -\tfrac{1}{2}\boldsymbol{F}x + \boldsymbol{F}(x - \tfrac{1}{2}l) + \boldsymbol{M}_{z2}(x) = 0 \rightarrow M_{z2}(x) = \tfrac{1}{2}\boldsymbol{F}l - \tfrac{1}{2}\boldsymbol{F}x \tag{m}$$

The results in Eqs. (j)–(m) are the same as for expressions in Eqs. (c)–(h). It should be noted that the free body diagrams were drawn by mind's eye inspection from Figure 6.5a since the figure is simple. For more involved loading cases, the drawing of free body diagrams is a must for more reliable results.

It is left to the reader to examine the case exactly at $x = l/2$. The *singularity function* of Section 6.1.3.3 ***C*** would help in that regard.

<u>*Slope and deflection:*</u> To determine the slope and deflection, one needs to apply Eqs. (6.25) and (6.26) for both zones $x = 0$ to $l/2$ and $x = l/2$ to l. In addition to the boundary conditions at the supports, a continuity condition should be observed at the connection between the two zones.

The *slope diagram* is obtained by Eq. (6.25) for both regions of the beam such that

$$\theta_1(x) = \int_0^x \frac{\boldsymbol{M}_{z1}(x)}{EI_z} dx + C_3 = \frac{1}{EI_z}\int_0^x \tfrac{1}{2}\boldsymbol{F}x\, dx + C_3 = \frac{\tfrac{1}{2}\boldsymbol{F}}{EI_z}\left(\tfrac{1}{2}x^2\right) + C_3$$

$$\text{At } x = \tfrac{1}{2}l,\ \theta_1\left(\tfrac{1}{2}l\right) = 0 \rightarrow C_3 = -\frac{\boldsymbol{F}\, l^2/4}{4EI_z} \rightarrow \theta_1(x) = \frac{4\boldsymbol{F}x^2}{16EI_z} - \frac{\boldsymbol{F}l^2}{16EI_z} = \frac{\boldsymbol{F}}{16EI_z}(4x^2 - l^2) \tag{n}$$

$$\theta_1(x) = \frac{\boldsymbol{F}\ l^2}{16EI_z}\left(\frac{4x^2}{l^2} - 1\right)$$

$$\theta_2(x) = \int_{l/2}^x \frac{\boldsymbol{M}_{z1}(x)}{EI_z} dx + C_3 = \frac{1}{EI_z}\int_{l/2}^x \left(-\frac{\boldsymbol{F}x}{2} + \frac{\boldsymbol{F}l}{2}\right) dx + C_3 = \frac{\boldsymbol{F}}{2EI_z}\left(-\frac{x^2}{2} + lx\right) + C_3$$

$$\text{At } x = \frac{1}{2}l,\ \theta_2\left(\frac{1}{2}l\right) = 0 \rightarrow C_3 = -\left(\frac{\boldsymbol{F}}{2EI_z}\right)\left(-\frac{l^2}{8} + \frac{l^2}{2}\right) = -\frac{\boldsymbol{F}l^2}{2EI_z}\left(-\frac{1}{8} + \frac{4}{8}\right) = -\frac{3\boldsymbol{F}l^2}{16EI_z} \tag{o}$$

$$\theta_2(x) = \frac{\boldsymbol{F}}{2EI_z}\left(-\frac{x^2}{2} + lx\right) - \frac{3\boldsymbol{F}l^2}{16EI_z} = \frac{\boldsymbol{F}l^2}{16EI_z}\left(\frac{8x}{l} - \frac{4x^2}{l^2} - 3\right)$$

From Eqs. (n) and (o), the *maximum slope* is found at $x = 0$ and $x = l$ such that

$$\theta_{1,\max}(0) = \frac{\boldsymbol{F}x^2}{16EI_z} - \frac{\boldsymbol{F}l^2}{16EI_z} = -\frac{\boldsymbol{F}l^2}{16EI_z}$$

$$\theta_{2,\max}(l) = \frac{\boldsymbol{F}l^2}{16EI_z}\left(\frac{8l}{l} - \frac{4l^2}{l^2} - 3\right) = \frac{\boldsymbol{F}l^2}{16EI_z}(8 - 4 - 3) = \frac{\boldsymbol{F}l^2}{16EI_z} \tag{p}$$

The *displacement* or *deflection diagram v(x)* is obtained by Eq. (6.26) for both regions of the beam such that

$$v_1(x) = \int_0^x \theta(x)\,dx + C_4 = \int_0^x \frac{\boldsymbol{F}}{16EI_z}(4x^2 - l^2)\,dx + C_4 = \frac{\boldsymbol{F}}{16EI_z}\left(\frac{4x^3}{3} - l^2x\right) + C_4$$
$$\text{At } x = 0,\ v_1(0) = 0 \ \rightarrow\ C_4 = 0 \ \rightarrow\ v_1(x) = \frac{\boldsymbol{F}}{48EI_z}(4x^3 - 3l^2x) \tag{q}$$

$$v_2(x) = \int_{l/2}^x \theta(x)\,dx + C_4 = \int_0^x \frac{\boldsymbol{F}l^2}{16EI_z}\left(\frac{8x}{l} - \frac{4x^2}{l^2} - 3\right)dx + C_4 = \frac{\boldsymbol{F}l^2}{16EI_z}\left(\frac{8x^2}{2l} - \frac{4x^3}{3l^2} - 3x\right) + C_4$$
$$\text{At } x = l,\ v_2(l) = 0 \ \rightarrow\ C_4 = -\frac{\boldsymbol{F}l^2}{16EI_z}\left(\frac{8l^2}{2l} - \frac{4l^3}{3l^2} - 3l\right) = -\frac{\boldsymbol{F}l^2}{16EI_z}\left(\frac{8l}{2} - \frac{4l}{3} - 3l\right)$$
$$= -\frac{\boldsymbol{F}l^3}{16EI_z}\left(\frac{24}{6} - \frac{8}{6} - \frac{18}{6}\right) = \frac{\boldsymbol{F}l^3}{48EI_z} \tag{r}$$
$$v_2(x) = \frac{\boldsymbol{F}l^2}{16EI_z}\left(\frac{8x^2}{2l} - \frac{4x^3}{3l^2} - 3x\right) + \frac{\boldsymbol{F}l^3}{48EI_z} = \frac{\boldsymbol{F}l^3}{48EI_z}\left(\frac{24x^2}{2l^2} - \frac{12x^3}{3l^3} - \frac{9x}{l} + 1\right)$$
$$= \frac{\boldsymbol{F}l^3}{48EI_z}\left(-\frac{4x^3}{l^3} + \frac{12x^2}{l^2} - \frac{9x}{l} + 1\right)$$

From Eqs. (q) and (r), the *maximum deflection* is found at $x = l/2$ as checked from both zones $x = 0$ to $l/2$ and $x = l/2$ to l.

$$v_{1,\max}\left(\frac{1}{2}l\right) = \frac{\boldsymbol{F}}{48EI_z}(4x^3 - 3l^2x) = \frac{\boldsymbol{F}}{48EI_z}\left(\frac{4l^3}{2^3} - \frac{3l^3}{2}\right) = \frac{\boldsymbol{F}l^3}{48EI_z}\left(\frac{4}{8} - \frac{3}{2}\right) = -\frac{\boldsymbol{F}l^3}{48EI_z}$$
$$v_{2,\max}\left(\frac{1}{2}l\right) = -\frac{\boldsymbol{F}l^3}{48EI_z}\left(-\frac{4x^3}{l^3} + \frac{12x^2}{l^2} - \frac{9x}{l} + 1\right) = -\frac{\boldsymbol{F}l^3}{48EI_z}\left(-\frac{4l^3}{l^3} + \frac{12l^2}{l^2} - \frac{9l}{l} + 1\right) = -\frac{\boldsymbol{F}l^3}{48EI_z} \tag{s}$$

As a summary of the slope θ and deflection v, Eqs. (p)–(s) give

$$\theta_1 = -\theta_2 = -\frac{\boldsymbol{F}l^2}{16EI_z} \quad \text{(p)}, \tag{6.29}$$
$$v_1 = -\frac{\boldsymbol{F}}{48EI_z}(4x^3 - 3xl^2) \quad \text{and} \quad v_2 = -\frac{\boldsymbol{F}}{48EI_z}(4x^3 - 12x^2l + 9xl^2 - l^3) \quad \text{(q)}, \tag{6.30}$$
$$v_{\max}\left(\frac{1}{2}l\right) = -\frac{\boldsymbol{F}l^3}{48EI_z} \quad \text{at } \frac{1}{2}l \quad \text{(s)}, \tag{6.31}$$

The values of maximum slope, deflection, and stress are obtained by substituting the numerical values of $\boldsymbol{F}$, l, and d_C into Eqs. (p), (s), and (6.14) to get the following.

The second area moments are found as follows:

$$I_z = \pi d_C{}^4/64 = \pi(0.1)^4/64 = 4.91E-06 \ [\text{m}^4]$$
$$I_z = \pi d_C{}^4/64 = \pi(4)^4/64 = 12.566 \ [\text{in}^4] \tag{t}$$

The maximum stresses are evaluated by applying Eq. (6.14) as follows:

$$\sigma_{x,\max} = \pm\frac{\boldsymbol{M}_z\ (d_C/2)}{I_z} = \pm\frac{10(10^3)(0.1/2)}{4.91(10^{-6})} = \pm 101.8\ (10^6)\ [\text{Pa}]$$
$$\sigma_{x,\max} = \pm\frac{\boldsymbol{M}_z\ (d_C/2)}{I_z} = \pm\frac{90(10^3)(4/2)}{12.566} = 14.324\ (10^3)\ [\text{psi}] \tag{u}$$

$$\theta_{\max} = \theta_1 = -\theta_2 = -\frac{\boldsymbol{F}l^2}{16\ EI_z} = -\frac{40(10^3)(1)^2}{16(207)(10^9)(4.91)(10^{-6})} = -0.002\,459\,7\ [\text{rad}] = -0.1409\ [°]$$
$$\theta_{\max} = \theta_1 = -\theta_2 = -\frac{\boldsymbol{F}l^2}{16\ EI_z} = -\frac{9(10^3)(40)^2}{16(30)(10^6)(12.566)} = -0.002\,39\ [\text{rad}] = -0.1369\ [°] \tag{v}$$

$$v_{\max}\left(\frac{1}{2}l\right) = -\frac{Fl^3}{48\,EI} = -\frac{40(10^3)(1)^3}{48(207)(10^9)(4.91)(10^{-6})} = -8.1991(10^{-4})\ [\text{m}] = -0.819\,91\ [\text{mm}]$$
$$v_{\max}\left(\frac{1}{2}l\right) = -\frac{Fl^3}{48\,EI} = -\frac{9(10^3)(40)^3}{48(30)(10^6)(12.566)} = -0.0318\ [\text{in}] = -0.809\,[\text{mm}] \qquad \text{(w)}$$

Radius of curvature: Since the deflection is small, it is expected to have a very large radius of curvature. The values are calculated using Eq. (6.11), which give

$$\frac{1}{R_c} = \frac{\boldsymbol{M}_z}{E\ I_z}, \quad \text{or} \quad R_c = \frac{E\ I_z}{\boldsymbol{M}_z} = \frac{(207)(10^9)(4.91)(10^{-6})}{10(10^3)} = 101.637\,[\text{m}]$$
$$R_c = \frac{E\ I_z}{\boldsymbol{M}_z} = \frac{(30)(10^6)(12.556)}{90(10^3)} = 4185\,[\text{in}] \qquad \text{(x)}$$

This example demonstrates the integration procedure and the application of boundary conditions. The procedure can be used for other types of beam loading; however it needs consideration of many regions of the beam as the different loading types and distributed loading zones.

Other Beams

Similar to the simply supported beam with concentrated load at the mid-span shown in Figure 6.5a, another simply supported beam with a concentrated load off the mid-span (asymmetric load) is shown in Figure 6.5b. The same procedure of integration is used to find the following relations of slope and deflection:

$$\theta_1 = -\frac{\boldsymbol{F}\ ab(l+b)}{6\ EI_z\ l} \quad \text{and} \quad \theta_2 = \frac{\boldsymbol{F}\ ab(l+a)}{6\ EI_z\ l} \qquad (6.29)$$

$$v_1 = -\frac{\boldsymbol{F}bx}{6EI_z l}(l^2 - b^2 - x^2) \quad \text{and} \quad v_2 = -\frac{\boldsymbol{F}bz}{6EI_z l}(l^2 - a^2 - z^2) \qquad (6.30)$$

The vales of the shear force $\boldsymbol{V}$ and bending $\boldsymbol{M}$ moments are depicted in Figure 6.5b.

Cantilever beams subjected to *concentrated load* at the tip can be modeled as one-half of a simply supported beam fixed at the middle. The length of the cantilever should be $2l$ and the load should be $2\boldsymbol{F}$. This *equivalent beam* will provide the slope and deflection using Eqs. (p), (q), and (s) of Example 6.2, and considering the right side of the equivalent beam, one gets the following:

$$\theta_1 = 0 \quad \text{and} \quad \theta_2 = \frac{2\boldsymbol{F}(2l)^2}{16EI_z} = \frac{\boldsymbol{F}l^2}{2EI_z} \qquad (6.31)$$

$$v_{\max}(l) = \frac{2\boldsymbol{F}(2l)^3}{48EI_z} = \frac{\boldsymbol{F}l^3}{3EI_z} \qquad (6.32)$$

Beams subjected to *uniformly distributed loads* are given in Figure 6.5c,d. The vales of the shear force $\boldsymbol{V}$ and bending $\boldsymbol{M}$ moments are depicted in Figure 6.5c,d. The following relations of slope and deflection can be obtained similar to the procedure in Example 6.2. The relations for *uniformly distributed* and *simply supported beam* in Figure 6.5c are as follows:

$$\theta_1 = -\theta_2 = \frac{\boldsymbol{w}l^3}{24EI_z} \qquad (6.33)$$

$$v_1 = \frac{\boldsymbol{w}x}{24EI_z}(l^3 - 2lx + x^3), \quad \text{and} \quad v_{\max} = \frac{5\boldsymbol{w}l^4}{384EI_z} \qquad (6.33')$$

In Eqs. (6.33) and (6.33′), one should note that the values are in the $\boldsymbol{w}$ direction, which is negative. That is why the $\boldsymbol{w}$ is substituted as a vector with its direction considered. It is unfortunate that the common practice is to have the loads in the negative y direction due to the gravitational downward direction. However, one is usually advised to assume the positive rather than the negative so that no confusion would materialize.

The relations for *uniformly distributed load* but *cantilever* beam in Figure 6.5d are as follows;

$$\theta_2 = -\frac{wl^3}{6EI_z} \tag{6.34}$$

$$v_2 = -\frac{wx^3}{24EI_z}(6l^2 - 4lx + x^2), \quad \text{and} \quad v_{max} = -\frac{wl^4}{8EI_z} \tag{6.34'}$$

These relations of Eq. (6.29)–(6.34) are important in determining the selection of the mathematical model of the real design defining the underlying differences in model adoption.

Beams subjected to other loads and different types of supports are provided in handbooks such as Young and Budynas (2002).

6.1.3.2 Deflection by Superposition

If there are several loads acting on a beam, the solution outcome is found by the summation of the contribution of each individual load. This summation process is called *superposition*. This is valid provided that the solution of each case such as slope, deflection, or stress is linear in terms of the load, be it a force or a moment. This can be very useful in complex loading conditions, where the solution of each load is known in a closed form terms. Superposition can be applied to all known cases such as beams found in several handbooks. To demonstrate the process, we use two of the known solutions for beams under *uniformly distributed* and *concentrated* loads of *simply supported* beams.

Example 6.3 A simply supported beam of length l is subjected to a mid-span load $-\boldsymbol{F}_i$ in addition to a uniformly distributed load $-\boldsymbol{w}_i$. It is required to find the reactions. Use the superposition method to find the shear force and bending moment diagrams. Find the maximum slope and deflection along the beam. If the force $\boldsymbol{F} = -40$ [kN] or -9 [klb], the uniformly distributed load $\boldsymbol{w}_i = -9$ [kN/m] or -50 [lb/in] the length $l = 1.0$ [m] or 40 [in], and the modulus of elasticity $E = 207$ [GPa] or 30 [Mpsi], what are the maximum slope, deflection, and normal stresses in the circular section profile of a diameter $d_C = 0.1$ [m] or4 [in]? What is the beam radius of curvature?

Solution

Data: $\boldsymbol{F} = -40$ [kN] or -9 [klb] and the length $l = 1.0$ [m] or 40 [in], $d_C = 0.1$ [m] or 4 [in], $\boldsymbol{w}_i = -9$ [kN/m] or -50 [lb/in] the length $l = 1.0$ [m] or 40 [in] and $E = 207$ [GPa] or 30 [Mpsi].

The solution is the addition of the solution of the concentrated force in Example 6.2 and the solution of the uniformly distributed load as defined by Figure 6.5c and Eqs. (6.33) and (6.33′). The second area moment $I_z = 4.91$ (10^{-6}) [m⁴] or 12.566 [in⁴] and the maximum stresses $\sigma_{x,max} = 101.8$ (109) [Pa] or 14.324 (103) [psi] as extracted from Example 6.2, Eqs. (t) and (u).

For the *concentrated force* in Example 6.2, we recall that

$$\begin{aligned} \theta_{max} &= \theta_1 = -\theta_2 = -0.002\,459\,7 \text{ [rad]} = -0.1409\,[°] \\ \theta_{max} &= \theta_1 = -\theta_2 = -0.002\,389\,3 \text{ [rad]} = -0.1369\,[°] \end{aligned} \tag{a}$$

$$\begin{aligned} v_{max}\left(\frac{1}{2}l\right) &= -8.1991(10^{-4})\,[\text{m}] = -0.8199\,[\text{mm}] \\ v_{max}\left(\frac{1}{2}l\right) &= -0.0318\,[\text{in}] = -0.809\,[\text{mm}] \end{aligned} \tag{b}$$

$$\sigma_{x,max} = \mp 101.8(10^6)\,[\text{Pa}] \quad \text{or} \quad \sigma_{x,max} = \mp 14.324(10^3)\,[\text{psi}] \tag{c}$$

For the *uniformly distributed load* and applying Eqs. (6.33), (6.33′), and (6.14), we get

$$\begin{aligned} \theta_1 &= -\theta_2 = \frac{wl^3}{24EI_z} = \frac{-9(10^3)(1)^3}{24(207)(10^9)(4.91)(10^{-6})} = -0.000\,369\,[\text{rad}] = -0.0211\,[°] \\ \theta_1 &= -\theta_2 = \frac{wl^3}{24EI_z} = \frac{-50(10^3)(40)^3}{24(30)(10^6)(12.566)} = -0.000\,354\,[\text{rad}] = -0.0203\,[°] \end{aligned} \tag{d}$$

$$v_{\max} = \frac{5\boldsymbol{w}l^4}{384EI_z} = \frac{5(-9)(10^3)(1)^4}{384(207)(10^9)(4.91)(10^{-6})} = -1.153(10^{-4})\,[\text{m}] = -0.1153\,[\text{mm}]$$
$$v_{\max} = \frac{5\boldsymbol{w}l^4}{384EI_z} = \frac{5(-50)(40)^4}{384(30)(10^6)(12.566)} = -0.004\,421\,[\text{in}] \tag{e}$$

$$\sigma_{x,\max} = \pm\frac{\boldsymbol{M}_z(d_C/2)}{I_z} = \pm\frac{\boldsymbol{w}_i(l^2/8)(d_C/2)}{I_z} = \pm\frac{-9(10^3)(1^2/8)(0.1/2)}{4.91(10^{-6})} = \mp 11.456(10^6)\,[\text{Pa}]$$
$$\sigma_{x,\max} = \pm\frac{\boldsymbol{M}_z(d_C/2)}{I_z} = \pm\frac{\boldsymbol{w}_i(l^2/8)(d_C/2)}{I_z} = \pm\frac{-50(40^2/8)(4/2)}{12.566} = \mp 1.592(10^3)\,[\text{psi}] \tag{f}$$

For the combination of loading, one just adds the corresponding quantities to consequently get

$$\theta_{\max} = \theta_1 = -\theta_2 = -0.002\,459\,7\,[\text{rad}] - 0.000\,369\,[\text{rad}] = -0.002\,828\,7\,[\text{rad}] = -0.1621\,[°]$$
$$\theta_{\max} = \theta_1 = -\theta_2 = -0.002\,389\,3\,[\text{rad}] - 0.000\,354\,[\text{rad}] = -0.002\,743\,3\,[\text{rad}] = -0.157\,[°] \tag{g}$$

$$v_{\max} = -8.1991(10^{-4})\,[\text{m}] - 1.153(10^{-4})\,[\text{m}] = -9.352(10^{-4})\,[\text{m}] = -0.9352\,[\text{mm}]$$
$$v_{\max} = -0.0318\,[\text{in}] - 0.004\,421\,[\text{in}] = -0.004\,421\,[\text{in}] = -0.036\,221\,[\text{in}] = -0.920\,[\text{mm}] \tag{h}$$

$$\sigma_{x,\max} = \mp 101.8(10^6)\,[\text{Pa}] \mp 11.456(10^6)\,[\text{Pa}] = \mp 113.256(10^6)\,[\text{Pa}]$$
$$\sigma_{x,\max} = \mp 14.324(10^3)\,[\text{psi}] \mp 1.592(10^3)\,[\text{psi}] = \mp 15.916(10^3)\,[\text{psi}] \tag{i}$$

6.1.3.3 Deflection by Singularity Function

Early in the analysis of beam deflections, Macaulay (1919) has developed a method to treat beam zones or sections as one continuous relation using one function instead of separate relations for each zone. This general treatment provides a functional relation that is valid along the full span of the beam. The general relation is called the *singularity function* since it uses singular *impulse* or *Dirac delta* functions (Dirac 1930) or *Heaviside step function* (Heaviside (1912), after Oliver Heaviside (1850–1925)) defined by the pointed brackets ⟨⟩, Macaulay (1919). The singularity functions for some of the mostly used and main loading cases are defined in Table 6.1. Others may be deduced by inspection or by especially dedicated references. The singularity function denotes that the relation is valid only at the concentrated load and zero otherwise. It also means that the distributed load is valid at the application point and after but zero before that.

The general form of the *singularity function* representing beam loading is governed by the following relations:

$$\langle x - x_0\rangle^n = \begin{cases} (x - x_0)^n & \text{for} \quad n > 0 \ \text{and} \ x \ge x_0 \\ 0 & \text{for} \quad n > 0 \ \text{and} \ x < x_0 \end{cases} \tag{6.35}$$

$$\langle x - x_0\rangle^0 = \begin{cases} 1 \ \text{for} & n = 0 \ \text{and} \ x \ge x_0 \\ 0 \ \text{for} & n = 0 \ \text{and} \ x < x_0 \end{cases} \tag{6.36}$$

$$\int \langle x - x_0\rangle^n = \frac{1}{n+1}\langle x - x_0\rangle^{n+1} + C, \quad n \ge 0 \tag{6.37}$$

$$\frac{d}{dx}\langle x - x_0\rangle^n = n\langle x - x_0\rangle^{n-1}, \quad n \ge 1 \tag{6.38}$$

where x_0 is the position at which the load is applied. Equation (6.38) is useful in finding the minimum or maximum quantity in question such as locating the position of the maximum bending or maximum deflection.

Different loading is then identified by the following relations that are also depicted in Table 6.1 with their integrals:

$\boldsymbol{q}_i(x) = \boldsymbol{w}_i\langle x - x_0\rangle^1$, linearly increaing distributed load starting at x_0

$$\boldsymbol{q}_i(x) = \boldsymbol{w}_i\langle x - x_0\rangle^0, \text{ uniformally distributed load starting at } x_0 \tag{6.39}$$

$$\boldsymbol{q}_i(x) = \boldsymbol{F}_i\langle x - x_0\rangle^{-1}, \text{ concentrated force at } x_0$$

$$\boldsymbol{q}_i(x) = M_i\langle x - x_0\rangle^{-2}, \text{ concentrated moment at } x_0$$

In this equation, the load is assumed positive as indicated in Table 6.1. If loads are negative (downward or clockwise moments), one needs to use the negative sign. A general beam having supports at $x = 0$ and $x = l$ with a loading case of several concentrated and distributed forces in addition to concentrated moments can take the following form of a *single function*:

$$\boldsymbol{q}_i(x) = \boldsymbol{R}_1\langle x - 0\rangle^{-1} + \sum \boldsymbol{w}_i\langle x - x_{\boldsymbol{w}_i}\rangle^0 + \sum \boldsymbol{F}_i\langle x - x_{\boldsymbol{F}_i}\rangle^{-1} + \sum \boldsymbol{M}_i\langle x - x_{\boldsymbol{M}_i}\rangle^{-2} + \boldsymbol{R}_2\langle x - l\rangle^{-1} \tag{6.40}$$

where the locations of the loads are within or outside the supports. One may also add support moments $\boldsymbol{M}_1$ and or $\boldsymbol{M}_2$ if the supports are restrained from rotation. If the distributed load ends at some location, we need to negate its effect by a negative value starting at that end. The inclusion of all reactions means that the *free body di*agram is the model used in that regard.

Employing Eqs. (6.40), one can write this *single function* of the beam loading with as many loads as those applied to the beam and for its supports. The reactions at the supports are included in that single beam loading equation. Performing successive integration of this equation defines the *shear* $\boldsymbol{V}$, the *bending moment* $\boldsymbol{M}$, the *slope* θ, and the *deflection* v. This alleviates the need of previous *method of direct integration* to divide the beam into regions or zones and thus should satisfy boundary conditions at the region's boundaries. If a beam has three different intermediate load types, one needs to consider four different regions of the beam between reactions and each loading on the beam in the previous classical section 6.1.3.1 of *deflection by integration*. Each region has two boundary conditions, which needs eight constants to evaluate. Even though, there are two reactions and 2×3 matching conditions between regions, one should then have to solve the eight equations simultaneously for the constants. It is taxing

Table 6.1 Singularity functions for some of the mostly used and main loading cases.

Loading	Form	Singularity function	Integral $\int q_i\, dx$
Linearly increasing distributed load	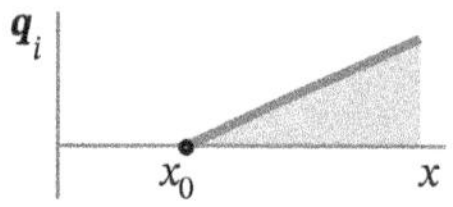	$\boldsymbol{q}_i = \boldsymbol{w}_i\langle x - x_0\rangle^1$	$((\boldsymbol{w}_i\langle x - x_0\rangle^2)/3) + C$
Uniformly distributed load	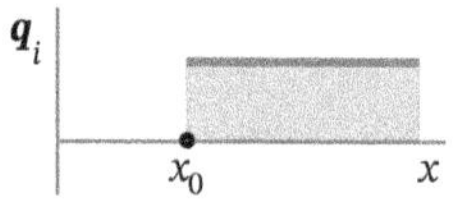	$\boldsymbol{q}_i = \boldsymbol{w}_i\langle x - x_0\rangle^0$	$((\boldsymbol{w}_i\langle x - x_0\rangle^1)/2) + C$
Concentrated force	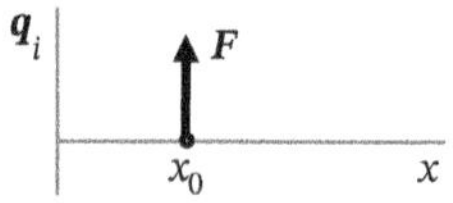	$\boldsymbol{q}_i = \boldsymbol{F}_i\,\langle x - x_0\rangle^{-1}$	$(\boldsymbol{F}_i\,\langle x - x_0\rangle^0) + C$
Concentrated moment	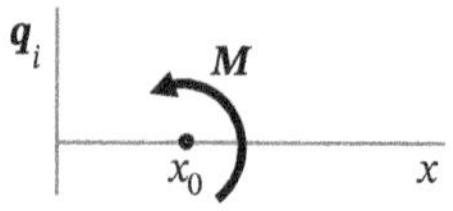	$\boldsymbol{q}_i = \boldsymbol{M}_i\langle x - x_0\rangle^{-2}$	$(\boldsymbol{M}_i\langle x - x_0\rangle^{-1}) + C$

to solve such a number of equations simultaneously. Therefore the singularity function approach is advantageous in that regard. The procedure is best clarified by the following example:

Example 6.4 A simply supported beam of length l is subjected to a mid-span load $\boldsymbol{F}_i$ in addition to a uniformly distributed load $\boldsymbol{w}_i$ as for the case of Example 6.3. It is required to find the reactions. Use the singularity function method to find the shear force, bending moment diagrams, the slope, and deflection along the beam. If the force $\boldsymbol{F} = -40$ [kN] or -9 [klb], the uniformly distributed load $\boldsymbol{w}_i = -9$ [kN/m] or -50 [lb/in] the length $l = 1.0$ [m] or 40 [in], and the modulus of elasticity $E = 207$ [GPa] or 30 [Mpsi], what are the maximum bending moment and slope in the uniform circular section profile of a diameter $d_C = 0.1$ [m] or4 [in]?

Solution

Data: $\boldsymbol{F} = -40$ [kN] or -9 [klb], $l = 1.0$ [m] or 40 [in], $d_C = 0.1$ [m] or 4 [in], and $\boldsymbol{w}_i = -9$ [kN/m] or -50 [lb/in]. From Example 6.3, the values of $E\,I_z$ are obtained from $E = 207$ [GPa] or 30 [Mpsi], the second area moment $I_z = 4.91\ (10^{-6})\ [\text{m}^4]$ or 12.566 [in^4], $E\,I_z = 1\,016\,370$ [Pa m^4], or 376 980 000 [psi in^4].

The reactions are calculated according to Figure 6.5a,c. This gives the reactions at the supports 1 and 2 as $\boldsymbol{R}_1 = \boldsymbol{R}_2 = -(-\tfrac{1}{2}\boldsymbol{F} - \tfrac{1}{2}\boldsymbol{w}l) = \tfrac{1}{2}40(10^3) + \tfrac{1}{2}\,9(10^3)\,(1) = 24.5\,(10^3)$ [N] $= 24.5$ [kN] or $\boldsymbol{R}_1 = \boldsymbol{R}_2 = -(-\tfrac{1}{2}\boldsymbol{F} - \tfrac{1}{2}\boldsymbol{w}l) = \tfrac{1}{2}\,9(10^3) + \tfrac{1}{2}\,50\,(40) = 5500$ [lb].

The solution is enacted by writing the general singularity loading ***including the reactions*** as defined in Eq. (6.40). The inclusion of all loadings and reactions means that the free body diagram is the model used in that regard. This gives the general loading as

$$\boldsymbol{q}_i(x) = \boldsymbol{R}_1\langle x-0\rangle^{-1} + \boldsymbol{w}_i\langle x-0\rangle^{0} + \boldsymbol{F}_i\langle x-\frac{1}{2}l\rangle^{-1} + \boldsymbol{R}_2\langle x-l\rangle^{-1} \tag{a}$$

Substituting the values of the loads (with their vectorial directions) and the previously found reactions gives

$$\begin{aligned} \boldsymbol{q}(x)|_{\text{SI}} &= 24\,500\langle x\rangle^{-1} - 9000\langle x\rangle^{0} - 40\,000\langle x-0.5\rangle^{-1} + 24\,500\langle x-1.0\rangle^{-1} \\ \boldsymbol{q}(x)|_{\text{US}} &= 5500\langle x\rangle^{-1} - 50\langle x\rangle^{0} - 9000\langle x-20\rangle^{-1} + 5500\langle x-40\rangle^{-1} \end{aligned} \tag{b}$$

Performing the integration of $\boldsymbol{q}_i\,(x)$ once (according to Eq. (6.23)) to get the shear $\boldsymbol{V}(x)$, which produces

$$\begin{aligned} \boldsymbol{V}(x)|_{\text{SI}} &= -(24\,500\langle x\rangle^{0} - 9000\langle x\rangle^{1} - 40\,000\langle x-0.5\rangle^{0} + 24\,500\langle x-1.0\rangle^{0}) \\ \boldsymbol{V}(x)|_{\text{SI}} &= -(5500\langle x\rangle^{0} - 50\langle x\rangle^{1} - 9000\langle x-20\rangle^{0} + 5500\langle x-40\rangle^{0}) \end{aligned} \tag{c}$$

Since we have included the reactions in the general loading $\boldsymbol{q}_i(x)$, there is no need to consider the constant of integration C (or C_1 in Eq. (6.23)) in the integration process. If it was included, it turns out to be zero. Also if we omit the reaction $\boldsymbol{R}_2$, it will not affect the solution, since any x will precede the right reaction, i.e. no contribution will present for $\boldsymbol{R}_2$.

Performing the integration of $\boldsymbol{V}(x)$ once (according to Eq. (6.24)) to get the moment $\boldsymbol{M}(x)$ such that

$$\begin{aligned} \boldsymbol{M}(x)|_{SI} &= 24\,500\langle x\rangle^{1} - \frac{9000}{2}\langle x\rangle^{2} - 40\,000\langle x-0.5\rangle^{1} + 24\,500\langle x-1.0\rangle^{1} \\ \boldsymbol{M}(x)|_{US} &- 5500\langle x-0\rangle^{1} - \frac{50}{2}\langle x\rangle^{2} - 9000\langle x-20\rangle^{1} + 5500\langle x-40\rangle^{1} \end{aligned} \tag{d}$$

Since we have included the reactions (if there are moments due to fixed supports) in the general loading $\boldsymbol{q}_i(x)$, there is no need to consider the constant of integration C (or C_2 in Eq. (6.24)) in the integration process. Note that in Eq. (d), the only term that is divided by $(n+1)$ is the distributed load. This is because it has an exponent $n \geq 0$ in the shear force expression $\boldsymbol{V}(x)$ of Eq. (c).

Performing the integration of $\boldsymbol{M}(x)$ once (according to Eq. (6.25)) to get the slope $\theta(x)$ in a modified form as

$$EI_z\ \theta(x)|_{SI} = \frac{24\,500}{2}\langle x\rangle^2 - \frac{4500}{3}\langle x\rangle^3 - \frac{40\,000}{2}\langle x-0.5\rangle^2 + \frac{24\,500}{2}\langle x-1.0\rangle^2 + C_3$$
$$EI_z\ \theta(x)|_{US} = \frac{5500}{2}\langle x\rangle^2 - \frac{25}{3}\langle x\rangle^3 - \frac{9000}{2}\langle x-20\rangle^2 + \frac{5500}{2}\langle x-40\rangle^2 + C_3 \tag{e}$$

Note that all terms are divided by $(n+1)$. All have exponents $n \geq 0$ in the bending moment expression $\boldsymbol{M}(x)$ in Eq. (d). The constant of integration C_3 is present to account for the boundary conditions that are defined latter on.

To get the deflection $v(x)$ for the full span, one integrates $\theta(x)$ by applying Eq. (6.26) once so that

$$EI_z v(x)|_{SI} = \frac{12\,250}{3}\langle x\rangle^3 - \frac{1500}{4}\langle x\rangle^4 - \frac{20\,000}{3}\langle x-0.5\rangle^3 + \frac{12\,250}{3}\langle x-1.0\rangle^3 + C_3x + C_4$$
$$EI_z v(x)|_{US} = \frac{2750}{3}\langle x\rangle^3 - \frac{8.333}{4}\langle x\rangle^4 - \frac{4500}{3}\langle x-20\rangle^3 + \frac{2750}{3}\langle x-40\rangle^3 + C_3x + C_4 \tag{f}$$

Next, one should apply the following boundary conditions to define the integration constants C_3 and C_4 such that

$$\text{At } x=0,\ v=0,\ \text{which gives } C_4|_{SI,US} = 0 \tag{g}$$

$$\text{At } x|_{SI}=1,\ v=0,\ \text{which gives } 0 = -(4083.3\langle 1\rangle^3 - 375\langle 1\rangle^4 - 6666.7\langle 1-0.5\rangle^3 + 4083.3\langle 0\rangle^3 + C_3(1))$$
$$C_3 = -(4083.3 - 375 - 833.375) = -2875 \tag{h}$$

$$\text{At } x|_{US}=40,\ v=0,\ \text{which gives } 0 = -(916.667\langle 40\rangle^3 - 2.083\,33\langle 40\rangle^4 - 1500\langle 40-20\rangle^3 + 916.67\langle 0\rangle^3 + C_3(40))$$
$$C_3 = -(58\,666\,666 - 5\,333\,333 - 12\,000\,000)/40 = -1\,033\,333 \tag{i}$$

The beam slope and deflection are then given by the following reiterated equations including the constants of integration:

$$EI_z\ \theta(x)|_{SI} = \frac{24500}{2}\langle x\rangle^2 - \frac{4500}{3}\langle x\rangle^3 - \frac{40\,000}{2}\langle x-0.5\rangle^2 + \frac{24\,500}{2}\langle x-1.0\rangle^2 - 2875$$
$$EI_z\ \theta(x)|_{US} = \frac{5500}{2}\langle x\rangle^2 - \frac{25}{3}\langle x\rangle^3 - \frac{9000}{2}\langle x-20\rangle^2 + \frac{5500}{2}\langle x-40\rangle^2 - 1\,033\,333 \tag{j}$$

$$EI_z\ v(x)|_{SI} = \frac{12\,250}{3}\langle x\rangle^3 - \frac{1500}{4}\langle x\rangle^4 - \frac{20\,000}{3}\langle x-0.5\rangle^3 + \frac{12\,250}{3}\langle x-1.0\rangle^3 - 2875x$$
$$EI_z\ v(x)|_{US} = \frac{2750}{3}\langle x\rangle^3 - \frac{8.3333}{4}\langle x\rangle^4 - \frac{4500}{3}\langle x-20\rangle^3 + \frac{2750}{3}\langle x-40\rangle^3 - 1\,033\,333x \tag{k}$$

As a special check and verification, it is apparent from symmetry that $\theta(l/2) = 0$. Substituting for $x = 0.5$ [m] or 20 [in] gives the following values:

$$EI_z\ \theta(0.5) = \frac{24\,500}{2}\langle 0.5\rangle^2 - \frac{4500}{3}\langle 0.5\rangle^3 - 2875 = 0$$
$$EI_z\ \theta(20) = \frac{5500}{2}\langle 20\rangle^2 - \frac{25}{3}\langle 20\rangle^3 - 1\,033\,333 = 0.3333 \approx 0 \tag{l}$$

The maximum bending moment can be obtained by differentiating Eq. (d) and equate to zero. Equation (6.38) can be used in that process. It is also apparent from Figure 6.6c or from Example 6.3 that the maximum bending moment occurs at mid-span. Substituting into Eq. (k), one finds

$$\boldsymbol{M}(0.5) = 24\,500\langle 0.5\rangle^1 - \frac{9000}{2}\langle 0.5\rangle^2 - 40\,000\langle 0.5-0.5\rangle^1 = 11\,125\ [\text{N m}]$$
$$\boldsymbol{M}(20) = 5500\langle 20\rangle^1 - \frac{50}{2}\langle 20\rangle^2 - 9000\langle 20-20\rangle^1 = 100\,000[\text{lb in}] \tag{m}$$

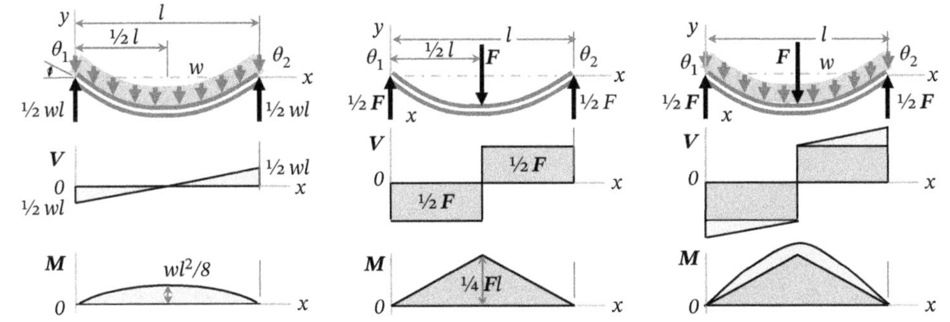

Figure 6.6 Beam solution by superposition. Uniformly distributed loaded beam in (a) is added to a concentrated loaded beam in (b) to get the end solution of both loading in (c).

Maximum slope θ_{max} is obtained from equation (j) at $x = 0$ or at $x = l$.

$$EI_z\ \theta(1) = \frac{24\,500}{2}\langle 1\rangle^2 - \frac{4500}{3}\langle 1\rangle^3 - \frac{40\,000}{2}\langle 1-0.5\rangle^2 - 2875 = 2875$$
$$EI_z\ \theta(40) = \frac{5500}{2}\langle 40\rangle^2 - \frac{25}{3}\langle 40\rangle^3 - \frac{9000}{2}\langle 40-20\rangle^2 - 1\,033\,333 = 1\,033\,333 \tag{n}$$

$$\theta(1) = \frac{2875}{EI_z} = \frac{2875}{1\,016\,370} = 0.002\,828\,7\ [\text{rad}] = 0.1621\ [°]$$
$$\theta(40) = \frac{1\,033\,333}{EI_z} = \frac{1\,033\,333}{376\,980\,000} = 0.002\,741\,1\ [\text{rad}] = 0.157\ [°] \tag{o}$$

These are almost the same results of Example 6.3, Eq. (g).

Example 6.5 A statically indeterminate beam shown in Figure 6.7 is fixed at both ends and loaded by an asymmetrical concentrated downward force $\boldsymbol{F}$. The beam of length l is subjected to the concentrated force $-\boldsymbol{F}$ at a distance a off the left support and a distance b off the right support. Use the singularity function method to find the shear force diagram, the bending moment diagram, the slope, and deflection along the beam. Find the reactions and moments at the supports.

Solution

Data: $\boldsymbol{F}$ is in the negative direction, and reactions are assumed positive.

The problem is statically indeterminate since the number of reactions is greater than the number of equilibrium equations. Therefore, additional deformation conditions or constraints are used to provide the additionally needed equations.

Applying the singularity function defined in Eq. (6.40) in addition to the inclusion of the moments at the supports gives

$$\boldsymbol{q}_i(x) = -\boldsymbol{M}_1\langle x\rangle^{-2} + \boldsymbol{R}_1\langle x\rangle^{-1} - \boldsymbol{F}_i\langle x-a\rangle^{-1} + \boldsymbol{R}_2\langle x-l\rangle^{-1} + \boldsymbol{M}_2\langle x-l\rangle^{-2} \tag{a}$$

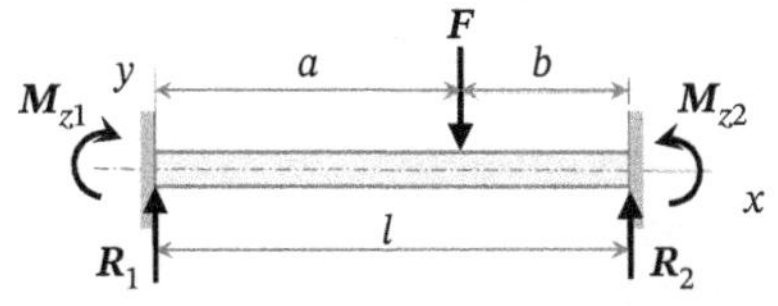

Figure 6.7 Statically indeterminate beam that is fixed at both ends and loaded by an asymmetrical concentrated force.

Performing the successive integration in Eqs. (6.23)–(6.26) and (6.37) in addition to considering C_1 and C_2 to have been taken care of due to the inclusion of all reactions, one gets

$$V(x) = \boldsymbol{M}_1\langle x\rangle^{-1} - \boldsymbol{R}_1\langle x\rangle^{0} + \boldsymbol{F}\langle x-a\rangle^{0} - \boldsymbol{R}_2\langle x-l\rangle^{0} - \boldsymbol{M}_2\langle x-l\rangle^{-1}$$

$$M(x) = -\boldsymbol{M}_1\langle x\rangle^{0} + \boldsymbol{R}_1\langle x\rangle^{1} - \boldsymbol{F}\langle x-a\rangle^{1} + \boldsymbol{R}_2\langle x-l\rangle^{1} + \boldsymbol{M}_2\langle x-l\rangle^{0}$$

$$\theta(x) = \left(-\boldsymbol{M}_1\langle x\rangle^{1} + \frac{\boldsymbol{R}_1}{2}\langle x\rangle^{2} - \frac{\boldsymbol{F}}{2}\langle x-a\rangle^{2} + \frac{\boldsymbol{R}_2}{2}\langle x-l\rangle^{2} + \boldsymbol{M}_2\langle x-l\rangle^{1} + C_3\right)/EI_z \tag{b}$$

$$v(x) = \left(-\frac{\boldsymbol{M}_1}{2}\langle x\rangle^{2} + \frac{\boldsymbol{R}_1}{6}\langle x\rangle^{3} - \frac{\boldsymbol{F}}{6}\langle x-a\rangle^{3} + \frac{\boldsymbol{R}_2}{6}\langle x-l\rangle^{3} + \frac{\boldsymbol{M}_2}{2}\langle x-l\rangle^{2} + C_3x + C_4\right)/EI_z$$

The boundary conditions at $x = 0$ gives the following:

$$\begin{aligned} \theta(0) &= 0 \quad \rightarrow \quad C_3 = 0 \\ v(0) &= 0 \quad \rightarrow \quad C_4 = 0 \end{aligned} \tag{c}$$

The boundary conditions at $x = l$ gives the following:

$$\theta(l) = 0 \quad \rightarrow \quad -\boldsymbol{M}_1 l + \frac{\boldsymbol{R}_1}{2}l^2 - \frac{\boldsymbol{F}}{2}b^2 = 0 \tag{d}$$

$$v(l) = 0 \quad \rightarrow \quad -\frac{\boldsymbol{M}_1}{2}l^2 + \frac{\boldsymbol{R}_1}{6}l^3 - \frac{\boldsymbol{F}}{6}b^3 = 0 \tag{e}$$

By solving Eq. (b) and Eq. (e) simultaneously and after some manipulations, one gets

$$\boldsymbol{M}_1 = -\frac{Fb^2a}{l^2} \quad \text{and} \quad \boldsymbol{R}_1 = \frac{Fb^2}{l^3}(3a+b) \tag{f}$$

Summing the forces in y direction and equate to zero gives

$$\boldsymbol{R}_2 = \frac{Fa^2}{l^3}(3b+a) \tag{g}$$

Summing the moments about either end and equate to zero gives

$$\boldsymbol{M}_2 = -\frac{Fa^2b}{l^2} \tag{h}$$

The process of solution by the singularity function is systematic and can be simply programmed on a computer. The extension of the beam beyond the right support would only need one to add whatever load type after that. One may also need to subtract a uniformly distributed load if it does not exist beyond the right support. Recalculations of the reactions and the integration constants are, however, needed to finalize the solution.

6.1.3.4 Deflection by Other Methods

Numerical integration is also a possible and attractive technique for finding the deflections by trapezoidal or *Simpson's* rule, Chapra (2006). Other methods have also been used to evaluate beam deflections such as the *moment-area method*, Beer et al. (2009). Nowadays, the readily available ***FE*** programs (see Section 6.11) supersedes these methods in efforts and time.

6.1.4 Simple Beam Synthesis

As a mathematical model representing the machine member to be designed, the simply supported beam with mid-span load is the more conservative model. Therefore the beam synthesis is more sensible to be implemented through such a model. Adjusting the design to be fairly evaluated rather than conservatively adapted is evidently attained. In this particular treatment, one may employ only the major loading such as pure bending rather than other loading conditions. This synthesis is considered as a jump-start for the intended final design that should account for other real loading conditions.

Consider a simply supported beam where a concentrated load is applied at the middle of the beam span l_B as shown in Figure 6.5a. As indicated previously, the maximum stress according to Eq. (6.13) for a cylindrical cross-section of a diameter d_C is given by

$$\sigma_{x,\max} = \pm\frac{\boldsymbol{M}_z c_S}{I_z} = \pm\frac{(\boldsymbol{F}l_B/4)(d_C/2)}{\pi d_C^{\,4}/64} = \pm\frac{8\boldsymbol{F}l_B}{\pi d_C^{\,3}} \tag{6.41}$$

If this maximum stress is permitted to reach an allowable stress limit such as $\sigma_{x,\max} = \sigma_{\text{all}}$, the synthesized diameter should then be

$$d_C = \sqrt[3]{\frac{8\boldsymbol{F}l_B}{\pi\sigma_{x,\max}}} = \sqrt[3]{\frac{8\boldsymbol{F}l_B}{\pi\sigma_{\text{all}}}} \tag{6.42}$$

If we assume that the *failure* of the beam material starts at the maximum stress reaching the material ***yield strength*** S_y, the ***allowable stress*** σ_{all} should be lower than that by a safety factor such that

$$\sigma_{x.\max} = \sigma_{\text{all}} = \frac{S_y}{K_{\text{SF}}} \tag{6.43}$$

In Eq. (6.43), K_{SF} is the ***safety factor*** to guard against reaching this elementary *failure mode;* see Section 7.7. The synthesized diameter in Eq. (6.42) becomes

$$d_C = \sqrt[3]{\frac{8K_{\text{SF}}\boldsymbol{F}l_B}{\pi S_y}} \tag{6.44}$$

If we conservatively assume a material of a yield strength $S_y = 200$ [MPa] = 200(10^6) [Pa] = 29 [kpsi] for a yield failure prevention, the synthesized beam diameter should then be

$$\begin{aligned} d_C &= \sqrt[3]{\frac{8K_{\text{SF}}\boldsymbol{F}l_B}{\pi 200(10^6)}}, \quad \text{SI} \\ d_C &= \sqrt[3]{\frac{8K_{\text{SF}}\boldsymbol{F}l_B}{\pi 29(10^3)}}, \quad \text{US} \end{aligned} \tag{6.45}$$

A chart is developed to synthesize the beam (or shaft) diameter d_C using Eq. (6.45) for a wide range of applied force $\boldsymbol{F}$ and beam length l_B. The chart is shown in Figure 6.6 for US and SI units. They define the diameter of the circular beam cross section for a maximum yield strength of 200 [MPs] or 29 [kpsi]. Beam material is usually a structural steel of designation such as (see Section 7.4.3) ASTM A283; AISI 1010; DIN St.37, St.42, C10; and ISO 630 Fe 37-A. The least of these materials has a yield strength of about 200 [MPs] or 29 [kpsi], such as DIN St. 37 or the equivalent ASTM A283 or so. Adjusting for a specific material is simply made into the evaluation of the proper safety factor K_{SF}.

Example 6.6 A simply supported beam of length l_B is subjected to a mid-span load $\boldsymbol{F}_i$ as for the case of Example 6.2. If the force $\boldsymbol{F} = -40$ [kN] or −9 [klb], the length $l = 1.0$ [m] or 40 [in], and the modulus of elasticity $E = 207$ [GPa] or 30 [Mpsi], what is the synthesized beam diameter for a safety factor K_{SF} of 2.0 and assuming a material of a yield strength $S_y = 200$ [MPa] = 200(10^6) [Pa] = 29 [kpsi] for a yield failure prevention? What is the safety factor if the selected material strength is doubled for the same beam diameter?

Solution

Data: $\boldsymbol{F} = -40$ [kN] or −9 [klb] and the length $l_B = 1.0$ [m] or 40 [in], $K_{\text{SF}} = 2.0$ and $S_y = 200$ [MPa] = 200(10^6) [Pa] = 29 [kpsi].

Using Eqs. (6.44) and (6.45) for the diameter syntheses, one gets

$$d_C = \sqrt[3]{\frac{8K_{\mathrm{SF}}\boldsymbol{F}l_B}{\pi 200(10^6)}} = \sqrt[3]{\frac{8(2.0)(40)(10^3)(1)}{\pi 200(10^6)}} = 0.1006\ [\mathrm{m}] \cong 100\ [\mathrm{mm}]$$

$$d_C = \sqrt[3]{\frac{8K_{\mathrm{SF}}\boldsymbol{F}l_B}{\pi 29(10^3)}} = \sqrt[3]{\frac{8(2.0)(9)(10^3)(40)}{\pi 29(10^3)}} = 3.984\ [\mathrm{in}] \cong 4.0\ [\mathrm{in}] \tag{a}$$

By consulting Figure 6.8 and using $K_{\mathrm{SF}}\,\boldsymbol{F} = 80\,000$ [N] or 18 000 [lb], one gets the synthesized beam diameter d_C at $l_B = 1$ [m] or 40 [in] as $d_C \approx 100$ [mm] and 4 [in]. These diameters are the same as the calculated values.

The chart is faster, but accuracy is an issue. However, diameters are rounded to the next standard diameter; see Appendix A.4.

For using a material that has double the strength and using Eqs. (6.41) and (6.43), we get the following relation for the safety factor:

$$\sigma_{x,\max} = \frac{8\boldsymbol{F}l_B}{\pi {d_C}^3} = \frac{S_y}{K_{\mathrm{SF}}} \quad \text{or} \quad K_{\mathrm{SF}} = S_y \frac{\pi {d_C}^3}{8\boldsymbol{F}l_B} \tag{b}$$

With all the terms to the right of S_y being constant, the doubling of material strength should double the safety factor. For our case the safety factor becomes $K_{\mathrm{SF}} = 4.0$ instead of $K_{\mathrm{SF}} = 2.0$. From Eq. (b), however, changing the diameter by 10% should increase the factor of safety by about 33% since $(1.1d_C)^3 = 1.331(d_C)^3$.

6.1.5 Comparing Stresses and Deflections in Beams

In this section, a comparison between beams with different loading and free, simply supported or fixed supports, i.e. different end conditions, is presented. A coefficient C_{EB} is deduced that represents an *equivalent beam* factor defining the comparable behavior to the simply supported beam with mid-span concentrated load and circular cross section (Figure 6.8a,b).

6.1.5.1 Beam Stresses

i. Consider a ***simply supported*** beam with *concentrated* load $\boldsymbol{F}$ at the middle of the span l_B. The maximum tensile stress $\sigma_{x\max}$ is at mid-span or at $l_B/2$ and is given by (see Eq. (6.41) and Figure 6.5a)

$$\sigma_{x,\max} = \frac{\boldsymbol{M}_z c_s}{I_z} = \frac{(\boldsymbol{F}l_B/4)(d_C/2)}{I_z} = \frac{\boldsymbol{F}l_B d_C}{8I_z} = \frac{\boldsymbol{F}l_B d_C}{8\pi d_C^4/64} = \frac{8\boldsymbol{F}l_B}{\pi d_C^3} \tag{6.46}$$

where $\boldsymbol{M}_z = \boldsymbol{F}l_B/4$ is the maximum bending moment, d_C is the diameter of the circular cross section, and the second area moment of the circular cross section is $I_z = \pi {d_C}^4/64$.

ii. For a ***fixed-free cantilever*** beam of the same length l_B, fixed at the left end and the *concentrated* load $\boldsymbol{F}$ is at the tip of the right end, the maximum stress at the fixed end is

$$\sigma_{x,\max} = \frac{\boldsymbol{M}_z d_C}{2I_z} = \frac{\boldsymbol{F}l_B d_c}{2I_z} = \frac{C_{\mathrm{EB}}\boldsymbol{F}l_B d_C}{8I_z} \tag{6.47}$$

where $C_{\mathrm{EB}} = 4.0$, which means that a cantilever beam develops four times the maximum stress in a simply supported beam of the same span considering the load at the end of the span and not at mid-span. However, if the location of the load with respect to the left support is kept at $l/2$, which is the usual load distance, the equivalent beam coefficient C_{EB} will be 2.0. This is the logically adapted case.

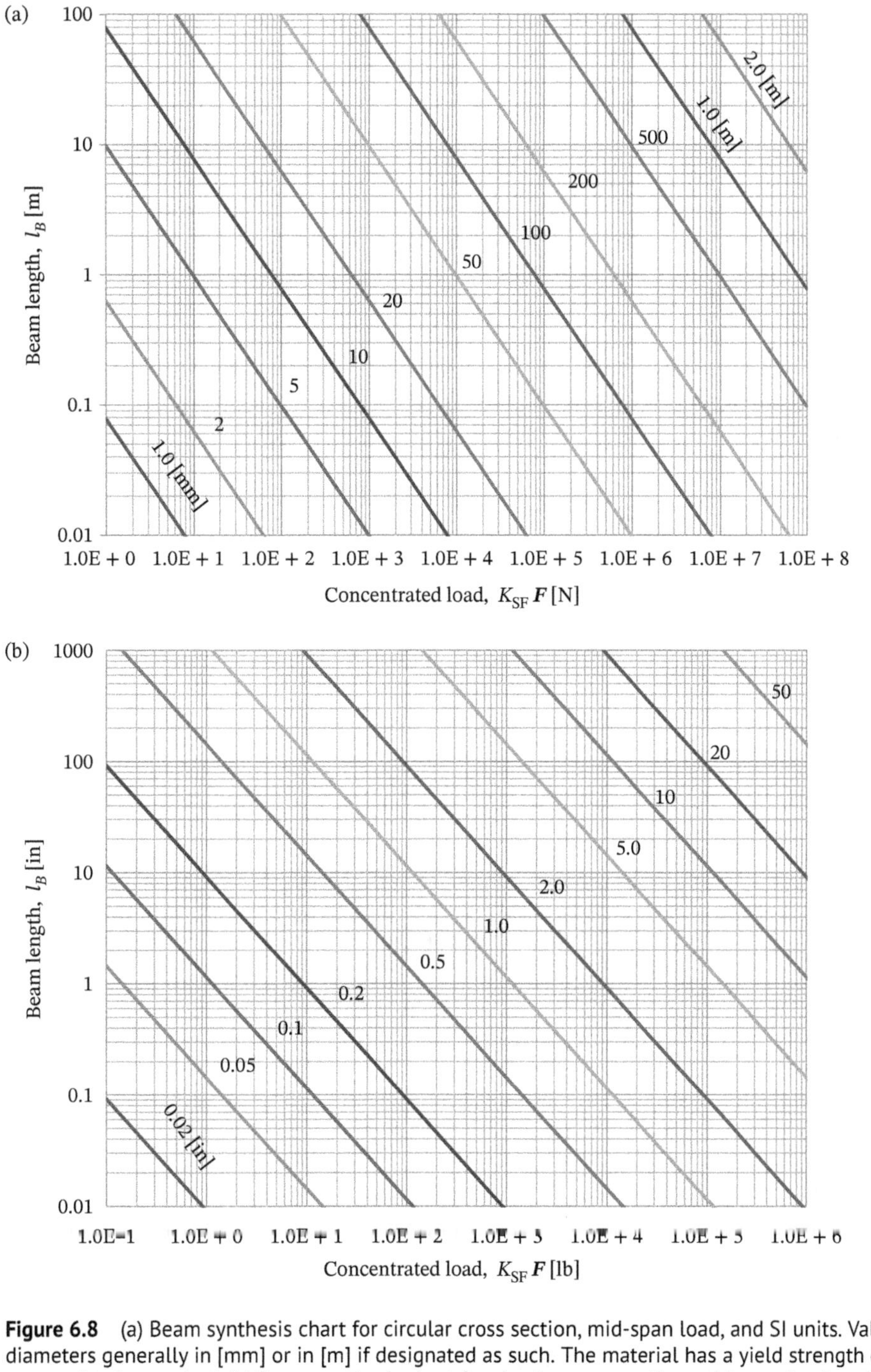

Figure 6.8 (a) Beam synthesis chart for circular cross section, mid-span load, and SI units. Values on the chart are for diameters generally in [mm] or in [m] if designated as such. The material has a yield strength of 200 [MPs], which is assumed to be reached. (b) Beam synthesis chart for circular cross section, mid-span load, and US units. Values on the chart are for diameters generally in [in] or in [in] if designated as such. The material has a yield strength of 29 [kpsi], which is assumed to be reached.

iii. When the load on a *simply supported* beam is *uniformly distributed* such that the distributed beam load $\boldsymbol{w} = \boldsymbol{F}/l_B$, the maximum stress is given by (see Figure 6.5c)

$$\sigma_{x,\max} = \frac{\boldsymbol{M}_z d_C}{2I_z} = \frac{\boldsymbol{w} l_B^2 d_C}{8(2I_z)} = \frac{\boldsymbol{F} l_B d_C}{16 I_z} = \frac{C_{EB} \boldsymbol{F} l_B d_C}{8 I_z} \tag{6.48}$$

where $C_{EB} = 1/2 = 0.5$, which indicates that a distributed load reduces the stresses by two times.

iv. For a *fixed-free cantilever* beam of the same length and the load is *uniformly distributed* such that the distributed beam load $\boldsymbol{w} = \boldsymbol{F}/l_B$,, the maximum stress is (see Figure 6.5d)

$$\sigma_{x,\max} = \frac{\boldsymbol{M}_z d_C}{2I_z} = \frac{\boldsymbol{w} l_B^2 d_C}{2(2I_z)} = \frac{\boldsymbol{F} l_B d_C}{4 I_z} = \frac{C_{EB} \boldsymbol{F} l_B d_C}{8 I_z} \tag{6.49}$$

where $C_{EB} = 2.0$, which indicates that a distributed load over the cantilever beam develops two times the maximum stress in a simply supported beam of the same span considering the load at the end of the span and not at mid-span.

6.1.5.2 Beam Deflection

i. Consider ***simply supported*** beam with *concentrated* load $\boldsymbol{F}$ at the middle of the span l_B. The maximum deflection $v_{\max}$ is at mid-span or at ½ l_B and is given by (see Example 6.2)

$$v_{\max} = \frac{\boldsymbol{F} l_B^3}{48 E I_z} \tag{6.50}$$

ii. For a ***fixed-free cantilever*** beam of the same length and the *concentrated* load $\boldsymbol{F}$ is at the tip, the maximum deflection $v_{\max}$, according to Eq. (6.32), is

$$v_{\max} = \frac{\boldsymbol{F} l_B^3}{3 E I_z} = \frac{C_{EB} \boldsymbol{F} l_B^3}{48 E I_z} \tag{6.51}$$

where $C_{EB} = (48)/3 = 16$, which is 16 times as for the deflection of the simply supported beam with the load at mid-span. However, if the location of the load with respect to the left support is kept at $l/2$, which is the usual load distance, the equivalent beam coefficient C_{EB} should be as follows:

$$v_{\max} = \frac{\boldsymbol{F}(l_B/2)^3}{3 E I_z} = \frac{\boldsymbol{F} l_B^3}{8(3 E I_z)} = \frac{C_{EB} \boldsymbol{F} l_B^3}{48 E I_z} \tag{6.52}$$

which gives $C_{EB} = (48)/24 = 2.0$. This is the logically adapted case.

iii. When the load on a *simply supported* beam is *uniformly distributed* such that the distributed beam load $\boldsymbol{w} = \boldsymbol{F}/l_B$, the maximum deflection is given by (see Eq. (6.32))

$$v_{\max} = \frac{5 \boldsymbol{w} l_B^4}{384 E I_z} = \frac{5 \boldsymbol{F} l_B^3}{384 E I_z} = \frac{C_{EB} \boldsymbol{F} l_B^3}{48 E I_z} \tag{6.53}$$

where $C_{EB} = 5(48)/384 = 0.625$. This indicates that the deflection is smaller by 0.626 times that for the equivalent concentrated load.

iv. For a *fixed-free cantilever* beam of the same length and the load is *uniformly distributed* such that the distributed beam load $\boldsymbol{w} = \boldsymbol{F}/l_B$,, the deflection as defined by Eq. (6.34) is given by

$$v_{\max} = \frac{\boldsymbol{w} l_B^4}{8 E I_z} = \frac{C_{EB} \boldsymbol{F} l_B^3}{48 E I_z} \tag{6.54}$$

where $C_{EB} = 48/8 = 6$, which indicates that a distributed load over the cantilever beam does increase the maximum deflection by a factor of 6 over the simply supported beam with a concentrated middle load $\boldsymbol{F}$.

6.1.5.3 Equivalent Loads on Simple Beams

Consider a simply supported beam where the load location is shifted toward the left support. The distance from the support is a as shown in Figure 6.5b. To get the same *maximum bending moment*, an *equivalent force* $\boldsymbol{F}_e$ at the mid-span can be identified as obtained in following Eq. (6.55):

$$\boldsymbol{M}_{z,\max} = \frac{\boldsymbol{F}ab}{l_B} = \frac{\boldsymbol{F}_e}{2}\frac{l_B}{2}, \quad \text{which gives} \quad \boldsymbol{F}_e = 4\boldsymbol{F}\left(\frac{ab}{l_B^2}\right) \tag{6.55}$$

It should be noted, however, that the two reactions in that equivalent case will be equal to $\boldsymbol{F}_e/2$, which is different from the values in Figure 6.5b. This process can only be useful in beam synthesis when loads are asymmetrically located.

Similar treatments can be used to develop equivalent simple force $\boldsymbol{F}_e$ for other distributed loading conditions to help in simpler but with an equivalent effect. The equivalence can be for the same maximum stresses or maximum deflections as previously developed for some cases in the preceding 6.1.5.1 and 6.1.5.2 parts of Section 6.1.5. The *equivalent beam* coefficient C_{EB} has been deduced accordingly for the previous cases. One can use C_{EB} to change the load for some other specific objectives.

6.2 Mathematical Model

As previously stated in Chapter 2, Figure 2.1 is redrawn as Figure 6.9a that shows the same assembly of a shaft carrying an element (such as a pully, a sprocket, or a connecting rod) through its hub and supported by two bearings. Recalling the same figure is more useful after the preceding presentation of beams or rods in this chapter. As indicated in Section 2.1 and reiterated herein, the clearance between the hub and the shaft is drawn extremely exaggerated to drive the point through. If the clearance is smaller than the shaft *deflection*, the distribution of the pressure between them might be as shown in the upper section of Figure 6.9a. Under load, the hub will first contact the shaft at the hub's right and left ends, and the deflection of the hub causes the interior to start transferring the load accordingly. That is why the pressure distribution between the hub and the shaft may be modeled as the depicted nonlinear form. To mathematically model the loading on the shaft, one can either assume the load $\boldsymbol{F}$ to be concentrated roughly about the mid-span of the hub or assume it as two loads each as $\boldsymbol{F}/2$ as shown in the

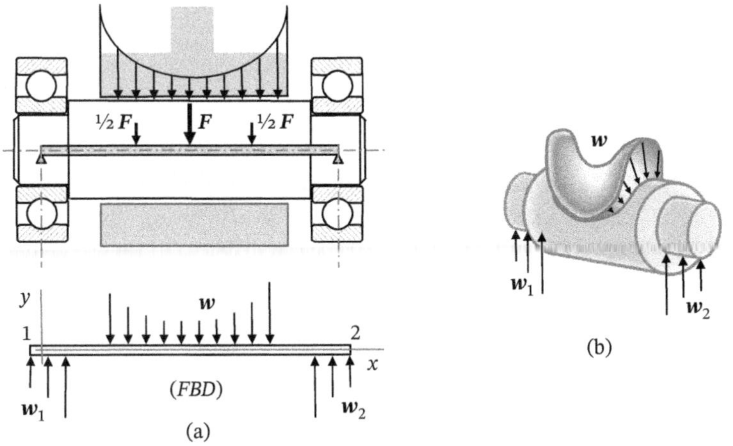

Figure 6.9 An assembly of a shaft carrying an element through its hub and supported by two bearings. One mathematical model is identified in the lower part of the design assembly in (a). The ***3D*** model in (b) shows an assumed pressure distribution w on top of the shaft without similar ***3D*** models for the bearing reactions w_1 and w_2.

upper section of Figure 6.9a. It is more realistic, however, to consider the nonlinear ***3D*** distributed load as shown in Figure 6.9b. The first assumption in Figure 6.9a is more conservative and may be adapted for the initial synthesis. The second assumption in Figure 6.9b might be valid if the clearance is little larger than the anticipated deflection of the shaft and hub.

The assumption that the reactions at the bearings are at the bearing centerline, as shown in the upper section of Figure 6.9a, is another simplified mathematical model. If the angular deflection of the shaft is not very small at the bearing, one might need to model the reaction as non-uniformly distributed load as indicated in the lower section of Figure 6.9a. Initially, the ***mathematical model*** of this assembly may then be defined as a *simply supported beam* with the load $\boldsymbol{F}$ at roughly about the middle as shown in upper part of Figure 6.9a as the grayish bar on the assembly axis. This assumption is approximate and may be adapted for the conservative treatment of this or similar problems. The other little more representative mathematical model is identified in the detached lower part of Figure 6.9a for this assembly. The distributed loads $\boldsymbol{w}_1$ and $\boldsymbol{w}_2$ can be linearly or nonlinearly varying for the right and left supports. The same simplification may be carried out for the middle load $\boldsymbol{w}$. Note that the differences and the equivalences have been discussed in Section 6.1. The needed accuracy should dictate the more appropriate mathematical model to adopt such as the complex loading case shown in Figure 6.9b. The complex ***3D*** model can also be applied to represent a more realistic representation of the reaction loads similar to the loading in Figure 6.9b. This loading is just an expectation for certain deformation and clearance conditions.

Simple models are initially used for synthesis. A more realistic model such as the expected ***3D*** pressure distribution in Figure 6.9b is usually difficult to evaluate. Some complex ***FE*** models may be generated to represent the more realistic ***3D*** model. This can be implemented in the final stages of the design to include all realistically modeled loads or complex contact models of mating elements; see, e.g. Metwalli and Moslehy (1983).

The random nature of the applied loads is also a factor to consider in the mathematical model of loading the design. The variation may be considered as sinusoidal with randomly varying magnitude and/or frequency. This dynamic consideration will be further discussed in Section 7.8.

For the design to be realistic, one should consider all of the aforementioned conditions. As the mathematical model is more sophisticated and considering all possible conditions, the safety factor can be as close to unity as the model approaches the exact representation of the real problem. This is usually the case in aviation industry. If the mathematical model is very conservative, the safety factor can be as close to unity as the conservatism increases. On the other hand, reliability is another form of statistical safety; see, e.g. Section 2.1.3. This reliability consideration is the main adapted theme and vigorously implemented in the field of *probabilistic design*, which is beyond the scope of this text.

6.3 Simple Stresses, Strains, and Deformations

Internal *normal* and *shear* stresses are generated by externally applied loads such as $\boldsymbol{w}_i$, $\boldsymbol{F}_i$, and $\boldsymbol{M}_i$. These stresses are produced by internal bending moments and shear forces as indicated previously. Additionally, the external loads such as axial forces $\boldsymbol{F}_i$ and torques $\boldsymbol{T}_i$ produce internal stresses. These stresses are treated in the following sections.

6.3.1 Uniform Tension and Compression

Uniform tension and compression stresses are produced when the applied force $\boldsymbol{F}_x$ is in the x direction in addition to being normal to the element cross section as indicated in Figure 6.10a for tension and Figure 6.10a,b for compression. The *normal stress* due to the externally applied load in the x direction is given by

$$\sigma_x = \pm\frac{\boldsymbol{F}_x}{A} \qquad (6.56)$$

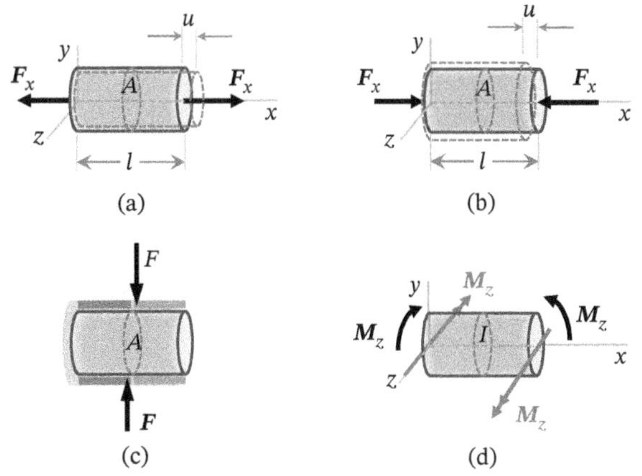

Figure 6.10 Simple cylindrical machine element subjected to (a) forces that cause tension and (b) forces that cause compression. The deformation u under these forces are depicted by the dashed cylinders relative to the original cylinders in both cases of (a) and (b). The loading in (c) is direct shear, and (d) is pure bending. The bending is designated by two methods: the usual angular arrows or the double-arrowed vector.

In Eq. (6.56), $\boldsymbol{\sigma}_x$ is the *normal stress vector* or *tensor* (see Section 6.4.2); $\boldsymbol{F}_x$ is the axial load or the *force* as a vector in the x direction; A is the cross-sectional area normal to the x direction. The stress is normal to what is labeled the x plane, i.e. in the direction x of the normal to the x plane. The area is the circular cross section of the cylinder as in Figure 6.10a or any other uniform shape along the element axis. In Eq. (6.56), the *tensile stress* is the positive sign. The *compressive stress* is the negative one. The direction of the stress vector is the same as the direction of the force vector, and it is usually in the same direction as the member axis or centerline.

The main assumptions in this treatment are as follows:

- The element has the same *uniform section* in the apparent x direction, which is the usual default direction.
- Loading is assumed along the *neutral axis* and not offset to cause an additional bending.
- The calculated stresses in Eq. (6.56) are away from loading so that the concentrated effect of the applied load is only affecting the local region at the load. This is what is called the *Saint-Venant's principle*, Barre de Saint-Venant (1797–1886).

The ***strain*** ε_x in the x direction is described as the deformation with respect to the original length, or

$$\varepsilon_x = \frac{du}{dx} = \pm\frac{u}{l} \tag{6.57}$$

where u is the constant *uniform* deformation in the x direction and l is the length of the member. It should be noted that a negative sign in Eq. (6.57) is denoting the fact that u is in the negative direction for compression, which is understandingly reducing the length l as shown in Figure 6.2b. Utilizing the ***Hook's law*** for materials, i.e. $\boldsymbol{\sigma}_x = E\varepsilon_x$ (see Section 7.1.3), and substituting Eq. (6.56) for the normal stress $\boldsymbol{\sigma}_x = \boldsymbol{F}/A$, one gets

$$\boldsymbol{\sigma}_x = E\varepsilon_x = \pm\frac{E\boldsymbol{u}}{l} = \frac{\boldsymbol{F}_x}{A}$$

or

$$\boldsymbol{u} = \pm\frac{\boldsymbol{F}_x l}{EA} \tag{6.58}$$

where E is the ***modulus of elasticity***, or *Young*'s modulus, $\boldsymbol{F}_x$ is the applied load in x direction, and l is the part length of constant cross-sectional area A. The modulus of elasticity E is the slope of the elastic part of the *stress–strain* curve for the material of the part. The stress–strain curve is developed from the *load-deflection* curve in material testing. The load is divided by the area to get the stress (Eq. (6.56)), and the deformation u is divided by the gauge length to get the strain (Eq. (6.57)). The positive sign in Eq. (6.58) is for tensile deformation and the negative sign is for the compressive deformation.

As observed from Figure 6.10 and to keep a constant volume of the part, the positive *strain* or elongation (stretching) in tension causes a lateral negative strain or contraction along the length of the part. The opposite will happen for compression. Poisson, *Siméon D. Poisson* (1781–1840), has observed that these longitudinal strains ε_x and transverse strains ε_y are proportional to each other. From that, the *Poisson's ratio* ν is defined as follows:

$$\nu = -\frac{\varepsilon_y}{\varepsilon_x} \tag{6.59}$$

One should note that the negative sign is stemming from the fact that when one is positive, the other should be negative. The value of the *Poisson's ratio* varies from 0.25 to 0.35 for most materials or a wider range of 0.1–0.5 for some other materials (e.g. 0.1 for some concretes and 0.5 for rubber).

6.3.2 Direct Uniform Shear

This stress is originally related to the production process of guillotine cutting, shearing, clipping, or nibbling (see Section 2.4.3) or similar arrangements where no cutting occurs. It is schematically shown in Figure 6.10c, where two close cylindrical casings bound the cylindrical element and force it, each in the opposite directions to the other. This is similar to the assembly in Figure 6.9, when the element (such as a connecting rod) occupies the full space between the bearings. The shaft will experience direct shear at its connection with the bearings. The developed stress throughout the cross section is

$$\tau = \frac{\boldsymbol{F}}{A} \tag{6.60}$$

where τ is the *shear stress* in the plane of the cross section, $\boldsymbol{F}$ is the load or shear force, and A is the cross-sectional area. If the element axis is in the x direction, the shear stress will be in the normal yz plane. Since this plane is normal to the x-axis, it is called the x plane due to the fact that its normal is in the x direction. The coordinates on this plane are the y and z coordinates. The shear stress on the x plane can either be in the y direction and thus called τ_{xy} or can be in the z direction and thus called τ_{xz}. In that case, Eq. (6.57) can be rewritten as

$$\tau_{xy} = \frac{\boldsymbol{F}_y}{A} \quad \text{and} \quad \tau_{xz} = \frac{\boldsymbol{F}_z}{A} \tag{6.61}$$

It should be noted that this *direct shear stress* is *uniform* across the section. The *transverse shear* stress in beams due to the internal shear force $\boldsymbol{V}$ is not uniform across the section; see Section 6.3.5.

6.3.3 Pure Bending

This type of loading generates a normal stress that has been discussed previously in Section 6.1.2 for beams or bars under the bending moment $\boldsymbol{M}_z$ as shown in Figure 6.10d. In that figure, an additional grayish double-arrowed vector is shown and can also represent the bending moment $\boldsymbol{M}_z$, which is also in a grayish font color. This representation is used in some references. The direction of the double-arrowed vector is in the z direction for the positive $\boldsymbol{M}_z$ bending moment. This bending moment is acting about the z-axis. In Section 6.1.2 the normal stress due to the bending moment has been found and was represented by Eq. (6.12), which is reiterated as follows:

$$\sigma_x = -\frac{\boldsymbol{M}_z y}{I_z} \tag{6.12}$$

where σ_x is the normal stress, $\boldsymbol{M}_z$ is the bending moment, and I_z is the second area moment about z coordinate. The maximum pure bending occurs at the outer fiber that is at $y = c_S = d_C/2$ for bars of circular cross sections. Observing that the second area moment for the circular cross section is $I_z = \pi d_C{}^4/64$, the maximum normal stress $\sigma_{x,\max}$ (see Eq. (6.13)) is given by

$$\sigma_{x,\max} = \pm\frac{\boldsymbol{M}_z y_{\max}}{I_z} = \pm\frac{\boldsymbol{M}_z c_S}{I_z} = \pm\frac{32\boldsymbol{M}_z}{\pi d_C^3} \tag{6.59}$$

where $I_z/y_{max} = I_z/c_S$ is the bar section modulus $Z_B = \pi d^3/32$. It should be reiterated also that for rectangular cross section of a width b_R and a depth h_R, the second area moment is $I_z = b_R h_R^3/12$, and its section modulus is $Z_R = b_R h_R^2/6$.

Beam curvature in pure bending has also been discussed previously in Section 6.1.2 for beams or bars under bending moment $\boldsymbol{M}_z$ only: see Figure 6.4. The curvature $\kappa = 1/R_c$ is the deformation outcome under the pure bending, and it is reiterated herein as a reminder.

$$\kappa = \frac{1}{R_c} = \frac{\boldsymbol{M}_z}{EI_z} \tag{6.11}$$

where R_C is the ***radius of curvature***, $\boldsymbol{M}_z$ is the bending moment, E is the modulus of elasticity, and I_z is the second area moment about the z-axis.

6.3.4 Shear Stress and Deformation Due to Torsion

This type of loading generates a shear stress in beams or bars under the torque $\boldsymbol{T}$, which can also be represented by the alternative moment $\boldsymbol{M}_x$ as shown in Figure 6.11a. In that figure, the additional grayish double-arrowed vector represents the alternative moment $\boldsymbol{M}_x$, which is written in a grayish font color. The direction of the double-arrowed vector is in the x direction for the positive $\boldsymbol{T}$ or $\boldsymbol{M}_x$ torsional moment. This torsional moment is acting about the x-axis. This alternative representation is used in some references. However, we will use the more familiar symbol $\boldsymbol{T}$ for the applied torque.

Figure 6.11a identifies the deformation in the cylindrical machine element due to the applied torque $\boldsymbol{T}$. The deformation is identified by the *angular rotation* ϕ of the cross section and the angle γ along the element length l. It also shows a deformed element dA on the surface, which was originally of a squared shape. The element dA is deformed due to shear stresses stemmed from the applied torque $\boldsymbol{T}$. Figure 6.11b shows the shear forces generated on the element dA due to the *twisting angle* ϕ caused by the applied torque $\boldsymbol{T}$. The following relations provide the development of the shear stresses and present the deformations resulting off that.

The integral of all internal *shear forces* generated by the *shear stress* τ in the circular cross section of the cylindrical machine member should balance with the applied torque $\boldsymbol{T}$. This gives the following equation:

$$\int_A \tau(r) dA\, r = \boldsymbol{T} \tag{6.63}$$

Substituting for $\tau(r) = r\,\tau_{max}/c_S$ from the triangle bound by the dashed line in Figure 6.12b gives

$$\int_A \frac{r}{c_S}\, \tau_{max}\, dA\, r = \boldsymbol{T} \tag{6.64}$$

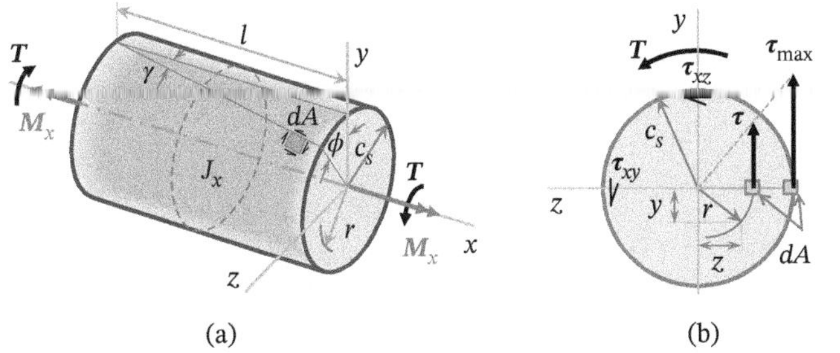

Figure 6.11 Simple cylindrical machine element subject to torsional loading: (a) deformation is identified by the angle ϕ of the cross section and the angle γ along the element length l and (b) shear forces generated on the element dA due to the twisting angle ϕ caused by the applied torque $\boldsymbol{T}$ or the double-arrowed vector $\boldsymbol{M}_x$.

where c_S is the radius of the outer fiber for the circular cross section. Carrying out the integration in Eq. (6.64) provides

$$\frac{\tau_{\max}}{c_S}\int_A r^2 dA = \boldsymbol{T} \tag{6.65}$$

or

$$\tau_{\max} = \frac{\boldsymbol{T}c_S}{J_x} \tag{6.66}$$

With reference to Figure 6.11b, it should be observed that the integral in Eq. (6.65) is the ***polar second area moment*** J_x of the cross-sectional area about the x axis, i.e.,

$$J_x = \int_A r^2\, dA = \int_A (y^2 + z^2)\; dA = I_y + I_z \tag{6.67}$$

For symmetrical solid circular cross sections, $I_y = I_z = \pi c_S{}^4/4 = \pi d_C{}^4/64$ and $J_x = 2I_y = 2I_z = \pi d_C{}^4/32$. The term d_C is the cylindrical cross-sectional diameter that is twice the maximum radius c_S of the outside fiber. The *polar second area moments* J_x of some other sections are available in mechanics textbooks and in handbooks such as Young and Budynas (2002). It should be noted, however, that the stresses developed in circular sections (*solid* or *hollow*) are different from shear stresses developed in noncircular sections. Equation (6.66) does not apply and other expressions are found in the literature; see Young and Budynas (2002). The maximum shear stress in those cases usually occurs at the narrowest part of the section rather than the thickest or the farthest or outermost point on the section.

The maximum shear stress $\tau_{\max}$ occurs at the outer fiber for the circular cross section and is given by Eq. (6.66). It can then be at c_S on the y-axis and thus should be in the z direction as represented by the one-sided arrow shown in Figure 6.11b. This shear should then be termed τ_{xz}, which means a shear stress on the x plane and in the z direction. On the other hand, the maximum shear stress can also be at c_S on the z-axis and thus should be in the -y direction as also evident from Figure 6.11b. The maximum shear will then be termed $-\tau_{xy}$. This gives the *maximum shear stress* $\tau_{\max}$ in the solid *circular cross section* due to external torque $\boldsymbol{T}$ as follows:

$$\tau_{\max} = \tau_{xy} = -\tau_{xz} = \frac{\boldsymbol{T}c_S}{J_x} = \frac{\boldsymbol{T}d_C/2}{J_x} = \frac{16\boldsymbol{T}}{\pi d_C^3} \tag{6.68}$$

As indicated previously and as shown in Figure 6.11a, the *angular deformation* due to the torque $\boldsymbol{T}$ is the rotation angle ϕ of the rotated cross section and the maximum angle γ along the outer fiber of the element length l. The following relation is for the outer arc of the cross section as evident from Figure 6.11a. From the outer arc relations, one gets

$$l\gamma = c_S\phi \tag{6.69}$$

By applying the *Hook's law* for linearly elastic materials, i.e. $\tau = G\gamma$, where G is the *shear modulus of elasticity* and γ is the *shear strain*, and utilizing Eqs. (6.68) and (6.69), one gets

$$\tau = G\gamma = G\left(\frac{c_S\phi}{l}\right) = \frac{\boldsymbol{T}c_S}{J_x}$$

or

$$\phi = \frac{\boldsymbol{T}l}{GJ_x} \tag{6.70}$$

where G is the ***shear modulus of elasticity*, or *modulus of rigidity***, which is related to the *modulus of elasticity* E and the *Poisson's ratio* ν by the following relation (see Section 6.6.1):

$$G = \frac{E}{2(1+\nu)} \tag{6.71}$$

This relation indicates that there are only two independent elastic material constants for any *isotropic material*, which is homogeneous with the same properties in all directions.

The main assumptions in this treatment are as follows:

- The element has the same *uniform section* in the apparent x direction, which is the usual default direction.
- The torque is assumed around the *neutral axis* and not offset to cause any other additional loadings.
- Plane circular sections of the machine element remains plane and perpendicular to the element axis.
- Shearing strains γ are linearly varying from the central axis of the element.

The applied torque on machine elements such as rotating shafts is often generated due to the power H transmitted by the shaft. The relation between power H, torque $\boldsymbol{T}$, and shaft speed ω is

$$\begin{aligned} H_W &= T_{N.m}\,\omega_{rad/s} = \frac{2\pi N_{rpm}}{60} T_{N.m} \\ H_{hp} &= \frac{T_{lb\,in}\,\omega_{rad/s}}{6600} = \frac{2\pi N_{rpm} T_{lb\,in}}{(12)(33\,000)} = \frac{N_{rpm} T_{lb\,in}}{63\,025} \end{aligned} \tag{6.72}$$

where $\omega_{rad/s}$ is rotational speed in [rad/s], N_{rpm} is the rotational speed in [rpm], $T_{N\,m}$ is the torque magnitude in [N m], $T_{lb\,in}$ is the torque magnitude in [lb in], and H is the power in [W] (for H_W) or [hp] (for H_{hp}); see Section 1.11.1. From Eq. (6.72), one can get the torque magnitude and thus use it in the *design synthesis* of the machine element. After synthesis, evaluate the stresses and deflections of the element under such loading; see, e.g. Section 8.4.2 for initial shaft synthesis.

One should note that the initial design synthesis could directly use a *shear yield strength,* which is usually deduced from the usual *tensile yield strength*. Usually *shear yield strength* is about one-half the *tensile yield strength*; see Section 7.7 concerning material *failure theories*. Shafts transmitting power are extensively treated in a separate chapter; see Chapter 17. In that chapter, more complex loading conditions are considered.

Example 6.7 A shaft is directly connected to an electric motor running at 1200 [rpm]. The shaft is transmitting the maximum power of 75 [kW] or 100.58 [hp] at that input speed of 1200 [rpm]. The shaft geometry and characteristics are defined as a length $l = 1.0$ [m] or 40 [in], a modulus of elasticity $E = 207$ [GPa] or 30 [Mpsi] and a modulus of rigidity $G = 79$ [GPa] or 11.5 [Mpsi]. What is the synthesized shaft diameter for a safety factor K_{SF} of 4.0? Assume a shaft material having a *shear yield strength* $S_{ys} = 200$ [MPa] $= 200\,(10^6)$ [Pa] $= 29$ [kpsi]. Find the shaft diameter to guard against a yield failure under torsional shear? At the maximum transmitted power from the motor, what is the angular deflection of the shaft?

Solution

Data: Shaft length $l = 1.0$ [m] or 40 [in], $H = 75\,000$ [W] or 100.58 [hp]. $N_{rpm} = 1200$ [rpm], $E = 207$ [GPa] or 30 [Mpsi] and $G = 79$ [GPa] or 11.5 [Mpsi], $K_{SF} = 4.0$, and $S_{ys} = 200$ [MPa] $= 200(10^6)$ [Pa] $= 29$ [kpsi].

By applying Eq. (6.43), the allowable shear stress $\boldsymbol{\tau}_{all} = S_{ys}/K_{SF} = 200/4 = 50$ [MPa] or $=29/4 = 7.25$ [kpsi].

Equation (6.72) is used to find the maximum applied torque such that

$$\begin{aligned} H_W &= \frac{2\pi\,N_{rpm}}{60} T_{N\,m}, \text{ or } T_{N\,m} = \frac{60 H_W}{2\pi\,N_{rpm}} = \frac{60(75\,000)}{2\pi(1200)} = 596.83\ [\text{N m}] \\ H_{hp} &= \frac{N_{rpm} T_{lb\,in}}{63\,025}, \text{ or } T_{lb\,in} = \frac{(63\,025) H_{hp}}{N_{rpm}} = \frac{(63\,025)(100.58)}{1200} = 5282.5\ [\text{lb in}] \end{aligned} \tag{a}$$

By using Eq. (6.68) and finding the diameter synthesis off that, one gets

$$\begin{aligned} d_C &= \sqrt[3]{\frac{16\boldsymbol{T}}{\pi\boldsymbol{\tau}_{all}}} = \sqrt[3]{\frac{16\,(596.83)}{\pi(50)(10^6)}} = 0.039\,32\ [\text{m}] = 39.32\ [\text{mm}] \\ d_C &= \sqrt[3]{\frac{16\boldsymbol{T}}{\pi\boldsymbol{\tau}_{all}}} = \sqrt[3]{\frac{16\,(5282.5)}{\pi(7.25)(10^3)}} = 1.5482\ [\text{in}] \end{aligned} \tag{b}$$

One would usually round the diameters to the closest preferred number (40 or 45 [mm]) or (1.6 or 1 3/4 [in]). However, we will use the 40 [mm] or the 1.6 [in].

To evaluate the angular deflection according to Eq. (6.70), one needs to evaluate the polar second area moment $J_x = \pi d_C{}^4/32$. Substituting for the selected rounded diameters, we have

$$J_x = \frac{\pi d_C{}^4}{32} = \frac{\pi(0.04)^4}{32} = 2.513\,27(10^{-7})\,[\mathrm{m}^4]$$
$$J_x = \frac{\pi d_C{}^4}{32} = \frac{\pi(1.6)^4}{32} = 0.6434\,[\mathrm{in}^4] \tag{c}$$

The angular deflection of the shaft at maximum transmitted power is obtained by Eq. (6.70) such that

$$\phi = \frac{Tl}{GJ_x} = \frac{596.83(1)}{79(10^9)(2.513\,27)(10^{-7})} = 0.030\,06\,[\mathrm{rad}] = 1.7223\,[^\circ]$$
$$\phi = \frac{Tl}{GJ_x} = \frac{5282.5(40)}{11.5(10^6)(0.6434)} = 0.028\,558\,[\mathrm{rad}] = 1.6362\,[^\circ] \tag{c}$$

It is expected to have close values of the deflections. The difference is due to the rounding of the diameter in the US system, which is little larger than that in the SI system.

6.3.5 Transverse Shear and Shear Flow

When the bending moment in a beam is not constant (i.e. $\boldsymbol{V} \neq 0$), transverse shear develops. Figure 6.12a shows a section in a simply supported beam loaded by a concentrated force $\boldsymbol{F}$. Along the beam, the bending moment is varying as evident from Figure 6.5a or b, where the shear forces $\boldsymbol{V} \neq 0$. An element A of the beam at a distance y from the centerline in Figure 6.12a is cut off the beam and its free body diagram is magnified as shown in Figure 6.12b. The balance of forces on that element A gives

$$\sum \boldsymbol{F}_x = \boldsymbol{F}_1 - d\boldsymbol{F} - \boldsymbol{F}_2 = 0$$
$$-d\boldsymbol{F} = \boldsymbol{F}_2 - \boldsymbol{F}_1 = \int_y^{c_S} \sigma_2 dA - \int_y^{c_S} \sigma_1 dA \tag{6.73}$$

where dA is the element area as shown in Figure 6.12b on side 1 and side 2 of the element and $\boldsymbol{\sigma}_1$ and $\boldsymbol{\sigma}_2$ are the compressive stresses on side 1 and side 2 of the element, respectively. Substituting for the bending stresses according to Eq. (6.12), $\{\boldsymbol{\sigma}_x = \boldsymbol{M}_z y/I_z\}$, and the *transverse shear stress* $\boldsymbol{\tau}_T$ over the area $b\,dx$ for the shear force $d\boldsymbol{F}$, one gets

$$-\tau_T b\,dx = \int_y^{c_S} \frac{\boldsymbol{M}_z + d\boldsymbol{M}_z}{I_z} y - \frac{\boldsymbol{M}_z}{I_z} y\,dA = \frac{d\boldsymbol{M}_z}{I_z}\int_y^{c_S} y\,dA = \frac{d\boldsymbol{M}_z}{I_z} Q_A \tag{6.74}$$

The term Q_A in Eq. (6.74) is the *first area moment* about the neutral axis. If the area dA is rectangular, the value of Q_A is equal to dA multiplied by its center of area distance y_c to the neutral axis. The variation of the bending moment $d\boldsymbol{M}_z$ can be substituted in Eq. (6.74) from Eq. (6.2) ($d\boldsymbol{M}_z = -\boldsymbol{V}\,dx$), to get

$$-\tau_T b\,dx = \frac{-\boldsymbol{V}dx}{I_z} Q_A$$
$$\tau_T = \frac{\boldsymbol{V} Q_A}{I_z b} \tag{6.75}$$

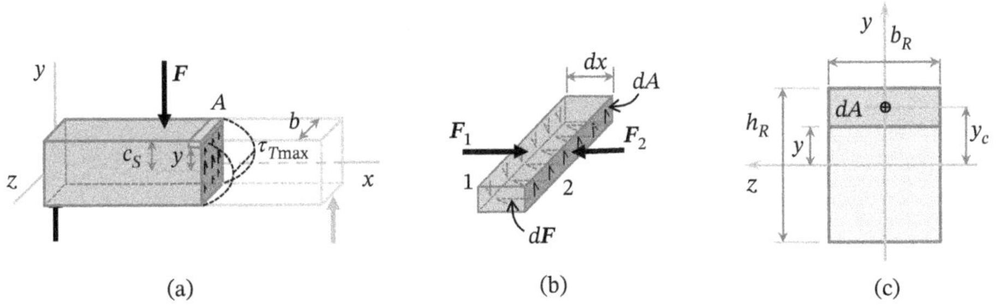

Figure 6.12 An element A of the beam at a distance y from the centerline in (a) is cut off the beam and its free body diagram is magnified in (b). The area dA above a distance y is shown in (c) with its area center at a distance y_c from the section neutral axis.

The ***shear flow*** q_S is defined as $\boldsymbol{q}_S = dF/dx$. With reference to Figure 6.12 and Eq. (6.74), one finds

$$\boldsymbol{q}_S = \frac{d\boldsymbol{F}}{dx} = \frac{\tau_T\ b\,dx}{dx} = \frac{\boldsymbol{V}dx}{I_z\,dx}Q_A = \frac{\boldsymbol{V}Q_A}{I_z} \tag{6.76}$$

From Eqs. (6.75) and (6.76), one can get

$$\tau_T = \frac{\boldsymbol{V}Q_A}{I_z\ b} = \frac{\boldsymbol{q}_S}{b}, \quad \text{or} \quad \boldsymbol{q}_S = \frac{\boldsymbol{V}Q_A}{I_z} = \tau_T b \tag{6.77}$$

This shear is on the face of the cross section in the direction of the internal shear force, i.e. $\tau_{xy} = \tau_T$. However, a complementary shear $-\tau_{yx}$ exists (notice $d\boldsymbol{F}$ in Figure 6.12b), which is in-between the horizontal layers of the beam. It is zero at the top and the bottom layers of the beam, and it has a maximum value $\boldsymbol{\tau}_{\max}$ at the neutral axis of the beam as shown in Figure 6.12a. Its average is naturally $\boldsymbol{V}/A$, which can be easily verified.

As evident from Figure 6.12 and Eqs. (6.75) and (6.77), the derivation is most suited for the *rectangular* cross sections. However, the derivation procedure can be applied to any other section as shown in Example 6.8. The transverse shear distribution for rectangular cross section is shown by the dashed parabolic surface in Figure 6.12a. The distribution is derived in Example 6.8.

Example 6.8 Develop the relation for the shear distribution over a rectangular cross section in Figure 6.12a,c. The section has a width of b_R (in place of b) and a depth of h_R (in place of $2c_S$). Find the maximum transverse shear for the beam of Example 6.2, if the rectangular section has a width $b_R = 0.05$ [m] or 2 [in] and a depth $h_R = 0.1$ [m] or 4 [in]. If the force $\boldsymbol{F} = -40$ [kN] or −9 [klb], the length $l = 1.0$ [m] or 40 [in], the modulus of elasticity $E = 207$ [GPa] or 30 [Mpsi], and the modulus of rigidity $G = 79$ [GPa] or 11.5 [Mpsi], what is the maximum transverse shear stress in rectangular section? Compare the result with a circular cross section beam of a diameter $d_C = 0.1$ [m] or 4 [in] utilizing information from Appendix A.4 or other references.

Solution

Data: Rectangle width = b_R and rectangle depth = h_R. Section $b_R = 0.05$ [m] or 2 [in], $h_R = 0.1$ [m] or4 [in], $d_C = 0.1$ [m] or 4 [in], $\boldsymbol{F} = -40$ [kN] or −9 [klb], $l = 1.0$ [m] or 40 [in], and $G = 79$ [GPa] or 11.5 [Mpsi].

Rectangular section area $A_R = b_R\, h_R = 0.05\,(0.1) = 0.005$ [m^2] or 2 (4) = 8 [in^2]. The circular section area $A_C = \pi\, d_C{}^2/4 = 0.007\,853$ [m^2] or 2 (4) = 12.57 [in^2].

The *first area moment* Q_A of the area A in Figure 6.12c is

$$Q_A = Ay_c = \left(\frac{h_R}{2} - y\right) b \left(\frac{1}{2}\left(\frac{h_R}{2} + y\right)\right)$$

or (a)

$$Q_A = \frac{b}{2}\left(\frac{h_R^2}{4} - y^2\right)$$

This is a parabolic equation that causes the transverse shear defined by Eq. (6.77) to be parabolic as shown in Figure 6.12a and identified by

$$\tau_T(y) = \frac{VQ_A}{I_z b} = \frac{V}{2I_z}\left(\frac{h_R^2}{4} - y^2\right) \tag{b}$$

As the second area moment I_x of the rectangular section is $I_z = b_R h_R{}^3/12$,

$$\tau_T(y) = \frac{6V}{b_R h_R{}^3}\left(\frac{h_R^2}{4} - y^2\right) \tag{c}$$

The maximum transverse shear $\tau_{T\max}$ occurs at $y = 0$, which is

$$\tau_{T\max} = \frac{3V}{2b_R h_R} = \frac{3V}{2A_R} \tag{d}$$

The term A_R in Eq. (d) is the area of the rectangular cross section.

The maximum transverse shear $\tau_{T\max}$ at $y = 0$ for the dimensions specified requires the calculation of the shear V. This can be calculated according to Figure 6.5a, where $V(\tfrac{1}{2}l) = -F = 40$ [kN] or 9 [klb]. The maximum transverse shear is calculated from Eq. (d), which gives

$$\begin{aligned} \tau_{T\max} &= \frac{3V}{2A_R} = \frac{3\,(40)(10^3)}{2\,(0.005)} = 12\,000\,000\ [\text{Pa}] = 12\ [\text{MPa}] \\ \tau_{T\max} &= \frac{3V}{2A_R} = \frac{3\,(9)(10^3)}{2\,(8)} = 1687.5\ [\text{psi}] \end{aligned} \tag{e}$$

For a circular cross section, the maximum transverse shear calculations need the evaluation of the center of area of the top half section, which gives (see Popov (1990))

$$\begin{aligned} \tau_{T\max} &= \frac{4V}{3A_C} = \frac{4(40)(10^3)}{3(0.007\,853)} = 6\,791\,460\ [\text{Pa}] = 6.791\ [\text{MPa}] \\ \tau_{T\max} &= \frac{4V}{3A_C} = \frac{4(9)(10^3)}{3(12.57)} = 954.66\ [\text{psi}] \end{aligned} \tag{f}$$

By comparing the values of the transverse shear to the values of the normal stresses in Example 6.2, one can find that the transverse shear is about 6.8% of the normal stress. This is mainly attributed to the large length of the beam, which affects the normal stress but does not affect the transverse shear stress.

6.3.5.1 Shear Center

When a beam is transversely loaded and the beam section is asymmetric as shown in Figure 6.13, the load has to be placed off the beam neutral axis as indicated by the *grayish load* $\boldsymbol{F}$ in part (a) of the figure. This condition is needed if the beam is not to be twisted. The condition is then necessitating to counteract the moment generated by the *shear flow* in the section due to the transverse shear as defined in Figure 6.13b. The location of the load $\boldsymbol{F}$ should be at the ***shear center*** located at a distance e_{Sc} off the cross section. Usually the distance of shear center

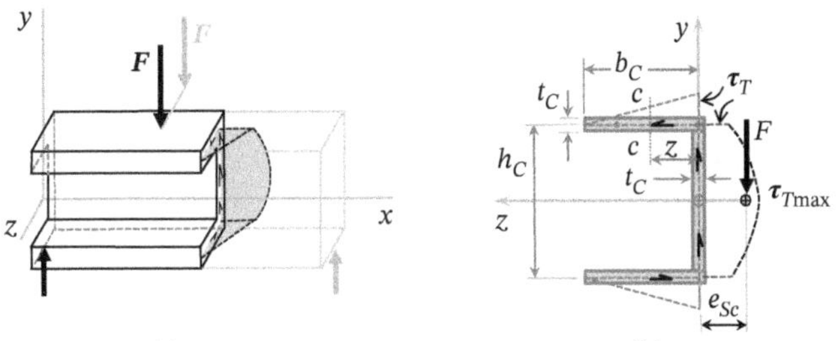

(a) (b)

Figure 6.13 A thin-section beam is transversely loaded and its section is asymmetric: (a) the load $\boldsymbol{F}$ has to be placed off the beam neutral axis as indicated by the grayish load and the shear distribution is the dashed garishly filled shape and (b) the section dimensions, shear flow distribution, and the offset force at the shear center distance e_{Sc}.

location is measured from the section center of gravity on a symmetric axis, even though it is seen in Figure 6.13b as measured from the center of the section side.

The calculation of the shear flow in the section is used to find the shear center of the section. The balance of the shear moment effect with the counteracting moment ($e_{Sc} \times \boldsymbol{F}$) produced by the offset force $\boldsymbol{F}$ defines the section shear center e_{Sc}. A cut at the section topside such as c–c in Figure 6.13b is usually made. The shear flow at that cut is calculated to get the shear flow along the topside as shown by the dashed shear flow distribution along the topside. The shear forces generated on these topside and bottom-side cause a moment about the circled web center of section, which is balanced by the countering moment $e_{Sc} \times \boldsymbol{F}$. The procedure is best demonstrated next by Example 6.9 and can be exploited with any other asymmetric cross section. For simplicity, however, all calculations in Example 6.9 uses the assumption of thin-walled sections, where all measures are defined from the centerline of the section thickness. Usually the shear flow is assumed to be running along that thickness centerline.

Example 6.9 Develop the relation for the shear center location of a channel cross section (C-section) of the loaded beam in Figure 6.13. The section has a width of b_C (off section thickness centerline) and a depth of h_C (off section thickness centerline). Find the transverse shear flow in the section. As for the beam of Example 6.2, the channel section has a width $b_C = 0.05$ [m] or 2 [in], a depth $h_C = 0.1$ [m] or 4 [in] and channel outer thickness $t_C = 0.01$ [m] or 0.4 [in]. The force is $\boldsymbol{F} = -40$ [kN] or -9 [klb], the length $l = 1.0$ [m] or 40 [in], the modulus of elasticity $E = 207$ [GPa] or 30 [Mpsi] and the modulus of rigidity $G = 79$ [GPa] or 11.5 [Mpsi]. Where is the shear center located in the channel section?

Solution

Data: Channel width = b_C, depth = h_C and outer thickness = t_C. Section $b_C = 0.05$ [m] or 2 [in], $h_C = 0.1$ [m] or 4 [in], $t_C = 0.01$ [m] or 0.4 [in], $\boldsymbol{F} = -40$ [kN] or -9 [klb], $l = 1.0$ [m] or 40 [in], and $G = 79$ [GPa] or 11.5 [Mpsi]. The shear at the force location $\boldsymbol{V}(\tfrac{1}{2}l) = -\boldsymbol{F} = 40$ [kN] or 9 [klb].

Channel section areas: flanges or sides $A_S = b_C\, t_C = 0.05\ (0.01) = 0.0005$ [m^2] or 2 (0.4) = 0.8 [in^2] and web $A_W = h_C\, t_C = 0.1\ (0.01) = 0.001$ [m^2] or 4 (0.4) = 1.6 [in^2].

A segment of the side at c–c of the section located at z in Figure 6.13b is selected. The shear flow at that location z is calculated. It is clear that the segment area from z (with its center) is linearly varying up to the connection of the section side with the section web at $z = 0$ as shown in Figure 6.13b. The value of the shear flow $\boldsymbol{q}_s$ (or shear stress τ_T) at z or at that intersection (where $z = 0$) is calculated according to Eq. (6.77) such that

$$\tau_T(z) = \frac{V(b_C - z)(t_C)(h_C/2)}{I_z t_C} = \frac{\boldsymbol{q}_s}{t_C}, \quad \text{or} \quad \boldsymbol{q}_s(z) = \frac{V(b_C - z)(t_C)(h_C/2)}{I_z} = \tau_T(z)t_C$$
$$\tau_T(0) = \frac{Vb_C h_C}{2I_z} = \frac{\boldsymbol{q}_s}{t_C}, \quad \text{or} \quad \boldsymbol{q}_s(0) = \frac{Vb_C h_C t_C}{2I_z} = \tau_T(0)t_C \tag{a}$$

The shear stress is varying from zero at the edge ($z = b_C$) to the value in equation (a) at $z = 0$. The average is ½ the value at $z = 0$. The total shearing force $\boldsymbol{F}_1$ on the top side (or $\boldsymbol{F}_2$ on the bottom side) of the channel is the average shear force $\boldsymbol{\tau}_{T,\text{avg}}$ on the side multiplied by the area of the top or bottom channel side or

$$\boldsymbol{F}_1 = -\boldsymbol{F}_2 = \boldsymbol{\tau}_{T,\text{avg}} \times (\text{side area}) = \frac{\boldsymbol{V} b_C h_C}{2(2)I_z}(b_C t_C) = \frac{\boldsymbol{V} b_C^2 h_C t_C}{4I_z} \tag{b}$$

The couple generated by the forces $\boldsymbol{F}_1$ and $\boldsymbol{F}_2$ should be balanced by the moment of the applied load $\boldsymbol{F}$ about the web center of the channel if the beam is not to be twisted. This gives the shear center location such that

$$\begin{aligned} \boldsymbol{F} e_{Sc} &= h_C \boldsymbol{F}_1 = h_C \frac{\boldsymbol{V} b_C^2 h_C t_C}{4I_z} = \frac{\boldsymbol{F} b_C^2 h_C^2 t_C}{4I_z} \\ e_{Sc} &= \frac{b_C^2 h_C^2 t_C}{4I_z} \end{aligned} \tag{c}$$

To calculate the shear center, we need to calculate the second area moment I_z of the channel in an approximate simplified form so that

$$\begin{aligned} I_z &= \frac{b_c(h_C + t_C)^3}{12} - \frac{(b_c - t_C)(h_C - t_C)^3}{12} = \frac{0.05(0.1 + 0.01)^3 - (0.05 - 0.01)(0.1 - 0.01)^3}{12} \\ &= 3.1158(10^{-6})\ [\text{m}^4] \\ I_z &= \frac{b_c(h_C + t_C)^3}{12} - \frac{(b_c - t_C)(h_C - t_C)^3}{12} = \frac{2(4 + 0.4)^3 - (2 - 0.4)(4 - 0.4)^3}{12} = 7.9765\ [\text{in}^4] \end{aligned} \tag{d}$$

The shear center is then calculated using Eq. (c) to get

$$\begin{aligned} e_{Sc} &= \frac{b_C^2 h_C^2 t_C}{4I_z} = \frac{0.05^2(0.1)^2(0.01)}{4(3.1158(10^{-6}))} = 0.020\,06\ [\text{m}] = 20.06\ [\text{mm}] \\ e_{Sc} &= \frac{b_C^2 h_C^2 t_C}{4I_z} = \frac{2^2(4)^2(0.4)}{4(7.9765)} = 0.802\,36\ [\text{in}] \end{aligned} \tag{e}$$

It should be noted that the derivation was considering a *very thin* thickness, which was not totally applied. The adapted thickness was 10% of the height h_C, which is not that thin enough. The results, however, may be considered as accurate within the acceptable engineering calculations. If the thickness is very thin, the second area moment I_z can be further approximated. The calculated shear center is also better defined from a measurable datum such as the surface of the channel section. If one uses a standard channel section (see Tables A.8.10 and A.8.11), the equations should be rechecked or derived to suit the variations in the dimensions of the section.

6.4 Combined Stresses

In Section 6.3, *simple stresses* are introduced where they are mainly in one dimension or direction. These stresses are due to loading of beams, bars, and simple loaded machine elements under a single defined force or moment. Generated stresses in those cases are unidirectional or *uniaxial*. However, the loading of real machine elements is usually in more than one direction and thus multi-axial. In this section we consider the more realistic combined stresses in two axes (or plane stress cases) and also the ***3D*** stress cases. The *plane stress* state is considered first and then followed by the *triaxial stress* state.

6.4.1 Plane Stress State

This case is one of the most common loading cases where the stress on one of the planes is not taking place or not present or insignificantly small. This is apparent from the beam loaded as shown in Figure 6.14a, where a

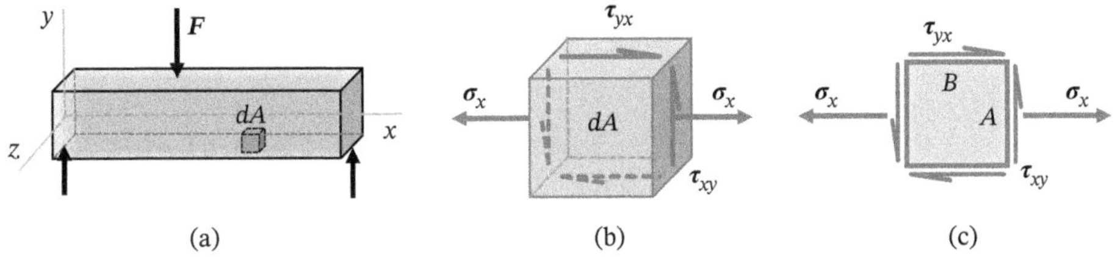

Figure 6.14 A simple beam loaded by a transverse force in (a). A small element *dA* is identified inside the beam and close to its bottom. The small element is taken away from the beam and its free body diagram is indicated in (b). The plane projection of *dA* on the *xy* plane is shown in (c) as a plane stress state.

small element *dA* is identified inside the beam and close to its bottom. This small element is taken away from the beam and its free body diagram is indicated in Figure 6.14b, where the generated *normal stress* $\boldsymbol{\sigma}_x$ due to bending and the transverse shear $\boldsymbol{\tau}_{xy}$ are in effect. These stresses are functions of the location of the element *dA* inside the beam. At the outer fiber, the normal stress is at its maximum value as indicated in Section 6.3.3. The transverse shear is zero at the outer fiber as defined in Section 6.3.5. That is why the selected location of the small element *dA* is not at the outer fiber, but close to it for this case. It is clear that element *dA* in Figure 6.14b does not have stresses in the *z* direction or on the *z* plane. That is why we can effectively consider the plane projection of *dA* on the *xy* plane as shown in Figure 6.14c and thus we have a *plane stress* state. On the side *A* of the *plane element* in Figure 6.14c, we have the normal stress $\boldsymbol{\sigma}_x$ and the shear stress $\boldsymbol{\tau}_{xy}$. One should also observe the complementary shear stress $\boldsymbol{\tau}_{yx}$ on the normal plane or face *B* of the *plane stress* element. The analysis of this case is best handled by the well-known *Mohr's circle*, after *Christian Otto Mohr* (1835–1918). The *Mohr's circle* gives a two-dimensional (***2D***) representation of the *stress state* and helps in defining the *maximum normal stress* values called the *principal stresses* and their orientations, which are the *principal directions*. The relations between the original plane stress state and the principal stress state are also called the *stress transformation equations*. These concepts are clarified in the following postulation.

It should be noted, however, that if there is an applied torsion to a circular machine element in addition to the bending, a similar small element to *dA* can be developed with the existence of plan stress case as defined earlier. The torsion generates the shear stress as depicted in Section 6.3.4. The shear stress direction is about the same as for the transverse shear depending on the direction of the torsion and the location of the element *dA* as shown in Figure 6.11. The following treatment then follows in the same way.

6.4.1.1 Mohr's Circle

To develop the *Mohr's circle*, Figure 6.14c has been reproduced in Figure 6.15a but with the addition of $\boldsymbol{\sigma}_y$ for generality. In addition, a wedge cut was operated on the element to be normal to the transformed coordinate x_T y_T. The *coordinate transformation* was through a rotation of an angle γ as shown in Figure 6.15a. It is required then to get the stresses on the wedge side or plane of normal $\boldsymbol{N}_T$. On the wedge surface, the *traction forces* $\boldsymbol{T}_x$ and $\boldsymbol{T}_y$ should balance with the forces generated from the stresses $\boldsymbol{\sigma}_x$ and $\boldsymbol{\sigma}_y$. The generated forces are functions of the lengths of the wedge sides relative to the unit inclined wedge length. These side lengths are then proportional to $\cos\gamma$ and $\sin\gamma$ for $\boldsymbol{\sigma}_x$, $\boldsymbol{\sigma}_y$, and $\boldsymbol{\tau}_{xy}$, each on its side with an assumed constant unit wedge thickness. This gives the following equation:

$$\begin{aligned} \boldsymbol{T}_x &= \sigma_x \cos\gamma + \tau_{xy} \sin\gamma \\ \boldsymbol{T}_y &= \tau_{xy} \cos\gamma + \sigma_y \sin\gamma \end{aligned} \tag{6.78}$$

or

$$\begin{bmatrix} \boldsymbol{T}_x \\ \boldsymbol{T}_y \end{bmatrix} = \begin{bmatrix} \sigma_x & \tau_{xy} \\ \tau_{xy} & \sigma_y \end{bmatrix} \begin{bmatrix} \cos\gamma \\ \sin\gamma \end{bmatrix} = \boldsymbol{\sigma N}_T \tag{6.79}$$

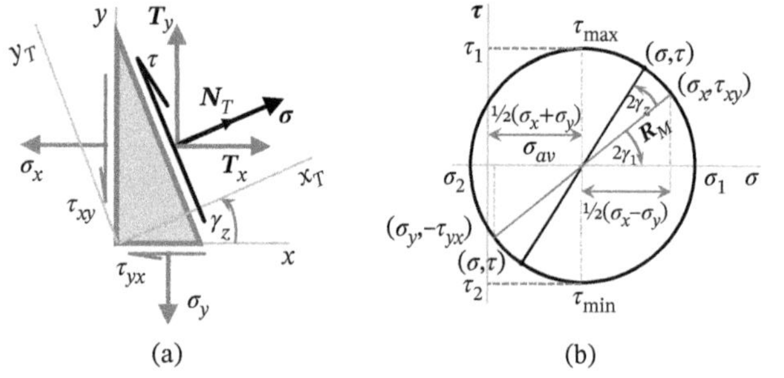

Figure 6.15 Figure 6.14c is reproduced in (a) but sectioned by a plane at an angle γ_z to form a wedge with the addition of σ_y for generality. The point (σ, τ) on the *Mohr's circle* is shown in (b) where $2\gamma_z$ is the angle from (σ_x, τ_{xy}) to (σ, τ).

where $\boldsymbol{\sigma}$ is the *stress tensor* (or *stress matrix*) and $\boldsymbol{N}_T$ is the direction cosine of the inclined plane of the wedge. The *traction forces* $\boldsymbol{T}_x$ and $\boldsymbol{T}_y$ are the components of the stress resultant $\boldsymbol{\sigma}$ and $\boldsymbol{\tau}$ on the inclined wedge length. The components of the stress resultant $\boldsymbol{\sigma}_T$ and $\boldsymbol{\tau}_T$ are then obtained by the rotational transformation $\boldsymbol{R}_z$ of Section 4.1.2 withholding the homogeneous coordinates and using ***2D*** only instead of the ***3D*** case. The stress resultants are then

$$\begin{bmatrix}\sigma\\ \tau\end{bmatrix} = \boldsymbol{R}_z\begin{bmatrix}\boldsymbol{T}_x\\ \boldsymbol{T}_y\end{bmatrix} = \begin{bmatrix}\cos\gamma_z & \sin\gamma_z\\ -\sin\gamma_z & \cos\gamma_z\end{bmatrix}\begin{bmatrix}\boldsymbol{T}_x\\ \boldsymbol{T}_y\end{bmatrix} = \begin{bmatrix}\cos\gamma_z & \sin\gamma_z\\ -\sin\gamma_z & \cos\gamma_z\end{bmatrix}\begin{bmatrix}\sigma_x\cos\gamma_z + \tau_{xy}\sin\gamma_z\\ \tau_{xy}\cos\gamma_z + \sigma_y\sin\gamma_z\end{bmatrix}$$
$$\begin{bmatrix}\sigma\\ \tau\end{bmatrix} = \begin{bmatrix}\sigma_x\cos^2\gamma_z + \tau_{xy}\sin\gamma_z\cos\gamma_z + \tau_{xy}\cos\gamma_z\sin\gamma_z + \sigma_y\sin^2\gamma_z\\ -\sigma_x\sin\gamma_z\cos\gamma_z - \tau_{xy}\sin^2\gamma_z + \tau_{xy}\cos^2\gamma_z + \sigma_y\cos\gamma_z\sin\gamma_z\end{bmatrix} \tag{6.80}$$

This equation gives

$$\begin{bmatrix}\sigma\\ \tau\end{bmatrix} = \begin{bmatrix}\sigma_x\cos^2\gamma_z + \sigma_y\sin^2\gamma_z + 2\tau_{xy}\sin\gamma_z\cos\gamma_z\\ \tau_{xy}(\cos^2\gamma_z - \sin^2\gamma_z) + (\sigma_y - \sigma_x)\sin\gamma_z\cos\gamma_z\end{bmatrix} \tag{6.81}$$

However, to reduce Eq. (6.81) to that in Eq. (6.83), one can use the following trigonometric functions:

$$\cos^2\gamma_z = \frac{1}{2}(1+\cos 2\gamma_z),\quad \sin^2\gamma_z = \frac{1}{2}(1-\cos 2\gamma_z)$$
$$\sin\gamma_z\cos\gamma_z = \frac{1}{2}(\sin 2\gamma_z) \tag{6.82}$$

Upon the application of Eq. (6.82), Eq. (6.81) becomes

$$\begin{bmatrix}\sigma\\ \tau\end{bmatrix} = \begin{bmatrix}\frac{1}{2}(\sigma_x+\sigma_y) + \frac{1}{2}(\sigma_x-\sigma_y)\cos 2\gamma_z + \tau_{xy}\sin 2\gamma_z\\ -\frac{1}{2}(\sigma_x-\sigma_y)\sin 2\gamma_z + \tau_{xy}\cos 2\gamma_z\end{bmatrix} \tag{6.83}$$

Equation (6.83) represents the point (σ, τ) on the *Mohr's circle* as shown in Figure 6.15b where 2γ is the angle from (σ_x, τ_{xy}) to (σ, τ). It should be noted that on the *Mohr's circle* the angle between σ_x and σ_y is 180° rather than the original 90° in the stress state. This means that the angles on the *Mohr's circle* are twice those in the original stress state.

6.4.1.2 Principal Stresses and Principal Directions

In order to find the maximum and minimum stresses in the preceding plane stress state, one should apply the condition $d\sigma/d\gamma = 0$ to Eq. (6.83). This yields the following relation:

$$-(\sigma_x - \sigma_y)\sin 2\gamma_z + 2\tau_{xy}\cos 2\gamma_z = 0 \tag{6.84}$$

or

$$\tan 2\gamma_1 = \frac{2\tau_{xy}}{(\sigma_x - \sigma_y)} \tag{6.85}$$

As $\tan 2\gamma = \tan(2\gamma + \pi)$, two perpendicular directions should satisfy Eq. (6.84). These directions are the ***principal directions***, along which the ***principal stresses*** $\boldsymbol{\sigma}_1$ and $\boldsymbol{\sigma}_2$ (i.e. the maximum and minimum stresses) will act. The principal directions are the normals of the ***principal planes***, i.e. the principal stress is normal to its principal plane. Comparing Eq. (6.84) with the shear stress in Eq. (6.83), it is clear that $\tau = 0$. This means that the perpendicular *principal planes* have no shear stresses. Each of the principal planes has only a principal stress, the principal stress σ_1 alone or the principal stress σ_2 alone. Implementing Eq. (6.85) on Figure 6.15b, the counterclockwise angle $2\gamma_1$ (i.e. $-2\gamma_1$) is the angle of the *first principal stress* $\boldsymbol{\sigma}_1$ from the stress state (σ_x, τ_{xy}). The *second principal stress* $\boldsymbol{\sigma}_2$ is perpendicular to $\boldsymbol{\sigma}_1$. The maximum shear stress $\boldsymbol{\tau}_{\max}$ is normal to the principal stress on the *Mohr's* circle; therefore it makes a 45° from the direction of the principal stress in the original stress state. Also implementing the conditions of Eq. (6.85) into Eq. (6.83) confirms that the last two terms of the principal stress values in Eq. (6.83) is the radius R_M of the *Mohr's* circle or

$$R_M = \frac{1}{2}(\sigma_x - \sigma_y)\cos 2\gamma_z + \tau_{xy}\ \sin 2\gamma_z = \sqrt{\left(\frac{1}{2}(\sigma_x - \sigma_y)\right)^2 + \tau_{xy}^2} \tag{6.86}$$

$$\sigma_{1,2} = \sigma_{\text{av}} \pm R_M = \frac{\sigma_x + \sigma_y}{2} \pm \sqrt{\left(\frac{\sigma_x - \sigma_y}{2}\right)^2 + \tau_{xy}^2} \tag{6.87}$$

where σ_{av} is the average stress, the value of the *first principal stress* $\boldsymbol{\sigma}_1$ is the larger with the positive (+) sign, and the value of the *second principal stress* $\boldsymbol{\sigma}_2$ is the smaller with the negative (−) sign in Eq. (6.87).

The maximum and minimum shear stresses $\boldsymbol{\tau}_{\max}$ and $\boldsymbol{\tau}_{\min}$ (or τ_1 and τ_2) can also be found by applying the condition $d\tau/d\gamma = 0$ to Eq. (6.83). This would result in a complementing angle to the angle in Eq. (6.85). Utilizing the *Mohr's circle* confirms that the values of the maximum and minimum shear stresses $\boldsymbol{\tau}_{\max}$ and $\boldsymbol{\tau}_{\min}$ (or τ_1 and τ_2) are simply the radii $\pm R_M$ of the *Mohr's circle*, i.e.,

$$\tau_{1,2} = \tau_{\max,\min} = \pm R_M = \pm\sqrt{\left(\frac{1}{2}(\sigma_x - \sigma_y)\right)^2 + \tau_{xy}^2} \tag{6.88}$$

6.4.1.3 Vector Space and Eigenvalue Problem

The problem of finding the maximum and minimum stresses for a stress state can be handled by transformation in *vector space* similar to the aforementioned rotation transformation. With reference to Eq. (6.78), the transformation is dealt with similarly for all previous relations of the stress state and its outcome. By rewriting Eq. (6.79) for the traction components of traction vector $\boldsymbol{T}$ and also for stress components matrix (or stress *tensor* $\boldsymbol{\tau}_{ij}$, where $\sigma_x = \tau_{xx}$, $\sigma_y = \tau_{yy}$, and $\tau_{xy} = \tau_{yx}$), one can have

$$\boldsymbol{T} = \begin{bmatrix} T_x \\ T_y \end{bmatrix} = \begin{bmatrix} \sigma_x & \tau_{xy} \\ \tau_{xy} & \sigma_y \end{bmatrix} \begin{bmatrix} \cos\gamma \\ \sin\gamma \end{bmatrix} = \boldsymbol{\sigma N}_T \tag{6.89}$$

This relation is valid for any stress state at any angle γ. Therefore, for the *principal stresses* σ_1 and σ_2, one can write

$$\begin{aligned} \boldsymbol{T}_1 &= \sigma_1 \boldsymbol{N}_1 = \begin{bmatrix} \sigma_1 & 0 \\ 0 & 0 \end{bmatrix} \boldsymbol{N}_1 \\ \boldsymbol{T}_2 &= \sigma_2 \boldsymbol{N}_2 = \begin{bmatrix} 0 & 0 \\ 0 & \sigma_2 \end{bmatrix} \boldsymbol{N}_2 \end{aligned} \tag{6.89}$$

The direction cosines $\boldsymbol{N}_1$ and $\boldsymbol{N}_2$ or the *principal directions* have been obtained by transformation of coordinates through the angular rotations γ_1 and γ_2 for the stress state or *stress tensor* given by

$$\boldsymbol{\sigma} = \begin{bmatrix} \sigma_x & \tau_{xy} \\ \tau_{xy} & \sigma_y \end{bmatrix} \tag{6.90}$$

Equation (6.89) can then be rewritten in the following form:

$$\begin{aligned} \boldsymbol{T}_1 &= \boldsymbol{\sigma N}_1 \quad \text{or} \quad \sigma_1 \boldsymbol{N}_1 = \boldsymbol{\sigma N}_1 \\ \boldsymbol{T}_2 &= \boldsymbol{\sigma N}_2 \quad \text{or} \quad \sigma_2 \boldsymbol{N}_2 = \boldsymbol{\sigma N}_2 \end{aligned} \tag{6.91}$$

or in general,

$$\sigma_i \boldsymbol{N}_i = \boldsymbol{\sigma N}_i \tag{6.92}$$

Equation (6.92) is the well-known ***eigenvalue problem*** in *vector space* transformation or matrix characteristic, which is defined in vector space as

$$\lambda \boldsymbol{x} = \boldsymbol{A x} \tag{6.93}$$

where λ is the *eigenvalue*, and $\boldsymbol{x}$ is the *eigenvector* of the matrix $\boldsymbol{A}$. Comparing Eqs. (6.92) and (6.93), it is evident that the ***principal stresses*** σ_i*'s* are the **eigenvalues** and the ***principal directions*** $\boldsymbol{N}_i$*'s* are the ***eigenvectors*** of the *stress state matrix* or *stress tensor* $\boldsymbol{\sigma}$ defined by Eq. (6.90).

6.4.1.4 Stress Invariants I_i

Further manipulation of Eq. (6.92) provides the procedure of finding the characteristic equation of the eigenvalue problem. Equation (6.92) can be put in the following form:

$$[\boldsymbol{\sigma} - \sigma \boldsymbol{I}]\boldsymbol{N}_i = 0 \tag{6.94}$$

where the *identity matrix* $\boldsymbol{I}$ is a matrix with unity diagonal terms and all other terms are zeros. For this expression in Eq. (6.94) to be valid, the determinant of $(\boldsymbol{\sigma} - \sigma \boldsymbol{I})$ should vanish or

$$\begin{vmatrix} \sigma_x - \sigma & \tau_{xy} \\ \tau_{xy} & \sigma_y - \sigma \end{vmatrix} = 0 \tag{6.95}$$

This gives the following *characteristic equation*:

$$\sigma^2 - (\sigma_x + \sigma_y)\sigma + (\sigma_x \sigma_y - \tau_{xy}^2) = 0 \tag{6.96}$$

The two roots of this characteristic equation are the *principal stresses* or *eigenvalues*. The invariants I_1 and I_2 of Eq. (6.96) are then

$$I_1 = (\sigma_x + \sigma_y) \quad \text{and} \quad I_2 = (\sigma_x \sigma_y - \tau_{xy}^2) \tag{6.97}$$

These *stress invariants* are constant irrespective of the orientation of the coordinates of the stress state and thus invariant.

6.4.1.5 A Common Stress State

In plane stress state, a common case is that of elements subjected to bending and transverse shear similar to Figure 6.14 or elements subjected to bending and torsional shear that gives a similar loading to that. In these cases, only one normal stress σ_x and another shear stress τ_{xy} exist with the other normal stress $\sigma_y = 0$. Such a case is represented by the tiny element shown in Figure 6.16a. The *Mohr's* circle is the simplest way to get the principal stresses and the principal directions. One needs only to sketch the two stress states on the surfaces A and B of Figure 6.16a as the abscissa σ and the ordinate τ for each point A and B as shown in Figure 6.16b. From that, the

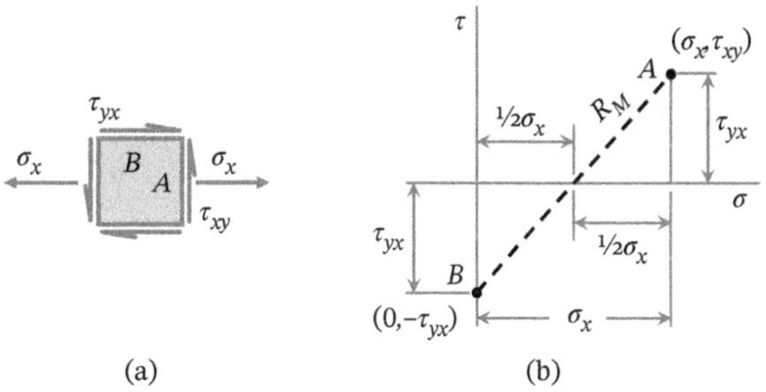

Figure 6.16 A common plane stress state shown in (a) and the basic construction points A and B of the *Mohr's circle* are given in (b).

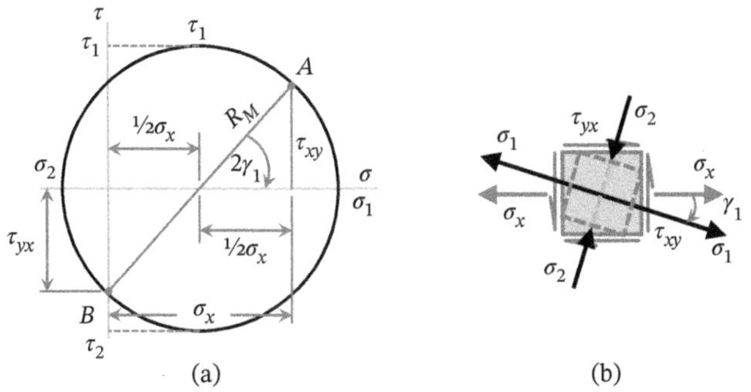

Figure 6.17 The *Mohr's circle* of Figure 6.16 is redrawn in (a) and the original tiny element is redrawn in (b), additionally showing the orientation γ_1 of the principal stress σ_1 and the principal stress σ_2 normal to it. The tiny principal element is dashed and reduced in size for clarity.

center of the circle is defined as $\frac{1}{2}\,\sigma_x$ and the *Mohr's* circle radius R_M is also defined from other lengths in the sketch. It is then easily stated that

$$\begin{aligned} R_M &= \sqrt{\left(\tfrac{1}{2}\sigma_x\right)^2 + \tau_{xy}^2} \\ \sigma_{1,2} &= \tfrac{1}{2}\sigma_x \pm R_M \\ \tau_{1,2} &= \pm R_M \\ \tan 2\gamma &= \frac{\tau_{xy}}{\frac{1}{2}\sigma_x} \end{aligned} \tag{6.98}$$

The *Mohr's* circle is drawn in Figure 6.17a and the original tiny element is also redrawn in Figure 6.17b additionally showing the orientation γ of the principal stress σ_1 and the principal stress σ_2 normal to it.

6.4.2 Triaxial Stress State

A general stress state where all stress components in ***3D*** may exist is called a *triaxial stress state*. Such a case is represented by an infinitesimal segment in Figure 6.18a for an inside fragment of a fully stressed machine element. The stress state in this case is represented by the following fully populated *stress matrix* or *stress tensor*:

$$\boldsymbol{\sigma} = \begin{bmatrix} \sigma_x & \tau_{xy} & \tau_{xz} \\ \tau_{yx} & \sigma_y & \tau_{yz} \\ \tau_{zx} & \tau_{zy} & \sigma_z \end{bmatrix} \tag{6.99}$$

A *stress tensor*, however, is usually defined by the *indicial* or *tensor notation* τ_{ij} rather than the adapted stress matrix $\boldsymbol{\sigma}$. The normal stress σ_x is the *tensor* component τ_{xx}. Similarly, the tensor component τ_{yy} is presumed for

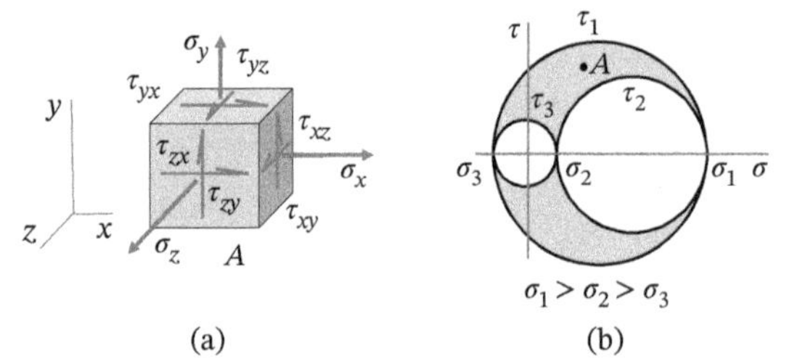

Figure 6.18 An infinitesimal segment A for an inside fragment of a fully stressed machine element is shown in (a). Point A representing the triaxial stress state in (a) is located inside the grayish region of the *Mohr's circles* in (b).

σ_y and the tensor component τ_{zz} is for σ_z. As indicated previously in Section 6.4.1, the shear stresses are mutually related such that the tensor components $\tau_{xy} = \tau_{yx}$, $\tau_{xz} = \tau_{zx}$, and $\tau_{yz} = \tau_{zy}$. The *stress matrix* or *stress tensor* is thus symmetric and given by

$$\boldsymbol{\sigma} = \begin{bmatrix} \sigma_x & \tau_{xy} & \tau_{xz} \\ \tau_{xy} & \sigma_y & \tau_{yz} \\ \tau_{xz} & \tau_{yz} & \sigma_z \end{bmatrix} \tag{6.100}$$

The eigenvalue problem in vector space transformation or matrix characteristic is used as for the case of the plane stress state to find the *principal stresses* for the triaxial state. These are formulated similarly as defined in the following form:

$$\sigma_i \mathbf{N}_i = \boldsymbol{\sigma} \mathbf{N}_i, i = 1, 2, 3 \qquad \textbf{\textit{(6.92)}}$$

where σ_i is the *principal stress*, $\mathbf{N}_i$ is the direction cosine or the *principal direction*, and the triaxial stress-matrix or *stress tensor* $\boldsymbol{\sigma}$ is defined by Eq. (6.99). For the triaxial stress state, the ***principal stresses*** σ_1, σ_2, and σ_3 are the ***eigenvalues*** and the **principal directions $\mathbf{N}_i$'s** are the ***eigenvectors***. The *Mohr's circles* defining this stress state are demonstrated in Figure 6.17b. However, point A inside the grayish region between the three *Mohr's circles* in Figure 6.17b defines the *triaxial stress state*. Point A would be located on one of the *Mohr's circles* in Figure 6.17b, if no stresses exist on any of the faces (or planes) of the infinitesimal segment in Figure 6.17a. Further proof of that can be found in the literature such as Reisman and Pawlik (1980, 1991).

6.4.2.1 Stress Invariants I_i

Similar to Section 6.4.1 and Eq. (6.95), the *characteristic equation* for the triaxial stress state can be obtained by the determinant of the stress matrix in Eq. (6.99), which gives

$$\begin{aligned} &\sigma^3 - (\sigma_x + \sigma_y + \sigma_z)\sigma^2 + (\sigma_x\sigma_y + \sigma_z\sigma_x + \sigma_y\sigma_z - \tau_{xy}^2 - \tau_{zx}^2 - \tau_{yz}^2)\sigma \\ &(\sigma_x\sigma_y\sigma_z - \sigma_x\tau_{yz}^2 - \sigma_y\tau_{zx}^2 - \sigma_z\tau_{xy}^2 + 2\tau_{xy}\tau_{zx}\tau_{yz}) = 0 \end{aligned} \tag{6.101}$$

The three roots of this characteristic equation are also the *principal stresses* or *eigenvalues*. The invariants I_1, I_2, and I_3 of Eq. (6.101) are then

$$\begin{aligned} I_1 &= (\sigma_x + \sigma_y + \sigma_z) \\ I_2 &= (\sigma_x\sigma_y + \sigma_z\sigma_x + \sigma_y\sigma_z - \tau_{xy}^2 - \tau_{zx}^2 - \tau_{yz}^2) \\ I_3 &= (\sigma_x\sigma_y\sigma_z - \sigma_x\tau_{yz}^2 - \sigma_y\tau_{zx}^2 - \sigma_z\tau_{xy}^2 + 2\tau_{xy}\tau_{zx}\tau_{yz}) \end{aligned} \tag{6.102}$$

The average *hydrostatic* stress σ_{av} and the *distortion stress* σ_d (or *deviatoric or von Mises* (1913), after *Richard E. von Mises (1883–1953)*), are then defined by the following equations:

$$\sigma_{av} = \frac{(\sigma_1 + \sigma_2 + \sigma_3)}{3} \tag{6.103}$$

$$\sigma_d = \left(\frac{(\sigma_1 - \sigma_2)^2 + (\sigma_2 - \sigma_3)^2 + (\sigma_3 - \sigma_1)^2}{2}\right)^{1/2} \tag{6.104}$$

The *distortion stress* is obtained by subtracting the *hydrostatic stress* from each *principal stress*. The *hydrostatic* stress σ_{av} and the *distortion* (or *deviatoric* or *von Mises*) stress σ_d are essential in the subject of failure theories discussed in Section 7.7.

Example 6.10 A shaft is transmitting a maximum power of 25 [kW] or 33.5 [hp] and running at 3000 [rpm] in the clockwise direction. It is also subjected to a maximum bending moment that is almost equal to the maximum applied torque. Find the internal stresses if the shaft cylindrical diameter is 45 [mm] or 1.75 [in]. Obtain the principal stresses and identify their directions or orientations. Demonstrate that the all the invariants $\boldsymbol{I}_i$ are constants. Find the average stress and the distortion or deviatoric stress for this state of stress.

Solution

Data: $H = 25$ [kW] or 33.5 [hp], $N_{rpm} = 3000$ [rpm], and $d_C = 0.045$ [m] or 1.75 [in].

The shaft is modeled similar to the generally loaded cylindrical element depicted in Figure 6.19a. The shaft, however, is not subjected to the axial load $\boldsymbol{F}$ and the torque $\boldsymbol{T}$ is in the opposite direction. Figure 6.19b shows the actual loaded shaft and an infinitesimal element in its lower fiber expanded to show the internal stresses on that element. These internal stresses are obtained according to the following procedure.

Equation (6.72) is used to evaluate the maximum applied torque or

$$H_W = \frac{2\pi N_{rpm}}{60} T_{N\,m}, \quad \text{or} \quad T_{N\,m} = \frac{60\, H_W}{2\pi\, N_{rpm}} = \frac{60(25\,000)}{2\pi(3000)} = 79.577\ [\text{N m}]$$
$$H_{hp} = \frac{N_{rpm} T_{lb\,in}}{63025}, \text{or} \quad T_{lb\,in} = \frac{(63\,025)H_{hp}}{N_{rpm}} = \frac{(63\,025)(33.5)}{3000} = 703.78\ [\text{lb in}] \tag{a}$$

Since the bending moment $\boldsymbol{M}_z$ is assumed to have the same value as the torque, then

$$\boldsymbol{M}_z = \boldsymbol{T}_{N\,m} = 79.577\ [\text{N m}]$$
$$\boldsymbol{M}_z = \boldsymbol{T}_{lb\,in} = 703.78\ [\text{lb in}] \tag{b}$$

The maximum shear stresses due to torque are obtained according to Eq. (6.68), which gives

$$\tau_{max} = \frac{16\,\boldsymbol{T}}{\pi d_C^3} = \frac{16(79.577)}{\pi(0.045)^3} = 4\,447\,542\ [\text{Pa}] = 4.4475\ [\text{MPa}]$$
$$\tau_{max} = \frac{16\,\boldsymbol{T}}{\pi d_C^3} = \frac{16(703.78)}{\pi(1.75)^3} = 668.79\ [\text{psi}] \tag{c}$$

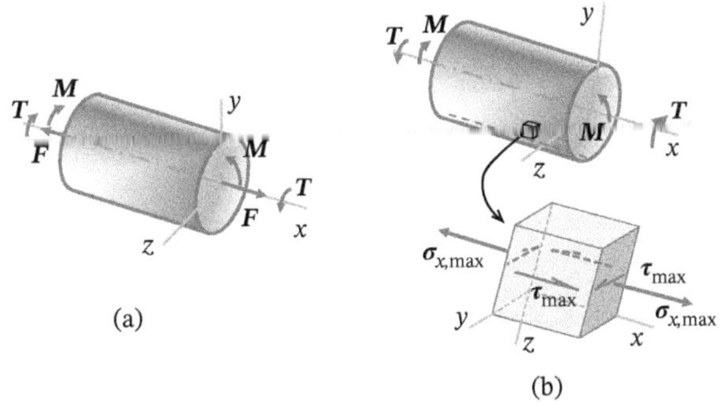

Figure 6.19 The shaft is modeled similar to cylindrical element depicted in (a), while (b) shows the actually loaded shaft and an infinitesimal element in its lower fiber expanded to show the internal stresses on that element.

The maximum normal stresses due to bending moment are obtained according to Eq. (6.68), which gives

$$\sigma_{x,\max} = \frac{32\boldsymbol{M}_z}{\pi d_C^3} = \frac{32(79.577)}{\pi(0.045)^3} = 8\,895\,084\ [\text{Pa}] = 8.8951\ [\text{MPa}]$$
$$\sigma_{x,\max} = \frac{32\boldsymbol{M}_z}{\pi d_C^3} = \frac{32(703.78)}{\pi(1.75)^3} = 1337.6\ [\text{psi}] \quad \text{(d)}$$

The principal stresses are obtained by the *Mohr's* circle defined in Figure 6.20a. The values of the stresses are reduced to lesser significant figures to allow for the *Mohr's* circle to be sketched. The line defining the *Mohr's* circle diameter is connecting the stresses on the x plane $(\sigma_{x\max}, \tau_{\max}) = (8.9, 4.45)$ and the stresses on the z plane $(\sigma_{z\max}, \tau_{\max}) = (0.0, -4.45)$ of the infinitesimal element in Figure 6.20b. The principal stresses according to Equation (6.98) are as follows:

$$R_M = \sqrt{\left(\tfrac{1}{2}\sigma_x\right)^2 + \tau_{xz}^2} = \sqrt{\left(\tfrac{1}{2}(8.8951)(10^6)\right)^2 + (4.4475(10^6))^2} = 6\,289\,715\ [\text{Pa}] = 6.2897\ [\text{MPa}]$$
$$R_M = \sqrt{\left(\tfrac{1}{2}\sigma_x\right)^2 + \tau_{xz}^2} = \sqrt{(\tfrac{1}{2}(1337.6))^2 + (668.79)^2} = 945.81\ [\text{psi}] \quad \text{(e)}$$

$$\sigma_1 = \tfrac{1}{2}\sigma_x + R_M = \tfrac{1}{2}(8.8951)(10^6) + 6\,289\,715 = 10\,737\,215\ [\text{Pa}] = 10.737\ [\text{MPa}]$$
$$\sigma_1 = \tfrac{1}{2}\sigma_x + R_M = \tfrac{1}{2}(1337.6) + 945.81 = 1614.6\ [\text{psi}] = 1.615\ [\text{kpsi}]$$
$$\sigma_2 = \tfrac{1}{2}\sigma_x - R_M = \tfrac{1}{2}(8.8951)(10^6) - 6\,289\,715 = -1\,842\,165\ [\text{Pa}] = -1.8422\ [\text{MPa}] \quad \text{(f)}$$
$$\sigma_2 = \tfrac{1}{2}\sigma_x - R_M = \tfrac{1}{2}(1337.6) - 945.81 = -277.01\ [\text{psi}] = -0.277\ [\text{kpsi}]$$

The directions of the principal stresses are defined by Eq. (6.98) for the first principal direction such that

$$\tan 2\gamma|_{SI} = \frac{\tau_{xz}}{\tfrac{1}{2}\sigma_x} = \frac{4.4475(10^6)}{\tfrac{1}{2}(8.8951)(10^6)} = 0.999\,99 \cong 1.0$$
$$\tan 2\gamma|_{US} = \frac{\tau_{xz}}{\tfrac{1}{2}\sigma_x} = \frac{668.79}{\tfrac{1}{2}(1337.6)} = 0.999\,99 \cong 1.0 \quad \text{(g)}$$

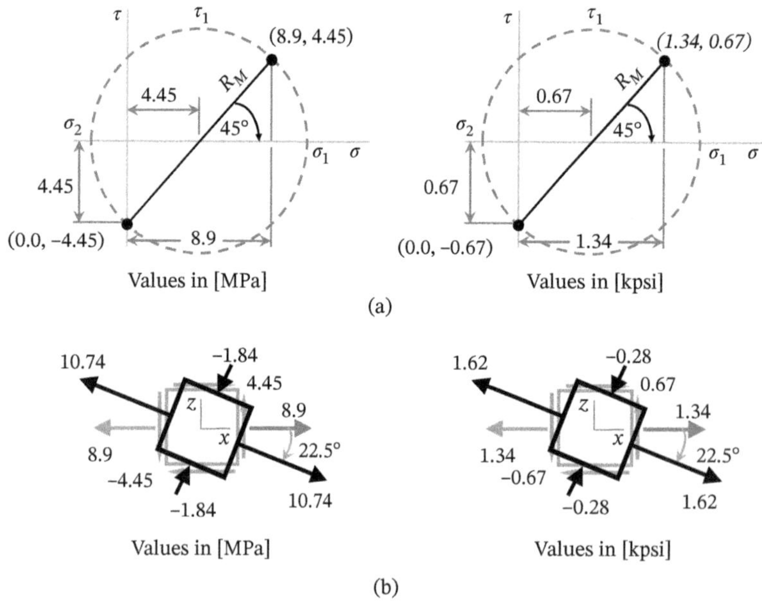

Figure 6.20 Principal stresses obtained by sketching the *Mohr's circle* defined in (a) for SI and US systems. The values of the stresses are reduced to lesser significant figures to allow for the *Mohr's circle* to be sketched. Stresses on the z plane and the principal stresses and directions are shown in (b) for SI and US systems.

Equation (g) states that $2\gamma = 45°$ or $\gamma = 22.5°$ in the clockwise direction. The other principal direction is normal to the first principal direction. Figure 6.20b illustrates the directions of the principal stresses σ_1 and σ_2 in addition to showing their abridged values. The figure also provides the original stresses on the infinitesimal element in a grayish color and the principal stresses in black color. It should be noted that the infinitesimal element projection on the xz plane (or the y plane) is the considered case in here. Same results are expected, if xy plane is used instead.

The maximum and minimum shear stresses are

$$\begin{aligned} \tau_{1,2} &= \pm R_M = \pm 6.2897\ [\text{MPa}] \\ \tau_{1,2} &= \pm R_M = \pm 945.81\ [\text{psi}] \end{aligned} \tag{h}$$

The invariants at the initial orientation of the infinitesimal element are calculated from Eq. (6.97) and bearing in mind that the coordinates are x and z instead of x and y with $\sigma_z = 0.0$ to give the following values:

$$\begin{aligned} I_1 &= (\sigma_x + \sigma_z) = \sigma_{x,\max} = 8.8951\ [\text{MPa}] \\ I_1 &= (\sigma_x + \sigma_z) = \sigma_{x,\max} = 1337.6\ [\text{psi}] \\ I_2 &= (\sigma_x\sigma_z - \tau_{xz}^2) = -\tau_{xz}^2 = -(4.4475)^2 = -19.780\ [\text{MPa}]^2 \\ I_2 &= (\sigma_x\sigma_z - \tau_{xz}^2) = -\tau_{xz}^2 = -(668.79)^2 = -447\,280\ [\text{psi}]^2 \end{aligned} \tag{i}$$

The invariants at the principal coordinates of the infinitesimal element are also calculated from Eq. (6.97) and taking into consideration that the shear stresses are zeros at this orientation, which gives

$$\begin{aligned} I_1 &= (\sigma_1 + \sigma_2) = 10.737 - 1.8422 = 8.895\ [\text{MPa}] \\ I_1 &= (\sigma_1 + \sigma_2) = 1614.6 - 277.01 = 1337.6\ [\text{psi}] \\ I_2 &= (\sigma_1\sigma_2 - 0.0^2) = 10.737(-1.8422) = -19.780\ [\text{MPa}]^2 \\ I_2 &= (\sigma_1\sigma_2 - 0.0^2) = 1614.6\,(-277.01) = -447\,260\ [\text{psi}]^2 \end{aligned} \tag{j}$$

The values in Eqs. (i) and (j) suggest that the invariants I_i are constants. The very small discrepancies are very much due to the rounding off in the respective values.

The average stress is definable by Eq. (6.103) that gives

$$\begin{aligned} \sigma_{av} &= \frac{(\sigma_1 + \sigma_2 + \sigma_3)}{3} = \frac{10.737 - 1.8422}{3} = 2.9649\ [\text{MPa}] \\ \sigma_{av} &= \frac{(\sigma_1 + \sigma_2 + \sigma_3)}{3} = \frac{1.615 - 0.277}{3} = 0.446\ [\text{kpsi}] \end{aligned} \tag{k}$$

The distortion or deviatoric stress is definable according to Eq. (6.104), which provides

$$\begin{aligned} \sigma_d &= \left(\frac{(\sigma_1 - \sigma_2)^2 + (\sigma_2 - \sigma_3)^2 + (\sigma_3 - \sigma_1)^2}{2}\right)^{1/2} \\ &= \sqrt{\frac{(10.737 + 1.8422)^2 + (-1.8422)^2 + (-10.737)^2}{2}} = 11.7668\ [\text{MPa}] \\ \sigma_d &= \left(\frac{(\sigma_1 - \sigma_2)^2 + (\sigma_2 - \sigma_3)^2 + (\sigma_3 - \sigma_1)^2}{2}\right)^{1/2} \\ &= \sqrt{\frac{(1.615 + 0.277)^2 + (-0.277)^2 + (-1.615)^2}{2}} = 1.7698\ [\text{kpsi}] \end{aligned} \tag{l}$$

Note that the principal stresses σ_1 and σ_2 are used with the assumption that σ_3 is zero, where σ_2 is the smallest negative eigenvalue. In ***3D***, one should have used $\sigma_3 = \sigma_2$ and setting $\sigma_2 = 0$ to have $\sigma_1 > \sigma_2 > \sigma_3$. However, this ought not to change the outcome. This is for the reader to qualify.

Figure 6.21 shows MATLAB codes to calculate the principal stresses and principal directions. The usual *xy* plane is used in place of the real case of the *xz* plane in Example 6.10. Figure 6.21a calculates the principal stresses and principal directions utilizing the *Mohr's* circle in ***2D***. Figure 6.21b calculates the principal stresses and principal directions utilizing the solution of the *eigenvalue problem* in ***2D***. Figure 6.21c calculates the principal stresses and principal directions utilizing the solution of the *eigenvalue problem* in ***3D***. The depicted values in Figure 6.21 are for the SI system of units, but values of stresses in the US system should be used instead to get the correct evaluation in that system of units. These MATLAB codes can, however, be used for any other problem by just plugging in the appropriate values of the stresses for the problem in question. The semicolon (;) at the end of the equation is to be omitted so as to print the required result. The programs are also available through **Wiley website**.

6.4.3 Applications in Plane Stress and Triaxial Stress States

Many applications such as thin-walled pressure cylinders and some boilers pertain to plane stress state and several others such as special hydraulic cylinders, high pressure pipes, gun barrels, particular boilers, and pressure vessels are modeled as triaxial stress states. The difference is in the acceptable assumption that one of the triaxial stress states can be assumed as negligibly small with respect to other stresses. The general case is the triaxial stress state and the exceptional is the abridged plane stress state. Some of these applications are the pressure cylinders upon which they may be categorized as thin or thick. If the pressure is small and the synthesized thickness is small relative to the cylinder diameter, the cylinder is considered a *thin pressure cylinder*. Otherwise, it is a *thick pressure cylinder*. In fact, it can be considered as a thick cylinder and the synthesis solution proclaims that it could have been assumed a thin cylinder to start with. Figure 6.22 present two configurations of assumed thin pressure cylinder or vessel (a) and a thick pressure cylinder or vessel (b); see ASME (2013) and Abd El-Aziz and Metwalli (2017). Each is considered in the following Sections 6.4.3.1 and 6.4.3.2.

6.4.3.1 Thin Pressure Cylinders

Figure 6.22a shows a thin pressure cylinder subjected to an internal pressure p_i. To calculate the internal stresses, the pressure cylinder is sectioned by an axial *x–y* plane at *A–A* running along the longitudinal *x*-axis of the cylinder. The section along the side of the cylinder is shown as a grayish area of a dashed boundary. The other section is similar but has not been extended along the back side. The free body diagram of the half cylinder is exposed to a force due to the bearing pressure of a magnitude equals to $p_i\ (d_c\ l_c)$, where d_c is the average cylinder diameter and l_c is the length of the cylinder. The forces generated by the internal *tangential stresses* σ_t have the magnitude of the stress σ_t on each side multiplied by the longitudinal cut area of $t_c\ l_c$, where t_c is the cylinder thickness. Equating these forces produce the *tangential stresses* σ_t acting tangentially around the cylinder as follows.

$$\begin{aligned} p_i(d_c l_c) &= 2\sigma_t(t_c l_c) \\ \sigma_t &= \frac{p_i d_c}{2t_c} \end{aligned} \tag{6.105}$$

This tangential stress σ_t is also called *hoop stress* since it is acting tangentially around the whole cylinder.

To evaluate the *longitudinal* or *axial stress* σ_a a similar procedure is applied; see Figure 6.22a. Consider that the two ends of the cylinder are to be eventually closed. The forces on the circular cross-sectional area amounting to $p_i\ (\pi d_c^{\,2}/4)$ should be balanced with the internal axial forces generated from the axial stress σ_a acting on the cylindrical cross-sectional area of $\pi\ d_c\ t_c$. This provides the relation to evaluate the axial stress such that

$$\begin{aligned} p_i(\pi d_c^{\,2}/4) &= \sigma_a(\pi d_c t_c) \\ \sigma_a &= \frac{p_i d_c}{4t_c} \end{aligned} \tag{6.106}$$

Equations (6.105) and (6.106) indicate that the tangential stress σ_t is twice as much as the axial stress σ_a.

(a)
```
clear all; clc; format compact
Sigma_x = 8.9                              % [MPa] or [psi]
Sigma_y = 0                                % [MPa] or [psi]
Taw_xy = 4.45                              % [MPa] or [psi]

Raduis = sqrt((Sigma_x/2-Sigma_y/2)^2 + Taw_xy^2);
Center = (Sigma_x + Sigma_y)/2;

Sigma_1 = (Center) + Raduis                % [MPa] or [psi] and No ; so as to print
Sigma_2 = (Center) - Raduis                % [MPa] or [psi] and No ; so as to print

Gama_1 = (-atan(Taw_xy/(Sigma_x/2-Sigma_y/2))/2) * 180/pi
Gama_2 = Gama_1 + 90
```

(b)
```
clear all; clc; format compact
Sigma_x = 8.9;                                        % [MPa] or [psi]
Sigma_y = 0 ;                                         % [MPa] or [psi]
Taw_xy = 4.45;                                        % [MPa] or [psi]
Stress_Tensor = [Sigma_x Taw_xy; Taw_xy Sigma_y]      % 2-D
[v e]=eig(Stress_Tensor);
Principal_Stress_Tensor = e                           % No ; so as to print
Principal_Dirctions = v                               % No ; so as to print

% MATLAB orders the smaller eigenvalue first, the order is reversed here.
Sigma_1 = e(2,2)                                      % [MPa] or [psi]
Sigma_2 = e(1,1)                                      % [MPa] or [psi]

Alpha_1 = -atan(v(2,2)/v(2,1))* 180/pi
Alpha_2 = -atan(v(1,2)/v(1,1))* 180/pi
```

(c)
```
clear all; clc; format compact

Sigma_x = 8.9; Sigma_y = 0 ; Sigma_z = 0 ;            % [MPa] or [psi]
Taw_xy = 0; Taw_yx = 4.45; Taw_yz = 0;                % [MPa] or [psi]
Stress_Tensor =     [Sigma_x Taw_xy Taw_zx;
          Taw_xy Sigma_y Taw_yz;
          Taw_zx Taw_yz Sigma_z]

[v e]=eig(Stress_Tensor);
Principal_Stress_Tensor = e                           % No ; so as to print
Principal_Dirctions = v                               % No ; so as to print

Sigma_1 = e(3,3)                                      % [MPa] or [psi]
Sigma_2 = e(2,2)                                      % [MPa] or [psi]
Sigma_3 = e(1,1)                                      % [MPa] or [psi]

% Assuming z-plane has zero stresses and Sigma_2 = 0. (Special Case Only)
Alpha_1 = atan(v(2,3)/v(1,3))* 180/pi
Alpha_2 = atan(v(3,2)/v(1,2))* 180/pi
Alpha_3 = atan(v(2,1)/v(1,1))* 180/pi
```

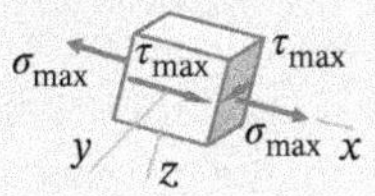

Figure 6.21 (a) MATLAB codes to calculate the principal stresses and principal directions: (a) utilizing the *Mohr's* circle in ***2D***, (b) utilizing the solution of the eigenvalue problem in ***2D***, and (c) using the solution of the eigenvalue problem in ***3D***.

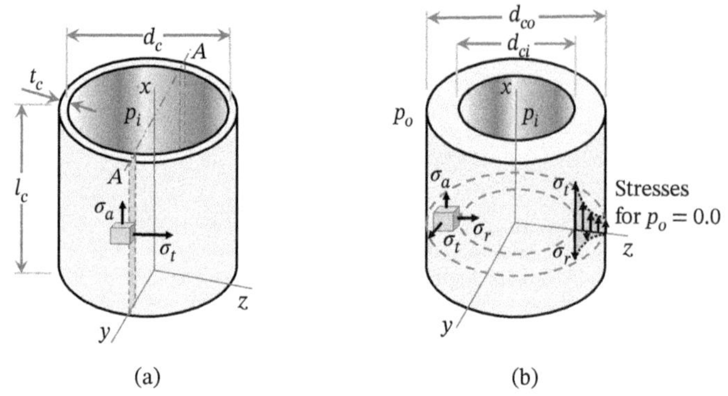

Figure 6.22 A configuration of assumed thin pressure cylinder or vessel subjected to an internal pressure in (a) and another thick pressure cylinder or vessel subjected to internal and external pressure in (b). Tangential and radial stress distributions are shown in (b) for internal pressure only.

The procedure of finding both *tangential* and *axial* stresses may suggest a more appropriate use of cylindrical coordinate axes rather than Cartesian coordinate axes. In that case, the radial stress across the cylinder thickness stemming from the internal pressure is very small relative to both *tangential* and *axial* stresses and the case is thus a *plane stress state.*

Since no shear exists in the *tangential* and *axial* stresses, the principal stresses are both *tangential* and *axial* stresses. This indicates that $\sigma_1 = \sigma_t$ and $\sigma_2 = \sigma_a = \sigma_1/2$. As previously indicated with this plane stress state $\sigma_3 = 0.0$ and the *Mohr's* circle is similar to Figure 6.18b adjusted for the values of $\sigma_1 = \sigma_t$, $\sigma_2 = \sigma_a = \sigma_1/2$ and $\sigma_3 = 0.0$ with two equal internal circles.

It should be noted that the ASME boiler and pressure vessel code (2013) uses similar equations to Eqs. (6.105) and (6.106), with adjustments to the longitudinal or circumferential welding process to manufacture the boiler or the pressure vessel; see Abd El-Aziz and Metwalli (2017).

6.4.3.2 Thick Pressure Cylinders

Figure 6.22b presents a thick pressure cylinder subjected to an internal pressure p_i and an outside pressure p_o. The internal cylindrical diameter is d_{ci} and the outside cylinder diameter is d_{co}. Assuming an eventually closed cylinder and that the case is away from the cylinder ends, the *tangential stresses* σ_t, the *radial stresses* σ_r, and the *axial stresses* σ_a at any cylindrical diameter d_c can be given by the following equations (see Reisman and Pawlik (1980, 1991)):

$$
\begin{aligned}
\sigma_t(d_c) &= \frac{p_i d_i^2 - p_o d_o^2}{d_o^2 - d_i^2} + \frac{(d_i^2 d_o^2)(p_i - p_o)}{(d_o^2 - d_i^2) d_c^2} \\
\sigma_r(d_c) &= \frac{p_i d_i^2 - p_o d_o^2}{d_o^2 - d_i^2} - \frac{(d_i^2 d_o^2)(p_i - p_o)}{(d_o^2 - d_i^2) d_c^2} \\
\sigma_a &= \frac{p_i d_i^2 - p_o d_o^2}{d_o^2 - d_i^2}
\end{aligned}
\tag{6.107}
$$

Note that the axial stress σ_a is constant at any diameter d_c as evident from Eq. (6.107).

When the outer pressure p_o is zero, Eq. (6.107) are reduced to the following expressions:

$$\sigma_t = \frac{p_i d_i^2}{d_o^2 - d_i^2} + \frac{(d_i^2 d_o^2)(p_i)}{(d_o^2 - d_i^2) d_c^2}$$

$$\sigma_r = \frac{p_i d_i^2}{d_o^2 - d_i^2} - \frac{(d_i^2 d_o^2)(p_i)}{(d_o^2 - d_i^2) d_c^2} \tag{6.108}$$

$$\sigma_a = \frac{p_i d_i^2}{d_o^2 - d_i^2}$$

The distributions of the tangential and radial stresses for $p_o = 0.0$ are depicted in Figure 6.22b. The axial stress σ_a is constant at any diameter d_c as also evident from Eq. (6.108).

The *principal stresses* are the *tangential*, *radial*, and *axial* stresses, which indicates that $\sigma_1 = \sigma_t$, $\sigma_2 = \sigma_a$, and $\sigma_3 = \sigma_r$ at any cylindrical diameter d_c for internal pressure exceeding the external pressure. As previously indicated with this triaxial stress state and usually for the inner diameter fibers, σ_r is negative (compressive), and σ_a is usually positive and smaller than σ_t. The *Mohr's* circle is similar to Figure 6.18b adjusted for the values of $\sigma_1 = \sigma_t$, $\sigma_2 = \sigma_a$, and $\sigma_3 = \sigma_r$. For the outer fibers of the cylinder, the order of the principal stresses depends on the magnitude of the outer and inner pressures such as a submarine hull in deeper depth below the sea surface.

6.4.3.3 Press and Shrink Fits

The equations of the thick pressure cylinders are used to advantage in the development of relations governing the press or shrink fit. When two cylinders are pressed or shrink-fitted together, the outside diameter of the inner cylinder 1 is compressed by the *shrink fitting pressure* p_F, whereas the inner diameter of the outer cylinder 2 is subjected to the same inner pressure p_F. Each of the cylinders will be deformed accordingly to have the same *interfacial shrink fit radius* r_F as shown in Figure 6.23. The outer radius of the inner cylinder 1 will be decreased in magnitude or deformed $-\delta_{1o}$. The inner diameter of the outer cylinder 2 will be increased in magnitude or deformed δ_{2i}. The total *interfacial shrink fit deformation* δ_F is then equal to $\delta_{2i} - \delta_{1o}$. These deformations are excessively exaggerated as partially shown in Figure 6.23. Each of these deformations can be obtained due to the *shrink fitting pressure* p_F such that (see Faupel and Fisher 1981 or Reisman and Pawlik 1980, 1991)

$$\delta_F = \delta_{2i} - \delta_{1o} = \frac{r_F\, p_F}{E_2}\left(\frac{r_{2o}^2 + r_F^2}{r_{2o}^2 - r_F^2} + \nu_2\right) + \frac{r_F\, p_F}{E_1}\left(\frac{r_F^2 + r_{1i}^2}{r_F^2 - r_{1i}^2} - \nu_1\right) \tag{6.109}$$

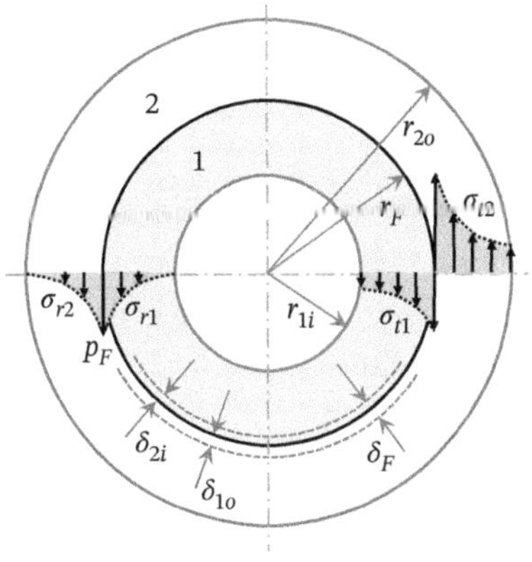

Figure 6.23 Two cylinders pressed or shrink-fitted together. The outside diameter of the inner cylinder 1 is compressed by the shrink fitting pressure p_F and the inner diameter of the outer cylinder 2 is subjected to the same inner pressure. Each cylinder is deformed to the same interfacial shrink fit radius r_F. Tangential and radial stress distributions are shown.

where E_1, E_2, and ν_1, ν_2 are the *moduli of elasticity* and *Poisson's* ratios for cylinder 1 and cylinder 2, respectively. Equation (6.17) is useful in defining the *fundamental deviation* δ_F of the *interference fit* to guarantee a required minimum shrink fit pressure p_F to withstand some maximum torque applied on cylinder 2, for instance. The maximum probable interference, however, is the *fundamental deviation* δ_F added to it, the tolerance zones of both cylinders; see Section 2.4.7 for the interference fit. This may be the case of a train wheel with a steel tier overlay ring shrunk-fitted onto it that should withstand the braking force applied onto it. On the other hand, if the total *interfacial shrink fit deformation* δ_F is known for a specific case, the *shrink fitting pressure* p_F can be evaluated from Eq. (6.109).

The tangential stresses generated due to the *shrink fitting pressure* p_F should be calculated to safeguard against cylindrical failure; see Section 7.7 for failure criterion. The tangential stresses due to the *shrink fitting pressure* p_F at the outer surface of the inner cylinder σ_{t1} and at the inner surface of the outer cylinder σ_{t2} are obtained by the application of Eq. (6.107), which gives

$$\begin{aligned} \sigma_{t1}(d_F) &= -p_i \frac{r_F^2 + r_{1i}^2}{r_F^2 - r_{1i}^2} \\ \sigma_{t2}(d_F) &= p_i \frac{r_{2o}^2 + r_F^2}{r_{2o}^2 - r_F^2} \end{aligned} \tag{6.110}$$

Since the cylinders are open, the axial stresses can be diminished particularly if the lengths of the cylinders are small relative to other dimensions. In that case, we would have a plane stress state. When the inner cylinder is a solid shaft, the relevant equations can be used with the inner radius r_{1i} assumed as zero. Stress distributions, however, should be adjusted accordingly.

The ensued tangential and radial stresses are normal to each other's and therefore both are the *principal stresses.* Note also the absence of shear stresses on these planes.

6.4.3.4 Contact Stresses

Contact stresses occur when two solid elastic bodies meet in contact under some loading force. Hertz (1882) developed the geometry and the *pressure distribution* at the contact area for such cases, and the stresses are dobbed *Hertzian stresses* in his honor, Heinrich Rudolf Hertz (1857–1894). The application of such cases is concerned with the mating of machine elements in contact such as ball bearings and gear teeth at the contact area. For ball bearings, the ball as a sphere is usually in contact with the bearing races in a torus like surface. At the spur gear contact region, it is assumed that *two cylindrical bodies* with different radii of curvature represent the contact case. The interest is to define the area of contact and maximum pressure developed between the two bodies in contact.

For ***spherical contact*** under radial force $\boldsymbol{F}$, the contact area is usually circular with a small circular diameter a_c as shown in Figure 6.24a. The pressure or *stress distribution* is not uniform as depicted in Figure 6.24a. The maximum pressure p_{max} (or *maximum compressive stress* σ_{max}) due to the normal radial force $\boldsymbol{F}$ is given by Hertz (1882).

$$p_{max} = \frac{6\boldsymbol{F}}{\pi a_c^2} = 1.5 p_{avg} = -\sigma_{max} \tag{6.111}$$

The average pressure p_{avg} is defined as the uniform stress over the area. The negative sign in front of the maximum stress σ_{max} indicates that the *Hertzian stress* is compressive. The *contact diameter* a_c of the circular contact area is defined by

$$a_c = \sqrt[3]{3\boldsymbol{F}\left(\frac{1-\nu_1^2}{E_1} + \frac{1-\nu_2^2}{E_2}\right)\left(\frac{d_{S1} d_{S2}}{d_{S1} + d_{S2}}\right)} \tag{6.112}$$

The symbols E_1, E_2, and ν_1, ν_2 are the *moduli of elasticity* and *Poisson's* ratios for the materials of sphere 1 and sphere 2, respectively. If one of the bodies is a concave spherical groove, its diameter can be substituted as a negative value in Eq. (6.112). If one of the bodies is a plane surface, its diameter can be substituted as an infinite value in Eq. (6.112).

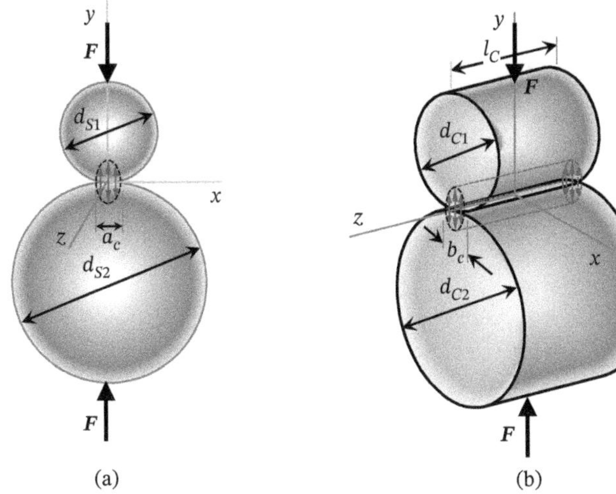

Figure 6.24 Contact zones and contact stress distribution for (a) two spherical bodies and (b) two cylindrical bodies; both under the radially applied forces $\boldsymbol{F}$ on each part of the bodies.

The maximum *tangential stress* σ_t in the contact area occurs at the perimeter of the contact area and its value is given by $\sigma_t = p_{max}(1 - 2\nu)/3$. The *maximum shear stress* τ_{max} occurs at a depth of 0.319 a_c. The value of the *maximum shear stress* is $\tau_{max} = p_{max}/3$.

For ***cylindrical contact***, the area is rectangular along the mating cylinders' lengths l_c and the small rectangular breadth b_c along the contact centerline. The maximum pressure p_{max} (or maximum compressive stress σ_{max}) at the centerline of the contact surface of length l_c due to the normal radial force $\boldsymbol{F}$ is given by Hertz (1882).

$$p_{max} = \frac{4\boldsymbol{F}}{\pi b_c l_c} = \frac{4}{\pi} p_{avg} = -\sigma_{max} \tag{6.113}$$

The average pressure p_{avg} is defined as the uniform stress over the rectangular contact area. The negative sign in front of the maximum stress σ_{max} indicates also that the *Hertzian* stress is compressive. The *breadth of contact area* b_c is defined by

$$b_c = \sqrt{\frac{16\boldsymbol{F}}{\pi l_c}\left(\frac{1-\nu_1^2}{E_1} + \frac{1-\nu_2^2}{E_2}\right)\left(\frac{d_{c1}d_{c2}}{d_{c1}+d_{c2}}\right)} \tag{6.114}$$

The symbols E_1, E_2, and ν_1, ν_2 are the *moduli of elasticity* and *Poisson's* ratios for the materials of cylinder 1 and cylinder 2, respectively. If one of the bodies is a concave cylindrical groove, its diameter can be substituted as a negative value in Eq. (6.114). If one of the bodies is a plane surface, its diameter can be substituted as an infinite value in Eq. (6.114).

The *maximum shear stress* τ_{max} occurs at a depth of 0.393 b_c. The value of the *maximum shear stress* is $\tau_{max} = 0.304$ p_{max} for a *Poisson's* ratio of 0.3.

6.5 Curved Beams

A curved beam of a symmetrical rectangular cross section and a center of curvature at 0 loaded by a pure bending $\boldsymbol{M}$ is shown in Figure 6.25. The rectangular cross section and its element of area dA along the z-axis (in the y–z plane) are rotated to facilitate defining the *neutral axis radius* r_n. The radius of the centroidal axis is R_c. At an angle θ and due to the moment $\boldsymbol{M}$, the cross section (grayish dashed line A–A) is rotated an extremely exaggerated angle $d\theta$ about the *neutral axis* to produce the blackish dashed line B–B. At the radius r, the grayish arch $r\theta$ is

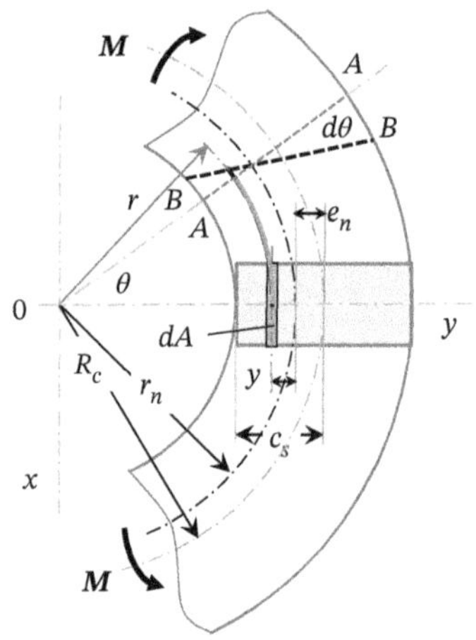

Figure 6.25 Curved beam of a symmetrical rectangular cross section and a center of curvature at 0 is loaded by a pure bending. The cross section and its element dA along the z-axis (in the y–z plane normal to the page) are rotated to facilitate defining the neutral axis of radius r_n. The radius of the beam centroidal axis is R_c.

therefore stretched by an amount $(r_n - r)\,d\theta$ that is shown as a small blackish arch at its end. The *strain* ε of that fiber is defined by its stretch (the small blackish arch) divided by its original length (the grayish arch); therefore

$$\varepsilon = \frac{(r_n - r)d\theta}{r\theta} \tag{6.115}$$

Applying *Hook's* law, one gets the normal stress σ due to the strain in Eq. (6.115) as

$$\sigma = \varepsilon E = \frac{E(r_n - r)d\theta}{r\theta} \tag{6.116}$$

where E is the *modulus of elasticity*. Since only the bending moment $\boldsymbol{M}$ is applied with no axial forces, the sum of the internal normal forces acting on the cross section should vanish. This is obtained by integrating the stresses σ over the area dA or

$$\int \sigma\, dA = \int \frac{E\,(r_n - r)\,d\theta}{r\theta} dA = E\frac{d\theta}{\theta}\left(r_n \int \frac{1}{r} dA - \int dA\right) = 0 \tag{6.117}$$

Solving the last term in parentheses in Eq. (6.117) gives the *neutral axis location* such that

$$r_n \int \frac{1}{r} dA = A_S$$

or (6.118)

$$r_n = \frac{A_S}{\int \frac{1}{r} dA}$$

where the parameter A_s is the cross-sectional area. The stress distribution is derived by equating the externally applied moment $\boldsymbol{M}$ with the internally developed counteracting moment. This gives – in lieu of Eq. (6.116) – the relation

$$\int (r_n - r)(\sigma dA) = E\frac{d\theta}{\theta}\int \frac{(r_n - r)^2}{r} dA = \boldsymbol{M} \tag{6.119}$$

Substituting for $(r_n - r)^2 = (r_n{}^2 - 2r_n r + r^2)$ and considering Eq. (6.117), one gets the following:

$$\boldsymbol{M} = E\frac{d\theta}{\theta}\left(r_n^2 \int \frac{1}{r} dA - r_n \int dA - r_n \int dA + \int r dA\right)$$

or (6.120)

$$\boldsymbol{M} = E\frac{d\theta}{\theta}\left(-r_n \int dA + \int r dA\right)$$

The first two terms in the first expression in Eq. (6.120) vanish due to Eq. (6.117). The second expression in Eq. (6.120) becomes

$$\boldsymbol{M} = E\frac{d\theta}{\theta}(-r_nA_s + R_cA_s) \tag{6.121}$$

or

$$\boldsymbol{M} = E\frac{d\theta}{\theta}e_nA_s \tag{6.122}$$

Here e_n is the *neutral axis eccentricity* of the centroidal axis as depicted in Figure 6.25, and A_s is the area of the cross section. Substituting Eq. (6.122) into Eq. (6.116) produces the following stress distribution at any point of a distant y off the neutral axis:

$$\sigma = \frac{\boldsymbol{M}y}{e_nA_c(r_n - y)} \tag{6.123}$$

This stress distribution has a *hyperbolic form* as shown in Figure 6.26a for the *rectangular section* as well as for the *circular* and the *"I"* sections shown in Figure 6.26b,c, respectively. The maximum stress occurs at the inner fiber of the curve, and the minimum stress occurs at the outer fiber of the curve. The values of the maximum and minimum stresses are then

$$\sigma_{\max} = \frac{\boldsymbol{M}c_i}{e_nA_cr_i} \quad \text{and} \quad \sigma_{\min} = \frac{\boldsymbol{M}c_o}{e_nA_cr_o} \tag{6.124}$$

For the rectangular section in Figure 6.26a, the *neutral axis* location can be obtained using Eq. (6.118), which gives

$$r_n = \frac{A_S}{\int \frac{1}{r}dA} = \frac{b_Rh_R}{\int_{r_i}^{r_o} \frac{b_R}{r}dr} = \frac{h_R}{\ln \frac{r_o}{r_i}} \tag{6.125}$$

For other sections, the location of the *neutral axis* can be obtained by Eq. (6.118) or a numerical integration of the equation that can take the following form:

$$r_n = \frac{\sum dA}{\sum \frac{dA}{r}} \tag{6.126}$$

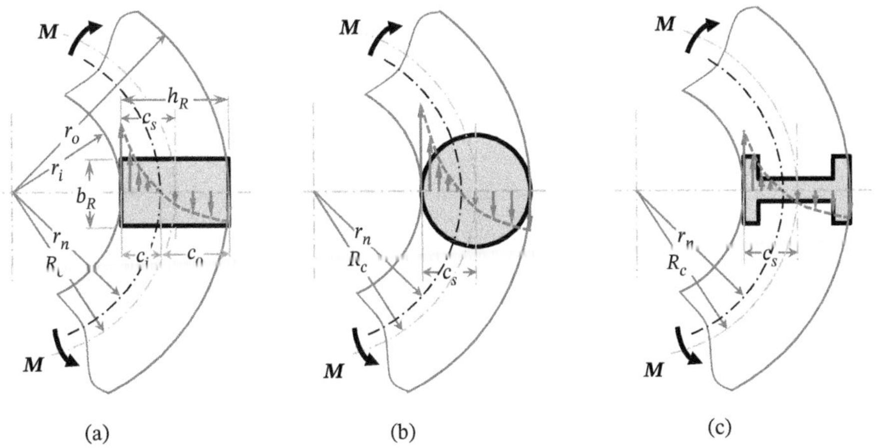

Figure 6.26 Curved beams under pure bending with symmetrical cross sections: (a) a rectangular section, (b) a circular cross section, and (c) an I section. The section inner fiber is at c_s distance from the curved beam radius R_c. The curvature ratio R_c/c_s in the figures is at 3.0. The stress distributions over the cross sections are also shown.

One can divide the section into 10–20 segments of dA and substitute of r for the center of each segment in Eq. (6.126) as depicted in Figure 6.25. As the section gets more complex, the number of segments is expected to increase so that the desired accuracy is achieved.

The symmetrical cross sections of a *rectangular*, a *circular*, and an *"I" section* are shown in Figure 6.26, where the stress distribution is approximately illustrated. When some defined proportions to these beams are set, one can get the graphs shown in Figure 6.27. These graphs define the *inner fiber factor* K_{ci} at different curvature ratio R_c/c_s which is multiplied directly in the following *normal stress* calculations representing Eq. (6.124):

$$\sigma_{x,\max} = \pm\frac{K_{ci}\boldsymbol{M}_z c_s}{I_z} \quad \text{and} \quad \sigma_{x,\min} = \pm\frac{K_{co}\boldsymbol{M}_z c_s}{I_z} \tag{6.127}$$

where K_{co} is the *outer fiber factor* in the minimum normal stress expression and c_s is the distance of the cross section's outer fiber from the centroidal axis as shown in Figures 6.25 and 6.26. Equation (6.127) is similar to the regular normal stress of Eq. (6.13) for straight bars under bending with the additional factor K_{ci} to account for the curvature ratio R_c/c_s. The positive or negative sign in Eq. (6.127) is for tensile or compressive stress, respectively, which is affected by the sign of the bending moment $\boldsymbol{M}_z$. In Figure 6.27, the rectangular cross section denoted by "R" is independent of dimensions or dimension ratios. The designated "I" section, however, is having a thickness that is one-third the width and one-sixth the height of the cross section. Figure 6.27a provides the values for the inner fiber factor K_{ci} at different curvature ratio R_c/c_s. The outer fiber factor K_{co} at different curvature ratio R_c/c_s is smaller than 1.0 and its effect can be deduced from the ratio of inner to outer factor K_{ci}/K_{co} at different curvature ratio R_c/c_s that is given in Figure 6.27b. This indicates the expected ratio of tensile to compressive strength of the material if that is available in some materials such as *gray cast iron*. From these figures, the effect of curvature on the normal stresses is pronounced for curvature ratios less than 2.0 or 3.0. Higher than that, the effect of curvature can be assimilated in the factor of safety, which should be modestly higher to consider the curvature effect.

It should be noted that many applications of curved machine elements such as crane hooks, clamps, and C frames usually experience normal forces in addition to the pure bending considered herein. In these cases superposition can be employed to account for the total loading case; see Nassef et al. (1999 and 2000) and Hegazi et al. (2002).

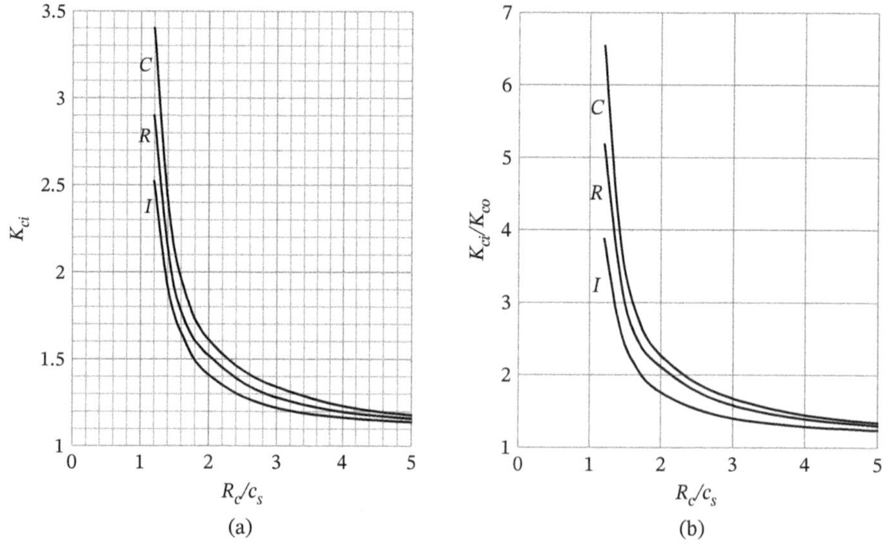

Figure 6.27 Curvature factors of circular –"C"-, rectangular –"R"-, and –"I"- sections: (a) for the inner fiber factor K_{ci} at different curvature ratio R_c/c_s and (b) for the ratio of inner to outer factors K_{ci}/K_{co} at different curvature ratio R_c/c_s.

6.6 Strain Energy and Deflection

The concept of strain has been introduced previously in Sections 6.1, 6.3, and 6.5. Here, we reiterate and further define the concept for completion.

6.6.1 Elastic Strain

The longitudinal or *normal elastic strain* ε_x in one direction x is the deformation u in the x direction per unit length l_x or it is the unit change in length given by

$$\varepsilon_x = \frac{du}{dx} \quad \text{or} \quad \varepsilon_x = \frac{u}{l_x} \tag{6.128}$$

The second expression is valid for a uniformly distributed deformation over the original length l_x; see Eq. (6.57).

For *plane strain* where the deformation in x direction is u and the deformation in y direction is v, the *longitudinal* or *normal stains* ε_x and ε_y become (see Figure 6.28a)

$$\varepsilon_x = \frac{\partial u}{\partial x}, \quad \varepsilon_y = \frac{\partial v}{\partial y} \tag{6.129}$$

The *shear strain* γ has also been discussed in Section 6.3.4 about torsion. The *shearing strain* is usually defined as the variation in the right angle of the infinitesimal element when subjected to shear loading as shown in Figure 6.28b, where normal strain is very small and shearing strain is exaggerated. The total change of the right angle between x and y axes in Figure 6.28b is the *shearing strain* γ_{xy} and thus

$$\gamma_{xy} = \gamma_x - \gamma_y = \frac{\partial u}{\partial y} + \frac{\partial v}{\partial x} \tag{6.130}$$

Since the angular change between y and x axes should be the same as between x and y axes, then $\gamma_{xy} = \gamma_{yx}$. The plane *strain tensor* ε (or *strain matrix*) is given by the following definition;

$$\varepsilon = \varepsilon_{ij} = \begin{bmatrix} \varepsilon_{xx} & \varepsilon_{xy} \\ \varepsilon_{yx} & \varepsilon_{yy} \end{bmatrix} = \begin{bmatrix} \varepsilon_x & \frac{1}{2}\gamma_{xy} \\ \frac{1}{2}\gamma_{xy} & \varepsilon_y \end{bmatrix} \tag{6.131}$$

where the ½ γ_{xy} is due to the fact that the strain ε_{xy} is ½ that shown in Figure 6.28b since we have the shearing on both sides rather than with respect to one side only as the shearing definition goes. This would be apparent if one rotates the deformed element clockwise till the side coincides with the x-axis or rotate it counterclockwise till the other side coincides with the y-axis. The shear angle is thus twice that of each of the shearing angles defined in Figure 6.28b, particularly if the sides are equal lengths and the strains are the same.

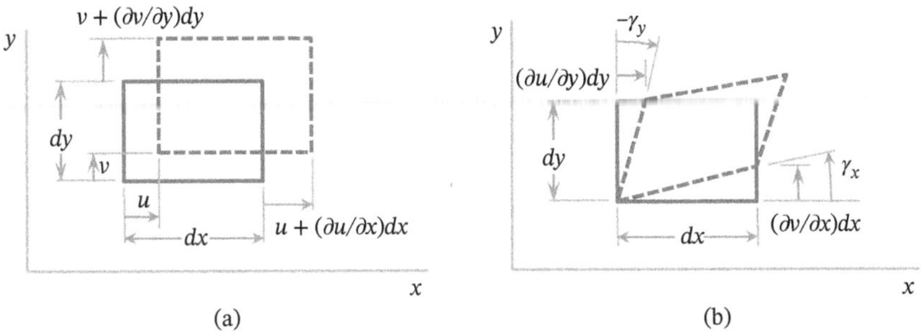

Figure 6.28 A plane strain where the deformation in x direction is u and the deformation in y direction is v and the longitudinal or normal strains ε_x and ε_y are defined from (a). The shear strain is the variation in the right angle of the infinitesimal element as shown in (b).

For ***3D*** or *triaxial strain state*, the *normal stains* ε_x, ε_y, and ε_z are then expected to be (see Eq. (6.129))

$$\varepsilon_x = \frac{\partial u}{\partial x}, \quad \varepsilon_y = \frac{\partial v}{\partial y}, \quad \varepsilon_z = \frac{\partial w}{\partial z} \tag{6.132}$$

For the ***3D*** of *shearing strain state*, the *shearing stains* γ_{xy}, γ_{yz}, and γ_{zx} are also expected to be similar to Eq. (6.130), or

$$\gamma_{xy} = \frac{\partial u}{\partial y} + \frac{\partial v}{\partial x} = \gamma_{yx}, \quad \gamma_{yz} = \frac{\partial v}{\partial z} + \frac{\partial w}{\partial y} = \gamma_{zy}, \quad \gamma_{zx} = \frac{\partial w}{\partial x} + \frac{\partial u}{\partial z} = \gamma_{xz} \tag{6.133}$$

Note that Eqs. (6.132) and (6.133) depend on the three displacements *u*, *v*, and *w*. Therefore, there can be only three independent equations in Eqs. (6.132) and (6.133). These are called the *equations of compatibility*; see Reisman and Pawlik (1980, 1991).

The ***strain tensor*** $\boldsymbol{\varepsilon}$, however, is usually defined by the *indicial* or *tensor notation* ε_{ij} having the following general equation:

$$\varepsilon_{xy} = \frac{1}{2}\left(\frac{\partial \delta_i}{\partial s_j} + \frac{\partial \delta_j}{\partial s_i}\right) \quad (i, j = x, y, z) \tag{6.134}$$

where $\delta_x = u$, $\delta_y = v$, $\delta_z = w$, $s_x = x$, $s_y = y$, and $s_z = z$. Equation (6.134) defines the relation between the shear components of the *strain tensor* $\boldsymbol{\varepsilon}$ and the *shearing strains* in Eq. (6.133) such that

$$\varepsilon = \varepsilon_{ij} = \begin{bmatrix} \varepsilon_{xx} & \varepsilon_{xy} & \varepsilon_{xz} \\ \varepsilon_{yx} & \varepsilon_{yy} & \varepsilon_{yz} \\ \varepsilon_{zx} & \varepsilon_{zy} & \varepsilon_{zz} \end{bmatrix} = \begin{bmatrix} \varepsilon_x & \frac{1}{2}\gamma_{xy} & \frac{1}{2}\gamma_{xz} \\ \frac{1}{2}\gamma_{xy} & \varepsilon_y & \frac{1}{2}\gamma_{yz} \\ \frac{1}{2}\gamma_{xz} & \frac{1}{2}\gamma_{yz} & \varepsilon_z \end{bmatrix} \tag{6.135}$$

The strain tensor (or matrix) has about the characteristics similar to the stress state tensor. Principal strains, principal directions, *Mohr's* circle of strain, and strain invariants are handled the same way as for the stress state.

6.6.2 Elastic Strain Energy

Linearly elastic materials store energy when loaded, if no loss of energy exists. If the material does not lose energy during or after loading, the work done by the external load will be equal to the strain energy stored. Figure 6.29 shows a general situation of an element loaded by a force $\boldsymbol{F}$ or a moment $\boldsymbol{M}$ or a stress σ. Figure 6.29b shows the case of an axially loaded element by the force $\boldsymbol{F}_x$. The response of the element is a deflection δ or θ or a strain ε. The energy stored for the case of the force $\boldsymbol{F}_x$ is the total work done by the force, i.e. the integral of the incremental work dW due to the incremental deformation $d\delta$ over the linear range of the load-deflection diagram shown in dark grayish color in Figure 6.29a. If the coordinates are the stress σ and the strain ε, the incremental energy is dU and the total energy stored is the *strain energy* U as indicated in Figure 6.29a. It is the dark grayish area under the straight line. The work done W_F by the force $\boldsymbol{F}_x$ to elongate the member in Figure 6.29b an amount of deformation δ is

$$W_F = \int_0^\delta \boldsymbol{F}_x \, d\delta \tag{6.136}$$

where $\boldsymbol{F}_x$ is a linear function of δ. This work done W_F should equal the energy stored in the strained material and is thus named *strain energy* U_ε. Substituting for the load $\boldsymbol{F}_x$ in terms of the stress ($\boldsymbol{F}_x = \sigma A$; see Eq. (6.56)) and for the deformation δ as a function of the strain ($\delta = l\varepsilon$, see Eq. (6.57)), one gets

$$W_F = U_\varepsilon = \int_0^\delta (\sigma A) l d\varepsilon = Al \int_0^\delta \sigma \, d\varepsilon \tag{6.137}$$

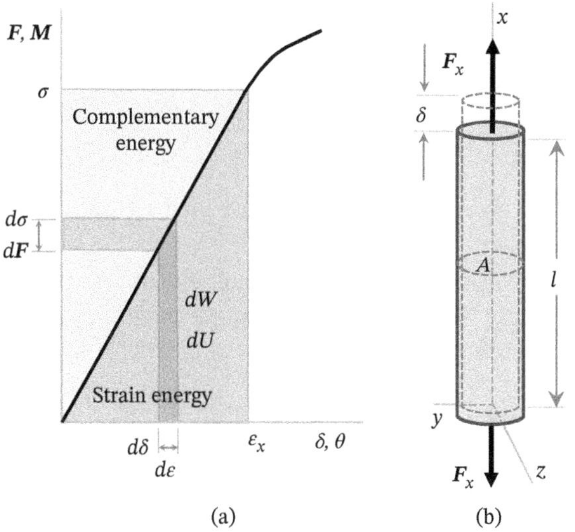

Figure 6.29 A general situation of an element loaded in (a) by a force $\boldsymbol{F}$ or a moment $\boldsymbol{M}$ or a stress σ, with the response of the element as a deflection δ or θ or a strain. The element in (b) shows only the case of an axially loaded element by the force $\boldsymbol{F}_x$. The energy stored is the total work done, i.e. the integral of the incremental work dW over the linear range of the diagram shown in dark grayish color in (a).

The stress σ, however, is a function of the strain ε through the *Hook's* law. This provides the following relations:

$$\varepsilon = \frac{\sigma}{E} \quad \text{and} \quad d\varepsilon = \frac{1}{E} d\sigma \tag{6.138}$$

Equation (6.137) becomes

$$U_\varepsilon = \int_0^\sigma \frac{Al}{E} \sigma \, d\sigma = \frac{Al}{E} \int_0^\sigma \sigma d\sigma = Al\left(\frac{\sigma^2}{2E}\right) = Alu_\varepsilon \tag{6.139}$$

The quantity in parentheses $(\sigma^2/2E)$ is the elastic strain energy per unit volume, which is called the *strain energy density* u_ε as depicted in Eq. (6.139). Employing the relation in (6.138), the *strain energy density* u_ε becomes

$$u_\varepsilon = \frac{\sigma^2}{2E} = \frac{1}{2}\sigma\varepsilon \tag{6.140}$$

This is expected since the area under the line in Figure 6.29a for the σ–ε plot is ½ $\sigma\varepsilon$. The complementary energy in the linear zone of Figure 6.29a has the same area as the *strain energy density* u_ε. The relation in Eq. (6.140) is also valid for any direction *x*, *y*, or *z*.

The *strain energies* of differently loaded components are similarly derived. By similarity, one can also substitute similar variables to get the relation needed just as the utility of *generalized variables* in Section 2.1.2.

Normal load ($\boldsymbol{F}_x$) has also been utilized in the preceding development of strain energy. The *strain energy density* u_ε is obtained by Eqs. (6.140), (6.56), and (6.57), which gives

$$u_\varepsilon = \frac{1}{2}\sigma\varepsilon = \frac{1}{2}\frac{\sigma^2}{E} = \frac{\boldsymbol{F}_x^2}{2EA^2} \tag{6.141}$$

The strain energy in normally loaded components is then

$$U_\varepsilon = \int_{\text{vol}} u_x \, dV = \int \frac{\boldsymbol{F}_x^2}{2EA^2}\left(\int dA\right) dx \tag{6.142}$$

Noting that the integral of dA is equal to A, the *strain energy* then becomes

$$U_\varepsilon = \int \frac{\boldsymbol{F}_x^2}{2EA} dx \tag{6.143}$$

Pure bending ($\boldsymbol{M}_z$) can be considered similar to normal loading, and utilizing Eqs. (6.140) and (6.12) provides

$$u_\varepsilon = \frac{1}{2}\frac{\sigma^2}{E} = \frac{\boldsymbol{M}_z^2 y^2}{2EI_z^2} \tag{6.144}$$

The strain energy in components loaded by pure bending is then

$$U_\varepsilon = \int_{\text{vol}} u_x \, dV = \int \frac{\boldsymbol{M}_z^2}{2EI_z^2} \left(\iint y^2 \, dy \, dz \right) dx \tag{6.145}$$

Noting that the double integral in Eq. (6.145) is the second area moment I_z, the *strain energy* then becomes

$$U_\varepsilon = \int \frac{\boldsymbol{M}_z^2}{2EI_z} \, dx \tag{6.146}$$

Torsion ($\boldsymbol{T}_x$) applied to machine members is treated as bending moment with the modulus of elasticity is replaced by the *shear modulus of elasticity* or *modulus of rigidity* G and the *polar second area moment* J_x replaces the second area moment I_z. Utilizing Eqs. (6.140) and (6.65) for circular cross sections gives

$$u_\varepsilon = \frac{1}{2} \frac{\tau^2}{G} = \frac{\boldsymbol{T}_x^2 r^2}{2GJ_x^2} \tag{6.147}$$

The strain energy in cylindrical components loaded by torsion is then

$$U_\varepsilon = \int_{\text{vol}} u_\varepsilon \, dV = \int \frac{\boldsymbol{T}_x^2}{2GJ_x^2} \left(\iint r^2 \, dy \, dz \right) dx \tag{6.148}$$

Noting that the double integral in Eq. (6.148) is the polar second area moment J_x, the *strain energy* becomes

$$U_\varepsilon = \int \frac{\boldsymbol{T}_x^2}{2GJ_x} \, dx \tag{6.149}$$

Since the strain energy is a scalar quantity, the total strain energy in an element subjected to a multitude of loadings can be obtained by the summation of the strain energy of each loading.

6.6.3 Castigliano's Theorem and Deflections

The method developed by *Castigliano* is used to advantage in the utilization of strain energy to calculate the displacements or deflections of complex machine members subjected to multitude of loads. *Castigliano's* theorem is named after *Carlo Alberto Castigliano* (1847–1884). He stated that "the partial derivative of the *strain energy*, considered as a function of the applied forces (or *loads)* acting on a linearly elastic structure, with respect to one of these forces (or *loads*), is equal to the displacement in the direction of the force (or *load)* of its point of application," *Castigliano* (1873). Mathematically stated, the *Castigliano's theorem* has the following known formula:

$$\delta_i = \frac{\partial U_\varepsilon}{\partial \boldsymbol{F}_i} \tag{6.150}$$

where δ_i is the displacement in the direction of the force $\boldsymbol{F}_i$ for any vectorial direction i.

This is also generalized to other *loads* such as moments $\boldsymbol{M}_i$ or torsions $\boldsymbol{T}_i$, which can provide

$$\theta_i = \frac{\partial U_\varepsilon}{\partial \boldsymbol{M}_i}, \quad \phi_i = \frac{\partial U_\varepsilon}{\partial \boldsymbol{T}_i} \tag{6.151}$$

These are the deformations θ_i and ϕ_i in the direction of the applied load $\boldsymbol{M}_i$ and $\boldsymbol{T}_i$ if any realistically exists.

The application of the theorem to find the displacement (δ, θ, or ϕ, etc.) at any point P of a machine element in, say, x direction, can be performed by applying a load $\boldsymbol{L}_P$ ($=\boldsymbol{F}_P$, $\boldsymbol{M}_P$, or $\boldsymbol{T}_P$) in the desired direction at the point and evaluating the strain energy of the element under all loads. The machine element oriented in the x direction is subjected to a multitude of loads such as $\boldsymbol{F}_i$, $\boldsymbol{M}_i$, and $\boldsymbol{T}_i$. As $\boldsymbol{L}_P$ may not be a function of the element direction x, the differentiation with respect to $\boldsymbol{L}_P$ can be performed under the integral using *Leibnitz* rule, named after *Godefroi Guillaume (von) Leibnitz* (1646–1716). The displacement Δ_L (δ, θ, *or* ϕ, *etc.*) in the direction of $\boldsymbol{L}_P$ is then given by

$$\Delta_L = \frac{\partial U_\varepsilon}{\partial \boldsymbol{L}_P} = \int \frac{\boldsymbol{F}_i}{EA}\frac{\partial \boldsymbol{F}_i}{\partial \boldsymbol{L}_P}dx + \int \frac{\boldsymbol{M}_i}{EI_i}\frac{\partial \boldsymbol{M}_i}{\partial \boldsymbol{L}_P}dx + \int \frac{\boldsymbol{T}_i}{GJ_i}\frac{\partial \boldsymbol{T}_i}{\partial \boldsymbol{L}_P}dx \tag{6.152}$$

Example 6.11 A uniformly distributed load $\boldsymbol{w}_i$ is applied to a simply supported beam of length l. Find the slope at the left reaction using the *Castigliano's* theorem. If the uniformly distributed load $\boldsymbol{w}_i = -9$ [kN/m] or -50 [lb/in], the length $l = 1.0$ [m] or 40 [in], and the modulus of elasticity $E = 207$ [GPa] or 30 [Mpsi], confirm the maximum slope where the circular section profile is of a diameter d_C. $= 0.1$ [m] or4 [in].

Solution

Data: The length $l = 1.0$ [m] or 40 [in], $d_C = 0.1$ [m] or 4 [in], and $\boldsymbol{w} = -9$ [kN/m] or -50 [lb/in]. $E = 207$ [GPa] or 30 [Mpsi], the second area moment $I_z = 4.91\,(10^{-6})$ [m^4] or 12.566 [in^4] as extracted from Example 6.2, Equation (t).

Since the slope is sought at the left support, one has to introduce a virtual or a fictitious positive moment $\boldsymbol{M}_P$ at the left support, Figure 6.30. The reactions at both ends are obtained by applying moment balance at each end as follows;

$$\begin{aligned} \boldsymbol{M}_{z1}\big|_{x=0} &= \boldsymbol{R}_2 l - \boldsymbol{w}l\,(l/2) + \boldsymbol{M}_P = 0 \quad \text{or} \quad \boldsymbol{R}_2 = {}^1\!/\!_2\,\boldsymbol{w}l - \boldsymbol{M}_P/l \\ \boldsymbol{M}_{z2}\big|_{x=l} &= -\boldsymbol{R}_1 l + \boldsymbol{w}l\,(l/2) + \boldsymbol{M}_P = 0 \quad \text{or} \quad \boldsymbol{R}_1 = {}^1\!/\!_2\,\boldsymbol{w}l + \boldsymbol{M}_P/l \end{aligned} \tag{a}$$

The bending moment along the beam is

$$\boldsymbol{M}(x) = -\left({}^1\!/\!_2\,\boldsymbol{w}l + \frac{\boldsymbol{M}_P}{l}\right)x + \frac{\boldsymbol{w}x^2}{2} + \boldsymbol{M}_P \quad (0 \le x \le l) \tag{b}$$

Applying Eq. (6.152), the slope θ_1 at the left support 1 can be obtained by the following expression:

$$\theta_1 = \frac{\partial U_\varepsilon}{\partial \boldsymbol{M}_P} = \int \frac{\boldsymbol{M}}{EI_z}\frac{\partial \boldsymbol{M}}{\partial \boldsymbol{M}_P}dx = \frac{1}{EI_z}\int_0^l \left(-\left({}^1\!/\!_2\,\boldsymbol{w}l + \frac{\boldsymbol{M}_P}{l}\right)x + \frac{\boldsymbol{w}x^2}{2} + \boldsymbol{M}_P\right)\left(-\frac{x}{l} + 1\right)dx \tag{c}$$

Integrating Eq. (c) and setting $\boldsymbol{M}_P = 0$, one obtains

$$\theta_1 = \frac{1}{EI_z}\int_0^l \left({}^1\!/\!_2\,\boldsymbol{w}x^2 - \frac{\boldsymbol{w}x^3}{2l}\right) + \left(-{}^1\!/\!_2\,\boldsymbol{w}lx + \frac{\boldsymbol{w}x^2}{2}\right)dx \tag{d}$$

$$\theta_1 = \frac{1}{EI_z}\left({}^1\!/\!_6\,\boldsymbol{w}x^3 - \frac{\boldsymbol{w}x^4}{8l} - {}^1\!/\!_4\,\boldsymbol{w}lx^2 + \frac{\boldsymbol{w}x^3}{6}\right)\Bigg|_0^l \tag{e}$$

$$\theta_1 = \frac{\boldsymbol{w}l^3}{EI_z}\left(\frac{4-3-6+4}{24}\right) = -\frac{\boldsymbol{w}l^3}{24EI_z} \tag{f}$$

This is the same as the value in Eq. (6.33)), but with the negative direction of $\boldsymbol{w}$ noted for.

Figure 6.30 Simply supported beam with a uniformly distributed load. A fictitious positive moment $\boldsymbol{M}_P$ at the left support is introduced to find the slope at that end. One may also add a load $\boldsymbol{F}_P$ (not shown) at point A or at the middle to get the deflections.

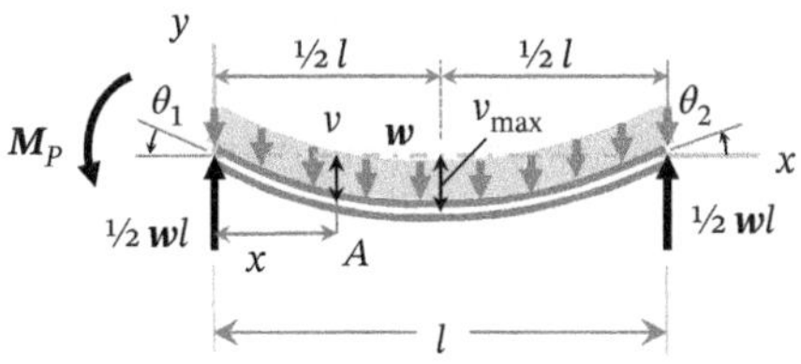

By substituting for the data of the loaded beam, one gets

$$\theta_1 = -\frac{wl^3}{24EI_z} = \frac{-9(10^3)(1)^3}{24(207)(10^9)(4.91)(10^{-6})} = -0.000\,369\ [\text{rad}] = -0.0211\ [^\circ]$$
$$\theta_1 = -\frac{wl^3}{24EI_z} = \frac{-50(10^3)(40)^3}{24(30)(10^6)(12.566)} = -0.000\,354\ [\text{rad}] = -0.0203\ [^\circ] \tag{g}$$

This confirms the value for both SI and US units in Example 6.3, Eq. (d).

6.7 Columns

Columns are usually long with respect to their cross-sectional size. When loaded in compression, they may deform laterally or buckle before reaching their maximum compressive strength. In that case their deformation is termed *buckling* or *instability*. If the buckling load is high, total collapse can occur where the lateral deformation increases till the two column ends meet. If the buckling load is reduced just before total collapse, the column may return to its original form. This can be easily demonstrated using an old *I.D.* card, credit card, or a similar piece of material. Hold the card off its longer side with one hand between the thumb on one end and the index, middle, and ring fingers on the other end. Squeezing the card increasingly, at one point the card will buckle into a half of a sine wave form. Usually the general reflex will reduce the squeezing force (or can control it) and the buckled card will not totally collapse. Releasing the forcing or squeezing action would usually return the card to its original shape. This return of the form indicates that the buckling was in the elastic zone and can be termed an *elastic buckling* case or *elastic stability or instability*. If the load is not controlled, the card would plastically have bent, and thus the card would be permanently damaged, and thus we experience *plastic instability*. This is the point of *instability* that cannot be controlled if the force is not controlled.

Many long machine elements are exposed to excessive compressive loading that they should withstand without buckling. Some of these are rods of hydraulic cylinders, long connecting rods, supporting columns, long drilling rods, long lifting screws or long power screws, compressive members of machine frames, members of ladders, long propeller shafts, posts of jib cranes, long compressive levers, compression links of lifting mechanisms, etc.

In this section, a study of this phenomenon is perused and the means to alleviate its occurrence by good synthesis is attained. An extensive work about the subject of buckling or *elastic stability*, which is beyond the scope of this text can be found in Timoshenko and Gere (1961), and Abd El Gawad et al. (2012).

6.7.1 Concentric Loading

Consider a simply supported column of length l_C compressively loaded with concentric force $\boldsymbol{P}$ acting along the centroidal axis at both ends, Figure 6.31a. Assuming the uniform-section column to be slim or slender and it will deform laterally an amount u in the x direction. The deformed lateral sections will experience a moment $\boldsymbol{M}_z$ of a $\boldsymbol{P}x$ value and will counteract with an internal moment $-\boldsymbol{P}x$. Utilizing Eq. (6.16) and observing that the displacement u is in the x rather than y direction, one gets

$$\frac{\boldsymbol{M}_z}{EI_z} = \frac{d^2u}{dy^2} = -\frac{\boldsymbol{P}x}{EI_z} \tag{6.153}$$

This equation is an ordinary differential equation in the form

$$\frac{d^2u}{dy^2} + \frac{\boldsymbol{P}x}{EI_z} = 0 \tag{6.154}$$

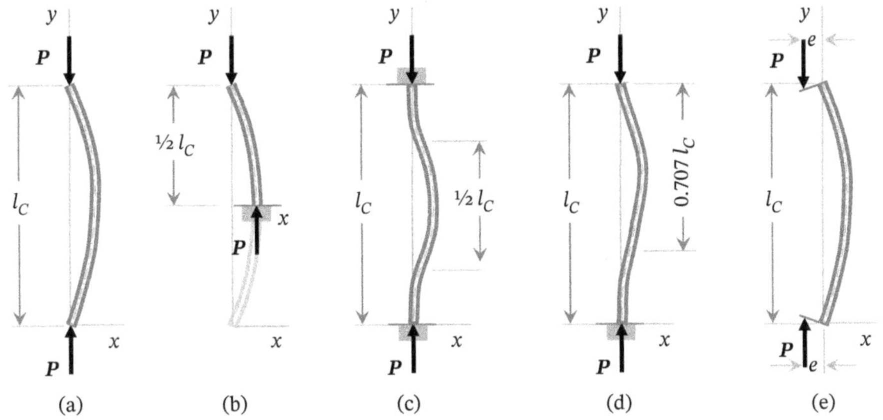

Figure 6.31 A column in (a) is simply supported case of an ideal end condition of rounded or pivoted form, compressively loaded with concentric force acting along the centroidal axis. In (b–d) other end conditions of free-fixed, fixed–fixed, and pinned–fixed are, respectively, shown. The case in (e) is a more realistic column with an exaggerated eccentric loading that is generally expected in practice.

The solution of this differential equation is usually assumed as a simple harmonic motion as follows:

$$x = C_1 \sin\sqrt{\frac{P}{EI_z}}x + C_2 \cos\sqrt{\frac{P}{EI_z}}x \tag{6.155}$$

where C_1 and C_2 are integration constants to be defined by the boundary conditions. These conditions are $x = 0$ at both $y = 0$ and $y = l_C$, which gives $C_2 = 0$. The solution becomes

$$0 = C_1\left(\sin\sqrt{\frac{P}{EI_z}}\,l_C\right) \tag{6.156}$$

This solution indicates that either $C_1 = 0$ or the term in parenthesis equals zero. Since $C_1 = 0$ provides a trivial and no solution, the acceptable nontrivial solution is

$$\left(\sin\sqrt{\frac{P}{EI_z}}\,l_C\right) = 0 \quad \text{or} \quad \sqrt{\frac{P}{EI_z}}\,l_C = n_B\pi, \quad n_B = 1, 2, 3, \ldots \tag{6.157}$$

where n_B is the buckling integer for critical buckling mode. For $n_B > 1$ the buckling mode or shape has n_B of ½ sine wave forms. Since for $n_B > 1$, the *critical buckling load* $\boldsymbol{P}_{cr} = \boldsymbol{P}$ will be higher than for $n_B = 1$, the least critical load will then be at $n_B = 1$ or

$$\boldsymbol{P}_{cr} = \frac{\pi^2 EI_z}{l_C^2} \tag{6.158}$$

This critical buckling load $\boldsymbol{P}_{cr}$ is termed ***Euler column*** formula (1744–1757) or *Euler critical load*, after *Leonhard Euler* (1707–1783).

The Euler critical buckling load is for an ideal end condition of rounded or pivoted form, Figure 6.31a. To account for other end conditions of free-fixed, fixed–fixed, and pinned–fixed, Figure 6.31b–d, respectively, an *end condition constant* C_E is used. The end condition is a theoretical account of the deformation curve being theoretically simulating an equivalent to a pinned–pinned *Euler* column. As shown in Figure 6.31b, the free-fixed column of ½ l_C in length has a buckling curve the same as an l_C long pinned–pinned column. Since the *Euler critical load* $\boldsymbol{P}_{cr}$ is a function of $(1/l^2)$, the *end condition constant* $C_E = 0.25$ should then be used for the free-fixed column of ½ l_C in length to have the same *Euler critical load* $\boldsymbol{P}_{cr}$ as the pinned–pinned column. Similarly, fixed–fixed column in Figure 6.31c should have an *end condition constant* $C_E = 4.0$ to have the same *Euler critical load* $\mathbf{P}_{cr}$ as the pinned–pinned column between the inflection points on the fixed–fixed column. The fixed-pivoted column in Figure 6.31d should have an *end condition constant* $C_E = 2.0$ to have the same *Euler critical load* $\mathbf{P}_{cr}$ as the pinned–pinned column between the inflection point and the pivoted end on the fixed-pivoted column. The *end condition constant* C_E is therefore added to Eq. (6.158) to account for the aforementioned conditions so that the *Euler critical load* $\mathbf{P}_{cr}$ becomes as follows:

$$\mathbf{P}_{cr} = \frac{C_E \pi^2 E I_z}{l_C^2}, \quad C_E = 1, 0.25, 4.0, 2.0 \tag{6.159}$$

These values of the *end condition constant* C_E are the theoretical values; however, some other practical values are available in some codes and standard, e.g. see AISC (2001). In this text, however, we use the theoretical values, and any other considerations should be included into the safety factor. A conservative designer always uses $C_E = 1.0$ or 1.2 at most, which can be included or accounted for in the safety factor K_{SF}. However, $C_E = 0.25$ should always be used for free-fixed columns, Figure 6.31b.

Figure 6.31d shows a more realistic column with a possible condition of *eccentric loading* that is generally expected in practice. This condition is treated in Section 6.7.2.

A convenient way to formulate the problem is by introducing the relation between the column *second area moment* I_C in place of I_z and the *cross-sectional area* A_C of the column. This allows more general consideration for the minimum value of the *second area moment* I_C for noncircular sections. The introduced relation is $I_C = A_C (r_{GC})^2$, where r_{GC} is the *radius of gyration* of the cross section. Therefore Equation (6.159) becomes

$$\mathbf{P}_{cr} = \frac{C_E \pi^2 E A_C}{(l_C / r_{GC})^2}, \quad \text{or} \quad \frac{\mathbf{P}_{cr}}{A_C} = \frac{C_E \pi^2 E}{(l_C / r_{GC})^2} = \frac{C_E \pi^2 E}{(s_r)^2} \tag{6.160}$$

where $s_r = l_C / r_{GC}$ is the designated ***slenderness ratio***. Even though the quotient $\mathbf{P}_{cr}/A_C$ has the dimension of a stress, it is called the *critical unit load*. As evident from Eq. (6.160), the *critical unit load* is dependent only on *slenderness ratio* s_r and the material *modulus of elasticity* E, and it is not related to any strength property of the material. If one would call the *critical unit load* as a stress, it might be confusingly compared with any strength property such as the yield strength. If that confusion is safeguarded against, one can have an enlightening relation between the *critical unit load ratio* ($P_{cr}/A_C\, E$ or the "critical stress ratio" σ_{cr}/E) and *slenderness ratio* s_r which is totally independent of whichever the material of the column is. This is shown in Figure 6.32, which obviously indicates that the relation is dimensionless and not depending on the material and thus valid for both SI and US units.

To *synthesize* a *Euler column*, usually the load $\mathbf{P}$ and the column length l_C are known. If we presumed that the required length is large, it can be expected that *Euler critical load* $\mathbf{P}_{cr}$ will be the limiting factor in design as shown in Figure 6.33. This figure indicates, however, that *Euler* column should be applied to long columns only since for shorter columns where *Euler* curve is a dashed line, the normal stress may quickly approach and can exceed the *compressive yield strength* S_{yc}. In the short column region of the figure another acceptable *parabolic equation* will be introduced later on.

To guarantee an acceptable synthesis in the *long column* or *Euler* range, at least the end condition C_E and the *buckling safety factor* K_{SF} should be known. If one assumes a higher *allowable critical load* $\mathbf{P}_{cr} = K_{SF}\, \mathbf{P}$ to use Eq. (6.159), the minimum acceptable synthesis of a cylindrical column diameter d_C is defined as follows:

$$\begin{aligned} I_C &= \frac{P_{cr} l_C^2}{\pi^2 E C_E} = \frac{\pi d_C^4}{64} \\ d_C^4 &= \frac{64 (K_{SF} \mathbf{P}) l_C^2}{\pi^3 E C_E} \end{aligned} \tag{6.161}$$

This relation is used to synthesize the *Euler's long cylindrical columns* as will be shown next where the *parabolic equation* is used to define the limit of *Euler* applicability zone as shown in Figure 6.33.

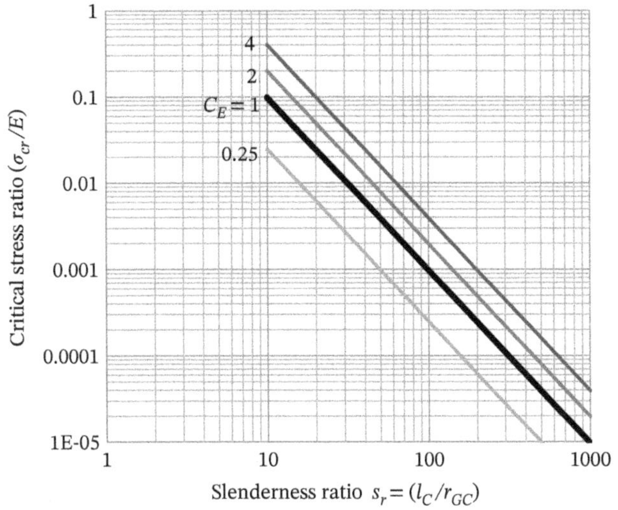

Figure 6.32 *Euler* critical buckling stresses ratio (σ_{cr}/E) versus slenderness ratio s_r for different end conditions C_E.

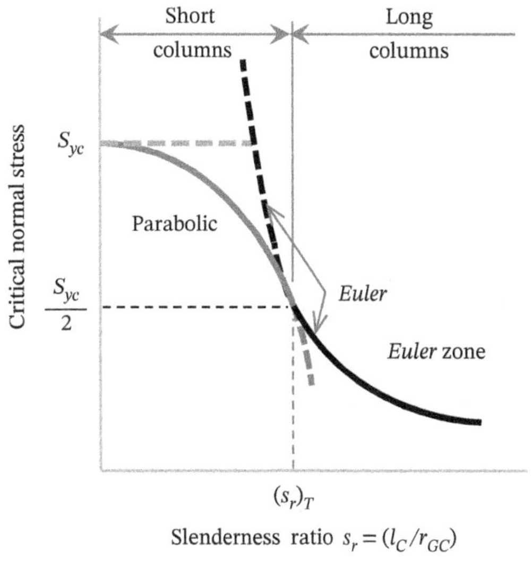

Figure 6.33 *Euler* critical buckling stress and parabolic equation versus slenderness ratio where both parabolic and *Euler* are tangent at the point separating short and long columns.

6.7.1.1 Johnson's Parabolic Equation

As the *Euler critical load* increases for smaller slenderness ratio, the normal stress quickly approaches the *compressive yield strength* S_{yc} as evident from Figure 6.33. In that case the design should be concerned with a compressive failure rather than instability failure or buckling. The extensive experiments show that the failure is lower than the yield line in the short column particularly above the parabolic curve in Figure 6.33, as sighted by Fauple and Fisher (1981). The parabolic equation is suggested by Johnson (1902), which is widely used in the short column zone. *Johnson's* parabolic equation can take the following form:

$$\frac{\boldsymbol{P}_{\mathrm{cr}}}{A_C} = a_P - b_P(s_r)^2 = a_P - b_P\left(\frac{l_C}{r_{\mathrm{GC}}}\right)^2 \tag{6.162}$$

where a_P and b_P are constants to be defined to fit experimental data and to be in harmony with *Euler's* equation and the end conditions of failure by yielding for very short columns as shown in Figure 6.33. This end condition defines that $a_P = S_{yc}$, which is the material *compressive yield strength*. The harmony or compatibility with *Euler's* equation suggests the wide spread condition to have both *Johnson's* parabola and *Euler's* equation be tangent at *critical unit load or critical stress* $P_{cr}/A_C = S_{yc}/2$. This condition provides the limit location of the point of *slenderness ratio tangent* $(s_r)_T$ or in short, the *slenderness ratio limit* $(s_r)_T$ as follows:

$$\frac{\boldsymbol{P}_{cr}}{A_C} = \frac{C_E\ \pi^2 E}{(s_r)_T^2} = \frac{S_{yc}}{2}$$

or **(6.163)**

$$(s_r)_T = \sqrt{\frac{2C_E\ \pi^2 E}{S_{yc}}}$$

Using this value in the parabolic equation defines the constant b_P from Eqs. (6.162) and (6.163) such that

$$\frac{\boldsymbol{P}_{cr}}{A_C} = a_P - b_P(s_r)^2 = S_{yc} - b_P\frac{2C_E\ \pi^2 E}{S_{yc}} = \frac{S_{yc}}{2}$$

or (6.164)

$$b_P = \frac{1}{C_E\,E}\left(\frac{S_{yc}}{2\pi}\right)^2 = \frac{S_{yc}^2}{4\pi^2 C_E\,E}$$

Substituting the values of a_P and b_P in Eq. (6.162) gives the *Johnson's parabolic equation* as follows:

$$\frac{\boldsymbol{P}_{cr}}{A_C} = S_{yc} - \frac{S_{yc}^2}{4\pi^2 C_E E}(s_r)^2, \quad \text{for} \quad (s_r) \leq (s_r)_T \qquad \textbf{(6.165)}$$

This relation is used to synthesize the *short cylindrical columns* where the *parabolic equation* is valid – off *Euler* applicability zone – for the *slenderness ratio* (s_r) less than the point of *slenderness ratio tangent* $(s_r)_T$ as shown in Figure 6.33.

6.7.2 Eccentric Loading

Figure 6.31e shows a more realistic column with an exaggerated *eccentric loading* that is generally expected in practice. In this case the loading axis is not concentric with the centroidal axis. The *eccentricity e* is the distance between the two axes. The eccentricity creates an initial moment of ***Pe***. The result of this moment suggested a solution, which is called the *secant formula*. The prospect of instability condition can then be accounted for by using the following *secant formula of the critical stress* σ_{cr} which has been developed for such a case; see Fauple and Fisher (1981):

$$\sigma_{cr} = \frac{P_{cr}}{A_C} = \frac{S_{yc}}{1 + (e\,C_S/r_{GC}^2)(\sec(s_r\sqrt{P/4A_C E}))} \qquad \textbf{(6.166)}$$

where C_S is the least distance from the *neutral axis* to the outer fiber of the column's section, which is $d_C/2$ for cylindrical columns. The *safety factor* K_{SF} is not included in Eq. (6.166) since it is typically included to get the *column synthesized diameter* d_C and thus included into the *column area* A_C. The quantity $(e\,C_S/(r_{GC})^2)$ in Eq. (6.166) is termed the ***eccentricity ratio***, and s_r is the *slenderness ratio*. It is to be noted, however, that Eq. (6.166) is an *implicit* equation where ***P*** exists on both sides of the equation. The solution is attained by iteration, which can be implemented in a computer code or in *Excel* as a *circular reference* or iterative calculation. The *secant* term can also be calculated as $\sec(\theta) = 1/\cos(\theta)$, if the *secant* function is missing. It is important to point out that the *secant formula* is valid for short or long columns. Figure 6.34 shows several conceptual plots of the *secant* curves at some levels of the *eccentricity ratio* $(e\,C_S/r_{GC}^2)$, indicating that as the eccentricity ratio increases, the curve is mainly reduced in size in the ordinate direction. The secant curves also show that the effect of eccentricity is not

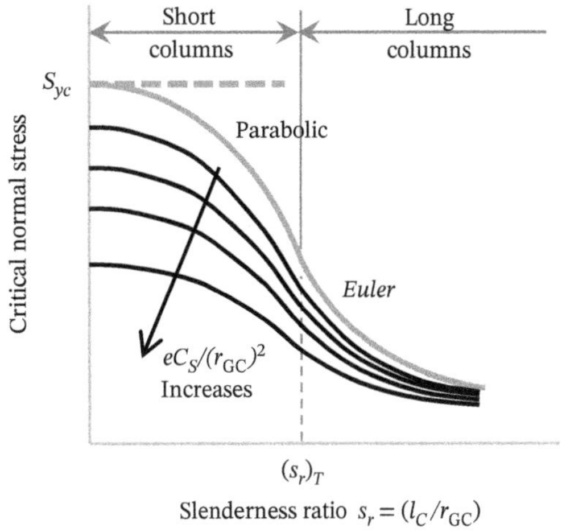

Figure 6.34 Critical buckling stresses for secant formula versus slenderness ratio. The conceptual secant curves in black are reduced in size in the ordinate direction as the eccentricity ratio increases. The secant formula is valid for short or long columns. *Euler* and parabolic curves (grayish line) are provided for comparison.

that pronounced at larger slenderness ratios. The parabolic and *Euler* curves (grayish line) are also provided in Figure 6.34 for relative comparison.

The usual eccentricity of loading in beams is taken as $l_B/400$. For columns, the same conceptual *eccentricity* $e = l_C/400$ might be an initial guess that can simulate inaccuracy of mounting, loading, and connecting columns or column construction or assembly. For more accurate consideration, the eccentricity should be calculated from fittings as *tolerance chain* or measured on actual assembly to have a more representative value.

Example 6.12 The rod of a hydraulic cylinder should withstand the maximum capacity of the compressive load $\boldsymbol{P} = -40$ [kN] or −9 [klb]. The maximum stroke or effective rod length is $l = 1.0$ [m] or 40 [in]. The modulus of elasticity of rod material is $E = 207$ [GPa] or 30 [Mpsi]. What is the synthesized rod diameter for a safety factor K_{SF} of 4.0 treating the rod as a column and assuming a material of a compressive yield strength $S_{yc} = 200$ [MPa] = $200(10^6)$ [Pa] = 29 [kpsi]? Check the applicability of *Euler* and *Johnson's* parabolic equations.

Solution

Data: $\boldsymbol{F} = -40$ [kN] or −9 [klb], rod or column $l_C = 1.0$ [m] or 40 [in], $E = 207$ [GPa] or 30 [Mpsi], $K_{SF} = 4.0$ and $S_{yc} = 200$ [MPa] = $200(10^6)$ [Pa] = 29 [kpsi].

Assuming that the rod-column to behave as an *Euler* column and with the hydraulic cylinder fully extended, the rod-column can be assumed as fixed-free configuration with the end condition constant $C_E = 0.25$. The free side is the pivot or pin at the end of the rod and fixed as being rigidly attached to the piston in the hydraulic cylinder and supported tightly at the rods exit sleeve bearing off the cylinder, see Figure 6.31b.

The expected minimum diameter for the *Euler* rod-column is obtained by the utilization of Eq. (6.161), which gives

$$d_C = \sqrt[4]{\frac{64(K_{SF}\boldsymbol{P})l_C^2}{\pi^3 E C_E}} = \sqrt[4]{\frac{64(4)(40)(10^3)(1)^2}{3.1416^3(207)(10^9)(0.25)}} = 0.050\,261\text{ [m]} = 50.261\text{ [mm]}$$

$$d_C = \sqrt[4]{\frac{64(K_{SF}\boldsymbol{P})l_C^2}{\pi^3 E C_E}} = \sqrt[4]{\frac{64(4)(9)(10^3)(40)^2}{3.1416^3(30)(10^6)(0.25)}} = 1.9954\text{ [in]} \tag{a}$$

The good design suggests the use of a round figure value for the rod diameter, which can be $d_C = 50$ [mm] or 2.0 [in].

To check the applicability of *Euler* or *Johnson's* parabolic equations, one needs to calculate the *slenderness ratio limit* $(s_r)_T$ using Eq. (6.163), which gives

$$(s_r)_T = \sqrt{\frac{2C_E\ \pi^2 E}{S_{yc}}} = \sqrt{\frac{2(0.25)(3.1416)^2(207)(10^9)}{200(10^6)}} = 71.467, \text{SI}$$
$$(s_r)_T = \sqrt{\frac{2C_E\ \pi^2 E}{S_{yc}}} = \sqrt{\frac{2(0.25)(3.1416)^2(30)(10^6)}{29(10^3)}} = 71.449, \text{US} \tag{b}$$

The actual *slenderness ratio* $s_r = l_C/r_{GC}$ requires the calculation of the column *radius of gyration* r_{GC}. For the circular section, this gives $r_{GC} = \sqrt{I_C/A_C} = \sqrt{(\pi d_C^4/64)/(\pi d_C^2/4)} = d_C/4$. The actual slenderness ratio is then

$$s_r = \frac{l_C}{r_{GC}} = \frac{l_C}{d_C/4} = \frac{1}{0.050/4} = 80, \ \text{SI}$$
$$s_r = \frac{l_C}{r_{GC}} = \frac{l_C}{d_C/4} = \frac{40}{2/4} = 80, \ \text{US} \tag{c}$$

This ratio is above the slenderness ratio limit $(s_r)_T$ calculated in Eq. (b). Therefore the *Euler* equation is valid, and the previously calculated rod diameters are the ones to be used.

Example 6.13 Redo Example 6.12, but for a shorter stroke or column length of $l_C = 0.5$ [m] or 20 [in] in place of Example 6.12 values. Assume a usual eccentricity e of $l_C/400$ of the original length of Example 6.12.

Solution

Data: $\boldsymbol{F} = -40$ [kN] or -9 [klb], rod or column $l_C = 0.5$ [m] or 20 [in], $E = 207$ [GPa] or 30 [Mpsi], $K_{SF} = 4.0$ and $S_{yc} = 200$ [MPa] $= 200(10^6)$ [Pa] $= 29$ [kpsi]. The eccentricity $e = 1/400 = 0.0025$ [m] $= 2.5$ [mm] or $40/400 = 0.1$ [in].

Again the assumptions of Example 6.12 are kept, and therefore the fixed-free configuration has the end condition constant $C_E = 0.25$. First, we assume an *Euler* column and get the expected diameter. If the slenderness ratio is lower than the limit between *Euler* and *Johnson's* parabola, one has to use *Johnson's* parabola to synthesize the column. It is also important to check the secant formula if the eccentricity is larger than expected.

The expected minimum diameter for the *Euler* rod-column is obtained by the utilization of Eq. (6.161), which gives

$$d_C = \sqrt[4]{\frac{64(K_{SF}\,\boldsymbol{P})\,l_C^2}{\pi^3 E C_E}} = \sqrt[4]{\frac{64(4)(40)(10^3)(0.5)^2}{3.1416^3(207)(10^9)(0.25)}} = 0.035\,540\ [\text{m}] = 35.54\ [\text{mm}]$$
$$d_C = \sqrt[4]{\frac{64(K_{SF}\,\boldsymbol{P})\,l_C^2}{\pi^3 E C_E}} = \sqrt[4]{\frac{64(4)(9)(10^3)(20)^2}{3.1416^3(30)(10^6)(0.25)}} = 1.4109\ [\text{in}] \tag{a}$$

Using a round figure value for the rod diameter as a good design suggests that the column diameters can be $d_C = 35$ [mm] or 1.4 [in].

The actual *slenderness ratio* $s_r = l_C/r_{GC}$ requires the calculation of the column *radius of gyration* r_{GC}. For the circular section, this gives $r_{GC} = d_C/4$ as noted before. The actual slenderness ratio is then

$$s_r = \frac{l_C}{r_{GC}} = \frac{l_C}{d_C/4} = \frac{0.5}{0.035/4} = 57.143, \quad \text{SI}$$
$$s_r = \frac{l_C}{r_{GC}} = \frac{l_C}{d_C/4} = \frac{20}{1.4/4} = 57.143, \quad \text{US} \tag{b}$$

To check the applicability of *Euler* or *Johnson's* parabolic equations, one needs to calculate the *slenderness ratio limit* $(s_r)_T$ using Eq. (6.163), which gives

$$(s_r)_T = \sqrt{\frac{2C_E\ \pi^2 E}{S_{yc}}} = \sqrt{\frac{2(0.25)(3.1416)^2(207)(10^9)}{200(10^6)}} = 71.467, \quad \text{SI}$$
$$(s_r)_T = \sqrt{\frac{2C_E\ \pi^2 E}{S_{yc}}} = \sqrt{\frac{2(0.25)(3.1416)^2(30)(10^6)}{29(10^3)}} = 71.449, \quad \text{US} \tag{c}$$

It is interesting to note that this *slenderness ratio limit* $(s_r)_T$ is exactly the same as Example 6.12. This should have been obvious since Eq. (6.163) is material dependent and independent of loading or column geometry except for end condition C_E, which did not change.

The actual *slenderness ratio* s_r ratio calculated in Eq. (b) is lower than the slenderness ratio limit $(s_r)_T$ calculated in Eq. (c). Therefore the *Johnson's parabolic* equation is valid, and the previously calculated rod diameters are either to be used or recalculations should be necessary. Assuming that the previously calculated rod diameters are to be used, the *Johnson's parabolic* Eq. (6.165) give

$$\frac{\boldsymbol{P}_{cr}}{A_C} = S_{yc} - \frac{S_{yc}^2}{4\pi^2 C_E\, E}(s_r)^2 = 200(10^6) - \frac{(200(10^6))^2}{4\pi^2(0.25)(207)(10^9)}(57.143)^2 = 136\,068\,340\ [\text{Pa}] = 136.07\ [\text{MPa}]$$
$$\frac{\boldsymbol{P}_{cr}}{A_C} = S_{yc} - \frac{S_{yc}^2}{4\pi^2 C_E\, E}(s_r)^2 = 29(10^3) - \frac{(29(10^3))^2}{4\pi^2(0.25)(30)(10^6)}(57.143)^2 = 19\,725\ [\text{psi}] = 19.725\ [\text{kpsi}] \tag{d}$$

Since there is an eccentricity taken as of the original length of Example 6.12 or $e = 1/400 = 0.0025$ [m] = 2.5 [mm] or $e = 40/400 = 0.1$ [in], we should calculate the secant critical stress formula. The secant formula of Eq. (6.166) requires the evaluation of the cross-sectional areas that gives $A_C = \pi\ d_C{}^2/4 = 9.6211(10^{-4})$ [m²] or $= 1.5394$ [in²]. The *secant critical stress* is then calculated by *iteration* in the *Excel* sheet of ***Column Synthesis Tablet*** developed to aid in column synthesis as discussed in Section 6.12.2, which gives

$$\sigma_{cr} = \frac{P_{cr}}{A_C} = \frac{S_{yc}}{1 + (e\,C_S/r_{GC}^2)(\sec(s_r\sqrt{P_{cr}/4A_C E}))}$$
$$= \frac{200(10^6)}{1 + (2.5(35/2)/(35/4)^2)(\sec(57.143\sqrt{P_{cr}/4(9.6211(10^{-4})(207)(10^9)}))} \Rightarrow 127.269\ [\text{MPa}] \tag{e}$$
$$= \frac{29(10^3)}{1 + (0.1(1.4/2)/(1.4/4)^2)(\sec(57.143\sqrt{P_{cr}/4(1.5394(30)(10^6)}))} \Rightarrow 18.454\ [\text{kpsi}]$$

From Eqs. (d) and (e), it is clear that the *secant critical stress* is lower than the *Johnsons parabolic* stress. In that case the *secant critical stress* is the controlling stability limit and from which the least factor of safety is defined. The prospective factor of safety is then as follows:

$$\sigma_{cr} = \frac{\boldsymbol{P}_{cr}}{A_C} = \frac{K_{SF}\boldsymbol{P}}{A_C}, \quad \text{or} \quad K_{SF} = \frac{\sigma_{cr}A_C}{\boldsymbol{P}} \tag{f}$$

For the *Johnson's parabolic stress*, the factors of safety are then

$$K_{SF} = \frac{\sigma_{cr}A_C}{\boldsymbol{P}} = \frac{136.07(10^6)(9.6211\ (10^{-4}))}{40(10^3)} = 3.273, \quad \text{SI}$$
$$K_{SF} = \frac{\sigma_{cr}A_C}{\boldsymbol{P}} = \frac{19\,725(1.5394)}{9(10^3)} = 3.374, \quad \text{US} \tag{g}$$

For the *secant critical stress*, the factors of safety are then

$$K_{SF} = \frac{\sigma_{cr} A_C}{\boldsymbol{P}} = \frac{127.269(10^6)(9.6211(10^{-4}))}{40(10^3)} = 3.0612, \quad \text{SI}$$

$$K_{SF} = \frac{\sigma_{cr} A_C}{\boldsymbol{P}} = \frac{18\,454(1.5394)}{9(10^3)} = 3.156, \quad \text{US} \tag{h}$$

From Eqs. (g) and (h), the least factor of safety K_{SF} is for the *secant critical stress* case. It is about 3 or so, and if one needs to have at least a factor of safety of 4, one should then need to change the column diameter d_C in slight iteration to attain the required safety. The provided *Excel* sheet of ***Column Synthesis Tablets*** can be used to advantage in that regard. The ***Tablets*** are accessible through **Wiley website** for *computer-aided synthesis* (CAS) of columns. This is discussed further in Section 6.12.2, *Computer-Aided Design and Optimization* of columns.

6.8 Equivalent Element

The equivalent machine element is that element which can behave the same as the original machine element. It is expected that the equivalence be in a simpler form or uncomplicated geometry than the original. It is usually difficult to get equivalence of all aspects of the original element due to the typical nonlinear relations of the behavioral variables. The equivalence can thus be for a specific behavior of the machine element. In that case, the equivalence can be used in that specific behavior.

The equivalence is ought to be normally used for the *preliminary stages* of design or synthesis to simplify a ballpark simpler configuration before more defined geometry is homed in. It is important to note that the mathematical model previously discussed in Section 6.2 is selected to be a simpler equivalence to the original. It was not a particular equivalence in some exacting way, but it is usually a conservative approximation and might approach equivalence in some cases.

In Section 6.1.5 a coefficient C_{EB} is deduced that represents an *equivalent beam* factor defining the comparable behavior to the simply supported beam with mid-span concentrated load and circular cross section. To indicate the same behavior, the equivalence coefficient C_{EB} is usually not equal to 1.0, but it differs to get a comparable behavior. It does not change the geometry or configuration. The equivalence in this section provides a changed simpler configuration that provides a comparable equivalence to the original one in some aspect or aspects.

Figure 6.35 presents an original machine element geometry in (a) and a sought much simpler equivalent to it in (b). The original machine element has three sections of diameters and lengths of (d_1, l_1), (d_2, l_2), and (d_3, l_3), respectively. The simpler equivalent in Figure 6.35b has one integral element of a diameter d_1 that has one equivalent length of three indistinctive zones of lengths l_1, l_2', and l_3'. It can, however, be either of the larger d_2 or d_3

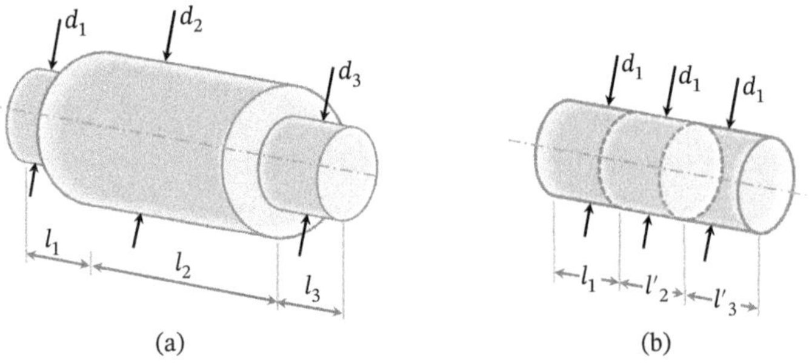

Figure 6.35 Machine element in (a) has different lengths and diameters. Equivalent machine element in (b) has the same diameter d_1 but different lengths to have the same torsional stiffness of the element in (a).

diameter instead of the selected d_1 diameter. The sought equivalence here is to get the simpler equivalence in the same *torsional rigidity* GJ_x, *torsional stiffness* $\boldsymbol{T}/\phi$, or *angular deformation* ϕ. For some other equivalent behavior, similar relations ought to be employed.

The *angular deformation* of the machine element sections of Figure 6.35a under torsion are defined by Eq. (6.70) such that

$$\phi_1 = \frac{Tl_1}{GJ_{x1}}, \quad \phi_2 = \frac{Tl_2}{GJ_{x2}}, \quad \phi_3 = \frac{Tl_3}{GJ_{x3}} \tag{6.167}$$

The total angular deformation is the summation of the deformations of these sections. Any equivalent zones (d_1, l_2') and (d_1, l_3') of Figure 6.35b should provide the same deformations ϕ_2 and ϕ_3 of the original sections in Figure 6.35a. For circular cross sections of *polar second area moments* $J_x = \pi d_C{}^4/32$, the equivalence gives

$$\begin{aligned} \frac{\phi_2 G}{T} &= \frac{l_2}{J_{x2}} = \frac{l_2'}{J_{x2}'}, \quad \text{or} \quad l_2' = l_2\frac{J_{x2}'}{J_{x2}} = l_2\frac{d_1^4}{d_2^4} \\ \frac{\phi_3 G}{T} &= \frac{l_3}{J_{x3}} = \frac{l_3'}{J_{x3}'}, \quad \text{or} \quad l_3' = l_3\frac{J_{x3}'}{J_{x3}} = l_3\frac{d_1^4}{d_3^4} \end{aligned} \tag{6.168}$$

This equivalent element gives exactly the same *torsional rigidity*, torsional stiffness, or angular deformation as the original machine element.

It should be pointed out that the torsional stresses in the equivalent element are severed due to the reduction in diameter. If the equivalent element had acquired the larger diameter, the maximum torsional stress would have been reserved. The equivalent in that case would be for the same torsional rigidity, torsional stiffness, or angular deformation in addition to keeping the maximum stress the same.

Other equivalence has been used to account for more complex situation of simulating ***3D*** area contact in ***FE*** model; see Metwalli and Moslehy (1983).

6.9 Thermal Effects

Thermal stresses are generated due to the constrained thermal expansion of material. If the material is constrained not to expand, the constraint rigidity will not allow the expansion to occur and thus the expected extension due to temperature will force the material to be compressed by that unallowed deformation. It is a case of restrained deformation or strain. This is usually a multidimensional effect since thermal expansion for isotropic materials is in all directions. Here, only a simple one-dimensional (*1D*) effect is considered. The *thermal strain* ε_{Tx} of a body in the x direction is given by

$$\varepsilon_{Tx} = \alpha_T(\Delta T_T) \tag{6.169}$$

where α_T is the *thermal expansion coefficient* and ΔT_T is the *temperature difference*. The *thermal deformation* $\boldsymbol{u}_T$ in the x direction of the machine element or bar of length l_B due to this temperature difference is obtained according to Eq. (6.57) such that

$$\boldsymbol{u}_T = \varepsilon_{Tx}\, l_B \tag{6.170}$$

where l_B is the bar or element length laid in the x direction. If the machine element is totally restrained in the x direction such as the member shown in Figure 6.36, a large longitudinal compressive thermal stress $\boldsymbol{\sigma}_{Tx}$ will be generated according to *Hook's* law such that

$$\boldsymbol{\sigma}_{Tx} = \varepsilon_{Tx} E = \alpha_T(\Delta T_T)E \tag{6.171}$$

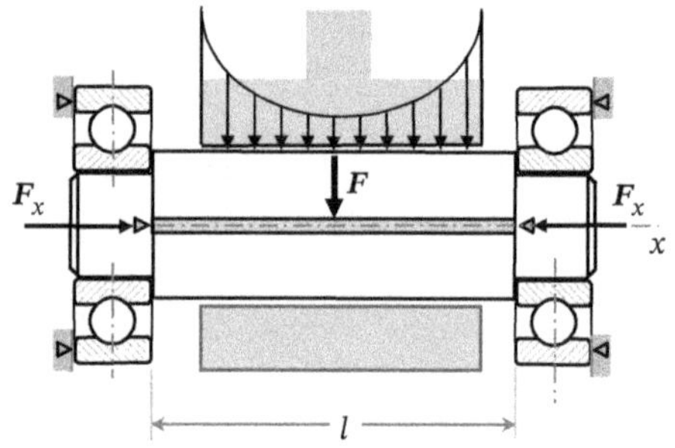

Figure 6.36 A suspended shaft with two rolling bearings supporting the shaft. It is assumed to be totally restrained in the x direction. A large longitudinal compressive thermal stress will be generated as if there are two axial forces $\boldsymbol{F}_x$ causing the compressive stress.

Table 6.2 Thermal expansion coefficients α_T and modulus of elasticity E for some engineering materials[a].

	Thermal expansion coefficient α_T, [SI] and [US]		Modulus of elasticity E, [SI] and [US]	
Material	**[10^{-6}/°C]**	**[10^{-6}/°F]**	**[GPa]**	**[Mpsi]**
Aluminum alloy	22	12	72	10.4
Brass, bronze	18.7	10.5	110	16
Gray cast iron	11.5	6.4	103	15
Magnesium alloy	26	14.5	45	6.5
Nickel alloy	13.1	7.0	207	30
Steel, alloy	10.8	6.3	207	30
Steel, carbon	12	6.7	207	30
Steel, stainless	17.3	9.6	190	27.5
Titanium, alloy	9	4.9	114	16.5
Tungsten	4.3	2.4	400	58

a) Values are representative and can vary according to composition.

where E is the *modulus of elasticity*. This compressive thermal stress σ_{Tx} will be added to (or subtracted from) the stresses induced by such other loading as of Figure 6.36. Even though the values of the *thermal expansion coefficient* α_T is small as shown in Table 6.2, the value of the modulus of elasticity E is large as also depicted in Table 6.2.

The construction of a suspended shaft in Figure 6.36 identifies two rolling bearings supporting the shaft. This arrangement is expected to be housed in a frame or a box. If the housing frame or box is of a heavy design and assumed to be rigidly attached to the ground and if the outer races of the bearings are thus tightly prevented from moving or allows any deformation, this deformation constraint will cause a very high *thermal compressive stresses*. This is as if there are two axial forces $\boldsymbol{F}_x$ on the shaft causing these compressive stress as demonstrated in Figure 6.36. The bearings will suffer from the same compressive force. This can cause a considerable damage to the bearings, the shaft and may be the housing. That is why it is recommended to leave a space between the outer races of the bearings and the housing of about 0.2–0.3 [mm] or 0.008–0.012 [in] (or appropriate *tolerance* calculation) to account for thermal expansion and other factors.

The ***bimetal element*** or strip is commonly used in thermal control systems and measurements. This application is not possibly damaging but instead very useful. A bimetal strip is shown in Figure 6.37 after experiencing a

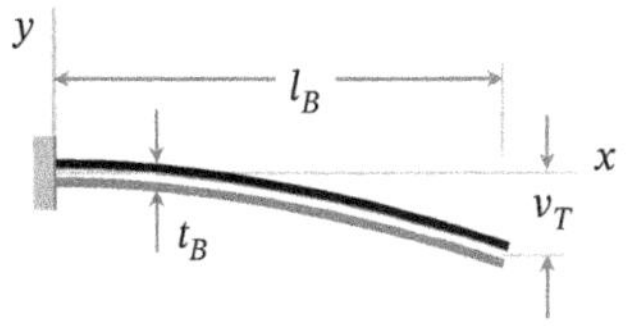

Figure 6.37 A *bimetal strip* of two different metals (usually steel and brass or copper) after undergoing a temperature rise of some degrees.

temperature rise of several degrees. Usually steel and brass or copper are the composition of bimetal strips. These two separate metals have different coefficients of thermal expansion. They are brazed, welded, bonded, or riveted together. The top black colored metal in Figure 6.37 has a higher thermal expansion coefficient than the lower grayish metal in the figure. Under the temperature increase ΔT_T, the strip will deflect a value of $\boldsymbol{v}_T$ in the y direction as exaggerated in Figure 6.37. It can be verified from curvature difference that the deflection $\boldsymbol{v}_T$ is then (Faupel and Fisher (1981))

$$\boldsymbol{v}_T = (3/4t_B)\ l_B^2(\Delta T_T)(\alpha_{T2} - \alpha_{T1}) \tag{6.172}$$

where t_B is the bimetal strip thickness. Other bimetal strip configurations are available and extensively used in thermostats, thermometers, clock balance wheel, circuit breakers, fluorescent lamp starters, etc.

Example 6.14 Find the equivalent force that would be generated in Figure 6.36 if the steel shaft length is 1 [m] or 40 [in], the circular diameter is 0.1 [m] or 4 [in], and the temperature rise is 50 [°C] or 90 [°F]. Note that the shaft is totally restrained in the x direction. The steel modulus of elasticity is $E = 207$ [GPa] or 30 [Mpsi] and the thermal expansion coefficient is $\alpha_T = 12\ [10^{-6}/°C]$ or $6.7\ [10^{-6}/°F]$.

Solution

Data: Shaft $d_C = 0.1$ [m] or 4 [in], $l_B = 1.0$ [m] or 40 [in], $E = 207$ [GPa] or 30 [Mpsi], $\alpha_T = 12\ [10^{-6}/°C]$ or $6.7\ [10^{-6}/°F]$, and $\Delta T_{T\cdot} = 50$ [°C] or 90 [°F].

The shaft circular cross-sectional area $A_C = \pi\, d_C{}^2/4 = 0.007\,853\ [m^2]$ or $2\,(4) = 12.57\ [in^2]$.

The *thermal strain* according to Eq. (6.169) is

$$\begin{aligned}\varepsilon_{Tx} &= \alpha_T(\Delta T_T) = 12(10^{-6})(50) = 600(10^{-6})\\ \varepsilon_{Tx} &= \alpha_T(\Delta T_T) = 6.7(10^{-6})(90) = 603(10^{-6})\end{aligned} \tag{a}$$

The values are expected to be about the same, but the rounding of the properties makes the difference. The thermal extension of the shaft $\boldsymbol{u}_T$, if it was not restricted from deformation is obtained according to Eq. (6.170) such that

$$\begin{aligned}\boldsymbol{u}_T &= \varepsilon_{Tx}\, l_B = 600(10^{-6})(1.0) = 600(10^{-6})\ [m] = 0.6\ [mm]\\ \boldsymbol{u}_T &= \varepsilon_{Tx}\, l_B = 603(10^{-6})\,(40.0) = 24\,120(10^{-6})\ [in] = 0.024\,12\ [in]\end{aligned} \tag{b}$$

This deformation is larger than the aforementioned recommended space between the outer races of the bearings and the housing of about 0.2–0.3 [mm] or 0.008–0.012 [in]. For such a case the assumptions of the problem should be further checked as to the rigidity of the construction, the temperature differences, the allowable clearances in the bearings, and the required *tolerance chain* for such a case.

If the housing is assumed rigid, the expected restraint compressive stress is obtained using the *Hooke's* law or Eq. (6.171). It should be noted, however, that the restraint should cause a negative or $-\varepsilon_{Tx}$ to return the shaft to its restraint length and thus get

$$\begin{aligned}\sigma_{Tx} &= -\varepsilon_{Tx}E = -600(10^{-6})(207)(10^9) = -124\,200(10^3)\ [Pa] = -124.2\ [MPa]\\ \sigma_{Tx} &= -\varepsilon_{Tx}E = -603(10^{-6})(30)(10^6) = -18\,090\ [psi] = -18.09\ [kpsi]\end{aligned} \tag{c}$$

This stress is very high and may cause failure to the shaft if the shaft *allowable stress* is lower than these values. The apparent force that causes this compressive stress is then

$$\begin{aligned} \boldsymbol{F} &= \sigma_{Tx}(A_C) = -124\,200(10^3)(0.007\,853) = -975\,343\ [\text{N}] = -975.34\ [\text{kN}] \\ \boldsymbol{F} &= \sigma_{Tx}(A_C) = -18\,090(12.57) = -227\,391\ [\text{lb}] = -227.39\ [\text{klb}] \end{aligned} \tag{d}$$

This is an excessive force that may damage the bearings if the restraining bodies are as rigid as assumed.

6.10 Stress Concentration Factors

The stress concentration stems from the geometry of some loaded machine elements that have holes, cavities, slots, slits, or sharp uneven distribution of material. These discontinuity or sharp uneven distribution of linearly elastic material causes the uneven and sharp distribution of internal stresses. These are called *stress raisers* or stress concentration generators or creators. At the boundaries of the stress raisers, the stresses are concentrated to generate higher values than the average stress in the neighborhood. A factor is devised to account of that, which is called the *stress concentration factor* K_{SC}.

The field of the *theory of elasticity* has the original analysis of the stress concentration problem; see e.g. Reisman and Pawlik (1980, 1991). The investigation of the stress distribution around a hole in a very wide or an infinite plate has been studied. The outcomes of *stress field calculations* show that the maximum stress σ_{max} at the major axis of an *elliptical hole* in a very wide plate (or an infinite plate) loaded as shown in Figure 6.38 is given by

$$\sigma_{max} = \sigma_{av}\left(1 + \frac{2b_h}{a_h}\right) \tag{6.173}$$

where σ_{av} is the *average stress*, and b_h and a_h are the principal dimensions of the elliptic hole. The stress distributions at the hole are also depicted in Figure 6.38.

Equation (6.173) indicates that when $b_h = a_h$, the ellipse turns into the grayish *circular hole* and the maximum stress becomes

$$\sigma_{max} = 3\sigma_{av} \tag{6.174}$$

This means that the stress has been raised to three times as much as the average stress and therefore the *stress concentration factor* $K_{SC} = 3$. Equation (6.173) indicates also that when $b_h \gg a_h$, the ellipse turns into a slit and the maximum stress becomes much larger than the average stress. The solution would not necessarily apply due to prospective *plastic failure* and a *fracture mechanics* approach is necessary; see Section 7.9 and Younan et al. (1983).

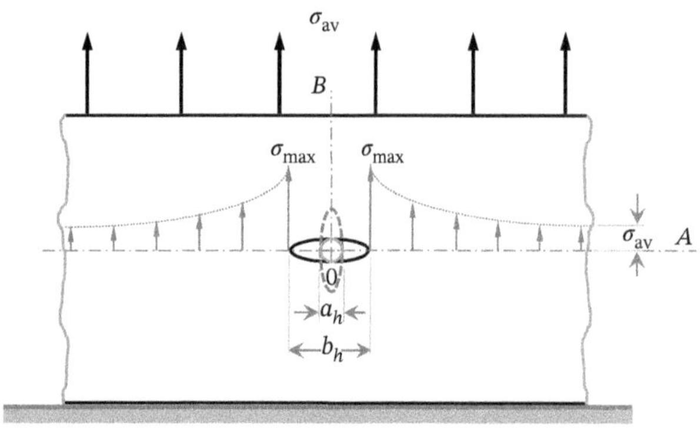

Figure 6.38 A very wide (infinite) plate with an elliptical hole uniformly loaded in tension and the stress distribution at the hole sides.

Equation (6.173) also indicates that when $b_h < a_h$, the ellipse turns around as shown in Figure 6.38 as the grayish dashed ellipse. The maximum stresses will approach the average stress and therefore the *stress concentration factor* K_{SC} approaches 1.0. Optimization studies confirm that the *optimum topology* of the circular hole turns the hole into an elongated slit in the B direction of Figure 6.38 similar to the rotated ellipse; see Section 6.12.3.

When the plate has a limited width w_P and a relatively small thickness t_P with a cylindrical hole of diameter d_h, the hole generates a stress distribution as depicted in Figure 6.39a. The solutions of this and other similar problems has been attained by several means such as experimental *photoelasticity* and fitting these and other extensively generated data; see, for example, Peterson (1974) and Young and Budynas (2002). One of these results for the relatively thin plate with a cylindrical hole is shown in Figure 6.39b. The stress concentration factor K_{SC} against the diameter to width ratio d_h/w_P is plotted in addition to the following curve fitting equation produced by using y for K_{SC} and x for d_h/w_P:

$$y = -1.7204\,x^3 + 3.8648\,x^2 - 3.1391\,x + 2.9965$$

or (6.175)

$$y = -1.72\,x^3 + 3.87\,x^2 - 3.14\,x + 3.0$$

The reduced expression in Eq. (6.175) may be accurate enough for engineering calculations. A similar equation found in Young and Budynas (2002) that gives slightly lower estimate of stress concentration factor K_{SC}, is as follows;

$$K_{SC} = -1.53\left(\frac{d_h}{w_P}\right)^3 + 3.66\left(\frac{d_h}{w_P}\right)^2 - 3.13\left(\frac{d_h}{w_P}\right) + 3.0 \tag{6.176}$$

The average stress in the finite width plate loaded by a total force $\boldsymbol{P}$ is easily obtained as

$$\sigma_{av} = \frac{\boldsymbol{P}}{(w_P - d_P)t_P} \tag{6.177}$$

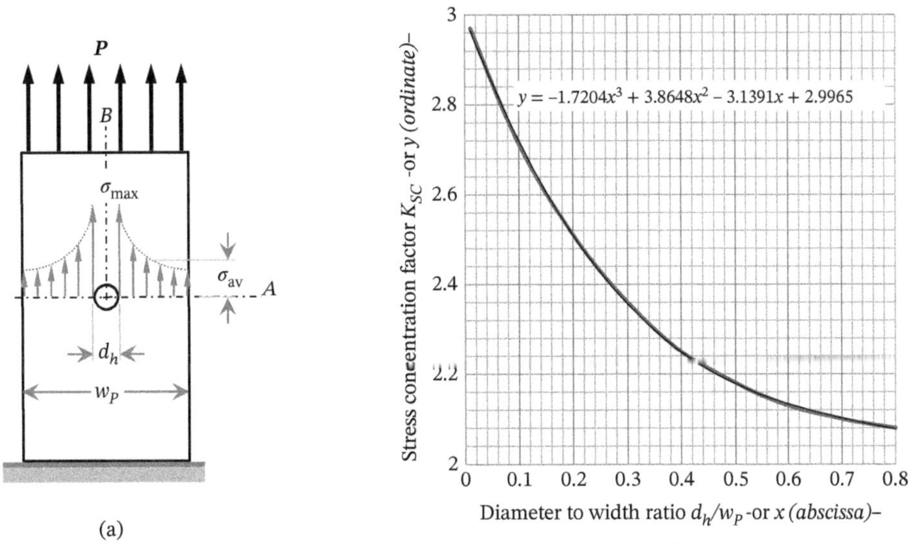

(a) (b)

Figure 6.39 A plate with a cylindrical hole uniformly loaded in tension and the stress distribution at the hole sides are shown in (a). The stress concentration factor K_{SC} against the diameter to width ratio is plotted in (b). The fitting equation is also shown in (b).

For an infinite plate width, both Eqs. (6.175) and (6.176) give the stress concentration factor $K_{SC} = 3.0$ as readily expected from Eq. (6.174).

For other numerous stress raisers, loading regimes and the associated stress concentration factors, refer to references such as Peterson (1974) and Young and Budynas (2002). The closely associated cases to the design of machine elements in this text are the circular bars with notches or with fillets between different diameters under axial, bending, or torsional loadings. These will be re-examined in due course.

6.11 Finite Element Method

The ***FE*** method is a *numerical technique* to solve for the response of complex continuum geometries to different loading conditions and deformation constraints. The technique approximates the formulation of the complex ***3D*** *partial differential equations* of the *continuum* domain to a much larger set of linear equations of the selected discreet elements to be accessibly solved. The number of discrete elements can be in the order of 10 000–100 000 or even much larger. Still, it should be noted that the ***FE*** technique is a numerical approximation that depends on the selected discrete *element type* and size. The smaller the element size, the more accuracy is expected if the resulting large number of elements can be accurately handled numerically in a reasonable time. The history of introducing the technique and its extensive development is beyond the scope of this text. There are usually some offered courses in some curricula that can be attended for broader coverage. There are also so many dedicated references that can be consulted such as Bathe (1982 or 2006), Zienkiewicz and Taylor (1991), Chandrupatla and Belegundu (2002), Logan (2007), and Bhatti (2005).

The typical process of achieving the mathematical modeling and the numerical solution is usually through the following general steps:

- Divide the ***2D*** or ***3D*** continuum of the design domain as illustrated in Figure 6.40a into discrete *elements* of different finite shape and size (with triangular and rectangular elements) as shown in Figure 6.40b. The hole in Figure 6.40a has been approximated by the eight sides of the *triangular elements* around it. Available ***FE*** software packages usually provide automatic division of the domain into different options of elements.
- The collection of elements is called a *mesh* connecting the elements at *nodes*, which are shown as circles in Figure 6.40b. Note that the loads on the design may be modeled as the black-arrowed loads, and the supports are modeled also as shown in Figure 6.40a. Further adjustments may also be implemented in Figure 6.40b. Available ***FE*** software packages may support some aid to introduce loads and support models.

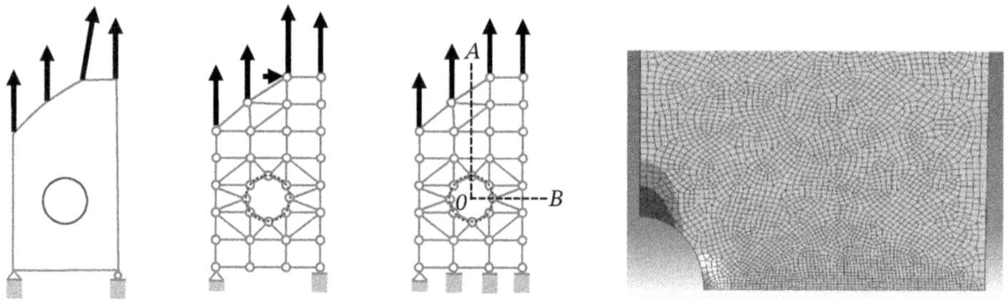

(a) (b) (c) (d)

Figure 6.40 A ***2D*** sketch of a small thickness ***3D*** design domain shown in (a). In (b), the domain is divided into elements of two different finite shapes and sizes. The domain is further adjusted in (c) to use the assumed symmetry of the geometry away from the load location and allow the use of only a quarter of the body sectioned by A–B partitions. The ***FE*** equations are solved and post-processing provides displays in a graphical form as in (d) for the stresses in part of the partition around the circular hole.

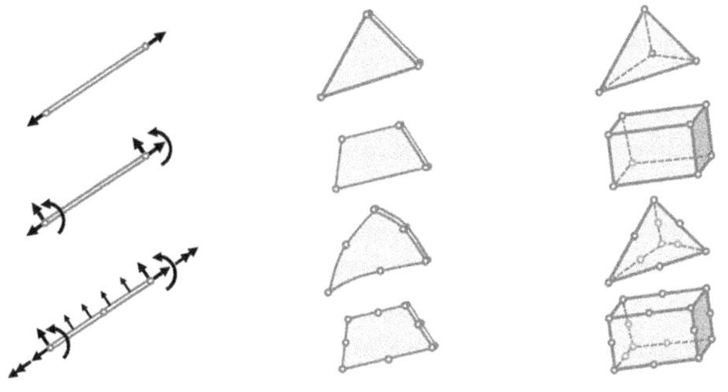

(a) (b) (c)

Figure 6.41 Typical elements in ***2D*** and ***3D*** that are available in most ***FE*** packages: (a) linear 2 node elements with optional extra intermediate node in the lower element, (b) surface elements with constant thickness and corner nodes such as triangular and quadrilateral in addition to optional extra intermediate nodes, and (c) ***3D*** elements with corner nodes such as tetrahedron and hexahedral with optional extra intermediate nodes.

- The system in Figure 6.40b may also be further reduced to that shown in Figure 6.40c to use the assumed symmetry of the geometry away from the load location. One can consider only the quarter of the body sectioned by *A–B* partitions, and new boundary conditions are set at these partitions. This may be used to reduce the total number of the elements in the model. Of course, re-meshing is needed.
- Governing equations such as *stiffness* of each *element* in its local coordinates are transformed into approximate algebraic *element* equations with assumed *shape or displacement function* solution. Equations of these elements are *assembled* based on the element connectivity at *nodes* to get the *global stiffness*. This is the usual *direct stiffness method* used by many ***FE*** software packages.
- Boundary conditions are imposed on the reduced domain such as the model in Figure 6.40b, and terms are shifted from one side of the equations to the other to allow for possible solution. Usually all of the governing equations, element assembly, and handling boundary conditions are provided in the available ***FE*** software packages.
- The system of equations is solved and *post-processing* provides displays in graphical form as shown in Figure 6.40d for part of the partition around the circular hole to show the stress distribution in different grayish level. Available ***FE*** software packages provide colors and color scale for values of the stresses and deformations.

As indicated previously, the accuracy of the results depends on the selected *element type* and size. The size of the element is expected to be smaller for the expected large *stress gradient* and larger for small stress gradient. This is also tied to the element type. A larger number of the simpler elements are required against a smaller number of more sophisticated element types. The element type is also dependent on the type of problem at hand. Figure 6.41 present a collection of typical element types in ***2D*** and ***3D*** that are available in most commercial ***FE*** packages. The first category of elements is the *line elements* with end nodes as shown in Figure 6.41a. The first linear type element is the linear ***2 node*** element that may be used in trusses where forces are only collinear with the *truss* element. The second type of 2 node elements may also represent *beams* that experience forces and moments at both ends. For more complex loading and more accuracy, the third linear element has the optional *extra intermediate node* in the element for higher order *shape* or *displacement function*.

The second category of elements is the *surface elements* with corner nodes as shown in Figure 6.41b. They are loaded in their own plane. These surface elements have constant thickness such as the *triangular* and *quadrilateral* shown in the two top elements. The third and the forth elements from the top are constructed with optional *extra intermediate nodes* for higher order *shape* or *displacement function*. These elements may be used for *plane stress* or *strain*, axisymmetry, thin flat plate, or shell in bending even though quads and extra nodes are preferred for bending.

The third category of elements is the *solid elements* with corner nodes as shown in Figure 6.41c. They are ***3D*** elements such as *tetrahedron* and *hexahedral* shown in the two top elements. The third and fourth elements from the top are developed with optional *extra intermediate nodes* for higher order shape or displacement functions. These elements are more suited for solid bodies, thick plates, and needed ***3D*** stress analyses.

Stresses can frequently vary in all directions; nodes can also be located in ***3D*** space and then the output displacements ought to be in three directions. Therefore, most elements should be prepared to handle these general requirements. If the real case is *1D* or ***2D***, the setting and solution would be a special case of the ***3D***.

Selecting the appropriate element type is essential in getting the proper solution. Choosing ***3D*** solid all the time entails much greater demand on computer resources and is harder to check for errors. The *boundary conditions* are also of a paramount importance in getting the right solution. Considering inappropriate or incorrect boundary conditions would create false solutions.

Even though the expected applications of the ***FE*** are in design for strength or rigidity (stresses and deformations), other engineering problems expressed by governing *partial differential equations* and experience *boundary conditions* may also be solved by ***FE***. For thermal problems, the property can be the *conductivity*, the behavior is the *temperature* and the load can be the *heat source*. For fluid problems, the property may be the *viscosity*, the behavior can be the *velocity*, and the load may be the *body force*. In general, one can then write the ***FE*** problem formulation such as

$$\boldsymbol{K\delta} = \boldsymbol{F} \qquad (6.178)$$

where $\boldsymbol{K}$ is the *property matrix* (e.g. stiffness), $\boldsymbol{\delta}$ is the *behavior vector* (e.g. displacements), and $\boldsymbol{F}$ is the *load* or action vector (e.g. forces). The solution is then obtained by the inverse of Eq. (6.178) such that

$$\boldsymbol{\delta} = \boldsymbol{K}^{-1}\boldsymbol{F} \qquad (6.179)$$

This process will be discussed in some details in the following sections.

6.11.1 Axially Loaded Elements

These elements are similar to truss members, but any member can be formed of several elements collinearly connected. Each of these elements can have different length and cross-sectional area. It may be tapered or other elements may be made of different material. All these options may be admissible in the total construction of the assembly of these elements to form the truss or similar machine assembly. The treatment can be unified if these options are considered in the procedure of formulation. The system or design model is composed of these *elements* connected to each other at connection points of junctions or *nodes*. The general procedure is to consider the *local* mathematical behavioral relation of properties for each element such as the *element stiffness* k_e and assemble these relations to reach the general or *global* form of the FE representing the assembly properties such as *global stiffness matrix* $\boldsymbol{K}$ relating the behaviors such as displacements $\boldsymbol{\delta}$ to the ensued actions or loads $\boldsymbol{F}$. The general form is then given by the previously provided relation:

$$\boldsymbol{K\delta} = \boldsymbol{F} \qquad (6.178)$$

where $\boldsymbol{K}$ is the system *assembly matrix* or the *global stiffness matrix* or the *global structural stiffness*, $\boldsymbol{\delta}$ is the *nodal displacement vector*, and $\boldsymbol{F}$ is the action or *load vector*. The aim of getting the stresses has to be dependent on the stain, which can be obtained from the displacements. From Eq. (6.178), the *displacement* vector is found by the inverse as follows:

$$\boldsymbol{\delta} = \boldsymbol{K}^{-1}\boldsymbol{F} \qquad (6.179)$$

As indicated previously in Section 6.3.1 and Eqs. (6.57) and (6.58), the *strains* ε_x and the *stresses* σ_x in the *local direction x* of the elements are obtained as

$$\varepsilon_x = \frac{d\delta}{dx} = \begin{bmatrix} u_1/l_1 \\ \vdots \\ u_{n_e}/l_{n_e} \end{bmatrix} \quad \text{and} \quad \sigma_x = E_i\varepsilon_x = \begin{bmatrix} E_1 u_1/l_1 \\ \vdots \\ E_{n_e} u_{n_e}/l_{n_e} \end{bmatrix} \tag{6.180}$$

where u_i is the uniform displacement in element i, l_i is the length of element i, n_e is the number of elements, and E_i is the modulus of elasticity of element i.

For such axially loaded elements, the process is best demonstrated by the following example.

Example 6.15 The simple system of a bar represented by three spring elements assembled in series is shown in Figure 6.42a as three connected flexible beams. The assembly is to be modeled by a system graph to define the system matrix $\boldsymbol{K}$. The system is acted upon by two flow driver or forces $\boldsymbol{F}_{\text{in1}}$ and $\boldsymbol{F}_{\text{in2}}$. Find the nodal displacements, strains, stresses, and forces in each element. The cross-sectional areas are $A_1 = 0.0004$ [m^2] or 0.62 [in]; $A_2 = 0.0005$ [m^2] or 0.775 [in^2]; and $A_3 = 0.0006$ [m^2] or 0.93 [in^2]. The element lengths are l_1, l_2, and $l_3 = 0.2$ [m] or 7.874 [in]. The input forces $\boldsymbol{F}_{\text{in1}} = 40\,000$ [N] or 8992 [lb]; $\boldsymbol{F}_{\text{in2}} = 10\,000$ [N] or 2248 [lb]; and the modulus of elasticity is $E = 207$ [GPa] or 29 [Mpsi].

Solution

The standard sign convention used in system graphs is depicted in Figure 6.42b. Any element assigned the direction exiting a node is positive (+), and it is negative (−) if the assigned direction is entering the node. The system graph is shown in Figure 6.42c indicating the links between the inputs at nodes (a) and (b). The spring element 1 is connecting (a) and (b), spring element 2 is connecting nodes (b) and (c), and spring element 3 is connecting nodes (c) and (g_t) which is the ground of translation g_t. Usually drivers are assigned a direction that is entering the node (a) and (b). If the driver is in the negative x direction as $\boldsymbol{F}_{\text{in2}}$, in may be assigned as exiting the node. The spring elements are assigned the direction from (a) to (b), from (b) to (c), and from (c) to the ground (g_t). The solution, however, is not affected by these assigned directions. Any assignment should work.

The node equations are written with reference to Eq. (2.8) as

$$\sum_{\text{at points of junction}} f_i = 0 \tag{a}$$

The node equations from Figure 6.42c at nodes (a), (b), and (c) are then

$$\begin{matrix} \text{at } a: \\ \text{at } b: \\ \text{at } c: \end{matrix} \begin{bmatrix} 1 & & & -1 & \\ -1 & 1 & & & 1 \\ & -1 & 1 & & \end{bmatrix} \begin{bmatrix} F_1 \\ F_2 \\ F_3 \\ F_{\text{in1}} \\ F_{\text{in2}} \end{bmatrix} = 0 \tag{b}$$

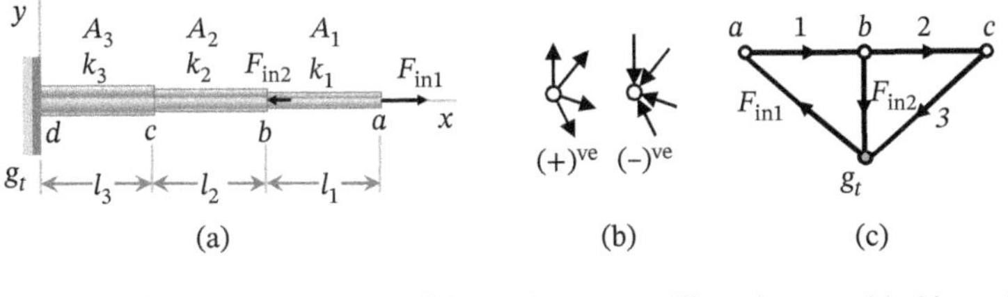

Figure 6.42 A simple system of three elements collinearly assembled in series: (a) schematic diagram, (b) the standard sign convention, and (c) is the system graph representation. The system graph model utilizes the generalized system formulation of Section 2.1.2.

or in a general *matrix* form,

$$\boldsymbol{A}_n \boldsymbol{F}_n = 0 \tag{b'}$$

where $\boldsymbol{A}_n$ is the *node matrix* $(n_{v-1} \times n_e)$ and $\boldsymbol{F}_n$ is the *flow vector* (n_e), with n_e as the number of components and n_{v-1} is the number of vertices or *nodes −1*. Partitioning Eq. (b) to separate the input driver term, one gets

$$\begin{bmatrix} 1 & & \\ -1 & 1 & \\ & -1 & 1 \end{bmatrix} \begin{bmatrix} F_1 \\ F_2 \\ F_3 \end{bmatrix} + \begin{bmatrix} -1 & \\ & 1 \\ & \end{bmatrix} \begin{bmatrix} F_{\text{in}1} \\ F_{\text{in}2} \end{bmatrix} = 0 \tag{c}$$

or in a general matrix form,

$$\boldsymbol{A}_1 \boldsymbol{F}_c + \boldsymbol{A}_{\text{in}} \boldsymbol{F}_{\text{in}} = 0 \tag{c'}$$

where $\boldsymbol{A}_1$ is the reduced node matrix (reduced by the input flow or across driver effect), $\boldsymbol{F}_c$ is the element flow vector, and $\boldsymbol{A}_{\text{in}}$ is the *input matrix*. The component equations (using energy rather than power in Section 2.1.2) are

$$\begin{bmatrix} F_1 \\ F_2 \\ F_3 \end{bmatrix} = \begin{bmatrix} k_1 & & \\ & k_2 & \\ & & k_3 \end{bmatrix} \begin{bmatrix} u_1 \\ u_2 \\ u_3 \end{bmatrix} \tag{d}$$

or in a matrix form,

$$\boldsymbol{F}_c = \boldsymbol{W}_c \ \boldsymbol{u}_c \quad \text{or} \quad \boldsymbol{F}_e = \boldsymbol{K}_e \ \boldsymbol{u}_e \tag{d'}$$

Here, $\boldsymbol{W}_c$ is the *components admittance matrix*, or $\boldsymbol{K}_e$ is the *element stiffness* matrix and $\boldsymbol{u}_c$ is the element across vector or $\boldsymbol{u}_e$ is the *element displacement vector*. For a prismatic or cylindrical element, we can find the stiffness k_i via the application of Eq. (6.58), which gives

$$\boldsymbol{u}_i = \frac{\boldsymbol{F}_i l_i}{EA_i} \quad \text{or} \quad k_i = \frac{F_i}{u_i} = \frac{EA_i}{l_i} \tag{e}$$

Substituting into the partitioned Eq. (c) gives

$$\begin{bmatrix} 1 & & \\ -1 & 1 & \\ & -1 & 1 \end{bmatrix} \boldsymbol{W}_c \begin{bmatrix} u_1 \\ u_2 \\ u_3 \end{bmatrix} + \begin{bmatrix} -1 & \\ & 1 \\ & \end{bmatrix} \begin{bmatrix} F_{in1} \\ F_{in2} \end{bmatrix} = 0 \tag{f}$$

or in a matrix form,

$$\boldsymbol{A}_1 \boldsymbol{W}_c \ \boldsymbol{u}_c + \boldsymbol{A}_{\text{in}} \boldsymbol{F}_{\text{in}} = 0 \tag{f'}$$

The *node variable transformation* relates the *element* displacements to the *node* displacements using the same sign convention of Figure 6.42b such that

$$\begin{bmatrix} u_1 \\ u_2 \\ u_3 \end{bmatrix} = \begin{bmatrix} 1 & -1 & \\ & 1 & -1 \\ & & 1 \end{bmatrix} \begin{bmatrix} u_a \\ u_b \\ u_c \end{bmatrix} \tag{g}$$

or in a matrix form,

$$\boldsymbol{u}_c = \boldsymbol{A}_1^T \boldsymbol{u}_n \tag{g'}$$

Again, $\boldsymbol{A}_1^{\mathrm{T}}$ is the *node variable transformation* matrix. This is known from graph theory that the node variable transformation matrix is the transpose of the reduced node matrix. Substituting Eq. (g) into Eq. (f) gives

$$\begin{bmatrix} 1 & & \\ -1 & 1 & \\ & -1 & 1 \end{bmatrix} \begin{bmatrix} k_1 & & \\ & k_2 & \\ & & k_3 \end{bmatrix} \begin{bmatrix} 1 & -1 & \\ & 1 & -1 \\ & & 1 \end{bmatrix} \begin{bmatrix} u_a \\ u_b \\ u_c \end{bmatrix} + \begin{bmatrix} -1 & \\ & 1 \\ & \end{bmatrix} \begin{bmatrix} F_{\mathrm{in1}} \\ F_{\mathrm{in2}} \end{bmatrix} = 0 \tag{h}$$

or in a matrix form,

$$\boldsymbol{A}_1 \, \boldsymbol{W}_c \, \boldsymbol{A}_1^T \, \boldsymbol{u}_n + \boldsymbol{A}_{\mathrm{in}} \, \boldsymbol{F}_{\mathrm{in}} = 0$$

$$\boldsymbol{K} \, \boldsymbol{u}_n + \boldsymbol{A}_{\mathrm{in}} \, \boldsymbol{F}_{\mathrm{in}} = 0 \tag{h$'$}$$

This is the ***general node equation*** that can be used irrespective of the number of elements and how they are connected in any system graph. In Eq. (h′), the term $\boldsymbol{K}$ is the *elements assembly stiffness matrix*. In the standard ***FE*** formulation, Eq. (h′) is

$$\boldsymbol{K} \boldsymbol{u}_n = -\boldsymbol{A}_{\mathrm{in}} \, \boldsymbol{F}_{\mathrm{in}} \tag{h$''$}$$

Multiplying the constituents of Eq. (g) provides the relation

$$\begin{bmatrix} k_1 & -k_1 & 0 \\ -k_1 & k_1 + k_2 & -k_2 \\ 0 & -k_2 & k_2 + k_3 \end{bmatrix} \begin{bmatrix} u_a \\ u_b \\ u_c \end{bmatrix} = - \begin{bmatrix} -1 & 0 \\ 0 & 1 \\ 0 & 0 \end{bmatrix} \begin{bmatrix} F_{\mathrm{in1}} \\ F_{\mathrm{in2}} \end{bmatrix} \tag{i}$$

This is the ***FE Equation*** of the system or design as a function of the *node displacement*. In ***FE***, the stiffness matrix $\boldsymbol{K}$ to the left of the node displacement vector $\boldsymbol{u}_n$ is the *assembled matrix* from each of the individual element. The generalized procedure generates this *assembly matrix* by the utilization of the node variable transformation, which is also known as the *elements connectivity* relations. It is observed that the ***stiffness matrix*** or the *assembled stiffness matrix* $\boldsymbol{K}$ on the left side of Eq. (i) is *always symmetric*. Setting the system equation into the ***FE*** form of *property-action* configuration, gives

$$\begin{bmatrix} k_1 & -k_1 & 0 \\ -k_1 & k_1 + k_2 & -k_2 \\ 0 & -k_2 & k_2 + k_3 \end{bmatrix} \begin{bmatrix} u_a \\ u_b \\ u_c \end{bmatrix} = \begin{bmatrix} 1 & 0 \\ 0 & -1 \\ 0 & 0 \end{bmatrix} \begin{bmatrix} F_{\mathrm{in1}} \\ F_{\mathrm{in2}} \end{bmatrix} = \begin{bmatrix} F_{\mathrm{in1}} \\ -F_{\mathrm{in2}} \\ 0 \end{bmatrix} \tag{j}$$

The solution of the ***FE*** problem is to get the node displacements $\boldsymbol{u}_n$. This is achieved by utilizing Eq. (i) to get the following:

$$\begin{bmatrix} u_a \\ u_b \\ u_c \end{bmatrix} = \begin{bmatrix} k_1 & -k_1 & 0 \\ -k_1 & k_1 + k_2 & -k_2 \\ 0 & -k_2 & k_2 + k_3 \end{bmatrix}^{-1} \begin{bmatrix} F_{\mathrm{in1}} \\ -F_{\mathrm{in2}} \\ 0 \end{bmatrix} \tag{k}$$

Substituting for the element stiffness from Eq. (e), one gets

$$\begin{bmatrix} u_a \\ u_b \\ u_c \end{bmatrix} = \frac{1}{E} \begin{bmatrix} \frac{A_1}{l_1} & -\frac{A_1}{l_1} & 0 \\ -\frac{A_1}{l_1} & \frac{A_1}{l_1} + \frac{A_2}{l_2} & -\frac{A_2}{l_2} \\ 0 & -\frac{A_2}{l_2} & \frac{A_2}{l_2} + \frac{A_3}{l_3} \end{bmatrix}^{-1} \begin{bmatrix} F_{\mathrm{in1}} \\ -F_{\mathrm{in2}} \\ 0 \end{bmatrix} \tag{l}$$

To get the stresses, one needs to get the stains and then use *Hook's* law. This process calculates the strain in each element ε_i using Eqs. (6.57) and (g) as follows:

$$\varepsilon\varepsilon_e = \begin{bmatrix} \varepsilon_1 \\ \varepsilon_2 \\ \varepsilon_3 \end{bmatrix} = \begin{bmatrix} u_1/l_1 \\ u_2/l_2 \\ u_3/l_3 \end{bmatrix} = \begin{bmatrix} 1/l_1 & -1/l_1 & \\ & 1/l_2 & -1/l_2 \\ & & 1/l_3 \end{bmatrix} \begin{bmatrix} u_a \\ u_b \\ u_c \end{bmatrix} \tag{m}$$

where $\boldsymbol{\varepsilon}_e$ is the *element strain vector*. The stresses in each element σ_i using Eqs. (6.58) and (m) to get

$$\boldsymbol{\sigma}_e = \begin{bmatrix} \sigma_1 \\ \sigma_2 \\ \sigma_2 \end{bmatrix} = E\boldsymbol{\varepsilon}_e = \begin{bmatrix} E\varepsilon_1 \\ E\varepsilon_2 \\ E\varepsilon_3 \end{bmatrix} \tag{n}$$

where $\boldsymbol{\sigma}_e$ is the *element stress vector*.

To find the reaction at *node d*, one can anticipate that it should be the same as the force in *element 3*. In that case, it is simple to find the element forces by either multiplying the stress by the area of the element or get the forces by the utilization of Eq. (d).

A MATLAB code to find the nodal displacements, element stresses, and forces for this simple system of three elements collinearly assembled is shown in Figure 6.43. The solution utilizes the formulation of generalized system modeling in Section 2.1.2, which has been given in Eqs. (e), (j), (k), (l), (m), (n), and (d). The code specifies the values of the design parameters given in the data section of Example 6.15 for the SI system. Changing the inputs to the US systems is simple to implement. Running the code produces the following results for SI or US data:

$$\begin{aligned} &\boldsymbol{u}_n = 1.0\text{e-}03 \begin{bmatrix} 0.2100 \\ 0.1100 \\ 0.0500 \end{bmatrix} [\text{m}], \quad \varepsilon_e = 1.0\text{e-}03 \begin{bmatrix} 0.5000 \\ 0.3000 \\ 0.2500 \end{bmatrix}, \quad \sigma_e = \begin{bmatrix} 100.000 \\ 60.000 \\ 50.000 \end{bmatrix} [\text{MPa}], \\ &\boldsymbol{F}_e = 1.0\text{e}+04 \begin{bmatrix} 8.4000 \\ 5.5000 \\ 3.0000 \end{bmatrix} [\text{N}] \\ &\boldsymbol{u}_n = \begin{bmatrix} 0.00827 \\ 0.00433 \\ 0.00197 \end{bmatrix} [\text{in}], \quad \varepsilon_e = 1.0\text{e-}03 \begin{bmatrix} 0.5000 \\ 0.3000 \\ 0.2500 \end{bmatrix}, \quad \sigma_e = 1.0\text{e}+04 \begin{bmatrix} 1.4503 \\ 0.8702 \\ 0.7252 \end{bmatrix} [\text{psi}], \\ &\boldsymbol{F}_e = 1.0\text{e}+04 \begin{bmatrix} 1.8883 \\ 1.2364 \\ 0.6744 \end{bmatrix} [\text{lb}] \end{aligned} \tag{o}$$

From Eq. (o), the reaction at the support *point d*, is $-\boldsymbol{F}_3$ or equals to −30 000 [N] or −674.4 [lb]. This value matches the balance of forces for a simple free body diagram, which also gives a negative sine to the force in element 3 as expected.

It should be noted that the foregoing procedure in Example 6.15 generates the global matrix directly. It does not employ the usual procedure in *regular* ***FE*** for generating the *element stiffness matrix*, which is usually formulated for element 1 as follows:

$$\begin{bmatrix} F_{1a} \\ F_{1b} \end{bmatrix} = \begin{bmatrix} k_1 & -k_1 \\ -k_1 & k_1 \end{bmatrix} \begin{bmatrix} u_a \\ u_b \end{bmatrix} \quad \text{or} \quad \boldsymbol{F}_{1n} = \boldsymbol{K}_e \boldsymbol{u}_n \tag{6.181}$$

```
clc; clear; clear all; format long; format compact
ESI=207*10^9; EUS=29*10^6;                         % Modulus of Elasticity SI&US
A1=0.0004; A2=0.0005; A3=0.0006;                   % Area in [m^2]
L1=0.2; L2=0.2; L3=0.2;                            % Length in [m]
Fin1=40000; Fin2=10000;                            % Force in [N]
Fin=[Fin1  -Fin2  0]';
k1=ESI*A1/L1; k2=ESI*A2/L2; k3=ESI*A3/L3;
K=[k1 -k1           0;
  -k1   k1+k2     -k2;
   0     -k2         k2+k3];
un=inv(K)*Fin                                      % Node Displacements
Nvt=[1/L1   -1/L1      0;
      0      1/L2    -1/L2;
      0       0         1/L3];                     % Node variable transformation
Strain=Nvt*un
Stress_MPa=ESI*Strain/10^6                         % Stress in [MPa] in each element
Ke= [k1  0  0; 0  k2  0; 0    0  k3];              % Stiffness of each element
Fe=Ke*Strain.* [L1; L2; L3]                        % Force in each element
```

Figure 6.43 MATLAB code to find the nodal displacements, element stresses, and forces for a simple system of three elements collinearly assembled. The solution employs the generalized modeling in Section 2.1.2 and the SI values of Example 6.15.

where $F_{1,a}$ and $F_{1,b}$ are the forces in element 1 at nodes a and b, $\boldsymbol{K}_e$ is the *local element stiffness matrix*, and $\boldsymbol{u}_n$ is the *node displacements* at a and b. Since $\boldsymbol{F}_1$ is the same at both ends a and b, Eq. (6.181) seems to have redundancies. These are taken care of in the assembly of *global stiffness matrix* $\boldsymbol{K}$ given by the following special summation;

$$\boldsymbol{K} = \sum_{1}^{n_e} \boldsymbol{K}_e \tag{6.182}$$

Note that the second redundant line in Eq. (6.181) will be algebraically added to the other element (say, *element 2*) connected at the *same node b* using the equation of *element 2* that is similar to Eq. (6.181). This has also been taken care of in the *node connectivity* in Eq. (h′), which has also included the boundary conditions from the start.

The *global stiffness matrix* is obtained by the assembly of element matrices; bearing in mind that elements are similarly connected according to Eq. (h) of Example 6.15, which gives

$$\begin{aligned}\boldsymbol{k}_1\boldsymbol{u}_e + \boldsymbol{k}_2\boldsymbol{u}_e + \boldsymbol{k}_3\boldsymbol{u}_e &= \begin{bmatrix} k_1 & -k_1 & \\ -k_1 & k_1 & \\ & & \end{bmatrix}\begin{bmatrix} u_a \\ u_b \\ u_c \end{bmatrix} + \begin{bmatrix} & & \\ & k_2 & -k_2 \\ & -k_2 & k_2 \end{bmatrix}\begin{bmatrix} u_a \\ u_b \\ u_c \end{bmatrix} + \begin{bmatrix} & & \\ & & \\ & & k_3 \end{bmatrix}\begin{bmatrix} u_a \\ u_b \\ u_c \end{bmatrix} \\ &= \begin{bmatrix} k_1 & -k_1 & 0 \\ -k_1 & k_1+k_2 & -k_2 \\ 0 & -k_2 & k_2+k_3 \end{bmatrix}\begin{bmatrix} u_a \\ u_b \\ u_c \end{bmatrix}\end{aligned} \tag{6.183}$$

which is exactly the same as the *assembly* or *global stiffness matrix* $\boldsymbol{K}$ in Eqs. (h″) and (i).

This procedure in regular *FE* will also be demonstrated in the following Section 6.11.2 for the solution of Example 6.16.

6.11.2 Prismatic Beam Element

Here a prismatic beam element is considered in x–y plane, even though this plane can be regarded in any ***3D*** space and the x–y plane is the *local plane* of the beam. In fact, the beam is laid in the x direction in its own local plane. If no load is assumed on the beam, $\boldsymbol{F}(x) = 0$ between nodes 1 and 2 as shown in Figure 6.44, displacement v, slope θ, moment $\boldsymbol{M}$, and force $\boldsymbol{F}$ at node 2 will be obtained from simple beam theory. No loads are present to consider the deformations in the z direction. Figure 6.44 is presenting the deformed beam and its displacement that is extremely

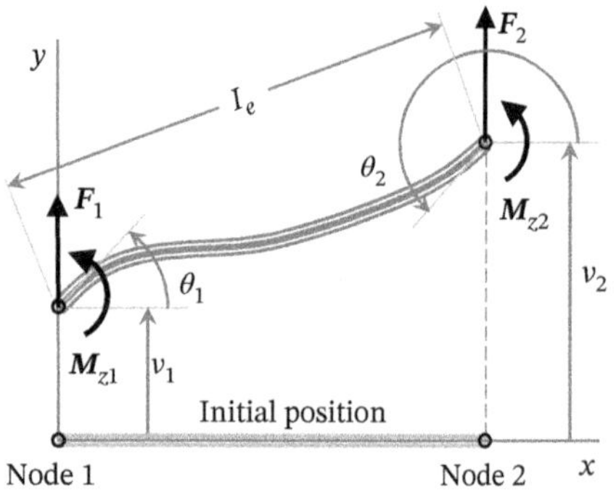

Figure 6.44 A deformed beam and its deflection extremely exaggerated to allow for the identification and definition of variables and to show initial and deformed states.

exaggerated to allow for clearer identification and definition of variables and to differentiate between the initial and the deformed states. The beam or *beam element* length l_e is assumed not to be changed as might seem in Figure 6.44. The deformation and displacements are assumed in the elastic range and thus very small to warrant the assumption of using the original l_e length. To formulate the ***FE*** relation of properties or *element stiffness matrix* $\boldsymbol{K}_e$, one needs to get the relation between the actions or *loads* and the element response or displacement. These actions or forces and moments loads are $\boldsymbol{F}_1$, $\boldsymbol{M}_1$, $\boldsymbol{F}_2$, and $\boldsymbol{M}_2$ at end *nodes* 1 and 2, while the behavior or beam response or *displacements* are v_1, θ_1, v_2, and θ_2 at the same *nodes* 1 and 2 as indicated in Figure 6.44. These loads and displacements in the local x–y coordinates are defined in Figure 6.44 between the initial position and the deformed one. Noting that $\boldsymbol{w}_i = 0$, $\boldsymbol{F}_i = \boldsymbol{F}(x) = 0$ and observing the load orientations regarding that they are at the nodes, the free body diagram in Figure 6.44 summing forces in the y direction and summing the moment about node 2 gives

$$\begin{aligned} \boldsymbol{F}_2 &= -\boldsymbol{F}_1 \\ \boldsymbol{M}_2 &= -\boldsymbol{M}_1 + l_e\boldsymbol{F}_1 \end{aligned} \tag{6.184}$$

It should be noted that the direction of $\boldsymbol{F}_1$ and $\boldsymbol{M}_1$ are opposite to $\boldsymbol{V}$ and $\boldsymbol{M}$ that are considered in Figure 6.1. $\boldsymbol{F}_1$ and $\boldsymbol{M}_1$ are forces and moments at node 1 and therefore different from the internal forces and moments on which Eqs. (6.27) and (6.28) depend. Eqs. (6.27) and (6.28) are representing the beam continuum and therefore depend on a free body diagram of an infinitesimal element of the beam in Figure 6.1. The loads $\boldsymbol{V}$ and $\boldsymbol{M}$ are in the negative global directions but positive on the negative plane to the left of the infinitesimal element of the beam in Figure 6.1. The *beam element*, however, is modeled as an entity between two nodes as defined in Figure 6.44. The direction of $\boldsymbol{F}_1$ and $\boldsymbol{M}_1$ are set in the positive local directions. This is because other beam elements may be connected at these nodes.

Substituting Eq. (6.184) into Eqs. (6.25) and applying Eq. (6.27) give

$$\begin{aligned} \theta_2 &= \int_{x_1}^{x_2} \frac{\boldsymbol{M}_z(x)}{EI_z}\,dx + C_3 = \frac{1}{EI_z}\int_0^{l_e}(-\boldsymbol{M}_1 + l_e\boldsymbol{F}_1)\,dx + \theta_1 = \theta_1 - \frac{l_e}{EI_z}\boldsymbol{M}_1 + \frac{l_e^2}{2EI_z}\boldsymbol{F}_1 \\ v_2 &= \int_0^{l_e}\theta(x)\,dx + C_4 = \int_0^{l_e}\left(\theta_1 - \frac{l_e}{EI_z}\boldsymbol{M}_1 + \frac{l_e^2}{2EI_z}\boldsymbol{F}_1\right)\,dx + v_1 = v_1 + l_e\,\theta_1 - \frac{l_e^2}{2EI_z}\boldsymbol{M}_1 + \frac{l_e^3}{6EI_z}\boldsymbol{F}_1 \end{aligned} \tag{6.185}$$

The constants C_3 and C_4 are the *boundary conditions* on the slope and displacement at $x = 0$ which obviously give θ_1 and v_1, respectively. To facilitate the derivation of the *element stiffness matrix* $\boldsymbol{K}_e$, Eqs. (6.184) and (6.185) are rewritten as follows:

$$v_2 = v_1 + l_e\,\theta_1 - \frac{l_e^2}{2EI_z}\boldsymbol{M}_1 + \frac{l_e^3}{6EI_z}\boldsymbol{F}_1 \tag{6.186}$$

$$\theta_2 = \theta_1 - \frac{l_e}{EI_z}\boldsymbol{M}_1 + \frac{l_e^2}{2EI_z}\boldsymbol{F}_1 \tag{6.187}$$

$$\boldsymbol{M}_2 = -\boldsymbol{M}_1 + l_e\,\boldsymbol{F}_1 \tag{6.188}$$

$$\boldsymbol{F}_2 = -\boldsymbol{F}_1 \tag{6.189}$$

Multiply Eq. (6.187) by ½ l_e and subtract that from Eq. (6.186), one gets – after the arrangement of terms – the expression of $\boldsymbol{F}_1$ in terms of the *displacements*. Using Eq. (6.189) produces the expression of $\boldsymbol{F}_2$ in terms of the *displacements*. Multiply Eq. (6.187) by $l_e/3$ and subtract that from Eq. (6.186), one gets – after the arrangement of terms – the expression of $\boldsymbol{M}_1$ in terms of the *displacements*. Using Eq. (6.188) afterward produces the expression of $\boldsymbol{M}_2$ in terms of the *displacements*. This process gives

$$\begin{bmatrix} \boldsymbol{F}_1 \\ \boldsymbol{M}_1 \\ \boldsymbol{F}_2 \\ \boldsymbol{M}_2 \end{bmatrix} = EI_z \begin{bmatrix} 12/l_e^3 & 6/l_e^2 & -12/l_e^3 & 6/l_e^2 \\ 6/l_e^2 & 4/l_e & -6/l_e^2 & 2/l_e \\ -12/l_e^3 & -6/l_e^2 & 12/l_e^3 & -6/l_e^2 \\ 6/l_e^2 & 2/l_e & -6/l_e^2 & 4/l_e \end{bmatrix} \begin{bmatrix} v_1 \\ \theta_1 \\ v_2 \\ \theta_2 \end{bmatrix} \tag{6.190}$$

or

$$\boldsymbol{K}_e = \frac{EI_z}{l_e^3} \begin{bmatrix} 12 & 6l_e & -12 & 6l_e \\ 6l_e & 4l_e^2 & -6l_e & 2l_e^2 \\ -12 & -6l_e & 12 & -6l_e \\ 6l_e & 2l_e^2 & -6l_e & 4l_e^2 \end{bmatrix} \tag{6.191}$$

The matrix $\boldsymbol{K}_e$ is the *element stiffness matrix* and its inverse is the *element flexibility matrix*. The element stiffness matrix $\boldsymbol{K}_e$ is a ***symmetric matrix***.

Now, it is desirable to get the displacements or deflections at any point x on the beam element. This can be accomplished by few methods. One method is to use Eqs. (6.23)–(6.26) and find the constants C_1–C_4 from the boundary conditions just like finding the deflections of beams by integration in Section 6.1.3. The other widely used method in **FE** field is to assume a function defining the element local deflection, which is called the ***shape function*** or the *displacement shape function* similar to the ***blending functions*** of Section 4.2. It is also necessary to find the coefficients of the function using also the boundary conditions. A function, which can be depicted from the integrated Eq. (6.26′) and resemble a *Hermite function* $\boldsymbol{h}_e(x)$ (after Charles Hermite (1822–1901)) is suggested. The assumed solution or interpolation function or blending function or *shape function* can then be assumed as

$$v(x) = a_3x^3 + a_2x^2 + a_1x + a_0 = \boldsymbol{h}_e(x) \tag{6.192}$$

where a_0, a_1, a_2, a_3, are the coefficients that satisfy nodal displacements and continuity. One can apply the boundary conditions and find the coefficients such that

$$\begin{aligned} &v(0) = v_1 = a_0 \quad \text{and} \quad v(l_e) = v_2 = a_3\,l_e^3 + a_2\,l_e^2 + a_1\,l_e x + a_0 \\ &\frac{dv(0)}{dx} = \theta_1 = a_1 \quad \text{and} \quad \frac{dv(l_e)}{dx} = \theta_2 = 3a_3\,l_e^2 + 2a_2\,l_e + a_1 \end{aligned} \tag{6.193}$$

Solving Eq. (6.193) for the coefficients a_0, a_1, a_2, and a_3 gives the *shape function* as

$$v(x) = \left(\frac{2}{l_e^3}(v_1 - v_2) + \frac{1}{l_e^2}(\theta_1 + \theta_2)\right)x^3 + \left(-\frac{3}{l_e^2}(v_1 - v_2) - \frac{1}{l_e}(2\theta_1 + \theta_2)\right)x^2 + \theta_1\,x + v_1 \tag{6.194}$$

In a vector form, Eq. (6.194) can be simply rearranged to have the usual ***FE*** form such that

$$v(x) = \begin{bmatrix} \frac{1}{l_e^3}(2x^3 - 3x^2 l_e + l_e^3) \\ \frac{1}{l_e^3}(x^3 l_e - 2x^2 l_e^2 + x l_e^3) \\ \frac{1}{l_e^3}(-2x^3 + 3x^2 l_e) \\ \frac{1}{l_e^3}(x^3 l_e - x^2 l_e^2) \end{bmatrix} \begin{bmatrix} v_1 \\ \theta_1 \\ v_2 \\ \theta_2 \end{bmatrix} = [N_1 \ N_2 \ N_3 \ N_4] \begin{bmatrix} v_1 \\ \theta_1 \\ v_2 \\ \theta_2 \end{bmatrix} = \boldsymbol{N}_e \boldsymbol{\delta}_e \tag{6.195}$$

where N_1, N_2, N_3, N_4 are the weight effect of each of the displacements $v_1, \theta_1, v_2, \theta_2$ on the displacement solution $v(x)$ along the beam element from 0 to l_e. The weights N_1 to N_4 are the components of the *shape function vector* set in a form of a raw vector $\boldsymbol{N}_e$. The displacements $v_1, \theta_1, v_2, \theta_2$ are the components of the element displacement vector $\boldsymbol{\delta}_e$. These can be used to find the bending moment, shear force and stresses along the beam element at any distance x.

To get the global stiffness matrix $\boldsymbol{K}$ of the model, one assembles all the element stiffness matrices $\boldsymbol{K}_e$'s for the entire beam elements of the model each written as in Eq. (6.191). However, the equations are written in the global coordinates that has all the displacements of all the nodes included similar to Eq. (6.183) . This gives the *global stiffness matrix* $\boldsymbol{K}$ of the model as will also be demonstrated in Example 6.16.

After solving for the reactions at the supports, one can get the bending moment diagram and the shear force diagram from singularity functions of section 6.1.3.3. Alternatively one can regenerate them using Eq. (6.194) for the deflection of each element then apply Eqs. (6.25), (6.23), and (6.24) to get the slope, bending moment diagram, and the shear force diagram by differentiating the deflection once, twice, and three times consecutively. This gives the diagrams of slope, bending moment, and shear force as follows:

$$\begin{aligned} \theta(x) &= v' \\ \boldsymbol{M}_z(x) &= EI_z v'' \\ \boldsymbol{V}(x) &= -EI_z v''' \end{aligned} \tag{6.23–6.25}$$

Performing the differentiations gives the following equations;

$$\theta(x) = 3\left(\frac{2}{l_e^3}(v_1 - v_2) + \frac{1}{l_e^2}(\theta_1 + \theta_2)\right) x^2 + 2\left(-\frac{3}{l_e^2}(v_1 - v_2) - \frac{1}{l_e}(2\theta_1 + \theta_2)\right) x + \theta_1 \tag{6.196}$$

$$\boldsymbol{M}(x) = 6EI_z\left(\frac{2}{l_e^3}(v_1 - v_2) + \frac{1}{l_e^2}(\theta_1 + \theta_2)\right) x + 2EI_z\left(-\frac{3}{l_e^2}(v_1 - v_2) - \frac{1}{l_e}(2\theta_1 + \theta_2)\right) \tag{6.197}$$

$$\boldsymbol{V}(x) = -6EI_z\left(\frac{2}{l_e^3}(v_1 - v_2) + \frac{1}{l_e^2}(\theta_1 + \theta_2)\right) \tag{6.198}$$

Alternatively, and in another form, the utility of Eqs. (6.23)–(6.25) applied to Eq. (6.195) gives

$$\theta(x) = v' = \boldsymbol{N}_e'\boldsymbol{\delta}_e = \left(\frac{d\boldsymbol{N}_e}{dx}\right)\left[v_1 \ \theta_1 \ v_2 \ \theta_2\right]^T \tag{6.199}$$

$$\boldsymbol{M}_z(x) = EI_z v'' = EI_z \boldsymbol{N}_e''\boldsymbol{\delta}_e = EI_z\left(\frac{d^2\boldsymbol{N}_e}{dx^2}\right)\left[v_1 \ \theta_1 \ v_2 \ \theta_2\right]^T \tag{6.200}$$

$$\boldsymbol{V}(x) = -EI_z v''' = -EI_z \boldsymbol{N}_e''' \boldsymbol{\delta}_e = -EI_z\left(\frac{d^3\boldsymbol{N}_e}{dx^3}\right)\left[v_1 \ \theta_1 \ v_2 \ \theta_2\right]^T \tag{6.201}$$

Note that from Eq. (6.195), the components of $\boldsymbol{N}_e$ are cubic functions of x that can be differentiated to get the components of the $\boldsymbol{N}_e$ derivatives in Eqs. (6.199)–(6.201). The first derivative of $\boldsymbol{N}_e$ is quadratic; hence $\theta(x)$ is quadratic. The second derivative of $\boldsymbol{N}_e$ is linear; therefore $\boldsymbol{M}_z(x)$ is linear. The third derivative of $\boldsymbol{N}_e$ is constant; then $\boldsymbol{V}(x)$ is constant.

The *normal stresses* and *shear stresses* in each beam can be calculated as regular beams that have the bending moment and shear force diagrams previously defined in Eqs. (6.197) and (6.198) or (6.200) and (6.201).

Frames are structures with rigidly connected members that are usually modeled as connected *beam elements*. These members will be similar to the beam elements except that axial loads and axial deformations are present. The beam elements of *frames* may typically have different orientations, and they are usually in ***3D*** space. Due to the different orientations of the beam elements in a frame, the local coordinates are different from the global coordinates and *coordinate transformation* would usually be necessary; see Section 4.1.3.

Example 6.16 A statically indeterminate beam of Example 6.5 is redrawn from Figure 6.7 into Figure 6.45. It is fixed at both ends and loaded by an asymmetrical concentrated force $\boldsymbol{F}$. The beam of length l is subjected to the concentrated force $\boldsymbol{F}$ at a distance a off the left support and a distance b off the right support. Use the ***FE*** method to find the global system matrix representing the beam. Find the reactions and moments at the supports. Indicate the procedure to find shear force diagram, the bending moment diagram, and the slope and deflection along the beam.

Solution

Data: $\boldsymbol{F}$ is in the negative direction and reactions are assumed positive in the previous Example 6.5. For the ***FE*** method, all loads and displacements are first assumed positive as shown in the lower part of Figure 6.45.

The problem is statically indeterminate since the number of reactions is greater than the number of equilibrium equations. Therefore additional deformation conditions are used to provide the additionally needed equations. These are found from the application of the boundary conditions into the developed ***FE*** model.

Applying the element stiffness matrices for elements 1 and 2 as in Eq. (6.191) and written in the global coordinates that has all the displacements $v_1, \theta_1, v_2, \theta_2, v_3, \theta_3$ gives the following global stiffness matrix:

$$K = K_1 + K_2 = \frac{EI_z}{l_1^3}\begin{bmatrix} 12 & 6l_1 & -12 & 6l_1 & 0 & 0 \\ & 4l_1^2 & -6l_1 & 2l_1^2 & & \\ & & 12 & -6l_1 & & \\ \text{Symmetric} & & & 4l_1^2 & & \\ 0 & & & & 0 & 0 \\ 0 & & & & 0 & 0 \end{bmatrix} + \frac{EI_z}{l_2^3}\begin{bmatrix} 0 & 0 & & & & 0 \\ 0 & 0 & & & & 0 \\ & & 12 & 6l_2 & -12 & 6l_1 \\ & & & 4l_2^2 & -6l_2 & 2l_2^2 \\ & & & & 12 & -6l_2 \\ 0 & 0 & \text{Symmetric} & & & 4l_2^2 \end{bmatrix} \tag{a}$$

The global coordinates XYZ are coincident with the local coordinates x, y, and z. The local coordinate of element 2, however, is shifted or translated by an amount of l_1 to the right with respect to XYZ. This should not

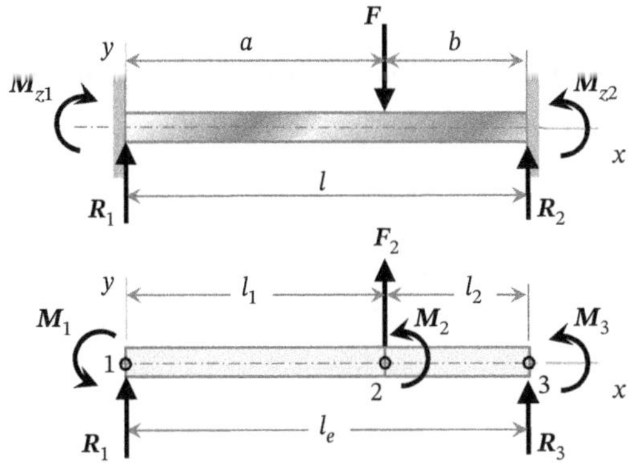

Figure 6.45 Statically indeterminate beam fixed at both ends, loaded by an asymmetrical concentrated force and its two beam element model of nodes 1–3 is given below it.

affect the results since the components of displacements are all in the y direction or angular displacements. If that was not the case, a *coordinate transformation* should have been applied.

Enacting the boundary conditions $v_1 = 0, \theta_1 = 0, v_3 = 0, \theta_3 = 0$ produces the following ***FE*** model:

$$\begin{bmatrix} F_1 \\ M_1 \\ F_2 \\ M_2 \\ F_3 \\ M_3 \end{bmatrix} = \frac{EI_z}{l_1^3}\begin{bmatrix} 12 & 6l_1 & -12 & 6l_1 & 0 & 0 \\ & 4l_1^2 & -6l_1 & 2l_1^2 & & \\ & & 12 & -6l_1 & & \\ & \text{Symmetric} & & 4l_1^2 & & \\ 0 & & & & 0 & 0 \\ 0 & & & & 0 & 0 \end{bmatrix}\begin{bmatrix} v_1 \\ \theta_1 \\ v_2 \\ \theta_2 \\ v_3 \\ \theta_3 \end{bmatrix} + \frac{EI_z}{l_2^3}\begin{bmatrix} 0 & 0 & & & & 0 \\ 0 & 0 & & & & 0 \\ & & 12 & 6l_2 & -12 & 6l_1 \\ & & & 4l_2^2 & -6l_2 & 2l_2^2 \\ & & \text{Symmetric} & & 12 & -6l_2 \\ 0 & 0 & & & & 4l_2^2 \end{bmatrix}\begin{bmatrix} v_1 \\ \theta_1 \\ v_2 \\ \theta_2 \\ v_3 \\ \theta_3 \end{bmatrix} \tag{b}$$

This gives

$$\begin{bmatrix} \boldsymbol{F}_2 \\ 0 \end{bmatrix} = \frac{EI_z}{l_1^3 l_2^3}\begin{bmatrix} 12l_2^3 & -6l_1 l_2^3 \\ -6l_1 l_2^3 & 4l_1^2 l_2^3 \end{bmatrix}\begin{bmatrix} v_2 \\ \theta_2 \end{bmatrix} + \frac{EI_z}{l_1^3 l_2^3}\begin{bmatrix} 12l_1^3 & 6l_2 l_1^3 \\ 6l_1 l_1^3 & 4l_2^2 l_1^3 \end{bmatrix}\begin{bmatrix} v_2 \\ \theta_2 \end{bmatrix}$$

$$\begin{bmatrix} \boldsymbol{F}_2 \\ 0 \end{bmatrix} = \frac{EI_z}{l_1^3 l_2^3}\begin{bmatrix} 12l_2^3 + 12l_1^3 & -6l_1 l_2^3 + 6l_2 l_1^3 \\ -6l_1 l_2^3 + 6l_1 l_1^3 & 4l_1^2 l_2^3 + 4l_2^2 l_1^3 \end{bmatrix}\begin{bmatrix} v_2 \\ \theta_2 \end{bmatrix} \tag{c}$$

Solving Eq. (c) provides the explicit expressions for v_2 and θ_2. If the load is at mid-span or $l_1 = l_2$, the angular deflection $\theta_2 = 0$, and the deflection $v_2 = -\boldsymbol{F}_2\, l_1^{\,3}/(24EI_z)$, where the negative sign is for the real direction of $\boldsymbol{F}_2$. If the load is not at mid-span, the substitution of the known values of the load $\boldsymbol{F}_2$, the beam dimensions, and the material modulus of elasticity E provides the values of v_2 and θ_2. With these values, one can substitute in Eq. (b) to get the reactions at the supports 1 and 3.

After solving for the reactions at both supports 1 and 3, one can get the bending moment diagram and the shear force diagram from Eq. (b) of Example 6.5. Alternatively one can regenerate them using Eq. (6.194) for the deflection of each element 1 and 2 and then apply Eqs. (6.25), (6.23), and (6.24) to get the slope, bending moment diagram, and the shear force diagram by differentiating the deflection once, twice, and three times consecutively. Equations (6.196)–(6.198) can be used for both elements 1 and 2 in tandem to get the total slope, bending moment, and shear force diagrams for the entire beam.

6.11.3 Constant Strain Triangle

Triangular elements are *constant thickness* and typically used for plane ***2D*** cases of relatively thin continuous body under *in-plane loading*. Figure 6.46 presents such a case of an abstract object subject to individual boundary in-plane loading $\boldsymbol{F}_i$ and constrained by a boundary where the deformation $\boldsymbol{\delta}$ is zero. The domain of the object is replaced by triangles as some depicted in Figure 6.46 with one highlighted triangle to be used as a general triangular element to study. The object is set in the global coordinate system XYZ. The triangular element is in an

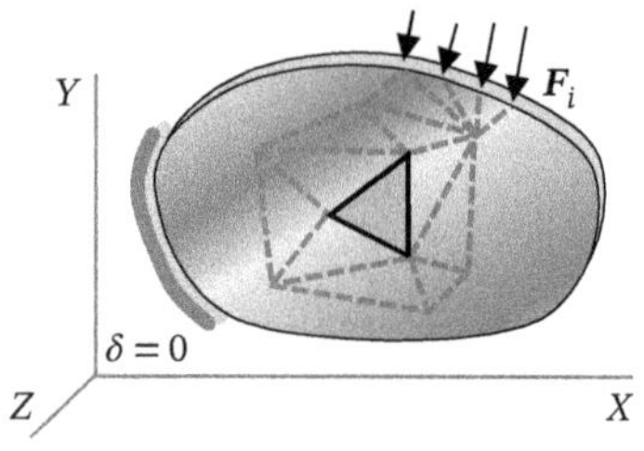

Figure 6.46 An object subject to individual in-plane boundary loading $\boldsymbol{F}_i$ and constrained by a boundary where the deformation δ is zero. The object is modeled by triangles, some of which are shown with one highlighted triangle to be used as a general element to study.

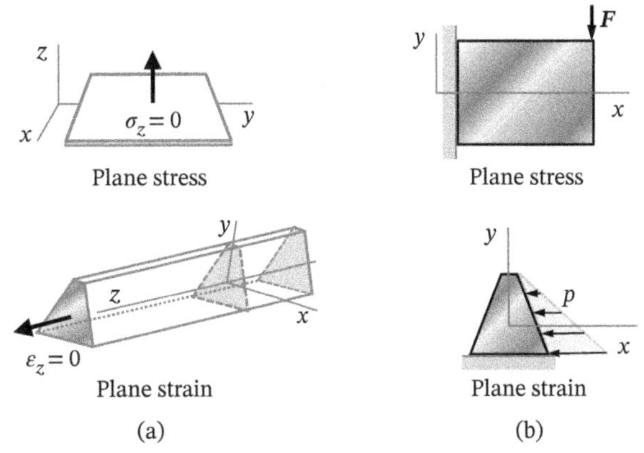

Figure 6.47 Cases of ***3D*** objects in (a) that may represent plane stress and plane strain problems. They can be modeled by ***2D*** cases shown in (b).

associated local coordinates *xyz*. To simplify the developments of ***FE*** relations the triangular element is assumed to be a *constant strain triangle* (***CST***). The assumption is an approximation that requires using so many triangular elements to warrant a reasonable representation of some expected high gradient strain and stress distributions.

The development of ***FE*** relations necessitates the differentiation between **2D** and **3D** consideration of *plane stress* and *plane strain* problems. Figure 6.47 demonstrates cases of ***3D*** objects that may reasonably represent the *plane stress* and *plane strain* problems that can be modeled by **2D** cases. In **3D**, the objects that have relatively thin continuous bodies under *in-plane loading* are *plane stress* cases where σ_z can be assumed zero as shown in Figure 6.47a. They can be modeled as ***2D*** objects, the same as the case shown at the top of Figure 6.47b. Another example is of a short ring press fitted on a shaft with a centerline in the *z* direction where the ring axial stress σ_z – in that simple case – is zero. The ***3D*** *plane strain* cases are sufficiently long objects loaded in such a way that one can assume a zero strain ε_z as defined in Figure 6.47b. Their ***2D*** model can be represented by an element in *xy* plane as shown at the bottom of Figure 6.47b. Another example is a long beam with a centerline in the *z* direction and transversely loaded uniformly along *z* that one can assume a zero strain ε_z. It should be noted, however, that the ***FE*** representation of the ***2D*** *plane stress* and *plane strain* elements should be differently handled. The ***CST*** is a ***plane stress*** problem.

In Section 6.6, some *plane strain* and ***3D*** *triaxial strain* relations have been discussed. Here, a revisit to some of these concepts is due to aid in the development of the constant strain triangular element of *plane stress* problems. With reference to Figure 6.48, at a local point (x_i, y_i) or any point $P(x,y)$, $\boldsymbol{\delta}$ is a vector of displacements $\boldsymbol{u},\boldsymbol{v}$ or any $\boldsymbol{u}_i,\boldsymbol{v}_i$ in the *x* and *y* directions, respectively. The displacement vector is thus given by

$$\delta = \begin{bmatrix} u \\ v \end{bmatrix} \tag{6.202}$$

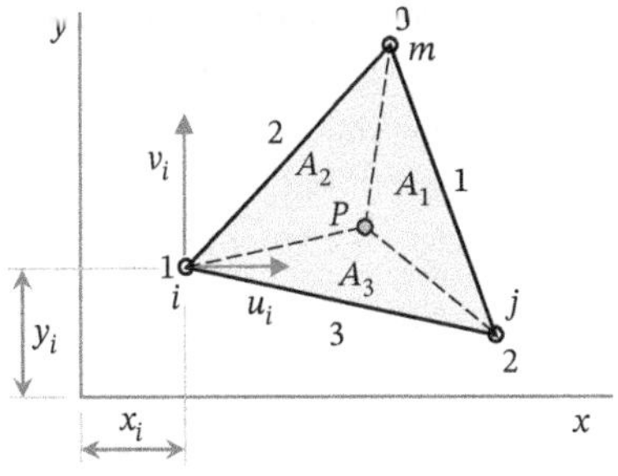

Figure 6.48 A constant strain triangle (***CST***) with corners or nodes marked by small circles. Nodes have the numbers 1, 2, 3 (or *i*, *j*, *m*) ordered in the positive counterclockwise direction. The triangle sides are numbered as the numbers of their opposite nodes. The subtriangle areas A_1, A_2, and A_3 are numbered as their edges are.

The strain ε_i at (x_i,y_i) or $\boldsymbol{\varepsilon}$ at any (x,y) is therefore

$$\begin{aligned}\boldsymbol{\varepsilon} &= \begin{bmatrix}\varepsilon_x & \varepsilon_y & \varepsilon_{xy}\end{bmatrix}^T = \begin{bmatrix}\dfrac{\partial u}{\partial x} & \dfrac{\partial \mathbf{v}}{\partial y} & \left(\dfrac{\partial u}{\partial y} + \dfrac{\partial \mathbf{v}}{\partial x}\right)\end{bmatrix}^T = \nabla\delta \\ \boldsymbol{\varepsilon} &= \mathbf{B}\,\delta\end{aligned} \tag{6.203}$$

where $\boldsymbol{B}$ is the strain-displacement matrix to be defined from the $\boldsymbol{u}$ and $\mathbf{v}$ latter on. The stress $\boldsymbol{\sigma}_i$ at (x_i,y_i) or $\boldsymbol{\sigma}$ at any (x,y) is

$$\boldsymbol{\sigma} = \begin{bmatrix}\sigma_x & \sigma_y & \sigma_{xy}\end{bmatrix}^T \tag{6.204}$$

For **plane stress**, the *constitutive stress–strain relations* (or the generalized *Hook's* law) for ***2D*** cases of homogeneous isotropic linear elastic materials are given by

$$\begin{bmatrix}\sigma_x \\ \sigma_y \\ \tau_{xy}\end{bmatrix} = \frac{E}{1-\nu^2}\begin{bmatrix}1 & \nu & 0 \\ \nu & 1 & 0 \\ 0 & 0 & (1-\nu)/2\end{bmatrix}\begin{bmatrix}\varepsilon_x \\ \varepsilon_y \\ \gamma_{xy}\end{bmatrix} \tag{6.205}$$

where E is the modulus of elasticity and ν is the *Poisson's* ratio. In the usual ***FE*** formulation that can be generalized, the vector–matrix form is given by

$$\boldsymbol{\sigma} = \boldsymbol{D}_E\boldsymbol{\varepsilon} \tag{6.206}$$

where for ***2D*** the stresses $\boldsymbol{\sigma}$, strains $\boldsymbol{\varepsilon}$, and the *stress–strain constitutive matrix* or the *elasticity matrix* $\mathbf{D}_E$ are defined according to the particulars in Eq. (6.205). The $\boldsymbol{D}_E$ matrix is also called the *material property matrix*.

For initial or additional *thermal strain* $\boldsymbol{\varepsilon}_T$, Eq. (6.206) becomes (see Section 6.9):

$$\boldsymbol{\sigma} = \boldsymbol{D}_E(\boldsymbol{\varepsilon} - \boldsymbol{\varepsilon}_T) \tag{6.207}$$

The thermal strain $\boldsymbol{\varepsilon}_T$ is given by

$$\boldsymbol{\varepsilon}_T = \begin{bmatrix}\alpha\Delta T_T & \alpha\Delta T_T & 0\end{bmatrix}^T \tag{6.208}$$

where α is *coefficient of thermal expansion* and ΔT_T *is the temperature difference.*

For **plane strain**, the *elasticity matrix* $\mathbf{D}_E$ is defined by the following from;

$$\begin{bmatrix}\sigma_x \\ \sigma_y \\ \tau_{xy}\end{bmatrix} = \frac{E}{(1+\nu)(1-2\nu)}\begin{bmatrix}1-\nu & \nu & 0 \\ \nu & 1-\nu & 0 \\ 0 & 0 & 1/2-\nu\end{bmatrix}\begin{bmatrix}\varepsilon_x \\ \varepsilon_y \\ \gamma_{xy}\end{bmatrix} \tag{6.209}$$

It should be noted that only the form of $\mathbf{D}_E$ is different in the *plane strain* than *plane stress* cases.

A *triangular element* in the *xy* plane shown in Figure 6.48 has a constant thickness t_e. The triangle corners or *nodes* are marked by small circles. Nodes have the numbers 1, 2, 3 ordered in the positive counterclockwise direction. The triangle sides are numbered as the numbers of their opposite nodes. It is assumed that this is a very small triangle that experiences a *constant strain* over its confined space. This ***CST*** in Figure 6.48 provides *linear displacements* through the element as can be deduced from Eq. (6.203). The *element nodal displacement* matrix $\boldsymbol{\delta}_e$ of the triangular element is (see Figure 6.48a)

$$\boldsymbol{\delta}_e = \begin{bmatrix}\boldsymbol{u}_1 & \mathbf{v}_1 & \boldsymbol{u}_2 & \mathbf{v}_2 & \boldsymbol{u}_3 & \mathbf{v}_3\end{bmatrix}^T \tag{6.210}$$

where $\boldsymbol{\delta}_e$ is a vector of displacements $\boldsymbol{u}$,$\mathbf{v}$ at any node or $\boldsymbol{u}_i$,$\mathbf{v}_i$ in the x and y directions, respectively. The nodes can also be given the counters i, j, and m as shown in Figure 6.48 to generalize the connections of other triangles in the assembly.

The *internal local displacements* at any point *P(x,y)* – shown in Figure 6.48 – are related to the *nodal displacements* $\boldsymbol{\delta}_e$ through the following *shape function:*

$$\boldsymbol{h}_e(x,y) = \boldsymbol{N}_e\boldsymbol{\delta}_e \quad \text{or} \quad \boldsymbol{h}_e(x,y) = \begin{bmatrix} \boldsymbol{u}(x,y) \\ \boldsymbol{v}(x,y) \end{bmatrix} \tag{6.211}$$

where $\boldsymbol{N}_e$ is the element *shape function matrix,* which is a function of nodal positions and obtained through the following development depending on the concept of *area coordinates.* Figure 6.48 defines the triangle edge or side number according to the triangle node opposite or facing it. The subtriangle areas A_1, A_2, A_3 are numbered as the edges are.

The concept of area coordinates uses the ratio of areas A_1, A_2, A_3 to the total area of the triangle A_e to locate *P(x,y)* as depicted in Figure 6.48a and thus represent the area coordinates as follows:

$$\begin{aligned} L_1 &= \frac{A_1}{A_e}, L_2 = \frac{A_2}{A_e}, L_3 = \frac{A_3}{A_e} \\ L_1 &+ L_2 + L_3 = 1 \end{aligned} \tag{6.212}$$

where L_1, L_2, L_3 are the *area coordinates.* However, from the second expression of Eq. (6.212), only two of area coordinates are independent. One also notes that $L_n = 0$ on side *n*, where side *n* faces node *n*. At node 1, one has $L_1 = 1$, $L_2 = L_3 = 0$; at node 2, $L_2 = 1$, $L_1 = L_3 = 0$; at node 3, $L_3 = 1$, $L_2 = L_1 = 0$. The areas A_1, A_2, A_3 and triangle area A_e are obtained from the node locations to get the area coordinates by the *cross product* of any 2 sides of the triangle such that (see Section 1.11)

$$\begin{aligned} 2A_e &= \begin{bmatrix} & 0 & (y_2 - y_1) \\ 0 & & -(x_2 - x_1) \\ -(y_2 - y_1) & (x_2 - x_1) & \end{bmatrix} \begin{bmatrix} x_3 - x_1 \\ y_3 - y_1 \\ 0 \end{bmatrix} = (x_2 - x_1)(y_3 - y_1) - (x_1 - x_3)(y_1 - y_2) \\ &= a_{x3}b_{y2} - a_{x2}b_{y3} \end{aligned} \tag{6.213}$$

Similarly, for the other two sides, one can get

$$2A_e = (x_1 - x_3)(y_2 - y_3) - (x_3 - x_2)(y_3 - y_1) = a_{x2}b_{y1} - a_{x1}b_{y2} \tag{6.213$'$}$$

where

$$\begin{aligned} a_{x1} &= (x_3 - x_2), \quad b_{y1} = (y_2 - y_3) \\ a_{x2} &= (x_1 - x_3), \quad b_{y2} = (y_3 - y_1) \\ a_{x3} &= (x_2 - x_1), \quad b_{y3} = (y_1 - y_2) \end{aligned} \tag{6.214}$$

The subscript numbering is in the sequence 32132... for the expression to get a_{x1}, a_{x2}, a_{x3} and the sequence 123123... for the expression to get b_{x1}, b_{x2}, b_{x3}. This sequencing order is very useful in writing computer code, which is not the objective of this text. However, understanding the technique helps in the appreciation of the developed software and understanding of many factors affecting the results.

Areas A_1, A_2, A_3 are obtained by replacing (x_1,y_1), (x_2,y_1), (x_3,y_3), by (x,y) for each, and therefore one gets

$$\begin{bmatrix} L_1 \\ L_2 \\ L_3 \end{bmatrix} = \frac{1}{2A_e} \begin{bmatrix} 2c_{23} & b_{y1} & a_{x1} \\ 2c_{31} & b_{y2} & a_{x2} \\ 2c_{12} & b_{y3} & a_{x3} \end{bmatrix} \begin{bmatrix} 1 \\ x \\ y \end{bmatrix} \tag{6.215}$$

where *a's* and *b's* are defined by Eq. (6.214) and the *c's* are as follows:

$$c_{23} = x_2y_3 - x_3y_2$$

$$c_{31} = x_3y_1 - x_1y_3 \tag{6.216}$$

$$c_{12} = x_1y_2 - x_2y_1$$

For the ***CST***, it is clear that assuming a *linear relation* or *linear blending function* for the displacement throughout the element provides

$$\boldsymbol{h}_e(x,y) = \begin{bmatrix} \boldsymbol{u}_1 & \boldsymbol{u}_2 & \boldsymbol{u}_3 \\ \boldsymbol{v}_1 & \boldsymbol{v}_2 & \boldsymbol{v}_3 \end{bmatrix} \begin{bmatrix} L_1 \\ L_2 \\ L_3 \end{bmatrix} \tag{6.217}$$

Note that $\boldsymbol{h}_e$ at point $P(x,y)$ is a function of $\boldsymbol{u}_e$ and $\boldsymbol{v}_e$. This relation can also be interpreted as $\boldsymbol{u}_e(x,y) = a_1x + a_2y + a_0$ and $\boldsymbol{v}_e(x,y) = b_1x + b_2y + b_0$, where a_0, a_1, a_2, and b_0, b_1, b_2 are related to the nodes and defined off Eqs. (6.217) and (6.215). These coefficients are different than the previous *a's* and *b's* defined in Eqs. (6.214) and (6.216).

Rearranging Eq. (6.217) provides the following:

$$\boldsymbol{h}_e = \begin{bmatrix} L_1 & 0 & L_2 & 0 & L_3 & 0 \\ 0 & L_1 & 0 & L_2 & 0 & L_3 \end{bmatrix} \begin{bmatrix} \boldsymbol{u}_1 & \boldsymbol{v}_1 & \boldsymbol{u}_2 & \boldsymbol{v}_2 & \boldsymbol{u}_3 & \boldsymbol{v}_3 \end{bmatrix}^T \tag{6.218}$$

or

$$\boldsymbol{h}_e = \boldsymbol{N}_e\,\boldsymbol{\delta}_e \tag{6.219}$$

From Eq. (6.203), and substitute Eqs. (6.218), (6.214), and (6.215), one gets the *strain matrix* at point *P(x,y)*:

$$\begin{bmatrix} \varepsilon_x \\ \varepsilon_y \\ \gamma_{xy} \end{bmatrix}_e = \frac{1}{2A_e} \begin{bmatrix} b_{y1} & & b_{y2} & & b_{y3} & \\ & a_{x1} & & a_{x2} & & a_{x3} \\ a_{x1} & b_{y1} & a_{x2} & b_{y2} & a_{x3} & b_{y3} \end{bmatrix} \begin{bmatrix} u_1 \\ v_1 \\ u_2 \\ v_2 \\ u_3 \\ v_3 \end{bmatrix}_e \tag{6.220}$$

$$\boldsymbol{\varepsilon}_e = \boldsymbol{B}_e\boldsymbol{\delta}_e \tag{6.221}$$

Since *a's* and *b's* are constants, the *strain* $\boldsymbol{\varepsilon}_e$ throughout the element at any point *P(x,y)* is constant. Thus the same is true for the ***CST***. The *stress* in the element is then obtained from Eq. (6.206) or (6.207) as

$$\boldsymbol{\sigma}_e = \boldsymbol{D}_E\,\boldsymbol{\varepsilon}_e \tag{6.222}$$

$$\boldsymbol{\sigma}_e = \boldsymbol{D}_E\,(\boldsymbol{\varepsilon}_e - \boldsymbol{\varepsilon}_T) \tag{6.223}$$

It should be noted that the material matrix $\boldsymbol{D}_E$ is usually valid for all elements in the object. Commonly the object would be made of one material and $\boldsymbol{D}_E$ is then constant.

Element stiffness matrix is formulated through the concept of *strain energy* that has been discussed in Section 6.6. The *element strain energy* U_e is obtained by integrating the element strain energy density u_e, which is $\frac{1}{2}\sigma_e\varepsilon_e$ for simple normal loads as defined previously in Eq. (6.140). For multidimensional case, the *element strain energy* U_e can be easily generalized to

$$U_e = \frac{1}{2}\int_e \boldsymbol{\sigma}_e^T\boldsymbol{\varepsilon}_e\; t_e dA \tag{6.224}$$

where t_e is the constant *element thickness*. Substituting for $\boldsymbol{\sigma}_e$ from Eq. (6.222) into Eq. (6.224), one gets

$$U_e = \frac{1}{2}\int_e \boldsymbol{\varepsilon}_e^T\boldsymbol{D}_E\,\boldsymbol{\varepsilon}_e\; t_e dA \tag{6.225}$$

Using the expression ε_e from Eq. (6.221), the element strain energy becomes

$$U_e = \frac{1}{2}\int_e \delta_e^T \mathbf{B}_e \mathbf{D}_E \mathbf{B}_e \delta_e t_e dA \tag{6.226}$$

As the element thickness t_e is constant, the terms of $\mathbf{B}_e$ and $\mathbf{D}_E$ are also constants. Equation (6.226) then reduces to

$$U_e = \frac{1}{2}\delta_e^T \mathbf{B}_e^T \mathbf{D}_E \mathbf{B}_e \delta_e t_e \int_e dA = \frac{1}{2}\delta_e^T\ t_e A_e \mathbf{B}_e^T \mathbf{D}_E \mathbf{B}_e\ \delta_e \tag{6.227}$$

where the integral of dA over the element is the element area A_e. Observing that the potential energy U_e is simply defined as $\frac{1}{2}\boldsymbol{F}\delta$ and similar to Eq. (6.181), $\boldsymbol{F}_e = \boldsymbol{K}_e\ \delta_e$, then in general $U_e = \frac{1}{2}\delta_e{}^T \boldsymbol{K}_e \delta_e$. Equation (6.227) indicates that the element potential energy U_e becomes

$$U_e = \frac{1}{2}\delta_e^T \mathbf{K}_e \delta_e \tag{6.228}$$

From Eqs. (6.227) and (6.228), the triangular *element stiffness* $\mathbf{K}_e$ should then be as follows:

$$\mathbf{K}_e = t_e A_e \mathbf{B}_e^T \mathbf{D}_E \mathbf{B}_e \tag{6.229}$$

Since the *elasticity matrix* $\mathbf{D}_E$ is symmetric, the *triangular-element stiffness* $\mathbf{K}_e$ is ***symmetric***. The element of the stiffness matrix can be simply obtained by matrix multiplication of its constituent matrices as in Eq. (6.229).

Similar to the beam element, to get the *global stiffness matrix* $\boldsymbol{K}$ of the object or the entire body, one must assemble all the element stiffness matrices $\boldsymbol{K}_e$*'s* for the entire triangular elements of the model each written as in Eq. (6.229). However, the equations are written in the global coordinates that has all the displacements of all the nodes included. This gives the *global stiffness matrix* $\boldsymbol{K}$ of the model as has also been demonstrated in Eq. (6.182) and Example 6.16 such that

$$\boldsymbol{K} = \sum_{1}^{n_e} \boldsymbol{K}_e \tag{6.230}$$

The *global stiffness matrix* $\boldsymbol{K}$ of the body or object is symmetric. The final ***FE*** model of the entire body is then similar to the general form of Eq. (6.178) or

$$\boldsymbol{K}\boldsymbol{\delta} = \boldsymbol{F} \tag{6.231}$$

where $\boldsymbol{K}$ is the global stiffness *matrix*, $\boldsymbol{\delta}$ is the *global displacement vector* for all nodes, and $\boldsymbol{F}$ is the *global load vector* or forces. For external or boundary forces, they would be distributed at the *nodes* depending on their distribution form on the boundary. Using the known *loadings* $\boldsymbol{F}_i$ and *displacement* boundary conditions $\boldsymbol{\delta}_i$ at some nodes i, the solution of the adjusted equations in Eq. (6.231) is then obtained by the inverse of the adjusted Eq. (6.231) such that

$$\boldsymbol{\delta} = \boldsymbol{K}^{-1}\boldsymbol{F} \tag{6.232}$$

This provides the displacements $\boldsymbol{\delta}$ at all the nodes. At the boundaries, one can get the reactions by applying Eq. (6.231). The stress σ_e at each element can be evaluated from Eqs. (6.221) and (6.222).

For external boundary forces or ***traction*** forces $\boldsymbol{T}$, they can be applied at the *nodes* depending on their specific value or their distribution. For external linearly distributed load or traction of the form $\boldsymbol{T}(x,y)$ per unit area on the boundary shown in Figure 6.49, the share of each node may be assigned as follows:

$$\begin{aligned} \boldsymbol{T}_{2e} &= \frac{l_1 t_e}{6}(2\boldsymbol{T}_2 + \boldsymbol{T}_3) \\ \boldsymbol{T}_{3e} &= \frac{l_1 t_e}{6}(2\boldsymbol{T}_3 + \boldsymbol{T}_2) \end{aligned} \tag{6.233}$$

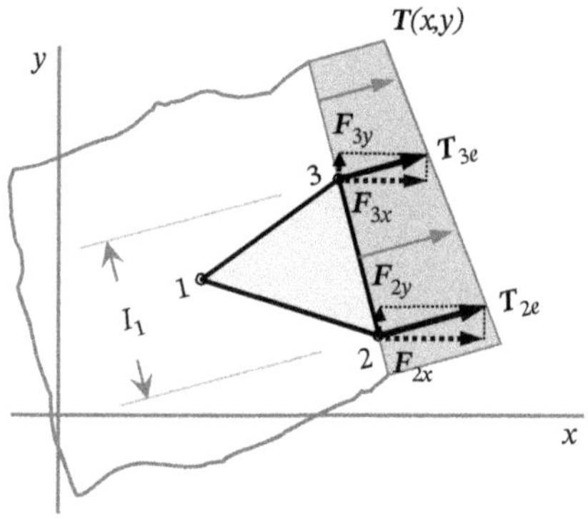

Figure 6.49 External linearly distributed load or traction $T(x,y)$ per unit area on the boundary are modeled as shares at each node. Their components in the x and y directions represent the loading vector components F_{2x}, F_{2y} and F_{3x}, F_{3y} at nodes 2 and 3, respectively.

where l_1 is the length of the triangle element edge or side 1 between nodes 2 and 3, t_e is the element thickness, $\boldsymbol{T}_2$ is the value of the distributed traction load at node 2, and $\boldsymbol{T}_3$ is the value of the distributed traction load at node 3. The distributed traction loads on the element at nodes 2 and 3 are $\boldsymbol{T}_{2e}$ and $\boldsymbol{T}_{3e}$, respectively. These representative shares are projected to their components in the x and y directions to represent the forcing load vector components $\boldsymbol{F}_{2x}$, $\boldsymbol{F}_{2y}$ and $\boldsymbol{F}_{3x}$, $\boldsymbol{F}_{3y}$ at nodes 2 and 3, respectively, as demonstrated by the dashed components of each vector in Figure 6.49.

6.11.4 General *3D* State: Linear Elasticity Problem

In this section, the main interest is to solve the *boundary value problem* (partial differential equation; ***PDE***) to find stresses and displacements in an elastic body under its own forces (body forces), external loading, and boundary conditions. This is the *theory of elasticity* field, which a quick overview relating to ***FE*** is given herein. For more involved treatment, refer to any theory of elasticity text such as Reisman and Pawlik (1980, 1991). The problem is characteristically having the following *unknowns*:

- Three displacements $\boldsymbol{\delta} \equiv (u, v, \text{and } w)$, where u, v, and w are the *displacements* in the x, y, and z directions, respectively.
- Six stress components (σ_{ij}), where the *normal stresses* and *shear stresses* are in the x, y, and z directions.
- Six strain components (ε_{ij}), where the *normal strains* and the *shear strains* are in the x, y, and z directions.

The available equations are as follows:

- Three ***equilibrium*** *equations* (partial differential equation; ***PDE***) relating the stresses σ to the applied loads or body forces $\boldsymbol{F}$ including *dynamics* due to *acceleration* $\ddot{\boldsymbol{\delta}}$ ($\sigma_{ij} \Leftrightarrow \boldsymbol{F}$)

$$\nabla\sigma + \boldsymbol{F} = \rho\ddot{\boldsymbol{\delta}} \tag{6.234}$$

where ρ is the density.

- Six strain-displacement ***compatibility*** *equations* relating the strains to the displacements ($\varepsilon_{ij} \Leftrightarrow \delta$)

$$\varepsilon = \frac{1}{2}(\nabla\delta + \nabla\delta^T) \tag{6.235}$$

- Six stress–strain ***constitutive*** *equations* relating the stresses to the strains, which represent the *elasticity matrix* or the *material property matrix* $\boldsymbol{D}_E$ ($\sigma_{ij} \Leftrightarrow \varepsilon_{ij}$)

$$\sigma = \boldsymbol{D}_E\varepsilon \tag{6.236}$$

The basic relations in ***3D*** for strain-displacement *compatibility equations* are as follows:

$$\varepsilon_x = \frac{\partial u}{\partial x} \quad \gamma_{xy} = \frac{\partial u}{\partial y} + \frac{\partial v}{\partial x}$$

$$\varepsilon_y = \frac{\partial v}{\partial y} \quad \gamma_{yz} = \frac{\partial v}{\partial z} + \frac{\partial w}{\partial y} \tag{6.237}$$

$$\varepsilon_y = \frac{\partial w}{\partial z} \quad \gamma_{zx} = \frac{\partial w}{\partial x} + \frac{\partial u}{\partial z}$$

The stress–strain relations in ***3D*** or the *constitutive equations* defining the material properties are as follows:

$$\begin{bmatrix} \sigma_x \\ \sigma_y \\ \sigma_z \\ \tau_{xy} \\ \tau_{yz} \\ \tau_{zx} \end{bmatrix} = \begin{bmatrix} (1-\nu)c_\lambda & \nu c_\lambda & \nu c_\lambda & 0 & 0 & 0 \\ & (1-\nu)c_\lambda & \nu c_\lambda & 0 & 0 & 0 \\ & & (1-\nu)c_\lambda & 0 & 0 & 0 \\ & & & G & 0 & 0 \\ & \text{Symetric} & & & G & 0 \\ & & & & & G \end{bmatrix} \begin{bmatrix} \varepsilon_x \\ \varepsilon_y \\ \varepsilon_z \\ \gamma_{xy} \\ \gamma_{yz} \\ \gamma_{zx} \end{bmatrix} \tag{6.238}$$

where

$$c_\lambda = \frac{E}{(1+\nu)(1-2\nu)} \quad \text{and} \quad G = \frac{E}{2(1+\nu)} \tag{6.239}$$

and where E is the modulus of elasticity, ν is the *Poisson's* ratio, and G is the shear modulus of elasticity. The symmetric matrix in Eq. (6.238) is the ***3D*** *material property matrix* $\mathbf{D}_E$.

The displacements inside the ***3D*** element are usually given by

$$\begin{bmatrix} u \\ v \\ w \end{bmatrix} = \begin{bmatrix} N_{31} & & & N_{32} & & & \cdots \\ & N_{31} & & & N_{32} & & \cdots \\ & & N_{31} & & & N_{32} & \cdots \end{bmatrix} \begin{bmatrix} u_1 \\ v_1 \\ w_1 \\ u_2 \\ v_2 \\ w_2 \\ \vdots \end{bmatrix} \tag{6.240}$$

where $N_{31}, N_{32}\ldots$ are the components of the ***3D*** *shape function*, which are defined specifically for each ***3D*** element of these shown in Figure 6.41c.

6.11.5 General *3D FE* Procedure

The ***FE*** procedure for general ***3D*** models or objects are dependent on the field of interest to do analysis or design. ***FE*** involves several fields that the commercial software can accommodate. Some of these are structural static analysis, dynamics and modal analysis, transient dynamic analysis, buckling analysis, contact problems, steady-state thermal analysis, transient thermal analysis, etc. With the availability of ***3D*** solid modeling, several software packages have the capabilities to do solid modeling. They are usually able to import solid models from other ***3D*** geometric modeling or CAD software. Some software packages of ***3D*** geometric modeling have the ***FE*** capabilities imbedded.

Nowadays, the general ***FE*** procedure is implemented mainly inside the program package as shown in Figure 6.50. The user should either provide a ***3D*** model for his object or build one inside the ***FE*** package as schematically defined in Figure 6.50a. The procedure should then allow the user to select element type either ***2D*** or ***3D***; linear or quadratic shape function; or truss, beam, shell, plate, solid, etc. The user should also select material properties in the form of E, υ, ρ, α, etc. The user or the ***FE*** package generates nodes either manual or in most cases automatically through the package internal code as envisioned in a ***2D*** projection of Figure 6.50b. It is necessary to build elements through mesh generation (internal code) and assign the element connectivity as

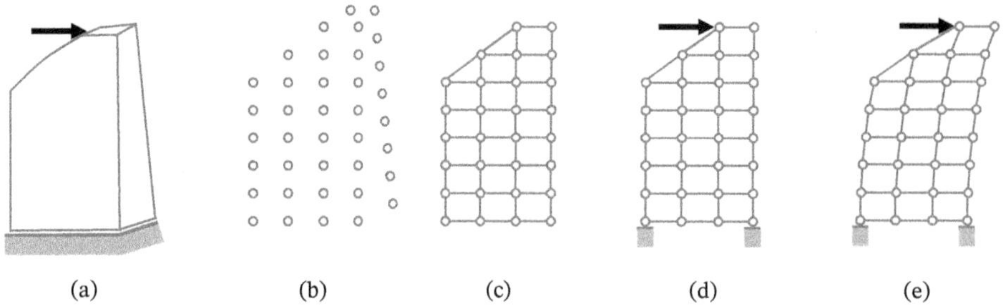

Figure 6.50 Finite element procedure to find deflections and stresses in an object. A ***3D*** model for an object schematically defined in (a). Generated nodes manually or automatically as in (b). Built elements through mesh generation as the ***2D*** projection in (c). Defined boundary conditions of displacement constraints and loads as in (d). Postprocessor provides deflections in (e) but extremely exaggerated for clarity.

shown in Figure 6.50c. Either before or after the previous process, one needs to apply the boundary conditions for displacement constraints and loads as suggested in Figure 6.50d. The ***FE*** package is generally ready for the process to solve the boundary value problem with codes that identify all elements stiffness matrices, assemble the global stiffness matrix, invert the stiffness matrix, and calculate the displacements and the internal stresses.

The ***FE*** package has a *postprocessor* that visualizes the calculated results in ***3D*** for displacements, stresses, strains, or natural frequencies and mode shapes for dynamic loading, or temperature distribution for thermal analysis, or time history for any of the available output options. The deformation output shown in Figure 6.50e is a possible output of the deflections but extremely exaggerated for clarity.

6.11.6 Errors in *FE* Modeling and Solution

The ***FE*** practice in analysis and design is after all a *numerical technique* to solve partial differential equation of complex continuum geometries subject to different loading conditions and deformation constraints. Numerical techniques are approximate and errors can develop due to inappropriate mathematical model of the object or the assessment of constraint boundary conditions and loading. Other main factor relates to the ***FE*** details concerning the approximation of the object domain by several ***FE*** *type* of some sort or another. The other detail is the *polynomial approximations* of the deformations that may not necessarily represent the true deformation. The critical factor is the simple *integration technique* commonly used in the solution codes. For large number of elements, the *calculation errors* may obscure the accurate attainment of reasonable results. Dynamic loading of a simple object has demonstrated erroneous results out of some reputable ***FE*** codes sometimes ago, Cook (1995). Therefore, *validation* and *verification* by other means such as experimentation is greatly encouraged.

6.11.7 Some Classical *FE* Packages

Some ***FE*** packages consider other special cases such as *contact* of components, existing *cracks*, nonlinear problems, and others. Several ***3D*** geometric modeling (CAD) packages have a selected ***FE*** package as an integral part to perform the analysis and allow geometry modification in a repetitive process to generate more suitable design. Some even have ***FE*** iterations or even some optimization in the ***FE*** package. Available commercial ***FE*** packages are imbedded with the ***3D*** modeling packages or standing alone such as the classical *ANSYS*, *Nastran*, *Abaqus*, *ADINA*, etc. Other ***FE*** packages are *free* and may be available to download such as *FreeFem++*, *Deal.II*, *Elmer*, *GetFEM++*, etc. However, general *verification* and *validation* of these might be confirmed and should be needed to allow for accurate use of the software. In all cases, the successful use of any ***FE*** package is dependent on the user defining the right model, load, and boundary conditions. Any misrepresentation of these would produce a solution that can be totally unrepresentative and can be grossly inaccurate and misleading. Several available commercial

FE packages are in a fast changing and maybe volatile field that some of these packages my change ownership, merge into other software, or change name, or go out of business. The following is an abridged list of some classical ***FE*** packages with few details and references at the end of the chapter:

- **Abaqus**: Franco-American software from SIMULIA, owned by Dassault Systemes SIMULIA, [Internet 2].
- **ANSYS**: American software, ANSYS, Inc., [Internet 3].
- **Nastran**: American software, The MacNeal-Schwendler Corporation (MSC), [Internet 4].
- ADINA: American software, ADINA R & D, Inc., [Internet 5].
- SOLIDWORKS Simulation, previously COSMOSWorks: A SolidWorks module, owned by Dassault Systemes COSMOS. In Solidworks 2009, COSMOSWorks was replaced/renamed to Solidworks Simulation, [Internet 6].
- COMSOL Multiphysics, COMSOL Inc., [Internet 7].

6.12 Computer-Aided Design and Optimization

In previous sections and examples, several means are available to provide synthesis tools for some machine elements such as the synthesis charts in Figure 6.8 for beam synthesis. The MATLAB codes in Figure 6.21 are available tools to find principal stresses and directions for stress states. Figure 6.43 provides a code that can be modified to solve simple collinearly connected members by ***FE*** method. The code can also be modified to develop another code for the synthesis of such simple problems by iterating on different member areas or lengths to have some constant stress or other deflection constraint; see Problems 6.78 and 6.79. These MATLAB codes are accessible through **Wiley website** for CAS of such similar components and cases.

Figures 6.51 and 6.52 provide the ***Beam Synthesis Tablets*** and the ***Column Synthesis Tablets*** developed from previously defined procedure and equations in Sections 6.1.4 and 6.7. These Tablets are CAS or *bona fide CAD* tools. The ***Tablets*** are accessible through **Wiley website** for CAS of beams. They provide direct synthesis without any appreciable need for repeated analysis. The iterative synthesis, however, is available for fine-tuning of required standards or satisfying other constraints.

In the ***Synthesis Tablets,*** the ***Default*** column is an example given as a reference and check. This column is not to be changed. No red (or dark grayish) or small font values *without* aqua color-or-grayish background are supposed to be changeable. These are due to predetermined and embedded mathematical relations such that they should not be accessible for input of any explicit variable. Any input to these cells will wipe out the solution procedure. Only cells with *aqua* (or *grayish*) background are allowable for *Inputs*. The synthesis ***Input*** column is to insert your case values only in place of the *blue-colored* values (or *greyish* larger font) with *aqua* or *light grey* background (as in Figures 6.51 and 6.52). All inputs are only allowable in cells with the blue colored (or dark grayish large) font with aqua (or light grayish) background. The red colored (or dark grayish smaller) font values with no colored background are the synthesized design parameters calculated as synthesized suggestions. The ***Analysis*** column is for design iterations to insert your case values only in place of the *blue colored* values (or *dark grayish larger font*) with *aqua* (or *light grayish*) background (as in Figures 6.51 and 6.52). The iteration is necessary only to account for existing standard geometrical dimensions or other constraints on space geometry. The red colored values (or dark grayish smaller font) with no colored background are also other design parameters calculated as behavioral outputs that are only changeable by entering in other cells of the *blue colored* values (or *dark grayish larger font*) with *aqua* (or *light grayish*) background. The ***Analysis*** column is suitable for quick and effective iterations in combination with the ***Input*** synthesis column. Change of parameters to accommodate specific strength values or necessary safety factor should need some little iteration. The *blue colored* values (or *dark grayish larger font*) with *aqua* (or *light grayish*) background in either ***Input*** or ***Analysis*** column are interactively exploitable to accommodate any other design parameter values.

The available ***Beam Synthesis Tablet*** and ***Column Synthesis Tablet*** assist in designing beams and columns using the procedure given herein particularly for *steel*. Any other material of known properties of the yield strength,

Beam Synthesis–US units			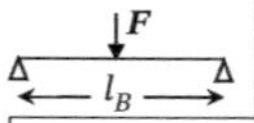
Inputs	Default	Input	Analysis
Load, F [lb]	225	-9000	-9000
Beam Length, l_B [in]	40	40	40
Torque, T [lb.in]	22.5	-900	-900
End Condition, C_{EB}	1	1	1
Safety Factor , K_{SF}	4	2	1.82
Modulus of Elasticity, E [Mpsi]	30	30	30
Output			
Diameter or Depth, d_B [in]	1.4058	3.9837	4
Cross-section Area, A_B [in^2]	1.55206	1.2464E+1	12.56637061
Second Area Moment, I_B [in^4]	1.9169E-1	1.2363E+1	12.56637061
Span to Depth Ratio, $r_{SD}=(l_B/d_B)$	28.454	10.041	10.000
Maximum Deflection, v_{max} [in]	0.052	-3.2354E-2	-3.1831E-2
Yield Strength, S_y [kpsi]	33	29	29
Ultimate Tensile Strength, S_{ut} [kpsi]	52	52	52
Endurance Limit, S_e [kpsi]	26	26	26
Stresses			
Normal Stress, σ_y [kpsi]	8.250	14.500	14.324
Shear Stress , τ_{xy} [kpsi]	0.072	-0.361	-0.358
Shear Stress due to Torque, τ_{xy} [kpsi]	0.041	-0.073	-0.072
von Mises Stress, σ_v [kpsi]	8.250	14.501	14.324
Factors of Safety			
Static Safety Factor, K_{SF}	4.00	2.00	2.02
Worst Static Safety Factor, K_{SFm}	NA	NA	2.02
Dynamic Safety Factor, K_{SFD}	3.15	1.79	1.82
For Other Sections			
Circular Section Modulus, Z_B [in^3]	2.7273E-1	6.2069E+0	6.2832E+0

(a)

Beam Synthesis–SI units			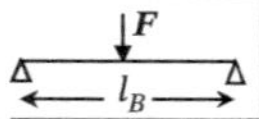
Inputs	Default	Input	Analysis
Load, F [N]	1000	-40000	-40000
Beam Length, l_B [m]	1	1	1
Torque, T [N.m]	2.5	-100	100
End Condition, C_{EB}	1	1	1
Safety Factor , K_{SF}	4	2	1.82
Modulus of Elasticity, E [GPa]	207	207	207
Output			
Diameter or Depth, d_B [m]	3.5380E-2	1.0062E-1	0.1
Cross-section Area, A_B [m^2]	9.8312E-4	7.9510E-3	0.007853982
Second Area Moment, I_B [m^4]	7.6913E-8	5.0308E-6	4.90874E-06
Span to Depth Ratio, $r_{SD}=(l_B/d_B)$	28.265	9.939	10.000
Maximum Deflection, v_{max} [mm]	1.309	-8.0022E-1	-8.2012E-1
Yield Strength, S_y [MPa]	230	200	200
Ultimate Tesile Strength, S_{ut} [MPa]	370	370	370
Endurance Limit, S_e [MPa]	185	185	185
Stresses			
Normal Stress, σ_y [MPa]	57.500	100.000	101.859
Shear Stress , τ_{xy} [MPa]	0.509	-2.515	-2.546
Shear Stress due to Torque, τ_{xy} [MPa]	0.288	-0.500	0.509
von Mises Stress, σ_v [MPa]	57.502	100.004	101.863
Factors of Safety			
Static Safety Factor, K_{SF}	4.00	2.00	1.96
Worst Static Safety Factor, K_{SFm}	NA	NA	1.96
Dynamic Safety Factor, K_{SFD}	3.22	1.85	1.82
For Other Sections			
Circular Section Modulus, Z_B [m^3]	4.3478E-6	1.0000E-4	9.8175E-5

(b)

Figure 6.51 Beam Synthesis Tablets: (a) US system of units and (b) SI system of units, for circular cross section. For other sections, you need to enter the depth, the area, the second area moment and the section modulus. This will eliminate these circular section relations.

Column Synthesis–US units	Default	Input	Analysis
Inputs			
Load, P [lb]	220	9000	9000
Column Length, l_C [in]	40	40	20
Load Eccentricity Offset, e [in]	0.1	0.1	0.1
End Condition, C_E	1	0.25	0.25
Buckling Safety Factor, K_{SF}	4	4	3.16
Modulus of Elasticity, E [Mpsi]	30	30	30
Output			
Diameter, d_C [in]	0.5579	1.9954	1.4
Cross-section Area, A_C [in^2]	0.2445	3.127056076	1.5393804
Second Area Moment, I_C [in^4]	0.0048	0.77814669	0.188574099
Radius of Gyration, $r_{GC} = (I_C/A_C)^{1/2}$[in]	0.1395	0.499	0.350
Slenderness Ratio, $sr = (l_C/r_{GC})$	286.79	80.19	57.14
Compressive Yield Strength, S_{yc} [kpsi]	33.5	29	29
Slenderness Ratio Limit, $(s_r)_T$	133	71	71
Euler Critical Load, P_{cr} [lb]	880	36000	34897
Parabolic Critical Load, P_{cr} [lb]	NA	NA	30365
Secant Critical Load, P_{cr} [lb]	3364	64730	28408
Stresses			
Normal Stress, σ_y [kpsi]	0.900	2.878	5.847
Euler Critical Stress, P_{cr}/A_C [kpsi]	3.600	46.050	90.677
Parabolic Critical Stress, P_{cr}/A_C [kpsi]	NA	NA	19.725
Secant Critical Stress, P_{cr}/A_C [kpsi]	13.763	20.700	18.454
Factors of Safety			
Euler Safety Factor, K_{SFE}	4.00	4.00	3.88
Parabolic Safety Factor, K_{SFP}	NA	NA	3.37
Secant Safety Factor, K_{SFS}	15.29	7.19	3.16
For Other Sections			
Circular Section Modulus, Z_C [in^3]	0.0170	0.779953258	0.26939157

(a)

Column Synthesis–SI units	Default	Input	Analysis
Inputs			
Load, P [N]	1000	40000	40000
Column Length, l_C [m]	1	1	0.5
Load Eccentricity Offset, e [mm]	2.5	2.5	2.5
End Condition, C_E	1	0.25	0.25
Buckling Safety Factor, K_{SF}	4	4	3.06
Modulus of Elasticity, E [GPa]	207	207	207
Output			
Diameter, d_C [mm]	14.1320	50.2614	35
Cross-section Area, A_C [mm^2]	156.856	1984.083172	962.1127502
Second Area Moment, I_C [mm^4]	1957.897	313263.563	73661.75743
Radius of Gyration, $r_{GC} = (I_C/A_C)1/2$[mm]	3.533	12.565	8.750
Slenderness Ratio, $s_r = (l_C/r_{GC})$	283.04	79.58	57.14
Compressive Yield Strength, S_{yc} [MPa]	230	200	200
Slenderness Ratio Limit, $(s_r)_T$	133	71	71
Euler Critical Load, P_{cr} [N]	4000	160000	150492
Parabolic Critical Load, P_{cr} [N]	NA	NA	130913
Secant Critical Load, P_{cr} [N]	14937	283855	122447
Stresses			
Normal Stress, σ_y [MPa]	6.375	20.160	41.575
Euler Critical Stress, P_{cr}/A_C [MPa]	25.501	322.567	625.671
Parabolic Critical Stress, P_{cr}/A_C [MPa]	NA	NA	136.069
Secant Critical Stress, P_{cr}/A_C [MPa]	95.226	143.066	127.269
Factors of Safety			
Euler Safety Factor, K_{SFE}	4.00	4.00	3.76
Parabolic Safety Factor, K_{SFP}	NA	NA	3.27
Secant Safety Factor, K_{SFS}	14.94	7.10	3.06
For Other Sections			
Circular Section Modulus, Z_C [mm^3]	277.09	12465.36	4209.24

(b)

Figure 6.52 Column Synthesis Tablets: (a) US system of units and (b) SI system of units.

ultimate strength, and the modulus of elasticity may be synthesized. The instantly shown results should be close to the allowable strengths. This provides a mean of optimum decision making by comparing results in the fine tuning of design.

6.12.1 Beam Synthesis Tablet

The ***Beam Synthesis Tablet*** requires the input of the specific design case either for SI or US system; see Figure 6.51. The ***Input*** includes the *load* $\boldsymbol{F}$ $[\mathrm{N}]_{\mathrm{SI}}$ or $[\mathrm{lb}]_{\mathrm{US}}$, the *beam length* l_B in [m] or [in], the *torque* $\boldsymbol{T}$ in [N m] or [lb in], the *end condition* C_{EB} (1 for simply supported or 2 for fixed-free cantilever), the *safety factor* K_{SF}, and the *modulus of elasticity* E_e [GPa] or [Mpsi]. The *yield strength* S_y, however, is a suggested *input* and if any of other preferred values is preferable, one should enter the value beforehand. The intended ***Output*** *synthesis* beam *diameter or depth* d_B is dependent on the assumption of a circular beam section. The *cross-sectional area* A_B and the *second area moment* I_B are calculated accordingly. If the cross section is not circular, the cross-sectional area A_B and the second area moment I_B should be calculated separately and entered in the ***Analysis*** column. These inputs will wipe out the equations to calculate the circular cross-sectional area A_B and the second area moment I_B. To return to circular cross sections, one needs to use a fresh *Synthesis Tablet*. The adapted material to be used is initially having a yield strength of 200 [MPa] or 29 [kpsi] (or 230 [MPa] or 33 [kpsi]) as a regular material property for common structural applications. For other materials, the *ultimate tensile strength* and the *endurance limit* should be entered as well in the ***Analysis*** column.

The design is made in two stages. The first stage is the *synthesis estimate* of *optimum* design that includes the beam solid *diameter* d_B, the circular *cross-sectional area* A_B, the circular cross-sectional *second area moment* I_B, the *span to depth ratio* r_{SD}, and the *maximum deflection* v_{max}. Major geometry and performance parameters are calculated accordingly as depicted in Figure 6.51. The normal stresses, shear stresses, shear stresses due to torsion, and the *von Mises* or distortion stresses are given. The static factor of safety, the worst static safety factor, and the dynamic safety factor are also provided. The worst static safety factor is calculated by considering the maximum normal stress and the maximum transverse shear are existing at the same point. This is seldom the case. The dynamic safety factor uses the endurance limit as the failure case; see Section 7.8. The *circular section modulus* Z_C is given so as to use it as a guide for the synthesis of other *noncircular* cross sections. These parameters and behavioral variables are obtained using the equations presented in this chapter. The geometry and performance parameters and variables are shown in Figure 6.51. Some of these variables are given as a check. Material designation is not provided since the material strength is used instead. The selection of a specific material designation is dependent on the available standards and the available manufacturing and treatment capabilities.

After the design synthesis is obtained in the first stage by introducing the case into the ***Input*** column, the user starts the second design stage. This entails the fine-tuning iterations to be performed into the ***Analysis*** column. The tuning may necessitate the change of some design parameters particularly *standard diameters* or use of noncircular cross section and may also include material strength refinement. Changing parameters in the *Input* column, the output values appear immediately in the *Output* locations in the ***Tablet***. Together with the input, the synthesized parameters provide necessary and sufficient information to calculate all other design parameters and performance or behavioral variables. If any of these are not satisfactory, iterations are possible through the utilization of the ***Analysis*** column in conjunction with the ***Input*** column to home into any needed design parameter or performance constraint. The design tuning process can be repeated until the user is satisfied with the design parameters. One can verify and possibly further adjust values in the available ***Beam Synthesis Tablets*** (for either *SI* or *US* system) through **Wiley website**.

After the beam is synthesized, the *principal stresses* at any point on the beam cross section can be obtained. The synthesis, however, has used the *von Mises* or distortion energy to evaluate the static safety factor; see failure theories in Section 7.7. The MATLAB codes of Figure 6.21 can be used to find principal stresses and principal directions.

These programs are available through **Wiley website**. This process is essential for thin thickness sections such as the channel or C-section and the L-section. One would consult Appendix A.8 for different standard sections and Example 6.9 where the transverse shear stresses at the top and bottom corners of the channel section are high combined with high normal stresses at the top and bottom flanges or sides.

6.12.2 Column Synthesis Tablet

The *Excel* sheet of ***Column Synthesis Tablet*** requires the input of the specific design case either for SI or US system; see Figure 6.52. The ***Inputs*** include the *compressive load* $\boldsymbol{P}$ $[\text{N}]_{\text{SI}}$ or $[\text{lb}]_{\text{US}}$ (***input as a positive value***); the *column length* l_B in [m] or [in]; the *load eccentricity offset e* [mm] or [in]; the *end condition* C_E (1 for pivoted ends, 0.25 for fixed-free cantilever, 2 for fixed-pivoted, and 4 for fixed–fixed); the *safety factor* K_{SF}, and the *modulus of elasticity* E_e in [GPa] or in [Mpsi]. The *compressive yield strength* S_{yc}, however, is suggested and if any of other preferred values is known, one should enter the value beforehand. The intended ***Output*** *synthesis* column *diameter* d_C is dependent on a circular section for the column. The *cross-sectional area* A_B and the *second area moment* I_B are calculated accordingly. If the cross section is not circular, the cross-sectional area A_B and the second area moment I_B should be calculated separately and may be entered in the ***Analysis*** column. These inputs will wipe out the equations to calculate the circular cross-sectional area A_B and the second area moment I_B. All calculations, however, use the diameter in their equations and thus utmost care should be exercised with needed verification. To return to circular cross sections, one needs to use a fresh Synthesis Tablet. The adapted material to be used is initially having yield strength of 200 [MPa] or 29 [kpsi] (or 230 [MPa] or 33 [kpsi]) as a regular material property for common structural applications. For other materials, the compressive yield strength should be entered as well in the ***Analysis*** column.

The design is made in two stages. The first stage is the *synthesis estimate* of *optimum* design, which includes the column solid *diameter* d_C, the circular *cross-sectional area* A_C, the circular cross-sectional *second area moment* I_C, the *radius of gyration* r_{GC}, and the *slenderness ratio* s_r. Major geometry and performance parameters are calculated accordingly as depicted in Figure 6.52. The *slenderness ratio limit* $(sr)_T$, *Euler critical load* $\boldsymbol{P}_{\text{cr}}$, *parabolic critical load* P_{cr}, and the *secant critical load* P_{cr}, are calculated. The secant critical load is iteratively calculated internally by using the secant critical stress, which is using the secant critical load. The *normal stress, Euler critical stress, parabolic critical stress, and secant critical stress* are determined. The *Euler* safety factor, *parabolic* safety factor, and *secant* safety factor are also provided. The *circular section modulus* Z_C is given so as to use it as a guide for the synthesis of other noncircular cross sections. These parameters and behavioral variables are obtained using the equations presented in this chapter. The geometry and performance parameters and variables are shown in Figure 6.52. Some of these variables are given as a check. Material designation is not provided since the material strength is used instead. The selection of a specific material designation is dependent on the available standards and the available manufacturing and treatment capabilities.

After the design synthesis is obtained in the first stage by introducing the case into the ***Input*** column, the user starts the second design stage. This entails fine tuning iterations to be performed into the ***Analysis*** column. The tuning may necessitate the change of some design parameters particularly *standard diameters* or use of noncircular cross section and may also include material strength refinement. Changing parameters in the *Input* column, the output values appear immediately in the *Output* locations in the ***Tablet***. Together with the input, the synthesized parameters provide necessary and sufficient information to calculate all other design parameters and performance or behavioral variables. If any of these are not satisfactory, iterations are possible through the utilization of the ***Analysis*** column in conjunction with the ***Input*** column to home into any needed design parameter or performance constraint. The design tuning process can be repeated until the user is satisfied with the design parameters. One can verify and possibly further adjust values in the available ***Column Synthesis Tablets*** (for either *SI* or *US* system) through **Wiley website**.

6.12.3 Optimum Stress Concentration

It is obvious that the global optimum stress concentration is having a minimum of a diminishing or insignificant effect that tends to compel the stress concentration value close to 1.0. Figure 6.53 presents a plate with an elliptical hole uniformly loaded in tension and the stress distribution at each side of the hole shown in (a) for an elongated ellipse in the direction of the load. The *stress concentration factor* K_{SC} against the major to minor ratio is plotted in Figure 6.53b. It is common, instead, to have the worst case of a circular hole that was considered in Figure 6.39 with the stress concentration factor shown in Figure 6.39b. The original derivation, however, is treating the elliptical hole shown in Figure 6.38 and provides Eq. (6.173) to estimate the maximum stress. The relation is still valid if the elliptical hole is rotated 90° as shown in Figure 6.38 (dashed grayish ellipse) or Figure 6.53a. The *stress concentration factor* K_{SC} against the minor to major ratio is a plot of Eq. (6.173), which suggests a much smaller stress concentration factor if the elliptical hole is rotated 90° as shown Figure 6.53a. If there is no compelling reason to have a circular hole, the optimum is seemingly an elliptical cavity rotated 90° with the minor to major axis as small as possible conforming to Figure 6.53b.

The optimum configuration or topology of a circular hole in a plate has been obtained by finding the optimum thickness of the plate around the circular hole to have a constant stress throughout the whole plate using the ***HGP*** similar to Saad and Metwalli (2012, 2013); see also Section 5.5.4. Figure 6.54 demonstrates a plate with an original hole of circular section after being optimized by ***FE*** to the shown parts of the ***3D*** model. A representative thickness values at nodes of a half plate is given in Figure 6.54a. The isometric view in Figure 6.54b demonstrates the thickness variations along the half plate replacing each half of the constant thickness plate with the circular hole. The optimized thicknesses have been interpolated by two spline surfaces that hold the plate material between them. The result affirms the notion of having an elliptical hole that is rotated 90° but with different ***3D*** thicknesses that even alters the topology of the rotated elliptical hole.

6.12.4 Optimum *FE* Prismatic Beams

This segment presents a simple optimization of the prismatic *1D* beam addressed previously in Example 6.15 utilizing the MATLAB code given in Figure 6.43. The procedure in Figure 6.55 is utilizing the ***HGP*** optimization of Section 5.5.4, which provides a derived *recursive relation* to home into the stress constraints generating equal

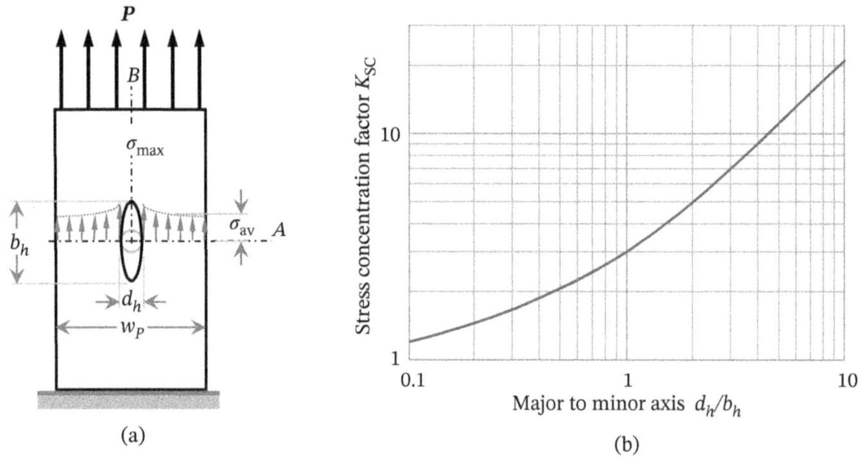

Figure 6.53 A plate with an elliptical hole uniformly loaded in tension and the stress distribution at the hole sides shown in (a) for an elongated ellipse in the direction of the load. The stress concentration factor K_{SC} against the major-to-minor ratio is plotted in (b).

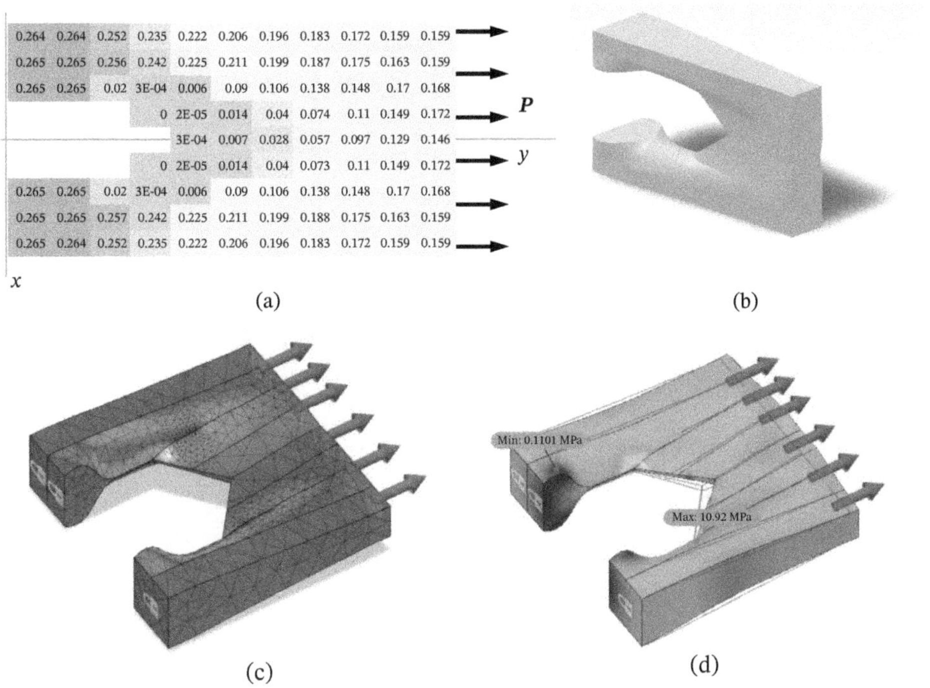

Figure 6.54 An optimum plate loaded in tension that had an initial circular hole is optimized by ***FE*** to the shown particulars: (a) representative optimum thickness values at nodes of the half plate and (b) an isometric view of optimum thickness variations along half the plate. This optimum replaces each half of the constant thickness plate that had a circular hole. The shown thicknesses are exaggerated.

allowable stresses σ_{all} at all of the selected points or regions of the design. The selected points are those having diameters d_i on the beam or bar in Figure 6.42. For simplified assumption that the three beam elements of areas A_1, A_2, and A_3 are to be optimized (i.e. minimized) for minimum weight. This is equivalently having equal maximum allowable stresses σ_{all} that cause the maximum utilization of the material.

The recursive formula of the ***HGP*** optimization is obtained as depicted in Section 5.5.4; see also Abd El-Rahman et al. (2017). If one uses the beam areas as the design vector, this gives the following relation:

$$A_i = A_{i,\text{old}}(\sigma_i/\sigma_{\text{all}}) \quad i = 1, 2, 3, \ldots \tag{6.241}$$

This relation is linear since the global stiffness matrix is linear in A_i. The current stress is σ_i and the target or allowable stress is σ_{all}. If one uses the diameters d_i, the stiffness matrix is nonlinear in d_i and the relation becomes as follows (see Section 5.5.4):

$$d_i = d_{i,\text{old}}(\sqrt{\sigma_i/\sigma_{\text{all}}}) \quad i = 1, 2, 3, \ldots \tag{6.242}$$

The current stress is σ_i and the target stress is σ_{all}. The old diameters $d_{i,\text{old}}$ are the previously calculated diameters.

Figure 6.55 shows the MATLAB code to find the optimum ***FE 1D*** beam of Example 6.15, where the diameters d_i's are calculated from the initial areas A_i's. Since the relations of the ***HGP*** are developed to reach the solution, the loop in the code should only take a single iteration. However, the loop is made to run for two iterations to demonstrate and guarantee total conversion.

Staring with cross-sectional areas of A_1 =0.0004 [m^2] or 0.62 [in^2], A_2 =0.0005 [m^2] or 0.775 [in^2], and A_3 =0.0006 [m^2] or 0.93 [in^2], the optimum areas are found to be A_1 =0.0004 [m^2] or 0.62 [in^2], A_2 =0.0003 [m^2]

```
clc; clear; clear all; format compact; format long ; format short;
ESI=200*10^9; EUS=29*10^6;                                          % Modulus of Elasticity SI&US
A1=0.0004; A2=0.0005; A3=0.0006;                                    % Area in [m^2]
d1=sqrt(A1*4/pi); d2=sqrt(A2*4/pi); d3=sqrt(A3*4/pi); di=[d1 d2 d3] % Diameter in [m]
d1old=d1; d2old=d2 ; d3old=d3;
Sallow=100;                                                         % Allowable stress [MPa]
L1=0.2; L2=0.2; L3=0.2;                                             % Length in [m]
Volume=A1*L1+A2*L2+A3*L3
Fin1=40000; Fin2=10000;  Fin=[Fin1 -Fin2 0]'                        % Force in [N]
for i=1:2
k1=ESI*A1/L1; k2=ESI*A2/L2; k3=ESI*A3/L3;
K=[k1  -k1   0; -k1  k1+k2  -k2;  0   -k2  k2+k3];
un=inv(K)*Fin;                                                      % Node Displacement
Nvt=[1/L1 -1/L1   0; 0   1/L2 -1/L2; 0   0   1/L3];                 % Node variable transformation
Strain=Nvt*un;
Stress_MPa=ESI*Strain/10^6                                          % Stress in [MPa]
Ke= [k1  0  0; 0  k2  0; 0  0  k3];                                 % Stiffness of each element
Fe=Ke*Strain.* [L1; L2; L3]                                         % Force in each element
d1=d1old*sqrt(Stress_MPa(1)/Sallow);                                % HGP
d2=d2old*sqrt(Stress_MPa(2)/Sallow);                                % HGP
d3=d3old*sqrt(Stress_MPa(3)/Sallow);                                % HGP
di=[d1 d2 d3]                                                       % New diameters
A1=pi*d1^2/4; A2=pi*d2^2/4; A3=pi*d3^2/4;
d1old=d1; d2old=d2 ; d3old=d3;
end
Volume=A1*L1+A2*L2+A3*L3
Fe=Ke*Strain.* [L1; L2; L3]                                         % Force in each element
```

Figure 6.55 MATLAB code to find the optimum three elements collinearly assembled for the bar of Example 6.15. The solution employs the ***FE*** modeling in Example 6.15 and its SI values. The optimization loop uses the ***HGP*** recursive relations inside the cited loop.

or 0.465 [in^2], and A_3 =0.0003 [m^2] or 0.465 [in^2]. The diameters are initially d_1 =0.0226 [m] or 0.89 [in], d_2 =0.0252 [m] or 0.993 [in], and d_3 =0.0276 [m] or 1.088 [in]. The optimum diameters are d_1 =0.0226 [m] or 0.89 [in], d_2 =0.0195 [m] or 0.77 [in], and d_3 =0.0195 [m] or 0.77 [in]. The stresses are all equal to the allowable stress σ_{all} 100 [MPa] or 14.5 [kpsi]. The optimum volume is reduced by 50% relative to the original volume.

One can verify and possibly further adjust values of Example 6.15 and Figure 6.55 through the available MATLAB code of ***Finite_Element_Example_6_15_Optimum.m*** (for either *SI* or *US* system) from **Wiley website**. If the total length is preserved, the optimum lengths can also be found by adjusting the code to include other recursive formulae for changing the lengths l_i.

6.12.5 Optimum *FE* Cantilever Beams

The optimization of beams has been the concern of many research and classical work in *geometry* and *topology* optimization of continuums, structural, and machine elements; see. e.g. Keveh et al. (2008). In this section a glimpse of an optimization alternative is presented as a sample of ***FE*** optimization, Saad and Metwalli (2012, 2013). It should be noted, however, that the optimum in such cases is usually multimodal with several geometries and topologies. The optima are also close to each other in terms of similar objectives and outcome shapes. Many commercial ***FE*** codes have some optimization capabilities that may be employed to observe the multimodal presence in many cases. This is not a curse, but it gives many optimum alternatives to pick from which the one that satisfy additional desires.

Figure 6.56 defines the design domain, nodes, and elements of a cantilever beam. The number closer to the node is the node number. The element number is designated as one of *e1–e31*, Saad and Metwalli (2012). Figure 6.57 presents the optimum design of the cantilever beam. In Figure 6.57a, the representative optimum thickness values at nodes are provided. Figure 6.57b shows an isometric view of the optimum thickness variations along the cantilever, Saad and Metwalli (2013). The optimum solutions utilize the ***HGP*** procedure that employs a ***FE*** definition

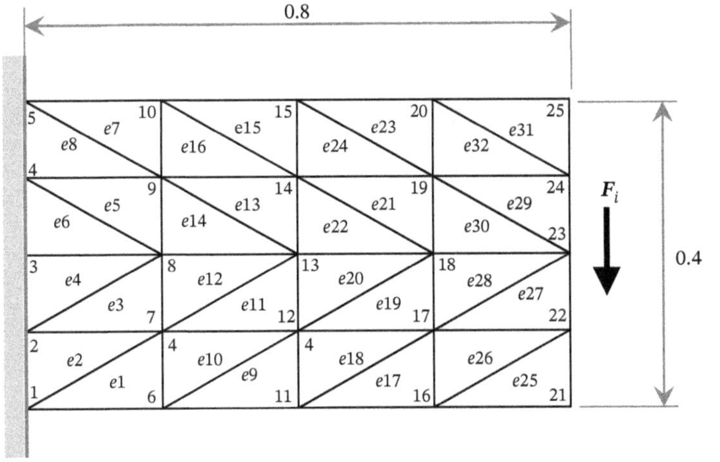

Figure 6.56 Design domain, nodes, and elements of a cantilever beam. The number closer to the node is the node number. The element number is designated as one of $e1-e31$, Saad and Metwalli (2012).

0.3434	0.3207	0.3222	0.3251	0.3231	0.2819	0.1153	0	0
0.353	0.3368	0.3219	0.3241	0.3297	0.3102	0.3412	0.0058	0
0.0076	0.0073	0.0069	0.0074	0.0117	0.0152	0.0796	0.2549	0.0132
0	0	0	0	0	0.0001	0.0131	0.2307	0.154
0	0	0	0	0	0.0001	0.0143	0.2977	0.1991
0	0	0	0	0	0.0001	0.0131	0.2307	0.154
0.0076	0.0073	0.0069	0.0074	0.0117	0.0152	0.0796	0.2549	0.0132
0.353	0.3368	0.3219	0.3241	0.3297	0.3102	0.3412	0.0058	0
0.3434	0.3207	0.3222	0.3251	0.3231	0.2819	0.1153	0	0

(a)

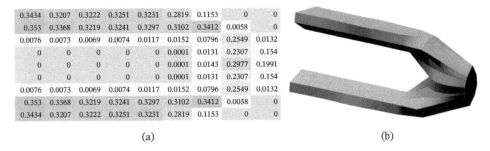

(b)

Figure 6.57 Optimum design of the cantilever beam: (a) representative optimum thickness values at nodes and (b) an isometric view of optimum thickness variations along the cantilever, Saad and Metwalli (2013).

of constant and linear strain triangles with variable thickness at each of the nodes and with extra intermediate nodes; see Saad and Metwalli (2012, 2013).

Figure 6.58 presents two various but similar geometries and objectives for the cantilever beam of Figure 6.56. The first optimum topology shows stresses in Figure 6.58a for an extracted geometry approximation to the work of Keveh et al. (2008). The stresses in the extracted geometry approximation is solved by ***FE*** (Abaqus 6.13); see Abd El-Rahman (2017). Figure 6.58b shows another similar optimum geometry of the same cantilever beam problem, which is optimized by ***FE*** (Abaqus 6.13). It is then clear that these types of problems are multimodal. By changing techniques and methods in solving such problems, different outputs may be obtained; see also Figure 6.57b. Moreover, usual optimization in ***FE*** codes may not take into account the same matching value of the stress through the model.

In classical beam treatment, optimization of a cantilever that has a rectangular cross section of a constant breadth b_B and loaded at the tip, is shown in Figure 6.59a. To have a constant stress beam along the x direction, the variable height h_B as a function of x is obtained by the application of Eq. (6.14) such that

$$\sigma_{x,\max} = \mp \frac{\boldsymbol{M}_z(x)}{Z_z} = \frac{6\boldsymbol{F}_y\,(l_B - x)}{b_B(h_B(x))^2} = \frac{6\boldsymbol{F}_y\, l_B}{b_B(h_B(0))^2} = \text{constant} \tag{6.243}$$

$$(h_B(x))^2 = (h_B(0))^2\frac{(l_B - x)}{l_B} \quad \text{i.e.} \quad h_B(x) = h_B(0)\sqrt{1 - \frac{x}{l_B}} \tag{6.244}$$

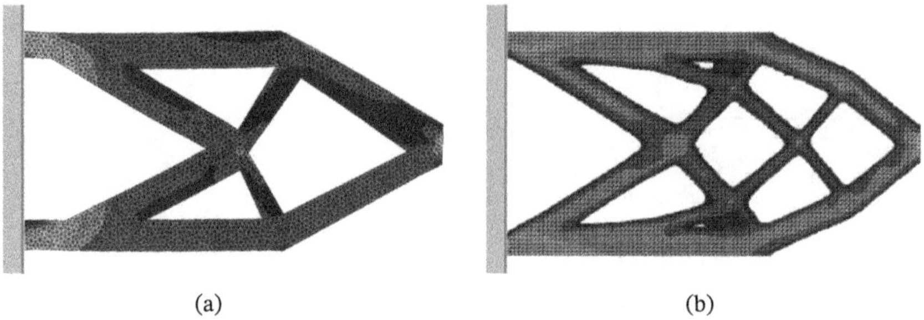

Figure 6.58 Optimum design of the cantilever beam: (a) stresses in an extracted geometry-approximation to the work of Keveh et al. (2008) that is solved by ***FE*** (Abaqus 6.13) and (b) stresses in the same cantilever beam optimized by ***FE*** (Abaqus 6.13). Source: Keveh et al. (2008). © Elsevier.

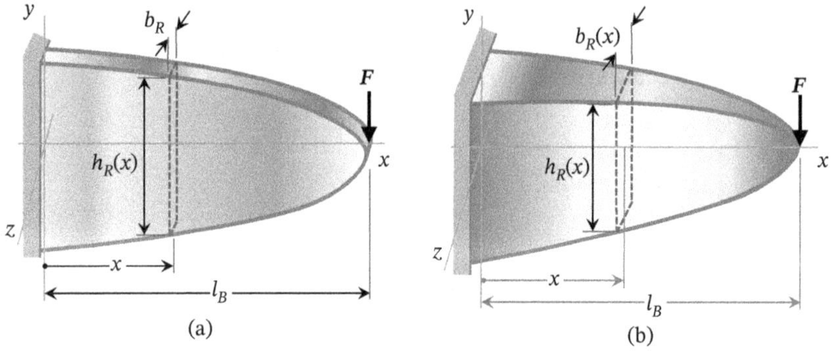

Figure 6.59 Optimum design of the classical continuum cantilever beam of a rectangular cross section of constant breadth or width b_R subject to a tip load in (a). In (b), variable breadth b_R and height h_R along x.

Equation (6.244) indicates that h_B is parabolic as depicted in Figure 6.59a. The proper value of $h_B(0)$ at $x = 0$ is defined by the needed or desired constant value of $\sigma_{x,\max}$; see Faupel and Fisher (1981).

If the width is also varied linearly along x as shown in Figure 6.59b, Eq. (6.243) is changed accordingly to give

$$b_B(x) = b_B(0)\left(\frac{h_B(0)}{h_B(x)}\right)^3 \tag{6.245}$$

This is the variation of the height of the *optimum beam* as shown in Figure 6.59b. For additional details see Faupel and Fisher (1981).

Now, the optimums in both cases of Eqs. (6.244) and (6.245) might be further examinable. Which one gives the minimum volume? Let alone to find an optimum by changing the thickness in the y direction. Fully fledged optimization for other related objectives is the realm of research in specific *geometry* and *topology* fields such as shape and topology optimization work by the author and so many others, for example, Keveh et al. (2008) and Abd El-Rahman et al. (2017).

It should be noted, however, that the previously presented results are dependent on the different adapted processes. For a specific material, however, one should investigate the correct values of properties and adopt values available in codes and standards. The synthesized and iterated parameters can be construed as initial values to be checked, verified and validated by available standard codes and adjusted to other conditions, loadings, and manufacturing processes.

6.13 Summary

The design of a machine is not only the synthesis of the component for its appropriate geometry and material. It also involves the interaction between components, which includes modeling of component-to-component interaction. The type of connection affects the loading, boundary condition, and the mathematical modeling of each component. A basic review of the necessary background of related fields has been introduced to facilitate the reasonable synthesis of basic machine elements. Considering basic and straightforward loading models and matching mathematical model for geometry help in machine element synthesis. Stresses generated in machine elements consequently obtained through the mechanics of materials, solid mechanics, stress analysis, theory of elasticity, or numerical evaluation by ***FE***. Basics of these fields are given in this chapter to facilitate machine element synthesis.

Externally applied loads and their effect on the internally ensued loads and stresses are presented. Simplified and combined loadings such as normal, transverse, bending, and torsion are included. From that, basic elements are synthesized considering basic allowable material strength considerations. Commonly used machine element material is adapted. Element behaviors such as deformation, deflection, buckling, and strains are defined. Different means to do that are presented. Specific applications of machine elements such as beams, columns, pressure vessels, press and shrink fits, contact stresses, curved beams, thermal effects, and stress concentration are treated. Means of synthesis such as computer-aided design (CAD) or CAS, ***FE*** method, and optimization are utilized.

CAD and optimization of some machine elements are obtained by utility of ***Synthesis Tablets*** or MATLAB codes. Beam Synthesis Tablet and column Synthesis Tablet provide real computer-aided design (real CAD) means to synthesize and optimize beams and columns under customary concentrated loads. Optimum stress concentration studies attempt to give insight into the proper configuration of holes and slots for minimum stresses and best utilization of material. Code to optimize ***FE*** prismatic beam presents a tool for further utility of larger optimum configurations. The study of optimum ***FE*** cantilever beams offers a guide that can help in the understanding of optimizing elements that are more complex.

Problems

6.1 What are the differences among deformations, displacements, and deflections? Identify the application of each of them in a machine element such as a spring, a lever, or a shaft under different loading conditions.

6.2 Acquire a straight piece of thick but very low-density sponge or foam rubber. The straight parallelepiped sponge of a rectangular section is to be equipped by two hefty straws or balsa-wood rods glued to the sponge similar to the one shown in Figure 6.4b but with each extended at both ends. With a permanent marker, draw centered straight lines along the centerline of each longitudinal face before any applications of load. Each line can be a 25 [mm] or 1.0 [in] of length. Apply loads at the end of the straws to cause bending of the sponge. Measure each of the drawn lines and observe the sponge shape change. Calculate the deformations, displacements, and deflections.

6.3 Draw a normal line at the end of each line in the kit of Problem 6.2. Observe the behavior of each of the normal lines under bending. Are the lines straight after loading is applied? What conclusions can be extracted from their performance?

6.4 Use the experimental apparatus or kit in Problem 6.2 to apply different loading types to the device. Observe the reaction of the device to each type of loading such as twisting, compression, or torsion.

6.5 In Example 6.1, what should be the width of the rectangular section b_R to have the same second area moment and maximum stresses of the circular cross section? Compare the areas of the resulting rectangular section to the circular cross section. Use either SI or US systems of units.

6.6 In Example 6.1, what should be the width of the T-section b_T to have the same second area moment and/or maximum stresses of the circular cross section? Compare the areas of the resulting T-sections to the circular cross section. Note to consider both tension and compression stresses for comparison of maximum normal stresses. Use either SI or US systems of units.

6.7 In Example 6.1, what should be the thickness of the T-section t_T and the width to height ratio b_T/h_T to have the same second area moment and maximum stresses of the circular cross section? Compare the areas of the resulting T-sections to the circular cross section. Note to consider both tension and compression stresses for comparison of maximum normal stresses. Use either SI or US systems of units.

6.8 If a hollow circular section of a thickness $t_C = 0.1\ d_C$ is used in place of the solid circular section in Example 6.1, what should be the diameter of the section d_C to have the same second area moment and/or maximum stresses of the circular cross section? Compare the resulting area to the initial circular area.

6.9 If a hollow rectangular section of a thickness t_R is used in place of the solid rectangular section in Example 6.1, what should be the thickness t_R of the section to have the same second area moment and/or maximum stresses of the circular cross section? Compare the resulting area to the initial circular area.

6.10 Use a 12 [in] or 0.30 [m] clear plastic ruler and two round pencils to simulate a simply supported beam under mid-span concentrated force. Measure the constant thickness and width of the ruler cross section. On a flat surface, place the two pencils parallel to each other and at a distance of 12 [in] or 0.30 [m] apart of their centerline. Use your figure to apply a load at mid-span to cause the ruler to touch the flat surface. With the known deflection to be the diameter of the pencils, find the stresses in the outer fiber of the ruler. Use the modulus of elasticity of the plastic material defined by the manufacturer. Find the load that has been applied by your finger.

6.11 Use the same ruler of previous problem to evaluate the modulus of elasticity of the plastic material by the following process. Weigh the plastic ruler by an accurate scale. Under the ruler uniformly distributed weight between the two parallel pencils at a distance of 12 [in] or 0.30 [m] apart, measure the sagging deflection. Use the mid-upper and lower surface of the ruler to get the neutral plane deflection at mid-span. With the assumed uniformly distributed load get an evaluation of the modulus of elasticity of the plastic material. Compare result to the modulus of elasticity of the manufacturer plastic material.

6.12 If a long steel ruler of 36 [in] or 1.0 [m] is available, apply the method of the previous problem to evaluate the modulus of elasticity of the ruler material. Compare the evaluated modulus of elasticity to that of steel or stainless steel material.

6.13 Redo Example 6.2 for the same beam but shorter with a length $l = 0.1$ [m].

6.14 Redo Example 6.2 for the same beam but longer with a length $l = 4$ [in].

6.15 If the beam in Example 6.2 has a length $l = 10$ [m], what should the diameter be to subject the beam to the same stresses and what should it be to get the same deflection?

6.16 If the beam in Example 6.2 has a length $l = 400$ [in], what should the diameter be to subject the beam to the same stresses and what should it be to get the same deflection?

6.17 What should be the distributed load in Example 6.3 to get equal contribution of concentrated and distributed load in stresses, deflection, and slope? Use either SI or US system of units.

6.18 If the beam length in Example 6.2 is twice the value or $l = 2.0$ [m], what should the diameter be to subject the beam to the same maximum stress and what should it be to get the same deflection?

6.19 If the beam length in Example 6.2 is twice the value or $l = 80$ [in], what should the diameter be to subject the beam to the same maximum stress and what should it be to get the same deflection?

6.20 Use the singularity function to find the slope and deflection of the case stated in Problem 6.17.

6.21 Use a uniformly distributed load over the length of the beam shown in Figure 6.7 in place of the concentrated load and solve the statically indeterminate beam for the reactions and moments at the supports. Use the singularity function method to find the shear force diagram, the bending moment diagram, and the slope and deflection along the beam.

6.22 If the left support of the statically indeterminate beam of Figure 6.7 is a simple support instead of a fixed support, use singularity function to find the reactions, shear force, and bending moment diagrams. In addition, find the slope and deflection along the beam.

6.23 Solve Problem 6.22 when a uniformly distributed load replaces the concentrated load.

6.24 Resynthesize the beam in the case stated in Example 6.6 for a shorter beam of $l_B = 0.5$ [m] and a longer beam of $l_B = 2.0$ [m]. Use both of the closed form expression and the synthesis chart defined by Figure 6.8a.

6.25 Resynthesize the beam in the case stated in Example 6.6 for a shorter beam of $l_B = 20$ [in] and a longer beam of $l_B = 80$ [in]. Use both of the closed form expression and the synthesis chart defined by Figure 6.8b.

6.26 Find the maximum slope and maximum deflection of the synthesized beams in Problem 6.24.

6.27 Find the maximum slope and maximum deflection of the synthesized beams in Problem 6.25.

6.28 Use the singularity function to derive Eqs. (6.33) and (6.33′) of the uniformly distributed and simply supported beam in Figure 6.5c.

6.29 Prove the relation to find the *equivalent force* $\boldsymbol{F}_e$ defined by Eq. (6.55).

6.30 Synthesize the circular beams of length $l_B = 0.1$ [m] if each is subjected to a mid-span concentrated load $\boldsymbol{F}$ of 5, 500, 50 000, and 500 000 [N] when the safety factor K_{SF} is assumed to be 2.0 and material S_y of 200 [MPa].

6.31 Synthesize the circular beams of length $l_B = 4$ [in] if each is subjected to a mid-span concentrated load $\boldsymbol{F}$ of 1, 100, 10 000, and 100 000 [lb] when the safety factor K_{SF} is assumed to be 2.0 and material S_y of 29 [kpsi].

6.32 Synthesize the circular beams of length $l_B = 1.0$ [m] if each is subjected to a mid-span concentrated load $\boldsymbol{F}$ of 5, 500, 50 000, and 500 000 [N] when the safety factor K_{SF} is assumed to be 2.0 and material S_y of 200 [MPa].

6.33 Synthesize the circular beams of length $l_B = 40$ [in] if each is subjected to a mid-span concentrated load $\boldsymbol{F}$ of 1, 100, 10 000, and 100 000 [lb] when the safety factor K_{SF} is assumed to be 2.0 and material S_y of 29 [kpsi].

6.34 Synthesize the circular beams of length $l_B = 10.0$ [m] if each is subjected to a mid-span concentrated load $\boldsymbol{F}$ of 5, 500, 50 000, and 500 000 [N] when the safety factor K_{SF} is assumed to be 2.0 and material S_y of 200 [MPa].

6.35 Synthesize the circular beams of length $l_B = 400$ [in] if each is subjected to a mid-span concentrated load $\boldsymbol{F}$ of 1, 100, 10 000, and 100 000 [lb] when the safety factor K_{SF} is assumed to be 2.0 and material S_y of 29 [kpsi].

6.36 Synthesize the circular beams of length $l_B = 50.0$ [m] if each is subjected to a mid-span concentrated load $\boldsymbol{F}$ of 5, 500, 50 000, and 500 000 [N] when the safety factor K_{SF} is assumed to be 2.0 and material S_y of 200 [MPa].

6.37 Synthesize the circular beams of length $l_B = 2000$ [in] if each is subjected to a mid-span concentrated load $\boldsymbol{F}$ of 1, 100, 10 000, and 100 000 [lb] when the safety factor K_{SF} is assumed to be 2.0 and material S_y of 29 [kpsi].

6.38 If a standard *I-Beam* is used in place of the circular cross section in any of the previous problems, identify the suitable standard *I-Beam* replacement that can be used. Note to use the same *section modulus* Z_z.

6.39 Inspect any highway bridge that is using *I-Beams* in their construction. Count the number of the I-Beams used in each span and estimate the length of the span. With the usual stated bridge capacity, identify the adapted safety factor of the bridge design. If the maximum desired deflection is to be less than $l_B/400$, what is the safety factor for that?

6.40 Inspect any available overhead crane of an *I-Beam* design or a built-up section and obtain its cross-sectional measurements and properties. From the stated load capacity, find the designer safety factor considering a mid-span maximum capacity loading.

6.41 For the previous problems of the synthesized circular beams, find the stress safety factor if the load is modeled as a uniformly distributed rather than concentrated at mid-span. Consider the shortest and the longest beam lengths.

6.42 A machine element of cylindrical form is to transmit a maximum of 5 [kW] at a speed of 200 [rpm]. The machine element has the length $l = 1.0$ [m], the modulus of elasticity $E = 207$ [GPa], and the modulus of rigidity $G = 79$ [GPa]. What is the synthesized element diameter for a safety factor K_{SF} of 4.0 and assuming a material of a shear yield strength $S_{ys} = 200$ [MPa] for a yield failure prevention due to torsional shear? What is the angular deflection of the element at maximum transmitted power?

6.43 A machine element of cylindrical form is to transmit a maximum of 6.7 [hp] at a speed of 200 [rpm]. The machine element has the length $l = 40$ [in], the modulus of elasticity $E = 30$ [Mpsi], and the modulus of rigidity $G = 11.5$ [Mpsi]. What is the synthesized element diameter for a safety factor K_{SF} of 4.0 and assuming a material of a shear yield strength $S_{ys} = 29$ [kpsi] for a yield failure prevention due to torsional shear? What is the angular deflection of the element at maximum transmitted power?

6.44 Develop the relation for the transverse shear distribution over a circular cross section.

6.45 Find the shear center of a channel cross section (C-section) of the loaded beam in Figure 6.13 for the following C-section dimensions. The channel section has a width $b_C = 0.1$ [m], a depth $h_C = 0.2$ [m] and channel outer thickness $t_C = 0.01$ [m].

6.46 Find the shear center of a channel cross section (C-section) of the loaded beam in Figure 6.13 for the following C-section dimensions. The channel section has a width $b_C = 4$ [in], a depth $h_C = 8$ [in], and channel outer thickness $t_C = 0.4$ [in].

6.47 Find the shear center of a standard channel cross section (C-section) of loaded beams similar to that shown in Figure 6.13.

6.48 Substitute Eq. (6.82) into Eq. (6.81) and apply the necessary mathematical manipulations to get Eq. (6.83).

6.49 Apply the condition $d\sigma/d\gamma = 0$ to Eq. (6.83) to yield the relations of Eqs. (6.84) and (6.85).

6.50 Implement the conditions of Eq. (6.85) into Eq. (6.83) to confirm that the last two terms of the principal stress values in Eq. (6.83) is the radius R_M of the *Mohr's circle*.

6.51 A cylindrical machine element is transmitting a maximum of 5 [kW] and running at 200 [rpm] in the clockwise direction. It is subjected to a maximum bending moment that is equal to the maximum applied torque. Find the internal stresses if the cylindrical diameter is 10 [mm]. Obtain the principal stresses and identify their directions or orientations. Use both *Mohr's* circle and the solution of the eigenvalue problem. Demonstrate that the invariants I_i are constants. Find the average stress and the distortion or deviatoric stress.

6.52 A cylindrical machine element is transmitting a maximum of 6.7 [hp] and running at 200 [rpm] in the clockwise direction. It is subjected to a maximum bending moment that is equal to the maximum applied torque. Find the internal stresses if the cylindrical diameter is 0.4 [in]. Obtain the principal stresses and identify their directions or orientations. Use both *Mohr's* circle and the solution of the eigenvalue problem. Demonstrate that the invariants I_i are constants. Find the average stress and the distortion or deviatoric stress.

6.53 Find the synthesized diameter for the element in Problem 6.51 or 6.52, assuming torsional load only. Compare the results of the principal stresses with and without the applied bending moment.

6.54 In the calculation of the distortion or deviatoric stress of Eq. (6.104), prove that using σ_2 as the smallest negative eigenvalue with the knowledge that σ_3 is zero is the same as using $\sigma_3 = \sigma_2$ and setting $\sigma_2 = 0$ to have $\sigma_1 > \sigma_2 > \sigma_3$.

6.55 A cylindrical pressure vessel with a hemispherical ends of 2.0 [m] or 79 [in] diameter and 4.0 [m] or 157 [in] in length is to be designed using a material of 50 [MPa] or 7.25 [kpsi] allowable stress. The maximum internal pressure is 1.5 [MPa] or 0.2175 [kpsi]. Find the suitable thicknesses for the cylindrical and the hemispherical parts of the cylinder. Draw the ***3D*** *Mohr's* circle circle and show the points of applied stresses. Compare your results with the ASME boiler and pressure vessel code assuming a similar material and a weld efficiency of 0.8 or 80%.

6.56 A small cylindrical pressure vessel with a hemispherical ends of 0.2 [m] or 7.9 [in] diameter and 0.4 [m] or 15.7 [in] in length is subjected to a maximum internal pressure of 15.0 [MPa] or 2.175 [Mpsi] in normal external atmospheric pressure. Apply the thick cylinder equations to design the pressure cylinder using a material of 50 [MPa] or 7.25 [kpsi] allowable stress. Compare your results with those obtained using thin cylinder equations.

6.57 A small submarine of a 2.0 [m] or 79 [in] spherical diameter is to be designed for deep sea exploration. The intended depth is 500 [m] or about 1640 [ft]. What should be the thickness if the internal pressure is to be maintained at atmospheric level? Assume that possible spherical buckling is to be handled in a separate way.

6.58 Search for the dimensions and allowable load of a train wheel with a steel tier overlay ring shrunk-fitted onto it that should withstand the braking force applied onto it. Assuming a shrink fit of H7/u6, find the shrink fitting pressure p_F. Find the maximum stresses on the steel tire and wheel. If the coefficient of friction between the wheel and rail is assumed as 1.0, what is the maximum breaking force that can be exerted not to have any sliding or failure of the shrink fitting?

6.59 Derive the shrink fit equations of a hub shrunk fitted over a solid shaft.

6.60 Select a roller bearing and find its dimensions and the maximum static capacity from a manufacturer catalogue. Use the cylindrical contact equations to get the maximum compressive stress.

6.61 Design a hook of a crane to carry a maximum load of 20 [kN] or 4.5 [klb] if the inner diameter of the hook is 0.1 [m] or 4 [in]. Use an allowable stress of 200 [MPa] or 29 [kpsi]. Assume a circular or a rectangular hook cross section. Start with at least the stem cross section that should carry the load in tension. The hook cross section should be larger since it should carry both tension and bending. Since iterations are needed, it is recommended to write a MATLAB or an Excel code to interactively enter the curvature factors off Figure 6.27.

6.62 Synthesize the door handle members for the case stated in Example 2.1, which is defined in Chapter 2.

6.63 For the synthesized door handle of previous problem, use Castigliano's method to find the deflection at the point of load application. Consider the maximum operating conditions.

6.64 Use Castigliano's method to solve Example 2.2 of two springs assembled in parallel as shown in Figure 2.10a.

6.65 Use Castigliano's method to solve Example 2.3 of two springs assembled in series as shown in Figure 2.11a.

6.66 Use Castigliano's method to solve Example 2.4 of the hoist assembly as shown in Figure 2.12a.

6.67 Calculate the critical stress ratio σ_{cr}/E in Example 6.12 and check the applicability of *Euler* buckling from Figure 6.32.

6.68 How high is the load in Example 6.12 to be, for the case of parabolic equation to be active?

6.69 If the load in Example 6.12 is doubled, tripled, or quadrupled, what should be the synthesized diameters?

6.70 A passenger elevator is operated by a hydraulic cylinder, which is usually buried underground below the elevator booth. The rod of the hydraulic cylinder lifts the elevator booth to the desired level. If the maximum strokes of such a system are 5, 10, 20, and 30 [m] or 200, 400, 800, and 1200 [in] that can carry a maximum load of 10 [kN] or 2200 [lb], what are the diameters of the hydraulic cylinder rods to guard against buckling with a safety factor of 7.5? Assume the case of a free-fixed rod construction.

6.71 Use the available Excel *Column Synthesis Tablet* to iterate for the maximum allowable eccentricity ratio the design in Example 6.12 can permit.

6.72 If the equivalence in Section 6.8 is used in a similar beam to that of Example 6.13 but with a stepped column at half the length with two different diameters where one is half the other, what should be the diameters for the bar not to buckle?

6.73 Apply the buckling consideration to the shaft in Example 6.14 when the length is three times as long.

6.74 What is the deflection of a bimetal strip of 25 [mm] or 1.0 [in] length, 1.0 [mm] or 0.04 [in] thick, and made of steel and brass material if the temperature difference is 28 [°C] or 50 [°F].

6.75 Compare the error difference in evaluating the stress concentration factor K_{SC} using Eq. (6.175) and Eq. (6.176) at different diameter to width ratios of 0.1, 0.3, 0.5, and 0.7.

6.76 Manipulate Eqs. (6.186)–(6.189) to derive the stiffness matrix of prismatic beam element in Eq. (6.191) using the indicated procedure in Section 6.11.2.

6.77 Derive the stiffness matrix of axially loaded elements for the general number of elements n_e instead of the three elements of Figure 6.42 and Example 6.15.

6.78 Adjust the MATLAB code of Figure 6.43 to consider 10 elements.

6.79 Use the generalized stiffness matrix of axially loaded elements in a MATLAB code to design a very large horizontal axis windmill tower of 100 [m] or 4000 [in] height with a main top axial vertical load of 1000 [kN] or 220 [klb]. Assuming a heavy tower, divide the tower into 10 sections with the weight of each section acting on the top of the lower section underneath it. The tower is made of steel and is assumed to stand an allowable stress of 100 [MPa] or 14.5 [kpsi], and its modulus of elasticity is $E = 207$ [GPa] or 29 [Mpsi]. Solve the top section first for the synthesized diameter of a solid or a hollow section.

6.80 For the statically indeterminate beam of Example 6.16, the force $\boldsymbol{F} = -40$ [kN] or -9 [klb], the length $l = 1.0$ [m] or 40 [in], and the modulus of elasticity $E = 207$ [GPa] or 30 [Mpsi]. What are the reactions, the maximum bending moment and the maximum slope in the uniform circular section profile of a diameter $d_{C\cdot} = 0.1$ [m] or4 [in] if $b = 0.5a$? What can be done to solve the problem if the diameter of each section is different?

6.81 Use any available ***FE*** package to solve the previous Problem 6.80 using a ***3D*** model. Does the ***FE*** solution verify the analytical solution? Discuss the attained results. Is there an error?

6.82 Use any available ***FE*** package to solve the door handle of Problem 6.62 using a ***3D*** model. Does the ***FE*** solution verify the analytical solution? Discuss the attained results. Is there an error?

6.83 In the ***FE*** package, change the model of the load application and the boundary constraints to be distributed rather than concentrated and resolve Problem 6.81 or 6.82. Discuss the attained results. Is there an error?

6.84 Search the Internet for the current active ***FE*** packages, their market share, their application scope, their connectivity to other geometric modeling (or CAD) packages, etc.

6.85 Use the available *Beam Synthesis Tablet* to synthesize selected cases in Problems 6.30–6.37. Compare results.

6.86 Solve some of the other previous beam problems using the *Beam Synthesis Tablets.*

6.87 Change the default circular cross section of beams to other cross sections, and solve these beams utilizing *Beam Synthesis Tablets.* Use standard beam sections to compare sought optimum solutions.

6.88 Use *Beam Synthesis Tablet* to synthesize each component of the door handle defined for the case stated in Example 2.1, which is defined in Chapter 2. Compare results to other methods of solution.

6.89 Use a ***FE*** package to verify results obtained by the *Beam Synthesis Tablets* in previous problems. Use suitable representative loading and boundary conditions. Compare results.

6.90 Use the available *Column Synthesis Tablet* to synthesize selected cases in Problems 6.67–6.73. Compare results with manual calculations.

6.91 Use the available *Column Synthesis Tablets* to resynthesize the rods of the hydraulic cylinders that operates the passenger elevators in Problem 6.70 considering fixed–fixed rod constructions. Sketch the details of the construction of the system and its critical connections.

6.92 Find the critical longitudinal load of the 12 [in] or 0.30 [m] clear plastic ruler cited in problem 6.10. Use *Column Synthesis Tablets* to check your calculations. Use utmost care in using the Tablet since its development is mainly for circular and not for rectangular cross sections. Check your calculations with some manual experimentation.

6.93 Redo the previous Problem 6.92 for the long steel ruler of 36 [in] or 1.0 [m].

6.94 A uniformly distributed load loads a simply supported beam of length l_B. The supports are placed symmetrically inward so that the distance between them is less than l_B. Where should the supports be placed to produce the minimum bending moments and normal stresses?

6.95 Use any ***FE*** package to optimize the plate with an elliptical and/or circular hole uniformly loaded in tension as shown in Figure 6.53a. Assume a suitable load, a cylindrical hole diameter to plate width ratios of 0.2, 0.4, and 0.6. Assume suitable dimensions of the rectangular cross-sectional area.

6.96 Create a ***3D*** solid plate with thicknesses as depicted in Figure 6.54. Find the stresses at the specified nodes due to selected values of the uniformly distributed load using any ***FE*** package. Compare the results with an optimized plate generated by the ***FE*** package.

6.97 Use an adjusted MATLAB code of Figure 6.43 to consider 10 elements. Modify the code similar to that in Figure 6.55 to perform optimization of the windmill tower of Problem 6.79. What should you do to consider the wind horizontal forces?

6.98 Create a ***3D*** solid plate with thicknesses as depicted in Figure 6.57. Find the stresses at the specified nodes due to selected values of the load shown in Figure 6.56 using any ***FE*** package. Compare the results with an optimized plate generated by the ***FE*** package similar to those in Figure 6.58.

References

Abd El-Rahman, M.S. (2017). *A generalization of the heuristic gradient projection for 2D and 3D frame optimization*, MS thesis, Cairo University.

Abd El-Rahman, M.S., Abd El-Aziz, K.M. and Metwalli, S.M. (2017). A generalization of the heuristic gradient projection for 2D and 3D frames. *Proceedings of the ASME IMICE2017*, Florida, paper # IMECE2017-70560.

Abd El-Aziz, K.M. and Metwalli, S.M. (2017). Optimum design of pressure vessels using Hybrid HGP and Genetic Algorithm, *Proceedings of the ASME 2017 Pressure Vessels and Piping Conference, PVP2017*, July 16–20, Waikoloa, Hawaii, United States, Paper No. PVP2017-65538

Abd El Gawad, M.A.M., Hegazi, H.A., and Metwalli, S.M. (2012). Optimum design of columns under elastic buckling, *Proceedings of IDETC/CIE 2012, ASME International Design Engineering Technical Conferences & Computers and Information in Engineering Conference*, August 12–15, Chicago, Illinois, paper # DETC2012–70601.

AISC (2001). *Manual of Steel Construction, LRED*, 3e. American Institute of Steel Construction.

ASME (2013). *ASME Boiler and Pressure Vessel Code*. New York: ASME.

Bathe, K.J. (1982). *Finite Element Procedures in Engineering Analysis*. Prentice-Hall.

Bathe, K.J. (2006). *Finite Element Procedures*. K-J Bathe.

Beer, F.P., Johnston, E.R. Jr., and DeWolf, J.T. (2009). *Mechanics of Materials*, 5e. McGraw-Hill.

Bhatti, M.A. (2005). *Finite Element Analysis with Mathematica and MATLAB Computations and Practical Applications: Fundamental Concepts*. Wiley.

Castigliano, C.A. (1873). *Intorno ai sistemi elastici*, PhD dissertation, Polytechnic of Turin.

Chandrupatla, T.R. and Belegundu, A.D. (2002). *Introduction to Finite Elements in Engineering*, 3e. Prentice-Hall.

Chapra, S.C. (2006). *Numerical Methods for Engineers*, 5e. McGraw-Hill.

Cook, R.D. (1995). *Finite Element Modeling for Stress Analysis*. Wiley.

Dirac, P. (1930). *The Principles of Quantum Mechanics*. Oxford University Press.

Faupel, J.H. and Fisher, F.E. (1981). *Engineering Design, a Synthesis of Stress Analysis and Material Engineering*, 2e. Wiley.

Heaviside, O. (1912). *Electromagnetic Theory*, vol. 3. The Electrician Printing and Publishing Co.

Hegazi, H.A., Nassef, A.O., and Metwalli, S.M. (2002). Shape optimization of NURBS modeled 3D C-frames using hybrid genetic algorithm, *Proceedings of the 2002 ASME Design Engineering Technical Conferences & Computers and Information in Engineering Conference*, Montreal, Canada, September 29–October 2, Paper # DETC2002/DAC-34107.

Hertz, H. (1882). The contact of solid elastic bodies. *Journal of Pure and Applied Mathematics (in German)* 92: 156–171.

Johnson, J.B. (1902). *The Theory and Practice of Modern Framed Structures*. Wiley.

Kaveh, A., Hassani, B., Shojaee, S., and Tavakkoli, S.M. (2008). Structural topology optimization using ant colony methodology. *Engineering Structures* 30 (9): 2559–2565.

Logan, D.L. (2007). *A First Course in the Finite Element Method*, 4 Nelson, a division of Thomson Canada
Macaulay, W.H. (1919). Note on the deflection of beams. *Messenger of Mathematics* 48: 129–130.
Metwalli, S.M. and Moslehy, F. (1983). *Development of Distributed Breach for the Conical Shock Tube*, Final Report, DOD/NRL, USA, July, ADA142210, 102 pages
von Mises, R. (1913). Mechanik der festen Körper im plastisch deformablen Zustand. *Göttin. Nachr. Math. Phys.* 1: 582–592.
Nassef, A.O., Hegazi, H.A., and Metwalli, S.M. (1999). Design of C-frame using real-coded genetic optimization algorithm and NURBS. *Proceedings of the 1999 ASME Design Engineering Technical Conference*, Las Vegas, Nevada (12–15 September). Paper No. DETC99/CIE-9138.
Nassef, A.O., Hegazi, H.A., and Metwalli, S.M. (2000). A hybrid genetic-direct search algorithm for the shape optimization of solid C-frame cross-section. *Proceedings of the ASME 2000 Design Engineering Technical Conference*, Baltimore, Maryland (10–14 September). Paper No. DETC2000/DAC-14295.
Peterson, R.E. (1974). *Stress Concentration Factors*. Wiley.
Popov, E.P. (1968). *Introduction to Mechanics of Solids*. Prentice-Hall.
Popov, E.P. (1990). *Engineering Mechanics of Solids*. Prentice-Hall.
Reisman, H. and Pawlik, P.S. (1980, 1991). *Elasticity Theory and Applications*. Wiley and Krieger Publishing Company.
Saad, M.R. and Metwalli, S.M. (2012). Heuristic GP finite element method for structural thickness optimization. *Proceedings of the ASME International Mechanical Engineering Congress & Exposition IMECE 2013* (15–21 November). San Diego, California, USA, paper # IMECE2013-66364
Saad, M.R. and Metwalli, S.M. (2013). Finite element synthesis for design optimization, *Proceedings of IDETC/CIE 2012*, ASME International Design Engineering Technical Conferences & Computers and Information in Engineering Conference, (12–15 August), Chicago, Illinois, paper # DETC2012-70582
Timoshenko, S.P. and Gere, J.M. (1961). *Theory of Elastic Stability*, 2e. McGraw-Hill.
Younan, M., Metwalli, S.M., and El-Zoughby, A. (1983). Fracture mechanics analysis of a fire tube boiler. *Journal of Engineering Fracture Mechanics* 17 (4): 335–338.
Young, W.C. and Budynas, R.G. (2002). *Roark's Formulas for Stress and Strain*, 7e. McGraw-Hill.
Zienkiewicz, O.C. and Taylor, R.L. (1991). *The Finite Element Method*, 4e. McGraw-Hill.

Internet Links

https://www.azom.com/articles.aspx, Materials Science industry.
https://www.3ds.com/products-services/simulia/products/abaqus/ or www.simulia.com, Abaqus ***FE*** software.
http://www.ansys.com/, ANSYS ***FE*** software.
www.mscsoftware.com/product/msc-nastran, NASTRAN ***FE*** software.
http://www.adina.com/, ADINA ***FE*** software.
http://www.solidworks.com/sw/products/simulation/structural-analysis.htm, Solidworks Simulation ***EF*** software.
https://www.comsol.com/, COMSOL Multiphysics, ***FE*** software.

7

Materials Static and Dynamic Strength

Abundant materials are existing and accessible in nature. Some are naturally available and may be usable as such or with little processing such as plants, trees, soil, and stones. They do need basic procedures to suit their direct employment in products. Other materials would need involved processing of ores to extract raw metal elements and generate alloys of elements in ingots or in different other forms such as casting, molding, hot or cold working, powder metallurgy, etc. Such abundance generated many materials of which over 80 000 are currently useful metallic alloys. In addition, about the same number or more are of nonmetallic or composite materials. The designer has thus ample possibilities to adapt a material for a machine element. Improper material selection causes *failure* or higher *cost*. The optimum material provides finest *service performance* and best *manufacturing processing*. Thus, possible combinations are almost infinite and optimization would be required. Historically adapted materials for some machine elements may provide a close *optimum material set* due to evolving experience in improving selection with time.

Normally, designers base their material selection on experience. These materials may not necessarily be the utmost optimum for machine elements. In many aerospace applications, however, materials are subject to allowable limits close to their strength properties. In consumer-oriented applications, the objective is frequently to decrease cost. In automotive industry, the objective is prospectively to increase energy efficiency through weight reduction. This required a revolution of material development, selection, and manufacturing.

In manufacturing businesses, cost of materials may amount to more than half of the total cost. In automotive industry, material cost is more than 70% of manufacturing cost. In shipbuilding, material cost is about 45% of total cost. As the automation degree and the production size increase, the percentage of material cost to the total cost increases.

In mechanical design, the *failures* of the components by *yielding* the material are not acceptable, let alone *fracture*. For overloading, yielding constitutes a possible permanent deformation that would surely prevent the component from performing its function. This permanent deformation is possible when the material is *ductile* such as *carbon steel*. *Brittle* materials such as *cast iron* usually fail by *fracture* with infinitesimal or no permanent deformation, which is even more catastrophic for the design. Brittle material *strength* in compression is usually higher than their strength in tension. Therefore, brittle materials are more suited for such compressive applications. Ductile materials *strength* in compression is about the same as in tension, and thus they can be more suitable for such loading conditions. If the design function is sensitive to deformation, a *stiff* material may be more suitable than a *strong* material. Stiffer materials need higher loads to deform, and stronger materials need higher loads to fail.

Symbols and Abbreviations

The adapted units are [in, lb, psi] or [m, kg, N, Pa], others given at each symbol definition ([k…] is 10^3, [M…] is 10^6 and [G…] is 10^9).

Machine Design with CAD and Optimization, First Edition. Sayed M. Metwalli.

Companion website: www.wiley.com/go/metwalli/machine

Symbol	Quantity, Units (adapted)
Elements	
Al	*Aluminum*
C	*Carbon*
Co	*Cobalt*
Cr	*Chromium*
Cu	*Copper*
Fe	*Iron*
H	*Hydrogen*
Mg	*Magnesium*
Mo	*Molybdenum*
Ni	*Nickel*
O	*Oxygen*
P	*Phosphorous*
S	*Sulfur*
Si	*Silicon*
Sn	*Tin*
Ti	*Titanium*
V	*Vanadium*
W	*Tungsten*
Zn	*Zinc*
Abbreviations	
AA	Aluminum Association
AISI	American Iron and Steel Institute
ASM	American Society for Materials (now the Materials Information Society)
ASTM	American Society for Testing and Materials
DIN	Deutsches Institut für Normung e. V.
EN	European Norm
ISO	International Organization for Standardization
SAE	Society of Automotive Engineers
UNS	Unified Numbering System
Symbols	
$2a_C$	Principal dimension of the elliptic hole or crack
$2b_C$	Principal dimension of the elliptic hole
$2N_{\mathrm{Cf}}$	Fatigue strain-life in reversals to failure
A_0	Original cross-sectional area
a_1, a_2	Constants of fatigue failure equation
a_{Cf}	Failure crack size
a_{Ci}	Initial crack size

Symbol	Quantity, Units (adapted)
$a_N^{1/2}$	*Neuber* constant
b_E	Elastic fatigue strength exponent
b_R	Breadth of rectangular section
C_C	Percentage of carbon content
CD	Cold drawn
c_P	Plastic fatigue ductility exponent
C_P	Crack propagation factor
CVN	Standard *Charpy* V-notch impact value
d_0	Original specimen *diameter*
da_C	Rate of crack growth
D_B	*Brinell* tester ball diameter in [mm]
d_C	Shaft cylindrical diameter
d_{HB}	*Brinell* indentation diameter in [mm]
dN_C	Rate of loading cycle
dN_C/da_C	Crack fatigue growth rate
d_P	Part diameter
$d_{P,e}$	Equivalent part diameter
E	Modulus of elasticity
fr	Failure ratio S_f/S_u
$\boldsymbol{F}_i$	Current load i
$\boldsymbol{F}_{\text{kg}}$	Applied force in "kg-force"
G	Shear modulus of elasticity or modulus of rigidity
H	Transmitted power
HB	*Brinell* hardness number
HR	Hot rolled
h_R	Height of rectangular section
HRB	*Rockwell* hardness for B scale
HRC	*Rockwell* hardness for C scale
HSLA	High-strength low-alloy
K_{conc}	Fatigue concentration factor
K_f	Fatigue stress concentration factor
K_{I}	Elastic stress intensity factor of mode I loading
$K_{\text{I,Th}}$	Threshold stress intensity factor
K_{IC}	Plane strain fracture toughness of material
K_{II}	Elastic stress intensity factor of mode II loading
K_{III}	Elastic stress intensity factor of mode III loading
K_{load}	Load factor
K_{miscel}	Miscellaneous factor
K_{reliab}	Reliability factor

Symbol	**Quantity, Units (adapted)**
K_S	Stress concentration factor
K_{SF}	Safety factor
$K_{SF,CM}$	Safety factor for *Coulomb–Mohr* theory
$K_{SF,F}$	Fatigue safety factor *f*
$K_{SF,FM}$	Safety factor for fracture mechanics
$K_{SF,MM}$	Safety factor for *modified Mohr* theory
$K_{SF,MNS}$	Safety factor for *max normal stress* theory
K_{size}	Size factor
K_{surf}	Surface factor
K_{temp}	Temperature factor
l_0	Original specimen *gauge length*
l_i	Deformed gauge length at load *i*
m_P	Crack propagation exponent
N_C	Number of loading cycles
N_{Cf}	Number of loading cycles to failure
$N_{Cf,i}$	Number of failure cycles at loading stress *i*
$N_{load,i}$	Number of loading cycles of loading stress *i*
N_{rpm}	Rotational speed in rpm
Q	Quenching
$Q\&T$	Quenching and tempering
q_N	Notch-sensitivity factor
r_C	Crack tip radius
r_{load}	Fluctuating loading ratio S_a/S_m
r_N	Notch radius
S	Strength
S_2	Failure strength projection (on the σ_2 axis)
S_{2CM}	Failure strength projection for *Coulomb–Mohr*
S_{2MM}	Failure strength projection for *modified Mohr*
S_{2MNS}	Failure strength projection for *max normal stress*
S_a	Alternating fatigue failure strength
S_e	Endurance limit of material
$S_{e,part}$	Endurance limit of machine part
S_{eN}	Endurance limit of notched specimen
S_{es}	Endurance limit in torsion (shear endurance)
S_f	Fatigue strength at a number of cycles to failure
S_{fb}	Fatigue strength of brittle material

Symbol	Quantity, Units (adapted)
S_m	Mean fatigue failure strength
S–N	Strength rotation diagram
S_u	Ultimate tensile strength (S_{ut})
S_{uc}	Ultimate compressive strength
S_{us}	Ultimate strength in shear
S_{ut}	Ultimate tensile strength (S_u)
S_y	Yield strength
S_{yc}	Compressive yield strength
S_{yl}	Lower yield strength
S_{ys}	Yield strength in shear
S_{yt}	Tensile yield strength
S_{yu}	Upper yield strength
T_M	Melting temperature
t_O	Operating time
T_O	Operating temperatures
u_ε	Strain energy density
u_{123}	Total strain energy density
u_d	Deviatoric or distortion strain energy density
u_h	Hydrostatic strain energy density
x	Property in abscissa coordinate
y	Property in ordinate coordinate
β_C	Correction factor for K_I
$\Delta\varepsilon_E$	Elastic strain range
$\Delta\varepsilon_F$	Strain change for fatigue cyclic stressing
$\Delta\varepsilon_F/2$	Strain amplitude
$\Delta\varepsilon_P$	Plastic strain range
ΔK_I	Range of stress intensity factor
$\Delta\sigma$	Stress range
$\varepsilon_1, \varepsilon_2, \varepsilon_3$	Principal strains
ε_C	Creep strain
ε_E	Elastic strain limit
ε_F	Failure strain
ε_f	Fatigue ductility coefficient
ε_i	Strain at load i
ε_{iT}	True strain at current load i

Symbol	Quantity, Units (adapted)
ε_P	Strain at proportional limit
ε_u	Strain at ultimate stress
ε_x	Normal strain in x direction
ν	*Poisson's* ratio
σ	Stress
$\sigma_1, \sigma_2, \sigma_3$	Principal stresses in principal directions
σ_a	Alternating stress
σ_{av}	Average or hydrostatic stress
σ_d	Deviatoric or distortion stresses
σ_E	Elastic stress limit
σ_f	Fatigue strength coefficient
σ_{FE}	Engineering stress at fracture
σ_{FT}	True failure stress
σ_i	Engineering stress at load i
$\sigma_{\text{load},i}$	Constant amplitude stress of loading stress i
σ_m	Mean stress
$\sigma_{\max}$	Maximum stress
$\sigma_{\min}$	Minimum stress
σ_O	Operating stresses
σ_P	Stress at proportional limit
σ_{vM}	Equivalent *von Mises* stress
$\sigma_{\text{vM},a}$	Equivalent alternating *von Mises* stress
$\sigma_{\text{vM},m}$	Equivalent mean *von Mises* stress
σ_x	Normal stress in x direction
$\sigma_x, \sigma_y, \sigma_z$	Stresses in x, y, and z directions
$\sigma_{x,a}, \sigma_{y,a}$	Alternating normal stresses in x and y directions
$\sigma_{x,m}, \sigma_{y,m}$	Mean normal stresses in x and y directions
τ_1	Principal shear stress
$\tau_{\max}$	Maximum shear stress
$\tau_{xy}, \tau_{yz}, \tau_{zx}$	Shear stresses on x, y, and z planes in x, y, and z directions
$\tau_{xy,a}$	Alternating shear stresses on x in y directions
$\tau_{xy,m}$	Mean shear stresses on x in y directions

7.1 Material Structure and Failure Modes

Material structure is composed of atoms of elements. Thus atomic bonding, crystal arrangements, and nano or micro or macro assemblies affect the structural geometry and behavior of material. This scale of *material science* affects the material properties as depicted in Figure 7.1. The goal of material science is to improve the properties of engineering materials by controlling various aspects of microstructure. Material microstructure can vary from

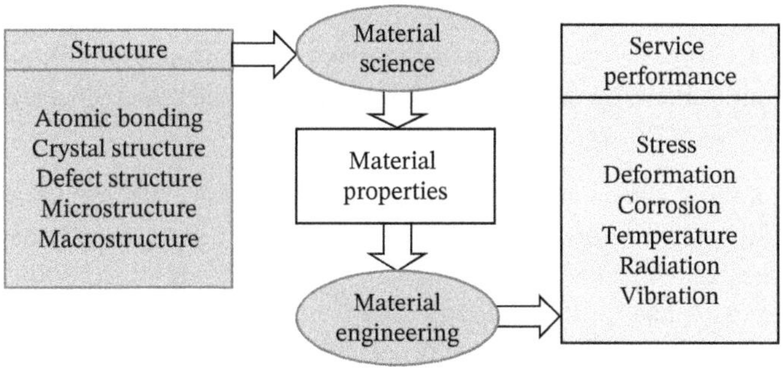

Figure 7.1 Material properties, the link between structure and service performance.

atomic dimensions to the dimensions of a macroscopic crack in weld. Chief methods of altering microstructure are through composition control (alloying) and heat treatment. Other *material engineering* operations such as deformation processing affect the material properties with respect to strength and service performance as shown in Figure 7.1. The performance or functional requirements of materials are usually in terms of properties in the associated fields of physical, mechanical, thermal, electrical, chemical, etc. The concern and purpose of material engineering is to define the means to evaluate the service and performance parameters and properties regarding stresses, deformation, corrosion, temperature, radiation, vibration, etc.

Machine design can usually span a large dimensional scale larger than 1.0 [m] or 40 [in] to less than 1.0 [mm] or 0.04 [in] for regular machine elements. The materials in these cases would be regular or customary engineering materials. If one is to design a *micro or nano elements*, the dimensional scale may vary from less than 0.1 [mm] or 0.004 [in] to 10^{-8} [m] or 4.0 (10^{-7}) [in] or less. The material behaviors in these cases would be different, and one has to consult material science or even solid-state physics and use dislocation, quantum, and wave mechanics in the design process particularly for *nano elements*.

This text dedicates the design and synthesis process to the usual large span of the large dimensional scale or for larger than 1.0 [m] or 40 [in] to less than 1.0 [mm] or 0.04 [in]. The main concern in that dimensional scale will be for the mechanical properties of the material. Machine component design of the *micro or nano* scale is not the scope of this text; see Senousy et al. (2008) as an example of micro scale. In that case and since the sizes were larger than 100 [nm], the procedures tolerated the use of engineering materials rather than quantum and wave mechanics.

7.1.1 Basic Elements of Material

The use of materials depends on the basic "atomic" *elements* involved in their composition. Previous background of chemistry and physics is essential. Material structure is composed of basic *material elements* that are more essential than others in regular machine component design. Categories and *symbols* of selected material elements interrelated to regular machine design are such as:

- *Common elements*: *Aluminum* (Al), *carbon* (*C*), *chromium* (Cr), *copper* (Cu), *iron* (Fe), *magnesium* (Mg), *nickel* (Ni), *titanium* (Ti), and *zinc* (Zn).
- *Alloying elements*: *Carbon* (C), *chromium* (Cr), *copper* (Cu), *iron* (Fe), *magnesium* (Mg), *manganese* (Mn), *molybdenum* (Mo), *nickel* (Ni), *phosphorous* (P), *silicon* (Si), *sulfur* (S), *tin* (Sn), *titanium* (Ti), *vanadium* (V), and *tungsten* (W). These elements are essential in producing *alloy steels* and other *alloys*. Some of alloying elements are present in the *common elements*, because they can be a major constituent in the named *alloys* such as *chromium, copper, magnesium, nickel, titanium, and zinc alloys*. In addition, they can be present in smaller quantities in other alloys such as *alloy steels, stainless steel, tool steels, and carbides*. Design of major machine components uses *carbon steels* and *alloy steels* extensively in addition to some other alloys.

- *Related affecting elements*: *Hydrogen* (H) and *oxygen* (O). The hydrogen element affects the material in an *embrittlement* action of hydrogen during manufacturing processes such as arc welding, electroplating, phosphating, or pickling. Oxygen, of course, is very important as an oxidizing element of most of the materials used in machine components.

Material structure is composed of the previous *material elements* and depends on the composition or alloying and the manufacturing processes. Usually metals start after extraction as casting ingots where alloying or composition control occurs. Manufacturing processes produce the desired plates, sheets, rods, or sections from ingots according to required standards. These standard material shapes are used as initial bodies to be formed, joined, or machined to produce the machine element. After the machine element is assembled with others into a design, it operates under load to perform the intended function. Good designs should not fail before their intended life if the design is good, and no overload anomaly occurs. This overloading condition may cause material failure depending on the loading type and the appropriateness of the design.

7.1.2 Material Failure Modes and Properties

Material *failure properties*, characteristics, modes, and limits are obtainable by standardized tests using standardized specimens. Some specific or standard machine elements are tested to define their load rating or failure limits such as threaded fasteners, couplings, seals, rolling bearings, belts, chains, wires, and ropes. However, usually the basic induced stresses are defined and accounted for, and the special characters of the compound element are taken into account.

Material testing is a field covered usually through the subjects of strength of materials or materials engineering. These subjects are usually prerequisite studies to the capstone courses of machine design. A short refreshing review gives a summary of some *failure modes* and their main associated material properties in Table 7.1. *Buckling* failure mode is a component failure affected greatly by material properties as discussed in Section 6.7.

As depicted in Table 7.1, the familiar properties in many prerequisite courses are the *yield strength*, *modulus of elasticity*, *ultimate strength*, and *hardness*. *Fatigue*, *creep*, *impact*, *brittle fracture*, and *corrosion* properties are subjects that may not receive sufficient details due additionally to their time dependency. These time-dependent properties and other characteristics may be of paramount significance in special machine design cases. In these cases, detailed considerations are in order. Major properties are revisited later; see Sections 7.1.3, 7.4, and 7.8. Mechanical properties of materials and their main failure modes, however, relate normally to their properties in *tensile* loadings; see dedicated references such as Callister and Rethwisch (2011) or Courtney (2000) and Ashby (1999).

Table 7.1 Some failure modes and their main associated material properties.

Failure mode	Material property	
Yielding	Yield strength	Shear yield strength
Buckling	Compressive yield strength	Modulus of elasticity
Fatigue	Endurance limit	Ultimate tensile strength
Contact fatigue	Compressive yield strength	Shear strength
Creep	Creep rate	
Brittle fracture	Impact energy, K_{1C}	Transition temperature
Corrosion	Electrochemical potential	Ultimate tensile strength
Wear	Hardness	Compressive strength

7.1.3 Tensile Properties

The *tension test* uses standard *test specimens* and qualified testing machines to define the behavior and several characteristics of a material. A typical round test specimen in Figure 7.2a shows essential standard dimensions. Different standards define the standard dimensions according to material type, utility, and loading. The set of some specimen *diameter* d_0 is 2.5, 6.25, 10, and 12.5 [mm] or 0.5 [in]. The set of some specimen *gauge length* l_0 is 10, 25, and 50 [mm] or 1 or 2 [in]. The two cylindrical ends are to be snuggly gripped by the testing machine. They can be machined to form a thread for testing brittle materials. A measuring tool (such as an extensometer) bridges between the two ends of the gauge length to measure the deformation. For more details about the *standard tensile test* and other forms of specimens such as flat ones, refer to ASTM E8/E8M-13 (2013) or ISO 6892-1 (2009) and ISO 6892-2 (2011). Other non-contact means of measuring deformations are endeavored; see e.g. Metwalli (1983, 1986), Ragab et al. (1986), Metwalli et al. (1983, 1987), and Ashry and Metwalli (2012).

Gripping the tensile test specimen in a material *testing machine*, the loading on the specimen starts at a defined standard rate until the *fracture* occurs. The *stress–strain* diagram in Figure 7.2b captures the response of a *ductile material* specimen to the applied load. The *elastic* segment of the diagram is exaggerated in the strain section of the line $(0 - \varepsilon_E)$ to distinctively show the different characteristics. In addition, it is better to distinguish between the *stress* and the *strength* by using σ for the *stress* and S for *strength* as a specific material property. The main properties and characteristics of the ductile material response are as follows (Figure 7.2b):

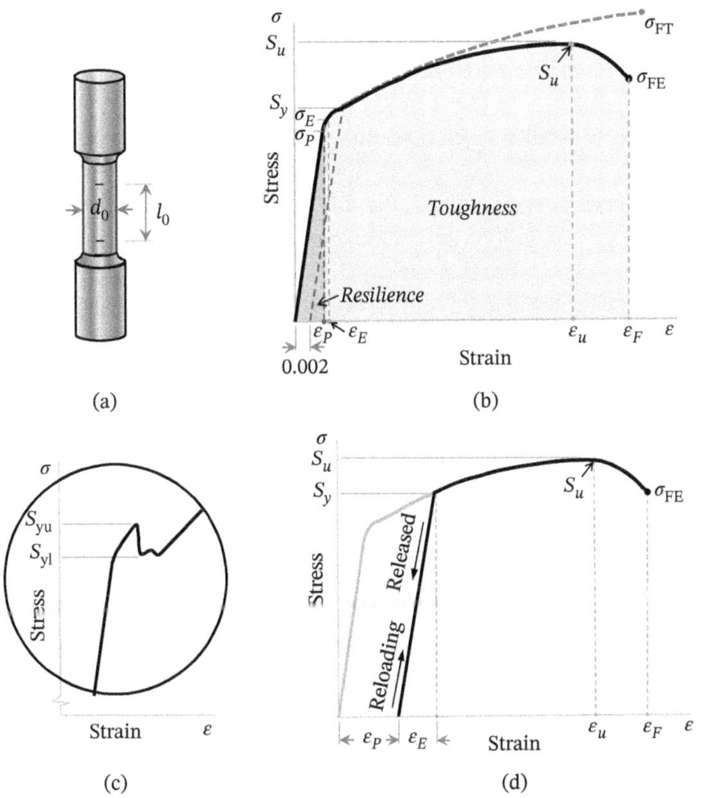

Figure 7.2 Tensile test: (a) essential standard dimensions of the test specimen, (b) stress–strain diagram of a representative ductile material, (c) upper yield strength point S_{yu} and lower yield strength point S_{yl} roughly magnified, and (d) plastic loading and reloading shifting the yield strength S_y to a higher strain hardening level.

- ***Proportional limit*** is the point defined by the proportional stress σ_P and characterizes the end of the linear portion of the diagram. The slope of the line is the *modulus of elasticity E* and *Hooke's law* applies up to that point. If the load is released at that point, the element should return to its original dimensions following the same straight loading line.
- ***Modulus of elasticity*** E is the slope of the line to the proportional limit or the tangent to the line if the line is not straight.
- ***Elastic limit*** is the point defined by the elastic stress σ_E and characterizes the end of the elastic portion of the diagram. The strain at this elastic stress is depicted as ε_E in Figure 7.2b. If the load is released at that point, however, the element should return to its original dimensions. Beyond the elastic limit the strain increases rapidly, and *permanent set* or permanent plastic deformation occurs.
- ***Yield strength*** S_y is the point at which *plastic deformation* starts. If it is difficult to identify, it is usually taken as the 0.002 offset strain as defined by the intersection with the material response of a parallel line offset by 0.002 (0.2%) of strain as shown by the dashed line in Figure 7.2b. This accounts for accepting a 0.2% of *permanent deformation* to the element for stating that the element has yielded. For many ductile materials the yield experiences an upper yield strength point S_{yu} and a lower yield strength point S_{yl} as the roughly and extremely magnified section in Figure 7.2c. The upper yield point is usually specified as the yield strength S_y.
- ***Ultimate tensile strength*** S_u is the maximum *engineering stress* on the σ–ε diagram after which *necking* in the specimen starts. The strain at this maximum stress is depicted as ε_u in Figure 7.2b. In many cases, one can use an alternative symbol of S_{ut} as an *ultimate* ***tensile*** *strength* to differentiate it from the *ultimate* ***compressive*** *strength* S_{uc}.
- ***Failure stress*** σ_{FE} is the terminal point on the *engineering stress–strain* (σ–ε) diagram at which *fracture* occurs. It is obvious that one does not consider it as strength. The strain at this failure stress is depicted as ε_F in Figure 7.2b.
- ***Resilience*** is the elastic energy stored in the element. It is the area under the line up to the elastic limit as defined by the grayish area in Figure 7.2b. The unit of resilience (or modulus of resilience) is then energy per unit volume, which comes to [Pa] or [psi].
- ***Toughness*** is the total energy imparted to the element by the load. It is the total area under the line up to the failure limit as defined by the light grayish area in Figure 7.2b. The unit of toughness (or modulus of toughness) is also energy per unit volume, which comes to [Pa] or [psi]. Tougher material is stronger and more ductile.
- ***Ductility*** is a property that indicates the ability of the material to be cold worked and absorb overloads or stand impact loads. Ductility can be expressed as the *percent elongation* or *reduction in area*. A high ductility or a large strain as in Figure 7.2b indicates a high ductility and thus a large ability of the material to be cold worked. A very small ductility indicates a large brittleness of the material.
- ***True failure stress*** σ_{FT} is the stress calculated using the actual smaller *fractured area* at the fractured section of the *necking*.
- ***Strain hardening*** occurs when the material is stressed beyond its yield strength and the load is then released. When the material is reloaded, it continues on the released line to the higher stress previously experienced. The yield point S_y is, therefore, the new maximum plastic stress previously attained as shown in Figure 7.2d. The permanent plastic strain ε_P is not recoverable, and the elastic strain ε_E is shifted and enlarged as depicted in Figure 7.2d. These are exaggerated to demonstrate the concept. The previous stress–strain curve is in grayish color. The new *stress–strain hardened* curve is the blackish color. If the load is higher than the new yield strength, a further plastic strain occurs and a newer stress–strain hardened curve develops with even higher yield strength.

For other no so ductile materials (or *brittle materials*) such as some of *cast iron* and very *high strength steels*, the properties and characteristics differ than the ductile materials. The distinction between the ultimate strength location and the failure point decreases, and the maximum failure strain gets closer to the proportional strain point. The ductility is much smaller, and the yield point may not be identifiable except for the assumed 0.2% offset rule, if achievable. The initial elastic loading line may not necessarily be straight.

The *engineering stress–strain* diagram is the solid black line, and the *true stress–strain* diagram is the grayish dashed line in Figure 7.2b from the yield point. The true *stress–strain* line may not be that far from the engineering *stress–strain* line up to the *ultimate tensile strength*. It is more distinguishable after the ultimate tensile strength due to *necking*. The ***engineering stress*** and ***engineering strain*** are defined with respect to the original dimensions such that

$$\sigma_i = \frac{\boldsymbol{F}_i}{A_0} \quad \text{and} \quad \varepsilon_i = \frac{l_i - l_0}{l_0} \tag{7.1}$$

where σ_i is the engineering stress at any current load $\boldsymbol{F}_i$ and A_0 is the original cross-sectional area. The strain ε_i is at any load $\boldsymbol{F}_i$ and deformed gauge length l_i, where l_0 is the original gauge length.

The ***true stress*** and ***true strain*** (or *logarithmic strain*) are defined with respect to the current dimensions such that

$$\sigma_{iT} = \frac{\boldsymbol{F}_i}{A_i} \quad \text{and} \quad \varepsilon_{iT} = \int_{l_0}^{l_i} \frac{dl}{l} = \ln \frac{l_i}{l_0} \tag{7.2}$$

where σ_{iT} is the true stress at any current load $\boldsymbol{F}_i$ and A_i is the current cross-sectional area. The true strain ε_{iT} is at any current load $\boldsymbol{F}_i$ and corresponding deformed gauge length l_i, where l_0 is the original gauge length. For more details, use dedicated materials references such as Callister and Rethwisch (2011).

The *true stress* and *true strain* concept is more important in the plastic zone. This area is useful in failure analysis and other scientific fields. In machine design, the expected synthesis is assumed to guard against any failure in function or components. That is why the plastic properties and characteristics are given a little coverage in this text as only to define constraints that should not be experienced or even approached in good designs. The *engineering stress* and *engineering strain* are thus the regularly adapted ones in most of the analysis, design, and synthesis of machine components.

The modulus of elasticity E is the tangent to the elastic straight line of the stress–strain diagram of Figure 7.2, i.e. it is the slope of this line. It is for the material obeying the ***Hooke's*** law as in Chapter 6 and reiterated theoretically as

$$\sigma_x = E\varepsilon_x, \quad \text{or} \quad E = \frac{\sigma_x}{\varepsilon_x} \tag{7.3}$$

Experimentally for the material obeying the ***Hooke's*** law, the modulus of elasticity E is then (Figure 7.2b) as

$$E = \frac{\sigma_x}{\varepsilon_x} = \frac{\sigma_P}{\varepsilon_P} \tag{7.4}$$

where σ_P is the stress at the proportional limit and ε_P is the strain at the proportional limit as shown in Figure 7.2b. This relation is dependent on a linear material response up to the proportional limit.

Ranges of basic *tensile property* spans of some materials indicate general trend as depicted in Figure 7.3. The lines show the spans of the *ultimate strength* and the *elastic modulus* for each category of materials. Properties can be off the line but within the span's limits. This general range for different material categories provides informative relative capacities of strength and stiffness. Materials of machine elements are usually defined as smaller subsets of some specific category, and may not even have the wide span of any category. These *material sets* are used to advantage in the synthesis of different machine elements. Section 7.5 provides such recognizable material sets for different machine elements. For the adaption of a *brittle material* rather than a *ductile material*, the *compressive strength* may be the deciding factor rather than the *tensile strength*. In such a case, the compressive strength is to be the major consideration in design.

For comparative purposes, the relevant and approximate strength ranges, modulus of elasticity, and specific gravity of some ***selected engineering materials*** are as follows (check ranges in Appendix A.7 for definite properties of particularly selected materials):

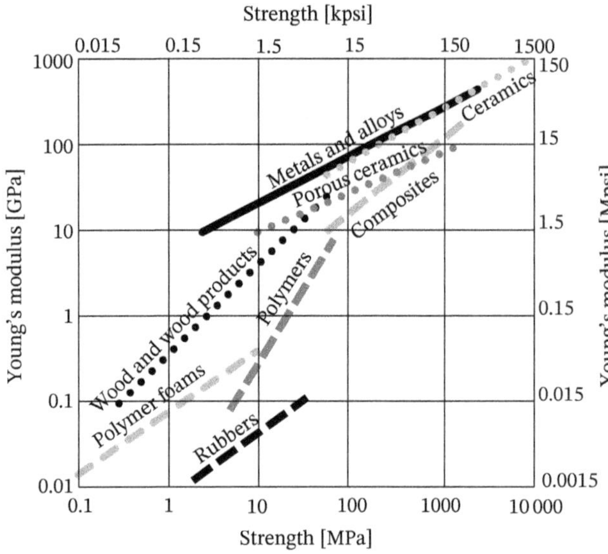

Figure 7.3 Spans of tensile properties of some general engineering materials. The lines show the extents of the strength and the elasticity modulus. Properties can be off the line but within the span's limits.

- *Gray cast iron*: The compressive strength (570–1290) [MPa] or (80–190) [kpsi] and the tensile strength (150–430) [MPa] or (22–63) [kpsi], modulus of elasticity (78–151) [GPa] or (11.3–21.9) [Mpsi], and specific gravity (6.95–7.35).
- *Ductile cast iron*: The yield strength (280–620) [MPa] or (40–90) [kpsi], modulus of elasticity (164–169) [GPa] or (23.8–24.5) [Mpsi], and specific gravity (7.10–7.30).
- *Plain and low-alloy steel*: The yield strength (165–2048) [MPa] or (24–297) [kpsi], modulus of elasticity (207) [GPa] or (30) [Mpsi], and specific gravity (7.83–7.87).
- *Stainless steels*: The yield strength (195–700) [MPa] or (28–102) [kpsi], modulus of elasticity (192–193) [GPa] or (27.6–28) [Mpsi], and specific gravity (7.53–8.00).
- *Aluminum alloys*: The yield strength (35–172) [MPa] or (5–27) [kpsi], modulus of elasticity (66–73) [GPa] or (9.6) [Mpsi], and specific gravity (2.71–2.8).
- *Copper alloys*: The yield strength (70–150) [MPa] or (10–22) [kpsi], modulus of elasticity (97–150) [GPa] or (14–22) [Mpsi], and specific gravity (8.25–8.96).
- *Magnesium alloys*: The yield strength (97–220) [MPa] or (14–32) [kpsi], modulus of elasticity (44–46) [GPa] or (6.4–6.7) [Mpsi], and specific gravity (1.76–1.84).
- *Titanium alloys*: The yield strength (220–1620) [MPa] or (32–235) [kpsi], modulus of elasticity (103–114) [GPa] or (15–16.5) [Mpsi], and specific gravity (4.43–4.71).
- *Polymers*: The yield strength (9–90) [MPa] or (1.3–132) [kpsi], modulus of elasticity (0.4–4.8) [GPa] or (0.06–0.7) [Mpsi], and specific gravity about (0.9–2.2).
- *Fibers*: The tensile strength (2500–6350) [MPa] or (360–920) [kpsi], modulus of elasticity (72–400) [GPa] or (10–58) [Mpsi], and specific gravity about (1.4–2.6).
- *Composites*: The longitudinal tensile strength (108–1380) [MPa] or (15.6–200) [kpsi], modulus of elasticity (45–220) [GPa] or (6.5–32) [Mpsi], and specific gravity about (1.4–2.1).

In the previous list, the tensile or compressive strength means the *ultimate* tensile or compressive strength. The low value of the range can be used for normal undemanding applications. The higher value of the strength range can be employed in demanding applications that requires or necessitates the upper attainable value. Some ranges are wide due to the extensive use of the alloying elements. Few of these ranges may even be extended due to high or special alloying or heat treating. For finer details of specific alloy or material in each category, one can consult later Sections 7.3 and 7.4, relevant standards, references, or handbooks such as ASM (1998). Other particular handbooks or texts books are useful such as Callister and Rethwisch (2011).

7.1.4 Other Static Properties

Loading type affects the categories of properties of material. Tension, compression, and torsion are loadings that might generate unique material properties for different materials. For ductile materials, compression properties are about the same as tensile properties. For brittle materials, however, compression properties may be different than tensile properties. Therefore, when the internal stresses are not predominantly in a tensile state, one should be careful particularly for brittle materials. Torsional behavior of material differs also from tensile properties particularly for yielding. Failure theories of Section 7.7 can shed some light on the difference. This is generally applicable for static loadings. For dynamic loadings, Section 7.8 is treating these cases. The following is a short assessment of some material properties under constant *static conditions* for different loading types:

Compression properties are similar to tensile properties, but compression test is more difficult to perform. If the specimen is relatively long, it may tend to buckle as indicated in Section 6.7. If the specimen is short, it can bulge as the end friction affects the behavior. In compression, even though properties are similar for ductile materials, the mode of failure is different from tension since there is no *necking*. Some *brittle materials* such as *cast iron* possess much higher compressive strength in the order of about 2–5 times their tensile strength.

Torsional and shear properties are relating and corresponding to the tensile properties as seen in Section 7.7 of the failure theories. However, conducting different tests finds the material properties under these different torsion or direct shear conditions. In torsion test, twisting standard specimens generates the *torque-twist* diagrams. The mechanism of deformation is different since the yield starts at the outer fiber while the inner fibers are still elastic. The yield progresses to the inner fibers until it reaches the central specimen axis, and then the entire cross section starts to yield. The recorded angular deformation defines *shear strain* as demonstrated in Section 6.3.4. The calculated *shear stress–shear strain* produces similar diagrams to Figure 7.2b for *ductile materials*. The *modulus of rupture* or *shear modulus of elasticity G* replaces the normal *modulus of elasticity E*. For *brittle materials*, failure occurs at a 45° that the reader can demonstrate by twisting a cylindrical piece of chalk until it breaks. For *direct shear properties*, however, similar tests to the direct shear in Figure 6.10c produce shear properties that can be useful in *direct shear* applications and *guillotine* manufacturing processes. Shear failure is revisited in Section 7.7.

Hardness properties are useful in wear applications. It is a property of how elements perform plastically under localized compressive loading. Performing hardness tests is through some familiar indentation hardness such as *Brinell (HB)*, *Vickers (HV)*, *Rockwell* (*HRB* or *HRC*, and *HRA* or *HRF*), and *Shore (HS)*, where values in parenthesis are for identification symbols of measured values. Other hardness tests exist such as *scratch* and *rebound* tests. The extent of the hardness can be greatly affected by heat treatment as viewed in Section 7.3. Hardness can also be correlated to ultimate strength as discussed in Section 7.6.

7.1.5 Other Time-Dependent Properties

When the material is loaded in a dynamic mode such as an impact for a short time or alternating in a deterministic or a random form, the material may behave differently. Specific properties are available to handle these time-dependent loading conditions. As depicted in Table 7.1, there are these other properties dedicated to *fatigue*, *creep*, *impact*, *corrosion*, and *wear*. A good synthesis of a design should consider these properties whenever the loading case matches.

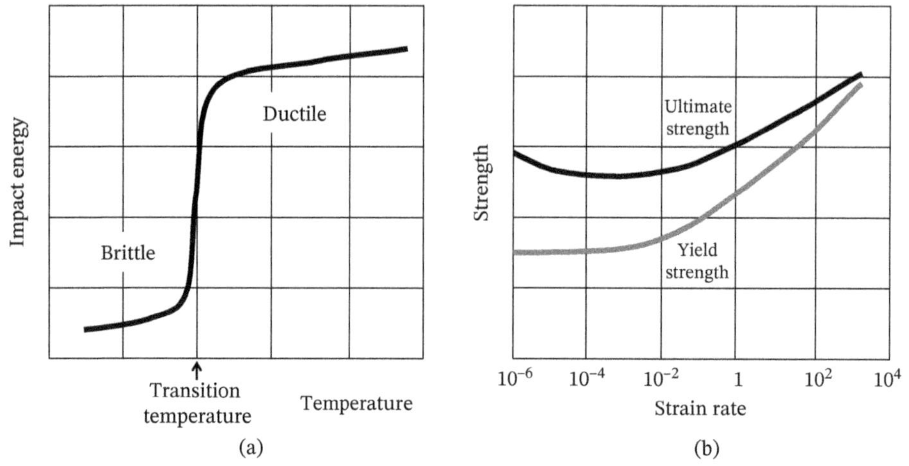

Figure 7.4 Some material properties under time varying load: (a) impact energy variation with temperature for ductile material and (b) strain rate effect on ductile material strength.

Impact characteristics are useful properties to design safeguarding against *low temperatures* under *dynamic shock loading*. The ***transition temperature*** of a material separates brittle behavior region from ductile behavior region (Figure 7.4a). For lower temperatures than the transition temperature, the material appears as *brittle*. The material is *ductile* for higher temperatures than the transition temperature. Around the transition temperature, the impact value increases sharply from the brittle to the ductile behavior. Usual tests are the *Charpy* and *Izod* impact outputs, which provide toughness information under dynamic loading. High impact *strain rates* cause the yield strength to approach ultimate strength for ductile materials such as in Figure 7.4b. Impact properties are essential in several applications in dynamically loaded machine elements with gaps or large clearances such as bearings, mating gears, dampers, rivets, shear blades, punches, dies, etc. or produced by drop forging, pressing, stamping, blanking, coining, etc.

Fatigue properties are of paramount importance in the design of machine elements subjected to repeated *dynamic loadings*. The usual test is using a standard rotating specimen loaded by pure bending to produce internal normal stresses due to fully *reversed cyclic loading*. At so many different loading stresses, recording the number of cycles to fatigue produces sufficient data to plot the average S_f–N_C diagram shown in Figure 7.5 (usually named *S*–*N* diagram). The ordinate axis of the *fatigue strength* S_f is logarithmic as well as the abscissa axis of the *number of cycles* N_C. *Ductile* materials usually have constant fatigue failure strength between 10^6 and 10^7 cycles and beyond as observed in Figure 7.5. This constant limit of the fatigue failure strength is denoted by the ***endurance limit*** S_e; see Section 7.8. The *low-cycle fatigue* range is between 10^0 and 10^3 cycles as observed in Figure 7.5. The *high-cycle fatigue* range is beyond 10^3 cycles. If the number of the loading cycles through the life of the component is between 10^3 and 10^6 cycles (such as the dashed arrow), one can obtain the *fatigue strength* S_f (or *brittle material fatigue strength* S_{fb}) by interpolation as indicated in Figure 7.5 or by a fitting equation as in Section 7.8. The ordinate values of the curves in Figure 7.5, however, are only representative of some illustrative ductile or brittle materials and just to demonstrate the concepts.

Fracture toughness characteristics are useful properties to design safeguarding against ***critical cracks*** existence, initiation, and propagation. The *brittle fracture* and *fracture toughness* are properties that get some coverage in Section 7.9. One of the main fracture toughness parameters is the ***stress intensity factor*** K_{1C}, which the subject of *fracture mechanics* covers in Section 7.9. It considers the instability of *cracks* as inducing sudden *brittle fracture* as the case of instability of *columns* in *buckling*.

Creep properties are valuable properties to design safeguarding against *high temperatures* and *excessive loading* for a *long time*. This can happen for turbine blades in power stations or gas turbines. Figure 7.6 shows general cases of *creep strain* ε_C versus operating time t_O at elevated operating temperatures T_O and elevated operating stresses

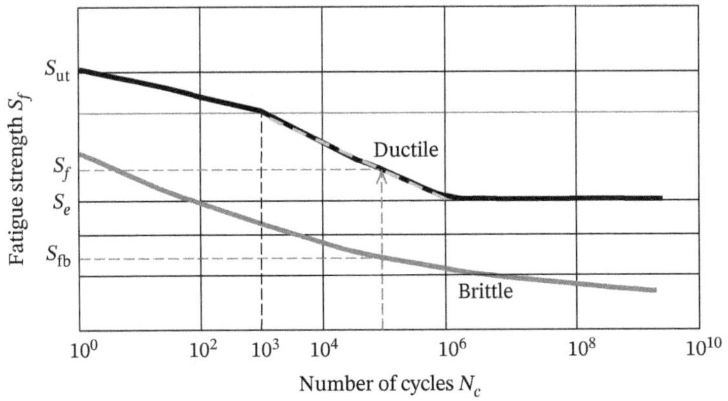

Figure 7.5 S–N (stress–number of cycles) diagram of fatigue strength versus the number of loading cycles for a representative of a ductile material and one of the brittle materials.

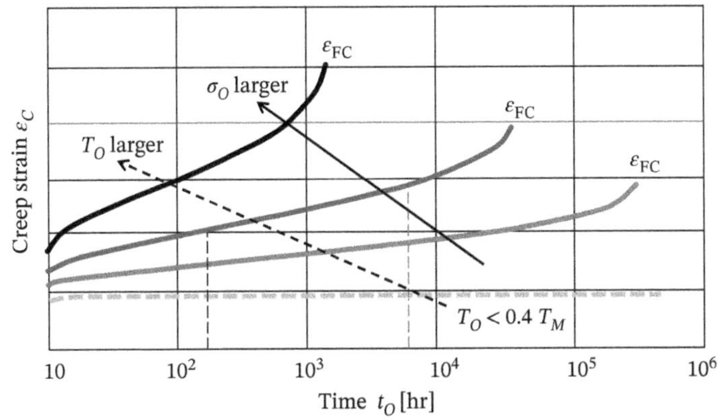

Figure 7.6 Diagram of creep rate versus time at different temperatures and high stress loading. For a middle range of the curves, the creep strain rate is constant as shown for the middle curve between the dashed lines amid operating time of 10^2 and 10^4 [hr].

σ_O (Callister and Rethwisch 2011). For a middle range of the curves, the ***creep strain rate*** is constant as shown for the middle curve between the dashed lines amid operating time of 10^2 and 10^4 hours [hr]. At longer time, the rate increases until the creep failure strain ε_{FC}. When the operating temperature T_O is less than 0.4 of the melting temperature T_M, the material does not undergo the creep phenomenon as depicted by the grayish dashed line in Figure 7.6. The values on the curves, however, are only representative of some illustrative materials prone to creep and just to demonstrate the concepts. For a specific material under prospective creep and elevated temperature, the material behavior should be obtained by a standard test such as ASTM Standard E139 (2003) or ISO 6892-2 (2011). Creep properties are essential in several applications in machine elements such as turbine blades, bolts, screws, springs, boiler vessels, etc.

Corrosion and wear characteristics should be of utmost consideration if the design is subjected to harsh *corrosive* environmental conditions, *friction*, or *abrasive* contacts. Some further discussions are accessible in Sections 7.2 and 7.3. For more information on corrosion and degradation of materials, refer to Callister and Rethwisch (2011). A well-known corrosion-resisting material is the *stainless steel*. Wear properties are essential in several machine elements such as clutches, breaks, chains, gears, crank shafts, wires, power screws, seals, spline shafts, universal joints, cams, camshafts, piston rings, pins, valves, cylinders for engines, piston pins, hay rake teeth, turbine blades, impellers, etc. Some of these are exposed to corrosive environments in addition.

7.2 Numbering Systems and Designations

Several numbering systems are present to designate materials particularly steels among others for so many standards of organizations or international bodies or countries such as the *US* ASTM, AISI, SAE, UNS, the international ISO, the European EN, the German DIN, the British BS, the French AFNOR, the Russian GOST, the Japanese JIS, etc., just to name a few. Here, the focus is on metallic materials inherently or traditionally more suitable for machine components such as *steel*, *alloy steel*, *cast iron*, and *aluminum alloys*. Other materials such as *copper*, *brass*, *bronze*, *magnesium alloys*, etc. are also considered. The *nickel alloys*, *titanium alloys*, and *zinc alloys* are useful and may have coverage in some enclosed applications. The focus is also on the standards, which are widely accessible and applicable. The standards, however, are evolving with the changes in the world. The United States has been trying to switch to the worldwide SI units, but the addressees are also trying to keep the familiar units; see ASM (1998). The emerging European Union has enticed many European nations to conform to or persuade the European standards *EN* to follow, such as the *DIN* of Germany. For world communities, they need an international standard for trade and globalization such as the ISO standards. The manufacturers also have their traditions and the need to abide by their global customer's needs. The evolving of technology also caused some standard organizations to abandon some of its standards for newer ones. The users, however, have been used to some legacy information, and it is not easy to switch to newer ones. The standards presented here are trying to bridge that with adherence to familiar standards. The needy designer, as always, has to keep up with the chase. The equivalent material data may be useful in that regard such as the updates of ASM (2006) or Bringas (2004).

To search for a suitable material for a design, one has the options of defining the *mechanical properties*, the *chemical composition*, or some *physical properties*. This is particularly necessary for *optimum selection* or synthesis of the material such as the work of Abdel Meguid (1999). Several standards embed some of these properties in their standard numbering designations. In cases where composition is very small relative to the major element, usually the main mechanical property is used, namely, the *minimum yield strength*. For alloys, the chemical composition may be preferable. For insufficient distinction, the standards may just use letters and numbers. The letters are for major (or *common*) elements, and the numbers for other variations.

For the *US* organizations concerned about standards for steel, the AISI and SAE numbering systems are very similar. The SAE and ASTM developed a Unified Numbering System (UNS) for metals and alloys. The UNS numbering system uses a prefix letter to identify the materials including steels. The rest of the numbers are four similar or identical to the AISI/SAE with an additional fifth number as zero or others for special considerations. The *prefix letters* are as follows:

- F for *cast irons*, G for *carbon steel* or *low-alloy steels*, J for *cast steels*, K for *special* or *heat-resisting steels*, S for *stainless steels*, and T for *tool steels*.
- A for *aluminum*, C for *copper base alloys*, N for *nickel alloys*, and Z for *zinc alloys*.
- M for miscellaneous *nonferrous metals* such as *magnesium*, *manganese*, and *silicon* alloys.

Few other prefix letters are in use for other materials that may rarely be used in regular machine component design; see ASM (1993). For the UNS G prefix for carbon or low-alloy steels, the four preceding numbers are the same as for AISI/SAE. The prefix letter in the UNS is usually having five preceding numbers in the range (00001–99999). These standards are in broad use by this text for the US material selection.

7.2.1 Carbon and Alloy Steels

The AISI/SAE major steel designation is defined in Figure 7.7a with two sample steels in Figure 7.7b. The first position of the number X in Figure 7.7a is denoting the major alloying element from 1 to 9 as indicated. The number 1 is important, since it is for the widely used *plain carbon steel*. Other numbers from 2 to 9 are for the major

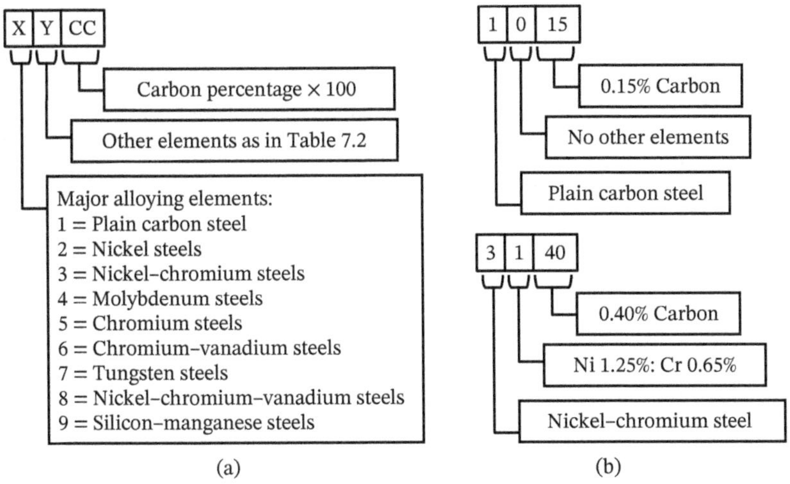

Figure 7.7 Selected AISI/SAE low-alloy steel designations: (a) major X and other alloying elements Y with carbon contents multiplied by 100 as CC and (b) two sample steels indicating alloy steel type and composition.

alloying elements stated in Figure 7.7a. The second position of the number Y in Figure 7.7a is in dedication for the other *low-alloying* elements in the composition, which partly defines the major alloying elements specifications and composition as the description in Table 7.2; see, e.g. ASM (1998). The third and fourth position of the number CC in Figure 7.7a is denoting the *carbon percentage* multiplied by 100. The examples in Figure 7.7b identify AISI 1015 steel and AISI 3140 steel. The AISI 1015 indicates that the first digit of the number, i.e. *one* states that it is a *plain carbon steel*; see Table 7.2. The second digit of the number, i.e. *zero*, states that there is no other alloying element. The last two digits of the number, i.e. 15, state that the material has a content of 0.15% carbon. The AISI 3140 indicates that the first digit of the number, which is *three*, states that it is a nickel–chromium *low-alloy steel*; see Table 7.2. The second digit of the number, i.e. *one*, states that the nickel–chromium alloying elements have a composition of 1.25% Ni and 0.65% Cr. The last two digits of the number, i.e. 40, state that the material has a content of 0.40% carbon.

Plain carbon steel and *low-alloy steel* designations for other numbering systems such as the *European standard* EN or the *German standards* DIN are similar or comparable to AISI/SAE. The EN and DIN standards, however, use two basic groups. The first group uses a *prefix letter* followed by usually the *minimum yield strength* in [MPa] for *plain carbon steel* and optionally trailed by other characteristics as in Figure 7.8a, upper part. The *prefix letter* is empty or C for carbon steel and may be S for structural steel or others; see EN 10027-1 (2005). The second group uses a *prefix letter* followed by the *percent carbon* content multiplied by 100, trailed by the composition of *alloying elements* and their *percentages* as in Figure 7.8a, lower part. For compositions less than 5%, the alloying elements percentages are multiplied by a factor to get round numbers; see EN 10027-1 (2005). There may also be additional trailing symbols or numbers for added information such as applications or manufacturing; see EN 10027-1 (2005) for more details. The examples in Figure 7.8b identify S355M steel and X10CrNi 18-8 steel according to EN 10027. The S355M steel indicates that the first letter of the designation, i.e. S states that it is a *structural steel*. The following digits of the number, i.e. 355 state that the yield strength is 355 [MPa]. The last letter of the designation, i.e. M, states that the material production is by mechanical rolling. Other *plain carbon steels* can have a simpler designation such as C35 for plain carbon steel with 0.35% carbon. The X10CrNi 18-8 indicates that the first letter of the designation, which is X, states that it is an *alloy steel*; see Figure 7.8b, lower part. The following number, i.e. 10, states that the material has a content of 0.10% carbon. The following symbols and numbers of the designation, i.e. CrNi18-8, state that the alloying elements are chromium and nickel having a composition of 18% Cr and 8% Ni.

Table 7.2 Selected AISI/SAE low-alloy steel designations and the associated material specifications and composition where the numbers after the element is its percentage with the balance as ferrite.

AISI steel	Designation	Specifications (numbers are element (%))
Carbon steel	10xx, 11xx, 12xx, 15xx	(Plain carbon steel, Mn 1.00 max), (resulfurized free cutting), (resulfurized–rephosphorized free cutting), (carbon steel, Mn 1.00–1.65)
Manganese steel	13xx	(Mn 1.75)
Nickel steel	23xx, 25xx	(Ni 3.50), (Ni 5.00)
Nickel chromium steel	31xx, 32xx, 33xx, 34xx	(Ni 1.25, Cr 0.65-0.80), (Ni 1.75, Cr 1.07), (Ni 3.50, Cr 1.50–1.57), (Ni 3.00, Cr 0.77)
Molybdenum steel	40xx, 44xx	(Mo 0.20–0.25), (Mo 0.40–0.52)
Chromium molybdenum steel	41xx	(Cr 0.50–0.95, Mo 0.12–0.30)
Nickel chromium molybdenum	43xx, 47xx	(Ni 1.82, Cr 0.50–0.80, Mo 0.25), (Ni 1.05, Cr 0.45, Mo 0.20–0.35)
Nickel molybdenum steel	46xx, 48xx	(Ni 0.85–1.82, Mo 0.20–0.25), (Ni 3.50, Mo 0.25)
Chromium steel	50xx, 51xx, 50xxx, 51xxx, 52xxx	(Cr 0.27–0.65), (Cr 0.80–1.05), (Cr 0.50, C 1.00 min), (Cr 1.02, C 1.00 min), (Cr 1.45, C 1.00 min)
Chromium vanadium steel	61xx	(Cr 0.60–0.95, V 0.10–0.15)
Tungsten chromium steel	72xx	(W 1.75, Cr 0.75)
Nickel chromium molybdenum	81xx, 86xx, 87xx, 88xx	(Ni 0.30, Cr 0.40, Mo 0.12), (Ni 0.55, Cr 0.50, Mo 0.20), (Ni 0.55, Cr 0.50, Mo 0.25), (Ni 0.55 Cr 0.50 Mo 0.35)
Silicon manganese steel	92xx	(Si 1.40–2.00, Mn 0.65–0.85 Cr 0.65)
Nickel chromium molybdenum	93xx, 94xx, 97xx, 98xx	(Ni 3.25, Cr 1.20, Mo 0.12), (Ni 0.45, Cr 0.40, Mo 0.12), (Ni 0.55, Cr 0.20, Mo 0.20), (Ni 1.00, Cr 0.80, Mo 0.25)

7.2.2 Aluminum and Aluminum Alloys

Aluminum and *aluminum alloys* are extensively in use for applications that require low specific gravity (2.7 versus 7.9 for steel), high thermal or electrical conductivity, and a decent specific strength (strength/density). They possess relatively high corrosion resistance in ordinary environment, but they have lower melting temperatures than steels.

The UNS numbering system for aluminum depends on the production process and the alloying elements. After the letter A for *aluminum* or *aluminum* alloys, the UNS numbering system is having five numbers. The *first number* is indicating the production method, which is ***9*** for a *wrought aluminum* or a ***0*** for a *cast aluminum alloys*. Other numbers may define casting methods such as die casting, permanent mold or sand casting, etc.; see ASM (1998) for more depiction. The *second number* defines the *principal alloying element*. The number for each *principal alloying element* is ***1*** for aluminum of at least 99.00%, ***2*** for *copper* alloys, ***3*** for *manganese* alloys, ***4*** for *silicon* alloys, ***5*** for *magnesium* alloys, ***6*** for *magnesium–silicon* alloys, and ***7*** for *zinc* alloys. The *third number* is to indicate a modification of original alloying element or defining limits (ASM 1998). The *fourth* and *fifth* numbers indicate other additional alloying elements.

The international standards ISO specifies the composition of the alloying elements in its designation, where the percentage is adjusted to the nearest 0.5% for the *main alloying element* and rounded off to the nearest 0.1% for the

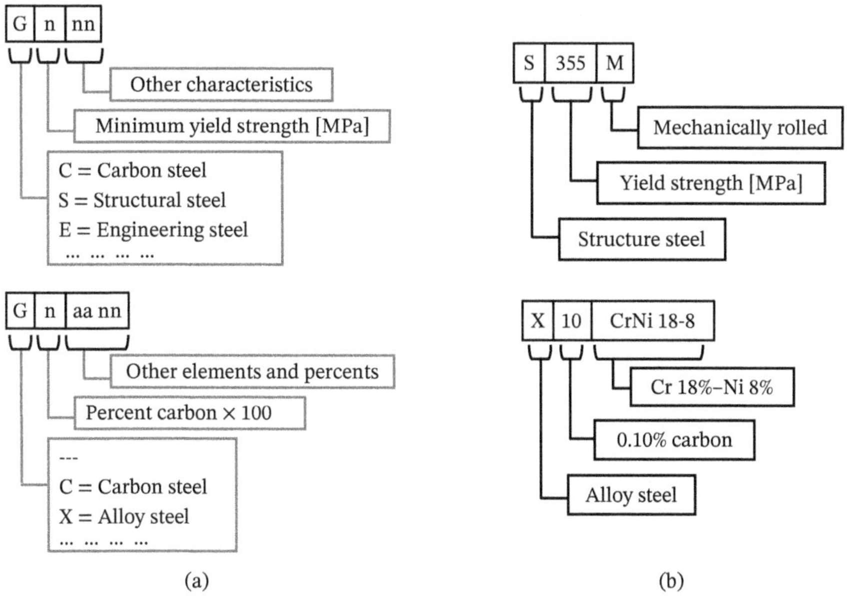

Figure 7.8 Selected EN alloy steel designations: (a) the top is for plain carbon steel and the bottom is for alloy steel and (b) two sample steels indicating steel type and alloy composition.

secondary alloying elements; see ISO 209-1 (1989). The numbers of alloying elements are restricted to four. As an example, the ISO designation for *aluminum alloy* Al Si1Mg0.8 indicates that the aluminum alloy has about 1.0% *silicon* and about 0.8% *magnesium*. The equivalent number for this ISO *aluminum alloy* Al Si1Mg0.8 is the UNS A96181. The *Aluminum Association* (AA) equivalent number is the *aluminum alloy* AA6181; see also ASM (1998).

7.2.3 Other Alloys

Numbering systems for several alloys are available, and some are previously stated in this Section 7.2. This section covers two widely used alloys, namely, copper and magnesium alloys. The other alloys infrequently used in this text are such as *zinc*, *titanium*, and *nickel* alloys. These and other alloys are for the reader to find more details off references such as ASM Handbook (1998).

7.2.3.1 Copper and Copper Alloys

Copper is one of the oldest metals in history. It is the first metal to be alloyed with other metal, tin (Sn), to produce bronze about 3500 BC. After steel and aluminum, copper and its alloys rank third in commercial metal consumption. They have excellent thermal and electrical conductivities, in addition to good corrosion resistance, good manufacturability, and relatively high strength; see Section 7.1.3. They have ample applications from wires, tubes for heat exchangers, valves, etc. to the especially distinctive components such as *self-lubricated bearings* or *impregnated bearings*.

The UNS numbering system identifies copper and copper alloys by the prefix letter "C" preceding five-digit numbers. UNS designation codes are then C00001–C99999. The codes C10000–C79999 indicates wrought alloys, and the codes C80000–C99999 are denoting cast alloys. The first number after the C letter is as follows: ***1*** for up to 96% coppers, ***2*** for brasses, ***3*** for leaded brasses, ***4*** for tin brasses, ***5*** for phosphor bronzes, ***6*** for aluminum and silicon bronzes, and ***7*** for copper-nickels. The following four numbers allude to indicate different alloy composition and other information; see ASM (1998).

The ISO designation for copper and copper alloys is using the element composition system (ISO 1190-1 1982). The ISO designation uses element symbols in descending percentage order after the main copper symbol (Cu). The ISO copper alloy CuZn38Pb2 has 60% Cu, 38% Zn, and 2% Pb. This designation, however, ignores whether the alloy metal is wrought or cast. Cross-reference between the ISO and other designations is available in references such as ASM (1998).

7.2.3.2 Magnesium and Magnesium Alloys

Magnesium is about the lightest commercial material with a specific gravity of 1.8. This is in comparison to the aluminum 2.7 specific gravity and 7.9 for steel. Its alloys competitive *specific strength* (strength/density) makes it very attractive for aviation and similar industries. They cannot, however, put up with high temperatures. Their strength is also low; see Section 7.1.3.

The designation of *magnesium alloys* are not well standardized but usually characterized by two prefix letters trailed by two numbers. The two letters denote the major alloying elements such as A for *aluminum*, M for *manganese*, S for *Silicon*, or Z for *Zinc*. After the two letters, the two numbers designate the approximate percentage of each alloying element. The magnesium alloy AZ63 has weight percentage of about 6 of aluminum and about 3 of zinc. More defined composition values are available in references such as ASTM B275 (2013). The equivalent UNS alloy to another AZ61 alloy is M11610, which is similar to the ASTM number in percentage values. The M11 in the UNS number would stand for magnesium, aluminum, and zinc.

7.3 Heat Treatment and Alloying Elements

Heat treatments and alloying are major strengthening mechanisms of materials used in machine elements. They can produce parts that have any of or combinations of higher strengths, higher toughness, higher hardness, or other improved properties. Other strengthening and hardening mechanisms of work hardening or strain hardening are accessible such as *hot rolling* or *cold rolling*, *drawing* and surface *shot peening*, or other similar plastic pre-stressing methods. Each of these manufacturing processes changes or improves some material properties of mostly ductile materials. The selection of the raw material of a machine element as originally cast, hot rolled (HR), or cold drawn (CD) affects the allowable strength value. A widespread practice is the utility of *heat treatment* and *alloying* to alter material properties to a better state of strength and utility. This practice is understandably more costly. The trade-off is then by optimizing cost or specific cost (cost/weight) of the part material and its manufacturing processes.

Heat treatment is very important in strengthening and hardening of *plain carbon steel*. Heat treatment can also affect the surface and thus used for *surface hardening*. Heat treatment is improving effectively and extensively the strength, ductility, and hardness of other *alloys*. It is critical is some applications, which defines the usable and suitable set of materials in those particular applications. Here, the text procedures focus on *steels*, but the main heat treatment processes are similar for other alloys.

7.3.1 Heat Treatment

Material heat treatment mainly includes *annealing*, *normalizing*, *case or surface hardening*, *precipitation strengthening*, *quenching*, and *tempering Q&T*. These typical topics are having some coverage in a detailed summary in this section particularly aimed at cast iron and steel. Heat treatments of different alloys, however, are typically having ample coverage in prerequisite materials courses and other related references such as Callister and Rethwisch (2011) or ASM (1998), and DIN EN 10083 (2006).

During the original production of a crude machine element and before machining, a sort of a heat treatment is in effect. The cooling rates in the ***casting*** or the ***forming*** processes affect the characteristics of the produced part.

Gray cast iron results when the cooling rate is relatively slow, which allows the high percent carbon (>1.7 and usually >2.0%) to be in the form of *pearlite*, thin flakes *graphite*, and *ferrite*. The material is brittle. Its compressive strength is three to four times the tensile strength (Section 7.1.3). When the cooling rate of cast iron is high, the *white cast iron* occurs due to the formation of a large fraction of *cementite* (iron-carbide). White cast iron is very brittle, wear resisting, and difficult to machine. A very slow *annealing* of the white cast iron (up to six days for large parts) produces the *malleable cast iron* or *nodular cast iron*. This process results in the formation of *nodular graphite* known as *temper carbon*. Malleable cast iron is relatively ductile with large elongation up to about 18%. *Ductile* and *nodular cast irons* also spawn by alloying introduction of magnesium and cesium.

Forming processes of hot working, cold working, or drawing such as hot rolling, heading, cold rolling, spinning, roll threading, stamping, extrusion, and forging produce parts and sections under the plastic deformation and strain hardening process. These affects the ensued properties of the original material used. Hot working produces properties depending on the cooling rates during or after forming. The material *grains* can be finer and thus stronger. On the other hand, the material *grains* have a deformed shape subsequent to the cold forming processes. The strength is usually higher, and the ductility is usually lower.

During the manufacture processing of a machine element, the nature of the internal microstructure *grain size* and its *composition* determines the general mechanical behavior of the element. The grain size and composition are functions of the *alloying elements* and the *cooling rate* in addition to the heating and cooling cycle or cycles. Therefore, the initial selection of the *alloy* affects the mechanical properties of the element by the heat treatment. The heat treatment of each alloy is particular to the specific alloy so that to deliver the intended property of the machine element. Therefore, care should be exercised with regard to a particular alloy, its original properties, and treatment procedures to get the ensued characteristics by consulting specific details of composition.

Carbon steel is widely used for many machine elements. The carbon content greatly affects the steel strength and hardness, particularly when heat treated. The heat treatment depends on the carbon percentage in the *plain carbon steel*, its composition, as depicted in the *phase diagram* of Figure 7.9, and the cooling rate scenarios. The outcome of the heat treatment is contingent upon the rate of cooling at a carbon percent and from the starting high temperature defining the initial composition as characterized in Figure 7.9. Generally, steels containing 0.05–0.3% carbon are *low-carbon steel*. Steels containing about 0.3–0.8% carbon are *medium-carbon steel*. Other steels containing about 0.8–2.0% are *high-carbon steel*. The main heat treatment processes of carbon steels are *annealing*, *normalizing*, or *quenching* and *tempering Q&T*. Other processes such as *spheroidizing*, *martempering* or *marquenching*, and *austempering* are applicable; however, the focus here is on the main treatment processes due to their major and general pronounced effects and utility.

Annealing is a stress-relieving process. The process starts at a temperature above the phase lines shown roughly by the grayish area in Figure 7.9 with a range of about 40 [°C] or 70 [°F]. The temperature is fixed at that level for enough time (about one hour) to guarantee the full transformation of the material to *austenite* or *austenite and cementite* as indicated in Figure 7.9. The cooling of the part is then so slow such as cooling in the air (or in the oven) to produce coarse or thick bands of *pearlite* formations. The part is thus ductile and soft, which is suitable for forming or cutting processes. *Process annealing* is for *low-carbon* steels of less than 0.3% carbon and for a temperature less than the *eutectic* temperature (727 [°C] or 1341 [°F] at about 0.76% carbon) as defined in Figure 7.9.

Normalizing is a process to set the material in a uniform microstructure. The process is similar to annealing but with initial heating to about 55 [°C] or 100 [°F] higher temperatures than the phase lines as roughly shown by the light grayish area in Figure 7.9. The cooling of the part is slow as to produce more uniform *fine pearlite* microstructure. This results in a higher strength and ductility than annealing.

Quenching *Q* is a process of producing a very high strength steel. The result is, however, a brittle steel of a very diminutive ductility. The process is not usually suitable to *low-carbon* steels. The carbon content is usually at least 0.4%. The process starts at temperatures similar to normalizing and kept at that level for enough time to guarantee the transformation of the material to *austenite* as indicated in Figure 7.9. The cooling of the part is then

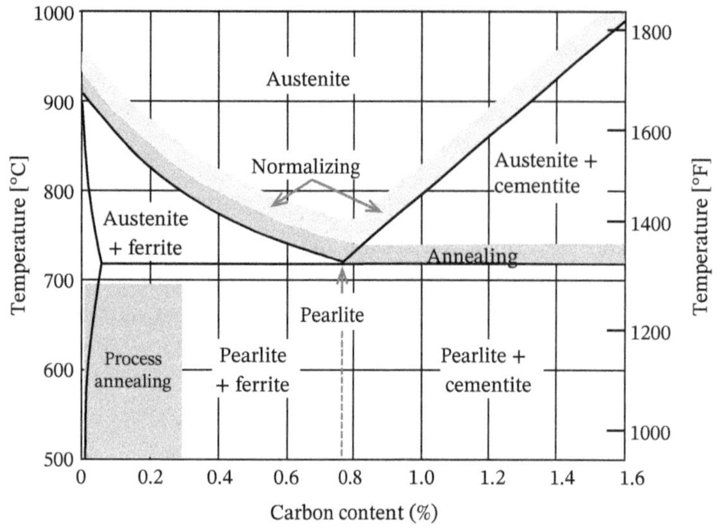

Figure 7.9 Iron–carbon phase diagram for plain carbon steel. The heat treatment zones of annealing, process annealing, and normalizing are roughly defined by the grayish areas.

swift such as dipping (or *quenching*) in *water*, *brine*, or *oil*. Water is the fastest quenching that gives the highest strength and brittleness, particularly at the surface. The fast quenching process traps the carbon within the iron to generate a *martensitic* microstructure possessing very high *internal stresses*. This gives the material its very high tensile strength (three to four times the normalized steel). The high internal stresses, however, may cause surface cracking, and therefore a post treatment is greatly advisable. This post treatment is the tempering process after quenching.

Tempering process is performed after quenching or *Q&T* to remedy (or temper) the extreme brittleness caused by quenching. The process starts at a *tempering temperatures* lower than the *eutectic* level and kept at that level for some time to guarantee the tempering process. The tempering temperature and time of cooling affect the reduction of brittleness and strength with gains to ductility. These can be tailored to achieve the desired level of *strength*, *hardness*, and *ductility* that suits the intended application.

7.3.2 Case Hardening

Part of case (or surface) hardening process is performed on steels as a heat treatment to impart specific hardness to the surface. Other case hardening techniques are also available to other materials and alloys to harden the surface. The aim is to treat the outer surface to attain certain hardness and/or wear resistance while keeping the core ductile and tough. The thickness of the outer surface layer or "case" is a function of machine component size. The smaller the component, the thinner is the surface layer or "case."

Flame or induction hardening heats the surface – of the carbon steel part – to the normalizing temperature and then followed by water quenching to harden the surface. These processes generate a very hard and wear-resisting *martensitic* microstructure layer on the surface. The depth of this surface layer or "case" is a function of time and part size. The carbon content, however, should be in the range of about 0.4–0.6%. If the material does not have this carbon content, a *carburization* is necessary.

Carburizing is a process of diffusing carbon into the surface layer of low-carbon steel (about 0.1–0.3% carbon) to increase the carbon percentage for a suitable heat treatment of the surface. The *diffusion* process involves exposing the part to a rich carbon environment (gas, liquid, or solid such as charcoal or carbon monoxide) under high set temperature and extended set time. The depth of the surface layer is dependent on temperature and time. Layer depth can be in the range of about 1–6 [mm] or 0.04–0.25 [in]. At the high temperature, austenite forms and then transforms to *martensitic* microstructure due to subsequent *quenching*. This creates a very hard layer or "case."

Nitriding, cyaniding, or carbonitriding are different surface treatment processes that each generates a different hard layer or case. Each would heat the part to a suitable temperature in a specific chemical environment. *Nitriding* exposes the part to an ammonia environment at about 480–620 [°C] or 900–1150 [°F] for a length of time needed to develop the required *nitride* case depth. To form hard nitrides, elements such as chromium, molybdenum, and aluminum ought to be present. However, no post heat treatment is necessary. *Cyaniding* immerses the part into a sodium cyanide bath at about 870–950 [°C] or 1600–1750 [°F] and then quenched to develop a very hard thin case of about 0.25–0.75 [mm] or 0.01–0.03 [in] deep. However, the part should be carefully washed to get rid of the toxic cyanide. The cyaniding process takes much less time (about 20–30 minutes) than carburizing (numerous hours). *Carbonitriding* subjects the part to ammonia and hydrocarbon environment at about 850 [°C] or 1550 [°F] for a short time needed to diffuse carbon and nitrogen required to form the case depth. The small layer depth can be in the range of 0.07–0.5 [mm] or 0.003–0.02 [in]. Quenching can then be in oil or gas for a harder undistorted surface.

Other alloy steels that need heat treatments benfit from alloying since quenching of carbon steel does not reach the inside for large machine elements. Alloying elements remedy these situations and prevents distortions in large elements. Procedures of heat treatments of alloy steels are similar to carbon steel.

7.3.3 Effect of Alloying Elements

The effect of *carbon* on the steel had a focused attention in the previous Sections 7.2.1, 7.3.1, and 7.3.2. This section gives a short focused attention to the other main alloying elements affecting steel properties. Alloys usually contain more than one alloying element, and thus the properties are not precisely extrapolated. Specific alloy steels should have properties in about the same trend as the effect of each alloying element. Specific values of the properties are available, and their utility and suitable precedent applications are the paramount target. Comparing properties and effects may lead to the identification of each alloying element responsibility and reasons for a particular property (ASM 1998). However, the general effects of each alloying element are as follows (Table 7.2):

Manganese (Mn) helps in hot working, machinability, and hardenability. In quantities more than 12%, it greatly increases wear resistance particularly for impact stressing. Steels containing more than 19% are suitable for low-temperature applications.

Nickel (Ni) improves impact strength, toughness, and corrosion resistance. Nickel stimulates resistance to reducing chemical and oxidizing attacks. It eases heat treatment of alloy steels. Nickel also enhances strength and toughness at low temperatures.

Chromium (Cr) enhances resistance to corrosion, chemical reactions, and oxidation. It improves hardenability, high-temperature strength, and abrasion resistance for high-carbon contents. Chromium forms carbides and is causing hardening properties. In existence with nickel, they provide higher hardenability, impact, and fatigue strength. In quantities more than 13%, it greatly increases corrosion resistance.

Molybdenum (Mo) maintains hardenability and increases tensile strength and creep resistance at high temperatures. It retards temper brittleness and promotes fine-grain structure. It forms carbides and reduces vulnerability to pitting.

Tungsten (W) enhances hardness, encourages fine-grain formation, and greatly improves heat resistance. It forms very hard and stable tungsten carbide that is valuable in high-temperature applications. Its employment in high-speed tooling is therefore extensive.

Silicon (Si) is a deoxidizer. It slightly improves strength and wear resistance with little reduction of ductility. Silicon also raises the elastic limit. In large percentage, it reduces scaling and acid attacks but hard to process.

7.4 Material Propertied and General Applications

In this section, considerations are only for some relevant properties of some suitable engineering materials applicable to machine design. Not all properties of all available materials are possibly important or significant to mechanical components. The span of *tensile strength* and *stiffness* (in the form of the *modulus of elasticity* or *Young's* modulus) of selected materials is show in Figure 7.3. The approximate strength ranges and some properties of ***selected engineering materials*** are also listed in Section 7.1.3. However, these properties are not coupled to their *standard designations* in previous citations to be accessibly adapted by designers. Here, the intension is to give some *standard designation* coverage to some properties of potentially applicable engineering materials to machine component design. For expanded and more extensive coverage, appropriate standards and relevant handbooks can provide more data and wider exposure; see standards and handbooks in References section.

The quoted *material properties* in this text are frequently the *estimated minimum* values for strength and average values for other properties. Few selected values are adapting trends consistent with previously developed relations. For a design to be conforming to a norm, material *specifications* should adhere to the standards of the intended market of the product. This should include the definition of the designation, composition, and properties of these characteristics either as defined or verified by accepted tests to conform. Material properties can thus deviate from values defined herein. The estimation of the *safety factor* and *reliability* should accommodate such indefinite limited variations.

7.4.1 Cast Iron

Cast iron is an iron alloy containing more than 2% carbon. It is usually the utilized material for applications in *machine frames, cylinder blocks, gearbox cases, cylinder heads, housings, brake drums, clutch plates, flywheels, pistons, valves, cylinder liners*, high strength castings, and even some *camshafts*. This list of selected applications progressively uses *gray cast iron* tensile strength of about 150–400 [MPa] or 20–60 [kpsi] and compressive strength of about 570–1290 [MPa] or 80–185 [kpsi]. Lower strength values are for the start of the application list, and highest strength values are for the last part of the application list. The designations and representative mechanical properties are in Table A.7.1, Appendix A.7. The designations cite the minimum value of the tensile strength, ASTM A48 or ISO 185 (1988). One should note that *gray cast iron* is a *brittle material* (with diminished ductility), and one should not expose it to shocks. For applications subjected to such a dynamic loading, *ductile or nodular cast iron* is the material to use (Table A.7.1, Appendix A.7). These materials are ductile with suitable ductility. As the strength gets higher, the ductility becomes lower. The ASTM designation (ASTM A536) cites the minimum value of the tensile strength, minimum value of the yield strength, and the ductility (elongation percent) with a dash between each. The ISO designation (ISO 1083) cites the minimum value of the tensile strength and the ductility with a dash between each as shown in Table A.7.1, Appendix A.7. Applications of the *ductile or nodular cast iron* are such as *fittings* and *valves* for steam or chemicals, components subject to shocks, even *crankshafts, gears, pistons, rollers*, and *slides*. *Malleable cast iron* is another type of cast iron that is useful in some thin thickness applications requiring strength, tolerance to elevated temperatures, and dynamic loading similar to *ductile iron*. The material is alloyed, heat treated, and more costly than ductile iron. If needed, the properties are obtainable from handbooks or standards such as ASM (1998) or ASTM A47, A220, and A897, and ISO S922 (1981).

Mechanical applications for each class of cast iron are relating to the following practical purposes. *Gray cast iron* is usually the preferred material for brake disks and drums, machine bases, engine blocks, and some gears. *Malleable cast iron* is mainly the selection material for railroad rolling stock, construction and farm machinery, and some heavy-duty bearings. *Nodular iron* is the choice for some heavy-duty gears, rollers, and some crankshafts. White cast iron is more appropriate for railroad brake shoes, shot-blasting nozzles, crushers, and ball mill liners.

Table 7.3 General uses of plain carbon steels as a function of carbon percentage.

Carbon (%)	Usual machine element usage	Manufacturing
0.05–0.1	Rivets, wires, wire nails, cold-drawn parts, tubing, sheets, strips, etc.	Stamping, deep drawing, free-cutting machining
0.1–0.2	Machine parts, structural shapes, screws, case-hardened parts, etc.	Rolling, carbonization, machining
0.2–0.3	Gears, shafts, camshafts forgings, levers, plates, tubing, structural profiles and bars, etc.	Forging, carbonization, machining
0.3–0.4	Shafts, axels, gears, connecting rods, hooks, etc.	Forging, tubing, heat treatment, machining
0.4–0.5	Gears, crankshafts, large forgings, etc.	Forging, heat treatment, machining
0.5–0.6	Gears, shafts, screws, spring wires, etc.	Forging, heat treatment, machining
0.6–0.7	Springs, lock washers, setscrews, rails, train tires, etc.	Hard-drawing, heat treatment, machining
0.7–0.9	Leaf springs, plows, shovels, hand tools, shear blades, chisels, hammers, band saws, etc.	Tooling, heat treatment
0.9–1.2	Springs, knives, drills, taps, reamers, cutters, dies, etc.	Tooling, heat treatment
1.2–1.4	Files, knives, razors, saws, drawing dies, etc.	Tooling, heat treatment

7.4.2 Plain and Low-Alloyed Carbon Steels

Plain and low-alloyed carbon steel is a widespread type of materials that is extensively used in the material selection of numerous or almost the majority of machine elements of this text. *Plain carbon steel* has mainly carbon as the alloying element with percentages about 0.05–0.8% carbon or little higher to 0.95% carbon. It is about the range of *low-carbon* to *medium-carbon* steels. A general usage of plain carbon steels is for a wide variety of machine elements and applications such as some of those pointed out in Table 7.3 with manufacturing processes indicated. For higher carbon contents of about 0.9–1.4%, many tools are made of *high-carbon steel*, and in many cases, *low-alloyed carbon steels* (or *tool steels*) are preferable. To be manufactured, many machine elements are usually employing plain carbon steel as HR or CD manufactured in basic cross sections such as cylindrical bars. The subsequent forming or machining generates the necessary geometry of many machine components such as shafts from the hot or cold rolled bars. If high strength is necessary, either higher carbon content or CD bars may be the alternatives. The other alternative is to use high-alloy steel. Chemical compositions of plain and low-alloy carbon steels are available in standards such as ASTM A29 (2005). Other comparative standards for composition and properties are also available in handbooks such as ASM (1998) and Bringas (2004).

Mechanical application samples for each class of carbon steel are relating to the following practical purposes. *Carbon and low-alloy steels* are typically useful for heavy-duty gear blanks, railroad wheels, and die blocks. *High-alloy steels* are the choice for applications such as rock crusher jaws, pump and valve components, and gas turbine housings.

7.4.2.1 Hot Rolled and Cold Drawn Plain-Carbon Steels

Table A.7.2 in Appendix A.7 defines typical plain carbon steel properties for HR and CD compiled from different sources; see references such as ASM (1998), ISO 683 (1987), and EN 10083 (2006). Figure 7.10 shows the variation of selected *plain carbon steel* properties as functions of the percentage of carbon contents. The black lines are for the HR plain carbon steels. The gray solid lines are for the CD steels. Dotted lines are for reported values of the ultimate tensile strength S_u and the yield strength S_y. The dashed line is for the estimation of the endurance limit S_e; see Section 7.8. These selected properties are the main ones since they are of major concerns in the design

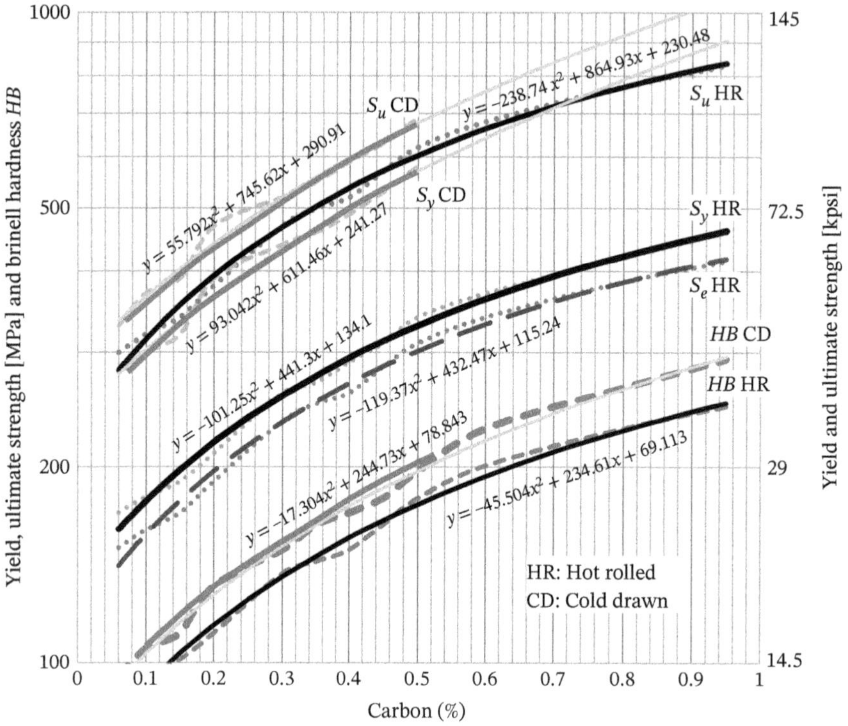

Figure 7.10 Selected plain carbon steel properties as functions of carbon contents. The black lines are for the hot rolled (HR) steels. The gray solid lines are for the cold drawn (CD) steels. Dotted lines are for reported values, and long dashed line is for endurance limit estimation.

process. The selection also highlights their proportions and relations with respect to the percentage of carbon content. As observed from Figure 7.10, the CD materials have higher yield strength S_y than the HR materials of the same carbon content. The CD yield strength S_y is also very close to the CD ultimate tensile strength S_u, which is a sign of more brittleness and less ductility. Carbon steels are also not CD beyond 0.5% carbon. The indicated trend relations are for the SI system, which help in approximating the required material for a certainly needed strength (Figure 7.10). The relations use the variable y to represent the property cited in the left ordinate and the variable x for the percentage of the carbon content cited in the abscissa of Figure 7.10. The symbol of the percentage of carbon content C_C is then replacing the variable x, and the different strength symbols of S_u, S_y, HB, and S_e are replacing the variable y. The trend relations for the *HR* plain carbon steels in Figure 7.10 are then approximately having the following equations for SI and the US systems:

$$\begin{aligned} S_u &= -238.74\, C_C^2 + 864.93\, C_C + 230.48 \text{ [MPa]} \\ S_u &= 0.145\left(-238.74\, C_C^2 + 864.93\, C_C + 230.48\right) \text{ [kpsi]} \end{aligned} \tag{7.5}$$

$$\begin{aligned} S_y &= -101.25\, C_C^2 + 441.3\, C_C + 134.1 \text{ [MPa]} \\ S_y &= 0.145\left(-101.25\, C_C^2 + 441.3\, C_C + 134.1\right) \text{ [kpsi]} \\ HB &= -45.504\, C_C^2 + 234.61\, C_C + 69.113 \end{aligned} \tag{7.6}$$

$$\begin{aligned} S_e &= -119.37\, C_C^2 + 432.47\, C_C + 115.24 \text{ [MPa]} \\ S_e &= 0.145\left(-119.37\, C_C^2 + 432.47\, C_C + 115.24\right) \text{ [kpsi]} \end{aligned} \tag{7.7}$$

These estimated equations are valid for the carbon content of 0.06–0.95%. They are useful in computer-aided selection and optimization of material. The other trend relations for the CD plain carbon steels in Figure 7.10 are then approximately having the following equations for SI and the US systems:

$$S_u = 55.792\,C_C^2 + 745.62\,C_C + 290.91\ [\text{MPa}]$$
$$S_u = 0.145\left(55.792\,C_C^2 + 745.62\,C_C + 290.91\right)\ [\text{kpsi}] \tag{7.8}$$

$$S_y = 93.042\,C_C^2 + 611.46\,C_C + 241.27\ [\text{MPa}]$$
$$S_y = 0.145\left(93.042\,C_C^2 + 611.46\,C_C + 241.27\right)\ [\text{kpsi}]$$
$$HB = -17.304\,C_C^2 + 244.73\,C_C + 78.843 \tag{7.9}$$

These estimated equations are valid for the carbon content of 0.1% to a maximum of 0.5%. The values for CD steels are higher than for HR steels due to the strain hardening associated with the cold drawing processing. No trend relation for the endurance limit is present for the CD steels. This is because the material is a little more brittle, and the assumed estimation may be in error due to the strain hardening effects. As observed from Figure 7.10, the CD material has higher yield strength S_y than the HR material of the same carbon content. The yield strength S_y is also very close to the ultimate tensile strength S_u, which is a sign of more brittleness and less ductility as shown in Table A.7.2 in Appendix A.7. Carbon steels are also not CD beyond 0.5% carbon.

7.4.2.2 Strength and Hardness of Annealed and Normalized Plain Carbon Steels

Figure 7.11 demonstrates the effect of carbon percent on the basic properties of plain carbon steels. These are, namely, the *yield strength* S_y, the *ultimate tensile strength* S_u, and the *Brinell hardness HB* for *annealed* and *normalized* heat treatments. The dashed lines are for smooth plotting of the values reported in Table A.7.2 in Appendix A.7. The plots noticeably indicate scattered values. This is also due to introducing an extra material of AISI 1080 hoping to improve the fit, but it was even more scattered than the original values in Table A.7.2 in Appendix A.7; see ASM (1998). Therefore, one must be cautious in adapting any information found in handbooks or other sources

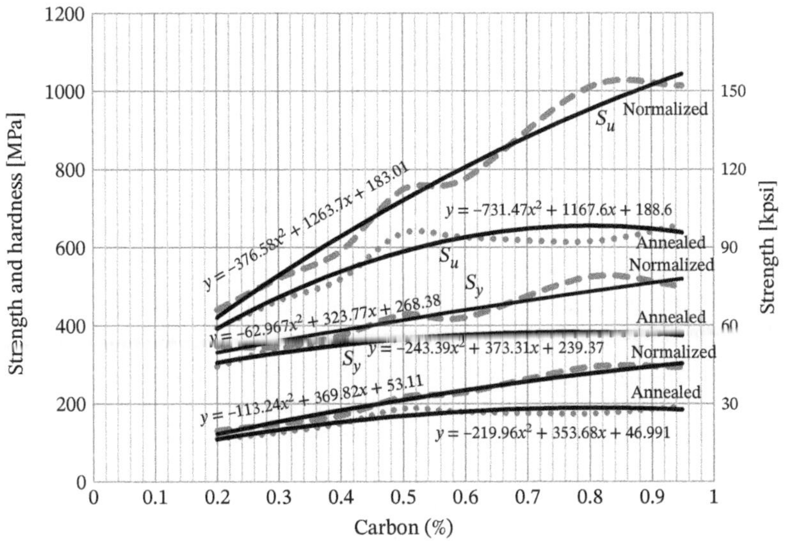

Figure 7.11 The effect of carbon percent on the ultimate strength, yield strength, and *Brinell* hardness of annealed and normalized carbon steel. Dashed grayish lines are for smooth connections of reported values. Solid black lines are for trend estimations of properties. Note that extra data values for 0.8% carbon are present in the grayish lines.

without checking. One must be willing to verify and to exert efforts to validate. The scale on the right-hand side of Figure 7.11 is for the approximate position of the US units of the adjacent strength label. The trend lines in Figure 7.11 present approximations to the scattered data assumingly the properties are reasonably behaving. To some extent, the *annealed HB* values are comparable to *HR* values, and the *normalized HB* values are also comparable to the *CD* values. The indicated trend relations of the *Brinell* hardness in Figure 7.11 are for both SI and the US systems. The others are *only* for the SI system of units. These relations should help in approximating the required material for a certainly needed hardness or carbon content. The trend relations use the variable y to represent the property cited on the ordinate axis and the variable x for the percentage of the carbon content cited in the abscissa of Figure 7.11. The symbol of the percentage of carbon content C_C is again replacing the variable x in the trend line equations. The *yield strength* S_y, the *ultimate tensile strength* S_u, and the *Brinell hardness HB number* are replacing the variable y. The trend relations for the *annealed* plain carbon steels in Figure 7.11 are then approximately having the following equation for SI in [MPa]:

$$\begin{aligned} S_u &= -731.47\ C_C^2 + 1167.6\ C_C + 188.6 \\ S_y &= -243.39\ C_C^2 + 373.31\ C_C + 239.37 \\ HB &= -219.96\ C_C^2 + 353.68\ C_C + 46.991 \end{aligned} \tag{7.10}$$

To get the value of the US system in [kpsi], multiply the right-hand side (in Eq. (7.10)) by 0.145 for either the *ultimate tensile strength* S_u or the *yield strength* S_y only.

The trend relations for the *normalized* plain carbon steels in Figure 7.11 are also approximately having the following equation for SI in [MPa]:

$$\begin{aligned} S_u &= -376.58\ C_C^2 + 1263.7\ C_C + 183.01 \\ S_y &= -62.967\ C_C^2 + 323.77\ C_C + 268.38 \\ HB &= -113.24\ C_C^2 + 369.82\ C_C + 53.11 \end{aligned} \tag{7.11}$$

To get its value in the US system in [kpsi], multiply the right-hand side (in Eq. (7.11)) by 0.145 for either the *ultimate tensile strength* S_u or the *yield strength* S_y only.

Equations (7.10) and (7.11) are applicable approximations for the carbon content in the range of 0.20–0.95%. If the strength and hardness information of the 0.8% plain carbon steel is absent, different equations than Eqs. (7.10) and (7.11) should materialize.

7.4.2.3 Quenched and Tempered Plain Carbon Steels

Figure 7.12 displays the effect of carbon percent on the yield strength, ultimate strength, and *Brinell* hardness number of water quenched and tempered (at 205 [°C] or 400 [°F]) carbon steel. The black lines are for the *SI* system. Dotted lines are for the reported SI values of the ultimate tensile strength S_u and the yield strength S_y. The *Brinell* hardness values are for the steels as quenched and tempered to a low temperature to express about the maximum useful hardness for both *SI* and the *US* systems. The grayish solid lines are for the *US* system values identified through the right ordinate scale. The grayish dashed lines are for the reported *US* values of the ultimate tensile strength S_u and the yield strength S_y. The lowest tempering temperature is adapted since it gives the highest attainable strength and a somewhat acceptable low ductility. The dotted and dashed lines are smooth plotting of the values presented in Juvinall (1983) after Bethlehem (1971) or Bethlehem (1980). The indicated trend relations in Figure 7.12 are for either *SI* or the *US* systems as marked. These should help in approximating the required material for a certainly needed maximum value of strength or hardness. The trend relations use the variable y to represent the property cited on the ordinate axis and the variable x for the percentage of the carbon content cited in the abscissa of Figure 7.12. The symbol of the percentage of carbon content C_C is again replacing the variable x, and the different strength symbols of S_u, S_y, and the *Brinell hardness HB* are replacing the variable y. The trend

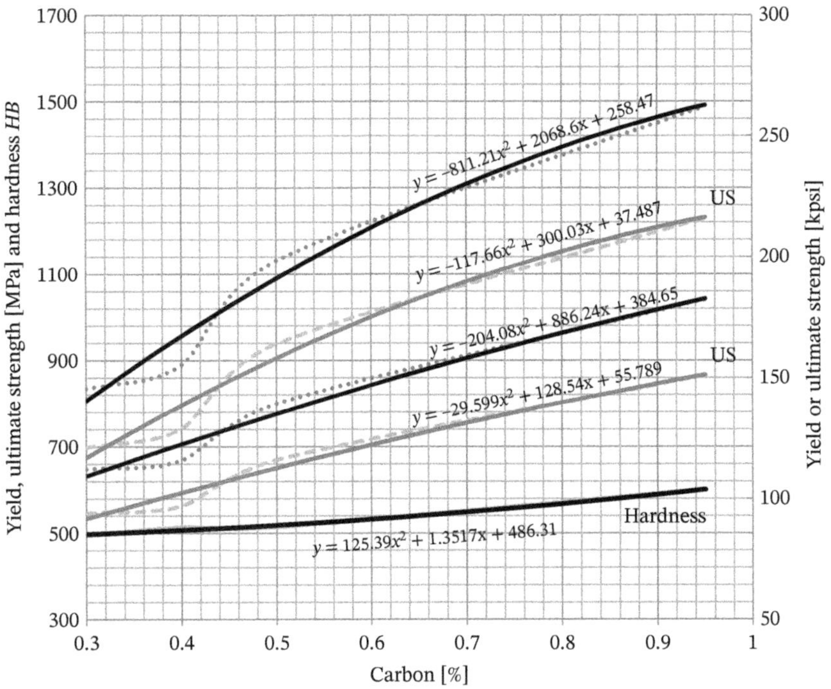

Figure 7.12 The effect of carbon percent on the yield strength, ultimate strength, and *Brinell* hardness of water quenched and tempered (to 205 [°C] or 400 [°F]) carbon steel. The *Brinell* hardness values are for the steel as quenched. The grayish solid lines are for the US system identified through the right ordinate scale.

relations for the *water quenched and tempered* plain carbon steels in Figure 7.12 are then approximately having the following equation for both SI and the US systems:

$$S_u = -811.21\, C_C^2 + 2068.6\, C_C + 258.47\ [\text{MPa}]$$
$$S_u = -117.66\, C_C^2 + 300.03\, C_C + 37.487\ [\text{kpsi}] \tag{7.12}$$

$$S_y = -204.08\, C_C^2 + 886.24\, C_C + 384.65\ [\text{MPa}]$$
$$S_y = -29.599\, C_C^2 + 128.54\, C_C + 55.789\ [\text{kpsi}] \tag{7.13}$$

$$HB = 125.39\, C_C^2 + 1.3517\, C_C + 486.31 \tag{7.14}$$

These estimated equations are only valid for the carbon content of the range from 0.3% to a maximum of 0.95%. The values for steels of higher carbon content are higher than for steels of lower carbon content.

Figure 7.13 best demonstrates the effect of the tempering temperature on the properties of *heat treated plain carbon steel*. Plots in Figure 7.13a are for AISI/SAE 1040 or (ISO C40) as an example. Plots in Figure 7.13b are for AISI/SAE 1095 or (DIN {Ck101}) as the other example. Tensile strength, yield strength, and hardness are reduced from high values to lower values, as the tempering temperature gets higher. The US temperature scale is on the top axis of the plot. The plots are mainly for the SI units just to demonstrate the effect. The US values are quoted in parenthesis beneath or above the SI units. Ductility (and toughness) behaves in an opposite manner, and the plotted elongation percentage is identifiable through the right ordinate scale. Considering the scales in Figure 7.13a,b, the effect of the tempering temperature on the properties of the higher carbon content of steel AISI 1095 is more

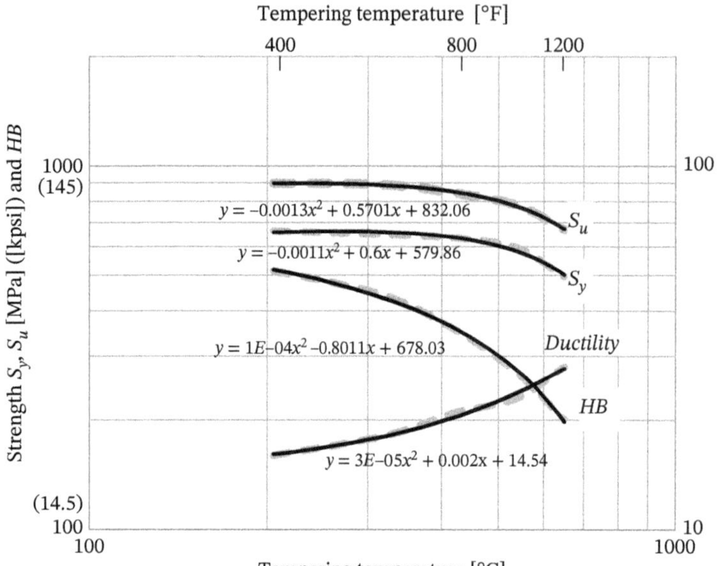

(a)

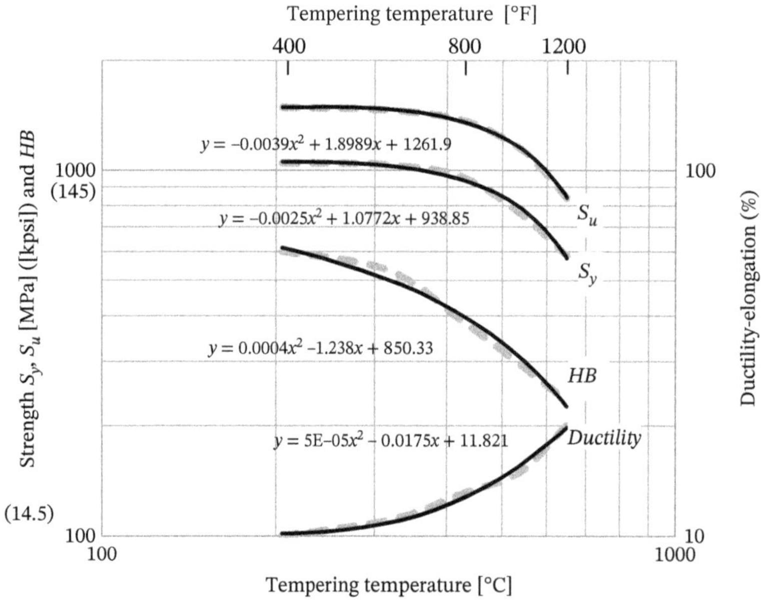

(b)

Figure 7.13 The effect of tempering temperatures on heat treated plain carbon steel properties. Tensile strength, yield strength, and hardness are decreased from high values to lower values as the tempering temperature gets higher. Ductility (and toughness) behaves in an opposite manner. (a) Plots are for AISI/SAE 1040 or (ISO C40) as an example. (b) Plots are for AISI/SAE 1095 or (EN (C98D2), DIN {Ck101}) as an example.

than for the lower carbon content of the steel AISI 1040. The quenched and tempered properties get closer to the normalized steel properties as the tempering temperature gets higher.

Table A.7.3 in Appendix A.7 represents the properties of selected *heat treated plain carbon steels*, which are quenched and tempered at 205 [°C] or (400) [°F]. Designation similarity between standards is realizable after careful assessment of individual case in terms of composition and properties. Given similarity or property is only a selected close correspondence. Cross-check and scrutiny are therefore necessary. Quoted properties belong to the AISI/SAE (UNS) designations. Unless noted otherwise, the quenching is in oil. The specified treatment of letter Q is oil quenched unless starred as WQ&T, which is water quenched WQ and then tempered. Values are compiled from different sources such as ASM (1993, 1998) and ASTM A1011 (2007). Oil quenching typically gives lower values for strength than water quenching.

To compare the quenched and tempered Q&T properties to the annealed steel properties, Figure 7.14 combines the effect of carbon percent on the *yield strength* and *Brinell* hardness of both annealed carbon steel and water quenched and tempered (to 205 [°C] or 400 [°F]) carbon steel. The yield strength is the main choice due to its importance to design. The lower three lines are for annealed carbon steels. The upper *Brinell* hardness line is for the steel as water quenched and tempered to 205 [°C] or 400 [°F]. The grayish solid lines are for the US system identified through the right ordinate scale. This demonstrates the increase in the strength due to quenching and tempering. Ductility is, however, lower as depicted earlier in Figure 7.13 and through comparing the data in Table A.7.3 in Appendix A.7. Trend line equations in Figure 7.14 are the same as defined in Figure 7.12 for the quenched and tempered steels. The extrapolated dashed lines from 0.3% to 0.1% carbon are only for comparison. No quenching and tempering are expected below 0.3% carbon. The lower three trend line equations in Figure 7.14 for the annealed steel are close to those in Figure 7.10, but, here, the carbon percentage range is for 0.1–0.95% carbon rather than 0.06–0.95% carbon.

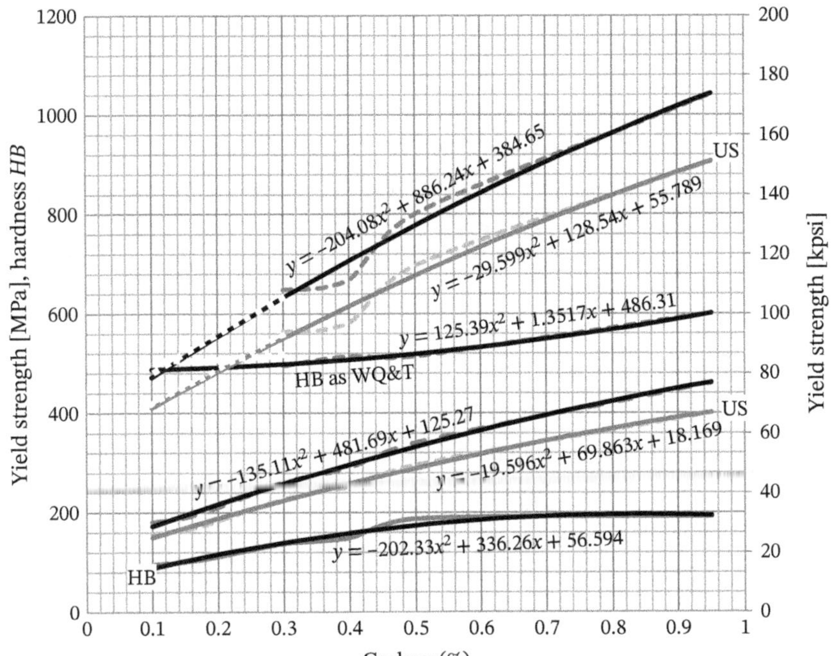

Figure 7.14 The effect of carbon percent on the yield strength and *Brinell* hardness of both annealed carbon steel and water quenched and tempered (to 205 [°C] or 400 [°F]) carbon steel. The lower three lines are for annealed carbon steels. The upper *Brinell* hardness line is for the steel as quenched. The grayish solid lines are for the US system identified through the right ordinate scale.

Example 7.1 A *mass spectrometer* testing the composition of a machine element indicated the carbon content is around 0.5%. Other elements exist, but they are within the admissible ranges of plain carbon steel. If the measurement of hardness indicates a value of about *HB* 195, what is the expected heat treatment of the steel? Estimate the mechanical properties of the steel, and compare the values to the AISI 1050 or the ISO C50. If the material is to be water quenched and tempered to 205 [°C] or (400) [°F], what are the expected properties? Compare results with available information about the heat treated plain carbon steels.

Solution

Data: $C_C = 0.5$, $HB = 195$ [rpm], material AISI 1050 or the ISO C50. In addition, water quenched and tempered.

For the *HR* plain carbon steels, the hardness in *HB* has estimation according to Eq. (7.10) to give

$$HB = -45.504\,(0.5)^2 + 234.61\ (0.5) + 69.113 = 175.042 \tag{a}$$

For the *CD* plain carbon steels, the hardness in HB has an evaluation according to Eq. (7.10) to give

$$HB = -17.304\,(0.5)^2 + 244.73\ (0.5) + 78.843 = 196.882 \tag{b}$$

From Eqs. (a) and (b), it is obvious that the material is CD. The properties are then definable by the applications of Eqs. (7.8) and (7.9). This gives the following results:

$$S_u = 55.792\,(0.5)^2 + 745.62\ (0.5) + 290.91 = 677.67\ [\text{MPa}]$$
$$S_u = 0.145\,(677.668) = 98.262\ [\text{kpsi}] \tag{c}$$

$$S_y = 93.042\,(0.5)^2 + 611.46\ (0.5) + 241.27 = 570.26\ [\text{MPa}]$$
$$S_y = 0.145\,(570.26) = 82.688\ [\text{kpsi}] \tag{d}$$

Comparing these values with the properties of AISI 1050 or the ISO C50 in Table A.7.2 in Appendix A.7, one gets the following:

- Calculated $HB = 196.882$ and the tabulated $HB = 197$.
- Calculated $S_u = 677.67$ [MPa] or 98.262 [kpsi] and the tabulated $S_u = 690$ [MPa] or 100 [kpsi].
- Calculated $S_y = 570.26$ [MPa] or 82.688 [kpsi] and the tabulated $S_y = 580$ [MPa] or 84 [kpsi].

These values are different by a maximum error of less than 2%, which is close enough for engineering and design calculations.

For water quenched and tempered steel to 205 [°C] or (400) [°F], the expected properties have estimations according to Eqs. (7.12)–(7.14). The estimated values are as follows:

$$S_u = -811.21\,(0.5)^2 + 2068.6\ (0.5) + 258.47 = 1090\ [\text{MPa}]$$
$$S_u = -117.66\,(0.5)^2 + 300.03\ (0.5) + 37.487 = 158.09\ [\text{kpsi}] \tag{e}$$

$$S_y = -204.08\,(0.5)^2 + 886.24\ (0.5) + 384.65 = 776.75\ [\text{MPa}]$$
$$S_y = -29.599\,(0.5)^2 + 128.54\ (0.5) + 55.789 = 112.66\ [\text{kpsi}] \tag{f}$$

$$HB = 125.39\,(0.5)^2 + 1.3517\,(0.5)^2 + 486.31 = 518 \tag{g}$$

Comparing these values with the properties of AISI 1050 or the ISO C50 in Table A.7.3 in Appendix A.7, one gets the following:

- Calculated $S_u = 1090$ [MPa] or 158.09 [kpsi] and the tabulated $S_u = 1120$ [MPa] or 163 [kpsi].
- Calculated $S_y = 776.75$ [MPa] or 112.66 [kpsi] and the tabulated $S_y = 807$ [MPa] or 117 [kpsi].
- Calculated $HB = 518$ and the tabulated $HB = 514$.

These values are different by a maximum error of less than 4%, which is still close enough for engineering and design calculations.

7.4.2.4 Quenched and Tempered Low-Alloy Steels

Table A.7.4 in Appendix A.7 lists properties of selected heat treated low-alloy steel for oil quenched and tempered at 205 [°C] or (400) [°F]. Designation similarity between standards is again realizable after careful assessment of individual case in terms of composition and properties. Given similarity or property is only a selected close correspondence. Cross-check and scrutiny are therefore necessary. The properties of the quenched and tempered, the normalized, and the annealed conditions are a choice for the typical average of the US standards. Property values and compositions are available from different sources and references such as ASM (1993, 1998), ASTM A29 (2005), and EN 10083 (2006). UNS designation is composed by the form defined by G(AISI/SAE)0. AISI/SAE 4130 is therefore the UNS G41300. In many cases, the ISO designation is similar to EN or DIN designation. Values in braces are for a selected close {DIN} equivalent. The sign of ~ in front of the designation is for approximate designation from unsubstantiated or past source.

As observed from Table A.7.4 in Appendix A.7, the quenched and tempered strength is higher than those of the plain carbon steel of the same carbon percentages in Table A.7.3 in Appendix A.7. For example, comparing the yield strength of the *quenched and tempered Q&T* AISI 1040 (593 [MPa] or 86 [kpsi]) to the yield strength of the quenched and tempered *Q&T* AISI 4140 (1640 [MPa] or 238 [kpsi]), it is clear that the low-alloying steel yield strength is about 2.75 times the strength of the plain carbon steel. The ductility is, however, lower by about 2.4 times from 19% to 8% elongation percentage. The low alloying is therefore useful and not that expensive means to get higher strength with some penalty loss of ductility. Cost optimization and size constraint would help in deciding which of the material to use. Cost should include material and manufacturing cost, which should be higher for higher strength material. A rough list for the *cost of material* usage from least to high cost is as follows: low-carbon steels, simple high-carbon steels, directly hardened carbon or low-alloy steels, low-carbon or low-alloy steels, medium-carbon chromium or chromium–aluminum steels, and directly hardened high-alloy steels.

7.4.3 Structural Steel

Frequently used structural components such as *standard sections, plates, bars, pipes*, and *rivets* are made of this type of steel. It is usually a *low-carbon steel* or *high-strength low-alloy* (HSLA) *steel* containing about 0.15–0.25% carbon. The yield strength S_y is in the range of 185–460 [MPa] or 27–66 [kpsi]. The HSLA structural steel has typically higher yield strength than the usual carbon steel. The HSLA structural steel contains small amounts of vanadium (V) and/or niobium (Nb) or small amounts of other alloying elements. Other structural steels are typically carbon–manganese or carbon–manganese–silicon steels. Material properties of selected *structural steels* are presented in Table A.7.5 in Appendix A.7. The compositions of different low-alloying elements are available in standards and handbooks. The related standards are ASTM A36/A36M (2014), A 283, A572/A672M (2007), A992/A992M (2011), A1011/A1011M (2007), ISO 630 (1995), EN 10025 (2005), and DIN 17100 (1986). Other information is also available in ASM (1998) or ASM (2006).

The structural type of steel is ductile, which allows easy manufacturing by rolling, drawing, or forming of these standard *structural profiles* or sections. Selected *standard sections* made of structural steel are defined in Tables A.8.1–A.8.14 in Appendix A.8. Designation similarity of structural steel in Table A.7.5 in Appendix A.7 is again realizable only after careful assessment of individual case. Given similarity is only a selected close correspondence. Values are also a compilation from different sources. Unless the same value or range in any category is quoted, the upper property line is for ASTM, and the lower property line is for ISO, EN, and DIN. Endurance limit is estimated as 0.5 of the ultimate tensile strength, if not available; see Section 7.8. Some values may differ with some citations. Cross-check is, therefore, recommendable. ASTM designation is different for different structural steel grade, class, and composition. ASTM A 1011/A1011M (2007) is mainly for sheet and strip, but is selected here for comparison. ISO designation is for ISO 630 (1995), and EN designation is for EN 10025 (2005). Values in brackets or braces are for selected close equivalent [EN] or classical {DIN} standard. Strength values are thickness dependent. As thickness increases the strength decreases. Quoted values are minimum cited or a range for general design purposes. The *Brinell* hardness values are an average or range citations. Values are according to the older DIN 1611 and 1612. For ductility, values are also an average or range of available information when designations provide different values. Check is advisable when citing a specific designation.

Mechanical application samples utilizing structural steel are relating to the following practical uses. Welded machine frames, crane booms and components, and rolls and components of material handling equipment can be made of structural steel plates, bars, and sections. *HSLA* structural steels are for the required higher strength applications such as truck frames, crane girders, bridges, noteworthy buildings, rail cars, and similar demanding usages and purposes.

7.4.4 Stainless Steel

Stainless steel is a material that resists oxidation particularly in normal ambient environment. Therefore, it does not seem to stain. This was where the name originates. Some of it can resist moderate acidic and bases or alkaline environment. Some of these stainless steels may also resist oxidation at elevated temperature and are then definable as *heat-resisting steels*. These properties are due to the alloying with a minimum of about 10.5% chromium (Cr) while carbon is in the range of about 0.02–1.2%. The chromium forms an external film or shield that prevents penetration or advancement of oxidation. The addition of nickel (Ni), molybdenum (Mo), manganese (Mn), or some other elements improves or sustains these properties. About 70% of utilized and popular stainless steels are of the *austenitic* type and mainly of ***304*** and ***316*** common designations. The 304 stainless steel is the most commonly used stainless steel. Many *stainless steels* are available in different forms such as *plates, sheets, strips, bars, pipes*, etc. They are covered by several standards and handbooks such as ASTM A176 (2004), ASTM A240/A240M (2004), ASTM A276 (2006), EN 10088 (2014), ISO 15510 (2003), ASM (1998), and Bringas (2004). Properties are varying due to different form, composition, treatment, and processing. Quoted values in Table A.7.6 in Appendix A.7 are minimum expected or average. Designation similarity between standards is realizable after careful assessment of individual case. Given similarity or property is only a selected close correspondence. Cross-check and scrutiny are therefore necessary.

The main categories of stainless steel are the *austenitic*, the *martensitic*, and the *ferritic* types, which are definable as follows:

Austenitic stainless steels contains a maximum of 0.15% carbon (C), at least 16% chromium (Cr), and enough nickel (Ni) or manganese (Mn) to sustain the *austenitic* microstructure over a very wide range of temperatures (cryogenic to melting). The widely known designations of austenitic stainless steels are the AISI 300 and the 200 series. The popular ***304*** (18-8, i.e. about 18% Cr and 8% Ni) and 316 (18-10, i.e. about 18% Cr and 10% Ni) stainless steel types are austenitic. The other austenitic 200 series of stainless steel has less nickel, more of the cheaper manganese and less corrosion resistance. Austenitic stainless steels are hardenable and strengthable by cold working, but are not heat treatable. See Table A.7.6 in Appendix A.7 for mechanical properties of some of the selected austenitic stainless steels in their annealed state. The equivalent ISO, EN, or DIN are close approximate selections. Therefore, properties are varying due to different composition, treatment, and processing. Quoted values are minimum expected or average.

Martensitic stainless steels have higher carbon contents (about 0.15–1.2%) and therefore can be heat treated. They can be stronger and tougher than other stainless steel types, but less corrosion-resistant. Martensitic stainless steels contain chromium (10.5–18%), nickel (up to 2%), and molybdenum (0.2–1.0%). They are exceedingly machinable that provides more manufacturing attractiveness. The widely known designations of martensitic stainless steels are some elements of the AISI 400 series. One of the most commonly used in this series is the ***410*** (mainly 12% Cr and 0.1% C). The annealed properties are shown in Table A.7.6 in Appendix A.7. When heat treated by quenching in oil from 980 [°C] or 1800 [°F] and tempered at 40 [°C] or 104 [°F], the yield strength of a bar can reach 1225 [MPa] or 178 [kpsi]. The ultimate tensile strength would reach 1525 [MPa] or 221 [kpsi]; the ductility is reduced to about 15% with an increased Rockwell hardness to about 45 HRB. Table A.7.6 in Appendix A.7 gives basic mechanical properties of some of the selected martensitic stainless steels in their annealed state. Heat treatment can provide higher values for strength and hardness with decreased ductility (ASM 1998).

Ferritic stainless steels contain a small amount of carbon (about 0.02–0.2%). They also contain chromium in the range of 11–30%. Ferritic stainless steels can be hardenable and strengthable by cold working, but they usually not often cold worked. They are also not heat treatable. The widely known designations of ferritic stainless steels are some other elements of the AISI 400 series. One of the most widely used in this series is the ***409*** (about 11% Cr, 0.08% C, 0.5% Ni, and up to 0.75% Ti). Table A.7.6 in Appendix A.7 presents mechanical properties of some of the selected ferritic stainless steels in their annealed state.

Other stainless steel types such as the *duplex* or the *precipitation-hardening* steels are available in related stainless steel references and standard handbooks such as ASM (1998).

Mechanical application samples for stainless steel are relating to the following practical products. Special bolts, nuts and fasteners, springs, gears, shafts, rollers, bearings, wires, heat exchanger coils, special pressure vessels, furnace and boiler parts, steam or gas turbine blades, jet engines parts, food and beverage components, domestic appliances or utensils, and other chemical, corrosion, or heat-resisting components would be made of stainless steel.

7.4.5 Tool Steel

Steels intended for tool manufacturing are usually medium- to high-carbon steels, low-alloy steels, or high-alloy steels. Designers of dies, manufacturing tools, and machine tools as such, or those concerned with the manufacturing of machine parts would be concerned about tool steel. Depending on the tool application, the important property or properties would be *hardness*, abrasion resistant, high stiffness or rigidity, steady cutting edge, and cracking abatement during heat treatment. Tool steel usually contains *carbides* to facilitate most of these properties. The alloying elements suitable for that are tungsten (W), molybdenum (Mo), chromium, (Cr), vanadium (V), and manganese (Mn). Other alloying elements can also exist in tool steels. The following is an ***AISI/SAE*** list of selected tool steel grades according to manufacturing processing, applications, and hardness citation range using the Rockwell hardness scale of *HRC* (see Section 7.6.):

- *Cold working* tools are usually for thread rolling dies, abrasion-resistant liners, forming dies, molds, gauges, intricate die shapes, slitters, short-run dies, coining, cold heading dies, and cutlery:
  - A: *Air hardening* medium alloy steel with a hardness range of about 58–60 *HRC*
  - O: *Oil hardening* with a hardness range of about 56–62 *HRC*
  - D: *High carbon–high chromium* (Cr) with a hardness range of about 54–62 *HRC*
  - W: *Water hardening* with a hardness range of about 50–64 *HRC*, should not be confused with other tungsten (W) alloy steel.
- *Hot working* tools are generally for dies designed for extrusion or forging. They are also suitable for highly stresses aircraft parts and die casting equipment:
  - H: *Chromium* (Cr) base with a hardness range of about 40–55 *HRC*
  - H: *Tungsten* (W) base with a hardness range of about 45–55 *HRC*
  - H: *Molybdenum* (Mo) base with a hardness range of about 50–60 *HRC*
- *High-speed* tools are mainly for cutting tools such as drills, taps, milling cutters, reamers, broaches, hobs, and other varieties of cutters;
  - T: *Tungsten* (W) base with a hardness range of about 60–68 *HRC*
  - M: *Molybdenum* (Mo) base with a hardness range of about 64–70 *HRC*
- *Shearing and drawing* tools are primarily for paper-cutting knives and some wiredrawing dies:
  - F: *Carbon–tungsten* with a hardness range of about 50–64 *HRC* and more wear resistance than W tool steel
- *Plastic mold* tools are suitable mainly for plastic molds and low-temperature die casting:
  - P: Mold tool steel with a hardness range of about 30–60 *HRC*

- *Special purpose* tools are for applications such as chisels, impact-loaded tools, hammers, feed fingers, chuck parts, and clutch plates;
  - S: Shock resistant with a hardness range of about 40–60 *HRC*
  - L: Low-alloy with a hardness range of about 45–64 *HRC*

For more information about *tool steel* composition, application appropriateness, properties, and processing, one can consult the following standards and references: ASTM A681 (2008), A600, A686, ISO 4957 (1999), ASM (1998), and Robert et al. (1998).

7.4.6 Other Nonferrous Metals

Many other nonferrous metals are in extensive use in many components, elements, and parts. Here, the focus has been on metallic materials inherently or traditionally more suitable for machine components such as *steel*, *alloy steel*, and *cast iron*. This section mainly covers the mechanical properties of *aluminum alloys*. Properties of other materials such as *copper*, *brass*, *bronze*, *magnesium alloys*, etc. are also cited. The *nickel alloys*, *titanium alloys*, and *zinc alloys* are useful and may have coverage in some enclosed applications.

7.4.6.1 Aluminum and Aluminum Alloys

Sections 7.2 and 7.2.2 previously introduced aluminum and aluminum alloys. Section 7.2.2 gives the numbering system for both UNS as well as the international ISO designations. The designations and representative mechanical properties of selected aluminum alloys are in Table A.7.7 in Appendix A.7 for wrought and cast aluminum. Values are compiled from different sources. Property line is for AA (or UNS), where AA stands for *Aluminum Association*. Designation is mainly for similar ISO and EN or DIN if needed. Designation similarity is only realizable after careful assessment of individual case. Given similarity is a selected close correspondence. Values of the fatigue strength are for 500 million cycles of full reversed stresses. Cross-check is necessary since endurance limit may not definitely exist. UNS designation is composed by the form defined by A9(AA) for wrought and A0(AA) representing cast with no decimal. AA 2024 is the UNS A92024 and AA 319.0 is UNS A03190. ISO designation is for ISO R209 (1989) and ISO 3522 (2007). Values in brackets or braces are for selected close equivalent [EN] or {DIN}, if needed. In so many cases, EN and DIN designations also embraced the ISO standards. The two numbers for ductility or elongation percentage are, respectively, for 1.6 [mm] or 1/16 [in] specimen thickness and – in parenthesis – for (13 [mm] or 1/2 [in]) diameter specimen. As observed from Table A.7.7 in Appendix A.7, the range of properties varies due to the alloying elements. The specific alloy is usually for targeted group of applications and rough material form such as billet, sheet, strip, plate, or rod. For more information about composition, available material form, suitable applications, temper designation, and other properties, one needs to consult with standards and references such as AA (2015), ISO 209 (1989), ISO 3522 (2007), and Davis (2001).

Mechanical application samples for aluminum alloys are relating to the following practical products. For *wrought aluminum alloys* the selected sample products are special bolts, nuts, screws, rivets and fasteners, light gears, automotive parts, appliance parts or trims, machine parts, railroad parts, trucks or trailers, recreational vehicles, wheels, fan blades, cylinders or pistons, containers, sheet metal works, kitchen utensils, pipes, extruded profiles or sections, and aircraft structure or parts. *Cast aluminum alloys* are suitable for products such as heavy-duty pistons, generators or motors housings, air-cooled cylinder heads or housing, engine crankcases or parts, carburetor bodies, tire molds, pipefittings, and cookware.

7.4.6.2 Copper and Magnesium Alloys

Section 7.2.3 previously introduced copper and magnesium alloys. The representative mechanical properties of selected alloys are given in Section 7.1.3 for copper and magnesium. Property values for specific alloy designation can be obtained from different standards and references such as ASTM B107/B107M (2007), EN 12163 (1998), CDA (2004), and ASM (1998).

7.4.7 Other Materials

So far, the text did not cover so many other materials extensively used in several highly developed industries. Of these are space, renewable energy, electronics, etc. They broadly use newly developed materials, composites, plastics, smart materials, and newly developed manufacturing processes such as ***3D*** printing. For that, special designing requirements, modeling tools, and procedures should be the way to synthesize components and systems. This section covers an introduction to some properties of the other intended materials for use in some component design. Various extensive works are available in dedicated research publications, references, and handbooks such as Schechtling (1987), Harper (2000), and Osswald et al. (2006).

7.4.7.1 Plastics

Plastics are mainly man-composed materials. They are usually solid polymers chemically composed of modified or growing molecules. Plastics are usable as such or with some included additives embedded to improve properties. In that case, one may call these as composite materials.

Plastics are usually set in two main categories. The first category is utilizing heating for softening processing. When cooled they keep the shape, but can reprocessed by heating. These materials are ***thermoplastics*** and manufactured by injection or blow molding, extrusion, calendar laminating, or similar processing. The second category requires heat and high pressure to cure or form into shapes. They are not further fusible or soluble by heat. Such materials are ***thermosets*** and usually manufactured by molding powders or similar processing.

Another classification for plastics depends on the elastic modulus E of the material. *Rigid plastics* have modulus of elasticity larger than 700 [MPa] or 100 [kpsi]. *Semi-rigid plastics* have modulus of elasticity range between 70 and 700 [MPa] or 10 and 100 [kpsi]. *Non-rigid plastics* have modulus of elasticity smaller than 70 [MPa] or 10 [kpsi].

Elastomers are mostly non-rigid plastics, and some are semi-rigid plastics. They return to their initial form when the load drops after considerable deformation. Therefore, they can be useful as soft *springs*; see Chapter 10. Natural and synthetic rubbers are elastomeric. The modulus of elasticity can be as low as about 3 [MPa] or 0.4 [kpsi]. Some are *thermoplastics* and most others are *thermosets*.

Properties of plastics are widespread due to their different types of rigid, semi-rigid, and non-rigid. Usually, they are, however, much softer and weaker than metals; see Figure 7.3 and Section 7.1.3. The range of the ultimate tensile strength S_{ut} is about 28–152 [MPa] or 4–22 [kpsi] including glass-reinforced polymers. The wider range of the modulus of elasticity E is 0.07–2.8 [GPa] or 10–400 [kpsi]. The range of the specific gravity is about 0.95–1.8. Table A.7.8 in Appendix A.7 lists the properties of some few *selected plastics*.

The main *utility* of these selected plastics is their main usage in machine components provided the adherence to constraints on their *lower strength*, *lower hardness*, *low service temperature* limits, and diminishing properties with temperature. Some are usable for *common mechanical* elements such as gears, springs, bearings, fasteners, cams, rollers, impellers, fan blades, rotors, valves, wash machine stirrer, etc. Some other plastics are utilizable in *light-duty mechanical* components such as pipefittings, steering wheels, camera cases, eyeglass frames, handles, knobs, etc. Some plastics are very useful in parts for *low-wear* functions such as gears, bearings, bushings, tracks, roller skates wheels, wear liners or strips, etc. So many further plastics are suitable for small and large housings and hollow shapes such as cases, helmets, small appliances, tanks, tubes, pipes, ducts, insulation linings, boat hulls, etc. Some special plastics are applicable in optical and transparent parts such as lenses, glasses, windshields, shelves, signs, etc.

Plastics have a high dispersion in their properties due to manufacturing processing and some other auxiliary factors. Usually quoted values are either a range or an approximate average. One should consult manufacturer for the properties and practical utilities of a specific product under the type of the plastic category selected for the intended application. The few selected plastics and their approximate properties in Table A.7.8 in Appendix A.7 use the symbolic name for the plastic chemical composition and some common, familiar, or trade names in parentheses as follows:

Thermoplastics:

- ABS is acrylonitrile-butadiene-styrene, used in light-duty mechanical applications.
- CA is cellulose acetate (cellulosic), used in light-duty applications.
- PA is polyamide (nylon), used in common mechanical and low-wear applications.
- PC is polycarbonate, used in common and light-duty applications.
- PE is polyethylene, used in light-duty applications.
- PMMA is poly-methyl-meth-acrylate (Acrylic), (Lucite), (Plexiglas), used in optical and transparent applications.
- POM is poly-oxy-methylene (acetal), used in common mechanical and light-duty applications.
- PP is polypropylene, used in common mechanical applications.
- PS is polystyrene, used in light-duty applications.
- PTFE is poly-tetra-fluoro-ethylene (Teflon), used in low-wear applications.
- PVC is polyvinylchloride, used in light-duty applications.

Thermosets:

- EP is epoxy, used in common mechanical bonding applications.
- PF is phenol-formaldehyde (Bakelite), used in light-duty, wear, and other applications.
- PU is polyurethane, used in light-duty applications.
- UP is polyester unsaturated risen, used in low-wear applications.

Elastomeric:

- NBR is butadiene-acrylonitrile (nitrile), used in mechanical application the same as rubber.
- SBR is styrene-butadiene, used in mechanical application the same as rubber.
- NR is polyisoprene (natural rubber), used in common rubber applications.
- SiR is polysiloxane (silicone rubber), used in mechanical application the same as rubber.

Many plastics are reinforced with fibers to improve their properties and they are then as composites. The improvement can be in the range of about 80–120% in strength for regular reinforced plastics. Mixing plastics is also another form of many plastic compounds. For further compounding, processing, and joining, refer to dedicated books such as Klein (2011). For more information on plastics and reinforced plastics, one may refer to handbooks such as Schechtling (1987), Harper (2000), Osswald et al. (2006), and Rosato and Rosato (2004).

7.4.7.2 Composites

A composite material is purposely made of two or more materials to improve one or few properties. Alloys in multiphase metals are denotable as composites. Compounds made by mixing of polymers are also composites. Adding *particles* or *fibers* to some materials (such as polymers) forms composites. Forming *laminates* or *sandwiches* of contiguous materials produces composite constructions such as *plywood*, *corrugated fiberboard* or *cardboard*, and *honeycomb* structures. The strength of these structures can be numerically addressed by finite element (***FE***) solutions such as laminate blades structure in Ismaiel et al. (2017). Some theories, however, exist to treat some of these composite structures, such as Tsai and Wu (1971). The scope of these treatments is beyond the intent of this text.

Composites are composed of a ***matrix*** as the surrounding portion of the added dispersed ***reinforcements*** by ***particles*** or ***fibers***, excluding the laminates or sandwich structures. Properties of composites depend on the properties of the aggregates or constituents. Added *particles* to the matrix change properties as a function of different sizes of particles. For small size particles (about 0.01–0.1 [μm] or 0.4–4 [μin]), they may increase strength due to retarding the dislocation movements. For larger particles, they would restrain the matrix movement around the particles. The properties of the composites are also a function of inserted *fibers* shapes and geometry or sizes (as in length and length to breadth ratio), the concentration extent (or volume fractions of fibers and matrix), and the orientation of fibers.

Fiber reinforcement composites are much stronger than the matrix, and the strength is a function of the extremely high *fiber strength* and other factors such as *volume fraction* as shown in Table A.7.9 in Appendix A.7. The table includes *carbon nanotubes (CNTs)* and *graphene* for reference and comparison. The *CNT* as a fiber is not yet acting as a regular fiber for composites, since it is relatively expensive. *Graphene* is also not utilizable as composites reinforcement. Other carbon fibers such as *graphite fibers* (***1D*** or ***2D*** fabric) are instead still less expensive and thus effectively more usable. Table A.7.9 in Appendix A.7 gives properties of some selected reinforcement fibers. Values of properties are approximate ranges compiled from different sources. However, necessary check should be with manufacturer for the properties of a specific product under the name or type selected for intended application. Strength depends on diameter. Strength is higher for smaller diameter. Very high strength values are, however, kept at the usual [MPa] ([Kpsi]) units to allow for comparison with most strength values of previous materials. The quoted diameters or thicknesses are only representative. Different values are also available. Higher values can usually be for multilayer or wires.

The longitudinal *composite strength* modeling may be simply as of a linear ratio of the *volume fractions*. The *composite strength* in the fiber direction is $S_{\mathrm{ut},C} = S_{\mathrm{ut},f}\ (V_f) + S_{\mathrm{ut},m}\ (1 - V_f)$, where $S_{\mathrm{ut},C}$ is the ultimate tensile strength of the composite, $S_{\mathrm{ut},f}$ is the ultimate tensile strength of the fiber, V_f is the volume fraction of the fiber material, and $S_{\mathrm{ut},m}$ is the ultimate tensile strength of the matrix. The same model also applies for the modulus of elasticity of composite material as a function of elasticity moduli of fiber and matrix.

Composites applications are widespread and expanding with a high rate. They are applicable in numerous regular and demanding products. So many conducted research work is hoping to advance the field and still going on to improve or develop products such as sample attempts by Elmoselhy et al. (2006), Hassan et al. (2014), and Ismaiel et al. (2017).

Tool material inserts are a form of ceramic composites that are usually producible as bits cemented to the tool to be the cutting or other processing edge. The *cutting tool inserts* are extremely hard, abrasion, and wear-resistant even at high temperatures, and maintain steady cutting edge. They are much better than *tool steels* but more expensive. The most common of the cutting inserts is a *tungsten carbide* WC in a binder metal of *cobalt* Co. The binder is supposedly the *matrix* with a very low volume fraction of about 4–20%. The tungsten carbide is the major part of the composite at a volume fraction of about 80–96%. Some inserts use mixtures of carbides such as *tungsten carbide* WC, *titanium carbide* TiC, and others in a *cobalt* Co binder. Other inserts use a *nickel* Ni binder and mixtures of carbides such as *titanium carbide* TiC and *chromium carbide* Cr_3C_2. For processing of these ceramics and other *tool inserts*, one needs to consult dedicated references such as Groover (2010).

7.5 Particular Materials for Machine Elements

Abundant knowledge is readily available for selecting particular material used for every machine element. This data exists in references, handbooks, and manufacturer of many machine elements such as the recommended *carbon steels* for numerous applications in McCauley (2012). In many other references, tables of material properties may have an application column to indicate the typical usage of the particular material. For steel, which is abundantly utilizable, Table 7.3 provides the common uses of *plain carbon steels* as a function of carbon percentage. If the material sought is a plain carbon steel, the designation of the steel is automatically definable as the AISI 1020 or 1030, ISO C20, or C30 for elements such as some gears, shafts, levers, camshafts forgings, plates, tubing, structural profiles, or similar elements (Table 7.3). For structural plates, bars, and standard profiles such as in Table A.7.5 in Appendix A.7 (and Tables A.8.1–A.8.14 in Appendix A.8), a structural steel of about the same strength and carbon percentage such as ASTM A36 (2014) or ISO E235 is more appropriate to conform with available manufacturers of structural profiles. These profiles are standard machine components suitable for frames, brackets, supports, etc.

7.5.1 Standard Machine Elements

Standard machine elements conform to specific standards that manufacturers and suppliers abide by. These standard machine elements are bolts, screws studs and fasteners, clips, snap rings, couplings, keys, pins, rivets, wires, wire nails, CD sections, tubing, sheets, strips, structural shapes, spring wires, lock washers, rolling bearings, etc. Material properties of these elements are usually set in these standards. One should use the available standards for these standard elements. Suppliers and manufacturers usually provide the *standard designations* for the materials of the standard machine elements they provide. Table 7.3 provides the common uses of *plain carbon steels* as a function of carbon percentage that generally conforms to these standard machine elements.

Several of the standard machine elements are also made of *cast iron*, *aluminum*, *magnesium*, or *copper* alloys and made of plastics or other materials. Some of these lists of applications as samples are in Section 7.4.6. Suppliers and manufacturers also provide the *standard designations* of these materials for the standard machine elements they provide.

7.5.2 Synthesized or Designed Machine Elements

Synthesized machine elements such as gears, shafts, springs, flywheels, and disks for brakes or clutches are of main concern of this text. The adaption of a particular material is the intended optimal goal. Optimization can be a procedure for adaption depending on the objective as for minimum cost or weight and many other objectives and constraints. Since the text is dedicated to design with CAD and optimization, initial good selection of appropriate material reduces the iterations to achieve the optimum. If the initial selection stemmed from vast experience, the

Table 7.4 Initially selected materials for some basic machine elements.

Machine element groups		Typical material AISI, ASTM, ISO, (EN, {DIN})[a]	Optional materials for special or crucial applications[a]
1	Bolts, screws, studs, fasteners, hooks	AISI 1030, ASTM A 307, ISO C30	AISI 1035, 1040, ISO C35, C40,
2	Bushings, cams	PA, PTFE, PU, UP Plastics	Phosphor bronze
3	Clips, snap rings, lock-washers	AISI 1060, ISO C60	AISI 1070, {DIN Ck67}
4	Couplings	ASTM 30, ISO 200	ASTM 35, ISO 300
5	Discs for brakes or clutches, housings	ASTM 30, ISO 200, AISI 1060, ISO C60	ASTM 35, ISO 300
6	Flywheels	ASTM 30, ISO 200	ASTM 35, ISO 300
7	Frames, housings, cases	ASTM 20, ISO 150	ASTM 30, ISO 200
8	Gears	AISI 1045, ISO C45	AISI 4340, 5140, EN X6CrMo17-1, ISO 41CrS4
9	Keys, pins	AISI 1020, ISO C20	AISI 1030, ISO C30
10	Rivets, wires	AISI 1010, {C10}	AA 1100, ISO Al99.0Cu
11	Shafts, camshafts, pinion shafts	AISI 1040, ISO C40, C45	AISI 1340, 3140, 4140, 4340, 5140, and 8650. ISO 42CrMoS4, 41CrS4, {50NiCrMo2} Carburized AISI 1020, ISO C20
12	Springs	AISI 1080, 1095, {Ck75}	AISI 6150, 9255 ISO 51CrV4, {55Si7}

a) See Appendix A.7 for designations and properties. Consult standards, references, and handbooks for missing designations in Appendix A.7.

initial selection might be an optimum estimate by knowledge and experience. Further iteration, however, is the optimum such as the attempts by Abdel Meguid (1999) and many others.

The initial material selection shown in Table 7.4 provides a reasonable start, which is stemming from previous knowledge and experience. The *machine elements* set in groups are of like materials, function, and utility. The element heading of each group or set is of basic concern to the subjects studied in this text. Several of these machine elements are *standard components* of direct interest. The table order is alphabetical with respect to the group head. The group head is according to the more familiar or simple name of the *set* rather than the subject name. Optional material for each group or set may be similar to other sets merely as an alternative selection. The bounded choice of the typical material is also a constraint selection from the limited list of materials in the appendix of the text to have available properties ready for synthesis. The distinction between the groups is due to different function and possible dissimilar alternative materials for the set group. The main categories of the initial material set are ***cast iron, plain carbon steel***, and ***plastics***. The machine elements categories by the order of materials and *strength* are as follows:

- ***Cast iron*** is suitable for groups: **7**-frames…, **4**-couplings…, **5**-disks for brakes… and **6**-flywheels…
- ***Plain carbon steel*** is suitable for groups: **10**-rivets…, **9**-keys…, **1**-bolts…, **11**-shafts…, **8**-gears…, **3**-clips…, **12**-springs…
- ***Plastic*** is suitable for group: **2**-bushings… and some of the above for light applications.

Other materials are utilizable such as *stainless steel, aluminum, or copper alloys* of about at least the same strength and comparable or close ductility. The other ***optional materials*** in Table 7.4 are also useful with more alternatives defined in each of the chapters covering these machine elements.

7.6 Hardness and Strength

Previously, Section 7.1.4 introduced the different types of hardness measurements. *Brinell* hardness (after *Johan A. Brinell* [1849–1925]) is a measure related to the diameter d_{HB} of the material plastic indentation under the 10 [mm] or 0.39 [in] steel ball (or a *tungsten carbide* ball) and a force of 29.42 [kN] (3000 kg-force) or 6614 [lb]. The *Brinell hardness number* is calculated as $HB = (2\boldsymbol{F}_{kg}/(\pi D_B(D_B - (D_B^2 - d_{HB}^2)^{1/2})))$, where $\boldsymbol{F}_{kg}$ is the applied force in "kg-force," D_B is the ball diameter in [mm], and d_{HB} is the indentation diameter in [mm]. The *Rockwell* hardness (after *Hugh M. Rockwell* (1890–1957) and *Stanley P. Rockwell* (1886–1940)) is a direct measure of penetration representing the hardness on the scale of a dial. The smaller penetration is indicating a harder material. The types of *Rockwell* hardness are *HRA, HRB, HRC* to *HRG*, which are using different ball diameters and loads. The commonly used are the *Rockwell HRB* and *HRC*. The *Rockwell HRB* uses 1/16 [in] (or 1.588 [mm]) diameter of a spherical steel ball and a force of 980.665 [N] (100 kg-force), which is suitable for softer materials such as aluminum, brass, and soft steels. The *Rockwell HRC* uses a 120° diamond *spheroconical* and a force of 1470.9975 [N] (150 kg-force), which is suitable for harder materials such as heat treated *alloys* and *alloy steels*. For more information on hardness testing methods, relations, and conversions, one can consult standards, references, and handbooks such as ASTM E140 (2002), ISO 18265 (2003), ISO 18265 (2013), Callister and Rethwisch (2011), and ASM (1998).

Figure 7.15 presents the variation of the ultimate tensile strength, *Rockwell* hardness (*HRB, HRC*), and *Brinell* hardness indentation diameter d_{HB} in [mm] as functions of *Brinell* hardness number *HB*. The values of the upper line of the *US* ultimate tensile strength S_u in [kpsi] and the lower line of *Brinell* hardness indentation diameter d_{HB} in [mm] can be identified through the right ordinate scale of Figure 7.15. The black dashed lines are for the reported values of the *SI* ultimate tensile strength S_u in Table A.7.4 in Appendix A.7 (and extension thereof), and the available associated values of the *Rockwell* hardness *HRB* and *HRC*. The black dotted lines are for the reported values of the *US* ultimate tensile strength S_u in Table A.7.4 in Appendix A.7 (and extension thereof), and the available associated values of the *Brinell* hardness indentation diameter d_{HB} in [mm]. The gray solid lines are

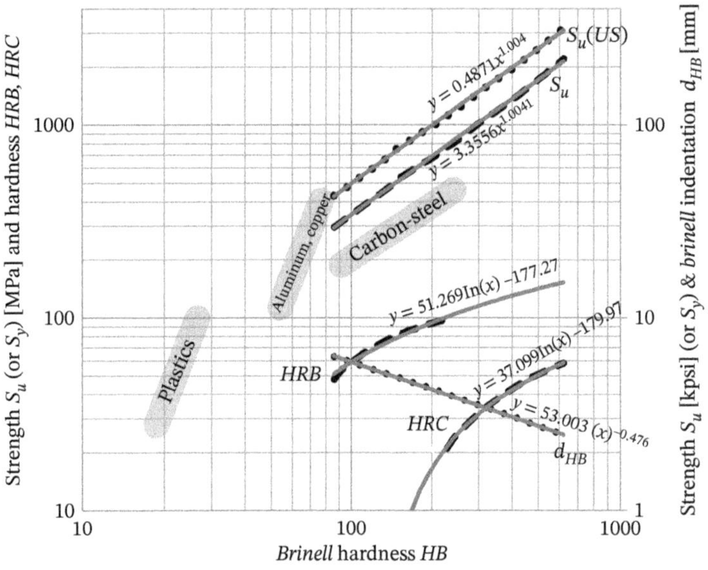

Figure 7.15 Ultimate tensile strength, *Rockwell* hardness (*HRB*, *HRC*), and *Brinell* hardness indentation diameter in [mm] as functions of *Brinell* hardness number *HB*. The values of the line of the *US* ultimate tensile strength S_u in [kpsi] and the line of *Brinell* hardness indentation diameter d_{HB} in [mm] are identified through the right ordinate scale. Very thick light gray lines are crude selection of different material ranges for yield strength S_y.

for the trend relations of these material behavioral responses *y* as functions of the *Brinell* hardness number *HB* as *x*. The dashed extensions of the *HRB* and *HRC* gray lines are just to indicate the trend and to show that they are beyond their normal valid regions of the covered *HB* scope. Very thick light gray lines are crude selection of different material ranges, just as a rough comparison. Strength of *aluminum*, *copper*, and *just carbon steel* refers to the ***yield strength*** S_y rather than the ultimate tensile strength S_u. This is due to the importance of the yield strength in design.

The indicated trend relations in Figure 7.15 should help in approximating the material behavior for a certainly needed strength if the hardness is possibly accessible. The trend relations are also useful to define the relations between the *Brinell* hardness and the *Rockwell* hardness. The range of applicability is also identifiable from Figure 7.15. The relations use the variable *y* to represent the property cited in the left (or right) ordinate and the variable *x* for the *Brinell* hardness number *HB* cited in the abscissa of Figure 7.15. The symbol of the *Brinell* hardness *HB* is then replacing the variable *x*, and the different strength symbols of S_u, S_u (*US*), *HRB*, and *HRC* are replacing the variable *y*. The trend relations for both ultimate tensile strengths S_u and S_u (*US*) in Figure 7.15 are then approximately having the following equations for *SI* and the *US* systems:

$$S_u = 3.3556\,HB^{1.0041}\ \text{[MPa]}$$
$$S_u = 0.4871\,HB^{1.004}\ \left[\text{kpsi}\right] \tag{7.15}$$

These estimated relations are only valid for steel and may be cautiously usable due to the actual scattered values of these properties. They are, however, suitable for engineering calculations of property estimates. In some various references, the relations are approximately definable as

$$S_u = 3.45\,HB \quad \text{or} \quad S_u = 3.4\,HB\ \text{[MPa]}$$
$$S_u = 0.5\,HB \quad \text{or} \quad S_u = 0.49\,HB\ \left[\text{kpsi}\right] \tag{7.16}$$

Equation (7.15) are little more accurate than Eq. (7.16). For a quick estimation, Eq. (7.16) can be useful. Similar relations can be accessible for other classes of materials in dedicated references and handbooks. Similar approximate relations applied to *cast iron* are as follows:

$$S_u = 1.58\,HB - 86 \;\; [\text{MPa}]$$
$$S_u = 0.23\,HB - 12.5 \;\; [\text{kpsi}] \qquad (7.17)$$

The trend relations for both *Rockwell hardness numbers HRB* and *HRC* in Figure 7.15 are then approximately having the following equations:

$$HRB = 51.269 \;\; \ln(HB) - 177.27$$
$$HRC = 37.099 \;\; \ln(HB) - 179.97 \qquad (7.18)$$

These estimated equations are only valid for the *HB* ranges of approximately (80–240) and (230–630) for calculating equivalent *HRB* and *HRC*, respectively.

The trend relations for the *Brinell* hardness indentation diameter d_{HB} in [mm] are then approximately having the following equations:

$$d_{HB} = 53.003\,HB^{-0.476} \;\; [\text{mm}] \qquad (7.19)$$

This equation is valid for the *Brinell* hardness span of 85–620 *HB*.

Example 7.2 With reference to Example 7.1, compare the ultimate strength properties to the *Brinell* hardness numbers utilizing the previously developed equations in Section 7.6.

Solution
Data: $C_C = 0.5$, $HB = 195$, material AISI 1050 or the ISO C50. In addition, water quenched and tempered.

The calculations in Example 7.1 deduced that the material is normalized or CD. For AISI 1050 or the ISO C50 the properties are as follows:

- Tabulated $HB = 197$
- Tabulated $S_u = 690$ [MPa] or 100 [kpsi]

According to Eq. (7.15), the ultimate tensile strengths S_u (*SI*) and S_u (*US*) are as follows:

$$S_u = 3.3556\,(197)^{1.0041} = 676 \;\; [\text{MPa}]$$
$$S_u = 0.4871\,(197)^{1.004} = 98 \;\; [\text{kpsi}] \qquad (\text{a})$$

The simpler Eq. (7.16) provide the estimate of the ultimate tensile strengths S_u (*SI*) and S_u (*US*) as follows:

$$S_u = 3.45\;(197) = 680 \quad \text{or} \quad S_u = 3.4\;(197) = 670 \;\; [\text{MPa}]$$
$$S_u = 0.5\;(197) = 98.5 \quad \text{or} \quad S_u = 0.49\;(197) = 96.5 \;\; [\text{kpsi}] \qquad (\text{b})$$

These results in Eqs. (a) and (b) favor the first of the simplified Eq. (7.16) for *SI* units and the first of the simplified Eq. (7.16) for the *US* units. This result may not necessarily be general for all other cases. The estimation of the *ultimate tensile strengths* S_u is, however, close enough for engineering or design calculations by any of Eq. (7.15) or (7.16).

7.7 Failure and Static Failure Theories

In previous Section 7.7, the treatment focused on the key properties of material stemming from common failure test means. That has mainly the consideration of the static tensile or compressive properties in addition to other simple one-directional properties, where failure occurs. The real ***3D*** loading of ***3D*** bodies is, however, more complex in character, form, and time. The *static failure* of components should realistically include the ***3D*** loading state and may include the microstructure of the material. The failure under these more complex loading situations is usually in a close relation to the unidirectional failures associated with the main static tensile or compressive properties of materials. The *static failure* should also account for instability in *buckling* and instability of *cracks* and its relation to *fracture toughness*; see Section 7.9. The *dynamic failure* includes *fatigue* (low-cycle or high-cycle) and cumulative fatigue for unsteady loadings; see Section 7.8. The other types of failures to worry about and should be considered are excessive wear, surface damage, stress corrosion, annealing due to excessive temperature, and local softening or hardening. Here, addressing static failure theories is a suitable initial introduction to failure.

The ***static*** *failure theories* attempt to define the complex ***3D*** failure occurrence with the knowledge of these available unidirectional static tensile or compressive properties. Failure in that sense is usually the *yielding* of ductile material or the *ultimate strength* of the brittle material if no definite yield is adaptable. The failure theories covered in this section are, chronologically, the *maximum normal stress*, the *maximum shear stress* (or Tresca 1864), and the *maximum distortion energy* or von Mises (1913), after *Richard Edler von Mises* (1883–1953). Other theories of failures are of use particularly for *brittle materials* such as the *Coulomb–Mohr* and *modified Mohr*, which are also helpful in materials with different strength values in tension and compression.

7.7.1 Maximum Normal Stress Theory

The *maximum normal stress* theory (attributed to Rankine (1857), after William J.M. Rankine (1820–1872)) postulates that failure in ***3D*** stress states occurs when the maximum principal stress reaches the *failure strength*. The failure in that case can be defined as the *tensile yield strength* S_{yt} or the *ultimate tensile strength* S_{ut}, for a *ductile* or a *brittle* material, respectively. If the minimum principal stress σ_3 reaches the *compressive strength* before the tensile failure, the *maximum normal stress* theory postulates failure occurrence. Figure 7.16a shows the *Mohr's* circle of the loading of an inside fragment point ***A*** in a ***3D*** stressed machine element as previously shown in Figure 6.18a. Figure 7.16a is the same as Figure 6.18b. The triaxial stress state in Figure 7.16a is located inside the grayish region of the *Mohr's circles* as point ***A***. The *Mohr's circle* in Figure 7.16b represents the simple tension. The *Mohr's circles* in Figure 7.16c represent pure torsion. The postulation of failures represented by the *tensile yield strength* S_{yt}, the *ultimate tensile strength* S_{ut}, the *compressive yield strength* S_{yc}, and the *ultimate compressive strength* S_{yc} is represented in Figure 7.16b,c.

The *maximum normal stress* theory postulates that *failure of* ***ductile*** *materials* would occur when reaching the following conditions:

$$\sigma_1 = S_{yt} \quad \text{or} \quad \sigma_3 = S_{yc} \tag{7.20}$$

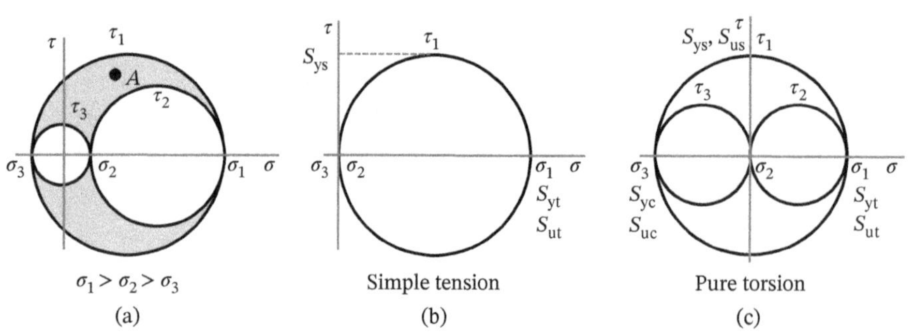

Figure 7.16 Point A in (a) representing an inside fragment of a fully stressed triaxial stress state in a machine element. Point A is located inside the grayish region of the *Mohr's circles*. The *Mohr's circle* for simple uniaxial tension is drawn in (b). The *Mohr's circle* for pure torsion is given in (c). Different expected failure strengths are pointed out.

where S_{yt} is the *tensile yield strength* and S_{yc} is the *compressive yield strength*. The compressive *principal stress* σ_3 is negative as well as the *compressive yield strength* S_{yc}.

The *maximum normal stress* theory postulates that *failure of* ***brittle*** *materials* would occur when reaching the following conditions:

$$\sigma_1 = S_{ut} \quad \text{or} \quad \sigma_3 = S_{uc} \tag{7.21}$$

where S_{ut} is the *ultimate tensile strength* and S_{uc} is the *ultimate compressive strength*. The compressive *principal stress* σ_3 is negative as well as the *ultimate compressive strength* S_{uc}.

This theory, however, is not valid for many cases that render its general utility as objectionable. It may be applicable for *brittle materials* under some loading conditions such as the failure of a bar of chalk under bending in Figure 7.17a. This results by gripping the two ends of the chalk bar by the fingers of the left and right hands and applying increasing moment until fracture. The failure is a fracture normal to the chalk centerline as a testament of the normal bending stress reaching the ultimate tensile strength of the chalk material. On the other hand, one of the obvious cases of the *maximum normal stress* theory failure is the pure torsion defined by the *Mohr's* circle in Figure 7.16c. This results by gripping the two ends of the chalk bar by the fingers of the left and right hands and applying increasing torsion until fracture. The failure of a bar of chalk under this pure torsion in Figure 7.16b demonstrates that the failure occurs at a 45° off the centerline of the chalk bar. This would entail the identification of the plane of the normal principal stress. The *maximum normal stress* theory, however, suggests that the failure strength in shear S_{us} is the same as the failure strength in tension S_{ut} or $S_{us} = S_{ut}$ as defined in Figure 7.16c. Similarly, the *maximum normal stress* theory suggests that the yield strength in shear S_{ys} is the same as the yield strength in tension S_{yt} or $S_{ys} = S_{yt}$. It should be observable, though, that the effort exerted to cause the torsional failure is less than that causing the failure in bending of Figure 7.17a. Other conducted experiments suggest that the *ultimate strength in shear* S_{us} is about 60% of the *ultimate tensile strength* S_{ut} particularly for brittle materials. The prediction of the *maximum normal stress* theory, therefore, fails in that regard.

For brittle materials, the *ultimate tensile strength* S_{ut} is usually less than the *ultimate compressive strength* S_{uc}. In ***2D*** stress state, the brittle failure in the *maximum normal stress* theory should occur outside the grayish region of Figure 7.18a. In ***3D***, however, the brittle failure in the *maximum normal stress* theory should occur outside the grayish cube of Figure 7.18b. It is observed, though, that the hydrostatic loading causing the ***3D*** stress state ($\sigma_1 = \sigma_2 = \sigma_3$) does not cause failure as defined in Figure 7.18b outside the cubic region. This is also more obvious for compression cases than tension cases. The *maximum normal stress* theory, consequently, fails as well in that prediction. The theory is presented here just for the record, and it is not globally applicable.

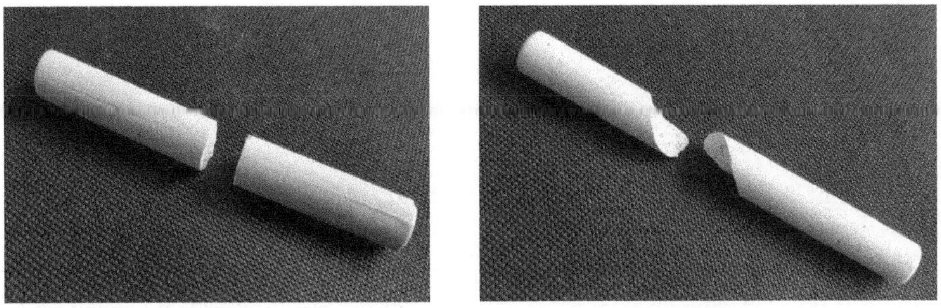

(a) (b)

Figure 7.17 The failure of a bar of chalk under manual bending in (a). The failure is normal to the bar centerline. The failure of a bar of chalk under manual pure torsion in (b) demonstrates that the failure occurs at a 45° off the centerline of the chalk-bar.

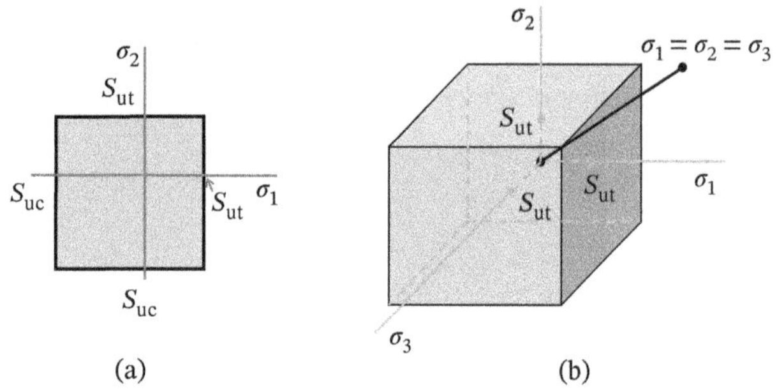

Figure 7.18 The *maximum normal stress* theory in ***2D*** stress state, where the brittle failure should occur outside the grayish region in (a). In ***3D*** stress state, the brittle failure should occur outside the grayish cube in (b). A hydrostatic loading causing the ***3D*** stress state ($\sigma_1 = \sigma_2 = \sigma_3$) is shown outside the cubic region in (b). Ultimate tensile strengths S_{ut} are taken less than the ultimate compressive strengths S_{uc}.

7.7.2 Maximum Shear Stress Theory

This theory has been suggested long ago by *Coulomb* (1773/1776), after *Charles-Augustin de Coulomb* (1736–1806). The theory, however, is usually attributed to Tresca (1864), after *Henri Édouard Tresca* (1814–885). The theory postulates that failure occurs when the maximum shear stress τ_{max} in a component reaches the maximum shear strength S_{ys} in the tensile test. The postulation stems from the failure observation in tension tests where the failure direction occurred at 45° from specimen centerline, which is the maximum shear direction. The *Mohr's* circle of simple tension in Figure 7.17b shows that the maximum shear stress $\tau_{max} = \tau_1$ and therefore the maximum *yield strength in shear* S_{ys} is equal to $S_{yt}/2$. The *maximum shear stress* theory states that failure occurs when the maximum shear stress τ_{max} is

$$\tau_{max} = S_{ys} = 0.5\, S_{yt} \tag{7.22}$$

where S_{ys} is the *yield strength in shear* and S_{yt} is the *tensile yield strength.*

The maximum shear stress theory in ***2D*** stress state is shown in Figure 7.19, where the *Mohr's* circle in Figure 7.19a shows σ_2 as compressive, σ_3 = *zero* and the maximum shear stress is the shear strength in simple tension S_{ys}. The failure should then occur outside the grayish region in Figure 7.19b, where the compressive yield strength is equal to the tensile yield strength as assumed for ductile materials. In the ***2D*** stress state, the convention of assuming that $\sigma_1 > \sigma_2 > \sigma_3$ is not used since σ_3 is always zero irrespective of the values of σ_1 and σ_2. Therefore in Figure 7.19b (or Figure 7.19a), σ_1 and σ_2 can take any positive or negative value with no restriction that $\sigma_1 > \sigma_2$. For the ***2D*** *stress state*, the *maximum shear stress* theory limit is then

$$\frac{\sigma_1 - \sigma_2}{2} = \frac{S_{yt}}{2}$$
$$\frac{\sigma_1}{S_{yt}} - \frac{\sigma_2}{S_{yt}} = 1 \tag{7.23}$$

Equation (7.23) represents the straight line in both the second and the fourth quadrants in Figure 7.19b. In the first quadrant, if a positive σ_2 is less than σ_1, one needs to consider that σ_3 = zero and the maximum shear stress τ_{max} is then equal to $\sigma_1/2 = S_{yt}$. This is the right limit of the *hexagon* in Figure 7.19b. The same applies for a positive σ_2, which is larger than the positive σ_1. One should then use σ_3 = zero, and the maximum shear stress τ_{max} is then equal to $\sigma_2/2 = S_{yt}$. This is the top limit of the *hexagon* in Figure 7.19b. The same logic applies for the third quadrant of Figure 7.19b, where both σ_1 and σ_2 are negative or compressive.

To use the *maximum shear stress* theory, one should compare the maximum shear stress in the machine element to the shear stress in tension as Eq. (7.23) suggests. To be on the safe side, a use of *safety factor* K_{SF} is advisable. In

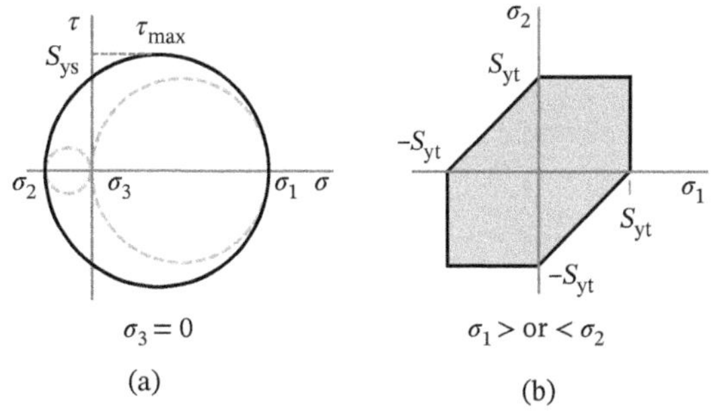

Figure 7.19 (a) The *maximum shear stress* theory in ***2D*** stress state, where the *Mohr's* circle shows σ_2 as compressive, $\sigma_3 = 0$ and the maximum shear stress is the shear strength in simple tension S_{ys}. The failure should occur outside the grayish region in (b), where the compressive yield strength magnitude is equal to the tensile yield strength as assumed for ductile materials.

that case, Eq. (7.23) simply converges to the following form:

$$\frac{\sigma_1 - \sigma_2}{2} = \frac{S_{yt}}{2K_{SF}} \tag{7.24}$$

For the ***3D*** *stress state* as shown in the *Mohr's* circle of Figure 7.16a, adding or subtracting a hydrostatic stress $\sigma_{av} = (\sigma_1 + \sigma_2 + \sigma_3)/3$ from the stress state or to the stress state will only shift the *Mohr's* circle without changing the maximum shear stress $\tau_{max} = \tau_1$. The failure limit of the ***3D*** stress state for the *maximum shear stress* theory is then

$$\frac{\sigma_1 - \sigma_3}{2} = \frac{S_{yt}}{2} \tag{7.25}$$

To be on the safe side again, a use of *safety factor* K_{SF} is once more advisable. In that case, Eq. (7.25) simply converges to a similar form to Eq. (7.24) replacing σ_2 by σ_3.

In Eq. (7.25), the failure consideration is set as the tensile yield strength S_{yt}, which is the most suitable application of the *maximum shear stress* theory for *ductile materials.* For a brittle material, using the ultimate tensile strength S_{ut} may be useful if the ultimate compressive strength S_{uc} is the same as ultimate tensile strength S_{ut}. However, so many brittle materials have their ultimate tensile strength S_{ut} different from their ultimate compressive strength S_{uc}. In these cases of *brittle materials*, different failure criterion such as the *Coulomb–Mohr* theory would be more applicable as indicated latter in Section 7.7.4.

7.7.3 Maximum Distortion Energy Theory (*von Mises*)

J.C. Maxwell in 1856, E. Beltrami in 1885, and M.T. Huber in 1904 proposed the use of energy in yielding criterion. The explanation and the main development are usually attributable to von Mises (1913), after *Richard Edler von Mises* (1883–1953). Hencky in 1925 has about the same contribution. That is why some would refer to the theory as *von Mises–Hencky*. Since *von Mises* was earlier than *Hencky*, it is regularly more attributable to *von Mises.*

The *maximum distortion energy* theory postulates that failure occurs when the *maximum distortion energy* in a machine member reaches the *maximum distortion energy* in the *tensile test*. The development of the maximum distortion energy in a ***3D loaded*** machine element that is made of a certain material in addition to that developed in the tensile test of the same material is definable as follows:

As stated before, for the ***3D*** *stress state* as shown in the *Mohr's* circle of Figure 7.16a, adding or subtracting a *hydrostatic stress* $\sigma_{av} = (\sigma_1 + \sigma_2 + \sigma_3)/3$ (see Eq. 6.103) from the stress state or to the stress state will only shift the *Mohr's* circle without changing the geometry of the *Mohr's* circle. In doing so, the developed stresses are the *deviatoric or distortion stresses* σ_d given by the same equation of (6.104), i.e.,

$$\sigma_d = \left(\frac{(\sigma_1 - \sigma_2)^2 + (\sigma_2 - \sigma_3)^2 + (\sigma_3 - \sigma_1)^2}{2} \right)^{1/2} \tag{6.104}$$

or

$$\sigma_d = \left(\sigma_1^2 + \sigma_2^2 + \sigma_3^2 - \left(\sigma_1\sigma_2 + \sigma_2\sigma_3 + \sigma_3\sigma_1\right)\right)^{1/2} \tag{7.26}$$

With reference to Chapter 6 and Section 6.6, the *strain energy density* u_ε is simply definable by (see Eq. 6.140)

$$u_\varepsilon = \frac{1}{2}\sigma\varepsilon = \frac{\sigma^2}{2E} \tag{7.27}$$

This can apply for any uniaxial stress state. To consider the lateral strains associated with the uniaxial strain, the *Poisson* effect is present (see Eq. 6.62) and for any principal stress, say, σ_1, one should then have the following principal strains:

$$\varepsilon_1 = \frac{\sigma_1}{E}, \quad \varepsilon_2 = -\nu\varepsilon_1 = -\frac{\nu\sigma_1}{E}, \quad \varepsilon_3 = -\nu\varepsilon_1 = -\frac{\nu\sigma_1}{E} \tag{7.28}$$

where ν is the *Poisson's* ratio. Considering the three principal stresses (σ_1, σ_2, σ_3) and using superposition, one gets the following *principal strains* (ε_1, ε_2, ε_3) due to the triaxial stresses.

$$\begin{aligned} \varepsilon_1 &= \frac{\sigma_1}{E} - \frac{\nu\sigma_2}{E} - \frac{\nu\sigma_3}{E} \\ \varepsilon_2 &= \frac{\sigma_2}{E} - \frac{\nu\sigma_1}{E} - \frac{\nu\sigma_3}{E} \\ \varepsilon_3 &= \frac{\sigma_3}{E} - \frac{\nu\sigma_1}{E} - \frac{\nu\sigma_2}{E} \end{aligned} \tag{7.29}$$

Applying Eqs. (7.29) to Eq. (7.27), the total *strain energy density* u_{123} is thus

$$u_{123} = u_{\varepsilon_1} + u_{\varepsilon_2} + u_{\varepsilon_3} = \frac{1}{2E}\left[\sigma_1^2 + \sigma_2^2 + \sigma_3^2 - 2\nu\left(\sigma_1\sigma_2 + \sigma_2\sigma_3 + \sigma_3\sigma_1\right)\right] \tag{7.30}$$

Now, the total strain energy density is the sum of the *hydrostatic strain energy density* u_h due to the average or hydrostatic stress σ_{av} and the deviatoric or *distortion strain energy density* u_d, i.e.,

$$u_{123} = u_h + u_d \tag{7.31}$$

Applying Eqs. (7.27) and (7.28) for the average or *hydrostatic stress* $\sigma_h = \sigma_{av} = (\sigma_1 + \sigma_2 + \sigma_3)/3$, or substituting σ_h for each of the principal stresses in Eq. (7.30), gives

$$\begin{aligned} u_h &= \frac{1}{2E}\left[\sigma_h^2 + \sigma_h^2 + \sigma_h^2 - 2\nu\left(\sigma_h\sigma_h + \sigma_h\sigma_h + \sigma_h\sigma_h\right)\right] \\ u_h &= \frac{1}{2E}\left[3\sigma_h^2 - 2\nu\left(3\sigma_h^2\right)\right] = \frac{3(1-2\nu)}{2E}\left[\sigma_h^2\right] \end{aligned} \tag{7.32}$$

In terms of the principal stresses, Eq. (7.32) becomes

$$\begin{aligned} u_h &= \frac{3(1-2\nu)}{2E}\left[\frac{\sigma_1 + \sigma_2 + \sigma_3}{3}\right]^2 = \frac{(1-2\nu)}{6E}\left[\sigma_1 + \sigma_2 + \sigma_3\right]^2 \\ u_h &= \frac{(1-2\nu)}{6E}\left[\sigma_1^2 + \sigma_2^2 + \sigma_3^2 + 2\left(\sigma_1\sigma_2 + \sigma_2\sigma_3 + \sigma_3\sigma_1\right)\right] \end{aligned} \tag{7.33}$$

The *distortion energy density* u_d is then obtained by Eq. (7.31) or by subtracting Eq. (7.33) from Eq. (7.30) to get

$$\begin{aligned} u_d &= \frac{(1+\nu)}{3E}\left[\sigma_1^2 + \sigma_2^2 + \sigma_3^2 - \left(\sigma_1\sigma_2 + \sigma_2\sigma_3 + \sigma_3\sigma_1\right)\right] \\ u_d &= \frac{(1+\nu)}{3E}\left[\frac{\left(\sigma_1 - \sigma_2\right)^2 + \left(\sigma_2 - \sigma_3\right)^2 + \left(\sigma_3 - \sigma_1\right)^2}{2}\right] \\ u_d &= \frac{1}{12G}\left[\left(\sigma_1 - \sigma_2\right)^2 + \left(\sigma_2 - \sigma_3\right)^2 + \left(\sigma_3 - \sigma_1\right)^2\right] \end{aligned} \tag{7.34}$$

where the *shear modulus of elasticity G* or the *modulus of rigidity* $G = E/2(1 + \nu)$. This clearly demonstrates that the distortion stemming from shear is the fundamental effect in the *distortion energy density* u_d.

The *distortion energy density* u_d in a tensile test specimen failing at the *tensile yielding strength* S_{yt} is then defined by Eq. (7.34). In this case, only the principal stress σ_1 reaching S_{yt} is active and $\sigma_2 = \sigma_3 =$ zero. This gives the following relation for the tensile test:

$$u_d = \frac{(1+\nu)}{3E} S_{yt}^2 = \frac{1}{6\,G} S_{yt}^2 \tag{7.35}$$

The *distortion energy* theory is then defining failure occurrence when the *distortion energy density* u_d in a machine member reaches the failing *distortion energy density* u_d in the tensile test of the same material. This is obtainable by equating the values in Eq. (7.34) to those in Eq. (7.35), i.e.,

$$2S_{yt}^2 = \left(\sigma_1 - \sigma_2\right)^2 + \left(\sigma_2 - \sigma_3\right)^2 + \left(\sigma_3 - \sigma_1\right)^2 \tag{7.36}$$

or

$$S_{yt}^2 = \sigma_1^2 + \sigma_2^2 + \sigma_3^2 - \left(\sigma_1\sigma_2 + \sigma_2\sigma_3 + \sigma_3\sigma_1\right) \tag{7.37}$$

This is the *distortion energy* theory for the failure criterion in a ***3D*** stress state.

In ***2D*** stress state, the *distortion energy* theory is therefore

$$S_{yt}^2 = \sigma_1^2 + \sigma_2^2 - \sigma_1\sigma_2 \tag{7.38}$$

or

$$\frac{\sigma_1^2}{S_{yt}^2} - \frac{\sigma_1\sigma_2}{S_{yt}^2} + \frac{\sigma_2^2}{S_{yt}^2} = 1 \tag{7.39}$$

Equation (7.39) represents an *ellipse*, which is plotted in Figure 7.20a on top of the *hexagon* of the maximum shear stress theory. The area inside the ellipse is the non-yield region. The outside is the failure region.

For the failure in pure shear as defined by the *Mohr's* circle in Figure 7.16c, $\sigma_2 = -\sigma_1$ and the *yield strength in shear* $S_{ys} = \tau_{max} = \sigma_1$. Substituting in Eq. (7.38) gives the following relation of the *yield strength in shear* S_{ys} to the tensile yield strength S_{yt}:

$$S_{yt}^2 = S_{ys}^2 + S_{ys}^2 + S_{ys}S_{ys} = 3S_{ys}^2 \tag{7.40}$$

or

$$S_{ys} = 0.577\,S_{yt} \tag{7.41}$$

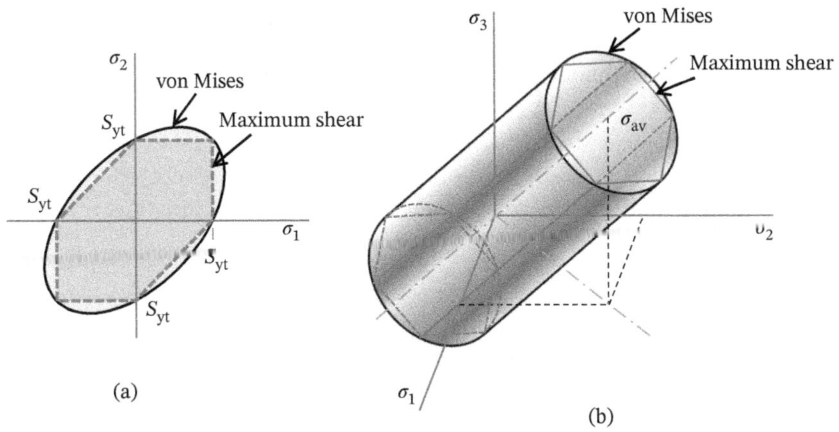

Figure 7.20 The ***2D*** ellipse of the *von Mises* distortion energy theory is plotted on top of the dashed maximum shear stress hexagon in (a). The ***3D*** stress state in (b) shows the cylindrical failure surface of *von Mises* and the inside hexagonal surface of the maximum shear stress theory.

This indicates that the *yield strength in shear* S_{ys} anticipated by the *maximum distortion energy* is more than the anticipated *maximum shear stress* theory identifiable by $S_{ys} = 0.5S_{yt}$ in Eq. (7.22).

In **3D**, the addition or subtraction of the *hydrostatic stress* σ_h or *average stress* σ_{av} causes the **2D** ellipse in the $\sigma_1\ \sigma_2$ plane to transform into the cylindrical failure surface in Figure 7.20b. The cylinder centerline makes equal angles to all σ_1, σ_2, and σ_3 axes. The intersection of the cylinder with the $\sigma_1\ \sigma_2$ plane is the *von Mises* ellipse. The *hexagonal* failure surface of the *maximum shear stress* theory is also shown in Figure 7.20b inside the *von Mises* cylinder.

For ***synthesis*** or ***design*** purposes in **3D**, one can define an *equivalent von Mises stress* σ_{vM}, which is as follows (see Eq. (7.37)):

$$\sigma_{vM} = \sqrt{\sigma_1^2 + \sigma_2^2 + \sigma_3^2 - (\sigma_1\sigma_2 + \sigma_2\sigma_3 + \sigma_3\sigma_1)} \tag{7.42}$$

Comparing the *equivalent von Mises stress* σ_{vM} to the material yield strength S_{yt} provides a good estimate of the *static safety factor* K_{SF}, or in that case, one can state that

$$K_{SF} = \frac{S_{yt}}{\sigma_{vM}} \tag{7.43}$$

In terms of the direct **3D** stress state, Eq. (7.42) is then expandable to the following form:

$$\sigma_{vM} = \sqrt{\frac{(\sigma_x - \sigma_y)^2 + (\sigma_y - \sigma_z)^2 + (\sigma_z - \sigma_x)^2 + 6(\tau_{xy}^2 + \tau_{yz}^2 + \tau_{zx}^2)}{2}} \tag{7.44}$$

For a **2D** stress state, Eq. (7.44) reduces to the following form:

$$\sigma_{vM} = \sqrt{\sigma_x^2 + \sigma_y^2 - \sigma_x\sigma_y + 3\tau_{xy}^2} \tag{7.45}$$

In so many cases, the **2D** state of stress has σ_y = zero, which renders Eq. (7.45) to the following simpler form:

$$\sigma_{vM} = \sqrt{\sigma_x^2 + 3\tau_{xy}^2} \tag{7.46}$$

The *equivalent von Mises stress* σ_{vM} in each of these cases should be compared to the material yield strength S_{yt} to provide a good estimate of the *static safety factor* K_{SF} for each of the design or synthesis case.

7.7.4 Other Failure Theories

The other failure theories are primarily for materials that have unequal failure strength in tension and compression. These are primarily brittle materials. In these cases, the failure is concerned with ultimate strength rather than yield strength. If any other material would have yield strength in tension S_{yt} different from the yield strength in compression S_{yc}, these theories can also be applicable. These yield values should then be replacing the ultimate tensile strength S_{ut} and the ultimate compressive strength S_{uc}, respectively, in all of the following failure theories:

i) ***Mohr's theory*** is dependent on the concept of *Mohr's* circle. For different failure strength in tension, compression, and torsion, three different constructed *Mohr's* circles are as exhibited in Figure 7.21a. The adapted failures are the ultimate tensile strength S_{ut}, ultimate shear strength S_{us}, and the ultimate compressive strength S_{uc}. The *Mohr's* circle of torsion is drawn using a dashed grayish line. The tangent envelopes or loci encompassing the three *Mohr's* circles are the two dashed grayish curves in Figure 7.21a. According to Mohr's theory, failures should then occur beyond these two bounding curves. No failure is possible within these two bounding curves. The bounds can be concave or convex depending on the magnitude of the ultimate shear strength S_{us} relative to the magnitudes of the ultimate tensile strength S_{ut} and the ultimate compressive strength S_{uc} of the material in question.

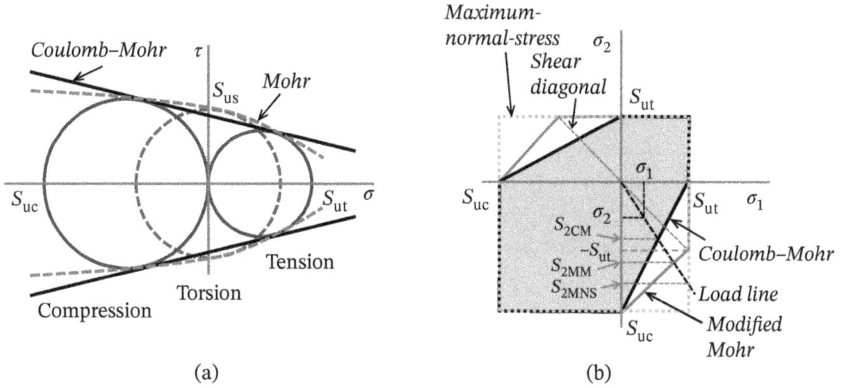

Figure 7.21 Different failure strength in tension, compression, and torsion provide three *Mohr's* circles in (a), where the two dashed grayish envelopes are the bounds of *Mohr's* theory. The two solid black lines in (a) represent the bounds of the *Coulomb–Mohr* theory. The black lines define the *Coulomb–Mohr* theory region filled with grayish color in (b). The grayish solid lines define the *Modified Mohr* theory in (b).

ii) ***Coulomb–Mohr theory*** is originally dependent on the internal friction of the material that affects its failure. The implementation produced the adaption of a linear bounding locus between the tensile and compressive *Mohr's* circles as depicted in Figure 7.21a. The two solid black lines in (a) represent the bounds of the *Coulomb–Mohr* failure theory. This modifies the *maximum normal stress* theory in the second and fourth quadrants of the ***2D*** stress state in Figure 7.21b. As for the *maximum shear stress* theory in the ***2D*** stress state, the convention of assuming that $\sigma_1 > \sigma_2 > \sigma_3$ is not used since σ_3 is always zero irrespective of the values of σ_1 and σ_2. Therefore in Figure 7.21b, σ_1 and σ_2 can take any positive or negative value with no restriction that $\sigma_1 > \sigma_2$. For the ***2D*** *stress state*, the linear relation between σ_1 and σ_2 in the second and fourth quadrant for the *Coulomb–Mohr* theory limit is then

$$\frac{\sigma_1}{S_{ut}} - \frac{\sigma_2}{S_{uc}} = 1 \qquad (7.47)$$

In the first quadrant of Figure 7.21b, if a positive σ_2 is less than σ_1, one needs to consider that σ_3 = zero and the maximum stress σ_1 is then equal to S_{ut}. This is the right grayish filled bound of the *Coulomb–Mohr* theory in the first quadrant in Figure 7.21b. The same applies for a positive σ_2, which is larger than the positive σ_1. One should then use σ_3 = zero and the maximum stress σ_2 is then equal S_{ut}. This is the top limit of the *Coulomb–Mohr* theory in the first quadrant in Figure 7.21b. The same logic applies for the third quadrant of Figure 7.21b, where both σ_1 and σ_2 are negative or compressive. The black lines define the *Coulomb–Mohr* theory region filled with grayish color in Figure 7.21b.

The liming maximum shear stresses in the second and fourth quadrants in Figure 7.21b are derivable by setting σ_1 and σ_2 as the one negative to the other, i.e. $\sigma_1 = -\sigma_2$ or $\sigma_2 = -\sigma_2$.

In Figure 7.21b, the *maximum normal stress* theory bounds are superimposed on the figure for reference indicating the same treatment in the first and third quadrant. The boundary lines of the *maximum normal stress* theory are having the dotted light grayish color.

iii) ***Modified Mohr theory*** is an adjustment to the *Coulomb–Mohr* theory limits in the second and fourth quadrants. It is less conservative than the *Coulomb–Mohr* theory, but would be more representative to the experimental date in the fourth quadrant. The adjusted boundaries are as shown by the solid grayish lines of Figure 7.21b in the second and fourth quadrants. The intersection of the grayish dotted *shear diagonal* line with the constant S_{ut} lines is the corner of the *modified Mohr* theory as defined in Figure 7.21b. Closing these intersections with the *ultimate compressive strength* S_{uc} completes the boundaries of the *modified Mohr* theory. These linear relations are derivable similar to the *Coulomb–Mohr* theory limits.

It is very essential to note that when the larger *ultimate compressive strength* S_{uc} approaches the value of the *ultimate tensile strength* S_{ut}, the *modified Mohr* theory converges to the *maximum shear stress* theory. It is also very important to note that when the *ultimate compressive strength* S_{uc} is much larger than the value of the *ultimate tensile strength* S_{ut}, the *modified Mohr* theory converges to the *Coulomb–Mohr* theory. As the ratio of the *ultimate compressive strength* S_{uc} to the *ultimate tensile strength* S_{ut} approaches unity, the difference between the *modified Mohr* theory and the *Coulomb–Mohr* theory becomes more pronounced. The ratio S_{uc}/S_{ut} in Figure 7.21b is close to 2, which intended to show a pronounced difference between the *modified Mohr* theory and the *Coulomb–Mohr* theory. In this and similar cases, the *modified Mohr* theory is thus much more conservative than the *Coulomb–Mohr* theory.

7.7.5 Comparison and Applications of Failure Theories

A good comparison of failure theories is to observe their expectation of failure in shear due to torsion. As the expected values of a theory gets closer to the experimental results, the more the acceptance and the generality of the theory. The *maximum normal stress* theory predicts that the shear failure due to pure torsion occurs when (see Section 7.7.1 and Figure 7.16c)

$$S_{us} = S_{ut} \quad \text{and} \quad S_{ys} = S_{yt} \tag{7.48}$$

The ultimate tensile and compressive strengths S_{ut} and S_{uc} are conventionally for *brittle* materials. These results, however, do not agree well with many experiments. For ductile materials, experiments show that the torsional yield strength is about 0.5–0.6 of that found from tension test.

The *maximum shear stress* theory predicts that the shear failure due to pure torsion occurs when (see Section 7.7.2 and Eq. (7.22)

$$S_{ys} = 0.5S_{yt} \quad \text{and} \quad S_{us} = 0.5S_{ut} \tag{7.49}$$

The ultimate strengths S_{us} and S_{ut} are foreseeable for *brittle* materials.

The *maximum distortion energy* theory predicts that the shear failure due to pure torsion occurs when (see Section 7.7.3 and Eq. (7.41))

$$S_{ys} = 0.577S_{yt} \tag{7.41}$$

These results do agree well with many experiments particularly for *ductile* materials.

The applicability of the aforementioned theories is then dependent on the type and properties of the intended material. The suggested adaption of the theory is as follows:

- ***Ductile materials*** are better selectable by the *maximum distortion energy* theory. The next choice is the *maximum shear stress* theory due to simplicity and inclination that is more conservative.
- ***Brittle materials*** are better selectable by the *Coulomb–Mohr* theory particularly for unequal tensile and compressive strengths. It is simple and on the conservative side. The next choice is the *maximum normal stress* theory or may be the *modified Mohr* theory.

The application of different failure theory in ***2D*** stress state depends on the loading ratio σ_1/σ_2, which defines the loading line as demonstrated in Figure 7.21b. A simple way to find the *safety factor* K_{SF} in the fourth quadrant is to calculate the *failure strength projection* S_2 (on the σ_2 axis) for each theory by the intersection of the *load line* with the failure boundaries of each theory. The *load line* is the dashed black line in Figure 7.21b. Comparing the failure strength projections S_2's with the *principal stress* σ_2 defines the *safety factor* K_{SF}. With reference to Figure 7.21b, the following is the failure strength S_2 for each failure theory and the associated safety factor for the fourth quadrant.

For *maximum normal stress* theory and after some handling, the *failure strength projection* S_{2MNS} is possible to get that

$$S_{2MNS} = \frac{\sigma_2}{\sigma_1} S_{ut}, \quad \sigma_2 \geq S_{uc} \tag{7.50}$$

The safety factor $K_{SF,MNS}$ is then

$$K_{SF,MNS} = \frac{S_{2MNS}}{\sigma_2} = \frac{S_{ut}}{\sigma_1}, \quad \sigma_2 \geq S_{uc} \tag{7.51}$$

For *Coulomb–Mohr* theory and after some manipulation, the *failure strength projection* S_{2CM} is possible to get that

$$S_{2CM} = \frac{S_{uc}}{\frac{S_{uc}}{S_{ut}} \frac{\sigma_1}{\sigma_2} - 1} \tag{7.52}$$

The safety factor $K_{SF,CM}$ is then

$$K_{SF,CM} = \frac{S_{2CM}}{\sigma_2} = \frac{S_{uc}}{\frac{S_{uc}\sigma_1}{S_{ut}} - \sigma_2} = \frac{S_{uc}S_{ut}}{S_{uc}\sigma_1 - S_{ut}\sigma_2} \tag{7.53}$$

For the *modified Mohr* theory and after some manipulation, the *failure strength projection* S_{2MM} is possible to obtain that

$$\begin{aligned} S_{2MM} &= \frac{\sigma_2}{\sigma_1} S_{ut}, \quad \sigma_2 \geq -S_{ut} \\ S_{2MM} &= \frac{S_{uc}}{\frac{S_{uc} - S_{ut}}{S_{ut}} \frac{\sigma_1}{\sigma_2} - 1}, \quad \sigma_2 \leq -S_{ut} \end{aligned} \tag{7.54}$$

The safety factor $K_{SF,MM}$ is then

$$\begin{aligned} K_{SF,MM} &= \frac{S_{2MM}}{\sigma_2} = \frac{S_{ut}}{\sigma_1}, \quad \sigma_2 \geq -S_{ut} \\ K_{SF,MM} &= \frac{S_{2MM}}{\sigma_2} = \frac{S_{uc}}{\frac{(S_{uc} - S_{ut})\,\sigma_1}{S_{ut}} - \sigma_2} = \frac{S_{uc}S_{ut}}{S_{uc}\sigma_1 - S_{ut}(\sigma_1 + \sigma_2)}, \quad \sigma_2 \leq -S_{ut} \end{aligned} \tag{7.55}$$

The top of Eq. (7.55) indicates that when $\sigma_2 > -S_{ut}$, the case is the same as that of the *maximum normal stress* theory; see Figure 7.21b for verification.

For the ***3D*** stress state, every term of σ_2 in Eqs. (7.50)–(7.55) is readily replaceable by σ_3. The σ_2 coordinate in Figure 7.21b is also replaceable by σ_3 coordinate.

Example 7.3 A shaft is transmitting a maximum of 25 [kW] or 33.5 [hp] and running at 3000 [rpm] in the clockwise direction. It is subject to a maximum bending moment that is equal to the maximum applied torque. Find the internal stresses if the shaft has a cylindrical diameter of 45 [mm] or 1.75 [in]. The selected material is set as HR AISI 1040 or ISO C40. Find the safety factor for each applicable theory of failure.

If this shaft function is as a camshaft made of gray cast iron, find the safety factor for each applicable theory of failure. This camshaft can be made of the gray cast iron designation ASTM A 48 class 40 or ISO 185 300.

Solution

Data: $H = 25$ [kW] or 33.5 [hp], $N_{rpm} = 3000$ [rpm], and $d_C = 0.045$ [m] or 1.75 [in]. *Ductile* material is HR AISI 1040 or ISO C40. *Brittle* material is ASTM A 48 class 40 or ISO 185 300.

Ductile material properties are yield strength of 42 [kpsi] or 290 [MPa] and ultimate tensile strength of 76 [kpsi] or 525 [MPa]; see Table A.7.2 in Appendix A.7.

Brittle material properties are ultimate tensile strength of 40 [kpsi] or 300 [MPa] and ultimate compressive strength of 140 [kpsi] or 965 [MPa]; see Table A.7.1 in Appendix A.7.

The shaft is similar to Example 6.10. The calculated results in Example 6.10 that can be applicable to this example are as follows:

The maximum shear stresses due to torque τ_{xz} are 4.4475 [MPa] or 0.668 79 [kpsi].

The maximum normal stresses due to bending moment σ_x are 8.8951 [MPa] or 1.3376 [kpsi].

The principal stresses σ_1, σ_2 are 10.737 [MPa] or 1.615 [kpsi] and −1.8422 [MPa] or −0.277 [kpsi], respectively.

The maximum shear stress τ_{max} is 6.2897 [MPa] or 0.945 81 [psi].

The distortion or deviatoric stress is 11.7668 [MPa] or 1.7698 [kpsi].

For the ***ductile shaft*** the applicable theories are as follows:

(i) The *maximum distortion energy* theory requires the calculation of the *equivalent von Mises stress* σ_{vM} noting that it is a ***2D*** *stress state* with σ_3 = zero. Eq. (7.42) becomes

$$\sigma_{vM} = \sqrt{\sigma_1^2 + \sigma_2^2 - \sigma_1\sigma_2} = \sqrt{(10.737)^2 + (-1.8422)^2 - (10.737 \times -1.8422)} = 11.767 \text{ [MPa]}$$

$$\sigma_{vM} = \sqrt{\sigma_1^2 + \sigma_2^2 - \sigma_1\sigma_2} = \sqrt{(1.615)^2 + (-0.277)^2 - (1.615 \times -0.277)} = 1.7698 \text{ [kpsi]} \tag{a}$$

Note that these *equivalent von Mises stresses* σ_{vM} are the same as the previously calculated distortion or deviatoric stress as 11.7668 [MPa] or 1.7698 [kpsi] from Example 6.10. Comparing the *equivalent von Mises stress* σ_{vM} to the material yield strength S_{yt} provides the *static safety factor* K_{SF} from Eq. (7.43) such that

$$K_{SF}\big|_{SI} = \frac{S_{yt}}{\sigma_{vM}} = \frac{290}{11.767} = 24.645$$

$$K_{SF}\big|_{US} = \frac{S_{yt}}{\sigma_{vM}} = \frac{42}{1.7698} = 23.731 \tag{b}$$

These values are extremely high, which indicates a very inefficient design as based on static calculations. However, later on the evaluation will be dependent on dynamic failure and not only on static failure.

(ii) The *maximum shear stress* theory compares the maximum shear stress to the shear stress in tension. Knowing that the maximum shear stress τ_{max} is equal to $(\sigma_1 - \sigma_2)/2$, using Eq. (7.24) gives the following safety factor K_{SF}:

$$K_{SF}\big|_{SI} = \frac{S_{yt}}{2\left((\sigma_1 - \sigma_2)/2\right)} = \frac{290}{2(6.2897)} = 23.05$$

$$K_{SF}\big|_{US} = \frac{S_{yt}}{2\left((\sigma_1 - \sigma_2)/2\right)} = \frac{42}{2(945.81/1000)} = 22.20 \tag{c}$$

These values are still extremely high. However, they are little more conservative than the values obtained by the *maximum distortion energy* theory in Eq. (a).

For the ***brittle shaft,*** the applicable theories are as follows:

(iii) The *Coulomb–Mohr* theory needs the application of Eq. (7.53) or applying Eq. (7.47) including the safety factor similar to Eq. (7.24), which gives

$$K_{SF}\big|_{SI} = \frac{1}{(\sigma_1/S_{ut}) - (\sigma_2/S_{uc})} = \frac{1}{(10.737/300) - (-1.8422/-965)} = 29.515$$

$$K_{SF}\big|_{US} = \frac{1}{(\sigma_1/S_{ut}) - (\sigma_2/S_{uc})} = \frac{1}{(1.615/40) - (-0.277/-140)} = 26.044 \tag{d}$$

These values are again very high. The discrepancy between them (29.515 against 26.044) is because the quoted value of the yield strength in *SI* is 300 [MPa], i.e. 43.5 [kpsi], which is higher than the quoted value of the yield strength in the *US* as 40 [kpsi].

(iv) The *maximum normal stress* theory involves the application of Eq. (7.51) for the safety factor K_{SF}, which gives

$$K_{\text{SF,MNS}}\big|_{\text{SI}} = \frac{S_{\text{ut}}}{\sigma_1} = \frac{300}{10.737} = 27.94, \quad \sigma_2\,(= -1.8422) \geq S_{\text{uc}}\,(= -965)\ [\text{MPa}]$$
$$K_{\text{SF,MNS}}\big|_{\text{US}} = \frac{S_{\text{ut}}}{\sigma_1} = \frac{40}{1.615} = 24.768, \quad \sigma_2\,(= -0.277) \geq S_{\text{uc}}\,(= -140)\ [\text{kpsi}] \tag{e}$$

These values are still very high. However, they are more conservative than the values obtained by the *Coulomb–Mohr* theory in Eq. (d).

(v) The *modified Mohr* theory entails the application of Eq. (7.55) for the safety factor $K_{SF,MM}$, which indicates that as $\sigma_2 > -S_{ut}$, the case is the same as that of the *maximum normal stress* theory. The safety factor $K_{SF,MM}$ is the same as defined by Eq. (e).

7.8 Fatigue Strength and Factors Affecting Fatigue

Realistic loading subjects machine elements to random and dynamic loading that causes an anticipated typical internal loading as shown in Figure 7.22a. Such a dynamically random loading may have a simplified internal stressing shown in Figure 7.22b. This model of generated internal stressing, however, is an acceptable representation of loading due to bending moments and axial forces of a rotating shaft. This is why the *fatigue strength* is obtained by testing a standard specimen loaded by pure bending that causes an internal stressing similar to Figure 7.22b but with zero *mean stress*, i.e. σ_m = zero. The existence of a loading that has a mean stress σ_m is an additional fatigue condition to tackle afterward. That case of Figure 7.22b is representable as the *fluctuating stresses* covered in Section 7.8. The values of the *mean stress* σ_m and *alternating stress* σ_a as functions of the *maximum stress* σ_{max} and the *minimum stress* σ_{min} are simply as follows:

$$\sigma_m = \frac{1}{2}\left(\sigma_{\max} + \sigma_{\min}\right) \tag{7.56}$$

$$\sigma_a = \frac{1}{2}\left(\sigma_{\max} - \sigma_{\min}\right) \tag{7.57}$$

These values are essential in dealing with *fluctuating stresses* and in recognizing the condition of σ_m = zero in the determination of the *endurance limit* S_e.

The concept of *fatigue strength* depends on the mechanism and stages of its occurrence. It is contingent upon internal phenomena of *crack initiation*, *crack propagation*, and *crack instability* to sudden fracture when the material is subjected to a repeated high loading for many cycles to failure. ***Crack initiation*** stage occurs usually at

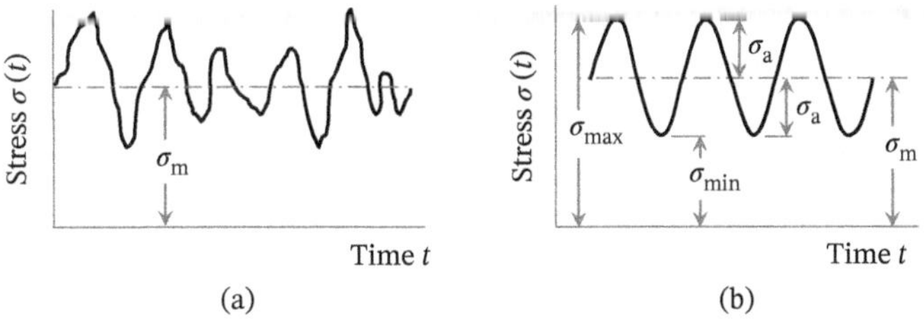

Figure 7.22 Anticipated typical internal stress loading shown in (a) due to a realistic external random input. The simplified modeling of internal loading shown in (b) due to the random external input or the model of the internal loading due to bending and axial load of a rotating shaft.

stress raisers or stress concentrations such as geometrical discontinuities, high contact stresses, fabrication faults, and microscopic material inhomogeneities. It is accelerated under corrosive environment and residual stresses. The ***crack propagation*** stage starts from the initiated crack, and the high stress concentration at the tip of the crack causes a failure that will arrest the crack, then when the stress goes in a cycle, the failed arrested tip will break causing the crack to propagate. This process may be similar to the *strain hardening* that causes the *embrittlement* of the material at the crack tip; see Figure 7.2d. As soon as the propagated crack reaches a critical size, *instability* and a sudden *brittle fracture* occur; see Section 7.9 about *fracture mechanics* and *toughness*.

Several factors accelerate the advance of crack initiation or its propagation. The fatigue life process is therefore affected by these factors. The factors affecting fatigue are discussed in Section 7.8.2.

A few fatigue life evaluation methods are available. The most suitable one depends on the loading type being applied to machine elements. The regular fatigue ***stress-life*** method, mostly used in this text, is suitable for high cyclic and infinite cyclic loading. The long-established use of the *endurance limit* is the traditional *stress-life* technique sustained by a wide range of supporting data. The ***strain-life*** method is expected to be suitable for *low-cycle* fatigue. It is possibly appropriate for high plastic strain due to the high level of stressing; see Section 7.8.5. However, the limited supporting data is hindering the widespread effectiveness of this method. The ***fracture mechanics*** approach is dedicated to the estimate of the crack propagation rate. This would relate the *stress intensity* and *fracture toughness* of the material; see Section 7.9. The fracture toughness is useful when some inspection, monitoring, and computer modeling procedures are present.

In this section, the dedication and focus of treatment is on the conventional fatigue *stress-life* method.

7.8.1 Fatigue Strength

Fatigue strength of a material has been introduced in Section 7.1.5. Fatigue properties of different materials in terms of their *endurance limit* S_e are available in Appendix A.7. The *endurance limit* S_e as an infinite life property is obtainable for lives in the range of (10^6)–(10^7) cycles for steels and similar ductile materials. Figure 7.23 is the same *S–N* diagram as Figure 7.5 but further identifying regions of *low-cycle fatigue, high-cycle fatigue, finite fatigue life*, and *endurance limit* for ductile materials. It is very useful in defining the *fatigue strength* S_f for *finite-life* between 10^0 and 10^6 loading cycles.

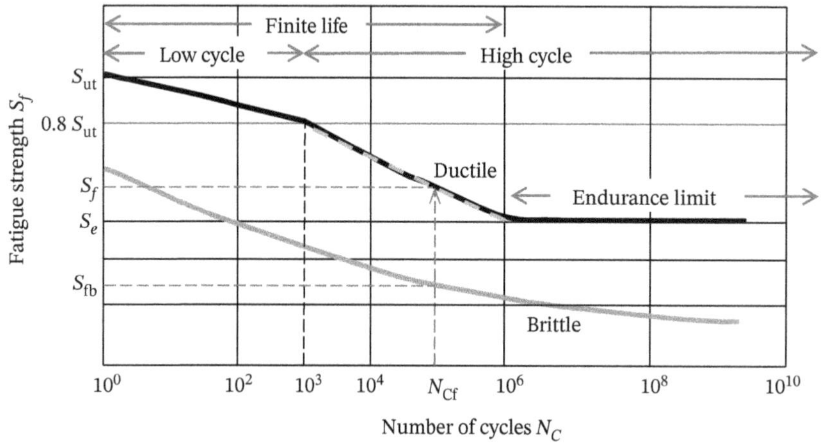

Figure 7.23 *S–N* diagram of fatigue strength versus the number of loading cycles for a representative of a ductile material and one of the brittle materials. Regions of low-cycle fatigue, high-cycle fatigue, finite life, and endurance limit are shown for ductile materials.

It is always advisable that the *designer* gets the *endurance limit* S_e of his selected material through the manufacturer, or the standards or the thorough and sufficient number of fatigue testing that provides the mean and standard deviation of S_e of his material or his machine element. If that is too costly for the machine element in question, some of the following selected empirical formulas may be cautiously usable with adaption of extra conservative safety factor. The dispersion of the fatigue strength data, particularly endurance limit, is usually large and the standard deviation of data can reach a 10%, which is high (Stulen et al. 1961). For engineering *students*, the following relations are useful in estimation and comparison of alternative material selection. They should, however, be handled with care.

7.8.1.1 Estimation of Endurance Limit

For ***wrought steel,*** the conventional relation between the *ultimate tensile strength* S_{ut} and the *endurance limit* S_e is as follows:

$$S_e \cong 0.5S_{ut},\ S_{ut} \le 1380\ [\text{MPa}]\quad \text{or}\quad 200\ [\text{kpsi}]$$
$$S_e \cong 690\ [\text{MPa}]\quad \text{or}\quad 100\ [\text{kpsi}],\ S_{ut} \ge 1380\ [\text{MPa}]\quad \text{or}\quad 200\ [\text{kpsi}] \tag{7.58}$$

The 0.5 constant in Eq. (7.58) is usually given the name of the *endurance ratio* S_{ut}/S_e. This 0.5 constant in Eq. (7.58) may be reduced to 0.4, and the 690 [MPa] or 100 [kpsi] may be reduced to 550 [MPa] or 80 [kpsi] for more safety. It should be critical to note that the *microstructure* of steel affects the 0.5 constant in Eq. (7.58); see Sors (1971). The 0.5 constant can take a value in the range of 0.57–0.63 for *ferritic* carbon steel. For *pearlite* carbon steels, the 0.5 constant would be in the range of 0.38–0.41. For *martensitic* carbon steels, the 0.5 constant can have an average of 0.25. For *martensitic alloy steels*, the 0.5 constant can have a range of 0.23–0.47.

For ***wrought iron*** or ***cast iron*** and ***cast steel,*** the estimated relation between the *ultimate tensile strength* S_{ut} and the *endurance limit* S_e is as follows:

$$S_e \cong 0.4S_{ut},\ S_{ut} \le 410\ [\text{MPa}]\quad \text{or}\quad 60\ [\text{kpsi}]$$
$$S_e \cong 165\ [\text{MPa}]\quad \text{or}\quad 24\ [\text{kpsi}],\ S_{ut} \ge 410\ [\text{MPa}]\quad \text{or}\quad 60\ [\text{kpsi}] \tag{7.59}$$

Equation (7.59) is very conservative. Some alternative *endurance ratio* S_{ut}/S_e is set as 0.45 rather than 0.4. The applicability region may also be extended to 600 [MPa] or 85 [kpsi], and the *endurance limit* beyond that is set at 275 [MPa] or 40 [kpsi].

For ***aluminum alloys,*** the expected relation between the *ultimate tensile strength* S_{ut} and the *endurance limit* S_e is as follows:

$$S_e|_{5\times10^8} \cong 0.4S_{ut},\ S_{ut} \le 330\ [\text{MPa}]\quad \text{or}\quad 48\ [\text{kpsi}]$$
$$S_e|_{5\times10^8} \cong 130\ [\text{MPa}]\quad \text{or}\quad 19\ [\text{kpsi}],\ S_{ut} \ge 330\ [\text{MPa}]\quad \text{or}\quad 48\ [\text{kpsi}] \tag{7.60}$$

Since the endurance limit of *aluminum alloys* is usually monotonically decreasing, the endurance limit in Eq. (7.60) is defined at $5(10^8)$ cycles rather than (10^6) to (10^7) cycles; see Figure 7.23.

For ***copper alloys,*** the likely relation between the *ultimate tensile strength* S_{ut} and the *endurance limit* S_e is set as follows:

$$S_e|_{5\times10^8} \cong 0.4S_{ut},\ S_{ut} \le 280\ [\text{MPa}]\quad \text{or}\quad 40\ [\text{kpsi}]$$
$$S_e|_{5\times10^8} \cong 97\ [\text{MPa}]\quad \text{or}\quad 14\ [\text{kpsi}],\ S_{ut} \ge 280\ [\text{MPa}]\quad \text{or}\quad 40\ [\text{kpsi}] \tag{7.61}$$

Again, since the endurance limit of *copper alloys* could be monotonically decreasing, the endurance limit in Eq. (7.61) is defined at $5(10^8)$ cycles rather than (10^6) to (10^7) cycles; see Figure 7.23.

For ***plastics*** and ***composites***, the *endurance ratio* S_{ut}/S_e beyond (10^7) cycles is in the range of 0.18–0.43.

7.8.1.2 Estimation of Fatigue Strength

This estimation of *fatigue strength* S_f is for a number of cycles (10^0–10^6) that is lower than the usual (10^6) to (10^7) cycles for the *endurance limit* S_e. The number of cycle range of (10^0–10^3) is the *low-cycle fatigue*, and number of cycle range of (10^3–10^6) is for the *high-cycle fatigue* (Figure 7.23). First, we develop the usual *high-cycle fatigue*, since the *low-cycle fatigue* may be treated by different means such as strain-life method of overstressing.

At one cycle (or less) of loading, the expected failure strength would be the *ultimate tensile strength* S_{ut} as shown in Figure 7.23 at 10^0 cycles. At loading of 10^3 cycles, the *fatigue strength* is about to be 0.8 of the *ultimate tensile strength* S_{ut}. This ratio of 0.8 is a *failure ratio* (*fr*) of the fatigue strength at 10^3 cycles to the ultimate tensile strength. Using this observation and the information that the *endurance limits* S_e are at 10^6 cycles provides the means to estimate the S_f in-between. Using a linear relation in *log–log* scale of *S–N* curves such as

$$\log S_f = a_1 \log N_{Cf} + \log a_2 \tag{7.62}$$

Substituting the boundary conditions at 10^3 cycles and 10^6 cycles and solving gives the a_1 and a_2 constants as

$$a_1 = \frac{\log \frac{0.8S_{ut}}{S_e}}{\left(\log 10^3 - \log 10^6\right)} \tag{7.63}$$

and

$$a_2 = \frac{\left(0.8S_{ut}\right)^2}{S_e} \tag{7.64}$$

By substituting Eqs. (7.63) and (7.64) into Eq. (7.62) and after minor manipulation, the *fatigue strength* S_f is then

$$S_f = a_2\ N_{Cf}^{a_1}$$

or

$$N_{Cf} = \left(\frac{S_f}{a_2}\right)^{1/a_1} \tag{7.65}$$

The value of ($0.8S_{ut}$) in Eqs. (7.63) and (7.64) can be different depending on the value of S_{ut}. In place of the 0.8 *failure ratio* (*fr*), another constant can be set in the range of 0.9–0.75 depending on S_{ut}. The high value of 0.9 *failure ratio* is for smaller S_{ut}, and the low value of 0.75 *failure ratio* is for higher value of S_{ut}. The value of the *failure ratio* can be estimated by the relation ($fr = 6\text{E}{-}06S_{ut}^2 - 0.0026S_{ut} + 1.0494$), where S_{ut} is only in [kpsi]. Equation (7.65) is, however, very useful in several machine element applications that the number of loading cycles are much less than the expected 10^6 during the total life span of the element to warrant using the endurance limit as the limiting strength of the material. One of these applications is a boiler that is operational daily at its maximum pressure, but is shutdown during the night shift. All associated mechanical components are exposed to much less life cycles than 10^6.

Similar expression to Eq. (7.65) is derivable for the range of *low-cycle fatigue* if the linear relation on the log–log scale is an acceptable representation of fatigue failure at the low-cycle zone. Several professionals, however, prefer to use *strain-life* method for such a zone.

7.8.2 Factors Affecting Fatigue Strength

So many factors affect the fatigue strength since the properly conducted tests are under some explicit conditions such as mirror-polished surface, specimen size, fully reversed bending, and definite room temperature. The machine component under consideration would not necessarily perform under these explicit constraints. The component is possibly made of different material composition and variability. The component design would have a different size, shapc, stress raisers, and speed. The manufacturing of the machine component can have different surface finish, microstructure, surface treatment, and internal stress concentration or pre-stressing. The operational environment is having different temperature, combined loading, extended time loading, and corrosive surrounding. All these factors would affect the *actual fatigue strength* of the machine component different from the constrained condition of the *endurance limit* of the material specimen. Some of the main factors affecting the fatigue strength are covered or discussed in this section. All other factors not being covered herein are considered under some miscellaneous factors.

When a component or machine part is subjected to dynamic loading, the *fatigue strength of the part* $S_{e,\text{part}}$ may then be relatable to the material fatigue strength or *endurance limit* S_e by the following amended relation of Marin (1962):

$$S_{e,\text{part}} = K_{\text{surf}} K_{\text{size}} K_{\text{load}} K_{\text{reliab}} K_{\text{temp}} K_{\text{conc}} K_{\text{miscel}} S_e \qquad (7.66)$$

where K_{surf} is surface factor, K_{size} is size factor, K_{load} is the load factor, K_{reliab} is a reliability factor, K_{temp} is a temperature factor, K_{conc} is related to the previously discussed *stress concentration factor* K_{SC}, and K_{miscel} is for other miscellaneous factors. If the numbers of cycles are less than that of the *endurance limit* S_e, the *fatigue strength* S_f as in Eq. (7.65) would replace S_e in Eq. (7.66).

7.8.2.1 *Surface Factor, K_{surf}*

This factor is mainly considering the surface finish produced by the manufacturing processing of the part or machine component. It has been under consideration for a long time; see Karpov (1939) and Lipson et al. (1950). It is also relevant to *notch sensitivity* of the material when notches are close to surface roughness size. Many experiments have been available to estimate the effect of surface condition on the endurance limit. Existing charts have been also available to identify the *surface factor* K_{surf} dependent on these experimental data (Shigley 1986; Juvinall and Marshek 2012). The charts, however, are not identical with some noticeable differences. The estimation of *surface factor* K_{surf} for each manufacturing processing was also available (Budynas and Nisbett 2015). Plotted in Figure 7.24 are the closer data to each other. The information in the existing chart, their trend lines, and the available estimator's equations reveal that there are some discrepancies. The grayish dotted lines and their black solid trend lines in Figure 7.24 are for the chart data in Shigley (1986). The estimated *SI* power equations of the trend lines are next to each line in Figure 7.24. The black dashed lines are for the estimated power equations of *surface factors* K_{surf} available in Budynas and Nisbett (2015) for each displayed manufacturing processing as referenced to Noll and Lipson (1946). The discrepancies in the estimated power equations for *surface factors* K_{surf} are apparent among each of the manufacturing processing. The displayed trend line equations are more conservative for lower tensile strength but more offbeat for the higher tensile strength values. Advisable, therefore, that one might need to conduct tests for the real part of the design under production. The recommended equations with the first as Shigley's (1986) trend lines and the second as of Budynas and Nisbett (2015) power equations are as follows:

For ***ground*** products:

$$\begin{aligned} K_{\text{surf}} &= 0.89\, S_u^{-16} = 0.89 \quad \text{for SI and US} \\ K_{\text{surf}}\big|_{\text{SI}} &= 1.58\, S_u^{-0.085} \text{ and } K_{\text{surf}}\big|_{\text{US}} = 1.34\, S_u^{-0.085} \end{aligned} \qquad (7.67)$$

For ***machined or CD*** products:

$$\begin{aligned} K_{\text{surf}}\big|_{\text{SI}} &= 2.9458\, S_u^{-0.212} \text{ and } K_{\text{surf}}\big|_{\text{US}} = 1.9572\, S_u^{-0.212} \\ K_{\text{surf}}\big|_{\text{SI}} &= 4.51\, S_u^{-0.265} \text{ and } K_{\text{surf}}\big|_{\text{US}} = 2.70\, S_u^{-0.265} \end{aligned} \qquad (7.68)$$

For ***HR*** products:

$$\begin{aligned} K_{\text{surf}}\big|_{\text{SI}} &= 19.878\, S_u^{-0.553} \text{ and } K_{\text{surf}}\big|_{\text{US}} = 6.8297\, S_u^{-0.553} \\ K_{\text{surf}}\big|_{\text{SI}} &= 57.7\, S_u^{-0.718} \text{ and } K_{\text{surf}}\big|_{\text{US}} = 14.4\, S_u^{-0.718} \end{aligned} \qquad (7.69)$$

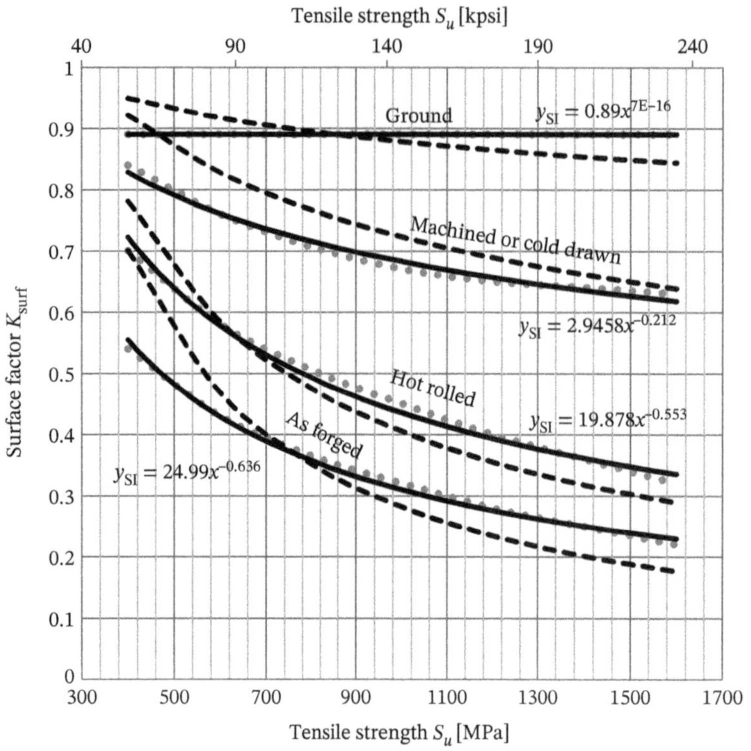

Figure 7.24 Values of surface factor K_{surf} for different manufacturing processing against the ultimate tensile strength. The grayish dotted lines and their black solid trend lines are for chart data in Shigley (1986). *SI* trend line equations are next to each line. The black dashed lines are for power equations of surface factors in Budynas and Nisbett (2015). The top horizontal *US* scale is an approximation to the values.

For ***as forged*** products:

$$K_{surf}|_{SI} = 24.99\, S_u^{-0.636} \text{ and } K_{surf}|_{US} = 7.3247\, S_u^{-0.636}$$
$$K_{surf}|_{SI} = 272\, S_u^{-0.995} \text{ and } K_{surf}|_{US} = 39.9\, S_u^{-0.995} \tag{7.70}$$

Again, for low tensile strength one may use the first line of each of Eqs. (7.67)–(7.70) since they are more conservative. Except for machined or CD processing and for high tensile strength, one may use the second line of each of Eqs. (7.67)–(7.70) since they are more conservative.

Ranges of *surface roughness* values for the forgoing manufacturing processes are obtainable as defined in Section 2.4.7 of the text.

7.8.2.2 *Size Factor, K_{size}*

The fitting of fatigue data by Mischke (1987), as reported by Budynas and Nisbett (2015), gives the following expressions for the size factor under bending or torsion:

For relatively small sizes of 2.8–50 [mm] or 0.11–2 [in], the size factor for a machine *part diameter* d_P is

$$K_{size}|_{SI} = 1.24 d_P^{-0.107} \quad \text{or} \quad K_{size}|_{US} = 0.879 d_P^{-0.107} \tag{7.71}$$

For larger size range of 50–254 [mm] or 2–10 [in], the size factor for a part diameter d_P is

$$K_{size}|_{SI} = 1.51 d_P^{-0.157} \quad \text{or} \quad K_{size}|_{US} = 0.91 d_P^{-0.157} \tag{7.72}$$

One should observe that the parts in Eqs. (7.71) and (7.72) are solid cylindrical shapes. For other shapes, one should account for the variations by the utilization of an *equivalent part diameter* $d_{P,e}$ according to some condition of equal volume to the 95% of maximum stressing such as suggested by Kuguel (1961). Using such a condition provides an *equivalent part diameter* $d_{P,e} = 0.37\, d_P$ for a hollow round section and $d_{P,e} = 0.808\,(h_R b_R)^{1/2}$ for rectangular section height h_R and breadth b_R, respectively.

7.8.2.3 *Loading Factor, K_{load}*

Since the values of the endurance limit are generated from fully reversed bending tests, therefore when the loading is bending, the *loading factor* should be one or $K_{load} = 1.0$. For axial or torsion, the experiments show lower values of the endurance limit than the regular endurance limit evaluated from fully reversed bending tests (Grover et al. 1960). From their tests on the ultimate strength variation due to different loading, it can extrapolate to the same relation between different endurance limits. This can lead to the following estimated *loading factors* for each of the individual *axial* and *torsional* loading:

$$K_{load}\big|_{axial} = 0.85$$
$$K_{load}\big|_{torsion} = 0.59 \qquad (7.73)$$

One can observe that the loading factor for torsion is very close to the *von Mises* ratio of 0.577 between the shear and tensile strength failures as depicted in Equation (7.41) for failure by ductile yielding.

7.8.2.4 *Reliability Factor, K_{reliab}*

Due to the scatter of experimental data in the evaluation of the *endurance limit* as depicted in *Grover* et al. (1960), a confidence in the adaption of a value for the *endurance limit* or a *reliability factor* K_{reliab} is in need. The reliability or the probability of unfailing is defined by the area under the probability density function of the *Gaussian* or *normal distribution* beyond the *mean* or average value by units of standard deviations. Beyond the *mean* value, the area under the *normal distribution* is 0.5, which indicates a 0.5 (or 50%) *probability of failure* and thus a *reliability* of (1 − *probability of failure*) 0.5, i.e. 50% (Haugen 1980). Therefore, if one accepts a 50% reliability in the endurance limit value (or 50% of lower value occurrences), the *reliability factor* K_{reliab} should be one, or $K_{reliab} = 1.0$. By the same logic, the *reliability factor* K_{reliab} at some accepted conventional *reliability percentages* is obtainable as depicted in Table 7.5. For computer usage, one may use the following equations: from 50% to 99% reliability, $K_{reliab} = 0.805\,25(100 - \text{reliability})^{0.052\,45}$ and from 99% to 99.999 999% reliability, $K_{reliab} = 0.804\,83(100 - \text{reliability})^{0.027\,97}$. The calculated the *reliability factor* $K_{reliab}|_{calc}$ is shown in Table 7.5 for comparison and error assessment.

7.8.2.5 *Temperature Factor, K_{temp}*

The effect of temperature on the endurance limit can be similar to the same effect on the ultimate strength of some steels (Brandes and Brook 1992). From that, the following empirical relation is available to apply for the

Table 7.5 Reliability factor K_{reliab} at some conventional reliability percentages.

Reliability (%)	50	90	95	99	99.9	99.99	99.999	99.9999	99.99999	99.999999
K_{reliab}	1	0.897	0.868	0.814	0.753	0.702	0.659	0.62	0.584	0.551
$K_{reliab}\|_{calc}$ [a)]	0.989	0.909	0.876	0.805	0.755	0.708	0.663	0.622	0.583	0.547

a) From 50% to 99% reliability, one may use $K_{reliab} = 0.805\,25\,(100 - \text{reliability})^{0.052\,45}$. From 99% to 99.999 999% reliability, one may use $K_{reliab} = 0.804\,83\,(100 - \text{reliability})^{0.027\,97}$.

temperature factor:

$$K_{\text{temp}}\big|_{\text{SI}} = -2.325\,(10)^{-6}T_O^2 + 0.0007178\;T_O + 0.9762, \quad T_O \text{ in } [°\text{C}]$$
$$K_{\text{temp}}\big|_{\text{US}} = -7.024\,(10)^{-7}T_O^2 + 0.0004273\;T_O + 0.9664, \quad T_O \text{ in } [°\text{F}] \tag{7.74}$$

where T_O is the operating temperature.

7.8.2.6 Fatigue Concentration Factor, K_{conc}

The *fatigue concentration factor* K_{conc} that affects fatigue strength in Eq. (7.66), and the *stress concentration factors* K_{SC} discussed in Section 6.10 are interdependent. Traditionally, another *fatigue stress concentration factor* K_f is defined as the ratio of the endurance limit of regular specimen S_e to the *endurance limit of notched specimen* S_{eN}.

$$K_f = \frac{S_e}{S_{\text{eN}}} \tag{7.75}$$

This means that

$$K_{\text{conc}} = \frac{1}{K_f} \tag{7.76}$$

The fatigue stress concentration factor K_f is a function of the existing *notches* just as the stress concentration factor K_{SC} is a function of the presence of stress raisers. This gave rise to the *notch-sensitivity factor* q_N as defined in Peterson (1974);

$$q_N = \frac{K_f - 1}{K_{\text{SC}} - 1} = \frac{\left(1/K_{\text{conc}}\right) - 1}{K_{\text{SC}} - 1} \tag{7.77}$$

Experiments on different materials developed some charts of the *notch-sensitivity factor* q_N as functions of material strength and *notch radius* r_N; see Sines and Waisman (1959). From these relations, the *fatigue stress concentration factor* K_f or the *fatigue concentration factor* K_{conc} are then obtainable by Eq. (7.77), which can be rewritten as

$$K_f = 1 + q_N\left(K_{\text{SC}} - 1\right), \quad \text{or} \quad K_{\text{conc}} = \frac{1}{1 + q_N\left(K_{\text{SC}} - 1\right)} \tag{7.78}$$

As stated before, the *notch sensitivity* q_N is dependent on the material property and the *notch radius* r_N such that

$$q_N = \frac{1}{1 + \sqrt{a_N/r_N}} \tag{7.79}$$

where $a_N^{1/2}$ is the *Neuber constant* and r_N is the *notch radius* (Neuber 1946). The *Neuber constant* depends on the material only. From Eqs. (7.78) and (7.79) the *fatigue stress concentration factor* K_f or the *fatigue concentration factor* K_{conc} are then obtainable. For our application, the *fatigue concentration factor* K_{conc} is then

$$K_{\text{conc}} = \frac{1 + \sqrt{a_N/r_N}}{K_{\text{SC}} + \sqrt{a_N/r_N}} \tag{7.80}$$

The value of the *Neuber constant* $a_N^{1/2}$ is a material dependent. It is obtainable from Eq. (7.79) when q_N is available from charts. On the other hand and under reversed ***bending***, Budynas and Nisbett (2015) suggest that the *Neuber constant* can take the following correlated expression as a function of the *ultimate tensile strength* S_{ut} for steels:

$$\sqrt{a_N} = 0.246 - 3.08(10)^{-3}S_{\text{ut}} + 1.51(10)^{-5}S_{\text{ut}}^2 - 2.67(10)^{-8}S_{\text{ut}}^3 \tag{7.81}$$

where the *ultimate tensile strength* S_{ut} is in [kpsi]. The value of the *Neuber constant* $a_N^{1/2}$ is in [in$^{1/2}$]. The equation is valid to a maximum of $S_{\text{ut}} = 200$ [kpsi] (i.e. about 1380 [MPa]).

For reversed **torsion**, Budynas and Nisbett (2015) suggest that the *Neuber constant* can take the following correlated expression as a function of the *ultimate tensile strength* S_{ut} for steels:

$$\sqrt{a_N} = 0.190 - 2.51(10)^{-3}S_{ut} + 1.35(10)^{-5}S_{ut}^2 - 2.67(10)^{-8}S_{ut}^3 \tag{7.82}$$

where the *ultimate tensile strength* S_{ut} is in [kpsi]. The value of the *Neuber constant* $a_N^{1/2}$ is in [in$^{1/2}$]. This equation is also valid to a maximum of $S_{ut} = 200$ [kpsi] (i.e. about 1380 [MPa]).

Knowing the material strength, the value of the *Neuber constant* $a_N^{1/2}$ is definable by substituting the value of S_{ut} into Eq. (7.81) for bending or Eq. (7.82) for torsion. This is convenient if a computer is available. For hand calculations or a quick check, it might be simpler to use Figure 7.25. The solid black line is for bending and the solid gray line is for torsion. Solid lines are plots of steel relations for bending and torsion (Budynas and Nisbett 2015). With the knowledge of the *notch radius* r_N and after getting the *Neuber constant* $a_N^{1/2}$, it is convenient to use Eq. (7.80) to evaluate the *fatigue concentration factor* K_{conc} to use it in Eq. (7.66).

For *aluminum*, Figure 7.26 provides the *Neuber constant* $a_N^{1/2}$ for annealed and hardened treatments. Dotted grayish line is for the annealed aluminum, and dashed grayish line is for hardened aluminum (Kuhn 1964). Again, with the knowledge of the *notch radius* r_N and after getting the *Neuber constant* $a_N^{1/2}$, it is convenient to use Eq. (7.80) to evaluate the *fatigue concentration factor* K_{conc} to use it in Eq. (7.66).

Inspecting Eq. (7.80) one can conclude that if a_N/r_N is much smaller than 1.0, the *fatigue concentration factor* K_{conc} tends to be the reciprocal of the *stress concentration factor* K_{SC} or

$$K_{conc} = \frac{1}{K_{SC}}, \quad \text{for } \left(a_N/r_N\right) \ll 1.0 \tag{7.83}$$

This can occur at high ultimate tensile strength as can be deduced from Figure 7.26. Larger notch radius also provides similar conclusion.

It is important to note that some designers prefer to use the *fatigue stress concentration factor* K_f rather than the *fatigue concentration factor* K_{conc}. The *fatigue stress concentration factor* K_f increases the stresses applied to the machine member, and thus Eq. (7.66) should not include the *fatigue concentration factor* K_{conc}. The *fatigue concentration factor* K_{conc} reduces the fatigue strength of the machine member. For combined loading, the first option of using the *fatigue stress concentration factor* K_f may be better. On the other hand, the existence of notches

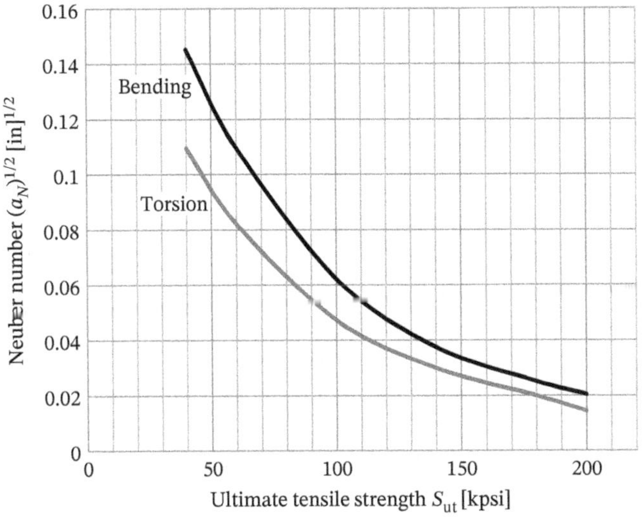

Figure 7.25 Values of *Neuber* number $(a_N)^{1/2}$ as a function of the ultimate tensile strength in [kpsi]. The black line is for bending and the gray line is for torsion. These are plots of relations in Budynas and Nisbett (2015).

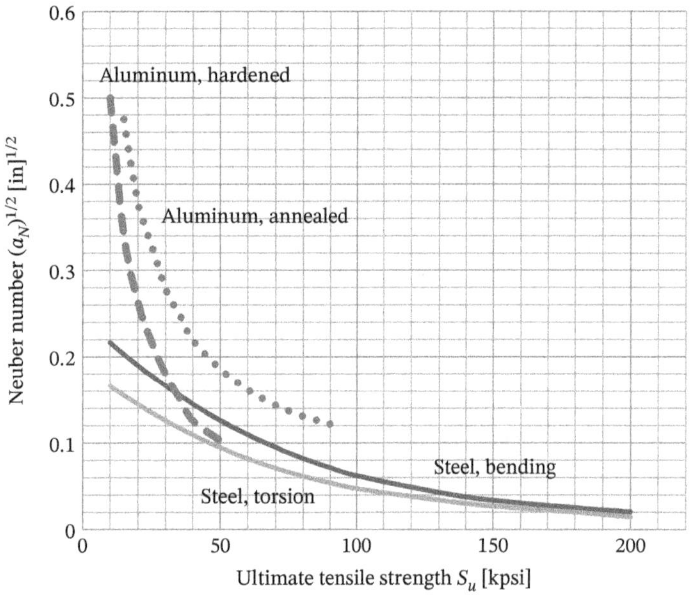

Figure 7.26 Values of *Neuber* number $(a_N)^{1/2}$ as a function of the ultimate tensile strength in [kpsi]. The solid black line is for bending and the solid gray line is for torsion. Solid lines are plots of steel relations for bending and torsion (Budynas and Nisbett 2015). Dotted and dashed grayish lines are for aluminum.

does reduce the fatigue strength. Then to employ the second option of using the *fatigue concentration factor* K_{conc}, a weighting factor of the ratio between bending and torsion may be used to get the *fatigue concentration factor* K_{conc} or use the higher value among both as being conservative. With the wide dispersion of the involved data, a conservative approach may be advisable.

7.8.2.7 Miscellaneous Factor, K_{miscel}

This factor should account for other effects that were not having prior specific coverage such as different or unusual manufacturing, operational, environment conditions, and other extraordinary factors. Some of these factors are dealt with in an introductory manner. If any is of a paramount existence, it should separately undergo thorough and extensive tests to have a reasonable account for its effect. Some of these factors are as follows:

- *Residual stresses* and *directional characteristics* of materials due to manufacturing processing have definite effects on the endurance limit. Existing compressive residual stresses due to shot peening or hammering, for instance, can improve fatigue strength on the surface. The core, however, may be the weaker part of the section. Directional production such as rolling or drawing generates fatigue properties in that direction, which can be better by 10–20% than the transverse direction.
- *Corrosion* is an important role to account for that effect. When a component is subjected to corrosive atmosphere, the miscellaneous factor k_f plays. Parts operating in corrosive atmosphere have lower fatigue strength due to roughening or pitting of the surface by that atmosphere. There might be no fatigue limit due to that.
- *Frequency* of cyclic loading affects strength, particularly in the presence of corrosion or high temperature. The lower the frequency of the cyclic rate and at higher temperature, the faster the crack propagates and shorter life occurs.
- *Electrolytic plating* such as metallic coatings (chromium, nickel, or cadmium plating) may reduce the endurance limit by 50%, i.e. k_f would be about 0.5. Zinc plating does not reduce fatigue strength. Anodic oxidation reduces endurance limit by about 39%.

- *Metal spraying* can cause surface imperfections that initiate cracks and lower endurance limit by about 14%.
- *Fretting* corrosion due to microscopic motion between tightly fitted parts involves surface pitting and fatigue. The miscellaneous factor k_f can be 0.24–0.9 depending on mating parts materials. Examples of that are bolted joints, bearing race fits, wheel hubs, or other tightly fitted parts.

7.8.3 Cumulative Fatigue Strength

The loading of a machine member may not necessarily be at one stress for a specific (or infinite) number of cycles. The machine member may be loaded at many stress levels with each stress applied for a different specific number of cycles. This cumulative fatigue loading is a problem that had the attention of both Palmgren (1924) and Miner (1945). Their finding is usually called *Miner's rule* or *Palmgren–Miner's rule*. In its simplified form, the rule postulates that cumulative fatigue damage happens when the following expression is satisfied:

$$\frac{N_{\text{load},1}}{N_{\text{Cf},1}} + \frac{N_{\text{load},2}}{N_{\text{Cf},2}} + \cdots + \frac{N_{\text{load},i}}{N_{\text{Cf},i}} + \cdots = 1 \quad \text{or} \quad \sum_{i=1}^{k} \frac{N_{\text{load},i}}{N_{\text{Cf},i}} = 1 \tag{7.84}$$

where $N_{\text{load},i}$ is the number of loading cycles under the exposure of the constant amplitude stress $\sigma_{\text{load},i}$ and $N_{\text{Cf},i}$ is the number of failure cycles at $\sigma_{\text{load},i}$, knowing that fatigue failure occurs after k of these stress loadings. This means that if the loadings were only one stress load of constant amplitude at the level of the endurance limit S_e, the number of cycles to failure would be about 10^6 cycles or 10^6 to 10^7 cycles. It also means that if the number of loading cycles under the exposure of one stress $\sigma_{\text{load},1}$ is equal to the number of failure cycles $N_{\text{Cf},1}$ at $\sigma_{\text{load},1}$, one cannot expose the member to any additional number of loading cycles. That is why equating the summation of load damage to the value of one in Eq. (7.84) is originally set at a value between 0.7 and 2.2. The common engineering communities use this *linear damage model* and the value of 1.0 as defined in Eq. (7.84) instead of the range (0.7–2.2). The evaluation of the $N_{\text{Cf},i}$ (the number of failure cycles at $\sigma_{\text{load},i}$) is through the application of Eq. (7.65).

The use of *Miner's linear damage model* with the value of 1.0 as in Eq. (7.84) facilitates the evaluation of cumulative loading safety in a simplified way. If the total loading damage is less than 1.0, it might be safe to assume that the cumulative loading is safe. For engineering students, this might be sufficient. However, for engineering design of products, one should conduct some extensive testing to get evaluation that is more definite.

7.8.4 Fluctuating Stresses

As indicated earlier in Section 7.8, the simplified model of a realistic loading may be as a simple fluctuating form as in Figure 7.22. The values of the *mean stress* σ_m and *alternating stress* σ_a as functions of the *maximum stress* $\sigma_{\max}$ and the *minimum stress* $\sigma_{\min}$ are simply as in Eqs. (7.56) and (7.57) that are rewritten as follows:

$$\sigma_m = \frac{1}{2}\left(\sigma_{\max} + \sigma_{\min}\right) \tag{7.56}$$

$$\sigma_a = \frac{1}{2}\left(\sigma_{\max} - \sigma_{\min}\right) \tag{7.57}$$

It is apparent that for the fully reversed bending tests to get the endurance limit, the mean stress σ_m = zero.

One of very useful methods of handling the fluctuating form of loading for ductile materials is to use the *von Mises distortion energy* theory introduced in Section 7.7.3. The procedure calls for developing two *Mohr's circles*, one for the mean stresses and the other for the alternating stresses. The two *Mohr's circles* would use the mean and alternating components of the normal and shear stresses (σ_x, σ_y, σ_z, τ_{xy}, τ_{yz}, τ_{xz}) at any internal point of the machine member.

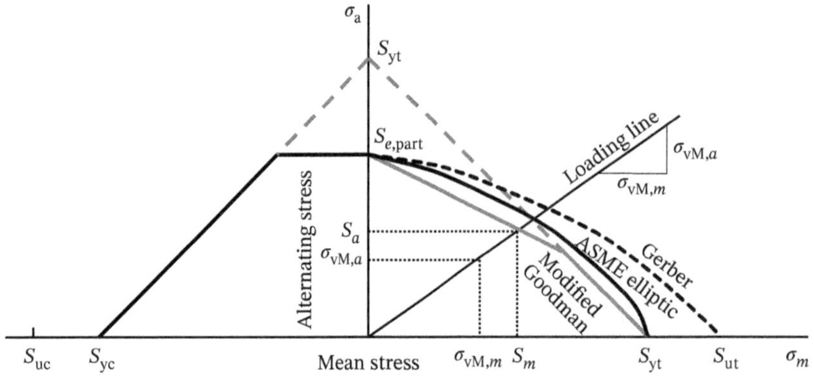

Figure 7.27 The fatigue diagram showing the failure limiting criteria of modified Goodman, ASME elliptic, and Gerber parabola. The limiting failure points in tension (S_{ut}, S_{yt}), compression (S_{uc}, S_{yc}), and machine part fatigue strength ($S_{e,part}$) are marked.

For ***2D*** stress state and from the two *Mohr's circles* of mean and alternating stresses, the *equivalent von Mises stresses*, as reduced from Eq. (7.42) for ***3D***, are as follows:

$$\sigma_{vM,m} = \sqrt{\sigma_{1,m}^2 - \sigma_{1,m}\ \sigma_{2,m} + \sigma_{2,m}^2}$$
$$\sigma_{vM,a} = \sqrt{\sigma_{1,a}^2 - \sigma_{1,a}\ \sigma_{2,a} + \sigma_{2,a}^2} \qquad \textbf{(7.85)}$$

The procedure then uses the equivalent stresses $\sigma_{vM,m}$ and $\sigma_{vM,\,a}$ in a *fatigue diagram* such as *Gerber*, *modified Goodman*, *ASME elliptic* as shown in Figure 7.27, or some other limiting lines (not shown). The fatigue diagram uses the *mean stress* σ_m as the abscissa axis and the *alternating stress* σ_a as the ordinate axis. The limiting failure points in tension (S_{ut}, S_{yt}), compression (S_{uc}, S_{yc}), and the machine part fatigue strength ($S_{e,part}$) are marked. The selected limiting lines are those shown in Figure 7.27. The other available limiting criteria are not preferable in this text since they are either historical or not totally in compliance with major tests in some regions. Some of these are the *Soderberg* and the other *Goodman* diagrams. The different relations for the selected fatigue-limiting lines are simply derivable from the boundary conditions of each line. These equations are as follows.

The ***Goodman*** diagram is connecting the two end points of the *part endurance limit* $S_{e,part}$ of Eq. (7.66) and the *ultimate tensile strength* S_{ut} (not shown in Figure 7.27). The last *modified Goodman* diagram in Figure 7.27 is the intersecting two lines with the first given as

$$\frac{S_a}{S_{e,part}} + \frac{S_m}{S_{ut}} = 1 \qquad (7.86)$$

where S_a is the *alternating failure strength* at failure limit and S_m is the *mean failure strength* at failure limit. The second intersecting line of the *modified Goodman* diagram is simply

$$\frac{S_a}{S_{yt}} + \frac{S_m}{S_{yt}} = 1 \qquad (7.87)$$

The *intersection* of the two lines in Eqs. (7.86) and (7.87) obtained by equating the equations of the lines gives

$$\frac{S_a}{S_m} = S_{e,part}\left(\frac{1}{S_{yt}} - \frac{1}{S_{ut}} + \frac{S_a}{S_m S_{yt}}\right) \qquad (7.88)$$

The *fluctuating loading ratio* r_{load} is the quotient of the *alternating failure strength* S_a at failure limit and the *mean failure strength* at failure limit S_m, i.e. $r_{\text{load}} = S_a/S_m$. This reinstates Eq. (7.88) as follows:

$$r_{\text{load}}\big|_{\text{intersect}} = \frac{(1/S_{\text{yt}}) - (1/S_{\text{ut}})}{(1/S_{e,\text{part}}) - (1/S_{\text{yt}})} \tag{7.89}$$

If the *fluctuating loading ratio* r_{load} is larger than the value of Eq. (7.89), Eq. (7.86) is used. If the *fluctuating loading ratio* r_{load} is lower than the value of Eq. (7.89), Eq. (7.87) is used.

The *modified Goodman* is linear, but its implementation is involved as indicated previously. With a computer program, the procedure is possibly implementable.

The ***Gerber parabolic*** criterion defines the fatigue failure parabola by the following equation and as in Figure 7.27:

$$\frac{S_a}{S_{e,\text{part}}} + \left(\frac{S_m}{S_{\text{ut}}}\right)^2 = 1 \tag{7.90}$$

The ***ASME elliptic*** criterion defines the fatigue failure elliptic line by the following equation and shown in Figure 7.27:

$$\left(\frac{S_a}{S_{e,\text{part}}}\right)^2 + \left(\frac{S_m}{S_{\text{yt}}}\right)^2 = 1 \tag{7.91}$$

These lines in Eqs. (7.90) and (7.91) are closer to some experimental results. The *ASME ellipse* is more conservative; however, the *modified Goodman* is even more conservative.

For design against *fluctuating fatigue failure* and assuming the same *fatigue safety factor* $K_{\text{SF},F}$ for both *mean loading* and *alternating loading*, this sets the relations $S_m = K_{\text{SF},F}\,\sigma_m$ and $S_a = K_{\text{SF},F}\,\sigma_a$. The *Gerber* and *ASME* criteria in Eqs. (7.90) and (7.91) become as follows:

$$\frac{K_{\text{SF},F}\;\sigma_a}{S_{e,\text{part}}} + \left(\frac{K_{\text{SF},F}\;\sigma_m}{S_{\text{ut}}}\right)^2 = 1 \tag{7.92}$$

$$\left(\frac{K_{SF,F}\;\sigma_a}{S_{e,part}}\right)^2 + \left(\frac{K_{SF,F}\;\sigma_m}{S_{yt}}\right)^2 = 1 \tag{7.93}$$

For the more conservative ***ASME*** *criterion*, the *fatigue safety factor* $K_{\text{SF},F}$ is simply obtainable from Eq. (7.93) such that

$$K_{\text{SF},F} = \frac{1}{\sqrt{(\sigma_a/S_{e,\text{part}})^2 + (\sigma_m/S_{\text{yt}})^2}} \tag{7.94}$$

For manual calculations, this *ASME* criterion is simpler than the *Gerber* criterion and more conservative. *ASME elliptic* is more recommendable in this text. For *synthesis*, however, the selected safety factor is employable at the start to guarantee securing appropriate or optimum geometry and material.

For a regularly encountered **2D** stress state, the general *distortion energy theory* of Eq. (7.85) is obtainable directly by using σ_x, σ_y, and τ_{xy} without the need for the *Mohr's circles* to define the principal stresses. This gives the following *equivalent von Mises* expressions:

$$\begin{aligned}\sigma_{\text{vM},m} &= \sqrt{\sigma_{x,m}^2 - \sigma_{x,m}\,\sigma_{y,m} + \sigma_{y,m}^2 + 3\tau_{xy,m}^2}\\ \sigma_{\text{vM},a} &= \sqrt{\sigma_{x,a}^2 - \sigma_{x,a}\sigma_{y,a} + \sigma_{y,a}^2 + 3\tau_{xy,a}^2}\end{aligned} \tag{7.95}$$

If the normal stress $\sigma_y = 0$, Eq. (7.95) is reduced even more to become as simple as follows:

$$\sigma_{\text{vM},m} = \sqrt{\sigma_{x,m}^2 + 3\tau_{xy,m}^2}$$
$$\sigma_{\text{vM},a} = \sqrt{\sigma_{x,a}^2 + 3\tau_{xy,a}^2} \tag{7.96}$$

This situation is frequently recurring than the full ***2D*** stress state as handled by Eq. (7.95).

Fluctuating ***compressive stressing*** with negative mean stress fails differently from the fatigue with tensile mean stress. Figure 7.27 indicates the fatigue failure limit in the second quadrant, which is just about representing different tests. The fatigue failure limit is indicating that failure occurs when $\sigma_a = S_{e,\text{part}}$ (machine part fatigue strength) or when $\sigma_m = S_{\text{yc}}$ (compressive yield strength). Therefore, one does not need any different failure criterion as for the tensile side of mean stressing.

In early experimental investigations, ***torsional fatigue stressing*** demonstrated different behavior from the fluctuating normal stressing. Later on, experiments show a reduction in fatigue strength as the mean torsional stress increases particularly for non-polished specimens. Fatigue failure criteria similar to those for tensile loading should then apply with consideration of coordinate transformation to shear. The relation between the yield strength in shear S_{ys} and the yield strength in tension S_{yt} is defined by the distortion energy theory according to Eq. (7.41), which gives $S_{\text{ys}} = 0.577\ S_{\text{yt}}$. The same might apply to the endurance limit in torsion S_{es} to get $S_{\text{es}} = 0.577\ S_e$. On the other hand the relation between the ultimate shear strength S_{us} and the ultimate tensile strength S_{ut} is predicted by the maximum shear stress theory according to Eq. (7.49), which gives $S_{\text{us}} = 0.5\ S_{\text{ut}}$. This does not match the test results, and the reasonable value might be $S_{\text{us}} = 0.67\ S_{\text{ut}}$, if no definite information is available. Some researchers roughly use $S_{\text{us}} = 0.8\ S_{\text{ut}}$ for steel and $S_{\text{us}} = 0.7\ S_{\text{ut}}$ for other ductile materials. With this information, a fatigue diagram similar to that in Figure 7.27 is achievable. Safety against *torsional fatigue* is definable similar to fatigue failure criterion of *ASME elliptic* as suggested earlier. Other criteria may also be employable.

Brittle materials may not be the appropriate material for fatigue loading. If there is a pressing need, the ultimate tensile strength S_{ut} should replace the yield strength S_{yt}. The *modified Goodman* diagram reduces to the *Goodman* diagram that may be used for synthesis or design check. Higher safety factor is advisable in that case, since not so much work is available in fatigue of brittle material. Other failure criterion such as *Smith-Dolan* is also applicable in the first quarter and takes the following form:

$$\frac{S_a}{S_{e,\text{part}}} = \left(\frac{1 - (S_m/S_{\text{ut}})}{1 + (S_m/S_{\text{ut}})}\right) \tag{7.97}$$

Assuming the same *fatigue safety factor* $K_{\text{SF,F}}$ for both *mean loading* and *alternating loading*, this provides the relations $S_m = K_{\text{SF,F}}\ \sigma_m$ and $S_a = K_{\text{SF,F}}\ \sigma_a$ to implement in Eq. (7.97).

7.8.5 Fatigue Failure Criteria

In the previous Sections 7.8.1–7.8.4, the *stress-life* technique had some sufficient coverage to allow the designer to reasonably guard against *high-cycle fatigue* failure. When the machine part is subject to *low-cyclic fatigue* loading, the *strain-life* technique could be a comparable or acceptable mean to consider. The fatigue generated by *crack propagation* to reach crack instability causing instantaneous brittle fracture is best understandable and alternatively treatable by *fracture mechanics* (Section 7.9). This section is then devotedly covering the necessary understanding of *stress-life* technique.

Many researchers dedicated their studies to *low-cycle fatigue* and the development of relations for handling fatigue from the standpoint of deformation and strain. The prediction of low-cycle fatigue equation has experimental support by Tavernelli and Coffin (1962) with a discussion by Manson. This developed the Manson–Coffin equation as defined later on. The low-cycle fatigue was also the target of experimental studies by Landgraf (1968) using many cyclic elastic plastic stress–strain tests.

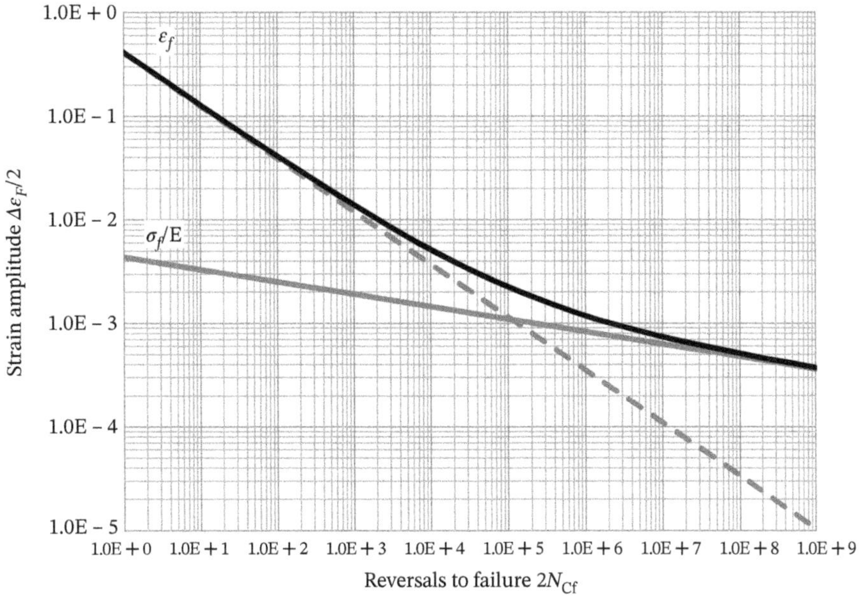

Figure 7.28 The fatigue *strain life* in reversals to failure ($2N_{Cf}$) is related to the strain amplitude $\Delta\varepsilon_F/2$ for AISI 1020. The grayish solid line is for the elastic strain, the grayish dashed line is for the plastic strain, and the black solid line is for the total fatigue strain-life equation.

The fatigue *strain-life* in reversals to failure ($2N_{Cf}$) is related to the strain amplitude $\Delta\varepsilon_F/2$ as shown in Figure 7.28 for AISI 1020. The $\Delta\varepsilon_F$ is the *strain change for fatigue* cyclic stressing, which is the summation of the *elastic strain range* $\Delta\varepsilon_E$ and the *plastic strain range* $\Delta\varepsilon_P$. The Manson–Coffin equation representing the *fatigue strain-life relation* is then

$$\frac{\Delta\varepsilon_F}{2} = \frac{\Delta\varepsilon_E}{2} + \frac{\Delta\varepsilon_P}{2} = \frac{\sigma_f}{E}\left(2N_{Cf}\right)^{b_E} + \varepsilon_f\left(2N_{Cf}\right)^{c_P} \tag{7.98}$$

where σ_f is the *fatigue strength coefficient*, b_E is the elastic *fatigue strength exponent*, ε_f is the *fatigue ductility coefficient*, c_P is the plastic *fatigue ductility exponent*, and N_{Cf} is the number of cycles to failure. The values of the empirical constants (σ_f, b_E, ε_f, c_P) in Eq. (7.98) are available in SAE (1977) for numerous materials. The values of these empirical constants for HR AISI 1020 (ISO C20) are used to plot Eq. (7.98) as shown in Figure 7.28. The empirical constants used for the *HR* ASI 1020 with $E = 205$ [GPa] are $\varepsilon_F = 0.96$, $\sigma_f = 895$ [MPa], $b_E = -0.12$, $\varepsilon_f = 0.41$, and $c_P = -0.51$. The grayish solid line is for the elastic strain, the grayish dashed line is for the plastic strain, and the black solid line is for the total fatigue strain-life equation.

The *strain-life* model of Eq. (7.98) does not give an asymptotic behavior representing a settling of life without failure beyond 10^6 or 10^7 as expected in Figure 7.28. At high cycles, the elastic strain dominates and the expectation is for the stress to decrease monotonically. No *endurance limit* is possibly to materialize. Some alterations to Eq. (7.98) provide means to approach the regular *S–N* fatigue diagram (Figure 7.23) such as the *four-point criteria* (ASM 1998). That is why the *strain-life* method would be more useful to apply in the *low-cycle* fatigue loading.

Example 7.4 Continuing on Example 7.2, the shaft is transmitting a maximum of 25 [kW] or 33.5 [hp] and running at 3000 [rpm] in the clockwise direction. It is subjected to a maximum bending moment that is equal to the maximum applied torque. Find the internal dynamic stresses if the shaft has a cylindrical diameter of 45 [mm] or 1.75 [in]. The selected material is set as HR AISI 1040 or ISO C40. Estimate the factors affecting fatigue knowing that the estimated life span is 10 years and the shaft operates eight hours a day, the reliability should be 99.99%, the operating temperature is about 380 [°C] or 716 [°F], the evaluated stress concentration factor K_{SC} is 2.1, and the notch radius is 2.5 [mm] or 0.1 [in]. Find the safety factor for ASME elliptic dynamic fatigue theory of failure.

Solution

Data: $H = 25$ [kW] or 33.5 [hp], $N_{rpm} = 3000$ [rpm], and $d_C = 0.045$ [m] or 1.75 [in]. *Ductile* material is HR AISI 1040 or ISO C40.

Ductile material properties are yield strength of 42 [kpsi] or 290 [MPa], ultimate tensile strength of 76 [kpsi] or 525 [MPa], and endurance limit of 37.7 [kpsi] or 260 [MPa]; see Table A.7.2 in Appendix A.7.

The shaft is similar to Example 6.10. The calculated results in Example 6.10 that can be applicable to this example are as follows:

The maximum shear stresses due to torque τ_{xz} are 4.4475 [MPa] or 0.668 79 [kpsi].
The maximum normal stresses due to bending moment σ_x are 8.8951 [MPa] or 1.3376 [kpsi].
The principal stresses σ_1, σ_2 are 10.737 [MPa] or 1.615 [kpsi] and −1.8422 [MPa] or −0.277 [kpsi], respectively.
The maximum shear stress τ_{max} is 6.2897 [MPa] or 0.945 81 [kpsi].
The distortion or deviatoric stress is 11.7668 [MPa] or 1.7698 [kpsi].

Information check: Endurance limit of AISI 1040 or ISO C40 is 37.7 [kpsi] or 260 [MPa]; see Table A.7.2 in Appendix A.7. Eq. (7.58) suggests $S_e = 0.5\, S_{ut} = 0.5(525) = 262.5$ [MPa] or $0.5(76) = 38$ [kpsi]. The two values are close, and one can then use the values in Table A.7.2 in Appendix A.7.

Finding the total number of cycles the shaft is subjected to by using the rotational speed 3000 [rpm], the operating hours in a day 8 [hr] per day, and the expected life span of 10 [years]. Note that the [hr] and [years] are not the regular units that are adapted by this text. The total number of cycles N_C is then

$$N_C = 3000 \times 8\,(60) \times 365 \times 10 = 5.256\text{e}+9\ [\text{cycles}] \tag{a}$$

This is far beyond the 10^6–10^7 cycles of the endurance limit. Therefore, one should calculate the case as a fatigue problem.

Estimating the factors affecting the fatigue strength provides the following values:

Surface factor K_{surf} for machined shaft (top Eq. (7.67)), $K_{surf}|_{SI} = 2.9458(525)^{-0.212} = 0.781$, and $K_{surf}|_{US} = 1.9572(76)^{-0.212} = 0.781$, which should be about the same.

Size factor K_{size} for d_p of 45 [mm] or 1.75 [in] (top Eq. (7.71)), $K_{size}|_{SI} = 1.24\,(45)^{-0.107} = 0.825$, and $K_{size}|_{US} = 0.879\,(1.75)^{-0.107} = 0.828$, which is about the same.

Loading factor K_{load} for reversed bending and steady torsion should be one, or $K_{load} = 1.0$.

Reliability factor K_{reliab} for a reliability of 99.99% (Table 7.5), $K_{reliab} = 0.702$. If calculated, $K_{reliab} = 0.7076$.

Temperature factor K_{temp} for such a high operating temperature (Eq. (7.74)), $K_{temp}|_{SI} = -2.325(10)^{-6}(380)^2 + 0.000\,717\,8\,(380) + 0.9762 = 0.913$ and $K_{temp}|_{US} = -7.024(10)^{-7}\,(716)^2 + 0.000\,427\,3\,(716) + 0.9664 = 0.912$, which are close enough.

Fatigue concentration factor K_{conc} is for a stress concentration factor K_{SC} of 2.1 and notch radius of 2.5 [mm] or 0.1 [in]. For the ultimate tensile strength of 76 [kpsi], the *Neuber constant* takes the value (Eq. (7.81))

$$\sqrt{a_N} = 0.246 - 3.08(10)^{-3}\,(76) + 1.51(10)^{-5}(76)^2 - 2.67(10)^{-8}(76)^3 = 0.0874\ \left[\text{in}^{1/2}\right] \tag{b}$$

Or $a_N = 0.007\,64$ [in] or 0.193 [mm] with $a_N/r_N = 0.007\,64/0.1 = 0.0764$. Using Eq. (7.80), the

$$K_{conc} = \frac{1 + \sqrt{a_N/r_N}}{K_{SC} + \sqrt{a_N/r_N}} = \frac{1 + \sqrt{0.0764}}{2.1 + \sqrt{0.076}} = \frac{1.2764}{2.3764} = 0.537 \tag{c}$$

Compare this to the approximate Eq. (7.83) or $K_{conc} = 1/K_{SC} = 1/2.1 = 0.476$.

Since no miscellaneous factor K_{miscel} is defined, one takes $K_{miscel} = 1.0$.

The *fatigue strength of the part* $S_{e,part}$ is then (Eq. (7.66))

$$S_{e,part} = (0.781)\,(0.825)\,(1.0)\,(0.702)\,(0.913)\,(0.537)\,(1.0)\ \ S_e = 0.2218\ \ (260) = 57.668\ [\text{MPa}]$$

$$S_{e,part} = (0.781)\,(0.828)\,(1.0)\,(0.702)\,(0.912)\,(0.537)\,(1.0)\ \ S_e = 0.2224\ \ (37.7) = 8.3845\ [\text{kpsi}] \tag{d}$$

Note that the *fluctuating stresses* are due to fiber rotation one-half a revolution causing the maximum normal stress (as tension) to change sign and become compressive. Calculating the mean and alternating stresses using Eqs. (7.56) and (7.57), one gets

$$\sigma_m = \frac{1}{2}\left(\sigma_{\max} + \sigma_{\min}\right) = \frac{1}{2}(8.8951 - 8.8951) = 0.0\ [\text{MPa}]$$
$$\sigma_m = \frac{1}{2}\left(\sigma_{\max} + \sigma_{\min}\right) = \frac{1}{2}(1.3376 - 1.3376) = 0.0\ [\text{kpsi}] \tag{e}$$

$$\sigma_a = \frac{1}{2}\left(\sigma_{\max} - \sigma_{\min}\right) = \frac{1}{2}(8.8951 + 8.8951) = 8.8951\ [\text{MPa}]$$
$$\sigma_a = \frac{1}{2}\left(\sigma_{\max} - \sigma_{\min}\right) = \frac{1}{2}(1.3376 + 1.3376) = 1.3376\ [\text{kpsi}] \tag{f}$$

The shear stress due to torsion τ_{xz} does not change with time, and therefore the mean shear stress is the maximum shear stress and the alternating shear stress is zero. This gives the following values:

$$\tau_m = \tau_{\max} = 4.4475\ [\text{MPa}]$$
$$\tau_m = \tau_{\max} = 0.66879\ [\text{kpsi}] \tag{g}$$

$$\tau_a = 0.0\ [\text{MPa}]$$
$$\tau_a = 0.0\ [\text{kpsi}] \tag{h}$$

Instead of using two *Mohr's circles* for mean and alternating stresses to get the equivalent *von Mises* stresses, one can use Eq. (7.96) for both mean and alternating components such that

$$\sigma_{\text{vM},m} = \sqrt{\sigma_m^2 + 3\tau_m^2} = \sqrt{0.0 + 3(4.4475)^2} = 7.7033\ [\text{MPa}]$$
$$\sigma_{\text{vM},m} = \sqrt{\sigma_m^2 + 3\tau_m^2} = \sqrt{0.0 + 3(0.668\,79)^2} = 1.1584\ [\text{kpsi}] \tag{i}$$

$$\sigma_{\text{vM},a} = \sqrt{\sigma_a^2 + 3\tau_a^2} = \sqrt{(8.8951)^2 + 0.0} = 8.8951\ [\text{MPa}]$$
$$\sigma_{\text{vM},a} = \sqrt{\sigma_a^2 + 3\tau_a^2} = \sqrt{(1.3376)^2 + 0.0} = 1.3376\ [\text{kpsi}] \tag{j}$$

The *ASME* criterion is simpler than the *Gerber* criterion and more conservative. The *fatigue safety factor* $K_{\text{SF},F}$ is due to Eq. (7.94), which gives

$$K_{\text{SF},F}\big|_{\text{SI}} = \frac{1}{\sqrt{\left(\sigma_{\text{vM},a}/S_{e,\text{part}}\right)^2 + \left(\sigma_{\text{vM},m}/S_{\text{yt}}\right)^2}} = \frac{1}{\sqrt{(8.8951/57.668)^2 + (7.7033/290)^2}} = 6.389$$
$$K_{\text{SF},F}\big|_{\text{US}} = \frac{1}{\sqrt{\left(\sigma_{\text{vM},a}/S_{e,\text{part}}\right)^2 + \left(\sigma_{\text{vM},m}/S_{\text{yt}}\right)^2}} = \frac{1}{\sqrt{(1.3376/8.3845)^2 + (1.1584/42)^2}} = 6.177 \tag{k}$$

These values of safety factor are high, which indicates an extremely conservative design. The diameter of the shaft should be reduced to a smaller value for either *SI* or the *US* system of units. Note that, so far, the previous Example 7.4 is an analysis process, and not a synthesis or a real design process.

7.9 Fracture Mechanics and Fracture Toughness

From the name of fracture mechanics, this field studies the mechanics of fracture theoretically and experimentally. Early sections of this chapter such as 7.1.2, 7.1.3, 7.1.5, 7.7, and 7.8 introduced many fracture concepts that relate to sudden failures such as fracture of brittle materials represented by their ultimate strength, brittle fracture of ductile materials under fatigue, and even failure of ductile materials under their ultimate strength. *Fracture mechanics*, however, is generally addressing the designer concern and understanding of the mechanisms of *crack stability*, initiation, propagation, and instability. These are the bases of fractures in ductile and brittle materials under static or dynamic loadings.

The crack can be internally or externally present in the component material. The internal cracks can be microscopic due to microstructure imperfections, nonmetallic inclusions, weld defects, flaws, and grain irregular connective boundaries. The external crack existence can be due to surface production imperfection, quenching cracks, grinding cracks, weld surface defects, assembly scratches or gouges, and even environmental corrosion during operational performance. The seriousness of these cracks was evident in the 1943 case of the splitting in half of the just built *SS Schenectady* tanker, 159 [m] or 523 [ft] in length, US Coast Guard (1944).

Since the crack is the main interest, approximately it may be mathematically modeled as an elliptical hole as previously discussed on the stress concentration factors or *stress raisers* in Section 6.10. The elliptical hole of Figure 6.38 is reproduced as Figure 7.29 with minor modifications to suit the analysis and development of the common fracture mechanics concepts. The stress distribution around a hole in an infinite plate has been introduced in Section 6.10. The maximum stress σ_{max} at the major axis of an *elliptical hole* in an infinite plate (or a very wide plate) loaded as shown in Figure 7.29a is given by (modification of Eq. (6.173) of Section 6.10)

$$\sigma_{max} = \sigma_{av}\left(1 + \frac{2a_C}{b_C}\right) \tag{7.99}$$

where σ_{av} is the *average stress*, $2a_C$ and $2b_C$ are the principal dimensions of the elliptic hole or crack. The stress distributions at the hole are also depicted in Figure 7.29a with the maximum at the edging of the hole or the crack tip. The crack tip radius r_C is much smaller than the radius of the ellipse at the edge. The maximum stress σ_{max} at the crack tip is thus much more than the stress in Eq. (7.99), which is plotted as shown in Figure 7.29b as a function of the ratio of major to minor elliptical axes a_C/b_C. The model of a crack as an elliptical shape will then have a very large ratio of major to minor elliptical axes a_C/b_C. It is expected to be a much greater value than the 100 in Figure 7.29b. The *stress concentration factor* $K_{SC} = \sigma_{max}/\sigma_{av}$ is then more than the 200 of Figure 7.29b. In fact, it may theoretically approach an infinite value beyond the tensile strength. Therefore, failure is expectable at the crack tip. The maximum stress σ_{max} at the crack tip of radius r_C may be approximately given by the following equation (see Inglis 1913; Griffith 1921):

$$\sigma_{max} = 2\sigma_{av}\sqrt{\frac{a_C}{r_C}} \tag{7.100}$$

where σ_{av} is the average applied stress, r_C is the *crack tip radius*, and a_c is *half the crack length.*

7.9.1 Stress Intensity Factor K_I

To calculate the maximum stress at the crack tip according to Eq. (7.100), it is very difficult to evaluate the real crack tip radius r_C. Therefore, the *linear elastic fracture mechanics* developed an alternative way by finding a quantity that can only be a function of the crack size $2a_C$. The quantity is the ***elastic stress intensity factor*** K_I for plane strain condition of mode I (tensile crack opening mode). The ***elastic stress intensity factor*** K_I has the following expression (Irwin 1957, 1960):

$$K_I = \sigma_{av}\sqrt{\pi a_C} \tag{7.101}$$

where a_C is the crack half size. This might suggest that the crack radius assumption in Eq. (7.100) is figuratively less than 4% of the crack length. From Eq. (7.101), the units of the *stress intensity factor* K_I is [MPa m$^{1/2}$] or [kpsi in$^{1/2}$]. To convert one to the other, one writes [MPa m$^{1/2}$] = [MPa × (0.145 kpsi/MPa) m$^{1/2}$ × (39.37 in/m)$^{1/2}$] ≈ 0.91 [kpsi in$^{1/2}$].

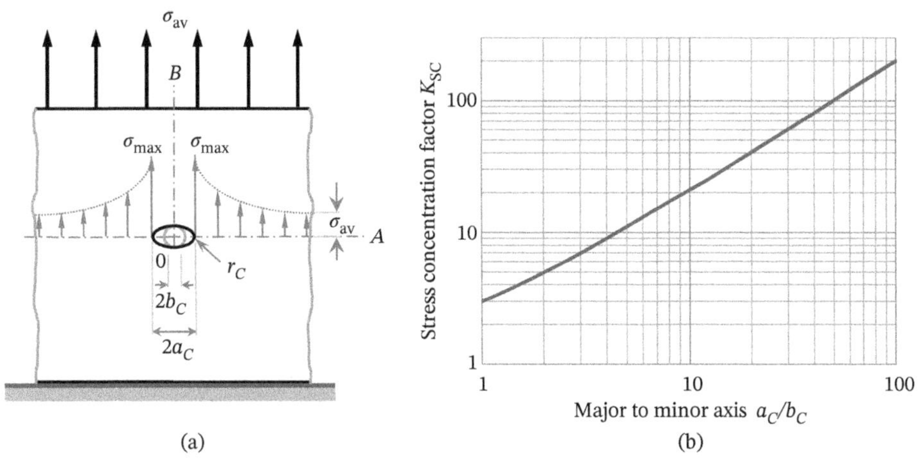

Figure 7.29 The elliptical hole of Figure 6.38 is reproduced in (a) with minor modifications to suit the analysis and development of fracture mechanics concepts. The stress concentration factor K_{SC} at the ellipse tip of radius r_C is plotted in (b) as a function of the ratio of major to minor elliptical axes a_C/b_C.

The other modes of crack opening are the *shear or sliding mode* II and the *tearing mode* III. Each of these modes would have the ***stress intensity factors*** K_{II} and K_{III}, respectively. The tensile crack opening mode I is the one mostly used.

It is important to note that Eq. (7.101) is for an infinite and relatively thin plate size. To account for other geometries of the plate and crack location, an additional *correction factor* β_C is adjusting Eq. (7.101) to be

$$K_I = \beta_C \; \sigma_{av} \sqrt{\pi a_C} \tag{7.102}$$

The factor β_C is usually close to 1.0 for relatively small size cracks. For a crack of length a_C at the edge of the plate, the value of the correction factor β_C is 1.12 for small cracks in a very wide and relatively thin plate. Other correction factor values for different geometries and loadings may be obtainable from dedicated handbooks such as Miedlar et al. (2002).

7.9.2 Fracture Toughness: Critical Stress Intensity Factor K_{IC}

Fracture toughness is a property of a material. It indicates how the material withstands cracks without sudden fracture, i.e. having unstable crack propagation. The limiting *critical stress intensity factor* K_{IC} a material can stand is the *plane strain fracture toughness* K_{IC} of the material. It is obtainable by standardized tests that have specific fatigue precracked specimens under slow varying static test and other extensive geometrical and operational constraints, e.g. ASTM E399 (2013). This means that when loading a cracked machine part to a high value of K_I (according to Eq. (7.101) or (7.102)), it will unstably fracture abruptly when K_I reaches K_{IC}. This brittle fracture can be catastrophic if the failing element is a member of a large structure or design such as a blade in a turbine. This text uses *fracture toughness* K_{IC} to mean the widely targeted *plane strain fracture toughness* K_{IC}.

The *fracture mechanics safety factor* $K_{SF,FM}$, considering fracture mechanics in design, would be as follows:

$$K_{SF,FM} = \frac{K_{IC}}{K_I} \tag{7.103}$$

As long as the *stress intensity factor* K_I is lower than the *fracture toughness* K_{IC}, there should be no danger of catastrophic brittle fracture since the crack is still in a stable mode. If the crack grows (i.e. propagates) due to increasing or fluctuating stresses, the brittle fracture will occur in time. If the crack size is observable and measurable, the expected life or cycles to failure may be possible to estimate as discussed in Section 7.9.3.

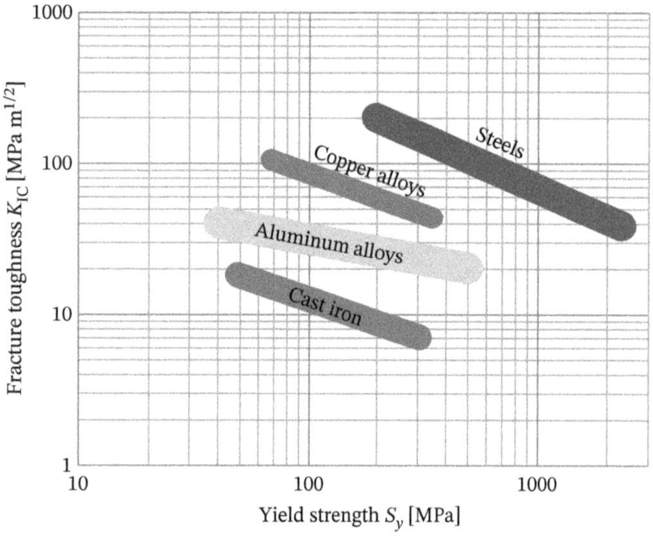

Figure 7.30 General approximate spans of fracture toughness against tensile yield strength of some common engineering materials. The lines show the extents of the strength and the toughness. Properties can be off the line but approximately within the span's limits; see Ashby (1999).

The designer finds from standards, manufacturer, and handbooks or by extensive standard tests the value of the *fracture toughness* K_{IC} of a material, if cost permits. For initial design and synthesis, one can roughly estimate the *fracture toughness* K_{IC} using the trend lines shown in Figure 7.30 relating K_{IC} to the *yield strength Sy*. The trend lines show the extents of the strength and toughness. Properties can be off the line but approximately within the span's limits; see Ashby (1999). Figure 7.30 indicates that as the yield strength increases, the fracture toughness decreases. Specially produced alloys are usually off the trend lines since they target additional *fracture toughness* (ASM 1996). Usually *martensitic* steels with meta-stable *austenite* microstructure would have additional *fracture toughness*, and steels dominated by *ferritic* and *pearlitic* microstructures would have lower *fracture toughness* (ASM 1998).

In general, engineering ***metals*** have fracture toughness convenient range of about 10–200 [MPa m$^{1/2}$] or 1.5–29 [kpsi in$^{1/2}$]. Engineering ***polymers*** have a fracture toughness range of about 0.5–5 [MPa m$^{1/2}$] or 0.07–0.7 [kpsi in$^{1/2}$]. Engineering ***composites*** have a fracture toughness range of about 15–75 [MPa m$^{1/2}$] or 2–11 [kpsi in$^{1/2}$]. Engineering ***ceramics*** have a fracture toughness range of about 2–10 [MPa m$^{1/2}$] or 0.3–1.5 [kpsi in$^{1/2}$]. For other materials, one can consult Ashby (1999). For a specific material, one should find from standards, manufacturer, and handbooks or by extensive standard tests the specific value of the *fracture toughness* K_{IC}.

Other mean of estimating the *fracture toughness* is to use the developed relation to the *Charpy* V-notch test (after *George A. Charpy* (1865–1945)), Barsom and Rolfe (1969). If the *Charpy* V-notch test values are accessible, the following approximate estimate relation may be usable (Faupel and Fisher 1981):

$$K_{IC} = \left(5\ S_y \left| \text{CVN} - \frac{S_y}{20} \right| \right)^{1/2} \tag{7.104}$$

where CVN is the standard *Charpy* V-notch impact value in [ft lb] and S_y is the yield strength in [kpsi]. The estimated K_{IC} is in [kpsi in$^{1/2}$]. This estimate has been effective in comparing results to the special fracture mechanics ***FE*** solutions (Younan et al. 1983).

7.9.3 Crack Propagation and Life

If the crack size a_C is observable and measurable, the expected life or cycles to failure N_f may be possible to estimate. The crack growth or propagation would be due to increasing of fluctuating stresses between σ_{max} and σ_{min} in each loading cycle or being under the *stress range* $\Delta\sigma = \sigma_{max} - \sigma_{min}$. The variation in the *stress intensity factor range* ΔK_I between a maximum K_{Imax} and a minimum K_{Imin} is then given by the application of Eq. (7.101) or (7.102), which gives

$$\Delta K_I = K_{Imax} - K_{Imin} = \beta_C\ \sigma_{max}\sqrt{\pi a_C} - \beta_C\ \sigma_{min}\sqrt{\pi a_C}$$
$$\Delta K_I = \beta_C\ \sqrt{\pi a_C}\left(\sigma_{max} - \sigma_{min}\right) = \beta_C\ \Delta\sigma\sqrt{\pi a_C} \tag{7.105}$$

With the advancing of loading cycle dN_C, the *crack size* a_C would propagate an increment da_C. In the crack propagation region, the *rate of crack growth* da_C as a function of *loading cycle rated* N_C is obtainable according to a simple empirical law by Paris and Erdogan (1963) as follows:

$$\frac{da_C}{dN_C} = C_P\left(\Delta K_I\right)^{m_P} \tag{7.106}$$

where dN_C/da_C is the *crack fatigue growth rate*, ΔK_I is variation or *range of the stress intensity factor* and the *propagation factors* C_P and m_P are approximate empirical constants of the material (Barsom 1971; Barsom and Rolfe 1987). For *martensitic steels*, C_P and m_P are 1.36(10^{-10}) [m/((MPa m$^{1/2}$)^{mP} cycle)] or 6.6(10^{-9}) [m/((kpsi in$^{1/2}$)^{mP} cycle)] and 2.25, respectively. For *ferritic–pearlitic steels*, C_P and m_P are 6.89(10^{-12}) [m/((MPa m$^{1/2}$)^{mP} cycle)] or 3.6(10^{-10}) [m/((kpsi in$^{1/2}$)^{mP} cycle)] and 2.25, respectively. For *austenitic stainless steels*, C_P and m_P are 5.61(10^{-12}) [m/((MPa m$^{1/2}$)^{mP} cycle)] or 3.00 10^{-10}) [m/((kpsi in$^{1/2}$)^{mP} cycle)] and 2.25, respectively. Figure 7.31 plots Eq. (7.106) in the propagation zone only and without the specific minimum *threshold stress intensity*

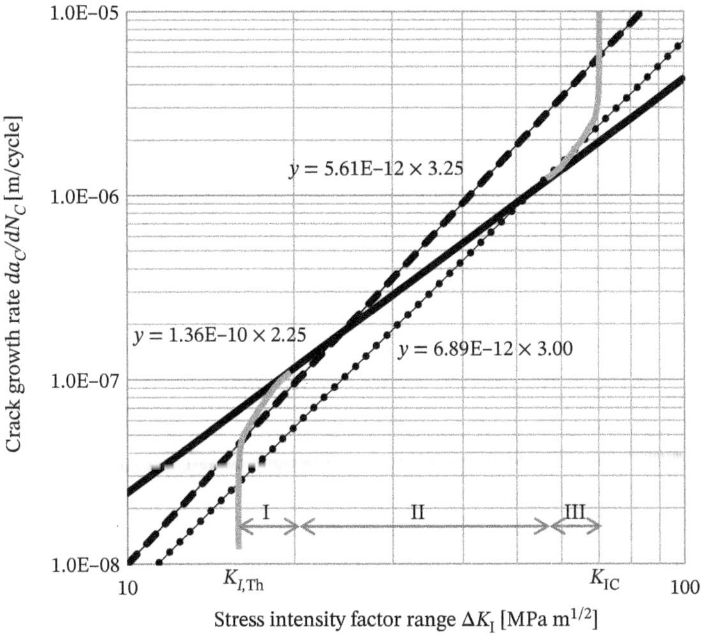

Figure 7.31 Plots of crack growth rates in the propagation zone only without the specific minimum or maximum graph ends. Fictitious representations of threshold stress intensity factor $K_{I,Th}$ and the fracture toughness K_{IC} maximum limit are plotted as a grayish solid line sprouting off the black solid line for martensitic steel. The dotted and dashed black lines are for the ferritic–pearlitic steels and the austenitic stainless steels, respectively.

factor $K_{I,Th}$ or the *fracture toughness* K_{IC} maximum limit, since these depend on the particular steel. Fictitious representations of these are plotted each as grayish solid line sprouting off the black solid line for *martensitic steels*. The dotted and dashed black lines are for the *ferritic–pearlitic steels* and the *austenitic stainless steels*, respectively. The black solid line with the two ends of grayish solid lines represents the usual sigmoid curve representing the usual crack growth rate as a function of the stress intensity factor range ΔK_I. This sigmoid curve has the zones I, II, and III shown at the bottom of the plot area of Figure 7.31. The second zone II is the crack propagation zone governed by Eq. (7.106). Zone I is the *crack initiation* stage, and zone III is the *crack instability* stage to fracture.

To approximately estimate the minimum *threshold stress intensity factor* $K_{I,Th}$ of the material, Rolfe and Barsom (1977) suggested the following relation:

$$1.5\left(10^{-4}\right) \leq \frac{K_{I,Th}}{E} \leq 1.8\left(10^{-4}\right)\ \left[\text{in}^{1/2}\right] \tag{7.107}$$

where E is the modulus of elasticity in [kpsi]. This estimate appears to be valid for various materials. However, qualification, validation, and verification are necessary for any particular material to be usable in design. Equation (7.107) might still be useful as an initial estimate. Other options state that the *threshold stress intensity factor* $K_{I,Th}$ of the material may be taken at an arbitrary 10^{-10} [m/cycle] of the fatigue crack growth rate or in the range of about 10^{-8}–10^{-13} [m/cycle] or about $4(10^{-7})$–$4(10^{-12})$ [in/cycle] of the fatigue crack growth rate. For this and other evaluations and considerations, one can consult dedicated references and handbooks such as ASM (1996).

It is stressed again that the *propagation factors* C_P and m_P are approximate empirical constants of the material and one needs to verify and validate the values for the particular material prospectively to be used.

The plot of the crack growth rate of Figure 7.31 is figuratively about a 90° clockwise rotation of the *S–N* fatigue curve as of Figure 7.23. The *threshold stress intensity factor* $K_{I,Th}$ is representing the *endurance limit* S_e in the sense that below the threshold stress intensity factor, there should be no crack propagation. Below the *endurance limit* S_e, there should be no fatigue failure. The same applies for the *fracture toughness* K_{IC}, which is corresponding to the low-cycle brittle fracture limit of a cracked element where the number of cycles to failure N_{Cf} is only one-half cycle and the *stress intensity factor* K_I reaches K_{IC}.

Consider the crack propagation stage under the *stress range* $\Delta\sigma$. To get the expected life or cycles to failure N_{Cf} from an *initial crack size* a_{Ci} to a *failure crack size* a_{Cf}, one can integrate Eq. (7.106) with the substitution of Eq. (7.105) to get the following expression:

$$N_{Cf} = \int_0^{N_{Cf}} dN_C = \frac{1}{C_P}\int_{a_{Ci}}^{a_{Cf}} \frac{da_C}{\left(\beta_C \Delta\sigma \sqrt{\pi a_C}\right)^{m_P}} \tag{7.108}$$

or

$$N_{Cf} = \frac{a_{Cf}^{1-(m_P/2)} - a_{Ci}^{1-(m_P/2)}}{C_P \beta_C^{m_P} \Delta\sigma^{m_P} \pi^{m_P/2}\left(1-(m_P/2)\right)} \tag{7.109}$$

This value of N_{Cf} may be compared to the value of N_{Cf} obtained from Eq. (7.65) with the fatigue failure strength S_f representing the *stress range* $\Delta\sigma$. The values of the *propagation factors* C_P and m_P are approximate empirical constants of the material and one needs to verify and validate the values for the particular material prospectively to be usable.

7.9.4 Crack Propagation and Real Case Study

Figure 7.32 pictures a seriously propagated crack and a *composite material repair* of an arm of a microwave coupler drive for the turntable. This image is after the fatigue-cracked arm is mended by a composite of *gauze fabric* (or flannel in other cases) in a matrix of *cyanoacrylate* adhesive bonding. The crack has propagated all the way to other

Figure 7.32 Seriously cracked arm of a microwave turntable coupler drive is mended by a composite of gauze fabric in a matrix of cyanoacrylate adhesive bonding. The crack has propagated all the way to other side flange as pointed at above the mending.

side flange of the arm section as pointed at above the mending. The fracture toughness of the material allowed a crack to propagate almost all the way without breakage or fracture. The *composite material repair* is surviving for years without further crack propagation.

7.10 Computer-Aided Selection and Optimization

This section is aiming at the productive use of previously developed relations of material properties and their interrelated relations. Static and fatigue strength and the associated safety factors are set in computer codes to aid in the evaluation of these factors and material properties. This allows carrying out easy iterations to reach part size optimization. The developed codes are computer-aided means for material selection and optimization. The following topics demonstrate the utility and applicability of these codes:

7.10.1 Material Properties: Carbon Steel

The selection of carbon steel properties is necessary for the synthesis of machine elements. The developed relations in the Section 7.4 are useful in constructing a computer code to aid in that process. One can use the input as trial values to converge into the desired property or use it as measured or needed values for the synthesis. The developed code in Figure 7.33 is valuable in that regard. The code intends to find the material properties of plain carbon steels as a function of carbon content as a percentage. Estimated properties are the yield strength (`Sy` for S_y), the ultimate tensile strength (`Su` for S_u), and the *Brinell* hardness number (`HB` for *HB*). Properties are for HR or CD carbon steel. Other properties are also for heat treatment by water quenching and tempering to 205 [°C] or 400 [°F], which is marked as (`WQT`). Figure 7.33 is a MATLAB code to calculate the material properties for carbon steel when both carbon percent and Brinell hardness *HB* number are measurable (or sought) as present in Example 7.1. Comparing both calculated HR hardness and CD hardness with measured *HB* value defines if the material is HR or CD. If the measured *HB* is closer to the calculated HR hardness, the material is then HR. If the measured *HB* is closer to the calculated CD hardness, the material is CD. Properties are then defined accordingly. The variables in the MATLAB code of Figure 7.33 are *intuitive*, as a CAD program should. They are combinations of the previous symbols. The name of each variable is a grouping of the hardness, strength, and processing such as `HB_HR` is the *Brinell* hardness number *HB* of HR steel. The strength such as `Su_HR_MPa` is a composition of `Su` = S_u, `HR` for HR and `MPa` for the SI unit [MPa]. The term `ksi` is for the US unit [ksi]. The rest of the variables are simply deductible by similar logic using terms as previously noted.

```
clear all; clc; format compact; format short      % Material_Properties_Carbon_Steel_HR_CD.m
          % (inputs need change or commenting)
Carbon_Percent=0.5                                % INPUT Measured Carbon Percentage
HB=195                                            % INPUT Measured BrinellHardness
CC=Carbon_Percent;                                % CC=0.06-0.95 (HR), 0.1-0.5 (CD)
HB_HR= -45.504*CC^2+234.61*CC+69.113              % HR Brinell Hardness (hot rolled)
HB_CD= -17.304*CC^2+244.73*CC+78.843              % CD Brinell Hardness (cold drawn)
          %Hot Rolled Properties (HR)             % MPa is [MPa] and kpsi is [kpsi]
Su_HR_MPa= -238.74*CC^2+864.93*CC+230.48          % Su is ultimate tensile strength
Sy_HR_MPa= -101.25*CC^2+441.3*CC+134.1            % Sy is yield strength
Se_HR_MPa= -119.37*CC^2+432.47*CC+115.24
Su_HR_kpsi= 0.145*Su_HR_MPa
Sy_HR_kpsi= 0.145*Sy_HR_MPa
Se_HR_kpsi= 0.145*Se_HR_MPa
          % Cold Drawn Properties (CD)
Su_CD_MPa= 55.792*CC^2+745.62*CC+290.91
Sy_CD_MPa= 93.042*CC^2+611.46*CC+241.27
Su_CD_kpsi= 0.145*Su_CD_MPa
Sy_CD_kpsi= 0.145*Sy_CD_MPa
          % Hear Treated, water quenched and tempered to 205 [oC] or (400) [oF]
Su_WQT_MPa= -811.21*CC^2+2068.6*CC+258.47         % WQT is water quenched and tempered
Sy_WQT_MPa= -204.08*CC^2+886.24*CC+384.65
Su_WQT_kpsi= 0.145*Su_WQT_MPa
Sy_WQT_kpsi= 0.145*Sy_WQT_MPa
HB_WQT= 125.39*CC^2+1.3517*CC+486.31
```

Figure 7.33 MATLAB code to calculate the material properties for carbon steel when both carbon percent and *Brinell* hardness *HB* number are measured (Example 7.1). Comparing both calculated hot rolled (HR) hardness and cold drawn (CD) hardness with measured *HB* value defines if the material is HR or CD. Properties are estimated accordingly.

```
clear all; clc; format compact; format short      % Material_Properties_Carbon_Steel_Ann_Norm.m
          % (inputs may need change or commenting)
Carbon_Percent=0.3                                % INPUT Sought Carbon Percentage
HB=495                                            % INPUT Sought HB as Brinell Hardness
CC=Carbon_Percent;                                % CC=0.2-0.95(Annealed or Normalized)
HB_HR= -45.504*CC^2+234.61*CC+69.113              % HR is hot rolled
HB_CD= -17.304*CC^2+244.73*CC+78.843              % CD is cold drawn
HB_Ann= -219.96*CC^2 + 353.68*CC + 46.99          % Ann is Annealed
HB_Norm= -113.24*CC^2 + 369.82*CC + 53.11         % Norm is Normalized
          % Annealed Properties                       % MPa is [MPa] and kpsi is [kpsi]
Su_Ann_MPa= -731.47*CC^2+1167.6*CC+188.6          % Su is ultimate tensile strength
Sy_Ann_MPa= -243.39*CC^2+373.31*CC+239.37         % Sy is yield strength
Su_Ann_kpsi= 0.145*Su_Ann_MPa
Sy_Ann_kpsi= 0.145*Sy_Ann_MPa
          % Normalized Properties
Su_Norm_MPa= -376.58*CC^2+1263.7*CC+183.01
Sy_Norm_MPa= -62.967*CC^2+323.77*CC+268.38
Su_Norm_kpsi= 0.145*Su_Norm_MPa
Sy_Norm_kpsi= 0.145*Sy_Norm_MPa
          % Hear Treated, water quenched and tempered to 205 [oC] or (400) [oF]
Su_WQT_MPa= -811.21*CC^2+2068.6*CC+258.47         % WQT is water quenched and tempered
Sy_WQT_MPa= -204.08*CC^2+886.24*CC+384.65
Su_WQT_kpsi= 0.145*Su_WQT_Mpa
Sy_WQT_kpsi= 0.145*Sy_WQT_MPa
HB_WQT= 125.39*CC^2+1.3517*CC+486.31
```

Figure 7.34 MATLAB code to calculate the material properties for carbon steel when both carbon percent and *Brinell* hardness *HB* number are sought (Example 7.5). Comparing calculated hardness of HR, CD, annealed, normalized, and water quenched and tempered with sought *HB* value defines the material. Properties are estimated accordingly.

Figure 7.34 is a MATLAB code to calculate the material properties for carbon steel when both carbon percent and Brinell hardness *HB* number are sought as present in Example 7.1. Comparing calculated hot rolled (`HR`) hardness, cold drawn (`CD`) hardness, annealed (`Ann`) hardness, normalized (`Norm`) hardness, and water quenched and tempered (`WQT`) hardness with measured *HB* value defines if the material is `HR`, `CD`, `Ann`, `Norm`, or `WQT`. Properties are then defined accordingly. The variables in the MATLAB code of Figure 7.34 are *intuitive*, as a CAD program should and as previously adapted. They are combinations of the previous symbols as before.

The available MATLAB code at ***Wiley website*** is a combination of the code in Figure 7.33 with the code in Figure 7.34. This code at ***Wiley website*** is under the name **`Material_Properties_Carbon_Steel.m`**. The code uses Eqs. (7.5)–(7.14) to evaluate the strength and hardness of plain carbon steel for different processing cases.

Example 7.5 Continuation on a similar example to the previous Example 7.1, the sought composition of the machine element indicated the need of a carbon content of around 0.3%. Other elements may exist, but they can be within the admissible ranges of plain carbon steel. If the sought hardness value is about *HB* 495, what is the expected material and heat treatment of the steel? Estimate the mechanical properties of the steel using a MATLAB code similar to that in Figures 7.33 and 7.34. Compare results with available information. If the material is to be annealed or normalized, what are the expected properties?

Solution

Data: $C_C = 0.3$, $HB = 495$.

This data is sufficient to apply into the MATLAB code in Figure 7.34.

Running the MATLAB code (`Material_Properties_Carbon_Steel.m`) gives the following *input* and *output* results.

Input: `Carbon_Percent = 0.3000 and HB = 495`

Output:

HB_HR = 135.4006 and HB_CD = 150.7046 (for *Brinell* hardness if the material is HR or CD) (a)

HB_Ann = 133.2976 and HB_Norm = 153.8644 (for *Brinell* hardness, annealed, or normalized) (b)

Su_HR_MPa = 468.4724, Sy_HR_MPa = 257.3775, Se_HR_MPa = 234.2377 (for hot rolled – SI) (c)

Su_HR_kpsi = 67.9285, Sy_HR_kpsi = 37.3197, Se_HR_kpsi = 33.9645 (for hot rolled – US) (d)

Su_CD_MPa = 519.6173, Sy_CD_MPa = 433.0818 (for cold drawn – SI) (e)

Su_CD_kpsi = 75.3445, Sy_CD_kpsi = 62.7969 (for cold drawn – US) (f)

Su_Ann_MPa = 473.0477 and Sy_Ann_MPa = 329.4579 (for annealed – SI) (g)

Su_Ann_kpsi = 68.5919 and Sy_Ann_kpsi = 47.7714 (for annealed – US) (h)

Su_Norm_MPa = 528.2278 and Sy_Norm_MPa = 359.8440 (for normalized – SI) (i)

Su_Norm_kpsi = 76.5930 and Sy_Norm_kpsi = 52.1774 (for normalized – US) (j)

Su_WQT_MPa = 806.0411, Sy_WQT_MPa = 632.1548 (for heat treated – SI) (k)

Su_WQT_kpsi = 116.8760, Sy_WQT_kpsi = 91.6624 (for heat treated – US) (l)

HB_WQT = 498.0006 (for heat treated – *Brinell* hardness number) (m)

Scrutinizing the previous values indicates that the material underwent water quenching and tempering to 205[°C] or 400 [°F], which gave the material *Brinell* hardness number of 498 close to the required 495 *HB*. The material is then AISI 1030 or ISO C30, water quenched, and tempered to 205[°C] or 400 [°F]. The estimated properties are as follows:

$$S_u = 806.0411\ [\text{MPa}]\ \text{or}\ 116.8760\ [\text{kpsi}]\ (\texttt{Su\_WQT\_MPa}\ \text{and}\ \texttt{Su\_WQT\_kpsi},\ \text{respectively}) \tag{n}$$

$$S_y = 632.1548\ [\text{MPa}]\ \text{or}\ 91.6624\ [\text{kpsi}]\ (\texttt{Sy\_WQT\_MPa}\ \text{and}\ \texttt{Sy\_WQT\_kpsi},\ \text{respectively}) \tag{o}$$

$$HB = 498.0006\ (\text{for}\ \texttt{HB\_WQT}) \tag{p}$$

The rest of the output properties do not apply. Finding the properties of AISI 1030 or ISO C30 according to Table A.7.3 in Appendix A.7 provides the following values:

$$S_u|_{\text{WQ\&T}} = 848\ [\text{MPa}]\ \text{or}\ 123\ [\text{kpsi}]\ (\text{compared to}\ 806.0411\ [\text{MPa}]\ \text{or}\ 116.8760\ [\text{kpsi}]) \tag{q}$$

$$S_y|_{\text{WQ\&T}} = 648\ [\text{MPa}]\ \text{or}\ 94\ [\text{kpsi}]\ (\text{compared to}\ 632.1548\ [\text{MPa}]\ \text{or}\ 91.6624\ [\text{kpsi}]) \tag{r}$$

The discrepancy is due to applying the equations of the water quenching and tempering excluding the data for 0.8% carbon steel as noted in Figure 7.12. The rest of the properties are in Table A.7.3 in Appendix A.7 and used for the *annealed* and *normalized* properties as follows:

$$S_u|_{\text{Annealed}} = 464\ [\text{MPa}]\ \text{or}\ 67\ [\text{kpsi}]\ (\text{compared to}\ 473.0477\ [\text{MPa}]\ \text{or}\ 68.5919\ [\text{kpsi}]) \tag{s}$$

$$S_y|_{\text{Annealed}} = 341\ [\text{MPa}]\ \text{or}\ 50\ [\text{kpsi}]\ (\text{compared to}\ 329.4579\ [\text{MPa}]\ 47.7714\ [\text{kpsi}]) \tag{t}$$

$$S_u|_{\text{Normalized}} = 521\ [\text{MPa}]\ \text{or}\ 76\ [\text{kpsi}]\ (\text{compared to}\ 528.2278\ [\text{MPa}]\ \text{or}\ 76.5930\ [\text{kpsi}]) \tag{u}$$

$$S_y|_{\text{Normalized}} = 345\ [\text{MPa}]\ \text{or}\ 50\ [\text{kpsi}]\ (\text{compared to}\ 359.8440\ [\text{MPa}]\ 52.1774\ [\text{kpsi}]) \tag{v}$$

The discrepancy is due to the use of Eqs. (7.10) and (7.11) including the data for 0.8% carbon steel as noted in Figure 7.11. This is a further indication of the wide data scattering than what the trend lines approximating character. The difference in values is in the range of 1–4%, which is within the character of initial engineering design calculations. This, however, renders the synthesis process more efficient and the optimization tuning much easier particularly with the use of more definite material properties.

7.10.2 Fatigue Strength and Factors Affecting Fatigue: Carbon Steel

A code to evaluate the *fatigue strength* S_f at a life span number of *fatigue failure cycles* N_{Cf} is useful in the design of any machine member that has a finite life span between 10^3 and 10^6 cycles. Figure 7.35 is a MATLAB code that calculates the SI and US *fatigue strength* S_f at any expected failure life or *cycles to failure* N_{Cf} for plain carbon steel machine elements. The code in Figure 7.35 is an opening part of the combined codes in some following figures. However, it can stand alone as an independent code. The code employs Eqs. (7.63) and (7.64) to calculate the constants a_1 and a_2 to use in Eq. (7.65) for the *fatigue strength* S_f. The variables used in the code are intuitive such as `Sut` for the *ultimate tensile strength* S_u or S_{ut} with attached SI or US for these systems of units. The same is done for the *endurance limit* S_e, the *yield strength* S_y, and *fatigue strength* S_f by using the variables `Se`, `Sy`, and `Sf` with attaching SI or US. The variable `NCf` stands for the number of loading cycles to failure N_{Cf}. The constants a_1 and

```
clear all; clc; format compact; format short % Fatigue_Strength_and_Factors_Affecting.m
          % (several inputs need change or commenting)
Sut=525; Ultimate_Strength_MPa=Sut          % INPUT Ultimate Strength [MPa]
SutUS= Sut*0.145;                               % [kpsi]
SutUS=76 ;                                  % INPUT
Ultimate_Strength_kpsi=SutUS                    % Ultimate Strength [kpsi]
SeSI=0.5*SutSI; SeSI=262.5;                 % INPUT [MPa]
Endurance_limit_MPa=SeSI                        % Endurance Limit [MPa]<1380
SeUS= 0.145*SeSI;                               % [kpsi]
SeUS= 37.7 ;                                % INPUT
Endurance_limit_kpsi=SeUS                       % Endurance Limit [kpsi]<200
SySI=290;                                   % INPUT [MPa]
Yield_Strength_MPa= SySI                        % Yield Strength [MPa]
SyUS=42;                                    % INPUT [kpsi]
Yield_Strength_kpsi= SyUS                       % Yield Strength [kpsi]
NCf=10^6;                                   % INPUT
Number_of_Cycles= NCf                           % Number of loading cycles to failure
fr=0.8 ;                                    % INPUT Failure ratio at NCf=1000, Change 0.8 if needed
Failure_ratio=fr
a1_SI=(log(0.8*SutSI)-log(SeSI))/(log(10^3)-log(10^6));
a2_SI=((fr*SutSI)^2)/SeSI;
a1_US=(log(0.8*SutUS)-log(SeUS))/(log(10^3)-log(10^6));
a2_US=((fr*SutUS)^2)/SeUS;
Sf_SI=a2_SI*NCf^a1_SI;
Sf_US=a2_US*NCf^a1_US;
Fatigue_Failure_MPa=Sf_SI
Fatigue_Failure_kpsi=Sf_US
```

Figure 7.35 MATLAB code to calculate the SI and US fatigue strength S_f of plain carbon steel at any failure life or cycles to failure N_{Cf} (Example 7.6). This is an opening part of the code in Figure 7.36. The final code includes the codes in Figures 7.35–7.37 as one code.

a_2 are the terms `a1` and `a2` with attaching underscores before the SI or US to indicate these systems. The terms `MPa` or `kpsi` are attached to the calculated values to mean [MPa] and [kpsi] units.

Figure 7.36 is a MATLAB code to calculate the factors affecting the *fatigue strength* $S_{e,\text{part}}$ of a machine part. This code is part of a unified code including the codes of Figures 7.35–7.37. The code employs Eqs. (7.67)–(7.82) to calculate most of the factors affecting fatigue and with few inputs calculates the *fatigue strength* $S_{e,\text{part}}$ of a machine part using Eq. (7.66). The variables used in the code are intuitive such as `dp` for part diameter in [mm] and `dpin` for part diameter in [in]. To and ToF are the operating temperature in [°C] and in [°F], respectively. The variable `rN` and `rNmm` are the *notch radius* r_N in [in] and [mm], respectively. The other variables are also intuitive as defined in Figure 7.36 such as the factors affecting fatigue as `K_` and attaching the factor type or the Neuber number as aN that stands for a_N. `Se_part` stands for the *part endurance limit* $S_{e,\text{part}}$. Attaching SI or the US stands for these systems of units.

Figure 7.37 is a MATLAB code to calculate the equivalent fatigue stresses. It also calculates the SI and US dynamic fatigue safety factors $K_{SF,F}$ according to ASME elliptic failure criterion. This code is part of a unified code including the codes of Figures 7.35–7.37. The code employs Eqs. (7.56) and (7.57) to calculate the mean and alternating values of the stresses given their maximum and minimum values as input. The variable `Sigma` is denoting the normal stress σ, and the variable `Tau` is the shear stress τ. The abbreviations max and min are for the maximum and the minimum, respectively. They are attached with a preceding underscore for each abbreviation. The variables `_m` and `_a` stand for the mean and alternating values calculated by similar equations to (7.56) and (7.57). The term `_vM` stands for the *von Mises* equivalent. Equations (7.95) are used to calculate the *equivalent von Mises* mean and alternating stresses. Variables or abbreviations are clearly intuitive, similar to the previously adapted symbols, and present in the code as comments preceded by the % character.

The previous MATLAB codes are set in one code under the name ***`Fatigue_Strength_and_Factors_Affecting.m`***. This combined MATLAB code is available at ***Wiley website.***

```
                                              % Fatigue_Strength_and_Factors_Affecting.m
        % Factors affecting fatigue (inputs may need change or commenting)
dp=45; Part_diameter_mm=dp                              % INPUT Part_diameter[mm]
dpin=1.75; Part_diam_in=dpin                            % INPUT Part_diameter[in]
To=380; Operating_Temperature_C=To                      % INPUT Operating_Temperature[oC]
ToF=716; Operating_Temperature_F=ToF                    % INPUT Operating_Temperature[oF]
rN=0.1; Notch_radius_in=rN                              % INPUT Notch_radius[in]
rNmm=rN*25.4; Notch_radius_mm=rNmm                      % INPUT Notch_radius[mm]
Reliab=99.99; Reliability =Reliab                       % INPUT Reliability
KSC=2.1; Stress_Concentration_Factor_KSC=KSC            % INPUT Stress Concentration Factor
K_surf=0.89;                                            % INPUT Surface Factor (ground)
K_surf=2.9458*SutSI^-0.212                              % INPUT Surface Factor (machined)
K_size_mm=1.51*dp^-0.157; K_size_mm=1.24*dp^-0.107          % Size > 50 [mm] and Size < 50 [mm]
K_size_in=0.91*dpin^-0.157; K_size_in=0.879*dpin^-0.107     % Size > 2 [in] and Size < 2 [in]
K_load=1.0                                              % INPUT0.85(axial), 0.59(torsion)
K_reliab= 0.80525 *(100-Reliab)^0.05245;                    % Reliab<= 99%
K_reliab= 0.80483*(100-Reliab)^0.02797                      % Reliab> 99%
K_temp_C=-2.325*10^-6*To^2+0.0007178*To+0.9762              % Operating Temperature [oC]
K_temp_F=-7.024*10^-7*ToF^2+0.0004273*ToF+0.9664            % Operating Temperature [oF]
        % Neuber constant aN in [in]
aN=(0.246-3.08*10^-3*SutUS+1.51*10^-5*SutUS^2-2.67*10^-8*SutUS^3)^2;
Neuber_constant=aN
K_conc=(1+sqrt(aN/rN))/(KSC+sqrt(aN/rN))                    % Concentration factor
K_miscel=1.0                                            % INPUT Miscellaneous factor
K_total_SI=K_surf*K_size_mm*K_load*K_reliab*K_temp_C*K_conc*K_miscel
K_total_US=K_surf*K_size_in*K_load*K_reliab*K_temp_F*K_conc*K_miscel
Se_part_SI=K_total_SI*Sf_SI                             % For Cycles NC<=10^6 (Sf_SIfrom previous code)
Se_part_US=K_total_US*Sf_US                             % For Cycles NC<=10^6 (Sf_USfrom previous code)
```

Figure 7.36 MATLAB code to calculate the factors affecting fatigue strength (Example 7.6). This code is part of a unified code including the codes of Figures 7.35–7.37.

```
                                              % Fatigue_Strength_and_Factors_Affecting.m
        % Fluctuating Stresses and ASME fatigue diagram (for Safety Factor [MPa])
        % (several inputs need change)
Sigma_max_MPa=8.8951                                    % INPUT[MPa] maximum stress
Sigma_min_MPa=-8.8951                                   % INPUT[MPa] minimum stress
Tau_max_MPa=4.4475                                      % INPUT[MPa] maximum stress
Tau_min_MPa=4.4475                                      % INPUT[MPa] minimum stress
Sigma_m_MPa= 0.5*(Sigma_max_MPa+Sigma_min_MPa)              % _m for mean
Sigma_a_MPa= 0.5*(Sigma_max_MPa-Sigma_min_MPa)              % _a for alternating
Tau_m_MPa= 0.5*(Tau_max_MPa+Tau_min_MPa)
Tau_a_MPa= 0.5*(Tau_max_MPa-Tau_min_MPa)
Sigma_vMm_MPa=sqrt(Sigma_m_MPa^2+3*Tau_m_MPa^2)             % vMvon Mises equivalent
Sigma_vMa_MPa=sqrt(Sigma_a_MPa^2+3*Tau_a_MPa^2)             % vMvon Mises equivalent
KSF_F_SI=1/sqrt((Sigma_vMa_MPa/Se_part_SI)^2+(Sigma_vMm_MPa/SySI)^2)    % ASME elliptic
        % Fluctuating Stresses and ASME fatigue diagram (for Safety Factor [kpsi])
        % (several inputs need change)
Sigma_max_kpsi=1.3376                                   % INPUT[kpsi] maximum stress
Sigma_min_kpsi=-1.3376                                  % INPUT[kpsi] minimum stress
Tau_max_kpsi=0.66879                                    % INPUT[kpsi] maximum stress
Tau_min_kpsi=0.66879                                    % INPUT[kpsi] minimum stress
Sigma_m_kpsi= 0.5*(Sigma_max_kpsi+Sigma_min_kpsi)           % _m for mean
Sigma_a_kpsi= 0.5*(Sigma_max_kpsi-Sigma_min_kpsi)           % _a for alternating
Tau_m_kpsi= 0.5*(Tau_max_kpsi+Tau_min_kpsi)
Tau_a_kpsi= 0.5*(Tau_max_kpsi-Tau_min_kpsi)
Sigma_vMm_kpsi=sqrt(Sigma_m_kpsi^2+3*Tau_m_kpsi^2)          % vMvon Mises equivalent
Sigma_vMa_kpsi=sqrt(Sigma_a_kpsi^2+3*Tau_a_kpsi^2)          % vMvon Mises equivalent
KSF_F_US=1/sqrt((Sigma_vMa_kpsi/Se_part_US)^2+(Sigma_vMm_kpsi/SyUS)^2)  % ASME elliptic
```

Figure 7.37 MATLAB code to calculate the equivalent fatigue stresses. It also calculates the SI and US dynamic safety factors according to ASME elliptic failure criterion (Example 7.6). This code is part of a unified code including the codes of Figures 7.35–7.37.

Example 7.6 Continue on the previous Examples 7.3 and 7.4, it is required to find the solution using a MATLAB code similar to that in Figures 7.35–7.37. Again, the shaft is transmitting a maximum of 25 [kW] or 33.5 [hp] and running at 3000 [rpm] in the clockwise direction. It is subject to a maximum bending moment that is equal to the maximum applied torque. For the internal stresses previously calculated and if the shaft has a cylindrical diameter of 45 [mm] or 1.75 [in], estimate the factors affecting fatigue knowing that the reliability should be 99.99%, the operating temperature is about 380 [°C] or 716 [°F], the evaluated stress concentration factor K_{SC} is 2.1 and the notch radius is 2.5 [mm] or 0.1 [in]. Find the safety factor for the distortion energy theory and the elliptic ASME fatigue failure criterion. The selected material is again set as HR AISI 1040 or ISO C40. Compare results to those of Examples 7.3 and 7.4.

Solution

Data: $H = 25$ [kW] or 33.5 [hp], $N_{rpm} = 3000$ [rpm], and $d_C = 0.045$ [m] or 1.75 [in]. *Ductile* material is HR AISI 1040 or ISO C40. The material properties are yield strength of 42 [kpsi] or 290 [MPa], ultimate tensile strength of 76 [kpsi] or 525 [MPa], and the endurance limit of 37.7 [kpsi] or 260 [MPa]; see Table A.7.2 in Appendix A.7. Equation (7.58) suggests $S_e = 0.5\,S_{ut} = 0.5(525) = 262.5$ [MPa] or $0.5(76) = 38$ [kpsi]. The two values are close, and one can then use the values in Table A.7.2 in Appendix A.7, which is the same as in Example 7.4. The data in Example 7.4 that can be applicable to this example is as follows:

1) The maximum shear stresses due to torque τ_{xz} are 4.4475 [MPa] or 0.668 79 [kpsi].
2) The maximum normal stresses due to bending moment σ_x are 8.8951 [MPa] or 1.3376 [kpsi].

All this data is sufficient to apply into the MATLAB codes in Figures 7.35–7.37.

Running the MATLAB code (`Fatigue_Strength_and_Factors_Affecting.m`) gives the following *input* and *output* results with minor editing.

Input: (The following bullets are for each of the codes in Figures 7.35–7.37.)

- `Ultimate_Strength_MPa = 525 [MPa], Ultimate_Strength_kpsi = 76 [kpsi], Endurance_limit_MPa = 262.5 [MPa], Endurance_limit_kpsi = 37.7 [kpsi], Yield_Strength_MPa = 290 [MPa], Yield_Strength_kpsi = 42 [kpsi], Number_of_Cycles = 1000000 cycles (`10^6` to check)`
- `Part_diameter_mm=45 [mm], Part_diam_in=1.7717 [in], Operating_Temperature_C = 380 [oC], Operating_Temperature_F = 716 [oF], Notch_radius_in = 0.1 [in], Notch_radius_mm = 2.54 [mm], Reliability = 99.99, Stress_Concentration_Factor_KSC = 2.1`
- `Sigma_max_MPa = 8.8951 [MPa], Sigma_min_MPa = -8.8951 [MPa], Tau_max_MPa = 4.4475 [MPa], Tau_min_MPa = 4.4475 [MPa], Sigma_max_kpsi = 1.3376 [kpsi], Sigma_min_kpsi = -1.3376 [kpsi], Tau_max_kpsi = 0.6688 [kpsi], Tau_min_kpsi = 0.6688 [kpsi]`

Output: (The following bullets are for each of the codes in Figures 7.35–7.37.)

- `Failure_ratio = 0.8, Fatigue_Failure_MPa = 262.5 [MPa], Fatigue_Failure_kpsi = 38.0625 [kpsi]`
- `K_surf = 0.7808, K_size_mm = 0.8251, K_size_in = 0.8268, K_load = 1, K_reliab = 0.7076, K_temp_C = 0.9132, K_temp_F = 0.9123, Neuber_constant = 0.0076 [in`$^{1/2}$`], K_conc = 0.5371, K_miscel = 1, K_total_SI = 0.2236, K_total_US = 0.2238, Se_part_SI = 58.6957 [MPa], Se_part_US = 8.4380 [kpsi]`
- `Sigma_m_MPa = 0 [MPa], Sigma_a_MPa = 8.8951 [MPa], Tau_m_MPa = 4.4475 [MPa], Tau_a_MPa = 0 [MPa], Sigma_vMm_MPa = 7.7033 [MPa], Sigma_vMa_MPa = 8.8951 [MPa], KSF_F_SI = 6.4996, Sigma_m_kpsi = 0 [kpsi], Sigma_a_kpsi = 1.3376 [kpsi], Tau_m_kpsi = 0.6688 [kpsi], Tau_a_kpsi = 0 [kpsi], Sigma_vMm_kpsi = 1.1584 [kpsi], Sigma_vMa_kpsi = 1.3376 [kpsi], KSF_F_US = 6.2149`

These values are very close to those found in Examples 7.4. The maximum difference is acceptable for engineering design calculations. The final result of each fatigue safety factor $K_{SF,F}$ is also close. The computer code gives

$K_{SF,F} = 6.4395$ and 6.2229 for SI units and the US units, respectively. The hand calculations in Example 7.4 gives $K_{SF,F} = 6.389$ and 6.177 for SI units and the US units, respectively, which is close enough for manual engineering design calculations. Variables defined by only three digits are present in some hand calculations, which may result in only two reliable digits in successive calculations.

7.10.3 Static Strength and Factors of Safety: Carbon Steel

For static loading or no appreciable variation in loads, the MATLAB fatigue strength code available as (`Fatigue_Strength_and_Factors_Affecting.m`) at ***Wiley website*** can be just as useful in static strength and the associated factors of safety. All one needs to do is to have the input of the maximum stresses be equal to the values of the minimum stresses. This causes all mean stresses to be stationary with each being at its maximum value with zero alternating stress components, i.e. no dynamic variations of loads. This is simply doable by resolving Example 7.3 using the fatigue strength MATLAB code as just suggested. Extra calculations such as the factors affecting fatigue do not influence computations.

Example 7.7 Resolve Example 7.3 by using the MATLAB fatigue strength code available at ***Wiley website*** and named as (`Fatigue_Strength_and_Factors_Affecting.m`). Use only the plain carbon steel material and the *von Mises* failure theory.

Solution

Data: $H = 25$ [kW] or 33.5 [hp], $N_{\text{rpm}} = 3000$ [rpm], and $d_C = 0.045$ [m] or 1.75 [in]. The material is HR AISI 1040 or ISO C40.

The material properties are the yield strength of 42 [kpsi] or 290 [MPa] and the ultimate tensile strength of 76 [kpsi] or 525 [MPa]; see Table A.7.2 in Appendix A.7.

From Example 7.3, the data that can be applicable to this example are as follows:

1) The maximum shear stress due to torque τ_{xz} is 4.4475 [MPa] or 0.668 79 [kpsi].
2) The maximum normal stress due to bending moment σ_x is 8.8951 [MPa] or 1.3376 [kpsi].

These are the same as previous Examples 7.5 and 7.6, which used the MATLAB code.

Since the assumed load is to be static in Example 7.3, the normal stress due to bending somehow should not change sign. All one should do is to have the minimum normal stress due to bending moment σ_x be 8.8951 [MPa] or 1.3376 [kpsi]. This changes the sign of the `Sigma_min_MPa` in Figure 7.37 to a positive value, i.e. `Sigma_min_MPa` = 8.8951 [MPa]. This should also change the sign of the `Sigma_min_kpsi` in Figure 7.37 to a positive value, i.e. `Sigma_min_kpsi` = 1.3376 [kpsi]. Running the MATLAB code accordingly, gives the following relevant outputs:

$$\texttt{Sigma\_vMm\_MPa} = \texttt{11.7671 [MPa]}, \texttt{Sigma\_vMa\_MPa} = \texttt{0 [MPa]}, \text{and } \texttt{KSF\_F\_SI} = \texttt{24.6451} \tag{a}$$

$$\texttt{Sigma\_vMm\_kpsi} = \texttt{1.7695 [kpsi]}, \texttt{Sigma\_vMa\_kpsi} = \texttt{0 [kpsi]}, \text{and } \texttt{KSF\_F\_US} = \texttt{23.7360} \tag{b}$$

These values are almost the same as the equivalent values in Example 7.3 such that

$$\sigma_{\text{vM}} = 11.767 \text{ [MPa] or } 1.7698 \text{ [kpsi] (compared to } 11.7671 \text{ [MPa] or } 1.7695 \text{ [kpsi])} \tag{c}$$

$$K_{\text{SF}}\big|_{\text{SI}} = 24.645 \text{ (compared to 24.6451)} \tag{d}$$

$$K_{\text{SF}}\big|_{\text{US}} = 23.731 \text{ (compared to 23.7360)} \tag{e}$$

```
                                                  % Fatigue_Strength_and_Factors_Affecting_Optimum.m
         % Factors affecting fatigue (one input needs change) % One input needs an iterative change
dp=45.0; Part_diameter_mm=dp                                  % INPUTPart_diameter[mm] (Input, iterative)
dpin=dp/25.4; Part_diam_in=dpin                               % INPUTPart_diameter[in]
         % Fluctuating Stresses and ASME fatigue diagram (for Safety Factor [MPa])
Bending = 79.577, Sigma = Bending*32/(pi*(dp/1000)^3);        % INPUT[N.m]  Bending
Torque = 79.577, Tau = Torque*16/(pi*(dp/1000)^3);            % INPUT[N.m]  Torque
Sigma_MPa=Sigma/10^6; Tau_MPa=Tau/10^6;                                 % [MPa]  Stresses
Sigma_max_MPa=Sigma_Mpa; Sigma_min_MPa=--Sigma_MPa                      % [MPa] maximum & minimum stress
Tau_max_MPa= Tau_MPa; Tau_min_MPa= Tau_MPa                              % [MPa] maximum  & minimum stress
Sigma_m_MPa= 0.5*(Sigma_max_MPa+Sigma_min_MPa) ;  Sigma_a_MPa= 0.5*(Sigma_max_MPa-Sigma_min_MPa)
Tau_m_MPa= 0.5*(Tau_max_MPa+Tau_min_MPa);  Tau_a_MPa= 0.5*(Tau_max_MPa-Tau_min_MPa)
Sigma_vMm_MPa=sqrt(Sigma_m_MPa^2+3*Tau_m_MPa^2)                         % vMvon Mises equivalent
Sigma_vMa_MPa=sqrt(Sigma_a_MPa^2+3*Tau_a_MPa^2)                         % vMvon Mises equivalent
KSF_F_SI=1/sqrt((Sigma_vMa_MPa/Se_part_SI)^2+(Sigma_vMm_MPa/SySI)^2) % ASME elliptic
         % Fluctuating Stresses and ASME fatigue diagram (for Safety Factor [kpsi])
Bending = 703.78, Sigma = Bending*32/(pi*(dpin)^3)            % INPUT [lb.in]  Bending
Torque = 703.78, Tau = Torque*16/(pi*(dpin)^3)                % INPUT [lb.in]  Torque
Sigma_kpsi=Sigma/10^3, Tau_kpsi=Tau/10^3,                               % [kpsi] Stresses
Sigma_max_kpsi= Sigma_kpsi; Sigma_min_kpsi= -Sigma_kpsi                 % [kpsi] maximum & minimum stress
Tau_max_kpsi= Tau_kpsi; Tau_min_kpsi= Tau_kpsi                          % [kpsi] maximum & minimum stress
         % Optimization for minimum volume or i.e. min. diameter for Safety Factor=2
Optimum_diam_SI=dp
Optimum_diam_US=dpin
Optimum_Vol_Percent_SI=100*dp^2/45^2
Optimum_Vol_Percent_US=100*dpin^2/(45/25.4)^2
Dynamic_Safety_Factor_SI=KSF_F_SI
Dynamic_Safety_Factor_US=KSF_F_US
         % HGP Optimization step
KSF_F_A = 2.0;                                                % INPUT Safety factor Aim (Input)
dp_new=dp*(KSF_F_A/KSF_F_SI)^(1/3)                                      % HGP suggested value
```

Figure 7.38 MATLAB code to iteratively optimize the part diameter for a needed value of SI or the US safety factor according to ASME elliptic failure criterion (Example 7.8). This code shows only the SI changes made in the unified code of Figures 7.35–7.37.

7.10.4 Optimization for a Specific Factor of Safety: Carbon Steel

A simple optimization is possible by utilizing the MATLAB code of (`Fatigue_Strength_and_Factors_Affecting.m`) with few little changes to allow for direct iteration. These changes are simply possible by adjusting few lines as indicated in Figure 7.37 with the code renamed as (**`Fatigue_Strength_and_Factors_Affecting_Optimum.m`**). The first change in the code is to iteratively input a different value of the part diameter (`dp`) in [mm] while observing the output dynamic safety factor. The second change is to tie the part diameter in [in] to the part diameter in [mm], so that one would only change one of them. One may alternatively change the diameter in [in] and tie the diameter in [mm] to it. The important change is to adjust the stresses according to the variations in diameters. This is achievable by the input of the bending moment (`Bending`) and the torque (`Torque`) in [N m] or [lb in], respectively. The normal stresses (`Sigma`) and the shear stresses (`Tau`) are then functions of the part diameter with results in [Pa] or [psi]. The stresses are then set in [MPa] and [kpsi]. These changes are evident in Figure 7.38 for both SI and the US systems of units. The last change is to rewrite the input diameters and the output safety factors to ease the comparison in the MATLAB command window and to plan the next iterative action of decreasing or increasing the part diameter dp. In addition, one needs the percentage savings in volume as a function of the iterative diameter raised to the power 2 divided by the original diameter raised to the power 2. These are the ratios of the areas considering a unit length to compute the volume.

The optimization can be set in a loop to automatically vary the diameter until the safety factor converges to the required safety factor aim. Utilizing the *HGP* optimization method for a 2.0 *safety factor aim* provides the new part diameter $d_{p,\mathrm{new}} = d_p(2.0/K_{\mathrm{SF},F})^{(1/3)}$ or in the code: `dpnew = dp*(2.0/KSF_F_SI)`$^{(1/3)}$. This relation is stemming from the fact that due to stresses, the safety factor $K_{\mathrm{SF},F}$ is a function of dp^3; see Section 5.6. When this relation is added to the optimization code, the optimum is reached in four iterations. For the next Example 7.8, the initial diameter $d_p = 45$ [mm] and the safety factor $K_{\mathrm{SF},F}$ for this manual iterations are as follows:

Part diameter d_p ($K_{\mathrm{SF},F}$): 45.00 (6.4395), 30.4746 (2.0825), 30.0667 (2.0028), 30.0527 (2.0001), 30.0522 (2.0000).

Example 7.8 Optimize Example 7.7 by using the adjusted MATLAB fatigue strength code, which is available at ***Wiley website*** as (`Fatigue_Strength_and_Factors_Affecting_Optimum.m`). These adjustments are evident in Figure 7.38. Use only the iteration of diameter to have a dynamic safety factor of about 2.0. Use a plain carbon steel material and the *von Mises* failure theory.

Solution

Data: $H = 25$ [kW] or 33.5 [hp], $N_{\text{rpm}} = 3000$ [rpm], and $d_C = 0.045$ [m] or 1.75 [in]. The material is HR AISI 1040 or ISO C40.

The material properties are the yield strength of 42 [kpsi] or 290 [MPa] and the ultimate tensile strength of 76 [kpsi] or 525 [MPa]; see Table A.7.2 in Appendix A.7.

From Example 6.10 and Example 7.6, the data that is of relevance to this example are as follows:

1) The bending moment is 79.577 [N m] or 703.78 [lb in] and the torque is 79.577 [N m] or 703.78 [lb in].
2) The computer code gave $K_{SF,F} = 6.4395$ and 6.2229 for SI units and the US units, respectively.

Running the MATLAB code of (`Fatigue_Strength_and_Factors_Affecting_Optimum.m`) by iteratively and manually changing the part diameter from the original 45 [mm], one can home into a part diameter of 30.053 [mm] or 1.1832 [in] and a safety factor close to 2.0. The other particular results are as follows:

```
Optimum_Vol_Percent_SI = 44.6016% and Optimum_Vol_Percent_US = 44.6016%
Dynamic_Safety_Factor_SI = 2.0002 and Dynamic_Safety_Factor_US = 2.0029
```

This indicates that one is able to reduce the volume to about 44.6% of the original volume and keep the safety factor at a reasonable value of about 2.0. The part diameter, however, ought to be in a round figure form. This would further adjust the part diameter to 30.0 [mm], which would reduce the safety factor to 1.9900. The US system is a little more conservative since the closest round figure may be 1.2 [in], which renders a higher safety factor of 2.0865. The adjusted factors of safety are obtained by MATLAB code for the adjusted diameters.

7.11 Summary

This chapter is extremely vital to the process of synthesis of machine elements. The understanding of material static and dynamic strength is indispensable in the selection of the most suitable material for a particular element. This process starts with the general material structure, its effect on properties, failure modes, and service performance. Failure modes have also association with different material properties that are necessary in the synthesis process. Static loading cases are utilizing some specific strength properties that are usually different from other strength properties needed for dynamic loading considerations. For static loading, the yield strength is typically the limit for ductile materials. The ultimate strength is generally the limit for brittle materials. For dynamic loading cases, the restraining property is normally the endurance limit. Even though the economics of the design figures heavily in the synthesis process, this is further set and left for an advanced optimization process.

The size of the design affects its material properties. Components of very large dimensions usually have lower strength than components of smaller dimensions made of the same material. Much smaller dimensions would have much higher strength for the same material used. Machine elements of close to *nano* dimensions have design methods that should differ from regular strength of material, elasticity, or continuum mechanics techniques used in this text. Such *nano* scale elements are then beyond the scope of this text.

Many machine elements are typically made of steel, cast iron, aluminum, and their alloys. Copper, magnesium, and their alloys are still having some market share for some machine elements. Plastics are gaining in their utility and replacing some traditional materials. The environmental implications are of concern and ought to be part of design optimization constraints or even objectives. Some research suggests that the optimum material for the environment is the one to generate the most *durable* machine element and thus most durable product design.

The properties of some of the most suitable set of materials are of significance to the synthesis of machine elements. Properties of some of these encompassed material sets are in this text. Focus is on reasonable selection for materials that are generally applicable to machine elements (Table 7.4). These materials have reasonable strength that can be adaptable for the majority of such elements. Higher strength and typically more expensive material would need optimum justification to use, which is a plan for more involved *cost optimization* process. The commonly adapted material is the usual selection in synthesis. The present synthesis process allows the selection of different materials. Properties and designation of some of these materials are available in this text for such a selection. The material designation completes the selection process and makes it accessible, acquirable, and deliverable.

The tensile test properties are the basis of many material strength properties used in the synthesis process. Fundamental strength properties, procedures, and constraints are in some link to these tensile properties. Heat treatment of carbon steel changes the properties and can be useful in synthesis. These properties and treatments are set in equations to be exploitable in performing the synthesis and optimization. Several computer codes are available to aid in such endeavors. The codes are verifiable by comparing their results with the provided material data. Some examples of verification are on hand in this chapter. These codes are also accessible through the **Wiley website**.

Problems

7.1 For the general tensile properties of materials stated in Section 7.1.3, search, verify, and validate the property ranges and narrow down the property range to an average stated in the majority of references or standards. Are there statistics about some of these properties that specify the average and standard deviation? What is the definition of the minimum value of the property in standards or manufacturers data?

7.2 If you had a curriculum course in strength of materials or material testing, collect the properties of materials in a data file similar to those in the Appendix A.7 of the text.

7.3 What is the difference between engineering stress–strain behavior and true stress–true strain behavior of brittle material?

7.4 Compare the engineering stress–strain and the true stress–true strain at failure of any of the ductile materials in property tables in the Appendix A.7 of the text using Eq. (7.2). Find the reduction in area of the material from other references or standards.

7.5 Chemical analysis of the surface of a machine element indicates a carbon content of 0.4% and the *Brinell* hardness *HB* is about 260. The chemical analysis of the core (at the end) of a machine element indicates a carbon content of 0.15% and the *Brinell* hardness *HB* is about 150. Indicate the designation of the material and estimate its original and final production processing.

7.6 Use Eqs. (7.8)–(7.14) to estimate the strength and the hardness of the original or core material and the surface strength and hardness of the machine element in Problem 7.5. Compare the estimated values with the properties of the material designation in Problem 7.5.

7.7 A machine element application suggests a carbon content of 0.3%. It also needs a high *Brinell* hardness of about 490. Indicate the type and the standard designation of the material and its manufacturing processing that is appropriate for this machine element. Use trend line equations of different properties to estimate the values and compare these values with the selected standard material.

7.8 The probable carbon content of a machine element is about 0.35%. Estimate its strength and hardness for hot rolled and cold drawn processing using trend line equations. What specified standard designation should be set for the material? Find the material properties in the selected standards for this designation.

7.9 The probable carbon content of a machine element is about 0.45%. Estimate its strength and hardness for processing of water quenched and tempered to 205 [°C] or (400) [°F]. What specified standard designation should be set for the material? Find the material properties in the selected standards for this designation.

7.10 Find the general mechanical usage of some particular *plain carbon steels* from other references.

7.11 Find the general mechanical usage of some particular *cast iron* from other references.

7.12 Find the general mechanical usage of some particular *low-alloy steels* from other references.

7.13 Find the general mechanical usage of some particular *aluminum alloys* from other references.

7.14 Find the general mechanical usage of some particular *copper alloys* from other references.

7.15 Find the general mechanical usage of some particular *magnesium alloys* from other references.

7.16 A suggested composite is made of 60% Kevlar reinforced fibers and 40% epoxy binding matrix. What is the expected strength of the composite? If carbon fibers are to be a replacement of Kevlar fibers, what should be the change in properties? Search for the current cost of each of these components and estimate the expected final cost of the material.

7.17 Find the mechanical applications of some reinforced plastics and the strength expected.

7.18 Find the mechanical applications of some reinforced composites and the strength expected.

7.19 Search for appropriate plastic materials employed in commercial and standard bolts, screws studs and fasteners, clips, snap rings, couplings, keys, pins, wires, sections, tubing, sheets, strips, structural shapes, springs, lock washers, journal, or rolling bearings, if any.

7.20 Use the approximate relation between ultimate tensile strength and the *Brinell* hardness number to check on some values provided in the appendix for plain carbon steel.

7.21 Use the approximate relation between ultimate tensile strength and the *Brinell* hardness number to check on some values provided in the appendix for heat treated plain carbon steel.

7.22 Use the approximate relation between ultimate tensile strength and the *Brinell* hardness number to check on some values provided in the appendix for low-alloy steel.

7.23 Use the approximate relation between ultimate tensile strength and the *Brinell* hardness number to check on some values provided in the appendix about cast iron.

7.24 Develop an approximate relation between ultimate tensile strength and the *Brinell* hardness number using some values provided in the appendix for aluminum alloys.

7.25 Use Eq. (7.31) to develop the distortion energy density defined by Eq. (7.34).

7.26 Use Eqs. (7.30) and (7.33) to develop the distortion energy density defined by Eq. (7.34).

7.27 Expand Eq. (7.42) to develop the distortion energy theory defined by Eq. (7.44) in terms of the normal and shear stresses instead of the principal stresses.

7.28 Develop the linear relation between σ_1 and σ_2 in the second and fourth quadrant for the *Coulomb–Mohr* theory limit.

7.29 Find the liming maximum shear stresses in the second and fourth quadrants for the *Coulomb–Mohr* theory.

7.30 Develop the linear relations for the boundaries of the *modified Mohr* theory for ***2D*** stress state.

7.31 An axel is subject to a bending moment and a torsional load of 80 [N m] or 708 [lb in]. The axel diameter is 30 [mm] or 1.2 [in]. The selected material is set as hot rolled AISI 1030 or ISO C30. Find the safety factor for each applicable theory of failure. Is it possible to use hot rolled AISI 1020 or ISO C20? Compare results.

7.32 A cylindrical supporting beam is loaded in bending and torsion. The maximum bending is 120 [N m] or 1060 [lb in]. The maximum torsion is 100 [N m] or 885 [lb in]. The beam diameter is 30 [mm] or 1.2 [in]. The selected material is set as gray cast iron ASTM A 48 class 30 or ISO 185 200. Find the safety factor for each applicable theory of failure. Is it possible to use gray cast iron ASTM A 48 class 20 or ISO 185 150? Compare results.

7.33 Very slow rotating roller transports heavy cargo containers with other adjacent parallel rollers. The maximum bending is 1200 [N m] or 10 600 [lb in]. The maximum torsion is 50 [N m] or 440 [lb in]. The roller diameter is 60 [mm] or 2.4 [in]. The selected material is set as ductile cast iron ASTM A 536 80-55-6 or ISO 1083 500-7. Find the safety factor for an applicable static theory of failure. Is it possible to use ductile cast iron ASTM A 536 65-45-12 or ISO 1083 450-10?

7.34 A propeller shaft of a large cruise ship rotates very slowly. The maximum bending is 25 [kN m] or 221 [klb in]. The maximum torsion is 2 [MN m] or 17.7 [Mlb in]. The propeller shaft diameter is 610 [mm] or 24 [in]. Is it safe to use a propeller shaft material as hot rolled AISI 1030 or ISO C30? Find the safety factor for an appropriate static theory of failure. What is the safety factor if one uses hot rolled AISI 1020 or ISO C20?

7.35 The shaft of a drone motor is rotating very fast and subject to very little bending moments from the propeller due to maneuvering. The maximum torsional moment is 0.05 [N m] or 0.44 [lb in] and the maximum bending is 0.002 [N m] or 0.018 [lb in]. The propeller shaft diameter is 2.0 [mm] or 0.079 [in]. Is it safe to use a shaft material as cold drawn AISI 1030 or ISO C30? Find the safety factor for an applicable static theory of failure. Can one use hot rolled AISI 1020 or ISO C20?

7.36 Check the relation between the endurance limit and the ultimate tensile strength for cast iron depicted in Eq. (7.59) with the data in the appendix of the text.

7.37 Check the relation between the endurance limit and the ultimate tensile strength for aluminum alloy depicted in Eq. (7.60) with the data in the appendix of the text.

7.38 Derive Eqs. (7.63) and (7.64) to evaluate the fatigue strength between the boundary conditions at 10^3 cycles and 10^6 cycles assuming that a failure ratio (*fr*) of the fatigue strength at 10^3 cycles to the ultimate tensile strength is 0.8. Assume that the boundary condition at 10^6 cycles is the endurance limit.

7.39 Estimate the loading number of cycles for a boiler that is operational daily at its maximum pressure, but is shutdown during the night shift if its expected life span is 20 years.

7.40 Use Eqs. (7.63)–(7.65) to find the fatigue strength of AISI 1030 or ISO C30 at life span of 10^4 and 10^5 cycles. Use water quenched and tempered heat treatment in addition to the normalized steel condition. If the failure ratio (*fr*) is 0.87 or 0.82 instead of 0.8, what will the fatigue strength be?

7.41 Derive a similar expression to Eq. (7.65) for the range of *low-cycle fatigue* if the linear relation on the log–log scale is an acceptable representation of fatigue failure at the low-cycle zone. Can one compare that to the use of *strain-life* method for such a zone?

7.42 Use Eqs. (7.78) and (7.79) of the *fatigue stress concentration factor* K_f to find the *fatigue concentration factor* K_{conc} as defined by Eq. (7.80).

7.43 Use the *Miner's linear damage model* defined by Eq. (7.84) to estimate the number of cycles and stress to assumed fatigue, if the machine element is heavily loaded to a stress level of 0.8 S_u for the first 10^3 cycles.

7.44 Develop the intersection relation of Eqs. (7.88) and (7.89) for the two *modified Goodman* diagram lines.

7.45 Solve Example 7.4 using the principal stresses in Eq. (7.85) instead of the normal and shear stresses and Eq. (7.95). Compare results.

7.46 Find the fatigue safety factors of Example 7.4 when using the *Gerber* and the *modified Goodman* limits. Compare results with the *ASME ellipse* failure limit.

7.47 Resolve Problem 7.33 to consider the variation of bending due to the roller rotation and the variation of torsion due to the gap between cargo containers where the torque vanishes. What is the fatigue safety factor in that case? Assume a machined condition, a reliability of 99.0%, stress concentration factor of 1.75, and ambient temperature. Should one change the material? What is the suitable material to use? Find the fatigue safety factor for this change.

7.48 Revisit Problem 7.34 to account for a 10% variation in torque and the alternating variation in bending due to rotation. Assume a machined condition, a reliability of 99.99%, stress concentration factor of 2.0, and ambient temperature. What is the fatigue safety factor for this new case?

7.49 A polished cylindrical bar of an unknown material in the storeroom of a workshop is reverse engineered to define the material designation. The mass spectrometer indicates a 0.35% carbon with some other elements of low percentages and the majority is ferrite. The hardness measure indicates a *Brinell* hardness value of 165 HB. Use the MATLAB code (`Material_Properties_Carbon Steel.m`) to define the material properties. Suggest the designation of the material for the US or SI systems. Check the result with the material data in the appendix of the text.

7.50 Use the MATLAB code (`Material_Properties_Carbon_Steel.m`) to define the material properties of a 0.45% carbon and a hardness of about 150 HB at the end core and about 510 HB at the surface.

7.51 Use the MATLAB code (`Material_Properties_Carbon_Steel.m`) to define the material properties of the material defined in Problems 7.5 and 7.6.

7.52 Use the MATLAB code (`Material_Properties_Carbon_Steel.m`) to define the material properties of the material defined in Problem 7.7.

7.53 Use the MATLAB code (`Material_Properties_Carbon_Steel.m`) to define the material properties of the material defined in Problem 7.8.

7.54 Use the MATLAB code (`Material_Properties_Carbon_Steel.m`) to define the material properties of the material defined in Problem 7.9.

7.55 Utilize the MATLAB code (`Fatigue_Strength_and_Factors_Affecting.m`) to solve Problems 7.33 and 7.47.

7.56 Utilize the MATLAB code (`Fatigue_Strength_and_Factors_Affecting.m`) to solve Problems 7.34 and 7.48.

7.57 Utilize the MATLAB code (`Fatigue_Strength_and_Factors_Affecting.m`) to solve Problem 7.35.

7.58 Utilize the MATLAB code (`Fatigue_Strength_and_Factors_Affecting_Optimum.m`) to optimize Problems 7.33 and 7.47.

7.59 Utilize the MATLAB code (`Fatigue_Strength_and_Factors_Affecting_Optimum.m`) to optimize Problems 7.34 and 7.48.

7.60 Utilize the MATLAB code (`Fatigue_Strength_and_Factors_Affecting_Optimum.m`) to optimize Problem 7.35.

7.61 Adjust the MATLAB code of Figure 7.36 to consider different relations for the surface factor for ground and machined surfaces. What is the implication of using such relations?

7.62 Adjust the MATLAB code of Figure 7.37 to include the more conservative *modified Goodman* failure criterion.

7.63 Adjust the MATLAB code of Figure 7.38 to include the search for an additional optimum material by including the current cost of different materials.

References

AA (2015). *International Alloy Designations and Chemical Composition Limits for Wrought Aluminum and Wrought Aluminum Alloys*. The Aluminum Association.

Abdel Meguid, M.R.M. (1999). *Computer aided material selection optimization*. M.S. thesis. Cairo University.

Ashby, M.F. (1999). *Materials Selection in Mechanical Engineering*, 2e. Butterworth-Heinemann.

Ashry, A.S. and Metwalli, S.M. (2012). Measuring elastic strain using radon transform-based image processing for tensile testing. *Proceedings of ASME 45196*, Volume 3, Houston, TX, USA (09–15 November), pp. 1193–1204, IMECE2012-86909.

ASM (1993). *Properties and Selection: Irons, Steels, and High-Performance Alloys: Metals Handbook*, 10e. ASM International.

ASM (1996). *ASM Handbook: Fatigue and Fracture*, vol. 19. ASM International.

ASM (1998). *Metals Handbook Desk Edition*, 2e. ASM International.

ASM (2006). *Worldwide Guide to Equivalent Irons & Steels*, 5e. ASTM International.

ASTM A1011/A1011M (2007). *Standard specification for steel, sheet and strip, hot-rolled, carbon, structural, high-strength low-alloy, high-strength low-alloy with improved formability, and ultra-high strength*. ASTM International.

ASTM A176 (2004). *Standard specification for stainless and heat-resisting chromium steel plate, sheet, and strip*. ASTM International.

ASTM A240/A240M (2004). *Standard specification for chromium and chromium–nickel stainless steel plate, sheet, and strip for pressure vessels and for general applications*. ASTM International.

ASTM A276 (2006). *Standard specification for stainless steel bars and shapes*. ASTM International.

ASTM A29/A29M (2005). *Standard specification for steel bars, carbon and alloy, hot-wrought*, General Requirements for ASTM International.

ASTM A681 (2008). *Standard specification for tool steels alloy*. ASTM International.

ASTM A36/A36M (2014). *Standard specification for carbon structural steel*. ASTM International.

ASTM A572/A672M (2007). *Standard specification for high-strength low-alloy columbium–vanadium structural steel*. ASTM International.

ASTM A992/A992M (2011). *Standard specification for structural steel shapes*. ASTM International.

ASTM B107/B107M (2007). *Standard specification for magnesium-alloy extruded bars, rods, profiles, tubes, and wire*. ASTM International.

ASTM B275 (2013). *Practice for codification of certain nonferrous metals and alloys, cast and wrought*. ASTM International.

ASTM E140 (2002). *Standard hardness conversion tables for metals relationship among Brinell hardness, Vickers hardness, Rockwell hardness, superficial hardness, Knoop hardness, and Scleroscope hardness*. ASTM International.

ASTM E399 (2013). *Standard test method for plane-strain fracture toughness of metallic materials*. ASTM International.

ASTM E8/E8M-13 (2013). *Standard test methods for tension testing of metallic materials*. ASTM International.

ASTM Standard E139 (2003). *Standard test methods for conducting creep, creep-rupture, and stress rupture tests of metallic materials*. ASTM International.

Barsom, J.M. (1971). Fatigue crack propagation in steels of various yield strengths. *Transactions of the ASME, Journal of Engineering for Industry* 93 (4): 1190–1196.

Barsom, J.M. and Rolfe, S.T. (1969). *Correlation between K_{IC} and Charpy V-notch test results in the transition-temperature range ASTM STP 446*, pp. 281–302.

Barsom, J.M. and Rolfe, S.T. (1987). *Fatigue and Fracture Control in Structures*, 2e. Prentice Hall, Copyright ASTM International.

Bethlehem (1971). *Modern Steels and their Properties*. Bethlehem Steel Corporation.

Bethlehem (1980). *Modern Steels and their Properties*. Bethlehem Steel Corporation.

Brandes, E.A. and Brook, G.B. (eds.) (1992). *Smithells Metal Reference Book*, 7e, 22–134 to 22–139. Butterworth.

Bringas, J.E. (ed.) (2004). *Handbook of Comparative World Steel Standards*, 3e, ASTM DS67B. ASTM International.

Budynas, R.G. and Nisbett, J.K. (2015). *Shigley's Mechanical Engineering Design*, 10e. McGraw-Hill.

Callister, W.D. Jr. and Rethwisch, D.G. (2011). *Material Science and Engineering*, 8e. Wiley.

CDA (2004). *Copper and Copper Alloys*, Publication No. 120. Copper Development Association.

Coulomb, C.A. (1773/1776). Essai sur une application des regles des maximis et minimis a quelquels problemesde statique relatifs, a la architecture. *Memoires Académie Royale des Sciences Divers Savants* 7: 343–387.

Courtney, T.H. (2000). *Mechanical Behavior of Materials*, 2e. McGraw-Hill.

Davis, J.R. (2001). *Aluminum and Aluminum Alloys. Alloying: Understanding the Basics* ASM International, pp. 351–416.

DIN 17100 (1986). *Steels for general structural purposes* Deutsches institut für Normung e. V. (German Institute for Standardization).

DIN EN 10083 (2006). *Steels for quenching and tempering –Part 3: Technical delivery conditions for alloy steels* Deutsches Institut für Normung e. V. (German Institute for Standardization).

Elmoselhy, S.A., Azzam, B.S., Metwalli, S.M., and Dadoura, H.H. (2006). Hybrid shape optimization and failure analysis of laminated fibrous composite E-springs for vehicle suspension. *SAE Transactions, Journal of Commercial Vehicles* 115 (2): 226–236.

EN 10025 (2005). *Hot rolled products of non-alloy structural steels*. European Committee for Standardization CEN.

EN 10027-1 (2005). *Designation systems for steels-Part 1: Steel names*. European Committee for Standardization CEN.

EN 10083 (2006). *Steels for quenching and tempering*. European Committee for Standardization CEN.

EN 10088 (2014). *Stainless steels*. European Committee for Standardization CEN.

EN 12163 (1998). *Copper and copper alloys rods for general purpose*. European Committee for Standardization CEN.

Faupel, J.H. and Fisher, F.E. (1981). *Engineering Design, A Synthesis of Stress Analysis and Material Engineering*, 2e. Wiley.

Griffith, A.A. (1921). The phenomena of rupture and flow in solids. *Philosophical Transactions of the Royal Society A: Mathematical, Physical and Engineering Sciences* 221: 163–198.

Groover, M.P. (2010). *Fundamentals of Modern Manufacturing, Materials, Processes, and Systems*, 4e. Wiley.

Grover, H.J., Gordon, S.A., and Jackson, L.R. (1960). *Fatigue of Metals and Structures*. Bureau of Naval Weapons, *Document NAVWEPS 00-2500435*.

Harper, C.A. (ed.) (2000). *Modern Plastics Handbook*. McGraw-Hill.

Hassan, M.T.Z., Esawi, A.M.K., and Metwalli, S. (2014). Effect of carbon nanotube damage on the mechanical properties of aluminium–carbon nanotube composites. *Journal of Alloys and Compounds* 607: 215–222.

Haugen, W.B. (1980). *Probabilistic Mechanical Design*. Wiley.

Inglis, C.E. (1913). Stresses in a plate due to the presence of cracks and sharp corners. *Proceedings Institution of Naval Architects* 55 (1): 219–230.

Irwin, G.R. (1957). Analysis of stresses and strains near the end of a crack traversing a plate. *Journal of Applied Mechanics* 24: 361–364.

Irwin, G.R. (1960). *Fracture Mechanics. Structural Mechanics*, Proceedings of First Naval Symposium. Pergamon Press.

Ismaiel, A.M.M., Metwalli, S.M., Elhadidi, B.M.N., and Yoshida, S. (2017). Fatigue analysis of an optimized HAWT composite blade. *EVERGREEN Joint Journal of Novel Carbon Resource Sciences & Green Asia Strategy* 04 (02/03): 1–6.

ISO 1190-1 (1982). *Copper and copper alloys – code of designation – Part 1: Designation of materials*. International Organization for Standardization.

ISO 18265 (2003). *Metallic materials – conversion of hardness values*. International Organization for Standardization.

ISO 18265 (2013). *Metallic materials conversion of hardness values*. International Organization for Standardization.

ISO 185 (1988). *Grey cast iron – classification*. International Organization for Standardization.

ISO 209 (1989). *Wrought aluminum and aluminum alloys – chemical composition and forms of products*. International Organization for Standardization.

ISO 209-1 (1989). *Wrought aluminum and aluminum alloys-chemical composition and forms of products – Part 1: Chemical composition*. International Organization for Standardization.

ISO 3522 (2007). *Aluminium and aluminium alloys — castings — chemical composition and mechanical properties*. International Organization for Standardization.

ISO 4957 (1999). *Tool steels*. International Organization for Standardization.

ISO 630 (1995). *Structural steel – plates, wide flats, bars, sections and profiles*. International Organization for Standardization.

ISO 683 (1987). *Heat-treatable steels, alloy steels and free-cutting steels*. International Organization for Standardization.

ISO 6892-1 (2009). *Metallic materials – tensile testing – method of test at ambient temperature*. International Organization for Standardization.

ISO 6892-2 (2011). *Metallic materials – tensile testing – method of test at elevated temperature*. International Organization for Standardization.

ISO S922 (1981). *Malleable cast iron*. International Organization for Standardization.

ISO-TS 15510 (2003). *Stainless steel – chemical composition*. International Organization for Standardization.

Juvinall, R.C. (1983) *Fundamentals of Machine Component Design* Wiley.

Juvinall, R.C. and Marshek, K.M. (2012). *Fundamentals of Machine Component Design*, 5e. Wiley.

Karpov, A.V. (1939). Fatigue problems in structural design, *Metals and Alloys*. 10 (11): 346–352.

Klein, R. (2011). *Laser Welding of Plastics*, 1e. Wiley-VCH.

Kuguel, R. (1961). A relation between theoretical stress-concentration factor and fatigue notch factor deduced from the concept of highly stressed volume. *Proceedings of ASTM* 61: 732–748.

Kuhn, P. (1964). The prediction of notch and crack strength under static and fatigue loading, *ASME* Paper 843c, April.

Landgraf, R.W. (1968). *Cyclic Deformation and Fatigue Behavior of Hardened Steels. Report No. 320*. Urbana: Department of Theoretical and Applied Mechanics, University of Illinois, pp. 84–90.

Lipson, C., Noll, G.C., and Clock, L.S. (1950). *Stress and Strength of Manufactured Parts*. McGraw-Hill.

Marin, J. (1962). *Mechanical Behavior of Engineering Materials*. Prentice Hall.

McCauley, C.J. (ed.) (Senior Editor) (2012). *Machinery's Handbook*, 29e. Industrial Press.

Metwalli, S.M. (1983). Direct strain measurements through laser speckle spectral density. *Proceedings of SESA 1983 Spring Meeting*, Cleveland, OH, USA (15–20 May), pp. 119–124

Metwalli, S.M. (1986). Industrial applications of computer image processing. *Proceedings of the 8th Annual Conference for Computer and Industrial Engineering*, Orlando, FL, USA (19–21 March). *Computer and Industrial Engineering*, Volume II, No. 1–4, pp. 608–612.

Metwalli, S.M., Moslehy, F.A., and Sheikhrezai, R. (1983). Application of spectral density in laser speckle strain measurement. *Proceedings of the 19th Midwestern Mechanics Conference, Developments in Mechanics*, Volume 12. The University of Iowa, Iowa City, USA (16–18 May), pp. 47–49.

Metwalli, S.M., Ragab, A.R., Kamel, A.H., and Saheb, A.A. (1987). Determination of plastic stress–strain behavior by digital image processing techniques. *Experimental Mechanics* 27 (4): 414–422.

Miedlar, P.C., Berens, A.P., Gunderson, A., and Gallagher, J.P. (2002). *USAF Damage Tolerant Design Handbook: Guidelines for the Analysis and Design of Damage Tolerant Aircraft Structures*, Analysis and Support Initiative for Structural Technology (ASIST) AFRL-VA-WP-TR-2003-3002.

Miner, M.A. (1945). Cumulative damage in fatigue. *Journal of Applied Mechanics* 12: A159–A164.

Mischke, C.R. (1987). Prediction of stochastic endurance strength. *Transactions of ASME, Journal of Vibration, Acoustics, Stress, and Reliability in Design* 109 (1): 113–122, Table 3.

Neuber, H. (1946). *Theory of Notch Stress-Principles for Exact Stress Calculation*. Ann Arbor, MI: J. W. Edwards Brothers, Inc.

Noll, C.J. and Lipson, C. (1946). Allowable working stresses. *Society for Experimental Stress Analysis* 3 (2): 29.

Osswald, T.A., Baur, E., Brinkmann, S. Oberbach, K. and Schmachtenberg, E. (2006). *International Plastics Handbook, The Resource for Plastics Engineers*, 4e. Hanser Publishers.

Palmgren, A. (1924). Die Lebensdauer von Kugellagern. *ZVDI* 68: 339–341.

Paris, P.C. and Erdogan, F. (1963). A critical analysis of crack propagation laws. *Transactions of the ASME, Journal of Basic Engineering* 85 (4): 528–533.

Peterson, R.E. (1974) *Stress Concentration Factors.* Wiley.

Ragab, A.A., Metwalli, S.M., and Reuada, J. (1986). Preliminary assessment of image processing application to large deformation measurement. In: *Current Advances in Mechanical Design and Production*, Proceedings of the 3rd Cairo University MDP Conference (26–29 December 1985) (eds. S.E.A. Bayoumi and M.Y.A. Younan), 53–38. Pergamon Press.

Rankine, W. (1857). On the stability of loose earth. *Philosophical Transactions of the Royal Society of London* 147: 9–27.

Robert, G., Krauss, G., and Kennedy, R. (1998). *Tool-Steels*, 5e. ASM International.

Rolfe, S.T. and Barsom, J.M. (1977). *Fracture and Fatigue Control in Structures.* Prentice Hall.

Rosato, D.V. and Rosato, D.V. (2004). *Reinforced Plastics Handbook*, 3e. Elsevier.

SAE (1977). Fatigue Properties. Technical Report, *SAE J1099*. Society of Automotive Engineers

Schechtling, H. (1987). *International Plastics Handbook, for Technologist, Engineer and User*, 2e. Hanser Publishers.

Senousy, M.S., Hegazi, H.A., and Metwalli, S.M. (2008). Topology optimization of a MEMS resonator using hybrid fuzzy techniques. *Proceedings of the ASME 2008 IDETC/CIE 2008 Conference*, Brooklyn, New York, USA (3–6 August), Paper # DETC2008-49563.

Shigley, J.E. (1986). *Mechanical Engineering Design*, First Metric Edition. McGraw-Hill.

Sines, G. and Waisman, J.L. (eds.) (1959). *Metal Fatigue*. McGraw-Hill.

Sors, L. (1971). *Fatigue Design of Machine Components*. Pergamon Press.

Stulen, F.B., Cummings, H.N., and Schulte, W.C. (1961). Preventing fatigue failures, Part 5. *Machine Design* 33: 161.

Tavernelli, J.F. and Coffin, L.F. Jr. (1962). Experimental support for generalized equation predicting low cycle fatigue, and S. S. Manson, discussion. *Transactions of the ASME, Journal of Basic Engineering* 84 (4): 533–537.

Tresca, H. (1864). Mémoire sur l'écoulement des corps solides soumis à de fortes pressions. *Comptes rendus de l'Académie des sciences Paris* 59: 754.

Tsai, S.W. and Wu, E.M. (1971). A general theory of strength in anisotropic materials. *Composite Material Journal* 5: 58–80.

US Coast Guard (1944). Report of Structural Failure of Inspected Vessel Schenectady, *Form CG796*, 1 April 1976.

von Mises, R. (1913). Mechanik der festen Körper im plastisch deformablen Zustand. *Göttingen, Mathematisch-Physikalische, Klasse 4* 1: 582–592.

Younan, M., Metwalli, S.M., and El-Zoughby, A. (1983). Fracture mechanics analysis of a fire tube boiler. *Journal of Engineering Fracture Mechanics* 17 (4): 335–338.

Internet Links

Material Selection

www.asminternational.org ASM American Society for Materials (now the Materials Information Society?)

http://www-materials.eng.cam.ac.uk/mpsite University of Cambridge, Material selection and Processing

http://www.matweb.com/index.asp Material Property Data Source for Material Information

http://www.materials.ac.uk/index.asp

https://www.secowarwick.com/wp-content/uploads/2017/03/HeatTreatingDataEBook.2011.pdf Heat treating data download.

www.plusmetals.net Aluminum alloys trader.

www.materialdatacenter.com Plastics material database.

Material Standards

www.asm-intl.org ASM International.
www.astm.org ASTM Website.
www.din.de DIN Deutsches Institut für Normung e.V. (German Institute for Standardization).
www.iso.chl International Standards Organization.
www.nist.gov National Institute for Standards and Technology (USA).
www.sae.org Society of Automotive Engineers (SAE).

8

Introduction to Elements and System Synthesis

Machine design involves the design or synthesis of machine elements and their assembly into the system design assemblage or the product. The design assembly can be a subsystem of a larger product. The design of the subsystem entails some standard components and other nonstandard (or non-on-the-shelve) components that need synthesis. The standard components that have some dedicated manufacturers produce them in a mass-production fashion. These components are screws, bolts, and rivets; ball screws; keys, pins, and retaining rings; couplings; seals; rolling bearings; etc. These standard components are seldom in need of synthesis except for special designs such as satellites, space gear, and some aerospace or special industries. The standard components are thus selected from the available set to withstand the design requirements and constraints. Most designs acquire such standard elements from specialized manufacturers. This chapter dedicates some sections to cover some of these standard elements, their fundamentals, and specifications. Other chapters are dedicated to specific standard machine elements such as fasteners, rolling bearings, belts, wire ropes, and chains to have the best or optimum selection thereof.

Symbols and Abbreviations

The adapted units are [in, lb, psi] or [m, kg, N, Pa], others given at each symbol definition ([k...] is 10^3, [M...] is 10^6, and [G...] is 10^9).

Symbols/Abbreviations	Quantity, Units (adapted)
Abbreviations	
AI	Artificial intelligence
AWG	American Wire Gauge or Brown & Sharpe gauge
FE	Finite element method
PF	Permanent fit for splines
SuL	Slide under load fit for splines
Symbols	
d_{Gn}	Wire diameter related to *gauge number Gn*
d_H	Hub diameter of the coupling
d_H	Hole diameter
d_i	Inner diameter of hollow shaft
d_m	Minor diameter of spline
d_m	Minor diameter of shaft

Machine Design with CAD and Optimization, First Edition. Sayed M. Metwalli.

Companion website: www.wiley.com/go/metwalli/machine

Symbols/Abbreviations	Quantity, Units (adapted)
d_O	Outside coupling diameter
d_P	Pin diameter
d_p	Spline pitch diameter
d_R	Diameter of ring bearing
d_S	Shaft diameter
$d_{S[\text{in}]}$	Shaft diameter in [in]
$d_{S[\text{m}]}$	Shaft diameter in [m]
$\boldsymbol{F}_a$	Axial force
$\boldsymbol{F}_i$	Applied forces, $i = 1, 2, \ldots$
$\boldsymbol{F}_s$	Shear force acting on the key section
$\boldsymbol{F}_t$	Tangential force on key side
Gn	Gauge number
H	Power
h_B	Boom height of *Jib crane*
H_{hp}	Power in [hp]
h_K	Key height
H_{kW}	Power in [kW]
H_{max}	Maximum failure power
h_S	Spline height or depth
H_{W}	Power in [W]
K_M	Additional moment factor
K_{SC}	Stress concentration factor
K_{SF}	Safety factor
l_B	Boom length of *Jib crane*
l_C	Clamp length of coupling
l_{DC}	Double-cone length of coupling
l_F	Flange length of the coupling
l_H	Hub length of the coupling
l_H	Hood length of *Jib crane*
l_K	Key length
l_M	Mast length of *Jib crane*
l_P	Pin length
l_R	Length or width of ring bearing
l_S	Spline length
$\boldsymbol{M}_1$	Moment at support 1
$\boldsymbol{M}_j$	Applied moments, $j = 1, 2, \ldots$
$\boldsymbol{M}_z$	Moment about z at support
N_{rpm}	Rotational speed in [rpm]
N_S	Number of splines
$\boldsymbol{R}_1, \boldsymbol{R}_2$	Reactions at supports

Symbols/Abbreviations	Quantity, Units (adapted)
$\boldsymbol{r}_i$	Position vector of applied force i, $i = 1, 2, \ldots$
S_{Ba}	Allowable bearing strength
S_e	Endurance limit
S_{es}	Endurance shear strength
S_u	Ultimate tensile strength
S_{yc}	Compressive yield strength
S_{ys}	Shear yield strength
S_{yt}	Tensile yield strength
$T_{\mathrm{lb\,in}}$	Torque magnitude in [lb in]
$T_{\mathrm{N\,m}}$	Torque magnitude in [N m]
t_S	Depth of keyway in the shaft
$\boldsymbol{T}_S$	Transmitted torque between shaft and mating hub
$\boldsymbol{T}_S$	Spline torque
T_S	Magnitude of shaft torque
w_k	Key width
w_R	Ring width of retaining ring
Z_z	Section modulus about z
σ_{vM}	Equivalent *von Mises* stress
$\sigma_{x,\max}$	Maximum normal stresses in x
$\omega_{\mathrm{rad/s}}$	Rotational speed in [rad/s]
τ_a	Axial shear stress
$\tau_{\max}$	Maximum shear stress
τ_{xy}	Direct shear stress on longitudinal key section

8.1 Introduction

The most critical factor of design is to define an estimate of the load. That load depends on the character of the design in terms of its function and if it is a consumer intensive or defined as a restrictive use as an industrial product. The load of the consumer-intensive products such as appliances, vehicles, etc. is varying more in magnitudes and durations. It is more dynamic and random. The estimation of load and the design process is then more probabilistic than deterministic. Consider designing a suspension spring for a vehicle (automobile). Most of the time the car sits at the curb or in the garage with a static load of its curb weight. The next mostly used in time is with the load of one passenger (the driver) travelling on paved roads causing little load variation. The less loading duration is usually with higher load of two, three, or four passengers. Seldom you have higher loads with five or six passengers with luggage and off-terrain travelling. The estimation of the load value and duration for the life of the vehicle is then more dynamic and random. Probabilistic treatment to design the vehicle spring is therefore advisable. One may, however, use averages with fatigue treatment and higher *safety factor*.

The load of the industrial products such as cranes, gearboxes, conveyers, motors, etc. is more defined in magnitude and loading durations. Its variation in magnitude and duration is in more defined control. These industrial products have defined capacities that are seldom overloaded and usually and efficiently operated at their limits of capabilities. Consider designing fasteners of the power station coupling between the turbine and generator. The turbine is controlled to have a fixed speed so that the generated power frequency is constant (usually 60 or

50 hertz [Hz]). The load, however, depends on the electric current variation due to consumption during the day. This current variation causes torque variation and thus force variation on the fasteners. The variation is kept at a minimum by the control system that reduces or increases power generation so as not to cause any variation to the power frequency. The fastener design cases are then usually designed as more static and deterministic than dynamic and probabilistic. At most the dynamics are deterministic. The safety factor is then not as high except to have an extremely high reliability to sustain the fastener for the life of the power station.

The *safety factor* is therefore greatly influenced by the requirements of the product. The safety factor is high for an indefinable load limits and lower for definite characterization of the load and the accuracy of the mathematical treatment with its resemblance to the real physical conditions, design, and loading.

A better design is the one that safely satisfies the function and can be competitive in cost. That is usually achievable by the maximum utilization of standard components and optimization. This can reduce geometry requirements and allows the utilization of most suitable material and components. Computer-aided design (*CAD*), synthesis, and optimization are well utilized in that regard. These generate tools (charts, codes, and programs) to define geometry (dimensions) and appropriate selection of material. Suitable numerical simulation such as finite element (***FE***) and lab testing for verification and validation are imperative to get a better design. The reverse engineering process gives many examples to aid in the design and redesign of components, subsystems, and systems.

It is imperative, then, to have a command of selecting the basic and common machine elements. The following sections attempt to cover such a necessary conventional set of information.

8.2 Basic and Common Machine Elements

Vital building blocks of machine element design include rods or wires and plates. So many machine elements have an initial raw form as a rod or a plate such as shafts being formed or machined out of rods. Screws, bolts, rivets, and pins are formed or manufactured out of rods. Helical springs are usually formed out of wires. Gears are usually machined out of rods or cut plates. Leaf springs are formed out of plates or strips. Pipes are formed out of plates or rolled sheets. Keys and retaining rings are manufactured out of plates or strips. Many frames, enclosures, chassis, and structures are built out of plates or formed sections off plates and strips. The basic or *preferred dimensions* of rods, wires, and plates are then extremely useful. The generally preferred sizes and dimensions are standards that are shown in Table A.3.1.

Standard ***bars*** or ***rods*** are usually following the preferred dimensions and sizes. Table A.4.1 gives the properties of some selected *round solid* steel bars or rods of mainly preferred dimensions. Few other dimensions are obvious additions due to the vast utility of bars in producing many components of narrower dimensional limits. The SI and US standards are side by side, but not the exact equivalent. They are the close alternative values for the other system. The *weight per length* is the only property that is material dependent. For other materials, a simple factor of (material specific gravity/steel specific gravity) produces the other material weight per [m] or per [ft]. General properties of common materials including specific gravities are provided in Chapter 7 (material strength) in Section 7.1.3.

Wires are thinner rods of sizes less than about 4 [mm] or 0.15 [in] and might be up to 6 [mm] or 0.25 [in]. Wires can be as small as 0.1 [mm] or 0.004 [in]. For the smaller range and smaller than that (0.0025–0.13 [mm] or 0.0001–0.005 [in]), one would be dealing with *filaments*, *fibers*, or *whiskers* (about 0.001 [mm] or 0.000 04 [in]), where some are woven in thin threads, fabric, or knits. *Wires and sheet metal* thicknesses for the US system are usually characterized by *gauges*. The larger the gauge the smaller the thickness. This is the case for the *American Wire Gauge* (AWG) or *Brown & Sharpe* wire gauge, where the diameter d_{Gn} is related to the *gauge number* Gn by the relation $d_{\mathrm{Gn}} = 0.127\ (92^{(36\text{-Gn})/39})$ [mm] and $d_{\mathrm{Gn}} = 0.005\ (92^{(36\text{-Gn})/39})$ [in]. This relation is also equivalent to $d_{\mathrm{Gn}} = e^{(2.1104 - 0.11954\mathrm{Gn})}$ [mm] and $d_{\mathrm{Gn}} = e^{(-1.124\,36 - 0.115\,94\mathrm{Gn})}$ [in]. The range of diameters is from about 8.251 [mm] or 0.3249 [in] for 0 gauge to about 0.0799 [mm] or 0.00314 [in] for 40 gauge.

Plates in small thicknesses would follow the *gauge numbers* for the US sheet metal thicknesses.

In general, *plates* have thicknesses with more intermediate values than the preferred sizes. The typical available *plate thicknesses* for the US inch system are in steps as follows:

- From 3/16 [in] to 7/8 [in] a step of 1/16 [in].
- From 1 [in] to 2¼ [in] a step of 1/8 [in].
- From 2½ [in] to 4½ [in] a step of ¼ [in].
- From 5 [in] to 8 [in] a step of ½ [in].
- From 8 [in] to 10 [in] a step of 1 [in].

The usual US *plate sizes* (width or length) are 36, 48, 60, 72, 84, 96, 108, 120, 144, and 240 [in].

For the SI *plate thicknesses*, *hot rolled sheets* or *plates* have an addition to the preferred sizes from 1.6 to 250 [mm] and some of the second preferred ones (35, 45, 55, 70, 90, 110, 180 [mm]); the other available thicknesses are 3.2, 12.5, 15, 32, 65, 75, 130, and 150 [mm]. These added thicknesses are to accommodate the equivalent US inch thicknesses and to provide more options. Other thicknesses would also be available upon agreements with manufacturer.

Thinner *cold rolled plates* of thicknesses in the range from 0.4 to 3.2 [mm] have thicknesses every 0.5 [mm] from 0.4 to 1.1 [mm]. From 1.1 to 2.2 [mm], the step is 1 [mm]. From 2.4 to 3.2 [mm], the step is 2 [mm] in addition to 2.5 [mm]. *Galvanized sheets* have the same thicknesses set as the cold rolled plates with the addition of the 3.4, 3.5, 3.6, 3.8, and 4 [mm] thicknesses. Other thicknesses may also be available.

The usual SI *plate sizes* (width or length) are 1000, 1250, 1500, 1800, 2000, 2400, 3000, 3600, and 6000 [mm]. Some of these are close to the equivalent US inch sizes.

Following the basic and vital building blocks rods, wires, and plates, there are other common standard components that are also essential. These are keys, pins, retaining rings, and seals. These components are available as standard joining and guarding parts. The couplings between shafts and the occasional need for splines to join such elements are also standardized. Even though couplings are not standardized, their abundant existence with exact fitting to shafts renders them as an essential common and elementary machine construction. Another important machine construction are the housings, enclosures, frames, and chassis that the design is needing to encompass, hold, and assemble the other design components.

Drills are also essential tools in producing holes that many components are housed in or through. Their available sizes are crucial to defining hole dimension. Drills have diameters with more intermediate values than the preferred sizes. Preferred and second choice sizes are surely available. More sizes are available because when mating parts are preferred sizes, a somewhat larger size hole is needed to accommodate the clearance fit between them. The typical available *drill sizes* for the US inch system are in the range of 0.006–2.0 [in] or 0.15–50 [mm] and in steps of 0.0004–1/16 [in] or 0.01–2.0 [mm]. Smaller steps are for smaller sizes, and large steps are for large sizes. Larger sizes to about 3.5 [in] or 100 [mm] are also available. For specific details, one can consult standards and handbooks such as ANSI/ASME B94.11M (1993) and Oberg et al. (2012). For large diameters, machining over a *lathe* or a *boring* machine can produce the specific sizes needed.

This section covers several vital and essential common machine elements, constructions, and assembly components.

8.2.1 Couplings

A coupling is a member that joins two shafts together and transmits power from one to the other. The shafts are usually collinear, but some couplings allow joining two shafts that are at an angle such as universal joints as to be described later on. Fitting of couplings to shafts is vital particularly for high rotational speed. The *alignment* of shafts is critical not to be a pronounced cause of vibrations. The *misalignment* of shafts occurs when shafts are not collinear, *parallel*, and/or *angular*, i.e. positioned offset or at an angle to each other. Rigid couplings should not tolerate any type of misalignment. Flexible couplings tolerate some types of misalignment but particularly should allow and smooth the torsional variations and impulses.

The coupling purpose is to connect shafts and thus fit their diameters. The shaft diameters usually follow the standard or *preferred sizes*, which are suggested as Table A.3.1. Manufacturers of couplings typically produce a set fitting the preferred shaft dimensions and normally made out of *steel*, *aluminum*, *cast iron*, and *stainless steel* or some other materials. These manufacturers are also commonly specialized in producing *rigid* and *flexible* or *universal* couplings at some specified set of dimensions. Some manufacturers produce customized sizes or materials according to order. One can get information and specific data from these manufacturers or according to some specific standards for each type of couplings as defined next.

8.2.1.1 Rigid Couplings

When the two connected shafts intended to be firmly joined, the rigid coupling is the most suitable. Alignment is, however, the most essential in joining the shafts and obviously in joining the two systems that the shafts are part of. Rigid couplings have several shapes depending on how large the system or power transmission and thus depending on shaft size. As the power and size are relatively high, the rigid coupling of Figure 8.1 is suitable. As the power and size drop, the couplings may use clamping means rather than positive means of power transmission. These clamping are to be discussed later.

Figure 8.1a shows a sectional view and the main dimension of a rigid coupling. The coupling is known also as a safety flange coupling. The side view in Figure 8.1b shows the key and keyway main dimensions. The keyway depth in the hub is larger than the part of the key placed in the hub. The number of bolts is commonly from 5 to 10 depending on size, but the holes in Figure 8.1b are for 6 bolts. This is suitable for shafts of around 7 [in] or 175 [mm]. Trend lines of the main dimensions of this type lay in Figure 8.2. The most accessible size of the shaft in this design is around 12 [in] or about 300 [mm]. Larger sizes may be possible, but other designs may be preferable.

Figure 8.2 shows the variation of selected dimensions as functions of the *shaft diameter* d_S in [in]. The black lines are for the coupling dimensions of outside diameter (OD) d_O, hub diameter d_H, hub length l_H, and flange length l_F along the coupling centerline. The gray dashed lines are some available dimensions. The indicated trend relations are for the US system, which help in approximating the required coupling dimensions for a certainly needed shaft diameter. The relations use the variable y to represent the dimension cited in the left ordinate, and the variable x for the shaft diameter cited in the abscissa of Figure 8.2 in [in]. The symbol of the shaft diameter d_S

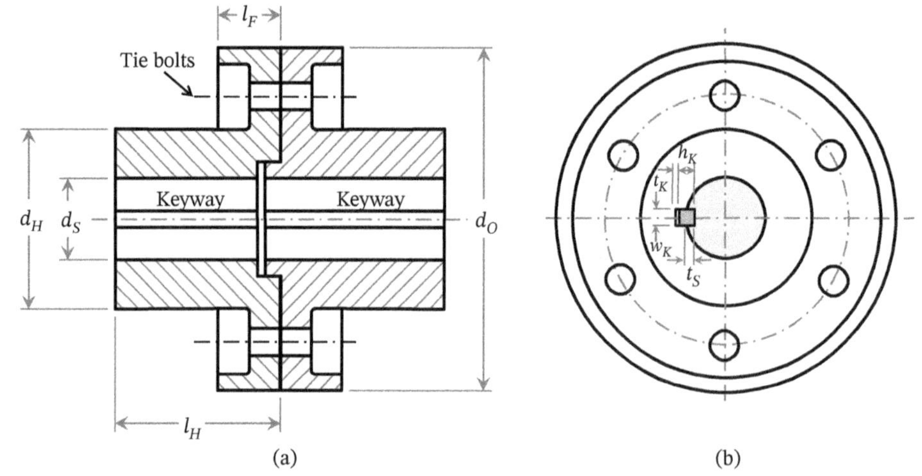

Figure 8.1 Sketches of a rigid coupling main dimensions on sectional view in (a) with tie bolts removed. Side view sketch in (b) indicates the key and keyway dimensions. The number of bolts is from 5 to 10 depending on size. The shown views are of holes for 6 bolts, which are usually for shafts of around 7 [in] or 175 [mm].

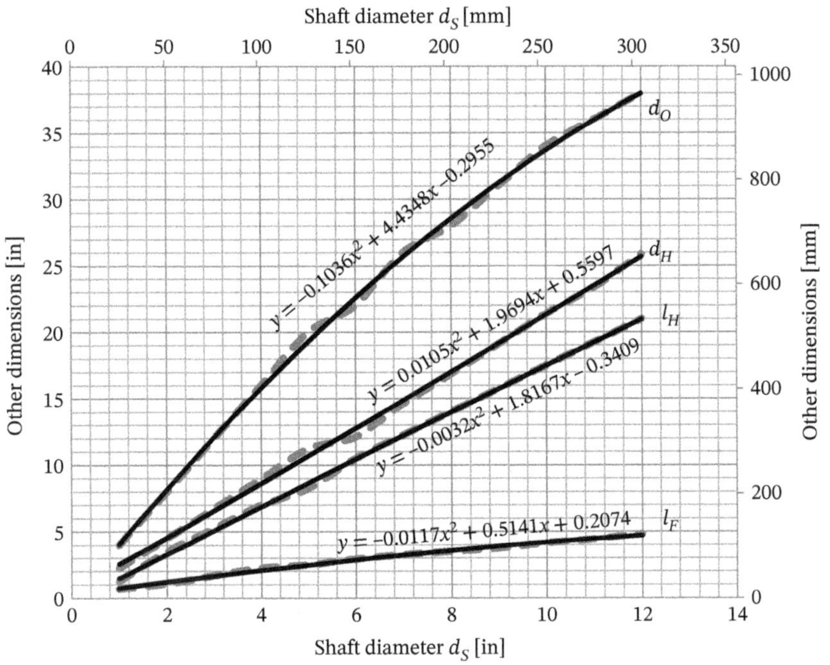

Figure 8.2 Rigid coupling main dimensions as functions of the shaft diameter. The trend equations are for the US units. Equivalent SI dimensions and captions are in the right and top secondary coordinates.

is then replacing the variable x, and the different dimension symbols of l_F, l_H, d_H, and d_O are replacing the variable y. The trend relations for the main dimensions of the *rigid coupling* in Figure 8.1 are then approximately having the following equations for the US system:

$$\begin{aligned} l_F &= -0.0117d_S^2 + 0.5141d_S + 0.2074 \\ l_H &= -0.0032d_S^2 + 1.8167d_S - 0.3409 \\ d_H &= 0.0105d_S^2 + 1.9694d_S + 0.5597 \\ d_O &= -0.1036d_S^2 + 4.4348d_S - 0.2955 \end{aligned} \tag{8.1}$$

The shaft diameters d_S are in [in], and flange length l_F, hub length l_H, hub diameter d_H, and coupling outer diameter d_O are all in [in]. It is simple to transform these dimensions to [mm] by a simple substitution of 25.4 [mm]/[in] for each term of the shaft diameter d_S. One can also use the top scale of Figure 8.2 for the shaft diameter in [mm] and get the main coupling dimensions off the right ordinate scale in [mm]. Anyway, all values should be rounded off to some reasonable significant figures for the US or SI system of units. The available equations, however, give an insight into the main geometric configuration of the coupling compared with other components of the assembly. This can be useful in a real CAD system before buying any specific coupling from a supplier or a manufacturer. Some of these manufacturers give ***3D*** geometric files to incorporate directly into the ***3D*** design after deciding on the appropriate coupling.

A very rough estimation of these trend relations may also be useful. It can be observable that the ranges of the *ratios* of the *main dimensions* to the *shaft diameter* d_S are as the following ***rule of thumb (conventional wisdom)***. The OD ratio d_O/d_S is about (3.2–4.0) with the larger ratio for the smaller diameters. The hub diameter ratio d_H/d_S is about (2.0–2.5) with the larger ratio for the smaller diameters. The hub length ratio l_H/d_S is (1.25–1.75) with the larger ratio for the larger diameters. The flange length ratio l_F/d_S is about (0.4–0.7) with the larger ratio for the smaller diameters.

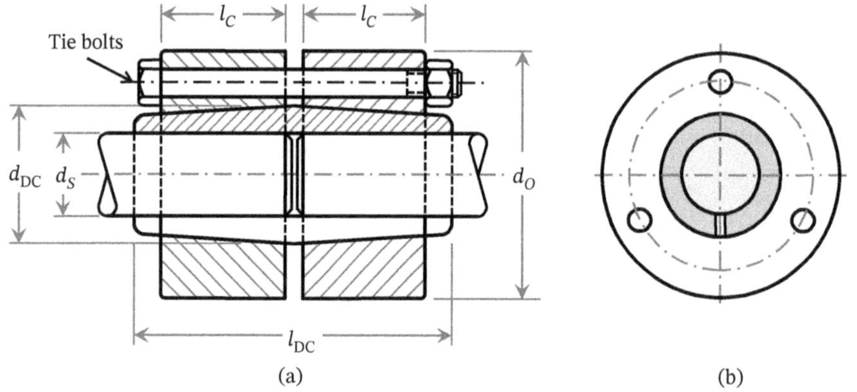

Figure 8.3 Sketches of a double-cone clamping coupling showing the main dimensions in (a). The side view in (b) indicating the holes of the tie bolts, when removed. The number of bolts is from 3 to 4 depending on size. The shown holes in (b) are for 3 bolts, which are suitable for shafts of 3 [in] or around 75 [mm].

Another coupling configuration that depends on gripping friction is shown in Figure 8.3. This type is suitable for shaft diameters up to about 6 [in] or 150 [mm]. This design has the designation of a *double-cone clamping coupling*. The *tie bolts* squeeze the *slit double cone* onto the shaft to cause a gripping friction that is enough to transmit the torque and the power. Figure 8.3a gives the main dimensions of the coupling. The side view in Figure 8.3b shows the holes of the tie bolts, when the coupling half is removed. The number of tie bolts is normally from 3 to 4 depending on size. The shown holes in (b) are for 3 bolts, which are suitable for shafts of 3 [in] or around 75 [mm]. The most accessible size of shaft diameter is around 6 [in] or about 150 [mm].

The trend relations for the main dimensions of this *rigid double-cone coupling* may approximately have the following equations for the US system:

$$\begin{aligned} l_C &= -0.1268d_S^2 + 2.0909d_S - 0.7518 \\ d_{DC} &= -0.1096d_S^2 + 2.5373d_S - 1.19 \\ d_O &= -0.1736d_S^2 + 3.8595d_S - 0.5899 \\ l_{DC} &= -0.2805d_S^2 + 5.0048d_S - 1.6012 \end{aligned} \tag{8.2}$$

In these relations, the shaft diameter d_S is in [in], and clamp length l_C, double-cone length l_{DC}, double-cone diameter d_{DC}, and coupling outer diameter d_O are all in [in]. It is simple to transform these dimensions to [mm] by a simple substitution of 25.4 [mm]/[in] for each term of the shaft diameter d_S.

A very rough estimation of these trend relations may also be useful. It can be discernible that the ranges of the *ratios* of the *main dimensions* to the *shaft diameter* d_S are as the following ***rule of thumb (conventional wisdom)***. The OD ratio d_O/d_S is about (2.7–3.3) with the larger ratio for the smaller diameters. The double-cone diameter ratio d_{DC}/d_S is about (1.6–1.8). The double-cone length ratio l_{DC}/d_S is (3.0–3.7) with the larger ratio for the smaller diameters. The double-cone clamp length ratio l_C/d_S is about (1.2–1.5) with the larger ratio for the smaller diameters.

Other rigid couplings are also available. Two of these are shown in Figure 8.4. The simple *sleeve coupling* in Figure 8.4a is shown with the *dowel pins* as principal means of transmitting torque. The *parallel* and *taper dowel pins* are only there for presenting other connecting alternatives that are given later. Shafts can also be without shoulders in Figure 8.4a. The coupling in Figure 8.4b is the *clamp coupling* with two drawing views. The elevation view is shown with one-half of the coupling removed. The sectional side view points to the two halves, a key, and two of the staggered clamping bolts. The outside semi-half circle in the side view represents the four webs shown

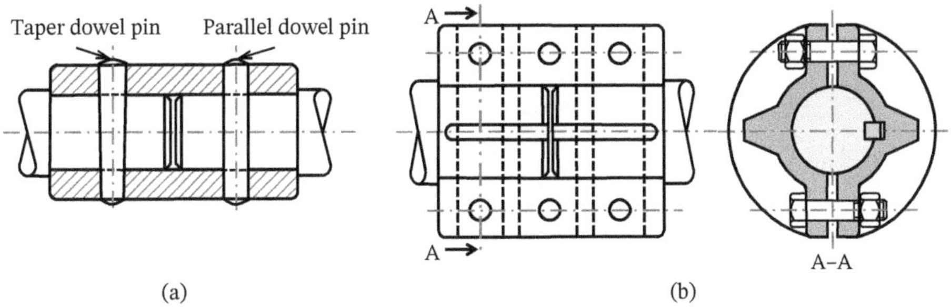

Figure 8.4 Sketches of two other rigid couplings. (a) Simple sleeve coupling with pins. Shafts can be without shoulders. (b) Clamp coupling with two views. The elevation view with one half removed. The A–A sectional side view indicates the two halves and two of the staggered clamping bolts. The number of bolts is from 4 to 8 depending on size.

as dashed lines in the elevation view. The number of bolts is from 4 to 8 depending on size. The shown holes in the elevation view in Figure 8.4b are for a 6-bolt configuration.

8.2.1.2 Flexible Couplings

In contrast to rigid couplings, the flexible couplings tolerate some axial, parallel, torsional, and angular misalignment. This is due to their inherent flexibility in their construction. They, however, are not to connect shafts with angular deviations of more than about 3° or for purposely offset shafts. Several flexible coupling types are available to accommodate different applications. Some are flexible in some directions, but not in other directions. The amount of flexibility is dependent on the construction. This flexibility helps in enduring impulses, damping vibrations, and reducing noise.

Many flexible coupling constructions are available through several manufacturers and suppliers. Only basic few give the general ideas that are differently configurable. Figure 8.5 presents sketches of two basic flexible couplings. The first flexible coupling in Figure 8.5a presents a *bush pin flanged coupling*. A local flexibility results at the connecting bolts between the two sides of the coupling. The rubber or elastomeric bush provides this flexibility between each special bolt and the flange. Flexibility is then limited by the bushes geometry, material, and number. In this case, permissible flexibility is for lesser amounts in parallel, angular, torsional, or axial misalignment. The configuration in Figure 8.5a is not symmetric. If one flips one of the special bolts, horizontally, a symmetric construction results for any even number of bolts. For this arrangement, the right half of the coupling is an exact replica of the left half. Manufacturers are thus fabricating only twice the numbers of the same part. This type of couplings usually couples electric motors to different machines.

The second flexible coupling in Figure 8.5b shows an *elastic disk flanged coupling*. The flexibility depends on the elastic disk construction and material. A special bolt ties the elastic or rubber disk to one side of the coupling. The other flipped bolt ties the other side of the coupling to the elastic disk. This results in a symmetric construction for any even number of bolts. This construction gives more flexibility arrangement than the *bush pin flanged coupling* in Figure 8.5a. The form, material, and shape of the special bolts and the disk generate so many other flexible couplings. The specific construction affects flexibility in parallel, angular, torsional, or axial misalignment.

If the special bolts are integral shaped jaws to the coupling side and the elastic disk is fitting between these jaws, the coupling turns to be a *jaw coupling*. When the jaws are straight, and the elastic disk is inelastic solid with slots or straight jaws to fit, the outcome is the *Oldham coupling*, after its inventor John Oldham (1779–1840). If the form of the jaws is as peripheral teeth and the elastic disk is shaped as a winding spring strip between the teeth, one gets the *serpentine coupling*. When the teeth are in an external spherical gear form and the elastic disk is a firm internal gear, one gets a *gear coupling*. If the elastic disk is the shape of a tire, a bellow, or a helical spring tied to the coupling sides or an integral part thereof, a *diaphragm coupling*, a *bellow coupling*, a *helical coupling*, or a *beam*

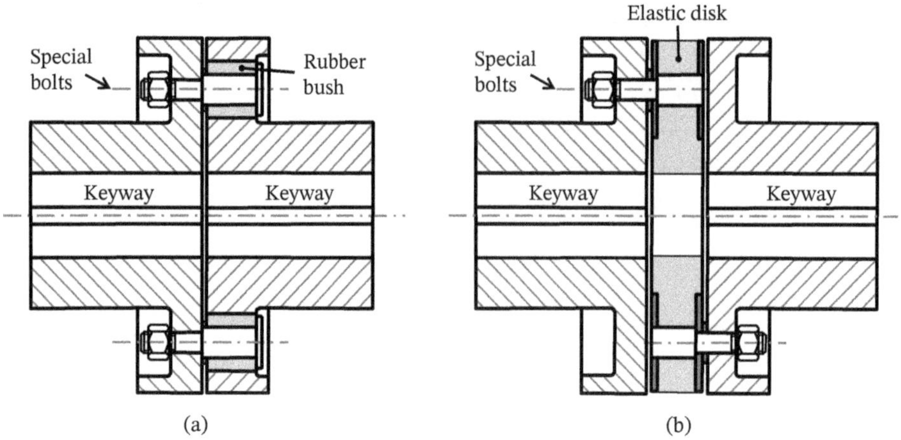

Figure 8.5 Sketches of two flexible couplings: (a) bush pin flanged coupling and (b) elastic disk flanged coupling.

coupling is the outcome. These types give different stiffness requirements for parallel, angular, torsional, or axial misalignment. Many manufacturers produce these and other different flexible (or rigid) couplings. Some Internet links are available in the references section of this chapter.

Magnetic couplings have the elastic disk in Figure 8.5b as a magnetic field between the two opposite pole magnets replacing the special bolts. One side of the coupling can have a distributed magnet of one pole, and the other side has the distributed opposite pole magnet. The attraction between the two sides is standing apart by thrust bearings on both sides of the coupling. Another means to keep the two sides apart is by using same pole magnets along the centerline of the coupling. There can be a physical barrier between the two sides of the coupling that does not diminish the magnetic field. It is then possible to have one side immersed in a fluid, while the other side is in a dry atmosphere. Many steering apparatuses in labs use the principle to mix in flasks.

A *fluid coupling* or a *hydromantic transmission* coupling (or clutch) is a flexible coupling with the elastic disk as a fluid moving between the two sides of the coupling. The transformed special bolts in Figure 8.5b are impeller blades fixed to one side, and the bolts on the other side of the coupling are blades of a turbine fixed to that side. One side is working as a pump driving the opposite turbine or motor side.

8.2.1.3 Universal Joints

The *universal joints* are also known as *Cardan* or *Hooke's* couplings. The two joined axes are not collinear but positioned at an angle with their centerline intersecting at a point as noted in the perspective view of Figure 8.6. A yoke extends from each hub, and through a hinge in the cross piece, connection to the other side occurs in an analogous way. The cross piece axes of the hinges are perpendicular to each other's. At the upper right corner of Figure 8.6, a simplified outline drawing demonstrates the concept. To further clarify the inlayed drawing, the cross piece is set in thicker black lines. Each hub and yoke is in double thin lines. The two hubs are obviously inserted in the shafts, which have their axes inclined to each other's.

Kinematically, using only one universal coupling between two shafts causes the output shaft to have unsteady rotational speed to a constant input rotational speed. This means that the velocity ratio of the output to the input shafts is not constant. Using two tandem universal couplings with intermediate shaft, however, remedy this problem; see, e.g. Mabie and Reinholtz (1987). This arrangement is continually applicable in the rear wheel drives of automotive vehicles with front engines. Other constant velocity ratio couplings that join inclined shafts have been devised and used in the front wheel drives of automotive vehicles with front engines. Some of these devised couplings use bales in grooves to cause the intermediate shaft length to be zero in length, where they work in a plane that bisects the angle between the two connected shafts; see, e.g. Mabie and Reinholtz (1987).

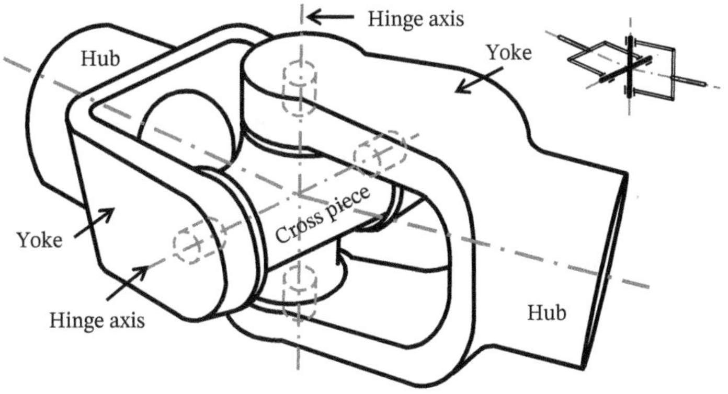

Figure 8.6 Outline of a universal joint coupling in a perspective view. The upper left corner shows the coupling sketch with the cross piece in black color.

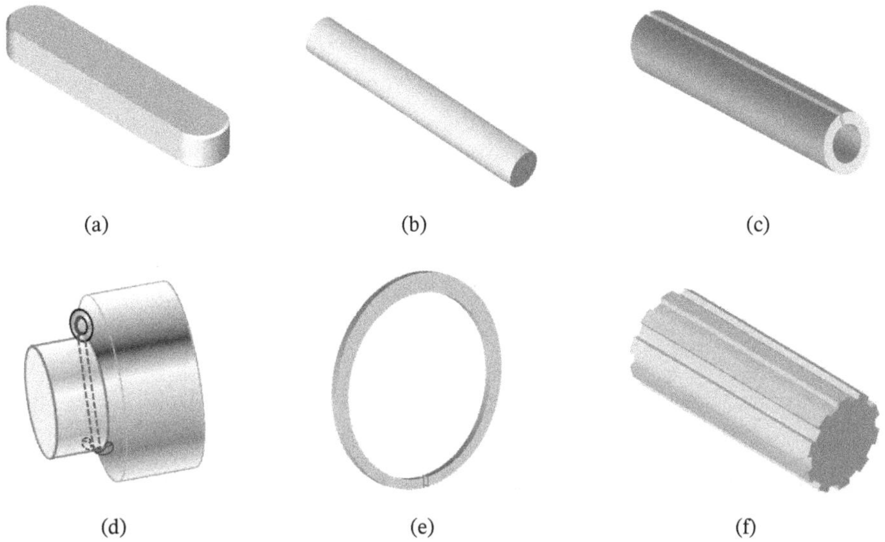

Figure 8.7 Images of some *3D* joining and fastening elements: (a) a parallel key with two round ends, (b) a parallel dowel pin, (c) slotted-type spring pin, (d) split pin (or a cotter pin or a cotter key) inserted in a shaft to prevent mating hub movement, (e) retaining ring without special attaching ends, and (f) a spline shaft with 10 straight splines.

8.2.2 Keys, Pins, Retaining Rings, and Splines

Keys, *pins*, *retaining rings*, and *splines* are joining parts between several machine elements and components. These fastening devices are presented in this section as nearly standard elements. Their dimensions are usually standardized. Their materials are usually predefined but may be selected off a limited set of choices. When a *spline* is generated into the machine element such as a *shaft* or a *hub*, the material is that of the shaft or the hub. Other joining or fastening parts such as *rivets*, *bolts*, and *fasteners* are covered in Chapter 9.

Figure 8.7 shows images of some *3D* joining and fastening elements. Figure 8.7a depicts a *parallel key* with two round ends. A parallel *dowel pin* is shown in Figure 8.7b, while Figure 8.7c depicts a slotted type *spring pin*. Figure 8.7d illustrates a *split pin (cotter pin or cotter key)* inserted in a shaft. *Retaining ring* without any special attaching ends is shown in Figure 8.7e. A *spline shaft* with 10 straight splines in Figure 8.7f should join an *internal-splined* hub of an element that is not shown.

8.2.2.1 Keys

Figure 8.7a shows a *key* with two round ends, and Figure 8.1b shows a section indicating the main dimensions of the *key* and *keyway* in the shaft and the joined hub of a coupling. Table A.5.1 presents some square and rectangular key dimensions as also defined in Figure 8.1b. The value of t_S is the depth of the keyway in the shaft. The depth of the keyway in the hub is having the value of $h_K + t_K - t_S$, where h_K is the key height and t_K is the clearance of the key in the mating hub. This relation is essential for manufacturing and thus for working drawings or ***3D*** models of mating components. Key length is calculated according to the required torque transmission. The area subjected to direct shear is considered as the key length l_K multiplied by the key width w_k. The key length l_K can be simply evaluated from the following simplified equation:

$$\tau_{xy} = \frac{\boldsymbol{F}_s}{l_K w_K} = \frac{\boldsymbol{T}_S/(d_S/2)}{l_K w_K} = \frac{S_{ys}}{K_{SF}}$$
$$l_K = \frac{2\boldsymbol{T}_S\, K_{SF}}{d_S w_K S_{ys}} \tag{8.3}$$

where τ_{xy} is the direct shear on the longitudinal section of the key ($l_K\, w_K$), $\boldsymbol{F}_s$ is the shear force acting on the key longitudinal section ($l_K\, w_K$), l_K is the key length, w_k is the key width, $\boldsymbol{T}_S$ is the transmitted torque from the shaft to the hub of the mounted element, d_S is the shaft diameter, K_{SF} is the safety factor, and S_{ys} is the shear yield strength of the key material. If the shaft is subject to alternating torque, the endurance shear strength S_{es} is to be used in place of the shear yield strength S_{ys}.

Bearing pressure or bearing stress on the key and slot sides needs also to be safe. Either the safe length is calculated or the key (or slot) length l_K is to be used to check the safety factor. The check is for the *bearing strength* S_B, which is assumingly like the compression yield strength S_{yc}. The subjected area to the pressure on the side of the key or *keyway* is approximately $l_K\, h_K/2$. Like Eq. (8.3), the calculated safe length l_K or the safety factor K_{SF} are then

$$\sigma_B = \frac{\boldsymbol{F}_t}{l_K h_K/2} = \frac{\boldsymbol{T}_S/(d_S/2)}{l_K h_K/2} = \frac{S_{yc}}{K_{SF}}$$
$$l_K = \frac{4\boldsymbol{T}_S\, K_{SF}}{d_S\, h_K S_{yc}}, \quad \text{or} \quad K_{SF} = \frac{d_S\; l_K h_K S_{yc}}{4\boldsymbol{T}_S} \tag{8.4}$$

where σ_B is the bearing stress on the side of the key or *keyway* ($l_K\, h_K/2$), $\boldsymbol{F}_t$ is the tangential force acting on the key side or *keyway*, l_K is the key length, h_k is the key height, $\boldsymbol{T}_S$ is the transmitted torque from the shaft to the hub of the mounted element, d_S is the shaft diameter, K_{SF} is the safety factor, and S_{yc} is the compressive yield strength of the key, shaft, or hub materials. If the shaft is subject to alternating torque, the compressive strength is also applicable.

Keys with round ends are secured in the shaft keyway and would be stationary in the shaft. Keys are also available with either or both flat (square) or round ends for some other applications. Flat end does not usually have axial constraints and should be handled accordingly through a setscrew fastening or any other construction restraining requirement. Manufacturing of flat end keyways in shafts is typically problematic.

For *keys* and as a ***rule of thumb (conventional wisdom)***, one may assume $w_K = 0.2$–$0.3\, d_S$ (or roughly $0.25\, d_S$), $h_K = 0.1\, d_S$–$0.3\, d_S$ (or roughly $0.2\, d_S$). For small sizes one can usually take $w_K = h_K$. It is usually safe to apply the couplings proportions to the keys. This gives the key length ratio as equal to the coupling hub length ratio l_H/d_S of (1.25–1.75) or $l_K/d_S = (1.25$–$1.75)$, with the *larger* ratio for the *larger* diameters. Eq. (8.3) is to be checked for the appropriate key material.

8.2.2.2 Pins and Cotter Pins

Pins shown in Figure 8.7b,c are suitable for location characterization between two parts such as fixed positioning and alignment. In some cases, they can carry loads and may be used as safety pins to fail before any of the connected components would fail. The *dowel pin* in Figure 8.7b is cylindrical, while the *slotted hollow pin* in Figure 8.7c is a spring that is held in the hole by contracting the pin diameter. The cylindrical dowel pin in Figure 8.7b is stuck in place inside the hole by the locational transition or interference fit such as H7/n6, H7/m6, or H7/p6; see Section 2.4.7.

A *taper dowel pin* (a taper of 1 : 48 for the US and 1 : 50 for SI) would be better in fitting two parts, but more difficult to produce the hole. The taper pin is missing in Figure 8.7, but it is simply a little tapered cylinder of Figure 8.7b. Producing a taper hole needs the use of number of drills. Each would drill to some specific depth before a taper *reamer* is to finish the tapered hole; see references such as Oberg et al. (2012).

Cotter pins shown in Figure 8.7d are manly used to prevent components from dismantling during operation. On the shaft in Figure 8.7d, the cotter pin restricts the motion of the collar to the left. Figure 8.7e shows a *retaining ring* without any special attaching ends at the space between the split ends. Figure 8.7f displays a part of a *spline* shaft with 10 *straight splines*.

The common ***dowel pin*** diameters and lengths are set according to the series of preferred dimensions; see Table A.3.1. The cylindrical pin has rounded corners, chamfered, or domed ends. Diameters are regularly smaller than lengths. There are more available lengths for each diameter. Common diameter ranges are (1/16–1) [in] or (1–25) [mm]. Common lengths are in the range of (3/16–5) [in] or (4–120) [mm]. For each diameter, the range is much smaller than that. The least pin length l_P is usually about 1.5–2 of the pin diameter d_P. The largest pin length l_P is usually about eight times the pin diameter d_P. A sample application of a dowel pin is in Figure 8.4a to fix the sleeve coupling to the shaft. The size of the pin is about 0.25 the shaft diameter. However, one should, check the pin diameter d_P for double shear on the pin section with like Eq. (8.3). The double the cross-sectional area of the pin $(2\pi d_P{}^2/4)$ under shear replaces the key section area of $l_K w_k$ in Eq. (8.3). For more information on available dimensions and properties, one can consult the standards such as ANSI/ASME B18.8.2 (2000), or ISO 2339 (1986).

The common ***taper pin*** is having a diameter and a length according to the series of preferred dimensions; see Table A3. The slightly conical pin has usually domed ends and designation according to the small end diameter. Diameters are regularly smaller than lengths. There are more available lengths for each diameter. Common diameter and length ranges are similar to the dowel pins, but regular ranges are smaller in span. Diameter limit is about 1.4 [in] or 35 [mm] and lengths are up to about 6 [in] or 140 [mm]. An application of a taper pin is in Figure 8.4a to fix the sleeve coupling to the shaft. The size of the pin is about 0.25 the shaft diameter. However, one should, check the pin size for double shear on the pin section with like Eq. (8.3). For more information, one can consult the standards such as ANSI/ASME B18.8.2 (2000), or ISO 2338 (1997).

A ***split pin*** or ***cotter pin*** or *cotter key* is a *fastener* made of a wire with half circular cross section (ASME B18.8.1 2014 or ISO 1234 1997, ASME B18.8.200M (2000)). It has an eye and two tines that form a split cylindrical extent; see Figure 8.7d. After insertion in a hole, the bent tines are as in Figure 8.7d. It is a fastener for light duties. It prevents a part on a shaft or a nut on a bolt from moving out under loosening conditions. The split or cotter pin has relatively smaller sizes of about 1/32–¾ [in] or 0.8–16 [mm]. Length range is also smaller than other pins. The common range of lengths is about 0.4–6 [in] or 10–160 [mm].

8.2.2.3 Retaining Rings

A *retaining ring* without any special attaching ends is shown in Figure 8.7e. The main purpose of a retaining ring is to hold an element such as a shaft from wondering along a hole. It should not purposely carry axial loads, even though it can. Figure 8.8 shows some retaining rings, mounting dimensions, and applications. Figure 8.8a presents the external retaining ring and shaft slot dimensions alongside an assembly of keeping a hub of an outer element from moving to the right of the shaft. Figure 8.8b depicts an internal retaining ring and hole slot dimensions and the assembly to constrain an inner element (shaft) from moving to the left (or right). Dimensions of *retaining rings* depend on the nominal diameters (NDs) of the shaft d_S or the hole d_H. These and other dimensions of the slot and shaft or hole recesses are in Tables A.5.2 and A.5.3. The material of retaining rings is like that of spring steel AISI

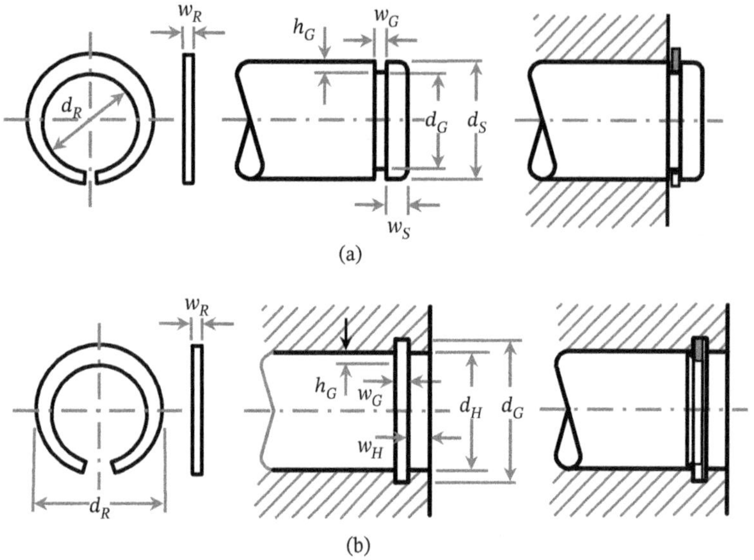

Figure 8.8 Retaining rings mounting dimensions: (a) external retaining ring and shaft slot dimensions alongside an assembly of retaining an outer element and (b) internal retaining ring and hole slot dimensions alongside an assembly of retaining an inner element.

1060-1095 or similar. For further information one can refer to ASME B27.7-1977 (R2017), ANSI B 27.7 (1977), MIL-DTL-27426B (1997), or ISO 464(2015).

The external retaining ring stretches in diameter to allow the ring to go into the shaft groove as shown in Figure 8.8a. For that, retaining rings can have some special ends and special tool to mount them in the groove. Usually larger ends with small holes and special pliers with small pins at the ends can do the job. Special end geometry and a special end pliers to fit can also do the job. The internal retaining rings contracts in diameter into the hole to go into the groove as shown in Figure 8.8b. The usual is to have the retaining ring in the hole and acts as a shaft stopper to the right. Special ends and/or special pliers are to do the mounting like the external retaining rings. These are available from the manufacturers and suppliers.

If a slight incidental axial force $\boldsymbol{F}_a$ exists, the retaining ring is then subject to direct shear on the circumferential section of the ring. This area is simply equal to $\pi d_S w_R$, where d_S is the shaft diameter and w_R is the ring width. When the ring is constraining a shaft in a hole, the hole diameter d_H substitutes the shaft diameter. Like Eq. (8.3), the calculated safety factor K_{SF} is then simply

$$\tau_a = \frac{\boldsymbol{F}_a}{\pi d_S w_R} = \frac{S_{\text{ys}}}{K_{\text{SF}}}$$

$$K_{\text{SF}} = \frac{\pi d_S w_R \, S_{\text{ys}}}{F_a} \quad \text{or} \quad K_{\text{SF}} = \frac{\pi d_H w_R \, S_{\text{ys}}}{F_a} \tag{8.5}$$

where τ_a is the axial shear on the circumferential section of the ring ($\pi d_S w_R$), $\boldsymbol{F}_a$ is the incidental axial force acting on the section, d_S is the shaft diameter (or hole diameter d_H), w_R is the ring width, K_{SF} is the safety factor, and S_{ys} is the shear yield strength of the ring material. The *bearing strength* can also be a factor to consider, which is evaluable like Eq. (8.4).

8.2.2.4 Splines

Splines are somewhat like keys (or *feathers*) that are integral part of the shaft. The cut mating part is compatibly in grooves to fit the spline shaft. Long ago, SAE has standardized the straight-sided splines; see SAE J499a (1936).

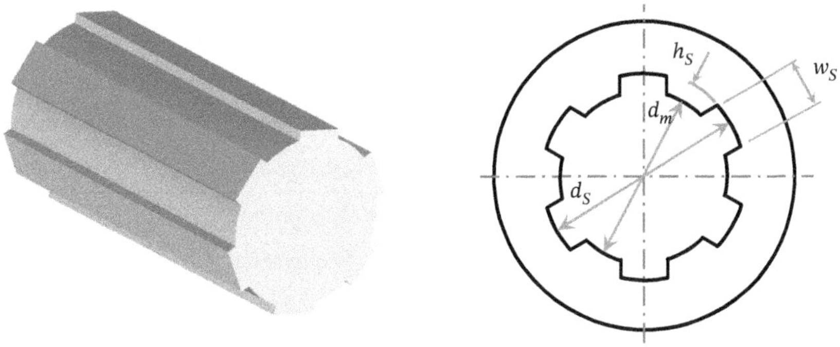

(a) (b)

Figure 8.9 Spline shaft and mating part side view: (a) ***3D*** shaft with 6 splines and (b) general spline dimensions defined on exaggerated section of 6-spline part grooves that mate the shaft.

Many straight-sided splines in numerous applications are still effectively in practice; see ISO 14 (1982). Involute splines, with involute rather than straight sides, are also having some widespread applications due to their higher load carrying capacities and standardized manufacturing. The design of involute spline is like the design of spur gear components, which are available in Chapter 14 of the text.

Straight splines

A straight-sided *spline shaft* shown with 10 *straight splines* is in Figure 8.7f. Figure 8.9 shows another 6-spline shaft and a mating part side view. The shaft with 6 splines in Figure 8.9a rather than the 10 splines in Figure 8.7f has exaggerated mating grooves in Figure 8.9b to clearly demonstrate the general spline dimensions.

SAE based the design of splines on the *allowable bearing strength* S_{Ba} of the softer mating part of 1000 [psi] or about 6.8948 [MPa]. This assumption is used on the spline faces to result in an allowable maximum spline torque $\boldsymbol{T}_S$ as follows:

$$\boldsymbol{T}_S = 1000 N_S \left(\frac{d_S + d_m}{4}\right) h_S \; l_S, \quad [\text{lb in}] \tag{8.6}$$

where N_S is the number of splines, d_S is the shaft or spline diameter, d_m is the spline minor diameter, $(d_S + d_m)/4$ is the mean spline radius, h_S is the spline height or depth, and l_S is the spline length. SAE suggests the maximum spline dimension ratios as defined in Table 8.1. All ratios are relative to the shaft diameter d_S. With these ratios and recognizing that the spline height h_S is equal to $(d_S - d_m)/2$, the allowable spline torque $\boldsymbol{T}_S$ of Eq. (8.6) becomes as follows:

$$\boldsymbol{T}_S = 1000 N_S \left(\frac{d_S + d_m}{4}\right)\left(\frac{d_S - d_m}{2}\right) l_S = 1000 N_S \left(\frac{d_S^2 - d_m^2}{8}\right) l_S, \quad [\text{lb in}] \tag{8.7}$$

Employing the spline dimension ratios according to Table 8.1 for one of the spline cases such as the 6 splines and a *slide under load (SuL)* fit, one gets the following relation:

$$\boldsymbol{T}_S = 1000 N_S \left(\frac{d_S^2 - d_m^2}{8}\right) l_S = 1000(6) d_S^2 \left(\frac{1 - 0.8^2}{8}\right) l_S = 270 \; d_S^2 \; l_S, \quad [\text{lb in}] \tag{8.8}$$

This process is useful for the other cases in Table 8.1 that carry load. The constant 270 in Eq. (8.8) changes according to the d_m/d_S ratio in Table 8.1.

The design of spline shaft depends on the shaft diameter d_S, which is to carry the torque $\boldsymbol{T}_S$. The common case is to have the transmitted torque as usually a known value. The shaft diameter is therefore a known value.

Table 8.1 Spline dimension ratios according to the dimensions defined in Figure 8.9.

Number of splines	All splines	Permanent fit		Slide under load		Slide without load	
	Spline width w_S/d_S	Minor diameter d_m/d_S	Spline depth h_S/d_S	Minor diameter d_m/d_S	Spline depth h_S/d_S	Minor diameter d_m/d_S	Spline depth h_S/d_S
4	0.241	0.85	0.075	—	—	0.75	0.125
6	0.25	0.9	0.05	0.8	0.1	0.85	0.075
10	0.156	0.91	0.045	0.81	0.095	0.86	0.07
16	0.098	0.91	0.045	0.81	0.095	0.86	0.07

Accordingly, all appropriate cases in Table 8.1 can then define the *spline lengths* l_S as functions of the ratio of the torque to the diameter squared; see Eq. (8.8). The spline lengths l_S as functions of the ratio of the torque to the diameter squared ($T_{max}/d_S{}^2$) for 4, 6, 10, and 16 splines due to *permanent fit* (PF) or SuL are then attainable as in Figure 8.10. Trend relations for the US system are thus derivable for each case as in Figure 8.10. The SI system values are definable off the secondary right and top axes with trend relations as given later. The US relations use the variable y to represent the dimension cited in the left ordinate, i.e. the spline length l_S in [in]. The variable x for the maximum magnitude of shaft torque T_S divided by the shaft diameter squared $d_S{}^2$ is the abscissa of Figure 8.10 in [lb/in]. The symbol of $T_S/d_S{}^2$ is then replacing the variable x, and the different spline lengths l_S are replacing the variable y.

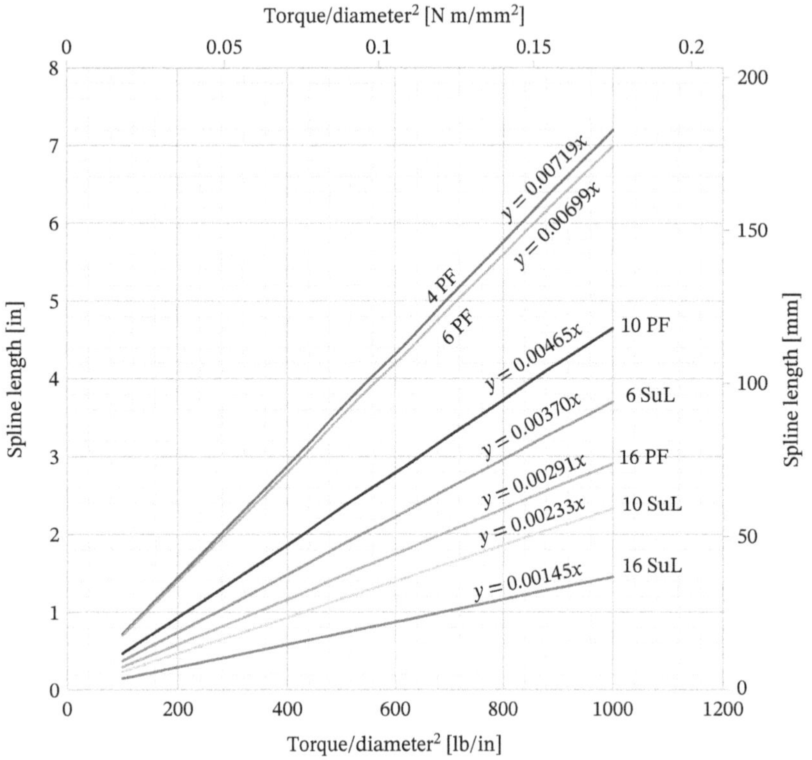

Figure 8.10 Spline shaft lengths as functions of the ratio of the torque to the diameter squared for 4, 6, 10, and 16 splines due to PF or a fit for SuL. Trend relations for the US system are given for each case. SI system values are found off the secondary axes with trend relations in the text.

As seen in Figure 8.10, the *spline lengths* l_S are functions of the torque and shaft diameter. For 4, 6, 10, and 16 splines PF or a fit for a SuL, the spline lengths l_S are as follows:

$$
\begin{aligned}
l_S &= 0.00719\ (T_S/d_S^2),\ [\text{in}]\ \text{or}\ l_S = 1043.4\ (T_S/d_S^2),\ [\text{mm}]\quad (4\ \text{Splines, PF})\\
l_S &= 0.00699\ (T_S/d_S^2),\ [\text{in}]\ \text{or}\ l_S = 1014.2\ (T_S/d_S^2),\ [\text{mm}]\quad (6\ \text{Splines, PF})\\
l_S &= 0.00465\ (T_S/d_S^2),\ [\text{in}]\ \text{or}\ l_S = 674.59\ (T_S/d_S^2),\ [\text{mm}]\quad (10\ \text{Splines, PF})\\
l_S &= 0.00370\ (T_S/d_S^2),\ [\text{in}]\ \text{or}\ l_S = 537.18\ (T_S/d_S^2),\ [\text{mm}]\quad (6\ \text{Splines, SuL})\\
l_S &= 0.00291\ (T_S/d_S^2),\ [\text{in}]\ \text{or}\ l_S = 421.62\ (T_S/d_S^2),\ [\text{mm}]\quad (16\ \text{Splines, PF})\\
l_S &= 0.00233\ (T_S/d_S^2),\ [\text{in}]\ \text{or}\ l_S = 337.3\ (T_S/d_S^2),\ [\text{mm}]\quad (10\ \text{Splines, SuL})\\
l_S &= 0.00145\ (T_S/d_S^2),\ [\text{in}]\ \text{or}\ l_S = 210.81\ (T_S/d_S^2),\ [\text{mm}]\quad (16\ \text{Splines, SuL})
\end{aligned}
\tag{8.9}
$$

where T_S is in [lb in] or [N m] and d_S is in [in] or [mm]. In SI system, the ratio $(T_S/d_S{}^2)$ has the dimension of [N m/mm^2]. The torque in the SI system is usually in [N m], therefore one **must not** change it to [N mm] before substituting into Eq. (8.9). It is also clear that the constants in Eq. (8.6) are the reciprocal of the constants derived like that in Eq. (8.8). The reciprocal of 270 in Eq. (8.8) is 0.0037 as for the constant of the US-[in] 6 splines (SuL) in Eq. (8.9).

Involute splines

Sketches of involute spline shafts of 30° pressure angle and mating or internal parts are shown in Figure 8.11. The shaft splines are in grayish color for clarity. Figure 8.11a is an involute with side fit between the external shaft spline and the internal hub spline. Figure 8.11b is an involute with major diameter fit of the shaft with the mating diameter of the hub. The major diameter fit, or the side fit, should suit the spline function. Ranges of common location and clearance fits are H/h, H/f, H/e, and H/d; see Section 2.4.7. Some interference fits are applicable for some PF applications. The *tolerance grade* is in the range IT4–IT7. In addition to the 30° pressure angle, there are other pressure angles such as 37.5° and 45°. Even number of spline teeth in the range of 6–60 teeth is the customary practice. For more details, one can consult handbooks or standards such as Oberg et al. (2012), ANSI B92.1 (1996), or ISO 4156 (2005). The capacity of the involute splines is about (1.6–5) times that of the straight-sided splines. Major involute spline diameters can be about (75–40%) smaller than the major diameter of the straight-sided splines. The design of involute spline is like the design of spur gears; see Chapter 14 of the text.

A preliminary design calculation is possible by approximating the strength to the shear at the root of the teeth. SAE suggests that only 25% of the area is eligible, since only this ratio of teeth is in contact. With these assumptions, the shear stress is like Eq. (8.3) to have

$$
\begin{aligned}
\tau_{\max} &= \frac{4\boldsymbol{F}_S}{(\pi d_m) l_S} = \frac{4T_S/(d_m/2)}{\pi d_m\, l_S} = \frac{S_{ys}}{K_{SF}}\\
l_S &= \frac{2T_S K_{SF}}{\pi d_m^2\, S_{ys}}
\end{aligned}
\tag{8.10}
$$

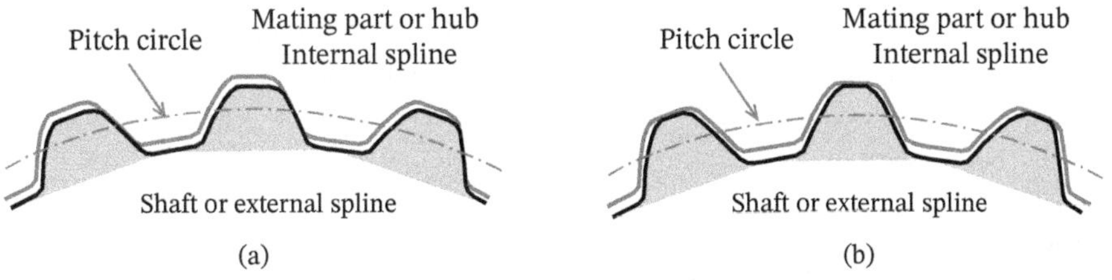

Figure 8.11 Sketches of involute spline shafts of 30° pressure angle and mating or internal parts. The shaft splines are filled with grayish color for clarity: (a) side involute fit and (b) involute major diameter fit.

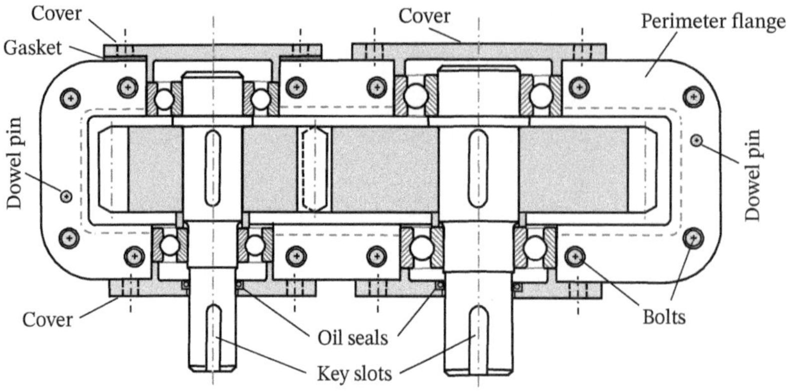

Figure 8.12 A sketch of a sectional view for a gearbox showing one gasket fitted between the upper left cover just to demonstrate possible application of cover sealing in a single stage *gearbox*. The view also demonstrates the utility of oil seals, dowel pins, keys, key slots, and other constructional details.

where τ_{max} is the maximum shear stress on 25% of the circumferential section of the spline, i.e. $(\pi d_m l_S/4)$, d_m is the shaft minor diameter, l_S is the spline length, $\boldsymbol{F}_s$ is the incidental shear force acting on the section, K_{SF} is the safety factor, and S_{ys} is the shear yield strength of the spline material. The second line of Eq. (8.10) presents the spline length l_S as a function of $T_S/d_m{}^2$ like the straight spline treatment of Eq. (8.9). This is to allow for comparison. SAE suggests that the length of the hollow spline shaft l_S can have the following value:

$$l_S \cong \frac{d_m^3(1 - d_m^4/d_i^4)}{d_p^2} \tag{8.11}$$

where d_m is the shaft minor diameter, d_i is the hollow shaft inner diameter, and d_p is the spline pitch diameter. When the number of splines is large, one can approximate Eq. (8.11) to get the involute spline length $l_S \approx d_m$. Customary practice suggests the spline length ratio to the shaft diameter to be in the range of 0.75–1.25.

The involute profile is useful in gears due to its rolling characteristics with no sliding at the pitch circle; see Section 14.2. Since there is no rolling motion in splines, inclined straight sides may just be as efficient as involute sides. Optimum inclination of sides is necessary to generate more efficient splines. An interactive ***FE*** procedure is useful in that regard. This research task is beyond the scope of this text.

8.2.3 Seals

When designing systems that involve fluids such as oils, water, gas, etc., one needs to prevent the fluids or any particles from leaking to outside or penetrating from outside or even seeping between system compartments. To do that, some types of *seals* are utilized to stop the leaking, penetration, or seepage. A *gearbox* is a system of gears with lubricating oil inside; see Figure 8.12. The sketch of a sectional view of the gearbox in Figure 8.12 is showing a *gasket* fitted between the upper left cover and the gearbox. This is just to demonstrate possible application of cover sealing in a single stage gearbox. Usually no such gaskets are exploitable in most gearbox covers. The usual cover-neck location-fit of H/h with the hole and the surface finish of the cover contact with the gearbox are sufficiently sealing the connection. In addition, the view in Figure 8.12 demonstrates the utility of *oil seals*, *dowel pins*, round end *keys*, key slots, and other constructional details.

Gaskets are *static seals* where no relative motion exists between components. Figure 8.13 is showing some sketches of some common static seals. In Figure 8.13a, the oil or fluid side is to the left as the dotted pattern. Each of the sketches in Figures 8.13–8.15 has the oil or fluid side similarly to the left. Figure 8.13a shows a *gasket* between an enclosure and a hole cover like the gearbox cover in Figure 8.12. The *gasket* thickness is extremely

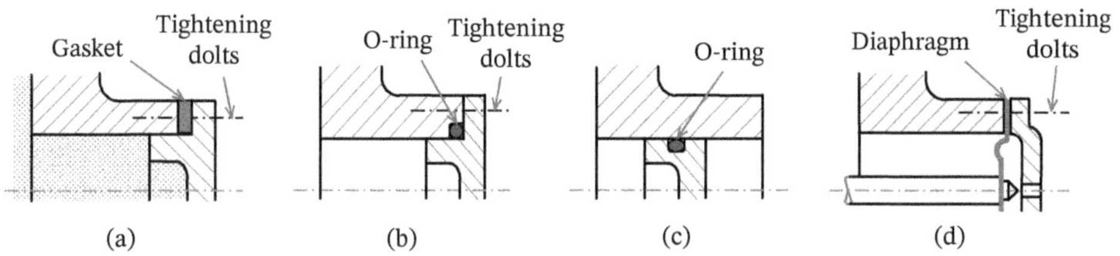

Figure 8.13 Sketches of some common static seals. (a) *Gasket* between an enclosure and a hole cover. The gasket thickness is extremely exaggerated to be visible. (b) An *O-ring* seal between the cover and an enclosure. (c) An *O-ring* seal between a component that can move slowly in a cylindrical hole. The component may be thought of as stationary. (d) Diaphragm pushed by the rod to close the hole in the cover.

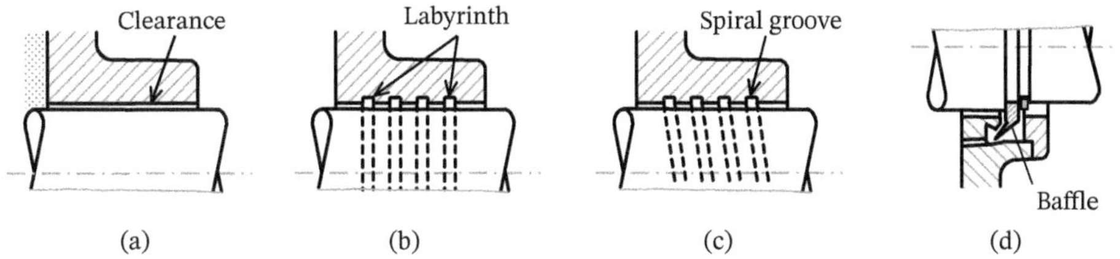

Figure 8.14 Sketches of some common non-rubbing seals. (a) *Close clearance* (gap type) seal. (b) Seal with *annular groove* or *labyrinth* type. (c) Seal with *spiral groove*. (d) *Flinger* ring or *baffle* plate type. The oil or fluid side is to the left of each sketch as shown by the dotted pattern in (a). The spiral grooves are in a direction to pump the oil or fluid back depending on the shaft rotating direction. The *flinger* in (d) is held on the shaft by a retaining ring. The drain of fluid in (d) is to be above the static fluid level. That is why the sketch is flipped vertically in (d).

exaggerated to be visible. The typical materials of conventional gaskets can be a cardboard or elastomeric of small thickness or some temperature-resistant material enclosed in a soft aluminum or copper skin to stand high temperatures such as the gasket between the engine block and the cylinder head in an internal combustion engine. With the tightening bolts between the two parts, the soft aluminum or copper skin or gasket material is somewhat yielding to fill the variable gap in between to provide sealing.

Figure 8.13b shows a *static seal* in the form of an *O-ring* between the cover and an enclosure or housing. The face of the cover compresses the O-ring sufficiently to guarantee sealing. The compression is enacted by the tightening bolts of the cover. The standard dimensions of the *O-ring* and grooves are as shown in Table A.6.2. These dimensions can be useful in defining the necessary space to affect sealing. Contact stresses and O-ring section material elasticity in addition to ***FE*** code are also helpful in that regard. O-rings are usually made of NBR elastomeric materials or PTFE or composite PTFE; see properties of selected plastics in Appendix A.7. Figure 8.13c represents another O-ring seal between a component that can slowly move axially or rotationally in a cylindrical hole. The component may be thought of as stationary. The surfaces are mirror polished and suitably lubricated to facilitate the slow motion without any damage to the O-ring. Figure 8.13d represents another sealing by means of a *diaphragm* that is pushed by the rod to close the hole in the cover. The left compartment of the assembly is totally sealed – by the diaphragm – from the right compartment that can have a fluid passing through the hole in the cover.

Non-rubbing seals depicted in Figure 8.14 show sketches of some common types of these *dynamic seals*. They involve designs with relative motion between their components. Figure 8.14a shows a close clearance (gap type) seal that can be useful in low speeds, little pressure differential, high viscosity fluids, or similar fluids such as air on both sides. Figure 8.14b presents a seal with annular groove or labyrinth type that causes successive pressure drops between succeeding labyrinth pockets. That inhibits the ability of the flow and achieves sealing. Figure 8.14c shows a seal with spiral groove that forces the fluid to flow inward and thus prevents it from seeping. Obviously, it

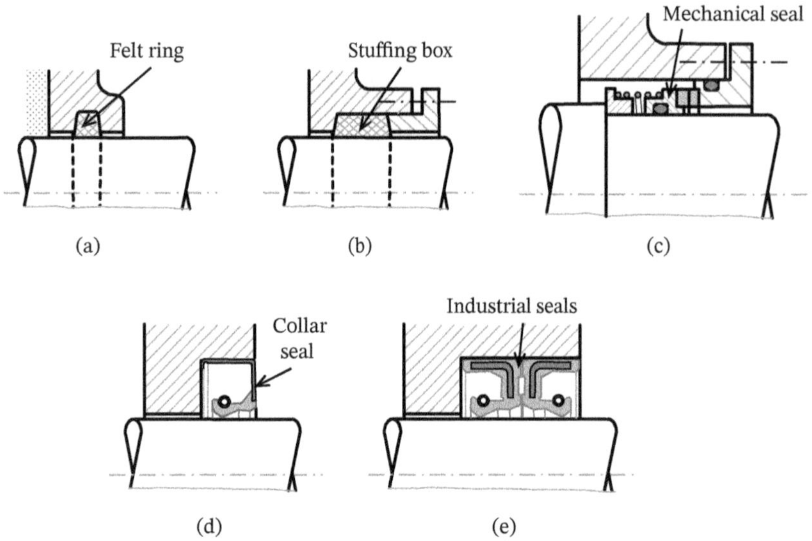

Figure 8.15 Sketches of some common rubbing seals. (a) *Felt ring* (checkered section) seal. (b) *Stuffing box* with an adjusting gland to take the slackness wear in the felt stuffing (checkered section). This is the basis of the *mechanical seal.* (c) A mechanical seal. (d) Widely used *collar seal* with metal jacket and special elastomeric body hugging the shaft through a soft spiral torus spring. (e) Two industrial collar seals arranged back to back. The oil or fluid side is to the left of each sketch as shown by the dotted pattern in Figure 8.15a.

depends on the direction of rotation. The spiral grooves are in a direction to pump the oil or fluid back when the shaft rotating direction is opposite to the spiral. Figure 8.14d presents the *flinger ring* or *baffle plate* type seal. With high rotational speeds, the fluid is forced to flow peripherally on the ring or plate and splashes into the groove to go back to the left compartment. The flinger in Figure 8.14d is held on the shaft by a retaining ring. The drain of fluid in Figure 8.14d is to be above the static fluid level. That is why the sketch is flipped vertically in Figure 8.14d. The oil or fluid side is to the left of each sketch as shown by the dotted pattern in Figure 8.14a.

Rubbing seals have direct contact between the seal and the moving part it seals. Figure 8.15 shows sketches of some common rubbing seals. Figure 8.15a presents the classical heritage *felt ring* seal as the part with the checkered section. It has been used effectively for long time with designs for low and medium speeds that are lubricated with grease or oil. The felt is absorbent and holds the oil or grease to reduce friction between the felt and the rotating element. Its effectiveness depends on the amount of compression it holds to keep the fluid from seeping out. Due to wear after some operating time, it needs to be replaced with a new felt ring. A *stuffing box* with an adjusting gland to take the slackness wear in the felt stuffing (checkered section) is shown in Figure 8.15b. This classical seal is also suitable for low and medium speeds, but the *felt* replacement cycle is stretched. This stuffing box philosophy has been the basis of the *mechanical seal* (Figure 8.15c). In the mechanical seal, the stuffing box in Figure 8.15b is replaced with another *small box*, which is attached to the shaft shoulder or just fixed on the shaft. The small box has an end disk or ring (checkered section), which rubs against another mating disk or ring (checkered section). The mating disk is fixed to the fixed gland (or housing). These disks are made of special materials that are having very low friction coefficient, very hard, extremely mirror polished, and wear resistant. The two disks are under some sufficiently low pressure to prevent seeping and to reduce in-between wear and power consumption. The low pressure is generated by a spring in the small box fixed to the shaft with an O-ring to prevent fluid from seeping outside. The small box can alternatively be fixed to the housing rather than the shaft. Many constructions are available for many applications; see, e.g. Internet references in this chapter.

Other widely used *rubbing seals* are shown in Figure 8.15d,e. They are known as *collar seals*. These have a metal jacket and special elastomeric body hugging the shaft by a soft spiral torus spring (ISO 6194 2007 or SAE J946

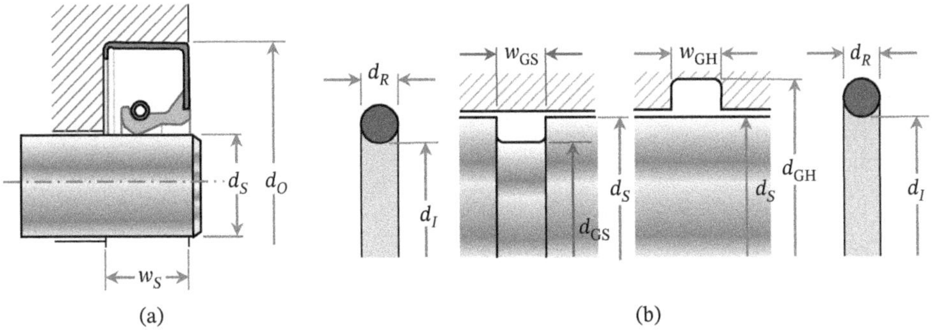

Figure 8.16 Collar seal and O-ring mounting dimensions: (a) collar seal and hole slot dimensions and (b) Internal and external O-ring and slot or groove dimensions.

2002). The elastomeric body is usually made of thermoplastics PU or UP, NBR elastomeric materials or PTFE, or composite PTFE; see properties of selected plastics in Table A.7.8. Simpler designs without the soft spiral torus spring are called *lip seals* (not shown). Tables A.6.1 and A.6.2 shows selected main dimensions of seals as defined in Figure 8.16 for metric and inch series. Figure 8.15e represents two industrial collar seals arranged back to back for attaining more sealing.

Figure 8.16 shows the selected *collar seal* and *O-ring* mounting dimensions. Figure 8.16a presents the selected seal and hole slot dimensions. Figure 8.16b gives the internal and external O-ring and their slot or groove dimensions. In Table A.6.1 selected metric and inch series of collar seal main dimensions as defined in Figure 8.16a; see ISO 6194 (2007) and SAE J946 (2002). Table A.6.2 gives selected metric and inch series of O-rings main dimensions as defined in Figure 8.16b; see ISO 3601-1 (2012), ISO 3601-2 (2014), SAE AS568 (2014), and SAE AS4716 (R) (2017). Smaller O-ring cross section (C-S) and smaller inside diameter (ID) are for shaft groove. Larger O-ring cross section (C-S) and larger ID are for hole groove.

8.2.4 Housings, Enclosures, Frames, and Chassis

Housings, enclosures, frames, chassis, etc. are structures made of casting materials or out of built-up sections made from standard plates or beams joined together by welding, riveting, bonding, and other joining processes. They may be treated individually as separate components with each having connection boundary conditions depending on the joints in-between. One can also design these by numerical techniques such as ***FE*** or any other suitable or special codes. To acquire these, the plates and beams subcomponents need to be accessible to the designer at reasonable cost. Usually plates and beams are structural components made of structural steel. They are manufactured to some specific standard dimensions. The designer ought to select out these standard dimensions to benefit from the lower mass-production cost. Other dimensions may be acquired by building up from basics or incurring the higher cost of custom production.

To facilitate building up housings, enclosures, frames, chassis, etc. from plates or beams, several of these standards are made available in Appendix A.8. Most standard dimensions use the preferred sizes, but more options are also available. Selected standard dimensions of *I-beams*, *channels*, *equal angles*, *round tubes*, *round pipes*, and *rectangular tubes* are provided in Appendix A.8. Other standard dimensions and thicknesses are available from standards, manufacturers, and publications such as ANSI B92.2M (1980), ISO 657 (1989), Oberg et al. (2012), ASTM A992/A992M (2015), and some available manufacturers or suppliers in the Internet References at the end of the chapter. Some section properties such as section area and second area moments are listed. The SI designations are usually including the section type and the main dimensions. The US designation customary includes the section type, main dimension, and the weight per length in [lb/ft]. For other related codes and standards, one may consult ANSI/AISC 303 (2016).

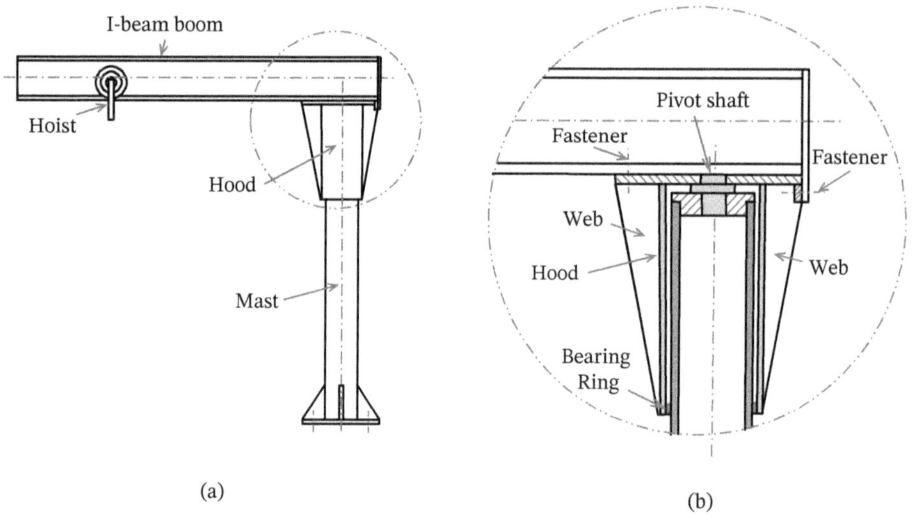

Figure 8.17 A conceptual *Jib crane* demonstrating the possible extensive use of standard sections such as wide flange I-beam, cylindrical tubes, plates, etc. The sections are welded or bolted together. (a) A suggested view of the crane. (b) Sectional view of a suggested construction of the main crane section (with I-beam not sectioned).

Most of the standard components have defined dimensions such as the sketches attached to the tables in some appendices at the end of the text. The round tubes and pipes would have either OD designation or ID designation in addition to the wall thickness. Pipe standards have been using the designations of the ID due to their past historical use in carrying fluids such as water. The main dimensions are usually the US or imperial inch system. The nominal pipe size (NPS) is utilizing the bore or ID in an abbreviated form close to the real value in [in]. Usually, the OD is larger to ensure that the IDs are close to the NPS. For larger pipes than 12 [in] in diameter, the OD is the same as the NPS. For the SI system, the ND is used to be close to the NPS equivalent in [in]; see ISO 6708 (1995). For specific dimensions and availability, one can consult ANSI/ASME B36.10M(1995), ISO 4200 (1991), and manufacturers or suppliers of pipes and tubes; see Internet references at the end of the chapter.

Many designs utilize the standard sections to gain from the lower cost of these mass-produced sections. This is particularly true for products of limited production size. An example of that is the *Jib crane* shown in Figure 8.17. This conceptual Jib crane demonstrates the possible extensive use of standard sections such as an *I-beam* or a wide flange *I-beam*, cylindrical or round tubes, plates, fasteners, etc. The standard sections are welded or fastened together. Figure 8.17a suggests a typical view of the crane. It utilizes an *I-beam* for the main hoist boom, which swivels around the mast through a hood acting as the swivel hinge. Figure 8.17b gives a sectional view of a suggested construction of the main crane section at the hood junction with the mast. The hood and mast are using round tubes that are shaded in grayish colors for their sections. The round tube of the hood is strengthened by webs that are welded to the tube and topped with a welded plate to be fastened to the *I-beam*. This top plate has a pivot shaft (dark grayish color) welded to it. The pivot shaft transmits the thrust load and side couple load to the mast. The bottom part of the hood tube is provided with one of the bearing rings attached to it. This bearing ring transmits the other couple load to the mast tube. The second bearing ring is attached to the mast tube. Both bearing rigs are shown in dark grayish sections between the hood and the mast tubes. The mast tube is topped with a thick part that the pivot shaft swivels are in. The dimensions of the *I-beam*, hood tube, mast tube, the webs, and the pivot shaft bearing are determined by the load capacity and the *Jib-crane* size. The bearings can be sleeve or rolling bearings that should need lubrication and sealings. An attempt to design one of these Jib cranes is accessible in Example 8.2.

Another example of a housing, enclosure, frame, and a chassis in one is the *gearbox* construction shown in Figure 8.12. The gearbox houses the shafts, gears, bearings, and other essential components in an enclosure, frame, and chassis. The frame of the gearbox is made of cast iron. It can alternatively be built up by welded plates and machined to hold other components. The welded construction might be more attractive than the casting alternative. This is particularly true for large sizes and few needed gearboxes. The gearbox base and top cover are built up separately with extra thickness for the perimeter flange. The interface flanges of the base and cover are milled and joined with two location dowel pins. Both base and cover are then fastened by the bolts in the suitably drilled holes as in Figure 8.12 (with cover removed). This forms the rough outer frame of the gearbox. The frame is then machined to accommodate the bearings and bearing covers. The other finished components are then assembled in the gearbox base as shown in Figure 8.12 before fastening the gearbox cover to the base.

Example 8.1 Continuation of the previous Example 7.6. Again, the cylindrical shaft is transmitting a maximum of 25 [kW] or 33.5 [hp] and running at 3000 [rpm] in the clockwise direction. It is subject to a maximum bending moment that is equal to the maximum applied torque at its critical location. For the internal stresses previously calculated and if the shaft has a cylindrical diameter of 45 [mm] or 1.75 [in] at a location of much smaller bending moment, estimate the necessary key, pin, and spline to join the hub of a coupling to the shaft. Consider a safety factor K_{SF} of 3.5. Find the joining element geometry or the safety factor for each of the elements used for the joint. The selected material is again set as hot rolled AISI 1040 or ISO C40. Compare results of the different joining elements.

Solution

Data: $H = 25$ [kW] or 33.5 [hp], $N_{rpm} = 3000$ [rpm], and $d_S = d_C = 0.045$ [m] or 1.75 [in]. *Ductile* material is hot-rolled AISI 1040 or ISO C40. The material properties are tensile yield strength $S_{yt} = 42$ [kpsi] or 290 [MPa], ultimate tensile strength $S_{ut} = 76$ [kpsi] or 525 [MPa], and the endurance limit $S_e = 37.7$ [kpsi] or 260 [MPa]; see Table A.7.2. Equation (7.58) suggests $S_e = 0.5\ S_{ut} = 0.5(525) = 262.5$ [MPa] or 0.5(76) = 38 [kpsi]. The two values are close, and one can then use the values in Table A.7.2, which is the same as in Example 7.4. The value is useful if the torque is fluctuating. The data in Examples 6.10 and 7.4 that can be applicable to this example is as follows:

The maximum torque on the shaft is $\boldsymbol{T}_S = 79.577$ [N m] or 703.78 [lb in].
The maximum shear stresses in the shaft due to torque τ_{xz} are 4.4475 [MPa] or 0.668 79 [kpsi].

The results of Example 7.6 are very close to those found in Example 7.4. The final result of each fatigue safety factor $K_{SF,F}$ is also close. The computer code in Example 7.6 gives $K_{SF,F} = 6.4395$ and 6.2229 for SI units and the US units, respectively. This is close enough to hand calculations for manual engineering design calculations. The safety factors are for the shaft at a maximum bending moment that is equal to the maximum applied torque at its critical location. In this example the stresses are in the key, pin, or spline to join the hub of a coupling to the shaft. The torque or the shear stress due to torque is of paramount importance.

Key design is effectively finding the length of the key. The dimensions of the key section for the 45 [mm] and 1.75 [in] shaft diameters are in Table A.5.1. The key width w_K can be 14 [mm] or 3/8 [in]. The key height h_K can be 9 [mm] or 3/8 [in]. The distortion energy theory suggests the shear yield strength $S_{ys} = 0.58$ of the tensile yield strength S_{yt}. Assuming the material of the key is about the same as the shaft, the length of the key for a safety factor of 3.5 is from Eq. (8.3) as follows:

$$l_K = \frac{2\boldsymbol{T}_S K_{SF}}{d_S w_K S_{ys}} = \frac{2(79.577)(3.5)}{(45)(10^{-3})(14)(10^{-3})(0.58 \times 290)(10^6)} = 0.005\,256\,8\ [\text{m}] = 5.2568\ [\text{mm}]$$

$$l_K = \frac{2\boldsymbol{T}_S K_{SF}}{d_S w_K S_{ys}} = \frac{2(703.78)(3.5)}{(1.75)(3/8)(0.58 \times 42)(10^3)} = 0.30817\ [\text{in}] = 7.8275\ [\text{mm}] \tag{a}$$

This value is very small and not realistic relative to the rule of thumb (conventional wisdom) defining l_K to be at least equal to d_S or $l_K = 45$ [mm] or 1.75 [in]. This is since the shaft diameter is much larger than usual and that the material of the key can have a lower strength than the assumed one. Take $l_K = 45$ [mm] or 1.75 [in].

Calculating the key safety factor with the selected length from the bearing strength, Eq. (8.4) gives

$$K_{SF} = \frac{d_S\ l_K h_K S_{yc}}{4T_S} = \frac{(45)(10^{-3})(45)(10^{-3})(9)(10^{-3})(290)(10^6)}{4(79.577)} = 16.604$$

$$K_{SF} = \frac{d_S\ l_K h_K S_{yc}}{T_S} = \frac{(1.75)(1.75)(3/8)(42)(10^3)}{4(703.78)} = 17.13 \quad \text{(b)}$$

These values are still high. This confirms again that the diameter is much higher that what is supposed to be. Counting the stress concentration factor K_{SC} of 3.0 at the contacts (*alleged*) gives the factors of safety as 5.5 or 5.7. The fatigue safety factors were also high in Example 7.6 that gave $K_{SF,F} = 6.4395$ and 6.2229 for SI units and the US units, respectively. This is an optimization case to get the best utilization of material and satisfy the function requirement.

Pin design is effectively finding the pin diameter or the safety factor of the selected pin diameter. The pin diameter d_P is about 0.25 the shaft diameter d_S. Assuming the same material as the shaft, the calculated pin diameter d_P is like Eq. (8.3).

$$\tau_{xy} = \frac{\boldsymbol{F}_S}{2(\pi d_P^2/4)} = \frac{\boldsymbol{T}_S/(d_S/2)}{2(\pi d_P^2/4)} = \frac{S_{ys}}{K_{SF}}$$

$$d_P = \sqrt{\frac{4\boldsymbol{T}_S K_{SF}}{\pi d_S\ S_{ys}}} \quad \text{(c)}$$

or

$$d_P = \sqrt{\frac{4\boldsymbol{T}_S K_{SF}}{\pi d_S S_{ys}}} = \sqrt{\frac{4(79.577)(3.5)}{\pi(45)(10^{-3})(0.58 \times 290)(10^6)}} = 0.006\,844\,8\ [\text{m}] = 6.8448\ [\text{mm}]$$

$$d_P = \sqrt{\frac{4\boldsymbol{T}_S K_{SF}}{\pi d_S S_{ys}}} = \sqrt{\frac{4(703.78)(3.5)}{\pi(1.75)(0.58 \times 42)(10^3)}} = 0.271\,24\ [\text{in}] = 6.8895\ [\text{mm}] \quad \text{(d)}$$

These calculated diameters are less than the suggested pin diameters of 0.25(45) = 11.25 [mm] and 0.25(1.75) = 0.4375 [in]. To use larger preferred sizes as defined in Table A.3, the pin sizes are then 8 [mm] and 0.3 or 5/16 [in]. These are about 0.18 or 0.16 d_S rather than 0.25 d_S. The safety factors of these pins are thus more than the assumed 3.5. Counting the stress concentration factor K_{SC} of 3.0 at the contacts (*alleged*) gives the pin diameters as about 11.86 [mm] and 0.47 [in]. These are close to the suggested pin diameters of 0.25(45) = 11.25 [mm] and 0.25(1.75) = 0.4375 [in].

Spline design is effectively finding the number of splines N_S and the spline length l_S. Considering *straight splines* and 6 PF splines produces the necessary relation to find the spline length l_S. Equation (8.9) gives the spline length l_S as follows:

$$l_S = 0.006\,99\ \ (T_S/d_S^2) = 0.006\,99\ \left(\frac{703.78}{(1.75)^2}\right) = 1.6063\ [\text{in}]$$

$$l_S = 0.031\,11\ \ (T_S/d_S^2) = 1014.2\ \left(\frac{79.577}{(45)^2}\right) = 39.855\ [\text{mm}] \quad \text{(e)}$$

These values are reasonable, and the spline length is about 0.92–0.89 of the shaft diameter d_S. This is because the relations of SAE use the allowable bearing strength of the softer mating part S_{Ba} of 1000 [psi] or about 6.8948 [MPa]. For higher number of splines, the length decreases as the number of spline increases. To use Figure 8.10, the ratio $(703.78/(1.75)^2)$ is ≈230 [lb/in]. This value gives the spline length as about 1.6 [in]. For SI system, the ratio $(79.577/(45)^2)$ is ≈0.039 [N m/mm^2]. This value gives the spline length as about 40 [mm] in Figure 8.10.

Considering *involute spline*, the spline length l_S is not affected by the number of splines except in reducing the minor diameter as the number of splines increases; see Eq. (8.10). This is also because of the simple assumption that the shear strength is only over 25% of the minor circumference. Approximating the minor diameter as the shaft diameter (for very large N_S), the spline length l_S according to Eq. (8.10) is then as follows:

$$l_S = \frac{2T_S K_{SF}}{\pi d_m^2 S_{ys}} = \frac{2(79.577)(3.5)}{\pi((45)(10^{-3}))^2(0.58 \times 290)(10^6)} = 5.2058\text{e-}4\ [\text{m}] = 0.520\,58\ [\text{mm}]$$

$$l_S = \frac{2T_S K_{SF}}{\pi d_m^2 S_{ys}} = \frac{2(703.78)(3.5)}{\pi(1.75)^2(0.58 \times 42)(10^3)} = 0.021\,02\ [\text{in}] = 0.5339\ [\text{mm}] \qquad \text{(f)}$$

This value is very small because the involute spline usually carries more torque and we also used the minor diameter as the shaft diameter. Counting the stress concentration factor K_{SC} of 3.0 at the contacts (*alleged*) gives the spline lengths as about 1.56 [mm] and 0.063 [in]. These involute spline lengths can be adjustable further, through the safety factor or substituting for the correct minor diameter and using smaller shaft diameter.

Example 8.2 A *Jib crane* like the one in Figure 8.17 is to be designed for a maximum load of 10 [kN] or 2.2 [klb]. The crane reach or active boom length is 3.3 [m] or 130 [in]. The height of the boom is to be about 3.3 [m] or 130 [in] above ground. It is required to investigate the possible use of *I-beam* or wide *I-beam* and round tubes like those in Figure 8.17. The *I-beam*, tubes, and plates are made of common structural steel ISO E235 or ASTM A36.

Solution

Data: $\boldsymbol{F} = 10$ [kN] or 2.2 [klb], $l_B = 3.3$ [m] or 130 [in], and $h_B = 3.3$ [m] or 130 [in]. *Ductile* structural material is ISO E235 or ASTM A36. The material properties are yield strength S_y of 36 [kpsi] or 250 [MPa], minimum ultimate tensile strength of 58 [kpsi] or 4200 [MPa], and the endurance limit of 29 [kpsi] or 200 [MPa]; see Table A.7.5 in Appendix A.7. Equation (7.58) suggests $S_e = 0.5\ S_{ut} = 0.5(400) = 200$ [MPa] or $0.5(59) = 29$ [kpsi]. The two values are the same as suggested by the quoted properties. The endurance value is useful if the load is fluctuating for many cycles.

Figure 8.18 shows free body diagrams of the Jib crane main modules with local *x*–*y* coordinates specified. Figure 8.18a is the main I-beam *boom* module. Figure 8.18b is the *hood* module. Figure 8.18c is the *mast* module. Dimensions are as specified as for the given data or as a suggested solution. Reactions are assumed positive and they are transferred as opposite direction forces to the connected modules.

Using Eqs. (2.4)–(2.6) for each of the *Jib-crane* modules in Figure 8.18 gives the following matrix–vector form to define the forces on each module. From that the maximum bending moment can be definable in magnitude and location. A possible synthesis of the main section is thus possible.

(a) Boom I-Beam Module

Assume all forces and reaction to be positive vectors (Figure 8.18a). If the applied force is negative in one direction, the magnitude in that direction is then negative. This is applicable to $\boldsymbol{F}_1$, where if one considers the real vector as negative, it may confuse the vector equations. If the bearing at $\boldsymbol{R}_1$ does not carry any bending and no external moments exist, Eqs. (2.4)–(2.6) give the following relations for moments:

$$\cancel{\sum_1^k \boldsymbol{M}_j} + \left(\sum_1^n \boldsymbol{r}_i \times \boldsymbol{F}_i\right) + \left(\boldsymbol{r}_{R2} \times \boldsymbol{R}_2\right) + \cancel{\boldsymbol{M}_1} = 0$$

$$\text{or} \quad \left(\sum_1^n [\boldsymbol{r}_i \times] \boldsymbol{F}_i\right) + [\boldsymbol{r}_{R2} \times] \boldsymbol{R}_2 = 0$$

$$\begin{bmatrix} 0 & 0 & 0 \\ 0 & 0 & -(-3.3) \\ 0 & (-3.3) & 0 \end{bmatrix} \begin{bmatrix} 0 \\ -10 \\ 0 \end{bmatrix} + \begin{bmatrix} 0 & 0 & 0 \\ 0 & 0 & -1.0 \\ 0 & 1.0 & 0 \end{bmatrix} \begin{bmatrix} R_{2x} \\ R_{2y} \\ R_{2z} \end{bmatrix} = 0 \rightarrow \begin{bmatrix} R_{2x} \\ R_{2y} \\ R_{2z} \end{bmatrix} = \begin{bmatrix} 0 \\ -33 \\ 0 \end{bmatrix} [\text{kN}]$$

$$\text{i.e. } R_{2y} = -33\ [\text{kN}] \qquad \text{(a)}$$

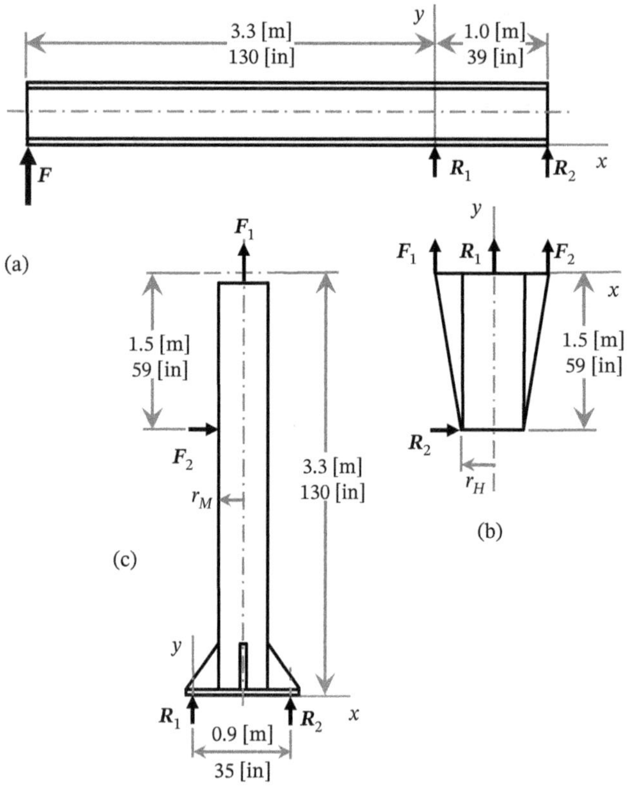

Figure 8.18 Free body diagrams of the Jib-crane main modules sketches with local x–y coordinates specified (not to scale): (a) main beam module, (b) hood module, and (c) stem module. Dimensions are as specified in Example 8.2 or as a suggested solution. Reactions are assumed positive and transferred as opposite direction forces to the connected modules.

or

$$\begin{bmatrix} 0 & 0 & 0 \\ 0 & 0 & -(-130) \\ 0 & (-130) & 0 \end{bmatrix}\begin{bmatrix} 0 \\ -2.2 \\ 0 \end{bmatrix} + \begin{bmatrix} 0 & 0 & 0 \\ 0 & 0 & -39 \\ 0 & 39 & 0 \end{bmatrix}\begin{bmatrix} R_{2x} \\ R_{2y} \\ R_{2z} \end{bmatrix} = 0 \quad \rightarrow \begin{bmatrix} R_{2x} \\ R_{2y} \\ R_{2z} \end{bmatrix} = \begin{bmatrix} 0 \\ -7.333 \\ 0 \end{bmatrix} \ [\text{klb}]$$

$$\text{i.e.} \quad R_{2y} = -7.333 \ [\text{klb}] \qquad \text{(a}')$$

For forces, one gets

$$\boldsymbol{F}_1 + \boldsymbol{R}_1 + \boldsymbol{R}_2 = 0 \quad \rightarrow \boldsymbol{R}_1 = -\boldsymbol{R}_2 - \boldsymbol{F}_1$$

$$\boldsymbol{R}_1 = -\begin{bmatrix} 0 \\ -33 \\ 0 \end{bmatrix} - \begin{bmatrix} 0 \\ -10 \\ 0 \end{bmatrix} = \begin{bmatrix} 0 \\ 43 \\ 0 \end{bmatrix} \ [\text{kN}] \quad \text{and} \quad \boldsymbol{R}_1 = -\begin{bmatrix} 0 \\ -7.33 \\ 0 \end{bmatrix} - \begin{bmatrix} 0 \\ -2.2 \\ 0 \end{bmatrix} = \begin{bmatrix} 0 \\ 9.533 \\ 0 \end{bmatrix} \ [\text{klb}]$$

$$\text{or} \ R_{1y} = 43 \ [\text{kN}] \ \text{and} \ R_{1y} = 9.533 \ [\text{klb}] \qquad \text{(a}'')$$

Note that Eqs. (a) and (a′) are equivalent to simplified moments about the origin of the local x–y coordinates. The second line of Eq. (a″) is simply $\sum F_y = 0$.

The maximum bending moment occurs at the location of $\boldsymbol{R}_1$. Assuming the model for the main beam is subject to bending moments only. The simplified mathematical model is simply a supported beam at $\boldsymbol{R}_1$ and $\boldsymbol{R}_2$. The magnitude of this bending moment is then as follows (see Section 6.1.2):

$$\boldsymbol{M}_z = \boldsymbol{F}_1 l_B = 10(10^3)(3.3) = 330\,00\ [\text{N m}] = 33\ [\text{kN m}]$$
$$\boldsymbol{M}_z = \boldsymbol{F}_1 l_B = 2.2(10^3)(130) = 286\,000\ [\text{lb in}] = 286\ [\text{k lb in}] \tag{b}$$

The needed section modulus Z_z according to Eq. (6.16) is obtainable such that

$$Z_z = \frac{\boldsymbol{M}_z}{\sigma_{x,\max}} = \frac{\boldsymbol{M}_z K_{\text{SF}}}{S_y} = \frac{33(10^3)(3.5)}{(250)(10^6)} = 0.000\,462\ [\text{m}^3] = 462\ [10^3\ \text{mm}^3]$$
$$Z_z = \frac{\boldsymbol{M}_z}{\sigma_{x,\max}} = \frac{\boldsymbol{M}_z K_{\text{SF}}}{S_y} = \frac{286(10^3)(3.5)}{(36)(10^3)} = 27.81\ [\text{in}^3] \tag{c}$$

For SI metric series *I-beam*, the section I-280 with a 280 [mm] depth and a 119 [mm] width has a section modulus $Z_z = 595\ [10^3\ \text{mm}^3]$; see Table A.8.7. For SI metric series wide flange *I-beam*, the section I-200 with a 200 [mm] depth and a 200 [mm] width has a section modulus $Z_z = 542\ [10^3\ \text{mm}^3]$; see Table A8.9.

For the US inch series *S-section I-beam*, the section S 10×35 with a 10 [in] depth and a 4.994 [in] width has a section modulus $Z_z = 29.4\ [10^3\ \text{mm}^3]$; see Table A.8.8. For the US inch series wide flange *I-beam*, the section W 16×26 with a 15.69 [in] depth and a 5.5 [in] width has a section modulus $Z_z = 38.4\ [\text{in}^3]$; see Table A.8.10. Different other sizes might be useful to reduce the deflection of the boom under load and ease the load travel along the boom. This depends on the hoist design, operational requirements, and other numerous available *I-beams*; see AISC (1980) and Oberg et al. (2012). Many other *I-beams* are available to have smaller deflections and less material. This is an optimization problem to be envisioned.

The other elements involved are the fasteners that are covered in Chapter 9 in the text. Initially, however, the two fasteners (one on each *I-beam* side) at the $\boldsymbol{R}_1$ reaction are not carrying a tensile load. They are working as a positioning and tightening. The two fasteners (one on each side) at $\boldsymbol{R}_2$ reaction are carrying a tensile load of 33 [kN] or 7.333 [klb] as shown in Eq. (a). Each fastener would carry 16.5 [kN]. The construction of Figure 8.17b indicates that the fasteners can carry the load in direct shear. A quick estimate with a safety factor of about 2.0 gives an *estimated bolt diameter* of about 16 [mm] (M16) diameter or about 3/4 [in]; see Section 8.4, Example 8.3, and Figure 8.19.

(b) Hood Module

The reactions in the boom module are transferred to the hood module such that $\boldsymbol{F}_1 = -\boldsymbol{R}_1$ and $\boldsymbol{F}_2 = -\boldsymbol{R}_2$. The reactions in the hood module are assumed positive (Figure 8.18b). If the bearing at $\boldsymbol{R}_1$ does not carry any bending moment, Eqs. (2.4)–(2.6) give the following for the moments:

$$\left(\sum_1^2 [\boldsymbol{r}_i \times] \boldsymbol{F}_i\right) + [\boldsymbol{r}_{R_2} \times] \boldsymbol{R}_2 = 0$$

$$\begin{bmatrix} 0 & 0 & 0 \\ 0 & 0 & -(-0.5) \\ 0 & -0.5 & 0 \end{bmatrix}\begin{bmatrix} 0 \\ -43 \\ 0 \end{bmatrix} + \begin{bmatrix} 0 & 0 & 0 \\ 0 & 0 & -0.5 \\ 0 & 0.5 & 0 \end{bmatrix}\begin{bmatrix} 0 \\ 33 \\ 0 \end{bmatrix} + \begin{bmatrix} 0 & 0 & -1.5 \\ 0 & 0 & 0 \\ 1.5 & 0 & 0 \end{bmatrix}\begin{bmatrix} R_{2x} \\ R_{2y} \\ R_{2z} \end{bmatrix} = 0$$

$$\text{Or}\quad \begin{bmatrix} R_{2x} \\ R_{2y} \\ R_{2z} \end{bmatrix} = -\begin{bmatrix} (21.5 + 16.5)/1.5 \\ 0 \\ 0 \end{bmatrix} = \begin{bmatrix} -25.33 \\ 0 \\ 0 \end{bmatrix}\ [\text{kN}] \to \text{i.e.}\quad R_{2x} = -25.33\ [\text{kN}] \tag{d}$$

or

$$\begin{bmatrix} 0 & 0 & 0 \\ 0 & 0 & -(-19.5) \\ 0 & -19.5 & 0 \end{bmatrix}\begin{bmatrix} 0 \\ -9.533 \\ 0 \end{bmatrix} + \begin{bmatrix} 0 & 0 & 0 \\ 0 & 0 & -19.5 \\ 0 & 19.5 & 0 \end{bmatrix}\begin{bmatrix} 0 \\ 7.333 \\ 0 \end{bmatrix} + \begin{bmatrix} 0 & 0 & -59 \\ 0 & 0 & 0 \\ 59 & 0 & 0 \end{bmatrix}\begin{bmatrix} R_{2x} \\ R_{2y} \\ R_{2z} \end{bmatrix} = 0$$

$$\text{Or}\quad \begin{bmatrix} R_{2x} \\ R_{2y} \\ R_{2z} \end{bmatrix} = -\begin{bmatrix} (19.5)(9.533 + 7.333)/59 \\ 0 \\ 0 \end{bmatrix} = \begin{bmatrix} -5.574 \\ 0 \\ 0 \end{bmatrix}\ [\text{klb}] \to \text{i.e.}\quad R_{2x} = -5.574\ [\text{klb}] \tag{d'}$$

For forces, one gets

$$\boldsymbol{F}_1 + \boldsymbol{F}_2 + \boldsymbol{R}_1 + \boldsymbol{R}_2 = 0 \quad \rightarrow \boldsymbol{R}_1 = -\boldsymbol{R}_2 - \boldsymbol{F}_1 - \boldsymbol{F}_2$$

$$\boldsymbol{R}_1 = -\begin{bmatrix} -25.33 \\ 0 \\ 0 \end{bmatrix} - \begin{bmatrix} 0 \\ -43 \\ 0 \end{bmatrix} - \begin{bmatrix} 0 \\ 33 \\ 0 \end{bmatrix} = \begin{bmatrix} 25.33 \\ 10 \\ 0 \end{bmatrix} \text{ [kN]}$$

$$\boldsymbol{R}_1 = -\begin{bmatrix} -5.574 \\ 0 \\ 0 \end{bmatrix} - \begin{bmatrix} 0 \\ -9.533 \\ 0 \end{bmatrix} - \begin{bmatrix} 0 \\ 7.333 \\ 0 \end{bmatrix} = \begin{bmatrix} 5.574 \\ 2.2 \\ 0 \end{bmatrix} \text{ [klb]} \qquad \text{(d'')}$$

The maximum bending moment occurs at the location of $\boldsymbol{R}_1$. Assuming the model for the main hood is subjected to bending moments only. The simplified mathematical model is simply a supported beam at $\boldsymbol{R}_1$ and $\boldsymbol{R}_2$ and to ignore the strength added webs. The magnitude of this bending moment is then as follows; see Section 6.1.2.

$$\boldsymbol{M}_z = \boldsymbol{R}_2\, l_H = 25.33(10^3)(1.5) = 37.995 \text{ [N m]} = 37.995 \text{ [kN m]}$$

$$\boldsymbol{M}_z = \boldsymbol{R}_2\, l_B = 5.574(10^3)(59) = 328\,866 \text{ [lb in]} = 328.87 \text{ [klb in]} \qquad \text{(e)}$$

The needed section modulus Z_z according to Eq. (6.16) is obtainable such that

$$Z_z = \frac{\boldsymbol{M}_z}{\sigma_{x,\max}} = \frac{\boldsymbol{M}_z K_{SF}}{S_y} = \frac{37.995(10^3)(3.5)}{(250)(10^6)} = 0.00053293 \text{ [m}^3\text{]} = 532.93 \text{ [}10^3\text{mm}^3\text{]}$$

$$Z_z = \frac{\boldsymbol{M}_z}{\sigma_{x,\max}} = \frac{\boldsymbol{M}_z K_{SF}}{S_y} = \frac{328.87(10^3)(3.5)}{(36)(10^3)} = 31.973 \text{ [in}^3\text{]} \qquad \text{(f)}$$

For SI metric series round pipes or tubes, the section DN-300 with a 323.8 [mm] OD and a 9.53 [mm] thickness has a section modulus $Z_z = 718.15$ [10^3 mm^3]; see Table A.8.1. For the US inch series round pipes or tubes,

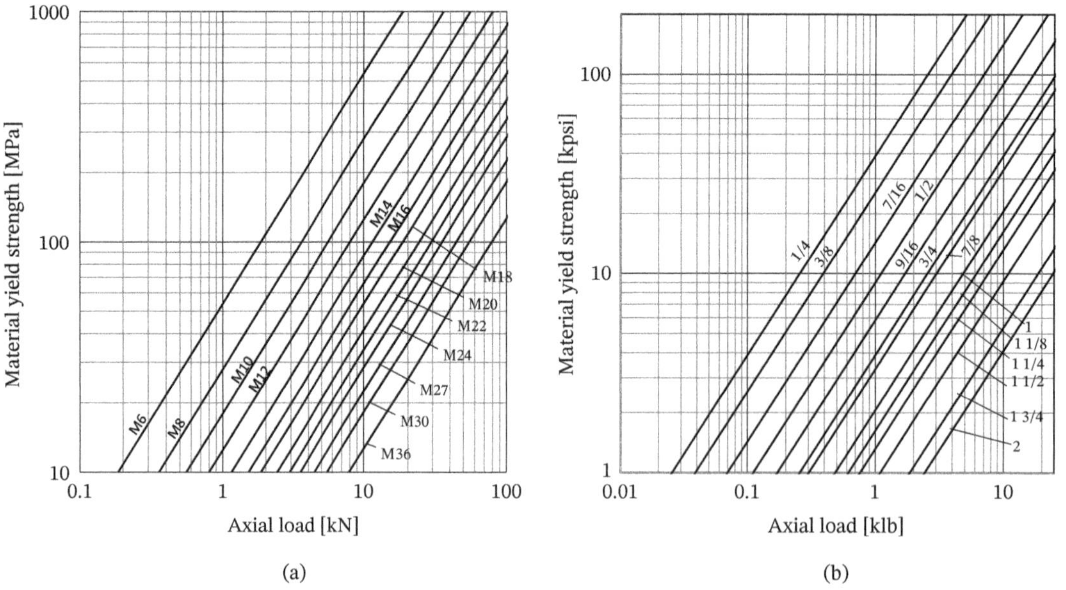

Figure 8.19 Preliminary bolt size estimation as a function of the tensile load and the yield strength of bolt material: (a) SI system of units and (b) the US system of units. The axial load is the load close to causing material yielding. The applied load should be multiplied by a proper safety factor before using the charts.

the section NSP-12 with a 12.75 [in] OD and a 0.375 [in] thickness has a section modulus $Z_z = 43.817$ [in^3]; see Table A.8.2. A larger size might be useful to accommodate the bearing at the contact with the crane mast module. This depends on the needed size of the mast tube.

The other elements involved are the fasteners at the $\boldsymbol{F}_1$ and $\boldsymbol{F}_2$ force, the stiffening webs, and the bearings. The fasteners have the dedicated Chapter 9 of the text. Initially, however, the two fasteners (one on each *I-beam* side) at the $\boldsymbol{F}_1$ load are not carrying a tensile load. They are working as positioning and tightening components. The two fasteners (one on each side) at $\boldsymbol{F}_2$ load are carrying a tensile load of 33 [kN] or 7.333 [klb] as shown in Eq. (a). Again, each fastener would carry about 16.5 [kN] or 3.67 [klb]. The construction of Figure 8.17b indicates that these fasteners are carrying the load in direct shear. A quick approximation with a safety factor of about 2.0 gave an *estimated bolt diameter* of about 16 [mm] (M16) diameter or about 3/4 [in]; see Section 8.4, Example 8.3, and Figure 8.19.

The bearings at the pivot shaft and the bearing ring at the lower section of the hood are best designed via the dedicated Chapters 11 and 12 of the text. The type of bearing necessitates a geometry commensurate with the selected bearing. The space between the hood and the mast provides the constraints in that regard. Viewing other existing products would help in reducing the possible alternatives. This will be addressable in Chapters 11 and 12 with an initial attempt in Example 8.5.

(c) Mast Module

The reactions in the hood module are transferred to the mast module such that $\boldsymbol{F}_1 = -\boldsymbol{R}_1$ and $\boldsymbol{F}_2 = -\boldsymbol{R}_2$. The reactions in the mast module are assumed positive (Figure 8.18c). If the load at $\boldsymbol{F}_1$ does not have any bending moment, Eqs. (2.4)–(2.6) give the following for moments:

$$\left(\sum_1^n [\boldsymbol{r}_i \times]\boldsymbol{F}_i\right) + [\boldsymbol{r}_{R_2}\times]\boldsymbol{R}_2 = 0$$

$$\begin{bmatrix} 0 & 0 & 3.3 \\ 0 & 0 & -(0.45) \\ -3.3 & 0.45 & 0 \end{bmatrix}\begin{bmatrix} -25.33 \\ -10 \\ 0 \end{bmatrix} + \begin{bmatrix} 0 & 0 & 1.8 \\ 0 & 0 & -(0.45 - r_S) \\ -1.8 & 0.45 - r_S & 0 \end{bmatrix}\begin{bmatrix} 25.33 \\ 0 \\ 0 \end{bmatrix} + \begin{bmatrix} 0 & 0 & 0 \\ 0 & 0 & -0.9 \\ 0 & 0.9 & 0 \end{bmatrix}\begin{bmatrix} R_{2x} \\ R_{2y} \\ R_{2z} \end{bmatrix} = 0$$

$$\begin{bmatrix} R_{2x} \\ R_{2y} \\ R_{2z} \end{bmatrix} = -\begin{bmatrix} 0 \\ 3.3(25.33) - 0.45(10) - 1.8(25.33) \\ 0 \end{bmatrix} /0.9 = \begin{bmatrix} 0 \\ -37.22 \\ 0 \end{bmatrix} \text{ [kN]} \tag{g}$$

or

$$\begin{bmatrix} 0 & 0 & 130 \\ 0 & 0 & -(17.5) \\ -130 & 17.5 & 0 \end{bmatrix}\begin{bmatrix} -5.574 \\ -2.2 \\ 0 \end{bmatrix} + \begin{bmatrix} 0 & 0 & 71 \\ 0 & 0 & -(17.5 - r_S) \\ -71 & 17.5 - r_S & 0 \end{bmatrix}\begin{bmatrix} 5.574 \\ 0 \\ 0 \end{bmatrix} + \begin{bmatrix} 0 & 0 & 0 \\ 0 & 0 & -35 \\ 0 & 35 & 0 \end{bmatrix}\begin{bmatrix} R_{2x} \\ R_{2y} \\ R_{2z} \end{bmatrix} = 0$$

$$\begin{bmatrix} R_{2x} \\ R_{2y} \\ R_{2z} \end{bmatrix} = -\begin{bmatrix} 0 \\ 130(5.574) - 17.5(2.2) - 71(5.574) \\ 0 \end{bmatrix} /35 = \begin{bmatrix} 0 \\ -8.296 \\ 0 \end{bmatrix} \text{ [klb]} \tag{g'}$$

For forces, one gets

$$\boldsymbol{F}_1 + \boldsymbol{F}_2 + \boldsymbol{R}_1 + \boldsymbol{R}_2 = 0 \quad \rightarrow \boldsymbol{R}_1 = -\boldsymbol{R}_2 - \boldsymbol{F}_1 - \boldsymbol{F}_2$$

$$\boldsymbol{R}_1 = -\begin{bmatrix} 0 \\ -37.22 \\ 0 \end{bmatrix} - \begin{bmatrix} -25.33 \\ -10 \\ 0 \end{bmatrix} - \begin{bmatrix} 25.33 \\ 0 \\ 0 \end{bmatrix} = \begin{bmatrix} 0 \\ 47.22 \\ 0 \end{bmatrix} \text{ [kN]}$$

$$\boldsymbol{R}_1 = -\begin{bmatrix} 0 \\ -8.296 \\ 0 \end{bmatrix} - \begin{bmatrix} -5.574 \\ -2.2 \\ 0 \end{bmatrix} - \begin{bmatrix} 5.574 \\ 0 \\ 0 \end{bmatrix} = \begin{bmatrix} 0 \\ 10.496 \\ 0 \end{bmatrix} \text{ [klb]} \tag{g''}$$

If the reaction at $\boldsymbol{R}_1$ is shifted to the mast centerline in addition to a bending moment (with no $\boldsymbol{R}_2$), Eqs. (2.4)(2.6) give the following for moments:

$$\sum_1^k \boldsymbol{M}_j + \left(\sum_1^n \boldsymbol{r}_i \times \boldsymbol{F}_i\right) + \left(\boldsymbol{r}_{R2} \times \boldsymbol{R}_2\right) + \boldsymbol{M}_1 = 0$$

$$\text{or} \quad \left(\sum_1^n [\boldsymbol{r}_i \times] \boldsymbol{F}_i\right) + \boldsymbol{M}_1 = 0$$

$$\begin{bmatrix} 0 & 0 & 3.3 \\ 0 & 0 & -0.0 \\ -3.3 & 0.0 & 0 \end{bmatrix} \begin{bmatrix} -25.33 \\ -10 \\ 0 \end{bmatrix} + \begin{bmatrix} 0 & 0 & 1.8 \\ 0 & 0 & -(-r_S) \\ -1.8 & -r_S & 0 \end{bmatrix} \begin{bmatrix} 25.33 \\ 0 \\ 0 \end{bmatrix} + \boldsymbol{M}_1 = 0$$

$$\boldsymbol{M}_1 = -\begin{bmatrix} 0 \\ 0 \\ 3.3(25.33) - 1.8(25.33) \end{bmatrix} = \begin{bmatrix} 0 \\ 0 \\ -37.995 \end{bmatrix} [\text{kN m}] \tag{h}$$

or

$$\begin{bmatrix} 0 & 0 & 130 \\ 0 & 0 & -0.0 \\ -130 & 0.0 & 0 \end{bmatrix} \begin{bmatrix} -5.574 \\ -10 \\ 0 \end{bmatrix} + \begin{bmatrix} 0 & 0 & 71 \\ 0 & 0 & -(-r_S) \\ -71 & -r_S & 0 \end{bmatrix} \begin{bmatrix} 5.574 \\ 0 \\ 0 \end{bmatrix} + \boldsymbol{M}_1 = 0$$

$$\boldsymbol{M}_1 = -\begin{bmatrix} 0 \\ 0 \\ 130(5.574) - 71(5.574) \end{bmatrix} = \begin{bmatrix} 0 \\ 0 \\ -328.87 \end{bmatrix} [\text{klb in}] \tag{h'}$$

To check the reaction $\boldsymbol{R}_1$, Eq. (2.4) becomes

$$\boldsymbol{F}_1 + \boldsymbol{F}_2 + \boldsymbol{R}_1 = 0 \quad \rightarrow \boldsymbol{R}_1 = -\boldsymbol{F}_1 - \boldsymbol{F}_2$$

$$\boldsymbol{R}_1 = -\begin{bmatrix} -25.33 \\ -10 \\ 0 \end{bmatrix} - \begin{bmatrix} 25.33 \\ 0 \\ 0 \end{bmatrix} = \begin{bmatrix} 0 \\ 10 \\ 0 \end{bmatrix} [\text{kN}]$$

$$\boldsymbol{R}_1 = -\begin{bmatrix} -5.574 \\ -2.2 \\ 0 \end{bmatrix} - \begin{bmatrix} 5.574 \\ 0 \\ 0 \end{bmatrix} = \begin{bmatrix} 0 \\ 2.2 \\ 0 \end{bmatrix} [\text{klb}] \tag{i}$$

To further check and considering the "fixed" character of the mast base, the maximum moment right before or at the base of the mast is simply

$$\boldsymbol{M}_z = \boldsymbol{F}_{1x} l_M + \boldsymbol{F}_{2x}(l_M - l_H) = -25.33(10^3)(3.3) + 25.33(10^3)(3.3 - 1.5) = -37\,995\ [\text{N m}] = -37.995\ [\text{kN m}]$$

$$\boldsymbol{M}_z = \boldsymbol{F}_{1x} l_M + \boldsymbol{F}_{2x}(l_M - l_H) = -5.574(10^3)(130) + 5.574(10^3)(130 - 59) = 328\,866\ [\text{lb in}] = -328.87\ [\text{klb in}] \tag{j}$$

where l_M is the mast length and l_H is the hood length. These values in Eqs. (j) are the same as the values in Eqs. (h) and (e). These results should have been expected. Why? Please reason.

The needed section modulus Z_z according to Eq. (6.16) is obtainable such that

$$Z_z = \frac{\boldsymbol{M}_z}{\sigma_{x,\max}} = \frac{\boldsymbol{M}_z K_{SF}}{S_y} = \frac{37.995(10^3)(3.5)}{(250)(10^6)} = 0.000\,532\,93\ [\text{m}^3] = 532.93\ [10^3\ \text{mm}^3]$$

$$Z_z = \frac{\boldsymbol{M}_z}{\sigma_{x,\max}} = \frac{\boldsymbol{M}_z K_{SF}}{S_y} = \frac{328.87(10^3)(3.5)}{(36)(10^3)} = 31.973\ [\text{in}^3] \tag{k}$$

These values are the same as the values in Eq. (f).

As for the hood and for SI metric series round pipes or tubes, the section DN-300 with a 323.8 [mm] OD and a 9.53 [mm] thickness has a section modulus $Z_z = 718.15\ [10^3\ \text{mm}^3]$; see Table A.8.1. For the US inch series round pipes or tubes, the section NSP-12 with a 12.75 [in] OD and a 0.375 [in] thickness has a section modulus $Z_z = 43.817$ [in^3]; see Table A.8.2. A larger size should then be used for the hood to accommodate the bearing at the contact with the mast module.

The suitable size of the hood tube can then be DN-350 with a 355.6 [mm] OD and a 9.53 [mm] thickness and has a section modulus $Z_z = 873.06$ [10^3 mm^3]; see Table A.8.1. For the US inch series round pipes or tubes, the section NSP-14 with a 14 [in] OD and a 0.375 [in] thickness has a section modulus $Z_z = 53.251$ [in^3]; see Table A.8.2. This gives an annular space of $(355.6-(2(9.53)-323.8)/2 = 12.75/2 = 6.375$ [mm] or $(14-(2(0.375)-12.75)/2 = 0.98/2 = 0.49$ [in] between the hood and the mast tubes. These spaces might be tight to have an annular *ring bearing* welded to each pipe. Rings of 5 [mm] or 0.3 [in] thickness can be welded to the tubes (pipes) and then finely finished to the required inner diameters to produce a close running fit such as $H8/f7$. The tolerance limits can be calculated for the *hood hole* and the *mast shaft* diameters; see Section 2.4.7. The detailed bearing-ring design is the subject of Chapter 12 of the text. A larger pipe can be nominated for the hood module to have more annular space if other bearings are necessary.

The stiffening plates at the end of the I-beam, the hood top and webs, and the mast base and webs (Figures 8.17 and 8.18) are of preferred thicknesses close to the thickness of the mating welded element; see Section 8.2 and Table A.3. This gives plates of thicknesses 10 [mm] or 3/8 [in] (or 0.4 [in]) for most cases.

8.3 Reverse Engineering

Element and product redesign or development should benefit from reverse engineering. Reverse engineering is not about copying products. In that, the experience of other similar products presents a wealth of examples to learn from. If the design is conventional such as a shaft of an electric motor, the diameter should be within the size set of other motors delivering the same maximum power at the same angular velocity. One can, therefore, gather a large sample of information about that from many sources including manufacturers. Due to open markets and competition, one finds that most samples are almost very similar in dimension and type of material used. This is an exercise the designers and manufacturers do, so at least would not be an oddball within the market and thus lose in the market share. The usual requirement of the motor shaft is to transmit the power at some rotational speed. Usually a bending moment is not recommended in such cases. This should thus be the same of all similar products. Most designers would get about the same answer for this requirement. This defines the knowledge base for this and other products and machine elements. The standards and codes formalize and unify such a knowledge. Another testament is that shaft *key* dimensions along other basic elements are standardized.

Reverse engineering of systems or products depends on knowledge of function, form, and properties of similar products. This is also intended for gaining knowledge of and learn from alternatives. If assembly is ambiguous, disassembly, measuring, and testing might be necessary for the learning process. This needs acquiring of other similar products to be at hand. One might need ***3D*** scanning of free form parts generating cloud point data to develop *CAD* modeling features and reconstruction using *CAD* software (e.g. geometric modeling) in addition to material identification; see Section 2.3. This learning process is being wider in scope by studying as many samples as possible. The knowledge base is thus extensive in scale, and the knowledge of alternatives is more reliable to warrant the selection of the best details to adopt for the targeted design, product, or system assembly.

The *Jib crane* of Figure 8.17 is just one of the product alternatives. It represents a conceptual form that needs more knowledge to acquire and develop. The Internet provides some information about so many Jib-crane alternatives; then one should search the net for *Jib-crane images* or *design drawings*. Even the possible connections and forms of the Jib-crane modules are available. It is necessary, however, to scrutinize and study the available figures and details to let these divulge the knowledge. The *hood* module can have a variety of forms different from a tube or a pipe in Figure 8.17b. The bearing ring can be a wheel or more. The pivot shaft can be attached differently in the *boom* and the *mast* instead of the hood. Different systems have diverse advantages and disadvantages. More reverse engineering analysis is advisable to home into the configuration that suits the objectives, possible manufacturing capabilities, and cost.

8.4 Sample Applications

In Section 8.2, some synthesis applications are available. Few component details are synthesized particularly shaft joining elements and *Jib-crane* construction. To synthesize the rest of components, charts, simple codes, and *CAD* tools are necessary. The rest of the text attempts to address many of these components. However, some initial synthesis of some basic components should be useful to set a general geometry and interaction of sizes at the onset of the design. Of these, initial synthesis of *bolt size* and number is valuable. Initial *shaft* synthesis is also essential. Further detailed treatments of these elements are available in Chapters 9 and 17. The shaft detailed synthesis is set at that late chapter, since most of the mounted elements on the shaft have forces affecting the shaft. The proper detailed evaluation of these forces is needed before the appropriate or optimum shaft synthesis is to materialize. These mounted elements are some bearings, gears, belts, chains, springs, etc.

8.4.1 Initial Bolt Synthesis

The means for the *initial synthesis* of ***bolts*** is available in Figure 8.19. Preliminary bolt size estimation as a function of the tensile load and the yield strength of bolt material is obtainable from Figure 8.19a for the SI system of units and from Figure 8.19b for the US system of units. The specified axial load is the load close to causing material yielding. The applied load should then be multiplied by a proper safety factor before using the charts. The information in Figure 8.19 is a ballpark legacy information that can be useful for initial estimate. The detailed calculations are available and dedicated in Chapter 9.

Example 8.3 The *Jib crane* of Example 8.2 has proposed bolts to connect the boom I-beam with the hood; see Figures 8.17 and 8.18. It is required to initially synthesize the bolts sizes for that connection and for tying the Jib-crane mast to the ground. The assumed bolt material for this application is like hot-rolled AISI 1030, or ISO C30. Assume a safety factor of 2 for now. (Such applications, however, would use more than 2 for safety factor; see Chapter 16.)

Solution

Data: The maximum tensile load between the boom I-beam with the hood is $\boldsymbol{F}_2 = 33$ [kN] or 7.333 [klb]. The maximum tensile load between the mast and the ground is $\boldsymbol{R}_2 = 37.22$ [kN] or 8.296 [klb]. The safety factor $K_{SF} = 2$. The assumed material of AISI 1030 or ISO C30 has the only needed property of yield strength S_y of 260 [MPa] or 37.5 [kpsi]; see Table A.7.2.

If only 2 bolts are carrying the load for both connections and the factor of safety is 2, one can just use the stated maximum loads to initially synthesize each one of the bolts. Figure 8.19 is then used to find the suitable bolts at the stated maximum loads. Figure 8.19a gives an M16 [mm] at about $\boldsymbol{R}_2 = 33$ [kN] on the abscissa and 260 [MPa] on the ordinate. Figure 8.19b gives a ¾ [in] at about $\boldsymbol{R}_2 = 7.3$ [klb] on the abscissa and 38 [kpsi] on the ordinate. The selected sizes are the ones higher than where the intersections occur.

The bolts sizes for tying the Jib-crane mast to the ground are to stand the maximum tensile load of $\boldsymbol{R}_2 = 37.22$ [kN] or 8.296 [klb]. Figure 8.19a gives an M16 [mm] at about $\boldsymbol{R}_2 = 37$ [kN] on the abscissa and 260 [MPa] on the ordinate. Figure 8.19b gives a ¾ [in] at about $\boldsymbol{R}_2 = 8.3$ [klb] on the abscissa and about 38 [kpsi] on the ordinate. The selected sizes are also the ones higher than where the intersections occur. Usually, however, more than 4 bolts (i.e. 6 bolts) are tying the crane to the ground. Since the crane rotates 360°, only two bolts would carry the load.

For a usual higher safety factor in such applications, this simple procedure is still the same. The solution here is just an initial synthesis of bolts. More detailed treatment is available in Chapter 9.

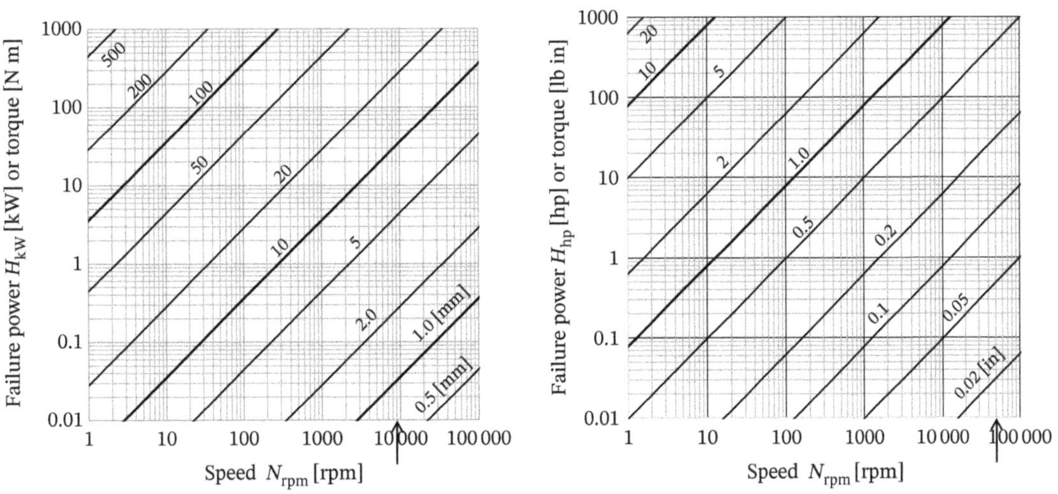

(a) (b)

Figure 8.20 Estimate of shaft size synthesis as a function of rotational speed and failure power: (a) SI system of units and (b) the US system of units. The torsional load would cause material failure for the employed material of a 300 [MPa] or 43.5 [kpsi] tensile yield strength. The arrow at about 10 000 [rpm] or about 60 000 [rpm] can be used to find the SI or the US diameter for the torques applied to stationary shafts.

8.4.2 Initial Shaft Synthesis

The implementation for the *initial synthesis* of ***shafts*** is available in Figure 8.20. Estimation of shaft size synthesis as a function of rotational speed and failure power is possible from Figure 8.20a for SI system of units and from Figure 8.20b for the US system of units. The ensuing torsional load due to power transmission would cause material failure for the employed material of a 300 [MPa] or 43.5 [kpsi] tensile yield strength. The transmitted power should be multiplied by a proper safety factor before using the charts to get the safe shaft diameter. The estimation of shaft size in Figure 8.20 is depending on *static distortion energy theory* (*von Mises*) of failure. The proper factor of safety K_{SF} can thus range from 3.5 to 5 due to prospective fatigue and its factors affecting the shaft as previously depicted in Chapter 7. The development of Figure 8.20 is established as follows.

One may consider applied torque only (for relatively short shafts) or consider bending moment only (for long shafts).

The relation between the transmitted power H, the torque $\boldsymbol{T}$, and the shaft speed ω for either system of units is obtained from (see Section 6.3.4 and Eq. (6.71))

$$H_W = \boldsymbol{T}_{N\,m}\,\omega_{rad/s} = \frac{2\pi\,N_{rpm}}{60}\,\boldsymbol{T}_{N\,m} = 0.10473\ N_{rpm}\,\boldsymbol{T}_{N\,m}$$

$$H_{hp} = \frac{\boldsymbol{T}_{lb\,in}\,\omega_{rad/s}}{6600} = \frac{2\pi\,N_{rpm}\boldsymbol{T}_{lb\,in}}{(12)(33\,000)} = \frac{N_{rpm}\boldsymbol{T}_{lb\,in}}{63\,025} = 1.5867(10^{-5})\,N_{rpm}\,\boldsymbol{T}_{lb\,in} \tag{8.12}$$

where $\omega_{rad/s}$ is the rotational speed in [rad/s], N_{rpm} is the rotational speed in [rpm], $\boldsymbol{T}_{N\,m}$ is the torque magnitude in [N m], $\boldsymbol{T}_{lb\,in}$ is the torque magnitude in [lb in], and H is the power in [W] (for H_W) or [hp] (for H_{hp}); see Section 1.11.1. From Eq. (6.71), one can get the torque magnitude and thus use it in the *design synthesis* of the machine element. After synthesis, evaluate the stresses and deflections of the element under such loading; see Chapter 17 for shaft synthesis.

$$\boldsymbol{T}_{N\,m} = 9549.3\,\frac{H_{kW}}{N_{rpm}} \approx 10\,000\,\frac{H_{kW}}{N_{rpm}}$$

$$\boldsymbol{T}_{lb\,in} = 63\,025\frac{H_{hp}}{N_{rpm}} \approx 65\,000\frac{H_{hp}}{N_{rpm}} \tag{8.13}$$

where N_{rpm} is in [rpm], $\boldsymbol{T}_{\text{N m}}$ is in [N m], $\boldsymbol{T}_{\text{lb in}}$ is in [lb in], H_{kW} is in [kW], and H_{hp} is in [hp]. The approximations in Eq. (8.13) may only be convenient in quick estimations. For other approximation of static loading, the normal and shear stresses are obtained from Eqs. (6.61) and (6.67) such that

$$\sigma_{x,\max} = \frac{32\boldsymbol{M}}{\pi d_S^3} \quad \text{and} \quad \tau_{\max} = \frac{16\,\boldsymbol{T}}{\pi d_S^3} \tag{8.14}$$

where M and T are the magnitudes of the *internal moments* (*bending* and *torsion*) obtained previously in Chapter 6 and d_S is the shaft diameter. The preliminary synthesis of shaft diameter d_S can be realized by the application of the *maximum distortion energy* (*von Mises*) theory, which gives (see Eq. 7.46)

$$\sigma_{vM} = S_{yt} = \sqrt{\sigma_{x,\max}^2 + 3\tau_{\max}^2} = \sqrt{\left(\frac{32\boldsymbol{M}}{\pi d_S^3}\right)^2 + 3\left(\frac{16\,\boldsymbol{T}}{\pi d_S^3}\right)^2}$$

$$d_S^3 = \left(\frac{32}{\pi\, S_{yt}}\right)(\boldsymbol{M}^2 + 0.75\boldsymbol{T}^2)^{1/2} \tag{8.15}$$

where the equivalent *von Mises* stress σ_{vM} is equated to the yield strength in tension S_{yt} and d_S is the shaft diameter. Assuming the bending $\boldsymbol{M}$ to be initially negligible relative to the torque $\boldsymbol{T}$. For an initial or approximate *hot-rolled* shaft material of $S_{yt} = 300$ [MPa] or 43.5 [kpsi] (AISI 1040 or ISO C45), Eqs. (8.15) and (8.13) gives the following:

$$\begin{aligned}
\boldsymbol{T}_{\text{N m}} &= \frac{d_{S[m]}^3}{\left(\dfrac{32}{\pi\, 300(10^6)}\right)(0.75)^{1/2}} = \frac{d_{S[m]}^3}{0.029\,40(10^{-6})} = 9549.3\,\frac{H_{kW}}{N_{rpm}} \\
\boldsymbol{T}_{\text{lb in.}} &= \frac{d_{S[in.]}^3}{\left(\dfrac{32}{\pi\, 43.5(10^3)}\right)(0.75)^{1/2}} = \frac{d_{S[in.]}^3}{0.202\,79(10^{-3})} = 63\,025\frac{H_{hp}}{N_{rpm}}
\end{aligned} \tag{8.16}$$

where the shaft diameters d_S have been qualified in SI unit as $d_{S[m]}$ in [m] and in the US units as $d_{S[in]}$ in [in]. From that, the value of the power as a function of the diameter is as follows:

$$\begin{aligned}
H_{kW} &= \frac{N_{rpm} d_{S[m]}^3}{0.02940(10^{-6})(9549.3)} = \frac{N_{rpm} d_{S[m]}^3}{280.75(10^{-6})} \\
H_{hp} &= \frac{N_{rpm} d_{S[in]}^3}{0.20279(10^{-3})(63\,025)} = \frac{N_{rpm} d_{S[in]}^3}{12\,781(10^{-3})}
\end{aligned} \tag{8.17}$$

These relations develop the charts in Figure 8.20 for the initial estimation of *shaft size synthesis* as a function of rotational speed and failure power. In that case, the factor of safety K_{SF} is assumed as unity. For hand calculations, the failure diameter is alternatively given by the following equations, which are extracted from Eq. (8.17):

$$\begin{aligned}
d_{S[m]} &= \left(280.749(10^{-6})\frac{H_{kW}}{N_{rpm}}\right)^{1/3} \ \text{[m]} \\
d_{S[in]} &= \left(12\,780.8(10^{-3})\frac{H_{hp}}{N_{rpm}}\right)^{1/3} \ \text{[in]}
\end{aligned} \tag{8.18}$$

Again, this is for approximate shaft material of $S_y = 300$ [MPa] or 43.5 [kpsi] (AISI 1040 or ISO C45).

To account for a different material of tensile strength $S_y \neq 300$ [MPa] or 43.5 [kpsi] and for a different safety than $K_{SF} = 1$, one should use a higher *maximum failure power* rating as follows:

$$H_{\max} = H_{kW}\frac{K_{SF}(300)}{S_{yt[MPa]}} \ \text{[kW]} \quad \text{or} \quad H_{\max} = H_{hp}\frac{K_{SF}(43.5)}{S_{yt[kpsi]}} \ \text{[hp]} \tag{8.19}$$

To account for the moment $\boldsymbol{M}$, (i.e. $\boldsymbol{M} \neq 0$), the shaft diameter obtained from Figure 8.20 should be multiplied by an additional moment factor K_M of the following form:

$$K_M = \left(\left(\frac{(\boldsymbol{M}/\boldsymbol{T})^2 + 0.75}{0.75}\right)^{1/2}\right)^{1/3} \tag{8.20}$$

The relation is apparent from Eq. (8.15). The additional moment factor K_M is pronounced when $\boldsymbol{M}/\boldsymbol{T}$ is more than 1. For $\boldsymbol{M}/\boldsymbol{T}$ of 1, the additional moment factor K_M is about 1.1517. For $\boldsymbol{M}/\boldsymbol{T}$ of 2, the additional moment factor K_M is about 1.3602. Usually, the construction of the shaft that transmits power should have as little bending moment as possible.

It can be easily shown that the *arrow* at about 10 000 [rpm] in Figure 8.20a can be used to find the estimated SI shaft diameter for the applied torque that causes the said material failure of *stationary shafts* (or *axels*). Rewriting Eq. (8.13), one gets the following:

$$\begin{aligned} \boldsymbol{T}_{\text{N.m}} &= \frac{9549.3}{N_{\text{rpm}}} H_{\text{kW}} \\ \boldsymbol{T}_{\text{lb.in}} &= \frac{63025}{N_{\text{rpm}}} H_{\text{hp}} \end{aligned} \tag{8.21}$$

It is clear from Eq. (8.21) that when N_{rpm} is equal to $9549.3 \approx 10\,000$ [rpm], the torque $\boldsymbol{T}_{\text{Nm}}$ is numerically equal to the power H_{kW}. Also, for the US system of units, Eq. (8.21) indicates that when N_{rpm} is numerically equal to $63\,025 \approx 60\,000$ [rpm], the torque $\boldsymbol{T}_{\text{lb in}}$ is numerically equal to the power H_{hp}. The arrow at about 60 000 [rpm] in Figure 18.20b can then be used to find the estimated US diameter for the applied torque that causes the said material failure of *stationary shafts* (or *axels*).

Example 8.4 A motor is running at 3000 [rpm] with a maximum power of 25 [kW] or 33.5 [hp]. The power is transmitted from the motor through a shaft. The shaft is also subjected to a bending moment. The bending moment is two times the maximum torque value. Find an initial synthesis of shaft diameter using a safety factor of 4. It is necessary to use a shaft material of AISI 1040 or ISO C40. The material must also be quenched and tempered at 205 [°C] or 400 [°F].

Solution

Data: $H = 25$ [kW] or 33.5 [hp], $N_{\text{rpm}} = 3000$ [rpm], $\boldsymbol{M}/\boldsymbol{T} = 2$, and $K_{\text{SF}} = 4$. *Heat treated* material is AISI 1040 or ISO C40 quenched and tempered at 205 [°C] or 400 [°F]. The main needed material properties for this problem is the yield strength S_y of 86 [kpsi] or 593 [MPa]; see Table A.7.3.

According to Eq. (8.19), the calculated failure power gives the following values:

$$\begin{aligned} H_{\max} &= H_{\text{kW}} \frac{K_{\text{SF}}(300)}{S_{\text{yt[MPa]}}} = 25\frac{4(300)}{593} = 50.59 \text{ [kW]} \\ H_{\max} &= H_{\text{hp}} \frac{K_{\text{SF}}(43.5)}{S_{\text{yt[kpsi]}}} = 33.5\frac{4(43.5)}{86} = 67.78 \text{ [hp]} \end{aligned} \tag{a}$$

Using Figure 8.20 at 3000 [rpm] and 50 [kW] or 68 [hp] and noting the logarithmic scales, the estimated synthesis diameters are about 15 [mm] and 0.65 [in], respectively. To check, one can use Eq. (8.18) to get

$$\begin{aligned} d_{S[\text{m}]} &= \left(280.749(10^{-6})\frac{H_{\text{kW}}}{N_{\text{rpm}}}\right)^{1/3} = \left(280.749(10^{-6})\frac{50.59}{3000}\right)^{1/3} = 0.016\,79 \text{ [m]} = 16.79 \text{ [mm]} \\ d_{S[\text{in}]} &= \left(12\,780.8(10^{-3})\frac{H_{\text{hp}}}{N_{\text{rpm}}}\right)^{1/3} = \left(12\,780.8(10^{-3})\frac{67.78}{3000}\right)^{1/3} = 0.660\,97 \text{ [in]} \end{aligned} \tag{b}$$

The calculated values are close to the values obtained from Figure 8.20. It should be noted that the obtained values from Figure 8.20 depend on the logarithmic estimation in-between two lines. This would need some expertise in estimations. That is why the values are off the calculated values by some maximum error of about 10%. As an initial preliminary estimate, this might be acceptable in design. The value of the selected shaft design is usually rounded off to a higher preferred size, which can be 16 or 20 [mm] and ¾ or 0.8 [in]; see Table A.4.

To account for the bending moment, the moment to torque ratio $\boldsymbol{M}/\boldsymbol{T}$ of 2 is used to find the diameter multiplier K_M from Eq. (8.20). This is $K_M = 1.3602$ as indicated after Eq. (8.20). The preliminary synthesis of shaft diameter should then be about $16\,(1.36) = 21.76$ [mm] or $0.75\,(1.36) = 1.02$ [in]. The preferred diameters can then be 25 [mm] or 1.0 [in].

8.4.3 Initial Bearing Synthesis

To design a system or a product, one might need a preliminary or *initial synthesis* of bearings required for that system or product. Which kind of bearing and its size is the necessary decision? After the gained experience from this text, the selection should be easier. As a beginner, however, it is useful to have an overview of different bearings and their suitability for intended loads and rotational speed. Bearing dimensions are essential outcomes to produce the system or product assembly. Figure 8.21 presents means to synthesis an estimate of *radial bearing* size as a function of bearing diameter, rotational speed, and usual load. Figure 8.21a is for SI system of units. Figure 8.21b is for the US system of units. The solid lines are for *rolling bearings* at a life of 10 000 hours [hr]; see Neale (1973). Each line is for specifically marked rotational speed. The dotted extension of some high rotational speed indicates a general unacceptable region. This is due to limitations of maximum speeds of some available commercial products.

The thicker dashed lines in Figure 8.21a,b are for *fluid-film journal bearings* at their maximum capacities and at different higher speeds. Journal bearings operate at much higher rotational speeds than rolling bearings. Their maximum load capacities are higher than rolling bearing for larger diameters and higher than 50 [mm] or 2 [in] as might be apparent from Figure 8.21. The maximum load capacity for some diameters – with the width same as the diameter – are as follows. For shaft diameter of 5 [mm] or 0.2 [in], the maximum load occurs at about 700 [rps]. For shaft diameter of 500 [mm] or 20 [in], the maximum load occurs at about 10 [rps]. For more details,

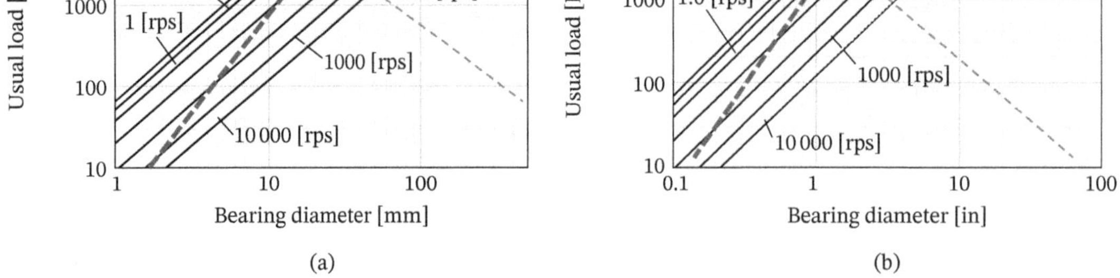

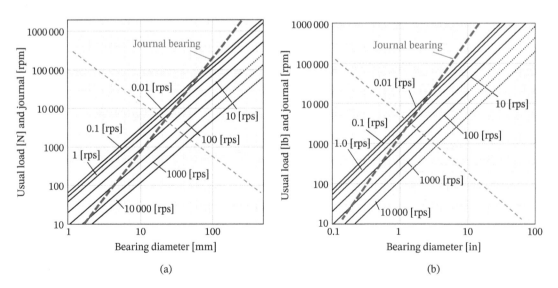

Figure 8.21 Estimate of bearing synthesis as a function of nominal bearing diameter and usual load: (a) SI units and (b) the US units. The solid lines are for rolling bearings at a life of 10 000 hours. Each line is for specifically marked rotational speed. The heavy dashed line is for fluid-film journal bearings at their maximum capacities and at certain much higher rotational speeds [rpm] (left scale) obtained from the dashed grayish line.

one can consult Chapter 12 of the text. The approximate pairs of other diameters and speeds at maximum load are as follows: (10 [mm] or 0.4 [in], 400 [rps]); (25 [mm] or 1 [in], 150 [rps]); (50 [mm] or 2 [in], 70 [rps]); (100 [mm] or 4 [in], 40 [rps]); and (250 [mm] or 10 [in], 13 [rps]); see Neale (1973).

The applied load onto a bearing is usually known. Also, the rotational speed is usually known. At the value of that load on the *abscissa*, a horizontal line can be beneficial to reach the line or the two encompassing lines of the rotational speed value. The suitable diameter is read off the horizontal *ordinate* axis at that rotational speed.

Other *plain bearings* can be like rolling bearings in the load carrying capacities. They are rubbing bearings between the relatively slow-moving parts relative to each other's. They can be direct contact of the shaft with the hole or contact through an interfacing *sleeve*, a *bush*, or a *collar*. Such bearings, however, should be useful in slower angular velocity applications than rolling bearings. Load capacities decrease with the increase of angular velocity. The values of diameters in these cases are for bearings with width or length equals to the diameter. Details and other considerations are available in Chapter 12 of the text.

Example 8.5 It is required to synthesize the bearing of the pivot shaft between the hood and the mast of the *Jib crane* in Example 8.2. Find the type and dimensions of the bearing. Assume that the maximum rotational speed of the boom is about 0.1 [rps].

Solution

Data: The maximum load between the pivot shaft of the hood and the mast is the mast force $\boldsymbol{F}_1 = (-25.33, -10, 0)$ [kN] or $(-5.574, -2.2, 0)$ [klb]. At the bearing ring of Figure 8.17b, the maximum load between the hood and the mast is the mast force $\boldsymbol{F}_2 = (25.33, 0, 0)$ [kN] or $(5.574, 0, 0)$ [klb].

The maximum load on the pivot shaft is acting as a thrust force in y direction and a radial force in the x direction. The radial force of $\boldsymbol{F}_{1x} = -25.33$ [kN] or -5.574 [klb] should be carried by the bearing between the pivot shaft of the hood and the top plate of the mast; see Figure 8.17b. The thrust force of $\boldsymbol{F}_{1y} = -10$ [kN] or -2.2 [klb] should also be carried by the bearing between the pivot shaft of the hood and the top plate of the mast; see Figure 8.17b.

The synthesis of the radial bearing is possible by utilizing Figure 8.21 to get the diameter of the pivot shaft. In Figure 8.21a and at about 25.0 [kN] on the ordinate axis, the diameter at 0.1 [rps] is about 40 [mm] read on the abscissa axis. In Figure 8.21b and at about 5.6 [klb] on the ordinate axis, the diameter at 0.1 [rps] is about 1.6 [in] read on the abscissa axis.

The synthesis of the thrust bearing may be approximately possible by utilizing Figure 8.21 to get the diameter and thus the area of the pivot shaft thrust bearing. In Figure 8.21a and at 10.0 [kN] on the ordinate axis, the diameter is about 22 [mm] read on the abscissa axis. In Figure 8.21b and at about 2.2 [klb] on the ordinate axis, the diameter is about 0.9 [in] read on the abscissa axis. The bearing projected area is the one carrying the load. With the length equal to the diameter, the projected area is $22(22) = 484$ [mm^2] or $0.9(0.9) = 0.81$ [in^2]. This area should be the annular area of the thrust bearing. As the shaft diameter is 40 [mm] or 1.6 [in], the outer diameter of the thrust bearing can be obtained from the following relation of the bearing area:

$$484 = \frac{\pi}{4}(d_{\mathrm{Bo}}^2 - d_S^2) = \frac{\pi}{4}(d_{\mathrm{Bo}}^2 - 40^2) \quad \text{or} \quad d_{\mathrm{Bo}} = \sqrt{\frac{484(4)}{\pi} + 40^2} = 47.077\ [\mathrm{mm}]$$

$$0.81 = \frac{\pi}{4}(d_{\mathrm{Bo}}^2 - d_S^2) = \frac{\pi}{4}(d_{\mathrm{Bo}}^2 - 1.6^2) \quad \text{or} \quad d_{\mathrm{Bo}} = \sqrt{\frac{0.81(4)}{\pi} + 1.6^2} = 1.895\ [\mathrm{in}] \tag{a}$$

Therefore, the collar outer diameter of the thrust bearing can be 50 [mm] or 2 [in]. This would be suitable as an initial rough estimation of the thrust bearing. Details and other considerations are available in Chapter 12 of the text.

One may, instead, use *rolling bearing* that can carry both radial and thrust loads. These details and other considerations are available in Chapter 11 of the text.

The maximum load on the bearing ring in Figure 8.17b is acting as a radial force in the x direction. The radial force of $\boldsymbol{F}_{1x} = 25.33$ [kN] or 5.574 [klb] should be carried by the bearing between the hood and the mast; see Figure 8.17b. In Figure 8.21a and at about 25.0 [kN] on the ordinate axis, the diameter is about 40 [mm] read on the abscissa axis. In Figure 8.21b and at about 5.6 [klb] on the ordinate axis, the diameter is about 1.6 [in] read on the abscissa axis. These are the same values as the bearing of the pivot shaft. Since the diameter of the bearing ring is much larger, the width or length of the bearing is the one to find. The bearing projected area is the one carrying the load. With the length equal to the diameter, the projected area is again 22(22) = 484 [mm^2] or 0.9(0.9) = 0.81 [in^2]. This area should be the annular area of the ring bearing. As the diameter of the ring bearing is about 340 [mm] or 13.4 [in], see Example 8.2, the width or length of the ring bearing l_R can be obtained from the following relation of the bearing area:

$$484 = d_R l_R \quad \text{or} \quad l_R = \frac{484}{340} = 1.42\,[\text{mm}]$$
$$0.81 = d_R l_R \quad \text{or} \quad l_R = \frac{0.81}{13.4} = 0.06\,[\text{in}] \qquad \text{(b)}$$

These values are very small, and the limits would be the possible width of the rings that are welded to the hood and the mast. Usually, the ring size should be about the thickness of the tubes for the mast or the hood. That would be about 10 [mm] or 0.4 [in].

8.5 Computer-Aided Design

The synthesis of machine elements is easier through the utility of computer codes. Many geometric modeling codes (usually called CAD packages) provide means to insert many standard or commercially available machine elements. The synthesis of a machine element, however, requires a priori identification of geometry and material for the elements. This task is achievable via the utility of previous Sections 8.2–8.4.

Many joining elements depend on the synthesis of the shaft diameter. Of these are the couplings, keys, pins, retaining rings, splines, and seals. The main dimensions of couplings are available in Section 8.2, and the need is usually for the ***rule of thumb (conventional wisdom)*** provided as the dimensional ratios of geometry relative to the shaft diameter. This provides enough information to construct the assembly of several machine elements. Inserting such elements in "CAD" packages also helps in that regard.

To support ***initial synthesis*** of shafts and joining elements, a computer-aided synthesis of connected elements is available through the code in Figure 8.22 written for MATLAB© (MathWorks 2013). One MATLAB© code includes Figure 8.22a–e. The text in the comment statement highlights the word **input** in the input data line. Figure 8.22a allows the **input** of power, rotational speed, and material properties of joined elements (initially shaft material). Transmitted torques and other parameters are the output to be ready for initial synthesis of shaft and other joining elements. Some inputs should need editing or commenting particularly for numerical or for defined variables rather than numerical values. Shaft synthesis is obtainable as in Figure 8.22b giving an initial approximate estimate. Shaft synthesis is obtained via **input** safety factor, stress concentration factor, and bending moment to torque ratio. Calculated shaft diameter is synthesized as a suggestion. **Input** a selected one of the close *preferred shaft diameters* to further calculate the joining elements.

Some joining elements synthesis is accessible via the rest of the MATLAB© codes in Figure 8.22c–e. All of these depend on the shaft diameter previously defined in Figure 8.22b. Figure 8.22c allows the synthesis of a *key* joining the shaft to a mating element's hub. One should input key width, key height, and a suggested key length as the output. Some **inputs** should need editing or commenting particularly for numerical or for defined variables rather than numerical values. When one inputs a key length, the new safety factor is the output. Only one key is considered. Other joining elements are synthesized just in case one needs to alternatively use any of these. Figure 8.22d provides a suggested *pin* diameter and a calculated pin diameter as the output. Stress concentration factor gives another calculated pin diameter. Again, some **inputs** should need editing or commenting particularly for numerical or for defined variables rather than numerical values. One should input a preferred pin diameter and the safety factor is then the output. Figure 8.22e synthesizes straight SAE splines or involute splines. One must **input** the

```
clear all; clc; format compact; format short   % Joining_Elements.m
 % Power Transmission and Applied Loads
        disp ('-Power Transmission and Applied Loads')
HkW = 25;                                      % INPUT-Transmitted Power [kW]
Hhp = 33.5;                                    % INPUT-Transmitted Power [hp]
Nrpm = 3000;                                   % INPUT-Rotational Speed [rpm]
        % Transmitted Torque
HW = HkW*1000;                                         % Transmitted Power in [W]
TNm = (60*HW)/(2*pi*Nrpm);                             % Transmitted Torque [N.m]
Tlbin = (63025*Hhp)/(Nrpm);                            % Transmitted Torque [lb.in]
 % Material Properties of Joined Elements
SutSI = 525;                                   % INPUT-Ultimate Strength [MPa]
Ultimate_Strength_MPa = SutSI
SutUS = SutSI*0.145;                                   % Calculated Ultimate Strength [kpsi]
SutUS = 76;                                    % INPUT-Ultimate Strength [kpsi]
SeSI = 0.5*SutSI;                                      % Calculated Endurance Limit [MPa]
SeSI = 260;                                    % INPUT-Endurance Limit [MPa]
Endurance_limit_MPa = SeSI                             %  Endurance Limit [MPa]<1380
SeUS = 0.145*SeSI;                                     % CalculatedEnduranceLimit[kpsi]
SeUS = 37.7;                                   % INPUT-Endurance Limit [kpsi]
Endurance_limit_kpsi = SeUS                            %  Endurance Limit [kpsi]<200
SySI = 290;                                    % INPUTYield Strength [MPa]
Yield_Strength_MPa = SySI                              %  Yield Strength [MPa]
SyUS = 42;                                     % INPUTYield Strength [kpsi]
Yield_Strength_kpsi = SyUS                             %  Yield Strength [kpsi]
```

(a)

```
 % Shaft Diameter Synthesis (initial) (inputs need change or commenting), Joining_Elements.m
        disp ('-Shaft Diameter Synthesis (initial) (inputs need change or commenting)')
KSF = 3.5;                                             % INPUT-Safety Factor
Input_Safety_Factor= KSF
KSC=3.0;                                               % INPUT-Stress Concentration Factor
Input_Stress_Concentration_Factor_KSC=KSC
MtoT= 2 ;                                              % INPUT-Moment/Torque Ratio
Input_Bending_Moment_to_Torque_Ratio= MtoT
dS=(((16*TNm)*(KSF*KSC)/(pi*0.58*SySI*10^6))^(1/3))*10^3;
KM=(((MtoT^2+0.75)/0.75)^0.5)^(1/3) ;
Calculated_Shaft_Multiplier_Factor= KM
Calculated_Shaft_Shaft_Diameter_mm= dS*KM                       % Calculated (Approx.) [mm]
dS=45;                                                 % INPUT-Preferred Shaft diameter [mm]
Input_Shaft_diameter_mm = dS
dSin=(((16*Tlbin)*(KSF*KSC)/(pi*0.58*SyUS*10^3))^(1/3));
Calculated_Shaft_Diameter_in= dSin*KM                           % Calculated (Approx.) [in]
dSin=1.75;                                             % INPUT-Preferred Shaft diameter [in]
Input_Shaft_diam_in=dSin
```

(b)

Figure 8.22 Computer-aided synthesis of connected elements. One code includes (a)–(e). (a) Input power, rotational speed, and material properties of joined elements (initially shaft material). Transmitted torque and other parameters are the output. (b) Shaft synthesis via input safety factor, stress concentration factor, and bending moment to torque ratio. Calculated shaft diameter is synthesized as a suggestion. Input a selected one of the preferred shaft diameters to further calculate the joining elements. (c) Input key width, key height, and a suggested key length is the output. Input key length and the new safety factor is the output. Only one key is considered. (d) A suggested pin diameter and a calculated pin diameter are the output. Stress concentration factor gives another calculated pin diameter. Input preferred pin diameter and the safety factor is the output. (e) Input trend line constant for other splines. Calculated SAE straight spline length is synthesized as a suggestion. Input one of the preferred shaft diameters and the calculated involute spline is one output with an additional stress concentration factor.

trend line constant for needed splines other than the 6-spline PF defined in Eq. (8.9) of Section 8.2.2.4. Calculated SAE straight spline length is synthesized as a suggestion. One must also **input** one of the preferred shaft diameters, and the calculated involute spline is one output with an additional stress concentration factor.

The MATLAB© code is available through the ***Wiley website*** under the name: ***Joining_Elements.m***. It solves Examples 8.1, 8.3, and 8.4. By appropriate editing or rewriting, any other case should be easily solved.

```
% Joining Components (Keys, for one key only), % Joining_Elements.m
        disp ('-Joining Components (Keys)')
wKmm = 14 ;                                        % INPUT-Key width in [mm]
hKmm = 9 ;                                         % INPUT-Key height in [mm]
wKin = 3/8 ;                                       % INPUT-Key width in [in]
hKin = 3/8 ;                                       % INPUT-Key height in [in]
Input_Key_width_mm=wKmm, Input_Key_height_mm=hKmm
Input_Key_width_in=wKin, Input_Key_height_in=hKin
lKmm = 2*TNm*KSF/(dS*wKmm*(0.58*SySI))*10^3;              % Key length in [mm] (Shear)
lKin = 2*Tlbin*KSF/(dSin*wKin*(0.58*SyUS)*10^3);          % Key length in [in] (shear)
Calculated_Key_Length_mm = lKmm
Calculated_Key_Length_in = lKin
lKmm = dS;                                                % Assumed Key Length [mm], dS
lKmm = dS;                                         % INPUT-Key Length [mm], dS -(Change)
lKin = dSin;                                              % Assumed Key Length [in], dSin
lKin = dSin;                                       % INPUT-Key Length [in], dSin-(Change)
Input_Key_Length_mm = lKmm
Input_Key_length_in = lKin
KSF_Key_SI = (dS*10^-3*lKmm*10^-3*hKmm*10^-3*SySI*10^6)/(4*TNm)
KSF_Key_US= (dSin*lKin*hKin*SyUS*10^3)/(4*Tlbin)
Key_Safety_Factor_counting_Stress_Concentration_SI = KSF_Key_SI/KSC
Key_Safety_Factor_counting_Stress_Concentration_US = KSF_Key_US/KSC
```

(c)

```
% Joining Components (Pins),                                        % Joining_Elements.m
        disp ('-Joining Components (Pins)')
dPmm = 0.25*dS ;                                   % Suggested Pin diameter [mm] (assumed)
dPin = 0.25*dSin;                                  % Suggested Pin diameter [in] (assumed)
Assumed_Pin_Diameter_mm = dPmm
Assumed_Pin_Diameter_in = dPin
dP_SI = (((4*TNm*KSF/2)/(pi*dS*10^-3*(0.58*SySI*10^6)))^(1/2))*10^3 ;    % Two sections loaded
dP_US = ((4*Tlbin*KSF/2)/(pi*dSin*(0.58*SyUS*10^3)))^(1/2) ;             % Two sections loaded
Calculated_Pin_Diameter_mm = dP_SI
Calculated_Pin_Diameter_in = dP_US
Pin_Diameter_counting_Stress_Concentration_SI = dP_SI*(KSC)^0.5
Pin_Diameter_counting_Stress_Concentration_US = dP_US*(KSC)^0.5
Input_Pin_Diameter_SI = dPmm                  % INPUT Preferred Pin diameter [mm] -(Change)
Input_Pin_Diameter_US = dPin                  % INPUT Preferred Pin diameter [in] -(Change)
Pin_Safety_Factor_SI = Input_Pin_Diameter_SI^2/(dP_SI*(KSC)^0.5)^2
Pin_Safety_Factor_US = Input_Pin_Diameter_US^2/(dP_US*(KSC)^0.5)^2
```

(d)

```
% Joining Component (Straight SAE Splines),          % Joining_Elements.m
        disp ('-Joining Component (Straight SAE Splines)')
Trend_Constant_US = 0.00699;                          % INPUT-6-Spiles PF (Change)
Trend_Constant_SI = 1014.2;                           % INPUT-6-Spiles PF (Change)
Trend_Line_Constant_SI = Trend_Constant_SI
Trend_Line_Constant_US = Trend_Constant_US
lS_US = Trend_Constant_US*(Tlbin/dSin^2);
lS_SI = Trend_Constant_SI*(TNm/dS^2);
Spline_Length_SAE_in = lS_US
Spline_Length_SAE_mm = lS_SI
% Joining Component (Involute Splines)
        disp ('-Joining Component (Involute Splines)')
dmSI = dS;                                            % INPUT assumption (Change)
dmSU = dSin;                                          % INPUT assumption (Change)
lS_Inv_SI=(2*TNm*KSF/(pi*(dmSI*10^-3)^2*(0.58*SySI*10^6)))*10^3;
lS_Inv_US=2*Tlbin*KSF/(pi*dmSU^2*(0.58*SyUS)*10^3);
Involute_Spline_Length_mm = lS_Inv_SI                          % Spline length [mm] (Shear)
Involute_Spline_Length_in = lS_Inv_US                          % Spline length [in] (Shear)
Involute_Spline_Length_counting_Stress_Concentration_mm = lS_Inv_SI*KSC
Involute_Spline_Length_counting_Stress_Concentration_in = lS_Inv_US*KSC
```

(e)

Figure 8.22 *(Continued)*

For *initial synthesis* or *CAD* of bearings, one can develop a code to help rather than using Figure 8.21. However, distinct assumptions characterize this figure so that a more rigorous treatment is required as defined in Chapters 11 and 12.

8.6 System Synthesis

Assembly of elements to generate products or systems depends on the way elements are synthesized. Firstly, each element location and orientation are globally set while the element synthesis is in the development stage. Secondly, the system synthesis is then automatically possible by adding these elements to that one product or system entity. As an example, this is possible if one is to design a gearbox while designing the shafts; the locations of the gears, keys, bearings, seals, fasteners, etc. are set in their locations and orientations; see Figure 8.12.

When detailed information and composition of the system are not definite, reverse engineering may be used for similar products; see the *Jib crane* in Section 8.2.4, Figure 8.17, and Example 8.2. It is difficult yet to employ *artificial intelligence (AI)* to produce the configuration; see Nassef (1990). This might entail a large knowledge-based library of elements and subsystems to fit together like a complex puzzle. In our stage of development and the level of technology used, this is beyond the scope of this text.

An attempt to synthesize a rigid coupling in terms of different concepts has generated the coupling in Figure 8.23. The concept is to reduce the rigid coupling geometry of Figure 8.1 to a smaller size with a smaller number of components. The first useful concept is to have the two coupling halves symmetric. The second useful concept is to let the torque transmit through slots and tongs like the *Oldham couplings* without an intermediate disk. The width of the slots and tongs may not be as large as in Figure 8.23b. The third useful concept is to achieve the centering through a small ring or a *bush* to assure the symmetry of the coupling halves. The tie bolts are not necessarily needed except for keeping the two coupling halves inseparable. The tie bolts can be 4 of very small bolts or maybe they can just be replaced by another fastener regime. All dimensions should go through optimization for minimum material volume and production cost; a task is left as an advanced exercise. Several manufactures have other designs of rigid couplings that are competitively optimized. A search over the Internet should provide several examples of those to compare with.

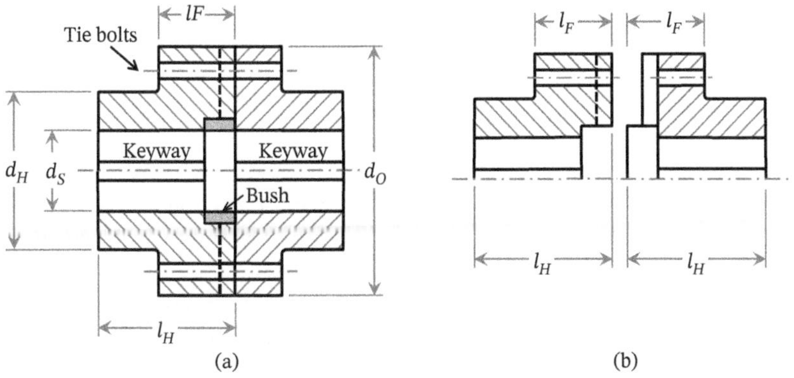

Figure 8.23 Redesign of a rigid coupling: main dimension in (a) and expanded half view in (b). The keyways are shown. The number of bolts is 4 to have symmetry. The shown holes for the bolts are through the jaws and the facing slots. They can be very small.

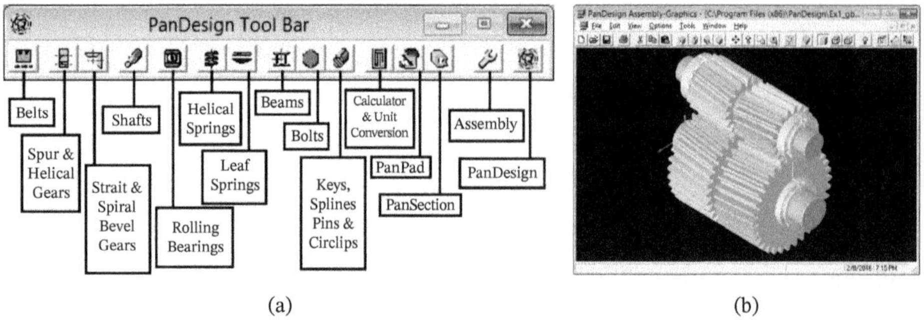

Figure 8.24 A legacy *PanDesign CAD* software interfaces: (a) components of PanDesign software for synthesis of shown machine elements and assembly of synthesized constructions. Source: Sayed Metwalli. (b) PanDesign assembly of some 15 [MW] or 20 000 [hp] gearbox components. Source: Sayed Metwalli.

8.7 Computer-Aided Assembly

As indicated in Section 8.6, the computer-aided assembly of products or systems is possible as a system synthesis. This is possible, if each element location and orientation are globally set while the element synthesis is in the development stage. Alteration of linked dimensions (parametric variables), *interference* and *collision* check, etc. is accessible in many *geometric modeling* or CAD packages; see Chapters 3 and 4. Assembly simulation, animation, and expanded views might also be possible through some *geometric modeling* or CAD packages.

PanDesign is a CAD and assembly of some machine elements. It was available for about the last three decades to undergraduate students. However, it is a legacy program under 32-bit system, which may need tweaking to run under the 64-bit Windows© system. Its construction was under the programming suite of MicrosoftVisual Studio© 6.0 (1998). It is available through the ***Wiley website*** under a possible download and the acceptance of the author to provide a license to the users of the textbook. It is mainly an SI system with few components that can solve the US system components or the ability to converge input and output. Figure 8.24 presents *PanDesign* interfaces. Figure 8.24a provides the components of *PanDesign* software for synthesis of shown machine elements and assembly of synthesized constructions. Figure 8.24b displays the *PanDesign* assembly of a 15 [MW] gearbox components.

8.8 Summary

This chapter is an introduction to common elements and system synthesis. It is concerned with reasonable load estimation, processes to define geometry and appropriate material selection, and tools for computer-aided synthesis and optimization. It intends to generate tools such as synthesis charts, codes, and programs to define geometry (basic dimensions) and initial selection of material or safety factor for assumed material.

Basic common machine elements are an initial set or building blocks for synthesis of machines. These elements are common basic and joining essentials to machine constructions. Their consideration includes the geometry and material of commercial rods, bars, plates, standard sections, and wires that are the raw stock to produce machine elements such as shafts, screws, bolts, keys, rivets, pins, pipes, tubes, springs, gears, and some machine frames or structures. Preferred sizes are emphasized for both the US and SI systems of units. Other sizes are also accessible to facilitate the production of several machine elements. Producing holes for housing of some of these elements is essential in assembly of designs. Drill dimensions are therefore in need to produce these holes. Drill dimensions are then vital and are made aware of.

Joining elements or components are available in conventional or standard form such as rigid, flexible, and universal couplings. These are close to be standardized particularly for preferred shaft diameters. Their geometries are set in formalized relations that would be useful in system or assembly of constructions or system designs. Other standardized joining components are keys, pins, cotter pins, retaining rings, and splines. These are available in specific dimensions and geometries to be applicable with little needed calculations to safeguard against failures. The *rule of thumb* or *conventional wisdom* about their dimensions usually provide some enough reliability.

Seals are needed for housing constructions where fluids are present in housings such as gearboxes. Their types and standard dimensions are made available to facilitate the proper assembly of these and similar housings. Detailed constructions utilizing such elements and standard components give some implicit sample applications.

Reverse engineering for components, assemblies, and redesigns is valuable as a synthesis tool. Elements in product redesign and development use reverse engineering as a knowledge of function, form, and properties of similar or legacy designs. One might need procedures of disassembly, measure, and test including ***3D*** scanning of parts to generate cloud point data. CAD modeling then features reconstruction using applicable software to generate ***3D*** geometric models. Possible material identification defines the suitability of part for intended loading. With this knowledge base, better designs are realizable.

Sample applications demonstrated examples of using developed charts, simple code, and *CAD* and optimization programs to find initial design synthesis of some common and basic machine elements for both the US and SI units. The extensively used elements such as *bolts*, *shafts*, and *bearings* are initially made possible to synthesize using developed charts and simple codes. This should be useful to set a general geometry and interaction of sizes at the onset of the design. Developed *CAD* codes provide tools to initially synthesize shafts and their joining elements such as keys, pins, and splines. Other *CAD* tools are possibly and similarly developable for other components in this chapter.

System synthesis and computer-aided assembly are possible through some formalized procedures or extensive use of ***AI***. Some of these are beyond the scope of this text. Assembly of elements to generate products or systems needs that each element location and orientation are globally set while the element synthesis is in the development stage. This might only be possible if each element size is predetermined a priori. The initial synthesis of each element is very useful in that regard. Some following adjustments may be necessary to reach an optimum system assembly. The concepts in this chapter can be helpful to reduce tuning iterations. In addition, some geometric modeling softwares provide linked dimensions (*parametric variables*), *interference* and *collision* check, assembly *simulation*, *animation,* and *expanded views*. Utilizing these tools helps in computer-aided assembly of systems and products.

Problems

8.1 Use the tolerances expected in the drilling production of holes in Chapter 2 to expect the clearance fit and required larger sizes for preferred diameters of drills.

8.2 Use Eq. (8.1) to estimate the coupling dimensions for shafts of diameters 1, 2, 4, 8, and 10 [in]. Compare values with those estimated from Figure 8.2 and those from the rule of thumb. Compare results with available couplings over the Internet.

8.3 Derive similar equations to those in Eq. (8.1) for the SI system.

8.4 Use the derived equations in Problem 8.3 to estimate the coupling dimensions for shafts of diameters 20, 50, 100, 200, and 250 [mm]. Compare values with those estimated from Figure 8.2 and those from the rule of thumb. Compare results with available couplings over the Internet.

8.5 Assume a double-cone coupling is to be used for Problem 8.2, what should be the overall dimensions and limitations. Employ Eq. (8.2) in the estimation process and compare with rule of thumb and available Internet options.

8.6 Derive similar equations to those in Eq. (8.2) for the SI system.

8.7 Assume a double-cone coupling is to be used for Problem 8.4, what should be the overall dimensions and limitations? Employ the derived equations in Problem 8.6 in the estimation process and compare with rule of thumb and available Internet options.

8.8 What are the expected overall dimensions of flexible coupling in Figure 8.5 compared to the rigid coupling in Figure 8.1? Compare results with available couplings over the Internet for the US and SI systems of units.

8.9 Explore the available dimensions and power transmission characteristics of universal joints.

8.10 Estimate the maximum torque and the maximum power transmitted for the shafts in Example 8.1, if the rotational speeds are 10, 100, 1000, or 10 000 [rpm]. Calculate the suitable key lengths for these shafts if usual material is used.

8.11 Estimate the maximum power transmitted for the shafts in Example 8.4, if the rotational speeds are 10, 100, 1000, or 10 000 [rpm]. Calculate the suitable key lengths for these shafts if usual material is used.

8.12 If a pin is used to join the shaft to a coupling for the cases stated in Problem 8.2, what is the conventional size and the safety factor for each of these pins. Are pins suitable for all shaft diameters?

8.13 If a pin is used to join the shaft to a coupling for the cases stated in Problem 8.4, what are the conventional size and the safety factor for each of these pins? Are pins suitable for all shaft diameters?

8.14 What are the dimensions of the retaining rings and their grooves for the shafts stated in Problem 8.2 and their mating holes? What is the maximum expected axial force that can be tolerated if the safety factor is 3.5? Use a suitable material.

8.15 What are the dimensions of the retaining rings and their grooves for the shafts stated in Problem 8.4 and their mating holes? What is the maximum expected axial force that can be tolerated if the safety factor is 3.5? Use a suitable material.

8.16 Find the constant in Eq. (8.8) for other spline cases than the 6 splines and sliding under load fit. Compare values to those in Eq. (8.9) for the US or SI systems of units.

8.17 Find the SAE straight spline lengths to be produced for the shaft diameters in Problem 8.10. Use 6, 10, and 16 SuL splines. Discuss the suitability of results.

8.18 Find the SAE straight spline lengths to be produced for the shaft diameters in Problem 8.10. Use 4, 6, 10, and 16 PF splines. Discuss the suitability of results.

8.19 Find the SAE straight spline lengths to be produced for the shaft diameters in Problem 8.11. Use 6, 10, and 16 SuL splines. Discuss the suitability of results.

8.20 Find the SAE straight spline lengths to be produced for the shaft diameters in Problem 8.11. Use 4, 6, 10, and 16 PF splines. Discuss the suitability of results.

8.21 Find the involute spline lengths to be produced for the shaft diameters in Problem 8.10. Use 6, 10, and 16 SuL splines. Discuss the suitability of results.

8.22 Find the involute spline lengths to be produced for the shaft diameters in Problem 8.10. Use 4, 6, 10, and 16 PF splines. Discuss the suitability of results.

8.23 Find the involute spline lengths to be produced for the shaft diameters in Problem 8.11. Use 6, 10, and 16 SuL splines. Discuss the suitability of results.

8.24 Find the involute spline lengths to be produced for the shaft diameters in Problem 8.11. Use 4, 6, 10, and 16 PF splines. Discuss the suitability of results.

8.25 Compare Eq. (8.10) of the involute spline length with the equations of the straight splines of Eq. (8.9).

8.26 Define the dimensions of the rubbing seals for the shafts depicted in Problem 8.2 and for the collar seals of Figure 8.15d.

8.27 Define the dimensions of the rubbing seals for the shafts depicted in Problem 8.4 and specifically for the collar seals of Figure 8.15d.

8.28 If one needs to use O-rings for shafts or mating holes of Problem 8.2, what should be the dimensions of the O-ring and its groove in the shaft or the mating hole?

8.29 If one needs to use O-rings for shafts or mating holes of Problem 8.4, what should be the dimensions of the O-ring and its groove in the shaft or the mating hole?

8.30 Redesign the Jib crane for the same dimensions as Example 8.2, but for a maximum load of 30 [kN] or 6.6 [klb].

8.31 If the crane reach or the active boom length is 5 [m] or 200 [in] for the Jib crane of Example 8.2, what should be the suitable synthesis of the crane?

8.32 Redesign the Jib crane of Problem 8.30, but for the crane reach or the active boom length of 5 [m] or 200 [in].

8.33 Explore different constructional details of Jib cranes and synthesize the attractive alternative for the load and the reach stated in Problems 8.30 and 8.32.

8.34 Derive Eq. (8.20) utilizing the treatment stated in Eq. (8.15).

8.35 Define the suitable dimensions of the ring bearing in Example 8.5.

8.36 Employ the code in the *Joining_Elements.m* to synthesize the joining components in Problems 8.10–8.24.

8.37 Synthesize the rigid coupling in Figure 8.23 by reducing the length of the hub l_H and the length of the flange l_F to a minimum space for one or two keys or a spline. The dimensions of the slots and the tongs are to be minimized. The tie bolts are to be minimized and the hub diameter d_H reduced. Devise some alternatives to replace the tie bolts by other simpler means of keeping the two coupling halves inseparable.

References

AISC (1980). *Manual of Steel Construction*, 8e. American Institute of Steel Construction.

ANSI/AISC 303 (2016). *Code of standard practice for steel buildings and bridges*. American Institute of Steel Construction

ANSI B92.1 (1996). *Involute spline and inspection*. Society of Automotive Engineers.

ANSI B92.2M (1980) (R1989). *Straight cylindrical involute splines, metric module, side fit – generalities, dimensions and inspection*. American National Standards Institute.

ANSI/ASME B18.8.2 (2000). *Taper pins, dowel pins, straight pins, grooved pins and spring pins (inch series)*. American Society of Mechanical Engineers.

ANSI/ASME B36.10M (1995). *Welded and seamless wrought steel pipe*. American Society of Mechanical Engineers

ANSI/ASME B94.11M (1993). *Twist drills*. American Society of Mechanical Engineers

ASME B18.8.1 (2014). *Clevis pins and cotter pins (inch series)*. American Society of Mechanical Engineers

ASME B18.8.200M (2000). *Cotter pins, headless clevis pins, and headed clevis pins (metric series)*. American Society of Mechanical Engineers

ASME B27.7-1977 (R2017), ANSI B 27.7 (1977). *General purpose tapered and reduced cross section retaining rings (metric)*. American Society of Mechanical Engineers

ASTM A992/A992M (2015). *Standard specification for structural steel shapes*. American Society of Mechanical Engineers

ISO 1234 (1997). *Split pins*. International Organization for Standardization.

ISO 14 (1982). *Straight-sided splines for cylindrical shafts with internal centering – dimensions, tolerances and verification*. International Organization for Standardization

ISO 2338 (1997). *Parallel pin, of unhardened steel and austenitic stainless steel*. International Organization for Standardization

ISO 2339 (1986). *Taper pin, of unhardened steel and austenitic stainless steel*. International Organization for Standardization

ISO 3601-1 (2012). *Fluid power systems – O-rings – Part 1: Inside diameters, cross-sections, tolerances and designation codes*. International Organization for Standardization

ISO 3601-2 (2014). *Fluid power systems – O-rings – Part 2: Housing dimensions for general applications*. International Organization for Standardization

ISO 4156 (2005). *Straight cylindrical involute splines – Metric module, side fit*. International Organization for Standardization

ISO 4200 (1991). *Plain end steel tubes, welded and seamless – General tables of dimensions and masses per unit length*. International Organization for Standardization

ISO 464 (2015). *Rolling bearings – Radial bearings with locating snap ring – dimensions, geometrical product specifications (GPS) and tolerance values*.International Organization for Standardization.

ISO 6194 (2007). *Rotary shaft lip-type seals incorporating elastomeric sealing elements*. International Organization for Standardization

ISO 657-1,2,5,11,14,15 … (1989). *Hot-rolled steel sections*. International Organization for Standardization

ISO 6708 (1995). *Pipework components – Definition and selection of DN (nominal size)*. International Organization for Standardization

Mabie, H.H. and Reinholtz, C.F. (1987). *Mechanisms and Dynamics of Machinery*, 4e. John Wiley & Sons.

MathWorks (2013). *MATLAB*©. The MathWorks, Inc.

Microsoft (1998). *Visual Studio*© *6.0*. Microsoft.

MIL-DTL-27426B (1997). *Rings, Retaining, Spiral (Uniform Cross Section)*. USA Department of Defense

Nassef, A.M.O.A. (1990). *Application of artificial intelligence and knowledge base to mechanical design*, MS thesis. Cairo University.

Neale, M.J. (ed.) (1973). *Tribology Handbook*. Butterworths.

Oberg, E., Jones, F.D., Horton, H.L., and Ryffel, H.H. (2012). *Machinery's Handbook*, *29*[t]e. Industrial Press.

SAE AS4716 (R) (2017). *Gland Design, O-ring and Other Seals*. Society of Automotive Engineers

SAE AS568 (2014). *Aerospace Size Standard for O-rings*. Society of Automotive Engineers

SAE J499a (1936). *Parallel Side Splines for Soft Broached Holes in Fittings*. Society of Automotive Engineers

SAE J946 (2002). *Application Guide to Radial Lip Seals*. Society of Automotive Engineers

Internet Links

Producers and Providers

www.daemar.com DMR Seals: retaining rings.
www.alro.com Alro Steel: metals, industrial supplies, and plastics.
www.chathamsteel.com Chatham Steel: steel plates and sheets.
www.skf.com SKF: flexible and rigid couplings, rolling bearings, and many others.
www.usseal.com American Seal and Packing: mechanical seals and gasket supply.
www.partsolutions.com CADENAS PARTsolutions: 3D CAD catalogs.
www.ahrinternational.com INA seals: snap rings and rollers.
www.ametric.com American Metric Corporation: retaining rings.
www.comintec.com/en ComInTec: flexible and rigid couplings, *3D* downloads.
www.lawsonproducts.com Lawson products: retaining rings.
www.metalsdepot.com ELCO Corporation: metals depot, hollow tubes supplier.
www.metricmetal.com Parker Steel Company: metric structural sections, rods, bars sheets, plates, and metal stock.
www.pspring.com Peterson Spring: springs and retaining rings.
www.ruland.com Ruland Manufacturing: flexible and rigid couplings.
www.smalley.com Smalley steel ring company: retaining rings, wave springs, and laminar seal rings labyrinth.
www.spirol.com Spirol: Ssotted spring pins, pin, and disk springs.
www.upstatesteel.com Upstate Steel: supplier of plates, sheets, sections and round bars, wear resistant steel, and custom steel fabrication.
www.corusgroup.com Corus Construction and Industrial: structural sections.
www.fennerdrives.com Fenner: fexible and rigid couplings, belts, etc.
www.onesteel.com One Steel: hot rolled bars, sections, and structural steel products.

Standards and Codes

www.aisc.org American Institute of Steel Construction.
www.ansi.org American National Standards Institute (ANSI).
www.asme.org American Society of Mechanical Engineers.
www.astm.org/Standards ASTM International.
www.iso.org International Organization for Standardization.
www.sae.org Society of Automotive Engineers.

Part III

Detailed Design of Machine Elements

Section A

Basic Joints and Machine Elements

9

Screws, Fasteners, and Permanent Joints

This chapter intends to present screws, fasteners, and permanent joints. Standards and types of threads including their geometry, designation, and internal stresses are presented. Modeling bolted connections joining few members are suggested to allow the evaluation of preloading due to tightening and stresses generated in the bolt and members under loading. As an application of threaded members, power screws are useful in mechanical applications such as jacks, vices, and presses. Torques that can be developed by power screws and the subsequent internal stresses are developed. Permanent joints that connect components without any dismantling option such as rivets, welding, and bonding are important tools of lifelong joining. These techniques are available for some machine design applications. Computer-aided design (*CAD*) of these types of fasteners and permanent joining are made accessible to support the synthesis and optimization of some of these components and tools. The coverage is geared toward the applications in machine design to develop a good selection of the appropriate joining of machine elements and components.

Symbols and Abbreviations

Symbols	Quantity, Units (adopted)
A_b	Cross-sectional area of solid unthreaded bolt
A_{CS}	Stripping shear area of the collar
A_m	Member or element area
A_T	Thread tensile stress area
A_{TS}	Thread-stripping shear area
C_R	Carrying ratio share due to stiffness ratio
d_a	Average major and minor bolt diameter
d_C	Average major and minor collar diameters
d_m	Major or nominal thread diameter
d_p	Pitch diameter of thread
d_r	Root or minor thread diameter
d_W	Washer outer diameter
d_{Wa}	Average washer contact diameter

Machine Design with CAD and Optimization, First Edition. Sayed M. Metwalli.

Companion website: www.wiley.com/go/metwalli/machine

Symbols	Quantity, Units (adopted)
dx	Element thickness
$d\delta$	Element contraction
E_b	Modulus of elasticity of bolt material
E_m	Member or element modulus of elasticity
$\boldsymbol{F}_b$	Force in the bolt
$\boldsymbol{F}_{\text{bmax}}$	Bolt maximum force
$\boldsymbol{F}_{\text{bmin}}$	Bolt minimum force
$\boldsymbol{F}_i$	Initial tightening force in bolt
$\boldsymbol{F}_L$	Force for lowering the load
$\boldsymbol{F}_m$	Force in members of a joint
$\boldsymbol{F}_n$	Internal normal force
$\boldsymbol{F}_R$	Force for raising the load
$\boldsymbol{F}_S$	Force that causes shear stripping
$\boldsymbol{F}_s$	Internal shear force
$\boldsymbol{F}_{\text{yt}}$	Force that cause tensile yielding
k_b	Bolt stiffness
k_m	Joint members stiffness
k_{m1} to k_{mi}	Stiffnesses of members number one to i
K_p	Pitch fraction factor of material
k_R	Stiffness ratio of the members to the bolt
K_{SF}	Safety factor
K_{SFb}	Bolt safety factor
K_{SFbf}	Fatigue safety factor of bolt
K_{SFbt}	Tensile safety factors of bolts
K_T	Tightening torque coefficient
l_b	Solid bolt length
l_m	Length of joint members
l_N	Nut engaging length or nut thread height
l_T	Thread lead
l_W	Weld length
$\boldsymbol{N}_h$	Force normal to thread helix
N_T	Number of threads per inch
n_T	Multiple start or multiple thread number
p	Thread pitch
$\boldsymbol{P}_l$	Externally applied load
$\boldsymbol{P}_m$	Joint members tightening load
$\boldsymbol{P}_T$	Tensile thread load or thread load

Symbols	Quantity, Units (adopted)
$\boldsymbol{P}_x$	Load in x direction
$\boldsymbol{P}_y$	Load in y direction
$\boldsymbol{R}_2$	Reaction force
S_a	Alternating load strength
S_e	Modified endurance strength of bolt material
S_{eb}	Bolt endurance limit
S_m	Mean or steady load strength
S_p	Proof strength of bolt material
S_{ut}	Ultimate tensile strength of bolt material
S_y	Yield strength of bolt material
$\boldsymbol{T}_C$	Collar friction torque
$\boldsymbol{T}_L$	Power screw lowering torque
t_m	Member thickness
$\boldsymbol{T}_P$	Torque applied to power screw
$\boldsymbol{T}_P$	Total power screw torque
t_P	Plate thickness
$\boldsymbol{T}_R$	Power screw raising torque
$\boldsymbol{T}_T$	Tightening torque
t_W	Weld throat thickness
α	Half of members cone angle
$\Delta\boldsymbol{P}_l$	Increased load on the bolt
$\Delta\boldsymbol{P}_m$	Decreased load on the joint members
$\Delta\delta_b$	Bolt deformation under external load
$\Delta\delta_m$	Joint members deformation
δ_b	Deflection of the bolt
δ_m	Compressive deflection of joint members
η_P	Power screw efficiency
γ_T	Half the thread tooth angle
λ_T	Mean helix angle or lead angle of thread
μ_C	Collar friction coefficient
μ_f	Friction coefficient in bolt thread
μ_T	Thread friction coefficient
μ_W	Friction coefficient in washer
σ_1	Maximum principal stress
σ_{ba}	Alternating stress in bolt
σ_{bm}	Mean or steady stress in bolt
σ_n	Normal stress

Symbols	Quantity, Units (adopted)
σ_t	Tensile stress
σ_{Tx}	Normal stress due to tooth bending
σ_y	Normal compression stress due to load
τ_{xy}	Stripping shear stress
τ_{xz}	Torsional shear stress
τ_{max}	Maximum shear stress
τ_s	Shear stress
τ_T	Tightening shear stress

9.1 Standards and Types

To have thread standards, the different types of threads are to be defined. The most used types are of more interest than others. This section defines thread terminology for commonly used threads. It defines their standards and particular designation of each standard. The joining alternative details are also the objectives of this section. The US and SI systems are provided. Basic relations are also specified.

9.1.1 Thread Terminology and Designation

So many machine components are threaded externally such as bolts or internally such as nuts. Figure 9.1 shows screw fastener as a bolt and basic standard thread. Figure 9.1a displays the end of a short bolt, its terminology, and main dimensions. It defines the *major diameter* d_m, the minor or *root diameter* d_r, and the *pitch diameter* d_p. It also shows the chamfer at the bolt end, the crest, and the root of the thread. Figure 9.1b defines the basic standard profile of the thread, which is defined in terms of the *thread pitch* p. The sketch of the *external* thread (bolt) is in light gray color, and the sketch of *internal* thread (nut) is in a darker gray color. There are fillets in the roots of the external and internal threads for the sake of manufacturing and reducing stresses. This standard profile is essentially the same for the US Unified National Standard (UNS) and the International Organization for Standardization (ISO) standards. However, the US is in [in], and the ISO is in [mm]. They are not interchangeable. The detailed specifications are available in ASME/ANSI B1.1-1989 (R2008) and ISO 68-1 (1998).

The usual designations for the US and the ISO standards are limited to the *nominal* or *major diameter* d_m, and the pitch p or the *number of threads per inch* N_T. The standard *nominal diameter* is the *major diameter* d_m. Usually preferred numbers for the nominal diameters are the most available ones; see Appendix A.3. Other diameters are also available. The preferred US *numbers of threads per inch* are 80, 72, 64, 56, 48, 40, 32, 24, 20, 18, 16, 14, 13, 12, 11, 10, 9, 8, 7, 6, 5, 4.5, and 4. These are for *fine* or *coarse* threads. Other numbers of threads per inch are also available. Usually large numbers of threads per inch are for small diameters, and small numbers of threads per inch are for large diameters. The selected ISO preferred *fine pitches* in [mm] are 0.2, 0.25, 0.3, 0.35, 0.5, 0.75, 1, 1.25, 1.5, 2, 3, and 4. The selected ISO preferred *coarse pitches* in [mm] are 0.25, 0.3, 0.35, 0.4, 0.45, 0.5, 0.6, 0.7, 0.8, 1, 1.25, 1.5, 1.75, 2, 2.5, 3, 3.5, 4, 4.5, 5, 5.5, 6, and 8 (ISO 262 (1998)). Other pitches are also available. Usually small pitches are for small diameters and vice versa.

The screw *thread lead* l_T is the distance moved for one turn by the nut or bolt when one is stationary. For *single thread*, the *thread lead* l_T is equal to the pitch or $l_T = p$. For multiple start thread or *number of multiple threads* of

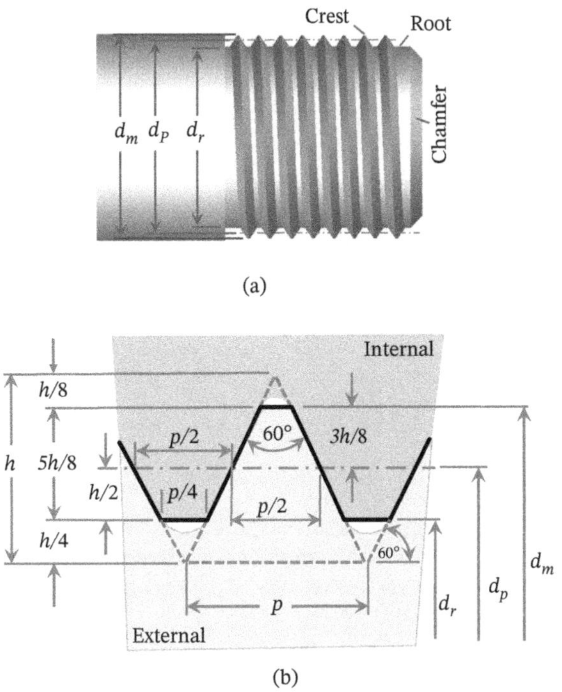

Figure 9.1 Screw fastener and thread. (a) The end of a short-threaded bolt and its terminology. (b) Standard basic profile of the thread. The sketch of the external thread (bolt) is in light gray, and the sketch of the internal thread (nut) is in a darker gray.

n_T, the lead is then

$$l_T = n_T p \tag{9.1}$$

The *double thread* has then a lead of twice the pitch. The triple thread has a lead of three times the pitch. Most screws have a single thread, i.e. a single start thread. For a single thread, the mean *helix angle,* or the *lead angle* λ_T of the thread volute, is then

$$\tan \lambda_T = \frac{l_T}{\pi d_m} \tag{9.2}$$

The lead angle is the smallest for single start threads, which ensures *self-locking*. The lead angle is then below the *friction angle* in almost all cases. Fine threads are then more self-locking than coarse threads since they have smaller pitches.

Most threads are *right hand* and thus operating according to the *right-hand rule*. This means that for a stationary internal thread (such as a nut or a piece of wood), when one turns the external screw (such as a bolt) in the *clockwise* direction (i.e. negative rotation), the bolt advances more inside the internal thread, i.e. in the negative direction following the *right-hand rule*. To untighten or unwind a right-hand bolt, one moves the bolt in the clockwise direction to retract the bolt in the positive or away from the fixed nut. The *left-hand* system acts in the opposite directions in the previous characterizations.

The ***designation*** for the US threads is given as the *diameter-number of threads per inch*. As an example of the US thread designation, 1/4–20 means that the nominal or major diameter is 1/4 [in] and the number of threads per inch is 20, i.e. the pitch is 1/20 or 0.05 [in]. The designation for the metric ISO threads is given as *M nominal or major diameter* [mm] × *pitch* [mm]. M stands for metric and the diameter is in [mm]. As an example of ISO thread designation, M10 × 1.5 means that the nominal or major diameter is 10 [mm] and the pitch is 1.5 [mm]. Table 9.1

Table 9.1 Selected screw threads.

US coarse threads	US fine threads	ISO coarse threads	ISO fine threads
(1–64)[a], (2–56)[a], (3–48)[a], (4–40)[a], (5–40)[a]	(0–80)[a], (1–72)[a], (2–64)[a], (3–56)[a], (4–48)[a], (5–44)[a]	(M1 × 0.25), (M1.2 × 0.25), (M1.6 × 0.35), (M2 × 0.4)	(M1 × 0.2), M1.2 × 0.2), (M1.6 × 0.2), (M2 × 0.25),
(6–32)[a], (8–32) [a], (10–24)[a], (12–24)[a]	(6–40)[a], (8–36) [a], (10–32)[a], (12-28)[a]	(M2.5 × 0.45), (M3 × 0.5), (M4 × 0.7), (M5 × 0.8)	(M2.5 × 0.35), (M3 × 0.35), (M4 × 0.5), (M5 × 0.5)
(¼–20), (5/16–18), (3/8–16), (7/16–14), (½–13)	(¼–28), (5/16–24), (3/8–24), (7/16–20), (½–20)	(M6 × 1), (M8 × 1.25), (M10 × 1.5), (M12 × 1.75)	(M6 × 0.75), (M8 × 1), (M10 × 1.25), (M12 × 1.5)
(9/16–12), (5/8–11), (¾–10), (7/8–9), (1–8)	(9/16–18), (5/8–18), (¾–16), (7/8–14), (1–12)	(M16 × 2), (M20 × 2.5), (M24 × 3)	(M16 × 1.5), (M20 × 2), (M24 × 2)
(1¼–7), (1½–6), (1¾–5)	(1¼–12), (1½–12)	(M30 × 3.5), (M36 × 4), (M42 × 4.5), (M48 × 5)	(M30 × 2), (M36 × 3), (M42 × 3), (M48 × 3)
(2–4½), (3–4), (4–4)		(M56 × 5.5), (M64 × 6), (M80 × 6), (M100 × 6)	(M56 × 2), (M64 × 2), (M80 × 1.5), (M100 × 2)

The designation for the US threads is (diameter-number of threads per inch) (ASME B18.2.1 (2010)). The designation for metric ISO threads is (M nominal or major diameter [mm] × pitch [mm]) (ISO 262 (1998)).
a) The nominal diameter for 0 to 12 is 0.06, 0.073, 0.086, 0.099, 0.112, 0.125, 0.138, 0.164, 0.19, 0.216 [in], respectively.

presents some selected screw threads with their designations (ASME B18.2.1 (2010) and ISO 262 (1998)). Other screw threads are also available in the standards or handbooks such as Oberg et al. (2012).

The fastener or *bolt lengths* and the *thread lengths* are also selected from standard preferred lengths. The available options are wide ranging, and one can refer to the standards or handbooks, or the manufacturers, for accessibility. The nominal hexagonal *nut heights* are approximately around *0.8* d_m, i.e. 0.8 of the nominal bolt diameters. Narrow or *jam nut* (*tightening nut*) height is roughly about 0.5 d_m. Thick or slotted nuts have a height of roughly around d_m or about the nominal bolt diameter. The specific nut height values are defined in the standards.

Other types of threads are available such as the *square* thread and *Acme* or trapezoidal thread, Figure 9.2. These types are typically used for power screws as covered in Section 9.5. The standard pitches and other dimensions are similar to preferred numbers and are accessible in standards or handbooks. More information is provided in Section 9.5.

9.1.2 Joining Alternative Details

In the previous section, bolts and nuts have been the main concern of threaded joining of components and connections. Other *detachable joining* alternatives are *cap screws*, *setscrews*, *locking devices*, pins, keys, and spring

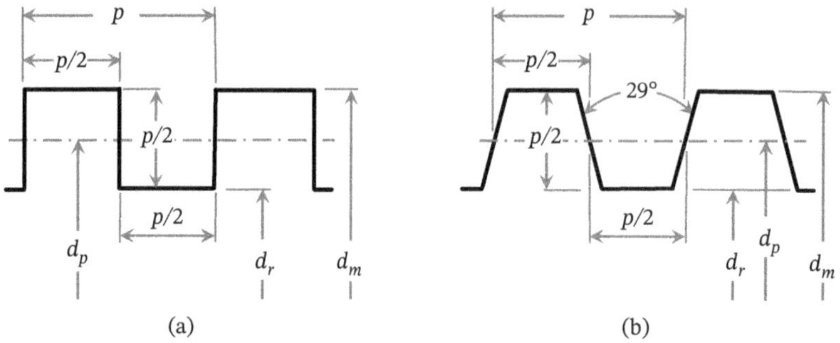

Figure 9.2 The square threads are shown in (a), and the Acme or trapezoidal threads is shown in (b).

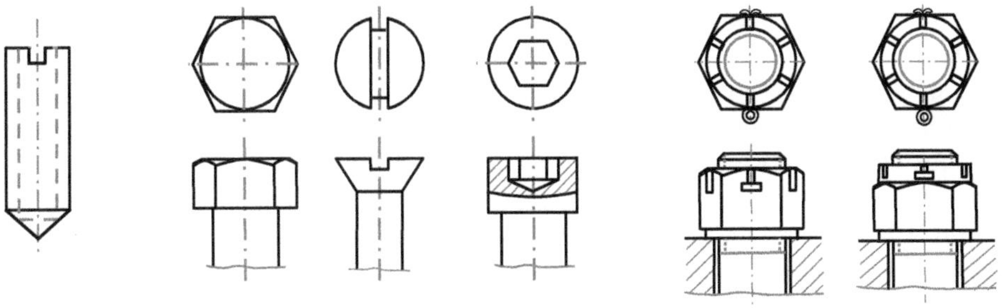

Figure 9.3 Few joining alternatives such as a set screw in (a), different screw heads and tightening means in (b), and a locking device particularly for nuts in (c).

retainers; see Section 8.2.2. The cap screws and set screws are threaded with the cap screw threaded all the way to the screw head. Cap screws are the same as bolts, but they are produced to tighten tolerances. Set screws are usually without a head, and they hold components when they are tightened against such components as for holding keys.

Figure 9.3 shows few joining alternatives such as a set screw, different screw heads and tightening means, and locking devices particularly for nuts. In Figure 9.3a a slotted headless set screw a cone point end is shown. Other ends are also available. Instead of the slot, the tightening means can be a hexagon or spline socket in a headless or even a square, hexagonal, or a round head. Different alternatives with standard dimensions are obtainable from standards or handbooks such as Oberg et al. (2012). Figure 9.3b presents few bolt heads such as the usual hexagon, the flat with countersunk slotted head, and the hexagonal socket round head. Most of the shown alternative tightening means may be used with different heads that may have fillister, round, or oval form. Figure 9.3c shows a locking device particularly for nuts, where a split or cotter pin is used through the slots of the locknut and the hole in the bolt or stud. Two different locknuts are shown in Figure 9.3c. Other locknuts have *nylon insert* rings that cause friction that grips the threads of the bolt and thus locks it.

9.2 Stresses in Threads

Threads and threaded screws and bolts have been available for a long time. Very early since the start of the nineteenth century, their standards have been evolving to enable compatibility between producers and users. When standards are entrenched, it is difficult to change slightly or drastically. The standard basic profile in Figure 9.1b is universally used and very difficult to change. Billions of threaded screws and joints are made. To change the profile, compatibility would not materialize. The optimization of threaded fasteners is constrained by these limits. The stresses generated by the joint geometry and loading must be accommodated as is. The possible change can only be applicable to special cases where interchangeability is not allowed such as unique or special products of very low lot sizes.

To be able to calculate the stresses in the threaded joints, mathematical models are simplified. The original loading is complex. It is stemming from elastic elements in complex contact configuration. If one considers the interaction between the threads in a bolt and a nut, it is a contact between two inclined helical surfaces in ***3D***. The surfaces are elastic due to the material existing behind each helix surface and its attachment to the root internal body of the bolt and the external body of the nut. This problem might effectively be handled numerically by *finite element (**FE**)*. The comprehensive solution cannot be performed for each and every case to design. In addition, the manufacturing process of the thread is important in defining the material state under stress. The production of the thread by *cutting* may provide a thread with little internally introduced stresses. The production of the thread by metal *forming* (such as *rolling*) provides a stronger thread that has internally deformed structure with strain hardening and residual stresses that may be adjusted by heat treatment. For all these deliberations, the conventional approach is to use a more simplified mathematical model that can provide reasonable consideration of the real case and simple enough to utilize. This is done by separately considering each simple loading case and may be combining some cases to get a closer representation. This approach is adopted in this text. The main simple loading cases are discussed separately as follows and covered in some details in Section 9.3.

The most apparently generated stress in the thread members (bolts or nuts) is the ***simple tensile stress***. This might be generated due to an external load in a free joint or an induced load due to tightening, in addition to the externally applied load. The tensile stresses in threads under simple tension are discussed in Section 9.3.1. The tightening load, however, introduces other stresses in addition to simple tension on the bolt cross section as discussed in Section 9.3.2. The simple *tensile stress* on the core cross section of the bolt or nut generates a ***shear stress*** into the root of the thread. This is called a ***stripping shear*** since it tends to strip the teeth off the base – root of the bolt or the nut element. This is discussed further in Section 9.3.1. This stripping shear stress can be combined with the tensile stress at the thread root.

Torsional stress is generated when screws are subjected to initial tightening. This tightening torsion is subjecting the root area to a shear stress. This is further considered and discussed in Section 9.3.3. The ***shear stress*** due to tightening torque can also be combined with the tensile stresses at the root section of the thread. These combined considerations may present a closeness to the physical problem of thread loading and the thread state of stress.

There is also a ***bearing stress*** between the helix surfaces of the bolt and nut. This stress is a function of the thread along the length of the contact between the bolt and the nut. That length is usually enough to safeguard against failure due to bearing stress. This is usually the case since the *stripping shear* requires a minimum length that is satisfied in nut thicknesses; see Section 9.3.1.

Stress concentration occurs at the root profile of the thread due to the small fillets at the roots as indicated by the standard profiles in Figure 9.1b. These stress-raisers affect the behavior of the screws particularly in the fatigue strength consideration.

Uneven stressing occurs due to the uniform spacing between mating thread teeth. The first mating teeth should carry more load, thus having more stresses than the following teeth. Calculations are usually made on this initial cross section. If even teeth loading would occur, more teeth would carry the load, and the stresses would be distributed among more teeth instead of having the full stress at the core section of the first tooth. In the *elastic* zone, this does not occur. In the *plastic* zone of yielding, however, the first mating teeth would be plastically deformed at the root for the following teeth to carry more load. Increasing the load further would let all the teeth carry the same share of stress if yielding deformation permits. That is why it may be recommended that tightening stress should approach the yield strength of the material or 0.9 of *proof strength* for permanent joints or 0.75 of *proof strength* for reusable joints (*proof strength* is about 0.9 the *yield strength*). This yielding approach should be carefully reexamined.

9.3 Bolted Connections

The simplest bolted connection is the one that loads the thread in tension only. Many other connections joint two or more members together by threads or screws. That usually causes or requires tightening, which generates preloading of fasteners and joined members. Tightening torque ensues stresses in the fastening bolt and the joined members. These are the topics of this section.

9.3.1 Threads Under Simple Tensile Load

For a loose connection with no initial tightening on the screw, it would be loaded under pure tension. It might be assumed that the area of the minimum or root diameter d_r should carry the maximum stress. Testing experience, however, has found that the area of the average pitch and root diameters is the resultant sectional area carrying the load. This area has been called the thread *tensile stress area* A_T. This is then given by (see Figure 9.1)

$$A_T = \frac{\pi}{4}\left(\frac{d_p + d_r}{2}\right)^2 \tag{9.3}$$

The ***tensile stress*** σ_t is then simply

$$\sigma_t = \frac{\boldsymbol{P}_T}{A_T} \tag{9.4}$$

where $\boldsymbol{P}_T$ is the *tensile thread load* and A_T is the *tensile stress area* given by Eq. (9.3). The expected *tensile safety factor* K_{SFbt} of a bolt only under tensile load is the yield strength S_y divided by the tensile stress σ_t.

The *tensile stress area* A_T and the minor diameter area are usually tabulated in standards to provide a quick way of calculating the expected maximum stress on the external thread (bolt). From the US (UNC) standards of basic profile, one can find that

$$d_p = d_m - 0.649\,519/N_T, \quad \text{and} \quad d_r = d_m - 1.299\,038/N_T \tag{9.5}$$

From the metric (ISO) standards of basic profile, one can find that

$$d_p = d_m - 0.649\,519p, \quad \text{and} \quad d_r = d_m - 1.299\,038p \tag{9.6}$$

These values can be used in a computer code for the design process instead of consulting tables in the standards.

For initial synthesis in Chapter 8, Figure 8.19 can be used to find the suitable bolts at the stated maximum loads. This is further scrutinized in Example 9.1.

The ***shear stress*** due to tension attempts to strip the teeth off the bolt. This shear is called the *stripping shear* under the tensile load $\boldsymbol{P}_T$. The *stripping shear area* A_{TS} is the area of the teeth roots at the root diameter d_r. The stripping shear area can be assessed as follows:

$$A_{\text{TS}} = \pi d_r\left(K_p\ l_N\right) \tag{9.7}$$

where K_p is the *pitch fraction factor* of material at the root diameter d_r and l_N is the *nut engaging length* with the bolt. The *pitch fraction factor* K_p is a function of thread type. For UNS/ISO threads, the *pitch fraction factor* K_p is 0.8. For square threads, the *pitch fraction factor* K_p is clearly 0.5. For the *Acme* threads, the *pitch fraction factor* K_p is 0.77. The nut engaging length with the bolt l_N may be the *nut length* or *nut height*. But, if there is some chamfering at the ends, the length should be less. Conventionally there should be at least three threads in fully engaging state between bolts and nuts to safeguard against *stripping shear*. The standard nut heights are higher than that and should be safe against shear stripping before tensile failure. Equating the force that causes tensile yielding $\boldsymbol{F}_{\text{yt}} \approx (\pi/4)(0.9d_m)^2 S_y$ with the force that causes shear stripping $\boldsymbol{F}_S \approx \pi d_m(0.8l_N)(0.58S_y)$, one gets the nut length $l_N \approx 0.44d_m$. That is why the *small* or *jam nuts* are having a height l_N of about 0.5 d_m.

Example 9.1 Redo Example 8.3 assuming only tensile loading, and calculate only the bolts between the boom I-beam with the hood. Example 8.3 states that the *jib crane* of Example 8.2 has proposed bolts to connect the boom I-beam with the hood; see Figures 8.17 and 8.18. It is required to initially synthesize the bolt sizes for that connection and for tying the jib-crane mast to the ground. The assumed bolt material for this application is like hot-rolled AISI 1030 or ISO C30. Assume a safety factor of 2 for now. (Such applications, however, would use more than 2 for safety factor; see Chapter 16.)

Solution

Data: The maximum tensile load between the boom I-beam with the hood is $\boldsymbol{F}_2 = 33$ [kN] or 7.333 [klb]. The safety factor $K_{SF} = 2$. The assumed material of AISI 1030 or ISO C30 has the only needed property of yield strength S_y of 260 [MPa] or 37.5 [kpsi]; see Table A.7.2.

In Example 8.3, only two bolts are used between the boom I-beam with the hood. Carrying the load for both connections and the factor of safety is 2, one can just use the stated maximum loads to initially synthesize the bolts. Figure 8.19a gives an M16 [mm] at about $\boldsymbol{R}_2 = 33$ [kN] on the abscissa and 260 [MPa] on the ordinate. Figure 8.19b gives a ¾ [in] at about $\boldsymbol{R}_2 = 7.3$ [klb] on the abscissa and 38 [kpsi] on the ordinate.

The bolt sizes for tying the *jib-crane* mast to the ground are found from Figure 8.19a as M16 [mm] at about $\boldsymbol{R}_2 = 37$ [kN] on the abscissa and 260 [MPa] on the ordinate. Figure 8.19b gives a ¾ [in] at about $\boldsymbol{R}_2 = 8.3$ [klb] on the abscissa and about 38 [kpsi] on the ordinate. These can be calculated similarly as defined next.

Between the boom I-beam with the hood, the two ISO M16s are selected with the designation M16 × 2 for coarse thread as depicted in Table 9.1. This defines the major diameter d_m as 16 [mm] and the pitch p as 2 [mm]. The two US bolts of ¾ [in] have the designation ¾ – 10 for coarse thread as depicted in Table 9.1. This defines the major diameter d_m as ¾ [in] and the number of teeth per inch as 10, i.e. a pitch of 0.1 [in]. From Eqs. (9.5) and (9.6), one gets

$$\begin{aligned} d_p &= 3/4 - 0.649\,519/10 = 0.685\,048\,1 \text{ [in]} \\ d_r &= 3/4 - 1.299\,038/10 = 0.620\,096\,2 \text{ [in]} \end{aligned} \tag{a}$$

and

$$\begin{aligned} d_p &= 16 - 0.649\,519\ (2) = 14.700\,962 \text{ [mm]} \\ d_r &= 16 - 1.299\,038\ (2) = 13.401\,924 \text{ [mm]} \end{aligned} \tag{b}$$

The tensile stress area A_T, according to Equation (9.3), is then

$$\begin{aligned} A_T &= \frac{\pi}{4}\left(\frac{d_p + d_r}{2}\right)^2 = \frac{\pi}{4}\left(\frac{0.685\,048\,1 + 0.620\,096\,2}{2}\right)^2 = 0.334\,46 \left[\text{in}^2\right] \\ A_T &= \frac{\pi}{4}\left(\frac{d_p + d_r}{2}\right)^2 = \frac{\pi}{4}\left(\frac{14.700\,962 + 13.401\,924}{2}\right)^2 = 155.07 \left[\text{mm}^2\right] \end{aligned} \tag{c}$$

The tensile stresses according to Equation (9.4) are then

$$\begin{aligned} \sigma_t &= \frac{\boldsymbol{P}_T}{A_T} = \frac{7.333/2}{0.334\,46} = 10.957 \left[\text{kpsi}\right] \\ \sigma_t &= \frac{\boldsymbol{P}_T}{A_T} = \frac{33\,000/2}{155.07} = 106.40 \text{ [MPa]} \end{aligned} \tag{d}$$

Note that these are the actual stresses applied on the bolts since we divided the load by two (the number of bolts). The expected tensile safety factors K_{SFbt} of bolts are as follows:

$$\begin{aligned} K_{SFbt}|_{US} &= \frac{S_y}{\sigma_t} = \frac{37.5}{10.957} = 3.42 \\ K_{SFbt}|_{SI} &= \frac{S_y}{\sigma_t} = \frac{260}{106.4} = 2.44 \end{aligned} \tag{e}$$

These values are more than the assumed safety factor of 2.0 while synthesizing the bolts in Example 8.3. The value for the US bolt is high because the closest bolt was much larger than needed; see Example 8.3. Note that the yield strength, and not the proof strength, is used.

9.3.2 Preloading Due to Tightening

When a fastener (a bolt) is used to join two members, a tightening is expected to keep the members together. It is recommended to keep the *initial tightening force* $\boldsymbol{F}_i$ as high as recommended to keep the members from separating under load $\boldsymbol{P}_l$. The tightening creates friction between the joined members that resists shear. This preloading improves the strength of the connection as demonstrated later. Figure 9.4 is showing a bolted joint tying two members in Figure 9.4a. The joint has been subjected to the initial tightening $\boldsymbol{F}_i$ first, and then the *externally applied load* $\boldsymbol{P}_l$ is administered. The load-deflection diagram of this joint is shown in Figure 9.4b. It is assumed that the total *length of joint members* l_m is about the same as the *solid bolt length* l_b that has its major diameter d_m along that whole length. To calculate this *statically indeterminate* joint, one needs to consider the stiffness and deflection of the bolt and the members.

The *deflection of the bolt* δ_b under initial tightening force $\boldsymbol{F}_i$ is given by (see Section 6.3, Eq. (6.60))

$$\delta_b = \frac{\boldsymbol{F}_i l_b}{E_b A_b} \tag{9.8}$$

where l_b is the solid bolt length, E_b is the *modulus of elasticity of bolt material*, and A_b is the *cross-sectional area of solid unthreaded bolt*. The stiffness of the bolt k_b in this joint is then

$$k_b = \frac{\boldsymbol{F}_i}{\delta_b} = \frac{E_b A_b}{l_b} \tag{9.9}$$

The *compressive deflection of joint members* δ_m is under the initial compression of *members tightening load* $\boldsymbol{P}_m$, which has the same magnitude as the initial tightening force $\boldsymbol{F}_i$. Since the members area is much larger than the bolt cross-sectional area, the expected *stiffness of joint members* k_m is much larger than the *bolt stiffness* k_b. Figure 9.4b shows the bolt and members stiffnesses, accordingly, but the mirror images of the members load deflection are also shown. Note that members load and deflection are both compressive, i.e. negative, otherwise separation occurs.

When the preloaded joint is externally loaded, the bolt is extended further by the following bolt deformation:

$$\Delta\delta_b = \frac{\Delta \boldsymbol{P}_b}{k_b} \tag{9.10}$$

where $\Delta\delta_b$ is *bolt deformation under external load*. Without members separation, the joined members experience a decrease in compression. The decrease in *member deformation* $\Delta\delta_m$ is then given by

$$\Delta\delta_m = \frac{\Delta \boldsymbol{P}_m}{k_m} \tag{9.11}$$

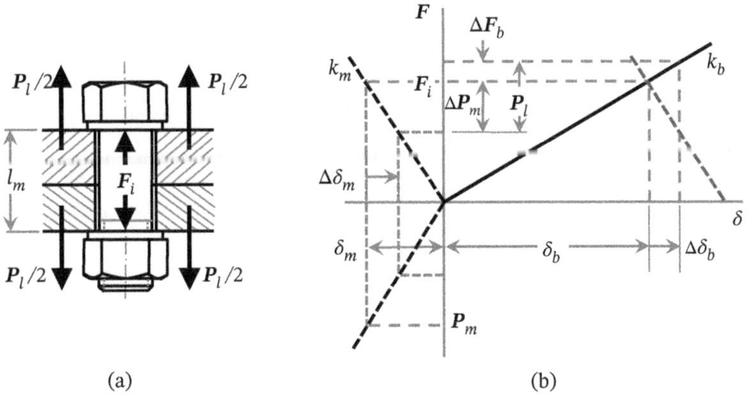

Figure 9.4 A bolted joint tying two members in (a). The load deflection diagram of the joint with the mirror images of the members load deflection are shown in (b).

This decrease in deformation of the joint members should be equal to the bolt extension in Eq. (9.10). This gives

$$\frac{\Delta \boldsymbol{P}_b}{k_b} = \frac{\Delta \boldsymbol{P}_m}{k_m} \tag{9.12}$$

The applied load $\boldsymbol{P}_l$ is the summation of the *increased load on the bolt* $\Delta \boldsymbol{P}_l$ and the *decreased load on the members* $\Delta \boldsymbol{P}_m$ or $\boldsymbol{P}_l = \Delta \boldsymbol{P}_l + \Delta \boldsymbol{P}_m$. Substituting Eqs. (9.10) and (9.11) into that summation gives

$$\boldsymbol{P}_l = \Delta \boldsymbol{P}_b + \Delta \boldsymbol{P}_m = k_b \Delta \delta_b + k_m \Delta \delta_m = \Delta \delta_b \left(k_b + k_m\right)$$

$$\Delta \delta_b = \frac{\boldsymbol{P}_l}{k_b + k_m} \tag{9.13}$$

From Eq. (9.10), (9.11), and (9.13), one also gets

$$\Delta \boldsymbol{P}_b = \frac{k_b \boldsymbol{P}_l}{k_b + k_m}, \quad \text{and} \quad \Delta \boldsymbol{P}_m = \frac{k_m \boldsymbol{P}_l}{k_b + k_m} \tag{9.14}$$

The *force in the bolt* $\boldsymbol{F}_b$ is then

$$\boldsymbol{F}_b = \boldsymbol{F}_i + \Delta \boldsymbol{P}_b = \boldsymbol{F}_i + \frac{k_b \boldsymbol{P}_l}{k_b + k_m} = \boldsymbol{F}_i + \frac{\boldsymbol{P}_l}{1 + \left(k_m / k_b\right)}$$

$$= \boldsymbol{F}_i + \frac{\boldsymbol{P}_l}{1 + k_R} = \boldsymbol{F}_i + C_R \boldsymbol{P}_l \tag{9.15}$$

where k_R is the *stiffness ratio* of the members to the bolt, i.e. $k_R = (k_m/k_b)$ and C_R is the *carrying ratio* due to stiffness ratio effect on the share of the external load carried by the bolt. As developed in Eq. (9.15), $C_R = (1/(1 + k_R))$.

The *force in the joint members* $\boldsymbol{F}_m$ is then

$$\boldsymbol{F}_m = \boldsymbol{F}_i - \Delta \boldsymbol{P}_m = \boldsymbol{F}_i - \frac{k_m \boldsymbol{P}_l}{k_b + k_m} = \boldsymbol{F}_i - \frac{\left(k_m / k_b\right) \boldsymbol{P}_l}{1 + \left(k_m / k_b\right)}$$

$$= \boldsymbol{F}_i - \frac{k_R \boldsymbol{P}_l}{1 + k_R} = \boldsymbol{F}_i - k_R C_R \boldsymbol{P}_l \tag{9.16}$$

where k_R is the *stiffness ratio* of the members to the bolt, i.e. $k_R = (k_m/k_b)$ and C_R is the *carrying ratio* due to stiffness ratio effect on the share of the external load carried by the bolt. As the joint members force $\boldsymbol{F}_m = \boldsymbol{P}_l - \boldsymbol{P}_b$, substituting Eq. (9.15) into Eq. (9.16) gives

$$\mathbf{F}_m = \left(1 - C_R\right) \mathbf{P}_l \tag{9.16}$$

For reusable joints, the recommended preloading for static or dynamic cases is about 75–80% of minimum yield strength or the *proof load* of bolt material. For permanent joints, the recommended preloading can be increased to about 90% of minimum yield strength or the *proof load* of bolt material. The *proof strength* S_p of bolt material is related to the *proportional limit* and may be estimated as about 0.9 of the *yield strength* S_y. The difficulty, however, is to estimate the members stiffness of the joint.

Estimating the ***members stiffness*** is rather involved since the number of members can vary and may have different thicknesses and different materials. However, members should be in series, and the total stiffness k_m ought to be

$$\frac{1}{k_m} = \frac{1}{k_{m1}} + \frac{1}{k_{m2}} + \frac{1}{k_{m3}} + \cdots + \frac{1}{k_{mi}} \tag{9.17}$$

where k_{m1} to k_{mi} are the stiffnesses of members number *one* to member number *i*. The problem is the effective area of each member since its thickness is supposedly known. One way of calculating the member stiffness is by considering a cone frustum starting at the washer contact with the members and having a *cone angle* of 2α as shown in Figure 9.5. The two frusta are symmetrical between the end washers as shown in Figure 9.5a. An element

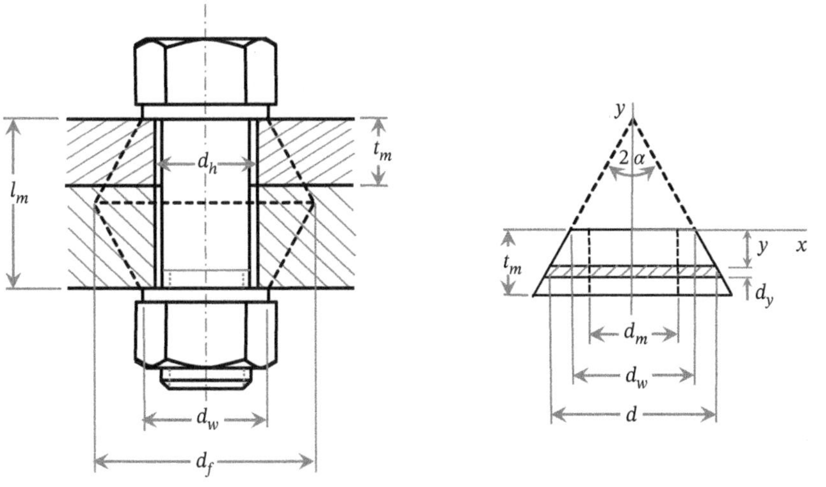

Figure 9.5 Member stiffness by using a cone frustum starting at the washer contact with the members and having a cone angle of 2α. The frusta are symmetrical between the washers as shown in (a). An element of thickness dx in (b) is used to evaluate the stiffness.

of thickness dy in Figure 9.5b is used to evaluate the stiffness of the *member thickness* t_m. The element contraction $d\delta$ of the element dx under the load is given by

$$d\delta = \frac{\boldsymbol{F}_m dy}{E_m A_m} \tag{9.18}$$

where $\boldsymbol{F}_m$ is the element force, E_m is the element modulus of elasticity, and A_m is the element area.

The area of the element A_m is given by (see Figure 9.5b)

$$\begin{aligned} A_m &= \pi\left((d/2)^2 - (d_m/2)^2\right) = \pi\left((y\ \tan\alpha + (d_w/2))^2 - (d_m/2)^2\right) \\ &= \pi\left(\left(y\ \tan\alpha + \frac{d_w + d_m}{2}\right)^2\left(\frac{d_w - d_m}{2}\right)^2\right) \end{aligned} \tag{9.19}$$

where d_m is the major bolt or approximately *hole diameter* and d_w is the *washer outer diameter*. The total contraction of the member δ_m is obtained by substituting Eq. (9.19) into Eq. (9.18) and integrating to get

$$\begin{aligned} \delta_m &= \frac{\boldsymbol{F}_m}{\pi E_m}\int_0^{t_m}\frac{dy}{\left((y\ \tan\alpha + (d_w + d_m)/2)^2((d_w - d_m)/2)^2\right)} \\ &= \frac{\boldsymbol{F}_m}{\pi E_m d_m \tan\alpha}\ln\frac{(2t_m \tan\alpha + d_w - d_m)(d_w + d_m)}{(2t_m \tan\alpha + d_w + d_m)(d_w - d_m)} \end{aligned} \tag{9.20}$$

The *member stiffness* k_m of *member thickness* t_m is then

$$k_m = \frac{\boldsymbol{F}_m}{\delta_m} = \frac{\pi E_m d_m \tan\alpha}{\ln\dfrac{(2t_m \tan\alpha + d_w - d_m)(d_w + d_m)}{(2t_m \tan\alpha + d_w + d_m)(d_w - d_m)}} \tag{9.21}$$

This equation is best utilized by a computer code particularly for several tightened members in a joint. Each member is calculated using Eq. (9.21) and the stiffness of the joined members is evaluated using Eq. (9.17). This could save time and can evaluate the assumption that the *cone angle* 2α is 60° as an average suggestion by Osgood

(1979). Several other researchers have used ***FE*** to evaluate the members stiffness; however, Eq. (9.21) may still be used and an ***FE*** package can verify or adjust the result.

If all the members are made of the same material, the length of the joint l_m is equal to $2t_m$, and Eq. (9.21) is used to calculate the *members stiffness* such that

$$k_m = \frac{\pi E_m d_m \tan\alpha}{2\ln\dfrac{\left(l_m \tan\alpha + d_w - d_m\right)\left(d_w + d_m\right)}{\left(l_m \tan\alpha + d_w + d_m\right)\left(d_w - d_m\right)}} \tag{9.22}$$

Note that the joint has two frusta; see Figure 9.5a. The washer diameter is typically related to the *nominal* or *major diameter* d_m and is usually taken as 1.5 d_m or $d_w = 1.5\ d_m$. Equation (9.22) then becomes

$$k_m = \frac{\pi E_m d_m \tan\alpha}{2\ln\dfrac{5\left(l_m \tan\alpha + 0.5 d_m\right)}{\left(l_m \tan\alpha + 2.5 d_m\right)}} \tag{9.23}$$

For a cone angle with $\alpha = 30°$, $\tan\alpha = 0.577\,35$. For an old assumption of a cone angle with $\alpha = 45°$, $\tan\alpha = 1.0$.

Example 9.2 Define the effect of the preloading on the bolts between the boom I-beam with the hood of the jib crane in Example 9.1. Assume the thickness of the member of the hood plate like the I-beam flange thickness. Find the stiffnesses, stresses, and safety factors of the bolts and members assuming a ½ cone angle of 30°.

Solution

Data: From Example 8.3, the maximum tensile load between the boom I-beam with the hood is $\boldsymbol{F}_2 = 33$ [kN] or 7.333 [klb] on two bolts. The safety factor $K_{SF} = 2$. The assumed material of AISI 1030 or ISO C30 has the needed properties of yield strength S_y of 260 [MPa] or 37.5 [kpsi]; see Table A.7.2. The modulus of elasticity E of steel is (207) [GPa] or (30) [Mpsi]; see Section 7.1.3.

From Example 8.2, a selection is made for SI metric series wide flange *I-beam*, the section I-200 with a 200 [mm] depth and a 200 [mm] width has a flange thickness of 16 [mm]; see Table A.8.9. For the US inch series wide flange *I-beam*, the section W 16x26 with a 15.69 [in] depth and a 5.5 [in] width has a flange thickness of 0.345 [in]; see Table A.8.10.

As before (Examples 8.3 and 9.1), the bolt sizes are M16 [mm] and a ¾ [in]. The members thicknesses or lengths are the flange thicknesses of 16 [mm] and 0.345 [in] in addition to the hood top plates taken at about the same thicknesses of 16 [mm] and 0.4 [in] as the close preferred sizes; see Table A.3.1. The member lengths are then 32 [mm] and 0.8 [in]. The bolt lengths are taken the same as member lengths.

The stiffnesses of the bolts are calculated according to Eqs. (9.9) and (9.10) that gives

$$k_b = \frac{E_b A_b}{l_b} = \frac{30\left(10^6\right)\left(\pi(0.75)^2/4\right)}{0.8} = 16.567\left(10^6\right)\ [\text{lb/in}]$$

$$k_b = \frac{E_b A_b}{l_b} = \frac{207\left(10^9\right)\left(\pi\left(16\left(10^{-3}\right)\right)^2/4\right)}{32\left(10^{-3}\right)} = 1300.6\left(10^6\right)\ [\text{N/m}] \tag{a}$$

The stiffnesses of the members are obtained as defined by Eq. (9.23) and calculated by a MATLAB code to give

$$k_m = 3.5028\left(10^7\right)\ [\text{lb/in}]$$
$$k_m = 3.6762\left(10^9\right)\ [\text{N/m}] \tag{b}$$

The initial preloading stress is estimated from the yield strength as 0.75 (0.9) S_y. Note that the *proof strength* is about 0.9 S_y. This gives the initial tightening forces $\boldsymbol{F}_i$ as follows:

$$F_i = 0.75\,(0.9)\left(S_y\right)\left(A_T\right) = 0.75\,(0.9)\left(37.5\left(10^3\right)\right)(0.334\,46) = 8.4660\left(10^3\right)\ [\mathrm{lb}]$$
$$F_i = 0.75\,(0.9)\left(S_y\right)\left(A_T\right) = 0.75\,(0.9)\left(260\left(10^6\right)\right)\left(155.07\left(10^{-6}\right)\right) = 27.215\left(10^3\right)\ [\mathrm{N}] \tag{c}$$

where A_T is the *tensile areas* calculated in Example 9.1. Note that these initial tightening forces are close to the applied loads. In fact, one may use 0.9 ratio of the proof strength rather than 0.75 for a permanent joint.

The *stiffness ratio*k_m/k_b is then given by

$$\frac{k_m}{k_b}\Big|_{\mathrm{US}} = \frac{3.5028\left(10^7\right)}{16.567\left(10^6\right)} = 2.1143$$
$$\frac{k_m}{k_b}\Big|_{\mathrm{SI}} = \frac{3.6762\left(10^9\right)}{1.3006\left(10^9\right)} = 2.8265 \tag{d}$$

The force in the bolt F_b is obtained from Eq. (9.15) and needs the *stiffness ratio* such that

$$F_b = F_i + \frac{P_l}{1+\left(k_m/k_b\right)} \tag{9.15}$$

Substituting for the values, one gets

$$F_b = F_i + \frac{P_l}{1+2.1143} = 8.4660\left(10^3\right) + \frac{7.333\left(10^3\right)/2}{3.1143} = 9.6433\left(10^3\right)\ [\mathrm{lb}]$$
$$F_b = F_i + \frac{P_l}{1+2.8265} = 27.215\left(10^3\right) + \frac{33\left(10^3\right)/2}{3.8265} = 31.527\left(10^3\right)\ [\mathrm{N}] \tag{e}$$

The stresses in the bolts including the preloading and the applied load is then given by

$$\sigma_t = \frac{F_b}{A_T} = \frac{9.6433\left(10^3\right)}{0.33446} = 28.832\ [\mathrm{kpsi}]$$
$$\sigma_t = \frac{F_b}{A_T} = \frac{31.527\left(10^3\right)}{155.07} = 203.31\ [\mathrm{MPa}] \tag{f}$$

The expected safety factors K_{SFbt} of the bolts are as follows:

$$K_{\mathrm{SFbt}}\big|_{\mathrm{US}} = \frac{S_y}{\sigma_t} = \frac{37.5}{28.832} = 1.30$$
$$K_{\mathrm{SFbt}}\big|_{\mathrm{SI}} = \frac{S_y}{\sigma_t} = \frac{260}{203.31} = 1.28 \tag{g}$$

These factors of safety might not be satisfactory to some applications, and larger initial safety factor might be recommended in the initial synthesis of the bolts.

9.3.3 Tightening Torque

To generate the initial preloading, a *tightening torque* on the bolts or nuts is needed. It is dependent on the desirable value of initial tightening force and other factors such as friction, lubrication, material type, surface treatment, etc. The tightening torque is also used to define the amount of initial preloading force F_i. This is because it is easier to control the tightening torque while preloading than measuring the force or its effects on the bolt. The relations between the tightening force and tightening torque is best derived by the analysis given later in Section 9.5 of *power screws*. The equation of the *tightening torque* T_T including the initial preloading force F_i and the friction in the washer is then (see Eqs. (9.45)–(9.47))

$$T_T = \frac{F_i d_a}{2}\left(\frac{1+\pi\mu_f d_a \sec\gamma_T}{\pi d_a - \mu_f l_T \sec\gamma_T}\right) + \frac{F_i \mu_W d_{Wa}}{2} \tag{9.24}$$

where d_a is the *average major and minor bolt diameter*, μ_f is the *friction coefficient in the bolt thread*, μ_W is the *friction coefficient in washer*, d_{Wa} is the average washer contact diameter, and γ_T is the *half thread tooth angle*. Considering the values of the variables in the relations for $\tan \lambda_T = l_T/\pi d_a$, and $d_{wa} = (d_m + 1.5d_m)/2$ and after some manipulations, one gets the following simplified relation:

$$T_T = K_T F_i d_m \cong 0.2\, F_i d_m \tag{9.25}$$

The value of the *tightening torque coefficient* K_T is about 0.2 for the friction coefficients of μ_f in the bolt thread and μ_W in the washer of about 0.15. This is valid for other ranges of variables in Eq. (9.24). Several relations other than Eq. (9.24) or (9.25) have been developed; however, Eq. (9.25) will still be used. The accuracy of bolt preload application, though, is about ±25% when using a torque wrench (Oberg et al. 2012). Other more accurate methods of adjusting the preloading or *tightening force and torque* are yet not practical such as using strain gages or ultrasonic sensing. Bolt elongation sensing might be practical for very critical and demanding cases with accuracy of about ±3–5% (Oberg et al. 2012).

The ***tightening shear stress*** τ_T generated due to the tightening torque T_T is obtained simply on the area of root diameter d_r such that

$$\tau_T = \frac{16\left(T_T/2\right)}{\pi d_r^3} \tag{9.26}$$

where the tightening torque is divided by 2.0 since not all torque reaches the root diameter d_r due to friction in the washers; see Eq. (9.24).

One can use the *von Mises* stress to combine the normal stress in Eq. (9.4) and the shear stress in Eq. (9.26) to evaluate the combined safety factor. One should note, however, that the preloading would usually cause yielding and the stress state will be different due to *prestressing*, *residual stresses*, and *strain hardening*. These cases are beyond the scope of this text.

Example 9.3 Use a MATLAB code to find the solution of Example 9.2. Find also the values of the tightening torque assuming the coefficients of friction to be about 0.15.

Solution

With reference to Example 9.1 and Eq. (9.25), a MATLAB code is developed to calculate all necessary requirements. The code is available in the ***Wiley website*** under the name of CAD_Bolts_Preload.m.

Data: $F_l = 16\,500$ [N] or 3.6665e+03 [lb], $d_m = 16$ [mm] or 0.75 [in], $E = 207$ [GPa] or 30 [Mpsi], $l_b = l_m = 32$ [mm] or 0.8 [in], $p = 2$ [mm] or $N_T = 10$ [/in], $S_y = 260$ [MPa] or 37.5 [kpsi], and $\alpha = 30°$.

Using the MATLAB code, the following results are obtained with minor editing:

- dp_mm = 14.7010 [mm], dr_mm = 13.4019 [mm] and dp_in = 0.6850 [in], dr_in = 0.6201 [in]
- AT_mm2 = 155.0714 [mm^2], AT_in2 = 0.3345 [in^2]
- km_SI = 3.6762e+09 [N/m], km_US = 3.5028e+07 [lb/in]
- Ab_mm2 = 201.0619 [mm^2], Ab_in2 = 0.4418 [in^2]
- kb_SI = 1.3006e+09 [N/m], kb_US = 1.6567e+07 [lb/in]
- Stiffness_Ratio_SI = 2.8265, Stiffness_Ratio_US = 2.1143
- Fi_SI = 2.7215e+04 [N], Fi_US = 8.4661e+03 [lb]
- Fb_SI = 3.1527e+04 [N], Fb_US = 9.6434e+03 [lb]
- sigma_MPa = 203.3065 [MPa], sigma_kpsi = 28.8325 [kpsi]
- KSF_cal_SI = 1.2789, KSF_cal_US = 1.3006
- T_torque_SI = 87.0881 [N.m], T_torque_US = 1.2699e+03 [lb.in]

The variables are kept as defined in the MATLAB code. Results are the same as calculated in Example 9.2. The tightening torque can be easily checked from Eq. (9.25).

9.4 Bolt Strength in Static and Fatigue

Bolt strength has been the concern of the standards as bolts can be produced out of so many materials. The standardization is important for design, assembly, and maintenance. The standards specify strength and enforce that it is marked on the bolts (SAE J429 (2001) and ISO 262 (1998)). It defines the minimum strength properties particularly the proof strength, the yield strength, and the ultimate tensile strength. Table 9.2 shows classes of material strength grades of screws and bolts. The properties are the minimum where at least 99% of the specimens have strength values at or above the stated minimum. The ISO classes are well spread over the property span, and the US classes in SAE or ASTM have some that closely resembles the ISO classes. These are indicated in Table 9.2 with *the grayish strength* values for the missing standards (). The classes in SAE or ASTM have more options that are not shown in Table 9.2. Those can be retrieved from these standards. The head marking that are stamped on the bolt head are shown for the ISO and the US with the US marking mainly for the SAE standard. The ASTM head marking for ASTM 449 has the same marking as the SAE 5.2. It is not shown to reduce confusion. One should consult the SAE and ASTM standards for their other grade designations and head marking. The SAE metric standard is closely resembling the ISO standards.

Table 9.2 Classes of material strength grades of screws and bolts.

Designation ISO class (SAE or {ASTM})	Size Range ISO (US)	Proof Strength [MPa] ([kpsi])	Yield Strength [MPa] ([kpsi])	Tensile Strength [MPa] ([kpsi])	Head Marking	
					ISO	US
4.6 (1,2 {A307})	M5-M36 (¼–1 ½)	225 (33)	240 (36)	400 (60)	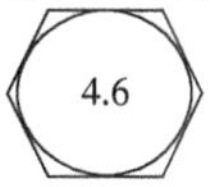 4.6	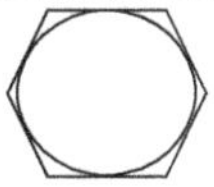
4.8 ( )	M5-M36 ( )	310 (*45*)	340 (*49*)	420 (*61*)	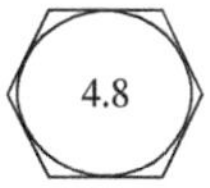 4.8	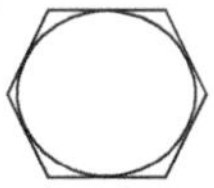
5.8 (2 {A449})	M5-M36 (¼–¾ , 1 ¾–3)	380 (55)	420 (57, {58})	520 (74, {90})	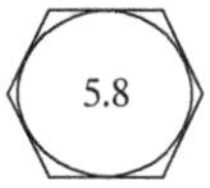 5.8	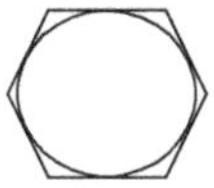
8.8 (5.2 {A325})	M5-M36 (1.4–1)	600 (85)	660 (92)	830 (120)	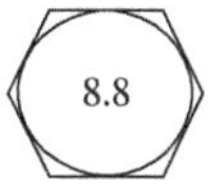 8.8	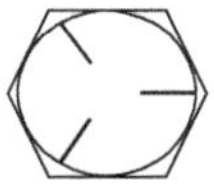
9.8 ({A354})	M5-M36 (2¾–4)	650 (95)	720 ({99}[a])	900 ({115}[a])	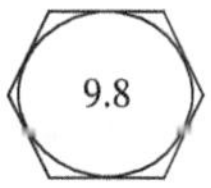 9.8	 BC
10.9 (8 {A490})	M5-M36 (¼–1½)	830 (120)	940 (130)	1040 (150)	 10.9	
12.9 ( )	M5-M36 ( )	970 (*141*)	1100 (*160*)	1220 (*177*)	 12.9	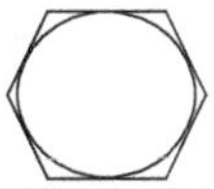

The SAE or ASTM classes that closely resemble the ISO classes are in brackets. The head marking of the US are mainly for the SAE standards.

a) The values are not close to the ISO values.

The classes of ISO standards have relations to the stated minimum strength. The numeral before the decimal point is approximately 0.01 of the ultimate tensile strength, i.e. the numeral 4 in the 4.6 class is 0.01 of the 400 [MPa] ultimate tensile strength. The numeral after the decimal point along with the decimal point is the approximate ratio of the yield strength to the ultimate tensile strength of the material. The ISO class of 4.6 indicates that the yield strength of 240 [MPa] is about 0.6 of the ultimate tensile strength or 0.6(400) = 240 [MPa]. The *proof strength* is the one to be used in design, and it is about 0.9 of yield strength. The yield strength is usually causing 0.2% permanent deformation, while the proof strength would not cause a permanent set. If the proof strength is available, the 0.9 of yield strength would not be used.

The ISO classes of 4.6, 4.8, and 5.8 are for low- or medium-carbon steel. They are ductile and to be used for simple connections with slight loads. The 8.8 and 9.8 classes are medium-carbon steel *quenched* and *tempered*. The 10.9 class is for low-carbon *martensitic* steel quenched and tempered. The 12.9 class is for *alloy steel* quenched and tempered.

The ***static strength*** of the bolted connection should be calculated according to specific selection of bolts according to ISO class and SAE grade. The *proof strength* S_p is to be used in place of $0.9S_y$ previously used. This renders the *bolt safety factor* K_{SFb} as follows:

$$K_{\mathrm{SFb}} = \frac{S_p}{\sigma_t} = \frac{S_p}{\boldsymbol{F}_b/A_T} = \frac{S_p A_T}{\boldsymbol{F}_i + C_R \boldsymbol{P}_l} \tag{9.27}$$

The ***fatigue strength*** is a property that can be evaluated as discussed before in Chapter 7 with factors affecting fatigue. The *stress concentration* factor for classes or grades of ISO 4.6–5.8 and SAE 1–2 is 2.2 for rolled threads, 2.8 for cut threads, and 2.1 for bolt fillets. The *stress concentration* factor for classes or grades of ISO 8.8–12.9, and SAE 5.2 –8 is 3.0 for rolled threads, 3.8 for cut threads, and 2.3 for bolt fillets. For cut threads, these factors are used in the same way discussed in Chapter 7. When these factors are considered with others, the *modified endurance strength* S_e is proposed to be as follows for *rolled* threads:

- SAE 5 grade, the endurance strength is 18.6 [kpsi] for size range of ¼–1 [in] and 16.3 [kpsi] for size range of 1⅛–1½ [in].
- SAE 8 grade, the endurance strength is 23.2 [kpsi] for size range of ¼–1½ [in]
- ISO 8.8 class, the endurance strength is 129 [MPa] for size range of M16–M36.
- ISO 9.8 class, the endurance strength is 140 [MPa] for size range of M1.6–M16.
- ISO 10.9 class, the endurance strength is 162 [MPa] for size range of M5–M36.
- ISO 12.9 class, the endurance strength is 190 [MPa] for size range of M1.6–M36.

These values are used directly for rolled threads without the factors affecting fatigue in Chapter 7. The cut threads can be calculated as in Chapter 7 with the *stress concentration factor* as specified earlier in addition to other factors affecting fatigue strength.

The ***dynamic strength*** of the bolted connection should be calculated according to specific selection of bolts according to ISO class and SAE grade. One must observe that the tightening load improves the dynamic strength as the load $\boldsymbol{F}_l$ is smaller as shown in Figure 9.4. With the load fluctuating between $\boldsymbol{F}_{l\max}$ and $\boldsymbol{F}_{l\min}$, the force on the bolt fluctuates according to Equation (9.15) such that the *bolt maximum force* $\boldsymbol{F}_{b\max}$ and the *bolt minimum force* $\boldsymbol{F}_{b\min}$ are as follows:

$$\begin{aligned} \boldsymbol{F}_{b\max} &= \boldsymbol{F}_i + \frac{\boldsymbol{P}_{l\max}}{1+k_R} = \boldsymbol{F}_i + C_R \boldsymbol{P}_{l\max} \\ \boldsymbol{F}_{b\min} &= \boldsymbol{F}_i + \frac{\boldsymbol{P}_{l\min}}{1+k_R} = \boldsymbol{F}_i + C_R \boldsymbol{P}_{l\min} \end{aligned} \tag{9.28}$$

where k_R is the *stiffness ratio* of the members to the bolt, i.e. $k_R = (k_m/k_b)$, and C_R is the *carrying ratio* due to stiffness ratio effect on the share of the external load carried by the bolt. As developed in Eq. (9.15), $C_R = (1/(1 + k_R))$.

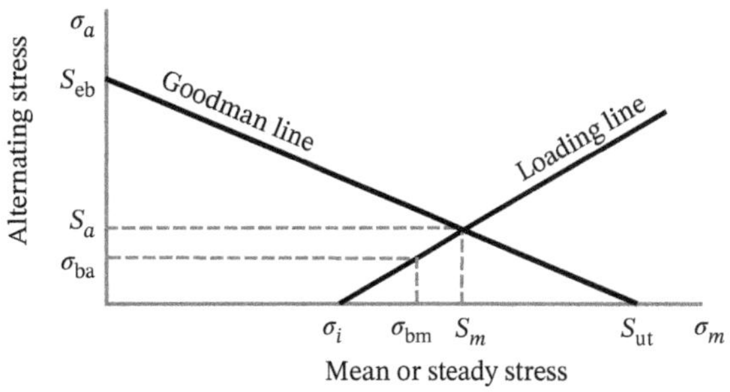

Figure 9.6 Fatigue diagram for a constant preload and a fluctuating load. The limit line of *Goodman* criteria is shown.

The *alternating stress* in the bolt σ_{ba} is then

$$\begin{aligned}\sigma_{ba} &= \frac{\boldsymbol{F}_{b\max} - \boldsymbol{F}_{b\min}}{2A_T} = \frac{(\boldsymbol{F}_i + C_R\boldsymbol{P}_{l\max}) - (\boldsymbol{F}_i + C_R\boldsymbol{P}_{l\min})}{2A_T} \\ &= \frac{\boldsymbol{P}_{l\max} - \boldsymbol{P}_{l\min}}{2A_T(1+k_R)} = \frac{C_R(\boldsymbol{P}_{l\max} - \boldsymbol{P}_{l\min})}{2A_T}\end{aligned} \qquad \textbf{(9.29)}$$

Again, k_R is the *stiffness ratio* of the members to the bolt, i.e. $k_R = (k_m/k_b)$, and C_R is the *carrying ratio* due to stiffness ratio effect on the share of the external load carried by the bolt. As developed in Eq. (9.15), $C_R = (1/(1+k_R))$.

The *mean* or *steady stress* in the bolt σ_{bm} is

$$\begin{aligned}\sigma_{bm} &= \frac{\boldsymbol{F}_{b\max} + \boldsymbol{F}_{b\min}}{2A_T} = \left(\left(\boldsymbol{F}_i + \frac{\boldsymbol{P}_{l\max}}{1+k_R}\right) + \left(\boldsymbol{F}_i + \frac{\boldsymbol{P}_{l\min}}{1+k_R}\right)\right)/2A_T \\ &= \frac{\boldsymbol{P}_{l\max} + \boldsymbol{P}_{l\min}}{2A_T(1+k_R)} + \frac{\boldsymbol{F}_i}{A_T} = \frac{C_R(\boldsymbol{P}_{l\max} + \boldsymbol{P}_{l\min})}{2A_T} + \frac{\boldsymbol{F}_i}{A_T}\end{aligned} \qquad \textbf{(9.30)}$$

Again, k_R is the *stiffness ratio* of the members to the bolt, i.e. $k_R = (k_m/k_b)$, and C_R is the *carrying ratio* due to stiffness ratio effect on the share of the external load carried by the bolt. As developed in Eq. (9.15), $C_R = (1/(1+k_R))$.

A plot of the bolt loading in the dynamic state is shown in Figure 9.6. The *loading line slope* is $\sigma_{ba}/(\sigma_{bm} - \sigma_i)$ as shown in Figure 9.6. The equation of the load line is then

$$S_a = \frac{\sigma_{ba}}{(\sigma_{bm} - \sigma_i)}(S_m - \sigma_i) \qquad (9.31)$$

where S_a is the *alternating load strength* and S_m is the *mean or steady load strength*. From Figure 9.6, the equation of the *Goodman* line is

$$S_a = S_{eb} - \frac{S_{eb}}{S_{ut}}S_m \qquad (9.32)$$

where S_{eb} is the *bolt endurance limit* and S_{ut} is the *ultimate tensile strength of bolt material*. The intersection of the two lines is obtained by equating Eqs. (9.31) and (9.32), which gives

$$S_m = \frac{S_{ut}((\sigma_{bm} - \sigma_i)S_{eb} + \sigma_{ba}\sigma_i)}{(\sigma_{bm} - \sigma_i)S_{eb} + \sigma_{ba}S_{ut}} \qquad (9.33)$$

Substituting Eq. (9.33) into Eq. (9.32) gives

$$S_a = \frac{S_{eb}\sigma_{ba}(S_{ut} - \sigma_i)}{(\sigma_{bm} - \sigma_i)S_{eb} + \sigma_{ba}S_{ut}} \qquad (9.34)$$

The *fatigue safety factor* K_{SFbf} of the bolt is then

$$K_{SFbf} = \frac{S_a}{\sigma_{ba}} = \frac{S_{eb}(S_{ut} - \sigma_i)}{(\sigma_{bm} - \sigma_i)S_{eb} + \sigma_{ba}S_{ut}} \qquad \textbf{(9.35)}$$

The *Goodman* criteria is conservative. If the case in question is designed for critical situations such as when minimum weight is an objective, other failure criterion such as *ASME*, modified *Goodman*, or *Gerber* would be used; see Section 7.8.5.

The procedure to find the fatigue behavior and safety from the previous equations is best utilized by a computer code particularly for several *CAD* iterations or optimization. MATLAB might be useful in that regard.

Example 9.4 Use a MATLAB code to find the solution of Example 9.3 with specific selection of bolts according to ISO class and SAE grade. Find also the effect of loading and unloading the crane on its fatigue strength if the maximum load is lifted and released so many times to warrant the use of endurance strength. Assume a *Goodman* fatigue criteria, and consider the coefficients of friction to be about 0.15.

Solution

With reference to Example 9.3 and the relevant equations, a MATLAB code is developed to calculate all necessary requirements.

Data: Initially $\boldsymbol{F}_l = 16\,500$ [N] or 3.6665e+03 [lb], $d_m = 16$ [mm] or 0.75 [in], $E = 207$ [GPa] or 30 [Mpsi], $l_b = l_m = 32$ [mm] or 0.8 [in], $p = 2$ [mm] or $N_T = 10$ [/in], $S_y = 260$ [MPa] or 37.5 [kpsi], and $\alpha = 30°$.

Since the jib crane is usually heavily employed, and not a product used with simple connections and little loads, the next class or grade of bolts should be used. These are the ISO 8.8 and the SAE 5.2 according to Table 9.2. The proof strength is 600 [MPa] or 85 [kpsi], the yield strength is 660 [MPa] or 92 [kpsi], and the ultimate strength is 830 [MPa] or 120 [kpsi]. From Section 9.4, the corrected endurance limit is 129 [MPa] or 18.6 [kpsi]. These values are the new inputs to the computer code to calculate all other variables.

Using the MATLAB code, the results are obtained as follows. The units are added and the output rearranged in single lines for each category:

- Sp_MPa = 600 [MPa], Sp_kpsi = 85 [kpsi]- Input proof strength
- Seb_MPa =129 [MPa], Seb_kpsi = 18.6000 [kpsi]- Input endurance strength
- Sut_MPa = 830 [MPa], Sut_kpsi = 120 [kpsi]- Input ultimate strength
- Pmax_N = 16500 [N], Pmin_N = 0 [N]- Input load limits [N]
- Pmax_lb =3.6665e+03 [lb], Pmin_lb = 0 [lb- Input load limits [lb]
- Fi_SI = 6.9782e+04 [N], Fi_US = 2.1322e+04 [lb]
- Fb_SI = 7.4094e+04 [N], Fb_US = 2.2499e+04 [lb]
- Fm_SI = 5.7594e+04 [N], Fm_US = 1.8833e+04 [lb]
- PS_N = 9.4470e+04 [N], PS_lb = 3.1407e+04 [lb]
- sigma_MPa = 477.8065 [MPa], sigma_kpsi = 67.2700 [kpsi]
- KSFb_cal_SI = 1.2557, KSFb_cal_US = 1.2636
- Pa_N = 8250 [N], Pa_lb = 1.8333e+03 [lb]
- Pm_N = 8250 [N], Pm_lb = 1.8333e+03 [lb]
- sigmaba_MPa = 13.9033 [MPa], sigmaba_kpsi = 1.7600 [kpsi]
- sigmabm_MPa = 463.9033 [MPa], sigmabm_kpsi = 65.5100 [kpsi]
- sigmai_MPa = 450.0000 [MPa], sigmai_kpsi = 63.7500 [kpsi]
- KSFbf_SI = 3.6765, KSFbf_kpsi = 4.2890

The variables are kept as defined in the MATLAB code. These values are not verified by hand calculations. It is left as an exercise to validate the MATLAB code.

9.5 Power Screws

A *power screw* is an element that changes the angular rotation into a rectilinear motion and transmits power. It is used in the lath lead screw, vices, presses, jacks, and similar arrangements. The torque is generated by the driver due to the force needed at the screw end. A square thread is usually used even though Acme profile may also be employed. The usual threads use preferred nominal or major diameters and pitches or number of teeth per inch. The larger power screw pitches can be about 0.125 or $1/8^{th}$ of the nominal diameter, and other *standard coarse pitches* can be used. The torque $\boldsymbol{T}_p$ is applied to the power screw end driving against the linear load $\boldsymbol{P}_l$.

9.5.1 Torque Requirements

Figure 9.7a shows a sketch of a portion of the power screw with a load $\boldsymbol{P}_l$. The force balance for raising the load is shown in the upper diagram of Figure 9.7b. The force balance for lowering the load is shown in the lower diagram of Figure 9.7b. Figure 9.7b is considering only a single turn of the thread.

The equilibrium equations for the force balance for *raising* the load is as follows; see top Figure 9.7b.

$$\begin{aligned} \sum \boldsymbol{F}_x &= \boldsymbol{F}_R - \mu_T \boldsymbol{N}_h \cos\lambda_T - \boldsymbol{N}_h \sin\lambda_T = 0 \\ \sum \boldsymbol{F}_y &= -\boldsymbol{P}_l - \mu_T \boldsymbol{N}_h \sin\lambda_T + \boldsymbol{N}_h \cos\lambda_T = 0 \end{aligned} \tag{9.36}$$

where $\boldsymbol{F}_R$ is the force for raising the load, μ_T is the *thread friction coefficient*, $\boldsymbol{N}_h$ is the force normal to the thread helix, λ_T is the thread lead angle, and $\boldsymbol{P}_l$ is the load to be raised. Eliminating $\boldsymbol{N}_h$ from Eq. (9.36) gives the following:

$$\boldsymbol{F}_R = \frac{\boldsymbol{P}_l \left(\sin\lambda_T + \mu_T \cos\lambda_T\right)}{\cos\lambda_T - \mu_T \sin\lambda_T} \tag{9.37}$$

Dividing the numerator and the denominator by $\cos\lambda_T$ and using $\tan\lambda_T = l_T/\pi d_p$, one gets

$$\boldsymbol{F}_R = \frac{\boldsymbol{P}_l \left(\sin\lambda_T + \mu_T \cos\lambda_T\right) / \cos\lambda_T}{\left(\cos\lambda_T - \mu_T \sin\lambda_T\right) / \cos\lambda_T} = \frac{\boldsymbol{P}_l \left(\left(l_T/\pi d_p\right) + \mu_T\right)}{\left(1 - \mu_T \left(l_T/\pi d_p\right)\right)} \tag{9.38}$$

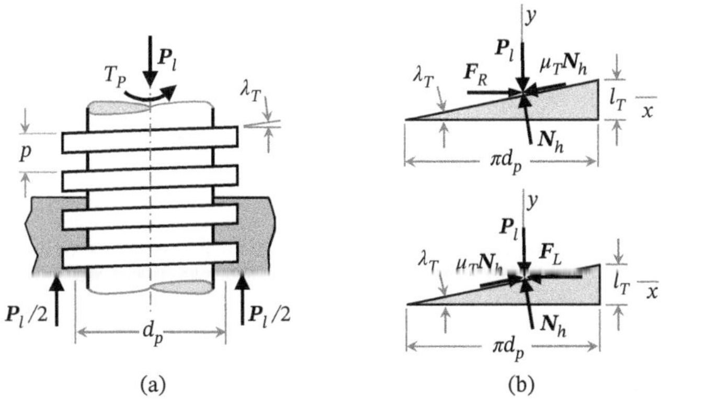

Figure 9.7 Sketch of a portion of power screw in (a). The force balance for raising the load in the upper diagram in (b) and the force balance for lowering the load in the lower diagram in (b).

Since the power screw *raising torque* $\boldsymbol{T}_R$ is simply given by $\boldsymbol{T}_R = \boldsymbol{F}_R\,(d_p/2)$, the torque to raise the load and overcome thread friction is then given by

$$\boldsymbol{T}_R = \frac{\boldsymbol{P}_l\left(\left(l_T/\pi d_p\right) + \mu_T\right)\left(d_p/2\right)}{\left(1 - \mu_T\left(l_T/\pi d_p\right)\right)} = \frac{\boldsymbol{P}_l d_p\left(l_T + \pi\mu_T d_p\right)}{2\left(\pi d_p - \mu_T l_T\right)} \tag{9.39}$$

The equilibrium equations for the force balance for *lowering* the load is as follows (see Figure 9.7b):

$$\begin{aligned}\sum \boldsymbol{F}_x &= -\boldsymbol{F}_L + \mu_T\boldsymbol{N}_h\cos\lambda_T - \boldsymbol{N}_h\sin\lambda_T = 0\\ \sum \boldsymbol{F}_y &= -\boldsymbol{P}_l + \mu_T\boldsymbol{N}_h\sin\lambda_T + \boldsymbol{N}_h\cos\lambda_T = 0\end{aligned} \tag{9.40}$$

where $\boldsymbol{F}_L$ is the force for lowering the load, μ_T is the thread friction coefficient, $\boldsymbol{N}_h$ is the force normal to the thread helix, λ_T is the lead angle, and $\boldsymbol{P}_l$ is the load to be lowered. Similar to the load raising procedure, eliminating $\boldsymbol{N}_h$ from Eq. (9.40) gives the following:

$$\boldsymbol{F}_L = \frac{\boldsymbol{P}_l\left(\mu_T\cos\lambda_T - \sin\lambda_T\right)}{\cos\lambda_T + \mu_T\sin\lambda_T} \tag{9.41}$$

Dividing the numerator and the denominator by $\cos\lambda_T$ and using $\tan\lambda_T = l_T/\pi d_p$, one gets

$$\boldsymbol{F}_L = \frac{\boldsymbol{P}_l\left(\mu_T\cos\lambda_T - \sin\lambda_T\right)/\cos\lambda_T}{\left(\cos\lambda_T + \mu_T\sin\lambda_T\right)/\cos\lambda_T} = \frac{\boldsymbol{P}_l\left(\mu_T - \left(l_T/\pi d_p\right)\right)}{\left(1 + \mu_T\left(l_T/\pi d_p\right)\right)} \tag{9.42}$$

Since the power screw *lowering torque* $\boldsymbol{T}_L$ is simply given by $\boldsymbol{T}_L = \boldsymbol{F}_R\,(d_p/2)$, the torque to lower the load and overcome thread friction is then given by

$$\boldsymbol{T}_L = \frac{\boldsymbol{P}_l\left(\mu_T - \left(l_T/\pi d_p\right)\right)\left(d_p/2\right)}{\left(1 + \mu_T\left(l_T/\pi d_p\right)\right)} = \frac{\boldsymbol{P}_l d_p\left(\pi\mu_T d_p - l_T\right)}{2\left(\pi d_p + \mu_T l_T\right)} \tag{9.43}$$

In some cases, however, the load may lower itself without any need for the external load. This can occur when the friction is slight and/or the helix angle is large. Observing Eq. (9.43), this will happen if the torque is negative or zero. The screw in that case is not ***self-locking***, and self-lowering the load happens. Self-locking would then be needed. From Eq. (9.43), the condition of *self-locking* is then $\pi\mu_T d_p > l_T$. Dividing that by πd_p and observing that $\tan\lambda_T = l_T/\pi d_p$, one notes that *self-locking* is assured when

$$\mu_T > \tan\lambda_T \tag{9.44}$$

Self-locking is then guaranteed if the thread friction coefficient is larger than the tangent of the *lead angle*. Self-locking may not be desired in some cases, when it might be more advantageous to let the head mass drive the system in some presses or other applications.

While the torque is applied, a collar or a thrust bearing is needed to take up the axial load. Figure 9.8a provides a sketch of the power screw with a collar to handle the thrust. The force diagram in Figure 9.8b considers an *Acme* or other thread for a more general treatment and potential applications in other screw threads. The *thread angle* is therefore given the symbol of $2\gamma_T$ as shown in Figure 9.8b. The load $\boldsymbol{P}_l$ is then inclined by an angle γ_T rather than being parallel to the screw axis. Its value should be divided by $\cos\gamma_T$. This inclination causes a wedging action that affects the friction term more than others, and that is why the friction terms in Eqs. (9.39) and (9.43) should be multiplied by $\sec\gamma_T$. The power screw raising torque $\boldsymbol{T}_R$ to *raise* the load and overcome thread friction is then

$$\boldsymbol{T}_R = \frac{\boldsymbol{P}_l d_p\left(l_T + \pi\mu_T d_p\sec\gamma\right)}{2\left(\pi d_p - \mu_T l_T\sec\gamma\right)} \tag{9.45}$$

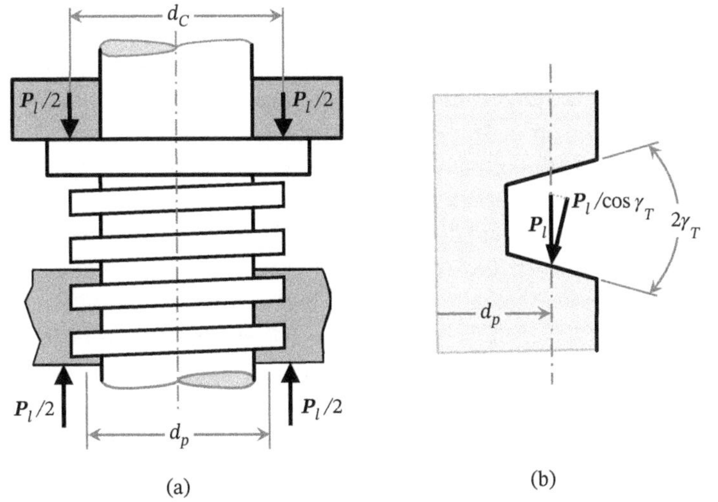

Figure 9.8 A sketch of the power screw with a collar to handle the thrust in (a). The force diagram in (b) considers an *Acme* or other thread for a more general treatment.

In addition to this power screw torque to *raise* the load, there is another torque to overcome *collar friction*; see Figure 9.8a. The load $\boldsymbol{P}_l$ on the collar causes a *collar friction torque* $\boldsymbol{T}_C$ when turning the power screw. This toque is then given by

$$\boldsymbol{T}_C = \frac{\boldsymbol{P}_l \mu_C d_C}{2} \tag{9.46}$$

where d_C is the average major and minor collar diameter and μ_C is the *collar friction coefficient* between the collar and the support. The *total power screw torque* $\boldsymbol{T}_P$ is the summation of Eqs. (9.45) and (9.46), which gives

$$\boldsymbol{T}_P = \frac{\boldsymbol{P}_l d_p \left(l_T + \pi\mu_T d_p \sec\gamma\right)}{2\left(\pi d_p - \mu_T l_T \sec\gamma\right)} + \frac{\boldsymbol{P}_l \mu_C d_C}{2} \tag{9.47}$$

The *coefficient of friction* of screw threads varies with the nut materials, and it is not usually definite in value. For lubricated steel on steel or cast iron, the friction coefficient range is about 0.11–0.17. For steel on brass or bronze, it is about 0.1–0.15. For thrust collar friction, the start friction coefficient is on the high side of that range, and the running friction coefficient is a little lower than the low side of that range, i.e. about (0.06–0.09). For specific materials, one should conduct experiments or consult dedicated references or handbooks to define the friction coefficient of the pair of materials used in the design; see, e.g. Rothbart and Brown (2006).

With the designer attempting to have as low value of the coefficient of friction as possible, the self-locking should always be checked.

9.5.2 Power Screw Efficiency

For evaluating power utility, one can let $\mu_T = 0$. This will define the torque, and thus the power, to raise the load without any friction. From Eq. (9.45), the torque needed to raise the load only without friction is $\boldsymbol{T}_0 = \boldsymbol{P}_l l_T/2\pi$, where l_T is the thread lead. The *power screw efficiency* η_P is therefore $\eta_P = \boldsymbol{T}_0/\boldsymbol{T}_R$ or

$$\eta_P = \frac{\boldsymbol{P}_l l_T}{2\pi \boldsymbol{T}_R} \tag{9.48}$$

This should demonstrate that the *square thread* would be somewhat more efficient than the *Acme* or trapezoidal threads. For other construction needs, *Acme* threads are still being used as power screw threads.

9.5.3 Stresses in Power Screws

The stresses developed in power screws are about the same as the stresses developed in threads as in Sections 9.3.1 and 9.3.3 with the *tightening torque* $\boldsymbol{T}_T$ replaced with the *raising torque* $\boldsymbol{T}_R$ of Eqs. (9.43) and (9.46). The collar stresses are similarly treated. Several locations on the power screw should be checked for safety factor. The collar root is subjected to normal stress due to external load, normal stress due to collar bending, shear stress due to torsion, and direct stripping shear stress. The thread root is subject to normal stress due to external load, normal stress due to tooth bending, shear stress due to torsion, and direct stripping shear stress as for the root area defined by Eq. (9.7). These are usually in different planes of the ***3D*** coordinates. The values of these stresses are evaluated similar to previously indicated. The assumption that only one thread is carrying the load can cause unrealistic conclusions, and numerical analysis such as ***FE*** is highly recommended. The external load is usually compressive, and buckling check over the full power screw length is in order. The bearing stress over the thread surface can also be checked.

Figure 9.9 shows a sketch of a power screw sections in Figure 9.9a with A and B locations highlighted, and their stress state in Figure 9.9b for the two locations of A and B is demonstrated. The first location at (A) is the connection at the root of the thread, and the second location at (B) is the point between the collar and the screw stem as demonstrated in Figure 9.9a. Stresses in A and B are caused by the loads on the power screw at these locations. The *raising torque* $\boldsymbol{T}_R$ and *collar friction torque* $\boldsymbol{T}_C$ of Eqs. (9.43) and (9.46) generate a shear stress in the thread root location at A due to raising torque $\boldsymbol{T}_R$ and a shear stress in the collar root location at B due to both raising torque $\boldsymbol{T}_R$ and collar friction torque $\boldsymbol{T}_C$. This is assuming that the driver is at the top of the power screw, which is usually the case. If the driver is the nut, both torques will act on both locations. The *shear stress* τ_{xz} due to torsion at location A may then be assumed as follows; see Eq. (9.43) and Eq. (6.67) of Chapter 6.

$$\tau_{xz}|_A = \frac{\boldsymbol{T}_R\left(d_r/2\right)}{J_y} = \frac{16\boldsymbol{T}_R}{\pi d_r^3} = \frac{8}{\pi d_r^3}\frac{\boldsymbol{P}_l d_p\left(l_T + \pi\mu_T d_p\right)}{\left(\pi d_p - \mu_T l_T\right)} \tag{9.49}$$

where d_r is the root diameter, d_p is the pitch diameter, μ_T is the thread friction coefficient, and l_T is the thread length.

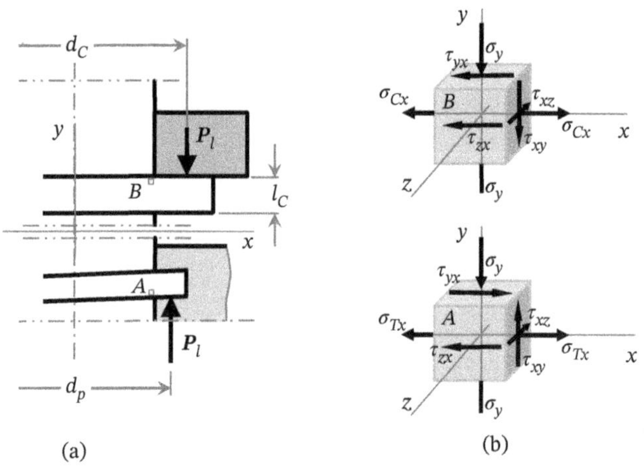

Figure 9.9 A sketch of a power screw sections in (a) and their stress state in (b) for two locations to be checked. The first location at (A) is the connection at the root of the thread, and the second location at (B) is the point between the collar and the screw stem.

The *shear stress* τ_{xz} at location *B* due to torsion is then (see Eqs. (9.43) and (9.46) and Eq. (6.67)):

$$\tau_{xz}|_B = \frac{(\boldsymbol{T}_R + \boldsymbol{T}_C)(d_r/2)}{J_y} = \frac{16(\boldsymbol{T}_R + \boldsymbol{T}_C)}{\pi d_r^3} = \frac{8}{\pi d_r^3}\left(\frac{\boldsymbol{P}_l d_p (l_T + \pi \mu_T d_p)}{(\pi d_p - \mu_T l_T)} + \boldsymbol{P}_l \mu_C d_C\right) \tag{9.50}$$

where d_r is the root diameter, d_p is the pitch diameter, l_T is the thread length, μ_T is the thread friction coefficient, and μ_C is the collar friction coefficient.

The simple *normal compressive stress* σ_y is due to the applied load $\boldsymbol{P}_l$ on the root area. This provides the same stresses at locations *A* and *B*, which gives the following value:

$$\sigma_y = -\frac{4\boldsymbol{P}_l}{\pi d_r^2} \tag{9.51}$$

where d_r is the root diameter.

The *normal stress* caused by *bending* of the tooth σ_{Tx} at location *A* along its *full nut length* l_N is given by (see Eq. (6.61))

$$\sigma_{Tx}|_A = \frac{12\boldsymbol{M}_z(p/4)}{(p/2)^3(\pi d_r N_T)} = \frac{12\boldsymbol{P}_l(p/4)^2}{(p/2)^3(\pi d_r N_T)} = \frac{6\boldsymbol{P}_l}{\pi d_r N_T p} = \frac{6\boldsymbol{P}_l}{\pi d_r l_N} \tag{9.52}$$

where d_r is the root diameter, N_T is the *number of engaging teeth*, l_N is the nut length, and *p* is the pitch.

The *normal stress* caused by *bending* of the collar σ_{Cx} at location *B* along its full *length* l_C is given by (see Eq. (6.61))

$$\sigma_{Cx}|_B = \frac{12\boldsymbol{M}_z(l_C/2)}{(l_C)^3(\pi d_r)} = \frac{12\boldsymbol{P}_l((d_C - d_r)/2)}{2(l_C)^2(\pi d_r)} = \frac{3\boldsymbol{P}_l(d_C - d_r)}{\pi d_r l_C^2} \tag{9.53}$$

where d_r is the root diameter, d_C is the collar diameter, and l_C is the *collar length* or thickness.

Due to the applied load and according to Eq. (9.7), the *stripping shear stress* τ_{xy} at location *A* over the root area of all teeth at location *A* is given by

$$\tau_{xy}|_A = \frac{\boldsymbol{P}_l}{A_{TS}} = \frac{\boldsymbol{P}_l}{\pi d_r (0.5\, l_N)} \tag{9.54}$$

where A_{TS} is the thread-stripping shear area and l_N is the nut length.

The *stripping shear* τ_{xy} over the root area of the collar at location *B* is given by

$$\tau_{xy}|_B = \frac{\boldsymbol{P}_l}{A_{CS}} = \frac{\boldsymbol{P}_l}{\pi d_r (l_C)} \tag{9.55}$$

where A_{CS} is the stripping shear area of the collar and l_C is the collar length or thickness.

The procedure to find the different ***3D*** stresses (Figure 9.9b), power screw performance, and safety from the previous situations is best utilized by a computer code. This should be useful particularly for several *CAD* iterations or optimization. The *von Mises* in ***3D*** is best utilized with the evaluation of the *principal stresses* or *eigenvalues* as defined by Eq. (6.101). The ***eigenvalues*** are of the ***3D*** *stress matrix* or *stress tensor* defined by Eq. (6.97). MATLAB is useful in that regard to develop such a code as indicated in Section 6.4.2.

9.5.4 Ball Screws

Regular power screws have threads with clearances to allow for sliding. This is not recommended for many applications such as lead screws of numerically controlled machine tools. Ball screws can solve this problem by eliminating the backlash through preloading means such as two half nuts, split, shifted, preloaded nuts, or oversized balls. The loads are transferred through the ball set from the screw shaft to the nut or vice versa (Figure 9.10). This

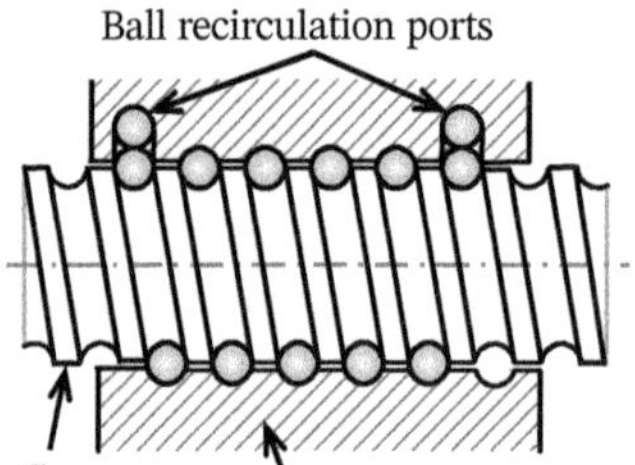

Figure 9.10 A sketch of a ball screw. The ball recirculation is usually off the nut. The balls exit the screw contact and move to the other end to engage the screw contact. When the screw rotates clockwise, the exit is on the right and the inlet on the left.

minimizes friction and thus maximizes efficiency. The preloading to eliminate backlash, however, causes some friction or preload torque, which is usually given by manufacturer; see the internet segment of the references of this chapter. The screw shaft is usually rolled with precision and may be ground. The ball recirculation is usually through the inside or off the nuts as shown in Figure 9.10. The balls exit the screw contact and move to the other end to engage the screw contact. If the screw rotates clockwise, the ball exit would be on the right, and the ball inlet on the left as shown in Figure 9.10, off-page circulating passage or track, is not shown. Many alternative designs are available for the ball recirculation schemes.

The design procedure of ball screws is similar to the rolling bearings in Chapter 11 of the text. The ball screw life under axial load is usually calculated for one million revolutions. The number of ball turns supporting the load affects the dynamic and static load carrying capacity. The usual number of ball turns is about 3–6 or more. Figure 9.10 shows five ball turns. The initial selection can be secured through the manufacturer's availability of ball screw type that fits the design.

The standards concerned with ball screw dimensions, loads, and tests are available in ASME B5.48 (1977) or ISO 3408-2 (1991), ISO 3408-3 (2006), ISO 3408-4 (2006), and ISO 3408-5 (2006). Ball screw manufacturers provide information on specific dimensions, allowable loads, speed rating, preloading torque, service life, and lubrication requirements for available types and dimensions; see the Internet segment of the references of this chapter.

Example 9.5 Use a MATLAB code to find the proper design and safety of a single thread power screw jack that should lift a load of 50 [kN] or 11 [klb]. Since the design is in a critical application, use a high initial safety factor of 4.5. Start with initial synthesis as defined in Section 8.4.1. Use a nut length equals to the thread nominal or major diameter and a general friction coefficient of 0.15. With reference to Figure 9.9, use an average collar diameter d_C equals to 1.1 the nominal or major thread diameter d_m. Use a collar length or thickness l_C equals to twice the thread pitch p.

Solution

With reference to Example 9.3, the initial synthesis used a simple material with $S_y = 260$ [MPa] or 37.5 [kpsi]. With relevant equations, a MATLAB code is developed to calculate all necessary requirements.

Data: Initially $\boldsymbol{F}_l = 50$ [kN] or 11 [klb], $p = 2$ [mm] or $N_T = 10$ [/in], $\mu_T = \mu_T = 0.15$, $l_N = d_m$, $d_C = 1.1\ d_m$, and $l_C = p$.

Initial synthesis uses material with $E = 207$ [GPa] or 30 [Mpsi], and $S_y = 260$ [MPa] or 37.5 [kpsi]. Using a safety factor of 4.5 generates a chart lookup material with a yield strength of about $S_y = 60$ [MPa] or 8.3 [kpsi] instead of $S_y = 260$ [MPa] or 37.5 [kpsi]. According to Figure 8.19, this suggests a bolt with ISO M36 or closer to the US 1½ [in] diameter. From Table 9.1, the standard coarse pitch p for M36 bolt is 4. The standard number of teeth per inch for the US 1½ [in] bolt is 6. Initially we may use these for the square or *Acme* thread.

Assume a minimum material of ISO 8.8 and SAE 5.2 according to Table 9.2 with a proof strength of 600 [MPa] or 85 [kpsi]. The MATLAB code is used to calculate all pertinent output as follows. The units are added, and the output is rearranged in single lines for each emboldened category. Safety factors for ***3D*** *von Mises* are also emboldened.

Input Parameters

- P_N = 50000 [N], P_lb = 11000 [lb], dm_mm = 36 [mm], dm_in = 1.5000 [in], lN_mm = 36 [mm], lN_in = 1.5000 [in], p = 4, NT = 6, dC_mm = 39.6000 [mm], dC_in = 1.6500 [in], lC_mm = 8 [mm], lC_in = 0.3333 [in].

Calculated Geometry

- dp_mm = 34 [mm], dr_mm = 32 [mm], dp_in = 1.4167 [in], dr_in = 1.3333 [in], AT_mm2 = 855.2986 [mm^2], AT_in2 = 1.4849 [in^2]

Input Material Properties (Same for Thread and Nut)

- Sp_MPa = 600 [MPa], Sp_kpsi = 85 [kpsi]

Normal Stress due to Loading

- sigma_MPa = 58.4591 [MPa], sigma_kpsi = 7.4079 [kpsi]

Raising Torque and Its Direct Shear Stresses on Thread and Collar

- mu = 0.1500, T_torqueS_SI = 160.2310 [N.m], T_torqueS_US = 1.4688e+03 [lb.in], T_torqueC_SI = 127.5000 [N.m], T_torqueC_US = 1.1688e+03 [lb.in], T_torque_SI = 287.7310 [N.m], T_torque_US = 2.6375e+03 [lb.in], tau_MPa = 44.7205 [MPa], tau_kpsi = 5.6670 [kpsi], tau_strip_MPa = 27.6311 [MPa], tau_strip_kpsi = 3.5014 [kpsi], tau_stripC_MPa = 62.1699 [MPa], tau_stripC_kpsi = 7.8782 [kpsi]

Stresses [MPa] and Von Mises Safety at Thread Location A for SI

- Sigma_x = 82.8932 [MPa], Sigma_y = −58.4591 [MPa], Sigma_z = 0 [MPa], Tau_xy = 27.6311 [MPa], Tau_zx = −44.7205 [MPa], Tau_yz = 0 [MPa]
- Principal_Stresses_A_MPa = (106.3341 −16.9207 −64.9793) [MPa], vonMises_MPa = 153.0517 [MPa], KSF_3D_A_SI = ***3.9202***

Stresses [kpsi] and Von Mises Safety at Thread Location A for the US

- Sigma_x = 15.7563 [kpsi], Sigma_y = −7.4079 [kpsi], Sigma_z = 0 [kpsi], Tau_xy = 3.5014 [kpsi], Tau_zx = 5.6670 [kpsi], Tau_yz = 0 [kpsi],
- Principal_Stresses_A_kpsi = (18.0206 −1.6445 −8.0276) [kpsi], vonMises_kpsi = 23.5156 [kpsi], KSF_3D_A_US = ***3.6146***

Stresses [MPa] and Von Mises Safety at Collar Location B for SI

- Sigma_x = 177.1842 [MPa], Sigma_y = −58.4591 [MPa], Sigma_z = 0 [MPa], Tau_xy = 62.1699 [MPa], Tau_zx = 44.7205 [MPa], Tau_yz = 0 [MPa]
- Principal_Stresses_B_MPa = (201.9316 −7.6643 −75.5422) [MPa], vonMises_MPa = 250.5291 [MPa], KSF_3D_A_SI = ***2.3949***

Stresses [kpsi] and Von Mises Safety at Collar Location B for the US

- Sigma_x = 22.4528 [kpsi], Sigma_y = −7.4079 [kpsi], Sigma_z = 0 [kpsi], Tau_xy = 7.8782 [kpsi], Tau_zx = 5.6670 [kpsi], Tau_yz = 0 [kpsi]
- Principal_Stresses_B_kpsi = (25.5888 −0.9712 −9.5727) [kpsi], vonMises_kpsi = 31.7470 [kpsi], KSF_3D_A_US = ***2.6774***

Thread Efficiency Percent

- Efficiency_SI = 19.8657 %, Efficiency_US = 19.8657 %

The variables are kept as defined in the MATLAB code. These values are not verified by hand calculations. It is left as an exercise to validate the MATLAB code.

The results of this example suggest several options to improve the design particularly assumed configuration and geometry. Some optimization is also possible. This is discussed in Section 9.7.

9.6 Permanent Joints

Riveted, soldered, brazed, welded, and adhesive bonded joints are permanently joining the machine members. These members are usually sheet metal components that form more complex ***3D*** members such as brackets, bearing bases, and frames or boxes housing gears or other components. Some of these are originally made of cast iron, but permanent joining of sheet metal parts would be more attractive for small number of produced housings. Rivets are still used in large structures such as bridges or ever airplanes. Recently, plane structures are produced by bonding of composite materials.

Rivets are permanent fasteners that can be treated as screws under very high tightening forces. These forces are generated from the residual stresses left after the plastic deformation the rivets undergo to join the two sheet metal members. The rivet can also be riveted under very high temperature forging so that to generate thermal stresses when it is cold. The rivets stems are calculated as reaching the plastic yield point and more. The rivets cross sections are usually subject to direct shear in the connection design. The rivet would replace the bolt in Figure 9.4a with loads rotated to be transverse and in opposite directions such that the two sheet members of the joint are directly shearing the rivet. The tightening force between the two sheet metal members should reduce the shear on the rivet. This is a statically indeterminate connection that might be approached by numerical analysis if the computer code can handle the nonlinearity of the plastic deformation and the contact problem requirement. Simplified analysis might help, but the conventional perception created a common configuration of using so many rivets to be safe with the simplified mathematical model. Many codes are available to facilitate such solution to be ultrasafe particularly for bridges, buildings, boilers, and similar structures. One should refer to codes and dedicated handbooks such as the *American Railway Engineering Association*, the boiler construction code of the ASME, or the *American Institute of Steel Construction Handbook*.

Soldering and brazing are other permanent means of joining components that are mainly of light sheet metal form that does not need any undue mechanical strength. The *soft soldering* material is usually composed of *tin* and *lead*. The hardest and lowest melting temperature (180 [°C] or 356 [°F]) alloy consists of about 66% tin and 34% lead. More lead would be a softer, cheaper, and having a higher melting temperature for a soldering material. Fluxes are also used with soft soldering such as chloride of zinc, rosin, or chloride of ammonia. Hard soldering of iron, brass, copper, aluminum, silver, etc. is called *brazing*. Red heat is needed for hard soldering or brazing. The soldering material is a thin film or filler of brass or aluminum silicon, nickel, silver, or others. A flux of *borax* is also needed to protect from oxides and oxidation. The composition of brazing brass is in the range of about 58–34% copper and 42–66% zinc, the higher copper percentage for a stronger brazing material. Tin and lead may be added to improve fusibility or provide a different white or grayish color from the usual reddish yellow. For more information, one may consult standards or handbooks such as AWS A5.8M/A5.8 (2011), or Oberg et al. (2012).

9.6.1 Welding

With the availability of various hot-rolled plates and shapes of different profiles, a designer should be able to construct lightweight and strong weldments with the utility of holding fixtures. A small sample is the *jib crane* discussed in Chapter 8. A hot-rolled section (I-beam) and pipes were used with some plates welded to form the main components of the jib crane. One can notice that most welds are fillet welds; however, butt welds shown in Figure 9.11 are extensively used in some applications such as pressure vessels. When sections or plates are welded together, there should be enough clearance between them for proper welding operation.

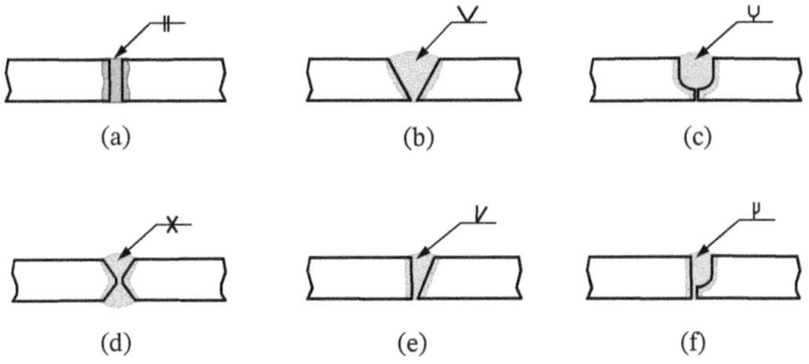

Figure 9.11 Sketches of frequently used butt welding types and simplified welding symbols (arrow to the right). Different end preparations are shown, which are mainly functions of plate thicknesses. No need to show end preparations with welding symbols. As the thickness increases, the end preparation is more pronounced as given in (b), (c), and (d). End preparations in (e) and (f) are for intermediate thicknesses.

Welding uses fillers and electrodes that would melt with the welded parts under extensively introduced or generated heating. A great opportunity of metallurgical changes in the parent metal occurs in the vicinity of the weld, which is called the *heat affected zone (HAZ)*. In addition, residual stresses result due to clamping or the order of welding passes and after cooling. Heat treatment may be needed after welding to have stress relieving. If the weldments are thick, preheating should be administered to reduce these problems.

9.6.1.1 Welding Types and Symbols

The main welding types are the *butt welds* and the *fillet welds*. Figures 9.11 and 9.12 show sketches of some of these types. The black lines are for the plate edges with or without end preparation. The grayish area is for the molten weld zone, indicating that the plates are melted to some extent where the filler material fuses with the base material.

Figure 9.11 displays sketches of frequently used ***butt welding*** types and simplified welding symbols (arrow to the right). Different end preparations are shown, which are mainly functions of *plate thicknesses*. No need to show end preparations with adopting welding symbols. Figure 9.11a indicates a *square butt weld*. This is usually suitable for plate thicknesses of up to 6 [mm] or 1/4 [in] for regular *shielded metal arc welding* and can go up to 16 [mm] or 5/8 [in] for *submerged arc welding*. Figure 9.11b shows a *single V weld*. This is usually suitable for thicknesses less than 20 [mm] or 3/4 [in]. Unlimited plate thicknesses are possible if standards such as groove preparation are followed. Figure 9.11c specifies a *U weld*. This is usually suitable for thicknesses more than 40 [mm] or 1 1/2 [in]. Figure 9.11d shows an *X weld* or *double V welds* on each side. This is usually suitable for thicknesses more than 20 [mm] or 3/4 [in]. Figure 9.11e indicates a single *bevel* weld, while Figure 9.11f defines a *J* weld type. These are usually suitable for thicknesses more than 40 [mm] or 1 1/2 [in] particularly *J* weld type. As the thickness increases, the end preparation is more pronounced as given in (b), (c), and (d), while (e) and (f) are for intermediate thicknesses. For high thicknesses *multipasses* are employed with each pass of a thickness, not usually more than about 8 [mm] or 1/3 [in]. There are *supplementary symbols* that can be drawn onto the welding symbol. Some of these are such as the *flat finish flush* "-" of Figure 9.11a, where the supplementary symbol will be on top of the *square butt* symbol "||." The other supplementary symbol is for *convex* weld top "⌒" of Figure 9.11b–f, where the convex symbol will be on top of the *V*, *U*, *X*, *bevel*, and *J* symbols. For more details on welding process preparation and allowances, one should consult standards such as AWS D1.5M-D1.5 (2002) and ISO 9692-1 (2013).

Figure 9.12 shows sketches of some ***fillet welding*** types and simplified welding symbols. Different end preparations are also possible for higher plate thicknesses. *Lap welding* is in Figure 9.12a, *T joint* is in Figure 9.12b, and *corner weld* is in Figure 9.12c. The end preparation or fillet welding groove is like the butt welding in Figure 9.11. There are also supplementary symbols that can be drawn onto the welding symbol like the *butt welding* types. One of these is the *concave* weld top "⌣" which can be on top of the weld symbol and mainly for some fillet welds such

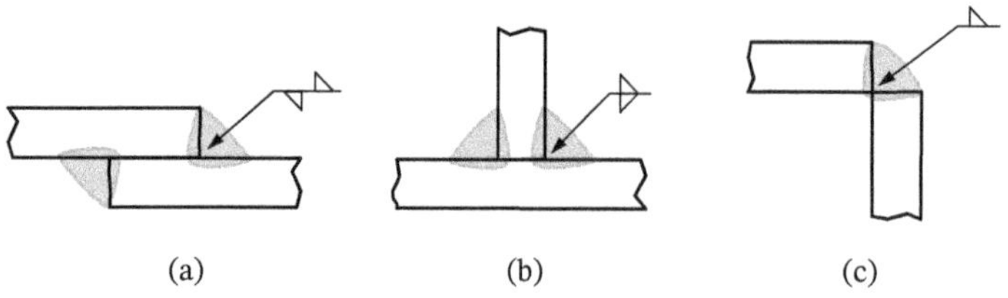

Figure 9.12 Sketches of some fillet welding types and simplified welding symbols. Different end preparations are also possible for higher plate thicknesses. Lap welding is in (a), T joint is in (b), and corner weld is in (c). The end preparation or groove is like the butt welding in Figure 9.11.

as on top of the *fillet-welds symbols* in Figure 9.12 (but not shown here). Complete expressive information can be displayed on the welding symbols and defined in standards. For more details, one should consult welding standards such as AWS A2.4 (1998) and ISO 2553 (2013). Both *American Welding Society* (AWS) and the ISO provide an extensive array of more extensive standards defining specifications for different welding codes.

Other organizations provide other welding codes for different application such as ASME for boilers and pressure vessels, American Petroleum Institute (API) for pipelines and related facilities, and other standards of some countries.

Welding process can also be categorized as *fusion welding* and *solid-state welding*. The process and manufacturing details are useful; however, the main interest of this text is on the design aspects of welding.

Fusion welding is the process where heat melts *the base metal*, and a *filler* metal is added to the molten junction. The molten junction belongs mainly to the base metal. Fusion welding methods are *arc welding* (AW), *resistance* welding (RW), *oxyfuel gas* welding (OFW), and others like *electron beam* welding and *laser beam* welding. The common welding processes are shielded metal arc welding (SMAW), gas metal arc welding (GMAW), flux-cored metal arc welding (FCAW), and submerged arc welding (SAW). Electrodes, fluxes, and processes are different for GMAW, FCAW, SMAW, SAW, gas tungsten AW, and plasma AW. For details of these different production methods, more dedicated references are to be consulted such as Oberg et al. (2012) and Groover (2010).

Solid-state welding is a process where pressure and/or heat diffuse the two base metal parts with no filler material added. Solid-state welding methods include diffusion welding (DFW), friction welding (FRW), and ultrasonic welding (USW). For more details of these manufacturing methods, more dedicated references are also to be consulted such as Oberg et al. (2012) and Groover (2010).

9.6.1.2 Stresses in Welded Joints

Welded joint is produced by melting the joint boundaries with or without a filler material. The homogeneity and form of the junction is somewhat randomly shaped. Stresses are then not uniformly generated (Norris 1945; Salakian and Claussen 1937). Simplified modeling and assumptions are pursued to be able to reasonably design the joint. The first assumption is the consideration that the *welding throat thickness* t_W is the main sidewise leg of the area carrying the load. The other side is the welding length l_W. Other extra materials are bypassed. The other assumption is that the generated stress is uniform. Figure 9.13 is a sketch of loaded *butt* and *fillet* welds showing assumed dimensions and loading, where $\boldsymbol{P}_x$ is the *load* in x direction and $\boldsymbol{P}_y$ is the load in the y direction.

Butt weld in Figure 9.13a has the assumed weld area of the *weld throat thickness* t_W multiplied by the *weld length* l_W. The generated normal stress σ_x is assumed uniform and has the following basic value:

$$\sigma_x = \frac{\boldsymbol{P}_x}{t_W l_W} \tag{9.56}$$

The weld throat does not include the *concave* top or bottom of the weld in Figure 9.13, and it is usually taken as the sheet metal or *plate thickness* t_P. If the load is *dynamic*, the intersection of the concave surface with the plate

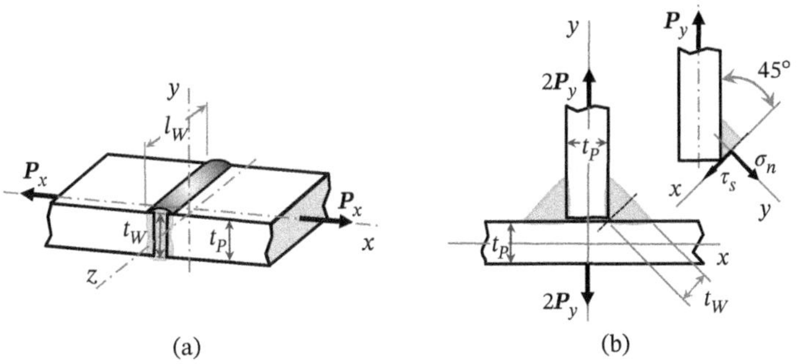

Figure 9.13 Sketches of loaded butt and fillet welds showing assumed dimensions and loading. In (a), a butt weld is shown. In (b), an example of a *T* joint with a free body diagram of the 45° cut is shown aside. The load in the cut is half the total load of $2P_x$.

causes stress concentration. It would then be advisable to grind the concave portion of the weld to make it flush with the plate. This would greatly reduce or eliminate the stress concentration.

Fillet weld in Figure 9.13b is an example of a *T* joint that can be applied to other similar weld cases. The *free body* diagram of the 45° cut is shown aside. The load in the cut is half the total load of $\boldsymbol{2P}_y$. The *internal normal force* $\boldsymbol{F}_n$ and *internal shear force* $\boldsymbol{F}_s$ are inclined by 45°. Neglecting the bending moment, the force analysis of this *free body* diagram gives the force in the local *y* direction normal to the cut as $\boldsymbol{F}_n = \boldsymbol{P}_y \sin 45°$ and the shear force in the local *x* direction as $\boldsymbol{F}_s = \boldsymbol{P}_y \cos 45°$. The internal stresses, namely, the *normal stresses* σ_n and the *shear stresses* τ_s are then obtained from the following relations:

$$\sigma_n = \sigma_y = \frac{\boldsymbol{P}_y \sin 45°}{t_W l_W} = \frac{\boldsymbol{P}_y (0.7071)}{t_W l_W}$$
$$\tau_s = \tau_{yx} = \frac{\boldsymbol{P}_y \cos 45°}{t_W l_W} = \frac{\boldsymbol{P}_y (0.7071)}{t_W l_W} \tag{9.57}$$

If the weld leg with the plate is equal to the *plate thickness* t_P, the *weld throat* thickness is then $t_W = t_P \cos 45°$ or $t_W = 0.707 t_P$. Using the *Mohr's circle* and Eq. (9.57), the *maximum principal stress* σ_1 is obtained by Eq. (6.85) such that

$$\sigma_1 = \frac{\sigma_y}{2} + \sqrt{\left(\frac{-\sigma_y}{2}\right)^2 + \tau_{xy}^2} = \frac{\boldsymbol{P}_y \sin 45°}{2 t_W l_W} + \sqrt{\left(\frac{\boldsymbol{P}_y \sin 45°}{2 t_W l_W}\right)^2 + \left(\frac{\boldsymbol{P}_y \cos 45°}{t_W l_W}\right)^2}$$
$$= \frac{\boldsymbol{P}_y}{2 t_P l_W} + \sqrt{\left(\frac{\boldsymbol{P}_y}{2 t_P l_W}\right)^2 + \left(\frac{\boldsymbol{P}_y}{t_P l_W}\right)^2} = 1.6180 \frac{\boldsymbol{P}_y}{t_P l_W} \tag{9.58}$$

The *maximum shear stress* τ_{max} is obtained by *Mohr's circle* according to Eq. (6.86) and Eq. (9.57) such that

$$\tau_{max} = \sqrt{\left(\frac{\sigma_y}{2}\right)^2 + \tau_{xy}^2} = \sqrt{\left(\frac{\boldsymbol{P}_y}{2 t_P l_W}\right)^2 + \left(\frac{\boldsymbol{P}_y}{t_P l_W}\right)^2} = 1.1180 \frac{\boldsymbol{P}_y}{t_P l_W} \tag{9.59}$$

The classical weld design depends on the throat thickness t_W and considers shear stresses only. The maximum shear stress of Eq. (9.59) becomes

$$\tau_{max} = 1.1180 \frac{\boldsymbol{P}_y}{t_P l_W} = 1.1180 \frac{\boldsymbol{P}_y}{0.7071\, t_W l_W} = 1.5811 \frac{\boldsymbol{P}_y}{t_W l_W} \tag{9.60}$$

This gives a higher shear stress for fillet welds, which should be considered.

Welds under bending and torsion are treated the same way other sections are subject to. The weldment section properties under these loads are defined such as the throat area, throat second area moment, and throat second polar moment of area. With large structures or thin sections, the welds are considered as thin areas with approximating section properties applied accordingly. The weld junctions are considered as lines or arcs with a weld throat thickness. The regular stress equations generated in direct transverse shear, bending moment, and torsion are then used on the weld junction. These are classical treatments that can be used as initial synthesis of welded joints. For more complex loadings, the numerical technique of ***FE*** can be used to a better estimation of the stress distribution in the welded joint. Available codes, and other numerical techniques, can be used for verification, redesign, and validation.

Design codes for weldments are existing to be followed for general or specific applications. Products are sometimes required to obey these codes for many products such as *boilers*, *presser vessels*, *automotive*, *sheet metal*, *structures*, *pipelines*, *buildings*, *bridges*, etc. The design *codes* usually utilize the classical analysis provided previously. They, however, specifically address the relevant junctions and joints in the application with concentration on the specific loading state, specific materials needed, production requirements, *weld joint efficiency*, environment, and the necessary safety. To design such joints under code requirements, the code should be followed. One of these is the ASME (2013) boiler and pressure vessel code (BPVC). Using this code, some optimization of pressure vessels parameters was conducted within the rules of the code; see, e.g. Abd El Aziz and Metwalli (2017). In this work, cylindrical pressure vessels with hemispherical ends are considered. They are required to hold a definite volume under a specific pressure. The optimum thicknesses of each hemispherical part and the cylindrical part satisfy the recommended ASME code. The optimum design also satisfied allowable stress constraints of potentially used materials. Other design or preparation codes can also be followed for other applications such as AWS D1.3/D1.3M (2008) and ISO 9692 (2013).

9.6.1.3 Welding Strength

The welding strength depends on the parent material, electrodes material, and the welding processes. The electrode or weld material is usually matching or stronger than the parent material. One should consult standards or manufacturers for selecting appropriate electrode or weld metal properties. Unless heat or forming treatment, the welding of cold-drawn material ends up as a hot-rolled material near the weld due to welding heat. Recommended minimum *allowable stresses* for butt welds in tension, bending, and compression is about 0.6 the *yield strength* S_y. of weld material or parent material. For butt or fillet welds subject to shear the suggested minimum *allowable stress* is about 0.3 the *yield strength* S_y. of weld material or parent material.

For *fatigue strength*, the regular procedure of machine elements can be used with the consideration of special stress concentration factor or fatigue strength reduction factor due to weldment. This factor ranges from about 1.2 to about 2.7. For butt weld, the factor is around 1.2 and for T-butt, weld it is approximately 2.0. For transverse fillet weld, it is about 1.5. For parallel fillet weld, it is around 2.7. The calculations are valid for both parent and weld materials.

9.6.1.4 Resistance Welding

Spot welding and *seam welding* are examples of resistance welding. This occurs when high electric current passes through the contact between the two parts and heats up due to the resistance. The two parts reach the point of melting that fuses the two parts together at the contact spot or seam. The seam is considered as many consecutive spots since the electrodes are rotating disks, and the electric current is fluctuating or pulsating. The welding region of the spot or seam depends on the current and pressure levels at the contact between the parts. Usually the design of such welding should construct the geometry so that the weld would be subjected to shear stress with no appreciable tension generated on the spot or seam. The shear stress is then simply the load divided by the spot or seam area. The application of this type of welding is generally applied to sheet metal work with relatively thin thicknesses. A relatively high safety factor is recommended.

9.6.2 Bonded Joints

Adhesives are becoming more attractive to use for permanently joining or *bonding* many components. The strength of these bonding materials is improving to warrant their utility in somewhat loaded components or relatively light constructions. The adhesive bonding has previously been used in very lightly loaded connections. With the combination of adhesives and composite materials, many components can be designed to save processing time and expensive fastening alternatives.

Bonded joints are similar in construction to brazing and soldering in some features. They utilize a filler material (an *adhesive*) between the two closely joined *adherend* surfaces to be bonded. Adhesives are, however, applied at room or somewhat higher temperatures, and they are not metallic. The adhesives are usually polymeric materials with a wide range of properties and strength. They range from simple glues, starch-based, or gums to cements, acrylics, or polymeric epoxy resins. The interest here is on *structural adhesives* that produce strong bond between more rigid components. The *liquid adhesive* needs *curing* by a chemical reaction to transform the liquid adhesive into a solid bonding medium that joins the parts. The curing or hardening may involve some low heat and/or a catalyst depending on the adhesive type. This can take time that may constitute a disadvantage.

The adhesion between joint *adherend* components should be strong to withstand the applied load. The adhesion is much better for cleaner and slightly roughened joint surfaces. The construction of the joint should capitalize on the advantages of the adhesive and should avoid the disadvantages. The adhesive *lap shear strength* as the load divided by the bond area (Figure 9.14) is usually more than the *cleavage* or *peel strength*, which is measured as the transverse load per *unit width* of the bonded joint. Combining formed geometry with the bonding can produce a stronger joint where a spot weld, a lip form, or a tongue shares the load with the bonding while the adhesive keeps the parts attached together. This is particularly useful in joints subject to bending. These *bend-loaded* joints should advisably be constructed differently to transfer bending into shear loading and eliminating peel stresses. Sketches of different types of ***lap joints*** with grayish and exaggerated adhesive thicknesses are shown in Figure 9.14. Shown in Figure 9.14(a–f) are several cases where (a) is a single lap, (b) is bevel ends preparations, (c) is step ends preparations, (d) is bent ends, (e) is double butt strap, and (f) is a double lap with half the load for each side of the lap. Except for the single lap in Figure 9.14a, all other types attempt to eliminate the *peeling* possibility by minimizing or eliminating the moment along the adhesive joint.

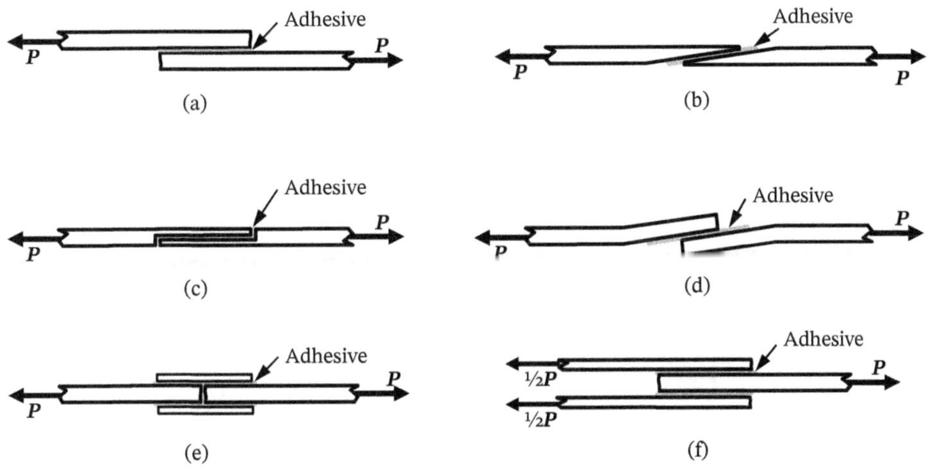

Figure 9.14 Sketches of different types of lap joints with grayish and *exaggerated adhesive thicknesses*: (a) single lap, (b) bevel ends preparation, (c) step ends preparation, (d) bent ends, (e) double butt strap, and (f) double lap with half the load for each side of the lap.

The main bonding adhesives that are used in machine design applications are the *structural adhesives*. In the order of *lap shear strength*, the adhesives are mainly ***epoxies***, ***phenolic***, ***acrylics***, ***polyurethane***, ***polyamides***, ***proteins***, etc.; see, e.g. Pizzi and Mittal (2003) and Pocius (2012). The *lap shear strength* of structural adhesives ranges from a high of about 40 [MPa] or 5.8 [kpsi] to a low of about 6.9 [MPa] or 1.0 [kpsi]. Other lower strength adhesives are available with ranges from a high of about 6.9 [MPa] or 1.0 [kpsi] to a low of about 0.01 [MPa] or 0.002 [kpsi] such as white glue, hot melts, rubber based, cellulosic, starch based, or pressure sensitive (Pocius (2012)). Some of the higher *lap shear strength* structural adhesives are such as follows (Pocius (2012)):

- Higher strength such as *modified epoxies* with strength of about 20–40 [MPa] or 3–6 [kpsi].
- Medium strength such as *polyimides, bis-maleimides,* or *modified phenolics* with strength of about 14–28 [MPa] or 2–4 [kpsi]. Also *unmodified epoxies* with strength of about 10–28 [MPa] or 1.5–4 [kpsi] and *rubber-modified acrylics* of about 14–24 [MPa] or 2–3.5 [kpsi].
- Lower strength such as *urethane, anaerobic acrylic, protein based,* or *cyanoacrylate* with strength of about 7–17 [MPa] or 1–2.5 [kpsi].

Some of these higher *lap shear strength* structural adhesives have *peel strength* measured in force per unit width of the joint, and the usual units are then [kN/m] or [lb/in]. The previous categories of higher lap shear strength have the *peel strength* in the range of 4.4–14 [kN/m] or 25–80 [lb/in], 0.18–8.8 [kN/m] or 1–50 [lb/in], or 0.18–3.5 [kN/m] or 1–20 [lb/in], respectively. The *peel strength* for each adhesive would vary within the previously stated ranges (Pocius (2012)).

Epoxies applications are in glass fiber-reinforced composites, adhesives for bonding aluminum structure components, aircraft honeycomb panels, sheet metal reinforcements for vehicles, bonding of heat exchanger, etc. *Phenolics* are used in bonding brake linings, abrasive wheels, etc. *Aerobic acrylics* and others are used in joining metal, plastic, glass, ceramics, and other substrates. They are also applied to lock fasteners and to bond a hub to a shaft and in gaskets.

Example 9.6 In addition to hand calculations, use a MATLAB code to check the weld between the hood top plate and the hood body pipe. Neglect the contribution of webs that are welded to strengthen the hood as shown in Figure 8.17 and discussed in Example 8.2. Consider the weld leg to be equal to the smaller of either plate or pipe thickness. Assume the strength of weld material to be the same as the parent materials and use the permissible or allowable welding strength.

Solution

With reference to Example 8.2 and the relevant equations, a MATLAB code is developed to calculate all necessary requirements.

Data: From Example 8.2, the material properties of the pipes are yield strength S_y of 36 [kpsi] or 250 [MPa], minimum ultimate tensile strength of 58 [kpsi] or 400 [MPa], and the endurance limit of 29 [kpsi] or 200 [MPa]; see Table A.7.5. Equation (7.58) suggests $S_e = 0.5\ S_{ut} = 0.5(400) = 200$ [MPa] or $0.5(59) = 29$ [kpsi]. The two values are the same as suggested by the quoted properties. The endurance value is useful if the load is fluctuating for many cycles.

According to section (b) of hood module in Example 8.2, assume the hood is subjected to bending moments only. The magnitude of this bending moment has been $\boldsymbol{M}_z = 37.995$ [kN m] or 327.87 [klb in]. The size of the hood tube is the DN-350 with a 355.6 [mm] outside diameter and a 9.53 [mm] thickness. For the US inch series round pipes or tubes, the section NSP-14 with a 14 [in] outside diameter and a 0.375 [in] thickness is selected.

With the known pipes thickness, this suggests a weld leg of 9.5 [mm] or 0.375 [in]. If the weld leg with the pipe is equal to the *pipe thickness* t_P, the *weld throat* thickness is then $t_W = t_P \cos 45^{s\circ}$ or $t_W = 0.707\ (9.5) = 6.72$ [mm] or $t_W = 0.707\ (0.375) = 0.265$ [in]. The second area moment of the circular weld about the z-axis is $I_z = \pi t_W (d_P/2)^3$. This is an approximation considering a thin hollow circular section. The values of the second area moment of the circular weld about the z-axis are then

$$I_z = \pi t_W \left(d_P/2\right)^3 = \pi\,(6.72)\,(355.6/2)^3 = 1.1866(10)^8\ [\text{mm}^4]$$
$$I_z = \pi t_W \left(d_P/2\right)^3 = \pi\,(0.265)\,(14/2)^3 = 2.8556(10)^2\ [\text{in}^4] \qquad \text{(a)}$$

The normals stresses in the weldment are therefore as follows:

$$\sigma_{y,\max} = \frac{\boldsymbol{M}_z\left(d_P/2\right)}{I_z} = \frac{37.995\left(10^3\right)(355.6/2)\,(10)^{-3}}{0.118\,66(10)^9(10)^{-12}} = 569\,39(10)^3\ [\text{Pa}] = 56.939\ [\text{MPa}]$$
$$\sigma_{y,\max} = \frac{\boldsymbol{M}_z\left(d_P/2\right)}{I_z} = \frac{327.87\left(10^3\right)(14/2)}{2.8556\,(10)^2} = 8037.2\ [\text{psi}] = 8.0372\ [\text{kpsi}] \qquad \text{(b)}$$

Depending on yield criterion only, the safety factor of the weldment is therefore as follows:

$$K_{\text{SF}}|_{\text{SI}} = \frac{S_y}{\sigma_{x,\max}} = \frac{250\left(10^6\right)}{(56.939)\left(10^6\right)} = 4.39$$
$$K_{\text{SF}}|_{\text{US}} = \frac{S_y}{\sigma_{x,\max}} = \frac{36\left(10^3\right)}{(8.0372)\left(10^3\right)} = 4.48 \qquad \text{(c)}$$

Using the MATLAB code, the output results are conforming the previous hand calculations. The MATLAB code includes a detailed calculation for the second area moment (outer and inner diameters), which gives the second area moment of the circular weld about the z-axis as 1.2549 (10^8) [mm^4] and 302.33 [in^4]. These cause higher safety factors of 4.6441 for SI and 4.7423 for the US cases, respectively.

Example 9.7 Use a MATLAB code to check the weld between the hood top plate and the hood body pipe considering the applied tensile load and the contribution of webs, which are welded to strengthen the hood as shown in Figure 8.17 and discussed in Example 8.2. Consider the weld leg to be equal to the smaller of either plate or pipe thickness. Assume the strength of weld material to be the same as the parent materials, and use the permissible or allowable welding strength.

Solution

With reference to Example 8.2 and the relevant equations, a MATLAB code is developed to calculate all necessary requirements.

Data: From Example 8.2, the material properties of the pipes are yield strength S_y of 36 [kpsi] or 250 [MPa], minimum ultimate tensile strength of 58 [kpsi] or 4200 [MPa], and the endurance limit of 29 [kpsi] or 200 [MPa]; see Table A.7.5. Equation (7.58) suggests $S_e = 0.5\,S_{\text{ut}} = 0.5(400) = 200$ [MPa] or 0.5(59) = 29 [kpsi]. The two values are the same as suggested by the quoted properties. The endurance value is useful if the load is fluctuating for many cycles.

According to Section (b) of hood module in Example 8.2, assume the hood is subjected to bending moments and normal force. The magnitude of the bending moment has been $\boldsymbol{M}_z = 37.995$ [kN m] or 327.87 [klb in]. The magnitude of the force has been $\boldsymbol{F}_y = 10$ [kN] or 2.2 [klb]. The size of the hood tube is the DN-350 with a 355.6 [mm] outside diameter and a 9.53 [mm] thickness. For the US inch series round pipes or tubes, the section NSP-14 with a 14 [in] outside diameter and a 0.375 [in] thickness is selected.

With the known pipes thickness, this suggests a weld leg of 9.5 [mm] or 0.375 [in]. If the weld leg with the pipe is equal to the *pipe thickness* t_P, the *weld throat* thickness is then $t_W = t_P \cos 45°$ or $t_W = 0.707\,(9.5) = 6.72$ [mm] or $t_W = 0.707\,(0.375) = 0.265$ [in]. The strengthening webs are four with a length of 1.5 [m] or 59 [in] and extend from the pipe diameter to the length of the top plate of the hood 1.0 [m] or 39 [in]; see Figure 8.18. Each welding area subject to normal or shear load is calculated. Stresses in these fillet welds are then calculated.

Circular weld under bending was considered in Example 9.6. The circular weld is also subjected to normal load $\boldsymbol{F}_y = 10$ [kN] or 2.2 [klb]. According to Eqs. (9.58) and (9.59) for fillet welds, the normal principal and maximum shear stresses are

$$\sigma_1 = 1.6180 \frac{\boldsymbol{P}_y}{t_P l_W} = 1.6180 \frac{10\,000}{9.5\,(\pi\,(355.6))} = 1.5246\ [\text{MPa}]$$
$$\sigma_1 = 1.6180 \frac{\boldsymbol{P}_y}{t_P l_W} = 1.6180 \frac{2\,200}{0.375\,(\pi\,(14))} = 215.82\ [\text{psi}] = 0.215\,82\ [\text{kpsi}] \tag{a}$$

and

$$\tau_{\text{max}} = 1.1180 \frac{\boldsymbol{P}_y}{t_P l_W} = 1.1180 \frac{10\,000}{9.5\,(\pi\,(355.6))} = 1.0535\ [\text{MPa}]$$
$$\tau_{\text{max}} = 1.1180 \frac{\boldsymbol{P}_y}{t_P l_W} = 1.1180 \frac{2\,200}{0.375\,(\pi\,(14))} = 149.13\ [\text{psi}] = 0.149\,13\ [\text{kpsi}] \tag{b}$$

Assume one may add the normal principal stress due to normal force to the normal stress due to bending as calculated in Example 9.6, one gets the estimated maximum normal stresses σ_{max} as follows:

$$\sigma_{\text{max}} = \sigma_{y,\text{max}} + \sigma_1 = 56.939 + 1.5246 = 58.464\ [\text{MPa}]$$
$$\sigma_{\text{max}} = \sigma_{y,\text{max}} + \sigma_1 = 8.0372 + 0.215\,82 = 8.2530\ [\text{kpsi}] \tag{c}$$

Depending on yield criterion only, the safety factor of the weldment is therefore as follows:

$$K_{SF}|_{SI} = \frac{S_y}{\sigma_{\text{max}}} = \frac{250\left(10^6\right)}{(58.464)\left(10^6\right)} = 4.28$$
$$K_{SF}|_{US} = \frac{S_y}{\sigma_{\text{max}}} = \frac{36\left(10^3\right)}{(8.2530)\left(10^3\right)} = 4.36 \tag{d}$$

Using the MATLAB code, the output results verified the previous hand calculations.

The safety factors in Eq. (d) produce less than the recommended minimum *allowable stresses* for butt welds in tension, bending, and compression of about 0.6 the *yield strength* S_y. of weld material or parent material. For butt or fillet welds subject to shear, the suggested minimum *allowable stress* is about 0.3 the *yield strength* S_y. of weld material or parent material.

Weld of strengthening webs is under shear. They were not considered in Example 9.6. The strengthening webs are subjected to shear force that is equal to the normal load $\boldsymbol{F}_y = 10$ [kN] or 2.2 [klb]. The welding length of the strengthening webs is then eight with each of a 1.5 [m] or 59 [in] length plus four of the extension from the pipe diameter to the 1.0 [m] or 39 [in] length of the top plate of the hood, or the total weld length is 14 578 [mm] or 572 [in]. Approximating the shear stress to be according to Eq. (9.59) for fillet welds, the maximum shear stresses τ_{max} are then

$$\tau_{\text{max}} = 1.1180 \frac{\boldsymbol{P}_y}{t_P l_W} = 1.1180 \frac{100\,00}{9.5\,(145\,78)} = 0.080\,727\ [\text{MPa}]$$
$$\tau_{\text{max}} = 1.1180 \frac{\boldsymbol{P}_y}{t_P l_W} = 1.1180 \frac{2200}{0.375\,(572)} = 11.467\ [\text{psi}] = 0.011\,467\ [\text{kpsi}] \tag{e}$$

The distortion energy theory suggests the shear yield strength $S_{ys} = (0.58)$ of the tensile yield strength S_{yt}. The associated safety factors are then

$$K_{SF}|_{SI} = \frac{0.58 S_y}{\tau_{\text{max}}} = \frac{0.58\left(250\left(10^6\right)\right)}{(0.080\,727)\left(10^6\right)} = 1796$$
$$K_{SF}|_{US} = \frac{0.58 S_y}{\tau_{\text{max}}} = \frac{0.58\left(36\left(10^3\right)\right)}{(0.011\,467)\left(10^3\right)} = 1821 \tag{f}$$

Using the MATLAB code, the output results verified the previous hand calculations. However, these safety factors in Eq. (f) are extremely large, and the design is expected to be extensively modified. The webs can be reduced to two instead of four. The webs can be drastically reduced in length and thickness; reduce the welding to sporadic dots rather than a continuous weld or even eliminating the webs. A good numerical analysis such as ***FE*** can be used for such a process.

9.7 Computer-Aided Design and Optimization

In this section, computer-aided programs are presented to calculate several design cases that have been introduced in this chapter. MATLAB codes for threads under simple tensile loads, preloading of bolts due to tightening, tightening torque, static, and fatigue safety factors are the main targets of bolted joint synthesis. MATLAB code is also developed to help in the design of power screws. The MATLAB code to help in the synthesis of welded joints is also introduced. These codes can be a starting platform to further development, synthesis, and optimization.

9.7.1 Threads Under Simple Tensile Load

For simply loaded threads under tension, there may be no need to consider tightening or fatigue. The simple calculations of Example 9.1 are thus the design process. Figure 9.15 is a MATLAB code to calculate the needed bolt material yield strength and the safety factor under tensile load only. This is for a specified yield strength of bolt material, and it also suggests the needed material yield strength according to the given safety factor as an initial input. The default values in the MATLAB code are for Example 9.1. The full code is available in the ***Wiley website*** under the name of ***CAD_Bolts.m***. This code can easily be used in the *CAD* of bolts for applications under simple loading on pure tension with no appreciable preloading or tightening. By few iterations, a better synthesis or an optimum can be achieved.

The input parameters in the code are highlighted in Figure 9.15 by the comment statements with the word **Input** in bold letters. The output parameters or results are displayed in the command window of MATLAB package. This is enacted by clearing the usual semicolon (;) at the end of the variable calculation statement or using a colon (,)

```
clc; clear all; format compact; format short; % CAD_Bolts.m
P_N = 33000/2                                 % Input load/bolt [N]
P_lb = 7333/2                                 % Input load/bolt [lb]
KSF = 2                                       % Input safety factor
dm_mm = 16                                    % Input synthesized diameter [mm]
dm_in = 3/4                                   % Input synthesized diameter [in]
p = 2                                         % Input the pitch [mm]
NT = 10                                       % Input number of teeth per inch
Material_Sy_MPa = 260                         % Input selected Material Sy [MPa]
Material_Sy_kpsi = 37.5                       % Input selected Material Sy [kpsi]

dp_mm = dm_mm - 0.649519*p                        % Pitch diameter [mm]
dr_mm = dm_mm - 1.299038*p                        % Root diameter [mm]
dp_in = dm_in - 0.649519/NT                       % Pitch diameter [in]
dr_in = dm_in - 1.299038/NT                       % Root diameter [in]
AT_mm2 = (pi/4)*((dp_mm+dr_mm)/2)^2               % Tensile stress area [mm^2]
AT_in2 = (pi/4)*((dp_in+dr_in)/2)^2               % Tensile stress area [in^2]
sigma_MPa = P_N/AT_mm2
sigma_kpsi = P_lb/AT_in2/1000
Syield_MPa = sigma_MPa*KSF                        % Needed yield strength
Syield_kpsi = sigma_kpsi*KSF                      % Needed yield strength
KSF_cal_SI = Material_Sy_MPa/ sigma_MPa           % Calculated safety factor
KSF_cal_US = Material_Sy_kpsi/ sigma_kpsi         % Calculated safety factor
```

Figure 9.15 MATLAB code to calculate the needed bolt material yield strength and the safety factor under tensile load only, for a specified yield strength of bolt material.

to indicate continuation with the next statement, and allowing the display of the result. The variables are usually defined by the comment statements at the right side of the code in Figure 9.15.

9.7.2 Preloading Due to Bolt Tightening

Usually, loaded threads are tightening two or more joint members. In that case, there is a need to consider tightening forces and stresses. The calculations of Examples 9.2 and 9.3 are then the design process. Figure 9.16 shows a MATLAB code to calculate the bolt and the joint members stiffnesses, the initial tightening, bolt force, and joint members force. This code is a continuation to the code in Figure 9.15. It also calculates the tightening force, the shear stress, and *von Mises* safety factor. The solution of Example 9.3 is an organized printout of the code results. The full code is available in the ***Wiley website*** under the name of ***CAD_Bolts_Preload.m***. This code can easily be used in the *CAD* of bolts for applications under preloading or tightening of two members of the same thickness. By few modifications, iterations, a better synthesis, or an optimum can be achieved. For more than two members in the joint, some modifications are needed.

The input parameters in the code are highlighted in Figure 9.16 by the comment statements with the word **Input** in bold letters. The output parameters or results are displayed in the command window of MATLAB package. This is enacted by clearing the usual semicolon (;) at the end of the variable calculation statement or using a colon (,) to indicate continuation with the next statement and allowing the display of the result. The variables are usually defined by the comment statements at the right side of the code in Figure 9.16.

9.7.3 Preloading, Bolt Tightening, and Fatigue Strength

Dynamically loaded threads may also be tightening two or more joint members. In that case, there is a need to consider tightening forces, stresses, and fatigue strength. The calculations of Examples 9.2 and 9.3 are then the design process. Figure 9.17.displays the MATLAB code to calculate the fatigue loading and stresses of the bolt in a joint with members stiffnesses and initial tightening. This code is a continuation to the code in Figures 9.15

```
                                                                 % CAD_Bolts_Preload.m
lb_mm = 32                                                  % Input bolt length [mm]
lb_in = 0.8                                                 % Input bolt length [in]
Alfa = 30                                                   % Input 1/2 cone angle
TA = tan(Alfa*(pi)/180);
ln_mm = log(5*(lb_mm*TA+0.5*dm_mm)/(lb_mm*TA+2.5*dm_mm));
km_SI = (pi*Material_E_GPa*10^9*dm_mm*(10^-3)*TA)/(2*ln_mm)            % Member stiffness [N/m]
ln_in = log(5*(lb_in*TA+0.5*dm_in)/(lb_in*TA+2.5*dm_in));
km_US = (pi*Material_E_Mpsi*(10^6)*dm_in*TA)/(2*ln_in)                 % Member stiffness [lb/in]
Ab_mm2 = pi*(dm_mm^2)/4;  Ab_in2 = pi*(dm_in^2)/4                      % Bolt area [mm^2] and [in^2]
kb_SI = Material_E_GPa*(10^9)*Ab_mm2*(10^-6)/(lb_mm*(10^-3))           % Bolt stiffness [N/m]
kb_US = Material_E_Mpsi* Ab_in2 * (10^6) / lb_in                       % Bolt stiffness [lb/in]
Stiffness_Ratio_SI = km_SI/ kb_SI
Stiffness_Ratio_US = km_US/ kb_US
Fi_SI= 0.75 * 0.9 * Material_Sy_MPa* AT_mm2                            % Initial tightening force [N]
Fi_US= 0.75 * 0.9 * Material_Sy_kpsi*(10^3) * AT_in2                   % Initial tightening force [lb]
Fb_SI= Fi_SI+ P_N/(1+Stiffness_Ratio_SI); Fb_US = Fi_US+ P_lb/(1+Stiffness_Ratio_US)% Bolt force [N] and [lb]
Fm_SI= Fi_SI-P_N/(1+(1/Stiffness_Ratio_SI)); Fm_US= Fi_US-P_lb/(1+(1/Stiffness_Ratio_US)) % Members force [N], [lb]
PS_N = Fi_SI* (1+(1/Stiffness_Ratio_SI)); PS_lb= Fi_US* (1+(1/Stiffness_Ratio_US))% Separation force [N],  [lb]
sigma_MPa = Fb_SI/AT_mm2 ; sigma_kpsi= Fb_US/AT_in2/1000               % With Preloading [MPa], [kpsi]
T_torque_SI = 0.2 * Fi_SI* dm_mm*(10^-3)                               % Tightening torque [N.m]
T_torque_US = 0.2 * Fi_US* dm_in                                       % Tightening torque [lb.in]
tau_MPa = 16 * T_torque_SI*10^3 /(2*pi*dr_mm^3)                        % [MPa] with 1/2 the torque
tau_kpsi = 16 * T_torque_US/(2*pi*dr_in^3* 1000)                       % [kpsi] with 1/2 the torque
vMises_MPa = sqrt(sigma_MPa^2 + 3*tau_MPa^2)                           % [MPa]
vMises_kpsi = sqrt(sigma_kpsi^2 + 3*tau_kpsi^2)                        % [kpsi]
KSF_vMises_SI = Material_Sy_MPa/ vMises_MPa                            % von Mises SI safety factor
KSF_vMises_US = Material_Sy_kpsi/ vMises_kpsi                          % von Mises US safety factor
```

Figure 9.16 MATLAB code to calculate the bolt and the joint members stiffnesses, initial tightening, bolt force, and joint members force. This code is a continuation to the code in Figure 9.15. It also calculates the tightening force, the shear stress, and von Mises safety factor.

and 9.16. It also calculates the carrying ratio due to the stiffness ratio, and the fatigue safety factor of the bolt. The solution of Example 9.4 is an organized printout of the code results. The full code is available in the ***Wiley website*** under the name of ***CAD_Bolts_Preload_Fatigue.m***. This code can easily be used in the *CAD* of bolts for applications exposed to dynamic loading, and the preloading or tightening of two members of the same thickness. By few modifications, iterations, a better synthesis, or an optimum can be achieved. For more than two members in the joint, some modifications are needed.

The input parameters in the code are highlighted in Figure 9.17 by the comment statements with the word **Input** in bold letters. The output parameters or results are displayed in the command window of MATLAB package. This is enacted by clearing the usual semicolon (;) at the end of the variable calculation statement or using a colon (,) to indicate continuation with the next statement and allowing the display of the result. The variables are usually defined by the comment statements at the right side of the code in Figure 9.17.

9.7.4 Power Screws

Loaded power screws have a higher torque to raise the load than to lower it. With that condition, there is a need to consider forces, stresses, at few locations to design the power screw. The calculations of Example 9.5 are, therefore, used as the design guide for usual cases of power screws. Figure 9.18 is a MATLAB code to calculate power screws inputs and outputs. It is split into two parts: (a) and (b). Figure 9.18 is the MATLAB code to input parameters and calculate power screw geometry, torques, and shear stresses due to torques, safety factors, and thread efficiency at A and B location. The code in Figure 9.18a is a pre-section to the code in Figure 9.18b. Figure 9.18b is the MATLAB code to calculate other stresses, safety factor, and thread efficiency at A and B locations. This code is a continuation to the code in Figure 9.18a. The solution of Example 9.5 is an organized printout of the code results. The full code is available in the ***Wiley website*** under the name of ***CAD_Power_Screw.m***. This code can easily be used in the *CAD* of power screws for applications similar to Example 9.5. By few modifications, iterations, a better synthesis, or an optimum can be achieved.

```
                                                    % CAD_Bolts_Preload_Fatigue.m
Sp_MPa = 600                                        % Input Material Sp[MPa] -ISO 8.8
Sp_kpsi = 85                                        % Input Material Sp[MPa] -SAE 5.2
Seb_MPa = 129                                       % Input Material Se [MPa] -ISO 8.8
Seb_kpsi = 18.6                                     % Input Material Se [MPa] -SAE 5.2
Sut_MPa = 830                                       % Input Material Sut [MPa] -ISO 8.8
Sut_kpsi = 120                                      % Input Material Sut [MPa] -SAE 5.2
Pmax_N = P_N                                        % Input maximum load [N]
Pmin_N = 0.0                                        % Input maximum load [N]
Pmax_lb = P_lb                                      % Input maximum load [lb]
Pmin_lb = 0.0                                       % Input maximum load [N]

CR_SI = 1/(1+Stiffness_Ratio_SI)                         % Carrying ratio
CR_US = 1/(1+Stiffness_Ratio_US)                         % Carrying ratio
Pa_N = (Pmax_N-Pmin_N)/2                                 % Alternating force [N]
Pa_lb = (Pmax_lb-Pmin_lb)/2                              % Alternating force [lb]
Pm_N = (Pmax_N+Pmin_N)/2                                 % Mean force [N]
Pm_lb = (Pmax_lb+Pmin_lb)/2                              % Mean force [lb]
sigmaba_MPa = CR_SI*(Pa_N)/(AT_mm2)                      % Alternating stress [MPa]
sigmaba_kpsi = (CR_US*(Pa_lb)/(AT_in2))/1000             % Alternating stress [kpsi]
sigmabm_MPa = CR_SI*Pm_N/AT_mm2+(Fi_SI/AT_mm2)           % Mean stress [MPa]
sigmabm_kpsi = (CR_US*Pm_lb/AT_in2+(Fi_US/AT_in2))/1000  % Mean stress [kpsi]
sigmai_MPa =Fi_SI/AT_mm2                                 % Initial stress [MPa]
sigmai_kpsi =Fi_US/(AT_in2*1000)                         % Initial stress [kpsi]
KSFbf_SI = Seb_MPa*(Sut_MPa-sigmai_MPa)/((sigmabm_MPa-sigmai_MPa)*Seb_MPa+(sigmaba_MPa*Sut_MPa))
KSFbf_kpsi = Seb_kpsi*(Sut_kpsi-sigmai_kpsi)/((sigmabm_kpsi-sigmai_kpsi)*Seb_kpsi+(sigmaba_kpsi*Sut_kpsi))
```

Figure 9.17 MATLAB code to calculate the fatigue loading and stresses of the bolt in a joint with members stiffnesses and initial tightening. This code is a continuation to the code in Figures 9.15 and 9.16. It also calculates the carrying ratio due to the stiffness ratio and the fatigue safety factor of the bolt.

```
clear all; clc; format compact; format short                    % CAD_Power_Screw.m
P_N = 50000                                                     % Input load [N]
P_lb = 11000                                                    % Input load [lb]
dm_mm = 36                                                      % Input synthesized diameter [mm]
dm_in = 1.5                                                     % Input synthesized diameter [in]
lN_mm = dm_mm%*1.5                                              % Input Nut length [mm]
lN_in = dm_in%*1.5                                              % Input Nut length [in]
p = 4                                                           % Input the pitch [mm]
NT = 6                                                          % Input number of teeth per inch
dC_mm = 1.1*dm_mm                                               % Input average Collar diameter [mm]
dC_in = 1.1*dm_in                                               % Input average Collar diameter [in]
lC_mm = 2.0*p                                                   % Input Collar length (thickness [mm]
lC_in = 2.0/NT                                                  % Input Collar length (thickness [in]
dp_mm = dm_mm-0.5 * p  ,  dr_mm= dm_mm-p                             % Pitch diameter and Root diameter [mm]
dp_in = dm_in-0.5/NT ,  dr_in= dm_in-1/NT                            % Pitch diameter and Root diameter [in]
AT_mm2 = (pi/4)*((dp_mm+dr_mm)/2)^2                                  % Average thread tooth area [mm^2]
AT_in2 = (pi/4)*((dp_in+dr_in)/2)^2                                  % Average thread tooth area [in^2]
Sp_MPa = 600                                                    % Input Material Sp[MPa] - ISO 8.8
Sp_kpsi = 85                                                    % Input Material Sp[MPa] - SAE 5.2
sigma_MPa = P_N/(AT_mm2), sigma_kpsi = P_lb/(AT_in2*1000)            % Normal Stress [MPa] and [kpsi]
mu = 0.15                                                       % Input Friction coefficient
T_torqueS_SI = P_N*dp_mm*10^-3*((p*10^-3)+pi*mu*dp_mm*10^-3)/(2*(pi*(dp_mm*10^-3)-mu*p*10^-3))% [N.m]
T_torqueS_US = P_lb*dp_in*((1/NT)+pi*mu*dp_in)/(2*(pi*(dp_in)-mu*(1/NT))) % [lb.in]
T_torqueC_SI = P_N*dp_mm*10^-3*mu/2                                   % [N.m]
T_torqueC_US = P_lb*dp_in*mu/2                                        % [lb.in]
T_torque_SI = T_torqueS_SI+ T_torqueC_SI                              % [N.m]
T_torque_US = T_torqueS_US+ T_torqueC_US                              % [lb.in]
tau_MPa = 16 * T_torque_SI*10^3 /(pi*dr_mm^3)                         % [MPa]
tau_kpsi = 16 * T_torque_US/(pi*dr_in^3* 1000)                        % [kpsi]
tau_strip_MPa = P_N / (0.5*pi*dr_mm*lN_mm)                            % [MPa]
tau_strip_kpsi = P_lb/ (0.5*pi*dr_in*lN_in*1000)                      % [kpsi]
tau_stripC_MPa = P_N / (pi*dr_mm*lC_mm)                               % [MPa]
tau_stripC_kpsi = P_lb/ (pi*dr_in*lC_in*1000)                         % [kpsi]
```

(a)

```
                                                                             % CAD_Power_Screw.m
Sigma_x= 6*P_N/(pi*dr_mm*lN_mm),  Sigma_y= -sigma_MPa,  Sigma_z = 0          % Normal stresses at A [MPa]
Tau_xy= tau_strip_MPa, Tau_zx= -tau_MPa, Tau_yz= 0                           % Shear stresses at A [MPa]
Stress_Tensor_A_MPa= [Sigma_xTau_xyTau_zx;  Tau_xySigma_yTau_yz;  Tau_zxTau_yzSigma_z];
[v e]=eig(Stress_Tensor_A_MPa); Principal_Stress_Tensor = e;                 % Principal_Dirctions = v
Sigma_1=e(3,3); Sigma_2=e(2,2);Sigma_3=e(1,1);                               % [MPa]
Principal_Stresses_A_MPa= [e(3,3), e(2,2), e(1,1)], s1 = Sigma_1; s2 = Sigma_2; s3 = Sigma_3;
vonMises_MPa= sqrt (((s1-s2)^2+(s2-s3)^2+(s3-s1)^2)/2)
KSF_3D_A_SI = Sp_MPa/vonMises_MPa                                            % Safety factor at A - SI units
Sigma_x= 6*P_lb/(pi*dr_in*NT*(1/NT)*1000), Sigma_y= -sigma_kpsi, Sigma_z = 0    % Normal stresses at A [kpsi]
Tau_xy= tau_strip_kpsi, Tau_zx= tau_kpsi, Tau_yz = 0                         % Shear stresses at A [kpsi]
Stress_Tensor_A_kpsi= [Sigma_xTau_xyTau_zx; Tau_xySigma_yTau_yz; Tau_zxTau_yzSigma_z];
[v e]=eig(Stress_Tensor_A_kpsi); Principal_Stress_Tensor = e;                % Principal_Dirctions = v
Sigma_1=e(3,3); Sigma_2=e(2,2);Sigma_3=e(1,1);                               % [kpsi]
Principal_Stresses_A_kpsi= [e(3,3), e(2,2), e(1,1)], s1 = Sigma_1; s2 = Sigma_2; s3 = Sigma_3;
vonMises_kpsi= sqrt (((s1-s2)^2+(s2-s3)^2+(s3-s1)^2)/2)
KSF_3D_A_US = Sp_kpsi/vonMises_kpsi                                          % Safety factor at A - US units
Sigma_x= 3*P_N*(dC_mm-dr_mm)/(pi*dr_mm*lC_mm^2), Sigma_y= -sigma_MPa, Sigma_z= 0          % Normal stress at B [MPa]
Tau_xy= tau_stripC_MPa, Tau_zx= tau_MPa, Tau_yz= 0                                        % Shear stresses at B [MPa]
Stress_Tensor_B_MPa= [Sigma_xTau_xyTau_zx; Tau_xySigma_yTau_yz; Tau_zxTau_yzSigma_z];
[v e]=eig(Stress_Tensor_B_MPa);Principal_Stress_Tensor= e;                   % Principal_Dirctions = v
Sigma_1=e(3,3); Sigma_2=e(2,2);Sigma_3=e(1,1);  % [MPa] or [kpsi]
Principal_Stresses_B_MPa= [e(3,3), e(2,2), e(1,1)], s1 = Sigma_1; s2 = Sigma_2; s3 = Sigma_3;
vonMises_MPa= sqrt (((s1-s2)^2+(s2-s3)^2+(s3-s1)^2)/2)
KSF_3D_B_SI = Sp_MPa/vonMises_MPa                                            % Safety factor at B - SI units
Sigma_x= 3*P_lb*(dC_in-dr_in)/(pi*dr_in*lC_in^2*1000), Sigma_y= -sigma_kpsi, Sigma_z= 0    % Normal stress at B [kpsi]
Tau_xy= tau_stripC_kpsi, Tau_zx= tau_kpsi, Tau_yz = 0                    % Shear stresses at B [kpsi]
Stress_Tensor_B_kpsi= [Sigma_xTau_xyTau_zx; Tau_xySigma_yTau_yz; Tau_zxTau_yzSigma_z];
[v e]=eig(Stress_Tensor_B_kpsi); Principal_Stress_Tensor= e;                 % Principal_Dirctions = v
Sigma_1=e(3,3); Sigma_2=e(2,2);Sigma_3=e(1,1);  % [MPa] or [kpsi]
Principal_Stresses_B_kpsi= [e(3,3), e(2,2), e(1,1)], s1 = Sigma_1; s2 = Sigma_2; s3 = Sigma_3;
vonMises_kpsi= sqrt (((s1-s2)^2+(s2-s3)^2+(s3-s1)^2)/2)
KSF_3D_B_US = Sp_kpsi/vonMises_kpsi                             % Safety factor at B - US units
Efficiency_SI= P_N*p*(10^-3)*(100) /(2*pi*T_torqueS_SI)         % 'Thread efficiency percent SI units, Equation (9.48)
Efficiency_US= P_lb*(1/NT)*(100) /(2*pi*T_torqueS_US)           % 'Thread efficiency percent US units, Equation (9.48)
```

(b)

Figure 9.18 (a) MATLAB code to calculate power screw geometry, loads, stresses, safety factors, and thread efficiency at *A* and *B* location. The code in (a) is a pre-section to the code in (b).

The input parameters in the code are highlighted in Figure 9.18 by the comment statements with the word **Input** in bold letters. The output parameters or results are displayed in the command window of MATLAB package. This is enacted by clearing the usual semicolon (;) at the end of the variable calculation statement or using a colon (,) to indicate continuation with the next statement and allowing the display of the result. The variables are usually defined by the comment statements at the right side of the code in Figure 9.18.

9.7.5 Permanent Weldment Joints

Welded joints can be treated as regular cross sections of machine elements. The weldment is usually a thin section attaching the welded elements. The regular treatment is by considering the thin section properties as

```
clc; clear all; format compact; format short;        % Jib_Crane_or_Similar_Weld.m
disp('-Circular Weld Safety SI-US for Bending Only')
Moment_Nm= 37995                                 % Input Bending Moment on the weld [N.m]
Moment_lbin= 327870                              % Input Bending Moment on the weld [lb.in]
Sy_MPa= 250                                      % Input weld material strength [MPa]
Sy_kpsi= 36                                      % Input weld material strength [kpsi]
tp_mm= 9.5                                       % Input pipe thickness [mm]
tp_in= 0.375                                     % Input pipe thickness [in]
dp_mm= 355.6                                     % Input pipe diameter [mm]
dp_in= 14                                        % Input pipe diameter [in]

tw_mm= 0.707 * tp_mm                                 % Weld throat thickness [mm]
tw_in= 0.707 * tp_in                                 % Weld throat thickness [in]
Iz_mm4 = pi*tw_mm*(dp_mm/2)^3                        % Weld second area moment [mm^4]
Iz_in4 = pi*tw_in*(dp_in/2)^3                        % Weld second area moment [in^4]
Iz_mm4 = pi*((dp_mm+2*tw_mm)^4-(dp_mm)^4)/64         % Calculated I_z-OR Comment
Iz_in4 = pi*((dp_in+2*tw_in)^4-(dp_in)^4)/64         % Calculated I_z-OR Comment

Sigma_MPa= Moment_Nm*(dp_mm/2)*(10^3)/(Iz_mm4)       % Normal stress [MPa]
Sigma_kpsi= Moment_lbin*(dp_in/2)*(10^-3)/(Iz_in4)   % Normal stress [kpsi]
KSF_SI = Sy_MPa/Sigma_MPa                            % Safety factor -SI
KSF_US = Sy_kpsi/Sigma_kpsi                          % Safety factor -US
```

(a)

```
                                                     % Jib_Crane_or_Similar_Weld.m(Cont.)
disp('-Circular Weld Safety SI-US for Bending and Normal Load')
Force_N= 10000                                      % Input Normal force on the weld [N]
Force_lb= 2200                                      % Input Normal force on the weld [lb]
lW_mm= 1500                                         % Input Web length [mm]
lW_in= 59                                           % Input Web length [in]
NW = 4                                              % Input Number of webs
lw_mm= pi*dp_mm                                       % Circular Weld length [mm]
lw_in= pi*dp_in                                       % Circular Weld length [in]
Sigma1_MPa = 1.618*Force_N/(tp_mm*lw_mm)              % Principal stress [MPa]
Sigma1_kpsi = 1.618*Force_lb/(tp_in*lw_in*1000)       % Principal stress [kpsi]
Sigma_max_MPa= Sigma_MPa+ Sigma1_MPa
Sigma_max_kpsi= Sigma_kpsi+ Sigma1_kpsi
KSF_SI = Sy_MPa/Sigma_max_MPa
KSF_US = Sy_kpsi/Sigma_max_kpsi

disp('-Webs Weld Safety SI-US for Normal Load')
lWw_mm= (8*lW_mm+4*(1000-dp_mm))                    % Strengthening Webs -Weld length [mm]
lWw_in= (8*lW_in+4*(39-dp_in))                      % Strengthening Webs -Weld length [in]
Tau_MPa= 1.118*Force_N/(tp_mm*lWw_mm)               % Shear stress [MPa]
Tau_kpsi= 1.118*Force_lb/(tp_in*lWw_in*1000)        % Shear stress [kpsi]
KSF_SI = 0.58*Sy_MPa/Tau_MPa
KSF_US = 0.58*Sy_kpsi/Tau_kpsi
```

(b)

Figure 9.19 (a) A MATLAB code for the calculation of weldment in *jib-crane hood* of Examples 9.6 and 9.7. Panel (b) is a continuation to panel (a).

an approximation considering the sections to be formed of lines or arcs and each having a small weld throat thickness. The treatment is just simpler that way. With that condition, there is then a need to consider forces and moments applied to such cross sections and the stresses generated thereof. The *jib*-crane weld calculations of Example 9.6 are used as a design guide for typical cases of simply loaded weldments. Figure 9.19 is a MATLAB code for the calculation of weldment in *jib-crane hood* of Examples 9.6 and 9.7. Figure 9.19b is a continuation to Figure 9.19a. The code results verified the solution of Examples 9.6 and 9.7.

The full welding code is available in the ***Wiley website*** under the name of ***Jib_Crane_or_Similar_Weld.m***. This code can easily be used in the *CAD* of weldments of similar applications to Examples 9.6 and 9.7. By some additions, modifications, iterations, a better synthesis, or an optimum can be achieved.

The input parameters in the code are highlighted in Figure 9.19 by the comment statements with the word **Input** in bold letters. The output parameters or results are displayed in the command window of MATLAB package. This is enacted by clearing the usual semicolon (;) at the end of the variable calculation-statement or using a colon (,) to indicate continuation with the next statement and allowing the display of the result. The variables are usually defined by the comment statements at the right side of the code in Figure 9.19.

9.7.6 Optimization

Any of the previously developed *CAD* codes can be utilized to perform design optimization. The objective function can be formulated off the variables in those codes and inserted into the code. A loop or more can be added to change the design variable or variables and achieve a minimum or a maximum of the stated objective. If the objective is simple, few runs with changing a variable or a parameter can realize the optimum. As an example of weldment applications, a study was referred to in Section 9.6.1, which used ASME pressure vessels code to optimize some design parameters. In that work, welded cylindrical pressure vessels with hemispherical ends have been considered. They were required to hold some definite volume under some specific pressure. The optimum thicknesses of each hemispherical part and the cylindrical part satisfied the recommended ASME code. The optimum design also satisfied allowable stress constraints of potentially used materials (Abd El Aziz and Metwalli (2017)).

9.8 Summary

This chapter was mainly dedicated to screws, fasteners, and permanent joins with emphasis on the synthesis of bolts, power screws, and welded joints. The types, standards, and designations are provided for usual utility and selected set of components or elements. A realistic application of a *jib crane* is used to demonstrate the synthesis suitability check of bolts and weldments. A jack is used as an application of power screw synthesis. The chief equations for these component design cases are highlighted by italic bold numbers. *CAD* codes have been developed to help in the further iteration or optimization of design. The optimization is left to the reader for code modification according to the application or individual favorite objectives and constraints.

Problems

9.1 Search for the latest thread standards ASME/ANSI B1.1 or ISO 68-1. Review terminology and designation, and acquire a copy for you reference. Compare the standard details to the information provided in the text sections and appendix particularly available diameters, pitches, or number of threads per inch, thread lengths associated with particular diameters, etc. Note the additional information for future reference and utility.

9.2 Can you find standard bolts and nuts with two or more multiple starts? Are any standard components available for right- or left-hand threads?

9.3 Study some applications of screws with both right-hand and left-hand threads at both ends or any other geometry.

9.4 Sketch different screw fasteners heads, end, or tip form and the different drivers to turn or tighten these fasteners.

9.5 Find different locking devices such as *spring washers*, and define their sizes relative to the fastener they lock.

9.6 Redesign the *jib crane* and its fasteners for the same dimensions as Example 8.2, but for a maximum load of 30 [kN] or 6.6 [klb].

9.7 If the crane reach or the active boom length is 5 [m] or 200 [in] for the *jib crane* of Example 8.2, what should be the suitable synthesis of the crane and its fasteners?

9.8 Redesign the *jib crane* and its fasteners of problem 9.6, but for the crane reach or the active boom length of 5 [m] or 200 [in].

9.9 Use two prismatic ***3D*** models of the x–y thread profile shown in Figure 9.1b including 5 or 6 teeth for the external and the internal threads with a constant thickness in the z direction to use in an ***FE*** program. Use more material in the transverse direction to the teeth of about 5 times the tooth height for a selected standard thread. With some assumptions of material and load, find the stresses in the external and internal threads considering concentric loads at the ends. Observe the uneven stressing in the thread roots.

9.10 Use the previous ***3D*** model in a nonlinear plastic ***FE*** program (elastic-perfectly plastic) to reach the yield in most of the thread roots of external and internal threads by increasing the load successively.

9.11 Recalculate the safety factors of Example 9.1 using the proof strength instead of the yield strength.

9.12 Calculate the safety factor of the bolts fixing the *jib-crane* mast to the ground in Example 9.1 assuming a base plate of the same size as the mast thickness.

9.13 Define the effect of the preloading on the bolts between the boom I-beam with the hood of the *jib crane* in Problem 9.6. Find the stiffnesses, stresses, and safety factors of the bolts and members assuming a ½ cone angle of 30°.

9.14 Define the effect of the preloading on the bolts between the boom I-beam with the hood of the *jib crane* in Problem 9.7. Find the stiffnesses, stresses, and safety factors of the bolts and members assuming a ½ cone angle of 30°.

9.15 Define the effect of the preloading on the bolts between the boom I-beam with the hood of the *jib crane* in Problem 9.8. Find the stiffnesses, stresses, and safety factors of the bolts and members assuming a ½ cone angle of 30°.

9.16 Use a MATLAB code to find the solution of Problem 9.6 with specific selection of bolts according to ISO class and SAE grade. Find also the values of the tightening torque assuming the coefficients of friction to be about 0.15. Define the effect of loading and unloading the crane on its fatigue strength if the maximum load is lifted and released so many times to warrant the use of endurance strength. Assume a *Goodman* fatigue criteria. What would be the tightening torque if the coefficients of friction are about 0.11 or 0.17? Would the fastener be self-locking for any of these values?

9.17 Use a MATLAB code to find the solution of Problem 9.7 with specific selection of bolts according to ISO class and SAE grade. Find also the values of the tightening torque assuming the coefficients of friction to be about 0.15. Define the effect of loading and unloading the crane on its fatigue strength if the maximum load is lifted and released so many times to warrant the use of endurance strength. Assume a *Goodman* fatigue criteria. What would be the results if the coefficients of friction are about 0.11 or 0.17?

9.18 Use a MATLAB code to find the solution of Problem 9.8 with specific selection of bolts according to ISO class and SAE grade. Find also the values of the tightening torque assuming the coefficients of friction to be about 0.15. Define the effect of loading and unloading the crane on its fatigue strength if the maximum load is lifted and released so many times to warrant the use of endurance strength. Assume a *Goodman* fatigue criteria. What would be the results if the coefficients of friction are about 0.11 or 0.17?

9.19 Use hand calculation to verify the solution of one of the Problems 9.16, 9.17, or 9.18. Suggest any changes in the MATLAB code to further improve the synthesis or the solution.

9.20 Use a MATLAB code to justify the use of Eq. (9.25) in place of Eq. (9.24).

9.21 Derive Eqs. (9.31)–(9.34) by means of the suggested procedure in the text.

9.22 Find the fatigue safety factor for other failure criterion such as *ASME* and modified *Goodman* or *Gerber* in the case of dynamic loading of bolts.

9.23 Redo Example 9.4 using the coefficients of friction of about 0.11 or 0.17. Compare results and indicate which is better.

9.24 Redo Example 9.4 using other failure criterion such as *ASME* and modified *Goodman* or *Gerber*. Compare results and indicate which is more conservative.

9.25 Modify the available MATLAB code (CAD_Bolts_Preload_Fatigue.m) to include the other failure criterion such as *ASME*, modified *Goodman*, or *Gerber*, and resolve Example 9.4 comparing results and indicate which is more suitable to use.

9.26 Use hand calculations to verify the results of Example 9.4.

9.27 Find the relation between the usual pitches and the nominal diameters of power screws.

9.28 Derive Eq. (9.37) from Eq. (9.36) by eliminating the force normal to the thread helix.

9.29 Derive Eq. (9.41) from Eq. (9.40) by eliminating the force normal to the thread helix.

9.30 Find the applications where self-locking may not be desirable and multi threads would be more desirable.

9.31 Derive Eq. (9.45) by developing the equilibrium equations for the force balance in Figure 9.8b. Also develop similar equation for lowering the load.

9.32 Demonstrate that the square thread would be more efficient than the *Acme* or trapezoidal threads.

9.33 Use hand calculations to verify the power screw results obtained in Example 9.5.

9.34 Design a single thread power screw jack that should lift a load of 5 [kN] or 1.1 [klb]. Use the same characteristics and proportions as in Example 9.5.

9.35 Design a single thread power screw jack that should lift a load of 500 [kN] or 110 [klb]. Use the same characteristics and proportions as in Example 9.5.

9.36 Define a better construction details than in Figure 9.9a to reduce the collar outer diameter. Can the collar outer diameter be as small as the screw outer diameter? Check stresses and construction acceptability and compatibility.

9.37 Transform the casting construction of the gearbox in Figure 8.12 to a welded construction including the base, box cover, and the side covers. Assume proportional dimensions to those shown in the figure.

9.38 Redo Examples 9.6 and 9.7 with a detailed calculation for the second area moments (outer and inner diameters), which gives a higher second area moment of the circular weld about the z-axis. Evaluate the difference in the calculated results relative to the simplified calculations.

9.39 In addition to hand calculations, use a MATLAB code to check the weld between the mast pipe and the base plate. Neglect or consider the contribution of webs that are welded to strengthen the mast as shown in Figure 8.17 and discussed in Example 8.2. Consider the weld leg to be equal to the smaller of either plate or pipe thickness. Assume the strength of weld material to be the same as the parent materials, and use the permissible or allowable welding strength.

9.40 Find the effect of loading and unloading the crane in Example 9.6 on its fatigue strength if the maximum load is lifted and released so many times to warrant the use of endurance strength. Assume a *Goodman* fatigue criteria.

9.41 Find the effect of loading and unloading the crane in Example 9.7 on its fatigue strength if the maximum load is lifted and released so many times to warrant the use of endurance strength. Assume a *Goodman* fatigue criteria.

9.42 Reconstruct the *jib crane* in Examples 9.6 and 9.7 to a better synthesis and more uniform distribution of stresses considering other factors such as cost or minimum weight. Adjust the MATLAB code to recalculate the safety of the welds.

9.43 Modify any of the available MATLAB codes to perform better synthesis or optimization of the fastener, the power screw, or the weldment of an application.

References

Abd El Aziz, K.M. and Metwalli, S.M. (2017). Optimum design of pressure vessels using hybrid HGP and genetic algorithm, *Proceedings of the ASME 2017 Pressure Vessels and Piping Conference, PVP2017* (July 16–20, 2017), Waikoloa, Hawaii, United States, Paper No. PVP2017-65538.

ASME (2013). *2013 ASME Boiler & Pressure Vessel Code*. American Society for Mechanical Engineers.

ASME B18.2.1 (2010). *Square, hex, heavy hex, and askew head bolts and hex, heavy hex, hex flange, lobed head, and lag screws (inch series)*. American Society of Mechanical Engineers.

ASME B5.48 (1977). *Ball screws*. American Society of Mechanical Engineers.

ASME/ANSI B1.1-(1989) (R2008). *Unified inch screw threads*. American Society of Mechanical Engineers/American National Standard Institute

AWS A2.4 (1998). *Standard symbols for welding, brazing, and nondestructive testing*. American Welding Society

AWS A5.8M/A5.8 (2011). *Specification for filler metals for brazing and braze welding*. American Welding Society an American National Standard.

AWS D1.3/D1.3M (2008). *Structural welding code – sheet steel american welding society/⅓*. American National Standard Institute.

AWS D1.5M/D1.5 (2002). *Design of Welded Connections*. American Welding Society (AWS).

Groover, M.P. (2010). *Fundamentals of Modern Manufacturing, Materials, Processing, and Systems*, 4e. Wiley.

ISO 2553 (2013). *Welding and Allied Processes – Symbolic Representation on Drawings – Welded Joints*, 4e. International Organization for Standardization.

ISO 262 (1998). *ISO general purpose metric screw threads – Selected sizes for screws, bolts and nuts*. International Organization for Standardization

ISO 3408-2 (1991). *Ball screws – Part 2: Nominal diameters and nominal leads – metric series*. International Organization for Standardization

ISO 3408-3 (2006). *Ball screws – Part 3: Acceptance Conditions and Acceptance Tests*. International Organization for Standardization.

ISO 3408-4 (2006). *Ball screws – Part 4: Static axial rigidity*. International Organization for Standardization.

ISO 3408-5 (2006). *Ball screws – Part 5: Static and dynamic axial load ratings and operational life*. International Organization for Standardization

ISO 68-1 (1998). *ISO general purpose screw threads – Basic profile – Part 1: Metric Screw Threads*. International Organization for Standardization.

ISO 9692-1 (2013). *Welding and allied processes – Types of joint preparation – Part 1: Manual metal arc welding, gas-shielded metal arc welding, gas welding, TIG welding and beam welding of steels*. International Organization for Standardization.

Norris, C.H. (1945). Photoelastic investigation of stress distribution in transverse fillet welds. *Welding J.* 24: 557s.

Oberg, E., Jones, F.D., Horton, H.L., and Ryffel, H.H. (2012). *Machinery's Handbook*, 29e. Industrial Press.

Osgood, C.C. (1979). Saving Weight on Bolted Joints. *Machine Design*, October 25.

Pizzi, A. and Mittal, K.L. (eds.) (2003). *Handbook of Adhesive Technology*, 2e, *Revised and Expanded*. Marcel Dekker.

Pocius, A.V. (2012). *Adhesion and Adhesives Technology an Introduction*, 3e. Carl HanserVerlag.

Rothbart, H.A. and Brown, T.H. Jr. (2006). *Mechanical Design Handbook*, 2e. McGraw-Hill.

SAE J429 (2001). Mechanical and Material Requirements for Externally Threaded Fasteners. *SAE Handbook*.

Salakian, A.G. and Claussen, G.E. (1937). Stress distribution in fillet welds: a review of the literature. *Welding J.* 16: 1–24.

Internet Links

http://aws.org/ AWS: Welding standards.
http://files.aws.org/technical/errata/A2.4errata.pdf AWS: Welding, brazing, and testing symbols.
http://www.asme.org/ ASME: Standards.
http://www.astm.org/ ASTM: Standards.
http://www.dupont.com/transportation/structural-adhesives.html DuPont; Adhesives.
http://www.toco.tw/en/products/ballscrew Toco: Transmission ball screw.
https://ampg.com/ AMPG: Screws, bolts, nuts, etc.
https://discoautomotive.com Disco: Automotive hardware supplier, fasteners, screws, etc.
https://webstore.ansi.org/ ANSI: Standards.
https://www.3m.com/3M/en_US/bonding-and-assembly-us/structural-adhesives/ 3M: Adhesives.
https://www.ball-screws.net/ Ball-screws manufacturers.
https://www.bigbolt.net/products.html Big Bolt: Bolts, screw rods and nuts.
https://www.fastenersuperstore.com/ Fastener superstore: Fasteners supplier.
https://www.skf.com/group/knowledge-centre/media-library/index.html#tcm:12-149715 SKF ball screws.

10

Springs

A spring is any element, member, body, or structure that deflects or deforms under load and subsequently rebounds when the load is removed. The deformation is a linear or a nonlinear function relating the deformation to the applied load. For simplification, however, one might consider a linear relation for the mathematical model of the spring. The scope of this chapter is on a limited number of mechanical springs that engineers usually and extensively employ in machine design. These springs store and dispense potential energy in the form of load–deformation interaction. They usually function in several modes such as compression, tension, bending, and torsion. The loads in those cases are usually dynamic; however, for slow variations one can effectively synthesize them using static equivalence. Other springs use different modes of energy storage such as compliant structures and pneumatic, hydraulic, or magnetic springs. The treatment of these springs is beyond the scope of this text.

Symbols and Abbreviations

The adapted units are [in, lb, psi] or [m, kg, N, Pa], others given at each symbol definition. [k…] is 10^3, [M…] is 10^6 and [G…] is 10^9.

Symbols	Quantity, units (adopted)
A	Intercept of fitted data line with wire diameter of 1.0, [psi in^b] or [MPa mm^b]
A_w	Wire cross-sectional area
b	Leaf spring width, [in] or [mm]
b_x	Triangular leaf spring width at x
C	Helical spring index
d_c	Mean coil diameter, [in] or [mm]
d_i	Inner diameter of spring, [in] or [mm]
d_o	Outer diameter of spring, [in] or [mm]
d_w	Wire diameter, [in] or [mm]
E	Modulus of elasticity, [Mpsi] or [GPa]
$\boldsymbol{F}_a$	Alternating value of the forces
$\boldsymbol{F}_i$	Initial loading force
$\boldsymbol{F}_{it}$	Initial tension force
$\boldsymbol{F}_m$	Mean value of applied forces
$\boldsymbol{F}_{max}$	Maximum applied force

Machine Design with CAD and Optimization, First Edition. Sayed M. Metwalli.

Companion website: www.wiley.com/go/metwalli/machine

Symbols	Quantity, units (adopted)
$\boldsymbol{F}_{\min}$	Minimum applied force
$\boldsymbol{F}_n$	Normal component of force
f_n	Fundamental surge frequency of spring resonance, [Hz]
$\boldsymbol{F}_o$	Operating force applied on the spring
f_o	Frequency of the dynamic operating force, [Hz]
$\boldsymbol{F}_r$	Radial component of force
$\boldsymbol{F}_s$	Solid force that sets the spring to be solid
G	Shear modulus of elasticity of wire material, [Mpsi] or [GPa]
h	Thickness of each leaf of leaf spring, [in] or [mm]
h_D	Disc spring height [in] or [mm]
i	Total number of leafs
I_L	Second area moment of leaf cross section
I_w	Wire second-area moment
J_w	Wire second polar moment of area
K_B	Bending factor due to wire curvature
K_c	Critical buckling factor
K_d	Direct shear factor
k_{Es}	Extension spring stiffness or spring rate
K_{FSF}	Fatigue safety factor
k_{Ls}	Leaf spring stiffness
k_s	Spring stiffness or spring rate
K_{SF}	Safety factor
k_{Ts}	Torsion spring stiffness or spring rate
K_T	Torsion coefficient that accounts for curvature
K_W	Wahl factor
L_a	Free length of the active part of spring
ℓ_a	Length of active coils along spring helix
L_f	Total free spring length
L_i	Initial spring length, [in] or [mm]
L_L	Leaf spring length
L_o	Operating spring length
L_s	Solid length of spring
L_{Ts}	Total torsion spring length along its wire
$\boldsymbol{M}$	Generated bending moment
m_a	Mass of active coils of spring, [$\mathrm{lb_m}$] or [kg]
m_s	Total spring mass, [$\mathrm{lb_m}$] or [kg]
$\boldsymbol{M}_o$	Operating bending moments
N_a	Number of active turns
N_e	Number of “dead” or idle ends of the spring
N_s	Total number of spring turns
N_{ta}	Number of active turns of torsion spring
p_h	Helix pitch of active turns

Symbols	Quantity, units (adopted)
p_i	Initial spring pitch
p_o	Spring pitch at operating load
s	Distance along the spring wire length
S_e	Endurance limit or fatigue strength, [psi] or [MPa]
S_{se}	Fatigue shear strength or endurance shear limit, [psi] or [MPa]
S_{se}	Fatigue shear strength of the spring material, [psi] or [MPa]
S_{su}	Ultimate shear strength, [psi] or [MPa]
S_{sy}	Shear yield strength, [psi], [MPa]
S_{ut}	Ultimate tensile strength, [psi] or [MPa]
$\boldsymbol{T}$	Torque on the spring
t	Time
t_D	Disc spring thickness [in] or [mm]
U_o	Total operating strain energy
v	Longitudinal wave velocity along the wire
w	Leaf spring total section widths at triangle base
w_x	Leaf spring section width at a distance x
x	Variable in fitted regression function $y(x)$
y	Ordinate of fitted regression function of a variable x
$\Delta \boldsymbol{F}$	Incremental change in spring force
ΔL	Incremental change in spring length
δ	Spring deflection
δ_w	Wave displacement at distance s along the spring wire
δ_{cr}	Critical deflection
δ_i	Initial imbedded or "stored" deflection
δ_{Lo}	Leaf spring deflection at the operating force $\boldsymbol{F}_o$
θ_h	Helix angle of helical spring
θ_t	Rotation of torsional spring, [rad]
ρ	Material density
σ_1, σ_2	Principal stresses
σ_{all}	Maximum allowable stress, [psi], [MPa]
σ_b	Maximum bending stress, [psi], [MPa]
σ_L	Maximum stress in leaf spring, [psi], [MPa]
τ_a	Maximum alternating shear stress, [psi], [MPa]
τ_{all}	Allowable shear stress, [psi], [MPa]
τ_d	Direct shear stress, [psi], [MPa]
τ_i	Initial torsional prestress, [psi], [MPa]
τ_m	Mean of the maximum shear stress, [psi], [MPa]
τ_{max}	Maximum shear stress, [psi], [MPa]
τ_T	Shear stress due to torsion, [psi], [MPa]
τ_w	Total shear stress in the wire, [psi], [MPa]
ϕ	Spring wire angle of twist
ω_n	Natural surge frequency of spring resonance, [rad/s]

10.1 Types of Springs

In this chapter compression, tension, torsion, and bending types of springs are considered. Some of these come in different shapes and forms as shown in Figure 10.1. Figure 10.1a shows typical common real-life applications of springs such as paper clips, clip pins, rubber bands, etc. They act mainly in tension, bending, and torsion modes or combinations of that. Some segments of these act as a *leaf spring* in a bending mode that is treated later in Section 10.3. Figure 10.1b presents some types of compression, extension, and torsion springs in a form of a helical cylindrical or other shapes. In this figure, the bottom spring is a non-cylindrical form that acts mainly in a bending function. The three cylindrical springs above the bottom spring on the left are *compression springs*. The upper-right corner spring is an extension spring with two different ends to apply the tension load in a simple hook-like form. The two springs below that are *torsion springs*. *Extension springs* have so many different ends than the ones shown in Figure 10.1b. These ends can be especially custom-made to the requirements of holding the tension applicator to the spring. However, several specific ends are available from the manufacturers. Several types of torsion springs can also have different ends to apply the torsion generating loads as shown in Figure 10.1b. Some other types of springs combine few loading regimes such as *Bellville springs*, elastomeric, and other spring types that are presented in Sections 10.4 and 10.5. Images of springs are available in many websites such as those found in the Internet images segment of the references in this chapter. One can also combine springs in series or parallel groups for extended function, utility, or space limitations to withstand more forces or to possess more or less stiffness. Usually springs in sires have the same load on each spring, and those in parallel would have the same deformation for each one. With this concept, one can easily derive relations of combined stiffness or load for such groups of springs.

10.2 Helical Springs

10.2.1 Geometry, Definitions, and Configurations

A helical spring is usually made of a wire that has a diameter d_w. Coiling the wire around a cylinder produces some spacious helix with a mean coil diameter of d_c, initial active *helix pitch* of p_h, and a *number of active turns*

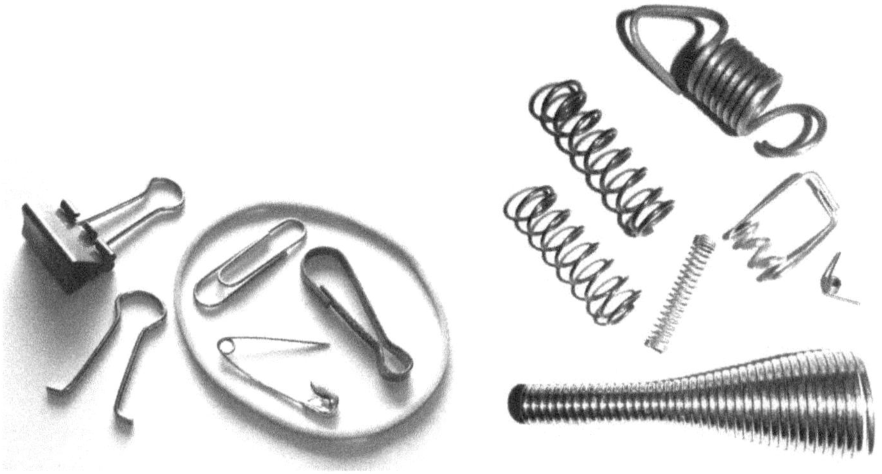

(a) (b)

Figure 10.1 Common springs showing (a) frequently encountered springs and (b) helical compression, torsion, and bending springs.

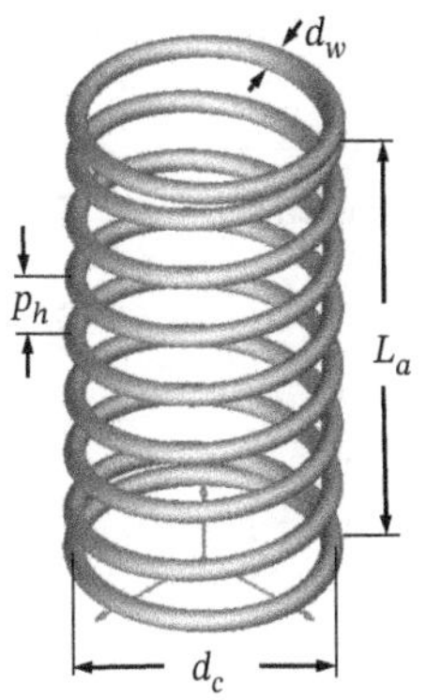

Figure 10.2 Common cylindrical helical spring showing basic geometrical parameters.

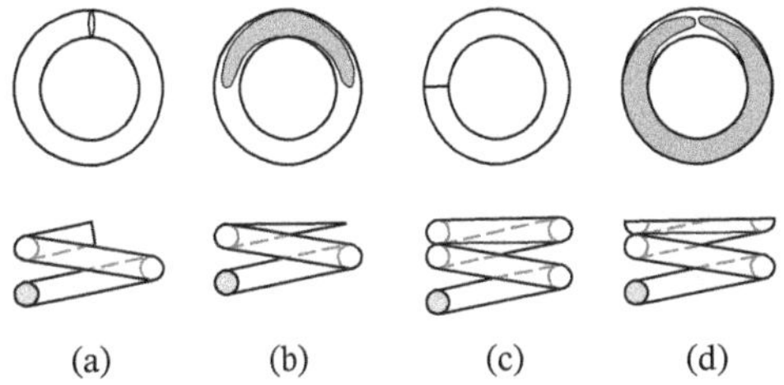

Figure 10.3 Spring end configurations: (a) plain end, (b) plain ground end, (c) square end, and (d) square and ground end.

of N_a as shown in Figure 10.2. The active turns are the ones that deflect upon load application and then rebound when the load diminishes. Figure 10.2 shows about seven active turns N_a. One can calculate the free length of the active part of the spring L_a as

$$L_a = N_a\, p_h + d_w \tag{10.1}$$

Note that the quantity of $N_a\, p_h$ is the length for the centerline of the helix. The added d_w is to account for the part of the wire off that helix centerline. The top and bottom "dead" or idle ends of the spring can have different configurations with their number counted as N_e. One of these that is shown in Figure 10.2 is a spring end diffused in a closed torus. More realistically produced ends affect the total number of spring turns N_s.

When nothing is added to both ends, it is called plain ends, Figure 10.3a. In that case, the number of end turns are counted as zero or $N_e = 0$. If both plain ends are ground as shown in Figure 10.3b, the number of end turns are counted as $N_e = -1$ because the part of the wire off the helix centerline is omitted by grinding. This adapted logical condition fits the developed equations for calculating spring lengths at different produced ends. Producing ends by a snuggly packed helix turns creates zero helix angle at both ends, which are called squared ends as shown in Figure 10.3c. If the squared ends are used, it is usually counted as two idle turns or $N_e = 2$. Frequently manufacturers ground these ends to have better contacts at the mating bodies; see Figure 10.3d. If full ground ends are used, one should count that as 1/2 extra turn for each end or $N_e = 1$. On the other hand, the numbers of active turns N_a are the engaged and sought ones in the synthesis and design procedure to satisfy the spring function. Adding the number of end turns to the active turns generates the total number of spring turns N_s.

The total *free length* of the spring under no load is obtained from the active turns N_a multiplied by the helix pitch p_h added to the length of the idle turns, which is N_e multiplied by the wire diameter d_w. The total free spring length L_f under no load is therefore

$$L_f = L_a + L_e = N_a p_h + d_w + N_e d_w \tag{10.2}$$

The *solid length* L_s of the spring is

$$L_s = (N_a + 1)d_w + N_e d_w \tag{10.3}$$

Table 10.1 The adapted number of inactive end turns N_e, the free length L_f, and the solid length L_s for different spring ends.

Type of spring ends	End turns, N_e	Free length, L_f	Solid length, L_s
Plain ends	0	$N_a p_h + d_w$	$(N_a + 1)d_w$
Ground plain ends	−1	$N_a p_h$	$N_a d_w$
Squared ends	2	$N_a p_h + 3d_w$	$(N_a + 3)d_w$
Ground squared ends	1	$N_a p_h + 2d_w$	$(N_a + 2)d_w$

The number 1 added to the active turns account for the spring material off the helix centerline. Table 10.1 gives the number of end turns N_e, the free length L_f, and the solid length L_s for different spring ends.

The outer spring diameter d_o and the inner spring diameter d_i are simply

$$d_o = d_c + d_w \tag{10.4}$$

$$d_i = d_c - d_w \tag{10.5}$$

These are useful when the spring is to be in a hole or in a recess for location or otherwise housed over a cylindrical rod for a definite function. The hole or the recess diameter is usually larger than the outer diameter d_o by a certain fit. The fit also accommodates spring deformation under load. When the spring is housed over a rod, one should select a suitable clearance or a fit for the inside diameter d_i depending on the function requirement.

The wire diameter d_w is usually specified from standards. Few standards are there depending on the adapted system of units and the material used for the spring. For the US system of units, many standard gauges are available. One such a gauge is the British SWG (standard wire gauge BS 6722, 1986). The ISO or metric wire standard is the usually using preferred numbers in a geometric series (European Standard, EN 10270-1 (2001), European Standard, EN 10270-3 (2001)). Wider selections of diameters are available by several manufacturers as given in the Internet segment of the references in this chapter.

The spring index C is a useful nondimensional parameter defined by

$$C = \frac{d_c}{d_w} \tag{10.6}$$

where d_c is the coil mean diameter and d_w is the wire diameter. It indicates the packing of the wire material in the spring. A small C indicates a dense spring with close wire coiling around the middle hollow core. A large C represents a lighter density spring with larger coiling around the middle hollow core. Figure 10.4 shows three helical springs with $C = 5$, 10, and 15, respectively, from the left. Usually one would use springs with lower C for heavier applications or tighter space constraint. The range of spring index C is usually 4–12. For both ground ends, the spring index C is usually 5–10. Usually C is >12 for light springs, $C = (6–12)$ for medium springs, and $C < 6$ for heavy springs. This knowledge is useful in spring synthesis.

The free length of the helical spring is reduced for installation purposes to accommodate a specific function requirement or under initial loading force $\boldsymbol{F}_i$ as an installation preloading as shown in Figure 10.5. If one initially specifies the spring length L_i, Eq. (10.2) generates the initial spring pitch p_i by substituting L_i for L_f and p_i for p_h. This, in addition to considering the deflections of Figure 10.5, gives the initial pitch p_i as

$$p_i = \frac{L_i - (1 + N_e)d_w}{N_a} = \frac{\delta_s - \delta_i}{N_a} + d_w \tag{10.7}$$

where δ_s is the solid deflection from the free length L_f and δ_i is the initial deflection for the initial load F_i as shown in Figure 10.5. Equation (10.7) is also used to get the free helix pitch p_h by replacing the subscript i by the subscript

Figure 10.4 Helical springs with each having a different spring index C of 5, 10, and 15, respectively, from the left.

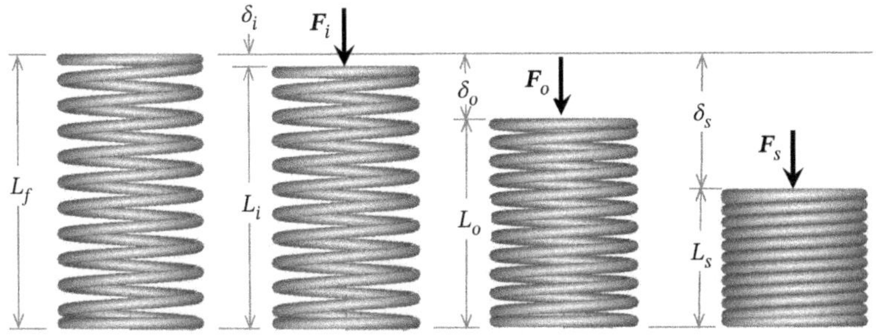

Figure 10.5 Helical springs at free length, initially loaded length, operating length, and solid length with the associated loading force for each case.

h and considering $\delta_i = 0$ at free length. This gives

$$p_h = \left(\frac{\delta_s}{N_a} + d_w\right) \tag{10.7'}$$

The same process of substitution is done for the operating condition at any operating force $\boldsymbol{F}_o$, where the operating spring length L_o and the operating pitch p_o are substituted for L_f and p_h, respectively in Eq. (10.2). This gives the operating spring pitch p_o as follows:

$$p_o = \frac{L_o - (1 + N_e)\, d_w}{N_a} \tag{10.8}$$

The operating length L_o is dependent on the operating force $\boldsymbol{F}_o$ and the spring stiffness (or spring rate). These different loading lengths, deflections, and associated forces are evident in Figure 10.5.

As usually defined, the spring stiffness is

$$k_s = \frac{\Delta \boldsymbol{F}}{\Delta L} = \frac{\boldsymbol{F}_o}{L_f - L_o} = \frac{\boldsymbol{F}_o}{\delta_o} \tag{10.9}$$

where $\Delta \boldsymbol{F}$ is the change in force, ΔL is the change in spring length from the free unloaded point to the operating position, $\boldsymbol{F}_o$ is the operating force, and δ_o is the operating deflection. Assuming a linear spring range allows the use of both $\Delta \boldsymbol{F}$ and ΔL at any loading point.

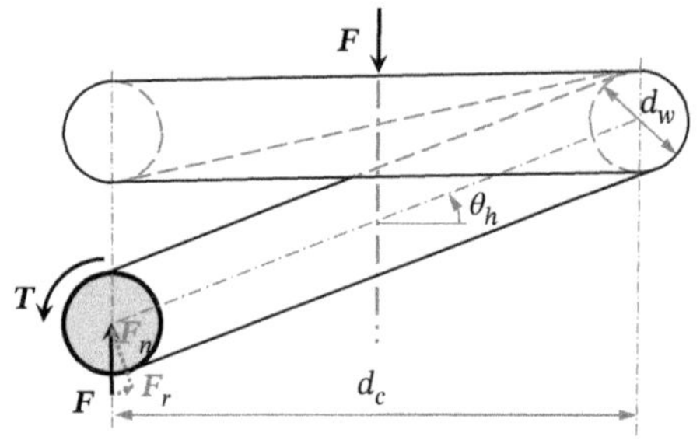

Figure 10.6 Free body diagram of a spring showing the torque $\boldsymbol{T}$ and the direct shear force $\boldsymbol{F}$ on the circular cross section of the cut.

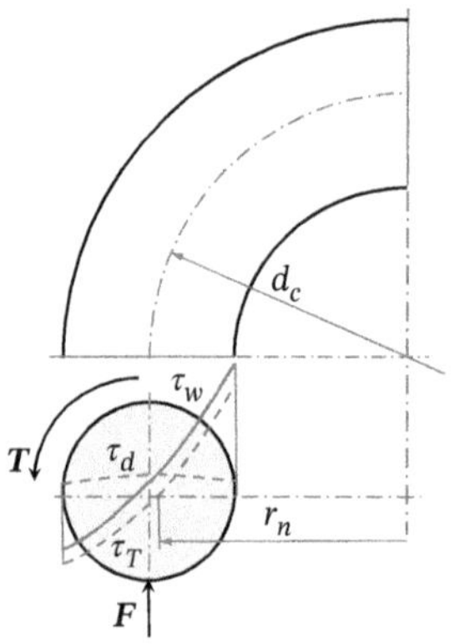

Figure 10.7 Distribution of shear stresses on the circular cross section of a low spring index showing the effect of the wire curvature in reducing the radius of curvature r_n of the neutral axes.

10.2.2 Stresses and Deflections

Figure 10.6 shows a free body diagram of a helical spring with a cut at the left cross section of the coil with dashed hidden lines for the hidden parts of the spring. The equivalent of the applied distributed load on the top contact area of the spring is the resultant force $\boldsymbol{F}$. This force is the intended axially applied force on the spring. It can be any of the initial force $\boldsymbol{F}_i$, the operating force $\boldsymbol{F}_o$, the solid force $\boldsymbol{F}_s$, or any force in-between. For the present analysis, assume that this axially applied force $\boldsymbol{F}$ is acting through the central y-axis of the spring with no load offset. The reactions on the cross-sectional area are the force $\boldsymbol{F}$ and the torque $\boldsymbol{T}$, which is equals to $\boldsymbol{F}\,(d_c/2)\cos\theta_h$. This value is obtained from the effect of the normal component of $\boldsymbol{F}_n$ of the force $\boldsymbol{F}$. This component is normal to the helix and thus is responsible for the generated torque. The radial component $\boldsymbol{F}_r$ of the force $\boldsymbol{F}$ should cause bending to the curved wire of the spring. The radial component $\boldsymbol{F}_r$ is equal to $\boldsymbol{F}\sin\theta_h$. The generated bending moment $\boldsymbol{M}$ is equal to $\boldsymbol{F}\,(d_c/2)\sin\theta_h$. The force $\boldsymbol{F}$ also causes a direct shear τ_d on the section and the torque $\boldsymbol{T}$ causes another shear stress due to torsion τ_T. Figure 10.7 shows the respective stresses generated as follows.

The *direct shear stress* τ_d on the wire cross-sectional area A_w due to the force $\boldsymbol{F}$ is (see Section 6.3.2)

$$\tau_d = \frac{K_d F}{A_w} \tag{10.10}$$

The factor K_d is the direct shear factor that depends on the cross-sectional geometry. Strength of material approach defines the factor to be 4/3 (or $\cong$1.33) for a circular cross section and a parabolic stress distribution rather than 1 for a uniform stress distribution (see Section 6.3.5 for details). The theory of elasticity, however, suggests a variation across the transverse or radial direction as shown by the dotted line in Figure 10.7. This reduces the factor K_d to 1.23 at the outer and inner surfaces, rather than the 4/3 (or $\cong$1.33) for the strength of materials approach. Strength of materials assumes a non-varying distribution in the transverse or radial direction. For a circular cross section, the factor is then (see Wahl 1963)

$$K_d = \frac{1.23}{C} \tag{10.11}$$

The shear stress on the wire cross section with a second polar area moment of J_w due to torsion $\boldsymbol{T}$ is

$$\tau_T = \pm\frac{K_T\boldsymbol{T}(d_w/2)\cos\theta_h}{J_w} \tag{10.12}$$

The positive sign (+) is for the inner surface, and the negative sign (−) is for the outer surface as shown in Figure 10.7. The factor K_T is a torsion coefficient that accounts for the curvature effect on the torsional shear stresses distribution across the section of the wire. For a circular cross section, the wire curvature causes the distribution to be nonlinear and shifts the neutral axes toward the centerline of the spring coil with higher shear stress at the inner surface than the shear stress at the outer surface as shown by the dashed line in Figure 10.7. The neutral axes become at a radius r_n, which is smaller than the average coil radius of $d_c/2$. The factor K_T is dependent on the curvature, which is a function of the spring index C and the cross-sectional geometry. When the wire cross section is circular, the torsion factor K_T – according to Wahl (1963) – is

$$K_T = \frac{4C-1}{4C-4} \tag{10.13}$$

The total shear stress in the wire τ_w due to the torque $\boldsymbol{T}$ and the direct shear force $\boldsymbol{F}$ is $\tau_w = \tau_d + \tau_T$ or

$$\tau_w = \pm\frac{K_T\boldsymbol{T}(d_w/2)\cos\theta_h}{J_w} + \frac{K_d\boldsymbol{F}}{A_w} \tag{10.14}$$

where J_w is the wire second polar moment of area and A_w is the wire cross-sectional area. For a wire of a circular cross section, $J_w = \pi(d_w)^4/32$ and $A_w = \pi(d_w)^2/4$. Substituting these values and the value of the torque $\boldsymbol{T}$ as $\boldsymbol{F}$ $(d_c/2)\cos\theta_h$, the maximum shear stress τ_{max} is then

$$\tau_{max} = \frac{16K_T\boldsymbol{F}(d_c/2)\cos\theta_h}{\pi d_w^3} + \frac{4K_d\boldsymbol{F}}{\pi d_w^2} \tag{10.15}$$

The maximum bending stress σ_b due to the bending moment $\boldsymbol{M}$ is

$$\sigma_b = \frac{K_B\boldsymbol{M}(d_w/2)}{I_w} \tag{10.16}$$

where I_w is the wire second-area moment and K_B is the bending factor due to the wire curvature, which can be evaluated as defined in the curved beam treatment of stresses (see Section 10.2.7). For a circular cross section, $I_w = \pi(d_w)^4/64$ and substituting for the moment $\boldsymbol{M}$ as $\boldsymbol{F}$ $(d_c/2)\sin\theta_h$, the normal stress becomes

$$\sigma_b = \frac{64\,K_B\;\boldsymbol{F}\sin\theta_h\;(d_c/2)(d_w/2)}{\pi d_w^4} = \frac{16\,K_B\,\boldsymbol{F}\,d_c\sin\theta_h}{\pi d_w^3} \tag{10.17}$$

The *principal stresses* for this stress state are (see Section 6.4)

$$\sigma_1, \sigma_2 = \frac{\sigma_b}{2} \pm \sqrt{\left(\frac{\sigma_b}{2}\right)^2 + \tau_{max}^2} \tag{10.18}$$

Substituting for the stresses of σ_b and τ_{max} gives the principal stresses. As σ_b approaches zero, σ_1 and $\sigma_2 \cong \pm\,\tau_{max}$.

Usually helical compression springs have small helix angle θ_h. Since it is required to design the spring at its maximum loading condition, the stresses attain the maximum values at very near or almost at the solid length. In that case, the helix angle is at its minimum and cos θ_h becomes close to unity and sine θ_h is close to zero. The bending stress becomes very small or near zero and the principal stresses are very close to $\pm\tau_{max}$. From Eq. (10.15) and utilizing the definition of C in Eq. (10.6), the maximum shear stress becomes

$$\tau_{max} = \left(\frac{8K_T\boldsymbol{F}(d_c)}{\pi d_w^3}\left(\frac{C}{(d_c/d_w)}\right) + \frac{4K_d\boldsymbol{F}}{\pi d_w^2}\right) = \left(\frac{8K_T\boldsymbol{F}C}{\pi d_w^2} + \frac{4K_d\boldsymbol{F}}{\pi d_w^2}\right) \tag{10.19}$$

Substituting for the stress factors K_T and K_d, the maximum shear stress τ_{ma} turns out to be

$$\tau_{max} = K_W\left(\frac{8\boldsymbol{F}C}{\pi d_w^2}\right) \tag{10.20}$$

where the *Wahl factor* K_W is equal to $K_T + (K_d/2)$ or from Eqs. (10.11) and (10.13) (Wahl 1963),

$$K_W = \frac{4C-1}{4C-4} + \frac{0.615}{C} \tag{10.21}$$

The deflection δ of the applied load $\boldsymbol{F}$ is obtainable from equating the work done by the force $\boldsymbol{F}$ to the energy stored in the spring due to torsion $\boldsymbol{T}$ and the wire twisting angle ϕ. This provides the following relation:

$$\frac{1}{2}\boldsymbol{F}\delta = \frac{1}{2}\boldsymbol{T}\phi \tag{10.22}$$

where ϕ is the angle, which the spring wire twists under the torque $\boldsymbol{T}$. For a circular cross section, the angle of twist ϕ depends on the length of the twisted wire, which is a function of the coil diameter d_c, and the number of active turns N_a. Referring to Section (6.3.4) about twisting of bars, this angle of twist is defined by

$$\phi = \frac{TN_a\pi d_c}{J_w G} \tag{10.23}$$

where $J_w = \pi\,(d_w)^4/32$ is the second polar area moment of the wire and G is the shear modulus of elasticity of wire material. Substituting for $\boldsymbol{T}$ as $\boldsymbol{F}\,(d_c/2)$ for small helix angle, using Eqs. (10.22), (10.23), and (10.6), and substituting for J_w, gives the deflection δ of the spring in the direction of the force $\boldsymbol{F}$ as

$$\delta = \frac{\boldsymbol{T}}{\boldsymbol{F}}\phi = \frac{\boldsymbol{T}}{\boldsymbol{F}}\frac{TN_a\pi d_c}{J_w G} = \frac{F^2(d_c/2)^2 N_a\pi d_c}{F(\pi d_w^4/32)G} = \frac{8Fd_c{}^3N_a}{d_w^4 G} \tag{10.24}$$

or

$$\delta = \frac{8FC^3N_a}{d_w G} \tag{10.25}$$

The spring stiffness k_s or spring rate is similar to Eq. (10.9) (or $k_s = \boldsymbol{F}/\delta$), which from Eq. (10.25) gives

$$k_s = \frac{d_w G}{8C^3 N_a} \tag{10.26}$$

10.2.2.1 Static Loading

The relations previously developed are valid for static and dynamic loading of the spring. Dynamically loaded spring, however, fails differently from statically loaded spring. While yield or fracture would occur in static loading, fatigue is the usual failure mode in dynamically loaded springs. This depends on the loading regime and the material strength properties. For the static case, the maximum operating force $\boldsymbol{F}_o$ is constant and no variation exists in that force.

10.2.2.2 Dynamic Loading

Dynamically loaded springs, however, need to take into account the variation of applied force with time. This force variation is usually characterizable as a function of the maximum applied force $\boldsymbol{F}_{\max}$ and the minimum applied force $\boldsymbol{F}_{\min}$ during the span of operation. From these values, one can estimate the mean value of the applied forces $\boldsymbol{F}_m$ and the alternating value of the forces $\boldsymbol{F}_a$ as follows:

$$\boldsymbol{F}_m = \frac{\boldsymbol{F}_{\max} + \boldsymbol{F}_{\min}}{2} \tag{10.27}$$

$$\boldsymbol{F}_a = \frac{\boldsymbol{F}_{\max} - \boldsymbol{F}_{\min}}{2} \tag{10.28}$$

According to Eq. (10.20), the maximum stresses generated due to these forces are

$$\tau_m = K_W\left(\frac{8\boldsymbol{F}_m C}{\pi d_w^2}\right) \tag{10.29}$$

$$\tau_a = K_W\left(\frac{8\boldsymbol{F}_a C}{\pi d_w^2}\right) \tag{10.30}$$

where τ_m is the mean of the maximum shear stresses and τ_a is the maximum alternating shear stress. These values are useful in the evaluation of the spring synthesis for the case of a fatigue expectation due to the applied dynamic loading. Later on, Section 10.2.5 handles this dynamic and the previous static loading considerations to synthesize the spring under each of the loading case.

10.2.3 Buckling

Similar to the bucking of columns (Section 6.7), helical springs will buckle when the free length L_f is large relative to the coil diameter d_c. The critical load $\boldsymbol{F}_{\mathrm{cr}}$ for *spring buckling* to occur is (Wahl (1963)

$$\boldsymbol{F}_{\mathrm{cr}} = K_{\mathrm{cr}}(k_s L_a) \tag{10.31}$$

where k_s is the stiffness, L_a is the active free length, and K_{cr} is the critical buckling factor, which depends on the free length to coil diameter ratio L_f/d_c. The critical buckling factor K_{cr} is plotted against the free length to coil diameter ratio (L_f/d_c) in Figure 10.8, Wahl (1963). The end conditions affect the critical buckling factor. Curve (a) is for fixed ends, which usually suits squared and ground ends. Curve (b), however, is for hinged ends that can model springs seated on spherical ends that allow free rotational movements at those ends. Substituting for the spring stiffness k_s in Eq. (10.31), the maximum critical deflection is then

$$\delta_{\mathrm{cr}} = K_{\mathrm{cr}}\, L_a \tag{10.32}$$

This maximum critical deflection δ_{cr} is the available deflection for the spring to move from the free length up to the solid length. The critical force $\boldsymbol{F}_{\mathrm{cr}}$ is also the maximum force that should act on the spring, which is the solid force. Fitting the curves of Figure 10.8 to third-order polynomials with reasonable errors gives

$$y_a = 0.0042x^3 - 0.0702x^2 + 0.2661x + 0.4186 \tag{10.33}$$

$$y_b = -0.0036x^3 + 0.0792x^2 - 0.587x + 1.5186 \tag{10.34}$$

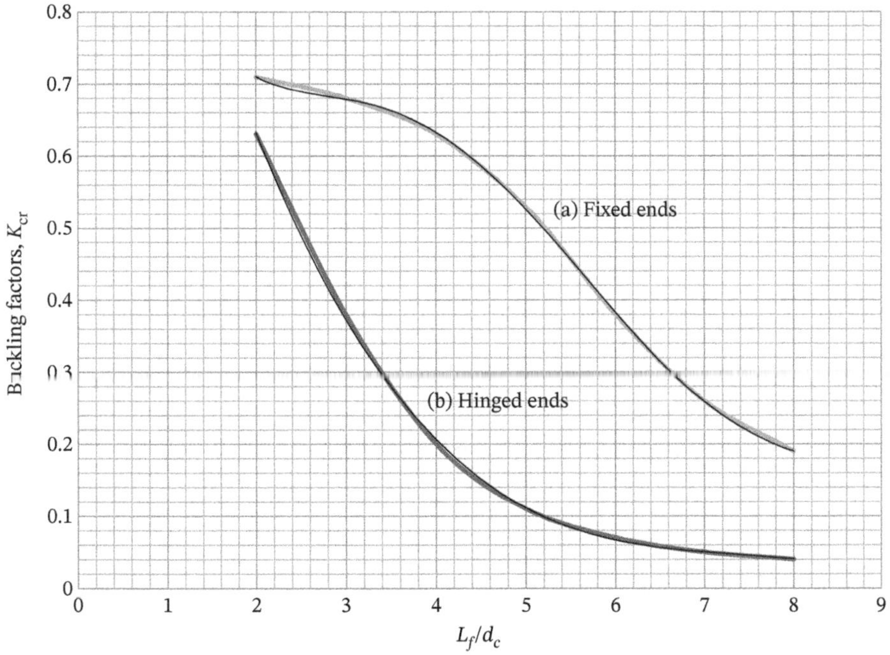

Figure 10.8 The critical buckling factor K_{cr} against the free length to coil diameter ratio (L_f/d_c). Curve (a) is for fixed ends and curve (b) is for hinged ends.

where x represents the free length to coil diameter ratio L_f/d_c, which is constrained to the range 2–8, y_a is the maximum critical fixed end deflection δ_{cr} for curve (a) of Figure 10.8, and y_b is the maximum critical free end deflection δ_{cr} for curve (b) of Figure 10.8. It is clear that the free end is more critical than the fixed end. Equations (10.33) and (10.34) are used to advantage in computer codes to check the stability of long springs. The maximum forces or deflections should be less than the critical ones or

$$\boldsymbol{F}_s \leq \boldsymbol{F}_{cr} \text{ or } \delta_s \leq \delta_{cr} \tag{10.35}$$

When design considerations necessitate using a critical longer spring, one can restrain the inner or outer spring space by a rod or a hollow cylinder. Clearance should be necessary to allow the coils to move freely in the restrained space up to its maximum travel. Spring outer and inner diameters increase with the application of loads. Under the maximum loading condition, or solid load, the outer diameter of the spring becomes

$$d_o = \sqrt{d_c^2 + \frac{p_h^2 - d_w^2}{\pi^2}} + d_w \tag{10.36}$$

where d_c is the coil diameter, d_w is the wire diameter, and p_h is the initial helix pitch of the spring obtained by Eqs. (10.1) or (10.2). It is usually recommended to have an extra 0.1 d_w added to the outer restraining body or a reduction of the same amount in the inner restraining body. This allows coil movements without rubbing.

10.2.4 Resonance

The spring wire helix acts as a string when excited by a load with varying frequency of oscillation. A longitudinal wave with a certain velocity v travels along the wire. The wave equation for this string wire assuming no internal damping is (Wahl, 1963)

$$\frac{\partial^2 \delta_w}{\partial t^2} = v^2 \frac{\partial^2 \delta_w}{\partial s^2} \tag{10.37}$$

where v is the velocity of the traveling wave along the wire and δ_w is the wave displacement at any distance s along the spring wire at time t. The velocity v of the wave depends on the spring stiffness k_s, spring mass m_a, and the length ℓ_a of its active coils along spring helix. The development of Eq. (10.37) suggests that the wave velocity v is (see Wahl 1963):

$$v = \ell_a \sqrt{\frac{k_s}{m_a}} \tag{10.38}$$

Considering the usual boundary conditions of the spring wire, the solution of Eq. (10.37) should satisfy

$$\sin \frac{\omega \ell_a}{v} = 0 \tag{10.38$'$}$$

This relation is satisfied, if

$$\frac{\omega_n \ell_a}{v} = \frac{2\pi f_n \ell_a}{v} = n\pi, \quad n = 1, 2, 3, \ldots \tag{10.39}$$

The fundamental *surge frequency* f_n of *spring resonance* is at $n = 1$ or

$$f_n = \frac{v}{2\ell_a} = \frac{1}{2}\sqrt{\frac{k_s}{m_a}} \tag{10.40}$$

This f_n is in cycles per second or [Hz]. The spring stiffness k_s is defined by Eq. (10.26), and the mass m_a of the active spring coils is

$$m_a = \frac{\pi}{4} d_w^2 (\pi d_c) N_a \rho = \frac{\pi^2}{4} d_w^2 d_c N_a \rho \tag{10.41}$$

where ρ is the material density. For ***spring steel wire*** (0.6–0.7% carbon), one can use ρ as 7860 [kg/m^3], or 0.282 [lb/in^3], and $G = 79.3\,(10^9)$ [Pa] or $11.5\,(10^6)$ [psi]. In the SI units, the fundamental surge frequency for steel is then

$$f_n = \frac{1}{2}\sqrt{\frac{k_s}{m_a}} = \frac{1}{2}\sqrt{\frac{d_w G}{\frac{\pi^2}{4} d_w^2 d_c N_a \rho (8C^3 N_a)}} = \frac{1}{2}\sqrt{\frac{G}{2\pi^2 d_w d_c N_a^2 \rho C^3}}$$

or

$$f_n = \frac{1}{2\sqrt{2}\ \pi}\sqrt{\frac{G}{\rho\ d_w^2\ N_a^2\ C^4}} \tag{10.42}$$

In SI units,

$$f_n = \frac{1}{2\sqrt{2}\pi}\sqrt{\frac{G}{\rho d_w^2 N_a^2 C^4}} = \frac{1}{8.885\,77}\sqrt{\frac{79.3\,(10^9)}{7860\, d_w^2 (10^{-6}) N_a^2 C^4}}$$

$$f_n = \frac{1}{8.885\,77}\sqrt{\frac{0.001\,008\,9\,(10^{16})}{d_w^2 N_a^2 C^4}} = 357\,462\sqrt{\frac{1}{d_w^2 N_a^2 C^4}} \tag{10.42$'$}$$

where d_w is in [mm] for SI units and f_n is in [Hz]. For US units, we substitute in Eq. (10.42) to get

$$f_n = \frac{1}{2\sqrt{2}\pi}\sqrt{\frac{G}{\rho d_w^2 N_a^2 C^4}} = \frac{1}{8.885\,77}\sqrt{\frac{11.5\,(10^6)}{(0.282/386) d_w^2 N_a^2 C^4}}$$

or

$$f_n = \frac{1}{8.885\,77}\sqrt{\frac{11.5\,(10^6)}{(0.282/386) d_w^2 N_a^2 C^4}} = 14\,120\sqrt{\frac{1}{d_w^2 N_a^2 C^4}} \tag{10.42$''$}$$

where d_w is in [in] for US units and f_n is in [Hz]. It is important to note that for the US units, we should divide the specific gravity ρ [lb/in^3] by the gravitational acceleration of 386 [in/s^2].

The operating frequency f_o for the operating dynamic force $\boldsymbol{F}_o$ (or any other kind of excitation) should be much less than the critical resonance frequency of the spring. A good range for that is $f_o < (1/15\text{–}1/20)\,f_n$

$$f_o < (1/15 \text{ to } 1/20)\ f_n, \quad [\text{Hz}] \tag{10.43}$$

10.2.5 Design Procedure

In this section, we introduce the initial synthesis that is mainly dependent on the static consideration of loading and an adapted material, which is a common inexpensive general-purpose spring steel. The developed synthesized design can be further refined to accommodate the dynamic loading if needed. If the case in question is a definite dynamic loading from the beginning, fatigue shear strength S_{se} should replace the static yield strength in shear S_{sy}. One may also use a larger factor of safety to account for the dynamic loading with an initial suitable safety factor of at least S_{sy}/S_{se} and still use the initial synthesis procedure provided herein.

10.2.5.1 Initial Synthesis

If it is unacceptable to have permanent deformation, one can consider failure ensued when the maximum shear stress reaches the yield strength in shear S_{sy}. To guard against that, one should use a safety factor K_{SF} to force the maximum shear stress τ_{max} to happen below the shear strength S_{sy}. This suggests that

$$\tau_{max} = S_{sy}/K_{SF} \tag{10.44}$$

The allowable shear stress τ_{all} is then

$$\tau_{all} = S_{sy}/K_{SF} \tag{10.45}$$

The maximum shear stress τ_{max} can then be equal to the allowable shear stress τ_{all}.

On the other hand, one can suggest a maximum acceptable condition indicating that failure may start when the maximum shear stress at the solid length just reaches the yield strength in shear S_{sy} with a safety factor $K_{SF} = 1.0$. This would allow the optimum maximum utilization of material without allowing permanent deformation or setting to occur. The spring should bounce back with no permanent alteration to its form or having internal distortion or *residual stresses*. This assumption depends on the accurate definition of the yield strength in shear S_{sy}, which may not necessarily be possible. To safeguard against that, one can use another larger safety factor K_{SF} that assumes the operating force $\boldsymbol{F}_o$ is less than the solid force by 10–20%. This small range depends on the confidence level of keeping $\boldsymbol{F}_o$ at its definite value and the confidence in the definite value of the yield strength of the spring material in shear S_{sy}. In that case, the solid force $\boldsymbol{F}_s$ is 1.1 $\boldsymbol{F}_o$ or 1.2 $\boldsymbol{F}_o$. The safety factor K_{SF} is then only 1.0 if synthesis is best at solid length assumption with synthesis according to solid force $\boldsymbol{F}_s$ of say 1.15 $\boldsymbol{F}_o$ rather than at operating force $\boldsymbol{F}_o$. The optimum synthesis, in that case, accounts for solid force $\boldsymbol{F}_s$ ensuing material yield in shear, which according to Eqs. (10.20) and (10.44) gives

$$S_{sy} = K_W\left(\frac{8\boldsymbol{F}_s C}{\pi d_w^2}\right) \tag{10.46}$$

From this equation and for any spring index C, one can find the wire diameter as

$$d_w = \sqrt{K_W\left(\frac{8\boldsymbol{F}_s C}{\pi S_{sy}}\right)} \tag{10.47}$$

From Eq. (10.21), one gets K_w as a function of the spring index C and again $\boldsymbol{F}_s$ is 1.1–1.2 $\boldsymbol{F}_o$. The selected material defines the value of the yield strength in shear S_{sy}. For initial synthesis, the initial material adapted is the **hard-drawn wire** (0.6–0.7% carbon), which is a conservative general-purpose spring steel. This is a widely used material for wide range of spring wire diameters. The ultimate tensile strength S_{ut} of this material depends on the wire diameter d_w with the assumed data fitting relation $S_{ut} = A/(d_w)^b$ such that

$$S_{ut} = \frac{142\,000}{d_w^{0.177}},\ [\text{psi in}^b] \quad \text{or} \quad S_{ut} = \frac{1734}{d_w^{0.177}},\ [\text{MPa mm}^b] \tag{10.48}$$

A is the intercept of the fitted line of data with the value of the wire diameter d_w of 1.0. One can find the *yield strength in shear* S_{sy} from the ultimate tensile strength S_{ut} by the usual application of failure theories for the type of material used. For initial synthesis, the widely used ***distortion energy*** relation for shear to tension strength is (see Section 7.7.3)

$$S_{sy} = 0.577\ S_y = 0.577\,(0.75\,S_{ut}) = 0.433\ S_{ut} \tag{10.49}$$

Here, the assumption is that the yield strength in tension S_y is about 0.75 the ultimate tensile strength S_{ut}. For detailed design, one can use the real yield strength in tension, if available. From Eqs. (10.47)–(10.49), one can find the spring wire diameter d_w for any spring index C. The value is as follows for the SI units giving d_w in meters [m]. One should multiply by 1000 to get d_w in [mm], if A is used as [Pa mm^b] rather than as [MPa mm^b]:

$$\begin{aligned} d_w &= \left(K_W\left(\frac{8\boldsymbol{F}_s C}{\pi(0.433\,S_{ut})}\right)\right)^{0.5} = \left(K_W\left(\frac{8\boldsymbol{F}_s C d_w^b}{\pi(0.433(A))}\right)\right)^{0.5} \\ &= d_w^{0.5b}\left(K_W\left(\frac{8\boldsymbol{F}_s C}{0.433\pi(A)}\right)\right)^{0.5} \end{aligned} \tag{10.50}$$

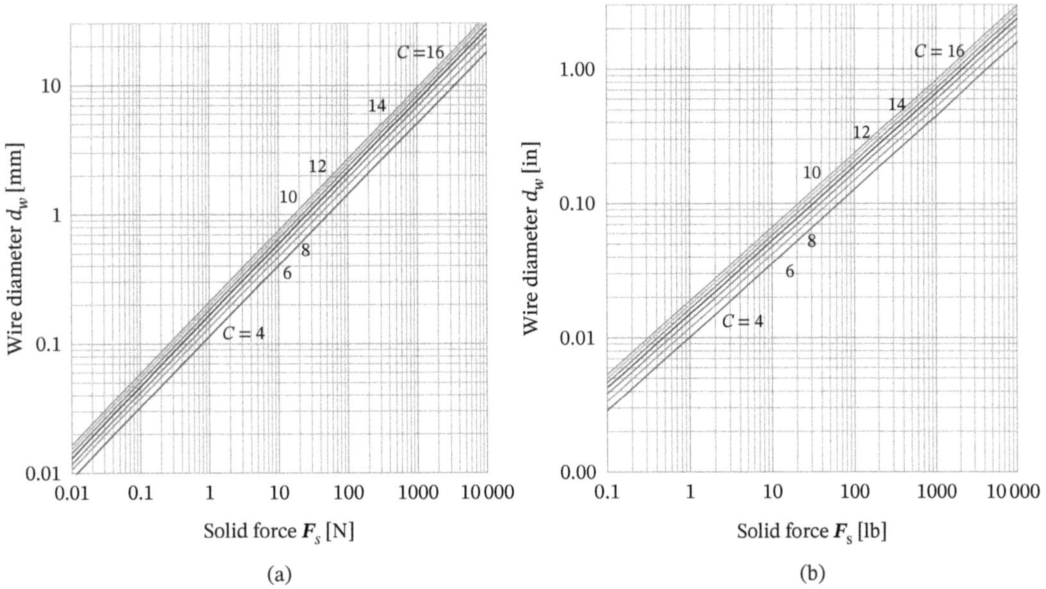

Figure 10.9 The wire diameter d_w for the maximum applied solid force $\boldsymbol{F}_s$ for different spring index $C = 4, 6, 8, 10, 12, 14$, and 16. (a) SI system and d_w is in [mm]. (b) US system and d_w is in [in].

or

$$d_w^{1-0.5\,b} = \left(K_W\left(\frac{8\boldsymbol{F}_s C}{0.433\pi(A)}\right)\right)^{0.5} \tag{10.51}$$

Then,

$$d_w = \left(\left(K_W\left(\frac{8\boldsymbol{F}_s C}{0.433\pi(A)}\right)\right)^{0.5}\right)^{\frac{1}{1-0.5b}} \tag{10.52}$$

This process develops the curves shown in Figure 10.9a, which defines the *wire diameter* d_w for the maximum applied *solid force* $\boldsymbol{F}_s$ in the SI system. Usually one selects the closest standard wire diameter from those shown in Table 10.2. Other diameters are also available from manufacturers of spring wires or particular springs (see Internet sites at references of this chapter). For a ***hard-drawn wire***, the fitted values are $A = 1734$ and $b = 0.177$. Later on, we will discuss other suitable materials and their properties in the following detailed design segment. The results obtained herein are not greatly affected and are open for tuning in synthesis refinement cycle. The developed ***Helical Spring Synthesis Tablets for US and SI*** systems provide a tool to do the necessary synthesis tuning. These are available through the ***Wiley website*** of this textbook.

Applying the same process for the US units gives the wire diameter in units of [in] same as Eq. (10.52) and substituting for the same $b = 0.177$ with A having a different value as defined by Eq. (10.48). The fitted value of A for the US units is then 142 000.

Again,

$$d_w = \left(\left(K_W\left(\frac{8\boldsymbol{F}_s C}{0.433\pi(A)}\right)\right)^{0.5}\right)^{\frac{1}{1-0.5b}} \tag{10.52}$$

Table 10.2 Selected spring wire diameters for US units [in] and SI units [mm].

US [in]	0.001, 0.0012, 0.0016, 0.0020, 0.0024, 0.0028, 0.0032, 0.0036, 0.004, 0.0044, 0.0048, 0.0052, 0.0060, 0.0068, 0.0076, 0.0084, 0.0092	0.0100, 0.0108, 0.0116, 0.0124, 0.0136, 0.0148, 0.0164, 0.0180, 0.020, 0.022, 0.024, 0.028, 0.032, 0.036, 0.040, 0.048, 0.056, 0.064, 0.072 0.080, 0.092	0.104, 0.116, 0.128, 0.144, 0.160, 0.176, 0.192, 0.212, 0.232, 0.252, 0.276, 0.300, 0.324, 0.348, 0.372, 0.400, 0.432. 0.464, 0.500	
SI[a)] [mm]	0.010, 0.0125, 0.016, 0.020, 0.025, 0.032, 0.040, 0.050, 0.063, 0.080	0.100, 0.125, 0.160, 0.200, 0.250, 0.320, 0.400, 0.500, 0.630, 0.800	1.000, 1.250, 1.600, 2.00, 2.50, 3.20, 4.00, 5.00, 6.30, 8.00	10.00, 12.50, 16.00, 20.00

a) More diameters are available e.g. Spring Wire (http://www.springwire.com/(range of more diameters)).

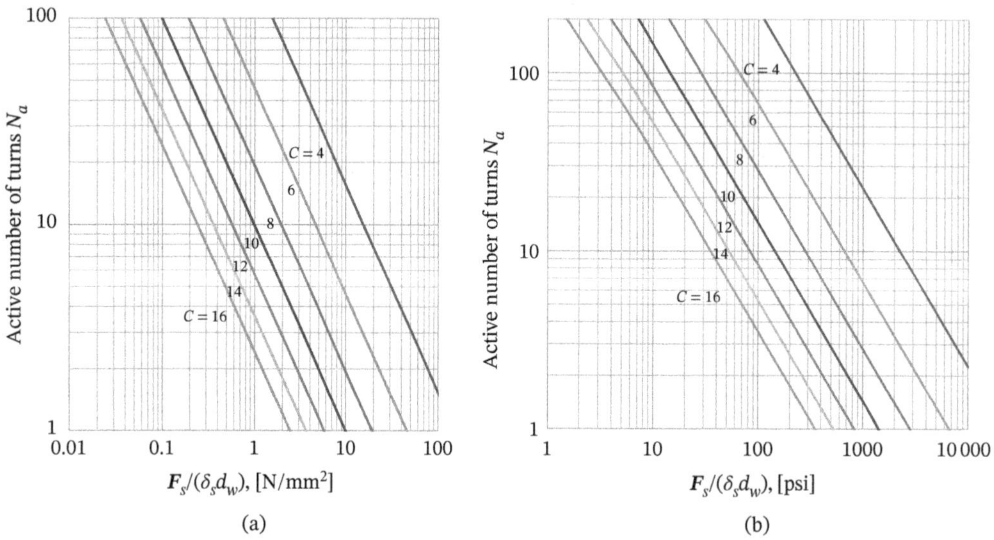

Figure 10.10 The Number of active turns N_a for different spring index C against the maximum solid force $\boldsymbol{F}_s$, solid deflection δ_a in [mm] or [in], and wire diameter d_w in [mm] or [in] for SI units in (a) and US units in (b).

The number of active turn N_a depends on the previously calculated wire diameter d_w and the modulus of rigidity G as depicted in Eq. (10.25), which is rearranged and rephrased as

$$N_a = \frac{\delta_s d_w G}{8F_s C^3} = \frac{\delta_o d_w G}{8F_o C^3} = \frac{d_w G}{8k_s C^3} \tag{10.53}$$

The shear modulus of rigidity G is for the selected ***hard-drawn steel*** with $G = 79.3\,(10^9)$ [Pa] or $11.5\,(10^6)$ [psi]. Substituting into Eq. (10.53) and using the US units of [lb] for $\boldsymbol{F}_s$ and [in] for both δ_s and d_w give

$$N_a = 1.4375(10^6)\frac{\delta_s d_w}{F_s C^3} \tag{10.54}$$

Substituting into Eq. (10.53) and using the SI units of [N] for $\boldsymbol{F}_s$ and [mm] for both δ_s and d_w give

$$N_a = 9.9125(10^3)\frac{\delta_s d_w}{F_s C^3} \tag{10.55}$$

The deflection and the wire diameter should be in [mm] or [in] for the units and constants to match. Using Eqs. (10.54) and (10.55) produces the curves shown in Figure 10.10 for the SI and the US units. This figure defines the *active number of turns* N_a for the maximum limit of the *solid force* $\boldsymbol{F}_s$, the *wire diameter* d_w, and the maximum *solid deflection* δ_s. One should note that the maximum solid deflection is obviously 1.1–1.2 δ_o.

Example 10.1 It is required to design a general-purpose spring that can carry a maximum operating load of 1000 [N] or 224.81 [lb] and a maximum operating deflection of 100 [mm] or 3.937 [in] assuming ground squared ends.

Solution

Data: $\boldsymbol{F}_o = 1000$ [N] or 224.81 [lb], $\delta_o = 100$ [mm] or 3.937 [in].

The initial synthesis suggests using a ***hard-drawn steel*** with $A = 1734$ [MPa mm^b] or 142 000 [psi in^b] and $b = 0.177$. Since the load is of a medium nature, we can select a spring index $C = 10$. For safety consideration, we can select the solid force $\boldsymbol{F}_s$ to be 15% larger than the maximum operating force $\boldsymbol{F}_o$ such that $\boldsymbol{F}_s = 1150$ [N] or 258.53 [lb] and the solid deflection $\delta_s = 100\,(1.15) = 115$ [mm] or $\delta_s = 3.937\,(1.15) = 4.5276$ [in].

Using Eq. (10.21) to evaluate the Wahl factor gives

$$K_W = \frac{4C-1}{4C-4} + \frac{0.615}{C} = \frac{4(10)-1}{4(10)-4} + \frac{0.615}{10} = 1.0833 + 0.0615 = 1.1448 \tag{a}$$

This value is valid for both US and SI systems of units. Using Eq. (10.52) gives

$$d_w = \left(\left(1.1448\left(\frac{8(1150)(10)}{0.433\pi(1734)}\right)\right)^{0.5}\right)^{\frac{1}{1-0.5(0.177)}} = (6.682\,14)^{1.097\,09} = 8.035\ [\text{mm}] \tag{b}$$

or

$$d_w = \left(\left(1.1448\left(\frac{8(258.53)(10)}{0.433\pi(142\,000)}\right)\right)^{0.5}\right)^{\frac{1}{1-0.5(0.177)}} = (0.350\,11)^{1.097\,09} = 0.3162\ [\text{in.}] \tag{b$'$}$$

One can simply convert the diameter in [mm] to that in [in] directly to get the value of the calculated diameter in the US units from that in the SI units or vice versa. The closest standard wire diameter, however, is 8 [mm] or 0.324 [in] as shown in the selected wire diameters of Table 10.2. Note that the selected wire diameter for the SI system is smaller than the calculated value, while the one for the US system is larger than the calculated value. In that case, the maximum shear stress at the maximum operating load is (see Eq. (10.20))

$$\tau_s = K_W\left(\frac{8\boldsymbol{F}_s C}{\pi d_w^2}\right) = 1.1448\left(\frac{8(1150)(10)}{\pi(8)^2}\right) = 523.83\ [\text{MPa}] \tag{c}$$

or

$$\tau_s = K_W\left(\frac{8\boldsymbol{F}_s C}{\pi d_w^2}\right) = 1.1448\left(\frac{8(258.53)\,(10)}{\pi(0.324)^2}\right) = 71\,794\ [\text{psi}] \tag{c$'$}$$

Note that the shear stress in the US system is less than that in the SI system. This is because the selected wire diameter is larger for the US system than the wire diameter in the SI system. The minimum ultimate tensile strength from Eq. (10.48) is

$$S_{ut} = \frac{1734}{d_w^{0.177}} = \frac{1734}{8^{0.177}} = \frac{1734}{1.4449} = 1200\ [\text{MPa}] \tag{d}$$

or

$$S_{ut} = \frac{142\,000}{d_w^{0.177}} = \frac{142\,000}{0.324^{0.177}} = \frac{142\,000}{0.81916} = 173\,348\ [\text{psi}] \tag{d$'$}$$

The shear yield strength from Eq. (10.49) is

$$S_{sy} = 0.433\ S_{ut} = 0.433\,(1200) = 519.6\ [\text{MPa}] \tag{e}$$

or

$$S_{sy} = 0.433\ S_{ut} = 0.433\,(173\,348) = 75\,060\ [\text{psi}] \tag{e$'$}$$

The safety factor from Eq. (10.44) is

$$K_{SF}\big|_{SI} = \frac{S_{sy}}{\tau_s} = \frac{519.6}{523.8} = 0.992 \tag{f}$$

or

$$K_{\mathrm{SF}}\big|_{\mathrm{US}} = \frac{S_{sy}}{\tau_s} = \frac{75\,060}{71\,794} = 1.046 \tag{f'}$$

One should expect to have a safety factor K_{SF} of 1.0 instead of 0.992 or 1.046 for the SI system or the US system, respectively. However, the strength at 8 [mm] diameter is little less than at 8.0354 [mm] as obtained by Eq. (d) and the stress at 0.324 [in] diameter is less than at 0.3162 [in] as obtained by Eq. (c′). If any of these safety factors were not acceptable, iterations would be required such as changing wire diameter to another preferred one. However, one should reiterate that a 1.15 of the maximum operating force defined the solid force, which is a safety in disguise.

The number of active coils as in Eqs. (10.55) and (10.54) is

$$N_a\big|_{\mathrm{SI}} = 9.9125(10^3)\frac{\delta_s d_w}{\boldsymbol{F}_s C^3} = 9.9125(10^3)\ \frac{100(1.15)(8)}{1150(10)^3} = 7.93 \tag{g}$$

and

$$N_a\big|_{\mathrm{US}} = 1.4375\ \frac{\delta_s\ d_w}{\boldsymbol{F}_s\ C^3} = 1.4375\,(10^6)\frac{3.934\,(1.15)\,(0.324)}{258.53\,(10)^3} = 8.16 \tag{g'}$$

The number of turns in the US system is larger than the SI system by about 2.9% due to the fact of a larger US wire diameter than the SI system and needing that little extra fractional turn to deliver the same required deflection.

This process of defining the geometry and material selection is also possible by the utilization of Figures 10.9 and 10.10. From Figure 10.9 and at $\boldsymbol{F}_s$ of 1150 [N] or 258.53 [lb], the wire diameter at $C = 10$ is about 8 [mm] or about 0.33 [in]. For Figure 10.10, calculate $\boldsymbol{F}_s/\delta_s\ d_w$ as 1150/(115(8)) = 1.25 [N/mm²] or 258.53/(4.5276 (0.324)) = 176 [psi]. From the Figure 10.10 and at $\boldsymbol{F}_s/\delta_s\ d_w = 1.25$ for SI or 176 for US and at $C = 10$, the number of active turns N_a is about 8.

It is clear that Figures 10.9 and 10.10 provide quick tools to produce the helical spring synthesis in a very short time.

Check *spring buckling* by assuming ground squared ends ($N_e = 1$ from Table 10.1) and calculating L_a/d_c using Eqs. (10.1), (10.6), and (10.7) to get

$$p_h = \left(\frac{\delta_s}{N_a} + d_w\right) = \left(\frac{115}{7.93} + 8\right) = 22.5\ [\mathrm{mm}] \tag{h}$$

and

$$p_h = \left(\frac{\delta_s}{N_a} + d_w\right) = \left(\frac{4.5276}{8.16} + 0.324\right) = 0.879\ [\mathrm{in}] \tag{h'}$$

$$L_a = N_a p_h + d_w = 7.93(22.5) + 8 = 186.43\ [\mathrm{mm}] \tag{i}$$

and

$$L_a = N_a p_h + d_w = 8.16(0.879) + 0.324 = 7.4966\ [\mathrm{in}] \tag{i'}$$

$$L_a/d_c = (N_a p_h + d_w)/C d_w = (186.43)/10(8) = 2.33 \tag{j}$$

and

$$L_a/d_c = (N_a p_h + d_w)/C d_w = (7.4967)/10(0.324) = 2.34 \tag{j'}$$

From Figure 10.8, the buckling factor K_{cr} at L_a/d_c of 2.33 or 2.34 is about 0.7. The critical solid deflection from Eq. (10.32) is

$$\delta_{\mathrm{cr}} = K_{\mathrm{cr}} L_a = 0.7(186.43) = 130.5\ [\mathrm{mm}] \tag{k}$$

and

$$\delta_{cr} = K_{cr}L_a = 0.7(7.5864) = 5.31 \text{ [in]} \qquad (k')$$

which are more than the available solid deflections of 115 [mm] or 4.5276 [in]. Then, buckling will not occur and no need for internal or external restraints.

From Eq. (10.42), the *surge frequency* is

$$f_n|_{SI} = 357\,462\sqrt{\frac{1}{d_w^2 N_a^2 C^4}} = \frac{357\,462}{(8)(7.93)(10)^2} = 56.35 \text{ [Hz]} \qquad (1)$$

and

$$f_n|_{US} = 14\,120\sqrt{\frac{1}{d_w^2 N_a^2 C^4}} = \frac{14\,120}{(0.324)(8.16)(10)^2} = 53.41 \text{ [Hz]} \qquad (1')$$

The surge frequencies are close with the difference attributed to the variation in geometry values of the wire diameters and the number of active turns. The driving or external excitation frequency should not be larger than 5–7% of the resonance or surge frequency f_n according to Eq. (10.43). This suggests that the driving or external excitation frequency has to be less than 2.8–3.7 [Hz] or about 3.0 [Hz].

The solutions to this example are in the ***Helical Spring Synthesis Tablets*** shown in Figures 10.11 and 10.12 for both SI and US systems, respectively. Entering the specified input data of the problem in the **Input** column generates the solution at any movement to any other location in the tablet. One then enters the preferred wire diameter and the expected coil diameter in the **Analysis** column to have the same spring index C in both columns if needed. This is attainable by iteratively changing the coil diameter. One should also enter the same ultimate tensile strength or other definitely known value in the **Analysis** column. If the same strength is used, iterate to get the same value of intercept *A* in both columns. If one uses other definitely known ultimate tensile strength value, no iteration is needed to get the same *A* value in both columns particularly if one uses another vale of the constant *b*. This process was not automatically pursuit to allow for other definitely known value or possible change of material selection. One should also iterate the number of active turns to get the same operating deflection in both columns if the value of operating deflection is inflexible. The results of these ***Helical Spring Synthesis Tablets*** are very close to the values calculated herein. To increase the static safety factor, one can use a different material that has a higher ultimate tensile strength as shown in Table 10.3 such as oil Q and tempered steel or music wire. This type of iteration is useful in the design optimization subsequently given in Section 10.6.

The ***Helical Spring Synthesis Tablets*** give other spring parameters and behavior variables as shown in Figures 10.11 and 10.12. Some of these parameters and behavior variables are spring stiffness, spring stresses, and other parameters such as the ones to employ for the abscissa of synthesis Figures 10.9 and 10.10.

10.2.5.2 Detailed Design

Detailed helical spring design depends on the specific material selected for the explicit application that may differ from the design synthesis material adapted in this text. The previously adapted relations apply with some variations on parameters such as *A* and *b* for the specific material requirements. Table 10.3 gives the values of *A* and *b* for different materials including the previously adapted ***hard-drawn steel***. Fitting the ultimate tensile strength variation with wire diameter data available in Associated Spring (1987) produced Table 10.3 in a new form. *A* is the intercept of the fitted line with the value of the wire diameter d_w of one. The value of *b* indicates the steepness of variation of the strength with the diameter. The variation of the strength is higher with higher values of the parameter *b*. The new values of the parameters *A* and *b* in Table 10.3 are not the same as observed in different other references on the subject. The difference, however, does not seriously affect the sought ultimate tensile strength S_{ut}. One, however, should use the accessible ultimate tensile strength at the design wire diameter in refining the

Helical Spring Synthesis - SI units			
Inputs	**Default**	**Input**	**Analysis**
Maximum Operating Force, F_o [N]	1000	1000	1000
Solid Force, F_s [N]	1150	1150	1150
Minimum Operating Force, F_{min} [N]	100	100	100
Safety Factor, K_{SF}	1	1	1.000
Spring Index, C	10	10	10.000
Operating Deflection, δ_o [mm]	100	100	100.000
Inactive End Turns, N_e	2	1	1
Output			**Input/Output**
Wire Diameter, d_w [mm]	8.036	8.036	8
Coil Diameter, d_c [mm]	80.36	80.355	80
Number of Active Turns, N_a	7.97	7.965	7.93
Material Constant, A [MPa mm^b]	1734	1734	1748.37
Material Exponent, b	0.177	0.177	0.177
Modulus of Rigidity, G [GPa]	79.30	79.3	79.3
Ultimate Tensile Strength, S_{ut} [MPa]	1199.12	1199.12	1210
Wahl Factor, K_W	1.145	1.145	1.145
Active Length, L_a [mm]	187.04	187.04	186.44
Free Length, L_f [mm]	203.11	195.08	194.44
Operating Length, L_o [mm]	103.11	95.08	94.44
Solid Length, L_s [mm]	88.11	80.08	79.44
Spring Stiffness, k_s [N/mm]	10.00	10.00	10.00
Free Helix Pitch, p_h [mm]	22.473	22.473	22.502
Free Helix Angle, θ_h [°]	5.08722	5.08722	5.11617
Solid Stress, τ_s [MPa]	519.217	519.217	523.842
Mean Stress, τ_m [MPa]	248.321	248.321	250.533
Alternating Stress, τ_a [MPa]	203.17	203.17	204.98
Fatigue Shear Strength, S_{se} [MPa]	345.95	345.95	465
Fatigue Safety Factor, K_{FSF}	1.116	1.116	1.334
Available values			
Critical Buckling Force, F_{cr} [N]	1443.34	1386.24	1381.76
Active Length to Coil Diameter, L_a/d_c	2.33	2.33	2.33
Buckling Factor, K_{cr}	0.71	0.71	0.71
Fundamental Surge Frequency, f_n [Hz]	55.849	55.849	56.346
Parameter, $F_s/(\delta_s\, d_w)$ [N/mm^2]	1.2445	1.2445	1.2500

Notes: *Examples 10.1 and 10.2 of the Textbook. The Analysis column is used to enter the preferred wire diameter instead of the synthesised one.*

Figure 10.11 *Helical Spring Synthesis Tablet* providing the solution to Examples 10.1 and 10.2 for SI system of units.

safety factor to a more accurate value. The listing of the spring materials in Table 10.3 is sorted with the relative cost ascending from a relative cost of 1 to a relative cost of about 8. Relative market values not specified since the market value for each material is not constant with time. Table 10.3 also provides the US units with the SI units shown in brackets.

The detailed design is also dependent on the loading dynamics. If the applied load does not wary with time, one should use the static maximum solid stress in the detailed design. This is the same as the ***initial synthesis*** procedure previously given. The detailed design utilizes Eqs. (10.20) and (10.44) to find the safety factor and Eqs. (10.25) and (10.26) to check the solid deflection and the spring stiffness. One can also determine other spring parameters and dimensions as a final definition of the spring. The ***Helical Spring Synthesis Tablets*** provide useful tools to determine most of the necessary spring parameters and behavioral variables. They also allow detailed

Helical Spring Synthesis - US units			
Inputs	**Default**	**Input**	**Analysis**
Maximum Operating Force, F_o [lb]	100	224.81	224.81
Solid Force, F_s [lb]	115	258.53	258.53
Minimum Operating Force, F_{min} [lb]	100	22.481	22.481
Safety Factor, K_{SF}	1	1	1.045
Spring Index, C	10	10	10.000
Operating Deflection, δ_o [in]	4	3.937	3.939
Inactive End Turns, N_e	2	1	1
Output			**Input/Output**
Wire Diameter, d_w [in]	0.203	0.316	0.324
Coil Diameter, d_c [in]	2.03	3.162	3.24
Number of Active Turns, N_a	11.66	7.960	8.16
Material Constant, A [psi in^b]	142000	142000	141999
Material Exponent, b	0.177	0.177	0.177
Modulus of Rigidity, G [Mpsi]	11.50	11.5	11.5
Ultimate Tensile Strength, S_{ut} [psi]	188345	174099	173348
Wahl Factor, K_W	1.145	1.145	1.145
Active Length, L_a [in]	7.17	7.361	7.497
Free Length, L_f [in]	7.57	7.677	7.821
Operating Length, L_o [in]	3.57	3.740	3.883
Solid Length, L_s [in]	2.97	3.149	3.292
Spring Stiffness, k_s [lb/in]	25.00	57.102	57.077
Free Helix Pitch, p_h [in]	0.597	0.885	0.879
Free Helix Angle, θ_h [°]	5.35727	5.09104	4.93607
Solid Stress, τ_s [psi]	81554	75385	71797
Mean Stress, τ_m [psi]	70916	36054	34338
Alternating Stress, τ_a [psi]	0.00	29499	28094
Fatigue Shear Strength, S_{se} [psi]	54338	50228	67500
Fatigue Safety Factor, K_{FSF}	1.78	1.116	1.405
Available values			
Critical Buckling Force, F_{cr} [lb]	126.38	311.51	317.18
Active Length to Coil Diameter, L_a/d_c	3.53	2.33	2.31
Buckling Factor, K_{cr}	0.67	0.71	0.71
Fundamental Surge Frequency, f_n [Hz]	59.7	56.101	53.407
Parameter, $F_s/(\delta_s\, d_w)$ [lb/in^2]	123.30	180.5908	176.1642

Notes: *Examples* 10.1 *and* 10.2 *of the Textbook. The Analysis column is used to enter the preferred wire diameter instead of the synthesised one.*

Figure 10.12 *Helical Spring Synthesis Tablet* providing the solution to Examples 10.1 and 10.2 for US system of units.

design iterations and possible means for optimization. These Tablets are available through the ***Wiley website*** of this textbook.

The detailed design process of dynamically loaded spring employs Eqs. (10.27) and (10.28) to define the mean and alternating forces $\boldsymbol{F}_m$ and $\boldsymbol{F}_a$ that describe the dynamic character of the load. Substituting these forces in Eqs. (10.29) and (10.30) obtains the mean of the shear stresses τ_m and the maximum alternating shear stress τ_a. This evaluates the spring synthesis for the case of a fatigue expectation due to the applied dynamic loading such as valve springs in the internal combustion engines. The fatigue safety factor K_{SFS} defines the condition of withstanding the dynamic load without possible fatigue failure. This process depends on the fatigue shear strength of the spring material S_{se} and the applied dynamic shear stresses τ_m and τ_a. Obtaining the fatigue shear strength of the spring material S_{se} is not very well assured or confidently defined for all materials. If the value is not accessible,

Table 10.3 Some spring materials, their composition, standards, size range, exponent parameter b, and intercept parameter A to evaluate the minimum ultimate tensile strength from the equation $S_{ut} = A/(d_w)^b$.

Spring material	Size range [in] or ([mm])	Exponent b	Intercept parameter A [psi in^b] ([MPa mm^b])
Hard-drawn steel wire (0.6–0.7% C, 0.6–1.3% Mn), ASTM A227, AISI 1066, JIS 3521, DIN 2076 17223, BS5216, ISO 8458, EN 10270-1	0.028–0.5 (0.7–12)	0.177	142 000 (1734)
Oil Q & tempered steel wire (0.6–0.7% C, 0.6–1.3% Mn), ASTM 229, AISI 1065, JIS G3560, DIN 2076 17223, BS2803, ISO 8458, EN 10270-2	0.02–0.5 (0.5–12)	0.176	149 700 (1824)
Music wire (0.8–0.95% C, 0.2–0.6% Mn), ASTM A228, AISI 1085, JIS G3522, DIN 2076 17223, BS5216, EN 10270-1	0.004–0.256 (0.1–6.5)	0.161	185 000 (2145)
Chrome vanadium wire (0.48–0.53% C, 0.8–1.1% Cr, 0.15% min. V), ASTM A232, AISI 6150, JIS G3561, BS2803, EN 10270-2	0.032–0.437 (0.8–11)	0.147	172 600 (1914)
Chrome silicon wire (0.51–0.59% C, 0.6–0.8% Cr, 1.2–1.6% Si), ASTM A401, AISI 9254, JIS G3561, DIN 2076 17223, BS2803, EN 10270-2	0.063–0.375 (1.6–9.5)	0.099	215 500 (2050)
Stainless steel wire (17–19% Cr, 8–10% Ni), ASTM A313, AISI 302, JIS G4314, DIN 17224, BS2056, ISO 6931, EN 10270-3	0.013–0.4 (0.3–10)	0.221	128 000 (1801)
Phosphor-bronze wire (94–96% Cu, 4–6% Sn), ASTM B159, JIS H3751 PBP1, BS2873	0.004–0.3 (0.1–7.5)	0.074	106 000 (926)

one should estimate its value. A reasonable estimate is by applying the distortion energy theory to first relate the normal and shear stress failures. If that applies, the fatigue shear strength of the spring material S_{se} is then

$$S_{se} = \frac{S_e}{\sqrt{3}} = 0.577\ S_e \tag{10.56}$$

where S_{se} is the endurance limit or fatigue strength in the reversed normal loading or reversed bending test. Again, S_e may not be accessible for the material at the anticipated diameter, and one therefore needs to estimate its value. A usual common relation between the ultimate tensile strength and the fatigue strength is (see Section 7.5)

$$S_e = 0.5\ S_{ut} \tag{10.57}$$

This relation is not very accurate for several material microstructures. The 0.5 factor can take another value in the range of 0.25–0.63 for carbon steel depending on ferrite, pearlite, or martensite microstructure. The range for alloy steel is 0.23–0.47, Sors (1971) (see Section 7.8).

Another way of estimating the fatigue shear strength is to adopt the results of Zemmerli (1957). He reported that the fatigue shear strength S_{se} levels out similar to cases of the endurance limit of the high tensile strength. His results show that

$$\begin{aligned} S_{se} &= 45\,000\ [\text{psi}]\ (310\ [\text{MPa}]) \quad \text{for } \textit{unpeened} \text{ springs} \\ S_{se} &= 67\,500\ [\text{psi}]\ (465\ [\text{MPa}]) \quad \text{for } \textit{shot peened} \text{ springs} \end{aligned} \tag{10.58}$$

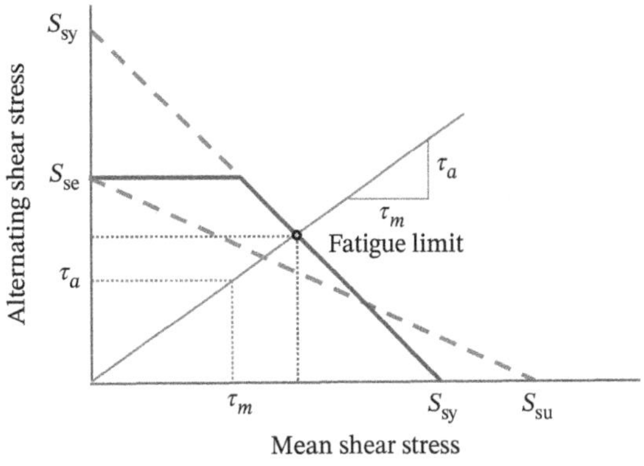

Figure 10.13 The torsional fatigue and the *modified Goodman* diagram for fatigue failure in shear.

Table 10.4 Some spring materials, size range, density ρ, and average shear modulus of elasticity or modulus of rigidity G.

Spring material	Size range [in] or ([mm])	Density ρ [lb/in^3] or ([kg/m^3])	Modulus of rigidity G [Mpsi] or ([GPa])
Hard-drawn steel wire (0.6–0.7% C, 0.6–1.3% Mn)	0.028–0.5 (0.7–12)	0.282 (7860)	11.5 (79.3)
Oil Q & tempered steel wire (0.6–0.7% C, 0.6–1.3% Mn)	0.02–0.5 (0.5–12)	0.284 (7915)	11.2 (77.2)
Music wire (0.8–0.95% C, 0.2–0.6% Mn)	0.004–0.256 (0.1–6.5)	0.281 (7830)	11.75 (81)
Chrome vanadium wire (0.48–0.53% C, 0.8–1.1% Cr, 0.15%min. V),	0.032–0.437 (0.8–11)	0.282 (7860)	11.2 (77.2)
Chrome silicon wire (0.51–0.59% C, 0.6–0.8% Cr, 1.2–1.6% Si)	0.063–0.375 (1.6–9.5)	0.282 (7860)	11.2 (77.2)
Stainless steel wire (17–19% Cr, 8–10% Ni)	0.013–0.4 (0.3–10)	0.286 (7920)	10.3 (71.0)
Phosphor-bronze wire (94–96% Cu, 4–6% Sn)	0.004–0.3 (0.1–7.5)	0.320 (8860)	6.25 (44.8)

With the existence of the mean shear stresses for the pulsating torsion, the *modified Goodman* criterion is usually used, Smith (1942). For that, one needs the value of the ultimate shear strength S_{su}. This value is estimated as (Joerres, 1981)

$$S_{su} = 0.67\ S_{ut} \tag{10.59}$$

where S_{ut} is the ultimate tensile strength of the wire material at the design wire diameter d_w (see Eq. (10.48)). The modified Goodman criterion requires the shear yield strength S_{sy}. Eq. (10.49) defines an applicable relation to estimate the shear yield strength S_{sy}. The dynamic failure criterion according to *modified Goodman* and the associated fatigue safety factor K_{FSF} is (see Section 7.4 and Figure 10.13)

$$\frac{\tau_a}{S_{se}} + \frac{\tau_m}{S_{su}} = \frac{1}{K_{FSF}} \quad \text{or} \quad K_{FSF} = \frac{S_{se}\ S_{su}}{\tau_a\ S_{su} + \tau_m\ S_{se}} \tag{10.60}$$

To evaluate the number of turns of the helical spring, one needs the shear modulus of rigidity G of the specific material selected for the explicit application. Table 10.4 provides the shear modulus of rigidity G for different materials used for springs. It also provides the density ρ that is essential in the calculations of spring weight or mass and fundamental resonance or surge frequency. Table 10.4 also provides the US units with the SI units shown in brackets for both quantities.

Example 10.2 Consider the design of the general-purpose spring of Example 10.1 that can carry a maximum operating load of 1000 [N] or 224.81 [lb], a maximum operating deflection of 100 [mm] or 3.937 [in], and assuming ground squared ends. Determine the needed difference in the design if the dynamic load varies between the specified maximum and a minimum of 100 [N] or 22.451 [lb]. Find the suitable fatigue safety factor K_{FSF}.

Solution

Data: $\boldsymbol{F}_{\max} = 1000$ [N] or 224.81 [lb], $\delta_{\max} = 100$ [mm] or 3.937 [in], $F_{\min} = 100$ [N] or 22.481 [lb], $\delta_{\min} = 10$ [mm] or 0.3937 [in].

The initial synthesis suggests the use of a ***hard-drawn steel*** with $A = 1734$ [MPa mm^b] or 142 000 [psi in^b] and $b = 0.177$. Even though the load is of a dynamic nature, we may keep the spring index $C = 10$. For safety consideration, we also keep the solid force $\boldsymbol{F}_s$ to be 15% larger than the maximum operating force or $\boldsymbol{F}_s = 1150$ [N] or 258.53 [lb] and the solid deflection $\delta_s = 100\ (1.15) = 115$ [mm] or $\delta_s = 3.937\ (1.15) = 4.5276$ [in]. The initial design found in Example 10.1 is of a geometry defining $d_w = 8$ [mm] or 0.324 [in] and $N_a = 7.93$ or 8.19 for SI and US systems, respectively.

Utilize the values of needed parameters previously calculated in Example 10.1 and calculate the mean and the alternating dynamic forces and stresses using Eqs. (10.26)–(10.30):

$$\boldsymbol{F}_m = \frac{\boldsymbol{F}_{\max} + \boldsymbol{F}_{\min}}{2} = \frac{1000 + 100}{2} = 550\ [\text{N}] \tag{a}$$

$$\boldsymbol{F}_a = \frac{\boldsymbol{F}_{\max} - \boldsymbol{F}_{\min}}{2} = \frac{1000 - 100}{2} = 450\ [\text{N}] \tag{b}$$

$$\tau_m = K_W\left(\frac{8\boldsymbol{F}_m C}{\pi d_w^2}\right) = 1.1448\left(\frac{8(550)(10)}{\pi(8^2)}\right) = 250.53\ [\text{Pa}] \tag{c}$$

$$\tau_a = K_W\left(\frac{8\boldsymbol{F}_a C}{\pi d_w^2}\right) = 1.1448\left(\frac{8(450)(10)}{\pi(8^2)}\right) = 204.96\ [\text{Pa}] \tag{d}$$

or

$$\boldsymbol{F}_m = \frac{\boldsymbol{F}_{\max} + \boldsymbol{F}_{\min}}{2} = \frac{224.81 + 22.481}{2} = 123.65\ [\text{lb}] \tag{a$'$}$$

$$\boldsymbol{F}_a = \frac{\boldsymbol{F}_{\max} - \boldsymbol{F}_{\min}}{2} = \frac{224.81 - 22.481}{2} = 101.15\ [\text{lb}] \tag{b$'$}$$

$$\tau_m = K_W\left(\frac{8\boldsymbol{F}_m C}{\pi d_w^2}\right) = 1.1448\left(\frac{8(123.65)(10)}{\pi(0.324^2)}\right) = 34\,338\ [\text{psi}] \tag{c$'$}$$

$$\tau_a = K_W\left(\frac{8\boldsymbol{F}_a C}{\pi d_w^2}\right) = 1.1448\left(\frac{8(101.15)(10)}{\pi(0.324^2)}\right) = 28\,090\ [\text{psi}] \tag{d$'$}$$

One could accurately convert the forces for the SI to the US system but cannot convert the stresses due to variations in wire diameters. The wire diameter in SI system is smaller than that for the US system.

Adopting the fatigue shear strength according to Eq. (10.58) for unpeened spring materials gives

$$S_{\text{se}} = 45\,000\ [\text{psi}]\quad (310\ [\text{MPa}]) \tag{e}$$

The ultimate shear strength of the adapted **hard-drawn wire** according to Eqs. (10.48) and (10.60) are

$$S_{\text{su}} = 0.67\frac{142\,000}{d_w^{0.177}},\ [\text{psi in}^b]\quad \text{or}\quad S_{\text{su}} = 0.67\frac{1734}{d_w^{0.177}},\ [\text{MPa mm}^b] \tag{f}$$

Using the same values as Example 10.1 gives

$$S_{\text{su}} = 0.67(1200) = 804\ [\text{MPa}] \tag{g}$$

$$S_{\text{ut}} = 0.67\ \ (173\,348) = 116\,143\ [\text{psi}] \tag{g$'$}$$

The fatigue safety factor K_{FSF} according to Eq. (10.60) is

$$K_{FSF}\big|_{SI} = \frac{S_{se}S_{su}}{\tau_a S_{su} + \tau_m S_{se}} = \frac{310(804)}{204.96(804) + 250.53(310)} = \frac{249\,240}{242\,452} = 1.028 \quad \text{(h)}$$

$$K_{FSF}\big|_{US} = \frac{S_{se}S_{su}}{\tau_a S_{su} + \tau_m S_{se}} = \frac{45\,000\ (116\,143)}{28\,090(116\,143) + 34\,338\ (45\,000)} = \frac{5226.4(10^6)}{4807.7(10^6)} = 1.087 \quad \text{(h')}$$

The fatigue safety factor is according to the *modified Goodman* criterion, which is a conservative estimate. The small fraction over 1.0 in the estimated value of the fatigue safety factor may be acceptable for some applications. If it is not acceptable, one can use *shot peened* material to increase the fatigue safety factor.

The solutions to this example are in the ***Helical Spring Synthesis Tablets*** shown in Figures 10.11 and 10.12 for SI and US systems, respectively. The results of these ***Helical Spring Synthesis Tablets*** are very close to the values provided herein. These Tablets are useful in the process of detailed redesign iterations to redefine other geometries or materials.

Using ***Helical Spring Synthesis Tablets*** and keeping the geometry intact, one can change the material processing to a *shot peened* material. The result is the increase in the fatigue safety factor K_{FSF} to higher values of 1.329 for SI system and 1.405 for the US system. This accounts for an increase of about 29% in the fatigue safety factor for either system but would entail extra cost. One can also use other types of iterations in the design optimization subsequently specified in Section 10.6.

Several springs and different constructional details of compression springs, their ends, and their connections are available through the Internet images of extension and compression springs.

10.2.6 Extension Springs

Extension springs usually have ends to help in exerting the applied tension load (see Figure 10.1b). The ends have different shapes to accommodate the application and connectivity to other elements that acts on the spring (see Internet images at the references of this chapter). Manufacturers of extension springs usually induce initial torsion while coiling the spring. This generates a solid spring (Figure 10.1b) with coils snuggled to each other. This prestressing generates initial internal force that needs an initial tension force $\boldsymbol{F}_{it}$ before starting the spring function. Figure 10.14 shows the expected load deflection characteristic for the active turns. The spring does not deflect until the applied force exceeds this initial tension force $\boldsymbol{F}_{it}$. This behavior does not include the effect of the shape of the spring ends and their deformation.

All previously developed equations for the compression springs apply. The number of end coils N_e is zero with the stresses in the spring ends treated separately as curved members for their synthesis. The only difference is that the solid force is now a tension force and treated as the maximum force $\boldsymbol{F}_{max}$. That force should not induce a torsional yield stress; otherwise the spring fails with permanent damage to its form and the initial tension diminishes. A construction constraint should present means to prevent that from happening. $\boldsymbol{F}_{max}$ should be 15–20%

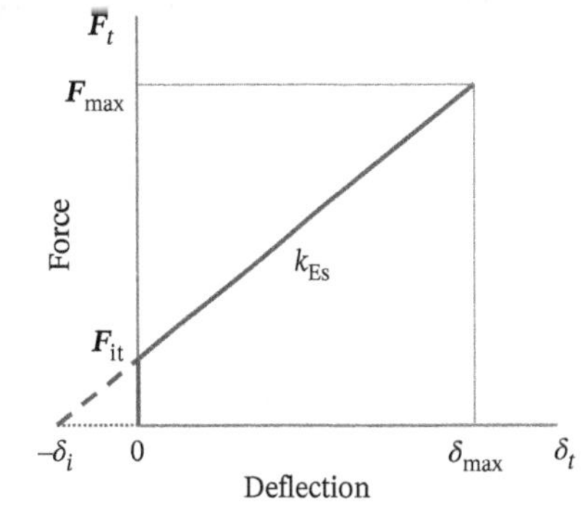

Figure 10.14 Extension spring characteristics indicating initial tension $\boldsymbol{F}_{it}$ of active coils and not including the effect of end configurations. The initial imbedded or "stored" deflection is δ_i.

more than the maximum operating force F_o so that the operating force F_o would not cause a torsional yield stress. This maximum force F_{max} is equivalent to the value of the solid force F_s in the compression spring.

The spring manufacturers, Associated Springs (1987), suggest an average preferred induced initial torsional stress that one should overcome. Using regression, this stress has the following fitting expression:

$$y = 38.69x^2 - 2141.7x + 29\,857 \quad \text{for US system} \tag{10.61}$$

$$y = 0.2668x^2 - 14.77x + 205.91 \quad \text{for SI system} \tag{10.62}$$

where y is the average preferred initial torsional prestress τ_i [psi] or [MPa] due to initial imbedded or "stored" tension F_{it} [lb] or [N] and x is the spring index C. The average initial torsional prestress is about 10–25% of the maximum torsional stress. The variation of this average is about ±20% for a preferred region between a maximum and a minimum. Check this value with the one provided by the intended manufacturer of the synthesized spring. For some specific and special applications, one may define the necessary initial torsional stress and ask the manufacturer to provide the spring with this particular specification.

We can carry out extension spring synthesis exactly as the helical compression springs with the previous considerations taken care of. The maximum operating force is the same as for compression springs. The solid stress in the compression spring is the same as the stress that is 10–20% more than the maximum stress due to the maximum operating tension force F_o. Alternatively, the maximum tension is the same as that for the solid force F_s in compression springs. The maximum operating deflection δ_o is the maximum deflection added to the initial "stored" deflection δ_i due to the initial internal tension F_{it} as in Figure 10.14. Adapted inactive end turns is zero. Initial spring length is the same as the solid length L_s of the compression spring with zero end turns. The maximum spring length after extension is the solid length added to the extension generated by the maximum force F_{max} minus the initial tension F_{it}, i.e. the maximum deflection δ_{max} as in Figure 10.14. The minimum operating force in the compression spring may act as the initial tension for the extension spring. Use the ***Helical Spring Synthesis Tablet*** to synthesize the extension spring. However, one should apply previous conditions to recalculate L_a, L_f, L_o, and L_s for extension spring from values calculated for compression spring. Buckling is not admissible in extension springs, surge, or resonance frequency behavior; however, it occurs. The evaluation of the extension-spring stiffness k_{Es} is the same as for the compression spring with the application of Eq. (10.9). With reference to Figure 10.14, one may also include the initial imbedded "stored" deflection in the calculation of the value of extension-spring stiffness k_{Es}.

Several different constructional details for extension springs and their connections are available through the Internet images Extension springs and Extension spring ends.

Example 10.3 Consider the design of the general-purpose spring of Example 10.1 that can carry a maximum operating tension load of 1000 [N] or 224.81 [lb], a maximum operating deflection of 100 [mm] or 3.937 [in] (including the initial deflection due to initial internal force), and assuming no concern for the ends. Determine the difference between this design of the extension spring and its similar compression spring of Examples 10.1 and 10.2. Consider a minimum initial tension of 100 [N] or 22.481 [lb] and the inception of yield at an extra 15% over the maximum operating tension load. Find the suitable fatigue safety factor K_{FSF} if the load varies between the minimum initial tension and the maximum applied force.

Solution

Data: $F_o = 1000$ [N] or 224.81 [lb], $\delta_o = 100$ [mm] or 3.937 [in] $- \delta_{min}$, $F_{min} = 100$ [N] or 22.481 [lb], $\delta_{min} = 10$ [mm] or 0.3937 [in], deflection at inception of yield in shear $\delta_{sy} = (115$ [mm] or 4.5276 [in]$) - \delta_{min}$.

Consider the synthesized spring to be of the same geometry similar to that of Example 10.1. Therefore, initially adopt the solutions depicted in the ***Helical Spring Synthesis Tablets*** of Figures 10.11 and 10.12. The initial torsional stress according to Eqs. (10.61) and (10.62) are

$$\tau_i = 38.69C^2 - 2141.7C + 29\,857 = 38.69(10)^2 - 2141.7(10) + 29\,857 = 12\,309 \text{ [psi]} \tag{a}$$

$$\tau_i = 0.2668C^2 - 14.77C + 205.91 = 0.2668(10)^2 - 14.77(10) + 205.91 = 84.89 \text{ [MPa]} \tag{b}$$

This indicates that the suggested manufacturer initial force is about 18.8% of the maximum force rather than the 10% adapted in this problem. Initial stresses due to the minimum initial force of 100 [N] or 22.481 [lb] are $(\tau_m - \tau_a)$ or 41.15 [MPa] or 6555 [psi], respectively. This, however, should not affect the synthesis main geometry of at least the wire diameter and the number of active coils. Entering the number of inactive coils as zero in the ***Helical Spring Synthesis Tablets*** gives the results we need to compare with.

The main variations in geometry for the SI system are as follows:

- The initial length in extension spring L_i (or the free length L_f) is the same as the solid length in compression spring L_s, which is $L_i = L_f = N_a d_w + d_w = 7.93\ (8) + 8 = 71.44$ [mm] (checks with Tablet).
- The length at maximum operating force is $L_o = L_i + (\delta_{max} - \delta_{min}) = 71.44 + 90 = 161.44$ [mm].
- The maximum length at the inception of torsional yield at 1.15 $\boldsymbol{F}_o$ is $L_{max} = L_i + (115 - 10) = 176.44$ [mm], which is the same as the free length of the compression spring minus the initial "stored" deflection δ_i (10 [mm]) in the extension spring as in Figure 10.14.

The main variations in geometry for the US system are as follows:

- The initial length in extension spring L_i (or the free length L_f) is the same as the solid length in compression spring L_s, which is $L_i = N_a\ d_w + d_w = 8.16\ (0.324) + 0.324 = 2.968$ [in] (checks with Tablet).
- The length at maximum operating force is $L_o = L_i + (\delta_{max} - \delta_{min}) = 2.968 + 0.3937 = 2.5741$ [mm].
- The maximum length at the inception of torsional yield at 1.15 $\boldsymbol{F}_o$ is $L_{max} = L_i + (4.5276 - 0.3937) = 7.102$ [in], which is the same as the free length of the compression spring minus the initial imbedded or "stored" deflection δ_i (0.3937 [in]) in the extension spring as in Figure 10.14.

The stiffness of the extension spring k_{Es} is the same as the compression spring, and using Eq. (10.9) gives

$$k_{Es} = \frac{\Delta \boldsymbol{F}}{\Delta L} = \frac{\boldsymbol{F}_{max} - \boldsymbol{F}_{it}}{\delta_{max} - \delta_{it}} = \frac{1000 - 100}{100 - 10} = 10 \text{ [N/mm]}$$

or

$$k_{Es} = \frac{\Delta \boldsymbol{F}}{\Delta L} = \frac{\boldsymbol{F}_{max} - \boldsymbol{F}_{it}}{\delta_{max} - \delta_{it}} = \frac{224.81 - 22.481}{3.937 - 0.3937} = 57.1 \text{ [lb/in]} \tag{c}$$

These are the same stiffness values of compression springs of Example 10.1.

10.2.7 Torsion Springs

Helical springs act as *torsional springs* when the ends are twisted by operating moments $\boldsymbol{M}_o$ about their central axis instead of the compression or tension forces along the axis. The internal stresses are due to the *twisting moments* $\boldsymbol{M}_o$ on the spring ends. The bending moment is usually acting in the same direction as the original coiling of the spring. If no stress relieving performed on the spring after coiling, there exists *residual stresses* that reduce the effect of applied moments. This would cause a reduction in the maximum stresses due to moment application. No consideration of that is included in the following analysis. If no stress relieving is performable, residual stresses can act as an extra safety.

The bending stresses due to moment $\boldsymbol{M}_o$ are similar to those of the curved-beam theory giving (see Section 6.5)

$$\sigma_t = K_i \frac{\boldsymbol{M}_o(d_w/2)}{I_w} \tag{10.63}$$

where K_i is the curvature factor for the inner diameter since that represents the maximum occurrence of the stress distribution due to curvature. The wire diameter is d_w and the second area moment of the circular wire I_w is equal to $\pi(d_w)^4/64$. The curvature factor K_i is analytically obtainable as (Wahl, 1963).

$$K_i = \frac{4C^2 - C - 1}{4C(C-1)} \tag{10.64}$$

where C is the spring index defined by Eq. (10.6). As the operating force $\boldsymbol{F}_o$ acts tangential to the spring mean coil diameter, the generated moment $\boldsymbol{M}_o$ becomes equal to $\boldsymbol{F}_o\, d_c/2$. Substituting for that in Eq. (10.63) gives the stress in the circular wire as

$$\sigma_t = K_i \frac{\boldsymbol{M}_o(d_w/2)}{I_w} = K_i \frac{\boldsymbol{F}_o\,(d_c/2)\ (d_w/2)}{\pi\,(d_w)^4/64} = K_i \frac{16\,\boldsymbol{F}_o\,(d_c)}{\pi\,(d_w)^3} = K_i \frac{16\,\boldsymbol{F}_o\,C}{\pi\,(d_w)^2} \tag{10.65}$$

Similar to helical compression springs, assume the inception of yield to happen 10–20% higher than the maximum operating force $\boldsymbol{F}_o$. Considering an average of 15% overload to cause yielding gives the wire diameter from Eq. (16.63) as

$$d_w = \sqrt{K_i \frac{16\,(1.15)\boldsymbol{F}_o\,C}{\pi S_y}} \tag{10.66}$$

where S_y is the tensile yield strength of the spring material. For initial synthesis, the initial material adapted is the **hard-drawn wire** (0.6–0.7% carbon), which is a conservative least expensive general-purpose spring steel. One can assume that the relation between the tensile yield strength S_y and the ultimate tensile strength S_{ut} ranges between 0.6 and 0.9. To have an average for our spring steel, the following conservative relation is acceptable (see Section 7.7):

$$S_y = 0.75\ S_{ut} \tag{10.67}$$

The values of the ultimate tensile strength are as those in Eq. (10.48) for the adapted material in US and SI systems. One can, however, use a different material that has a higher ultimate tensile strength as shown in Table 10.3.

The number of active coils N_{ta} in the torsion spring depends on the required rotation θ_t at the end of the spring. According to *Castigliano's* theorem, the deflection $((d_c/2)\,\theta_t)$ of the operating force $\boldsymbol{F}_o$ is obtained by

$$(d_c/2)\,\theta_t = \frac{\partial U_o}{\partial \boldsymbol{F}_o} \tag{10.68}$$

where U_o is the total operating strain energy due to the operating force $\boldsymbol{F}_o$. The operating strain energy U_o due to the bending $\boldsymbol{M}_o$ generated by the force $\boldsymbol{F}_o$ is

$$U_o = \int \frac{\boldsymbol{M}_o^2}{2EI_w}\,ds = \int \frac{(\boldsymbol{F}_o(d_c/2))^2}{2E\ (\pi\ d_w^4/64)}\,ds \tag{10.69}$$

where s is the distance along the wire and E is the modulus of elasticity of the spring material. Performing the differentiation of Eq. (10.68) on the strain energy defined by Eq. (10.69) and considering the length s of the spring to be $\pi\ d_c\ N_{Ta}$, where d_c is coil diameter and N_{Ta} is the *number of active turns*, give

$$(d_c/2)\theta_t = \int_0^{\pi d_c N_{Ta}} \frac{\partial}{\partial \boldsymbol{F}_o}\left(\frac{(\boldsymbol{F}_o(d_c/2))^2}{2E(\pi d_w^4/64)}\right) ds = \int_0^{\pi d_c N_{Ta}} \frac{16\boldsymbol{F}_o d_c^2}{E(\pi d_w^4)}\,ds \tag{10.70}$$

This gives θ_t in [rad] as

$$\theta_t = \frac{64\boldsymbol{F}_o(d_c/2)\pi d_c N_{Ta}}{\pi E d_w^4} = \frac{32\boldsymbol{F}_o d_c^2 N_{Ta}}{E d_w^4} = \frac{32\boldsymbol{F}_o C^2 N_{Ta}}{E d_w^2} \tag{10.71}$$

From this equation, one can find the torsion spring stiffness or spring rate k_{Ts} as

$$k_{Ts} = \frac{\boldsymbol{M}_o}{\theta_t} = \frac{\boldsymbol{F}_o(d_c/2)}{\theta_t} = \frac{E\, d_w^4}{64\, d_c\, N_{Ta}} = \frac{E\, d_w}{64\;\, C\, N_{Ta}} \tag{10.72}$$

From Eq. (10.71), the number of active turns N_{Ta} is then

$$N_{Ta} = \frac{E\, d_w^4}{32\boldsymbol{F}_o d_c^2}\theta_t = \frac{E\, d_w^2}{32\boldsymbol{F}_o C^2}\theta_t \tag{10.73}$$

Alternatively, from Eq. (10.72), the *number of active turns* is also

$$N_{Ta} = \frac{E\, d_w^4}{64\, d_c\, k_{Ts}} = \frac{E\, d_w}{64\;\, C\, k_{Ts}} \tag{10.74}$$

This is only the number of active turns with no consideration for the deflections of the ends that apply the operating force $\boldsymbol{F}_o$ to be tangent to the spring. These deflections depend on different end configurations and can be separately calculated and added to produce total deflection.

The derivation of Eq. (10.71) is easily produced by considering the torsion spring as a bar of a diameter d_w and a length L_{Ts} of $\pi\, d_c\, N_{Ta}$ subjected to the moment $\boldsymbol{M}_o$. The rotation θ_t is simply

$$\theta_t = \frac{\boldsymbol{M}_o L_s}{EI_w} = \frac{\boldsymbol{F}_o(d_c/2)\,(\pi d_c N_{Ta})}{E(\pi\, d_w^4/64)} = \frac{32\boldsymbol{F}_o C^2\, N_{Ta}}{E\, d_w^2} \tag{10.71}$$

The convincing logic of this simple procedure might be harder to grasp than the formalized *Castigliano's* theorem. The imagination that the bar is originally bendable in a helix might have been an issue to accept the applicability of this simple postulation. The rigorous procedure, however, verified the simpler postulation.

One can simply generate similar graphs to those of Figures 10.9 and 10.10 for torsion springs using Eqs. (10.66), (10.67), and (10.72). Developing such an aid to *torsion spring* synthesis is an exercise that is simply attainable similar to the previous ***Spring Synthesis Tablets***.

Several different constructional details for torsion spring ends and their connections are available through the Internet images Torsion spring ends.

10.3 Leaf Springs

Usually *leaf springs* are made of constant thickness strips of spring material. In a compound form, regular leaf springs utilize the concept of uniform beam strength in their construction. As shown in Figure 10.15a, a constant strength–constant thickness beam must have the width w_x varying linearly with the length L_l. To utilize this configuration, one should divide the triangular form into equal strips (Figure 10.15b) and rearrange these strips into *multilayer leaf spring* as shown in Figure 10.15c. This is one configuration of leaf sprigs acting as a cantilever. Other configurations include popular simply supported leaf springs and the over-hanged ones. Figure 10.16 shows these forms with similarity equations that can provide the equivalent cantilever for each. This process is useful since the development herein will use the cantilever spring as the one to synthesize.

10.3.1 Stresses and Deflections

With reference to Figure 10.14c, the maximum *normal stress* in the leaf spring σ_L in each one of the leafs i due to the operating load $\boldsymbol{F}_o$ is

$$\sigma_L = \frac{\boldsymbol{M}_o(h/2)}{i\; I_L} = \frac{\boldsymbol{F}_o\, L_L\;\, (h/2)}{i\;\, (bh^3/12)} = \frac{6\;\, \boldsymbol{F}_o\, L_L}{ibh^2} \tag{10.75}$$

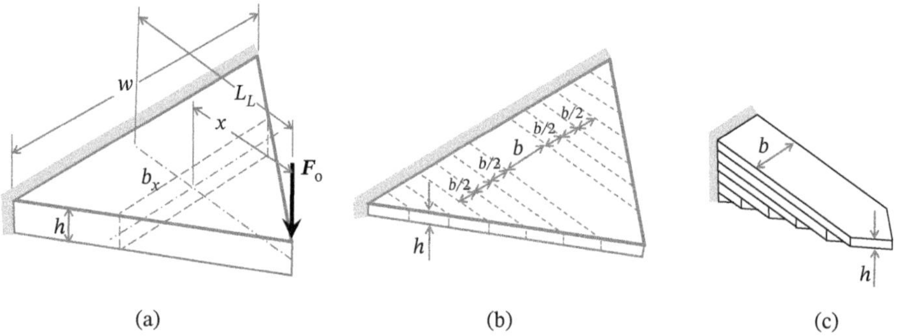

Figure 10.15 Constant strength–constant thickness triangular beam (a), similarly divided into equal strips (b) and rearrange strips into multilayer leaf spring (c).

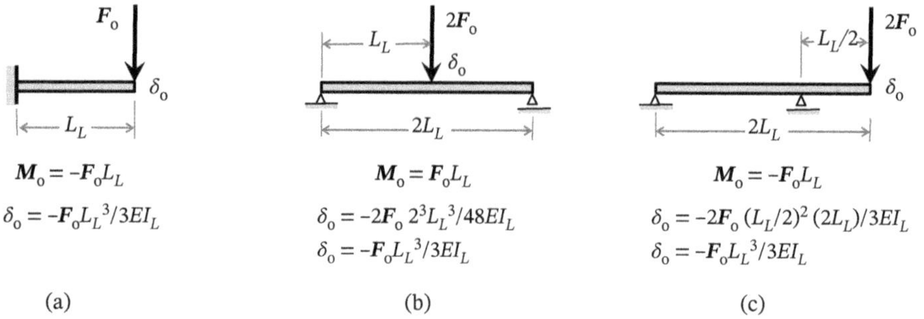

Figure 10.16 An equivalent leaf spring configuration relative to the original leaf spring shown in (a), where (b) is the equivalent simply supported spring, and (c) is the equivalent overhang spring.

where L_L is the spring length, i is the total number of leafs, I_L is the second-area moment of each leaf, b is the spring width, and h is the thickness of each leaf. Note that the second area moment of the spring section is the total second area moments of sections of all layers or equals to $i\,I_L$. For each rectangular section, the second area moment L_L is $(bh^3/12)$.

If the operating force is dynamic, one can use the same procedure to evaluate the mean operating force $\boldsymbol{F}_m$ and the alternating operating forces $\boldsymbol{F}_a$ similar to Eqs. (10.27) and (10.28). The mean stresses σ_m and the alternating stresses σ_a are consequently defined by repeated application of Eq. (10.75) using $\boldsymbol{F}_m$ and $\boldsymbol{F}_a$, respectively, in place of $\boldsymbol{F}_o$.

Since the operating moment $\boldsymbol{M}_o$ and the second area moment I_L are changing along the length of the constant strength triangular spring (Figure 10.14a), the better way to get the leaf spring deflection δ_{Lo} at the operating force $\boldsymbol{F}_o$ is to use the *Castigliano's theorem*. The application of the theorem to the operating strain energy U_o gives (see Section 6.6)

$$\delta_{Lo} = \frac{\partial U_o}{\partial \boldsymbol{F}_o} = \frac{\partial}{\partial \boldsymbol{F}_o}\left(\int \frac{\boldsymbol{M}_o^2}{2EI_L}dx\right) = \int \frac{1}{EI_L}\left(\boldsymbol{M}_o \frac{\partial \boldsymbol{M}_o}{\partial \boldsymbol{F}_o}\right) dx \tag{10.76}$$

where E is the modulus of elasticity of spring material. Substituting for $\boldsymbol{M}_o$ as $-\boldsymbol{F}_o\,x$ and I_L as $(wx\,h^3/12\,L_L)$ where w is the total width at the base (see Figure 10.15a with the width at x is $b_x = wx/L_L$), the leaf spring deflection δ_{Lo} at the operating load $\boldsymbol{F}_o$ is then

$$\delta_{Lo} = \int \frac{1}{EI_L}\left(\boldsymbol{M}_o \frac{\partial \boldsymbol{M}_o}{\partial \boldsymbol{F}_o}\right) dx = \int_0^{L_L} \frac{1}{E\,(wx\,h^3/12L_L)}\left(-\boldsymbol{F}_o x \frac{\partial(-\boldsymbol{F}_o x)}{\partial \boldsymbol{F}_o}\right) dx$$

$$= \int_0^{L_L} \frac{1}{E\,(wx\,h^3/12L_L)}(\boldsymbol{F}_o x^2) dx$$

or

$$\delta_{\mathrm{Lo}} = \frac{12L_L\boldsymbol{F}_o}{E\,w\,h^3}\int_0^{L_L} x\,dx = \frac{6L_L^3\boldsymbol{F}_o}{E\,w\,h^3} \tag{10.77}$$

Since the total width at the base $w = ib$, the leaf spring deflection δ_{Lo} is then

$$\delta_{\mathrm{Lo}} = \frac{6L_L^3\boldsymbol{F}_o}{i\,b\,h^3E} \tag{10.78}$$

Using Eq. (10.78), the leaf spring stiffness k_{Ls} (or the spring rate) is

$$k_{\mathrm{Ls}} = \frac{\boldsymbol{F}_o}{\delta_{\mathrm{Lo}}} = \frac{ibh^3E}{6\,L_L^3} \tag{10.79}$$

Figure 10.16 shows the other configurations of leaf springs. It includes; the cantilever spring (a), a simply supported spring (b), and an overhang spring (c). Figure 10.16b,c indicates dimensions and applied forces to produce the exact equivalence to the cantilever spring of Figure 10.16a. This can provide the exact same behavior as for the previous performance of cantilever spring. Relations are the same if the equivalent dimensions and forces are used. The equations derived in Figure 10.16 verify the equivalence. The procedure for either case of (b) or (c) is to divide the applied force by two and the length of the simply supported spring by two. For the overhang spring, the equivalent is valid if the extended overhang distance is 0.5 L_L and the distance between the two supports should be 1.5 L_L. If these ratios are not satisfiable, the equivalence does not apply. This is because the maximum stress and the deflection should be the same as the original cantilever spring. The specified dimensions of the overhang spring in Figure 10.16c are the only configuration that satisfies these requirements.

10.3.2 Design Procedure

Leaf spring synthesis depends on the postulation that usually one has the known operating force $\boldsymbol{F}_o$, the operating deflection δ_{Lo}, and the length of the leaf spring L_L. The initial synthesis defines the leaf thickness h, the leaf width b, the number of leafs i, and the spring material. Detailed design performs iterations for refinements and further optimization. Changing leaf spring length might also be an iteration objective.

10.3.2.1 Initial Synthesis

Materials used for leaf springs are similar to those used for helical springs. The usual thicknesses are not as small as for helical springs. The effect of smaller dimensions on the strength is therefore within the safety factor considerations. Other **spring steels** (0.6–0.8% carbon) are more general-purpose widely used material for a wide range of leaf spring thicknesses. Table 10.4 provides selected *leaf spring material* standards, designation numbers, ultimate tensile strength S_{ut}, yield strength S_y, and endurance limits S_e. One can also use the density ρ as 7860 [kg/m^3], or 0.282 [lb/in^3] for the basic spring steel material to evaluate the spring mass and natural frequency.

The next step is to find the appropriate synthesized geometry for the leaf spring. By dividing Eqs. (10.75) and (10.76), the *leaf thickness h* is implicitly obtainable from

$$\frac{\sigma_L}{\delta_{\mathrm{Lo}}} = \frac{E\ h}{L_L^2} \tag{10.80}$$

Explicitly,

$$h = \frac{L_L^2\ \sigma_L}{E\ \delta_{\mathrm{Lo}}} \tag{10.81}$$

One should observe the SI and US units in the direct application of Eq. (10.81), if one wishes to use different units than the ones adapted herein. Consideration of yield strength in [MPa], the modulus of elasticity in [Mpsi] or [GPa], the leaf spring thickness in [mm], and the use of a leaf spring *safety factor* K_{LSF} provide the following relations;

$$h = \frac{L_L^2 \sigma_{all}}{E\delta_{Lo}} = \frac{L_L^2 (S_y/K_{LSF})}{E\delta_{Lo}} = \frac{L_L^2 (S_y/K_{LSF})}{E_{Mpsi}(10^6)\delta_{Lo}} \text{ [in]} \tag{10.81'}$$

$$h = \frac{L_L^2 \sigma_{all}}{E\delta_{Lo}} = \frac{L_{L[mm]}^2 (10^{-6})(S_{y[MPa]}(10^6)/K_{LSF})}{E_{[GPa]}(10^9)\delta_{Lo[mm]}(10^{-3})}(10^3) = \frac{L_{L[mm]}^2 (S_{y[MPa]}/K_{LSF})}{E_{[GPa]}\delta_{Lo[mm]}(10^3)} \text{ [mm]} \tag{10.81''}$$

In these equations, the leaf spring safety factor K_{LSF} is equal to (S_y/σ_{all}) or equivalently considering the allowable stress σ_{all} as (S_y/K_{LSF}). The value of h should withstand the maximum *allowable stress* or σ_{all}. If we use the maximum force with a deflection intercepting stopper at an extra 10–20% of the maximum force and assuming the maximum stress to reach the yield strength, this would be the optimum utilization of the material with a margin of 10–20% to failure. One can further use an extra safety factor if the value of the maximum operating force is not definite or the maximum deflection stopper is not possible to implement. This initial synthesis is therefore dependent on the assumption of static loading. Finding a synthesized design for dynamic loading requires the application of the mean and alternating stresses coupled with a *modified Goodman* fatigue limit as defined latter in the detailed design. Alternatively, one can initiate the dynamic synthesis by utilizing an extra safety factor over and above the usual 10–20%. This can eliminate the iterations in the detailed design process.

After obtaining the leaf spring thickness h and if the deflection is known, one can use Eq. (10.81) and obtain the quantity $i\,b$. On the other hand if the stiffness is required, one can use Eq. (10.82) to get the quantity $i\,b$. Selecting a practical *leaf numbers* i, one can then find the *leaf width* b. Usually the range of i is from three to five. Any other reasonable value is admissible. Using Eqs. (10.78) or (10.79), the value defining $i\,b$ is

$$i\,b = \frac{6L_L^3 \boldsymbol{F}_o}{h^3 E\,\delta_{Lo}} \tag{10.82}$$

or

$$ib = \frac{6L_L^3 k_{Ls}}{h^3 E} \tag{10.82'}$$

One should observe the SI and US units in the direct application of Eqs. (10.82) or (10.82′), if one wishes to use different units than the ones adapted herein. Consideration of the modulus of elasticity in [Mpsi] or [GPa] and the possible use of [mm] for leaf thickness provide the following relations for Eq. (10.82′):

$$ib = \frac{6L_{L[mm]}^3 (10^{-3})^3 \boldsymbol{F}_o}{h_{[mm]}^3 (10^{-3})^3 E_{[GPa]}(10^9)\,\delta_{Lo[mm]}(10^{-3})}(10^3) = \frac{6L_{L[mm]}^3 \boldsymbol{F}_o}{h_{[mm]}^3 E_{[GPa]}\delta_{Lo[mm]}(10^3)} \text{ [mm]} \tag{10.82''}$$

$$ib = \frac{6L_L^3 \boldsymbol{F}_o}{h^3 E\delta_{Lo}} = \frac{6L_L^3 \boldsymbol{F}_o}{h^3 E_{[Mpsi]}(10^6)\delta_{Lo}} \text{ [in]} \tag{10.82'''}$$

As indicated previously, select a practical leaf numbers i (usually 3–5) and find the leaf width b. If the width is low, decrease the number of leafs and vice versa.

Example 10.4 Consider the design of a leaf spring of the same form as in Figure 10.16b that can carry a maximum operating load of 2000 [N] or 449.62 [lb] and a maximum operating deflection of 150 [mm] or 5.906 [in]. The length of the spring is 1.0 [m] or 39.36 [in]. Determine the thickness of each leaf, the number of leafs, and the width of the spring. Consider the inception of yield at an extra 20% over the maximum operating load.

Table 10.5 Selected spring material standards, numbers, ultimate tensile strength S_{ut}, yield strength S_y, endurance limits S_e, and modulus of elasticity E.

Standards (numbers)	S_{ut} [psi] ([MPa])	S_y [psi] ([MPa])	S_e [psi] ([MPa])	E [Mpsi] ([GPa])
AISI (1060)	142 000 (980)	87 000 (600)	58 000 (400)	29.0 (200)
AISI (1074,1080), DIN (D75-2),	171 000 (1180)	128 000 (880)	—	29.0 (200)
DIN (46Si7)	184 000 (1270)	156 000 (1080)	—	29.0 (200)
AISI (9255), DIN (51Si7, 55Si7), ISO (55SiCr63)	191 000 (1320)	164 000 (1130)	79 000 (550)	29.4 (203)
AISI (9260, 5155, 5160, 4161, 6150), DIN (60Cr7, 55Cr3, 50CrV4, 51CrMoV4), ISO (59Si7, 55Cr3, 51CrV4, 60CrMo33)	198 000 (1370)	171 000 (1180)	94 000 (650)	29.4, 29.0, 28.4 (203, 200, 196)

Solution

Data: To use the developed procedure for the cantilever spring, the equivalent data from Figure 10.16a,b is obtainable as $\boldsymbol{F}_o = 1000$ [N] or 224.81 [lb], $\delta_o = 150$ [mm] or 5.906 [in], and the maximum load and deflection at inception of yield are $F_{max} = 1200$ [N] or 269.77 [lb]. Equivalent cantilever spring length $L_L = 500$ [mm] or 19.69 [in].

Consider a more widely used **spring steel** (0.6% carbon) that is a general-purpose spring material, which is the first line in Table 10.5. It is more common to use as the least expensive. The yield strength S_y is (600) [MPa] or 87 000 [psi] and the modulus of elasticity E is (200) [GPa] or 29.0 [Mpsi]. Using Eq. (10.81) and a leaf spring safety factor $K_{LSF} = 1.2$, the initial value of the leaf height h is

$$h = \frac{L^2_{L[mm]}\ (S_{y[MPa]}/K_{LSF})}{E_{[GPa]}\ \delta_{Lo[mm]}(10^3)} = \frac{(500)^2(600/1.20)}{200(150)(10^3)} = 4.17\ [mm] \tag{a}$$

or

$$h = \frac{L^2_L\ (S_y/K_{LSF})}{E_{[Mpsi]}(10^6)\ \delta_{Lo}} = \frac{(19.69)^2(87\,000/1.2)}{29.0(10^6)\ (5.906)} = 0.1641\ [in] \tag{a$'$}$$

One might use a usual standard thickness of 4.0 [mm] or 0.160 [in]. The SI standard thickness is smaller than the calculated one, and the US thickness is smaller than the calculated thickness. This affects the stresses and other spring geometry and parameters.

According to Eq. (10.82), the parameter ib is

$$ib = \frac{6L^3_L\boldsymbol{F}_o}{h^3E_{[GPa]}(10)^9\ \delta_{Lo}(10^{-3})}(10^3) = \frac{6(500)^3(1000)}{4^3(200)(150)(10^3)} = 390\ [mm] \tag{b}$$

or

$$ib = \frac{6L^3_L\boldsymbol{F}_o}{h^3E_{[Mpsi]}(10^6)\delta_{Lo}} = \frac{6(19.69)^3(224.81)}{0.160^3(29.0)(10^6)(5.906)} = 14.678\ [in] \tag{b$'$}$$

Taking the number of leafs as 4, the width of ach leaf for SI or US system is then

$$b = \frac{ib}{i} = \frac{390}{4} = 97.5\ [mm] \tag{c}$$

or

$$b = \frac{ib}{i} = \frac{14.678}{4} = 3.6695\ [\text{in}] \tag{c$'$}$$

With the same number of leafs, note that the width of the US spring is smaller than the width of the SI spring. This is because the thickness is little higher for US spring than the SI one. The width of leafs in either SI or US does not need to be standardized or significantly rounded. The deflection due to the selected geometry is

$$\delta_{\text{Lo}} = \frac{6L_L^3 \boldsymbol{F}_o}{ibh^3E} = \frac{6(500)^3(10^{-3})^3(1000)}{4(97.5)(10^{-3})(4)^3(10^{-3})^3(200)(10^9)}(10)^3 = 150.24\ [\text{mm}] \tag{d}$$

$$\delta_{\text{Lo}} = \frac{6L_L^3 \boldsymbol{F}_o}{ibh^3E} = \frac{6(19.69)^3(224.81)}{4(3.6695)(0.160)^3(29)(10^6)} = 5.9058\ [\text{in}] \tag{d$'$}$$

Calculating the maximum leaf stresses σ_L at the maximum operating force, Eq. (10.75) gives

$$\sigma_L = \frac{6\boldsymbol{F}_o L_L}{ibh^2} = \frac{6(1000)(500)(10^{-3})}{4(97.5)(10^{-3})(4)^2(10^{-3})^2} = 480\ (10^6)\ [\text{Pa}] \tag{e}$$

or

$$\sigma_L = \frac{6\boldsymbol{F}_o L_L}{ibh^2} = \frac{6(224.81)(19.69)}{4(3.6695)(0.160)^2} = 70\,681\ [\text{psi}] \tag{e$'$}$$

Relative to the yield strength S_y, the synthesized leaf spring safety factor K_{SF} for SI system is

$$K_{\text{SF}} = \frac{S_y}{\sigma_L} = \frac{600(10)^6}{480(10)^6} = 1.25 \tag{f}$$

And for US system

$$K_{\text{SF}} = \frac{S_y}{\sigma_L} = \frac{87\,000}{70\,681} = 1.23 \tag{f$'$}$$

These calculated safety factors are close to the initial value of 1.2, which is the adapted safety factor at the start of the synthesis process. The difference is due to the selection of the thicknesses from the preferred ones.

10.3.2.2 Detailed Design

Detailed leaf spring design depends on the specifically selected material for the explicit application that may differ from the adapted material of the design synthesis in the previous initial synthesis section. The previously adapted relations apply with some variations on parameters such as the yield strength S_y, the modulus of elasticity E, and the endurance limit S_e for different selected material. Table 10.4 gives the values of S_y, E, and S_e for different materials including the previously adapted ***spring steel***. Table 10.4 also provides the US units with the SI units shown in brackets.

The detailed design is also dependent on the dynamics of loading conditions. If the applied load does not wary with time, one should use the static maximum force in the detailed design. This is the same as the previous ***initial synthesis*** procedure with varying the material properties to those for the selected one.

The detailed design process of dynamically loaded leaf springs is about the same as helical springs. The procedure employs Eqs. (10.27) and (10.28) to define the mean and alternating forces $\boldsymbol{F}_m$ and $\boldsymbol{F}_a$ that describe the dynamic character of the load. Substituting these $\boldsymbol{F}_m$ and $\boldsymbol{F}_a$ forces in place of $\boldsymbol{F}_o$ in Eq. (10.75) obtains the *mean normal stress* σ_m and the maximum *alternating stress* σ_a as follows:

$$\sigma_m = \frac{6\boldsymbol{F}_m L_L}{ibh^2} \tag{10.83}$$

$$\sigma_a = \frac{6\boldsymbol{F}_a L_L}{ibh^2} \tag{10.84}$$

These equations evaluate the spring synthesis for the case of a fatigue expectation due to the applied dynamic loading such as springs in vehicle suspensions. The *fatigue safety factor* K_{FSF} defines the condition of withstanding the dynamic load without failure. This depends on the fatigue strength or *endurance limit* of the spring material S_e and the applied dynamic normal stresses σ_m and σ_a. The fatigue strength of the spring material S_e is in Table 10.5 for some materials. If the value is not accessible, one should estimate its value. A usual common relation between the ultimate tensile strength and the fatigue strength is reiterated as (see Section 10.2.5)

$$S_e = 0.5\ S_{ut} \tag{10.85}$$

Again, this relation is not very accurate for several material microstructures. The 0.5 factor can take another value in the range of 0.25–0.63 for carbon steel depending on ferrite, pearlite, or martensite microstructure. The range for alloy steel is 0.23–0.47, Sors (1971). One should identify the endurance limit of the material used if the fatigue safety factor K_{FSF} would need to be more accurate. The conservative dynamic failure criterion according to *modified Goodman* and the associated fatigue safety factor K_{FSF} is then (see Section 7.8)

$$\frac{\sigma_a}{S_e} + \frac{\sigma_m}{S_{ut}} = \frac{1}{K_{FSF}} \quad \text{or} \quad K_{FSF} = \frac{S_e\ S_{ut}}{\sigma_a\ S_{ut} + \sigma_m\ S_e} \tag{10.86}$$

Example 10.5 Consider the detailed design of the equivalent leaf spring of the same form as in Example 10.4 that can carry a maximum operating load of 1000 [N] or 224.81 [lb] and a maximum operating deflection of 150 [mm] or 5.906 [in]. The length of the spring is 0.50 [m] or 19.68 [in]. Check the thickness of each leaf, the number of leafs, and the width of the spring if the material needs changing due to dynamic loading. Find the suitable fatigue safety factor K_{FSF} if the load varies between 20% and 100% of the maximum applied force.

Solution

Data: The equivalent spring has $\boldsymbol{F}_{max} = \boldsymbol{F}_o = 1000$ [N] or 224.81 [lb], $\delta_{max} = \delta_o = 150$ [mm] or 5.906 [in], $\boldsymbol{F}_{min} = 200$ [N] or 44.962 [lb], and $\delta_{min} = 30$ [mm] or 1.181 [in]. Spring length $L_L = 500$ [mm] or 19.69 [in].

Again, consider the initially adapted **spring steel** (0.6% carbon) that is a general-purpose spring material, which is the first line in Table 10.5. It is more common to use as the least expensive. The ultimate tensile strength S_{ut} is (980) [MPa] or 142 000 [psi], the endurance limit is 400 [MPa] or 58 000 [psi] and the modulus of elasticity E is (200) [GPa] or 29.0 [Mpsi]. Use the initially synthesized leaf spring defined by Example 10.4 such that the standard thicknesses $h = 4.0$ [mm] or 0.160 [in], the number of leafs as 4, and the width of each leaf $b = 97.5$ [mm] or 3.6695 [in].

Use Eqs. (10.27) and (10.28) to get the mean and alternating forces as

$$\boldsymbol{F}_m = \frac{\boldsymbol{F}_{max} + \boldsymbol{F}_{min}}{2} = \frac{1000 + 200}{2} = 600\ [\text{N}]$$

$$\boldsymbol{F}_m = \frac{\boldsymbol{F}_{max} + \boldsymbol{F}_{min}}{2} = \frac{224.81 + 44.962}{2} = 134.886\ [\text{lb}] \tag{a}$$

or

$$\boldsymbol{F}_a = \frac{\boldsymbol{F}_{max} - \boldsymbol{F}_{min}}{2} = \frac{1000 - 200}{2} = 400\ [\text{N}]$$

$$\boldsymbol{F}_a = \frac{\boldsymbol{F}_{max} - \boldsymbol{F}_{min}}{2} = \frac{224.81 - 44.962}{2} = 89.924\ [\text{lb}] \tag{b}$$

Obtain the mean of the normal stress σ_m and the alternating stress σ_a by using Eqs. (10.83) and (10.84) as follows:

$$\sigma_m = \frac{6\boldsymbol{F}_m L_L}{ibh^2} = \frac{6(600)(500)(10^{-3})}{4(97.5)(10^{-3})(4.0)^2(10^{-3})^2} = 288.46(10^6)\ [\text{Pa}]$$

$$\sigma_m = \frac{6\boldsymbol{F}_m L_L}{ibh^2} = \frac{6(134.89)(19.69)}{4(3.6695)(0.160)^2} = 42\,410\ [\text{psi}] \tag{c}$$

$$\sigma_a = \frac{6F_a L_L}{ibh^2} = \frac{6(400)(500)(10^{-3})}{4(97.5)(10^{-3})(4.0)^2(10^{-3})^2} = 192.31(10^6)\ [\text{Pa}]$$

$$\sigma_m = \frac{6F_m L_L}{ibh^2} = \frac{6(134.89)(19.69)}{4(3.6695)(0.160)^2} = 28\,272\ [\text{psi}] \tag{d}$$

Substituting these values into Eq. (10.85) gives

$$K_{\text{FSF}}|_{\text{SI}} = \frac{S_e\ S_{\text{ut}}}{\sigma_a\ S_{\text{ut}} + \sigma_m\ S_e} = \frac{400(10^6)\ (980)(10^6)}{192.31(10^6)(980)(10^6) + 188.46(10^6)(400)(10^6)} = 1.49$$

$$K_{\text{FSF}}|_{\text{US}} = \frac{S_e\ S_{\text{ut}}}{\sigma_a\ S_{\text{ut}} + \sigma_m\ S_e} = \frac{58\,000\ (142\,000)}{28\,272(142\,000) + 42\,410(58\,000)} = 1.27 \tag{e}$$

Both values are higher than the safety factors for static loading (1.25 and 1.23, respectively) as found in Example 10.4 and thus no need to change material to accommodate this dynamic loading condition.

Many leaf springs, particularly in vehicle suspensions, have initial camber in their unloaded manufactured condition. This camber is to allow more space to be available for the maximum deflection not to interfere with other connected elements of the design. The maximum loading condition usually causes the spring to be flat as our adapted mathematical model suggests. Existence of different camber should not affect the synthesized design in any appreciable way. Construction of the spring end connections should be accommodating the extensions or contractions of the spring. Several different constructional details are available through the Internet images for Leaf springs.

10.4 Belleville Springs

Belleville springs have conical disc shape with a relatively small thickness. The French inventor Julien Belleville patented this type of springs in 1867. Conical spring washer is another name for these springs. The form and the cross section are shown in Figure 10.17, with basic dimensions indicated. This type of *disc springs* provides relatively higher load carrying capacity for relatively smaller deflections. If *Belleville* springs are stacked in series or in parallel, Figure 10.18, they can produce a spring combination of high compactness and provide either a larger deflection for the same force as in Figure 10.18a or a much larger force for a small deflection as in Figure 10.18b. These options are very useful in applications such as suspension of heavy moving equipment, metal forming dies, and gun recoil systems. Other series-parallel combinations of these springs can provide the needed characteristics for other applications that require compact construction.

At some dimensional proportions, disc springs also demonstrate nonlinear behavior as observed in Figure 10.19. The figure presents representative characteristics of *Belleville* springs in nondimensional form. Nondimensional load is the load divided by the thickness squared and the modulus of elasticity of the material. The negative spring rate occurs with higher h_D/t_D where snap action happens, which can be useful in some applications. At some h_D/t_D, the spring demonstrates a range of almost zero stiffness at some deflection range as the case for $h_D/t_D \approx 1.41$ in Figure 10.19. This characteristic can be very useful in applications such as clutches, seals, and nonlinear suspension systems, Metwalli (1986). In these applications, nearly constant load is possibly available at some varying range of displacement with almost zero sprig rates. Most applications, however, utilize the relatively linear part of *Belleville* spring characteristics at lower values of h_D/t_D.

The synthesis of *Belleville* springs is beyond the scope of this text. The mathematical model, treatment, and optimization are available in several references, which are accessible in Shawki et al. (1977, 1981) and Metwalli et al. (1985). Some software programs are available for the selection of these springs such as the one provided through the Internet in http://www.mubea.com/(disc springs).

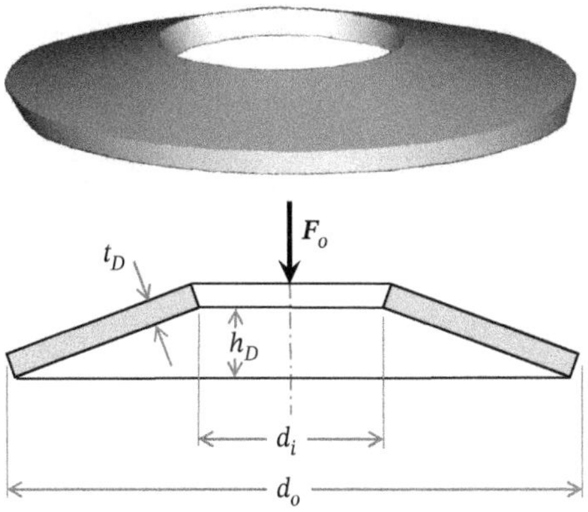

Figure 10.17 *Belleville* spring and its sectional view showing main geometry in dimensional form.

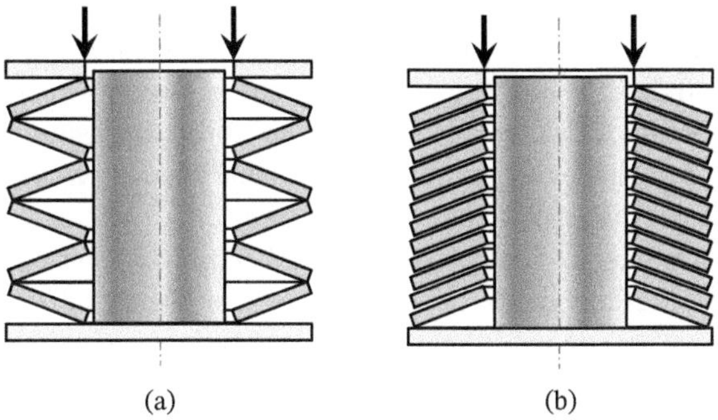

Figure 10.18 *Belleville* springs stacked in combined form: (a) for the same force, the combination provides more deflection (6 of a single disk), and (b) for the same deflection of a single disk, the combination provides more force (12 of a single disk).

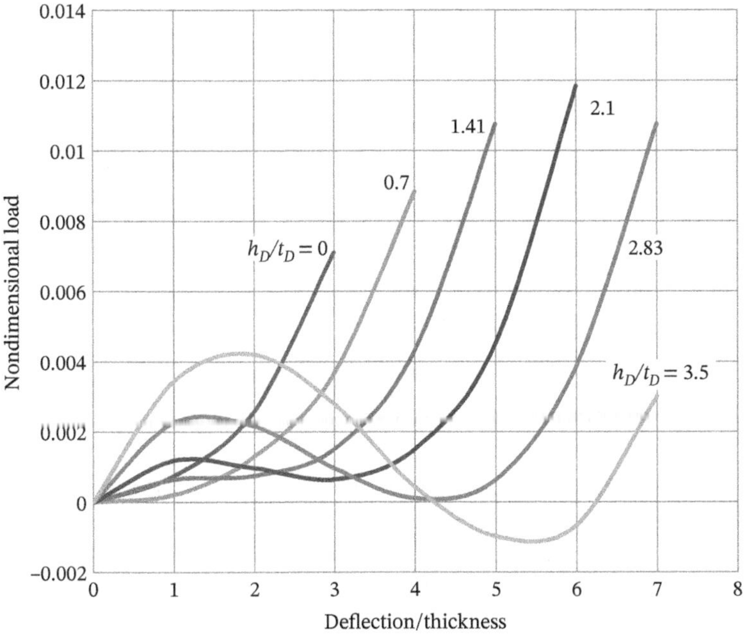

Figure 10.19 Representative characteristics of *Belleville* springs in nondimensional form. Nondimensional load is the load divided by the thickness squared and the modulus of elasticity of the material.

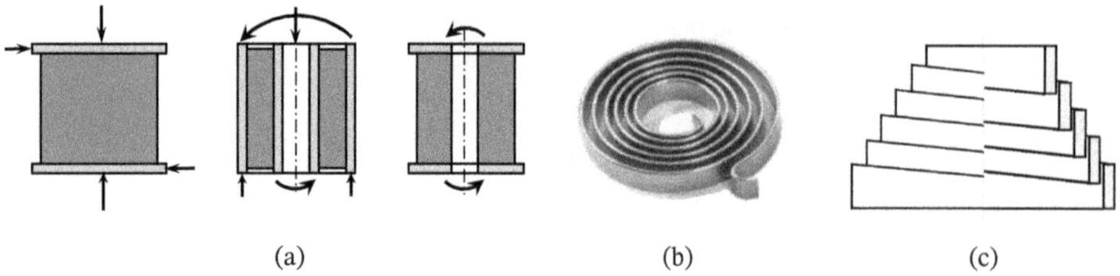

Figure 10.20 Other springs: (a) three different elastomeric spring blocks, (b) constant load spiral clock spring, and (c) volute spring.

10.5 Elastomeric and Other Springs

These elastomerics are rubber or similar polymer material shaped in blocks, bands, or strings. They act as springs when loaded. Rubber bands are one type of these springs. An elastomeric spring is usually acting as a compliant solid body or block under the applied load. Elastomeric springs store high energy per unit weight than several other materials. They are usually having low weights and high reliability that attract wide applications in aerospace industry. Figure 10.20a shows some forms of elastomeric springs. These are simple forms as cubes, cylinders, hollow cylinders, or alike bound by metal plates or cylinders. The metal bounds in Figure 10.20a are lighter in shade than the darker elastomeric material. The blocks are treatable as bars or short beams under different loading conditions of compression, shear, torsion, or bending. Stresses are calculated as usual machine elements, but the deflections are much larger due to their usually lower modulus of elasticity. Relations that account for large deformations under different loading conditions are more appropriately adapted to attain more representative mathematical model of these springs. Regular relations of deformations can, however apply for initial rough synthesis.

Constant force springs are useful in applications that need maintaining of uniform loads over a relatively large working range. One of these is the spiral clock spring of Figure 10.20b. Another spring type is the volute spring, which is in a half-sectional view in Figure 10.20c. It is made of thin rectangular section of a long strip wound to form the spring. The solid length of this spring is the width of the rectangular section. As the spring is loaded, more coils touch the base and fewer coils remain active. This provides a variable spring rate with the increase of deflection, which produces a nonlinear spring that has a stiffening behavior. Similar springs are available with a circular cross section rather than a rectangular one.

Several other spring types are available such as wave springs (http://www.smalley.com/(wave springs)), serpentine springs, and air springs. Their form and applications are available through the Internet images. Other springs of different shapes and made of composite materials such as laminated fibrous composite E-springs can find applications in automotive suspensions (Elmoselhy 2005; Elmoselhy et al. 2004, Elmoselhy et al. 2005a, 2005b, 2005c, 2006. Compliant mechanisms are also useful devices that act as springs (Shalaby et al., 2003).

10.6 Computer-Aided Design and Optimization

In previous sections and examples, several means are available to provide synthesis tools for helical compression or extension springs as the synthesis charts in Figures 10.9 and 10.10 or Figures 10.11 and 10.12 of the developed ***Spring Synthesis Tablets***. These *Excel* sheet Tablets are computer-aided synthesis (CAS) rather than computer-aided design (CAD) tools. Usually available CAD tools may generate designs by repeated analysis. The tablets are accessible through ***Wiley website*** for CAS or real *CAD* of springs. They provide direct synthesis without any need for repeated analysis. The iterative synthesis, however, is available for fine-tuning of required standards or satisfying other constraints. Developing such an aid to torsion spring or leaf springs synthesis is an exercise similar to these ***Spring Synthesis Tablets***.

In the ***Spring Synthesis Tablets***, the **Default** column is an example given as a reference and check. This column is not to be changed. No red or small dark gray font values without aqua color or light grayish background are changeable. These are due to predetermined and embedded mathematical relations such that they should not be accessible for input of any explicit variable. Any input to these cells will wipe out the solution procedure. Only cells with aqua (or light grayish) background are allowable for Inputs. The synthesis **Input** column is to insert your case values only in place of the dark gray colored values (or larger font) and aqua or light grayish background (as in Figures 10.11 and 10.12). All inputs are only allowable in cells with the blue colored (or large dark gray) font and aqua (or light grayish) background. The red colored (or smaller dark gray) font values and no colored background are the synthesized design parameters calculated as synthesized suggestions. The **Analysis** column is for design iterations to insert your case values only in place of the blue colored values (or larger dark gray font) and aqua (or light grayish) background (as in Figures 10.11 and 10.12). The iteration is necessary only to account for existing standard wire diameters or other constraint on space geometry such as smaller outer diameter and spring length restrictions. The red colored values (or dark gray smaller font) with no colored background are also other design parameters calculated as behavioral outputs that are only changeable by entering the blue colored values (or dark gray larger font) and aqua (or light grayish) background. The **Analysis** column is suitable for quick and effective iterations in combination with the **Input** synthesis column. Change of material to accommodate specific strength values or necessary safety factor should need some other iteration. The blue colored values (or dark gray larger font) and aqua (or light grayish) background in either **Input** or **Analysis** column are interactively exploitable to accommodate any other design parameter values.

For the helical compression springs, the ***Tablets*** allow the usual inputs of spring design in the **Input** column. These inputs are; the maximum operating force $\boldsymbol{F}_o$, the desired solid force $\boldsymbol{F}_s$ for crucial optimum safety, the minimum operating force $\boldsymbol{F}_{\min}$ for dynamic consideration, an extra safety factor K_{SF} for extra contemplation, the proposed spring index C, the maximum operating deflection δ_o, and the number N_e of the inactive coils of spring ends. These provide sufficient information to synthesize the spring geometry, namely, the wire diameter d_w, the coil diameter d_c, the number of active coils N_a, and the suggested material properties. These values appear immediately in the **Output** section of the ***Tablet***. Together with the input, the synthesized parameters provide necessary and sufficient information to calculate all other design parameters and performance or behavioral variables. If any of these are not satisfactory, iterations are possible through the utilization of the **Analysis** column in conjunction with the **Input** column to home into any design parameter or performance constraint.

With some extensions to the ***Spring Synthesis Tablets,*** iterations for different optimization are possible. Additional calculations are necessary for intended definition of objectives such as minimum volume or weight, maximum energy or resilience, and specific surge frequency. Iterations for optimization can follow either manually or by activating the optimization options in the Excel tools. This process is a simple exercise for anyone looking for further interest or solution to specific optimization problem. Other existing software is also available for such a task through the Internet.

Employing the ***Spring Synthesis Tablets*** in optimization is achievable as carried out in the following example.

Example 10.6 Use the ***Spring Synthesis Tablet*** to find the *optimum spring index* C for the spring specified in Example 10.1. The optimum is targeting a minimum mass of the spring.

Solution

Data: Instead of spring ends of ground squared, one should only use squared spring ends because one needs further grinding on top of that for ground squared ends. This adds two end coils to the intended spring. The results of running the ***Spring Synthesis Tablet*** at six different C values of 6, 8, 10, 12, 14, and 16 are copied to Table 10.6. The results are graphically available in Figure 10.21 through Excel. Iteration results are of optimum spring index C for the mass m_a of active coils and the total spring mass m_s. Total spring mass includes the squared ends as mentioned before. With higher spring index, the spring helix angle is higher. The length of the helix is then larger and its effect on the active coils mass m_a is therefore

$$m_a = (\rho\,\pi^2 d_w^2 d_c N_a)/(\cos\theta_h) \qquad \text{(a)}$$

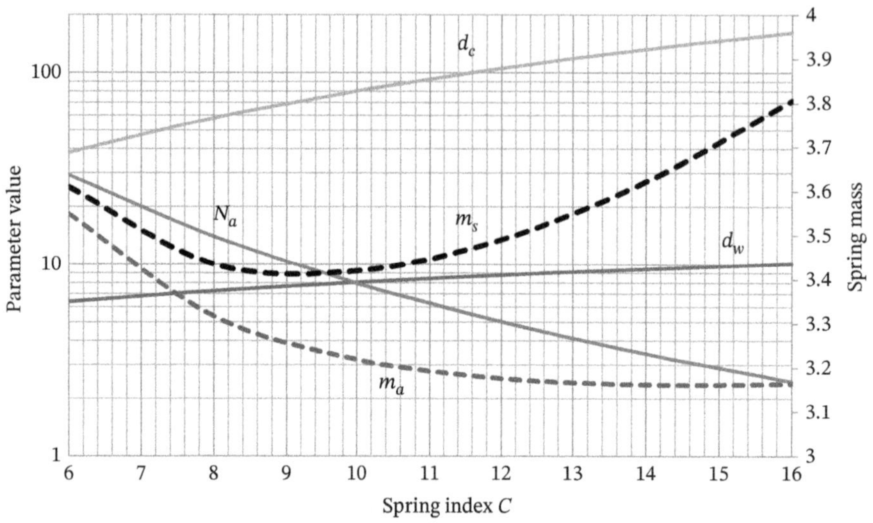

Figure 10.21 Iteration results of optimum spring index C for active coils mass m_a and total spring mass m_s by employing the *Spring Synthesis Tablet*. Total spring mass includes squared ends. Spring masses [kg] are dashed with values off the right secondary coordinate.

Adding the mass of the inactive end coils to the active coils mass, one gets the total sprig mass m_s as

$$m_s = (\rho\pi^2 d_w^2 d_c N_a)/(\cos\theta_h) + (\rho N_e \pi^2 d_w^2 d_c) \tag{b}$$

where the density ρ for steel is 7860 [kg/m^3] as previously noted. Using continuous values of wire diameter makes it easier to calculate the optimum, but preferred values are also possible with even lower number of iterations. The optimization is for the SI system and the same results are similarly doable for the US system. This is a simple verification exercise using the US ***Spring Synthesis Tablet*** instead of the SI ***Spring Synthesis Tablet***.

In Figure 10.21, the spring masses are the dashed lines with their values off the right secondary coordinate. The *optimum spring index C* for the total spring mass is about 9.216–9.25, which is iteratively reachable by using the ***Spring Synthesis Tablet***. Changing the C value in the **Input** synthesis column and defining the design parameters in the **Analysis** column gives a chance to monitor the spring mass variations. Iterating C from the value of 10 with steps of 0.5, 0.1, and 0.05 until the final value of 9.25 indicated the smaller total spring mass. However, using single quadratic interpolation (see Section 5.2) over the range 9–11 produced an optimum C of 9.216. Table 10.6 and its Figure 10.21 indicate that the minimum is between 8 and 10 but closer to 10. If the spring has no idle or inactive end turns, the optimum C for the active turns is higher than 14 as deduced from Table 10.6 with the mass at 14 given in bold characters.

Note that the optimum C for the objective of minimizing the *total spring mass* is depending on the loading force and the deflection specifications of the intended spring. Other loading cases should apparently define another optimum C. The ***Spring Synthesis Tablet*** is accessible to perform such iterations for the optimization of other loading cases. As derived by Wahl (1963), the expected optimum C was between 4 and 5 or even lower for the maximum space efficiency. The objective and the loading case were different from the case of our example, and one should then expect the results to be different.

Elmeligy (1992) and Metwalli et al. (1993, 1994) covered the general case of computer-aided design and optimization of helical compression and torsion springs. Nondimensional parameters and several objective functions defined optimum springs for a wide range of design requirements. Several optimum figures are available to get the optimum spring parameters and behavioral variables directly from the figures. One can consult these results for a wider treatment of optimum helical springs.

Table 10.6 Iteration results of optimum spring index C for active coils and total spring by employing the *Spring Synthesis Tablet*.

C	6	8	10	12	14	16
d_w [mm]	6.379	7.242	8.036	8.772	9.463	10.115
d_c [mm]	38.272	57.938	80.355	105.265	132.475	161.834
N_a	29.273	14.021	7.965	5.032	3.418	2.448
Spring mass m_s [kg]	3.6096	3.4351	**3.4200**	3.4906	3.6223	3.8058
Active mass m_a [kg]	3.5492	3.3172	3.2187	3.1764	**3.1622**	3.1636

Total spring mass includes squared ends. The minimum in the Table is in bold font.

Optimum designs of disk springs are available in the work of Shawki et al. (1977 and 1981) and Metwalli et al. (1985). One main result is that the optimum ratio of the outer diameter d_o to the inner diameter d_i ought to be in the range of 0.59–0.62 instead of the usually widely used value of 0.5. The other result is indicating that the optimum ratio h_D/t_D ought to be in the range of 0.4–1.4. These results are useful in providing guide to the appropriate synthesis of *Belleville* springs. Some software programs are available for the selection of these springs such as the one provided through the Internet in www.diamondwire.com.

10.7 Summary

In this chapter, we developed the mathematical tools and generated procedures for the synthesis of helical and leaf springs. Helical springs occupied the main thrust of the treatment particularly compression springs, which is the main type in that group. Extension and torsion springs are included but in a more confined way since they are similar to compression springs. Real design procedure utilizing CAS was the focus of attention since direct synthesis would be better than design by repeated analysis. This allowed the generation of tools to provide the definition of geometry and the selection of the suitable material. The ***Spring Synthesis Tablets*** are the simple tools generated to perform the task of the CAS. These tables provide the tools to specify the usual inputs for US or SI systems and they generate the appropriate geometry and material selection. The tablets allow the efficient iterations to optimize or tailor the design to some specific constraints.

The ***Tablets*** for helical compression springs allow the usual inputs of spring design. These inputs include the maximum operating force, the preferred solid force, the minimum operating force, an extra safety factor, the proposed spring index, the maximum operating deflection, and the number of inactive coils of the spring ends. These are sufficient to produce the synthesized spring geometry, namely; the wire diameter, the coil diameter, and the number of active coils, in addition to the suggested material properties. The ***Tablets*** also provide all other dependent design parameters and performance or behavioral variables. If any of these are not satisfactory, iterations are possible. Moreover, the ***Tablets*** facilitate the optimization of any parameter such as the spring index for the minimum spring mass. Several examples to that effect are available in this chapter to demonstrate the utility of the developed procedures and tools. Iterations on the cost of utilizing other materials are possible and can be useful if material and manufacturing cost data are available.

Similar procedures are available in this chapter to generate similar tools as Tablets or computer codes to handle extension, torsion and leaf springs. The treatments provided herein for these springs allow the development of these similar synthesis tools.

Problems

10.1 Use Eqs. (10.2) and (10.3) to derive the relations defined in Table 10.1 for the free length L_f and the solid length L_s for different spring ends.

10.2 Prove Eq. (10.7) and define the free spring helix pitch from the operating spring pitch and the solid.

10.3 Show the reason for considering the end turns of the ground plain ends to be −1.

10.4 What is the percentage inaccuracy for considering $\cos\theta_h$ equals to 1.0 for helix angles of 5°, 10°, and 15°. What should be the helix angle at the solid length of the spring?

10.5 Derive the equation for the outer diameter of the spring d_o at solid length as a function of the initial coil diameter d_c, the wire diameter d_w, and the initial helix pitch p_h (Eq. (10.36)). What is the inner diameter value at solid length and the variation in its value from the initial outside diameter?

10.6 Compare the values used in the procedure by assuming the operating force $\boldsymbol{F}_o$ is less than the solid force by 10–20% to the values used by the spring manufacturers. What is the difference between the value use for the yield strength in shear S_{sy} from the ultimate tensile strength S_{ut}, and those used by the spring manufacturers?

10.7 What is the effect of changing the material on the wire diameter estimation? Use the range of A and b of the materials to define the effect of the material on the wire diameter value.

10.8 Compare the values of the material intercept values of A and the exponent b from different sources and calculates the differences in the evaluation of the wire diameters due to these variations.

10.9 What is the relation between the values of the material intercept parameter A for both US and SI system of units and justify that the value of b is the same?

10.10 In Example 10.1, change the material to a music wire and compare all relevant results. Check the cost of both springs using a current estimate of material prices. Consider the manufacturing cost to be the same for both materials and mass production consideration.

10.11 Perform the same tasks in Problem 10.10 on the particulars of Example 10.2.

10.12–10.23 Design helical compression springs by initial synthesis procedure and verify results using the ***Spring Synthesis Tablet – US and SI*** for the required specifications in Table 10.P1. The SI specification may not relate to the US specification. Select different end conditions to observe the difference. Find the solid stress, other properties, and the safety factor for each spring. Compare results with the values possibly extracted from Figures 10.9 and 10.10. Find the fatigue safety factor for each case if the minimum force is 10% of the maximum operating one.

10.24 Design a front suspension spring for a quarter suspension of a vehicle of total curb weight of 10 000 [N] or 2205 [lb] that should stand a maximum of twice the curb weight and a maximum deflection of 200 [mm] or 7.87 [in]. Assuming the dynamics of loading is between 80% and 125% of curb weight.

Table 10.P1 The required specifications in SI and US specifications for different springs to be synthesized.

Problem number	Max. operating force F_o [lb] ([N])	Max. operating deflection δ_o [in] ([mm])	Spring index C
10.12	1 (1)	1 (1)	4
10.13	1 (1)	1 (1)	8
10.14	1 (1)	1 (1)	12
10.15	1 (1)	1 (1)	16
10.16	100 (100)	10 (100)	4
10.17	100 (100)	10 (100)	8
10.18	100 (100)	10 (100)	12
10.19	100 (100)	10 (100)	16
10.20	10 000 (10 000)	20 (500)	4
10.21	10 000 (10 000)	20 (500)	8
10.22	10 000 (10 000)	20 (500)	12
10.23	10 000 (10 000)	20 (500)	16

10.25 Design front and rear suspension springs of an adult bicycle. Assume an average worldwide weight of an adult and average dimensions and specifications of the bicycle. Assume the dynamics of loading is between 50% and 150% of the weight.

10.26 The seat of the bicycle is supported by two rear springs with a front pivot. Design the two springs for the average loading of the previous problem and the average dimensions of available seats. Select a reasonable deflection anticipated under maximum loading conditions. If the front pivot of the seat is turned into a torsional spring, find the appropriate design for it.

10.27 Design a tension spring needed to support a jumper diving a distance of 1000 [in] or 25.4 [m] down before the spring goes into effect. Assume that the initial tension is zero, and find the deflection when the jumper reaches a zero velocity. The spring need not be the same length as the distant of the dive.

10.28 The nub handle of a door locker operates under a torsional spring. Design a suitable torsional spring for the locker shaft that is not hard to turn and not too loose for a young child to turn. Define average loads and dimensions from ergonomics and human strength references or personal environment.

10.29 Develop an aid to torsion spring synthesis similar to the ***Spring Synthesis Tablets*** of compression springs for US or SI system of units. Use the solution of Problem 10.28 to verify your developed tablet. Validate the tablet results against other problems in other textbooks.

10.30 Develop an aid to leaf spring synthesis similar to the ***Spring Synthesis Tablets*** of compression springs for US or SI system of units. Use the solution of Example 10.5 to verify your developed tablet. Validate the tablet results against other problems in other textbooks.

10.31 Design a leaf spring board for a swimming pool to aid adult jumpers performing Olympic jumps. Find the necessary specifications for that from relevant references.

10.32 Use the ***Spring Synthesis Tablet (US)*** to find the optimum spring index for the same Examples 10.1 and 10.6. Compare results with the ***Spring Synthesis Tablet (SI)*** using tablet iteration and quadratic interpolation optimization.

10.33 Find the optimum spring index *C* for the synthesized springs of some of the Problems 10.12–10.22.

10.34 Find the optimum torsion spring for Problem 10.28 using the tablet developed in Problem 10.29.

10.35 Find the optimum leaf spring for Example 10.4 using the tablet developed in Problem 10.30.

10.36 Compare the optimum result of Example 10.6 with the optimum results of Metwalli et al. (1993), which suggests optimum solutions for different objectives.

10.37 Compare the optimum result of Problem 10.33 with the optimum results of Metwalli et al. (1994), which suggests optimum solutions for different objectives.

References

Associated Spring (1987). *Design Handbook: Engineering Guide to Spring Design*. Associated Spring, Barnes Group Inc.

Elmeligy, A.M.A. (1992). *Computer aided design and optimization of helical springs*. MS Thesis, Cairo University.

Elmoselhy, S.A.M. (2005). *Design, shape optimization and experimental analysis of micro-composite E-springs for vehicle suspension systems*. MS Thesis, Cairo University.

Elmoselhy, S.A., Azzam, B.S., and Metwalli, S.M. (2004). Theoretical and numerical analyses of laminated fibrous composite E-springs for vehicle suspension systems. *SME Proceedings of The 14th International Conference on Composite Materials (ICCM-14)*, San Diego, California, USA, 14–18 July, paper # TP04PUB77.

Elmoselhy, S.A., Azzam, B.S., and Metwalli, S.M. (2005a). Conceptual design, analysis and optimization of laminated hybrid composite E-springs for vehicle suspension systems. *Proceedings of the 3rd International CIRP Conference on Reconfigurable Manufacturing (RM) of University of Michigan* Ann Arbor, 10–12 May, University of Michigan, Ann Arbor, MI, USA, A03-RMS05, pp. 25–32.

Elmoselhy, S.A., Azzam, B.S., and Metwalli, S.M. (2005b). Experimental analysis of laminated fibrous micro-composite E-springs for vehicle suspension systems. *Proceedings of ASME IMECE2005*, 5–11 November, Orlando, Florida, USA, Paper # IMECE2005-80780.

Elmoselhy, S.A., Azzam, B.S., Metwalli, S.M., Dadoura, H.H. (2005c). Transmissibility and experimental analyses of laminated fibrous micro-composite E-springs for vehicle suspension systems. *Proceedings of the 2005 SAE Commercial Vehicle Engineering Congress*, Illinois, USA, SAE Paper # 2005-01-3607, pp. 10–18.

Elmoselhy, S.A., Azzam, B.S., Metwalli, S.M., Dadoura, H.H. (2006). Hybrid shape optimization and failure analysis of laminated fibrous composite E-springs for vehicle suspension. *SAE Transactions, Journal of Commercial Vehicles* 115 (2): 226–236.

European Standard, EN 10270-1 (2001). *Steel wire for mechanical springs Part 1: Patented cold drawn unalloyed spring steel wire*. European Committee for Standardization.

European Standard, EN 10270-3 (2001). *Steel wire for mechanical springs Part 3: Stainless spring steel wire*. European Committee for Standardization.

Joerres, R.E. (1981). Springs," Chapter 24 in: *Standard Handbook of Machine Design* (eds. J.E. Shigley and C.R. Mischke). New York, NY: McGraw-Hill.

Metwalli, S.M. (1986). Optimum nonlinear suspension systems. *Journal of Mechanical Design* 108 (2): 197–203.

Metwalli, S.M., Radwan, M.A., and Elmeligy, A.A. (1993). Optimization of helical compression and tension springs. *Proceeding of the 19th ASME Design Automation Conference*, Albuquerque, N.M., 19–22 September.

Metwalli, S.M., Radwan, M. A., and Elmeligy, A.A.M. (1994). CAD and optimization of helical torsion springs. *Proceedings of ASME International Computers in Engineering Conference*, Minneapolis, Minnesota, USA, 11–14 September, p. 767–773.

Metwalli, S.M., Shawki, G.S.A., and Elzoghby, A. (1985). Optimal design of disk springs. *Journal of Mechanical Design* 107 (4): 477–481.

Shalaby, M.M., Hegazi, H.A., Nassef, A.O., Metwalli, S.M. (2003). Topology optimization of a compliant gripper using hybrid simulated annealing and direct search. *Proceeding of ASME IDETC/CIEC, (DETC'03)*, 2–6 September, Chicago, Illinois, Paper # DETC2003/DAC-48770.

Shawki, G.S.A., Metwalli, S.M. and Elzoghby, A.A. (1977). Optimum design of uniform-section disk spring under various constraints. *Proceedings of ASME Design Engineering Conference*, Winter Annual Meeting, Atlanta, GA, ASME Paper No. 77-WA/DE-22, December.

Shawki, G.S.A., Metwalli, S.M., and El-Zoghby, A.A. (1981). Optimization of radially-tapered disk spring. In: *Current Advances in Mechanical Design and Production* (eds. G.S.A. Shawki and S.M. Metwalli), 21–32. Pergamon Press.

Smith, J.O. (1942). The effect of range of stress on the fatigue strength of metals. *University of Illinois Bulletin* 334: 1–52.

Sors, L. (1971). *Fatigue Design of Machine Components*. Pergamon Press.

Wahl, A.M. (1963). *Mechanical Springs*. McGraw Hill.

Zemmerli, F.P. (1957). *Human Failures in Spring Applications. The Mainspring*, no.17. Associated Spring Corporation.

Internet: Information and Some Manufacturer

www.diamondwire.com
www.leespring.com
http://www.mubea.com/(disc springs)
www.peninsulaspring.com
http://www.springwire.com/(range of more diameters)
www.springworksutah.com
www.centuryspring.com
www.smalley.com/(wave springs)

Internet: Images

Springs:
https://www.google.com/search?q=constant+force+springs&tbm=isch&tbo=u&source=univ&sa=X&ved=0ahUKEwiW_YyX_czKAhWGyT4KHZM2C9sQsAQILw&biw=1037&bih=546&dpr=2.5#tbm=isch&q=springs

Compression springs:
https://www.google.com/search?q=compression+springs&biw=1037&bih=546&source=lnms&tbm=isch&sa=X&ved=0ahUKEwiu_4_Xq83KAhUFWT4KHX2OAZ8Q_AUIBygC

Extension springs:
https://www.google.com/search?tbm=isch&q=extension+springs+ends&ei=2qV7VtDJAYW3UbvyiOAD#tbm=isch&q=tension+springs
Extension spring ends:
https://www.google.com/search?tbm=isch&q=extension+springs+ends&ei=2qV7VtDJAYW3UbvyiOAD
Torsion spring ends:
https://www.google.com/search?q=torsion+springs&source=lnms&tbm=isch&sa=X&ved=0ahUKEwjX8cu2sfXKAhVK7iYKHQ3fBn0Q_AUICCgC&biw=1101&bih=479#tbm=isch&q=torsion+spring+ends
Leaf springs:
https://www.google.com/search?q=leaf+spring+suspension&source=lnms&tbm=isch&sa=X&ved=0ahUKEwik75rLmvXKAhVDFj4KHQZLDzMQ_AUIBygB&biw=1101&bih=479
Disk springs:
https://www.google.com/search?q=compression+springs&biw=1037&bih=546&source=lnms&tbm=isch&sa=X&ved=0ahUKEwiu_4_Xq83KAhUFWT4KHX2OAZ8Q_AUIBygC#tbm=isch&q=disk+springs

11

Rolling Bearings

Rolling has been anciently discovered and used as wheels in howling carts and chariots. Rolling bearing or *rolling elements bearing* stands for some rolling components inside the bearing such as the *ball bearing* or the *roller bearing* in Figure 11.1. In some cases, the rolling components (such as spheres, cylinders, or needles) may be used directly between the two elements such as a shaft and a hub. Rolling bearings are common machine elements that allow relative movement between the components of machines. They also provide some type of location between machine components. The conception that rolling eliminates friction is not true, and rolling does not mean zero friction. There is always a *rolling friction* under load. It is usually very small to a point that these bearings, some time ago, were given the erroneous name as *anti-friction* bearings. Rolling resistance is due to deformation at contact region under load. Coefficient of rolling resistance is a function of elastic and plastic properties of mating materials, roughness, lubricant, and others. The *friction coefficient* of rolling is mostly in the range of about 0.0011–0.006.

Symbols and Abbreviations

Symbol	Quantity, units (adapted)
[hr]	Hours
[Mrev]	Million revolutions
[rev]	Revolutions
[rpm]	Revolutions per minute
[rps]	Revolution per second
a_B	Bearing load-life exponent
ABMA	American Bearing Manufacturers Association
ANSI	American National Standards Institute
a_T	Effective load center distance
B	Bearing width *in various literature*
b_B	Bearing breadth or width
b_W	*Weibull* shape parameter
C_D	Dynamic load rating capacity

Machine Design with CAD and Optimization, First Edition. Sayed M. Metwalli.

Companion website: www.wiley.com/go/metwalli/machine

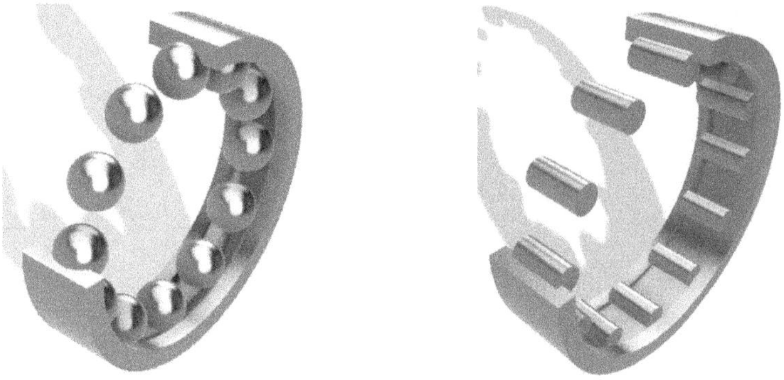

(a) (b)

Figure 11.1 Images of a simple ***3D*** rolling bearings with cages and inner rings removed and cut outer rings for clarity. Part of the shadow is shown where (a) is a deep groove ball bearing and (b) is a cylindrical roller bearing with reduced roller numbers for clarity. The ***3D*** modeling has been generated by Fusion 360 of Autodesk.

Symbol	Quantity, units (adapted)
C_{D90}	Bearing dynamic load rating at 90 [Mrev]
C_{DS}	Synthesis dynamic load rating
C_o	Static load rating capacity
D	Bearing outer diameter *in various literature*
d	Bearing inner diameter *in various literature*
d_b	Ball diameter
d_i	Bearing inner diameter
d_o	Bearing outer diameter
e_B	Load ratio limit
$\boldsymbol{F}_1, \boldsymbol{F}_2$	Side element loads
$\boldsymbol{F}_{a1}, \boldsymbol{F}_{a2}$	Thrust or axial components for $\boldsymbol{P}_{l1}$ and $\boldsymbol{P}_{l2}$
$\boldsymbol{F}_i$	Load on rolling element (i)
$\boldsymbol{F}_o$	Lower rolling element load
ISO	International Organization for Standardization
k_B	Integer half of number of balls
K_g	Bearing geometry and material factor
K_{SF0}	Static safety factor
K_{SFD}	Dynamic safety factor
K_{T1}, K_{T2}	Ratios of radial rating to the thrust rating
L_{10}	Bearing rating life (10% failures)
L_{10S}	Desired synthesis rating life of bearing
L_B, L_{Brev}	Bearing life in revolutions [rev]

Symbol	Quantity, units (adapted)
L_{B1}, L_{B2}	Respective rating lives [rev]
L_{B90}	Bearing rating life at 90 [Mrev]
L_{Bhr}	Bearing life in number of hours [hr]
L_{BM}	Bearing life in million revolutions [Mrev]
L_{BS}	Desired synthesis rating life of bearing
n_b	Number of balls in bearing
$N_{i,\mathrm{rpm}}$	Rotational speed in [rpm] during $\boldsymbol{P}_i$ loading
n_P	Number of fluctuating or unsteady loads
n_r	Number of rows in a bearing
N_{rpm}	Rotational speed in [rpm]
N_{rpm10}	Rotational speed to match L_{10}
$\boldsymbol{P}_1, \boldsymbol{P}_2, \ldots, \boldsymbol{P}_n$	Fluctuating or unsteady loads
$\boldsymbol{P}_{a0}$	Static axial load
$p_B(x)$	Bearing failure probability
$\boldsymbol{P}_{e0}$	Equivalent static load
$\boldsymbol{P}_{e1}, \boldsymbol{P}_{e2}$	Equivalent dynamic loads
$\boldsymbol{P}_l$	Bearing radial load
$\boldsymbol{P}_{l1}, \boldsymbol{P}_{l2}$	Different radial loads
$\boldsymbol{P}_{\mathrm{le}}$	Equivalent load
$\boldsymbol{P}_{\mathrm{le}}, \boldsymbol{P}_e$	Equivalent dynamic load
$\boldsymbol{P}_{\mathrm{lS}}$	Synthesis load (at L_{10})
$\boldsymbol{P}_p$	Axial preload
$\boldsymbol{P}_{r0}$	Static radial load
$\boldsymbol{P}_{r1}, \boldsymbol{P}_{r2}$	Radial loads
$R_B(x)$	*Weibull* bearing reliability
T	Tapered roller bearing width *in various literature*
$t_1, t_2, \ldots, t_n$	Load durations in revolutions [rev]
t_B	Main tapered roller bearing thickness
$t_{i,\mathrm{hr}}$	Duration time of loading $\boldsymbol{P}_{\mathrm{i}}$ in hours [hr]
VWXYZ	Generic bearing designation number
x	Bearing life measure $x = L_B/L_{10}$
λ_0	Guaranteed bearing life
X_0	Static radial load factor
Y_0	Static axial load factor
α	Angle of angular contact bearing
δ_i	Deflection of rolling element (i)
θ_b	Angle between rolling elements
θ_W	*Weibull* characteristic value

11.1 Bearing Types and Selection

Rolling bearings are mainly classified into two main groups. The *ball bearing* type is the first group as shown in Figure 11.2. The second group is the *roller bearing* type as shown in Figure 11.3. Both types can be loaded mainly in the radial directions. Some can carry combined radial and thrust loads. Others would mainly carry thrust loading.

Figure 11.2 presents some of the main types of ball bearings with simple sketches of retainers. ***Deep groove*** ball bearing is the most commonly used. Its main terminologies and dimensions are shown in Figure 11.2. The inner ring with the nominal or an *inner diameter* d_i is to tightly fit onto a shaft or an axle by a location fit; see Section 2.4.7 of the text. The outer ring with an *external outer ring diameter* d_o is to snugly fit into the housing or a hub of the

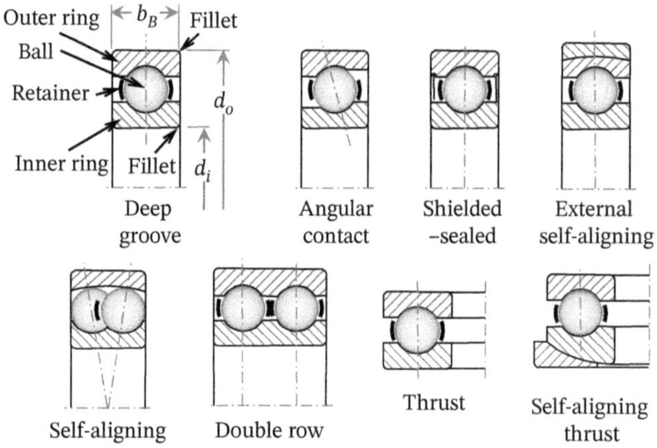

Figure 11.2 Some of main types of *ball* bearings. Deep groove ball bearing is commonly used, and the main terminologies and dimensions are shown. The balls roll in ring tracks that are grooved in the inner and outer rings. The transverse radius of the track is little larger than the radii of the rolling balls.

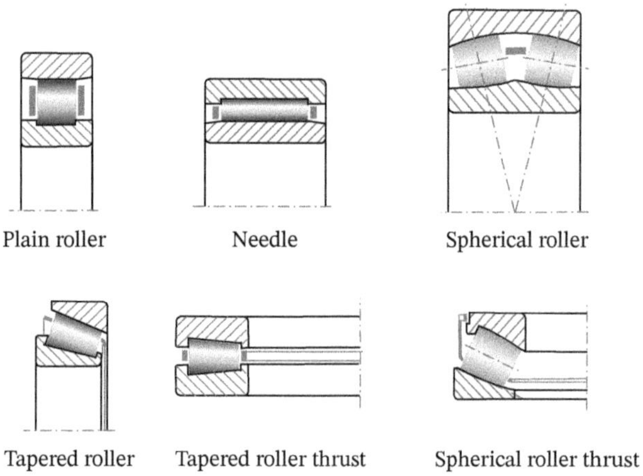

Figure 11.3 Some of the main types of *roller* bearings showing simple sketches of retainers. The roller bearings are in plain form, needle shape, tapered roller profile, or spherical roller type. The roller bearings can carry radial load only, thrust load, or can carry both loads particularly for spherical and tapered roller types.

mating element. The balls roll in ring tracks that are grooved in the inner and outer rings. The transverse radius of the track is little larger than the radii of the rolling balls. The balls are equally held apart by a retainer, which comes in different simple forms and materials to allow for ball spacing with a minimum rubbing friction. The designer can select the suitable retainer for his application. If the load is mainly radial, it is handled by selecting deep groove ball bearing. However, deep groove ball bearings can accommodate light to medium axial load.

Other types of ball bearings shown in Figure 11.2 are configured to handle different combinations of radial and thrust components. The ***angular contact*** ball bearings are most suitable for loads that are having the same angle as the contact angle of the bearing as shown by the inclined ball centerline in Figure 11.2. Most bearings can also be *shielded* or being *sealed* by contact or noncontact covers on one side as in Figure 11.2 or on both sides. Some ball bearings may be constructed as ***external self-aligning*** or double row inherently ***self-aligning*** to be able to accommodate angular displacements between the shaft and the housing. There are also ***double row*** ball bearings that may handle higher loads and may be some moments. Ball bearings are also constructed as ***thrust*** bearings to mainly withstand thrust loadings. They may also be configured as ***self-aligning thrust*** bearings as shown in Figure 11.2. Other types of ball bearings are offered by some manufacturers to combine some of the previous characteristics.

Figure 11.3 provides some of the main types of roller bearings showing simple sketches of retainers. The roller bearings are in straight or ***plain roller*** form, ***needle*** shape, ***tapered roller*** profile, or ***spherical roller*** type. The roller bearings can carry radial load and thrust load or can carry both loads particularly for spherical and tapered roller types. The plain roller bearing in Figure 11.3 would carry pure radial load. The inner ring cannot stand any axial load but should allow some axial displacement. Similarly, the shown needle bearing would behave similar to the plain roller bearing. Other configurations exist for plain and needle bearings to restrict axial displacements in single or both directions. These, however, should not carry appreciable axial loads. The spherical roller bearings can carry radial and some axial loading. Taper roller bearings can carry combined radial and axial loadings. Spherical roller thrust can carry mainly thrust and combine some radial loads.

For selecting the suitable *radial* bearing and no other types of bearings, Chapter 8 presents a simple way to define the appropriate bearing to adopt and provides its initial synthesis; see Section 8.4.3 and Figure 8.21. Other types of rolling bearings can be used if loads other than radial loads exist such as axial loads or if only axial loads exist.

Relative to radial ***deep groove*** *ball* bearings and considering the radial load as assigned a *reference capacity* of ***1.0***, the following is an estimate of *relative capacities* to carry radial and axial loads for common bearings:

- *Deep groove* ball bearing radial capacity of ***1.0*** and axial capacity of about ***0.7***.
- *Angular contact* ball radial capacity of about ***1.0–1.15*** and axial capacity of about ***1.5–2.3,*** angle dependent.
- *Self-aligning* ball radial capacity of about ***0.7*** and axial capacity of about ***0.2***.
- *Double row* ball radial capacity of about ***1.5*** and axial capacity of about ***1.4***.
- *Thrust* ball *axial* capacity of about ***1.5*** and radial capacity of about ***0.0***.
- *Plain roller* radial capacity of about ***1.55*** and axial capacity of about ***0.0***.
- *Needle* roller radial capacity of about ***1.4*** and axial capacity of about ***0.0***.
- *Spherical roller* radial capacity of about ***2.4*** and axial capacity of about ***0.7***.
- *Tapered roller* radial capacity of about ***1.9*** and axial capacity of about ***0.7***.
- *Tapered roller thrust* axial capacity of about ***1.7*** and radial capacity of about ***0.0***.
- *Spherical roller thrust* axial capacity of about ***1.8*** and radial capacity of about ***0.1***.

This information provides guidance to initially select the type of rolling bearing according to the combined loading in a design. Detailed calculations are necessary to reach the proper class of bearing. The following is additional information to reach the appropriate selection.

11.2 Standard Dimension Series

This section provides information about the standard boundary dimensions of rolling bearings. The designations of boundary dimensions are an integral part of bearing designation numbers that additionally defines the type of bearing and other added features such as shields, seals, cadge type, and many others.

11.2.1 Boundary Dimensions

Boundary dimensions are standardized; see ISO 104, 15, 355 (2015, 2017, 2019) for thrust, radial, and tapered roller bearings, respectively. For metric radial ball bearings, the standard 2-digit ***dimension series*** are combinations of ***width series*** (8, 0, 1, 2, 3, 4, 5, 6) with ***diameter series*** (8, 9, 0, 1, 2, 3, 4) in relation to the same bore or inner diameter. Figure 11.4 shows the commonly used dimension series for radial bearings of same bore diameter. It is composed of the width series followed by the outside diameter series. The 02 and 03 series are most commonly used. The numeral 0 to the left is usually omitted, and the *dimension series* is left to be defined by the *diameter series* only.

For the *radial* bearing and the *diameter series* of 0, the bearing *outer diameter* d_o is related to the bearing *inner diameter* d_i by the following projected relation:

$$d_o = d_i + 0.84d_i^{0.9} \tag{11.1}$$

For other diameter series, the 0.84 constant would be 1.12, 1.48, 1.92, and 2.56 for the diameter series of 1, 2, 3, and 4, respectively. The subsequent values of the outside diameters would be rounded to the closest preferred numbers and advisably retrieved from intended manufacturer. For the *width series* of 0, the *bearing width* b_B is related to the outer diameter d_o and the inner diameter d_i by the following projected relation:

$$b_B = 0.64(d_o - d_i)/2 \tag{11.2}$$

where b_B is the bearing width or breadth. For other width series, the 0.64 constant would be 0.88, 1.15, 1.5, and 2.0 for the width series of 1, 2, 3, and 4, respectively. The subsequent values of the bearing width or breadth would be rounded to the closest preferred numbers and advisably retrieved from intended manufacturer.

Other bearing types would have similar, but different, relations than Eqs. (11.1) and (11.2). The actual values are to be used according to the bearing supplier that conforms to the adapted standard.

Figure 11.5 is an updated continuation of Figure 8.12. It shows a sketch of a sectional view for a *gearbox* employing *deep groove ball* bearings supporting the two shafts of the gearbox. The construction indicates the special character of bearing *mounting* and shoulder dimension constraints. The outer shoulder constraint defines the

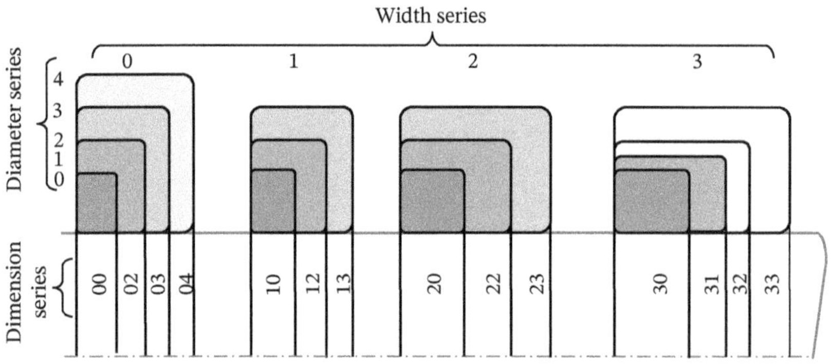

Figure 11.4 Dimension series for radial bearings of same bore diameter. It is composed of the *width series* followed by the outside *diameter series*.

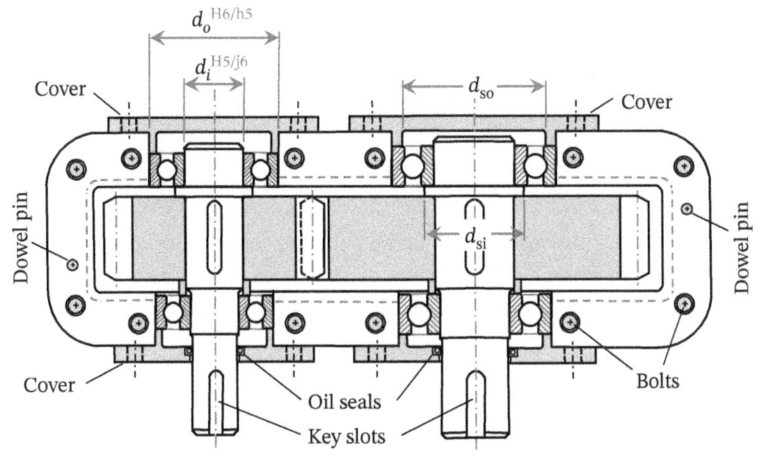

Figure 11.5 A sketch of a sectional view for a *gearbox* showing deep groove *ball bearings* supporting the *shafts* of the gearbox. Indicated is the special character of bearing mounting and *shoulder* dimension constraints. The bearing dimension fits with the shaft and housing are suggested in grayish color.

limiting outer ring shoulder diameter d_{so}, which is allowed for the mating cover. The inner shoulder constraint defines the limiting inner ring shoulder diameter d_{si}, which is allowed for the mating shaft collar. These are noted on the gearbox in Figure 11.5.

It should be noted that many references and manufacturers use "*D*" for outer bearing diameter d_o, use only "*d*" for bearing inner diameter d_i, and use "*B*" for the *bearing width* b_B. Since this text is not only dedicated to just rolling bearings, it was reasonable to usually define lower case for dimensions with qualified subscript to indicate what the dimension is for.

11.2.2 Bearing Designation Number

Designation of *bearing number* may take a form such as VWXYZ. The YZ digits denote the shaft diameter d_S, which is the bearing nominal or *inner diameter* d_i (i.e. bearing bore diameter). The digits YZ $= d_i/5$ [mm] for shaft diameters larger than 17 [mm]. For smaller diameters, the YZ = 00 for the inner or bore diameter $d_i = 10$ [mm]. YZ = 01 for inner diameter $d_i = 12$ [mm], YZ = 02 for $d_i = 15$ [mm], and YZ = 03 for $d_i = 17$ [mm]. For smaller than $d_i = 10$ [mm], the numeral is equal to the value of the diameter in [mm]. These can be part of *miniature bearing* types.

The X digit (in VWXYZ) usually represents the *width series* only, since any *zero* in the dimension series is also omitted. The VW digits indicate the *bearing type* and whether it may include any special features. The special features can have more digit's *prefixes* or *suffixes* to the bearing designation such as the cage type, materials, shields, seals, tolerances, internal fit, preload, lubrication, heat treatment, design modification, retaining ring groove, etc. These are defined by manufacturers for different types of bearings they provide. For ball bearings, few manufacturers use 6 for *deep groove* ball bearings, 7 for the *angular contact*, 1 for *self-aligning*, 4 for *double row deep groove*, and 5 for *thrust ball* bearings; see Figure 11.2. For roller bearings, few manufacturers use 2 for *spherical roller*, 3 or T for *tapered roller*, 8 for *cylindrical roller thrust*, and N for *plain or needle roller*; see Figure 11.3. Other designations are provided by manufacturers. The designer is advised to consult the intended manufacturer by using the *bearing type* rather than the number designation of the bearing.

The *bearing* ***designation*** numbers for different types of bearings are not standardized. Manufacturers are using their own numbers. The radial deep groove ball bearing is usually given the *bearing type* designation of 6. As an example, the widely used deep groove ball bearing of *dimension series* 02 with an *inner diameter* 35 [mm] would have a designation of 6207 instead of 60207. The numeral 6 stands for the *deep groove ball bearing* in this designation. The numeral 2 stands for the dimension series (only the diameter part), and the numeral 07 stands for $d_i/5$ in [mm].

Rolling bearings *accuracy* and ***tolerances*** are extremely important for the proper performance of the bearing and the design. Fine tolerances such as ISO Normal, P5, P6, CL2, CL3, CL0, and others are then defined by standards and provided by manufacturers for different types of bearing's inner diameter, outer diameters, width, and chamfer; see, e.g. ANSI/ABMA 20 (2011) (1987), ANSI/ABMA 19.2 (2013), ISO 492 (2014), or ISO 199 (2014). To get some feel of that, the usual tolerances of inner bearing diameter d_i ranges from 0 to −8 or 0 to −200 [μm] for diameters of 1–2000 [mm]. The usual tolerances of outer bearing diameter d_o ranges also from 0 to −8 or 0 to −200 [μm] for diameters of 6–2500 [mm]. Suggested tolerances for mating shaft and housing are also recommended (g6 or h6, or IT 5, 6, or 7 with fit depending on *load* and *size* in the range of f to r and F to P); see, e.g. ANSI/ABMA 7-(S2013) (1995), ISO 492 (2014), ISO 286-2 (2010), and Section 2.4.7 about tolerances, surface finish, and fits. For light loads, the shaft tolerance can range from h5 for smaller sizes (<20 [mm]) to p6 for very large sizes (>500 [mm]. For normal loads, the shaft tolerance can range from j5 for smaller sizes (<20 [mm]) to r7 for very large sizes (>500 [mm]. For heavy loads, the shaft tolerance can range from k5 for smaller sizes (<20 [mm]) to r7 for very large sizes (>500 [mm]). For housings, the tolerance can range from G7 to P6 depending on loading, rotational conditions, displacement limitation, and other additional conditions (ANSI/ABMA 7-(S2013) (1995)).

There are also *internal clearances* to allow the rolling elements to revolve without unwarranted resistance. These are small and can be acceptable for normal applications or can be reduced by preloading for more accurate movements of mating parts in precision applications. Internal clearances may also be added to the bearing number as a suffix or a prefix.

Bearing ***materials*** may also be added to the bearing number as a suffix or usually as a prefix. The usual materials of rolling bearings are of alloy steels with a high *carbon* content of about (0.95–1.2% C). Other alloying elements can be *chromium* of about (0.4–1.65% Cr), some *manganese* of about (0.4–1.2% Mn), and *silicone* of about (0.15–0.65% Si). Other alloys are also used by manufacturers, and heat treatment, case hardening, and surface coating are indicated. Some rolling bearings are made of stainless steels, high-temperature steels, and balls, or rollers made of ceramics. Retaining rings may be made of stamped steel, stainless steel, brass sheets, or reinforced polymers. Standards such as AISI/SAE indicate steels of 51100 and 52100 designations for rolling bearings. These are *chromium* steels with (1.0% C) and (1.02–1.45% Cr), respectively. The ultimate strength is about 1150 [MPa] or 165 [kpsi], and the yield strength is about 900 [MPa] or 130 [kpsi] in the annealed form. It can be heat treated and case hardened.

Again, Figure 11.5 is an updated continuation of Figure 8.12. It suggests the bearing ***fits*** with the shaft and housing. The fit is shown in grayish color since bearing clearances are only considered as anticipated regular, and not a precision clearance system. It is assumed that the bearing outer ring external diameter d_o has a clearance class of h5. It is also assumed that the bearing inner ring bore diameter d_i has a clearance class of H5. The fit should be adjusted to accommodate the load variation and gearbox application and utilization conditions.

11.3 Initial Design and Selection

For selecting the proper bearing, Chapter 8 presents a simple way to define the appropriate ***radial bearing*** to adopt and provides its initial synthesis; see Section 8.4.3 and Figure 8.21. Figure 11.6 is similar to Figure 8.21 for the estimation of radial bearing synthesis as a function of inner bearing diameter d_i and *usual loading*. Figure 11.6a is for SI units. Figure 11.6b is for the US units. The solid lines are for *ball bearings* at a *life* of 10 000 hour. Each line is for specifically marked rotational speed in revolutions per second [rps]. The heavy dashed line is for fluid-film journal bearings at their maximum capacities and at certain much higher rotational speeds [rpm] (left scale) read off the dashed grayish line. The journal bearings are presented in Chapter 12, but its preliminary synthesis is useful in asserting the suitability of selecting rolling bearings.

The applied load onto the radial bearing is usually known. Also, the rotational speed is usually known. At the value of that load on the *abscissa*, a horizontal line can be beneficial to reach the line or the two encompassing

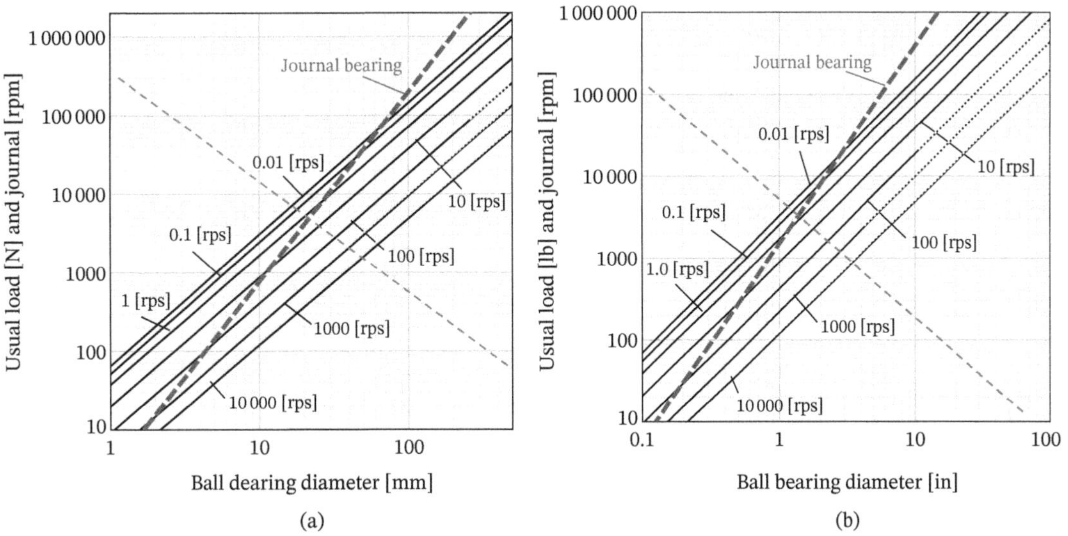

Figure 11.6 (Similar to Figure 8.21) Estimate of radial bearing synthesis as a function of nominal bearing diameter and usual load. (a) SI units. (b) US units. The solid lines are for rolling bearings at a life of 10 000 hours. Each line is for specifically marked rotational speed. The heavy dashed line is for fluid-film journal bearings at their maximum capacities and at certain much higher rotational speeds [rpm] (left scale) read off the dashed grayish line.

lines of the rotational speed value in revolutions per second [rps]. The suitable diameter is read off the horizontal *ordinate* axis at that rotational speed.

Thrust bearings can be treated as radial bearings, and initial synthesis graphs can also be developed similarly. Figure 11.7 presents means to synthesize an estimate of *thrust ball bearing* size as a function of bearing diameter, rotational speed, and *usual load*. Figure 11.7a is for SI system of units. Figure 11.7b is for the US system of units. The solid lines are for *bearings* at a life of 10 000 hours [hr]; see Neale (1973). Each line is for specifically marked rotational speed. The dotted extension of some high rotational speed indicates a general unacceptable region. This is due to limitations of maximum speeds of some available commercial products.

The applied load onto a bearing is usually known. Also, the rotational speed is usually known. At the value of that load on the *abscissa*, a horizontal line can be beneficial to reach the line or the two encompassing lines of the rotational speed value in revolutions per second [rps]. The suitable diameter is read off the horizontal *ordinate* axis at that rotational speed.

For ***other types*** of bearings, the *relative capacities* of common bearings defined in Section 1.1 can be used to get an approximate estimate of the nominal or bore diameter of any of these bearings. The applied load is scaled according to the relative capacity to ball bearings and the approximate diameter obtained accordingly. Note that the synthesis process is an approximate estimation and should be qualified by detailed design calculations as will be defined later in Section 11.5.

Example 11.1 It is required to synthesize the bearing of the pivot shaft between the hood and the mast of the *Jib crane* in Example 8.2. Find the type and dimensions of the bearing. Assume that the maximum rotational speed of the boom is about 0.1 [rps] and the life is about 10 000 hours.

Solution

Data: The maximum load between the pivot shaft of the hood and the mast is the mast force $\boldsymbol{F}_1 = (-25.33, -10, 0)$ [kN] or $(-5.574, -2.2, 0)$ [klb]. Rotational speed of the boom is 0.1 [rps]. At the lower bearing ring of Figure 8.17b, the maximum load between the hood and the mast is the mast force $\boldsymbol{F}_2 = (25.33, 0, 0)$ [kN] or $(5.574, 0, 0)$ [klb].

If we assume that a construction can accommodate a thrust bearing and a radial bearing at the top pivot shaft, the diameters of the *radial* bearing can be obtained from Figure 11.6. In Figure 11.6a and at the intersection of the

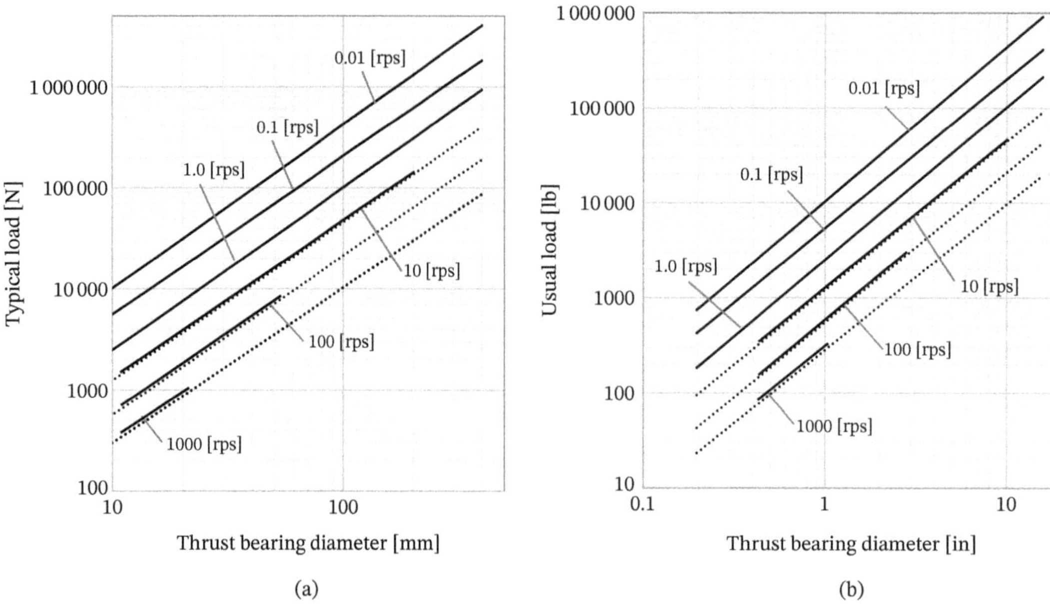

Figure 11.7 Estimate of thrust bearing synthesis as a function of nominal bearing diameter and usual load, (a) for SI units, and (b) for the US units. The solid lines are for thrust bearings at a life of 10 000 hours. Each line is for specifically marked rotational speed with invalid dotted extensions for large diameters.

usual load of 25 [kN] with the 0.1 [rps] line, the nominal or inner radial bearing diameter d_i is found to be about

$$d_i = 40\,[\text{mm}] \quad \text{(a)}$$

In Figure 11.6b and at the intersection of the usual load of 5.6 [klb] with the 0.1 [rps] line, the nominal or inner bearing diameter d_i is found to be about

$$d_i = 1.5\,[\text{in}] \quad \text{(b)}$$

One must adjust these diameters to the closest or larger *standard* diameters that can be found with the supplier or manufacturer; see the Internet references.

The diameters of the *thrust* bearing can be obtained from Figure 11.7. In Figure 11.7a and at the intersection of the usual load of 10 [kN] with the 0.1 [rps] line, the nominal or inner thrust bearing diameter d_i is found to be about

$$d_i = 13\,[\text{mm}] \quad \text{(c)}$$

In Figure 11.7b and at the intersection of the usual load of 2.2 [klb] with the 0.1 [rps] line, the nominal or inner bearing diameter d_i is found to be about

$$d_i = 0.55\,[\text{in}] \quad \text{(d)}$$

One must adjust these diameters to the closest or larger *standard* diameters that can be found with the supplier or manufacturer; see the Internet references. The 13 [mm] thrust ball bearing diameter would be 15 [mm] or 20 [mm], whichever available. The 0.55 [in] thrust ball bearing can be 0.6 [in] or 5/8 [in] (0.625 [in]), whichever available.

Maybe one bearing can carry both radial and axial load. This can be examined for deep groove or angular contact bearings. It might be suggested for the text in Section 11.5 about detailed design and selection.

11.4 Bearing Load

Bearing load involves load direction either radial or axial, or combined, rotational speed, and duration of operating the bearing during its intended life. This affects the selection of the bearing and its life reliability. This section provides information about bearing life, reliability, load distribution, and bearing load rating for the proper selection and assessment of bearings. See rolling bearings standards about load ratings such as ISO 281 (2007), ISO 76 (2006), ANSI/ABMA 9 (2015), and ANSI/ABMA 11 (2014),

11.4.1 Bearing Life and Reliability

From experiments, the minimum or ***rating life*** L_{10} is defined by the life at which 90% of bearings will not fail, i.e. having a 90% reliability. *Weibull* probability distribution can represent bearing *failure probability* $p_B(x)$, where the *life measure* $x = L_B/L_{10}$ and L_B is the *bearing life* in revolutions [rev]. These experiments have the estimated mathematical model of the *Weibull* distribution for bearing *reliability* $R_B(x)$ against bearing life L_B as shown in Figure 11.8. The figure is expanded around the low *guaranteed bearing life* x_0 for clarity. The *bearing reliability* $R_B(x) = 1 - p_B(x)$ gives the *three-parameter Weibull* distribution as follows:

$$R_B = \exp\left(-\left(\frac{x - x_0}{\theta_W - x_0}\right)^{b_W}\right) \tag{11.3}$$

where x_0 is the guaranteed minimum value of the life measure x, θ_W is the *Weibull characteristic value*, and b_W is the *Weibull shape parameter*. The usual estimate of the guaranteed minimum value of *life measure* x is $x_0 = 0.02$; see Figure 11.8. The values of θ_W and b_W are varying according to several factors of estimation. The set of (θ_W, b_W) can be (4.459, 1.483) or (6.86, 1.17), which are validated at about 90% reliability. The first set (4.459, 1.483), however, is more valid for the data of Harris (1963) at higher reliabilities and therefore is adapted in this text. Other values may exist that can provide similar estimates.

When the minimum value of x (i.e. x_0) is very small, the simpler *two-parameter Weibull* distribution becomes

$$R_B = \exp\left(-\left(\frac{L_B}{\theta_W L_{10}}\right)^{b_W}\right) \tag{11.4}$$

where b_W is the *Weibull* shape parameter and θ_W is the *adjusted Weibull characteristic value*. The *Weibull* parameter and characteristic value are dependent on the extensive bearing tests and robust regression estimation.

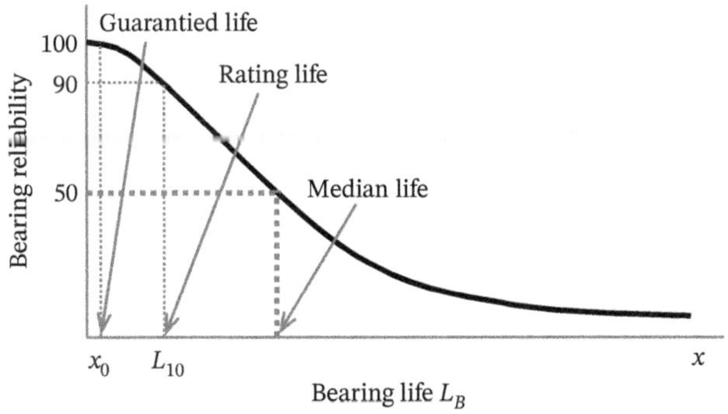

Figure 11.8 A sketch of expanded bearing reliability at low guarantied bearing life x_0 and showing an estimated mathematical model for a *Weibull* distribution of reliability against bearing life.

Taking the natural logarithm of Eq. (11.3), one can get the *rating life*L_{10} for any desired *bearing life* L_B in terms of x as follows:

$$\ln\left(\frac{1}{R_B}\right) = \left(\frac{x - x_0}{\theta_W - x_0}\right)^{b_W}, \quad \text{or} \quad \left(\ln\left(\frac{1}{R_B}\right)\right)^{1/b_W} = \frac{x - x_0}{\theta_W - x_0} \tag{11.5}$$

which gives

$$x = x_0 + (\theta_W - x_0)\left(\ln\left(\frac{1}{R_B}\right)\right)^{1/b_W} \tag{11.6}$$

For the simpler two-parameter *Weibull* distribution of Eq. (11.4), the *synthesis rating life* L_{10S} for any desired *bearing life* L_B can be obtained simply from Eq. (11.6) as

$$\frac{L_B}{L_{10S}} = x_0 + (\theta_W - x_0)\left(\ln\left(\frac{1}{R_B}\right)\right)^{1/b_W} \tag{11.7}$$

or

$$L_{10S} = \frac{L_B}{(x_0) + (\theta_W - x_0)\left(\ln\left(\frac{1}{R_B}\right)\right)^{1/b_W}} \tag{11.8}$$

When the bearing reliability R_B is 0.9 (or 90%) and L_B is 10^6 [rev], the expected L_{10S} should be equal to L_{10}. For the simpler two-parameter *Weibull* distribution and the adapted set of (θ_W, b_W) can be (4.459, 1.483), Eq. (11.8) reduces to the following:

$$L_{10S} = \frac{L_B}{4.439\left(\ln\left(\frac{1}{R_B}\right)\right)^{1/1.483}} \tag{11.9}$$

This relation is easier to use for hand calculations to evaluate the desired *synthesis rating life* L_{10S} for the baring life L_B of the design. However, this produces a little larger L_{10S} than using Eq. (11.8).

For ***tapered roller bearings***, Eq. (11.4) is reduced to the following expression (Mischke (1986)):

$$R_B = \exp\left(-\left(\frac{L_B}{\theta_W L_{10S}}\right)^{b_W}\right) = \exp\left(-\left(\frac{L_B}{4.48 L_{10S}}\right)^{1.5}\right) \tag{11.10}$$

This relation is simpler to use than those equations including the guaranteed minimum value of the life measure x_0. The set of new (θ_W, b_W) is then (4.48, 1.5) for the estimation of reliability in tapered roller bearings calculation of reliability.

For the simpler two-parameter *Weibull* distribution of Eq. (11.10) for *tapered roller bearing*, the *synthesis rating life* L_{10S} for any desired *bearing life* L_B can be obtained similar to Eq. (11.6) as

$$L_{10S} = \frac{L_B}{4.48\left(\ln\left(\frac{1}{R_B}\right)\right)^{1/1.5}} \tag{11.11}$$

This relation is easier to use for hand calculations to evaluate the desired synthesis rating life L_{10S} for the baring life L_B of the design. However, this relation indicates that x_0 is presumed as zero.

It should be noted that the *rating life* L_{10} of manufactured bearings is usually 10^6 revolutions. This necessitates that the calculated *synthesis rating life* L_{10S} should scale the reliability carve to coincide with rating life, i.e. the loads should be adjusted to a higher capacity bearing for the calculated L_{10S} to be scaled to 10^6 rating life. This is also done by using the new calculated L_{10S} in place of L_B or L_{BS} = new L_{10S} for the scaling to materialize.

Example 11.2 It is required to estimate the behavior of a ball bearing rotating at 6 rpm and can last for 10 000 hours with a reliability of 99%. What should be the rated life of bearing in that case? If the duration of rotation in hours is only 2778 [hr], what is the effect of the 99% reliability on the rated life of the bearing?

Solution

Data: $L_B = 10\,000$ [hr](6[rpm]) (60) [min/hr] = 3 600 000 [rev], $R_B = 99\% = 0.99$. The second case $L_B = 2778$ [hr] (6[rpm]) (60) [min/hr] = 1 000 080 [rev] or about = 10^6 [rev].

Substituting in Eq. (11.8) gives the following:

$$L_{10S} = \frac{L_B}{(x_0) + (\theta_W - x_0)\left(\ln\left(\frac{1}{R_B}\right)\right)^{1/b_W}} = \frac{3\,600\,000}{(0.02) + (4.439)\left(\ln\left(\frac{1}{0.99}\right)\right)^{1/1.483}} = \frac{3\,600\,000}{0.219\,59} = 16\,394\,220\ [\text{rev}] \tag{a}$$

Using Eq. (11.9) gives

$$L_{10S} = \frac{L_B}{4.439\left(\ln\left(\frac{1}{R_B}\right)\right)^{1/1.483}} = \frac{3\,600\,000}{4.439(0.010\ 05)^{0.674\ 31}} = \frac{3\,600\,000}{0.199\,59} = 18\,037\,420\ [\text{rev}] = 18.04\,(10^6)\ [\text{rev}] \tag{b}$$

This value is about 18 times the rated life, which is higher than the 16.4 times in Eq. (a).

The second case needs to replace the 3 600 000 [rev] in Eqs. (a) and (b) by 10^6 [rev], which renders Eqs. (a) and (b) to be as follows:

$$L_{10S} = \frac{1\,000\,000}{0.219\,59} = 4\,553\,941\ [\text{rev}] = 4.5539\,(10^6)\ [\text{rev}] \tag{c}$$

$$L_{10S} = \frac{1\,000\,000}{0.199\,59} = 5\,010\,271\ [\text{rev}] = 5.0103\,(10^6)\ [\text{rev}] \tag{d}$$

The approximate two-parameter *Weibull* distribution gives about 10% more rated value than Eq. (11.8). This is how the reliability curve should be scaled to adjust for the desired L_{10S}.

11.4.2 Load Distribution

The rolling elements transfer the load from the inner to the outer rings or vice versa through the intermediate balls or rollers. The load is divided among the rolling elements depending on bearing geometry. For radial bearings, only half of the balls will carry the load unevenly distributed. For thrust bearing, it is expected to have all the balls or rollers carrying about an equal share of the load. The process of load distribution, contact stresses, rolling friction, material, and lubrication properties is of a paramount importance to the designers and manufacturers of the rolling element bearings. The simplified treatment of load distribution shades light on some of these processes. For engineers that use the rolling bearing in the construction of a product design, the standard dimensions and load rating are the main parameters needed for a successful design. These are provided by the standards and the manufacturers.

For *radial* bearings, load is not equally distributed among rolling elements; see Figure 11.9. Lower rolling elements carry heavier load, and side and upper ones carry almost no load. The lower element load is assigned the symbol $\boldsymbol{F}_o$. The side element loads are $\boldsymbol{F}_1$, and $\boldsymbol{F}_2$. It is assumed that approach between ball elements is proportional to $\boldsymbol{F}_i^{2/3}$. The load sharing is a function of the *relative deflection* ($\delta_i \propto \boldsymbol{F}_i^{2/3}$) and the geometry ($\delta_i = \delta_o \cos i\theta_b$), which gives

$$\left(\frac{\boldsymbol{F}_i}{\boldsymbol{F}_o}\right)^{2/3} = \frac{\delta_i}{\delta_o} = \cos i\theta_b \tag{11.12}$$

where θ_b is the angle between rolling elements. The resulting total load $\boldsymbol{P}_l$ is the sum of all force components, which is

$$\boldsymbol{P}_l = \boldsymbol{F}_o + 2\boldsymbol{F}_1 \cos\theta_b + 2\boldsymbol{F}_2 \cos 2\theta_b + \cdots \tag{11.13}$$

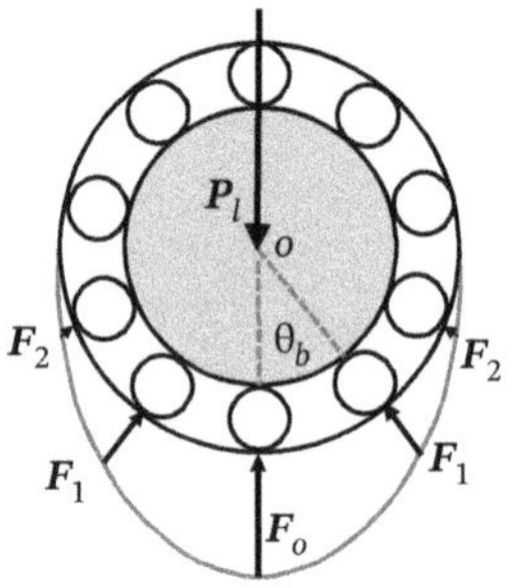

Figure 11.9 A sketch of load distribution among rolling elements.

Substituting the relations of Eq. (11.12) into Eq. (11.13) generates the following relation:

$$P_l = F_o\left(1 + 2\sum_{i=1}^{k_B}(\cos\, i\theta_b)^{5/2}\right) \tag{11.14}$$

In previous equation, k_B is the integer half of the number of balls or the number of balls in the lower half of the bearing. From Eq. (11.14), one can get the maximum force F_o on a ball if the number of balls is known. With that force, the *Hertzian* contact stresses are evaluated and the fatigue or endurance limits defined. The contact stresses are also affected by the present lubricant in the contact area. This is usually treated by the *elastohydrodynamics* field, which is beyond the scope of this text; see, e.g. Gohar (2003), Harrisand Kotzalas (2006), or Abbas et al. (2010), and Abbas and Metwalli (2011).

In relation to preceding development, the *dynamic load rating capacity* C_D can be represented by the following expression for ball diameters less than or equal to 25.4 [mm] or 1.0 [in] (ISO 281 (2007) and Oberg et al. (2012)).

$$C_D = K_g(n_r \;\cos\, \alpha)^{0.7} n_b^{2/3} d_b^{1.8} \tag{11.15}$$

where d_b is the *ball diameter* in [mm] or [in], n_b is the *number of balls*, α is the *angle* for the *angular contact* bearing, n_r is the *number of rows*, and K_g is a *geometry and material factor* that is a function of the ball to the mean diameter ratio. The value of K_g is about 44–59 for SI, 3350–4550 for the US system, and bearings of single row radial contact, single and double row angular contact, and double row radial contact. For self-aligning bearings, K_g is about 17–41 for SI and 1300–3100 for the US system (ISO 281 (2007) and Oberg et al. (2012)).

When the ball diameter is larger than 25.4 [mm] or 1.0 [in], the *dynamic load rating capacity* C_D can be represented by the following expressions:

$$\begin{aligned} C_D|_{SI} &= 3.647\, K_g(n_r \;\cos\, \alpha)^{0.7} n_b^{2/3} d_b^{1.4} \\ C_D|_{US} &= K_g(n_r \;\cos\, \alpha)^{0.7} n_b^{2/3} d_b^{1.4} \end{aligned} \tag{11.16}$$

where d_b is the *ball diameter* in [mm] or [in], n_b is the *number of balls*, α is the *angle* for the *angular contact* bearing, n_r is the *number of rows*, and K_g is a *geometry and material factor* that is a function of the ball to the mean diameter ratio. Values of K_g are the same as for Eq. (11.15).

Other similar equations are available for thrust bearings, roller bearings, and others (Oberg et al. (2012)). Again, these equations are of a paramount importance to the designers and manufacturers of the rolling element bearings. The information about material, ball diameter, and the number of balls is therefore known. Usually, manufacturers may not provide such information in their catalogs for users to apply Eq. (11.15) or (11.16) or any other relation. For users or practicing engineers, the standard *dimensions*, *dynamic load rating capacity* C_D, and *static load rating capacity* C_o are the main parameters needed for a successful design. The detailed design in Section 11.5 and other following Sections 11.4.3, 11.5.1–11.5.4 provide means to do that.

11.4.3 Bearing Load Rating

From experiments, two groups of identical bearings have been tested under different *radial loads* $\boldsymbol{P}_{l1}$ and $\boldsymbol{P}_{l2}$. They have respective *rating lives* of L_{B1} and L_{B2} in revolutions [rev]. After numerous tests, the following expression has been observed to represent the proportional *bearing load-life* relation:

$$\left(\frac{L_{B1}}{L_{B2}}\right) = \left(\frac{\boldsymbol{P}_{l2}}{\boldsymbol{P}_{l1}}\right)^{a_B} \tag{11.17}$$

where the bearing load-life exponent $a_B = 3$ is for ball bearings and $a_B = 10/3$ is for roller bearings.

Since the *basic dynamic load rating* C_D is evaluated at rating life L_{10} of 10^6 revolutions, the bearing life L_B in [rev] should be specified as bearing life in million revolutions L_{BM}, i.e. divide L_B in [rev] by 10^6 to get L_{BM} in [Mrev]. Due to Eq. (11.17), the required *bearing life* L_{BM} *in million revolutions* [Mrev] under *radial load* $\boldsymbol{P}_l$ is then related to *synthesis dynamic load rating* C_{DS} or the *synthesis load* $\boldsymbol{P}_{\mathrm{lS}}$ (at L_{10}) such that

$$\left(\frac{L_B}{L_{10}}\right) = \left(\frac{\boldsymbol{P}_{\mathrm{lS}}}{\boldsymbol{P}_l}\right)^{a_B}, \quad \text{or} \quad L_{\mathrm{BM}} = \left(\frac{\boldsymbol{P}_{\mathrm{lS}}}{\boldsymbol{P}_l}\right)^{a_B} = \left(\frac{C_{\mathrm{DS}}}{\boldsymbol{P}_l}\right)^{a_B}, \quad \text{in } 10^6 \text{ [rev] or in [Mrev]} \tag{11.18}$$

or

$$C_{\mathrm{DS}} = \boldsymbol{P}_{\mathrm{lS}} = \boldsymbol{P}_l\left(\frac{L_B}{10^6}\right)^{1/a_B} = \boldsymbol{P}_l(L_{\mathrm{BM}})^{1/a_B} \tag{11.19}$$

where, yet again, $a_B = 3$ is for ball bearings and $a_B = 10/3$ is for roller bearings. Again, note that in Eq. (11.19), the required bearing synthesis life L_{BS} should be in million revolutions to get the proper value of *synthesis load* $\boldsymbol{P}_{\mathrm{lS}}$ (at L_{10}), i.e. the desired *synthesis dynamic load rating* C_{DS}.

Usually, the bearing would be running at a certain *rotational speed* in revolutions per minute N_{rpm} [rpm] for a certain number of hours L_{Bhr} [hr]. To translate that into million revolutions, the following relation is obvious:

$$L_{\mathrm{BM}} = \frac{L_{\mathrm{Bhr}} N_{\mathrm{rpm}}(60)}{10^6}, \text{ in } 10^6 \text{ [rev] or in [Mrev]} \tag{11.20}$$

Bearing life L_{Bhr} in terms of operating *hours* under rotational speed N_{rpm}, and certain *synthesis dynamic load rating* C_{DS} evaluated at a rating life L_{10} of 10^6 [rev], is then given by

$$L_{\mathrm{Bhr}} = \left(\frac{C_{\mathrm{DS}}}{\boldsymbol{P}_l}\right)^{a_B}\left(\frac{10^6}{60N_{\mathrm{rpm}}}\right) \tag{11.21}$$

where $\boldsymbol{P}_l$ can also be the *equivalent dynamic force* $\boldsymbol{P}_{\mathrm{le}}$, which is the *bearing load* $\boldsymbol{P}_l$ for steady operations.

Most manufacturers use *rating life* L_{10} of one million revolutions for evaluating the *basic dynamic load rating* C_D. Some other manufacturers additionally define the *basic dynamic load rating* C_{D90} at rotational speed N_{rpm} of 500 [rpm] for 3000 [hr] or 90 million revolutions [Mrev], Timken (2014a, 2014b). To account for that, one must evaluate the required bearing life in 90 million revolutions, which renders Eqs.(11.19) and (11.20) as follows:

$$\begin{aligned} L_{B90} &= \frac{L_{\mathrm{Bhr}} N_{\mathrm{rpm}}(60)}{90(10^6)}, \text{ in } 90(10^6) \text{ [rev]} \\ C_{D90} &= \boldsymbol{P}_l(L_{B90})^{1/a_B} \end{aligned} \tag{11.22}$$

where L_{B90} is the *bearing rating life* at 90 [Mrev] and C_{D90} is the *dynamic load rating* at 90 [Mrev]. Timken (2014a, b), as well, provides the *basic dynamic load rating* C_D for *rating life* L_{10} of 1 million revolutions. Equation (11.22), therefore, may not be needed in that case.

Example 11.3 Find the synthesis dynamic load rating C_{DS} for a roller bearing, if a bearing life L_B of 36 million revolutions is required. If the bearing is changed to a deep groove ball bearing, find the new synthesis dynamic load rating C_{DS}.

Solution

Data: $L_B = 36\ (10^6)$ [rev], $L_{BM} = 36$ [Mrev], $a_B = 3$ (ball bearings), and $a_B = 10/3$ (roller bearings).

For roller bearing, the basic dynamic load rating for a roller bearing for a required life of 36 million revolutions is then given according to Eq. (11.18) with $a_B = 10/3$. The basic dynamic load rating C_D or the required *synthesis load* $\boldsymbol{P}_{lS}$ should then be calculated as follows:

$$C_{DS} = \boldsymbol{P}_{lS} = \boldsymbol{P}_l(36)^{3/10} = 2.9302\ \boldsymbol{P}_l \tag{a}$$

or about 2.93 times the actual force to account for the needed higher bearing life. One would use this basic dynamic load rating C_D in a manufacturers catalog for the selected bearing to be close to the synthesis dynamic load rating C_{DS}.

For deep groove ball bearing, the basic dynamic load rating for a required life of 36 million revolutions is also given according to Eq. (11.18) with $a_B = 3$. The synthesis dynamic load rating C_{DS} or the required *synthesis load* $\boldsymbol{P}_{lS}$ should then be calculated as follows:

$$C_{DS} = \boldsymbol{P}_{lS} = \boldsymbol{P}_l(36)^{1/3} = 3.3019\ \boldsymbol{P}_l \tag{b}$$

or about 3.3 times the actual force to account for the needed higher bearing life. One would use this basic dynamic load rating C_D in a manufacturers catalog for the selected bearing to be close to the synthesis dynamic load rating C_{DS}.

Example 11.4 Prepare for the selection of a deep groove ball bearing to withstand a radial force of 25.33 [kN] or 5.574 [klb] with a life of 10 000 hours [hr] and rotating at a speed of 0.1 [rps]. Find the synthesis dynamic load rating you should prepare to select the bearing from the manufacturer catalog. Find, also, the synthesis dynamic load rating you should prepare to select the bearing from the manufacturer catalog if reliability should be 99%.

Solution

Data: Initially $\boldsymbol{P}_l = 25.33$ [kN], $L_B = 10\,000$ [hr], and $N_{\text{rpm}} = 0.1\ (60) = 6$ [rpm].

Using Eq. (11.21), and $a_B = 3$, gives

$$(C_{DS})^3 = L_{Bhr}(\boldsymbol{P}_l)^3 \left(\frac{60N_{\text{rpm}}}{10^6}\right) \tag{a}$$

or

$$C_{DS} = \boldsymbol{P}_l \left(L_{Bhr}\frac{60N_{\text{rpm}}}{10^6}\right)^{1/3} \tag{b}$$

Then, with $L_{Bhr} = 10\,000$ [hr] and to have a 90% reliability, one gets

$$\begin{aligned} C_{DS} &= \boldsymbol{P}_{lS} = \boldsymbol{P}_l \left(L_B\frac{60N_{\text{rpm}}}{10^6}\right)^{1/3} = 25.33\left(10\,000\left(\frac{60(6)}{10^6}\right)\right)^{1/3} = 25.33(3.6)^{1/3} = 25.33(1.5326) \\ &= 38.821\ [\text{kN}] \\ C_{DS} &= \boldsymbol{P}_{lS} = \boldsymbol{P}_l \left(L_B\frac{60N_{\text{rpm}}}{10^6}\right)^{1/3} = 5.574\left(10\,000\left(\frac{60(6)}{10^6}\right)\right)^{1/3} = 5.574(3.6)^{1/3} \\ &= 5.574(1.5326) = 8.5428\ [\text{klb}] \end{aligned} \tag{c}$$

These values can be used to identify the bearing in any supplier or manufacturer catalog that can be downloaded such as SKF (2018), Timken (2014a, b), NSK (2017), and NTN (2009), just to name a few; see the Internet references.

In Example 11.1, the selected bearing diameter was 40 [mm]. The SKF bearing designation number 6308 ($C_D = 42.4$ [kN]) may be adapted to accommodate the synthesis dynamic load rating. A usual series of SKF number 6208 ($C_D = 32.5$ [kN]) can accommodate a little lower life rating than the 3.6 million revolutions needed in this application.

To have a 99% reliability as examined in Example 11.2, then one can use $L_B = L_{10S} = 16.394\ (10^6)$ [rev] in place of the 10 000 [hr]. With that, one can get the following expected synthesis dynamic load rating:

$$\begin{aligned} C_{\mathrm{DS}} &= \boldsymbol{P}_l\left(\frac{L_B}{10^6}\right)^{1/3} = 25.33\left(\frac{16.394(10^6)}{10^6}\right)^{1/3} = 25.33(2.5404) = 64.348\ [\mathrm{kN}] \\ C_{\mathrm{DS}} &= \boldsymbol{P}_l\left(\frac{L_B}{10^6}\right)^{1/3} = 5.574\left(\frac{16.394(10^6)}{10^6}\right)^{1/3} = 5.574(2.5404) = 14.160\ [\mathrm{klb}] \end{aligned} \tag{d}$$

These values are higher for the application like a *Jib crane*. Such reliability may not be needed in that case, because the failure of the bearing is not catastrophic in any sense depending on the appropriate design.

11.5 Detailed Design and Selection

If the bearing design has a known rotational speed N_{rpm} and the number of operating hours L_{Bhr} in its expected life in years is also known, the number of million revolutions of the bearing life L_B should be known. The suggested life of a rolling bearing in hours for several applications is useful to select the appropriate bearing for an application with a known rotational speed. The available experience suggests the following life of a rolling bearing in *hours* L_{Bhr} for several machinery applications:

• Instruments and infrequently used devices	500 [hr]
• Aircraft engines	500–2000 [hr]
• Equipment intermittently used for short durations	4000–8000 [hr]
• Appliances intermittently used but requiring high reliability	8000–14 000 [hr]
• Products operating 8 [hr] a day, but not continuously	14 000–20 000 [hr]
• Machines continuously operating 8 [hr] a day	20 000–30 000 [hr]
• Machineries continuously operating 24 [hr] a day	40 000–60 000 [hr]
• Machinery continuously operating 24 [hr] a day and requiring extreme reliability	100 000–200 000 [hr]

The bearing rotational speed N_{rpm} should be used to evaluate the total revolutions of bearing life L_B in [rev] as indicated in Eq. (11.20).

Some bearings can have a combined *radial load* $\boldsymbol{P}_r$ and an *axial load* $\boldsymbol{P}_a$. An equivalent load $\boldsymbol{P}_e$ may be calculated to accommodate such a combined loading. The initial synthesis load $\boldsymbol{P}_l$ for ball bearings should at least be the resultant of the combined load to find the bearing diameter as defined in Section 11.3. The initial *bearing load* $\boldsymbol{P}_l$ or the *equivalent load* $\boldsymbol{P}_e$ could then be

$$\boldsymbol{P}_l = \boldsymbol{P}_e = \sqrt{\boldsymbol{P}_r^2 + \boldsymbol{P}_a^2} \tag{11.23}$$

Loads would also be *static* or *unsteady*, and a way should be used to account for that by using an equivalent representative value of the load. These will be touched next.

It is also common to have a *load application factor* K_l that increases the *equivalent load* $\boldsymbol{P}_e$ as a safety and to account of different application characters. This factor depends on the type of application as follows:

- Precision sets or equipment under steady load would employ $K_l = 1.0$–1.1.
- Commercial equipment with light shocks would employ $K_l = 1.1$–1.3.
- Equipment operating under moderate shock loading would employ $K_l = 1.3$–1.8.
- Equipment operating under heavy shock loading would employ $K_l = 2.0$–3.0.

The general procedure of detailed design and selection starts with the review of initial design and adjusting that usual *synthesis* to the real application life other than the 10 000 hours and the load application factor previously indicated. Another ***initial synthesis*** might be necessary to reduce detailed design iterations required to account for *static* loading, *combined* loading, *unsteady* loading, adoption of *tapered roller bearings*, checking speed limits, and suitable lubrication effects. These are treated in the following segments and sections with more defined procedure recapped in Section 11.5.5.

11.5.1 Static Loading

In rolling bearings, static loading generates *Hertzian* stresses that should be within acceptable limits; see Section 6.4.3 about contact stresses. Due to the special case of application, the limit of static loading has been defined as ***basic static load rating*** C_0 that causes a permanent deformation of approximately 0.0001 of the rolling element diameters (ISO 76 (2006)). The *contact stresses* in these cases would have values of 4.2 [GPa] for all ball bearings except a 4.6 [GPa] for the self-aligning ball bearings, and 4.0 [GPa] for all roller bearings. This plastic contact stress problem is in the field of *plasticity*, which is beyond the scope of the text. One can, therefore, use the values provided by the manufacturers for the specifically selected bearing. The bearing should stand the *equivalent static load* $\boldsymbol{P}_{e0}$ and a *static safety factor* K_{SF0} such that

$$K_{\text{SF0}} = \frac{C_0}{\boldsymbol{P}_{e0}} \tag{11.24}$$

The static safety factor K_{SF0} is a function of the certainty of load level and the motion if continuous or intermittent. The value of K_{SF0} ranges from 0.4 to 2.0.

An important variable in bearing selection is the *load ratio limit* e_B, which is used in combined loading. The factor of *load ratio limit* e_B is proportional to the ratio of the *axial load* $\boldsymbol{P}_a$ to the *basic static load rating* C_0 in the form

$$e_B \propto \frac{\boldsymbol{P}_a}{C_0} \tag{11.25}$$

For available information about *radial load ball bearings*, Eq. (11.25) is estimated as follows:

$$e_B = 0.5065\left(\frac{\boldsymbol{P}_a}{C_0}\right)^{0.2320} \tag{11.26}$$

When static applied loading is composed of *static radial load* $\boldsymbol{P}_{r0}$ and *static axial load* $\boldsymbol{P}_{a0}$, an *equivalent static load* $\boldsymbol{P}_{e0}$ can be found for radial ball bearing as the greater of:

$$\begin{aligned} \boldsymbol{P}_{e0} &= X_0\boldsymbol{P}_{r0} + Y_0\boldsymbol{P}_{a0} = 0.6\boldsymbol{P}_{r0} + 0.5\boldsymbol{P}_{a0} \\ \boldsymbol{P}_{e0} &= \boldsymbol{P}_{r0} \end{aligned} \tag{11.27}$$

where X_0 is the static *radial load factor* and Y_0 is the static *axial load factor*. For other types of bearings, the static *radial load factor* X_0 and the static *axial load factor* Y_0 are dependent on the *load ratio limit* e_B and can be obtained from manufacturers catalogs that can be downloaded from their websites; see the Internet references. Note that for radial ball bearings, $X_0 = 0.6$ and $Y_0 = 0.5$ as depicted in Eq. (11.27).

Example 11.5 Select a deep groove ball bearing to withstand a radial force of 25.33 [kN] or 5.574 [klb] and axial force of 10.0 [kN] or 2.2 [klb] with a life of 10 000 hours [hr] and rotating at a speed of 0.1 [rps]. If these loads are assumed as the static loads due to possible long hesitation in operations, find an estimate of equivalent load on the deep groove ball bearing assuming a light shock application. For an adjusted synthesis of the selected bearing in Example 11.1, obtain the basic dynamic load rating C_D, the basic static load rating C_0, and load ratio limit e_B

from the manufacture catalog, and compare with Eq. (11.26). You should select the bearing from the manufacturer catalog by defining the bearing designation number. Find, also, the static safety factor.

Solution

Data: Initially $\boldsymbol{P}_r = 25.33$ [kN] or 5.574 [klb], $\boldsymbol{P}_a = 10.00$ [kN] or 2.2 [klb], $L_B = 10\,000$ [hr], and $N_{\text{rpm}} = 0.1$ (60) = 6 [rpm]. For light shocks, take $K_l = 1.3$. Bearing life in revolution $L_B = 3.6\,(10^6)$ [rev].

Using Eq. (11.23), the equivalent load is then

$$\begin{aligned}\boldsymbol{P}_{e0} &= \sqrt{\boldsymbol{P}_{r0}^2 + \boldsymbol{P}_{a0}^2} = \sqrt{25.33^2 + 10^2} = 27.233\ [\text{kN}]\\ \boldsymbol{P}_{e0} &= \sqrt{\boldsymbol{P}_{r0}^2 + \boldsymbol{P}_{a0}^2} = \sqrt{5.574^2 + 2.2^2} = 5.9925\ [\text{klb}]\end{aligned} \tag{a}$$

By multiplying these values by the load application factor $K_l = 1.3$, one gets

$$\begin{aligned}\boldsymbol{P}_{e0} &= K_l\boldsymbol{P}_{e0} = 1.3(27.233) = 35.403\ [\text{kN}]\\ \boldsymbol{P}_{e0} &= K_l\boldsymbol{P}_{e0} = 1.3(5.9925) = 7.7902\ [\text{klb}]\end{aligned} \tag{b}$$

For an adjusted synthesis of the selected bearing in Example 11.1, use Figure 11.6a, and at the intersection of the usual load of about 35 [kN] with the 0.1 [rps] line, the nominal or inner radial bearing diameter d_i is found to be 50 [mm]. In Figure 11.6b and at the intersection of the usual load of 7.8 [klb] with the 0.1 [rps] line, the nominal or inner bearing diameter d_i is found to be 2.0 [in].

Adjusting the synthesis dynamic load rating similar to Example 11.4, Eq. (c), the new values should be

$$\begin{aligned}C_{\text{DS}} &= K_l\boldsymbol{P}_l\left(L_B\frac{60N_{\text{rpm}}}{10^6}\right)^{1/3} = 1.3(25.33)\left(10^4\left(\frac{60(6)}{10^6}\right)\right)^{1/3} = 32.929(3.6)^{1/3} = 32.929(1.5326)\\ &= 50.467\ [\text{kN}]\\ C_{\text{DS}} &= K_l\boldsymbol{P}_l\left(L_B\frac{60N_{\text{rpm}}}{10^6}\right)^{1/3} = 1.3(5.574)\left(10^4\left(\frac{60(6)}{10^6}\right)\right)^{1/3} = 7.2462(3.6)^{1/3} = 7.2462(1.5326)\\ &= 11.106\ [\text{klb}]\end{aligned} \tag{c}$$

It should be noted that these values are high due to the use of the equivalent load instead of the radial load. Using SKF catalog, a deep groove bearing designation at 50 [mm] or 2.0 [in] can be 6310 ($C_D = 65$ [kN], $C_0 = 38$ [kN], $f_0 = 13$, with e_B (named *e* by SKF) *depending on* $f_0\ P_a/C_0$).

To calculate load ratio limit e_B from the manufacture catalog, one needs to calculate the radial load and axial load without the previous equivalent load utility. This gives the following values:

$$\begin{aligned}\boldsymbol{P}_r &= K_l\boldsymbol{P}_r = 1.3(25.33) = 32.929\ [\text{kN}], \quad \boldsymbol{P}_a = K_l\boldsymbol{P}_a = 1.3(10.0) = 13.0\ [\text{kN}]\\ \boldsymbol{P}_r &= K_l\boldsymbol{P}_r = 1.3(5.574) = 7.2462\ [\text{klb}], \quad \boldsymbol{P}_a = K_l\boldsymbol{P}_a = 1.3(2.20) = 2.86\ [\text{klb}]\end{aligned} \tag{d}$$

The value of the load ratio limit e_B from the manufacturer (SKF) catalog is then depending on $f_0\ P_a/C_0 = 4.4474$, which gives e_B about 0.39. According to Eq. (11.26), the load ratio limit e_B is then

$$e_B = 0.5065\left(\frac{\boldsymbol{P}_a}{C_0}\right)^{0.2320} = 0.5065\left(\frac{13.0}{38}\right)^{0.2320} = 0.5065(0.7797) = 0.3949 \tag{e}$$

which is close to the *e* of the manufacturer catalog.

The static safety factor K_{SF0} is obtained from Eq. (11.24) so that

$$K_{\text{SF0}} = \frac{C_0}{\boldsymbol{P}_{e0}} = \frac{38}{35.403} = 1.0734 \tag{f}$$

To account for the combined static radial and axial loads, one applies Eqs. (11.27) and (d) to get

$$\begin{aligned} \boldsymbol{P}_{e0} &= 0.6\boldsymbol{P}_{r0} + 0.5\boldsymbol{P}_{a0} = 0.6(32.929) + 0.5(7.2462) = 23.381\ [\text{kN}] \\ \boldsymbol{P}_{e0} &= \boldsymbol{P}_{r0} = 32.929\ [\text{kN}] \end{aligned} \tag{g}$$

The equivalent static load is the largest of the two values in Eq. (g) or $\boldsymbol{P}_{r0} = 32.929$ [kN].

The static safety factor K_{SF0} is then obtained from Eq. (11.24) so that

$$K_{\text{SF0}} = \frac{C_0}{\boldsymbol{P}_{e0}} = \frac{38}{32.929} = 1.154 \tag{h}$$

This value might seem small, but an application factor of 1.3 was used ahead of the calculations. The load application factor is considered part of the safety consideration.

11.5.2 Combined Loading

Combining radial load $\boldsymbol{P}_r$ and axial load $\boldsymbol{P}_a$ for radial, angular contact ball bearings, and roller bearings is treated using an equivalent load $\boldsymbol{P}_e$ as the maximum of the following two values:

$$\begin{aligned} \boldsymbol{P}_e &= XV_B\boldsymbol{P}_r + Y\boldsymbol{P}_a \\ \boldsymbol{P}_e &= V_B\boldsymbol{P}_r \end{aligned} \tag{11.28}$$

where X is the *radial load factor*, Y is the *axial load factor*, and V_B is a *bearing rotation factor* that equals 1.0 for inner ring rotation and 1.2 for outer ring rotation. For rotating outer ring of self-aligning bearings, the *rotation factor* V equals 1.0. The load factors X and Y depend on the type of bearing and the ratio $\boldsymbol{P}_a/V\boldsymbol{P}_r$. as being less or equal to e_B, or larger than e_B; see Eq. (11.26) for radial load ball bearings. Also, the value of the radial load factor X is *constant* at ***0.56***. The value of the axial load factor Y is estimated by the following relation:

$$Y = 0.8632\left(\frac{\boldsymbol{P}_a}{C_0}\right)^{-0.2349} \tag{11.29}$$

where this equation is for radial load ball bearings and should be valid for both the US and SI systems since it is nondimensional. For other types of bearings, one should consult manufacturers catalogs or develop similar relation as Eq. (11.29).

Since the previously obtained equivalent load $\boldsymbol{P}_e$ does not account for shock or impact loading, it is multiplied by the appropriate service or *load application factor* K_l as indicated previously at the beginning of Section 11.5.

The *dynamic safety factor* K_{SFD} is expected to be as follows:

$$K_{\text{SFD}} = \frac{C_D}{\boldsymbol{P}_e} \tag{11.30}$$

This factor can be limited in value larger than 1.0, particularly when a load application factor K_l is used as safety too.

Example 11.6 Consider the selected deep groove ball bearing in Example 11.5 to withstand a radial force of 25.33 [kN] or 5.574 [klb] and axial force of 10.0 [kN] or 2.2 [klb] with a life of 10 000 hours [hr] and the outer ring rotating at a speed of 0.1 [rps]. If these loads are assumed as the combined dynamic loads, find an estimate of equivalent dynamic load on the deep groove ball bearing assuming a light shock application. Obtain the basic dynamic load rating C_D, the basic static load rating C_0, load ratio limit e_B, and the axial load factor Y from the manufacture catalog, and compare with Eq. (11.29). Use the previously selected bearing from the manufacturer catalog by defining the bearing designation number. Find, also, the dynamic safety factor.

Solution

Data: Initially $\boldsymbol{P}_r = 25.33$ [kN] or 5.574 [klb], $\boldsymbol{P}_a = 10.00$ [kN] or 2.2 [klb], $L_B = 10\,000$ [hr], and $N_{\text{rpm}} = 0.1$ (60) = 6 [rpm]. For light shocks, take $K_l = 1.3$. Bearing life in revolution $L_B = 3.6\,(10^6)$ [rev].

Example 11.5 suggested an SKF deep groove bearing designation of 6310 ($C_D = 65$ [kN], $C_0 = 38$ [kN], $f_0 = 13$, with e_B *depending on* $f_0\,P_a/C_0$) with calculated $e_B = 0.3949$, $\boldsymbol{P}_r = 32.929$ [kN], or/and $\boldsymbol{P}_a = 13$ [kN]. The SKF catalog gives Y between 1.15 and 1.04 with in-between interpolation required. $V_B = 1.2$ (rotating outer ring). For the US system, only unit conversion will be applied. No bearing designation in the US system is searched for.

Using Eq. (11.29), the value of the axial load factor Y is then

$$Y = 0.8632\left(\frac{\boldsymbol{P}_a}{C_0}\right)^{-0.2349} = 0.8632\left(\frac{13}{38}\right)^{-0.2349} = 0.8632(1.2865) = 1.1105 \qquad \text{(a)}$$

This value is within the rage of the catalog value. It should be noted that this value should be valid for both the US and SI systems.

To get the equivalent load $\boldsymbol{P}_e$ as the maximum of the following two values using Eq. (11.28) and $X = 0.56$, one thus gets

$$\begin{aligned}
\boldsymbol{P}_e &= XV_B\boldsymbol{P}_r + Y\boldsymbol{P}_a = 0.56(1.2)(32.929) + 1.1105(13) = 36.565\,[\text{kN}] \\
\boldsymbol{P}_e &= V_B\boldsymbol{P}_r = 1.2(32.929) = 39.515\,[\text{kN}] \\
\boldsymbol{P}_e &= XV_B\boldsymbol{P}_r + Y\boldsymbol{P}_a = 0.56(1.2)(7.2462) + 1.1105(2.86) = 8.0455\,[\text{klb}] \\
\boldsymbol{P}_e &= V_B\boldsymbol{P}_r = 1.2(7.2462) = 8.6954\,[\text{klb}]
\end{aligned} \qquad \text{(b)}$$

The value of the equivalent load is then $\boldsymbol{P}_e = 39.515$ [kN], or $\boldsymbol{P}_e = 8.6954$ [klb].

The dynamic safety factors K_{SFD} are then obtained from Eq. (11.30) so that

$$\begin{aligned}
K_{\text{SFD}}\big|_{\text{SI}} &= \frac{C_D}{\boldsymbol{P}_e} = \frac{65}{39.515} = 1.645 \\
K_{\text{SFD}}\big|_{\text{US}} &= \frac{C_D}{\boldsymbol{P}_e} = \frac{65(0.224\,81)}{8.6954} = 1.681
\end{aligned} \qquad \text{(c)}$$

The value seems more than 1.0, which may not need design adjustment. A different smaller bearing may not be safe, since it has much lower C_D.

11.5.3 Tapered Roller Bearings

Tapered roller bearings were introduced as a type of rolling bearings in Section 11.1 and Figure 11.3. It was invented by J.L. Scott in 1895 and by H. Timken in 1898. The *Timken*© company has been producing *tapered roller bearings* soon after the invention; see, e.g. Timken (2014a, b). Figure 11.10 shows a sketch of main terminologies and dimensions of a tapered roller bearing. The main dimensions are the *outer diameter* d_o, the nominal or *inner diameter* d_i, the main *bearing* width, or *thickness* t_B, and the *effective load center distance* a_T defines the location at which the tapered radial and axial components meet. It should be noted, however, that many references and manufacturers use "D" for *outer bearing diameter* d_o, also use only "d" for bearing nominal or *inner diameter* d_i, and use "T"or "B" for the width or main *tapered roller bearing thickness* t_B. Since this text is not only dedicated to just rolling bearings, it was reasonable to define lower case for dimensions with qualified subscript to indicate what the dimension is belonging to.

Here, only the single-raw bearings are considered. The *Timken* category of this most widely used bearing is *type TS*. More types are available and consulting the manufacturer's catalog is necessary; see, e.g. Timken (2014a, b). The treatment given herein is for users to get *initial synthesis* that should be qualified and checked by the manufacturer procedure of design for qualified bearing designation number. Several factors are given by the manufacturer that may not be considered in here.

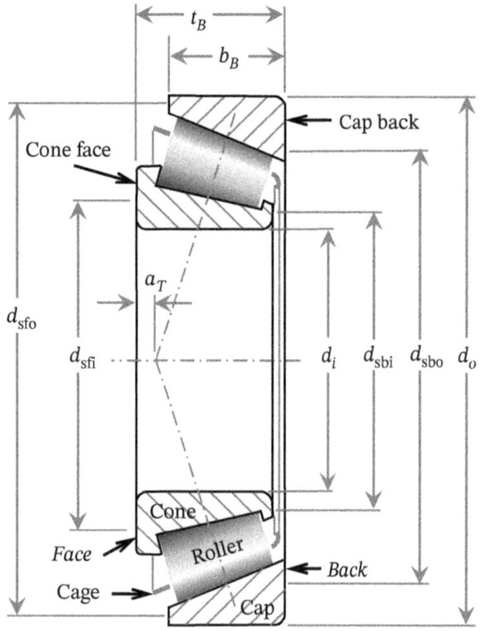

Figure 11.10 Sketch of main terminologies and dimensions of a tapered roller bearing.

With the existence of the tapered roller, an axial load is generated for the applied radial load to have the resultant normal to the roller. The axial load must be supported by any *mounting* arrangement. The usual utilization of tapered roller bearings is by the employment of two in tandem bearings in line or collinear, one opposite to the other so as to cancel the axial load. The two bearings will share the radial load but has to be brought together by eliminating the internal clearance through a probable and slight *axial preload* $\boldsymbol{P}_p$, which would be axially caused by any means. Figure 11.11 shows the two possible mounting arrangements. They are arranged *back-to-back*, which is also called *direct mounting*, as in Figure 11.11a or arranged *face-to-face*, which is called *indirect mounting* as in Figure 11.11b. Figure 11.10 indicates the ***back*** and the ***face*** of the bearing in bold italic font. The mounting of the two tapered roller bearings has to be confined or limited to move as a set in the axial direction. Figure 11.11a shows the housing confining the bearing set to move to the right or to the left. The slackness elimination or restraining is performed on the shaft between its shoulder and the left end part of the restraining mechanism on the shaft as a screw, a shim, etc. This restraining is performed against the housing restrictive protrusion between the two bearings. The slight axial forces are balanced on the housing restraining protrusion (opposite squeezing forces not shown). Figure 11.11b shows the housing confining the bearing set to move to the right or to the left. The restraining is performed on the housing between its shoulder and the left end part of the restraining mechanism in the housing such as a screw, a shim, etc. This restraining is performed against the shaft shoulder between the two bearings. The slight axial forces are balanced on the shaft shoulder (opposite squeezing forces not shown).

In Figure 11.11, no external bending moments are considered. The only externally applied loads are the *radial loads* ($\boldsymbol{P}_{l1}$ and $\boldsymbol{P}_{l2}$) and the externally applied *axial load* $\boldsymbol{P}_a$ in addition to the imaginable *axial preload* $\boldsymbol{P}_p$. The best possible axial preloading $\boldsymbol{P}_p$ is near zero to a slight preload. One only needs to eliminate end play or axial clearance within a mounted tapered roller bearing. Any externally applied axial load $\boldsymbol{P}_a$ should be handled by considering the different direction from the assumed one in Figure 11.11 as negative. The span length l_s between the *effective load centers* is different for the two arrangements and has to be used for the shaft calculations.

The thrust or axial components $\boldsymbol{F}_{a1}$ and $\boldsymbol{F}_{a2}$ produced by the radial loads $\boldsymbol{P}_{l1}$ and $\boldsymbol{P}_{l2}$ can be estimated as follows (Timken (2017)):

$$\boldsymbol{F}_{a1} = \frac{0.47\boldsymbol{P}_{l1}}{K_{T1}} \quad \text{and} \quad \boldsymbol{F}_{a2} = \frac{0.47\boldsymbol{P}_{l2}}{K_{T2}} \tag{11.31}$$

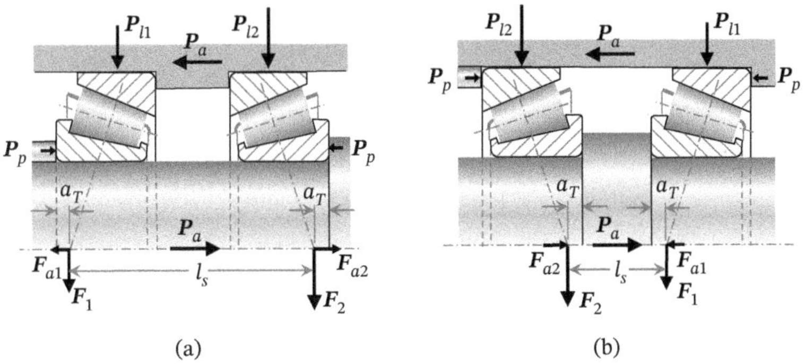

Figure 11.11 Sketch of tapered roller bearing mountings and applied loads. Forces generated due to tapered rollers are shown particularly induced axial forces and shifted radial forces. Bearings are arranged back-to-back in (a) and arranged face-to-face in (b).

where K_{T1} and K_{T2} are the ratios of radial rating to the thrust rating of the tapered roller bearings 1 and 2. Suggested initial iteration values of K_{T1} and K_{T2} are 1.5 for purely radial bearing and 0.75 for steep roller angle. The real value depends on the tapered roller angle. They are defined by the manufacturer for each individual bearing; see, e.g. Timken (2014a, b) for tapered roller bearings. Other similar equations to (11.31) for the evaluation of the thrust or axial components $\boldsymbol{F}_{a1}$ and $\boldsymbol{F}_{a2}$ produced by the radial loads $\boldsymbol{P}_{l1}$ and $\boldsymbol{P}_{l2}$ are estimated differently by employing a 0.5 instead of 0.47 and factors Y_1 and Y_2 in place of K_{T1} and K_{T2}; see, e.g. SKF (2017). Therefore, one is advised to follow the procedure suggested by the prospective manufacturer. The procedure presented herein is useful as an initial synthesis and should be verified by the process given by the bearing's source of acquisition.

To select the tapered roller bearing, an equivalent load is calculated similar to Eq. (11.28) for each bearing under the combined radial and axial loads. The *bearing rotation factor* V_B in Eq. (11.28) is not used with this type of bearings. Figure 11.11 assumes the external axial load $\boldsymbol{P}_a$ in the same direction for part (a) and part (b). The bearing that carries the axial load is always given the numeral label 1. The other bearing is always given numeral 2. The estimated *equivalent dynamic loads* $\boldsymbol{P}_{e1}$ and $\boldsymbol{P}_{e2}$ are then given by the maximum of the two values in the following relations for bearings 1 and 2 in Figure 11.11 (see, e.g. Timken (2017)):

$$
\begin{aligned}
\boldsymbol{P}_{e1} &= 0.4\boldsymbol{P}_{r1} + K_{T1}\left(\frac{0.47\boldsymbol{P}_{r2}}{K_{T2}} + \boldsymbol{P}_a\right), \quad \text{or} \quad \boldsymbol{P}_{e1} = \boldsymbol{P}_{r1} = \boldsymbol{P}_{l1} \\
\boldsymbol{P}_{e2} &= 0.4\boldsymbol{P}_{r2} + K_{T2}\left(\frac{0.47\boldsymbol{P}_{r1}}{K_{T1}} - \boldsymbol{P}_a\right), \quad \text{or} \quad \boldsymbol{P}_{e2} = \boldsymbol{P}_{r2} = \boldsymbol{P}_{l2}
\end{aligned}
\tag{11.32}
$$

where K_{T1} and K_{T2} are the values defined by the manufacturer for each individual bearing. Also, the radial loads $\boldsymbol{P}_{r1}$ and $\boldsymbol{P}_{r2}$ are the same as the applied loads $\boldsymbol{P}_{l1}$ and $\boldsymbol{P}_{l2}$ and the same as the radial loads $\boldsymbol{F}_1$ and $\boldsymbol{F}_2$ on the bearings in Figure 11.11. The equivalent loads should be less than the basic dynamic load ratings of the selected bearings in the *manufacturer catalog*.

Other similar equations to (11.32) for the evaluation of the equivalent dynamic loads produced by the axial components $\boldsymbol{F}_{a1}$ and $\boldsymbol{F}_{a2}$ and by the radial loads $\boldsymbol{P}_{l1}$ and $\boldsymbol{P}_{l2}$ are estimated differently by employing a 0.5 instead of 0.47 and factors Y_1 and Y_2 in place of K_{T1} and K_{T2}, in addition to the parameter e_B (named e by SKF); see, e.g. SKF (2017). Therefore, one is advised to follow the procedure suggested by the manufacturer. The procedure presented herein is useful as an initial synthesis and should be ensured by the process given by the bearings source of acquisition.

It should be noted that one would be advised to employ the mounting, which sets the bearing carrying the external axial load as the one that carries the least radial load as shown in Figure 11.11. This might cause the two bearing to be about the same size. The relative weight of the applied axial load $\boldsymbol{P}_a$ to the thrust or axial components

$\boldsymbol{F}_{a1}$ and $\boldsymbol{F}_{a2}$ produced by the radial loads $\boldsymbol{P}_{l1}$ and $\boldsymbol{P}_{l2}$ can play a role in that mounting decision. Computer-aided design (*CAD)* would provide means to iterate and select the best or optimum selection.

11.5.4 Unsteady Loading

When loads are fluctuating or unsteady in magnitudes ($\boldsymbol{P}_1, \boldsymbol{P}_2, \ldots, \boldsymbol{P}_n$) and during respective durations ($t_1, t_2, \ldots, t_n$ in revolutions [rev]), a value of the equivalent load $\boldsymbol{P}_{\text{le}}$ is useful to simulate the total effect of these loads. This is a mathematical modeling process that would hopefully represent the real effect of the physical conditions. The equivalent load $\boldsymbol{P}_{\text{le}}$ may then be simulated similar to bearing load life in Eqs. (11.17) and (11.20) as follows:

$$\boldsymbol{P}_{\text{le}} = \left(\frac{\sum_{i=1}^{n_P} t_i \boldsymbol{P}_i^a}{\sum_{i=1}^{n_P} t_i} \right)^{1/a_B}, \quad t_i = (t_{i,\text{hr}}\ N_{i,\text{rpm}}(60))\ [\text{rev}] \tag{11.33}$$

where n_P is the number of fluctuating or unsteady loads $\boldsymbol{P}_i$, t_i is the duration time of applied load $\boldsymbol{P}_i$ in revolutions [rev], $a = a_B$, $a_B = 3$ for ball bearings and $a_B = 10/3$ for roller bearings, $t_{i,\text{hr}}$ is the duration time of loading $\boldsymbol{P}_i$ in hours [hr], and $N_{i,\text{rpm}}$ is the rotational speed in [rpm] during the loading $\boldsymbol{P}_i$.

This equivalent load can be a radial, axial, or combined depending on the direction of the original loads $\boldsymbol{P}_1$, $\boldsymbol{P}_2, \ldots,$ and $\boldsymbol{P}_n$. The equivalent $\boldsymbol{P}_{\text{le}}$ is treated as the bearing load $\boldsymbol{P}_l$ in all previous relations.

11.5.5 Detailed Design Procedure

The procedure of rolling bearings synthesis and selection is suggested as follows:

1. From the design of mating elements, obtain the geometry and values of the loads applied to the bearing. Adjust the loads to the equivalent one $\boldsymbol{P}_e$ including different previously considered factors such as load application factor K_l (as a safety factor), expected bearings life and reliability, static loading, combined loading, unsteady loading, etc. Consider, also, the following factors such as lubrication, friction, temperature, heat dissipation, speed limit, etc., if any. These and other factors may be assessed or obtained from the manufacturer. If in doubt, use the larger safety factor range.
2. Use the estimated shaft synthesis and *bearing selection synthesis* to define the bearing nominal or *inner diameter* d_i to be safe for the equivalent load and rotational speed.
3. If the bearing is expected to be a *deep groove* ball bearing (most commonly used), calculate the load ratio limit e_B, $\boldsymbol{P}_a/V_B\boldsymbol{P}_r$, the factors X and Y, and the equivalent load $\boldsymbol{P}_e$ using given relations. If the bearing is not a deep groove ball bearing, calculate the same values, but one has to use the *manufacturer catalog*. Calculate the required *synthesis dynamic load rating* C_{DS}.
4. To speed up the selection process, the estimated *catalog* dynamic load rating C_D and the static load rating C_0 for ***02*** series of common ***deep groove*** *ball bearings* are simply given by the following projected relations:

$$\begin{aligned} C_D &= 0.2342(d_i)^{1.3128}\ [\text{kN}] \\ C_0 &= 0.0665(d_i)^{1.4922}\ [\text{kN}] \end{aligned} \tag{11.34}$$

For common ***angular contact*** *ball bearings*, the estimated *catalog* dynamic load rating C_D and the static load rating C_0 for 02 series are simply given by the following projected relations:

$$\begin{aligned} C_D &= 0.2075(d_i)^{1.3651}\ [\text{kN}] \\ C_0 &= 0.0509(d_i)^{1.5977}\ [\text{kN}] \end{aligned} \tag{11.35}$$

Real catalog values can be higher or different due to possible improvements on design, manufacturing, materials, or other factors. These relations might also be used inversely to find the expected inner diameter for the required dynamic load rating C_D and the static load rating C_0 of the ***02*** series.
Other *dimension series* should have different relations than Eqs. (11.34, 11.35).

5. Select a *bearing designation number* from the *manufacturer catalog* that satisfies the calculated C_{DS} and the static load rating C_0. Find the static and dynamic load ratings C_D and C_0 for this bearing. Calculate the extra static and dynamic safety factors or the extra reliability factor using this bearing.

This procedure is amenable to a computer code development. Some manufacturers provide similar codes at their website to aid in this selection or design process of their products. Developed MATLAB code is available at ***Wiley website***, which implements the procedure given in this text; see Section 11.9. The selection process of a *bearing designation number* from the manufacturer catalog has to be completed in any way as defined previously in step 5.

11.6 Speed Limits

Relatively high speeds in general applications can be achieved without employing any special measures. The following information can be used as a guide to determine the speed limits for some types of bearings and assuming adequate control of internal clearance, lubrication, etc.

The maximum speed a bearing should not exceed depends on several factors such as bearing design, internal clearance, lubrication, temperature, etc. Usually, these speed limits are much higher than regular applications. The regular limits are defined by the bearing inner diameter d_i in [mm] multiplied by the rotational speed N_{rpm}. This value depends on the bearing accuracy class. For regular accuracy, the speed limits are given as follows:

- *Radial ball bearings:* The range of d_i (N_{rpm}) is (250 000–400 000) depending on cadge and lubricant types. More accurate bearings can reach higher values.
- *Angular contact bearings:* The range of d_i (N_{rpm}) is (200 000–400 000) depending on cadge and lubricant types. More accurate bearings can reach higher values.
- *Tapered roller bearings:* The range of d_i (N_{rpm}) is (25 000–80 000) depending on cadge and lubricant types. Special high speed types can reach about seven times these limits.

When the application speed is close to the previous speed limit range, the individual bearing speed limit should be checked and observed. Usually ball and roller thrust bearings for high speed applications must continually be preloaded.

11.7 Lubrication and Friction

Grease is generally used at normal speed, loads, and temperature conditions. Bearings and housing are typically filled with grease up to 30–50% of the free space. Oil lubrication is used when operating conditions preclude the use of grease. Suitable oil viscosity is essential in bearing operation to reduce rolling *friction*, starting and *running torques*, induced high temperature, and power loss. The oil viscosity increases with pressure and provides a film to withstand transmitted force at contact. The rolling friction stems from the rolling sliding interface, which is the field of *elastohydrodynamics* as discussed briefly in Section 11.9. Based on these aspects, the power dissipated by lubrication and generated rolling friction in bearings can be estimated. According to conventional considerations, the selection of grease or oil is according to the following situations:

- ***Grease*** is used for low speeds, temperature below 100 [°C] or 200 [°F], and infrequent attention or worry about contamination.
- ***Oil*** is used for high speeds, higher temperatures, sealing existence, or continuous supply of lubrication.

Bearings of larger diameters and higher rotational speeds require lower viscosity lubricants. The suggested lubrication needed by the manufacturer is useful to accommodate the specific bearing type used. It provides means to evaluate starting and *running torques*, power loss, and expected temperature rise. Manufacturers may provide information about these starting and running torques, power loss, and expected temperature rise with some special bearings that have lower friction.

11.8 Mounting and Constructional Details

A typical *mounting* and *assembly* of rolling bearings is shown in a gearbox sketch of Figure 11.5. Important considerations are, however, needed for such a construction arrangement. A small clearance of about 0.25 [mm] or 0.01 [in] is needed in the outer ring track to allow for heat expansion. The typical mounting of bearings on shafts assembled in the gearbox housing is shown in Figure 11.12. Shown are two sketches of basic mountings of bearings on shafts housed in the gearbox. The simplest is the one in Figure 11.12a, where the inner rings are confined to the right and to the left by the shaft shoulders. The outer rings are confined from opposite sides by the housing covers leaving a small gap of 0.2–0.3 [mm] or 0.008–0.012 [in] for possible expansion. The second mounting in Figure 11.12b is firmly holding the inner and outer rings of the right bearing. The outer ring is tightly confined between the housing and the cover cap, where the cover flange is left agape to effective tightening. The inner ring is firmly held to shaft by a special nut in Figure 11.12b. This nut is a special *locknut* with a special *lock washer* accessible as a standard for each bearing. The left bearing is relatively free. The inner ring may be restricted to further movement using a retaining ring on the shaft; see Section 8.2.2 and Appendix A.5. This is not shown in Figure 11.12b since the outer ring is confining the bearing to any further move. No essential need for duplicate arrangements.

Figure 11.13 shows few sketches of bearing *mounting* and *assembly* needs. Assembly of bearing *locknut* and *lock washer* is in (a) with a shaft neck or a recess to accommodate bearing fillet. The neck or the recess can be replaced by a smaller fillet than the bearing fillet, which is usually 0.6–2.0 [mm] (0.02–0.08 [in]) for bearings inner diameter range of 10–100 [mm] (0.4–4 [in]). Figure 11.13b depicts some details of a quarter locknut and lock washer. The inner tongue of the lock washer goes in the specially milled shaft slot as seen in Figure 11.13a. The outer tongues can be in much more replicas than in Figure 11.13b. One of these outer tongues is bent in one of the encountered outer slots of the locknut after sufficiently tighten the locknut. The necessary tightening options of *tapered roller*

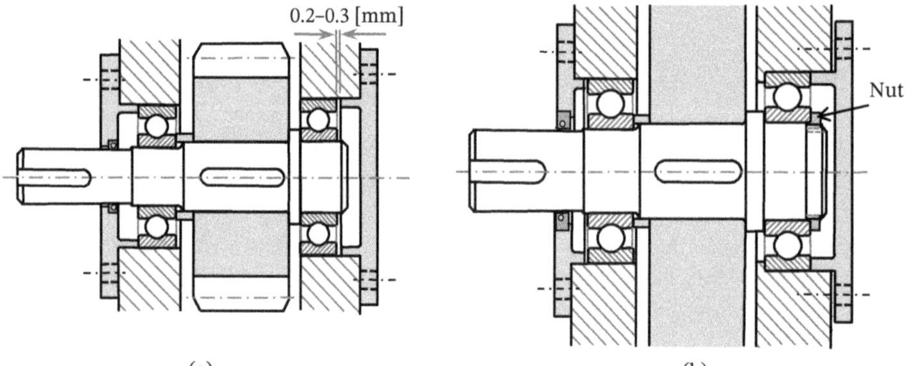

Figure 11.12 Sketches of basic mountings of bearings on shafts in housings. The simplest is the one in (a), where the inner rings are confined by shaft shoulders. The outer rings are confined by the housing covers leaving a small gap for possible expansion. The second mounting in (b) is firmly holding the inner and outer rings of the right bearing. The left bearing is relatively free.

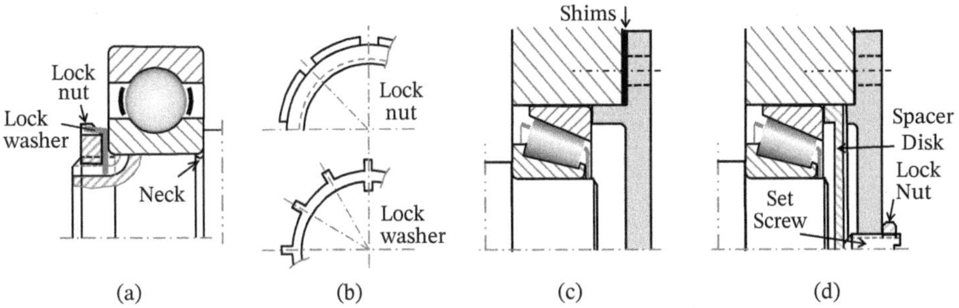

Figure 11.13 Sketches of bearing mounting or assembly needs. Assembly of bearing locknut and lock washer is in (a) with a shaft neck or a recess to accommodate bearing fillet. The details of a quarter locknut and lock washer are in (b). The necessary tight holding options of tapered roller bearings are shown in (c,d) for the right bearing of Figure 11.12b.

bearings are shown in Figure 11.13c,d for the right bearing of Figure 11.12b. The first option in Figure 11.13c is using measured shims to eliminate the internal *clearances* of the right and left bearings of Figure 11.12b or to produce a slight preloading. The second option in Figure 11.13d is to use a loose *spacer disk* between the bearing and the housing cover without being pushed by the cover. A set screw is attached to the cover and pushes the spacer disk to produce the necessary elimination of the bearings clearances or produce a slight preloading. The set screw is locked with the thin *regular locknut* (not those specifically for bearings); see Section 9.1.2.

Other mounting arrangements are available in numerous assemblies of products. The ones presented herein are simple few alternatives. It would be advisable to find other alternatives in available designs.

11.9 Computer-Aided Design and Optimization

Rolling bearings are special machine components made of rolling elements between two races. The rolling bearings are to be selected, and not to be designed in the sense of defining materials, numbers, and dimensions of internal rolling parts or groove forms. The design and manufacturing of these are the responsibilities of the manufacturers, and the users are to select from the rolling bearings offered by the manufacturer. The dimensions are almost always the same due to the standards. The load ratings are usually dependent on the manufacturer, since they use different materials and internal designs. The differences are not much, but they cannot be identical, particularly, if they use different evaluation procedures and calculation coefficients. The ***CAD*** is then restricted to getting close to whatever load ratings, which the manufacturers list. Following the calculation procedure of the manufacturer is necessary for selection. Checking the further safety factors between the designer values and the manufacturer values provides some extra reliabilities in rolling bearing performance. The main task of the ***CAD*** code is then to provide extensive assistance in that regard and allow for possible iterations toward optimization.

The *CAD* codes in this text use the same characters of the parameters employed in the text. The subscripts, however, are not present in some coding tools, and the subscripts are replaced by regular font. This might ease the identification of parameters. To differentiate between the US and SI systems, either the units or either letters of the US or SI are attached to the parameter after an underscore character.

11.9.1 Initial Ball Bearing Synthesis

Figure 11.14 is a MATLAB code for initial synthesis of rolling ball bearings. The input is emphasized by bold capital letters in the comment statements of the code (% **INPUT**). Few more inputs are entered more than directly needed in this figure. This is just to have some completeness in the entered information set. In addition to the radial load $\boldsymbol{P}_l$, the load application factor K_l, the rotational speed N_{rpm}, the bearing life L_{Bhr} in hours, and the rating life L_{10},

```
clc; clear all; format compact; format short;      % CAD_Rolling_Bearings_Selection.m
Pl_kN = 25.330                                   % INPUT Bearing radial load [kN]
Pl_klb = 5.574                                   % INPUT Bearing radial load [klb]
Pla_kN = 10.0                                    % INPUT Bearing axial load [kN]
Pla_klb = 2.2                                    % INPUT Bearing axial load [klb]
Kl = 1.3                                         % INPUT Load application factor (safety)
Nrpm = 6                                         % INPUT Rotational speed [rpm]
LBhr = 10000                                     % INPUT Required bearing life [hr]
L10 = 10^6                                       % INPUT Rating life [rev] 0.9 Reliability or (90%)

LBrev = LBhr*Nrpm*60                                 % Bearing life in revolutions [rev]
Nrpm10 = Nrpm/(LBrev/L10)                            % Adjust rotational speed [rpm] to 10^6 [rev]
         disp('Estimated Bearing diameter for L10, di_mm [mm], di_in [in]')
di_mm = (Pl_kN*1000/(76.19085*Nrpm10^-0.24026))^(1/1.64674)       % Scope 0.6-60000 [rpm]
di_in = di_mm/ 25.4
Pl_kN = Pl_kN* Kl                                    % Adjust load for application factor
         disp('Estimated Bearing diameter with Application Factor Kl, di_mm[mm], di_in[in]')
di_mm = (Pl_kN*1000/(76.19085*Nrpm10^-0.24026))^(1/1.64674)       % Scope 0.6-60000 [rpm]
di_in = di_mm/ 25.4

% Pl_kN = sqrt(Pl_kN^2+Pla_kN^2)* Kl                 % Adjust for resultant and load application factor
% di_mm = (Pl_kN*1000/(76.19085*Nrpm10^-0.24026))^(1/1.64674)     % Scope 0.6-60000 [rpm]
% di_in = di_mm/ 25.4
```

Figure 11.14 MATLAB code for initial synthesis of rolling bearings. The input is emphasized by bold capital letters in the comment statements. The output is the estimated synthesized ball bearing inner diameter. It is adjusted for the input load application factor.

the axial load $\boldsymbol{P}_a$ is added to the input. For these variables, the computer code uses `Pl`, `Kl`, `Nrpm`, `LBhr`, `L10`, and `Pa`, respectively. The output is the estimated synthesized ball bearing inner diameter d_i. It is adjusted for the input load application factor K_l. The estimated inner diameter is calculated first for the rating life L_{10} only. The designated bearing life is adjusted to L_{10} by adjusting the rotational speed N_{rpm} to an adjusted new L_{10} rotational speed of $N_{\text{rpm10}} = N_{\text{rpm}}/(L_{\text{Brev}}/L_{10})$. The estimated inner diameter equation is based on a life of 10 000 hours. The estimated *inner diameter* d_i in [mm] (`di_mm` in the code) is calculated as a function of the usual applied *radial load* $\boldsymbol{P}_l$ and the L_{10} *rotational speed* N_{rpm10}, i.e. by a *hypersurface* in the following form:

$$d_i = \frac{\boldsymbol{P}_l}{(76.190\,85(N_{\text{rpm10}})^{-0.240\,26})^{1/1.646\,74}} \text{ [mm]} \tag{11.36}$$

This function, hypersurface, or meta-model is approximating the usual bearing performance of Figure 11.6a. The trend lines in this figure have about the same inner diameter exponent, which is averaged to 1.646 74. The effect of the rotational speeds is estimated by the exponential function of N_{rpm10}. The scope or rotational speed N_{rpm} is considered in the range 0.6–60 000 [rpm]. This is a larger scope than many bearing diameters can stand; see Figure 11.6. The accuracy is within the design engineering range. It gives an approximate initial estimation of the inner diameter that can be beneficial in a computer code instead of using Figure 11.6.

Then the procedure includes the load application factor and adjusting the estimated synthesis of the bearing inner diameter; see code in Figure 11.14. This diameter should be changed to the higher round figure diameter, which is provided in the standards and offered by the manufacturer. An option to include the load resultant is allowed by uncommenting the code lines for that; see the end of Figure 11.14. This might be useful in angular contact ball bearings. The name of the MATLAB code is ***`CAD_Rolling_Bearings_Selection.m`***. Figure 11.14 is a part of the code, and Figures 11.15–11.17 are continuing parts of the same code.

Figure 11.15 is a continuation part of the MATLAB code for initial synthesis of rolling ball bearings. Again, the input is emphasized by bold capital letters in the comment statement. The output is the estimated synthesized ball bearing inner diameter d_i using Eq. (11.36). It is adjusted for speed, reliability, and application factor. The estimated inner diameter is calculated for the *Weibull* three-parameter valuation of reliability. The basic *Weibull* distribution parameters x_0, b_W, and θ_W (`x0`, `bW`, and `thetaW` in the code) need to be entered as inputs. The bearing reliability R_B (`RB`) is calculated for the supplied or entered bearing life in hours L_{Bhr} and rotational speed N_{rpm}. These are used in evaluating bearing life L_{Brev} (`LBrev`) in revolutions [rev]. The *Weibull* distribution uses the

```
        disp('Estimated Bearing Reliability for LBM [Mrev]') % CAD_Rolling_Bearings_Selection.m
x0 = 0.02                                        % INPUT Guaranteed minimum value of x
bW = 1.483                                       % INPUT Weibull shape parameter (or 1.17)
thetaW = 4.459                                   % INPUT Weibull characteristic value (or 6.86)
x = LBrev/L10                                         % Weibull life measure (multiplier)
RB = exp(-((x-x0)/(thetaW-x0))^bW);                   % Calculated Bearing reliability
Reliability = RB
        disp('Estimated bearing synthesis for adjusted life of L10S, di_mm[mm], di_in[in]')
LBhr = 10000                                     % INPUT Required bearing life [hr]
RB = 0.9;                                        % INPUT Required Reliability
% RB = 0.99;
LB = LBhr*Nrpm*60                                     % Bearing life [rev]
Reliability = RB
L10S = LB/(x0+(thetaW-x0)*(log(1/RB))^(1/bW))         % Required Life for input Reliability
% L10S = LB/(4.439*(log(1/RB))^(1/bW))                % Approximate L10S
        % disp('Adjust Initial Synthesis For Application Factor, Reliability, and Life')
if RB==0.9; LBS = LB; else; LBS = L10S; end           % Adjust bearing life for reliability factor
aB = 3                                           % INPUT aB = 3 (Ball), aB=10/3 (Roller)
P1S_kN = P1_kN* (LBS/10^6)^(1/aB)                     % Adjust load for reliability factor
P1S_klb = P1_klb* (LBS/10^6)^(1/aB)                   % Adjust load for reliability factor

di_mm = (P1S_kN*1000/(76.19085*Nrpm10^-0.24026))^(1/1.64674); % Scope 0.6-60000 [rpm]
Initial_Bearing_Inner_Diameter_mm = di_mm
Initial_ Bearing_Inner_Diameter_in_Reliability = di_mm/25.4
```

Figure 11.15 MATLAB code for initial synthesis of rolling bearings. The input is emphasized by bold capital letters in the comment statement. The output is the estimated synthesized ball bearing inner diameter. It is adjusted for speed, reliability, and application factor.

```
                                                      % CAD_Rolling_Bearings_Selection.m
        disp('Bearing Synthesis for Required Synthesis Dynamic Load Rating CDS')
% P1_kN = 25.330                                 % INPUT Bearing radial load [kN]
% P1_klb = 5.574                                 % INPUT Bearing radial load [klb]
% P1a_kN = 10.0                                  % INPUT Bearing axial load [kN]
% P1a_klb = 2.2                                  % INPUT Bearing axial load [klb]
aB = 3                                           % INPUT aB = 3 (Ball), aB=10/3 (Roller)
VB = 1.0                                         % INPUT VB = 1.0 (inner ring rotation)
VB = 1.2                                         % INPUT VB = 1.2 (outer ring rotation)
RB = 0.9;                                        % INPUT Required Reliability
L10S = LB/(x0+(thetaW-x0)*(log(1/RB))^(1/bW))         % Required Life for a Reliability
if RB==0.9; LB = LB;
else; LB = L10S; end                                  % Scaling LB to Reliability change
P1a_kN = K1*P1a_kN                                    % Adjust Bearing axial load [kN]
P1a_klb = K1*P1a_klb                                  % Adjust Bearing axial load [klb]
P1_kN = K1*P1_kN                                      % Adjust Bearing radial load [kN]
P1_klb = K1*P1_klb                                    % Adjust Bearing radial load [klb]
P1R_kN = sqrt(P1_kN^2 + P1a_kN^2)                       % Resultant load [kN]
P1R_klb = sqrt(P1_klb^2 + P1a_klb^2)                    % Resultant load [klb]

CDS_kN = P1_kN*(LB/10^6)^(1/aB)                         % Adjust Synthesis load rating [kN]
CDS_klb = P1_klb*(LB/10^6)^(1/aB)                       % Adjust Synthesis load rating [klb]
        disp('Considering Resultant Load for Angular Contact Bearing Synthesis CDRS')
CDRS_kN = P1R_kN*(LB/10^6)^(1/aB)                       % Adjust Synthesis load rating [kN] Resultant
CDRS_klb = P1R_klb*(LB/10^6)^(1/aB)                     % Adjust Synthesis load rating [klb] Resultant
```

Figure 11.16 MATLAB code for required synthesis of rolling bearings. The input is emphasized by bold capital letters in the comment statement. The output is the estimated synthesis ball bearing dynamic load rating. It is adjusted for speed, reliability, and application factor.

entered parameters and the calculated life measure $x = L_{Brev}/L_{10}$ to calculate the expected bearing reliability R_B. This is just to know the expected failure if no attention to life and rotational speed are considered. The effect of life and speed are then taken into consideration by calculating the needed synthesis rating life L_{10S} (`L10S`) to scale the *Weibull* distribution for any input reliability needed as given by Eq. (11.8). The estimated inner diameter is then calculated for the *Weibull* three-parameter reliability, speed, and application factor.

This progression of the calculated estimated inner diameter presents a useful tool to evaluate the effect of each parameter on the estimated inner diameter. It affects the understanding of the developed synthesis dynamic load rating C_{DS} (`CDS` in the code) needed to select the specific ball bearing off the *manufacturer catalog*. This process is to be defined next.

```
                                                              % CAD_Rolling_Bearings_Selection.m
          disp('Bearing Selection for Basic Dynamic and Static Load Ratings CD and C0')
Selected_Bearing='   SKF 6310'            % INPUT Selected Bearing Number 'SKF 6310'
C0 = 38                                   % INPUT Static load rating [kN] SKF 6310
CD = 65                                   % INPUT Dynamic load rating [kN] SKF 6310
C0_US = C0*0.2248089                      % INPUT Static load rating [klb] SKF 6310
CD_US = CD*0.2248089                      % INPUT Dynamic load rating [klb] SKF 6310
eB = 0.5065*(Pla_kN/C0)^0.232                       % Load ratio limit (radial ball bearing)
eB_US = 0.5065*(Pla_klb/C0_US)^0.232                % Load ratio limit (radial ball bearing)
          disp('Resultant Static Safety Factor of Selected Bearing with Static Load Rating C0')
KSF0R_SI = C0/P1R_kN                                % Static safety factor (Resultant load)
KSF0R_US = C0_US/P1R_klb                            % Static safety factor (Resultant load
          disp('Combined Static Safety Factor of Selected Bearing with Static Load Rating C0')
Pe0_kN = 0.6*Pl_kN+ 0.5*Pla_kN
if Pe0_kN < Pl_kN; Pe0_kN = Pl_kN; end
Pe0_klb = 0.6*Pl_klb + 0.5*Pla_klb
if Pe0_klb < Pl_klb; Pe0_klb = Pl_klb; end
KSF0_SI = C0/Pe0_kN                                 % Static safety factor (SI)
KSF0_US = C0_US/Pe0_klb                             % Static safety factor (US)
          disp('Combined Safety Factor for Selected Bearing with Dynamic Load Rating CD')
X = 0.56                                  % INPUT Radial load factor
Y = 0.8632*(Pla_kN/C0)^-0.2349;                     % Estimated Axial load factor
Y = Y                                     % INPUT Axial load factor
Pe_kN = X*VB*PlS_kN+ Y*Pla_kN
if Pe_kN< VB*PlS_kN; Pe_kN = VB*PlS_kN; end
Pe_klb= 0.6*PlS_klb+ 0.5*Pla_klb
if Pe_klb< VB*PlS_klb; Pe_klb = VB*PlS_klb; end
KSFD_SI = CD/Pe_kN                                  % Combined load safety factor (SI)
KSFD_US = CD_US/Pe_klb                              % Combined load safety factor (US)
```

Figure 11.17 MATLAB code for rolling bearings selection safety. The input is emphasized by bold capital letters in the comment statement. The output is the estimated ball bearing safety. It is for static and dynamic rating of selected bearing.

11.9.2 Dynamic Load Rating Estimate

Figure 11.16 presents part of the MATLAB code for required synthesis of rolling bearings. Also, the input is emphasized by bold capital letters in the comment statement. The bearing load-life exponent a_B (aB) is an input of 3 for ball bearings and $a_B = 10/3$ for roller bearings. The input bearing rotation factor V_B (VB) is equal to 1.0 for inner ring rotation and 1.2 for outer ring rotation. The recalculation of the loads with the consideration of a different radiality is important to the better definition of the synthesis dynamic load rating C_{DS} (CDS). The load is adjusted for speed, reliability, and application factor, which causes the synthesis dynamic load rating C_{DS} (CDS) to be used for the selection of the specific bearing designation number from a definite manufacturer. The output is that synthesis dynamic load rating (CDS) for adjusted radial loading and (CDSR) for the resultant loading that may be useful in angular contact ball bearings.

11.9.3 Ball Bearing Selection

Figure 11.17 provides the part of the MATLAB code for rolling bearings selection safety. The input is also emphasized by bold capital letters in the comment statement. The input pertains to the selection of the specific bearing designation number from a definite manufacturer. This entails the definition of the dynamic load rating C_D (CD) and the static load rating C_0 (C0), which to be the input variables to the ***CAD*** program for check. The other inputs are the radial and axial load factors X and Y, even though the X is usually constant at 0.56 and Y has been approximated as given by Eq. (11.29) for radial load ball bearings. One can also change the value of X or Y to the values given in the *manufacturer catalog*. The load ratio limit e_B (eB) has also been estimated by Eq. (11.26), but one can change it from the manufacturer catalog. The output variables are the equivalent load $\boldsymbol{P}_e$ (Pe) to the applied radial and axial loading and the estimated ball bearing safety factors K_{SF0} (KSF0) and K_{SFD} (KSFD). These are for static and dynamic ratings of the selected bearing, respectively. Safety factors are expected to be around or little over the value 1.0. This is an extra safety over the load application factor, which is considered as a safety factor. The selected bearing should have, alternatively, a little more reliability than the required input value.

11.9.4 Rolling Bearing Optimization

Optimization of rolling bearing selection is different from optimization of rolling bearing design. The *selection optimization* would be concerned about bearing size, cost, friction, accuracy, and bearing type. The selection is the main consideration of the user designing machines or products that employ rolling bearings. The rolling bearing design optimization is the burden on the manufacturers. Herein, the concern is about selection optimization.

In previous examples, few bearing alternatives are selected to accommodate loading requirements. The size or the inner diameter values and dimension series were the main parameters. The bearing volume or cost can be introduced in the optimization to get a better selection. With the same bearing diameter, using a higher series than 02, i.e. 03, causes an increase in basic dynamic rating C_D and boundary volume due to increases in width and outer diameter; see Figure 11.4. With about the same acceptable basic dynamic load rating C_D, using series 02 necessitates the change of the inner diameter to 70 [mm] with a C_D of about 64 [kN] instead of 50 [mm] diameter of series 03 with a C_D of 65 [kN]. The dimensions of the 02 series are $d_i = 70$ [mm], $d_o = 125$ [mm], and the width $b_B = 24$ [mm]. The dimensions of the 03 series with a C_D of 65 [kN] are $d_i = 50$ [mm], $d_o = 110$ [mm], and $b_B = 27$ [mm]. Going to a higher 04 series with C_D about 64 [kN], the dimensions become $d_i = 40$ [mm], $d_o = 110$ [mm], and the width $b_B = 27$ [mm]. It seems that series 03 might have a smaller volume and maybe at a lower cost, which must be verified. A computer program can be simply devised to get the optimum. However, it requires the input of many properties of bearings that necessitates the link to the *manufacturer catalog*. The other option is resorting to some estimation functions of the basic dynamic load rating C_D with the inner diameter such as Eqs. (11.34, 11.35) and Eqs. (11.1, 11.2) for outer diameter and width. The parameters would be rounded to some integers, which might need an integer programming optimization process. This is outside the scope of this text.

Even though, the rolling bearing design is a burden on the manufacturer, many academic researchers have attempted optimization. The usual field is the elastohydrodynamics. In that, the most important bearing parameters that have the major and final decision concerning the bearing selection is the bearing's fatigue life. In these research efforts, the wear and fatigue using the elastohydrodynamic theory maximizes life of the bearing. In addition, adequate amount of oil can provide damping to the moving parts under cyclic load and leads to complete separation between the moving parts. These efforts include factors that can address these issues; see, e.g. Abbas and Metwalli (2011).

Example 11.7 Redo Examples 11.1, 11.2, and 11.4–11.6 using the MATLAB code for bearing synthesis and selection. Select a deep groove ball bearing to withstand a radial force of 25.33 [kN] or 5.574 [klb] and axial force of 10.0 [kN] or 2.2 [klb] with a life of 10 000 hours [hr] and rotating at a speed of 0.1 [rps]. Find the basic dynamic load rating you should use to select the bearing from the manufacturer catalog. Find, also, the basic dynamic load rating you should use to select the bearing from the manufacturer catalog if reliability should be 99%.

Solution

Data: Initially $\boldsymbol{P}_l = 25.33$ [kN] or 5.574 [klb], $\boldsymbol{P}_a = 10.0$ [kN] or 2.2 [klb], $L_B = 10\,000$ [hr], $N_{rpm} = 0.1\,(60) = 6$ [rpm], and $R_B = 0.9$ or 0.99.

Using the MATLAB code, the said loads, the rotating at a speed, and reliability, the code output for each example is given as follows:

Example 11.1:

- *Input*: `Pl_kN =25.3300 [kN], Pl_klb = 5.5740 [klb], Pla_kN = 10 [kN], Pla_klb = 2.2000 [klb], Kl = 1.3000 Nrpm = 6 [rpm], LBhr = 10000 [hr], L10 = 1000000 [rev].`
- *Output*: `LBrev = 3600000 [rev], LBM = 3.6000 [Mrev], Nrpm10 = 1.6667 [rpm].` The change in the rotational speed is done to get the bearing rating life at 10^6, i.e. L_{10} revolutions as usual bearings.

The inner diameters without including `K1` are as follows:

`di_mm = 36.6198` [mm], `di_in = 1.4417` [in]

(The loads and the inner diameters with the inclusion of the load application factor `K1` gives `P1S_kN = 32.9290 [kN], P1S_klb = 7.2462 [klb], and di_mm = 42.9447 [mm], di_in = 1.6907 [in]`. These were not required by the example.)

These values are close to the inner diameters found from Figure 11.6 and reported as $d_i = 40$ [mm] or 1.5 [in]. The difference is due to estimate Eq. (11.36) and the resolution of Figure 11.6. The output diameters of the computer code should be rounded off to the higher values reported by Example 11.1. The computer code does not offer the synthesis of the thrust bearings. It is left to the interested to do that or approximate the thrust bearing as a radial one, which is not that accurate.

Example 11.2: Note that the input has to be changed by uncommenting both statements of reliability `RB = 0.99` and the next approximate equation for `L10S`.

- *Input*: `LBhr = 10000 [hr], RB = 0.9900`
- *Output*: `LB = 3600000 [rev], L10S = 1.6394e+07 [rev]`
  For approximate Eq. (11.9), `L10S = 1.8037e+07 [rev]`

These values are the same as those of Example 11.2. Note that the input was changed by uncommenting the reliability `RB = 0.99` and the approximate equation for `L10S`. They are recommented after this run of the code:

- *Input*: Changing the bearing life to `LBhr =2778 [hr]` in the code
- *Output*: `L10S = 4.5510e+06 [rev]`, and the approximate `L10S = 5.0071e+06`

These are close to the values of Example 11.2. Note that the input was changed by changing `LBhr`, uncommenting the reliability `RB = 0.99`, and uncommenting the approximate equation for `L10S`. They are changed back to `LBhr = 10000 [hr]` recommented after this run of the code.

Example 11.4:

- *Input*: Same as Example11.1 and `aB = 3.`
- *Output*: `P1S_kN = 38.8212 [kN], P1S_klb = 8.5428 [klb]` (which are very close to results of Example 11.3). The suggested inner diameters are as follows: `Initial_Bearing_Inner_Diameter_mm = 47.4595 [mm], Initial_Bearing_Inner_Diameter_in = 1.8685 [in]` (which are larger inner diameters than discussed in Example 11.4. The example suggests that the option of using a heavier bearing series 03 and keeping the inner diameter might be the solution.)
  Radiality of 0.99 was not attempted in this run. It is left for the user to check.

Example 11.5:

- *Inputs*: Same as Example 11.1 and `aB = 3.`
- *Output*: `P1R_kN = 35.4022 [kN], P1R_klb = 7.7902 [klb], CDS_kN = 50.4676 [kN], CDS_klb = 11.1057 [klb]` (which are about the same as Example 11.5. The example suggests the increase in inner diameter to 50 [mm] or 2.0 [in] as evident with Example 11.4. The code did not suggest inner diameters due to variation in diameter estimation conditions.) The example selects an SKF 6310 bearing for this synthesis dynamic load rating `CDS`.
  Continue with the selection of the SKF 6310 bearing, the following was the code requirement and results:
- *Input*: Continue from before and `aB = 3, VB = 1.2000`
  `Selected_Bearing = SKF 6310, C0 = 38 [kN], CD = 65 [kN]`

- *Output*: `C0_US = 8.5427 [klb], CD_US = 14.6126 [klb], Pla_kN = 13 [kN], Pla_klb = 2.8600 [klb], Pl_kN = 32.9290 [kN], Pl_klb = 7.2462 [klb], PlR_kN = 35.4022 [kN], PlR_klb = 7.7902 [klb]`
  `eB = 0.3949, eB_US =0.3929, KSFOR_SI = 1.0734, KSFOR_US = 1.0966`
  The static safety factor and its needed variables have been given as follows:
  `Pe0_kN = 26.2574 [kN], Pe0_klb = 5.7777 [klb], KSF0_SI = 1.1540, KSF0_US = 1.1789`
  All these are almost the same as Example 11.5.

Example 11.6:

- *Input*: Same as before, and `X = 0.5600, Y = 1.1105` (it is also an output)
- *Output*: `Y = 1.1105, Pe_kN = 36.5654 [kN], Pe_klb = 5.7777 [klb],KSFD_SI = 1.6450, KSFD_US = 1.6805`
  All these are almost the same as Example 11.6.

Developed MATLAB code is available at ***Wiley website***, which implements the procedure given in this text. The name of the program is ***`CAD_Rolling_Bearings_Selection.m`***. It proceeds about the same as Example 11.7, with the inputs and outputs result of running the program. Anyway, the selection of a *bearing designation number* from the *manufacturer catalog* has to be completed as demonstrated previously. Verification can be performed by the software of the manufacturer, if possibly available in their website or downloadable.

11.10 Summary

This chapter was mainly dedicated to rolling bearings synthesis and selection. Initial synthesis of common rolling bearing is provided by synthesis charts. Detailed design is performed by a *CAD* code that would also need the selection of a bearing designation number from a manufacturers catalog to calculate the selected bearing safety factor.

Problems

11.1 Can you find and download some manufacturers catalogs for ball and roller bearings? Use the Internet references or any other search options.

11.2 Search for the possible construction to accommodate a deep groove ball bearing only and both radial and thrust ball bearings in the *Jib crane* of Example 11.1. Draw a proportional sketch or a ***3D*** model of each selected construction. What should be the construction, if an angular contact bearing is used? What angle should be selected for the ball contact with the races or rings? Compare the standard details with the information provided by a manufacturer.

11.3 Justify the use of L_{10S} in place of L_{10} in Eq. (11.7).

11.4 Find the three- or two-parameter *Weibull* distribution set (x_0, θ_W, b_W) that are adapted by different manufacturers, if any. How close is that set to those used in the text? Compare results of the life measure x for reliabilities of 0.9, 0.95, 0.99, 0.995, and 0.999.

11.5 Calculate the reliability of the selected radial ball bearing of Example 11.1. What should be the assumed load and synthesis diameter to have a 90%, 99%, 99.5%, and 99.9% reliability?

11.6 Recalculate Example 11.2 if the reliability is 99.5%.

11.7 Prove Eq. (11.14) by substituting the relations of Eq. (11.2) into Eq. (11.13).

11.8 Calculate the expected dynamic load rating C_D of a deep groove ball bearing having inner diameters of 40, 50, and 60 [mm] for different number of balls using Eq. (11.16). Can you find a manufacturer that defines the number of balls in his catalog of deep groove ball bearings? Compare results or find the expected number of balls in a manufacturer that does not define the number of balls in his catalog.

11.9 If you use Timken load rating method, what should be the rated life you search for to solve Example 11.2?

11.10 If Example 11.4 is a Jib crane working in a workshop, what is the estimation of more suitable life and usual rotating speed than assumed in Example 11.4? Use suggested life ranges stated in Section 11.5.

11.11 Estimate the bearing inner diameter for the cases of Example 11.4.

11.12 Recalculate Example 11.5 if reliability should be 99%. Select more appropriate bearing designation number. Continue with recalculations to include Example 11.6.

11.13 Redesign the bearing in Example 11.5 assuming angular contact bearing in place of the deep grove ball bearing. What is the difference in procedure, parameters, and equations suggested by the manufacturer than those for deep grove ball bearing?

11.14 For the redesign of the Jib crane with the same dimensions as Example 8.2, but for a maximum load of 30 [kN] or 6.6 [klb], synthesize a radial ball bearing in the hood to carry the load. Consider a life of 20 000 hours [hr] and rotating at a speed of 0.1 [rps].

11.15 If the crane reach or the active boom length is 5 [m] or 200 [in] for the Jib crane of Example 8.2, what should be the suitable synthesis of the ball bearing? Consider a life of 30 000 hours [hr] and rotating at a speed of 0.1 [rps].

11.16 Redesign the bearing of the Jib crane of Problem 11.14, but for the crane reach or the active boom length of 5 [m] or 200 [in].

11.17 Use few bearings in a manufacturers catalog to estimate the axial load factor Y relation similar to Eq. (11.29) for some different types of bearings or dimension series.

11.18 If reliability should be 99% in Example 11.6, what should be the bearing designation number of a selected manufacturer catalog?

11.19 Estimate the *catalog* dynamic load rating C_D and the static load rating C_0 similar to Eq. (11.34) for 01, 02, 03, or 04 series of common deep groove ball bearings from a manufacturer catalog using few preferred inner diameters and covering a reasonable range of dimensions.

11.20 Estimate the *catalog* dynamic load rating C_D and the static load rating C_0 similar to Eq. (11.35) for the available 01, 02, 03, or 04 series of angular contact ball bearings of different angles from a manufacturer catalog using few preferred inner diameters and covering a reasonable range of dimensions.

11.21 Estimate the *catalog* dynamic load rating C_D and the static load rating C_0 similar to Eqs. (11.34, 11.35) for any of the available 01, 02, 03, or 04 series of any of the rolling bearings of interest from a manufacturer catalog using few preferred inner diameters and covering a reasonable range of dimensions.

11.22 Explore different constructional details of bearing mountings similar to those shown in Figures 11.12 and 11.13. Draw proportional sketches or ***3D*** models of each selected construction.

11.23 Download available constructional details of bearing mounting applications similar to those shown in Figures 11.12 and 11.13 provided by some rolling bearing manufacturer.

11.24 Explore different constructional details of Jib-crane bearing mounting options for the hood deep groove ball bearing, angular contact bearing, or tapered roller bearings. Synthesize the attractive alternative for the load and the reach stated in any one of the previous problems.

11.25 Adjust the MATLAB code of Figure 11.17 to synthesize the tapered roller bearing sets of Figure 11.11. Use any necessary part of Figures 11.14–11.16.

11.26 Use the available MATLAB code `CAD_Rolling_Bearings_Selection.m` to verify any of the previously applicable problems.

11.27 Use the available MATLAB code `CAD_Rolling_Bearings_Selection.m` to solve Example 11.4 for 99% reliability.

11.28 Adjust the available MATLAB code `CAD_Rolling_Bearings_Selection.m` to account for unsteady loading defined in Section 11.5.4.

11.29 Adjust the available MATLAB code `CAD_Rolling_Bearings_Selection.m` to optimize the selection of radial ball bearings.

11.30 Find the rolling bearings manufacturer that provides software in their website to help in the selection of rolling bearing designation number. Check some of the previous problems utilizing available software.

References

Abbas, M.H., Youssef, A.M.A., and Metwalli, S.M. (2010). Ball bearing fatigue and wear life optimization using elastohydrodynamic and genetic algorithm. *Proceedings of the ASME 2010 International Design Engineering Technical Conferences & Computers and Information in Engineering Conference IDETC/CIE 2010*, Montreal, QC, Canada (15–18 August), DETC2010-28849. American Society of Mechanical Engineers.

Abbas, M.H. and Metwalli, S.M. (2011). Elastohydrodynamic ball bearing optimization using genetic algorithm and heuristic gradient projection. *Proceeding of the ASME 2011 IDETC/CIE*, Washington, DC, USA (29–31 August), Paper # DETC2011-47624. American Society of Mechanical Engineers.

ANSI/ABMA 11 (2014). *Load ratings and fatigue life for roller bearings*. American National Standards Institute/American Bearing Manufacturers Association.

ANSI/ABMA 19.2 (2013). *Tapered roller bearings, radial inch design*. American National Standards Institute/American Bearing Manufacturers Association.

ANSI/ABMA 20 (2011) (1987). *Radial bearings of ball cylindrical roller and spherical roller types metric design*. American National Standards Institute/American Bearing Manufacturers Association.

ANSI/ABMA 7-(S2013)(1995). *Shaft and housing fits for metric radial ball and roller bearings (except tapered roller bearings) conforming to basic boundary plan*. American National Standards Institute/American Bearing Manufacturers Association.

ANSI/ABMA 9 (2015). *Load ratings and fatigue life for ball bearings*. American National Standards Institute/American Bearing Manufacturers Association.

Gohar, R. (2003). *Elastohydrodynamics*, 2e. Imperial College Press.

Harris, T. and Kotzalas, M. (2006). *Rolling Bearing Analysis, Advanced Concepts of Bearing Technology*, 5e. Taylor & Francis, CRC Press.

Harris, T.A. (1963). *Predicting Bearing Reliability. Machine Design* 35 (1): 129–132.

ISO 104 (2015). *Rolling bearings – thrust bearings – boundary dimensions, general plan*. International Organization for Standardization.

ISO 15 (2017). *Rolling bearings – radial bearings – boundary dimensions, general plan*. International Organization for Standardization.

ISO 199 (2014). *Rolling bearings – thrust bearings – geometrical product specification (GPS) and tolerance values*. International Organization for Standardization.

ISO 281 (2007). *Rolling bearings – dynamic load ratings and rating life*. International Organization for Standardization.

ISO 286-2 (2010). *Geometrical product specifications (GPS) – ISO code system for tolerances on linear sizes – part 2: tables of standard tolerance classes and limit deviations for holes and shafts*. International Organization for Standardization.

ISO 355 (2019). *Rolling bearings – tapered roller bearings – boundary dimensions and series designations*. International Organization for Standardization.

ISO 492 (2014). *Rolling bearings – radial bearings – geometrical product specifications (GPS) and tolerance values*. International Organization for Standardization.

ISO 76 (2006). *Rolling bearings – static load ratings*. International Organization for Standardization.

Mischke, C.R. (1986). *Rolling contact bearings*. In: *Standard Handbook of Machine Design* (eds. J.E. Shigley and C.R. Mischke), 27.12. McGraw Hill.

Neale, M.J. (ed.) (1973). *Tribology Handbook*. Butterworths.

NSK (2017). *Rolling Bearings for Industrial Machinery Cat. No. E1103 NSK*.

NTN (2009). Ball and Roller Bearings Cat. A-1000-XI NTN.

Oberg, E., Jones, F.D., Horton, H.L., and Ryffel, H.H. (2012). *Machinery's Handbook*, 29e. Industrial Press.

SKF (2018). *Rolling Bearings Catalogue 17000 SKF*.

Timken (2014a). *Timken Ball Bearings Catalog*. The Timken Company.

Timken (2014b). *Timken Tapered Roller Bearing Catalog*. The Timken Company.

Timken (2017). *Timken-Engineering-Manual 10424*. The Timken Company.

Internet

http://www.ntnamericas.com/en NTN: Manufacturer of rolling bearings.

https://www.skf.com/group/splash/index.html SKF: Manufacturer of rolling bearings, and others, etc.

www.timken.com Timken: Manufacturer of rolling bearings, chains, seals.

http://www.nsk.com/index.html NSK: Manufacturer of rolling bearings, ball screws, etc.

https://www.schaeffler.de/content.schaeffler.de/en/index.jsp Schaeffler: Manufacturer of FAG rolling bearings, etc.

12

Journal Bearings

Bearings are vital machine components that should carry load in static or dynamic modes. Sliding bearings or plain surface bearings usually have two surfaces, which move relative to each other. The suitable shapes of surfaces that permit the relative motion are the flat surfaces and the concentric cylinders. Various types of sliding or journal bearings are used in mechanical products such as in static supports, simple hinges, axels, shafts, etc. For the bearing on a rotating shaft, the portion of the shaft mating with the bearing is called the *journal*. The stationary part supporting the shaft is called the *bearing*. That is where the name of *journal bearing* came from.

For the dynamic modes of bearings, a widespread use of rolling bearings has been introduced in Chapter 11. On the other hand, for static supports or basis, the bearing area should directly stand the pressure of the weights and applied loads. Simple allowable *bearing pressure* or *bearing stress* is utilized to define the adequate bearing area in the design; see Section 12.5. As the function of the machine element involves some movements, this simple design process is not enough. In that case, *friction* becomes a major factor in the design procedure.

As known from previous mechanics background, the friction phenomenon is apparent from the cases shown in Figure 12.1. The blocks having the *weight* $\boldsymbol{W}$ shown in Figure 12.1 rest on the ground surface with extremely exaggerated space between them to clearly define the mutual interaction. The *normal force* to the ground is given the bold symbol $\boldsymbol{N}_n$ since major references use N. The component of the weight normal to the ground surface $\boldsymbol{W}_n$ should be equal to $\boldsymbol{N}_n$. The component of the weight tangent to the ground surface $\boldsymbol{W}_t$ should be equal to the friction force $\boldsymbol{F}_f$. The ground inclined surface in Figure 12.1a is tilted by an angle θ_f that was increased to the shown inception of sliding or an impending motion of the block. If no substance exists in between the block and the ground surface, the friction is considered as a *Coulomb, dry,* or *static friction*. The contacts are usually between asperities in the block and the ground surface. The *wet friction* occurs when a substance such as a *lubricant* exists between the block and the ground surface. The increase of the angle θ to a maximum till the impending motion defines the *friction angle* θ_f. The *Coulomb* or *static friction coefficient* μ_S is defined as tan θ_f. The coefficient of friction is evaluated for dry and wet cases by the simple inclination of ground surface. The values of the static friction coefficients are usually having a relatively large dispersion range depending on so many factors such as *mating material* types, surface roughness, dryness or wetness, type of wetting substance or *lubricant*, humidity, temperature, pressure, contamination, oxide films, chemical conditions, etc. That is why care should be exercised in adapting values to represent the case at hand.

The main concern of machine design is the dynamic or ***kinetic friction*** due to expected motion between components. It can be characterized by the motion of the block in Figure 12.1b under the horizontal applied load $\boldsymbol{P}$, which is balanced with the kinetic friction force $\boldsymbol{F}_K$. Figure 12.2 shows simplified friction force development from *Coulomb*, dry, or *static coefficient of friction* μ_S for impending motion to the kinetic coefficient of friction μ_K for sliding or dynamic motion. Usually, kinetic coefficient of friction μ_K is smaller than static coefficient of friction μ_S, and μ_K is almost constant at moderate rubbing speeds. In Figure 12.1b, the kinetic friction force $\boldsymbol{F}_K$ is defined by the *kinetic friction coefficient* μ_K and the normal force $\boldsymbol{N}_n$. The friction force is then having the form $\boldsymbol{F}_K = \mu_K \boldsymbol{N}_n = \mu_K \boldsymbol{W} = \boldsymbol{P}$ as evident from Figure 12.1b. The *kinetic friction coefficient* μ_K can then be evaluated as

Machine Design with CAD and Optimization, First Edition. Sayed M. Metwalli.

Companion website: www.wiley.com/go/metwalli/machine

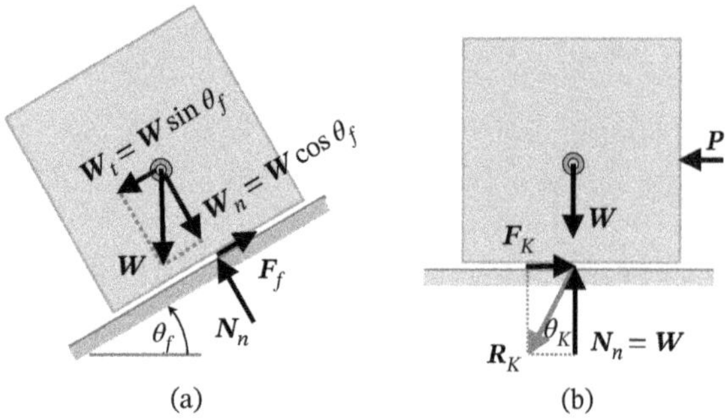

Figure 12.1 Friction phenomenon and force development. Dry or *Coulomb* coefficient of friction μ_C for impending motion in (a). Kinetic coefficient of friction μ_K for sliding or dynamic motion and the force needed as shown in (b).

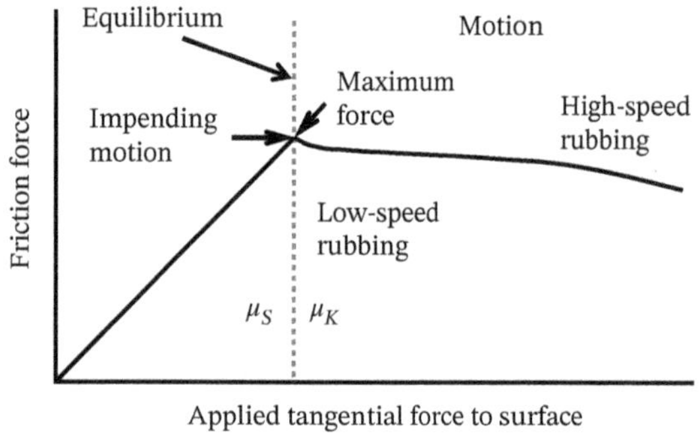

Figure 12.2 Simplified friction force development. *Coulomb*, dry, or static coefficient of friction μ_S for impending motion. Kinetic coefficient of friction μ_K for sliding or dynamic motion.

$\mu_K = \boldsymbol{F}_K/\boldsymbol{W}$. Usually, a *lubricant* is introduced between the block and the ground surface to reduce the friction and thus the applied load $\boldsymbol{P}$, unless the materials of the block and the ground have a very low mutual kinetic friction coefficient μ_K. The interest in having a reduced friction and thus smaller applied load $\boldsymbol{P}$ is one of the main objectives of *lubrication* and efficient journal bearings.

Symbols and Abbreviations

Symbol	Quantity, units (adapted)
$\mathfrak{F}$	Functional relation
(c_D/d_j)	Diametral clearance ratio
(c_J/r_J)	Radial clearance ratio
(d_J/c_D)	Inverse diametral clearance ratio
$(p_p\, \boldsymbol{V}_J)_{\max}$	Maximum allowable pressure velocity limit
$(p_p)_{\max}$	Maximum projected pressure
(r_J/c_J)	Inverse radial clearance ratio
$(T_J)_{\max}$	Maximum allowable temperature
$(\boldsymbol{V}_J)_{\max}$	Maximum allowable speed
[cP]	*Centipoise* = [0.001 Pa s]
[cSt]	Centistokes = [0.01 St] = [0.01 cm^2/s]
[Je]	Joulean heat equivalent = 9336 [in lb/Btu]
[P]	*Poise* = [0.1 Pa s], or {Pa s} = [10 P]
[Pa s]	*Pascal* seconds = [Ns/m^2], *dynamic viscosity*

Symbol	Quantity, units (adapted)
[R]	*Reynolds* [reyn] = [lb s/in^2] = [psi s]
[St]	*Stokes* = [cm^2/s], [m^2/s] = [100 cSt]
[μR]	*Micro Reynolds* = [10^{-6} R] = [10^{-6} reyn]
[μR]	*Micro Reynolds* = 6.895 [cP]
A_J	Journal area exposed to lubricant
a_{TC}, b_{TC}	Constants for temperatures in [°C]
a_{TF}, b_{TF}	Constants for temperatures in [°F]
C_1, C_2	Constants of integration
CAD	Computer-aided design
c_D	Diametral clearance between journal bearing
C_{Hf}	Specific heat of fluid lubricant
c_J	Radial clearance amid journal and bearing
c_o	Fundamental deviation of tolerance
d_C	Cylindrical shaft or hole diameter
d_J	Journal diameter
$dx\,dy$	Small element in x- and y-coordinates
e	Eccentricity
$\boldsymbol{F}$	Force to move the object
$\boldsymbol{F}_f$	Friction force
$\boldsymbol{F}_K$	Kinetic friction force
h_0	Minimum film thickness
h_f	Lubricant film height or thickness
H_f	Power loss in fluid
h_{max}	Maximum film thickness
i_g	Tolerance grade number (... 7, 8, …)
ISO	International Organization for Standardization
l_J	Journal bearing length
$\boldsymbol{N}_n$	Normal force to ground
N_{rpm}	Rotational speed in [rpm]
N_{rps}	Journal rotational speed in [rps]
$\boldsymbol{P}$	Applied load
p	Pressure
p_0	External supply pressure at any zero p_p
p_m	Maximum pressure
p_p	Projected pressure, load per unit are
Q_x	Fluid flow in the x direction
Q_z	Side leakage in z direction
r_J	Journal radius
SAE	Society of automotive engineers

Symbol	Quantity, units (adapted)
S_N	*Sommerfeld* number
T_{av}	Average temperature of lubricant film
T_C, T_F	Temperature in [°C] or [°F], respectively
$\boldsymbol{T_f}$	Friction torque
$\mathrm{T_{id}}$	Tolerance value for $\mathrm{IT}i_g$ and diameter
T_{in}	Inlet temperature of lubricant
$\boldsymbol{v}$	Velocity (italic lower case for $\boldsymbol{v}$)
$\boldsymbol{V_J}$	*Journal* velocity at lubricant contact
$\boldsymbol{W}$	Block weight or light load
$\boldsymbol{W_n}$	Weight component normal to ground
$\boldsymbol{W_t}$	Weight component tangent to ground
Greek	
β_B	Partial bearing sector
ΔT_f	Temperature rise
ε	Eccentricity ratio
γ_f	Fluid mass per unit volume
μ_K	Kinetic friction coefficient
μ_S	*Coulomb* or static friction coefficient
ν	Lubricant viscosity ((μ) in some references)
$\nu_C{}^*$	Nondimensional optimum viscosity criterion
ν_D	Dynamic or absolute viscosity
$\nu_D{}^*$	Optimum dynamic or absolute viscosity
ν_K	Kinematic viscosity
ρ	Fluid density
τ_f	Shear stress between lubricant film layers
θ	Bearing contact angle
θ_f	Friction angle

12.1 Lubricants

Lubrication, lubricants, friction, and wear are the main interest of the *tribology* field. The name of tribology is stemming from the Greek word "*tribos*," which means rubbing. The rest of the word "logy" is an indicative of a field of study. The applications span the areas of maintenance, reliability, and design. Lubrication objectives are to reduce friction losses, heat, and *wear* such as for friction drives, slip grip, adhesion, antiskid, damping, manufacturing processes, etc. Wear is a progressive and accumulative loss of matter from the surface of a body subjected to mechanical interaction with other mating surfaces. Many types of wear regimes may occur such as corrosion, abrasion, pitting, scoring, scuffing, and spalling; see, e.g. Neale (1973). These are caused by some intruding substances or higher surface stresses initiating surface failure or fatigue. A lubricant is used to reduce such an immanent wear.

Lubricants are substances introduced between mating surfaces to reduce rubbing friction, heat, and wear. They function and act as one of *boundary*, *hydrodynamic*, *hydrostatic*, *elastohydrodynamic*, or as a *solid film*, depending on the relative motion, *speed*, lubricant *viscosity*, and other physical properties of lubricants. Very early in history, Egyptians used a mixed lubrication of *silt and water* in moving stones to build the pyramids and temples. The *Nile* silt, as very fine particles, produces a very slippery substance when mixed with little water. This has been used as a lubricant under heavy stone blocks on sleds to move it on dry silty tracks by just pouring some water onto it. Moving heavy blocks does not, therefore, need so many persons to pull these objects even on some inclined ramps. Portraits of that can be found on the wall of Egyptian temples.

On a micro level, resistance to sliding is through the interface of asperities or protrusions between the two surfaces. As the surfaces encounter a lubricant, a film of that lubricant will adhere to these surfaces and reduce the effect of asperities or protrusions interaction. Incursion of some abrasive particles (grit) between surfaces contributes to excessive wear. That is why a finer surface roughness and a filtered lubricant should reduce wear rates. The lubrication type affects or supports such a postulation as discussed in the following brief explanations:

- ***Hydrodynamic lubrication*** occurs when surfaces are separated by a relatively thick film of lubricant that prevents surface contacts. The film is created by developing a wedging pressure due to lubricant high velocity progression in between the wedged surfaces. The developed pressure balances the loads on the mating surfaces. Hydrodynamic lubrication is also given the names of *full film* or fluid lubrication. This type of lubrication techniques is widespread and is extensively used in many designs particularly for higher rotational speeds as discussed in Chapter 8 of the text and in the following Sections 12.2 and 12.3.
- ***Hydrostatic lubrication*** is provided by introducing lubricant between the mating surfaces at a relatively high pressure to separate these surfaces. The developed pressure balances the applied loads on the mating surfaces. Lubricants can be a fluid of a low friction or viscosity such as air or water. This does not need the motion of one surface relative to the other surface to generate the wedging pressure. It is effectively used in hover crafts.
- ***Elastohydrodynamic lubrication*** occurs when lubricants are present between surfaces in rolling contact. Rolling bearings and mating gears are examples of such applications; see Chapters 11, 14, and 15. Combination of *Hertzian* theory of contact stresses and fluid dynamics is used to model and treat such cases. This type of lubrication is beyond the scope of this text and left to higher level courses.
- ***Boundary lubrication*** happens when asperities of the two mating surfaces are separated by lubrication film of several molecular dimensions in thickness. Viscosity by itself may not be an important factor in the treatment of such a case. Further discussion is to follow in Section 12.4.
- ***Solid film lubrication*** happens when the asperities of the two mating surfaces are separated by a solid lubricant film composed of small size particles that can roll on each other and on the surface asperities. Lubricants such as *graphite* or molybdenum disulfide can act that way.

12.1.1 Lubricant Viscosity

One of the major factors in journal bearing performance and design is the *lubricant viscosity* ν. This fluid property is usually introduced in the subject of *fluid mechanics*. In some machine design references, the symbol used for absolute lubricant viscosity is (μ), which conflicts with the same symbol for the coefficient of friction. The symbols in this text are selected with the minimum conflict, and the use of subscripts should also help in that regard.

Dynamic viscosity ν_D or *absolute viscosity* of the *lubricant* is perceived as the resistance of the liquid substance to flow. It is the amount of friction or resistance between molecules when contacting each other during their motion. In a simplified model of Figure 12.3, the lubricant layers between the stationary base or *bearing* and the moving object should have different velocities up to the velocity $\boldsymbol{V}_J$ of the moving object or *journal*. This variation develops shear stress between layers. The lubricant ***absolute*** or ***dynamic viscosity*** ν_D is defined by the viscosity law of *Newton*, which postulates that the shear stress is proportional to the rate of change of velocity or velocity

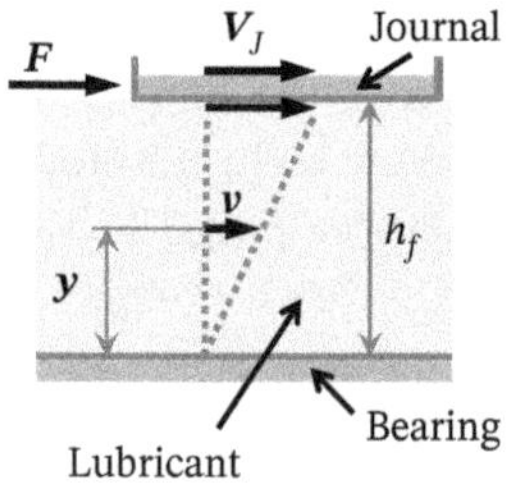

Figure 12.3 Lubricant film between the stationary base or *bearing* and the moving object or *journal* should have different velocities up to the velocity $\boldsymbol{V}_J$ of the moving journal.

gradient. This produced the following *viscosity law* of *Newton*:

$$\tau_f = \frac{\boldsymbol{F}}{A_J} = \nu_D \frac{d\boldsymbol{v}}{dy} \tag{12.1}$$

where τ_f is the shear stress between the layers of the lubricant film, $\boldsymbol{F}$ is the force needed to move the object or journal, A_J is the area of the journal exposed to lubricant, ν_D is the lubricant *absolute* or *dynamic viscosity*, and $\boldsymbol{v}$ is the lubricant velocity at distant y from the stationary bearing surface. Note that the viscosity symbol ν_D (Greek lowercase letter Nu) is close in shape to the velocity $\boldsymbol{v}$ (italic **v**), but it is not closely the same. Please make sure that the difference is clear to you. The top right part of the letter is different. The velocity of the lubricant layer $\boldsymbol{v}$ is zero at the stationary bearing surface and is $\boldsymbol{V}_J$ of the lubricant layer at the journal surface as shown in Figure 12.3. The distribution or variation in layers speed is assumed linear. The rate of change of velocity or velocity gradient is then constant, and *Newton's* law becomes as follows:

$$\tau_f = \frac{\boldsymbol{F}}{A_J} = \nu_D \frac{\boldsymbol{V}_J}{h_f} \tag{12.2}$$

where h_f is the lubricant film height or thickness and ν_D is the lubricant *absolute* or *dynamic viscosity*. The lubricant *absolute* or *dynamic viscosity* is then given by the following relation:

$$\nu_D = \frac{\tau_f}{\boldsymbol{V}_J / h_f} = \frac{\tau_f h_f}{\boldsymbol{V}_J} \tag{12.3}$$

Equation (12.3) indicates that the dimension of the *absolute* or *dynamic viscosity* ν_D is [N s/m^2] = [Pa s] or [lb s/in^2] = [psi s]. The standard, common, or conventional ***units*** of the *absolute* or *dynamic viscosity* ν_D and their equivalence are as follows:

- ***SI units***: Basically [N s/m^2] = [Pa s]. For reasonable values, one uses ***centipoise*** [cP] = [0.01 P]. The *poise* [P] = [0.1 Pa s], or [Pa s] = [10 P]. 1 [cP] = [0.001 Pa s].
- ***US units***: Basically [lb s/in^2] = [psi s] = *Reynolds* [reyn] = [R]. For reasonable values, one uses ***micro Reynolds*** [μR] = [10^{-6} R] = [10^{-6} reyn].
- ***Unit conversion***: The relation between the usually used SI and US units is given by the following relation:

$$1 \text{ micro-reyn } [\mu R] = 6.895 \text{ centipoise } [cP] \approx 7\,[cP] \tag{12.3'}$$

Other units can be obtained from this relation.

The most commonly used lubricants are the SAE engine oils (petroleum or mineral). The SAE engine oil grades are 10, 20, 30, 40, 50, 60, and 70 (SAE J-300 1997). The equivalents for ISO VG (viscosity grades) are 46, 68, 100, 150, 220, 320, 460, and 680 (ISO 3448 1992). These standards may be revised, and information may be somewhat different. The basics are usually the same.

Dynamic viscosities of some liquid at 20[°C] or 68[°F] are given next for comparison:

- Dynamic viscosity ν_D of *water* is about 1 *centipoise* [cP] or 0.0145 [μR].
- Dynamic viscosity ν_D of *SAE 30* engine oil is about 393 *centipoise* [cP] or 57 [μR].
- Dynamic viscosity ν_D of *air* is about 0.0179 *centipoise* [cP] or 0.0026 [μR].

Kinematic viscosity ν_K is an easier concept to grasp since it is defined by measuring the time required for a specific liquid quantity to flow through a narrow pipe under gravity, Gupta (2014) and ISO 3104 (1994). It is commonly used in some references as ν (without the subscript) for defining the *fluid viscosity*, and it is related to the *absolute* or *dynamic viscosity* ν_D by the following relation:

$$\nu_K = \frac{\nu_D}{\rho} \tag{12.4}$$

where ν_K is the *kinematic viscosity*, ν_D is the *absolute* or *dynamic viscosity*, and ρ is the *fluid density*. For the selected units in this text, the *fluid density* ρ is [kg/m^3] or [lb$_m$/in^3]; see Section 1.11. Therefore, Eq. (12.4) indicates that the dimension of the *kinematic viscosity* ν_K is [m^2/s] or [in^2/s]. The standard, common, or conventional ***units*** of the *absolute* or *dynamic viscosity* ν_D and their equivalence are as follows:

- ***SI units***: Basically [m^2/s]. For reasonable values, one uses *Stokes* [St] = [cm^2/s] or ***centistokes*** [cSt] = [0.01 St]. 1 [mm^2/s] = [10^{-6} m^2/s] = 1 [cSt]. The *Stokes* [St] is named after *Sir George G. Stokes* (1819–1903). The basic SI unit is then [m^2/s] = [10^4 St] or [m^2/s] = [100 cSt].
- ***US units***: Basically [in^2/s]. In the US usage, ***centistokes*** [cSt] is often used. At one time, the unit *Saybolt* Universal Seconds (SUS) and other acronyms for alternatives have been employed. The unit is in seconds t_S spent for a specific volume to go through a calibrated tube, and viscosity ν_K in [cSt] is obtained thereof as ν_K [cSt] = (0.22 t_s–(180/t_s)) ASTM D2161 (2019) and ASTM D445 (2019).
- ***Unit conversion***: The relation between the usually used SI and US units is not necessary since the US unit is usually the same ***centistokes*** [cSt]. The conversion from the occasionally cited SUS or t_S to [cSt] is given in the previous paragraph of the *US units*.

Kinematic viscosities of some oils at 100 [°C] or 212[°F] and other fluids at 20[°C] or 68[°F] are given next for data comparison:

- Kinematic viscosities ν_K of SAE engine oil grades 20, 30, 40, 50, and 60, the minimum viscosities in [mm^2/s], i.e. ***centistokes*** [cSt], are 5.6, 9.3, 12.5, 16.3, and 21.9, respectively.
- Kinematic viscosity ν_K of *water* is about 1 *centistoke* [cSt] at 20[°C] or 68[°F]. One should check the value at 100 [°C] or 212[°F] for comparison.
- Kinematic viscosity ν_K of *air* is about 15.2 *centistoke* [cSt] at 20[°C] or 68[°F]. Note that the density is very small. One should check values for comparison.

12.1.2 Lubricant Selection

Lubricants are evolving, and research is adding further improvements that make lubrication more efficient and closer to optimum in many applications. The main factors are the *stability* of lubrication film, *viscosity* reduction with *temperature*, and lubricant *degradation*. The selected lubricant would be of a certain required ***viscosity*** at an operating temperature and pressure. These are needed to hold for the operating life or scheduled maintenance time. The change of viscosity with temperature and pressure is usually available from the selected manufacturer and standards. The variation of lubricant viscosity with pressure is only noticeable at higher pressures, and it is improved with *extreme pressure* additives. The usual reduction of viscosity with temperature is used in the design process. However, the variation of viscosity with temperature depends on the lubricant type such as the *multi-viscosity* oils and some synthetic oils.

For SAE lubricating oils, the change of the absolute or *dynamic viscosity* ν_D with *temperature* T_C[°C] or T_F [°F] is approximately estimated according to the following relations (see Seireg and Dandage 1982):

$$\begin{aligned} \nu_D &= a_{\mathrm{TC}}\left(T_C\right)^{b_{\mathrm{TC}}}, \quad [\mathrm{cP}] \\ \nu_D &= a_{\mathrm{TF}}\ e^{(b_{\mathrm{TF}}/(T_F+95))}, \quad [\mu\mathrm{R}] \end{aligned} \tag{12.5}$$

Table 12.1 Constants for SAE lubricating oils to change the absolute or dynamic viscosity ν_D with temperature.

Oil grade SAE	a_{TC} [cP]	b_{TC}	a_{TF} [μ reyn]	b_{TF}, [°F]
10	1.2706E+05	−2.2455	0.0158	1157.5
20	3.2147E+05	−2.3899	0.0136	1271.6
30	6.5105E+05	−2.4876	0.0141	1360
40	4.3141E+06	−2.8496	0.0121	1474.4
50	1.0842E+07	−2.9563	0.0170	1509.6
60	2.2069E+07	−3.0545	0.0187	1564
70	6.7185E+07	−3.2492	—	—

The values of the constants are given for the SI and US systems of units. The constants of the first two columns are for SI system and are defined by regression. The constants of the following two columns are for the US System.
Source: Seireg and Dandage (1982).

where ν_D is the absolute or *dynamic viscosity*, T_C is the temperature in [°C], the constants a_{TC} and b_{TC} are for temperatures in [°C], the T_F is the temperature in [°F], and the constants a_{TF} and b_{TF} are for temperatures in [°F]. The values of the constants in Eq. (12.5) are given in Table 12.1, where the constants of the first equation are obtained by regression, and the constants of the second equation are due to Seireg and Dandage (1982). The viscosity generally decreases with the increase of temperature. A lower rate of decreasing viscosity with temperature is considered advantageous for the utility of the oil in low-temperature environments. This rate of decreasing viscosity with temperature is usually defined as the *viscosity index*. That is why manufacturers provide *multigrade oils* for such cases of lower *viscosity index* and for other special applications.

12.1.2.1 Stable Lubrication

To have a stable full film bearing, the bearing characteristics should be kept above some values (Figure 12.4). The *bearing characteristic* $\nu_D N_{\mathrm{rpm}}/p_p$ defines the stability of lubricant film (McKee and McKee 1932). The bearing characteristic is composed of the *dynamic viscosity* ν_D, the *rotational speed* N_{rpm}, and the *projected pressure* p_p on the bearing, which is the load divided by the projected area. Figure 12.4 shows the coefficient of friction variation as a function of bearing characteristic value. At low values, the inception of motion occurs at the kinetic coefficient of friction. The first thin film region to the left is a *boundary lubrication*. The next is the unstable region of a *mixed lubrication* phenomenon to the left of the right dashed line in Figure 12.4. In that unstable region, a decrease of viscosity will increase the friction and thus the temperature that would reduce the viscosity furthermore. Beyond that, a stable full or *thick film lubrication* occurs. The characteristics of these regions are as follows:

- ***Boundary lubrication*** causes actual contact between the moving and stationary surfaces of the journal bearing. This occurs despite the presence of a film of lubricant even though a film of lubricant is present.
- ***Mixed film lubrication*** causes some instability of transition between boundary and full film lubrication.
- ***Full or thick film lubrication*** produces a full lubricant film between the moving and stationary parts of the journal bearing. This generates a separating film of lubricant that carries the load. This type is also termed a ***hydrodynamic lubrication*** since the moving lubrication film hydrodynamically develops a squeezing pressure that carries the load.

The mixed film lubrication should be avoided because it is not possible to predict the bearing performance. The boundary lubrication should be expected for slow speed operation, low viscosity lubricant, and high projected

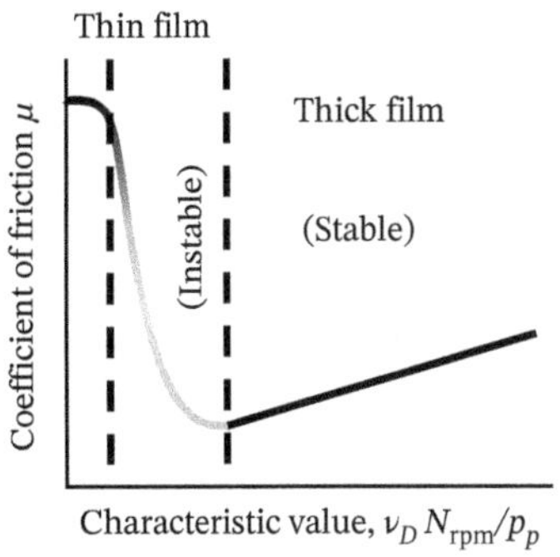

Figure 12.4 Coefficient of friction variation as a function of bearing characteristic value. At low values, the inception of motion occurs at the kinetic coefficient of friction. The first thin film region is a boundary lubrication. The next unstable region is a mixed lubrication. Source: Based on McKee and McKee (1932).

pressure of the load; see Figure 12.4. The selection of lubricant for journal bearing operation should then consider the main factors of operation type (boundary or thick film), surface speed, and bearing load.

12.2 Hydrodynamic Lubrication

The generation of full or *thick film* lubricant characterizes the hydrodynamic development of pressure that carries the load. The development of the hydrodynamic lubrication subject started early on by researchers working on friction and lubrication (Petrov 1883; Towers 1883; Reynolds 1886; Sommerfeld 1904). The following is a brief review of the development in that field, which should help and provide the foundation in the design of journal bearings.

12.2.1 Petroff's Equation

Petrov (1883) and Towers (1883) suggested about the same formulation of the concentric or lightly loaded bearing shown in Figure 12.5. Petroff's equation is named after *Nikolai Petrov* (1836–1920). The light load $\boldsymbol{W}$ was not supposed to cause eccentricity between the journal and the bearing. The lubricant film has a constant thickness of h_f, which is extremely exaggerated in Figure 12.5. The velocity distribution in the film has a linear form shown in Figure 12.5 with a maximum $\boldsymbol{V}_J$ at the journal surface and zero at the stationary surface of the bearing. The friction can be found from the circumferential shear τ_f generated from the velocity gradient in the film as follows (see Eq. (12.2)):

$$\tau_f = \nu_D \frac{\boldsymbol{V}_J}{h_f} = \nu_D \frac{2\pi r_J N_{\text{rps}}}{c_J} \tag{12.6}$$

where h_f is the lubricant film height, which is the *radial clearance* c_J between the journal and the bearing, ν_D is the lubricant absolute or *dynamic viscosity*, $\boldsymbol{V}_J$ is the velocity of the lubricant layer at the journal surface $\boldsymbol{V}_J = 2\pi r_J N_{\text{rps}}$, N_{rps} is the journal *rotational speed* in [rps], and r_J is the *journal radius*. The journal radius r_J is equal to $d_J/2$, where d_J is the journal diameter. The *friction torque* $\boldsymbol{T}_f$ due to friction force $(\tau_f\, A_f)$ generated by the film shear τ_f is given from Eq. (12.6) as

$$\boldsymbol{T}_f = \left(\tau_f A_f\right)\left(\frac{d_J}{2}\right) = \nu_D \frac{2\pi r_J N_{\text{rps}}}{c_J}\left(2\pi r_J l_J\right)\left(r_J\right) = \nu_D \frac{4\pi^2 r_J^3 l_J N_{\text{rps}}}{c_J} \tag{12.7}$$

where A_f is the film area, l_J is the bearing length, and the other terms are defined in Eq. (12.6). The very light load or weight $\boldsymbol{W}$ is distributed over the projected bearing area of $d_J l_J$ or $2r_J l_J$ to generate a *projected pressure* p_p. Under this small load $\boldsymbol{W}$, the torque in Eq. (12.7) is then given by

$$\boldsymbol{T}_f = \mu_K \boldsymbol{W}\left(r_J\right) = \mu_K\left(2r_J l_J p_p\right)\left(r_J\right) = \nu_D \frac{4\pi^2 r_J^3 l_J N_{\text{rps}}}{c_J} \tag{12.8}$$

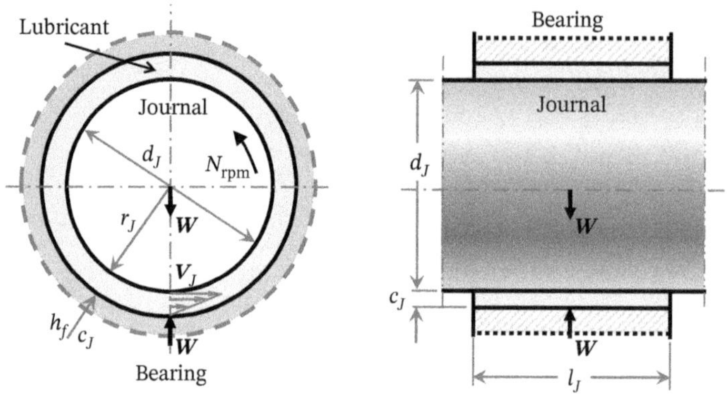

Figure 12.5 A concentric lightly loaded bearing used to derive *Petroff's* equation. The linear velocity distribution is shown in the left cross-sectional view. The clearance c_f between the journal and bearing is extremely exaggerated.

where μ_K is the *kinematic coefficient of friction*. From Eq. (12.8), the kinematic coefficient of friction μ_K is then given by the following relation:

$$\mu_K = \nu_D \frac{4\pi^2 r_j^3 l_J N_{rps}}{c_J\left(2r_J^2 l_J p_p\right)} = \nu_D \frac{2\pi^2 r_J N_{rps}}{c_J p_p} = 2\pi^2 \frac{\nu_D N_{rps}}{p_p}\left(\frac{r_J}{c_J}\right) \tag{12.9}$$

where (r_J/c_J) is the ***inverse radial clearance ratio***, which should be the same as the *inverse diametral clearance ratio* (d_J/c_D). The expression in Eq. (12.9) is the *Petroff's equation*, which is named after *Petrov* (1883) as previously indicated. Note that the term $\nu_D N_{rpm}/P_p$ in Eq. (12.9) is the same as the *bearing characteristic* defined in Figure 12.4 by McKee and McKee (1932).

In his later work, Sommerfeld (1904) suggested a nondimensional number S_N, which is later named after him, *Arnold Sommerfeld* (1868–1951). The *Sommerfeld number* S_N has been extensively used in journal bearing analysis and design. The nondimensional *Sommerfeld number* S_N is similar to a basic part of *Petroff's Equation*(12.9). The nondimensional *Sommerfeld number* S_N is given by

$$S_N = \frac{\nu_D N_{rps}}{p_p}\left(\frac{r_J}{c_J}\right)^2 = \frac{\nu_D l_J d_J N_{rps}}{\boldsymbol{P}}\left(\frac{d_J}{c_D}\right)^2 \tag{12.10}$$

where ν_D is the lubricant absolute or *dynamic viscosity*, $p_p = \boldsymbol{P}/l_J d_J = \boldsymbol{P}/2l_J r_J$, $r_J = (d_J/2)$, $c_J = (c_D/2)$, and c_D is the diametral clearance. In Eq. (12.10), the light load $\boldsymbol{W}$ is replaced by the applied load $\boldsymbol{P}$. In many references, the symbol S is used as the *Sommerfeld* number in place of our symbol S_N.

From Eqs. (12.9) and (12.10), one can observe that $\mu_K\,(r_J/c_J) = 2\pi^2 S_N$.

12.2.2 Journal Bearings

Due to the load $\boldsymbol{P}$ on the journal bearing, the lubricant film in between the journal and the bearing can have the exaggerated eccentric form shown as a very light gray color in Figure 12.6. The load is assumed to be carried by the partial bearing sector β_B in Figure 12.6a. Under the load $\boldsymbol{P}$, the journal center is shifted off the bearing center by an *eccentricity e* causing the *minimum film thickness* h_0 to develop along its direction. The N_{rps} rotational speed of the journal (i.e. shaft) drives the lubricant film by a velocity $\boldsymbol{v}$ through the narrowing passage. The partial bearing sector β_B in Figure 12.6a is assumed smaller than the usual 180° used in most cases, which is shown in Figure 12.6b. Figure 12.6b shows the expected *pressure distribution p* due to the hydrodynamic effect of squeezing the lubricant film through the narrowing passage. The pressure distribution produces an equal and opposite resultant force $\boldsymbol{P}$ to support the applied load $\boldsymbol{P}$. A small element $(dx\ dy)$ within the film in Figure 12.6 is used to develop the

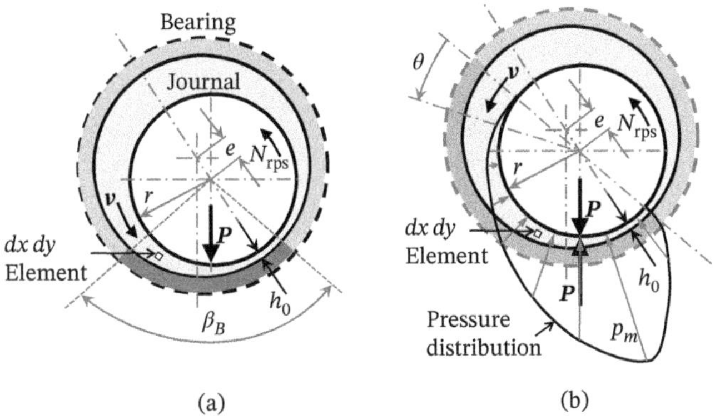

Figure 12.6 The lubricant film in between the journal and the bearing has an exaggerated eccentric form with a minimum film thickness h_0 along the direction of the eccentricity e. The partial bearing is shown in (a) and the pressure distribution sketch of 180° bearing is in (b).

Figure 12.7 Small element ($dx\,dy$) in the lubricant film with forces due to pressure variation on the right and left sides of the element. The shear force variation acts on the top and bottom sides of the element. The slope of the moving surface of the journal is exaggerated.

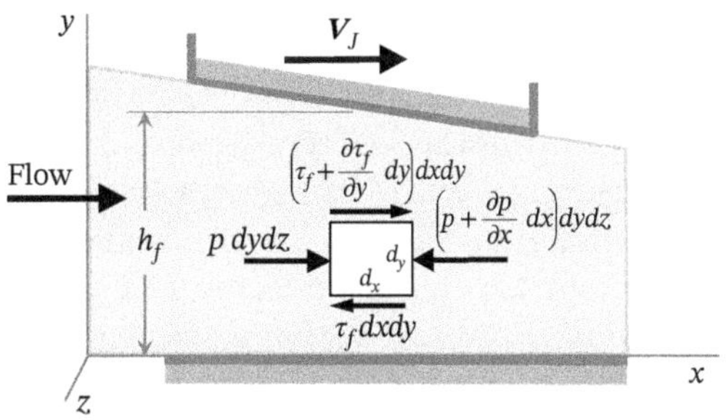

mathematical model of the journal bearing. The model is valid for the small sector β_B or the 180° sector, and the curvature of the film is assumed very minor in the small ($dx\,dy$) element model in Figure 12.7. For simplicity, the Cartesian coordinates are used in Figure 12.7 instead of the expected cylindrical coordinates. The pressure p is assumed constant along the journal bearing length l_J in the z direction, which is assumed long enough (or infinite) to warrant the assumption of no fluid flow in the z direction. The pressure p is also assumed constant along the y direction. The mathematical model is developed next (Reynolds 1886).

With the previous simplified assumptions, the forces on the small ($dx\,dy$) element are observed in Figure 12.7. In this figure, the small element ($dx\,dy$) in the lubricant film has forces due to *pressure* p on the left side and its variation on the right side of the element. The shear force components act on the top and bottom sides of the element ($dx\,dy$) due to the *shear stress* τ_f and its variation along the y direction. The slope of the moving surface of the journal is exaggerated. The summation of forces in the x direction gives the following expression:

$$\sum \boldsymbol{F}_x = p\,dy\,dz - \left(p + \frac{\partial p}{\partial x}dx\right)dy\,dz - \tau_f\,dx\,dy + \left(\tau_f + \frac{\partial \tau_f}{\partial y}dy\right)dx\,dy = 0 \tag{12.11}$$

where p is the *pressure* and τ_f is the *shear stress* in the lubricant film. This gives

$$\frac{dp}{dx} = \frac{\partial \tau_f}{\partial y} \tag{12.12}$$

Upon the consideration of Eq. (12.1), one gets

$$\frac{dp}{dx} = \frac{\partial}{\partial y}\left(v_D \frac{dv}{dy}\right) = v_D \frac{\partial^2 \mathbf{v}}{\partial y^2} \tag{12.13}$$

where ν_D is the dynamic viscosity of the lubricant. The partial derivative is kept for $\boldsymbol{v}$ in Eq. (12.13), since it should depend on x and y. Integrating Eq. (12.13) twice and keeping x constant gives

$$\frac{\partial \boldsymbol{v}}{\partial y} = \frac{1}{\nu_D}\frac{dp}{dx}y + C_1$$

$$\boldsymbol{v} = \frac{1}{2\nu_D}\frac{dp}{dx}y^2 + C_1 y + C_2 \tag{12.14}$$

where C_1 and C_2 are constants of integration that are obtained from boundary conditions. It is obvious that at $y = 0$, the velocity $\boldsymbol{v} = 0$, and at $y = h_f$, the velocity $\boldsymbol{v} = \boldsymbol{V}_J$. By substituting these conditions in Eq. (12.14), one gets the following:

$$C_1 = \frac{\boldsymbol{V}_J}{h_f} - \frac{h_f}{2\nu_D}\frac{dp}{dx}, \quad \text{and} \quad C_2 = 0 \tag{12.15}$$

By substituting the values of C_1 and C_2 from Eq. (12.15) into Eq. (12.14), one gets the velocity distribution as follows:

$$\boldsymbol{v} = \frac{1}{2\nu_D}\frac{dp}{dx}(y^2 - h_f y) - \frac{\boldsymbol{V}_J}{h_f}y \tag{12.16}$$

Equation (12.16) indicates that the pressure distribution across the film thickness h_f is nonlinear, and a parabolic term exists. This parabolic term depends on the rate of change of pressure with x, which is dp/dx. At the maximum pressure p_m, the rate of change of pressure with x is zero, and one gets the linear variation of velocity across the film, which is $\boldsymbol{v} = (\boldsymbol{V}_J/h_f)y$. The maximum pressure p_m is approximately as shown in Figure 12.6b.

The fluid flow Q_x along the x direction is obtained by integrating the velocity $\boldsymbol{v}$ of Eq. (12.16) in the y direction from 0 to h_f. Upon considering an incompressible fluid (i.e. $dQ_x/dx = 0$), one gets the following *one-dimensional Reynolds' equation*:

$$\frac{d}{dx}\left(\frac{h_f^3}{\nu_D}\frac{dp}{dx}\right) = 6\boldsymbol{V}_J\frac{dh_f}{dx} \tag{12.17}$$

Equation (12.17) neglects the side leakage Q_z in the z direction. If the side leakage in the z direction is not neglected, one gets the following *Reynolds'* equation (see Reynolds 1886):

$$\frac{\partial}{\partial x}\left(\frac{h_f^3}{\nu_D}\frac{\partial p}{\partial x}\right) + \frac{\partial}{\partial z}\left(\frac{h_f^3}{\nu_D}\frac{\partial p}{\partial z}\right) = 6\boldsymbol{V}_J\frac{\partial h_f}{\partial x} \tag{12.18}$$

Equation (12.18) does not have a closed form analytical solution. Sommerfeld (1904), however, suggested a solution as a function of the *Sommerfeld number* S_N such that

$$\frac{r_J}{c_J}\mu_K = \mathfrak{F}(S_N) = \mathfrak{F}\left(\frac{\nu_D N_{\text{rps}}}{p_P}\left(\frac{r_J}{c_J}\right)^2\right) \tag{12.19}$$

where $\mathfrak{F}$ is indicating a functional relation, ν_D is the lubricant absolute or *dynamic viscosity*, $p_p = \boldsymbol{P}/l_J d_J = \boldsymbol{P}/2l_J r_J$, $r_J = (d_J/2)$, $c_J = (c_D/2)$, and c_D is the diametral clearance.

From the preceding mathematical model, it is apparent that the journal bearing is affected by the two main sets of parameters, namely, the geometry and the operation:

- *Geometry parameters*: These are the journal *bearing diameter* d_J, the *diametral clearance* c_D, and the *journal length* l_J. The bearing or *journal radius* r_J is half the bearing diameter, i.e. $r_J = (d_J/2)$. The ***radial clearance*** c_J is half the diametral clearance or $c_J = (c_D/2)$. The partial bearing sector or effective bearing angle β_B is usually assumed as 180°, which is named half bearing. The other useful derived parameters are as follows:

Length to diameter ratio is l_J/d_J and the ***clearance ratio*** is $c_J./r_J = c_D/d_J$.
Eccentricity ratio ε is given as the following ratio of eccentricity e to *radial clearance* c_J, i.e.,

$$\varepsilon = \frac{e}{c_J} \tag{12.20}$$

Film thickness ratio is defined by h_f/c_J, and the minimum film thickness ratio h_0/c_J is then given as

$$h_0 = c_J\,(1-\varepsilon),$$
$$\frac{h_0}{c_J} = (1-\varepsilon) \tag{12.21}$$

where h_0 is the minimum film thickness; see Figure 12.6. The maximum film thickness h_{max} around the 360° is given by

$$h_{max} = c_J\,(1+\varepsilon) \tag{12.22}$$

Operational parameters: These are the parameters of running the journal function such as rotational ***speed*** N_{rps}, bearing ***load P*** or *projected bearing* ***pressure*** $p_p = (P/l_J d_J)$, and *lubricant* ***viscosity*** ν_D.

Performance parameters: These are the journal bearing functioning behavioral variables. The performance characteristics of interest are the ***friction*** *coefficient* μ_K or friction *torque* T_f, side leakage or *side lubricant* ***flow*** Q_z, the ***temperature*** *rise* ΔT_f, maximum *film* ***pressure*** p_m, the ***eccentricity*** *e* or *eccentricity ratio* ε, and *minimum film thickness ratio* (h_0/c_J) or the *minimum* ***film thickness*** h_0.
It should be noted that the eccentricity e and the film thickness h_f or minimum film thickness h_0 are behavioral or performance variables. The geometrical parameter of journal clearance c_J is, however, the one variable, which allowed the minimum film thickness to be considered as geometry parameter. This is because its minimum and maximum (without operating oil) is equal to zero and $2c_J$ or c_D, respectively; see Eqs. (12.21) and (12.22) and Figures 12.5 and 12.6.

Since Eq. (12.18) does not have a general closed form analytical solution, the two bounds of infinite or extremely *long bearing* and the *short bearing* may be used effectively to check the expected behavior of bearing design. In fact, short bearing formulation may be used efficiently to solve other bearings.

12.2.2.1 Long Bearing

The *Reynolds'* equation (12.18) does not have a closed form analytical solution. Sommerfeld (1904), however, found a solution for *infinite* or very *long bearing* where the pressure distribution along *z* direction is constant. This renders $\partial p/\partial z$ to be zero, and Eq. (12.18) becomes as follows:

$$\frac{\partial}{\partial x}\left(h_f^3\frac{\partial p}{\partial x}\right) = 6\nu_D \boldsymbol{V}_J \frac{\partial h_f}{\partial x} \tag{12.23}$$

which is about the same as Eq. (12.17). The solution of this equation is given by (Sommerfeld 1904)

$$p_p = \frac{\nu_D \boldsymbol{V}_J r_J}{c_J^2}\left(\frac{6\varepsilon\,(\sin\theta)\,(2+\varepsilon\cos\theta)}{\left(2+\varepsilon^2\right)(1+\varepsilon\cos\theta)^2}\right) + p_0 \tag{12.24}$$

where p_p is the *projected pressure* around the journal bearing with angle θ from zero to π (Figure 12.6b), ν_D is the lubricant absolute or *dynamic viscosity*, $\boldsymbol{V}_J$ is the *velocity* of the lubricant layer at the journal surface, N_{rps} is the journal *rotational speed* in [rps], r_J is the *journal radius*, c_J is the *radial clearance* or half the diametral clearance $c_J = (c_D/2)$, ε is the *eccentricity ratio*, and p_0 is the *external supply pressure* at any zero p_p location (say at $\theta < 0$). Note that the pressure distribution for the projected pressure p_p in Figure 12.6b is for no external supply of pressure (i.e. $p_0 = 0$).

The total load $\boldsymbol{P}$ is also given by the following relation (Sommerfeld 1904):

$$\boldsymbol{P} = \frac{\nu_D \boldsymbol{V}_J l_J r_J^2}{c_J^2}\left(\frac{12\pi\varepsilon}{(2+\varepsilon^2)(1+\varepsilon)^{1/2}}\right) \tag{12.25}$$

where most variables are defined after Eq. (12.24) and l_J is the journal bearing length. Note that $\boldsymbol{V}_J = 2\pi r_J\, N_{\text{rps}}$, $p_p = \boldsymbol{P}/l_J d_J = \boldsymbol{P}/2l_J r_J$, and $S_N = (\nu_D\, N_{\text{rps}}/p_p)(r_J/c_J)^2$ from Eq. (12.10).

The torque $\boldsymbol{T}_f$ can also be obtained as related to *Petroff's* equation such that (see Eq. (12.7))

$$\boldsymbol{T}_f = \left(\nu_D \frac{4\pi^2 r_J^3 l_J N_{\text{rps}}}{c_J}\right)\left(\frac{(1+2\varepsilon)^2}{(2+\varepsilon^2)(1-\varepsilon^2)^{1/2}}\right) \tag{12.26}$$

where ν_D is the lubricant absolute or *dynamic viscosity*, $r_J = (d_J/2)$, $c_J = (c_D/2)$, c_D is the diametral clearance, N_{rps} is the journal *rotational speed* in [rps], and l_J is the *journal length*. This torque $\boldsymbol{T}_f$ in Eq. (12.26) is expected to converge to *Petroff's* equation when ε is zero.

From this torque, the coefficient of friction would be evaluated similar to Eq. (12.8).

12.2.2.2 Short Bearing

As indicated previously, the *Reynolds'* equation (12.18) does not have a closed form analytical solution. Ocvirk (1952) and DuBois and Ocvirk (1955) have provided a solution for short bearing application for full journal bearings. This is reviewed next.

With short bearings, the side or end leakage is significant and not zero as for long bearings. To consider the end leakage, *Reynolds'* equation (12.18) attention should be to the following part:

$$\frac{\partial}{\partial z}\left(\frac{h_f^3}{\nu_D}\frac{\partial p}{\partial z}\right) = 6\boldsymbol{V}_J \frac{\partial h_f}{\partial x} \tag{12.27}$$

This reduces the importance of the circumferential flow of lubricant in the bearing relative to the end or side leakage in z direction. Integrating Eq. (12.27) gives the following *pressure distribution* in z and x directions (Ocvirk solution).

$$p_p = \frac{\nu_D \boldsymbol{V}_J}{r_J c_J^2}\left(\frac{l_J^2}{4} - z^2\right)\left(\frac{3\varepsilon(\sin\theta)}{(1+\varepsilon\cos\theta)^3}\right) \tag{12.28}$$

where p_p is the *projected pressure* distribution around the journal bearing with angle θ from zero to π (Figure 12.8a) and its distribution in the z direction from zero to l_J (Figure 12.8b), l_J is the journal length, ν_D is the lubricant absolute or *dynamic viscosity*, $\boldsymbol{V}_J$ is the *velocity* of the lubricant layer at the journal surface, N_{rps} is the journal *rotational speed* in [rps], r_J is the *journal radius*, c_J is the *radial clearance* or half the diametral clearance $c_J = (c_D/2)$, and ε is the *eccentricity ratio*. In Figure 12.8 the space between the journal and the bearing is extremely exaggerated. The *projected pressure* distribution in Figure 12.8a is only an approximate sketch. The other *projected pressure* distribution in Figure 12.8b is also an approximate sketch, and it is not at the maximum pressure location. It should be scaled around the journal so that its maximum matches the value in Figure 12.8a from 0 to π.

The total load $\boldsymbol{P}$ is also given by the following relation (Ocvirk 1952):

$$\boldsymbol{P} = \frac{\nu_D \boldsymbol{V}_J l_J^3}{c_J^2}\left(\frac{\varepsilon\left(\pi^2(1-\varepsilon^2)+16\varepsilon^2\right)^{1/2}}{4(1-\varepsilon^2)^2}\right) \tag{12.29}$$

where the variables are defined after Eq. (12.28).

The torque $\boldsymbol{T}_f$ can also be obtained as related to *Petroff's* equation such that (see Eq. (12.7))

$$\boldsymbol{T}_f = \left(\nu_D \frac{4\pi^2 r_J^3 l_J N_{\text{rps}}}{c_J}\right)\left(\frac{1}{(1-\varepsilon^2)^{1/2}}\right) \tag{12.30}$$

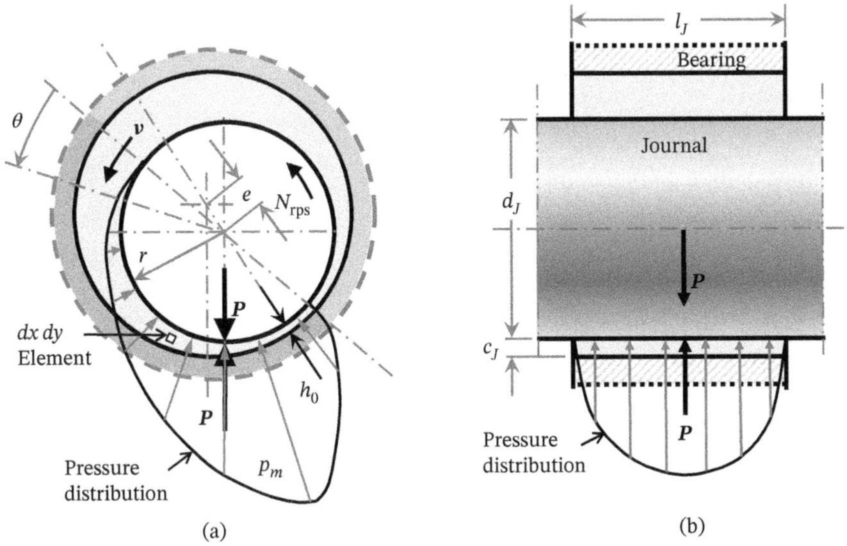

Figure 12.8 Short bearing showing the projected pressure distribution sketches around the journal bearing with angle θ from zero to π shown in (a) and in the z direction from zero to l_J shown in (b). Space between the journal and the bearing is extremely exaggerated.

This torque $\boldsymbol{T}_f$ is expected to converge to *Petroff's* equation when ε is zero.

From this torque, the coefficient of friction would be evaluated similar to Eq. (12.8).

12.2.2.3 Finite Length Bearing

As indicated previously, the *Reynolds'* equation (12.18) does not have a closed form analytical solution. Using computers, Raimondi and Boyd (1958) presented a *numerical solution* for finite length bearings particularly for *length to diameter ratio* l_J/d_J of 1.0, 0.5, and 0.25. The l_J/d_J value of ∞ (very *long bearing*) has also been included. Their efforts have been utilized effectively in the analysis, design, and optimization of journal bearings. The synthesis procedure used herein utilized their *numerical solution* particularly for the 180° bearings (half bearings). Other 360° (full bearings) and 60° partial bearings can be handled similarly using other tables and graphs of their works, such as Raimondi and Boyd (1958) and Boyd and Raimondi (1951). Their information is henceforward referred to as the *numerical solution*. The nondimensionalized performance parameters of their *numerical solution* allow the evaluation of bearing performances for both the US and SI systems of units. They have also utilized the nondimensional *Sommerfeld number* S_N as a main abscissa or primary variable (say x) to define any other performance ratio (say y) against that S_N.

One of the main performance parameters is the ***minimum film thickness ratio*** defined as h_0/c_J, where h_0 is the *minimum film thickness* and c_J is the *radial* clearance of the journal bearing ($c_J = c_D/2$). This ratio is $(1-\varepsilon)$, where ε is the eccentricity ratio defined as (e/c_J) in Eq. (12.20); see Eq. (12.21). As the design process is the inverse of the analysis procedure, this inverse suggests that the output ought to be the *Sommerfeld* number S_N for a good minimum film thickness or minimum film thickness ratio (h_0/c_J). The *numerical solution* of the 180° bearing performance at $l_J/d_J = 1.0$ is used to get the following regression formula:

$$y = 42.765x^6 - 87.052x^5 + 66.687x^4$$
$$-23.165x^3 + 4.0453x^2 - 0.0795x + 0.0041 \qquad (12.31)$$

where y is the *Sommerfeld* number S_N at $(l_J/d_J = 1)$ and x is the *minimum film thickness ratio* (h_0/c_J). In a substituted form, Eq. (12.31) is then

$$S_N = 42.765\left(\frac{h_0}{c_J}\right)^6 - 87.052\left(\frac{h_0}{c_J}\right)^5 + 66.687\left(\frac{h_0}{c_J}\right)^4 \\ -23.165\left(\frac{h_0}{c_J}\right)^3 + 4.0453\left(\frac{h_0}{c_J}\right)^2 - 0.0795\left(\frac{h_0}{c_J}\right) + 0.0041 \qquad (12.31)$$

The design of the journal bearing is easier to define if the *Sommerfeld* number S_N is known. One can also use the *Sommerfeld* number S_N at $(l_J/d_J = 1.0)$ to evaluate all other behavioral parameters from the *numerical solution*. Other l_J/d_J values can be treated the same way. The available l_J/d_J ratios are at limited values of ∞, 1, 0.5, and 0.25. Therefore, interpolation is necessary. The very *long* or l_J/d_J of ∞ is not usually admissible. Figure 12.9 shows the regression curves of the *Sommerfeld* number S_N for l_J/d_J ratios of ∞, 1, 0.5, and 0.25 against the minimum film thickness ratio (h_0/c_J) for 180° bearings. The range of (h_0/c_J) is 0.03–0.9 as available in the *numerical solution*. The values of the *numerical solution* are cited on each curve. The dotted curves are very rough lines for the *maximum load* (upper dotted curve) and the *minimum power loss* (lower dotted curve). The design space is expected to be between those two dotted lines in Figure 12.9. The typical, general, and practice range of l_J/d_J ratio is about 0.5–2.4 or so. For length to diameter ratio $l_J/d_J = 1.0$, the minimum film thickness ratio (h_0/c_J) is expected to be between 0.3 and 0.55. For some larger l_J/d_J than 1.0, the range of (h_0/c_J) is expected to be between 0.4 and 0.6. For some smaller l_J/d_J than 1.0, the range of (h_0/c_J) is expected to be between 0.2 and 0.45; see Figure 12.9. These thickness ratios (h_0/c_J) are more feasible and viable to select than the *Sommerfeld* numbers, which are functions of many variables; some should be known but others are design variables to be synthesized. The range of the *Sommerfeld* numbers S_N in the previous design space is about 0.04–0.4, which is much wider to pick an estimate.

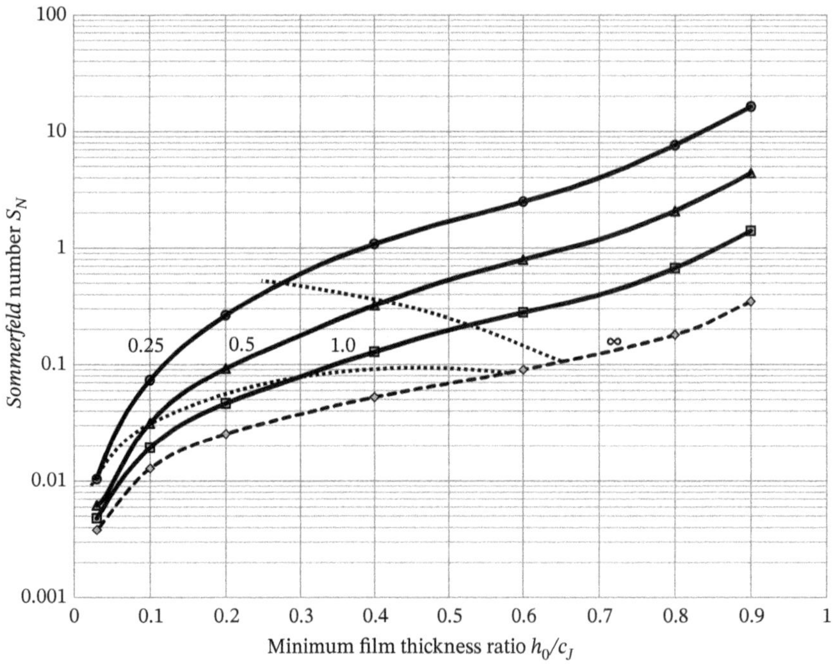

Figure 12.9 The regression curves of the Sommerfeld number S_N for l_J/d_J ratios of ∞, 1.0, 0.5, and 0.25 against the minimum film thickness ratio (h_0/c_J) for 180° bearings. The range of (h_0/c_J) is 0.03–0.9. The values of the *numerical solution* are cited on each curve.

Another important performance parameter is the ***coefficient of friction variable*** defined as $\mu_K(r_J/c_J)$, where μ_K is the *kinetic coefficient of friction*, r_J is the journal *bearing radius* ($r_J = d_J/2$), and c_J is the *radial clearance* of the journal bearing ($c_J = c_D/2$). The *numerical solution* of the 180° bearing performance at $l_J/d_J = 1.0$ is used to get the following regression formula:

$$y = 0.090x^2 + 19.188x + 0.737 \tag{12.32}$$

where x is the *Sommerfeld* number S_N at ($l_J/d_J = 1$) and y is the *coefficient of friction variable* $\mu_K\ (r_J/c_J)$. In a substituted form, Eq. (12.32) is then

$$\mu_K\left(\frac{r_J}{c_J}\right) = 0.090\ S_N^2 + 19.188\ S_N + 0.737 \tag{12.32}$$

The kinetic coefficient of friction μ_K can then be approximately evaluated from Eq. (12.32). The friction torque and power loss are simply calculated from the kinetic coefficient of friction μ_K, the load $\boldsymbol{P}$, the journal radius r_J, and rotational speed N_{rpm} or N_{rps}.

The other important performance parameter is the ***temperature rise variable*** defined as $\Delta T_f\ (\gamma_f C_{Hf}/p_p)$, where ΔT_f is the *temperature rise* of the lubricant film [°C] or [°F], γ_f is the fluid *mass per unit volume* [kg/m³] or [lb$_m$/in³], C_{Hf} is the *specific heat* of the fluid lubricant [J/kg °C], and p_p is the *projected pressure* ($\boldsymbol{P}/d_J l_J$). The *numerical solution* of the 180° bearing performance at $l_J/d_J = 1.0$ is used to get the following regression formula:

$$y = -0.3459x^2 + 77.394x + 2.9807 \tag{12.33}$$

where x is the *Sommerfeld* number S_N at ($l_J/d_J = 1$) and y is the *temperature rise variable* $\Delta T_f\ (\gamma_f C_{Hf}/p_p)$. In a substituted form, Eq. (12.33) is then

$$\Delta T_f\left(\frac{\gamma_f\ C_{Hf}}{p_p}\right) = -0.3459\ S_N^2 + 77.394\ S_N + 2.9807 \tag{12.33}$$

The *temperature rise* ΔT_f can then be approximately evaluated from Eq. (12.33). This temperature rise is for the lubricating oil, and does not consider the flow of heat to the journal bearing or the surrounding. It does not include any oil cooling process from the bearing to the sump and back to the bearing. It does not consider the quantity of oil pumped into the bearing. The temperature rise depends only on the work done by the bearing on the oil and the friction power dissipated into heating of the oil film. The average temperature of the lubricant film T_{av} can then be approximately estimated as follows:

$$T_{\text{av}} = T_{\text{in}} + \frac{\Delta T_f}{2} \tag{12.34}$$

where T_{in} is the inlet temperature of the lubricant flowing into the circumferential lubricant film at zero p_p location before pressure development and before the friction power is dissipated or transferred into heat to the lubricant; see Figure 12.6b. The average temperature rise between the lubricant input and output is taken as the total *temperature rise* $\Delta T_f/2$.

Initial average values of the lubricant parameters (SAE engine oil) are the fluid *mass per unit volume* $\gamma_f = 861$ [kg/m³] or 0.0311 [lb$_m$/in³], the *specific heat* of the fluid lubricant $C_{Hf} = 1760$ [J/kg °C] or 0.42 [Btu/(lb$_m$ °F)], and *Joulean* heat equivalent [Je] = 9336 [in lb/Btu].

Other performance parameters such as *circumferential flow, side leakage, maximum film pressure,* position of minimum film thickness, and position of film termination can be derived from the *numerical solution* the same way as the previous performance parameters.

The expected ***friction torque*** $\boldsymbol{T_f}$ and the **power loss** H_f in the bearing film are basically given by the following simple expressions:

$$\boldsymbol{T}_f = \mu_K \boldsymbol{P}\ r_J \tag{12.35}$$

and

$$H_f = 2\pi \boldsymbol{T}_f N_{\text{rps}} \tag{12.36}$$

where μ_K is the kinetic coefficient of friction, $\boldsymbol{P}$ is the externally applied load, r_J is the bearing radius, and N_{rps} is the rotational speed in [rps]. These are basic expressions that can be used to simply account for the expected friction and power loss in the bearing.

For a full 360° journal bearing, other data can be extracted from the *numerical solution* of Raimondi and Boyd (1958) and Boyd and Raimondi (1951). The differences between half bearing and full bearing solutions are expected to be within the design engineering range. For initial design synthesis, the previously introduced information may be used. Further computer simulation means such as *computational fluid dynamics (CFD)* should be used to define the performance of a specific bearing design. This is beyond the scope of this text.

Example 12.1 It is required to find a selected design of the hydrodynamic journal bearing and estimate its behavioral parameters values. The bearing design calls for a journal bearing diameter d_J of 40 [mm] or 1.5 [in], a bearing length $l_J = 40$ [mm] or 1.5 [in], and a diametral clearance c_D of 0.064 [mm] or 0.003 [in], running at N_{rpm} of 1800 [rpm], i.e. an N_{rps} of 30 [rps], and should carry a load $\boldsymbol{P}$ of 2.5 [kN] or 500 [lb]. It is expected to use an SAE 30 lubricating fluid at an inlet temperature T_{in} of 60 [°C] or 140 [°F]. Find the appropriateness of the lubricating oil, the expected temperature rise, the coefficient of friction, the friction torque, and the power loss in the bearing. Assume a simplified 180° half bearing.

Solution

Data: $d_J = 40$ [mm] or 1.5 [in], $l_J = 40$ [mm] or 1.5 [in], $c_D = 0.064$ [mm] or 0.003 [in], $N_{\text{rps}} = 30$ [rps], $\boldsymbol{P} = 2.5$ [kN] or 500 [lb], and T_{in} of 60 [°C] or 140 [°F]. The length to diameter ratio is $l_J/d_J = 40/40$ or $1.5/1.5 = 1.0$.

The projected pressure p_p is defined as follows:

$$p_p = \frac{\boldsymbol{P}}{d_J l_J} = \frac{2.5\,(10)^6}{40\,(40)} = 1562.5\ [\text{kPa}]$$
$$p_p = \frac{\boldsymbol{P}}{d_J l_J} = \frac{500}{1.5\,(1.5)} = 222.22\ [\text{psi}] \tag{a}$$

The inverse clearance ratio (r_J/c_J) is obtained as

$$\left.\frac{r_J}{c_J}\right|_{\text{SI}} = \left(\frac{d_J}{c_D}\right) = \frac{40}{0.064} = 625$$
$$\left.\frac{r_J}{c_J}\right|_{\text{SI}} = \left(\frac{d_J}{c_D}\right) = \frac{1.5}{0.003} = 500 \tag{b}$$

The two calculated inverse clearance ratios are not the same, and the solutions of the SI and US systems should then be independently calculated. The clearance ratio (c_J/r_J) is then 0.0016 for SI or 0.002 for the US system.

First, the evaluation of nondimensional parameters should apply to both the US and SI systems. The initially given diametral clearance c_D of 0.064 [mm] or 0.003 [in], length to diameter ratio of 1.0, and the average minimum film thickness ratio (h_0/c_J) is about 0.42 for $l_J/d_J = 1.0$ (the range of h_0/c_J is about 0.3–0.55). This is easily selected and changed. According to Eq. (12.31), the following nondimensional *Sommerfeld* number is evaluated as follows:

$$S_N = 42.765\left(\frac{h_0}{c_J}\right)^6 - 87.052\left(\frac{h_0}{c_J}\right)^5 + 66.687\left(\frac{h_0}{c_J}\right)^4$$
$$-23.165\left(\frac{h_0}{c_J}\right)^3 + 4.0453\left(\frac{h_0}{c_J}\right)^2 - 0.0795\left(\frac{h_0}{c_J}\right) + 0.0041$$
$$= 42.765(0.42)^6 - 87.052(0.42)^5 + 66.687(0.42)^4$$
$$-23.165(0.42)^3 + 4.0453(0.42)^2 - 0.0795\,(0.42) + 0.0041 = 0.1402 \tag{c}$$

With this value of S_N, and according to Eqs. (12.32) and (12.33), the following nondimensional parameters are calculated:

$$\mu_K\left(\frac{r_J}{c_J}\right) = 0.090\ S_N^2 + 19.188\ S_N + 0.737 = 0.090(0.1402)^2 + 19.188\,(0.1402) + 0.737 = 3.4288 \tag{d}$$

$$\Delta T_f\left(\frac{\gamma_f\, C_{Hf}}{p_p}\right) = -0.3459\ S_N^2 + 77.394\ S_N + 2.9807 = -0.3459\ (0.1402)^2 + 77.394\ (0.1402) + 2.9807$$
$$= 13.8241 \tag{12.33}$$

Second, the SI system variables at operating conditions are evaluated as follows:

$$\nu_D = S_N\frac{p_P}{N_{\text{rps}}}\left(\frac{c_J}{r_J}\right)^2 = 0.1402\frac{1562.5(10)^3}{30}\left(\frac{0.064}{40}\right)^2 = 0.018\,693\ \ [\text{Pa s}] = 18.693\ \ [\text{cP}] \tag{f}$$

$$\Delta T_f = 13.8241\left(\frac{p_P}{\gamma_f\, C_{Hf}}\right) = 13.8241\left(\frac{1562.5(10)^3}{861\,(1760)}\right) = 14.254\ [^\circ\text{C}] \tag{g}$$

The average temperature T_{av} of the lubricant is evaluated according to Eq. (12.34), which gives

$$T_{\text{av}} = T_{\text{in}} + \frac{\Delta T_f}{2} = 60 + \frac{14.254}{2} = 67.127\ \ [^\circ\text{C}] \tag{h}$$

It is necessary to find an oil with the calculated viscosity of 18.693 [cP] as in Eq. (f) at the operating temperature of 67.127 [°C]. Using Equation (12.5) and Table 12.1 for an SAE 30 should give the following viscosity:

$$\nu_D = a_{\text{TC}}\left(T_C\right)^{b_{\text{TC}}} = 6.5105(10)^5(67.127)^{-2.4876} = 18.579\ [\text{cP}] \tag{i}$$

This value is close to the 18.693 [cP] in Eq. (f), which indicates that SAE 30 would be a suitable lubrication oil. If the value in Eq. (i) is not close to that in Eq. (f), an iteration is needed to select another SAE oil or use other means to cool the inlet oil or increase the flow by pumping the oil.

The coefficient of friction is defined from Eqs. (d) and (b) such that

$$\mu_K = 3.4288\left(\frac{c_J}{r_J}\right) = 3.4288\left(\frac{1}{625}\right) = 0.005\,486\,1 \tag{j}$$

The expected friction torque $\mathbf{T}_f$ and the power loss H_f in the bearing film are calculated according to Eqs. (12.35) and (12.36) to give

$$\mathbf{T}_f = \mu_K \mathbf{P}\ r_J = 0.005\,486\,1\,\left(2.5(10)^3\right)\left(20(10)^{-3}\right) = 0.274\ \ 30\ \ [\text{N m}] \tag{k}$$

$$H_f = 2\pi \mathbf{T}_f N_{\text{rps}} = 2\pi\,(0.274\,30)\,(30) = 51.704\ \ [\text{W}] \tag{l}$$

The minimum film thickness, the eccentricity ratio, and the eccentricity are very useful geometric parameters to observe. These are evaluated due to Eqs. (12.20) and (12.21) so that

$$h_0 = \left(\frac{h_0}{c_J}\right)c_J = \left(\frac{h_0}{c_J}\right)\left(\frac{c_D}{2}\right) = 0.42\left(\frac{0.064}{2}\right) = 0.013\,44\ \ [\text{mm}] \tag{m}$$

$$\varepsilon = 1 - \frac{h_0}{c_J} = 1 - 0.42 = 0.58 \tag{n}$$

$$e = \varepsilon c_J = \varepsilon\left(\frac{c_D}{2}\right) = 0.58\left(\frac{0.064}{2}\right) = 0.018\,56\ [\text{mm}] \tag{o}$$

It should be expected that the *minimum film thickness* h_0 should be larger than the surface roughness of the journal or bearing surfaces. This should be checked.

Third, the US system variables at operating conditions are evaluated as follows:

$$\nu_D = S_N \frac{p_P}{N_{\text{rps}}} \left(\frac{c_J}{r_J}\right)^2 = 0.1402 \frac{222.22}{30} \left(\frac{0.003}{1.5}\right)^2 = 4.1540(10)^6 \,[\text{R}] = 4.1540 \quad [\mu\text{R}] \tag{p}$$

The coefficient of friction is defined from Eqs. (d) and (b) such that

$$\mu_K = 3.4288 \left(\frac{c_J}{r_J}\right) = 3.4288 \left(\frac{1}{500}\right) = 0.006\,857\,6 \cong 0.007 \tag{q}$$

The expected friction torque $\boldsymbol{T}_f$ and the power loss H_f in the bearing film are calculated according to Eqs. (12.35) and (12.36) to give (see Section 1.11)

$$\boldsymbol{T}_f = \mu_K \boldsymbol{P}\ r_J = 0.006\,857\,6\,(500)\,(0.75) = 2.5716 \ [\text{lb in}] \tag{r}$$

$$H_f = 2\pi \boldsymbol{T}_f N_{\text{rps}} = 2\pi\,(2.5716)\,(30) = 484.74 \quad [\text{lb in/s}]$$

$$H_f = \frac{\boldsymbol{T}_f N_{\text{rps}}}{1050} = \frac{2.5716\,(30)}{1050} = 0.073474 \quad [\text{hp}]$$

$$H_f = \frac{2\pi \boldsymbol{T}_f N_{\text{rps}}}{778\,(12)} = \frac{484.74}{9336} = 0.051922 \quad [\text{Btu/s}] \tag{s}$$

The temperature rise is obtained from Eq. (e) with some needed *Joulean* heat equivalent [Je] units' conversion between heat and work (778(12) = 9336 [in lb/Btu]) as follows:

$$\Delta T_f = 13.8241 \left(\frac{p_P}{\gamma_f\, C_{Hf}}\right) = 13.8241 \left(\frac{222.22}{0.0311\,(0.42)\ [778\,(12)]}\right) = 25.191 \ [^\circ\text{F}] \tag{t}$$

The average temperature T_{av} of the lubricant is evaluated according to Eq. (12.34), which gives

$$T_{\text{av}} = T_{\text{in}} + \frac{\Delta T_f}{2} = 140 + \frac{25.191}{2} = 152.60 \quad [^\circ\text{F}] \tag{u}$$

It is necessary to find an oil with the calculated viscosity of 4.154 [μR] as in Eq. (p) at 152.6 [°F]. Using Eq. (12.5) and Table 12.1 for an SAE 30 should give the following viscosity:

$$\nu_D = a_{\text{TF}} \ e^{(b_{\text{TF}}/(T_F+95))} = 0.0141 e^{(1360/(152.6+95))} = 3.4252 \quad [\mu\text{R}] \tag{v}$$

This value is not very close to the 4.154 [μR] in Eq. (p), which indicates that SAE 30 might not be a very suitable lubrication oil for this design configuration. If the value in Eq. (v) is not close to that in Eq. (p), an iteration is needed to select another SAE oil, or use other means to cool the oil or increase the oil flow by pumping, etc. The viscosity of the SAE 40 at 152.6 [°F] is obtained by Eq. (12.5) as 4.6656 [μR], which is more than the needed value of 4.154 [μR] in Eq. (p). Some iteration is then needed to change this design configuration to be closer to the SI design or iterate the assumed average minimum film thickness ratio (h_0/c_J) or have a mix of SAE 30 and SAE 40, which is not common to do.

The minimum film thickness, the eccentricity ratio, and the eccentricity are very useful geometric parameters to observe. These are evaluated due to Eqs. (12.20) and (12.21) so that

$$h_0 = \left(\frac{h_0}{c_J}\right) c_J = \left(\frac{h_0}{c_J}\right)\left(\frac{c_D}{2}\right) = 0.42 \left(\frac{0.003}{2}\right) = 0.000\,63 \ [\text{in}] \tag{x}$$

$$\varepsilon = 1 - \frac{h_0}{c_J} = 1 - 0.42 = 0.58 \tag{y}$$

$$e = \varepsilon c_J = \varepsilon \left(\frac{c_D}{2}\right) = 0.58 \left(\frac{0.003}{2}\right) = 0.00087 \ [\text{in}] \tag{z}$$

Again, it is expected that the *minimum film thickness* h_0 should be larger than the surface roughness of the journal or bearing surfaces. This should be checked.

Even though the approach herein might be different, the results of this example agree with some of the sample calculations or examples in Shigley (1986) and Budynas and Nisbett (2015).

12.3 Journal Bearing Design Procedure

In Chapters 8 and 11 and particularly for the initial synthesis of rolling bearings, Figures 8.21 and 11.6 are introduced to aid in the synthesis process. Figure 12.10 is similar to these two figures. The emphasis here is on the initial synthesis of journal bearings. Figure 12.10 provides some estimate of special journal bearing design as a function of nominal bearing diameter and *maximum load*. Figure 12.10a is for SI units and Figure 12.10b is for the US units. The heavy dashed line is for the fluid film journal bearings at their *maximum capacities* operating at certain *high rotational speeds* [rpm] that are read off the left scale for the dashed light grayish line. The solid lines are for rolling bearings at a life of 10 000 hours. Each rolling bearing line is for specifically marked rotational speed. Figure 12.10 can be used to define the maximum capacity of a journal bearing running at defined higher rotational speeds. The estimate of maximum load journal bearing design is restricted to the associated distinct rotational speed. Therefore, its utility is confined to these conditions. The bearing diameter for maximum load can be used as a reference since it usually gives a smaller size bearing than usual but should be running at higher rotational speeds. This smaller size and its parametric requirements may be associated with the upper dotted curve in Figure 12.9 defining the *maximum load*. If need be, the associated minimum film thickness ratio (h_0/c_J) for 180° bearings may be used to finalize the synthesis of the *maximum load* bearing at the distinct higher rotational speed. This is not usually the general design process to synthesize the practical journal bearing for some other rotational speeds.

The *practical journal* bearing designs are associated with Figures 12.11 and 12.12. Figure 12.11 presents estimates of practical journal bearing synthesis as functions of nominal bearing diameter. Figure 12.11a is for SI units, and Figure 12.11b is for the US units. Each line is for a marked specific length to diameter ratio (l_J/d_J).

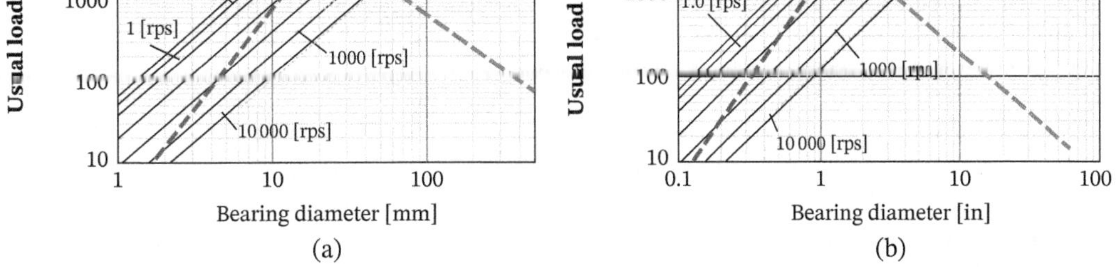

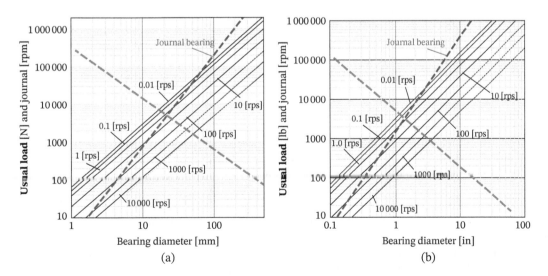

Figure 12.10 (Similar to Figures 8.21 and 11.6) Estimate of practical journal bearing synthesis as a function of nominal bearing diameter and maximum load. (a) SI units and (b) US units. The heavy dashed line is for fluid film journal bearings at their maximum capacities and operating at certain rotational speeds [rpm] (left scale) read off the dashed grayish line. The solid lines are for rolling bearings at a life of 10 000 hours and for specifically marked rolling bearing speed.

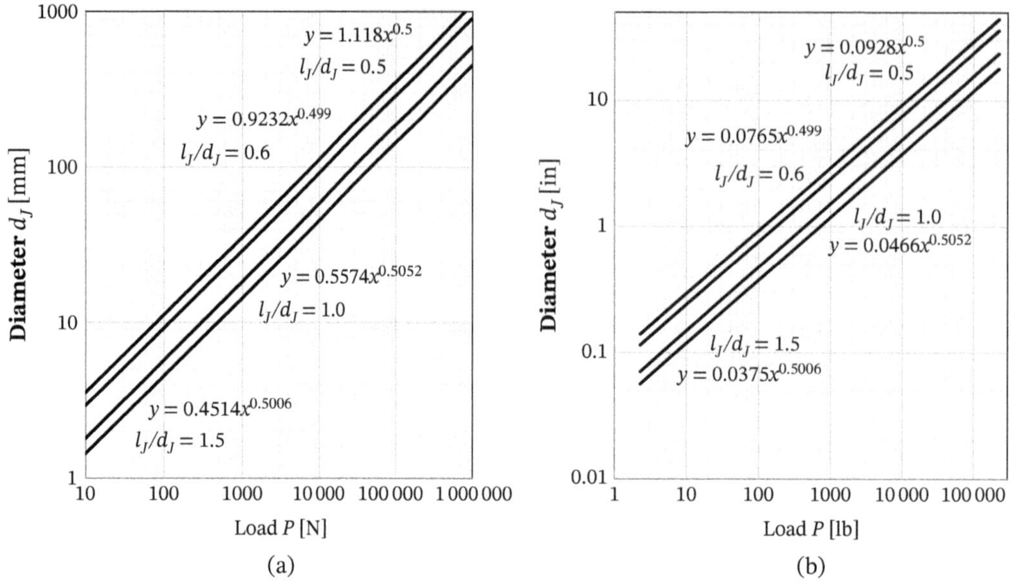

Figure 12.11 Estimate of practical journal bearing synthesis as a function of nominal bearing diameter, where (a) the SI units and (b) the US units. Each line is for a specific length to diameter ratio. The design should fulfill definite clearance and viscosity depending on the rotational speeds as in Figure 12.12.

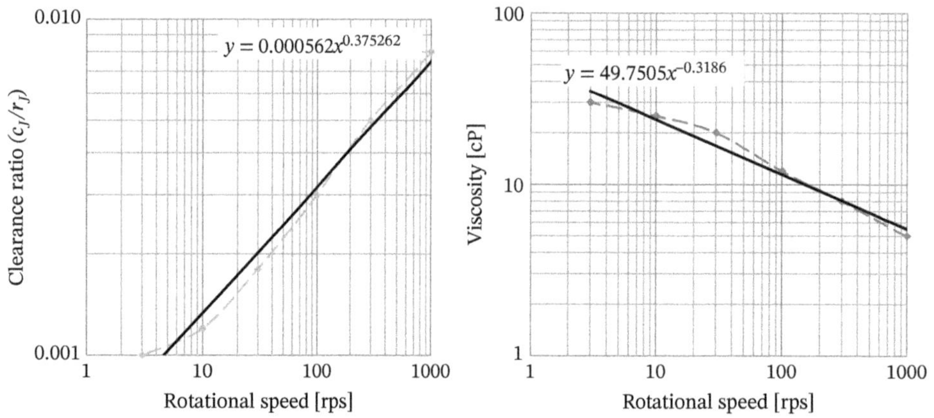

Figure 12.12 The practical journal bearing designs are characterized by clearances and oil viscosities dependent on the shown rotational speeds. Clearance ratio limit is about 0.001. Viscosity limit is about 30 [cP]. The dashed lines are for practical cases. Source: Neale (1973). Tribology Handbook Butterworths.

The design should fulfill definite clearance and viscosity depending on the rotational speeds as in depicted in Figure 12.12. Figure 12.12 provides the practical journal bearing designs, which are characterized by clearances and oil viscosities dependent on the rotational speeds shown on the abscissa coordinate. The ***clearance ratio*** ($c_J/r_J = c_D/d_J$) has usually a lower limit of about 0.001, which indicates a maximum limit of the *inverse clearance ratio* ($r_J./c_J = d_J/c_D.$) of 1000. Viscosity preferred limit is about 30 [cP]. The dashed lines are for practical cases (Neale 1973). From these figures, one should find the practical bearing synthesis as follows:

i. The bearing design is usually for a defined radial load $\boldsymbol{P}$ and rotating speed N_{rps}. In some few cases, the bearing diameter d_J is restrained not to be less than a certain value due to shaft stresses. If not, one can find the practical bearing diameter d_J from Figure 12.11 for the radial load $\boldsymbol{P}$ at the assumed specific length to diameter ratio (l_J/d_J) of 1.5, 1.0, 0.6, or 0.5. The common length to diameter ratio (l_J/d_J) is 1.0. An extra safety factor K_{SF} may be used to account for overloads and ensuring the avoidance of metal to metal contact in those cases. The regression relations to get the practical synthesis of the diameters are as follows (see Figure 12.11):

$$\begin{aligned} d_J &= 1.118\ \boldsymbol{P}^{0.5}\ [\text{mm}], \quad (l_J/d_J) = 0.5 \\ d_J &= 0.9232\ \boldsymbol{P}^{0.499}\ [\text{mm}], \quad (l_J/d_J) = 0.6 \\ d_J &= 0.5574\ \boldsymbol{P}^{0.5052}\ [\text{mm}], \quad (l_J/d_J) = 1.0 \\ d_J &= 0.4514\ \boldsymbol{P}^{0.5006}\ [\text{mm}], \quad (l_J/d_J) = 1.5 \end{aligned} \tag{12.37}$$

$$\begin{aligned} d_J &= 0.0928\,\boldsymbol{P}^{0.5}\ [\text{in}], \quad (l_J/d_J) = 0.5 \\ d_J &= 0.0765\ \boldsymbol{P}^{0.499}\ [\text{in}], \quad (l_J/d_J) = 0.6 \\ d_J &= 0.0466\ \boldsymbol{P}^{0.5052}\ [\text{in}], \quad (l_J/d_J) = 1.0 \\ d_J &= 0.0375\ \boldsymbol{P}^{0.5006}\ [\text{in}], \quad (l_J/d_J) = 1.5 \end{aligned} \tag{12.38}$$

The bearing geometrical and some performance parameters are then readily available to be calculated such as bearing length l_J (from (l_J/d_J)), load per unit area $p_p = \boldsymbol{P}/l_J d_J$, and bearing performance parameter $p_p\ \boldsymbol{V}_J$, where $\boldsymbol{V}_J$ is the peripheral or circumferential journal velocity; see Section 12.4.

ii. At the known rotating speed N_{rps} of the journal bearing, the needed diametral clearance c_D and the lubricant viscosity ν_D are obtained.

$$\begin{aligned} \frac{c_J}{r_J} &= \frac{c_D}{d_J} = 0.000\,562\, N_{\text{rps}}^{0.375262}, \geq 0.001 \\ c_D &= 0.000\,562\, N_{\text{rps}}^{0.375262}\,(d_J) \end{aligned} \tag{12.39}$$

$$\nu_D = 49.751\, N_{\text{rps}}^{-0.3186}, \qquad [\text{cP}] \tag{12.40}$$

The bearing geometrical and some performance parameters are then readily available to be calculated such as clearance c_D (from (c_D/d_J)) and *Sommerfeld* number S_N using Eq. (12.10).

iii. The minimum film thickness ratio h_0/c_J is *iteratively* changed to have the iterative *Sommerfeld* number S_N as close as possible to the calculated *Sommerfeld* number S_N.

The bearing performance parameters are then readily available to be calculated such as the nondimensional friction coefficient variable $(d_J/c_D)\mu_K$, and temperature rise variable T_{var}, given by Eqs. (12.32) and (12.33).

The other performance parameters are also obtainable such as minimum film thickness h_0, (from h_0/c_J), eccentricity ratio e (from Eqs. (12.20) to (12.21)), friction coefficient μ_K from friction coefficient variable $(d_J/c_D)\mu_K$, lubricant temperature rise ΔT from temperature rise variable T_{var}, operating temperature T_{av} from inlet lubricant temperature T_{in} and lubricant temperature rise ΔT (Eq. (12.34)), resisting friction torque T_f (Eq. (12.35)), friction power loss H_f (Eq. (12.36)), and bearing performance parameter $p_p\ \boldsymbol{V}_J$, where $\boldsymbol{V}_J$ is the peripheral or circumferential journal velocity.

iv. The *lubricant definition* is enacted by selecting an SAE lubricant grade and *iteratively* has its viscosity close to the operating viscosity at the operating temperature as defined by Eq. (12.5).

Example 12.2 It is required to design the hydrodynamic journal bearing of Example 12.1 according to practice or knowledge base information and estimate its characteristic values for the geometrical and behavioral parameters. The bearing design calls for a journal bearing running at N_{rpm} of 1800 [rpm], i.e. N_{rps} of 30 [rps], and should carry a load $\boldsymbol{P}$ of 2.5 [kN] or 500 [lb]. The length to diameter ratio l_J/d_J is assumed as 1.0. Find the bearing diameter, one of the expected SAE lubricating fluids at an inlet temperature T_{in} of 60 [°C] or 140 [°F]. Find the expected temperature rise of the lubricating oil, the coefficient of friction, the friction torque, and the power loss in the bearing. Assume a simplified 180° half bearing, and compare results with Example 12.1.

Solution

Data: $N_{\text{rps}} = 30$ [rps], $\boldsymbol{P} = 2.5$ [kN] or 500 [lb], and T_{in} of 60 [°C] or 140 [°F]. Example 12.1 input $d_J = 40$ [mm] or 1.5 [in], $l_J = 40$ [mm] or 1.5 [in], $c_D = 0.064$ [mm] or 0.003 [in]. The length to diameter ratio was $l_J/d_J = 1.0$, and inverse clearance ratio is $r_J/c_J. = d_J/c_D = 625$ for SI or 500 for the US.

To compare results with Example 12.1, a length to diameter ratio l_J/d_J is assumed as 1.0. The rest of the parameters are to be synthesized. Safety factor is also assumed as 1.0. The following is the design procedure as previously stated:

i. The practical synthesis of the journal bearing diameter for $l_J/d_J = 1.0$ is given according to Eq. (12.37) as

$$d_J = 0.5574\ \boldsymbol{P}^{0.5052} = 0.5574\ (2500)^{0.5052} = 29.027\ \ [\text{mm}]$$

$$d_J = 0.0466\ \boldsymbol{P}^{0.5052} = 0.0466\ (500)^{0.5052} = 1.0762\ \ [\text{in}] \tag{a}$$

These values are expected to be rounded to either 29 [mm] and 1.1 [in] or 30 [mm] and 1.2 [in]. In this example the values are kept, so that computer values can be compared directly without interaction. The differences should not be alarming. Then, the bearing length $l_J = 29.027$ [mm] or 1.0762 [in], and the load per unit area for each system is given by the following expression:

$$p_P = \frac{\boldsymbol{P}}{l_J d_J} = \frac{2500}{(29.027)^2} = 2.9671\ \ [\text{MPa}]$$

$$p_P = \frac{\boldsymbol{P}}{l_J d_J} = \frac{500}{(1.0762)^2} = 431.70\ \ [\text{psi}] = 0.4317\ \ [\text{kpsi}] \tag{b}$$

ii. At the rotating speed $N_{\text{rps}} = 30$, the needed diametral clearance c_D and the lubricant viscosity ν_D are obtained from Eqs. (12.39) and (12.40) such that

$$c_D = 0.000\,562\,N_{\text{rps}}^{0.375262}\left(d_J\right) = 0.000\,562(30)^{0.375262}\,(29.027) = 0.058\,458\ \ [\text{mm}]$$

$$c_D = 0.000\,562\,N_{\text{rps}}^{0.375262}\left(d_J\right) = 0.000\,562(30)^{0.375262}\,(1.0762) = 0.002\,167\,4\ \ [\text{in}] \tag{c}$$

$$\nu_D = 49.751\,N_{\text{rps}}^{-0.3186} = 49.751\ \ (30)^{-0.3186} = 16.834\ \ [\text{cP}]$$

$$\nu_D = 16.834\ \ [\text{cP}]\left(1\ \ [\mu\text{R}]\,/6.895\ \ [\text{cP}]\right) = 2.4415\ \ [\mu\text{R}] \tag{d}$$

Since the US is commonly using [cP], there might be no need for the unit conversion.
The *Sommerfeld* number S_N using Eq. (12.10) gives the following value:

$$S_N = \frac{\nu_D N_{\text{rps}}}{p_P}\left(\frac{d_J}{c_D}\right)^2 = \frac{(16.834/1000)\,(30)}{2.9671(10)^6}\left(\frac{29.027}{0.058\,458}\right)^2 = 0.041\,966$$

$$S_N = \frac{\nu_D N_{\text{rps}}}{p_P}\left(\frac{d_J}{c_D}\right)^2 = \frac{\left(2.4414/10^6\right)(30)}{0.4317(10)^3}\left(\frac{1.0762}{0.002\,167\,4}\right)^2 = 0.041\,83 \tag{e}$$

which are very close to each other, and each may be approximated as 0.042. This is, however, different from the 0.1402 of Example 12.1. With these values of S_N and according to Eqs. (12.32), and (12.33), the following nondimensional parameters are calculated;

$$\mu_K\left(\frac{r_J}{c_J}\right) = 0.090\ \ S_N^2 + 19.188\ \ S_N + 0.737 = 0.090(0.042)^2 + 19.188\,(0.042) + 0.737 = 1.5431 \tag{f}$$

$$\Delta T_f \left(\frac{\gamma_f C_{Hf}}{p_P} \right) = -0.3459 \ S_N^2 + 77.394 \ S_N + 2.9807 = -0.3459 \ (0.042)^2 + 77.394 \ (0.042) + 2.9807$$

$$= 6.2306 \tag{g}$$

iii. Using an Excel sheet, the minimum film thickness ratio h_0/c_J is iteratively changed to have the iterative *Sommerfeld* number S_N as close as possible to the calculated *Sommerfeld* number S_N. This gives the following approximate minimum film thickness ratio h_0/c_J and minimum film thickness h_0, (from h_0/c_J):

$$\frac{h_0}{c_J} \cong 0.1844, \ \ h_0 \cong 0.1844 \left(\frac{c_D}{2}\right)$$

$$h_0 \cong 0.1844 \left(\frac{0.058\,458}{2}\right) = 0.005\,389\,8 \ [\text{mm}], \quad \text{and}$$

$$h_0 \cong 0.1844 \left(\frac{0.002\,167\,4}{2}\right) = 0.000\,199\,83 \ [\text{in}] \tag{h}$$

These are realistically defining the minimum film thicknesses $h_0 = 0.0054$ [mm] and 0.0002 [in]. From Eqs. (12.20) and (12.21), the eccentricity ratio ε and the eccentricity e are given by

$$\varepsilon = 1 - \frac{h_0}{c_J} = 1 - 0.1844 = 0.8156$$

$$e = \varepsilon c_J = 0.8156 \left(\frac{0.058\,458}{2}\right) = 0.02384 \ [\text{mm}], \quad \text{and}$$

$$e = \varepsilon c_J = 0.8156 \left(\frac{0.002\,167\,4}{2}\right) = 0.000\,883\,9 \ [\text{in}] \tag{i}$$

The friction coefficient μ_K is obtained from friction coefficient variable $(d_J/c_D)\mu_K$ of Eq. (f) such that

$$\mu_K|_{\text{SI}} = 1.5431 \left(\frac{c_D}{d_J}\right) = 1.5431 \left(\frac{0.058\,458}{29.027}\right) = 0.003\,108$$

$$\mu_K|_{\text{US}} = 1.5431 \left(\frac{c_D}{d_J}\right) = 1.5431 \left(\frac{0.002\,167\,4}{1.0762}\right) = 0.003\,108 \tag{j}$$

The resisting friction torque $\boldsymbol{T}_f$ (Eq. (12.35)) is then given by

$$\boldsymbol{T}_f = \mu_K \boldsymbol{P} \ r_J = 0.003\,108 \left(2.5(10)^3\right) \left(29.027(10)^{-3}/2\right) = 0.112\,77 \ [\text{N.m}]$$

$$\boldsymbol{T}_f = \mu_K \boldsymbol{P} \ r_J = 0.003\,108\,(500)\,(1.0762/2) = 0.836\,21 \ [\text{lb.in}] \tag{k}$$

The friction power loss H_f (Eq. (12.36)) are then

$$H_f = 2\pi \boldsymbol{T}_f N_{\text{rps}} = 2\pi\,(0.112\,77)\,(30) = 21.257 \ \ [\text{W}] \tag{l}$$

and

$$H_f = 2\pi \boldsymbol{T}_f N_{\text{rps}} = 2\pi\,(0.836\,21)\,(30) = 78.811 \ \ [\text{lb in/s}]$$

$$H_f = \frac{\boldsymbol{T}_f N_{\text{rps}}}{1050} = \frac{0.836\,21\,(30)}{1050} = 0.023\,892 \ \ [\text{hp}]$$

$$H_f = \frac{2\pi \boldsymbol{T}_f N_{\text{rps}}}{778\,(12)} = \frac{78.811}{9336} = 0.008\,441\,6 \ \ [\text{Btu/s}] \tag{m}$$

The lubricant temperature rise ΔT is calculated from temperature rise variable T_{var} in Eq. (g). The operating temperature T_{av} is calculated from inlet lubricant temperature T_{in} and lubricant temperature rise ΔT (Eq. (12.34)). This gives the following values:

$$\Delta T_f = 6.2306 \left(\frac{p_P}{\gamma_f C_{Hf}} \right) = 6.2306 \left(\frac{2.9671\,(10^6)}{861\,(1760)} \right) = 12.1996 \ [°\text{C}]$$

$$\Delta T_f = 6.2306\left(\frac{p_p}{\gamma_f\, C_{Hf}}\right) = 6.2306\left(\frac{0.4317\,(10^3)}{0.0311\,(0.42)\;[778\,(12)]}\right) = 22.057\;[^\circ\mathrm{F}] \tag{n}$$

The average temperature T_{av} of the lubricant is evaluated according to Eq. (12.34), which gives

$$T_{\mathrm{av}} = T_{\mathrm{in}} + \frac{\Delta T_f}{2} = 60 + \frac{12.1996}{2} = 66.0998\;[^\circ\mathrm{C}]$$
$$T_{\mathrm{av}} = T_{\mathrm{in}} + \frac{\Delta T_f}{2} = 140 + \frac{22.057}{2} = 151.03\;[^\circ\mathrm{F}] \tag{o}$$

iv. It is necessary to find an oil with the calculated viscosity of 18.384 [cP] as in Eq. (d) at 66.1 [°C]. Using Eq. (12.5) and Table 12.1 for an SAE 30 should give the following viscosity:

$$\nu_D = a_{\mathrm{TC}}\left(T_C\right)^{b_{TC}} = 6.5105(10)^5(66.1)^{-2.4876} = 19.306\;[\mathrm{cP}]$$
$$\nu_D = a_{\mathrm{TF}}\;e^{(b_{\mathrm{TF}}/(T_F+95))} = 0.0141e^{(1360/(151.03+95))} = 3.5473\;[\mu\mathrm{R}] \tag{p}$$

These values are not very close to the 16.834 [cP] or 2.4415 [μR] in Eq. (d), which indicates that SAE 30 might not be a very suitable lubricating oil for this design configuration. If the values in Eq. (d) is not close to that in Eq. (p), an iteration is needed to select another SAE oil, or use other means to cool the oil or change the design parameters or increase the lubricant flow by pumping, etc.

To compare results with Example 12.1, it is apparent that this example allows the synthesis of the bearing, and it is not just an analysis task. The output diameter, clearance, and lubricant viscosity are generated from the practical relation as a knowledge base in the design synthesis. This synthesis in Example 12.2 may not be the optimum, but it produces less friction coefficient (0.003 11) and power (21.26 [W] or 0.024 [hp]) than Example 12.1, where friction coefficient is about (0.007) and power is about (51.7 [W] or 0.073 [hp]). The diameter in Example 12.1 is, however, smaller (about 29 [mm] or about 1.1 [in]), while in Example 12.1 it was 40 [mm] or 1.5 [in]. The minimum film thickness in Example 12.1 is, however, about half of that in Example 12.1. An optimization would improve such discrepancy if a maximum–minimum film thickness is one of the objectives.

12.4 Boundary and Mixed Lubrication

The boundary and mixed film lubrication occur at lower rotational speeds and low viscosity. This causes the critical bearing characteristic value ($\nu_D\, N_{\mathrm{rpm}}/p_p$) to be lower than the instability limit as shown in Figure 12.4. This condition is usually encountered at the starting and stopping of journal bearing operation or when lubricant supply diminishes. The lubricant film thickness is small, and not enough full film materializes to cause a hydrodynamic lubrication as defined in Section 12.2. The journal surface would directly contact the bearing surface unless there is a separating ***thin film*** of lubricant in between. The *thin film* is usually adhering to the surfaces of the journal bearing allowing a *boundary lubrication*. This adhesion is mainly dependent on the lubricant and bearing type. The thin film may, however, be squeezed out, and solid material contacting occurs if the applied *projected pressure* p_p is high enough at still or at very low rotational speeds. The friction is consequently high, and wear of surface increases. To reduce that risk, an *extreme pressure EP* additive to lubricants may be used and a lower *projected pressure* p_p (or bearing *unit load*) would also be recommended. The commonly used values of the projected pressure p_p in *sleeve bearings* for some applications and their projected pressure ranges are given in Table 12.2. The steady loads are for relatively steady rather than purely static. These may be checked for the bearing design of some similar applications.

Most of fluctuating load applications would inherently have a forced flow of lubricant to reduce the boundary or thin film lubrication pitfalls. That is why these fluctuating load applications are operating at a higher projected pressure p_p than anticipated; see Table 12.2. They are usually running at higher speed to form a proper *full film*

Table 12.2 *Sleeve bearings* for some applications and their projected pressure range.

Applications		**Pressure p_p[MPa]**	**Pressure p_p[kpsi]**
Steady loading	Light shafting	0.1–0.2	0.015–0.03
	Marine line shaft bearings	0.2–0.3	0.03–0.04
	Centrifugal pump	0.6–1.2	0.08–0.15
	Electric generators and motors	0.7–1.5	0.1–0.2
	Turbines and gearboxes	0.8–1.7	0.1–0.25
	Heavy shafting	0.8–1.0	0.1–0.15
	Air compressors	1.0–2.0	0.15–0.3
Fluctuating loading	Marine engine bearings	2.8–4.1	0.4–0.6
	Automotive main bearings	3.5–4.8	0.5–0.7
	Aircraft engine bearings	4.8–20.7	0.7–3
	Diesel main bearings	5.5–10.3	0.8–1.5
	Diesel connecting rod	6.9–13.8	1–2
	Automotive connecting rod	10.3–17.2	1.5–2.5
	Automotive wrist pins	12.4–13.8	1.8–2

hydrodynamic bearing character in steady state operation. In starting and stopping of journal bearing operation or when lubricant supply weakens, a *mixed film lubrication* between the thin film and the full film occurs. In addition to low speeds, mixed film lubrication would also occur at low viscosity, high overloads, tight clearances, or misalignment of journal and bearing elements. The use of proper sleeve or bearing material would reduce the effects of mixed film lubrication problems. A useful measure in that regard is the quantity of projected pressure p_p times the journal circumferential speed $\boldsymbol{V}_J$, which can be observed not to exceed the maximum allowable limit $(p_p\ \boldsymbol{V}_J)_{max}$ for the sleeve or bearing material. These materials, however, should not be loaded beyond their maximum projected pressure $(p_p)_{max}$, maximum allowable speed $(\boldsymbol{V}_J)_{max}$, or maximum allowable temperature $(T_J)_{max}$.

The projected pressure p_p in the journal bearing is not usually uniform over the 180° of the full half of the bearing. Its distribution can be assumed as a sinusoidal from the beginning of half the bearing at $\theta = 0$ to the end of that half at $\theta = 180°$ as shown in Figure 12.13. The clearance between the journal shaft and the bearing sleeve or bush is assumed very small to seriously affect that assumption. The maximum projected pressure p_p is not uniform at the value of $\boldsymbol{P}/l_J d_J$ but can be mathematically obtained as the sinusoidal $p_p = (p_p)_{max} sin\theta$ from $\theta = 0$ to $\theta = 180°$, *i.e.* $\theta = \pi$. The force component in the y direction is $p_p dA\ \sin\theta$. The total load $\boldsymbol{P}$ is then obtained as follows:

$$\boldsymbol{P} = \int_0^{\pi} p_p dA \sin\theta = \int_0^{\pi} p_p \left(l_J r_J\ d\theta\right) \sin\theta = \int_0^{\pi} \left(p_p\right)_{\max} \sin\theta \left(l_J r_J\ d\theta\right) \sin\theta$$
$$= \int_0^{\pi} \left(p_p\right)_{\max} \left(l_J r_J\right) \sin\theta^2\ d\theta = \frac{\pi}{2} \left(p_p\right)_{\max} \left(l_J r_J\right) \tag{12.41}$$

or

$$\left(p_p\right)_{\max} = \frac{2}{\pi\left(l_J d_J/2\right)} \boldsymbol{P} = \frac{4}{\pi} \frac{\boldsymbol{P}}{\left(l_J d_J\right)} = \frac{4}{\pi} p_p \cong 1.2732\ p_p \tag{12.42}$$

Equation (12.42) indicates that it is more advisable to use the maximum projected pressure $(p_p)_{max}$ as about $(1.2732\ p_p)$ rather than only p_p; in the evaluation of maximum pressure, the bearing sleeve should stand.

The sleeve or bush material should be able to withstand the *maximum projected pressure* $(p_p)_{max}$ and have other qualifications as will be discussed next.

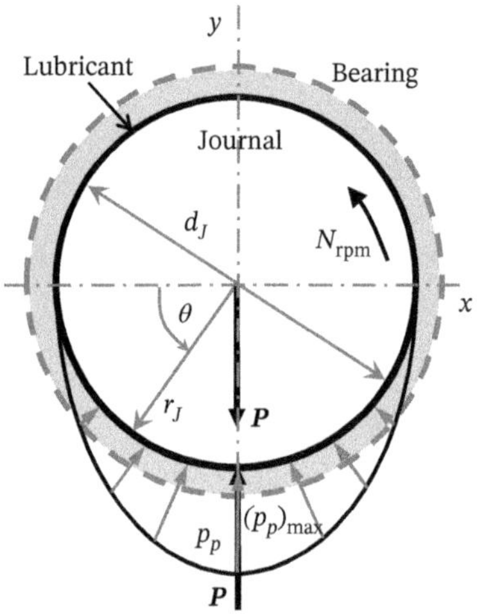

Figure 12.13 A more realistic projected pressure distribution is not usually uniform over the 180° of the full half of the bearing. Its assumed distribution can have a sinusoidal form starting from the beginning of half the bearing at $\theta = 0$ to the end of the half at $\theta = 180°$.

12.5 Plain Bearing Materials

The desirable properties of materials used for plain bearings are somewhat unique, and compromises must be made among needed and available materials. The general *preferability* requirements that bearing materials ought to satisfy include static or fatigue strength, embeddability of invading particles, corrosion resistance, and cost. Some available materials for such bearings with their expected maximum properties and requirements are shown in Table 12.3. Some metallic bearings are made of alloys that are characterized by low friction with other materials in dry or wet states. Nonmetallic materials are also characterized by low friction in dry or some wet states, but they are usually suitable for lower pressure-speed values. Some other materials are self-lubricated porous bearings made typically of sintered materials. The sintering process uses particles of few different materials that are bonded together under measured compression and heat. This process leaves voids between particles that are typically filled with lubricant, and thus the porous assemblage becomes self-lubricating. Some might purposely have cavities that are filled with lubricant to replenish the bearing with.

The general characteristic ranges of clearance ratios (c_J/r_j), layer thicknesses t_L, and relative *preferability* (best, very good, good, and fair) for some conventional bearing alloys are as follows:

- *Tin or lead base babbitt*: $(c_J/r_j) = 0.001$–0.0017, $t_L = 0.1$–0.6 [mm] or 0.004–0.022 [in], and best–very good.
- *Silver plus overlay*: $(c_J/r_j) = 0.001$–0.0017, $t_L = 0.33$ [mm] or 0.013 [in], and best–good.
- *Leaded bronze*: $(c_J/r_j) = 0.001$–0.002, $t_L =$ solid, and very good–good.
- *Copper–lead*: $(c_J/r_j) = 0.001$–0.002, $t_L = 0.6$ [mm] or 0.022 [in], and very good–fair.
- *Aluminum alloy*: $(c_J/r_j) = 0.002$–0.0025, solid, and very good–fair.
- *Cadmium-1.5%Ni*: $(c_J/r_j) = 0.002$–0.0025, $t_L = 0.6$ [mm] or 0.022 [in], and good–fair.

The layer thickness facing the journal is deposited or cast as an overlay on the metal backing of the sleeve or bush. Some layer thicknesses can be higher than the ones previously stated due to large sizes and required applications. The solid layer indicates that the whole sleeve or bush is made of the said alloy. Some typical *dimensions* and *materials* of sleeves and bushes are provided by manufacturers and suppliers; see, e.g. the Internet Section of References.

The main approach to the design of boundary lubrication bearings is to have a bearing pressure p_p less than the maximum allowable pressure of selected bearing material in Table 12.3. The preliminary guide of Figure 12.10 can also be approximately employed for *sleeve bearings* in place of the rolling bearing lines at low speeds (Neale 1973).

Table 12.3 Some available materials for bearings with their expected maximum properties.

Material		Pressure $(p_p)_{max}$ [MPa] ([kpsi])	Temperature $(T_J)_{max}$ [°C] ([°F])	Speed $(V_J)_{max}$ [m/s] ([fpm])	$(p_p\ V_J)_{max}$ [MPa m/s] ([kpsi fpm])
Metallic, and porous, self-impregnated bearings	Copper–iron (hardenable)	345 (50)	180 (350)	0.2 (35)	2.6 (75)
	Copper–iron	138 (20)	180 (350)	1.1 (225)	1.2 (35)
	Cast bronze	31 (4.5)	160 (325)	7.6 (1500)	1.7 (50)
	Porous bronze	32 (4.5)	70 (160)	7.6 (1500)	1.7 (50)
	Porous iron	55 (8)	70 (160)	4 (800)	1.7 (50)
	Aluminum	28 (4)	150 (300)	6.1 (1200)	1.7 (50)
	Lead–iron	28 (4)	65 (150)	4 (800)	1.7 (50)
	Bronze–iron	72 (10.5)	65 (150)	4 (800)	1.2 (35)
	Iron	69 (10)	65 (150)	2 (400)	1.0 (30)
Nonmetallic bearings	Teflon fabric	410 (60)	260 (500)	0.25 (50)	0.8 (25)
	Phenolics	6	93 (200)	12.7 (2500)	0.5 (15)
	Carbon graphite	6	400 (750)	12.7 (2500)	0.5 (15)
	Reinforced Teflon	2.5	260 (500)	5 (1000)	0.3 (10)
	Nylon	1	93 (200)	5 (1000)	0.1 (3)
	Delrin	1	82 (180)	5 (1000)	0.1 (3)
	Teflon	0.5	260 (500)	0.5 (100)	0.03 (1)
	Rubber	0.05	65 (150)	20 (4000)	0.5 (15)

The coefficient of friction may range between 0.03 and 0.05 for some boundary lubricated sleeves or bushes. This might also be helpful in assessing the design of the bearing.

Example 12.3 It is required to synthesize the bearings of the pivot shaft between the hood and the mast of the *Jib crane* in Examples 8.2, 8.5, 11.1, and 11.4–11.7. The sketch of the involved *Jib-crane* section is shown in Figure 12.14. It is a modified form of Figure 8.17b. Find the types and dimensions of the suggested journal bearings. Assume that the maximum rotational speed of the boom is about 0.1 [rps].

Solution

Data: The maximum load between the pivot shaft of the hood and the mast is the mast force $\boldsymbol{F}_1 = (-25.33, -10, 0)$ [kN] or $(-5.574, -2.2, 0)$ [klb] (Figure 8.18b and Example 8.2). At the bearing ring of Figures 8.17b and 11.17, the maximum load between the hood and the mast is the mast force $\boldsymbol{F}_2 = (25.33, 0, 0)$ [kN] or $(5.574, 0, 0)$ [klb] as calculated previously in Example 8.2. The calculations of Example 8.5 are repeated herein to allow for the considerations of the boundary lubrication rules to take over the previous results.

From Example 8.5, the following pertinent information is imported as cited:

The maximum load on the pivot shaft is acting as a thrust force in y direction and a radial force in the x direction. The radial force of $\boldsymbol{F}_{1x} = -25.33$ [kN] or -5.574 [klb] should be carried by the bearing between the pivot shaft of the hood and the top plate of the mast; see Figure 8.18b. The thrust force of $\boldsymbol{F}_{1y} = -10$ [kN] or -2.2 [klb] should also be carried by the bearing between the pivot shaft of the hood and the journal bearing in the top plate of the mast; see Figure 11.17.

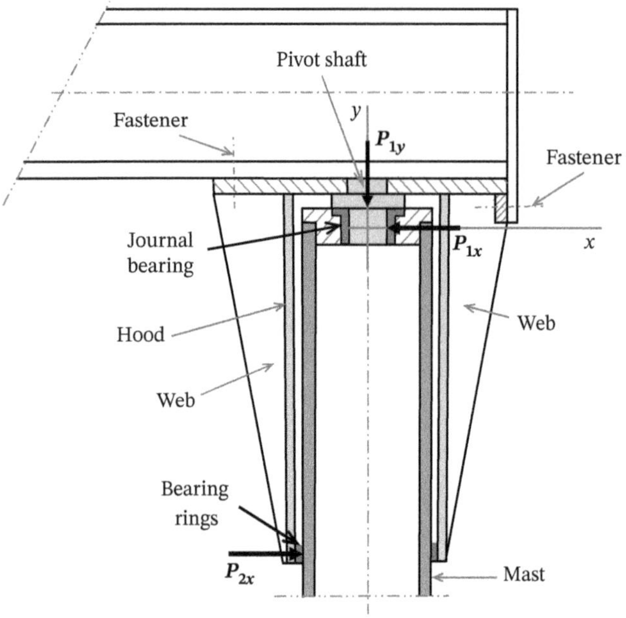

Figure 12.14 (Similar to Figure 8.17b) The conceptual *Jib crane* uses standard sections such as wide flange I-beam, cylindrical tubes, plates, etc. A sectional view of a suggested construction of the main crane section (with I-beam not sectioned) showing a journal bearing sketch.

The synthesis of the radial bearing is possible by utilizing Figure 8.21 to get the diameter of the pivot shaft. In Figure 8.21a and at about 25.0 [kN] on the ordinate axis, the diameter is about 40 [mm] read on the abscissa axis. In Figure 8.21b and at about 5.6 [klb] on the ordinate axis, the diameter is about 1.6 [in] read on the abscissa axis.

The synthesis of the thrust bearing may be approximately possible by utilizing Figure 8.21 to get the diameter and thus the area of the pivot shaft thrust bearing. In Figure 8.21a and at 10.0 [kN] on the ordinate axis, the diameter is about 22 [mm] read on the abscissa axis. In Figure 8.21b and at about 2.2 [klb] on the ordinate axis, the diameter is about 0.9 [in] read on the abscissa axis. The bearing projected area is the one carrying the load. With the length equal to the diameter, the projected area is 22(22) = 484 [mm^2] or 0.9(0.9) = 0.81 [in^2]. This area should be the annular area of the thrust bearing. As the shaft diameter d_S is 40 [mm] or 1.6 [in], the outer diameter of the thrust bearing d_{Bo} can be obtained from the following relation of the bearing area:

$$484 = \frac{\pi}{4}\left(d_{Bo}^2 - d_S^2\right) = \frac{\pi}{4}\left(d_{Bo}^2 - 40^2\right) \quad \text{or} \quad d_{Bo} = \sqrt{\frac{484\,(4)}{\pi} + 40^2} = 47.077 \quad [\text{mm}]$$

$$0.81 = \frac{\pi}{4}\left(d_{Bo}^2 - d_S^2\right) = \frac{\pi}{4}\left(d_{Bo}^2 - 1.6^2\right) \quad \text{or} \quad d_{Bo} = \sqrt{\frac{0.81\,(4)}{\pi} + 1.6^2} = 1.895 \quad [\text{in}] \qquad \text{(a)}$$

Therefore, the sleeve collar outer diameter d_{Bo} of the thrust bearing can be 50 [mm] or 2 [in]. This would be suitable as an initial rough estimation of the thrust bearing.

The maximum load on the bearing ring in Figures 8.18b and 12.13 is acting as a radial force in the x direction. The radial force of $\boldsymbol{F}_{1x} = 25.33$ [kN] or 5.574 [klb] should be carried by the bearing between the hood and the mast; see Figure 8.17b. In Figure 8.21a and at about 25.0 [kN] on the ordinate axis, the diameter is about 40 [mm] read on the abscissa axis. In Figure 8.21b and at about 5.6 [klb] on the ordinate axis, the diameter is about 1.6 [in] read on the abscissa axis. These are the same values as the bearing of the pivot shaft. Since the diameter of the bearing ring is much larger, the width or length of the bearing is the one to find. The bearing projected area is the one carrying the load. With the length equal to the diameter, the projected area is again 22(22) = 484 [mm^2] or

0.9(0.9) = 0.81 [in^2]. This area should be the annular area of the ring bearing. As the diameter of the ring bearing is about 340 [mm] or 13.4 [in] (see Example 8.2), the width or length of the ring bearing l_R can be obtained from the following relation of the bearing area:

$$484 = d_R l_R \quad \text{or} \quad l_R = \frac{484}{340} = 1.42 \ [\text{mm}]$$
$$0.81 = d_R l_R \quad \text{or} \quad l_R = \frac{0.81}{13.4} = 0.06 \ [\text{in}] \tag{b}$$

These values are very small, and the limits would be the possible width of the rings that are welded to the hood and the mast. Usually, the ring size should be about the thickness of the tubes for the mast or the hood. That would be about 10 [mm] or0.4 [in].

To check on the findings of Example 8.5 and to prepare for the information needed for boundary lubrication and noting that the loads are given the symbol $\boldsymbol{P}$ rather than $\boldsymbol{F}$, the subsequent data is calculated for the transverse loading $\boldsymbol{P}_{1x}$ of the upper bearing in Figure 12.14 as follows:

- The circumferential journal speed $\boldsymbol{V}_J$ (assuming the pivot shaft diameter to be 40 [mm] or 1.6 [in]) is

$$\boldsymbol{V}_J = 2\pi \, r_J N_{\text{rps}} = 2\pi \left(\frac{40 \left(10^{-3}\right)}{2} \right) (0.1) = 0.012566 \ \left[\text{m/s}\right]$$
$$\boldsymbol{V}_J = 2\pi \, r_J N_{\text{rps}} = 2\pi \left(\frac{1.6}{2\,(12)} \right) (0.1) = 0.041888 \ \left[\text{ft/s}\right] = 2.5133 \ \left[\text{fpm}\right] \tag{c}$$

- The projected pressure p_p on the journal bearing in the x direction of Figure 12.14 is given by

$$p_p = \frac{\boldsymbol{P}_{1x}}{l_J d_J} = \frac{25.33 \left(10^3\right)}{(40)\,(40)\left(10^{-6}\right)} = 15.831 \ [\text{MPa}]$$
$$p_p = \frac{\boldsymbol{P}_{1x}}{l_J d_J} = \frac{5.574 \left(10^3\right)}{(1.6)\,(1.6)} = 2.1773 \ \left[\text{kpsi}\right] \tag{d}$$

- The maximum projected pressure $(p_p)_{\max}$ on each bearing, which is calculated according to Eq. (12.42), is given by

$$\left(p_p\right)_{\max} = 1.2732 p_p = 1.2732\,(15.831) = 20.156 \ [\text{MPa}]$$
$$\left(p_p\right)_{\max} = 1.2732 p_p = 1.2732\,(2.1773) = 2.7721 \ \left[\text{kpsi}\right] \tag{e}$$

- The maximum $p_p \ \boldsymbol{V}_J$ is then obtained as follows:

$$p_p.\boldsymbol{V}_J = p_p \left(\boldsymbol{V}_J\right) = 20.156\,(0.012566) = 0.25328 \ \left[\text{MPa.m/s}\right]$$
$$p_p.\boldsymbol{V}_J = p_p \left(\boldsymbol{V}_J\right) = 2.7721\,(2.5133) = 6.9671 \ \left[\text{kpsi.fpm}\right] \tag{f}$$

The proper selection of the *journal bearing* for transverse loading $\boldsymbol{P}_{1x}$ can be a porous bronze according to Table 12.3. All its properties are below the calculated ones in Eqs. (c)–(f). The final selection is contingent on its ability to withstand the axial loading $\boldsymbol{P}_{1y}$; see Figure 12.14.

To check on the findings of Example 8.5 for the axial load $\boldsymbol{P}_{1y}$, the subsequent data is calculated for the axial loading $\boldsymbol{P}_{1y}$ of the upper bearing in Figure 12.14 as follows:

- The maximum circumferential thrust journal speed $\boldsymbol{V}_J$ (assuming the pivot shaft diameter to be 40 [mm] or 1.6 [in], and the sleeve collar outer diameter of 50 [mm] or 2 [in] as the thrust journal bearing outer diameter) is

$$\boldsymbol{V}_J = 2\pi \, r_J N_{\text{rps}} = 2\pi \left(\frac{50 \left(10^{-3}\right)}{2} \right) (0.1) = 0.015\,708 \ \left[\text{m/s}\right]$$
$$\boldsymbol{V}_J = 2\pi \, r_J N_{\text{rps}} = 2\pi \left(\frac{2}{2\,(12)} \right) (0.1) = 0.052\,36 \ \left[\text{ft/s}\right] = 3.1416 \ \left[\text{fpm}\right] \tag{g}$$

- The projected pressure p_p on the journal bearing in the y direction of Figure 12.14 (where d_{BC} is the sleeve collar diameter in Eq. (a)) is given by

$$p_p = \frac{4\boldsymbol{P}_{1y}}{\pi\left(d_{Bc}^2 - d_J^2\right)} = \frac{4\left(10.0\left(10^3\right)\right)}{\pi\left(50^2 - 40^2\right)\left(10^{-6}\right)} = 14.147 \text{ [MPa]}$$
$$p_p = \frac{4\boldsymbol{P}_{1y}}{\pi\left(d_{Bc}^2 - d_J^2\right)} = \frac{4\left(2.2\left(10^3\right)\right)}{\pi\left(2^2 - 1.6^2\right)} = 1.9452 \text{ [kpsi]} \tag{h}$$

- The maximum projected pressure $(p_p)_{max}$ on each bearing, which is calculated according to Eq. (12.42), is given by

$$\left(p_p\right)_{max} = 1.2732 p_p = 1.2732\,(14.147) = 18.012 \text{ [MPa]}$$
$$\left(p_p\right)_{max} = 1.2732 p_p = 1.2732\,(1.9452) = 2.4767 \text{ [kpsi]} \tag{i}$$

- The maximum $p_p\ \boldsymbol{V}_J$ is then obtained as follows:

$$p_p.\boldsymbol{V}_J = p_p\left(\boldsymbol{V}_J\right) = 18.012\,(0.015\,708) = 0.282\,93 \text{ [MPa.m/s]}$$
$$p_p.\boldsymbol{V}_J = p_p\left(\boldsymbol{V}_J\right) = 2.4767\,(3.1416) = 7.7808 \text{ [kpsi.fpm]} \tag{j}$$

The proper selection of the *journal bearing* for axial loading $\boldsymbol{P}_{1y}$ can then be the porous bronze according to Table 12.3. All its properties are below the calculated ones in Eqs. (g)–(j). The final selection is then the porous bronze, which was able to withstand the axial loading $\boldsymbol{P}_{1y}$; see Figure 12.14. The safety factor is obviously high, and the porous bronze is usually self-lubricated or impregnated with lubricant for a long life. Supplying lubricant can be intermittent, and the means of delivery should be accounted for.

To check on the findings of Example 8.5 for the radial load $\boldsymbol{P}_{2x}$, the subsequent data is calculated for this radial loading of the lower bearing ring in Figure 12.14 as follows:

- The circumferential journal speed $\boldsymbol{V}_J$ (assuming the inside diameter of the hood tube to be 330 [mm] or 13 [in] for the DN-350 pipe or tube with a 355.6 [mm] outside diameter and an 9.53 [mm] thickness or NSP-14 with a 14 [in] outside diameter and an 0.375 [in] thickness; see Example 8.2) is

$$\boldsymbol{V}_J = 2\pi\, r_J N_{rps} = 2\pi\left(\frac{330\left(10^{-3}\right)}{2}\right)(0.1) = 0.10367 \text{ [m/s]}$$
$$\boldsymbol{V}_J = 2\pi\, r_J N_{rps} = 2\pi\left(\frac{13}{2\,(12)}\right)(0.1) = 0.34034 \text{ [ft/s]} = 20.420 \text{ [fpm]} \tag{k}$$

- The projected pressure p_p on the journal bearing in the x direction of Figure 12.14 is given by

$$p_p = \frac{\boldsymbol{P}_{1x}}{l_J d_J} = \frac{25.33\left(10^3\right)}{(10)\,(330)\left(10^{-6}\right)} = 7.6758 \text{ [MPa]}$$
$$p_p = \frac{\boldsymbol{P}_{1x}}{l_J d_J} = \frac{5.574\left(10^3\right)}{(0.4)\,(13)} = 1.0719 \text{ [kpsi]} \tag{l}$$

- The maximum projected pressure $(p_p)_{max}$ on each bearing, which is calculated according to Eq. (12.42), is given by

$$\left(p_p\right)_{max} = 1.2732 p_p = 1.2732\,(7.6758) = 9.7728 \text{ [MPa]}$$
$$\left(p_p\right)_{max} = 1.2732 p_p = 1.2732\,(1.0719) = 1.3647 \text{ [kpsi]} \tag{m}$$

- The maximum $p_p\ \boldsymbol{V}_J$ is then obtained as follows:

$$p_p.\boldsymbol{V}_J = p_p\left(\boldsymbol{V}_J\right) = 9.7728\,(0.103\,67) = 1.0131 \text{ [MPa m/s]}$$
$$p_p.\boldsymbol{V}_J = p_p\left(\boldsymbol{V}_J\right) = 1.3647\,(20.420) = 27.868 \text{ [kpsi fpm]} \tag{n}$$

The proper selection of the *bearing ring* for transverse loading $\boldsymbol{P}_{2x}$ can be a bronze–iron according to Table 12.3. All its properties are below the calculated ones in Eqs. (k)–(n). The safety factor should be acceptable enough, and lubricant (grease) can be easily admitted. Supplying lubricant can be intermittent, and the means of delivery should also be accounted for. Note also that this ring should be attached or welded to the hood or the mast of the *Jib crane*. A use of just iron bearing may be desirable. The bearing length can then be much more than 10 [mm] or 0.4 [in] to satisfy the maximum properties.

12.6 CAD and Optimization

In Section 12.2.2 of journal bearings, the behavior of finite length bearing has been considered. The *numerical solution* of the 180° bearing performance is used to get the *Sommerfeld* number S_N at $(l_J/d_J = 1)$ as a function of the *minimum film thickness ratio* (h_0/c_J); see Eq. (12.31). With the known *Sommerfeld* number S_N, one can then get the main nondimensional bearing performance variables such as the *coefficient of friction variable* $\mu_K\ (r_J/c_J)$ and the *temperature rise variable* $\Delta T_f\ (\gamma_f C_{Hf}/p_p)$. From these nondimensional performance variables, one can get the coefficient of friction, the expected temperature rise, friction torque, and the power loss for the usual known design requirement of the applied load $\boldsymbol{P}$, rotational speed N_{rps}, and under some diametral clearance c_D. It was postulated that it is easier to assume the *minimum film thickness ratio* (h_0/c_J) and design the bearing with its required performance accordingly. The knowledge base or practice experience helps in the selection of the unknown parameters such as the journal bearing diameter d_J, the lubricant viscosity ν_D, other performance variables, and the lubricant grade definition. This process has been clarified in Section 12.3 of journal bearing design procedure. Herein, this knowledge base or practice process and an optimization approach to bearing synthesis are to be presented next.

12.6.1 CAD of Bearing Synthesis Using Knowledge Base Practice

The design procedure under the knowledge base practice has been suggested in Section 12.3 and implemented in Example 12.2. The process has some requirements such as the assumed length to diameter ratio (l_J/d_J) of 1.0, and the proper minimum film thickness ratio h_0/c_J to calculate the *Sommerfeld* number S_N. Also, there are assumed relations defining the bearing diameters as functions of the load $\boldsymbol{P}$ in Eqs. (12.37) and (12.38). Other relations define or suggest the needed diametral clearance c_D and the lubricant viscosity ν_D as functions of the known journal bearing's rotating speed N_{rps} in Eqs. (12.39) and (12.40). With an assumed lubrication oil grade and initial lubricant temperature, the final temperature is calculated where the lubricant viscosity ν_D should be close to the suggested viscosity. The oil viscosity change with temperature is defined by Eq. (12.5) and Table 12.1.

Figure 12.15 shows the MATLAB code for journal bearings – 180° bearing under the name `CAD_Journal_Bearings.m`. All components of Figure 12.15a–c are parts of the unified code. Figure 12.15a presents the applied loads and geometry with the inputs clearly identified as comments `%` **INPUT** in bold letters, and variables are defined as close as possible to the symbols used in this text. The inputs are the loads (`P_kN P_lb`), rotational speed (`Nrpm`) [rpm}, assumed journal bearing diameters and lengths with l_J/d_J of 1.0 (`dJ_mm=lJ_mm, dJ_in=lJ_in`), diametral clearances (`cD_mm, cD_in`), and minimum film thickness ratio h_0/c_J as (`h0_cJ`) for both SI and the US systems. Figure 12.15b calculates the *Sommerfeld number* S_N as (`SN`), coefficient of friction variable, and temperature rise variable as functions of the assumed minimum film thickness ratio h_0/c_J. The inputs are the oil inlet temperature (`Temp_in_C, Temp_in_F`) and oil properties. Figure 12.15c calculates the lubricant viscosity at operating temperature (`nu_cPoise, nu_micoReyn`), kinetic coefficient of friction (`mu_K_SI, mu_K_US`), and the temperature rise (`dlt_T_C, dlt_T_F`) as functions of the *Sommerfeld number*. The input is the SAE oil grade as (`SAE`). The hand calculations of Example 12.2 verified the results of the MATLAB code and vice versa.

```
clear all; clc; format compact; format short      % CAD_Journal_Bearings.m
 disp('-Journal Bearings Applied Loads and Geometry -180 bearing')
P_kN = 2.5 , P_lb= 500                      % INPUT Load [kN] and [lb]
Nrpm = 1800                                 % INPUT Rotational speed [rpm]
Nrps = Nrpm/60
dJ_mm = 40 , lJ_mm = 40                     % INPUT Journal bearing diameter and length [mm]
dJ_in = 1.5 ,  lJ_in = 1.5                  % INPUT Journal bearing diameter and length [in]
cD_mm = 0.064 , cD_in = 0.003               % INPUT Diametral clearance [mm] and [in]
rJ_mm = dJ_mm/2   , rJ_in = dJ_in/2 % Radial clearance [mm] and [in]
lJ_dJ = lJ_mm/dJ_mm                                   % Length to diameter ratio
pp_kPa = P_kN*10^6/(dJ_mm*lJ_mm)                      % Projected pressure [kPa]
pp_psi = P_lb/(dJ_in*lJ_in)                           % Projected pressure [psi]
cJ_mm =  cD_mm/2                                      % Radial clearance [mm]
cJ_in = cD_in/2                                       % Radial clearance [in]
rJ_cJ_SI = rJ_mm/ cJ_mm                               % Radius to radial clearance ratio SI
rJ_cJ_US = rJ_in/ cJ_in                               % Radius to radial clearance ratio US

 disp('-Journal Bearings Assumed Solution (lJ/dJ=1.0) -180 bearing')
h0_cJ = 0.42                                % INPUT Minimum film thickness ratio h0/cJ
h0_mm = h0_cJ*cJ_mm                                   % Minimum film thickness h0 [mm]
h0_in = h0_cJ*cJ_in                                   % Minimum film thickness h0 [in]
epsilon = 1-h0_cJ                                     % Eccentricity ratio
e_mm= epsilon*cJ_mm                                   % Eccentricity [mm]
e_in= epsilon*cJ_in                                   % Eccentricity [in]
```

(a)

```
 disp('-Lubrication Oil -Assumed Data ')                    %CAD_Journal_Bearings.m
Temp_in_C = 60                                   % INPUT Inlet temperature [oC]
Temp_in_F = 140                                  % INPUT Inlet temperature [oF]
Gamaf_SI = 861                                   % INPUT Mass per unit volume [kg/m3]
CHf_SI = 1760                                    % INPUT Specific heat [J/kg oC]
Gamaf_US = 0.0311                                % INPUT Mass per unit volume [lbm/in3]
CHf_US = 0.42                                    % INPUT Specific heat [Btu/lbm oF]
Energy_Conv_US = 778*12

 disp('-Journal Bearings Nondimensional and Basic Solution (lJ/dJ=1.0) -180 bearing')
 disp('-Sommerfeld number at the minimum film thickness ratio -180 bearing')
x = h0_cJ;
y = 42.765*x^6 -87.052*x^5 + 66.687*x^4 -23.165*x^3 + 4.0453*x^2 -0.0795*x + 0.0041;
SN = y

 disp('-Coefficient of friction variable as a function of Sommerfeld number -180 bearing')
x = SN;
y = 0.090*x^2 + 19.188*x + 0.737;
mu_var = y

 disp('-Temperature rise variable as a function of Sommerfeld number -180 bearing')
x = SN;
y = -0.3459*x^2 + 77.394*x + 2.9807;
dlt_T_var = y
```

(b)

Figure 12.15 (a) MATLAB code for journal bearings – 180 bearing; (a) applied loads and geometry; (b) Sommerfeld number, coefficient of friction variable, and temperature rise variable; and (c) calculates coefficient of friction, temperature rise, and lubricant viscosity.

The code in Figure 12.15 can be used iteratively to change any of the assumed values of diameters, minimum film thickness ratio, inlet lubricant temperature, and the selected SAE oil grade to satisfy the same calculated lubricant viscosity. SAE oil grades other than SAE 30 and SAE 40 need to be formulated into the code using Eq. (12.5) and Table 12.1; see Figure 12.15c.

The code is available at the ***Wiley website*** under the name ***`CAD_Journal_Bearings.m`***.

```
 disp('-Calculated Lubricant Viscosity at Operating Conditions') %CAD_Journal_Bearings.m
nu_cPoise = 1000*(SN*pp_kPa*1000/(Nrps*(rJ_mm/cJ_mm)^2))
nu_micoReyn = 1000000*(SN*pp_psi/(Nrps*(rJ_in/cJ_in)^2))

 disp('-Calculated Coefficient of Friction -180 bearing')
mu_K_SI = mu_var/ rJ_cJ_SI
mu_K_US = mu_var/ rJ_cJ_US

 disp('-Calculated Temperature Rise and Average -180 bearing')
dlt_T_C = dlt_T_var*(pp_kPa*1000/(Gamaf_SI*CHf_SI))
Temp_av_C = Temp_in_C + (dlt_T_C/2)
dlt_T_F = dlt_T_var*(pp_psi/((Gamaf_US*CHf_US)*Energy_Conv_US))
Temp_av_F = Temp_in_F + (dlt_T_F/2)

 disp('-Absolute dynamic viscosity as a Function of Temperature')
SAE = 30                                  % INPUT SAE oil grade
% SAE = 40                                % INPUT SAE oil grade
if SAE == 30 aTC = 6.5105E+05; bTC = -2.4876; else end
if SAE == 30 aTF = 0.0141; bTF = 1360; else end
if SAE == 40 aTC = 4.3141E+06; bTC = -2.8496; else end
if SAE == 40 aTF = 0.0121; bTF = 1474.4; else end
                                                             % Add other SAE oil grades
nu_Tav_cPoise = aTC*(Temp_av_C)^bTC
nu_Tav_micoReyn = aTF*exp(bTF/(Temp_av_F+95))
```

(c)

Figure 12.15 *(Continued)*

12.6.2 CAD of Bearing Synthesis Using an Optimization Approach

Optimization of journal bearings has been studied by many researchers. Different approaches have been used to optimize the bearing design for different objectives and some design parameters; see, e.g. Seireg and Ezzat (1969) and Metwalli et al. (1984). The interest here is to use some of the previous optimization results to aid in the synthesis of the journal bearings. The previous knowledge base, or part of it, is also utilized in this effort.

The results of one of the multi-objective optimizations concerning the minimum power loss, maximum–minimum film thickness, and minimum bearing size is used to synthesize a better bearing; see Metwalli et al. (1984). The results are for a full 360° bearing, and using these for half 180° bearings might not be accurate. The adapted procedure should be checked against other procedures and should be cautiously utilized as an approximate solution. The optimization results gave an optimum length to diameter ratio l_J/d_J of 0.6 rather than 1.0. The outcome also suggests an optimum relation between the clearance ratio (c_J/r_J) and the optimum viscosity criterion as defined in Figure 12.16, where the clearance ratio (c_J/r_J) is in the range of about 0.001–0.01. The dashed line is for the optimum output (Metwalli et al. 1986). The assumed objective is selected for minimum power loss, maximum–minimum film thickness, and minimum bearing size, all having the same weight or importance. The nondimensional *optimum viscosity criterion* ν_C^* is related to the lubricant viscosity ν_D by the following relation:

$$\nu_C^* = \frac{\nu_D N_{\text{rps}} d_J^2}{\boldsymbol{P}}$$

$$\nu_D = \nu_C^* \frac{\boldsymbol{P}}{N_{\text{rps}} d_J^2} \tag{12.43}$$

where ν_D is the lubricant dynamic or absolute viscosity in [N s/m^2] or [Pa s], N_{rps} is the rotational speed in [rps], d_J is the journal bearing diameter in [m], and $\boldsymbol{P}$ is the applied load in [N]. The regression estimation of the optimum relation in Figure 12.16 is given by

$$\nu_C^* = 0.0866\left(\frac{c_J}{r_J}\right)^{1.9497} \tag{12.44}$$

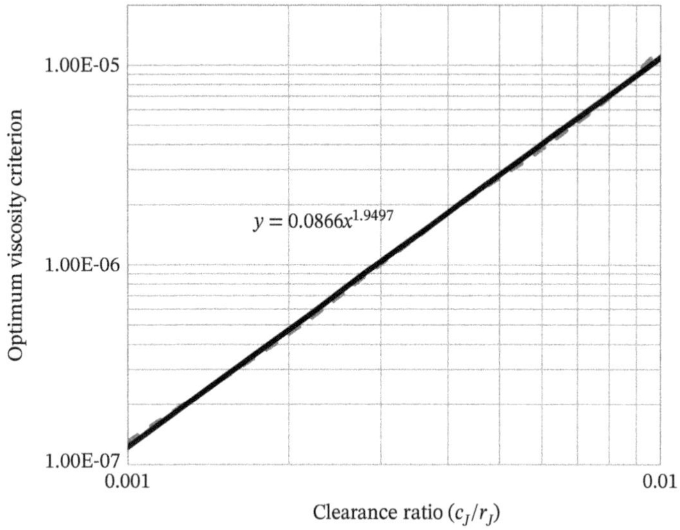

Figure 12.16 The optimum viscosity criterion as a function of the clearance ratio for a range of about 0.001–0.01. The dashed line is for optimum output. The assumed objective is for minimum power loss, maximum–minimum film thickness, and minimum bearing size, all having the same objective weight. Source: Metwalli et al. (1986).

where c_J is the radial clearance and r_J is the bearing radius. Eqs. (12.43) and (12.44) produce the lubricant optimum dynamic or absolute viscosity $v_D{}^*$ in [N s/m^2] or [Pa s] as follows:

$$v_D^* = v_C^* \frac{\boldsymbol{P}}{N_{\text{rps}} d_J^2} = 0.0866 \left(\frac{c_J}{r_J}\right)^{1.9497} \left(\frac{\boldsymbol{P}}{N_{\text{rps}} d_J^2}\right) \tag{12.45}$$

Therefore, when the clearance ratio (c_J/r_J) is known, the optimum lubricant viscosity $v_D{}^*$ is then defined in [Pa s]. Adjusting to the more common centipoise [cP] or micro Reynolds [μR] should be a simple conversion factor multiplication; see Section 12.1.1 and Eq. (12.3′).

The proper tolerances and fits (Section 2.4.7 of the text) are an appropriate knowledge base to use in the definition of the proper expected clearance. Equation (2.29) states that the value of the tolerance T_{id} is dependent on the *grade number* i_g in (IT*i*) and the size or diameter d_c of the cylindrical surface.

$$\text{T}_{\text{id}} = 10^{0.2(i_g-1)} \left(0.45\sqrt[3]{d_C} + 0.001\, d_C\right) = 10^{0.2(i_g-1)} \left(0.45\sqrt[3]{d_J} + 0.001\, d_J\right) \tag{12.46}$$

where the tolerance zone T_{id} is in [μm], i_g is the *tolerance grade number*, and d_C is the diameter of the cylindrical shaft or hole in [mm]. Note that the cylindrical surface diameter d_C is our journal bearing diameter d_J. For fits, the deviation letters d, e, f, and g have the *fundamental deviations* c_o of $16\, d_J{}^{0.44}$, $11\, d_J{}^{0.41}$, $5.5\, d_J{}^{0.41}$, and $2.5\, d_J{}^{0.34}$, respectively; see Section 2.4.7. The expected proper clearance of a *running* ***fit*** of H8/f7 or H8/e7 can then be obtained from the relation of the *fundamental deviation* c_o plus the expected half of each tolerance zone IT8 and IT7. The H8/f7 fit is a *close running* fit for sensible speed and pressure; see Table 2.6. The free running fit in Table 2.6 is H9/d8. For usual journal bearing applications, the reasonable or proper ***running fit*** may be selected as H8/e7. The expression for the expected clearance of that proper fit of H8/e7 (noting that the fundamental deviation of H is zero) is then given by (see Section 2.4.7)

$$\begin{aligned} c_D &= c_O + \frac{(\text{IT8} + \text{IT7})}{2} = 11 d_J^{0.41} + \frac{25 i_g + 16 i_g}{2} \\ &= 11 d_J^{0.41} + 20.5\left(0.45 d_J^{1/3} + 0.001 d_J\right) \end{aligned} \tag{12.47}$$

where c_D is in [μm], i_g is the tolerance grade number, and d_J is the diameter of the cylindrical shaft or hole in [mm]. This equation can be used to estimate the expected diametral clearance c_D for the journal bearing.

With these parameters the optimum bearing for the minimum power loss, maximum–minimum film thickness, and minimum bearing size is estimated for any journal bearing diameter.

The optimization did not suggest an optimum journal bearing diameter, but it gives an optimum length to diameter ratio of 0.6. The diameter should then be obtained from other constraints. The knowledge base can be utilized in that regard, and the relations in Eqs. (12.37) and (12.38) may be used as guide.

To use a similar procedure to Section 12.3, the relations for the length to diameter ratio of 0.6 should be available. For the length to diameter ratio of 0.6, the *Sommerfeld* number S_N as a function of (h_0/c_J) can be estimated by the following regression formula:

$$y = 100.85\,x^6 - 206.79\,x^5 + 159.38\,x^4 - 55.571\,x^3 + 9.8489\,x^2 - 0.4052\,x + 0.0104 \qquad (12.48)$$

where y is the *Sommerfeld* number S_N at $(l_J/d_J = 0.6)$ and x is the *minimum film thickness ratio* (h_0/c_J). In a substituted form, Eq. (12.48) in then

$$S_N = 100.85\left(\frac{h_0}{c_J}\right)^6 - 206.79\left(\frac{h_0}{c_J}\right)^5 + 159.38\left(\frac{h_0}{c_J}\right)^4 - 55.571\left(\frac{h_0}{c_J}\right)^3 + 9.8489\left(\frac{h_0}{c_J}\right)^2 - 0.4052\left(\frac{h_0}{c_J}\right) + 0.0104 \qquad (\mathbf{\mathit{12.48}})$$

One can then use the *Sommerfeld* number S_N at $(l_J/d_J = 0.6)$ to evaluate all other behavioral parameters from the *numerical solution* as done before in Section 12.3. Eq. (12.48) is obtained by interpolation of data for l_J/d_J of ∞, 1.0, 0.5, and 0.25, which are available in Raimondi and Boyd (1958) and Shigley (1986). The interpolation formula is the one suggested by Raimondi and Boyd(1958).

12.6.3 Journal Bearing Synthesis Tablet

To facilitate the availability of a *CAD* tool to synthesize both practice and optimization knowledge base options, a ***tablet*** in Figure 12.17 has been developed in an *Excel*© datasheet. Allowed inputs are in cells with a light gray color background in Figure 12.17 (and light blue background color in the tablet file). The tablet provides side by side bearing synthesis for both ***practice*** and some ***optimum*** procedures previously discussed in Sections 12.6.1 and 12.6.2. In addition, the tablet offers an ***Analysis*** column to allow the definition of more design parameters and obtain the bearing behavioral performance variables for comparison. The SAE oil selection is done through clicking the dropdown list in the Suggest Oil Type cells – in the Input or Analysis columns – or just type the grade number. The viscosity of that oil will be promptly calculated for the operating temperature. The matching of the needed lubricant viscosity and the selected oil viscosity can be attained by changing the inlet oil temperature; see Figure 12.17.

The main ***inputs*** in the synthesis section are the applied load, rotational speed in [rpm]. length to diameter ratios of only 0.6 or 1.0, inlet oil temperature, and a safety factor that would only increase the load. Other inputs in the synthesis section are the iterative *minimum film thickness ratio* and the bearing diameter variation. The iterative *minimum film thickness ratio* is necessary to have matching *Sommerfeld* numbers. The bearing diameter variation is only available for the optimum columns of synthesis section. The selected optimum solution does not specify an optimum diameter. One may use the suggested values of practice procedure as shown in Figure 12.17 or enter any other diameter. Small diameters, however, may result in much higher viscosity than usual oil type can accommodate. The outputs are most of the useful bearing performance variables and parameters that have been previously considered; see Figure 12.17.

The ***tablet*** uses the ***practice*** procedure in Sections 12.3 and 12.6.1 with all associated equations for length to diameter ratio of 1.0 in the first **Input** column. It also uses the *Sommerfeld* number S_N as a function of (h_0/c_J) for

Journal Bearing Synthesis - SI units

Inputs	Default 1	Input	Default 2	Input	Analysis
Applied Load P [N]	2500	2500	2500	2500	2500
Rotational Speed N_{rpm} [rpm]	1800	1800	1800	1800	1800
Length to Diameter Ratio l_J/d_J	0.6	1	0.6	1	1
Inlet Oil Temperature T_{in} [°C]	60	60	47.7	52.3	60
Safety Factor K_{SF}	1	1	1	1	1.00
Output (Knowledge Base)	Practice	Practice	Optimum	Optimum	
Diameter d_J [mm]	45.800	29.027	45.8	40	40
Lubricant Viscosity ν_D [c_P] or [mPa.s]	16.834	16.834	16.9031	26.0543	18.693
Diametral Clearance c_D [mm]	0.09224	0.05846	0.0867	0.0823	0.064
Bearing Length l_J [mm]	27.480	29.027	27.480	40.000	40
Clearance Ratio $c_D/d_J = c_J/r_J$	0.00201	0.00201	0.00189	0.00206	0.00160
Load per unit Area $p_p = P/l_Jd_J$ [MPa]	1.9863	2.9671	1.9864	1.5625	1.5625
Performance					
Sommerfeld Number S_N	0.06269	0.04197	0.07122	0.11821	0.14019
Sommerfeld Number (*iterated*) S_N	0.06269	0.04197	0.07118	0.11811	0.14019
Film Thickness Ratio h_0/c_J	0.1770	0.1844	0.19430	0.3830	0.4200
Minimum Film Thickness h_0 [mm]	0.00816	0.00539	0.00842	0.01576	0.0134
Eccentricity e [mm]	0.0380	0.0238	0.0349	0.0254	0.0186
Friction Coefficient Variable $(d_J/c_D)\mu_K$	1.9403	1.5424	2.1041	3.0065	3.4288
Friction Coefficient μ_K	0.00391	0.00311	0.00398	0.00618	0.00549
Temperature Rise Variable T_{var}	7.831	6.228	8.488	12.117	13.824
Lubricant Temperature Rise Δ_T [°C]	10.265	12.195	11.126	12.494	14.254
Operating Temperature T_{av} [°C]	65.133	66.098	53.263	58.547	67.127
Resisting Torque T_f [N.m]	0.2237	0.1127	0.228	0.309	0.274
Power Loss H_f [W]	42.17	21.24	42.988	58.290	51.705
Performance Parameter p_pV_J [kPa.m/s]	0.00857	0.00812	0.00857	0.00589	0.00589
Lubricant Definition					
Lubricant Viscosity ν_D [cP]	16.834	16.834	16.903	26.054	18.693
Suggest Oil Type SAE 10–60	30	30	10	30	30
Calculated Viscosity of Oil Type at T_{av}	19.993	19.275	16.911	26.065	18.579

Journal Bearing Synthesis - SI units

Inputs	Default 1	Input	Default 2	Input	Analysis
Applied Load P [N]	2500	2500	2500	2500	2500
Rotational Speed N_{rpm} [rpm]	1800	1800	1800	1800	1800
Length to Diameter Ratio l_J/d_J	0.6	1	0.6	1	1
Inlet Oil Temperature T_{in} [°C]	60	60	67.5	63	60
Safety Factor K_{SF}	1	1	1	1	1.00
Output (Knowledge Base)	Practice	Practice	Optimum	Optimum	
Diameter d_J [mm]	45.800	29.027	45.8	40	40
Lubricant Viscosity ν_D [cP] or [mPa.s]	16.834	16.834	8.3326	12.8781	18.693
Diametral Clearance c_D [mm]	0.09224	0.05846	0.0603	0.0573	0.064
Bearing Length l_J [mm]	27.480	29.027	27.480	40.000	40
Clearance Ratio $c_D/d_J = c_J/r_J$	0.00201	0.00201	0.00132	0.00143	0.00160
Load per unit Area $p_p = P/l_Jd_J$ [MPa]	1.9863	2.9671	1.9864	1.5625	1.5625
Performance					
Sommerfeld Number S_N	0.06269	0.04197	0.07254	0.12038	0.14019
Sommerfeld Number (*iterated*) S_N	0.06269	0.04197	0.07253	0.12094	0.14019
Film Thickness Ratio h_0/c_J	0.1770	0.1844	0.19700	0.3880	0.4200
Minimum Film Thickness h_0 [mm]	0.00816	0.00539	0.00594	0.01112	0.0134
Eccentricity e [mm]	0.0380	0.0238	0.0242	0.0175	0.0186
Friction Coefficient Variable $(d_J/c_D)\mu_K$	1.9403	1.5424	2.1293	3.0482	3.4288
Friction Coefficient μ_K	0.00391	0.00311	0.00280	0.00437	0.00549
Temperature Rise Variable T_{var}	7.831	6.228	8.592	12.336	13.824
Lubricant Temperature Rise ΔT [°C]	10.265	12.195	11.263	12.719	14.254
Operating Temperature T_{av} [°C]	65.133	66.098	73.131	69.360	67.127
Resisting Torque T_f [N.m]	0.2237	0.1127	0.161	0.218	0.274
Power Loss H_f [W]	42.17	21.24	30.266	41.173	51.705
Performance Parameter p_pV_J [kPa.m/s]	0.00857	0.00812	0.00857	0.00589	0.00589
Lubricant Definition					
Lubricant Viscosity ν_D [cP]	16.834	16.834	8.333	12.878	18.693
Suggest Oil Type SAE 10–60	30	30	10	20	30
Calculated Viscosity of Oil Type at T_{av}	19.993	19.275	8.300	12.791	18.579

Figure 12.17 *CAD* tablet to synthesize both *practice* and *optimum* knowledge base bearings. An *Analysis* column allows more design parameters to define and obtain the bearing behavioral performance variables for comparison. Right Tablet is for tight tolerance.

the length to diameter ratio of 0.6 defined in Eq. (12.48). This is implemented in the first Default 1 column. The procedures, equations, and results are the same as of Examples 12.1 and 12.2. One may change the values in the Default 1 column for the cells of light grayish background (or light bluish background color in the tablet file).

The ***tablet*** also uses an ***optimum*** approach given in Section 12.6.2 with all associated equations for length to diameter ratios of 0.6 and 1.0, each one handled in a separate column. It utilizes the *Sommerfeld* number S_N as a function of (h_0/c_J) for the length to diameter ratios of 0.6 and 1.0 defined in Eqs. (12.48) and (12.31). Another input for the bearing diameter is needed. The practice solution may be used as a guide for diameter selection. Again, small diameters would result in higher viscosities. The iteration of the minimum film thickness ratio, however, is slight in these optimum approach columns. One may also change the values in the Default 2 column for the cells of light grayish background (or light bluish background color in the tablet file).

The utility of the *synthesis* ***tablet*** is best demonstrated by the next example, observing the tablet outcomes in Figure 12.17, and running the tablet for different inputs. The tablet is available at ***Wiley website*** under the name ***Journal Bearing Synthesis-SI.xlsx***. It should be noted that the tablet provides simplified and approximate solution depending on the assumed cases. They are not intended to be suitable or accurate for other cases. The intension herein is for simplified half 180° bearing. Discrepancies are expected for cases where the consideration is for full 360° bearing. The *elastohydrodynamic* approach is not considered. Further computer simulation means such as *CFD* should be used to define the numerical performance of a specific bearing design.

Example 12.4 It is required to design the hydrodynamic journal bearing of Examples 12.1 and 12.2 according to optimum, or knowledge base information, and estimate its characteristic values for the geometrical and behavioral parameters. The bearing design calls for a journal bearing running at N_{rpm} of 1800 [rpm], i.e. N_{rps} of 30 [rps], and should carry a load $\boldsymbol{P}$ of 2.5 [kN] or 500 [lb]. Find one of the expected SAE lubricating fluids at an inlet temperature T_{in} of 60 [°C] or 140 [°F]. Find the expected temperature rise of the lubricating oil, the coefficient of friction, the friction torque, and the power loss in the bearing. Assume a simplified 180° half bearing, and compare results with Examples 12.1and 12.2.

Solution

Data: $N_{rps} = 30$ [rps], $\boldsymbol{P} = 2.5$ [kN] or 500 [lb], and T_{in} of 60 [°C] or 140 [°F]. Example 12.1 input $d_J = 40$ [mm] or 1.5 [in], $l_J = 40$ [mm] or 1.5 [in], and $c_D = 0.064$ [mm] or 0.003 [in]. The length to diameter ratio was $l_J/d_J = 1.0$, and inverse clearance ratio is $r_J/c_J. = d_J/c_D = 625$ for SI or 500 for the US. The clearance ratios were then $c_J/r_J. = c_D/d_J = 0.0016$ for SI or 0.002 for the US.

The optimization suggests an optimum $l_J/d_J = 0.6$. It does not suggest a bearing diameter or diametral clearance. The practice synthesis suggests a diameter according to Eq. (12.27) for $l_J/d_J = 0.6$ to get

$$d_J = 0.9232\ \boldsymbol{P}^{0.499} = 0.9232\ (2500)^{0.499} = 45.800\ \ [\text{mm}] \tag{a}$$

Then, the bearing length $l_J = 0.6\,(45.8) = 27.480$ [mm].

The clearance of a proper fit of H8/e7 is then given due to Eq. (12.47) by

$$\begin{aligned} c_D &= 11 d_J^{0.41} + 20.5\left(0.45 d_J^{1/3} + 0.001 d_J\right) \\ &= 11(45.8)^{0.41} + 20.5\left(0.45(45.8)^{1/3} + 0.001\,(45.8)\right) = 86.7097\ [\mu\text{m}] = 0.08671\ [\text{mm}] \end{aligned} \tag{b}$$

The clearance ratios are then

$$\frac{c_J}{r_J} = \frac{c_D}{d_J} = \frac{0.086\,709\,7}{45.8} = 0.001\,893\,2 \tag{c}$$

The inverse clearance ratio is then $r_J/c_J. = d_J/c_D = 1/0.001\,875\,1 = 528.2$, which is different from the previous values of 625 in Example 12.2.

The optimization results suggest the lubricant optimum dynamic or absolute viscosity $\nu_D{}^*$ in [N s/m^2] or [Pa s] to be given by Eq. (12.45), which produces the following:

$$\nu_D^* = 0.0866\left(\frac{c_J}{r_J}\right)^{1.9497}\left(\frac{\boldsymbol{P}}{N_{\text{rps}}d_J^2}\right) = 0.0866(0.001\,893\,2)^{1.9497}\left(\frac{2.5\left(10^3\right)}{30\left(45.8\left(10^{-3}\right)\right)^2}\right) = 0.016\,903\ [\text{Pa s}]$$
$$= 16.903\ [\text{cP}] \tag{d}$$

The load per unit area p_p is given by the following expression:

$$p_p = \frac{\boldsymbol{P}}{l_J d_J} = \frac{2500}{27.480\,(45.8)} = 1.9864\ \ [\text{MPa}] \tag{e}$$

The *Sommerfeld* number S_N using Eq. (12.10) gives the following value:

$$S_N = \frac{\nu_D N_{\text{rps}}}{p_p}\left(\frac{d_J}{c_D}\right)^2 = \frac{(0.016\,903)\,(30)}{1.9864(10)^6}\left(\frac{45.8}{0.086\,71}\right)^2 = 0.071\,221 \tag{f}$$

Using the developed Excel tablet, the minimum film thickness ratio h_0/c_J is iteratively changed to 0.1943 with the purpose of having the *iterative Sommerfeld* number S_N as close as possible to the calculated *Sommerfeld* number S_N; see Figure 12.17. This gives the following approximate minimum film thickness ratio h_0/c_J and minimum film thickness h_0 (from h_0/c_J):

$$\frac{h_0}{c_J} = 0.1943,\ \ h_0 = 0.1943\left(\frac{c_D}{2}\right)$$
$$h_0 = 0.1943\left(\frac{0.086\,71}{2}\right) = 0.008\,423\,9\ [\text{mm}] \tag{g}$$

This is realistically defining the minimum film thicknesses $h_0 = 0.0084$ [mm], which is a little more than 8.4 [μm]. As stated in Section 2.4.7, "In manufacturing processes, excellent *surface finish* has a roughness up to about 0.2 [μm] or 8 [μin]." Therefore, *excellent surface finish* is required for this journal bearing; see Table 2.5. The surface roughness should be much smaller than the minimum film thicknesses.

The rest of the journal bearing parameters are calculated in the same manner as performed in Example 12.2, which are done through the synthesis tablet ***Journal Bearing Synthesis-SI.xlsx***.

The developed tablet has the Default 2 Optimum column set at the design information of this Example 12.3. The tablet output compares extremely well with the results in this example. The input column provided the ability to change the diameter as shown in Figure 12.17, with the restriction of having the length to diameter ratio at 1.0. For the length to diameter ratio at 0.6, one can change the value of the diameter in the Default 2 Optimum column as apparent from Figure 12.17. The ability to change the inlet oil temperature allowed the lubricant viscosity to be very close to the calculated oil viscosity at operating temperature as shown in Figure 12.17.

The developed *tablet* has the **Analysis** column set at the design information of Example 12.1. The tablet output compares extremely well with the results in Example 12.1.

The developed *tablet* has the **Input** column of practice knowledge base set at the design information of Example 12.2. The *tablet* output compares extremely well with the results in Example 12.2.

Comparing the results of Examples 12.1, 12.2, and 12.4, the synthesis *tablet* provides the following interesting results for similar input parameters:

- The optimum gives about the same results for the minimum power loss and better results maximum–minimum film thickness ratio for the same bearing size of the optimum $l_J/d_J = 0.6$. The minimum film thickness ratio is 0.19703 for optimum against 0.1770 for practice. The minimum film thickness is 0.00842 for optimum against 0.00816 for practice.
- The optimum gives about the same results for the friction coefficient.

- The optimum uses a lower viscosity oil of SAE 10 against SAE 30 for practice but required a lower inlet oil temperature.
- The current optimum procedure, however, selects an expected tighter cleanse resulting from the fit (H8/e7) than the practice. If an even *tighter fit* of H8/f7 is used, the optimum produces lower power loss, lower friction coefficient, lower viscosity oil, and ***lower*** minimum film thickness, which requires a much finer surface roughness and machining constraints. The ***tighter fit*** is shown as the right tablet in Figure 12.17. The tighter fit tablet is provided as the Excel sheet ***Journal Bearing Synthesis-SI-small tolerance.xlsx***.
- For the same length to diameter ratio $l_J/d_J = 1.0$, the practice and optimum procedures provide overall advantages over the analysis column of Example 12.1, except for the *input* maximum–minimum film thickness ratio.

Without changing the selected oil, the inlet temperature can be reduced by cooling to have the calculated viscosity the same as the needed viscosity at operating temperature as shown in Figure 12.17.

Further trials and iterations through the *tablet* are necessary to achieve other better requirements or constraints. The solutions, however, are restricted to the length to diameter ratios of 0.6 and 1.0. The ratio of 1.0 for the optimum tablet column is using the optimum viscosity relation of the length to diameter ratios of 0.6. Other optimum viscosity relations might be developed from the multi-objective optimization or other optimization efforts. Therefore, care must be exercised, and the *Analysis* column might be an acceptable alternative to adjust for that.

12.7 Summary

The underlying principles are presented to facilitate the understanding of the behavior of journal bearings. Some basic relations are introduced or derived, which can be used for estimating the bearing synthesis. The mathematical models provide basic relations that have also been numerically solved some time ago for *finite length bearings*. The data of the *numerical solution* has been subsequently used in the design procedures of journal bearings. This data is also used herein for the synthesis of journal bearings in addition to the implementation of knowledge base practice and optimization. Several knowledge base data provide means to estimate the design parameters for better synthesis. The numerical solution has also been utilized in the evaluation of bearing performance and adjusting parameter for better or near optimum synthesis of journal bearings.

Simple *CAD* tools have been developed in MATLAB and Excel to help in the journal bearing synthesis. These tools can be used or further developed to accommodate more options, special requirements, or additional optimization objectives and constraints.

Short bearings or infinitely long bearing relations can be used in some cases where the bearing fluid is supplied with lubricant under pressure or the bearing ends are limiting side flow. The process presented herein should be altered to accommodate these cases. Further computer simulation means such as *CFD* would also be used thereafter to define the performance of a specific bearing design.

Problems

12.1 Find the history of journal bearings and the early implementation of their principles. What kind of applications journal bearings has been used? Was it operating under hydrodynamic, hydrostatic, boundary, or solid film lubrication?

12.2 What real applications use viscosity to judge the quality of fluids? What types of viscometers are used in these tests? What is the accuracy and how would the accuracy be guaranteed?

12.3 Have you employed a simple viscosity test to check on a product quality such as honey?

12.4 Prove the conversion relation between micro *Reynolds* and centipoise. Is the centipoise widely used in the US? Why is the US the usual unit of the kinematic viscosity the same centistokes as the SI units? What did you use in Fluid Mechanics course?

12.5 Find the viscosities of different SAE engine oil grades and the effect of density and temperature variation on their values. Can you check on the accuracy of the change of the absolute or dynamic viscosity with temperature given by Eq. (12.5)? What is the viscosity index of different oil grades you can find information on?

12.6 Do the dynamic and kinematic viscosities of water hold for higher temperatures? Check the values at about 100 [°C] or 212 [°F] and compare the effect of temperature.

12.7 Can you justify the relatively high value of the kinematic viscosity of air? What should be the effect of temperature and pressure to use in an air bearing?

12.8 Find the viscosities of some synthetic oils and the viscosity index of each. Can they have values of viscosity similar to SAE oils? Are they more expensive?

12.9 Find the relation between the kinematic coefficient of friction and the *Sommerfeld* number in the *Petroff's* equation.

12.10 Provide the development of Eq. (12.25) as a function of the *Sommerfeld* number.

12.11 Verify that the torque in Eq. (12.26) is expected to converge to *Petroff's* equation when ε is zero.

12.12 Evaluate the coefficient of friction in long bearings similar to Eq. (12.8). Does it converge *Petroff's* equation when ε is zero?

12.13 Evaluate the coefficient of friction in short bearings similar to Eq. (12.8). Does it converge *Petroff's* equation when ε is zero?

12.14 Develop a similar curves or curve to those in Figure 12.9 for the full or 360° bearings. Find the information from the same source used in this text. You may use other available texts that provide the information.

12.15 Find the practical range of the length to diameter ratios used in different applications. Was the usual range expected to lie in between 0.3 and 2.4? What kind of applications operates outside that range? Can you justify the reason for being out of that range?

12.16 Find the expected surface roughness of the machining process for the bearing in Example 12.1 to be much less than the minimum film thickness.

12.17 Redo Example 12.1 using diametral clearance of 0.057 [mm] or.0027 [in] instead of 0.064 [mm] or 0.003 [in]. What do you expect the minimum film thickness ratio to be? Can you simply scale it? Can you use Figure 12.9?

12.18 A journal bearing is carrying 8 [kN] or 1800 [lb] and running at 3600 [rpm]. It has a diameter of 40 [mm] or 1.6 [in], a bearing length $l_J = 40$ [mm] or 1.6 [in], and a diametral clearance of 0.09 [mm] or 0.0035 [in].

It is expected to use an SAE 30 lubricating fluid at an inlet temperature T_{in} of 57 [°C] or 135 [°F]. Find the appropriateness of the lubricating oil, the expected temperature rise, the coefficient of friction, the friction torque, and the power loss in the bearing. Assume a simplified 180° half bearing.

12.19 Synthesize a bearing according to practice, which can carry 8 [kN] or 1800 [lb] and run at 3600 [rpm]. The length to diameter ratio is assumed as 1.0. Find the bearing diameter, one of the expected SAE lubricating fluids at an inlet temperature T_{in} of 70 [°C] or 160 [°F]. Find the expected temperature rise of the lubricating oil, the coefficient of friction, the friction torque, and the power loss in the bearing. Assume a simplified 180° half bearing.

12.20 The ring bearing in the *Jib crane* of Example 12.3 is assumed as bronze–iron. Change the material to iron bearing in two parts. One is welded to the mast, and the other is welded to the hood. Find the appropriate safe length of the new ring bearing. Draw a ***3D*** model of the ring bearing connection in any ***3D*** modeling and simulating package. Can you find the stresses between the two parts assuming no in-between clearance?

12.21 For the *Jib crane* in Example 12.3, find the torque necessary to rotate the crane at its full load and maximum load position. Can an operator move the crane at this condition? What is the force and power exerted by the operator in that case? Note the possible angles at which the operator forces the crane to move around.

12.22 For the *Jib crane* in Example 12.3, sketch the bearings construction to scale and check its assembly, disassembly, and lubrication means. Change sleeve and bearing ring dimensions to construction requirements, and use suitably needed fits and tolerances.

12.23 For the *Jib crane* in Example 12.3, would a semi-spherical bearing be suitable, and what dimensions should it have to be safe? Sketch the construction requirement to scale.

12.24 Develop a simple MATLAB code to help in the selection of sleeve bearings like the *Jib-crane* example and can be used in similar applications.

12.25 Produce a ***3D*** model for the *Jib-crane* bearing, and find the stresses on its contact with the pivot shaft using any ***3D*** modeling and simulating package. Compare the maximum stress with the calculated maximum projected pressure value. Assume no clearance between the sleeve and the shaft.

12.26 Use a tighter fit of H8/f7 to solve Example 12.4 and compare results. Use the same journal bearing diameter. What are the advantages and disadvantages of using a tighter fit? Note that the *fundamental deviations* c_o is 5.5 $d_J^{0.41}$ rather than 11 $d_J^{0.41}$.

12.27 Use the available MATLAB code `CAD_Journal_Bearings.m` in Figure 12.15a to solve Problem 12.17. Use some iterations to adjust any bearing parameters.

12.28 Use the available MATLAB code `CAD_Journal_Bearings.m` in Figure 12.15b to solve Problem 12.18. Use some iterations to adjust any bearing parameters.

12.29 Use the available MATLAB code `CAD_Journal_Bearings.m` in Figure 12.15c to solve Problem 12.19. Use some iterations to adjust any bearing parameters.

12.30 Synthesize a bearing according to an optimum design, which can carry 8 [kN] or 1800 [lb] and runs at 3600 [rpm]. The length to diameter ratio is expected to be 0.6. Use a bearing diameter of 40 [mm] and one of the expected SAE lubricating fluids at an inlet temperature T_{in} of 40 [°C] or 104 [°F]. Find the expected minimum film thickness, temperature rise of the lubricating oil, the coefficient of friction, the friction torque, and the power loss in the bearing. Assume a simplified 180° half bearing.

12.31 Use the available bearing synthesis tablet to compare the assumed optimum solutions with a bearing carrying 8 [kN] and running at 3600 [rpm]. The bearing has a 50 [mm] diameter, a 0.6 length to diameter ratio, and a radial clearance of 0.0375 [mm]. The initial oil temperature is 40 [°C]. Find the expected minimum film thickness, temperature rise of the lubricating oil, the coefficient of friction, the friction torque, and the power loss in the bearing.

12.32 Use the available bearing synthesis tablet to compare different designs for a journal bearing of your choice. Compare the results of your choice to the results of the source of information.

12.33 Develop a MATLAB code to help in the synthesis of journal bearings. The code can use the practice and/or the optimum procedures. Compare results with the available tablet or any available source of information. Notice the difference between the full 360° bearings and the half 180° bearing assumptions.

12.34 Compare the results of any of the previous problems with the solutions of the same bearing using the short and long bearing assumptions.

References

ASTM D2161 (2019). *Standard practice for conversion of kinematic viscosity to Saybolt universal viscosity or to Saybolt Furol viscosity*. ASTM International.

ASTM D445 (2019). *Standard test method for kinematic viscosity of transparent and opaque liquids (and calculation of dynamic viscosity)*. ASTM International.

Boyd, J. and Raimondi, A.A. (1951). Applying bearing theory to the analysis and design of journal bearings, Parts I and II. *Journal of Applied Mechanics* 73: 298–316.

Budynas, R.G. and Nisbett, J.K. (2015). *Shigley's Mechanical Engineering Design*, 10e. McGraw Hill.

DuBois, G.B. and Ocvirk, F.W. (1955). The short bearing application for full journal bearings. *Trans. ASME* 77: 1173–1178.

Gupta, S.V. (2014). *Viscometry for Liquids*. Springer Series in Materials Science 194. Springer International Publishing Switzerland.

ISO 3104 (1994). *Petroleum products – Transparent and opaque liquids – Determination of kinematic viscosity and calculation of dynamic viscosity*. International Organization for Standardization.

ISO 3448 (1992). *Industrial liquid lubricants – ISO viscosity classification*. International Organization for Standardization.

McKee, S.A. and McKee, T.R. (1932). Journal bearing friction in the region of thin film lubrication. *SAE Journal* 31: 371–377.

Metwalli, S.M., Shawki, G.S.A., Mokhtar, M.O.A., and Seif, M.A.A. (1984). Multiple design objectives in hydrodynamic journal bearing optimization (ASME Paper No. 83-DET-41). *ASME Journal of Mechanisms, Transmissions and Automation in Design* 106 (1): 54–61.

Metwalli, S.M., Shawki, G.S.A., Mokhtar, M.O.A., and Seif, M.A.A. (1986). Direct approach to the evaluation of hydrodynamic journal bearing characteristics. In: *Current Advances in Mechanical Design and Production* (eds. S.E.A. Bayoumi and M.Y.A. Younan), 69–78. Pergamon Press.

Neale, M.J. (ed.) (1973). *Tribology Handbook*. Butterworths.

Ocvirk, F.W. (1952). *Short-bearing approximation for full journal bearings*. TN 2808 NACA.

Petrov, N.P. (1883). Friction in machines and the effect of lubricant. *Inzh. Zh. St. Petersburg* 1: 71–140. (vol. 2, pp. 227–279; vol. 3, pp, 377–463; vol. 4, pp. 535–564).

Raimondi, A.A. and Boyd, J. (1958). A solution for the finite journal bearing and its application to analysis and design, Parts I, II, and III. *A S L E Transactions* 1 (1): 159–209.

Reynolds, O. (1886). On the theory of lubrication and its application to Mr. Beauchamp Tower's experiment, including an experimental determination of the viscosity of olive oil, Part I. *Philosophical Transactions of the Royal Society* 52: 228–310.

SAE J-300 (1997). *Engine oil viscosity classification*. Society of Automotive Engineers International.

Seireg, A. and Ezzat, H. (1969). Optimum design of hydrodynamic journal bearings. *ASME Journal of Lubrication Technology* 91: 516–523.

Seireg, A.S. and Dandage, S. (1982). Empirical design procedure for the thermodynamic behavior of journal bearings. *Journal of Lubrication Technology* 104: 135–148.

Shigley, E.J. (1986). *Mechanical Engineering Design*, First Metrice. McGraw Hill.

Sommerfeld, A. (1904, 1904). Zur Hydrodynamischen Theorie der Schmiermittel-Reibung [On the hydrodynamic theory of lubrication]. *Zeitschrift für Mathematik und Physik* 50: 97–155.

Towers, B. (1883). First report on friction experiments (friction of lubricated bearings). *Proceedings of the Institution of Mechanical Engineers* 34: 632–659.

Internet Link

https://www.oilesglobal.com/america/en/products/ Oils: Metallic, multi-layer, plastic, and air bearings.

Section B

Power Transmitting and Controlling Elements

13

Introduction to Power Transmission and Control

Various machine elements are power transmission elements. They transmit power from a prime mover or an engine (e.g. electric motor, internal combustion engine, steam turbine, hydraulic turbine, a pump, etc.) to a machine that consumes power or transforms it. These machines (e.g. hoisting and conveyers, vehicles, hydraulic cylinders, machine tools, etc.) would usually use a *shaft* to pass the power through another power transmission element to the machine (Figure 13.1). The transmission can be directly through collinear shafts or indirectly through a noncollinear shafts. Figure 13.1 shows a flowchart of power transmission between a prime mover and a machine. The first power transmission element is the shaft directly or indirectly connecting the prime mover to other power transmission element and usually another shaft from the power transmission element to the machine. Direct power transmission would usually have the same input and output rotational speed and torques. If the power transmission is rectilinear, the input and output forces and velocities would be the same. This indicates that no *power transformation* is involved and the *input power* $\boldsymbol{T}_1 \cdot \boldsymbol{\omega}_1$ is closely equal to the *output power* $\boldsymbol{T}_2 \cdot \boldsymbol{\omega}_2$ and the *input torque* $\boldsymbol{T}_1$ is closely equal to the *output torque* $\boldsymbol{T}_2$. This should be obvious, since the *input rotational speed* $\boldsymbol{\omega}_1$ is the same as the *output rotational speed* $\boldsymbol{\omega}_2$.

Indirect power transmission would not usually have the same input and output rotational speeds and torques. If the power transmission is noncollinear, the input and output forces and velocities may not be the same. This indicates that a *power transformation* is involved and the *input power* $\boldsymbol{T}_1 \cdot \boldsymbol{\omega}_1$ is not usually equals to the *output power* $\boldsymbol{T}_2 \cdot \boldsymbol{\omega}_2$. The relation depends on the loss of power during transformation, which cause a power transmission efficiency of less than 100%. The *input torque* $\boldsymbol{T}_1$ is usually not equal to the *output torque* $\boldsymbol{T}_2$. This should be obvious, since the *input rotational speed* $\boldsymbol{\omega}_1$ may not be the same as the *output rotational speed* $\boldsymbol{\omega}_2$, particularly if the *reduction ratio* r_R is not 1.0. The definition of the *reduction ratio* is the ratio of the *output rotational speed* $\boldsymbol{\omega}_2$ to the *input rotational speed* $\boldsymbol{\omega}_1$.

Symbols and Abbreviations

Symbols	Quantity, units (adapted)
CAD	Computer-aided design
d_S	Shaft diameter
$\boldsymbol{F}$	Force
$\boldsymbol{F}_1$	Input force
$\boldsymbol{F}_2$	Output force
H	Power

Machine Design with CAD and Optimization, First Edition. Sayed M. Metwalli.

Companion website: www.wiley.com/go/metwalli/machine

Symbols	Quantity, units (adapted)
H_1	Input power
H_2	Output power
K_{SC}	Stress concentration factor
K_{SFF}	Fatigue safety factor
$\boldsymbol{M}$	Moment
$\boldsymbol{M/T}$	Moment to torque ratio
N_{rpm}	Rotational speed in [rpm]
r_R	Reduction ratio
S_e	Endurance strength of material
S_{eP}	Endurance strength of part
$\boldsymbol{T}_1$	Input torque
$\boldsymbol{T}_2$	Output torque
$\boldsymbol{v}$	Velocity
$\boldsymbol{v}_1$	Input velocity
$\boldsymbol{v}_2$	Output velocity
v_1, v_2	Magnitudes of $\boldsymbol{v}_1$ and $\boldsymbol{v}_2$
η	Efficiency
σ_a	Alternating normal stress
σ_m	Mean normal stress
σ_{max}	Maximum normal stress
σ_{min}	Minimum normal stress
σ_{vMa}	*von Mises* alternating stress
σ_{vMm}	*von Mises* mean stress
τ_{max}	Maximum shear stress
τ_{min}	Minimum shear stress
$\boldsymbol{\omega}_1$	Input rotational speed [rad/s]
$\boldsymbol{\omega}_2$	Output rotational speed [rad/s]
ω_1, ω_2	Magnitudes of $\boldsymbol{\omega}_1$ and $\boldsymbol{\omega}_2$

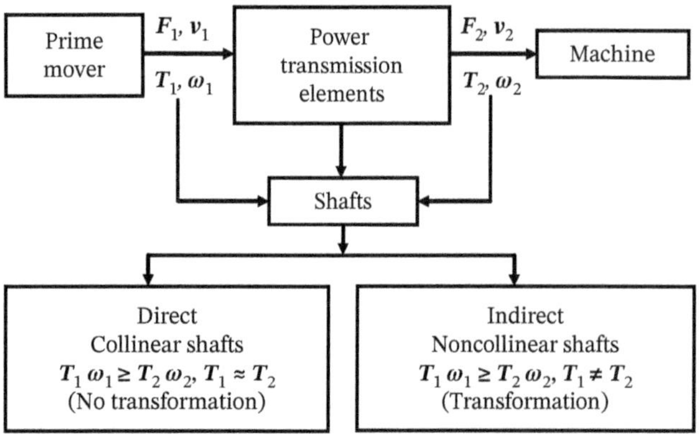

Figure 13.1 Power transmission elements between a prime mover and a machine. The first element is a shaft directly or indirectly connecting the prime mover to the power transmission element and usually another shaft from the power transmission element to the machine.

13.1 Prime Movers and Machines

Prime movers and machines may be considered as power transformation components. The prime movers usually transform chemical energy to mechanical one in internal combustion engines or transform electric power into mechanical power in the case of electric motors. Other prime movers act similarly. Machines would also consume mechanical power into frictional heat or into forming or cutting energy in machine tools.

Prime movers such as combustion engines, electric motors, steam turbines, hydraulic turbines, pumps, etc. provide power and is considered as power sources or drivers; see Section 2.1. These prime movers are also power transformation elements that transform one energy source to another energy or power form. The combustion engines transform the chemical energy into mechanical energy. The power is typically unsteady during each cycle of the engine and is usually smoothed up by a *flywheel*. The electric motors transform the electric power to the mechanical power. The power flow is usually steady. Steam, pneumatic, and hydraulic turbines transform the heat or fluid energy into other mechanical energy, usually rotational. A pump as a prime mover generates fluid power from an electric or an engine source. The mechanical interest herein is mainly related to power in terms of torque and rotational speed, force and velocity, pressure and flow rate, or temperature and heat flux; see Section 2.1. Machine element design is primarily concerned with rotational and translational functions. Some others may also be considered.

Machines consumes prime movers' energy to generally other transformed and usually irretrievable states such as the vehicle consumes the power of the engine to move from one position or state to the other. Usually, one needs to consume more engine power to get back to the original position or state. This is if the vehicle was not to climb a mountain and returns back without using any engine power. A hydraulic cylinder works as an actuator and consumes the hydraulic power to lift or move objects. Hoisting and conveyers consume power to transport objects and materials from one location to the other. Machine tools use power to cut, form, or deform materials into other shapes and components.

In most machines, the best utilization of power is to employ the dynamic matching of the prime mover, power transmission element, and machine consumption. This might need some optimization that may be used in some power transmission elements; see, e.g. Elmaghraby et al. (1979) and Section 13.4.

13.2 Collinear and Noncollinear Transmission Elements

For ***collinear*** shafts, the centerlines of input and output shafts coincide. Examples of the collinear transmission elements are couplings, clutches, brakes, flywheels, power screws, etc.; see Figure 13.2. Some of these act as control elements particularly during transit time of power transmission. Figure 13.2 displays a simplified flowchart of power transmission elements carried by input and output shafts. Shafts must be radially and axially supported. Direct or collinear elements, control elements, and indirect or noncollinear elements are cited. Figure 13.3 shows simplified outlines of some power transmission elements. From Figure 13.3a–d, some collinear elements are sketched such as a *coupling* (a), a *clutch* (b), a *brake* (c), and a *flywheel* (d). Couplings join two shafts together and have been addressed in Section 8.2.1. Clutches, brakes, and flywheels control the transient motion between shafts or between a shaft and a ground. The clutch outlined in Figure 13.3b shows two end disks with one attached to the input shaft and the other attached to the output shaft. Between the two disks, there is a friction wheel, and upon the application of the squeezing force from the two disks, it starts the transmission of torque from the input disk to the output disk. This transit state ends when the two disks are having no relative motion with both input and output shafts rotating at the same speed. The brake outlined in Figure 13.3c shows a disk attached to the input shaft with a two-sided caliper shoes forcing the input shaft to decelerate under friction. The caliper is attached to a ground. The outline of a flywheel in Figure 13.3d indicates a massive disk attached to the shaft, which stores the

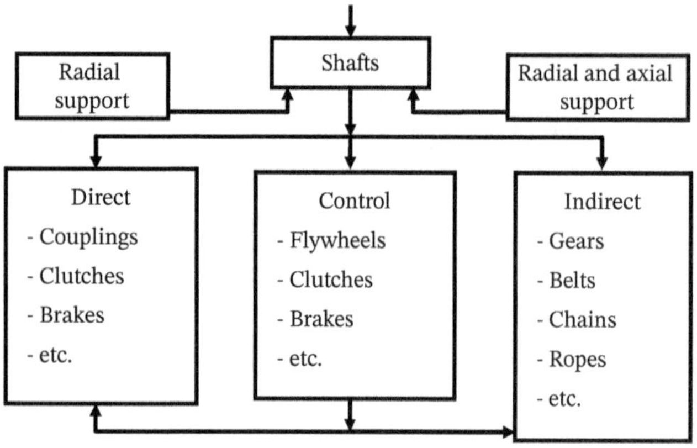

Figure 13.2 Flowchart of power transmission elements carried by input and output shafts. Shafts must be radially and axially supported. Direct or collinear elements, control elements, and indirect or noncollinear elements are cited.

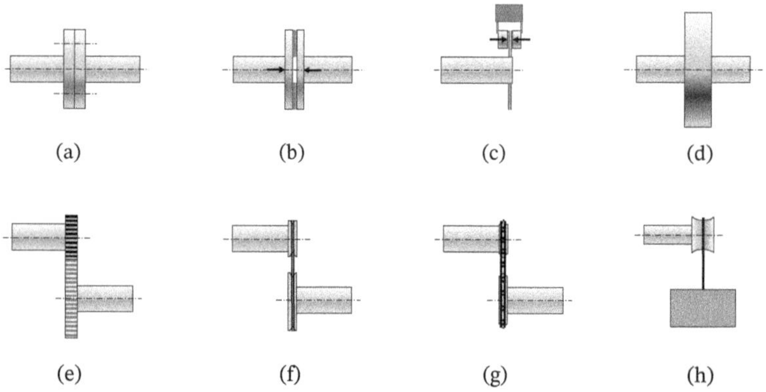

Figure 13.3 Simplified outlines of some power transmission elements. From (a) to (d) are collinear elements as a coupling (a), a clutch (b), a brake (c), and a flywheel (d). From (e) to (h) are noncollinear elements such as a gear set (e), a belt (f), a chain (g), and a rope (h).

variation of input power in its massive inertia to deliver a smoothed power to the output shaft. Clutches, brakes, and flywheels are covered in Chapter 18.

Control elements adjust the flow of power from input shaft to the output. They are usually arranged as collinear, even though some control elements are arranged in a noncollinear form. Clutches, brakes, and flywheels are examples of control elements. They are usually collinear with the input shaft.

Noncollinear power transmission elements would usually connect two noncollinear shafts as illustrated in Figure 13.3e–g. From Figure 13.3e–h, noncollinear elements are outlined such as a *gear set* (e), a *belt* (f), a *chain* (g), and a *rope* (h). They usually transform power particularly if the *reduction ratio* r_R is not equal to 1.0. Gears are generally used as speed reducers and they are covered in Chapter 14. In their simplest form shown in Figure 13.3e, they are formed off cylinders with peripheral teeth that engaged together, delivering the needed reduction ratio as a function of the number of teeth in each gear. Belts and chains are mostly used also as speed reducers, and Chapter 16 discusses these elements. The belt in Figure 13.3f is a flexible-rubber torus of a usually small V-shaped cross section that is wedged between the two pulleys attached on the input and output shafts. The chain in Figure 13.3g is a linked torus of some flexibly connected small links, which goes between the teeth of the two sprockets attached on the input and output shafts. Ropes are flexible elements that are utilized in several capacities and they are presented in Chapter 16. Figure 13.3h illustrates the connection of a shaft to a noncollinear object such as a load lifted by a rope wound on a drum that is attached to the shaft.

As indicated previously, indirect transmission may also incorporate power transformation, i.e. $\boldsymbol{T}_1 \neq \boldsymbol{T}_2$ and $\boldsymbol{\omega}_1 \neq \boldsymbol{\omega}_2$ with $\boldsymbol{T}_1 \cdot \boldsymbol{\omega}_1 \approx \boldsymbol{T}_2 \cdot \boldsymbol{\omega}_2$. Transformation is needed for higher torques with lower rotational speed or higher rotational speed with lower torques. The same applies in force and velocity transformation. The transformations are useful in speed reducers, leverages, hoistings, or hydraulic amplifiers. A *power screw* transforms rotational power to a translational power.

The relation between the power H, the force $\boldsymbol{F}$, and the velocity $\boldsymbol{v}$ or the power H, the torque $\boldsymbol{T}$, and the rotational speed $\boldsymbol{\omega}$ is as follows:

$$
\begin{aligned}
&H_1 = \boldsymbol{F}_1 \cdot \boldsymbol{v}_1, \quad H_2 = \boldsymbol{F}_2 \cdot \boldsymbol{v}_2 \\
&H_1 = \boldsymbol{T}_1 \cdot \boldsymbol{\omega}_1, \quad H_2 = \boldsymbol{T}_2 \cdot \boldsymbol{\omega}_2
\end{aligned}
\tag{13.1}
$$

where the subscript numerals 1 and 2 stand for input and output, respectively. For rotational systems and SI or US systems of units, one gets the following relations (see Eq. (8.12)):

$$
\begin{aligned}
H_1\big|_{\mathrm{SI}} &= \frac{2\pi N_{\mathrm{rpm1}}}{60}\boldsymbol{T}_1, \quad H_1\big|_{\mathrm{SI}} = \frac{2\pi N_{\mathrm{rpm2}}}{60}\boldsymbol{T}_2 \\
H_1\big|_{\mathrm{US}} &= \frac{2\pi N_{\mathrm{rpm1}}}{(12)(33\,000)}\boldsymbol{T}_1, \quad H_1\big|_{\mathrm{US}} = \frac{2\pi N_{\mathrm{rpm2}}}{(12)(33\,000)}\boldsymbol{T}_2
\end{aligned}
\tag{13.2}
$$

where N_{rpm} is the rotational speed in [rpm], $\boldsymbol{T}$ is the torque in [N m] for SI or [lb in] for US, H is the power in [W] for SI or [hp] for US, and the subscript numerals 1 and 2 stand for input and output, respectively.

The *collinear* shafts typically have the same power H, the same force $\boldsymbol{F}$, or torque $\boldsymbol{T}$, and the same shaft speed $\boldsymbol{v}$, $\boldsymbol{\omega}$, or N_{rpm}. Equation (13.1) becomes as follows:

$$
H_1 = \boldsymbol{F}_1 v_1 = H_2 = \boldsymbol{F}_2 v_2 \quad \text{or} \quad H_1 = \boldsymbol{T}_1 \omega_1 = H_2 = \boldsymbol{T}_2 \omega_2 \tag{13.3}
$$

where ω_1, ω_2 are the magnitudes of $\boldsymbol{\omega}_1$ and $\boldsymbol{\omega}_2$, and v_1, v_2 are the magnitudes of $\boldsymbol{v}_1$ and $\boldsymbol{v}_2$.The *noncollinear* shafts usually transmit most but not necessarily all the power. This is due to friction losses or the geometrical configuration of the element such as the *power screw*; see Section 9.5.2. The *efficiency* η is used to account for that. For about the same power H or an efficiency close to a 100%, and for rotational systems, the torque $\boldsymbol{T}$ decreases as the shaft speed $\boldsymbol{\omega}$ increases. Equations (13.1) and (13.2) become as follows with the introduction of the *reduction ratio* r_R:

$$
\begin{aligned}
H_1 &= \boldsymbol{T}_1 \omega_1 = \frac{H_2}{\eta} \cong H_2 = \boldsymbol{T}_2 \omega_2 \\
r_R &= \frac{\omega_1}{\omega_2} = \frac{\boldsymbol{T}_2}{\boldsymbol{T}_1} \\
\boldsymbol{T}_2 &= \frac{\omega_1 \boldsymbol{T}_1}{\omega_2} = r_R \boldsymbol{T}_1 \\
\omega_2 &= \frac{\boldsymbol{T}_1 \omega_1}{\boldsymbol{T}_2} = \frac{\omega_1}{r_R}
\end{aligned}
\tag{13.4}
$$

where the *efficiency* η is considered as about 100%. If the *efficiency* η is not close to a 100%, Eq. (13.4) must be adjusted using the first equation in (13.4) without the approximation.

The *efficiency* η of a power transmission element is a characteristic that depends on the design of the element. The shaft is usually transmitting all the power or having a 100% efficiency with only infinitesimal heat losses inside the shaft material under power variations. Collinear elements should have very close to a 100% efficiency. For clutches and brakes as controlling elements shown in Figure 13.2, the efficiency changes with time during engagements. The clutches should have about 100% efficiency after both sides reach the same rotational speed. Even though the brakes in Figure 13.3c does not seem to be collinear with the ground, two symmetric calipers and the shoe-type brakes are. Noncollinear elements are usually having a very high efficiency but not at a full 100%. The losses are usually due to friction and geometry that may not transfer the full power. Examples are just as *power screws* (Section 9.5.2) and similarly worm gears. The following is a tally of the power transmission elements with their main conventional characteristics for one-stage configurations. These characteristics provide the usual bounds and the extreme limits that may be used. The main characteristics of concern are the ranges of bounds and extreme limits of *reduction ratio* r_R, ranges of *efficiency* η relative to the reduction ratio, and the usual *peripheral velocity* bounds for only one-stage configurations:

- ***Gears*** *(spur and helical)* have usual reduction ratios up to 8 and 20 as a limit, efficiency of 99% to 96%, and peripheral velocity up to 200 [m/s] or 8000 [in/s].
- ***Planetary gears*** have usual reduction ratios up to 8 and 12 as a limit, efficiency of 99% to 98%, and peripheral velocity up to 50 [m/s] or 2000 [in/s].
- ***Worm gears*** have usual reduction ratios up to 60 and 100 as a limit, efficiency of 97% to 45%, and peripheral velocity up to 70 [m/s] or 2800 [in/s].
- ***Belts*** (V belts) have usual reduction ratios up to 8 and 15 as a limit, efficiency of 97% to 94%, and peripheral velocity up to 25 [m/s] or 980 [in/s].
- ***Chains*** have usual reduction ratios up to 6 and 10 as a limit, efficiency of 97% to 98%, and peripheral velocity up to 20 [m/s] or 790 [in/s].
- ***Friction gears*** have usual reduction ratios up to 6 and 10 as a limit, efficiency of 98% to 95%, and peripheral velocity up to 90 [m/s] or 3500 [in/s].

The previously mentioned characteristics are provided just as a general reference and approximate guide. Other values may be available for special or tailored cases. The power transmission elements should, however, stand internal stresses due to torques, moments, and other loadings. This provides the real characteristics supportive or disagreeing with the previous bounds.

Example 13.1 The cylindrical shaft in Example 8.1 is transmitting a maximum of 25 [kW] or 33.5 [hp] and running at 3000 [rpm] in the clockwise direction. It is subject to a maximum bending moment that is half of the maximum applied torque at its critical location. It is connected to a noncollinear shaft that runs at 1000 [rpm] with an efficiency close to a 100%. Find the reduction ratio, the input torque, and the output torque. Consider an initial safety factor K_{SF} of 3.5. For shafts, a selected material is set as hot rolled AISI 1040 or ISO C40. Find the initial diameter synthesis of both input and output shafts.

Solution

Data: $H = 25$ [kW] or 33.5 [hp], $N_{rpm} = 3000$ [rpm], the output $N_{rpm\_out} = 1000$ [rpm], $\boldsymbol{M/T} = 0.5$, and $K_{SF} = 3.5$. Material properties are $S_{yt} = 42$ [kpsi] or 290 [MPa], $S_{ut} = 76$ [kpsi] or 525 [MPa], and $S_e = 37.7$ [kpsi] or 260 [MPa]; see Table A.7.2. Equation (7.58) suggests $S_e = 0.5\ S_{ut} = 0.5(525) = 262.5$ [MPa] or $0.5(76) = 38$ [kpsi].

The data in Examples 6.10, 7.4, and 8.1 that can be applicable to this example is the maximum torque on the shaft $\boldsymbol{T}_S = 79.577$ [N m] or 703.78 [lb in].

The reduction ratio r_R is found according to Eq. (13.4) as follows:

$$r_R = \frac{\omega_1}{\omega_2} = \frac{\boldsymbol{T}_2}{\boldsymbol{T}_1} = \frac{2\pi N_{rpm\_in}(60)}{60(2\pi N_{rpm\_out})} = \frac{N_{rpm\_in}}{N_{rpm\_out}} = \frac{3000}{1000} = 3.0 \tag{a}$$

The maximum torque on the input shaft according to Eqs. (13.2), (13.4), and (8.13) is as follows:

$$\boldsymbol{T}_1 = 9549.3\frac{H_{kW}}{N_{rpm}} = 9549.3\frac{25}{3000} = 79.578 \text{ [N m]}$$

$$\boldsymbol{T}_1 = 63\,025\frac{H_{hp}}{N_{rpm}} = 63\,025\frac{33.5}{3000} = 703.78 \text{ [lb in]} \tag{b}$$

These values are the same as Examples 6.10 and 7.4.

The maximum torque on the output shaft according to Eqs. (13.2), (13.4), and (8.13) is as follows:

$$\boldsymbol{T}_2 = 9549.3\frac{H_{\text{kW}}}{N_{\text{rpm}}} = 9549.3\frac{25}{1000} = r_R\boldsymbol{T}_1 = 3(79.578) = 238.73\ [\text{N m}]$$

$$\boldsymbol{T}_2 = 63\,025\frac{H_{\text{hp}}}{N_{\text{rpm}}} = 63\,025\frac{33.5}{1000} = r_R\boldsymbol{T}_1 = 3(703.78) = 2111.3\ [\text{lb in}] \tag{c}$$

From Eqs. (b) and (c), and as the rotational speed goes down, the torque goes up for the same power. It is also assumed that the *efficiency* of power transmission is approximately very close to a 100% or $\eta = 1.0$.

The diameters of the input or output shafts are obtained from Eqs. (8.15)–(8.20) such that

$$d_S = K_M\left(\frac{32\boldsymbol{T}(K_{\text{SF}})}{\pi S_{\text{yt}}}\right)^{1/3}$$

$$K_M = \left(\left(\frac{(\boldsymbol{M}/\boldsymbol{T})^2 + 0.75}{0.75}\right)^{1/2}\right)^{1/3} \tag{d}$$

From Eq. (d) and with $\boldsymbol{M}/\boldsymbol{T} = 0.5$, the factor K_M is then

$$K_M = \left(\left(\frac{(\boldsymbol{M}/\boldsymbol{T})^2 + 0.75}{0.75}\right)^{1/2}\right)^{1/3} = \left(\left(\frac{0.5^2 + 0.75}{0.75}\right)^{1/2}\right)^{1/3} = (1.1547)^{1/3} = 1.0491 \tag{e}$$

Substituting into Eq. (d) gives the following input diameters:

$$d_{S1} = K_M\left(\frac{32\boldsymbol{T}_1(K_{\text{SF}})}{\pi S_{\text{yt}}}\right)^{1/3} = 1.0491\left(\frac{32(79.578)(3.5)}{\pi(290)(10)^6}\right)^{1/3} = 0.022\,437\ [\text{m}] = 22.437\ [\text{mm}]$$

$$d_{S1} = K_M\left(\frac{32\boldsymbol{T}_1(K_{\text{SF}})}{\pi S_{\text{yt}}}\right)^{1/3} = 1.0491\left(\frac{32(703.78)(3.5)}{\pi(42)(10)^3}\right)^{1/3} = 0.883\,56\ [\text{in}] \tag{f}$$

Substituting into Eq. (d) gives the following output diameters:

$$d_{S2} = K_M\left(\frac{32\boldsymbol{T}_2(K_{\text{SF}})}{\pi S_{\text{yt}}}\right)^{1/3} = 1.0491\left(\frac{32(238.73)(3.5)}{\pi(290)(10)^6}\right)^{1/3} = 0.032\,360\ [\text{m}] = 32.36\ [\text{mm}]$$

$$d_{S1} = K_M\left(\frac{32\boldsymbol{T}_2(K_{\text{SF}})}{\pi S_{\text{yt}}}\right)^{1/3} = 1.0491\left(\frac{32(2111.3)(3.5)}{\pi(42)(10)^3}\right)^{1/3} = 1.2743\ [\text{in}] \tag{g}$$

From Eqs. (f) and (g), and as the rotational speed went down, the synthesized diameter went up for the same power. The selected diameters would be the closest round figure of each value obtained in Eqs. (f) and (g). It might be much safer, however, if one selects the higher round figure or preferred value for the diameter. This would give much higher safety factor than intended.

13.3 Power Control Elements

The power control elements in this book are passively controlling the power flow from the input shaft to the output shaft. The controls are not having feedback features that characterize the regular *control systems*. *Clutches*, *brakes*, and *flywheels* are examples of control elements discussed in this book; see Figure 13.3c,d. Some present designs of clutches and brakes, however, would have feedback controls for activation or deactivation such as hydra-mantic clutches, continuously **v**ariable **t**ransmission (CVT), and **a**nti-lock **b**raking **s**ystem (ABS brakes). These systems and capabilities are beyond the scope of this book. The book is focusing on the basic principles of these elements and their traditional applications as indicated in Chapter 18. The feedback controls can be attached thereafter.

The transmission power control utilizing a flywheel, or a clutch, is intended to smooth or absorb excess energy. The transmission power control employing a brake is projected to dissipate energy (e.g. to stop the machine or the system). These elements are discussed further in Chapter 18.

The power control elements should be designed to withstand internal stresses due to torque variations and other loading conditions. For flywheels, evaluating stresses and implementing optimization are extremely useful in that regard; see, e.g. Bazaj and Metwalli (1971), Shawki et al. (1984), and Metwalli et al. (1983). Disk brake optimization improves the design and performance of this control element (Metwalli and Hegazi 1999, 2001).

13.4 Computer-Aided Design of a Power Transmission System

In this section, concern is about two noncollinear shafts. More shafts can be handled consecutively the same way. The input first shaft and the second shaft are solved. The second shaft is the input to the third and so on. Each step has its own reduction ratio and the power input should be the same if no split power occurs on the same shaft. Computer-aided design (***CAD***) codes are developed to help in the synthesis of any two noncollinear shafts.

A MATLAB© code is developed to help in the design of such cases as given in Figures 13.4–13.6. The initial synthesis of the *input shaft* and *output shaft* is the outcome of Figure 13.4. For both *shafts*, the same procedure is used with few changes in values particularly the output torque and its effects on the normal and shear stresses due to the *reduction ratio*. The MATLAB© code is the concatenation of all three Figures 13.4–13.6 with some added descriptive statements (not shown). The code helps in the design of power transmission elements particularly initial synthesis of shafts. The code name is ***CAD_Power_Transmission.m***. In the code, the input shaft is not assigned with a subscript of 1 or the word input. The output shaft is assigned with the word *output* in the variable definition in place of the numeral 2 in the previously given Eqs. (13.1)–(13.4).

```
clear all; clc; format compact; format short     % CAD_Power_Transmission.m
          disp('-Power Transmission and Applied Loads')
HkW = 25;                                        % INPUT-Transmitted Power [kW]
Hhp = 33.5;                                      % INPUT-Transmitted Power [hp]
Nrpm = 3000;                                     % INPUT-Input Rotational Speed [rpm]
Nrpm_out = 1000;                                 % INPUT-Output rotational Speed [rpm]
MtoT = 0.5 ;  MtoT_out = 0.5 ;                   % INPUT-Moment/Torque Ratio
rR = Nrpm/Nrpm_out;                                          % Reduction ratio
          disp('-Power Transmission and Torques')
HW = HkW*1000;                                               % Transmitted Power in [W]
TNm  = (60*HW)/(2*pi*Nrpm);                            % Transmitted Torque [N.m]
Tlbin  = (63025*Hhp)/(Nrpm);                           % Transmitted Torque [lb.in]
TNm_out  = TNm*rR; Tlbin_out = Tlbin*rR;
KM=(((MtoT^2+0.75)/0.75)^0.5)^(1/3) ; KM_out=(((MtoT_out^2+0.75)/0.75)^0.5)^(1/3)a;
          disp('-Material Properties of Element')
SeSI = 260;                                      % INPUT-Endurance Limit [MPa]
SeUS = 37.7;                                     % INPUT-Endurance Limit [kpsi]
SySI = 290;                                      % INPUT-Yield Strength [MPa]
SyUS = 42;                                       % INPUT-Yield Strength [kpsi]
KSF = 3.0;                                       % INPUT-Safety Factor
          disp('-Shaft Diameter Synthesis (Initial) (inputs need change or commenting)')
dS =KM* (((32*TNm)*(KSF)/(pi*SySI*10^6))^(1/3))*10^3;                  % [mm]
dS_out = KM_out* ( ((32*TNm_out)*(KSF)/(pi*SySI*10^6))^(1/3))*10^3;   % [mm]
dSin = KM* (((32*Tlbin)*(KSF)/(pi*SyUS*10^3))^(1/3));                  % [in]
dSin_out = KM_out* (((32*Tlbin_out)*(KSF)/(pi*SyUS*10^3))^(1/3));      % [in]
```

Figure 13.4 Torque and preliminary shaft synthesis for power transmission elements. The main inputs are the power, rotational speed of input and output shafts, material properties, and substantial initial safety factor.

```
                                              %CAD_Power_Transmission.m
        disp('-Fluctuating Stresses and ASME fatigue diagram (for Safety Factor SI and US)')
dS=20;                                      % INPUT-Shaft diameter [mm]
dS_out=30;                                  % INPUT-Output-Shaft diameter [mm]
% dS = dS_out                               % Uncomment to calculate
SeSI=260;                                   % INPUT-Endurance Limit [MPa]
KSC=3.0;                                    % INPUT-Stress Concentration Factor
        disp('-Fluctuating Stresses and ASME fatigue diagram (for Safety Factor SI)')
Tau_max_MPa=(16*TNm/(pi*(dS/10^3)^3))/10^6          % [MPa] Maximum shear stress
Tau_min_MPa=(16*TNm/(pi*(dS/10^3)^3))/10^6          % [MPa] Minimum shear stress
Sigma_max_MPa= (32*TNm*MtoT/(pi*(dS_out/10^3)^3))/10^6  % [MPa] maximum stress
Sigma_min_MPa= -(32*TNm*MtoT/(pi*(dS_out/10^3)^3))/10^6 % [MPa] minimum stress
Sigma_m_MPa= 0.5*(Sigma_max_MPa+Sigma_min_MPa)          % _m for mean
Sigma_a_MPa= 0.5*(Sigma_max_MPa-Sigma_min_MPa)          % _a for alternating
Tau_m_MPa= 0.5*(Tau_max_MPa+Tau_min_MPa)
Tau_a_MPa= 0.5*(Tau_max_MPa-Tau_min_MPa)
Sigma_vMm_MPa=sqrt(Sigma_m_MPa^2+3*Tau_m_MPa^2)     % vM von Mises equivalent
Sigma_vMa_MPa=sqrt(Sigma_a_MPa^2+3*Tau_a_MPa^2)     % vM von Mises equivalent
K_total_SI=1/KSC                                    % Approximate (Change) (INPUT)
Se_part_SI= K_total_SI*SeSI
KSF_F_SI=1/sqrt((Sigma_vMa_MPa/Se_part_SI)^2+(Sigma_vMm_MPa/SySI)^2) % ASME elliptic
```

Figure 13.5 Dynamic shaft synthesis process for power transmission of SI system. The main inputs are in Figure 13.4, in addition to endurance limit, selected shaft diameters (both input and output), and stress concentration factor. The output is the fatigue safety factor.

```
                                              %CAD_Power_Transmission.m
        disp('-Fluctuating Stresses and ASME fatigue diagram (for Safety Factor SI and US)')
dSin=0.75;                                  % INPUT-Shaft diameter [in]
dSin_out=1.0;                               % INPUT-Output Shaft diameter [in]
% dSin = dSin_out                                   % Uncomment to calculate
SeUS= 37.7;                                 % INPUT-Endurance Limit [kpsi]
SeUS= 0.145*SeSI;                                   % Calculated Endurance Limit [kpsi]
KSC=3.0;                                    % INPUT-Stress Concentration Factor
        disp('-Fluctuating Stresses and ASME fatigue diagram (for Safety Factor US)')
Sigma_max_kpsi=(32*Tlbin*MtoT/(pi*dSin^3))/10^3     % [kpsi] Maximum normal stress
Sigma_min_kpsi=-(32*Tlbin*MtoT/(pi*dSin^3))/10^3    % [kpsi] Minimum normal stress
Tau_max_kpsi= (16*Tlbin/(pi*dSin^3))/10^3           % [kpsi] Maximum shear stress
Tau_min_kpsi= (16*Tlbin/(pi*dSin^3))/10^3           % [kpsi] Minimum shear stress
Sigma_m_kpsi= 0.5*(Sigma_max_kpsi+Sigma_min_kpsi)           % _m for mean
Sigma_a_kpsi= 0.5*(Sigma_max_kpsi-Sigma_min_kpsi)           % _a for alternating
Tau_m_kpsi= 0.5*(Tau_max_kpsi+Tau_min_kpsi)
Tau_a_kpsi= 0.5*(Tau_max_kpsi-Tau_min_kpsi)
Sigma_vMm_kpsi=sqrt(Sigma_m_kpsi^2+3*Tau_m_kpsi^2)          % vM von Mises equivalent
Sigma_vMa_kpsi=sqrt(Sigma_a_kpsi^2+3*Tau_a_kpsi^2)          % vM von Mises equivalent
K_total_US=1/KSC                                    % Approximate (Change) (INPUT)
Se_part_US= K_total_US*SeUS
KSF_F_US=1/sqrt((Sigma_vMa_kpsi/Se_part_US)^2+(Sigma_vMm_kpsi/SyUS)^2) % ASME elliptic
```

Figure 13.6 Dynamic shaft synthesis process for power transmission of US system. The main inputs are in Figure 13.4, in addition to endurance limit, selected shaft diameters (both input and output), and stress concentration factor. The output is the fatique safety factor.

Figure 13.4 presents the code for torque and preliminary shaft synthesis of power transmission elements. The main *inputs* are the power [kW] or [hp], rotational speed of input and output shafts [rpm], moment to torque ratio, material properties in SI and US units, and a substantial initial safety factor of 3.5. The inputs are highlighted by the bold uppercase font for the word **INPUT**. The *outputs* are the initial shaft diameter synthesis for input and output shafts in the adapted SI and US systems of units.

Figure 13.5 shows the code for dynamic shaft synthesis process for power transmission of SI system. The main *inputs* are in Figure 13.4, in addition to endurance limit [MPa], selected shaft diameters (both input and output shafts) [mm], and stress concentration factor. The *output* is the fatigue safety factor for the *input shaft* in the SI system of units. The code for dynamic shaft synthesis process for power transmission of *output shaft* is similarly defined in the MATLAB code of `CAD_Power_Transmission.m`. The main differences are in the reduction ratio effect on the output variables. The output is defined as a word *out* attached to the variables.

Figure 13.6 shows the code of dynamic shaft synthesis process for power transmission of US system. The main *inputs* are in Figure 13.4, in addition to endurance limit [kpsi], selected shaft diameters (both input and output shafts) [in], and stress concentration factor. The *output* is the fatigue safety factor for the *input shaft* in the US system of units. The code for dynamic shaft synthesis process for power transmission of *output shaft* is similarly defined in the MATLAB code of `CAD_Power_Transmission.m`. The main differences are in the reduction ratio effect on the output variables. The output is defined as a word *out* attached to the variables.

The MATLAB code `CAD_Power_Transmission.m` provides the same outputs for the *output shafts* in both SI and US systems of units (not shown in Figures 13.5 and 13.6).

Example 13.2 The cylindrical shaft in Example 13.1 is transmitting a maximum of 25 [kW] or 33.5 [hp] and running at 3000 [rpm]. The power transmission is steady with no appreciable variation in values. Again, it is subject to a maximum bending moment that is half of the maximum applied torque at its critical location. It is connected to a noncollinear shaft that runs at 1000 [rpm] with an efficiency close to a 100%. With an initial safety factor K_{SF} of 3.5, a selected material of hot rolled AISI 1040 or ISO C40, and if the stress concentration factor at the critical location is found to be 3.0, find the fatigue safety factor for some close diameters to the initially found ones in Example 13.1.

Solution

Data: $H = 25$ [kW] or 33.5 [hp], $N_{\text{rpm}} = 3000$ [rpm], the output $N_{\text{rpm\_out}} = 1000$ [rpm], $K_{SF} = 3.5$, and $K_{SC} = 3.0$. Material properties are $S_{yt} = 42$ [kpsi] or 290 [MPa], $S_{ut} = 76$ [kpsi] or 525 [MPa], and $S_e = 37.7$ [kpsi] or 260 [MPa]; see Table A.7.2. Equation (7.58) suggests $S_e = 0.5S_{ut} = 0.5(525) = 262.5$ [MPa] or $0.5(76) = 38$ [kpsi].

The data in Example 13.1 that can be applicable to this example is the torque on each shaft and the initial diameter synthesis of both input and output shafts. These torques are $\boldsymbol{T}_1 = 79.578$ [N m] or 703.78 [lb in] and $\boldsymbol{T}_2 = 238.73$ [N m] or 2111.3 [lb in]. The initial synthesized diameters are $d_{S1} = 22.437$ [mm] or 0.883 56 [in] and $d_{S2} = 32.36$ [mm] or 1.2743 [in].

The usual process is to select some higher preferred diameters for the input and output shafts. However, with the synthesis safety factor at 3.5, it might be acceptable to select the closest preferred diameters for the input and output shafts. The selected diameters are then $d_{S1} = 20$ [mm] or 0.8 [in] and $d_{S2} = 30$ [mm] or 1.2 [in] for the input and output shafts, respectively. These are about 10% less than the synthesized initial diameters of input and output shafts. If the fatigue safety factors are not satisfactory, one can go to the higher preferred diameters. With these values, the calculations are ensuing the fluctuating stresses and ASME fatigue diagram to develop the fatigue safety factor as previously depicted in Chapter 7.

Since the power transmission is steady with no variation in values, the minimum torque is assumed to be the same as the maximum torque. The bending moment, however, causes the normal stress to switch signs from tension to compression as the shafts rotate. This gives the following normal and shear stress values for *input shaft* employing Eq. (8.14) and given a subscript of 1. The maximum normal stresses are then

$$\sigma_{\max 1} = \frac{32\boldsymbol{M}}{\pi d_S^3} = \frac{32(0.5\boldsymbol{T}_1)}{\pi d_S^3} = \frac{32(0.5(79.578))}{\pi(20(10)^{-3})^3} = 50\,660\,928\ [\text{Pa}] = 50.661\ [\text{MPa}]$$

$$\sigma_{\max 1} = \frac{32\boldsymbol{M}}{\pi d_S^3} = \frac{32(0.5\boldsymbol{T}_1)}{\pi d_S^3} = \frac{32(0.5(703.78))}{\pi(0.8)^3} = 7000.6\ [\text{psi}] = 7.0006\ [\text{kpsi}] \qquad \text{(a)}$$

The minimum normal stresses are as follows:

$$\sigma_{\min1} = -\frac{32\boldsymbol{M}}{\pi d_S^3} = -50.661 \text{ [MPa]}$$
$$\sigma_{\min1} = -\frac{32\boldsymbol{M}}{\pi d_S^3} = -7.0006 \text{ [kpsi]} \tag{b}$$

The maximum shear stresses are then

$$\tau_{\max1} = \frac{16\boldsymbol{T}_1}{\pi d_S^3} = \frac{16(79.578)}{\pi(20(10)^{-3})^3} = 50\,660\,928 \text{ [Pa]} = 50.661 \text{ [MPa]}$$
$$\tau_{\max1} = \frac{16\boldsymbol{T}_1}{\pi d_S^3} = \frac{16(703.78)}{\pi(0.8)^3} = 7000.6 \text{ [psi]} = 7.0006 \text{ [kpsi]} \tag{c}$$

The minimum shear stresses are as follows:

$$\tau_{\min1} = \frac{16\boldsymbol{T}_1}{\pi d_S^3} = 50.661 \text{ [MPa]}$$
$$\tau_{\min1} = \frac{16\boldsymbol{T}_1}{\pi d_S^3} = 7.0006 \text{ [kpsi]} \tag{d}$$

The mean and alternating normal and shear stresses are defined as follows:

$$\sigma_{m1} = 0.5(\sigma_{\max1} + \sigma_{\min1}) = 0.5(50.661 - 50.661) = 0.0 \text{ [MPa]}$$
$$\sigma_{a1} = 0.5(\sigma_{\max1} - \sigma_{\min1}) = 0.5(50.661 + 50.661) = 50.661 \text{ [MPa]}$$
$$\sigma_{m1} = 0.5(\sigma_{\max1} + \sigma_{\min1}) = 0.5(7.0006 - 7.0006) = 0.0 \text{ [kpsi]}$$
$$\sigma_{a1} = 0.5(\sigma_{\max1} - \sigma_{\min1}) = 0.5(7.0006 + 7.0006) = 7.0006 \text{ [kpsi]} \tag{e}$$

and

$$\tau_{m1} = 0.5(\tau_{\max1} + \tau_{\min1}) = 0.5(50.661 + 50.661) = 50.661 \text{ [MPa]}$$
$$\tau_{a1} = 0.5(\tau_{\max1} - \tau_{\min1}) = 0.5(50.661 - 50.661) = 0.0 \text{ [MPa]}$$
$$\tau_{m1} = 0.5(\tau_{\max1} + \tau_{\min1}) = 0.5(7.0006 + 7.0006) = 7.0006 \text{ [kpsi]}$$
$$\tau_{a1} = 0.5(\tau_{\max1} - \tau_{\min1}) = 0.5(7.0006 - 7.0006) = 0.0 \text{ [kpsi]} \tag{f}$$

The *maximum distortion energy (von Mises) theory* gives (see Eq. (7.46))

$$\sigma_{\text{vMm1}} = \sqrt{\sigma_{m1}^2 + 3\tau_{m1}^2} = \sqrt{(0.0)^2 + 3(50.661)^2} = 87.747 \text{ [MPa]}$$
$$\sigma_{\text{vMa1}} = \sqrt{\sigma_{a1}^2 + 3\tau_{a1}^2} = \sqrt{(50.661)^2 + 3(0.0)^2} = 50.661 \text{ [MPa]}$$
$$\sigma_{\text{vMm1}} = \sqrt{\sigma_{m1}^2 + 3\tau_{m1}^2} = \sqrt{(0.0)^2 + 3(7.0006)^2} = 12.125 \text{ [kpsi]}$$
$$\sigma_{\text{vMa1}} = \sqrt{\sigma_{a1}^2 + 3\tau_{a1}^2} = \sqrt{(7.0006)^2 + 3(0)^2} = 7.0006 \text{ [kpsi]} \tag{g}$$

The endurance strength of the material S_e is adjusted to endurance strength S_{eP} of the part or element by the stress concentration factor K_{SC} such that

$$S_{eP} = \frac{S_e}{K_{SC}} = \frac{260}{3.0} = 86.667\ [\text{MPa}]$$

$$S_{eP} = \frac{S_e}{K_{SC}} = \frac{37.7}{3.0} = 12.567\ [\text{kpsi}] \tag{h}$$

The ASME fatigue safety factor K_{SFF} as previously depicted in Chapter 7 is then given by the following:

$$K_{SFF}|_{SI} = \left(\sqrt{\left(\frac{\sigma_{vMa1}}{S_{eP}}\right)^2 + \left(\frac{\sigma_{vMm1}}{S_{yt}}\right)^2}\right)^{-1} = \left(\sqrt{\left(\frac{50.661}{86.667}\right)^2 + \left(\frac{87.747}{290}\right)^2}\right)^{-1} = \frac{1}{0.658\,22} = 1.5193$$

$$K_{SFF}|_{US} = \left(\sqrt{\left(\frac{\sigma_{vMa1}}{S_{eP}}\right)^2 + \left(\frac{\sigma_{vMm1}}{S_{yt}}\right)^2}\right)^{-1} = \left(\sqrt{\left(\frac{7.0006}{12.567}\right)^2 + \left(\frac{12.125}{42}\right)^2}\right)^{-1} = \frac{1}{0.627\,42} = 1.5938 \tag{i}$$

For the *output shaft*, the same procedure is used with few changes in values particularly the output torque and its effects on the normal and shear stresses. A MATLAB© code developed to help in the design of such cases is given in Figures 13.4–13.6. The code is a concatenation of all three figures with some added descriptive statements. The code helps in the design of power transmission elements particularly initial synthesis of shafts and dynamic safety factors for selected diameters. The code name is `CAD_Power_Transmission.m`. It verified all previous results in Eqs. (a)–(i). In the code, the *input* shaft is not assigned with a subscript 1 or the word input. The *output* shaft is assigned with the word out in the variable definition rather than the subscript of 2. For the *output* shaft, the code gives the following results:

The ***input***:

```
dS_out= 30 [mm], SeSI= 260 [MPa], KSC=3.0,
dSin_out=1.2 [in], SeUS= 37.7 [kpsi], KSC=3.0
```

The ***output***:

```
    Tau_max_out_MPa = 45.0316 [MPa], Tau_min_out_MPa = 45.0316 [MPa],
Sigma_max_out_MPa = 45.0316 [MPa], Sigma_min_out_MPa = -45.0316 [MPa],
Sigma_m_out_MPa = 0 [MPa], Sigma_a_out_MPa = 45.0316 [MPa], Tau_m_out_MPa =
45.0316 [MPa], Tau_a_out_MPa = 0 [MPa], Sigma_vMm_out_MPa = 77.9971 [MPa],
Sigma_vMa_out_MPa = 45.0316 [MPa], K_total_SI = 0.3333, Se_part_SI =
86.6667 [MPa],
    KSF_F_out_SI = 1.7092.
    Tau_max_out_kpsi = 6.2228 [kpsi], Tau_min_out_kpsi = 6.2228 [kpsi],
Sigma_max_out_kpsi = 6.2228 [kpsi], Sigma_min_out_kpsi = -6.2228 [kpsi],
Sigma_m_out_kpsi = 0 [kpsi], Sigma_a_out_kpsi = 6.2228 [kpsi],
Tau_m_out_kpsi = 6.2228 [kpsi], Tau_a_out_kpsi = 0 [kpsi],
Sigma_vMm_out_kpsi = 10.7782 [kpsi], Sigma_vMa_out_kpsi = 6.2228 [kpsi],
K_total_US = 0.3333, Se_part_US = 12.5667 [kpsi],
    KSF_F_out_US = 1.7930.
```

These values can be easily verified by hand calculations.

The MATLAB© code is available through the ***Wiley website*** under the name ***CAD_Power_Transmission.m***. It solves Examples 13.1 and 13.2. By appropriate editing or rewriting, any other case should be easily solvable.

13.5 Summary

This chapter presents an introduction to power transmission elements. It emphasizes the connectivity between the prime movers that provide power and the machines that consumes or transforms this power. The connections are usually through shafts as power transmission elements on which the other power transmission elements are mounted. The initial design of shafts is highlighted due to their importance in connecting power transmission elements. Main collinear and noncollinear power transmission elements are underlined, and necessary equations to calculate their main parameters are presented. For noncollinear shafts, the effect of the reduction ratio on the synthesis of input and output shafts is demonstrated. Fluctuating torques and moments, material properties, and stress concentration factors are initially used to check the basic fatigue safety factor of the input and output shafts. These can be achieved at any critical section thereof. Fluctuating stresses and ASME fatigue diagram are used to evaluate the fatigue safety factors at any of the shaft's sections. A ***CAD*** code is, therefore, developed to help in this synthesis process of shafts. Rounding values of synthesized shaft diameters are to be designated into the ***CAD*** program at each of these sections. By few iterations, the most suitable initial shaft diameter synthesis is obtainable at any section under fluctuating normal and shear stresses.

Problems

13.1 Find other power transmission elements not cited by this chapter. Can you find and download some of these? Present sketches and references or any other search options that cover these different elements.

13.2 Select one of the noncollinear power transmission elements in Section 13.2 (gears, belts, chains, etc.), and search for the maximum available reduction ratio (for one stage). Is the efficiency cited or quoted? Evaluate or find the maximum peripheral speed.

13.3 Resolve Example 13.1 for a bending to torque ratios of 0.2, 1.0, 2.0, 5.0, and 10.0. Use the same material and safety factor for input and output shafts. Compare shaft diameters.

13.4 Resolve Example 13.1 for the reduction ratios of 2.0, 5.0, and 10.0. Use the same material and safety factor for input and output shafts. Compare shaft diameters.

13.5 Resolve Example 13.1 for the reduction ratios of 0.5, 0.2, and 0.1. Use the same material and safety factor for input and output shafts. Compare shaft diameters. Can you justify the *reduction* ratio of less than 1.0?

13.6 How can the value of the initial safety factor affect the initial synthesis of shaft diameter? Select one of the previous problems to study the effect of initial safety factor of 5.0, 4.0, and 3.0.

13.7 Find the general principles of hydra-mantic or automatic clutches. Sketch the diagram indicating the components and actions to perform the automatic clutching.

13.8 Find the general principles of continuously variable transmission (CVT). Sketch the diagram indicating the components and actions to perform the power transmission and defining the minimum and maximum reduction ratios.

13.9 Search for the general principles of anti-lock braking system (ABS brakes). Sketch the diagram indicating the components and actions to perform the braking controls.

13.10 Identify several applications of flywheels and calculate their properties.

13.11 Redo Example 13.2 for a bending to torque ratios of 0.2, 1.0, 2.0, 5.0, and 10.0. Use the same material and safety factor for input and output shafts. Compare the fatigue safety factor for different cases.

13.12 Redo Example 13.2 for the reduction ratios of 2.0, 5.0, and 10.0. Use the same material and safety factor for input and output shafts. Compare the fatigue safety factor for different cases.

13.13 Redo Example 13.2 for the reduction ratios of 0.5, 0.2, and 0.1. Use the same material and safety factor for input and output shafts. Compare the fatigue safety factor for different cases.

13.14 Use MATLAB code *CAD_Power_Transmission.m* to solve Problems 13.3 and 13.11, and compare results with hand calculations. What should be the change in diameter synthesis if the torque should vary from zero to the maximum?

13.15 Use MATLAB code *CAD_Power_Transmission.m* to solve Problems 13.4 and 13.12 and compare results with hand calculations. What should be the change in diameter synthesis if the torque should vary from zero to the maximum?

13.16 Use MATLAB code *CAD_Power_Transmission.m* to solve Problems 13.5 and 13.13, and compare results with hand calculations. What should be the change in diameter synthesis if the torque should vary from zero to the maximum?

References

Bazaj, D.K. and Metwalli, S.M. (1971). Stress analysis of compounded rotating disks. *Journal of the Franklin Institute* 292 (4): 265–275.

Elmaghraby, S.E., Metwalli, S.M., and Zorowski, C.F. (1979). Some operations research approaches to automobile gear train design. In: *Engineering Optimization*, Mathematical Programming Study, vol. 11 (eds. M. Avriel and R.S. Dembo), 150–175. The Mathematical Programming Society – Springer.

Metwalli, S.M. and Hegazi, H.A. (1999). CAD of disc brakes by multi-objective optimization. *Proceedings of the 1999 ASME DETC*, Las Vegas, Nevada (12–15 September). Paper No. DETC99/CIE-9139, American Society of Mechanical Engineers.

Metwalli, S.M. and Hegazi, H.A. (2001). Computer-based design of disc brakes by multi-objective form optimization. *Proceedings of the ASME DETC/CIE*, Pittsburgh, PA (9–12 September). Paper No. DETC2001/CIE-21680, American society of mechanical engineers.

Metwalli, S.M., Shawki, G.S.A., and Sharobeam, M.H. (1983). Optimum design of variable-material flywheels, ASME Paper No. 82-DET-99. *Journal of Mechanical Design* 105 (2): 249–253.

Shawki, G.S.A., Metwalli, S.M., and Sharobeam, M.H. (1984). Optimum configuration for an isotropic rotor. *Journal of Mechanical Design* 106 (3): 376–379.

14

Spur Gears

Gear drives are the most commonly used of all types of drives. One of these gear drives is the spur gear that has a cylindrical shape with teeth parallel to the cylinder center line as shown in Figure 14.1. Spur gears are used for *parallel shafts*. The geometry of the teeth, kinematic definitions, and usual types and standards are introduced in the chapter. Forces generated during meshing and power transmission are identified. These forces are acting on the teeth of the meshing gears, and their effect in developing stresses on the teeth and gear design is defined. Synthesis procedure utilizes computer-aided design (*CAD*), and optimization tools to generate better gearing design are provided. Constructional details for gear set assembly of gearboxes including planetary, or epicyclic, gear sets are also presented.

Symbol and Abbreviations

The adapted units are [in, lb, psi] or [m, kg, N, Pa], others given at each symbol definition ([k…] is 10^3, [M…] is 10^6, and [G…] is 10^9).

Symbol	Quantity, Units (adopted)
AGMA	American Gear Manufacturers Association
B_{BC}	Material property factor (ratio of S_{ab}/S_{ac})
b_g	Breadth of the *Hertzian* contact area
B_v	Quality constant for velocity factor equation
C_g	Center distance between the mating gears
C_p	Elastic coefficient, $[\text{MPa}]^{1/2}$, $[\text{psi}]^{1/2}$
d_b	Base circle diameter
d_{bG}	Base circle diameter of the gear
d_{bP}	Base circle diameter of the pinion
d_G	Pitch diameter for gear [mm], [in]
d_g	Gear pitch diameter for any pinion or gear
d_o	Gear outside diameter
d_{oG}	Gear outside diameter for the gear
d_{oP}	Gear outside diameter for the pinion
d_P	Pitch diameter for pinion [mm], [in]

Machine Design with CAD and Optimization, First Edition. Sayed M. Metwalli.

Companion website: www.wiley.com/go/metwalli/machine

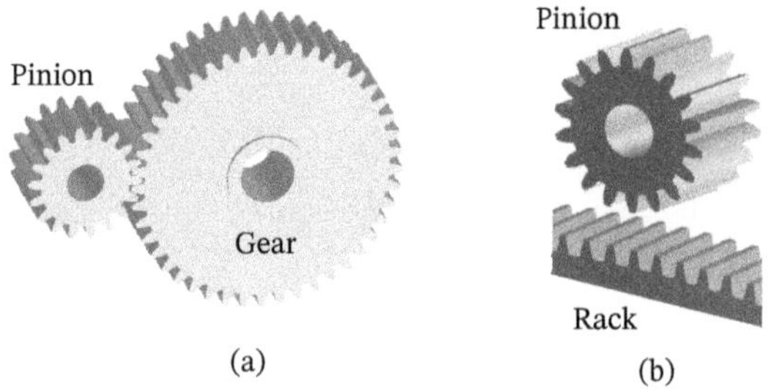

Figure 14.1 Gears showing (a) spur gear and pinion that have cylindrical shape with teeth parallel to the cylinder center line and (b) a rack and pinion where the rack having infinite diameter but a definite straight length.

Symbol	Quantity, Units (adopted)
d_r	Gear root diameter
d_{rP}	Gear root diameter for pinion
E_G	Elasticity modulus of gear material [Mpsi], [GPa]
E_P	Elasticity modulus of pinion material [Mpsi], [GPa]
f_g	Face width of the gear
f_g^*	Optimum face width
$f_i, f_{1,2,3}$	Flow variables in system model
$\boldsymbol{F}_{\mathrm{in}}$	Generated force between teeth
$\boldsymbol{F}_n$	Normal force to tooth profile
f_n	Natural frequency, [Hz]
f_T	Tooth frequency of gear resonance, [Hz]
$\boldsymbol{F}_r$	Radial force
$\boldsymbol{F}_t$	Tangential force
H	Transmitted power
HB	Brinell hardness
HRC	Hardness Rockwell C
H_{hp}	Horse power, [hp]
H_{kW}	Power in kilo watt, [kW]
I_g	Contact geometry factor
J_g	Bending geometry factor
k_a	Surface finish factor
k_b	Size factor
k_c	Reliability factor
k_d	Temperature factor
k_e	Stress concentration factor
K_f	Stress concentration factor
k_f	One-way or two-way bending factor
K_{IN}	Correction factor for contact geometry factor I_g
k_m	Load distribution factor
k_o	Load application factor

Symbol	Quantity, Units (adopted)
K_{MSF}	Material safety factor
K_R	Reliability factor
K_{SF}	Safety factor
$K_{\text{SF}}\|_b$	Bending safety factor
$K_{\text{SF}}\|_c$	Contact safety factor
K_{SFE}	Extra safety factor
K_v	Velocity or dynamic factor
l	Tooth height from the tip to the tooth root
L_N	Life cycle number
m_1	Mass of gear 1 in contact with gear 2
m_2	Mass of gear 2 in contact with gear 1
m_e	Equivalent mass of gear 1 and gear 2
m_n^*	Optimum module, [mm]
m_n	Module or normal module [mm]
m_p	Contact ratio
n_g	Gear ratio or gear set velocity ratio
N_G	Number of teeth for gear
N_g	Number of gear teeth for any pinion or gear
N_{in}	Input rotational speed
$N_{\text{in,rpm}}$	Input rotational speed in revolutions per minute
N_{inP}	Input pinion rotational speed
N_c	Number of carrier teeth in planetary set
N_r	Number of ring teeth in planetary set
N_s	Number of sun teeth in planetary set
N_{out}	Output rotational speed
$N_{\text{rpm},G}$	Gear rotational speed in revolution per minute
$N_{\text{rpm},P}$	Pinion rotational speed in revolution per minute
N_P	Number of teeth for pinion
N_p	Number of teeth for the planet gear in planetary set
N_P^*	Optimum number of pinion teeth
$N_P\|_{\text{min}}$	Minimum number of teeth to ensure no interference
p_c	Circular pitch
p_d	Diametral pitch (US) [teeth/in]
p_d^*	Optimum diametral pitch (US) [teeth/in]
p_{max}	Maximum pressure (or stress) for *Hertzian* contact
Q_v	Quality number
r_b	Base circle radius for either pinion or gear
r_{bG}	Base circle radius for the gear

Symbol	Quantity, Units (adopted)
r_{bP}	Base circle radius for the pinion
r_{cG}	Curvature radius of gear tooth at contact centerline
r_{cP}	Curvature radius of pinion tooth at contact centerline
R_e	Reliability (e.g. 0.9 for 90%)
r_f	Fillet radius at tooth root
r_G	Pitch radius of gear
r_g	Gear pitch circle radius for any pinion or gear
r_P	Pitch radius of pinion
S_{ab}	Allowable bending fatigue strength [MPa], [kpsi]
S_{ac}	Allowable contact fatigue strength [MPa], [kpsi]
S_c	Allowable contact fatigue strength
S_{cf}	Contact fatigue strength
S_e	Endurance limit of gear material
S_{ec}	Contact fatigue endurance strength limit
S_{ecg}	Contact endurance strength of a gear
S_{eg}	Endurance strength limit of the gear
S_f	Fatigue strength
S_t	Allowable bending stress number
S_{ut}	Ultimate tensile strength
S_y	Yield strength in tension
$\boldsymbol{T}_{in}$	Input torque, [N m], or [lb in]
t_p	Tooth thickness at pitch diameter
$\boldsymbol{T}_P$	Pinion torque, [N m], or [lb in]
t_r	Tooth thickness at the root
$\boldsymbol{v}_G$	Gear normal velocity
V_G	Gear pitch cylinder volume
$\boldsymbol{v}_P$	Pinion normal velocity
V_P	Pinion pitch cylinder volume
v_P, v_G	Magnitude *of pinion and gear velocities*
V_S	Gear set pitch cylinders volume
$\boldsymbol{v}_t$	Pitch point velocity, [m/s], or [ft/min]
$\boldsymbol{v}_{t,max}$	Limiting speed of a quality number, [m/s], or [ft/min]
y_L	*Lewis* form factor
Y_L	Modified *Lewis* form factor
Y_N	Life cycle factor
Z_N	Life cycle factor
$\delta_{1,2,3}$	Deflections in components 1, 2, and 3
δ_a, δ_b	Deflections of nodes *a* and *b* in system model

Symbol	Quantity, Units (adopted)
ν_G	Poisson's ratios for gear material
ν_P	Poisson's ratio for pinion material
$\sigma_{ab}\rvert_{max}$	Maximum bending fatigue stress
σ_{ac}	Contact fatigue stress
$\sigma_{ac}\rvert_{max}$	Maximum contact fatigue stress
σ_b	Normal stress due to bending
σ_c	*Hertzian* contact stress
ϕ	Pressure angle
ω_c	Angular velocity of carrier in planetary set
ω_r	Angular velocity of ring gear in planetary set
ω_s	Angular velocity of sun gear in planetary set
ω_G	Angular velocity of gear
ω_n	Natural frequency, [rad/s]
ω_P	Angular velocity of pinion
ω_P, ω_G	Magnitude of *angular velocities of pinion and gear*

14.1 Types and Utility

As indicated before, a *spur gear* has a cylindrical shape with *teeth* parallel to the center line of the cylinder as shown in Figure 14.1a, where usually the small gear is named the ***pinion*** and the large one is named the ***gear***. A ***rack and pinion*** is shown in Figure 14.1b where the *rack* is a mating gear with infinite diameter (or curvature) but has a definite straight length. It has been used in steering of some motor vehicles. The rack, however, is utilized as a cutting tool in the generation of gears, but it should be made of a *cutting tool* material. The spur gear is the simplest possible gear type. Gear trains are sets of linked spur or other gears for high speed ratios such as the one depicted in Figure 14.2a. Epicyclic gears or planetary gear trains are also sets of gears such as the one depicted in Figure 14.2b, which has a sun in the middle, three planets, and an external ring. Epicyclic gears are used in several applications such as automatic transmission gearboxes and some hoisting equipment.

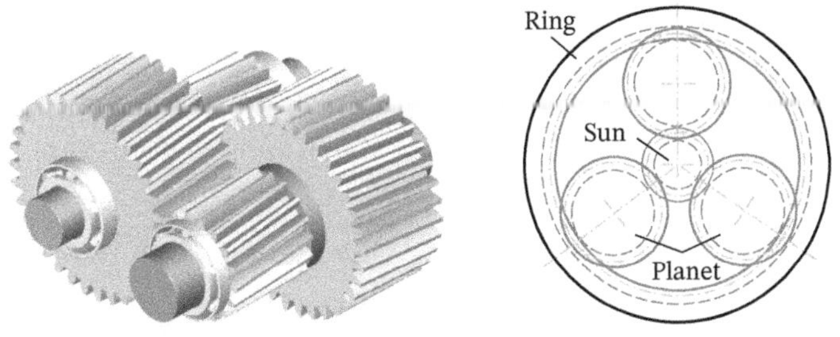

(a) (b)

Figure 14.2 Gear train sets of linked spur or other gears (a) for high speed ratios and epicyclic or planetary gear train (b), which has a sun in the middle, three planets, and an external ring.

Other non-spur types of gears such as helical and double-helical gears are also used for *parallel shafts*. Helical and double-helical gears are treated in Chapter 15 of this text, but an assembly of helical gears is viewed in Figure 14.2a. For shafts that are not parallel and with nonintersecting center lines, crossed helical gears, worm gears, and hypoid gears are employed. Straight bevel gears and spiral bevel gears are utilized when shafts are not parallel but with intersecting center lines. Straight bevel gears, spiral bevel gears, and worm gears are also presented in Chapter 15.

In this chapter definitions and kinematics of spur gears are presented, which are essential to the synthesis and design process. Forces generated on mating gears contact are defined in the sense of synthesis rather than analysis. The forces exerting efforts and transmitting the power are the first to attend to and used in gear synthesis. The synthesis and design procedures are to be specified next. The effects of gear operation on the dynamics of the gear set are considered. A simple system model is introduced to define the critical speed that should be avoided to reduce vibration and noise close to or at resonance. Sample applications provide insight into the design and specifications of spur gears. *CAD* and optimization tools are also provided in the course of the chapter. Constructional details are also presented to be able to physically build gearboxes and other gear sets.

14.2 Definitions, Kinematics, and Standards

Involute tooth profile is used for spur gears (Figure 14.3a). This ensures rolling of teeth in contact at *pitch point*, where the two mating gears rolling as two cylinders meeting at *pitch point* as shown in Figure 14.3b. It produces constant angular velocity ratio during meshing as developed next. Usually the small gear is given the name ***pinion***, and the larger gear is dobbed as the ***gear***. Subscripts are used as uppercase letters where P is for pinion and G is for gear. The subscript g is used to represent a gear be it any pinion or gear. For the following definitions and relations, refer to Figures 14.3–14.6 and Tables 14.1–14.4.

The *involute* curve is the path of a string end that is wrapped on a circular cylinder when the string is unwrapped from the **base circle** as presented in Figure 14.3a. The **base circle** is then the base for the involute profile generation. Involute profile provides a *line of action* tangent to both base circles shown in Figure 14.3b. The *line of action* is inclined with an angle named the *pressure angle* ϕ. Components along the line of action of the normal velocities $\mathbf{v}_P$ and $\mathbf{v}_G$ are the same *pitch line velocity* $\mathbf{v}_t$ as demonstrated in Figure 14.4a,b. *Pitch point velocity* $\mathbf{v}_t$ is at the pitch

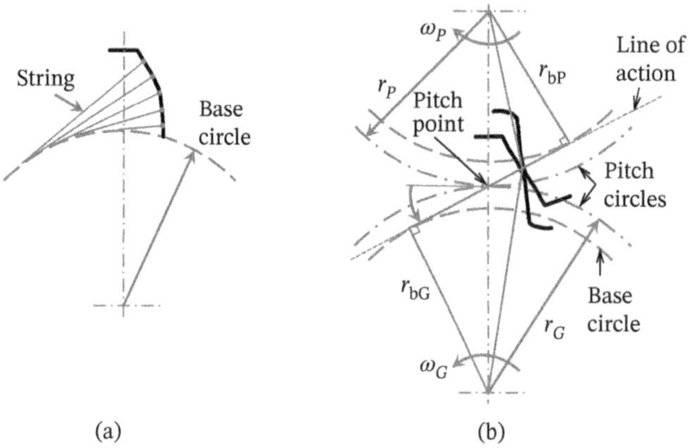

(a) (b)

Figure 14.3 Involute tooth profile for spur gears (a) and (b) shows the two mating gear rolling as two pitch cylinders meeting at pitch point.

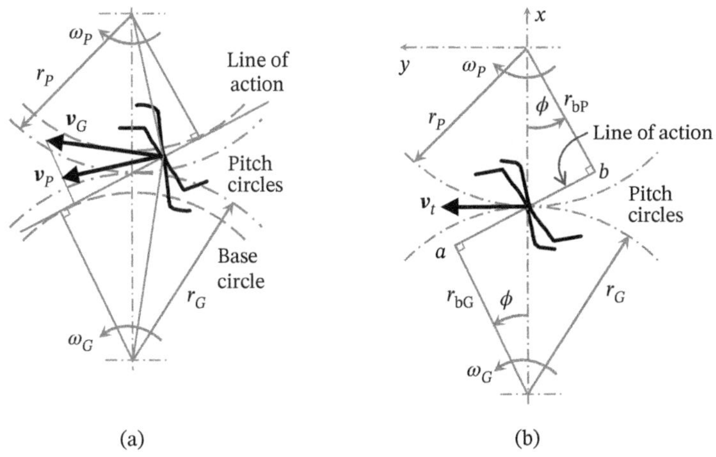

Figure 14.4 Components along the line of action of the normal velocities $\boldsymbol{v}_P$ and $\boldsymbol{v}_G$ in (a) are the same pitch point velocity $\boldsymbol{v}_t$ as demonstrated in (b).

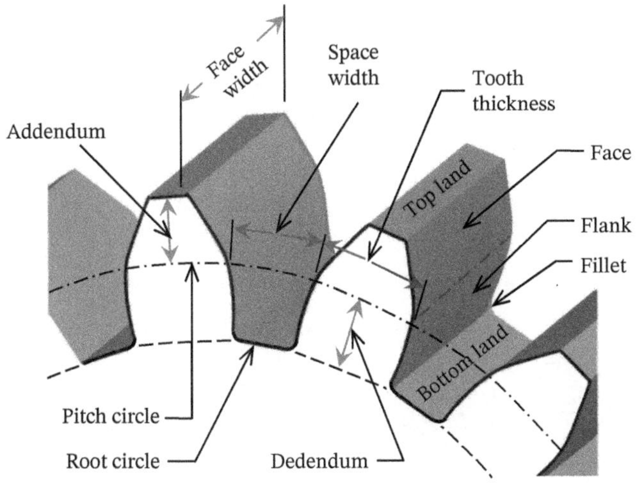

Figure 14.5 A part of a spur gear sketch where some basic gear features and definitions are demonstrated.

point of contact between both gears on the ***pitch circle*** of the gears as shown in Figure 14.4b. It is then tangent to both *pitch circles* that are touching at the pitch point. The two gears as if they are rolling on each other's as two pitch cylinders of radii r_P and r_G with same contact *pitch point velocity* $\boldsymbol{v}_t$. The involute profile then ensures rolling with no sliding of teeth in contact at the pitch point. The *pitch point velocity* $\boldsymbol{v}_t$ is tangent to both pitch circles, and noting that the pinion angular velocity vector $\boldsymbol{\omega}_P$ is in the negative z direction, the vector representing the pinion radius $\boldsymbol{r}_P$ is in the negative x direction, the gear angular velocity vector $\boldsymbol{\omega}_G$ is in the positive z direction, and the vector representing the pinion radius $\boldsymbol{r}_P$ is in the positive x direction, therefore according to the cross product (Eq. (1.22)),

$$\boldsymbol{v}_t = \boldsymbol{\omega}_P \times \boldsymbol{r}_P = \omega_P r_P = \boldsymbol{\omega}_G \times \boldsymbol{r}_G = \omega_G r_G \tag{14.1}$$

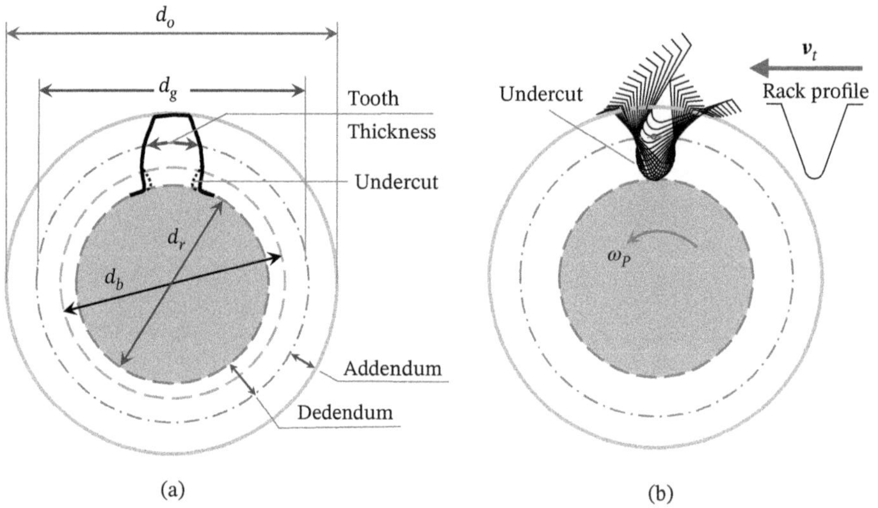

Figure 14.6 The gear main geometrical parameters, dimensions, and tooth form are shown in (a). The undercut is generated as shown in (b) due to teeth manufacturing by the rack cutter.

where ω_P and ω_G are the magnitude of *angular velocities* of the pinion and gear, respectively, and r_P and r_G are the magnitude of *vectors* representing radii of the pinion and gear, respectively. From this relation and previous characterizations, important definitions are distinctively developed in scalar form as follows:

i. *Velocity ratio*

   The velocity ratio is obtained from Eq. (14.1) such that

$$\frac{\omega_P}{\omega_G} = \frac{r_G}{r_P} = \frac{d_G}{d_P} = n_g \tag{14.2}$$

   where d_G and d_P are the *pitch diameters* for gear and pinion, respectively. This velocity ratio is also used at the same time as the **gear ratio** n_g as noted in Eq. (14.2).

ii. *Module*

   The basic standard gear parameter in the SI systems is the ***module*** or *normal module* m_n, in [mm], which is defined by

$$m_n = \frac{d_P}{N_P} = \frac{d_G}{N_G} \tag{14.3}$$

   where d_P is the *pinion pitch diameter* in [mm], N_P is the *number of pinion teeth*, d_G is the *gear pitch diameter* in [mm], and N_G is the *number of gear teeth*. From Eq. (14.1), one can also get the relation for the gear set ***velocity ratio*** or ***gear ratio*** n_g by substituting Eq. (14.3) into Eq. (14.1) to obtain

$$n_g = \frac{\omega_P}{\omega_G} = \frac{N_G}{N_P} \tag{14.4}$$

iii. *Circular pitch*

   The ***circular pitch*** p_c is the distance between similar points on the profiles along the ***pitch circle*** or

$$p_c = \frac{\pi d_g}{N_g} = \frac{\pi d_P}{N_P} = \frac{\pi d_G}{N_G} \tag{14.5}$$

   The notation d_g and N_g are the gear pitch diameter and number of gear teeth, respectively, for any pinion or gear. From Eqs. (14.3) and (14.5), one gets a relation between the *circular pitch* p_c [mm] and the *module*

[mm]:

$$p_c = \pi m_n \tag{14.6}$$

iv. *Pressure angle*
The standard term of a ***pressure angle*** defines the tangent to both base circles of meshing gears as shown in Figure 14.3b. It is also the direction of the *resultant force* on a gear tooth, which should act normal to both teeth profiles of the pinion and the gear. From standards, the old standard value of the pressure angle is $\phi = 14.5°$. This standard is not used nowadays. The new standard is for the pressure angle ϕ to be 20° or 25°. The value of $\phi = 20°$ is much more in use than 25°. Therefore, the adapted *pressure angle* ϕ in this text is emphasized here as

$$\phi = 20° \tag{14.7}$$

v. *Pitch circle*
The ***pitch circle*** and ***module*** m_n are basic standards of gears in a gear set. Even through the pitch circle has been introduced before, one can use a special distinction to that. When one states a radius or a diameter of a gear, it means distinctively defining the pitch circle radius r_g or *pitch circle diameter* d_g of the gear. A pinion or gear radius can be identified as r_P or r_G, and both are for the pitch circle, as depicted in Figures 14.3b and 14.4. In some cases, as indicated before, the gear diameter or gear pitch circle diameter of any gear can be identified as d_g without any need to a specific designation of a pinion or a gear.

vi. *Diametral pitch*
In the US system, another parameter that has been used is the ***diametral pitch*** p_d. It has been mainly used in inch systems and is defined by

$$p_d = \frac{N_g}{d_g\big|_{[\text{in}]}}\ [\text{teeth/in}] \text{ for US}, \quad p_d\big|_{[\text{teeth/in}]} = \frac{25.4}{m_n\big|_{[\text{mm}]}} \tag{14.8}$$

The notations of d_g and N_g are the gear pitch diameter and number of gear teeth, respectively, for any pinion or gear. The gear diameter d_g is in [in] for the US traditional system. The relationship between the diametral pitch and the module is obtained by converting the module to inches. This gives the diametral pitch $p_d = 25.4/m_n$ as depicted in Eq. (14.8). The relation between the *circular pitch* p_c in [in] and the *diametral pitch* p_d is just that $p_c = \pi/p_d$ as defined by Eqs. (14.5) and (14.8), for the US system of units.

vii. *Gear features*
Figure 14.5 shows a part of a spur gear sketch where some basic *gear features* and definitions are demonstrated with others shown in Figure 14.6. The *pitch circle* and the *base circle* are pointed out in Figure 14.6. The *circular pitch* is equally divided into the *space width* between teeth and the ***tooth thickness***, see Figure 14.5. The ***addendum*** is the part of the tooth from the pitch circle to the outer diameter. The ***dedendum*** is the part of the tooth from the pitch circle to the *root circle*. The *face width* f_g is the width of the gear over the teeth region as shown in Figure 14.5. The *top land* is the outer cylindrical tip surface of the tooth. The *bottom land* is the cylindrical surface on the root circle between two teeth. The tooth *face* is the upper surface of the tooth profile above the pitch circle. The tooth *flank* is the lower surface of the tooth profile below the pitch circle. The *tooth fillet* is a cylindrical surface connecting the tooth profile to the base cylindrical surface of the root circle.

viii. *Gear standards*
Basic standards of gears include the ***standard module*** m_n [mm] for both SI and the US systems particularly adapted by the American Gear Manufacturers Association (ANSI/AGMA 2004 or ISO 54 1977 or 1996). The main tooth dimensions are the *addendum* and the *dedendum*. For regular SI tooth dimensions, the value of the addendum is m_n, and the value of the dedendum is $1.25m_n$. The other specifications of *tooth dimensions* for *full depth gears* are defined in Table 14.1 for both SI and the US systems. *Standard modules* in general use

Table 14.1 Tooth specifications for full depth spur gears.

Quantity	Value, SI	Value, US
Addendum	m_n	$1.000/p_d$
Dedendum	$1.25m_n$	$1.250/p_d$
Working depth	$2m_n$	$2.000/p_d$
Whole depth (min)	$2.25m_n$	$2.250/p_d$[a)]
Tooth thickness	$\pi m_n/2$	$1.571/p_d$
Fillet radius of basic rack	$0.3m_n$	$0.300/p_d$
Width of top land (min)	$0.25m_n$	$0.250/p_d$
Clearance (min)	$0.25m_n$	$0.250/p_d$[a)]
Clearance (shaved or ground teeth)	$0.35m_n$	$2.250/p_d$[a)]

a) For fine pitch, whole depth = $2.2/p_d + 0.002$ [in], clearance = $0.2/p_d + 0.002$ [in], and clearance (shaved or ground teeth) = $0.35/p_d + 0.002$ [in].

Table 14.2 Modules [mm] in general use.

Preferred	1	1.25	1.5	2	2.5	3	4	5	6	8
	10	12	16	20	25	32	40	50	60	70
Next choice	1.125	1.375	1.75	2.25	2.75	3.5	4.5	5.5	7	9
	11	14	18	22	28	36	45	55	65	75

Table 14.3 Standard diametral pitches [teeth/in].

Coarse pitch ($p_d < 20$)	1	1.25	1.5	1.75	2	2.5	3	4	5	6	8	10	12	14	16	18
Fine pitch ($p_d > 20$)	20		24		32		48		64		72		80		96	120

are given in Table 14.2 for widely used *preferred modules* and the next choice if need necessitates. *Standard diametral pitches* in general use are shown in Table 14.3 for coarse and fine pitches.

ix. *Interference*

Interference occurs if gears are not produced by *generation of teeth* for pinions with small number of teeth. Generation produces *undercuts* below base circle as the indicated dashed lines inside the tooth profile in Figure 14.6a. The main gear geometrical parameters and dimensions are also indicated in Figure 14.6a. The undercut is produced by cutting as shown in Figure 14.6b due to the process of manufacturing gear teeth by the rack cutter generation. The rack cutter moves to the left with the same peripheral velocity $\boldsymbol{v}_t$ of the pinion at the pitch circle and cuts in a perpendicular direction to the horizontal profile plane. Each cutting frame is frozen with the pinion and rotates with it in the cutting process, which generates the teeth shown in Figure 14.6b. These undercuts weaken the tooth but allow for smaller number of teeth that might be needed in some applications. The minimum number of teeth $N_P|_{\min}$ to ensure no such undercut and possible teeth interference is given by the following relation:

$$N_P\big|_{\min} = 2/\sin 2\phi \tag{14.9}$$

This gives a minimum number of pinion teeth $N_P|_{\min}$ as 18 for a 20° pressure angle and 12 for 25° pressure angle if manufacturing is done by traditional *milling machines* without any undercut. The acceptable range of the number of teeth a *milling cutter* can produce is inscribed on the cutter to ensure accurate tooth cutting. Minimum number of teeth per pair of mating gears is then twice as much for each case. If manufacturing is performed by teeth generation such as *gear hobbing*, *broaching*, or *shaping*, the generated gear can have a lower number of teeth, and no interference will occur when gears are meshed (Townsend 1992).

As indicated previously, the standards for gear *modules* in general use are given in Table 14.2. The preferred values are more accessible and easier to find cutting tools to produce the gears with. The next choice modules are less accessible and might be more expensive to acquire. If it is a critical situation in space or weight, the next choice modules can be used. The standard *diametral pitches* p_d [teeth/in] are provided in Table 14.3. Coarse pitches ($p_d < 20$) generate gears with larger teeth than the fine pitches ($p_d > 20$), which has more teeth per inch. Table 14.4 presents a comparison among modules m_n, their equivalent diametral pitches, and the closest standard diametral pitch p_d for each module. This demonstrates that one should be careful to differentiate – for instance – between a gear made in the module system with $m_n = 5$ [mm] and a gear made with a diametral pitch $p_d = 5$ [teeth/in] as evident from Table 14.4. This is the case when one has such a gear and wants to define its specifications. Since the equivalent diametral pitch to the module m_n of 5 [mm] is very close (p_d of 5.080 [teeth/in]), very accurate measurements should be performed to identify the system, which the gear has been produced under.

x. *Gear geometry formulas*

The gear has several geometry parameters that can be defined in addition to the main ones given before. Some of these are derived using the previous defining specifications. Figure 14.6a provides a drawing of a general

Table 14.4 Equivalent diametral pitches to modules and the closest standard diametral pitches to modules.

Modules m_n [mm]	Equivalent p_d	Standard p_d [teeth/in]
0.3	84.667	80
0.4	63.500	64
0.5	50.800	48
0.8	31.750	32
1	25.400	24
1.25	20.320	20
1.5	16.933	16
2	12.700	12
2.5	10.160	10
3	8.466	8
4	6.350	6
5	5.080	5
6	4.233	4
8	3.175	3
10	2.540	2.5
12	2.117	2
16	1.587	1.5
20	1.270	1.25
25	1.016	1

gear that can also be considered as a pinion. The gear ***outside diameter*** d_o is easily obtained as the gear pitch diameter d_g with the addition of two addendums (Figure 14.6a and Table 14.1)

$$d_o = d_g + 2m_n = m_n(N_g + 2) = d_g\Big|_{[\text{in}]} + 2.000/p_d \tag{14.10}$$

It should be noted that d_g is the pitch diameter of a pinion d_P or the gear d_G. N_g is the number of teeth in a gear, be it a pinion P or a gear G with these replacing the subscript g. The use of the upper case subscripts P and G identifies the relation as applied to the pinion P or the gear G. The *gear outside diameter* of the pinion can then be defined as $d_{oP} = d_P + 2m_n = m_n(N_P + 2)$, which Eq. (14.10) is a simpler identifier as a general form. The same applies to the gear so that $d_{oG} = d_G + 2m_n = m_n(N_G + 2)$. The same applies also to the equation with the diametral pitch p_d.

The gear ***root diameter*** d_r is the pitch diameter d_g minus twice the dedendum (Figure 14.6a and Table 14.1)

$$d_r = d_g - 2.5m_n = d_g - 2.500/p_d \tag{14.11}$$

The ***base circle diameter*** d_b is defined relative to the pitch circle diameter of a gear d_g by

$$d_b = d_g \cos\phi \tag{14.12}$$

The ***tooth thickness*** t_p *at pitch diameter* d_g is

$$t_p = \frac{\pi}{2}m_n = \frac{\pi}{2p_d} \tag{14.13}$$

This is the theoretical tooth thickness along the pitch circle. To have proper meshing and mating, clearance should be available that may usually reduce the tooth thickness as discussed in the *backlash* section below.

The ***center distance*** C_g between the mating gears centerlines is

$$C_g = \frac{1}{2}(d_P + d_G) = \frac{1}{2}(m_n(N_P + N_G))\Big|_{\text{SI}} \tag{14.14}$$

The ***contact ratio*** m_p is the apparent number of teeth in full contact between the two mating gears. This is found through the length on the line of action between the two gears divided by the projection of the circular pitch on the line of action (Figure 14.4b). This results in the following expression:

$$m_p = \frac{\sqrt{r_{\text{oP}}^2 - r_{\text{bP}}^2} + \sqrt{r_{\text{oG}}^2 - r_{\text{bG}}^2} - C_g \sin\phi}{\pi m_n \cos\phi} \tag{14.15}$$

In Eq. (14.15), r_o is the outside radius, and r_b is the *base circle radius* for either pinion P or gear G as subscripts. For the US system, one can replace m_n by $1/p_d$ in Eq. (14.15) provided all other parameters are in [in]. It is recommended that the contact ratio would be more than 1.2 for smoother initial engagement of teeth at the beginning of contact.

xi. *Backlash*

Gear backlash is recommended to provide space between teeth. It indicates that the space between teeth is little larger than the tooth thickness, which needs the tooth width to be little smaller than the regular thickness. The extra space is needed for proper lubrication between teeth, teeth deflection, thermal expansion, etc. The recommended minimum backlash for SI and the US system are shown in Figures 14.7 and 14.8, respectively. The heavy thick lines are the available information in references. The fitted lines with equations close to each one are extended to cover larger range of modules or diametral pitch. The relations in the SI units are linear, while in the US units are of a power form. Each curve represents specific center distance between the two mating gears. Higher values of backlash are for larger center distances. The SI curves are for center distances C_g of 50, 100, 200, 400, and 800 [mm] (Figure 14.7). The US curves are for center distances C_g of 2, 4, 8, 16, and 32 [in] (Figure 14.8). The value of the backlash is usually used in reducing the thickness of the teeth.

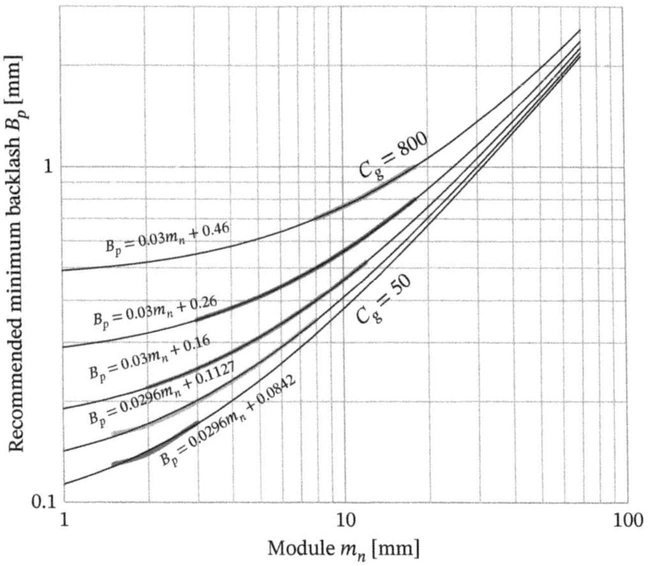

Figure 14.7 The recommended minimum backlash for SI system indicating the fitted equation for each specified center distance C_g in [mm] as a function of the module m_n.

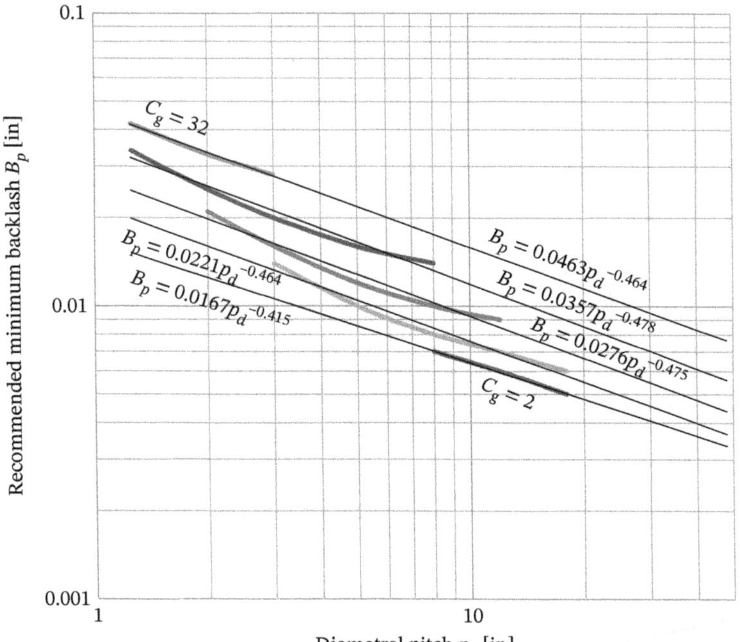

Figure 14.8 The recommended minimum backlash for the US system indicating the fitted equation for each specified center distance C_g in [in] as a function of the diametral pitch p_d.

xii. *Quality*

Gear quality is a very important factor in the design and in the selection of gear manufacturing process. It affects the dynamic performance and requires consideration in selecting appropriate manufacturing process and has to be accounted for in design. High quality gears are requiring higher finishing procedures such as *high precision shaving* or *grinding*. Producing gears by *hobbing* or *shaping* generates quality less than shaving or grinding. *Milling* gears are also lower quality than hobbing. Manufacturing gears by *casting* only should produce much lower quality than previous production means. As indicated in Section 2.4.7, the manufacturing process affects the outcome of *tolerances* and *surface finish* of products. Smaller tolerances and surface finish values are the characteristics of higher quality products for gearing. Smaller tolerances and surface finish values give less variations of dimensions and surface undulations, which generate much less dynamic variations when gears are running. Larger tolerances and surface finish values introduce high variations of dimensions and surface undulations, which generate much higher dynamic disturbances when gears are running. That is why high quality gearing has lower dynamic effect to be considered in design. Measurement of total composite tolerances and expected values are essential in ensuring the intended gear quality control. AGMA 2000-A88 (1988) suggested a scale of *quality numbers* Q_v that indicates the range of quality for some applications (Table 14.5). Table 14.5 also provides the quality numbers of ISO 1328 (2013) but shown in brackets. The ISO quality numbers are in harmony with the *IT* numbers of *tolerances* (see Section 2.4.7) and the higher values are for larger tolerances. The low Q_v quality numbers are, however, for less sensitive applications, and the high quality numbers are for more accuracy expectedly needed applications. Other approaches for quality, however, have been lately suggested. American Gear Manufacturing Association (AGMA) has adapted the quality numbers of ISO 1328 (2013) but using the letter *A* as a prefix. The approach of using the traditional *AGMA quality numbers* Q_v is the one adapted herein.

Example 14.1 ***SI System***

A reduction gear set or a gearbox transmits a maximum of 75 [kW] with 1200 [rpm] input speed. The gear reduction ratio is 4. The standard gearing pressure angle is 20°. Standard module is 8 [mm], and the number of input pinion teeth is 18. Determine (a) the number of teeth of output gear, (b) the output speed of the reducer gearbox, and (c) the input and output gear geometry specifications.

Solution

Data: $H_{\text{kW}} = 75$ [kW] = 75 000 [W], $N_{\text{in,rpm}} = N_{\text{in}} = 1200$ [rpm], $\omega_{\text{in}} = \omega_P = 2\pi N_{\text{in}}/60 = 125.66$ [rad/s], the gear ratio $n_g = \omega_P/\omega_G = 4$, $\phi = 20°$, $m_n = 8$ [mm], and input pinion teeth $N_P = 18$.

Table 14.5 Recommended quality numbers Q_v for some applications (AGMA 2000-A88 (1988)) also the quality number of ISO 1328 (2013) are shown in brackets.

Application	Quality number[a)]	Application	Quality number[a)]
Cement mixer	3–5(14–12)	Washing m/c	8–10 (9–7)
Cement kiln, steel mill	5–6 (12–11)	Printing press	9–11 (8–6)
Corn picker, cranes, punch press, mining conveyer	5–7 (12–10)	Computing mechanism, automotive transmission	10–11 (7–6)
Paper box making machine	6–8 (11–9)	Radar antenna, marine drive	10–12 (7–5)
Gas meter mechanism, small power drills	7–9 (10–8)	Aircraft engine	10–13 (7–4)
		Gyroscope	12–14 (5–3)

a) The sum of AGMA and ISO quality is always 17.

(a) The number of teeth of output gear is obtained from Eq. (14.4), which gives

$$N_G = n_g(N_P) = 4(18) = 72 \text{ teeth} \tag{a}$$

(b) The output speed of reducer gearbox is defined also by Eq. (14.4) $\omega_P/\omega_G = 4$ or

$$\omega_G = \frac{\omega_P}{n_g} = \frac{125.66}{4} = 31.415 \text{ [rad/s]} \tag{b}$$

The output speed in revolution per minute is $N_{\text{out}} = \omega_G(60/2\pi) = 299.99 = 300$ [rpm]. The rounding was not necessary because the result is exactly $N_{\text{out}} = 300$ [rpm] when applying Eq. (14.4) such that

$$n_g = \frac{N_{\text{in,rpm}}}{N_{\text{out,rpm}}} \quad \text{or} \quad N_{\text{out,rpm}} = \frac{N_{\text{in,rpm}}}{n_g} = \frac{1200}{4} = 300 \text{ [rpm]} \tag{c}$$

(c) The input and output gear geometry specifications are obtained from the applicable equations from (14.3) to (14.15).
Pinion and gear diameters are exactly the same as circular pitch diameters for pinion and gear (Eqs. (14.3) and (a)):

$$\begin{aligned} d_P &= m_n N_P = 8(18) = 144 \text{ [mm]} = 0.144 \text{ [m]} \\ d_G &= m_n N_G = 8(72) = 576 \text{ [mm]} = 0.576 \text{ [m]} \end{aligned} \tag{d}$$

The circular pitch p_c is defined from Eqs. (14.5) and (14.6):

$$p_c = \frac{\pi d_P}{N_P} = \frac{\pi d_G}{N_G} = \pi m_n = \pi(8) = 25.133 \text{ [mm]} \tag{e}$$

The outside diameters of the pinion and the gear are obtained from Eq. (14.10):

$$\begin{aligned} d_{\text{oP}} &= d_P + 2m_n = m_n(N_P + 2) = 144 + 2(8) = 160 \text{ [mm]} \\ d_{\text{oG}} &= d_G + 2m_n = m_n(N_G + 2) = 576 + 2(8) = 592 \text{ [mm]} \end{aligned} \tag{f}$$

The root diameters of the pinion and the gear are defined from Eq. (14.11):

$$\begin{aligned} d_{\text{rP}} &= d_r = d_p - 2.5m_n = 144 - 2.5(8) = 124 \text{ [mm]} \\ d_{\text{rG}} &= d_r = d_p - 2.5m_n = 576 - 2.5(8) = 556 \text{ [mm]} \end{aligned} \tag{g}$$

The pinion and the gear base circles are obtained from Eq. (14.12):

$$\begin{aligned} d_{\text{bP}} &= d_g \cos\phi = d_P \cos\phi = 144(\cos 20^\circ) = 135.316 \text{ [mm]} \\ d_{\text{bG}} &= d_g \cos\phi = d_G \cos\phi = 576(\cos 20^\circ) = 541.263 \text{ [mm]} \end{aligned} \tag{h}$$

The tooth thickness at pitch diameter is found in Eq. (14.13):

$$t_p = \frac{\pi}{2} m_n = \frac{\pi}{2}(8) = 12.6637 \text{ [mm]} \tag{i}$$

This value is having more accuracy since when rounded it becomes 12.664 [mm]. For a necessary backlash, the value of the thickness should be reduced from 12.6637 [mm], rather than from 12.664 [mm]. The amount of backlash might be much more than the rounding of 0.001 [mm], but for much smaller gears, it might be necessary to use higher accuracy.
The center distance C_g between the mating gears centerlines is found from Eq. (14.14):

$$C_g = \frac{1}{2}(d_P + d_G) = \frac{1}{2}(m_n(N_P + N_G)) = \frac{1}{2}(8(18 + 72)) = 360 \text{ [mm]} \tag{j}$$

At that center distance, the minimum backlash is obtained from Figure 14.7 as about 0.470 [mm]. Using the more conservative estimate from the equation of the 400 [mm] center distance, one finds the backlash

$B_p = 0.03m_n + 0.26 = 0.03(8) + 0.26 = 0.500$ [mm]. The value is close to the one obtained from Figure 14.7 but more than the suggested minimum, which is acceptably conservative.
The contact ratio m_p is defined by Eq. (14.15), which gives

$$\begin{aligned} m_p &= \frac{\sqrt{r_{\text{oP}}^2 - r_{\text{bP}}^2} + \sqrt{r_{\text{oG}}^2 - r_{\text{bG}}^2} - C_g \sin\phi}{\pi m_n \cos\phi} \\ &= \frac{\sqrt{(160/2)^2 - (135.316/2)^2} + \sqrt{(592/2)^2 - (541.263/2)^2} - 360\sin(20°)}{\pi(8)\cos(20°)} \\ &= \frac{42.6895 + 119.894 - 123.127}{23.617} = 1.671 \end{aligned} \tag{k}$$

Example 14.2 ***US System***
A reduction gear set or a gearbox transmits a maximum of 100.5 [hp] with 1200 [rpm] input speed. The gear reduction ratio is 4. The standard gearing pressure angle is 20°. Standard diametral pitch is 3 [teeth/in], and the number of input pinion teeth is 18. Determine (a) the number of teeth of output gear, (b) the output speed of the reducer gearbox, and (c) the input and output gear geometry specifications.

Solution
Data: $H_{\text{kW}} = 100.5$ [hp] $\approx 75\,000$ [W], $N_{\text{in,rpm}} = N_{\text{in}} = 1200$ [rpm], $\omega_{\text{in}} = \omega_P = 2\pi N_{\text{in}}/60 = 125.66$ [rad/s], the gear ratio $n_g = \omega_P/\omega_G = 4$, $\phi = 20°$, $p_d = 3$ [tooth/in], and input pinion teeth $N_P = 18$.

(a) The number of teeth of output gear is obtained from Eq. (14.4), which gives exactly the same value of $N_G = 72$ teeth as Example 14.1 in Eq. (a). The relation is dimension independent.
(b) The output speed of reducer gearbox is defined also by Eq. (14.4) $\omega_P/\omega_G = 4$ or $\omega_G = 31.415$ [rad/s], which is exactly the same as Example 14.1 in Eq. (b). The relation is also dimension independent.
The output speed in revolution per minute is $N_{\text{out}} = \omega_G(60/2\pi) = 299.99 = 300$. The rounding was not necessary because the result is exactly $N_{\text{out}} = 300$ when applying Eq. (14.4), which is exactly the same value as Example 14.1 in Eq. (c). The relation is also dimension independent.
(c) The input and output gear geometry specifications are obtained from the applicable equations from (14.3) to (14.15).
Pinion and gear diameters are exactly the same as circular pitch diameters for pinion and gear (Eq. (14.8)):

$$\begin{aligned} d_P &= \frac{N_P}{p_d} = \frac{18}{3} = 6 \text{ [in]} \\ d_G &= \frac{N_G}{p_d} = \frac{72}{3} = 24 \text{ [in]} \end{aligned} \tag{a}$$

The circular pitch p_c is defined from Eqs. (14.5) and (14.8):

$$p_c = \frac{\pi d_g}{N_g} = \frac{\pi}{p_d} = \frac{\pi}{3} = 1.0472 \text{ [in]} \tag{b}$$

The outside diameters of the pinion and the gear are obtained from Eq. (14.10):

$$\begin{aligned} d_{\text{oP}} &= d_g + 2.000/p_d = 6 + 2/3 = 6.6667 \text{ [in]} \\ d_{\text{oG}} &= d_g + 2.000/p_d = 24 + 2/3 = 24.667 \text{ [in]} \end{aligned} \tag{c}$$

The root diameters of the pinion and the gear are defined from Eq. (14.11):

$$\begin{aligned} d_{\text{rP}} &= d_P - 2.500/p_d = 6 - 2.5/3 = 5.1667 \text{ [in]} \\ d_{\text{rG}} &= d_G - 2.500/p_d = 24 - 2.5/3 = 23.1667 \text{ [in]} \end{aligned} \tag{d}$$

The base circles of the pinion and the gear are obtained from Eq. (14.12):

$$\begin{aligned} d_{\text{bP}} &= d_P \cos\phi = 6(\cos 20^\circ) = 5.6382\ [\text{in}] \\ d_{\text{bG}} &= d_G \cos\phi = 24(\cos 20^\circ) = 22.5526\ [\text{in}] \end{aligned} \tag{e}$$

The tooth thickness at pitch diameter is found by Eq. (14.13):

$$t_p = \frac{\pi}{2p_d} = \frac{\pi}{2(3)} = 0.523\,599\ [\text{in}] \tag{f}$$

This value is having more accuracy than usual since when rounded it becomes 0.5236 [in]. For a necessary backlash, the value of the thickness should be reduced from 0.523 599 [in]. The amount of backlash might be much more than the rounding, but for much smaller gears, it might be necessary to use higher accuracy.

The center distance C_g between the mating gears centerlines is found from Eq. (14.14):

$$C_g = \frac{1}{2}(d_P + d_G) = \frac{1}{2}(6 + 24) = 15\ [\text{in}] \tag{g}$$

At that center distance, the minimum backlash is obtained from Figure 14.8 as about 0.02 [in]. Using the more conservative estimate from the equation of the 16 [in] center distance, one finds the backlash $B_p = 0.0357 p_d^{-0.478} = 0.0357(3)^{-0.478} = 0.0211$ [in]. The value is close to the one obtained from Figure 14.7 but more than the suggested minimum, which is acceptably conservative.

The contact ratio m_p is defined by Eq. (14.15), and replacing m_n by $1/p_d$ gives

$$\begin{aligned} m_p &= \frac{\sqrt{r_{\text{oP}}^2 - r_{\text{bP}}^2} + \sqrt{r_{\text{oG}}^2 - r_{\text{bG}}^2} - C_g \sin\phi}{(\pi(\cos\phi)/p_d)} \\ &= \frac{\sqrt{(6.6667/2)^2 - (5.6382/2)^2} + \sqrt{(24.667/2)^2 - (22.5526/2)^2} - 15\sin(20^\circ)}{\pi\cos(20^\circ)/3} \\ &= \frac{1.7787 + 4.9960 - 5.1303}{0.9840} = 1.671 \end{aligned} \tag{h}$$

This is the same value of contact ratio for the SI system of Example 14.1. It is expected to get about the same result since dimensions, geometry, and specifications are very close.

14.3 Force Analysis and Power Transmission

The simple gear set consists of a pinion and a gear mounted on two parallel shafts using *keys* or *splines* as discussed in Section 8.2.2. When the gear set is housed in a box, it is called a gearbox. The gear set is a ***power transmission*** device from one shaft to the other. It can transform power such that the input torque and rotational speed are not equal to each of the output torque and rotational speed, if the gear ratio is different from one. The power, however, is about the same with the output power equals the input power multiplied by the gear set efficiency. The efficiency is usually in the range of 96–99%, which is very close to 100%. This is because involute teeth roll on each other at pitch points with very little friction loses around that. In addition, the set is supported by a common use of very efficient bearings such as rolling bearings.

Forces develop between teeth of mating gears as shown in Figure 14.9. The main resultant force $\boldsymbol{F}_n$ is a *normal force* to both teeth profiles, where contact between the two teeth is at the pitch point (Figure 14.9). The normal force $\boldsymbol{F}_n$ acts in the direction of the line of action, which is inclined by the pressure angle ϕ. The components of this force are the *tangential force* $\boldsymbol{F}_t$ and the *radial force* $\boldsymbol{F}_r$. The tangential force is tangent to both gears at the pitch point. It is the only force that exerts effort or work. Therefore the *transmitted power H* from the driver or

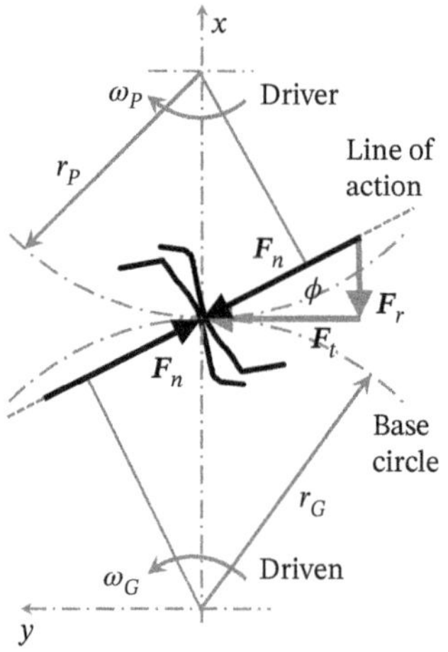

Figure 14.9 Forces developed between teeth of mating gears. The main force $\boldsymbol{F}_n$ is normal to both teeth profiles at the pitch point. The components of $\boldsymbol{F}_n$ are the tangential force $\boldsymbol{F}_t$ and the radial force $\boldsymbol{F}_r$.

pinion is enacted only by the tangent force $\boldsymbol{F}_t$. The known relation of the power as a function of the transmitted torque and the rotational speed is

$$H = \boldsymbol{T}_P \cdot \boldsymbol{\omega}_P = (\boldsymbol{r}_P \times \boldsymbol{F}_t) \cdot \boldsymbol{\omega}_P = \boldsymbol{F}_t \cdot \boldsymbol{v}_t \tag{14.16}$$

where $\boldsymbol{T}_P$ is the *pinion torque*, ω_P is the magnitude of the *angular velocity* of the pinion in [rad/s], $\boldsymbol{r}_P$ is the *pitch circle* radius of the pinion, and $\boldsymbol{v}_t$ is the tangential *pitch point velocity*. From Eq. (14.16), one can get

$$\boldsymbol{F}_t = \frac{H}{r_P \omega_P} = \frac{H}{\boldsymbol{v}_t} \tag{14.17}$$

Figure 14.9 indicates the vector composition of the *total normal force* $\boldsymbol{F}_n$, and from that one can get the total force as

$$\boldsymbol{F}_n = \frac{\boldsymbol{F}_t}{\cos\phi} \tag{14.18}$$

The radial component of the total normal force $\boldsymbol{F}_n$ is then derived from the total force or from the tangential component of the force, which gives

$$\begin{aligned} \boldsymbol{F}_r &= \boldsymbol{F}_n \sin\phi \\ \boldsymbol{F}_r &= \boldsymbol{F}_t \tan\phi \end{aligned} \tag{14.19}$$

This is the useful way to proceed with the evaluation of forces in the design process. First to evaluate the tangential force $\boldsymbol{F}_t$, find the total normal force $\boldsymbol{F}_n$, and then calculate the radial force $\boldsymbol{F}_r$. These forces are typically needed to design the shaft on which the gear is mounted.

Example 14.3 Again, a gearbox transmits a maximum power of 75 [kW] or 100.58 [hp] at 1200 [rpm] input speed. The gear reduction ratio is 4. The standard gearing pressure angle is 20°. Standard module is 8 [mm], or the standard diametral pitch of 3 [teeth/in] is used, and the number of input pinion teeth is 18. Determine the forces between gears in SI and the US systems.

Solution

Data: $H_{\text{kW}} = 75$ [kW] = 75 000 [W] or $H_{\text{hp}} = 100.58$ [hp], $N_{\text{in,rpm}} = N_{\text{in}} = 1200$ [rpm], $\omega_P = 2\pi N_{\text{in}}/60 = 125.66$ [rad/s], $n_g = \omega_P/\omega_G = 4$, $\phi = 20°$, $m_n = 8$ [mm] or $p_d = 3$ [teeth/in], and $N_P = 18$.

Forces between gears, and noting that speed should be in [rad/s] or again $\omega_P = 2\pi(1200)/60 = 125.66$ [rad/s], the tangential force between the pinion and the gear is found through Eq. (14.17) such that

$$\boldsymbol{F}_t = \frac{H}{r_p\omega_p} = \frac{H}{\boldsymbol{v}_t} = \frac{75\,000}{(0.144/2)(125.66)} = \frac{75\,000}{9.047\,52} = 8289.56\ [\mathrm{N}] = 8.2896\ [\mathrm{kN}] \tag{a}$$

Note that the pitch line velocity is calculated in Eq. (a) as $\boldsymbol{v}_t = 9.047\,52$ [m/s]. The total normal force is obtained from Eq. (14.18) to get

$$\boldsymbol{F}_n = \frac{\boldsymbol{F}_t}{\cos\phi} = \frac{8289.56}{\cos 20^\circ} = \frac{8289.56}{0.939\,69} = 8821.63\ [\mathrm{N}] = 8.8216\ [\mathrm{kN}] \tag{b}$$

The rounding of the value suggests that it is sufficient to carry five digits for now. The radial force is obtained by the use of Eq. (14.19) to give

$$\boldsymbol{F}_r = \boldsymbol{F}_n \sin\phi = 8821.63 \sin 20^\circ = 3017.18\ [\mathrm{N}] = 3.0172\ [\mathrm{kN}] \tag{c}$$

For the US system the power transmitted is 100.58 [hp], which is the same as 75 [kW]. One can, therefore, just convert the values of the forces instead of carrying the same calculations over again. This gives

$$\begin{aligned} \boldsymbol{F}_r &= 8289.6\,(0.224\,81\ [\mathrm{lb}]/[\mathrm{N}]) = 1863.6\ [\mathrm{lb}] \\ \boldsymbol{F}_n &= 8821.6(0.224\,81\ [\mathrm{lb}]/[\mathrm{N}]) = 1983.2\ [\mathrm{lb}] \\ \boldsymbol{F}_r &= 3017.2\,(0.224\,81\ [\mathrm{lb}]/[\mathrm{N}]) = 678.29\ [\mathrm{lb}] \end{aligned} \tag{d}$$

The same equations of (14.17)–(14.19) should give the same values. However, one should use the appropriate evaluation of the power H_{hp} in relation to the torque $\boldsymbol{T}_P$ for the US system. This process has been treated in Section 1.11.

14.4 Design Procedure

The design procedure is divided into three sections. The first section is about the classical approach. It is a conventional procedure related to the basic approach of gear design considering the tooth as a cantilever fixed to the gear rim. The second section is concerned with the gear synthesis approach, which depends on gear optimization results. The gear optimization utilized the process defined in the third section of the design procedure, namely, detailed design.

14.4.1 Classical Procedure

This design procedure depends on an early work by Wilfred Lewis (1854–1927). He introduced and discussed his equation during 1892–1910 (Lewis 1893, 1910). The *Lewis equation* has been continually used and modified by introducing several factors to account for geometry, loading, and other conditions. A modified form is presented in this section, and other considerations are introduced in Section 14.4.3. The derivation of *Lewis equation* depends on the tooth considered in Figure 14.10. The tooth is loaded at the tip by the total normal force $\boldsymbol{F}_n$ where the tangential component $\boldsymbol{F}_t$ is the main force considered for calculating the loading effect. The tooth is assumed as a cantilever beam simply modeled as in Figure 14.10b with the cross section flipped about y to show the gear face width f_g. The main dimensions in Figure 14.10a show l as the tooth height from the tip to the tooth root, t_r as the tooth thickness at the root, and r_f as the root fillet radius. The bending of the modeled beam generates the usual *normal stress* σ_b due to bending in the customary form

$$\sigma_b = \frac{\boldsymbol{M}_z y}{I_z} = \frac{(\boldsymbol{F}_t\, l)(t_r/2)}{\frac{1}{12}t_r^3 f_g} = \frac{6\boldsymbol{F}_t l}{t_r^2 f_g} \tag{14.20}$$

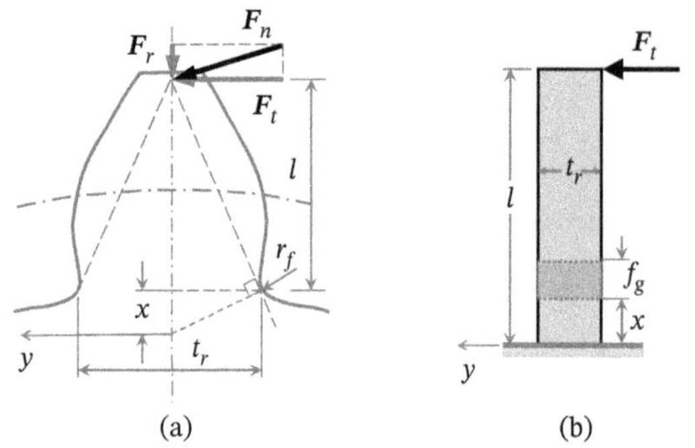

Figure 14.10 For the derivation of *Lewis* equation, (a) the tooth is loaded at the tip by $\boldsymbol{F}_n$ where the tangential component $\boldsymbol{F}_t$ is the main bending force in the cantilever beam in (b). Main dimensions are shown.

where $\boldsymbol{M}_z$ is the moment about the z-axes, y is the distance from the neutral axes of the section, and I_z is the second area moment of the section; see Section 6.1. The term f_g is the gear *face width* (Figure 14.10b), and $\boldsymbol{F}_t$ is the *tangential force*. From the right triangle geometry in Figure 14.10a, the following relation can be deduced:

$$\frac{t_r/2}{x} = \frac{l}{t_r/2} \quad \text{or} \quad x = \frac{t_r^2}{4l} \tag{14.21}$$

Substituting Eq. (14.21) into Eq. (14.20), one gets

$$\sigma_b = \frac{6\boldsymbol{F}_t l}{t^2 f_g} = \frac{6\boldsymbol{F}_t l}{(4xl)f_g} = \frac{\boldsymbol{F}_t}{\frac{2}{3}xf_g} \tag{14.22}$$

Lewis introduced his form factor y_L as a function of the *circular pitch* p_c such that the coined *Lewis form factor* y_L is

$$y_L = \frac{2}{3}\frac{x}{p_c} \quad \text{or} \quad \frac{2}{3}x = p_c y_L \tag{14.23}$$

Substituting Eq. (14.23) into Eq. (14.22), the developed *Lewis equation* becomes

$$\sigma_b = \frac{\boldsymbol{F}_t}{\frac{2}{3}xf_g} = \frac{\boldsymbol{F}_t}{f_g p_c y_L} \tag{14.24}$$

Substituting for the relation between the module m_n and the circular pitch p_c (Eq. (14.6)), or $m_n = p_c/\pi$ and introducing a *modified Lewis form factor* $Y_L = \pi y_L$, one gets the *Lewis equation* as

$$\sigma_b = \frac{\boldsymbol{F}_t}{f_g m_n Y_L} \quad \text{for SI,} \quad \sigma_b = \frac{\boldsymbol{F}_t p_d}{f_g Y_L} \quad \text{for US} \tag{14.25}$$

The *modified Lewis form factor* Y_L is shown in Figure 14.11 with an indicated approximate fitting relation of the following form:

$$Y_L = 0.0619 \ln(N_P) + 0.1243 \tag{14.26}$$

The error in Eq. (14.26) is lower than 5% in the usual span of teeth numbers in use. This is sufficient for the type of calculations utilizing the *Lewis equation.*

Dynamic factors such as K_v have also been introduced to get the *modified Lewis equation*. Introducing a dynamic, or a *velocity factor*, K_v accounts for the effect of gear *quality* and *pitch line velocity* on the dynamics affecting the bending stress and endurance. The *modified Lewis equation* becomes

$$\sigma_b = \frac{\boldsymbol{F}_t K_v}{f_g m_n Y_L} \tag{14.27}$$

where Y_L is the *modified Lewis factor*, K_v is the *velocity factor*, $\boldsymbol{F}_t$ is the transmitted (or tangential) force, f_g is the *face width* of the gear, and m_n is the *module*. The value of the *modified Lewis factor* Y_L is obtained as defined in

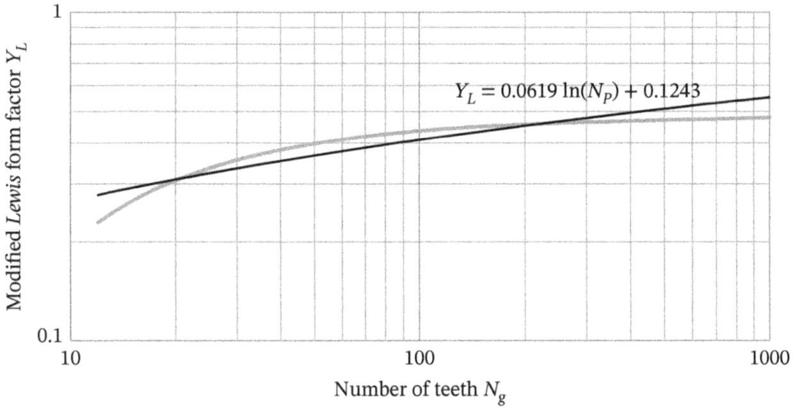

Figure 14.11 The modified *Lewis* form factor Y_L is shown with an indicated approximate fitting relation as a function of the number of teeth of a gear N_g.

Eq. (14.26), which is valid for both SI and the US systems. The dynamic or velocity factor K_v has been adapted for *pitch line velocities* $\mathbf{v}_t$ in [m/s] or [ft/min] as

$$
\begin{aligned}
K_v\big|_{SI} &= \frac{3+v_t}{3},\quad K_v\big|_{US} = \frac{600+v_t\big|_{[\text{ft/min}]}}{600} \quad \text{(for cast iron with cast teeth)}\\
K_v\big|_{SI} &= \frac{6+v_t}{6},\quad K_v\big|_{US} = \frac{1200+v_t\big|_{[\text{ft/min}]}}{1200} \quad \text{(for milled or not carefully generated gears)}\\
K_v\big|_{SI} &= \frac{50+\sqrt{200v_t}}{50},\quad K_v\big|_{US} = \frac{50+\sqrt{v_t\big|_{[\text{ft/min}]}}}{50} \quad \text{(for hobbing or shaping gears)}\\
K_v\big|_{SI} &= \sqrt{\frac{78+\sqrt{200v_t}}{78}},\quad K_v\big|_{US} = \sqrt{\frac{78+\sqrt{v_t\big|_{[\text{ft/min}]}}}{78}} \quad \text{(for high precision shaved or ground teeth)}\\
K_v\big|_{\text{SIçUS}} &= 1,\quad \text{(for high-precision and no appreciable dynamic load)}
\end{aligned}
\tag{14.28}
$$

This indicates clearly the pronounced effect of manufacturing process on the generated dynamic loading and thus the stresses. The last three expressions in Eq. (14.28) have been adapted earlier in a reciprocal form by AGMA (1959). These *velocity factors* may, however, still be used for initial estimate to select gear materials. The *velocity factors* have been further modified to other expressions that also include the *quality numbers* as defined in Section 14.4.3. Equation (14.28) may still be used as initial estimates in the classical procedure.

For a ***classical estimate*** of selecting gear materials, one can initially start with estimating the standard *module* m_n or the *diametral pitch* p_d; use the *minimum number of teeth* to safeguard against interference (i.e. start with $N_P = 18$); apply a *velocity factor* K_v of cast, milled, or not carefully generated gears; and adapt the conventional *face width* of gearing $f_g = (3\text{–}5)p_d = (3\text{–}5)\pi m_n$ (or a simpler 10–15m_n). One can then select the material based on its *yield strength* in tension S_y and using a high *safety factor* $K_{SF} = 3\text{–}5$. This initially high safety factor might be a safeguard against other factors, not initially included. Again this safety factor is an ignorance factor concerning the estimate of the module. If some better information is available in that regard, the safety factor can be greatly reduced. This problem is significantly resolved by introducing the *gear synthesis* procedure of the following initial synthesis (Section 14.4.2).

14.4.2 Initial Synthesis

Instead of using the classical treatment of the previous presentation of modified *Lewis* equation, one can rely on the *optimum design* of gears provided in the literature such as Metwalli and El Danaf (1996). The optimum design postulation depends on attaining the maximum allowable bending fatigue strength and the maximum allowable contact fatigue strength at the same time. This provides the maximum utilization of material capacities and thus results in the minimum gear set volume for this material. The optimum geometry has been attained for different material properties. The main *material property factor* is the ratio of the *allowable bending fatigue strength* to the *allowable contact fatigue strength*, which was given the symbol B. It was found that the material property factor B ranges mainly from about 0.2 to 0.6 (Metwalli and El Danaf 1996). Here and to have more representation, the symbol B_{BC} is used instead of symbol B in the paper. The subscript BC refers to the bending and contact fatigue strength ratios.

As an initial synthesis, the results of the optimization are used as an opening jump start position of synthesis. The initial material property value is set at 0.3; the allowable bending fatigue strength is selected in the range of 100–300 [MPa], and other parameters are conservatively selected off the optimum cases. The chart of the optimum module m_n^* in [mm] is reduced to the one shown in Figure 14.12 for a large range of input torque in [N m]. The optimum number of pinion teeth N_P^* is given in Figure 14.13 against the allowable bending fatigue strength σ_{ab} in [MPa]. The optimum face width f_g^* is defined in Figure 14.14 against the input torque in [N m]. It should be noted that these charts are generated for an average value of rotational speed, gear ratio, and gear quality for steel materials. The charts are for a *gear ratio* $n_g = 4$, pinion *angular velocity* $N_{inP} = 1000$ [rpm], the AGMA *quality number* $Q_v = 8$, and *elastic coefficient* $C_P = 191$ $[\text{MPa}]^{1/2}$; see Section 14.4.3. The effect of shifting away from these averages is not appreciably pronounced and can be examined in Metwalli and El Danaf (1996). The consideration of different other cases should be taken care of in the detailed calculations for the particular case of the design. This should greatly reduce iterations toward the *optimum design* of those different cases as discussed in Section 14.6 about *CAD* and optimization.

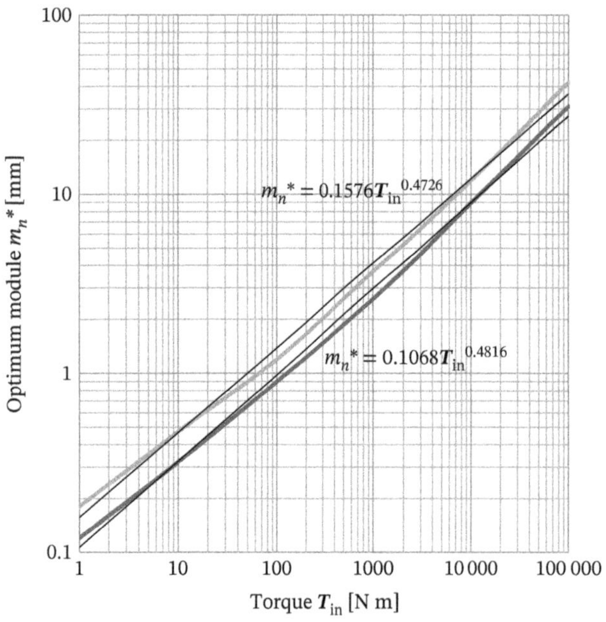

Figure 14.12 The optimum module m_n^* in [mm] for a large span of input torque T_{in} in [N m]. The upper curve is for an allowable bending fatigue of 300 [MPa], and the lower curve is for an allowable bending fatigue of 100 [MPa].

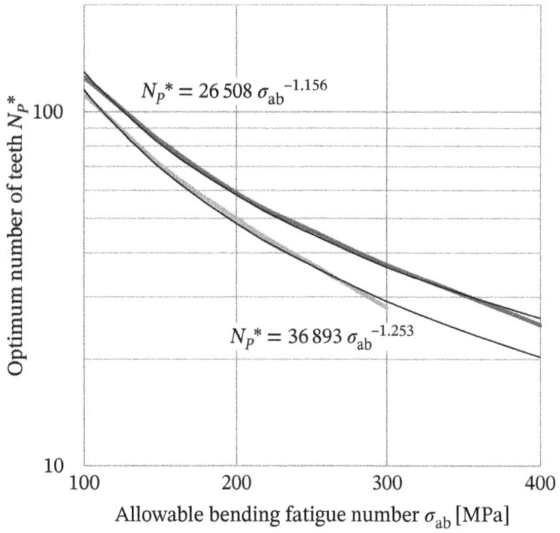

Figure 14.13 The optimum number of pinion teeth N_P^* against the allowable bending fatigue strength σ_{ab} in [MPa]. The upper curve is for a low torque of 10 [N m], and the lower curve is for a much higher torque of 10^5 [N m].

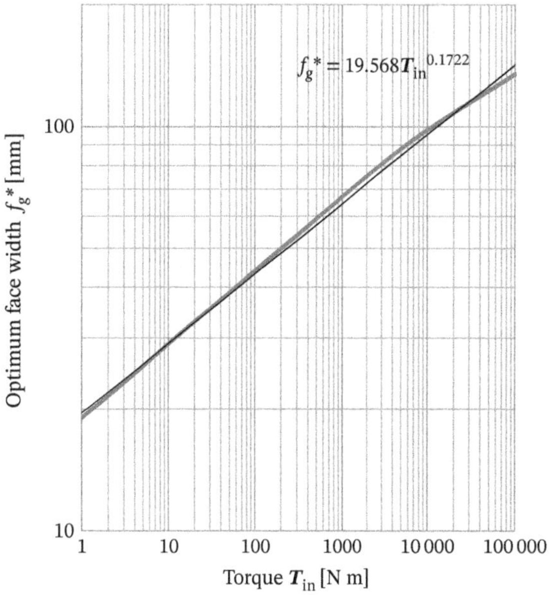

Figure 14.14 The optimum face width f_g^* in [mm] against the input torque T_{in} in [N m].

To facilitate the utility of the adapted optimum design, the optimum curves of Figures 14.12–14.14 are used as the optimum synthesis *meta-model* or *surrogate functions* with the following relations. The relations are also adapted to the US system of units as shown next.

The *optimum module* m_n^* in [mm] is modeled as a function of the input torque T_{in} [N m] (Figure 14.12) such that the optimum module and the *optimum diametral pitch* are

$$\begin{aligned} m_n^*|_{SI} &= (0.1576)\boldsymbol{T}_{in}^{0.4726}\ \text{[mm]} \\ p_d^*|_{US} &= 25.4/(0.1576)\,(0.113\,(\boldsymbol{T}_{in}))^{0.4726}\ \text{[teeth/in]} \end{aligned} \tag{14.29}$$

In the equation of the optimum diametral pitch (US), the input torque T_{in} is in [lb in]. This value of module or diametral pitch is for the higher *allowable bending fatigue strength* of ***300*** [MPa] or ***43.5*** [kpsi]. It is adapted for initial synthesis since it gives a higher module value and thus supposedly more conservative. The selected material represents average *through-hardened steel* commonly used or introductory *case hardened steel* used for ***typical*** gears. For lower allowable bending fatigue strength of 100 [MPa] or 14.5 [kpsi], the following relation gives the optimum module and the optimum diametral pitch as:

$$\begin{aligned} m_n^*|_{SI} &= (0.1068)\boldsymbol{T}_{\text{in}}^{0.4816}\ [\text{mm}] \\ p_d^*|_{US} &= 25.4/(0.1068)\,(0.113(\boldsymbol{T}_{\text{in}}))^{0.4816}\ [\text{teeth/in}] \end{aligned} \qquad (14.30)$$

The smaller optimum module for the lower allowable contact fatigue strength is compensated for by the larger increase in the optimum number of teeth for the lower allowable contact fatigue strength as subsequently noted.

The *optimum number of pinion teeth* N_P^* is modeled as a function of the *allowable bending fatigue strength* S_{ab} (100–400) [MPa] or (14.4–58) [kpsi] (Figure 14.13) such that

$$\begin{aligned} N_P^*|_{SI} &= (26\,508)S_{ab}^{-1.156} \\ N_P^*|_{US} &= (2846.5)S_{ab}^{-1.156} \end{aligned} \qquad (14.31)$$

This number of teeth is for the lower applied torque $\boldsymbol{T}_{\text{in}}$ of 10 [N m] (or 88.5 [lb in]) and thus higher speed for the same power. It is also adapted, since it should be conservatively inclined. For a much higher applied torque of 10^5 [N m] (or 885 [klb in]) and much lower speed for the same power, the relation is given by

$$\begin{aligned} N_P^*|_{SI} &= (36\,893)S_{ab}^{-1.253} \\ N_P^*|_{US} &= (3284.4)S_{ab}^{-1.253} \end{aligned} \qquad (14.32)$$

It should be noted that the number of teeth should be an integer. An adjustment to that effect is necessary. The difference in optimum number of teeth due to different transmitted torque is tolerable. At an allowable bending fatigue strength $S_{b,\text{all}}$ of 300 [MPa] or (43.5 [kpsi]), the optimum number of pinion teeth is about 37 for the lower applied torque of 10 [N m] (or 88.5 [lb in]) and 29 for the much higher applied torque of 10^5 [N m] (or 885 [klb in]). Adapting the lower value of torque would thus be for higher speeds at the same power. This would produce a smaller yet safe gear. Adjusting this value can be performed in the provided ***Spur Gear Synthesis Tablet***, which is available through the ***Wiley website*** of this textbook.

The optimum face width f_g^* in [mm] (or in [in]) (Figure 14.14) is modeled as a function of the input torque $\boldsymbol{T}_{\text{in}}$ in [N m] (or in [lb in]) such that

$$\begin{aligned} f_g^*\big|_{SI} &= (19.568)\boldsymbol{T}_{\text{in}}^{0.1722} \\ f_g^*\big|_{US} &= (0.5292)\boldsymbol{T}_{\text{in}}^{0.1722} \end{aligned} \qquad (14.33)$$

These optimum values are used for the adapted *material property factor* B_{BC} of 0.3. This selection of B_{BC} is for median or regular ductile gear material; see gear materials in Section 14.4.3.

For different values off the adapted optimum or parameters, fine-tuning iteration is necessary to satisfy other objectives as discussed in Section 14.6. The optimum equations (14.29)–(14.33) are used for the US system provided that appropriate dimensions are used and substitution of $1/p_c$ for m_n. Safeguard verification is important in that regard.

Defined relations in Eqs. (14.29)–(14.33) are for selected case of optimum design. Other cases are provided in Metwalli and El Danaf (1996). The general optimum trends, however, suggest some interesting observations. The one concerning the *optimum number of teeth* indicates that as the material strength diminishes, the number of the teeth approaches infinity (Figure 14.13). This is reasonable since one can use a shaft with no teeth (infinite number of teeth with infinitesimal module) driving a rubber cylinder (infinite number of teeth with even smaller module). Note that the involute tooth form does not need to apply. The mechanism in that case is different, but the trend is expected to be reserved.

Synthesis general trend departing off the adapted value incites the *optimum module* m_n^* to decrease a little with the reduction of *allowable bending fatigue strength* S_{ab}. The *diametral pitch* p_d would do the opposite. The optimum module m_n^*, however, would be higher for a lower value of the *material property factor* B_{BC}. The *optimum number of teeth* N_P^* decreases with the decrease of the material property factor B_{BC}. The *optimum face width* f_g^* decreases a little with the decrease of the material property factor B_{BC}. These observations about the optimum values are deduced from the trends depicted in Metwalli and El Danaf (1996).

Example 14.4 Again, a reduction gear in a crane driven by a maximum power of 75 [kW] or 100.58 [hp] electric motor running at 1200 [rpm] is to be designed for a gear ratio of 4 : 1. Gears are to be of 20° full depth. *Synthesize* a gear set for such a type of application.

Solution

Data: Power transmitted $H_{kW} = 75$ [kW] or 100.58 [hp], rotational speed $N_{in} = 1200$ [rpm], and the gear ratio $n_g = 4$. The rotation speed $\omega_P = (2\pi N_{in}/60) = 125.66$ [rad/s].

The value of the maximum input torque $\boldsymbol{T}_{in}$ is obtained from Eq. (14.16) such that

$$\begin{aligned} \boldsymbol{T}_{in} &= \frac{H_{kW}(10)^3}{\omega_P} = \frac{75(10)^3}{125.66} = 596.85\ [\text{N m}] \\ \boldsymbol{T}_{in} &= \frac{H_{hp}(6600)}{\omega_P} = \frac{100.58(6600)}{125.66} = 5282.7\ [\text{lb in}] \end{aligned} \tag{a}$$

According to Table 14.5, the quality number of a crane type of applications has the range of 5–7. The results of optimum design were based on a quality number of 8. This quality is higher than the required 5–7. It will be, however, more conservative to use better quality gears or accept higher safety factor out of the selected optimum solution.

Initial synthesis: From Figures 14.12–14.14, the synthesized geometry of the gear can be defined by the optimum of $m_n \approx 3.2$ [mm], $N_P \approx 37$, and $f_g \approx 60$ [mm] at $\boldsymbol{T}_{in} \approx 600$ [N m] and $S_{ab} = 300$ [MPa] or 43.5 [kpsi]. To have more specific values, one could use Eqs. (14.29), (14.31), and (14.32). Using these initial synthesis equations, the expected optimum gear module m_n^* or diametral pitch p_d are calculated such that

$$\begin{aligned} m_n^* &= (0.1576)T_{in}^{0.4726} = (0.1576)(596.85)^{0.4726} = 3.232\ [\text{mm}] \\ p_d^*|_{US} &= 25.4/(0.1576)\,(0.113\,(\boldsymbol{T}_{in}))^{0.4726} = 7.859\ [\text{teeth/in}] \end{aligned} \tag{b}$$

The standard module and diametral pitch closer to these values are $m_n = 3.25$ [mm] and $p_d = 8$ [teeth/in]. The optimum number of pinion teeth N_P^* is calculated from the optimum at allowable bending fatigue strength $S_{ab} = 300$ [MPa] or 43.5 [kpsi] such that

$$\begin{aligned} N_P^* &= (26\,508)S_{ab}^{-1.156} = (26\,508)(300)^{-1.156} = 36.29 = 36 \\ N_P^*|_{US} &= (2846.5)S_{ab}^{-1.156} = (2846.6)(43.5)^{-1.156} = 36.33 = 36 \end{aligned} \tag{c}$$

It should be apparent that the two values of the number of pinion teeth should be very close to each others. The optimum face width f_g^* is calculated from the optimum to get

$$\begin{aligned} f_g^*\big|_{SI} &= (19.368)T_{in}^{0.1722} = (19.368)(596.85)^{0.1722} = 58.82\ [\text{mm}] \\ f_g^*\big|_{US} &= (0.5292)T_{in}^{0.1722} = (0.5292)(5282.7)^{0.1722} = 2.315\ [\text{in}] \end{aligned} \tag{d}$$

The value of the face width in the SI system and the value of the face width in the US system are very closely the same in length (58.82 [mm] ≅ 2.3157 [in]).

It is clear from Eqs. (a)–(c) that all the values are close to those obtained from Figures 14.12–14.14. It is also feasible to just work in SI or the US system and then covert the final results to the other system.

The pinion diameter d_P is evaluated from Eq. (14.3), i.e. $d_P = N_P m_n = 36(3.25) = 117(10)^{-3}$ [m] or $d_P = N_P/p_d = 36/8 = 4.5$ [in]. The pitch line velocity $\boldsymbol{v}_t$ is obtained from Eq. (14.1), i.e. $\boldsymbol{v}_t = \omega_P(d_P/2) = 125.66((117/2)/1000) =$

7.35 [m/s] or $\boldsymbol{v}_t = \omega_P(d_P/2) = 125.66(4.5/2) = 282.74$ [in/s] = 1413.7 [ft/min]. This will change the tangential force on the teeth but will also change the dynamic effect.

The suggested material would have allowable bending fatigue strength S_{ab} of 300 [MPa] or 43.5 [kpsi]. This will be defined in the detailed design in Section 14.4.3, which considers the other unspecified factors in this problem.

Example 14.5 For the problem in Example 14.4, the reduction gear set in the crane is driven by a maximum power of 75 [kW] or 100.58 [hp] electric motor running at a rotational speed of 1200 [rpm]. The gear set is to be designed for a gear ratio of 4 : 1. Gears are also of 20° full depth. *Conventionally* or classically design the gear set for such a type of application, and compare it to the synthesized set of Example 14.4.

Solution

Data: (same as Example 14.4) Power transmitted $H_{\text{kW}} = 75$ [kW] or 100.58 [hp], rotational speed $N_{\text{in}} = 1200$ [rpm], and the gear ratio $n_g = 4$. The rotation speed $\omega_P = (2\pi N_{\text{in}}/60) = 125.66$ [rad/s].

The value of the maximum input torque $\boldsymbol{T}_{\text{in}}$ is obtained from Eq. (14.16) such that

$$\begin{aligned} \boldsymbol{T}_{\text{in}} &= \frac{H_{\text{kW}}(10)^3}{\omega_P} = \frac{75(10)^3}{125.66} = 596.85\ [\text{N m}] \\ \boldsymbol{T}_{\text{in}} &= \frac{H_{\text{hp}}(6600)}{\omega_P} = \frac{100.58(6600)}{125.66} = 5282.7\ [\text{lb in}] \end{aligned} \tag{a}$$

Again and according to Table 14.5, the quality number of a crane type of applications has the range of 5–7. To compare conventional results to the results of optimum design, the same quality number of 8 is used. This quality is higher than the required 5–7. It will be, however, more conservative to use better quality gears and also to compare conventional solution to optimum solution on equal footing.

Conventional procedure: The conventional, or classical, procedure is a more conservative procedure that assumes the number of pinion teeth, to ensure no interference, as $N_P = 18$. Therefore, the number of gear teeth is $N_G = n_g(N_P) = 4(18) = 72$.

The module m_n and the diametral pitch p_d may be taken as 8 [mm] and 3 [teeth/in] consecutively at 75 [kW] or 100.58 [hp] and 1200 [rpm]. These values are estimated based on some very conservative consideration of lower quality steel (Mott 1999). The ***face width*** f_g is usually taken as $(3–5)\pi m_n \approx (10–15)m_n$ or $(3–5)\pi/p_d \approx (8/p_d$ to $16/p_d)$. This gives the face width f_g as 75–126 [mm] with an average of 100 [mm] or 2.7–5.3 [in] with an average of 4 [in] as a selected rounding.

To substitute into the *modified Lewis equation*, one needs to evaluate few other parameters and factors. According to Eq. (14.3), the pinion diameter $d_P = N_P m_n = 18(8) = 144$ [mm] $= 144(10)^{-3}$ [m] or $d_P = N_P/p_d = 18/(3) = 6$ [in]. Note that the pinion diameter in the US system is little larger than the pinion diameter in the SI system. The pitch line velocity $\boldsymbol{v}_t$ is obtained from Eq. (14.1) as follows:

$$\begin{aligned} \boldsymbol{v}_t &= \omega_P r_P = \omega_P(d_P/2) = 125.66(144(10)^{-3}/2) = 9.0475\ [\text{m/s}] \\ \boldsymbol{v}_t &= \omega_P r_P = \omega_P(d_P/2) = 125.66(6/2) = 376.98\ [\text{in/s}] = 1884.9\ [\text{ft/min}] \end{aligned} \tag{b}$$

The maximum tangential force applied at the tip of the pinion tooth (Figure 14.10) is

$$\begin{aligned} \boldsymbol{F}_t &= \frac{H_{\text{kW}}(10)^3}{\boldsymbol{v}_t} = \frac{75(1000)}{(9.0475)} = 8.29\ [\text{kN}] \\ \boldsymbol{F}_t &= \frac{H_{\text{hp}}(6600)}{\boldsymbol{v}_t} = \frac{100.58(6600)}{(376.98)} = 1761\ [\text{lb}] \end{aligned} \tag{c}$$

The modified *Lewis* form factor Y_L can be evaluated by Eq. (14.26) to get

$$Y_L = 0.0619\ \ln(N_P) + 0.1243 = 0.0619\ \ln(18) + 0.1243 = 0.3032 \tag{d}$$

The dynamic or velocity factor K_v is chosen for the most conservative selection (Eq. (14.28)), which gives

$$K_v|_{SI} = \frac{3+v_t}{3} = \frac{3+9.0475}{3} = 4.016$$
$$K_v|_{US} = \frac{600+v_t|_{[ft/min]}}{600} = \frac{600+1761}{600} = 3.935 \tag{e}$$

Substituting in the modified *Lewis* equation (14.25) gives the bending stress as

$$\sigma_b = \frac{F_t K_v}{f_g m_n Y_L} = \frac{8.29(1000)(4.016)}{(100)(8)(0.3032)} = 137.5\ [\text{MPa}]$$
$$\sigma_b = \frac{F_t K_v p_d}{f_g Y_L} = \frac{1761(3.935)(3)}{(4)(0.3032)} = 17\,141\ [\text{psi}] = 17.141\ [\text{kpsi}] \tag{f}$$

If the *allowable bending strength* S_{ab} is 300 [MPa] or 43.5 [kpsi], the *extra safety factor* K_{SFE} may be obtained as

$$K_{SFE} = \frac{S_{ab}}{\sigma_b} = \frac{300}{137.5} = 2.18 \quad \text{for SI}$$
$$K_{SFE} = \frac{S_{ab}}{\sigma_b} = \frac{43.5}{17.141} = 2.54 \quad \text{for US} \tag{g}$$

These *extra safety factors* may be considered acceptable since most design parameters are conservatively selected particularly the velocity factor and a high *allowable strength*. The crane application of this example has quality number in the range of 5–7. This would allow a selection of the velocity factor K_v for milled rather than cast gears. In that case, the value of the velocity factor will be $K_v = (6 + \mathbf{v}_t)/6 = (6 + 9.0475)/6 = 2.51$ or $K_v = (1200 + \mathbf{v}_t)/1200 = (1200 + 1884.9)/1200 = 2.57$, which gives a bending stress of 85.81 [MPa] or 11.195 [kpsi]. The extra safety factor becomes 3.49 or 3.89 instead of 2.18 or 2.54. It should be noted, however, that no consideration has been taken for all other factors, which are considered in the detailed design section.

The conventional *material selection* depends on the consideration of a relatively high safety factor K_{SF} of, say, 4 without the use of the allowable strength value. The sought material should have a yield strength of about $K_{SF}(\sigma_b) = 4(137.5) = 550$ [MPa] or $4(17.141) = 68.6$ [kpsi]. Assuming a material AISI 3140 *normalized* that has a yield strength $S_y = 599.8$ [MPa] $= 87$ [kpsi], Brinell hardness HB = 262, and ultimate tensile strength $S_{ut} = 891.5$ [MPa] $= 129.3$ [kpsi], the safety factor becomes as follows:

$$K_{SF} = \frac{S_y}{\sigma_b} = \frac{599.8(10^6)}{137.5(10^6)} = 4.36$$
$$K_{SF} = \frac{S_y}{\sigma_b} = \frac{87(10^3)}{17.141(10^3)} = 5.08 \tag{h}$$

Again, note that no consideration has been taken for all other factors, which are discussed in detailed design in Section 14.4.3.

Applying the same values for the initial synthesis gives a different modified *Lewis* stress (Eq. (14.27)). To substitute in Eq. (14.27), one needs to reevaluate the parameters. As calculated before, the synthesized pinion diameter $d_P = 117(10)^{-3}$ [m] or 4.5 [in], the number of teeth $N_P = 36$, and the face width $f_g = 59$ [mm] or 2.135 [in]. The pitch line velocity $\mathbf{v}_t = 7.35$ [m/s] or 282.74 [in/s] = 1413.7 [ft/min]. One can use a higher quality $k_v = [(50 + (200\mathbf{v}_t)^{1/2})/50] = 1.767$ or $= [(50 + (\mathbf{v}_t)^{1/2})/50] = 1.752$. Alternatively, a more precision dynamic factor $k_v = [(78 + (200\mathbf{v}_t)^{1/2})/78]^{1/2} = 1.221$ or $= [(78 + (\mathbf{v}_t)^{1/2})/78]^{1/2} = 1.217$ can be used. The modified *Lewis* form factor (Eq. (14.26)) is $Y_L = 0.0619 \ln(36) + 0.1243 = 0.3461$. The tangential force is, however, different than the conventional process such that $\boldsymbol{F}_t = H_{kW}(1000)/\mathbf{v}_t = 75\,000/7.35 = 10.2041$ [kN] or $= H_{hp}(6600)/\boldsymbol{v}_t = 100.58(6600)/282.74 = 2347.8$ [lb]. The modified *Lewis* equation (14.27) gives the bending stress as

$$\sigma_b = \frac{F_t K_v}{f_g m_n Y_L} = \frac{10.2041(1000)(1.221)}{(59)(3.25)(0.3461)} = 187.7\ [\text{MPa}]$$
$$\sigma_b = \frac{F_t K_v p_d}{f_g Y_L} = \frac{2347.8(1.217)(8)}{(2.135)(0.3461)} = 30\,934\ [\text{psi}] = 30.934\ [\text{kpsi}] \qquad \text{(i)}$$

If the velocity factor K_v of 1.767 or 1.752 is used instead of 1.221 or 1.217, the bending stress becomes 271.6 [MPa] or 44.5 [kpsi]. These values are higher than the conventional calculations in Eq. (i) due to the fact that the optimum design is considering different factors and conditions. These factors and conditions are such as calculation for a better quality gears ($Q_v = 8$) and more accurate K_v, accounting for other factors, and it should also be closer to the allowable bending fatigue strength of 300 [MPa] or 43.5 [kpsi]. More precise computations are also covered in the following detailed design in Section 14.4.3. Adjusting these values can be performed through the provided ***Spur Gear Synthesis Tablet***, which is available through the ***Wiley website*** of this textbook. It should be noted, however, that the ***Spur Gear Synthesis Tablet*** is using the detailed design calculations, which should closely approach more realistic conditions. More on the capabilities and utility of the ***Spur Gear Synthesis Tablet*** is covered in the *CAD* and optimization in Section 14.6.

14.4.3 Detailed Design

This section provides the *detailed design* of gears in the sense of repeated analysis. However the usual start of the iteration is the initial synthesis of Section 14.4.2. The number of cycles of repeated analysis to reach a satisfactory or an alternative optimum design goal is thus reduced. The design entails the definition of the geometry and the material. The initial synthesis adapted a material of *allowable bending fatigue strength* S_{ab} of 300 [MPa] or 43.5 [kpsi] and *allowable contact fatigue strength* S_{ac} of $(S_{ab})/B_{BC} = 300/0.3 = 1000$ [MPa] or 145 [kpsi]. These values might not be satisfactory for some applications. It might be low for extensively loaded cases or very high for very insubstantial, delicate, inexpensive, or undemanding products. The choice of the material, therefore, necessitates some iteration with regard to these cases. This and other details of loading specifics, severity, life cycling, reliability, etc. are covered in the following *detailed design* process. The material set that is usually employed in gear design is presented in section 14.4.3.1. The bending and the contact fatigue cases are covered in sufficient details in the following sections of 14.4.3.2 and 14.4.3.3. The aim of the detailed design is to specifically define the material designation that conforms to the requirements. It should necessitate specifying the heat treatment, or the case hardening details, to end up with the required allowable strength and safety conditions.

14.4.3.1 Material Set

Numerous materials can be used for gears; however, special considerations are narrowing down the appropriate set of materials that can be used. Remember that centuries ago the early gears were made of wood or even processed trees. The teeth were sort of cylindrical pigs or beams depending on construction; see http://www.hellenicaworld.com/Greece/Technology/en/ArchimedesGears.html and others for other gear material and information. Exploiting previous knowledge is important since some crude optimization has been performed with the previous results as the experience mature and outcome established. Experience is an optimization process that ends up with better selections or sort of close to the best solutions. The previous experience is a ***knowledge base*** that provides the set of materials that are effectively employed in gears.

As indicated previously (see Eq. (14.28) and Table 14.5), the quality of the gear is dependent on the application and necessitate some defined manufacturing processes. The main categories of materials and some selected representative samples are considered the ***material set*** for gears. The selected gear material set is as follows:

- *Steel:*
  - *Through hardened*: Carbon steel AISI: 1040, 1050 and alloy steel AISI: 3140, 3150, 4140, 4340, 5150, 6150.
  - *Case hardened*: Flame and induction hardening, carburizing, cyaniding, and nitriding: AISI 1020, 1320, 2317, 3115, 3310, 4119, 4320, 4620, 4815, 5120, 8620, 9310, 5140.
- *Cast iron:* Gray cast iron ASTM A48, ductile (nodular) ASTM A536, and malleable ASTM A602.
- *Nonferrous:* Aluminum bronze ASTM 148 UNS C95400, manganese bronze UNS C67500, 86200, and aluminum ASM 2017 T4, 2024 T4, 6061 T6.
- *Nonmetallic:*
  - *Thermoplastics*: Nylon, Teflon, acetal copolymer, and polycarbonate.
  - *Thermosets*: Phenolic, polyurethane, and composite laminate.

The *heat treatments* of steels depend on the range of carbon content. *Carburizing* is performed on steels with carbon content range of about 0.1–0.25%. *Flame hardening* or *induction hardening* is suitable for carbon content range of approximately 0.3–0.55%. Total *quenching* is applied to steels of carbon content starting about 0.4%. Nitriding is used for steels with carbon content range of about 0.25–0.55%. Nitriding, however, necessitates that the material contains one or more alloy elements, such as Al, Cr, Mo, or V; see Section 7.3.

Some property ranges of classes of *material set* are given in Table 14.6. The property range, according to AGMA (2001), helps in identifying the different categories of materials to be employed as alternative diverse range other than the usual average ***through-hardened steel*** previously used in the initial synthesis; see also ISO 6336 (2006) and (2016). The procedure presented herein with the *material property factor* $B_{BC} = 0.3$ would apply to the *material*

Table 14.6 Recommended ranges of hardness, allowable bending and contact fatigue strength S_{ab} and S_{ac}, and average property factor B_{BC} for some ranges of gear material set (AGMA 2001-B88).

Material	Hardness HB (HRC)	Allowable bending fatigue S_{ab} [MPa] ([kpsi])	Allowable contact fatigue S_{ac} [MPa] ([kpsi])	Average property factor B_{BC}
Steel				
Through hardened	180–400	170–390 (25–56)	590–1200 (85–175)	0.29–0.33
Case hardened	(50–64)	310–520 (45–75)	1200–1550 (170–225)	0.26–0.34
Cast Iron				
Gray	175–200	70–90 (8–13)	340–590 (50–85)	0.172
Ductile (nodular)	140–230	150–280 (22–40)	530–870 (77–126)	0.28–0.32
Malleable	165–240	70–125 (10–21)	500–650 (72–94)	0.170
Nonferrous				
Bronze	70–210	40–160 (6–24)	200–450 (30–65)	0.2–0.36
Nonmetallic				
Thermoplastics Thermosets	8–30	2–15 (0.3–2.1)[a)]	—	—

a) Dupont (2000).

set with the exception of *brittle* and *nonmetallic* materials; see Table 14.6 for average B_{BC} values of all material set. Other materials can also be used through their specific standard properties according to their material designation (Davis 2005). The following procedure is helpful in that regard and can be used as a specific material designation to be tuned for specific requirements. Plastic gears, however, are designed in a similar way for allowable bending fatigue (Table 14.6), but other conditions are more pronounced than those covered in this text such as temperature and the limit of peripheral velocity to be less than 5 [m/s]. For more details on the design of *plastic gears*, one should consult with material manufacturers such as DuPont (2000). The *optimum design* suggests a smaller module m_n and an exceptionally large number of teeth N_P for especially low strength materials.

14.4.3.2 Bending Fatigue

The evaluation of the bending fatigue stress is performed by a remodified *Lewis equation* (Eq. (14.27)). The ***bending fatigue stress*** σ_{ab} becomes

$$\begin{aligned} \sigma_{ab} &= \frac{\mathbf{F}_t K_v}{f_g m_n J_g} \quad \text{for SI} \\ \sigma_{ab} &= \frac{\mathbf{F}_t K_v p_d}{f_g J_g} \quad \text{for US} \end{aligned} \tag{14.34}$$

where J_g is the *remodified Lewis factor*, K_v is the *velocity factor*, $\mathbf{F}_t$ is the tangential force, f_g is the *face width* of the gear, and m_n is the *module*. The remodified *Lewis* factor J_g is the *bending geometry factor* that replaces the modified *Lewis* factor Y_L. It includes the *fatigue stress concentration factor* K_f for gear root effect. The *bending geometry factor* is then defined as $J_g = Y_L/K_f$. The *stress concentration factor* K_f for gear root is obtained by Dolan and Broghamer (1942) using *photoelasticity* and still applicable. An equation for 20° pressure angle has the following estimated form:

$$K_f = 0.18 + \left(\frac{t_r}{r_f}\right)^{0.15} + \left(\frac{t_r}{l}\right)^{0.45} \tag{14.35}$$

The thickness at tooth root t_r, the fillet radius r_f, and the tooth height l are defined in Eqs. (14.21) and (14.23) and Figure 14.10. The height of the load location at l can be either the tooth height or the highest point of single tooth contact, which can be lower than the tooth tip. Taking l as the tooth height of $2.25m_n$ or $2.25/p_d$ is a conservative estimate. The fillet radius r_f can be the standard $0.3m_n$ or $0.3/p_d$ as a conservative estimate. The tooth thickness t_r at the tooth root may then be evaluated from Eqs. (14.21) and (14.23).

Considering the load at the tooth tip of a gear g, the *bending geometry factor* J_g may, however, be approximated for $\phi = 20°$ according to the following relation as a function of the number of teeth N_g:

$$J_g = 0.32(1 - (1.14/N_g^{0.546})) \tag{14.36}$$

This relation has been suggested for *regular gears* (Dimargonas 1989). For *high-accuracy gears*, he suggested including the effect of mating gears to find the following expression:

$$J_{g1} = 0.56(1 - 0.38/N_{g2} - (0.88/N_{g2}))(1 - (0.26/N_{g1} - 5.5/N_{g1})) \tag{14.37}$$

It is supposed that the subscript g1 is for the gear to which the geometry factor is desired and the subscript g2 is for the mating gear. The values obtained by Eq. (14.37), however, have not been close enough compared with the values suggested by AGMA 218.01 (1982) and AGMA 908-B89 (1999) for loads applied at the *highest point of a tooth contact*. An attempt to produce fitting results for loads applied at the *highest point of a tooth contact* suggests the following relation:

$$J_{g1} = (0.0685\ \ln(N_{g1}) + 0.1652)(0.0408\ \ln(N_{g2}) + 0.7668) \tag{14.38}$$

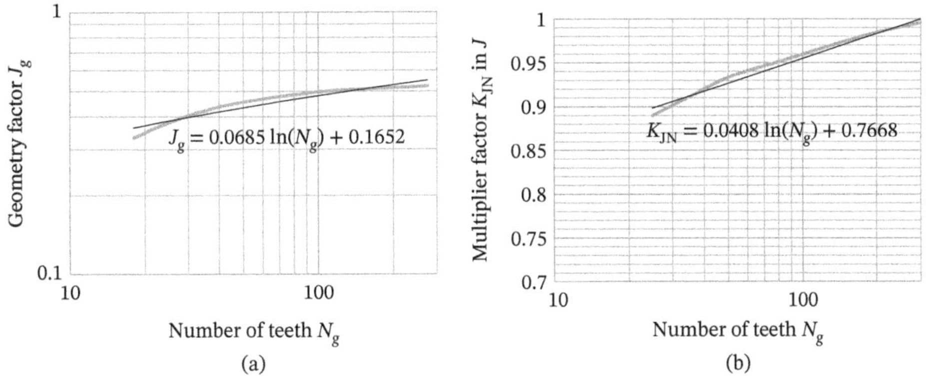

Figure 14.15 Bending geometry factor evaluation: (a) value of geometry factor J_{g1} for a gear mating with a rack and (b) the reduction multiplier K_{JN} of the J_{g1} value when smaller number of teeth is mating with the gear.

The first right-hand part of Eq. (14.38) represents the value of geometry factor J_{g1} for a gear mating with a rack (Figure 14.15a). The second right-hand part is the reduction multiplier in the value of J_{g1} when smaller number of teeth is mating with the gear (Figure 14.15b). This relation has been verified and found to be within engineering accuracy for loads applied at the *highest point of a tooth contact*. The limit of N_{g2} in Eq. (14.38) is about 300 teeth for the value of J_g to be within representative engineering accuracy (Figure 14.15b). The *bending geometry factor* J_g may have a constant value of 0.54 for the number of teeth around and beyond 300.

The dynamic or *velocity factor* K_v that is more accurately defined as a function of *pitch line velocity* $\mathbf{v}_t$ and the AGMA gear *quality number* Q_v is shown in Figure 14.16. Each line in Figure 14.16 is an approximate functional fit of information in AGMA 2001-C95 (1995). The appropriate and more accurate values of K_v for $Q_v = 6$–11 are obtained from the following AGMA relations (ANSI/AGMA 2001-D04):

$$K_v|_{SI} = \left(\frac{50 + 56(1 - B_v) + \sqrt{200(v_t)}}{50 + 56(1 - B_v)}\right)^{B_v}$$

$$K_v|_{US} = \left(\frac{50 + 56(1 - B_v) + \sqrt{(v_t)_{ft/min}}}{50 + 56(1 - B_v)}\right)^{B_v} \qquad (\mathbf{\mathit{14.39}})$$

$$B_v = 0.25(12 - Q_v)^{2/3}$$

It should also be noted that each adapted quality number has a limiting speed associated with it. The *speed limit* $\mathbf{v}_{t,\max}$ is suggested by AGMA to have the following values (ANSI/AGMA 2001-D04):

$$\mathbf{v}_{t,\max}\Big|_{SI} = (50 + 56(1 - B_v) + (Q_v - 3))^2/200$$

$$\mathbf{v}_{t,\max}\Big|_{US} = (50 + 56(1 - B_v) + (Q_v - 3))^2 \qquad (14.40)$$

The term B_v in Eq. (14.40) is defined in Eq. (14.39). The speed limit values $\mathbf{v}_{t,\max}$ have been approximately terminating the curves shown in Figure 14.16. If the value in a problem is close to these values in Figure 14.16, the limits of Eq. (14.40) ought to be observed.

The *bending fatigue stress* σ_{ab} of Eq. (14.34) is now possible to evaluate since all components have been defined in Eqs. (14.38)–(14.40).

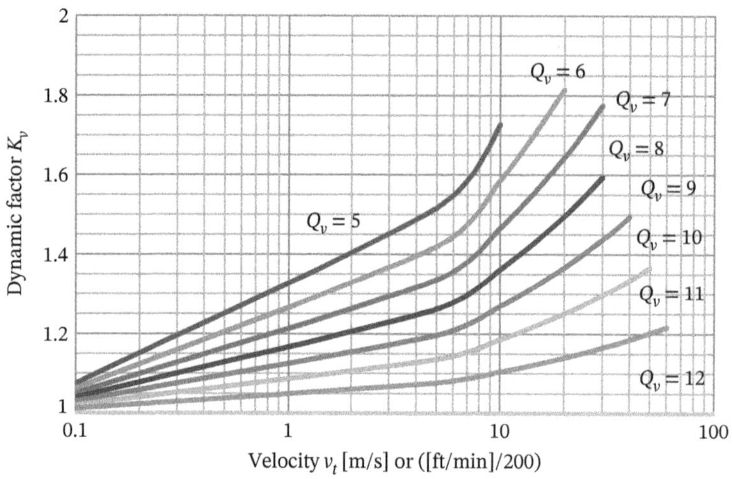

Figure 14.16 The dynamic or velocity factor K_v as function of pitch line velocity v_t and AGMA gear quality number Q_v. The velocity axis is to be multiplied by 200 if velocity is in [ft/min].

Fatigue Strength

To compare the *bending fatigue stress* σ_{ab} with the *endurance strength limit* of the gear S_{eg}, the factors affecting the *endurance limit* S_e are to be considered in addition to a gear safety factor K_{SF}. This is the same procedure used for other machine components, which has been defined in Section 7.8. The *endurance limit* S_e of materials is not readily available and may, however, be estimated from other properties such as the *ultimate tensile strength* S_{ut}. The factors affecting the endurance limit defined specifically for gears that include the *surface finish factor* k_a, the *size factor* k_b, the *reliability factor* k_c, the *temperature factor* k_d, the stress concentration factor k_e (defined previously as K_f), and the *miscellaneous factor* k_f are taken here as the *one-way or two-way bending*. The one-way bending is defined when the gear set is rotating in one direction with no reversing of rotation. The two-way bending is when the gear set rotates in both directions about an equal amount of time. The different factors affecting fatigue strength are defined from the available information in the literature, and the fitted relations are as follows.

The *surface finish factor* k_a decreases with the increase of the *ultimate tensile strength* S_{ut} and can be obtained from the following relation:

$$\begin{aligned} k_a &= 3.0597\,(S_{\text{ut}}|_{[\text{MPa}]})^{-0.219} \\ k_a &= 2.0039(S_{\text{ut}}|_{[\text{kpsi}]})^{-0.219} \end{aligned} \tag{14.41}$$

The effect of tooth size on the *size factor* k_b is estimated according to the module m_n [mm] or the diametral pitch p_d [teeth/in] according to the fitted relation as follows:

$$\begin{aligned} k_b &= -0.0721\ \ln(m_n) + 1.0133 \\ k_b &= 0.0721\ \ln(p_d) + 0.7793 \end{aligned} \tag{14.42}$$

The *reliability factor* k_c is estimated according to the usual practice of having the *reliability* R_e of the material property in the range of 0.99 (or 99%) (ANSI/AGMA 2001-D04). The reliability factor k_c can then be estimated as follows:

$$k_c = 1/(0.6735(1 - R_e)^{-0.088}) \tag{14.43}$$

For the *temperature factor* k_d and since gear sets are usually lubricated, the temperature is kept as low as possible for good operation. This would require the temperature to be below the value of 120 [°C] (250 [°F]) to warrant a consideration of k_d different than 1. Therefore, the adapted value of $k_d = 1$ is reasonable for most cases.

The *stress concentration factor* k_e is taken care of in the *bending geometry factor* J_g as indicated in the discussion leading to Eq. (14.35). As discussed, the *bending geometry factor* J_g includes the stress concentration factor. The stress concentration factor, therefore, should be $k_e = 1.0$. There is no need to consider it twice.

One-way bending factor k_f is dependent on the *ultimate tensile strength* of the gear. The fitted relation is given by

$$\begin{aligned} k_f\Big|_{SI} &= -7(10)^{-8} S^2_{ut,[MPa]} + 0.0005 S_{ut,[MPa]} + 0.7874 \\ k_f\Big|_{US} &= -3(10)^{-6} S^2_{ut,[kpsi]} + 0.0034 S_{ut,[kpsi]} + 0.7874 \end{aligned} \tag{14.44}$$

If the *ultimate tensile strength* S_{ut} of the gear material is lower than 1400 [MPa], the *one-way bending* factor k_f should always be 1.33. For *two-way bending* the factor $k_f = 1.0$.

Equations (14.41)–(14.44) conclude the factors affecting fatigue strength for gear material. These factors in addition to the effect of *expected life* beyond or below the endurance limit should be taken care of by *adjusting the endurance limit* different from the conventional 10^6–10^7 cycles. This can be resolved by using the *fatigue strength* S_f rather than the endurance limit S_e; see Section 7.8. To compare the *bending fatigue stress* σ_{ab} to the *endurance strength limit* of the gear S_{eg}, the factors affecting the endurance limit S_e are considered in addition to a gear safety factor K_{SF}. The factors affecting the *endurance strength limit* of the gear S_{eg} or the *fatigue strength* S_f for specific number of *life cycles* L_c give the following equation:

$$S_{eg} = k_a k_b k_c k_d k_e k_f S_e\Big|_{10^6,10^7} = k_a k_b k_c k_d k_e k_f S_f\Big|_{Cycles} \tag{14.45}$$

This represents the ***material fatigue strength*** for the specific application to the gear in question. Note that the number of loading cycles for the smaller pinion is surely higher than the number of loading cycles for the larger mating gear.

Bending Fatigue Stress

The applied bending fatigue stress is not only the stress defined by Eq. (14.34). This relation does not account for the service loading conditions such as the steadiness of the power source (electric motor, combustion engines of multi or single cylinders) or the type of driven machines. Table 14.7 suggests a *load application factor* k_o to account for these conditions of different classes of power sources and driven machines. This factor should be included in the evaluation of the applied stress since it accounts for the dynamic changes of the applied loads. Another factor that should also be included is the *load distribution factor* k_m. This factor accounts for the inevitable variation of the way the load is distributed over the face of the teeth. Equation (14.34) assumes perfect uniform distribution over the whole face width. Due to deflections of the shafts on which gears are mounted and teeth deflections accordingly, the load distribution is not expected to be uniform or centered over the teeth. That nonuniform distribution is clearly a function of each gear face width f_g and should be included in the evaluation of the applied stress since

Table 14.7 Recommended load application factor k_o for different classes of power sources and driven machines.

Power source	Driven machine		
	Uniform	Moderate shock	Heavy shock
Uniform (electric motor, ...)	1.00	1.25	1.75 or more
Light shock (I.C. Engines,...)	1.25	1.50	2.00 or more
Medium shock (Single-cyl.,...)	1.50	1.75	2.25 or more

some portion of the tooth will have higher loads than the rest of the tooth. A suggested fitted relation to account for the *load distribution factor* k_m effect is given as defined in Metwalli and El Danaf (1996).

$$\begin{aligned} k_m\big|_{SI} &= -4(10)^{-7} f_g^2\Big|_{[mm]} + 0.0011 f_g\Big|_{[mm]} + 1.544 \\ k_m\big|_{US} &= -6.28(10)^{-4} f_g^2\Big|_{[in]} + 0.0364\, f_g\Big|_{[in]} + 1.523 \end{aligned} \qquad (14.46)$$

The maximum value of k_m is considered as 2.0 for the face width of more than 500 [mm] or 20 [in].

The applied bending fatigue stress is not only the stress defined by Eq. (14.34). It should then be adjusted to include the *load application factor* k_o and the *load distribution factor* k_m. It is then more reasonable to incorporates the k_o and k_m factors into the probable maximum loading and thus into the ***bending fatigue stress*** σ_{ab}. A *maximum* ***bending fatigue stress*** $\sigma_{ab}|_{max}$ may then be rewritten as

$$\begin{aligned} \sigma_{ab}\big|_{max} &= \frac{F_t K_v k_o k_m}{f_g m_n J_g} \quad \text{for SI} \\ \sigma_{ab}\big|_{max} &= \frac{F_t K_v p_d k_o k_m}{f_g J_g} \quad \text{for US} \end{aligned} \qquad (14.47)$$

For the SI system, usually the module m_n is substituted in [mm], and the tooth face width f_g is also substituted in [mm] to calculate the stress directly in [MPa]. It should also be noted that the *load distribution factor* k_m is included in the *optimum* data used in the *synthesis*. In the synthesis procedure, however, the *load application factor* k_o should be incorporated into the maximum input power, which materializes directly into the tangential force $\boldsymbol{F}_t$. The adjusted tangential force will then incorporate the *load application factor* k_o and the newly adjusted $\boldsymbol{F}_t = k_o \boldsymbol{F}_t$.

Bending Fatigue Safety Factor

To get the safety factor, one should include the *load application factor* k_o and the *load distribution factor* k_m into the stress defined by Eq. (14.34) and as shown in Eq. (14.47). Considering the *endurance strength limit* of the gear S_{eg} or the *fatigue strength* S_f defined by Eq. (14.45) and introducing a *material safety factor* K_{MSF}, the final ***bending fatigue safety factor*** $K_{SF}|_b$ is defined as follows:

$$K_{SF}\big|_b = \frac{(S_{eg}/K_{MSF})}{k_o k_m \sigma_{ab}} = \frac{(S_{eg}/K_{MSF})}{\sigma_{ab}\big|_{max}} = \frac{S_{ab}}{\sigma_{ab}\big|_{max}} \qquad (14.48)$$

As indicated in Eq. (14.48) where the *allowable fatigue strength* S_{ab} is used in place of the *endurance strength limit* of the gear S_{eg} divided by the *material safety factor* K_{MSF}, the ***bending fatigue safety factor*** $K_{SF}|_b$ should be about 1.0. As defined in Eq. (14.47), the k_o and k_m factors have been incorporated into the *bending fatigue stress* σ_{ab} defined by Eq. (14.34) to have the ***maximum bending fatigue stress*** $\sigma_{ab}|_{max}$ defined by Eq. (14.47).

It should also be noted that the *load distribution factor* k_m is included in the *optimum* data used in the *synthesis*. The available ***Gear Synthesis Tablet*** through the ***Wiley website*** incorporates the load distribution factor k_m into the calculations of the bending fatigue stress σ_{ab}. The *load application factor* k_o should, however, be incorporated into the input power to get the ***maximum bending fatigue stress*** $\sigma_{ab}|_{max}$. More on the capabilities and utilities of the ***Spur Gear Synthesis Tablet*** is covered in the *CAD* and optimization in Section 14.6.

AGMA Bending Stress Number

Extensive work has been done by the AGMA in developing codes and standards. They have also been using the technical information developed by researchers in addition to experimental work. The *Lewis equation* and its modifications have been the bases of the developed codes and standards for the bending stresses and gear strength. Equation (14.34) is the same as that adapted expression by AGMA. The particular development of the geometry and other factors are of paramount importance as evident in the preceding developments. For other cases not included in the previous treatment, one should consult the AGMA standards.

The main difference in the treatment of AGMA is their utility of ***allowable bending stress number*** S_t, which is not necessarily coincident with the treatment of the ***fatigue strength*** previously discussed in this section. These numbers are literally the *allowable bending* ***fatigue*** *strength*, and they have been quoted in most of Table 14.6 in ranges rather than for specific materials. The selection of a specific material is therefore dependant on its hardness number (see Table 14.6) since there are no accounted factors directly associated with it. The AGMA standards, however, specify relations between the *hardness* and the *allowable bending stress number* S_t (ANSI/AGMA 2001-D04). These relations are also available in the form of graphs that can be viewed in these standards if need be. Herein, it is recommended to find the suitable *allowable bending stress number* S_t (***allowable bending fatigue strength***) and then specify the suitable fitting material or the heat treatment that provides the ***hardness*** to fit. This would narrow down the selection of a specific *material designation* standard. The available relations of the ***Brinell hardness HB*** and ***allowable bending stress number*** S_t for some selected materials are as follows:

- ***Steel:***
  - *Through hardened*:

$$\begin{aligned}
\text{HB} &= 1.876S_t - 165.7, \quad \text{or} \quad S_t = 0.533\text{HB} + 88.3\ [\text{MPa}], \quad \text{grade 1} \\
\text{HB} &= 0.0129S_t - 165.6, \quad \text{or} \quad S_t = 77.3\text{HB} + 12\,800\ [\text{psi}], \quad \text{grade 1} \\
\text{HB} &= 1.422S_t - 160.7, \quad \text{or} \quad S_t = 0.703\text{HB} + 113\ [\text{MPa}], \quad \text{grade 2} \\
\text{HB} &= 0.0098S_t - 160.8, \quad \text{or} \quad S_t = 102\text{HB} + 16\,400\ [\text{psi}], \quad \text{grade 2}
\end{aligned} \tag{14.49}$$

  - *Case hardened*: Nitride:

$$\begin{aligned}
\text{HB} &= 1.76S_t - 147.5, \quad \text{or} \quad S_t = 0.568\text{HB} + 83.8\ [\text{MPa}], \quad \text{grade 1} \\
\text{HB} &= 0.0122S_t - 147.6, \quad \text{or} \quad S_t = 82.3\text{HB} + 12\,150\ [\text{psi}], \quad \text{grade 1} \\
\text{HB} &= 1.335S_t - 146.9, \quad \text{or} \quad S_t = 0.749\text{HB} + 110\ [\text{MPa}], \quad \text{grade 2} \\
\text{HB} &= 0.0092S_t - 146.3, \quad \text{or} \quad S_t = 108.6\text{HB} + 15\,890\ [\text{psi}], \quad \text{grade 2}
\end{aligned} \tag{14.50}$$

- ***Cast iron*** and some ***nonferrous*** that are used for gears; one may use the following allowable strength relations, *previously available*, which have some embedded safety:

$$\begin{aligned}
S_t &= 35,\ \ 59,\ \ 90\ [\text{MPa}]\ \ \text{Cast Iron grade 20, 30, 40} \\
S_t &= 5,\ \ 8,\ \ 13\ [\text{kpsi}]\ \ \text{Cast Iron grade 20, 30, 40} \\
S_t &= 160\ \ [\text{MPa}],\ \ 90\ [\text{kpsi}]\ \ \text{Aluminum bronze (ASTM B-148-52)}
\end{aligned} \tag{14.50$'$}$$

  For these and other materials, it is advisable to consult the current AGMA standards for the allowable bending stress number.

The AGMA factors affecting the allowable bending stress number S_t are the *reliability factor* K_R and the *bending life cycle factor* Y_N. The other factors are for special cases such as the *rim thickness factor* that is equal to 1.0 for rims that are more than the tooth height by a factor of 1.2, which should generally be the case. The reliability factor K_R is similar to that factor k_c given by Eq. (14.43) and thus can be used to represent K_R. Therefore, one can state that

$$K_R = 1/k_c = 0.6735\,(1 - R_e)^{-0.088} \tag{14.43$'$}$$

The *life cycle factor* Y_N accounts for stressing the gear to a higher or a lower number of cycles than the assumed endurance of 10^7 cycles. The stressing at higher cycles than the 10^7 is the critical one and can be assumed conservatively as AGMA suggests as follows:

$$Y_N = 1.6831N_L - 0.0323 \geq 10^7 \text{ cycles} \tag{14.51}$$

If the number of life cycles is less than 10^7, this factor will be more than 1.0. The value of Y_N will then depend on the material type, and it is recommended to consult AGMA standard for this value (ANSI/AGMA 2001-D04).

The *bending safety factor* $K_{SF}|_b$ previously defined by Eq. (14.47) as K_{SF} will be adjusted to the AGMA bending stress number S_t so that

$$K_{SF}|_b = \frac{Y_N S_t / K_R}{(k_o k_m \sigma_{ab})} \tag{14.52}$$

This equation suggests relating S_t to S_{ab} with the same consideration for the fatigue strength factors indicated before and noting that K_R has been included in S_{eg}. Note that the life cycle factor Y_N was previously alluded to by considering *the fatigue strength* S_f rather than the *endurance limit* S_e. The *fatigue strength* should include Y_N for higher life cycles of more than 10^7 and should also include consideration for life cycles less than 10^7 just as Y_N. Therefore, S_t is expected to be equal to S_{eg}/K_{MSF}. That is may be why AGMA did not go for the conventional S_e and preferred to just use the term of allowable bending stress number instead. The *allowable stress number* S_t, however, has been evaluated with safety factor inclusion or assuming $S_t \approx S_{ab}$ the *allowable bending fatigue strength*; see Table 14.6. Comparing the *allowable stress number* S_t and the Brinell hardness HB with the *endurance limit* S_e of some materials, it is apparent that S_t has an embedded safety factor of about 2, which is also about the regular safety factor. Evaluating another safety factor $K_{SF}|_b$ in Eq. (14.52) is considered as an *extra safety factor* K_{SFE} for fear of missing any factor or misrepresenting some of the previous factors. In other words, Eq. (14.52) should be such that

$$Y_N S_t / K_R \geq (k_o k_m \sigma_{ab}) = \sigma_{ab}|_{max} \tag{14.53}$$

Equation (14.53) suggests that the *extra safety factor* can be 1.0 or little more than 1.0. The right-hand term of the equation represents the *expected maximum loading* of the gear, and the left-hand term represents the *expected allowable strength* of the material.

Example 14.6 For the problem in Example 14.4, the reduction gear set in the crane is driven by a maximum power of 75 [kW] or 100.58 [hp] electric motor running at a rotational speed of 1200 [rpm]. The gear set is to be designed for a gear ratio of 4 : 1. Gears are also of 20° full depth. Present a detailed design depending on the bending fatigue for the previously synthesized gear set.

Solution

Data: (same as Example 14.4) Power transmitted $H_{kW} = 75$ [kW] or 100.58 [hp], rotational speed $N_{in} = 1200$ [rpm], and the gear ratio $n_g = 4$. The rotation speed $\omega_P = (2\pi N_{in}/60) = 125.66$ [rad/s].

From Examples 14.4 and 14.5, the synthesized geometry provides $m_n = 3.25$ [mm], $N_P = 36$, and $f_g = 59$ [mm] for SI system or $p_d = 8$ [teeth/in], $N_P = 36$, and $f_g = 2.315$ [in] for the US system. The pinion diameter $d_P = N_P m_n = 117(10)^{-3}$ [m] or $d_P = N_P/p_d = 4.5$ [in] and the pitch line velocity $\boldsymbol{v}_t = \omega_P(d_P/2) = 7.35$ [m/s] or 282.74 [in/s] = 1413.7 [ft/min]. The tangential force is $\boldsymbol{F}_t = H_{kW}(1000)/\boldsymbol{v}_t = 10.2041$ [kN] or $= H_{hp}(6600)/\boldsymbol{v}_t = 100.58(6600)/282.74 = 2347.8$ [lb]. The gear quality number $Q_v = 8$, and the suggested material has allowable bending fatigue strength S_{ab} of 300 [MPa] or 43.5 [kpsi].

To find the *bending fatigue stress* σ_{ab} by applying Eq. (14.47), one needs to calculate the geometry factor J_g, the velocity factor K_v, and the load distribution factor K_m using Eqs. (14.38), (14.39), and (14.46). The load application factor k_o is assumed 1.0 for smooth operation. To get J_P for the pinion P (subscript $g1$), the gear G is subscript $g2$. To get J_G for the gear G (subscript $g1$), the pinion P is subscript $g2$.

$$\begin{aligned} J_P &= (0.0685\ \ln(N_P) + 0.1652)(0.0408\ \ln(N_G) + 0.7668) \\ &= (0.0685\ \ln(36) + 0.1652)(0.0408\ \ln(144) + 0.7668) = 0.398 \end{aligned} \tag{a}$$

$$\begin{aligned} J_G &= (0.0685\ \ln(N_G) + 0.1652)(0.0408\ \ln(N_P) + 0.7668) \\ &= (0.0685\ \ln(144) + 0.1652)(0.0408\ \ln(36) + 0.7668) = 0.462 \end{aligned}$$

Since the maximum allowable velocity for $Q_v = 8$ is higher than 7.35 [m/s] (see Figure 14.16), the velocity factor K_v (Eq. (14.39)) is

$$B_v = 0.25(12 - Q_v)^{2/3} = 0.25(12 - 8)^{2/3} = 0.63$$

$$K_v|_{SI} = \left(\frac{50 + 56\,(1 - B_v) + \sqrt{200(v_t)}}{50 + 56(1 - B_v)}\right)^{B_v} = \left(\frac{50 + 56(0.37) + \sqrt{200(7.35)}}{50 + 56(0.37)}\right)^{0.63} = 1.3138 \tag{b}$$

$$K_v|_{US} = \left(\frac{50 + 56\,(1 - B_v) + \sqrt{v_t}}{50 + 56(1 - B_v)}\right)^{B_v} = \left(\frac{50 + 56(0.37) + \sqrt{1413.7}}{50 + 56(0.37)}\right)^{0.63} = 1.3039$$

The load distribution factor k_m is obtained by applying Eq. (14.46) to get

$$\begin{aligned} k_m|_{SI} &= -4(10)^{-7} f_P^2\Big|_{[mm]} + 0.0011 f_P|_{[mm]} + 1.544 = -4(10)^{-7}(59)_P^2 + 0.0011(59)_P + 1.544 \\ &= 1.6075 \\ k_m|_{US} &= -6.28(10)^{-4} f_g^2\Big|_{[in]} + 0.0364 f_g\Big|_{[in]} + 1.523 = -6.28(10)^{-4}(2.315)^2 + 0.0364(2.315) + 1.523 \\ &= 1.6039 \end{aligned} \tag{c}$$

Applying the pinion variables and the values in Eqs. (a), (b), and (c) into Eq. (14.47) gives

$$\begin{aligned} \sigma_{ab}|_P &= \frac{F_t K_v k_o k_m}{f_P m_n J_P} = \frac{10.2041(1000)(1.3138)(1.0)(1.6075)}{59(3.25)(0.398)} = 282.39\ [\text{MPa}] \\ \sigma_{ab}|_P &= \frac{F_t K_v p_d k_o k_m}{f_P J_P} = \frac{2347.8(1.3039)(8)(1.0)(1.6039)}{2.315(0.398)} = 42\,633[\text{psi}] = 42.633\ [\text{kpsi}] \end{aligned} \tag{d}$$

These values of the pinion *bending fatigue stress* σ_{ab} were expected to be 300 [MPa] and 43.5 [kpsi], respectively. It is less than the two expected values by about 6% and 2%, respectively. This is due to the rounding of all geometry values and the inexactness of fitting equations as can be seen in the ***Default*** column of Figure 14.17a,b where $m_n = 3.23$ [mm] and $f_P = 58.8$ [mm] for SI system or $p_d = 8$ [teeth/in] and $f_g = 2.315$ [in] for the US system. The number of teeth has also been forcefully rounded to integer values.

The detailed design depends on the selection of material. However, the adapted material by synthesis has allowable *bending fatigue strength* S_{ab} of 300 [MPa] or 43.5 [kpsi]. Therefore either calculate the expected endurance limit S_e or the fatigue strength S_f and select the material accordingly or find the Brinell hardness HB number from AGMA and select the material accordingly. Assuming a usual life cycle of 10^6–10^7, one can then use the endurance limit S_e and apply Eq. (14.45) to get the required endurance limit of the material. Applying Eq. (14.45) requires the evaluation of the factors affecting fatigue using Eqs. (14.41)–(14.44). Assume a 99% reliability (i.e. $k_c = 1.0$), no excess temperature (i.e. $k_d = 1.0$), and the operation is two-way bending (i.e. $k_f = 1.0$) since no additional

(a)

Spur gear synthesis – SI units			
Inputs	**Default**	**Input**	**Analysis**
Maximum power, H_{kW} [kW]	75	75	75
Input rotational speed, $N_{rpm,P}$ [rpm]	1200	1200	1200
Output rotational speed, $N_{rpm,G}$ [rpm]	300	300	300.000
Safety factor, K_{SF}	1.000	1.000	1.222
Quality number, Q_v	8	8	8
Operating direction, (*2-way or 1-way*)	2	2	2
Reliability, R_e	0.99	0.99	0.99
Output			**Input/Output**
Module, m_n [mm]	3.23	3.232	8
Number of pinion teeth, N_P	36	57	18
Number of gear teeth, N_G	144	228	72
Face width, $f_{g\,(P,G)}$ [mm]	58.8	58.823	100
Allowable bending strength, S_{ab} [MPa]	300	200	300
Allowable contact strength, S_{ac} [MPa]	1000	667	1000
Gear ratio, n_G	3.968	4.000	4.000
Pinion pitch diameter, d_P [mm]	117.29	184.204	144.000
Gear pitch diameter, d_G [mm]	465.36	736.815	576.000
Center distance, C_g [mm]	291.32	460.509	360.00
Maximum torque, $T_{g,max}$ [N.m]	596.83	596.83	596.83
Tangential force, F_t [N]	10 177.31	6480.12	8289.32
Radial force, F_r [N]	3704.24	2358.57	3017.06
Pinion volume, V_P [m^3]	0.000 708	0.001 679 535	0.002 010 619
Gear volume, V_G [m^3]	0.010 285	0.025 523 532	0.027 525 378
Pitch point velocity, v_t [m/s]	7.369	11.574	9.048
Dynamic factor, K_v	1.314 12	1.386 71	1.345 37
Bending geometry factor, J_g	0.398 71	0.436 98	0.341 87
Surface geometry factor, I_g	0.119 74	0.128 55	0.103 84
Bending stress, σ_b [MPa]	283.622	173.873	67.282
Contact stress, σ_c [MPa]	974.332	615.046	670.016
Bending safety factor, $K_{SF}\vert_b$	1.058	1.150	4.459
Contact safety factor, $K_{SF}\vert_c$	1.013	1.041	1.222
Other values			
Tooth frequency, f_T [Hz]	725.86	1140.00	360.00
Contact ratio, m_p	1.791	1.848	1.671
Circular pitch, p_c [mm]	10.153	10.153	25.133
Outer pinion diameter, d_{oP} [mm]	123.750	190.667	160.000
Outer gear diameter, d_{oG} [mm]	471.820	743.278	592.000
Total gear set volume, V_S [m^3]	0.010 992	0.027 203 067	0.029 535 997

Figure 14.17 The image of the *Spur Gear Synthesis Tablet*. It also shows a synthesized gear for an allowable bending fatigue strength of (a) 200 [MPa] and (b) 29 [kpsi] in the *Input* column. The *Analysis* column presents the conventional or classical solution for the *Default* problem.

(b)

Spur gear synthesis – US units			
Inputs	**Default**	**Input**	**Analysis**
Maximum power, H_{hp} [hp]	100.5	100.5	100.5
Input rotational speed, $N_{rpm,P}$ [rpm]	1200	1200	1200
Output rotational speed, $N_{rpm,G}$ [rpm]	300	300	300.000
Safety Factor, K_{SF}	1.000	1.000	1.245
Quality Number, Q_v	8	8	8
Operating direction, (*2-way* or *1-way*)	2	2	2
Reliability, R_e	0.99	0.99	0.99
Output			**Input/Output**
Diametral pitch, p_c [teeth/in]	7.862	7.862	3
Number of pinion teeth, N_P	36	58	18
Number of gear teeth, N_G	144	232	72
Face width, $f_{g(P,G)}$ [in]	2.315	2.315	4
Allowable bending strength, S_{ab} [kpsi]	43.50	29	43.5
Allowable contact strength, S_{ac} [kpsi]	145.00	96.67	145
Gear ratio, n_G	4.000	4.000	4.000
Pinion pitch diameter, d_P [in]	4.579	7.377	6.000
Gear pitch diameter, d_G [in]	18.316	29.509	24.000
Center distance, C_g [in]	11.447	18.443	15.000
Maximum torque, $T_{g,max}$ [lb.in]	5278.4	5278.374	5278.374
Tangential force, F_t [lb]	2305.51	1431.005	1759.458
Radial force, F_r [lb]	839.14	520.843	640.390
Pinion volume, V_P [in^3]	42.483	105.9128	139.6263
Gear volume, V_G [in^3]	627.122	1610.9343	1911.4846
Pitch point velocity, v_t [ft/min]	1438.5	2317.60	1884.956
Dynamic factor, K_v	1.310 60	1.386 92	1.351 93
Bending geometry factor, J_g	0.398 17	0.438 48	0.341 87
Surface geometry factor, I_g	0.119 81	0.128 55	0.103 84
Bending stress, σ_b [kpsi]	42.85	25.560	9.031
Contact stress, σ_c [kpsi]	144.66	89.173	93.476
Bending safety factor, $K_{SF}\|b$	1.015	1.135	4.817
Contact safety factor, $K_{SF}\|c$	1.001	1.041	1.245
Other values			
Tooth frequency, f_T [Hz]	720.00	1160.00	360.00
Contact ratio, m_p	1.790	1.850	1.671
Circular pitch, p_c [in]	0.400	0.400	1.047
Outer pinion diameter, d_{oP} [in]	4.833	7.632	6.667
Outer gear diameter, d_{oG} [in]	18.570	29.763	24.667
Total gear set volume, V_S [in^3]	669.60	1716.8471	2051.1109

Figure 14.17 (*Continued*)

specification has been required. The other factors affecting fatigue are then (Eqs. (14.41)–(14.44))

$$\begin{aligned} k_a &= 3.0597(S_{ut}|_{[MPa]})^{-0.219} = 3.0597((2)(2)(300)|_{[MPa]})^{-0.219} = 0.647\,65 \\ k_a &= 2.0039(S_{ut}|_{[kpsi]})^{-0.219} = 2.0039((2)(2)(43.5)|_{[kpsi]})^{-0.219} = 0.6474 \\ k_b &= -0.0721\ \ln(m_n) + 1.0133 = -0.0721\ \ln(3.25) + 1.0133 = 0.928\,32 \\ k_b &= 0.0721\ \ln(p_d) + 0.7793 = 0.0721\ \ln(8) + 0.7793 = 0.929\,23 \end{aligned} \tag{e}$$

It should be noted that the sought ultimate tensile strength S_{ut} in Eq. (e) has been grossly approximated as twice the endurance, which had about an additional 2.0 factor of safety relative to the allowable of 300 [MPa] and 43.5 [kpsi].

$$\begin{aligned} S_e|_{10^6,10^7} &= \frac{S_{eg}/K_{MSF}}{k_a k_b k_c k_d k_e k_f} = \frac{300 \text{ or } (282.39)}{(0.647\,65)(0.928\,32)(1)(1)(1)(1)} = 498.98 \text{ (or } 469.69) \text{ [MPa]} \\ S_e|_{10^6,10^7} &= \frac{S_{eg}/K_{MSF}}{k_a k_b k_c k_d k_e k_f} = \frac{43.5 \text{ or } (42.633)}{(0.6474)(0.929\,23)(1)(1)(1)(1)} = 72.31 \text{ (or } 70.87) \text{ [kpsi]} \end{aligned} \tag{f}$$

Previously in Example 14.5, the material was assumed as AISI 3140 *normalized* that has a yield strength $S_y = 599.8$ [MPa] = 87 [kpsi], Brinell hardness HB = 262, and ultimate tensile strength $S_{ut} = 891.5$ [MPa] = 129.3 [kpsi]. This material will fall short to the required endurance limit of at least 470 [MPa] or 71 [kpsi] since a rough estimate for fatigue is $S_{ut}/2$ = 891.5/2 = 445.75 [MPa] or 129.3/2 = 64.65 [kpsi]. A different, more suitable, and widely used material would be *through-hardening* steel ***AISI 4340*** that also has defined endurance limits of 489–668 [MPa] or 71–97 [kpsi] (Stulen et al. 1961). The AISI 4340 *normalized* has a yield strength $S_y = 861.8$ [MPa] = 125.0 [kpsi], Brinell hardness HB = 363, and ultimate tensile strength $S_{ut} = 1279.0$ [MPa] = 185.5 [kpsi]. These values will change when a *through hardening* and different *tempering* temperature occurs.

The AGMA procedure can suggest finding the Brinell hardness HB number from AGMA allowable *bending fatigue number* S_t of 300 [MPa] (43.5 [kpsi]) and select the material accordingly. Equation (14.53) applies and utilizing Eq. (14.49) for *through-hardened* steel (grade 1 or 2), one gets the Brinell hardness HB number as

$$\begin{aligned} \text{HB} &= 1.876S_t - 165.7 = 1.876(300) - 165.7 = 397.1 \quad \text{grade 1, SI} \\ \text{HB} &= 0.0129S_t - 165.6 = 0.0129(43\,500) - 165.6 = 395.6 \quad \text{grade 1, US} \end{aligned} \tag{g}$$

The AISI 4340 *normalized* has a Brinell hardness number HB = 363. Therefore for grade 1, *through hardening* and a suitable *tempering* temperature is needed to obtain a little higher Brinell hardness number of about 400.

Again and if necessary, change the material or iterations of few design parameters such as face width to safeguard against bending fatigue, if need be. Iterations can be easily accomplished by utilizing the available ***Spur Gear Synthesis Tablet*** through ***Wiley website***. It should be noted, however, that the foregoing design procedure has been very close to the requirements.

14.4.3.3 Surface Fatigue

When two solid elastic bodies meet in contact, Hertz (1857–1894) theory applies. Hertz (1881) developed the geometry and the pressure distribution at the contact area; see Section 6.4.3. The applicable part herein is concerned with the mating of gear teeth as the contact area. At the spur gear contact region, it is assumed that *two cylindrical bodies* with different radii of curvature represent the case. The interest is to define the area of contact and maximum pressure developed between the two teeth. For cylindrical contact, the area is rectangular along the mating gears for the width f_g and the small breadth b_g along the contact centerline. The maximum pressure p_{max} (or stress) at the contact length due to the normal gear force $\boldsymbol{F}_n$ is given by (see Section 6.4.3)

$$p_{max} = \frac{2F_n}{\pi b_g f_g} \tag{14.54}$$

The *breadth of contact area* b_g is defined by

$$b_g = \sqrt{\frac{4\boldsymbol{F}_n(r_{cP}r_{cG})}{\pi f_g(r_{cP}+r_{cG})}\left(\frac{1-\nu_P^2}{E_P}+\frac{1-\nu_G^2}{E_G}\right)} \tag{14.55}$$

The symbols ν_P and ν_G are the Poisson's ratios for pinion and gear materials, respectively. E_P is the *modulus of elasticity* of pinion material, and the E_G is the *modulus of elasticity* of gear material. The terms r_{cP} and r_{cG} are the instantaneous *curvature radii* of pinion tooth and gear tooth at the centerline of contact. These radii are obtained from the *involute teeth* surfaces along the contact centerline at the *pitch point*, which gives (see Figure 14.4b)

$$r_{cP} = \frac{d_{bP}\sin\phi}{2} \quad \text{and} \quad r_{cG} = \frac{d_{bG}\sin\phi}{2} \tag{14.56}$$

The two terms of d_{bP} and d_{bG} are the *base circle diameters* for the pinion and the gear, respectively. The angle ϕ is the pressure angle.

The maximum pressure p_{max} is the maximum *compressive contact stress* σ_c at the centerline of the contact surface, which is also named *Hertzian stress*. The *Hertzian contact stress* σ_c is then given by Eq. (14.54) with the substitution of Eqs. (14.55) and (14.56) to obtain the following:

$$\sigma_c = -\frac{2\boldsymbol{F}_n}{\pi f_g}\sqrt{\frac{\pi f_g(r_{cP}+r_{cG})}{4\boldsymbol{F}_n(r_{cP}r_{cG})\left(\frac{1-\nu_P^2}{E_P}+\frac{1-\nu_G^2}{E_G}\right)}} \tag{14.57}$$

The negative sign in Eq. (14.57) indicates that the *Hertzian stress* is compressive. After some manipulations and substituting from Eq. (14.56), Eq. (14.57) becomes

$$\sigma_c = C_p\sqrt{\frac{\boldsymbol{F}_n}{f_g d_P I_g}} \tag{14.58}$$

The symbol C_p in Eq. (14.58) is the *elastic coefficient* given by

$$C_p = \sqrt{\frac{1}{\pi\left(\frac{1-\nu_P^2}{E_P}+\frac{1-\nu_G^2}{E_G}\right)}} \tag{\textbf{14.59}}$$

Table 14.8 displays the calculated values of the *elastic coefficients* C_p $[\text{MPa}]^{1/2}$ $([\text{psi}]^{1/2})$ for different mating gear materials. The symbol I_g in Eq. (14.58) is the *contact geometry factor* given by

$$I_g = \frac{\cos\phi\sin\phi\, n_g}{2(n_g\pm 1)} \quad \begin{cases} + \text{ external gear} \\ - \text{ internal gear} \end{cases} \tag{14.60}$$

Note that n_g in Eq. (14.60) is the *gear ratio* (N_G/N_P) as defined by Eqs. (14.2) and (14.4). Equations (14.55) and (14.57) are for external contact. When the contact is internal, the gear sign changes to negative, and that is why the negative sign in Eq. (14.60) appears for internal gear contacting a pinion. It is also important to note that the derivation of I_g has been performed when the contact is at the *pitch point* between the two gears (Figure 14.4b). At other contact points, the radii of curvatures are different with one smaller at the root of the involute and the other is larger at the tip of the tooth. That is why a correction multiplier K_{IN} is developed and the calculated value of I_g is multiplied by the correction factor K_{IN} to account for contact with different number of teeth (Figure 14.18). The adjusted *contact geometry factor* I_P is then

$$I_P = K_{IN}\frac{\cos\phi\sin\phi\, n_g}{2(n_g\pm 1)} \quad \begin{cases} + \text{ external gear} \\ - \text{ internal gear} \end{cases} \tag{\textbf{14.61}}$$

Table 14.8 Elastic coefficients for different mating gear materials, C_p [MPa]$^{1/2}$ ([psi]$^{1/2}$).

Pinion material elasticity modules E_p [Mpa] ([psi])		Gear material					
		Steel	Mall. iron	Nod. iron	Cast iron	Al. bronze	Tin bronze
Steel	$2\times10^5(30\times10^6)$	191(2300)	181(2300)	179(2160)	174(2100)	162(1950)	158(1900)
Malleable iron	$1.7\times10^5(25\times10^6)$	181(2180)	174(2090)	172(2070)	168(2020)	158(1900)	154(1850)
Nodular iron	$1.7\times10^5(24\times10^6)$	179(2160)	172(2070)	170(2050)	166(2000)	156(1880)	152(1830)
Cast iron	$1.5\times10^5(22\times10^6)$	174(2100)	168(2020)	166(2000)	163(1960)	154(1850)	149(1800)
Aluminum bronze	$1.2\times10^5(25\times10^6)$	162(1950)	158(1900)	156(1880)	154(1850)	145(1750)	141(1700)
Tin bronze	$1.1\times10^5(25\times10^6)$	158(1900)	154(1850)	152(1830)	149(1800)	141(1700)	137(1550)

Poisson's ratio $\nu = 0.3$, Mall. is malleable, Nod. is nodular, and Al. is aluminum.

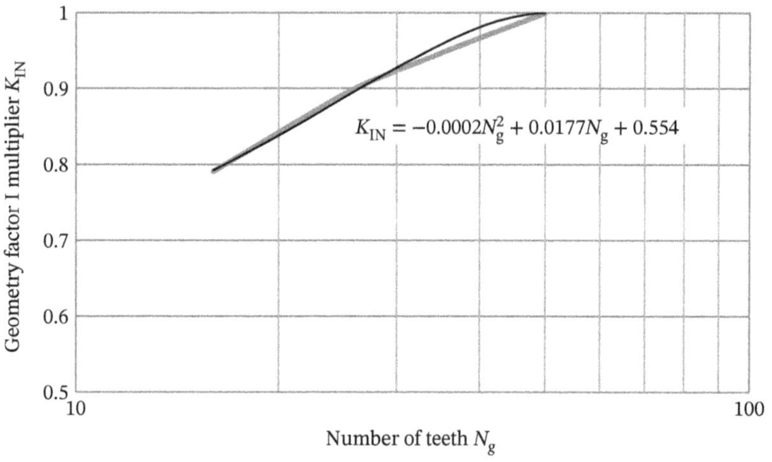

Figure 14.18 The correction factor K_{IN} to be multiplied by geometry factor I_g to account for contact with different number of teeth, for $\phi = 20°$ and standard center distance.

where

$$K_{IN} = -0.0002N_P^2 + 0.0177N_P + 0.554 \tag{14.62}$$

The multiplier factor K_{IN} is calculated as a correction for the smaller number of teeth, which is usually the pinion. This correction is for $\phi = 20°$ and standard center distance. It is confirmed by the different number of teeth for the lowest point of single tooth contact on the pinion (AGMA 1982, 1999).

Similar to bending fatigue, the inclusion of the dynamic or *velocity factor* K_v is necessary to account for the effect of gear *quality* and *pitch line velocity* on the dynamics affecting the contact stress. The same K_v factor defined by Eq. (14.39) is used herein to include AGMA quality number Q_v in the calculations. This would also account for the stresses causing fatigue due to the dynamic loading action. The evaluated stress is thus dubbed the *contact fatigue stress* σ_{ac}, which is given by

$$\sigma_{ac} = C_p\sqrt{\frac{F_nK_v}{f_gd_pI_g}} \tag{14.63}$$

As for the bending fatigue case, the relation in (14.63) does not account for the service loading conditions such as the steadiness of power source (electric motor, combustion engines of multi or single cylinders) or the type of driven machines. Table 14.7 suggests a *load application factor* k_o to account for different classes of power sources and driven machines, which has previously been used for the bending fatigue case. This factor should be included in the evaluation of the applied contact stress since it accounts for the dynamic changes of the applied loads. Another factor that should also be included is the *load distribution factor* k_m as was done for the bending case. This factor accounts for the inevitable variation of the way the load is distributed over the face of the teeth. Equation (14.63) assumes perfect uniform distribution over the whole face width. Due to deflections of the shafts on which gears are mounted and teeth deflections accordingly, the load distribution is not expected to be uniform or centered over the teeth. That nonuniform distribution is clearly a function of the gear face width f_g. The *load distribution factor* k_m should then be included in the evaluation of the applied contact stress since some portion of the tooth will have higher contact loads than the rest of the tooth. A previously suggested fitted relation (Eq. (14.46)) was used in bending fatigue for the same reason. It can also be used for contact fatigue stress. Therefore, including the *load application factor* k_o of Table 14.7 and the *load distribution factor* k_m as given by Eq. (14.46) into Eq. (14.63), the *maximum contact fatigue stress* σ_{ac} becomes

$$\sigma_{ac}\big|_{max} = C_P \sqrt{\frac{k_o k_m \boldsymbol{F}_n K_v}{f_g d_P I_g}} \tag{14.64}$$

It should be noted that the two factors k_o and k_m are affecting the applied normal force $\boldsymbol{F}_n$, and that is why they have been included under the radical. Usually for the SI system, the pinion diameter d_P is substituted in [mm], and the tooth face width f_g is also substituted in [mm] to calculate the stresses in [MPa]. No need to do similar shortcuts for the US system.

Contact Fatigue Strength

To compare the *contact fatigue stress* σ_{ac} with the *contact endurance strength* of the gear S_{ecg}, the factors affecting the contact endurance limit S_{ec} are to be considered in addition to a *material safety factor* K_{MSF}. The *contact endurance strength* S_{ec}, however, is not readily available for regular materials, and it may be evaluated from failure mechanics. Contact fatigue for most cases starts below the surface undergoing *Hertzian* stresses. The maximum *Hertzian* shear stress is about 0.32 the normal compressive stress, and it is located at a distance of about 0.5 the breadth of the contact area b_g below the surface; see Eq. (14.55) and Section 6.4.3. A crack is usually initiated below the surface at the maximum shear stress location and then propagates to the surface forming a chip. The chip size that is bound by the propagated crack is dislodged, and thus pitting occurs. This is one of the main mechanisms of contact fatigue.

Since the stresses are compressive, the *contact fatigue endurance limit* S_{ec} has been empirically correlated to the Brinell hardness number HB implying the following relation for steels at 10^8 cycles:

$$\begin{aligned} S_{ec} &= 2.76\text{HB} - 70\ [\text{MPa}], \quad \text{HB for softer surface} \\ S_{ec} &= 0.4\text{HB} - 10\ [\text{kpsi}], \quad \text{HB for softer surface} \\ S_{ec} &= 0.95(S_{ec})_{Steel}, \quad \text{for Nodular steel} \end{aligned} \tag{14.65}$$

These relations have been in common use for some time. It applies only to steels, and other materials may have similar empirical relations that can be found in the literature. One may alternatively use the previous *allowable contact strength* relations that have some embedded safety.

$$S_{ec} = 379, 482, 551\ [\text{MPa}] \quad \text{Cast Iron grade } 20, 30, 40$$

$$S_{ec} = 55, 70, 80\ [\text{kpsi}] \quad \text{Cast Iron grade } 20, 30, 40$$

$$S_{ec} = 207\ [\text{MPa}],\ 30\ [\text{kpsi}]\ \ \text{Tin Bronze}(11\%\ \ \text{tin})$$

$$S_{ec} = 448\ [\text{MPa}],\ 65\ [\text{kpsi}]\ \ \text{Aluminum bronze (ASTM B-148-52)} \tag{14.65'}$$

The *contact fatigue* has been the concern of AGMA. Similar relation to those in Eq. (14.65) has been used by AGMA for steels at 10^7 cycles and 99% confidence. This relation has the following form:

$$S_{ec} = 0.327\text{HB} + 26\ [\text{kpsi}] \tag{14.66}$$

In this text, we adapt the ratio between allowable bending and contact fatigue strength or ***material property factor*** B_{BC} as 0.3. This value is about the average property ratio for steel, ductile or nodular iron, and for bronze as shown in Table 14.6. This may provisionally solve the uncertainty of the contact fatigue strength value for the adapted procedure herein. For a specific material, however, one should either investigate the correct value or adapt values available in codes and standards such as AGMA. The adapted value of B_{BC} can be an initial synthesis that can be adjusted in the available ***Spur Gear Synthesis Tablet*** in ***Wiley website***; see Figure 14.17a,b for sample images of different cases covered herein. In fact, one can use any other value for allowable bending and contact fatigue strengths in the ***Input*** and ***Analysis*** column as shown in Figure 14.17. The *synthesis* solution, however, is more accurate for allowable bending fatigue strength of 300 [MPa] or 43.5 [kpsi] and strength ratio or *material property factor* $B_{BC} = 0.3$. The adapted *allowable contact fatigue strength* S_{ac} is then defined herein as follows:

$$S_{ac} = \frac{S_{ab}}{B_{BC}} \tag{14.67}$$

If the fatigue is not known, this relation may be useful as an initial attempt to synthesize the gear. Consult Table 14.6 for the ranges of *allowable fatigue strengths* of different materials.

Similar to bending case, factors that affect the contact fatigue strength S_{ec} should also be included. These are related to the reliability of the material contact fatigue endurance value and the life cycle of the gear set different from 10^6 to 10^8 cycles. As for the bending case, the *reliability factor* k_c defined by Eq. (14.43) may be use. As for the *life factor*, it is necessary to find the relation between the fatigue strength and the contact endurance limit for different life cycles. Conventional relations of the *life factor* or the factor similar to the contact *AGMA life factor* may be used. For contact fatigue, however, the relation at higher cycles than the 10^7 endurance is the critical one and can be assumed conservatively as ANSI/AGMA (2004) for steel. This *life factor* Z_N may then take the following relation:

$$Z_N = 2.466 N_L^{-0.056} \geq 10^7\ \text{ life cycles} \tag{14.68}$$

If the number of *life cycles* N_L is less than 10^7, this factor will be more than 1.0. The value of Z_N will then depend on the material type, and it is recommended to consult AGMA standard for this value (ANSI/AGMA 2004). The *contact fatigue strength* S_{cf} may then take the following form:

$$S_{cf} = k_c Z_N S_{ec} \tag{14.69}$$

Other applicable factors affecting contact fatigue similar to those for bending fatigue such as *surface finish factor* k_a, *the temperature factor* k_d, and a *material safety factor* K_{MSF} may also be included to get the *allowable contact fatigue strength* S_{ac}.

Contact Safety Factor

To get the contact safety factor, one should include the *load application factor* k_o and the *load distribution factor* k_m into the stress as defined by Eq. (14.64). Considering the *reliability factor* k_c (Eq. (14.43)), the *material safety factor* K_{MSF}, and the *allowable contact strength* of S_{ac} (Eq. (14.67)), the ***contact safety factor*** $K_{SF}|_c$ is defined as follows:

$$K_{SF}|_c = \frac{k_c Z_N (S_{ec}/K_{MSF})}{\sigma_{ac}|_{max}} = \frac{k_c Z_N S_{ac}}{\sigma_{ac}|_{max}} \tag{14.70}$$

If the *allowable contact fatigue strength* S_{ac} is used in place of the *contact fatigue endurance limit* S_{ec} or the *contact fatigue strength* S_{cf}, and the reliability is 99% ($k_c = 1.0$), the ***contact safety factor*** $K_{SF}|_c$ should be about 1.0 and $Z_N = 1.0$ if S_{cf} is used. It should be noted that the load distribution factor k_m is included in the optimum data used in the ***Gear Synthesis Tablet***. The load application factor k_o should, however, be incorporated into the input power to get the ***maximum contact fatigue stress*** $\sigma_{ac}|_{max}$. More on the capabilities and utilities of the ***Spur Gear Synthesis Tablet*** is covered in the *CAD* and optimization in Section 14.6.

The images of the ***Spur Gear Synthesis Tablet*** are shown in Figure 14.17. It presents the solution of Examples 14.4 and 14.5 in the ***Default*** column as indicated previously. It also gives the synthesized gear using an allowable bending fatigue strength of 200 [MPa] in the ***Input*** column. The ***Analysis*** column presents the *conventional* or *classical* solution of Example 14.5. Figure 14.17a presents the adjusted solution for the synthesized gear according to bending fatigue strength of 200 [MPa]. The adjustment is performed to account for the standard module and rounding of some parameters such as the face width. The adjustment tries also to have equal safety factor of about 1.0 for both bending and contact. The calculated geometry values in Example 14.5 are very close to the values shown in the ***Synthesis Tablet***. All values, however, are calculated according to the ***detailed design*** procedure given herein including ***contact fatigue.*** More *geometry* and *performance parameters* and variables are shown in Figure 14.19a,b for the two cases shown in Figure 14.17a,b. Some of these variables are given as a check, and they would necessitate the input of the *ultimate tensile strength* S_{ut} and the *number of life cycles* N_L beyond the usual endurance limit. The calculated values are, however, concerned with the check of the normal reversed cycle fatigue strength, and not the contact fatigue strength.

AGMA Contact Fatigue Number

As for the AGMA bending stress number, extensive work has also been done by AGMA in developing *allowable contact fatigue numbers* S_{ac}, which are for most of the available gear materials shown in Table 14.6. The main difference in the treatment of AGMA is their utility of ***allowable contact stress number*** S_c, which is not necessarily coincident with the treatment of the ***contact fatigue endurance limit*** S_{ac} or the ***contact fatigue strength*** S_{cf} previously discussed in this section. These numbers may be considered literally as the *allowable contact* ***fatigue*** *strength*, and they have been quoted for most of the materials in Table 14.6 in ranges rather than for specific materials. Similar to bending case, the selection of a specific material is therefore dependant on its hardness number; see Table 14.6. This is due to the probability that there are no factors directly associated with it. The AGMA standards also specify relations between the *hardness* and the *allowable contact stress number* (ANSI/AGMA 2001-D04). These relations are also available in the form of graphs that can be viewed in these standards if need be. Herein, it is recommended to find the suitable *allowable contact stress number* (***allowable contact strength***) and then specify or verify the suitably selected material or the heat treatment that provides the ***hardness*** to fit. This would reduce the iterations for the selection of a specific material standard. The available relations of the ***Brinell Hardness HB*** and ***allowable contact stress number*** for through-hardened steels are as follows:

$$\begin{aligned} &\text{HB} = 0.4505S_c - 90.09, \quad \text{or} \quad S_c = 2.22\text{HB} + 200\ [\text{MPa}], \quad \text{grade 1} \\ &\text{HB} = 0.003\,11S_c - 90.37, \quad \text{or} \quad S_c = 322\text{HB} + 29\,100\ [\text{psi}], \quad \text{grade 1} \\ &\text{HB} = 0.4149S_c - 93.34, \quad \text{or} \quad S_c = 2.41\text{HB} + 237\ [\text{MPa}], \quad \text{grade 2} \\ &\text{HB} = 0.002\,87S_c - 98.28, \quad \text{or} \quad S_c = 349\text{HB} + 34\,300\ [\text{psi}], \quad \text{grade 2} \end{aligned} \tag{14.71}$$

For other materials, it is advisable to consult the current AGMA standards for the allowable contact stress number. If that is not available, one may use previous values such as the ones defined in Eq. (14.65′) for the allowable contact stress numbers.

The AGMA factors affecting the allowable contact stress number S_c are mainly the reliability factor K_R and the life cycle factor Z_N. Other factors are either not defined or their effect is only present in defined conditions such as the hardness ratio factor, which is not effective unless the hardness of the pinion is 1.2 times or more than the

(a)

Spur gear synthesis – SI units			Page 2
Output geometry			
Addendum [mm]	3.232	3.232	8.000
Dedendum [mm]	4.040	4.040	10.000
Working depth [mm]	6.463	6.463	16.000
Whole depth [mm]	7.271	7.271	18.000
Tooth thickness [mm]	5.076	5.076	12.566
Fillet radius of dasic rack [mm]	0.969	0.969	2.400
Width of top land (min.) [mm]	0.808	0.808	2.000
Clearance (min.) [mm]	0.808	0.808	2.000
Clearance, (shaved or ground) [mm]	1.131	1.131	2.800
Diametral pitch (US), p_c [teeth/in]	2.502	2.502	1.011
Pinion outside diameter, d_{oP} [mm]	123.750	190.667	160.000
Pinion root diameter, d_{rP} [mm]	109.207	176.125	124.000
Pinion base circle diameter, d_{bP} [mm]	110.213	173.095	135.316
Gear outside diameter, d_{oG} [mm]	479.899	751.357	612.000
Gear root diameter, d_{rG} [mm]	455.204	726.662	550.867
Gear base circle diameter, d_{bG} [mm]	437.292	692.380	541.263
Other output			
Normal force, F_n [N]	10 830.47	6896.00	8821.31
Gear torque, T_G [N.m]	150.42	149.21	149.21
Speed limit for a quality, $v_{t,\max}$ [m/s]	28.669	28.669	28.669
Reliability factor , K_R	1.000	1.000	1.000
Load distribution factor, k_m	1.607	1.607	1.650
Load application factor, k_o	1.000	1.000	1.222
Other values			
Ultimate tensile strength, S_{ut} [MPa]	1279	1279	1279
Number of life cycles, $L_N \times 10^6$	10.000	10.000	10.000
Expected fatigue strength, S_f [MPa]	379.33	379.33	352.63
Bending life cycle factor, Y_N	1.000	1.000	1.000
Contact life cycle factor, Z_N	1.000	1.000	1.000
Surface finish factor, k_a	0.639	0.639	0.639
Size factor, k_b	0.929	0.929	0.863
Temperature factor , k_d	1.000	1.000	1.000

Notes: *Examples* 14.1, 14.3–14.6 *of the textbook. The Analysis column is used to enter the preferred module instead of the synthesized one and round design variables.*

Figure 14.19 More geometry and performance parameters and variables are shown in page 2 of (a) the SI Tablet and (b) the US Tablet. Some are given as a check. One may input the ultimate tensile strength S_{ut} and the number of life cycles N_L beyond the usual endurance limit.

(b)

Spur gear synthesis – US units			Page 2
Output geometry			
Addendum [in]	0.127	0.1272	0.3333
Dedendum [in]	0.159	0.1590	0.4167
Working depth [in]	0.254	0.2544	0.6667
Whole depth [in]	0.286	0.2862	0.7500
Tooth thickness [in]	0.200	0.1998	0.5236
Fillet radius of basic rack [in]	0.038	0.0382	0.1000
Width of top land (min.), [in]	0.032	0.0318	0.0833
Clearance (min.) [in]	0.032	0.0318	0.0833
Clearance, (shaved or ground), [in]	0.045	0.0445	0.1167
Module (SI), m_n [mm]	3.231	3.2307	8.4667
Pinion outside diameter, d_{oP} [in]	4.833	7.632	6.667
Pinion root diameter, d_{rP} [in]	4.261	7.059	5.167
Pinion base circle diameter, d_{bP} [in]	4.303	6.932	5.638
Gear outside diameter, d_{oG} [in]	18.888	30.081	25.500
Gear root diameter, d_{rG} [in]	17.916	29.109	22.953
Gear base circle diameter, d_{bG} [in]	17.211	27.729	22.553
Other output			
Normal force, F_n [lb]	2453.47	1522.84	1872.38
Gear torque, T_G [lb.in]	1319.59	1319.59	1319.59
Speed limit for a quality, $v_{t,\max}$ [ft/min]	5733.9	5733.85	5733.853
Reliability factor, K_R	1.000	1.000	1.000
Load distribution factor, k_m	1.663	1.663	1.731
Load application factor, k_o	1.000	1.000	1.245
Other values			
Ultimate tensile strength, S_{ut} [kpsi]	185.50	185.5	185.5
Number of life cycles, $L_N \times 10^6$	10.000	10.000	10.000
Expected fatigue strength, S_f [kpsi]	54.95	54.95	50.84
Bending life cycle factor, Y_N	1.000	1.000	1.000
Contact life cycle factor, Z_N	1.000	1.000	1.000
Surface finish factor, k_a	0.638	0.638	0.638
Size factor, k_b	0.928	0.928	0.859
Temperature factor , k_d	1.000	1.000	1.000

Notes: *Examples 14.1, 14.3–14.6 of the textbook. The Analysis column is used to enter the preferred module instead of the synthesized one and round design variables.*

Figure 14.19 (*Continued*)

hardness of the gear. Since that may not be the case all the time, the factor is assumed as 1.0. For these special cases, one should consult the AGMA standards (ANSI/AGMA 2001-D04).

The reliability factor K_R is similar to the factor k_c given by Eq. (14.43) and thus can be used to represent K_R. Therefore, on can also state that

$$K_R = 1/k_c = 0.6735\,(1 - R_e) - 0.088 \tag{14.34'}$$

The life cycle factor Z_N accounts for stressing the gear to a higher or a lower number of cycles than the assumed endurance of 10^7 cycles. The stressing at higher cycles than the 10^7 endurance is the critical one and can be assumed conservatively as AGMA suggests (Eq. (14.68)):

$$Z_N = 2.466 N_L^{-0.056} \geq 10^7 \text{ life cycles} \tag{14.68}$$

If the number of life cycles is less than 10^7, this factor will be more than 1.0. The value of Z_N will then depend on the material type, and it is recommended to consult AGMA standard for this value (ANSI/AGMA 2001-D04).

The *contact safety factor* $K_{SF}|_c$ previously defined as $K_{SF}|_c$ by Eq. (14.70) will be adjusted to the AGMA *contact stress number* S_c rather than the *contact fatigue endurance limit* S_{ec} so that

$$K_{SF}\big|_c = \frac{Z_N S_c / K_R}{\sigma_{ac}\big|_{max}} \tag{14.72}$$

This relation is very similar to Eq. (14.70), where S_c is the AGMA *allowable contact stress number*, which may not necessarily equal to the *allowable contact fatigue strength* S_{ac} used herein. They should, however, be very closely equal. Note also that in Eq. (14.72), the reliability factor K_R is used in place of the reliability factor k_c since they are reciprocal; see Eq. (14.34′). The safety factor in Eq. (14.72) should be close to 1.0 since we are using the allowable contact stress number, which has embedded factor of safety. The calculated safety factor of Eq. (14.72) may then be considered as an *extra safety factor* K_{SFE} for fear of missing any factor or misrepresenting some of the previous factors. In other words, Eq. (14.72) should be such that

$$Z_N S_c / K_R \geq \sigma_{ac}\big|_{max} \tag{14.73}$$

Equation (14.73) suggests that the *extra safety factor* can be 1.0 or little more than 1.0. The right-hand term of the equation represents the *expected maximum loading* of the gear, and the left-hand term represents the *expected allowable strength* of the material.

Example 14.7 For the problem in Example 14.6, again the reduction gear set in the crane is driven by a maximum power of 75 [kW] or 100.58 [hp] electric motor running at a rotational speed of 1200 [rpm]. The same gear set is to be designed for a gear ratio of 4 : 1. Gears are also of 20° full depth. Present a detailed design to find the contact fatigue safety for the previously synthesized gear set.

Solution

Data: (same as Examples 14.4 and 14.6) Power transmitted $H_{kW} = 75$ [kW] or 100.58 [hp], rotational speed $N_{in} = 1200$ [rpm], and the gear ratio $n_g = 4$. The rotation speed $\omega_P = (2\pi N_{in}/60) = 125.66$ [rad/s].

From Examples 14.4 and 14.5, the synthesized geometry provides $m_n = 3.25$ [mm], $N_P = 36$, and $f_g = 59$ [mm] for SI system or $p_d = 8$ [teeth/in], $N_P = 36$, and $f_g = 2.315$ [in] for the US system. The pinion diameter $d_P = N_P m_n = 117(10)^{-3}$ [m] or $d_P = N_P/p_d = 4.5$ [in] and the pitch line velocity $\boldsymbol{v}_t = \omega_P(d_P/2) = 7.35$ [m/s] or 282.74 [in/s] = 1413.7 [ft/min]. The tangential force is $\boldsymbol{F}_t = H_{kW}(1000)/\boldsymbol{v}_t = 10.2041$ [kN] or $= H_{hp}(6600)/\boldsymbol{v}_t =$ 100.58(6600)/282.74 = 2347.8 [lb]. The gear quality number $Q_v = 8$, and the suggested material has allowable bending fatigue strength S_{ab} of 300 [MPa] or 43.5 [kpsi]. The dynamic or velocity factor K_v, load distribution factor k_m, and load application factor k_o have been evaluated in Example 14.6 as $K_v = 1.3138$, $k_m = 1.6075$, and $k_o = 1.0$ for SI system and $K_v = 1.3039$, $k_m = 1.6039$, and $k_o = 1.0$ for the US system. Also assume a 99% reliability (i.e. $k_c = 1.0$), and no excess temperature (i.e. $k_d = 1.0$), since no additional specification has been required.

From the *bending fatigue* results of Example 14.5, a more suitable and widely used material was the through-hardening steel *AISI 4340*. It also has defined endurance limits of 489–668 [MPa] or 71–97 [kpsi] (Stulen et al. 1961). The AISI 4340 *normalized* has a yield strength $S_y = 861.8$ [MPa] = 125.0 [kpsi], Brinell hardness HB = 363, and ultimate tensile strength $S_{ut} = 1279.0$ [MPa] = 185.5 [kpsi]. These values can change when a through hardening and different *tempering* temperature occurs. The surface hardness can also change if case hardening is performed if need be.

Assuming the gear material is also steel, the maximum *Hertzian* contact fatigue stress defined by Eq. (14.64) needs the evaluation of several factors. The *elastic coefficients* C_p for steel pinion in contact with steel gear is calculated according to Eq. (14.58) and the value in Table 14.8 as 191 [MPa]$^{1/2}$ or 2300 [psi]$^{1/2}$. The *contact geometry factor* I_g is obtained from Eqs. (14.61) and (14.62) such that

$$\begin{aligned} I_P &= (-0.0002N_P^2 + 0.0177N_P + 0.554)\left(\frac{\cos\phi\,\sin\phi\, n_g}{2(n_g \pm 1)}\right) \\ &= (-0.0002(36)^2 + 0.0177(36) + 0.554)\left(\frac{\cos 20°\,\sin 20°(4)}{2(4+1)}\right) \\ &= (0.932)\left(\frac{0.9397(0.3420)(4)}{10}\right) = 0.119\,81 \end{aligned} \tag{a}$$

The maximum *Hertzian* contact fatigue stress defined by Eq. (14.64) is then

$$\begin{aligned} \sigma_{ac}\big|_{max} &= C_p\sqrt{\frac{k_o k_m \boldsymbol{F}_n K_v}{f_g d_P I_g}} = 191\sqrt{\frac{(1.0)(1.6075)(10.2041)(1000)(1.3138)}{(59)(117)(0.119\,81)}} = 974.98\text{ [MPa]} \\ \sigma_{ac}\big|_{max} &= C_p\sqrt{\frac{k_o k_m \boldsymbol{F}_n K_v}{f_g d_P I_g}} = 2300\sqrt{\frac{(1.0)(1.6039)(2347.8)(1.3039)}{(2.315)(4.5)(0.119\,81)}} = 144\,268\text{ [psi]} = 144.268\text{ [kpsi]} \end{aligned} \tag{b}$$

This value of the pinion *contact fatigue stress* σ_{ac} was expected to be 1000 [MPa] or 145 [kpsi], respectively. It is less than expected by about 2.5% and 0.5%, respectively. This is due to the rounding of geometry values and the inexactness of fitting equations as can be seen in the ***Default*** column of Figure 14.17a,b where $m_n = 3.23$ [mm] and $f_P = 58.8$ [mm] for SI system or $p_d = 8$ [teeth/in] and $d_P = 4.579$ [in] for the US system. The number of teeth has also been forcefully rounded to integer values. Figure 14.17b presents the adjusted solution for the synthesized gear according to bending fatigue strength of 29 [kpsi]. The adjustment should also be performed to account for the standard diametral pitch and rounding of some parameters such as the number of teeth. The adjustment tries also to have equal safety factor of about 1.0 for both bending and contact.

The AGMA procedure would suggest finding the Brinell hardness HB number from AGMA allowable *contact fatigue number* S_t of 1000 [MPa] or 145 [kpsi] and select the material accordingly. Equation (14.73) applies and utilizing Eq. (14.71) for *through-hardened* steel (grade 1 or 2), one gets the Brinell hardness HB number as

$$\begin{aligned} \text{HB} &= 0.4505\,(1000) - 90.09 = 360.41, \quad \text{SI grade 1} \\ \text{HB} &= 0.003\,11(145\,000) - 90.37 = 360.58, \quad \text{US grade 1} \\ \text{HB} &= 0.4149(1000) - 93.34 = 321.56, \quad \text{SI grade 2} \\ \text{HB} &= 0.002\,87(145\,000) - 98.28 = 317.87, \quad \text{US grade 2} \end{aligned} \tag{c}$$

The AISI 4340 *normalized* has a Brinell hardness number HB = 363. Therefore for grade 1 or 2, no case hardening or a *tempering* temperature is needed to obtain the expected Brinell hardness number for contact fatigue. This might have been unexpected, but with the previous synthesis, the result might be acceptable.

Again and if necessary, change the material or iterations of few design parameters such as face width to safeguard against contact fatigue, if need be. Iterations can be easily accomplished by utilizing the available ***Spur Gear Synthesis Tablet*** through ***Wiley website***. It should be noted, however, that the foregoing design procedure has been very close to the requirements.

14.5 Critical Speed

Gears can be a major source of noise and large vibration amplitudes when the frequency of tooth engagement coincides practically with any of the natural frequencies of the system. The *tooth frequency* ω_T [rad/s] or f_T [Hz] is the product of the rotational frequency of a gear ω_g [rad/s] or f_g [Hz] and the number of teeth on the gear N_g or

$$\omega_T = \omega_P N_P = \omega_G N_G \quad \text{or} \quad f_T = f_P N_P = f_G N_G \tag{14.74}$$

For a forced vibration, we notice the high response magnification at $\omega_T = \omega_n$ [rad/s] or $f_T = f_n$ [Hz], where ω_n or f_n is the *natural frequency* of the gear system. Noise and large amplitude vibration would result in this case. Operating speeds should then totally avoid coincidence with such natural frequencies.

Consider the connection between two gears in contact as the main source of vibrations due to teeth engagements and the imperfection of teeth that generates inherent dynamic force $\boldsymbol{F}_{\text{in}}$ which changes at the tooth frequency rate. This disturbance is a self-excitation force between teeth of the two mating gears. Figure 14.20a shows a schematic diagram of the two gears of masses (m_1, m_2) and represents the teeth contact flexibility as a spring k_3. A representative free-body diagram (***FBD***) and the system graph of a mathematical model for the two masses m_1 and m_2 representing gears in contact through their flexible teeth of stiffness k_3 and the in-between generated input force $\boldsymbol{F}_{\text{in}}$ are shown in Figure 14.20b. The two masses and the connecting spring shown in Figure 14.20c are reduced to the equivalent mass m_e and the spring k_3 as shown in Figure 14.20c. Assuming the connection stiffness as the most in the gear set system at its connection to other components, the mathematical model may be an acceptable subset of the total machine system. If that is not acceptable, more components should be included in the mathematical model of the total machine system. Using the *system graph* model of such systems is systematic, and does not depend much on the number of components in the system; see Section 2.1.2. Only the size of matrices would be larger. The node equations are written for Figure 14.20b with reference to Eq. (2.8) as

$$\sum_{\text{at points of junction}} f_i = 0 \tag{14.75}$$

The term f_i in Eq. (14.75) is the flow variable in each of the components. The node equations are then

$$\begin{bmatrix} 1 & 1 & 1 & \\ & 1 & -1 & -1 \end{bmatrix} \begin{bmatrix} f_1 \\ f_2 \\ f_3 \\ \boldsymbol{F}_{\text{in}} \end{bmatrix} = \boldsymbol{0} \tag{14.76}$$

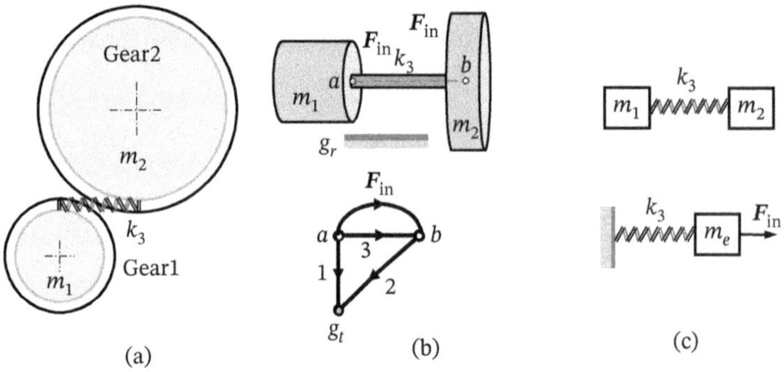

Figure 14.20 A schematic diagram (a) and ***FBD*** and system graph (b) of a mathematical model for two masses m_1 and m_2 representing gears in contact through their flexible teeth of stiffness k_3. The two masses and the connecting spring (c) are reduced to the equivalent mass m_e and the spring k_3 as also in (c).

Partitioning Eq. (14.76) to separate the input driver term, one gets

$$\begin{bmatrix} 1 & & 1 \\ & 1 & -1 \end{bmatrix} \begin{bmatrix} f_1 \\ f_2 \\ f_3 \end{bmatrix} + \begin{bmatrix} -1 \\ 1 \end{bmatrix} \boldsymbol{F}_{\text{in}} = \boldsymbol{0} \tag{14.77}$$

The component equations are given in the differential and integral form as

$$\begin{bmatrix} f_1 \\ f_2 \\ f_3 \end{bmatrix} = \begin{bmatrix} m_1(d/dt) & & \\ & m_2(d/dt) & \\ & & k_3 \int dt \end{bmatrix} \begin{bmatrix} \dot{\delta}_1 \\ \dot{\delta}_2 \\ \dot{\delta}_3 \end{bmatrix} \tag{14.78}$$

The node variable transformation relates the *element* displacements to the *node* displacements such that

$$\begin{bmatrix} \dot{\delta}_1 \\ \dot{\delta}_2 \\ \dot{\delta}_3 \end{bmatrix} = \begin{bmatrix} 1 & \\ & 1 \\ 1 & -1 \end{bmatrix} \begin{bmatrix} \dot{\delta}_a \\ \dot{\delta}_b \end{bmatrix} \tag{14.79}$$

One should then substitute the component equation (14.78) and the node variable transformation of Eq. (14.79) into Eq. (14.77). After matrix multiplications, we get

$$\begin{bmatrix} m_1(d/dt) + k_3 \int dt & -k_3 \int dt \\ -k_3 \int dt & m_2(d/dt) + k_3 \int dt \end{bmatrix} \begin{bmatrix} \dot{\delta}_a \\ \dot{\delta}_b \end{bmatrix} = \begin{bmatrix} -1 \\ 1 \end{bmatrix} \boldsymbol{F}_{\text{in}} \tag{14.80}$$

Multiplying the first equation by $\{m_2/(m_1 + m_2)\}$ and the second equation by $\{m_1/(m_1 + m_2)\}$ and subtracting, one gets

$$\left(\left(\frac{m_1 m_2}{m_1 + m_2} \right) (d/dt) + \left(k_3 \int dt \right) \right) [\dot{\delta}_b - \dot{\delta}_a] = \boldsymbol{F}_{\text{in}} \tag{14.81}$$

This is the same differential equation for the equivalent single degree of freedom models with the equivalent mass $m_e = \{m_1 m_2/(m_1 + m_2)\}$ and the spring k_3 shown in Figure 14.20c. The displacement in the spring k_3 is the differential $(\delta_b - \delta_a)$ with the *equivalent mass* m_e defined obviously as follows:

$$m_e = \left(\frac{m_1 m_2}{m_1 + m_2} \right) \tag{14.82}$$

The *natural frequency* ω_n [rad/s] or f_n [Hz] of this equivalent system is given by

$$\omega_n = \sqrt{\frac{k_3}{m_e}} \text{ [rad/s]} \quad \text{or} \quad f_n = 2\pi \sqrt{\frac{k_3}{m_e}} \text{ [Hz]} \tag{14.83}$$

Evaluating appropriate k_3 and m_e is left as an exercise for interested individuals. Approximate value for the tooth stiffness modeled as a cantilever suggests $k_3 \approx 11\,500 f_g$ [N/mm] for steel gears, where f_g is the face width in [mm]. The masses can be approximated as the gear volume multiplied by the density. It should be noted, however, that the simplified model presented herein is only to develop a rough picture of the problem. The masses m_1 and m_2 should have included the apparent masses of the system on both sides of the gears and the *gear ratio* effect. A more involved model can be derived with the help of the *transformer* given in Section 2.1.2. Also, rotational model is more realistic than translation model, even though both are analogous to each other. The results are therefore only an indicative ballpark estimate. If the tooth frequency ω_T or f_T, nonetheless, coincides with the natural frequency ω_n or $f_n = 2\pi\omega_n$, resonance vibrations would occur. Operating speeds should then totally avoid coincidence with such natural frequency. Also, any other fundamental frequency of the system should be far from the tooth frequency ω_T or f_n.

The *tooth frequency* f_T [Hz] is also calculated in the available ***Spur Gear Synthesis Tablets*** through ***Wiley website***.

Example 14.8 For the problem in Example 14.6, again the reduction gear set in the crane is driven by a maximum power of 75 [kW] or $H_{hp} = 100.5$ [hp] electric motor running at a rotational speed of 1200 [rpm]. The same gear set has a gear ratio of 4 : 1. Gears are also of 20° full depth. Define the frequencies that should be avoided to reduce resonant vibrations. Calculate the tooth frequency for the output gear. Find a rough estimate of the natural frequency of the gear set.

Solution

Data: (same as Examples 14.4 and 14.6) Power transmitted $H_{kW} = 75$ [kW] or $H_{hp} = 100.5$ [hp], rotational speed $N_{in} = 1200$ [rpm], and the gear ratio $n_g = 4$. The rotation speed $\omega_P = (N_{in}/60) = 20$ [Hz] and the angular velocity of the gear $\omega_G = (N_{out}/60) = (N_{in}/60n_g) = 5$ [Hz]. The number of pinion teeth $N_P = 36$ and the number of gear teeth $N_G = 144$.

The tooth frequency is obtained using Eq. (14.74) to get

$$\omega_T = \omega_P N_P = 20(36) = 720 \text{ [Hz]} \tag{a}$$

This frequency should be far from any natural frequency of the system in which the gear set is part of.

The tooth frequency for the output gear is obtained using Eq. (14.74) to get

$$\omega_T = \omega_G N_G = 5(144) = 720 \text{ [Hz]} \tag{b}$$

It is clear that the tooth frequency should be the same.

The rough estimate of the natural frequency of the gear set is obtained by first evaluating the equivalent gear set mass m_e according to Eq. (14.82) and which is shown in Figure 14.20.

$$m_e = \left(\frac{m_1 m_2}{m_1 + m_2}\right) = \rho\left(\frac{V_P V_G}{V_P + V_G}\right) = (7860)\left(\frac{0.001\,68(0.025\,52)}{0.001\,68 + 0.025\,52}\right) = 12.389 \text{ [kg]} \tag{c}$$

The gear volume is calculated as a cylinder with an outer diameter of the gear pitch circle. The rough estimate of the natural frequency of the gear set is then

$$f_n = 2\pi\sqrt{\frac{k_3}{m_e}} \approx 2\pi\sqrt{\frac{11\,500(10^3)(58.8)}{12.389}} \approx 46\,420 \text{ [Hz]} \tag{d}$$

The tooth frequency of this gear set has been calculated as 720 [Hz], which is far lower than the approximate natural frequency of 46 420 [Hz]. This indicates that the design would be suitable for the objective of reducing noise radiation from gears. Variations of the design variables are then not needed to shift the tooth frequency and the natural frequency further apart. The synthesized gears in the *Input* columns of Figure 14.17a,b have the tooth frequency of 1140 [Hz], which is also away from the natural frequency. Design iterations to have different tooth frequency and other natural frequencies are highly recommended with the inclusion of more components to the dynamic system particularly shafts, gearbox casing, and other flexible masses in the total mechanical system. Using the *system graph* model of such a system is systematic, and does not depend on the number of components in the system. Only the size of matrices would be larger, and one needs to get the natural frequencies of the larger system. The tooth frequency may thus activate other lower natural frequencies of the system that is close to it. A better estimation of tooth stiffness may also be necessary. A full *finite element* model may also be needed to perform the full dynamic analysis of the total mechanical system and generates the natural frequencies of the system. This is beyond the scope of this text.

14.6 CAD and Optimization

In Sections 14.4.2 and 14.4.3 and Examples 14.4 and 14.5, several means are available to provide synthesis tools for spur gears such as the optimum synthesis charts in Figures 14.12–14.14. Figures 14.17 and 14.19 are snap shots of the developed ***Spur Gear Synthesis Tablets***. These Tablets are *Computer-Aided Synthesis* (CAS) rather than *CAD* tools. Usually available *CAD* tools may generate designs by repeated analysis. The ***Tablets*** are accessible through **Wiley website** for CAS of spur gears. They provide direct synthesis without any appreciable need for repeated analysis. The iterative synthesis, however, is available for fine tuning of required standards or satisfying other constraints.

In the ***Spur Gear Synthesis Tablets***, the ***Default*** column is an example given as a reference and check. This column is not to be changed. No red (or small grayish) font values *without* aqua color (or light grayish) background are changeable. These are due to predetermined and embedded mathematical relations such that they should not be accessible for input of any explicit variable. Any input to these cells will wipe out the solution procedure. Only cells with *aqua* (or light *grayish*) background are allowable for inputs. The synthesis ***Input*** column is to insert your case values only in place of the *blue-colored* values (or grayish *larger font*) and *aqua* or *grayish* background (as in Figures 14.17 and 14.19). All inputs are only allowable in cells with the blue-colored (or grayish large) font and aqua (or light grayish) background. The red-colored (or grayish smaller) font values and no colored background are the synthesized design parameters calculated as synthesized suggestions. The ***Analysis*** column is for design iterations to insert your case values only in place of the *blue-colored* values (or grayish *larger font*) and *aqua* (or light *grayish*) background (as in Figures 14.17 and 14.19). The iteration is necessary only to account for existing standard modules or diametral pitch and other constraints on space geometry such as smaller face width. The red-colored values (or grayish smaller font) with no colored background are also other design parameters calculated as behavioral outputs that are only changeable by entering the *blue colored* values (or grayish *larger font*) and *aqua* (or light *grayish*) background. The ***Analysis*** column is suitable for quick and effective iterations in combination with the ***Input*** synthesis column. Change of parameters to accommodate specific strength values or necessary safety factor should need some iteration. The *blue-colored* values (or grayish *larger font*) and *aqua* (or light *grayish*) background in either ***Input*** or ***Analysis*** column are interactively exploitable to accommodate any other design parameter values.

The available ***Spur Gear Synthesis Tablet*** assists in designing spur gears using the procedure given herein particularly for *steel* gears. Any other material such as cast iron, aluminum, or bronze of known properties of the ultimate strength, the allowable bending fatigue strength and the allowable contact fatigue strength may be synthesized if the material property ratio B_{BC} is around 0.3. If the B_{BC} is not around 0.3, the synthesis can only be approximate, and iteration is necessary to find a better design. The detailed design calculations consider bending fatigue stress and contact fatigue stress. The results are shown and should be close to the allowable fatigue strengths in bending and contact. Pinion, gear, and total volume of the gear set are also calculated as solid cylinders each with pitch circle diameter as the outer cylinder diameter and the face width as the cylinder length. This provides a mean of optimum decision making by comparing results in the fine tuning of design.

The ***Spur Gear Synthesis Tablet*** requires the input of the specific design case either for SI or the US system. The input includes the power H_{kW} [kW]$_{SI}$ or H_{hp} [hp]$_{US}$, the input rotational speed of the pinion $N_{rpm,P}$ in [rpm], the required output rotational speed of the gear $N_{rpm,G}$ in [rpm], the safety factor K_{SF} as a service or load application factor k_o, the gear quality Q_v, the operating direction such as two-way or one-way rotational direction (2 or 1), and the intended reliability R_e if over 99%. The gear ratio n_G, however, is calculated, and if any of the angular velocities of the pinion or gear is not known, one should calculate it from the required gear ratio beforehand. The intended gear ratio is dependent on the integer number of teeth for the pinion and the gear, where each one is rounded to the closest integer during calculations. The adapted material to be used is initially having allowable bending fatigue strength of 300 [MPa] or 43.5 [kpsi] and a material property ratio B_{BC} of 0.3.

The design is made in two stages. The first stage is the *synthesis* estimate of *optimum* design, which includes the *module* m_n, or the diametral pitch p_d, the *number of teeth* for both pinion and gear N_P and N_G, and the *face width* f_g. Major geometry and performance parameters are calculated accordingly as depicted in Figure 14.17. These parameters are obtained using the equations presented herein. More *geometry* and *performance parameters* and variables are shown in Figure 14.19a,b. Some of these variables are given as a check, and they would necessitate the input of the *ultimate tensile strength* S_{ut} and the *number of life cycles* L_N beyond the usual endurance limit. Material designation is not provided since the allowable fatigue strength is used instead. The selection of a specific material designation is dependent on the available standards and the available manufacturing and treatment capabilities.

After the design synthesis is obtained in the first stage by introducing the case into the ***Input*** column, the user starts the second design stage. This entails fine tuning iterations to be performed into the ***Analysis*** column. The tuning may necessitate the change of some design parameters particularly *standard module* or *standard diametral pitch* and rounded face width, minor adjustment of number of teeth, and may also include material strength refinement. It is regularly recommended to have the number of teeth at odd so that no tooth would meet another tooth intentionally more times than others. Changing parameters in the *Input* column, the output values appear immediately in the *Output* locations in the ***Tablet***. Together with the input, the synthesized parameters provide necessary and sufficient information to calculate all other design parameters and performance or behavioral variables. If any of these are not satisfactory, iterations are possible through the utilization of the ***Analysis*** column in conjunction with the ***Input*** column to home into any design parameter or performance constraint. The design tuning process can be repeated until the user is satisfied with the design parameters.

Even though the optimum case utilized in the ***Spur Gear Synthesis Tablet*** depends on an allowable bending fatigue strength of 300 [MPa] or 43.5 [kpsi] and a material property ratio B_{BC} of 0.3, the results for other allowable bending fatigue strength and material property ratio can be synthesized. One can reach an optimum by observing the trends in Figures 14.12–14.14 and adjust design parameters accordingly. Figure 14.17 changes the allowable bending fatigue strength to 200 [MPa] or 29 [kpsi] in the ***Input*** column. The material property ratio B_{BC} of 0.3 is kept in this particular process. The calculated module m_n or diametral pitch p_d did not change because the optimum is dependent on S_{ab} of 300 [MPa] or 43.5 [kpsi], which was embedded in the procedure. Observing Figure 14.12, the optimum module is expected to be a little lower for a lower S_{ab}. This can be enacted in the ***Analysis*** column. Observing Figure 14.14, the optimum number of teeth is expected to be higher for a lower S_{ab}, which is observed as shown in Figure 14.17. This higher value in the ***Input*** column would be relatively kept in the ***Analysis*** column during iteration. The variation in the face width is expected to be different for lower S_{ab}. This can be enacted in the ***Analysis*** column and varies to get better results. This process is iterated in the ***Analysis*** column, and the outcome of the safety factors is observed so as to approach 1.0 by adjusting the number of teeth and/or the face width. This is the process of the *CAD* and *optimization* for allowable bending fatigue strength of 200 [MPa] and 29 [kpsi] rather than the adapted 300 [MPa] and 43.5 [kpsi]. The module and the diametral pitch were adjusted to the standard 3.25 [mm] and 8 [teeth/in], the number of teeth was kept at 57, but the face width was iterated to 51 [mm] or 2.1 [in]. It is obvious that the adjustments were not much, but the gear set volume V_s was reduced from 0.027 203 [m^3] or 1716.8 [in^3] in the bottom of the ***Input*** column to 0.023 853 9 [m^3] or 1503.9 [in^3] in the bottom of the ***Analysis*** column. This is a further reduction of about 14% for both SI and the US systems. One can verify and possibly further adjust values in the available ***Tablets*** through **Wiley website**.

Note that the conventional approach of Example 14.5 and Figure 14.17 with the higher allowable bending fatigue strength of 300 [MPa] and 43.5 [kpsi] was not optimum and the gear set had a volume of 0.029 536 [m^3] or 2051.1109 [in^3]. This is about 269% or 306% higher than the synthesized gear that has a volume of 0.010 981 [m^3] or 669.6 [in^3] as shown in Figure 14.17. A comparison between optimization results for allowable bending fatigue strength of 200 [MPa] or 29 [kpsi] and a *tuned* conventional solution was carried out. The tuning was performed by increasing the face width to 105 [mm] or 3.9 [in] so as to assure a safety factor of about 1.0. The volumes of the tuned conventional designs were 0.031 013 [m^3] and 1999.8 [in^3], which are about 30% and 33% higher than the tuned optimum solutions of 0.023 853 9 [m^3] or 1503.9 [in^3]. The difference is not as pronounced because some tuning (sort of an optimization) has been performed. It gave a lower bending safety factor and a very close contact safety factor to 1.0, which is utilizing the material closer to the optimum contact case. If further optimum tuning is implemented by reducing the module, by increasing the number of teeth and adjusting the face width, one would reach the optimum values suggested herein.

With some extensions to the ***Spur Gear Synthesis Tablets,*** iterations for different optimization are possible. Additional calculations may be necessary for intended definition of other objectives such as cost, which is a function of the volume, material cost, and quality of production. If the objective is the tooth frequency f_n, one should observe this specific output in the *Tablet*. In that case, iterations for optimization can follow either manually or by activating the optimization options in the *Excel* spreadsheet tools. This process is a simple exercise for anyone looking for further interest or solution to specific optimization problem. Other existing software is also available through the Internet, which may be utilized for such a task.

The ***synthesis general trend*** departing off the adapted value indicates that the *module* m_n would decrease a little with the reduction of *allowable bending fatigue strength* S_{ab}. The *diametral pitch* p_d would do the opposite. The module m_n, however, would be higher for a lower value of the *material property factor* B_{BC}. The *number of teeth* N_P should decrease with the decrease of the material property factor B_{BC}. The *face width* f_g may decrease a little with the decrease of the material property factor B_{BC}. These observations are deduced for the optimum values used herein as depicted in Metwalli and El Danaf (1996). Instead of performing optimization iterations with the utility of the ***Spur Gear Synthesis Tablets,*** it is advisable to consult the optimum values in the published paper first to reduce iterations.

Employing the ***Spur Gear Synthesis Tablets*** in optimization is achievable as carried out in previously calculated cases of the initial synthesized spur gear design and the other conventional or classical case as shown in Figures 14.21–14.24. The case in Figure 14.21 reduces the allowable bending fatigue strength to 100 [MPa] in the ***Input*** column. To obtain about the same 1.0 safety factors for both bending and contact, the module is initially reduced a little to be 3 [mm] in the ***Analysis*** column instead of 3.232 [mm] in the ***Input*** synthesis column. For the US system, the diametral pitch is initially increased a little to be 8 [teeth/in] in the ***Analysis*** column instead of 7.862 [teeth/in] in the ***Input*** synthesis column as shown in Figure 14.22. The number of teeth is reduced iteratively in the **Analysis** column to 120 for SI or 118 for the US system as observed in Figures 14.21 and 14.22. The face width is rounded to 59 [mm] or 2.2 [in] in the ***Analysis*** column. With about the same safety factors close to 1.0, the volume is reduced in the iterated ***Analysis*** column relative to the ***Input*** column by about 33% and 30% as shown in Figures 14.21 and 14.22, respectively. Similar initial iterations are shown in Figures 14.23 and 14.24 when reducing both the allowable bending fatigue strength to 100 [MPa] or 14.5 [kpsi] and the material property factor B_{BC} to 0.20 for a brittle material. The module is reduced a little to be 3 [mm], and the diametral pitch is increased a little to 8 [teeth/in] in the ***Analysis*** column. The number of teeth is *iterated* in the **Analysis** column to 75 for SI and 60 for the US. The face width is increased to 90 [mm] and 4.3 [in] in the ***Analysis*** column. The volume is reduced in the iterated ***Analysis*** column relative to the ***Input*** column by about 23% and 55% as shown in Figures 14.23 and 14.24. This reduction is relatively large due to the fact that the synthesis output is dependent on $B_{BC} = 0.3$ rather than 0.2 and for SI system iterations could have been attempted for further safety factor reduction closer to 1.0. This means that further optimization is still achievable. Other cases can also be optimized with the *Tablet* utilization. Full-fledged optimization for other objectives is the realm of research in specific gear fields such as optimization and other related works by the author and others (Elmaghraby et al. 1979; Metwalli 1982; Hegazi and Metwalli 1999; Townsend 1992).

It should be noted, however, that the previously presented results are dependent on the current adapted process. For a specific material, however, one should investigate the correct values of properties and adapt values available in codes and standards of gears such as AGMA. The synthesized and iterated parameters can be construed as initial values to be checked, verified, and validated by the standard codes and adjusted to other conditions, loadings, and manufacturing processes.

Spur gear synthesis – SI units			
Inputs	**Default**	**Input**	**Analysis**
Maximum power, H_{kW} [kW]	75	75	75
Input rotational speed, $N_{rpm,P}$ [rpm]	1200	1200	1200
Output rotational speed, $N_{rpm,G}$ [rpm]	300	300	300.000
Safety factor, K_{SF}	1.000	1.000	1.005
Quality number, Q_v	8	8	8
Operating direction, (2-*way* or 1-*way*)	2	2	2
Reliability, R_e	0.99	0.99	0.99
Output			**Input/Output**
Module, m_n [mm]	3.23	3.232	3
Number of pinion teeth, N_P	36	129	120
Number of gear teeth, N_G	144	516	480
Face width, $f_{g(P,G)}$ [mm]	58.8	58.823	59
Allowable dending strength, S_{ab} [MPa]	300	100	100
Allowable contact strength, S_{ac} [MPa]	1000	333	333
Gear ratio, n_G	3.968	4.000	4.000
Pinion pitch diameter, d_P [mm]	117.29	416.882	360.000
Gear pitch diameter, d_G [mm]	465.36	1667.529	1440.000
Center distance, C_g [mm]	291.32	1042.205	900.00
Maximum torque, $T_{g,\max}$ [N.m]	596.83	596.83	596.83
Tangential force, F_t [N]	10177.31	2863.31	3315.73
Radial force, F_r [N]	3704.24	1042.16	1206.83
Pinion volume, V_P [m^3]	0.000 708	0.008 279 95	0.006 207 319
Gear volume, V_G [m^3]	0.010 285	0.129 462 688	0.096 889 894
Pitch Point velocity, v_t [m/s]	7.369	26.193	22.619
Dynamic factor, K_v	1.314 12	1.558 92	1.523 55
Bending geometry factor, J_g	0.398 71	0.498 10	0.493 14
Surface geometry factor, I_g	0.119 74	0.128 55	0.128 55
Bending stress, σ_b [MPa]	283.622	75.772	93.034
Contact stress, σ_c [MPa]	974.332	288.145	329.392
Bending safety factor, $K_{SF}\|b$	1.058	1.320	1.075
Contact safety factor, $K_{SF}\|c$	1.013	1.075	1.005
Other values			
Tooth frequency, f_T [Hz]	725.86	2580.00	2400.00
Contact ratio, m_p	1.791	1.915	1.911
Circular pitch, p_c [mm]	10.153	10.153	9.425
Outer pinion diameter, d_{oP} [mm]	123.750	423.345	366.000
Outer gear diameter, d_{oG} [mm]	471.820	1673.992	1446.000
Total gear set volume, V_S [m3]	0.010 992	0.137 742 638	0.103 097 213

Figure 14.21 Optimization iteration for initially synthesized case to reduce the allowable bending fatigue strength to 100 [MPa] in the *Input* column. The module is reduced to 3 [mm] in the *Analysis* column. The number of teeth is also iterated in the *Analysis* column to 120.

Spur gear synthesis – US units			
Inputs	Default	Input	Analysis
Maximum power, H_{hp} [hp]	100.5	100.5	100.5
Input rotational speed, $N_{rpm,P}$ [rpm]	1200	1200	1200
Output rotational speed, $N_{rpm,G}$ [rpm]	300	300	300.000
Safety factor, K_{SF}	1.000	1.000	1.004
Quality number, Q_v	8	8	8
Operating direction, (2-*way* or 1-*way*)	2	2	2
Reliability, R_e	0.99	0.99	0.99
Output			Input/Output
Diametral pitch, p_c [teeth/in]	7.862	7.862	8
Number of pinion teeth, N_P	36	129	118
Number of gear teeth, N_G	144	516	472
Face width, $f_{g(P,G)}$ [in]	2.315	2.315	2.2
Allowable bending strength, S_{ab} [kpsi]	43.50	14.5	14.5
Allowable contact strength, S_{ac} [kpsi]	145.00	48.33	48.33
Gear ratio, n_G	4.000	4.000	4.000
Pinion pitch diameter, d_P [in]	4.579	16.408	14.750
Gear pitch diameter, d_G [in]	18.316	65.631	59.000
Center distance, C_g [in]	11.447	41.020	36.875
Maximum torque, $T_{g,\max}$ [lb.in]	5278.4	5278.374	5278.374
Tangential force, F_t [lb]	2305.51	643.398	715.712
Radial force, F_r [lb]	839.14	234.177	260.498
Pinion volume, V_P [in^3]	42.483	504.8806	388.7721
Gear volume, V_G [in^3]	627.122	7894.1545	6065.8165
Pitch point velocity, v_t [ft/min]	1438.5	5154.67	4633.849
Dynamic factor, K_v	1.310 60	1.554 91	1.529 20
Bending geometry factor, J_g	0.398 17	0.498 10	0.491 99
Surface geometry factor, I_g	0.119 81	0.128 55	0.128 55
Bending stress, σ_b [kpsi]	42.85	11.342	13.402
Contact stress, σ_c [kpsi]	144.66	42.452	47.953
Bending safety factor, $K_{SF}\|^b$	1.015	1.278	1.082
Contact safety factor, $K_{SF}\|c$	1.001	1.067	1.004
Other values			
Tooth frequency, f_T [Hz]	720.00	2580.00	2360.00
Contact ratio, m_p	1.790	1.915	1.910
Circular pitch, p_c [in]	0.400	0.400	0.393
Outer pinion diameter, d_{oP} [in]	4.833	16.662	15.000
Outer gear diameter, d_{oG} [in]	18.570	65.886	59.250
Total gear set volume, V_S [in^3]	669.60	8399.0351	6454.5886

Figure 14.22 Optimization iteration for initially synthesized case to reduce the allowable bending fatigue strength to 14.5 [kpsi] in the *Input* column. The diametral pitch is increased to 8 [teeth/in] in the *Analysis* column. The number of teeth is also iterated in the *Analysis* column to 118.

Spur gear synthesis – SI units			
Inputs	**Default**	**Input**	**Analysis**
Maximum power, H_{kW} [kW]	75	75	75
Input rotational speed, $N_{rpm,P}$ [rpm]	1200	1200	1200
Output rotational speed, $N_{rpm,G}$ [rpm]	300	300	300.000
Safety factor, K_{SF}	1.000	1.000	1.004
Quality number, Q_v	8	8	8
Operating direction, (2-*way* or 1-*way*)	2	2	2
Reliability, R_e	0.99	0.99	0.99
Output			**Input/Output**
Module, m_n [mm]	3.23	3.232	3
Number of pinion teeth, N_P	36	129	75
Number of gear teeth, N_G	144	516	300
Face width, $f_{g(P,G)}$ [mm]	58.8	58.823	90
Allowable bending strength, S_{ab} [MPa]	300	100	100
Allowable contact strength, S_{ac} [MPa]	1000	500	500
Gear ratio, n_G	3.968	4.000	4.000
Pinion pitch diameter, d_P [mm]	117.29	416.882	225.000
Gear pitch diameter, d_G [mm]	465.36	1667.529	900.000
Center distance, C_g [mm]	291.32	1042.205	562.50
Maximum torque, $T_{g,max}$ [N m]	596.83	596.83	596.83
Tangential force, F_t [N]	10177.31	2863.31	5305.16
Radial force, F_r [N]	3704.24	1042.16	1930.92
Pinion volume, V_P [m^3]	0.000 708	0.008 27 995	0.003 771 867
Gear volume, V_G [m^3]	0.010 285	0.129 462 688	0.058 021 478
Pitch point velocity, v_t [m/s]	7.369	26.193	14.137
Dynamic factor, K_v	1.314 12	1.558 92	1.423 63
Bending geometry factor, J_g	0.398 71	0.498 10	0.460 72
Surface geometry factor, I_g	0.119 74	0.128 55	0.128 55
Bending stress, σ_b [MPa]	283.622	75.772	99.557
Contact stress, σ_c [MPa]	974.332	288.145	416.601
Bending safety factor, $K_{SF}\|b$	1.058	1.320	1.004
Contact safety factor, $K_{SF}\|c$	1.013	1.317	1.096
Other values			
Tooth frequency, f_T [Hz]	725.86	2580.00	1500.00
Contact ratio, m_p	1.791	1.915	1.875
Circular pitch, p_c [mm]	10.153	10.153	9.425
Outer pinion diameter, d_{oP} [mm]	123.750	423.345	231.000
Outer gear diameter, d_{oG} [mm]	471.820	1673.992	906.000
Total gear set volume, V_S [m^3]	0.010992	0.137742638	0.061793345

Figure 14.23 Initial iteration of reducing both the allowable bending fatigue strength to 100 [MPa] and the material property factor B_{BC} to 0.20 for *a brittle material*. In the *Analysis* column, the module is reduced to the standard 3 [mm], the number of teeth is iterated to 75, and the face width is iterated to 90 [mm].

Spur gear synthesis – US units			
Inputs	**Default**	**Input**	**Analysis**
Maximum power, H_{hp} [hp]	100.5	100.5	100.5
Input rotational speed, $N_{rpm,P}$ [rpm]	1200	1200	1200
Output rotational speed, $N_{rpm,G}$ [rpm]	300	300	300.000
Safety factor, K_{SF}	1.000	1.000	1.011
Quality number, Q_v	8	8	8
Operating direction, (2-*way* or 1-*way*)	2	2	2
Reliability, R_e	0.99	0.99	0.99
Output			**Input/Output**
Diametral pitch, p_c [teeth/in]	7.862	7.862	8
Number of pinion teeth, N_P	36	129	60
Number of gear teeth, N_G	144	516	240
Face width, $f_{g(P,G)}$ [in]	2.315	2.315	4.3
Allowable bending strength, S_{ab} [kpsi]	43.50	14.5	14.5
Allowable contact strength, S_{ac} [kpsi]	145.00	72.5	72.5
Gear ratio, n_G	4.000	4.000	4.000
Pinion pitch diameter, d_P [in]	4.579	16.408	7.500
Gear pitch diameter, d_G [in]	18.316	65.631	30.000
Center distance, C_g [in]	11.447	41.020	18.750
Maximum torque, $T_{g,\max}$ [lb in]	5278.4	5278.374	5278.374
Tangential force, F_t [lb]	2305.51	643.398	1407.566
Radial force, F_r [lb]	839.14	234.177	512.312
Pinion volume, V_P [in^3]	42.483	504.8806	202.8438
Gear volume, V_G [in^3]	627.122	7894.1545	3090.3601
Pitch point velocity, v_t [ft/min]	1438.5	5154.67	2356.194
Dynamic factor, K_v	1.310 60	1.554 91	1.389 85
Bending geometry factor, J_g	0.398 17	0.498 10	0.441 39
Surface geometry factor, I_g	0.119 81	0.128 55	0.128 55
Bending stress, σ_b [kpsi]	42.85	11.342	14.344
Contact stress, σ_c [kpsi]	144.66	42.452	65.896
Bending safety factor, K_{SF} $\vert b$	1.015	1.278	1.011
Contact safety factor, K_{SF} $\vert c$	1.001	1.307	1.049
Other values			
Tooth frequency, f_T [Hz]	720.00	2580.00	1200.00
Contact ratio, m_p	1.790	1.915	1.854
Circular pitch, p_c [in]	0.400	0.400	0.393
Outer pinion diameter, d_{oP} [in]	4.833	16.662	7.750
Outer gear diameter, d_{oG} [in]	18.570	65.886	30.250
Total gear set volume, V_S [in^3]	669.60	8399.0351	3293.2040

Figure 14.24 Initial iteration of reducing both the allowable bending fatigue strength to 14.5 [kpsi] and the material property factor B_{BC} to 0.20 for *a brittle material.* In the *Analysis* column, the module is increased to the standard 8 [teeth/in], the number of teeth is iterated to 60, and the face width is iterated to 4.3 [in].

14.7 Constructional Details

The design of a gear set is not only the synthesis of the pinion and the gear for their appropriate geometry and material. It is essential, as well, to assemble these components with the shafts on which the gears are mounted and by which the power is transmitted. The mounting is implemented by the proper *tolerances and fits*. The *power transmission* is positively guaranteed trough *keys, feathers,* or *splines*; see Section 8.2.2. The shafts should also be supported by proper bearings and a ready connection to the prime mover or the machine. These connections are usually selected as keys or splines. Figure 14.25 defines the main requirements and constraints for the proper *assembly*.

14.7.1 Gearboxes

The gears in Figure 14.25 are mounted on two *shafts* with a center distance specifically defined to have the two pitch circles (cylinders) tangent to each other at the pitch point. The pinion and gear are keyed to the two shafts to transmit the power from one shaft to the other. Each shaft is supported by two bearings that are housed in a frame or a box. Each shaft is also equipped with a keyway for *joining* the shaft with a coupling to the prime mover and to the machine. The main constructional requirement is to locate the gear on the shaft allowing no movement along the shaft. This is guaranteed by the shaft *shoulder* on one side and a *spacer sleeve* between the gear and the retained bearing on the other side. To hold the gear, the shaft and the end bearings prevent unnecessary movement along the shaft centerline by the two *retainers* that harness the movement of the outer races of the two end bearings. The two end *retainers* are to be fastened to the frame. Usually the end *retainers* are not flush with the outer race of the bearing, and there is about 0.2–0.3 [mm] or 0.008–0.01 [in] space at either end. This allows a breathing space for thermal expansion and deformations. In addition, the *fits, surface finish, and tolerances* between components are basic requirements for proper *location*, *operation*, and *maintenance*. For suggested *fits* and proper *tolerances*, consult Section 2.4.7 and specifications by suppliers such as tolerances of bearings from manufacturers; see Figure 11.5. This would complete the constructional constraints of the gear set. It should be noted, however, that the previously defined constructional details are only the concepts of the simplest way of satisfying these constraints. Other means are also available that may serve more requirements and complexities.

Figure 14.26 is a material implementation of the previous constructional constraints and requirements. Figure 14.26 shows one view of the *assembly* of a single stage gear set or a *gearbox*. The *housing frame* is shown as the boundary of the two gears in the form of the flange of the lower or the upper part of the gearbox housing. The two housing halves of the gearbox are partitioned at the plane of the two centerlines of the gear shafts. The two housing halves are produced by, say, casting. The bottom of the lower half and the partition between the two

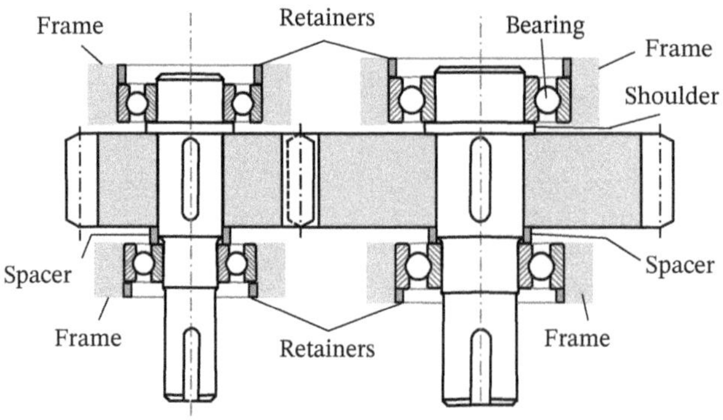

Figure 14.25 An assembly drawing that defines the main requirements and constraints for the constructional details of a gearbox.

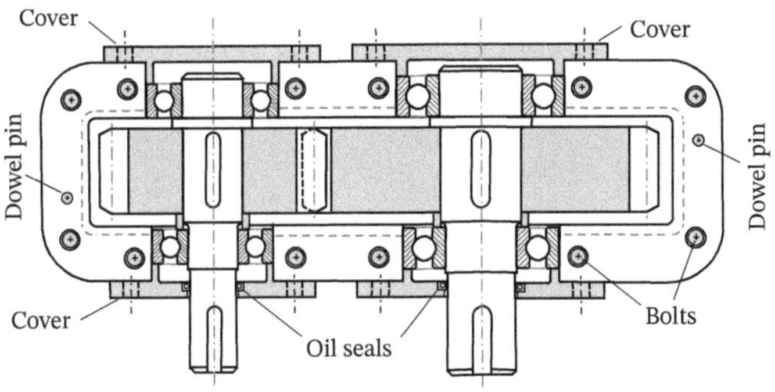

Figure 14.26 A single sectional view for the physical substance implementation of the necessary constructional constraints and some other requirements for the assembly of a single stage gear set or *a gearbox*.

halves are usually surface milled after raw casting. The two housing halves are then assembled while empty, and two tapered holes are drilled for the two tapered dowel pins to assure accurate location between the two halves. Holes for bolts are then drilled, and the two housing halves are bolted together before further machining. Other machining operations are then carried out such as through holes for the bearings, facing sides for the covers, drilling holes, and taping of threaded holes. The dashed lines on the flange in Figure 14.26 represent the outer surface of bottom or top half of the gearbox housing. The covers have the shown dashed holes for the cover fixation bolts. The bolts are staggered not to interfere with the fixation bolts of the two gearbox housing halves. Oil seals are also housed into the covers of the through shafts to prevent oil from seeping out of the gearbox.

The construction in Figure 14.26 is only a *geometric model*. No real dimensions are adhered to. The dimensions and associated material adaption is the *real design* or *synthesis* for that construction. The dimensions and materials should have been generated or synthesized before the construction details are defined. That has been the process developed herein to synthesize before *geometric modeling* is settled upon. The construction in Figure 14.26 is considered only as a *sketch* to support decision on the relative positioning among components that need to be synthesized. After synthesis, the proper geometries of all components are defined, and the connectivity or *joining* between any two components is then well assured. In Figure 14.26, for instance, the center distance between the two gears might be smaller than the space required by the bearings or the covers. This is highly unlikely, but one would not decide off hand particularly when optimization is the objective to get the smallest design possible or minimizing cost. Optimization might be the priority, and the necessary construction detail is to adapt to the optimum geometry. This is the challenge of a good designer.

Other constructional details that should be considered are, for instance, oil drainage plugs, oil supply caps, oil level meters, venting means, and lifting means or attachments for positioning heavy gearboxes. The reduction of gear volume by considering webs and holes between the *gear hub* and the *teeth rim* might be necessary in some applications and dynamic performance. Dynamic *balancing* of high speed components should also be essential. Means of reducing clearances might also be needed in some applications. Means of reducing *noise radiation* off gearboxes can also be useful in many applications.

14.7.2 Gear Trains

Gear trains are frequently multistages with more than two shafts or a single stage as the one shown in Figure 14.2. The gear train shown in Figure 14.2a is represented schematically as shown in Figure 14.27a. Even though the set in Figure 14.2a is a helical gear set, the schematic diagram in Figure 14.27a can equally apply to spur gear trains also. Figure 14.2b is a schematic diagram for a part of *planetary* or *epicyclic* gear train. One of the simplest ways to get the relation between input and output angular velocities is to use the velocity triangles as shown in

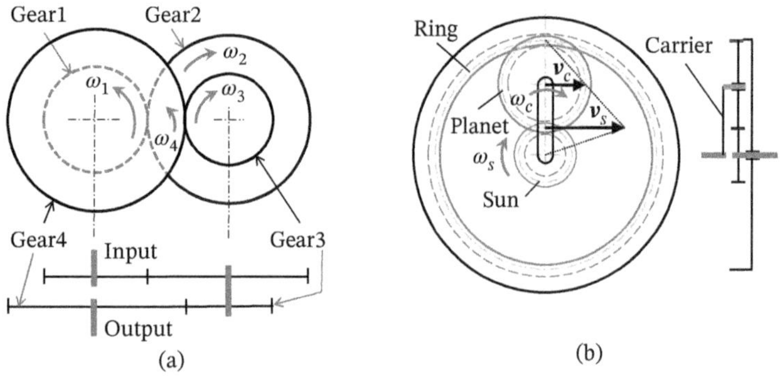

Figure 14.27 A two-stage gear train schematically represented in (a) with a connection diagram below it. The schematic in (b) represents a part of *planetary*, or *epicyclic*, gear train with a connection diagram to the right-hand side.

Figure 14.27b. Similar diagrams can be erected in Figure 14.27a at each contact or pitch point between each pair of gears. One may, however, just use the relationship between any pair of gears to get the total gearbox reduction ratio. For the gear train in Figure 14.27a, gear1 is the input gear, which runs on a separate shaft other than the output gear4, and gear1 and gear2 are behind gear4 and gear3. Both gear2 and gear3 are attached to the same shaft. The lower part of Figure 14.27a is a connection diagram below the schematic drawing. The first reduction is achieved between gear1 and gear2, where r_{g1} is the pitch radius of gear1 and r_{g2} is the pitch radius of gear2. Since the velocity of the pitch point is the same, the angular velocity ω_2 of gear2 is then less than the angular velocity ω_1 of gear1. Note that r_{g1} is less than r_{g2} (Figure 14.27a). The angular velocity ω_3 of gear3 is the same as the angular velocity ω_2 of gear2, since both are attached or keyed to the same shaft. The second reduction is achieved between gear3 and gear4, where r_{g3} is the pitch radius of gear3 and r_{g4} is the pitch radius of gear4. The angular velocity ω_4 of gear4 is then less than the angular velocity ω_3 of gear3, since r_{g3} is less than r_{g4} (Figure 14.27a). Using Eq. (14.2) for each reduction gear ratio and realizing that $\omega_2 = \omega_3$, one gets the following gear train ratio or gearbox reduction ratio n_g:

$$\frac{\omega_1}{\omega_2} = \frac{r_{g2}}{r_{g1}} \quad \text{and} \quad \frac{\omega_3}{\omega_4} = \frac{r_{g4}}{r_{g3}}$$
$$n_g = \frac{\omega_1}{\omega_4} = \frac{\omega_1}{(\omega_2)} \frac{(\omega_3)}{\omega_4} = \frac{r_{g2}}{r_{g1}} \frac{r_{g4}}{r_{g3}} \tag{14.84}$$

Employing Eq. (14.4) into Eq. (14.84), the gear train ratio or *gearbox* reduction ratio n_g is then

$$n_g = \frac{\omega_1}{\omega_4} = \frac{r_{g2}}{r_{g1}} \frac{r_{g4}}{r_{g3}} = \frac{N_{g2}N_{g4}}{N_{g1}N_{g3}} = \frac{\prod \text{drivens}}{\prod \text{drivers}} \tag{14.85}$$

where N_{g1} is the number of teeth of gear1, N_{g2} is the number of teeth of gear2, N_{g3} is the number of teeth of gear3, and N_{g4} is the number of teeth of gear4. For the gear set to have the configuration in Figure 14.27a, the input and output shafts should be *in-line*. This adds a constraint that the sum of radii of gear1 and gear2 should be the same as the sum of the radii of gear3 and gear4. This constraint is also set in number of teeth if the modules or diametral pitches are the same. Employing Eq. (14.4), in addition, gives the following relation:

$$r_{g1} + r_{g2} = r_{g3} + r_{g4} \quad \text{or} \quad N_{g1} + N_{g2} = N_{g3} + N_{g4} \tag{14.86}$$

This concept and similar relations can easily be developed for the more complex gear sets of multi axis. The requirements of *in-line* configuration of some of the more complex gear sets can also be derived similarly.

14.7.3 Planetary or Epicyclic Gear Trains

Planetary or *epicyclic* gear train is shown in Figure 14.2, where only a part of it is copied in Figure 14.27b to allow for analysis. The rest of the planets and carrier will follow the same developed rules. The added link between the sun and the planet in Figure 14.27b is the housing arm or the *carrier*. The right-hand part of Figure 14.27b is a connection diagram to the side of the schematic drawing to clarify the interconnections. One can get the relation between input and output angular velocities through the use of the velocity triangle depicted in Figure 14.27b. In this particular assembly, the sun rotates with an angular velocity of ω_s, and the outer ring is stationary. The three basic components of the assembly (sun, carrier, and ring), however, can be stationary and the two others rotating as input or output. Planets cannot be stationary in this arrangement. The carrier housing is representing the accommodation cluster of the three planets, which cause all planets' axis to rotate as one unit around the sun with angular velocity ω_c. The planets, however, rotate inside the *carrier* or housing cluster with angular velocity ω_p. From the velocity triangle of Figure 14.27b and using Eq. (14.1), the *carrier* angular velocity ω_c is obtained relative to the sun velocity ω_s as follows:

$$\begin{aligned}
&\boldsymbol{v}_s = r_s\omega_s \quad \text{and} \quad \boldsymbol{v}_c = v_s/2 = r_s\omega_s/2 \\
&\boldsymbol{v}_c = r_c\omega_c = (r_s + r_p)\omega_c = r_s\omega_s/2 \\
&\text{or} \quad \frac{\omega_c}{\omega_s} = \frac{r_s}{2(r_s + r_p)}
\end{aligned} \tag{14.87}$$

It is obvious that for meshing, the module or the diametral pitch should be the same for all mating gears. Substituting for the radii of the sun and the planets as defined by Eq. (14.3) or (14.8) into Eq. (14.87), one gets

$$\frac{\omega_c}{\omega_s} = \frac{m_n N_s}{2m_n(N_s + N_p)} = \frac{N_s}{2(N_s + N_p)} \tag{\textbf{\textit{14.88}}}$$

where N_s is the number of teeth of the sun gear and N_p is the number of teeth of the planet gear. The planet angular velocity ω_p is obtained from the general relation of planetary gear train (Meirovitch 1986). The two equations defining the sun–planet and the planet–ring relations are as follows:

$$\begin{aligned}
&N_s\omega_s + N_p\omega_p - (N_s + N_p)\omega_c = 0 \\
&N_r\omega_r - N_p\omega_p - (N_r - N_p)\omega_c = 0
\end{aligned} \tag{14.89}$$

Adding these two equations, one gets

$$\begin{aligned}
&N_s\omega_s + N_r\omega_r = (N_s + N_r)\omega_c \\
&-\frac{N_r}{N_s} = \frac{\omega_s - \omega_c}{\omega_r - \omega_c}
\end{aligned} \tag{14.90}$$

The quotient $-N_r/N_s$ in the second expression of Eq. (14.90) is frequently called the ***train ratio***. From Eq. (14.90) and for the case of $\omega_r = 0$ (stationary ring), one gets

$$\omega_c = \frac{N_s\omega_s + N_r\omega_r}{(N_s + N_r)} = \frac{N_s\omega_s}{(N_s + N_r)} \tag{\textbf{\textit{14.91}}}$$

For the case of $\omega_r = 0$, the second equation in (14.89) is reduced to

$$\omega_p = -\frac{(N_r - N_p)}{N_p}\omega_c \tag{\textbf{\textit{14.92}}}$$

The velocity triangle and Eq. (14.3) gives a relation, which should be *conforming* to Eq. (14.89) such that

$$\begin{aligned}
&\boldsymbol{v}_c = \boldsymbol{v}_s/2 = r_s\omega_s/2 = -r_p\omega_p \\
&\omega_p = -\frac{r_s\omega_s}{2r_p} = -\frac{N_s\omega_s}{2N_p}
\end{aligned} \tag{14.93}$$

This relation does represent the angular velocity of the planet. It is important to note this derivation of Eq. (14.93) so that no oversight would occur. The velocity triangle is representing the relation between the sun, the carrier, and the ring. It should also be noted that the rotation of the planet gear is in the opposite direction relative to the sun or the carrier.

As indicated previously, any of the three basic components of the *planetary* or *epicyclic* gear train can be stationary. The other two components can then rotate either as an input or as an output. The carrier can be stationary while the sun gear can be an input, and the ring can then be the output or vice versa. On the other hand, the sun can be stationary while the carrier can be an input, and the ring can then be the output or vice versa. These cases can easily be modeled using the appropriate velocity triangle. A stationary arm is defined by a zero velocity at its pivot with the planet. A stationary sun gear is defined by a zero velocity at its outer radius.

It is important to note that the power transmitted is divided by the number of planets when designing the sun, planet, and the ring. It is obvious that as many teeth as the number of planets are transmitting the load at the same time. However, the teeth are subjected to the loading cycles at multiple times as the number of planets. In addition and similar to the relation of the gear train to guarantee *in-line* axis, the sum of the sun radius and the planet diameter should be the same as the ring radius. This constraint is also set in the form of number of teeth if the modules or diametral pitches are the same, which should be the case in this simple planetary gear train form. In addition, employing Eq. (14.4) gives the following relation:

$$r_s + 2r_p = r_r \quad \text{or} \quad N_s + 2N_p = N_r \tag{14.94}$$

It should also be noted that there are constraints on the specific number of teeth for the sun, the planets, and the ring to allow *assembly* where the planet gears should simultaneously engage the sun and the ring gear at opposite ends of each planet. The rule to guarantee equal spacing assembly of this simple planetary gear trains is given by (Townsend 1992)

$$\frac{N_s + N_r}{\text{number of planets}} = \text{integer} \tag{14.95}$$

This problem has also been considered, and a solution is suggested for multiple planets and compound planetary gear assembly; see Metwalli (1982).

Example 14.9 An epicyclic or a planetary or gear train is shown in Figure 14.2. It has a stationary ring. The rotational speed of the input sun gear is 1200 [rpm]. The number of teeth of the planet gear is 30, the number of teeth of the sun gear is 20, and the number of teeth of the ring gear is 80. Gears are of 20° pressure angle and all of full depth spurs. The carrier houses the planets. Define the carrier output angular velocity. What is the gear set ratio of the planetary gear? Find the planets angular velocity. Define the number of planets that can be assembled between the sun and the ring.

Solution

Data: Input rotational speed of the sun gear $N_{\text{in}} = 1200$ [rpm]. The input sun gear angular velocity $\omega_1 = (2\pi N_{\text{in}}/60) = 125.664$ [rad/s]. The number of sun gear teeth $N_s = 20$, the number of planet gear teeth $N_p = 30$, and the number of ring gear teeth $N_r = 80$. These values of teeth numbers agree with the relation in Eq. (14.94).

The output angular velocity of the carrier ω_c that houses the planets is obtained from Eq. (14.88) as follows:

$$\omega_c = \frac{\omega_s N_s}{2(N_s + N_p)} = \frac{125.664(20)}{2(20 + 30)} = 25.133\ [\text{rad/s}] = 240\ [\text{rpm}] \tag{a}$$

The carrier angular velocity ω_c is obtained again from Eq. (14.91) as follows:

$$\omega_c = \frac{N_s\omega_s}{(N_s + N_r)} = \frac{20(125.664)}{(20 + 80)} = 25.133\,[\text{rad/s}] = 240\,[\text{rpm}] \tag{b}$$

This is the same as the value obtained from the velocity triangle of Eq. (14.88). The planetary gear ratio is obtained from Eq. (14.88) as follows:

$$\frac{\omega_c}{\omega_s} = \frac{N_s}{2(N_s + N_p)} = \frac{25.133}{125.664} = \frac{20}{2(20 + 30)} = 0.2 \tag{c}$$

The planet angular velocity ω_p that are housed in the carrier is obtained from Eq. (14.92) as follows:

$$\omega_p = -\frac{(N_r - N_p)}{N_p}\omega_c = -\frac{80 - 30}{30}(25.133) = -41.888\,[\text{rad/s}] = -400\,[\text{rpm}] \tag{d}$$

From Eq. (14.93), the planet angular velocity ω_p is

$$\omega_p = \left(-\frac{N_s}{2N_p}\right)\omega_s = \left(-\frac{20}{2(30)}\right)(125.664) = -41.888\,[\text{rad/s}] = -400\,[\text{rpm}] \tag{e}$$

This value in Eq. (e) is the same as the value in Eq. (d) obtained from the general relation of Eq. (14.92).

Assembly of this planetary gear set requires the satisfaction of Eq. (14.95). From this equation, the number of equal spacing planets should then be as follows:

$$\text{number of planets} = \frac{N_s + N_r}{\text{integer}} = \frac{20 + 80}{\text{integer}} = \frac{100}{\text{integer}} \tag{f}$$

This means that only 2, 4, 5, etc. planets can be assembled. Therefore, three planets cannot be assembled at equal spacing of 120°. With the available space between the sun and the ring gears, four planets may barely be accommodated at equal spacing of 90°. Five planets would not have enough space to fit every 72° without teeth interfering with each others.

14.8 Summary

In this chapter spur gears are presented. They, with other types of gears, are the most commonly used of all types of transmission drives. The geometry of the teeth, kinematic definitions, and usual types and standards of spur gears are introduced in the chapter. Forces generated during meshing and power transmission are identified. These forces are acting on the teeth of the meshing gears, and their effect in developing stresses on the teeth and gear design is defined. The relations defining the bending fatigue and contact fatigue stresses are derived. Material sets usually used in gears are presented and their strength properties introduced. The material property factor identified by the bending fatigue strength to the contact fatigue strength is utilized in the synthesis of spur gears. Synthesis procedure utilizes *CAD*, and optimization tools and results to generate better gearing design are provided. Constructional details for gear set assembly of gearboxes including planetary, or epicyclic, gear sets are also presented.

The CAS process is provided in a real *CAD* tablet using Excel spreadsheet. This is available as a ***Spur Gear Synthesis Tablets*** through **Wiley website**. The utility of these SI and the US tables has been demonstrated to effectively synthesize spur gears near the required optimum. Little iteration is used to tune the synthesized spur gear design to satisfy multitude of conditions, cases, and possible optimum objectives. If need be, the synthesized and iterated parameters can be construed as good values to be checked, verified, and validated by standard codes and adjusted to other conditions, loadings, and manufacturing processes that have not been included herein.

Problems

14.1 Place a can on a sheet of paper, and draw a circle around its end on the sheet. Wrap a string around the can close to one end with a pencil tied at the end of the string. Fit the can on the drawn circle on the sheet with the cord rapped end close to the sheet. With the pencil snuggled to the can, unwrap the string with the pencil marking on the sheet while unwrapping. The can should be held tight on top of the sheet. Observe the curve generated by the pencil marking. Is it an involute? How can you tell?

14.2 Investigate the reasons behind the abandonment of the 14.5° pressure angle for recent gears and the wide spread use of the 20° pressure angle. Why is the 25° pressure angle not favored? Are there other options for pressure angles?

14.3 Backlash is very important to gearing operations. Can you tie that to tolerances in manufacturing processes, and what is called total composite tolerances?

14.4 If one is interested in having the least number of teeth on about 1.0 [in] or 25 [mm] outside diameter gear, what reasonable options does he have? Consider the gear is generated to have no interference problem, i.e. the number of teeth can be much less than 18. What is the base circle diameter? What is the module or the diametral pitch for each case? Use standard module or diametral pitch values.

14.5 A gear set of a reduction ratio of 8 is to be geometrically defined. It has an input speed of 3600 [rpm]. The pinion teeth number is 36, the module is 16 [mm], and the pressure angle is 20°. Determine (a) the number of teeth of the output gear, (b) the output speed of the reducer gearbox, (c) the input and output gear geometry specifications, and (d) the recommended backlash for proper operation.

14.6 A gear set of a reduction ratio of 8 is to be geometrically defined. It has an input speed of 3600 [rpm]. The pinion teeth number is 36, the diametral pitch is 1.5 [teeth/in], and the pressure angle is 20°. Determine (a) the number of teeth of the output gear, (b) the output speed of the reducer gearbox, (c) the input and output gear geometry specifications, and (d) the recommended backlash for proper operation.

14.7 A small gear set has an input speed of 120 [rpm] and a reduction ratio of 4. The 20° pressure angle pinion has a module of 1.0 [mm], and the number of teeth is 57. Determine (a) the number of teeth of the output gear, (b) the output speed of the reducer gearbox, (c) the input and output gear geometry specifications, and (d) the recommended backlash for proper operation.

14.8 A small gear set has an input speed of 120 [rpm] and a reduction ratio of 4. The 20° pressure angle pinion has a diametral pitch of 24 [teeth/in], and the number of teeth is 57. Determine (a) the number of teeth of the output gear, (b) the output speed of the reducer gearbox, (c) the input and output gear geometry specifications, and (d) the recommended backlash for proper operation.

14.9 The gear set of Problem 14.5 can transmit a maximum power of 7500 [kW]. Find the maximum transmitted torque. Determine the forces between gears and the pitch point velocity.

14.10 The gear set of Problem 14.6 can transmit a maximum power of 10 058 [hp]. Find the maximum transmitted torque. Determine the forces between gears and the pitch point velocity.

14.11 The small gear set of Problem 14.7 transmits a maximum power of 0.5 [kW]. Find the maximum transmitted torque. Determine the forces between gears and the pitch point velocity.

14.12 The small gear set of Problem 14.8 can transmit a maximum power of 0.67 [hp]. Find the maximum transmitted torque. Determine the forces between gears and the pitch point velocity.

14.13 Synthesize the gear set defined in Problem 14.9, and compare the results with the conventional procedure of design. How can one handle the selection of quality if the pitch line speed is high? What quality number should be suggested, and what are the consequences in the design procedure?

14.14 Synthesize the gear set defined in Problem 14.10, and compare the results with the conventional procedure of design. How can one handle the selection of quality if the pitch line speed is high? What quality number should be suggested, and what are the consequences in the design procedure?

14.15 Synthesize the gear set defined in Problem 14.11, and compare the results with the conventional procedure of design. How can one handle the selection of quality if the pitch line speed is very low? What quality number should be suggested, and what are the consequences in the design procedure?

14.16 Synthesize the gear set defined in Problem 14.12, and compare the results with the conventional procedure of design. How can one handle the selection of quality if the pitch line speed is very low? What quality number should be suggested, and what are the consequences in the design procedure?

14.17 Recommend a better material than the adapted AISI 4340, and resynthesize the gear set of Problem 14.13.

14.18 Suggest a better material than the adapted AISI 4340, and resynthesize the gear set of Problem 14.14.

14.19 Propose a lower grade material than the adapted AISI 4340, and resynthesize the gear set of Problem 14.15.

14.20 Select a lower grade material than the adapted AISI 4340, and resynthesize the gear set of Problem 14.16.

14.21 Compare the value of the bending geometry factor J_g with the modified *Lewis* factor Y_L and the calculated stress concentration factor K_f at teeth numbers of 18, 30, 60, and 100 for full form standard 20° pressure angle. Use the procedure defined to get K_f from Eq. (14.35).

14.22 Derive the following equations by applying the following means: (a) Use Figure 14.4b to derive Eq. (14.15). (b) Use the maximum *Hertzian* contact pressure of Eq. (14.54), the breadth of contact

Table 14-P1 The required specifications for Problems 14.23–14.34.

Problem number	Power H_{kW} or (H_{hp}) [kW] ([hp])	Input and (output) angular velocities [rpm]	Quality Q_v	Allowable bending fatigue strength S_{ab} [MPa] ([kpsi])	Property factor B_{BC}
14.23	1 (1)	750 (150)	6	100 (14.5)	0.2
14.24	1 (1)	1200 (300)	8	200 (29)	0.25
14.25	1 (1)	2000 (500)	10	300 (43.5)	0.3
14.26	1 (1)	3000 (1000)	12	400 (58)	0.35
14.27	100 (100)	750 (150)	6	100 (14.5)	0.2
14.28	100 (100)	1200 (300)	8	200 (29)	0.25
14.29	100 (100)	2000 (500)	10	300 (43.5)	0.3
14.30	100 (100)	3000 (1000)	12	400 (58)	0.35
14.31	10 000 (10 000)	750 (150)	6	100 (14.5)	0.2
14.32	10 000 (10 000)	1200 (300)	8	200 (29)	0.25
14.33	10 000 (10 000)	2000 (500)	10	300 (43.5)	0.3
14.34	10 000 (10 000)	3000 (1000)	12	400 (58)	0.35

area of Eq. (14.55), and the radii of teeth curvature of Eq. (14.56) to develop the *Hertzian* stress given by Eqs. (14.58)–(14.60).

14.23–14.34 Design some of the spur gear sets by initial synthesis procedure and verify results using the ***Spur Gear Synthesis Tablet – SI and US*** for the required specifications in Table 14-P1. The SI specification may not relate to the US specification. Select different ***material designations*** to observe the difference. Find the bending and contact safety factors and other properties and parameters for each gear set. Compare results with the values possibly extracted from Figures 14.12–14.14. Find the fatigue safety factor for each case if the reliability is still 0.99 and the load application factor is increased to 1.5 of the maximum operating one. The power, angular velocities in [rpm], the quality, allowable bending fatigue strength, and property factor are given for each problem in Table 14-P1.

14.35–14.46 Use some of the problems stated in Table 14-P1 as a start to minimize the gear set volume by attempting to have the bending and contact fatigue safety factors closer to 1.0. Find the tooth frequency for each case, and compare it to the tooth frequency at the iteration start point. Calculate the natural frequency of each case, and compare it to the tooth frequency.

14.47 Attempt to design a multistage plastic gear set for a small toy that uses two 1.5 AA type batteries to drive a small car that has a weight of 5 [N] or 1.12 [lb]. The DC motor can have a maximum power of 0.6 [W], a maximum speed of 12 000 [rpm], and a maximum stall torque of 0.005 [N m].

14.48 Design a gear set to drive a bicycle using a small 0.2 [kW] DC motor with an output pinion mounted directly on top of the rear wheel tire. Assume a maximum bicycle speed of 6 [m/s]. Estimate the maximum angular velocity of the motor if the estimated pinion diameter is about 8 [mm] and the outer wheel diameter is 0.6 [m]. What is the maximum force applied by the pinion on the tire? What problems you might have with such a design, and can you find solutions to these problems?

14.49 Design a gearbox for a small one occupant solar car that can generate 1.0 [kW] of power from solar cells. Use four DC motors such that each one drives a wheel. For a very high gear ratio, design the gear set at each wheel. Check if you can drive each wheel the same way as the bicycle in Problem 14.48.

14.50 Multistage gearboxes similar to the one that is shown in Figure 14.27a are to be designed for a total gear set ratio of about 10, 12, 15, 30, and 50. Determine the gear ratio for each stage of any selected two of these gearboxes. Define the number of teeth of each gear to come up with an in-line configuration. Is there an optimum value for the gear ratio of each stage to produce the minimum volume? Design the gearboxes to transmit one of the powers defined in Problems 14.23 to 14–34. Suggest a construction for the gearbox similar to that shown in Figure 14.26.

14.51 Develop a spreadsheet to aid in the synthesis of similar gearboxes to those defined in Problem 14.50. The spreadsheet should consider the satisfaction of relation (14.86) and integer number of teeth.

14.52 Find the mathematical development of the general equations defining the sun–planet and the planet–ring relation of Eq. (14.89). Is it possible to develop the same relations with the help of velocity triangles?

14.53 Use the velocity triangles to develop the kinematic relations for other planetary gear trains. The trains should allow the fixation of the sun gear and the carrier consecutively while the other members are used as the input and the output. Can you utilize Eqs. (14.87)–(14.93) to find the relations for the case of assuming the carrier as the input and the sun as the output? Check your derived expression with the general relations of Eq. (14.89). Which of these arrangements provides the maximum possible gear train ratio?

14.54 Planetary gear trains of train ratios of about 3, 5, 7, and 9 are to be designed. Select one of these, and define the number of teeth and pitch diameters for the sun, planets, and the ring. How many planets can you fit in the space between the sun and the ring? Check the assembly requirement for three planets.

14.55 Develop a spreadsheet to aid in the synthesis of planetary gearboxes. The spreadsheet should consider the satisfaction of relation (14.94) and integer number of teeth.

References

AGMA 225.01 (1959). *Strength of spur, helical, herringbone and bevel gear teeth.* American Gear Manufacturers Association, Alexandria, Virginia, United States.

AGMA 218.01 (1982). *Rating the pitting resistance and bending strength of spur and helical gear teeth.* AGMA 218.01, replaced by AGMA 2001-D04 and AGMA 908-B89. American Gear Manufacturers Association, Alexandria, Virginia, United States.

AGMA 2000-A88 (1988). *Gear classification and inspection handbook: tolerances and measuring methods for unassembled spur and helical gears (including metric equivalents).* AGMA 2000-A88, replaced by 915-1-A02, 915-2-A05, 2015-1-A01, and 2015-2-A06. American Gear Manufacturers Association, Alexandria, Virginia, United States.

ANSI/AGMA 2101-C95 (1995). *Fundamental, rating factors and calculation methods for involute spur and helical gear teeth.* ANSI/AGMA 2101-C95, Metric Edition of ANSI/AGMA 2001-C95. American Gear Manufacturers Association, Alexandria, Virginia, United States.

AGMA 908-B89 (1999). *Geometry factors for determining the pitting resistance and bending strength of spur, helical and herringbone gear teeth.* AGMA 908-B89, April 1989, (Revision of AGMA 226.01), (Reaffirmed August 1999). American Gear Manufacturers Association, Alexandria, Virginia, United States.

ANSI/AGMA 2001-D04 (2004). *Fundamental rating factors and calculation methods for involute spur and helical gear teeth.* ANSI/AGMA 2001-D04 (revised AGMA 2001-C95) and ANSI/AGMA 2101-D04 (metric edition of ANSI/AGMA 2001-D04). American Gear Manufacturers Association, Alexandria, Virginia, United States.

Davis, J.R. (ed.) (2005). *Gear Materials, Properties, and Manufacture.* ASM International.

Dimargonas, A.D. (1989). *Computer Aided Machine Design.* Prentice Hall.

Dolan, T.J. and Broghamer, E.L. (1942). *A Photoelastic Study of the Stresses in Gear Tooth Fillets.* University of Illinois, Engineering Experiment Station, Bulletin No. 335.

DuPont (2000). *General Design Principals for DuPont Engineering Polymers.* E.I. du Pont de Nemours and Company.

Elmaghraby, S.E., Metwalli, S.M., and Zorowski, C.F. (1979). Some operations research approaches to automobile gear train design. *Mathematical Programming Study* 11: 150–175.

Hegazi, H.A. and Metwalli, S.M. (1999). Reverse engineering of standard mechanical elements. *Proceedings of the 1999 ASME Design Engineering Technical Conference*, Las Vegas, Nevada (12–15 September). Paper No. DETC99/CIE-9137, ASME.

Hertz, H. (1881). The contact of solid elastic bodies. *Journal of Pure and Applied Mathematics* (in German) 91: 156–171.

ISO 54 (1977). *Cylindrical gears for general engineering and for heavy engineering – Modules and diametral pitches.* International Organization for Standardization, Geneva, Switzerland.

ISO 54 (1996). *Cylindrical gears for general engineering and for heavy engineering – Modules.* International Organization for Standardization, Geneva, Switzerland.

ISO 1328 (2013). *Cylindrical gears – ISO system of accuracy – Part 1: Definitions and allowable values of deviations relevant to corresponding flanks of gear teeth.* International Organization for Standardization, Geneva, Switzerland.

ISO 6336 1-3 (2006). *Calculation of load capacity of spur and helical gears – Part 1: Basic principles, introduction and general influence factors, Part 2: Calculation of surface durability (pitting), Part 3: Calculation of tooth bending strength.* International Organization for Standardization, Geneva, Switzerland.

ISO 6336 5 (2016). *Calculation of load capacity of spur and helical gears – Part 5: Strength and quality of materials.* International Organization for Standardization, Geneva, Switzerland.

Lewis, W. (1893). Investigation of the strength of gear teeth. *An address to the Engineers Club of Philadelphia*, October 15, reprinted in *Gear Technology*, vol 9, no. 6, November/December (1992), pp 19-23, American Gear Manufacturers Association.

Lewis, W. (1910). Interchangeable involute gearing. *Journal of ASME* October 10: 1631.

Metwalli, S.M. (1982). Multiple cluster spacing of compound epicyclic gear trains. In: *Proceedings of Second Cairo University Conference on Mechanical Design and Production MDP-2,* Cairo, Egypt (ed. M.A. El-Salamoni), 97–108. Cairo University.

Metwalli, S.M. and El Danaf, E.A. (1996). CAD and optimization of spur and helical gear sets. *Proceeding of the 21st ASME Design Automation Conference*, Irvine, California (18–22 August). Paper No:96-DETC/DAC-1433. ASME, New York, NY.

Meirovitch, L. (1986). *Elements of Vibration Analysis.* McGraw-Hill.

Mott, R.L. (1999). *Machine Elements in Mechanical Design*, 3e. Prentice Hall.

Stulen, F.B., Cummings, H.N., and Schulte, W.C. (1961). Preventing fatigue failures, part 5. *Machine Design* 33: 161.

Townsend, D.P. (ed.) (1992). *Dudley's Gear Handbook*, 2e. McGraw-Hill.

Internet Links

http://khkgears.net/ KHK stock gears: Gears.
http://www.bostongear.com/ Boston Gear: Gears, speed reduces, and others.
http://www.dupont.com/ Dupont: Plastic gears.
http://www.hellenicaworld.com/Greece/Technology/en/ArchimedesGears.html History: Archimedes Gears.
http://www.slideshare.net/palanivendhan/gear-manufacturing-process Slide Share: Gear manufacturing.
https://www.agma.org/standards/ AGMA, American Gear Manufacturers Association: Gear standards.

15

Helical, Bevel, and Worm Gears

In Chapter 14, spur gear drives have been introduced as power transmission and transformation elements. Other gear drives are also commonly used for power transmission and transformation such as helical, bevel, and worm gears. The spur gears have a cylindrical shape with teeth parallel to the cylinder centerline. ***Helical gears*** are also cylindrical, but the teeth are inclined to the cylinder centerline as shown in the diagrammatic sketch of Figure 15.1a. The teeth are represented by the inclined lines on the cylindrical surface. ***Bevel gears*** have a conical shape with teeth in the same plane as the cone centerline for straight bevel gears as shown in Figure 15.1b. The pinion axis is inclined to the bevel gear axis, and it is usually normal to it in most cases. ***Worm gears*** can be cylindrical, but the *worm* is like a screw thread tangentially driving the worm gear over its peripheral teeth with the worm axis normal to the worm gear axis as shown in the simplified diagrammatic sketch of Figure 15.1c.

The geometry of the teeth, kinematic definitions, and usual types and standards of these gears are introduced in this chapter. Forces generated during meshing and power transmission are identified. These forces that are acting on the teeth of the meshing gears and their effects in developing stresses on the teeth and gear design are discussed. Synthesis procedure utilizes computer-aided design (*CAD*), and optimization tools to generate better gearing design are projected. Constructional details for gear set assembly of sample gearboxes are also expected.

Symbols and Abbreviations

The adapted units are [in, lb, psi] or [m, kg, N, Pa], others given at each symbol definition ([k...] is 10^3 [M...] is 10^6 and [G...] is 10^9).

Symbol	Quantity, Units (adopted)
A_G	Gear cone distance
A_P	Pinion cone distance
B_{BC}	Material property factor (ratio of S_{ab}/S_{ac})
B_v	Quality constant for velocity factor equation
C_g	Gear center distance
C_P	Elastic coefficient $[MPa]^{1/2}$, $[psi]^{1/2}$
C_W	Center distance of worm gear set
d_g	A gear pitch diameter
d_p	Pinion pitch diameter
d_W	Worm pitch diameter

Machine Design with CAD and Optimization, First Edition. Sayed M. Metwalli.

Companion website: www.wiley.com/go/metwalli/machine

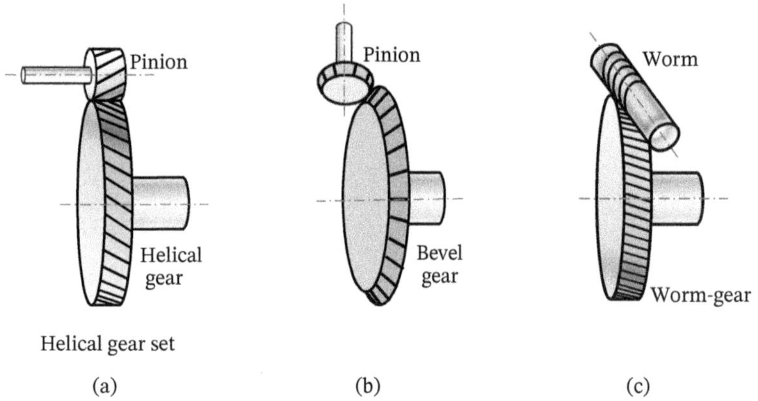

Figure 15.1 Simple diagrammatic sketches of helical, bevel, and worm gears: (a) helical gear set, (b) bevel gear set, and (c) worm gear set.

Symbol	Quantity, Units (adopted)
$\boldsymbol{F}_a$	Axial force
f_g	Face width of a gear
f_G	Gear face width
$f_g{}^*$	Optimum face width
$\boldsymbol{F}_n$	Normal force to tooth profile
f_P	Pinion face width
$\boldsymbol{F}_r$	Radial force
$\boldsymbol{F}_t$	Tangential force
H	Transmitted power
h_w	Worm whole tooth *working depth*
I_g	Contact geometry factor
J_g	Bending geometry factor
k_c	Reliability factor
K_f	Fatigue stress concentration factor
k_m	Load distribution factor
k_{mb}	Load distribution constant
K_{MSF}	Material safety factor
k_o	Load application factor
$K_{\mathrm{SF}}\|_b$	Bending fatigue safety factor
$K_{\mathrm{SF}}\|_c$	Contact fatigue safety factor
K_{SFE}	Extra safety factors
K_v	Dynamic or velocity factor
l_W	Worm lead
L_W	Warm length (minimum)

Symbol	Quantity, Units (adopted)
m_n	Normal module
m_N	Load sharing ratio
$m_n{}^*$	Optimum normal module
m_t	Transverse module
N_g	Number of teeth on a gear
n_g	Gear ratio (N_G/N_P)
$N_g{}^v$	Virtual number of teeth
$N_P{}^*$	Optimum number of pinion teeth
$N_P{}^v, N_G{}^v$	Virtual number of teeth for pinion and gear
N_W	Number of starts or threads of worm
p_d	Diametral pitch [teeth/in]
$p_d{}^*$	Optimum diametral pitch [teeth/in]
p_{dt}	Transverse diametral pitch
p_n	Normal circular pitch
p_t	Transverse pitch
p_x	Axial pitch
Q_V	AGMA quality number
r_{bG}	Gear back cone radius
r_{bP}	Pinion back cone radius
r_{bP}, r_{bG}	Base circle radii for pinion and gear
$\boldsymbol{r}_P$	Pinion pitch circle radius
r_p, r_G	Pitch radii magnitudes for pinion and gear
r_W	Worm radius
S_{ab}	Allowable bending fatigue strength
S_{ac}	Allowable contact fatigue strength
S_{cf}	Contact fatigue strength
S_{eg}	Endurance strength limit of gear
S_f	Fatigue strength
t_{av}	Average gear radius
$\boldsymbol{T}_{in}$	Input torque
$\boldsymbol{T}_P$	Pinion torque
$\boldsymbol{T}_W$	Worm input torque
$\boldsymbol{v}_G$	Worm tangential pitch point velocity
$\boldsymbol{v}_t$	Tangential pitch point velocity
v_t	Pitch point velocity magnitude
$v_{t,max}$	Speed limit
$\boldsymbol{v}_W$	Worm tangential pitch point velocity
Y_L	Modified *Lewis* factor
Z	Length of line of action
Z_N	Life factor

Symbol	Quantity, Units (adopted)
Greek	
η_W	Worm gear efficiency
γ	Pinion pitch cone angle
Γ	Gear pitch cone angle
λ_W	Worm lead angle
μ_k	Kinetic coefficient of friction
ω_G	Angular velocity of the worm gear
ω_P	Pinion angular velocity
ω_W	Angular velocity of the worm
ϕ_n	Normal pressure angle
ψ	Helix angle
ψ^*	Optimum helix angle
ϕ_t	Transverse pressure angle
σ_{ab}	Allowable bending fatigue strength
σ_{ab}	Bending fatigue stress
$\sigma_{ab}\vert_{max}$	Maximum bending fatigue stress
$\sigma_{ab}\vert_{Synth.}$	Synthesis contact fatigue stress
σ_{ac}	Contact surface fatigue stress
$\sigma_{ac}\vert_{max}$	Maximum contact fatigue stress

15.1 Helical Gears

The ***3D*** models of some helical gear sets are shown in Figure 15.2. Figure 15.2a is an isometric view of frequently encountered *helical gear* set. Figure 15.2b is a *crossed helical* gear set that is utilized in some applications. Figure 15.3 is showing other compound helical gears in ***3D*** isometric views. Figure 15.3a is a *double helical* gear set that can be treated as two common helical gear sets side by side. Figure 15.3b shows a *herringbone* gear set. It can also be treated as two common helical gear sets side by side without in between space. These are usually employed to eliminate the axial forces generated upon gear engaging and operation. Details of force analysis are given in Section 15.1.3. Further details of the frequently encountered helical gear sets are the focus of this section. Crossed helical and other types of helical gears can be similarly treated and can be obtained from handbooks, standards, and codes dedicated to these special gears; see, e.g. AGMA (2016).

15.1.1 Types and Utility

Some of the different types of helical gear sets are given in Figures 15.2 and 15.3. The common helical gear sets of Figure 15.2a, the double helical gear sets, and the herringbone gear sets of Figure 15.3 are the usual helical gears that have the same treatment in this section.

Helical gears, double helical gears, and herringbone gears are used for parallel shafts. For nonparallel shafts and with nonintersecting centerlines, crossed helical gears can be used.

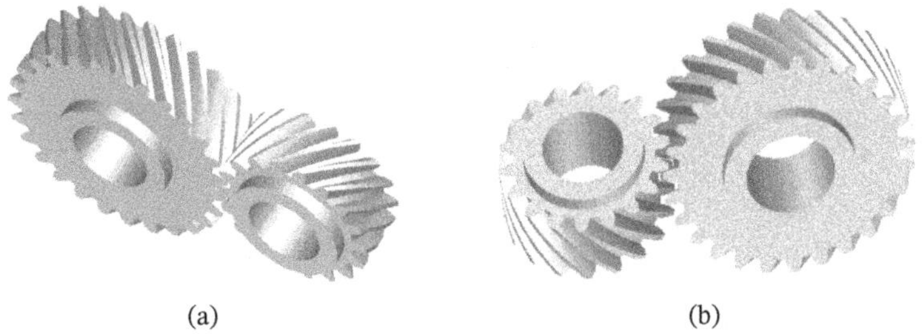

Figure 15.2 Helical gears in ***3D*** isometric view: (a) frequently encountered helical gear set and (b) crossed helical gear set that are utilized in some applications.

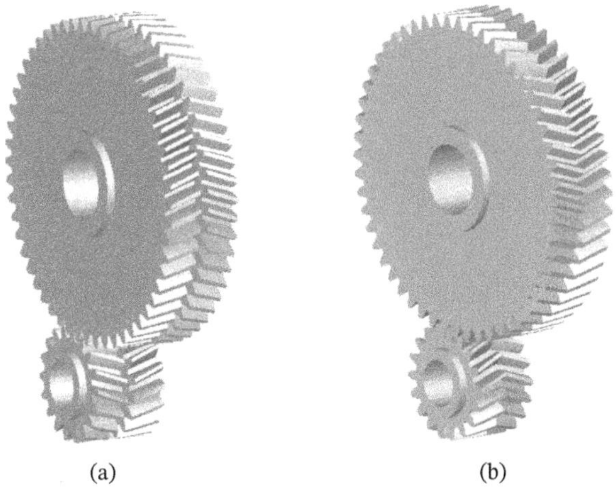

Figure 15.3 Other compound helical gears in ***3D*** isometric view: (a) double helical gear set and (b) herringbone gear set.

Helical gears usually run quieter than spur gears due to longer and gradual engagement along the spiral teeth. In addition, helical gears carry more loads and transmit more power than spur gears of the same size. On the average, helical gears transmit about 1.5–1.6 times the transmitted power of the same size spur gears. They, however, are more expensive to produce due to needed higher accuracy and surface finish. They also produce axial forces, while spur gears do not.

15.1.2 Definitions, Kinematics, and Standards

The form of a helical gear is normally cylindrical, but the teeth are of spiral formation; see Figure 15.2. *Involute helicoid* tooth profile is generated during helical gear manufacturing. This ensures longer teeth in contact with each other. It also produces more teeth in contact during pinion and gear meshing. Figure 15.4 demonstrates that an *involute helicoid* is generated from the simulated path of the grayish sheet having an inclined or slanted edge with the helix angle ψ (Figure 15.4a). When the grayish sheet is wrapped around the base cylinder and unwinds, the slanted edge generates the involute helicoid as shown in Figure 15.4b. Each tooth of the helical gear is produced by a process that generates such an involute helicoid. *Hobbing* or milling machines can produce such a tooth form; see, e.g. Townsend (1992).

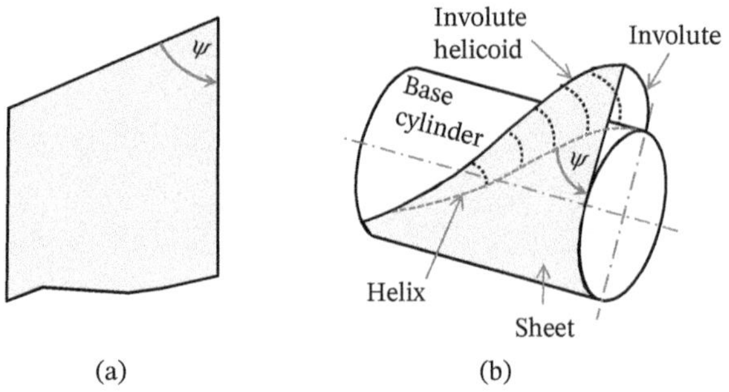

Figure 15.4 *Involute helicoid* simulated generation. (a) The grayish sheet having an inclined or slanted edge with the helix angle ψ. (b) The grayish sheet is wrapped around the base cylinder and unwinds to generate the involute helicoid.

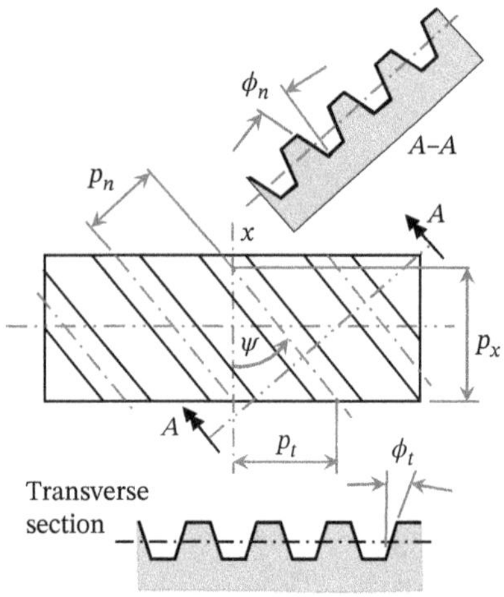

Figure 15.5 Helical gear parameters for the normal (n) section A–A, transverse section (t), and axial (x) directions. The variables in parenthesis are used as subscripts.

As for the spur gears, the involute and involute helicoid surfaces have the rolling properties at the pitch circle and thus a much lower *sliding friction* occurrence between such mating surfaces. The surfaces roll on each other's at pitch points intersection.

Figure 15.5 presents different helical gear parameters for normal (n) section (A–A section), transverse (t) section, and axial (x) directions. The gear axis is set in the x direction. The helical gear view shows the teeth and their centerlines inclined with the gear centerline at an angle, which is the *helix angle* ψ. The normal distance between the teeth centerlines is the *normal pitch* p_n of the helical gear. This is an important parameter that is coinciding with the standard spur gears particularly in the normal section A–A in Figure 15.5. The *normal pressure angle* ϕ_n is the standard pressure angle of spur gears, namely, 20° and 25°. This normal pressure angle is the one used in specifications and manufacturing tools such as milling cutters. The helical gear view in Figure 15.5 also shows the *axial pitch* p_x, which is the distance between the teeth centerlines in the x direction. The transverse section in Figure 15.5 defines the *transverse pitch* p_t and the *transverse pressure angle* ϕ_t. The relations between these transverse properties should hold with those of the spur gears as will be given later.

From Figure 15.5, the relation between the *normal pitch* p_n and the *transverse pitch* p_t is given as follows:

$$p_n = p_t \cos \psi \tag{15.1}$$

The relation between the transverse pitch p_t and the normal pitch p_n is then given by

$$p_t = \frac{p_n}{\cos \psi} \tag{15.2}$$

The relation between the *transverse pressure angle* ϕ_t and the *normal pressure angle* ϕ_n is

$$\cos\psi = \frac{\tan\phi_n}{\tan\phi_t} \tag{15.3}$$

Since the normal section of the helical gear is the important one in manufacturing, the gear module in the normal direction is the *standard normal module* m_n just as the standard *normal pressure angle* ϕ_n; see Figure 15.5. The normal module standards are given in Chapter 14, and they are given in millimeters [mm]; see Table 14.2. The important relation between the ***normal module*** m_n and the *transverse module* m_t is given by

$$m_n = m_t\cos\psi \tag{15.4}$$

For the transverse plane in Figure 15.5, the spur gear relations hold for the helical gears, and thus one has the following Eqs. (15.5) and (15.6):

$$m_t = \frac{d_g}{N_g} \tag{15.5}$$

$$p_t = \frac{\pi d_g}{N_g} \tag{15.6}$$

where the *transverse module* m_t is the module that has been used in the spur gears as the normal module, d_g is the *gear diameter* (pitch diameter in the transverse plane), and N_g is the *number of teeth on a gear*. One should be careful to note that the *transverse module* m_t is the module that has been used in the spur gears and it was given the symbol m_n because it must be the *standard normal module*. The *transverse module* m_t for the helical gears, however, does not have to be standard, and Eq. (15.5) would render m_t as a possible decimal number. This is apparent from the following development. From Eqs. (15.4) and (15.5), one gets

$$d_g = N_g\, m_t = \frac{N_g m_n}{\cos\psi} \tag{15.7}$$

As the *number of teeth* N_g is an integer and the normal module m_n is a whole or an integer number, then the transverse module m_t may not be an integer or a whole number. The gear or *pinion diameter* d_p with the mating *gear diameter* d_g can both provide an integer *center distance* by adjusting the helix angle ψ in Eq. (15.7) to satisfy that goal; see Example 15.1.

From Eq. (15.7) the following relation should then hold:

$$p_t = \pi\ m_t \tag{15.8}$$

The ***diametral pitch*** p_d used in the US system is the number of teeth per inch of gear diameter in the normal plane, and p_{dt} is the *transverse diametral pitch* in the rotation plane or (see Eq. (14.8))

$$p_{dt} = \frac{N_g}{d_g|_{[in]}}\ \left[\text{teeth/in}\right], \qquad p_{pt} = p_d\cos\psi\ \left[\text{teeth/in}\right], \qquad p_d|_{[\text{teeth/in}]} = \frac{25.4}{m_n|_{[\text{mm}]}} \tag{15.9}$$

The standard diametral pitch values are given in Table 14.3. These are also measured in the normal plane as for the modules.

The *virtual number of teeth* $N_g^{\,v}$ is useful for manufacturing and is given by the following relation:

$$N_g^v = \frac{N_g}{\cos^3\psi} \tag{15.10}$$

Since cutting of teeth is performed perpendicular to the normal plane, the space between teeth is larger for large helix angles; the milling cutter will have this to account for. That is why gear milling cutters have the range of the suitable teeth number itched on their sides. The virtual number of teeth should be within that suitable range itched on the *gear milling cutter*. For spur gears, the virtual number of teeth is the same as the number of teeth.

15.1.3 Force Analysis

Due to power transmission, the contact between mating gears experiences a *transmission force* $\boldsymbol{F}_n$ normal to the contact line. It is distributed along that line, but herein it is treated as a concentrated force at the line midpoint as shown in Figure 15.6. The *resultant normal force* $\boldsymbol{F}_n$ has components as a *radial force* $\boldsymbol{F}_r$ in the radial direction, an *axial force* $\boldsymbol{F}_a$ in the axial direction, and a *tangential force* $\boldsymbol{F}_t$ tangent to the helical gear pitch cylinder. The axial force $\boldsymbol{F}_a$ and the radial force are exaggerated just to clearly show the angles, where the helix angle ψ and other angles are also exaggerated than usual.

The power transmission is occurring only through the *tangential force* $\boldsymbol{F}_t$, which is the only force exerting effort or doing work. It is the only moving force that is applying a *gear* or *pinion torque* $\boldsymbol{T}_P$ on the helical pinion or gear. If the tangential force is known, the radial and axial forces are then obtainable from the *transverse pressure angle* ϕ_t, the *normal pressure angle* ϕ_n, and the *helix angle* ψ, as shown in Figure 15.6. The *tangential force* $\boldsymbol{F}_t$ is then obtained from the power transmission H as follows:

$$H = \boldsymbol{T}_P \cdot \boldsymbol{\omega}_P = \left(\boldsymbol{r}_P \times \boldsymbol{F}_t\right) \cdot \boldsymbol{\omega}_P = \boldsymbol{F}_t \cdot \boldsymbol{v}_t \tag{15.11}$$

where $\boldsymbol{\omega}_P$ is the *angular velocity* of the pinion in [rad/s], $\boldsymbol{r}_P$ is the *pitch circle* radius of the pinion, and $\boldsymbol{v}_t$ is the tangential *pitch point velocity* $\boldsymbol{v}_t = \omega_P\,(d_P/2) = \omega_P r_P$. From Eq. (15.11), one can then get

$$\boldsymbol{F}_t = \frac{H}{r_P \omega_P} = \frac{H}{\boldsymbol{v}_t} \tag{15.12}$$

From Figure 15.6, the resultant *normal force* $\boldsymbol{F}_n$, the *radial force* $\boldsymbol{F}_r$, and the *axial force* $\boldsymbol{F}_a$ are then given by

$$\boldsymbol{F}_n = \frac{\boldsymbol{F}_t}{\cos\psi \cos\phi_n} \tag{15.13}$$

$$\boldsymbol{F}_r = \boldsymbol{F}_n \sin\phi_n \tag{15.14}$$

$$\boldsymbol{F}_a = \boldsymbol{F}_n \cos\phi_n \sin\psi \tag{15.15}$$

And for check,

$$\boldsymbol{F}_n = \sqrt{\boldsymbol{F}_t^2 + \boldsymbol{F}_r^2 + \boldsymbol{F}_a^2} \tag{15.16}$$

This is the useful way to proceed with the evaluation of forces in the design process. First to evaluate the tangential force $\boldsymbol{F}_t$, find the resultant transmission normal force $\boldsymbol{F}_n$, and then calculate the radial force $\boldsymbol{F}_r$ and the axial force $\boldsymbol{F}_a$. These forces are typically needed to design the shaft on which the gear is mounted.

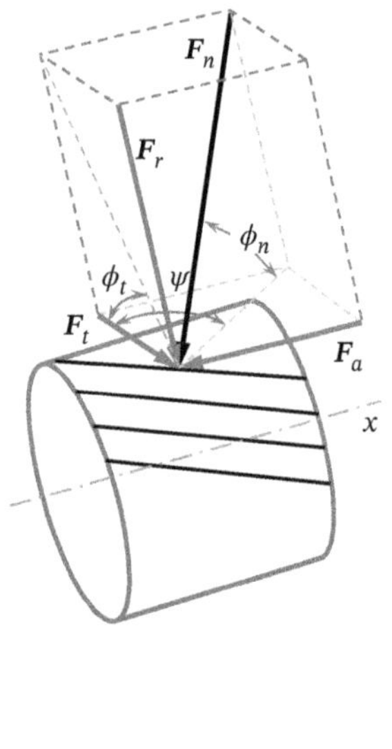

Figure 15.6 Helical gear forces. The grayish cylinder lines represent the bevel gear pitch cylinder. The black lines are some tooth pitch lines, and the resultant force $\boldsymbol{F}_n$ is normal to the tooth surface at the pitch line contact with mating gear.

Example 15.1 A gearbox transmits a maximum power of 75 [kW] or 100.58 [hp] at 1200 [rpm] input speed. The gear reduction ratio is 4. The standard helical gearing pressure angle is 20°. Standard module is 5 [mm], or the standard diametral pitch of 5 [teeth/in] is used, and the number of input pinion teeth is 23. The gear set must have a round figure center distance of 300 [mm] or 12 [in] since the set is part of transmission with more sets to engage at different gear shifts on the same two shafts. Determine the forces between gears in SI and the US systems.

Solution

Data: $H_{\text{kW}} = 75$ [kW] = 75 000 [W] or $H_{\text{hp}} = 100.58$ [hp], $N_{\text{in,rpm}} = N_{\text{in}} = 1200$ [rpm], and $\omega_P = 2\pi\, N_{\text{in}}/60 =$ 125.66 [rad/s].

The gear ratio $n_g = \omega_P/\omega_G = 4$, $\phi_n = 20°$, $m_n = 5$ [mm] or $p_d = 5$ [teeth/in], and $N_P = 23$. The number of gear teeth $N_G = 4(23) = 92$. The center distance $C_g = ½\,(d_P + d_G) = 300$ [mm] or 12 [in].

The constraint center distance is obtained by changing the helix angle ψ. With the gear ratio of 4, the gear diameter d_g is defined as $d_G = 4d_P$. From the center distance C_g, the pinion diameter is defined as follows:

$$C_g = 0.5\left(d_P + d_G\right) = 0.5\left(d_P + 4d_P\right) = 2.5\ d_P$$
$$d_P = 0.4C_g = 0.4\,(300) = 120\ \ [\text{mm}]$$
$$d_P = 0.4C_g = 0.4\,(12) = 4.8\ \ [\text{in}] \tag{a}$$

The gear diameters are then $d_G = 480$ [mm] or 19.2 [in]. Using Eq. (15.7), one gets

$$\psi = \cos^{-1}\left(\frac{N_g m_n}{d_g}\right) = \cos^{-1}\left(\frac{23\,(5)}{120}\right) = \cos^{-1}(0.958\,333\,33) = 16.597\,842° \tag{b}$$

The number of digits should be enough to get the helix angle set as accurate as the machine tool can operate. One might need to specify the angle into degrees, minutes, and seconds. This might give 16° 35′ 52.23″. One should note that using the US system and using Eqs. (15.7) and (15.9) and using the diameter *dg* in [in], and not in [mm] would get the following:

$$\psi = \cos^{-1}\left(\frac{N_g m_n}{d_g}\right) = \cos^{-1}\left(\frac{N_g}{d_g\,(p_d)}\right) = \cos^{-1}\left(\frac{(23)}{4.8\,(5)}\right) = \cos^{-1}(0.958\,333\,333) = 16.597\,842° \tag{c}$$

which is the same as Eq. (b).

The forces are affected by this angle as given next.

Forces between gears, and noting that speed should be in [rad/s] or $\omega_P = 2\pi(1200)/60 = 125.66$ [rad/s], the tangential force between the pinion and the gear is found through Eq. (15.12) such that

$$\boldsymbol{F}_t = \frac{H}{r_P\omega_P} = \frac{H}{v_P} = \frac{75\left(10^3\right)}{(0.120/2)\,(125.66)} = 9947.477\ [\text{N}] = 9.9475\ [\text{kN}]$$
$$\boldsymbol{F}_t = \frac{H_{hp}\ (6600)}{r_P\omega_P} = \frac{100.58\,(6600)}{(4.8/2)\,(125.66)} = 2201.1\quad [\text{lb}] \tag{d}$$

The pitch line velocity $v_t = \omega_P\ (d_P/2)$ is calculated in Eq. (d) as $v_t = 7.5396$ [m/s] or 301.58 [in/s]. The resultant transmission normal force $\boldsymbol{F}_n$, the radial force $\boldsymbol{F}_r$, and the axial force $\boldsymbol{F}_a$ are then given from Equs. (15.13), (15.14), and (15.15) to get

$$\boldsymbol{F}_n = \frac{\boldsymbol{F}_t}{\cos\psi\cos\phi_n} = \frac{9.9475}{\cos(16.597\,842)\cos(20)} = 11.046\ [\text{kN}]$$
$$\boldsymbol{F}_n = \frac{\boldsymbol{F}_t}{\cos\psi\cos\phi_n} = \frac{2201.1}{\cos(16.597\,842)\cos(20)} = 2444.2\ [\text{lb}]$$
$$\boldsymbol{F}_r = \boldsymbol{F}_n\sin\phi_n = 11.046\ \ \sin(20) = 3.77795\ [\text{kN}] = 3.778\ [\text{kN}]$$
$$\boldsymbol{F}_r = \boldsymbol{F}_n\sin\phi_n = 2444.2\quad \sin(20) = 835.97\ [\text{lb}]$$

$$F_a = F_n \cos\phi_n \sin\psi = 11.046 \quad \cos(20) \quad \sin(16.597\,842) = 2.9650 \text{ [kN]}$$

$$F_a = F_n \cos\phi_n \sin\psi = 2444.2 \quad \cos(20) \quad \sin(16.597\,842) = 656.09 \text{ [kN]} \tag{e}$$

As a check, one can use Eq. (15.16) to get

$$F = \sqrt{F_t^2 + F_r^2 + F_a^2} = \sqrt{(9.9475)^2 + (3.778)^2 + (2.965)^2} = 11.046 \text{ [kN]}$$

$$F = \sqrt{F_t^2 + F_r^2 + F_a^2} = \sqrt{(2201.1)^2 + (835.97)^2 + (656.09)^2} = 2444.2 \text{ [lb]} \tag{f}$$

which are the same values as that in Eq. (e) for SI and the US units.

15.1.4 Design Procedure

The procedure of helical gear synthesis is similar to that of spur gears. Only few factors are having different charts or relations to use. That is why one can just use the spur gear synthesis for classical helical gear design and simply reduce the transmitted power by 1.6–1.5. All previous information about materials and most factors are applicable. Tooth proportions for helical gears are the same as for spur gears; see Table 14.1. However, other geometrical relations are not necessarily the same, and the relations governing that should be used.

15.1.4.1 Initial Synthesis

For the initial synthesis of helical gears, one can rely on the *optimum design* of gears provided in the literature. The optimum design postulation depends on attaining the maximum allowable bending fatigue strength and the maximum allowable contact fatigue strength at the same time. This provides the maximum utilization of material capacities and thus results in the minimum gear set volume for this material. The optimum geometry has been attained for different material properties. The main *material property factor* is the ratio of the *allowable bending fatigue strength* to the *allowable contact fatigue strength*, which was given the symbol "*B*" in the applied reference. It was found that the material property factor "*B*" ranges mainly from about 0.2–0.6 (Metwalli and El Danaf 1996). In most cases, however, the range is about 0.27–0.35. Here and to have a less confusing representation, the symbol B_{BC} is used instead of symbol "*B*" in the reference. The subscript *BC* refers to the *bending* and *contact* fatigue strength ratios; see Section 14.4.2.

As an initial synthesis, the results of the optimization are used as an opening startup position of synthesis. The initial material property value is set at 0.35; the allowable bending fatigue strength is selected in the range of 100–400 [MPa], and other parameters are conservatively selected off the optimum cases. The chart of the optimum module m_n* in [mm] is shown in Figure 15.7 for a large range of input torques in [N m]. The grayish markers represent the optimum results for allowable bending stresses of 100 and 400 [MPa]. The lines represent the trend relations for the same allowable bending stresses. The optimum number of pinion teeth N_P* is given in Figure 15.8 against the allowable bending fatigue stress σ_{ab} in [MPa]. The grayish markers represent the optimum results. The lines represent the trend relations for the torques of 100, 1000, 10 000, and 100 000 [N m] from top to bottom. The optimum face width f_g* is defined in Figure 15.9 against the input torque in [N m]. The grayish markers represent the optimum results for allowable bending fatigue of 100 and 400 [MPa]. These markers are shown to indicate the dispersion. The lines represent the trend relations for the same allowable bending fatigue. It should be noted that these charts are generated for an average value of rotational speed, gear ratio, and gear quality for steel materials. The charts are for a *gear ratio* $n_g = 4$, pinion *angular velocity* $N_{rpm} = 1000$ [rpm], the AGMA *quality number* $Q_V = 8$, and *elastic coefficient* $C_P = 191$ [MPa]$^{1/2}$; see Section 14.4.3. The effect of shifting away from these averages is not appreciably pronounced and can be reviewed in Metwalli and El Danaf (1996). The consideration of different other cases should be taken care of in the detailed calculations for the specific case of the design. This should greatly reduce iterations toward the *optimum design* of those different cases as previously discussed in Section 14.6 in the case of *CAD* and optimization of spur gears.

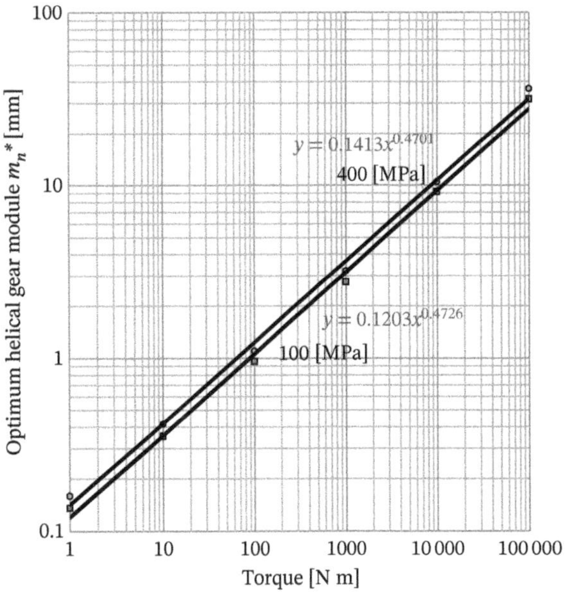

Figure 15.7 The optimum module m_n^* for helical gears. The grayish markers represent the optimum results for allowable bending stresses of 100 and 400 [MPa]. The lines represent the trend relations for the same allowable bending stresses.

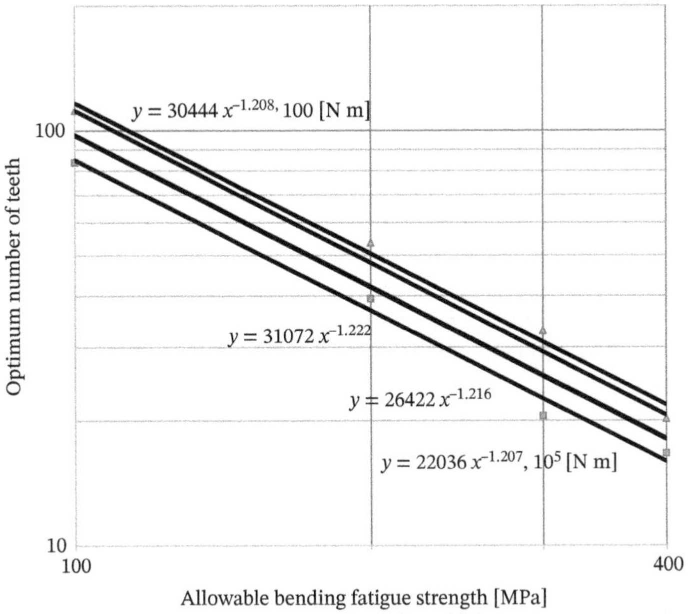

Figure 15.8 The optimum number of teeth N_p^* for helical gears. The grayish markers represent the optimum results. The lines represent the trend relations for the torques of 100, 1000, 10 000, and 100 000 [N m] from top to bottom.

To facilitate the utility of the adapted optimum design, the optimum curves of Figures 15.7–15.9 are used as the optimum synthesis *meta-model* or *surrogate functions* with the following relations. The relations are also adapted to the US system of units as shown next.

The *optimum module* m_n* in [mm] is modeled as a function of the input torque $\boldsymbol{T}_{\text{in}}$ [N m] (Figure 15.7) such that the optimum module and the *optimum diametral pitch* p_d* are

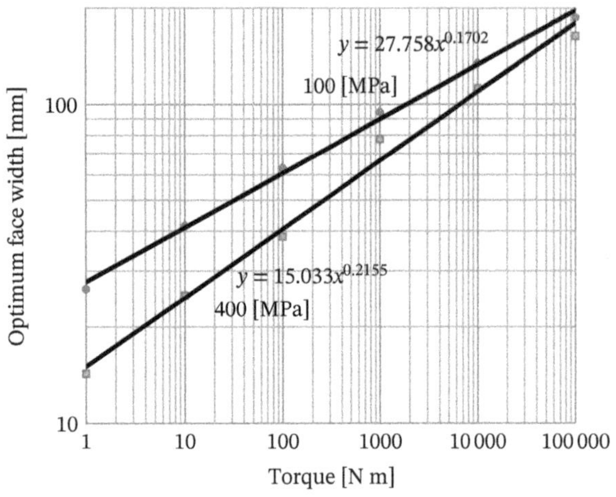

Figure 15.9 The optimum face width $f_g{}^*$ for helical gears. The grayish markers represent the optimum results for allowable bending fatigue of 100 and 400 [MPa]. The lines represent the trend relations for the same allowable bending fatigue.

$$m_n^* = (0.1413)\ \boldsymbol{T}_{\text{in}}^{0.4701} \qquad [\text{mm}]$$
$$p_d^* = 25.4/(0.1413)\left(0.113\ \boldsymbol{T}_{\text{in}}\right)^{0.4701} \qquad [\text{teeth/in}] \tag{15.17}$$

In the equation of the optimum diametral pitch (US), the input torque $\boldsymbol{T}_{\text{in}}$ is in [lb in]. This value of module or diametral pitch is for the higher *allowable bending fatigue strength* of *400* [MPa] or *58* [kpsi]. It is adapted for initial synthesis since it gives a higher module value and thus supposedly more conservative. The selected material represents average *through-hardened steel* commonly used or introductory *case-hardened steel* used for *typical helical* gears. For lower allowable bending fatigue strength of 100 [MPa] or 14.5 [kpsi], the following relation gives the optimum module and the optimum diametral pitch as:

$$m_n^* = (0.1203)\ \boldsymbol{T}_{\text{in}}^{0.4726} \qquad [\text{mm}]$$
$$p_d^* = 25.4/(0.1203)\left(0.113\ \boldsymbol{T}_{\text{in}}\right)^{0.4726} \qquad [\text{teeth/in}] \tag{15.18}$$

The smaller optimum module for the lower allowable contact fatigue strength is compensated for by the larger increase in the optimum number of teeth for the lower allowable contact fatigue strength as subsequently noted.

The *optimum number of pinion teeth* $N_P{}^*$ is modeled as a function of the *allowable bending fatigue strength* S_{ab} (100–400) [MPa] or (14.5–58) [kpsi] (Figure 15.8) such that

$$N_P^*|_{\text{SI}} = (30\,444)\ S_{\text{ab}}^{-1.208}$$
$$N_P^*|_{\text{US}} = (2954.16)\ S_{\text{ab}}^{-1.208} \tag{15.19}$$

This number of teeth is for the lower applied torque $\boldsymbol{T}_{\text{in}}$ of 100 [N m] or 885 [lb in] and thus higher speed for the same power. It is also adapted, since it should be conservatively inclined. For a much higher applied torque of 10^5 [N m] (or 885 [klb in]) and much lower speed for the same power, the relation is given by

$$N_P^*|_{\text{SI}} = (22\,036)\ S_{\text{ab}}^{-1.207}$$
$$N_P^*|_{\text{US}} = (2142.11)\ S_{\text{ab}}^{-1.207} \tag{15.20}$$

It should be noted that the number of teeth should be an integer. An adjustment to that effect is necessary. The difference in optimum number of teeth due to different transmitted torque is tolerable. At an allowable bending fatigue strength S_{ab} of 300 [MPa] or (43.5 [kpsi]), the optimum number of pinion teeth is about 31 for the lower applied torque of 100 [N m] (or 885 [lb in]) and 22 for the much higher applied torque of 10^5 [N m] (or 885 [klb in]). Using Figure 15.8 would help in the selection of the number of teeth at the applied torque. Adopting the lower value of torque would thus be for higher speeds at the same power. This would produce a smaller yet safe gear.

Adjusting this value can be performed in the provided ***CAD_Helical_Gears.m***, which is available through the ***Wiley website*** of this textbook.

The optimum face width f_g* in [mm] (Figure 15.9) is modeled as a function of the input torque T_{in} in [N m] or in [lb in] such that

$$f_g^*|_{\text{SI}} = (27.758)\ \boldsymbol{T}_{\text{in}}^{0.1702}$$
$$f_g^*|_{\text{US}} = (0.754\,07)\ \boldsymbol{T}_{\text{in}}^{0.1702} \tag{15.21}$$

where f_g* is in [mm] or in [in] for SI or the US, respectively. These optimum values are used for the allowable bending fatigue strength S_{ab} of 100 [MPa] or (14.5 [kpsi]) and *material property factor* B_{BC} of 0.35. This selection of B_{BC} is for median or regular ductile gear material; see gear materials in Section 14.4.3. This face width is for the lower fatigue strength and thus higher for higher fatigue strength; see Figure 15.9. It may be adapted, since it should be conservatively inclined. For a much higher allowable bending fatigue strength S_{ab} of 400 [MPa] or (58 [kpsi]), the optimum face width f_g* in [mm] or in [in] is given by

$$f_g^*|_{\text{SI}} = (15.033)\ \boldsymbol{T}_{\text{in}}^{0.2155}$$
$$f_g^*|_{\text{US}} = (0.369\,98)\ \boldsymbol{T}_{\text{in}}^{0.2155} \tag{15.22}$$

where f_g* is in [mm] or in [in] for SI or the US, respectively. This gives a smaller face width as shown in Figure 15.9.

The optimum helix angle ψ* is also dependent on the allowable bending fatigue strength S_{ab}. At a constant value of allowable bending fatigue strength, the optimum helix angle ψ* is nearly constant. For allowable bending fatigue strength S_{ab} of 100 [MPa] or 14.5 [kpsi], the optimum helix angle ψ* is about 11.2° for low torque to 13.6° for a very high torque. For allowable bending fatigue strength S_{ab} of 200 [MPa] or 29 [kpsi], the optimum helix angle ψ* is about 17.6° for low torque to 20.2° for a very high torque. For allowable bending fatigue strength S_{ab} of 300 [MPa] or 43.5 [kpsi], the optimum helix angle ψ* is about 24.6° for low torque to 34.6° for a very high torque. For allowable bending fatigue strength S_{ab} of 400 [MPa] or 58 [kpsi], the optimum helix angle ψ* is constant at 34.8° for a low to a very high torque.

For different values than the adapted optimum or the assumed parameters, fine-tuning iteration is necessary to satisfy other objectives. The optimum Eqs. (15.17)–(15.21) are used for the US system provided that appropriate dimensions are used and substitution of $1/p_c$ for m_n while using dimensions in [in] instead of [mm]. Safeguard verification is important in that regard.

Defined relations in Eqs. (15.17)–(15.21) are for selected case of optimum design. Other cases are provided in Metwalli and El Danaf (1996). The general optimum trends, however, suggest some interesting observations. The one concerning the *optimum number of teeth* indicates that as the material strength diminishes the number of the teeth approaches infinity (Figure 15.8). This is reasonable since one can use a shaft with no teeth (infinite number of teeth with infinitesimal module) driving a rubber cylinder (infinite number of teeth with even smaller module). Note that the involute tooth form does not need to apply. The mechanism in that case is different, but the trend is reserved.

Synthesis general trend departing away from the adapted value incites the *optimum module* m_n* to decrease a little with the reduction of *allowable bending fatigue strength* S_{ab}. The *diametral pitch* p_d would do the opposite. The optimum module m_n*, however, would be higher for a lower value of the *material property factor* B_{BC}. The *optimum number of teeth* N_p* decreases with the decrease of the material property factor B_{BC}. The *optimum face width* f_g* decreases a little with the decrease of the material property factor B_{BC}. These observations about the optimum values are deduced from the trends depicted in Metwalli and El Danaf (1996).

Example 15.2 A reduction gear in a crane driven by electric motor at a maximum power of 75 [kW] or 100.58 [hp] and running at 1200 [rpm] is to be designed for a gear ratio of 4 : 1. Gears are to be of 20° full-depth helical type. Synthesize an initial helical gear set for such a type of application.

Solution

Data: Power transmitted $H_{kW} = 75$ [kW] or 100.58 [hp], rotational speed $N_{in} = 1200$ [rpm], and the gear ratio $n_g = 4$. The rotation speed $\omega_P = (2\pi\ N_{in}/60) = 125.66$ [rad/s].

The value of the maximum input torque T_{in} is obtained from Eq. (15.11) such that

$$\boldsymbol{T}_{in} = \frac{H_{kW}(10)^3}{\omega_P} = \frac{75(10)^3}{125.66} = 596.85 \quad [\text{N m}]$$

$$\boldsymbol{T}_{in} = \frac{H_{hp}\,(6600)}{\omega_P} = \frac{100.58\,(6600)}{125.66} = 5282.7 \quad [\text{lb in}] \tag{a}$$

According to Table 14.5, the quality number of a crane type of applications has the range of 5–7. The results of optimum design were based on a quality number of 8. This quality is higher than the required 5–7. It will be, however, more conservative to use better quality gears or accept higher safety factor out of the selected optimum solution.

From Figures 15.7–15.9, the synthesized geometry of the gear can be defined by the optimum of $m_n \approx 3$ [mm], $N_P \approx 30$, and $f_g \approx 70$ [mm] at $T_{in} \approx 600$ [N m] and $S_{ab} = 300$ [MPa] or 43.5 [kpsi]. To have more specific values, one could use Eqs. (15.17), (15.19), and (15.20). Using these initial synthesis equations, the expected optimum gear module m_n* or diametral pitch p_d* for a higher $S_{ab} = 400$ [MPa] or 58 [kpsi] are calculated such that

$$m_n^* = (0.1413)\ \boldsymbol{T}_{in}^{0.4701} = (0.1413)\,(596.85)^{0.4701} = 2.8515\ [\text{mm}]$$

$$p_d^* = 25.4/\,(0.1413)\,(0.113\ \ \boldsymbol{T}_{in})^{0.4701} = 25.4/\,(0.1413)\,(0.113\ \ (5282.7))^{0.4701} = 8.907\ [\text{teeth/in}] \tag{b}$$

For lower $S_{ab} = 100$ [MPa] or 14.5 [kpsi], Eq. (15.18) gives

$$m_n^* = (0.1203)\ \boldsymbol{T}_{in}^{0.4726} = (0.1203)\,(596.85)^{0.4726} = 2.4668\ [\text{mm}]$$

$$p_d^* = 25.4/\,(0.1203)\,(0.113\ \ \boldsymbol{T}_{in})^{0.4726} = 25.4/\,(0.1203)\,(0.113\ \ (5282.7))^{0.4726} = 10.296\ [\text{teeth/in}] \tag{c}$$

The smaller optimum module or larger diametral pitch for the lower allowable contact fatigue strength of 100 [MPa] or 14.5 [kpsi] in Eq. (c) may not be used for extra safety. The optimum module and optimum diametral pitch should be changed to the closest available standards of, say, 2.5 [mm] or 10 [teeth/in]; see Tables 14.3 and 14.4.

The optimum number of pinion teeth N_P* is calculated from the optimum at allowable bending fatigue strength $S_{ab} = 300$ [MPa] or 43.5 [kpsi] in Eqs. (15.19) and (15.20) for torques of 100 [N m] or 885 [lb in] and 10^5 [N m] or 885 [klb in], respectively, such that

$$N_P^*|_{SI} = (30\,444)\ S_{ab}^{-1.208} = (30\,444)\ (300)^{-1.208} = 30.984$$

$$N_P^*|_{US} = (2954.16)\ S_{ab}^{-1.208} = (2954.16)\ (43.5)^{-1.208} = 30.984$$

$$N_P^*|_{SI} = (22\,036)\ S_{ab}^{-1.207} = (22\,036)\ (300)^{-1.207} = 22.555$$

$$N_P^*|_{US} = (2142.11)\ S_{ab}^{-1.207} = (2142.11)\ (43.5)^{-1.207} = 22.552 \tag{d}$$

The values of optimum number of teeth for the SI and US systems are the same for the same torques. The values at different torques are different, and interpolation for the applied torque should be estimated at integer value only. An optimum number of pinion teeth at 30 might be acceptable. The optimum number of gear teeth is then $4(30) = 120$.

The optimum face width f_g* is calculated from Eqs. (15.21) and (15.22) for 100 [MPa] or 14.5 [kpsi] and 400 [MPa] or 58 [kpsi], respectively, to get

$$f_g^*|_{SI} = (27.758)\ \boldsymbol{T}_{in}^{0.1702} = (27.758)\,(596.85)^{0.1702} = 82.384\ [\text{mm}]$$

$$f_g^*|_{US} = (0.754\,07)\ \boldsymbol{T}_{in}^{0.1702} = (0.754\,07)\ (5282.7)^{0.1702} = 3.2437\ [\text{in,}]$$

$$f_g^*|_{SI} = (15.033)\ \boldsymbol{T}_{in}^{0.2155} = (15.033)\,(596.85)^{0.2155} = 59.60\ [\text{mm}]$$

$$f_g^*|_{US} = (0.36998)\ \boldsymbol{T}_{in}^{0.2155} = (0.369\,98)\ (5282.7)^{0.2155} = 2.346\,67\ [\text{in,}] \tag{e}$$

The value of the face width in the SI system and the value of the face width in the US system are very closely the same. Interpolation for the applied torque should be estimated at round figure values. Optimum face width of 65 [mm] and 2.6 [in] might be acceptable.

The optimum helix angle ψ^* is about 24.6° for low torque to 34.6° for a very high torque. The optimum helix angle ψ^* may be selected as 27°.

Larger values for the design parameters (except inversely p_d) would be safer, but would not necessarily be optimum. For realistic design, however, values should be integers or round figures anyway. The resulting stresses and safety factors are left to the detailed design calculations given hereafter.

15.1.4.2 Detailed Design

This section provides the *detailed design* of helical gears in the sense of repeated analysis. However, the usual start of the iteration is the initial synthesis Section 15.1.4.1. The number of cycles of repeated analysis to reach a satisfactory or an alternative optimum design objective is thus reduced. The design entails the definition of the geometry and the material. The initial synthesis adapted a material of *allowable bending fatigue strength* S_{ab} of 300 [MPa] or 43.5 [kpsi] and *allowable contact fatigue strength* S_{ac} of about $(S_{ab})/B_{BC} = 300/0.35 = 860$ [MPa] or 125 [kpsi]. These values might not be satisfactory for some applications. It might be low for extensively loaded cases or very high for very insubstantial, rough, inexpensive, or undemanding products. The choice of the material, therefore, necessitates some iteration about these cases. This and other details of loading specifics, severity, life cycling, reliability, etc. are covered in the following *detailed design* process. The *material set* that is usually employed in helical gear design is presented in Section 14.4.3 for spur gears. The materials for helical gears are usually the higher ends of these sets with contact strength to warrant a value of B_{BC} of about 0.35. The bending and the contact fatigue cases are covered in enough details in the following parts. The aim of the detailed design is to specifically define the material designation that conforms to the requirements. It should necessitate specifying the heat treatment, or the case-hardening details to end up with the required allowable strength and safety conditions.

The detailed design process is almost exactly like that for spur gear. Only few parameters are defined through a different figure or table. These are emphasized in the next bending and contact fatigue considerations. The detailed design process is tailored according to classical information and AGMA procedures starting at AGMA (1982) and AGMA (1999). AGMA, however, is constantly improving on that process to approach the real situations. To follow a standard, the current codes and procedures are to be followed. The discrepancies may be tolerable within design considerations.

Bending Fatigue

Conforming to the spur gears, the ***bending fatigue stress*** σ_{ab} is given by (see Eq. (14.34))

$$\sigma_{ab} = \frac{\mathbf{F}_t K_v}{f_g\ m_t J_g} \quad \text{for SI}$$

$$\sigma_{ab} = \frac{\mathbf{F}_t K_v\ p_{dt}}{f_g\ J_g} \quad \text{for US} \tag{15.23}$$

where J_g is the remodified *Lewis* factor for helical gears, K_v is the *velocity factor* as for spur gears, $\mathbf{F}_t$ is the tangential force, f_g is the *face width* of the gear, and m_t is the *transverse module*. For the US system, the p_{dt} is the transverse diametral pitch as defined in Eq. (15.9). The remodified *Lewis* factor J_g is the *bending geometry factor*, which replaces the modified *Lewis* factor Y_L; see Section 14.4. This form factor J_g includes the *fatigue stress concentration factor* K_f for gear root effect. The *bending geometry factor* is then defined as $J_g = Y_L/K_f$. The *stress concentration factor* K_f for gear root is obtained by Dolan and Broghamer (1942) using *photoelasticity* and would still be applicable. An equation of J_g for 20° pressure angle has the following estimated form.

The geometry factor J_g is a function of the helix angle ψ, the pinion number of teeth N_{g1}, and the mating gear number of teeth N_{g2}. The geometry factor J_g as a function of the helix angle ψ, and pinion number of teeth N_P or

N_{g1}, is given by

$$J_g = -\left(0.000\,145\ N_g^{0.161597}\right)\psi^2 + \left(0.007\,237\ N_g^{0.05324}\right)\psi + 0.342\,121\ N_g^{0.100154} \tag{15.24}$$

To account for the number of teeth of mating gear different from 75, the adjusting multiplier factor K_{g2} affects Eq. (15.24) so that one has the following average relation:

$$J_g = \left\{-\left(0.000\,145\ N_{g1}^{0.161597}\right)\psi^2 + \left(0.007\,237\ N_{g1}^{0.05324}\right)\psi + 0.342\,121 N_{g1}^{0.100154}\right\}\left(0.8691 N_{g2}^{0.0295}\right) \tag{15.25}$$

It is supposed that the subscript g1 is for the gear to which the geometry factor is desired and the subscript g2 is for the mating gear. This relation is within design accuracy for loads applied at the *highest point of a tooth contact.* The limit of N_{g2} in Eq. (15.25) is about 500 teeth for the value of J_g to be within representative design accuracy. The values obtained by Eq. (15.25), however, would be close enough compared to the values suggested by AGMA 218.01 (1982) and AGMA (1999) for loads applied at the *highest point of a tooth contact.* Other relations exist such as those given by Mitchener and Mabie (1982). The relation in Eq. (15.25) is, however, more convenient to use with helical gears when representative design accuracy is acceptable. The proper value of J_g can be attained by the standard code followed by the designer such as the current AGMA and ISO standards and codes such as ISO 6336 (2006) and (2016).

The dynamic or *velocity factor* K_v that is more accurately defined as a function of *pitch line velocity* v_t and the AGMA gear *quality number* Q_v are obtained from the following AGMA relations (see ANSI/AGMA 2001, 2004, 2001-D4) as given in the spur gear treatment as Eq. (14.39) and reiterated herein such that

$$K_v|_{SI} = \left(\frac{50 + 56\left(1 - B_v\right) + \sqrt{200\left(v_t\right)}}{50 + 56\left(1 - B_v\right)}\right)^{B_v}$$

$$K_v|_{US} = \left(\frac{50 + 56\ \left(1 - B_v\right) + \sqrt{\left(v_t\right)_{ft/min}}}{50 + 56\left(1 - B_v\right)}\right)^{B_v}$$

$$B_v = 0.25\left(12 - Q_v\right)^{2/3} \tag{15.26}$$

It should also be noted that each adapted quality number has a limiting speed associated with it. The *speed limit* $v_{t,max}$ is suggested by AGMA to have the following values, which are the same as Eq. (14.40) of spur gears (ANSI/AGMA 2004, 2001-D4):

$$v_{t,max}|_{SI} = \left(50 + 56\left(1 - B_v\right) + \left(Q_v - 3\right)\right)^2/200$$

$$v_{t,max}|_{US} = \left(50 + 56\left(1 - B_v\right) + \left(Q_v - 3\right)\right)^2 \tag{15.27}$$

The term B_v in Eq. (15.27) is defined in Eq. (15.26). If the value in a problem is close to these values in Figure 14.16, the limits of Eq. (15.27) ought to be observed.

The *bending fatigue stress* σ_{ab} of Eq. (15.23) is now possible to evaluate since all components have been defined in Eqs. (15.25)–(15.27).

The applied bending fatigue stress is not only the stress defined by Eq. (15.23). This relation does not account for the service loading conditions such as the steadiness or not of the power source (electric motor, combustion engines of multi or single cylinders) and the type of driven machines. Table 14.7 for spur gears suggests a *load application factor* k_o to account for these conditions of different classes of power sources and driven machines. This factor should be included in the evaluation of the applied stress since it accounts for the dynamic changes of the applied loads. Another factor that should also be included is the *load distribution factor* k_m. This factor accounts for the inevitable variation of the way the load is distributed over the face of the teeth. Eq. (15.23) assumes perfect uniform distribution over the whole face width. Due to deflections of the shafts on which gears are mounted and teeth deflections accordingly, the load distribution is not expected to be uniform or centered over the teeth. That nonuniform distribution is clearly a function of each gear face width f_g and should be included in the evaluation of the applied stress. This is since some portion of the tooth will have higher loads than the rest of the tooth. A suggested fitted relation to account for the effect of *load distribution factor* k_m is given as defined in Metwalli and El Danaf (1996) and as previously used for spur gears so that

$$k_m|_{SI} = -4(10)^{-7} f_g^2|_{[mm]} + 0.0011 f_g|_{[mm]} + 1.544$$
$$k_m|_{US} = -6.28(10)^{-4} f_g^2|_{[in]} + 0.0364 f_g|_{[in]} + 1.523 \quad (15.28)$$

The maximum value of k_m is considered as 2.0 for the face width of more than 500 [mm] or 20 [in].

The applied bending fatigue stress is not only the stress defined by Eq. (15.23). It should then be adjusted to include the *load application factor* k_o and the *load distribution factor* k_m. It is then more reasonable to incorporate the k_o and k_m factors into the probable maximum loading and thus into the ***bending fatigue stress*** σ_{ab}. A *maximum* ***bending fatigue stress*** $\sigma_{ab}|_{max}$ may then be rewritten as

$$\sigma_{ab}|_{max} = \frac{\boldsymbol{F}_t\, K_v k_o k_m}{f_g\ m_t\, J_g} \quad \text{for SI}$$
$$\sigma_{ab}|_{max} = \frac{\boldsymbol{F}_t\, K_v\ p_{dt}\, k_o k_m}{f_g\ J_g} \quad \text{for US} \quad (15.29)$$

For the SI system, usually the module m_n is substituted in [mm], and the tooth face width f_g is also substituted in [mm] to calculate the stress directly in [MPa]. It should also be noted that the *load distribution factor* k_m is included in the *optimum* data used in the *synthesis*. In the synthesis procedure, however, the *load application factor* k_o should be incorporated into the maximum input power, which materializes directly into the tangential force $\boldsymbol{F}_t$. The adjusted tangential force will then incorporate the *load application factor* k_o and the newly adjusted $\boldsymbol{F}_t = k_o \boldsymbol{F}_t$.

Contact Surface Fatigue

The *Hertzian* stress developed due contact is typically the same as for spur gears. The equations are as given in Section 14.4.3.3.

Conforming to the spur gears, the contact *surface fatigue stress* σ_{ac} is given by (see Eqs. (14.57), (14.58), (14.59), and (14.63))

$$\sigma_{ac} = C_p \sqrt{\frac{\boldsymbol{F}_n}{f_g\ d_P\ I_g}} \quad (15.30)$$

where $\boldsymbol{F}_n$ is the *normal gear force* or resultant force, C_p is the *elastic coefficient* (Eqs. (14.59) and (15.31)), d_P is the *pitch circle diameter* of the pinion, f_g is the gear *face width*, and I_g is the *contact geometry factor*. The elastic coefficient is given by (see Eq. (14.59))

$$C_p = \sqrt{\frac{1}{\pi \left(\frac{1-\nu_P^2}{E_P} + \frac{1-\nu_G^2}{E_G} \right)}} \quad (15.31)$$

As for spur gears, Table 14.8 displays the calculated values of the *elastic coefficients* C_p [MPa]$^{1/2}$ ([psi]$^{1/2}$) for different mating gear materials. The symbol I_g in Eq. (15.30) is the *contact geometry factor* given by

$$I_g = \frac{\cos\phi\ \sin\phi}{2m_N} \frac{n_g}{(n_g \pm 1)} \begin{cases} +\text{external gear} \\ -\text{internal gear} \end{cases} \quad (15.32)$$

where n_g is the *gear ratio* (N_G/N_P) as defined by Eqs. (14.2) and (14.4). Equation (15.32) is like Eq. (14.60) of spur gears except for the addition of the *load sharing ratio* m_N, where the load sharing ratio for spur gears m_N is considered as 1.0. The load sharing ratio m_N for helical gears is given by

$$m_N = \frac{p_n \cos \phi_n}{0.95Z} \tag{15.33}$$

where p_n is the normal circular pitch, ϕ_n is the normal pressure angle, and Z is the *length of line of action* in the transverse plane, which is given by (see Figures 14.3 and 14.4 of spur gears)

$$Z = \sqrt{(r_P + a)^2 - r_{bP}^2} + \sqrt{(r_G + a)^2 - r_{bG}^2} - (r_P + r_G) \sin \phi_t \tag{15.34}$$

where r_p and r_G are the *pitch radii* for pinion and gear, respectively, a is the *addendum*, and r_{bP}, r_{bG} are the *base circle radii* for pinion and gear, respectively. Note that the base circle radii are given by

$$r_{bP} = r_P \cos \phi_t, \quad r_{bG} = r_G \cos \phi_t \tag{15.35}$$

In applying Eq. (15.34), any of the radicals should be replaced by the third term if it is smaller than that term.

Like bending fatigue, the inclusion of the dynamic or *velocity factor* K_v is necessary to account for the effect of gear *quality* and *pitch line velocity* on the dynamics affecting the contact stress. The same K_v factor defined by Eq. (15.26) is used herein to include AGMA quality number Q_v in the calculations. This would also account for the stresses causing fatigue due to the dynamic loading action. The evaluated stress is thus dubbed the *contact fatigue stress* σ_{ac}, which is given by

$$\sigma_{ac} = C_P \sqrt{\frac{\boldsymbol{F}_n K_v}{f_g\ d_P\ I_g}} \tag{15.36}$$

As for the bending fatigue case, the relation in Eq. (15.36) does not account for the service loading conditions such as the steadiness or not of power source (electric motor, combustion engines of multi- or single cylinders) and the type of driven machines. Table 14.7 (of spur gears) suggests the *load application factor* k_o to account for different classes of power sources and driven machines, which has previously been used for the bending fatigue case. This factor should be included in the evaluation of the applied contact stress since it accounts for the dynamic changes of the applied loads. Another factor that should also be included is the *load distribution factor* k_m as was done for the bending case. This factor accounts for the inevitable variation of the way the load is distributed over the face of the teeth. Equation (15.36) assumes perfect uniform distribution over the whole face width. Due to deflections of the shafts on which gears are mounted and teeth deflections accordingly, the load distribution is not expected to be uniform or centered over the teeth. That nonuniform distribution is clearly a function of the gear face width f_g. The *load distribution factor* k_m should then be included in the evaluation of the applied contact stress since some portion of the tooth will have higher contact loads than the rest of the tooth. A previously suggested fitted relation (Eq. (15.28)) was used in bending fatigue for the same reason. It can also be used for contact fatigue stress. Therefore, including the *load application factor* k_o of Table 14.7 and the *load distribution factor* k_m as given by Eq. (15.28) into Eq. (15.36), the *maximum contact fatigue stress* $\sigma_{ac}|_{max}$ becomes

$$\sigma_{ac}|_{max} = C_P \sqrt{\frac{k_o k_m\ \boldsymbol{F}_n K_v}{f_g\ d_P\ I_g}} \tag{15.37}$$

It should be noted that the two factors k_o and k_m are affecting the applied normal force $\boldsymbol{F}_n$ and that is why they have been included under the radical. The *load application factor* k_o may alternatively be used to increase the applied transmission load $\boldsymbol{F}_n$ or as part of the safety factor. Usually for the SI system, the pinion diameter d_p is substituted in [mm], and the tooth face width f_g is also substituted in [mm] to calculate the stresses in [MPa]. No need to do similar shortcuts for the US system.

Material Set and Safety Factor

The material set and properties introduced in the spur gear chapter should also apply for the helical gears; see Section 14.4.3 and Table 14.6. Usually, however, the helical gears are used in demanding applications that their higher quality necessitates the better materials. Also employing the higher quality manufacturing is expected; see Davis (2005).

The same treatment of safety factor in spur gears is also applicable in helical gear treatment; see Section 14.4.3. In that sense, there is a *bending fatigue safety factor* $K_{SF}|_b$ as Eq. (14.48) and contact fatigue safety factor $K_{SF}|_c$ as Eq. (14.70).

As for spur gears, the bending fatigue safety factor $K_{SF}|_b$ can be as follows (see Eq. (14.48)):

$$K_{SF}|_b = \frac{\left(S_{eg}/K_{MSF}\right)}{k_o k_m \sigma_{ab}} = \frac{\left(S_{eg}/K_{MSF}\right)}{\sigma_{ab}|_{max}} = \frac{S_{ab}}{\sigma_{ab}|_{max}} \tag{15.38}$$

where S_{eg} is the gear *endurance limit* or *endurance strength* (or instead using the *fatigue strength* S_f defined by Eq. (14.45)), K_{MSF} is a *material safety factor*, $\sigma_{ab}|_{max}$ is the *maximum* ***bending fatigue stress*** given by Eq. (15.29), and S_{ab} is the *allowable fatigue strength*.

As for spur gears and considering the *reliability factor* k_c (Eq. (14.43)), the *material safety factor* K_{MSF}, and the *allowable contact fatigue strength* S_{ac} (Eq. (14.67)), the ***contact safety factor*** $K_{SF}|_c$ is defined as follows (see Eqs. (14.70) and (14.72)):

$$K_{SF}|_c = \frac{k_c Z_N \left(S_{ec}/K_{MSF}\right)}{\sqrt{k_o k_m}\ \sigma_{ac}} = \frac{k_c Z_N S_{ac}}{\sigma_{ac}|_{max}} \triangleright \frac{S_{ac}}{\sigma_{ac}|_{max}} \tag{15.39}$$

where the result in Eq. (15.39) is occurring at $k_c = 1$ for 99% reliability and the *life factor* $Z_N = 1$ if the *contact fatigue strength* S_{cf} is used instead of S_{ac}. These are the usual cases and if the case is different, the values of k_c and Z_N can be evaluated by Eqs. (14.43) and (14.68) for spur gears or consult standards such as AGMA.

If the proper *allowable fatigue strength* S_{ab} and *allowable contact fatigue strength* S_{ac} are used, the safety factors are expected to be about 1.0. If not, the safety factor would be an *extra safety factors* K_{SFE} over the allowable strength values.

Since the synthesized optimum contact stress includes the *load distribution factor* k_m (see Eqs. (15.36), (15.37), and (15.39)), one can get the following relation for the safety factor (not including the *load application factor* k_o):

$$K_{SF}|_c = \frac{S_{ac}}{\sigma_{ac}|_{max}}, \quad \text{and} \quad \sigma_{ac}|_{Synth.} = \sigma_{ac}\sqrt{k_m}$$

$$K_{SF}|_c = \frac{S_{ac}}{\sigma_{ac}} = \frac{S_{ac}\sqrt{k_m}}{\sigma_{ac}|_{Synth.}} \tag{15.40}$$

That is because the *synthesis contact fatigue stress* $\sigma_{ab}|_{Synth.}$ is including the *load distribution factor* k_m but not the *load application factor* k_o. The synthesis depends on applying the *load application factor* k_o directly to the load as sort of a direct safety factor.

Example 15.3 A reduction gear in a crane driven by electric motor at a maximum power of 75 [kW] or 100.58 [hp] running at 1200 [rpm] and for a gear ratio of 4 : 1 has been synthesized in Example 15.2. Gears are 20° full-depth helical type. Check the synthesized initial helical gear set for stresses and safety of such an application.

Solution

Data: Power transmitted $H_{kW} = 75$ [kW] or 100.58 [hp], rotational speed $N_{in} = 1200$ [rpm], and the gear ratio $n_g = 4$. The rotation speed $\omega_P = (2\pi\ N_{in}/60) = 125.66$ [rad/s]. Results of Example 15.2 with a value of B_{BC} of 0.35 are as follows:

- Input torque $\boldsymbol{T}_{in} = 596.85$ [N m] and 5282.7 [lb in].

- Initial synthesis for the optimum module of $m_n \approx 3$ [mm], $N_P \approx 30$, and $f_g \approx 70$ [mm] at $T_{in} \approx 600$ [N m] and $S_{ab} = 300$ [MPa] or 43.5 [kpsi].
- Calculated synthesis for optimum module and optimum diametral pitch are changed to the closest available standards of, say, 2.5 [mm] or 10 [teeth/in].
- Calculated synthesis for optimum number of pinion teeth N_P* is suggested as 30. One can start with N_P* = 30 and N_G* = 4(30) = 120.
- Calculated synthesis for optimum face width f_g* is 65 [mm] and 2.6 [in]. One can start with f_g* = 65 [mm] or 2.6 [in].
- Calculated synthesis for optimum helix angle ψ* is about 24.6° for low torque to 34.6° for a very high torque. It may be selected as 27°.

With the previous results of Example 15.2, one needs to calculate gear geometry and other parameters needed for calculating stresses.

With the normal module m_n and the diametral pitch p_d, the transverse module m_t and the transverse diametral pitch p_{dt} are obtained from Eqs. (15.4) and (15.9) such that

$$m_t = \frac{m_n}{\cos\psi} = \frac{2.5}{\cos 27} = 2.8058 \text{ [mm]}$$
$$p_{dt} = p_d \cos\psi = 10\cos 27 = 8.9101 \text{ [teeth/in]} \quad \text{(a)}$$

The pinion and gear diameters are defined from Eqs. (15.7) and (15.9) as

$$d_P = N_P\, m_t = 30\,(2.8058) = 84.174 \text{ [mm]}$$
$$d_G = N_G\, m_t = N_P\,(n_g)\, m_t = 30\,(4)\,(2.8058) = 336.70 \text{ [mm]}$$
$$d_P = \frac{N_P}{p_{dt}} = \frac{30}{8.9101} = 3.3670 \text{ [}in\text{]}$$
$$d_G = \frac{N_G}{p_{dt}} = \frac{N_G\,(n_g)}{p_{dt}} = \frac{30\,(4)}{8.9101} = 13.468 \text{ [in]} \quad \text{(b)}$$

The pitch line velocities are simply calculated as follows:

$$v_t = \omega_P r_P = 125.66\left(84.174\left(10^{-3}\right)/2\right) = 5.2887 \text{ [m/s]}$$
$$v_t = \omega_P r_P = 125.66\,(3.3670/2) = 211.55 \text{ [in/s]} = 1057.7 \text{ [ft/ min]} \quad \text{(c)}$$

For forces between gears, the tangential force between the pinion and the gear is found through Eq. (15.12) such that

$$\boldsymbol{F}_t = \frac{H}{r_P\omega_P} = \frac{H}{v_t} = \frac{75\left(10^3\right)}{(5.2887)} = 14\,181 \text{ [N]} = 14.181 \text{ [kN]}$$
$$\boldsymbol{F}_t = \frac{H_{hp}\,(6600)}{r_P\omega_P} = \frac{100.58\,(6600)}{(211.55)} = 3137.9 \quad \text{[lb]} \quad \text{(d)}$$

The resultant transmission normal force $\boldsymbol{F}_n$, the radial force $\boldsymbol{F}_r$, and the axial force $\boldsymbol{F}_a$ are then given from Eqs. (15.13), (15.14), and (15.15) to get

$$\boldsymbol{F}_n = \frac{\boldsymbol{F}_t}{\cos\psi\cos\phi_n} = \frac{14.181}{\cos(27)\cos(20)} = 16.937 \text{ [kN]}$$
$$\boldsymbol{F}_n = \frac{\boldsymbol{F}_t}{\cos\psi\cos\phi_n} = \frac{3137.9}{\cos(27)\cos(20)} = 3747.8 \text{ [lb]}$$
$$\boldsymbol{F}_r = \boldsymbol{F}_n \sin\phi_n = 16.937 \quad \sin(20) = 5.7928 \text{ [kN]}$$
$$\boldsymbol{F}_r = \boldsymbol{F}_n \sin\phi_n = 3747.8 \quad \sin(20) = 1281.8 \text{ [lb]}$$

$$\boldsymbol{F}_a = \boldsymbol{F}_n \cos\phi_n \sin\psi = 16.937 \ \cos(20) \ \sin(27) = 7.2255 \ [\text{kN}]$$
$$\boldsymbol{F}_a = \boldsymbol{F}_n \cos\phi_n \sin\psi = 3747.8 \ \cos(20) \ \sin(27) = 1598.9 \ [\text{lb}] \quad \text{(e)}$$

To evaluate the bending fatigue stress, the following parameters need to be calculated:
The geometry factor is obtained from Eq. (15.25) such that

$$\begin{aligned} J_P &= \left\{-\left(0.000\,145 \ N_P^{0.161597}\right)\psi^2 + \left(0.007\,237 \ N_P^{0.05324}\right)\psi + 0.342\,121 N_P^{0.100154}\right\}\left(0.8691 N_G^{0.0295}\right) \\ &= \left\{-\left(0.000\,145 \ (30)^{0.161597}\right)(27)^2 + \left(0.007\,237 \ (30)^{0.05324}\right)(27) + 0.342\,121(30)^{0.100154}\right\} \\ &\quad \left(0.8691(120)^{0.0295}\right) = 0.5325 \end{aligned} \quad \text{(f)}$$

The dynamic or velocity factor is given by Eq. (15.26) so as

$$B_v = 0.25\left(12 - Q_v\right)^{2/3} = 0.25(12-8)^{2/3} = 0.629\,96$$

$$K_v|_{\text{SI}} = \left(\frac{50 + 56\left(1 - B_v\right) + \sqrt{200\left(v_t\right)}}{50 + 56\left(1 - B_v\right)}\right)^{B_v} = \left(\frac{50 + 56\,(0.370\,04) + \sqrt{200\,(5.2887)}}{50 + 56\,(0.370\,04)}\right)^{0.62996} = 1.2691$$

$$K_v|_{\text{US}} = \left(\frac{50 + 56\left(1 - B_v\right) + \sqrt{\left(v_t\right)_{\text{ft/min}}}}{50 + 56\left(1 - B_v\right)}\right)^{B_v} = \left(\frac{50 + 56\,(0.370\,04) + \sqrt{1057.7}}{50 + 56\,(0.370\,04)}\right)^{0.62996} = 1.2691 \quad \text{(g)}$$

The bending fatigue stress is then calculated according to Eqs (15.28) and (15.29) to get

$$k_m|_{\text{SI}} = -4(10)^{-7} f_g^2|_{[\text{mm}]} + 0.0011 f_g|_{[\text{mm}]} + 1.544 = -4(10)^{-7}(65)^2 + 0.0011\,(65) + 1.544 = 1.6138$$
$$k_m|_{\text{US}} = -6.28(10)^{-4} f_g^2|_{[\text{in}]} + 0.0364 f_g|_{[\text{in}]} + 1.523 = -6.28(10)^{-4}(2.6)^2 + 0.0364\,(2.6) + 1.523 = 1.6134 \quad \text{(h)}$$

$$\sigma_{\text{ab}} = \frac{\boldsymbol{F}_t K_v k_o k_m}{f_g \ m_t J_g} = \frac{14.181\left(10^3\right)(1.2691)(1)(1.6138)}{65\,(2.8058)\,(0.5325)} = 299.06 \ [\text{MPa}]$$
$$\sigma_{\text{ab}} = \frac{\boldsymbol{F}_t K_v \ p_{\text{dt}} \, k_o k_m}{f_g \ J_g} = \frac{3137.9\,(1.2691)\,(8.9101)\,(1)\,(1.6134)}{2.6\,(0.5325)} = 41318 \ [\text{psi}] = 41.318 \ [\text{kpsi}] \quad \text{(i)}$$

To evaluate the contact fatigue stress from Eq. (15.37) and for the elastic coefficient $C_p = 191 \ [\text{MPa}]^{1/2}$ and 2300 $[\text{psi}]^{1/2}$, the following parameters need to be calculated according to Eqs. (15.32)–(15.35):

$$\phi_t = \tan^{-1}\left(\frac{\tan\phi_n}{\cos\psi}\right) = \tan^{-1}\left(\frac{\tan 20}{\cos 27}\right) = \tan^{-1}\left(\frac{0.363\,97}{0.891\,01}\right) = 22.2197^\circ$$

$$\begin{aligned} r_{bP} &= r_P \cos\phi_t = 42.087\,(0.925\,74) = 38.962 \ [\text{mm}], \quad r_{bG} = r_G \cos\phi_t \\ &= 168.35\,(0.925\,74) = 155.85 \ [\text{mm}] \end{aligned}$$
$$\begin{aligned} r_{bP} &= r_P \cos\phi_t = 1.6835\,(0.925\,74) = 1.5585 \ [\text{in}], \quad r_{bG} = r_G \cos\phi_t \\ &= 6.734\,(0.925\,74) = 6.2339 \ [\text{in}] \end{aligned} \quad \text{(j)}$$

$$\begin{aligned} Z_{\text{mm}} &= \sqrt{\left(r_P + a\right)^2 - r_{bP}^2} + \sqrt{\left(r_G + a\right)^2 - r_{bG}^2} - \left(r_P + r_G\right)\sin\phi_t \\ &= \sqrt{(42.087 + 2.5)^2 - 38.962^2} + \sqrt{(168.35 + 2.5)^2 - 155.85^2} \\ &\quad - (42.087 + 168.35)\sin 22.2197 = 12.108 \ [\text{mm}] \end{aligned}$$

$$\begin{aligned} Z_{\text{in}} &= \sqrt{\left(r_P + 1/p_d\right)^2 - r_{bP}^2} + \sqrt{\left(r_G + 1/p_d\right)^2 - r_{bG}^2} - \left(r_P + r_G\right)\sin\phi_t \\ &= \sqrt{(1.6835 + 0.1)^2 - 1.5585^2} + \sqrt{(6.734 + 0.1)^2 - 6.2339^2} - (1.6835 + 6.734)\sin 22.2197 = 0.4843 \ [\text{in}] \end{aligned} \quad \text{(k)}$$

$$m_N = \frac{p_n \cos \phi_n}{0.95 Z_{mm}} = \frac{\pi m_n \cos \phi_n}{0.95 Z_{mm}} = \frac{\pi\,(2.5) \cos 20}{0.95\,(12.108)} = 0.641\,62$$

$$m_N = \frac{\left(\pi / P_d\right) \cos \phi_n}{0.95 Z_{in}} = \frac{(\pi/10) \cos 20}{0.95\,(0.4843)} = 0.641\,65 \tag{l}$$

$$I_g = \frac{\cos \phi \, \sin \phi}{2 m_N} \frac{n_g}{\left(n_g \pm 1\right)} = \frac{\cos 20 \, \sin 20}{2\,(0.6416)} \frac{4}{(4+1)} = 0.200\,37 \tag{m}$$

The contact fatigue stresses are then (Eq. (15.37))

$$\sigma_{ac}|_{max} = C_p \sqrt{\frac{k_o k_m}{f_g} \frac{\boldsymbol{F}_n K_v}{d_P\; I_g}} = 191 \sqrt{\frac{(1)\,(1.6138)}{65\,(84.174)} \frac{(16.937)\,(1.2691)}{(0.200\,37)}} = 1074.4 \text{ [MPa]}$$

$$\sigma_{ac}|_{max} = C_p \sqrt{\frac{k_o k_m}{f_g} \frac{\boldsymbol{F}_n K_v}{d_P\; I_g}} = 2300 \sqrt{\frac{(1)\,(1.6134)}{2.6\,(3.367)} \frac{(3747.8)\,(1.2691)}{(0.200\,37)}} = 152128 \; [\text{psi}] = 152.13 \; [\text{kpsi}] \tag{n}$$

The allowable bending fatigue strength $S_{ab} = 300$ [MPa] or 43.5 [kpsi] and the bending to contact fatigue strength ratio B_{BC} of 0.35. The safety factors are calculated according to Eqs. (15.38) and (15.40) to get

$$K_{SFb}|_{SI} = \frac{S_{ab}}{\sigma_{ab}|_{max}} \frac{300}{299.06} = 1.0031$$

$$K_{SFc}|_{SI} = \frac{S_{ac} \sqrt{k_m}}{\sigma_{ac}|_{Synth.}} = \frac{(300/0.35)\,\sqrt{1.6138}}{1074.4} = 1.0135$$

$$K_{SFb}|_{US} = \frac{S_{ab}}{\sigma_{ab}|_{max}} \frac{43.5}{41.318} = 1.0528$$

$$K_{SFc}|_{US} = \frac{S_{ac} \sqrt{k_m}}{\sigma_{ac}|_{Synth.}} = \frac{(43.5/0.35)\,\sqrt{1.6138}}{152.13} = 1.0378 \tag{o}$$

Since the allowable stresses are used in the design process, the safety factors are expected to be around 1.0. The little difference is due to the selection of the design variables in a little conservative way during the selection of the design variables rounding. If more factors are considered in the design process, the load should be increased, or the material is changed to have higher allowable stresses and/or adjusting the design variables. This is doable if there is a *CAD* program to be used; see Section 15.6.

15.2 Bevel Gears

Bevel gear sets accommodate connections of inclined intersecting or nonintersecting axes. The usual usage is for intersecting normal axes. Figure 15.10 shows two main bevel gear sets. The most usual is the *straight bevel gear* set with intersecting perpendicular centerline as shown in Figure 15.10a. The teeth are straightly formed with their centerlines falling in the same plane as the gear's axis. The second helical gear set in Figure 15.10b is the *spiral bevel gear* set with intersecting perpendicular centerline. The teeth are spirally formed with their centerlines inclined to the gear's axis at a specific helix angle. The straight bevel gears are the ones that this section is devoted to. Other types are covered in dedicated handbooks and standards such as Townsend (1992), AGMA (1988), AGMA (2003), AGMA ISO 22849 (2011), and ISO 23509 (2016).

15.2.1 Definitions, Kinematics, and Standards

This section considers straight bevel gears only, where the general gear form is a truncated cone with straight teeth as shown in Figure 15.10a. The gear set shafts can be intersecting at any angle from a few degrees to 90°. The

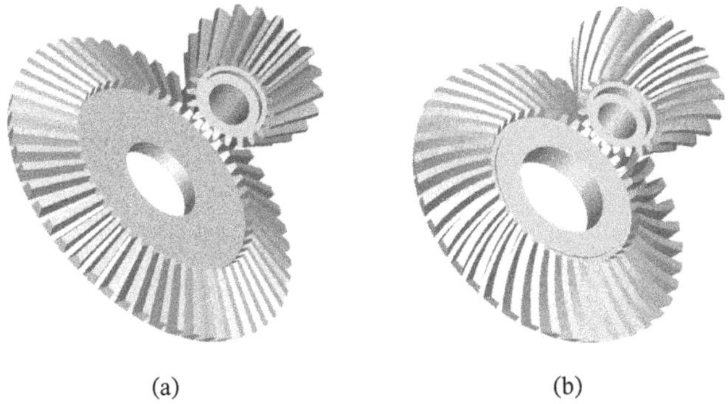

Figure 15.10 Two main types of bevel gears: (a) straight bevel gear set and (b) spiral bevel gear.

normal intersecting shafts are the usual cases that are considered in the following treatment. Figure 15.11 shows a sketch of the sectional view for bevel gear set indicating the main geometrical parameters and basic designations of such a straight bevel gear set. The pinion is the smaller gear, and its main parameters are marked where similar ones for the gear are not shown. The basic angles are the *pinion pitch cone angle* γ and the *gear pitch cone angle* Γ. The summation of both angles is the 90° angle between their two intersecting shafts. The *pinion pitch diameter* d_P and the *gear pitch diameter* d_G are measured at the large ends of both truncated cones. The *pinion face width* f_P is shown with the *pinion cone distance* A_P. Similarly, the *gear face width* f_G would usually be the same as for the pinion. The *gear cone distance* A_G is not shown but can be similarly defined. The *gear back cone radius* r_{bG} is shown, and the *pinion back cone radius* r_{bP} is similarly defined for the pinion. The relations between the main parameters are similar to spur gears and are particularly as follows:

$$m_n = \frac{d_P}{N_P} = \frac{d_G}{N_G}, \quad [\text{mm}] \tag{15.41}$$

where for SI system, m_n is the *normal module* in [mm], d_P and d_G are the *pinion and gear large end diameters*, respectively, in [mm], N_P is the *pinion number of teeth*, and N_G is the *gear number of teeth*. For the US system, the *diametral pitch* p_d is given in [teeth/in] as follows:

$$p_d = \frac{N_P}{d_P} = \frac{N_G}{d_G}, \quad [\text{teeth/in}] \tag{15.42}$$

where d_P and d_G are the pinion and gear large end diameters, respectively, in [in]. The *normal circular pitch* p_n is given by

$$p_n = \frac{\pi d_G}{N_G} = \frac{\pi d_P}{N_P} \tag{15.43}$$

From the geometry and previous relations, one can find the following expressions:

$$\tan\gamma = \frac{N_P}{N_G}, \quad \tan\Gamma = \frac{N_G}{N_P} \tag{15.44}$$

The *virtual number of teeth* $N_P{}^v$ or $N_G{}^v$ for cutting of pinions and gears, respectively, are as follows:

$$N_P^v = \frac{2\pi r_{bP}}{p_n}, \quad N_G^v = \frac{2\pi r_{bG}}{p_n} \tag{15.45}$$

where r_{bP} is the pinion back cone radius, r_{bG} is the gear back cone radius (see Figure 15.11), and p_n is the circular pitch at the large teeth end.

Table 15.1 is presenting some of the proportions of straight bevel gear parameters.

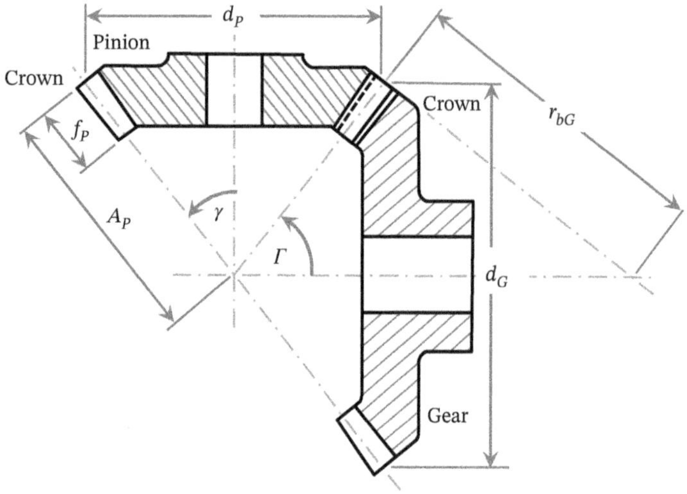

Figure 15.11 Main geometrical parameters and basic designations of straight bevel gear set.

Table 15.1 Proportions of the main straight bevel gear parameters.

Item	Formula for SI and (US)				
Working depth	$h_w = 2.0\ m_n$, $(h_w = 2.0\ p_d)$				
Clearance	$c_g = 1.88\ m_n + 0.05$, $(c_g = 1.88/p_d + 0.002)$				
Addendum of gear	$a_G = m_n\ [0.54 + 0.460/(m_{90})^2]$, $(a_G = 0.54/p_d + 0.460/p_d\ (m_{90})^2)$				
Gear ratio	$n_g = N_G/N_P$				
Cone distance	$A_P = d_P/2\sin\gamma$, $A_G = d_G/\ 2\sin\Gamma$				
Equivalent 90° ratio	$n_{90} = [n_G\ (\cos\gamma/\cos\Gamma)]^{1/2}$ when $\sum(\gamma + \Gamma) \neq 90°$				
Face width	$f_g = 0.3\ A_g$ or $f_g = 10\ m_n$ or $(f_g = 10/p_d)$				
Minimum number of teeth	Pinion	16	15	14	13
	Gear	16	17	20	30

15.2.2 Force Analysis

The bevel gear is transmitting power, usually, from the pinion to the gear. The force that does the work is the *tangential force component* $\boldsymbol{F}_t$ of the *resultant normal force* $\boldsymbol{F}_n$ to the tooth at the pitch line as shown in Figure 15.12. The figure shows the forces on a straight bevel gear tooth. The *radial force component* $\boldsymbol{F}_r$ and the *axial force component* $\boldsymbol{F}_a$ are in the plane of the gear section. The tangential component $\boldsymbol{F}_t$ is normal to the gear section, and the resultant force $\boldsymbol{F}_n$ is normal to the tooth at the pitch line. The angle between the tangential component $\boldsymbol{F}_t$ and the resultant normal force $\boldsymbol{F}_n$ is the *normal pressure angle* ϕ_n. As in Eqs. (15.11) and (15.12), the tangential component $\boldsymbol{F}_t$ is obtained from power transmission such that

$$\boldsymbol{F}_t = \frac{H}{r_{\text{av}}\omega_P} = \frac{H}{v_t} \tag{15.46}$$

where t_{av} is the *average gear radius* used as an approximation for the distributed loads over the length of the tooth; see Figure 15.12. The rest of the terms in Eq. (15.46) are defined at Eq. (15.12). The *pitch point velocity* v_t is evaluated

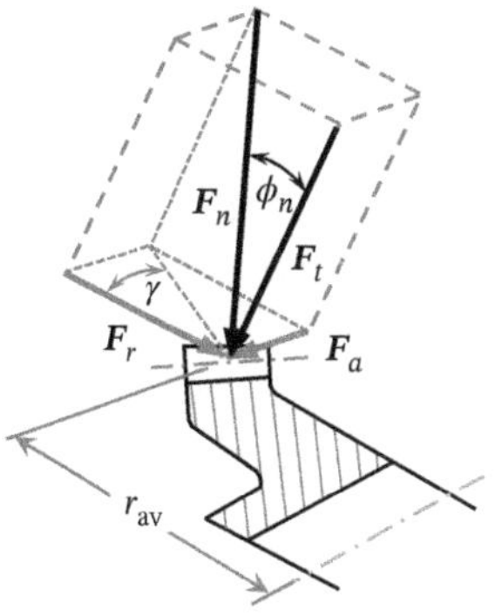

Figure 15.12 Forces on straight bevel gear tooth. The radial and axial components are in the plane of the gear section. The tangential component is normal to the gear section, and the resultant force is normal to the tooth at the pitch line.

at the average gear radius as shown in Eq. (15.46). From Figure 15.12, the resultant *normal force* $\boldsymbol{F}_n$, the *radial force* $\boldsymbol{F}_r$, and the *axial force* $\boldsymbol{F}_a$ are then given by

$$\boldsymbol{F}_n = \frac{\boldsymbol{F}_t}{\cos \phi_n} \tag{15.47}$$

$$\boldsymbol{F}_r = \boldsymbol{F}_n \tan \phi_n \cos \gamma \tag{15.48}$$

$$\boldsymbol{F}_a = \boldsymbol{F}_n \tan \phi_n \sin \gamma \tag{15.49}$$

The calculation check is that the *normal force* $\boldsymbol{F}_n$ is the resultant of the *tangential component* $\boldsymbol{F}_t$, the *radial force* $\boldsymbol{F}_r$, and the *axial force* $\boldsymbol{F}_a$ identical to Eq. (15.16).

These forces are typically needed to design the shaft on which the gear is mounted.

15.2.3 Design Procedure

The design procedure of bevel gears is equivalent to that of the spur gears. The initial synthesis of spur gears can be used for the *initial synthesis* of bevel gears. Both gears of the same materials and specifications can transmit the same power. The bevel gears geometry at the large ends is the same as for spur gears.

15.2.3.1 Initial Design

The *initial synthesis* procedure of Section 14.4.2 can be used for bevel gears, where the module or the diametral pitch, the number of teeth, and the face width can be adapted. The main constraint, however, is the face width that should be less than the pinion cone distance A_P. The suggested value should be within that shown in Table 15.1, i.e. about one third the pinion cone distance or smaller. These were not the cases shown in Figure 15.10 since the objective was only for the demonstration of the types.

15.2.3.2 Detailed Design

The detailed design procedure is like the spur gears with most parameters having the same guide for evaluation. Only few parameters are defined in a different way. These are given in the following stress evaluation expressions.

The *maximum* ***bending fatigue stress*** $\sigma_{ab}|_{max}$ may then be rewritten as

$$\begin{aligned} \sigma_{ab}|_{max} &= \frac{\boldsymbol{F}_t K_v k_o k_m}{f_g\ m_n J_g} \quad \text{for SI} \\ \sigma_{ab}|_{max} &= \frac{\boldsymbol{F}_t K_v\ p_d\ k_o k_m}{f_g\ J_g} \quad \text{for US} \end{aligned} \tag{15.50}$$

where K_v is the *velocity factor* as spur gears, $\boldsymbol{F}_t$ is the tangential force (usually at the large end of gear and smaller than at the average radius), f_g is the *face width* of the gear, m_n is the *normal module*, p_d is the diametral pitch, k_m is the *load distribution factor* (different from spur gears; see Eq. (15.51)), the *load application factor* k_o should be incorporated into the maximum input, and J_g is a *bending geometry factor* for bevel gears (different from spur gears) and is to be given later.

The *load distribution factor* k_m is different from spur gears and is concerned about mounting and face width f_g and is given by the following expression (see ANSI/AGMA 2003-B97 (2003)):

$$\begin{aligned} k_m &= k_{mb} + 0.0036 f_g^2 \quad \text{(US)} \\ k_m &= k_{mb} + 5.6(10)^{-6} f_g \quad \text{(SI)} \end{aligned} \tag{15.51}$$

where k_{mb} is the *load distribution constant*, which is 1.0 for both saddle-mounted pinion and gear, 1.10 for one saddle-mounted gear, and 1.25 for neither saddle-mounted gears.

The *bending geometry factor* for bevel gears J_g is given by the following regression equation:

$$J_g = \left(0.161\,93\left(N_{g2}^{-0.00294}\right)\right) N_{g1}^{\left(0.00703\left(N_{g2}^{0.72653}\right)\right)} \tag{15.52}$$

where the subscript g1 is for the gear to which the bending factor is desired and the subscript g2 is for the mating gear. The limit of N_{g1} and N_{g2} in Eq. (15.52) is about 100 teeth for the value of J_g to be within representative design accuracy. The values obtained by Eq. (15.52), however, would be close enough compared with the values suggested by ANSI/AGMA 2003-B97 (2003). Equation (15.52) is a simplified expression to be used effectively in any *CAD* code. The complete design calculations, however, should be rechecked against the design code that one should use in any product development.

The *maximum contact fatigue stress* $\sigma_{ac}|_{max}$ becomes

$$\sigma_{ac}|_{max} = C_p \sqrt{\frac{k_o k_m}{f_g} \frac{\boldsymbol{F}_n K_v}{d_P\; I_g}} \tag{15.53}$$

where $\boldsymbol{F}_n$ is the *normal gear force* or resultant force at large end of the gear (different from Eqs. (15.47) and (15.46)), K_v is the *velocity factor* as spur gears, k_m is the *load distribution factor* (different from spur gears; see Eq. (15.51)), the *load application factor* k_o should be incorporated into the maximum input, C_p is the *elastic coefficient* (close to Eqs. (14.59) and (15.31)), d_P is the *pitch circle diameter* of the pinion at the large end, f_g is the gear *face width*, and I_g is the *contact geometry factor*. The elastic coefficient C_p is given by the values in Table 14.8 as spur gears.

The *contact geometry factor* I_g is given by the following regression equation:

$$\begin{aligned} I_{g1} = &\left(4.12(10)^{-6} N_{g2}^{-0,290}\right) N_{g1}^3 - \left(4.84(10)^{-6} N_{g2}^{0,776}\right) N_{g1}^2 \\ &+ \left(1.92(10)^{-4} N_{g2}^{0.774}\right) N_{g1} + \left(3.63(10)^{-1} N_{g2}^{-0.612}\right) \end{aligned} \tag{15.54}$$

where the subscript g1 is for the gear to which the contact geometry factor is desired and the subscript g2 is for the mating gear. The limit of N_{g1} and N_{g2} in Eq. (15.54) is about 50 teeth for the value of I_g to be within representative design accuracy. The values obtained by Eq. (15.54), however, would be close enough compared with the values suggested by ANSI/AGMA 2003-B97 (2003). Equation (15.54) is a simplified expression to be used effectively in any *CAD* code. The complete design calculations, however, should be rechecked against the design code that one should use in any product development.

15.2.3.3 Material Set and Safety Factor

The material set and properties introduced in the spur gear chapter should also apply for the helical gears; see Section 14.4.3 and Table 14.6. Usually the straight bevel gears are used in regular applications. The spiral bevel gears are usually used in demanding applications that their higher quality necessitates the better materials. Also employing the higher quality manufacturing is also expected in spiral bevel gears (Davis 2005).

The same treatment of safety factor in spur gears is also applicable in bevel gear treatment; see Section 14.4.3. In that sense, there is a bending fatigue safety factor $K_{SF}|_b$ as Eqs. (14.48) and (15.38) and contact fatigue safety factor $K_{SF}|_c$ as Eqs. (14.70) and (15.39).

As for spur gears, the bending fatigue safety factor $K_{SF}|_b$ can be as Eqs. (14.48) and (15.38), where S_{eg} is the *endurance strength limit* of the gear or the *fatigue strength* S_f defined by Eq. (14.45), K_{MSF} is the *material safety factor*, $\sigma_{ab}|_{max}$ is the *maximum* ***bending fatigue stress*** given by Eq. (15.50), and S_{ab} is the *allowable fatigue strength*.

As for spur gears and considering the *reliability factor* k_c (Eq. (14.43)), the *material safety factor* K_{MSF} and the *allowable contact fatigue strength* of S_{ac} (Eq. (14.67)), the ***contact safety factor*** $K_{SF}|_c$ is defined as in Eq. (15.39), and the *maximum* ***contact fatigue stress*** is given by Eq. (15.53). The particular result in Eq. (15.39) is occurring at $k_c = 1$ for 99% reliability and the *life factor* $Z_N = 1$ if the *contact fatigue strength* S_{cf} is used. These are the usual cases, and if the case is different, the values of k_c and Z_N can be evaluated as Eqs. (14.43) and (14.68) for spur gears or consult standards such as AGMA.

If the proper *allowable fatigue strength* S_{ab} and *allowable contact fatigue strength* S_{ac} are used, the safety factors are expected to be about 1.0. If not, the safety factor would be an *extra safety factors* K_{SFE} over the allowable strength values.

Since the synthesized optimum contact stress includes the *load distribution factor* k_m; see Eqs. (15.50), (15.53), and (15.39), one can get the same relation for the safety factor (not including the *load application factor* k_o) as Eq. (15.40). This is because the *synthesis contact fatigue stress* $\sigma_{ab}|_{Synth.}$ is including the *load distribution factor* k_m, but not the *load application factor* k_o. The synthesis depends on applying the *load application factor* k_o directly to the load as sort of a direct safety factor. See a similar application in Example 15.3.

15.3 Worm Gears

The *worm* has a mating worm wheel or *worm gear*; see Figures 15.1c and 15.13. The set is usually called worm gears. The worm can be considered a helical gear with a helix angle exceeding 45°, i.e. as a screw thread. The *worm wheel* or worm gear is usually with axis at a right (90°) angle with the worm axis. The axes, however, are not intersecting. The rim of the worm gear is machined to match or *envelop* the root cylinder of the worm as shown in Figure 15.13. The machining of the worm is like machining a screw thread. The machining of the worm gear or worm wheel is by *hobbing*, which employs a hob cutter in the form of the worm screw thread.

Due to lower efficiency of worm gear systems, they are mainly used for lower power and torque transmission. The usual speed ratio ranges from 8 to 70 or may be 100.

15.3.1 Definitions, Kinematics, and Standards

Figure 15.13 shows a sketch of a *single-enveloping* worm gear set viewing the worm, the worm gear, and the main geometry parameters and nomenclatures. The displayed worm has two starts or two leads as for some threads. The *double-enveloping* worm gear has the worm enveloping the worm gear also, which is not shown in Figure 15.13. The main concern of this section is focused on the common right (90°) angle between the worm and worm gear axes. The following are the main definitions and fundamental relations for the single-enveloping worm gear set.

The ***axial pitch*** p_x of the worm and the *transverse circular pitch* p_t of the worm gear are equal, i.e.,

$$p_x = p_t \tag{15.55}$$

Again, this is for angles of worm and worm gear shafts at 90°; see Figure 15.13.

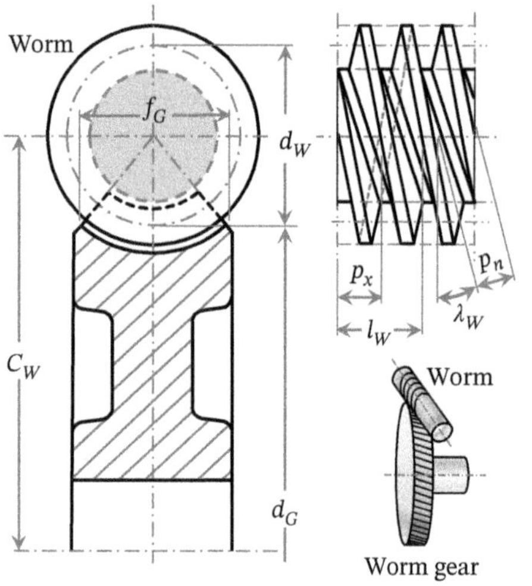

Figure 15.13 A sketch of a single-enveloping worm gear set showing the worm, the worm gear, and the main geometry parameters and nomenclatures. The shown worm has two starts.

The ***worm lead*** l_W is given by (see Figure 15.13)

$$l_W = p_x N_W \tag{15.56}$$

where N_W is the *number of starts* or *threads* of the worm. Figure 15.13 shows a worm having two starts for clarity of the sketch, where the whole tooth *working depth* h_w is also large. The number of starts is usually 1, but can be 2 or 4 and may be 3, 6, 8, etc. The *single thread* or *single start* ($N_W = 1$) is adapted to obtain a high ratio, even though, single threads are usually less efficient.

The ***gear ratio*** n_G of the worm gear set is defined as follows:

$$n_G = \frac{N_G}{N_W} = \frac{\omega_W}{\omega_G} \tag{15.57}$$

where N_G is the *number of teeth on worm gear*, N_w is the *number of starts* or *threads* of the *worm*, ω_W is the rotational speed of the worm, and ω_G is the rotational speed of the worm gear. Usually N_W can be 1 for getting a large gear ratio, but the efficiency is expected to be lower than the efficiency of a higher number of starts.

The *worm* ***lead angle*** λ_W is obtained as follows (see Figure 15.13 and Eq. (15.56)):

or

$$\lambda_W = \tan^{-1}\frac{l_W}{\pi d_W} = \tan^{-1}\frac{p_x N_W}{\pi d_W}$$

$$d_W = \frac{p_x N_W}{\pi \tan \lambda_W} \tag{15.58}$$

where d_W is the **worm pitch diameter** and l_W is the worm lead. The common lead angles are usually dependent on the number of threads or worm starts N_W. For single start or $N_W = 1$, the lead angle is usually less than 6°. For $N_W = 2$, the lead angle is usually between 3° and 12°. For $N_W = 3$, the lead angle is usually between 6° and 18°. For $N_W = 4$, the lead angle is usually between 12° and 24°. For $N_W = 5$, the lead angle is usually between 15° and 30°. For $N_W = 6$, the lead angle is usually between 18° and 36°. For $N_W = 7$ and larger, the lead angle is usually not more than $6N_W$°.

The ***worm gear pitch diameter*** d_G is given by

$$d_G = \frac{N_G p_t}{\pi} \tag{15.59}$$

where p_t is the *transverse circular pitch* as defined by Eq. (15.55).

The ***center distance*** C_W of worm gear set is simply found from Figure 15.13 as

$$C_W = \frac{1}{2}\left(d_G + d_W\right) = \frac{p_t}{2\pi}\left(N_G + \frac{N_W}{\tan\lambda_W}\right) = \frac{p_n}{2\pi}\left(\frac{N_G}{\cos\lambda_W} + \frac{N_W}{\sin\lambda_W}\right) \quad \textbf{(15.60)}$$

where p_n is the *normal circular pitch,* which is used to develop the hob profile to manufacture the worm. The relation between the normal circular pitch p_n, and the axial pitch p_x is a function of the worm lead angle λ_W. The normal circular pitch p_n, the normal module m_n, and the diametral pitch p_d are then given by (see Figure 15.13)

$$p_n = p_t \cos\lambda_W = \frac{\pi d_W}{N_W}\sin\lambda_W$$

$$m_n = \frac{p_n}{\pi} \quad [\text{mm}], \quad p_d = \frac{\pi}{p_n} \quad [\text{teeth/in}] \quad \textbf{(15.61)}$$

where the normal circular pitch p_n is in [mm] or in [in] for the SI system or the US system, respectively.

The usual values for the *normal circular pitch* p_n are 0.030, 0.040, 0.050, 0.065, 0.080, 0.100, 0.130, and 0.160 [in] for the US fine pitch system. The usual values for the normal circular pitch p_n are 1.0, 1.25, 1.5, 2.0, 2.5, 3.0, 3.5, and 4.0 [mm] for SI fine pitch system. For course pitch system, the usual values for normal circular pitch p_n are 0.2, 0.25, 0.3, 0.4, 0.5, 0.100, 0.625, 0.75, 1.0, 1.25, and 1.5 [in] for the US system. For course pitch system, the usual values for normal circular pitch p_n are 5.0, 6.5, 8.0, 10.0, 12.0, 15.0, 20.0, 25.0, 30.0, and 40.0 [mm] for SI system; see Townsend (1992).

The suggested ***worm pitch diameter*** d_W for workable average is given by the following relation (see Dudley 1984):

$$d_W = \frac{C_W^{0.875}}{2.2} \quad [\text{in}], \text{ for US}$$

$$d_W = \frac{C_W^{0.875}}{1.4683} \quad [\text{mm}], \text{ for SI} \quad (15.62)$$

where the center distance C_W is in [in] or [mm], and d_W is in [in] or [mm] for the US and SI systems, respectively. The constant 2.2 in Eq. (15.62) for the US system has the range of 3.0–1.6 with an average of about 2.2. For SI system the constant 1.4683 in Eq. (15.62) has the range of 2.002–1.0679. These values can be rounded off.

The ***normal pressure angle*** ϕ_n is usually adapted as 20°. The worm gearing parameters in that case would be similar to other previously discussed gears. The worm gear face width f_G is shown in Figure 15.13. Table 15.2 provides the usual proportions of the main worm gear parameters. Some gear parameters depend on the type of thread profile for gear cutting process. The profiles are either *Archimedean* basic rack for straight-sided worms, or *involute helicoidal*, or concave thread profile (Dimargonas 1989). It is useful to note that the fine pitches may also have different properties than the course pitches (Townsend 1992). The current standards are to be consulted in that regard.

15.3.2 Force Analysis

The force analysis of worm gear set is involved due to the inclined ***3D*** tooth form and the presence of friction. Figure 15.14 presents a schematic diagram of the main worm forces, where forces are not proportional for sake of clarity and the friction components are moved offset to the right on the tangent to the tooth. The grayish cylinder lines represent the worm pitch cylinder. The black lines on the cylinder surface are some tooth pitch lines, and the resultant force $\boldsymbol{F}_n$ is normal to the tooth surface at the pitch line contact with the mating worm gear. The *x, y,* and *z* coordinates are coincident with the orthogonal components of the resultant *normal force* $\boldsymbol{F}_n$. The *x*-coordinate is parallel to the worm axis. For clarity, the friction components are moved offset to the right on the same line of the tangent to the worm tooth at the contact point with the mating worm gear. The orthogonal components of the

Table 15.2 The usual proportions of the main worm gear parameters.

Item	Formula for SI and (US)
Addendum	$a_g = m_n, (a_g = 0.3183\, p_n, \text{ or } = 0.3683\, p_x)$
Dedendum	$b_g = 1.2\, m_n, (b_g = 0.3683\, p_n, \text{ or } = 0.3820\, p_n)$
Working depth	$h_g = 2.0\, m_n, (h_g = 0.6366\, p_n)$
Whole depth	$h_w = 2.2\, m_n, (h_w = 0.7043\, p_n, \text{ or } = 0.7003\, p_n + 0.002)$
Tooth thickness	$t_W = \pi m_n/2, (t_W = 0.5\, p_n)$
Clearance	$c_g = h_w - h_g$
Gear ratio	$n_g = N_G/N_W$
Outer diameter	$d_{Wo} = d_W + 2a_g, d_{Go} = 2C_W - d_W + 2a_g$
Worm length (minimum)	$L_W = (d_{Go}{}^2 - d_G{}^2)^{1/2}$
Face width (minimum)	$f_G = 1.125\, ((d_{Wo} + 2C_W)^2 - (d_{Wo} - 2a_g)^2)^{1/2}$

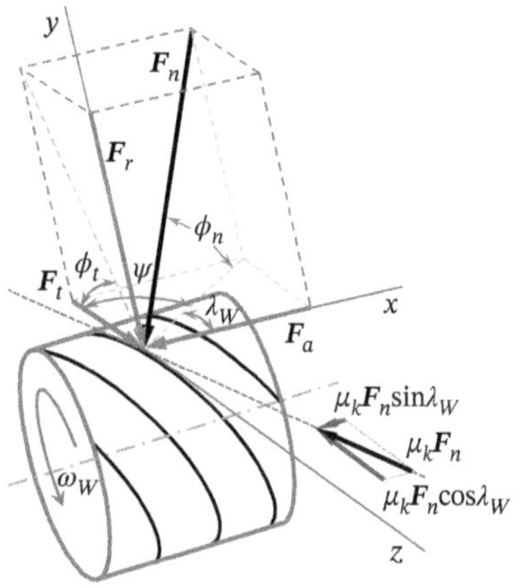

Figure 15.14 A schematic of worm forces, where forces are not proportional for sake of clarity. The grayish cylinder lines represent the worm pitch cylinder. The black lines on the cylinder surface are some tooth pitch lines, and the resultant force $\boldsymbol{F}_n$ is normal to the tooth surface at the pitch line contact with mating worm gear.

normal force $\boldsymbol{F}_n$ are then given by the following relations (see Figure 15.14):

$$\boldsymbol{F}_t = \boldsymbol{F}_n \left(\cos\phi_n \sin\lambda_W + \mu_k \cos\lambda_W\right) \tag{15.63}$$

$$\boldsymbol{F}_r = \boldsymbol{F}_n \sin\phi_n \tag{15.64}$$

$$\boldsymbol{F}_a = \boldsymbol{F}_n \left(\cos\phi_n \cos\lambda_W - \mu_k \sin\lambda_W\right) \tag{15.65}$$

where $\boldsymbol{F}_t$ is the *tangential force*, $\boldsymbol{F}_r$ is the *radial force*, $\boldsymbol{F}_a$ is the *axial force*, μ_k is the *kinetic coefficient of friction*, ϕ_n is the *normal pressure angle*, and λ_W is the worm *lead angle*.

The power transmission is occurring only through the *tangential force* $\boldsymbol{F}_t$, which is the only force exerting effort or doing work. It is the only moving force that is applying a *gear* or *pinion torque* $\boldsymbol{T}_P$ on the worm or the worm gear. If the tangential force is known, the radial and axial forces are then obtainable from the *normal pressure angle* ϕ_n, and the *worm lead angle* λ_W, as shown in Figure 15.14 and Eqs. (15.63)–(15.65). The *tangential force* $\boldsymbol{F}_t$ is then obtained from the power transmission H as in Eqs. (15.63)–(15.65), which gives

$$\boldsymbol{F}_t = \frac{H}{r_W \omega_W} = \frac{H}{v_W} \tag{15.66}$$

where the power transmission H can be either H_{kW} [kW] or H_{hp} [hp], r_W is the worm radius, ω_W is the magnitude of the *angular velocity* of the pinion in [rad/s], and v_W is the magnitude of the tangential pitch point velocity of the worm, *i.e.* $v_W = \omega_W (d_W/2) = \omega_W r_W$. Knowing the tangential force on the worm from Eq. (15.66) provides the way to calculate the normal force $\boldsymbol{F}_n$ from Eq. (15.63), the radial force $\boldsymbol{F}_r$ from Eq. (15.64), and the axial force $\boldsymbol{F}_a$ from Eq. (15.65). These forces are essential to calculate the reactions on the worm shaft and the shaft of the worm gear.

It is important to note that there is sliding between the worm and worm gear. The sliding velocity is on the line of the tangent to the worm tooth at the contact pitch point with the mating worm gear. It is in the same direction as the friction force ($\mu_k \boldsymbol{F}_n$) in Figure 15.14. The velocity diagram at the pitch point is like the offset force diagram in Figure 15.4. The pitch line velocity of the worm gear $\boldsymbol{v}_G$ is then related to the pitch line velocity of the worm $\boldsymbol{v}_W$ by the following relation:

$$\boldsymbol{v}_G = \frac{\boldsymbol{v}_W}{\cos \lambda_W} \tag{15.67}$$

where λ_W is the worm lead angle. That is why the velocities of the worm and worm gear are not a direct quotient function of both pitch diameters as other previously presented gears.

The ***worm gear efficiency*** η_W is obtained by the following development:

$$\eta_W = \frac{\boldsymbol{F}_t|_{\text{without friction}}}{\boldsymbol{F}_t|_{\text{with friction}}} = \frac{(\cos \phi_n \sin \lambda_W)}{(\cos \phi_n \sin \lambda_W + \mu \ \cos \lambda_W)} \tag{15.68}$$

As an enlightening example and for a usual friction coefficient of 0.05, and for different lead angles λ_W, the approximate efficiencies are given as the pairs (λ_W, η_W) such as (1°, 25%), (2.5°, 46%), (5°, 63%), (7.5°, 71%), (10°, 77%), (15°, 83%), (20°, 87), and (30°, 89%). Therefore, if the worm gear unit is continuously operating and heat generation due to inefficiency is of concern, a worm gear unit design with higher lead angle would be the objective. This would mean a higher number of threads or worm starts N_W of more than one.

15.3.3 Design Procedure

The procedure of worm gear set design is following the same process as other gear sets. The *initial synthesis* depends on the available knowledge base, but not on formal optimization as other previous gear sets. The detailed design has other concerns in addition to previous gear sets process. The material set is concerned about lower friction in addition to strength; see Davis (2005). The safety factor is also concerned about the wear rate due to sliding friction. These factors are simply perused as follows:

15.3.3.1 Initial Synthesis

The initial synthesis of worm gear sets depends on the available information about manufactured worm gear sets. The knowledge base can guide the synthesis in terms of getting the usual center distance C_W for the needed power transmission H and speed ratio or gear ratio n_G. The geometry of the worm gear set is then found by the utilization of the previous relations and suggested parameters. Some iterations would be needed to adjust for standards or preferred values and manufacturing requirements.

Figure 15.15 provides the worm gearbox approximate characteristic torque versus speed ratio at different centerline distances. The worm rotational speed N_{rpm} is the usual 1750 [rpm] of various electric motors. This value of rotational speed produces the maximum power transmission than other lower rotational speeds. At lower rotational speeds, the transmitted torque is less and the detailed design procedure or rounding values can adjust parameters to cope with the difference.

The worm *input torque* $\boldsymbol{T}_W$ is simply given by the following relation (see Eqs. (15.11) and (15.12)):

$$\boldsymbol{T}_W = \frac{H}{\omega_W} \tag{15.69}$$

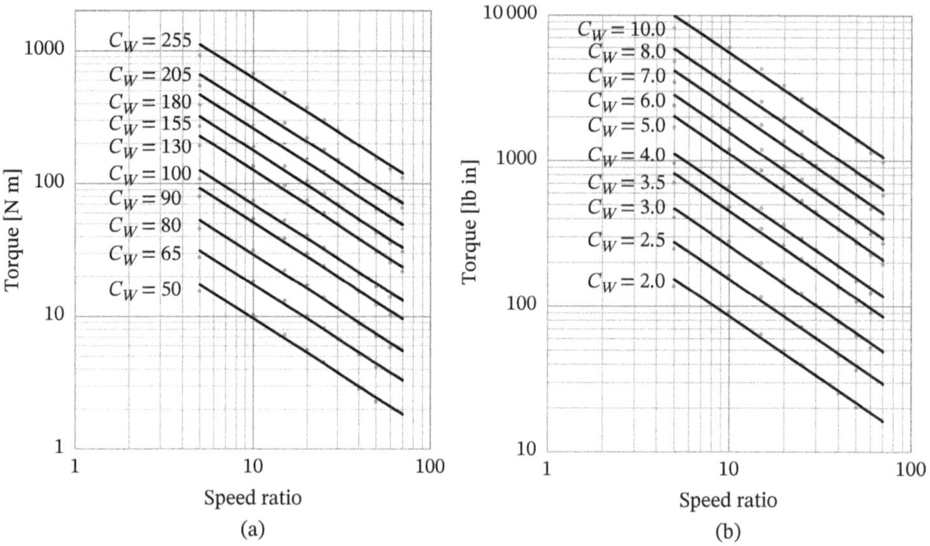

Figure 15.15 Worm gearbox approximate characteristic torque versus speed ratio at different centerline distances. The input rotational speed is the usual 1750 [rpm]. (a) SI system of units, and (b) US system of units.

where H is the maximum transmitted power and ω_W is the *angular velocity* of the worm in [rad/s]. If the rotational speed is 1750 [rpm], the initial synthesis is very close to the knowledge base. If not, the initial synthesis is just an approximate value for the expected center distance C_W of the worm gear set. A larger round figure value for the center distance C_W of the worm gear set would then be projected.

The initial synthesis procedure would then call for the utility of Figure 15.14 to get the expected center distance C_W of the worm gear set for the desired speed ratio. In Figure 15.14, the line of the maximum input torque $\boldsymbol{T}_W$ according to Eq. (15.69) should intersect the line of the desired speed ratio n_G at the expected center distance C_W of the worm gear set. With this center distance C_W, the worm gear set would then be synthesized using Eqs. (15.62), 15.57–15.59, and (15.60) to get the worm and worm gear pitch diameters for an acceptable lead angle and pitch.

15.3.3.2 Detailed Design

The detailed design procedure is like the spur gears with most parameters having the same guide for evaluation. Only few parameters are defined in a different way. These are given in the standards following stress evaluation expressions. The *bending fatigue* is the stress of concern. The other factor of interest is the *Buckingham* allowable *wear load*. This load depends on a wear factor, the worm gear diameter, and face width. The wear factor depends on the material pairs of worm and worm gear (Buckingham 1981). The main concern is the worm gear since the worm is usually safe. Detailed design procedures and factors depend on the standards, and it is advisable to follow the adapted standard for the specific case at hand; see ANSI/AGMA 6022-C93 R(2014) and ISO 10347 (1999).

15.3.3.3 Material Set and Safety Factor

The important factors in selecting material pairs are the strength and the lower friction coefficient to reduce heat generation and power losses. The power screw material pairs would be adequate for worm gearing applications; see Section 9.5. In addition, consult dedicated texts and standards for admissible gear materials and safety factors for allowable strength and wear coefficients; see Davis (2005).

15.4 Gear Failure Regimes and Remedies

Gear failure occurs due to bad design, which does not consider or expect overloads or adopt a sustaining proper operating condition. A perfect design supposedly should not fail during the lifetime of the product. A good design would fail after the expected or required lifetime. The variability of diverse loading conditions should be predictable, and the design is expected to accommodate that. A very large safety factor might be an escape from failure or a delayed time to failure. This policy has been adapted at the inception of technology due to lack of knowledge. But this is not the optimum solution to failure. As indicated, the safety factor is a factor of ignorance.

Wear as a failure regime happens due to several conditions with operating time progress. It is defined as a progressive loss of matter from a component surface by a mechanical encounter. Gear wear involves some incidences such as corrosion, abrasion, fretting, pitting, scoring, scuffing, and spalling. The following short description indicates the failure regime, its cause, and the apparent way to diminish its effect in gear sets.

Corrosion is a loss of surface matter by a corrosive medium present in the set. *Abrasion* causes wear by fine solid particles existing in the lubricant. *Fretting* corrosion introduces destructions of surfaces by sliding and corrosive action. *Pitting* is a localized removal of material spigots from surface due to high contact fatigue. *Scoring* forms scratches across rubbing surfaces. *Scuffing* generates gross damage by local welds between surfaces. *Spalling* is a local separation of flakes of metal from surfaces.

Breakage and ***fatigue*** are direct symptoms of a bad gear design. Higher or unaccounted for loading variation could cause either breakage or brittle fatigue failures. *Breakage* is caused by high fatigue stressing on the tooth root or tip that generates eventual separation of material continuum. The tooth loses its geometry, and contact would be erratic. Vibration and noise are generated with interruption of continuous power transmission. Overloading and impact prevention means, improving root strength, increasing material toughness, diminishing misalignment of axes, and using tooth crowning would reduce the probability of such failures. *Surface failures* or *wear* are caused by high surface or contact stresses than the surface fatigue strength. Improving contact load distribution should be part of the remedy. This would also require the minimization of friction forces using high nominal oil viscosity and extreme pressure lubricants that does not cause corrosion, the use of shorter addenda with proper tip relief, and reducing lubricant contamination. In short, a better design would safeguard against these failures.

15.5 Computer-Aided Design and Optimization

The design procedures introduced for helical, bevel, and worm gears are amenable to computer programming. Just as other previously introduced machine elements, the objective is to have a computer-aided synthesis for the machine elements presented in this chapter. Elements such as bevel gears are left to be synthesized as spur gears in Chapter 14. The variations of some parameters in the detailed design can be considered independently, and necessary changes in the load or the safety factor can account for that. The *CAD* of helical and worm gears are treated in Sections 15.5.1 and 15.5.3.

Some codes are available to help in the design of gears particularly those provided by AGMA such as AGMA Gear (2016) and AGMA Bevel (2016). The software packages should help to abide by the needed or necessary standards in product design.

15.5.1 Helical Gears Synthesis

MATLAB code is used to implement the synthesis procedure of helical gears as developed in Section 15.1.4. The initial synthesis, the detailed design, and the safety factors are developed for a selected material property factor B_{BC} of about 0.35. The allowable fatigue strength, however, can be changed. This distinct material property factor is selected since optimum parameters are available for this material property factor as discussed in Section 15.1.4.1.

```
clear all; clc; format compact; format short          % CAD_Helical_Gears.m
H_kW = 75                                         % INPUT Power [kW]
H_hp = 100.58                                     % INPUT Power [hp]
Nrpm = 1200                                       % INPUT Rotational speed [rpm]
ng = 4                                            % INPUT Gear ratio
Sab_MPa = 300                                     % INPUT Allowable Fatigue Strength [MPa]
Sab_kpsi = 43.5                                   % INPUT Allowable Fatigue Strength [kpsi
Omega =2*pi*Nrpm/60                                   % Rotational speed [rad/s]
Tin_Nm = H_kW*10^3/Omega                              % Torque input [N.m]
Tin_lbin = H_hp*6600/Omega                            % Torque input [lb.in]
Nrpm_out = Nrpm/ng                                    % Out rotational speed [rpm]
Omega_out = Omega/ng                                  % Out rotational speed [rad/s]
Tout_Nm = Tin_Nm*ng                                   % Torque out [N.m]
Tout_lbin = Tin_lbin*ng                               % Torque out [lb.in]
mn_o_400MPa = 0.1413 * Tin_Nm^0.4701                  % Module* [mm] at 400 [MPa]
pd_o_58_lb = 25.4/(0.1413*(0.113*Tin_lbin)^0.4701)    % Diametral pitch* [teeth/in] at 58 [kpsi]
mn_o_100MPa = 0.1203 * Tin_Nm^0.4726                  % Module* [mm] at 100 [MPa]
pd_o_15_lb = 25.4/(0.1203*(0.113*Tin_lbin)^0.4726)    % Diametral pitch* [teeth/in] at 14.5 [kpsi]
Np_SI_100 = 30444*Sab_MPa^-1.208                      % Pinion teeth number at 100 [N.m]
Np_US_885 = 2954.16*Sab_kpsi^-1.208                   % Pinion teeth number at 885 [lb.in]
Np_SI_10_5 = 22036*Sab_MPa^-1.207                     % Pinion teeth number at 105[N.m]
Np_US_885_3 = 2142.11*Sab_kpsi^-1.207                 % Pinion teeth number at 885 [klb.in]
fg_mm100 = 27.758*Tin_Nm^0.1702                       % Face width at 100 [MPa]
fg_in15 = 0.75407*Tin_lbin^0.1702                     % Face width at 14.5 [kpsi]
fg_mm400 = 15.033*Tin_Nm^0.2155                       % Face width at 400 [MPa]
fg_in58 = 0.36998*Tin_lbin^0.2155                     % Face width at 58 [kpsi]
```

Figure 15.16 MATLAB code for synthesis of single stage helical gearbox. The optimum is mainly calculated for the pinion. The full code is the combination of Figures 15.15–15.18.

Figure 15.16 provides the start MATLAB code for the synthesis of a single stage helical gearbox. The optimum is mainly calculated for the pinion. The full code is the combination of Figures 15.16–15.19. The input variables are highlighted by the bold **INPUT** comment statement (`%` ...) at each variable. The start inputs are the usual design need of transmitting power H (`H_kW or H_hp`) in [kW] or [hp], the input rotational speed of the pinion N_{rpm} (`Nrpm`) in [rpm], the gear ratio n_g (`ng`), and the allowable fatigue strength S_{ab} (`Sab_MPa or Sab_kpsi`) in [MPa] or [kpsi]. This section of the code calculates all the parameters defined in the comment statement (`%` ...) at each one. These include the input and output torques; the optimum module, diametral pitch, pinion and gear diameters, pinion number of teeth, and face width. The optimum values are at the bounds of the material strength or the applied torques. These have been previously discussed in Section 15.1.4.1 and implemented in Example 15.2.

Figure 15.17 is a continuation of the synthesis process in the MATLAB code, but it allows the definite choice of the standard or round figure values of the main design parameters. In addition to selecting synthesis parameters, this part of the code calculates geometry and some other variables. The calculated optimum parameters had the ranges defined in Figure 15.15 and should be used as guide to input the parameters. The needed input parameters are highlighted by the bold **INPUT** comment statement (`%` ...) at each parameter. These parameters are the normal module m_n (`mn`) in [mm], the diametral pitch p_d (`pd`) in [teeth/in], number of pinion teeth N_P (`Np`), the face width f_g (`fg_mm_SI or fg_in_US`) in [mm] or [in], the helix angle ψ (`psi`), the normal pressure angle ϕ_n (`phi`), and the gear quality number Q_v (`Qv`). This section of the code calculates all the variables defined in the comment statement (`%` ...) at each one. These include the transverse module, transverse diametral pitch, pinion and gear diameters, and tangential velocities.

Figure 15.18 is the third part of the MATLAB code for synthesis of the single stage helical gearbox. It calculates forces between gears, other factors needed to calculate stresses, and the maximum bending fatigue stress (Eq. (15.29)). Also, the code needs the input of some other parameters for contact fatigue stress calculation. The needed input parameters are highlighted by the bold **INPUT** comment statement (`%` ...) at each parameter. These are the elastic coefficients C_p (`Cp_SI or Cp_US`) and the bending to contact fatigue strength ratio or material property factor B_{BC} (`BBC`). Few other variables are calculated and continued into Figure 15.19.

Figure 15.19 is the last part of the MATLAB code for synthesis of the single stage helical gearbox. It calculates the other needed factors to estimate the maximum contact fatigue stress in Eq. (15.30) and the extra safety factors for bending and contact in both SI and the US systems as in Eqs. (15.38) and (15.40). The full synthesis code is the combination of Figures 15.16–15.19.

```
                                              % CAD_Helical_Gears.m
Optimum_Module_Range = [mn_o_100MPa mn_o_400MPa]
mn = 2.5;                                     % INPUT module [mm]
Optimum_Diametral_pitch_Range = [pd_o_15_lb pd_o_58_lb]
pd = 10;                                      % INPUT Diametral pitch [teeth/in]
Optimum_Number_of_Teeth_Range = [Np_SI_100 Np_SI_10_5]
Np = 30;                                      % INPUT Pinion teeth number
Optimum_Face_Width_Range_mm = [fg_mm100 fg_mm400]
fg_mm_SI = 65;                                % INPUT Face width [mm]
Optimum_Face_Width_Range_in = [fg_in15 fg_in58]
fg_in_US = 2.6;                               % INPUT Face width [in]
Ng = Np*ng                                            % Number of gear teeth
psi = 27                                      % INPUT the Optimum Helix Angle
phi = 20                                      % INPUT Pressure Angle
Qv = 8                                        % INPUT Quality number
mt = mn/cos(pi*psi/180)                               % Transverse module [mm]
pdt = pd*cos(pi*psi/180)                              % Transverse diametral pitch [teeth/in]
dp_mm = Np*mn/cos(pi*psi/180)                         % Pinion diameter [mm]
dg_mm = dp_mm*ng                                      % Gear diameter [mm]
dp_in = Np/pdt                                        % Pinion diameter [in]
dg_in = dp_in*ng                                      % Gear diameter [in]
vt_SI = Omega*dp_mm/(2*1000)                          % Tangential Velocity [m/s]
vt_US = Omega*dp_in/2                                 % Tangential Velocity [in/s]
vt_US_ftmin = Omega*dp_in*60/(2*12)                   % Tangential Velocity [ft/min]
```

Figure 15.17 MATLAB code for synthesis of single stage helical gearbox. The optimum is mainly calculated for the pinion. This part is for selecting synthesis parameters and calculating geometry and some other parameters.

```
                                                  % CAD_Helical_Gears.m
Ft_kN = H_kW/vt_SI                                % Tangential force [kN]
Ft_lb = H_hp*6600/vt_US                           % Tangential force [lb]
Fn_kN = Ft_kN/(cos(pi*psi/180)*cos(pi*phi/180))   % Normar or resultant force [kN]
Fn_lb = Ft_lb/(cos(pi*psi/180)*cos(pi*phi/180))   % Normar or resultant force [lb]
Fr_kN = Fn_kN* sin(pi*phi/180)                    % Radial force [kN]
Fr_lb = Fn_lb* sin(pi*phi/180)                    % Radial force [lb]
Fa_kN = Fn_kN*(sin(pi*psi/180)*cos(pi*phi/180))   % Axial force [kN]
Fa_lb = Fn_lb*(sin(pi*psi/180)*cos(pi*phi/180))   % Axial force [lb]
        disp('-Maximum Bending Fatigue Stress for Input Allowable Fatigue Strength ')
Kg2 = 0.8691*Ng^0.0295;                           % Geometry Factor Multiplier of Jg
Jg = -(0.000145*Np^0.161597)*psi^2 + (0.007237*Np^0.05324)*psi + 0.342121*Np^0.100154;
Jg = Jg * Kg2                                     % Geometry Factor Jg
Bv = 0.25*(12-Qv)^(2/3)
Bv_f = (50+ 56*(1.0-Bv));
Kv_SI = ((Bv_f+sqrt(200*vt_SI))/Bv_f)^(Bv)        % Dynamic or velocity factor SI
Kv_US = ((Bv_f+sqrt(vt_US_ftmin))/Bv_f)^(Bv)      % Dynamic or velocity factor US
km_SI = -4*10^-7*fg_mm_SI^2+0.0011*fg_mm_SI+1.544 % Load distribution factor SI
km_US = -6.28*10^-4*fg_in_US^2+0.0364*fg_in_US+1.523  % Load distribution factor US
sigma_ab_MPa = (Ft_kN*1000*Kv_SI*km_SI)/(fg_mm_SI*mt*Jg)      % Maximum Bending Fatigue Stress [Mpa]
sigma_ab_kpsi = (Ft_lb*Kv_US*km_US*pdt)/(1000*fg_in_US*Jg)    % Maximum Bending Fatigue Stress [kpsi]
        disp('-Maximum Contact Fatigue Stress for Input Allowable Fatigue Strength')
Cp_SI = 191                                   % INPUT Elastic Coefficient [MPa]^0.5
Cp_US = 2300                                  % INPUT Elastic Coefficient [psi]^0.5
BBC = 0.35                                    % INPUT Bending to Contact Strength ratio
pt_mm = (pi*mt)                                   % Transverse circular pitch [mm]
pn_mm = pt_mm*cos(pi*psi/180)                     % Normal circular pitch [mm]
```

Figure 15.18 MATLAB code for synthesis of single stage helical gearbox. The optimum is mainly calculated for the pinion. Calculating forces between gears, other factors, and maximum bending fatigue stress. Also input some parameters for contact fatigue.

This MATLAB code in Figures 15.16–15.19 is available in the ***Wiley* website** for *computer-aided synthesis* or real *CAD* of helical gears under the name of ***CAD_Helical_Gears.m***. This provides a direct synthesis close to an optimum without any appreciable need for repeated analysis. The iterative synthesis, however, is available for fine tuning of required standards or satisfying other constraints. This is possible by iterating some of the input geometry parameters to reach the desired goal.

Example 15.4 This example is redoing Example 15.3, which is about the reduction gear in a crane driven by an electric motor at a maximum power of 75 [kW] or 100.58 [hp], running at 1200 [rpm], and for a gear ratio of 4 : 1.

```
                                                          % CAD_Helical_Gears.m
tan_phit = tan(pi*phi/180)/cos(pi*psi/180)
phit = atan(tan_phit)*180/pi
rp_mm = dp_mm/2
rg_mm = dg_mm/2
rbP_mm = rp_mm*cos(pi*phit/180)
rbG_mm = rg_mm*cos(pi*phit/180)
Z = sqrt((rp_mm+mn)^2-rbP_mm^2)+sqrt((rg_mm+mn)^2-rbG_mm^2)-(rp_mm+rg_mm)*sin(pi*phit/180)
mN = pn_mm*cos(pi*phi/180)/(0.95*Z)
Ig =(cos(pi*phi/180)*sin(pi*phi/180)/(2*mN))*(ng/(ng+1))          % Geometry factor SI
rp_in = dp_in/2
rg_in = dg_in/2
rbP_in = rp_in*cos(pi*phit/180)
rbG_in = rg_in*cos(pi*phit/180)
Z = sqrt((rp_in+1/pd)^2-rbP_in^2)+sqrt((rg_in+1/pd)^2-rbG_in^2)-(rp_in+rg_in)*sin(pi*phit/180)
mN_US = (pi/pd)*cos(pi*phi/180)/(0.95*Z)
Ig_US =(cos(pi*phi/180)*sin(pi*phi/180)/(2*mN_US))*(ng/(ng+1))    % Geometry factor US
sigma_ac_MPa = Cp_SI*sqrt(Fn_kN*1000*Kv_SI*km_SI/(fg_mm_SI*dp_mm*Ig))       % Maximum Contact Fatigue Stress [Mpa]
sigma_ac_kpsi = Cp_US*sqrt(Fn_lb*Kv_US*km_US/(fg_in_US*dp_in*Ig_US))/1000 % Maximum Contact Fatigue Stress [kpsi]
Sac_MPa = Sab_MPa/BBC                                             % Maximum Contact Fatigue Strength [Mpa]
Sac_kpsi = Sab_kpsi/BBC                                           % Maximum Contact Fatigue Strength [kpsi]
      disp('-Extra Safety Factors for Bending and Contact Allowable Fatigue')
KSF_b_SI = ((Sab_MPa)/sigma_ab_MPa)
KSF_b_US = ((Sab_kpsi)/sigma_ab_kpsi)
KSF_c_SI = (Sac_MPa)*sqrt(km_SI)/sigma_ac_MPa
KSF_c_US = (Sac_kpsi)*sqrt(km_US)/sigma_ac_kpsi
```

Figure 15.19 MATLAB code for synthesis of single stage helical gearbox. The optimum is mainly calculated for the pinion. Calculating factors, maximum contact fatigue stress, and the extra safety factors. The full code is the combination of Figures 15.15–15.19.

This has been synthesized in Example 15.2. Gears are 20° full-depth helical type. Check the synthesized initial helical gear set for stresses and safety using the MATLAB code of `CAD_Helical_Gears.m`.

Solution

Data: Again, power transmitted $H_{\text{kW}} = 75$ [kW] or $H_{\text{hp}} = 100.58$ [hp], rotational speed $N_{\text{in}} = 1200$ [rpm], and the gear ratio $n_g = 4$. The rotation speed $\omega_P = (2\pi\, N_{\text{in}}/60) = 125.66$ [rad/s]. Results of Examples 15.2 and 15.3 are with a value of B_{BC} of 0.35.

Using the MATLAB code of `CAD_Helical_Gears.m`, the following input and output are obtained with some minor presentation editing (see Figures 15.16–15.19):

Input:

`H_kW = 75, H_hp = 100.5800, Nrpm = 1200, ng = 4, Sab_MPa = 300, Sab_kpsi = 43.5000`

Output:

- *Input shaft*: `Omega = 125.6637, Tin_Nm = 596.8310, Tin_lbin = 5.2826e+03`
- *Output shaft*: `Nrpm_out = 300, Omega_out = 31.4159, Tout_Nm = 2.3873e+03, Tout_lbin = 2.1130e+04`
- *Helical gears optimum synthesis*: `mn_o_400MPa = 2.8515, pd_o_58_lb = 8.9070, mn_o_100MPa = 2.4668, pd_o_15_lb = 10.2959, Np_SI_100 = 30.9841, Np_US_885 = 30.9841, Np_SI_10_5 = 22.5552, Np_US_885_3 = 22.5520, fg_mm100 = 82.3831, fg_in15 = 3.2437, fg_mm400 = 59.5994, fg_in58 = 2.3467, Optimum_Module_Range = 2.4668 2.8515, Optimum Module Range is for 100 [MPa] to 400 [MPa]`
- *Some standard modules* = `0.3,0.4,0.5,0.8,1,1.25,1.5,2,2.5,3,4,5,6,8,10,12,16,20,25`
`Optimum_Diametral_pitch_Range = 10.2959 8.9070`
- *Some diametral pitches* = `80,64,48,32,24,20,16,12,10,8,6,5,4,3,2.5,2,1.5,1.25,1,25`
`Input_module = 2.5000, Input_diametral_pitch = 10`
`Optimum_Number_of_Teeth_Range = 30.9841 22.5552, Optimum Number of Teeth Range is for 100 [Nm] to 10^5 [Nm]`
`Input_number_of_teeth = 30`

```
Optimum_Face_Width_Range_mm = 82.3831 59.5994, - Optimum Face Width Range is
for 100 [MPa] to 400 [MPa]
Input_face_width_mm = 65
Optimum_Face_Width_Range_in = 3.2437 2.3467
Input_face_width_in = 2.6000
Ng = 120
```

Input:

```
psi = 27, phi = 20, Qv = 8
```

Output:

```
mt = 2.8058, pdt = 8.9101
dp_mm = 84.1745, dg_mm = 336.6979, dp_in = 3.3670, dg_in = 13.4679,
vt_SI = 5.2888, vt_US = 211.5535, vt_US_ftmin = 1.0578e+03
Ft_kN = 14.1808, Ft_lb = 3.1379e+03, Fn_kN = 16.9369, Fn_lb = 3.7477e+03,
Fr_kN = 5.7928, Fr_lb = 1.2818e+03, Fa_kN = 7.2255, Fa_lb = 1.5988e+03
```

- *Maximum bending fatigue stress for input allowable fatigue strength*

```
Jg = 0.5325, Bv = 0.6300, Kv_SI = 1.2691, Kv_US = 1.2691, km_SI = 1.6138,
km_US = 1.6134
sigma_ab_MPa = 299.0639, sigma_ab_kpsi = 41.3491
```

- *Maximum contact fatigue stress for input allowable fatigue strength*

```
Cp_SI = 191, Cp_US = 2300, BBC = 0.3500
pt_mm = 8.8147, pn_mm = 7.8540, tan_phit = 0.4085, phit = 22.2197,
rp_mm = 42.0872, rg_mm = 168.3489, rbP_mm = 38.9619, rbG_mm = 155.8475,
Z = 12.1077, mN = 0.6416
Ig = 0.2004
rp_in = 1.6835, rg_in = 6.7340, rbP_in = 1.5585, rbG_in = 6.2339,
Z = 0.4843, mN_US = 0.6416
Ig_US = 0.2004
sigma_ac_MPa = 1.0744e+03, sigma_ac_kpsi = 152.1349
Sac_MPa = 857.1429, Sac_kpsi = 124.2857
```

- *Extra safety factors for bending and contact allowable fatigue*

```
KSF_b_SI = 1.0031, KSF_b_US = 1.0520
KSF_c_SI = 1.0134, KSF_c_US = 1.0377
```

These values conform with the hand calculation results of Example 15.3.

15.5.2 Bevel Gears Synthesis

The initial synthesis of spur gears can be used for the *initial synthesis* of bevel gears. The *bevel gear geometry* at the large end is the same as for spur gear geometry. Therefore, procedure of Section 14.4.2 can be used for bevel gears, where the module or the diametral pitch, the number of teeth, and the face width can be adapted. The main constraint, however, is the face width that should be less than the pinion cone distance A_P. The suggested value should be within that shown in Table 15.1, i.e. about one-third the pinion cone distance or smaller. A computer code can be developed to check on that. In addition, the detailed design factors that are different from spur gears can be included in the same design code to evaluate the multipliers needed to adjust the results for bevel gears. Both spur and bevel gears of the same materials and specifications can transmit about the same power, and the differences can then be defined.

The different factors are the *load distribution factor* k_m in Eq. (15.51), the *bending factor* for bevel gears J_g in Eq. (15.52), and the *contact geometry factor* I_g in Eq. (15.54). The variations of these parameters from those of spur gears can be considered independently, and necessary changes in the load, material, or the safety factor can account for that.

The development of a computer code to adjust the synthesis of bevel gears relative to the spur gears is left to the reader. The complete design calculations, however, should be rechecked against the design code that one should use in any product development.

15.5.3 Worm Gears Synthesis

MATLAB code is used to implement the synthesis procedure of worm gears as developed in Section 15.3.3. The initial synthesis is initiated by the utility of Figure 15.15 to estimate the expected center distance C_W as indicated in Section 15.3.3.1. The MATLAB code (`CAD_Worm_Gears.m`) in Figures 15.20 and 15.21 is then used to proceed with the synthesis of the worm gears geometry. The synthesis is mainly calculated for geometry and contact forces. Figure 15.20 gives the worm and worm gear diameters. Figure 15.21 gives the rest of worm and worm gear geometry and contact forces. The full code is the combination of Figures 15.20 and 15.21. The input variables are highlighted by the bold **INPUT** comment statement (`%` ...) at each variable. The start inputs are the usual design need of transmitting power H (`H_kW or H_hp`) in [kW] or [hp], the input rotational speed of the pinion N_{rpm} (`Nrpm`) in [rpm], and the gear ratio n_g (`ng`). The other inputs are the expected center distance C_W (`CW`) and the worm thread umber N_W (`NW`). One can start with $N_W = 1$ and then advance the value to adjust design parameters. The output geometry is dependent on the utility of Eq. (15.62). The range of the constants 2.2 and 1.4683 can be used to adjust the design to any other acceptable one.

In addition to giving the worm and worm gear diameters, Figure 15.15 provides the calculated input and output torques and the output rotational speed. Figure 15.21 of the code allows the user to input the normal pressure angle ϕ_n, (`phi`), and the friction coefficient μ_k (`mu`). This section of the code calculates all the parameters defined in the comment statement (`%` ...) at each one. These include the number of worm gear teeth, transverse axial pitch, lead angle, normal circular pitch, normal module, diametral pitch, and tangential velocities. Also output forces include tangential forces, normal forces, radial forces, and axial forces. All are in SI [kN] and the US [lb].

The constant 2.2 in Eq. (15.62) has the range of 3.0–1.6 and can be examined in the MATLAB code to get different geometrical configuration or optimization. The smaller constant of 1.6 gives a smaller worm lead angle λ_W and

```
clear all; clc; format compact; format short                          % CAD_Worm_Gears.m
          disp('-Worm Gears Applied Loads and Geometry')
H_kW = 75                                                   % INPUT Power [kW]
H_hp = 100.58                                               % INPUT Power [hp]
% Nrpm = 1200                                               % INPUT Rotational speed [rpm]
Nrpm = 1750                                                 % INPUT Rotational speed [rpm]
ng = 15                                                     % INPUT Gear ratio
          disp('-Input Shaft')
Omega =2*pi*Nrpm/60
Tin_Nm = H_kW*10^3/Omega                                              % Torque input [N.m]
Tin_lbin = H_hp*6600/Omega                                            % Torque input [lb.in]
          disp('-Output Shaft')
Nrpm_out = Nrpm/ng                                                    % Out rotational speed [rpm]
Omega_out = Omega/ng                                                  % Out rotational speed [rad/s]
Tout_Nm = Tin_Nm*ng                                                   % Torque out [N.m]
Tout_lbin = Tin_lbin*ng                                               % Torque out [lb.in]
          disp('-Worm Gears Input Center Distance, Thread Number, and Get Geometry')
CW_mm = 254                                                 % INPUT Center distance [mm]
CW_in = 10                                                  % INPUT Center distance [in]
NW = 1                                                      % INPUT Worm thread number
NW = 6                                                      % INPUT Worm thread number
dW_mm = CW_mm^0.875/1.4683                                            % Worm diameter [mm]
dW_in = CW_in^0.875/2.2                                               % Worm diameter [in]
dG_mm = 2*CW_mm -dW_mm                                                % Worm-gear diameter [mm]
dG_in = 2*CW_in -dW_in                                                % Worm-gear diameter [in]
```

Figure 15.20 MATLAB code for synthesis of single stage worm gear set. The synthesis is mainly calculated for geometry and contact forces. This gives the worm and worm gear diameters. The full code is the combination of Figures 15.20 and 15.21.

```
clear all; clc; format compact; format short                    % CAD_Worm_Gears.m
NG = ng*NW                                                      % Number of teeth of worm-gear
pt_mm = pi*dG_mm/NG                                             % Transverse or Axial pitch
pt_in = pi*dG_in/NG                                             % Transverse or Axial pitch
lamdaW_SI = atan(pt_mm*NW/(pi*dW_mm))*180/pi                    % Lead angle (worm) SI
lamdaW_US = atan(pt_in*NW/(pi*dW_in))*180/pi                    % Lead angle (worm) US
pn_mm = pt_mm*cos(lamdaW_SI*pi/180)                             % Normal circular pitch
pn_in = pt_in*cos(lamdaW_US*pi/180)                             % Normal circular pitch
mn_mm = pn_mm/pi                                                % Normal module [mm]
pd = pi/pn_in                                                   % Diametral pitch [tooth/in]
phi = 20                                                   % INPUT Pressure Angle
mu = 0.05                                                  % INPUT Friction coefficient
eta = (cos(20*pi/180)*sin(lamdaW_SI*pi/180))/(cos(20*pi/180)*sin(lamdaW_SI*pi/180)+mu*cos(lamdaW_SI*pi/180))
                                                                       % Efficiency SI
          disp('-Worm Gear Forces')
vt_SI = Omega*dW_mm/(2*1000)                                    % Tangential velocity [m/s]
vt_US = Omega*dW_in/2                                           % Tangential velocity [in/s]
Ft_kN = H_kW/vt_SI                                                     % Tangential force [kN]
Ft_lb = H_hp*6600/vt_US                                                % Tangential force [lb]
Fn_kN = Ft_kN/(cos(pi*phi/180)*sin(pi*lamdaW_SI/180)+ mu*cos(pi*lamdaW_SI/180))    % Normal force [kN]
Fn_lb = Ft_lb/(cos(pi*phi/180)*sin(pi*lamdaW_US/180)+ mu*cos(pi*lamdaW_US/180))    % Normal force [lb]
Fr_kN = Fn_kN * sin(pi*phi/180)                                        % Radial force [kN]
Fr_lb = Fn_lb * sin(pi*phi/180)                                        % Radial force [lb]
Fa_kN = Fn_kN*(cos(pi*phi/180)*cos(pi*lamdaW_SI/180)-mu*sin(pi*lamdaW_SI/180))     % Axial force [kN]
Fa_lb = Fn_lb*(cos(pi*phi/180)*cos(pi*lamdaW_US/180)-mu*sin(pi*lamdaW_US/180))     % Axial force [lb]
```

Figure 15.21 MATLAB code for synthesis of single stage worm gear set. The synthesis is mainly calculated for geometry and contact forces. This gives the rest of worm and worm gear geometry and contact forces. The full code is the combination of Figures 15.20, 15.21.

lower efficiency. The larger number of worm threads gives a smaller tooth, a larger worm lead angle λ_W, higher efficiency, and unlocking condition.

The MATLAB code in Figures 15.20 and 15.21 is available in the ***Wiley* website** for *computer-aided synthesis* or real *CAD* of worm gears under the name of ***CAD_Worm_Gears.m***. This should be useful in calculating initial synthesis of worm gear sets and help in the understanding of their geometry and performance. The forces generated by the code can be used to design shafts and bearings needed for the construction of such gearboxes.

Example 15.5 A reduction gear in Example 15.3 of a crane driven by electric motor at a maximum power of 75 [kW] or 100.58 [hp] running at 1750 [rpm] and for a worm gear ratio of 15 : 1 has to be synthesized. Gears are 20° full-depth worm type. Use the available MATLAB code *CAD_Worm_Gears.m* to synthesize an initial worm gear set and find the interfacing forces between gears for such an application.

Solution

Data: Power transmitted $H_{kW} = 75$ [kW] or $H_{hp} = 100.58$ [hp], rotational speed $N_{in} = 1750$ [rpm], and the gear ratio $n_g = 15$. The rotation speed $\omega_P = (2\pi\, N_{in}/60) = 183.26$ [rad/s].

Using the MATLAB code of *CAD_Worm_Gears.m*, the following inputs and outputs are obtained with some minor presentation editing (see Figures 15.20 and 15.21):

Input:

H_kW = 75, H_hp = 100.5800, Nrpm = 1750, ng = 15

Output:

- *Input shaft*: Omega = 183.2596, Tin_Nm = 409.2556, Tin_lbin = 3.6223e+03
- *Output shaft*: Nrpm_out = 116.6667, Omega_out = 12.2173, Tout_Nm = 6.1388e+03, Tout_lbin = 5.4335e+04

Input:

- *Worm gears input center distance, thread number*, and *gear geometry*

CW_mm = 254, CW_in = 10, NW = 6

Output:

dW_mm = 86.5794, dW_in = 3.4086

dG_mm = 421.4206, dG_in = 16.5914, NG = 90

pt_mm = 14.7104, pt_in = 0.5791, lamdaW_SI = 17.9781, lamdaW_US = 17.9782,
pn_mm = 13.9921, pn_in = 0.5509, mn_mm = 4.4538, pd = 5.7030

Input:

phi = 20, mu = 0.0500

Output:

- *Worm gear efficiency*: eta = 0.8591
- *Worm gear forces*: vt_SI = 7.9333, vt_US = 312.3302

Ft_kN = 9.4539, Ft_lb = 2.1254e+03
Fn_kN = 28.0034, Fn_lb = 6.2956e+03
Fr_kN = 9.5777, Fr_lb = 2.1532e+03
Fa_kN = 24.5976, Fa_lb = 5.5299e+03

These values are to be checked by hand calculations and parameter variations to get a better design for other objectives.

15.6 Constructional Details

The helical, bevel, and worm gear sets experience ***3D*** forces generated from the teeth interaction to transmit power. The simple gear set consists of a pinion (or a worm) and a gear mounted on two shafts using *keys* or *splines* is discussed in Section 8.2.2. The forces are transmitted from the gears to the shafts, to the bearings, to the gearbox housing, and then to the ground support; see Section 14.7. Figure 15.22 shows a sectional view for the assembly of a single stage helical gearbox similar to Figure 14.26. The gears are shown in a simplified half sectional view to indicate the angle of the helix on each gear. The helix angle is identified by the slanted line on each gear view. The helix angle generates the axial force that one of the bearings should carry. It is assumed that the *end bearings* will carry the axial force in addition to part of the radial force (about half) depending on bearing and gear locations on the shaft. That is why, the end bearings on both shafts are larger than the front bearings (Figure 15.22). The output shaft is also larger than the input shaft due to the larger output torque. The gears are keyed to the shafts to transmit the input and output torques. The fits between the gears and the shafts can be a location fit of H7/h6 or H7/j6 (not shown in Figure 15.22). See Figures 11.15 and 11.12 for additional details about mounting of bearings.

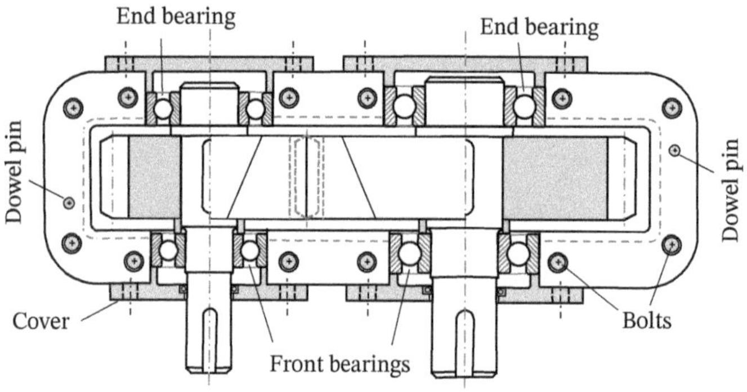

Figure 15.22 A sectional view for the assembly of a single stage helical gear set. The gears are shown in a simplified half sectional view to indicate the angle of the helix on each gear. The helix angle is identified by the slanted line on each gear view.

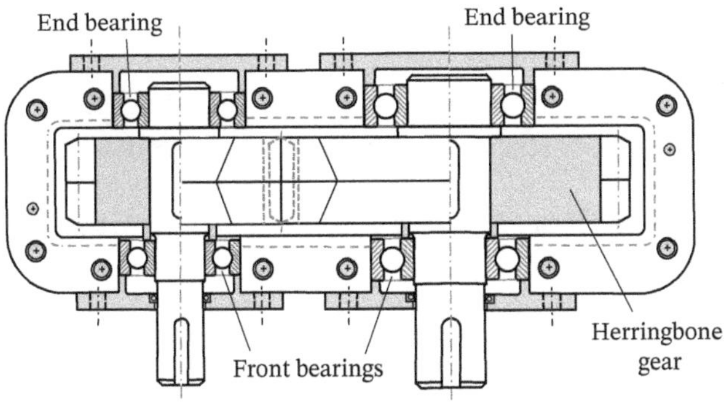

Figure 15.23 A sectional view for the assembly of a single stage herringbone gear set. The gears are shown in a simplified half sectional view to indicate the angles of the helix on each gear. The helix angles are identified by the two slanted lines on each gear view.

The other option of bearing selection is to use *tapered roller bearings* to carry a higher axial load. The end bearings should then have the configuration in Figure 11.11b with the axial tight holding options shown in Figure 11.13c or d for the right bearing of Figure 11.12b.

Figure 15.23 shows a sectional view for the assembly of a single stage *herringbone* gear set. The gears are shown in a simplified half sectional view to indicate the angles of the helix on each gear. The helix angles are identified by the two slanted lines on each gear view. In this case, the larger end bearings can be reduced in size (not shown) since the axial forces are canceled out among the two halves of the *herringbone* gears. This other solution to handle the axial force would be more expensive since the herringbone gears are more expensive to produce than helical gears. The cost comparison or trade-off is between herringbone gears, larger ball bearings, or tapered rolling bearings in Figure 15.22. Other factors may also be affecting the design, which can favor either option such as size, heavy loading, reliability, quieter operations, etc.

15.7 Summary

In this chapter helical, bevel, and worm gears are presented. The geometry of the teeth, kinematic definitions, and usual types and standards are introduced in the chapter. Forces generated during meshing and power transmission are identified. These forces that are acting on the teeth of the meshing gears and their effects in developing stresses on the teeth and gear design are discussed. The relations defining the bending fatigue and contact fatigue stresses are reported. Material set usually used in gears are utilized. The material property factor identified by the bending fatigue strength to the contact fatigue strength is utilized in the synthesis of helical and bevel gears. Synthesis procedure utilizes *CAD* and optimization tools, and their results to generate better gearing design are employed. Constructional details for gear set assembly of gearboxes are also presented.

The computer-aided synthesis process is provided in a real *CAD* codes using MATLAB. These are available as ***CAD_Helical_Gears.m*** and ***CAD_Worm_Gears.m*** through **Wiley website**. The utility of these SI and US codes has been demonstrated to effectively synthesize helical and worm gears near the required optimum or the available feasible solutions. Little iteration is needed to tune the synthesized gear design to satisfy multitude of conditions, cases, and possible optimum objectives. If need be, the synthesized and iterated parameters can be construed as good values to be checked, verified, and validated by standard codes and adjusted to other conditions, loadings, and manufacturing processes that have not been included herein.

Problems

15.1 Identify few applications or products that utilize helical, or bevel, or worm gearing. Use the Internet to collect images of such applications. Why these applications are not using spur gears instead?

15.2 Define other helical gear sets than the double helical, the crossed helical, and the herringbone gear sets. Sketch those other types and indicate their utility.

15.3 Place a can on a sheet of paper having an inclined or slanted edge with an angle ψ. Wrap the paper around the can close to one end. Unwrap the sheet of paper and observe while unwrapping. Observe the curved surface generated by the inclined edge of the sheet.

15.4 Search the Internet for *hobbing* or milling machines that can produce helical gears. Download any video that would describe the manufacturing process. Describe the technical specifications of the hob that is provided by the hobbing machine manufacturer.

15.5 Prove that the transverse pressure angle ϕ_t is related to the normal pressure angle ϕ_n by Eq. (15.3).

15.6 Find other helical gear parameters in addition to those in Table 14.1.

15.7 Find an image over the Internet of a gear milling cutter with a visible range of the virtual number of teeth itched on the side.

15.8 Prove that the virtual number of teeth is the same as the number of teeth for spur gears.

15.9 How small can you make the module in Example 15.1 to be able to reduce the round figure center distance to 200 [mm]? What should be the helix angle in that case? Find the forces developed between the teeth of the gear set.

15.10 How large can you make the diametral pitch in Example 15.1 to be able to reduce the round figure center distance to 8 [in]? What should be the helix angle in that case? Find the forces developed between the teeth of the gear set.

15.11 Find the range of the ratio B_{BC} between the allowable bending fatigue strength and the allowable contact fatigue strength for several materials used in gears.

15.12 Use any available information to verify the acceptability of Eq. (15.25). AGMA or ISO standard should be a good source.

15.13 A helical gear set is needed to transmit a maximum power of 7500 [kW] and has an input speed of 3600 [rpm]. Find the maximum transmitted torque. Determine the forces between gears and the pitch point velocity. Synthesize the gear set, and compare the results to the set having an input speed of 1750 [rpm]. How can one handle the selection of quality if the pitch line speed is high? What quality number should be suggested and what are the consequences in the design procedure?

15.14 A helical gear set is needed to transmit a maximum power of 10 058 [hp] and has an input speed of 3600 [rpm]. Find the maximum transmitted torque. Determine the forces between gears and the pitch point velocity. Synthesize the gear set, and compare the results to the set having an input speed of 1750 [rpm]. How can one handle the selection of quality if the pitch line speed is high? What quality number should be suggested and what are the consequences in the design procedure?

15.15 A small helical gear set has an input speed of 120 [rpm] and a reduction ratio of 4 and transmits a maximum power of 0.5 [kW]. Find the maximum transmitted torque. Determine the forces between gears and the pitch point velocity. Synthesize the gear set, and compare the results to the set having an input speed of 1750 [rpm].

15.16 A small helical gear set has an input speed of 120 [rpm] and a reduction ratio of 4 and transmits a maximum power of 0.67 [hp]. Find the maximum transmitted torque. Determine the forces between gears and the pitch point velocity. Synthesize the gear set, and compare the results to the set having an input speed of 1750 [rpm].

15.17–15.28 Design helical gear sets by initial synthesis procedure, and verify results using the `CAD_Helical_Gears.m` for the required specifications in Table 15.P1. The SI specification may not relate to the US specification. Select different *material designations* to observe the difference. Find the bending and contact safety factors, other properties, and parameters for each gear set. Compare results with the values possibly extracted from Figures 15.7–15.9. The power, angular velocities in [rpm], the quality, allowable bending fatigue strength, and property factor are given for each problem in Table 15.P1.

Table 15.P1 Data for Problems 15.17–15.28.

Problem number	Power H_{kW} or (H_{hp}) [kW] ([hp])	Input and (output) angular velocities [rpm]	Quality Q_v	Allowable bending fatigue strength S_{ab} [MPa] ([kpsi])	Property factor B_{BC}
15.17	1 (1)	750 (150)	8	100 (14.5)	0.3
15.18	1 (1)	1200 (300)	8	200 (29)	0.3
15.19	1 (1)	1750 (500)	10	300 (43.5)	0.35
15.20	1 (1)	3000 (1000)	12	400 (58)	0.35
15.21	100 (100)	750 (150)	8	100 (14.5)	0.3
15.22	100 (100)	1200 (300)	8	200 (29)	0.3
15.23	100 (100)	1750 (500)	10	300 (43.5)	0.35
15.24	100 (100)	3000 (1000)	12	400 (58)	0.35
15.25	10 000 (10 000)	750 (150)	8	100 (14.5)	0.3
15.26	10 000 (10 000)	1200 (300)	8	200 (29)	0.3
15.27	10 000 (10 000)	1750 (500)	10	300 (43.5)	0.35
15.28	10 000 (10 000)	3000 (1000)	12	400 (58)	0.35

15.29 From the geometry and available relations, prove Eq. (15.44).

15.30 Find the factors that one can use to adjust the inputs in a spur gear design to be able to utilize the ***spur gear synthesis tablet – SI and US*** in the design of bevel gears. The detailed design factors that are different from spur gears should be included in evaluating the *difference* factor or factors. Note that forces on teeth have different relations. Also note that the load distribution factor, the bending geometry factor, and the contact geometry factor for bevel gears are different from spur gears.

15.31 Adjust the MATLAB code `CAD_Helical_Gears.m` to be able to use it in the synthesis of bevel gears and save as `CAD_Bevel_Gears.m`.

15.32 Resolve any three spur gear problems (of different power and angular velocity inputs) among Problems 14.23–14.34 of Chapter 14 as normal bevel gears, and compare results to spur gears solutions.

15.33 Reconstruct the gearbox of Figure 14.26 to accommodate a bevel gear set. Consult available constructions from the Internet or any other references to help in that regard. Change the requirements in Figure 14.25 to aid in the development of construction variations.

15.34 Find the different configurations of automobile differentials utilizing bevel gears. Sketch the differential, and find the input–output relations for straight travelling or turning.

15.35 From Figure 15.13, derive Eq. (15.58) of the worm lead angle. Use similarity of threads as in Chapter 9.

15.36 The constant 2.2 in Eq. (15.62) has the range of 3.0–1.6 with an average of about 2.2. For SI system the constant 1.4683 in Eq. (15.62) has the range of 2.002–1.0679. Find the effect of that range on the worm gear geometry and characteristics.

15.37 Resolve any three helical gear problems (of different power and angular velocity inputs) among Problems 15.17–15.28 using the available MATLAB code `CAD_Helical_Gears.m`, and compare results with hand calculations and relevant spur gears solutions. Use Example 15.4 as a guide.

15.38 Use hand calculations to verify the results of Example 15.5. Change the design variables to standard or round figure values for a more producible worm gear set, and recalculate other parameters and variables.

15.39 Use the available MATLAB code (`CAD_Worm_Gears.m`) to study the synthesis effect of changing the constant 2.2 in Eq. (15.62) for the range of 3.0–1.6 or the constant 1.4683 in Eq. (15.62) for the range of 2.002–1.0679. Which bound is more reasonable in the re-evaluation of Example 15.5?

15.40 Use standard or round figure values for the design variables in an adjusted `CAD_Worm_Gears.m` MATLAB code to redesign the worm gear set of Example 15.5.

15.41 Reconstruct the gearbox of Figure 14.26 to accommodate the worm gear set of Example 15.5. Consult available constructions from the Internet or any other references to help in that regard. Change the requirements in Figure 14.25 to aid in the development of construction variations.

15.42 Reconstruct the gearbox of Figure 15.22 to accommodate the helical gear set of Example 15.4 using tapered roller bearings for both shafts.

References

AGMA (1982). *Rating the pitting resistance and bending strength of spur and helical gear teeth.* AGMA 218.01 Replaced by AGMA 2001-D04 and AGMA 908-B89, American Gear Manufacturers Association.

AGMA (1988). *Gear classification and inspection handbook: Tolerances and measuring methods for unassembled spur and helical gears (Including metric equivalents).* AGMA 2000-A88, Replaced by 915-1-A02, 915-2-A05, 2015-1-A01, and 2015-2-A06. American Gear Manufacturers Association.

AGMA (1999). *Geometry factors for determining the pitting resistance and bending strength of spur, helical and herringbone gear teeth.* (AGMA 908-B89, April 1989, (Revision of AGMA 226.01), (Reaffirmed August 1999), American Gear Manufacturers Association.

AGMA (2003). *Rating the pitting resistance and bending strength of generated straight bevel.* (R2015). Zerol Bevel, and Spiral Bevel Gear Teeth 2003-C10, American Gear Manufacturers Association.

AGMA (2016). *Technical Publications Catalog.* American Gear Manufacturers Association.

AGMA Bevel (2016). *Bevel Gear Rating Suite v.1.3.* American Gear Manufacturers Association.

AGMA Gear (2016). *Gear Rating Suite v.3.1.* American Gear Manufacturers Association.

AGMA ISO 22849 (2011). *Design Recommendations for Bevel Gears.* American Gear Manufacturers Association.

ANSI/AGMA (2001). *Fundamental rating factors and calculation methods for involute spur and helical gear teeth.* ANSI/AGMA 2001-D04 (revised AGMA 2001-C95) and ANSI/AGMA 2101-D04 (metric edition of ANSI/AGMA 2001-D04). American Gear Manufacturers Association.

ANSI/AGMA (2004). *Fundamental rating factors and calculation methods for involute spur and helical gear teeth.* ANSI/AGMA 2001-D04 (revised AGMA 2001-C95) and ANSI/AGMA 2101-D04 (metric edition of ANSI/AGMA 2001-D04). American Gear Manufacturers Association.

ANSI/AGMA 2003-B97 (2003). *Rating the pitting resistance and bending strength of generated straight bevel, zerol bevel and spiral bevel gear teeth.* American Gear Manufacturers Association.

ANSI/AGMA 6022-C93 (2014). *Design manual for cylindrical wormgearing.* American Gear Manufacturers Association.

Buckingham, E. (1981). *Design of Worm and Spiral Gears.* Industrial Press.

Davis, J.R. (ed.) (2005). *Gear Materials, Properties, and Manufacture.* ASM International.

Dimargonas, A. (1989). *Computer Aided Machine Design.* Prentice Hall.

Dolan, T.J. and Broghamer, E.L. (1942). *A Photoelastic Study of the Stresses in Gear Tooth Fillets.* University of Illinois, Engineering Experiment Station, Bulletin No. 335.

Dudley, D.W. (1984). *Handbook of Practical Gear Design.* McGraw Hill.

ISO 10347 (1999). *Worm gears – Geometry of worms – Name plates for worm gear units, center distances, information to be supplied to gear manufacturer by the purchaser.* International Organization for Standardization.

ISO 23509 (2016). *Bevel and hypoid gear geometry.* International Organization for Standardization.

ISO 6336 1-3 (2006). *Calculation of load capacity of spur and helical gears – Part 1: Basic principles, introduction and general influence factors, Part 2: Calculation of surface durability (pitting), Part 3: Calculation of tooth bending strength.* International Organization for Standardization.

ISO 6336 5 (2016). *Calculation of load capacity of spur and helical gears – Part 5: Strength and quality of materials.* International Organization for Standardization.

Metwalli, S.M., El Danaf, E. A. (1996). CAD and optimization of spur and helical gear sets. *Proceeding of the 21st ASME Design Automation Conference*, Irvine, CA (18–22 August). Paper # 96-DETC/DAC-1433.

Mitchiner, R.G. and Mabie, H.H. (1982). Determination of the Lewis form factor and the AGMA geometry factor J for external spur gear teeth. *ASME Journal of Mechanisms, Transmissions and Automation in Design* 104: 148–158.

Townsend, D.P. (ed.) (1992). *Dudley's Gear Handbook, The Design, Manufacture, and Application of Gears*, 2e. McGraw-Hill.

Internet Links

http://agma.org/ AGMA: Gears standard and design codes.
http://ansi.org ANSI: Standards.
http://iso.org ISO: Standards.
http://khkgears.net/ KHK: Gears provider.
http://www.bostongear.com/ Boston Gear: Gears provider.
http://www.slideshare.net/palanivendhan/gear-manufacturing-process Slide Share: Gear manufacturing.

16

Flexible Elements

Flexible power transmission elements have advantages over other elements. They are used between shafts that are relatively a long distance apart. In that case, they cause cost reduction over other options such as gears. Some are elastic which absorb shock loads and dampen vibrations. *V-belts*, *link V-belts*, *flat belts*, *round belts*, *timing belts*, *ropes*, and *roller chains* are some of these elements that are used in conveying systems and *power transmission*. *Friction drives* may also be considered as *flexible elements* due to slip possibility and high elasticity under higher loading. Belts and other flexible elements may also allow slipping to the point that the output my not necessarily has the exact intended velocity ration relative to the input. The timing belt is an exception to that since it engages through their built-in teeth with the input and output sprockets or toothed wheels that should guarantee the intended velocity ratio.

In this chapter, focus will be on the design and selection of *V-belts*, *flat belts*, *rope drives*, and *roller chains*.

Symbols and Abbreviations

The adopted units are [in, lb, psi] or [m, kg, N, Pa], others given at each symbol definition, ([k…] is 10^3 [M…] is 10^6 and [G…] is 10^9).

Symbols	Quantity, units (adopted)
a_{KL}	Regression constant for length factor
A_{MR}	Metal area in a wire rope section
a_V	V-belt cross-sectional width
b_{KL}	Regression exponent constant for length factor
b_V	V-belt cross-sectional depth
$C\#_{ANSI}$	ANSI chain number or size
C_C	Center distance between chain sprockets
C_P	Center distance between two pulleys
$d\boldsymbol{N}$	Normal force on V-belt sides
d_P	Pulley pitch diameter
d_r	Diameter of round belt
d_R	Rope diameter
d_S	Sprocket pitch diameter
d_W	Wire diameter

Machine Design with CAD and Optimization, First Edition. Sayed M. Metwalli.

Companion website: www.wiley.com/go/metwalli/machine

Symbols	Quantity, units (adopted)
$d\phi$	Small wedge-element of belt
E_B	Belt modules of elasticity
E_R	Wire rope modulus of elasticity
E_W	Wire modulus of elasticity
$\boldsymbol{F}$	Longitudinal force on V-belt section
$\boldsymbol{F}_1, \boldsymbol{F}_2$	Force in tight-side and force in loose-side
$\boldsymbol{F}_c$	Centrifugal force
$\boldsymbol{F}_C$	Chain tight side force
$\boldsymbol{F}_D$	Design force
$\boldsymbol{F}_f$	Ultimate fatigue strength force
$\boldsymbol{F}_i$	Initial tension
$\boldsymbol{F}_{\text{Rb}}$	Equivalent bending load
$\boldsymbol{F}_{\text{ut}}$	Ultimate tensile force
g	Gravitational acceleration
H	Power, applied or transmitted
H_D	Design power
H_R	Chain fatigue rated power
H_{RT}	Total power rating
H_V	V-belt rated power per belt
I_z	Rope second area moment of section
K_θ	Arc of contact factor
K_{ASF}	Application service factor
K_{CS}	Chain speed factor
K_{HR}	Power ratio factor
K_L	Length factor
K_{SF}	Safety or service factor
K_{SFE}	Extra safety factor
K_{SL}	Strength loss factor
K_{SR}	Strength reduction factor
L_C	Chain length
L_i	Belt inside circumference length
L_p	Pitch length of belt
m_V	Mass per unit length of V-belt
$\boldsymbol{M}_z$	Bending moment
n_C	Chain speed ratio
N_{CS}	Number of chain strands
n_F	Flat belt speed ratio

Symbols	Quantity, units (adopted)
N_R	Number of ropes
N_S	Sprocket teeth number
N_S	Sprocket number of teeth
N_{S1}, N_{S2}	Number of input and output sprocket teeth
n_V	V-belt speed reduction ratio or speed ratio
N_{VB}	Number of V-belts
p_C	Chain pitch
p_P	Pulley bearing pressure
r_1, r_2	Small and large pulleys radii
R_c	Radius of curvature of pulley
r_S	Sprocket pitch radius
S_S	Allowable service strength
S_{ut}	Ultimate tensile strength
t_F	Flat belt thickness
$\boldsymbol{T}_P$	Pulley torque
$\boldsymbol{T}_S$	Sprocket torque
v_B	Tangential belt velocity
v_C	Longitudinal chain velocity
$v_{C\max}$	Maximum longitudinal chain velocity
$v_{C\min}$	Minimum longitudinal chain velocity
v_F	Flat belt speed
w_F	Flat belt width
$\boldsymbol{w}_R$	Rope weight-per-unit-length
β_V	V-belt section ½ angle
$\Delta \boldsymbol{F}$	Needed variation of force
Δv_C	Chordal speed variation
θ_1, θ_2	Wrapping angle for driver or driven pulley
θ_S	Sprocket pitch angle
μ	Friction coefficient
μ_V	Apparent coefficient of fiction for V-belts
σ_b	Bending stresses
σ_c	Centrifugal stress
σ_{F1}, σ_{F2}	Tension stresses due to $\boldsymbol{F}_1$ and $\boldsymbol{F}_2$
ω_1, ω_2	Small and large pulleys rotational speeds
ω_P	Pulley rotational speed
ω_S	Sprocket rotational speed
ϕ	Belt exit contact-angle

16.1 V-belts

The V-belt is an elastic closed rubber loop internally fortified by a woven fabric to increase its tensile strength. Figure 16.1 shows isometric sketches of one of V-belt longitudinal sectioned-part to show its cross-section and a ***3D*** view of the pulley (sheave) on which the belts would be wrapped around. Figure 16.1a demonstrates a sample of a V-belt section which is internally fortified by a woven fabric near the top. The fabric is shown as dark grayish dots. There are also fabrics wrapped around the rubber core with a thicker top fabric, which are shown in Figure 16.1a as triple grayish lines. The V-belt section has an angle $2\beta_V$ as shown in Figure 16.1a,b. Figure 16.1a shows the classical V-belt section and Figure 16.1b shows a similar V-belt section, but with a transverse cogged form longitudinally repeated and with only a woven fabric near the top that reaches the sides. Figure 16.1c shows a ***3D*** image of the V-belt pulley viewing the grooves for 6 V-belts, where belts are not shown for viewing clarity.

Many types of V-belts exist to suite the applications. The commonly used ones are the ***classical V-belts*** (mainly A, B, C, D, E types) as the section shown in Figure 16.1a, ANSI/RMA IP-20 (1988). These are the most commonly used and they are herein considered as the adopted V-belts. Another V-belt type is the *narrow V-belts* or *wedge V-belts* that are used for lighter and more compact applications, see ANSI/RMA IP-22 (1984). They are given the designations of $3V$, $5V$, and $8V$. The number preceding the letter V is the belt cross-sectional width a_V divided by 8 and the pairs (a_V, b_V) of width a_V and depth b_V are (3/8, 5/16), (5/8, 17/32), and (1, 7/8) in [in], or (9, 8), (15, 13), and (15, 23) in [mm], respectively. Since the depth of these belts is relatively large, an option is available to have molded *cogs* in the narrow part of the section to allow higher bendability and durability of the belt, Figure 16.1b. These looks like transverse cogged teeth in the inner belt circumference as displayed in Figure 16.1b where the transverse cogged cut off and no fabric wrapping around the core are shown. This option is usually indicated by adding a letter X to the section designation. These *cogged belts* can be used with lower minimum pully diameters and can generally transmit more power than the classical V-belts.

The *double V-belt* is another belt which is composed of two integral top-to-top V-belts to facilitate operation on both sides of the cross section, ANSI/RMA IP-21 (1991), and Oberg et al. (2012). The *light duty V-belts* are different belts used for fractional powers. They employ small diameter sheaves with a usual single belt accommodation. They are given the designations of $2L$, $3L$, $4L$, and $5L$. The number preceding the letter L is about the belt cross-sectional width divided by 8, ANSI/RMA IP-23 (1968). Other similar V-belts are used in automotive industries as given by the Society of Automobile Engineers (SAE), SAE J637 (2012). Another type of V-belts that are used in automotive applications are the *V-Ribbed belts*. It has a flat cross-section with small internally protruding V-shaped ribs on the usual internal driving-side. These belts are more suitable for high-speed and serpentine applications, ANSI/RMA IP-26 (1977), ISO 9982 (1998). The *variable speed-belts* are like the V-belts, but they are much

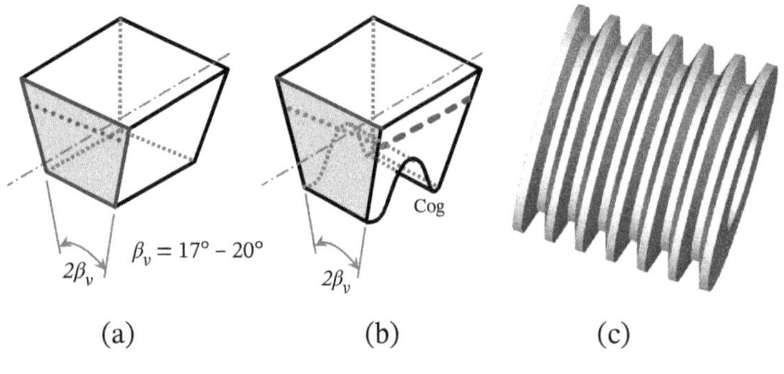

Figure 16.1 Isometric sketches of V-belt sections and a pulley on which the belts would be wrapped about. (a) Belt section internally fortified by a woven fabric near the top and fabrics wrapping around the core. (b) V-belt section with transverse cogged form. (c) V-belt pulley with grooves for 6 V-belts (belts not shown).

wider to accommodate the variation of pulley groove width while changing the speed ratio in some applications such as continuous variable transmission (CVT), ANSI/RMA IP-25 (1982), ISO 1604 (1989).

The *timing-belts* or *synchronous belts* are sort of V-belts, but the V cross-section is a tooth lateral to the belt longitudinal centerline. They are used to synchronize the rotation of the output to the rotation of the input. They do not depend on the possible slip wedging action of V-belts, but a positive no-slip act of the teeth fitting in the slotted pulley or sprocket. The V-shaped lateral teeth should withstand the needed torque and the belt section should withstand the tension needed for that. The timing belt size is following specific pitches between the teeth, such as 1/5, 3/8, 1/2, 7/8, or 1 1/4 [in]. Timing-belt widths can be 0.25, 0.5, 0.75, 1.5, 2, 3, 4, and 5 [in], see ANSI/RMA IP-24 (1983).

Belt lengths are standardized, and the ***pulleys*** (***sheaves***) center distances should accommodate these values. The standard or available lengths may be the inside circumferential length, the pitch length, or the outside circumference belt length. One must find these from standards, manufacturers, or handbooks, see Oberg et al. (2012).

The *standard V-belts* are treated in this text and other belt types may be handled similarly, or one should consult standards, manufacturers, or handbooks for these other belt types.

16.1.1 V-belt Drive Relations

Figure 16.2 presents simple diagrammatic sketches of a belt connecting two pulleys. Figure 16.2a shows the driver and the driven pulleys with some defined parameters such as the *pulleys radii* r_1 and r_2, the *rotational speeds* ω_1 and ω_2, and the *center distance* c_P between the two pulleys. In that form, the *driver speed* ω_1 is higher than the *driven speed* ω_2, and the speed *reduction ratio* (ω_1/ω_2) or the *speed ratio* n_V of V-belt system is then defined as $n_V = (\omega_1/\omega_2)$. In practice, the common range of the speed ratios n_V is about 1–6 and may rarely reach 8–15. Figure 16.2b presents two separated free body diagrams of driver and driven pulleys to allow for the analysis of the internal forces $\boldsymbol{F}_1$ and $\boldsymbol{F}_2$. These forces are generated due to power transmission. For the analysis, the two pulleys are connected by one belt only, as also shown in Figure 16.2a. *Driver pulley 1* is driving pulley *2*. Force $\boldsymbol{F}_1$ is thus larger on the tight-side than the force $\boldsymbol{F}_2$ in the loose-side. Only one free body diagram allows the analysis of forces generated due to power transmission. The other free body diagram would give the same result with the wrapping angle θ representing either θ_1 for the driver or θ_2 for the driven pulley.

Consider a small wedge-element $d\phi$ of belt in Figure 16.3 showing simple diagrammatic sketches of free body diagrams of a single pully and the wrapped belt section. Figure 16.3a defines the wrapped belt of *driven pulley* with the small wedge-element $d\phi$ of the belt before its exit contact-angle ϕ. Figure 16.3b presents a free body diagram of the V-belt wedge-element showing the boundary forces. The following is the force analysis of the V-belt wedge-element in the vertical and horizontal directions.

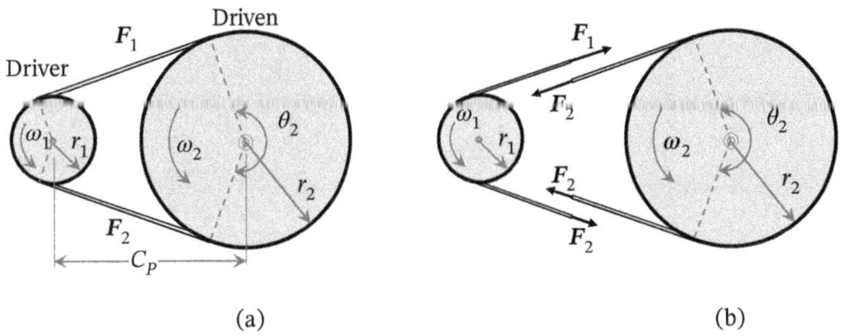

Figure 16.2 Simple diagrammatic sketches of a belt connecting two pulleys. (a) Driver and driven pulleys with defined parameters. (b) Free body diagrams of driver and driven pulleys to allow for the analysis of forces generated due to power transmission.

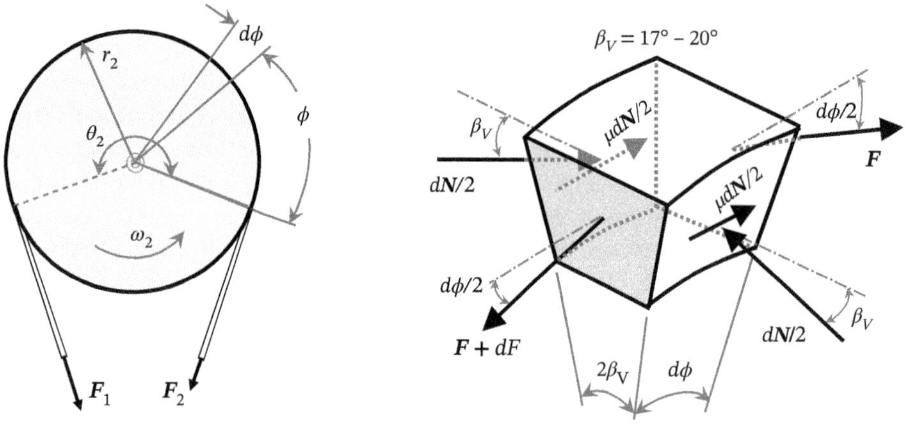

Figure 16.3 Simple diagrammatic sketches of free body diagrams of a single pully and the wrapped belt section. (a) Wrapped belt of driven pulley showing a small wedge-element $d\phi$ of the belt before its exit angle ϕ. (b) A free body diagram of the V-belt wedge-element.

Summing forces in the vertical direction in Figure 16.3b gives the following relation.

$$\sum \boldsymbol{F}_{\text{Vertical}} = \frac{\boldsymbol{dN}}{2}\sin\beta_V + \frac{\boldsymbol{dN}}{2}\sin\beta_V - \boldsymbol{F}\sin\frac{d\phi}{2} - (\boldsymbol{F} + \boldsymbol{dF})\sin\frac{d\phi}{2} = 0 \tag{16.1}$$

where $\boldsymbol{dN}$ is the normal force on the sides of the V-belt, $2\beta_V$ is the angle of the V-belt section, $\boldsymbol{F}$ is the longitudinal force on V-belt section, and ϕ is the contact exit-angle to the small wedge-element, as shown in Figure 16.3. From Eq. (16.1) and since for small angles, $\sin(d\phi/2)$ can be closely equal to $d\phi/2$, one gets

$$\boldsymbol{dN}\sin\beta_V = \boldsymbol{F}d\phi \tag{16.2}$$

Summing forces in the horizontal direction in Figure 16.3b gives the following.

$$\sum \boldsymbol{F}_{\text{Horizontal}} = \mu\frac{\boldsymbol{dN}}{2} + \mu\frac{\boldsymbol{dN}}{2} + \boldsymbol{F}\cos\frac{d\phi}{2} - (\boldsymbol{F} + \boldsymbol{dF})\cos\frac{d\phi}{2} = 0 \tag{16.3}$$

where μ is the coefficient of friction, $\boldsymbol{dN}$ is the normal force on the sides of the V-belt, $2\beta_V$ is the angle of the V-belt section, $\boldsymbol{F}$ is the longitudinal force on V-belt section, and ϕ is the contact exit-angle to the small wedge-element. From Eq. (16.3) and since for small angles, $\cos(d\phi/2)$ can be closely equal to *1*, one gets

$$\mu\ \boldsymbol{dN} = \boldsymbol{dF} \tag{16.4}$$

Substituting Eq. (16.2) into Eq. (16.4) gives the following relation.

$$\boldsymbol{dF} = \mu\frac{\boldsymbol{F}}{\sin\beta_V}d\phi \tag{16.5}$$

Integrating Eq. (16.5), the following expression is obtained.

$$\int_{\boldsymbol{F}_2}^{\boldsymbol{F}_1}\frac{\boldsymbol{dF}}{\boldsymbol{F}} = \int_0^{\theta}\frac{\mu}{\sin\beta_V}d\phi \tag{16.6}$$

where ϕ goes from *0* to θ. From Eq. (16.6), it is obvious to get the following important relation.

$$\frac{\boldsymbol{F}_1}{\boldsymbol{F}_2} = e^{\frac{\mu}{\sin\beta_V}\theta} \tag{16.7}$$

where F_1 is the *tight side force* and F_2 is the slack or *loose side force*. One should note that the *wrapping angle* θ is in radians [rad] and the smaller wrapping angle for the driving or driven pulley is to be used. It should also be noted that the forgoing analysis is bases on simplified assumption that the full wrapping is happening with no sliding zones, see Hussein et al. (2010, 2013).

The power H is simply given by the following relation.

$$H = T_P \cdot \omega_P = \left(F_1 - F_2\right) r_P \omega_P \tag{16.8}$$

where T_P is the pulley torque, ω_P is the magnitude of pulley rotational speed, r_P is the pulley radius, F_1 is the tight side force, and F_2 is the loose side force.

Substituting Eq. (16.7) into Eq. (16.8) and performing minor manipulation, one gets an expression defining the force F_1 on the tight side of the belt as follows.

$$H = \left(F_1 - F_2\right) r_P \omega_P = \left(F_1 - \frac{F_1}{e^{\frac{\mu}{\sin \beta_V}\theta}} \right) r_P \omega_P \tag{16.9}$$

or

$$F_1 = \frac{H}{\left(1 - \frac{1}{e^{\frac{\mu}{\sin \beta_V}\theta}} \right) r_1 \omega_1} \tag{16.10}$$

This force can be checked against the maximum tensile force the belt can withstand. It should also be noted that the loose side force F_2 ought not be zero or negative and it is obtained from Eq. (16.7). Belts are flexible elements that would not carry compression.

To approach some realistic consideration, the belt centrifugal force is added to the analysis. The *centrifugal force* F_c is also considered as follows.

$$F_c = m_V \; v_B^2 / g \tag{16.11}$$

where m_V is the *mass per unit length* of the V-belt, v_B is the tangential velocity of the belt, and g is the gravitational acceleration, see Section 1.11.1. For a sample consideration, the ***mass per unit length*** m_V for some common V-belt sections are as follows: for section-type A, $m_V = 0.094$ [kg/m]; for section-type B, $m_V = 0.163$ [kg/m]; for section-type C, $m_V = 0.284$ [kg/m]; and for section-type D, $m_V = 0.56$ [kg/m]. The specific value of the belt type is advisably obtained from the manufacturer or provider. The centrifugal force F_c reduces the contact force between the belt and the pulley, since the belt tends to leave the pulley. Equation (16.7) then becomes as follows.

$$\frac{F_1 - F_c}{F_2 - F_c} = e^{\frac{\mu}{\sin \beta_V}\theta} \tag{16.12}$$

Without an *initial tension* F_i, however, F_2 becomes a compression force. This is not allowed. As power is transmitted with the existence of the initial force F_i, the following initial force variation should occur.

$$\begin{aligned} F_1 &= F_i + \Delta F \\ F_2 &= F_i - \Delta F \end{aligned} \tag{16.13}$$

where F_i is the force due to initial tension, and ΔF is the needed variation of force to transmit the power. Solving Eq. (16.13) for the initial tension F_i, one gets

$$F_i = \frac{F_1 + F_2}{2} \tag{16.14}$$

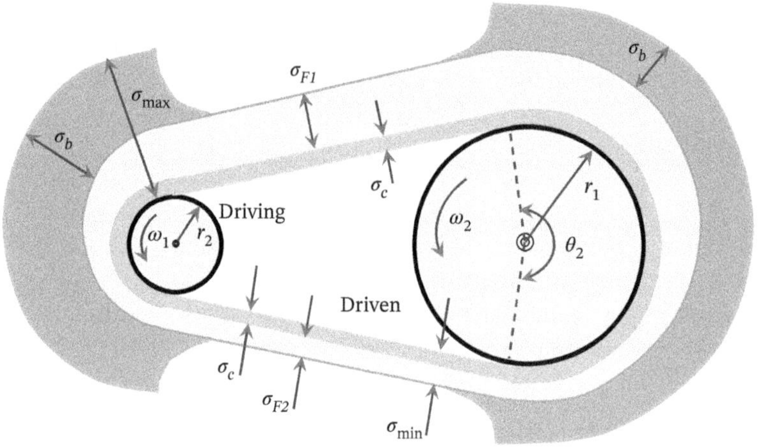

Figure 16.4 Simple sketch of the stress diagram of the wrapped V-belt. Tension stresses σ_{F1} and σ_{F2} due to $\boldsymbol{F}_1$ and $\boldsymbol{F}_2$, centrifugal stress σ_c, and bending stresses σ_b are demonstrated.

where the limit minimum is for the initial tension $\boldsymbol{F}_i$ to be equals to $\boldsymbol{F}_1/2$ with the minimum of $\boldsymbol{F}_2 = 0$. The maximum condition is then $\boldsymbol{F}_2 = 0$ and $\boldsymbol{F}_1 = 2\ \boldsymbol{F}_i$. Solving Eq. (16.14) for $\boldsymbol{F}_2$ gives $\boldsymbol{F}_2 = 2\ \boldsymbol{F}_i - \boldsymbol{F}_1$. Substitution these values into Eq. (16.8) gives the following relation.

$$H = (\boldsymbol{F}_1 - \boldsymbol{F}_2)\, r_P\omega_P = 2\,(\boldsymbol{F}_1 - \boldsymbol{F}_i)\, v_B \tag{16.15}$$

where v_B is the *tangential velocity* of the belt.

Stresses in belts are as shown in the diagrammatic sketch of Figure 16.4. It roughly demonstrates the stresses generated during the wrapped V-belt operation. The expected stresses are the *tension stresses* σ_{F1} and σ_{F2} due to $\boldsymbol{F}_1$ and $\boldsymbol{F}_2$, the *centrifugal stress* σ_c due to belt velocity, and the *bending stresses* σ_b due to bending of belt around pulleys. The maximum stress in the belt is then given by, see Figure 16.4

$$\sigma_{\max} = \sigma_{F1} + \sigma_b + \sigma_c \tag{16.16}$$

The bending stress $\sigma_b = E_B b_V / d_P$, where b_V is the *V-belt thickness*, E_B is the belt *modules of elasticity*, and d_P is the *pulley diameter*, which is equal to d_1 when it is the smaller pulley. The bending stress at the smaller pulley is therefore larger than at the larger pulley, see Figure 16.4. The *tension stresses* σ_{F1} due to $\boldsymbol{F}_1$ is obtained by dividing the force $\boldsymbol{F}_1$ by the belt sectional area.

16.1.2 Standards and Geometric Relations

Table 16.1 presents the selected common V-belt types, dimensions of belt sections as given in the top figure, minimum pulley pitch diameters, suggested power ranges per belt, the advised belt speed ranges, the maximum tension a belt can generally withstand, and expected top rotational speed of the smaller pulley. The usual angle of the V-belt section $2\beta_V$ ranges between 34° and 40° as indicated in Figure 16.1a. For the selected common V-belt types in Table 16.1, the angle of the V-belt section $2\beta_V$ is 40°. The suggested power range is given for the advised V-belt speed v_V of ***4000*** feet per minute [ft/min] or ***20*** [m/s]. The start of that range is for the minimum pulley diameter and the end of the rage is for the usual top pulley diameter after which the transmitted power does not increase. These top pulley diameters are 5 [in] or 125 [mm] for section-type A, 7 [in] or 180 [mm] for section-type B, 12 [in] or 315 [mm] for section-type C, 17 [in] or 450 [mm] for section-type D, and 28 [in] or 710 [mm] for section-type E. Other V-belt speeds (other than the stated range in Table 16.1) would transmit lower power. That is why the advised V-belt speed is adopted in this text for the design of V-belt drives.

The usual standards for ***pulley diameters*** are covering wide span of diameters to accommodate various applications. There are smaller pulley diameters than the recommended minimum puller diameters in Table 16.1.

Table 16.1 Selected common V-belt types, dimensions of belt sections, minimum pulley pitch diameters, suggested power ranges, the advised belt speed ranges, the maximum tension the belt can generally withstand, and top rotational speed.

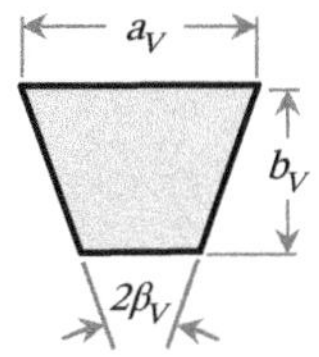

Belt section type	Dimensions of belt section				Minimum pulley diameter		Suggested power range		Belt speed range (advised)		Maximum tension		Top rotational speed
	a_V		b_V		d_P	d_P	H_{kW}	H_{hp}					
	[mm]	[in]	[mm]	[in]	[mm]	[in]	[kW]	[hp]	[m/s]	[ft/min]	[N]	[lb]	[rpm]
A	13	1/2	8	11/32	70	2.6	0.7–1.9	0.9–2.5	15–20	3000–4000	175	40	5000
B	16	21/32	10	7/16	112	4.6	1.9–4.1	2.5–5.7	15–20	3000–4000	290	65	3300
C	22	7/8	13	17/32	180	7.0	4.1–7.2	5.7–10	15–25	3000–5000	555	125	2000
D	32	1 1/4	19	3/4	250	10	7.2–17	10–24	15–25	3000–5000	1110	250	1400
E	38	1 1/2	25	29/32	450	17	17–27	24–37	20–25	4000–5000	1780	400	800

However, the minimum is reported in Table 16.1 as the start of pulley diameter set to implement the suggested lowest diameter recommendations. The suggested pulley diameters for US and SI systems are as follows.

- The suggested set of the pulley *pitch diameters* in [in] for US system is as follows.
  2.6, 3, 3.2, 3.4, 3.6, 4.2, 4.6, 5, 5.4, 5.8, 6.2, 6.6, 7, 8, 9, 10, 11, 12, 13, 14, 15, 16, 17, 18, 20, 22, 24, 26, 28, 30, 35, 40, 45, 50, 55, 60, 65, 70, 80, 90, 100, 110, 125, 150, 180, 200
- The suggested set of the pulley *pitch diameters* in [mm] for SI system is as follows.
  56, 75, 80, 85, 90, 100, 112, 125, 140, 160, 180, 200, 224, 250, 280, 300, 335, 375, 400, 450, 500, 630, 710, 800, 900, 1000, 1120,1250, 1400, 1600, 1800, 2000, 2240, 2500, 2800, 3150, 3550, 4000, 4500, 5000
- For A-type belts, the US pulley diameters start at 3.0 [in] and increment by 0.2 [in]. For SI pulleys, the start is 75 [mm] and increment by 5 [mm].
- For B-type belts, the US pulley diameters start at 5.4 [in] and increment by 0.2 [in]. For SI pulleys, the start is 140 [mm] and increment by 5 [mm].
- For C-type belts, the US pulley diameters start at 9.0 [in] and increment by 0.5 [in]. For SI pulleys, the start is 224 [mm] and increment by 10 [mm].
- For D-type belts, the US pulley diameters start at 13.0 [in] and increment by 0.5 [in]. For SI pulleys, the start is 330 [mm] and increment by 20 [mm].
- For E-type belts, the US pulley diameters start at 21.0 [in] and increment by 1.0 [in]. For SI pulleys, the start is 540 [mm] and increment by 25 [mm].

The standard ***belt lengths*** are selected as subsets as defined in the following general set of belt lengths in [in] and the subset range for each belt type.

- The suggested standard set of the belt ***inside circumference length*** L_i in [in] for US system is as follows.
  26, 31, 33, 35, 38, 42, 46, 48, 51, 53, 55, 57, 60, 62, 64, 66, 68, 71, 75, 78, 80, 85, 90, 96, 105, 112, 120, 128, 131, 136, 144, 158, 173, 180, 195, 210, 240, 270, 300, 330, 360, 390, 420, 480, 540, 600. 660. Note that other intermediate lengths may be available.

- Type A belt length subset range is 26–128 [in] and one would add 1.3 [in] to get the pitch length L_p.
- Type B belt length subset range is 35–300 [in] and one would add 1.8 [in] (for 35–240) or 2.1 [in] (for 240 up) to get the pitch length L_p.
- Type C belt length subset range is 51–420 [in] and one would add 2.9 [in] (for 51–210) or 3.8 (for 210 up) to get the pitch length L_p.
- Type D belt length subset range is 120–660 [in] and one would add 3.3 [in] (for 120–210) or 4.1 (for 210 up) to get the pitch length L_p.
- Type E belt length subset range is 180–660 [in] and one would add 4.5 [in] (for 180–240) or 5.5 (for 240 up) to get the pitch length L_p.
- The suggested set of the belt ***pitch length*** L_p in [mm] for SI system can approximate the US system converted from inch [in] to [mm] and adjusting the length to a pitch length L_p instead of *inside circumferential length* L_i. **630**, 700, **710**, **800**, **900**, **1000**, 1100, **1120**, **1250**, 1300, 1200, **1600**, 1700, **1800**, 1900, **2000**, 2100, 2200, **2240**, 2300, 2400, **2500**, 2600, 2700, **2800**, 2900, 3000, 3100, **3150**, 3200, 3400, 3500, **3550**, 3600, 3700, 3800, 3900, **4000**, **4500**, 4600, **5000**, 5400, **5600**, 6100, **6300**, **7100**, 7600, 7700, **8000**, **9000**, 9100, **10000**, 10700, **11200**, **11500**, 12200, 13700, where the bold numbers are ISO standards, ISO 4184 (1992). Note that other numerous intermediate lengths are available through manufacturers for different types of belt sections and for every 5 or 10 [mm] multiple increments. One should use the available *pitch length* L_p instead of the previously quoted round figure pitch lengths, see available standards and manufacturers catalogues downloaded through the Internet of references in this chapter.

  The suggested set of round figure *pitch lengths* (others available) in [mm] for certain belt types are as follows.
- Type A belt length subset range is 520–3950 [mm] and one would subtract 34 [mm] to get the inside circumference length L_i.
- Type B belt length subset range is 700–6070 [mm] and one would subtract 45 [mm] to get the inside circumference length L_i.
- Type C belt length subset range is 1300–11 330 [mm] and one would subtract 58 [mm] to get the inside circumference length L_i.
- Type D belt length subset range is 2740–13 700 [mm] and one would subtract 85 [mm] to get the inside circumference length L_i.
- Type E belt length subset range is 2540–11 430 [mm] and one would subtract 115 [mm] to get the inside circumference length L_i.

The standard *belt length* is used to calculate the needed center distance between the small and large pulleys (sheaves) as shown in Figure 16.2a. The center distance is assumed from constructional details and the needed belt length is then estimated. The belt length is usually the pitch length. A close standard belt length is selected, and the needed *center distance* C_P is then calculated. An iteration might be needed to satisfy the construction need and the available belt lengths.

From Figure 16.2a, the length of the belt between the small and large pulleys (sheaves) contact is given by the triangle of the *center distance* C_P and the difference between pully diameters. The total belt *pitch length* L_p is then given by the following relation.

$$L_p = \sqrt{4C_P^2 - (d_2 - d_1)^2} + \frac{1}{2}(d_2\theta_2 + d_1\theta_1) = 2C_P + \frac{\pi}{2}(d_2 + d_1) + \frac{(d_2 - d_1)^2}{4C_P} \quad \textbf{(16.17)}$$

where θ_1 and θ_2 are the *belt wrapping angles* or *angles of contact* for the small and large pulleys, respectively, and C_P is the center distance between the two pulleys. From this relation, the needed *center distance* C_P for a specific *pitch length* L_p is then

$$C_P = \frac{1}{4}\left\{\left(L_p - \frac{\pi}{2}(d_2 + d_1)\right) + \sqrt{\left(L_p - \frac{\pi}{2}(d_2 + d_1)\right)^2 - 2(d_2 - d_1)^2}\right\} \quad \textbf{(16.18)}$$

The usual center distance C_P range is recommended as follows.

$$d_2 < C_P < 3\left(d_2 + d_1\right) \tag{16.19}$$

The *wrap contact angles* of θ_1 and θ_2 are given by the following relations, see Figure 16.2a.

$$\begin{aligned}\theta_1 &= \pi - 2\sin^{-1}\frac{d_2 - d_1}{2C_P} \cong \pi - \frac{d_2 - d_1}{C_P}\\ \theta_2 &= \pi + 2\sin^{-1}\frac{d_2 - d_1}{2C_P} \cong \pi + \frac{d_2 - d_1}{C_P}\end{aligned} \tag{16.20}$$

where the wrap angles are in [rad] and the approximation is for reasonable center distances and sensible reduction or speed ratios.

16.1.3 Design Procedure

The main interest in V-belt design is the *power transmission*. For that goal, the following *initial synthesis* and detailed design process are consequently adopted in this text.

16.1.3.1 Initial Synthesis

The design for proper V-belt selection involves the proper definition of belt and pulley geometry that should be used. The operational range of belt speed is usually 1000–5000 [ft/min] or 5–25 [m/s]. The conventional experience calls for the belt speed adoption of 4000 [ft/min] or 20 [m/s], where the speed range is narrowed to 3000–5000 [ft/min] or 15–25 [m/s] as depicted in Table 16.1. The optimum belt speed is expected to be varying with belt type, size, pulley diameter, constraints and some other few objectives. As an initial synthesis, the adopted belt speed is an expected better choice for a large span of cases. A formal optimization is needed to tune the selection of the belt speed.

With the adopted belt speed v_V of 4000 [ft/min] or 20 [m/s], the initial synthesis of pulley diameter is derived from the knowledge of the conservative *rated power* H_V per belt for the classical V-belts associated with an acceptable life and a speed ratio of 1.0. The estimated *pulley pitch diameter* d_1 is then given by the following regression relations, Figure 16.5.

$$\begin{aligned}d_1 &= 2.561\,66 H_{V,\text{hp}}^{0.62932}\ [\text{in}]\\ d_1 &= 78.242\,57 H_{V,\text{kW}}^{0.62932}\ [\text{mm}]\end{aligned} \tag{16.21}$$

Figure 16.5 provides the conservative V-belt rated power H_V per belt against the pulley pitch diameter d_1 for both US and SI systems. The small grayish circles are the selected knowledge base for different belt types in the suggested power range per belt as given in Table 16.1. The estimated trendline is given by the black solid line and the regression relation is displayed in the plot area, where x is for power rating and y is for the pulley diameter. The relations in Eq. (16.21) can be uses to find an estimate of the pulley diameter if the number of belts is expected or some iteration is alternatively anticipated. This is possible by increasing the *number of belts* N_{VB}. Divide the power by that number of belts and the calculated pulley diameter from Eq. (16.21) should fall within the range of the belt type in Table 16.1. Other constraints are expected to be observed in that process particularly the speed limit and suggested power ratings of each section-type of the belt, see e.g. ISO 5292 (1995).

The adopted belt speed v_V of 4000 [ft/min] or 20 [m/s] necessitates that the initial synthesis of pulley diameter should satisfy that constraint and thus the diameter would be defined. The procedure of initial synthesis for the belt system design is then suggested as follows.

- Usually the transmitted or *applied power* H, the input rotational speed ω_1, and the speed *reduction ratio* (ω_1/ω_2) or *speed ratio* n_V of the V-belt system are given, see Figure 16.2.

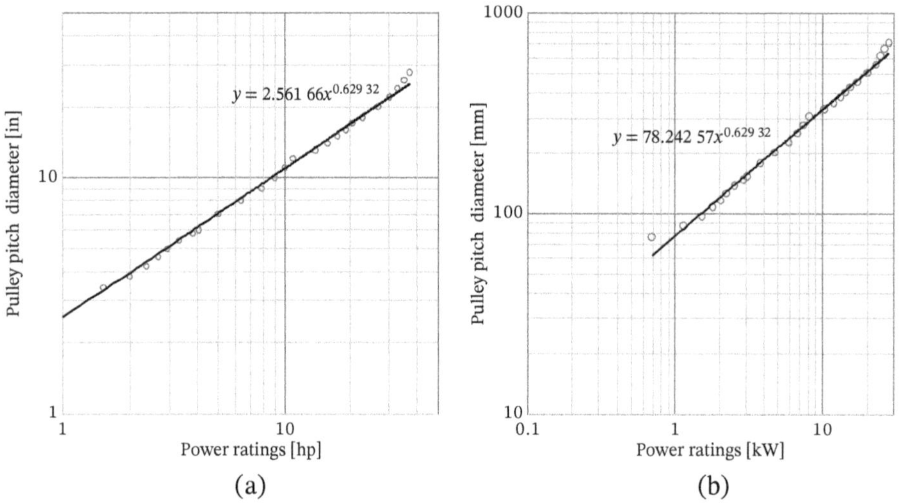

Figure 16.5 Pulley diameter estimated from the knowledge of the *rated power* H_V for classical V-belt sections with an acceptable life and at a speed ratio of 1.0 for US and SI systems.

- With an adopted belt speed v_V, the initial synthesis of pulley diameter is then defined as follows.

$$d_1 = \frac{2\ v_V}{\omega_1} = 2\ v_V\left(\frac{60}{2\pi N_{\text{rpm}}}\right) \tag{16.22}$$

- For the belt speed adoption of 4000 [ft/min] or 20 [m/s], the pulley diameter synthesis from Eq. (16.22) should then be as follows.

$$d_1 = 2\ v_V\left(\frac{60}{2\pi N_{\text{rpm}}}\right) = 4000\left(\frac{12}{60}\right)\left(\frac{60}{\pi N_{\text{rpm}}}\right) \cong \frac{15\,279}{N_{\text{rpm}}}\ \text{[in]}$$

$$d_1 = 2\ v_V\left(\frac{60}{2\pi N_{\text{rpm}}}\right) = 20\,(1000)\left(\frac{60}{\pi N_{\text{rpm}}}\right) \cong \frac{381\,972}{N_{\text{rpm}}}\ \text{[mm]} \tag{16.23}$$

And the driven pully diameters due to the *speed ratio* $n_V = (\omega_1/\omega_2) = (d_2/d_1)$ are then obtained from:

$$d_2 = d_1 n_V, \quad \text{[in] or [mm]} \tag{16.24}$$

- From the top rotational speed of each belt as given in Table 16.1, the belt type section is selected by having the input speed in [rpm] less than the belt-section top-speed in [rpm] and higher than the next belt-section top-speed in [rpm].
- From Figure 16.5, the estimated V-belt rated power H_V of each belt type-section is obtained.
- The *total number of V-belts* N_{VB} depends on the design power H_D including a safety or a service factor K_{SF} affecting the maximum transmitted power H and the V-belt rated power H_V. The design power is then given by

$$H_D = K_{SF} H \tag{16.25}$$

Get the estimated number of V-belts N_{VB} by dividing the design power H_D by the rated power H_V of the selected belt type-section, which gives

$$N_{VB} \approx \frac{H_D}{H_V} \approx \frac{K_{SF}\ H}{H_V} \tag{16.26}$$

The integer number of V-belts N_{VB} is then estimated by the rounding the value of $(N_{VB} + 0.5)$.

This initial synthesis procedure defined the initial pully diameters, the V-belt section-type, and the number of belts. These are the main parameters of the V-belt system that are refined with other additional geometric parameters given through the next detailed design procedure.

16.1.3.2 Detailed Design Process

This process refines the selected parameters in the preceding section of the initial synthesis. The following steps are to finalize the selection of the defined parameters and other geometrical requirements of the design.

- It is recommended at this point to select standard pulley diameters for the driver and driven sides. If the speed and speed ratio must be specifically required, the pully diameters can be produced to the needed diameters according to a standard groove profile. Otherwise the driver pulley can be standard, and the driven pulley is manufactured. If the two pulleys are standard, the speed ratio n_V and the output rotational speed might be different. Selected pulley diameters are given in Section 16.1.2. These or others might be available from suppliers or manufacturers.
- The expected average center distance C_P between the pulley's centerlines are then obtained by Eq. (16.19).
- The expected average belt length L_P is, therefore, found by Eq. (16.17). One should then select the closest pitch length L_P that is available from a supplier or a manufacturer. Selected belt lengths are given in Section 16.1.2. These and many others should be available from suppliers or manufacturers.
- The needed center distance C_P between the pulley's centerlines are then obtained by Eq. (16.18). The two belt-contact wrap-angles θ_1 and θ_2 on both pulleys are then calculated utilizing Eq. (16.20). The center distance would be adjusted to generate the initial tightening force $\boldsymbol{F}_i$ defined by Eqs. (16.7) and (16.14).
- The selection of the number of belts is reconfirmed from the total design power H_D and the V-belt rated power H_V of the selected belt section type. This would be simply defined by the top speed of each belt section type in [rpm] as defined in Table 16.1 and as indicated in the initial synthesis. The design power H_D should include the service factor or safety factor K_{SF} that depends on the prime mover type, application severity, and the duration of operation or duty hours per day, Table 16.2. The relation is reiterated as follows.

$$H_D = K_{SF} H \tag{16.25}$$

where H is the transmitted or *applied power* through the belt system and K_{SF} is the service factor.
- As the pulley diameter d_1 is defined from Eq. (16.21) and/or standardized, the V-belt rated power H_V of the selected belt section type is obtained by the inverse of Eq. (16.21), which gives

$$\begin{aligned} H_V|_{\text{hp}} &= 0.230\,11\ d_1^{1.5778}, \quad d_1 \text{ in [in]} \\ H_V|_{\text{kW}} &= 0.001\,04\ d_1^{1.5778}, \quad d_1 \text{ in [mm]} \end{aligned} \tag{16.27}$$

where the V-belt section-type is selected from the rotational speed in [rpm] such that it should be within the top suggested rotational speeds of the belt sections defined in Table 16.1. This is feasible since the high belt speed v_V of 4000 [ft/min] or 20 [m/s] is adopted. The use of a lower belt speed would transmit lower power at the same pulley diameter. Another check is that the power rating obtained by Eq. (16.26) would also be within the suggested power range of belt section-type depicted in Table 16.1.
- The integer number of V-belts N_{VB} is then defined by rounding the value of $(N_{VB} + 0.5)$ in the following refined relation.

$$N_{VB} = \frac{H_D}{(K_L K_\theta)\, H_V} = \frac{K_{SF}\ H}{K_L K_\theta\ H_V} \tag{16.28}$$

where the calculated power H_V is adjusted by the *length factor* K_L and for having a speed ratio other than 1.0 using the *arc of contact factor* K_θ. The length factor K_L has the range of 0.85–1.2 and is given by the following regression equation.

$$K_L = a_{KL} (L_V)^{b_{KL}} \tag{16.29}$$

Table 16.2 Proposed service factor K_{SF} for common V-belts for different prime movers and driven machines. The range of values is for the range of duty hours per day from under 10 [hr] to over 16 [hr].

Prime mover	Driven machine				
	Light duty	Medium duty	Heavy duty	Extra heavy duty	Extreme duty
Soft start	1–1.2	1.1–1.3	1.2–1.4	1.3–1.5	2
Heavy start	1.1–1.3	1.2–1.4	1.4–1.6	1.5–1.8	2

Table 16.3 The values of the coefficients and exponents for the evaluation of the length factor K_L for different V-belt section types.

Coefficients and exponents	V-belt section type				
	A	B	C	D	E
a_{KL} US	0.372	0.3588	0.3281	0.2848	0.2783
a_{KL} SI	0.1739	0.1714	0.1592	0.1363	0.1345
b_{KL} US or SI	0.235	0.2284	0.2236	0.2278	0.2249

where the constant a_{KL} is dependent on the V-belt section-type, the exponent b_{KL} is almost constant, and the belt length L_V is in [in] for US system and in [mm] for SI system. The values of a_{KL} and b_{KL} are given in Table 16.3 for the common V-belt section-types. The arc of contact factor K_θ is given by the following curve fit expression cited by Budynas and Nisbett (2015).

$$K_\theta = 0.143\,543 + 0.007\,468\;\theta_1 - 0.000\,015\,052\;\theta_1^2 \tag{16.30}$$

where the wrap angle θ_1 is in the range of 90°–180° and the contact factor K_θ is then in the range of 0.7–1.0. Other trend line equations are available. Equation (16.30), however, is usable with reasonable accuracy in the 90°–180° range.

- If needed, iterate the previous steps to get a better or optimum design.
- Find the force $\boldsymbol{F}_1$ of the belt tight side, the force $\boldsymbol{F}_2$ of the belt slack side, the initial tension force $\boldsymbol{F}_i$, and the centrifugal force $\boldsymbol{F}_c$ acting on the belt. These are obtained through Eqs. (16.10), (16.7), (16.14), and (16.11). This action, however, necessitates an a priori knowledge of the friction coefficient μ, the belt angle $2\beta_V$, and the mass per unit length m_V of the belt section. These forces can be used to check on the expected V-belt rated power per belt using Eq. (16.15).
- Determine the final dimensions of a standard ***V-belt pulley*** and groove dimensions as the one suggested in Table 16.4, e.g. see ANSI/RMA IP-20 (1988). Standards may vary and one should consult the one that is applicable in the product design and dissemination.
- Verify the design suitability by following the suggested manufacturer's design procedure, from which the belts are to be acquired.

Example 16.1 A V-belt system transmits a maximum power of 8.9 [kW] or 11.93 [hp] at an input rotational speed of 3000 [rpm] and a needed speed reduction ratio of 2.0. The common classical V-belts are to be employed. The system is driven by a steady prime mover and is driving a heavy-duty load for about 12 hours. Synthesize

Table 16.4 Suggested pulley and grooves dimensions for different V-belt section types.

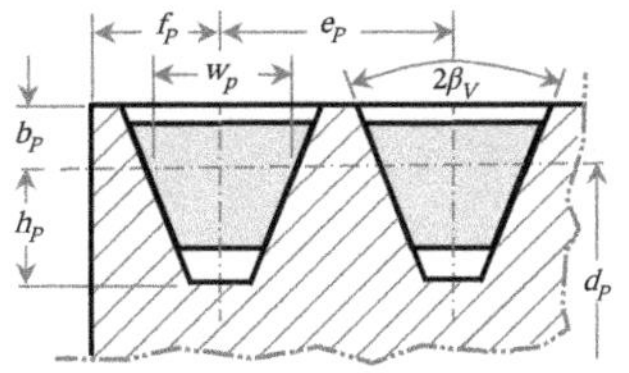

Belt section type	Pulley pitch diameter, d_p [mm] ([in])	Angle, $2\beta_V$ [°]	Pitch width, w_p [mm] ([in])	b_P [mm] ([in])	h_p [mm] ([in])	e_P [mm] ([in])	f_P [mm] ([in])
A	≤75 (3.0)	32	11 (0.418)	3.3 (0.130)	8.7 (0.335)	15 (0.625)	10 (0.375)
	75–125 (3.0–4.9)	34					
	>125 (4.9)	38					
B	<125 (4.9)	32	14 (0.530)	4.2 (0.165)	10.8 (0.375)	19 (0.750)	12.5 (0.500)
	125–200 (4.9–7.87)	34					
	>200 (7.87)	38					
C	<200 (7.87)	32	19 (0.757)	5.7 (0.224)	14.3 (0.412)	25.5 (1.00)	17 (0.688)
	200–300 (7.87–11.8)	34					
	>300 (11.8)	38					
D	<355 (14)	32	27 (1.076)	8.1 (0.319)	19.9 (0.550)	37 (1.438)	24 (0.875)
	355–500 (14–19.7)	34					
	>500 (19.7)	38					
E	≤630 (24.8)	36	32 (1.260)	9.6 (0.378)	23.4 (0.921)	44.5 (1.752)	29 (1.142)
	>630 (24.8)	38					

the V-belt system defining the belt section type, pulley diameters, and the number of belts. Determine the forces between the V-belt system in SI and US systems.

Solution

Data: $H_{kW} = 8.9$ [kW] = 8900 [W] or $H_{hp} = 11.95$ [hp], $N_{in,rpm} = N_{in} = 3000$ [rpm], $\omega_1 = 2\pi N_{in}/60 = 314.16$ [rad/s]. The speed ratio $n_V = \omega_1/\omega_2 = 2$, and $\beta_V = 20°$. Service factor $K_{SF} = 1.3$ for steady prime mover driving a heavy-duty load 12 [hr], (Table 16.2).

A. Initial Synthesis

The initial synthesis of the driver-pulley pitch-diameter is given according to Eqs. (16.22) and (16.23) such that

$$d_1 = 2\ v_V \left(\frac{60}{2\pi N_{\text{rpm}}} \right) \cong \frac{15\,279}{N_{\text{rpm}}} = \frac{15\,279}{3000} = 5.093\ [\text{in}]$$

$$d_1 = 2\ v_V \left(\frac{60}{2\pi N_{\text{rpm}}} \right) \cong \frac{381\,972}{N_{\text{rpm}}} = \frac{381\,972}{3000} = 127.32\ [\text{mm}] \tag{a}$$

From that and from Section 16.1.2, one can assume a close standard pulley pitch-diameter of 5 [in] or 125 [mm]. With the speed reduction ratio $n_V = d_2/d_1 = 2$ (see Eq. (16.24)), the output pulley diameters are then 10 [in] and 250 [mm]. These are also standard pulley pitch-diameter. If not, the speed reduction ratio would be different.

From Figure 16.5, the estimated V-belt rated power H_V at pulley pitch-diameter of 5 [in] or 125 [mm] are about 2.9 [hp] and 2.1 [kW].

According to Eq. (16.25), the design power H_D is given by

$$H_D = K_{SF} H_{hp} = 1.3\,(11.95) = 15.535\ [\text{hp}]$$
$$H_D = K_{SF} H_{kW} = 1.3\,(8.9) = 11.57\ [\text{kW}] \tag{b}$$

According to Eq. (16.26), the number of needed belts are then

$$N_{VB}|_{US} \approx \frac{H_D}{H_V} \approx \frac{15.535}{2.9} = 5.357$$
$$N_{VB}|_{SI} \approx \frac{H_D}{H_V} \approx \frac{11.57}{2.1} = 5.5095 \tag{c}$$

The round figure number of belts are then ($5.357 + 0.5 = 5.857 = 6$) for US system and ($5.5095 + 0.5 = 6.095 = 6$) for SI system.

The belt section-type is B due to the rotational speed of 3000 [rpm], which is in between the 3300 [rpm] top speed of section-type B and the 2000 [rpm] top speed of section-type C.

Definitively, then, the initial synthesis procedure defined the initial pully diameters, the V-belt section type, and the number of belts.

B. Detailed Design Process

The procedure suggests the selection of standard pulley diameters for the driver and driven sides, which has been adopted in the initial synthesis.

The expected average center distance C_P between the pulley's centerlines are obtained by Eq. (16.19), which gives the following average values.

$$C_P = (3\,(d_2 + d_1) + d_2)\,/2 = (3\,(10 + 5) + 10)\,/2 = 27.5\ [\text{in}]$$
$$C_P = (3\,(d_2 + d_1) + d_2)\,/2 = (3\,(250 + 125) + 250)\,/2 = 687.5\ [\text{mm}] \tag{d}$$

According to Eq. (16.17), the expected belt lengths are

$$L_p = 2C_P + \frac{\pi}{2}(d_2 + d_1) + \frac{(d_2 - d_1)^2}{4C_P} = 2\,(27.5) + \frac{\pi}{2}(10 + 5) + \frac{(10 - 5)^2}{4\,(27.5)} = 78.789\ [\text{in}]$$
$$L_p = 2C_P + \frac{\pi}{2}(d_2 + d_1) + \frac{(d_2 - d_1)^2}{4C_P} = 2\,(687.5) + \frac{\pi}{2}(250 + 125) + \frac{(250 - 125)^2}{4\,(687.5)} = 1969.7\ [\text{mm}] \tag{e}$$

Selecting the closest pitch length L_P that is available from the list in Section 16.1.2 or from a supplier or a manufacturer, the pitch length L_P can then be 80 [in] and 2000 [mm]. The adjusted center distance C_P between the pulley's centerlines are then obtained by Eq. (16.18) such that

$$C_P = \frac{1}{4}\left\{\left(L_p - \frac{\pi}{2}(d_2 + d_1)\right) + \sqrt{\left(L_p - \frac{\pi}{2}(d_2 + d_1)\right)^2 - 2(d_2 - d_1)^2}\right\}$$
$$= \frac{1}{4}\left\{\left(80 - \frac{\pi}{2}(10 + 5)\right) + \sqrt{\left(80 - \frac{\pi}{2}(10 + 5)\right)^2 - 2(10 - 5)^2}\right\} = 28.108\ [\text{in}]$$
$$C_P = \frac{1}{4}\left\{\left(L_p - \frac{\pi}{2}(d_2 + d_1)\right) + \sqrt{\left(L_p - \frac{\pi}{2}(d_2 + d_1)\right)^2 - 2(d_2 - d_1)^2}\right\}$$
$$= \frac{1}{4}\left\{\left(2000 - \frac{\pi}{2}(250 + 125)\right) + \sqrt{\left(2000 - \frac{\pi}{2}(250 + 125)\right)^2 - 2(250 - 125)^2}\right\} = 702.7\ [\text{mm}] \tag{f}$$

These center distances would be adjusted to generate the initial tightening force $\boldsymbol{F}_i$ defined by Eqs. (16.7) and (16.14). The construction of the belt system should provide bolting slots on one side or any other mean to accomplish this needed requirement.

The two belt-contact wrap angles θ_1 and θ_2 on both pulleys are then calculated utilizing Eq. (16.20).

$$\theta_1|_{\text{US}} \cong \pi - \frac{d_2 - d_1}{C_P} = \pi - \frac{10 - 5}{28.108} = 2.9637 \text{ [rad]} = 169.81^\circ$$
$$\theta_2|_{\text{US}} \cong \pi + \frac{d_2 - d_1}{C_P} = \pi + \frac{10 - 5}{28.108} = 3.3195 \text{ [rad]} = 190.19^\circ \quad \text{(g)}$$

And

$$\theta_1|_{\text{SI}} \cong \pi - \frac{d_2 - d_1}{C_P} = \pi - \frac{250 - 125}{702.7} = 2.9637 \text{ [rad]} = 169.81^\circ$$
$$\theta_2|_{\text{SI}} \cong \pi + \frac{d_2 - d_1}{C_P} = \pi + \frac{250 - 125}{702.7} = 3.3195 \text{ [rad]} = 190.19^\circ \quad \text{(h)}$$

The same design powers H_D calculated in Eq. (b) are reused to get the number of belts N_B. The V-belt rated power H_V of the selected belt section-type is obtained by Eq. (16.27), which gives

$$H_V|_{\text{hp}} = 0.230\,11\ d_1^{1.5778} = 0.230\,11(5)^{1.5778} = 2.9159\ [\text{hp}]$$
$$H_V|_{\text{kW}} = 0.001\,04\ d_1^{1.5778} = 0.001\,04(125)^{1.5778} = 2.1161\ [\text{kW}] \quad \text{(i)}$$

These values are close to the estimated values in the previous initial synthesis.

To get the integer number of V-belts N_{VB} according to Eq. (16.28), one needs to define the length factor K_L from Eq. (16.29) and the contact factor K_θ from Eq. (16.30). These give the following values.

$$K_L|_{\text{US}} = a_{\text{KL}}\left(L_V\right)^{b_{\text{KL}}} = 0.3588(80)^{0.2284} = 0.976\,15$$
$$K_L|_{\text{SI}} = a_{\text{KL}}\left(L_V\right)^{b_{\text{KL}}} = 0.1714(2000)^{0.2284} = 0.972\,67 \quad \text{(j)}$$

And

$$K_\theta = 0.143\,543 + 0.007\,468\ \theta_1 - 0.000\,015\,052\ \theta_1^2$$
$$= 0.143\,543 + 0.007\,468\ (169.81) - 0.000\,015\,052\ (169.81)^2 = 0.9776 \quad \text{(k)}$$

The number of V-belts N_{VB} according to Eq. (16.28) is then

$$N_{\text{VB}}|_{\text{US}} = \frac{H_D}{\left(K_L K_\theta\right) H_V} = \frac{15.535}{(0.976\,15)\,(0.9776)\,(2.9159)} = 5.5829\ [\text{hp}]$$
$$N_{\text{VB}}|_{\text{SI}} = \frac{H_D}{\left(K_L K_\theta\right) H_V} = \frac{11.57}{(0.972\,67)\,(0.9776)\,(2.1161)} = 5.7500\ [\text{kW}] \quad \text{(l)}$$

These are little larger than the values in Eq. (c). The round figure number of belts are then $(5.5829 + 0.5 = 6.0829 = 6)$ for US system and $(5.75 + 0.5 = 6.25 = 6)$ for SI system.

The following is a check on the induced forces in one belt using its rated power H_V as the expected maximum transmitted power per belt. To find the force $\boldsymbol{F}_1$ of the belt tight side, the force $\boldsymbol{F}_2$ of the belt slack side, the initial tension force $\boldsymbol{F}_i$, and the centrifugal force $\boldsymbol{F}_c$ acting on the belt, assume the friction coefficient $\mu = 0.3$, the belt angle $2\beta_V = 40^\circ$, and the mass per unit length $m_V = 0.163$ [kg/m] for the belt section-type B, see Section 16.1.1. This check is carried out for the SI system just as a check. Using Eqs. (16.10), (16.7), (16.14), and (16.11) gives the following values.

$$\boldsymbol{F}_1 = \frac{H_V}{\left(1 - \left(1/e^{\frac{\mu}{\sin \beta_V}\theta}\right)\right) r_1 \omega_1} = \frac{2.1161}{\left(1 - \left(1/e^{\frac{0.3}{\sin 20}(169.81\,(\pi/180))}\right)\right)\left(\frac{125\,(314.16)}{2\,(1000)}\right)} = 116.42\ [\text{N}] \quad \text{(m)}$$

$$F_2 = \frac{F_1}{e^{\frac{\mu}{\sin \beta_V}\theta}} = \frac{116.42}{e^{\frac{0.3}{\sin 20}(169.81\,(\pi/180))}} = 8.6507\ [\text{N}] \tag{n}$$

$$F_i = \frac{F_1 + F_2}{2} = \frac{116.42 + 8.6507}{2} = 62.535\ [\text{N}] \tag{o}$$

$$F_c = m_V\ v_B^2/g = 0.163(20)^2/9.806\,65 = 6.6485\ [\text{N}] \tag{p}$$

The total belt tight side force F_t and belt slack side force F_s are given by the following expressions.

$$F_t = F_1 + F_c = 116.42 + 6.6485 = 123.07\ [\text{N}]$$
$$F_s = F_2 + F_c = 8.6507 + 6.6485 = 15.299\ [\text{N}] \tag{q}$$

These are the forces in each of the 6 belts. The forces on the shafts that carry the pulleys are the vectorial sum of F_1 and F_2 generated from all 6 belts. The resultant force F_R has the magnitude given by the following simplified value.

$$F_R = N_{\text{VB}}\sqrt{F_1^2 + F_2^2} = 6\sqrt{123.07^2 + 15.299^2} = 700.45\ [\text{N}] \tag{r}$$

where N_{VB} is the number of V-belts. This force affects the shaft on which the 6-belt pulley is mounted.

16.2 Flat Belts

Traditional flat belts have been used almost since the start of power transmission. The prime movers were then expensive, and they used to drive many machines through several elevated shafts and connecting belts. The prime mover drives these shafts through flat belts. Other flat belts drive the machines off these elevated shafts. At that time, most belts were made of leather strips that are joined lengthwise by different means such as hand stitching or clipper lacing. Some of the flat belts are still employed in different capacities, applications, and materials, particularly in conveyor belts and others, see ISO 14890 (2013), and ISO 22 (1991). Flat belts are manufactured in different materials such as rubber-impregnated composites, polyamide, urethane and can be reinforced by nylon cords or steel wires. With advanced manufacturing, very thin (0.002 [in] or 0.05 [mm]) and very narrow (up to 0.025 [in] or 0.6 [mm]) metallic belts are potentially useful in special applications such as sterile, pharmaceutical and food industries or abrasive or highly corrosive production environment.

Figure 16.6 presents a simple diagrammatic sketches of flat belt drives that are usually having larger *centerline distance* c_P than the displayed reduced spaces. Figure 16.6a shows the conventional and usual open belt, which

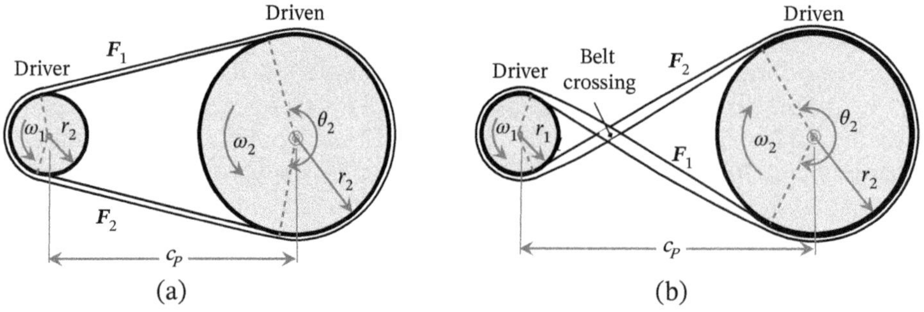

Figure 16.6 Simple diagrammatic sketches of flat belt drives. (a) Regular open belt. (b) Reversing crossed belt, where the in between rubbing ought to be prevented by some separation.

runs over the circumferentially crowned and smoothed periphery of the pully. Figure 16.6b introducing a rough draft of a reversing crossed belt. The belt is twisted to reverse the rotation direction. This causes an eventual in between rubbing of the belt faces crossing each other's. This ought to be prevented by some separation means. Between these two configurations, there can be an intermediate position of one shaft relative to the other since the crossed belt can be viewed as a 180° rotation of one shaft out-of-plane i.e. in the normal plane to the two pulley's centerline. One shaft of the two pulleys can be rotated 90° in the normal plane of the two pulley's centerline. This would form a half-twist belt drive. These and other configurations have been extensively used in heritage production plants and may still be used in conveying belts.

16.2.1 Drive Relations

Proceeding with the analysis just like the V-belts in Section 16.1.1 and with reference to Figure 16.3, one can get the flat belt drive relations by setting the wedge angle β_V to 90°. This causes the V-belt section to be flat with friction forces right on the peripheral of the flat circumference pulley. The force analysis of the flat belt is thus identical to the V-belt wedge-element in the vertical and horizontal directions with the wedge angle $\beta_V = 90°$. This renders Eqs. (16.2) and (16.4) to have the following forms.

$$d\boldsymbol{N} \sin \beta_V = d\boldsymbol{N} = \boldsymbol{F} d\phi \tag{16.31}$$

$$\mu \ d\boldsymbol{N} = d\boldsymbol{F} \tag{16.32}$$

Substituting Eq. (16.31) into Eq. (16.32), integrating and solving gives the following important relation.

$$\frac{\boldsymbol{F}_1}{\boldsymbol{F}_2} = e^{\mu\theta} \tag{16.33}$$

where $\boldsymbol{F}_1$ is the *tight side force* and $\boldsymbol{F}_2$ is the slack or *loose side force*, see Figures 16.2 and 16.6. One should note that the wrapping angle θ is in radians [rad] and the smaller wrapping angle for the driving or driven pulley is used. It should also be noted that the forgoing analysis is bases on simplified assumption that the full wrapping is happening with no sliding zones, see Hussein et al. (2010, 2013).

Again, the power H is simply given by the same V-belt relation of Eq. (16.8). Substituting Eq. (16.33) into Eq. (16.8) and performing minor manipulation, one gets an expression defining the force $\boldsymbol{F}_1$ on the tight side of the belt as follows.

$$\boldsymbol{F}_1 = \frac{H}{\left(1 - \frac{1}{e^{\mu\theta}}\right) r_1 \omega_1} \tag{16.34}$$

This force can be checked against the maximum tensile force the belt can withstand. It should also be noted that the loose side force $\boldsymbol{F}_2$ ought not be zero or negative and is obtained from Eq. (16.33). Flat belts are flexible elements that would not carry compression.

As for V-belts, without an *initial tension* $\boldsymbol{F}_i$, though, $\boldsymbol{F}_2$ becomes a compression force. This is not allowed and the initial tension $\boldsymbol{F}_i$, is obtained exactly as V-belts to be as follows.

$$\boldsymbol{F}_i = \frac{\boldsymbol{F}_1 + \boldsymbol{F}_2}{2} \tag{16.35}$$

where the limit minimum is for the initial tension $\boldsymbol{F}_i$ to be equals to $\boldsymbol{F}_1/2$ with a minimum of $\boldsymbol{F}_2 = 0$. The maximum condition is then $\boldsymbol{F}_2 = 0$ and $\boldsymbol{F}_1 = 2\,\boldsymbol{F}_i$. Solving Eq. (16.5) for $\boldsymbol{F}_2$ gives $\boldsymbol{F}_2 = 2\,\boldsymbol{F}_i - \boldsymbol{F}_1$. Substitution of these values into Eq. (16.8) gives the following relation.

$$H = \left(\boldsymbol{F}_1 - \boldsymbol{F}_2\right) r_P \omega_P = 2\left(\boldsymbol{F}_1 - \boldsymbol{F}_i\right) v_B \tag{16.36}$$

where v_B is the *tangential velocity* of the flat belt.

Stresses in flat belts are as in the V-belt presented in the diagrammatic sketch of Figure 16.4. This roughly demonstrates the stresses generated during the wrapped flat belt action. The expected stresses are the same *tension stresses* σ_{F1} and σ_{F2} due to $\boldsymbol{F}_1$ and $\boldsymbol{F}_2$, the *centrifugal stress* σ_c due to belt velocity, and the *bending stresses* σ_b due to bending of belt around pulleys. The maximum stress in a flat belt is then given by Eq. (16.16), see Section 16.1.1 and Figure 16.4. Since the flat belt thickness is usually much smaller than V-belts, the *bending stress* σ_b and the *centrifugal stress* σ_c due to belt velocity are relatively small compared to the *tension stresses* σ_{F1} due to $\boldsymbol{F}_1$. This also allows flat belts to run at higher speeds than V-belts.

Example 16.2 With reference to Eqs. (16.7) and (16.33), find the apparent friction of the V-belt relative to the flat belt for usual V-belt wedge angles. Find the limiting ratio F_1/F_2 when the belt is just about to slip for a V-bet of wedge angle $2\beta_V$ of 40° and a flat belt. The small wrap angle θ_1 is 170° and the coefficient of friction μ is 0.3.

Solution

Data: $2\beta_V = 40°$, $\beta_V = 20°$, $\theta_1 = 170° = (170\pi/180) = 2.9671$ [rad], and μ is 0.3.

Comparing Eq. (16.7) to Eq. (16.33), the *apparent coefficient of fiction* for the V-belts μ_V relative to the coefficient of fiction for flat belt μ can be

$$\mu_V = \frac{\mu}{\sin \beta_V} = \frac{\mu}{\sin 20} = 2.9238\,\mu \tag{a}$$

or the apparent coefficient of fiction for the V-belts is about three times that of flat belts, particularly if the wedge angle $2\beta_V$ of the V-belt is about 38°. For a wedge angle $2\beta_V$ of 34°, or 32°, the apparent coefficient of fiction for the V-belts is about 3.42 or 3.63 times that of flat belts, respectively.

Using Eq. (16.7) for V-belt, one gets

$$\frac{\boldsymbol{F}_1}{\boldsymbol{F}_2} = e^{\frac{\mu}{\sin \beta_V}\theta} = e^{\frac{0.3}{\sin 20}2.9671} = e^{0.877\,14(2.9671)} = 13.498 \tag{b}$$

Using Eq. (16.33) for flat belt, gives the following.

$$\frac{\boldsymbol{F}_1}{\boldsymbol{F}_2} = e^{\mu\theta} = e^{0.3(2.9671)} = 2.4354 \tag{c}$$

This shows that the material is more efficiently utilized for V-belts than for flat belts.

16.2.2 Standards and Geometry Relations

Table 16.5 presents some selected flat-belts and round-belts materials, and the suggested ranges of some basic properties. Smaller property values are generally for smaller thicknesses or diameters. The minimum pulley diameters are given as recommendations, but other values are available from manufacturers or suppliers, see Internet part of the references. The allowable service strength range in Table 16.5 is provided to allow the proper calculation of an initial synthesis for the needed belt section. The proper belt should be selected from the manufacturers or supplier's dada for a specific material, thickness, and width. The allowable maximum tension is usually provided in these data and a verification of the selected belt is then required. The suggested belt speed is provided to indicate an advisable value as a design parameter. Other higher values are admissible, and available data verification is therefore required.

It should be noted that the supposedly flat circumference of the pulley is normally crowned to prevent the belt from slipping off the pulley. The normal round crowning is small such that the flat character of the belt is still valid. Crowning values are small and in the range of about 0.012–0.10 [in] or 0.3–2.5 [mm] for pulley diameters in the range of 1.5–80 [in] or 38–2030 [mm]. Crowning can also be about 0.01 [in] per inch of the pulley width or 10 [mm] per meter of the pulley width. Values can also be available in the standards; flat pulley manufactures or providers.

Table 16.5 Some selected flat-belts and round-belts materials and suggested ranges of some basic properties. Smaller property values are generally for smaller thicknesses or diameters.

Material	Thickness, w_F or Diameter, d_r [mm] ([in])	Minimum pulley diameter, d_1 [mm] ([in])	Allowable service strength, S_S [MPa] ([kpsi])	Belt speed 'suggested', v_F [m/s] ([ft/min])	Friction coefficient, μ
Polyamide	0.75–6.4 (0.03–0.25)	15–340 (0.6–13.5)	2.3–7.5 (0.33–1.1)	50 (10 000)	0.5–0.7
Urethane	1.5–2.5 (0.06–0.1)	10–19 (0.38–0.75)	1.15–1.17 (0.17)	50 (10 000)	0.7
Rubber cotton duck	3–7 (0.1–0.28)	20–300 (0.8–12)	2.5–4.5 (0.36–0.65)	40 (8000)	0.3–0.5
Leather	4.4–9 (11/64–23/64)	180–230 (7–9)	1.0–1.2 (1.5–1.75)	30 (6000)	0.4
Urethane-round	6.5–20 (0.25–0.77)	40–180 (1.5–7.1)	1.16–1.2 (0.17)	50 (10 000)	0.7

The standard pitch diameters for flat belt pulleys are expected to be about the same as the set for V-belts in Section 16.1.2. Some selected standard set of the belt lengths is also about the same as for V-belts in Section 16.1.2. Available values from manufactures or providers should be used unless some appropriate joining process is utilized to get the desired length.

The standard *belt length* is used to calculate the needed center distance between the small and large pulleys (sheaves) as shown in Figure 16.6. For regular open belt arrangement in Figure 16.6a, the same equations used in V-belt geometry are used for flat belts, i.e. Eqs. (16.17), (16.18), and (16.20). Equation (16.19) for the center distance C_P range may be used, even though a larger center distance C_P is normally admissible. The ***inside circumference length*** L_i of the belt is used in place of the belt ***pitch length*** L_p. However, since the thickness is usually small, the difference should be taken care of while adjusting the belt tension.

Flat belt thicknesses are generally smaller than V-belts, see Tables 16.1 and 16.5. For few belt materials, some selected thicknesses w_F or diameters d_r for round belts are as follows, see Table 16.5.

- Polyamide thicknesses are possibly 0.03, 0.05, 0.07, 0.11, 0.13, 0.2, 0.25 [in] or 0.75, 1.25, 1.8, 2.8, 3.3, 5, 6.4 [mm].
- Urethane thicknesses are possibly 0.062, 0.078, 0.09 [in] or 1.6, 2, 2.3 [mm]
- Rubber cotton duck thicknesses are possibly 0.11, 0.13, 0.15, 0.16, 0.17, 0.2, 0.22, 0.23, 0.26 [in] or 2.8, 3.3, 3.8, 4, 4.3, 5, 5.6, 5.8, 6.6 [mm]
- Leather thicknesses are possibly 11/64, 13/64, 18/64, 20/64, 23/64 [in] or 4.4, 5.2, 7.1, 7.9, 9.1 [mm]
- Urethane-round diameters are possibly 0.25, 0.375, 0.5, 0.75 [in] or 6.4, 9.5, 12.7, 19 [mm]

Many other materials, thicknesses, and a variety of widths can also be offered or provided by manufacturers.

16.2.3 Design Procedure

The design procedure of flat belts is about the same as for V-belts in Section 16.1.3, particularly if the interest is for power transmission. For other applications such as conveyor belts, the objectives might be different and adjustments to that effect are therefore necessary.

16.2.3.1 Initial Synthesis

Particularly if the interest is for power transmission, the initial synthesis is suggested as follows.

- Usually the transmitted or *applied power* H, the input rotational speed ω_1, and the flat belt *reduction ratio* (ω_1/ω_2) or flat belt *speed ratio* n_F of the belt system are given.

- With an adopted *flat belt speed* v_F, the initial synthesis of pulley diameter is then defined as follows.

$$d_1 = \frac{2\ v_F}{\omega_1} = 2\ v_F\left(\frac{60}{2\pi N_{\text{rpm}}}\right) \quad (16.37)$$

- With an adopted belt speed as suggested in Table 16.5, the pulley diameter synthesis from Eq. (16.37) should then be obtained. Note that the range of the adopted belt speed in Table 16.5 is 6000–10 000 [ft/min] or 30–50 [m/s] for different materials. Excluding the leather belts, the range of the adopted belt speed would be 8000–10 000 [ft/min] or 40–50 [m/s] for other materials. If one uses 8000 [ft/min] or 40 [m/s], Eq. (16.37) becomes

$$d_1 = 2\ v_F\left(\frac{60}{2\pi N_{\text{rpm}}}\right) = 8000\left(\frac{12}{60}\right)\left(\frac{60}{\pi \text{Nm}}\right) \cong \frac{30\,558}{N_{\text{rpm}}}\ [\text{in}]$$

$$d_1 = 2\ v_F\left(\frac{60}{2\pi \text{Nm}}\right) = 40\,(1000)\left(\frac{60}{\pi N_{\text{rpm}}}\right) \cong \frac{763\,940}{N_{\text{rpm}}}\ [\text{mm}] \quad (16.38)$$

And the driven pully diameters due to the *speed ratio* $n_F = (\omega_1/\omega_2) = (d_2/d_1)$ are then

$$d_2 = d_1\ n_F, \quad [\text{in}] \text{ or } [\text{mm}] \quad (16.39)$$

- From the application needs and design constraints, the belt material is adopted from options in Table 16.5. The geometry and strength ranges are then defined. These are the ranges of minimum pulley diameter, allowable service strength, and friction coefficient.
  If the calculated small pulley diameter is larger than the ranges of the minimum pulley diameter, a smaller belt speed than the suggested one can be used (say 3000 [ft/min] or 15 [m/s] or less). This would potentially transmit lower power, but at a smaller pulley diameter.
- With a mean value of the friction coefficient, the *applied power H* (or the design power H_D including a safety or a service factor K_{SF}), a service factor K_{SF} (Table 16.2), and the input rotational speed ω_1, the force $\boldsymbol{F}_1$ on the tight side of the belt is estimated according to Eqs. (16.34) and (16.25) to include the service factor. A wrap angle θ_1 of 180°, i.e. (π), may firstly be used for the initial estimated synthesis. The actual wrap angle θ_1 would be used in the detailed design process.
- Use the estimated tight-side force $\boldsymbol{F}_1$ to find the expected cross-sectional area of belt from an allowable service strength S_S picked-out of the range available from Table 16.5. An average service strength S_S may be used. The estimated tight-side force $\boldsymbol{F}_1$ can also be used directly to select an available belt thickness (or diameter) and the belt width from available manufacturer or belt provider catalogue. Usually the tensile force per unit of width is provided in the catalogues.
- Select an available *flat belt thickness* t_F (or *round diameter* d_R) and calculate the *flat belt width* w_F to satisfy the needed cross-sectional area of belt. The selection would be from available manufacturer or belt provider catalogues.

This initial synthesis procedure defined the initial pully diameters, initial belt material, and the belt cross-sectional dimensions. These are the main parameters of the flat belt system that are refined with other geometric parameters given by the next detailed design procedure.

16.2.3.2 Detailed Design Process

As for V-belts, the detailed design process refines the selected parameters in the preceding section of the initial synthesis. The following steps are to finalize the selection of the defined parameters and other geometrical requirements of the design.

- It might be recommended at this point to select standard pulley diameters for the driver and driven sides if possible. The selected set of the pulley diameters for US and SI systems of the V-belts in Section 16.1.2 may be used as the admissible diameters of the flat belt pulleys.
- One can use the expected average center distance C_P between the pulley's centerlines as obtained by Eq. (16.19). If the needed center distance C_P between the pulley's centerlines is larger, one should use the needed one.
- The expected belt length L_P is, therefore, found by Eq. (16.17). One should then select the closest length that is available from a supplier or a manufacturer.
- The needed *center distance* C_P between the pulley's centerlines are then obtained by Eq. (16.18). The two belt-contact *wrap angles* θ_1 and θ_2 on both pulleys are also calculated utilizing Eq. (16.20). The center distance would be adjusted to generate the *initial tightening force* $\boldsymbol{F}_i$ defined by Eqs. (16.33) and (16.35).
- The maximum *design power* H_D should include the service factor or *safety factor* K_{SF} which depends on the prime mover type, application severity, and the duration of operation or duty hours per day, Table 16.2. The applied relation is Eq. (16.25) to also be used in flat belts. Consequently, the maximum force the belt is subjected to or the maximum design force $\boldsymbol{F}_D$ should then include the service factor or safety factor K_{SF}. This gives the following expression.

$$\boldsymbol{F}_D = K_{\text{SF}}\ \boldsymbol{F}_1 \tag{16.40}$$

  where $\boldsymbol{F}_1$ is the tight-side force given by Eq. (16.34).
- The *width* of the *flat belt* w_F is then defined by the following relation.

$$w_F = \frac{\boldsymbol{F}_D}{(K_\theta)\ t_F S_S} = \frac{K_{\text{SF}}\ \boldsymbol{F}_1}{K_\theta\ t_F S_S} = \frac{K_{\text{SF}}\ \boldsymbol{F}_1}{K_\theta\ \boldsymbol{F}_S} \tag{16.41}$$

  where w_F is the flat belt width, t_F is the selected flat belt thickness in the initial synthesis, K_θ is the contact arc factor given by Eq. (16.30), S_S is the allowable service strength (Table 16.5), and $\boldsymbol{F}_S$ is the allowable service tension per unit length. This allowable service tension per unit length is usually provided by the manufacturers as a maximum tension.
- One must select the larger available belt thickness w_F from the manufacturer or provider's catalogue.
- If needed, change belt thickness t_F, or iterate the previous steps to get a better or optimum design.
- Verify the design suitability by following the suggested manufacturer's design procedure, from which the belts are to be acquired.

16.3 Ropes

Ropes can be classified as *fiber ropes* and *wire ropes*. This section is mainly dedicated to *wire ropes* since their utility is larger in mechanical applications such as hoists, cranes, winches, haulage, elevators, conveyor cords, cable cars, gondola lift, bridge suspension structure, … etc. The ***fiber ropes*** are also used in manual construction, marine, areal, theater, climbing, and other sports, etc., see S9086-UU-STM-010 (1995) and TM 3-34.86 (2012). Figure 16.7 shows some fiber ropes with different sizes, structure, and materials, particularly fibers, or cords. Similar wire ropes are using metal wires in place of the fibers. The fibers, yarns, or plies are twisted, wound, spun, or braided together to form the rope as those shown in Figure 16.7, where the upper two ropes are *braided*. Materials used in ropes are widespread and depend on the applications. Table 16.6 presents some selected fiber materials for ropes and some expected basic properties. Fibers, lines, threads, and ropes made of these materials can have a wide range of tensile strength that would be suitable for composites to accommodate many demanding applications. When fibers are spun to get yarns, threads, strings, or other forms, the strength of the outcome would be different from the fiber strength.

Figure 16.7 Some ropes with different sizes, structure, and materials.

Table 16.6 Some selected fiber materials for ropes and some expected basic properties.

Material	Tensile strength [MPa] ([kpsi])	Modulus of elasticity [GPa] ([Mpsi])	Specific gravity
Aramid fiber	3930 (570)	124 (18.0)	1.45
Carbon fiber	3430 (497)	300 (43.5)	1.77
Alloy steel	1330 (193)	210 (30.5)	7.8
Polyester fiber	784 (114)	13.2 (1.9)	1.39
Nylon fiber	616 (89.3)	3.9 (0.57)	1.14
Hemp	300 (43.5)	32 (4.64)	1.49
Cotton	225 (32.6)	7.9 (1.15)	1.54
Bulk nylon	63 (9.14)	2.5 (0.36)	1.09
Bulk polyester	50 (7.25)	2.9 (0.42)	1.3

Wire ropes are wound, coiled, or twisted wires forming strands that are wound, coiled, or twisted into intertwined helixes to form the wire rope as shown in Figure 16.8. Wires are usually made of steel, alloy steel, stainless steel, plow steel (PS), bridge rope steel, galvanized, marlin clad, nonferrous phosphor bronze, or others. Figure 16.8 presents sketches of a wire rope with different structure and *lays*. Figure 16.8a displays a cress-sectional vies of a wire rope type (6 × 7), where 7 wires are wound into 6 strands and the strands are wound around a fiber core (or other cores) to form the wire rope. Figure 16.8b shows a sketch of the *regular lay* (left-hand winding). Figure 16.8c shows a sketch of the *lang lay* (also left-hand winding). *Lays* can be either right-handed or left-handed, which is the direction of twisting. The wires are laid, wound, or twisted in a right-hand or left-hand direction. For steel wire standards, one can review ISO 2408 (2017), ISO 3108 (2017), ISO 17893 (2004), and ASTM A1023/A1023M (2009). The left-hand lays are usually given the symbol (S), while the right-hand lays are usually given the symbol (Z), ASTM A1023/A1023M (2009). The types of lay windings are given as follows.

- ***Regular lay*** is produced when wires in strands are twisted in a direction and the strands are twisted in an opposite direction to form the rope as seen in Figure 16.8b. In this case, wire ropes do not kink and twist. This lay-type is the most common.
- ***Lang-lay*** is produced when the strands are twisted in the same direction of wires to form the rope as seen in Figure 16.8c. In this case, wire ropes do kink and twist. However, this lay-type is usually better for wear and fatigue due to several lengths of exposed wires, see Figure 16.8c.

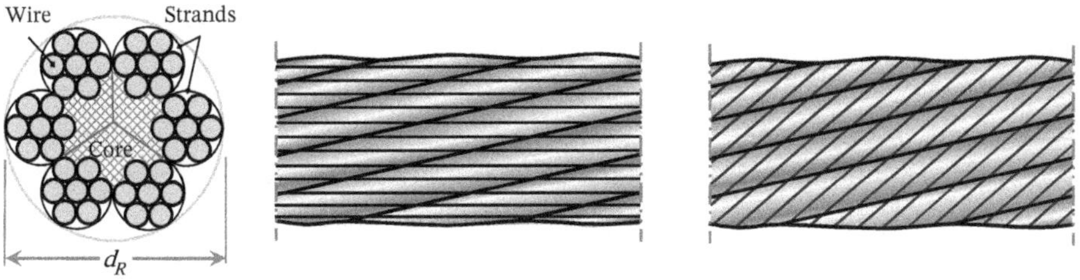

(a) (b) (c)

Figure 16.8 Wire rope sketches with different structure and lays. (a) Cress-sectional vies of a wire rope type (6 × 7), where 7 wires are wound in 6 strands and the strands are wound around a core. (b) Regular lay left. (c) Lang lay left. Lays can be either right or left-handed.

- ***Alternate-lay*** or *combined-lay* is produced when alternately regular lay and lang lay are laid to form the wire rope. The rope, thus, combined both regular and lang lays side by side. Its structure is made by combining a strand from Figure 16.8b with an adjacent strand from Figure 16.8c. This type is usually having limited use in some conveyors.

The ***wire rope core*** can be of a nonmetallic *fibrous core* (FC) or a *wire-strand core* (WSC), or *independent wire rope core* (IWRC). WSC is generally less flexible than FC or IWRC. Wire ropes with metallic cores are usually stronger than those with nonmetallic FC.

The ***wire rope structure*** can be more than the type (6 × 7) cross-section (6 strands each made of 7 wires) viewed in Figure 16.8a. In other structures, a wire rope can have 7, 9, 16, 17, 19, 25, 27, 29, 30, 35, 36, 41, or 46 wires in a strand, and 6, 7, 8, or 19 strands in the wire rope, see Oberg et al. (2012). Wires in strand may not necessarily be of the same size and not all combinations of number of wires in a strand and number of strands in a rope are available. Some of the typical combinations are (**6 × 7**), (6 × 12), (6 × 17), (**6 × 19**), (6 × 21), (6 × 25), (6 × 29), (6 × 36), (**6 × 37**), (6 × 41), (7 × 19), (7 × 36), (**8 × 19**), (8 × 21), (8 × 25), (8 × 36), (18 × 7), (19 × 7), and (34 × 7), where the bold-labeled ones are more common. There are also other forms such as flat, flattened strand, spiral strands, rotation resistant, locked clad, etc. that can be useful in special applications.

16.3.1 Sizes and Properties

Wire rope ***designations*** can be given by the rope diameter and other construction attributes. The wire rope designation can then be provided by the ***wire rope diameter*** d_R as measured in Figure 16.8a and the designation of the form such as the (6 × 7) lay-type. The rope of $\frac{1}{4}$ [in] or 6.5 [mm] diameter d_R with a nonmetallic FC may be quoted as $\frac{1}{4}''$ 6 × 7 (FC) or 6.5 6 × 7 (FC) wire. The wires in the wire ropes are commonly following the standard wire gauges and preferred diameters. Unless distinctive requirements are needed, no specific reference to the wires is needed in the rope designation. The supplier or manufacturer, however, is responsible for these details to provide a rope with a specific strength and service characteristics. Additionally, needed properties might be specified in wire rope order such as material requirement, lay-type and left or right-handed formation.

Wire rope ***sizes*** are generally adhering to the preferred sizes. Some of the available wire ***rope diameters*** d_R are selected as follows.

- US system in [in] are $\frac{1}{4}$, $\frac{5}{16}$, $\frac{3}{8}$, $\frac{7}{16}$, $\frac{1}{2}$, $\frac{9}{16}$, $\frac{5}{8}$, $\frac{3}{4}$, $\frac{7}{8}$, 1, $1\frac{1}{8}$, $1\frac{1}{4}$, $1\frac{3}{8}$, $1\frac{1}{2}$, $1\frac{5}{8}$, $1\frac{3}{4}$, $1\frac{7}{8}$, 2, $2\frac{1}{8}$, $2\frac{1}{4}$, $2\frac{1}{2}$, $2\frac{3}{4}$, 3, $3\frac{1}{4}$, $3\frac{1}{2}$, $3\frac{3}{4}$, 4, $4\frac{1}{4}$, $4\frac{1}{2}$, $4\frac{3}{4}$, 5.
- SI system in [mm] are 6.5, 8, 9.5, 11.5, 13, 14.5, 16, 19, 22, 26, 29, 32, 35, 38, 42, 45, 48, 51, 54, 58, 60, 64, 67, 71, 74, 77, 80, 83, 87, 90, 96, 103, 109, 115, 122, 128.

Other sizes might be available from manufacturers. However, not all rope sizes may be available for small or large rope structures. For example, the rope diameter of the (6 × 7) lay-type is usually not more than 1 ½ [in] or 38 [mm].

The ***wire diameter*** d_W in the strands of the wire rope are usually of a size depending on the wire ***rope diameter*** d_R. The size of the outer wire and the minimum ***pulley diameter*** d_P for some selected wire ropes are as follows.

- For the common (6 × 7) lay-type, the wire diameters d_W is equal to the rope diameter $d_R/9$. The minimum pulley diameter d_P is 42 d_R.
- For the (6 × 19) lay-type, the wire diameters d_W is in the range of the rope diameter $d_R/13$ to $d_R/16$. The minimum pulley diameter d_P is in the range of 26–34 d_R.
- For the (6 × 37) lay-type, the wire diameters d_W is equal to the rope diameter $d_R/22$. The minimum pulley diameter d_P is in the range of 18–23 d_R.
- For the (8 × 19) lay-type, the wire diameters d_W is in the range of the rope diameter $d_R/15$ to $d_R/19$. The minimum pulley diameter d_P is in the range of 21–27 d_R.

The smaller outer wire diameter d_W is normally preferable for lower bending and better fatigue strength.

16.3.1.1 Wire Rope Strength

The rope strength is not the same as the wire strength. The construction of the rope is a ***3D*** structure and interaction of wires in the strands, strands with the core and with other strands provides a rather complex system to determine a closed form expression for the strength. Some empirical relations are then obtained from available data or the maximum loads suggested by the standards or manufacturers are to be adopted. The strength of the selected wire rope should be verified by the information provided by the manufacturer or supplier catalogue.

The ***metal area*** A_{MR} in a wire rope section can be empirically estimated as follows.

$$A_{MR} = 0.38\ d_R^2 \cong 0.40\ d_R^2 \tag{16.42}$$

where d_R is the wire rope diameter measured as in Figure 16.8a and the rope type is mostly for *hauling* and *hoisting*. This relation suggests that the metal in that wire rope is about 48% of the solid area having a d_R diameter. It is not suitable to use this area in the strength calculation of the rope. The rope is a complex ***3D*** structure that has more reactions of the wound wires and interaction between the wires and the strands. The wires and the strands are in spiral or helical forms that would try to unwind and to twist the rope under the longitudinal load. Also, this increases the longitudinal extension and thus really reduces the rope modulus of elasticity relative to the wire modulus of elasticity.

The ***breaking-strength*** of the wire rope is defined as the *minimum ultimate tensile* ***force*** $\boldsymbol{F}_{ut}$ causing the wire rope to break. The stress is not adopted in this case due to the complex stress-state in the wire rope as a ***3D*** system under longitudinal loading. Figure 16.9 presents the *ultimate wire rope breaking-strength force* $\boldsymbol{F}_{ut}$ and the rope *weight-per-unit-length* $\boldsymbol{w}_R$ for few common wires of lay-types (6 × 7), (6 × 19), (6 × 37), and (8 × 19) for the available *wire rope diameters*, see ASTM A1023/A1023M (2009). The material of the conventional wire ropes is the *PS* (AISI 1080 or equivalent, EN C86D), which is a relatively common material in the lower range of strength used in the wire ropes. Figure 16.9a provides the wire ropes for US units where available data are shown as grayish markers and the trend line of the averages with the estimated trend line equations are also shown. Figure 16.9b provides wire ropes for SI units where available data are also shown as grayish markers and the trend line of the averages with the estimated trend line equations are also shown. It is apparent that the variations of the strength and the weight-per-unit-length are small between these common wires at any wire diameter. That is why the average is used to represent the estimated variation of the strength and weight-per-unit-length with wire *rope diameter* d_R for both US and SI units. The regression relations for the *ultimate breaking-strength force* $\boldsymbol{F}_{ut}$ are given by the following trend line equations.

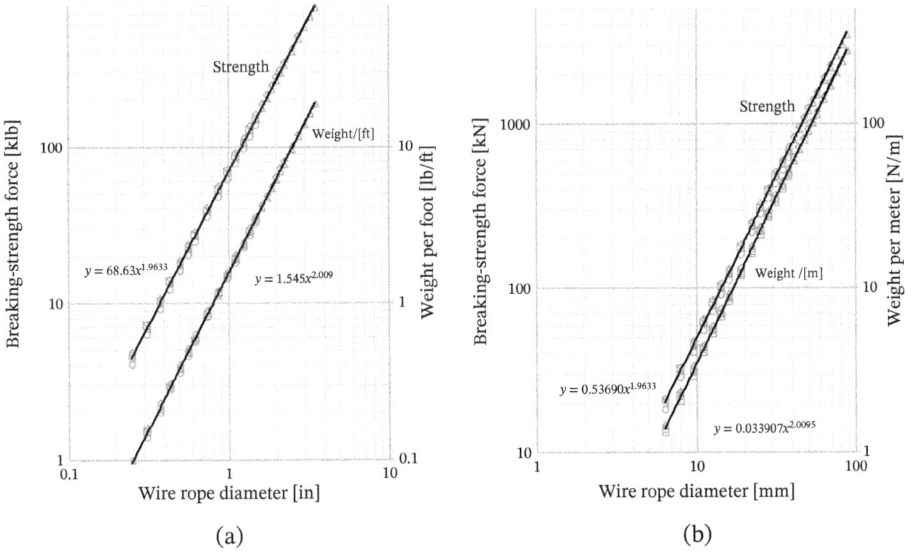

Figure 16.9 Wire rope breaking-strength force and weight per unit length. (a) US system of units. (b) SI system of units. Available data are shown as grayish markers.

$$\begin{aligned} \boldsymbol{F}_{\text{ut}}|_{\text{US}} &= 68.636\ d_R^{1.9633} \quad [\text{klb}], d_R \text{ in } [\text{in}] \\ \boldsymbol{F}_{\text{ut}}|_{\text{SI}} &= 0.53690\ d_R^{1.9633} \quad [\text{kN}], d_R \text{ in } [\text{mm}] \end{aligned} \qquad \textbf{(16.43)}$$

These relations can be used to initially synthesize the needed wire rope diameter for a given loaded wire rope or ropes. Other similar expressions to Eq. (16.43) exist with slight variations of the coefficients and the exponents, see Shigley (1986). The calculation difference is within the design engineering variation that is accommodated by the usual high safety factor in the utility of the wire ropes as discussed later. For higher strength materials than the *PS* (AISI 1080 or equivalent, EN C86D), the ultimate breaking-strength force $\boldsymbol{F}_{\text{ut}}$ are higher than the values obtained by Eq. (16.43). One should use the value of the ultimate breaking-strength force $\boldsymbol{F}_{\text{ut}}$ quoted by the manufacturer or supplier catalogue to verify the selection of the adopted wire rope.

The regression relations for the wire *rope weight-per-unit-length* $\boldsymbol{w}_R$ are given by the following trend line equations.

$$\begin{aligned} \boldsymbol{w}_R|_{\text{US}} &= 1.5459\ d_R^{2.0095} \quad [\text{lb/ft}], d_R \text{ in } [\text{in}] \\ \boldsymbol{w}_R|_{\text{SI}} &= 0.033907\ d_R^{2.0095} \quad [\text{N/m}], d_R \text{ in } [\text{mm}] \end{aligned} \qquad \textbf{(16.44)}$$

These relations can be used to initially synthesize the needed wire rope weight for a given loaded wire rope or ropes. This is particularly useful when the length of rope is large, and its weight would affect the applied load.

The wire ***rope modulus of elasticity*** E_R is less than the *wire modulus of elasticity* E_W due to the **3D** effect of its structure. The extension of the rope is much more than the wire-material due to the spiral form of the wires in the rope. The wire rope modulus of elasticity is useful in evaluating the bending characteristic of the rope (Section 16.3.2). For our common rope lay-types, the wire rope modulus of elasticity E_R is in the range of 10–14 [Mpsi] or 69–97 [GPa] compared to 30.5 [Mpsi] or 210 [GPa] for the usual steel-wire modulus of elasticity E_W, see Table 16.6.

The bending of wire ropes and its fatigue failure are considered later in the detailed design section.

16.3.1.2 Other Wire Rope Properties

Some wire rope properties such as *flexibility*, *wear*, distortion, and crushing resistance are useful in some applications. *Flexibility* is needed when the rope is repeatedly moving over small sheaves, drums, guys or rods that develop excessive bending and potential fatigue. *Wear resistance* is important when the rope must move over a stationary rod or guy surface. Lang lay ropes or flattened strands with proper material selection can be a solution. *Distortion* and *crushing* may occur when the rope is exposed to transverse loading. A rope with larger outer wires and ropes with IWRC can help in resisting distortion and crushing.

16.3.2 Design Procedure

The design procedure entails an initial synthesis and then a detailed design process to guarantee the acceptability of the initial synthesis and perform any needed adjustments or optimization. The assurance of safety and intended life of the wire rope is another possible objective of the detailed design.

16.3.2.1 Initial Synthesis

The procedure of initial synthesis to identify the rope type, rope diameter, number of ropes, the pulley diameter, and the required rope material are the needed objectives of the design. The process of initial synthesis is suggested as follows.

- Usually the *maximum rope system load* $\boldsymbol{F}_R$ (one or more ropes) is known. The type of application, rope speed, and estimated length are also expected to be known. The projected safety factor K_{SF} is then identified from Table 16.7. The higher the speed or the rope length, the higher the adopted safety factor.
- The *minimum ultimate tensile force* $\boldsymbol{F}_{ut}$ can then be simply defined from the load on the rope system $\boldsymbol{F}_R$ multiplied by the safety factor K_{SF}, or $\boldsymbol{F}_{ut} = K_{SF}\ (\boldsymbol{F}_R)$.
- The *rope diameter* d_R of a rope-type, say (6 × 19), is obtained from Eq. (16.43) for each of an assumed *number of ropes* N_R from 1 to say 6 and the standard rope diameter is selected. The weight of the rope system can be obtained from Eq. (16.44) multiplied by the number of ropes N_R and the expected rope length L_R. Comparing the different weights of the rope system, the number of ropes can be identified as the number which would render the minimum weight of the rope system for a rope type, say (6 × 19). More than one rope may be advisable because one must have a redundant rope if one is to fail. A rope replacement is then immediately required.
- The pulley diameter d_P is then defined from the selected rope type out of say (6 × 7), (6 × 19), (6 × 37), or (8 × 19) and the minimum pulley diameter d_P associated with the selected type. This selection of (d_P/d_W) is needed as an input to define the pulley diameter. The relatively simple structure rope-type such as (6 × 7) would need a larger pulley diameter than the more dense-structured ropes such as (6 × 19). If information is available, other rope types may be selected, see Section 16.3.1.
- Iteration is possible on the different rope types to select an acceptable pulley diameter that suits the design more than others.

Table 16.7 Suggested safety factors for some applications operating wire ropes. High values for high speed or distance (rope length).

Application	Safety factor	Application	Safety factor
Guys, track cables	3.2–3.5	Electric hoists	7.0
Hand elevators	4.5–5.0	Freight elevators	7.0–11.0
Hoisting, cranes	5.0	Grain elevators	7.5
Mine conduits	5.0–8.0	Private elevators	7.5
Haulage	6.0	Passenger elevators	8.0–12.0

16.3.2.2 Detailed Design Process

The detailed design process is important in satisfying the safe operational character of the wire rope for the intended life. An utmost concern affecting safety factor is the *bending moments* in the wires when the rope passes over a small diameter pulley or guy. The repeated occurrences of bending would cause *fatigue*, which can trigger snapping of one or more wires and eventually rope rupturing. Therefore, the minimum pulley, sheave, guy or drum diameter has been suggested to considerably reduce that effect. The type of loading variation, the rope speed, and rope type affect the severity of bending and fatigue. Smaller wire diameter in the strands and larger pulley, sheave, guy or drum diameter reduce the bending effect and improves the fatigue resistance of the wire rope.

The ***bending** moment* in the wire rope is developed due to the rope traversing around the pulley diameter. As per the strength of material discussed in Chapter 6, particularly Sections 6.1.2, on can use Eqs. (6.11) and (6.12) to develop the following relation.

$$\boldsymbol{M}_z = \frac{E_R\ I_z}{R_C} \quad \text{and} \quad \sigma_x = \frac{\boldsymbol{M}_z\ y}{I_z} \quad \text{or} \quad \sigma_x = \frac{E_R\ y}{R_C} \tag{16.45}$$

where $\boldsymbol{M}_z$ is the bending moment, E_R is the *rope elasticity modulus*, I_z is the rope second area moment, R_c is the radius of curvature of the pulley or else the *pulley radius* $(d_P/2)$, y is the distance from the neutral axis of the wire in the rope or the *wire radius* $(d_W/2)$, and σ_x is the tensile stress in the outer fiber of the wire. Equation (16.45) assumes the longitudinal axis of the rope to be in the x direction. Substituting these parameters and the first expression for the bending moment $\boldsymbol{M}_z$ in the second of Eq. (16.45), one gets the bending stress in the wire as follows.

$$\sigma_x = \frac{E_R\ d_W/2}{d_P/2} = \frac{E_R\ d_W}{d_P} \tag{16.46}$$

where E_R is, again, the *rope elasticity modulus* and not the wire elasticity modulus. The *rope elasticity modulus* E_R is dependent on the rope lay type and is given the value in the range of 10–14 [Mpsi] or 69–97 [GPa]. This value is useful as a check, but the proper value of E_R should be certified by the manufacturer or supplier and verified by tests if need be.

The *equivalent bending load* $\boldsymbol{F}_{\text{Rb}}$ representing the effect of bending on the rope can simply be calculated from Eq. (16.46) to give the following expression.

$$\boldsymbol{F}_{\text{Rb}} = \sigma_x A_{\text{MR}} = \frac{E_R\ d_W A_{\text{MR}}}{d_P} \tag{16.47}$$

where A_{MR} is the metal area of the rope approximately defined by Eq. (16.42).

As evident from Eq. (16.46), the bending stress is greatly reduced for smaller wire diameters and larger pulley, guy, or drum diameters. If the ratio of the pulley diameter to wire diameter (d_P/d_W) is large enough, the bending stress will have a little impact on the loading of the wire rope. A check is, however, advisable as part of the detailed design process. The *strength loss factor* K_{SL} is about 40% when the pulley diameter to rope diameter (d_P/d_R) is about 2 for ropes of lay-type (6 × 17) and (6 × 19). However, the strength loss factor K_{SL} is about 5% when the pulley diameter to rope diameter (d_P/d_R) is about 40 for ropes of lay-type (6 × 17) and (6 × 19). This factor has been the reason for suggesting the minimum pulley diameter d_P with reference to the rope diameter d_R as given in Section 16.3.1 for different rope types. The high pulley diameter to wire diameter (d_P/d_W) can be attained since the wire diameter to the rope diameter is small ($d_W = d_R/9$, for 6 × 7 rope) and the minimum pulley diameter is large relative to the rope diameter ($d_P/d_R = 42$, for 6 × 7 rope), i.e. $d_P/d_W = 9 \times 42 = 378$, see Section 16.3.1 for other rope lay-types. The bending stress is then equal to $E_R/378$ or about 14/378 = 0.037 [Mpsi] = 37 [kpsi] or 97/378 = 0.2570 [GPa] = 257 [MPa] which is relatively slight with respect to the wire strength of *improved plow steel* (IPS) of about 228 [kpsi] or 1570 [MPa]. The *strength reduction factor* K_{SR} is equal to $(1 - K_{SL})$. It can be approximately evaluated for ropes of lay-type (6 × 17) and (6 × 19) by the following regression interpretation.

$$K_{\mathrm{SR}} = 0.5495\left(\frac{d_P}{d_R}\right)^{0.2035}, \quad \left\{\frac{d_P}{d_R}\right\} < 10$$

$$K_{\mathrm{SR}} = 0.839\,57\left(\frac{d_P}{d_R}\right)^{0.031\,33}, \quad \left\{\frac{d_P}{d_R}\right\} \geq 10 \tag{14.48}$$

This value is important only if the ratio of pulley diameter to rope diameter is much smaller than the suggested minimum pulley diameter. This, however, can be the case if the rope is passing over a pin or a guy with a small diameter. In that case, the strength reduction factor K_{SR} should be considered in the detailed design process.

The ***wear*** and ***fatigue*** go hand in hand. Wear occurs due to high *bearing pressure* on the pulley, sheave, drum, guy, and the rope wires traversing on any of them. This high and repeated stress eventually would cause fatigue, particularly on the outer wires in the rope. The pulley ***bearing pressure*** p_P is basically defined as the load divided by the projected area. This gives the following simple relation.

$$p_P = \frac{2\boldsymbol{F}_R}{d_R d_P} \tag{16.49}$$

where $\boldsymbol{F}_R$ is the tension in the rope, d_R is the rope diameter, and d_P is the pulley (sheave, drum, or guy) diameter.

The minimum *allowable bearing pressure* depends on the pulley or drum material and the rope type. The suggested ranges of the minimum allowable bearing pressure of some pulley materials for some selected rope types ((6 × 7), (6 × 19), (6 × 37), (8 × 19)) are as follows.

- *Cast iron* allowable bearing pressure range is about 0.3–0.7 [kpsi] or 2–5 [MPa].
- *Cast steel* allowable bearing pressure range is about 0.6–1.3 [kpsi] or 4–9 [MPa].
- *Manganese steel* allowable bearing pressure range is about 1.5–3.5 [kpsi] or 10–24 [MPa].

The lower value of the allowable bearing pressure range is for the common lower strength rope type. These values are just as a legacy practice and should be used as a background knowledgebase. More proper values are advisably defined by the recommendation of manufacturer or vendor.

Fatigue of wire ropes is somewhat more complex and is affected by many factors including the pulley or drum diameter, groove dimension and shape, material, the load variation cycle, speed and acceleration, and rope construction. The lower the bearing pressure, however, the longer life the rope is expected to have. Experimentally, the lower the ratio of the bearing pressure p_P to the *ultimate tensile strength* S_{ut} of the rope (which is not that definite), the longer the rope life is expected. A suggested smaller ration than 0.001 p_P/S_{ut} is projected to achieve a close to an asymptotic endurance strength at 10^6 bends. The ultimate *fatigue strength force* $\boldsymbol{F}_f$ may then be obtained from that small ratio of 0.001 and Eq. (16.49) to get the following suggested limit.

$$\boldsymbol{F}_f = \frac{\left({}^{p_P}/_{S_{\mathrm{ut}}}\right) S_{\mathrm{ut}} d_R d_P}{2} \leq \frac{0.001\left(S_{\mathrm{ut}} d_R d_P\right)}{2} \tag{16.50}$$

Alternatively, to expect a longer rope life before fatigue failure, the material of the wire ought to have an ultimate tensile strength S_{ut} of at least the following.

$$S_{\mathrm{ut}} \geq \frac{2000\boldsymbol{F}_f}{d_R d_P} = \frac{2000\boldsymbol{F}_{Rb}}{d_R d_P} \tag{16.51}$$

where $\boldsymbol{F}_f$ is simply taken as the equivalent bending load $\boldsymbol{F}_{\mathrm{Rb}}$ since the fatigue is assumingly due to the moment variation with the other tensile force $\boldsymbol{F}_R$ as a mean load value. This would help in adopting a mild PS (about 180 [kpsi] or 1240 [MPa]), or a PS (about 210 [kpsi] or 1448 [MPa]), or IPS (about 228 [kpsi] or ISO 1570 [MPa]), or extra improved plow steel (EIPS) (about 257 [kpsi] or ISO 1770 [MPa]), or extra–extra improved plow steel (EEIPS) (about 284 [kpsi], or ISO 1960 [MPa]), or even extra–extra–extra improved plow steel (EEEIPS) (about 313 [kpsi] or ISO 2160 [MPa]). One should also abide by the strength of the material used by the manufacturer or vendor.

The procedure for the detailed design process is, therefore, as follows.

- Usually the load on the rope system (one or more ropes) is known. The type of application, rope speed, and the expected rope length are also known. The expected *safety factor* is then reaffirmed from Table 16.7. The higher the speed, the higher the safety factor.
- With the initial synthesis of rope type, rope diameter, the number of ropes, the pulley diameter, and the normal stresses; the *bending stresses* can be estimated from Eq. (16.46).
- The bearing pressure on pulley, drum, or guy is calculated from Eq. (16.49). The suitable material of the pulley, drum, or guy is then adopted as suggested after Eq. (16.49).
- The suggested *ultimate tensile strength* S_{ut} for the rope material to have a longer rope life before fatigue failure is calculated according to Eq. (16.51). The proper wire rope material is then adopted among the different plow steels or IPSs. The manufacturer designation is then defined.
- Iteration might be needed to have a more extended rope life by increasing the rope diameter, or change its type, increasing the pulley, drum, or guy diameter. Optimization might also be needed to reduce cost or increase life.

16.4 Chains

So many types of chains exist such as the necklaces, lockets, and pendants. Industrial chains such as closed-linked hoisting, slings, and crane chains are employed in several applications. The most used chains are the common roller chains, which are the focus of this section.

Roller chains are extensively used in bicycles and motorcycles in addition to so many other power transmission applications. The common roller chain components and assembly are shown in Figure 16.10. Figure 16.10a presents the components needed to produce the *roller link* shown in Figure 16.10b. The two bushings are securely attached to the top plate in Figure 16.10a. The two rollers are inserted onto the bushings with a running clearance and the bottom plate is then attached securely into the protruding bushing ends. This generates the roller link of Figure 16.10b. Figure 16.10c shows the two pins securely attached to the top link plate to form the *pin link*. Then the two pins of the pin link assemble two of the roller links (Figure 16.10d) by the insertion of the pins into the bushing holes of each of the two-roller links. The bottom link plate is then attached securely into the protruding pin ends, Figure 16.10d. By repeating the process, a full roller chain length is produced by attaching the last link-plate to connect both loose ends of the chain. This produces an endless chain with specific number of roller links with each having a specific standard chain pitch p_C. The length of the chain is the number of roller-links multiplied by twice the chain pitch for even number of chain links, or the chain length is the number of chain

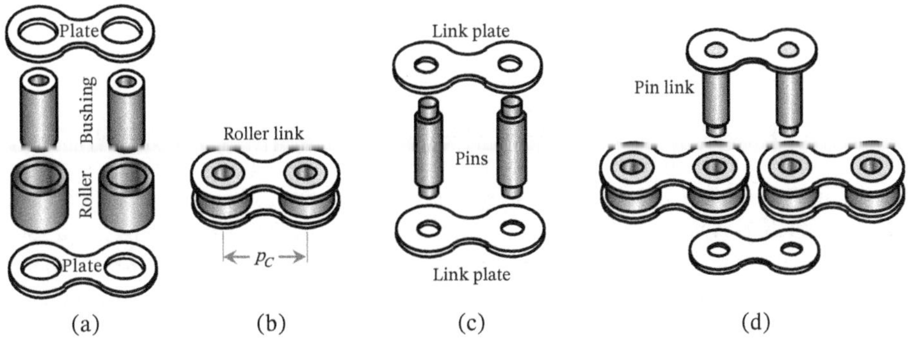

Figure 16.10 Simple sketches of the common roller chain components and assembly. (a) Components needed to produce the roller-link. (b) The roller-link having a standard chain pitch p_C. (c) Two pins to be attached to the top link plate forming the pin-link. (d) Assembly of two roller-links by the pin-link and securely attaching the bottom link-plate to pin-link.

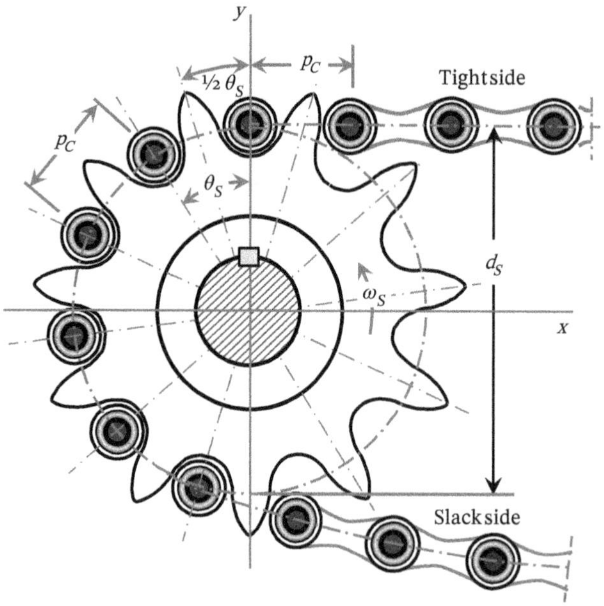

Figure 16.11 A draft of the roller chain engaging the driving sprocket and indicating the main dimensions, geometry and rotational direction.

links multiplied by the chain pitch. Note that even number of links is preferred so that no *offset link* (not shown in Figure 16.10) would be needed. The offset link is having one roller and one pin with two offset link plates to accommodate the longer pin.

Figure 16.11 shows a draft of the roller chain engaging the driving ***sprocket*** and indicating the main dimensions, geometry, and rotational direction. The tight side of the chain is the part of the chain under the driving tension. The slack side of the chain is the chain return to the driven sprocket. The slack side has a *catenary* geometry between the driving and driven sprockets. The sprocket in Figure 16.11 has an odd *number of teeth* (11). The main parameters are the *chain pitch* p_C, the sprocket *pitch diameter* d_S, and the sprocket *pitch angle* θ_S. The transverse centerlines of the chain links coincide with the pitch-diameter circle. The base or bottom of the sprocket teeth fits the radius of the chain roller, see Figures 16.10 and 16.11. Further geometric relations are given in Sections 16.4.2 and 16.4.3.

16.4.1 Standards

There are many types of chains that are made of steel, cast steel, cast iron, forged, and plastics. The keen interest in this section is on the common steel roller chain made of the components shown in Figure 16.10, which is also engaged with the sprocket shown in Figure 16.11. The emphasis herein is on single-strand chains and the sprockets associated with that, see standards such as ANSI/ASME B29.1 (2011), ISO 606 (2015), ISO 10823 (2004), and ISO 487 (1998). Other standards are available for other types of chains such as ISO 9633 (2001), ISO 10190 (2008) for bicycles and motorcycles, and BSR/ASME B29.200 (1995) for welded steel type mill chains. The initial roller chain inventions have been attributed to *Hans Renold* from 1880–1915, even though *Leonardo Da Vinci* (1452–1519) had previous sketches showing a chain with roller bearing. Since the initial invention, the chains have been mainly in imperial or [in] dimensions. The SI chains are just converting the inches into millimeters for interchangeability particularly for maintenance.

Table 16.8 The common roller chain standard sizes or numbers, main dimensions, and ultimate breaking loads, see ANSI/ASME B29.1 (2011), and ISO Standards.

Chain number	Pitch, p_C		Maximum roller diameter		Roller width		Ultimate breaking load	
ANSI (ISO)	[in]	[mm]	[in]	[mm]	[in]	[mm]	[kN]	[klb]
25[a)](04C-1)	0.25	6.35	*0.13*	*3.30*	0.125	3.18	3.5	0.78
35[a)](06C-1)	0.375	9.53	*0.2*	*5.08*	0.188	4.78	7.8	1.76
40 (08A-1)	0.5	12.70	0.312	7.92	0.312	7.92	13.9	3.13
41 (085-1)	0.5	12.70	0.306	7.77	0.25	6.35	6.7	1.50
50 (010A-1)	0.625	15.88	0.4	10.16	0.375	9.53	21.7	4.88
60 (12A-1)	0.75	19.05	0.469	11.91	0.5	12.70	31.3	7.03
80 (16A-1)	1	25.40	0.625	15.88	0.625	15.88	55.6	12.50
100 (20A-1)	1.25	31.75	0.75	19.05	0.75	19.05	86.9	19.53
120 (24A-1)	1.5	38.10	0.875	22.23	1	25.40	125.1	28.13
140 (28A-1)	1.75	44.45	1	25.40	1	25.40	170.3	38.28
160 (32A-1)	2	50.80	1.125	28.58	1.25	31.75	222.4	50.00
180 (36A-1)	2.25	57.15	1.406	35.71	1.406	35.71	281.5	63.28
200 (40A-1)	2.5	63.50	1.562	39.67	1.5	38.10	347.7	78.18
240 (48A-1)	3	76.20	1.875	47.63	1.875	47.63	500.4	112.50

a) Chain is rollerless. Italic dimension is *bushing diameter*.

16.4.1.1 Chain Size or Number

Table 16.8 presents the common roller *chain standard sizes or numbers*, the main dimensions, and ultimate breaking loads. The main chain standard dimension is the *chain pitch* p_C. The US standard pitches in [in] are the decimal values of the usual fractions of ¼, ⅜, ½, ⅝, ¾, …. The SI values in [mm] are just the conversion of these inch fractions to millimeters with some little rounding as shown in Table 16.8. The maximum roller diameter of the chain is to allow enough tooth size in the sprocket as shown in Figure 16.11 and enough space and thickness for the roller to rotate around the bushing, see Figure 16.10. There should be a lubricant in between the roller and the bushing to allow free rotation with minimum friction. The lubricant is usually grease or other more suitable oil for higher speed applications. The roller width in Table 16.8 allows for the suitable rotation of the roller under the maximum applied load as a reasonable safe fraction of the indicated *ultimate breaking load*. The *minimum ultimate tensile strength* is defined as the ultimate breaking load in pounds [lb] and it is around or greater than (12 500 (p_C in [in])^2). This is for a single strand chain.

For *multiple strand chains*, the ultimate breaking loads are less than the multiple breaking loads of a single strand. The two-strands can carry about 1.7 times the single strand and not twice as much. The three-strands can carry about 2.5 times the single strand. The four-strands, the five-strands, and the six-strands can carry about 3.3, 3.9, and 4.6 times the single strand, respectively.

Chain numbers or size such as ANSI 25 or 40 or 41 in Table 16.8 are given such that the left-hand digit (**2** or **4**) is the number of eighths of an inch that make up the chain pitch p_C such as (**2**/8 or **4**/8, i.e. a pitch of ¼ [in] or ½ [in]). The right-hand digit of the chain number or size (2**5** or 4**0** or 4**1**) denotes *rollerless* bushing chain for the digit **5**, normal chain for the digit **0**, and lightweight chain for the digit **1**, ANSI/ASME B29.1 (2011). The SI chain number uses the roller diameter in a round-figure form or so in [mm] as part of the chain number, see Table 16.8,

ISO 606 (2015), and ISO 487 (1998). With each chain number or size, there is an associated sprocket to fit with it, see Figure 16.11.

16.4.1.2 Chain Sprockets

Sprocket *teeth number* N_S in the US and SI systems are usually the following preferred sizes.

- US sprocket *number of teeth*: 9, 10, 11, 12, 13, 14, 15, 16, 17, 18, 19, 20, 28, 30, 32, 35, 36, 40, 42, 45, 48, 52, 54, 60, 70, 72, 80, 84, 96, 112.
- SI sprocket *number of teeth*: 11, 12, 13, 15, 17, 19, 20, 21, 23, 25, 27, 30, 38, 45, 57, 76, 95, 114.

An odd number of teeth in the drive side is also preferred to ensure a uniform wear all over the numerous teeth in the chain system. It is also suggested not to use sprockets below 19 teeth, particularly for better *articulation*, lower loads on sprocket teeth, better life, and to have a more cost-effective selection.

Sprocket teeth form and width should engage through the chain roller with enough defined clearances. There are plain-plate sprockets, one-side, or two-side, integral or detachable hubs, and other sprocket constructions. The detailed alternative designs, basic dimensions, clearances, and other details for single or multiple-strand chains can be found in the standards, handbooks, and sprocket manufacturer catalogues, see references and internet websites in this chapter.

16.4.2 Drive Relations

Geometric relations of the chain and the sprocket can be obtained by examining Figure 16.11. Noting that the links are straight lines between the rollers, the trigonometric relation for the half sprocket angle (½ θ_S) gives the following.

$$\sin\left(\frac{1}{2}\theta_S\right) = \frac{\frac{1}{2}p_C}{\frac{1}{2}d_S}, \text{ or } \quad d_S = \frac{p_C}{\sin\left(\frac{1}{2}\theta_S\right)} \tag{16.52}$$

where p_C is the *chain pitch*, d_S is the sprocket *pitch diameter*, and θ_S is the sprocket *pitch angle*, see Figure 16.11. However, the sprocket *pitch angle* θ_S is related to the *sprocket teeth number* N_S by the simple relation such that $\theta_S = 360°/N_S$. This renders Eq. (16.52) into the following form.

$$d_S = \frac{p_C}{\sin\left(180°/N_S\right)} \tag{16.53}$$

Chain articulation occurs due to the chain links forming a tangential connection between sprocket teeth, Figure 16.11. When the roller at the top of the *y*-axis starts to move in the −*x* direction with the sprocket rotation, it will also move in the −*y* direction pulling the chain down. It travels one half the sprocket angle (½θ_S) till the incoming chain roller starts moving up in the positive *y* direction. The half sprocket angle (½θ_S) is called the *articulation angle.* The more this angle, the more it causes the articulation of the incoming chain into the sprocket triggering vibration and impact between the rollers and the teeth of the sprocket. This angle (½θ_S) is a function of the *sprocket teeth number* N_S and it is equal to $360/N_S$. As the number of sprocket teeth increases, the articulation angle decreases and consequently the vibration and impact reduction occur.

With reference to Figure 16.11, the *longitudinal chain velocity* v_C can be defined as follows.

$$v_C|_{US} = \frac{\pi d_S\ N_{\text{rpm}}}{12} = \frac{N_S\ p_C\ N_{\text{rpm}}}{12}\ [\text{ft/min}]$$
$$v_C|_{SI} = \frac{\pi d_S\ N_{\text{rpm}}}{60} = \frac{N_S\ p_C\ N_{\text{rpm}}}{60}\ [\text{m/s}] \tag{16.54}$$

The *maximum longitudinal chain velocity* $v_{C\max}$ of the chain and the minimum longitudinal chain velocity $v_{C\min}$ of the chain depend on the sprocket velocity at the pitch circle diameter d_S and at the lower position of the rollers at

($\frac{1}{2}\theta_S$) where the diameter is equal to $d_S \cos(\frac{1}{2}\theta_S)$. The maximum chain velocity $v_{C\max}$ and the minimum velocity $v_{C\min}$ are then given by

$$v_{C\max}|_{US} = \frac{\pi d_S\, N_{\text{rpm}}}{12} = \frac{N_S\, p_C\, N_{\text{rpm}}}{12\sin\left(\frac{1}{2}\theta_S\right)}, \quad v_{C\max}|_{SI} = \frac{\pi d_S\, N_{\text{rpm}}}{60} = \frac{N_S\, p_C\, N_{\text{rpm}}}{60\sin\left(\frac{1}{2}\theta_S\right)}$$

$$v_{C\min}|_{US} = \frac{\pi d_S \cos\left(\frac{1}{2}\theta_S\right) N_{\text{rpm}}}{12} = \frac{N_S\, p_C \cos\left(\frac{1}{2}\theta_S\right) N_{\text{rpm}}}{12\sin\left(\frac{1}{2}\theta_S\right)},$$

$$v_{C\min}|_{SI} = \frac{\pi d_S \cos\left(\frac{1}{2}\theta_S\right) N_{\text{rpm}}}{60} = \frac{N_S\, p_C \cos\left(\frac{1}{2}\theta_S\right) N_{\text{rpm}}}{60\sin\left(\frac{1}{2}\theta_S\right)} \tag{16.55}$$

Noting that $\frac{1}{2}\theta_S = 180^\circ/N_S$, the variation of the chain velocity can then be found by minor manipulation as

$$\frac{\Delta v_C}{v_C} = \frac{v_{C\max} - v_{C\min}}{v_C} = \frac{\pi}{N_S}\left\{\csc\left(\frac{180^\circ}{N_S}\right) - \cot\left(\frac{180^\circ}{N_S}\right)\right\} \tag{16.56}$$

Equation (16.56) represents the *chordal speed variation* Δv_C which has the approximate values of about 20.4%, 4.98%, 2.2%, 1.24%, 0.79%, 0.55%, 0.40%, and 0.31% for the sprocket teeth number N_S of 5, 10, 15, 20, 25, 30, 35, and 40, respectively.

Therefore, it is recommended to have at least 19 or 25 teeth sprocket for the higher speed axis and low chain noise. For more precision drives, the chordal speed variation should be carefully considered.

Power transmission is a characteristic which depends on the chain size or number, and the rotational speed of the sprocket. Each chain size has an *optimum speed* where the power is maximum. These optimum conditions are preferably used to get the best utilization of the chain. A lower or higher speed than the optimum would transmit a lower power. This would give an inferior alternative unless an abiding constraint is mandatory. Figure 16.12 is a plot of the roller chains characteristics and trend lines for maximum power rating H, ultimate strength force $\boldsymbol{F}_{\text{ut}}$, and marks of chain numbers (ANSI chain numbers 25–160) for US and SI systems. The range of rated power and rotational speed are limited to the bounds of the plots and Table 16.8. The *maximum power ratings* H_{hp} and H_{kW} are the grayish marks around the black solid trend-lines. The *ultimate tensile strength forces* $\boldsymbol{F}_{\text{ut}}$ [klb] and $\boldsymbol{F}_{\text{ut}}$ [kN] are the grayish marks around the black dashed trend-lines. Values are read off the left ordinate axis for the SI units and off the right ordinate axis for the US units. The *chain number* markers are read off the right ordinate axis for the ANSI standards. The plots are for common chains running over 17 tooth sprockets for a projected service-life of 15 000 hours [hr].

The trend lines of the *maximum power ratings* H_{hp} and H_{kW} have the following estimated trend formulas.

$$H_{\text{hp}} = 8.8586\left(10^6\right) N_{\text{rpm}}^{-1.6852}$$
$$H_{\text{kW}} = 6.6109\left(10^6\right) N_{\text{rpm}}^{-1.6852} \tag{16.57}$$

where the power is a function of the sprocket rotational speed N_{rpm}. It should be mainly valid at the optimum rotational speed of each chain size or number. The supposed optimum rotational speeds are shown in Figure 16.12 for each chain size or number. An approximate estimate for the *sprocket rotational speed* N_{rpm} as a function of the *ANSI chain number* $C\#_{\text{ANSI}}$ is projected as follows.

$$N_{\text{rpm}} \approx 1.6842\left(10^5\right)\left(C\#_{\text{ANSI}}\right)^{-1.1543} \tag{16.58}$$

where this optimum rotational speed is a rough estimate and is assumed usually as a round-figure approximation, as depicted in Figure 16.12. This projected rough rotational speed is to be used cautiously as an initial guide in the design process and should be verified or adjusted to the actual values provided by the manufacturer.

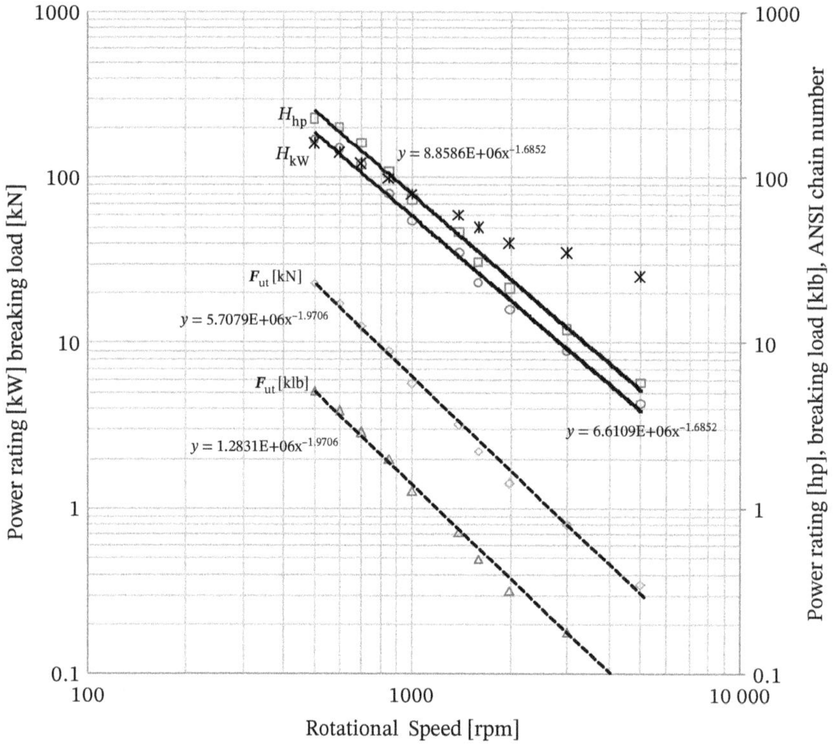

Figure 16.12 Roller chains characteristics and trend lines for maximum power rating, strength, and chain number (as black markers of ANSI chain numbers 25–160) for US and SI systems.

Other estimates of rated power are present in dedicated references, which additionally include the number of teeth in the sprocket and the chain pitch, ACA (2006). The rated power is, however, within the standard values in Table 16.8 and within design engineering accuracy shown in Figure 16.12 for the optimum rotational speed. If the rotational speed N_{rpm} is different from optimum and much lower, the link or roller-plate limited *fatigue rated power* H_R is given by the following expression, ACA (2006).

$$H_R|_{US} = 0.004 N_S^{1.08} N_{rpm}^{0.9}\ p_C^{(3-0.07p_C)} \quad [\text{hp}], \quad p_C \text{ in [in]}$$
$$H_R|_{SI} = 0.746\left(0.004 N_S^{1.08} N_{rpm}^{0.9}\ p_C^{(3-0.07p_C)}\right) \quad [\text{kW}], \quad p_C \text{ in [in]} \tag{16.59}$$

where N_S is the *sprocket number of teeth*, p_C is the chain pitch in [in]. A straight conversion to the SI system is shown in the equation. If the chain pitch is in [mm], one can just divide by 25.4 [mm]/[in] for every p_C in the equation.

The trend lines of the *ultimate breaking strength* forces $\boldsymbol{F}_{ut}$ [klb] and $\boldsymbol{F}_{ut}$ [kN] have the following estimated trend expressions.

$$\boldsymbol{F}_{ut} = 1.2831\left(10^6\right) N_{rpm}^{-1.9706} \quad [\text{klb}]$$
$$\boldsymbol{F}_{ut} = 5.7079\left(10^6\right) N_{rpm}^{-1.9706} \quad [\text{kN}] \tag{16.60}$$

where these forces should not be reached otherwise the chain will break. Suitable factors such as safety factor and other factors are used to guarantee unexpected failure of the chain. This is discussed in the following Section 16.4.4 for the design procedure.

16.4.3 Set Dimensions and Constraints

Unlike the regular V-belts, the chains are made of rigid links. The *chain length* L_C is then a multiple of the chain pitch p_C. This means that L_C/p_C must be an integer. In a chain system, the *center distance* C_C between sprockets is usually more or less a known value. Also, the center distance C_C is usually recommended to be about 30–50 times the chain pitch p_C. From that approximate C_C, one can get an estimate of the needed chain length L_C as given by the following relation.

$$L_C = 2C_C + p_C\left(\frac{N_{S1}+N_{S2}}{2}\right) + p_C^2\left(\frac{(-N_{S1}+N_{S2})^2}{4\pi^2 C_C}\right) \tag{16.61}$$

where N_{S1} and N_{S2} are the *number of sprocket teeth* for the input and the output sprockets. This should render the chain length L_C more than 60–100 the chain pitch p_C. As calculated from Eq. (16.60), an adjustment to the real chain length as an integer multiple of p_C should then be implemented. The center distance C_C should then be readjusted using the following expression.

$$C_C = \frac{p_C}{4}\left(\left(\frac{L_C}{p_C} - \frac{N_{S1}+N_{S2}}{2}\right) + \sqrt{\left(\frac{N_{S1}+N_{S2}}{2} - \frac{L_C}{p_C}\right)^2 - 8\left(\frac{N_{S2}+N_{S1}}{2\pi}\right)^2}\right) \tag{16.62}$$

The construction of the chain system should include adjustment means to accommodate the chain assembly, appropriate initial chain tightening, chain wear, and sagging. An *idle sprocket* may be used to advantage in that regard.

16.4.4 Design Procedure

The main interest in roller chains is the *power transmission*. To get the design procedure in the usual manner, an initial synthesis followed by detailed design is adopted.

16.4.4.1 Initial Synthesis

The design of chains for power transmission involves the selection of chain size or number and the associated sprockets for the input and output sides. The main interest is for single chain or multi-stand chains unless constraining factors exist to necessitate specific number of multiple-strand chains. The usual requirements are the power transmitted at some initial rotational speed and the output rotational speed or speed reduction ratio. The following is the expected useful initial synthesis process.

For *optimum rotational speed* (within 300–5000 [rpm]) and *optimum power*, the procedure is as follows.

- Usually the transmitted or *applied power* H, the input rotational speed ω_1, and the *reduction ratio* (ω_1/ω_2) or the *speed ratio* n_C of the chain system are given, see Figure 16.2 for a similar system with the chain replacing the belt and the sprockets replacing the pulleys.
- The power H is simply given by the following relation.

$$H = \boldsymbol{T}_S \cdot \boldsymbol{\omega}_S = \boldsymbol{F}_C r_S \omega_S \tag{16.63}$$

  where $\boldsymbol{T}_S$ is the *sprocket torque*, ω_S is the *sprocket rotational speed*, r_S is the *sprocket pitch radius*, and $\boldsymbol{F}_C$ is the *chain tight side force*. It is assumed that the loose slack side chain does not contribute to the force polygon of the chain system.
- The initial synthesis depends on the design state conditions such as the *safety factor* K_{SF} and *application service factor* K_{ASF} affecting the maximum transmitted *design power* H_D. The *design power* is then given by

$$H_D = K_{SF}K_{ASF}\ H \tag{16.64}$$

where the application service factor K_{ASF} can be taken like the service factor for V-belts shown in Table 16.2 or any other available information defining the service differently. The safety factor is considered to account for anticipated higher service life than the 15 000 hours [hr] or any other abnormal conditions in the sought design. Other factors such as *temperature* or *high-speed factor* would be included in the detailed design process (16.4.4.2). Note that one would seldom uses a longitudinal chain speed higher than 200 [ft/min] or 1 [m/s].
- The chain number or size is obtained to satisfy Eq. (16.58) and rounding off the value. The chain number or size, in this case, is dependent on the optimum rotational speed only and the number of chain strands is one.
- With the known chain number, the chain pitch p_C is known, and the chain characteristics are as defined later.

For *nonoptimal rotational speeds* and *different rated powers*, the process for *multiple-strand chains* is proposed as follows.

- If the driving rotational speed N_{rpm} is lower than the optimum rotational speed (about 500–5000 [rpm] assumed range), the chain number or size would need to be changed. The chain number or size is altered to conform with Table 16.8 and Figure 16.12. Using the design power H_D, the optimum rotational speed is obtained by the utility of Eq. (16.57). The chain number $C\#_{\text{ANSI}}$ is then found by Eq. (16.58) and by rounding off the value. This also defines the chain pitch p_C as given in Table 16.8. The *rated power* H_R at the driving speed is then evaluated by Eq. (16.59).
- From the effect of the *number of chain strands* N_{CS} on the power, a regression expression is found to define the *power ratio factor* K_{HR}. This process gives the power ratio factor K_{HR} as a function of the *chain strands number* N_{CS} such as follows.

$$K_{\text{HR}} = 0.97625 \ N_{\text{CS}}^{0.86212} \tag{16.65}$$

This is for chain strands number N_{CS} higher than one. From this relation, one can select or find the appropriate chain size or number that fits the power rating of the chain size from Eq. (16.59) and the number of chain strands N_{CS} to accommodate the *design power* H_D. The number of chain strands N_{CS} is then given by rounding the following equation

$$N_{\text{CS}} = \text{round}\left(\frac{H_D}{K_{\text{HR}}} + 0.5\right) \tag{16.66}$$

where K_{HR} is power ratio factor given by Eq. (16.65).
- The available *total power* rating H_{RT} is calculated from the following expression.

$$H_{\text{RT}} = K_{\text{HR}} \ H_R \tag{16.67}$$

where K_{HR} is the power ratio factor, and H_R is the rated power. The *extra safety factor* K_{SFE} can then be obtained by realizing that $K_{\text{SFE}} = H_{\text{RT}}/H_D$.

The chain system characteristics are then found from the following.

- The chain and sprocket system characteristics are obtained by applying Eq. (16.53) to get the input and output sprocket diameters. The chain velocity is obtained from Eq. (16.54). The chain length L_C and center distance C_C are found from Eqs. (16.61) and (16.62).

This process completely defines the chain system as an initial synthesis. Adjustments and possible optimization are defined through the following detailed design process.

16.4.4.2 Detailed Design Process

The previous initial synthesis included so many details particularly for the multi-strand chains. The detailed design process should include some other specified factors such as speed factor, temperature factor and would consider different number of small sprocket teeth than 17, different chain length than 100 p_C, and different expected life span than 15 000 [hr]. These factors can also be obtained from dedicated references such as ACA (2006).

The *rated power* H_R in Eq. (16.59) can be adjusted for different *sprocket teeth number* N_S such that

$$H_R|_{US} = 0.004\left(\frac{N_S}{17}\right)^{1.08} N_S^{1.08}\, N_{rpm}^{0.9}\, p_C^{(3-0.07p_C)} \quad [\text{hp}], \quad p_C \text{ in } [\text{in}]$$

$$H_R|_{SI} = 0.746\left(0.004\left(\frac{N_S}{17}\right)^{1.08} N_S^{1.08}\, N_{rpm}^{0.9}\, p_C^{(3-0.07p_C)}\right) \quad [\text{kW}], \quad p_C \text{ in } [\text{in}] \tag{16.68}$$

The *chain speed factor* K_{CS} can be considered as a multiplier to the design power H_D as to adjust Eq. (16.64). Using approximate estimates, the chain speed factor K_{CS} can be given as follows.

$$K_{CS}|_{US} = 0.965\,39\; e^{0.002\,565 v_C}, \quad v_C \text{ in } [\text{ft/min}]$$

$$K_{CS}|_{SI} = 0.965\,39\; e^{0.504\,97 v_C}, \quad v_C \text{ in } [\text{m/s}] \tag{16.69}$$

where v_C is the longitudinal chain speed given by Eq. (16.54). The chain speed factor K_{CS} ranges from 1.0 for $v_C < 20$ [ft/min] or 0.1 [m/s] to about 1.6 for v_C about 200 [ft/min] or 1.0 [m/s]. Usually longitudinal chain speed may not go beyond that upper speed value.

The temperature effect is not pronounced for the service temperatures in the range of 10–300 [°F] or −10–150 [°C]. Lower or higher than that, the power rating of the chain deteriorates at a fast rate.

As noted, some of these factors affect the power rating and some affect the design power. For a critical design, these factors should adjust the calculations of the chain system to achieve the intended life without failure. Chain providers should be consulted to realize this objective.

16.5 Friction Drives

The opportunity of using different friction drives is pioneered by vehicle tires. The movement of the vehicle is dependent on the friction between the tires and the ground. If the ground is very slippery, the vehicle would not move, and the wheels would spin in-place as for the case of ice on the ground. The classical friction disc clutch moves the vehicle from standstill till the clutch speed is the same as the input rotational speed, see Chapter 18.

Other types of friction drives are close in form to the gearboxes. The driven side is like a drum or a disk. The driver side is a member such as a cylinder, a V-shaped disk, a bevel, or a spherical shaped member. The friction between the two members is the driving force limit. The contact between members can be external, internal, or on a frontal face. Intermediate members such as balls, rings, or disks may also exist to allow for more alternatives. Surfaces may be treated or made of different materials to increase friction coefficient and thus more power transmission. A sample friction drive can be constructed by a reasonably large thin disk with a central shaft and an O-ring housed in a perimetral groove. When the shaft is inserted in a hand drill chuck, the disk upon its rotation can drive another small shaft at very high speeds. Note that the power is limited by the O-ring tolerance of the friction force in between.

16.6 Flexible Shafts

Flexible shafts are power transmission elements when the two ends are not necessarily colinear or parallel. The shaft should then have as high a transverse and angular flexibility as needed along the length. A spring member in Figure 16.13 is having a very large number of coils and a small wire diameter that gives it a large transverse

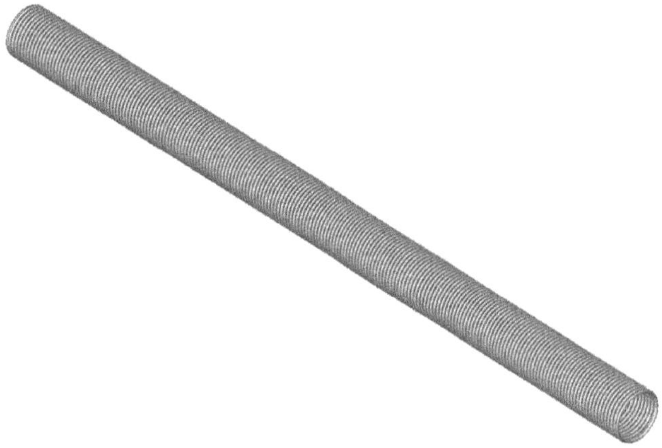

Figure 16.13 A spring member having a very large number of coils and a small wire diameter that gives it a large transverse flexibility but a torsional stiffness to act as a flexible shaft.

flexibility but has enough torsional stiffness to act as a flexible shaft. If one end is twisted, the torque is transmitted to the other end with limited torsional deflection, particularly if the internal diameter is constrained by a subject such as a wire rope with a rope diameter about the same inside diameter of the spring. If the spring wire diameter is much smaller than the coil diameter, the wire rope would also transmit power in a torsional mode. The bounding spring is then the constraint of the rope and the rope can transmit all the power, if the outer spring is stationary and formed to the needed bends. This outer spring can even be a flexible tube rather than a spring. The mechanics of the outer spring must be developed differently from the helical spring development in Chapter 10.

Some of the common applications of flexible shafts are the dental rotary tools, plumber's snakes, jewelry tools, remote control of instrument knobs, and *Bowden* cables for bicycle brakes or automotive applications, where axial forces are transmitted rather than torsional loads, Bowden (1898). Other constructions exist and can be viewed with patent office's such as the US patent office.

16.7 Computer-Aided Design and Optimization

In this chapter, the design procedures presented for V-belts, flat belts, wire ropes, and roller chains are amenable for computer implementation. Some of these are presented next and the others are left as exercises, see e.g. Shalaby et al. (2003).

16.7.1 V-belts Synthesis

MATLAB code is used to implement the synthesis procedure of V-belts as developed in Section 16.1.3. The initial synthesis begins by the utility of Figure 16.14 to synthesize the pulley diameter for SI and US systems as indicated in Section 16.1.3.1. The MATLAB code (*CAD_V_Belts.m*) in Figures 16.14–16.16 is then used to proceed with the synthesis of the V-belt system. The synthesis is mainly calculated for geometry and forces. Figure 16.14 is part 1 of the MATLAB code for the synthesis of V-belts. The main inputs needed to synthesize the pulley diameter for SI and US systems are defined with interactive assistance to specify pulley diameters. Figure 16.15 gives the center distance and belt length calculations to help in their input. In addition, wrap angle, belt type and rated power are defined for SI and US systems. Figure 16.16 gives the calculated number of belts, and belt forces. The full code is the combination of Figures 16.14–16.16. The input variables are highlighted by the bold **INPUT** comment

```
clear all; clc; format compact; format short                          % CAD_V_Belts.m
        disp('- V-belts Applied Power and Speed ')
H_kW = 8.9                                              % INPUT Power [kW]
H_hp = 11.93                                            % INPUT Power [hp]
Nrpm = 3000                                             % INPUT Rotational speed [rpm]
nV = 2                                                  % INPUT Speed Reduction Ratio
KSF = 1.3                                               % INPUT Safety and Service factor
Omega =2*pi*Nrpm/60
        disp('- Input Shaft')
Tin_Nm = H_kW*10^3/Omega , Tin_lbin = H_hp*6600/Omega          % Torque input [N.m], and [lb.in]
        disp('- Output Shaft')
Nrpm_out = Nrpm/nV, Omega_out = Omega/nV                       % Out rotational speed [rpm], and [rad/s]
Tout_Nm = Tin_Nm*nV, Tout_lbin = Tin_lbin*nV                   % Torque out [N.m], and [lb.in]
        disp('- V-Belt CALCULATED Pulley Diameters and Center Distance ')
        disp('- Assume V-Belt Speed is 4000 [Ft/min] or 20 [m/s] ')
vV_SI = 20                                              % INPUT V-Belt Speed 20 [m/s]
vV_US = 4000                                            % INPUT V-Belt Speed 4000 [ft/min]
d1_mm = vV_SI*2*1000/Omega,                                    % Pulley 1 diameter [mm]
d2_mm = (vV_SI*2*1000/Omega)*nV                                % Pulley 2 diameter [mm]
        disp('- Some Standard Pulley Diameters [mm] = ')
        disp(' 56,75,80,85,90,100,112,125,140,160,180,200,224,250,280,300,335,375,400,450,500 ')
d1_mm = 125                                             % INPUT Pulley 1 Diameter [mm]
d2_mm = d1_mm * nV
d1_in = (vV_US*2/Omega)/(60/12)                                % Pulley 1 diameter [in]
d2_in = ((vV_US*2/Omega)*nV)/(60/12)                           % Pulley 2 diameter [in]
        disp('- Some Standard Pulley Diameters [in] = 3, 3.2, 3.4, 3.6, 4, 4.5, 5, 5.5, 6, 7, 8, 9, 10,')
        disp(' 11, 12, 13, 15, 16, 18, 20, 25, 28, 30, 35, 0, 45, 50, 55, 60, 65, 70, 80, 90, 100, 110, 125, 150, 180, 200')
d1_in = 5                                               % INPUT Pulley 1 Diameter [in]
d2_in = d1_in * nV
```

Figure 16.14 MATLAB code for the synthesis of V-belts (part 1). The main inputs needed to synthesize the pulley diameter for SI and US systems.

```
                                      % CAD_V_Belts.m
        disp('- Calculated Center Distance and Suggested Range ')
CPmin_mm = d2_mm;                                         % Minimum Center Distance [mm]
CPmax_mm = 3*(d2_mm+d1_mm);                               % Maximum Center Distance [mm]
CPmin_in = d2_in ;                                        % Minimum Center Distance [in]
CPmax_in = 3*(d2_in+d1_in);                               % Maximum Center Distance [in]
CP_range_mm =[CPmin_mm CPmax_mm]; CP_range_in =[CPmin_in CPmax_in];
        disp('- Calculated, Suggested, and Input Center Distance ')
CP_mm =(CPmin_mm+CPmax_mm)/2                              % Average center distance [mm]
CP_in =(CPmin_in+CPmax_in)/2                              % Average center distance [in]
CP_mm = 690 , CP_in = 28                             % INPUT Center distance [mm], and [in]
        disp('- Calculated Belt Length and Input the Standard Length ')
Lp_mm = 2*CP_mm+pi*(d2_mm+d1_mm)/2+((d2_mm-d1_mm)^2)/(4*CP_mm)
Lp_in = 2*CP_in+pi*(d2_in+d1_in)/2+((d2_in-d1_in)^2)/(4*CP_in)
Lp_mm = 2000, Lp_in = 80                             % INPUT Belt Length [mm], and [in]
CP_mm =0.25*((Lp_mm-(pi/2)*(d2_mm+d1_mm))+sqrt((Lp_mm-(pi/2)*(d2_mm+d1_mm))^2 -2*(d2_mm-d1_mm)^2))
CP_in =0.25*((Lp_in-(pi/2)*(d2_in+d1_in))+sqrt((Lp_in-(pi/2)*(d2_in+d1_in))^2 -2*(d2_in-d1_in)^2))
        disp('- Calculated Belt Wrap Angles ')
theta1_SI = (pi - (d2_mm-d1_mm)/CP_mm)*(180/pi)           % Belt 1 Wrap Angle (SI)
theta2_SI = (pi + (d2_mm-d1_mm)/CP_mm)*(180/pi)           % Belt 2 Wrap Angle (SI)
theta1_US = (pi - (d2_in-d1_in)/CP_in)*(180/pi)           % Belt 1 Wrap Angle (US)
theta2_US = (pi + (d2_in-d1_in)/CP_in)*(180/pi)           % Belt 2 Wrap Angle (US)
        disp('- Estimating, Belt Section Type, Rated Power, and Belt Numbers ')
if Nrpm >3300, if Nrpm <=5000, disp('Belt type is A'), Belt_type = 1, end, end
if Nrpm >2000; if Nrpm <=3300, disp('Belt type is B'), Belt_type = 2, end, end
if Nrpm >1400; if Nrpm <=2000, disp('Belt type is C'), Belt_type = 3, end, end
if Nrpm >800; if Nrpm <=1400, disp('Belt type is D'), Belt_type = 4, end, end
if Nrpm <=800; disp('Belt type is E'), Belt_type = 5, end;
HV_kW = 0.00104* d1_mm^1.57780, HV_hp = 0.23011* d1_in^1.57780
```

Figure 16.15 MATLAB code for the synthesis of V-belts (part 2). Center distance, belt length, wrap angle, belt type and power calculated for SI and US systems.

```
                    % CAD_V_Belts.m
Belt_type = input('Input Belt Section Type (ANSI A=1,B=2,C=3,D=4,E=5) = '),     % INPUT Belt section-type
      if Belt_type ==1, aKL_US=0.372 , aKL_SI=0.1739 , bKL= 0.235, end
      if Belt_type ==2, aKL_US=0.3588 , aKL_SI=0.1714 , bKL= 0.2284, end
      if Belt_type ==3, aKL_US=0.3281 , aKL_SI=0.1592 , bKL= 0.2236, end
      if Belt_type ==4, aKL_US=0.2848 , aKL_SI=0.1363 , bKL= 0.2278, end
      if Belt_type ==5, aKL_US=0.2783 , aKL_SI=0.1345 , bKL= 0.2249, end;
KL_US = aKL_US*Lp_in^bKL
KL_SI = aKL_SI*Lp_mm^bKL
Ktheta1_US = 0.143543+0.007468*theta1_US - 0.000015052*theta1_US^2
Ktheta1_SI = 0.143543+0.007468*theta1_SI - 0.000015052*theta1_SI^2
NVB_SI = H_kW * KSF / (KL_SI*Ktheta1_SI* HV_kW)
NVB_US = H_hp * KSF / (KL_US*Ktheta1_US* HV_hp)
NVB_SI = round(NVB_SI+0.5)                                      % Number of V-belts -SI
NVB_US = round(NVB_US+0.5)                                      % Number of V-belts -US
      disp('- Estimating Belt Forces at V-belt Top Speed for SI ')
mu = 0.3                                    % INPUT Coefficient of friction (0.3, 0.5, to 0.8)
beta = 20 *(pi/180)                         % INPUT Belt angle = 2 beta [rad]
mV = 0.163                                  % INPUT Belt mass/length [kg/m]
muV =1/sin(beta)
Fc_N = mV*vV_SI^2/9.80665                                       % Centrifugal force [N]
expV = exp(mu*theta1_SI*(pi/180)/sin(beta))
F1_N = HV_kW*1000*1000/((1-1/expV)*d1_mm*Omega/2)               % 10^6 for [kW] & [mm]
F1_N = H_kW*KSF*1000*1000/(NVB_SI*(1-1/expV)*d1_mm*Omega/2) % Force for design power [N]
F2_N = F1_N/expV
Fi_N = (F1_N+F2_N)/2                                            % Initial tightening force [N]
HV_kW_Calc = 2*(F1_N-Fi_N)*vV_SI/1000                           % Calculated rated power/belt [kW]
Ft_N = F1_N                                                     % Tight side force [N]
Fs_N = F2_N                                                     % Slack side force [N]
```

Figure 16.16 MATLAB code for the synthesis of V-belts (part 3). Calculating number of belts, and belt system forces.

statement (% …) at each variable. The start inputs are the usual design need of transmitting power H (H_kW or H_hp) in [kW] or [hp], the input rotational speed of the small pulley N_{rpm} (Nrpm) in [rpm], and the V-belt speed reduction ratio n_V (nV). The other inputs are the safety and service factor K_{SF} (KSF), belt speed v_V (vV_SI and vV_US), and the pulley diameters d_1 and d_2 for SI and US (d1_mm and d1_in). These diameters are selected from a provided standard values. The expected range of pulley center distance C_p (CP_SI and CP_US), and the selected input pulley center distances are needed for SI and US systems. The expected output belt pitch lengths L_p (Lp_mm and Lp_in) are calculated, and the selected belt lengths are required inputs for SI and US systems. The corrected output center distances and wrap angles are then calculated. The output number of V-belts N_{VB} (NVB_SI and NVB_US), the expected rated power, and the belt forces are then obtained.

The MATLAB code in Figures 16.14–16.16 is available as one code in the ***Wiley website*** for *Computer-Aided Synthesis* or real *Computer-Aided Design* (*CAD*) of V-belts under the name of ***CAD_V_Belts.m***. This should be useful in calculating initial synthesis of V-belt systems and help in the understanding of their geometry and performance. The forces generated by the code can be used to design shafts and bearings needed for the construction of such belt systems.

Example 16.3 A V-belt system of Example 16.2 transmits a maximum power of 8.9 [kW] or 11.93 [hp] at an input rotational speed of 3000 [rpm] and a needed speed reduction ratio of 2.0. The common classical V-belts are to be employed. The system is also driven by a steady prime mover and is driving a heavy-duty load for about 12 [hr]. Synthesize the V-belt system using the MATLAB code (*CAD_V_Belts.m*) to define the belt section type, pulley diameters, and the number of belts. Also, determine the forces between the V-belt system in SI and US systems.

Solution

Data: $\boldsymbol{H}_{\text{kW}} = 8.9$ [kW] = 8900 [W] or $H_{\text{hp}} = 11.95$ [hp], $N_{\text{in,rpm}} = N_{\text{in}} = 3000$ [rpm], $\omega_1 = 2\pi\ N_{\text{in}}/60 = 314.16$ [rad/s]. The speed ratio $n_V = \omega_1/\omega_2 = 2$, and $\beta_V = 20°$. Service factor $K_{SF} = 1.3$ for steady prime mover driving a heavy-duty load 12 [hr], (Table 16.2).

Using the MATLAB code of (*CAD_V_Belts.m*), the following inputs and outputs are obtained with some minor presentation editing, see Figures 16.14–16.16.

Input:
H_kW = 8.9000, H_hp = 11.93, Nrpm = 3000, nV = 2, KSF = 1.3000

Output:
Omega = 314.1593,
- *Input Shaft*: Tin_Nm = 28.3296, Tin_lbin = 250.6308,
- *Output Shaft*: Nrpm_out = 1500, Omega_out = 157.0796, Tout_Nm = 56.6592, Tout_lbin = 501.2617
- *V-Belt Calculated Pulley Diameters and Center Distance,*
- *Assume V-Belt Speed* is 4000 [Ft/min] or 20 [m/s]: vV_SI = 20, vV_US = 4000
d1_mm = 127.3240, d2_mm = 254.6479
- *Some Standard Pulley Diameters* [mm] = 56,75,80,85,90,100,112,125,140,160,180,200,224,250,280, 300,335,375,400,450,500

Input: d1_mm = 125, d2_mm = 250,

Output:
d1_in = 5.0930, d2_in = 10.1859
- *Some Standard Pulley Diameters* [in] =3,3.2,3.4,3.6,4,4.5,5,5.5,6,7,8,9,10,11,12,13,15,16,18,20,25,28, 30,35,40,45,50,55,60,65,70,80,90,100,110,125,150,180,200

Input: d1_in = 5, d2_in = 10

Output:
- *Calculated Center Distance and Suggested Range*
- *Calculated, Suggested, and Input Center Distance*
CP_mm = 687.5000, CP_in = 27.5000

Input: CP_mm = 690, CP_in = 28

Output:
- *Calculated Belt Length and Input the Standard Length*
Lp_mm = 1.9747e+03, Lp_in = 79.7852

Input: Lp_mm = 2000, Lp_in = 80

Output:
CP_mm = 702.6962, CP_in = 28.1078
- *Calculated Belt Wrap Angles*
theta1_SI = 169.8079, theta2_SI = 190.1921, theta1_US = 169.8079, theta2_US = 190.1921
- *Estimating, Belt Section Type, Rated Power, and Belt Numbers*
Belt type is B, Belt_type = 2
HV_kW = 2.1161, HV_hp = 2.9159
aKL_US = 0.3588, aKL_SI = 0.1714, bKL = 0.2284, KL_US = 0.9762, KL_SI = 0.9727, Ktheta1_US = 0.9776, Ktheta1_SI = 0.9776, NVB_SI = 5.7497, NVB_US = 5.5733, NVB_SI = 6, NVB_US = 6
- *Estimating Belt Forces at V-belt Top Speed for SI*

Input: mu = 0.3000, beta = 0.3491, mV = 0.1630,

Output:
muV = 2.9238
Fc_N = 6.6485, expV = 13.4582,
F1R_N = 116.4233, F2_N = 8.6507, Fi_N = 62.5370, HV_kW_Calc = 2.1555, Ft_N = 116.4233, Fs_N = 8.6507
F1_N = 106.0923, F2_N = 7.8831, Fi_N = 56.9877, HV_kW_Calc = 1.9642, Ft_N = 106.0923, Fs_N = 7.8831

These values are verifying the results of Example 16.1 obtained by hand calculations for the rated power. The forces are recalculated from the design power. The differences are within design expectations. The MATLAB code can be used for parameter variations to get a better design or optimization for other objectives.

```
clear all; clc; format compact; format short                     % CAD_Wire_Ropes.m
         disp('- Wire Ropes Applied Loads, and Properties ')
FR_klb = 2.200 ,  FR_kN = 9.800                                   % INPUT Force on rope system [klb] and [kN]
ER_US = 12*10^6 , ER_SI = 83*10^9                                 % INPUT Rope elasticity modulus [psi] and [Pa]
LR_US = 590 , LR_SI = 180                                         % INPUT Rope length [ft] and [m]
KSF = 6.5                                                         % INPUT Safety and Service factor
dP_to_dR = 34                                                     % INPUT (Pulley diameter/Rope diameter) (6x19)
dR_to_dW = 13                                                     % INPUT (Rope diameter/Wire diameter) (6x19)
dP_to_dW = dP_to_dR*dR_to_dW , dW_to_dP =1/dP_to_dW
Fut_klb = FR_klb * KSF , Fut_kN = FR_kN * KSF                          % Ultimate strength force [klb] and [kN]
         disp('- Estimating, Rope Diameters and Rope Weight for Rope Numbers - US ')
for NR_US = 1:6 ;  Fb_klb = 0.0 ; KSL_US=1;
   dR_US = ((1/68.636)*((Fut_klb*KSL_US) - Fb_klb)/NR_US)^(1/1.9633);
 if dR_US <0.25, dR_US= 0.25;
   elseif dR_US <0.3125, dR_US= 0.3125; elseif dR_US <0.375, dR_US= 0.375; elseif dR_US <0.4375, dR_US= 0.4375;
   elseif dR_US <0.5, dR_US= 0.5; elseif dR_US <0.5625, dR_US= 0.5625; elseif dR_US <0.625,dR_US=0.625;
   elseif dR_US <0.75,dR_US=0.75; elseif dR_US <0.875,dR_US=0.875; elseif dR_US <1.0,dR_US=1.0;
   elseif dR_US <1.125,dR_US=1.125; elseif dR_US <1.25,dR_US=1.25; elseif dR_US <1.5,dR_US=1.5;
 end;
   dw_US = dR_US/16 ;                                                  % 6x19
   KSL_US =0.83957*dP_to_dR^0.03133;
   Fb_klb = ER_US*dw_US*0.4*dw_US^2/(1000*dP_to_dR*dw_US);
   dR_US_V(NR_US) = dR_US;
   wR_US = 1.5459*dR_US^2.0095*NR_US*LR_US/1000;                       % Rope weight [klb]
   wR_US_V(NR_US)= wR_US;
end
Maximum_Number_of_Ropes = NR_US                                        % Number of ropes considered
Wire_Rope_Diameters_in = dR_US_V                                       % Rope diameters vector [in]
Wire_Rope_Weight_klb =wR_US_V                                          % Rope weight vector [klb]
```

Figure 16.17 MATLAB code for the synthesis of wire ropes (part 1). The main inputs needed to synthesize the rope diameter for SI and US systems are shown.

16.7.2 Wire Rope Synthesis

MATLAB code is used to implement the synthesis procedure of wire ropes as developed in Section 16.3.2. The initial synthesis is started by the utility of Figure 16.17 to synthesize the rope diameter for SI and US systems as indicated in Section 16.3.2.1. The MATLAB code (***CAD_Wire_Ropes.m***) in Figures 16.17–16.19 is then used to proceed with the synthesis of the wire rope system. The synthesis is mainly calculated for geometry, forces, and expected strength of a rope type such as (6 × 19). For other types, some entries identified by code comments for (6 × 19) should be adjusted.

Figure 16.17 presents the MATLAB code for the synthesis of wire ropes. The main inputs needed to synthesize the rope diameter for SI and US systems are shown by the bold **INPUT** comment statement (% …) at each variable. Figure 16.18 is the second part of the MATLAB code. The main inputs are the rope diameter and the number of ropes needed for SI and US systems. These are optimally selected from the calculated array of the 1–6 ropes and their weights displayed for US system as row vectors shown at the ending of Figure 16.17 and at the middle of Figure 16.18 for the SI system. Figure 16.19 is the last part of the MATLAB code for the synthesis of wire ropes. The main synthesize outputs are the tensile stress, bearing pressure, and ultimate strength for SI and US systems.

Example 16.4 A wire rope system is to carry a maximum load of 2.0 [klb] or 8.9 [kN]. The wire rope length of 590 [ft] or 180 [m] is used in a mine hoist. Synthesize the wire rope system using the MATLAB code (*CAD_Wire_Ropes.m*) to define the wire rope diameter, number of ropes, pulley, or drum diameter, bearing pressure, and the expected ultimate strength of wire material.

Solution

Data: $F_R = 2.0$ [klb] or 8.9 [kN], $L_R = 590$ [ft] or 180 [m]. Safety factor $K_{SF} = 6.5$ for average mine hoist applications (Table 16.7).

Using the MATLAB code of (*CAD_Wire_Ropes.m*), the following inputs and outputs are obtained with some minor presentation editing, see Figures 16.17–16.19.

```
% CAD_Wire_Ropes.m
        % NR_US_Input = input('Input Number of Ropes (US) = ');
NR_US_Input = 3                                          % INPUT Number of Ropes (US)
dR_US = 0.25                                             % INPUT Rope diameter [in]
dP_US = dR_US*dP_to_dR                                   % Pulley or drum diameter [in]
        disp('- Estimating, Rope Diameters and Rope Weight for Rope Numbers - SI ')
for NR_SI = 1:6
  Fb_klb = 0.0 ; KSL_US=1;
  dR_SI = ((1/0.5369)*Fut_kN/NR_SI)^(1/1.9633);
if dR_SI <6.5, dR_SI= 6.5;
  elseif dR_SI <8, dR_SI= 8; elseif dR_SI <9.5, dR_SI= 9.5; elseif dR_SI <11.5, dR_SI= 11.5;
  elseif dR_SI <13,dR_SI=13; elseif dR_SI <14.5, dR_SI= 14.5; elseif dR_SI <16,dR_SI=16;
  elseif dR_SI <19,dR_SI=19; elseif dR_SI <22,dR_SI=22; elseif dR_SI <26,dR_SI=26;
  elseif dR_SI <29, dR_SI= 29; elseif dR_SI <32,dR_SI=32; elseif dR_SI <38,dR_SI=38;
end;
  dR_SI_V(NR_SI) = dR_SI;
  wR_SI = 0.033907*dR_SI^2.0095*NR_SI*LR_SI/1000;              % Rope weight [kN]
  wR_SI_V(NR_SI)= wR_SI;
end
Maximum_Number_of_Ropes = NR_SI                          % Number of ropes considered
Wire_Rope_Diameters_mm = dR_SI_V                         % Rope diameters vector [mm]
Wire_Rope_Weight_kN =wR_SI_V                             % Rope weight vector [kN]
        % NR_SI_Input = input('Input Number of Ropes (SI) = ');
NR_SI_Input = 4                                          % INPUT Number of Ropes (SI)
dR_SI = 6.5                                              % INPUT Rope diameter [mm]
dP_SI = dR_SI*dP_to_dR                                   % Pulley or drum diameter [mm]
```

Figure 16.18 MATLAB code for the synthesis of wire ropes (part 2). The main inputs are the needed rope diameter and number of ropes for SI and US systems.

```
% CAD_Wire_Ropes.m
        disp('- Estimating Normal Stresses ')
AMR_US = 0.4*dR_US^2                                     % Metal Area of Rope [in^2]
AMR_SI = 0.4*dR_SI^2                                     % Metal Area of Rope [mm^2]
sigma_t_kpsi = (FR_klb/AMR_US)/NR_US_Input               % Tensile stress [kpsi]
sigma_t_MPa = ((FR_kN*1000)/ AMR_SI)/NR_SI_Input         % Tensile stress [MPa]
        disp('- Estimating, Bending Moments, and Fatigue Stresses ')
sigma_b_kpsi = ER_US*dW_to_dP / 10^3                     % [kpsi]
sigma_b_MPa = ER_SI*dW_to_dP / 10^6                      % [MPa]
FRb_klb = sigma_b_kpsi*AMR_US                            % Equivalent bending load [klb]
FRb_kN = sigma_b_MPa*AMR_SI /10^3                        % Equivalent bending load [kN]
        disp('- Estimating Bearing Pressure for Lower Fatigue Expectation ')
pP_kpsi = (2*FR_klb/(dR_US*dP_US))/NR_US_Input           % Bearing pressure [kpsi]
pP_MPa = (2*FR_kN*1000/(dR_SI*dP_SI))/NR_SI_Input        % Bearing pressure [MPa]

        disp('- Estimating Ultimate Strength using Wire Rope and Bending Forces ')
sigma_ut_kpsi = (2000*(FR_klb+FRb_klb)/(dR_US*dP_US))/NR_US_Input     % Ultimate Strength [kpsi]
sigma_ut_MPa = (2000*(FR_kN+FRb_kN)*1000/(dR_SI*dP_SI))/NR_SI_Input   % Ultimate Strength [MPa]

        % Note that the bending force is used in place of the tensile force
        disp('- Estimating Ultimate Strength using Bending Force Only ')
sigma_ut_kpsi = (2000*FRb_klb/(dR_US*dP_US))/NR_US_Input          % Ultimate Strength [kpsi]
sigma_ut_MPa = (2000*FRb_kN*1000/(dR_SI*dP_SI))/NR_SI_Input       % Ultimate Strength [MPa]
```

Figure 16.19 MATLAB code for the synthesis of wire ropes (part 3). The main synthesize output are the tensile stress, bearing pressure, and ultimate strength for SI and US systems.

Input:

FR_klb = 2, FR_kN = 8.9000, ER_US = 12000000, ER_SI = 8.3000e+10, LR_US = 590, LR_SI = 180, KSF = 6.5000, dP_to_dR = 34, dR_to_dW = 13

Output:

dP_to_dW = 442, dW_to_dP = 0.0023

Fut_klb = 13, Fut_kN = 57.8500

- Estimating, Rope Diameters and Rope Weight for Rope Numbers - US

Maximum_Number_of_Ropes = 6
Wire_Rope_Diameters_in = [0.4375 0.3125 0.2500 0.2500 0.2500 0.2500]
Wire_Rope_Weight_klb = [0.1732 0.1762 0.1688 0.2250 0.2813 0.3376]
Input:
NR_US_Input = 3, dR_US = 0.2500,
Output:
dP_US = 8.5000
- *Estimating, Rope Diameters and Rope Weight for Rope Numbers – SI*
Maximum_Number_of_Ropes = 6
Wire_Rope_Diameters_mm = [11.500 8.0000 6.5000 6.5000 6.5000 6.5000]
Wire_Rope_Weight_kN = [0.8261 0.7968 0.7875 1.0500 1.3124 1.5749]
Input:
NR_SI_Input = 3, dR_SI = 6.5000,
Output:
dP_SI = 221
- *Estimating Normal Stresses*
AMR_US = 0.0250, AMR_SI = 16.9000
sigma_t_kpsi = 26.6667, sigma_t_MPa = 175.5424
- *Estimating, Bending Moments, and Fatigue Stresses*
sigma_b_kpsi = 27.1493, sigma_b_MPa = 187.7828
FRb_klb = 0.6787, FRb_kN = 3.1735
- *Estimating Bearing Pressure for Lower Fatigue Expectation*
pP_kpsi = 0.6275, pP_MPa = 4.1304
- *Estimating Ultimate Strength using Wire Rope and Bending Force*
sigma_ut_kpsi = 840.3868, sigma_ut_MPa = 5.6032e+03
- Estimating Ultimate Strength using Bending Force Only
sigma_ut_kpsi = 212.9359, sigma_ut_MPa = 1.4728e+03

These values must be verified by hand calculations and using other sample applications in other references or manufacturer's catalogues. The MATLAB code, however, can be used for parameter variations to get a better design or optimization for other objectives.

16.7.3 Roller Chains Synthesis

MATLAB code is used to implement the synthesis procedure of roller chains as developed in Section 16.4.4. The initial synthesis is started by the utility of Figure 16.20 to synthesize the chain system including chain size, number of chain strands, and sprocket diameters for SI and US systems as indicated in Section 16.3.2.1. The MATLAB code (***CAD_Chains.m***) in Figures 16.20–16.22 is then used to proceed with the synthesis of the roller chains system. The synthesis is mainly calculated for geometry and expected power rating of a roller chains. For other consideration of ultimate breaking loads, similar procedure can be implemented.

Figure 16.20 presents the MATLAB code for the synthesis of roller chains (part 1). The main synthesize needed to synthesize the chain size or number for SI and US systems are shown by the bold **INPUT** comment statement (% …) at each variable. Figure 16.21 is the second part of the MATLAB code. The main syntheses outputs are the single chain size and sprocket diameters for US and SI units. This is the option for the optimum rotational speed and single strand chain. Figure 16.22 is the last part of the MATLAB code for the synthesis of roller chains. The main syntheses outputs are the chain size, the number of chain strands, and the sprocket diameters for SI and US systems.

```
clear all; clc; format compact; format short                        % CAD_Chains.m
            disp('- Chain Applied Loads, and Properties ')
H_hp = 90                                                  % INPUT Power [hp]
H_kW = 67                                                  % INPUT Power [kW]
Nrpm = 300                                                 % INPUT Rotational speed [rpm]
nR = 2                                                     % INPUT Speed Reduction Ratio
KSF = 1.5                                                  % INPUT Safety factor
KASF = 1.3                                                 % INPUT Application Service factor
NS1 = 19                                                   % INPUT Small sprocket teeth number
% NS1 = 17                                                 % INPUT Small sprocket teeth number
NS2 = NS1 * nR                                                   % Large sprocket teeth number
            disp('- Input Shaft')
Omega =2*pi*Nrpm/60                                              % Rotational speed of input [rad]
Tin_Nm = H_kW*10^3/Omega                                         % Torque input [N.m]
Tin_lbin = H_hp*6600/Omega                                       % Torque input [lb.in]
            disp('- Output Shaft')
Nrpm_out = Nrpm/nR                                               % Out rotational speed [rpm]
Omega_out = Omega/nR                                             % Out rotational speed [rad/s]
Tout_Nm = Tin_Nm*nR                                              % Torque out [N.m]
Tout_lbin = Tin_lbin*nR                                          % Torque out [lb.in]
            disp('- Chain Design Power ')
HD_hp = H_hp * KSF * KASF                                        % Design power [hp]
HD_kW = H_kW * KSF * KASF                                        % Design power [kW]
```

Figure 16.20 MATLAB code for the synthesis of roller chains (part 1). The main inputs needed to synthesize the chain size or number for SI and US systems are shown.

```
                              % CAD_Chains.m
            disp('- Chain Number Calculation (ANSI) for Optimum Rotational Speed and Power')
C_No = round((Nrpm/(1.6842*10^5))^(-1/1.1543))                      % Not a function of Strand Number
if C_No <= 25, pc = 0.375; C_No_S =25; end                          % Chain number or size
if C_No > 25,if C_No <= 35 end, pc = 0.375; C_No_S =35; end
if C_No > 35, if C_No <= 40 end, pc = 0.5; C_No_S =40; end
if C_No > 40, if C_No <= 50 end, pc = 0.625; C_No_S =50; end
if C_No > 50, if C_No <= 60 end, pc = 0.75; C_No_S =60; end
if C_No > 60, if C_No <= 80 end, pc = 1.0; C_No_S =80; end
if C_No > 80, if C_No <= 100 end, pc = 1.25; C_No_S =100; end
if C_No > 100, if C_No <= 120 end, pc = 1.5; C_No_S =120; end
if C_No > 120, if C_No <= 140 end, pc = 1.75; C_No_S =140; end
if C_No > 140, if C_No <= 160 end, pc = 2.0; C_No_S =160; end
if C_No > 160, if C_No <= 180 end, pc = 2.25; C_No_S =180; end
if C_No > 180, if C_No <= 200 end, pc = 2.5; C_No_S =200; end
if C_No > 200, if C_No <= 240 end, pc = 3.0; C_No_S =240; end
Chain_No_ANSI = C_No_S                                              % Chain size (ANSI)
Chain_Pitch_in = pc , Chain_Pitch_mm = pc*25.4                      % Chain pitch [in] and [mm]
HR_hp =0.004*NS1^1.08*Nrpm^0.9*pc^(3-0.07*pc), HR_kW=0.746*HR_hp    % Rated power [hp] and [kW]
NCS = round((HD_hp / HR_hp)+0.5)                                    % Number of chain strands
KHR = 0.97625*NCS^0.86212;                                          % For strand # higher than 1
if NCS == 1, KHR = 1; end
HRT_hp = HR_hp*KHR                                                  % Total power rating [hp]
            disp('- Check Chain Power Calculation (ANSI) and the Extra Safety ')
HR_hp = 0.004*NS1^1.08*Nrpm^0.9*pc^(3-0.07*pc)
KSF_E = HRT_hp/ (HD_hp)                                             % Extra safety factor
            disp('- Basic Chain and Sprocket Geometry (ANSI) ')
dS1_in = pc/sin(pi/NS1) , dS1_mm = dS1_in * 25.4                    % Small sprocket diameter [in] and [mm]
dS2_in = pc/sin(pi/NS2) , dS2_mm = dS2_in * 25.4                    % Large sprocket diameter [in] and [mm]
```

Figure 16.21 MATLAB code for the synthesis of wire ropes (part 2). The main syntheses outputs are the single chain size and sprocket diameters for US and SI units.

Example 16.5 A roller chain system is to transmit a maximum power of 90 [hp] or 67 [kW] at an input rotational speed of 300 [rpm] and a needed speed reduction ratio of 2.0. The system is driven by a steady prime mover and is driving a heavy-duty load for about 12 hours. The environment is harsh, and the design safety factor ought to be not less than 1.5. Synthesize the roller chain system using the MATLAB code (*CAD_Chains.m*) to define

```
% CAD_Chains.m
        disp('- Chain Number Calculation (ANSI), for Nonoptimal Speed or Power')
Nrpm_C_US = (HD_hp/(8.8586*10^6))^(-1/1.6852)
C_No = round((Nrpm_C_US/(1.6842*10^5))^(-1/1.1543))
if C_No <= 25, pc = 0.375; C_No_S =25; end
if C_No > 25,if C_No <= 35 end, pc = 0.375; C_No_S =35; end
if C_No > 35, if C_No <= 40 end, pc = 0.5; C_No_S =40; end
if C_No > 40, if C_No <= 50 end, pc = 0.625; C_No_S =50; end
if C_No > 50, if C_No <= 60 end, pc = 0.75; C_No_S =60; end
if C_No > 60, if C_No <= 80 end, pc = 1.0; C_No_S =80; end
if C_No > 80, if C_No <= 100 end, pc = 1.25; C_No_S =100; end
if C_No > 100, if C_No <= 120 end, pc = 1.5; C_No_S =120; end
if C_No > 120, if C_No <= 140 end, pc = 1.75; C_No_S =140; end
if C_No > 140, if C_No <= 160 end, pc = 2.0; C_No_S =160; end
if C_No > 160, if C_No <= 180 end, pc = 2.25; C_No_S =180; end
if C_No > 180, if C_No <= 200 end, pc = 2.5; C_No_S =200; end
if C_No > 200, if C_No <= 240 end, pc = 3.0; C_No_S =240; end
Chain_No_ANSI = C_No_S                                          % Chain size (ANSI)
Chain_Pitch_in = pc , Chain_Pitch_mm = pc*25.4                  % Chain pitch [in] and [mm]
HR_hp=0.004*NS1^1.08*Nrpm^0.9*pc^(3-0.07*pc), HR_kW=0.746*HR_hp % Rated power [hp] and [kW]
NCS = round((HD_hp / HR_hp)+0.5)                                % Number of chain strands
KHR = 0.97625*NCS^0.86212                                       % For strand # higher than 1
HRT_hp = HR_hp*KHR                                              % Total power rating [hp]
        disp('- Check Chain Power Calculation (ANSI) and the Extra Safety ')
HR_hp = 0.004*NS1^1.08*Nrpm^0.9*pc^(3-0.07*pc)                  % Rated power [hp]
KSF_E = HRT_hp/ (HD_hp)                                         % Extra safety factor
        disp('- Basic Chain and Sprocket Geometry (ANSI) ')
dS1_in = pc/sin(pi/NS1), dS1_mm = dS1_in * 25.4                 % Small sprocket diameter [in] and [mm]
dS2_in = pc/sin(pi/NS2), dS2_mm = dS2_in * 25.4                 % Large sprocket diameter [in] and [mm]
```

Figure 16.22 MATLAB code for the synthesis of wire ropes (part 3). The main syntheses outputs are the chain size, the number of strands, and sprocket diameters for SI and US units.

the roller chain size or number, sprocket diameters, number of roller chain strands, the expected total projected power rating and safety factor.

Solution

Data: $H_{hp} = 90$ [hp], or $H_{kW} = 67$ [kW], $N_{rpm} = 300$ [rpm], $\omega_1 = 2\pi\ N_{in}/60 = 31.416$ [rad/s]. The speed ratio $n_C = \omega_1/\omega_2 = 2$. Service factor $K_{SF} = 1.3$ for steady prime mover driving a heavy-duty load 12 [hr], (Table 16.2).

Using the MATLAB code of (*CAD_Chains.m*), the following inputs and outputs are obtained with some minor presentation editing, see Figures 16.20–16.22.

Input:

- *Roller Chain Applied Loads, and Properties*: H_hp = 90, H_kW = 67, Nrpm = 300, nR = 2, KSF = 1.5000, KASF = 1.3000, NS1 = 19, NS2 = 38

Output:

- *Input Shaft*: Omega = 31.4159, Tin_Nm = 2.1327e+03, Tin_lbin = 1.8908e+04
- *Output Shaft*: Nrpm_out = 150, Omega_out = 15.7080, Tout_Nm = 4.2654e+03, Tout_lbin = 3.7815e+04
- Roller Chain Design Power: HD_hp = 175.5000, HD_kW = 130.6500
- *Chain Number Calculation (ANSI) for Optimum Rotational Speed and Power:*

C_No = 241, Chain_No_ANSI = 240, Chain_Pitch_in = 3, Chain_Pitch_mm = 76.2000
HR_hp = 349.6965, HR_kW = 260.8736, NCS = 1
HRT_hp = 349.6965,

- *Check Chain Power Calculation (ANSI) and the Extra Safety:* HR_hp = 349.6965, KSF_E = 1.9926
- *Basic Chain and Sprocket Geometry* (ANSI)

dS1_in = 18.2266, dS1_mm = 462.9557, dS2_in = 36.3287, dS2_mm = 922.7489

Output 2:

- *Chain Number Calculation (ANSI), for Nonoptimal Speed or Power*

Nrpm_C_US = 617.7587, C_No = 129,
Chain_No_ANSI = 140, Chain_Pitch_in = 1.7500, Chain_Pitch_mm = 44.4500
HR_hp = 81.6329, HR_kW = 60.8981, NCS = 3,
KHR = 2.5171, HRT_hp = 205.4763
- *Check Chain Power Calculation (ANSI) and the Extra Safety*: HR_hp = 81.6329
KSF_E = 1.1708,
- *Basic Chain and Sprocket Geometry* (ANSI)
dS1_in = 10.6322, dS1_mm = 270.0575, dS2_in = 21.1917, dS2_mm = 538.2702

These values must be verified by hand calculations and using other sample applications in other references or manufacturer's catalogues. It should be noted that the first solution is a large ANSI 240 one strand chain with large sprockets and the second option is a small ANSI 140 three strands chain with smaller 3 strands sprockets. The MATLAB code, however, can be used for parameter variations to get a better design or optimization for other objectives.

16.8 Summary

In this chapter, V-belts, flat belts, wire ropes, and chains are presented in addition to a preview about friction drives and flexible shafts. The geometry, definitions, drive relations, and common types and standards are introduced in the chapter. Forces generated during operation and power transmission are identified. These forces are acting on the shafts, which these flexible elements are mounted. The effect of developing internal stresses affecting design is defined. The relations defining the ultimate strength and fatigue are considered. Material set usually used are reported and utilized. Developed synthesis procedures attracting *CAD* and optimization tools and their findings are exploited to generate better designs.

The computer-aided synthesis process is provided in a real *CAD* codes using MATLAB. These are available as ***CAD_V_Belts.m***, ***CAD_Wire_Ropes.m***, and ***CAD_Chains.m*** through **Wiley website**. The utility of these SI and US codes has been demonstrated to effectively synthesize V-belts, wire ropes, and chains near the operational optimum or the available feasible solutions. Little iteration may be needed to tune the synthesized design to satisfy multitude of conditions, cases, and possible optimum objectives. If need be, the synthesized and iterated parameters can be construed as good values to be checked, verified, and validated by standard codes and adjusted to other conditions, loadings and manufacturing processes that have not been included herein.

Problems

16.1 Identify few applications or products that utilize V-belts, flat belts, wire ropes, or roller chains. Use the net to collect images of such applications. Why these applications are not using spur gears instead?

16.2 Define other V-belt types than the ones covered in the chapter. Sketch those other types and indicate their utility.

16.3 Download a standard about V-belt and study the differences between the material presented in Section 16.1 and that indicated in the standard.

16.4 Download a standard about timing belts and study the procedure for design process and selection. Find information from some manufacturer of timing belts about standard dimensions, and design procedure and selection. Are there differences between the standards and the manufacturer's material and design procedure or selection?

16.5 What are the differences between the classical V-belts (A–E) and the narrow V-belts (3V, 5V, and 8V) in terms of utility and power rating?

16.6 Are cogged V-belts having more strength, more life, and transmit more power, or are they the same? Compare values from a manufacturer catalogue.

16.7 Download a manufacturer catalogue for belts, V-belts or other belts. Use the Internet section of this chapter for some of available manufacturers or use the internet to define other available manufacturers.

16.8 Find the metric standard for the timing belts similar or equivalent to the US standard.

16.9 Check the effect of reducing the wrapping angle on the power transmission if the assumption of no sliding zones is not holding which reduces the wrapping angle.

16.10 It is suggested that the V-belt speed v_V of 4000 [ft/min] or 20 [m/s] is advised. From other information you can get, is that advised speed an optimum one or in some cases it should be higher or lower? Specify the cases if any and justify the results.

16.11 From a manufacturer catalogue, identify the difference in the set of the pulley pitch diameters for US and SI systems for a different belt type such as A–E.

16.12 Is the standard belt lengths available for belt types existing or more options are available as well?

16.13 Check some manufacturers catalogues for the way they quote the belt length either as the pitch or the inside circumference length. Do they give both?

16.14 From Figure 16.2a, derive Eq. (16.17).

16.15 Derive Eq. (16.18) from Eq. (16.17).

16.16 Why is the operational range of belt speed usually 1000–5000 [ft/min] or 5–25 [m/s]? What happens if the operational belt speed is lower than 1000 [ft/min] or 5 [m/s] or higher than 5000 [ft/min] or 25 [m/s]?

16.17 Redo Example16.1 for a soft start and light duty driven machine. What would be the design for heavy prime mover starts and extreme duty for the driven machine.

16.18 A V-belt system transmits a maximum power of 20 [kW] or 27 [hp] at an input rotational speed of 850 [rpm] and a needed speed reduction ratio of 2.0. The common classical V-belts are to be employed. The system is driven by a steady prime mover and is driving a heavy-duty load for about 12 hours. Synthesize the V-belt system defining the belt section type, pulley diameters, and the number of belts. Determine the forces between the V-belt system in SI or US systems.

16.19 A V-belt system transmits a maximum power of 1.0 [kW] or 1.34 [hp] at an input rotational speed of 3600 [rpm] and a needed speed reduction ratio of 2.0. The common classical V-belts are to be employed.

The system is driven by a steady prime mover and is driving a heavy-duty load for about 12 hours. Synthesize the V-belt system defining the belt section type, pulley diameters, and the number of belts. Determine the forces between the V-belt system in SI or US systems.

16.20 Find a construction details for the design of a bolting slots on one side of the pulley system or any other mean to accomplish the adjustment of the center distances between the two V-belt pulleys to generate the needed initial tightening force.

16.21 A V-belt system transmits a maximum power of 20 [kW] or 27 [hp] at an input rotational speed of 3000 [rpm] and a needed speed reduction ratio of 2.0. The common classical V-belts are to be employed. The system is driven by a steady prime mover and is driving a heavy-duty load for about 12 hours. Synthesize the V-belt system defining the belt section type, pulley diameters, and the number of belts. Determine the forces between the V-belt system in SI or US systems.

16.22 A V-belt system transmits a maximum power of 1.0 [kW] or 1.34 [hp] at an input rotational speed of 850 [rpm] and a needed speed reduction ratio of 2.0. The common classical V-belts are to be employed. The system is driven by a steady prime mover and is driving a heavy-duty load for about 12 hours. Synthesize the V-belt system defining the belt section type, pulley diameters, and the number of belts. Determine the forces between the V-belt system in SI or US systems.

16.23 Search for other flat belt configurations than the ones shown in Figure 16.6. Consider conveyor belts for news paper production. Sketch the different construction configurations and define their operations. Observe the objectives of idler and reversing pulleys.

16.24 Resolve Example 16.1 using flat belt instead of V-belts. Select rubber cotton duck in Table 16.5 and the procedure in Section 16.2.3. Compare results with the V-belts.

16.25 Use a similar relation to Eq. (16.8), substituting Eq. (16.33) into Eq. (16.8) and performing minor manipulation to get an expression defining the force $\boldsymbol{F}_1$ on the tight side of the belt as given by Eq. (16.34).

16.26 Find the property values quoted in Table 16.5 for available belt materials in some manufacturer's catalogues. Compare values and suggest the reasons for the differences.

16.27 Design a flat belt system to transmit a 15 [hp] or 11 [kW] from a steady source running at 1750 [rpm] to a light shock driven machine to run at about 583 [rpm]. The distance between the two pulleys is 96 [in] or 2440 [mm]. Use a polyamide material with thickness of 0.13 [in] or 3.3 [mm] and find other properties from Table 16.5. Apply the design methodology of Section 16.2.3 and compare results with similar material in any available manufacturer's catalogue.

16.28 For wire ropes, find the strength reduction ratio for the minimum pulley diameter of each rope lay-type given in Section 16.3.1.

16.29 Use the wire rope design procedure in Section 16.3.2 to synthesize a wire rope that can carry a maximum load of 2.0 [klb] or 8.9 [kN]. The wire rope length of 590 [ft] or 180 [m] is used in a mine hoist. Define the wire rope diameter, number of ropes, pulley, or drum diameter, bearing pressure, and the expected ultimate strength of wire material. Compare the results for different rope lay-types such as (6 × 7), (6 × 19,

6 × 37). What is the effect of increasing or decreasing the wire rope length on the selection of the rope lay-type?

16.30 Justify using the equivalent bending load only in the estimation of the needed ultimate tensile strength of the wire in Eq. (16.51) to achieve a close to an asymptotic endurance strength at 10^6 bends under the ultimate fatigue strength force.

16.31 Probe the different chain types different from the one commonly used chain in Section 16.4. Sketch the different types and suggest means to synthesize or select a suitable dimension under different loading cases, particularly for the closed-linked hoisting.

16.32 From Eq. (16.55), derive Eq. (16.56) and plot the variation of the chain velocity against the sprocket teeth number.

16.33 Derive Eq. (16.61) using a similar diagram to Figures 16.6a and 16.11.

16.34 Adjust the design procedure in Section 16.4.4 to account for the maximum or ultimate breaking load in addition to the power transmission.

16.35 Use the roller chain design procedure in Section 16.4.4 to synthesize a roller chain that can transmit a maximum power of 90 [hp] or 67 [kW] at an input rotational speed of 300 [rpm] and a needed speed reduction ratio of 2.0. The system is driven by a steady prime mover and is driving a heavy-duty load for about 12 hours. The environment is harsh, and the design safety factor ought to be not less than 1.5.

16.36 Adjust the MATLAB code of (*CAD_V_Belts.m*) to include any missing variables needed to complete the US system design of V-belts.

16.37 Use the MATLAB code of (*CAD_V_Belts.m*) to solve Problem 16.18 and compare results.

16.38 Use the MATLAB code of (*CAD_V_Belts.m*) to solve Problem 16.19 and compare results.

16.39 Use the MATLAB code of (*CAD_V_Belts.m*) to solve Problem 16.21 and compare results.

16.40 Use the MATLAB code of (*CAD_V_Belts.m*) to solve Problem 16.22 and compare results.

16.41 Adjust the MATLAB code of (*CAD_V_Belts.m*) to solve flat belts by assuming the different belt angle and any other needed relations. Check results by solving Problem 16.24 and compare results.

16.42 Use the adjusted MATLAB code of (*CAD_V_Belts.m*) to solve flat belts of Problem 16.27 and compare results.

16.43 Use the MATLAB code of (*CAD_Wire_Ropes.m*) to solve Problem 16.29 and compare results.

16.44 Use the MATLAB code of (*CAD_Wire_Ropes.m*) to resolve Problem 16.29 for a maximum load of 20 [klb] or 89 [kN] and wire rope length of 6000 [ft] or 1800 [m].

16.45 Use the MATLAB code of (*CAD_Wire_Ropes.m*) to resolve Problem 16.29 for a maximum load of 0.2 [klb] or 0.9 [kN] and wire rope length of 60 [ft] or 18 [m].

16.46 Use the MATLAB code of (*CAD_Chains.m*) to resolve Problem 16.35 for a maximum load of 9 [hp] or 6.7 [kW] at an input rotational speed of 1750 [rpm] and a needed speed reduction ratio of 3.0.

16.47 Use the MATLAB code of (*CAD_Chains.m*) to resolve Problem 16.35 for a maximum load of 0.9 [hp] or 0.67 [kW] at an input rotational speed of 60 [rpm] and a needed speed reduction ratio of 3.0. Note that the minimum rotational speed of the procedure is about 300 [rpm] and one should be careful to adjust variables to that effect.

16.48 Can you use the MATLAB code of (*CAD_Chains.m*) to design a suitable chain for a bicycle. Note that the sustained human power is about 0.25–0.3 [kW] or 0.3–0.4 [hp]. Estimate the pedaling speed of the driver and the speed reduction ratio between the pedal and the rear wheel. Consider using an electric motor of the same capacity to power the bicycle. Find different constructions to select from and give a ***3D*** model for your design.

References

ACA (2006). *Standard Handbook of Chains, Chains for Power Transmission and Material Handling*, 2e. American Chain Association: CRC Press, Taylor & Francis.

ANSI/ASME B29.1 (2011). *Precision Power Transmission Roller Chains, Attachments, and Sprockets.* American Society of Mechanical Engineers.

ANSI/RMA IP-20 (1988), (R2007). *Classical V-Belts and Sheaves - Specifications for Drives Using Classical V-Belts and Sheaves.* Rubber Manufacturers Association.

ANSI/RMA IP-21 (1991). *Specifications for Drives Using Double-V (Hexagonal) Belts - Part I: Metric (SI) Dimensions (13D, 16D, and 22D Cross Sections) - (AA, BB, and CC Cross Sections).* Rubber Manufacturers Association.

ANSI/RMA IP-22 (1984), (R1991). *Specifications for Drives Using Narrow V-Belts and Sheaves: Part I Metric (SI) Dimensions (9N/9NX, 15N/15NX and 25N Cross Sections), Part II Inch-Pound Dimensions (3V/3VX, 5V/5VX and 8V Cross Sections).* Rubber Manufacturers Association.

ANSI/RMA IP-23 (1968). *Light Duty V-Belt.* Rubber Manufacturers Association.

ANSI/RMA IP-24 (1983). *Synchronous Belts.* Rubber Manufacturers Association.

ANSI/RMA IP-25 (1982), (R1991). *Specifications for Drives Using Variable Speed V-Belts (12 Cross Sections).* Rubber Manufacturers Association.

ANSI/RMA IP-26 (1977). *V-Ribbed Belts.* Rubber Manufacturers Association.

ASTM A1023/A1023M (2009). *Standard Specification for Stranded Carbon Steel Wire Ropes for General Purposes.* American Society for Testing and Materials - ASTM International.

Bowden, E.M. (1898). *Mechanism for transmitting motion or power* US Patent No. 609,570.

BSR/ASME B29.200 (1995). *Welded Steel Type Mill Chains, Welded Steel Drag Chains, Attachments and Sprocket Teeth (revision, redesignation and consolidation of ANSI/ASME B29.16M-1995, and ANSI/ASME B29.18M-1993).* American Society of Mechanical Engineers.

Budynas, R.G. and Nisbett, J.K. (2015). *Shigley's Mechanical Engineering Design*, 10e. McGraw Hill.

Hussein, M.M., Hussein, B.A., and Metwalli, S.M. (2013). Optimization for slip analysis of belt drives with finite element verification. *Proceedings of the ASME International Mechanical Engineering Congress & Exposition IMECE 2013*, 15–21 November, San Diego, California, USA, paper # IMECE2013-62813.

Hussein, M.M., Metwalli, S.M., Mohamed, A.A., and Shabana, A.A. (2010). Belt drive mechanics with absolute nodal coordinate finite element verification. *Proceedings of the ASME 2010 International Mechanical Engineering Congress & Exposition IMECE 2010*, 12–18 November, Vancouver, British Colombia, Canada, IMECE2010-38449.

ISO 10190 (2008). *Motorcycle Chains - Characteristics and Test Methods*. International Organization for Standardization.

ISO 10823 (2004). *Guidelines for the Selection of Roller Chain Drives*. International Organization for Standardization.

ISO 14890 (2013), (R2018). *Conveyor Belts - Specification for Rubber- or Plastics-covered Conveyor Belts of Textile Construction for General Use*. International Organization for Standardization.

ISO 1604 (1989), (R2017). *Belt Drives - Endless Wide V-belts for Industrial Speed-changers and Groove Profiles for Corresponding Pulleys*. International Organization for Standardization.

ISO 17893 (2004). *Steel Wire Ropes - Vocabulary, Designation, and Classification*. International Organization for Standardization.

ISO 22 (1991), (R2015). *Belt Drives - Flat Transmission Belts and Corresponding Pulleys - Dimensions and Tolerances*. International Organization for Standardization.

ISO 2408 (2017). *Steel Wire Ropes - Requirements*. International Organization for Standardization.

ISO 3108 (2017). *Steel Wire Ropes for General Purposes - Determination of Actual Breaking Load*. International Organization for Standardization.

ISO 4184 (1992), (R2017). *Belt Drives - Classical and Narrow V-belts - Lengths in Datum System*. International Organization for Standardization.

ISO 487 (1998). *Steel Roller Chains, Types S and C, Attachments and Sprockets*. International Organization for Standardization.

ISO 5292 (1995), (R2015). *Belt Drives - V-belts and V-ribbed Belts - Calculation of Power Ratings*. International Organization for Standardization.

ISO 606 (2015). *Short-pitch Transmission Precision Roller and Bush Chains, Attachments and Associated Chain Sprockets*. International Organization for Standardization.

ISO 9633 (2001). *Cycle Chains - Characteristics and Test Methods*. International Organization for Standardization.

ISO 9982 (1998). *Belt Drives - Pulleys and V-ribbed Belts for Industrial Applications - PH, PJ, PK, PL and PM Profiles: Dimensions*. International Organization for Standardization.

Oberg, E., Jones, F.D., Horton, H.L., and Ryffel, H.H. (2012). *Machinery's Handbook, 29*e. Industrial Press.

S9086-UU-STM-010 (1995). *Wire and Fiber Rope and Rigging*. NSTM Chapter 613 Naval Sea Systems Command.

SAE J637 (2012). *Automotive V-Belt Drives*. Society of Automobile Engineers.

Shalaby, M.M., Hegazi, H.A., Nassef, A.O., Metwalli, S.M. (2003). Topology optimization of a compliant gripper using hybrid simulated annealing and direct search. *Proceedings of ASME 2003 DETC/CIE Conference (DETC'03)*, 2–6 September, Chicago, Illinois, Paper # DETC2003/DAC-48770.

Shigley, J.E. (1986). *Mechanical Engineering Design*, First Metric Edition. McGraw Hill.

TM 3-34.86 (2012). *Rigging Techniques, Procedures, and Applications*. US Department of The Army.

Internet Links

www.unirope.com – Unirope: Wire ropes, fiber ropes.
www.hitbelt.com – HIT: V-belts, ribbed, automotive, conveyor belts, timing, and others.
http://www.pixhose.com/-PIX: V-belts, ribbed, timing belts, hoses, and others.
www.ushamartin.com – Usha Martin: Wire ropes.
www.alliedlocke.com – Allied Locke: Chains, sprockets, and others.
www.baldor.com – ABB-Baldor: V-belts and others.
www.diamondchain.com – Diamond Chain: Roller chains.
www.fptgroup.com – Fenner: V-belts and others.

www.habasit.com – Habasit: Flat belts, timing, and others.
www.kettenwulf.com – KettenWulf: Special chains, sprockets, and others.
www.maskapulleys.com – Maska: Pulleys.
www.optibelt.com/us/home – Optibelt: V-belts, timing, other belts, and others.
www.renold.com – Renold: Chains, and others.
www.skf.com – SKF: Belts, Chains, sprockets, rolling bearings, and others.
www.unionrope.com – Union: Wire ropes.
www.ustsubaki.com – Tsubaki: Chains.
www.ansi.org – ANSI: Standards.
www.iso.org – ISO: Standards.
www.americanchainassn.org – ACA: American Chain Association.
www.asme.org – American Society of Mechanical Engineers: Standards.
www.steel.org – American Iron and Steel Institute: Standards.

17

Shafts

The shaft is one of the main power transmission elements, which connects the prime mover to other transmission elements and again retransmits power from these elements to the machine; see Chapter 13 and Figures 13.1 and 13.2. The shaft is then a primary component of power transmission elements. Any other power transmitting element is usually connected on either or both of its input and output ends by a shaft. In addition to transmitting power, shafts support rotating machine components and thus run in bearings. They provide a constant positioning of the axis of rotation of these components. They are, therefore, subjected to applied forces, moments, and torques and may be treated as rotating beams of bodies of revolution.

Symbols

The adapted units are [in, lb, psi] or [m, kg, N, Pa], others given at each symbol definition ([k…] is 10^3, [M…] is 10^6 and [G…] is 10^9).

Symbols	Quantity, units (adopted)
$(d_1, l_2'), (d_1, l_3')$	Equivalent zones on equivalent shaft
$[\boldsymbol{r}\times]$	Cross-product matrix of $\boldsymbol{r}$
A_S	Shaft cross-sectional area
d_1	Diameter of segment 1 of shaft
d_i	Diameter of segment i of shaft
$d_{p,\text{new}}$	New part diameter
d_S	Shaft diameter
E	Elasticity modulus of the material
$\boldsymbol{F}_1, \ldots, \boldsymbol{F}_n$	External n force vectors
$\boldsymbol{F}_a, \boldsymbol{M}_a$	Operating loads at point a
F_x, F_y, F_z	Components of the force vector $\boldsymbol{F}$
G	Shear modulus of elasticity or modulus of rigidity
g_a	Gravitational acceleration
H_{hp}	Transmitted power in [hp]

Machine Design with CAD and Optimization, First Edition. Sayed M. Metwalli.

Companion website: www.wiley.com/go/metwalli/machine

Symbols	Quantity, units (adopted)
H_{kW}	Transmitted power in [kW]
I_z	Second area moment of shaft cross section
J_i	Polar second area moment of shaft section i
K_{conc}	Related to stress concentration factor K_{SC}
K_{load}	Load factor
K_M	Moment factor
K_{miscel}	Miscellaneous factors
K_{relaib}	Reliability factor
k_{Sb}	Transverse shaft stiffness or rigidity
K_{SC}	Stress concentration factor
K_{SF}	Factor of safety
$K_{\mathrm{SF},F}$	Fatigue safety factor
K_{size}	Size factor
k_{St}	Total angular stiffness of the shaft
k_{Sti}	Angular stiffness of shaft segment i
K_{surf}	Surface factor
K_{temp}	Temperature factor
l	Length between bearings
l_i	Length of the ith segment of shaft
l_i'	Equivalent length of segment i relative to segment 1
l_S	Shaft length between bearings
$\boldsymbol{M}$	Bending moment
$\boldsymbol{M}_1, \ldots, \boldsymbol{M}_k$	External k moment vectors
$\boldsymbol{M}_a$	Bending moment at point a
m_i	Mass of body i attached to shaft
m_S	Mass per unit length of shaft
N_{rpm}	Rotational speed in [rpm]
$\boldsymbol{R}_1, \boldsymbol{M}_1$	Reaction vectors at support 1
$\boldsymbol{r}_1, \ldots \boldsymbol{r}_n$	Position vectors of external forces
$\boldsymbol{R}_1, \boldsymbol{R}_2$	Reaction vectors at supports
S_e	Endurance limit
$S_{e,\mathrm{part}}$	Fatigue strength of a machine part
$S_{e,S}$	Fatigue strength of the shaft
S_f	Fatigue strength
S_{yt}	Tensile yield strength of material
$\boldsymbol{T}$	Operating torque
$\boldsymbol{V}_a$	Shear force at point a
v_i	Transverse static deflection of body mass m_i
v_{max}	Maximum transverse deflection

Symbols	Quantity, units (adopted)
θ_{max}	Maximum slope
σ_a, τ_a	Alternating stresses
σ_m, τ_m	Mean stresses
σ_{max}	Maximum normal stress
σ_{min}	Minimum normal stress
$\sigma_{vM,m}, \sigma_{vM,a}$	Equivalent *von Mises* stresses
τ_{max}	Maximum shear stress
τ_{min}	Minimum shear stress
ρ	Density of shaft material
ω_n	Critical speed or fundamental natural frequency
ϕ_i	Angular deformation of segment i of shaft
ϕ_S	Total angular deformation of the shaft

17.1 Types of Shafts and Axles

Some different types of shafts and axles are displayed in Figure 17.1 as ***3D*** isometric views. Figure 17.1a shows a legacy railway axle where the two ends are recessed to house the two halves of each journal bearing. The two inner recesses are used to shrink fit the wheels. Figure 17.1b presents a stepped shaft with two keyways and several undercuts. The two keyways are used to connect two gears, or any other elements such as a pulley, a sprocket, a clutch, or a flywheel in a power transmission system. Figure 17.1c shows a machine tool spindle with a hefty right side to give sufficient rigidity needed for tight tolerance achievement. Figure 17.1d indicated a turbine shaft with many keyways to hold the turbine disks and two end keyways for connections to couplings or other components.

Other types of shafts and axles exist such as transmission shafts, crankshafts, and nonrotating axles. They have some different forms than those in Figure 17.1. The transmission shaft is usually unstepped cylindrical body to be

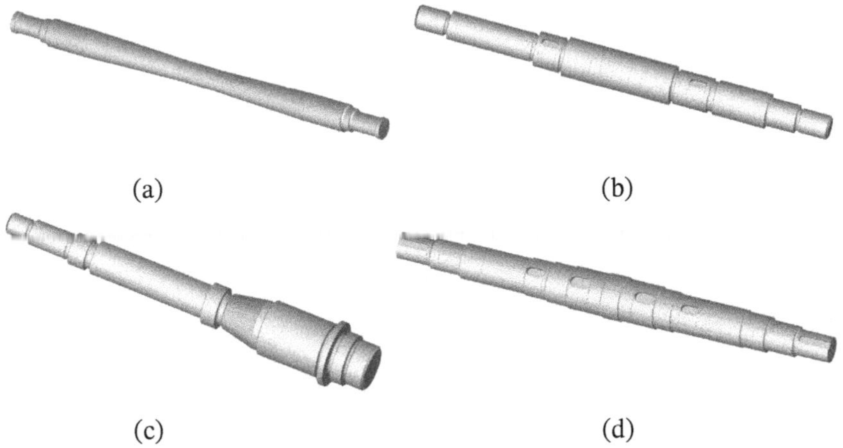

Figure 17.1 Isometric views of few different shafts: (a) legacy railway axle, (b) stepped shaft with two keyways, (c) machine tool spindle, and (d) turbine shaft with many keyways.

connected by special gripping couplings or other end preparations. The crankshaft has one or more offset sections to house the connecting rod or rods in addition to concentric sections to accommodate the axial or central bearings. Nonrotating axles are stationary beams that might have supports at some locations and certain rotating elements on other attached bearings.

17.2 Mathematical Model

A mathematical model is needed to synthesize the shaft. Usually, the shaft transmits power and might have more than one power transmission element. To evaluate external loads, internal loads, and stresses, select proper material, and define geometry, the mathematical modeling is necessary. Chapter 2 has a coverage of means to do that particularly Sections 2.2 and 2.3 and Figures 2.1–2.3. Chapter 6 discussed the mathematical models to simplify elements as beams with concentrated loads and reactions at approximate locations; see Section 6.2 and Figure 6.9. The simplified concentrated loads and reactions are conservative approximations to actual random variations. Chapter 8 presented many components connecting shafts to other elements in addition to keys and other features in shaft shaping to facilitate the connectivity of these components.

In this section, a simple mathematical model is adapted to initially synthesize the shafts. The treatment is mainly dependent on the *engineering solid mechanics* or *strength of materials* of Chapter 6 rather than the rigorous *theory* of *elasticity*. Improving the model representation of physical reality is achieved by load distribution rather than concentrated loadings or concentrated reactions at supports. This is doable if an efficient computer-aided program is utilized for the initial synthesis. After the initial synthesis, numerical finite element (***FE***) models can be used to verify or modify the design.

Figure 17.2 represents a simplified mathematical model for the external and internal *free body diagrams* (***FBD***) of a simple shaft. The modeled shaft is a simply supported cylindrical body under external loads at point or node

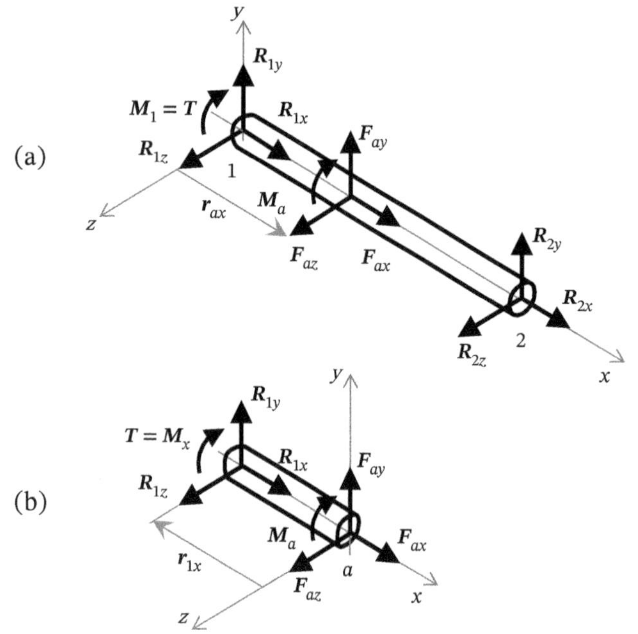

Figure 17.2 External and internal free body diagrams (***FBD***) of a simple shaft: (a) the ***FBD*** of the simple shaft under an operation loading $\boldsymbol{F}_a$, $\boldsymbol{T}$, and $\boldsymbol{M}_a$ and (b) ***FBD*** of cut segment of the shaft with coordinates transferred to the cut section at point or node *a*.

a and supports at locations 1 and 2. Figure 17.2a is the ***FBD*** of the simple shaft under the operation loading $\boldsymbol{F}_a$, $\boldsymbol{T}$, and $\boldsymbol{M}_a$. Figure 17.2b is the ***FBD*** of a cutout segment of the shaft with the coordinates transferred to the cut section at the point or node a. The transferred coordinates to the cutout section can help in finding the internal shear force and bending moment at any point or node on the shaft. All loads and reactions are assumed in the positive directions, and the negative ones are entered as such. The output negative values are then truly in the negative directions. Figure 17.2 demonstrates a situation of one loading location or node. Several other loading cases, however, are simply added to the mathematical model at the stipulated locations or node.

The procedure to complete a mathematical model before shaft synthesis is then as follows:

- Identify the external forces and moments generated due to power transmission elements or other sources. The forces and moments previously developed in spur, helical, bevel, and worm gears are transmitted to the shaft on which the gears are attached to. Similarly, belts, chains, and other power transmission elements develop forces and moments affecting transversed shafts at some defined locations or nodes.
- Generate a mathematical model to suit constructional considerations such as the one shown in Figure 17.2. Estimation of the spacing between elements is possible since the geometry of the attached power transmission elements has been previously generated during the synthesis process of these elements. The gear, pulley, bearings, and sprocket widths should be already known to a great extent.
- Perform ***external force analysis*** off the ***FBD*** in Figure 17.2a to find the reactions at the supports 1 and 2. Seldom that the shaft would have a single bearing, which can withstand radial loads and moments. As discussed in Section 2.1.1, the equilibrium of the ***FBD*** is then governed by the following reiterated relations:

$$\sum_{\text{FBD}} \boldsymbol{F} = 0 \tag{17.1}$$

$$\sum_{\text{About any Point}} \boldsymbol{M} = 0 \tag{17.2}$$

The procedure and tools to apply these equations are adapted as presented in Chapter 2. In a general ***FBD*** subjected to *external* n force vectors $\boldsymbol{F}_1, \boldsymbol{F}_2, \ldots, \boldsymbol{F}_n$ acting at position vectors $\boldsymbol{r}_1, \boldsymbol{r}_2, \ldots, \boldsymbol{r}_n$, respectively, and other *external* k moment vectors $\boldsymbol{M}_1, \boldsymbol{M}_2, \ldots, \boldsymbol{M}_k$ with reaction vectors at supports $\boldsymbol{R}_1$ and $\boldsymbol{R}_2$ or $\boldsymbol{R}_1$ and $\boldsymbol{M}_1$, the equilibrium equations of Eqs. (17.1) and (17.2) will be as follows:

$$\left(\sum_{1}^{n} \boldsymbol{F}_i\right) + \boldsymbol{R}_1 + \boldsymbol{R}_2 = 0 \tag{17.3}$$

$$\sum_{1}^{k} \boldsymbol{M}_j + \left(\sum_{1}^{n} \boldsymbol{r}_i \times \boldsymbol{F}_i\right) + (\boldsymbol{r}_{R_2} \times \boldsymbol{R}_2) + \boldsymbol{M}_1 = 0 \tag{17.4}$$

Locating the coordinate origin at support 1 and with moment $\boldsymbol{M}_1 = 0$, the moment Eq. (17.4) produces $\boldsymbol{R}_2$. Substitution of $\boldsymbol{R}_2$ into Eq. (17.3) allows the evaluation of $\boldsymbol{R}_1$. If support 1 carries a reaction moment $\boldsymbol{M}_1$, support 2 is assumed free with $\boldsymbol{R}_2 = 0$, and $\boldsymbol{r}_2 = 0$. Equation (17.4) will then produce the moment $\boldsymbol{M}_1$, and Eq. (17.3) generates $\boldsymbol{R}_1$ at support number 1. This is only applicable if the shaft is a cantilever, which is rarely utilized. The *cross product* in Eq. (17.4) is best handled by the following matrix product:

$$\boldsymbol{r} \times \boldsymbol{F} = \begin{bmatrix} 0 & -r_z & r_y \\ r_z & 0 & -r_x \\ -r_y & r_x & 0 \end{bmatrix} \begin{bmatrix} F_x \\ F_y \\ F_z \end{bmatrix} = [\boldsymbol{r}\times]\boldsymbol{F} \tag{17.5}$$

where (r_x, r_y, r_z) are the components of the position vector $\boldsymbol{r}$ that comprise the ***cross-product matrix*** $[\boldsymbol{r}\times]$ and (F_x, F_y, F_z) are the components of the force vector $\boldsymbol{F}$. This expression is the same as getting the solution of the cross product by any other means. It is, however, very useful as has been demonstrated in Example 2.1 and in computer coding;

- Perform analysis to produce the ***internal shear force and bending moment*** at any *critical location* or develop the *shear force and bending moment diagrams* along the shaft. This is attainable by generating ***FBD*** of a shaft segment from the left end to the anticipated critical location as specified in Figure 17.2b. The *external reactions* should have been previously defined for the entire shaft. Applying Eqs. (17.3) and (17.4) to the *cut segment* produces the *shear force* and *bending moment* in the form of the reaction force and moment vectors at that location. Repeating the process along the shaft generates the *shear force diagram* and *bending moment diagram*. If one can identify the crucial locations or potentially critical locations, there might be no need to develop the full shear force and bending moment diagrams. The ***crucial locations*** are usually where maximum bending moments, maximum torques, and high stress concentrations are expected.

After this process, one can synthesize the shaft and define stresses, slopes, and deflections at any location. This will be addressed later in this chapter.

Figure 17.3 represents part (a) of the MATLAB code to find the ***3D** external reactions* for a loaded shaft in SI system. External operating loads are f $= \boldsymbol{F}_a$, and m $= (\boldsymbol{T}$ and $\boldsymbol{M}_a)$ as of Figure 17.2b. Values are for Example 17.1. For other problems, the inputs should be changed. The inputs are identified by the bold **INPUT** comment statement (% …) at each variable. After getting the reactions, their values are used in part (b) of the MATLAB code in Figure 17.4. The values are also for Example 17.1. Figure 17.4 presents part (b) of the MATLAB code to find the *internal shear forces and moments* for the shaft in SI system. External reactions from Figure 17.3 and operating loads f $= \boldsymbol{F}_a$, m $= \boldsymbol{M}_a$ are inputs. The coordinates are transferred to the cut section of the shaft segment at point a as shown in Figure 17.2b. To reduce any anomaly, the coordinates are transferred an infinitesimal distance after

```
clear all; clc; format compact; format short                    % Force_Analysis_3D_Shafts_Short.m
            disp('- Analysis of 3D Body with n forces and 1 or 2 supports, R1 at [0 0 0]')
            disp ('- Default: Text Example 17.1 External Force Analysis')          % INPUT
n=1;                                            % INPUT Number of external forces
 rR2=[0.11; 0; 0];                              % INPUT Position of reaction R2 [m]
 f(1,:)=[7225.5; -15318; 0];                    % INPUT Force 1 components as a column vector [N]
 r(1,:)=[0.055; 0; 0];                          % INPUT Force 1 position as a column vector[m]
 m(1,:)=[596.83; 0; 0];                         % INPUT Moment 1 components as a column vector [N.m]
            disp('* Input External Forces and Moments ');
Reaction_R2_position = rR2 ; External_Forces = f' ; External_Forces_Locations = r'; External_Moments = m'
            disp('* Output External Forces and Moments ');
a=[0 -rR2(3) rR2(2);rR2(3) 0 -rR2(1); -rR2(2) rR2(1) 0];     % Cross Product Matrix for reaction position "rR2"
ainv=pinv(a);
for i=1:n                                                    % Calculate Moments of each f= (r X f)
  af=[0 -r(i,3) r(i,2); r(i,3) 0 -r(i,1); -r(i,2) r(i,1) 0];
  rXf(i,:)=af*f(i,:)';
end
sumM=[0; 0; 0]; sumf=sumM;                                   % Initialize the sum
for i=1:n                                                    % Calculate Sum of Moments (r X f) and Sum of forces
  sumM = sumM - rXf(i,:)' - m(i,:)';
  sumf = sumf - f(i,:)';
end
            disp(' - Reactions in Row Vector Form:')
R2= [ainv*[sumM]]'; disp(' R2 = '); disp(R2);                % Row vector for display [N]
R1= (sumf-R2')'; disp(' R1 = '); disp(R1);                   % Row vector for display [N]
M1= [sumM-a*R2']'; disp(' M1 = '); disp(M1);                 % Row vector for display [N.m]
```

Figure 17.3 MATLAB code to find the ***3D*** external reactions for a loaded shaft in SI system. External operating loading: f $= \boldsymbol{F}_a$ and m $= (T$ and $M_a)$. Values are for Example 17.1.

```
% Force_Analysis_3D_Shafts_Short.m
        disp ('- Default: Text Example 17.1 Internal Forces and Moments ')     % INPUT
n=2;                                          % INPUT Number of external forces including R1
 rR2=[0; 0; 0];                                          % Position of R2 as the shear force at the cut [m]
 fi(1,:)= R1 ;                                           % This is the external reaction R1
 ri(1,:)= - r(1,:);                                      % Position of external reaction R1 [m]
 mi(1,:)= M1 ;                                           % Moments at external reaction R1 [N.m]
 fi(2,:)=[7225.5; -15318; 0];                 % INPUT Applied load [N]
 ri(2,:)=[-0.0000001; 0; 0];                  % INPUT Location of applied load [m]
 mi(2,:)=[596.83; 0; 0];                      % INPUT Applied moment [N.m]
        disp('* Input Section Forces and Moments ');             % Reaction_R2_position = rR2, where coordinates moved
External_Forces_Including_R1 = fi'; External_Forces_Locations = ri' ; External_Moments = mi'
        disp('* Output Internal Shear Forces and Moments at the cut, where coordinates are transferred and x = 0.0. ');
ai=[0 -rR2(3) rR2(2);rR2(3) 0 -rR2(1); -rR2(2) rR2(1) 0];       % Cross Product Matrix for reaction position "rR2"
ainv=pinv(ai);
for i=1:n                                                % Calculate Moments of each fi = (ri X fi)
  afi=[0 -ri(i,3) ri(i,2); ri(i,3) 0 -ri(i,1); -ri(i,2) r(i,1) 0];
  rXfi(i,:)=afi*fi(i,:)';
end
sumM=[0; 0; 0]; sumf=sumM;                               % Initialize the sum
for i=1:n                                                % Calculate Sum of Moments (r X f) and Sum of forces
  sumM = sumM - rXfi(i,:)' - mi(i,:)';  sumf = sumf - fi(i,:)';
end
        % disp(' - Reactions in Row Vector Form:')
R1= (sumf -R2')'; disp('   Shear Forces = '); disp(R1);              % Row vector for display [N]
M1=[sumM - ai*R2']'; disp('   Internal Moments  = '); disp(M1);      % Row vector for display [N.m]
```

Figure 17.4 MATLAB code to find the internal shear forces and moments for the shaft in SI system. External reactions from Figure 17.3 and operating loads f$=\boldsymbol{F}_a$, m$=\boldsymbol{M}_a$ are inputs. The coordinates are transferred to the cut section of the shaft segment at point *a*.

the cut section of the shaft segment, i.e. at point *a+0.000 000 1* as depicted in Figure 17.4. For the US system of units, if one would use the standard units, the output will be standard, and no need to converge any units, i.e. forces in [lb], moments in [lb in], and distances in [in]. This was also the case for SI units, where forces are in [N], moments are in [N m], and distances are in [m].

The MATLAB code in Figures 17.3 and 17.4 is available as one code in the ***Wiley website*** under the name of ***Force_Analysis_3D_Shafts_Short.m***. This should be useful in calculating external and internal forces and moments for shafts loaded by other elements and transmitting power from and to other systems. The internal shear forces and moments generated by the code can be used to initially synthesize the shaft as will be defined later in Sections 17.2–17.4.

Example 17.1 A gearbox transmits a maximum power of 75 [kW] or 100.58 [hp] at 1200 [rpm] input speed. The gear reduction ratio is 4. The standard helical gears on the main input and output shafts have been synthesized through Examples 15.1–15.4. The gear width on the input shaft has been defined as 65 [mm] or 2.6 [in]. Assume deep groove ball bearings are used as supporting means to carry the reactions. The bearing width would need about 30 [mm] or 1.2 [in] on the shaft. Assume some spacer or shoulder on the shaft between the gear and the right and left bearings; see Figures 8.12, 11.5, and 15.21. It is required to start the synthesize process of the input shaft on which the gear is attached.

Solution

Data: $H_{\text{kW}} = 75$ [kW] or $H_{\text{hp}} = 100.58$ [hp], $N_{\text{in,rpm}} = N_{\text{in}} = 1200$ [rpm], and $\omega_P = 2\pi\, N_{\text{in}}/60 = 125.66$ [rad/s].

Gear width = 65 [mm] or 2.6 [in] and bearing width = 30 [mm] or 1.2 [in].

From Examples 15.3 and 15.4, the following relevant information are retrieved:

- Input torque T_{in} = 596.83 [N m] and 5282.6 [lb in].
- F_t = 14.1808 [kN], F_t = 3.1379e+03 [lb], F_n = 16.9369 [kN], F_n = 3.7477e+03 [lb], F_r = 5.7928 [kN], F_r = 1.2818e+03 [lb], F_a = 7.2255 [kN], and F_a = 1.5988e+03 [lb].

The tangential force F_t is moved to the shaft centerline and resists the input torque moment T_{in}. That tangential force and the radial force F_r have the following resultant:

$$F_R = \sqrt{F_t^2 + F_r^2} = \sqrt{(14.1808)^2 + (5.7928)^2} = 15.318\,[\text{kN}] = 15\,318\,[\text{N}]$$

$$F_R = \sqrt{F_t^2 + F_r^2} = \sqrt{(3137.9)^2 + (1281.8)^2} = 3389.6\,[\text{lb}] \tag{a}$$

The resultant force F_R is used with the axial force F_a in the x–y plane as the gear forces on the shaft $F_y = F_R$ at the location r_{1x} of 55 [mm] or 2.2 [in] as shown in Figure 17.2b. The axial force F_x in Figure 17.2b is equal to F_a.

The location r_{1x} is evaluated as more than half the gear face width of 65 [mm] or 2.6 [in] and the bearing width of about 30 [mm] or 1.2 [in], which is rounded to (65/2 + 30/2 = 47.5 [mm] or 2.6/2 + 1.2/2 = 1.9 [in]). Taking some spacer or shoulder between the gear and the right and left bearings would render the location r_{1x} as 55 [mm] = 0.055 [m] or 2.2 [in]. The full length of the shaft from the left to the right bearings is 110 [mm] = 0.11 [m] or 4.4 [in].

Using the MATLAB code of (***Force_Analysis_3D_Shafts_Short.m***), the following inputs and outputs are obtained with some presentation editing; see Figures 17.3 and 17.4. Distances are in [m], forces are in [N], and moments are in [N m].

Input: External forces and moments

$$Reaction\_R2\_position = [0.11\ \ 0\ \ 0]^T \tag{b}$$

$$External\_Forces = [7226\ \ 15\,318\ \ 0]^T \tag{c}$$

$$External\_Forces\_Locations = [0.0550\ \ 0\ \ 0]^T \tag{d}$$

$$External\_Moments = [596.83\ \ 0\ \ 0]^T \tag{e}$$

Output: External reactions in row vector form

$$R2 = [0\ \ 7659.0\ \ 0]^T \tag{f}$$

$$R1 = [7225.5\ \ \ 7659\ \ \ 0]^T \tag{g}$$

$$M1 = [-596.83\ \ 0\ \ 0]^T \tag{h}$$

Input: Input shaft segment forces and moments

$$External\_Forces\_Including\_R1 = \begin{bmatrix} -7226 & 7226 \\ 7659 & -15\,318 \\ 0 & 0 \end{bmatrix} \tag{i}$$

$$External\_Forces\_Locations = \begin{bmatrix} 0.0550 & -0.0000 \\ 0 & 0 \\ 0 & 0 \end{bmatrix} \tag{j}$$

$$External\_Moments = \begin{bmatrix} -596.83 & 596.83 \\ 0 & 0 \\ -0.0000 & 0 \end{bmatrix} \tag{k}$$

Output: Internal shear forces and moments at the cut, where coordinates are transferred and $\underline{x} = 0.0$. Vectors are in row form.

$$\text{Shear Forces} = \begin{bmatrix} 0 & 7659 & 0 \end{bmatrix}^T \tag{l}$$

$$\text{Internal Moments} = \begin{bmatrix} 0 & 0 & 421.2435 \end{bmatrix}^T \tag{m}$$

This shaft is modeled as a simply supported beam like that in Figure 6.5a, where $\boldsymbol{F}$ is the resultant force $\boldsymbol{F}_R$ in Eq. (a) and l is the full length of the shaft, i.e. 110 [mm] = 0.11 [m] or 4.4 [in]. The shear force $\boldsymbol{V}_a$ and bending moment $\boldsymbol{M}_a$ at point a (just after the load $\boldsymbol{F} = \boldsymbol{F}_R$) in Figure 17.2 are then given by the following values:

$$\boldsymbol{V}_a = \tfrac{1}{2}\boldsymbol{F}_R = \tfrac{1}{2}(15.318) = 7.659\ [\text{kN}] = 7659\ [\text{N}]$$

$$\boldsymbol{V}_a = \tfrac{1}{2}\boldsymbol{F}_R = \tfrac{1}{2}(3389.6) = 1694.8\ [\text{lb}] = 1.6948\ [\text{klb}] \tag{n}$$

$$\boldsymbol{M}_a = \tfrac{1}{4}\boldsymbol{F}_R\ l = \tfrac{1}{4}(15.318)(0.11) = 0.421245\ [\text{kN m}] = 421.245\ [\text{N m}]$$

$$\boldsymbol{M}_a = \tfrac{1}{4}\boldsymbol{F}_R\ l = \tfrac{1}{4}(3389.6)(4.4) = 3728.56\ [\text{lb in}] = 3.72856\ [\text{klb in}] \tag{o}$$

As expected, these values are exactly verifying Eqs. (l) and (m) for SI system. If the MATLAB code is run for the US values, it is expected to get the same results as for Eqs. (n) and (o).

After this procedure, one can define stresses at any location and continue with shaft design. The previously generated internal shear forces and moments are used to define stresses that can be used to start the initial synthesize of the shaft as defined next.

17.3 Initial Design Estimate

In Chapter 8, an *initial shaft synthesis* has been introduced in Section 8.4.2. Estimate of shaft size as a function of rotational speed and failure power is possible from Figure 8.20a for SI system of units and from Figure 8.20b for the US system of units. It was stipulated that the torsional load due to power transmission would cause material failure for the employed basic material of a 300 [MPa] or 43.5 [kpsi] tensile yield strength (AISI 1040 or ISO C45). The transmitted power should be multiplied by a proper *safety factor* K_{SF} before using the charts to get a safe shaft diameter. The estimation of shaft size in Figure 8.20 is depending on *static distortion energy theory* (*von Mises*) of failure (von Mises 1913). The proper *safety factor* K_{SF} can thus range from 3.5 to 5 due to prospective fatigue and its factors affecting the shaft as previously depicted in Chapter 7. The development of Figure 8.20 with its implementation process is established in Section 8.4.2. In this section, reiteration of the process highlights with concise specifics is presented. For detailed derivations, one is advised to review Section 8.4.2.

The *initial synthesis* of shaft diameter d_S can be realized by the application of the *maximum distortion energy* (*von Mises*) theory, which gives (see Eqs. (7.46) and (8.12)–(8.19))

$$d_S|_{\text{SI}} = \left(280.749(10^{-6})\frac{K_{\text{SF}}H_{\text{kW}}}{N_{\text{rpm}}}\frac{300}{S_{yt[\text{MPa}]}}\right)^{1/3}\ [\text{m}]$$

$$d_S|_{\text{US}} = \left(12\,780.8(10^{-3})\frac{K_{\text{SF}}H_{\text{hp}}}{N_{\text{rpm}}}\frac{43.5}{S_{yt[\text{kpsi}]}}\right)^{1/3}\ [\text{in}] \tag{17.6}$$

where S_{yt} is the tensile yield strength of the material in [MPa] or [kpsi], N_{rpm} is the rotational speed in [rpm], H_{kW} is the *transmitted power* in [kW], H_{hp} is the transmitted power in [hp], and K_{SF} is the factor of safety, which is in the range of 3.5–5 to account for fatigue and its factors affecting the shaft. If the torque $\boldsymbol{T}$ is known and due to Eq. (8.13), the $H_{\text{kW}}/N_{\text{rpm}}$ is equal to $\boldsymbol{T}_{N.m}/9549.3$, and the $H_{\text{hp}}/N_{\text{rpm}}$ is equal to $\boldsymbol{T}_{\text{lb.in}}/63\,025$. Eq. (17.6) is developed for shafts under torsion and transmitting power with no account for bending moments. This might be acceptable if the *bending moment* $\boldsymbol{M}$ to the *torque* $\boldsymbol{T}$ is much smaller than 1.0. To account for the moment $\boldsymbol{M}$ (i.e. $\boldsymbol{M} \neq 0$), the shaft diameter obtained from Eq. (17.6) should be multiplied by an additional *moment factor* K_M of the following form (see Eq. (8.15)):

$$K_M = \left(\left(\frac{(\boldsymbol{M}/\boldsymbol{T})^2 + 0.75}{0.75} \right)^{1/2} \right)^{1/3} \tag{17.7}$$

The additional moment factor K_M is pronounced when $\boldsymbol{M}/\boldsymbol{T}$ is more than 1. For $\boldsymbol{M}/\boldsymbol{T}$ of 1, the additional moment factor K_M is about 1.1517. For $\boldsymbol{M}/\boldsymbol{T}$ of 2, the additional moment factor K_M is about 1.3602. Usually, the construction of the shaft that transmits power should have as little bending moment as possible.

Eqs. (17.6) and (17.7) can be used directly to synthesize the power transmitting shaft, which is additionally subjected to bending moment. In Chapter 8, the computer-aided design (*CAD*) in Section 8.5 introduced a MATLAB code to help in the initial synthesis of connected elements including initial shaft synthesis in conjunction with joining components such as keys, pins, and splines. The code is called ***Joining_Elements.m*** and is presented in Figure 8.22. This code can be used to initially synthesize shafts in addition to the prospective joining components along the shaft. The procedure in that code is using the process in Section 8.4.2, which led to Eqs. (17.6) and (17.7).

Example 17.2 The gearbox of Example 17.1 transmits a maximum power of 75 [kW] or 100.58 [hp] at 1200 [rpm] input speed. The standard helical gears on the main input and output shafts have been synthesized through Examples 15.1–15.4. The gear width on the input shaft has been defined as 65 [mm] or 2.6 [in]. The deep groove ball bearing width is 30 [mm] or 1.2 [in] on the shaft. It is required to synthesize the input shaft on which the gear is attached employing a basic material of a 300 [MPa] or 43.5 [kpsi] tensile yield strength and a safety factor of 4.0.

Solution

Data: $H_{kW} = 75$ [kW] or $H_{hp} = 100.58$ [hp], $N_{in,rpm} = N_{in} = 1200$ [rpm], and $\omega_P = 2\pi\, N_{in}/60 = 125.66$ [rad/s].

Gear width = 65 [mm] or 2.6 [in], bearing width = 30 [mm] or 1.2 [in], and safety factor $K_{SF} = 4.0$.

From 17.1, the following relevant information are reiterated:

- Input torque $\boldsymbol{T}_a = 596.83$ [N m] and 5282.6 [lb in].
- Internal moments at location a, $\boldsymbol{M}_a = [0\,0\,421.2435]^T$ [N m], and $[0\,0\,3728.56]^T$ [lb in].

The moment to torque ratio $\boldsymbol{M}_a/\boldsymbol{T}_a = 421.2435/596.83 = 0.705\,80$ or $\boldsymbol{M}_a/\boldsymbol{T}_a = 3728.56/5282.6 = 0.705\,82$, which either is less than 1.0. The shaft diameter multiplier or the moment factor K_M given by Eq. (17.7) is then

$$K_M = \left(\left(\frac{(\boldsymbol{M}_a/\boldsymbol{T}_a)^2 + 0.75}{0.75} \right)^{1/2} \right)^{1/3} = \left(\left(\frac{(0.7058)^2 + 0.75}{0.75} \right)^{1/2} \right)^{1/3} = 1.0886 \tag{a}$$

This should be valid for both the US and SI system of units. This also indicates that the shaft diameter would increase by only 1.0886 or 8.86 percent due to the moment $\boldsymbol{M}_a$, when calculated solely under the torque $\boldsymbol{T}_a$.

Using Eq. (17.6), the *initial synthesis* of shaft diameter d_S under the torque $\boldsymbol{T}_a$ is then

$$d_S|_{SI} = \left(280.749(10^{-6}) \frac{K_{SF} H_{kW}}{N_{rpm}} \frac{300}{S_{yt[MPa]}} \right)^{1/3} = \left(280.749(10^{-6}) \frac{4(75)}{1200} \frac{300}{300} \right)^{1/3} = 0.041\,25[\text{m}] = 41.25\,[\text{mm}]$$

$$d_S|_{US} = \left(12780.8(10^{-3}) \frac{K_{SF} H_{hp}}{N_{rpm}} \frac{43.5}{S_{yt[kpsi]}} \right)^{1/3} = \left(12780.8(10^{-3}) \frac{4(100.58)}{1200} \frac{43.5}{43.5} \right)^{1/3} = 1.6242\,[\text{in}] \tag{b}$$

The diameter in Eq. (b) should be multiplied by K_M to give

$$d_S|_{SI} = K_{SF}\, d_S|_{SI} = 1.0886(0.041\,25) = 0.044\,90\ [\text{m}] = 44.90\ [\text{mm}]$$
$$d_S|_{US} = K_{SF}\, d_S|_{US} = 1.0886(1.6242) = 1.7681\ [\text{in}] \tag{c}$$

Rounding off the diameters to the nearest preferred dimensions suggests the values of 45 [mm] or 1.75 [in].

17.4 Detailed Design

The initial synthesis in the preceding Section 17.3 is using *von Mises* static failure theory, and that is why a high safety factor of 3.5–5 is needed for the process to reach a reasonable estimate. In detailed design procedure, the same theory is used but with dynamic considerations. This has been introduced in Section 7.8 about fatigue strength and factors affecting fatigue; see also Barsom and Rolfe (1987) and ASM (1996). In this section, reiteration of the process highlights with concise specifics is presented. For detailed derivations, one is advised to review Section 7.8.

Shafts are mainly transmitting power to mounted elements along their span or overhangs. Therefore, shafts are subjected to torsion and bending moments along some of their lengths. In some cases, their axial forces exist due to mounted elements on the shaft such as helical gears, bevel gears, and worm gears. These situations create ***3D*** combined case of stressing. In a better design, the axial forces are minimized by the utility of some means such as double helical gears, reducing their values or their location span.

Due to the shaft rotation under steady loading, the bending moment $\boldsymbol{M}$ produces alternating normal stresses. The torsional load $\boldsymbol{T}$ produces stresses that are usually steady. Fluctuation of shear stresses due to torsion might happen due to unsteady power transmission, stop-and-go, and the change of power demands in the driven machine or system. The transverse shear stress due to the transverse shear force $\boldsymbol{V}$ can affect the maximum shear stress due to torsion. The transverse shear will be left out of the following presentation.

For shafts under bending moment $\boldsymbol{M}$ and axial force $\boldsymbol{F}$, the *maximum normal stress* $\sigma_{\max}$ and *minimum normal stress* $\sigma_{\min}$ at the outer fiber of a cylindrical shaft can be written as follows (see Eqs. (6.62) and (6.56)):

$$\sigma_{\max} = \frac{32\,\boldsymbol{M}}{\pi\, d_S^3} + \frac{\boldsymbol{F}}{(\pi/4)\, d_S^2}, \qquad \sigma_{\min} = -\frac{32\boldsymbol{M}}{\pi\, d_S^3} \mp \frac{\boldsymbol{F}}{(\pi/4)\, d_S^2} \tag{17.8}$$

where d_S is the shaft diameter, and the *compression* of $\sigma_{\min}$ occurs when the shaft rotates in addition to an *axial compressive* or *tensile force* $\boldsymbol{F}$. The shaft under fluctuating torsion $\boldsymbol{T}$ experiences *maximum shear stress* $\tau_{\max}$ and *minimum shear stress* $\tau_{\min}$ at the outer fiber given by the following relations (see Equation (6.67)):

$$\tau_{\max} = \frac{16\,\boldsymbol{T}}{\pi d_S^3}, \qquad \tau_{\min} = -\frac{16\,\boldsymbol{T}}{\pi d_S^3} \tag{17.9}$$

where d_S is the shaft diameter, and the minimum shear stress happens only if the shaft reverses direction of rotation and power transmission, which is not usually happening.

For a simple fluctuating form as in Figure 7.22b, the values of the *mean stresses* σ_m, τ_m and the *alternating stresses* σ_a, τ_a as functions of the *maximum stresses* $\sigma_{\max}, \tau_{\max}$ and the *minimum stresses* $\sigma_{\min}, \tau_{\min}$ are simply as in Eqs. (7.56) and (7.57) that are rewritten as follows:

$$\sigma_m = \frac{1}{2}(\sigma_{\max} + \sigma_{\min}), \qquad \tau_m = \frac{1}{2}(\tau_{\max} + \tau_{\min})$$
$$\sigma_a = \frac{1}{2}(\sigma_{\max} - \sigma_{\min}), \qquad \tau_a = \frac{1}{2}(\tau_{\max} - \tau_{\min}) \tag{17.10}$$

For *steady torsion*, the minimum shear stress is the same as the maximum shear stress, which renders the values in Eq. (17.10) as follows:

$$\tau_m = \frac{1}{2}(\tau_{max} + \tau_{min}) = \tau_{max}$$
$$\tau_a = \frac{1}{2}(\tau_{max} - \tau_{min}) = 0 \tag{17.11}$$

For shafts under *bending moment* $\boldsymbol{M}$, and no axial force $\boldsymbol{F}$, the maximum normal stress σ_{max} and minimum normal stress σ_{min} at the outer fiber of a cylindrical shaft will be the same magnitude but with negative signs. The values of the *mean stress* σ_m and the *alternating stress* σ_a in Eq. (17.10) are then as follows:

$$\sigma_m = \frac{1}{2}(\sigma_{max} + \sigma_{min}) = 0$$
$$\sigma_a = \frac{1}{2}(\sigma_{max} - \sigma_{min}) = \sigma_{max} \tag{17.12}$$

A good way of handling the fluctuating form of loading for ductile materials is to use the *von Mises distortion energy* theory as demonstrated in Section 7.7.3 (von Mises 1913). The procedure develops two *Mohr's circles*, one for the *mean stresses* and the other for the *alternating stresses*. The two *Mohr's circles* would use the mean and alternating components of the normal and shear stresses (σ_x, σ_y, σ_z, τ_{xy}, τ_{yz}, and τ_{xz}) at any internal point of the shaft in a machine. For **2D** stress state and from the two *Mohr's circles* of mean and alternating stresses, the *equivalent von Mises stresses* $\sigma_{vM,m}$ and $\sigma_{vM,a}$ are as follows (see Section 7.8.4):

$$\sigma_{vM,m} = \sqrt{\sigma_{1,m}^2 - \sigma_{1,m}\ \sigma_{2,m} + \sigma_{2,m}^2}$$
$$\sigma_{vM,a} = \sqrt{\sigma_{1,a}^2 - \sigma_{1,a}\ \sigma_{2,a} + \sigma_{2,a}^2} \tag{17.13}$$

where σ_1 and σ_2 are the principal stresses for mean and alternating stresses as subscripts m and a, respectively. The procedure is to use the equivalent stresses $\sigma_{vM,m}$ and $\sigma_{vM,a}$ in a *fatigue diagram* such as *Gerber*, *modified Goodman*, and *ASME elliptic* as shown in Figure 7.26. The fatigue diagram uses the *mean stress* σ_m as the abscissa axis and the *alternating stress* σ_a as the ordinate axis. If the normal stress $\sigma_y = 0$, Eq. (17.13) is reduced to become as simple as follows (see Eq. (7.95)):

$$\sigma_{vM,m} = \sqrt{\sigma_m^2 + 3\tau_m^2}$$
$$\sigma_{vM,a} = \sqrt{\sigma_a^2 + 3\tau_a^2} \tag{17.14}$$

here, the *ASME elliptic* diagram is used since it includes the static yield failure criterion (ANSI/ASME B106.1M 1985). For the *ASME criterion*, the *fatigue safety factor* $K_{SF,F}$ is simply obtainable from Eq. (7.93) such that

$$K_{SF,F} = \frac{1}{\sqrt{(\sigma_{vM,a}/S_{e,S})^2 + (\sigma_{vM,m}/S_{yt})^2}} \tag{17.15}$$

where $S_{e,S}$ is the *endurance limit of shaft* material including the factors affecting fatigue as defined in Section 7.8.2 and S_{yt} is the *tensile yield strength* of shaft material.

The *fatigue strength of the shaft* $S_{e,S}$ is related to the material fatigue strength or *endurance limit* S_e by the following relation (see Eq. (7.66) and Marin (1962)):

$$S_{e,S} = K_{surf}\ K_{size}\ K_{load}\ K_{reliab}\ K_{temp}\ K_{conc}\ K_{miscel}\ S_e \tag{17.16}$$

where K_{surf} is the *surface factor*, K_{size} is the *size factor*, K_{load} is the *load factor*, K_{relaib} is a *reliability factor*, K_{temp} is a *temperature factor*, K_{conc} is related to the previously discussed *stress concentration factor* K_{SC}, and K_{miscel} is for other *miscellaneous factors*. All these factors are attainable from Eqs. (7.67)–(7.83). If the numbers of cycles are less than that of the *endurance limit* S_e, the *fatigue strength* S_f as in Eq. (7.65) would replace S_e in Equation (17.16).

A MATLAB code is used to calculate the factors affecting the *fatigue strength* $S_{e,\text{part}}$ of a machine part such as shafts ($S_{e,S} = S_{e,\text{part}}$). This code is part of a unified code including the codes of Figures 7.34–7.36. The code employs Eqs. (7.67)–(7.82) to calculate most of the factors affecting fatigue and with few inputs calculates the *fatigue strength* $S_{e,\text{part}}$ of a machine part using Eq. (7.66), i.e. Equation (17.16). These MATLAB codes are set in one code under the name ***Fatigue_Strength_and_Factors_Affecting.m***. This combined MATLAB code is available at ***Wiley website.***

The detailed design procedure is then as follows:

- With the initial shaft synthesis that defined the initial shaft diameter d_S and the material selection for a static safety factor, it is left to define details needed to evaluate dynamic or fatigue safety factor.
- Evaluate the maximum and the minimum normal stresses and the maximum and the minimum shear stresses as defined in Eqs. (17.8) and (17.9).
- Calculate the mean and the alternating normal stresses and the mean and the alternating shear stresses as defined in Eq. (17.10).
- Determine the mean and the alternating principal stresses according to Eqs. (17.13) or (17.14).
- Estimate the factors affecting fatigue and the shaft fatigue strength using Eqs. (7.67)–(7.83) and (17.16).
- Find the fatigue safety factor $K_{\text{SF},F}$ from Eq. (17.15).

As needed, some of the factors and parameters may be tuned to improve the shaft design. For a critical design, these factors should adjust the calculations to achieve the intended life or other objectives without failure.

After this process of design for stress is consummated, one can find the slope and deflection along the shaft to check clearances and suitability of bearings and other mounted elements. This will be discussed in Section 17.4 and can be checked by any available codes of ***FE***.

Example 17.3 The gearbox Examples of 17.1 and 17.2 transmits a maximum power of 75 [kW] or 100.58 [hp] at 1200 [rpm] input speed. It was required to synthesize the input shaft on which the gear is attached employing a basic material of a 300 [MPa] or 43.5 [kpsi] tensile yield strength and a static safety factor of 4.0. Find the internal dynamic stresses if the shaft is synthesized as a cylinder of a 45 [mm] or 1.75 [in] diameter. Use the estimated factors affecting fatigue in Example 7.4 for the estimated life span of 10 years and the shaft operating 8 hours a day; the reliability should be 99.99%, the operating temperature is about 380 [°C] or 716 [°F], the evaluated stress concentration factor K_{SC} is 2.1, and the notch radius is 2.5 [mm] or 0.1 [in]. Find the safety factor for ASME elliptic dynamic fatigue theory of failure.

Solution

Data: $H_{\text{kW}} = 75$ [kW] or $H_{\text{hp}} = 100.58$ [hp], $N_{\text{rpm}} = N_{\text{in}} = 1200$ [rpm], $\omega_P = 2\pi\ N_{\text{in}}/60 = 125.66$ [rad/s], and $d_C = 0.045$ [m] or 1.75 [in]. *Ductile* material is hot rolled AISI 1040 or ISO C40.

Ductile material properties are yield strength of 42 [kpsi] or 290 [MPa], ultimate tensile strength of 76 [kpsi] or 525 [MPa], and endurance limit of 37.7 [kpsi] or 260 [MPa]; see Table A.7.2. Equation (7.58) suggests $S_e = 0.5\ S_{\text{ut}} = 0.5(525) = 262.5$ [MPa] or $0.5(76) = 38$ [kpsi]. The two values are close, and one can then use the values in Table A.7.2.

From Examples 17.1 and 17.2, the following relevant information are reiterated:

- Input torque $\boldsymbol{T} = 596.83$ [N m] and 5282.6 [lb in].
- Internal moments at location a $\boldsymbol{M} = [0\,0\,421.2435]^T$ [N m] and $[0\,0\,3728.56]^T$ [lb in].
- The shaft diameters to the nearest preferred dimensions are 45 [mm] or 1.75 [in].

The shaft maximum and the minimum normal stresses and the maximum and the minimum shear stresses as defined in Eqs. (17.8) and (17.9) at location a are as follows:

$$\sigma_{max} = \frac{32\ \boldsymbol{M}}{\pi\ d_S^3} = \frac{32\ (421.2435)}{\pi\ (45/1000)^3} = 47\,086\,431\ [\text{Pa}] = 47.086\ [\text{MPa}], \quad \sigma_{min} = -\sigma_{max} = -47.086\ [\text{MPa}]$$

$$\tau_{max} = \frac{16\ \boldsymbol{T}}{\pi\ d_S^3} = \frac{16\ (596.83)}{\pi\ (45/1000)^3} = 33\,356\,710\ [\text{Pa}] = 33.357\ [\text{MPa}], \quad \tau_{min} = 0\ [\text{MPa}] \tag{a}$$

and

$$\sigma_{max} = \frac{32\ \boldsymbol{M}}{\pi\ d_S^3} = \frac{32\ (3728.56)}{\pi\ (1.75)^3} = 7086.4\ [\text{psi}] = 7.0864\ [\text{kpsi}], \quad \sigma_{min} = -\sigma_{max} = -7.0864\ [\text{kpsi}]$$

$$\tau_{max} = \frac{16\ \boldsymbol{T}}{\pi\ d_S^3} = \frac{16\ (5282.6)}{\pi\ (1.75)^3} = 5019.99\ [\text{psi}] = 5.020\ [\text{kpsi}], \quad \tau_{min} = 0\ [\text{MPa}] \tag{b}$$

The values of the mean stresses σ_m, τ_m and the alternating stresses σ_a, τ_a as in Eq. (17.11) and (17.12) are then as follows:

$$\sigma_m = \frac{1}{2}(\sigma_{max} + \sigma_{min}) = 0$$

$$\sigma_a = \frac{1}{2}(\sigma_{max} - \sigma_{min}) = \sigma_{max} = 47.086\ [\text{MPa}] \ \text{ or } = 7.0864\ [\text{kpsi}]$$

$$\tau_m = \frac{1}{2}(\tau_{max} + \tau_{min}) = \tau_{max} = 33.357\ [\text{MPa}] \ \text{ or } = 5.020\ [\text{kpsi}]$$

$$\tau_a = \frac{1}{2}(\tau_{max} - \tau_{min}) = 0 \tag{c}$$

The *equivalent von Mises stresses* $\sigma_{vM,m}$ and $\sigma_{vM,a}$ are as follows (see Eq. (17.14)):

$$\sigma_{vM,m} = \sqrt{\sigma_m^2 + 3\tau_m^2} = \sqrt{3\tau_m^2} = \sqrt{3}\,\tau_m = 1.7321(33.357) = 57.776\ [\text{MPa}] \ \text{ or } = 1.7321(5.02)$$
$$= 8.6951\ [\text{kpsi}]$$

$$\sigma_{vM,a} = \sqrt{\sigma_a^2 + 3\tau_a^2} = \sqrt{\sigma_a^2} = \sigma_a = 47.086\ [\text{MPa}] \ \text{ or } = 7.0864\ [\text{kpsi}] \tag{d}$$

To estimate the factors affecting fatigue and the shaft fatigue strength using Eqs. (7.67)–(7.83) and (17.16), one observes that the shaft dimension and the dynamic considerations are identical to that given in Example 7.4. From Example 7.4, Eq. (d) is rewritten as follows:

$$S_{e,S} = (0.781)(0.825)(1.0)(0.702)(0.913)(0.537)(1.0)\ S_e = 0.2218\ (260) = 57.668\ [\text{MPa}]$$

$$S_{e,S} = (0.781)(0.828)(1.0)(0.702)(0.912)(0.537)(1.0)\ S_e = 0.2224\ (37.7) = 8.3845\ [\text{kpsi}] \tag{e}$$

Equation (e) indicates that the cumulative factors affecting fatigue are 0.2218 or 0.2224. This reduces the endurance limit by (1/0.2218) = 4.5 times or (1/0.2224) = 4.5 times also. This can be used as the cumulative reduction factor affecting fatigue as a reciprocal value similar to a safety factor that is larger than 1.0.

For the *ASME criterion*, the fatigue safety factor is then obtained from Eq. (17.15) such that

$$K_{SF,F}\Big|_{SI} = \frac{1}{\sqrt{(\sigma_{vM,a}/S_{e,S})^2 + (\sigma_{vM,m}/S_{yt})^2}} = \frac{1}{\sqrt{\left(\frac{47.086}{57.668}\right)^2 + \left(\frac{57.776}{290}\right)^2}} = 1.1898$$

$$K_{SF,F}\Big|_{US} = \frac{1}{\sqrt{(\sigma_{vM,a}/S_{e,S})^2 + (\sigma_{vM,m}/S_{yt})^2}} = \frac{1}{\sqrt{\left(\frac{7.0864}{8.3845}\right)^2 + \left(\frac{8.6951}{42}\right)^2}} = 1.1492 \tag{f}$$

These values are small due to the harsh consideration of the factors affecting fatigue. The material can be changed to a material with a higher endurance limit and having a ground surface finish, or the shaft diameter and/or the notch radius should be increased. A shaft diameter of 50 [mm] or 2 [in] might be safe for the harsh factors affecting fatigue. The *quenched and tempered* process can improve the material fatigue strength to accommodate the harsh factors affecting fatigue.

17.5 Design for Rigidity

Shaft design for rigidity is important for special cases where shaft deflection would disturb proper operating function or cause potential damage to mounted elements. A gear mounted on a shaft would be affected by shaft transverse deflection, particularly if the deflection would eliminate the backlash or would cause larger separation between teeth beyond the allowable. The maximum range of allowable deflection for spur gears is between 0.003 [in] for $p_d < 50$ [teeth/in] or 0.08 [mm] for normal module $m_n > 0.5$ [mm] to a value of 0.01 [in] for diametral pitch $p_d < 10$ [teeth/in] or 0.25 [mm] for normal module $m_n > 2.5$ [mm]. The transverse angular deflection is also important for bearings particularly cylindrical roller bearings. The maximum slope range for deep groove ball bearings is 0.06°–0.17°. The maximum slope range for taper roller bearings is 0.03°–0.07°. These situations are particularly the issue when shafts are *long* and *slender*, which greatly amplify the deflections. A limiting case occurs if the shaft is considered as a simply supported model shown in Figures 17.2 and 6.5a. For a constant diameter shaft, the maximum *transverse deflection* v_{max} and the *maximum slope* θ_{max} are as follows (see Eqs. (6.30) and (6.28)):

$$v_{max}\left(\frac{1}{2}l\right) = -\frac{\boldsymbol{F}l^3}{48EI_z} \quad \text{and} \quad \theta_{max}\ (0 \text{ or } l) = \frac{\boldsymbol{F}l^2}{16EI_z} \tag{17.17}$$

where $\boldsymbol{F}$ is the force, l is the length between bearings, E is the elasticity modulus of the material, and I_z is the second area moment of the shaft cross section. Note that the slope is in [rad] with the maximum occurring at the bearings ($x = 0$ and $x = l$); see Figure 6.5a. Equation (17.17) indicates the greater effect of the length l on the deflection, since it is a function of l^3. The effect of shaft diameter is also great since it is a function of $1/d^4$, ($I_z = \pi d^4/64$). A larger shaft diameter greatly reduces the deflection and slope.

The transverse shaft *stiffness* or *rigidity* k_{Sb} of the simply supported model of the shaft at the applied central load is then given by the following expression:

$$k_{Sb} = \frac{\boldsymbol{F}}{v_{max}} = \frac{48\ EI_z}{l^3} = \frac{48\ E(\pi d_S^4/64)}{l^3} = \frac{0.75\pi d_S^4 E}{l^3} \tag{17.18}$$

where this is only valid for a constant shaft diameter d_S under the simply supported central force model.

Shafts are usually stepped and subjected to several loadings that require involved means to calculate the deflections at critical points. *Superposition* or *singularity functions* can be used to define the slopes and deflections along the variable diameter shaft as discussed in Section 6.1.3.

For a stepped shaft, a simplified equivalent may also be used to evaluate slopes and deflections; see Figure 6.35 and Section 6.8. The *angular deformation* and *angular stiffness* k_{Sti} of the shaft segments of Figure 6.35a under torsion are defined by Eq. (6.69) such that (see Eq. (6.163))

$$\phi_i = \frac{\boldsymbol{T}l_i}{GJ_i}, \quad k_{Sti} = \frac{\boldsymbol{T}}{\phi_i} = \frac{GJ_i}{l_i} \tag{17.19}$$

where ϕ_i is the angular deformation of segment i of the shaft, $\boldsymbol{T}$ is the applied torque, l_i is the length of the ith segment, G is the shear modulus of elasticity or modulus of rigidity, and J_i is the polar second area moment of the shaft section. Any equivalent zones (d_1, l_2') and (d_1, l_3') of Figure 6.35b should provide the same deformations ϕ_2 and ϕ_3 of the original sections in Figure 6.35a. The shaft in Figure 6.35b is then equivalent to the shaft in

Figure 6.35a. The equivalent length l_i' of segment i relative to segment 1 is given by the following expression (see Eq. (6.168));

$$l_i' = l_i \frac{d_1^4}{d_i^4} \tag{17.20}$$

where d_1 is the diameter of segment 1, which the equivalent shaft will have as a diameter, and d_i is the diameter of segment i. The total *angular deformation* ϕ_S and *angular stiffness* k_{St} of the shaft is then the summation of all shaft segments in Eq. (17.19), which gives the following relation:

$$\phi_S = \sum \frac{\boldsymbol{T} l_i}{G J_i} = \frac{\boldsymbol{T}}{G} \sum \frac{l_i}{J_i}, \quad k_{St} = \frac{\boldsymbol{T}}{\phi_S} = \frac{G}{\sum l_i / J_i} \tag{17.21}$$

where the final expression is assuming that the shaft is under a constant torque along the full length.

The readily available ***FE*** programs (see Section 6.11) supersede these methods in efforts and time for the comprehensive numerical evaluation of slopes and deflections. The simple model of a simply supported shaft may give useful limiting values before the extensive analysis of ***FE***.

17.6 Critical Speed

The critical speed of the shaft is a vital factor in the design process. This is the shaft rotational speed at which the shaft starts to vibrate in resonance with the drivers. This vibration may lead to instability, which causes damage to the shaft and the machine that houses the shaft. Shafts should then operate more than 20% above or below that speed. The following is a simplified explanation that gives some formulas for general-purpose evaluation of the critical speed, which may be used for preliminary estimation. For more analysis and considerations, one is advised to consult references in the theory of vibrations and applications such as Inman (2017), Gorman (1975), and Thomson and Dahleh (1998).

The *critical speed* or fundamental *natural frequency* ω_n of a constant diameter shaft, which is simply supported is given by the following equation (Thomson and Dahleh 1998):

$$\omega_n = \pi^2 \sqrt{\frac{E I_z}{m_S l_S^4}} = \frac{\pi^2}{l_S^2} \sqrt{\frac{E I_z}{\rho A_S}} \tag{17.22}$$

where the *critical speed* ω_n is in [rad/s], E is the elasticity modulus of the material, and I_z is the second area moment of the shaft cross section, m_S is the mass per unit length of the shaft, l_S is the shaft length between bearings, A_S is the shaft cross-sectional area, and ρ is the density of the shaft material. All of these parameters should be in consistent units.

For the critical speed of shafts with several bodies each of mass m_i attached to it, one may determine the first or fundamental natural (or critical) frequency ω_n [rad/s] as follows (see Thomson and Dahleh 1998 and ANSI/ASME B106.1M 1985):

$$\omega_n = \pi^2 \sqrt{\frac{g_a \sum m_i v_i}{\sum m_i v_i^2}} \tag{17.23}$$

where g_a is the gravitational acceleration (9.8 [m/s^2], or 386 [in/s^2]), and v_i is the transverse static deflection of the body mass m_i.

Some available ***FE*** programs (see Section 6.11) supersede these models in efforts and time for the comprehensive numerical evaluation of critical or fundamental frequency particularly for stepped shafts with several attached bodies. The simple model of a simply supported shaft may give useful limiting values before the extensive analysis of ***FE***.

17.7 Computer-Aided Design and Optimization

The initial design estimate of shafts in Section 17.3 employed a basic material of a 300 [MPa] or 43.5 [kpsi] tensile *yield strength* (AISI 1040 or ISO C45). The static synthesis of the shaft diameter, however, considers an optional different material in Eq. (17.6). The detailed design Section 17.4 presented added considerations to include the fluctuation of stresses in rotating shafts. The *endurance limit* S_e or the *fatigue strength* S_f of shaft material would be needed. Chapter 7 of this text introduced information to suitably define these values. Other shaft materials are potentially more useful in some applications or severe loading or size constraints. In these cases, different material selection is conceivable.

17.7.1 Shaft Materials

In Chapter 7.5.2 of synthesized or designed machine elements, the initial material selection in Table 7.4 provides a reasonable start, which is stemming from previous knowledge and experience. To search for a suitable material for a shaft design, one has the options of defining the *mechanical properties*, the *chemical composition*, or some *physical properties*. This is particularly necessary for *optimum selection* or synthesis of the material such as the work of Abdel Meguid (1999). Some standards embed some of these properties as guidance for shaft materials such as ANSI/ASME B106.1M (1985). An initial good selection of appropriate material reduces the iterations to achieve the optimum. If the initial selection stemmed from vast experience, the initial selection might be an optimum estimate by knowledge and experience. Further iteration, however, is the optimum such as the attempts by Abdel Meguid (1999) and many others. The costs of the material in addition to the cost of manufacturing are important factors for large or little number of shafts required. This optimization problem is needed particularly for very large or mass production of a specific shaft design. The optimization is also very crucial in the case of a small number of shafts if the minimum weight is an objective. Hollow, forged, and heat-treated shafts might be needed to satisfy the optimum objective. These optimization problems are beyond the scope of this text but are useful to contemplate; see Section 7.10 and Figure 7.37.

In addition to the basic material of a 300 [MPa] or 43.5 [kpsi] tensile *yield strength* (AISI 1040 or ISO C45), Table 7.4 indicates other optional materials for special or crucial applications. The limited list includes AISI 1340, 3140, 4140, 4340, 5140, and 8650; ISO 42CrMoS4, 41CrS4, and {50NiCrMo2}; and carburized AISI 1020 and ISO C20. The carbon steels can be hot rolled, cold drawn, or alloyed. The higher carbon content materials can be heat treated such as the employed basic material of a 300 [MPa] or 43.5 [kpsi] tensile *yield strength* (AISI 1040 or ISO C45). The materials with low carbon contents can be carburized and heat treated, particularly for pinion shafts and camshafts. Few other shafting steels are suggested in addition to those in Table 7.4 such as AISI 1045, 1050, and 8620; see ANSI/ASME B106.1M (1985). These are also higher carbon steels such as AISI 1045 and AISI 1050 or a lower carbon steel such as AISI 8620.

17.7.2 Computer-Aided Design of Shafts

Figure 17.5 presents the images of *Excel tablets* for shaft synthesis devoted to the US and SI systems of units. The **Default** columns of numbers are acting as a guide, and values are the input and solution of Example 8.3 in Section 8.4.2 of Chapter 8. It is not allowed to change any of the cell's entries. All cells with clear background should not be changed. They have formulas to evaluate dependent on the ***Input*** *cells* with light grayish or light blue background color. The column with **Input** heading is used to input data in the *Input cells* with light grayish or light blue background color. These are the only values needed to *initially synthesize* the shaft geometry in terms of the shaft diameter at the shaft location where the power is transmitted in addition to the existence of a bending moment. The needed inputs are then the power, the rotational speed, the bending moment, the initial static safety factor, the ultimate tensile strength, and the yield strength of shaft material. The ultimate tensile strength and the yield strength of shaft material are initially for the basic material of a 300 [MPa] or 43.5 [kpsi] tensile *yield strength* (AISI 1040 or ISO C45). There is also an option to enter the axial stress emanating from additional applied axial force either a positive tensile or a negative compressive force. This *mean normal stress* must be calculated outside the *Synthesis Tablet*. Optionally, the *alternating shear stress* due to prospective torque variation must also be calculated outside the *Synthesis Tablet* and entered in that cell with light grayish or light blue background color. This would only affect the dynamic safety factor without changing the normal and shear components of the static stresses. These optional cases are not usually present or seldom occurring.

The ***input*** shaft synthesis column provides the initial shaft diameter for torque only and for both torque and moment loadings. The synthesized diameter due to both torque and moment loadings is used to calculate the static and dynamic stresses and the dynamic safety factor according to ASME fatigue boundary criterion. The input values in this column are conforming to Example 17.3. The output values are close to the solution of Example 17.3. The differences are due to the synthesis dependent on the synthesized diameter, and not the selected round figure diameter. If the selected round figure diameter is entered in the **Analysis** column, the output would then be the same as the hand calculations of Example 17.3 (not shown in Figure 17.5). The ultimate tensile strength, the yield strength, and the endurance limit of shaft material are changed in the **Analysis** column to improve the dynamic safety factor according to ASME.

There are other software packages that can be used to analyze or design shafts. One may find some over the Internet that can be downloaded for free. There is no guarantee that any of these would be accurate or free of bugs. ***PanDesign*** is one of these *CAD* and assembly of some machine elements including shaft synthesis. It was available for about the last three decades to undergraduate students. However, it is a legacy program under 32bit system, which may need tweaking to run under the 64bit Windows© system. Its construction was under the Microsoft© programming suite of Visual Studio© 6.0 (1998). It is available through the ***Wiley website*** under a possible download and the acceptance of the author to provide a license to the users of the textbook. It is mainly an SI system with few components that can solve the US system components or the ability to converge input and output. One can see Figure 8.24, which presents *PanDesign* interfaces. Figure 8.24a provides the components of *PanDesign* software for synthesis of shown machine elements and assembly of synthesized constructions. Figure 8.24b displays the *PanDesign* assembly of 15 [MW] gearbox components.

Example 17.4 The gearbox of Examples 17.1–17.3 transmits a maximum power of 75 [kW] or 100.58 [hp] at 1200 [rpm] input speed. It was required to adjust the synthesized input shaft on which the gear is attached employing a different material than the one of a 300 [MPa] or 43.5 [kpsi] tensile yield strength to have a better fatigue safety factor higher than the value of Example 17.3. Find the internal dynamic stresses if the shaft is synthesized as a cylinder of a 45 [mm] or 1.75 [in] diameter. Use the estimated factors affecting fatigue in Example 7.4 for the estimated life span of 10 years and the shaft operating 8 hours a day; the reliability should be 99.99%, the operating temperature is about 380 [°C] or 716 [°F], the evaluated stress concentration factor K_{SC} is 2.1, and the notch radius is 2.5 [mm] or 0.1 [in]. Find the better safety factor for ASME elliptic dynamic fatigue theory of failure, if the material is quenched and tempered to 205 [°C] or 400 [°F]. Discuss other alternatives.

Solution

Data: $H_{kW} = 75$ [kW] or $H_{hp} = 100.58$ [hp], $N_{rpm} = N_{in} = 1200$ [rpm], $\omega_P = 2\pi\ N_{in}/60 = 125.66$ [rad/s], and $d_C = 0.045$ [m] or 1.75 [in]. *Ductile* material was hot rolled AISI 1040 or ISO C40, yield strength of 42 [kpsi] or 290 [MPa], ultimate tensile strength of 76 [kpsi] or 525 [MPa], and endurance limit of 37.7 [kpsi] or 260 [MPa]; see Table A.7.2. The material quenched and tempered to 205 [°C] or 400 [°F] has yield strength of *86* [kpsi] or *593* [MPa], ultimate tensile strength of 113 [kpsi] or 779 [MPa], and calculated endurance limit of *56.95* [kpsi] or *392.6* [MPa].

Shaft Synthesis - US units

Inputs	Default	Input	Analysis
Power, H [hp]	33.50	100.58	100.58
Rotational Speed, N_{rpm} [rpm]	3000	1200	1200
Moment to Torque Ratio, M/T	1	0.706	0.706
Bending Moment, M [lb.in]	703.779	3728.56	3728.560
Safety Factor, K_{SF}	4	4	7.67
Output			Input;Output
Diameter, d_S [in] ***(no bending)***	0.8296	1.6242	
Diameter, d_S [in]	0.9554	1.7681	1.75
Ultimate Strength, [kpsi]	76	76	113
Yield Strength, [kpsi]	43.5	43.5	86
Endurance Limit, [kpsi]	38.304	38.304	56.95
Torque, T_x [lb.in]	703.779	5282.545	5282.545
Stresses (Static)			
Normal Stress, σ_x [kpsi]	8.221	6.8705E+00	7.086
Shear Stress, τ_{xy} [kpsi]	4.110	4.8670E+00	5.020
Max. Principal Stress, σ_1 [kpsi]	9.923	9.3924E+00	9.688
Min. Principal Stress, σ_2 [kpsi]	−1.703	−2.5220E+00	−2.601
Max. Shear Stress, τ_{max} [kpsi]	5.813	5.9572E+00	6.144
von Mises Stress, σ_{vM} [kpsi]	10.875	1.0875E+01	11.217
Stresses (Dynamic)			
Mean Normal Stress, σ_m [kpsi]	0.000	0.0000E+00	0.000
Alternating Normal Stress, σ_a [kpsi]	8.221	6.8705E+00	7.086
Mean Shear Stress, τ_m [kpsi]	4.110	4.8670E+00	5.020
Alternating Shear Stress, τ_a [kpsi]	0.000	0.0000E+00	0.000
Equivalent Mean Stress, σ'_m [kpsi]	7.119	8.4298E+00	8.695
Equivalent Alternating Stress, σ'_a [kpsi]	8.221	6.8705E+00	7.086
Commulative Factors Affecting Fatigue	4	4.5	4.5
Dynamic Safety Factor (ASME)	1.14	1.20	1.76

Shaft Synthesis - SI units

Inputs	Default	Input	Analysis
Power, H [kW]	25.00	75	75
Rotational Speed, N_{rpm} [rpm]	3000	1200	1200
Moment to Torque Ratio, M/T	1	0.706	0.706
Bending Moment, M [N.m]	79.5775	421.2435	421.2435
Safety Factor, K_{SF}	4	4	7.96
Output			Input;Output
Diameter, d_S [mm] ***(no bending)***	21.0743	41.2516	
Diameter, d_S [mm]	24.2707	44.9064	45
Ultimate Strength, [MPa]	520	520	779
Yield Strength, [MPa]	300	300	593
Endurance Limit, [MPa]	262	262.08	392.6
Torque, T_x [N.m]	79.5775	596.83125	596.83125
Stresses (Static)			
Normal Stress, σ_x [MPa]	56.695	4.7382E+01	47.086
Shear Stress, τ_{xy} [MPa]	28.347	3.3566E+01	33.357
Max. Principal Stress, σ_1 [MPa]	68.437	6.4775E+01	64.372
Min. Principal Stress, σ_2 [MPa]	−11.742	−1.7394E+01	−17.285
Max. Shear Stress, τ_{max} [MPa]	40.089	4.1084E+01	40.828
von Mises Stress, σ_{vM} [MPa]	75.000	7.5000E+01	74.533
Stresses (Dynamic)			
Mean Normal Stress, σ_m [MPa]	0.000	0.0000E+00	0.000
Alternating Normal Stress, σ_a [MPa]	56.695	4.7382E+01	47.086
Mean Shear Stress, τ_m [MPa]	28.347	3.3566E+01	33.357
Alternating Shear Stress, τ_a [MPa]	0.000	0.0000E+00	0.000
Equivalent Mean Stress, σ'_m [MPa]	49.099	5.8138E+01	57.776
Equivalent Alternating Stress, σ'_a [MPa]	56.695	4.7382E+01	47.086
Commulative Factors Affecting Fatigue	4	4.5	4.5
Dynamic Safety Factor (ASME)	1.14	1.20	1.82

Figure 17.5 Shaft Synthesis Tablets for the US and SI systems of units.

From Examples 17.1–17.3, the following relevant information are reiterated:

- Input torque $\boldsymbol{T} = 596.83$ [N m] and 5282.6 [lb in].
- Internal moments at location a, $\boldsymbol{M} = [0\,0\,421.2435]^T$ [N m] and $[0\,0\,3728.56]^T$ [lb in].
- The shaft diameters to the nearest preferred dimensions are 45 [mm] and 1.75 [in].

The shaft maximum and the minimum normal stresses and the maximum and the minimum shear stresses as defined in Eqs. (17.8) and (17.9) at location a are as follows (same as Example 17.3):

$$\sigma_{\max} = \frac{32\boldsymbol{M}}{\pi d_S^3} = \frac{32(421.2435)}{\pi\,(45/1000)^3} = 47\,086\,431\text{ [Pa]} = 47.086\text{ [MPa]},\quad \sigma_{\min} = -\sigma_{\max} = -47.086\text{ [MPa]}$$

$$\tau_{\max} = \frac{16\boldsymbol{T}}{\pi d_S^3} = \frac{16(596.83)}{\pi\,(45/1000)^3} = 33\,356\,710\text{ [Pa]} = 33.357\text{ [MPa]},\quad \tau_{\min} = 33.357\text{ [MPa]} \tag{a}$$

and

$$\sigma_{\max} = \frac{32\boldsymbol{M}}{\pi d_S^3} = \frac{32\ (3728.56)}{\pi\ (1.75)^3} = 7086.4\text{ [psi]} = 7.0864\text{ [kpsi]},\quad \sigma_{\min} = -\sigma_{\max} = -7.0864\text{ [kpsi]}$$

$$\tau_{\max} = \frac{16\boldsymbol{T}}{\pi d_S^3} = \frac{16\ (5282.6)}{\pi\ (1.75)^3} = 5019.99\text{ [psi]} = 5.020\text{ [kpsi]},\quad \tau_{\min} = 5.020\text{ [MPa]} \tag{b}$$

The values of the mean stresses σ_m, τ_m and the alternating stresses σ_a, τ_a as in Eqs. (17.11) and (17.12) are then as follows (same as Example 17.3):

$$\sigma_m = \frac{1}{2}(\sigma_{\max} + \sigma_{\min}) = 0$$

$$\sigma_a = \frac{1}{2}(\sigma_{\max} - \sigma_{\min}) = \sigma_{\max} = 47.086\text{ [MPa] or } = 7.0864\text{ [kpsi]}$$

$$\tau_m = \frac{1}{2}(\tau_{\max} + \tau_{\min}) = \tau_{\max} = 33.357\text{ [MPa] or } = 5.020\text{ [kpsi]}$$

$$\tau_a = \frac{1}{2}(\tau_{\max} - \tau_{\min}) = 0 \tag{c}$$

The *equivalent von Mises stresses* $\sigma_{\text{vM},m}$ and $\sigma_{\text{vM},a}$ are as follows (same as Example 17.3; see Eq. (17.14)):

$$\sigma_{\text{vM},m} = \sqrt{\sigma_m^2 + 3\tau_m^2} = \sqrt{3\tau_m^2} = \sqrt{3}\ \tau_m = 1.7321(33.357) = 57.776\text{ [MPa]}\ \text{ or } = 1.7321(5.02)$$
$$= 8.6951\text{ [kpsi]}$$

$$\sigma_{\text{vM},a} = \sqrt{\sigma_a^2 + 3\tau_a^2} = \sqrt{\sigma_a^2} = \sigma_a = 47.086\text{ [MPa]}\ \text{ or } = 7.0864\text{ [kpsi]} \tag{d}$$

To estimate the factors affecting fatigue and the shaft fatigue strength using Eq. (7.67)–(7.83) and (17.16), one observes that the shaft dimension and the dynamic considerations are identical to that given in Example 7.4. From Example 7.4, Eq. (d) is recalculated as follows (***not*** the same as Example 17.3):

$$S_{e,S} = (0.781)(0.825)(1.0)(0.702)(0.913)(0.537)(1.0)\ S_e = 0.2218\ \ (392.6) = 87.079\ \text{ [MPa]}$$

$$S_{e,S} = (0.781)(0.828)(1.0)(0.702)(0.912)(0.537)(1.0)\ S_e = 0.2224\ \ (56.95) = 12.666\ \text{ [kpsi]} \tag{e}$$

Equation (e) indicates that the cumulative factors affecting fatigue are 0.2218 or 0.2224. This reduces the endurance limit by (1/0.2218) = 4.5 times or (1/0.2224) = 4.5 times also. This can be used as the cumulative reduction factor affecting fatigue as a reciprocal value similar to a safety factor that is larger than 1.0.

For the *ASME criterion*, the fatigue safety factor is then obtained from Eq. (17.15) such that

$$K_{SF,F}\Big|_{SI} = \frac{1}{\sqrt{(\sigma_{\text{vM},a}/S_{e,S})^2 + (\sigma_{\text{vM},m}/S_{yt})^2}} = \frac{1}{\sqrt{\left(\frac{47.086}{87.079}\right)^2 + \left(\frac{57.776}{593}\right)^2}} = 1.8201$$

$$K_{SF,F}\Big|_{US} = \frac{1}{\sqrt{(\sigma_{\text{vM},a}/S_{e,S})^2 + (\sigma_{\text{vM},m}/S_{yt})^2}} = \frac{1}{\sqrt{\left(\frac{7.0864}{12.666}\right)^2 + \left(\frac{8.6951}{86}\right)^2}} = 1.7589 \tag{f}$$

These values, confirming those in Figure 17.5, are not so small considering the harsh consideration of the factors affecting fatigue. The material can be further changed to another material with a higher endurance limit and having a ground surface finish, or the shaft diameter and/or the notch radius should be increased. A shaft diameter of 50 [mm] or 2 [in] might again be much safer for the harsh factors affecting fatigue. The *quenched and tempered* process did improve the material fatigue strength to accommodate these harsh factors affecting fatigue. However, the tempering temperature is not realistic with the much higher operating temperature. The properties should be changed to a tempering temperature higher than the operating temperature. The dynamic safety factor is expected to be less than the results in Eq. (f). The diameters may effectively be changed to 50 [mm] or 2 [in] to remedy this reduction in safety factor.

The *Excel tablets* in Figure 17.5 are available in the ***Wiley website*** for *computer-aided synthesis* or real *CAD* of shafts under the name of ***Shaft Synthesis Tablet – SI.xlsx*** and ***Shaft Synthesis Tablet – US.xlsx***. These should be useful in calculating synthesized shaft diameters at all points along the shaft.

17.7.3 Optimum Design of Shafts

As indicated previously in Section 17.7.1, an initial good selection of appropriate material reduces the iterations to achieve the optimum. Alternatively, the design can be changed to satisfy a specific safety factor as indicated in Section 7.10.4 about the optimization for a specific factor of safety for carbon steel shafts. The optimization code adjusted few lines in the MATLAB code of (*Fatigue_Strength_and_Factors_Affecting.m*) as indicated in Figure 7.36. This code is provided in this chapter with the code readjusted and renamed as (***CAD_Shaft_Fatigue_Optimum.m***). The code is used iteratively to input a different value of the shaft diameter as a part diameter (dp) in [mm] or [in] while observing the output dynamic safety factor and the output reiteration optimum shaft diameter. This is achievable by the input of the bending moment (Bending) and the torque (Torque) in [N m] or [lb in], respectively. The normal stresses (Sigma) and the shear stresses (Tau) are then functions of the part diameter with results in [Pa] or [psi]. The stresses are then set in [MPa] and [kpsi]. These changes are evident in Figure 7.37 for both SI and the US systems of units with only the SI fluctuating stresses shown. The code, however, includes both changes in the SI and US systems of units.

These MATLAB code under the name ***CAD_Shaft_Fatigue_Optimum.m*** is available at ***Wiley website.***

Alternatively, the ***Shaft Synthesis Tablet – SI.xlsx*** and ***Shaft Synthesis Tablet – US.xlsx*** can be used to perform the iteration. However, one needs the percentage savings in volume as a function of the iterative diameter raised to the power 2 divided by the original diameter raised to the power 2; see Figure 7.37. These are the ratios of the areas considering a unit length to compute the volume; see Section 7.10.4. The optimization conversion is extremely faster by the diameter relation used in that section. This relation is stemming from the fact that due to stresses, the safety factor $K_{SF,F}$ is a function of $(\text{dp})^3$. Utilizing the *HGP* optimization method for a 2.0 *safety factor aim* provides the new part diameter $d_{p,\text{new}} = d_p(2.0/K_{SF,F})^{(1/3)}$. For some sections of the shaft, the diameter is then rounded off to be able to produce and fit with other mounted elements.

Example 17.5 The gearbox of Examples 17.1–17.3 transmits a maximum power of 75 [kW] or 100.58 [hp] at 1200 [rpm] input speed. It is required to optimize the synthesized input shaft on which the gear is attached employing the original material of a 300 [MPa] or 43.5 [kpsi] tensile yield strength to satisfy a better fatigue safety factor of at least 2.0. Start with the synthesized shaft of a 45 [mm] or about 1.75 [in] diameter. Use the estimated factors affecting fatigue in Example 7.4 for the estimated life span of 10 years and the shaft operating 8 hours a day; the reliability should be 99.99%, the operating temperature is about 380 [°C] or 716 [°F], the evaluated stress concentration factor K_{SC} is 2.1, and the notch radius is 2.5 [mm] or 0.1 [in]. Satisfy the better dynamic safety factor for ASME elliptic dynamic fatigue theory of failure using the MATLAB code (*CAD_Shaft_Fatigue_Optimum.m*).

Solution

Data: $H_{kW} = 75$ [kW] or $H_{hp} = 100.58$ [hp], $N_{rpm} = N_{in} = 1200$ [rpm], $\omega_P = 2\pi\ N_{in}/60 = 125.66$ [rad/s], and $d_C = 0.045$ [m] or 1.75 [in]. *Ductile* material was hot rolled AISI 1040 or ISO C40, yield strength of 42 [kpsi] or

290 [MPa], ultimate tensile strength of 76 [kpsi] or 525 [MPa], and endurance limit of 37.7 [kpsi] or 260 [MPa]; see Table A.7.2.

From Examples 17.1–17.3, the following relevant information are reiterated:

- Input torque $\boldsymbol{T} = 596.83$ [N m] and 5282.6 [lb in].
- Internal moments at location a $\boldsymbol{M} = [0\,0\,421.2435]^T$ [N m] and $[0\,0\,3728.56]^T$ [lb in].
- The shaft diameters to the nearest preferred dimensions are 45 [mm] or 1.75 [in].
- The maximum and the minimum normal stresses and the maximum and the minimum shear stresses are as follows:

$$\sigma_{\max} = 47.086\ [\text{MPa}], \quad \sigma_{\min} = -\ \sigma_{\max} = -47.086\ [\text{MPa}]$$

$$\tau_{\max} = 33.357\ [\text{MPa}], \quad \tau_{\min} = 0\ [\text{MPa}] \tag{a}$$

$$\sigma_{\max} = 7.0864\ [\text{kpsi}], \quad \sigma_{\min} = -\ \sigma_{\max} = -7.0864\ [\text{kpsi}]$$

$$\tau_{\max} = 5.020\ [\text{kpsi}], \quad \tau_{\min} = 0\ [\text{MPa}] \tag{b}$$

- The mean stresses σ_m, τ_m and the alternating stresses σ_a, τ_a

$$\sigma_m = \frac{1}{2}(\sigma_{\max} + \sigma_{\min}) = 0, \quad \sigma_a = \frac{1}{2}(\sigma_{\max} - \sigma_{\min}) = \sigma_{\max} = 47.086\ [\text{MPa}] \text{ or } = 7.0864\ [\text{kpsi}]$$

$$\tau_m = \frac{1}{2}(\tau_{\max} + \tau_{\min}) = \tau_{\max} = 33.357\ [\text{MPa}] \ \text{ or } = 5.020\ [\text{kpsi}], \quad \tau_a = \frac{1}{2}(\tau_{\max} - \tau_{\min}) = 0 \tag{c}$$

- The *equivalent von Mises stresses* $\sigma_{\text{vM},m}$ and $\sigma_{\text{vM},a}$ are as follows (same as Example 17.3; see Eq. (17.14)):

$$\sigma_{\text{vM},m} = 57.776\ [\text{MPa}] \ \text{ or } = 8.6951\ [\text{kpsi}]$$

$$\sigma_{\text{vM},a} = 47.086\ [\text{MPa}] \ \text{ or } = 7.0864\ [\text{kpsi}] \tag{d}$$

- The shaft fatigue strength including factors affecting fatigue is

$$S_{e,S} = 57.668\ \ [\text{MPa}] \text{ or } S_{e,S} = 8.3845\ \ [\text{kpsi}] \tag{e}$$

- The *ASME* fatigue safety factor is then

$$K_{\text{SF},F}\Big|_{\text{SI}} = 1.1898, \text{ or } K_{\text{SF},F}\Big|_{\text{US}} = 1.1492 \tag{f}$$

These safety factors are small due to the harsh consideration of the factors affecting fatigue. A larger shaft diameter of 50 [mm] or 2 [in] or higher might be safer for the harsh factors affecting fatigue. The dynamic safety factor may be less than the required fatigue safety factor of at least 2.0. The diameters may effectively be changed by optimization to satisfy that. Utilizing the *HGP* optimization method for a 2.0 *safety factor aim* suggested the new part diameter $d_{p,\text{new}} = d_p(2.0/K_{\text{SF},F})^{(1/3)}$. This gives the following first iteration values:

$$d_{p,\text{new}} = d_p\sqrt[3]{\frac{2.0}{K_{\text{SF},F}}} = 45\ \sqrt[3]{\frac{2.0}{1.1898}} = 53.505\ [\text{mm}]$$

$$d_{p,\text{new}} = d_p\sqrt[3]{\frac{2.0}{K_{\text{SF},F}}} = 1.75\ \sqrt[3]{\frac{2.0}{1.1492}} = 2.104\,99\ [\text{in}] \tag{g}$$

Using the MATLAB code (***CAD_Shaft_Fatigue_Optimum.m***) with an initial shaft diameter of 45 [mm] gives a first iteration value of 53.3691 [mm]. Using this value for the next iteration gives the second iteration value of 53.6765 [mm]. Using this value for the next iteration gives the third iteration value of 53.6869 [mm]. Using this value for the next iteration gives the fourth iteration value of 53.6872 [mm] and the *ASME* fatigue safety factor of

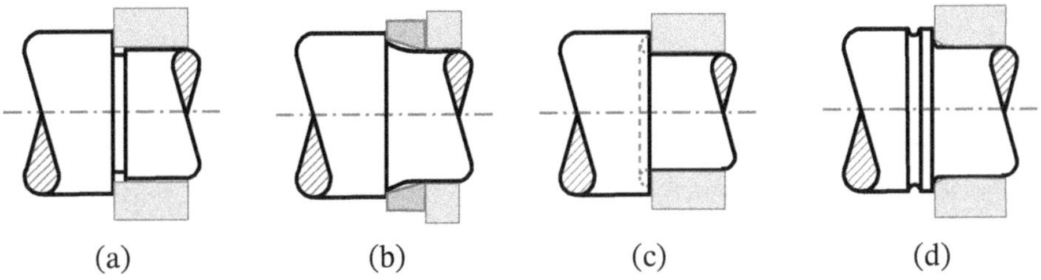

Figure 17.6 Some methods of improving the stress concentration factor and fatigue strength at shaft shoulders: (a) a recess or a neck, (b) a large elliptic fillet and an intermediate spacer ring, (c) a fillet and undercut shoulder face, and (d) stress relief groove and a fillet.

2.0000 for SI units. This conversion is good enough. The round off values of shaft diameter of 55 [mm] and 2.125 [in] (i.e. $2\frac{1}{8}$ [in]) give the *ASME* fatigue safety factors of 2.1451 and 2.0323 for SI and the US units, respectively.

17.8 Constructional Details

Many constructional details have been discussed in previous chapters, which involve machine elements mounted on shafts such as gears in gearboxes, rolling bearings, and other joining elements in Chapters 8 and 11. It is recommended to go over figures related to these methods of mounting such as Figures 8.4, 8.8, 8.12, 8.14, 8.15, 11.5, 11.12, and 11.13.

Of importance to shaft fatigue strength and factors affecting fatigue, Figure 17.6 presents some constructional methods of improving the stress concentration factor and fatigue strength at shaft shoulders. The figure shows the mounted elements as light grayish sections resting on the left shoulders of the shaft. Figure 17.6a shows a recess or a neck to allow for grinding wheel clearance to produce a better surface finish. Figure 17.6b defines a large elliptic fillet to reduce stress concentration factor and suggests an intermediate spacer ring in darker grayish color for proper mounting. Figure 17.6c defines a fillet and undercut made in the shoulder face. Figure 17.6d presents a stress relief groove and a fillet, which should reduce stress concentration factor. The stress concentration factor and the design of these methods can be attained and improved by numerical ***FE*** and geometric optimization.

Figure 17.7 presents a sketch of a sample left end assembly of a shaft. This end is supported by two bearings, and the V-belt pulley is overhung. The right end of the shaft does not need another support. Another mounted element such as a gear is attached to the right overhang end (not shown). It is clear that the inner rings of the bearings are tightly attached to the shaft right shoulder with a spacer and the pulley's hub keeping all of them squeezed to the shoulder by the locking nut at the left end of the shaft. The outer rings of the rolling bearings are fixed to the outer housing by a shoulder to the right and an inner snap or retaining ring to the left. This can stand a higher axial load pushing the shaft to the right and a slight axial load pushing the shaft assembly to the left against the snap ring. The shoulder and the snap or retaining ring can switch locations if the axial load is larger to the left. The left rolling bearing should have a self-sealing cover if the setting is not extremely imbedded in lubricant to the right of the bearings. In that case a proper seal is needed between the left bearing and the snap ring. The pulley is kept as close to the bearing as possible to reduce the shaft bending moment. Other arrangements can let the V-belt grooves on top of the left bearing to further reduce the bending. The pulley will be capping the outer housing of the two bearings and all around their housing assembly. The pulley, however, will have a bending moment on its hub, and that should be taken care of in pulley design. The shaft will also have a bending moment at its left end.

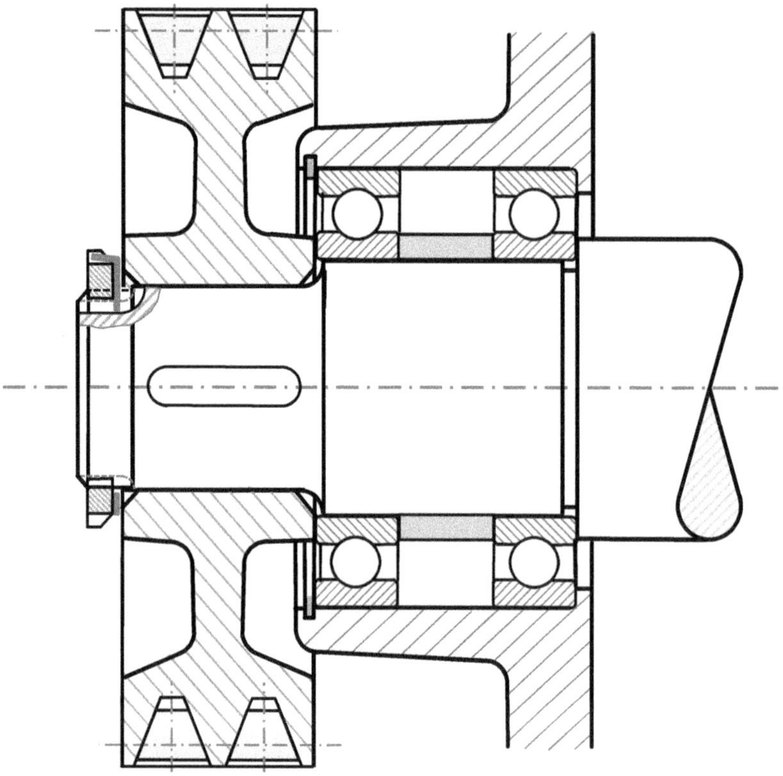

Figure 17.7 A sketch of a left end assembly of a shaft. This end is supported by two bearings, and the V-belt pulley is overhung. The right end of the shaft does not need another support. Another mounted element such as a gear is attached to the right overhang end (not shown).

17.9 Summary

The shaft is a very important machine element that connects other components in power transmission. The synthesis of a shaft is, consequently, dependent on other machine elements mounted on the shaft. Therefore, this chapter has been pushed back after so many other machine elements in this text. The initial synthesis of shafts depending mainly on power transmission has been presented in Chapter 8 to facilitate a better synthesis of other mounted machine elements.

This chapter introduces the different types of shafts and axles and presents the mathematical model to handle proper synthesis of shafts or axles. Computer code (MATLAB; *Force_Analysis_3D_Shafts_Short.m*) is provided to find the external force analysis and the internal shear force and bending moment at any location or any *critical location* along the shaft ***FBD***. One needs to input all applied external forces and moments stemming from the mounted elements on the shaft at their defined location vectors. After the process of finding the internal shear force and bending moment, one can initially synthesize the shaft and define stresses, slopes, and deflections at any location along the shaft. Another MATLAB code of *Joining_Elements.m* can be used to initially synthesize shafts according to *von Mises* static failure theory in addition to the prospective joining components along the shaft; see Section 8.4.2.

The detailed design of shafts involves the consideration of various dynamic loadings and the different factors affecting fatigue strength. The *von Mises* dynamic failure theory and the *ASME elliptic criterion* are used to obtain the *fatigue safety factor* $K_{SF,F}$ for the initially synthesized shafts. A MATLAB code is provided to perform this process under the name *Fatigue_Strength_and_Factors_Affecting.m*. The code is also used to calculate the factors affecting fatigue strength of shaft material. Considerations of shaft rigidity and deflection in addition to avoiding critical speed and excessive vibrations are introduced in this chapter.

For *CAD* and optimization of shafts, *Excel tablets* are developed and available in the ***Wiley website*** as *computer-aided synthesis* or real *CAD* of shafts under the name of *Shaft Synthesis Tablet – SI.xlsx* and *Shaft Synthesis Tablet – US.xlsx*. These should be useful in calculating synthesized shaft diameters at all points along the shaft. These may also be used, iteratively, to optimize the shaft. Another MATLAB code of *CAD_Shaft_Fatigue_Optimum.m* is developed to optimize the shaft for a set value of the dynamic safety factor. It might be adjusted to do other optimization objectives.

Previously provided tools are to be used along the shaft to provide a better or optimum design. A variation, verification, and validation would be performed on the ***3D*** model by any numerical ***FE*** procedure or other means.

Problems

17.1 Identify few applications or products that utilize shafts in their construction. Use the Internet to collect images of such applications. Why are these applications using shafts in that capacity?

17.2 Define more shaft types other than the ones covered in the chapter. Sketch those other types and indicate their utility.

17.3 Download or acquire a standard about shafts, and study the procedure for design and selection.

17.4 If the input forces in Example 17.1 are considered as uniformly distributed over the gear width, what would be the bending moment and shear force distributions along the shaft? What is the expected change in the initial synthesis of shaft diameter?

17.5 Do you need to adjust any code line in the MATLAB code of Figures 17.3 and 17.4 to be used in the US system of units? What are the changes needed in code lines or comments?

17.6 Adjust the MATLAB code in Figures 17.3 and 17.4 to develop the bending moment and shear force distributions along the shaft.

17.7 Download or acquire a computer code or a package that develops the bending moment and shear force distributions along the shaft. *Hint*: Add the distance x to the definition of the position vectors of the applied loads.

17.8 A propeller shaft of a large cruise ship rotates very slowly. The maximum bending is 25 [kN m] or 221 [klb in]. The maximum torsion is 2 [MN m] or 17.7 [Mlb in]. For an initial safety factor of 5, what should be the propeller shaft diameter if the propeller shaft material is a hot rolled AISI 1030 or ISO C30? Use the safety factor for an appropriate static theory of failure. What is the safety factor if one uses hot rolled AISI 1020 or ISO C20?

17.9 The shaft of a drone motor is rotating very fast and subject to very little bending moments from the propeller due to maneuvering. The maximum torsional moment is 0.05 [N m] or 0.44 [lb in], and the maximum bending is 0.002 [N m] or 0.018 [lb in]. For an initial safety factor of 4, what should be the propeller shaft diameter if the propeller shaft material is a hot rolled AISI 1030 or ISO C30? Use the safety factor for an appropriate static theory of failure. What is the safety factor if one uses hot rolled AISI 1020 or ISO C20?

17.10 Use the available MATLAB code of *Force_Analysis_3D_Shafts_Short.m* to resolve Example 17.1, but double the power transmission at the same input speed and different shaft length. Two helical gears of the same size are used face-to-face to accommodate the transmitted power to and from the shaft. The *input* shaft is longer due the two adjacent gears, and each bearing width should also be larger. Expect the bearing width is required to be about 1.5 that of the bearings in Example 17.1. Use an initial safety factor of 5 and a basic material of a 300 [MPa] or 43.5 [kpsi] tensile yield strength to initially synthesize the shaft diameter at the expected critical location.

17.11 The gearbox of Examples 17.1 and 17.2 transmits a maximum power of 75 [kW] or 100.58 [hp] at 1200 [rpm] input speed. The standard helical gears on the main input and output shafts have been synthesized through Examples 15.1–15.4. The gear width on the *output* shaft has been defined as 65 [mm] or 2.6 [in]. Presume the deep groove ball bearing width is 45 [mm] or 1.75 [in] on the shaft. It is required to synthesize the *output* shaft on which the gear is attached employing a basic material of a 300 [MPa] or 43.5 [kpsi] tensile yield strength and a safety factor of 4.0.

17.12 Use the available MATLAB code of *Force_Analysis_3D_Shafts_Short.m* to resolve Example 17.1, but double the power transmission at the same input speed and different shaft length. Two helical gears of the same size are used face-to-face to accommodate the transmitted power to and from the shaft. The *output* shaft is longer due the two adjacent gears, and each bearing width should also be larger. Expect the bearing width on the *output* shaft is required to be about 2.5 that of the bearings in Example 17.1. Use an initial safety factor of 5 and a basic material of a 300 [MPa] or 43.5 [kpsi] tensile yield strength to initially synthesize the shaft diameter at the expected critical location.

17.13 The forces on the shaft of the gearbox transmitting a maximum power of 108 [kW] or 145 [hp] at 1750 [rpm] input speed are assumed to be about the same as those in Example 17.1. It is required to initially synthesize the *input* shaft on which the gear is attached. Assume bearings of the same expected width.

17.14 The forces on the shaft of the gearbox transmitting a maximum power of 45 [kW] or 60 [hp] at 750 [rpm] input speed are assumed to be about the same as those in Example 17.1. It is required to initially synthesize the *input* shaft on which the gear is attached. Assume bearings of the same expected width.

17.15 The forces on the shaft of the gearbox transmitting a maximum power of 108 [kW] or 145 [hp] at 1750 [rpm] input speed are assumed to be about the same as those in Example 17.1. It is required to initially synthesize the *output* shaft on which the gear is attached. Expect the bearing width on the *output* shaft is required to be about 2.5 that of the bearings in Example 17.1.

17.16 The forces on the shaft of the gearbox transmitting a maximum power of 45 [kW] or 60 [hp] at 750 [rpm] input speed are assumed to be about the same as those in Example 17.1. It is required to initially synthesize the *output* shaft on which the gear is attached. Expect the bearing width on the *output* shaft is required to be about 2.5 that of the bearings in Example 17.1.

17.17 The forces on the shaft of the gearbox transmitting a maximum power of 745 [kW] or 1000 [hp] at 12 000 [rpm] input speed are assumed to be about the same as those in Example 17.1. It is required to initially synthesize the *input* shaft on which the gear is attached. Assume bearings of the same expected width.

17.18 The forces on the shaft of the gearbox transmitting a maximum power of 745 [kW] or 1000 [hp] at 12 000 [rpm] input speed are assumed to be about the same as those in Example 17.1. It is required to initially synthesize the *output* shaft on which the gear is attached. Expect the bearing width on the *output* shaft is required to be about 2.5 that of the bearings in Example 17.1.

17.19 Redo Example 17.3 considering the same factors affecting fatigue but with operating temperature of about 100 [°C] or 212 [°F], the evaluated stress concentration factor K_{SC} of 1.5, and the notch radius of 5 [mm] or 0.2 [in].

17.20 Calculate the maximum transverse deflection and slope for the shaft in Examples 17.1–17.3. Are they within the allowable values?

17.21 Use an ***FE*** program to estimate the maximum transverse deflection and slope for the shaft in Examples 17.1–17.3. Compare values with the simplified model in Section 17.5.

17.22 Use the ***FE*** program to estimate the critical speed for the shaft in Examples 17.1–17.3. Compare values with the simplified model in Section 17.6.

17.23 Search for the unit cost of different shaft materials. What is the relative cost of heat treatment of these materials?

17.24 Redo Example 17.4 with the same material and factors affecting fatigue but with quenched and tempered to 430 [°C] or 800 [°F]. Compare results.

17.25 Redo Example 17.4 with the same material and factors affecting fatigue but with different diameters to 50 [mm] or 2 [in], and no heat treatment. Compare results.

17.26 Use one of the available tablets *Shaft Synthesis Tablet – SI.xlsx* or *Shaft Synthesis Tablet – US.xlsx* to resolve Problem 17.8.

17.27 Use one of the available tablets *Shaft Synthesis Tablet – SI.xlsx* or *Shaft Synthesis Tablet – US.xlsx* to resolve Problem 17.9.

17.28 Use one of the available tablets *Shaft Synthesis Tablet – SI.xlsx* or *Shaft Synthesis Tablet – US.xlsx* to resolve Problem 17.10.

17.29 Use one of the available tablets *Shaft Synthesis Tablet – SI.xlsx* or *Shaft Synthesis Tablet – US.xlsx* to resolve Problem 17.11.

17.30 Use one of the available tablets *Shaft Synthesis Tablet – SI.xlsx* or *Shaft Synthesis Tablet – US.xlsx* to resolve Problem 17.12.

17.31 Use one of the available tablets *Shaft Synthesis Tablet – SI.xlsx* or *Shaft Synthesis Tablet – US.xlsx* to resolve Problem 17.13.

17.32 Use one of the available tablets *Shaft Synthesis Tablet – SI.xlsx* or *Shaft Synthesis Tablet – US.xlsx* to resolve Problem 17.14.

17.33 Use one of the available tablets *Shaft Synthesis Tablet – SI.xlsx* or *Shaft Synthesis Tablet – US.xlsx* to resolve Problem 17.15.

17.34 Use one of the available tablets *Shaft Synthesis Tablet – SI.xlsx* or *Shaft Synthesis Tablet – US.xlsx* to resolve Problem 17.16.

17.35 Use one of the available tablets *Shaft Synthesis Tablet – SI.xlsx* or *Shaft Synthesis Tablet – US.xlsx* to resolve Problem 17.17.

17.36 Use one of the available tablets *Shaft Synthesis Tablet – SI.xlsx* or *Shaft Synthesis Tablet – US.xlsx* to resolve Problem 17.18.

17.37 Use one of the available tablets *Shaft Synthesis Tablet – SI.xlsx* or *Shaft Synthesis Tablet – US.xlsx* to resolve Problem 17.19. Use the MATLAB code of *Fatigue_Strength_and_Factors_Affecting.m* to recalculate the factors affecting fatigue.

17.38 Use one of the available tablets *Shaft Synthesis Tablet – SI.xlsx* or *Shaft Synthesis Tablet – US.xlsx* to resolve Problem 17.25. Use the MATLAB code of *Fatigue_Strength_and_Factors_Affecting.m* to recalculate the factors affecting fatigue.

17.39 Use one of the available tablets *Shaft Synthesis Tablet – SI.xlsx* or *Shaft Synthesis Tablet – US.xlsx* to resolve Example 17.5. Use the MATLAB code of *Fatigue_Strength_and_Factors_Affecting.m* to recalculate the factors affecting fatigue.

17.40 Use one of the available tablets *Shaft Synthesis Tablet – SI.xlsx* or *Shaft Synthesis Tablet – US.xlsx* to resolve Example 17.5 for a basic material of a 300 [MPa] or 43.5 [kpsi] tensile yield strength. Use the MATLAB code of *Fatigue_Strength_and_Factors_Affecting.m* to recalculate the factors affecting fatigue.

17.41 Use the available MATLAB code *CAD_Shaft_Fatigue_Optimum.m* to resolve any of the previous problems for some acceptable safety factor.

17.42 It is required to design a power transmission unit. It is to use a 8.9 [kW] or 12 [hp] motor running at 3000 [rpm] and driving the shaft that runs at 1500 [rpm] through a belt system. The shaft transmits the power to a bevel gear set; see, e.g. Figure 17.7 for the shaft. The gear reduction ratio is about 4. The output speed of the bevel gear set should be about 375 [rpm]. The forces on the V-belt pulley are about $\boldsymbol{F}_y = -484$ [N] or -109 [lb], $\boldsymbol{F}_z = -44$ [N] or -10 [lb], and the torque $\boldsymbol{M}_x = 56.66$ [N m] or 502 [lb in]. The forces on the bevel gear are about $\boldsymbol{F}_x = -163$ [N] or -37 [lb], $\boldsymbol{F}_y = -651$ [N] or -147 [lb], and $\boldsymbol{F}_z = -1842$ [N] or -415 [lb], and the torque $\boldsymbol{M}_x = -56.66$ [N m] or -502 [lb in]. Assume the pulley to start at the left with 120 [mm] or 4.8 [in] (for 120 [mm] or 4.8 [in] width), the left bearing may need 60 [mm] or 2.4 [in] (including cover and seal), the right bearing centerline is about 100 [mm] or 4 [in] from the left bearing centerline, the right bearing may need 40 [mm] or 1.5 [in], and the pinion footprint on the shaft is about 40 [mm] or

1.5 [in]. This will require nodes (or positions and distances in x-direction) at 0 start, 60 [mm] or 2.4 [in] (middle of pulley), 120 [mm] or 4.8 [in] (end of pulley), 180 [mm] or 7.2 [in] (left bearing centerline), 280 [mm] or 11.2 [in] (right bearing centerline), 320 [mm] or 12.8 [in] (pinion centerline), and 340 [mm] or 13.6 [in] (end of shaft). It is required to synthesize the shaft.

References

Abdel Meguid, M.R.M. (1999). *Computer aided material selection optimization*. MS thesis. Cairo University.

ANSI/ASME B106.1M (1985). *Design of transmission shafting*. American Society of Mechanical Engineers.

ASM (1996). *ASM handbook volume 19 – fatigue and fracture*. ASM International.

Barsom, J.M. and Rolfe, S.T. (1987). *Fatigue and Fracture Control in Structures*, 2e. Prentice Hall, Copyright ASTM International.

Gorman, D.J. (1975). *Free Vibration Analysis of Beams and Shafts*. Wiley.

Inman, D.J. (2017). *Vibration with Control*. Wiley.

Marin, J. (1962). *Mechanical Behavior of Engineering Materials*. Prentice Hall.

Microsoft (1998) *Visual Studio© 6.0* Microsoft

Thomson, W.T. and Dahleh, M.D. (1998). *Theory of Vibration with Applications*, 5e. Prentice Hall.

von Mises, R. (1913). Mechanik der festen Körper im plastisch deformablen Zustand. *Göttin. Nachr. Math. Phys.* 1: 582–592.

Internet Links

www.ansi.org ANSI: Standards.

www.iso.org ISO: Standards.

www.asme.org American Society of Mechanical Engineers: Standards.

www.steel.org American Iron and Steel Institute: Standards.

18

Clutches, Brakes, and Flywheels

This chapter is dedicated to the elements that control power transmission, particularly clutches, brakes, and flywheels. Coupling elements are used in rotation transfer from one shaft to another, and if they are flexible, they will perform some control over some short time. The fluid coupling, torque converter, or a hydromantic has such a flexibility to be considered as close to a clutch than a coupling, if one needs to differentiate.

Clutches and brakes are used to control rotation between shafts for clutches or between shafts and the ground for brakes. Flywheels are used to smooth variations in rotational speeds due to generated or consumed energy variation with time.

Symbols

The adapted units are [in, lb, psi] or [m, kg, N, Pa], others given at each symbol definition ([k...] is 10^3, [M...] is 10^6, and [G...] is 10^9).

Symbol	Quantity, units (adopted)
$\dot{E}_C$	Rate of clenching, braking, or clutching energy
$\dot{E}_{\max}$	Maximum energy dissipation
$(2\beta_C/2\pi)$	Caliper subtended ratio
$(n_F\mu_K p_{\max})$	Multi-disk clutch-brake parameter
(pv_C)	Pressure times velocity
C_F	Coefficient of flywheel speed fluctuation
c_p	Specific heat of material
C_w	Wear constant
dA_C	Area of the conical element
d_{Fo}, d_{Fi}	Flywheel outer and inner diameters
d_i	Inner diameter
d_o	Outer diameter
dr	Radius variation at element
E_C	Total clenching, braking, or clutching energy
$E_{\min}$, $E_{\max}$	Minimum and maximum kinetic energies

Machine Design with CAD and Optimization, First Edition. Sayed M. Metwalli.

Companion website: www.wiley.com/go/metwalli/machine

Symbol	Quantity, units (adapted)
$\boldsymbol{F}_a$	Axial force
H	Transmitted power
J_F	Flywheel mass moment of inertia
J_i	Input mass moment of inertia
J_o	Output mass moment of inertia
K_{service}	Service factor
K_{SF}	Safety factor
K_{start}	Start factor
m_C	Clutch-brake mass of disk contact components
m_F	Flywheel mass
n_D	Number of disks
n_F	Number of friction surfaces
N_{rpm}	Rotational speed [rpm]
OEM	Original equipment manufacturer
p	Applied pressure or pressure intensity
$p_{\max}$	Maximum allowable uniform pressure
r	Radius
r_{Fo}, r_{Fi}	Flywheel outer and inner radii
$\boldsymbol{T}$	Transmitted torque
t	Time
$\boldsymbol{T}_{\text{av}}$	Average or nominal torque
t_C	Clenching, braking, or clutching time
v_C	Velocity at clutch-brake outer diameter
α_C	Cone clutch angle
β_C	Caliper disk angle [rad]
ΔE_F	Flywheel change in kinetic energy
ΔT_C	Contact temperature rise
μ_K	Kinetic friction coefficient
ρ	Density
ω_{av}	Average rotational speed or nominal speed
ω_i, $\dot{\theta}_i$	Angular velocity of the input
$\omega_{\max}$	Maximum rotational speed
$\omega_{\min}$	Minimum rotational speed
ω_o, $\dot{\theta}_o$	Angular velocity of the output
$\ddot{\theta}_i$	Angular acceleration of the input
$\ddot{\theta}_o$	Angular acceleration of the output
θ_i	Angular rotation of input
θ_o	Angular rotation of output

18.1 Classifications of Clutches and Brakes

The classification of clutches or brakes depends on the method of motion control and engagement activation. The motion control can be a positive mechanical method such as rigid teeth or jaws engagement or can be through friction interaction between engaging sides. The engagement activation method can be mechanical, pneumatic, hydraulic, electrical, magnetic, etc.

The classification of clutches and brakes can then be defined in the following categories:

- Positive mechanical engagement such as toothed or jaw clutches, where the control of motion is abrupt or in a very short duration.
- Friction-type engagement such as disk type, multi-disk type, cone type, radial type with external or internal shoes, band, or belt type. The motion control is gradual and taking some time to settle due to friction allowing the slipping action.
- Friction-positive engagement such as overrunning clutches or brakes, where slipping in one direction and a friction activated links causes a mechanical engagement such as bike overrunning drives. A *synchromesh* combines both, and then a friction clutch allows the two sides of positive jaws or teeth to smoothly engage.

Clutches such as classical auto disk clutches, torque limiting clutches, and multiple disk clutch-brake types are mostly frictionally operated. Brakes such as rotor caliper, drum, or bike pads are also frictionally operated.

If one is not familiar with any of the previously stated gadgets, it would be recommended to search over the Internet for the construction and operating principle or wait for the treatment presented in this chapter. Most of these gadgets or systems operate under some of the principles given next. In this chapter, some frictionally operated clutches and brakes are addressed in addition to other motion control elements such as flywheels.

18.2 Cone Clutches and Brakes

Figure 18.1 shows sketches of cone clutch or brake. It is out of scale and the *cone angle* α_C is larger than usual. The usual *cone angle* α_C is between 10° and 15°. Lower than 10° would cause possible welding during engagement. More than 15° would cause a larger value of the axial force needed for engagement as demonstrated later. The cone angle in Figure 18.1 is shown much larger for more clarity of demonstration. Figure 18.1a is a ***3D*** cone engaging the sectioned cup, which is keyed to the mating shaft. The cone is also attached to another mating shaft by a key or a spline. The spline is used if the cone is to move axially over the shaft for the cup engagement. Figure 18.1b defines the main cone dimensions and the *pressure intensity* or *applied pressure* p on the cone outer surface. An element of width $dr/\sin\alpha_C$ at a radius r is used to develop the mathematical model. The area dA_C of the element on the outer surface of the cone is simply equal to $2\pi r(dr/\sin\alpha_C)$. The *axial force* $\boldsymbol{F}_a$ is the integral of the axial component of force $p\,dA_C$ on the element dA_C; see Figure 18.1b. The axial component is equal to $p\,dA_C(\sin\alpha_C)$. This gives the following relation:

$$\boldsymbol{F}_a = \int_{d_i/2}^{d_o/2} p\ (dA_C \sin\alpha_C) = \int_{d_i/2}^{d_o/2} p\ \left(\frac{2\pi r\,dr}{\sin\alpha_C}\sin\alpha_C\right) = \int_{d_i/2}^{d_o/2} p\ \ (2\pi r\,dr) \tag{18.1}$$

where d_i is the *inner diameter* of the cone, d_o is the *outer diameter* of the cone, and the pressure p can be constant or variable over the cone's surface. The mathematical model of the *pressure intensity* or *applied pressure* p on the cone outer surface is considered later.

The transmitted torque $\boldsymbol{T}$ is the result of the kinetic friction force, which is normal to the page of Figure 18.1b. The kinetic friction force on the element of width $dr/\sin\alpha_C$ at a radius r is $\mu_K p\,dA_C$ normal to the page. The *transmitted torque* $\boldsymbol{T}$ is the integral of the product of the *kinetic friction* force $(\mu_K p\,dA_C)$ with the radius r. The

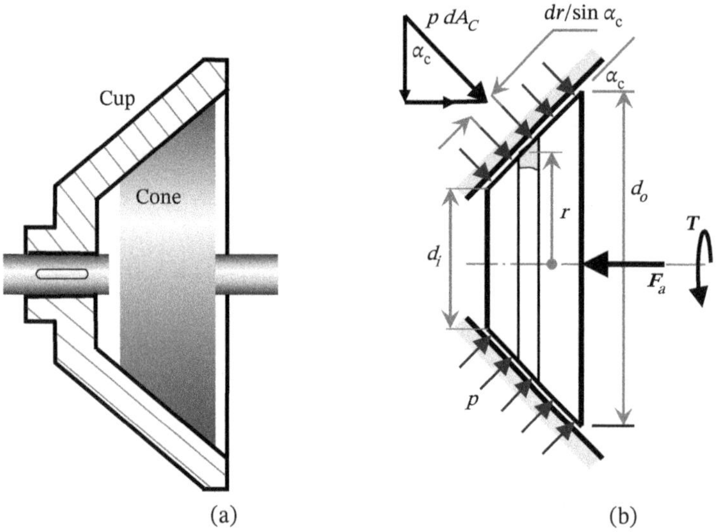

Figure 18.1 Out of scale sketches of a cone clutch or a cone brake. (a) ***3D*** cone engaging the sectioned cup. (b) Main cone dimensions and the applied pressure on cone outer surface.

friction torque of the force $\mu p \, dA_C$ on the element dA_C is then given by the following relation (see Figure 18.1b):

$$\boldsymbol{T} = \int_{d_i/2}^{d_o/2} \mu_K p \; (dA_C) r = \int_{d_i/2}^{d_o/2} \mu_K p \; \left(\frac{2\pi r \, dr}{\sin \alpha_C} \right) r = \int_{d_i/2}^{d_o/2} \mu_K p \; \left(\frac{2\pi r^2}{\sin \alpha_C} \right) dr \tag{18.2}$$

where d_i is the *inner diameter* of the cone, d_o is the *outer diameter* of the cone, p is the pressure, and μ_K is the *kinetic friction coefficient.*

The mathematical model of the *pressure intensity* or *applied pressure p* on the cone outer surface can be constant or variable over the cone's surface. The model affects the expressions for the *axial force* $\boldsymbol{F}_a$ in Eq. (18.1) and the transmitted torque $\boldsymbol{T}$ in Eq. (18.2). The two mathematical models of the *pressure intensity* or *applied pressure p* on the cone are assumed. The first model is a *uniform pressure* over the contact surface. The second model is assuming the pressure p stemming from a *uniform wear rate.* These depend on the contact flexibility of the mating parts and the transient time to steady-state operation. It is speculated that if the mating parts are flexible, the pressure distribution would be uniform. With operation time elapsing, the situation approaches a uniform wear condition. A more rigid mating part would approach a uniform wear condition faster. These different conditions and their fundamental relations are discussed next.

18.2.1 Uniform Pressure

Assuming a uniform pressure, one would employ the *maximum allowable uniform pressure* p_{max} in the mathematical model. According to Eq. (18.1), the maximum *axial force* $\boldsymbol{F}_a$ is then given by the following expression:

$$\boldsymbol{F}_a = \int_{d_i/2}^{d_o/2} p \quad (2\pi r \, dr) = 2\pi p_{max} \int_{d_i/2}^{d_o/2} r \; dr = \frac{\pi}{4} p_{max} (d_o^2 - d_i^2) \tag{18.3}$$

which indicates that the axial force $\boldsymbol{F}_a$ is simply the pressure p multiplied by the projected area of the mating surfaces; see Figure 18.1. If the maximum allowable *uniform pressure* p_{max} is used in Eq. (18.3), the axial force $\boldsymbol{F}_a$ is then the maximum *allowable force.*

According to Eq. (18.2), the expected *transmitted torque* $\boldsymbol{T}$ is then given by the following equation:

$$\boldsymbol{T} = \int_{d_i/2}^{d_o/2} \mu_K p \left(\frac{2\pi r^2}{\sin \alpha_C}\right) dr = \frac{2\pi \mu_K p_{\max}}{\sin \alpha_C} \int_{d_i/2}^{d_o/2} r^2 \, dr = \frac{\pi \mu_K p_{\max}}{12 \sin \alpha_C}(d_0^3 - d_i^3) \tag{18.4}$$

Substituting for the *uniform pressure* $p_{\max}$ from Eq. (18.3) into Eq. (18.4), one gets the torque as a function of the axial force, i.e.,

$$\boldsymbol{T} = \frac{\pi \mu_K p_{\max}}{12 \sin \alpha_C}(d_0^3 - d_i^3) = \frac{\mu_K}{3 \sin \alpha_C} \boldsymbol{F}_a \frac{(d_0^3 - d_i^3)}{(d_0^2 - d_i^2)} \tag{18.5}$$

Equation defines the maximum axial force due to reaching the maximum allowable surface pressure. If the axial force starts at a lower value, the applied torque would be at a lower value according to Eq. (18.5). This is the process of controlling the power transmission from one side to the other.

18.2.2 Uniform Wear Rate

For a uniform wear, it is stipulated that the work done in the wear of the material is the same along the contact radius from the inner to the outer boundary of contact. This suggests that the pressure p multiplied by the radius r is constant or $pr = C_w$, where C_w is the *wear constant*. The pressure p is then expected to be a maximum $p_{\max}$ at the inner diameter d_i. The relation of the pressure distribution is then given by

$$p = p_{\max} \frac{d_i}{2r} \tag{18.6}$$

which means that the pressure decreases as the radius r increases from r_i to r_o.

According to Eq. (18.1) and substituting for Eq. (18.6), the maximum *axial force* $\boldsymbol{F}_a$ is then given by the following expression:

$$\boldsymbol{F}_a = \int_{d_i/2}^{d_o/2} p \quad (2\pi r \, dr) = \pi p_{\max} d_i \int_{d_i/2}^{d_o/2} dr = \frac{\pi}{2} p_{\max} d_i (d_o - d_i) \tag{18.7}$$

According to Eq. (18.2) and substituting for Eq. (18.6), the expected *transmitted torque* $\boldsymbol{T}$ is then given by the following equation:

$$\boldsymbol{T} = \int_{d_i/2}^{d_o/2} \mu_K p \quad \left(\frac{2\pi r^2}{\sin \alpha_C}\right) dr = \frac{2\pi \mu_K}{\sin \alpha_C} p_{\max} \frac{d_i}{2} \int_{d_i/2}^{d_o/2} r \, dr = \frac{\pi \mu_K}{8 \sin \alpha_C} p_{\max} d_i (d_o^2 - d_i^2) \tag{18.8}$$

Equation (18.7) defines the maximum axial force due to attaining the maximum allowable surface pressure. If the axial force starts at a lower value, the applied torque would be at a lower value according to Eq. (18.8). This is the process of controlling the power transmission from one side to the other.

18.3 Disk Clutches and Brakes

The treatment of disk clutches and brakes are made simple by considering the cone angle α_C to be equals to 90°. This causes the cone surface to be a flat surface. Figure 18.2 presents the sketch of such a disk clutch or a disk brake with one friction surface. The main dimensions and the applied pressure on one friction surface are shown. To have a more efficient clutch, one would use both friction surfaces of a disk, not shown in Figure 18.2. Also, more than one disk and thus more friction surfaces can be employed. The treatment in this section is considering one friction surface first.

For one friction surface in Figure 18.2, setting the cone angle α_C to a 90° in the equations of the cone clutches or brakes in Section 18.2 provides the relations for the disk clutches and brakes. This is done for both conditions of uniform pressure and uniform wear.

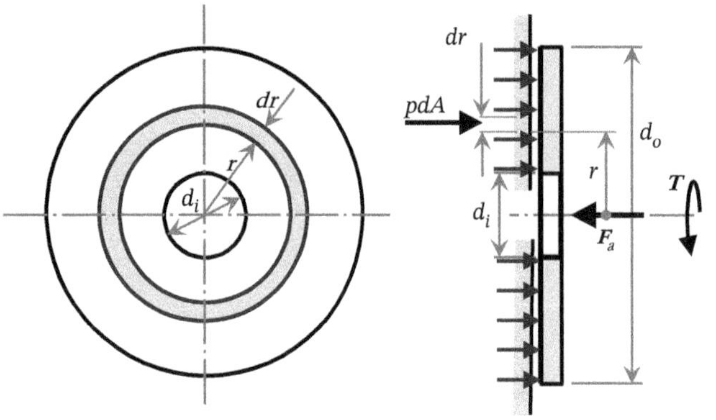

Figure 18.2 Sketch of a disk clutch or disk brake with one friction surface. The main dimensions and the applied pressure on one friction surface are shown.

18.3.1 Uniform Pressure

The maximum *axial force* $\boldsymbol{F}_a$ is given by the same expression for the cone clutch or brake in Eq. (18.3) since no cone angle appears in that equation. A fresh derivation as the cone clutch or brake produces the same results; see Figure 18.2. The maximum *axial force* $\boldsymbol{F}_a$ for one friction surface is then given by the following equation:

$$\boldsymbol{F}_a = \int_{d_i/2}^{d_o/2} p\,(2\pi r\,dr) = \frac{\pi}{4} p_{\max}\,(d_o^2 - d_i^2) \tag{18.9}$$

According to Eq. (18.4) and setting the cone angle $\alpha_C = 90°$, the expected *transmitted torque* $\boldsymbol{T}$ is then given by the following equation:

$$\boldsymbol{T} = \frac{\pi}{12} \mu_K p_{\max}\,(d_o^3 - d_i^3) \tag{18.10}$$

This relation is different from the cone clutches or cone brakes in Eq. (18.4). It provides a smaller torque for the same geometry. However, more disks and friction surfaces would make the difference as discussed later.

Substituting for the *uniform pressure* $p_{\max}$ from Eq. (18.9) into Eq. (18.10), one gets the torque as a function of the axial force, i.e.,

$$\boldsymbol{T} = \frac{\pi\,\mu_K p_{\max}}{12}(d_o^3 - d_i^3) = \frac{\mu_K \boldsymbol{F}_a}{3}\frac{(d_o^3 - d_i^3)}{(d_o^2 - d_i^2)} \tag{18.11}$$

Resembling cone clutches and cone brakes, Eq. (18.9) defines the maximum axial force due to attaining the maximum allowable surface pressure. If the axial force starts at a lower value, the applied torque would be at a lower value according to Eq. (18.11). Again, this is the process of controlling the power transmission from one side to the other.

18.3.2 Uniform Wear Rate

The maximum *axial force* $\boldsymbol{F}_a$ is given by the same expression of the cone clutch or brake in Eq. (18.7) since no cone angle appears in that equation. A fresh derivation as the cone clutch or cone brake produces the same results; see Figure 18.2. The maximum *axial force* $\boldsymbol{F}_a$ for one friction surface is then given by the following equation:

$$\boldsymbol{F}_a = \frac{\pi}{2} p_{\max}\, d_i(d_o - d_i) \tag{18.12}$$

According to Eq. (18.8) and setting the cone angle $\alpha_C = 90°$, the expected *transmitted torque* $\boldsymbol{T}$ is then given by the following equation:

$$\boldsymbol{T} = \frac{\pi}{8}\mu_K p_{\max} d_i(d_o^2 - d_i^2) \qquad \textbf{(18.13)}$$

Again, this relation is different from the cone clutches or brakes in Eq. (18.4). It provides a smaller torque for the same geometry. However, more disks and friction surfaces would make the difference as discussed later.

Substituting for the *uniform pressure* $p_{\max}$ from Eq. (18.12) into Eq. (18.13), one gets the torque as a function of the axial force, i.e.,

$$\boldsymbol{T} = \frac{\pi}{8}\mu_K p_{\max} d_i(d_o^2 - d_i^2) = \frac{\mu_K \boldsymbol{F}_a}{4}(d_o + d_i) \qquad \textbf{(18.14)}$$

Again, like the cone clutches and cone brakes, Eq. (18.13) defines the maximum axial force due to reaching the maximum allowable surface pressure. If the axial force starts at a lower value, the applied torque would be at a lower value according to Eq. (18.14). Again, this is the process of controlling the power transmission from one side to the other.

18.3.3 Multi-disk Clutch-Brake

The multi-disk clutch-brake arrangements are constructed to have multiple friction surfaces to increase the torque capabilities. With the same axial force $\boldsymbol{F}_a$, the multiple friction surfaces are capable of transmitting multiple torques more than a single friction surface. Figure 18.3 provides schematic diagrams of multiple friction surface clutch-brake. Figure 18.3a shows two friction surfaces in a single disk of a traditional manual auto-clutch. The clutch disk is splined with the takeoff shaft. This shaft runs freely in the flywheel, which is attached to the engine. The engagement and disengagement of the clutch is enacted by releasing or pushing the thrust bearing arrangement, which slides over the takeoff shaft. This would force the diaphragm spring to move the pressure plate away from the clutch disk, releasing it off the flywheel. The configuration in Figure 18.3 has the pressure plate pressing against the clutch disk in the default position. The pushing of the thrust bearing arrangement would disengage the clutch. Figure 18.3b shows a multi-disk clutch-brake sketch. The axial force does not increase with multiple friction surfaces. It is transmitted from the applied force through each disk to the next. A simple *free body diagram* of each disk attests to that.

The multi-disk clutch-brake performs as a function of the *number of friction surfaces* n_F, which may not always equal to twice the *number of disks* n_D. The construction may have a disk as an integral part of the construction, which may render the number of friction surfaces not necessarily twice the number of disks. It is then useful to use the *number of friction surfaces* n_F in the calculations of the transmitted torque.

The *axial forces* for multi-disk clutch-brake are the same as for a single friction surface. The *axial force* $\boldsymbol{F}_a$ for multi-disk clutch-brake for both the US and SI units are then given by Eq. (18.9) for *uniform pressure* case and by Eq. (18.13) for *uniform wear* situation. The transmitted torques, however, are different for either conditions of uniform pressure or uniform wear. These are as follows.

18.3.3.1 Uniform Pressure

The transmitted torques for every friction surface is added to the transmitted torques from every other friction surface. This generates the expected *transmitted torque* $\boldsymbol{T}$ as follows (see Eqs. (18.10) and (18.11)):

$$\boldsymbol{T} = (n_F)\ \frac{\pi}{12}\ \mu_K p_{\max}(d_0^3 - d_i^3) \qquad \textbf{(18.15)}$$

or

$$\boldsymbol{T} = (n_F)\ \frac{\mu_K \boldsymbol{F}_a}{3}\frac{(d_0^3 - d_i^3)}{(d_0^2 - d_i^2)} \qquad \textbf{(18.16)}$$

where n_F is the *numbers of friction surfaces* and μ_K is the kinetic friction coefficient.

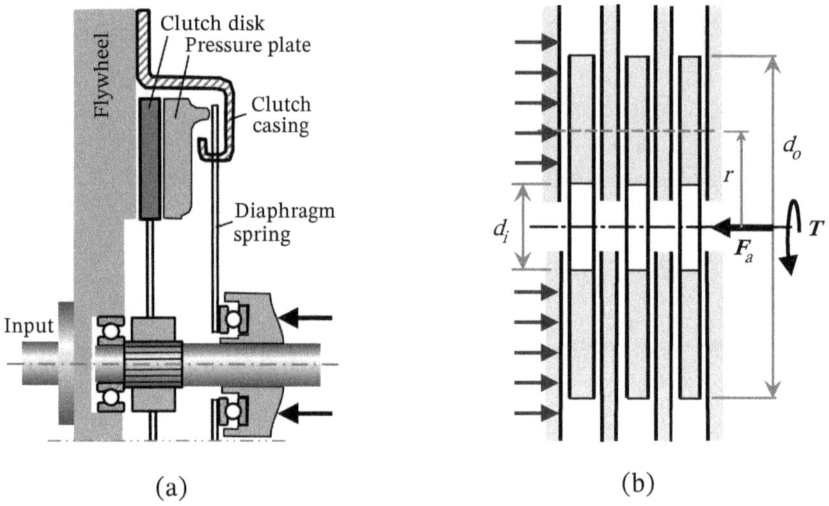

Figure 18.3 Schematic diagrams of multiple friction surface clutch-brake. (a) Two friction surfaces in a single disk traditional auto-clutch. (b) Multi-disk clutch-brake sketch.

18.3.3.2 Uniform Wear Rate

Again, the transmitted torques for every friction surface is added to the transmitted torques from every other friction surface. This generates the expected *transmitted torque* $\boldsymbol{T}$ as follows (see Eqs. (18.13) and (18.14)):

$$\boldsymbol{T} = (n_F)\ \frac{\pi}{8}\ \mu_K p_{\max} d_i (d_o^2 - d_i^2) \tag{18.17}$$

or

$$\boldsymbol{T} = (n_F)\ \frac{\mu_K \boldsymbol{F}_a}{4}(d_o + d_i) \tag{18.18}$$

where n_F is the *numbers of friction surfaces* and μ_K is the kinetic friction coefficient.

18.3.4 Initial Disk Clutch-Brake Synthesis

Usually, the transmitted power and the rotational speed of the system are known in most cases. From the previous treatment of disk clutches and brakes, the main parameters to consider in the synthesis are then the disk *outer diameter* d_o, the *inner diameter* d_i, and the *number of friction surfaces* n_F. The material selection is the main parameter, which depends on the application and operating environments. The friction between the mating parts is dependent on the friction material and the operating conditions with wide variation and uncertainty. Table 18.1 presents expected basic properties of some selected friction materials (for clutches or brakes). Table 18.2 provides some expected additional properties of these selected friction materials (for clutches or brakes). Values in Tables 18.1 and 18.2 are compiled from several sources and should be cautiously used; see, e.g. Neale (1973). The values adapted in a design should depend on quoted values from the manufacturer or provider and should also be verified. The operating conditions and cost would enter in the selection of material and its properties. Optimization might be essential for mass production and selection of a manufacturer or provider; see Internet references. With the material advancements, manufacturers are striving to provide better friction materials in terms of friction, maximum allowable pressure, maximum power rating, maximum temperature, and lower wear rate.

The synthesis is primarily picked for a *uniform pressure* condition. An optimum inner to outer diameter ratio d_i/d_o of 0.577 35 is selected; see, e.g. Hegazi (1997) and Metwalli et al. (1999), Metwalli and Hegazi (2001). The difference in synthesis between conditions of *uniform pressure* and *uniform wear* is estimated to be small at that

Table 18.1 Expected basic properties of some selected friction materials (for clutches or brakes).

	Coefficient of friction μ_k		Maximum pressure p_{max}	
Material type	**Dry**[a)]	**Wet**[a)]	**SI [MPa]**	**US [psi]**
Woven	0.25–0.6	0.1–0.2	0.3–0.7	40–100
Molded	0.25–0.5	0.04–0.12	0.3–1.0	40–150
Sintered	0.1–0.45	0.05–0.1	0.35–2.7	50–400
Cermet	0.3–0.4	0.06–0.08	0.7–1.4	100–200
Resin-graphite	0.25–0.3	0.05–0.1	0.35–2.1	50–300
Cast iron or steel	0.15–0.20	0.03–0.06	1.03–1.72	150–250

a) On smooth steel or cast iron.

Table 18.2 Some expected additional properties of some selected friction materials (for clutches or brakes).

	Power rating		Maximum temperature	
Material type	**SI [MN/m²]**	**US [hp/in²]**	°C	°F
Woven	0.3–1.8	0.25–1.6	95–260	200–500
Molded	0.6–1.2	0.5–1.0	260–400	500–750
Sintered	1.7–2.3	1.5–2	300–520	570–970
Cermet	3.5–4.0	3–3.5	400–800	750–1500
Resin-graphite	3–3.5	2.5–3	370–540	700–1000
Cast iron or steel	0.5–1.0	0.4–0.9	260–320	500–600

inner to outer diameter ratio (Budynas and Nisbett 2015). The uniform wear is the more conservative option. With possible rounding of design variables, either option might be acceptable.

Usually the *transmitted power H* and the *rotational speed* N_{rpm} are defined. The transmitted torques for SI and the US units are then given by the following familiar expressions:

$$\boldsymbol{T} = 1000\, H_{kW} \left(\frac{60}{2\,\pi N_{rpm}} \right) = 9549\, \frac{H_{kW}}{N_{rpm}} \;[\text{N m}] \tag{18.19}$$

$$\boldsymbol{T} = 6600 H_{hp} \left(\frac{60}{2\,\pi N_{rpm}} \right) = 63\,025 \frac{H_{hp}}{N_{rpm}} \;[\text{lb in}] \tag{18.20}$$

where H_{kW} is the power in [kW] and H_{hp} is the power in [hp]. With the known transmitted torque $\boldsymbol{T}$, the outer diameter d_o would be obtainable if other parameters are defined. From Eq. (18.15) and considering the optimum inner to outer diameter ratio d_i/d_o of 0.577 35, one can find the outer diameter d_o for *uniform pressure* as follows:

$$d_o = \left(\frac{12\boldsymbol{T}}{(0.807\,55)\pi n_F \mu_K p_{max}} \right)^{1/3} = \left(\frac{4.7300\boldsymbol{T}}{n_F \mu_K p_{max}} \right)^{1/3} \tag{18.21}$$

where n_F is the *numbers of friction surfaces*, μ_K is the *kinetic friction coefficient*, and p_{max} is the *maximum allowable uniform pressure*. Selecting the *friction material* (Tables 18.1 and 18.2) and assuming the number of friction surfaces would then define the outer diameter d_o, which would be usually rounded to an upper acceptable value.

Equation (18.21) is valid for both SI and the US units. If the torque $\boldsymbol{T}$ is in [N m], and the maximum pressure p_{max} is in [Pa], the outer diameter d_o would be in [m]. If the torque $\boldsymbol{T}$ is in [lb in], and the maximum pressure p_{max} is in [psi], the outer diameter d_o would be in [in].

To represent Eq. (18.21) in terms of the transmitted power H and the rotational speed N_{rpm}, one needs to use Eqs. (18.19) and (18.20) to get the following expressions for the outer diameter d_o:

$$d_o = \left(\frac{4.7300}{n_F \mu_K p_{max}} 9549 \frac{H_{kW}}{N_{rpm}} \right)^{1/3} = \left(\frac{45\,166.8}{n_F \mu_K p_{max}} \frac{H_{kW}}{N_{rpm}} \right)^{1/3}$$

$$d_o = \left(\frac{4.7300}{n_F \mu_K p_{max}} 63\,025 \frac{H_{hp}}{N_{rpm}} \right)^{1/3} = \left(\frac{296\,218}{n_F \mu_K p_{max}} \frac{H_{hp}}{N_{rpm}} \right)^{1/3} \tag{18.22}$$

Knowing the disk outer diameter d_o, the disk inner diameter d_i is then 0.577 35 d_o. The maximum *axial force* $\boldsymbol{F}_a$ is also defined from Eq. (18.9) for the assumed uniform pressure or from Eq. (18.13) for prospective uniform wear condition. Using the maximum allowable pressure gives the maximum capacity of the clutch or brake.

For the *uniform wear* and to evaluate the outer diameter d_o, one can use Eqs. (18.17)–(18.20) to get the following expression:

$$d_o = \left(\frac{8\boldsymbol{T}}{0.384\,90 \pi n_F \mu_K p_{max}} \right)^{1/3} = \left(\frac{6.6159}{n_F \mu_K p_{max}} 9549 \frac{H_{kW}}{N_{rpm}} \right)^{1/3} = \left(\frac{63\,175}{n_F \mu_K p_{max}} \frac{H_{kW}}{N_{rpm}} \right)^{1/3}$$

$$d_o = \left(\frac{8\boldsymbol{T}}{0.384\,90 \pi n_F \mu_K p_{max}} \right)^{1/3} = \left(\frac{6.6159}{n_F \mu_K p_{max}} 63\,025 \frac{H_{hp}}{N_{rpm}} \right)^{1/3} = \left(\frac{416\,967}{n_F \mu_K p_{max}} \frac{H_{hp}}{N_{rpm}} \right)^{1/3} \tag{18.23}$$

To implement the previous procedure for disk clutch-brake systems, *initial synthesis charts* are developed as shown in Figure 18.4. The service or *safety factor* K_{SF} is included in the abscissa to account for *starting* and *service* conditions. Eqs. (18.19)–(18.21) are used with the inclusion of the *safety factor* K_{SF} to apparently increase the failure or overrun power transmission. The safety factor K_{SF} may be assumed as $K_{SF} = K_{start}$ ($K_{service}$), where the *start factor* K_{start} = (1.2–2), and the *service factor* $K_{service}$ = (1–2.5). Figure 18.4a presents the chart for the SI units, where the maximum allowable pressure p_{max} is in [MPa]. Figure 18.4b shows the chart for the US units, where p_{max} is in [psi]. These units are used since these are the quoted values of the maximum allowable pressure p_{max} in Table 18.1. The plots are lines for a constant parameter ($n_F \mu_K p_{max}$) of 0.01–100 for the SI system and 0.1–1000 for the US system. The ranges in the charts of Figure 18.4 are very wide to include the vast variations for the prospective ranges of the friction coefficient, maximum allowed pressure, number of friction surfaces, and the general design space for power and rotational speed.

Example 18.1 A disk clutch is to be designed to transmit a maximum of 25 [kW] or 33.5 [hp] at a running speed of 1575 [rpm]. It has only one disk on the takeoff or driven shaft. Find the disk outside and inside diameters for the usual steel and woven or molded friction materials. Consider a usual coefficient of friction for dry engagement as 0.3 and the maximum axial pressure as 0.35 [MPa] or 51 [psi]. Take the safety factor for this design as 1 since no overload is allowed and slipping should occur at the limiting maximum transmitted torque at the maximum transmitted power.

Solution

Data: $H_{kW} = 25$ [kW] or $H_{hp} = 33.5$ [hp], $N_{rpm} = 1575$ [rpm], $n_F = 2$, $\mu_K = 0.3$, $p_{max} = 0.35$ [MPa] or 51 [psi], and $K_{SF} = 1$.

Prepare the variables to be used in the synthesis chart in Figure 18.4 as follows:

$H_{kW} K_{SF}/N_{rpm} = 25 \times 1/1575 = 0.015\,873$, which can be assumed as (0.016) for realistically possible implementation on the chart resolution.

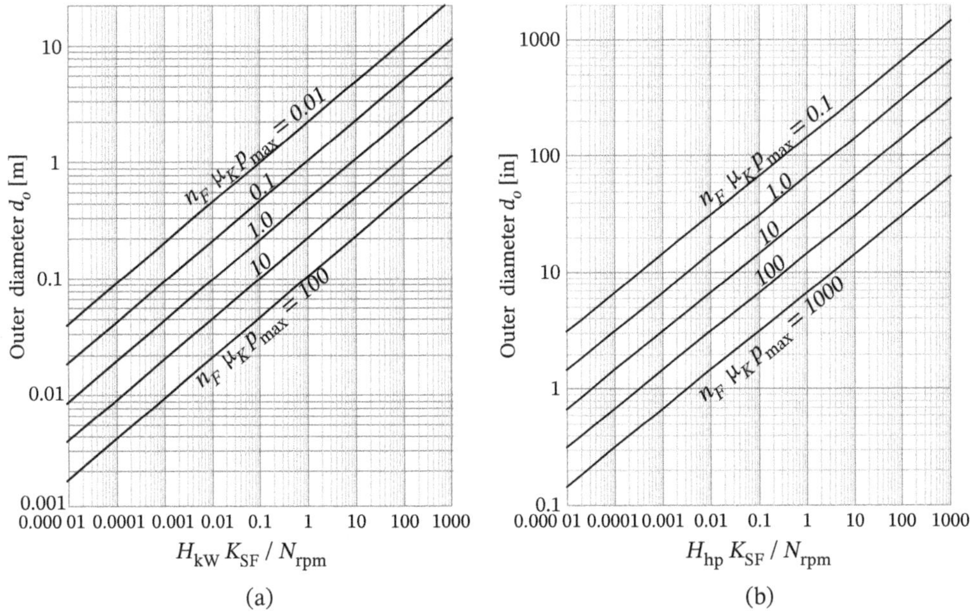

Figure 18.4 Charts of initial clutch-brake synthesis noting the safety factor inclusion in the abscissa. (a) SI units with p_{max} in [MPa]. (b) US units with p_{max} in [psi].

$H_{hp}K_{SF}/N_{rpm} = 33.5 \times 1/1575 = 0.021\,27$, which can be assumed as (0.02) for realistically possible implementation on the chart resolution.

The parameter for SI units ($n_F\mu_K p_{max}$) = 2 × 0.3 × 0.35 = 0.21, which can be assumed approximately as 0.2.

The parameter for the US units ($n_F\mu_K p_{max}$) = 2 × 0.3 × 51 = 30.6, which can be assumed approximately as 30.

From the charts in Figure 18.4, one can meticulously find the approximate outer diameter $d_o \cong 0.15$ [m] = 150 [mm] and 6 [in]. The inner diameters are then $d_i = 0.577\,35\, d_o = 86.60$ [mm] or 3.464 [in].

The charts in Figure 18.4 are not having the good resolution, and the application of Eq. (18.22) would be advisable. Equation (18.22) gives the following:

$$d_o = \left(\frac{45\,166.8}{n_F\mu_K p_{max}}\,\frac{H_{kW}}{N_{rpm}}\right)^{1/3} = \left(\frac{45\,166.8}{0.21(10^6)}\left(\frac{25}{1575}\right)\right)^{1/3} = 0.150\,58\,[\text{m}] = 151\,[\text{mm}]$$

$$d_o = \left(\frac{296\,218}{n_F\mu_K p_{max}}\,\frac{H_{hp}}{N_{rpm}}\right)^{1/3} = \left(\frac{296\,218}{30.6}\left(\frac{33.5}{1575}\right)\right)^{1/3} = 5.9050\,[\text{in}] \qquad \text{(a)}$$

These values are close to the values obtained from Figure 18.4. Selecting round figures for the *outer diameter* $d_o = 0.15$ [m] = 150 [mm] and 6 [in] would be acceptable. The inner round figure diameters may then be $d_i = 85$ [mm] or 3.5 [in]. The round figure values should affect the maximum axial force needed to transmit the power. The maximum axial force $\boldsymbol{F}_a$ can be found from Eq. (18.9) for uniform pressure as follows:

$$\boldsymbol{F}_a = \frac{\pi}{4} p_{max}(d_o^2 - d_i^2) = \frac{\pi}{4}(0.35(10^6))((0.15)^2 - (0.085)^2) = 4198.9\,[\text{N}]$$

$$\boldsymbol{F}_a = \frac{\pi}{4} p_{max}(d_o^2 - d_i^2) = \frac{\pi}{4}(51)((6)^2 - (3.5)^2) = 951.31\,[\text{lb}] \qquad \text{(b)}$$

The maximum axial force $\boldsymbol{F}_a$ can be found from Eq. (18.9) for uniform wear as follows:

$$\boldsymbol{F}_a = \frac{\pi}{2} p_{max}\, d_i(d_o - d_i) = \frac{\pi}{2}(0.35(10^6))(0.085)(0.15 - 0.085) = 3037.5\,[\text{N}]$$

$$\boldsymbol{F}_a = \frac{\pi}{2} p_{\max} d_i(d_o - d_i) = \frac{\pi}{2}(51)(3.5)(6 - 3.5) = 700.97\ [\text{lb}] \tag{c}$$

The difference between the values in Eqs. (b) and (c) is large.

For the *uniform wear* and to evaluate the outer diameter d_o, one can use Eq. (18.23) to get

$$d_o = \left(\frac{63175}{n_F \mu_K p_{\max}} \frac{H_{\text{kW}}}{N_{\text{rpm}}} \right)^{1/3} = \left(\frac{63\,175}{0.21(10^6)} \left(\frac{25}{1575} \right) \right)^{1/3} = 0.168\,39\ [\text{m}] = 168.39\ [\text{mm}]$$

$$d_o = \left(\frac{416967}{n_F \mu_K p_{\max}} \frac{H_{\text{hp}}}{N_{\text{rpm}}} \right)^{1/3} = \left(\frac{416\,967}{30.6} \left(\frac{33.5}{1575} \right) \right)^{1/3} = 6.6178\ [\text{in}] \tag{d}$$

The difference between the values in Eqs. (a) and (d) is large. Selecting round figures for the *outer diameter* $d_o = 0.17$ [m] = 170 [mm], and 6.6 [in] would be acceptable. The inner round figure diameters may then be $d_i = 100$ [mm] or 3.8 [in]. The maximum axial force $\boldsymbol{F}_a$ can then be found from Eq. (18.11) for *uniform wear* as follows:

$$\boldsymbol{F}_a = \frac{\pi}{2} p_{\max} d_i(d_o - d_i) = \frac{\pi}{2}(0.35(10^6))(0.1)(0.17 - 0.1) = 3848.5\ [\text{N}]$$

$$\boldsymbol{F}_a = \frac{\pi}{2} p_{\max} d_i(d_o - d_i) = \frac{\pi}{2}(51)(3.8)(6.6 - 3.8) = 852.38\ [\text{lb}] \tag{e}$$

These values are close to the maximum axial forces in Eq. (b). The dimensions of the uniform wear clutch are larger, which is considered as a more conservative design. If one uses the axial forces in Eq. (e), the initial power transmission would not be the expected maximum at the initial running according to the uniform pressure condition.

18.4 Caliper Disk Brakes

Currently, brake calipers with the disk brakes are extensively used in automotive and other applications. They are exceedingly competitive in vehicle brakes to the traditional *drum* or *shoe brakes*. They utilize *calipers* to squeeze mounted *pads* onto the disk, which is attached to the element that needs braking such as the wheel of a vehicle. Figure 18.5 provides a schematic outlines of caliper disk brake with the main components identified. Figure 18.5a defines the rotor of caliper disk brakes identifying the grayish area of the *caliper disk angle* β_C. Shoe pads on both sides of the rotor disk are shown in Figure 18.5b with the piston squeezing the shoe pads to the brake disk. For other configurations, one may search disk brake patents such as Lambert (1945, 1946).

The caliper disk brake is then a partial disk of a *caliper angle* β_C, which is specified in [rad] so that the previously developed relations would hold. Since part of the disk (angle β_C) is used on both sides of rotor, a $2\beta_C/2\pi$ would be used as the *apparent number of friction surfaces* n_F in the forgoing Eqs. (18.21)–(18.23). The parameter $(2\beta_C/2\pi)$ is labeled as *caliper subtended ratio*. It is noted that *uniform wear* should be a reasonable assumption and Eqs. (18.17), (18.18), and (18.23) become as follows:

$$\boldsymbol{T} = \left(\frac{2\beta_C}{2\pi} \right) \frac{\pi}{8} \mu_K p_{\max} d_i(d_o^2 - d_i^2) = \frac{\beta_C}{8} \mu_K p_{\max} d_i(d_o^2 - d_i^2) \tag{18.24}$$

$$\boldsymbol{T} = \left(\frac{2\beta_C}{2\pi} \right) \frac{\mu_K \boldsymbol{F}_a}{4}(d_o + d_i) = \beta_C \frac{\mu_K \boldsymbol{F}_a}{4\pi}(d_o + d_i) \tag{18.25}$$

As the *axial force* $\boldsymbol{F}_a$ in Eq. (18.12) is applied over the full disk area, the axial force on the caliper should only be on the sector $(\beta_C/2\pi)$ of the friction surface of the disk with an equal and opposite axial force on the other side of the disk. This gives the axial force $\boldsymbol{F}_a$ of Eq. (18.12) as follows:

$$\boldsymbol{F}_a = \left(\frac{\beta_C}{2\pi} \right) \frac{\pi}{2} p_{\max} d_i(d_o - d_i) = \left(\frac{\beta_C}{4} \right) p_{\max} d_i(d_o - d_i) \tag{18.26}$$

For the *uniform wear* and to evaluate the outer diameter d_o, one can use Eq. (18.23) and the *caliper subtended ratio* parameter $(2\beta_C/2\pi)$ in place of the number of friction surfaces n_F to get the following expression:

$$d_o = \left(\frac{8\boldsymbol{T}}{0.38490\pi(\beta_C/\pi)\ \mu_K p_{max}}\right)^{1/3} = \left(\frac{8\boldsymbol{T}}{0.38490(\beta_C)\ \mu_K p_{max}}\right)^{1/3} \qquad \textbf{(18.27)}$$

Equation (18.27) is valid for both SI and the US units. Note also that β_C is in [rad]. If the torque $\boldsymbol{T}$ is in [N m], and the maximum pressure p_{max} is in [Pa], the outer diameter d_o would be in [m]. If the torque $\boldsymbol{T}$ is in [lb in], and the maximum pressure p_{max} is in [psi], the outer diameter d_o would be in [in].

18.5 Energy Dissipation and Temperature Rise

In this section simplified assumptions are used to allow for an estimation of the temperature rise in the clutch or brake disk and the main contact components of the assembly. A simple joining model of clutch-brake is shown in Figure 18.6. The clutch or brake is connecting an input system having an *input mass moment of inertia* J_i to the output system having an *output mass moment of inertia* J_o. Assuming stiff links between the system inertias and the clutch-brake joint and assuming a constant joining torque $\boldsymbol{T}$, one can have the following simple mathematical model:

$$J_i\ddot{\theta}_i = -\boldsymbol{T} \quad \text{and} \quad J_o\ddot{\theta}_o = \boldsymbol{T} \qquad (18.28)$$

where $\ddot{\theta}_i$ is the *angular acceleration of the input* and $\ddot{\theta}_o$ is the *angular acceleration of the output*. Integrating Eq. (18.28) for a time t to get the instant angular velocities, one gets the following relations:

$$\dot{\theta}_i = -\frac{\boldsymbol{T}}{J_i}t + \omega_i \quad \text{and} \quad \dot{\theta}_o = \frac{\boldsymbol{T}}{J_o}t + \omega_o \qquad (18.29)$$

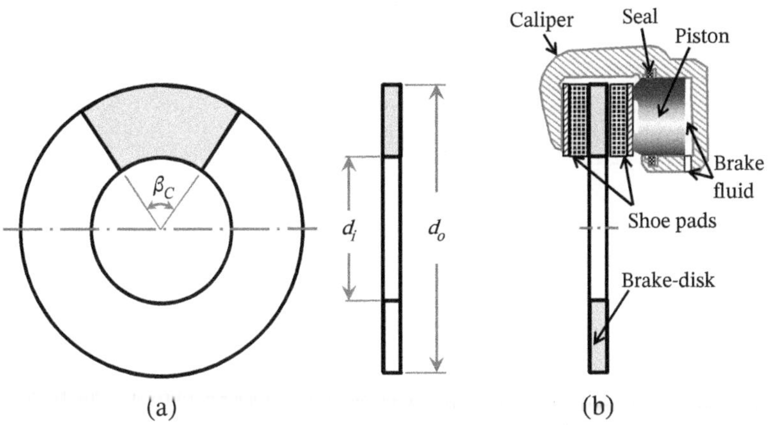

Figure 18.5 Outline of caliper disk brake with main components identified. (a) The rotor of caliper disk brakes defining the grayish area of the disk angle β_C. (b) Shoe pads on both sides of rotor disk are shown with piston squeezing shoe pads with brake disk.

Figure 18.6 A simple joining model of clutch-brake connecting an input inertia system to the output inertia system.

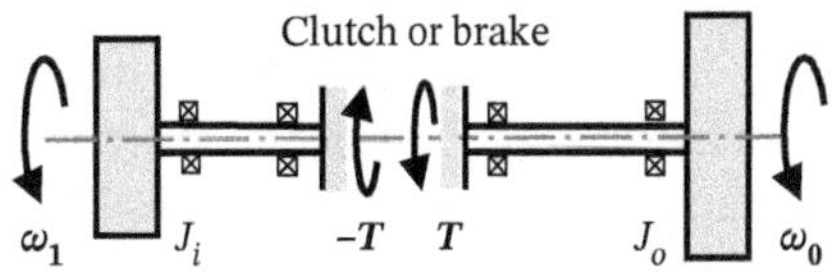

where $\dot{\theta}_i = \omega_i$ as the *angular velocity of the input* at $t = 0$ and $\dot{\theta}_o = \omega_o$ as the *angular velocity of the output* at $t = 0$. One should note that at the start where the angular velocities are usually different, one has $\omega_i \neq \omega_o$.

The relative velocity between the input and the output is then given by the following relation:

$$\dot{\theta} = \dot{\theta}_i - \dot{\theta}_o = -\left(\frac{\boldsymbol{T}}{J_i}\right) t + \omega_i - \left(\frac{\boldsymbol{T}}{J_o} t + \omega_o\right) = -\boldsymbol{T}\left(\frac{1}{J_i} + \frac{1}{J_o}\right) t + (\omega_i - \omega_o)$$

$$\dot{\theta} = -\boldsymbol{T}\left(\frac{J_i + J_o}{J_i\ J_o}\right) t + (\omega_i - \omega_o) \tag{18.30}$$

The clutching is concluded when the relative velocity $\dot{\theta}$ in Eq. (18.30) is zero. This *braking, clenching*, or *clutching time* t_C is then

$$t_C = \frac{(J_i\ J_o)}{\boldsymbol{T}(J_i + J_o)}(\omega_i - \omega_o) \tag{18.31}$$

18.5.1 Energy Dissipation

The rate of clutching–braking contact energy dissipation or rate of *clenching energy* $\dot{E}_C$ (i.e. the power loss) is then given by the following relation:

$$\dot{E}_C = \boldsymbol{T}\dot{\theta} = \boldsymbol{T}\left[-\boldsymbol{T}\left(\frac{1}{J_i} + \frac{1}{J_o}\right) t + (\omega_i - \omega_o)\right] \tag{18.32}$$

Equation (18.32) indicates that the *maximum energy dissipation* $\dot{E}_{\max}$ occurs when $t = 0$, i.e. at the start of clutching or braking. This gives the maximum energy dissipation $\dot{E}_{\max}$ as follows:

$$\dot{E}_{\max} = \boldsymbol{T}(\omega_i - \omega_o) \tag{18.33}$$

The total *clenching energy* E_C dissipated from $t = 0$ to $t = t_C$ is then given by

$$E_C = \int_0^{t_C} \dot{E}\ dt = \boldsymbol{T}\int_0^{t_C}\left[-\boldsymbol{T}\left(\frac{1}{J_i} + \frac{1}{J_o}\right) t + (\omega_i - \omega_o)\right] dt \tag{18.34}$$

After integration and substituting for t_C from Eq. (18.31), Eq. (18.34) reduces to the following:

$$E_C = \frac{(\omega_i - \omega_o)^2}{2\left(\frac{1}{J_i} + \frac{1}{J_o}\right)} = \frac{(J_i\ J_o)}{2(J_i + J_o)}(\omega_i - \omega_o)^2 \tag{18.35}$$

Equation (18.35) indicates that the total dissipated energy is not a function of the joining torque $\boldsymbol{T}$. For automotive braking to the ground ($\omega_o = 0, J_o = \infty$), the total dissipated energy in Eq. (18.35) is then provided by the following expected expression:

$$E_C = \frac{1}{2} J_i \omega_i^2 \tag{18.36}$$

It should be noted that the inertia of the mass traveling at a linear velocity should be transferred to the equivalent mass moment of inertia rotating at ω_i of the brake disk on the wheel shaft. This is supposed to be a simple dynamics problem; see, e.g. Figure 2.12 and Example 2.4.

18.5.2 Temperature Rise

The dissipated energy in friction during clutching or braking is expected to cause some temperature rise to the clutch disk or brake disk and the main contact components of the assembly. The *contact temperature rise* ΔT_C is rather a complex problem that has been a subject for extensive research; see, e.g. Metwalli et al. (1999, 2001). The temperature is high at the contact zone, and heat is conducted to the contact disk or components to be dissipated to the surrounding. To reduce the high contact temperature, the materials of the disks and pads should have a high heat conductivity, and disks may also have internal fins to cool the disk. For fluid immersed clutch-brake arrangements, the fluid should also cool the disks. The mathematical model for the contact components is presented next in a simplified way that may be used as an initial estimate.

The *temperature rise* ΔT_C can be obtained approximately as follows:

$$\Delta T_C = \frac{E_C}{c_p m_C} \tag{18.37}$$

where E_C is the total *dissipated energy* defined by Eq. (18.35), c_p is the *specific heat* of material, and m_C is the *clutch-brake mass* of the disk or contact components. When ΔT_C is the temperature rise in [°C], c_p is the specific heat, and m_C can be the clutch or brake disk mass in [kg]. Using a *cast iron* disk, the specific heat $c_p = 540$ [J/kg °C], and the density $\rho = 7200$ [kg/m^3]. For the adapted US system of units, the energy generated as heat in British thermal units of [Btu] is used in Eq. (18.37) such that

$$\Delta T_C = \frac{E_C/9336}{c_p m_C} \ [°\text{F}] \tag{18.38}$$

where ΔT_C is the temperature rise in [°F], c_p is the specific heat in [Btu/lb$_m$ °F], and m_C can be the clutch or brake disk mass in [lb$_m$]. Using a *cast iron* disk, the specific heat $c_p = 0.13$ [Btu/lb$_m$ °F], and the density $\rho = 0.26$ [lb/in^3].

The *disk thickness* might be one of the design variables to consider in the reduction of the temperature rise. Venting of the disk, proper material selection, and optimization can contribute to that effect (Metwalli et al. 1999, 2001).

The previous treatment did not indicate that the friction pad material can be a factor to the heat dissipation. The manufacturers would need to improve the composite materials to have higher friction, higher maximum allowable pressure and temperature, higher heat transfer, and lower wear or higher power rating; see Tables 18.1 and 18.2. For potential disk materials and composite friction materials, one can consult handbooks and dedicated material property references for possible material properties of higher heat transfer, higher temperature resistance, and higher specific heat and density; see, e.g. Neale (1973).

18.6 Design Process

Clutches and brakes are becoming one of the *original equipment manufacturer* (OEM) products with many manufacturers making these general or special products for others. The designer in those outfits should be capable of synthesizing clutches and brakes to satisfy specific demands. This job is more than design and development. It involves some research and innovation. The other regular designer should have the knowledge and tools to define, verify, and validate the selection of the available clutches and brakes. The process and development provided herein might support such an effort. A typical initial synthesis followed by a detailed design process is the suggested procedure in this text.

18.6.1 Initial Synthesis

In Section 18.3.4 an initial disk clutch and brake synthesis process was introduced. The process involves initial synthesis of *outer disk diameter* d_o and *inner disk diameter* d_i in addition to the option of selecting the *number of friction surfaces* n_F or disks and their materials. Simple expressions in Eqs. (18.21)–(18.23) and plots in Figure 18.4 provide tools to reasonably achieve initial clutch or brake synthesis. Example 18.1 applies these in an initial synthesis of a disk clutch application.

The procedure can be extended to *caliper brakes* if the *number of friction surfaces* n_F is replaced by the *caliper subtended ratio* $(2\beta_C/2\pi)$. The partial disk of a *caliper angle* β_C is specified in [rad].

18.6.2 Detailed Design Process

After the previous simplified initial synthesis, other details should be included in the design process. These may not necessarily change geometry, but it can adjust some parameters such as disk thickness, number of disks, and other material selection to account for energy dissipation, temperature rise, and other constraints. This would be easily performed if a suitable computer-aided design (*CAD*) tool is available for better synthesis and possible optimization.

Energy dissipation and temperature rise are affected by the constructional details such as the space needed for multi-disk, disk thickness, internal fins of a disk, heat conduction, convection or removal by ventilation, or fluid circulation. Equation (18.37) can be used to define the disk thickness or select a different material of better specific heat. Disk and pad materials of higher conductivity can better improve heat dissipation and temperature rise. Optimization would help in that regard.

Few other parameters are useful in complete design assessment. Some of these are given in Table 18.2 such as power rating and maximum temperature. Another useful parameter is the *pressure times the velocity* (pv_C), which is maximum at the *outer diameter*. The critical parameter of pressure times the velocity (pv_C) depends on the duty cycle of loading and the efficiency of heat transfer or dissipation. Usual values of (pv_C) are as follows:

- For continuous operation and poor heat dissipation, (pv_C) ranges are 30–40 [kpsi ft/min] or 1–1.3 [MPa m/s].
- For occasional operation and poor heat dissipation, (pv_C) ranges are 60–80 [kpsi ft/min] or 1.7–2.1 [MPa m/s].
- For continuous operation and good heat dissipation, (pv_C) ranges are 85–120 [kpsi ft/min] or 3–4.2 [MPa m/s].

These are ballpark values that should be scrutinized by manufacturer information or recommendations. The lower the value, however, the higher the life of the clutch-brake element.

18.7 Computer-Aided Design and Optimization

Figure 18.7 presents the images of *Excel tablets* for ***clutch-brake*** synthesis devoted to SI and the US systems of units, Excel© (2016). The **Default** column of numbers is acting as a guide, and values are the input and solution of Example 18.1. It is not allowed to change any of the cell's entries. All cells with clear background should not be changed. They have formulas to evaluate dependent on the ***Input*** *cells* with light greyish background (or light blue background color for the *Excel Tablet*). The column with **Input** heading is used to input data in the *Input cells* with light grayish or light blue background color. These are the only values needed to *initially synthesize* the clutch-brake geometry in terms of the disk outer diameter and the disk inner diameter. The needed inputs are the power, the rotational speed, the initial safety factor, and the expected number of friction surfaces. The kinetic friction coefficient and the maximum pressure are initially for the basic material of usual coefficient of friction for dry engagement as 0.3 and the maximum axial pressure as 0.35 [MPa] or 51 [psi]. There is also an option to enter the number of disks for your construction record. Its value does not affect results. Calculations depend on the number of friction surfaces and not the number of disks.

The ***Input*** clutch-brake synthesis column provides the initial clutch-brake disk outer diameter and disk inner diameter for optimum even or uniform pressure distribution at the maximum adapted value. There is an option to change the kinetic friction coefficient and the maximum pressure. The values are left as for the basic initial material of usual coefficient of friction for dry engagement as 0.3 and the maximum axial pressure as 0.35 [MPa] or 51 [Pa]. The input values in this column are conforming to Example 18.1. The ***output*** values are close to the solution of Example 18.1. The values are due to the synthesized diameters and not the selected round figure diameters. If the selected round figure diameters are entered in the **Analysis** column, the output would then be the same as the hand calculations of Example 18.1.

Other ***output*** variables confirming calculations or presented calculation variables for *uniform wear* are also provided. The *working pressure* and *power rating* are available to present some useful parameter for comparison;

(a)

Disc Clutch-Brake Synthesis - SI units			
Inputs	Default	Input	Analysis
Power, H_{kW} [kW]	25	25	25
Rotational Speed, N_{rpm} [rpm]	1575	1575	1575
Safety Factor, K_{SF}	1	1	1.42
Number of Friction Surfaces, n_F	2	2	2
Output	*Assuming even pressure*		Input/Output
Outer Diameter, d_o [m]	0.15058	0.15058	0.170
Inner Diameter, d_i [m]	0.08694	0.08694	0.100
Number of Disks, n_D	1	1	1
Number of Friction Surfaces, n_F	2	2	2
Inner to Outer Diameter, d_i/d_o	0.57735	0.57735	0.58824
Torque, T [N.m]	151.58	151.58	151.58
Friction Coefficient, μ_K	0.3	0.3	0.3
Working Pressure, p [MPa]	0.350	0.350	0.247
Maximum Pressure, p_{max} [MPa]	0.35	0.35	0.35
Power Rating, [MW/m^2]	1.053	1.053	0.842
Pressure * Velocity, pv_C [MPa.m/s]	4.346	4.346	3.457
Axial Force, F_a [N]	4155.11	4155.11	3660.60
Other Output	*Values for verification*		
Axial Force (even wear), F_a [N]	3041.75	3041.75	3848.45
Axial Force (even pressure), F_a [N]	4155.11	4155.11	5195.41
Torque (even wear), T [N m]	108.37	108.37	155.86
Torque (even pressure), T [N m]	151.58	151.58	215.13
Power*SF/ Speed, $(H_{kW}\,K_{SF}\,/N_{rpm})$	0.015873	0.015873016	0.022528195
Parameter, $n_F\,\mu_K\,p_{max}$ [MPa]	0.21	0.2100	0.2100
Outer Diameter (even wear), d_o [m]	0.16839	0.16839	0.18924

(b)

Disk Clutch-Brake Synthesis - US units			
Inputs	Default	Input	Analysis
Power, H_{hp} [hp]	33.5	33.5	33.5
Rotational Speed, N_{rpm} [rpm]	1575	1575	1575
Safety Factor, K_{SF}	1	1	1.39
Number of Friction Surfaces, n_F	2	2	2
Output	*Assuming even pressure*		Input/Output
Outer Diameter, d_o [in]	5.91752	5.91752	6.6
Inner Diameter, d_i [in]	3.41648	3.41648	3.8
Number of Disks, n_D	1	1	1
Number of Friction Surfaces, n_F	2	2	2
Inner to Outer Diameter, d_i/d_o	0.57735	0.57735	0.57576
Torque, T [lb in]	1340.54	1340.54	1340.54
Friction Coefficient, μ_K	0.3	0.3	0.3
Working Pressure, p [psi]	51.000	51.000	36.686
Maximum Pressure, p_{max} [psi]	51	51	51
Power Rating, [hp/in^2]	0.914	0.914	0.732
Pressure * Velocity, pv_C [psi.in/s]	24,888.0	24,888.0	19,967.7
Axial Force, F_a [lb]	935.08	935.08	839.05
Available values	*Values for verification*		
Axial Force (even wear), F_a [lb]	684.53	684.527	852.377
Axial Force(even pressure), F_a [lb]	935.08	935.081	1166.411
Torque (even wear), T [lb in]	958.41	958.406	1329.708
Torque (even pressure), T [lb in]	1340.54	1340.540	1863.565
Power*SF/ Speed, $(H_{hp}\,K_{SF}/N_{rpm})$	0.021270	0.021270	0.029568497
Parameter, $n_F\,\mu_K\,p_{max}$ [psi]	30.6	30.600	30.600
Outer Diameter Check, d_o [in]	6.61782	6.61782	7.38590

Figure 18.7 Clutch-brake synthesis tablets for (a) SI and (b) US systems of units.

see Table 18.2. The parameters (HK_{SF}/N_{rpm}) and ($n_F \mu_K p_{max}$) are given to help in the utility of Figure 18.4 as a visual verification of hand calculations. The pressure times the velocity (pv_C) at the outer diameter is given for further duty assessment, better selection of material, and heat dissipation.

The *Excel tablets* in Figure 18.7 are available in the ***Wiley website*** for *computer-aided synthesis* or real *CAD* of clutch-brake under the name of ***DiskClutch-Brake Synthesis Tablet – SI.xlsx*** and ***Disk Clutch-Brake Synthesis Tablet – US.xlsx***. These should be useful in calculating synthesized clutch-brake diameters for multitude of applications.

The use of these tablets can be extended to ***caliper brakes*** if the *number of friction surfaces* n_F is replaced by the *caliper subtended ratio* ($2\beta_C/2\pi$). The partial disk of a *caliper angle* β_C is specified in [rad]. The main assumption in these tablets is the consideration of uniform pressure rather than uniform wear. Values of both, however, are calculated.

Example 18.2 The disk clutch of Example 18.1 was required to be designed to transmit a maximum of 25 [kW] or 33.5 [hp] at a running speed of 1575 [rpm]. It had only one disk on the takeoff or driven shaft. The disk outside and inside diameters for the usual steel and woven or molded friction materials was defined. Use the available *Excel* files of *Disk Clutch-Brake Synthesis Tablet – SI.xlsx* or *Disk Clutch-Brake Synthesis Tablet – US.xlsx* to evaluate the results and for further assessment of the design.

Solution

Data: $H_{kW} = 25$ [kW] or $H_{hp} = 33.5$ [hp], $N_{rpm} = 1575$ [rpm], $n_F = 2$, $\mu_K = 0.3$, $p_{max} = 0.35$ [MPa] or 51 [psi], and $K_{SF} = 1$.

From Example 18.1, $d_o = 0.15$ [m] = 150 [mm] and 6 [in], $d_i = 85$ [mm] or 3.5 [in].

Using *Shaft Synthesis Tablet – SI.xlsx* and *Shaft Synthesis Tablet – US.xlsx*, the results are shown in Figure 18.7. Values in the **Input** column are the synthesized diameters and not the selected ones. The output parameters, however, are close to the calculated values in Example 18.1.

The *uniform wear* diameters are selected in Example 18.1 as $d_o = 0.17$ [m] = 170 [mm] and 6.6 [in] and $d_i = 100$ [mm] or 3.8 [in]. These values are used in the **Analysis** column in Figure 18.7. The output parameters are the same as the calculated values in Example 18.1. The safety factor has increased to 1.42 and 1.39 for SI and the US, respectively.

The power rating values are reduced to 0.824 [kW/m^2] and 0.732 [hp/in^2]; see Table 18.2. The pressure times the velocity (pv_C) at the outer diameter improved to 3.457 [MPa m/s] and 19 967.7 [psi in/s]; see previously suggested usual ranges in Section 18.6.2.

Optimum design of clutches and brakes can be performed by iteratively using the *Excel tablets* for *clutch-brake* synthesis in Figure 18.7. The optimum can be satisfying some minimum safety factor, a defined power rating values, or a specific pressure times the velocity for either SI or the US systems of units. The lower temperature rise may be separately calculated for some disk thickness according to Eqs. (18.37) and (18.38).

The optimization of a clutch or a brake involves more details and considerations such as thermal stresses that are beyond the scope of this text. For more details, one can investigate the area in the literature such as Hegazi (1997), Metwalli and Hegazi (2001), and beyond.

18.8 Flywheels

Flywheels are used to *control* the variation of torques or power in the sense of smoothing or averaging that variation. They are extensively used in engines that generate unsteady power like the internal combustion engine. The power is usually dependent on the engine cycle, ignition means, and slider-crank kinematics. Figure 18.8 shows an approximation of the torque fluctuation on crankshaft of a single-cylinder engine similar to Filipi and Assanis

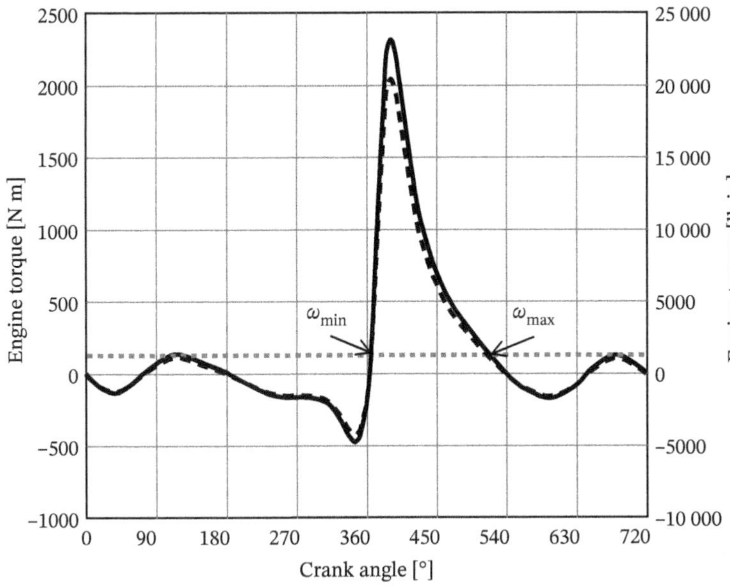

Figure 18.8 Torque fluctuation on crankshaft of a single-cylinder engine. SI curve is the solid line, and the US curve is the dashed line. The dotted grayish line is the fluctuation average.

(2001). The SI curve is the solid line, and the US curve is the dashed line. To smooth the output torque, a flywheel is used. The clutch in Figure 18.3a shows such a flywheel before the engagement of the clutch. The dotted grayish line in Figure 18.8 is the average of that varying torque. The size of the flywheel affects the steadiness of that average with a minimum variability. The objective is then to minimize the variation of the output torque about that average.

To develop a simple mathematical model representing the input system, flywheel, and the output system, one can start with the initial model shown in Figure 18.6. After full engagement or clutching, the system reduces to an input from the driver to the flywheel and an output from the flywheel to the driven system. A more involved system requires more detailed dynamics and vibration modeling; see, e.g. Inman (2017) or Thomson and Dahleh (1998). Assuming a simplified rigid connection from the driver to the flywheel and a rigid connection from the flywheel to the driven system, one can simply utilize the power and kinetic energy variation as follows.

Consider a temporary *steady state*, where the driver or engine is delivering a steady power of $\boldsymbol{T}_{av}\,\omega_{av}$. The average torque $\boldsymbol{T}_{av}$ is the *nominal torque*, and the average rotational speed ω_{av} is the *nominal speed* of the flywheel at that steady power. The average torque is resembling that in Figure 18.8 of the dotted grayish line defining the torque fluctuation average. At the start of this cycle, the engine demands energy or takes away power off the flywheel since the torque is below the average. The rotational speed ω_{av} is then reduced to the minimum speed ω_{min} when the torque crosses the average torque, as shown in Figure 18.8. At that defined point in Figure 18.8, the engine starts delivering positive power to the flywheel till the speed is increased to a maximum speed ω_{max}. At that defined point in Figure 18.8, the engine starts taking away power from the flywheel and so on. The minimum and the maximum kinetic energies E_{min} and E_{max} of the flywheel at these two switching points are then given by the following established relations:

$$E_{min} = \frac{1}{2}J_F\,\omega_{min}^2, \quad E_{max} = \frac{1}{2}J_F\,\omega_{max}^2 \tag{18.39}$$

where J_F is the *flywheel mass moment of inertia*. The flywheel *change in kinetic energy* ΔE_F is then given by

$$\Delta E_F = E_{max} - E_{min} = \frac{1}{2}J_F(\omega_{max}^2 - \omega_{min}^2) = \frac{1}{2}J_F(\omega_{max} + \omega_{min})(\omega_{max} - \omega_{min}) \tag{18.40}$$

One should also note that the *average rotational speed* ω_{av} is given by the following simple expression:

$$\omega_{av} = \frac{\omega_{max} - \omega_{min}}{2} \tag{18.41}$$

As a measure of flywheel performance, define the *coefficient of speed fluctuation* C_F such that

$$C_F = \frac{(\omega_{max} - \omega_{min})}{\omega_{av}} \tag{18.42}$$

From Eqs. (18.40)–(18.42), one can find the needed *flywheel mass moment of inertia* J_F as follows:

$$J_F = \frac{2\Delta E_F}{(\omega_{max} + \omega_{min})(\omega_{max} - \omega_{min})} = \frac{2\Delta E_F}{(2\omega_{av})(C_F\omega_{av})} = \frac{\Delta E_F}{C_F\omega_{av}^2} \tag{18.43}$$

Equation (18.43) assumes that the flywheel change in kinetic energy ΔE_F is handled by the flywheel only. This assumes that the driver or *input inertia* J_i and the *output system inertia* J_o as shown in Figure 18.6 are inconsequential. Considering these inertias should reduce the needed *flywheel mass moment of inertia* J_F by those two inertias. However, before clutching or at the *idle stage*, the *output system inertia* J_o should not be considered if the *coefficient of speed fluctuation* C_F at that stage is needed to be definitely and absolutely defined to manage driver idle vibration.

The design of flywheel as a simple uniform thickness disk having an *outer flywheel diameter* d_{Fo} and an *inner flywheel diameter* d_{Fi} should have a *mass moment of inertia* J_F given by the following relation:

$$J_F = \frac{1}{2}m_F(r_{Fo}^2 - r_{Fi}^2) = \frac{1}{8}m_F(d_{Fo}^2 - d_{Fi}^2) \tag{18.44}$$

where m_F is the *flywheel mass*, r_{Fo} is the flywheel outer radius, and r_{Fi} is the inner flywheel radius.

The traditional design of flywheels has been in the form of a large rim cast with spokes linking the hub and the rim. The spokes are usually replaced by a continuous disk of constant or variable thickness. Stresses are calculated in the form of a rotating disk (like turbine disks) particularly for high rotational speed. Flywheels running in vacuum at a very high speed are potentially used as energy storage; see, e.g. Alhneaish (2018) and Shaltout et al. (2020). A solid disk flywheel will have lower stresses than a disk with a central hole (Bazaj and Metwalli 1971). Stresses, design, and optimization of disks and flywheels can be found in the literature such as Metwalli et al. (1982), Shawki et al. (1984), and Metwalli (1986).

Example 18.3 Torque fluctuation on crankshaft of a single-cylinder engine is similar to that in Figure 18.8. The engine rotational speed is 900 [rpm], which is the idle phase. The SI curve is the solid line, and the US curve is the dashed line. The dotted grayish line is the torque fluctuation average of about 50.5 [N m] or 447 [lb in]. If the change in kinetic energy is about 750 [N m] or 6650 [lb in], what should be the flywheel mass moment of inertia to have a coefficient of speed fluctuation of 0.1? What is the expected mass of the solid flywheel, if the outer diameter is 0.5 [m]? Find the minimum and maximum rotational speed.

Solution

Data: $N_{rpm} = 900$ [rpm], $\omega_{av} = 2\pi N_{in}/60 = 94.248$ [rad/s], $\boldsymbol{T}_{av} = 50.5$ [N m] or 447 [lb in], $\Delta E = 750$ [N m] or 6650 [lb in], $C_F = 0.1$.

From Eq. (18.43), the flywheel mass moment of inertia is given by the following value:

$$J_F = \frac{\Delta E}{C_F\omega_{av}^2} = \frac{750}{0.1(94.248)^2} = 0.844\,34\ [\text{kg m}^2]$$

$$J_F = \frac{\Delta E}{C_F\omega_{av}^2} = \frac{6650}{0.1(94.248)^2} = 7.4865\ [\text{lb s}^2\ \text{in}] \tag{a}$$

The expected mass of the solid flywheel according to Eq. (18.44) is

$$m_F = \frac{8J_F}{(d_{Fo}^2 - d_{Fi}^2)} = \frac{8(0.844\,34)}{0.5^2} = 27.019\ [\text{kg}] \tag{b}$$

To find the minimum and maximum rotational speed, one uses Eqs. (18.41) and (18.43) to get

$$\omega_{\text{max}} = \frac{\omega_{\text{av}}}{2}(2 + C_F) = \frac{94.248}{2}(2 + 0.1) = 98.960\,[\text{rad/s}]$$
$$\omega_{\text{min}} = 2\omega_{\text{av}} - \omega_{\text{max}} = 2(94.248) - 98.960 = 89.536\,[\text{rad/s}] \qquad \text{(c)}$$

These two speeds in Eq. (c) are expected to occur at the crank angles shown in Figure 18.8.

18.9 Constructional Details

Few constructional details have been introduced as sketches in Figure 18.3a of a traditional auto-clutch and Figure 18.5b about caliper disk brake. The main components are presented, and constructional drawings can be obtained from these by few additional details that can be certainly deduced. Figure 18.3a requires the inclusion of the clutch disk details with its expected construction and dimensions from OEM manufacturer. The details may include internal springs (not shown) to cushion the impacting torque. The flywheel size and fixation to the input side should be synthesized. The embedded ball bearing should be sized, and its assembly requirements should be selected; see Chapter 11 for needed constructional details. The diaphragm spring should be synthesized, and its fulcrum in the clutch casing is detailed. The attachment of the clutch casing to the flywheel by screws and its centering must be constructed. The thrust bearing layout is to be synthesized and its operating mechanism detailed. Several alternative constructions may be reviewed over the Internet or the associated patents.

Figure 18.5b is about the caliper disk brake. It requires different detail. The caliper disk brake is usually supplied by an OEM manufacturer. Some details might be negotiated with the manufacturer about the fitting location means and space. The fixation of the brake disk to the wheel boss and its centering must be constructed; see similar constructions on the Internet. The fixation of the caliber to the wheel steering knuckle and its centering must also be constructed; see alternative constructions on the Internet.

For a simplified constructional details of multi-disk clutch-brake, Figure 18.9 provides a sample half-sectional view of a hydraulically operated multi-disk clutch-brake. Main components are displayed. The fluid can be either air or a special hydraulic liquid. The actuating piston is an annular construction with internal and external seal inside the similar cavity in the left casing. The actuating piston can, however, be three or four pistons inside the left casing squeezing the friction disks against the end disk (not advocated). The inner friction disks are splined over the left casing, and the alternate outer friction disks are splined with the right casing. The left casing is keyed to the input shaft, but it is not positively prevented from moving to the right. There can be some in between sliding interference employed near the left shoulder of the input shaft. However, some positive fixation can be made such as extending the tightening bolt of the end disk to an enlarged left shoulder (not shown). This option is needed if

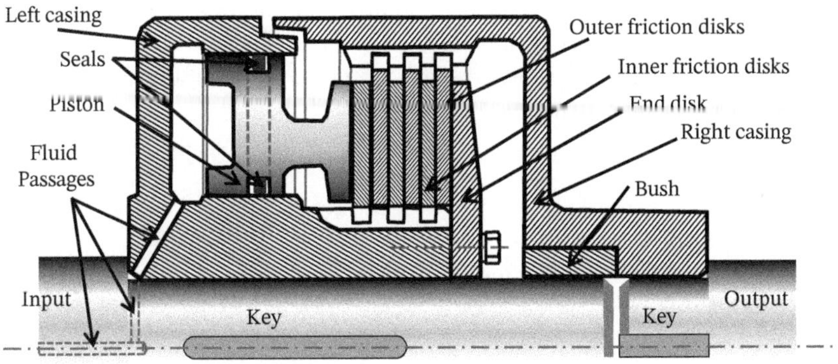

Figure 18.9 Half-sectional view of a hydraulically operated multi-disk clutch-brake. Main components are displayed.

the clutch-brake operating fluid is different from the surrounding environment. In such a case, an added locknut over the input shaft (see Figure 11.13a) and a seal between the left casing and the shaft are needed. The assembly of the right casing with the inner spline to the outer friction disk spline is to be jigged. The input shaft must also be inserted into the internal bush of the right casing for centering.

The construction details in Figure 18.9 might be altered in some previously indicated cases. If the clutch-brake is pneumatically operated and operating in air or operated by hydraulic oil and operating and immersed in hydraulic oil, the construction details might not change much of that in Figure 18.9. If the clutch-brake is operated and operating differently, the previously indicated seals and fixations must be considered, and fluid passages may also be relocated.

Several alternative constructions may be available over the Internet or the associated patents. In addition, several available manufacturers provide OEM clutch-brake products that can be acquired.

18.10 Summary

Clutches, brakes, and flywheels are important machine elements that control power transmission. The synthesis of these, consequently, is not necessarily dependent on other machine elements. Therefore, this chapter has been pushed back after so many other machine elements in this text. The initial synthesis of each of the clutches, brakes, and flywheels is dependent mainly on power transmission variation.

Classification of clutches and brakes as positive, friction, or mixed friction-positive indicates the importance of the friction in the design of these components. The basic treatment in this chapter has been on the friction activated clutches and brakes. Relations for developed torques and axial forces according to the assumptions of uniform wear and uniform pressure have been introduced. Applications in conical and plane disks provide needed information to synthesize single or multi-disk configurations for clutches and brakes.

The initial synthesis of disk clutches and brakes depends on optimum ratio of inner to outer diameters. Knowing the transmitted power and the rotational speed is enough to initially synthesize the single or multi-disk configurations for clutches and brakes using usual or selected materials in these applications. Information about these materials is provided to evaluate, modify, or verify the design. Wide ranging graphs provide tools to initially synthesize the single or multi-disk configurations for clutches and brakes. These would also be potentially useful in the initial designs of caliper disk brakes. Initial mathematical models for energy dissipation and temperature rise are provided to support the detailed design process of clutches and brakes. *Excel tablets* for clutch-brake synthesis are provided in both SI and the US system of units to help in calculating synthesized clutch-brake elements for multitude of applications. They are available in the ***Wiley website*** under the name of *DiskClutch-Brake Synthesis Tablet – SI.xlsx* and *Disk Clutch-Brake Synthesis Tablet – US.xlsx*. The use of these tablets can be extended to caliper brakes. Hopefully, these tools can be useful in computer-aided synthesis or a real *CAD* of such elements.

Flywheels are presented to control the variation of torques or power and to smooth or average that variation. The tools to achieve that are provided and their applications are demonstrated.

Some constructional details of clutches, brakes, and flywheels are provided to aid in the understanding and possible initiation of complete designs of these machine elements.

Problems

18.1 Find the constructional details and method of operation for the bike overrunning drive. Use the Internet, manufacturers, patents, etc.

18.2 Find the constructional details and method of operation for a synchromesh. Use the Internet, manufacturers, patents, etc.

18.3 Use the Internet, manufacturers, patents, etc. to obtain the construction of a manual transmission clutch of a vehicle. Is it different in type than a motorcycle clutch?

18.4 What applications do you find for a conical clutch? Sketch an application that uses a conical clutch and define the method of operation.

18.5 Why should the cone angle of a cone clutch be between 10° and 15°? How is that related to the friction angle? If the angle is small, should there be an axial force for disengagement?

18.6 Friction clutches start operating in a uniform pressure mode, and after some operational time, they turn to a uniform wear mode. Search for the principle behind that postulation.

18.7 Why is the design according to uniform wear assumption more conservative than the design according to uniform pressure assumption?

18.8 For the same conditions, is the axial force in Eq. (18.3) larger that the axial force in Eq. (18.7)?

18.9 For the same conditions, is the torque in Eq. (18.4) smaller that the torque in Eq. (18.8)?

18.10 From Eqs. (18.7) and (18.8), find the torque as a function of the axial force.

18.11 Construct a free body diagram of one of the disks in the multi-disk clutch-brake arrangement to show that the axial force is the same over each disk.

18.12 Define an arrangement of multi-disk clutch-brake where the number of friction surfaces is not the same as twice the number of disks.

18.13 Scrutinize some of the material property ranges given in Tables 18.1 and 18.2.

18.14 What are the implications of using a uniform pressure condition rather than the uniform wear conditions? Use Eqs. (18.22) and (18.23) to compare.

18.15 Redesign the clutch in Example 18.1 adapting a constant wear assumption.

18.16 Redo Example 18.1 using the same material, but with a safety factor of 2.0.

18.17 Submit an initial synthesis for a disk clutch-brake to transmit a maximum of 75 [kW] or 100.5 [hp] at a running speed of 525 [rpm]. Consider a usual coefficient of friction of 0.3 and a maximum axial pressure of 0.35 [MPa] or 51 [psi]. Take the safety factor for this design as 2.

18.18 What should be the implication if one uses the maximum force needed for the assumption of uniform pressure in the clutch designed for uniform wear assumption for the case in Example 18.1?

18.19 Study the effect of increasing the number of friction surfaces on the design in Example 18.1.

18.20 How many numbers of friction surfaces are needed in the same construction of Example 18.1 if the clutch is used in a wet environment?

18.21 Consider the case of a multi-disk clutch that transmits 25 [kW] or 33.5 [hp] at a rotational speed of 1575 [rpm]. It has three disks on the driving shaft and two disks on the other driven shaft. Disks outside and inside diameters are 200 [mm] or 8 [in] and 110 [mm] or 4.5 [in]. The coefficient of friction is 0.3. Find the maximum axial pressure assuming an even pressure distribution and an even wear condition.

18.22 Generate initial designs of clutches to transmit 0.5, 5, 50, 500, and 5000 [kW] running at 750, 1250, 1750, 2500, 3000, and 3600 [rpm]. Consider a usual coefficient of friction for dry engagement as 0.3 and the maximum axial pressure as 0.35 [MPa]. Take the safety factor for these cases as 2.

18.23 Generate initial designs of clutches to transmit 0.5, 5, 50, 500, and 5000 [hp] running at 750, 1250, 1750, 2500, 3000, and 3600 [rpm]. Consider a usual coefficient of friction for dry engagement as 0.3 and the maximum axial pressure as 51 [Pa]. Take the safety factor for these cases as 2.

18.24 Develop initial synthesis charts for the caliper disk brake similar to clutch-brake synthesis charts of Figure 18.4.

18.25 The system in Example 18.1 needs to be stopped. Use a caliper brake of a 120° caliper angle to stop the system right after the power shutdown. Use the power and the initial rotational speed to define the torque needed to stop the system.

18.26 Write the expression to transfer the inertia of the mass traveling at a linear velocity to the equivalent mass moment of inertia rotating at ω_i of the brake disk on the wheel shaft.

18.27 It is required to design a caliper disk brake set for a front wheel group of an airplane of about 1200 [kN] or 270 [klb] traveling at a speed of 85 [m/s] or 190 [mph] in case of aborting takeoff. Assume the front wheel diameter is 1.4 [m] or 55 [in], the load distribution is 20% at that wheel group, caliper pads subtended angle is 220° of the disk, and the maximum deceleration is $0.6g_a$. Find the expected temperature rise of the caliper disk brake set. Use the most suitable parameters for this critical case.

18.28 Use Eqs. (18.31) and (18.34) to derive Eq. (18.35).

18.29 Find the properties of the prospective brake materials concerning the specific heat and density.

18.30 Use either *Disk Clutch-Brake Synthesis Tablet – SI.xlsx* or *Disk Clutch-Brake Synthesis Tablet – US.xlsx* to solve Problems 18.17 or 18.21. Compare results with hand calculations.

18.31 Use either *Disk Clutch-Brake Synthesis Tablet – SI.xlsx* or *Disk Clutch-Brake Synthesis Tablet – US.xlsx* to solve Problem 18.20. Compare results with hand calculations.

18.32 Use either *Disk Clutch-Brake Synthesis Tablet – SI.xlsx* or *Disk Clutch-Brake Synthesis Tablet – US.xlsx* to solve Problem 18.22 or 18.23. Compare results with hand calculations.

18.33 What should you change in either *Disk Clutch-Brake Synthesis Tablet – SI.xlsx* or *Disk Clutch-Brake Synthesis Tablet – US.xlsx* to solve Problem 18.25.

18.34 Can you use either *Disk Clutch-Brake Synthesis Tablet – SI.xlsx* or *Disk Clutch-Brake Synthesis Tablet – US.xlsx* to solve Problem 18.25. Compare results with hand calculations.

18.35 For some clutch or brake manufacturers, can you evaluate or find their values of the pressure times the velocity (pv_C) for the products.

18.36 Use either *Disk Clutch-Brake Synthesis Tablet – SI.xlsx* or *Disk Clutch-Brake Synthesis Tablet – US.xlsx* to optimize any of the previous problems to satisfy some minimum safety factor or achieve defined power rating value or a specific pressure times the velocity value for either SI or the US systems of units.

18.37 Find the width of the flywheel in Example 18.3 if it is made of cast iron. If the outer diameter is increased to 0.6 [m], what is the needed moment of inertia and flywheel thickness.

18.38 Change the coefficient of speed fluctuation in Example 18.3 to 0.05. What should be the needed moment of inertia and flywheel thickness?

18.39 The needed energy in one cycle of a flywheel blanking press is about 7000 [N m] or 62 000 [lb in]. What should be the flywheel moment of inertia if the angular velocity variation should be between 50 and 70 [rpm]? If the diameter of this cast iron flywheel can be 2 [m] or 80 [in], what is its expected thickness?

18.40 Use relevant constructions of traditional auto-clutch to that in Figure 18.3a to produce a ***3D*** working design. Employ a reasonable OEM disk clutch in your design. What are the characteristics of that OEM disk clutch? Can you verify the design by the utilization of the tools in this chapter?

18.41 Select a caliper disk brake from a manufacturer, and present a detailed ***3D*** design and assembly for its detailed components. What are the characteristics of that caliper disk brake? Can you verify the design by the utilization of the tools in this chapter?

18.42 Adjust the constructional details of Figure 18.9 to include the indicated design requirements of seals and fixations. Present a ***3D*** construction and provide a ***2D*** assembly drawing.

18.43 Investigate the constructional details of any available manufacturers' OEM clutch-brake products. Can you identify any one like that in Figure 18.9? Provide a ***3D*** design and assembly for its detailed components. Can you verify the design by the utilization of the tools in this chapter?

References

Alhneaish, M.M. (2018). *Optimal design and control of integrated wind turbine and flywheel system*. MS thesis. Cairo University.

Bazaj, D.K. and Metwalli, S.M. (1971). Stress analysis of compounded rotating disks. *Journal of The Franklin Institute* 292 (4): 265–275.

Budynas, R.G. and Nisbett, J.K. (2015). *Shigley's Mechanical Engineering Design*, 10e. McGraw Hill.

Excel© (2016). *Microsoft Office Excel*. Microsoft Corporation.

Filipi, Z.S. and Assanis, D.N. (2001). A nonlinear, transient, single-cylinder diesel engine simulation for predictions of instantaneous engine speed and torque. *Journal of Engineering for Gas Turbines and Power* 123 (4): 951–959.

Hegazi, H.A.M. (1997). *Computer aided design and optimization of disc brakes*. MS thesis. Cairo University.

Inman, D.J. (2017). *Vibration with Control*. Wiley.

Lambert, H.T. (1945). *Multiple disk brake*. US Patent US2375855.

Lambert, H.T. (1946). *Disk brake*. US Patent US2405219.

Metwalli, S.M. (1986). Flywheel optimization under speed fluctuation effects. *The 11Th ASME Design Automation Conference*, Columbus, OH, USA (5–8 October), ASME Paper No. 86-DET-129. American Society of Mechanical Engineers.

Metwalli, S.M. and Hegazi, H.A. (2001). Computer-based design of disc brakes by multi-objective form optimization. *Proceedings of the ASME 2001 Design Engineering Technical Conference and Computers and Information in Engineering Conference*, Pittsburgh, PA, USA (9–12 September), Paper No. DETC2001/CIE-21680. American Society of Mechanical Engineers.

Metwalli, S.M., Shawki, G.S.A., and Sharobeam, M.H. (1982). Optimum design of variable-material flywheels, presented at *The ASME Eighth Design Automation Conference*, Washington, DC, 12–15 September, paper number 82-DET-99, American Society of Mechanical Engineers. *Journal of Mechanical Design* 105 (2): 249–253.

Metwalli, S.M. Hegazi, H.A., and Abdel-All, U.M. (1999). CAD of disc brakes by multi-objective optimization. *Proceedings of the 1999 ASME Design Engineering Technical Conference*, Las Vegas, NV, USA (12–15 September), Paper No. DETC99/CIE-9139. American Society of Mechanical Engineers.

Neale, M.J. (ed.) (1973). *Tribology Handbook*. Butterworths.

Shaltout, M., Alhneaish, M., and Metwalli, S. (2020). An economic model predictive control approach for wind power smoothing and tower load mitigation. *Journal of Dynamic Systems, Measurement, and Control* 142: 061005-1-10.

Shawki, G.S.A., Metwalli, S.M., and Sharobeam, M.H. (1984). Optimum configuration for an isotropic rotor. *Journal of Mechanical Design* 106 (3): 376–379.

Thomson, W.T. and Dahleh, M.D. (1998). *Theory of Vibration with Applications*, 5e. Prentice Hall.

Internet Links

www.twindisc.com Twin Disc: Clutches, power transmission, etc.
https://akebonobrakes.com Akebono: Brake pads.
https://loganclutch.com Logan Clutch Corporation: Clutches, brakes, etc.
https://nrsbrakes.com NRS: Brake pads.
www.ate-brakes.com ATE: Brake Calipers, disk brakes, drum brakes, clutches.
www.brembo.com Brembo: Brakes.
www.exedyusa.com EXEDY Corporation: Clutches, flywheels.
www.hilliardcorp.com Hilliard Corporation: Clutches, brakes, drive train, etc.
www.schaeffler.com Schaeffler (Luk): Clutch systems, etc.
www.valeoservice.com Valeo: Clutches, brakes, flywheels, etc.
www.warnerelectric.com Warner Electric: Clutches, brakes, electromagnetic clutches & brakes, etc.
www.zf.com ZF: Clutches, brakes, power train and mobility systems, etc.

Appendix A

Figures and Tables

A.1 Conversion Between US and SI Units

Table A.1.1 Conversion between basic US and SI units including conversion ratios.

Quantity	US set	SI set	Ratio SI/US or (conversion ratio = 1)
Length	Inch [in or ']	Meter [m]	39.370, 1 [m] = 39.37 [in] or (0.0254 [m]/[in])
Mass	[lb_m]	Kilogram [kg]	2.2046, 1 [kg] = 2.2046 [lb_m] or (0.4536 [kg]/[lb_m])
	slug [1 lb s^2/ft]		0.068 522, 1 [kg] = 0.0685 [slug] or (14.594 [kg]/[slug])
Time	Second [s]	Second [s]	1
Force	Pound [lb]	Newton [N]	0.224 81, 1 [N] = 0.224 81 [lb] or (4.4482 [N]/[lb])

Machine Design with CAD and Optimization, First Edition. Sayed M. Metwalli.

Companion website: www.wiley.com/go/metwalli/machine

Table A.1.2 Conversion ratios of selected basic quantities.

Quantity	US system	SI system	(Conversion ratio = 1)
Acceleration a	$[\mathrm{in/s^2}]$	$[\mathrm{m/s^2}]$	$(0.025\,40\ [\mathrm{m/s^2}]/[\mathrm{in/s^2}])$
Area second moment I_A	$[\mathrm{in^4}]$	$[\mathrm{m^4}]$	$(0.416\,23\times10^{-6}\ [\mathrm{m^4}]/[\mathrm{in^4}])$
Density ρ	$[\mathrm{lb}_m/\mathrm{in^3}]$	$[\mathrm{kg/m^3}]$	$(27\,680\ [\mathrm{kg/m^3}]/[\mathrm{lb}_m/\mathrm{in^3}])$
Heat energy E	[Btu]	$[\mathrm{N\,m}]\equiv[\mathrm{J}]$	(1055.056 [J]/[Btu])
Mass m	Slug $[\mathrm{lb\,s^2/ft}]$	[kg]	(14.594 [kg]/[slug])
Mass m	$[\mathrm{lb\,s^2/in}]$	[kg]	$(175\ [\mathrm{kg}]/[\mathrm{lb\,s^2/in}])$
Mass m	$[\mathrm{lb}_m]$	[kg]	$(0.453\,592\,37\ [\mathrm{kg}]/[\mathrm{lb}_m])$
Mass moment of inertia I_m	$[\mathrm{lb}_m\ \mathrm{in^2}]$	$[\mathrm{kg\,m^2}]$	$(2.9264\times10^{-4}\ [\mathrm{kg\,m^2}]/[\mathrm{lb}_m\ \mathrm{in^2}])$
Power H	[lb in/s] or [hp]	$[\mathrm{N\,m/s}]\equiv[\mathrm{W}]$	(745.7 [W]/[hp])
Pressure p or stress σ, τ	[psi]	[Pa]	(6894.8 [Pa]/[psi])
Section modulus Z	$[\mathrm{in^3}]$	$[\mathrm{m^3}]$	$(16.387\times10^{-6}\ [\mathrm{m^3}]/[\mathrm{in^3}])$
Spring rate or stiffness k_s	[lb/in]	[N/m]	(175.13 [N/m]/[lb/in])
Torque T	[in lb]	[m N]	(0.112 985 [m N]/[in lb])
Velocity v	[in/s]	[m/s]	(0.0254 [m/s]/[in/s])
Weight w	[lb]	[N]	(0.453 592 37 [N]/[lb])
Work W and energy E	[lb in]	$[\mathrm{N\,m}]\equiv[\mathrm{J}]$	(0.112 985 [J]/[lb in])

A.2 Standard SI Prefixes

Table A.2.1 Suggested prefix symbols for SI units and adapted for the US units.

Classification	Symbol	Name		Factor or decimal
Adapted basic	[G...]	giga	(billion)	$10^9 = 1\,000\,000\,000$
	[M...]	mega	(million)	$10^6 = 1\,000\,000$
	[k...]	kilo	(thousand)	$10^3 = 1000$
	[m...]	milli	(thousandth)	$10^{-3} = 0.001$
	[μ...]	micro	(millionth)	$10^{-6} = 0.000\,001$
Adapted occasionally	[T...]	tera	(trillion)	$10^{12} = 1\,000\,000\,000\,000$
	[c...]	centi	(hundredth)	$10^{-2} = 0.01$
	[n...]	nano	(billionth)	$10^{-9} = 0.000\,000\,001$
Rarely used	[h...]	hecto	(hundred)	$10^2 = 100$
	[da...]	deka	(ten)	$10^1 = 10$
	[d...]	deci	(tenth)	$10^{-1} = 0.1$
	[p...]	pico	(trillionth)	$10^{-12} = 0.000\,000\,000\,001$

A.3 Preferred Numbers and Sizes

Table A.3.1 Series of preferred dimensions (see ANSI B4.2 (2009), ANSI B4.1 (2009), ISO 3 (1973), and references in Chapter 2).

Preferred[a)][mm] ([in])	Second[b)][mm] ([in])
0.05, 0.06, 0.08 (0.002, 0.003)	0.055, 0.07, 0.09
0.1, 0.12, 0.16, 0.2, 0.25, 0.3, 0.4, 0.5, 0.6, 0.8 (0.004, 0.005, 0.006, 0.008, 0.010, 0.012, 0.016, 0.020, 0.025, 0.032) (1/64, 1/32)	0.11, 0.14, 0.18, 0.22, 0.28, 0.35, 0.45, 0.55, 0.7, 0.9
1.0, 1.2, 1.6, 2, 2.5, 3, 4, 5, 6, 8 (0.04, 0.05, 0.06, 0.08, 0.10, 0.12, 0.16, 0.20, 0.24, 0.30) (1/16, 3/32, 1/8, 3/16, 1/4, 5/16)	1.1, 1.4, 1.8, 2.2, 2.8, 3.5, 4.5, 5.5, 7, 9
10, 12, 16, 20, 25, 30, 40, 50, 60, 80 (0.40, 0.50, 0.60, 0.80, 1.00, 1.20, 1.6, 2.0, 2.4, 3.0) (3/8, 1/2, 5/8, 3/4, 1, 1 1/4, 1 1/2, 2, 2 1/2, 3)	11, 14, 18, 22, 28, 35, 45, 55, 70, 90 (1.40, 1.80, 2.40, 2.60, 2.80, 3.20) (7/16, 9/16, 11/16, 7/8, 1 3/4, 2 3/4, 3 1/4)
100, 120, 160, 200, 250, 300, 400, 500, 600, 800, 1000 (4.0, 5.0, 6.0, 8.0, 10.0, 12.0, 16.0, 20.0, 24.0, 30.0)	110, 140, 180, 220, 280, 350, 450, 550, 700, 900 (3.2, 3.4, 3.6, 3.8, 4.2, 4.4, 4.6, 4.8, 5.2, 5,4, 5.6, 5.8, 7.0, … 0.5 step to 20) (3 1/4, 3 1/2, 3 3/4, 4 1/4, 4 1/2, 4 3/4, 5 1/4, 5 1/2, 5 3/4, 6 1/2, … ½ step to 20)

The series of preferred ranges depend on metric decimal or base ten and their inch decimal and fractional equivalents in parentheses. The second column is available if deemed necessary or critically needed.

a) The preferred inch decimals and fractions are approximately selected.

b) The second inch choices are approximately selected.

A.4 Standard Rods, or Bars

Table A.4.1 Properties of selected round solid steel bars or rods of main preferred dimensions.

Diameter [mm]	Area [mm^2]	Weight [N/m]	Second area moment @z [10^6 mm^4]	Section modulus Z_z [10^3 mm^3]	Radius of gyration r_G [mm]	Diameter [in]	Area [in^2]	weight [lb/ft]	Second area moment @z [in^4]	Section modulus Z_z [in^3]	Radius of gyration r_G [in]
1	0.7854	0.061	0.049 09	0.098 17	0.25	0.031 25	7.670E−4	0.002 61	4.681E−8	2.996E−6	0.007 813
2	3.142	0.242	0.7854	0.7854	0.50	0.062 50	3.068E−3	0.0104	7.490E−7	2.397E−5	0.015 625
3	7.069	0.546	3.9761	2.651	0.75	0.12	0.011 31	0.0384	1.018E−5	1.696E−4	0.03
4	12.57	0.970	12.566	6.283	1.00	0.16	0.020 11	0.0684	3.217E−5	4.021E−4	0.04
5	19.63	1.52	30.680	12.27	1.25	0.2	0.031 42	0.1068	7.854E−5	7.854E−4	0.05
6	28.27	2.18	63.617	21.21	1.50	0.25	0.049 09	0.1669	1.917E−4	1.534E−3	0.0625
8	50.27	3.88	201.06	50.27	2.00	0.3	0.070 69	0.2403	3.976E−4	2.651E−3	0.075
10	78.54	6.06	490.87	98.17	2.50	0.4	0.1257	0.4272	1.257E−3	6.283E−3	0.1
12	113.1	8.73	1017.9	169.6	3.00	0.5	0.1963	0.6675	3.068E−3	0.012 27	0.125
15	176.7	13.6	2485.0	331.3	3.75	0.6	0.2827	0.9612	6.362E−3	0.021 21	0.15
16	201.1	15.5	3217.0	402.1	4.00	0.7	0.3848	1.308	0.011 79	0.033 67	0.175
20	314.2	24.2	7854.0	785.4	5.00	0.8	0.5027	1.709	0.020 11	0.050 27	0.2
25	490.9	37.9	19 175	1534	6.25	1	0.7854	2.670	0.049 09	0.098 17	0.25
30	706.9	54.6	39 761	2651	7.50	1.2	1.131	3.845	0.1018	0.1696	0.3
40	1257	97.0	125 664	6283	10.0	1.6	2.011	6.835	0.3217	0.4021	0.4
50	1963	152	306 796	12 272	12.5	2	3.142	10.68	0.7854	0.7854	0.5
60	2827	218	636 173	21 206	15.0	2.4	4.524	15.38	1.629	1.357	0.6
80	5027	388	2.011E+6	50 265	20.0	3	7.069	24.03	3.976	2.651	0.75
100	7854	606	4.909E+6	98 175	25.0	4	12.57	42.73	12.57	6.283	1
120	11 310	873	1.018E+7	169 646	30.0	5	19.63	66.76	30.68	12.27	1.25
150	17 671	1364	2.485E+7	331 340	37.5	6	28.27	96.13	63.62	21.21	1.5
160	20 106	1552	3.217E+7	402 124	40.0	7	38.48	130.9	117.9	33.67	1.75
200	31 416	2425	7.854E+7	785 398	50.0	8	50.27	170.9	201.1	50.27	2
250	49 087	3789	1.917E+8	1.534E+6	62.5	10	78.54	267.0	490.9	98.17	2.5
300	70 686	5456	3.976E+8	2.651E+6	75.0	12	113.1	384.5	1018	169.6	3
400	125 664	9699	1.257E+9	6.283E+6	100	16	201.1	683.5	3217	402.1	4
500	196 350	15 155	3.068E+9	1.227E+7	125	20	314.2	1068	7854	785.4	5
600	282 743	21 824	6.362E+9	2.121E+7	150	24	452.4	1538	16 286	1357	6
800	502 655	38 798	2.011E+10	5.027E+7	200	32	804.2	2734	51 472	3217	8
1000	785 398	60 621	4.909E+10	9.817E+7	250	40	1257	4272	125 664	6283	10

A.5 Standard Joining and Retaining Elements

Table A.5.1 Some square and rectangular key dimensions as defined in figure.

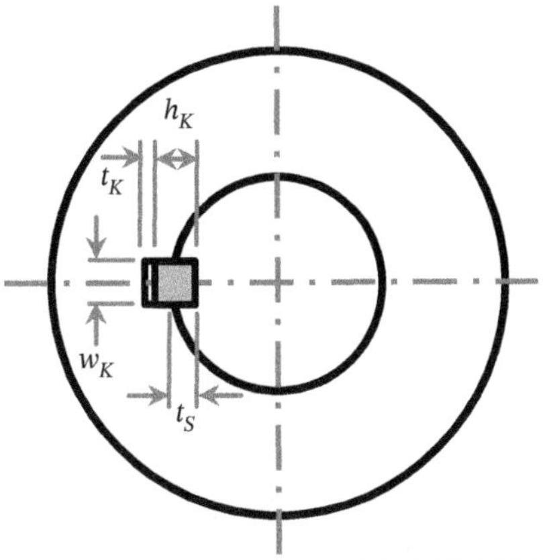

Shaft diameter d_S [mm]	**Shaft diameter** (d_S) ([in])	**Key width** w_K [mm]	**Key height** h_K (t_S)[a] [mm]	**Key width** (w_K) ([in])	**Key height** (h_K) ([in])	**Keyway clearance** t_K [mm]
6–8	(5/16–7/16)	2	2 (1.2)	(3/32)	(3/32)	0.2
9–10		3	3 (1.8)			0.2
11–12	(1/2–9/16)	4	4 (2.5)	(1/8)	(1/8)	0.3
13–17		5	5 (3)			0.3
18–22	(5/8–7/8)	6	6 (3.5)	(3/16)	(3/16)	0.3
23–30	(15/16–1.25)	8	7 (4)	(1/4)	(1/4)	0.3
31–38	(1 5/16–1 3/8)	10	8 (5)	(5/16)	(5/16)	0.3
39–44	(1 7/16–1.75)	12	8 (5)	(3/8)	(3/8)	0.3
45–50	(1 13/16–2.25)	14	9 (5.5)	(1/2)	(1/2)	0.3
51–58		16	10 (6)			0.3
59–65	(2 5/16–2.75)	18	11 (7)	(5/8)	(5/8)	0.4
66–75	(2 13/16–3.25)	20	12 (7.5)	(3/4)	(3/4)	0.4
76–85		22	14 (9)			0.4
86–95	(3 5/16–3.75)	25	14 (9)	(7/8)	(7/8)	0.4
96–110	(3 13/16–4.5)	28	16 (10)	(1)	(1)	0.4
111–130	(4 5/16–5.5)	32	18 (11)	(1.25)	(1.25)	0.4
131–150	(5 9/16–6.5)	36	20 (12)	(1.5)	(1.5)	0.4
151–170		40	22 (13)			0.4
171–200	(6 9/16–7.5)	45	25 (15)	(1.75)	(1.5)	0.4
201–230	(7 9/16–9)	50	28 (17)	(2)	(1.5)	0.4
231–260	—	56	32 (20)	—	—	0.4
261–290	—	63	32 (20)	—	—	0.4
291–330	—	70	36 (22)	—	—	0.4
331–380	—	80	40 (25)	—	—	0.4
381–440	—	90	45 (28)	—	—	0.4
441–500	—	100	50 (31)	—	—	0.5

a) The value of (t_S) is the depth of the keyway in the shaft. The depth of the keyway in the hub = $h_K + t_K - t_S$.

Table A.5.2 Selected inch series of internal and external retaining ring main dimensions as defined in figures.

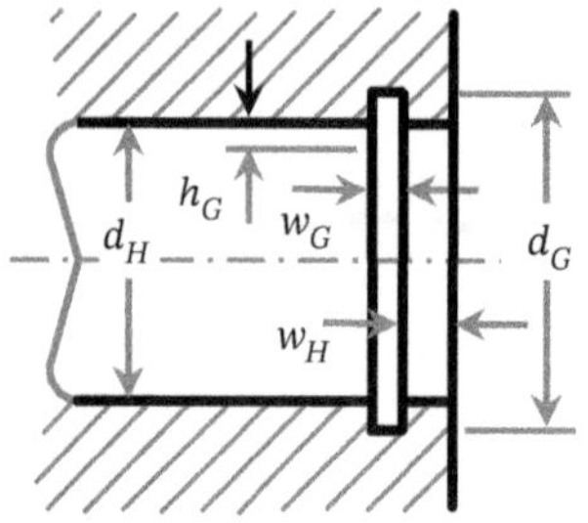

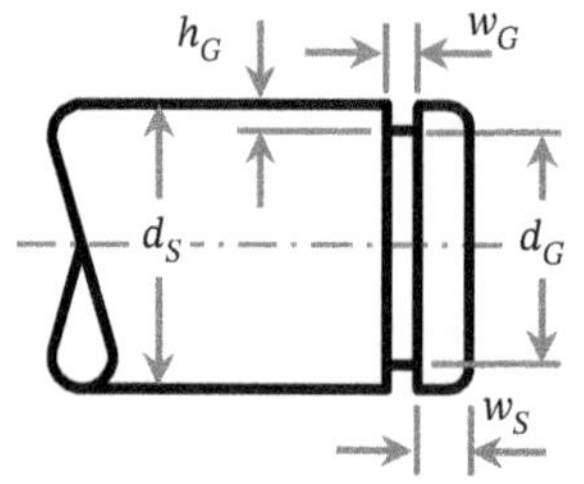

Housing	Ring		Groove		Shaft	Ring		Groove	
Diameter	Diameter	Thickness	Diameter	Width	Diameter	Diameter	Thickness	Diameter	Width
d_H [in]	d_R [in]	w_R [in]	d_G [in]	w_G [in]	d_S [in]	d_R [in]	w_R [in]	d_G [in]	w_G [in]
					0.125	0.112	0.01	0.117	0.012
0.25	0.28	0.015	0.268	0.018	0.25	0.225	0.025	0.23	0.029
0.375	0.415	0.025	0.397	0.029	0.375	0.338	0.025	0.352	0.029
0.5	0.548	0.035	0.53	0.039	0.5	0.461	0.035	0.468	0.039
0.625	0.694	0.035	0.665	0.039	0.625	0.579	0.035	0.588	0.039
0.75	0.831	0.035	0.796	0.039	0.75	0.693	0.042	0.704	0.046
0.875	0.971	0.042	0.931	0.046	0.875	0.81	0.042	0.821	0.046
1	1.111	0.042	1.066	0.046	1	0.925	0.042	0.94	0.046
1.125	1.249	0.05	1.197	0.056	1.125	1.098	0.05	1.118	0.056
1.25	1.388	0.05	1.33	0.056	1.25	1.156	0.05	1.176	0.056
1.375	1.526	0.05	1.461	0.056	1.375	1.272	0.05	1.291	0.056
1.5	1.66	0.05	1.594	0.056	1.5	1.387	0.05	1.406	0.056
1.625	1.804	0.062	1.725	0.068	1.625	1.503	0.062	1.529	0.068
1.75	1.942	0.062	1.858	0.068	1.75	1.618	0.062	1.65	0.068
1.875	2.054	0.062	1.989	0.068	1.875	1.735	0.062	1.769	0.068
2	2.21	0.062	2.122	0.068	2	1.85	0.062	1.896	0.068
2.125	2.35	0.078	2.251	0.086	2.125	1.964	0.078	2.003	0.086
2.25	2.49	0.078	2.382	0.086	2.25	2.081	0.078	2.12	0.086
2.375	2.63	0.078	2.517	0.086	2.375	2.197	0.078	2.239	0.086
2.5	2.775	0.078	2.648	0.086	2.5	2.313	0.078	2.36	0.086
2.75	3.05	0.093	2.914	0.103	2.75	2.543	0.083	2.602	0.103
3	3.325	0.093	3.182	0.103	3	2.775	0.093	2.838	0.103

Table A.5.3 Selected metric series of internal and external retaining ring main dimensions as defined in figures.

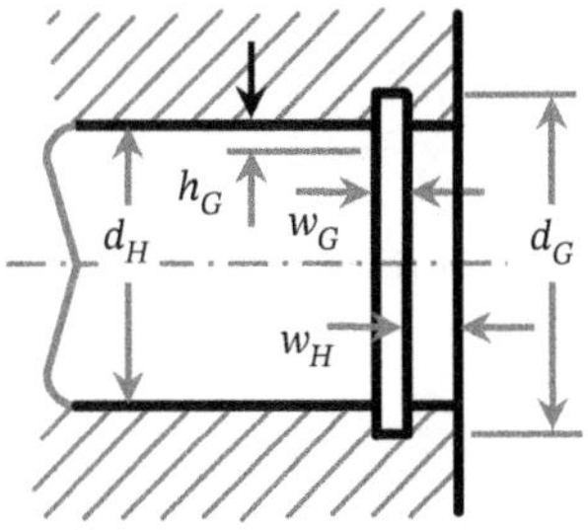

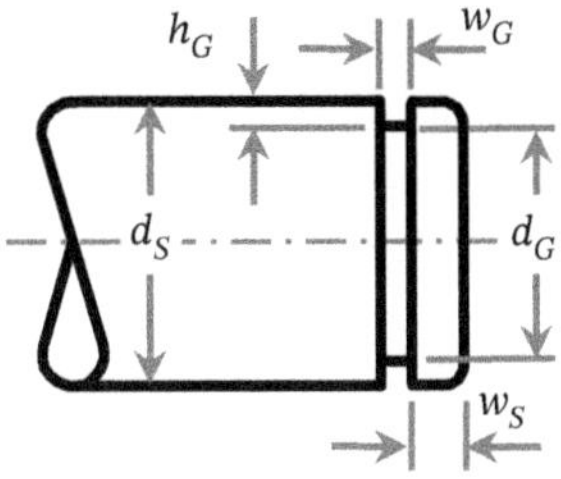

Housing	Groove		Ring		Shaft	Groove		Ring	
Diameter d_H [mm]	Diameter d_G [mm]	Width w_G [mm]	Thickness w_R [mm]	Diameter d_R [mm]	Diameter d_S [mm]	Diameter d_G [mm]	Width w_G [mm]	Thickness w_R [mm]	Diameter d_R [mm]
10	10.4	1.1	1	10.8	4	3.8	0.5	0.4	3.7
12	12.5	1.1	1	13	6	5.7	0.8	0.7	5.6
14	14.6	1.1	1	15.1	8	7.6	0.9	0.8	7.4
15	15.7	1.1	1	16.2	10	9.6	1.1	1	9.3
16	16.8	1.1	1	17.3	12	11.5	1.1	1	11
18	19	1.1	1	19.5	14	13.4	1.1	1	12.9
19	20	1.1	1	20.5	15	14.3	1.1	1	13.8
20	21	1.1	1	21.5	16	15.2	1.1	1	14.7
22	23	1.1	1	23.5	18	17	1.3	1.2	16.5
24	25.2	1.3	1.2	25.9	20	19	1.3	1.2	18.5
25	26.2	1.3	1.2	26.9	22	21	1.3	1.2	20.5
27	28.4	1.3	1.2	29.1	25	23.9	1.3	1.2	23.2
28	29.4	1.3	1.2	30.1	28	26.6	1.6	1.5	25.9
29	30.4	1.3	1.2	31.1	29	27.6	1.6	1.5	26.9
30	31.4	1.3	1.2	32.1	30	28.6	1.6	1.5	27.9
32	33.7	1.3	1.2	34.4	32	30.3	1.6	1.5	29.6
35	37	1.6	1.5	37.8	36	32	1.6	1.5	32.2
38	40	1.6	1.5	40.8	38	36	1.85	1.75	35.2
40	42.5	1.85	1.75	43.5	40	37.5	1.85	1.75	36.6

A.6 Standard Sealing Elements

Table A.6.1 Selected metric and inch series of seals main dimensions as defined in the figure (ISO 6194 and SAE J946).

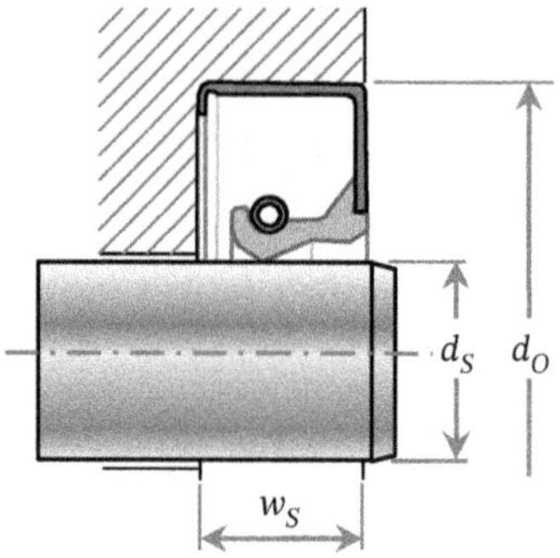

Shaft diameter d_S [mm]	Outer diameter d_O [mm]	Width w_S [mm]	Shaft diameter d_S [in]	Outer diameter d_O [in]	Width w_S [in]
			0.25	0.749	0.25
12	22, 28, 35	7	0.5	0.875, 1.375	0.25
16	25, 30, 35	7	0.625	1.25, 1.375	0.25
20	30, 35, 40	7	0.75	1.25, 1.375	0.25
25	35, 40, 45, 50	7	1	1.375, 2, 2.25	0.25
30	40, 45, 50, 60	7, 8	1.125	2	0.25, 0.375
40	55, 60, 70, 80, 90	7, 8	1.5	2, 2.25, 2.875	0.25, 0.5
50	65, 70, 80, 90	8	2	2.5, 3	0.375, 0.5
60	75, 80, 90, 110	8	2.5	3.5	0.375
80	100, 110, 140	10, 12	3	3.5, 4.5	0.375
100	120, 140	12	4	6.25	0.5
120	140, 160	12	5	6.25	0.5
160	190	15	6.5	7.5	0.5
280	320	20	11	13	0.625

Table A.6.2 Selected metric and inch series of O-rings main dimensions as defined in the figures (ISO 3601, SAE AS568, and SAE AS4716).

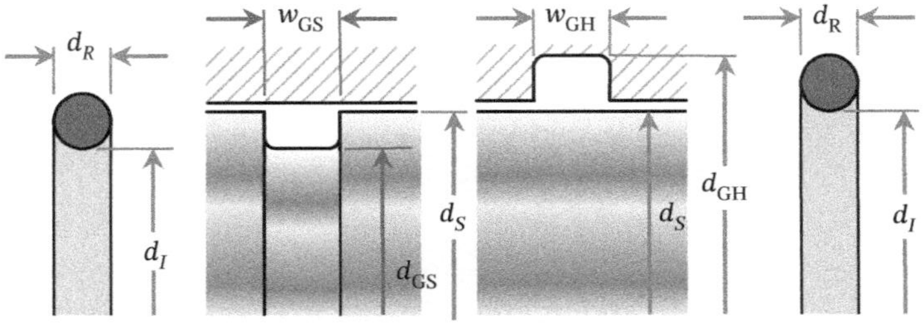

Shaft or hole diameter d_S [mm]	Hole groove diameter d_{GH} [mm]	Shaft groove diameter d_{CS} [mm]	Ring C-S diameter d_R [mm]	Ring inside diameter d_I [mm]	Groove width w_G [mm]	Shaft or hole diameter d_S [in]	Hole groove diameter d_{GH} [in]	Shaft groove diameter d_{GS} [in]	Ring C-S diameter d_R [in]	Ring inside diameter d_I [in]	Groove width w_G [in]
6	10	3.3	1.6, 2.4	3.1, 5.3	2.3, 3.2	0.25	0.35	0.15	0.07	0.114, 0.239	0.093
12	16	8.4	2.4	7.3, 11.3	3.2	0.5	0.6	0.4	0.07	0.364, 0.489	0.093
16	20	12.4	2.4	11.3, 15.3	3.2	0.625	0.725	0.525	0.07	0.489, 0.614	0.093
20	25	16.4	2.4, 3	15.3, 19.2	3.2, 4	0.75	0.85	0.65	0.07	0.614, 0.739	0.093
25	30	20.2	3	19.2, 24.2	4	1	1.1	0.9	0.07	0.864, 0.989	0.093
30	35	25.2	3	24.2, 29.2	4	1.125	1.225	1.025	0.07	0.989, 1.114	0.093
40	45	35.2	3	34.2, 39.2	4	1.5	1.6	1.4	0.07	1.364, 1.489	0.093
50	60	45.2	5.7	44.2, 49.2	7.5	2	2.1	1.9	0.07	1.864, 1.989	0.093
60	70	50.3	5.7	49.2, 59.2	7.5	2.5	2.6	2.4	0.07	2.364, 2.489	0.093
80	90	70.3	5.7	69.2, 79.2	7.5	3	3.162	2.838	0.103	2.8, 2.989	0.14
100	110	90.3	5.7	89.1, 99.1	7.5	4	4.222	3.778	0.139	3.734, 3.984	0.187
120	130	110.3	5.7	109.1, 119.3	7.5	5	5.222	4.778	0.139	4.734, 4.984	0.187
160	175	145	8.4	144.1, 159.1	11	6.5	6.722	6.278	0.139	6.234, 6.484	0.187
200	215	135	8.4	184.1, 199.1	11	8	8.222	7.778	0.139	7.734, 7.984	0.187
250	259.6	240.4	5.7	239.3, 249.3	7.5	10	10.222	9.778	0.139	9.734, 9.984	0.187
300	309.6	290.4	5.7	289.3, 299.3	7.5						

Smaller O-ring cross section (C-S) and inside diameter are for shaft groove. Larger O-ring cross section (C-S) and inside diameter are for hole groove.

A.7 Material Properties

Table A.7.1 Typical cast iron properties (gray and ductile or nodular) compiled from different sources.

	Designation[a)]		Tensile strength	Compressive or yield strength[b]	Modulus of elasticity	Endurance limit	Brinell hardness	Ductility (percent elongation)
	ASTM A48 or A536[c)]	ISO 185 or 1083[d)]	S_{ut} [MPs] ([kpsi])	S_{uc} or S_y [MPs] ([kpsi])	E [GPs] ([Mpsi])	S_e [MPs] ([kpsi])	HB	in 50 [mm] or 2 [in]
Gray cast iron	20	150	150 (20)	570 (83)	78 (11.3)	69 (10)	156	<1
	30	200	200 (30)	750 (109)	100 (14.7)	97 (14)	201	<1
	40	300	300 (40)	965 (140)	120 (17.5)	127 (18.5)	235	<1
	50	350	350 (50)	1130 (164)	143 (20.8)	148 (21.5)	262	<1
	60	—	— (60)	— (187.5)	— (21.9)	— (24.5)	302	<1
Ductile (nodular) iron	60-40-18	400-18	414 (60)	275 (40)	169 (24.5)	160 (23)	170	18
	65-45-12	450-10	448 (65)	310 (45)	168 (24.4)	170 (25)	180	12
	80-55-6	500-7	554 (80)	340 (55)	169 (24.4)	200 (29)	190	6
	100-70-03	700-3	690 (100)	480 (70)	168 (24.4)	260 (38)	220	3
	120-90-02	800-2	827 (120)	620 (90)	164 (23.8)	270 (39)	330	2

a) Designation similarity is realizable after careful assessment of individual case. Given similarity is a close correspondence.
b) Yield strength is for ductile (nodular) iron, and compressive strength is for gray cast iron. Strength values are the expected minimum; others are the average values.
c) ASTM A536 is for ductile (nodular) iron, and ASTM A48 is for gray cast iron.
d) ISO 1083 is for ductile (nodular) iron, and ISO 185 is for gray cast iron.

Table A.7.2 Typical plain carbon steel properties for hot rolled and cold drawn bars.

Designation[a)]		Yield strength	Tensile strength	Brinell hardness	Endurance limit[b)]	Ductility elongation %
AISI/SAE UNS[c)]	ISO[d)] [EN] {DIN}	S_y [e)] [MPs] ([kpsi])	S_u [e)] [MPs] ([kpsi])	HB[f)]	S_e [b)] [MPs] ([kpsi])	in 50 [mm] or in 2 ([in])
1006	(C6)	165, 285 (24, 41)	295, 330 (43, 48)	86 (95)	150 (21.8)	30 (20)
1010	{C10, Ck10}	180, 300 (26, 43)	325, 365 (47, 53)	95 (105)	160 (23.2)	28 (20)
1015	{Ck15}	190, 320 (27.5, 47)	345, 390 (50, 57)	101 (111)	170 (24.6)	28 (18)
1020	C20	210, 390 (30, 57)	380, 420 (55, 61)	111 (121)	190 (27.6)	25 (15)
1030	C30	260, 440 (37.5, 64)	470, 525 (68, 76)	137 (149)	235 (34)	20 (12)
1035	C35	270, 460 (39.2, 67)	495, 550 (72, 80)	143 (163)	250 (36.3)	18 (12)
1040	C40	290, 490 (42, 71)	525, 585 (76, 85)	149 (170)	260 (37.7)	18 (12)
1045	C45	310, 530 (45, 77)	565, 625 (82, 91)	163 (179)	285 (41.3)	16 (12)
1050	C50	340, 580 (49.3, 84)	620, 690 (90, 100)	179 (197)	310 (45)	15 (10)
1060	C60	370 (54)	675 (98)	201 (229)	340 (49.3)	12
1080	[C86D] {Ck75}	425 (61.6)	770 (112)	229 (265)	385 (55.8)	10
1095	[C98D2] {Ck101}	455 (66)	825 (120)	248 (293)	415 (60.2)	10

Values are estimated minimum and compiled from different sources.

a) Designation similarity is realizable after careful assessment of individual case. Given similarity is a close correspondence.
b) Estimated as 0.5 ultimate tensile strength. Some values may differ with some citations. Cross-check is recommended.
c) UNS designation is composed by the form defined by G(AISI/SAE)0. AISI/SAE 1045 is the UNS G10450.
d) ISO designation is about the same as DIN designation. Values in brackets or braces may not be in accessible ISO records, but the closest equivalent [EN] or {DIN} is given.
e) Values are for hot rolled HR and cold drawn CD for SI and for the US in parenthesis for (HR, CD) if CD is available. Otherwise, values are for hot rolled.
f) Brinell hardness are given as annealed (normalized), which are assumed similar to HR (CD).
g) Values are given as HR (CD) if CD is available.

Table A.7.3 Properties of selected heat treated plain carbon steel for quenched and tempered at 205 [°C] or (400) [°F], unless noted otherwise.

Designation[a)]		Treatment and temperature[b)]		Yield strength	Tensile strength	Elongation %	Brinell hardness
AISI/SAE (UNS)[c)]	ISO[d)] [EN] {DIN}	[°C] ([°F])		S_y [MPa] ([kpsi])	S_u [MPa] ([kpsi])	in 50 [mm] or 2 [in]	HB
1030	C30	WQ&T*	205 (400)	648 (94)	848 (123)	17	495
		Normalized	925 (1700)	345 (50)	521 (76)	32	149
		Annealed	870 (1600)	341 (49)	464 (67)	35	126
1040	C40	Q&T	205 (400)	593 (86)	779 (113)	19	262
		Normalized	900 (1650)	374 (54)	590 (86)	28	170
		Annealed	790 (1450)	353 (51)	519 (75)	30	149
1050	C50	WQ&T*	205 (400)	807 (117)	1120 (163)	9	514
		Normalized	900 (1650)	427 (62)	748 (108)	20	217
		Annealed	790 (1450)	365 (53)	636 (92)	24	187
1060	C60	Q&T	425 (800)	765 (111)	1080 (156)	14	311
		Normalized	900 (1650)	421 (61)	776 (112)	18	229
		Annealed	790 (1450)	372 (54)	626 (91)	22	179
1095	[EN C98D2] {Ck101}	Q&T	315 (600)	813 (118)	1260 (183)	10	375
		Normalized	900 (1650)	500 (72)	1014 (147)	9	293
		Annealed	790 (1450)	379 (55)	657 (95)	13	192

Values are compiled from different sources.

a) Designation similarity is realizable after careful assessment of individual case. Given similarity is a close correspondence.
b) Treatment Q is oil quenched unless starred as WQ&T*, which is water quenched WQ and tempered.
c) UNS designation is composed by the form defined by G(AISI/SAE)0. AISI/SAE 1045 is the UNS G10450.
d) ISO designation is about the same as DIN designation. Values in braces are not in accessible ISO records, but a close {DIN} equivalent is given.

Table A.7.4 Properties of selected heat treated low-alloy steel for quenched and tempered at 205 [°C] or (400) [°F].

Designation[a)]		Treatment and temperature		Yield strength	Tensile strength	Elongation %	Brinell hardness
AISI/SAE (UNS)[a)]	ISO[b)] [EN], {DIN}	[°C] ([°F])		S_y [MPa] ([kpsi])	S_u [MPa] ([kpsi])	in 50 [mm] or 2 [in]	HB
1144	44 SMn 28 {45 S 20}	Q&T	205 (400)	627 (91)	876 (127)	17	277
		Normalized	900 (1650)	400 (58)	667 (97)	21	197
		Annealed	790 (1450)	347 (50)	585 (85)	25	167
1340	~{40 Mn 5}	Q&T	205 (400)	1593 (231)	1806 (262)	11	505
		Normalized	870 (1600)	559 (81)	836 (121)	22	248
		Annealed	810 (1490)	472 (69)	745 (108)	22	217
4130	25 CrMo 4	Q&T	205 (400)	1460 (212)	1630 (236)	10	467
		Normalized	870 (1600)	436 (63)	671 (97)	25	197
		Annealed	865 (1585)	361 (52)	561 (81)	28	156
4140	42 CrMoS 4	Q&T	205 (400)	1640 (238)	1770 (257)	8	510
		Normalized	870 (1600)	655 (95)	1020 (148)	18	302
		Annealed	815 (1500)	417 (61)	656 (95)	26	197
4340	[X6CrMo 17-1]	Q&T	315 (600)	1590 (230)	1720 (250)	10	486
		Normalized	870 (1600)	862 (125)	1279 (186)	12	363
		Annealed	810 (1490)	472 (69)	745 (108)	22	217
5140	41 CrS 4	Q&T	205 (400)	1641 (238)	1793 (260)	9	490
		Normalized	870 (1600)	472 (69)	793 (115)	23	229
		Annealed	830 (1525)	293 (43)	572 (83)	29	167
5160	[60 Cr 3]	Q&T	205 (400)	1793 (260)	2220 (322)	4	627
		Normalized	855 (1575)	531 (77)	957 (139)	18	269
		Annealed	815 (1495)	276 (40)	723 (105)	17	197
6150	51 CrV 4	Q&T	205 (400)	1689 (245)	1931 (280)	8	538
		Normalized	870 (1600)	616 (89)	940 (136)	22	269
		Annealed	815 (1500)	412 (60)	667 (97)	23	197
8630	~{30 NiCrMo 2}	Q&T	205 (400)	1503 (218)	1641 (238)	9	465
		Normalized	870 (1600)	430 (62)	650 (94)	24	187
		Annealed	845 (1550)	372 (54)	564 (82)	29	156
8650	~{50 NiCrMo 2}	Q&T	205 (400)	1675 (243)	1937 (281)	10	525
		Normalized	870 (1600)	688 (100)	1024 (149)	14	302
		Annealed	795 (1465)	386 (56)	716 (104)	23	212
8740	~{40 NiCrMo 2 2}	Q&T	205 (400)	1655 (240)	1999 (290)	10	578
		Normalized	870 (1600)	607 (88)	929 (135)	16	269
		Annealed	815 (1500)	416 (60)	695 (101)	22	201
9255	~{55 Si 7}	Q&T	205 (400)	2048 (297)	2103 (305)	1	601
		Normalized	900 (1650)	579 (84)	933 (135)	20	269
		Annealed	845 (1550)	486 (71)	774 (112)	22	229

Values are compiled from different sources.

a) Designation similarity is realizable after careful assessment of individual case. Given similarity is a close correspondence.
b) UNS designation is composed by the form defined by G(AISI/SAE)0. AISI/SAE 4130 is the UNS G41300.
c) ISO designation is similar to EN or DIN designation. Values in braces are for a close {DIN} equivalent. The ~ is for unauthenticated source.

Table A.7.5 Properties of selected structural steel.

Designation[a)]		Yield strength	Tensile strength	Brinell hardness	Endurance limit[b)]	Ductility elongation %
ASTM[c)]	ISO[d)] [EN], {DIN}	S_y [e)] [MPs]([kpsi])	S_u [e)] [MPs] ([kpsi])	HB[f)]	S_e [b)] [MPs] ([kpsi])	in 50 [mm] or in 2 [in][g)]
A283 Grade B	E 185 [S 185], {St 37}	185 (27)	345–450 (50–65) 300–540 (44–78)	100–130	170–225 (25–33) 150–270 (22–39)	25 18
A 36	E 235 [S 235], {St 42}	250 (36) 235–175 (34–27)	400–550 (58–80) 340–470 (49–68)	115–145	200–275 (29–40) 170–235 (25–34)	21 26-21
A 572 Grade 42 [290]	E 275 [S 275], {St 46}	290 (42) 275–215 (34–27)	415 (60) 410–540 (59–78)	120–150	208 (30) 205–270 (22–39)	24 22–17
A 572 Grade 55 [380]	E 295 [E 295], {St 50}	380 (55) 295 (43)	485 (90) 470–610 (68–88)	135–170	242 (35) 235–305 (34–44)	20 20–22
A 572 Grade 65 [460]	E 355 [S 355], {St 60}	460 (65) 355–285 (51–41)	550 (80) 490–640 (71–93)	165–200	275 (40) 245–320 (35–46)	17 22–17
A 1011 Grade 90 [620]	[E 360], {St 70}	620 (90) 360 (52)	690 (100) 650–830 (94–120)	195–245	345 (50) 325–415 (47–60)	16–14 10-6

Values are compiled from different sources. Unless the same in any category, the upper property line is for ASTM, and the lower property line is for ISO, EN, and DIN.

a) Designation similarity is realizable after careful assessment of individual case. Given similarity is a selected close correspondence.
b) Estimated as 0.5 ultimate tensile strength if not available. Some values may differ with some citations. Cross-check is recommended.
c) ASTM designation is different for different steel grade, class, and composition. ASTM A 1011 is mainly for sheet and strip but is selected here for comparison.
d) ISO designation is for ISO 630, and EN designation is for EN 10025. Values in brackets or braces are for selected close equivalent [EN] or classical {DIN}.
e) Values are thickness dependant. As thickness increases the strength decreases. Quoted values are minimum cited or a range for general design purposes.
f) Brinell hardness values are given as average or range citations. Values are according to the older DIN 1611, 1612.
g) Values are given as average or range of available information when designations provide different values. Check is advisable when citing a specific designation.

Table A.7.6 Properties of selected stainless steel.

	Designation[a)]		Yield strength	Tensile strength	Brinell hardness	Endurance limit[b)]	Elongation %
	AISI/SAE (UNS)[c)]	ISO[d)] [EN], {DIN}	S_y [MPs] ([kpsi])	S_u [MPs] ([kpsi])	HB	S_e [b)] [MPs] ([kpsi])	in 50 [mm] or 2 [in]
Austenitic	304	X 5 CrNi 18 9 [X5CrNi18-10]	235 (34) 195 (28)	585 (85) 500–700 (72–101)	149	250–350 (36–51)	60 40
Austenitic	316	X3CrNiMo17-12 [X3CrNiMo17-12-2] {X5CrNiMo 18 12}	240 (35) 205–220 (30–32) 210 (30)	550 (80) 510–710 (74–103) 490 (71)	149	270 (39) 245–350 (36–51)	60 40
Austenitic	201	[X12CrMnNiN17-7-5]	380 (55) 275–260 (40–38)	760 (110) 515 (75)	172	258–380 (37–55)	52 40
Austenitic	314	—	345 (50)	689 (100)	180	345 (50)	45
Austenitic	347	X 6 CrNiNb 18 10 [X6CrNiNb1810]	240 (35) 210 (30)	620 (90) 490 (71)	160	245–310 (36–45)	50 35
Martensitic	410	{X10Cr13} {X12Cr13}	205 (30) 250 (36)	450 (65) 450–650 (65–94)	95	225–320 (33–46)	22 20
Martensitic	403	{X7Cr13}	205 (30)	485 (70)	88	243 (35)	25
Martensitic	420	{X20Cr13}	345 (50)	655 (95)	195	328 (48)	25
Martensitic	414	—	620 (90)	795 (115)	235	398 (58)	20
Martensitic	431	[X17CrNi16-2]	655 (95) 600–700 (87–102)	860 (125) 800–1050 (116–152)	260	400–525 (58–76)	20 14–10
Martensitic	440C	[X105CrMo17]	450 (65)	760 (110)	222	380 (55)	14
Ferritic	409	X6CrTi12	240 (35)	450 (65)	137	225 (33)	25
Ferritic	405	X6Cr13 [X6CrAl13]	276 (40) 230–210 (33–30)	483 (70) 400–600 (58–87)	150 200	200–300 (29–44)	30 17
Ferritic	430	X6Cr17 [X8Cr17]	310 (45) 260–240 (38–35)	517 (75) 430–630 (62–91)	155	215–315 (31–46)	30 20–18
Ferritic	442	[X2CrMoTi18-2]	310 (45)	515 (75)	160	258 (37)	30

Values are compiled from different sources. Unless the same in any category, the upper property line is for AISI/SAE UNS, and the lower property line is for close ISO, EN, or DIN.

a) Designation similarity is realizable after careful assessment of individual case. Given similarity is a selected close correspondence.
b) Estimated as 0.5 ultimate tensile strength if not available. For brittle martensitic steel the estimate endurance may be 0.35 the ultimate strength. Cross-check is therefore necessary.
c) UNS designation is composed by the form defined by S(AISI/SAE)00. AISI/SAE 304 is the UNS S30400.
d) ISO designation is for ISO 9329, and EN designation is for EN 10028. Values in brackets or braces are for selected close equivalent [EN] or {DIN}.

Table A.7.7 Properties of selected aluminum alloys.

	Designation[a)]		Yield strength	Tensile strength	Brinell hardness	Fatigue strength[b)]	Ductility elongation %
	AA (UNS)[c)]	ISO[d)] [EN], {DIN}	S_y [MPs] ([kpsi])	S_u [MPs] ([kpsi])	HB	S_f [MPs] ([kpsi])	in 50 [mm] or 2 ([in])[e)]
Wrought	1100	Al 99.0 Cu	35 (5)	90 (13)	23	35 (5)	37 (45)
	2024	Al Cu4Mg1	75 (11)	186 (27)	47	90 (13)	20 (22)
	3003	Al Mn1Cu	40 (6)	110 (16)	28	50 (7)	30 (40)
	5005	Al Mg1(B)	40 (6)	125 (18)	28	—	25 —
	5050	Al Mg1.5(C)	55 (8)	145 (21)	36	85 (12)	24 —
	5052	Al Mg2.5	90 (13)	195 (28)	47	110 (16)	25 (30)
	5154	Al Mg3.5	115 (17)	241 (35)	58	115 (17)	27 —
	6061	Al Mg1SiCu	55 (8)	125 (18)	30	60 (9)	25 (30)
	6063	Al Mg0.7Si	50 (7)	90 (13)	25	55 (8)	—
	7075	Al Zn5.5MgCu	105 (15)	230 (33)	60	—	17 (16)
Cast	242.0	AICu4Ni2Mg2	124 (18)	186 (27)	70	55 (8)	1
	319.0	AISi5Cu3	124 (18)	186 (27)	70	69 (10)	2
	443.0	AISi5	55 (8)	131 (19)	40	55 (8)	8
	514.0	A1Mg3	83 (12)	172 (25)	50	48 (7)	9
	710.0	—	172 (25)	241 (35)	75	55 (8)	5

Values are compiled from different sources. Property line is for AA or UNS. Designation is mainly for similar ISO and EN or DIN if needed.

a) Designation similarity is realizable after careful assessment of individual case. Given similarity is a selected close correspondence.
b) Values for 500 million cycles of totally reversed stresses. Cross-check is necessary since endurance limit may not exist.
c) UNS designation is composed by the form defined by A9(AA) for wrought and A0(AA) for cast with no decimal. AA 2024 is the UNS A92024, and AA 319.0 is UNS A03190.
d) ISO designation is for ISO R209 and 3522. Values in brackets or braces are for selected close equivalent [EN] or {DIN}, if needed.
e) The two numbers are for 1.6 [mm] or 1/16 [in] specimen thickness and in parenthesis for (13 [mm] or 1/2 [in]) diameter specimen.

Table A.7.8 Properties of selected plastics.

Type	Chemical name[a]	Trade or common name or (types)	Tensile strength	Compressive or {yield strength}	Tensile modulus	Elongation %	Specific gravity	Maximum operating temperature
	Symbol		S_u [MPa] ([kpsi])	S_{uc} $\{S_y\}$ [MPa] ([kpsi])	E [GPa] ([Mpsi])	in 50 [mm] or 2 [in]		[°C] ([°F])
Thermoplastics	ABS		38–41 (5.5–6)	{45–65 (6.5–9.4)}	1.7–3 (0.26–0.44)	5–25	1.02–1.07	90 (195)
	CA	Cellulosic	30–37 (4.5–5.5)	{25–30 (3.5–4.5)}	1–3 (0.15–0.45)	25–50	1.2–1.3	100 (210)
	PA	Nylon (6,6-6,12)	60–94 (8.8–13.7)	{45–83 (6.5–12)}	1.6–3.8 (0.23–0.55)	15–300	1.06–1.15	90 (194)
	PC		60–72 (8.8–10.5)	{62 (9)}	2.3 (0.33)	110–150	1.2	200 (390)
	PE	(Low–high density)	8–31 (2.6)	{9–33 (1.3–4.8)}	0.17–1.1 (0.2)	10–1200	0.92–0.97	105 (220)
	PMMA	Acrylic, Lucite, Plexiglas	48–72 (7–10.5)	{54–73 (7.8–10.6)}	2.2–3.2 (0.32–0.46)	2.0–6	1.17–1.2	100 (212)
	POM	Acetal	60–69 (8.8–10)	124 (18)	2.8–3.4 (0.4–0.5)	10–75	1.42	160 (320)
	PP		31–41 (4.5–6)	{31–37 (4.5–5.3)}	1.2–1.5 (0.17–0.22)	100–600	0.9–0.91	150 (300)
	PS		36–52 (5.2–7.5)	{25–69 (3.6–10)}	2.3–3.3 (0.33–0.48)	1.2–2.5	1.04–1.07	100 (212)
	PTFE	Teflon	21–35 (3–5)	{14–15 (2–2.2)}	0.4–0.55 (0.06–0.08)	200–400	2.14–2.2	285 (550)
	PVC		40.7–51.7	{41–45 (5.9–6.5)}	2.4–4.1	40–80	1.3–1.58	100 (212)
Thermosets	EP	Epoxy (resin)	69–138 (10–20)	100–170 (15–25)	2.4 (0.35)	3–6	1.1–1.4	170 (340)
	PF	Phenolic (resin), Bakelite	35–62 (5–9)	83–103 (12–15)	2.7–4.8 (0.4–0.7)	1.5–2.0	1.24–1.32	200 (390)
	PU	(Resin)	69–76 (10–11)	—	2.8 (0.4)	3–6	1.05	120 (250)
	UP	Polyester (resin)	30–55 (4.4–8)	—	14–20 (2–2.9)	0.6–1.2	1.07–1.31	170 (340)
Elastomers	NBR	Nitrile	24–28 (3.5–4)	—	(3.4 (0.5)) 10^{-3}	500–760	0.98	108 (226)
	SBR		18–21 (2.6–3)	—	(2–10 (0.3–1.5))10^{-3}	450–600	0.94	120 (393)
	NR	Natural rubber	28 (4)	—	—	700	0.93	90 (195)
	SiR	Silicone rubber	8 (1.2)	—	—	100–800	1.1–1.6	315 (588)

Values are approximate ranges compiled from different sources. Check with manufacturer for the properties of a specific product under the type of the plastic category selected for intended application.

a) ABS (acrylonitrile-butadiene-styrene), CA (cellulose acetate), PA (polyamide), PC (polycarbonate), PE (polyethylene), PMMA (poly-methyl-meth-acrylate), POM (poly-oxy-methylene), PP (polypropylene), PS (polystyrene), PTFE (poly-tetra-fluoro-ethylene), PVC (polyvinylchloride), EP (epoxy), PF (phenol-formaldehyde), PU (polyurethane), UP (polyester – unsaturated resin), NBR (butadiene-acrylonitrile), SBR (styrene-butadiene), NR (polyisoprene), and SiR (polysiloxane).

Table A.7.9 Properties of selected reinforcement fibers.

Material name	Common name or (type)	Tensile strength[a]	Tensile modulus	Diameter or {thickness}[b]	Specific gravity
		S_u [MPa] ([Kpsi])	E [GPa] ([Mpsi])	[μm] ([μin])	
Carbon	CNT (SWCNT-MWCNT)	(20–100 (3–14)) (10^3)	1000 (150)	(0.4–4 (0.02–0.16)) (10^{-3})	1.3–1.4
Carbon	Graphene (2D)	(130 (19)) (10^3)	1000 (150)	{(0.01 (0.4)) (10^{-3})}	0.4
Aramid	Kevlar	(3.6–4.1 (0.5–0.6)) (10^3)	130 (19)	13 (0.5)	1.44
Carbon	Graphite	(0.4–6 (0.06–0.85)) (10^3)	228–724 (33–105)	5–10 (0.2–0.4)	1.78–2.25
Glass	(E-glass)	(3.5–4.1 (0.5–0.6)) (10^3)	73.1 (10.6)	10 (0.4)	2.54–2.58
Ceramic	Aluminum oxide	(1.4–1.9 (0.2–0.28)) (10^3)	380 (55)	20 (0.8)	3.4–3.95
Ceramic	Silicon carbide	(3.28–3.9 (0.48–0.57)) (10^3)	400 (58)	130 (5)	3.0–3.1
Boron		(3.1–3.6 (0.45–0.52)) (10^3)	393 (57)	140 (5.5)	2.34–2.57
Steel	(High strength)	(1.0–2.4 (0.15–0.35)) (10^3)	209 (30.5)	130 (5)	7.9
Tungsten		(2.9–4.0 (0.42–0.58)) (10^3)	407 (59)	13 (0.5)	19.3

Values are approximate ranges compiled from different sources. Check with manufacturer for the properties of a specific product under the name or type selected for intended application.

a) Strength depends on diameter. Strength is higher for smaller diameter. Values are kept as the usual [MPa] ([Kpsi]) for comparison with most of previous values.

b) Diameters or thicknesses are representative. Different values are available. Higher values can be for multilayer or wires.

A.8 Standard Sections or Profiles and Section Properties

Table A.8.1 Selected metric series of round pipes main dimensions and some properties as defined in the figures.

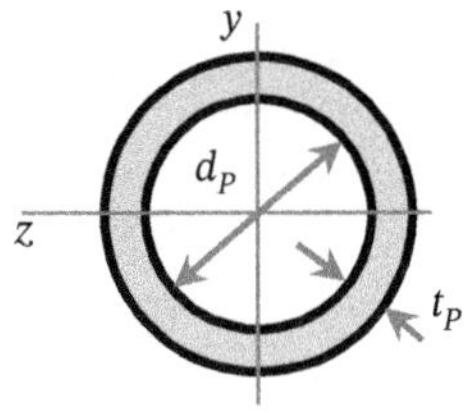

Nominal size DN	Outside diameter [mm]	Thickness t_P [mm]	Weight [N/m]	Area [mm^2]	Second area moment @z [10^3 mm^4]	Section modulus Z_z [10^3 mm^3]	Radius of gyration r_G [10^3 mm]
10	17.1	2.31	3.72	107	3.0064	0.351 62	0.167 36
15	21.3	2.77	5.62	161	7.0756	0.664 38	0.209 47
20	26.7	2.87	7.48	215	15.473	1.1590	0.268 35
25	33.4	3.38	11.1	319	36.365	2.1775	0.337 75
32	42.2	3.56	15.0	432	81.338	3.8549	0.433 84
40	48.3	3.68	17.9	516	129.25	5.3521	0.500 56
50	60.3	3.91	24.1	693	276.65	9.1757	0.631 97
65	73	5.16	38.2	1100	636.32	17.433	0.760 67
80	88.9	5.49	49.9	1439	1256.5	28.268	0.934 57
100	114.3	6.02	71.1	2048	3010.5	52.678	1.2125
125	141.3	6.55	96.3	2773	6308.3	89.289	1.5083
150	168.3	7.11	125	3600	11 716	139.23	1.8039
200	219.1	8.18	188	5420	30 187	275.55	2.3599
250	273	9.27	267	7680	66 858	489.80	2.9504
300	323.8	9.53	327	9409	116 268	718.15	3.5153
350	355.6	9.53	360	10 361	155 229	873.06	3.8706
400	406.4	9.53	413	11 882	234 071	1151.9	4.4384
450	457	9.53	465	13 397	335 461	1468.1	5.0040
500	508	9.53	518	14 924	463 691	1825.6	5.5741
600	610	9.53	624	17 978	810 468	2657.3	6.7143
700	711	9.53	729	21 002	1 291 993	3634.3	7.8434
750	813	9.53	835	24 055	1 941 435	4776.0	8.9837
800	864	9.53	886	25 582	2 335 056	5405.2	9.5539
900	914	9.53	940	27 079	2 769 386	6059.9	10.113
1000	1016	9.53	1046	30 133	3 815 876	7511.6	11.253
1200	1219	9.53	1257	36 211	6 621 632	10 864	13.523

Table A.8.2 Selected inch series of round pipes main dimensions and some properties as defined in the figures.

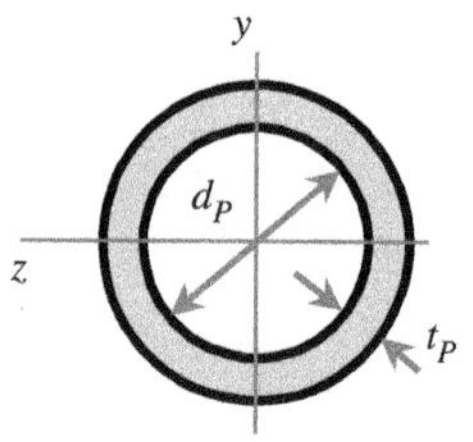

Nominal size NPS	Outside diameter [in]	Thickness t_P [in]	Weight [lb/ft]	Area [in²]	Second area moment @z [in⁴]	Section modulus Z_z [in³]	Radius of gyration r_G [in]
3/8	0.675	0.091	0.564	0.1670	0.007 291	0.021 602	0.208 97
1/2	0.84	0.109	0.853	0.2503	0.017 092	0.040 695	0.261 30
3/4	1.05	0.113	1.13	0.3326	0.037 036	0.070 545	0.333 68
1	1.315	0.133	1.68	0.4939	0.087 343	0.132 84	0.420 54
1 1/4	1.66	0.14	2.28	0.6685	0.194 71	0.234 59	0.539 68
1 1/2	1.9	0.145	2.72	0.7995	0.309 89	0.326 20	0.622 60
2	2.375	0.154	3.65	1.0745	0.665 75	0.560 63	0.787 13
2 1/2	2.875	0.203	5.80	1.7041	1.5296	1.0640	0.947 42
3	3.5	0.216	7.58	2.2285	3.0172	1.7241	1.1636
4	4.5	0.237	10.8	3.1740	7.2326	3.2145	1.5095
5	5.563	0.258	14.6	4.2999	15.162	5.4511	1.8778
6	6.625	0.28	19.0	5.5814	28.142	8.4958	2.2455
8	8.625	0.322	28.6	8.3993	72.489	16.809	2.9378
10	10.75	0.365	40.5	11.908	160.73	29.904	3.6739
12	12.75	0.375	49.6	14.579	279.34	43.817	4.3772
14	14	0.375	54.6	16.052	372.76	53.251	4.8190
16	16	0.375	62.6	18.408	562.08	70.261	5.5259
18	18	0.375	70.6	20.764	806.63	89.626	6.2328
20	20	0.375	78.7	23.120	1113.5	111.35	6.9398
24	24	0.375	94.8	27.833	1942.3	161.86	8.3538
28	28	0.375	111	32.545	3105.1	221.79	9.7678
32	32	0.375	127	37.257	4658.5	291.15	11.182
34	34	0.375	135	39.614	5599.3	329.37	11.889
36	36	0.375	143	41.970	6658.9	369.94	12.596
40	40	0.375	159	46.682	9163.0	458.15	14.010
48	48	0.375	191	56.107	15 908.3	662.84	16.839

Table A.8.3 Selected metric series of square tubes main dimensions and some properties.

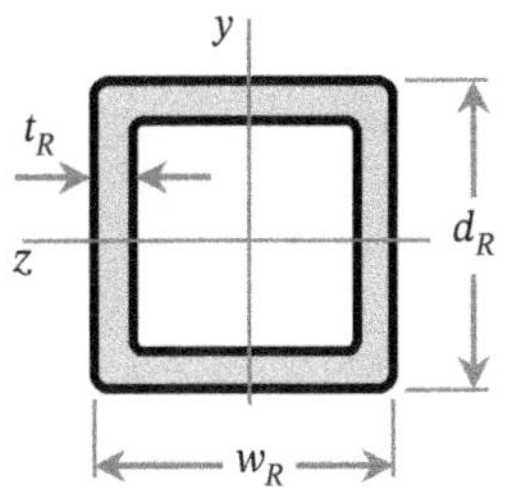

Section type	Depth d_R [mm]	Width w_R [mm]	Thickness t_R [mm]	Area [mm²]	Weight [N/m]	Second area moment @z or @y [10^6 mm⁴]	Section modulus Z_z or Z_y [10^3 mm³]
R-20	20	20	2	144	11	0.008	0.8
R-25	25	25	2	184	12	0.016	1.3
R-30	30	30	2	224	17.3	0.029	2
R-32	32	32	2	240	18.6	0.036	2.3
R-34	34	34	2	256	19.8	0.044	2.6
R-35	35	35	2	264	20.5	0.048	2.7
R-40	40	40	2	304	23.6	0.073	3.7
R-42	42	42	4	608	44.5	0.148	7
R-50	50	50	2	384	29.3	0.148	5.9
R-60	60	60	3	684	53.1	0.371	12.4
R-70	70	70	3	804	61.3	0.603	17.2
R-80	80	80	3	924	71.9	0.914	22.9
R-30	30	30	2.5	275	21.2	0.035	2.3
R-35	35	35	3	384	29.5	0.066	3.8
R-50	50	50	2.5	475	37.7	0.179	7.2
R-60	60	60	4	896	67.1	0.471	15.7
R-70	70	70	4	1056	79.7	0.769	22
R-80	80	80	4	1216	92.2	1.174	29.3

Section type is R for rectangular.

Table A.8.4 Selected inch series of square tubes main dimensions and some properties.

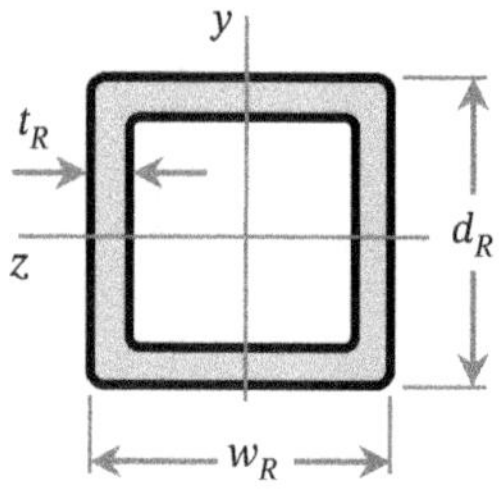

Section type[a)]	Depth d_R [in]	Width w_R [in]	Thickness t_R [in]	Area [in^2]	Weight [lb/ft]	Second area moment @z or @y [in^4]	Section modulus Z_z or Z_y [in^3]
R-3/4	6/8	6/8	1/16	0.223	0.123	0.0192	0.0192
R-1	1	1	1/16	0.285	0.134	0.0384	0.0384
R-1 3/16	1 3/16	1 3/16	1/16	0.347	0.193	0.0697	0.0697
R-1 1/4	1 4/16	1 4/16	1/16	0.372	0.208	0.0865	0.0865
R-1 5/16	1 5/16	1 5/16	1/16	0.397	0.221	0.1057	0.1057
R-1 3/8	1 6/16	1 6/16	1/16	0.409	0.229	0.1153	0.1153
R-1 9/16	1 9/16	1 9/16	1/16	0.471	0.264	0.1754	0.1754
R-1 10/16	1 10/16	1 10/16	3/16	0.942	0.497	0.3556	0.3556
R-1 15/16	1 15/16	1 15/16	1/16	0.595	0.327	0.3556	0.3556
R-2 3/8	2 6/16	2 6/16	2/16	1.060	0.593	0.8913	0.8913
R-2 3/4	2 12/16	2 12/16	2/16	1.246	0.685	1.4487	1.4487
R-3 1/8	3 2/16	3 2/16	2/16	1.432	0.803	2.1959	2.1959
R-1 3/16	1 3/16	1 3/16	2/16	0.426	0.237	0.0841	0.0841
R-1 3/8	1 6/16	1 6/16	2/16	0.595	0.329	0.1586	0.1586
R-1 15/16	1 15/16	1 15/16	2/16	0.736	0.421	0.4300	0.4300
R-2 3/8	2 6/16	2 6/16	3/16	1.389	0.749	1.1316	1.1316
R-2 3/4	2 12/16	2 12/16	3/16	1.637	0.890	1.8475	1.8475
R-3 1/8	3 2/16	3 2/16	3/16	1.885	1.030	2.8205	2.8205

a) R for rectangular. Approximate conversion from metric.

Table A.8.5 Selected metric series of rectangular tubes or hollow structural sections (HSS) main dimensions and some properties.

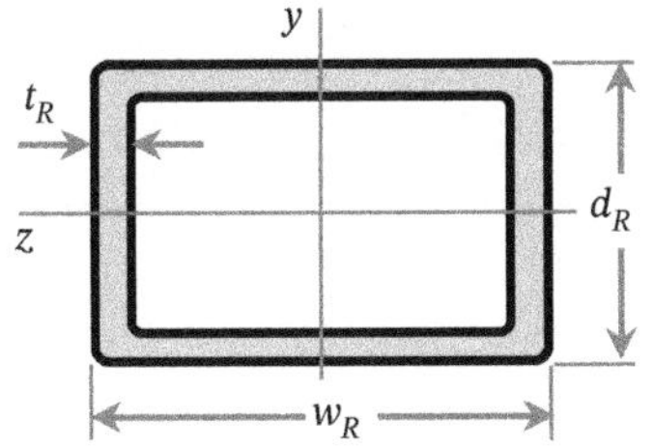

Section type[a)]	Depth d_R [mm]	Width w_R [mm]	Thickness t_R [mm]	Area [mm^2]	Weight [N/m]	Second area moment @z [10^6 mm^4]	Second area moment @y [10^6 mm^4]	Section modulus Z_z [10^3 mm^3]	Section modulus Z_y [10^3 mm^3]
R-28	20	28	2	176	1.36	0.018	0.01	1.3	1
R-28	20	28	2	176	1.3	0.018	0.01	1.3	1
R-30	15	30	2	164	1.26	0.018	0.006	1.2	0.7
R-30	20	30	2	184	1.42	0.022	0.011	1.4	1.1
R-30	20	30	2	184	1.36	0.022	0.011	1.4	1.1
R-35	20	35	2	204	1.58	0.032	0.013	1.8	1.3
R-40	20	40	2	224	1.73	0.044	0.014	2.2	1.4
R-40	25	40	2.5	300	1.89	0.062	0.029	3.1	2.3
R-40	28	40	2	256	1.98	0.056	0.032	2.8	2.3
R-40	30	40	2	264	2.05	0.059	0.037	2.9	2.5
R-40	35	40	2.5	350	2.71	0.079	0.064	4	3.7
R-45	30	45	4	536	3.88	0.135	0.068	6	4.6
R-50	20	50	2	264	2.05	0.079	0.018	3.1	1.8
R-50	25	50	2	284	2.2	0.09	0.03	3.6	2.4
R-50	30	50	2	304	2.36	0.102	0.045	4.1	3
R-50	34	50	2	320	2.49	0.111	0.06	4.4	3.5
R-50	35	50	2	324	2.52	0.113	0.064	4.5	3.7
R-50	40	50	2	344	2.67	0.125	0.088	5	4.4
R-55	34	35	2	260	0.64	0.047	0.045	2.7	2.6
R-60	20	60	2	304	2.36	0.126	0.021	4.2	2.1
R-60	30	60	2	344	2.67	0.159	0.053	5.3	3.5
R-60	35	60	2	364	2.83	0.176	0.075	5.9	4.3
R-60	40	60	2	384	2.99	0.193	0.102	6.4	5.1
R-65	40	65	3	594	4.6	0.334	0.153	10.3	7.7

(Continued)

Table A.8.5 (Continued)

Section type[a)]	Depth d_R [mm]	Width w_R [mm]	Thickness t_R [mm]	Area [mm^2]	Weight [N/m]	Second area moment @z [10^6 mm^4]	Second area moment @y [10^6 mm^4]	Section modulus Z_z [10^3 mm^3]	Section modulus Z_y [10^3 mm^3]
R-70	35	70	3	594	4.6	0.367	0.12	10.5	6.9
R-80	40	80	3	684	5.31	0.559	0.184	14	9.2
R-80	50	80	3	744	5.78	0.648	0.308	16.2	12.3
R-80	60	80	3	804	6.31	0.736	0.469	18.4	15.6
R-90	45	90	2.5	650	5.06	0.687	0.23	15.3	10.2
R-100	40	100	3	804	6.13	0.98	0.225	19.6	11.3
R-100	50	100	3	864	6.72	1.121	0.374	22.4	15
R-100	60	100	3	924	7.19	1.262	0.567	25.2	18.9
R-120	40	120	3	924	7.07	1.562	0.267	26	13.3
R-120	60	120	3	1044	8.01	1.973	0.664	32.9	22.1
R-35	35	35	3	384	2.95	0.066	0.066	3.8	3.8
R-40	20	40	2.5	275	2.03	0.053	0.017	2.7	1.7
R-40	25	40	4	456	3.26	0.087	0.039	4.3	3.1
R-40	28	40	2.5	315	2.35	0.067	0.038	3.4	2.7
R-40	30	40	2.5	325	2.43	0.071	0.044	3.5	3
R-50	20	50	2.5	325	2.51	0.094	0.021	3.8	2.1
R-50	25	50	2.5	350	2.71	0.109	0.035	4.3	2.8
R-50	30	50	2.5	375	2.9	0.123	0.054	4.9	3.6
R-50	35	50	2.5	400	3.01	0.137	0.077	5.5	4.4
R-50	40	50	2.5	425	3.21	0.151	0.106	6	5.3
R-60	30	60	2.5	425	3.21	0.193	0.063	6.4	4.2
R-60	35	60	3	534	4.01	0.249	0.105	8.3	6
R-60	40	60	2.5	475	3.6	0.235	0.123	7.8	6.2
R-70	40	70	4	816	6.08	0.508	0.204	14.5	10.2
R-80	40	80	4	896	6.71	0.711	0.23	17.8	11.5
R-80	50	80	4	976	7.34	0.827	0.389	20.7	15.6
R-80	60	80	4	1056	7.97	0.943	0.596	23.6	19.9
R-100	40	100	4	1056	7.97	1.257	0.282	25.1	14.1
R-100	60	100	4	1216	9.22	1.626	0.722	32.5	24.1
R-120	40	120	4	1216	9.22	2.014	0.334	33.6	16.7

a) R for hollow rectangular.

Table A.8.6 Selected inch series of rectangular tubes or hollow structural sections (HSS) main dimensions and some properties.

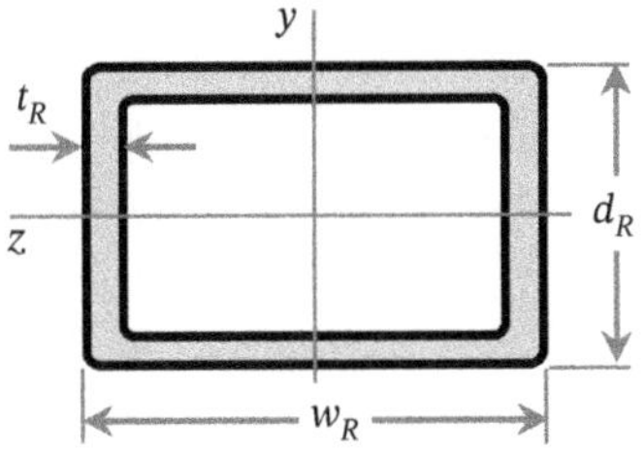

Section type[a)]	Depth d_R [in]	Width w_R [in]	Thickness t_R [in]	Area [in^2]	Weight [lb/ft]	Second area moment @z [in^4]	Second area moment @y [in^4]	Section modulus Z_z [in^3]	Section modulus Z_y [in^3]
R-1 1/8	13/16	1 1/8	1/16	0.273	0.152	0.0432	0.0240	0.1098	0.0436
R-1 3/16	9/16	1 3/16	1/16	0.254	0.141	0.0432	0.0144	0.1465	0.0244
R-1 3/16	13/16	1 3/16	1/16	0.285	0.159	0.0529	0.0264	0.1343	0.0448
R-1 3/16	13/16	1 3/16	1/16	0.285	0.152	0.0529	0.0264	0.1343	0.0448
R-1 3/4	13/16	1 3/8	1/16	0.316	0.176	0.0769	0.0312	0.1953	0.0453
R-1 9/16	13/16	1 9/16	1/16	0.347	0.193	0.1057	0.0336	0.2685	0.0427
R-1 9/16	1	1 9/16	2/16	0.465	0.211	0.1490	0.0697	0.3027	0.0885
R-1 9/16	1 1/8	1 9/16	1/16	0.397	0.221	0.1345	0.0769	0.2441	0.0976
R-1 9/16	1 3/16	1 9/16	1/16	0.409	0.229	0.1417	0.0889	0.2400	0.1129
R-1 9/16	1 3/8	1 9/16	2/16	0.543	0.303	0.1898	0.1538	0.2755	0.1953
R-1 12/16	1 3/16	1 3/4	3/16	0.831	0.433	0.3243	0.1634	0.5492	0.1844
R-1 15/16	13/16	1 15/16	1/16	0.409	0.229	0.1898	0.0432	0.4821	0.0439
R-1 15/16	1	1 15/16	1/16	0.440	0.246	0.2162	0.0721	0.4394	0.0732
R-1 15/16	1 3/16	1 15/16	1/16	0.471	0.264	0.2451	0.1081	0.4150	0.1098
R-1 15/16	1 3/8	1 15/16	1/16	0.502	0.281	0.2715	0.1538	0.3940	0.1562
R-1 15/16	1 9/16	1 15/16	1/16	0.533	0.298	0.3003	0.2114	0.3814	0.2148
R-1 3/8	1 5/16	1 3/8	1/16	0.403	0.071	0.1129	0.1081	0.1687	0.1569
R-2 3/8	13/16	2 3/8	1/16	0.471	0.264	0.3027	0.0505	0.7689	0.0427
R-2 3/8	1 3/16	2 3/8	1/16	0.533	0.298	0.3820	0.1273	0.6469	0.1078
R-2 3/8	1 3/8	2 3/8	1/16	0.564	0.316	0.4228	0.1802	0.6137	0.1526
R-2 3/8	1 9/16	2 3/8	1/16	0.595	0.334	0.4637	0.2451	0.5889	0.2075
R-2 3/8	1 9/16	2 9/16	2/16	0.921	0.514	0.8024	0.3676	1.0191	0.2873
R-2 3/4	1 3/8	2 3/4	2/16	0.921	0.514	0.8817	0.2883	1.2798	0.2092
R-2 3/4	1 9/16	2 3/4	2/16	0.967	0.541	0.9634	0.3940	1.2235	0.2859
R-2 3/4	1 15/16	2 3/4	3/16	1.389	0.749	1.4295	0.8313	1.4524	0.6033

(Continued)

Table A.8.6 (Continued)

Section type[a)]	Depth d_R [in]	Width w_R [in]	Thickness t_R [in]	Area [in²]	Weight [lb/ft]	Second area moment @z [in⁴]	Second area moment @y [in⁴]	Section modulus Z_z [in³]	Section modulus Z_y [in³]
R-3 1/8	1 9/16	3 1/8	2/16	1.060	0.593	1.3430	0.4421	1.7056	0.2807
R-3 1/8	1 15/16	3 1/8	2/16	1.153	0.646	1.5568	0.7400	1.5817	0.4699
R-3 1/8	2 3/8	3 1/8	2/16	1.246	0.705	1.7682	1.1268	1.4971	0.7155
R-3 9/16	1 3/4	3 9/16	2/16	1.008	0.565	1.6505	0.5526	1.8633	0.3119
R-3 15/16	1 9/16	3 15/16	2/16	1.246	0.685	2.3545	0.5406	2.9902	0.2746
R-3 15/16	1 15/16	3 15/16	2/16	1.339	0.751	2.6932	0.8985	2.7363	0.4565
R-3 15/16	2 3/8	3 15/16	2/16	1.432	0.803	3.0320	1.3622	2.5671	0.6920
R-4 3/4	1 9/16	4 3/4	2/16	1.432	0.790	3.7527	0.6415	4.7660	0.2716
R-4 3/4	2 3/8	4 3/4	2/16	1.618	0.895	4.7402	1.5953	4.0133	0.6753
R-1 9/16	13/16	1 9/16	2/16	0.426	0.227	0.1273	0.0408	0.3234	0.0519
R-1 9/16	1	1 9/16	3/16	0.707	0.364	0.2090	0.0937	0.4247	0.1190
R-1 9/16	1 1/8	1 9/16	2/16	0.488	0.262	0.1610	0.0913	0.2920	0.1159
R-1 9/16	1 3/16	1 9/16	2/16	0.504	0.271	0.1706	0.1057	0.2888	0.1343
R-1 15/16	13/16	1 15/16	2/16	0.504	0.280	0.2258	0.0505	0.5736	0.0513
R-1 15/16	1	1 15/16	2/16	0.543	0.303	0.2619	0.0841	0.5321	0.0854
R-1 15/16	1 3/16	1 15/16	2/16	0.581	0.324	0.2955	0.1297	0.5004	0.1318
R-1 15/16	1 3/8	1 15/16	2/16	0.620	0.336	0.3291	0.1850	0.4777	0.1880
R-1 15/16	1 9/16	1 15/16	2/16	0.659	0.359	0.3628	0.2547	0.4607	0.2587
R-2 3/8	1 3/16	2 3/8	2/16	0.659	0.359	0.4637	0.1514	0.7852	0.1281
R-2 3/8	1 3/8	2 3/8	2/16	0.828	0.448	0.5982	0.2523	0.8683	0.2136
R-2 3/8	1 9/16	2 3/8	2/16	0.736	0.402	0.5646	0.2955	0.7170	0.2502
R-2 3/4	1 9/16	2 3/4	3/16	1.265	0.679	1.2205	0.4901	1.5500	0.3557
R-3 1/8	1 9/16	3 1/8	3/16	1.389	0.749	1.7082	0.5526	2.1694	0.3509
R-3 1/8	1 15/16	3 1/8	3/16	1.513	0.820	1.9869	0.9346	2.0187	0.5935
R-3 1/8	2 3/8	3 1/8	3/16	1.637	0.890	2.2656	1.4319	1.9182	0.9093
R-3 15/16	1 9/16	3 15/16	3/16	1.637	0.890	3.0200	0.6775	3.8353	0.3442
R-3 15/16	2 3/8	3 15/16	3/16	1.885	1.030	3.9065	1.7346	3.3075	0.8812
R-4 3/4	1 9/16	4 3/4	3/16	1.885	1.030	4.8387	0.8024	6.1451	0.3397
R-4 3/4	2 3/8	4 3/4	3/16	2.133	1.170	6.1312	2.0373	5.1911	0.8625

a) R is rectangular. Approximate conversion from metric.

Table A.8.7 Selected metric series of I-section main dimensions and some properties as defined in the sketch.

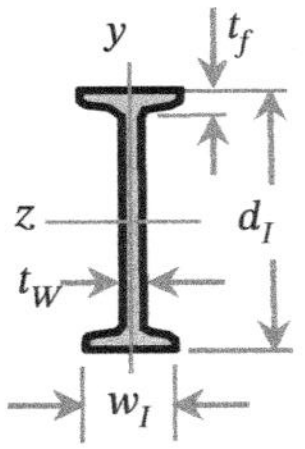

Section type	Depth d_I [mm]	Width w_I [mm]	Web thickness t_w [mm]	Flange thickness t_f [mm]	Area [mm^2]	Weight [N/m]	Second area moment @z [10^6 mm^4]	Section modulus Z_z [10^3 mm^3]	Radius of gyration in z [mm]	Second area moment @y [10^6 mm^4]	Section modulus Z_y [10^3 mm^3]	Radius of gyration in y [mm]
I-80	80	42	3.9	5.9	758	59.5	0.778	19.5	32	0.0629	3	9.1
I-100	100	50	4.5	6.8	1060	83.2	1.71	34.2	40.1	0.122	4.88	10.7
I-120	120	58	5.1	7.7	1420	112	3.28	54.7	48.1	0.215	7.41	12.3
I-140	140	66	5.7	8.6	1830	144	5.73	81.9	56.1	0.352	10.7	14
S-120	120	80	5	10	2100	112	5.27	87.83	48.1	0.8543	21.35	12.3
I-160	160	74	6.3	9.5	2280	179	9.35	117	64	0.547	14.8	15.5
I-180	180	82	6.9	10.4	2790	219	14.5	161	72	0.813	19.8	17.1
I-200	200	90	7.5	11.3	3350	263	21.4	214	80	1.17	26	18.7
I-220	220	98	8.1	12.2	3960	311	30.6	278	88	1.62	33.1	20.2
I-240	240	106	8.7	13.1	4610	362	42.5	354	95.9	2.21	41.7	22
I-260	260	113	9.4	14.1	5340	419	57.4	442	104	2.88	51	23.2
I-280	280	119	10.1	15.2	6110	480	75.9	542	111	3.64	51.2	24.5
I-300	300	125	10.8	16.2	6910	542	93	653	119	4.51	72.2	25.6
I-320	320	131	11.5	17.3	7780	611	125.1	782	127	5.55	84.7	26.7
I-340	340	137	12.2	18.3	8680	681	157	923	135	6.74	98.4	28
I-360	360	143	13	19.5	9710	762	196.1	1090	142	8.18	114	29
I-380	380	149	13.7	20.5	10 700	840	240.1	1260	158	9.75	131	30.2
I-400	400	155	14.4	21.6	11 800	926	292.1	1460	157	11.5	149	31.3
I-425	425	163	15.3	23	13 200	1040	369.7	1740	167	14.4	176	33
I-450	450	170	16.2	24.3	14 700	1150	485.5	2040	177	17.3	203	34.3
I-475	475	178	17.1	25.6	16 300	1280	564.8	2390	186	20.9	235	36
I-500	500	185	18	27	18 000	1410	687.4	2750	196	24.8	268	37.2
I-550	550	200	19	30	21 300	1670	991.8	3610	216	34.9	349	40.2
I-600	600	215	21.6	32.4	25 400	1990	1390	4630	234	46.7	434	43

Table A.8.8 Selected inch series of S-section I-beam main dimensions and some properties as defined in the sketch.

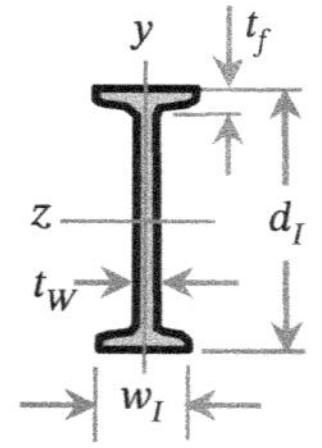

Section type[a)]	Depth d_I [in]	Width w_I [in]	Web thickness t_w [in]	Flange thickness t_f [in]	Area [in^2]	Weight [lb/ft]	Second area moment @z [in^4]	Section modulus Z_z [in^3]	Radius of gyration in z [in]	Second area moment @y [in^4]	Section modulus Z_y [in^3]	Radius of gyration in y [in]
S 3×5.7	3	2.33	0.26	0.17	1.67	5.7	2.52	1.68	1.23	0.455	0.39	0.522
S 3×7.5	3	2.509	0.26	0.349	2.21	7.5	2.93	1.95	1.15	0.586	0.468	0.516
S 4×7.7	4	2.663	0.293	0.193	2.26	7.7	6.08	3.04	1.64	0.764	0.574	0.581
S 4×9.5	4	2.796	0.293	0.326	2.79	9.5	6.79	3.39	1.56	0.903	0.646	0.569
S 5×10	5	3.004	0.326	0.214	2.94	10	12.3	4.92	2.05	1.22	0.809	0.643
S 5×14.75	5	3.284	0.326	0.494	4.34	14.75	15.2	6.09	1.87	1.67	1.01	0.62
S 6×12.5	6	3.332	0.359	0.232	3.67	12.5	22.1	7.37	2.45	1.82	1.09	0.705
S 6×17.25	6	3.565	0.359	0.465	5.07	17.25	26.3	8.77	2.28	2.31	1.3	0.675
S 7×15.3	7	3.662	0.392	0.252	4.5	15.3	36.7	10.5	2.86	2.64	1.44	0.766
S 7×20	7	3.86	0.392	0.45	5.88	20	42.4	12.1	2.69	3.17	1.64	0.734
S 8×18.4	8	4.001	0.426	0.271	5.41	18.4	57.6	14.4	3.26	3.73	1.86	0.831
S 8×23	8	4.171	0.426	0.441	6.77	23	64.9	16.2	3.1	4.31	2.07	0.798
S 10×25.4	10	4.661	0.491	0.311	7.46	25.4	124	24.7	4.07	6.79	2.91	0.954
S 10×35	10	4.944	0.491	0.594	10.3	35	147	29.4	3.78	8.36	3.38	0.901
S 12×31.8	12	5	0.544	0.35	9.35	31.8	218	36.4	4.83	9.36	3.74	1
S 12×50	12	5.477	0.659	0.687	14.7	50	305	50.8	4.55	15.7	5.74	1.03
S 15×42.9	15	5.501	0.622	0.411	12.6	42.9	447	59.6	5.95	14.4	5.23	1.07
S 15×50	15	5.64	0.622	0.55	14.7	50	486	64.8	5.75	15.7	5.57	1.03
S 18×54.7	18	6.001	0.691	0.461	16.1	54.7	804	89.4	7.07	20.8	6.94	1.14
S 18×70	18	6.251	0.691	0.711	20.6	70	926	103	6.71	24.1	7.72	1.08
S 20×66	20	6.255	0.795	0.505	19.4	66	1190	119	7.83	27.7	8.85	1.19
S 20×96	20.3	7.2	0.92	0.8	28.2	96	1670	165	7.71	50.2	13.9	1.33
S 24×80	24	7	0.87	0.5	23.5	80	2100	175	9.47	42.2	12.1	1.34
S 24×121	24.5	8.05	1.09	0.8	35.6	121	3160	258	9.43	83.3	20.7	1.53

a) Section type is (S depth × weight/foot).

Table A.8.9 Selected metric series of wide flange I-sections main dimensions and some properties as defined in the side sketch.

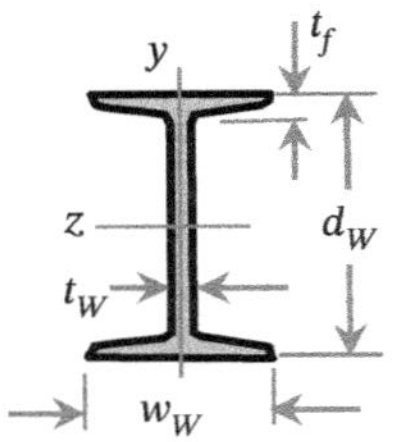

Section type	Depth d_W [mm]	Width w_W [mm]	Web thickness t_w [mm]	Flange thickness t_f [mm]	Area [mm²]	Weight [N/m]	Second area moment @z [10^6 mm⁴]	Section modulus Z_z [10^3 mm³]	Radius of gyration in z [mm]	Second area moment @y [10^6 mm⁴]	Section modulus Z_y [10^3 mm³]	Radius of gyration in y [mm]
WI-100	100	100	6.5	10	2610	20.5	4.47	89.3	41.4	1.67	33.4	25.3
WI-120	120	120	7	11	3430	26.9	8.64	144	5.02	3.17	53	30.4
WI-140	140	140	8	12	4410	34.6	15.2	217	5.87	5.5	79	35.4
WI-160	160	160	9	14	5840	45.8	26.3	329	67.2	9.58	120	40.5
WI-180	180	180	9	14	6530	51.6	38.3	426	76.3	13.6	151	45.5
WI-200	200	200	10	16	8270	64.9	59.5	595	84.8	21.4	214	50.8
WI-220	220	220	10	16	1910	71.5	80.5	732	93.7	28.4	258	55.9
WI-240	240	240	11	18	11 100	87.4	116.9	974	103	41.5	346	61.1
WI-260	260	260	11	18	12 100	94.8	150.5	1160	112	52.8	406	66.1
WI-280	280	280	12	20	14 400	113	207	1480	120	73.2	523	71.4
WI-300	300	300	12	20	15 400	121	257.6	1720	129	90.1	600	76.5
WI-320	320	300	13	22	17 100	135	322.5	2020	137	99.1	661	76
WI-340	340	300	13	22	17 400	137	369.4	2170	145	99.1	661	75.5
WI-360	360	300	14	24	19 200	150	451.2	2510	153	108.1	721	75.1
WI-380	380	300	14	24	19 400	153	509.5	2680	162	108.1	721	74.6
WI-400	400	300	14	26	20 900	164	606.4	3030	170	117.1	781	74.9
WI-450	450	300	15	28	23 200	182	842.2	3740	190	126.2	841.1	73.8
WI-500	500	300	16	30	25 500	200	1132	4530	210	135.3	902	72.8
WI-550	550	300	16	30	26 300	207	1403	5100	231	135.3	902	71.7
WI-600	600	300	17	32	28 900	227	1808	6030	250	144.4	962	70.7
WI-650	650	300	17	32	29 700	234	2168	6670	270	144.4	962	69.7
WI-700	700	300	18	34	32 400	254	2703	7720	289	153.5	1020	68.8
WI-750	750	300	18	34	33 300	261	3163	8430	308	153.5	1020	67.9
WI-800	800	300	18	34	34 200	268	3664	9160	327	153.5	1020	67
WI-850	850	300	19	36	37 200	292	4439	10 440	346	162.7	1080	66.1
WI-900	900	300	19	36	38 100	299	5060	11 250	364	162.7	1080	65.3
WI-950	950	300	19	36	39 100	307	5730	12 060	303	162.7	1080	64.5
WI-1000	1000	300	19	36	40 000	314	6447	12 900	401	162.8	1080	63.7

Table A.8.10 Selected inch series of wide flange I-sections main dimensions and some properties as defined in the side sketch.

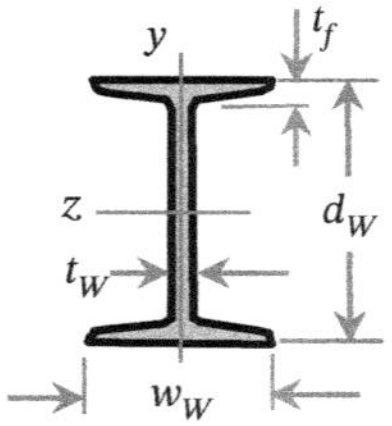

Section type	Depth d_W [in]	Width w_W [in]	Flange thickness t_f [in]	Web thickness t_w [in]	Area [in^2]	Second area moment @z [in^4]	Section modulus Z_z [in^3]	Radius of gyration r_{Gz} in z [in]	Second area moment @y [in^4]	Section modulus Z_y [in^3]	Radius of gyration r_{Gy} in y [in]
W 4×13	4.16	4.06	0.345	0.28	3.83	11.3	5.46	1.72	3.86	1.9	1
W 5×16	5.01	5	0.36	0.24	4.68	21.3	8.51	2.13	7.51	3	1.27
W 5×19	5.15	5.03	0.43	0.27	5.54	26.2	10.2	2.17	9.13	3.63	1.28
W 6×9	5.9	3.94	0.215	0.17	2.68	16.4	5.56	2.47	2.19	1.11	0.905
W 6×25	6.38	6.08	0.455	0.32	7.34	53.4	16.7	2.7	17.1	5.61	1.52
W 8×10	7.89	3.94	0.205	0.17	2.96	30.8	7.81	3.22	2.09	1.06	0.841
W 8×67	9	8.28	0.935	0.57	19.7	272	60.4	3.72	88.6	21.4	2.12
W 10×12	9.87	3.96	0.21	0.19	3.54	53.8	10.9	3.9	2.18	1.1	0.785
W 10×112	11.36	10.415	1.25	0.755	32.9	716	126	4.66	236	45.3	2.68
W 12×14	11.91	3.97	0.225	0.2	4.16	88.6	14.9	4.62	2.36	1.19	0.753
W 12×336	16.82	13.385	2.955	1.775	98.8	4060	483	6.41	1190	177	3.47
W 14×22	13.74	5	0.335	0.23	6.49	199	29	5.54	7	2.8	1.04
W 14×730	22.42	17.89	4.91	3.07	215	14 300	1280	8.17	4720	527	4.69
W 16×26	15.69	5.5	0.345	0.25	7.68	301	38.4	6.26	9.59	3.49	1.12
W 16×100	16.97	10.425	0.985	0.585	29.4	1490	175	7.1	186	35.7	2.51
W 18×35	17.7	6	0.425	0.3	10.3	510	57.6	7.04	15.3	5.12	1.22
W 18×119	18.97	11.265	1.06	0.655	35.1	2190	231	7.9	253	44.9	2.69
W 21×44	20.66	6.5	0.45	0.35	13	843	81.6	8.06	20.7	6.36	1.26
W 21×147	22.06	12.51	1.15	0.72	43.2	3630	329	9.17	376	60.1	2.95
W 24×55	23.57	7.005	0.505	0.395	16.2	1350	114	9.11	29.1	8.3	1.34
W 24×162	25	12.955	1.22	0.705	47.7	5170	414	10.4	443	68.4	3.05
W 27×84	26.71	9.96	0.64	0.46	24.8	2850	213	10.7	106	21.2	2.07
W 27×178	27.81	14.085	1.19	0.725	52.3	6990	502	11.6	555	78.8	3.26

Section type is (W depth × weight [lb/ft]).

Table A.8.11 Selected metric series of C-section or channel main dimensions and some properties as defined in the side sketch.

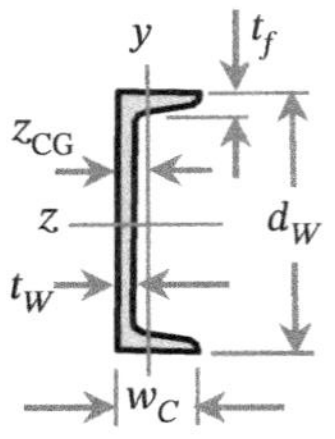

Section type	Depth d_C [mm]	Width w_C [mm]	Web thickness t_w [mm]	Flange thickness t_f [mm]	Area [mm^2]	Weight [N/m]	C.G. distance z_{CG} [mm]	Second area moment @z [10^6 mm^4]	Section modulus Z_z [10^3 mm^3]	Second area moment @y [10^6 mm^4]	Section modulus Z_y [10^3 mm^3]
C-30	30	33	5	7	544	42.7	13.1	0.0639	4.26	0.0533	2.68
C-40	40	35	5	7	621	48.7	13.3	0.141	7.05	0.0668	3.08
C-50	50	38	5	7	712	55.9	13.7	0.264	10.6	0.0912	3.75
C-65	65	42	5.5	7.5	903	70.9	14.2	0.375	17.7	0.141	5.07
C-80	80	45	6	8	1100	86.4	14.5	1.06	26.5	0.19	6.36
C-100	100	50	6	8.5	1350	106	15.5	2.06	41.2	0.293	8.49
C-120	120	50	7	9	1700	134	16	3.64	60.7	0.432	11.1
C-140	140	60	7	10	2040	160	17.5	6.05	86.4	0.627	14.8
C-160	160	60	75	10.5	2400	198	18.4	9.25	116	0.853	18.3
C-180	180	70	8	11	2800	220	19.2	13.5	150	1.14	22.4
C-200	200	75	8.5	11.5	3200	253	20.1	19.1	191	1.48	27
C-220	220	80	9	12.5	3740	294	21.4	26.9	245	1.97	33.6
C-240	240	85	9.5	13	4230	332	22.3	36	300	2.48	39.6
C-260	260	90	10	14	4830	379	23.6	48.2	371	3.17	47.7
C-280	280	95	10	15	5300	418	25.3	62.8	448	3.99	57.2
C-300	300	100	10	16	5880	462	27	80.3	535	4.95	67.8
C-320	320	100	14	17.5	7580	595	26	108.7	679	5.97	80.6
C-350	350	100	14	16	7730	606	24	128.4	734	5.7	75
C-380	380	102	13.34	16	7900	626	23.5	157.3	826	6.13	78.4
C-400	400	110	14	18	9150	718	26.5	203.5	1020	8.46	102

Table A.8.12 Selected inch series of C-section or channel main dimensions and some properties as defined in the side sketch.

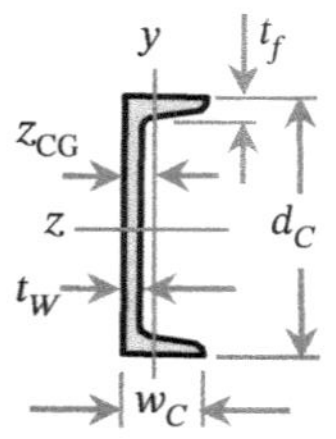

Section type	Depth d_C [in]	Width w_C [in]	Web thickness t_w [in]	Flange thickness t_f [in]	Area [in^2]	Weight [lb/ft]	C.G. distance z_{CG} [in]	Second area moment @z [in^4]	Section modulus Z_z [in^3]	Second area moment @y [in^4]	Section modulus Z_y [in^3]
C 3×4.1	3	1.41	0.273	0.17	1.21	4.1	0.436	1.66	1.1	0.197	0.202
C 3×6	3	1.596	0.273	0.356	1.76	6	0.455	2.07	1.38	0.305	0.268
C 4×5.4	4	1.584	0.296	0.184	1.59	5.4	0.457	3.85	1.93	0.319	0.283
C 4×7.25	4	1.721	0.296	0.321	2.13	7.25	0.459	4.59	2.29	0.433	0.343
C 5×6.7	5	1.75	0.32	0.19	1.97	6.7	0.484	7.49	3	0.479	0.378
C 5×9	5	1.885	0.32	0.325	2.64	9	0.478	8.9	3.56	0.632	0.45
C 6×8.2	6	1.92	0.343	0.2	2.4	8.2	0.511	13.1	4.38	0.693	0.492
C 6×13	6	2.157	0.343	0.437	3.83	13	0.514	17.4	5.8	1.05	0.642
C 7×9.8	7	2.09	0.366	0.21	2.87	9.8	0.54	21.3	6.08	0.968	0.625
C 7×14.75	7	2.299	0.366	0.419	4.33	14.75	0.532	27.2	7.78	1.38	0.779
C 8×11.5	8	2.26	0.39	0.22	3.38	11.5	0.571	32.6	8.14	1.32	0.781
C 8×18.75	8	2.527	0.39	0.487	5.51	18.75	0.565	44	11	1.98	1.01
C 9×13.4	9	2.433	0.413	0.233	3.94	13.4	0.601	47.9	10.6	1.76	0.962
C 9×20	9	2.648	0.413	0.448	5.88	20	0.583	60.9	13.5	2.42	1.17
C 10×20	10	2.739	0.436	0.379	5.88	20	0.606	78.9	15.8	2.81	1.32
C 10×30	10	3.033	0.436	0.673	8.82	30	0.649	103	20.7	3.94	1.65
C 12×20.7	12	2.942	0.501	0.282	6.09	20.7	0.698	129	21.5	3.88	1.73
C 12×30	12	3.17	0.501	0.51	8.82	30	0.674	162	27	5.14	2.06
C 15×33.9	15	3.4	0.65	0.4	9.96	33.9	0.787	315	42	8.13	3.11
C 15×50	15	3.716	0.65	0.716	14.7	50	0.798	404	53.8	11	3.78

Table A.8.13 Selected metric series of L-section or angle with equal legs main dimensions and some properties as defined in the side sketch.

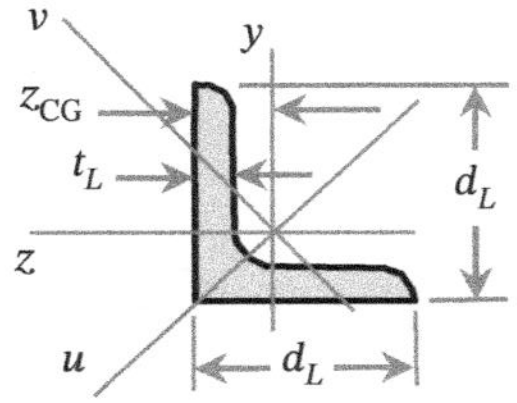

Section type	Depth d_L [mm]	Thickness t_L [mm]	Area [mm²]	Weight [N/m]	C.G. distance z_{CG} [mm]	Second area moment @y [10^6 mm⁴]	Second area moment @u [10^6 mm⁴]	Second area moment @v [10^6 mm⁴]	Mixed area moment [10^6 mm⁴]	Section modulus Z_y [10^3 mm³]
L 45×45×5	45	5	430	33.8	12.9	0.0783	0.12	0.03	0.0458	2.439
L 45×45×7	45	7	586	46	13.6	0.104	0.16	0.04	0.061	3.312
L 50×50×5	50	5	480	37.7	14	0.11	0.17	0.05	0.0641	3.05
L 50×50×9	50	9	824	64.7	15.6	0.179	0.28	0.08	0.1023	5.2
L 55×55×6	55	6	631	49.5	15.6	0.173	0.27	0.07	0.1006	4.39
L 55×55×10	55	10	1010	79	17.2	0.263	0.41	0.11	0.15	6.95
L 60×60×6	60	6	691	54.2	16.9	0.228	0.36	0.09	0.1337	5.29
L 60×60×10	60	10	1110	86.9	18.5	0.349	0.55	0.15	0.203	8.4
L 65×65×7	65	7	870	68.3	18.5	0.334	0.53	0.14	0.196	7.18
L 65×65×11	65	11	1320	103	20	0.488	0.77	0.21	0.2811	10.8
L 70×70×7	70	7	940	73.8	19.7	0.424	0.67	0.18	0.244	8.42
L 70×70×11	70	11	1430	112	21.3	0.618	0.98	0.26	0.358	12.6
L 75×75×8	75	8	1150	90.3	21.3	0.589	0.93	0.24	0.345	10.96
L 75×75×12	75	12	1670	131	22.9	0.824	1.3	0.35	0.477	15.81
L 80×80×8	80	8	1230	96.6	22.6	0.723	1.15	0.3	0.427	12.59
L 80×80×12	80	12	1790	141	24.1	1.02	1.61	0.43	0.59	18.24
L 90×90×9	90	9	1550	122	25.4	1.16	1.84	0.48	0.682	17.95
L 90×90×13	90	13	2180	171	27	1.58	2.5	0.66	0.921	25.07

(Continued)

Table A.8.13 (Continued)

Section type	Depth d_L [mm]	Thickness t_L [mm]	Area [mm^2]	Weight [N/m]	C.G. distance z_{CG} [mm]	Second area moment @y [10^6 mm^4]	Second area moment @u [10^6 mm^4]	Second area moment @v [10^6 mm^4]	Mixed area moment [10^6 mm^4]	Section modulus Z_y [10^3 mm^3]
L 100 × 100 × 10	100	10	1920	151	28.2	1.77	2.81	0.73	1.037	24.65
L 100 × 100 × 14	100	14	2620	206	29.8	2.35	3.72	0.98	1.367	33.47
L 110 × 110 × 10	110	10	2120	166	30.7	2.39	3.79	0.99	1.404	30.13
L 110 × 110 × 14	110	14	2900	228	32.1	3.19	5.05	1.33	1.86	40.94
L 120 × 120 × 11	120	11	2540	199	33.6	3.41	5.42	1.4	2.01	39.46
L 120 × 120 × 15	120	15	3390	266	35.1	4.46	7.06	1.86	2.6	52.5
L 130 × 130 × 12	130	12	3000	236	36.4	4.72	7.5	1.94	2.78	50.42
L 130 × 130 × 16	130	16	3930	309	38	6.05	9.59	2.51	3.54	65.76
L 140 × 140 × 13	140	13	3500	275	39.2	6.38	10.14	2.62	3.76	63.29
L 140 × 140 × 17	140	17	4500	353	40.8	8.05	12.76	3.34	4.71	81.14
L 150 × 150 × 14	150	14	4030	316	42.1	8.45	13.43	3.47	4.98	78.31
L 150 × 150 × 18	150	18	5100	401	43.6	10.5	16.62	4.38	6.12	98.68
L 160 × 160 × 15	160	15	4610	362	44.9	11	17.47	4.53	6.47	95.56
L 160 × 160 × 19	160	19	5750	451	46.5	13.5	21.42	5.58	7.92	118.94
L 180 × 180 × 16	180	16	5540	435	50.2	16.8	26.81	6.79	10.01	129.42
L 180 × 180 × 20	180	20	6840	537	51.8	20.4	32.5	8.3	12.1	159.12
L 200 × 200 × 16	200	16	6180	485	55.2	23.4	37.37	9.43	13.97	161.6
L 200 × 200 × 20	200	20	7640	599	56.8	28.5	45.4	11.6	16.9	199.02

Table A.8.14 Selected inch series of L-section or angles with equal legs main dimensions and some properties as defined in the sketch.

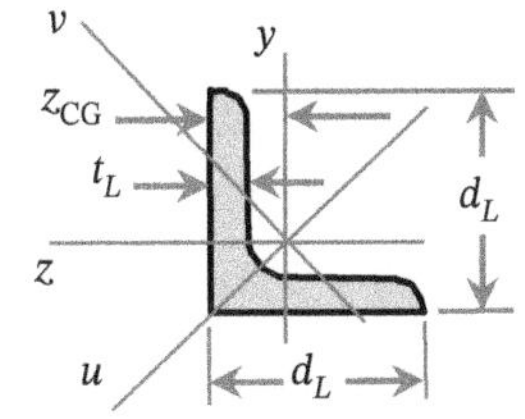

Section type	Depth d_L [in]	Thickness t_L [in]	Area [in^2]	Weight [lb/ft]	C.G. distance z_{CG} [in]	Second area moment @z [in^4]	Radius of gyration r_{Gz} in z [in]	Section modulus Z_y [in^4]	Radius of gyration r_{Gz} in v [in]
L 1×1×1/8	1	0.125	0.234	0.8	0.29	0.021	0.298	0.029	0.191
L 1×1×1/4	1	0.25	0.437	1.49	0.336	0.036	0.287	0.054	0.193
L 1½×1½×1/8	1.5	0.125	0.36	1.23	0.41	0.074	0.45	0.068	0.2
L 1½×1½×1/4	1.5	0.25	0.69	2.34	0.46	0.135	0.44	0.13	0.29
L 2×2×1/8	2	0.125	0.484	1.65	0.546	0.19	0.626	0.131	0.398
L 2×2×1/4	2	0.25	0.938	3.19	0.592	0.348	0.609	0.247	0.391
L 2×2×3/8	2	0.375	1.36	4.7	0.636	0.479	0.594	0.351	0.389
L 2½×2½×1/4	2.5	0.25	1.19	4.1	0.717	0.703	0.769	0.394	0.491
L 2½×2½×3/8	2.5	0.375	1.73	5.9	0.762	0.984	0.753	0.566	0.487
L 3×3×1/4	3	0.25	1.44	4.9	0.842	1.24	0.93	0.577	0.592
L 3×3×3/8	3	0.375	2.11	7.2	0.888	1.76	0.913	0.833	0.587
L 3×3×1/2	3	0.5	2.75	9.4	0.932	2.22	0.898	1.07	0.584
L 3½×3½×1/4	3.5	0.25	1.69	5.8	0.968	2.01	1.09	0.794	0.694
L 3½×3½×3/8	3.5	0.375	2.48	8.5	1.01	2.87	1.07	1.15	0.687
L 3½×3½×1/2	3.5	0.25	3.25	11.1	1.06	3.64	1.06	1.49	0.683
L 4×4×1/4	4	0.25	1.94	6.6	1.09	3.04	1.25	1.05	0.795
L 4×4×3/8	4	0.375	2.86	9.8	1.14	4.36	1.23	1.52	0.788
L 4×4×1/2	4	0.5	3.75	12.8	1.18	5.56	1.22	1.97	0.782
L 4×4×5/8	4	0.625	4.61	15.7	1.23	6.66	1.2	2.4	0.779
L 5×5×3/8	5	0.375	3.61	12.3	1.39	8.74	1.56	2.4	0.99
L 5×5×1/2	5	0.5	4.75	16.2	1.43	10	1.54	3.2	0.983
L 5×5×5/8	5	0.625	5.86	20	1.48	11.3	1.52	3.9	0.978
L 5×5×3/4	5	0.75	6.94	23.6	1.52	15.7	1.51	4.5	0.975
L 6×6×3/8	6	0.375	4.36	14.9	1.64	15.4	1.88	3.53	1.19
L 6×6×1/2	6	0.5	5.75	19.6	1.68	19.9	1.86	4.61	1.18
L 6×6×5/8	6	0.625	7.11	24.2	1.73	24.2	1.84	5.66	1.18
L 6×6×3/4	6	0.75	8.44	28.7	1.78	28.2	1.83	6.66	1.17

Index

Bold-italic numerals are for dedicated sections or variables in the text. *Italic* numerals are usually for variables in tables

Machine Design with CAD and Optimization, First Edition. Sayed M. Metwalli.

Companion website: www.wiley.com/go/metwalli/machine

b

C

d

e

f

i

q

r